Element	Symbol	Atomic Number	Average Atomic Mass[a]
Actinium	Ac	89	[227]
Aluminum	Al	13	26.982
Americium	Am	95	[243]
Antimony	Sb	51	121.76
Argon	Ar	18	39.948
Arsenic	As	33	74.922
Astatine	At	85	[210]
Barium	Ba	56	137.33
Berkelium	Bk	97	[247]
Beryllium	Be	4	9.0122
Bismuth	Bi	83	208.98
Bohrium	Bh	107	[270]
Boron	B	5	10.811
Bromine	Br	35	79.904
Cadmium	Cd	48	112.41
Calcium	Ca	20	40.078
Californium	Cf	98	[251]
Carbon	C	6	12.011
Cerium	Ce	58	140.12
Cesium	Cs	55	132.91
Chlorine	Cl	17	35.453
Chromium	Cr	24	51.996
Cobalt	Co	27	58.933
Copernicium	Cn	112	[285]
Copper	Cu	29	63.546
Curium	Cm	96	[247]
Darmstadtium	Ds	110	[281]
Dubnium	Db	105	[268]
Dysprosium	Dy	66	162.50
Einsteinium	Es	99	[252]
Erbium	Er	68	167.26
Europium	Eu	63	151.96
Fermium	Fm	100	[257]
Flerovium	Fl	114	[289]
Fluorine	F	9	18.998
Francium	Fr	87	[223]
Gadolinium	Gd	64	157.25
Gallium	Ga	31	69.723
Germanium	Ge	32	72.63
Gold	Au	79	196.97
Hafnium	Hf	72	178.49
Hassium	Hs	108	[277]
Helium	He	2	4.0026
Holmium	Ho	67	164.93
Hydrogen	H	1	1.0079
Indium	In	49	114.82
Iodine	I	53	126.90
Iridium	Ir	77	192.22
Iron	Fe	26	55.845
Krypton	Kr	36	83.798
Lanthanum	La	57	138.91
Lawrencium	Lr	103	[262]
Lead	Pb	82	207.2
Lithium	Li	3	6.941
Livermorium	Lv	116	[293]
Lutetium	Lu	71	174.97
Magnesium	Mg	12	24.305
Manganese	Mn	25	54.938
Meitnerium	Mt	109	[276]
Mendelevium	Md	101	[258]
Mercury	Hg	80	200.59
Molybdenum	Mo	42	95.96
Neodymium	Nd	60	144.24
Neon	Ne	10	20.180
Neptunium	Np	93	[237]
Nickel	Ni	28	58.693
Niobium	Nb	41	92.906
Nitrogen	N	7	14.007
Nobelium	No	102	[259]
Osmium	Os	76	190.23
Oxygen	O	8	15.999
Palladium	Pd	46	106.42
Phosphorus	P	15	30.974
Platinum	Pt	78	195.08
Plutonium	Pu	94	[244]
Polonium	Po	84	[209]
Potassium	K	19	39.098
Praseodymium	Pr	59	140.91
Promethium	Pm	61	[145]
Protactinium	Pa	91	231.04
Radium	Ra	88	[226]
Radon	Rn	86	[222]
Rhenium	Re	75	186.21
Rhodium	Rh	45	102.91
Roentgenium	Rg	111	[280]
Rubidium	Rb	37	85.468
Ruthenium	Ru	44	101.07
Rutherfordium	Rf	104	[265]
Samarium	Sm	62	150.36
Scandium	Sc	21	44.956
Seaborgium	Sg	106	[271]
Selenium	Se	34	78.96
Silicon	Si	14	28.086
Silver	Ag	47	107.87
Sodium	Na	11	22.990
Strontium	Sr	38	87.62
Sulfur	S	16	32.065
Tantalum	Ta	73	180.95
Technetium	Tc	43	[98]
Tellurium	Te	52	127.60
Terbium	Tb	65	158.93
Thallium	Tl	81	204.38
Thorium	Th	90	232.04
Thulium	Tm	69	168.93
Tin	Sn	50	118.71
Titanium	Ti	22	47.867
Tungsten	W	74	183.84
Ununoctium	Uuo	118	[294]
Ununpentium	Uup	115	[288]
Ununseptium	Uus	117	[294]
Ununtrium	Uut	113	[284]
Uranium	U	92	238.03
Vanadium	V	23	50.942
Xenon	Xe	54	131.29
Ytterbium	Yb	70	173.05
Yttrium	Y	39	88.906
Zinc	Zn	30	65.38
Zirconium	Zr	40	91.224

[a] Average atomic mass values for most elements are from *Pure Appl. Chem.* (2011) **83**, 359. Those for B, C, Cl, H, Li, N, O, Si, S, and Tl are from *Pure Appl. Chem.* (2009) **81**, 2131 and are within the ranges cited in the first reference. Atomic masses in brackets are the mass numbers of the longest-lived isotopes of elements with no stable isotopes.

CHEMISTRY

AN ATOMS-FOCUSED APPROACH

CHEMISTRY

AN ATOMS-FOCUSED APPROACH

Thomas R. Gilbert

NORTHEASTERN UNIVERSITY

Rein V. Kirss

NORTHEASTERN UNIVERSITY

Natalie Foster

LEHIGH UNIVERSITY

With Contributions by

Geoffrey Davies

NORTHEASTERN UNIVERSITY

W. W. NORTON & COMPANY

NEW YORK · LONDON

W. W. Norton & Company has been independent since its founding in 1923, when William Warder Norton and Mary D. Herter Norton first published lectures delivered at the People's Institute, the adult education division of New York City's Cooper Union. The Nortons soon expanded their program beyond the Institute, publishing books by celebrated academics from America and abroad. By mid-century, the two major pillars of Norton's publishing program—trade books and college texts—were firmly established. In the 1950s, the Norton family transferred control of the company to its employees, and today—with a staff of four hundred and a comparable number of trade, college, and professional titles published each year—W. W. Norton & Company stands as the largest and oldest publishing house owned wholly by its employees.

Editor: Erik Fahlgren
Developmental Editor: Andrew Sobel
Project Editor: Carla L. Talmadge
Assistant Editor: Renee Cotton
Production Manager: Eric Pier-Hocking
Marketing Manager: Stacy Loyal
Managing Editor, College: Marian Johnson
Design Director: Hope Miller Goodell
Book Designer: Lissi Sigillo
Photo Editor: Stephanie Romeo
Photo Researcher: Rona Tuccillo
Associate Media Editor: Jennifer Barnhardt
Assistant Media Editor: Paula Iborra
Science Media Editor: Robert Bellinger
Composition: Precision Graphics
Illustrations: Precision Graphics; Electrostatic Potential Surfaces: Daniel Zeroka
Manufacturing: Transcontinental Interglobe

Library of Congress Cataloging-in-Publication Data

Gilbert, Thomas R.
  Chemistry : an atoms-focused approach / Thomas R. Gilbert, Northeastern University, Rein V. Kirss, Northeastern University, Natalie Foster, Lehigh University ; with contributions by Geoffrey Davies, Northeastern University.
    pages cm
  Includes index.
  ISBN 978-0-393-91234-0 (hardcover)
  1. Chemistry—Textbooks. I. Kirss, Rein V. II. Foster, Natalie. III. Title.
  QA33.2.G54 2013
  540—dc23

                        2013016774

W. W. Norton & Company, Inc., 500 Fifth Avenue, New York, NY 10110
www.wwnorton.com

W. W. Norton & Company Ltd., Castle House, 75/76 Wells Street, London W1T 3QT

4 5 6 7 8 9 0

Brief Contents

Contents

Why does black ironwood sink in seawater? *(Chapter 1)*

What are atoms made of? *(Chapter 2)*

Why is magma red? *(Chapter 3)*

Why are greenhouse gases linked to climate change? *(Chapter 4)*

What is responsible for the shimmering, colorful display known as an aurora?
(Chapter 5)

Why do some liquids flow slowly?
(Chapter 6)

What processes control the composition of seawater? *(Chapter 8)*

What reactions occur when wood burns? *(Chapter 9)*

What allows hot air balloons to fly?
(Chapter 10)

What caused this ship to rust?
(Chapter 12)

How is chemical equilibrium manipulated
to produce the ammonia needed to
fertilize crops? *(Chapter 14)*

What is responsible for the color of
hydrangeas? *(Chapter 15)*

What makes aquamarine crystals blue?
(Chapter 16)

How do we power cars that do not rely on gasoline? *(Chapter 17)*

Why is Kevlar so strong? *(Chapter 19)*

How has genetic engineering helped
protect the Hawaiian papaya crop?
(Chapter 20)

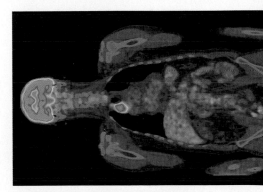

How does nuclear chemistry help
diagnose and treat diseases? *(Chapter 21)*

How do ancient living trees help us achieve accurate radiocarbon dating?
(Chapter 21)

Applications

▶‖ ChemTours

About the Authors

Thomas R. Gilbert has a BS in chemistry from Clarkson and a PhD in analytical chemistry from MIT. After 10 years with the Research Department of the New England Aquarium in Boston, he joined the faculty of Northeastern University, where he is currently associate professor of chemistry and chemical biology. His research interests are in chemical and science education. He teaches general chemistry and science education courses and conducts professional development workshops for K–12 teachers. He has won Northeastern's Excellence in Teaching Award and Outstanding Teacher of First-Year Engineering Students Award. He is a fellow of the American Chemical Society and in 2012 was elected to the ACS Board of Directors.

Rein V. Kirss received both a BS in chemistry and a BA in history as well as an MA in chemistry from SUNY Buffalo. He received his PhD in inorganic chemistry from the University of Wisconsin, Madison, where the seeds for this textbook were undoubtedly planted. After two years of postdoctoral study at the University of Rochester, he spent a year at Advanced Technology Materials, Inc., before returning to academics at Northeastern University in 1989. He is an associate professor of chemistry with an active research interest in organometallic chemistry.

Natalie Foster is emeritus professor of chemistry at Lehigh University in Bethlehem, Pennsylvania. She received a BS in chemistry from Muhlenberg College and MS, DA, and PhD degrees from Lehigh University. Her research interests included studying poly(vinyl alcohol) gels by NMR as part of a larger interest in porphyrins and phthalocyanines as candidate contrast enhancement agents for MRI. She taught both semesters of the introductory chemistry class to engineering, biology, and other nonchemistry majors and a spectral analysis course at the graduate level. She is the recipient of the Christian R. and Mary F. Lindback Foundation Award for distinguished teaching.

Preface

Dear Student,

An old adage says that you can't (or shouldn't) judge a book by its cover. However, the cover of this book—and its title—say a great deal about what's inside. The cover photo is an image of graphite, the most common form of a common element: carbon. The illustration shows what might happen if we could pull back an ultrathin surface layer of graphite and view it using an extremely powerful microscope. What we would see is a layer of carbon atoms, each chemically bonded to three others, forming a seemingly endless array of hexagons. Here's the cool part: peeling back and viewing a single layer of graphite is more than a theoretical possibility—it's actually happened! In addition, scientists have been able to study the properties of these monolayers of carbon atoms, and what they've discovered is pretty amazing. These layers represent the strongest material known—much stronger than the strongest steel—and they conduct electricity better than the most conductive metal. This remarkable material even has its own name: graphene.

Why does graphene have such remarkable properties? The answer to that question is contained in this book's title: *Chemistry: An Atoms-Focused Approach*. The properties of graphene are tied directly to the presence of those hexagonal arrays of carbon atoms and the nature of the chemical bonds that join them together. The geometry and strength of those bonds contribute to the overall strength of graphene, and the unusually high mobilities of the electrons that make up those bonds contribute to the material's outstanding conductivity.

Our cover illustrates a central message of this book: the properties of substances are directly linked to their atomic and molecular structures. In our book we start with the smallest particles of matter and assemble them into more elaborate structures: from subatomic particles to single atoms to monatomic ions and polyatomic ions, and from atoms to small molecules to bigger ones to truly gigantic polymers. By constructing this layered particulate view of matter, we hope our book helps you visualize the underlying chemistry of a wide range of substances and the changes they undergo. With this ability to visualize atoms and molecules, you won't have to resort to memorizing formulas and reactions as a strategy for *surviving* general chemistry. Instead, you will be able to understand *why* elements combine to form compounds with particular formulas and *why* substances react with each other the way they do. For example, you won't have to memorize the charges of the common ions that make seawater (and your blood plasma and tears) salty; instead, you will understand why the many billions of tons of sodium dissolved in the sea (and the 100 grams of it inside your body) exist entirely as Na^+ ions.

Context

While our primary learning goal is for you to be able to interpret and even predict the physical and chemical properties of substances based on their atomic and molecular structures, we would also like you to understand how chemistry is linked to other scientific disciplines. We illustrate these connections using contexts drawn from fields such as biology, environmental science, materials science, astronomy, geology, and medicine. We hope that this approach helps you better understand how scientists apply the principles of chemistry to treat and cure diseases, to make more-efficient use of natural resources, and to minimize the impact of human activity on our planet and its climate.

Problem-Solving Strategies

Another major goal of our book is to help you improve your problem-solving skills. To solve problems in chemistry, you first need to recognize the connections between the information provided in a problem and the answer you are asked to find. Sometimes the hardest part of solving a problem is distinguishing between information that is relevant and information that is not. Once you are clear on where you are starting and where you are going, planning for and carrying out a solution become much easier.

To help you hone your problem-solving skills, we have developed a framework that we introduce in Chapter 1. It is a four-step approach we call **COAST**, which is our acronym for (1) **C**ollect and **O**rganize, (2) **A**nalyze, (3) **S**olve, and (4) **T**hink about it. We use these four steps in *every* Sample Exercise and in the solutions to *odd* problems in the Student's Solutions Manual. They are also used in the hints and feedback embedded in the SmartWork online homework program. To summarize the four steps:

COLLECT AND ORGANIZE helps you understand where to begin to solve the problem. In this step we often rephrase the problem and the answer that is sought, and we identify the relevant information that is provided in the problem statement or available elsewhere in the book.

ANALYZE is where we map out a strategy for solving the problem. As part of that strategy we often estimate what a reasonable answer might be.

SOLVE applies our analysis of the problem from the second step to the information and relations from the first step to actually solve the problem. We walk you through each step in the solution so that you can follow the logic and the math.

THINK ABOUT IT reminds us that an answer is not the last step in solving a problem. We should check the accuracy of the solution and think about the value of a quantitative answer. Is it realistic? Are the units correct? Is the number of significant figures appropriate? Does it agree with our estimate from the Analyze step?

SAMPLE EXERCISE 10.5 Applying Amontons's Law **LO4**

Labels on aerosol cans caution against their incineration because the cans may explode when the pressure inside them exceeds 3.00 atm. At what temperature in degrees Celsius might an aerosol can burst if its internal pressure is 2.00 atm at 25°C?

COLLECT AND ORGANIZE We are given the temperature ($T_1 = 25$°C) and pressure ($P_1 = 2.00$ atm) of a gas and asked to determine the temperature (T_2) at which the pressure (P_2) reaches 3.00 atm.

ANALYZE Because the gas is isolated in a rigid aerosol can, we know that the quantity of gas and its volume are constant. Amontons's law (Equation 10.18) relates the pressures of a confined quantity of gas at two different temperatures. To estimate our answer, we note that the pressure in the can must increase by 50% to reach 3.00 atm. Pressure is directly proportional to absolute temperature, so it, too, must increase by 50%. The initial temperature of 25°C is nearly 300 K, so a 50% increase in absolute temperature corresponds to a final temperature near 450 K, or about 175°C.

SOLVE Rearranging Equation 10.18 to solve for T_2:

$$T_2 = \frac{T_1 P_2}{P_1}$$

and inserting the given T and P values:

$$T_2 = \frac{[(25 + 273)\ \text{K}](3.00\ \cancel{\text{atm}})}{2.00\ \cancel{\text{atm}}} = 447\ \text{K}$$

Converting T_2 to degrees Celsius:

$$T_2 = 447\ \text{K} - 273 = 174°\text{C}$$

THINK ABOUT IT This temperature is close to our estimated value. It is also well below the temperatures that solid waste experiences in the fires of an operating incinerator, which makes the warning label on the can all the more important.

Practice Exercise Air pressure in each of the tires of an automobile is adjusted to 34 psi at a gas station in San Diego, California, where the air temperature is 68°F. After a 3-hour drive along Interstate Highway 8, the car and driver are in Yuma, Arizona, where the temperature is 110°F. What is the pressure in the tires? ⚙

Many students use the **Sample Exercises** more than any other part of the book. Sample Exercises take the concept being discussed and illustrate how to apply it to solve a problem. We think that repeated application of the COAST framework will help you refine your problem-solving skills and hope that the approach becomes habit-forming for you. When you finish a Sample Exercise, you'll find a Practice Exercise to try on your own. If you have the ebook, the Practice Exercises are "live," meaning that you can solve them and receive hints and answer-specific feedback when you need help. The next few pages describe how to use the tools built into each chapter to gain a conceptual understanding of chemistry.

Chapter Structure

Each chapter begins with an **opening story**, which provides glimpses of how the chemistry in the chapter that follows connects to the world. We have used topics that should be familiar to you, but we place them in chemical contexts that may surprise you.

If you are trying to decide what is most important in a chapter, check the **Learning Outcomes** listed on the first page. Whether you are reading the chapter from first page to last or reviewing it for an exam, the Learning Outcomes should help you focus on the key information you need and the skills you should acquire. You will also see which Learning Outcomes are linked to which Sample Exercises in the chapter.

As you study each chapter, you will find **key terms** in boldface in the text and in a running glossary in the margin. We have deliberately duplicated these definitions so that you can continue reading without interruption but quickly find them when doing homework or reviewing for a test. All key terms are also defined in the Glossary in the back of the book.

Many concepts are related to others described earlier in the book. We point out these relationships with **Connection** icons in the margins. We hope they help you draw your own connections between major themes covered in the book.

To help you develop your own microscale view of matter, we use **molecular art** to enhance photos and figures, and to illustrate what is happening at the atomic and molecular levels.

If you're looking for additional help visualizing a concept, we have about 100 **ChemTours**, denoted by the ChemTour icon, available online at wwnpag.es/chemtours. ChemTours demonstrate dynamic processes and help you visualize events at the molecular level. Many of the ChemTours allow you to manipulate variables and observe the resulting changes. Questions at the end of the ChemTour tutorials offer step-by-step assistance in solving problems and provide useful feedback.

Concept Tests are short, conceptual questions that serve as a self-check by asking you to stop and answer a question relating to what you just read. We designed them to help you see for yourself whether you have grasped a key concept and can apply it. We have an average of one Concept Test per section and many have a visual component. You may find some Concept Tests challenging. We provide the answers to all Concept Tests in the back of the book.

CONNECTION In Chapter 9 we defined standard conditions of temperature and pressure as they apply to thermochemistry. Note that *STP* and *standard conditions* are not the same. STP applies strictly to calculations involving the gas laws, while standard conditions apply to thermochemical data.

▶❚❚ **CHEMTOUR**

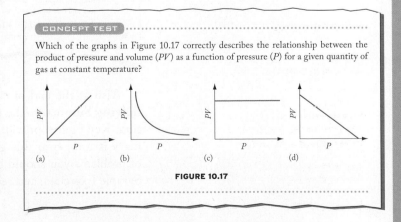

CONCEPT TEST

Which of the graphs in Figure 10.17 correctly describes the relationship between the product of pressure and volume (*PV*) as a function of pressure (*P*) for a given quantity of gas at constant temperature?

FIGURE 10.17

At the end of each chapter is a special Sample Exercise that draws on several key concepts from the chapter and occasionally others from preceding chapters to solve a problem that is framed in the context of a real-world scenario or incident. We call these **Integrated Sample Exercises**. You may find them more challenging than most of those that precede them in each chapter, but please invest your time in working through them because they represent authentic exercises that will enhance your problem-solving skills.

SAMPLE EXERCISE 10.16   Integrating Concepts: Air for a Jet Engine

The Boeing 767 (Figure 10.42) is one of the most popular wide-body commercial airliners (over 1000 have been built), and it's also one of the most fuel efficient. While cruising at 851 km/hr (530 mph) at an altitude of 11,000 m (36,000 ft), a 767-200ER (the extended-range version of the plane) consumes about 1720 U.S. gallons of jet fuel per hour.

We are given the following facts: (1) the density of jet fuel is 0.80 g/mL; (2) at an altitude of 11,000 m, $P_{atm} = 210$ mmHg and $T = -56°C$; (3) 1 U.S. gallon = 3.785 L; and (4) dodecane, $C_{12}H_{26}$, is considered an appropriate model hydrocarbon for jet fuel.

a. What volume of air, in liters, does a cruising 767 need so that it can completely burn an hour's worth of jet fuel?
b. Fuel efficiency is often based on number of passengers times distance traveled per volume of fuel consumed. In the United States, this value is typically expressed in units of passenger-miles per gallon. However, in much of the rest of the world, efficiency units are inverted and are typically expressed in liters per 100 km per passenger. Express the fuel efficiency of a full 767-200ER (which holds 224 passengers) in both sets of units.

COLLECT AND ORGANIZE We know the quantity of dodecane to be combusted, and we are asked to calculate the volume of air needed for complete combustion—that is, to convert its C and H content into CO_2 and H_2O. We know the pressure and temperature of the air. According to Table 10.1, dry air is 20.95% O_2.

ANALYZE This exercise involves a chemical reaction, so writing a balanced chemical equation describing it is a good place to start. Next we need to convert the volume of fuel into an equivalent number of moles of fuel, and then to use the stoichiometry of the reaction to convert that value to moles of O_2. We will then calculate the equivalent volume of O_2 using the ideal gas equation. (The pressure is below 1 atm, so there should be no need to correct for nonideal behavior.) Finally we will convert the volume of O_2 to the corresponding volume of air. The volume of air needed each hour by the engines of a 767 should be enormous.

SOLVE
a. Calculating volume of air needed in 1 hour:
1. Write the balanced chemical equation describing the combustion reaction. The reactants and products are

$$C_{12}H_{26}(\ell) + O_2(g) \rightarrow CO_2(g) + H_2O(g)$$

We first balance the numbers of C and H atoms:

$$C_{12}H_{26}(\ell) + O_2(g) \rightarrow 12\,CO_2(g) + 13\,H_2O(g)$$

This leaves us with an odd number of O atoms on the right, requiring that we multiply all the terms by 2 and then balance the number of O atoms:

$$2\,C_{12}H_{26}(\ell) + 37\,O_2(g) \rightarrow 24\,CO_2(g) + 26\,H_2O(g)$$

FIGURE 10.42 A Boeing 767-200ER.

2. Converting the volume of fuel consumed in 1 hour into an equivalent number of moles of $C_{12}H_{26}$:

$$1720\ \text{gal}\left(\frac{3.785\ \text{L}}{1\ \text{gal}}\right)\left(\frac{1000\ \text{mL}}{1\ \text{L}}\right)\left(\frac{0.80\ \text{g}}{\text{mL}}\right)\left(\frac{1\ \text{mol}}{170.33\ \text{g}}\right)$$

$$= 3.06 \times 10^4\ \text{mol}\ C_{12}H_{26}$$

3. Converting moles of $C_{12}H_{26}$ into moles of O_2:

$$3.06 \times 10^4\ \text{mol}\ C_{12}H_{26}\left(\frac{37\ \text{mol}\ O_2}{2\ \text{mol}\ C_{12}H_{26}}\right) = 5.66 \times 10^5\ \text{mol}\ O_2$$

4. In converting moles of O_2 into a volume of air using the ideal gas law, we need to convert pressure units from mmHg to atm and temperature to the Kelvin scale:

$$V = \frac{nRT}{P} = \frac{(5.66 \times 10^5\ \text{mol})\left(0.08206\ \frac{\text{L} \cdot \text{atm}}{\text{mol} \cdot \text{K}}\right)(273 - 56)\ \text{K}}{210\ \text{mmHg}\left(\frac{1\ \text{atm}}{760\ \text{mmHg}}\right)}$$

$$= 3.65 \times 10^7\ \text{L}\ O_2$$

5. Air is 20.95% O_2 by volume, so the volume of air the engines must take in each hour is

$$3.65 \times 10^7\ \text{L}\ O_2\left(\frac{100\ \text{L air}}{20.95\ \text{L}\ O_2}\right) = 1.74 \times 10^8\ \text{L air}$$

b. Calculating fuel efficiency in U.S. Customary units based on the distance traveled and fuel consumed in 1 hour:

$$\frac{224\ \text{passengers} \times 530\ \text{miles}}{1720\ \text{gal}} = 69.0\ \text{passenger-miles/gallon}$$

The corresponding efficiency value in liters per 100 km per passenger is

$$\frac{1720\ \text{gal} \times \frac{3.785\ \text{L}}{1\ \text{gal}}}{224\ \text{passengers} \times 851\ \text{km}} \times 100 = 3.41\ \text{L/100 km-passenger}$$

THINK ABOUT IT The calculated volume of air is 174 million liters. That is a lot of air, but then 1720 U.S. gallons of jet fuel is a lot of fuel. Perhaps a more interesting value is the 69.0 passenger-miles per gallon: better than that of most automobiles with two occupants, and you reach your destination much faster.

Also at the end of each chapter are a thematic **Summary** and a **Problem-Solving Summary**. The first is a brief synopsis of the chapter, organized by section. Key figures provide visual cues as you review. The Problem-Solving Summary is unique to this general chemistry book—it outlines the different types of problems you should be able to solve, where to find examples of them in the Sample Exercises, and reiterates relevant concepts and equations.

PROBLEM-SOLVING SUMMARY

TYPE OF PROBLEM	CONCEPTS AND EQUATIONS		SAMPLE EXERCISES
Calculating relative rates of effusion	$\dfrac{\text{effusion rate}_A}{\text{effusion rate}_B} = \sqrt{\dfrac{\mathcal{M}_B}{\mathcal{M}_A}}$	(10.7)	10.1
Measuring gas pressure with a manometer; converting pressure units	See the conversion factors inside the back cover and Table 10.2.		10.2
Applying Boyle's law	$P_1V_1 = P_2V_2$	(10.10)	10.3
Applying Charles's law	$\dfrac{V_1}{V_2} = \dfrac{T_1}{T_2}$	(10.14)	10.4
Applying Amontons's law	$\dfrac{P_1}{T_1} = \dfrac{P_2}{T_2}$	(10.18)	10.5
Applying the combined gas law	$\dfrac{P_1V_1}{T_1} = \dfrac{P_2V_2}{T_2}$	(10.19)	10.6
Applying the ideal gas law	$PV = nRT$	(10.20)	10.7–10.10
Calculating mole fractions and partial pressures	$x_i = \dfrac{n_i}{n_{\text{total}}}$ $P_i = x_i P_{\text{total}}$	(10.23) (10.24)	10.11
Calculating the quantity of a gas collected by water displacement	Calculate the partial pressure (P_i) of the collected gas using the equation $P_i = P_{\text{atm}} - P_{H_2O}$		10.12
Calculating gas solubility using Henry's law	$C_{\text{gas}} = k_H P_{\text{gas}}$	(10.26)	10.13
Calculating root-mean-square speeds	$u_{\text{rms}} = \sqrt{\dfrac{3RT}{\mathcal{M}}}$	(10.27)	10.14
Using the van der Waals equation	$\left[P + a\left(\dfrac{n}{V}\right)^2 \right](V - nb) = nRT$	(10.28)	10.15

Following the summaries are groups of questions and problems. The first group consists of **Visual Problems**. In many of them, you are asked to interpret a molecular view of a sample or a graph of experimental data.

Concept Review Questions and Problems come next, arranged by topic in the same order as they appear in the chapter. Concept Reviews are qualitative and often ask you to explain why or how something happens. Problems are paired and can be quantitative, conceptual, or a combination of both. **Contextual problems** have a title that describes the context in which the problem is placed. Finally, **Additional Problems** can come from any section or combination of sections in the chapter. Some of them incorporate concepts from previous chapters. Problems marked with an asterisk (*) are more challenging and often take multiple steps to solve.

We want you to have confidence in using the answers in the back of the book as well as the Student's Solutions Manual, so we used a rigorous triple-check accuracy program for this book. Each end-of-chapter question and problem was solved independently by the Solutions Manual author, Karen Brewer, and by two additional chemical educators. Karen compared her solutions to those from the two reviewers and resolved any discrepancies. This process is designed to ensure clearly written problems and accurate answers in the appendices and Solutions Manual.

10.99. A sample of oxygen is collected over water at 25°C and 1.00 atm. If the total sample volume is 0.480 L, how many moles of O_2 are collected?

10.100. Water vapor is removed from the O_2 sample in Problem 10.99. What is the volume of the dry O_2 at 25°C and 1.00 atm?

10.101. The following reactions are carried out in sealed containers. Will the total pressure after each reaction is complete be greater than, less than, or equal to the total pressure before the reaction? Assume all reactants and products are gases at the same temperature.
a. $N_2O_5(g) + NO_2(g) \rightarrow 3\,NO(g) + 2\,O_2(g)$
b. $2\,SO_2(g) + O_2(g) \rightarrow 2\,SO_3(g)$
c. $C_3H_8(g) + 5\,O_2(g) \rightarrow 3\,CO_2(g) + 4\,H_2O(g)$

10.102. In each of the following gas-phase reactions, determine whether the total pressure at the end of the reaction (carried out in a sealed, rigid vessel) will be greater than, less than, or equal to the total pressure at the beginning. Assume all reactants and products are gases at the same temperature.
a. $H_2(g) + Cl_2(g) \rightarrow 2\,HCl(g)$
b. $4\,NH_3(g) + 5\,O_2(g) \rightarrow 4\,NO(g) + 6\,H_2O(g)$
c. $2\,NO(g) + O_2(g) \rightarrow 2\,NO_2(g)$

*10.103. **High-Altitude Mountaineering** Most alpine climbers breathe pure oxygen near the summits of the world's highest mountains. How much more O_2 is there in a lungful of pure O_2 at an elevation where atmospheric pressure is 266 mmHg than in a lungful of air at sea level? Express your answer as a percentage.

10.104. **Scuba Diving** A scuba diver is at a depth of 50 m, where the pressure is 5.0 atm. What should be the mole fraction of O_2 in the gas mixture the diver breathes to achieve the same P_{O_2} as at sea level?

Solubilities of Gases and Henry's Law

CONCEPT REVIEW

10.109. Why is the Henry's law constant for CO_2 so much larger than those for N_2 and O_2 at the same temperature? *Hint*: Does CO_2 react with water?

10.110. As water in a beaker is heated, bubbles form inside the beaker at temperatures well below the boiling point of water. What gas is in the bubbles?

10.111. What type of intermolecular interaction accounts for the limited solubility of methane in water?

*10.112. Air is primarily a mixture of nitrogen and oxygen. Is the Henry's law constant for the solubility of air in water the sum of k_H for N_2 and k_H for O_2? Explain why or why not.

PROBLEMS

*10.113. **Arterial Blood** Arterial blood contains about 0.25 g of oxygen per liter at 37°C and standard atmospheric pressure. What is the Henry's law constant, in mol/(L · atm), for O_2 dissolution in blood?

10.114. The solubility of O_2 in water is 6.5 mg/L at an atmospheric pressure of 1 atm and a temperature of 40°C. Calculate the Henry's law constant of O_2 at 40°C.

*10.115. **Oxygen for Climbers and Divers** Use the Henry's law constant for O_2 dissolved in arterial blood from Problem 10.113 to calculate the solubility of O_2 in the blood of (a) a climber on Mt. Everest ($P_{\text{atm}} = 0.35$ atm) and (b) a scuba diver breathing air at a depth of 20 meters ($P \approx 3.0$ atm).

*10.116. The solubility of air in water is approximately 7.9×10^{-4} M at 20°C and 1.0 atm. Calculate the Henry's law constant for air.

Gas Diffusion: Molecules Moving Rapidly

PROBLEMS

Acknowledgments

As we launch this first edition of *Chemistry: An Atoms-Focused Approach*, our thanks go out to our publisher, W. W. Norton, for supporting us in writing a book that is written the way we much prefer to teach general chemistry. We especially wish to acknowledge the hard work and dedication of our editor/motivator/taskmaster, Erik Fahlgren. Erik has been an indefatigable source of guidance, perspective, persuasion, and inspiration to all of us. He has kept this project on track and on time over many months of conceptualization, development, and production. Erik is the single greatest reason the first copies of this book will roll off the presses in the summer of 2013, and, we hope, help change the way general chemistry is taught in many colleges and universities in the years that follow.

We are pleased to acknowledge the contributions of an outstanding developmental editor, Andrew Sobel. Andrew provided invaluable advice on how the chapters and topics of this book should be organized to achieve our learning goals, and he was also tenacious in reading and weighing every word of text and every illustration to ensure that we achieved those goals. Thanks to him, the following pages contain presentations and explanations that are more clear, more concise, and more likely to engage readers just setting out on their college careers.

Carla Talmadge is our project editor and doyenne of the many features in the book designed to help students visualize matter and the chemical changes it undergoes on an atomic and molecular scale. Assistant editor Renee Cotton kept information flowing in a timely fashion among all of the players involved in writing, reviewing, and writing some more. Thanks as well to Stephanie Romeo and Rona Tuccillo for finding great photos; production manager Eric Pier-Hocking for his work behind the scenes; Jennifer Barnhardt for managing the print ancillaries; Rob Bellinger for his diligence with the book's many electronic enhancements; and Stacy Loyal for her encouragement, fellowship, and marketing prowess. The entire Norton team was staffed by skilled, dedicated professionals who, as a bonus, were delightful people to work with and, on occasion, relax with.

Many reviewers, listed here, contributed to the development and production of this book. First among them are the Editorial Consultants: Lee Friedman, Jeremy Kua, Jeff Macedone, and Wayne Wesolowski. These insightful colleagues provided a continuous stream of useful suggestions on how to make a general chemistry textbook truly *atoms focused*. Essentially every section of every chapter in this book benefited from their input. We also owe an extra special thanks to Karen Brewer for her dedicated work on the Solutions Manuals and for her invaluable suggestions on how to improve the inventory and organization of problems and concept questions at the end of each chapter. She, along with Jordan Fantini and Amy Irwin, comprised the triple-check accuracy team who helped ensure the quality of the back-of-book answers and Solutions Manuals. Finally, we wish to acknowledge the care and thoroughness of Petia Bobadova-Parvanova, Tara Carpenter, Garry Crosson, Greg Domski, Doug English, Daniel Groh, Megan Grunert, Tracy Hamilton, Maria Kolber, Willem Leenstra, Douglas Magde, Gellert Mezei, Nancy Mullins, Sherine Obare, Edith Osborne, Robert Parson, James Patterson, Garry Pennycuff, and John Pollard for checking the accuracy of the myriad facts that frame the contexts and the science in the pages that follow.

Thomas R. Gilbert
Rein V. Kirss
Natalie Foster

Reviewers

Ioan Andricioaei, University of California, Irvine

Merritt Andrus, Brigham Young University

David Arnett, Northwestern College

Christopher Babayco, Columbia College

Carey Bagdassarian, University of Wisconsin, Madison

Craig Bayse, Old Dominion University

Vladimir Benin, University of Dayton

Philip Bevilacqua, Pennsylvania State University

Robert Blake, Glendale Community College

David Boatright, University of West Georgia

Petia Bobadova-Parvanova, Rockhurst University

Stephanie Bousscrt, DePaul University

Jasmine Bryant, University of Washington

Michael Bukowski, Pennsylvania State University

Charles Burns, Wake Technical Community College

Jon Camden, University of Tennessee, Knoxville

Tara Carpenter, University of Maryland, Baltimore County

David Carter, Angelo State University

Allison Caster, Colorado School of Mines

Colleen Craig, University of Washington

Gary Crosson, University of Dayton

Guy Dadson, Fullerton College

David Dearden, Brigham Young University

Danilo DeLaCruz, Southeast Missouri State University

Anthony Diaz, Central Washington University

Greg Domski, Augustana College

Jacqueline Drak, Bellevue Community College

Michael Ducey, Missouri Western State University

Lisa Dysleski, Colorado State University

Amina El-Ashmawy, Collin College

Doug English, Wichita State University

Jim Farrar, University of Rochester

MD Abul Fazal, College of Saint Benedict & Saint John's University

Anthony Fernandez, Merrimack College

Lee Friedman, University of Maryland

Arthur Glasfeld, Reed College

Daniel Groh, Grand Valley State University

Megan Grunert, Western Michigan University

Margaret Haak, Oregon State University

Tracy Hamilton, University of Alabama at Birmingham

David Hanson, Stony Brook University

Roger Harrison, Brigham Young University

David Henderson, Trinity College

Carl Hoeger, University of California, San Diego

Adam Jacoby, Southeast Missouri State University

James Jeitler, Marietta College

Christina Johnson, University of California, San Diego

Maria Kolber, University of Colorado, Boulder

Regis Komperda, Wright State University

Jeffrey Kovac, University of Tennessee, Knoxville

Jeremy Kua, University of California, San Diego

Robin Lammi, Winthrop University

Annie Lee, Rockhurst University

Willem Leenstra, University of Vermont

Ted Lorance, Vanguard University

Charity Lovitt, Bellevue Community College

Suzanne Lunsford, Wright State University

Jeffrey Macedone, Brigham Young University

Douglas Magde, University of California, San Diego

Rita Maher, Richland College

Heather McKechney, Monroe Community College

Anna McKenna, College of Saint Benedict & Saint John's University

Claude Mertzenich, Luther College

Gellert Mezei, Western Michigan University

Katie Mitchell-Koch, Emporia State University

Stephanie Morris, Pellissippi State Community College

Nancy Mullins, Florida State College at Jacksonville

Joseph Nguyen, Mount Mercy University

Sherine Obare, Western Michigan University

Edith Osborne, Angelo State University

Ruben Parra, DePaul University

Robert Parson, University of Colorado, Boulder

Brad Parsons, Creighton University

James Patterson, Brigham Young University

Garry Pennycuff, Pellissippi State Community College

Thomas Pentecost, Grand Valley State University

Sandra Peszek, DePaul University

John Pollard, University of Arizona

Gretchen Potts, University of Tennessee at Chattanooga

William Quintana, New Mexico State University

Cathrine Reck, Indiana University, Bloomington

Alan Richardson, Oregon State University

Dawn Richardson, Collin College

James Roach, Emporia State University

Jill Robinson, Indiana University

Perminder Sandhu, Bellevue Community College

James Silliman, Texas A&M University, Corpus Christi

Joseph Simard, University of New England

Kim Simons, Emporia State University

Sergei Smirnov, New Mexico State University

Justin Stace, Belmont University

Alyssa Thomas, Utica College

Jess Vickery, SUNY Adirondack

Wayne Wesolowski, University of Arizona

Thao Yang, University of Wisconsin, Eau Claire

Teaching and Learning Resources

SMARTWORK ONLINE HOMEWORK FOR GENERAL CHEMISTRY

wwnorton.com/smartwork

Created by chemistry educators, SmartWork is the most intuitive online tutorial and homework management system available for general chemistry. The many question types, including graded molecule drawing, math and chemical equations, ranking tasks, and interactive figures, help students develop and apply their understanding.

Every problem in SmartWork includes response-specific feedback and general hints using the steps in COAST. Links to the ebook version of *Chemistry: An Atoms-Focused Approach* take students to the specific place where the concept is explained. All problems in SmartWork use the same language and notation as the textbook.

SmartWork also features Tutorial Problems. If a student asks for help in a Tutorial Problem, SmartWork breaks the problem down into smaller steps, providing hints, answer-specific feedback, and probing questions within each step. At any point in a Tutorial, the student can return and answer the original problem.

Assigning, editing, and administering homework within SmartWork is easy. SmartWork allows the instructor to search for problems using both the text's Learning Objectives and Bloom's taxonomy. Instructors can use premade assignment sets provided by Norton authors, modify those assignments, or create their own. Instructors can also make changes at the question level. All instructors have access to our WYSIWYG (What You See Is What You Get) authoring tools—the same ones Norton authors use. Those intuitive tools make it easy to modify existing problems or to develop new content that meets the specific needs of your course.

Wherever possible, SmartWork makes use of algorithmic variables so that students see slightly different versions of the same problem. Assignments are graded automatically, and SmartWork includes sophisticated yet flexible tools for managing class data. Instructors can use the Item Analysis features to assess how students have done on specific problems within an assignment. Instructors can also review individual students' work on problems.

SmartWork for *Chemistry: An Atoms-Focused Approach* features the following problem types:

> ➤ Selected End-of-Chapter Problems. These problems, which use algorithmic variables when appropriate, coach students through mastering single- and multiconcept problems based on chapter content. They make use of all of SmartWork's answer-entry tools.

> ➤ Multistep Tutorials. These problems offer students who demonstrate a need for help a series of linked, step-by-step subproblems to work through. They are based on the Concept Review problems at the end of each chapter. Tutorials make use of student-focused artwork from the Student's Solutions Manual.

> ➤ Math Review Problems. These problems can be used by students for practice, or by instructors to diagnose the mathematical ability of their students.

> ➤ Ranking Task Problems. These problems ask students to make comparative judgments between items in a set.

➤ Visual and Graphing Problems. These problems challenge students to identify chemical phenomena and to interpret graphs. They use SmartWork's Drag-and-Drop and Hotspot functionality.

➤ Reaction Visualization Problems. Based on both static art and simulated reaction videos, these problems are designed to help students visualize what happens at the atomic level—and why.

➤ Nomenclature Problems. New matching and multiple-choice problems help students master course vocabulary.

➤ ChemTour Problems. These are based on Norton's popular animations.

EBOOK

An affordable and convenient alternative to the print text, the ebook retains the content and design of the print book and allows students to highlight and take notes with ease, print chapters as needed, and search the text.

www.nortonebooks.com

The online version of *Chemistry: An Atoms-Focused Approach* provides students with stand-alone Interactive Practice Exercises (which require Flash). These are self-grading SmartWork problems that allow students to practice solving problems, and receive hints and feedback, without being penalized. The online ebook also allows students one-click access to the 100 ChemTour animations.

The online ebook is available bundled with the print text and SmartWork at no extra cost, or it may be purchased bundled with SmartWork access.

Norton also offers a downloadable PDF version of the ebook as well as a version optimized for delivery to any device—laptop, tablet, or smartphone.

STUDENT'S SOLUTIONS MANUAL by Karen Brewer, Hamilton College

The Student's Solutions Manual provides students with fully worked solutions to select end-of-chapter problems using the COAST four-step method (Collect and Organize, Analyze, Solve, and Think about it). Each chapter in the Student's Solutions Manual contains several pieces of art designed to help students visualize ways to approach problems. This artwork is also used in the hints and feedback within SmartWork.

CLICKERS IN ACTION: INCREASING STUDENT PARTICIPATION IN GENERAL CHEMISTRY by Margaret Asirvatham, University of Colorado, Boulder

An instructor-oriented resource providing information on implementing clickers in general chemistry courses, *Clickers in Action* contains more than 250 class-tested, lecture-ready questions with histograms showing student responses, as well as insights and suggestions for implementation. Question types include macroscopic observation, symbolic representation, and atomic or molecular views of processes.

INSTRUCTOR'S SOLUTIONS MANUAL by Karen Brewer, Hamilton College

The Instructor's Solutions Manual provides instructors with fully worked solutions to every end-of-chapter Concept Review and Problem. Each solution uses the COAST four-step method (Collect and Organize, Analyze, Solve, and Think about it).

INSTRUCTOR'S RESOURCE MANUAL

Each chapter in this complete resource manual for instructors begins with a brief overview of the text chapter followed by tips for those teaching with the atoms-focused approach for the first time, suggestions for integrating the contexts featured in the book into a lecture, suggested sample lecture outlines, alternate contexts to use with each chapter, and instructor notes for suggested activities from the *ChemConnections* and *Calculations in Chemistry* workbooks. Summaries of the ChemTours and suggested laboratory exercises round out each chapter.

INSTRUCTOR'S RESOURCE DISC

This helpful classroom presentation tool features:

➤ Stepwise animations and classroom response questions. Developed by Jeffrey Macedone of Brigham Young University and his team, these animations, which use native PowerPoint functionality and textbook art, help instructors to "walk" students through over 100 chemical concepts and processes. Where appropriate, the slides contain two types of questions for students to answer in class: questions that ask them to predict what will happen next and why, and questions that ask them to apply knowledge gained from watching the animation. Self-contained notes help instructors adapt these materials to their own classrooms.

➤ Lecture PowerPoint slides that include integrated figures from the text, ChemTours, and stick-or-switch clicker questions. These are particularly helpful to first-time teachers of the introductory course.

➤ All ChemTours and multilevel visualizations.

➤ *Clickers in Action* clicker questions for each chapter provide instructors with class-tested questions they can integrate into their course.

➤ Photographs, drawn figures, and tables from the text available in PowerPoint and JPEG.

DOWNLOADABLE INSTRUCTOR'S RESOURCES

wwnorton.com/instructors · This password-protected site for instructors includes:

➤ Stepwise animations and classroom response questions. Developed by Jeffrey Macedone of Brigham Young University and his team, these animations, which use native PowerPoint functionality and textbook art, help instructors to "walk" students through over 100 chemical concepts and processes. Where appropriate, the slides contain two types of questions for students to answer in class: questions that ask them to predict what will happen next and why, and questions that ask them to apply knowledge gained from watching the animation. Self-contained notes help instructors adapt these materials to their own classrooms.

➤ Lecture PowerPoints with stick-or-switch clicker questions.

➤ All ChemTours and multilevel visualizations.

➤ Test bank in PDF, Word RTF, and *ExamView* Assessment Suite formats.

➤ Instructor's Solutions Manual in PDF and Word, so that instructors may edit solutions.

➤ All of the end-of-chapter questions and problems available in Word along with the key equations.

➤ Photographs, drawn figures, and tables from the text available in PowerPoint and JPEG.

➤ *Clickers in Action* clicker questions.

➤ BlackBoard and WebCT materials.

BLACKBOARD AND WEBCT COURSE CARTRIDGES

Course cartridges for BlackBoard and WebCT include access to the ChemTours, a Study Plan for each chapter, multiple-choice tests (Maryfran Barber, Wayne State University), and links to premium content in the ebook and SmartWork.

TEST BANK by Randa Roland, University of California, Santa Cruz

Norton uses an innovative, evidence-based model to help instructors create high-quality and pedagogically effective quizzes and testing materials. Each chapter of the Test Bank is structured around an expanded list of student learning objectives and evaluates student knowledge on six distinct levels based on Bloom's taxonomy: remembering, understanding, applying, analyzing, evaluating, and creating.

Questions are further classified by section and difficulty, making it easy to construct tests and quizzes that are meaningful and diagnostic according to instructor need. The more than 2100 questions are divided into multiple-choice and short answer.

The Test Bank is available in *ExamView* Assessment Suite, Word RTF, and PDF formats.

CHEMISTRY

AN ATOMS-FOCUSED APPROACH

1

Matter and Energy

An Atomic Perspective

Back in Time

The first tools were made of stone. Their use ushered in the Stone Age, which lasted for more than 2 million years and ended with the development of the first tools made of metal. In a region of the Middle East called the Fertile Crescent, prosperity and access to minerals containing copper and tin enabled groups of craftsmen to engage in one of the earliest applications of chemical technology: converting the minerals into free metals and then blending the metals to form an alloy—bronze. The Bronze Age began around the 4th millennium BCE, and weapons made of bronze were the strongest and most durable ones available for nearly 3000 years.

Making bronze was, for ancient metal workers, more craft than science. They knew *how*, for example, to build primitive smelters that produced hotter flames than typical campfires. They also knew how to fill the smelters with chunks of charcoal and copper ore, then ignite the charcoal, and later recover pure copper from the cooling embers. Yet they had no idea *why* their smelters produced metals, nor could they explain why adding another element such as tin or arsenic (a favorite additive of Incan metal workers) produced an alloy that was much stronger than copper alone.

Today we understand the chemistry of smelting metals because we know the chemical composition of metal ores, and we can explain why alloys are stronger than their parent elements because we know the structures of the materials at the atomic level. We know, for example, that the atoms in copper are arranged in three-dimensional arrays of tightly packed layers. A piece of copper wire or foil is easily bent because the layers of copper atoms can slide past each other when subjected to an external force. However, when slightly larger atoms of tin are also present in neighboring layers of copper atoms, they produce atomic-scale imperfections in the layers that inhibit their sliding past each other. As a result, an object made of bronze is much tougher to bend than if it were made of pure copper. This property translated into Bronze Age tools and weapons that held their shape better, stayed sharper longer, and, in the case of shields and body armor, provided better protection for warriors in battle.

Bronze Age Warriors This shield decoration, from Greece in the 6th century BCE, ▶ is made of bronze, an alloy of copper and tin. Tin atoms create irregularities in the layers of copper atoms in bronze. As a result, the layers do not pass each other as easily, making bronze objects harder than copper objects and less easily deformed.

In this chapter we begin our exploration of how the properties of substances are linked to their atomic-level structure. As we begin, we need to acknowledge the pioneering contributions of thoughtful Greek philosophers of the late Bronze Age who espoused *atomism*, a belief that all forms of matter are composed of extremely tiny, indestructible building blocks called atoms. Atomism is an example of a natural philosophy; it was *not* a scientific theory. The difference between them, as we discuss in this chapter, is that while both seek to explain natural phenomena, scientific theories are based on observations and experimental results, *and* they are testable. An important quality of a scientific theory is that it accurately predicts the results of future experiments and, indeed, serves as a guide in designing experiments. Alas, the ancient Greeks did not have the technology to test whether matter really is made of atoms—but we do.

1.1 States of Matter

All things that are physically real—from the air we breathe to the ground we walk on—are forms of matter. Scientists define **matter** as everything in the universe that has **mass (*m*)** and occupies space. **Chemistry** is the study of the composition, structure, and properties of matter. The science of chemistry has led to the synthesis of many new forms of matter that impact our lives, from the food we eat and clothes we wear to the technological innovations that surround us.

Matter exists in three phases or physical states: solid, liquid, or gas. You are probably familiar with their characteristic properties:

> ➤ A **solid** has a definite volume and shape.

> ➤ A **liquid** has a definite volume but not a definite shape. Instead, it takes the shape of the container it is in.

> ➤ A **gas** (or *vapor*) has neither a definite volume nor a definite shape. Rather, it expands to occupy the entire volume and shape of its container.

A gas, unlike a solid or a liquid, is highly compressible, which means it can be squeezed into a smaller volume if its container is not rigid and pressure is applied to it.

Let's try to explain why the states of matter have these properties. Why, for example, do solids have definite shapes whereas liquids do not, and why are gases highly compressible but liquids and solids are not? The answers to both questions are based on the particulate nature of matter and the behavior of the particles that make up these phases. Our notion that matter is made of tiny particles comes down to us from the writings of 5th century BCE Greek philosophers who believed that all matter is made of indivisible, extremely tiny building blocks they called **atoms**. Today their natural philosophy of *atomism* has been refined into the atomic theory of matter. According to atomic theory, all forms of matter are composed of microscopic particles. The particles may be individual atoms, or they may be groups of atoms called **molecules** that are held together in a characteristic pattern and proportion by forces called **chemical bonds**.

Consider the three familiar states of water, as shown with macroscopic views and particle-view magnification in Figure 1.1. All three states contain the same particles: molecules made of two atoms of hydrogen bonded to a central atom of

matter anything that has mass and occupies space.

mass (*m*) the property that defines the quantity of matter in an object.

chemistry the study of the composition, structure, and properties of matter and of the energy consumed or given off when matter undergoes a change.

solid a form of matter that has a definite shape and volume.

liquid a form of matter that occupies a definite volume but flows to assume the shape of its containers.

gas a form of matter that has neither definite volume nor shape, and that expands to fill its container; also called *vapor*.

atom the smallest particle of an element that retains the chemical characteristics of the element.

molecule a collection of atoms chemically bonded together.

chemical bond a force that holds two atoms in a molecule or a compound together.

(a) Solid   (b) Liquid   (c) Gas

FIGURE 1.1 The three states of water. (a) In the solid state, each particle in ice is held in place in a rigid, three-dimensional array. (b) In the liquid state, the particles are close together but free to tumble over one another. (c) In the gas state, the particles are far apart, largely independent of one another, and move freely. Water vapor is invisible, but we can see clouds and fog that form when atmospheric water vapor condenses to liquid water.

oxygen. In solid ice, each water molecule is surrounded by the same four others, locked in place in a hexagonal array of molecules that extends in all three dimensions as shown in Figure 1.1(a). Molecules in the array may vibrate a little depending on their temperature, but they are not free to move past the molecules that surround them. Thus, ice is rigid at both the microscopic and macroscopic levels. On the other hand, the molecules in liquid water are not arranged in a crystalline pattern but instead are more randomly ordered (Figure 1.1b) and can flow past one another. They are still in close proximity to each other, but their nearest neighbors change over time. Molecules of water vapor and other gases are widely separated and have much more freedom of motion (Figure 1.1c). The volume the particles occupy is negligible relative to the volume occupied by the vapor itself. Most of the volume is made up of nothing—just empty space between the particles of gas. This empty space between particles accounts for the compressibility of gases.

Matter can be transformed from one physical state to another as its temperature is raised or lowered. Water undergoes such changes in each of the processes illustrated in Figure 1.2: drops of water fall from *melting* icicles. A tray of ice cubes stored in a frost-free freezer slowly disappears as solid ice becomes water vapor in a process called **sublimation**. Bubbles of water vapor form in boiling water as the liquid undergoes *vaporization*. Water vapor from the air *condenses* as drops of liquid water form on a glass holding a cold drink on a humid summer day. A layer of ice forms in early winter on the surface of a pond as the surface layer of water *freezes*. And ice forms around the door of the freezer compartment of a refrigerator when water vapor undergoes **deposition**.

sublimation transformation of a solid directly into a vapor (gas).

deposition transformation of a vapor (gas) directly into a solid.

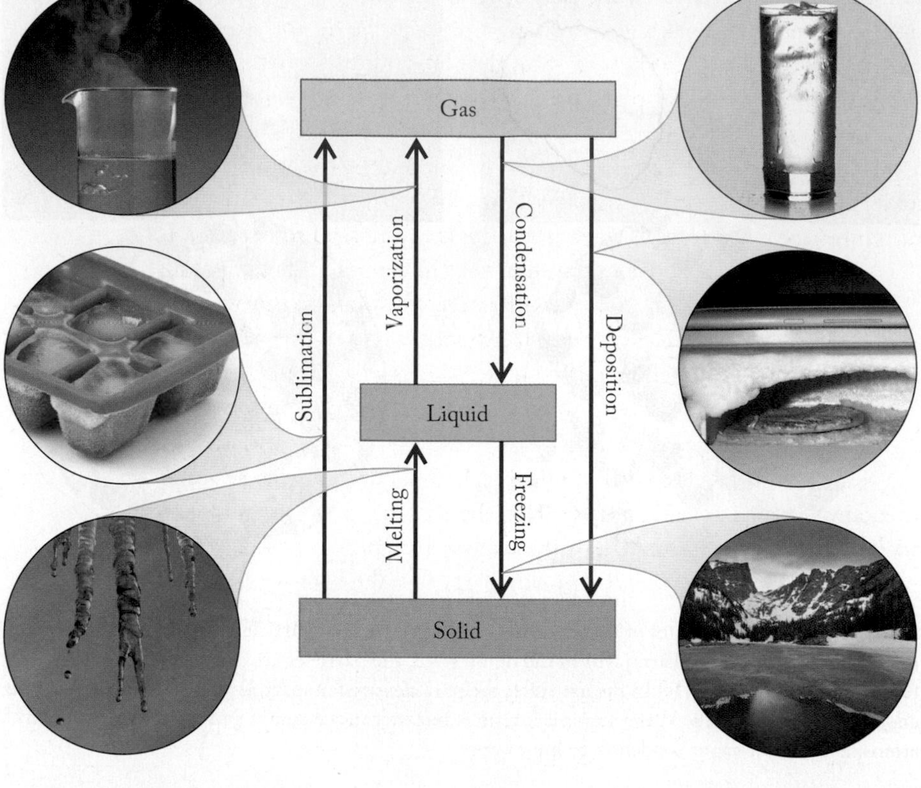

FIGURE 1.2 Matter changes from one state to another when energy is either added or removed. Arrows pointing upward represent transformations that require adding energy; arrows pointing downward represent transformations that release energy.

CONCEPT TEST

Ice cream vendors often use dry ice, which is solid carbon dioxide (CO_2), to keep their ice cream frozen. Over time, the dry ice disappears as solid CO_2 turns into CO_2 gas.

a. What is the name of this change in physical state?
b. What is the name of the reverse process in which dry ice is produced from CO_2 gas?

(Answers to Concept Tests are in the back of the book.)

energy the ability to do work.

work the energy required to move an object through a given distance.

heat a flow of energy from one object or place to another due to differences in the temperatures of the objects or places.

chemical reaction the conversion of one or more substances into one or more different substances.

potential energy (PE) the energy stored in an object because of its position or composition.

kinetic energy (KE) the energy of an object in motion due to its mass (m) and its speed (u): $KE = \frac{1}{2}mu^2$.

law of conservation of energy the principle that energy cannot be created or destroyed, but can be changed from one form to another.

1.2 Forms of Energy

All the changes of state shown in Figure 1.2 are accompanied by transfers of energy. The directions of the transfers are indicated by the arrows in the figure. The upward-pointing arrows represent changes that require the addition of energy; those represented by downward-pointing arrows release energy. In this section we explore the different forms of energy and the ways in which energy is transferred from one object or place to another. Let's begin by defining what exactly we mean by *energy*.

Energy is a word that has many meanings and is used in many different ways. However, in the physical sciences **energy** is narrowly defined as, simply, the ability to do work. In this context **work** (w) means the exertion of a force (F) through a distance (d). We can express this definition in the form of an equation:

$$w = F \times d \tag{1.1}$$

An example of physical work is shown in Figure 1.3, where a group of football linemen pushes a blocking sled down a practice field. The work they are doing is the product of the force they exert on the sled times the distance the linemen push the sled. Doing work is one of the processes by which energy is transferred—in this case, from the linemen to the sled. Another kind of energy transfer is the spontaneous flow of energy from a warm object to a cooler one, which is defined as **heat**. Thus, heat is energy that is moving from place to place because of a difference in temperatures. The term *thermal energy* is often used to describe that portion of the total energy of an object or system that increases as its temperature increases.

As this is a chemistry book, we will primarily focus on the energy that is released or absorbed during **chemical reactions**, that is, when one or more substances are converted into one or more different substances. When energy is produced in a chemical reaction, it may be lost via the transfer of heat, or it may be harnessed to do work. For example, the energy that powers our bodies is derived from chemical reactions fueled by glucose (blood sugar). The chemical energy stored in glucose is an example of **potential energy (PE)**, the energy stored in an object because of its position or composition. When the energy in glucose is released during vigorous exercise, like running the 100-meter dash (Figure 1.4), part of it is transformed into **kinetic energy (KE)**, which is the energy of motion. The remainder of it does other work, like pumping blood through the sprinter's circulation system, and some of it is given off as heat. These transformations are consistent with the **law of conservation of energy**, which states that energy cannot be created or destroyed. It can, however, be converted from one form to another.

The quantity of kinetic energy in the body of an Olympic sprinter (or any moving object) is proportional to the product of its mass (m) and the square of its speed (u):

$$KE = \tfrac{1}{2} m u^2 \tag{1.2}$$

This equation confirms something you probably already knew: a heavy object has more kinetic energy than a lighter one moving at the same speed. Similarly, two objects having the same mass but traveling at different speeds have different kinetic energies.

At the molecular level, particles of matter can have significant energies depending on their temperature and physical state. What they lack in mass they make up for in speed. For example, the tiny molecules of oxygen in the air you are breathing are moving around at supersonic speeds. The microscopic views of the three states of water in Figure 1.1 give us a clue about what happens when we add energy to ice by heating it: doing so increases the kinetic energy of the water molecules. The particles have limited kinetic energy when they are locked in place in ice, but they have more when they can tumble over each other in liquid water—and they have a lot more when they are free to move as individual particles of matter in gaseous steam.

$w = F \times d$

FIGURE 1.3 Football linemen doing work: pushing a blocking sled during practice.

FIGURE 1.4 Olympic sprinters converting chemical energy into kinetic energy.

CONCEPT TEST ···

If the speed of a vehicle doubles, by what factor does its kinetic energy increase?

(Answers to Concept Tests are in the back of the book.)

pure substance matter that cannot be separated into simpler matter by a physical process.

physical process a transformation of a sample of matter, such as a change in its physical state, that does not alter the chemical identity of any substance in the sample.

element a pure substance that cannot be separated into simpler substances by any chemical process.

compound a pure substance that is composed of two or more elements linked together in fixed proportions and that can be broken down into those elements by some chemical process.

chemical formula a notation for representing the elemental composition of a pure substance using the symbols of the elements; subscripts indicate the relative number of atoms of each element in the substance.

law of constant composition the principle that all samples of a particular compound always contain the same elements combined in the same proportions.

1.3 Classes of Matter

Matter in any of the three physical states is classified based on its composition (Figure 1.5). The two principal classes of matter are *pure substances* and *mixtures*. A **pure substance** has a constant composition that does not vary from one sample of it to another. For example, the composition of pure water is always the same, no matter what its source, because water molecules always have the same structure and composition: two atoms of hydrogen bonded to a single atom of oxygen. Pure substances cannot be separated into simpler forms of matter by any physical process. By **physical process**, we mean a transformation of a substance (or group of them) that does not change its chemical identity, that is, does not change the structure and composition of the particles that make it up.

Pure substances are further subdivided into two groups: *elements* and *compounds*. An **element** is a pure substance that cannot be separated into simpler substances. The periodic table inside this book's front cover includes all the known elements. They are the basic building blocks of matter. However, only a few elements including sulfur (S) and gold (Au) are found in nature in their elemental states, that is, as pure (or nearly pure) substances that contain only one element (Figure 1.6). Instead, most elements occur chemically combined with other elements in the form of compounds.

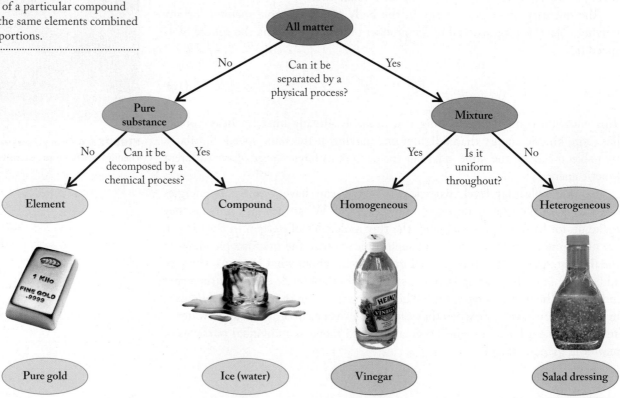

FIGURE 1.5 Matter is classified as shown in this diagram. The two principal categories are pure substances and mixtures. A substance may be a compound (such as water) or an element (such as gold). When the substances making up a mixture are distributed uniformly, as they are in vinegar (a mixture of acetic acid and water), the mixture is homogeneous. When the substances making up a mixture are not distributed uniformly, as when solids are suspended in a liquid and then settle to the bottom of the container, the mixture is heterogeneous.

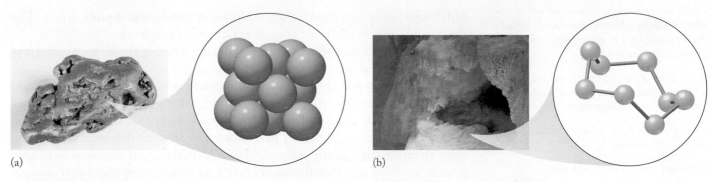

FIGURE 1.6 (a) Gold and (b) sulfur are among the few elements that occur in nature uncombined with other elements.

A **compound** is a pure substance that consists of two or more elements that cannot be physically separated from each other. The elements in a compound are present in characteristic and definite proportions that reflect the composition of the particles that make up the compound. A compound's composition is described by its **chemical formula**, which contains the symbols of the elements in the compound with subscripts to indicate the proportions of the elements. The chemical formula of water, H_2O, reflects the fact that its molecules each contain two atoms of hydrogen and one of oxygen. This atomic view of the composition of water explains a phenomenon that scientists first observed centuries ago: when hydrogen gas reacts with oxygen gas to form H_2O, the two gases always combine in the same proportion: two volumes of hydrogen for every one volume of oxygen. If we reverse the process and decompose water into hydrogen and oxygen, we always obtain two volumes of hydrogen gas for every one volume of oxygen gas (Figure 1.7). This consistency illustrates the **law of constant composition**: every sample of a compound is composed of the same elements combined in the same proportions.

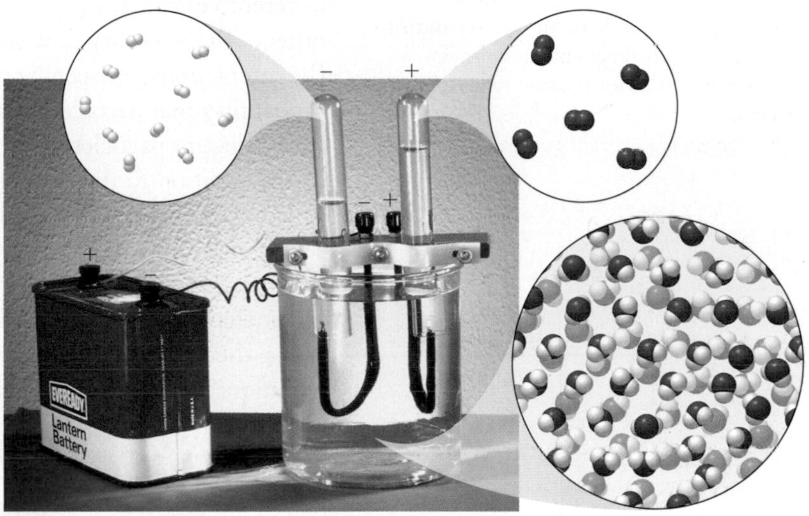

FIGURE 1.7 An electric current passed through water decomposes the water into oxygen gas and hydrogen gas. Two volumes of hydrogen gas collect above the electrode connected to the negative terminal of the battery for every one volume of oxygen above the positive electrode. Quantitative observations like this illustrate the law of constant composition.

CONCEPT TEST ••••••••••••••••••••••••••••••••

A compound with the formula NO may be present in the exhaust gases leaving a car's engine. As NO travels through the car's exhaust system, some of it decomposes into nitrogen and oxygen gas. What is the volume ratio of nitrogen to oxygen formed from NO?

(Answers to Concept Tests are in the back of the book.)

Mixtures are composed of two or more pure substances, and are classified as either *homogeneous* or *heterogeneous* mixtures. The substances in a **homogeneous mixture** are distributed uniformly, and the composition and appearance of the mixture are uniform throughout. Homogeneous mixtures are also called **solutions**, a term that scientists apply to homogeneous mixtures of gases and solids as well as liquids. In contrast, the substances in a **heterogeneous mixture** are not

mixture a combination of pure substances in variable proportions in which the individual substances retain their chemical identities and can be separated from one another by a physical process.

homogeneous mixture a mixture in which the components are distributed uniformly throughout and have no visible boundaries or regions.

solution another name for a *homogeneous mixture*. Solutions are often liquids, but they may also be solids or gases.

heterogeneous mixture a mixture in which the components are not distributed uniformly, so that the mixture contains distinct regions of different compositions.

immiscible liquids combinations of liquids that are incapable of mixing with, or dissolving in, each other.

distillation a process using evaporation and condensation to separate a mixture of substances with different volatilities.

volatility a measure of how readily a substance vaporizes.

filtration a process for separating solid particles from a liquid or gaseous sample by passing the sample through a porous material that retains the solid particles.

chromatography a process involving stationary and mobile phases for separating a mixture of substances based on their different affinities for the two types of phases.

distributed uniformly. They include mixtures of **immiscible liquids**, which, like the oil and water in salad dressing, do not dissolve in each other.

Nearly all the forms of matter we encounter, including the air we breathe and the food and drink we consume, are mixtures. The substances in mixtures can be separated from one another based on differences in their physical properties. The differences are closely linked to how strongly the particles in each substance interact with each other and with the particles of the other substances in the mixture. For example, the water in seawater can be made drinkable by separating it from the salts (mostly sodium chloride, NaCl) that are dissolved in it. One process for doing so is **distillation** (Figure 1.8), which incorporates two changes of physical state: a component (water in this case) of a mixture is separated by evaporation, and then the resulting vapor is recovered by condensation. Distillation works as a separation technique whenever the components of a mixture have different **volatilities**, that is, one or more of them vaporize more readily than the others. The volatility of a substance is inversely proportional to the strength of the interactions between its particles: the stronger the interactions, the lower the probability that particles of the substance will have enough energy to break away from adjacent particles in the liquid phase and become particles of vapor.

In addition to dissolved salts, seawater may contain tiny particles of single-celled plants called phytoplankton (Figure 1.9). Scientists who study these microscopic algae separate them from seawater by **filtration**, which works in this case because the phytoplankton cells, which are many micrometers in diameter, are larger than the pores in the filter (Figure 1.10). The cells are trapped on the filter while the water and dissolved salts readily pass through. Filtration is a useful technique for removing particles that are suspended in gases as well as liquids.

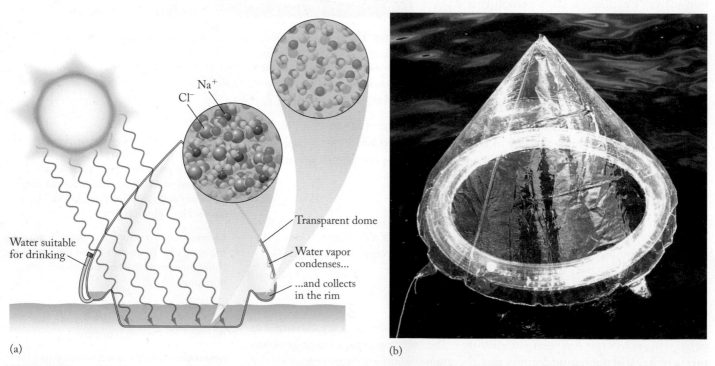

Na⁺

Cl⁻

Transparent dome

Water vapor condenses...

...and collects in the rim

Water suitable for drinking

(a)

(b)

FIGURE 1.8 (a) In a solar-powered distillation apparatus used in survival gear to provide freshwater from seawater, sunlight passes through the transparent dome and heats a pool of seawater. Water vapor rises from the pool, contacts the inside of the transparent dome (which is relatively cool), and condenses. The distilled water collects in the depression around the rim and then passes into the attached tube, from which one may drink it. (b) A solar-powered distillation apparatus in use.

(a)   (b)

FIGURE 1.9 (a) Very high concentrations (called "blooms") of phytoplankton known as *coccolithophores* in the Bering Sea were photographed by a NASA satellite. During intense blooms, coccolithophores turn the color of the ocean a milky aquamarine. The color comes from chlorophyll and other pigments in the phytoplankton; the milkiness comes from sunlight scattering off the organisms' textured exterior shells. (b) A single coccolithophore cell.

For example, the air in hospitals and in the clean rooms of laboratories is typically purified using HEPA (high-efficiency particulate air) filters to remove dust, bacteria, and even viruses.

The components of the phytoplankton caught on the filter in Figure 1.10 can be further separated by soaking the wet filter in acetone (Figure 1.11a), which dissolves some of the compounds present inside the phytoplankton cells. These compounds include chlorophyll and other pigments that have distinctive colors. Once dissolved in acetone, the pigments can be separated from each other using another separation technique called **chromatography** (Figure 1.11b). In a chromatographic separation, the components of a liquid or gaseous mixture are distributed between two phases: one of them is a stationary solid (or liquid-coated solid), and the other is a moving liquid or gas. The more strongly the particles in the mobile phase interact with the stationary solid, the slower they move. Those that interact weakly or not at all with the stationary solid move more rapidly.

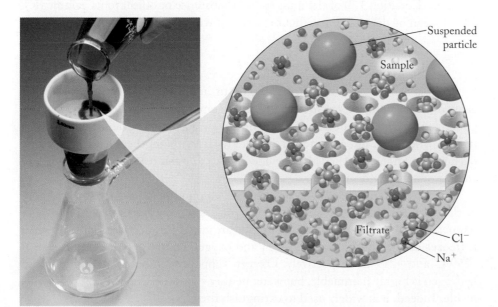

Suspended particle

Sample

Filtrate

Cl⁻

Na⁺

FIGURE 1.10 Particles suspended in a liquid, such as a culture of phytoplankton in seawater, can be separated from the liquid by filtration. (The suspended particles and filter pores are not drawn to scale; in fact, they are many thousands of times larger than the water molecules and ions in seawater.)

(a)

(b)

FIGURE 1.11 (a) Chlorophyll and other pigments can be extracted from a phytoplankton sample trapped on a filter using the solvent acetone. (b) Dissolved pigments can be separated from one another using chromatography, producing a characteristic pattern of colored bands.

intensive property a property that is independent of the amount of substance present.

extensive property a property that varies with the amount of substance present.

physical property a property of a substance that can be observed without changing the substance into another substance.

density (d) the ratio of the mass (m) of an object to its volume (V).

chemical property a property of a substance that can be observed only by reacting the substance chemically to form another substance.

CONCEPT TEST ••

Which physical process—distillation, filtration, or chromatography—would you use to perform each of the following tasks?

    a. Removing particles of rust from drinking water.
    b. Separating the different coloring agents in a sample of ink.
    c. Separating volatile compounds normally found in natural gas that have dissolved in a sample of crude oil.

(Answers to Concept Tests are in the back of the book.)

••

1.4 Properties of Matter

Pure substances have distinctive properties. Pure gold, for example, has a distinctive color, is soft for a metal, is malleable (it can be hammered into very thin sheets called gold leaf), is ductile (it can be drawn out into thin wires), and melts at 1064°C. Those are all examples of **intensive properties**, properties that characterize a pure substance independent of the quantity of the material present. On the other hand, an ingot of gold has a particular length, width, mass, and volume. Those properties of a particular sample of a pure substance, which depend on how much of the substance is present, are **extensive properties**.

CONCEPT TEST ••

Which of these properties of a sample of pure iron are intensive: (a) its mass, (b) its density, (c) its volume, or (d) its hardness?

(Answers to Concept Tests are in the back of the book.)

••

    The properties of substances fall into two other general categories: *physical* and *chemical*. **Physical properties** can be observed or measured without changing the substance into another substance. The properties of gold described in the previous paragraph are all physical properties. Another example is **density (d)**, which is the ratio (Equation 1.3) of the mass (m) of a substance or object to its volume (V). Density can be determined without reacting it with another substance.

$$d = \frac{m}{V} \tag{1.3}$$

    Gold is one of the few elements that occur in nature uncombined with other elements. The tendencies of most elements to combine with others and form compounds represent the **chemical properties** of the elements. Chemical properties include whether or not a particular element reacts with another element or with a particular compound. They also include *how rapidly* the reactions take place and what products are formed.

    The physical and chemical properties of a compound are different from those of the elements that combine to form it. For example, water is a liquid at room temperature, whereas hydrogen and oxygen are gases. Liquid water expands when it freezes at 0°C, but liquid hydrogen and oxygen contract when they freeze at −259°C and −219°C, respectively. Oxygen supports combustion reactions, and hydrogen is highly flammable, but water neither supports combustion nor is flammable. Indeed, it is widely used to extinguish fires.

Which of the following properties of gold are chemical and which are physical?

  a. Gold reacts with a mixture of nitric and hydrochloric acids known as aqua regia.
  b. Gold melts at 1064°C.
  c. Gold can be hammered into sheets so thin that light passes through them.
  d. Gold can be recovered from gold ore by treating the ore with a solution containing cyanide, which reacts with and dissolves gold.

(Answers to Concept Tests are in the back of the book.)

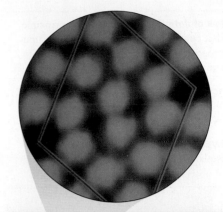

1.5 Atomic Theory: The Scientific Method in Action

The natural philosophers of ancient Greece believed in atoms but lacked the technology to test their belief. Today, we have that technology. Consider the images in Figure 1.12. On the bottom is a photograph of silicon (Si) wafers, the material used to make computer chips and photovoltaic cells. The magnified view above it is a photomicrograph of a Si wafer produced by an instrument called a scanning tunneling microscope, or STM.[1] The fuzzy spheres are individual atoms of silicon—the smallest representative particles of silicon. If, for example, you were to grind a sample of pure silicon into the finest dust imaginable, there would be a limit to how tiny a particle of the dust you could produce and still have a particle of silicon. That limit would be an atom of silicon. (Of course, no grinding tool exists that can produce particles anywhere near the size of individual atoms, but *if* such a grinder existed, the smallest particles it could make that would still be particles of elements would be atoms.)

FIGURE 1.12 Silicon wafers are widely used to make computer chips and photovoltaic cells for solar panels. Since the 1980s, scientists have been able to image individual atoms using an instrument called a scanning tunneling microscope (STM). In the STM image (top), the fuzzy spheres are individual silicon atoms. The radius of each atom is 117 picometers (pm), or 117 trillionths of a meter. Atoms are the tiniest particles of silicon that still retain the chemical characteristics of silicon.

Laws and Theories

Scanning tunneling microscopes have been used to image atoms since the early 1980s, but, as noted before, belief in the existence of atoms dates back at least to the 5th century BCE, and probably earlier. However, belief in atoms would not become a *scientific* theory for another 2000 years. To be a scientific theory, it needed to be systematically tested and, if necessary, refined, following an approach to investigating and understanding natural phenomena called the **scientific method**.

One of the earliest descriptions of the scientific method was published in 1620 by the English philosopher Francis Bacon (1561–1626). In his book *Novum Organum* (*New Organ* or *New Instrument*), Bacon described a way to systematically acquire knowledge and understanding of the natural world through a process that begins with observation and experimentation. The results of observations and experiments are then carefully examined. Scientists look for patterns and trends, trying to determine how different factors might have influenced the results. Sometimes the patterns and relationships can be generalized and used to express a fundamental scientific principle. When this expression, which can be verbal or in the form of an equation, is both concise and generally applicable, it is called a **scientific law**.

scientific method an approach to acquiring knowledge based on observation of phenomena, development of a testable hypothesis, and additional experiments that test the validity of the hypothesis.

scientific law a concise and generally applicable statement of a fundamental scientific principle.

[1] German physicist Gerd Binnig (b. 1947) and Swiss physicist Heinrich Rohrer (b. 1933) shared the 1986 Nobel Prize in Physics for their development of scanning tunneling microscopy.

law of definite proportions the principle that compounds always contain the same proportions of their component elements; equivalent to the *law of constant composition.*

scientific theory a general explanation of widely observed phenomena that has been extensively tested.

hypothesis a tentative and testable explanation for an observation or a series of observations.

Let's consider a scientific law that is linked to the notion that matter is composed of atoms. It is called the **law of definite proportions** and was first articulated by French chemist Joseph Louis Proust (1754–1826). Proust's research into the composition of compounds, particularly those containing metals and oxygen, led him to conclude that compounds always contain the same proportions of their component elements. Note the similarity between Proust's law of definite proportions and the law of constant composition (a compound always has the same *elemental composition*).

When Proust published his law of definite proportions, some of the leading chemists of the time refused to believe it. Their own experiments seemed to show, for example, that the compound that tin forms with oxygen could have variable tin content. These scientists did not realize that the samples they were synthesizing and analyzing were probably not *pure* compounds, but rather mixtures, which could have variable composition. Still, acceptance of Proust's law not only required corroborating results from other scientists but also needed a *theoretical* basis: a convincing argument that explained *why* the composition of a compound is always the same. Such an argument, in the form of a concise explanation of why a natural phenomenon is observed or why a scientific law is true, is called a **scientific theory**.

Note the complementary nature of scientific laws and theories: scientific laws describe natural phenomena and relationships; scientific theories explain *why* phenomena happen or *why* relationships are true. Theories usually start out as tentative explanations of why a set of experimental results was obtained, or why a particular phenomenon is consistently observed. Such a tentative explanation is called a **hypothesis**. An important feature of a hypothesis is that it can be tested through additional observations and experiments. Another feature is that it enables scientists to accurately predict the results of future experiments and observations. Further testing and observation might support a hypothesis or disprove it, or perhaps require that it be modified. A hypothesis that withstands the tests of many experiments and accurately predicts the results of further observations and experimentation may be elevated to the rank of a scientific theory.

A scientific theory explaining Proust's law of definite proportions was developed by an English contemporary of Proust, John Dalton (1766–1844). Whereas Proust studied the composition of the solid compounds formed by metals and oxygen, Dalton's own research focused on the behavior and composition of gases. Like Proust, Dalton observed that when two elements combine to form compounds, they may form two or more compounds with different compositions. For example, Proust discovered that tin (Sn) and oxygen combine to form two compounds: the composition by mass of one of them is 88.1% Sn and 11.9% O; the other is 78.7% Sn and 21.3% O. Dalton noted that the ratio of oxygen to tin in the first compound:

$$\frac{11.9\% \text{ O}}{88.1\% \text{ Sn}} = 0.135$$

is very close to half that in the second compound:

$$\frac{21.3\% \text{ O}}{78.7\% \text{ Sn}} = 0.271$$

Similar results were obtained with other sets of compounds formed by pairs of elements. Sometimes their compositions would differ by a factor of 2 as with oxygen and tin (and with oxygen and carbon). For other pairs of elements the compositions differed by other factors, but in all cases they differed *by ratios of small numbers*. This pattern led Dalton to formulate the **law of multiple proportions**: when two elements combine to make two (or more) compounds, the ratio of the masses of the first element, which combine with a given mass of the second element, is a ratio of small whole numbers. For example, 15 grams of oxygen may combine with 10 grams of sulfur; or, under different reaction conditions, only 10 grams of oxygen may combine with 10 grams of sulfur:

$$\frac{15 \text{ g oxygen}}{10 \text{ g oxygen}} = \frac{3}{2}$$

This is indeed the ratio of two small whole numbers.

The law of multiple proportions led Dalton to conclude that elements combine this way because *they are composed of atoms*. Thus, the factor-of-2 difference in the proportion of oxygen to tin in Proust's compounds can be explained this way: the compound with the O:Sn ratio of 0.135 contains 1 atom of oxygen for each atom of tin, but the compound with 2 times that ratio contains 2 atoms of O per atom of Sn. Those atomic ratios are reflected in the chemical formulas of the two compounds: SnO and SnO_2. As in all chemical formulas, the lack of subscripts after Sn and O in SnO, and after Sn in SnO_2, means 1 atom of each of those elements: there is 1 atom of Sn for every atom of O in SnO, and 1 atom of Sn for every 2 atoms of O in SnO_2. The same logic applies to the two compounds formed by oxygen and sulfur. For them, the ratio of 3:2 translates into the chemical formulas SO_3 and SO_2.

1.6 A Molecular View

As we noted in Section 1.1, a molecule is a collection of atoms held together in a characteristic pattern and proportion by chemical bonds. Figure 1.1 provided several molecular views of water molecules. Figure 1.13(a) features molecules of another common molecular compound, carbon dioxide gas. Note how each is composed of a central carbon atom, represented by the black spheres, that is bonded to two atoms of oxygen, the red spheres. (In the inside back cover of this book, our Atomic Color Palette shows the standard colors used to represent atoms of the elements we study most often.) Some pure elements also exist as molecules—molecules made of only one kind of atom. The molecular views of hydrogen, nitrogen, and oxygen gas in Figure 1.13(b) show that these elements exist as *diatomic* (two-atom) molecules: H_2, N_2, and O_2, respectively. The elements in column 17 of the periodic table also form diatomic molecules: F_2, Cl_2, Br_2, and I_2.

The chemical formulas of molecular compounds are also called *molecular formulas*. The element symbols and subscripts in a **molecular formula** indicate how many atoms of each element are present in one molecule of the compound. The molecular formulas of acetone, a common solvent, and acetic acid, the ingredient that gives vinegar its distinctive aroma and taste, are given in Figure 1.14(a). These two formulas provide information about the number of atoms of carbon, hydrogen, and oxygen in a molecule of each compound, but they do not tell us how the atoms are bonded together, nor do they tell us anything about the shape of the molecules. One way to show how they are connected is to draw a structure that includes the chemical bonds between atoms. Such a drawing is called a **structural formula** (Figure 1.14b).

law of multiple proportions the principle that, when two masses of one element react with a given mass of another element to form two different compounds, the two masses of the first element have a ratio of two small whole numbers.

molecular formula a chemical formula that shows how many atoms of each element are in one molecule of a pure substance.

structural formula a representation of a molecule that uses short lines between the symbols of elements to show chemical bonds between atoms.

(a)   (b)

FIGURE 1.13 Molecular views of (a) pure carbon dioxide and (b) a mixture of hydrogen molecules (pairs of white spheres), oxygen molecules (pairs of red spheres), and nitrogen molecules (pairs of blue spheres).

FIGURE 1.14 Five ways to represent the arrangement of atoms in molecules of acetone and acetic acid (a principal ingredient in vinegar): (a) molecular formulas; (b) structural formulas; (c) condensed structural formulas; (d) ball-and-stick models, where white spheres represent hydrogen atoms, black spheres represent carbon atoms, and red spheres represent oxygen atoms; (e) space-filling models.

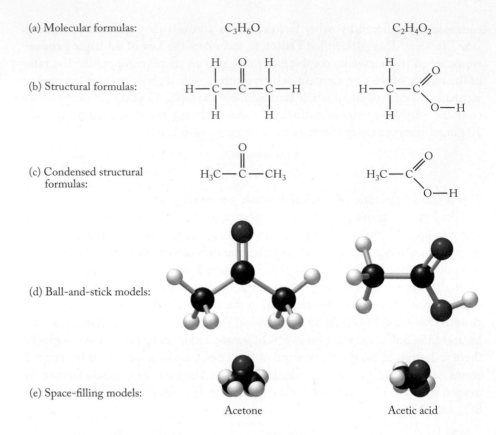

(a) Molecular formulas: C_3H_6O $C_2H_4O_2$

(b) Structural formulas:

(c) Condensed structural formulas:

(d) Ball-and-stick models:

(e) Space-filling models:

Acetone   Acetic acid

Sometimes the structures of molecules are represented using *condensed* structural formulas (Figure 1.14c) in which the symbols of elements appear in a pattern that shows how the atoms are arranged relative to one another, but some common structural components, such as C—H bonds and C—C bonds, are omitted. For example, the structural formula of acetic acid tells us that three of the four hydrogen atoms in each molecule are bonded to the same carbon atom. The same information is conveyed in the condensed structural formula by grouping three H atoms before the first C atom in the formula (H_3C–). Overall, the condensed structural formula of acetone more quickly tells us that there is a group of three H atoms bonded to each outer C atom, and that these outer carbon atoms are bonded to a central carbon atom that is also bonded to an oxygen atom.

Ball-and-stick models (Figure 1.14d) provide three-dimensional views of molecules. They use balls to represent atoms and sticks to represent chemical bonds. An advantage of ball-and-stick models is that they accurately show the angles between the bonds in a molecule. A disadvantage is that the sticks make the atoms seem far apart although they actually overlap each other. *Space-filling* models (see Figures 1.13 and 1.14e) more accurately show us how the atoms are arranged in a molecule and its overall three-dimensional shape. However, it is sometimes hard to see all the atoms and the angles between the bonds in space-filling models, especially in those with many atoms.

CONCEPT TEST ●

Figure 1.15 is the space-filling model of methanol (also called methyl alcohol and wood alcohol). What is its molecular formula?

(Answers to Concept Tests are in the back of the book.)

FIGURE 1.15 Space-filling model of methanol.

1.7 COAST: A Framework for Solving Problems

Throughout this book, you will find Sample Exercises designed to help you better understand chemical concepts and develop your problem-solving skills. Each Sample Exercise follows a systematic approach to problem solving that we encourage you to apply to the Practice Exercises that follow each Sample Exercise and to the end-of-chapter problems. We use the acronym COAST (**C**ollect and **O**rganize, **A**nalyze, **S**olve, and **T**hink about the answer) to represent the four steps in our approach to problem solving. As you read about it here and use it later, keep in mind that COAST is merely a *framework* for solving problems, not a recipe. Use it as a guide to developing your own approach to solve each problem.

COLLECT AND ORGANIZE First start by sorting through the information given in the problem and identifying other relevant information. These actions help you understand the problem, including the fundamental chemical principles on which it is based. As part of the collecting and organizing process, you:

- ➤ Identify the key concept of the problem.

- ➤ Identify and define the key terms used to express that concept. You may find it useful to restate the problem in your own words.

- ➤ Sort through the information given in the problem, separating what is pertinent from what is not.

- ➤ Assemble any supplemental information that may be needed, including equations, definitions, and constants.

ANALYZE The next step is to analyze the information you have collected to determine how to relate it to the answer you seek. Sometimes it is easier to work backward to create the relationships: consider the nature of the answer first, and think about how you might get to it from the information provided in the problem and other sources. If the problem is quantitative and requires a numerical answer, frequently the units of the initial values and the final answer will help you identify how they are connected and which equation (or equations) may be useful. This step may include rearranging equations to solve for an unknown or setting up conversion factors. For some problems, drawing a sketch based on molecular models or an experimental setup may help you visualize how the starting points and final answer are connected. You should also look at the numbers involved and estimate your answer. Having an order-of-magnitude ("ballpark") estimate of your final answer before entering numbers into your calculator provides a check on the accuracy of your calculated answer.

SOLVE For most conceptual questions, the solution flows directly from your analysis of the problem. To solve quantitative problems, you need to insert the starting values and appropriate constants into the relevant equations or conversion factors and calculate the answer. In the Solve step, make sure that units are consistent and cancel as needed, and that the certainty of the quantitative information is reflected in how many *significant figures* (see Section 1.8) you used.

THINK ABOUT IT Finally, you need to think about your result and answer such questions as: Does this answer make sense based on my own experience and based on what I have just learned? Is the value for a quantitative answer reasonable—is it close to my estimate from the Analyze step? Are the units correct and the number of significant figures appropriate? Then ask yourself how confident you are that you could solve another problem, perhaps drawn from another context but based on the same chemical concept. You may also think about how this problem relates to other observations you may have made about matter in your daily life.

The COAST approach should help you solve problems in a logical way and avoid certain pitfalls, such as grabbing an equation that seems to have the right variables and simply plugging numbers into it, or resorting to trial and error. As you study the steps in each Sample Exercise, try to answer these questions about each step:

> ➤ **What** is done in this step?

> ➤ **How** is it done?

> ➤ **Why** is it done?

After answering the questions, you will be ready to solve the Practice Exercises and end-of-chapter problems in a systematic way.

1.8 Making Measurements and Expressing the Results

Advances in scientific inquiry in the 18th century, including those which led to the atomic theory of matter, brought about a heightened awareness of the need for accurate measurements and an international system of units for expressing the results of those measurements. In 1791 French scientists proposed a standard unit of length, which they called the **meter (m)** after the Greek *metron*, which means "measure." They based the length of the meter on 1/10,000,000 of the distance along an imaginary line running from the North Pole to the equator. By 1794 hard work by teams of surveyors had established the length of the meter that is still used today.

The French scientists also settled on a decimal-based system for designating lengths that are multiples or fractions of a meter (Table 1.1). They chose Greek prefixes such as *kilo-* for lengths much greater than 1 meter (1 kilometer = 1000 m) and Latin prefixes such as *centi-* for lengths much smaller than a meter (1 centimeter = 0.01 m).

Since 1960 scientists have by international agreement used a modern version of the French *metric* system of units: the *Système International d'Unités*, commonly abbreviated SI. Table 1.2 lists six SI base units; many others are derived from them. For example, a common SI unit for volume, the cubic meter (m^3), is derived from the base unit for length, the meter. A common SI unit for speed, meters per second (m/s), is derived from the base units for length and time. The SI unit for energy is the **joule (J)**, which is equivalent to $1 \text{ kg} \cdot (\text{m/s})^2$. This equivalency makes sense given the relationship we saw in Section 1.2 between kinetic energy and mass and speed: $KE = \frac{1}{2}mu^2$ (Equation 1.2). When mass (m) is expressed in kilograms and speed (u) is in meters per second, then the units of KE are $\text{kg} \cdot (\text{m/s})^2$, or joules.

Table 1.3 contains some of the SI units and their equivalents in the U.S. Customary System of units. They include the volume corresponding to 1 cubic decimeter (a cube 1/10 meter on a side), which we call a liter (L). A more extensive list of units and their equivalents is on the inside back cover of this book.

Modern science requires that the length of the meter, as well as the dimensions of other SI units, be known or defined by quantities that are much more

meter (m) the standard unit of length, named after the Greek *metron*, which means "measure"; equivalent to 39.37 inches.

joule (J) the SI unit of energy, equivalent to $1 \text{ kg} \cdot (\text{m/s})^2$.

constant and precisely known than, for example, the distance from the North Pole to the equator. Two such quantities are the speed of light (c) and time. In 1983, 1 m was redefined as the distance traveled in 1/299,792,458 of a second in a vacuum by the light emitted from a helium–neon laser. This modern definition of the meter is consistent with the one adopted in France in 1794.

TABLE 1.1 Commonly Used Prefixes for SI Units

PREFIX		VALUE	
Name	Symbol	Numerical	Exponential
zetta	Z	1,000,000,000,000,000,000,000	10^{21}
exa	E	1,000,000,000,000,000,000	10^{18}
pcta	P	1,000,000,000,000,000	10^{15}
tera	T	1,000,000,000,000	10^{12}
giga	G	1,000,000,000	10^{9}
mega	M	1,000,000	10^{6}
kilo	k	1,000	10^{3}
hecto	h	100	10^{2}
deka	da	10	10^{1}
deci	d	0.1	10^{-1}
centi	c	0.01	10^{-2}
milli	m	0.001	10^{-3}
micro	μ	0.000001	10^{-6}
nano	n	0.000000001	10^{-9}
pico	p	0.000000000001	10^{-12}
femto	f	0.000000000000001	10^{-15}
atto	a	0.000000000000000001	10^{-18}
zepto	z	0.000000000000000000001	10^{-21}

TABLE 1.2 SI Base Units

Quantity or Dimension	Unit Name	Unit Abbreviation
Mass	kilogram	kg
Length	meter	m
Temperature	kelvin	K
Time	second	s
Electric current	ampere	A
Quantity of a substance	mole	mol

TABLE 1.3 Conversion Factors for SI and Other Commonly Used Units

Quantity or Dimension	Equivalent Units
Mass	1 kg = 2.205 pounds (lb); 1 lb = 0.4536 kg = 453.6 g
	1 g = 0.03527 ounce (oz); 1 oz = 28.35 g
Length (distance)	1 m = 1.094 yards (yd); 1 yd = 0.9144 m (exactly)
	1 m = 39.37 inches (in); 1 foot (ft) = 0.3048 m (exactly)
	1 in = 2.54 cm (exactly)
	1 km = 0.6214 miles (mi); 1 mi = 1.609 km
Volume	1 m^3 = 35.31 ft^3; 1 ft^3 = 0.02832 m^3
	1 m^3 = 1000 liters (L) (exactly)
	1 L = 0.2642 gallon (gal); 1 gal = 3.785 L
	1 L = 1.057 quarts (qt); 1 qt = 0.9464 L

precision the extent to which repeated measurements of the same variable agree.

accuracy agreement between an experimental value and the true value.

significant figures all the certain digits in a measured value plus one estimated digit. The greater the number of significant figures, the greater the certainty with which the value is known.

Precision and Accuracy

All scientific measurements have one thing in common: there is a limit to how well we can know the results. Nobody is perfect, and no analytical method is perfect either. Every method is inherently limited in its ability to produce accurate results. Two terms—precision and accuracy—are widely (though sometimes inappropriately) used to describe how well a measured quantity or a value calculated from a measured quantity is known. **Precision** indicates how repeatable a measurement is. Suppose we use the balance on the top in Figure 1.16 to determine the mass of a penny over and over again, and suppose the reading on the balance is always 2.53 g. These results tell us that the mass of the penny is precisely 2.53 g. We can also say that we are certain of its mass to the nearest 0.01 g.

Now suppose we use the balance on the bottom in Figure 1.16 to determine the mass of the same penny five more times and obtain these values:

Measurement	Mass (g)
1	2.5270
2	2.5271
3	2.5272
4	2.5271
5	2.5271

Note that there is a small variability in the last decimal place. Such variability is not unusual when using a balance that can report masses to the nearest 0.0001 g. These results are quite consistent with one another, so we can say that the balance is precise. Many other factors—particles of dust landing on the balance, vibration of the laboratory bench, or the transfer of moisture from our fingers to the penny as we handle it—could have produced a change in mass of 0.0001 g or more.

One way to express the precision of these results is to cite the range between the highest and lowest values—in this example, 2.5270 to 2.5272. Range can also be expressed using the average value (2.5271 g) and the range above (0.0001) and below (0.0001) the average that includes all the observed results. A convenient way to express the observed range in this case is 2.5271 ± 0.0001, where the symbol ± means "plus or minus" the value that follows it.

While precision relates to the agreement among repeated measurements, **accuracy** reflects how close the measured value is to the true value. Suppose the true mass of our penny is 2.5267 g. That means the average result obtained with the bottom balance in Figure 1.16—2.5271 g—is 0.0004 g too high. Thus the measurements made on this balance may be precise to within 0.0001 g, but they are not accurate to within 0.0001 g of the true value. A way to visualize the difference between accuracy and precision is presented in Figure 1.17.

How can we be sure that the results of a measurement are accurate? The accuracy of a balance can be checked by weighing objects of known mass. A thermometer can be calibrated by measuring the temperature at which a substance changes state. For example, an ice-water bath should have a temperature of 0.0°C. At sea level, liquid water boils at 100.0°C, which means an accurate thermometer dipped into boiling water reads that temperature. A measurement that is validated by calibration with an accepted standard material is considered accurate.

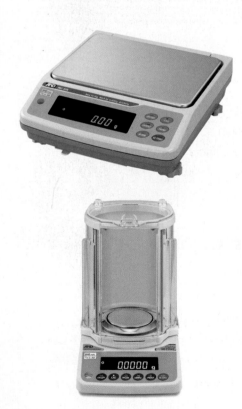

FIGURE 1.16 The mass of a penny can be measured to the nearest 0.01 gram with the balance on the top and to the nearest 0.0001 gram with the balance on the bottom.

Significant Figures

Let's now explore how we can express how well we know the results of measurements or calculated values derived from measurements. Our approach makes use of **significant figures**. The number of significant figures used to report a value indicates how certain we are of the value. For example, let's return to the two balances shown in Figure 1.16. Assuming both balances are working properly, we can use the one on the top to weigh a penny to the nearest 0.01 g, obtaining a mass value of 2.53 g. The balance on the bottom can measure the mass of the same penny to the nearest 0.0001 g, or 2.5271 g. The mass obtained with the top balance has three significant figures: the 2, 5, and 3 are considered *significant*, which means we are confident in their values. The mass obtained using the bottom balance has five significant figures (2, 5, 2, 7, and 1). The mass of the penny can be determined with greater certainty with the balance on the right.

Now consider this experimental result: an aspirin tablet is placed on the balance on the bottom in Figure 1.16, and the display reads 0.0810 g. How many significant figures are there in this value? You might be tempted to say five because that is the number of digits displayed. However, the first two zeros are not considered significant because they serve only to determine the location of the decimal point. Those two zeros function the way exponents do when we express values using scientific notation. Expressing 0.0810 g using scientific notation gives us 8.10×10^{-2} g. Only the three digits in the decimal part indicate how precisely we know the value; the "−2" in the exponent does not. (See Appendix 1 for a review of how to express values using scientific notation.)

Why is the rightmost zero in 0.0810 g significant? The answer is related to the ability of the balance to measure masses to the nearest 0.0001 g. If the balance is operating correctly, we may assume that the last digit is significant. If we dropped the zero and recorded a value of only 0.081 g, we would be implying that we knew the value to only the nearest 0.001 g, which is not the case. The following guidelines will help you handle zeros (highlighted in green) in deciding the number of significant figures in a value:

1. Zeros at the beginning of a value, as in **0.0**592, are never significant. In this example, they just set the decimal place.
2. Zeros after a decimal point and after a non-zero digit, as in $3.\mathbf{00} \times 10^8$, are always significant.
3. Zeros at the end of a value that contains no decimal point, as in 96,**500**, may or may not be significant.[2] They may be there only to set the decimal place. We can use scientific notation to indicate whether or not these terminal zeros are significant. For example, expressing the above value as 9.65×10^4 indicates that the value has three significant figures and that the terminal zeros are there only to set the decimal place.
4. Zeros between nonzero digits, as in 1**0**1.3, are always significant.

(a)

(b)

(c)

FIGURE 1.17 (a) Three dart throws meant to hit the center of the target are both accurate and precise. (b) The three throws meant to hit the center of the target are precise but not accurate. (c) This set of three throws is neither precise nor accurate.

[2] Some books add decimal points after terminal zeros to indicate that the zeros are significant. We do not follow that practice. One reason is that it does not work for values in which only some of the terminal zeros are significant.

▶❚❚ **CHEMTOUR** Significant Figures

▶❚❚ **CHEMTOUR** Scientific Notation

CONCEPT TEST ••

How many significant figures are there in the values used as examples in guidelines 1, 2, and 4?

(Answers to Concept Tests are in the back of the book.)

••

Significant Figures in Calculations

Now let's consider how significant figures are used to express the results of calculations involving measured quantities. An important rule to remember is that significant figure rules should be used only at *the end of a calculation*, never on intermediate results. A reasonable guideline to follow is that one digit to the right of the last significant digit should be carried forward in all intermediate steps.

Suppose we believe that a small nugget of yellow metal is pure gold. We could test our belief by determining the mass and volume of the nugget and then calculating its density. If its density matches that of gold (19.3 g/mL), chances are good that the nugget is pure gold because very few minerals are that dense. We find that the mass of the nugget is 4.72 g and its volume is 0.25 mL. What is the density of the nugget, expressed in the appropriate number of significant figures?

Using Equation 1.3 to calculate the density (*d*) from the mass (*m*) and volume (*V*) produces the following result:

$$d = \frac{m}{V} = \frac{4.72 \text{ g}}{0.25 \text{ mL}} = 18.88 \text{ g/mL}$$

This density value appears to be slightly less than that of pure gold. However, we need to answer the question, "How well do we know the result?" The mass value is known to three significant figures, but the volume value is known only to two. At this point we need to invoke the *weak-link principle*, which is based on the idea that a chain is only as strong as its weakest link. In calculations involving measured values, this principle means that we can know the answer of a calculation only as well as we know the least well-known value used in the calculation. In calculations involving multiplication or division, the weak link is the value with the fewest significant figures. In this example the weak link is the value of the volume, because it has only two significant figures. Our final answer cannot have more than two significant figures, so we must convert 18.88 g/mL to a value with two significant figures.

To round off 18.88 g/mL to two significant figures, we drop those to the right of the two that are significant and then round up the rightmost significant digit (the "8" in "18") to 9, giving us 19 g/mL. In this example we round up because the first insignificant digit (the "8" to the right of the decimal point) is greater than 5. If the digit to the right had been less than 5, we would have rounded down, which would have meant leaving the value to the left of the decimal unchanged at 18 g/mL. In cases where the first dropped digit is 5 and there are nonzero digits to the right of the 5, we round up. If there are no nonzero digits to the right of the 5, then a good rule to follow is to round to the nearest even number.

So, is the nugget made of gold? The density of pure gold expressed to two significant figures is 19 g/mL, so based on how well we know the measured values, we may conclude that the nugget *could* be pure gold. Keep in mind that knowing only the density of a substance rarely confirms its identity. On the other hand, if other properties such as color, sheen, and malleability also match, then chances are the nugget really is made of gold.

Does our measured density value prove that the nugget is pure gold, or does it only *fail* to prove that the nugget is *not* gold? Explain the difference between these two conclusions, and explain why you prefer one over the other.

(Answers to Concept Tests are in the back of the book.)

The weak-link principle for significant figures also applies to calculations requiring addition and subtraction. To illustrate, consider how the volume of a gold nugget might be determined. One approach is to measure the volume of water the nugget displaces. Suppose we partially fill a 100 mL graduated cylinder as shown in Figure 1.18. Note that the cylinder is graduated in milliliters. However, the space between the graduations allows us to estimate the volumes of samples to the nearest tenth of a milliliter. For example, the meniscus of the water in Figure 1.18 is aligned exactly with the 50 mL graduation. We may record the volume of water as 50.0 mL because we can read the "50" part of this value directly from the cylinder, and we can estimate that the next digit is ".0". Even though there is some uncertainty in the last digit—the true volume might actually be 49.9 mL, or perhaps 50.1—the last digit is still considered significant.

Then we place the nugget in the graduated cylinder, being careful not to splash any water out. The level of the water is now almost exactly halfway between the 58 and 59 mL graduations, and so we estimate the volume of the combined sample to be 58.5 mL. Taking the difference of the two values, we have the volume of our nugget:

$$
\begin{array}{r}
58.5 \text{ mL} \\
- 50.0 \text{ mL} \\
\hline
8.5 \text{ mL}
\end{array}
$$

The result has only two significant figures because the initial and final volumes are known to the nearest tenth of a milliliter. Therefore, we can know the difference between them to only the nearest tenth of a milliliter. In general, when measured numbers are added or subtracted, the result has the same number of digits to the right of the decimal as the measured number with the fewest digits to the right of the decimal.

Consider one more example. Suppose the average mass of a U.S. penny is 2.53 g and that a full roll of pennies contains exactly 50 of them. We use the balance on the top in Figure 1.16 to determine the mass of a roll of pennies. Using the tare feature, which enables us to correct for the mass of the wrapper, we find that the mass of just the pennies in the roll is 124.01 g. Dividing this value by the average mass of a penny and canceling out the gram units in the numerator and denominator, we get

$$
\frac{124.01 \text{ g}}{2.53 \text{ g/penny}} = 49.0 \text{ pennies}
$$

However, the roll contains only *whole numbers* of pennies. We may conclude that the wrapper holds exactly 49 pennies, and that we have to add one more to make the roll complete.

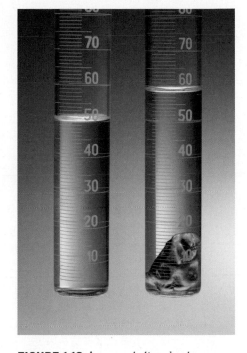

FIGURE 1.18 A nugget believed to be pure gold is placed in a graduated cylinder containing 50.0 mL of water. The volume rises to 58.5 mL, which means that the volume of the nugget is 8.5 mL.

SAMPLE EXERCISE 1.1 Using Significant Figures in Calculations LO7

Suppose we add a penny with a mass of 2.5271 grams to 49 other pennies with a combined mass of 124.01 grams. What is the combined mass of the 50 pennies?

COLLECT AND ORGANIZE We are asked to calculate the combined mass of 49 pennies that were weighed to the nearest 0.01 g and a single penny that was weighed to the nearest 0.0001 g. To express the results of this calculation we must follow the weak-link rule: we can know a combination of measured values only as well as we know the least well-known measured value.

ANALYZE We are summing two values in this example, so the weak-link value is the one with the fewest digits to the right of its decimal point, which is the combined mass of the 49 pennies.

SOLVE Adding the mass of the 50th penny to the mass of the other 49, we have

$$\begin{array}{r} 124.01\ \text{g} \\ +\ \ 2.5271\ \text{g} \\ \hline 126.53\mathbf{71}\ \text{g} = 126.54\ \text{g} \end{array}$$

THINK ABOUT IT We can know the value of the sum to only the nearest 0.01 g, so we round 126.5371 to 126.54. We round up because the first digit to be dropped is the "7" highlighted in red.

Practice Exercise According to the rules of golf, a golf ball cannot weigh more than 45.97 grams and it must be at least 4.267 cm in diameter. The volume of a sphere is $(4/3)\pi r^3$ where r is the radius (half the diameter).

a. What is the maximum density of a golf ball, expressed in g/cm^3 to the appropriate number of significant figures?
b. Is such a golf ball more dense or less dense than water? (You may need to look up the density of water—unless you are a golfer, in which case you already know the answer.)

(Answers to Practice Exercises are in the back of the book.)

CONCEPT TEST

An adventurous thru-hiker begins hiking the Appalachian Trail (Figure 1.19) at its southern terminus at Springer Mountain, Georgia, at 7:00 AM on April 3 and completes his journey at its northern terminus on the peak of Mt. Katahdin, Maine, at 4:47 PM on August 28. According to the Appalachian Trail Conservancy, the trail is 2175 miles long.

a. What is the hiker's average speed in miles/day?
b. Which do you think is the weak link in calculating the hiker's average speed: the actual distance hiked or the time that it took?

(Answers to Concept Tests are in the back of the book.)

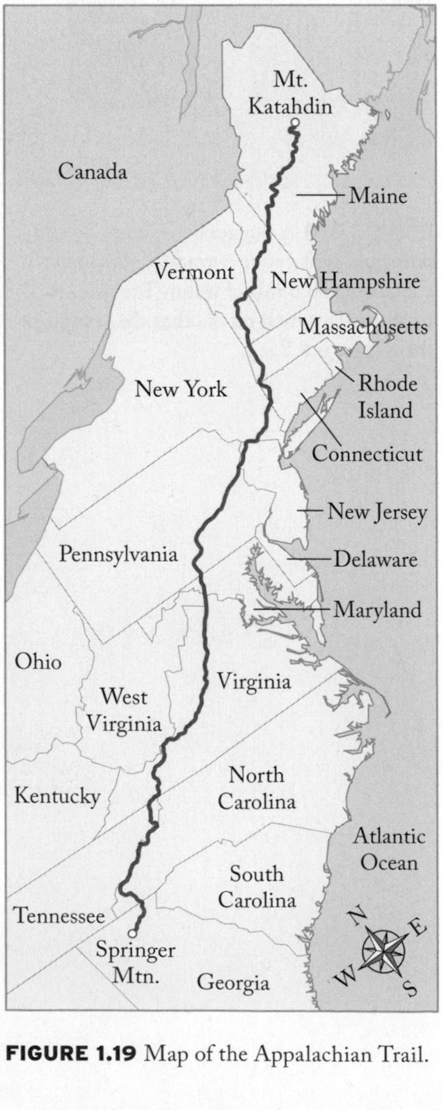

FIGURE 1.19 Map of the Appalachian Trail.

Measurements always have some degree of uncertainty, which limits the number of significant figures we can use to report any measurement. On the other hand, some quantities are known exactly because they can be determined by counting, such as the number of pennies in a roll or the number of eggs in a carton. Some other values are defined exactly, like 2.54 centimeters in an inch or 3 feet in a yard. Such values have no uncertainty and so do not influence the number of significant figures used to express the result of a calculation in which they appear.

**SAMPLE EXERCISE 1.2 Distinguishing Exact from LO7
Uncertain Values**

Which of the following quantities associated with the Washington Monument in Washington, DC (Figure 1.20), are exact values and which are not exact?

    a. The monument is made of 36,941 white marble blocks.
    b. The monument is 169 m tall.
    c. There are 893 stair steps to the top.
    d. The mass of the aluminum apex is 2.8 kg.
    e. The area of the foundation is 1487 m^2.

COLLECT AND ORGANIZE This exercise involves distinguishing between values based on an exact number, such as 12 eggs in a dozen, from those that are not exact numbers, such as the mass of an egg.

ANALYZE One way to distinguish exact from inexact values is to answer the question, "Which values represent quantities that can be counted?" Exact values can be counted.

SOLVE The number of marble blocks (a) and the number of stairs (c) are quantities we can count, so they are exact numbers. The other three quantities are based on measurements of length (b), mass (d), and area (e) and therefore are not exact.

THINK ABOUT IT The "you can count them" property of exact numbers was used in this exercise, which assumes that there is no uncertainty in a counted value. Can you think of a type of counting whose certainty is sometimes challenged?

Practice Exercise Which of the following statistics associated with the Golden Gate Bridge in San Francisco, California (Figure 1.21), are exact numbers and which have some inherent uncertainty?

    a. The roadway is six lanes wide.
    b. The width of the bridge is 27.4 m.
    c. The bridge has a mass of 3.808×10^8 kg.
    d. The length of the bridge is 2740 m.
    e. The toll for a car traveling south is $6.00 (as of July 1, 2012).

(Answers to Practice Exercises are in the back of the book.)

FIGURE 1.20 The Washington Monument.

FIGURE 1.21 The Golden Gate Bridge.

1.9 Unit Conversions and Dimensional Analysis

On July 23, 1983, Air Canada Flight 143, a Boeing 767 that had been in service only a few months, was near the halfway point on a flight from Montreal, Quebec, to Edmonton, Alberta, when it ran out of fuel. Without power the plane became a very large, very heavy glider. Fortunately, the pilot of Flight 143 was also an experienced glider pilot, and he was able to make a successful emergency landing at an abandoned airbase in Gimli, Manitoba. There were no serious injuries even though the nose gear collapsed during the landing (Figure 1.22). The plane was repaired and flown out of Gimli and back into service, where it remained until its retirement in 2008. During its 25 years of operation the plane became widely known in aviation circles as the *Gimli Glider*.

Why did Flight 143 run out of fuel? There were several reasons, including faulty communications between ground and flight crews about a malfunctioning fuel gauge. However, the most immediate reason was an error in calculating how much fuel the plane needed to fly from Montreal to Edmonton. The error occurred at a time when Canada and its national airline were in the process of

FIGURE 1.22 Air Canada Flight 143 after landing safely in Gimli, Manitoba.

conversion factor fraction in which the numerator is equivalent to the denominator but is expressed in different units, making the fraction equivalent to 1.

▶❚❚ **CHEMTOUR** Dimensional Analysis

converting from a unit system called the *Imperial System*, in which masses are expressed in ounces and pounds (as in the United States today), to the SI system in which the base unit of mass is the kilogram. The Boeing 767 that landed in Gimli was among the first in the Air Canada fleet to use the SI system, so its fuel capacity and consumption were expressed in kilograms. The plane was supposed to take off from Montreal with 22,300 kg of fuel in its tanks. It left with less than half that. Let's explore why.

To make sure the plane had enough fuel—but not too much to avoid transporting excess weight—the refueling crew in Montreal measured the volume of fuel already in its tanks and found that they contained 7700 liters. Next they converted the volume of fuel into an equivalent mass of fuel. To carry out this calculation they used an approach sometimes called the *unit factor method* but more often called *dimensional analysis*. The approach makes use of **conversion factors**, which are fractions in which the numerators and denominators have different units but represent equivalent quantities. This equivalency means that multiplying a quantity by a conversion factor is like multiplying by 1: the intrinsic value of the quantity does not change; it is simply expressed with different units. The key to using conversion factors is to set them up correctly, with the appropriate units in the numerator and denominator.

To calculate the mass of fuel onboard Flight 143, the refueling crew should have used the density of jet fuel, 0.80 kg/L, as a conversion factor. Had they done so they would have determined that the mass of fuel on board was

$$7700 \; \cancel{\text{L}} \times \frac{0.80 \; \text{kg}}{1 \; \cancel{\text{L}}} = 6200 \; \text{kg}$$

Notice how multiplying by density allows us to convert a volume into the equivalent mass when the initial volume units (liters in this case) match the denominator units of the density value. The volume units cancel, and we obtain a mass in the units of the numerator of the density term (kg in this case). The general equation for converting a value from one set of units to another is

$$\cancel{\text{initial units}} \times \frac{\text{desired units}}{\cancel{\text{initial units}}} = \text{desired units}$$

The refueling crew next had to calculate the mass of jet fuel to be added. They knew they needed 22,300 kg, so they should have subtracted 6200 kg from that value:

$$\begin{array}{r} 22{,}300 \text{ kg needed} \\ - \; 6{,}200 \text{ kg onboard} \\ \hline 16{,}100 \text{ kg to be added} \end{array}$$

Next, they calculated *the volume* of jet fuel to be added. Why? Because jet fuel is pumped into airplanes the way gasoline is pumped into automobile tanks: with a metering system based on volume. Gasoline is sold by the gallon in the United States, but by the liter in the rest of the industrialized world, including Canada. So, the refueling crew calculated the volume of jet fuel in liters corresponding to the mass of fuel needed. In this calculation the reciprocal of the density serves as the conversion factor so that mass units cancel out:

$$16{,}100 \; \cancel{\text{kg}} \times \frac{1 \; \text{L}}{0.80 \; \cancel{\text{kg}}} = 20{,}100 \; \text{L} \approx 20{,}000 \; \text{L}$$

Had this volume of fuel been added, Flight 143 would have made it to Edmonton. Unfortunately, the refueling crew added only 4900 L of fuel, which was nowhere near enough. Why so little? They used the wrong value of jet fuel density to convert volume to mass and back again. The value they used was 1.77 (instead of 0.80), which is the density of jet fuel expressed in *pounds* per liter, not kilograms per liter. Using the wrong conversion factor resulted in their overestimating the mass of fuel already onboard the airplane. Then, to make matters worse, they underestimated how much fuel to add by using the reciprocal of 1.77 instead of 0.80 to calculate the volume of fuel to pump into the plane's tanks. As a result of the faulty conversion factors that overestimated how much fuel was already onboard and then underestimated how much more to add, Flight 143 left Montreal with less than half as much fuel as it should have carried.

The flight of the *Gimli Glider* is not the only disaster (or near disaster) caused by mixing up units. In 1999 the *Mars Climate Orbiter* space probe crashed on the surface of the planet instead of orbiting it because the programmers who wrote the software that controlled the spacecraft used one set of units to express the thrust of its rocket engines, and the engineers who designed the engines used another. These examples remind us that we must take care to use the correct units in expressing quantities of substances. This section's goal is to help you become more skilled at converting measured and calculated values from one set of units to another. You will have an opportunity to develop this skill in the sample and practice exercises below.

SAMPLE EXERCISE 1.3 Converting Units   LO8

The largest cut diamond in the world is the yellow-brown Golden Jubilee Diamond (Figure 1.23). Its mass is 545.67 carats (1 carat = 200 milligrams). What is the mass of this diamond in grams?

COLLECT AND ORGANIZE We need to convert the mass of a diamond expressed in carats to an equivalent mass in grams (g). We know that 1 carat is equivalent to 200 milligrams (mg) and, from Table 1.1, that the prefix *milli-* represents 10^{-3} of the base unit, which is grams (g) in this case.

ANALYZE We need two conversion factors: one for converting carats to milligrams and another to convert milligrams to grams. The first conversion factor must have carats in its denominator to cancel out the initial units of the diamond's mass. This conversion step will give us a mass in milligrams, which means that milligrams must be in the denominator of the second conversion factor. Therefore, the two conversion factors (exact numbers) are

$$\frac{200 \text{ mg}}{1 \text{ carat}} \quad \text{and} \quad \frac{10^{-3} \text{ g}}{1 \text{ mg}}$$

The value of the two conversion factors when multiplied together is 0.2, which means that our final answer should be 0.2 times, or 1/5, the initial value—or a little more than 100 grams.

SOLVE We multiply the initial mass by the two conversion factors to obtain the final answer:

$$545.67 \text{ } \cancel{\text{carats}} \times \frac{200 \text{ } \cancel{\text{mg}}}{1 \text{ } \cancel{\text{carat}}} \times \frac{10^{-3} \text{ g}}{1 \text{ } \cancel{\text{mg}}} = 109.13 \text{ g}$$

THINK ABOUT IT The calculated result is reasonable given how close it is to our estimate. The Golden Jubilee Diamond is a really massive diamond!

Practice Exercise The Eiffel Tower in Paris is 324 m tall, including a 24 m television antenna that was not there when the tower was built in 1889. What is the overall height of the Eiffel Tower in kilometers and in centimeters? ⚙

(Answers to Practice Exercises are in the back of the book.)

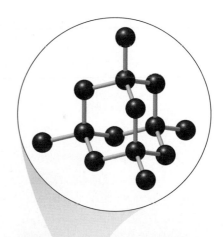

FIGURE 1.23 The Golden Jubilee Diamond. Like all diamonds, it consists of carbon atoms.

FIGURE 1.24 A warning to Mount Washington hikers from the U.S. Forest Service.

SAMPLE EXERCISE 1.4 Converting Customary U.S. Units and SI Units I   LO8

The summit of Mount Washington in New Hampshire is famous for its awful weather: it experiences hurricane-force winds an average of 110 days per year (Figure 1.24). On April 12, 1934, a wind gust of 231 miles per hour was recorded at the summit. What is this wind speed in meters per second?

COLLECT AND ORGANIZE Wind speed is expressed initially in miles per hour; the desired value has units of meters per second. Therefore, we need conversion factors to convert distance from miles to meters and time from hours to seconds. We know that 1 km = 0.6214 mi (from Table 1.3) and that there are 60 minutes in an hour and 60 seconds in a minute, which combine to give us the equality 1 hr = 3600 s.

ANALYZE The distance conversion factor based on (1 km = 0.6214 mi) must have miles in its denominator to cancel out the miles in the numerator of the initial value. Similarly, the time conversion factor (1 hr = 3600 s) must have hours in its numerator to cancel out the hours in the denominator of the initial value. We will also need to convert kilometers into meters. Therefore, we will need these three conversion factors:

$$\frac{1 \text{ km}}{0.6214 \text{ mi}}, \quad \frac{10^3 \text{ m}}{1 \text{ km}}, \quad \text{and} \quad \frac{1 \text{ hr}}{3600 \text{ s}}$$

The values of the three conversion factors, when multiplied together, give us a fraction with a numerator of 1000. Its denominator is a little less than 2/3 of 3600, or a little more than 2000. Our final answer should therefore be a little less than $\frac{1}{2}$ of 231, or about 100 m/s.

SOLVE

$$\frac{231 \text{ mi}}{\text{hr}} \times \frac{1 \text{ km}}{0.6214 \text{ mi}} \times \frac{10^3 \text{ m}}{1 \text{ km}} \times \frac{1 \text{ hr}}{3600 \text{ s}} = 103 \frac{\text{m}}{\text{s}}$$

THINK ABOUT IT The answer seems reasonable based on our estimate. Each conversion factor is an exact number. We rounded the result to three significant figures because that is the number of significant figures in the initial measurement of 231 mi/hr.

Practice Exercise The average distance from Earth to the moon is 238,857 miles.

   a. What is this distance in kilometers?
   b. On average, how much time does it take moonlight (reflected sunlight) to travel from the moon to Earth? The speed of light in space is 2.998×10^8 m/s.

(Answers to Practice Exercises are in the back of the book.)

SAMPLE EXERCISE 1.5 Converting Customary U.S. Units and SI Units II   LO8

A pediatrician prescribes the antibiotic amoxicillin to treat a 2-year-old child suffering from a recurring middle ear infection. The pediatrician decides that a therapeutic dose of amoxicillin in this case is 75 milligrams of the drug per kilogram of body mass of the patient per day. The drug is to be administered twice each day in the form of a flavored liquid that contains 125 milligrams of amoxicillin per milliliter of liquid. If the child weighs 28 pounds, how many milliliters of the liquid should be administered in each dose?

COLLECT AND ORGANIZE We need to calculate the volume of liquid amoxicillin in milliliters to be administered twice each day to provide a daily dose of 75 mg/kg patient. We know the patient's mass (28 pounds) and the concentration of liquid amoxicillin (125 mg/mL). We know from the conversion factors inside the back cover that 1 pound = 0.453592 kg.

ANALYZE There is usually more than one pathway to solving a problem like this one that involves several conversion steps. A good place to start is with a quantity that has a single unit, such as the mass of the patient: 28 pounds. Dosage is expressed in mg amoxicillin/kg patient, so we can use it as a conversion factor to turn patient mass into milligrams of amoxicillin, but only after we convert patient mass from pounds into kilograms. When we multiply patient mass in kilograms by dosage (mg amoxicillin/kg patient), the "/kg patient" denominator cancels out and we have mg amoxicillin to be given each day. To convert mg amoxicillin to an equivalent volume of the liquid it's dissolved in, we invert the concentration of the liquid from "mg amoxicillin/mL" to "mL/mg amoxicillin" so that the mass units cancel out. Finally, we need to divide the daily volume in two because the drug is administered twice a day. The resulting conversion factors are

$$\frac{0.453592 \text{ kg}}{1 \text{ lb}} \quad \frac{75 \text{ mg}}{\text{kg} \cdot \text{day}} \quad \frac{1 \text{ mL}}{125 \text{ mg}} \quad \frac{1 \text{ day}}{2 \text{ doses}}$$

To estimate the result of our calculation, we note that there is a little less than half a kilogram in one pound, so a 28-pound child weighs a little less than 14 kg—say, 12 kg. The product of 12 kg × 75 mg amoxicillin/kg is about 1000 mg amoxicillin. Dividing this value by the concentration of the liquid (125 mg/mL) gives us a volume of about 8 mL per day, or 4 mL twice a day.

SOLVE

$$28 \text{ lb} \times \frac{0.453592 \text{ kg}}{1 \text{ lb}} \times \frac{75 \text{ mg}}{\text{kg} \cdot \text{day}} \times \frac{1 \text{ mL}}{125 \text{ mg}} \times \frac{1 \text{ day}}{2 \text{ doses}} = 3.8 \frac{\text{mL}}{\text{dose}}$$

THINK ABOUT IT The answer nearly matches our estimate. We rounded the result to two significant figures because that is the number of significant figures in the child's weight and the therapeutic daily dose. Those values represent the weak links in our significant-figure chain. The number of doses per day (2) is an exact number and has no uncertainty (except when a forgetful parent misses a dose). Medicine droppers for dispensing liquid medications are often calibrated in milliliters, and also (in the United States) in teaspoons (1 tsp = 5 mL).

Practice Exercise A student planning a party has $20 to spend on her favorite soft drink. It is on sale at Store A for $1.29 for a 2-liter bottle (plus 10-cent deposit); at Store B the price of a 12 pack of 12-fluid-ounce cans is $2.99 (plus a 5-cent deposit per can). At which store can she buy the most of her favorite soft drink for no more than $20? (There are 29.57 mL in one U.S. fluid ounce, which is a unit of volume, not mass.)

(Answers to Practice Exercises are in the back of the book.)

1.10 Temperature Scales

Temperature is by far the most frequently measured quantity, and it has been since the first thermometers were developed nearly 500 years ago. In 1592 Galileo invented an air thermometer consisting of a glass bulb with a long tube attached. The end of the tube was immersed in a cold liquid and the air in the bulb was heated, causing it to expand and bubble out through the end of the tube. Then the heat was removed, the air inside cooled and contracted, and liquid was sucked up into the tube. The level of the liquid provided a measure of the temperature of the air in the bulb. This air thermometer was very sensitive to small changes in temperature, but unfortunately it was also sensitive to fluctuations in atmospheric pressure, functioning as a barometer as well as a thermometer.

absolute zero (0 K) zero point on Kelvin temperature scale; theoretically the lowest temperature possible.

▶Ⅱ **CHEMTOUR** Temperature Conversion

Modern thermometers based on the thermal expansion of liquids such as mercury and alcohol were introduced in the early 18th century by German scientist (and glass blower) Daniel Gabriel Fahrenheit (1686–1736). He later developed a temperature scale with a zero point corresponding to the freezing point of a concentrated salt solution and an upper value (100°) corresponding to the average internal temperature of the human body. The Fahrenheit scale widely used in the United States today is a slightly modified version of the one based on those two reference temperatures. Later in the 18th century Swedish astronomer Anders Celsius (1701–1744) proposed an alternative temperature scale based on the freezing and boiling points of pure water. Today's Celsius temperature scale is linked to these temperatures: zero Celsius (0°C) is the temperature at which water freezes and 100°C is the average temperature at which it boils (at sea level). In 1848 British scientist William Thomson (later Lord Kelvin, 1824–1907) proposed another temperature scale that was not based on the physical properties of any substance, but rather on the notion that there is a lower limit to temperature, called **absolute zero (0 K)**. That temperature is the zero point of the Kelvin scale named in his honor.

The Fahrenheit and Celsius scales differ from each other in two ways, as shown in Figure 1.25. First, as we have seen, their zero points are different. Zero degrees Celsius (0°C) is the temperature at which water freezes under normal conditions, but that temperature is 32 degrees on the Fahrenheit scale (32°F) because 0°F is

FIGURE 1.25 Three temperature scales are commonly used today, although the Fahrenheit scale is rarely used in scientific work.

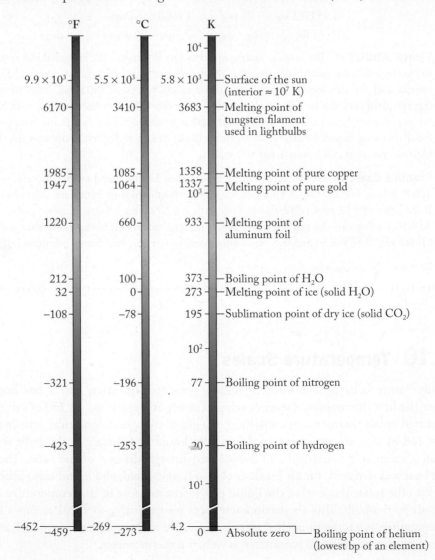

based on a lower temperature: the freezing point of concentrated salt solutions (used to make homemade ice cream). The other difference is in the size of the temperature change corresponding to 1 degree. The difference between the freezing and boiling points of water is $212 - 32 = 180$ degrees on the Fahrenheit scale but only $100 - 0 = 100$ degrees on the Celsius scale. This difference means that a Fahrenheit degree is 100/180, or $\frac{5}{9}$, as large as a Celsius degree.

To convert temperatures from Fahrenheit into Celsius, we need to account for the differences in zero point and in degree size. Equation 1.4 does both:

$$T(°C) = \frac{5}{9}[T(°F) - 32] \qquad (1.4)$$

The Kelvin scale is the basis for the SI unit of temperature, the **kelvin (K)**. (Note that although we speak of temperatures on the other scales in "degrees Fahrenheit" and "degrees Celsius," temperatures on the Kelvin scale are simply in "kelvin.") The zero point on the Kelvin scale, absolute zero, is equivalent to $-273.15°C$. No one has ever been able to chill matter to absolute zero, nor is it theoretically possible to do so, but scientists have come very close, cooling samples to less than 10^{-9} K.

The zero point on the Kelvin scale differs from that on the Celsius scale, but the size of 1 degree is the same on the two scales. For this reason, the conversion from a Celsius temperature to a Kelvin temperature is simply a matter of adding 273.15 to the Celsius value:

$$T(K) = T(°C) + 273.15 \qquad (1.5)$$

kelvin (K) the SI unit of temperature.

SAMPLE EXERCISE 1.6   Temperature Conversions   LO8

The temperature of interstellar space is 2.73 K. What is this temperature on the Celsius scale and on the Fahrenheit scale?

COLLECT AND ORGANIZE We are asked to convert a temperature from kelvin to degrees Celsius and degrees Fahrenheit. Equation 1.4 relates Celsius and Fahrenheit temperatures; Equation 1.5 relates Kelvin and Celsius temperatures.

ANALYZE We will first use Equation 1.5 to convert 2.73 K into an equivalent Celsius temperature and then use Equation 1.4 to calculate an equivalent Fahrenheit temperature. We can estimate that the value in degrees Celsius should be close to absolute zero (about $-273°C$). Because $1°F$ is about half the size of $1°C$, the temperature on the Fahrenheit scale should be a little less than twice the value of the temperature on the Celsius scale (around $-500°F$).

SOLVE To convert from kelvin to degrees Celsius, we have

$$T(K) = T(°C) + 273.15$$
$$T(°C) = T(K) - 273.15$$
$$= 2.73 - 273.15 = -270.42°C$$

To convert degrees Celsius to degrees Fahrenheit, we rearrange Equation 1.4,

$$T(°C) = \frac{5}{9}[T(°F) - 32]$$

to solve for degrees Fahrenheit. Multiplying both sides by 9 and dividing both by 5 gives us

$$\frac{9}{5}T(°C) = T(°F) - 32$$

$$T(°F) = \frac{9}{5}T(°C) + 32 = \frac{9}{5}(-270.42) + 32 = -454.76°F$$

The value of 32°F is considered a definition and so is not used in determining the number of significant figures in the answer. The number that determines the accuracy to which we can know this value is −270.42°C.

THINK ABOUT IT The calculated Celsius value of −270.42°C makes sense because it represents a temperature only a few degrees above absolute zero, just as we estimated. The Fahrenheit value is within 10% of our estimate, so it is reasonable too.

Practice Exercise The temperature of the moon's surface varies from −233°C at night to 123°C during the day. What are the temperatures on the Kelvin and Fahrenheit scales?

(Answers to Practice Exercises are in the back of the book.)

SAMPLE EXERCISE 1.7   Integrating Concepts: Searching for Cheaper Gas

When would it be worth it to make a special trip across town (or out of the country) to buy gas at a station where it is cheaper than nearby? Many motorists have been asking themselves this question lately. Let's examine some of the factors that might go into answering it.

Suppose a driver lives in Niagara Falls, Ontario, on a day when the price of gasoline at a local station is 1.299 Canadian dollars (CAD) per liter. On the same day the price of gasoline at a station in Niagara Falls, New York (4.0 km away), is 3.89 U.S. dollars (USD) per gallon, and the currency exchange rate is 1.000 USD = 0.980 CAD. How much money in Canadian dollars would the driver save by filling up with 60.0 liters of gasoline at the American station? In calculating the driver's savings, consider that his car can go 12.0 km on one liter of gasoline (about 26 miles per gallon), and that he has to make a round trip, which includes crossing the Rainbow Bridge over the Niagara River. The toll on the bridge is $3.25 (CAD or USD) for vehicles crossing to the Canadian side (there is no toll the other way).

COLLECT AND ORGANIZE The problem gives the gasoline prices at a nearby station and one that is 4.0 km away, and asks how much money would be saved by driving to the distant station to purchase 60.0 liters of gasoline. The prices are given in different currencies and volume units: liters and U.S. gallons. According to Table 1.3, there are 3.785 liters in one U.S. gallon. We are also given the distance to the American station and the fuel efficiency of the car. Other factors impacting the money saved are the cost of driving to the American station and back (we know the car's fuel efficiency and the distance to the American station), and the toll across the Rainbow Bridge.

ANALYZE The problem contains many pieces of information, and we need to sort them out. A logical starting point is the different prices of gasoline. We have to assume that gas is cheaper at the American station; otherwise, why make the trip? To compare prices we need to express both using the same set of units. Converting the price at the American station into Canadian dollars per liter makes sense because (1) the car's fuel tank capacity is given in liters, (2) its fuel efficiency is expressed in km/L, and (3) the distance to the American station is given in kilometers. We know the distance to the American station in km and the car's fuel efficiency in km/L, so we can calculate the volume of gas consumed by dividing the round-trip distance by the car's fuel efficiency.

To estimate the savings, we start with the approximate difference in the cost of 60.0 liters of gasoline locally and at the U.S. station. Locally the cost is about $1\frac{1}{3}$ CAD/L × 60 L = 80 CAD. At the American station the cost is about 4 USD/~~gal~~ × (1 ~~gal~~ / 4 ~~L~~) × 60 ~~L~~ = 60 USD, which is nearly the same as 60 CAD and represents a 20 CAD savings. The cost of the gasoline consumed in a round trip of 2 × 4.0 km = 8.0 km is about (8 ~~km~~ × (1 ~~L~~ / 12 ~~km~~) × $1\frac{1}{3}$ CAD/~~L~~) = ~1 CAD, and the bridge toll is about 3 CAD, so the savings should be about 20 − (1 + 3) = 16 CAD.

SOLVE

1. Cost of buying gasoline at the local station:

   $$60.0 \text{ L} \times (1.299 \text{ CAD/L}) = 77.94 \text{ CAD}$$

2. Cost of buying gasoline at the American station:

   $$60.0 \text{ L} \times (1 \text{ gal} / 3.785 \text{ L}) \times (3.89 \text{ USD/gal})$$
   $$\times (0.980 \text{ CAD} / 1.000 \text{ USD}) = 60.43 \text{ CAD}$$

3. Difference in gasoline costs:

   $$(77.94 - 60.43) \text{ CAD} = 17.51 \text{ CAD}$$

4. Cost of gasoline to drive to and from the American station:

   $$2 \times 4.0 \text{ km} \times (1 \text{ L} / 12.0 \text{ km}) \times 1.299 \text{ CAD/L}$$
   $$= 0.87 \text{ CAD}$$

5. Net savings factoring in the 3.25 CAD bridge toll:

   $$17.51 - (0.87 + 3.25) = 13.39 \approx 13.4 \text{ CAD}$$

THINK ABOUT IT The calculated savings are slightly less than we estimated, mostly because we based the estimate on a relatively small difference in two large estimated values. Would you be willing to drive to another country to save about $13 on a tankful of gasoline? How about driving the same distance within your own country to save only half as much? Finally, can you identify an assumption made solving the problem that you disagree with? Did it have a significant impact on the results of the calculation?

SUMMARY ·······························■

Section 1.1 **Matter** is everything in the universe that has **mass (*m*)** and occupies space, and all matter is made of particles. **Chemistry** is the study of matter and the changes that it undergoes. A **molecule** is a collection of **atoms** held together by **chemical bonds** in a characteristic pattern and proportion. The states (or phases) of matter include **solid**, in which the particles have an ordered structure; **liquid**, in which the particles are free to move past each other; and **gas** (or *vapor*), in which the particles have the most freedom and completely fill their container. Familiar phase changes include melting, freezing, vaporization, and condensation. The transformation of a solid directly into a gas is **sublimation**; the reverse process is **deposition**.

Section 1.2 **Energy** can be defined as the ability to do **work**. **Heat** is the flow of energy due to a difference in temperature. **Potential energy (PE)** is the energy in an object due to its position or composition. **Kinetic energy (KE)** is the energy of motion. The **law of conservation of energy** states that energy cannot be created or destroyed.

Section 1.3 The principal categories of matter are **mixtures** and **pure substances**, which may be either **elements** or **compounds** (elements chemically combined together according to the **law of constant composition**). A **chemical formula** indicates the proportion of elements in a substance. Mixtures may be **homogeneous** (such mixtures are also called **solutions**) or

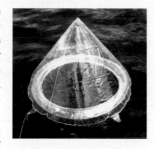

heterogeneous, and they can be separated by **physical processes** including **distillation**, **filtration**, and **chromatography**. Distillation separates substances of differing **volatility**.

Section 1.4 The properties of a substance are either **intensive properties**, which are independent of quantity, or **extensive properties**, which are related to the quantity of the substance. The **physical properties** of a substance can be observed without changing the substance into another one; the **chemical properties** of a substance (such as **flammability**) can be observed only through chemical reactions involving the substance. The **density (*d*)** of an object is the ratio of its mass to its volume.

Section 1.5 An atom is the smallest representative particle of an element. Dalton's atomic theory explains Proust's **law of definite proportions** and Dalton's **law of multiple proportions**. The **scientific method** is based on observations of natural phenomena and the results of laboratory experiments; developing a tentative explanation, or **hypothesis**, for the observations and results; testing the hypothesis through further experimentation; and then formulating a **scientific theory**. A **scientific law** is a comprehensive, succinct description of a phenomenon or process.

Section 1.6 **Structural formulas**, ball-and-stick models, and space-filling models are all used to show molecular structure, the three-dimensional arrangement of atoms in a molecule.

Section 1.7 The COAST framework used in this book to solve problems has four components: **C**ollect and **O**rganize information and ideas, **A**nalyze the information to determine how it can be used to obtain the answer, **S**olve the problem (often the math-intensive step), and **T**hink about the answer.

Section 1.8 The International System of Units (SI), in which the **meter (m)** is the standard unit of length, evolved from the metric system and is widely used in science to express the results of measurements. Prefixes naming powers of 10 are used with SI base units to express quantities much larger or much smaller than the base units. The appropriate number of **significant figures** is used to express the certainty in the result of a measurement or calculation. The **precision** of any set of measurements indicates how repeatable the measurement is; the **accuracy** of a measurement indicates how close to the true value the measured value is.

Section 1.9 Dimensional analysis uses **conversion factors** (fractions in which the numerators and denominators have different units but represent the same quantity) to convert a value from one unit into another unit.

Section 1.10 Three temperature scales are widely used today: Fahrenheit, Celsius, and Kelvin. Zero on the Kelvin scale is **absolute zero (0 K)**, the coldest possible temperature.

PROBLEM-SOLVING SUMMARY ·······················■

TYPE OF PROBLEM	CONCEPTS AND EQUATIONS	SAMPLE EXERCISES
Using significant figures in calculations	Apply the weak-link rule: the number of significant figures in a calculated quantity involving multiplication or division can be no greater than the number of significant figures in the least certain value used to calculate it.	1.1
Distinguishing exact from uncertain values	Quantities that can be counted are exact. Measured quantities or conversion factors that are not exact values are inherently uncertain.	1.2
Converting units using dimensional analysis; converting temperatures	Converting values from one set of units to another involves multiplication by one or more conversion factors, which are set up so that the original units cancel out. The following equations convert between the Fahrenheit, Celsius, and Kelvin temperature scales: $$T(^\circ C) = \tfrac{5}{9}[T(^\circ F) - 32] \qquad (1.4)$$ $$T(K) = T(^\circ C) + 273.15 \qquad (1.5)$$	1.3–1.6

VISUAL PROBLEMS ···■

(Answers to boldface end-of-chapter questions and problems are in the back of the book.)

1.1. For each image in Figure P1.1, identify what class of pure substance is depicted (an element or compound) and identify the physical state(s).

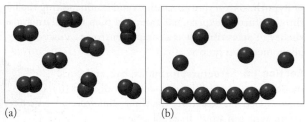

(a)   (b)

FIGURE P1.1

1.2. For each image in Figure P1.2, identify what class of matter is depicted (an element, a compound, a mixture of elements, or a mixture of compounds) and identify the physical state.

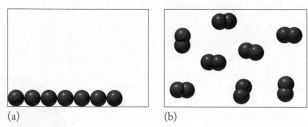

(a)   (b)

FIGURE P1.2

1.3. Which of the following statements best describes the change depicted in Figure P1.3?

FIGURE P1.3

a. A mixture of two gaseous elements undergoes a chemical reaction, forming a gaseous compound.
b. A mixture of two gaseous elements undergoes a chemical reaction, forming a solid compound.
c. A mixture of two gaseous elements undergoes deposition.
d. A mixture of two gaseous elements condenses.

1.4. Which of the following statements best describes the change depicted in Figure P1.4?

FIGURE P1.4

a. A mixture of two gaseous elements is cooled to a temperature at which one of them condenses.
b. A mixture of two gaseous compounds is heated to a temperature at which one of them decomposes.
c. A mixture of two gaseous elements undergoes deposition.
d. A mixture of two gaseous elements reacts to form two compounds, one of which is a liquid.

1.5. A space-filling model of formic acid is shown in Figure P1.5. What is the molecular formula of formic acid?

FIGURE P1.5

1.6. A ball-and-stick model of the compound in rubbing alcohol is shown in Figure P1.6.

FIGURE P1.6

a. What is the molecular formula of the compound?
b. Write a condensed structural formula for the compound.

QUESTIONS AND PROBLEMS ··························■

States of Matter

CONCEPT REVIEW

1.7. In what ways are the arrangement of water molecules in ice and liquid water similar and what ways are they different?
1.8. What is in the space in between the particles that make up a gas?

1.9. Substances have characteristic *triple points*: unique combinations of temperature and pressure at which substances can simultaneously exist as solids, liquids, and gases. In which of these three states do the particles of a substance at its triple point have the greatest motion and in which state do they have the least motion?

1.10. A pot of water on a stove is heated to a rapid boil. Identify the gas inside the bubbles that form in the boiling water.

1.11. A brief winter storm leaves a dusting of snow on the ground. During the sunny but very cold day after the storm the snow disappears even though air temperature never gets above freezing. If the snow didn't melt, where did it go?

1.12. Equal masses of water undergo condensation, deposition, evaporation, and sublimation.
  a. Which of the processes is accompanied by the *release* of the greatest amount of energy?
  b. In which of the above processes is the greatest amount of energy *absorbed*?

Forms of Energy

CONCEPT REVIEW

1.13. How are energy and work related?

1.14. Explain the difference between potential energy and kinetic energy.

1.15. Which of the following statements about heat are true?
  a. Heat is the transfer of energy from a warmer place to a cooler one.
  b. A thermos bottle has more heat inside when it is full of hot coffee than when it is half full. (Assume no heat is lost from the thermos bottle).
  c. The temperature of an object is a measure of its heat content.

1.16. Describe three examples of energy transfer that happen when you speak on a cell phone to a friend.

PROBLEMS

1.17. A subcompact car with a mass of 1400 kg and a loaded dump truck with a mass of 18,000 kg are traveling at the same speed. How many times more kinetic energy does the dump truck have than the car?

1.18. Speed of Baseball Pitches One of the reasons why Johan Santana (Figure P1.18) has been an effective major league pitcher is because the speed of his fastball is over 90 miles per hour (mph), while the speed of his changeup (a pitch that fools batters because it looks like a fastball leaving the pitcher's hand) is less than 80 mph. How much more kinetic energy does a 92 mph fastball have than a 78 mph changeup? Express your answer as a percentage of the kinetic energy of the changeup.

FIGURE P1.18

Classes of Matter

CONCEPT REVIEW

1.19. Which of the following foods is a heterogeneous mixture? (a) solid butter; (b) a Snickers bar; (c) grape juice; (d) an uncooked hamburger

1.20. Which of the following foods is a homogeneous mixture? (a) freshly brewed coffee; (b) vinegar; (c) a slice of white bread; (d) a slice of ham

1.21. Which of the following foods is a heterogeneous mixture? (a) apple juice; (b) cooking oil; (c) solid butter; (d) orange juice; (e) tomato juice

1.22. Which of the following is a homogeneous mixture? (a) a bronze sword from ancient Greece; (b) sweat; (c) Nile River water; (d) gasoline; (e) compressed air in a scuba tank

Properties of Matter

CONCEPT REVIEW

1.23. List one chemical and four physical properties of gold.

1.24. Describe three physical properties that gold and silver have in common, and three physical properties that distinguish them.

1.25. Give three properties that enable a person to distinguish between table sugar, water, and oxygen.

1.26. Give three properties that enable a person to distinguish between table salt, sand, and copper.

1.27. Indicate whether each of the following properties is a physical or a chemical property of sodium (Na):
  a. Its density is greater than that of kerosene and less than that of water.
  b. It has a lower melting point than most metals.
  c. It is an excellent conductor of heat and electricity.
  d. It is soft and can be easily cut with a knife.
  e. Freshly cut sodium is shiny, but it rapidly tarnishes in contact with air.
  f. It reacts very vigorously with water, releasing hydrogen gas (H_2).

1.28. Indicate whether each of the following is a physical or chemical property of hydrogen gas (H_2):
  a. At room temperature, its density is less than that of any other gas.
  b. It reacts vigorously with oxygen (O_2) to form water.
  c. Liquefied H_2 boils at a very low temperature (−253°C).
  d. H_2 gas does not conduct electricity.

1.29. Can an extensive property be used to identify a substance? Explain why or why not.

1.30. Which of these are intensive properties of a sample of a substance? (a) freezing point; (b) heat content; (c) temperature

Atomic Theory: The Scientific Method in Action

CONCEPT REVIEW

1.31. What kinds of information are needed to formulate a hypothesis?

1.32. How does a hypothesis become a theory?

1.33. Is it possible to disprove a scientific hypothesis?

1.34. Why was the belief that matter consists of atoms a philosophy in ancient Greece, but was considered a theory in the early 1800s?

1.35. How do people use the word *theory* in normal conversation?

1.36. Can a theory be proven?

Making Measurements and Expressing the Results; Unit Conversions and Dimensional Analysis

CONCEPT REVIEW

1.37. Describe in general terms how the SI and U.S. Customary systems of units differ.

1.38. Suggest two reasons why SI units are not more widely used in the United States.

PROBLEMS

NOTE: Some physical properties of the elements are listed in Appendix 3.

1.39. Olympic Mile An Olympic "mile" is actually 1500 m. What percentage is an Olympic mile of a U.S. mile (5280 feet)?

1.40. A sport-utility vehicle has an average mileage rating of 18 miles per gallon. How many gallons of gasoline are needed for a 389-mile trip?

1.41. A single strand of natural silk may be as long as 4.0×10^3 m. What is this length in miles?

1.42. The speed of light in a vacuum is 2.998×10^8 m/s. What is the speed of light in km/hr?

1.43. If a wheelchair-marathon racer moving at 13.1 miles per hour expends energy at a rate of 665 Calories per hour, how much energy in Calories would be required to complete a marathon race (26.2 miles) at that pace?

1.44. Boston Marathon To qualify to run in the 2012 Boston Marathon, a distance of 26.2 miles, an 18-year-old woman had to have completed another marathon in 3 hours and 35 minutes or less. Translate this qualifying time and distance into average speeds expressed in (a) in miles per hour and (b) in meters per second.

1.45. Nearest Star At a distance of 4.3 light-years, Proxima Centauri is the nearest star to our solar system. What is the distance to Proxima Centauri in kilometers? (The speed of light in space is 2.998×10^8 m/s.)

1.46. Sports Car The Porsche Boxster Spyder in Figure P1.46 is powered by a 320-horsepower gasoline engine. The electric motor in the Tesla Roadster (on the left in Figure P1.46) is rated at 215 kilowatts. Which is the more powerful sports car? (1 horsepower = 745.7 watts.)

FIGURE P1.46

*1.47. The level of water in an Olympic size swimming pool (50.0 meters long, 25.0 meters wide, and about 2 meters deep) needs to be lowered 3.0 cm. If water is pumped out at a rate of 5.2 liters per second, how long will it take to lower the water level 3.0 cm?

*1.48. The price of a popular soft drink is $1.00 for 24 fluid ounces (fl oz) or $0.75 for 0.50 L. Which is a better buy? (1 qt = 32 fl oz.)

1.49. Suppose a runner completes a 10K (10.0 km) road race in 41 minutes and 23 seconds. What is the runner's average speed in meters per second?

1.50. Kentucky Derby Record In 1973 a horse named Secretariat ran the fastest Kentucky Derby in history, taking 1 minute and 59 seconds to run 1.25 miles. What was Secretariat's average speed in (a) miles per hour and (b) meters per second?

1.51. What is the mass of a magnesium block that measures 2.5 cm × 3.5 cm × 1.5 cm?

1.52. What is the mass of an osmium block that measures 6.5 cm × 9.0 cm × 3.25 cm? Do you think you could lift it with one hand?

1.53. A chemist needs 35.0 g of concentrated sulfuric acid for an experiment. The density of concentrated sulfuric acid at room temperature is 1.84 g/mL. What volume of the acid is required?

1.54. What is the mass of 65.0 mL of ethanol? (Its density at room temperature is 0.789 g/mL.)

1.55. A brand new silver U.S. dollar weighs 0.934 ounces. Express this mass in grams and kilograms. (1 oz = 28.35 g.)

1.56. A U.S. dime weighs 2.5 g. What is the U.S. dollar value of exactly 1 kg of dimes?

1.57. What volume of gold would be equal in mass to a piece of copper with a volume of 125 cm³?

*1.58. A small hot-air balloon is filled with 1.00×10^6 L of air at a temperature at which the density of air is 1.18 g/L. As the air in the balloon is heated further, it expands and 9×10^4 L escapes out the open bottom of the balloon. What is the density of the heated air remaining inside the balloon?

1.59. What is the volume of 1.00 kg of mercury?

1.60. A student wonders whether a piece of jewelry is made of pure silver. She determines that its mass is 3.17 g. Then she drops it into a 10 mL graduated cylinder partially filled with water, and determines that its volume is 0.3 mL. Could the jewelry be made of pure silver?

*1.61. The average density of Earth is 5.5 g/cm³. The mass of Venus is 81.5% of Earth's mass, and the volume of Venus is 88% of Earth's volume. What is the density of Venus?

1.62. Earth has a mass of 6.0×10^{27} g and an average density of 5.5 g/cm³.
   a. What is the volume of Earth in cubic kilometers?
   *b. Geologists sometimes express the "natural" density of Earth after doing a calculation that corrects for gravitational squeezing (compression of the core because of high pressure). Should the natural density be more or less than 5.5 g/cm³?

*1.63. Utility Boats for the Navy A plastic material called high-density polyethylene (HDPE) was once evaluated for use in impact-resistant hulls of small utility boats for the U.S. Navy. A cube of this material measures 1.20×10^{-2} m on a side and has a mass of 1.70×10^{-3} kg. Seawater at the surface of the ocean has a density of 1.03 g/cm³. Will this cube float on water?

1.64. **Dimensions of the Sun** The sun is a sphere with an estimated mass of 2×10^{30} kg. If the radius of the sun is 7.0×10^5 km, what is the average density of the sun in units of grams per cubic centimeter? The volume of a sphere is $\frac{4}{3}\pi r^3$.

1.65. Diamonds are measured in carats, where 1 carat = 0.200 g. The density of diamond is 3.51 g/cm³. What is the volume of a 5.0-carat diamond?

1.66. If the concentration of mercury in the water of a polluted lake is 0.33 μg (micrograms) per liter of water, what is the total mass of mercury in the lake, in kilograms, if the lake has a surface area of 10.0 km² and an average depth of 15 m?

1.67. **Sodium in Candy Bars** Three different analytical techniques were used to determine the quantity of sodium in a Mars Milky Way candy bar. Each technique was used to analyze five portions of the same candy bar, with the following results (expressed in milligrams of sodium per candy bar):

mg of Na: Technique 1	mg of Na: Technique 2	mg of Na: Technique 3
109	110	114
111	115	115
110	120	116
109	116	115
110	113	115

The actual quantity of sodium in the candy bar was 115 mg. Which techniques would you describe as precise, which as accurate, and which as both? What is the range of the values (the difference between the highest and lowest measurements) for each technique?

*1.68. **Circuit Boards** The widths of copper lines in printed circuit boards must be close to a specified value. Three manufacturers were asked to prepare circuit boards with copper lines that are 0.500 μm (micrometers) wide. Each manufacturer's quality control department reported the following line widths on five sample circuit boards (given in micrometers):

Cu Line Width (μm): Manufacturer 1	Cu Line Width (μm): Manufacturer 2	Cu Line Width (μm): Manufacturer 3
0.512	0.514	0.500
0.508	0.513	0.501
0.516	0.514	0.502
0.504	0.514	0.502
0.513	0.512	0.501

a. What is the range of the data provided by each manufacturer?
b. Can any of the manufacturers justifiably advertise that they produce circuit boards with "high precision"?
c. Is there a data set for which this claim is misleading?

1.69. Which of the following numbers have just three significant figures? (a) 7.02; (b) 6.452; (c) 302; (d) 6.02×10^{23} (e) 12.77; (f) 3.43

1.70. Which of the following quantities have four significant figures? (a) 0.0592; (b) 0.08206; (c) 8.314; (d) 5420; (e) 5.4×10^3; (f) 3.752×10^{-5}

1.71. Perform each of the following calculations and express the answer with the correct number of significant figures:
a. $0.6274 \times 1.00 \times 10^3 / [2.205 \times (2.54)^3] =$
b. $6 \times 10^{-18} \times (1.00 \times 10^3) \times 17.4 =$
c. $(4.00 \times 58.69)/(6.02 \times 10^{23} \times 6.84) =$
d. $[26.0 \times 60.0)/43.53]/(1.000 \times 10^4) =$

1.72. Perform each of the following calculations, and express the answer with the correct number of significant figures:
a. $[(12 \times 60.0) + 55.3]/(5.000 \times 10^3) =$
b. $(2.00 \times 183.9)/[6.02 \times 10^{23} \times (1.61 \times 10^{-8})^3] =$
c. $0.8161/[2.205 \times (2.54)^3] =$
d. $(9.00 \times 60.0) + (50.0 \times 60.0) + (3.00 \times 10^1) =$

Temperature Scales

CONCEPT REVIEW

1.73. Both the Fahrenheit and Celsius scales are based on reference temperatures that are 100 degrees apart. Suggest a reason why the Celsius scale is preferred by scientists.

1.74. In what way are the Celsius and Kelvin scales similar and in what way are they different?

1.75. What is meant by an *absolute* temperature scale?

1.76. Can a temperature in °C ever have the same value in °F?

PROBLEMS

1.77. Liquid helium boils at 4.2 K. What is the boiling point of helium in degrees Celsius?

1.78. Liquid hydrogen boils at −253°C. What is the boiling point of H_2 on the Kelvin scale?

1.79. **Topical Anesthetic** Ethyl chloride acts as a mild topical anesthetic because it chills the skin when sprayed on it. It dulls the pain of injury and is sometimes used to make removing splinters easier. The boiling point of ethyl chloride is 12.3°C. What is its boiling point on the Fahrenheit and Kelvin scales?

1.80. **Dry Ice** The temperature of the dry ice (solid carbon dioxide) in ice cream vending carts is −78°C. What is this temperature on the Fahrenheit and Kelvin scales?

1.81. **Record Low** The lowest temperature measured on Earth is −128.6°F, recorded at Vostok, Antarctica, in July 1983. What is this temperature on the Celsius and Kelvin scales?

1.82. **Record High** The highest temperature ever recorded in the United States is 134°F at Greenland Ranch, Death Valley, California, on July 13, 1913. What is this temperature on the Celsius and Kelvin scales?

1.83. Critical Temperature The discovery of "high-temperature" superconducting materials in the mid-1980s spurred a race to prepare the material with the highest superconducting temperature. The *critical temperatures* (T_c)—the temperatures at which the material becomes superconducting—of $YBa_2Cu_3O_7$, Nb_3Ge, and $HgBa_2CaCu_2O_6$ are 93.0 K, −250.0°C, and −231.1°F, respectively. Convert these temperatures into a single temperature scale, and determine which superconductor has the highest T_c value.

1.84. The boiling point of O_2 is −183°C; the boiling point of N_2 is 77 K. As air is cooled, which gas condenses first?

Additional Problems

*1.85. Agricultural Runoff A farmer applies 1.50 metric tons of a fertilizer that contains 10% nitrogen to his fields each year (1 metric ton = 1000 kg). Fifteen percent of the fertilizer washes into a stream that runs through the farm. If the stream flows at an average rate of 1.4 cubic meters per minute, what is the additional concentration of nitrogen (expressed in milligrams of nitrogen per liter) in the stream water due to the farmer's yearly application of fertilizer?

1.86. Your laboratory instructor has given you two shiny, light-gray metal cylinders. Your assignment is to determine which one is made of aluminum ($d = 2.699$ g/mL) and which one is made of titanium ($d = 4.54$ g/mL). The mass of each cylinder was determined on a balance to five significant figures. The volume of each was determined by immersing it in a partially filled graduated cylinder as shown in Figure P1.86.

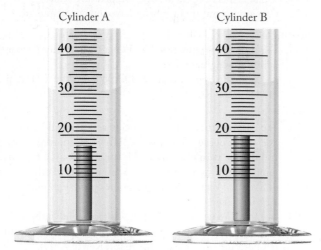

FIGURE P1.86

The initial volume of water was 25.0 mL in each graduated cylinder. The following data were collected:

	Mass (g)	Height (cm)	Diameter (cm)
Cylinder A	15.560	5.1	1.2
Cylinder B	35.536	5.9	1.3

a. Calculate the volume of each cylinder using the dimensions of the cylinder only.
b. Calculate the volume from the water displacement method.
c. Which volume measurement allows for the greater number of significant figures in the calculated densities?
d. Express the density of each cylinder to the appropriate number of significant figures.

*1.87. The road salt that is used in cold climates to melt ice and snow in the winter months contains 1.54 grams of chloride ions for every 1.00 grams of sodium ions. Which of the following mixtures would react to produce NaCl with no sodium or chlorine left over?
a. 11.0 grams of sodium and 17.0 grams of chlorine
b. 6.5 grams of sodium and 10.0 grams of chlorine
c. 6.5 grams of sodium and 12.0 grams of chlorine
d. 6.5 grams of sodium and 8.0 grams of chlorine

*1.88. The wood of the black ironwood tree (*Krugiodendron ferreum*, Figure P1.88), which grows in the West Indies and coastal areas of South Florida, is so dense that it sinks in seawater. Does it sink in freshwater, too? Explain your answer.

FIGURE P1.88

1.89. Manufacturers of trail mix have to control the distribution of ingredients in their products. Deviations of more than 2% from specifications cause production delays and supply problems. A favorite trail mix is supposed to contain 67% peanuts and 33% raisins. Bags of trail mix were sampled from the production line on different days with the following results:

Day	Number of Peanuts	Number of Raisins
1	50	32
11	56	26
21	48	34
31	52	30

On which day(s) did the product meet the specification of 65% to 69% peanuts?

*1.90. Gasoline and water are immiscible. Regular-grade (87 octane) gasoline has a lower density (0.73 g/mL) than water (1.00 g/mL). A 100 mL graduated cylinder with an inside diameter of 3.2 cm contains 34.0 g of gasoline and 34.0 g of water. What is the combined height of the two liquid layers in the cylinder? The volume of a cylinder is $\pi r^2 h$, where r is the radius and h is the height.

1.91. In 1999 a drug overdose incident occurred when a prescription that called for a patient to receive 0.5 *grains* of the powerful sedative phenobarbital each day was misread, and the patient was given 0.5 *grams* of the drug. Actually, four intravenous injections of 130 mg each were administered each day for 3 days. How many times as much phenobarbital was administered with respect to the prescribed amount? (1 grain = 64.79891 mg.)

*1.92. **Mercury in Dental Fillings** The controversy over human exposure to mercury from dental fillings (Figure P1.92) is linked to concerns that mercury may volatilize from fillings made of a combination of silver and mercury, and may then be breathed into the lungs and absorbed into the blood. In 1995 the U.S. Environmental Protection Agency set a safe exposure level for mercury vapor in air of 0.3 μg/m³. Typically, an adult breathes in 0.5 liters of air 15 times per minute.

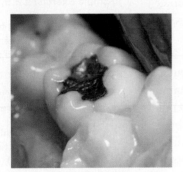

FIGURE P1.92

a. What rate of volatilization of mercury (in μg/minute) from dental fillings would create an exposure level of 0.3 μg Hg/m³ in the air entering the lungs of an adult?

b. The safe exposure level to inhaled mercury vapor adopted by Health Canada is only 0.06 μg/m³. What rate of volatilization of mercury (in μg/minute) from dental fillings would create this exposure level in air entering the lungs of a child who breathes in 0.35 liters of air 18 times per minute?

*1.93. The digital thermometers used in a hospital are evaluated by immersing them in an ice-water bath at 0.0°C and then in boiling water at 100.0°C. Evaluations of three thermometers yield the following results.

Thermometer	MEASURED TEMPERATURE	
	Ice Water	Boiling Water
A	−0.8	99.4
B	0.2	99.8
C	0.4	101.0

a. Which, if any, of the three thermometers would detect an increase of 0.1°C in the temperature of a patient?

b. Which, if any, of the three thermometers would accurately give a reading of 36.8°C (under-the-tongue temperature) for a patient without a fever?

1.94. **Deepwater Horizon Oil Spill** According to the U.S. government, 4.9 billion barrels of crude oil flowed into the Gulf of Mexico following the explosion that destroyed the Deepwater Horizon drilling rig in April, 2010. Express this volume of crude oil in liters and in cubic kilometers. (1 barrel of oil = 42 gallons.)

2

Atoms, Ions, and Molecules

The Building Blocks of Matter

When Artillery Shells Bounced off Tissue Paper

Philosophers in ancient Greece proposed that matter was composed of indestructible particles called atoms. For over 2000 years the concept of an *indestructible* atom also meant an *indivisible* atom. However, a series of discoveries made from the end of the 19th century through the 1930s showed that atoms are not indivisible. Instead, they are made of even tinier *sub*atomic particles: positively charged protons, negatively charged electrons, and particles with no electrical charge called neutrons.

By 1905, scientists had been able to accurately determine the mass and charge of electrons, but they were unclear on how electrons were distributed inside atoms. Nor were they clear on the location of the atom's positive charge that had to be there to neutralize the negative charges of the electrons. In one model of atomic structure from that time, electrons were thought to be embedded within diffuse clouds of positive charge. In 1909 students of Ernest Rutherford at the University of Manchester in England conducted experiments to test this model. They bombarded a thin piece of gold foil with a beam of positively charged alpha particles—and were shocked by what happened.

Most of the alpha particles went straight through the foil, as Rutherford had expected, because they should interact only weakly with the electrons and diffuse clouds of positive charge. However, a few of the particles were deflected well away from the incident beam, and a very few bounced right back at the beam's source. Rutherford later described his amazement at the result: "It is about as incredible as if you had fired a 15-inch shell[1] at a piece of tissue paper and it came back and hit you."

The 1905 model of the atom could not explain such large angles of deflection. Clearly, a new model was needed, and Rutherford proposed one in which every atom has at its center a positively charged nucleus that contains nearly all the atom's mass, and most of an atom's volume is occupied by a cloud of electrons that swirl around the nucleus. The nucleus–electron cloud model has

[1]A shell with a diameter of 15 inches (38 cm) was the largest projectile that could be fired by a British battleship in 1909.

The Large Hadron Collider Inside this particle accelerator located near Geneva, ▶ Switzerland, beams of protons traveling at near the speed of light collide with each other, releasing energy and subatomic particles never observed before.

subatomic particles the neutrons, protons, and electrons in an atom.

FIGURE 2.1 In 1897 J. J. Thomson discovered electrons as he studied how gases conduct electricity. The research earned him the 1906 Nobel Prize in Physics.

▶️‖ **CHEMTOUR** Cathode-Ray Tube

been extensively tested and is still used today. Later experiments in Rutherford's own lab led to the discovery of the proton, the subatomic carrier of positive charge in nuclei. In the 1930s the neutron was discovered and our modern view of the atom, linked closely to Rutherford's, was complete. And yet there is still much that we do not know about the particles in atoms. According to the theories of modern physics, neutrons and protons are made up of even tinier elementary particles called *quarks*. Quarks are held together by particles of force called *gluons*. In the 1960s another force particle, called the *Higgs boson*, was hypothesized to be responsible for the masses of elementary particles and matter itself. In July 2012, teams of scientists working with the Large Hadron Collider reported the discovery of this elusive component of the world we live in.

2.1 The Rutherford Model of Atomic Structure

By the end of the 19th century, many scientists realized that atoms were not the smallest particles of matter, but instead were made up of even smaller **subatomic particles**. The first subatomic particle to be discovered and studied was the electron. The scientist who discovered it was an Englishman, Joseph John (J. J.) Thomson (1856–1940, Figure 2.1).

Electrons

In the late 19th century J. J. Thomson set out to better understand how gases conduct electricity. In his research he used a device called a cathode-ray tube, or CRT (Figure 2.2), which consists of a glass tube from which most of the air has been removed. Electrodes within the tube are attached to the poles of a high-voltage power supply. An electrode called the cathode is connected to the negative

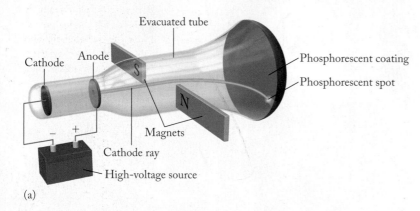

(a)

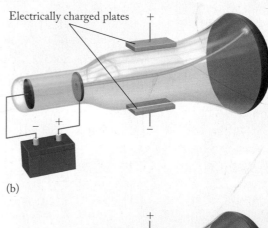

(b)

FIGURE 2.2 A cathode ray is generated when electricity is passed through a tube from which most of the air has been removed. Though invisible to the unaided eye, the path of the ray can be inferred by the bright spot it makes in a phosphorescent material coated on the end of the tube. (a) Cathode ray deflected in one direction by a magnetic field; (b) cathode ray deflected in the opposite direction by an electric field; (c) electric and magnetic fields tuned to balance out the deflections.

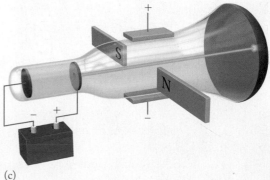

(c)

terminal of the power supply, and an electrode called the anode is connected to the positive terminal. When these connections are made, electricity may travel the length of the glass tube in the form of a beam of **cathode rays** that flows from the cathode toward the anode, passing through a hole cut into the center of the anode. Cathode rays are invisible to the naked eye, but when the end of the tube opposite the cathode is coated with a phosphorescent material, a glowing spot appears where the beam hits the coating.

Thomson discovered that cathode-ray beams can be deflected by magnetic (Figure 2.2a) and electric (Figure 2.2b) fields. This behavior told Thomson that cathode rays were not rays of energy, but rather charged particles of matter. The directions of the deflections told him that their charges were negative. By adjusting the strengths of the electric and magnetic fields, Thomson could balance out the deflections (Figure 2.2c) so that the particles passed straight through the CRT. From the strengths of the opposing electric and magnetic fields, he was able to calculate the mass-to-charge ratio of the particles. He also observed that the deflection pattern and calculated mass-to-charge ratio was always the same no matter what cathode material he used to generate the beam of particles. This observation convinced Thomson that these particles, which he called *corpuscles* but which we now know as **electrons**, are fundamental particles that occur in all forms of matter.

In 1909 American physicist Robert Millikan (1868–1953) advanced Thomson's work by determining the charge of an electron and, indirectly, its mass. Figure 2.3 illustrates Millikan's experimental apparatus. It consisted of two chambers filled with air. A fine spray of oil drops produced in the top chamber fell through a hole into the bottom chamber. Highly energetic X-rays also passed through the bottom chamber, colliding with and dislodging electrons from molecules of the air inside it, mostly N_2 and O_2.

When a molecule (or atom) loses an electron, it also loses the electron's negative charge, which means that it is left with a net positive charge. For example, when an X-ray removes an electron from a molecule of nitrogen, N_2, the product is a species called a molecular **ion** that has a positive charge:

$$N_2 \rightarrow N_2^+ + e^- \tag{2.1}$$

where the superscripts indicate the electrical charges on the nitrogen ion (1+) and electron (1−). Note that the charges on the right side of the reaction arrow add up to zero. They have to because their sum must match the zero charge of the neutral molecule on the left side if Equation 2.1 is to be balanced—as all equations must be.

An atom or molecule may lose more than one electron. Those that lose two, three, or four electrons, for example, also become positively charged ions, called **cations**, with corresponding charges of 2+, 3+, and 4+. Particles of matter (such as Millikan's drops of oil, as we will see) can also gain electrons. When atoms or molecules gain electrons, they form negatively charged ions called **anions** with charges proportional to the number of electrons gained: 1− from gaining one electron, 2− from gaining two electrons, and so on.

As Millikan's oil drops fell through the bottom chamber, they collided with and absorbed

▶❚❚ **CHEMTOUR** Millikan Oil-Drop Experiment

FIGURE 2.3 In Millikan's oil-drop experiment, X-rays ionized the air in the lower chamber, producing electrons that were absorbed by tiny drops of oil falling through the chamber. The descent of the electrically charged drops could be slowed, stopped, or even reversed by applying an electric field with charged plates above and below the chamber. A microscope allowed Millikan to follow the descent of the oil drops.

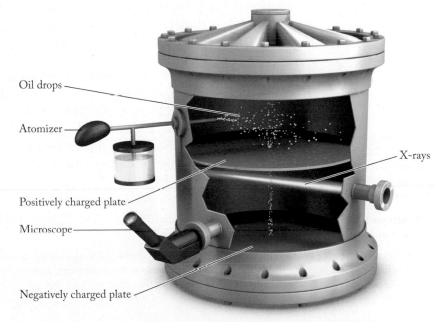

Oil drops

Atomizer

Positively charged plate

Microscope

Negatively charged plate

X-rays

free electrons, thereby acquiring the electrons' negative charges. He observed the rate of fall of the negatively charged oil drops with a microscope in the side of the bottom chamber (Figure 2.3). Because like charges repel one another and opposite charges attract, Millikan could adjust the drops' rate of fall using charged metal plates located above and below the bottom chamber, thus creating a vertical electric field within it. From the strength of the electric field and the rate of fall of the drops, he calculated the charges on them. Millikan discovered that the charge on each drop was always a whole-number multiple of a minimum charge. He concluded that this minimum charge must be the charge on one electron. Millikan's determination of the charge of an electron was within 1% of the modern value, -1.602×10^{-19} coulombs. (The coulomb, abbreviated C, is the SI unit of electric charge.) From Thomson's value of the electron's mass-to-charge ratio, Millikan calculated its mass.

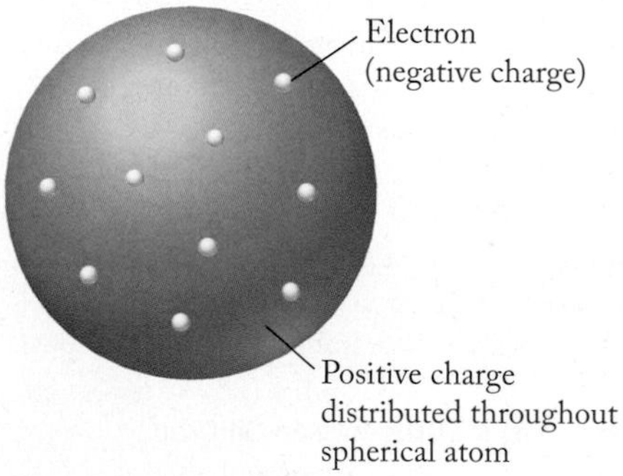

CONCEPT TEST ....

By adjusting the electrical charges on the top and bottom plates of the lower chamber of his apparatus, Millikan was able to slow, stop, and even reverse the fall of oil drops. Which of the two plates must have been positively charged?

(Answers to Concept Tests are in the back of the book.)
....

Electron
(negative charge)

Positive charge
distributed throughout
spherical atom

FIGURE 2.4 In Thomson's plum-pudding model, atoms consist of electrons distributed throughout a massive, positively charged, but very diffuse sphere. The plum-pudding model lasted only a few years before it was replaced by a model based on experiments carried out under the direction of Thomson's former student Ernest Rutherford.

The discovery of the electron raised questions about the possibility of other subatomic particles. Scientists knew that matter was electrically neutral, but they didn't know how the electrons and positive charges were arranged inside atoms. Thomson proposed what came to be called the "plum-pudding" model (Figure 2.4) in which electrons were distributed throughout the atom like raisins in an English plum pudding (or blueberries in a muffin). Thomson's plum-pudding model lasted only a few years. Its demise was linked to another scientific discovery of the 1890s: radioactivity.

Radioactivity

In 1896 French physicist Henri Becquerel (1852–1908) discovered that pitchblende, a brownish-black mineral that is the principal source of uranium, produces radiation that can be detected using photographic plates. Becquerel and his contemporaries initially thought that this radiation consisted of X-rays, which had just been discovered by German scientist Wilhelm Conrad Röntgen (1845–1923).[2] Additional experiments by Becquerel, by the Polish and French wife-and-husband team of Marie Curie (born Marie Skłodowska, 1867–1934) and Pierre Curie (1859–1906), and by British scientist Ernest Rutherford (1871–1937, Figure 2.5) showed that Becquerel's radiation was actually several types of **radioactivity**, a term used to describe the spontaneous emission of high-energy radiation and particles by radioactive materials such as pitchblende.

In studying the particles emitted by pitchblende, Rutherford found that one type, which he named **beta (β) particles**, penetrated materials better than another type, which he named **alpha (α) particles**. He knew that both types of particles

radioactivity the spontaneous emission of high-energy radiation and particles by materials.

beta (β) particle a radioactive emission that is a high-energy electron.

alpha (α) particle a radioactive emission with a charge of 2+ and a mass equivalent to that of a helium nucleus.

[2]Röntgen discovered X-rays in experiments with a cathode-ray tube much like the apparatus used by J. J. Thomson. After completely encasing the tube in a black carton, Röntgen discovered that invisible rays escaped the carton and were detected by a photographic plate. Because he knew so little about the rays, he called them X-rays.

could be deflected by magnetic fields, proving that they were electrically charged. How much they were deflected by fields of different strengths allowed him to calculate their mass-to-charge ratios. He found that this ratio for β particles exactly matched the mass-to-charge ratio of the electron that had been determined by Thomson, which led to the conclusion that β particles are simply high-energy electrons.

The direction in which α particles are deflected in a magnetic field is opposite that of β particles, which means α particles are positively charged. If the β particle (an electron) is assigned a relative charge of 1−, the corresponding charge of an α particle is 2+. Rutherford also discovered that α particles are about 10,000 times more massive than β particles.

FIGURE 2.5 Ernest Rutherford was born in New Zealand. He was awarded a scholarship in 1894 that enabled him to go to Trinity College in Cambridge, England, where he was a research assistant in the laboratory of J. J. Thomson. His contributions included characterizing the properties of α and β particles. By 1907 he was a professor at the University of Manchester, where his famous gold-foil experiments led to our modern view of atomic structure. He received the Nobel Prize in Chemistry in 1908.

Alpha particles emitted by pitchblende played a key role in the demise of the plum-pudding model of atomic structure. In 1909 Rutherford directed two of his students at Manchester University—Hans Geiger (1882–1945, for whom the Geiger counter was named) and Ernest Marsden (1889–1970)—to test the plum-pudding model by bombarding a thin foil of gold with a beam of α particles (Figure 2.6a). If the plum-pudding model were correct, then most of the particles would pass straight through the diffuse spheres of positive charge that made up the gold atoms, though a few might interact with the electrons (the "raisins" embedded in the pudding) enough to be deflected slightly (Figure 2.6b). This was the result Rutherford expected to see.

▶ ‖ **CHEMTOUR** Rutherford Experiment

What Geiger and Marsden observed was not what Rutherford expected. For the most part, the α particles did pass straight through the gold. However, about 1 in every 8000 particles was deflected by an average angle of 90 degrees (Figure 2.6c), and a very few bounced almost straight back at their source. These results ended the short life of the plum-pudding model because it could not account for such large angles of deflection. Rutherford concluded that the deflections

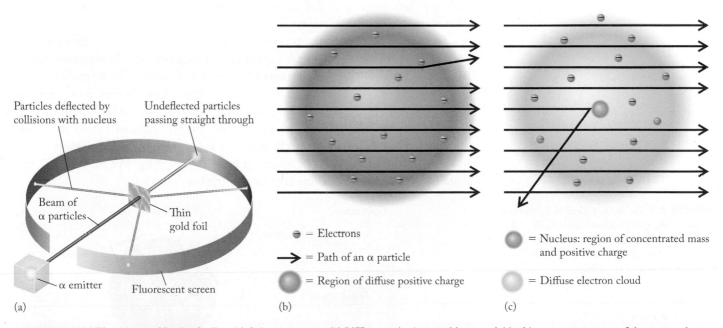

FIGURE 2.6 (a) The design of Rutherford's gold-foil experiments. (b) If Thomson's plum-pudding model had been correct, most of the α particles would have passed straight through thin gold foil, though a few might have been deflected slightly. (c) In fact, most did pass straight through, but a few were scattered widely as shown in (a). This unexpected result led to the theory that an atom has a small, positively charged nucleus that contains most of the mass of the atom.

nucleus (of an atom) the positively charged center of an atom that contains nearly all the atom's mass.

proton a subatomic particle, present in the nucleus of an atom, that has a relative charge of 1+ and a mass number of 1.

neutron an electrically neutral (uncharged) subatomic particle with a mass number of 1.

atomic mass unit (amu) unit used to express the relative masses of atoms and subatomic particles that is exactly 1/12 the mass of 1 atom of carbon with 6 protons and 6 neutrons in its nucleus.

dalton (Da) a unit of mass equal to 1 atomic mass unit.

occurred because the α particles occasionally encountered small regions of high positive charge and large mass. Based on the relative numbers of α particles that were deflected in this way, Rutherford determined that the diameter of a gold atom is over 10,000 times greater than the diameter of the region of positive charge at its center. Rutherford's model of the atom is the basis for our current understanding of atomic structure. It assumes that an atom consists of a tiny **nucleus** that contains the positive charge and most of the mass of the atom, and is surrounded by a diffuse cloud of negatively charged electrons.

CONCEPT TEST

If an α particle hits an electron in an atom of gold, why doesn't it bounce back the way it does when it hits the nucleus of a gold atom?

(Answers to Concept Tests are in the back of the book.)

The Nuclear Atom

In the decade following the gold-foil experiments, Rutherford and others observed that bombarding elements with α particles sometimes changed, or *transmuted*, the elements into other elements. They also discovered that hydrogen nuclei were frequently produced during transmutation reactions. By 1920 a consensus was growing that hydrogen nuclei, which Rutherford called **protons** (from the Greek *protos,* meaning "first"), were part of all nuclei. For example, to account for the mass and charge of an α particle, Rutherford assumed that it was made of four protons, two of which had combined with two electrons to form two electrically neutral particles, which he called **neutrons**. Repeated attempts to produce neutrons by neutralizing protons with electrons were unsuccessful. However, in 1932 one of Rutherford's former students, James Chadwick (1891–1974), was the first to successfully detect and characterize free neutrons. With the discovery of neutrons the current model of atomic structure was complete, as illustrated by the gold atom in Figure 2.7.

Table 2.1 summarizes the properties of neutrons, protons, and electrons. The masses of these tiny particles are expressed in kilograms (the SI standard unit of mass) and in **atomic mass units (amu)**. One amu is exactly 1/12 the mass of a carbon atom that has 6 protons and 6 neutrons in its nucleus. Atomic mass units are also called **daltons (Da)** or *unified atomic mass units* (u). The term "dalton" honors English chemist John Dalton, who published the first table of atomic masses in 1803. If we divide the mass in kilograms of any of the particles in Table 2.1 by its mass in amu, we find that 1 amu = 1.66054×10^{-27} kg. We make use of this equality in several calculations later in this chapter.

FIGURE 2.7 The modern view of Rutherford's model of the gold atom includes a nucleus that is about 1/10,000 the overall size of the atom. Note that the nucleus would be too small to see if drawn to scale in the left drawing.

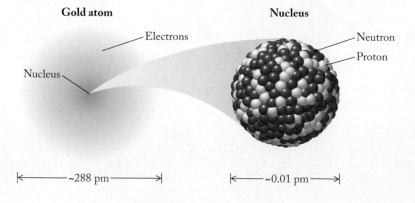

Gold atom

Electrons

Nucleus

Nucleus

Neutron

Proton

|← ~288 pm →|

|← ~0.01 pm →|

Particle	Symbol	Mass (amu)	Mass Number	Mass (kg)	Charge (relative value)	Charge (C)
Neutron	^1_0n	1.00867	1	1.67493×10^{-27}	0	0
Proton	^1_1p	1.00728	1	1.67262×10^{-27}	1+	$+1.602 \times 10^{-19}$
Electron	$^0_{-1}e$ or $^0_{-1}\beta$	5.485799×10^{-4}	0	9.10938×10^{-31}	1−	-1.602×10^{-19}

TABLE 2.1 Properties of Subatomic Particles

The data in Table 2.1 show that the masses of neutrons and protons are approximately equal to each other and much greater than the mass of an electron. The relative masses of these particles are reflected in their mass numbers, which are not real mass values, but rather a way of counting the number of subatomic particles in an atom. Neutrons and protons, which make up nearly all an atom's mass, each have mass numbers of 1, and the minimal contribution that electrons make to the mass of an atom is reflected in their mass number: zero.

2.2 Nuclides and Their Symbols

Around the time that Thomson was exploring the nature of cathode rays, he and other scientists were also using modified cathode-ray tubes to analyze the beams of positively charged particles that flowed from the anode toward the cathode when cathode rays (electrons) flowed in the opposite direction. A combination of electric and magnetic fields surrounded the tubes of these *positive-ray analyzers*, deflecting positively charged ions depending on their masses and charges. If all the ions in a particular beam had the same charge, say 1+, and different masses, the ions with the greatest mass would be deflected the least, and those with the smallest mass would be deflected the most.

In 1912 Thomson and his research assistant, Francis W. Aston (1877–1945), observed that when they passed an electric current through a tube that contained small quantities of neon gas, two bright patches formed on a photographic plate, as shown in Figure 2.8. They assumed that the patches were produced by neon atoms that had lost electrons in the tube, forming positively charged Ne^+ ions. The presence of two patches meant that Ne^+ ions and their parent Ne atoms can have two different masses. The one with lower mass produced the brighter patch, indicating that more of its ions hit the plate and that it must be the more abundant of the two.

Today we know that the Ne^+ ions detected by Thomson and Aston came from two different **isotopes** of neon. Isotopes are atoms of the same element that have the same number of protons (10 for neon) in their nuclei, but different numbers of neutrons. The lighter Ne isotope has 10 neutrons per nucleus, giving it a total mass of about 20 amu. Atoms of the less abundant isotope have 12 neutrons in their nuclei, giving them a mass of about 22 amu. The term **nuclide** is used to refer to any atom of any element that has a particular number of neutrons in its nucleus.

Since the time of John Dalton, scientists had defined an element as *matter composed of identical atoms, all of which have the same mass.* The work of Thomson, Aston, Chadwick, and others required a modification of this definition: henceforth an element was defined as *matter composed of atoms all having the same number of protons in their nuclei.* This number of protons is called the **atomic number (Z)** of

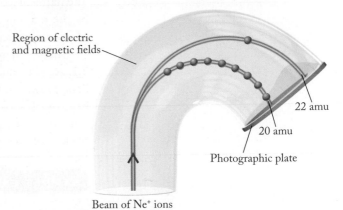

Region of electric and magnetic fields

22 amu
20 amu
Photographic plate

Beam of Ne^+ ions

FIGURE 2.8 Aston's positive-ray analyzer. A beam of Ne^+ ions passing through electric and magnetic fields separates into two beams. Ions with a mass of about 20 amu—90% of the total—are deflected more than the other 10%, which have a mass of about 22 amu. Aston's positive-ray analyzer was the forerunner of the modern mass spectrometer.

isotopes atoms of an element containing the same number of protons but different numbers of neutrons.

nuclide a specific isotope of an element.

atomic number (Z) the number of protons in the nucleus of an atom.

nucleon a proton or neutron in a nucleus.

mass number (A) the number of nucleons in an atom.

the element. The total number of **nucleons** (neutrons and protons) in the nucleus of an atom defines its **mass number (A)**. The isotopes of a given element thus all have the same atomic number Z but different mass numbers A.

The symbol we use to represent a particular nuclide has the generic form

$$^A_Z X$$

where X represents the one- or two-letter symbol for the element, Z is the element's atomic number, and A is the mass number of the particular nuclide. For example, two of the isotopes of oxygen (O) and two of the isotopes of lead (Pb) are written

$$^{16}_8 O \qquad ^{18}_8 O \qquad ^{206}_{82} Pb \qquad ^{208}_{82} Pb$$

Because Z and X provide the same information—each by itself identifies the element—the subscript Z may be omitted, so that the nuclide symbol may be simplified to just

$$^A X$$

The name of a nuclide may also be spelled out as the name of the element followed by the mass number of the nuclide. For example, the names of the two isotopes of neon detected by Thomson and Aston may be written *neon-20* and *neon-22*.

The $^A_Z X$ symbol can also be used to represent subatomic particles as in Table 2.1. As with nuclides, the superscripts of the symbols are the particles' mass numbers. However, the subscripts represent the relative charges of the particles: 0 for a neutron, 1 for a proton, and −1 for an electron. This notation for subscripts makes sense because an element's atomic number Z is simply the number of protons in the nucleus of an atom of that element, which is the positive charge of that nucleus.

SAMPLE EXERCISE 2.1   Writing Symbols of Nuclides   LO1

Write symbols in the form $^A_Z X$ for the nuclides that have (a) 6 protons and 6 neutrons, (b) 11 protons and 12 neutrons, and (c) 92 protons and 143 neutrons.

COLLECT AND ORGANIZE We know the number of protons and neutrons in the nuclei of three nuclides and are asked to write symbols of the form $^A_Z X$ where Z is the atomic number, A is the mass number, and X is the symbol of the element.

ANALYZE The number of protons in the nucleus of an atom defines its atomic number (Z), which in turn defines which element it is (X). The sum of the number of nucleons (protons plus neutrons) is the mass number (A) of the nuclide.

SOLVE

a. This nuclide has 6 protons, so $Z = 6$, which is the atomic number of carbon. Six protons plus six neutrons gives a mass number of 12. Therefore this nuclide is carbon-12, or $^{12}_6 C$.

b. This nuclide has 11 protons, so $Z = 11$, which is the atomic number of sodium. Eleven protons plus 12 neutrons gives a mass number of 23. This isotope is sodium-23, or $^{23}_{11} Na$.

c. This nuclide has 92 protons, so $Z = 92$, which is the atomic number of uranium. The mass number is $92 + 143 = 235$, so this isotope is uranium-235, or $^{235}_{92} U$.

THINK ABOUT IT In working through this exercise, did you use the periodic table of the elements inside the front cover to identify the symbol of the element once you knew its atomic number? It's an easy search because the elements in the table are arranged in order of increasing atomic number.

Practice Exercise Use the formats $^A_Z X$ and $^A X$ to write the symbols of the nuclides whose atoms each have (a) 26 protons and 30 neutrons, (b) 7 protons and 8 neutrons, (c) 17 protons and 20 neutrons, and (d) 19 protons and 20 neutrons. ⚙

(Answers to Practice Exercises are in the back of the book.)

SAMPLE EXERCISE 2.2   **Identifying Ions and Writing**   **LO1**
Their Isotopic Symbols

In March, 2011, an earthquake and tsunami off the coast of northeast Japan crippled nuclear reactors at a power station in Fukushima, Japan. The resulting explosions and fires released radiation into the atmosphere and ocean that included three radioactive, single-atom particles that had the following numbers of protons, neutrons, and electrons:

	Protons	Neutrons	Electrons
(a)	53	78	54
(b)	54	79	54
(c)	55	82	54

Identify which of the species are ions, and write the symbols of all of them in the form $^A X^Q$ where Q is the charge of the particles that are ions.

COLLECT AND ORGANIZE We know the numbers of protons, neutrons, and electrons in three particles that are either atoms or single-atom ions, and we are asked to write their symbols in the form $^A X^Q$ where Q is the charge of each ion.

ANALYZE The number of protons in the nucleus of an atom or ion defines its atomic number, which in turn defines which element it is (X). The sum of the number of nucleons (protons plus neutrons) is the mass number (A). The charge (Q) is the difference between the number of protons and the number of electrons. It will have a negative value in ions having more electrons than protons, and it will have a positive value in ions with more protons than electrons. Particles with equal numbers of protons and electrons have no charge.

SOLVE

a. This single-atom particle has 53 protons, so its atomic number is 53, which makes its parent element iodine (I). Its mass number (A) is $53 + 78 = 131$. It is an ion because it has unequal numbers of positively charged protons (53) and negatively charged electrons (54). The net charge on the ion is $53 - 54$, or $1-$. The charge symbol of an ion with a single negative charge is simply "$-$", so the symbol of this ion is $^{131}I^-$.

b. This particle has equal numbers of protons and electrons (54), which means it has no charge and, therefore, is an *atom* of xenon (Xe). The mass number (A) of this xenon isotope is $54 + 79 = 133$, so its symbol is simply ^{133}Xe.

c. This particle has 55 protons, which makes its parent element cesium (Cs). Its mass number (A) is $55 + 82 = 137$, and its charge is $55 - 54 = 1$. The charge symbol of an ion with a single positive charge is simply "$+$", so the symbol of the ion is $^{137}Cs^+$.

THINK ABOUT IT A difference between the ions in this Sample Exercise (I^- and Cs^+) and the atoms in this exercise and the previous one (Xe, C, Na, and U) is that each of the ions has unequal numbers of protons and electrons. In contrast, each atom has an equal number of protons and electrons, and is therefore electrically neutral.

Practice Exercise Write symbols of three of the major ions in seawater in the form $^AX^Q$ where Q is the symbol for the charge of the ion. The ions have the following numbers of protons, neutrons, and electrons:

	Protons	Neutrons	Electrons
(a)	19	20	18
(b)	12	12	10
(c)	17	18	18

(Answers to Practice Exercises are in the back of the book.)

2.3 Navigating the Periodic Table

Long before chemists knew about subatomic particles and the concept of atomic numbers they knew that groups of elements such as Li, Na, and K, or F, Cl, and Br had similar properties, and that when the elements were arranged by increasing atomic mass, there were repeating patterns of similarities in their properties. This *periodicity* in the properties of the elements inspired several 19th-century scientists to create tables of the elements in which the elements were arranged in patterns based on similarities in their chemical properties.

By far the most successful of these scientists was Russian chemist Dmitri Mendeleev (1834–1907). In 1872 he published a table (Figure 2.9a) that is widely considered the forerunner of the modern **periodic table of the elements** (Figure 2.9b). In addition to organizing all the elements that were known at the time, Mendeleev realized that there might be elements in nature that were yet to be discovered. This insight meant that he could leave empty cells in his table for unknown elements. Doing so allowed him to align the known elements so that those in each column had similar chemical properties. Based on the locations of the empty cells, Mendeleev was able to predict the chemical properties of the missing elements. These predictions greatly facilitated the subsequent discovery of these elements by other scientists. Note that Mendeleev arranged the elements in his periodic table in order of increasing atomic *mass*, but in the modern periodic table the elements appear in order of their atomic *numbers*.

CONCEPT TEST

Suggest a reason why the elements in Mendeleev's version of the periodic table are in order of atomic mass and not atomic number.

(Answers to Concept Tests are in the back of the book.)

periodic table of the elements a chart of the elements in order of their atomic numbers and in a pattern based on their physical and chemical properties.

period (of elements) all the elements in a row of the periodic table.

group or **family** (of elements) all the elements in a column of the periodic table.

radionuclide a radioactive (unstable) nuclide.

The elements in the modern periodic table are arranged in seven horizontal rows (also called **periods**) and 18 columns that are commonly called **groups** (and sometimes **families**). The rows are numbered at the far left. The group numbers appear at the top of each column. The periodic table inside the front cover has a second set of column headings consisting of a number followed by the letter A or B. These secondary headings were widely used in earlier versions of the table, and many scientists still find them useful because the number preceding A or B is linked to the charges on the single-atom, or *monatomic*, ions that the elements in that column have in ionic compounds (Figure 2.10), or, as we discover in Chapter 4, to the number of chemical bonds their atoms form in molecular compounds. The coincidence of new and old group numbers with the most common ionic charges is illustrated in the following table. Actually, *all* the cations of group 1 and 2 elements have charges of 1+ and 2+, respectively.

Group	Charges of the Most Common Monatomic Cations
1/1A	1+
2/2A	2+
3/3B	3+
4/4B	4+
13/3A	3+

Figure 2.10 also shows that the charges on the common monatomic anions of group 17 elements are always 1−, those in group 16 are always 2−, and those in group 15 are always 3−. Note how in all three groups the charges of the anions are equal to the group number minus 18 (the total number of groups in the table).

The first row of the periodic table contains only two elements—hydrogen and helium—and the second and third rows each contain only eight. Starting with the fourth row, all 18 columns are full. Actually, the sixth and seventh rows contain more elements than there is space for in an 18-column array. The additional elements appear in the two separate rows at the bottom of the main table. Elements in the row with atomic numbers from 58 to 71 are called the lanthanides (after element 57, lanthanum) and those with atomic numbers between 90 and 103 are called actinides (after element 89, actinium). Most of the nuclides of the actinide elements are radioactive, that is, they are **radionuclides**. Those with atomic numbers above 94 are highly radioactive, which means they spontaneously emit high-energy radiation and particles and are transformed into other nuclides. Therefore, they are not

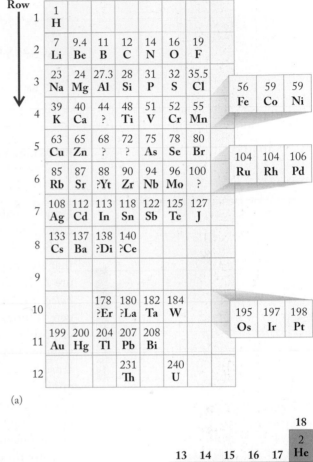

FIGURE 2.9 Two views of the periodic table. (a) Mendeleev organized his periodic table around similar properties and atomic masses. He assigned three elements with similar properties to group VIII in rows 4, 6, and 10. As a result, his rows 4 and 5 together contain spaces for 18 elements, corresponding to the 18 groups in the modern periodic table. (b) In the modern table elements are arranged in order of atomic number (*Z*). Those in tan are metals, those in blue are nonmetals, and those in green are metalloids (also called semimetals).

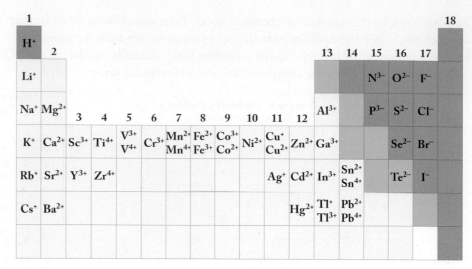

FIGURE 2.10 Periodic trends in the charges of common monatomic ions.

found in nature. They can, however, be synthesized in the laboratory, as we discuss in the last section of this chapter.

Several of the groups of elements have names in addition to numbers—names that are based on chemical properties common to the elements in that group (Figure 2.11a). For example, the elements of group 17 are called **halogens**. The word *halogen* is derived from the Greek for "salt former." Chlorine is a typical halogen. It forms 1:1 binary (two-element) ionic compounds (compounds that consist of positive and negative ions) with the group 1 elements; a familiar example is sodium chloride, NaCl (the principal ingredient in table salt). The 1:1 ratio makes sense because all the group 1 elements form cations that have a 1+ charge, and group 17 elements all form 1− anions (as we saw in Figure 2.10). Equal numbers of these anions and cations means that their charges cancel out, as they must in all ionic compounds so that, overall, they are neutral substances. For their part, the group 1 elements (except for hydrogen) are called **alkali metals**.

Group 2 elements are called **alkaline earth metals**; they form ionic compounds with halogens in which the ratio of cations to anions is 1:2. This ratio makes sense because all the group 2 cations have 2+ charges, so a neutral compound must have twice as many anions with 1− charges as alkaline earth cations. A common example of such a compound is calcium chloride, $CaCl_2$, which is widely used in regions with cold winters to melt ice and snow from sidewalks and roadways.

The elements in the periodic table are also divided into three broad categories highlighted by the three different cell colors in Figure 2.9(b). Elements in the tan cells are **metals**. They tend to conduct heat and electricity well, they're malleable (capable of being shaped by hammering) and ductile (capable of being drawn out in a wire), and all but mercury (Hg) are shiny solids at room temperature. Elements in the blue cells are **nonmetals**. They are poor conductors of heat and electricity. Most are gases at room temperature; the solids among them tend to be brittle, and bromine (Br) is a liquid at room temperature. The elements in the green cells are called **metalloids** or **semimetals**, so named because they tend to have the physical properties of metals but the chemical properties of nonmetals.

Groups 1, 2, and 13 through 18 are referred to collectively as **main group elements** or **representative elements** (Figure 2.11b). They include the most

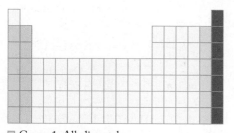

☐ Group 1: Alkali metals
☐ Group 2: Alkaline earth metals
☐ Group 17: Halogens
☐ Group 18: Noble gases

(a)

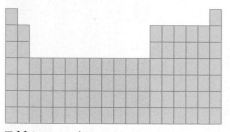

☐ Main group elements
  (representative elements)

☐ Transition metals

(b)

FIGURE 2.11 (a) The commonly used names of groups 1, 2, 17, and 18 of the periodic table. (b) The *main group* (or *representative*) elements are in groups 1, 2, and 13–18. They are separated by the *transition metals* in groups 3–12.

abundant elements in the solar system and many of the most abundant on Earth. Note that they are the "A" elements in the older group numbering system. The elements in groups 3 through 12 are called **transition metals**; they are the old "B" elements. All of them except mercury exhibit the classic physical properties of metals: they are hard, shiny, ductile, malleable solids, and excellent conductors of heat and electricity. The group 18 elements are called **noble gases** because for the most part they do not interact with other elements. In the next chapter, we will explore further connections between the chemical properties of the elements and the groups in which they reside.

SAMPLE EXERCISE 2.3   Navigating the Periodic Table   LO3

The elements described below are the major components of Portland cement, which is used in construction to make concrete and mortar. Which elements are they?

  a. The group 13 element in the third row (period)
  b. The group 16 element with the smallest atomic number
  c. The third-row metalloid
  d. The fourth-row alkaline earth

COLLECT AND ORGANIZE We are asked to identify elements based on the locations of their symbols in the periodic table. In parts a, c, and d we are given the row number; in b we are not provided the row number directly, but we know that the element has the lowest atomic number of all the elements in its group. Information about group locations comes from group number in a and b, the type of element in c, and the group name in d.

ANALYZE (a) The cell address is group 13, row 3; (b) group 16 starts with row 2; (c) the only metalloid in row 3 is in group 14; (d) all the alkaline earths, including the one in row 4, are in group 2.

SOLVE (a) Al, aluminum; (b) O, oxygen; (c) Si, silicon; (d) Ca, calcium

THINK ABOUT IT Each element has a unique location in the periodic table that is linked to both its atomic number and its chemical properties. Chemists usually describe the locations in terms of their group number or name (the column they're in) and their row number.

Practice Exercise What are the symbol and name of each of these elements?

  a. The metalloid in group 15 closest in mass to the noble gas krypton
  b. A representative element in the fourth row that is an alkaline earth metal
  c. A transition metal in the sixth row and in the same group as zinc (Zn)

(Answers to Practice Exercises are in the back of the book.)

---

halogen an element in group 17 of the periodic table.

alkali metal an element in group 1 of the periodic table.

alkaline earth metal an element in group 2 of the periodic table.

metals elements that are typically shiny, malleable, ductile solids that conduct heat and electricity well and tend to form positive ions.

nonmetals elements with properties opposite those of metals, including poor conductivity of heat and electricity.

metalloids or semimetals elements that tend to have the physical properties of metals and the chemical properties of nonmetals.

main group elements or representative elements the elements in groups 1, 2, and 13 through 18 of the periodic table.

transition metals the elements in groups 3 through 12 of the periodic table.

noble gases the elements in group 18 of the periodic table.

average atomic mass the weighted average of masses of all isotopes of an element, calculated by multiplying the natural abundance of each isotope by its mass in atomic mass units and then summing the products.

2.4 The Masses of Atoms, Ions, and Molecules

At the center of each of the cells in the periodic table inside the front cover is the symbol of an element. The number above the symbol is the element's atomic number Z; the number below the symbol is the element's **average atomic mass**. More precisely, the number below the symbol is the *weighted* average of the masses of all the isotopes of the element.

Isotope	Mass (amu)	Natural Abundance (%)
Neon-20	19.9924	90.4838
Neon-21	20.9940	0.2696
Neon-22	21.9914	9.2465

To understand the meaning of a weighted average, let's consider the masses and **natural abundances** of the three isotopes of neon in the table to the left. Natural abundances are usually expressed as percentages. Thus, 90.4838% of all neon atoms are neon-20, 9.2465% are neon-22, and only 0.2696% are neon-21. The natural abundance of neon-21 is so small that Thomson and Aston could not detect it with their positive-ray analyzers. However, modern mass spectrometers are much more sensitive, so we can obtain precise natural abundance values for even such low-abundance isotopes.

We can calculate the average atomic mass of neon by multiplying the mass of each of its isotopes by its natural abundance (in the language of mathematics, we *weight* the isotope's mass using natural abundance as the *weighting factor*) and then summing the three weighted masses. To simplify the calculation we first convert the percent abundance values into their decimal equivalents:

$$
\begin{aligned}
\text{Average atomic mass of neon} = \quad & (19.9924 \text{ amu} \times 0.904838) \\
+ \ & (20.9940 \text{ amu} \times 0.002696) \\
\underline{+ \ } & \underline{(21.9914 \text{ amu} \times 0.092465)} \\
& 20.1799 \text{ amu}
\end{aligned}
$$

It is important to note that no single atom of neon has this average atomic mass; every atom of neon in the universe has the mass of one of the three neon isotopes. The value we just calculated is the *average* mass of all neon atoms.

This method for calculating average atomic mass works for all elements. The general equation for doing the calculation is

$$m_X = a_1 m_1 + a_2 m_2 + a_3 m_3 + \cdots \tag{2.2}$$

where m_X is the average atomic mass of element X, which has isotopes with masses $m_1, m_2, m_3, \ldots$, for which the natural abundances expressed in decimal form are $a_1, a_2, a_3, \ldots$.

SAMPLE EXERCISE 2.4 Calculating Average Atomic Mass   LO4

The precious metal platinum ($Z = 78$) has six isotopes with these natural abundances:

Symbol	Mass (amu)	Natural Abundance (%)
^{190}Pt	189.96	0.014
^{192}Pt	191.96	0.782
^{194}Pt	193.96	32.967
^{195}Pt	194.97	33.832
^{196}Pt	195.97	25.242
^{198}Pt	197.97	7.163

Use the data to calculate the average atomic mass of platinum.

COLLECT AND ORGANIZE We know the masses and natural abundances of each of the six isotopes of platinum and need to calculate the average atomic mass. Equation 2.2 may be used to perform such a calculation.

ANALYZE Using Equation 2.2, we multiply the mass of each isotope by its natural abundance expressed as a decimal, and then add the products together.

natural abundance the proportion of a particular isotope, usually expressed as a percentage, relative to all the isotopes of that element in a natural sample.

molecular mass the mass in amu of one molecule of a molecular compound.

formula unit the smallest electrically neutral unit of an ionic compound.

formula mass the mass in amu of one formula unit of an ionic compound.

SOLVE

$$
\begin{aligned}
\text{Average atomic mass} = \ & (189.96 \text{ amu})(0.00014) \\
& + (191.96 \text{ amu})(0.00782) \\
& + (193.96 \text{ amu})(0.32967) \\
& + (194.97 \text{ amu})(0.33832) \\
& + (195.97 \text{ amu})(0.25242) \\
& + (197.97 \text{ amu})(0.07163) \\
\hline
& 195.08 \text{ amu}
\end{aligned}
$$

THINK ABOUT IT Note that the six values of natural abundances expressed as decimals should add up to 1.00000, and they do. Sometimes this is not the case (check the neon abundances in the previous table). Uncertainties in the last decimal place may be due to uncertainties in measured values or in rounding them off. Our calculated average atomic mass of platinum agrees with the value given inside the front cover.

Practice Exercise Silver (Ag) has two stable isotopes: ^{107}Ag, 106.90 amu, and ^{109}Ag, 108.90 amu. If the average atomic mass of silver is 107.87 amu, what is the natural abundance of each isotope? *Hint*: Let x be the natural abundance of one of the isotopes. Then $1 - x$ is the natural abundance of the other.

(Answers to Practice Exercises are in the back of the book.)

When we apply the concept of particle mass to molecular compounds, the particles involved are not individual atoms but rather groups of atoms chemically bonded together in molecules. Just as an atom of an element has an average atomic mass, a molecule of a molecular compound has a **molecular mass** that is the sum of the average atomic masses of the atoms in it. Like average atomic masses, molecular masses are expressed in amu. To calculate the molecular mass of a compound, we simply add up the average atomic masses of the atoms in each of its molecules. For example, the molecular mass of sulfur dioxide (SO_2), which is one of the gases released during volcanic eruptions (Figure 2.12), is the sum of the average atomic mass of one atom of sulfur and two atoms of oxygen:

$$
\left(\frac{1 \text{ atom S}}{\text{molecule } SO_2} \times \frac{32.065 \text{ amu}}{1 \text{ atom S}} \right) + \left(\frac{2 \text{ atoms O}}{\text{molecule } SO_2} \times \frac{15.999 \text{ amu}}{1 \text{ atom O}} \right) = \frac{64.063 \text{ amu}}{\text{molecule } SO_2}
$$

There are no molecules in ionic compounds such as sodium chloride (NaCl). Instead ionic compounds consist of three-dimensional arrays of positive and negative ions (Figure 2.13). The lack of molecules means that ionic compounds do not have *molecular* masses. However, the formulas of ionic compounds define quantities called **formula units**, the smallest electrically neutral unit in an ionic compound. A formula unit has a particular **formula mass**: the sum of the average atomic masses of the cations and anions that make up a neutral formula unit. Thus, the formula unit of NaCl consists of one Na^+ ion and one Cl^- ion. The formula of calcium chloride is $CaCl_2$, so its formula unit consists of one calcium ion, Ca^{2+}, and *two* chloride ions, Cl^-, and its formula mass is the combined mass of 1 Ca^{2+} ion and 2 Cl^- ions:

$$
\left(\frac{1 \text{ } Ca^{2+} \text{ ion}}{\text{formula unit}} \times \frac{40.078 \text{ amu}}{1 \text{ } Ca^{2+} \text{ ion}} \right) + \left(\frac{2 \text{ } Cl^- \text{ ions}}{\text{formula unit}} \times \frac{35.453 \text{ amu}}{1 \text{ } Cl^- \text{ ion}} \right) = \frac{110.984 \text{ amu}}{\text{formula unit}}
$$

Note that we used the average atomic masses of calcium and chlorine from the periodic table for the masses of the Ca^{2+} and Cl^- ions in this calculation. We can do that because losing or gaining an electron or two has very little impact on the

FIGURE 2.12 Volcanic eruptions such as this one of Kilauea in Hawaii in March, 2011, can introduce thousands of metric tons (1 ton = 1000 kg) of SO_2 into the atmosphere each day.

▶❙❙ **CHEMTOUR** NaCl Reaction

One formula unit

FIGURE 2.13 Crystals of sodium chloride. The cubic shape of the crystals mirrors the cubic array of Na^+ and Cl^- ions that make up its structure. The formula unit of NaCl (inside the dashed oval), or of any ionic compound, contains the same numbers of cations and anions as there are in the chemical formula of the compound—in this case, one of each.

mole (mol) an amount of a substance that contains Avogadro's number ($N_A = 6.0221 \times 10^{23}$) of particles (atoms, ions, molecules, or formula units).

Avogadro's number (N_A) the number of carbon atoms in exactly 12 grams of the carbon-12 isotope; $N_A = 6.0221 \times 10^{23}$. It is the number of particles in one mole.

molar mass ($\mathcal{M}$) the mass of 1 mole of a substance.

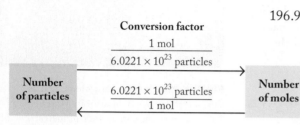

FIGURE 2.14 Converting between a number of particles and an equivalent number of moles (or vice versa) is a matter of dividing (or multiplying) by Avogadro's number.

▶❚❚ **CHEMTOUR** Avogadro's Number

FIGURE 2.15 The quantities shown represent 1 mole of each element: 4.0026 grams of helium gas fill the balloon and, left to right in front, 32.065 grams of solid sulfur, 63.546 grams of copper metal, and 200.59 grams of liquid mercury.

mass of an atom; nearly all of its mass is concentrated in its nucleus. Moreover, the masses of the electrons lost when atoms form cations are balanced by the masses gained when atoms form anions, so the overall mass of a formula unit is exactly the same as the sum of the average atomic masses of the elements in it.

2.5 Moles and Molar Masses

Individual atoms are very tiny particles, which is why we need very powerful instruments such as scanning tunneling microscopes (Figure 1.12) to see them. Atoms also have very little mass. The periodic table tells us that the average mass of a gold atom is 196.97 amu. That is heavy—for an atom—but what is 196.97 amu in grams? If one amu equals 1.66054×10^{-27} kg, then the mass of an average gold atom is only

$$196.97 \text{ amu Au} \times \frac{1.66054 \times 10^{-27} \text{ kg}}{1 \text{ amu}} \times \frac{1000 \text{ g}}{1 \text{ kg}} = 3.2708 \times 10^{-22} \text{ g Au}$$

In our macroscopic (visible) world, chemists usually work with quantities of substances that they can see, transfer from one container to another, and weigh on balances. Inevitably the quantities of substances contain enormous numbers of atoms, ions, or molecules. To deal with such very large numbers, chemists need a unit that relates macroscopic quantities of substances, such as their masses expressed in grams, to the number of particles they contain. That unit is the **mole (mol)**, the SI base unit for expressing quantities of substances (see Table 1.2).

One mole of a substance is defined as the quantity of the substance that contains the same number of particles as the number of carbon atoms in exactly 12 g of the isotope carbon-12. This number, to five significant figures, is 6.0221×10^{23}. This very large value is called **Avogadro's number (N_A)** after the Italian scientist Amedeo Avogadro (1776–1856), whose research enabled other scientists to accurately determine the atomic masses of the elements. To put a number of this magnitude in perspective, it would take 9.4 trillion computer flash drives, each capable of storing 64 gigabytes (6.4×10^{10} bytes) of data, to store a mole of bytes.

Dividing the number of particles in a sample by Avogadro's number yields the number of moles of those particles. For example, U.S. pennies minted before 1960 each contained about 2.4×10^{22} atoms of Cu. That number is equivalent to 4.0×10^{-2} moles of copper:

$$2.4 \times 10^{22} \text{ atoms Cu} \times \frac{1 \text{ mol}}{6.0221 \times 10^{23} \text{ atoms}} = 4.0 \times 10^{-2} \text{ mol Cu}$$

On the other hand, multiplying a number of moles by Avogadro's number gives us the number of particles in that many moles. For example, silicon wafers used in solar panels are typically 10–20 cm on a side and between 200 and 300 μm thick. A silicon wafer with dimensions 10.0 cm by 10.0 cm by 2.00×10^{-2} cm contains 0.166 moles of silicon. The number of silicon atoms in such a wafer is

$$0.166 \text{ mol Si} \times \frac{6.0221 \times 10^{23} \text{ atoms}}{1 \text{ mol}} = 1.00 \times 10^{23} \text{ atoms Si}$$

These atom-to-mole conversions are illustrated in Figure 2.14, and the quantities of some common elements equivalent to 1 mole of each are shown in Figure 2.15. While it is sometimes useful to know the number of atoms of an element in a sample, most of the time we'll focus on how many *moles* of an element are present.

How does a unit of measure such as 500 facial tissues in a box relate to the concept of the mole?

(Answers to Concept Tests are in the back of the book.)

SAMPLE EXERCISE 2.5 **Converting Number of Moles into Number of Particles** **LO5**

The silicon used to make computer chips has to be extremely pure. For example, it must contain less than 3×10^{-10} moles of phosphorus (a common impurity in silicon) per mole of silicon. What is this level of impurity expressed in atoms of phosphorus per mole of silicon?

COLLECT AND ORGANIZE The problem states the maximum number of moles of phosphorus allowed per mole of silicon and asks us to calculate the equivalent number of atoms of P per mole of Si. Avogadro's number defines the number of atoms in 1 mole of an element: 6.0221×10^{23}.

ANALYZE We can convert a number of moles into an equivalent number of atoms by multiplying by Avogadro's number because its denominator contains our initial unit (mol) and its numerator contains the units we seek (atoms). We can get a rough estimate of the correct answer from the exponents of the values in the calculation. The product of 10^{23} and 10^{-10} is $10^{(23-10)} = 10^{13}$.

SOLVE

$$\frac{3 \times 10^{-10} \text{ mol P}}{1 \text{ mol Si}} \times \frac{6.0221 \times 10^{23} \text{ atoms P}}{1 \text{ mol P}} = \frac{2 \times 10^{14} \text{ atoms P}}{\text{mol Si}}$$

THINK ABOUT IT The result reveals that even a very small number of moles of an impurity translates to a very large number of atoms. The exponent of our calculated value (14) is close to our estimate (13), so our answer is probably right.

Practice Exercise If 1.0 mL of seawater contains about 2.5×10^{-14} moles of dissolved gold, how many atoms of gold are in that volume of seawater?

(Answers to Practice Exercises are in the back of the book.)

Molar Mass

The mole provides an important link between the atomic mass values in the periodic table and measurable masses of elements and compounds. To see how this link works, let's convert the average mass of an atom of gold, 196.97 amu, into an equivalent mass expressed in grams per mole of gold:

$$\frac{196.97 \text{ amu Au}}{1 \text{ atom Au}} \times \frac{6.0221 \times 10^{23} \text{ atoms}}{1 \text{ mol}} \times \frac{1.66054 \times 10^{-27} \text{ kg}}{1 \text{ amu}} \times \frac{1000 \text{ g}}{1 \text{ kg}}$$

$$= \frac{196.97 \text{ g Au}}{\text{mol Au}}$$

The atomic mass values expressed in amu/atom and in g/mol *are exactly the same.* This equality holds for all elements: the mass in grams of 1 mole of an atom, a quantity called its **molar mass (M)**, has exactly the same numerical value as the mass of one atom of the element expressed in amu. Thus, the molar mass of gold is 196.97 g/mol (Figure 2.16a).

79
Au
196.97

Atomic mass of Au	196.97 amu/atom
Mass of 1 mol of Au	196.97 g
Molar mass of Au	196.97 g/mol

(a)

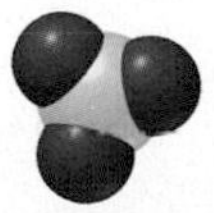

Molecular mass of SO_3	80.06 amu/molecule
Mass of 1 mol of SO_3	80.06 g
Molar mass of SO_3	80.06 g/mol

(b)

FIGURE 2.16 (a) The atomic mass (in amu/atom) and the molar mass (in g/mol) of gold have the same numerical value. (b) The molecular mass (in amu/molecule) and the molar mass (in g/mol) of SO_3 have the same numerical value.

The concept of molar mass applies to compounds as well as elements. Just as the average atomic mass in amu of an element translates exactly into its molar mass in g/mol, the molar mass of a molecular compound is equivalent to its molecular mass in amu, and the molar mass of an ionic compound is equivalent to its formula mass in amu. Thus, the molecular mass of SO_2 that we calculated in the previous section, 64.06 amu, is equivalent to a molar mass of 64.06 g/mol. Molecules of sulfur trioxide, SO_3, have one more atom of oxygen (mass = 15.999 amu) per molecule, giving them a mass of (64.06 + 15.999) = 80.06 amu and giving sulfur trioxide a molar mass of 80.06 g/mol (Figure 2.16b). Similarly, the formula masses of ionic compounds, expressed in amu, have the same values as the molar masses of the compounds expressed in g/mol. As we saw in the previous section, the formula mass of $CaCl_2$ is 110.984 amu, which means its molar mass is 110.984 g/mol. If we didn't already know the molecular mass or formula mass of a compound, we could calculate its molar mass directly from the molar masses of the elements in it. For example, the molar mass of carbon dioxide, CO_2, is the sum of the masses of 1 mole of carbon and 2 moles of oxygen:

$$\left(\frac{1 \ \cancel{mol \ C}}{mol \ CO_2} \times \frac{12.011 \ g}{1 \ \cancel{mol \ C}} \right) + \left(\frac{2 \ \cancel{mol \ O}}{mol \ CO_2} \times \frac{15.999 \ g}{1 \ \cancel{mol \ O}} \right) = \frac{44.010 \ g}{mol \ CO_2}$$

The mole enables us to know the number of particles in any sample of a given substance simply by knowing the mass of the sample. We can do this because the mole represents both a fixed number of particles (Avogadro's number) and a specific mass (the molar mass) of the substance. It may be useful to think about moles and mass using the following analogy: A box that contains 1 dozen golf balls and a box that contains 1 dozen baseballs have very different masses, but each contains 12 balls (Figure 2.17). Indeed, the mole is sometimes referred to as the "chemist's dozen." Figure 2.18 summarizes how to use Avogadro's number, chemical formulas, and molar masses to convert between the mass, the number of moles, and the number of particles in a given quantity of a substance.

FIGURE 2.17 A dozen golf balls weigh less than a dozen baseballs, but both quantities contain the same number of balls.

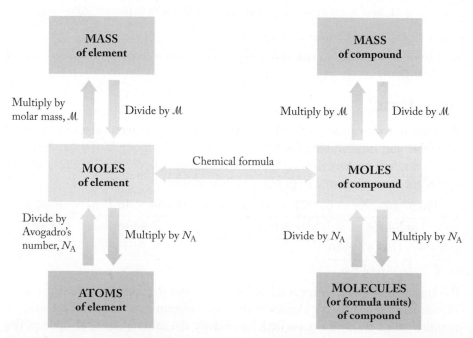

FIGURE 2.18 The mass of a pure substance can be converted into the equivalent number of moles or number of particles (atoms, ions, or molecules) and vice versa.

SAMPLE EXERCISE 2.6 **Converting Number of Moles into Mass** **LO5**

A helium balloon sold at an amusement park contains 0.462 moles of He. How many grams of He are in the balloon?

COLLECT AND ORGANIZE The number of grams of He in 0.462 moles represents the number of helium particles present in this quantity of He. We need the molar mass of helium to solve the problem: 4.0026 g/mol.

ANALYZE We need to convert moles of He into grams of He. Our initial value is in moles and our answer must be in grams, so if we multiply by molar mass, the mole units cancel out and we have our answer in grams. The balloon contains approximately $\frac{1}{2}$ mole of helium, so our answer should be about $\frac{1}{2}$ of 4, or 2 grams.

SOLVE

$$0.462 \text{ mol He} \times \frac{4.0026 \text{ g He}}{1 \text{ mol He}} = 1.85 \text{ g He}$$

THINK ABOUT IT Our calculated answer is reassuringly close to our estimate. In general, we multiply by molar mass to convert a number of moles into an equivalent mass in grams.

Practice Exercise A silicon wafer with dimensions 20 cm $\times$ 20 cm $\times$ 0.03 cm contains 0.0996 moles of Si. How many grams of Si does this represent?

(Answers to Practice Exercises are in the back of the book.)

SAMPLE EXERCISE 2.7 **Converting Mass into Number of Moles** **LO5**

If Geiger and Marsden used a piece of gold foil with dimensions 1.0 cm by 1.0 cm by 4.0×10^{-5} cm thick, how many moles of gold did the foil contain? (The density of gold is 19.3 g/cm^3.)

COLLECT AND ORGANIZE We are given the dimensions of a piece of gold foil and its density, which we can use to calculate the mass of gold in the foil. The mass of gold and the equivalent number of moles of gold are related by its molar mass: 196.97 g/mol.

ANALYZE To find the volume of the gold foil, we multiply the dimensions given to us in the problem. Density is mass per unit volume ($d = m/V$), so multiplying the volume of the gold foil by gold's density gives us the mass of the foil. To convert grams of Au into moles of Au, we divide by the molar mass of Au. The volume of the gold is $(1.0 \times 1.0 \times 4.0 \times 10^{-5})$ cm^3, or 4.0×10^{-5} cm^3. Multiplying this value by a density of about 20 g/cm^3 and then dividing by a molar mass of about 200 g/mol should give us a value around 1/10 that of the volume of the foil, or 4×10^{-6} mol.

SOLVE Volume of gold foil:

$$1.0 \text{ cm} \times 1.0 \text{ cm} \times 4.0 \times 10^{-5} \text{ cm} = 4.0 \times 10^{-5} \text{ cm}^3$$

Moles of gold:

$$4.0 \times 10^{-5} \text{ cm}^3 \text{ Au} \times \frac{19.3 \text{ g Au}}{1 \text{ cm}^3 \text{ Au}} \times \frac{1 \text{ mol Au}}{196.97 \text{ g Au}} = 3.9 \times 10^{-6} \text{ mol Au}$$

THINK ABOUT IT The foil sample is very thin, so we would expect it to contain a small number of moles of Au. The answer reflects that fact and agrees with our predicted value.

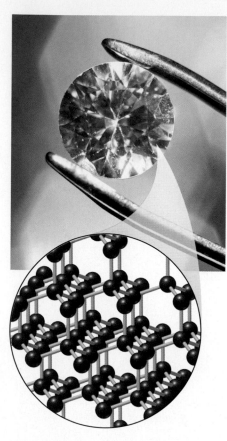

FIGURE 2.19 A 3.25-carat diamond.

Practice Exercise The mass of the diamond in Figure 2.19 is 3.25 carats (1 carat = 0.200 g). Diamonds are nearly pure carbon. How many moles and how many atoms of carbon are in the diamond?

(Answers to Practice Exercises are in the back of the book.)

■ •••••••••••••••••••••••••••••••••••••••

CONCEPT TEST ••••••••••••••••••••••••••••••••••

Which contains more atoms: 1 gram of gold (Au) or 1 gram of silver (Ag)?

(Answers to Concept Tests are in the back of the book.)

SAMPLE EXERCISE 2.8 **Calculating the Molar Mass of a Compound** **LO6**

When the SO_2 released during a volcanic eruption combines with the steam (H_2O vapor) that is also released, the two of them form sulfurous acid, H_2SO_3. What is the molar mass of this molecular compound?

COLLECT AND ORGANIZE In this problem we are asked to calculate the molar mass of H_2SO_3. The molar mass of this, or any, molecular compound is the sum of the molar masses of the elements in its formula, each multiplied by the number of atoms of that element in a molecule of the compound. The molar masses of the elements can be found in the periodic table: H is 1.0079 g/mol, S is 32.065 g/mol, and O is 15.999 g/mol.

ANALYZE The molecular formula of H_2SO_3 tells us that one mole of it contains 2 moles of H atoms, 1 mole of S atoms, and 3 moles of O atoms. Working with whole numbers, we can estimate contributions of $(2 \times 1 = 2)$ g/mol H, $(1 \times 32 = 32)$ g/mol S, and $(3 \times 16 = 48)$ g/mol O, for a total of $2 + 32 + 48 = 82$ g/mol.

SOLVE

$$\left(\frac{2 \text{ mol H}}{\text{mol H}_2\text{SO}_3} \times \frac{1.0079 \text{ g}}{1 \text{ mol H}} \right) + \left(\frac{1 \text{ mol S}}{\text{mol H}_2\text{SO}_3} \times \frac{32.065 \text{ g}}{1 \text{ mol S}} \right) + \left(\frac{3 \text{ mol O}}{\text{mol H}_2\text{SO}_3} \times \frac{15.999 \text{ g}}{1 \text{ mol O}} \right)$$
$$= 82.078 \text{ g/mol H}_2\text{SO}_3$$

THINK ABOUT IT The molar mass of a compound reflects the masses of the elements in the compound and the number of atoms of each element in one molecule of the compound. Most of the molar mass of H_2SO_3 comes from the sulfur and oxygen atoms in its molecules.

Practice Exercise During photosynthesis green plants convert water and carbon dioxide into glucose ($C_6H_{12}O_6$) and oxygen. What is the molar mass of glucose?

(Answers to Practice Exercises are in the back of the book.)

■ ••••••••••••••••••••••••••••••••

SAMPLE EXERCISE 2.9 **Interconverting Grams, Moles, Molecules, and Formula Units** **LO5**

The dosage of the active ingredient in a popular antacid tablet is 500 mg of the ionic compound calcium carbonate, $CaCO_3$.

    a. How many moles of $CaCO_3$ are in each tablet?
    b. How many formula units of $CaCO_3$ are in each tablet?

COLLECT AND ORGANIZE We are asked to convert a given mass of a compound into moles of the compound and the number of formula units that mass represents. Such a conversion requires dividing the given mass by the molar mass of the compound. To obtain a compound's molar mass, we need the molar masses of its elements: 40.078 g/mol Ca, 12.011 g/mol C, and 15.999 g/mol O. To convert moles of an ionic compound into an equivalent number of formula units, we multiply by Avogadro's number, 6.0221×10^{23}.

ANALYZE We need to calculate the molar mass of $CaCO_3$, which is the sum of the molar masses of Ca and C plus three times the molar mass of O. Because the units of molar masses are g/mol, we also need to convert the initial mass from milligrams to grams. To estimate our answers, we note that the starting mass is 500 mg, or 0.5 g, and the molar mass of $CaCO_3$ is approximately $40 + 12 + (3 \times 16)$ g/mol, or about 100 g/mol. Therefore the answer to part a should be about 0.5/100 = 0.005 moles, and the answer to part b should be about $0.005 \times (6 \times 10^{23})$, or about 3×10^{21} formula units.

SOLVE

a. The molar mass of $CaCO_3$ is

$$\left(\frac{1\ \cancel{mol\ Ca}}{mol\ CaCO_3} \times \frac{40.078\ g}{1\ \cancel{mol\ Ca}}\right) + \left(\frac{1\ \cancel{mol\ C}}{mol\ CaCO_3} \times \frac{12.011\ g}{1\ \cancel{mol\ C}}\right) + \left(\frac{3\ \cancel{mol\ O}}{mol\ CaCO_3} \times \frac{15.999\ g}{1\ \cancel{mol\ O}}\right)$$

$$= 100.086\ g/mol\ CaCO_3$$

Converting from milligrams $CaCO_3$ to moles $CaCO_3$:

$$5.00 \times 10^2\ \cancel{mg\ CaCO_3} \times \frac{1\ \cancel{g}}{1000\ \cancel{mg}} \times \frac{1\ mol\ CaCO_3}{100.086\ \cancel{g\ CaCO_3}}$$

$$= 0.00500\ mol\ CaCO_3$$

b. Multiplying by Avogadro's number:

$$0.00500\ \cancel{mol}\ CaCO_3 \times \frac{6.0221 \times 10^{23}\ formula\ units}{1\ \cancel{mol}}$$

$$= 3.01 \times 10^{21}\ formula\ units\ CaCO_3$$

THINK ABOUT IT Because atoms and ions are tiny, we expect to find a large number of formula units in a relatively small mass of $CaCO_3$. We can't "count up" individual formula units directly, but we can use molar masses to calculate how many of them are in a sample of known mass.

Practice Exercise A standard aspirin tablet contains 325 mg of aspirin, which has the molecular formula $C_9H_8O_4$. How many moles and how many molecules of aspirin are in one tablet? ⚙

(Answers to Practice Exercises are in the back of the book.)

2.6 Making Elements

We began this chapter by discussing the remarkable advances that took place in the early 20th century—advances that provided scientists with a clearer understanding of the structure of atoms. As those discoveries were taking place, scientists were also advancing our understanding of how atoms may have formed in the first place.

Theoretical physicists believe that our universe began about 13.7 billion years ago with an enormous release of energy that has come to be known as the Big Bang. Current theories suggest that within a few microseconds of the Big Bang

▶❚❚ **CHEMTOUR** Big Bang

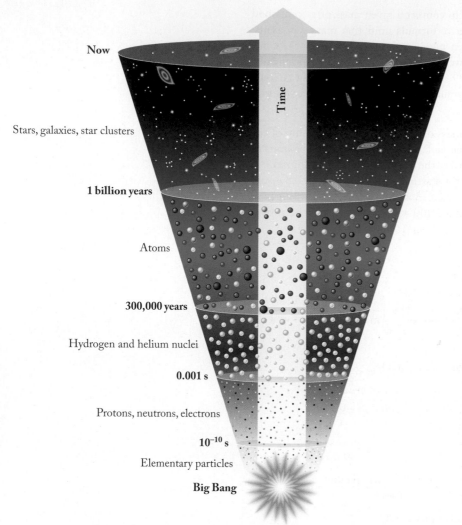

Now

Stars, galaxies, star clusters

1 billion years

Atoms

300,000 years

Hydrogen and helium nuclei

0.001 s

Protons, neutrons, electrons

10^{-10} s

Elementary particles

Big Bang

Time

FIGURE 2.20 Timeline for energy and matter transformations believed to have occurred since the universe began. In this model, protons, neutrons, and electrons were formed from quarks in the first millisecond after the Big Bang, followed by hydrogen and helium nuclei. Atoms of ^{1}H and ^{4}He did not form until after 300,000 years of expansion and cooling, and other elements did not form until the first galaxies appeared around 1 billion years after the Big Bang. According to this model, our solar system, our planet, and all life-forms on it are composed of elements synthesized in stars that were born, burned brightly, and then disappeared millions to billions of years after the Big Bang.

▶❚❚ **CHEMTOUR** Synthesis of Elements

much of its energy had transformed into matter made of elementary particles such as electrons and **quarks** (Figure 2.20). Less than a millisecond later, the universe had expanded and "cooled" to a mere 10^{12} K, which allowed quarks to combine with one another to form neutrons and protons. Thus, in less than a second, the universe contained the three types of subatomic particles that would eventually make up atoms.

Primordial Nucleosynthesis

By about 4 minutes after the Big Bang, the universe had expanded and cooled to about 10^{9} K. In this hot, dense subatomic "soup," neutrons and protons that collided with one another began to fuse together in a process called primordial **nucleosynthesis**. When a proton and a neutron collide and fuse together, they form a *deuteron* ($^{2}_{1}$D), which is the nucleus of the deuterium isotope of hydrogen, or hydrogen-2. We can describe this fusion process using the following *nuclear equation*:

$$^{1}_{1}\text{p} + ^{1}_{0}\text{n} \rightarrow ^{2}_{1}\text{D} \qquad (2.3)$$

In writing Equation 2.3 we follow the rules described in Section 2.2 for writing isotopic symbols: each species' superscript is its mass number, and the subscripts are the atomic numbers of nuclei, or the relative charges of subatomic particles.

Every nuclear equation must be balanced. This means the sum of the mass numbers (superscripts) of the particles to the left of the arrow must be equal to the sum of the mass numbers of the particles to the right of the arrow. Similarly, the sum of the charges (subscripts) of the particles on the left side must be equal to the sum of the charges of the particles on the right. A check of the superscripts and subscripts in Equation 2.3 reveals that it is balanced because the sum of the mass numbers (superscripts), $1 + 1 = 2$, is equal to the mass number of the deuteron on the right, and the sum of the relative charges (subscripts) on the left side, $0 + 1 = 1$, matches the atomic number of the deuteron on the right. In later chapters of this book we will write and balance *chemical equations* that describe how substances chemically react with each other.

Deuteron formation proceeded rapidly in the minutes following the Big Bang, consuming most of the free neutrons in the universe. However, no sooner did deuterons form than they too were consumed in collisions with one another, fusing to make ^4_2He nuclei (α particles):

$$2\,^2_1D \rightarrow\,^4_2He \qquad\qquad (2.4)$$

By about 5 minutes after the Big Bang the matter of the universe had become 75% (by mass) protons and 25% α particles. Nucleosynthesis then came to a screeching halt. Perhaps you are wondering why. Why didn't α particles and protons, for example, fuse together to make ^5Li:

$$^4_2He +\,^1_1p \overset{?}{\rightarrow}\,^5_3Li$$

or why didn't two α particles fuse together to make a nucleus of ^8Be?

$$2\,^4_2He \overset{?}{\rightarrow}\,^8_4Be$$

Neither process took place because neither ^5Li nor ^8Be is a stable nuclide. In fact, there are no stable nuclides with mass numbers of 5 or 8.

To understand why ^5Li and ^8Be are not stable, we need to explore the nature of the energy that keeps the nucleons in atomic nuclei together. By the 1930s, scientists had discovered that the mass of a stable nucleus is always less than the sum of the separate masses (Table 2.1) of its nucleons. For example, a helium-4 nucleus, which consists of 2 neutrons and 2 protons, has a mass of 6.64466×10^{-27} kg. The mass of a single neutron is 1.67493×10^{-27} kg, and the mass of a single proton is 1.67262×10^{-27} kg. Therefore the combined masses of the four nucleons in helium-4 are

$$
\begin{aligned}
\text{Mass of 2 neutrons} &= 2(1.67493 \times 10^{-27}\text{ kg}) \\
+\ \text{Mass of 2 protons} &= 2(1.67262 \times 10^{-27}\text{ kg}) \\
\hline
\text{Total mass of nucleons} &= 6.69510 \times 10^{-27}\text{ kg}
\end{aligned}
$$

The difference between this value and the actual mass of the ^4He nucleus (6.64466×10^{-27} kg) is

$$
\begin{aligned}
\text{Total mass of nucleons} &= 6.69510 \times 10^{-27}\text{ kg} \\
-\ \text{Mass of nucleus} &= 6.64466 \times 10^{-27}\text{ kg} \\
\hline
&\quad\ 0.05044 \times 10^{-27}\text{ kg}
\end{aligned}
$$

or 5.044×10^{-29} kg. This difference in mass is called the **mass defect (Δm)** of the ^4He nucleus. It represents the energy that holds the four nucleons together in the nucleus, that is, the **binding energy (BE)** of the nucleus. The binding energy of a nucleus is the energy that would be released if free nucleons were to fuse together to form the nucleus. It is also the energy needed to split the nucleus apart into free nucleons. How big is this energy? We can calculate it using Albert Einstein's famous equation that relates mass (m) and energy (E):

$$E = mc^2 \qquad\qquad (2.5)$$

where c is the speed of light in a vacuum, 2.998×10^8 m/s. Let's use it to calculate the binding energy of an α particle (^4He nucleus) based on its mass defect of 5.044×10^{-29} kg.

$$
\begin{aligned}
E &= mc^2 \\
&= 5.044 \times 10^{-29}\text{ kg} \times (2.998 \times 10^8\text{ m/s})^2 \\
&= 4.534 \times 10^{-12}\text{ kg} \cdot (\text{m/s})^2 = 4.534 \times 10^{-12}\text{ J}
\end{aligned}
$$

quarks elementary particles that combine to form neutrons and protons.

nucleosynthesis the natural formation of nuclei as a result of fusion and other nuclear processes.

mass defect (Δm) the difference between the mass of a stable nucleus and the masses of the individual nucleons that comprise it.

binding energy (BE) the energy that holds the nucleons together in a nucleus.

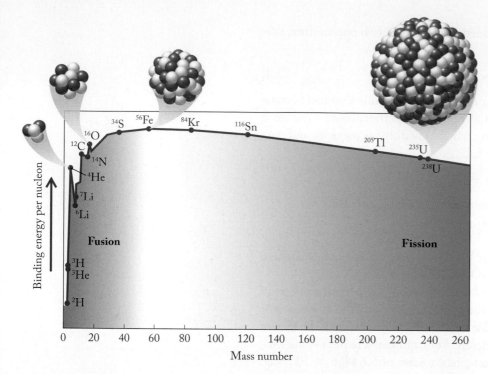

FIGURE 2.21 The stability of a nucleus is directly proportional to its binding energy per nucleon, which reaches a maximum at ^{56}Fe. For all nuclides up to ^{56}Fe, fusion reactions form more massive nuclei that have greater binding energy per nucleon than the particles that fused together. These fusion reactions release energy. However, fusion of nuclei with atomic numbers greater than 26 produces nuclei that have less binding energy per nucleon, and so consumes energy.

This value may seem very small, but keep in mind that it is based on the formation of only a single particle. If we multiply it by Avogadro's number, we find that it is equivalent to 2.730×10^{12} J per mol ^{4}He. This enormous energy means that ^{4}He is a stable nuclide.

Most of the stable nuclei with larger numbers of nucleons than ^{4}He have even larger binding energies. This makes sense because a nucleus with many protons in close proximity to each other must be held together by an enormous energy to overcome the electrostatic repulsion that all of the positively charged particles exert on each other. In such close proximity, nucleons come under the influence of a fundamental force of nature known as the **strong nuclear force**. It operates only over very small distances, such as the diameters of atomic nuclei, but it is 100 times stronger than the repulsions the protons experience. The strong nuclear force binds nucleons together and stabilizes atomic nuclei.

To make comparisons of nuclear binding energies fair, they are usually divided by the number of nucleons in each nucleus. Expressing binding energy on a per nucleon basis allows us to compare the relative stabilities of different nuclides. When we plot binding energy per nucleon values against atomic number, we get the curve shown in Figure 2.21.

Now let's reconsider the reason why ^{8}Be did not form during primordial nucleosynthesis. This nuclide does not exist because it has no more binding energy per nucleon than ^{4}He. In other words, no energy would be released if two ^{4}He nuclei were to fuse together to form ^{8}Be. Similarly, ^{8}Be nuclei require no energy to spontaneously decompose into ^{4}He nuclei, so they would immediately do so. For this reason, ^{8}Be nuclei never form in the first place.

As the early universe continued to expand and cool, the kinetic energies of the particles of matter in it (hydrogen and helium nuclei and free electrons) dropped and their motions slowed, allowing nuclei to combine with free electrons to produce neutral atoms. The result was a universe that was 75% hydrogen and 25% helium. It remained that way for millions of years—until the first galaxies formed.

Stellar Nucleosynthesis

Most of the matter detected and identified in the universe today is still hydrogen and helium, which is a strong piece of evidence supporting the Big Bang theory. But how did the other elements in the periodic table eventually form, including those that make up most of our planet? Scientists theorize that the synthesis of elements more massive than helium had to wait until nuclear fusion resumed in the first generation of stars. Inside the coalescing masses of hydrogen and helium that became the first stars, these gases underwent enormous compression heating, becoming hot enough to ignite the nuclear furnaces that are the source of the

strong nuclear force the fundamental force of nature that keeps quarks together in subatomic particles and nucleons together in atomic nuclei.

neutron capture the absorption of a neutron by a nucleus.

beta (β) decay the process by which a neutron in a neutron-rich nucleus decays into a proton and a β particle.

energy in all stars. Initially hydrogen nuclei served as the fuel for the furnaces, and their fusion to form more helium nuclei (α particles) provided the energy.[3]

Some of these stars, known as *red giants*, had cores so extraordinarily hot and dense that sometimes three α particles collided with each other simultaneously. When they did, they fused together to form a stable nucleus that contained 6 protons and 6 neutrons, which is the most common isotope of carbon:

$$3\,{}^{4}_{2}\text{He} \rightarrow {}^{12}_{6}\text{C} \tag{2.6}$$

With the formation of ${}^{12}\text{C}$, the barrier that had halted primordial nucleosynthesis was overcome. In the cores of giant stars, ${}^{12}\text{C}$ nuclei fuse with α particles to form ${}^{16}\text{O}$ (Figure 2.22). Then ${}^{16}\text{O}$ nuclei may fuse with more α particles to form ${}^{20}\text{Ne}$, and so on. Additional fusion reactions involving nuclei with increasingly greater positive charges are possible in intensely hot (10^9 K) stars because the nuclei have enough kinetic energy, and are moving fast enough, to overcome the electrostatic repulsion experienced by particles with large positive charges.

For billions of years fusion reactions of this sort have simultaneously fueled the nuclear furnaces of stars and produced isotopes as heavy as ${}^{56}\text{Fe}$ (Figure 2.23). However, once the core of a star turns into iron, the star is in trouble because fusion reactions involving iron nuclei do not release energy; instead they *consume* it. This happens because the binding energy per nucleon (and, therefore, the nuclear stability) reaches a maximum with ${}^{56}\text{Fe}$ (Figure 2.21). Thus, a star with an iron core has essentially run out of fuel. Its nuclear furnace goes out, and the star begins to cool and collapse into itself.

As the star collapses, compression reheats its core to above 10^9 K. At such temperatures, nuclei begin to disintegrate into free protons and neutrons. Free neutrons may collide and fuse with atomic nuclei in a process called **neutron capture**. If a stable nucleus captures enough neutrons, it becomes unstable. For example, if a nucleus of ${}^{56}\text{Fe}$ captures three neutrons it forms the unstable nuclide ${}^{59}\text{Fe}$:

$${}^{56}_{26}\text{Fe} + 3\,{}^{1}_{0}\text{n} \rightarrow {}^{59}_{26}\text{Fe}$$

The neutron *richness* of ${}^{59}\text{Fe}$ means that this radioactive nuclide spontaneously undergoes a kind of radioactive decay that reduces the ratio of neutrons to protons in its nucleus. This kind of decay is called **beta (β) decay** because it involves the ejection of a high-energy electron, or β particle. The decay process is summarized in the following nuclear equation:

$${}^{59}_{26}\text{Fe} \rightarrow {}^{59}_{27}\text{Co} + {}^{0}_{-1}\beta \tag{2.7}$$

Note that the nuclide formed in the reaction has an atomic number of 27, or one more than the radioactive isotope that produced it. This increase is the result of the decomposition of a neutron into a proton, which remains in the nucleus, and an electron (β particle) that is

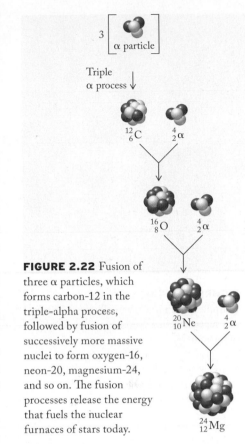

FIGURE 2.22 Fusion of three α particles, which forms carbon-12 in the triple-alpha process, followed by fusion of successively more massive nuclei to form oxygen-16, neon-20, magnesium-24, and so on. The fusion processes release the energy that fuels the nuclear furnaces of stars today.

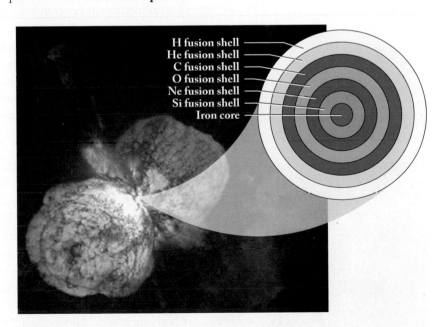

FIGURE 2.23 The star Eta Carinae is believed to be evolving toward an explosion. The outer regions are still fueled by energy released as hydrogen isotopes fuse, but the star is increasingly hotter and denser closer to the center. This central heating allows the fusion of larger nuclei and results in the production of ${}^{56}\text{Fe}$ in the core.

---

[3]The hydrogen–helium fusion process that takes place in stars such as our sun is not the same as in primordial nucleosynthesis because there are few free neutrons in the stars. Details of stellar hydrogen fusion are described in Chapter 21.

FIGURE 2.24 The Crab Nebula in the constellation Taurus is actually the debris field of a supernova that was observed on Earth in the year 1054.

ejected from it. The additional proton results in an increase in atomic number. The combination of repeated neutron capture and β decay events in the cores of collapsing stars produces the most massive stable nuclides in the periodic table.

Eventually, the enormous heating that occurs when a massive star collapses produces a gigantic explosion. Cosmologists call such an event a *supernova*. In addition to finishing the job of synthesizing the elemental building blocks found in the universe—up to and including the isotopes of uranium—a supernova serves as its own element-distribution system, blasting its inventory of elements throughout its galaxy (Figure 2.24). The legacies of supernovas are found in the elemental composition of later-generation stars like our sun and in the planets that orbit the stars. Indeed, all the matter in our solar system—and in us—is essentially demolition debris from ancient exploding stars.

2.7 Artificial Nuclides

Rutherford's experiments with gold foil led him to bombard other elements with alpha particles. In a 1919 experiment, Rutherford reported that passing α particles through nitrogen yielded two products: oxygen-17 and free protons. Bombardment of nuclei by α particles became a popular method for transmuting elements in the 1920s and 1930s. In 1933, the French chemists Irène (1897–1956) and Frédéric (1900–1958) Joliot-Curie synthesized the first radionuclide not found in nature, phosphorus-30, by bombarding aluminum-27 with α particles:

$$_{13}^{27}\text{Al} + _{2}^{4}\text{He} \rightarrow _{15}^{30}\text{P} + _{0}^{1}\text{n} \qquad (2.8)$$

In 1940 elements 93 (neptunium) and 94 (plutonium) were produced at the University of California, Berkeley, by bombarding uranium-238 with neutrons. Transmutation by neutron bombardment is easier than by α bombardment because both α particles and atomic nuclei have positive charges, and so repel each other. The α particles must be traveling at high speeds to overcome this repulsion in a successful bombardment. Because neutrons have no charge, however, they are more readily captured by nuclei. For example, when nuclei of uranium-238 are bombarded with neutrons, a nuclear reaction may occur in which two β particles are ejected, leading to the formation of plutonium-239:

$$_{92}^{238}\text{U} + _{0}^{1}\text{n} \rightarrow _{94}^{239}\text{Pu} + 2\,_{-1}^{0}\beta \qquad (2.9)$$

Between 1944 and 1961, the Berkeley research team synthesized elements through $Z = 103$, lawrencium, by using combinations of neutron and α-particle bombardment of actinide nuclei. The research team was led by American chemist Glenn T. Seaborg (1912–1999, Figure 2.25), who won the Nobel Prize in Chemistry in 1951 for his team's ability to first synthesize and then characterize the chemical properties of these *transuranium* elements.

Seaborg's team's methods did not work for synthesizing elements with $Z > 103$ because nuclides more massive than californium-249 are highly unstable and rapidly lose α or β particles, so they are not useful target materials for making even more massive nuclides. However, by bombarding californium-249 targets with carbon, nitrogen, and oxygen nuclei instead of α particles, scientists have been able to synthesize isotopes of rutherfordium ($Z = 104$), dubnium ($Z = 105$), and seaborgium ($Z = 106$), as described in these nuclear equations:

$$_{98}^{249}\text{Cf} + _{6}^{12}\text{C} \rightarrow _{104}^{257}\text{Rf} + 4\,_{0}^{1}\text{n}$$

$$_{98}^{249}\text{Cf} + _{7}^{15}\text{C} \rightarrow _{105}^{260}\text{Db} + 4\,_{0}^{1}\text{n}$$

$$_{98}^{249}\text{Cf} + _{8}^{18}\text{O} \rightarrow _{106}^{263}\text{Sg} + 4\,_{0}^{1}\text{n}$$

FIGURE 2.25 When element 106 was named seaborgium in 1994, Glenn T. Seaborg became the only living scientist to have an element named after him.

SAMPLE EXERCISE 2.10 Completing and Balancing Nuclear Equations LO8

In the summer of 1944 Seaborg's research team at the University of California, Berkeley, bombarded a target of plutonium-239 with α particles. One of the products of this bombardment was curium-242. Write a balanced nuclear equation describing this nuclear reaction. Use symbols of the form $_Z^A X$ to represent the nuclides and any subatomic particles that may also have formed.

COLLECT AND ORGANIZE The information contained in the periodic table of the elements (inside front cover) tells us that the symbol for plutonium is Pu and its atomic number is 94, and that the symbol for curium is Cm and its atomic number is 96. Alpha particles are helium-4 nuclei.

ANALYZE We can combine the element symbols and atomic numbers with the given mass numbers to write symbols of the nuclides involved in the reaction. To write a balanced nuclear equation, the sum of the subscripts (atomic numbers) of the nuclides on the left side of the reaction arrow must equal the sum of the subscripts on the right. The same is true for the mass numbers denoted by superscripts.

SOLVE The symbols of the nuclides involved in the reaction are

$$_{94}^{239}\text{Pu} + _2^4\text{He} \rightarrow _{96}^{242}\text{Cm}$$

To check whether the equation is balanced, we add the superscripts (239 + 4) and the subscripts (94 + 2) on the left side to see if they match the mass number (242) and atomic number (96) of the product. The atomic numbers match, but the mass numbers do not. To balance the equation we need to add another product: a particle with no atomic number (or charge) and a mass number of 1. A neutron is such a particle. Adding a neutron to the right side balances the nuclear equation:

$$_{94}^{239}\text{Pu} + _2^4\text{He} \rightarrow _{96}^{242}\text{Cm} + _0^1\text{n}$$

THINK ABOUT IT By bombarding plutonium with α particles, the Berkeley scientists synthesized an even more massive nuclide in much the way that lighter elements fuse with α particles to make heavier elements and release the energy that fuels giant stars.

Practice Exercise In March of 1945, Seaborg's research team synthesized another isotope of curium, ^{240}Cm, by bombarding a ^{239}Pu target with α particles. Write a balanced nuclear equation describing this reaction. ⚙

(Answers to Practice Exercises are in the back of the book.)

In recent years, scientists have reported synthesizing nuclides that have as many as 118 protons by bombarding targets as massive as californium-249 with medium-mass nuclei, such as calcium-48. For example, in January 1999 an atom with 114 protons and 175 neutrons was synthesized and lasted for 30 s (most supermassive nuclides have half-lives that are fractions of a second) before undergoing a series of α decay events that yielded isotopes of elements 112, 110, and 108. Some of these "supermassive" elements, and the bombarding ions and target nuclides used to make them, are listed in Table 2.2.

Why bother to make such short-lived elements? The answer is that their mere existence, no matter for how brief a time, can be a source of insight into the nature of nuclear structure. Their behavior illustrates the competition between the strong nuclear force that holds nucleons together and the electrostatic repulsion between protons. Supermassive elements are pieces of a puzzle that someday may tell us whether there is a limit to the size of atoms.

TABLE 2.2 Synthesis of Supermassive Nuclides

Bombarding Ion	Target	Nuclide Synthesized[a]	Year First Synthesized
^{62}Ni	^{208}Pb	$_{110}^{269}$Ds	1994
^{64}Ni	^{209}Bi	$_{111}^{272}$Rg	1994
^{69}Zn	^{208}Pb	$_{112}^{277}$Cn	1996
^{70}Zn	^{209}Bi	$_{113}^{278}$Uut	2003
^{48}Ca	^{244}Pu	$_{114}^{289}$Fl	1999
^{48}Ca	^{243}Am	$_{115}^{288}$Uup	2003
^{48}Ca	^{248}Cm	$_{116}^{293}$Lv	2000
^{48}Ca	^{249}Bk	$_{117}^{293}$Uus	2009
^{48}Ca	^{249}Cf	$_{118}^{294}$Uuo	2002

[a]No permanent names or symbols have yet been adopted for elements 113, 115, 117, and 118. They have temporary names and symbols, ununtrium (Uut), ununpentium (Uup), ununseptium (Uus), and ununoctium (Uuo), which are derived from the digits in their atomic numbers.

SAMPLE EXERCISE 2.11   **Integrating Concepts: Searching for Islands of Stability**

One of the reasons why scientists are trying to synthesize super-massive elements is to test the theory that certain combinations of neutrons and protons make unusually stable nuclides. Super-massive nuclides with these combinations are believed to represent "islands of stability" in a sea of highly radioactive short-lived nuclides. Theoretically, a nucleus that contains 120 protons and 182 neutrons should occupy one of these islands. As this book is being written, all attempts to synthesize that isotope, tentatively named unbinilium (Ubn) have been unsuccessful, but more are planned. All involve bombarding targets containing actinide elements with the nuclei of a much lighter element.

a. Complete and balance the following nuclear equations describing attempts to synthesize this isotope of Ubn. Assume that no neutron loss occurs in these reactions.

$$^{238}_{92}U + \, ^{?}_{?}? \rightarrow \, ^{?}_{?}Ubn$$

$$^{244}_{94}? + \, ^{?}_{?}? \rightarrow \, ^{?}_{?}Ubn$$

$$^{248}_{?}Cm + \, ^{?}_{?}? \rightarrow \, ^{?}_{?}Ubn$$

b. If element Ubn is ever synthesized:
  1. To which group in the periodic table will it belong?
  2. What will be the charge on its monatomic ions?
  3. What will be the formula of the neutral compound it forms with oxygen?
  4. What existing element would have the most similar properties?
  5. Will it have an average atomic mass and molar mass?

COLLECT AND ORGANIZE We are given the number of protons and neutrons in an isotope of an element and asked for its mass number and to complete nuclear reactions describing how the nuclide might be synthesized. We are also asked to place the new element in the periodic table and to describe some of its chemical properties.

ANALYZE The mass number of a nuclide is the sum of the number of protons and neutrons in its nucleus. In a balanced nuclear equation the sum of the atomic numbers (subscripts) and the sum of the mass numbers (superscripts) of the particles to the left of the reaction arrow must match the corresponding sums of the particles to the right. If we count two spaces over in the periodic table from the most massive element (Uuo, $Z = 118$) to get to 120, we wrap around the table to the second column of the next row, just below radium.

SOLVE

a. The mass number of the Ubn isotope is equal to $120 + 182 = 302$. Insert this mass number and the atomic number 120 next to the Ubn symbol in the nuclear equation:

$$^{238}_{92}U + ? \rightarrow \, ^{302}_{120}Ubn$$

To balance the equation the unknown particle must have a mass number of $302 - 238 = 64$ and an atomic

number of $120 - 92 = 28$, which makes it nickel-64. The complete equation is

$$^{238}_{92}U + \, ^{64}_{28}Ni \rightarrow \, ^{302}_{120}Ubn$$

Repeating this process for the next two syntheses, we start by completing the terms for product nuclides, which are all ^{302}Ubn, and the target isotopes. The atomic number of the target of the first one is 94, which makes it an isotope of plutonium:

$$^{244}_{94}Pu + ? \rightarrow \, ^{302}_{120}Ubn$$

The target of the second is curium, which means that its atomic number is 96:

$$^{248}_{96}Cm + ? \rightarrow \, ^{302}_{120}Ubn$$

In the first equation the unknown particle must have a mass number of $302 - 244 = 58$ and an atomic number of $120 - 94 = 26$, which makes it iron-58. In the second equation the unknown particle must have a mass number of $302 - 248 = 54$ and an atomic number of $120 - 96 = 24$, which makes it chromium-54. The complete equations are

$$^{244}_{94}Pu + \, ^{58}_{26}Fe \rightarrow \, ^{302}_{120}Ubn$$

$$^{248}_{96}Cm + \, ^{54}_{24}Cr \rightarrow \, ^{302}_{120}Ubn$$

b.

  1. The location of element 120 in row 8 and column 2 of the periodic table makes Ubn an alkaline earth metal.
  2. Like the other alkaline earths, Ubn should form 2+ cations, Ubn^{2+}.
  3. Oxygen forms 2− anions (O^{2-}), so Ubn^{2+} will combine with it in a 1:1 ratio, and the resulting neutral compound would have the formula UbnO.
  4. Elements in the same group have similar chemical properties, so those of Ubn should most closely match those of the element directly above it in the periodic table, radium.
  5. The atomic mass of an element is the weighted average mass of its naturally occurring isotopes. However, Ubn has no naturally occurring isotopes, so it can't have an atomic mass or a molar mass.

THINK ABOUT IT When Mendeleev first published his periodic table, he left open spaces for elements that were unknown at the time. He even gave the undiscovered elements names based on the elements above their places in the table. Thus, he named gallium "eka-aluminum" and germanium "eka-silicon" using a prefix derived from the Sanskrit word meaning "one" (as in *one* row down in the table). Following Mendeleev's naming convention, some of the scientists trying to make element 120 refer to it as eka-radium.

SUMMARY ·· ■

Section 2.1 Atoms are composed of negatively charged **electrons** surrounding a **nucleus**, which contains positively charged **protons** and electrically neutral **neutrons**. The values of the charge and mass of the electron were determined by J. J. Thomson's studies using cathode-ray tubes and by Robert Millikan's oil-drop experiments. The research of

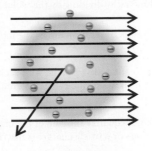

Ernest Rutherford's group, who bombarded thin gold foil with **alpha (α) particles** and determined how the particles were deflected, showed that all the positive charge and nearly all the mass of an atom are contained in its nucleus. An atom or group of atoms having a net charge is called an **ion**. If the charge is positive, it is a **cation**; if the charge is negative, it is an **anion**.

Section 2.2 The number of protons in the nucleus of an element defines its **atomic number (Z)**; the number of **nucleons** (protons and neutrons) in the nucleus defines the element's **mass number (A)**. The different **isotopes** of an element consist of atoms with the same number of protons per nucleus but different numbers of neutrons.

Section 2.3 Elements are arranged in the **periodic table of the elements** in order of increasing atomic number and in a pattern based on their chemical properties, including the charges of the monatomic ions they form. Elements in the same column are in the same **group** and have similar properties. Elements in groups 1, 2, and 13–18 are **main group** (or **representative**) **elements**. The **transition metals** are in groups 3–12. **Metals** are mostly malleable, ductile solids; they form cations and are good conductors of heat and electricity. **Nonmetals** include elements in all three physical states; they form anions and are poor conductors of heat and electricity. **Metalloids**, or **semimetals**, have the physical properties of metals and chemical properties of nonmetals.

Section 2.4 To calculate the **average atomic mass** of an element, multiply the mass of each of its stable isotopes by the **natural abundance** of that isotope, and then sum the products. The **molecular**

mass of a compound is the sum of the average atomic mass of each of the atoms in one of its molecules. The formula of an ionic compound defines the simplest combination of its ions that gives a neutral **formula unit** of the compound, which has a corresponding **formula mass**.

Section 2.5 The **mole (mol)** is the SI base unit for quantity of substances. One mole of a substance consists of **Avogadro's number** ($N_A = 6.0221 \times 10^{23}$) of particles of the substance. The mass of 1 mole of a substance is its **molar mass ($\mathcal{M}$)**.

Section 2.6 Neutrons, protons, and electrons formed within seconds of the Big Bang. During primordial **nucleosynthesis**, protons and neutrons fused to produce nuclei of helium. After galaxies formed, the nuclei of atoms with $Z \leq 26$ formed when the nuclei of lighter elements fused together in the cores of giant stars (stellar nucleosynthesis). The nuclei of elements with $Z > 26$ formed by a combination of **neutron capture**, **beta (β) decay**, and other nuclear reactions that occurred during supernovas (explosions of giant stars). As a result of these explosions, the elements produced were distributed throughout galaxies for possible inclusion in later-generation stars and in planets such as our own. Stellar nucleosynthesis continues today.

Section 2.7 Artificial nuclides are produced from high-speed collisions between atomic nuclei and subatomic particles and in collisions between two nuclei. The latter have produced supermassive elements with atomic numbers up to 118.

PROBLEM-SOLVING SUMMARY ··································· ■

TYPE OF PROBLEM	CONCEPTS AND EQUATIONS	SAMPLE EXERCISES
Writing symbols of nuclides and ions	To the left of the element symbol, place a superscript for the mass number (A) and a subscript for the atomic number (Z). If the particle is an ion, add its charge as a superscript following the symbol.	2.1, 2.2
Navigating the periodic table	Use row numbers to identify periods on the periodic table, and use column numbers to identify groups. Groups with special names include the *alkali metals* (group 1), *alkaline earth metals* (group 2), *halogens* (group 17), and *noble gases* (group 18).	2.3
Calculating the average atomic mass of an element	Multiply the mass (m) of each stable isotope of the element by the natural abundance (a) of that isotope; then sum the products: $$m_X = a_1m_1 + a_2m_2 + a_3m_3 + \cdots \qquad (2.2)$$	2.4
Converting number of particles into number of moles (or vice versa)	Convert number of particles or formula units to number of moles by dividing by Avogadro's number. Convert number of moles to number of particles or formula units by multiplying by Avogadro's number.	2.5, 2.9

TYPE OF PROBLEM	CONCEPTS AND EQUATIONS	SAMPLE EXERCISES
Converting mass of a substance into number of moles (or vice versa)	Convert mass of the substance to number of moles by dividing by the molar mass ($\mathcal{M}$) of the substance.	2.6, 2.7, 2.9
	Convert number of moles of the substance to mass by multiplying by the molar mass ($\mathcal{M}$) of the substance.	
Calculating the molar mass of a compound	Sum the molar masses of the elements in the compound's formula, with each element multiplied by the number of atoms of that element in one molecule or formula unit of the compound.	2.8, 2.9
Completing and balancing nuclear equations	Add nuclides or subatomic particles as needed to equalize the sum of the mass numbers (superscripts) and the sum of the atomic numbers or particle charges (subscripts) on the left and right sides of the equation.	2.10

VISUAL PROBLEMS

(Answers to boldface end-of-chapter questions and problems are in the back of the book.)

2.1. Which of the highlighted elements in Figure P2.1 formed, according to the theory of the Big Bang, before the first generation of galaxies formed?

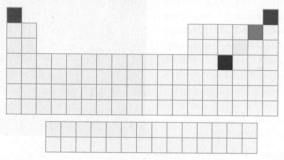

FIGURE P2.1

2.2. Which one of the highlighted elements in Figure P2.1 is *not* formed by fusion of lighter elements in the cores of giant stars?

2.3. Which one of the highlighted elements in Figure P2.1 has a stable isotope with no neutrons in its nucleus?

2.4. Which of the highlighted elements in Figure P2.4 is (a) a transition metal; (b) an alkali metal; (c) a halogen?

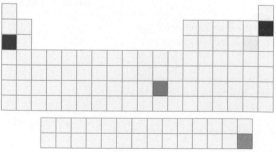

FIGURE P2.4

2.5. Which of the highlighted elements in Figure P2.4 is (a) a nonmetal; (b) a chemically inert gas; (c) a metal?

2.6. Which of the highlighted elements in Figure P2.4 has no stable isotopes?

2.7. Alpha and beta particles emitted by a sample of pitchblende escape through a narrow channel in the shielding surrounding the sample and into an electric field as shown in Figure P2.7. Identify which colored arrow corresponds to each of the two forms of radiation.

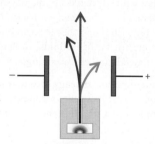

FIGURE P2.7

2.8. Does the radiation that follows the red path in Figure P2.7 penetrate solid objects better than the radiation following the green path?

2.9. Which of the highlighted elements in Figure P2.9 forms monatomic ions with a charge of (a) 1+; (b) 2+; (c) 3+; (d) 1−; (e) 2−?

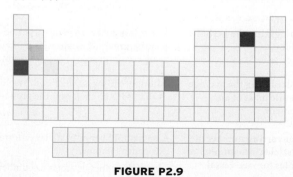

FIGURE P2.9

2.10. Which of the nuclear processes listed below do Figures P2.10(a) and P2.10(b) represent?
 a. primordial nucleosynthesis
 b. synthesis of a supermassive nuclide
 c. solar fusion
 d. fission
 e. β decay

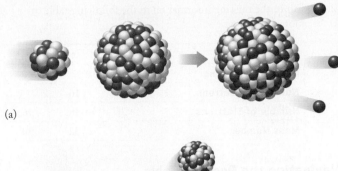

(a)

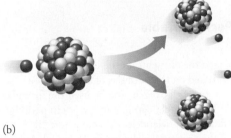

(b)

FIGURE P2.10

QUESTIONS AND PROBLEMS

The Rutherford Model of Atomic Structure

CONCEPT REVIEW

2.11. Explain how the results of the gold-foil experiment led Rutherford to dismiss the plum-pudding model of the atom and create his own model based on a nucleus surrounded by electrons.

2.12. Had the plum-pudding model been valid, how would the results of the gold-foil experiment have differed from what Geiger and Marsden actually observed?

2.13. What properties of cathode rays led Thomson to conclude that they were not rays of energy, but rather particles with an electric charge?

2.14. Describe two ways in which α particles and β particles differ.

***2.15.** Helium in Pitchblende The element helium was first discovered on Earth in a sample of pitchblende, an ore of radioactive uranium oxide. How did helium get in the ore?

2.16. How might using a thicker piece of gold foil have affected the scattering pattern of α particles observed by Rutherford's students?

Nuclides and Their Symbols

CONCEPT REVIEW

2.17. If the mass number of a nuclide is more than twice the atomic number, is the neutron-to-proton ratio less than, greater than, or equal to 1?

2.18. How are the mass number and atomic number of a nuclide related to the number of neutrons and protons in each of its nuclei?

2.19. Nearly all stable nuclides have at least as many neutrons as protons in their nuclei. Which very common nuclide is an exception?

2.20. Explain the inherent redundancy in the nuclide symbol $_Z^A X$.

PROBLEMS

2.21. How many protons, neutrons, and electrons are there in the following atoms? (a) ^{14}C; (b) ^{59}Fe; (c) ^{90}Sr; (d) ^{210}Pb

2.22. How many protons, neutrons, and electrons are there in the following atoms? (a) ^{11}B; (b) ^{19}F; (c) ^{131}I; (d) ^{222}Rn

2.23. Fill in the missing information about atoms of the four nuclides in the following table.

Symbol	^{23}Na	?	?	?
Number of Protons	?	39	?	79
Number of Neutrons	?	50	?	?
Number of Electrons	?	?	50	?
Mass Number	?	?	118	197

2.24. Fill in the missing information about atoms of the four nuclides in the following table.

Symbol	^{27}Al	?	?	?
Number of Protons	?	42	?	92
Number of Neutrons	?	56	?	?
Number of Electrons	?	?	60	?
Mass Number	?	?	143	238

2.25. Fill in the missing information in the following table of monatomic ions.

Symbol	$^{37}Cl^-$	?	?	?
Number of Protons	?	11	?	88
Number of Neutrons	?	12	46	?
Number of Electrons	?	10	36	86
Mass Number	?	?	81	226

2.26. Fill in the missing information in the following table of monatomic ions:

Symbol	$^{137}Ba^{2+}$	?	?	?
Number of Protons	?	30	?	40
Number of Neutrons	?	34	16	?
Number of Electrons	?	28	18	36
Mass Number	?	?	32	90

Navigating the Periodic Table

CONCEPT REVIEW

2.27. Mendeleev arranged the elements on the left side of his periodic table based on the formulas of the binary compounds they form with oxygen, and he used the formulas as column labels. For example, group 1 in a modern periodic table was labeled "R_2O" in Mendeleev's table where "R" represents one of the elements in the group. What labels did Mendeleev use for groups 2, 3, and 4 from the modern periodic table?

2.28. Mendeleev arranged the elements on the right side of his periodic table based on the formulas of the binary compounds they form with hydrogen and used these formulas as column labels. Which groups in the modern periodic table were labeled "HR," "H_2R," and "H_3R," where "R" represents one of the elements in the group?

2.29. Mendeleev left empty spaces in his periodic table for elements he suspected existed but that had yet to be discovered. However, he left no spaces for the noble gases (group 18 in the modern periodic table). Suggest a reason why he left no spaces for them.

2.30. Describe how the charges of the monatomic ions that elements form change as group number increases in a particular row of the periodic table and how ion charges change as the row number increases in a particular group.

PROBLEMS

2.31. **TNT** Molecules of the explosive TNT contain atoms of hydrogen and second-row elements in groups 14, 15, and 16. Which three elements are they?

2.32. **Phosgene** Phosgene was used as a chemical weapon during World War I. Despite the name, phosgene molecules contain no atoms of phosphorus. Instead, they contain atoms of carbon and the group 16 element in the second row of the periodic table and the group 17 element in the third row. What are the identities and atomic numbers of the two elements?

2.33. **Catalytic Converters** The catalytic converters used to remove pollutants from automobile exhaust contain the compounds of several fairly expensive elements including those described below. Which elements are they?
   a. The group 10 transition metal in the fifth row of the periodic table
   b. The transition metal whose symbol is to the left of your answer to part a
   c. The transition metal whose symbol is directly below your answer to part a

2.34. **Swimming Pool Chemistry** Compounds containing chlorine have long been used to disinfect the water in swimming pools, but in recent years a compound of a less corrosive halogen has become a popular alternative disinfectant. What is the name of this fourth-row element?

2.35. How many metallic elements are there in the third row of the periodic table?

2.36. Which third-row element in the periodic table has the chemical properties of a nonmetal but physical properties closer to those of metals?

The Masses of Atoms, Ions, and Molecules

CONCEPT REVIEW

2.37. What is meant by a *weighted average*?

2.38. Explain how percent natural abundances are used to calculate average atomic masses.

2.39. A hypothetical element consists of two isotopes (X and Y) with masses m_X and m_Y. If the natural abundance of the X isotope is exactly 50%, what is the average atomic mass of the element?

2.40. In calculating the formula masses of binary ionic compounds we use the average masses of neutral atoms, not ions. Why?

PROBLEMS

2.41. The vanadium in nature consists of two isotopes: ^{50}V and ^{51}V. Which one is the more abundant?

2.42. Manganese has only one stable isotope. How many neutrons are in each of its atoms?

2.43. Boron, lithium, and nitrogen each have two stable isotopes. Use the average atomic masses of the elements to determine which isotope in each of the following pairs of stable isotopes is the more abundant. (a) ^{10}B or ^{11}B; (b) ^6Li or ^7Li; (c) ^{14}N or ^{15}N

2.44. Neon, chlorine, and xenon have two stable isotopes. Use the average atomic masses of the elements to determine which isotope in each of the following pairs of stable isotopes is the more abundant. (a) ^{20}Ne or ^{22}Ne; (b) ^{35}Cl or ^{37}Cl; (c) ^{132}Xe or ^{134}Xe

2.45. Naturally occurring copper contains a mixture of 69.17% copper-63 (62.9296 amu) and 30.83% copper-65 (64.9278 amu). Use this information to calculate the average atomic mass of copper.

2.46. Naturally occurring sulfur consists of four isotopes: ^{32}S (31.9721 amu, 95.04%); ^{33}S (32.9715 amu, 0.75%); ^{34}S (33.9679 amu, 4.20%); and ^{36}S (35.9671 amu, 0.01%). Use this information to calculate the average atomic mass of sulfur.

2.47. **Chemistry of Mars** Chemical analyses conducted by the first Mars rover robotic vehicle in its 1997 mission produced the magnesium isotope data shown in the table below. Is the average atomic mass of magnesium in this Martian sample the same as on Earth (24.31 amu)?

Isotope	Mass (amu)	Natural Abundance (%)
^{24}Mg	23.9850	78.70
^{25}Mg	24.9858	10.13
^{26}Mg	25.9826	11.17

2.48. The natural abundances of the four isotopes of strontium are 0.56% ^{84}Sr (83.9134 amu), 9.86% ^{86}Sr (85.9094 amu), 7.00% ^{87}Sr (86.9089 amu), and 82.58% ^{88}Sr (87.9056 amu). Calculate the average atomic mass of strontium and compare it to the value in the periodic table inside the front cover.

2.49. Use the data in the following table of abundances and masses of the five naturally occurring titanium isotopes to calculate the atomic mass of ^{48}Ti.

Isotope	Mass (amu)	Natural Abundance (%)
^{46}Ti	45.9526	8.25
^{47}Ti	46.9518	7.44
^{48}Ti	?	73.72
^{49}Ti	48.94787	5.41
^{50}Ti	49.94479	5.18
Average	47.867	

2.50. Use the following table of abundances and masses of the three naturally occurring isotopes of argon to calculate the atomic mass of ^{40}Ar.

Isotope	Mass (amu)	Natural Abundance (%)
^{36}Ar	35.9675	0.337
^{38}Ar	37.9627	0.063
^{40}Ar	?	99.60
Average	39.948	

2.51. What are the masses of the formula units of each of the following ionic compounds? (a) CaF_2; (b) Na_2S; (c) Cr_2O_3

2.52. What are the masses of the formula units of each of the following ionic compounds? (a) KCl; (b) MgO; (c) Al_2O_3

2.53. How many carbon atoms are there in one molecule of each of the following compounds? (a) CH_4; (b) C_3H_8; (c) C_6H_6; (d) $C_6H_{12}O_6$

2.54. How many hydrogen atoms are there in each of the molecules in Problem 2.53?

2.55. Rank the following molecules based on increasing molecular mass. (a) CO; (b) Cl_2; (c) CO_2; (d) NH_3; (e) CH_4

2.56. Rank the following molecules based on decreasing molecular mass. (a) H_2; (b) Br_2; (c) NO_2; (d) C_2H_2; (e) BF_3

Moles and Molar Masses

CONCEPT REVIEW

2.57. In principle, we could use the more familiar unit *dozen* in place of mole when expressing the quantities of particles (atoms, ions, or molecules). What would be the disadvantage in doing so?

2.58. In what way is the molar mass of an ionic compound the same as its formula mass, and in what ways are they different?

PROBLEMS

2.59. Earth's atmosphere contains many volatile substances that are present in trace amounts. The following quantities of trace gases were found in a 1.0 mL sample of air. Calculate the number of moles of each gas in the sample.
 a. 4.4×10^{14} atoms of Ne
 b. 4.2×10^{13} molecules of CH_4
 c. 2.5×10^{12} molecules of O_3
 d. 4.9×10^9 molecules of NO_2

2.60. The following quantities of trace gases were found in a 1.0 mL sample of air. Calculate the number of moles of each compound in the sample.
 a. 1.4×10^{13} molecules of H_2
 b. 1.5×10^{14} atoms of He
 c. 7.7×10^{12} molecules of N_2O
 d. 3.0×10^{12} molecules of CO

2.61. How many moles of iron are there in 1 mole of the following compounds? (a) FeO; (b) Fe_2O_3; (c) $Fe(OH)_3$; (d) Fe_3O_4

2.62. How many moles of Na^+ ions are there in 1 mole of the following compounds? (a) $NaCl$; (b) Na_2SO_4; (c) Na_3PO_4; (d) $NaNO_3$

2.63. What is the mass of 0.122 mol of $MgCO_3$?

2.64. What is the volume of 1.00 mol of benzene (C_6H_6) at 20°C? The density of benzene at 20°C is 0.879 g/mL.

2.65. How many moles of titanium and how many atoms of titanium are there in 0.125 moles of each of the following? (a) ilmenite, $FeTiO_3$; (b) $TiCl_4$; (c) Ti_2O_3; (d) Ti_3O_5

2.66. How many moles of iron and how many atoms of iron are there in 2.5 moles of each of the following? (a) wolframite, $FeWO_4$; (b) pyrite, FeS_2; (c) magnetite, Fe_3O_4; (d) hematite, Fe_2O_3

2.67. Which substance in each of the following pairs of quantities contains more moles of oxygen?
 a. 1 mol Al_2O_3 or 1 mol Fe_2O_3
 b. 1 mol SiO_2 or 1 mol N_2O_4
 c. 3 mol CO or 2 mol CO_2

2.68. Which substance in each of the following pairs of quantities contains more moles of oxygen?
 a. 2 mol N_2O or 1 mol N_2O_5
 b. 1 mol NO or 1 mol $Ca(NO_3)_2$
 c. 2 mol NO_2 or 1 mol $NaNO_2$

2.69. **Elemental Composition of Minerals** Aluminum, silicon, and oxygen form minerals known as aluminosilicates. How many moles of aluminum are in 1.50 moles of the following?
 a. pyrophyllite, $Al_2Si_4O_{10}(OH)_2$
 b. mica, $KAl_3Si_3O_{10}(OH)_2$
 c. albite, $NaAlSi_3O_8$

2.70. **Radioactive Minerals** The uranium used for nuclear fuel exists in nature in several minerals. Calculate how many moles of uranium are in 1 mole of the following.
 a. carnotite, $K_2(UO_2)_2(VO_4)_2$
 b. uranophane, $CaU_2Si_2O_{11}$
 c. autunite, $Ca(UO_2)_2(PO_4)_2$

2.71. Calculate the molar masses of the following atmospheric molecules. (a) SO_2; (b) O_3; (c) CO_2; (d) N_2O_5

2.72. Determine the molar masses of the following minerals.
a. rhodonite, $MnSiO_3$
b. scheelite, $CaWO_4$
c. ilmenite, $FeTiO_3$
d. magnesite, $MgCO_3$

2.73. **Flavoring Additives** Calculate the molar masses of the following common flavors in food.
a. vanillin, $C_8H_8O_3$
b. oil of cloves, $C_{10}H_{12}O_2$
c. anise oil, $C_{10}H_{12}O$
d. oil of cinnamon, C_9H_8O

2.74. **Sweeteners** Calculate the molar masses of the following common sweeteners.
a. sucrose, $C_{12}H_{22}O_{11}$
b. saccharin, $C_7H_5O_3NS$
c. aspartame, $C_{14}H_{18}N_2O_5$
d. fructose, $C_6H_{12}O_6$

2.75. How many moles of carbon are there in 500.0 grams of carbon?

2.76. How many moles of gold are there in 2.00 ounces of gold?

2.77. How many moles of Ca^{2+} ions are in 0.25 mol of calcium titanate, $CaTiO_3$? What is the mass in grams of the Ca^{2+} ions?

2.78. How many moles of O^{2-} ions are in 0.55 mol of aluminum oxide, Al_2O_3? What is the mass in grams of the O^{2-} ions?

2.79. Suppose pairs of balloons are filled with 10.0 g of the following pairs of gases. Which balloon in each pair has the greater number of particles? (a) CO_2 or NO; (b) CO_2 or SO_2; (c) O_2 or Ar

2.80. If you had equal masses of the substances in the following pairs of compounds, which of the two would contain the greater number of ions? (a) NaBr or KCl; (b) NaCl or $MgCl_2$; (c) $CrCl_3$ or Na_2S

2.81. How many moles of SiO_2 are there in a quartz crystal (SiO_2) that has a mass of 45.2 g?

2.82. How many moles of NaCl are there in a crystal of halite that has a mass of 6.82 g?

2.83. The density of uranium (U; 19.05 g/cm^3) is more than five times as great as that of diamond (C; 3.514 g/cm^3). If you have a cube (1 cm on a side) of each element, which cube contains more atoms?

*2.84. Aluminum ($d = 2.70$ g/mL) and strontium ($d = 2.64$ g/mL) have nearly the same density. If we manufacture two cubes, each containing 1 mol of one element or the other, which cube will be smaller? What are the dimensions of this cube?

Making Elements

CONCEPT REVIEW

2.85. In the history of the universe, which of these particles formed first and which formed last? (a) deuteron; (b) neutron; (c) proton; (d) quark

2.86. Why did primordial nucleosynthesis last such a short time?

2.87. Why is energy released in a nuclear fusion process when the product is an element preceding iron in the periodic table?

2.88. **Components of Solar Wind** Most of the ions that flow out from the sun in the solar wind are hydrogen ions. The ions of which element should be next most abundant?

*2.89. **Nucleosynthesis in Giant Stars** A star needs a core temperature of about 10^7 K for hydrogen fusion to occur. Core temperatures above 10^8 K are needed for helium fusion. Why does helium fusion require much higher temperatures?

2.90. Why was the triple-alpha process unlikely to happen in a rapidly cooling universe soon after the Big Bang?

2.91. Early nucleosynthesis produced a universe that was more than 99% hydrogen and helium with less than 1% lithium. Why were the other elements not formed?

2.92. **Origins of the Elements** Our sun contains carbon even though its core is not hot or dense enough to sustain carbon synthesis through the triple-alpha process. Where could the carbon have come from?

2.93. What is the effect of β decay on the ratio of neutrons to protons in a nucleus?

2.94. Explain how the product of β decay has a higher atomic number than the radionuclide from which the product forms.

PROBLEMS

2.95. What nuclide is produced in the core of a giant star by each of the following fusion reactions? Assume there is only one product in each reaction.
a. $^{12}C + {}^4He \rightarrow$
b. $^{20}Ne + {}^4He \rightarrow$
c. $^{32}S + {}^4He \rightarrow$

2.96. What nuclide is produced in the core of a giant star by each of the following fusion reactions? Assume there is only one product in each reaction.
a. $^{28}Si + {}^4He \rightarrow$
b. $^{40}Ca + {}^4He \rightarrow$
c. $^{24}Mg + {}^4He \rightarrow$

2.97. What nuclide is produced in the core of a collapsing giant star by each of the following reactions?
a. $^{96}_{42}Mo + 3\,^1_0n \rightarrow ? + {}^0_{-1}\beta$
b. $^{118}_{50}Sn + 3\,^1_0n \rightarrow ? + {}^0_{-1}\beta$
c. $^{108}_{47}Ag + {}^1_0n \rightarrow ? + {}^0_{-1}\beta$

2.98. What nuclide is produced in the core of a collapsing giant star by each of the following reactions?
a. $^{65}_{29}Cu + 3\,^1_0n \rightarrow ? + {}^0_{-1}\beta$
b. $^{68}_{30}Zn + 2\,^1_0n \rightarrow ? + {}^0_{-1}\beta$
c. $^{88}_{38}Sr + {}^1_0n \rightarrow ? + {}^0_{-1}\beta$

2.99. Radioactive ^{137}I decays to ^{137}Xe, which is also radioactive and decays to ^{137}Cs. Do either, or both, of the decay processes involve emission of a β particle?

2.100. **Isotopes in Geochemistry** The relative abundances of the stable isotopes of the elements are not entirely constant. For example, in some geological samples (soils and rocks) the ratio of ^{87}Sr to ^{86}Sr is affected by the presence of a radioactive isotope of another element, which slowly undergoes β decay to produce more ^{87}Sr. What is this other isotope?

Artificial Nuclides

PROBLEMS

2.101. Write a balanced nuclear equation describing how a ^{209}Bi target might be bombarded with subatomic particles to form ^{211}At.

*__2.102.__ Bombardment of a ^{239}Pu target with α particles produces ^{242}Cm and another particle.
   a. Use a balanced nuclear equation to determine the identity of the missing particle.
   b. The synthesis of which other nuclide described in this chapter involves the same subatomic particles?

2.103. Complete the following nuclear equations describing the preparation of isotopes for nuclear medicine.
   a. ^{32}S + ^{1}n → ? + ^{1}H
   c. ^{75}As + ? → ^{77}Br
   b. ^{55}Mn + ^{1}H → ^{52}Fe + ?
   d. ^{124}Xe + ^{1}n → ? → ^{125}I + ?

2.104. Complete the following nuclear equations describing the preparation of isotopes for nuclear medicine.
   a. ^{6}Li + ^{1}n → ^{3}H + ?
   c. ^{56}Fe + ? → ^{57}Co + ^{1}n
   b. ^{16}O + ^{3}H → ^{18}F + ?
   d. ^{121}Sb + ^{4}He → ? + 2 ^{1}n

2.105. Complete the following nuclear equations.
   a. $^{131}_{52}$Te → $^{131}_{53}$I + ?
   c. ? + $^{4}_{2}$He → $^{13}_{7}$N + $^{1}_{0}$n
   b. ? → $^{122}_{54}$Xe + $^{0}_{-1}$β
   d. ? + $^{1}_{1}$H → $^{67}_{31}$Ga + 2 $^{1}_{0}$n

2.106. Complete the following nuclear equations.
   a. ^{210}Po → ^{206}Pb + ?
   c. $^{14}_{6}$C → ? + $^{0}_{-1}$β
   b. $^{3}_{1}$H → $^{3}_{2}$He + ?
   d. ? + $^{1}_{0}$n → $^{59}_{26}$Fe

Additional Problems

2.107. In April 1897, J. J. Thomson presented the results of his experiment with cathode-ray tubes in which he proposed that the rays were actually beams of negatively charged particles, which he called "corpuscles."
   a. What is the name we use for the particles today?
   b. Why did the beam deflect when passed between electrically charged plates?
   c. If the polarity of the plates were switched, how would the position of the light spot on the phosphorescent screen change?

2.108. Strontium has four isotopes: ^{84}Sr, ^{86}Sr, ^{87}Sr, and ^{88}Sr.
   a. How many neutrons are there in an atom of each isotope?
   b. Use the data in the table below to calculate the natural abundances of ^{87}Sr and ^{88}Sr.

Isotope	Mass (amu)	Natural Abundance (%)
^{84}Sr	83.9134	0.56
^{86}Sr	85.9094	9.86
^{87}Sr	86.9089	?
^{88}Sr	87.9056	?
Average	87.621	

*__2.109.__ There are three stable isotopes of magnesium. Their masses are 23.9850, 24.9858, and 25.9826 amu. If the average atomic mass of magnesium is 24.3050 amu and the natural abundance of the lightest isotope is 78.99%, what are the natural abundances of the other two isotopes?

2.110. In the summer of 2003 a team of American and Russian scientists reported that they had synthesized an isotope of element 115 with a mass number of 288 by bombarding a target of ^{243}Am with ^{48}Ca. Write a balanced nuclear equation describing the nuclear reaction that may have produced this nuclide.

*__2.111.__ The absorption of a neutron by ^{11}B produces a radioactive nuclide that decays by either α decay or β decay. Write balanced nuclear equations describing the decay reactions.

2.112. The following nuclear equations are based on successful attempts to synthesize supermassive elements. Complete each equation by writing the symbol of the supermassive nuclide that was synthesized.
   a. $^{58}_{26}$Fe + $^{209}_{83}$Bi → ? + $^{1}_{0}$n
   b. $^{64}_{28}$Ni + $^{209}_{83}$Bi → ? + $^{1}_{0}$n
   c. $^{62}_{28}$Ni + $^{208}_{82}$Pb → ? + $^{1}_{0}$n
   d. $^{22}_{10}$Ne + $^{249}_{97}$Bk → ? + 4 $^{1}_{0}$n
   e. $^{58}_{26}$Fe + $^{208}_{82}$Pb → ? + $^{1}_{0}$n

2.113. An atom of darmstadtium-269 was synthesized in 2003 by bombardment of a ^{208}Pb target with ^{62}Ni nuclei. Write a balanced nuclear equation describing the synthesis of ^{269}Ds.

*__2.114.__ The origins of the two naturally occurring isotopes of boron, ^{11}B and ^{10}B, are unknown. Both isotopes may be formed from collisions between protons and carbon, oxygen, or nitrogen in the aftermath of supernova explosions. Write balanced nuclear equations describing how ^{10}B might be formed from such collisions with ^{12}C and ^{14}N.

2.115. **Hope Diamond** The Hope Diamond (Figure P2.115) at the Smithsonian National Museum of Natural History has a mass of 45.52 carats.
   a. How many moles of carbon are in the Hope Diamond (1 carat = 200 mg)?
   b. How many carbon atoms are in the diamond?

FIGURE P2.115

2.116. Suppose we know the atomic mass of each of the three stable isotopes of an element to six significant figures, and we know the natural abundances of the isotopes to the nearest 0.01%. How well can we know the average atomic mass, that is, how many significant figures should be used to express its value?

3

Atomic Structure

Explaining the Properties of Elements

Can Nature Be as Absurd as It Seems?

By the end of the 19th century, many leading scientists believed that all the major scientific discoveries that *could* be made *had* been made. All that was left was explaining a few phenomena that seemed at odds with well-established laws. One of these outliers, called the photoelectric effect, was the observation that electrons are ejected from charged metal surfaces when those surfaces are illuminated. Another involved the colors of light given off by materials heated to incandescence, such as the magma ejected during volcanic eruptions. Attempts to explain these phenomena helped pave the way for a new theory of the structure of matter at the atomic level. We call this theory *quantum theory*. In addition to giving us powerful insights into the composition and behavior of atoms, quantum theory has forever altered our sense of what we can really know about the world around us.

According to quantum theory, our world at the atomic level is governed by rules that are different from those that apply to the macroscopic world. Two of the giants of quantum theory, Niels Bohr and Werner Heisenberg, wrestled with these differences and their implications. The two scientists briefly worked together at the University of Copenhagen, and after a particularly heated discussion about a feature of quantum theory, Heisenberg posed this question to himself: "Can nature possibly be as absurd as it seems?" The answer is probably yes.

In this chapter we explore the evolution of quantum theory and the interaction of matter and energy at the atomic level. We link the spectra of atoms to their atomic structure: in particular, we describe how electrons are distributed around the nuclei of their atoms. These distribution patterns provide a theoretical framework that explains the properties of the elements, including their tendencies to form ions with characteristic positive or negative charges, and their interactions to form compounds.

Hot Magma Volcanic magma (primarily silica) emits a broad range of electromagnetic radiation, which volcanologists can use to estimate its temperature. Efforts in the 20th century to explain properties of the radiation produced by hot material led to a whole new way of viewing matter at the atomic level.

LEARNING OUTCOMES

LO1 Interconvert the energies, wavelengths, and frequencies of electromagnetic radiation and link their values to the appropriate regions of the electromagnetic spectrum
Sample Exercises 3.1, 3.2

LO2 Explain the photoelectric effect using quantum theory
Sample Exercise 3.3

LO3 Relate the energies and wavelengths of photons absorbed and emitted by atoms to electron transitions between atomic energy levels
Sample Exercises 3.4, 3.5

LO4 Apply the Heisenberg uncertainty principle to particles in motion and calculate their de Broglie wavelengths
Sample Exercises 3.6, 3.7

LO5 Assign quantum numbers to orbitals and use their values to describe the sizes, energies, and orientations of orbitals
Sample Exercises 3.8, 3.9

LO6 Use the aufbau principle and Hund's rule to write electron configurations and draw orbital diagrams of atoms and monatomic ions
Sample Exercises 3.10, 3.11, 3.12

LO7 Use the concept of effective nuclear charge to explain differences in the energies of atomic orbitals and to predict the relative sizes of atoms and monatomic ions
Sample Exercise 3.13

LO8 Relate the ionization energies and electron affinities of the elements to their positions in the periodic table
Sample Exercise 3.14

electromagnetic radiation any form of radiant energy in the electromagnetic spectrum.

electromagnetic spectrum a continuous range of radiant energy that includes gamma rays, X-rays, ultraviolet radiation, visible light, infrared radiation, and radio waves.

CONNECTION We discussed Rutherford's gold-foil experiments and the model of the atom that evolved from them in Chapter 2.

CHEMTOUR Electromagnetic Radiation

3.1 Waves of Light

In Chapter 2 we explored the development of our modern view of atomic structure. Rutherford's gold-foil experiments established that atoms are mostly empty space: negatively charged electrons surround a tiny nucleus containing most of the atom's mass and all of its positive charge. Because they have opposite electrical charges, the electrons in an atom are attracted to its nucleus. What keeps them from falling into it? Rutherford thought that the electrons were in motion, revolving around the nucleus in orbits much as the planets revolve around the sun. However, classical physics predicts that negatively charged electrons orbiting a positively charged nucleus should lose energy and eventually spiral into it. The existence of stable atoms tells us that this does not happen. Why doesn't the electron crash into the nucleus?

To understand the stability of atoms structured according to Rutherford's model, we need to study how atoms interact with a form of energy called *radiant energy* or **electromagnetic radiation**. Visible light is the most familiar form of electromagnetic radiation, but there are several others you may have heard of. Together they make up the **electromagnetic spectrum** (Figure 3.1). Some forms, such as the gamma rays that accompany nuclear reactions and the X-rays used in medical imaging, have much more energy than visible light. Other forms have much less energy, like the infrared radiation we can feel as heat flowing from warm objects, the microwaves used in ovens and wireless networks, and radio waves.

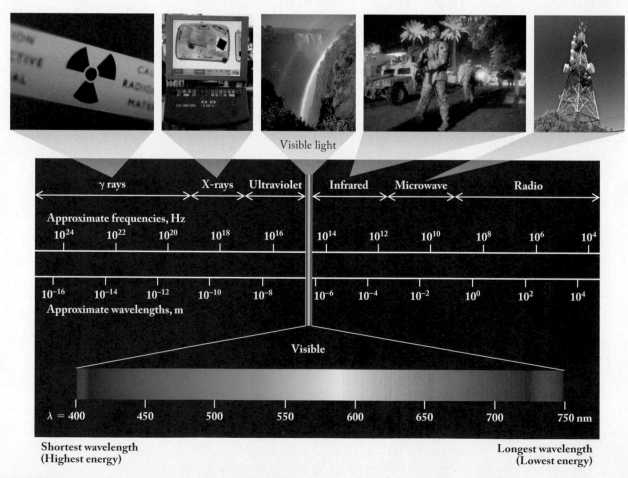

FIGURE 3.1 Visible light occupies a tiny fraction of the electromagnetic spectrum, which ranges from ultrashort-wavelength, high-frequency gamma (γ) rays to long-wavelength, low-frequency radio waves. Note that frequencies increase from right to left, and wavelengths increase from left to right.

All the forms of radiation in Figure 3.1 are considered *electromagnetic* because of a theory about their properties developed by Scottish scientist James Clerk Maxwell (1831–1879). According to Maxwell's theory, electromagnetic radiation moves through space (or through any transparent medium) as waves with two perpendicular components: an oscillating *electric* field and an oscillating *magnetic* field (Figure 3.2). Maxwell derived a set of equations based on his oscillating-wave model that accurately describes nearly all the observed properties of light and the other forms of radiant energy.

A wave of electromagnetic radiation, like any wave traveling through any medium, has a characteristic **wavelength (λ)**, which is the distance from one wave crest to the next, as shown in Figure 3.3. Note that wave A in Figure 3.3 has twice the wavelength of wave B. Each wave also has a characteristic **frequency (ν)**, which is the number of crests that pass a stationary point in space per second. Frequency has units of waves per second, or simply *per second* (s^{-1}). Scientists also use the frequency unit called the **hertz (Hz)**, which is equal to 1 wave per second ($1 \text{ Hz} = 1 \text{ s}^{-1}$).

Note that Figure 3.3 contains two complete cycles of wave A and four complete cycles of wave B, so the frequency of wave B is twice that of wave A. Also note that wave B has a wavelength half that of wave A. The wavelengths and frequencies of electromagnetic radiation are inversely related in exactly this way: the shorter the wavelength, the higher the frequency. This occurs because all forms of radiant energy travel at the same speed in the same transparent medium, and a wave's speed is the product of its wavelength and its frequency. Equation 3.1 shows this relationship for radiant energy traveling through a vacuum, where its speed is called the *speed of light, c*:

$$\lambda \nu = c \tag{3.1}$$

The value of c to four significant figures is 2.998×10^8 m/s. The reciprocal relationship between wavelength and frequency may be more evident if we rearrange the terms in Equation 3.1:

$$\nu = \frac{c}{\lambda} \tag{3.2}$$

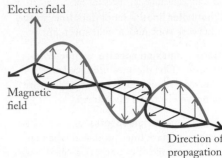

Electric field

Magnetic field

Direction of propagation

FIGURE 3.2 Electromagnetic waves consist of electric and magnetic fields that oscillate in planes oriented at right angles to each other.

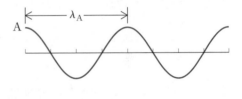

A

λ_A

B

λ_B

FIGURE 3.3 Every wave has a characteristic wavelength (λ) and frequency (ν). Wave A has a longer wavelength (and lower frequency) than wave B.

| SAMPLE EXERCISE 3.1 | **Calculating Frequency from Wavelength** | LO1 |

What is the frequency of the yellow-orange light (λ = 589 nm) produced by sodium-vapor streetlights?

COLLECT AND ORGANIZE We are given the wavelength of a color of visible light and asked to calculate its frequency. Frequency and wavelength are related by Equation 3.2:

$$\nu = \frac{c}{\lambda}$$

where c is 2.998×10^8 m/s.

ANALYZE The value c is in meters per second, but the wavelength is given in nanometers. Therefore, we need to convert nanometers to meters using the equality 1 nm = 10^{-9} m as a conversion factor. The unit labels in Figure 3.1 indicate that frequencies of visible light are around 10^{14} Hz, or 10^{14} s^{-1}, so we expect our answer to have an exponent in that range.

---

wavelength (λ) the distance from crest to crest or trough to trough on a wave.

frequency (ν) the number of crests of a wave that pass a stationary point of reference per second.

hertz (Hz) the SI unit of frequency with units of reciprocal seconds: $1 \text{ Hz} = 1 \text{ s}^{-1} = 1$ cycle per second (cps).

Fraunhofer lines a set of dark lines in the otherwise continuous solar spectrum.

atomic emission spectra characteristic patterns of bright lines produced when atoms are vaporized in high-temperature flames or electrical discharges.

atomic absorption spectra characteristic patterns of dark lines produced when an external source of radiation passes through free, gaseous atoms.

SOLVE Substituting the given values and appropriate unit conversion factor into Equation 3.2:

$$\nu = \frac{c}{\lambda} = \frac{2.998 \times 10^8 \text{ m/s}}{589 \text{ nm} \times \dfrac{10^{-9} \text{ m}}{1 \text{ nm}}}$$

$$= 5.09 \times 10^{14} \text{ s}^{-1}$$

THINK ABOUT IT The large value is in the range that we expected. Remember that λ and ν are reciprocal: as one increases, the other decreases.

Practice Exercise If the radio waves transmitted by a radio station have a frequency of 90.9 MHz, what is the wavelength of the waves, in meters?

(Answers to Practice Exercises are in the back of the book.)

CONCEPT TEST

The ultraviolet (UV) region of the electromagnetic spectrum contains waves with wavelengths from about 10^{-7} m to 10^{-9} m; the infrared (IR) region contains waves with wavelengths from about 10^{-4} m to 10^{-6} m. Are waves in the UV region higher in frequency or lower in frequency than waves in the IR region?

(Answers to Concept Tests are in the back of the book.)

3.2 Atomic Spectra

In this section we explore how atoms and electromagnetic radiation interact. We start with observations that were made over 200 years ago by English scientist William Hyde Wollaston (1766–1828). In 1800, Wollaston was studying the spectrum of sunlight using carefully ground glass prisms when he discovered that the spectrum was not completely continuous. Instead, it contained a series of very narrow dark lines. Using even better prisms, German physicist Joseph von Fraunhofer (1787–1826) resolved and mapped the wavelengths of over 500 of these lines, now called **Fraunhofer lines** (Figure 3.4).

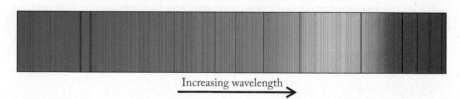

Increasing wavelength

FIGURE 3.4 The spectrum of sunlight is not continuous but contains numerous narrow gaps, which appear as dark lines called Fraunhofer lines.

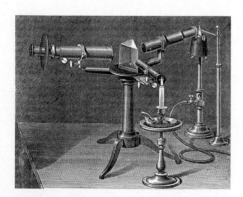

FIGURE 3.5 Kirchhoff and Bunsen used spectroscopes such as this one to observe the atomic emission spectra of elements introduced to the flame of the burner on the right.

Fraunhofer did not know *why* those narrow lines were present in the sun's spectrum. That understanding came nearly a half century later from discoveries made by two other Germans, Robert Wilhelm Bunsen (1811–1899) and physicist Gustav Robert Kirchhoff (1824–1887), using an instrument called a spectroscope (Figure 3.5). They discovered that the light given off, or *emitted*, by elements vaporized in the transparent flame of a gas burner developed by Bunsen produced spectra that were the opposite of Fraunhofer's. Whereas the sun's spectrum was nearly continuous except for a set of narrow dark lines, the spectra produced by the elements consisted of only a few narrow bright lines on a dark background. Bunsen and Kirchhoff

FIGURE 3.6 The atomic emission spectrum of sodium, when viewed through a spectroscope like that developed by Kirchhoff and Bunsen, consists of a single bright yellow-orange line with a wavelength of 589 nm.

discovered that the lines in the **atomic emission spectra** of certain elements *exactly matched the wavelengths* of some of the Fraunhofer lines in the spectrum of sunlight. For example, the Fraunhofer D line exactly matched the bright yellow-orange line in the emission spectrum produced by hot sodium vapor (Figure 3.6).

Those experiments and others employing light sources called gas-discharge tubes showed that at very high temperatures the atoms of each element *emit* a characteristic spectrum (Figure 3.7).

(a) Emission spectrum of hydrogen

(b) Emission spectrum of helium

(c) Emission spectrum of neon

Conversely, the atoms of elements in the gaseous state *absorb* electromagnetic radiation when illuminated by an external source of radiation. This absorption of radiation by atoms produces **atomic absorption spectra** of narrow dark lines in otherwise continuous spectra (Figure 3.8). For any given element, the dark lines in its absorption spectrum are at exactly the same wavelengths as the bright lines in its emission spectrum. The phenomenon of atomic absorption explains the Fraunhofer lines: gaseous atoms in the outer regions of the sun absorb characteristic wavelengths of the sunlight passing through them on its way to Earth.

FIGURE 3.7 The colored light emitted from gas-discharge tubes filled with various gaseous elements produces atomic emission spectra that are characteristic of the element: (a) hydrogen, (b) helium, and (c) neon.

▶❚❚ **CHEMTOUR** Light Emission and Absorption

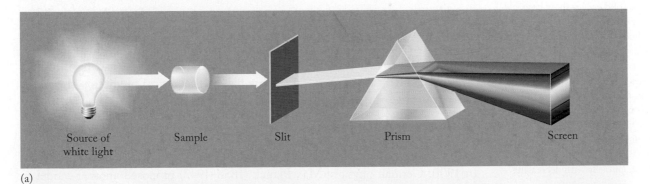

(a)

Absorption spectrum of hydrogen

Absorption spectrum of helium

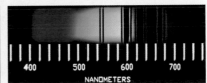

Absorption spectrum of neon

(b)

FIGURE 3.8 (a) When gaseous atoms of hydrogen, helium, and neon are illuminated by an external source of white light (containing all colors of the visible spectrum), the resultant atomic absorption spectra contain dark lines that are characteristic of the elements. (b) The dark lines have the same wavelengths as the bright lines in the elements' atomic emission spectra, shown in Figure 3.7.

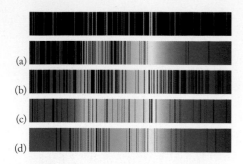

FIGURE 3.9 Atomic emission spectrum of mercury vapor (top) and four atomic absorption spectra (a–d).

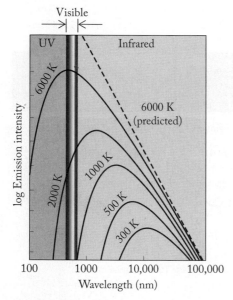

FIGURE 3.11 The solid curves show how the intensities and wavelengths of radiation emitted by blackbody radiators change with changing temperature. Logarithmic scales are used on both axes to include a wide range of intensities and wavelengths. The dashed line shows the predicted spectrum of a 6000 K radiator. It fits the experimental data at longer (infrared) wavelengths but not at shorter wavelengths, and especially not in the ultraviolet (UV) region, where emission intensity drops precipitously with decreasing wavelength.

quantum (plural *quanta*) the smallest discrete quantity of a particular form of energy.

The top spectrum in Figure 3.9 is the emission spectrum of mercury vapor. Select the absorption spectrum of mercury vapor from the other four spectra.

(Answers to Concept Tests are in the back of the book.)

3.3 Particles of Light: Quantum Theory

In addition to his studies of the narrow-line emission spectra of atoms, Kirchhoff also studied the continuous emission spectra produced by hot objects such as metals heated to incandescence (Figure 3.10).

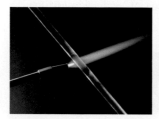 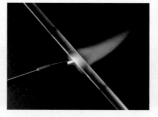 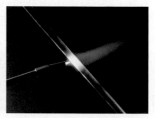

FIGURE 3.10 When a metal rod is heated it glows red at first, then orange, and finally becomes white-hot as its temperature increases and light of shorter wavelengths is emitted.

He and others recorded the spectra of sources of radiant energy that are called *blackbody* radiators because, when cold, their surfaces are jet black and they absorb all the light that strikes them. These perfect absorbers of light when cold become perfect radiators of light when hot. The spectra they emit depend only on their temperature: the hotter they are, the more radiation they emit, and the shorter the wavelengths of that radiation. However, the laws of classical physics could not explain the spectra. In fact, those laws predicted that emission brightness should increase with increasing frequency (decreasing wavelength) without limit, as shown by the dashed line in Figure 3.11 for $T = 6000$ K. This meant that at very short wavelengths brightness should approach infinity, which does not make sense: no source of radiation, no matter how hot it is, can be infinitely bright. Scientists at the time called this flawed theoretical prediction the *ultraviolet catastrophe*.

Photons of Energy

In 1900, German physicist Max Planck (1858–1947) proposed another model to describe the emission spectra of hot objects. Then a professor at the University of Berlin, Planck (Figure 3.12) had been commissioned by electric companies to maximize the output of the incandescent lightbulbs invented by Thomas Edison. His attempts at developing a theoretical model that accurately described their incandescence led him to discard classical physics and make a critical assumption about radiant energy: no matter what its source, it can never be truly continuous. Instead, Planck proposed that objects emit electromagnetic radiation only in integral multiples of an elementary unit, or **quantum**, of energy defined by the equation

$$E = h\nu \tag{3.3}$$

where ν is the frequency of the radiation and h is the **Planck constant**, 6.626×10^{-34} J · s. If we combine Equations 3.2 and 3.3, we obtain an equation that relates the energy of a quantum of electromagnetic radiation to its wavelength:

$$E = \frac{hc}{\lambda} \tag{3.4}$$

Because Planck's model is characterized by these quanta, it has become known as **quantum theory**.

To visualize the meaning of Planck's quantum of energy, consider two ways you might get from the sidewalk to the entrance of a building (Figure 3.13). If you walked up the steps, you would be able to stand at only discrete heights above the sidewalk. You could not stand at a height between adjacent steps because there would be nothing to stand on. If you walked up the ramp, however, you could stop at any height between the sidewalk and the entrance. On the steps, height is **quantized**: the discrete changes in height model Planck's hypothesis that energy is released (analogous to walking down the steps) or absorbed (walking up the steps) in discrete packets, or quanta, of energy.

> **CONCEPT TEST** ....
>
> Which of the following are quantized?
>
> a. The volume of water in the Atlantic Ocean
> b. The number of eggs remaining in a carton
> c. The time it typically takes you to get ready for class in the morning
> d. The number of red lights encountered when driving the length of Fifth Avenue in New York City
>
> *(Answers to Concept Tests are in the back of the book.)*
> ....

Given the extremely tiny value of the Planck constant, the energy values we obtain using Equation 3.4 tend to be very small because E represents the number of joules in a single quantum of energy. Today we call the tiny packets of radiant energy **photons**. They represent elementary building blocks of electromagnetic radiation in much the same way that atoms represent the building blocks of matter. The observed brightness of a source of radiant energy is the sum of the energies of the enormous number of photons it produces per unit of time.

> **SAMPLE EXERCISE 3.2** **Calculating the Energy of a Photon** **LO1**

What is the energy of a photon of red light that has a wavelength of 656 nm?

COLLECT AND ORGANIZE We need to calculate the energy of a photon of electromagnetic radiation starting with its wavelength. Equation 3.4,

$$E = \frac{hc}{\lambda}$$

relates the energy of a photon to its wavelength. The value of the Planck constant (h) is 6.626×10^{-34} J · s, and the speed of light (c) is 2.998×10^8 m/s.

ANALYZE The wavelength is given in nanometers, but the value of the speed of light has units of meters per second. Therefore, we need to convert nanometers to meters so the distance units cancel out. The value of h is extremely small, so it is likely the results of our calculation, even factoring in the speed of light, will be very small, too.

FIGURE 3.12 German scientist Max Karl Ernst Ludwig Planck is considered the father of quantum physics. He won the 1918 Nobel Prize in Physics for his pioneering work on the quantized nature of electromagnetic radiation. Planck was revered by his colleagues for his personal qualities as well as his scientific accomplishments.

FIGURE 3.13 Quantized and continuously varying heights. A flight of stairs exemplifies quantization: each step rises by a discrete height to the next step. In contrast, the heights on a ramp are not quantized.

....

Planck constant (h) the proportionality constant between the energy and frequency of electromagnetic radiation expressed in $E = h\nu$; $h = 6.626 \times 10^{-34}$ J · s.

quantum theory a model based on the idea that energy is absorbed and emitted in discrete quantities of energy called quanta.

quantized having values restricted to whole-number multiples of a specific base value.

photon a quantum of electromagnetic radiation.

....

photoelectric effect the release of electrons from a material as a result of electromagnetic radiation striking it.

threshold frequency (ν_0) the minimum frequency of light required to produce the photoelectric effect.

SOLVE

$$E = \frac{hc}{\lambda} = \frac{(6.626 \times 10^{-34}\,\text{J}\cdot\text{s})(2.998 \times 10^8\,\frac{\text{m}}{\text{s}})}{656\,\text{nm} \times \frac{10^{-9}\,\text{m}}{1\,\text{nm}}}$$

$$= 3.03 \times 10^{-19}\,\text{J}$$

THINK ABOUT IT This quantity of energy is indeed small, as it should be, because a photon is an atomic-level particle of radiant energy. If we multiply this energy value by Avogadro's number we obtain the energy content of a mole of the photons:

$$\left(3.03 \times 10^{-19}\,\frac{\text{J}}{\text{photon}}\right) \times \left(6.0221 \times 10^{23}\,\frac{\text{photons}}{\text{mol}}\right) = 1.82 \times 10^5\,\frac{\text{J}}{\text{mol}} = 182\,\frac{\text{kJ}}{\text{mol}}$$

Practice Exercise Some instruments differentiate individual quanta of electromagnetic radiation based on their energies. Assume such an instrument has been adjusted to detect photons that have 1.00×10^{-16} J of energy. What is the wavelength of the detected radiation? Give your answer in nanometers and in meters. ⚙

(Answers to Practice Exercises are in the back of the book.)

The Photoelectric Effect

Although Planck's quantum model explained the emission spectra of hot objects, there was no experimental evidence in 1900 to support the existence of quanta of energy. In 1905, Albert Einstein (1879–1955) supplied that evidence. It came from his studies of a phenomenon called the **photoelectric effect**, in which electrons are emitted from metals and semiconductor materials when they are illuminated by and absorb electromagnetic radiation. Because light releases these electrons, they are called *photoelectrons*, derived from the Greek *photo,* meaning "light."

Photoelectrons are emitted when the frequency of incident radiation is above some minimum **threshold frequency (ν_0)** (Figure 3.14). Radiation of frequencies less than the threshold value produces no photoelectrons, no matter how intense the radiation is. On the other hand, even a dim source of radiant energy produces at least a few photoelectrons when the frequencies it emits are equal to or greater than the threshold frequency.

〜〜〜 High-frequency light
〜〜 Low-frequency light
⊝ Electron

FIGURE 3.14 A phototube includes a positive electrode and a negative metal electrode. (a) If radiation of high enough frequency and energy (violet) illuminates the negative electrode, electrons are dislodged from the surface and flow toward the positive electrode. This flow of electrons completes the circuit and produces an electric current. The size of the current is proportional to the intensity of the radiation—to the number of photons per unit time striking the negative electrode. (b) Photons of lower frequency (red) and hence lower energy do not have sufficient energy to dislodge electrons and do not produce the photoelectric effect no matter how many of them (c) bombard the surface of the metal. The circuit is not complete, and there is no current.

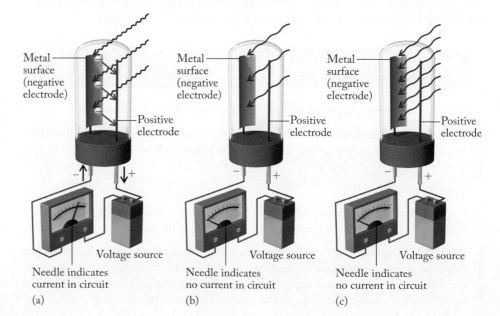

Metal surface (negative electrode)
Positive electrode
Voltage source
Needle indicates current in circuit
(a)

Metal surface (negative electrode)
Positive electrode
Voltage source
Needle indicates no current in circuit
(b)

Metal surface (negative electrode)
Positive electrode
Voltage source
Needle indicates no current in circuit
(c)

For example, some night vision goggles used by military and law enforcement personnel (Figure 3.15) can detect extremely low levels of light because they incorporate photoelectric sensors and amplifier circuits that produce over 10,000 electrons for each photoelectron initially emitted. When such large numbers of electrons strike a phosphorescent screen (like the ones used in cathode-ray tubes, as in Figure 2.2) they produce images that are tens of thousands of times brighter than the original.

Einstein used Planck's quantum model to explain the ability of the photoelectric materials to emit photoelectrons. He proposed that every photoelectric material has a characteristic threshold frequency ν_0 associated with the minimum quantum of absorbed energy needed to remove a single electron from the material's surface. The threshold frequency, for example, of a sensor used in night vision goggles is in the infrared region of the electromagnetic spectrum. This means that when photons of radiation with higher frequencies than this (including visible light) are focused on one of the sensors, photoelectrons are emitted.

The minimum quantity of energy needed to emit photoelectrons from a photoelectric material is called the material's **work function (Φ)**:

$$\Phi = h\nu_0 \tag{3.5}$$

The value of Φ is related to the strength of the attraction between the nuclei of the metal's atoms and the electrons surrounding those nuclei. In Einstein's model, a photoelectron is emitted when a quantum of radiant energy (a photon) provides an electron with enough energy to break free of the surface of a photoelectric material. If the incoming beam includes photons with frequencies above the threshold frequency ($\nu > \nu_0$), then each photon has more than enough energy to dislodge an electron. This extra energy in excess of the work function is imparted to each ejected electron as kinetic energy: the higher the frequency above the threshold frequency, the greater the kinetic energy of the ejected electron. The kinetic energies of photoelectrons ($KE_{electron}$) can be determined using instruments that measure their speeds. If we know $KE_{electron}$ and the energy of the incident photons ($h\nu$), we can calculate the value of the work function of the target metal using Equation 3.6:

$$\Phi = h\nu - KE_{electron} \tag{3.6}$$

FIGURE 3.15 Night vision goggles incorporate photoelectric sensors and electronic amplifiers called image intensifiers to produce images that are tens of thousands of times brighter than the original.

SAMPLE EXERCISE 3.3   **Using the Work Function**   **LO2**

The work function of mercury is 7.22×10^{-19} J.

  a. What is the minimum frequency of radiation required to eject photoelectrons from a mercury surface?
  b. Could visible light produce the photoelectric effect in mercury?

COLLECT AND ORGANIZE We are asked to use the value of the work function (Φ) of a metal to find its corresponding threshold (minimum) frequency (ν_0). Equation 3.5 relates the parameters. Figure 3.1 contains information about the frequencies and wavelengths of the different regions of the electromagnetic spectrum.

ANALYZE We need to rearrange the terms in Equation 3.5 to solve for ν_0. According to Figure 3.1, the frequencies of visible light are between 10^{14} and 10^{15} Hz.

SOLVE
  a. Rearranging Equation 3.5 to solve for threshold frequency,

$$\nu_0 = \frac{\Phi}{h} = \frac{7.22 \times 10^{-19} \text{ J}}{6.626 \times 10^{-34} \text{ J} \cdot \text{s}} = 1.09 \times 10^{15} \text{ s}^{-1}$$

work function (Φ) the amount of energy needed to dislodge an electron from the surface of a material.

b. The threshold frequency is greater than 10^{15} Hz and, according to Figure 3.1, is in the ultraviolet region of electromagnetic radiation. The frequencies of visible light are below the threshold frequency, so visible light cannot generate photoelectrons from mercury.

THINK ABOUT IT The calculated threshold frequency is close to, but still greater than, that of the highest frequency (violet) visible light. Photons of visible light simply lack the energy needed to dislodge electrons from atoms on the surface of a drop of mercury.

Practice Exercise The work function of silver is 7.59×10^{-19} J. What is the longest wavelength of electromagnetic radiation that can eject a photoelectron from the surface of a piece of silver? ⚙

(Answers to Practice Exercises are in the back of the book.)

3.4 The Hydrogen Spectrum and the Bohr Model

In formulating his quantum theory, Planck was influenced by the results of investigations of the emission spectra produced by free, gas-phase atoms (as described in Section 3.2). These line spectra led him to question whether any spectrum, even that of an incandescent lightbulb, was truly continuous. Among the early investigators of atomic emission spectra was a Swiss mathematician and schoolteacher named Johann Balmer (1825–1898). His work focused on the pattern of emission lines produced by high-temperature hydrogen atoms. In 1885 Balmer formulated an empirical equation that accurately predicted the wavelengths of the four brightest atomic emission lines in the visible spectrum of hydrogen. To express the wavelengths in nanometers, the units most often used for visible light, we can use this form of the Balmer equation:

$$\lambda \text{ (nm)} = \left(\frac{364.56\, m^2}{m^2 - n^2} \right) \tag{3.7}$$

where m is an integer greater than 2 (the values 3, 4, 5, and 6 predict the wavelengths of the four brightest hydrogen lines) and $n = 2$. If, for example, we let $m = 3$ and solve for λ, we get

$$\lambda \text{ (nm)} = \left(\frac{364.56 \times 3^2}{3^2 - 2^2} \right) = 364.56 \times \frac{9}{5}$$

$$= 656.21 \text{ nm}$$

which is the wavelength of the red line in the atomic emission spectrum of hydrogen (Figure 3.7a). By using m values of 4, 5, and 6 in Equation 3.7, we get the wavelengths of the blue-green (486.08 nm), blue (434.00 nm), and violet (410.13 nm) lines, respectively.

The Balmer equation is called an *empirical* equation because he derived it strictly from experimental data: the wavelengths of light in the visible atomic emission spectrum of hydrogen. It had no theoretical foundation. No one in 1885 could explain *why* his equation fit the hydrogen spectrum, though other scientists of the time used it successfully to search for hydrogen lines corresponding to m values greater than 6, which Balmer could not see because they are in the ultraviolet region of the electromagnetic spectrum.

In 1888 Swedish physicist Johannes Robert Rydberg (1854–1919) published a more general empirical equation for predicting the wavelengths of hydrogen's spectral lines:

$$\frac{1}{\lambda} = R_{\mathrm{H}}\left(\frac{1}{n_1{}^2} - \frac{1}{n_2{}^2}\right) \tag{3.8}$$

where n_1 and n_2 are any positive integers (though n_2 has to be greater than n_1) and R_{H} is the Rydberg constant. As you can see, the term on the left in Rydberg's equation is not wavelength, but rather the reciprocal of wavelength ($1/\lambda$). That parameter is called the *wavenumber* of a spectral line. To understand why Rydberg found wavenumbers useful in formulating his equation, remember that the energy of a photon is inversely proportional to its wavelength. Therefore, its energy is *directly* proportional to its wavenumber. Though Rydberg did not know it in 1888, the values of n_1 and n_2 on the right side of his equation *define energy levels inside hydrogen atoms*, and the energy of a photon that a hydrogen atom absorbs or emits is exactly the same as the difference in energy between a pair of energy levels.

The value of Rydberg's constant, R_{H}, depends on the units used to express wavelength. Here are three common options:

$$R_{\mathrm{H}} = 1.097 \times 10^7 \text{ m}^{-1}$$
$$= 1.097 \times 10^5 \text{ cm}^{-1}$$
$$= 1.097 \times 10^{-2} \text{ nm}^{-1}$$

When hydrogen lines are in the ultraviolet or visible regions of the electromagnetic spectrum (as happens when $n_1 = 1$ or 2), their wavelengths are expressed in nanometers, and the most convenient form of the Rydberg equation is:

$$\frac{1}{\lambda} = (1.097 \times 10^{-2} \text{ nm}^{-1})\left(\frac{1}{n_1{}^2} - \frac{1}{n_2{}^2}\right) \tag{3.9}$$

When we let $n_1 = 2$ and $n_2 = 3, 4, 5$, or 6 in Equation 3.9 and then solve for λ, we get four wavelength values that exactly match those we obtained with the Balmer equation when the value of his m parameter was 3, 4, 5, or 6. They are the wavelengths of the four lines in the visible emission spectrum of hydrogen. Actually, the Balmer equation represents a special case of the Rydberg equation for the hydrogen lines that correspond to $n_1 = 2$. An advantage of Rydberg's equation is that it allowed scientists to predict the wavelengths of other series of hydrogen emission lines for which $n_1 \neq 2$. None of the lines were in the visible region. In 1908 German physicist Friedrich Paschen (1865–1947) discovered a series of hydrogen emission lines in the infrared region with wavelengths corresponding to Rydberg equation values of $n_1 = 3$ and $n_2 = 4, 5, 6$, and so on. A few years later, Theodore Lyman (1874–1954) at Harvard University discovered another series of hydrogen emission lines in the ultraviolet region corresponding to $n_1 = 1$. By the 1920s the $n_1 = 4$ and $n_1 = 5$ series had been discovered. Like the $n_1 = 3$ series, they are also in the infrared region.

SAMPLE EXERCISE 3.4   Calculating the Wavelength of a   LO3
                       Line in the Hydrogen Spectrum

What is the wavelength in nanometers of the line in the hydrogen spectrum that corresponds to $m = 7$ in the Balmer equation (Equation 3.7)? Check your answer by also calculating this wavelength using the Rydberg equation.

COLLECT AND ORGANIZE We need to calculate the wavelength of a line in hydrogen's atomic emission spectrum from its m value in the Balmer equation. The n value in the Balmer equation is always 2. We will then use the Rydberg equation to check the accuracy of our calculation. The requested units of wavelength are nanometers, so Equation 3.9 should be the most useful form of the Rydberg equation for this calculation.

ANALYZE To calculate a wavelength using the Balmer equation we insert the appropriate values of m and n (7 and 2 in this case) and solve for λ. In the corresponding calculation based on the Rydberg equation, we let $n_2 = 7$ because it is the greater integer and $n_1 = 2$. Our answer should be near 400 nm because the wavelength corresponding to $m = 6$ in the Balmer equation is 410 nm, and the wavelengths generated by the Balmer equation become both smaller and closer to each other as m increases.

SOLVE Using the Balmer equation,

$$\lambda = 364.56 \text{ nm} \left(\frac{m^2}{m^2 - n^2} \right) = 364.56 \text{ nm} \left(\frac{7^2}{7^2 - 2^2} \right)$$

$$= 396.97 \text{ nm}$$

Using the Rydberg equation (Equation 3.9),

$$\frac{1}{\lambda} = [1.097 \times 10^{-2} \text{ (nm)}^{-1}] \left(\frac{1}{n_1^2} - \frac{1}{n_2^2} \right)$$

$$= 1.097 \times 10^{-2} \text{ (nm}^{-1}) \left(\frac{1}{2^2} - \frac{1}{7^2} \right)$$

$$= 1.097 \times 10^{-2} \text{ (nm}^{-1})(0.2296) = 2.519 \times 10^{-3} \text{ (nm}^{-1})$$

$$\lambda = 397.0 \text{ nm}$$

THINK ABOUT IT The results of the two calculations are consistent although the first one has one more significant figure. This is because there are five significant figures in the constant in Equation 3.7 and only four in the constant in Equation 3.9. The calculated values are also close to the wavelength that we estimated based on $m = 6$.

Practice Exercise What is the wavelength of the photon emitted by a hydrogen atom that corresponds to $m = 12$ in the Balmer equation? Would Balmer have been able to see this line?

(Answers to Practice Exercises are in the back of the book.)

The Bohr Model

When Balmer and Rydberg derived their empirical equations, they didn't know why their equations worked or what the integers in them physically represented. A few years later, Max Planck proposed that the discrete wavelengths of the lines meant that hydrogen atoms lost and gained only discrete quanta of energy, which indicated to him that there were discrete energy levels inside the atoms. However, classic (macroscopic-scale) physics could not explain the existence of quantized energy levels. A new model was needed to account for them.

In 1913 Danish scientist Niels Bohr (1885–1962) proposed such a model. With it he could explain (1) why hydrogen atoms lose and gain discrete quanta of energy and (2) why their electrons do not spiral into their nuclei. His explanation was based on Rutherford's planetary model, in which electrons orbit nuclei, combined with Planck's notion of quantized energy. In Bohr's model, the electron in a hydrogen atom revolves around the nucleus in one of an array of available orbits. Each orbit represents a discrete energy level inside the atom. Bohr assigned each orbit a number, n, starting with $n = 1$ for the orbit closest to the nucleus (Figure 3.16). In his model, orbits farther from the nucleus have larger values of n, and the electrons in them have higher energies based on the following equation:

$$E = -2.178 \times 10^{-18}\,\text{J}\left(\frac{1}{n^2}\right) \tag{3.10}$$

where $n = 1, 2, 3, \ldots, \infty$. According to Equation 3.10, an electron in the orbit closest to the nucleus ($n = 1$) has the lowest (most negative) energy:

$$E = -2.178 \times 10^{-18}\,\text{J}\left(\frac{1}{1^2}\right) = -2.178 \times 10^{-18}\,\text{J}$$

An electron in the next closest ($n = 2$) orbit has an energy of

$$E = -2.178 \times 10^{-18}\,\text{J}\left(\frac{1}{2^2}\right) = -5.445 \times 10^{-19}\,\text{J}$$

Note that this value is less negative than the value for the electron in the $n = 1$ orbit. The values of n for orbits farther and farther from the nucleus approach ∞, and as they do, E approaches zero:

$$E = -2.178 \times 10^{-18}\,\text{J}\left(\frac{1}{\infty^2}\right) = 0$$

Zero energy means that the electron at $n = \infty$ is no longer part of the hydrogen atom, and that the atom no longer exists as a single entity. Rather, it has become two separate particles: a H^+ ion and a free electron.

Perhaps you are wondering why there is a negative sign in front of the right side of Equation 3.10. What does it mean for an electron to have negative energy? To understand this, let's start with an assumption: the addition of energy is required to pry a negatively charged electron away from a positively charged nucleus. This is logical because oppositely charged particles are attracted to each other, and this attraction must be overcome to separate them. An electron that has been separated from an atom no longer interacts with the nucleus, and there is no energy of attraction between the particles ($E = 0$). So, if we had to add energy to get to zero energy, the initial energy of the electron in the atom must have been less than zero, as we would calculate using Equation 3.10.

An important feature of the Bohr model is that it provides a theoretical framework for explaining the experimental observations of Balmer, Rydberg, and others we have discussed. To see the connection, consider what happens when an electron moves between two allowed energy levels in Bohr's model. If we label the energy level where the electron starts n_{initial}, and the level where the electron ends up n_{final}, then the change in energy of the electron is

$$\Delta E = -2.178 \times 10^{-18}\,\text{J}\left(\frac{1}{n_{\text{final}}{}^2} - \frac{1}{n_{\text{initial}}{}^2}\right) \tag{3.11}$$

Here we use the capital Greek letter delta (Δ) to represent change (and we will do so many more times in this book).

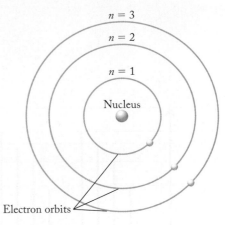

FIGURE 3.16 In the Bohr model of the hydrogen atom, its electron revolves around the nucleus in one of a series of concentric orbits. Each orbit represents an allowed energy level. An electron in the orbit closest to the nucleus ($n = 1$) has the lowest energy.

▶❙❙ **CHEMTOUR** Bohr Model of the Atom

If the electron moves to an orbit farther from the nucleus, then $n_{final} > n_{initial}$, and the overall value of the terms inside the parentheses in Equation 3.11 is negative because $1/n_{final}^2 < 1/n_{initial}^2$. This negative value multiplied by the negative coefficient gives us a positive ΔE and represents an increase in electron energy. On the other hand, if an electron moves from an outer orbit to one closer to the nucleus, then $n_{final} < n_{initial}$, and the sign of ΔE is negative. This means the electron loses energy.

When the electron in a hydrogen atom is in the lowest ($n = 1$) energy level, the atom is said to be in its **ground state**. According to the Bohr model, the electron cannot have any less energy than it has in the ground state, which means that it can't lose more energy and spiral into the nucleus.

If the electron in a hydrogen atom is in an energy level above $n = 1$, then the atom is said to be in an **excited state**. An electron can move from the $n = 1$ (ground state) orbit to a higher level (for example, $n = 3$) by absorbing a quantum of energy (ΔE) that exactly matches the energy difference between the two states. Similarly, an electron in an excited state can move to an even higher energy level by absorbing a quantum of energy that exactly matches the energy difference between those two states. On the other hand, an electron in an excited state can move to a lower energy excited state, or all the way down to the ground state, by emitting a quantum of energy that exactly matches the energy difference between those two states. The movement of an electron between any two energy levels is called an **electron transition**.

The energy-level diagram in Figure 3.17 shows some of the transitions that an electron in a hydrogen atom can make. The black arrow pointing upward represents absorption of sufficient energy to completely remove the electron from a hydrogen atom (ionization). The downward-pointing colored arrows represent decreases in the internal energy of the hydrogen atom that occur when photons are emitted as the electron moves from a higher energy level to a lower energy level. If the colored arrows pointed up, they would represent *absorption* of photons leading to *increases* in the internal energy of the atom. In every case the energy of the photon absorbed or emitted matches the absolute value of ΔE.

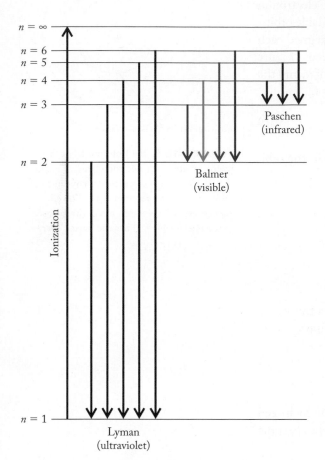

FIGURE 3.17 An energy-level diagram showing some of the possible electron transitions in a hydrogen atom. The arrow pointing up represents ionization. Arrows pointing down represent different electron transitions. During any one of them the atom loses a particular quantity of energy that may be emitted as a photon of electromagnetic radiation.

ground state the most stable, lowest energy state of a particle.

excited state any energy state above the ground state.

electron transition movement of an electron between energy levels.

CONCEPT TEST

Based on the lengths of the arrows in Figure 3.17, rank the following transitions in order of greatest change in electron energy to the smallest change:

a. $n = 4 \rightarrow n = 2$
b. $n = 3 \rightarrow n = 2$
c. $n = 2 \rightarrow n = 1$
d. $n = 4 \rightarrow n = 3$

(Answers to Concept Tests are in the back of the book.)

SAMPLE EXERCISE 3.5 **Calculating the Energy of a** **LO3**
Transition in a Hydrogen Atom

How much energy is required to ionize a ground-state hydrogen atom? Put another way, what is the *ionization energy* of hydrogen?

COLLECT AND ORGANIZE We are asked to determine the energy required to remove the electron from a hydrogen atom in its ground state. Equation 3.11 enables us to calculate the energy change associated with any electron transition.

ANALYZE To use Equation 3.11, we need to identify the initial ($n_{initial}$) and final (n_{final}) energy levels of the electron. The ground state of a H atom corresponds to the $n = 1$ energy level. If the atom is ionized, $n = \infty$ and the electron is no longer associated with the nucleus.

SOLVE

$$\Delta E = -2.178 \times 10^{-18}\,\text{J}\left(\frac{1}{n_{final}^2} - \frac{1}{n_{initial}^2}\right)$$

$$= -2.178 \times 10^{-18}\,\text{J}\left(\frac{1}{\infty^2} - \frac{1}{1^2}\right)$$

Dividing by ∞^2 yields zero, so the value inside the parentheses simplifies to -1, which gives us

$$\Delta E = 2.178 \times 10^{-18}\,\text{J}$$

THINK ABOUT IT Note how this energy is equal in magnitude but opposite in sign to the energy of an electron in the $n = 1$ orbit of a hydrogen atom. The sign of ΔE is positive because energy must be added to overcome the attraction between the negatively charged electron and the positively charged nucleus.

Practice Exercise Calculate the energy, in joules, required to ionize a hydrogen atom when its electron is initially in the $n = 3$ energy level. Before doing the calculation, predict whether this energy is greater than or less than the 2.178×10^{-18} J needed to ionize a ground-state hydrogen atom. ⚙

(Answers to Practice Exercises are in the back of the book.)

Before we end this section, let's compare two equations: Equation 3.11, which describes the differences in energy between pairs of energy levels in hydrogen atoms, and the Rydberg equation (Equation 3.9) that describes the wavenumbers for the lines in hydrogen's atomic emission and absorption spectra. Notice how much alike they are. The coefficients in them differ because of the different units used to express wavenumber and energy. The key point is that the empirical Rydberg equation has the same form as the theoretical equation developed by Bohr to explain the internal structure of the hydrogen atom. Thus, atomic emission and absorption spectra reveal the energies of electrons inside hydrogen atoms, and, as we will see, the atoms of other elements, too.

When we extend our exploration of atomic spectra and internal energies of atoms to other elements, we discover that the Bohr model of electrons revolving around nuclei in stable circular orbits works only for atoms or ions with only one electron. In multielectron atoms, the electrons interact with each other in ways that the Bohr model does not take into account. Thus, the picture of the atom provided by Bohr's model is limited. However, it was an important stepping stone to the development of more complete models of atomic structure based on quantum theory.

3.5 Electrons as Waves

A decade after Bohr published his model of the hydrogen atom, a French graduate student named Louis de Broglie (1892–1987) developed another explanation for the stability of electrons orbiting the nuclei of hydrogen atoms. De Broglie based his hypothesis on the assumption that electrons could behave like *waves of matter* as well as particles of matter. This *dual* nature of electrons was modeled after the behavior of electromagnetic radiation, which could be explained by assigning it wavelike properties, as James Maxwell and many others had done, but also by considering it to be made up of particles, or quanta, as Einstein had done to explain the photoelectric effect.

De Broglie Wavelengths

▶❚❚ **CHEMTOUR** De Broglie Wavelength

De Broglie calculated electron wavelengths using an equation he derived from Einstein's equation relating energy and mass (Equation 2.5):

$$E = mc^2$$

and the energy of a photon (Equation 3.4):

$$E = \frac{hc}{\lambda}$$

We can set the right-hand sides of the two equations equal to each other:

$$mc^2 = \frac{hc}{\lambda}$$

and solve for wavelength:

$$\lambda = \frac{h}{mc}$$

To apply this equation to an electron, de Broglie replaced c (the speed of light) with u, the speed of an orbiting electron in an atom:

$$\lambda = \frac{h}{mu} \tag{3.12}$$

where h is the Planck constant, m is the mass of the electron in kilograms, and u is its speed in meters per second. The wavelength of an electron calculated in this way is often called its *de Broglie wavelength*.

De Broglie's equation is not restricted to electrons. It tells us that *any moving particle* has wavelike properties. In other words, the particle behaves as a **matter wave**. De Broglie predicted that moving particles much bigger than electrons, such as atomic nuclei, molecules, and even tennis balls and airplanes, have characteristic wavelengths that can be calculated using Equation 3.12. The wavelengths of large objects are extremely small, given the tiny size of the Planck constant and their considerable mass, so we never notice the wave nature of large objects in motion.

SAMPLE EXERCISE 3.6 **Calculating the Wavelength of a Particle in Motion** **LO4**

Compare the de Broglie wavelength of a 142 g baseball thrown at 44 m/s (98 mi/hr) with the size (diameter) of the ball, which is 15.0 cm.

COLLECT AND ORGANIZE We know the mass and speed of a baseball and need to calculate its de Broglie wavelength so that we can compare its value to the size of the baseball. Equation 3.12 may be used to calculate this wavelength.

matter wave the wave associated with any moving particle.

ANALYZE Given the small value of h, it is likely that the wavelength of a pitched baseball is only a tiny fraction of the size of the baseball. The right side of Equation 3.12 has units of joule-seconds in the numerator and mass and speed in the denominator. To combine the units in a way that gives us a unit of length, we need to use this conversion factor:

$$1\,\text{J} = 1\,\text{kg} \cdot \text{m}^2/\text{s}^2$$

To use this equality, we must express the mass of the baseball in kilograms: 142 g = 0.142 kg.

SOLVE The de Broglie wavelength of the baseball is

$$\lambda = \frac{h}{mu} = \frac{6.626 \times 10^{-34}\,\cancel{\text{J}} \cdot \text{s}}{(0.142\,\cancel{\text{kg}})(44\,\cancel{\text{m/s}})} \times \frac{1\,\cancel{\text{kg}} \cdot \text{m}^2/\text{s}^2}{1\,\cancel{\text{J}}}$$

$$= 1.06 \times 10^{-34}\,\text{m}$$

This wavelength is

$$\frac{1.06 \times 10^{-34}\,\cancel{\text{m}}}{0.150\,\cancel{\text{m}}} \times 100\% = 7.1 \times 10^{-32}\%$$

of the ball's diameter.

THINK ABOUT IT The wavelength of the matter wave of the baseball is much too small to be observed, so its character contributes nothing to the behavior of the baseball. This is what we expected: large moving objects behave like large moving objects, not like waves.

Practice Exercise The speed of the electron in the ground state of the hydrogen atom is 2.2×10^6 m/s. What is the wavelength of the electron in meters? ⚙

(Answers to Practice Exercises are in the back of the book.)

De Broglie used matter waves to explain the stability of the electron levels in Bohr's model of the hydrogen atom. He proposed that the orbiting electron behaves like a circular wave oscillating around the nucleus. To understand the implications of this statement, we need to examine what is required to make a stable, circular wave. Consider the motion of a vibrating violin string of length L (Figure 3.18a). Because the string is fixed at both ends, there is no vibration at the ends and the vibration is at its maximum in the middle. The wave created by the combination of fixed ends and maximum vibration in the middle is called a **standing wave** because it oscillates back and forth within a fixed space rather than moving through space the way waves of light travel through space. The sound wave produced in this way on a violin string is called the *fundamental* of the string. The wavelength of the fundamental is $2L$ because a complete wave cycle, with both an upward and a downward deflection, requires two lengths of the string. It has the lowest frequency and longest wavelength possible for that string. The length of the string is one-half the wavelength of the fundamental ($L = \lambda/2$).

On a standing wave, any points that have zero displacement, such as the two ends of the string, are called **nodes**. If the string is held down in the middle (creating a third node) and plucked halfway between the middle and one end, a new, higher frequency wave called the *first harmonic* is produced. The wavelength of the first harmonic is equal to L, and the length of the string accommodates one complete wave cycle. The wavelengths of the fundamental, the first harmonic,

standing wave a wave confined to a given space with a wavelength λ related to the length L of the space by $L = n(\lambda/2)$, where n is a whole number.

node a location in a standing wave that experiences no displacement.

and higher-level harmonics of the violin string are all related to the length of the string by the equation

$$L = \frac{n\lambda}{2}$$

where *n* is a whole number equal to 1 for the fundamental, 2 for the first harmonic, 3 for the second harmonic, and so on.

CONCEPT TEST ••

Why are the various possible waves in a violin string examples of quantization?

(Answers to Concept Tests are in the back of the book.)
••

The standing-wave pattern for a circular wave generated by an electron differs slightly from that of a vibrating violin string in that the circular wave has no defined stationary ends. Instead, the electron vibrates in an endless series of waves, but only if, as shown in Figure 3.18(b), the circumference of the circle equals a whole-number multiple of the electron's wavelength:

$$\text{Circumference} = n\lambda \tag{3.13}$$

Equation 3.13 gives a new meaning to Bohr's orbit label *n*: it represents the number of matter-wave wavelengths in that orbit's circumference.

De Broglie's matter-wave hypothesis created a quandary for the graduate faculty at the University of Paris, where he studied. Bohr's model of electrons moving between allowed energy levels had been widely criticized as an arbitrary suspension of well-tested physical laws. De Broglie's rationalization of Bohr's model seemed even more outrageous to many scientists. Before the faculty would accept his doctoral thesis based on matter waves, they wanted another opinion, so they sent it to Albert Einstein for review. Einstein wrote back that he found de Broglie's work "quite interesting." That endorsement was good enough for the faculty: de Broglie's thesis was accepted in 1924 and immediately submitted for publication.

The Heisenberg Uncertainty Principle

After de Broglie proposed that electrons exhibited both particle-like and wave-like behavior, questions arose about the impact of wave behavior on our ability to locate the electron, given that a wave by its very nature is spread out in space. The question, "Where is the electron?" was addressed by German physicist Werner Heisenberg (1901–1976), who proposed the following thought experiment: What if we tried to watch an electron orbiting the nucleus of an atom? We would need an extremely powerful microscope to see a particle as tiny as an electron. No such microscope exists, but if it did, it would need to use gamma rays for illumination because they are the only part of the electromagnetic spectrum with wavelengths short enough to match the diminutive size of electrons. Longer wavelength radiation would pass right by an electron without being reflected by it. Unfortunately, the ultrashort wavelengths and high frequencies of gamma rays mean that they have enormous energies—so large that any gamma ray striking an electron would knock the electron off course. The only way not to affect the electron's motion would be to use a much lower energy, longer wavelength source of radiation to illuminate it, but then we would not be able to see the tiny electron clearly.

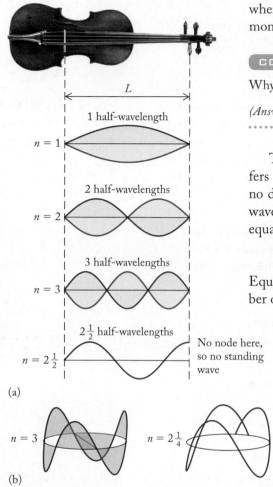

FIGURE 3.18 Linear and circular standing waves. (a) The wavelength of a standing wave in a violin string fixed at both ends is related to the distance *L* between the ends of the string by the equation $L = n(\lambda/2)$. In the standing waves shown, *n* = 1, 2, and 3. An *n* value of $2\frac{1}{2}$ does not produce a standing wave because there can be no string motion at either end. (b) The circular standing waves proposed by de Broglie account for the stability of the energy levels in Bohr's model of the hydrogen atom. Each stable wave must have a circumference equal to *nλ*, with *n* restricted to being an integer, such as the *n* = 3 circular standing wave shown here. If the circumference is not an exact multiple of *λ*, as shown in the $n = 2\frac{1}{4}$ image, there is no standing wave.

This situation presents an experimental dilemma. The only means for clearly observing an electron make it impossible to know the electron's motion or, more precisely, its momentum, which is defined as an object's velocity times its mass. Therefore we can never know exactly both the position and the momentum of the electron simultaneously. This conclusion is known as the **Heisenberg uncertainty principle** and is mathematically expressed by

$$\Delta x \cdot m\, \Delta u \geq \frac{h}{4\pi} \tag{3.14}$$

where Δx is the uncertainty in the position of the electron, m is its mass, Δu is the uncertainty in its velocity, and h is the Planck constant. To Heisenberg, this uncertainty was the essence of quantum theory. Its message for us is that there are limits to what we can observe, measure, and therefore know about particles the size of electrons.

> **Heisenberg uncertainty principle** the principle that one cannot simultaneously know the exact position and the exact momentum of an electron.

SAMPLE EXERCISE 3.7 Calculating Heisenberg Uncertainties **LO4**

Use the data in Sample Exercise 3.6 to compare the uncertainties in the speeds of a thrown baseball and an electron. Assume that the position of the baseball is known to within one wavelength of red light ($\Delta x_{\text{baseball}}$ = 680 nm) and the position of the electron is known to within the radius of the hydrogen atom (Δx_{e} = 5.3×10^{-11} m).

COLLECT AND ORGANIZE We are asked to calculate the uncertainty in the speeds of two particles and are given the uncertainties in their positions. Equation 3.14 provides a mathematical connection between the variables. From Sample Exercise 3.6 we know that the mass of a baseball is 0.142 kg. The mass of an electron is 9.109×10^{-31} kg.

ANALYZE According to Equation 3.14, the uncertainty in the speed and position of a particle is inversely proportional to its mass. Therefore, we can expect little uncertainty in the speed of the baseball but much greater uncertainty in the speed of the electron. We need to rearrange the terms in the equation to solve for the uncertainty in speed (Δu):

$$\Delta u \geq \frac{h}{4\pi\, \Delta x \cdot m}$$

SOLVE For the baseball,

$$\Delta u \geq \frac{6.626 \times 10^{-34}\ \cancel{\text{J}} \cdot \cancel{\text{s}}}{4\pi(6.80 \times 10^{-7}\ \cancel{\text{m}})(0.142\ \cancel{\text{kg}})} \times \frac{1\ \cancel{\text{kg}} \cdot \text{m}^2/\text{s}^2}{1\ \cancel{\text{J}}}$$

$$\geq 5.46 \times 10^{-28}\ \text{m/s}$$

For the electron,

$$\Delta u \geq \frac{6.626 \times 10^{-34}\ \cancel{\text{J}} \cdot \cancel{\text{s}}}{4\pi(5.3 \times 10^{-11}\ \cancel{\text{m}})(9.109 \times 10^{-31}\ \cancel{\text{kg}})} \times \frac{1\ \cancel{\text{kg}} \cdot \text{m}^2/\text{s}^2}{1\ \cancel{\text{J}}}$$

$$\geq 1.09 \times 10^{6}\ \text{m/s}$$

THINK ABOUT IT The uncertainty in the speed of the baseball is insignificant compared, for example, to the speed of a major league fastball (> 40 m/s). That result is expected for objects in the macroscopic world. The uncertainty in the speed of the electron is many orders of magnitude larger because its mass (which is in the denominator of the expression we derived from Equation 3.14) is many orders of magnitude smaller.

Practice Exercise What is the uncertainty, in meters, in the position of an electron moving near a nucleus at a speed of 8×10^{7} m/s? Assume the relative uncertainty in the speed of the electron is 1%, that is, 8×10^{5} m/s. ⚙

(Answers to Practice Exercises are in the back of the book.)

wave mechanics or **quantum mechanics** a mathematical description of the wavelike behavior of electrons and other particles.

Schrödinger wave equation a description of how the electron matter wave varies with location and time around the nucleus of a hydrogen atom.

wave function (ψ) a solution to the Schrödinger wave equation.

orbitals defined by the square of the wave function (ψ^2); regions in an atom where the probability of finding an electron is high.

▶❚❚ **CHEMTOUR** Quantum Numbers

FIGURE 3.19 The probability of finding an ink spot in the pattern produced by a source of ink spray decreases with increasing distance from the center of the pattern, in much the way that electron density in the 1*s* orbital decreases with increasing distance from the nucleus.

When Heisenberg proposed his uncertainty principle, he was working with Bohr at the University of Copenhagen. The two scientists had widely different views about the significance of the uncertainty principle and the idea that particles could behave like waves. To Heisenberg, uncertainty was a fundamental characteristic of nature. To Bohr, it was merely a mathematical consequence of the wave–particle duality of electrons; there was no physical meaning to an electron's position and path. The debate between the two gifted scientists was heated at times. Heisenberg later wrote about one particularly emotional debate:

> [A]t the end of the discussion I went alone for a walk in the neighboring park [and] repeated to myself again and again the question: "Can nature possibly be as absurd as it seems. . . ?"[1]

3.6 Quantum Numbers

Many of the leading scientists of the 1920s were unwilling to accept de Broglie's model of electron waves until it had a stronger theoretical foundation. They wanted a mathematical model that accurately described the behavior of matter waves and accounted for the atomic spectra of hydrogen. During a Christmas vacation in the Swiss Alps in 1925, Austrian physicist Erwin Schrödinger (1887–1961) created that mathematical foundation by developing what came to be called **wave mechanics** or **quantum mechanics**.

Schrödinger's mathematical description of electron waves is known as the **Schrödinger wave equation**. We do not examine it in detail in this book, but you should know that solutions to it are called **wave functions**: mathematical expressions represented by the Greek letter psi (**ψ**) that describe how the matter wave of an electron in an atom varies in both time and location inside the atom. Wave functions define the energy levels in the hydrogen atom. They can be simple trigonometric functions, such as sine or cosine waves, or they can be very complex.

What is the physical significance of a wave function? Schrödinger believed that a wave function depicted the "smearing" of an electron through three-dimensional space. However, this notion of subdividing a discrete particle was later rejected in favor of the model developed by German physicist Max Born (1882–1970), who proposed that the square of a wave function, ψ^2, defines an **orbital**: the space within an atom where the probability of finding an electron is high. Born later showed that his interpretation could be used to calculate the probability of electron transitions between orbitals, as happens when an atom absorbs or emits a quantum of energy.

To help visualize the probabilistic meaning of ψ^2, consider what happens when we spray ink onto a flat surface as in Figure 3.19. If we then draw a circle encompassing most of the ink spots, we are identifying the region of maximum probability for finding the spots.

It is important to understand that quantum mechanical orbitals are not two-dimensional concentric orbits as in Bohr's model of the hydrogen atom, or even two-dimensional circles as in the pattern of ink drops in Figure 3.19. Instead, they are three-dimensional regions of space with distinctive shapes, orientations, and average distances from the nucleus. Each orbital is a solution to Schrödinger's wave equation and is identified by a unique combination of three integers called

---

[1] Werner Heisenberg, *Physics and Philosophy: The Revolution in Modern Science* (Harper & Row, 1958), p. 42.

quantum numbers, whose values flow directly from the mathematical solutions to the wave equation. The quantum numbers are as follows:

➤ The **principal quantum number** n is like Bohr's n value for the hydrogen atom in that it is a positive integer that indicates the relative size and energy of an orbital or of a group of orbitals in an atom. Orbitals with the same value of n are in the same *shell*. Orbitals with larger values of n extend farther from the nucleus and, in the hydrogen atom, represent higher energy levels, consistent with Bohr's model. In multielectron atoms, the relationship between energy levels and orbitals is more complex, but increasing values of n generally represent higher energy levels.

➤ The **angular momentum quantum number** ℓ is an integer with a value ranging from zero to $(n-1)$ that defines the shape of an orbital. Orbitals with the same values of n and ℓ are in the same *subshell* and have the same energy. Orbitals with a given value of ℓ are identified with a letter according to the following scheme:

Value of ℓ	0	1	2	3	4
Letter Identifier	s	p	d	f	g

➤ The **magnetic quantum number** m_ℓ is an integer with a value from $-\ell$ to $+\ell$. It defines the orientation of an orbital in the space around the nucleus of an atom.

Each subshell has a two-part label that contains the appropriate value of n and a letter designation for ℓ. For example, orbitals with $n = 3$ and $\ell = 1$ are called $3p$ orbitals, and electrons in $3p$ orbitals are called $3p$ electrons. How many $3p$ orbitals are there? We can answer that question by finding all possible values of m_ℓ. Because p orbitals are those for which $\ell = 1$, they have m_ℓ values of -1, 0, and $+1$. The three values mean that there are three $3p$ orbitals, each with a unique combination of n, ℓ, and m_ℓ values. All the possible combinations of these three quantum numbers for the orbitals of the first four shells are given in Table 3.1.

quantum number one of four related numbers that specify the energy, shape, and orientation of orbitals in an atom and the spin orientation of electrons in the orbitals.

principal quantum number (_n_) a positive integer describing the relative size and energy of an atomic orbital or group of orbitals in an atom.

angular momentum quantum number (_ℓ_) an integer having any value from 0 to $(n-1)$ that defines the shape of an orbital.

magnetic quantum number (_m_ℓ_) defines the orientation of an orbital in space; an integer that may have any value from $-\ell$ to $+\ell$, where ℓ is the angular momentum quantum number.

TABLE 3.1   Quantum Numbers of the Orbitals in the First Four Shells

Value of n	Allowed Values of ℓ	Subshell Letter	Allowed Values of m_ℓ	NUMBER OF ORBITALS IN: Subshell	NUMBER OF ORBITALS IN: Shell
1	0	s	0	1	1
2	0	s	0	1	
	1	p	$-1, 0, +1$	3	4
3	0	s	0	1	
	1	p	$-1, 0, +1$	3	
	2	d	$-2, -1, 0, +1, +2$	5	9
4	0	s	0	1	
	1	p	$-1, 0, +1$	3	
	2	d	$-2, -1, 0, +1, +2$	5	
	3	f	$-3, -2, -1, 0, +1, +2, +3$	7	16

SAMPLE EXERCISE 3.8 **Identifying the Subshells and** **LO5**
Orbitals in an Energy Level

a. What are the names of all the subshells in the $n = 4$ shell?
b. How many orbitals are in all of the subshells of the $n = 4$ shell?

COLLECT AND ORGANIZE We are asked to describe the subshells in the fourth shell and to determine how many orbitals are in all of the subshells. Table 3.1 contains an inventory of all the subshells in the first four shells.

ANALYZE Subshell designations are based on the possible values of quantum numbers n and ℓ. The allowed values of ℓ depend on the value of n, in that ℓ is an integer between 0 and $(n - 1)$. The number of orbitals in a subshell depends on the number of possible values of m_ℓ, from $-\ell$ to $+\ell$.

SOLVE

a. The allowed values of ℓ for $n = 4$ range from 0 to $(n - 1)$—that is, 0 to 3—so they are 0, 1, 2, and 3. The ℓ values correspond to the subshell designations s, p, d, and f. The appropriate subshell names are thus $4s$, $4p$, $4d$, and $4f$.

b. The possible values of m_ℓ from $-\ell$ to $+\ell$ are

➤ $\ell = 0$; $m_\ell = 0$: This combination of ℓ and m_ℓ values for the $n = 4$ shell represents a single $4s$ orbital.

➤ $\ell = 1$; $m_\ell = -1, 0,$ or $+1$: These three combinations of ℓ and m_ℓ values for the $n = 4$ shell represent the three $4p$ orbitals.

➤ $\ell = 2$; $m_\ell = -2, -1, 0, +1,$ or $+2$: These five combinations of ℓ and m_ℓ values represent the five $4d$ orbitals.

➤ $\ell = 3$; $m_\ell = -3, -2, -1, 0, +1, +2,$ or $+3$: These seven combinations of ℓ and m_ℓ values represent the seven $4f$ orbitals.

Thus there are $1 + 3 + 5 + 7 = 16$ orbitals in the $n = 4$ shell.

THINK ABOUT IT We determined that there are 16 orbitals in the fourth shell. The number of orbitals in each shell is equal to the square of the principal quantum number of the shell.

Practice Exercise How many orbitals are there in the $n = 5$ shell? What are the names of all the subshells in the $n = 5$ shell? ⚙

(Answers to Practice Exercises are in the back of the book.)

Several relationships are worth noting in the quantum numbering system:

➤ There are n subshells in the nth shell: one subshell ($1s$) in the $n = 1$ shell, two subshells ($2s$ and $2p$) in the $n = 2$ shell, and so on.

➤ There are n^2 orbitals in the nth shell: $1^2 = 1$ in the $n = 1$ shell, $2^2 = 4$ in the $n = 2$ shell, and so on.

➤ There are $(2\ell + 1)$ orbitals in each subshell: one s orbital $(2 \times 0 + 1 = 1)$ in each s subshell, three p orbitals $(2 \times 1 + 1 = 3)$ in each p subshell, five d orbitals $(2 \times 2 + 1 = 5)$ in each d subshell, and so on.

The Schrödinger equation accounts for most, but not all, aspects of atomic spectra. The emission spectrum of hydrogen, for example, when viewed through a high-resolution spectrometer, contains a pair of red lines at 656 nm where Balmer saw only one (Figure 3.7a). There are also pairs of lines in the spectra of multielectron atoms that have a single electron in their outermost shells.

In 1925, two students at the University of Leiden in the Netherlands, Samuel Goudsmit (1902–1978) and George Uhlenbeck (1900–1988), proposed that the pairs of lines, called *doublets*, were caused by a property they called *electron spin*. In their model, electrons spin in one of two directions designated "spin up" and "spin down." A moving electron (or any charged particle) creates a magnetic field by virtue of its movement through space. The spinning motion produces a second magnetic field oriented up or down. To account for the two spin orientations, Goudsmit and Uhlenbeck proposed a fourth quantum number, the **spin quantum number**, m_s. There are two possible values of m_s: $+\frac{1}{2}$ for spin up and $-\frac{1}{2}$ for spin down.

Even before Goudsmit and Uhlenbeck proposed the electron-spin hypothesis, two other scientists, Otto Stern (1888–1969) and Walther Gerlach (1889–1979), observed the effect of electron spin when they shot a beam of silver ($Z = 47$) atoms through a magnetic field (Figure 3.20). Those atoms in which the net electron spin was "up" were deflected in one direction by the field; those in which the net electron spin was "down" were deflected in the opposite direction.

In 1925, Austrian physicist Wolfgang Pauli (1900–1958) proposed that no two electrons in a multielectron atom can have the same set of values for the four quantum numbers n, ℓ, m_ℓ, and m_s. This idea is known as the **Pauli exclusion principle**. We have seen how unique combinations of the three quantum numbers from Schrödinger's wave equation define each orbital in an atom. The fourth (spin) quantum number provides a unique address for each electron in each orbital.

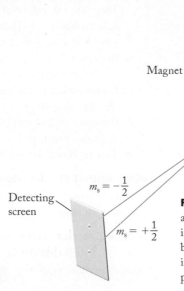

FIGURE 3.20 A narrow beam of silver atoms passed through a magnetic field is split into two beams because of the interactions between the field and the spinning electrons in the atoms. This observation led to proposing the fourth quantum number, m_s.

SAMPLE EXERCISE 3.9 **Identifying Valid Sets of Quantum Numbers** **LO5**

Which of these five combinations of quantum numbers are valid?

	n	ℓ	m_ℓ	m_s
(a)	1	0	−1	$+\frac{1}{2}$
(b)	3	2	−2	$+\frac{1}{2}$
(c)	2	2	0	0
(d)	2	0	0	$-\frac{1}{2}$
(e)	−3	−2	−1	$-\frac{1}{2}$

COLLECT AND ORGANIZE Table 3.1 contains valid combinations of quantum numbers for the first four shells. The rules for which values of n, ℓ, and m_ℓ are possible are given at the beginning of this section.

ANALYZE The principal quantum number (n) can be any positive integer. The valid values of ℓ in a given shell are integers from 0 to ($n-1$), and the values of m_ℓ in a given subshell include all integers from $-\ell$ to $+\ell$ including 0. The only two options for m_s are $+\frac{1}{2}$ or $-\frac{1}{2}$.

spin quantum number (m_s) either $+\frac{1}{2}$ or $-\frac{1}{2}$, indicating the spin orientation of an electron.

Pauli exclusion principle no two electrons in an atom can have the same set of four quantum numbers.

SOLVE

a. Because n is 1, the maximum (and only) value of ℓ is $(n-1) = 1-1 = 0$. Therefore the values of n and ℓ are valid. However, if $\ell = 0$, then m_ℓ must be 0; it cannot be -1. Therefore, this set is not valid. The spin quantum number is a possible one, however.

b. Because n is 3, ℓ can be 2 and m_ℓ can be -2. Also, $m_s = +\frac{1}{2}$ is a valid choice for the spin quantum number. This set is valid.

c. Because n is 2, ℓ cannot be 2, making this set invalid. In addition m_s has an invalid value (0).

d. Because n is 2, ℓ can be 0, and for that value of ℓ, m_ℓ must be 0. The value of m_s is also valid, and so is the set.

e. This set contains two impossible values, $n = -3$ and $\ell = -2$, so it is invalid.

THINK ABOUT IT The values of n, ℓ, and m_ℓ are related mathematically, and m_s can be either $+\frac{1}{2}$ or $-\frac{1}{2}$. Together the four numbers provide a unique address for every electron in an atom.

Practice Exercise Write all the possible sets of quantum numbers for an electron in the $n = 3$ shell that has an angular momentum quantum number $\ell = 1$ and a spin quantum number $m_s = +\frac{1}{2}$.

(Answers to Practice Exercises are in the back of the book.)

3.7 The Sizes and Shapes of Atomic Orbitals

We have learned that atomic orbitals have three-dimensional shapes that are graphical representations of ψ^2. In this section we examine how the shapes of orbitals impact the energies of the electrons in them.

s Orbitals

Figure 3.21 provides several representations of the 1s orbital of hydrogen. In Figure 3.21(a), electron density is plotted against distance from the nucleus and seems to show that density decreases with increasing distance. However, Figure 3.21(b) provides a more useful profile of electron distribution. To understand why, think of the hydrogen atom as a tiny onion, made of many concentric spherical layers all of the same thickness. A cross section of that image of the atom is shown in Figure 3.21(c). What is the probability of finding the electron in one of the spherical layers? A layer very close to the nucleus has a very small radius, so it accounts for only a small fraction of the total volume of the atom. A layer with a larger radius makes up a much larger fraction of the volume of the atom because the volume of the layers increases as a function of r^2. Even though electron densities are higher closer to the nucleus (as Figure 3.21a shows), the volumes of the spherical shells closest to the nucleus are so small that the chances of the electron being near the center of an atom are extremely low. This low probability is shown in Figure 3.21(b), where the curve starts off at essentially zero for electron distribution values at distances very close to the nucleus. Farther from the nucleus, electron densities are lower but the volumes of the layers are much larger, so the probability of the electron being in one of the layers is relatively high, represented by the peak in the curve of Figure 3.21(b). At greater distances, volumes of the

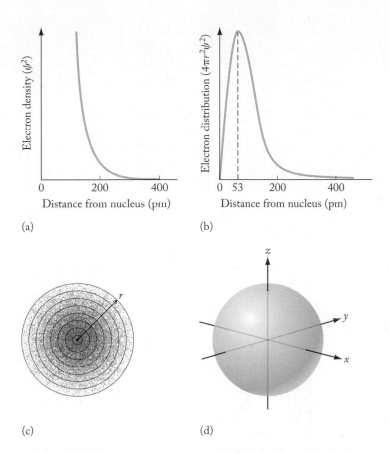

(a)

(b)

(c)

(d)

FIGURE 3.21 (a) Probable electron density in the 1s orbital of the hydrogen atom represented by a plot of electron density (ψ^2) versus distance from nucleus. (b) Electron distribution in the 1s orbital versus distance from nucleus. The distribution is essentially zero both for very short distances from the nucleus and for very long distances from the nucleus. (c) Cross section through the hydrogen atom, with the space surrounding the nucleus divided into an arbitrary number of thin, concentric hollow layers. Each layer has a unique value for radius r. The probability of finding an electron in a particular layer of radius r depends on the volume of the layer and the density of electrons in the layer. (d) Boundary–surface representation of a sphere within which the probability of finding a 1s electron is 90%.

layers are very large but ψ^2 drops to nearly zero (see Figure 3.21a); therefore, the chances of finding the electron in layers far from the nucleus are very small.

Thus Figure 3.21(b) represents a combination of two competing factors—increasing layer volume and decreasing probability of the electron being far from the nucleus. This combination produces a *radial distribution profile* for the electron. Figure 3.21(b) is a plot not of ψ^2 versus distance from the nucleus as in Figure 3.21(a), but rather of $4\pi r^2 \psi^2$ versus distance from the nucleus. In geometry, $4\pi r^2$ is the formula for the surface area of a sphere, but here it represents the volume of one of the very thin spherical layers in Figure 3.21(c). The maximum value of the curve in Figure 3.21(b), at 53 pm, corresponds to the most likely radial distance of a 1s electron from the nucleus.

Figure 3.21(d) provides a view of the spherical shape of this (or any other) *s* orbital. The surface of the sphere encloses the volume within which the probability of finding a 1s electron is 90%. This type of depiction, called a *boundary–surface representation*, is one of the most useful ways to view the relative sizes, shapes, and orientations of orbitals. All *s* orbitals are spheres, which means that they have no angular dependence on orientation.

Radial electron distribution profiles of hydrogen's 1s, 2s, and 3s orbitals are shown in Figure 3.22. Note that orbital size increases with increasing values of the principal quantum number n. Note also that in the quadrants above the profile curves there are bands in which the density of dots is high. The dots represent the probability of finding an electron at those locations, and each dark band represents a local maximum in electron distribution. In all three profiles, there is a local maximum close to the nucleus. This means that electrons in *s* orbitals, even *s* orbitals with high values of n, have some probability of being close to the nucleus.

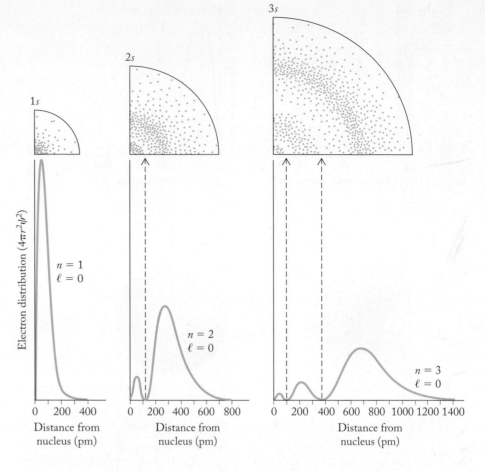

FIGURE 3.22 Radial distribution profiles of 1s, 2s, and 3s orbitals. Electrons in all these s orbitals have some probability of being close to the nucleus, but 3s electrons are more likely to be farther away from the nucleus than 2s electrons, which are more likely to be farther away than 1s electrons. The dashed lines connect two representations of the nodes in the radial distributions in the two diagrams.

p and d Orbitals

All shells with $n \geq 2$ have a subshell containing three p orbitals ($\ell = 1$; $m_\ell = -1$, 0, +1). Each of the three p orbitals has two lobes oriented on either side of the nucleus along one of the three perpendicular Cartesian axes x, y, z. The true shape of the lobes is squashed and roundish, like mushroom caps (Figure 3.23a), but we draw them in an elongated teardrop shape to more easily visualize their orientation (Figure 3.23b). The orbitals are designated p_x, p_y, and p_z, depending on the axis along which the lobes are aligned. The two lobes of a p orbital are sometimes labeled with plus or minus signs to indicate the sign of the wave function that defines each lobe (but don't confuse the signs with electrical charges—all electrons have negative charges).

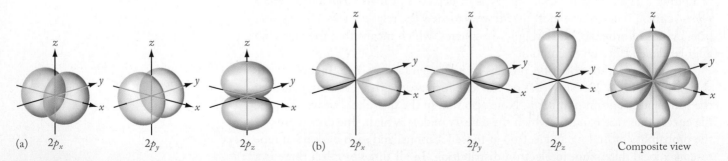

FIGURE 3.23 Boundary–surface views of the three 2p orbitals, showing their orientation along the x-, y-, and z-axes. (a) These views of $2p_x$, $2p_y$, and $2p_z$ are obtained from the wave functions of the orbitals. (b) We use an elongated version of the theoretical shapes of the orbitals throughout this book to make it easier to see the orientation of their lobes.

Shells with principal quantum numbers of 3 or higher have five d orbitals ($\ell = 2$, $m_\ell = -2, -1, 0, +1, +2$). Their shapes are shown in Figure 3.24. Four of them have teardrop-shaped lobes oriented like the leaves in a four-leaf clover. The lobes of three of the four, designated d_{xy}, d_{xz}, and d_{yz}, lie between the Cartesian axes. The lobes of the fourth orbital, $d_{x^2-y^2}$, are centered on the x- and y-axes. The fifth d orbital, d_{z^2}, is mathematically equivalent to the other four but has a much different shape, with two teardrop-shaped lobes oriented along the z-axis and a doughnut shape called a *torus* in the x–y plane that surrounds the middle of the two lobes. We will not address the shapes and geometries of f orbitals here because they are not included in our discussions of chemical bonding in the chapters to come.

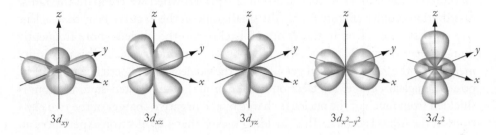

$3d_{xy}$ $3d_{xz}$ $3d_{yz}$ $3d_{x^2-y^2}$ $3d_{z^2}$

FIGURE 3.24 Boundary–surface views of the five $3d$ orbitals, showing their orientation relative to the x-, y-, and z-axes. As with the $2p$ orbitals in Figure 3.23(b), the theoretical boundary surfaces of the $3d$ orbitals are elongated to make it easier to see the orientation of their lobes. The d_{xy}, d_{xz}, and d_{yz} orbitals are not aligned along any axis; the $d_{x^2-y^2}$ orbital lies along the x and y axes; the d_{z^2} orbital consists of two teardrop-shaped lobes along the z axis with a donut-shaped torus ringing the point where the two lobes meet.

3.8 The Periodic Table and Filling Orbitals

In this section we explore how the sizes and shapes of orbitals determine the order in which they fill with electrons as we work our way down the periodic table, starting with hydrogen. In assigning electrons to orbitals we will follow the **aufbau principle** (German *aufbauen*, "to build up"), which states that the electrons are placed in the lowest energy orbitals available. Our only other restriction is that each orbital can have no more than two electrons.

Using these rules, let's begin by assigning the single electron in hydrogen ($Z = 1$) to the $1s$ orbital. We represent this arrangement with the **electron configuration** $1s^1$, where the first 1 indicates the principal quantum number (n) of the orbital, s indicates the subshell, and the superscript ^{1} indicates that there is *one* electron in the $1s$ orbital. Note that when we write the electron configurations of elements, we assume that their atoms are in their ground states.

The atomic number of helium is 2, which tells us there are two protons in the nucleus surrounded by two electrons in the neutral atom. Using the aufbau principle, we simply add another electron to the $1s$ orbital. The $1s$ orbital already contains one electron, so it has space for one more. The Pauli exclusion principle dictates that the spin quantum number for the second electron cannot be the same as the first: one must be $+\frac{1}{2}$ and the other must be $-\frac{1}{2}$. The two electrons with opposite spin quantum numbers are said to be *spin-paired*. Their presence gives helium a ground-state electron configuration of $1s^2$. With the two electrons the $1s$ orbital is filled to capacity and so is the $n = 1$ shell.

The concept of filled shells and subshells is critical to understanding the chemical properties of the elements. Helium and the other group 18 elements are composed of atoms that have filled s, or s and p, orbitals in their outermost shells. The group 18 elements, the noble gases, are chemically stable and generally unreactive. Other main group elements have partially filled sets of s and p

aufbau principle the method of building electron configurations of atoms by adding one electron at a time as atomic number increases across the rows of the periodic table.

electron configuration the distribution of electrons among the orbitals of an atom or ion.

⊚⊙ **CONNECTION** In Chapter 2 we defined the main group elements as those in groups 1, 2, and 13 through 18 in the periodic table.

▶❚❚ **CHEMTOUR** Electron Configuration

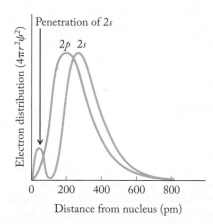

FIGURE 3.25 Radial distribution profiles of electrons in 2s and 2p orbitals. The 2s orbital is lower in energy because electrons in it penetrate more closely to the nucleus, as indicated by the local maximum in electron distribution about 50 pm from the nucleus. As a result, 2s electrons experience a greater effective nuclear charge than 2p electrons.

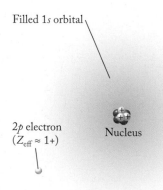

FIGURE 3.26 The effective nuclear charge (Z_{eff}) experienced by a 2p electron in an excited-state Li atom approximately equals the sum of the actual nuclear charge (3+) and the shielding effect of the negative charges of the two 1s electrons (2−). This shielding produces a net Z_{eff} of about 1+.

orbitals in their outermost shell. They engage in chemical reactions in which their atoms lose, gain, or share electrons in ways that produce filled outermost s and p subshells. In that way they acquire the stable electron configurations of group 18 elements.

The location of lithium ($Z = 3$) in the periodic table—the first element in the second period—is a signal that lithium has one electron in its $n = 2$ shell. The row numbers in the periodic table correspond to the n values of the outermost shells of the elements in the rows. The second shell has four orbitals: one 2s and three 2p, which can hold up to eight electrons. Lithium's third electron occupies the lowest energy orbital in the second shell, which is the 2s orbital, making the electron configuration of Li $1s^2 2s^1$.

Perhaps you are wondering why the 2s orbital is lower in energy than any of the 2p orbitals. The answer to that question is revealed when we compare the radial distribution profiles (Figure 3.25). The small peak on the 2s curve near the nucleus tells us that an electron in the 2s orbital is closer to the nucleus more frequently than an electron in a 2p orbital, which has no such secondary peak. This proximity means that the 2s electron can get closer to the nucleus and experience more of its positive nuclear charge. An electron in a 2p orbital of an excited-state Li atom is shielded from much of the nucleus's charge by the negative charges of the two electrons in the filled 1s orbital. This shielding means that a 2p electron experiences an **effective nuclear charge (Z_{eff})** that is only about $(3+) + (2-) = 1+$ (Figure 3.26). However, an electron in a 2s orbital can penetrate the cloud of 1s electrons better and experience a greater Z_{eff}. Therefore, a 2s electron has less energy than a 2p electron, which is why the 2s orbital fills first.

Condensed Electron Configurations

We can simplify the electron configurations of Li and all the elements that follow it in the periodic table by writing their *condensed* electron configurations. In this format the symbols representing all of the electrons in orbitals that were filled in the rows above the element of interest are replaced by the symbol of the group 18 element at the end of the row above the element. For example, the condensed electron configuration of Li is $[\text{He}]2s^1$. Condensed electron configurations are useful because they eliminate the symbols of the **core electrons** in filled shells and subshells. Core electrons are not involved in the chemistries of the elements, so they have less interest for us than those that reside in the outermost shells and subshells. The outermost electrons are the ones involved in bond formation and are called **valence electrons**. The outermost shell is called the **valence shell**. Notice that lithium has a single electron in its valence-shell s orbital, as does hydrogen, the element directly above it in the periodic table. Therefore, both elements have the same general valence-shell configuration, ns^1, where n represents both the number of the row in which the element is located and the principal quantum number of the valence shell.

Beryllium ($Z = 4$) is the fourth element in the periodic table and the first in group 2. Its electron configuration is $1s^2 2s^2$ or, in condensed form, $[\text{He}]2s^2$. The other elements in group 2 also have two spin-paired electrons in the s orbital of their valence shell. The second shell is not full at this point because it also has three p orbitals, which are all empty and fill as we move to the next elements in the periodic table.

Boron ($Z = 5$) is the first element in group 13. Its fifth electron is in one of its three $2p$ orbitals, resulting in the condensed electron configuration [He]$2s^2 2p^1$. It does not matter which of the three $2p$ orbitals contains the fifth electron because all three have the same energy. Chemists call orbitals with the same energy *degenerate* orbitals.

Hund's Rule and Orbital Diagrams

The next element is carbon ($Z = 6$). It has another electron in one of its $2p$ orbitals, giving it the condensed electron configuration [He]$2s^2 2p^2$. The second $2p$ electron resides in a different $2p$ orbital than the first. Why? Because electrons repel each other, and they will logically repel each other less if they are in separate orbitals rather than in the same orbital. This separation of the $2p$ electrons into different orbitals is an application of **Hund's rule**, named after German physicist Friedrich Hund (1896–1997), which states that the lowest energy electron configuration for degenerate orbitals, like the three in the $2p$ subshell, is the one with the maximum number of unpaired valence electrons, all of which have the same spin.

We use **orbital diagrams** to show in detail how electrons, represented by single-headed arrows, are distributed among orbitals, which are represented by boxes. A single-headed arrow pointing upward represents an electron with spin up ($m_s = +\frac{1}{2}$), and a downward-pointing single-headed arrow represents an electron with spin down ($m_s = -\frac{1}{2}$). To obey Hund's rule, the orbital diagram for carbon must be

Carbon: ⇅ ⇅ ↑ ↑ ☐
       $1s$ $2s$ $2p$

The two $2p$ electrons are unpaired (in separate orbitals). By convention, the spin arrows of the single electrons in those orbitals are drawn pointed up.

The condensed electron configuration of the next element, nitrogen ($Z = 7$), is [He]$2s^2 2p^3$. According to Hund's rule, the third $2p$ electron resides alone in the third $2p$ orbital, so that the electron distribution is

Nitrogen: ⇅ ⇅ ↑ ↑ ↑
        $1s$ $2s$ $2p$

with all three spin arrows pointed up. As we proceed across the second row to neon ($Z = 10$), we fill the $2p$ orbitals as shown in Figure 3.27. The last three $2p$ electrons added (in oxygen, fluorine, and neon) pair up with the first three so that the spin orientations of each pair have opposite directions. All three $2p$ orbitals are filled to capacity in an atom of neon, and so is the $n = 2$ shell.

Sodium ($Z = 11$) follows neon. It is the third element in group 1 and the first in the third row. Its position means that its outermost electron is in the third shell. There are three types of orbitals in the third shell: $3s$, $3p$, and $3d$. Which type gets sodium's 11th electron? Figure 3.28 provides a clue to the answer from the radial distribution profiles of electrons in the orbitals. Note the two peaks in the $3s$ profile and single peak in the $3p$ profile that are all close to the nucleus. The peaks mean that electrons in $3s$ and $3p$ orbitals penetrate through the filled orbitals of the first two shells and experience greater effective nuclear charge than do $3d$ electrons.

effective nuclear charge (Z_{eff}) the attraction toward the nucleus experienced by an electron in an atom; the positive charge on the nucleus reduced by the extent to which other electrons in the atom shield the electron from the nucleus.

core electrons electrons in the filled, inner shells in an atom or ion that are not involved in chemical reactions.

valence electrons electrons in the outermost occupied shell of an atom having the most influence on the atom's chemical behavior.

valence shell the outermost occupied shell of an atom.

Hund's rule the lowest energy electron configuration of an atom has the maximum number of unpaired electrons, all of which have the same spin, in degenerate orbitals.

orbital diagram depiction of the arrangement of electrons in an atom or ion using boxes to represent orbitals.

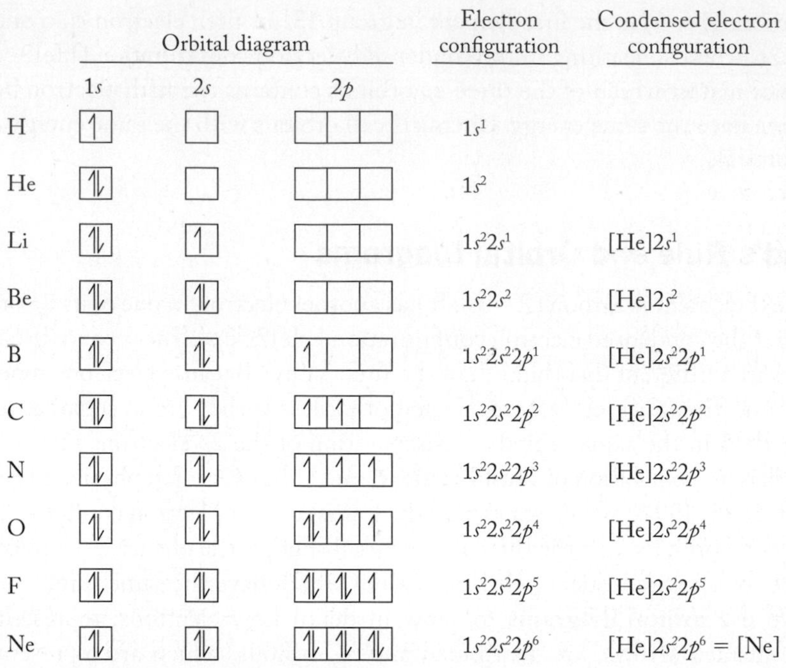

	Orbital diagram			Electron configuration	Condensed electron configuration
	1s	2s	2p		
H	↑			$1s^1$	
He	↑↓			$1s^2$	
Li	↑↓	↑		$1s^2 2s^1$	[He]$2s^1$
Be	↑↓	↑↓		$1s^2 2s^2$	[He]$2s^2$
B	↑↓	↑↓	↑	$1s^2 2s^2 2p^1$	[He]$2s^2 2p^1$
C	↑↓	↑↓	↑ ↑	$1s^2 2s^2 2p^2$	[He]$2s^2 2p^2$
N	↑↓	↑↓	↑ ↑ ↑	$1s^2 2s^2 2p^3$	[He]$2s^2 2p^3$
O	↑↓	↑↓	↑↓ ↑ ↑	$1s^2 2s^2 2p^4$	[He]$2s^2 2p^4$
F	↑↓	↑↓	↑↓ ↑↓ ↑	$1s^2 2s^2 2p^5$	[He]$2s^2 2p^5$
Ne	↑↓	↑↓	↑↓ ↑↓ ↑↓	$1s^2 2s^2 2p^6$	[He]$2s^2 2p^6$ = [Ne]

FIGURE 3.27 Orbital diagrams and condensed electron configurations for the first 10 elements. The squares in the orbital diagrams represent orbitals. Each can hold up to two electrons of opposite spin. Condensed electron configurations for all elements are given in Appendix 3.

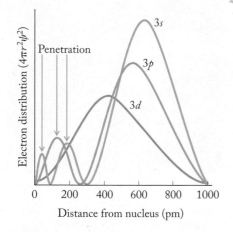

FIGURE 3.28 Radial distribution profiles of electrons in 3s, 3p, and 3d orbitals. The highlighted maxima in the 3s and 3p profiles mean that electrons in these orbitals penetrate closer to the nucleus, are less shielded by the electrons in filled inner shells, and have lower energy than 3d electrons.

Therefore, their relative energies are lower, with 3s the lowest of all. So, in the third shell, the 3s orbital fills first and the 3d orbitals fill last. The same sequence applies to all the shells with n values greater than 3. Shells for which $n \geq 4$ also contain f orbitals: those orbitals have higher energies than the d orbitals in the same shell, which means the f orbitals fill last.

Because the s orbital in a shell always fills first, the condensed electron configuration of Na is [Ne]$3s^1$. Just as we write condensed electron configurations that focus on the valence shell, we can draw condensed orbital diagrams that do the same thing. The one for sodium is

Sodium: [Ne] ↑
                   3s

This diagram reinforces the message that a sodium atom has a neon core of 10 electrons plus one more in its 3s orbital. Sodium has the same ns^1 valence-shell configuration as lithium and hydrogen, where n is the principal quantum number of the outermost shell. This pattern holds throughout the periodic table: all the elements in a given group have the same generic valence-shell electron configuration. For instance, the electron configuration of magnesium ($Z = 12$) is [Ne]$3s^2$, and the condensed electron configuration of every other element in group 2 consists of the immediately preceding noble gas core followed by ns^2.

CONCEPT TEST

What is the generic valence-shell configuration of the halogen (group 17) elements?

(Answers to Concept Tests are in the back of the book.)

The next six elements in the periodic table—aluminum, [Ne]$3s^2 3p^1$, through argon, [Ne]$3s^2 3p^6$—contain increasing numbers of $3p$ electrons until they achieve a filled $3p$ subshell in argon. As we noted when we discussed helium, the filled $3s$ and $3p$ subshells of argon impart a chemical stability that is in keeping with its membership in the noble gas family of elements.

Before leaving the third row, let's revisit the condensed electron configuration of sodium, [Ne]$3s^1$. This configuration represents a ground-state sodium atom because all of the electrons, and most importantly its valence electron, occupy the lowest energy orbitals available. Now think back to the discussion about atomic emission spectra in Section 3.2 and the distinctive yellow-orange glow that sodium makes in the flames of Bunsen burners, as illustrated back in Figure 3.6 and here in Figure 3.29(a). Sodium emits yellow-orange light after its atoms have absorbed quanta of energy that raise them from the ground state to an excited state (a transition represented by the black arrow in Figure 3.30). The easiest excited state to populate is the one with the smallest energy above the ground state. We have seen that the $3p$ orbitals fill after the $3s$ orbital because they have the next lowest energy. Therefore, it is logical that the lowest energy (or *first*) excited state of sodium is one in which its $3s$ electron has moved up to a $3p$ orbital. This excited state has the electron configuration [Ne]$3p^1$ (see Figure 3.30) and a very short lifetime. The electron typically takes less than a nanosecond to fall back to the ground state in a transition represented by the yellow-orange arrow in Figure 3.30. This transition releases a quantum of energy ($h\nu$) equal to the difference in energy between the $3p$ and $3s$ orbitals in a Na atom—the energy of a photon of yellow-orange light.

Potassium ($Z = 19$) is the first element of the fourth row. Like all group 1 elements, its generic valence-shell electron configuration is ns^1. For potassium, that translates into the condensed electron configuration [Ar]$4s^1$. Similarly, the condensed electron configuration of the next element, calcium ($Z = 20$), is [Ar]$4s^2$. At this point, the $4s$ orbital is filled. However, the $3d$ subshell *is still empty*. Why does the $4s$ orbital fill before the $3d$ subshell? The answer to that question is linked to the relative energies of electrons in the orbitals. While it is true that energy levels increase with increasing n values, the increases get smaller as n values get larger, as shown in Figure 3.31. The differences in energy between orbitals in the third and fourth shells are so small that the $3d$ subshell is slightly higher in energy than the $4s$ orbital. Therefore, the $4s$ orbital fills first.

> ### CONCEPT TEST
>
> The excited-state configuration [Ar]$3p^1$ for potassium emits yellow light. Which is possible for the emission from the [Ar]$4p^1$ excited state? (a) ultraviolet light; (b) the same yellow light; (c) red light; (d) infrared light

(Answers to Concept Tests are in the back of the book.)

(a)

(b)

FIGURE 3.29 The alkali metals produce characteristic colors in Bunsen burner flames because the high flame temperatures produce excited-state atoms of these elements. (a) The yellow-orange glow of Na atoms. (b) The lavender color of potassium atoms.

FIGURE 3.30 A ground-state Na atom absorbs a quantum of energy as its valence electron moves from the $3s$ orbital to a $3p$ orbital. This $3p$ electron in the excited-state atom spontaneously falls back to the empty $3s$ orbital, emitting a photon of yellow-orange light. The energy of the photon exactly matches the difference in energy between the $3p$ and $3s$ orbitals of Na atoms.

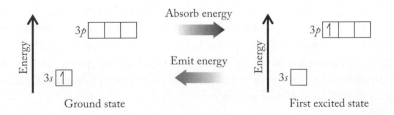

Absorb energy

Emit energy

$3p$

$3s$

Energy

Ground state

$3p$

$3s$

Energy

First excited state

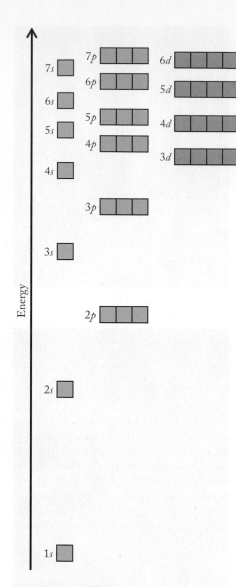

FIGURE 3.31 The energy levels in multielectron atoms increase with increasing values of n and with increasing values of ℓ within a shell. The difference in energy between adjacent shells decreases with increasing values of n, which may cause the energies of subshells in two adjacent shells to overlap. For example, electrons in $3d$ orbitals have slightly higher energy than those in the $4s$ orbital, resulting in the order of subshell filling $4s \rightarrow 3d \rightarrow 4p$.

Only after the $4s$ orbital is full do the $3d$ orbitals begin to fill, starting with scandium ($Z = 21$), the first transition metal in the fourth row. Scandium has the condensed electron configuration $[Ar]3d^1 4s^2$. Note how the valence-shell orbitals are arranged in order of increasing principal quantum number in that electron configuration, not in the order in which they were filled. The reason why we use such sequences will become clear in the next section, where we discuss which valence electrons are lost when transition metals form cations.

The $3d$ orbitals are filled in the fourth-row transition metals from scandium (group 3) to zinc (group 12). This pattern of filling the d orbitals of the shell whose principal quantum number is one less than the row number, abbreviated $(n-1)d$, is followed throughout the periodic table. Thus, the $4d$ orbitals are filled in the transition metals of the fifth row, the $5d$ orbitals in the sixth row, and so on, as shown in Figure 3.32.

Titanium ($Z = 22$) has one more $3d$ electron than scandium, so it has the condensed electron configuration $[Ar]3d^2 4s^2$. At this point, you may feel that you can accurately predict the electron configurations of the remaining transition metals in the fourth period. However, because the energies of the $3d$ and $4s$ orbitals are so close together (see Figure 3.31), the sequence of d-orbital filling deviates from the pattern you might expect. The first deviation appears in the electron configuration of chromium ($Z = 24$). You might expect it to be $[Ar]3d^4 4s^2$; however, it is actually $[Ar]3d^5 4s^1$. The reason for this difference is that the second configuration puts one electron in each of the five d orbitals:

Chromium: $\quad$ [Ar] $\boxed{\uparrow|\uparrow|\uparrow|\uparrow|\uparrow}$ $\boxed{\uparrow}$
$$\qquad\qquad\qquad\qquad 3d^5 \qquad 4s^1$$

This half-filled set of d orbitals represents a lower energy, more stable electron configuration than $[Ar]3d^4 4s^2$. Its stability compensates for the energy needed to raise a $4s$ electron to a $3d$ orbital.

Another deviation from the expected filling pattern happens near the end of each row of transition metals. For example, the electron configuration of copper ($Z = 29$) is $[Ar]3d^{10}4s^1$ instead of $[Ar]3d^9 4s^2$ because the electron configuration with a completely filled d subshell is more stable. By zinc ($Z = 30$), the $3d$ subshell is full and the next six electrons are added to $4p$ orbitals to reach the end of the fourth row, giving krypton ($Z = 36$) the condensed electron configuration $[Ar]3d^{10}4s^2 4p^6$. The pattern we have just described for the fourth row is repeated, though with additional deviations from the expected pattern, in the fifth row. In the fifth row the deviations are due to the similar energies of $5s$ and $4d$ orbitals.

Figure 3.32 illustrates the overall orbital filling pattern described above. It also shows how the periodic table can be used to predict the electron configurations of the elements. The color patterns and labels in Figure 3.32 indicate which type of orbital is filled going across each row from left to right. For example, groups 1 and 2 are called s block elements because their outermost electrons are in s orbitals. Groups 13–18 (except for helium) are p block elements because their outermost electrons are in p orbitals. Note how the principal quantum numbers of the outermost orbitals in the s and p blocks match their row numbers: $2s$ and $2p$ in row 2, $3s$ and $3p$ in row 3, and so on.

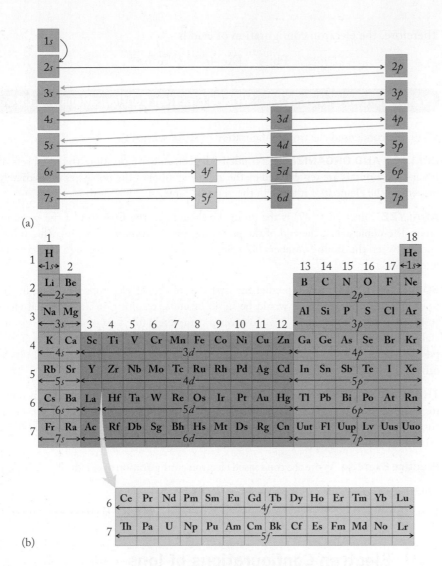

FIGURE 3.32 (a) This diagram shows the sequence in which atomic orbitals fill. (b) The same color coding is used in this version of the periodic table to highlight the four "blocks" of elements in which valence-shell s (green), p (blue), d (orange), and f (purple) orbitals are filled with increasing atomic number across a row of the table.

The transition metals in groups 3–12 make up the d block. The principal quantum number of the outermost d orbitals in each row, starting with row 4, is always one less than the row number. The lanthanides ($Z = 58$ through 71) and actinides ($Z = 90$ through 103) at the bottom of the periodic table make up the f block. The principal quantum number of the outermost f orbitals in each of these rows is always *two* less than the row number of the element that precedes them. This means, for example, that the $4f$ orbitals, starting with cerium ($Z = 58$), do not fill until after the $6s$ orbital is full. The bottom two rows are each 14 elements long, which makes sense because each f subshell contains seven orbitals ($\ell = 3$, $m_\ell = -3, -2, -1, 0, +1, +2,$ and $+3$) that can hold up to 14 electrons.

With these patterns in mind, let's write the condensed electron configuration of lead ($Z = 82$), which is the group 14 element in the sixth row. The nearest noble gas above it is xenon ($Z = 54$). The difference in atomic numbers means that we need to account for $82 - 54 = 28$ electrons in the electron configuration symbols. The block labels in Figure 3.32 tell us that the 28 electrons are distributed as follows:

2 electrons in $6s$	10 electrons in $5d$
14 electrons in $4f$	2 electrons in $6p$

Therefore, the electron configuration of lead is

$$\text{Pb:} \qquad [\text{Xe}]4f^{14}5d^{10}6s^26p^2$$

SAMPLE EXERCISE 3.10 **Writing Electron Configurations** **LO6**

Write the condensed electron configuration of silver ($Z = 47$).

COLLECT AND ORGANIZE In a condensed electron configuration, the filled sets of inner-shell orbitals are represented by the atomic symbol of the noble gas immediately preceding the element of interest in the periodic table.

ANALYZE Silver ($Z = 47$) is the group 11 element in the fifth row of the periodic table. The noble gas at the end of the preceding row is krypton ($Z = 36$). The difference between the atomic numbers ($47 - 36$) means that we need to assign 11 electrons to orbitals.

SOLVE We would initially predict the first two of the 11 electrons would be in the $5s$ orbital and the next nine would be in $4d$ orbitals, resulting in a condensed electron configuration of $[\text{Kr}]4d^95s^2$. However, a completely filled set of d orbitals is more stable than a partially filled set, as we saw for copper, the element just above Ag in the periodic table. Therefore, silver has 10 electrons in its $4d$ orbitals and only one in its $5s$ orbital, making its condensed electron configuration $[\text{Kr}]4d^{10}5s^1$.

THINK ABOUT IT We can generate a tentative electron configuration by simply moving across a row in Figure 3.32 until we come to the element of interest. However, in the transition metals, we have to remember the special stability of half-filled and full d orbitals and make the appropriate adjustments to our configuration.

Practice Exercise Write the condensed electron configuration of cobalt ($Z = 27$).

(Answers to Practice Exercises are in the back of the book.)

3.9 Electron Configurations of Ions

To write the electron configuration of monatomic ions, we begin with the electron configuration of the parent element. If the ion has a positive charge, we remove the appropriate number of electrons from the orbital(s) with the highest principal quantum number. If the ion has a negative charge, we add the appropriate number of electrons to one or more partially filled outer-shell orbitals.

Ions of the Main Group Elements

The s block elements (see Figure 3.32) form monatomic cations by losing all their outer-shell electrons, leaving their ions with the electron configuration of the noble gas immediately preceding them in the periodic table. For example, an atom of sodium forms a Na^+ ion by losing its single $3s$ electron:

$$\text{Na} \rightarrow \text{Na}^+ + \text{e}^-$$

$$[\text{Ne}]3s^1 \rightarrow [\text{Ne}] + \text{e}^-$$

An element of the p block that forms a monatomic anion does so by gaining enough electrons to completely fill its valence-shell p orbitals. The ion it forms has the electron configuration of the noble gas at the end of its row in the periodic table. For example, an atom of fluorine ($[\text{He}]2s^22p^5$) forms a fluoride (F^-) ion by gaining one electron, which fills its set of three $2p$ orbitals and gives it the electron configuration of neon:

$$F + e^- \rightarrow F^-$$

$$[He]2s^22p^5 + e^- \rightarrow [He]2s^22p^6 = [Ne]$$

Thus, a Na^+ ion and a F^- ion have the same electron configuration as an atom of Ne. We say that the three species, Na^+, F^-, and Ne, are **isoelectronic**, meaning that they have the same electron configuration.

> **isoelectronic** describes atoms or ions that have identical electron configurations.

SAMPLE EXERCISE 3.11   Determining Isoelectronic Species in Main Group Ions   LO6

    a. Determine the electron configuration of each of the following ions: Mg^{2+}, Cl^-, Ca^{2+}, and O^{2-}.

    b. Which ions in part a are isoelectronic with neon?

COLLECT AND ORGANIZE In part a, we are asked to determine the electron configurations of four ions. In part b, we are asked to identify which of the part a ions have the same electron configuration as neon, which has 10 electrons filling its $n = 1$ and $n = 2$ shells.

ANALYZE The ions include

➤ two from group 2, Mg and Ca, which form 2+ cations;

➤ one from group 16, O, which forms a 2− anion;

➤ one from group 17, Cl, which forms a 1− anion.

Let's arrange the atoms and ions in a table in order of increasing atomic number, remembering that atoms form cations by losing electrons and form anions by gaining them:

Element	Electron Configuration	Atomic Number (Z)	Charge on Ion	Electrons/Ion
O	$[He]2s^22p^4$	8	2−	10
Mg	$[Ne]3s^2$	12	2+	10
Cl	$[Ne]3s^23p^5$	17	1−	18
Ca	$[Ar]4s^2$	20	2+	18

SOLVE

    a. The electron configurations for the ions are

        O^{2-} $[He]2s^22p^6 = [Ne]$

        Mg^{2+} $[Ne]$

        Cl^- $[Ne]3s^23p^6 = [Ar]$

        Ca^{2+} $[Ar]$

    b. Two of the ions formed—O^{2-} and Mg^{2+}—are isoelectronic with Ne and with each other. The other two—Cl^- and Ca^{2+}—are isoelectronic with Ar.

THINK ABOUT IT Each of the ions has the stability we associate with the electron configuration of a noble gas.

Practice Exercise Write the electron configurations of K^+, I^-, Ba^{2+}, S^{2-}, and Al^{3+}. Which of the ions are isoelectronic with Ar? ⚙

(Answers to Practice Exercises are in the back of the book.)

Transition Metal Cations

As with the main group elements, writing the electron configurations of transition metal cations begins with the atoms from which the cations form. Nickel atoms, like those of many transition metals, form ions with 2+ charges by losing both electrons from their valence-shell *s* orbital:

$$Ni \rightarrow Ni^{2+} + 2\ e^-$$

$$[Ar]3d^84s^2 \rightarrow [Ar]3d^8 + 2\ e^-$$

We might have expected Ni atoms to form cations by losing 3*d* electrons, reasoning that the last orbitals to be filled are the highest in energy and so should be the first to be emptied. This does not happen for Ni and the other transition metals. Among the reasons are these:

> ➤ The differences in energy between the valence-shell *s* orbitals (*ns*) and the $(n-1)d$ orbitals of transition metals are very small.

> ➤ As the $(n-1)d$ orbitals fill with increasing atomic number, the effective nuclear charge (Z_{eff}) felt by their electrons increases more than the Z_{eff} felt by the *ns* electrons. This happens because the $(n-1)d$ electrons in a transition metal atom are shielded less than the *ns* electrons in the next higher shell. As a result the *ns* electrons have higher energy and ionize first.

The rule that the electrons in orbitals with the highest *n* value ionize first applies to transition metals, too. Preferential loss of outer-shell *s* electrons explains why the most frequently encountered charge on transition metal ions is 2+.

Many transition metals form ions with charges greater than 2+ by losing *d* electrons in addition to their valence-shell *s* electrons. Atoms of scandium, $[Ar]3d^14s^2$, for example, lose both 4*s* electrons *and* their single 3*d* electron as they form Sc^{3+} ions. The chemistry of titanium, $[Ar]3d^24s^2$, is dominated by its tendency to lose two 4*s* and two 3*d* electrons, forming Ti^{4+} ions. In general, transition metals form 1+ and 2+ cations by losing all of their valence-shell *s* electrons (some, such as Ag, have only one to lose). They form cations with charges greater than 2+ by also losing $(n-1)d$ electrons.

SAMPLE EXERCISE 3.12   **Writing Electron Configurations**   **LO6**
                                of Transition Metal Ions

What are the electron configurations of Fe^{2+} and Fe^{3+}?

COLLECT AND ORGANIZE We are asked to write the electron configuration of two ions formed by iron ($Z = 26$). Iron is the group 8 element of the fourth row of the periodic table. Transition metals preferentially lose their outermost *s* electrons when they form ions.

ANALYZE The location of iron on the periodic table tells us that it has two 4*s* and six 3*d* electrons built on an argon core: the electron configuration of Fe is

$$\text{Fe:} \qquad [Ar]3d^64s^2$$

SOLVE We remove the two 4*s* electrons to form Fe^{2+}. We remove the two 4*s* electrons and one of the 3*d* electrons to form Fe^{3+}:

$$\text{Fe}^{2+}\text{:} \qquad [Ar]3d^6 \qquad\qquad \text{Fe}^{3+}\text{:} \qquad [Ar]3d^5$$

THINK ABOUT IT The electron configuration of Fe^{3+} means that each of its $3d$ orbitals contains one electron. This half-filled set of $3d$ orbitals should have enhanced stability. Actually, both Fe^{2+} and Fe^{3+} are common in nature. Most of the iron in your blood and tissues is Fe^{2+}.

Practice Exercise Write the electron configurations for Mn^{3+} and Mn^{4+}.

(Answers to Practice Exercises are in the back of the book.)

CONCEPT TEST

The electron configuration of Eu ($Z = 63$) is $[Xe]4f^7 6s^2$, but the electron configuration of Gd ($Z = 64$) is $[Xe]4f^7 5d^1 6s^2$. Suggest a reason why the additional electron in Gd is not in a $4f$ orbital, which would result in the electron configuration $[Xe]4f^8 6s^2$.

(Answers to Concept Tests are in the back of the book.)

3.10 The Sizes of Atoms and Ions

In the previous sections we learned how electrons populate orbitals with higher values of n as we go down a column of elements in the periodic table. Earlier in the chapter we learned that electrons in orbitals with higher values of n are farther from the nucleus. For example, the radial distribution profiles in Figure 3.22 showed how the distances from the nucleus to electrons in s orbitals increase with increasing principal quantum number n. In this section we explore how those ideas and the concept of effective nuclear charge can explain the relative sizes of the atoms and the common monatomic ions of the elements.

We usually express the sizes of atoms in terms of their radii. Given the wavelike behavior of electrons, the "edge" of a single atom cannot be exactly defined, so we must look at interactions between atoms to determine their radii. The atomic radius of an element that exists as a diatomic molecule, such as N_2 or O_2, is simply half the distance between the nuclear centers in the molecule (Figure 3.33a). The atomic radius of a metal, also called its *metallic* radius, is half the distance between the nuclear centers in the solid metal (Figure 3.33b). The values of ionic radii are derived from the distances between nuclear centers in solid ionic compounds (Figure 3.33c).

Trends in Atomic Size

Periodic trends in the relative sizes of the atoms of the main group elements are shown by the sizes of the spheres in Figure 3.34. Note how the sizes of atoms *increase* as we go down a group of the elements in the table (and their atomic numbers increase). However, the sizes of the atoms also change as we move from left to right across a row, especially the second and third rows, where the elements are in sequential order based on atomic number. Within these periods atomic size appreciably *decreases* with increasing atomic number. Let's explore both of these trends.

We have noted that the value of the principal quantum number n of an orbital determines the most probable distance from the nucleus of the electrons in that orbital. As n increases, this distance increases. This factor supports the observed

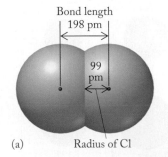

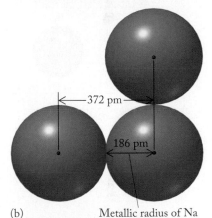

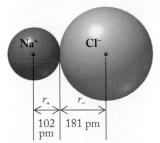

(a)   Radius of Cl

Bond length
198 pm

99 pm

(b)   Metallic radius of Na

372 pm

186 pm

Na^+ Cl^-

r_+ r_-

102 pm | 181 pm

(c) Ionic radii of Na^+ and Cl^-

FIGURE 3.33 A comparison of atomic, metallic, and ionic radii. (a) An atomic radius is half the distance between identical nuclei in a molecule, such as the distance between chlorine nuclei in Cl_2. (b) A metallic radius is based on the distance of closest approach of adjacent atoms in a solid metal. (c) An ionic radius is determined by a series of comparisons among ionic compounds containing the ion of interest.

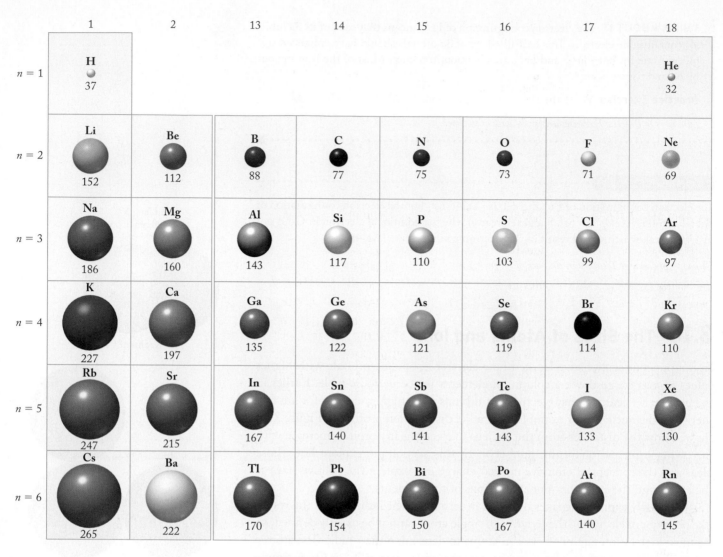

FIGURE 3.34 Atomic radii in picometers of the main group elements. Size generally increases from top to bottom in any group and generally decreases from left to right across any period.

increase in atomic size with increasing atomic number (Z) in a group of elements. There is, however, another factor to consider: as Z increases, the positive charge of the nucleus increases. If electrons experience a stronger nuclear charge, they should be pulled closer to the nucleus, shrinking the size of the atom. Clearly that does not happen as we go down a group of elements. Why not?

The answer is contained in the concept of effective nuclear charge. As we discussed in Section 3.8, the negative charges of electrons in the filled inner shells of an atom shield electrons in the outermost shell from much of the positive nuclear charge. Shielding means that the effective nuclear charge experienced by outer-shell (valence) electrons can be a tiny fraction of the total nuclear charge. Therefore, the valence electrons experience a relatively weak attraction to the nucleus and they keep their distance from it. Thus, the atomic radii of elements in the same group of the periodic table increase with increasing atomic number as we move from top to bottom. The atomic radii of the halogens in group 17 exemplify this trend:

Element	Atomic Number	n Value of Valence Shell	Atomic Radius (pm)
F	9	2	71
Cl	17	3	99
Br	35	4	114
I	53	5	133
At	85	6	140

As you can see in Figure 3.34, this pattern holds for all the other main group elements.

Now let's address why the sizes of atoms decrease with increasing atomic number across a row of the periodic table. This pattern makes sense if we focus on how increasing atomic number means more positive nuclear charge. We have seen that the orbitals that are filled as we go across a row are in the same shells. For example, the n value of the valence-shell s and p orbitals always matches the number of the row. Electrons in the s and p orbitals feel the full effect of the growing positive charge of their nuclei with increasing atomic number. Therefore, they are more strongly attracted to their nuclei and are pulled closer to them, and as a result, the sizes of atoms tend to decrease with increasing Z across a row. Of course, packing more electrons into the same shells and subshells means that there will be more repulsion between the electrons, which might tend to spread them out from each other. However, this repulsion is not strong enough to overcome the increasing Z_{eff} experienced by valence electrons with increasing atomic number.

Trends in Ionic Size

The cations of the main group elements are much smaller than their parent atoms, but the anions of the main group elements are much larger (Figure 3.35). To understand the trends, let's revisit what happens when a Na atom forms a Na^+ ion. The atom loses its only valence-shell ($3s$) electron, forming an ion with a much smaller neon core of $1s$, $2s$, and $2p$ electrons. On the other hand, when a Cl atom acquires an electron and forms a Cl^- ion, it contains more electrons than protons; hence, the attractive force per electron decreases as electron–electron repulsion increases. That is why Cl^- ions and all monatomic anions are larger than the atoms from which they form.

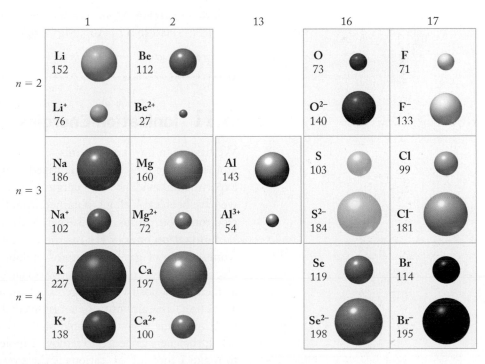

FIGURE 3.35 Comparison of atomic and ionic radii, in picometers.

Arrange each set by size, largest to smallest: (a) O, P, S; (b) Na^+, Na, K.

COLLECT AND ORGANIZE We are asked to rank a set of three atoms based on their size and a set of two atoms and a cation of one of the atoms based on their size. The location of elements in the periodic table can be used to determine the relative sizes of their atoms.

ANALYZE All of the atoms and ions are of main group elements. Atomic radii of the elements decrease with increasing atomic number across rows because their valence electrons experience increasing Z_{eff} with increasing atomic number. Atomic radii increase with increasing atomic number down groups of elements because their valence electrons are in orbitals farther from the nucleus. Main group cations are always smaller than their parent atoms because they have lost their valence-shell electrons.

SOLVE

a. Sulfur is below oxygen in group 16, so S atoms are larger than O atoms. Phosphorus is to the left of sulfur in row 3, so P atoms are larger than S atoms. Therefore, these elements in order of decreasing atomic size are P > S > O.

b. Cations are smaller than their parent atoms, so Na > Na^+. Size increases down a group of elements and K is below Na, so K > Na. Therefore, the size order is K > Na > Na^+.

THINK ABOUT IT The trend in the size of the atoms in set a reflects decreasing atomic size and increasing Z_{eff} across a row of elements in the periodic table, and increasing atomic size with increasing n values of their valence shells down a group. The relative sizes of the particles in set b are linked (1) to increasing atomic size down a group of elements in the periodic table and (2) to the smaller size of a cation relative to its parent atom.

Practice Exercise Arrange each set in order of increasing size (smallest to largest): (a) Cl^-, F^-, Na^+; (b) P^{3-}, Al^{3+}, Mg^{2+}. ⚙

(Answers to Practice Exercises are in the back of the book.)

3.11 Ionization Energies

We have seen how atomic emission spectra led to theories about the structure of atoms and the presence of quantized energy levels inside them. Another type of experimental evidence for energy levels in atoms and the electron configurations we have been exploring is obtained from the different energies required to remove electrons from atoms, that is, their ionization energies.

Ionization energy (IE) is the energy needed to remove 1 mole of electrons from 1 mole of gas-phase atoms or ions in their ground states. Removing the electrons always consumes energy because a negatively charged electron is always attracted to a positively charged nucleus, and energy is required to overcome that attraction. The amount of energy needed to remove 1 mole of electrons from 1 mole of atoms to make 1 mole of 1+ cations is called the *first ionization energy* (IE_1); the energy needed to remove 1 mole of electrons from 1 mole of 1+ cations to make 1 mole of 2+ cations is the *second ionization energy* (IE_2), and so forth. For example, for Mg atoms in the gas phase,

$$Mg \rightarrow Mg^+ + 1\ e^- \qquad IE_1 = 738\ kJ/mol$$

$$Mg^+ \rightarrow Mg^{2+} + 1\ e^- \qquad IE_2 = 1451\ kJ/mol$$

ionization energy (IE) the amount of energy needed to remove 1 mole of electrons from 1 mole of ground-state atoms or ions in the gas phase.

The total energy required to make 1 mole of Mg^{2+} cations from 1 mole of Mg atoms in the gas phase is the sum of the two ionization energies:

$$Mg \rightarrow Mg^{2+} + 2\ e^-$$

$$Total\ IE = (738 + 1451)\ kJ/mol = 2189\ kJ/mol$$

Figure 3.36 shows how the first ionization energies vary in the main group elements. The IE_1 of hydrogen is 1312 kJ/mol. The IE_1 of helium is nearly twice as big: 2372 kJ/mol. This difference seems reasonable because He atoms have two protons per nucleus, whereas H atoms have only one. In general, first ionization energies increase from left to right across a period; the easiest element to remove an electron from is in group 1, and the hardest is in group 18. This pattern makes sense because Z_{eff} increases with increasing atomic number across a row, and so do the strengths of attraction between nuclei and valence electrons.

Two anomalies occur in the general trend of increasing IE_1 across each row. One shows up between the group 2 and 13 elements that are next to each other in the second and third rows: There is a decrease in IE_1 between Be and B and between Mg and Al. The reason is that B and Al lose a p electron when they ionize, whereas Be and Mg must lose an s electron. Recall that electrons in an s orbital penetrate closer to the nucleus than electrons in the p orbitals in the same shell ($n > 2$). Therefore, a p electron experiences proportionately less effective nuclear charge than an s electron in its shell and requires less energy to be removed from the atom.

Another anomaly occurs in the values of IE_1 between the group 15 and 16 elements. The decrease makes sense when we recall that an electron lost by a group 15 element, such as nitrogen, was originally alone in a half-filled orbital:

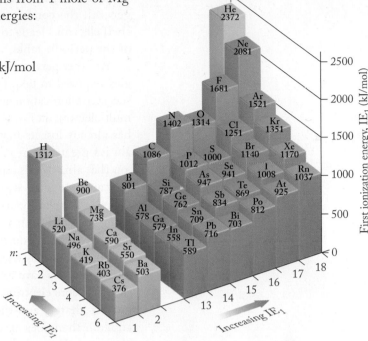

FIGURE 3.36 The first ionization energies of the main group elements generally increase from left to right in a period and decrease from top to bottom in a group.

N: ⬚ ⬚ ⬚⬚⬚ → N⁺: ⬚ ⬚ ⬚⬚⬚ + e⁻
　1s　2s　2p　　　　　1s　2s　2p

However, the electron lost by a group 16 element, such as oxygen, was paired with another in a filled orbital:

O: ⬚ ⬚ ⬚⬚⬚ → O⁺: ⬚ ⬚ ⬚⬚⬚ + e⁻
　1s　2s　2p　　　　　1s　2s　2p

Repulsion between the paired electrons raises their energy, which means less energy must be added to ionize one of them.

Now let's see how ionization energy changes as we go down a group of elements in the periodic table. We begin with group 1 and the IE_1 values of hydrogen and lithium. A lithium atom has three times the nuclear charge of a hydrogen atom, so we might expect its ionization energy to be about three times larger. However, its IE_1 value is only 520 kJ/mol, or less than half that of hydrogen, 1312 kJ/mol. Why? Because the $2s$ electron in a lithium atom is shielded from the nucleus by the two electrons in the $1s$ orbital, so the effective nuclear charge felt by the lithium valence electron is much less than 3+. In general, the combination of larger atomic size and more shielding by inner-shell electrons leads to decreasing IE_1 values from top to bottom in every group of the periodic table.

Another perspective on energy levels in atoms is provided by looking at energies required to take more than one electron away from an atom. Consider the successive ionization energies for the first 10 elements, as shown in Table 3.2. For multielectron atoms, the energy required to remove an electron from an ion that has already lost its first electron—that is, the second ionization energy, IE_2—is always greater than IE_1 because the second electron is being removed from an ion that already has a positive charge. Because electrons carry a negative charge, they are held more strongly in a cation than in the atom from which the cation is formed. The energy needed to remove a third electron is greater still because it is being removed from a 2+ ion.

Superimposed on this trend are much more dramatic increases in ionization energy (defined by the red line in Table 3.2) when all the valence electrons in ionizing atoms have been removed and the next electrons must come from inner shells. A core electron experiences much less shielding and much more effective nuclear charge, and therefore requires much more energy to be removed from an atom than does an outer-shell electron. The impact of this energy difference between valence-shell and inner-shell electrons is illustrated in the first and second ionization energies of Na and Mg:

	Na	Mg
IE_1 (kJ/mol) (orbital of lost electron)	496 (3s)	738 (3s)
IE_2 (kJ/mol) (orbital of lost electron)	4562 (2p)	1451 (3s)

Note that the IE_2 of Na is nearly 10 times its IE_1 because Na has only one 3s electron to lose. The second must come from an inner-shell 2p orbital. The IE_2 of Mg is only twice its IE_1 because the two electrons lost in forming a Mg^{2+} ion come from the same valence-shell 3s orbital.

TABLE 3.2   Successive Ionization Energies^a of the First 10 Elements

Element	Z	IE_1	IE_2	IE_3	IE_4	IE_5	IE_6	IE_7	IE_8	IE_9	IE_{10}
H	1	1312									
He	2	2372	5249								
Li	3	520	7296	12040							
Be	4	900	1758	15050	21070						
B	5	801	2426	3660	24682	32508					
C	6	1086	2348	4617	6201	37926	46956				
N	7	1402	2860	4581	7465	9391	52976	64414			
O	8	1314	3383	5298	7465	10956	13304	71036	84280		
F	9	1681	3371	6020	8428	11017	15170	17879	92106	106554	
Ne	10	2081	3949	6140	9391	12160	15231	19986	23057	115584	131236

^{a}In kJ/mol.

SAMPLE EXERCISE 3.14 **Recognizing Trends in Ionization Energies** **LO8**

Without referring to Figure 3.36, arrange argon, magnesium, and phosphorus in order of increasing first ionization energy, from lowest to highest value.

COLLECT AND ORGANIZE We are asked to order three elements on the basis of their IE_1 values. All three elements are in the third row of the periodic table.

ANALYZE First ionization energies generally increase from left to right across a row because the values of Z_{eff} increase.

SOLVE Assuming that increasing ionization energy with increasing atomic number is the dominant factor for these elements, their order of increasing first ionization energies should be

$$Mg < P < Ar$$

THINK ABOUT IT Magnesium forms stable 2+ cations, so we expect its ionization energy to be smaller than the values for phosphorus and argon, which do not form cations. Argon is a noble gas with a stable valence-shell electron configuration, so its first ionization energy would logically be the highest in the set. We can check our prediction against the IE_1 values in Figure 3.36: Mg, 738 kJ/mol; P, 1012 kJ/mol; Ar, 1521 kJ/mol.

Practice Exercise Without referring to Figure 3.36, arrange cesium, calcium, and neon in order of decreasing first ionization energy, from largest value to smallest value. ⚙

(Answers to Practice Exercises are in the back of the book.)

CONCEPT TEST

Why does ionization energy decrease with increasing atomic number down the group 18 elements?

(Answers to Concept Tests are in the back of the book.)

3.12 Electron Affinities

In the preceding section we examined the periodic nature of the energy required to remove electrons from atoms. Now we look at a complementary process and examine the change in energy when electrons are added to atoms to form monatomic anions. The energies involved are called **electron affinities (EA)**. They are the energy changes that occur when 1 mole of electrons is added to 1 mole of atoms or ions in the gas phase.

For example, the energy associated with adding 1 mole of electrons to 1 mole of chlorine atoms in the gas phase is

$$Cl + e^- \rightarrow Cl^- \qquad EA = -349 \text{ kJ/mol}$$

The negative sign tells us that energy is lost, or released, when chlorine forms chloride ions. The same is true for most of the main group elements, as shown in Figure 3.37. This release of energy makes sense because the association of a negatively charged electron with a

electron affinity (EA) the energy change that occurs when 1 mole of electrons combines with 1 mole of atoms or ions in the gas phase.

1							18
H −72.6	2	13	14	15	16	17	He (0.0)[a]
Li −59.6	Be >0	B −26.7	C −122	N +7	O −141	F −328	Ne (+29)[a]
Na −52.9	Mg >0	Al −42.5	Si −134	P −72.0	S −200	Cl −349	Ar (+35)[a]
K −48.4	Ca −2.4	Ga −28.9	Ge −119	As −78.2	Se −195	Br −325	Kr (+39)[a]
Rb −46.9	Sr −5.0	In −28.9	Sn −107	Sb −103	Te −190	I −295	Xe (+41)[a]
Cs −45.5	Ba −14	Tl −19.2	Pb −35.2	Bi −91.3	Po −183.3	At −270[a]	Rn (+41)[a]

[a]Calculated values.

FIGURE 3.37 Electron affinity (EA) values of main group elements are expressed in kilojoules per mole. The more negative the value, the more energy is released when 1 mole of atoms combines with 1 mole of electrons to form 1 mole of anions with a 1− charge. A greater release of energy reflects more attraction between the atoms of the elements and free electrons.

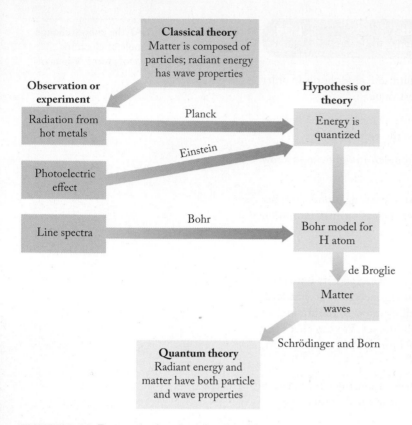

FIGURE 3.38 During the first three decades of the 20th century, quantum theory evolved from classical 19th-century theories of the nature of matter and energy. The arrows trace the development of modern quantum theory, which assumes that radiant energy has both wave properties and particle properties, and that mass (matter) also has both wave properties and particle properties.

positively charged nucleus should produce an ion that is more stable and has less energy than the free atom and electron had before they came together.

An examination of the EA values in Figure 3.37 reveals that the trends in EA are not as regular as the trends in size and ionization energy. Electron affinities generally increase (become less negative) with increasing atomic number among the group 1 and group 17 elements (except for F and Cl), but other groups do not display a clear trend. In general, electron affinity becomes more negative with increasing atomic number across a row, but there are exceptions to that trend, too. The halogens of group 17 have the most negative EA values of the elements in each of their rows because of the relatively high Z_{eff} experienced by electrons in their valence-shell p orbitals (including the electron each of their atoms acquires in becoming an anion), and because the anions they form have the stable electron configurations of noble gases.

On the other hand, adding an electron to an atom of a noble gas requires the addition of energy, which makes sense because noble gas atoms have stable electron configurations already. Beryllium and magnesium have positive electron affinities because the added electrons have to occupy outer-shell p orbitals that are significantly higher in energy than the outer-shell s orbitals. We can also rationalize the positive electron affinity of nitrogen by noting that adding an electron to an N atom means that the atom loses the stability associated with a half-filled set of $2p$ orbitals:

$$\text{N: } \begin{array}{ccc} \boxed{\uparrow\downarrow} & \boxed{\uparrow\downarrow} & \boxed{\uparrow|\uparrow|\uparrow} \\ 1s & 2s & 2p \end{array} + e^- \rightarrow \text{N}^-: \begin{array}{ccc} \boxed{\uparrow\downarrow} & \boxed{\uparrow\downarrow} & \boxed{\uparrow\downarrow|\uparrow|\uparrow} \\ 1s & 2s & 2p \end{array}$$

CONCEPT TEST

Describe at least one similarity and one difference in the periodic trends in first ionization energies and electron affinities among the main group elements.

(Answers to Concept Tests are in the back of the book.)

To end this chapter, let's return to the first three decades of the 20th century, which saw remarkable advances in our understanding of the structure of atoms and how nature works at the atomic and subatomic level. In this chapter we have tried to connect the advances together, showing how one led to another (Figure 3.38) while also conveying a sense of how profoundly unsettling these new ideas about the laws of nature were to the leading scientists of the time.

Consensus in the scientific community on the ideas of quantum theory did not come easily. For example, we have seen how excited-state sodium atoms emit photons of yellow-orange light as they fall to the ground state. Einstein puzzled over this phenomenon for several years before deciding that neither the exact moment when emission occurs nor the path of the emitted photon could be predicted exactly. He concluded that quantum theory allows us to calculate only the probability of a spontaneous electron transition; the details of the event are left to chance. In other words, no force of nature causes an excited-state sodium atom to emit a photon and fall to a lower energy level at a particular instant.

This probabilistic view of the interaction of matter and energy was very different from the familiar cause-and-effect interactions involving energy exchange among large objects. For example, a marble rolling off a table drops immediately to the floor, yet an electron remains in an excited state for an indeterminate time before falling to a lower energy level. This lack of determinacy bothered Einstein and many of his colleagues. Had they discovered an underlying theme of nature—that some processes cannot be described or known with certainty? Are there fundamental limits to how well we can know and understand our world and the events that change it?

Many scientists in the early decades of the 20th century did not agree with this uncertain view of nature. They preferred the Newtonian view, where events occur for a reason and where there are clearly understood causes and effects. They believed that the more they studied nature with ever more sophisticated tools, the more they would understand why things happen the way they do. Soon after Max Born published his probabilistic interpretation of Schrödinger's wave functions in 1926, Einstein wrote Born a letter in which he contrasted the new theories with the certainties many people find in religious beliefs:

> Quantum mechanics is very impressive. But an inner voice tells me that it is not yet the real thing. The theory produces a great deal but hardly brings us closer to the secret of the Old One. I am at all events convinced that He does not play dice.[2]

SAMPLE EXERCISE 3.15   Integrating Concepts: Red Fireworks

The brilliant red color of some fireworks (Figure 3.39) is produced by the presence of lithium carbonate (Li_2CO_3) in shells that are launched into the night sky and explode at just the right moment. The explosions produce enough energy to vaporize the lithium compound and produce excited-state lithium atoms. The atoms quickly lose energy as they transition from their excited states to their ground states, emitting photons of red light that have a wavelength of 670.8 nm.

FIGURE 3.39 The red color of some fireworks is produced by the transition of electrons in lithium atoms from their first excited state to their ground state.

a. How much energy is in each photon of red light emitted by an excited-state lithium atom?

b. Lithium atoms in the lowest energy (first) state above the ground state emit the red photons. What is the difference in energy between the ground state of a Li atom and its first excited state?

c. How much energy is needed to ionize a Li atom in its first excited state?

d. What are the electron configurations of (1) a ground-state Li atom, (2) a first-excited-state Li atom, and (3) a ground-state Li^+ ion?

COLLECT AND ORGANIZE We are given the wavelength of the red light produced by excited Li atoms and are asked to calculate the energy of the photons of this light, as well as the differences in energy between a ground-state and first-excited-state Li atom and between this excited-state Li atom and a ground-state Li^+ ion. We are also to write the electron configurations of the three states. The energy of a photon is related to its wavelength (λ) by Equation 3.4:

$$E = \frac{hc}{\lambda}$$

Lithium is the group 1 element in the second row of the periodic table. Its first ionization energy (Figure 3.36) is 520 kJ/mol.

---

[2]Letter to Max Born, 12 December 1926; quoted in R. W. Clark, *Einstein: The Life and Times* (New York: HarperCollins, 1984), p. 880.

ANALYZE The energy of a photon (E) emitted by an excited-state atom is the same as the difference in energies (ΔE) between the excited state and the lower energy state (in this case, the ground state) of the transition that produced the photon. It takes 520 kJ to ionize one mole of, or 6.0221×10^{23}, Li atoms, so dividing this 520 kJ by 6.0221×10^{23} gives us the energy required to ionize one Li atom. This should be greater than the energy needed to raise a Li atom to its first excited state, and the difference between the energies is that required to ionize the excited-state atom. We have seen in other calculations in this chapter that photons of visible light have energies in the 10^{-19} J range.

SOLVE

a. The energy of a photon with a wavelength of 670.8 nm is

$$E = \frac{hc}{\lambda} = \frac{(6.626 \times 10^{-34}\,\text{J} \cdot \text{s})(2.998 \times 10^8\,\text{m/s})}{670.8\,\text{nm}\left(\dfrac{10^{-9}\,\text{m}}{1\,\text{nm}}\right)}$$

$$= 2.961 \times 10^{-19}\,\text{J}$$

b. The difference in energies between the ground state of a Li atom and its lowest energy excited state must match exactly the energy of the photon emitted; therefore,

$$\Delta E = 2.961 \times 10^{-19}\,\text{J}$$

c. The energy required to ionize a ground-state Li atom (E_{ionize}) is

$$E_{\text{ionize}} = \left(520\,\frac{\text{kJ}}{\text{mol}}\right)\left(\frac{1000\,\text{J}}{1\,\text{kJ}}\right)\left(\frac{1\,\text{mol}}{6.0221 \times 10^{23}\,\text{atoms}}\right)$$

$$= 8.63 \times 10^{-19}\,\text{J}$$

The energy required to ionize a Li atom that is in the first excited state (and 2.961×10^{-19} J above the ground state) is E_{ionize} less 2.961×10^{-19} J:

$$\begin{array}{r} 8.63 \;\times 10^{-19}\,\text{J} \\ -2.961 \times 10^{-19}\,\text{J} \\ \hline 5.67 \;\times 10^{-19}\,\text{J} \end{array}$$

d. Li is the first atom in the second row of the periodic table, which means it has one electron in its 2s orbital. Therefore,

1. Ground-state Li atoms have the electron configuration $1s^2 2s^1$.
2. The first subshell (and electron energy level) above $2s$ is $2p$, so the valence electron in a Li atom in the lowest excited state is in a $2p$ orbital, giving it the electron configuration $1s^2 2p^1$.
3. The valence electron of a Li atom is lost when it forms a Li^+ ion, which means that the electron configuration of a ground-state Li^+ ion is simply $1s^2$.

THINK ABOUT IT You might expect that the colors that lithium and other elements make in fireworks explosions are the same ones they produce in Bunsen burner flames. You would be right. You may also know that flame tests, examples of which we saw in Figure 3.29, can be used to identify which elements are present in samples. Many of the colors are produced by transitions from first excited states to ground states, because the lowest energy excited state is the one most easily populated by the thermal energy available in a gas flame or exploding fireworks shell.

SUMMARY

Section 3.1 Visible light is one form of **electromagnetic radiation**. Like the other forms, it has wave properties described by characteristic **wavelengths (λ)** and **frequencies (ν)**, and it travels through a vacuum at the speed of light, $c = 2.998 \times 10^8$ m/s.

Section 3.2 Free atoms in flames and in gas-discharge tubes produce **atomic emission spectra** consisting of narrow bright lines at characteristic wavelengths. When continuous radiation passes through atomic gases, absorption produces the dark lines of **atomic absorption spectra**. The bright lines of an element's emission spectrum and the dark lines of its absorption spectrum are at the same wavelengths.

Section 3.3 According to **quantum theory** there are discrete energy levels in atoms, which means they absorb or emit discrete amounts of energy called **photons**. Planck used these energy **quanta** to explain the radiation emitted by incandescent objects, and Einstein used them to explain the **photoelectric effect**. The radiant energy required to dislodge a photoelectron from a metal surface is called the **work function (Φ)** of the metal.

Section 3.4 Balmer derived an equation that accounted for the bright lines in the visible emission spectrum of hydrogen and that predicted the existence of bright lines in the UV and IR regions of hydrogen's emission spectrum. Bohr proposed that the lines predicted by Balmer are related to energy levels occupied by electrons inside the hydrogen atom. A **ground-state** atom or ion has all its electrons in the lowest possible energy levels. Other arrangements are called **excited states**. **Electron transitions** from higher to lower energy levels in an atom cause the atom to emit particular frequencies of radiation; absorption of the same frequencies accompanies electron transitions from the same lower to higher energy levels.

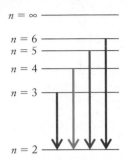

Section 3.5 De Broglie proposed that electrons in atoms, and all other moving particles, have wave properties and can be treated as **matter waves**. He explained the stability of the electron orbits in the Bohr hydrogen atom in terms of **standing waves**: the circumferences of the allowed orbits had to be whole-number multiples of the hydrogen electron's characteristic wavelength. The **Heisenberg uncertainty principle** states that the position and momentum of an electron cannot both be precisely known at the same time.

Section 3.6 The solutions to **Schrödinger's wave equation** are mathematical expressions called **wave functions (ψ)** that define the regions within an atom, called **orbitals**, where electron densities are high. Each orbital has a unique set of three **quantum numbers**: **principal quantum number n**, which defines orbital size and energy level; **angular momentum quantum number ℓ**, which defines orbital shape; and **magnetic quantum number m_ℓ**, which defines orbital orientation in space. Two electrons in the same orbital have opposite **spin quantum numbers m_s**: $+\frac{1}{2}$ and $-\frac{1}{2}$. The **Pauli exclusion principle** states that no two electrons in an atom can have the same four values of n, ℓ, m_ℓ, and m_s.

Section 3.7 Orbitals have characteristic three-dimensional sizes, shapes, and orientations that are depicted by boundary–surface representations. All s orbitals are spheres that increase in size with increasing values of n. Each of the three p orbitals in any $n \geq 2$ shell has two lobes aligned along the x-, y-, or z-axis. The five d orbitals in $n \geq 3$ shells come in two forms: four are shaped like a four-leaf clover, and the fifth has two lobes oriented along the z-axis and a torus surrounding the middle of the two lobes.

Section 3.8 According to the **aufbau principle**, electrons fill the lowest-energy atomic orbitals of a ground-state atom first. **Effective nuclear charge (Z_{eff})** is the net nuclear charge felt by outer-shell electrons when they are shielded from the full nuclear charge by inner-shell electrons. Greater Z_{eff} means lower energy. An **electron configuration** is a set of numbers and letters expressing the number of electrons in each occupied orbital in an atom. The electrons in the outermost occupied shell of an atom are **valence electrons**. They are lost, gained, or shared in chemical reactions. All the orbitals of a given p, d, or f subshell are said to be degenerate, which means they all have the same energy. **Hund's rule** states that in any set of degenerate orbitals, each orbital must contain one electron before any orbital in the set can accept a second electron. We use **orbital diagrams** to show in detail how electrons are distributed among orbitals, which are represented by boxes.

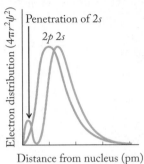

Section 3.9 Atoms of group 1 and group 2 elements tend to lose electrons and form 1+ and 2+ cations, respectively. In so doing they become **isoelectronic** with the noble gas in the preceding period. Atoms in groups 16 and 17 tend to gain electrons to form 2− and 1− anions, respectively, thereby becoming isoelectronic with the noble gas at the end of their period. When transition metals form ions, the electrons are removed from the shell of highest n until the charge on the ion is achieved.

Section 3.10 The sizes of atoms increase with increasing atomic number in a group of elements because valence-shell electrons with higher n values are, on average, farther from the nucleus. However, the sizes of atoms decrease with increasing number across a row of elements because the valence electrons experience higher effective nuclear charges. Anions are larger than their parent atoms due to additional electron–electron repulsion, but cations are smaller than their parent atoms—sometimes much smaller when all the electrons in the valence shell are lost.

Section 3.11 Ionization energy (IE) is the amount of energy needed to remove 1 mole of electrons from 1 mole of atoms or ions in the gas phase. IE values generally increase with increasing effective nuclear charge across a row and decrease with increasing atomic number down a group in the periodic table. The energy differences between the different shells and subshells of atoms are reflected in the values of successive ionization energies (IE_1, IE_2, IE_3, . . .).

Section 3.12 Electron affinity (EA) is the energy change that occurs when 1 mole of electrons combines with 1 mole of atoms or ions in the gas phase. The EA values of many main group elements are negative, indicating that energy is released when they acquire electrons.

PROBLEM-SOLVING SUMMARY

TYPE OF PROBLEM	CONCEPTS AND EQUATIONS		SAMPLE EXERCISES
Calculating frequency from wavelength	$\nu = \dfrac{c}{\lambda}$	(3.2)	3.1
Calculating the energy of a photon	$E = \dfrac{hc}{\lambda}$	(3.4)	3.2
Using the work function	$\Phi = h\nu_0 = h\nu - \mathrm{KE}_{electron}$	(3.5, 3.6)	3.3
Calculating the wavelength of a line in the hydrogen spectrum	$\lambda \text{ (nm)} = \left(\dfrac{364.56\ m^2}{m^2 - n^2}\right)$ $\dfrac{1}{\lambda} = R_H\left(\dfrac{1}{n_1{}^2} - \dfrac{1}{n_2{}^2}\right)$	(3.7) (3.8)	3.4
Calculating the energy of a transition in a hydrogen atom	$\Delta E = -2.178 \times 10^{-18}\,\mathrm{J}\left(\dfrac{1}{n_{final}{}^2} - \dfrac{1}{n_{initial}{}^2}\right)$	(3.11)	3.5
Calculating the wavelength of a particle in motion	$\lambda = \dfrac{h}{mu}$	(3.12)	3.6
Calculating Heisenberg uncertainties	$\Delta x \cdot m\,\Delta u \geq \dfrac{h}{4\pi}$	(3.14)	3.7
Identifying the subshells and orbitals in an energy level and valid quantum number sets	n is the shell number; ℓ defines both the subshell and the type of orbital; an orbital has a unique combination of allowed n, ℓ, and m_ℓ values. ℓ is any integer from 0 to $(n-1)$; m_ℓ is any integer from $-\ell$ to $+\ell$, including zero. m_s equals $+\frac{1}{2}$ or $-\frac{1}{2}$.		3.8, 3.9
Writing electron configurations of atoms and ions	Orbitals fill up in the following sequence: $1s^2$, $2s^2$, $2p^6$, $3s^2$, $3p^6$, $4s^2$, $3d^{10}$, $4p^6$, $5s^2$, $4d^{10}$, $5p^6$, $6s^2$, $4f^{14}$, $5d^{10}$, $6p^6$, $7s^2$, $5f^{14}$, $6d^{10}$, $7p^6$, (superscripts represent maximum numbers of electrons). Arrange orbitals in electron configuration based on increasing value of n. In forming transition metal ions, electrons are removed to maximize the number of d electrons; there is enhanced stability in half-filled and filled d subshells.		3.10–3.12
Ordering atoms and ions by size	In general, sizes decrease left to right across a row and increase down a column of the periodic table; cations are smaller and anions are larger than their parent atoms.		3.13
Recognizing trends in ionization energies	First ionization energies increase across a row and decrease down a column.		3.14

VISUAL PROBLEMS

(Answers to boldface end-of-chapter questions and problems are in the back of the book.)

3.1. Which of the elements highlighted in Figure P3.1 consists of atoms with
  a. a single s electron in their outermost shells? (More than one answer is possible.)
  b. filled sets of s and p orbitals in their outermost shells?
  c. filled sets of d orbitals?
  d. half-filled sets of d orbitals?
  e. two s electrons in their outermost shells?

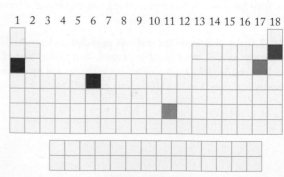

FIGURE P3.1

3.2. Which of the highlighted elements in Figure P3.1 has the greatest number of unpaired electrons per ground-state atom?

3.3. Which of the highlighted elements in Figure P3.1 form common monatomic ions that are (a) larger than their parent atoms and (b) smaller than their parent atoms?

3.4. Which of the highlighted elements in Figure P3.4 forms monatomic ions by
a. losing an s electron?
b. losing two s electrons?
c. losing two s electrons and a d electron?
d. adding an electron to a p orbital?
e. adding electrons to two p orbitals?

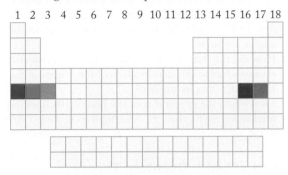

FIGURE P3.4

3.5. Which of the highlighted elements in Figure P3.4 form(s) common monatomic ions that are smaller than their parent atoms?

3.6. Rank the highlighted elements in Figure P3.4 based on (a) increasing size of their atoms and (b) increasing first ionization energy.

3.7. Suppose three beams of radiation are focused on a negatively charged metallic surface. The beam represented by the A waves in Figure P3.7 causes photoelectrons to be emitted from the surface. Which of the following statements accurately describes the abilities of the beams represented by waves B and C to also produce photoelectrons from this surface?
a. Neither the B nor the C wave can produce photoelectrons.
b. Both the B and C waves should produce photoelectrons if the sources of the waves are bright enough.
c. The B wave may or may not produce photoelectrons, but the C wave surely will.
d. The C wave may or may not produce photoelectrons, but the B wave surely will.

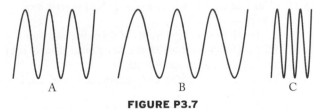

FIGURE P3.7

3.8. A group 2 element (M) and a group 17 element (X) have nearly the same atomic radius, as shown by the relative sizes of the blue and red spheres representing their atoms in Figure P3.8. Are the two elements in the same row of the periodic table? If not, which one is in the higher row?

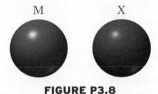

FIGURE P3.8

3.9. Which of the pairs of spheres in Figure P3.9 best depicts the relative sizes of the cation and anion formed by atoms of the elements M and X from Figure P3.8?

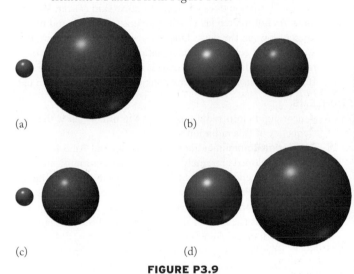

FIGURE P3.9

3.10. The arrows A, B, and C in Figure P3.10 show three transitions among the $n = 3$, 4, and 5 energy levels in a single-electron ion.
a. Assuming the transitions are accompanied by the loss or gain of electromagnetic energy, do the arrows depict absorption or emission of radiation?
b. The spectral lines of the transitions are, in order of increasing wavelength, 80, 117, and 253 nm. Which wavelength goes with which arrow?

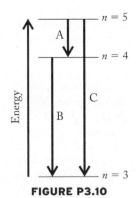

FIGURE P3.10

QUESTIONS AND PROBLEMS

Waves of Light

CONCEPT REVIEW

3.11. Why are the various forms of radiant energy called *electromagnetic* radiation?

3.12. Explain with a sketch why the frequencies of long-wavelength waves of electromagnetic radiation are lower than those of short-wavelength waves.

3.13. **Dental X-Rays** When X-ray images are taken of your teeth and gums in the dentist's office, your body is covered with a lead shield. Explain the need for this precaution.

3.14. **UV Radiation and Skin Cancer** Ultraviolet radiation causes skin damage that may lead to cancer, but exposure to infrared radiation does not seem to cause skin cancer. Why?

3.15. **Lava** As hot molten lava cools it begins to solidify and no longer glows in the dark. Does this mean it no longer emits any kind of electromagnetic radiation? If not, what kind of radiation is it likely to emit once it is no longer "red" hot?

*3.16. If light consists of waves, why don't objects look "wavy" to us?

PROBLEMS

3.17. A neon light emits radiation of $\lambda = 616$ nm. What is the frequency of this radiation?

3.18. **Submarine Communications** The Russian and American navies developed extremely low-frequency communications networks to send messages to submerged submarines. The frequency of the carrier wave of the Russian network was 82 Hz, while the Americans used 76 Hz.
   a. What was the ratio of the wavelengths of the Russian network to the American network?
   b. To calculate the actual underwater wavelength of the transmissions in either network, what additional information would you need?

3.19. **Broadcast Frequencies** FM radio stations broadcast in a band of frequencies between 88 and 108 megahertz (MHz). Calculate the wavelengths corresponding to the broadcast frequencies of the following radio stations:
   a. KRNU (Lincoln, NE), 90.3 MHz
   b. WBRU (Providence, RI) 95.5 MHz
   c. WYLD (New Orleans, LA), 98.5 MHz
   d. WAAF (Boston, MA), 107.3 MHz

3.20. Which radiation has the longer wavelength, (a) radio waves from an AM radio station broadcasting at 680 kHz or (b) infrared radiation emitted by the surface of Earth ($\lambda = 15$ μm)?

3.21. Which radiation has the lower frequency, (a) radio waves from an AM radio station broadcasting at 1090 kHz or (b) the green light ($\lambda = 550$ nm) from an LED (light-emitting diode) on a stereo system?

3.22. Which radiation has the higher frequency, (a) the red light on a bar-code reader at a grocery store or (b) the green light on the battery charger for a laptop computer?

3.23. **Speed of Light** How long does it take light to reach Earth from the sun when the distance between them is 149.6 million kilometers?

3.24. **Exploration of the Solar System** The Voyager 1 spacecraft was launched in 1977 to explore the outer Solar System and interstellar space. By December 2012 it was 1.85×10^{10} km from Earth and still sending and receiving data. How long did it take a signal from Earth to reach Voyager 1 over this distance?

Atomic Spectra

CONCEPT REVIEW

3.25. Describe the similarities and differences in the atomic emission and absorption spectra of an element.

3.26. Are the Fraunhofer lines the result of atomic emission or atomic absorption?

3.27. How did the study of the atomic emission spectra of elements lead to the identification of the origins of the Fraunhofer lines in sunlight?

*3.28. How would the appearance of the Fraunhofer lines in the solar spectrum be changed if sunlight were passed through a flame containing high-temperature sodium atoms?

Particles of Light: Quantum Theory

CONCEPT REVIEW

3.29. What is a quantum?

3.30. What is a photon?

3.31. If a piece of tungsten metal were heated to 1000 K, would it glow in the dark? If so, what color?

3.32. **Incandescent Lightbulbs** A variable power supply is connected to an incandescent lightbulb. At the lowest power setting, the bulb feels warm to the touch but produces no light. At medium power, the lightbulb filament emits a red glow. At the highest power, the lightbulb emits white light. Explain this emission pattern.

PROBLEMS

3.33. **Tanning Booths** Prolonged exposure to ultraviolet radiation in tanning booths significantly increases the risk of skin cancer. What is the energy of a photon of UV light with a wavelength of 3.00×10^{-7} m?

3.34. When dentists take X-rays, the patient is given a protective shield from excess radiation. How much more energy is there per photon from an X-ray ($\lambda = 71.2$ pm) than from UV radiation ($\lambda = 250$ nm)?

3.35. Which of the following have quantized values? Explain your selections.
   a. The elevation of the treads of a moving escalator
   b. The elevations at which the doors of an elevator open
   c. The speed of an automobile

3.36. Which of the following have quantized values? Explain your selections.
   a. The pitch of a note played on a slide trombone
   b. The pitch of a note played on a flute
   c. The wavelengths of light produced by the heating elements in a toaster
   d. The wind speed at the top of Mt. Everest

3.37. When a piece of metal is irradiated with UV radiation (λ = 162 nm), electrons are ejected with a kinetic energy of 5.34×10^{-19} J. What is the work function of the metal?

3.38. The first ionization energy of a gas-phase atom of a particular element is 6.24×10^{-19} J. What is the maximum wavelength of electromagnetic radiation that could ionize this atom?

3.39. Solar Power Photovoltaic cells convert solar energy into electricity. Could tantalum ($\Phi = 6.81 \times 10^{-19}$ J) be used to convert visible sunlight to electricity? Assume that most of the electromagnetic energy from the sun is in the visible region near 500 nm.

3.40. With reference to Problem 3.39, could tungsten ($\Phi = 7.20 \times 10^{-19}$ J) be used to construct solar cells?

*3.41. Pieces of potassium ($\Phi = 3.68 \times 10^{-19}$ J) and sodium ($\Phi = 4.41 \times 10^{-19}$ J) metal are exposed to radiation of wavelength 300 nm. Which metal emits electrons with the greater velocity? What is the velocity of the electrons?

3.42. Titanium ($\Phi = 6.94 \times 10^{-19}$ J) and silicon ($\Phi = 7.24 \times 10^{-19}$ J) surfaces are irradiated with UV radiation with a wavelength of 250 nm. Which surface emits electrons with the longer wavelength? What is the wavelength of the electrons emitted by the titanium surface?

3.43. Red Lasers The power of a red laser (λ = 630 nm) is 1.00 watt (abbreviated W, where 1 W = 1 J/s). How many photons per second does the laser emit?

*3.44. Starlight The energy density of starlight in interstellar space is 10^{-15} J/m^3. If the average wavelength of starlight is 500 nm, what is the corresponding density of photons per cubic meter of space?

The Hydrogen Spectrum and the Bohr Model

CONCEPT REVIEW

3.45. Why is the Balmer equation considered a special case of the Rydberg equation?

3.46. How does the value of n of an orbit in the Bohr model of hydrogen relate to the energy of an electron in that orbit?

3.47. Does the electromagnetic energy emitted by an excited-state H atom depend on the individual values of n_1 and n_2, or only on the difference between them $(n_1 - n_2)$?

3.48. Explain the difference between a ground-state H atom and an excited-state H atom.

3.49. Without calculating any wavelength values, predict which of these four electron transitions in the hydrogen atom is associated with radiation having the shortest wavelength.
a. $n = 1 \rightarrow n = 2$
b. $n = 2 \rightarrow n = 3$
c. $n = 3 \rightarrow n = 4$
d. $n = 4 \rightarrow n = 5$

3.50. Without calculating any frequency values, rank the following transitions in the hydrogen atom in order of increasing frequency of the electromagnetic radiation that could produce them.
a. $n = 4 \rightarrow n = 6$
b. $n = 6 \rightarrow n = 8$
c. $n = 9 \rightarrow n = 11$
d. $n - 11 \rightarrow n - 13$

3.51. Electron transitions from $n = 2$ to $n = 3, 4, 5,$ or 6 in hydrogen atoms are responsible for some of the Fraunhofer lines in the sun's spectrum. Are there any Fraunhofer lines due to transitions that start from the ground-state hydrogen atoms?

3.52. In the visible portion of the atomic emission spectrum of hydrogen, are there any bright lines due to electron transitions to the ground state?

3.53. Balmer observed a hydrogen emission line for the transition from $n = 6$ to $n = 2$, but not for the transition from $n = 7$ to $n = 2$. Why?

*3.54. In what ways should the emission spectra of H and He$^+$ be alike, and in what ways should they be different?

PROBLEMS

3.55. What is the wavelength of the photons emitted by hydrogen atoms when they undergo $n = 4 \rightarrow n = 3$ transitions? In which region of the electromagnetic spectrum does this radiation occur?

3.56. What is the frequency of the photons emitted by hydrogen atoms when they undergo $n = 5 \rightarrow n = 3$ transitions? In which region of the electromagnetic spectrum does this radiation occur?

*3.57. The energies of photons emitted by one-electron atoms and ions fit the equation

$$E = (2.18 \times 10^{-18} \text{ J})Z^2\left(\frac{1}{n_1^2} - \frac{1}{n_2^2}\right)$$

where Z is the atomic number, n_1 and n_2 are positive integers, and $n_2 > n_1$. Is the emission associated with the $n = 2 \rightarrow n = 1$ transition in a one-electron ion ever in the visible region? Why or why not?

*3.58. Can transitions from higher energy states to the $n = 2$ state in He$^+$ ever produce visible light? If so, for what values of n_2? Refer to the equation in Problem 3.57.

3.59. What is the wavelength of the $n = 3 \rightarrow n = 2$ transition in a Li^{2+} ion? Refer to the equation in Problem 3.57.

*3.60. The hydrogen atomic emission spectrum includes a UV line with a wavelength of 92.3 nm.
a. Is this line associated with a transition between different excited states, or between an excited state and the ground state?
b. What is the energy of the longest wavelength photon that a ground-state hydrogen atom can absorb?

Electrons as Waves

CONCEPT REVIEW

3.61. Identify the symbols in the de Broglie relation $\lambda = h/mu$, and explain how the relation links the properties of a particle to those of a wave.

3.62. Why do matter waves not add significantly to the challenge of hitting a baseball thrown at 99 mph (44 m/s)?

3.63. Would the density or shape of an object have an effect on its de Broglie wavelength?

3.64. How does de Broglie's hypothesis that electrons behave like waves explain the stability of the electron orbits in a hydrogen atom?

3.65. Two objects are moving at the same speed. Which (if any) of the following statements about them are true?
   a. The de Broglie wavelength of the heavier object is longer than that of the lighter one.
   b. If one object has twice as much mass as the other, its wavelength is one-half the wavelength of the other.
   c. Doubling the speed of one of the objects will have the same effect on its wavelength as doubling its mass.

3.66. Which (if any) of the following statements about the frequency of a particle is true?
   a. Heavy, fast-moving objects have lower frequencies than those of lighter, faster-moving objects.
   b. Only very light particles can have high frequencies.
   c. Doubling the mass of an object and halving its speed result in no change in its frequency.

PROBLEMS

3.67. Calculate the wavelengths of the following objects:
   a. A muon (a subatomic particle with a mass of 1.884×10^{-28} kg) traveling at 325 m/s
   b. Electrons ($m_e = 9.10938 \times 10^{-31}$ kg) moving at 4.05×10^6 m/s in an electron microscope
   c. An 82 kg sprinter running at 9.9 m/s
   d. Earth (mass $= 6.0 \times 10^{24}$ kg) moving through space at 3.0×10^4 m/s

3.68. How rapidly would each of the following particles be moving if they all had the same wavelength as a photon of red light ($\lambda = 750$ nm)?
   a. An electron of mass 9.10938×10^{-28} g
   b. A proton of mass 1.67262×10^{-24} g
   c. A neutron of mass 1.67493×10^{-24} g
   d. An α particle of mass 6.64×10^{-24} g

3.69. **Particles in a Cyclotron** The first cyclotron was built in 1930 at the University of California, Berkeley, and was used to accelerate molecular ions of hydrogen, H_2^+, to a velocity of 4×10^6 m/s. (Modern cyclotrons can accelerate particles to nearly the speed of light.) If the relative uncertainty in the velocity of the H_2^+ ion was 3%, what was the uncertainty of its position?

3.70. **Radiation Therapy** An effective treatment for some cancerous tumors involves irradiation with "fast" neutrons. The neutrons from one treatment source have an average velocity of 3.1×10^7 m/s. If the velocities of individual neutrons are known to within 2% of that value, what is the uncertainty in the position of one of them?

Quantum Numbers and the Sizes and Shapes of Atomic Orbitals

CONCEPT REVIEW

3.71. How does the concept of an orbit in the Bohr model of the hydrogen atom differ from the concept of an orbital in quantum theory?

3.72. What properties of an orbital are defined by each of the three quantum numbers n, ℓ, and m_ℓ?

3.73. How many quantum numbers are needed to identify an orbital?

3.74. How many quantum numbers are needed to identify an electron in an atom?

PROBLEMS

3.75. How many orbitals are there in an atom with each of the following principal quantum numbers? (a) 1; (b) 2; (c) 3; (d) 4; (e) 5

3.76. How many orbitals are there in an atom with the following combinations of quantum numbers?
   a. $n = 3$, $\ell = 2$
   b. $n = 3$, $\ell = 1$
   c. $n = 4$, $\ell = 2$, $m_\ell = 2$

3.77. What are the possible values of quantum number ℓ when $n = 4$?

3.78. Which are the possible values of m_ℓ when $\ell = 2$?

3.79. Which subshell corresponds to each of the following sets of quantum numbers?
   a. $n = 2$, $\ell = 0$
   b. $n = 3$, $\ell = 1$
   c. $n = 4$, $\ell = 2$
   d. $n = 1$, $\ell = 0$

3.80. Which subshell corresponds to each of the following sets of quantum numbers?
   a. $n = 2$, $\ell = 1$
   b. $n = 5$, $\ell = 3$
   c. $n = 3$, $\ell = 2$
   d. $n = 4$, $\ell = 3$

3.81. How many electrons could occupy orbitals with the following quantum numbers?
   a. $n = 2$, $\ell = 0$
   b. $n = 3$, $\ell = 1$, $m_\ell = 0$
   c. $n = 4$, $\ell = 2$
   d. $n = 1$, $\ell = 0$, $m_\ell = 0$

3.82. How many electrons could occupy orbitals with the following quantum numbers?
   a. $n = 3$, $\ell = 2$
   b. $n = 5$, $\ell = 4$
   c. $n = 3$, $\ell = 0$
   d. $n = 4$, $\ell = 1$, $m_\ell = 1$

3.83. Which of the following combinations of quantum numbers are allowed?

    a. $n = 1, \ell = 1, m_\ell = 0, m_s = +\frac{1}{2}$

    b. $n = 3, \ell = 0, m_\ell = 0, m_s = -\frac{1}{2}$

    c. $n = 1, \ell = 0, m_\ell = 1, m_s = -\frac{1}{2}$

    d. $n = 2, \ell = 1, m_\ell = 2, m_s = +\frac{1}{2}$

3.84. Which of the following combinations of quantum numbers are allowed?

    a. $n = 3, \ell = 2, m_\ell = 0, m_s = -\frac{1}{2}$

    b. $n = 5, \ell = 4, m_\ell = 4, m_s = +\frac{1}{2}$

    c. $n = 3, \ell = 0, m_\ell = 1, m_s = +\frac{1}{2}$

    d. $n = 4, \ell = 4, m_\ell = 1, m_s = -\frac{1}{2}$

The Periodic Table and Filling Orbitals; Electron Configurations of Ions

CONCEPT REVIEW

3.85. What is meant when two or more orbitals are said to be degenerate?

3.86. Explain how the electron configurations of the group 2 elements are linked to their location in the periodic table.

3.87. How do we know from examining the periodic table that the $4s$ orbital is filled before the $3d$ orbitals?

3.88. Why do so many transition metals form ions with a 2+ charge?

PROBLEMS

3.89. Identify the subshells with the following combinations of quantum numbers and arrange them in order of increasing energy in a multielectron atom:

    a. $n = 3, \ell = 2$

    b. $n = 7, \ell = 3$

    c. $n = 3, \ell = 0$

    d. $n = 4, \ell = 1, m_\ell = -1$

3.90. Identify the subshells with the following combinations of quantum numbers and arrange them in order of increasing energy in an atom of gold:

    a. $n = 2, \ell = 1$

    b. $n = 5, \ell = 0$

    c. $n = 3, \ell = 2$

    d. $n = 4, \ell = 3$

3.91. What are the electron configurations of Li^+, Ca, F^-, Mg^{2+}, and Al^{3+}?

3.92. Which species in Problem 3.91 are isoelectronic with Ne?

3.93. What are the condensed electron configurations of K, K^+, Ba, Ti^{4+}, and Ni?

3.94. In what way are the electron configurations of C, Si, and Ge similar?

3.95. Write the condensed electron configurations of the following species: Ra, I, In, Mn, and Mn^{2+}.

3.96. Write the condensed electron configurations of the following species: La, Se, Pb, Ti, and Ti^{3+}.

3.97. How many unpaired electrons are there in the following ground-state atoms and ions? (a) N; (b) O; (c) P^{3-}; (d) Na^+

3.98. How many unpaired electrons are there in the following ground-state atoms and ions? (a) Sc; (b) Ag^+; (c) Cd^{2+}; (d) Zr^{4+}

3.99. Identify the element whose condensed electron configuration is $[Ar]3d^2 4s^2$. How many unpaired electrons are there in the ground state of this atom?

3.100. Identify the element whose condensed electron configuration is $[Ne]3s^2 3p^3$. How many unpaired electrons are there in the ground state of this atom?

3.101. Which monatomic ion has a charge of 1− and the condensed electron configuration $[Ne]3s^2 3p^6$? How many unpaired electrons are there in the ground state of this ion?

3.102. Which monatomic ion has a charge of 1+ and the electron configuration $[Kr]4d^{10} 5s^2$? How many unpaired electrons are there in the ground state of this ion?

3.103. Which of the following condensed electron configurations represent an excited state?

    a. $[He]2s^1 2p^5$

    b. $[Kr]4d^{10} 5s^2 5p^1$

    c. $[Ar]3d^{10} 4s^2 4p^5$

    d. $[Ne]3s^2 3p^2 4s^1$

3.104. Which of the following condensed electron configurations represent an excited state?

    a. $[Ne]3s^2 3p^1$

    b. $[Ar]3d^{10} 4s^1 4p^2$

    c. $[Kr]4d^{10} 5s^1 5p^1$

    d. $[Ne]3s^2 3p^6 4s^1$

3.105. In which subshell are the highest energy electrons in a ground-state atom of the isotope ^{131}I? Are the electron configurations of ^{131}I and ^{127}I the same?

*__3.106.__ No known element contains electrons in an $\ell = 4$ subshell in the ground state. If such an element were synthesized, what is the minimum atomic number it would have to have?

The Sizes of Atoms and Ions

CONCEPT REVIEW

3.107. Sodium atoms are much larger than chlorine atoms, but sodium ions are much smaller than chloride ions. Why?

3.108. Why does atomic size tend to decrease with increasing atomic number across a row of the periodic table?

3.109. Which of the group 1 elements has the largest atoms? Explain your selection.

3.110. Which of the group 17 elements forms the largest monatomic ions? Explain your selection.

Ionization Energies

CONCEPT REVIEW

3.111. How do ionization energies change with increasing atomic number (a) down a group of elements in the periodic table and (b) from left to right across a row?

3.112. Explain the differences in ionization energy between (a) He and Li; (b) Li and Be; (c) Be and B; (d) N and O.

3.113. Explain why it is more difficult to ionize a fluorine atom than a boron atom.

3.114. Predict whether the ionization energies of anions of the group 17 elements are lower or higher than the ionization energies of their parent elements. Explain your prediction.

3.115. Which of the following elements should have the smallest *second* ionization energy? Br, Kr, Rb, Sr, Y

3.116. Why is the first ionization energy of Al less than that of Mg *and* less than that of Si?

Electron Affinities

CONCEPT REVIEW

3.117. An electron affinity (EA) value that is negative indicates that the free atoms of an element are less stable than the 1− anions they form by acquiring electrons. Does this mean that all of the elements with negative EA values exist in nature as anions? Give some examples to support your answer.

3.118. The electron affinities of the group 17 elements are all negative values, but the EA values of the group 18 noble gases are all positive. Explain this difference.

3.119. The electron affinities of the group 17 elements increase with increasing atomic number. Suggest a reason for this trend.

3.120. Ionization energies generally increase with increasing atomic number across the second row of the periodic table, but electron affinities generally decrease. Explain the opposing trends.

Additional Problems

3.121. **Interstellar Hydrogen** Astronomers have detected hydrogen atoms in interstellar space in the $n = 732$ excited state. Suppose an atom in that excited state undergoes a transition from $n = 732$ to $n = 731$.
  a. How much energy does the atom lose as a result of this transition?
  b. What is the wavelength of radiation corresponding to this transition?
  c. What kind of telescope would astronomers need to detect radiation of that wavelength? (*Hint*: It would not be a visible-light telescope.)

*3.122. When an atom absorbs an X-ray of sufficient energy, one of its $2s$ electrons may be ejected, creating a hole that can be spontaneously filled when an electron in a higher energy orbital—$2p$, for example—falls into it. A photon of electromagnetic radiation with an energy that matches the energy lost in the $2p \rightarrow 2s$ transition is emitted. Predict how the wavelengths of $2p \rightarrow 2s$ photons would differ between (a) different elements in the fourth row of the periodic table, and (b) different elements in the same column (for example, between the noble gases from Ne to Rn).

*3.123. Two helium ions (He$^+$) in the $n = 3$ excited state emit photons of radiation as they return to the ground state. One ion does so in a single transition from $n = 3$ to $n = 1$. The other does so in two steps: $n = 3$ to $n = 2$ and then $n = 2$ to $n = 1$. Which of the following statements about the two pathways is true?
  a. The sum of the energies lost in the two-step process is the same as the energy lost in the single transition from $n = 3$ to $n = 1$.
  b. The sum of the wavelengths of the two photons emitted in the two-step process is equal to the wavelength of the single photon emitted in the transition from $n = 3$ to $n = 1$.
  c. The sum of the frequencies of the two photons emitted in the two-step process is equal to the frequency of the single photon emitted in the transition from $n = 3$ to $n = 1$.
  d. The wavelength of the photon emitted by the He$^+$ ion in the $n = 3$ to $n = 1$ transition is shorter than the wavelength of a photon emitted by a H atom in an $n = 3$ to $n = 1$ transition.

*3.124. Use your knowledge of electron configurations to explain the following observations:
  a. Silver tends to form ions with a charge of 1+, but the elements to the left and right of silver in the periodic table tend to form ions with 2+ charges.
  b. The heavier group 13 elements (Ga, In, Tl) tend to form ions with charges of 1+ or 3+ but not 2+.
  c. The heavier elements of group 14 (Sn, Pb) and group 4 (Ti, Zr, Hf) tend to form ions with charges of 2+ or 4+.

3.125. Should the same trend in the first ionization energies for elements with atomic numbers $Z = 31$ through $Z = 36$ be observed for the second ionization energies of the same elements? Explain why or why not.

3.126. **Chemistry of Photo-Gray Glasses** "Photo-gray" lenses for eyeglasses darken in bright sunshine because the lenses contain tiny, transparent AgCl crystals. Exposure to light removes electrons from Cl$^-$ ions, forming a chlorine atom in an excited state (indicated below by the asterisk):

$$Cl^- \xrightarrow{h\nu} Cl^* + e^-$$

The electrons are transferred to Ag$^+$ ions, forming silver metal:

$$Ag^+ + e^- \rightarrow Ag$$

Silver metal is reflective, producing the photo-gray color. How might substitution of AgBr for AgCl affect the light sensitivity of photo-gray lenses? In answering this question, consider whether more energy, or less, is needed to remove an electron from a Br$^-$ ion than from a Cl$^-$ ion.

3.127. Tin (in group 14) forms both Sn^{2+} and Sn^{4+} ions, but magnesium (in group 2) forms only Mg^{2+} ions.
   a. Write condensed ground-state electron configurations for the ions Sn^{2+}, Sn^{4+}, and Mg^{2+}.
   b. Which neutral atoms have ground-state electron configurations identical to Sn^{2+} and Mg^{2+}?
   c. Which 2+ ion is isoelectronic with Sn^{4+}?

3.128. **Oxygen Ions in Space** Between 1999 and 2007 the *Far Ultraviolet Spectroscopic Explorer (FUSE)* satellite analyzed the spectra of emission sources within the Milky Way. Among the satellite's findings were interplanetary clouds containing oxygen atoms that have lost five electrons.
   a. Write an electron configuration for the highly ionized oxygen atoms.
   b. Which electrons have been removed from the neutral atoms?
   c. The ionization energies corresponding to removal of the third, fourth, and fifth electrons are 4581 kJ/mol, 7465 kJ/mol, and 9391 kJ/mol, respectively. Explain why removal of each additional electron requires more energy than removal of the previous one.
   d. What is the maximum wavelength of radiation that will remove the fifth electron from O^{4+}?

*3.129. Effective nuclear charge (Z_{eff}) is related to atomic number (Z) by a factor called the shielding parameter (σ) according to the equation $Z_{eff} = Z - \sigma$.
   a. Calculate Z_{eff} for the outermost s electrons of Ne and Ar given $\sigma = 4.24$ (for Ne) and $\sigma = 11.24$ (for Ar).
   b. Explain why the shielding parameter is much greater for Ar than for Ne.

3.130. **Fog Lamp Technology** Sodium fog lamps and street lamps contain gas-phase Na atoms and Na^+ ions. Sodium atoms emit yellow-orange light at 589 nm. Do Na^+ ions emit the same yellow-orange light? Explain why or why not.

*3.131. How can an electron get from one lobe of a p orbital to the other without going through the point of zero electron density between them?

3.132. Einstein did not fully accept the uncertainty principle, remarking that "He [God] does not play dice." What do you think Einstein meant? Niels Bohr allegedly responded by saying, "Albert, stop telling God what to do." What do you think Bohr meant?

4

Chemical Bonding
Understanding Climate Change

The Greenhouse Effect: Good News and Bad

In Chapter 3 we saw how scientific advances in the early decades of the 20th century produced a radically different, probabilistic view of the structure of matter and the changes it undergoes. This new perspective was inspired by studies of how matter and radiant energy interact, and it created the need for new ways to explain those interactions. In this chapter we continue our exploration of matter and energy, but we expand our focus to include interactions involving not only atoms, but also molecules and compounds.

Let's consider one example of how radiant energy—in this case, infrared radiation—interacts with molecules in Earth's atmosphere, particularly molecules of carbon dioxide (CO_2). You are probably aware of the debate that is raging over global warming and what should be done about it. Most scientists see a connection between recent increases in the average temperature of Earth's surface (about half a Celsius degree in the last half century) and increasing concentrations of CO_2 and other greenhouse gases in the atmosphere. The gases trap Earth's heat much like panes of glass trap heat in a greenhouse. Increases in atmospheric concentrations of greenhouse gases have been linked to human activity, particularly to increasing rates of fossil fuel combustion and the destruction of forests that would otherwise consume CO_2 during photosynthesis.

Greenhouse gases are transparent to visible radiation and allow sunlight to pass through the atmosphere and warm Earth's surface, but they trap heat that would otherwise flow from the surface into space. Were it not for the presence of greenhouse gases, Earth would be much colder than it is. Indeed, it would be too cold to be habitable.

Unfortunately, during the last 50 years, atmospheric concentrations of CO_2 have increased from about 315 parts per million to more than 390 ppm. The atmosphere hasn't contained so much CO_2 in over half a million years, long before our species evolved. In this chapter we do not address the social consequences of climate change, but we do answer the question: why does CO_2 in the atmosphere trap heat, while the two most abundant gases, O_2 and N_2, do not?

Power Production and Global Warming Coal-fired power plants, such as this one in Germany, add to the increasing concentration of carbon dioxide in the atmosphere. ▶

LEARNING OUTCOMES

LO1 Describe ways in which covalent, ionic, and metallic bonds are alike and ways in which they differ

LO2 Calculate the relative strengths of ion–ion interactions
Sample Exercise 4.1

LO3 Name molecular and ionic compounds and write their formulas
Sample Exercises 4.2, 4.3, 4.4, 4.5, 4.6, 4.7

LO4 Draw Lewis structures of ionic and molecular compounds and polyatomic ions
Sample Exercises 4.8, 4.9, 4.10, 4.11

LO5 Predict the polarity of covalent bonds based on differences in the electronegativities of the bonded elements
Sample Exercise 4.12

LO6 Draw resonance structures and use formal charges to evaluate their relative importance
Sample Exercises 4.13, 4.14, 4.15, 4.16

LO7 Describe how bond order, bond energy, and bond length are related
Sample Exercise 4.17

valence the capacity of the atoms of an element to form chemical bonds.

electrostatic potential energy (E_{el}) the energy a charged particle has because of its position relative to another charged particle; it is directly proportional to the product of the charges of the particles and inversely proportional to the distance between them; also called *coulombic attraction.*

ionic bond a bond resulting from the electrostatic attraction of a cation for an anion.

Carbon dioxide is good at trapping heat because it absorbs infrared radiation. To understand why it does, we need to consider the characteristics of the chemical bonds that hold atoms together in CO_2 and other molecular compounds. How do chemical bonds form when pairs of atoms share their outermost electrons? When do atoms share bonding electrons equally and when do they not? And how does *unequal* sharing make it possible for some compounds, such as CO_2, to absorb infrared radiation and contribute to climate change?

4.1 Chemical Bonds

There are fewer than 100 stable (nonradioactive) elements that make up all the matter in the universe. The elements combine to form over 50 million compounds that we know of, and the number grows daily as chemists synthesize new ones. This multitude of compounds exists because most of them are more stable than the free elements that make them up. They are more stable because atoms or ions that are linked together by chemical bonds tend to be lower in chemical energy than the free atoms or ions that form these bonds.

We learned in Chapter 3 that an atom's outermost electrons are the ones involved in forming chemical bonds and that these electrons are called *valence* electrons. Actually, chemists use the term **valence** all by itself to describe the capacity of the atoms of a particular element to form chemical bonds. Later in this chapter we will see how the atoms of some elements may form different numbers of chemical bonds and thus exhibit more than one *valency*. Before we explore how and why some elements display multiple valencies, let's first examine the three major categories of chemical bonds, beginning with the one involving the strong attraction between ions of opposite charge.

Ionic Bonds

CONNECTION In Chapter 2 we learned that ions (charged particles) are further classified as cations (positively charged) and anions (negatively charged).

In Chapter 2 we saw that binary (two-element) ionic compounds consist of combinations of cations formed from metallic elements and anions formed from nonmetals. These ions are held together by the electrostatic attraction that ions of opposite charge have for each other. The strength of this attraction is a form of potential energy called **electrostatic potential energy (E_{el})**. It is also called *coulombic attraction.*

CONNECTION The coulomb is the SI unit of electrical charge (see Chapter 2).

The value of E_{el} between a pair of ions is directly proportional to the product of their charges (Q_1 and Q_2) and inversely proportional to the distance (d) between their nuclei:

$$E_{el} \propto \frac{Q_1 \times Q_2}{d}$$

We can turn this expression into a real equation by replacing the "proportional to" symbol ($\propto$) with a constant of proportionality that gives us energy in joules when we express distance in nanometers:

$$E_{el} = 2.31 \times 10^{-19} \, \text{J} \cdot \text{nm} \left(\frac{Q_1 \times Q_2}{d} \right) \tag{4.1}$$

CONNECTION In Chapter 1 we defined potential energy as the energy of position or composition. Note how the electrostatic potential energy between two ions depends on both the distance between them (position) and their electrical charges (composition).

Two ions that have the same charge—that is, two positive ions or two negative ions—repel each other. Equation 4.1 tells us that this repulsion is accompanied by a positive energy because two positive or two negative values multiplied together in the numerator on the right side of the equation produce a positive E_{el} value

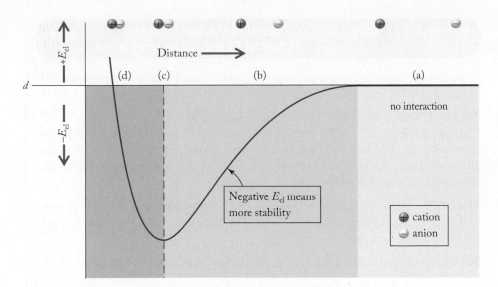

FIGURE 4.1 Changes in electrostatic potential energy (E_{el}) in the formation of an ionic bond. (a) A positive ion and a negative ion are so far apart that they do not interact. (b) As the ions move closer together the electrostatic potential energy between them becomes more negative. (c) At this distance, the ions have the lowest electrostatic potential energy and form an ionic bond. (d) If the ions are forced even closer together, they repel each other and E_{el} rises.

on the left side. Conversely, the attraction between a positive ion and a negative ion produces a negative E_{el}: the greater the attraction, the more negative the E_{el} value. More negative energy values correspond to more stable configurations of particles, as we saw in the energy states of atoms in Chapter 3: the ground state of an atom has the lowest (most negative) energy value and is the most stable state a free atom can have.

Figure 4.1 shows that E_{el} between two ions of opposite charge is zero when the ions are far apart (in zone a of Figure 4.1). As the ions approach each other in zone (b), E_{el} becomes negative, reaching a minimum when the distance between nuclei reaches the value represented by (c) in Figure 4.1. This energy minimum corresponds to a stability maximum and the formation of an **ionic bond** between the ions. If the ions involved are Na^+ and Cl^- ions, then their distance at point (c) is 0.282 nanometers. This distance of maximum stability exactly matches the distance between the nuclei of adjacent Na^+ and Cl^- ions in a crystal of table salt, sodium chloride. If the ions come closer together (zone d in Figure 4.1), E_{el} rises and the pair of ions becomes less stable. This instability results from repulsions between nuclei and between the ions' two sets of outer-shell electrons as they begin to overlap.

Equation 4.1 allows us to compare the relative strengths of attraction between the ions in different ionic compounds, assuming we know their charges and the distances between their nuclei. The charges of some of the most common monatomic ions are shown in Figure 2.10. The distance values are equivalent, within a picometer (0.001 nm) or so, to *the sum of the radii of the two ions*. We discussed in Chapter 3 how atomic radii increase with increasing atomic number down a group of the periodic table, but decrease with increasing atomic number across a row. We also discussed how the radii of monatomic anions are always larger than the atomic radii of their parent elements, and the radii of monatomic cations are always smaller. The ionic radii of the most common main group monatomic ions are given in Figure 3.35.

▶‖ **CHEMTOUR** Bonding

∞ **CONNECTION** We discussed how to predict the charges of monatomic ions from the positions of their parent elements in the periodic table in Chapter 2.

CONCEPT TEST •

Without doing any calculations, rank the pairs of ions in each group in order of decreasing (more negative) electrostatic potential energy: (a) NaCl, KCl, RbCl; (b) KF, MgO, CaO.

(Answers to Concept Tests are in the back of the book.)

SAMPLE EXERCISE 4.1 **Calculating the Electrostatic Potential Energy of an Ionic Bond** LO2

What is the electrostatic potential energy (E_{el}) of the ionic bond between a potassium ion and a chloride ion?

COLLECT AND ORGANIZE We are asked to calculate the E_{el} of an ionic bond. Equation 4.1 relates E_{el} to the charges on a pair of ions and the distance between them. An ionic bond forms when E_{el} reaches a minimum, and that happens when the distance between the nuclei of the ions is close to the sum of their ionic radii. Ionic radii of main group elements are given in Figure 3.35.

ANALYZE Potassium is in group 1, so it forms 1+ ions. Chlorine is in group 17, so it forms 1− ions. According to Figure 3.35, the ionic radii of K^+ and Cl^- ions are 138 and 181 picometers, respectively.

SOLVE The distance (d) between the nuclei of a pair of K^+ and Cl^- ions is

$$d = (K^+ \text{ ionic radius}) + (Cl^- \text{ ionic radius})$$

$$= (138 \text{ pm} + 181 \text{ pm}) = 319 \text{ pm}$$

The distance unit in Equation 4.1 is nm, so we need to convert this d value to nanometers as part of the calculation of E_{el}:

$$E_{el} = 2.31 \times 10^{-19} \text{ J} \cdot \text{nm} \left(\frac{Q_1 \times Q_2}{d} \right)$$

$$= 2.31 \times 10^{-19} \text{ J} \cdot \text{nm} \left(\frac{(1+) \times (1-)}{319 \text{ pm} \times (1 \text{ nm}/1000 \text{ pm})} \right)$$

$$= -7.24 \times 10^{-19} \text{ J}$$

THINK ABOUT IT The calculated energy has a negative value because the two ions are more stable when they are connected by an ionic bond than when they are apart. We would have to use 7.24×10^{-19} J of energy to break the bond and separate the ions.

Practice Exercise What is the electrostatic potential energy (E_{el}) of the ionic bond between a Ca^{2+} ion and a S^{2-} ion? Before doing the calculation, predict whether the result you get will be less than (more negative) or greater than the E_{el} value for KCl. ⚙

(Answers to Practice Exercises are in the back of the book.)

---

lattice energy (*U*) the energy released when 1 mole of an ionic compound forms from its free ions in the gas phase.

crystal lattice an ordered three-dimensional array of particles.

covalent bond a bond created by two atoms sharing one or more pairs of electrons.

bond length the distance between the nuclear centers of two atoms joined together in a bond.

In the preceding Sample Exercise we calculated the electrostatic potential energy in the ionic bond between a single pair of K^+ and Cl^- ions. However, the energy associated with bonding is more conventionally expressed per *mole* of ions. It is tempting to simply multiply the result from Sample Exercise 4.1 by Avogadro's number to calculate the energy that holds the ions together in a mole of KCl. The calculation produces this:

$$\frac{6.0221 \times 10^{23} \text{ ion pairs}}{1 \text{ mol}} \times \frac{-7.24 \times 10^{-19} \text{ J}}{\text{ion pair}} = -4.36 \times 10^5 \text{ J/mol, or } -436 \text{ kJ/mol}$$

The answer is in the ballpark of the true value, but it's not exactly right. The actual change in energy that occurs when 1 mole of K^+ ions in the gas phase combines with 1 mole of gaseous Cl^- ions to form 1 mole of solid KCl is −720 kJ/mol, not −436 kJ/mol. Why the difference? The answer is contained in the name we give the real value: the **lattice energy (*U*)** of KCl. Lattice energy is the term used to describe the energy released when free, gas-phase ions combine to form one mole of a crystalline solid. A crystalline solid consists of an ordered three-dimensional array of particles (atoms, ions, or molecules) called a **crystal lattice**. Figure 4.2 provides a view of the crystal lattice that makes up solid KCl. Note the tightly packed array of K^+ ions (purple spheres) and Cl^- ions (green spheres).

As a result of the tight packing of the ions, every K^+ ion touches *six* Cl^- ions, not just one, and every Cl^- ion touches *six* K^+ ions. All of the additional interactions between ions of opposite charge make E_{el} in solid KCl more negative and make the ionic bonds stronger. The tight packing also produces some repulsion between ions of the same charge, which do not touch but are still near each other. The net effect of the interactions is an overall decrease in E_{el}, corresponding to stronger ionic bonding than our calculation predicted based on individual pairs of ions.

Table 4.1 contains lattice energy values for several common ionic compounds. The values show trends that make sense based on the strengths of the ion-pair interactions predicted by Equation 4.1. For example, the only major difference between the crystal lattices of NaF, NaCl, and NaBr is the size of the anion (fluoride is the smallest anion in this group; bromide is the biggest). Differences in ionic radii translate into differences in the distance d between the nuclei of the pairs of ions. A large value of d in the denominator of Equation 4.1 means a weaker attraction between ions of opposite charge. Thus, the ions in NaF experience the strongest attractions, giving it the most negative lattice energy of the three compounds, and the ions in NaBr experience the weakest attraction, giving it the least negative U value.

The large difference between the lattice energies of NaF and MgO cannot be due to differences in d because the sums of the radii of the ions involved are nearly the same. However, the crystal lattice of MgO consists of ions with charges of $2+$ and $2-$, whereas NaF is made of $1+$ and $1-$ ions. Inserting these charges into the numerator of Equation 4.1 gives us Q_1Q_2 values of -4 for MgO and -1 for NaF. It is logical, then, that the lattice energy of MgO is about four times more negative than that of NaF.

Covalent Bonds

Now let's consider how the potential energy of two neutral atoms changes as they approach each other, as shown in Figure 4.3. We begin with two hydrogen atoms that are far enough apart that they do not interact with each other (in zone a in Figure 4.3) and have zero potential energy of interaction. When the two atoms approach each other in zone (b), the proton in the nucleus of one is attracted to the electron of the other, and vice versa. This mutual attraction of both electrons for both nuclei is accompanied by a drop in potential energy that reaches a minimum at $d = 74$ pm (point c in Figure 4.3). At this distance the two atoms share both of their electrons equally; they have formed a single **covalent bond**. There are no longer two separate H atoms in the system, but rather a more stable single particle: a molecule of H_2. If the atoms come any closer together, potential energy rises as the repulsion between the two positive nuclei increases (zone d in Figure 4.3).

The distance of minimum energy at 74 pm represents the **bond length** of the H—H bond. Covalent bond lengths have values close to the sum of the atomic radii of the atoms involved in making the bonds. The minimum potential energy of −432 kJ/mol in Figure 4.3 is the energy that is released when two moles of H atoms form one mole of H_2 molecules. That much energy would have to be added

FIGURE 4.2 Crystals of KCl contain within them an ordered three-dimensional array, or crystal lattice, of alternating K^+ ions (purple spheres) and Cl^- ions (green spheres).

▶❚❚ **CHEMTOUR** Lattice Energy

TABLE 4.1	Lattice Energies (*U*) of Some Common Ionic Compounds
Compound	**U (kJ/mol)**
LiF	−1049
LiCl	−864
NaF	−930
NaCl	786
KCl	−720
NaBr	−754
KBr	−691
MgO	−3791
MgCl₂	−2540

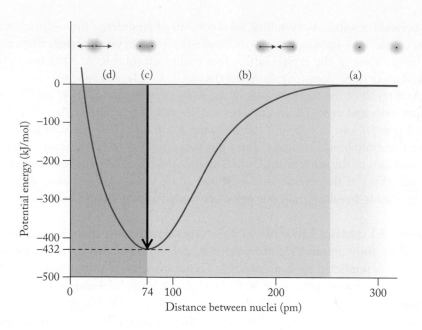

FIGURE 4.3 Changes in potential energy in the formation of a covalent bond. (a) Two hydrogen atoms that are far apart do not interact. (b) As the atoms move closer together, potential energy values decrease. (c) At an internuclear distance of 74 pm, potential energy reaches a minimum value, corresponding to a covalent bond between the atoms. (d) If the atoms are even closer together, repulsion between them causes potential energy to increase.

FIGURE 4.4 The atoms in copper metal are arranged in a crystal lattice in which each Cu atom touches and forms metallic bonds with 12 others. The electrons in metallic bonds are mobile, making copper and other metals good conductors of electricity.

to break up one mole of H_2 molecules into two moles of free H atoms. It is called the **bond energy** or *bond strength* of a H—H bond. We explore the significance of bond lengths and bond energies in more detail in Section 4.9.

Metallic Bonds

A third type of chemical bond holds the atoms together in metallic solids. As with atoms in molecules, the positive nucleus of each atom in a metallic solid is attracted to the electrons of the atoms that surround it. These attractions, coupled with overlapping valence-shell orbitals, result in the formation of **metallic bonds**. These bonds are different from covalent bonds in that they are not pairs of electrons shared by pairs of atoms. Instead, the shared electrons in metallic bonds form a "sea" of mobile electrons that flow freely among all the atoms in a piece of metal.

To understand the mobility of metallic bonds, let's focus on the bonding in a familiar metal, copper. The electron configuration of Cu ($Z = 29$) is $[Ar]3d^{10}4s^1$. If two free Cu atoms approach each other, we might expect them to share their $4s$ electrons and form a molecule of Cu_2—in much the way two hydrogen atoms form a covalent bond and a molecule of H_2 by sharing their $1s$ electrons. However, the molecular model does not correspond to the observed structure of copper: there are no Cu_2 molecules in copper metal. Instead, there is a crystal lattice of copper atoms, as shown in Figure 4.4. Inside this lattice each Cu atom is surrounded by *12* other Cu atoms. This means that each Cu atom shares its $4s$ electron with 12 other atoms, not just one, and those other 12 share their $4s$ electrons with 12 others, and so on. This pattern means that the bonding electrons are not confined to particular pairs of Cu atoms but instead can easily move from one atom to another throughout a piece of copper. Electron mobility means that copper (and metals in general) are excellent conductors of electricity.

Table 4.2 summarizes some of the similarities and differences between the three types of chemical bonds. As you make note of them, keep in mind that many bonds do not fall exclusively into just one category, but rather have some covalent, ionic, and even metallic character.

TABLE 4.2	**Types of Chemical Bonds**		
	Ionic	**Covalent**	**Metallic**
Elements Involved	Metals and nonmetals	Nonmetals and/or metalloids	Metals
Electron Distribution	Transferred	Shared	Pooled
Example	KCl	H_2	Cu

bond energy the energy needed to break 1 mole of a particular covalent bond in a molecule or polyatomic ion in the gas phase; also called *bond strength*.

metallic bond a bond consisting of the nuclei of metal atoms surrounded by a "sea" of shared electrons.

4.2 Naming Compounds and Writing Formulas

Most of the matter on our planet consists of ionic and molecular compounds. As we explore the particulate nature of some of these compounds in this chapter, we need to be clear on how to write their names and their chemical formulas. We focus on simple compounds, starting with binary ionic compounds formed by combinations of main group elements.

Binary Ionic Compounds of Main Group Elements

Nearly all binary ionic compounds formed by main group elements contain a cation formed by a group 1, 2, or 13 element and an anion formed by an element from groups 15–17. The compounds have two-word names, and there are only two rules for naming them:

1. The first word is the name of the cation, which is simply the name of its parent element.
2. The second word is the name of the anion, which is the name of its parent element, except that the ending is changed to *-ide*.

To translate the name of a binary ionic compound into a formula, we need to keep one fundamental concept in mind: the positive and negative charges of the cations and anions must be in balance. Sample Exercise 4.2 illustrates how to apply this concept and provides practice in naming ionic compounds.

> **SAMPLE EXERCISE 4.2** **Naming and Writing Formulas of Binary Ionic Compounds** **LO3**

What are the names and formulas of the compounds formed by ions of the following combinations of elements: (a) potassium and bromine, (b) calcium and oxygen, (c) sodium and sulfur, (d) magnesium and chlorine, and (e) aluminum and oxygen?

COLLECT AND ORGANIZE We are asked to name and write formulas of five binary (two-element) ionic compounds. The periodic table inside the front cover gives us the names and symbols of the elements. Figure 2.10 relates the charges of the monatomic ions of the elements of interest to their locations in the periodic table.

ANALYZE A scan of the periodic table and Figure 2.10 reveals that the five compounds contain the following combinations of ions: (a) K^+ and Br^-, (b) Ca^{2+} and O^{2-}, (c) Na^+ and S^{2-}, (d) Mg^{2+} and Cl^-, and (e) Al^{3+} and O^{2-}. The name of each compound will start with the name of the parent element of the cation followed by the name of the parent element of the anion in which the ending is changed to *-ide*. Writing chemical formulas of the compounds involves balancing the positive and negative charges on their ions.

SOLVE

a. The name of the compound is potassium bromide. The charges of its ions are 1+ and 1−. A 1:1 ratio of the ions provides electrical neutrality, making the formula KBr.

b. The name of the compound is calcium oxide. The charges of its ions are 2+ and 2−, so a 1:1 ratio of Ca^{2+} to O^{2-} ions balances the charges, making the formula CaO.

c. The name of the compound is sodium sulfide. The charges of its ions are 1+ and 2−, so a 2:1 ratio of Na^+ to S^{2-} is needed: Na_2S.

d. The name of the compound is magnesium chloride. The charges of its ions are 2+ and 1−, so a 1:2 ratio of Mg^{2+} to Cl^- is needed: $MgCl_2$.

e. The name of the compound is aluminum oxide. The charges of its ions are 3+ and 2−. A combination of two Al^{3+} ions and three O^{2-} ions balances the charges, so the formula is Al_2O_3.

THINK ABOUT IT Balancing a formula like Al_2O_3 with subscripts following both ions may be challenging. Here is a shortcut you may find useful: make the number of the charge of each ion the subscript for the other ion. Thus the 3+ charge on Al^{3+} becomes a subscript $_3$ after O, and the 2− charge on the oxide ion becomes a subscript $_2$ after Al. The result is Al_2O_3:

$$Al^{3+}O^{2-} \rightarrow Al_2O_3$$

In this exercise we used Figure 2.10 to identify the charges of the ions. Try using only the periodic table inside the front cover as you complete the following Practice Exercise.

Practice Exercise Without referring to Figure 2.10, write the chemical formulas for (a) strontium chloride, (b) magnesium oxide, (c) sodium fluoride, and (d) calcium bromide.

(Answers to Practice Exercises are in the back of the book.)

Binary Ionic Compounds of Transition Metals

Some metallic elements, including many of the transition metals, form cations with different charges. For example, most of the copper found in nature is present as Cu^{2+}; however, some copper compounds contain Cu^+ ions. Because the name *copper chloride* could apply to either $CuCl_2$ or $CuCl$, we need a naming system that distinguishes between the two compounds. One system uses a Roman numeral after the name of the transition metal. Its value matches the positive charge on the metal's ion. Thus copper(II) chloride is the chloride of Cu^{2+} ($CuCl_2$), and copper(I) chloride ($CuCl$) is the chloride of Cu^+. Roman numerals are used to indicate the charge of nearly all transition metal ions; Ag^+ and Zn^{2+} are exceptions because those ions are the only ones that silver and zinc typically form. For example, the compounds they form with Cl^- are simply called silver chloride ($AgCl$) and zinc chloride ($ZnCl_2$).

Chemists for many years have also used special names to identify different cations of the same element. In the older naming system, Cu^+ is called the *cuprous* ion and Cu^{2+} is called the *cupric* ion. Similarly, Fe^{2+} and Fe^{3+} are called *ferrous* and *ferric* ions, respectively. Note that, in both of the pairs of ions, the name of the ion with the lower charge ends in *-ous* and the name of the ion with the higher charge ends in *-ic*.

SAMPLE EXERCISE 4.3 **Writing Formulas of Transition** **LO3**
Metal Compounds

a. What are the chemical formulas of iron(II) sulfide and iron(III) oxide?
b. What are the alternative names of the compounds that do not use Roman numerals?

COLLECT AND ORGANIZE We are asked to write the chemical formulas for two binary ionic compounds of iron. We also need to write alternative names for them that do not use Roman numerals to indicate the charges on the iron ions.

ANALYZE The Roman numerals indicate the value of the positive charges on the iron cations: 2+ for iron(II) and 3+ for iron(III). Oxygen and sulfur are both in group 16. Therefore the charge on both the sulfide ion and oxide ion is 2−. In the alternate naming system, Fe^{2+} ions are ferrous ions and Fe^{3+} ions are ferric ions.

SOLVE

a. A charge balance in iron(II) sulfide is achieved with equal numbers of Fe^{2+} and S^{2-} ions, so the chemical formula is FeS. To balance the charges on Fe^{3+} and O^{2-}, we need three O^{2-} ions for every two Fe^{3+} ions. Thus, the formula of iron(III) oxide is Fe_2O_3.
b. The alternate names of FeS and Fe_2O_3 are ferrous sulfide and ferric oxide, respectively.

THINK ABOUT IT Though the Roman numeral system is the most widely used to indicate the charges on transition metal ions, you may encounter *-ous/-ic* endings in the names of compounds formed by iron, copper, and several other metals, including mercury, lead, and tin.

Practice Exercise Write the formulas of manganese(II) chloride and manganese(IV) oxide. 🔅

(Answers to Practice Exercises are in the back of the book.) ■

Polyatomic Ions

Table 4.3 lists the formulas and names of common **polyatomic ions**. Polyatomic ions consist of more than one kind of atom joined by covalent bonds. The ammonium ion (NH_4^+) is the most common cation among the polyatomic ions; all the others in the table are anions.

Many of the polyatomic anions in Table 4.3 have the generic formula XO_m^{n-}. They are called **oxoanions**. An oxoanion's name is based on the name of the element that appears before oxygen in its formula, with that element's ending changed to either *-ite* or *-ate*. The *-ate* oxoanion of an element has a greater number of oxygen atoms than its *-ite* oxoanion. For example, SO_4^{2-} is the sulfate ion and SO_3^{2-} is the sulfite ion.

If an element forms more than two kinds of oxoanions (for example, chlorine and the other group 17 elements form four), we add prefixes to distinguish between them. The oxoanion with the largest number of oxygen atoms has the prefix *per-* and ends in *-ate*, and the oxoanion with the smallest number of oxygen atoms has the prefix *hypo-* combined with an *-ite* ending. Table 4.3 contains the four oxoanions of chlorine that follow the naming rules: perchlorate (ClO_4^-), chlorate (ClO_3^-), chlorite (ClO_2^-), and hypochlorite (ClO^-).

While there are trends in the names and formulas of polyatomic ions, a name by itself does not tell you the chemical formula or the charge. Therefore, you need

polyatomic ion a charged group of more than one kind of atom joined together by covalent bonds.

oxoanion a polyatomic anion that contains at least one nonoxygen central atom bonded to one or more oxygen atoms.

TABLE 4.3	Names, Formulas, and Charges of Some Common Polyatomic Ions
Name	**Chemical Formula**
Acetate	CH_3COO^-
Ammonium	NH_4^+
Azide	N_3^-
Carbonate	CO_3^{2-}
Chlorate	ClO_3^-
Chlorite	ClO_2^-
Chromate	CrO_4^{2-}
Cyanide	CN^-
Dichromate	$Cr_2O_7^{2-}$
Dihydrogen phosphate	$H_2PO_4^-$
Disulfide	S_2^{2-}
Hydrogen carbonate or bicarbonate	HCO_3^-
Hydrogen phosphate	HPO_4^{2-}
Hydrogen sulfite or bisulfite	HSO_3^-
Hydroxide	OH^-
Hypochlorite	ClO^-
Nitrate	NO_3^-
Nitrite	NO_2^-
Perchlorate	ClO_4^-
Permanganate	MnO_4^-
Phosphate	PO_4^{3-}
Sulfate	SO_4^{2-}
Sulfite	SO_3^{2-}
Thiocyanate	SCN^-

to spend some time memorizing the formulas, charges, and names of the common polyatomic ions in Table 4.3. Note two important polyatomic ions that contain oxygen but are not considered oxoanions: the hydroxide (OH^-) ion and the acetate (CH_3COO^-) ion. You will encounter them frequently in this book.

SAMPLE EXERCISE 4.4   Writing the Formulas of Compounds Containing Oxoanions   LO3

What are the chemical formulas of (a) sodium sulfite and (b) magnesium phosphate?

COLLECT AND ORGANIZE We are given the names of two compounds containing oxoanions and are asked to write their chemical formulas. The cations in the compounds are Na^+ (a group 1 element) and Mg^{2+} (a group 2 element). The formulas and charges of common polyatomic ions are listed in Table 4.3.

ANALYZE To write the formulas of the compounds, we need to know the formulas and charges of the ions. The charges on their sodium and magnesium ions are 1+ and 2+, respectively. The sulfite ion is SO_3^{2-}, and phosphate is PO_4^{3-}.

SOLVE
a. To balance the charges on Na^+ and SO_3^{2-}, we need twice as many Na^+ ions as SO_3^{2-} ions. Therefore the formula is Na_2SO_3.
b. To balance the charges on Mg^{2+} and PO_4^{3-}, we need three Mg^{2+} ions for every two PO_4^{3-} ions, which gives us $Mg_3(PO_4)_2$.

THINK ABOUT IT In writing the formula of magnesium phosphate, we used parentheses around the phosphate ion to make it clear that the subscript 2 applies to all of the atoms in the preceding oxoanion.

Practice Exercise What are the chemical formulas of (a) strontium nitrate and (b) potassium bicarbonate?

(Answers to Practice Exercises are in the back of the book.)

SAMPLE EXERCISE 4.5   Naming Compounds Containing Oxoanions   LO3

What are the names of the compounds with these chemical formulas: (a) $CaCO_3$; (b) $LiNO_3$; (c) $MgSO_3$; (d) NH_4NO_2; (e) $KClO_3$; (f) $NaHCO_3$?

COLLECT AND ORGANIZE We are given the formulas of six compounds, each containing at least one oxoanion, and are to name them. The names of ionic compounds start with the name of the cation followed by the name of the anion.

ANALYZE The cations in five of the compounds are those formed by atoms of the elements (a) calcium, (b) lithium, (c) magnesium, (e) potassium, and (f) sodium. According to the list of polyatomic ion names in Table 4.3, the cation in formula (d) is the ammonium ion, and the oxoanions in the six compounds are (a) carbonate, (b) nitrate, (c) sulfite, (d) nitrite, (e) chlorate, and (f) hydrogen carbonate.

SOLVE Combining the names of the cations and anions, we get (a) calcium carbonate, (b) lithium nitrate, (c) magnesium sulfite, (d) ammonium nitrite, (e) potassium chlorate, and (f) sodium hydrogen carbonate.

THINK ABOUT IT Sodium hydrogen carbonate is often called sodium bicarbonate. The prefix *bi-* is widely used to indicate that there is a hydrogen ion (H^+) attached to an oxoanion.

Practice Exercise Name these compounds: (a) $Ca_3(PO_4)_2$; (b) $Mg(ClO_4)_2$; (c) $LiNO_2$; (d) $NaClO$; (e) $KMnO_4$.

(Answers to Practice Exercises are in the back of the book.)

Binary Molecular Compounds

When two nonmetallic elements combine to form compounds, they form binary molecular compounds. To translate the molecular formula of a binary molecular compound into its two-word name, proceed as follows:

1. The first word is the name of the first element in the formula.
2. For the second word, change the ending of the name of the second element to -*ide*.
3. Use prefixes (Table 4.4) to indicate the number of atoms of each type in the molecule. (However, do not use the prefix *mono-* with the first element in a name.)

For example, NO_2 is nitrogen dioxide (not *mono*nitrogen dioxide), SO_2 is sulfur dioxide, and SO_3 is sulfur trioxide. When prefixes ending in *o-* or *a-* (like *mono-* and *tetra-*) precede a name that begins with a vowel (such as *oxide*), the *o* or *a* at the end of the prefix is deleted to make the combination of prefix and name easier to pronounce. Thus, the name of NO is nitrogen monoxide, not nitrogen mon*o*oxide.

The order in which the elements are named and given in formulas corresponds to their relative positions in the periodic table: the element with the lower group number appears first. When the elements are in the same group—for example, sulfur and oxygen—the name of the element with the higher atomic number goes first. Note that the pattern follows the sequence of names in ionic compounds because the elements that form cations are to the left of and below those in the periodic table that form anions.

TABLE 4.4	Naming Prefixes for Molecular Compounds
one	*mono-*
two	*di-*
three	*tri-*
four	*tetra-*
five	*penta-*
six	*hexa-*
seven	*hepta-*
eight	*octa-*
nine	*nona-*
ten	*deca-*

SAMPLE EXERCISE 4.6   **Naming Binary Molecular Compounds**   **LO3**

What are the names of the compounds with these chemical formulas: (a) N_2O; (b) N_2O_4; (c) N_2O_5?

COLLECT AND ORGANIZE All three compounds are binary nonmetal oxides, so they are molecular compounds. Therefore, we use prefixes from Table 4.4 in the names to indicate the number of atoms of each element present in one molecule.

ANALYZE The first element in all three compounds is nitrogen, so the first word in each name is *nitrogen* with the appropriate prefix. There are two nitrogen atoms in each formula, so add the prefix *di-* to the name nitrogen. The second element in all three compounds is oxygen, so the second word in each name is *oxide* with the appropriate prefixes: *mono-* to indicate one O atom in (a); *tetra-* to indicate four O atoms in (b); *penta-* to indicate five O atoms in (c).

SOLVE
   a. dinitrogen monoxide
   b. dinitrogen tetroxide
   c. dinitrogen pentoxide

THINK ABOUT IT To avoid back-to-back vowels in the middle of the second terms in all three names, we deleted the last letter of the three prefixes before *oxide*.

Practice Exercise Name these compounds: (a) P_4O_{10}; (b) CO; (c) NCl_3.

(Answers to Practice Exercises are in the back of the book.)

Binary Acids

Some compounds have special names because of their particular chemical properties. Among these are acids. We will discuss the properties of acids in greater detail in later chapters, but for now we can consider an acid to be any compound whose molecules contain one or more *ionizable* hydrogen atoms—H atoms that are released as H^+ ions when the compound dissolves in water.

Hydrogen appears first in the formulas of acids in which *all* of the hydrogen atoms in each molecule could form H^+ ions. The simplest acids, called *binary acids*, have the generic formula HX where X is the symbol of a group 17 element. The names of the hydrogen halides follow the naming rules discussed above; thus, the name of HCl is hydrogen chloride. However, when HCl dissolves in water, it produces an acidic solution that is called *hydrochloric acid*. The rules for naming this solution and those formed by other binary acids are as follows:

1. Add the prefix *hydro-* to the name of the second element in the formula.
2. Replace the last syllable in the element's name with *-ic*, followed by the word *acid*.

CONCEPT TEST

What is the name of the solution that is produced when HF dissolves in water?

(Answers to Concept Tests are in the back of the book.)

Oxoacids

Oxoanions that have combined with H^+ ions form neutral, molecular **oxoacids**, as illustrated in Table 4.5. If the oxoanion name ends in *-ate*, the name of the corresponding oxoacid ends in *-ic*. If the oxoanion name ends in *-ite*, the name of the oxoacid ends in *-ous*. Thus, SO_4^{2-} is the sulf*ate* ion and H_2SO_4 is sulfur*ic* acid; NO_2^- is the nit*rite* ion and HNO_2 is nit*rous* acid.

TABLE 4.5 **Oxoanions of Chlorine and Their Corresponding Acids**

Ion	Formula	Name
	ClO^-	hypochlorite
	ClO_2^-	chlorite
	ClO_3^-	chlorate
	ClO_4^-	perchlorate
Acid	**Formula**	**Name**
	HClO	hypochlorous acid
	$HClO_2$	chlorous acid
	$HClO_3$	chloric acid
	$HClO_4$	perchloric acid

SAMPLE EXERCISE 4.7 **Naming Oxoacids** **LO3**

What are the names of the oxoacids formed by the following oxoanions: (a) SO_3^{2-}; (b) ClO_4^-; (c) NO_2^-?

COLLECT AND ORGANIZE We are given the formulas of three oxoanions and are asked to name the oxoacids they form when they combine with H^+ ions.

ANALYZE According to Table 4.3, the names of the oxoanions are (a) sulfite, (b) perchlorate, and (c) nitrite. When the oxoanion name ends in *-ite*, the corresponding oxoacid name ends in *-ous*. When the anion name ends in *-ate*, the oxoacid name ends in *-ic*.

SOLVE Making the appropriate changes to the endings of the oxoanion names and adding the word *acid*, we get (a) sulfurous acid, (b) perchloric acid, and (c) nitrous acid.

THINK ABOUT IT Once we know the names of the common oxoanions, naming the corresponding oxoacids is simply a matter of changing the ending of the oxoanion name from *-ate* to *-ic*, or from *-ite* to *-ous*, and then adding the word *acid*.

Practice Exercise Name these acids: (a) HBrO; (b) $HBrO_2$; (c) H_2CO_3.

(Answers to Practice Exercises are in the back of the book.)

Living organisms produce another class of acids called **carboxylic acids**. All of their names end in -ic followed by the word *acid*. Table 4.6 contains the formula and structure of one of the most common of them, acetic acid, which is the ingredient in vinegar that gives it its distinctive taste and pungent aroma. The shaded portion of its molecular structure is common to all carboxylic acids and is called the carboxylic acid **functional group**. We use that term to describe the parts of molecules that are largely responsible for the physical and chemical properties of molecular compounds.

When acetic acid dissolves in water, some of its molecules ionize, releasing the H atoms in their carboxylic acid groups (highlighted in red below) as H^+ ions. After a H^+ ion leaves a carboxylic acid, the part that remains behind is called a carboxylate anion. For example, when a molecule of acetic acid releases a H^+ ion, it forms the acetate ion, which has a charge of 1−:

Acetic acid
(carboxylic acid)

Acetate ion
(carboxylate anion)

TABLE 4.6	**Formula and Molecular Structure of Acetic Acid**
Name	Acetic acid
Formula	CH_3COOH
Molecular Structure	

4.3 Lewis Symbols and Lewis Structures

In 1916, American chemist Gilbert N. Lewis (1875–1946) proposed that atoms form chemical bonds by sharing pairs of electrons. He further suggested that through this sharing each atom acquires enough valence electrons to mimic the electron configuration of a noble gas. Today we associate such an electron configuration with the chemical stability that arises when the valence-shell *s* and *p* orbitals of atoms are completely full. Lewis's view of chemical bonding predated quantum mechanics and the notion of atomic orbitals, but it was consistent with what he called the **octet rule**: atoms tend to lose, gain, or share electrons so that each atom has eight valence electrons, or an *octet* of them.

Lewis Symbols

Lewis developed a system of symbols, called **Lewis symbols** or *Lewis dot symbols*, that reflected the number of chemical bonds each atom of an element typically forms to complete its octet, that is, its **bonding capacity**. A Lewis symbol consists of the symbol of an element surrounded by dots representing its valence electrons. The dots are placed on the four sides of the symbol (top, bottom, right, and left). The order in which they are placed does not matter *as long as one dot is placed on each side before any dots are paired*. The number of *unpaired* dots in a Lewis symbol indicates the typical bonding capacity of the atoms of that element.

Figure 4.5 shows the Lewis symbols of the main group elements. Because all elements in a family have the same number of valence electrons, they all have the same arrangement of dots in their Lewis symbols. For example, the Lewis symbols of carbon and all other group 14 elements have four unpaired electrons. Four unpaired electrons means that a carbon atom tends to form four chemical bonds.

oxoacid an acid whose molecules contain oxygen atoms, ionizable hydrogen atoms, and atoms of another element.

carboxylic acid a compound containing the −COOH functional group.

functional group a group of atoms in a molecule that is largely responsible for the physical and chemical properties of a molecular compound.

octet rule atoms of main group elements make bonds by gaining, losing, or sharing electrons to achieve a valence shell containing 8 electrons, or four electron pairs.

Lewis symbol the chemical symbol for an element surrounded by one or more dots representing valence electrons; also called *Lewis dot symbol*.

bonding capacity the number of covalent bonds an atom forms to have an octet of electrons in its valence shell.

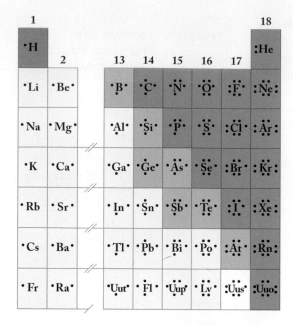

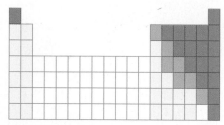

FIGURE 4.5 Lewis symbols of the main group elements of the periodic table. Because elements in a family have similar outer-shell electron configurations, they have the same arrangement of dots in their Lewis symbols, which represent valence electrons.

In doing so it completes its octet because the four bonds contain the carbon atom's original four valence electrons plus four more from the atoms with which it forms the bonds. Similarly, the Lewis symbols of nitrogen and all other group 15 elements have three unpaired electrons; a nitrogen atom typically forms three bonds to complete its octet.

There are some important exceptions to Lewis's octet rule. The atoms of a few elements typically form molecular compounds in which they have less than an octet of valence electrons. For example, a hydrogen atom has only one orbital ($1s$), so it can hold only two electrons. The one electron in the Lewis symbol of hydrogen indicates it has a bonding capacity of 1. Forming a single covalent bond gives it two electrons, or a *duet* rather than an octet, and the stable electron configuration of He.

The valence shell of beryllium is also underpopulated in some of its compounds. Its Lewis symbol, ·Be·, tells us that a Be atom has the capacity to form two bonds. However, adding two more electrons to the two it already has gives it only four electrons in its valence shell, not eight. Similarly, the three dots in the Lewis symbols of boron and aluminum indicate that their atoms have the capacity to form three bonds, but doing so gives them only six valence-shell electrons. Thus, according to Lewis theory, Be, B, and Al will be *electron deficient* in the molecular compounds they form. Another type of electron deficiency happens when a molecule has an odd number of total valence electrons. One of the atoms in such a molecule usually ends up with only seven valence electrons instead of eight. We examine exceptions to the octet rule in more detail in Section 4.8.

Before we start to use Lewis symbols to explore how elements form compounds, we need to keep in mind that the symbols indicate the number of bonds that the atoms of an element *typically* form. We will see in this chapter that atoms may exceed their nominal bonding capacities in some molecules and polyatomic ions, and fail to reach them in others. Bonding capacity is a useful concept, but we will treat it more like a guideline than a requirement.

CONCEPT TEST

Devise a formula that relates the bonding capacity (BC) of the atoms of the elements in groups 14–17 with their group number (GN).

(Answers to Concept Tests are in the back of the book.)

Lewis Structures of Ionic Compounds

A **Lewis structure** is a two-dimensional representation of a compound showing how its atoms are connected. Though Lewis structures are most often used to show the bonding in molecular compounds, they also provide insights into the composition of ions and ionic compounds. The Lewis structures of binary ionic compounds such as NaCl are some of the simplest to draw, so we start with them.

Crystals of sodium chloride are held together by the attraction between oppositely charged Na$^+$ and Cl$^-$ ions. We discussed in Chapter 3 how sodium and the other group 1 and 2 elements achieve noble gas electron configurations by losing their valence-shell s electrons and forming positively charged cations. Therefore, the ions of group 1 and 2 elements have no valence-shell electrons, and their

Lewis structures have no dots around them. On the other hand, nonmetals such as chlorine, which acquire electrons to achieve noble gas electron configurations, have completely filled s and p orbitals in their valence shells. Therefore, their monatomic ions have Lewis structures with four pairs of dots. We place brackets around the dots to emphasize that all eight valence electrons are associated with the anion, and the charge of the ion is placed outside the bracket to indicate the overall charge of everything inside. Applying this notation to NaCl gives us this Lewis structure:

$$\text{Na}\cdot + \cdot\ddot{\underset{\cdots}{\text{Cl}}}: \;\rightarrow\; \text{Na}^{+}\left[:\ddot{\underset{\cdots}{\text{Cl}}}:\right]^{-}$$

> **Lewis structure** a two-dimensional representation of the bonds and lone pairs of valence electrons in an ionic or molecular compound.
>
> **bonding pair** a pair of electrons shared between two atoms.

SAMPLE EXERCISE 4.8 Drawing the Lewis Structure of a Binary Ionic Compound **LO4**

Draw the Lewis structure of calcium oxide.

COLLECT AND ORGANIZE We are asked to draw the Lewis structure of the ionic compound formed by calcium and oxide ions. The charges of the ions are related to the group numbers and valence-shell electron configurations of the parent elements.

ANALYZE When a metal combines with a nonmetal, the metal forms a cation and the nonmetal forms an anion. The monatomic ion formed by calcium has a charge of 2+ because Ca is in group 2; it achieves a noble gas electron configuration by losing its two valence-shell electrons. The oxide ion has a charge of 2− because oxygen is in group 16; it acquires two valence-shell electrons to achieve the electron configuration of Ne.

SOLVE A Ca^{2+} ion has no valence electrons left, so its Lewis structure is simply

$$\text{Ca}^{2+}$$

After acquiring two electrons, an O^{2-} ion has a complete octet and this Lewis structure:

$$\left[:\ddot{\underset{\cdots}{\text{O}}}:\right]^{2-}$$

Combining the Lewis structures of the Ca^{2+} and O^{2-} ions, we have the Lewis structure of CaO:

$$\text{Ca}^{2+}\left[:\ddot{\underset{\cdots}{\text{O}}}:\right]^{2-}$$

THINK ABOUT IT The lack of dots in the structure of Ca^{2+} reinforces the fact that Ca atoms, like those of the other main group metals, lose all their valence-shell electrons when they form monatomic cations. In contrast, atoms of the nonmetals form monatomic anions with complete octets.

Practice Exercise Draw the Lewis structure of magnesium fluoride.

(Answers to Practice Exercises are in the back of the book.)

▶❚❚ **CHEMTOUR** Lewis Dot Structures

Lewis Structures of Molecular Compounds

Molecular compounds are held together by covalent bonds. Because covalent bonds are *pairs* of shared valence electrons, the Lewis structures of molecular compounds focus on how the shared pairs, also called **bonding pairs**, are distributed among the atoms in their molecules. The bonding pairs in Lewis structures

single bond a bond that results when two atoms share one pair of electrons.

lone pair a pair of electrons that is not shared.

are represented by dashes, as in the structural formulas we first saw in Chapter 1. A single bonding pair of electrons is called a **single bond**. Electron pairs that are not involved in bonds appear as pairs of dots on one atom. The unshared electron pairs are called **lone pairs**. Atoms with one or more pairs of dots in their Lewis symbols frequently (though not always) end up with that same number of lone pairs of electrons in the Lewis structures of the molecules they form.

A five-step process for drawing Lewis structures is described below. It is particularly useful for drawing the structures of small molecules and polyatomic ions that have a single central atom bonded to atoms with lower bonding capacities. To draw the Lewis structures of molecules with more than one "central" atom, we may need to break the structure into subunits, as in hydrogen peroxide:

$$\text{H—}\ddot{\text{O}}\text{—}\ddot{\text{O}}\text{—H}$$

We treat each oxygen atom in H_2O_2 as the "central atom" within a three-atom subunit, as shown by the O atoms highlighted in red:

$$\text{—}\ddot{\text{O}}\text{—}\ddot{\text{O}}\text{—H}$$

$$\text{H—}\ddot{\text{O}}\text{—}\ddot{\text{O}}\text{—}$$

Five Steps for Drawing Lewis Structures

1. *Determine the number of valence electrons.* For a neutral molecule, count the valence electrons in all the atoms in the molecule. For a polyatomic ion, count the valence electrons and then add or subtract the number of electrons needed to account for the charge on the ion.

2. *Arrange the symbols of the elements in a pattern that shows how their atoms are bonded together (a* skeletal structure*) and then connect them with single bonds (single pairs of bonding electrons).* Make the element with the greatest bonding capacity the central atom. (If two elements have the same bonding capacity, choose the one that is less *electronegative,* an atomic property we examine in Section 4.4.) Place the remaining atoms around the central atom and those that form the fewest bonds (for example, hydrogen) around the periphery. Sometimes the formulas of compounds and polyatomic ions, such as HNCO and SCN^-, indicate how their atoms are connected. For example, the skeletal structures of HNCO and SCN^- are H—N—C—O and S—C—N, respectively.

3. *Complete the octets of all the atoms (except hydrogen) bonded to the central atom by adding lone pairs of electrons.* Stop when all valence electrons are used.

4. *Compare the number of valence electrons in the Lewis structure to the number determined in step 1.* If there are electrons left over, add lone pairs of electrons to the central atom, even if doing so means giving it more than an octet of valence electrons.

5. *Complete the octet on the central atom.* If there is an octet on the central atom, the structure is complete. If there is less than an octet on the central atom, create additional bonds to it by converting one or more lone pairs of electrons on atoms surrounding the central atom into bonding pairs.

SAMPLE EXERCISE 4.9   **Drawing the Lewis Structure** **LO4**
                                              of a Small Molecule I

Chloroform ($CHCl_3$) is a volatile liquid that was once used as an anesthetic in surgery. Draw its Lewis structure.

COLLECT AND ORGANIZE   Chloroform has the molecular formula $CHCl_3$. We can use the five-step process described above to generate its Lewis structure.

ANALYZE   The formula $CHCl_3$ tells us that a chloroform molecule contains one carbon atom, one hydrogen atom, and three chlorine atoms. Because carbon is a group 14 element, the carbon atom has 4 valence electrons and needs 4 more to complete its octet. Hydrogen, in group 1, has 1 valence electron and needs 1 more to complete its duet. Chlorine, in group 17, has 7 valence electrons and needs 1 more to complete its octet.

SOLVE

1. The number of valence electrons in the $CHCl_3$ molecule is

Element:	C	+	H	+	3 Cl
Valence electrons:	4	+	1	+	$(3 \times 7) = 26$

2. The carbon atom has the most (4) unpaired electrons in its Lewis symbol, so it has the greatest bonding capacity. Therefore, C is the central atom. Each H atom and each Cl atom needs one more electron to achieve the electron configuration of a noble gas, so each forms one bond to the C atom:

$$
\begin{array}{c}
\text{H} \\
| \\
\text{Cl}\!-\!\text{C}\!-\!\text{Cl} \\
| \\
\text{Cl}
\end{array}
$$

3. In this structure the hydrogen atom has a complete duet because it shares a bonding pair of electrons with the central carbon atom. We need to add lone pairs of electrons to complete the octets on the three chlorine atoms:

$$
\begin{array}{c}
\text{H} \\
| \\
:\!\ddot{\text{Cl}}\!-\!\text{C}\!-\!\ddot{\text{Cl}}\!: \\
| \\
:\!\ddot{\text{Cl}}\!:
\end{array}
$$

4. This structure contains 4 pairs of bonding electrons and 9 lone pairs, for a total of

$$(4 \times 2) + (9 \times 2) = 26 \text{ electrons}$$

   which matches the total number of valence electrons in the molecule (see step 1).
5. The carbon atom is surrounded by four bonds, which means it has 8 valence electrons and a full octet. The structure is complete.

THINK ABOUT IT   In this example, there was no need to change the structure during steps 4 and 5 because all the electrons were accounted for and all the atoms had an octet (or duet in the case of the hydrogen atom).

Practice Exercise   Draw the Lewis structure of methane, CH_4.

(Answers to Practice Exercises are in the back of the book.)

SAMPLE EXERCISE 4.10 **Drawing the Lewis Structure of a Small Molecule II** **LO4**

Draw the Lewis structure of ammonia, NH_3.

COLLECT AND ORGANIZE Ammonia has the molecular formula NH_3. As in the previous exercise, we use the five-step process described above to draw its Lewis structure.

ANALYZE The chemical formula tells us that NH_3 contains one atom of nitrogen and three atoms of hydrogen per molecule. Nitrogen is a group 15 element with 5 valence electrons, 3 of which are unpaired, giving it a bonding capacity of 3. Hydrogen atoms have 1 valence electron each and a bonding capacity of 1.

SOLVE

1. The number of valence electrons in the NH_3 molecule is

Element:	N	+	3 H
Valence electrons:	5	+	$(3 \times 1) = 8$

2. The nitrogen atom has the greater bonding capacity and is the central atom. Connecting each H atom to the nitrogen atom with a covalent bond yields

   $$H-N-H$$
   $$|$$
   $$H$$

3. Each bonded H atom has a complete duet of electrons.

4 and 5. The three bonds in the structure represent $3 \times 2 = 6$ valence electrons, but there are supposed to be 8. Adding a lone pair of electrons to nitrogen gives us 8 valence electrons and completes the octet on nitrogen:

   $$H-\overset{\displaystyle ..}{N}-H$$
   $$|$$
   $$H$$

   The Lewis structure is complete.

THINK ABOUT IT This structure makes sense because the Lewis symbol of the nitrogen atom contains 2 electrons that are paired and 3 that are unpaired. Therefore, it is reasonable that the nitrogen atom in NH_3 has one lone pair and three bonding pairs of electrons.

Practice Exercise Draw the Lewis structure of phosphorus trichloride.

(Answers to Practice Exercises are in the back of the book.)

Lewis Structures of Molecules with Double and Triple Bonds

Lewis structures can be used to show the bonding in molecules in which two atoms share more than one pair of bonding electrons. A bond in which two atoms share two pairs of electrons is called a **double bond**. For example, the two oxygen atoms in a molecule of O_2 share two pairs of electrons, forming an $O{=}O$ double bond. When the two nitrogen atoms in a molecule of N_2 share three pairs of electrons, they form a $N{\equiv}N$ **triple bond**. In Section 4.9 we discuss the characteristics of multiple bonds and compare them with those of single bonds.

How do we know when a Lewis structure has a double or triple bond? Typically we find it out when we apply steps 3 through 5 in the guidelines. Suppose

double bond a bond that results when two atoms share two pairs of electrons.

triple bond a bond that results when two atoms share three pairs of electrons.

we fill the octets of all the atoms attached to the central atom in step 3, and in doing so we use all the valence electrons available. If we discover that the central atom does not have an octet, we can provide the needed electrons in step 5 by *converting one or more lone pairs of electrons* from the other atoms into bonding pairs. The following example illustrates an application of the guidelines.

Let's draw the Lewis structure of formaldehyde (H_2CO). We begin with the five-step guidelines:

1. The total number of valence electrons is

Element:	C	+	2 H	+	O
Valence electrons:	4	+	(2×1)	+	$6 = 12$

2. Of the three elements, carbon has the greatest bonding capacity (4) and is the central atom. Connecting it with single bonds to the other three atoms, we have

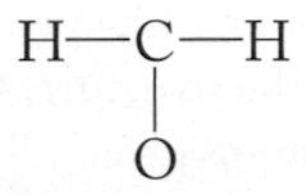

$$
\begin{array}{c}
\text{H—C—H} \\
| \\
\text{O}
\end{array}
$$

3. Each H atom has a single covalent bond (2 electrons) completing its valence shell. Oxygen needs 3 lone pairs of electrons added to complete its octet:

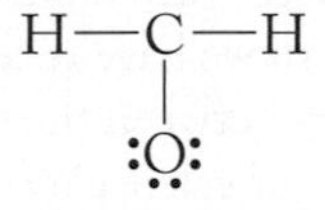

$$
\begin{array}{c}
\text{H—C—H} \\
| \\
\text{:\ddot{O}:}
\end{array}
$$

4. There are 12 valence electrons in this structure, which matches the number determined in step 1.
5. The central C atom has only 6 electrons. To provide the carbon atom with 2 more electrons so that it has an octet—without removing any electrons from oxygen, which already has an octet—we convert one of the lone pairs on the oxygen atom into a bonding pair between C and O:

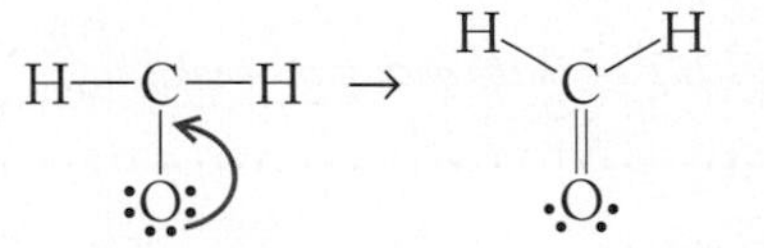

$$
\begin{array}{c}
\text{H—C—H} \\
| \\
\text{:\ddot{O}:}
\end{array}
\rightarrow
\begin{array}{c}
\text{H} \quad \text{H} \\
\diagdown \diagup \\
\text{C} \\
\| \\
\cdot\ddot{\text{O}}\cdot
\end{array}
$$

It does not matter which of the three lone pairs we move because they are equivalent. The central carbon atom now has a complete octet, and the oxygen atom still does. This structure makes sense because the four covalent bonds around carbon—two single bonds and one double bond—match its bonding capacity. The double bond to oxygen makes sense because oxygen is a group 16 element with a bonding capacity of 2, and it has two bonds in this structure.

Notice that we have drawn the double bond and the two single bonds on the central carbon atom at an angle of about 120° from each other in the final structure; we have done the same thing with the lone pairs of electrons on oxygen with respect to the double bond. Drawing them this way maximizes the separation between pairs of electrons and more closely represents their actual orientation in the molecules, as we will investigate in Chapter 5. For now, be assured there is nothing wrong with drawing a Lewis structure for formaldehyde in which the bond angles are, for example, 90° instead of 120°. The purpose of Lewis structures is to show how atoms are bonded to each other in molecules, not necessarily how the bonds are oriented.

polar covalent bond a bond resulting from unequal sharing of bonding pairs of electrons between atoms.

nonpolar covalent bond a bond characterized by an even distribution of charge; electrons in the bonds are shared equally by the two atoms; pure covalent bonds give rise to nonpolar diatomic molecules.

electronegativity a relative measure of the ability of an atom in a bond to attract electrons to itself when bonded to another atom.

SAMPLE EXERCISE 4.11   **Drawing Lewis Structures with Double and Triple Bonds**   **LO4**

Draw the Lewis structure of acetylene, C_2H_2, the fuel used in oxyacetylene torches for cutting steel and other metals.

COLLECT AND ORGANIZE We are asked to draw the Lewis structure of acetylene, which has the molecular formula C_2H_2. We follow the steps in the guidelines, using double or triple bonds as needed.

ANALYZE Each molecule contains two atoms of carbon and two atoms of hydrogen. Carbon is a group 14 element with a bonding capacity of 4.

SOLVE
1. The total number of valence electrons is

   | Element: | 2 C | + | 2 H |
   | Valence electrons: | (2×4) | + | $(2 \times 1) = 10$ |

2. Of the two elements, carbon has the greater bonding capacity. Therefore C atoms go in the middle of the structure. Connecting each carbon atom with single bonds to the other atoms gives

$$H—C—C—H$$

3. Each H atom has a single covalent bond (made up of 2 electrons) and a complete valence shell.

4 and 5. There are 6 valence electrons in our structure, but there are 10 valence electrons in the molecule, so we have to add 4 more. One way to do that is to add two bonding pairs between the carbon atoms. This gives the structure the right number of valence electrons, and it completes the octets of both carbon atoms:

$$H—C≡C—H$$

THINK ABOUT IT The structure makes sense because carbon has a bonding capacity of 4, and there are four covalent bonds around each carbon atom—1 single bond and 1 triple bond—giving both complete octets.

Practice Exercise Draw the Lewis structure of carbon dioxide. ⚙

(Answers to Practice Exercises are in the back of the book.)

4.4 Electronegativity, Unequal Sharing, and Polar Bonds

When Lewis proposed that atoms form chemical bonds by sharing electrons, he knew that electron sharing in covalent bonds did not necessarily mean *equal* sharing. For example, Lewis knew that molecules of HCl ionized, forming H^+ and Cl^- ions, when HCl dissolves in water. To explain this phenomenon Lewis proposed that the bonding pair of electrons in a molecule of HCl is closer to the chlorine end of the bond than the hydrogen end. This unequal sharing makes the H—Cl bond a **polar covalent bond**. It also means that the bond may break unequally so that both of the shared pair of electrons end up on the Cl atom, forming a Cl^- ion, and leaving the H atom with no electrons and a charge of 1+.

The polarity of the bonds in HCl means that each bond possesses a tiny electric *dipole*, meaning that there is a slightly positive pole at the H end of the bond and a slightly negative pole at the Cl end, analogous to the positive and negative ends of a battery as shown in Figure 4.6. This figure also shows two of the ways

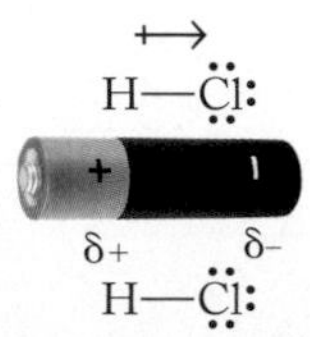

FIGURE 4.6 Just as a battery has positive and negative terminals, a polar bond such as the one in a molecule of HCl has positive and negative ends, represented here by the arrow with a positive tail above the H—Cl bond and by delta symbols ($\delta+$ and $\delta-$).

we represent unequal sharing of bonding pairs of electrons. One of them incorporates an arrow with a plus sign embedded in its tail. The arrow points toward the more negative, electron-rich end of the bond, and the plus sign represents the more positive, electron-poor end. Another way makes use of the lowercase Greek delta, δ, followed by a + or − sign. The deltas represent *partial* electrical charges, as opposed to full electrical charges that accompany the complete transfer of one or more electrons as atoms become ions.

Figure 4.7 shows examples of equal and unequal sharing of bonding pairs of electrons. It includes (a) Cl_2, in which the Cl atoms are connected by a **nonpolar covalent bond** in which the bonding electrons are shared equally between two identical atoms, (b) a polar covalent bond (H—Cl), and (c) an ionic bond between a pair of Na^+ and Cl^- ions in NaCl. An ionic bond represents unequal sharing taken to an extreme. In a sample of NaCl, the bonding pairs have been completely transferred from Na atoms to Cl atoms, resulting in *total*, rather than *partial*, separation of electrical charge: 1+ on Na and 1− on Cl. The degree of charge separation may be represented using color, as shown in the calibration bar at the top of Figure 4.7. The middle color (yellow-green) represents equal sharing and no partial charge, the red and violet ends represent full charge separation, and the colors in between represent various degrees of partial charge separation.

Another American chemist, Linus Pauling (1901–1994), explained bond polarity through the concept of **electronegativity**, which is represented by the Greek letter χ (chi) and defined as the tendency of an atom chemically bonded to another to attract electrons toward itself. Central to this concept is the assumption that the bonds between atoms of different elements are neither 100% covalent nor 100% ionic, but somewhere in between. To determine where, Pauling developed electronegativity values for the most common elements (Figure 4.8). The degree of ionic character of a bond depends on the difference ($\Delta\chi$) in the electronegativity values of the two elements the bond connects.

CHEMTOUR Partial Charges and Bond Dipoles

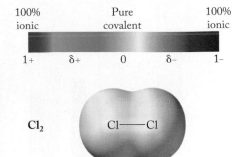

(a) Pure covalent: even charge distribution

(b) Polar covalent: uneven charge distribution

(c) Ionic: complete transfer of electron

FIGURE 4.7 Variations in valence electron distribution are represented using colored surfaces in these molecular models. (a) In the covalent bond in Cl_2, uniform electron density is represented by an evenly yellow-green surface, indicating that the two atoms share their bonding pair of electrons equally. (b) Unequal sharing of the bonding pair of electrons occurs in HCl, as shown by the orange-red color of the Cl (δ−) and the blue-green color around the H (δ+). (c) In ionic NaCl, the violet color on the surface of the sodium ion indicates that it has a full 1+ charge, and the red of the chloride ion reflects its charge of 1−.

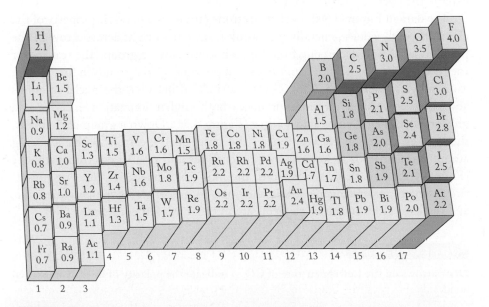

FIGURE 4.8 Electronegativity values of the elements. Electronegativity increases from left to right across a period and decreases from top to bottom down a group. The greater an element's electronegativity, the greater is that atom's ability to attract electrons in a bond.

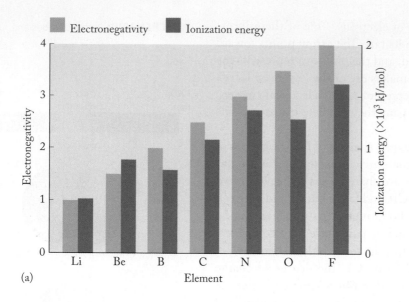

(a)

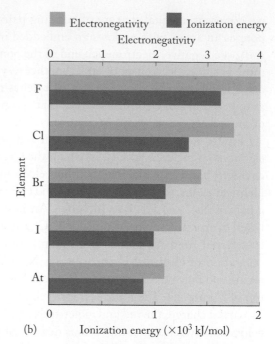

(b)

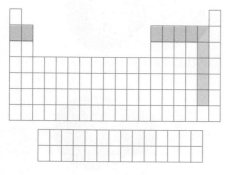

FIGURE 4.9 The trends in the electronegativities of the main group elements follow those of ionization energies: both tend to (a) increase with increasing atomic number across a row and (b) decrease with increasing atomic number within a group.

▶❚❚ **CHEMTOUR** The Periodic Table

When $\Delta\chi$ is less than or equal to 0.4, a covalent bond is considered essentially nonpolar. The bond between Cl ($\chi = 3.0$) and Br ($\chi = 2.8$) is in that category. When $\Delta\chi$ is between 0.4 and 2.0, bonds are considered polar covalent, and when $\Delta\chi$ is equal to or greater than 2.0, bonds are considered ionic. Thus, the bonds in HF and HCl are considered polar covalent, and a H—F bond is *more* polar than a H—Cl bond because the $\Delta\chi$ between H (2.1) and F (4.0) is greater than the $\Delta\chi$ between H (2.1) and Cl (3.0). Calcium oxide is considered an ionic compound because the $\Delta\chi$ between Ca (1.0) and O (3.5) is 2.5. Keep in mind that these cutoff values are not strict limits, but more like guidelines. We revisit the characterization of bonds as either ionic or polar covalent in Chapter 5, where we explore the overall polarity of entire molecules.

The data in Figure 4.8 show that electronegativity is a periodic property of the elements, with values generally increasing from left to right across a row in the periodic table, and decreasing from top to bottom down a group. The reasons for the trends are essentially the same ones that produce similar trends in first ionization energies (see Figure 3.36). Greater attraction between the nuclei of atoms and their outer-shell electrons produces both higher ionization energies and greater electronegativities across a row (Figure 4.9a). Down a group of elements, the weaker attraction between nuclei and outer-shell electrons with increasing atomic number leads to lower ionization energies and smaller electronegativities (Figure 4.9b). For those two reasons, the most electronegative elements—fluorine, oxygen, and nitrogen—are in the upper right corner of the periodic table, and the least electronegative elements are in the lower left corner.

CONCEPT TEST

Draw arrows on the Lewis structure of CO to indicate the polarity of its carbon–oxygen bond.

(Answers to Concept Tests are in the back of the book.)

SAMPLE EXERCISE 4.12 Comparing Bond Polarities   LO5

Rank, in order of increasing polarity, the bonds formed between O and C, Cl and Ca, N and S, O and Si. Are any of the bonds considered ionic?

COLLECT AND ORGANIZE We are given four pairs of elements and asked to rank the pairs according to the polarity of the bond each pair forms and to identify any ionic bonds in the set. We need to refer to the elements' electronegativities (Figure 4.8) to judge the relative polarity.

ANALYZE The polarity of a bond is related to the difference in electronegativity between the atoms in the bond. The guidelines we follow are

$$\Delta\chi \leq 0.4 \qquad \text{Nonpolar covalent}$$
$$0.4 < \Delta\chi < 2.0 \qquad \text{Polar covalent}$$
$$\Delta\chi \geq 2.0 \qquad \text{Ionic}$$

SOLVE Calculate the electronegativity difference between the atoms:

O and C: $\qquad \Delta\chi = 3.5 - 2.5 = 1.0$

Cl and Ca: $\qquad \Delta\chi = 3.0 - 1.0 = 2.0$

N and S: $\qquad \Delta\chi = 3.0 - 2.5 = 0.5$

O and Si: $\qquad \Delta\chi = 3.5 - 1.8 = 1.7$

The electronegativity differences are proportional to the polarity of the bonds formed between the pairs of atoms. Therefore, ranking them in order of increasing polarity we have:

$$\text{N—S} < \text{O—C} < \text{O—Si} < \text{Cl—Ca}$$

The bond between Cl and Ca is so polar it is considered ionic because $\Delta\chi = 2.0$.

THINK ABOUT IT Calcium is a metal and chlorine is a nonmetal, so the conclusion that the bond between them is ionic is reasonable. Ionic bonds tend to be formed between metals and nonmetals. Two of the other bonds, N—S and O—C, connect pairs of nonmetals, and the O—Si bond connects a nonmetal with a metalloid. We expect those three bonds to be covalent.

Practice Exercise Which pair forms the most polar bond: O and S, Be and Cl, N and H, or C and Br? Is the bond between that pair ionic? ⚙

(Answers to Practice Exercises are in the back of the book.)

4.5 Vibrating Bonds and the Greenhouse Effect

Covalent bonds are not rigid. They all vibrate a little, stretching and bending like tiny atomic-sized springs (Figure 4.10). As polar bonds vibrate, the strengths of tiny electrical fields produced by the partial separations of charge in the bonds fluctuate at the same frequencies as their vibrations. The natural frequencies of the vibrations correspond to frequencies of infrared radiation. As we described in Chapter 3, all forms of radiant energy, including infrared rays, travel through space as oscillating electrical and magnetic fields. Now, suppose a photon of infrared radiation traveling through Earth's atmosphere strikes a molecule containing a polar bond that is vibrating at exactly the same frequency as the photon. The fluctuating fields of the photon and the vibrating bond can interact. The molecule

▶❚❚ **CHEMTOUR** Greenhouse Effect

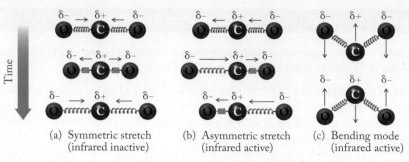

(a) Symmetric stretch (infrared inactive) (b) Asymmetric stretch (infrared active) (c) Bending mode (infrared active)

FIGURE 4.10 Three modes of bond vibration in a molecule of CO_2 include (a) symmetric stretching of the C=O bonds, which produces no overall change in the polarity of the molecule. (b) Asymmetric stretching does produce side-to-side fluctuations in polarity that may result in absorption of IR radiation. (c) The bending mode produces up-and-down fluctuations that may also absorb IR radiation.

▶❚❚ **CHEMTOUR** Vibrational Modes

∞ **CONNECTION** The average temperature of Earth's surface is 287 K, which means that it emits its peak intensity of electromagnetic radiation in the infrared region (see Figure 3.11).

may absorb that photon, temporarily raising its internal energy, and later emit a photon of the same energy as it returns to its ground state. This molecule–photon interaction is at the heart of the greenhouse effect.

To understand the connection between vibrating bonds and a warming atmosphere, we need to recall that infrared radiation is the part of the electromagnetic spectrum that we cannot see but can feel as heat. Any warm object, including Earth's surface, emits infrared radiation. When infrared photons strike atmospheric molecules such as CO_2 that contain polar bonds, the photons may be absorbed. When they are reemitted, they are just as likely to be moving back toward Earth's surface as upward toward space. In this process of absorption and reemission, a significant fraction of the heat flowing from Earth's surface is trapped in the atmosphere, much in the way that the glass covering a greenhouse traps heat inside it.

Not all polar bond vibrations result in absorption and emission of infrared radiation. For example, two kinds of stretching vibrations can occur in a molecule of CO_2, which has two C=O bonds. One is a *symmetric* stretching vibration (Figure 4.10a) in which the two C=O bonds stretch and then compress at the same time. In that case the two fluctuating electrical fields produced by the two C=O bonds cancel each other out, and no infrared absorption or emission is possible. This vibration is said to be *infrared inactive*. However, when the bonds stretch such that one gets shorter as the other gets longer (Figure 4.10b), the changes in charge separation do not cancel. This *asymmetric* stretch produces a fluctuating electrical field that enables CO_2 to absorb infrared radiation, so the vibration is *infrared active*. Molecules can also bend (Figure 4.10c) to produce fluctuating electrical fields that match the frequencies of other photons of infrared radiation.

CONCEPT TEST ●

Nitrogen and oxygen make up about 99% of the gases in the atmosphere. Is the stretching of the N≡N and O=O bonds in the molecules infrared active? Why or why not?

(Answers to Concept Tests are in the back of the book.)

4.6 Resonance

The atmosphere contains two kinds of oxygen molecules. Most of them are O_2, but a tiny fraction are O_3, which is a form of oxygen called ozone. Different molecular forms of the element are called **allotropes**, and they have different chemical and physical properties. Ozone, for example, is an acrid, pale blue gas that is toxic even at low concentrations, whereas O_2 is a colorless, odorless gas that is essential for most life-forms. Ozone is produced naturally by lightning (Figure 4.11) and is the source of the pungent odor you may have smelled after a severe thunderstorm. Ozone in the lower atmosphere is sometimes referred to as "bad ozone" because high levels in polluted air can damage crops, harm trees, and cause human health problems. Ozone is also present in the upper atmosphere, where exactly the same substance is considered "good ozone" because it shields life on Earth from potentially harmful ultraviolet radiation from the sun.

Let's draw the Lewis structure of ozone by following our five-step process from Section 4.3. Oxygen is a group 16 element and so has 6 valence electrons. Therefore the total number of valence electrons in an ozone molecule is $3 \times 6 = 18$ (step 1). Connecting the three O atoms with single bonds, we have (step 2)

$$O—O—O$$

Completing the octets of the atoms on the ends gives us (step 3)

$$:\ddot{O}—O—\ddot{O}:$$

The structure contains 16 electrons. We determined in step 1 that there are 18 valence electrons in the molecule, so we add a lone pair to the central oxygen atom (step 4):

$$:\ddot{O}—\ddot{O}—\ddot{O}:$$

This structure leaves the central atom 2 electrons short, so we convert one of the lone pairs on the O atom on the left end of the molecule into a bonding pair (step 5):

We could have used a lone pair from the O atom on the right, which would have given us

The two structures illustrate an important concept in Lewis theory called **resonance**: the existence of multiple Lewis structures, called **resonance structures**, which have the same arrangement of atoms but have bonding electrons and lone pairs of electrons in different positions.

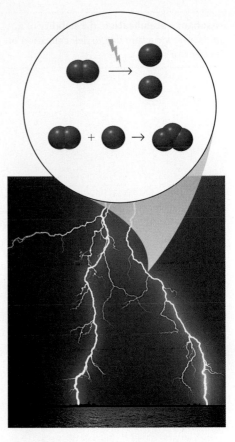

FIGURE 4.11 Lightning strikes contain sufficient energy to break oxygen–oxygen double bonds. The O atoms formed in this fashion collide with other O_2 molecules, forming ozone (O_3), an allotrope of oxygen.

▶❚❚ **CHEMTOUR** Resonance

allotropes different molecular forms of the same element, such as oxygen (O_2) and ozone (O_3).

resonance characteristic of electron distributions when two or more equivalent Lewis structures can be drawn for one compound.

resonance structure one of two or more Lewis structures with the same arrangement of atoms but different arrangements of bonding pairs of electrons.

resonance stabilization the stability of a molecular structure due to delocalization of its electrons.

Not only are resonance structures equivalent; they are interconvertible. To see how, consider what happens when the bonding electrons and the lone-pair electrons in structure (a) below move about as shown by the red arrows:

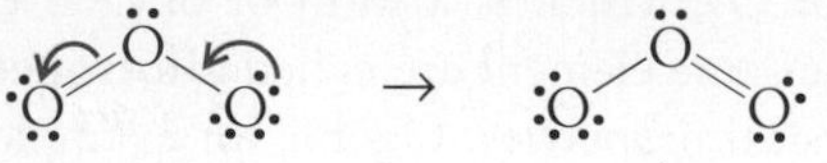

Note how this rearrangement of electron pairs produces structure (b). The process is completely reversible: the electron pairs in (b) can just as easily be rearranged to produce the pattern in (a):

We often represent the interconversion of resonance forms by drawing a double-headed arrow between them:

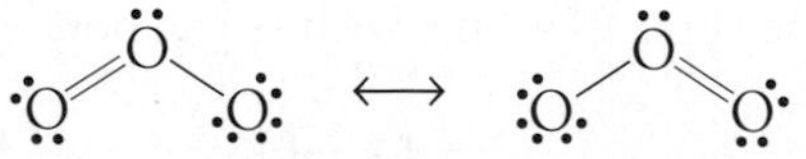

So, which of the two structures is correct? Experimental evidence indicates that, technically, neither is. Scientists have determined that the two bonds in ozone have exactly the same length, 128 pm. As we will see in Section 4.9, this value is about halfway between the length of an O—O single bond (148 pm) and an O=O double bond (121 pm). One way to explain this result is to assume that the actual bonding in an O_3 molecule is the average of the two resonance structures, which means identical bonds between the O atoms that are intermediate in length and strength between O—O single and O=O double bonds. It is as if two bonding pairs of electrons connect the center atom to the two others and then a third bonding pair is spread out across all three atoms.

Pairs of electrons that are spread out among atoms are said to be *delocalized*. A key point about delocalization is that it reduces the electrons' potential energy and makes the molecule more stable, a phenomenon called **resonance stabilization**. We will see in the chapters ahead that resonance can strongly influence the chemical properties of molecular substances.

Resonance may also occur in polyatomic ions, as illustrated in Sample Exercise 4.13. In drawing the Lewis structures of polyatomic ions, we need to account for their overall electrical charges. This means adding the appropriate number of valence electrons to polyatomic anions or subtracting the appropriate number from polyatomic cations.

SAMPLE EXERCISE 4.13 **Drawing Resonance Structures** **LO6**

Draw all the resonance structures of the nitrate ion, NO_3^-.

COLLECT AND ORGANIZE We are asked to draw the resonance structures of the NO_3^- ion, which contains one nitrogen atom and three oxygen atoms. The charge of the ion is 1−.

ANALYZE Nitrogen is a group 15 element and has 5 valence electrons per atom and a bonding capacity of 3. Oxygen is in group 16 and has 6 valence electrons and a bonding capacity of 2. The 1− charge means there is an additional valence electron in

t[...]ving resonance structures we expect to change positions of double and
si[...]

[...]ber of valence electrons is

	N	+	3 O
[...] electrons:	5	+	$(3 \times 6) = 23$
[...]al electron due to the 1− charge:			1
[...]lence electrons:			24

[...] has the higher bonding capacity, so N is the central atom. Connecting
[...]ingle bonds to the three O atoms gives us this skeletal structure:

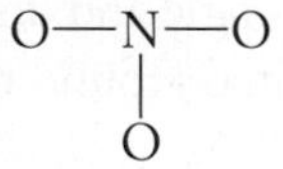

$$O\!-\!N\!-\!O$$
$$|$$
$$O$$

3. Each O atom needs three lone pairs of electrons to complete its octet:

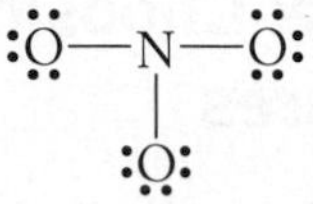

4. There are 24 valence electrons in this structure, which matches the number
determined in step 1.
5. The central N atom has only 6 electrons. To provide it with the 2 more that
it needs to complete its octet, we convert a lone pair on one of the oxygen
atoms into a bonding pair:

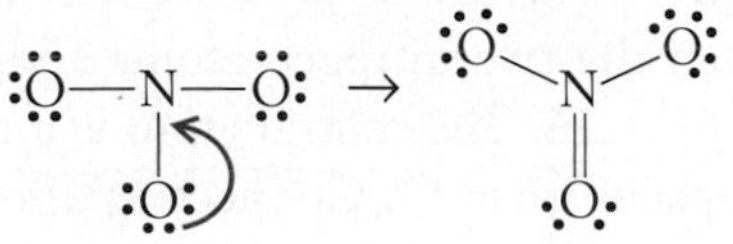

The nitrogen atom now has a complete octet. Adding brackets and the ionic
charge, we have a complete Lewis structure:

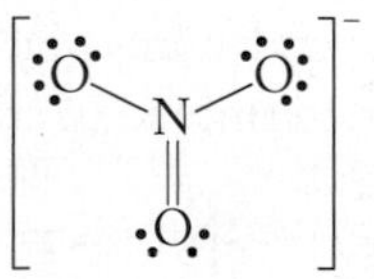

Using the O atom to the left or right of the central N atom to form the double bond
creates two additional resonance forms, or three in all:

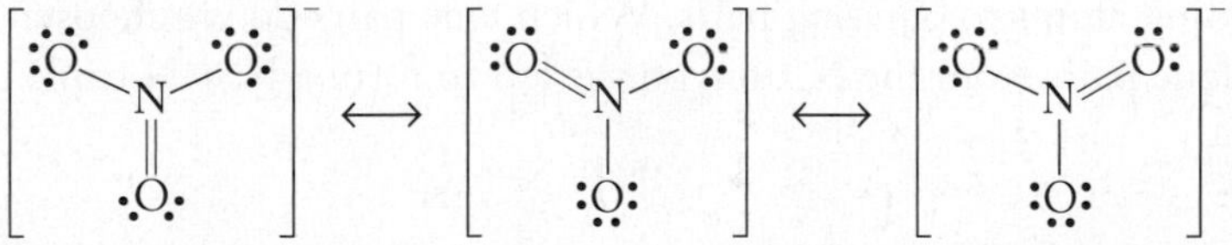

THINK ABOUT IT It makes sense that there are three resonance forms of the NO_3^-
ion because there are three equivalent O atoms bonded to the central N atom, and
any one of the O atoms could be the one with the double bond. The Lewis symbol
of nitrogen indicates that it has a bonding capacity of 3, and N atoms typically form
three bonds in neutral molecules such as NH_3 (see Sample Exercise 4.10). However,
the central N atom forms four bonds in a nitrate oxoanion.

Practice Exercise Draw all the resonance forms of the azide ion, N_3^-, and the nitronium ion, NO_2^+.

(Answers to Practice Exercises are in the back of the book.)

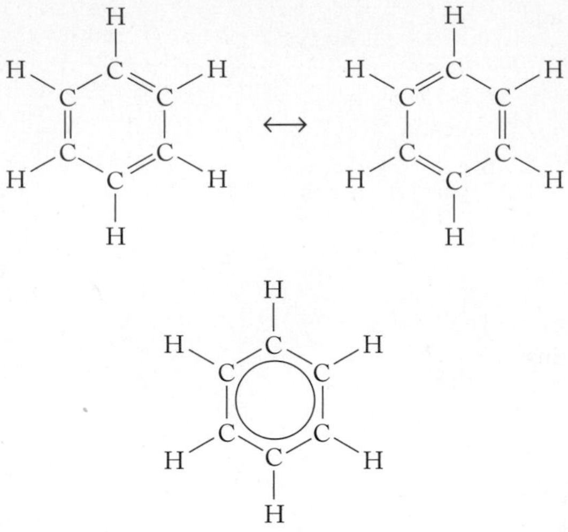

FIGURE 4.12 The molecular structure of benzene is an average of the two equivalent structures at the top. It is frequently represented by a circle inside the hexagonal ring indicating completely uniform distribution of the electrons in the bonds around the ring.

A test of whether a compound has multiple equivalent resonance forms—and therefore exhibits resonance stabilization—is the presence of one or more atoms having both single and double bonds to two or more atoms of another element, as in the ozone molecule and the nitrate ion. A molecule of benzene (C_6H_6) also has that property, as it contains a ring of six carbon atoms with alternating single and double bonds (Figure 4.12). Assuming the atoms don't move, there are two equivalent ways to draw the single and double bonds. Chemists frequently draw benzene molecules with circles in their centers, as shown in Figure 4.12, to represent an averaging of the two resonance structures. The symbol emphasizes that the six carbon–carbon bonds in the ring are all identical and intermediate in character between single and double bonds, and that bonding electrons are uniformly distributed around the ring.

4.7 Formal Charge: Choosing among Lewis Structures

Let's turn our attention to the molecular structure of another atmospheric gas, dinitrogen monoxide (N_2O), also known as nitrous oxide or, more commonly, laughing gas. Its common name is derived from the euphoria people feel when they inhale it; that reaction is closely linked to the use of N_2O as an anesthetic in dentistry and medicine.

To draw the Lewis structure of N_2O we first count the number of valence electrons: 5 each from the two nitrogen atoms and 6 from the oxygen atom for a total of $(2 \times 5 + 6) = 16$. The central atom is a nitrogen atom because N has a higher bonding capacity than O. Connecting the atoms with single bonds we have

$$N—N—O$$

Completing the octets of the end atoms gives us a structure with 16 valence electrons, which is the number determined in step 1:

$$:\ddot{N}—N—\ddot{O}:$$

However, there are only two bonds and thus only 4 valence electrons on the central N atom. To give it 4 more electrons, we need to convert lone pairs on the surrounding atoms to bonding pairs. Which lone pairs do we choose? We could use two lone pairs from the N atom on the left to form a N≡N triple bond:

$$:N≡N—\ddot{O}:$$

or we could use two lone pairs from the O atom to make a N≡O triple bond:

$$:\ddot{N}—N≡O:$$

or we could use one pair from each terminal atom to make two double bonds:

$$:\ddot{N}=N=\ddot{O}:$$

Which of the resonance structures best represents the actual bonding pattern inside N_2O? We have seen that in some sets of resonance structures, such as those of O_3 and NO_3^-, all the structures are equivalent, so no one of them is more important than another in giving us a sense of the actual bonding in the molecule. That is not the case with the nonequivalent resonance forms of N_2O. To help us decide which resonance form in a nonequivalent set is the most important in representing the actual bonding pattern in a molecule, we make use of the concept of formal charge.

A **formal charge (FC)** is not a real charge but rather a measure of the number of electrons *formally assigned* to an atom in a molecular structure. To understand what formal charge means, let's go through the process of calculating it for an atom in the Lewis structure of a molecule or polyatomic ion:

formal charge (FC) value calculated for an atom in a molecule or polyatomic ion by determining the difference between the number of valence electrons in the free atom and the sum of lone-pair electrons plus half of the electrons in the atom's bonding pairs.

Calculating Formal Charge

1. Determine the number of valence electrons in the free atom (the number of dots in its Lewis symbol).
2. Count the number of lone-pair electrons on the atom in the structure.
3. Count the number of electrons in bonds to the atom and divide that number by 2.
4. Sum the results of steps 2 and 3, and subtract that sum from the number determined in step 1.

Summarizing these steps in the form of an equation we have

$$FC = \left(\begin{array}{c} \text{number of} \\ \text{valence } e^- \end{array} \right) - \left[\begin{array}{c} \text{number of} \\ \text{unshared } e^- \end{array} + \frac{1}{2} \left(\begin{array}{c} \text{number of } e^- \\ \text{in bonding pairs} \end{array} \right) \right] \qquad (4.2)$$

The calculation of formal charge assumes that each atom is *formally assigned* (or can claim ownership of) all the electrons in its lone pairs and half of the electrons it shares in bonding pairs. If this number matches the number of valence electrons in a single atom of the element, as reflected in its Lewis symbol, then the formal charge on it is zero.

We can confirm that the three resonance forms of N_2O we have drawn are not equivalent by calculating the formal charges on the atoms in each structure. The steps in this calculation are highlighted in the table below, where we have colored the lone pairs of electrons red and the shared pairs green to make it easier to track the quantities of the electrons in the formal charge calculations. The numbers of valence electrons in free atoms of N (5) and O (6) are shown in blue.

Formal Charge Calculations for the Resonance Structures of N_2O									
Step	:N≡N—Ö:			N̈=N=Ö			:N̈—N≡O:		
1 Number of valence electrons	5	5	6	5	5	6	5	5	6
2 Number of lone pair electrons	2	0	6	4	0	4	6	0	2
3 Number of shared electrons	6	8	2	4	8	4	2	8	6
4 FC = valence − [lone pair + ½ (shared)]	0	+1	−1	−1	+1	0	−2	+1	+1

To illustrate one of the formal charge calculations in the table, consider the N atom at the left end of the left resonance structure. It has 2 electrons in a lone pair and 6 electrons in three shared (bonding) pairs. Using Equation 4.2 to calculate the formal charge on this N atom,

$$FC = 5 - [2 + \tfrac{1}{2}(6)] = 0$$

which is the first value in the bottom row of the table. The results of similar FC calculations for all the other atoms in the three resonance structures complete the row in the table. Note that the sum of the formal charges on the three atoms in all three structures is zero, as it should be for a neutral molecule. When we do an analysis of the formal charges of atoms in a polyatomic ion, the formal charges on its atoms must add up to the charge on the ion.

Having done the formal charge calculations, we now need to use the results to decide which of the three N_2O structures best represents the actual bonding pattern. To make this decision for the three N_2O structures, or for any set of nonequivalent resonance structures, we use the following three general criteria:

1. The best molecular structure is the one in which the formal charge on each atom is zero.
2. If no such structure can be drawn, or if the structure is that of a polyatomic ion, then the best structure is the one with the most formal charges equal to zero or closest to zero.
3. Any negative formal charges should be on the atom(s) of the more/most electronegative element.

Let's apply these criteria to the nonequivalent resonance structures of N_2O. First of all, none of them meet criterion (1) because each structure has at least two nonzero formal charge values. So we proceed to criterion (2) to find the structure with the most FC values that are closest to zero, such as −1 or +1. On this count we have a tie between the structure on the left (0, +1, −1) and the one in the middle (−1, +1, 0).

To break the tie, we invoke criterion (3) and answer the question, "In which structure is the negative formal charge on the more electronegative atom?" The answer is the structure on the left, in which the formal charge on the atom of the more electronegative element, O, is −1. We conclude that the structure on the left is the best of the three in representing the actual bonding in a molecule of N_2O.

In reality, it is known from experimental measurements that the middle structure also contributes to the bonding in N_2O. For example, the length of the bond between the two nitrogen atoms is between the length of a N=N bond and the length of a N≡N bond. This reality check is important as we interpret the results of formal charge analyses. Just because a resonance structure scores the best in an FC analysis *does not mean that other structures don't also influence the actual bonding pattern.*

CONCEPT TEST ··

What is the formal charge on a sulfur atom that has three lone pairs of electrons and one bonding pair?

(Answers to Concept Tests are in the back of the book.)

**SAMPLE EXERCISE 4.14   Selecting a Resonance Structure   LO6
Based on Formal Charges**

Which of these resonance forms best describes the actual bonding in a molecule of CO_2?

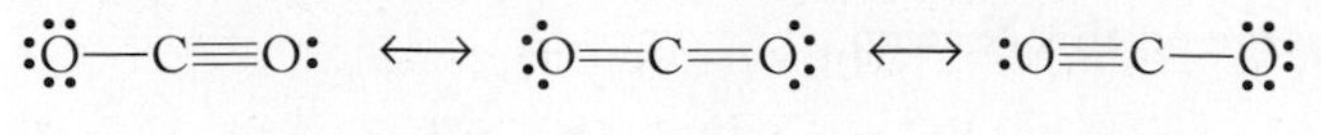

$$:\ddot{O}-C\equiv O: \quad \longleftrightarrow \quad :\ddot{O}=C=\ddot{O}: \quad \longleftrightarrow \quad :O\equiv C-\ddot{O}:$$

COLLECT AND ORGANIZE We are given three resonance forms for CO_2. Formal charges can be used to select the most representative structure.

ANALYZE The preferred structure is one in which the formal charges are closest to zero and any negative formal charges are on the more electronegative atom. In this case, oxygen is a group 16 element and is more electronegative than carbon, a group 14 element. Each free carbon atom has 4 valence electrons, and free oxygen atoms have 6 valence electrons each.

SOLVE We use Equation 4.2 to find the formal charge on each atom. We illustrate the results in a table, applying the same color coding scheme used for the earlier N_2O calculations:

Formal Charge Calculations for the Resonance Structures of CO_2									
Step	:Ö—C≡O:			:Ö=C=Ö:			:O≡C—Ö:		
1 Number of valence electrons	6	4	6	6	4	6	6	4	6
2 Number of lone pair electrons	6	0	2	4	0	4	2	0	6
3 Number of shared electrons	2	8	6	4	8	4	6	8	2
4 FC = valence − [lone pair + $\frac{1}{2}$ (shared)]	−1	0	+1	0	0	0	+1	0	−1

The formal charges are all zero on the atoms in the middle resonance form with the two double bonds. Therefore this structure best represents the actual bonding in a molecule of CO_2.

THINK ABOUT IT Notice that the sum of the formal charges is zero in all three resonance forms, as it should be for a neutral molecule. In this case we have a clear winner in our FC analysis of resonance structures, so we may conclude that the two losing structures do not significantly influence the actual bonding in CO_2.

Practice Exercise Which resonance forms of the azide ion, N_3^-, and the nitronium ion, NO_2^+, contribute the most to the actual bonding in each ion? ⚙

(Answers to Practice Exercises are in the back of the book.)

Let's take a final look at the resonance structures for N_2O. Our purpose here is to examine the link between the formal charge on an atom in a resonance structure and the bonding capacity of that atom. The Lewis symbol of nitrogen, which has 3 unpaired electrons, tells us that a nitrogen atom can complete its octet by forming 3 bonds. The 2 unpaired electrons in the Lewis symbol of oxygen tell us that the bonding capacity of an oxygen atom is 2. In the N_2O calculations, the nitrogen atom with 3 bonds has a formal charge of zero (left N in the table's first structure), and the oxygen atom with 2 bonds has a formal charge of zero (O in the second structure). As a general rule—and assuming the octet rule is obeyed—atoms have formal charges of zero in resonance structures in which the numbers of bonds they form match their bonding capacities. If the number of bonds an atom forms is one more than its bonding capacity, such as an oxygen atom with 3 bonds (O in the third structure), the formal charge is +1. If the number of bonds it forms is one fewer than the bonding capacity, such as an oxygen atom with 1 bond (O in the first structure), the formal charge is −1.

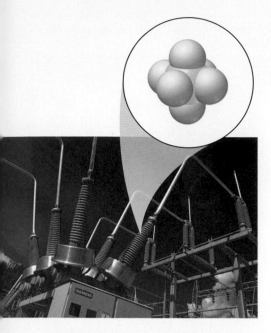

FIGURE 4.13 Electrical transformers use SF_6 as an insulator because it is thermally stable and does not react with water. When it leaks into the atmosphere it becomes a potent greenhouse gas. It absorbs much more infrared radiation than CO_2 and lasts for thousands of years in the atmosphere.

4.8 Exceptions to the Octet Rule

Earth's atmosphere contains trace concentrations of several compounds that illustrate the limitations of the octet rule. They include two nitrogen oxides, NO and NO_2, that contribute to photochemical smog formation in urban areas. Each has an odd number of valence electrons per molecule, which means that at least one atom in each molecule cannot have a complete octet. Another compound of environmental concern is sulfur hexafluoride (SF_6), which is the most potent greenhouse gas present in the atmosphere (Figure 4.13). Each of its molecules contains six sulfur–fluorine covalent bonds, which means that, according to the Lewis model of one pair of electrons per bond, each S atom is surrounded by 12 valence electrons, not 8.

Odd-Electron Molecules

Traces of nitrogen monoxide (NO) enter the atmosphere from automobile exhaust and then react with O_2, forming nitrogen dioxide, NO_2. Both of these oxides of nitrogen are odd-electron molecules. To understand what that means, let's draw their Lewis structures. NO has 11 valence electrons: nitrogen contributes 5 and oxygen 6. There is no central atom, so we start with a single bond between N and O and then complete the octet around O, which is the more electronegative element:

$$N\!\!-\!\!\ddot{\underset{\displaystyle ..}{O}}:$$

We then place the remaining 3 electrons around the N atom:

$$\underset{\displaystyle ..}{N}\!\!-\!\!\ddot{\underset{\displaystyle ..}{O}}:$$

That leaves the N atom short of valence electrons. We can increase its number by converting a lone pair on the O atom into a bonding pair:

$$\cdot\dot{N}\!\!=\!\!\ddot{O}:$$

The change has the added advantage of creating a double-bonded O atom, which gives it a formal charge of zero. The formal charge on nitrogen is also zero.

The only problem with the structure is that nitrogen does not have an octet: it has only 7 electrons. Is there a way to give it 8 electrons? Not really. We could use another of the O atom lone pairs to make a bonding pair:

$$\cdot\dot{N}\!\!\Rrightarrow\!\!O:$$

but the N atom in the structure has 9 valence-shell electrons, which is impossible for an atom with only four orbitals in its valence shell. Moving an electron from the O atom to the N atom does not solve the problem of having at least one atom with only 7 valence electrons:

$$:\dot{N}\!\!\Rrightarrow\!\!\dot{O}.$$

In fact, it makes it worse, because in this structure the atom of the more electronegative element (O), which attracts bonding electrons more strongly than the other (N), is the one deprived of a complete octet. Because nitrogen is less electronegative than oxygen, it is reasonable that we short-change it when there are not enough electrons to complete the octets of both. Compounds that have odd numbers of valence electrons are called **free radicals**. They are typically very reactive species because it is often energetically favorable for them to acquire or share an electron from another molecule or ion.

free radical an odd-electron atom, ion, or molecule.

**SAMPLE EXERCISE 4.15   Drawing the Lewis Structures   LO6
                          of an Odd-Electron Molecule**

Draw the resonance structures of nitrogen dioxide (NO_2) and assign formal charges to the atoms.

COLLECT AND ORGANIZE We need to draw the resonance forms of NO_2 and use formal charges to determine which structure best reflects the actual bonding in the molecule. Each molecule contains one atom of nitrogen (bonding capacity = 3) and two atoms of oxygen (bonding capacity = 2).

ANALYZE A molecule of NO_2 contains one atom of an element with an odd atomic number, so it is an odd-electron molecule, which means one of the atoms will not have a complete octet.

SOLVE The number of valence electrons is

Element:	N	+	2 O
Valence electrons:	5	+	$(2 \times 6) = 17$

Nitrogen has the greater bonding capacity, so it is the central atom:

$$O—N—O$$

Completing the octets on the O atoms gives

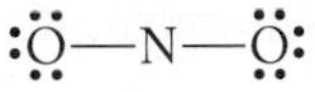

$$:\ddot{O}—N—\ddot{O}:$$

There are 16 valence electrons in the structure, but we need 17 to match the number available in the molecule. We add 1 more electron to the N atom:

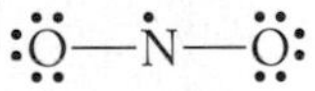

$$:\ddot{O}—\dot{N}—\ddot{O}:$$

There are only 5 valence electrons around the N atom, fewer than the 8 we need. We can increase the number by converting a lone pair on one of the O atoms to a bonding pair, giving the formal charges shown in red:

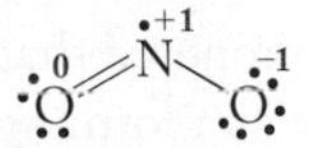

An equivalent resonance form can be drawn with the double bond on the right side:

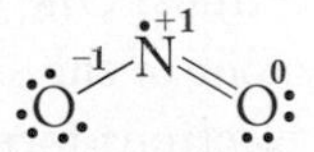

THINK ABOUT IT The two Lewis structures are equivalent because the two O atoms in each structure are equivalent. The structures do not satisfy the octet rule, but the formal charges of the atoms are close to zero, and the negative formal charge is on an atom of the more electronegative element. Both O atoms have complete octets, leaving the less electronegative N atom one electron short in the odd-electron molecule.

Practice Exercise Nitrogen trioxide (NO_3) may form in polluted air when NO_2 reacts with O_3. Draw its Lewis structure(s).

(Answers to Practice Exercises are in the back of the book.)

Expanded Octets

▶‖ **CHEMTOUR** Expanded Valence Shells

Atoms of the nonmetals in the third row and below in the periodic table may form molecules in which they appear to have more than an octet of valence electrons. Consider, for example, the Lewis structure of SF_6. Its valence electron inventory (step 1) is

Element:	S	+	6 F
Valence electrons:	6	+	$(6 \times 7) = 48$

Sulfur has a greater bonding capacity (2) than fluorine (1), so S is the central atom (step 2). However, connecting six fluorine atoms to the sulfur atom means that sulfur's Lewis bonding capacity must increase from 2 to 6:

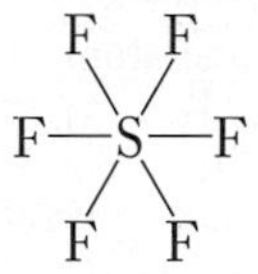

$$\begin{array}{c} \text{F} \quad \text{F} \\ \diagdown \diagup \\ \text{F} \!-\! \text{S} \!-\! \text{F} \\ \diagup \diagdown \\ \text{F} \quad \text{F} \end{array}$$

Completing the octets on the fluorine atoms (step 3),

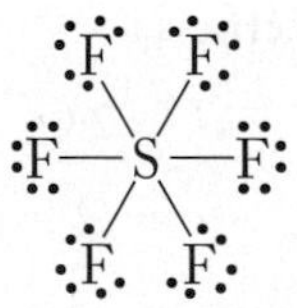

gives us a structure with 48 valence electrons (step 4), which matches the number the molecule should have. However, forming the six covalent bonds to six F atoms means that the S atom must have *expanded its valence shell* to accommodate a total of six pairs or 12 electrons. An SF_6 molecule with six covalent S—F bonds has zero formal charge on each atom:

$$\text{Formal charge of S} = 6 - [0 + \tfrac{1}{2}(12)] = 0$$

$$\text{Formal charge of F} = 7 - [6 + \tfrac{1}{2}(2)] = 0$$

Based on our criteria for judging Lewis structures, the structure is a perfectly acceptable one, but how can a sulfur atom have more than 8 valence electrons?

Over the years chemists have devised several explanations for what many have called the *hypervalency* of sulfur and other nonmetals in the third row and below in the periodic table. Some have proposed that these elements expand their octets by incorporating valence *d* orbitals in forming covalent bonds. Others have proposed that hypervalency is really an illusion and that atoms can form five or six bonds using *only eight* valence electrons. One of the ways they explain the bonding in SF_6 is to assume that only four of the six S—F bonds are covalent bonds. Those four bonds and 8 valence electrons form the polyatomic ion $SF_4{}^{2+}$, which has this Lewis structure:

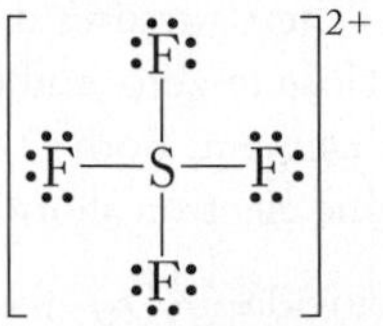

To obtain a particle with the formula SF_6, we combine the SF_4^{2+} ion with two F^- ions:

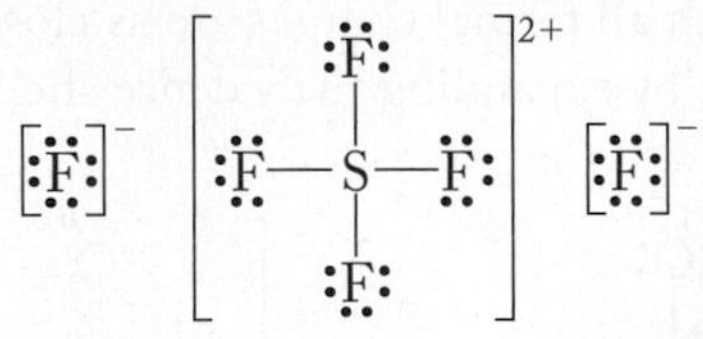

The Lewis structure may be misleading in that it casts SF_6 as an ionic compound when it's not. We know that because it has the properties of a molecular compound. For example, SF_6 is a gas at room temperature, not a solid (which all ionic compounds are). In addition, spectroscopic analyses indicate that all six S—F bonds in a molecule of SF_6 are exactly the same. How is that possible when two bonds are theoretically ionic? Resonance provides the answer. Consider the three resonance structures below. The four covalent bonds are blue and the two ionic bonds are red:

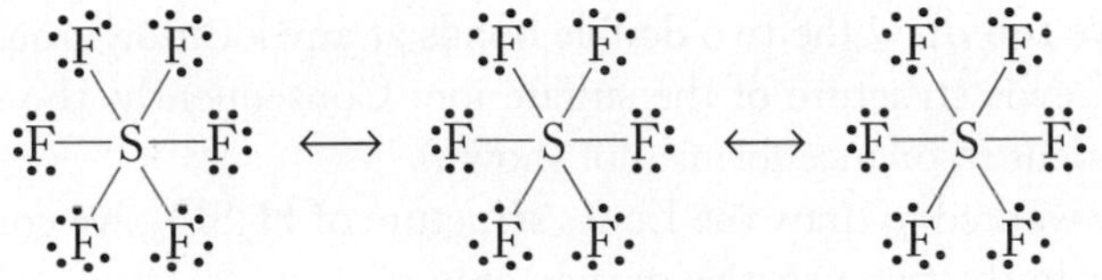

Note how the positions of the two kinds of bonds are not confined to particular pairs of S and F atoms. This means that each S—F bond has, on average, $\frac{1}{3}$ of the character of an ionic bond and $\frac{2}{3}$ of the character of a covalent bond. The proportions mean that the bonds are polar covalent, not ionic, and are consistent with the molecular properties of SF_6.

The fact that some atoms *appear* to have the ability to expand their valence shell does not mean the atoms always make use of the ability. Instead, they tend to do so in two situations:

1. When they form compounds with strongly electronegative elements, particularly F, O, and Cl.
2. When an expanded shell results in smaller formal charges on the atoms in a molecule.

Let's explore a case in which expanding the valence shell of a central atom (sulfur again) produces a Lewis structure with more favorable (closer-to-zero) formal charges. Our example is the sulfate ion, SO_4^{2-}, the oxoanion that is formed when sulfuric acid, H_2SO_4, a principal component of acidic precipitation in eastern North America and Europe, dissolves in water and releases two H^+ ions.

In the Lewis structure of SO_4^{2-}, the central S atom is bonded to four O atoms. Following the usual procedures for drawing Lewis structures and assigning formal charges, we get

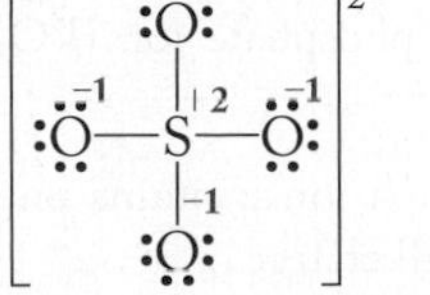

The sum of the formal charges on atoms in the ion is $1(+2) + 4(-1) = -2$. The calculation yields the correct ionic overall charge, but remember that the goal is a Lewis structure in which all formal charges are as close to zero as possible. We can draw such a structure by expanding the valence shell of sulfur:

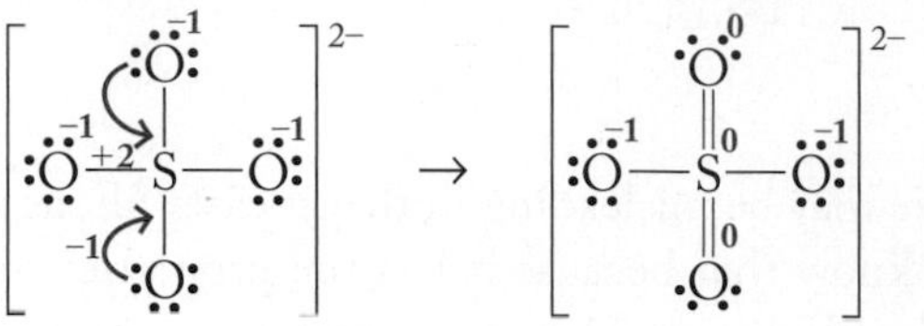

Note that each oxygen atom still has a complete octet, but now the sulfur has expanded its valence shell to accommodate 12 electrons. In this way, the formal charges change from +2 on sulfur and −1 on each of the four oxygen atoms to 0 on sulfur, 0 on two of the oxygen atoms, and −1 on the other two oxygen atoms. The values sum to −2, which is the value required to give the structure its overall 2− charge. We can draw the two double bonds at any location around the sulfur atom in the Lewis structure of the sulfate ion. Consequently, the structure has several equivalent resonance forms (not shown).

If we now wanted to draw the Lewis structure of H_2SO_4, we could bond two hydrogen ions to the two negative oxygen atoms:

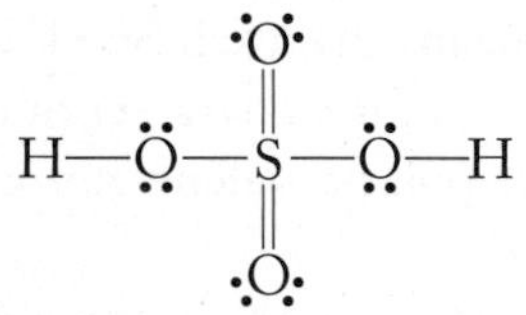

Each hydrogen atom has achieved a duet of electrons, each oxygen atom has an octet, and every atom has a formal charge of zero.

So, based on the above formal charge analyses, we might conclude that the preferred bonding pattern in SO_4^{2-} ions and in molecules of H_2SO_4 includes two S=O double bonds and zero formal charges all around. If only bonding were that simple. In reality, the two structures with zero formal charges do contribute to the bonding in the species, but structures that obey the octet rule and have no S=O double bonds seem to contribute as well. There is evidence that the actual bonding is an average of both the expanded octet and normal octet options.

SAMPLE EXERCISE 4.16 **Drawing a Lewis Structure Containing an Atom with an Expanded Valence Shell** **LO6**

Draw a Lewis structure for the phosphate ion (PO_4^{3-}) that minimizes the formal charges on its atoms.

COLLECT AND ORGANIZE Each ion contains one atom of phosphorus and four atoms of oxygen and has an overall charge of 3−.

ANALYZE Phosphorus and oxygen are in groups 15 and 16 and have bonding capacities of 3 and 2, respectively. Phosphorus is in row 3 ($Z = 15$), so we can also expand its octet if we need to.

SOLVE

The number of valence electrons is

Element:	P	+	4 O
Valence electrons:	5	+	$(4 \times 6) = 29$
Additional electrons due to the 3− charge:			3
Total valence electrons:			32

Phosphorus has the greater bonding capacity (3), so it is the central atom:

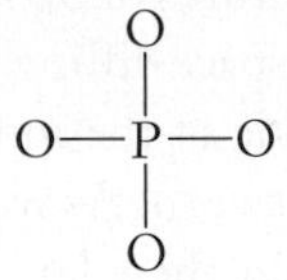

$$
\begin{array}{c}
\text{O} \\
| \\
\text{O—P—O} \\
| \\
\text{O}
\end{array}
$$

Each O atom needs three lone pairs of electrons to complete its octet:

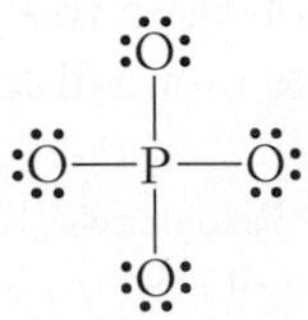

$$
\begin{array}{c}
:\ddot{\text{O}}: \\
| \\
:\ddot{\text{O}}\text{—P—}\ddot{\text{O}}: \\
| \\
:\ddot{\text{O}}:
\end{array}
$$

There are 32 valence electrons in the structure, which matches the number determined for the ion. Therefore, it is a complete Lewis structure of a polyatomic ion once we add the brackets and electrical charge:

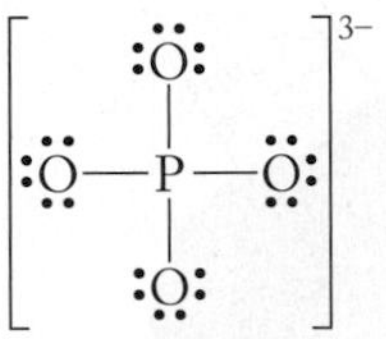

$$
\left[\begin{array}{c}
:\ddot{\text{O}}: \\
| \\
:\ddot{\text{O}}\text{—P—}\ddot{\text{O}}: \\
| \\
:\ddot{\text{O}}:
\end{array} \right]^{3-}
$$

Each O has a single bond and a formal charge of −1; the four bonds around the P atom are one more than its bonding capacity, so its formal charge is +1. The sum of the formal charges, $[+1 + 4(-1)]$, matches the charge on the ion, 3−.

We can reduce the formal charge on P by increasing the number of bonds to it, and we can do that by converting a lone pair on one of the O atoms into a bonding pair:

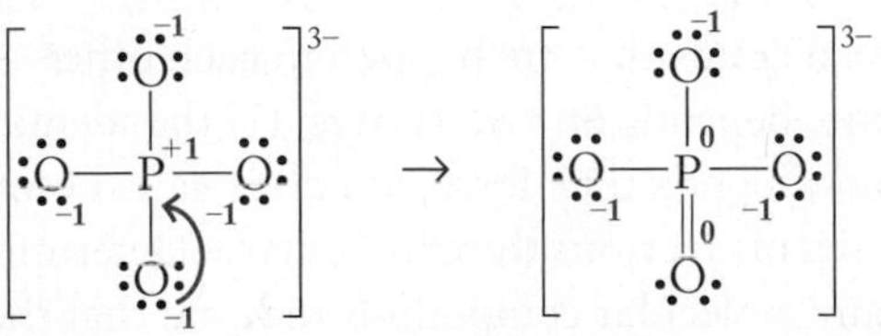

$$
\left[\begin{array}{c}
\overset{-1}{:\ddot{\text{O}}:} \\
| \\
:\ddot{\text{O}}\overset{+1}{\text{—P—}}\ddot{\text{O}}: \\
\underset{-1}{} \quad \underset{-1}{} \\
:\ddot{\text{O}}: \\
\underset{-1}{}
\end{array} \right]^{3-}
\quad \rightarrow \quad
\left[\begin{array}{c}
\overset{-1}{:\ddot{\text{O}}:} \\
| 0 \\
:\ddot{\text{O}}\text{—P—}\ddot{\text{O}}: \\
\underset{-1}{} \quad || 0 \quad \underset{-1}{} \\
\cdot\ddot{\text{O}}\cdot \\
0
\end{array} \right]^{3-}
$$

At the same time, we change a single-bonded O atom into a double-bonded O atom and thereby make its formal charge zero. Therefore, the structure on the right, in which the P atom has an expanded valence shell, is the best Lewis structure we can draw for the phosphate ion.

THINK ABOUT IT The phosphorus atom in the final structure has an expanded octet. It is also stabilized by resonance because the P—O double bond could link any of the four O atoms to the central P atom.

Practice Exercise Draw the resonance structures of the selenite ion (SeO_3^{2-}) that minimize the formal charges on the atoms. ⚙

(Answers to Practice Exercises are in the back of the book.)

bond order the number of bonds between atoms: 1 for a single bond, 2 for a double bond, and 3 for a triple bond.

PF$_5$ exists, but NF$_5$ does not. Suggest a reason why.

(Answers to Concept Tests are in the back of the book.)

4.9 The Lengths and Strengths of Covalent Bonds

Let's return to the resonance structures of ozone (Section 4.6), which result in two equivalent bonds as shown in the space-filling model in Figure 4.14(a). Each of the two equivalent bonds on either side of the molecule is neither a single bond nor a double bond, but something in between. As we noted in Section 4.6, the "tweener" nature of the bonds is reflected in their length, 128 pm, which is between the length of a typical O=O double bond (121 pm) and an O—O single bond (148 pm), as shown in Figure 4.14(b). In this section we explore the use of bond length, and also bond strength, to rationalize and validate molecular structures.

FIGURE 4.14 (a) The molecular structure of ozone is an average of the two resonance structures shown at the top. Both bonds in ozone are 128 pm long. (b) The value falls between the average length of an O=O double bond (121 pm) and the average length of an O—O single bond (148 pm). The intermediate value for the ozone bond length indicates that the bonds in ozone molecules are neither single bonds nor double bonds, but something in between.

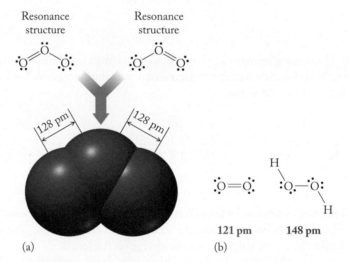

Bond Length

The distance between atoms that are bonded to each other—that is, the length of the bond they form—depends on two things: (1) the identities of the two atoms and (2) the number of bonds they form, which is called **bond order** and is equal to the number of pairs of electrons the atoms share. Determinations of the lengths of the bonds in many molecular compounds indicate that the length of any given bond varies by only a few picometers. For example, the C—H bond length in formaldehyde, CH$_2$O, is nearly the same as the C—H bond length in methane, CH$_4$. Similarly, the lengths of the C=O double bonds in formaldehyde and CO$_2$ are nearly identical (Figure 4.15).

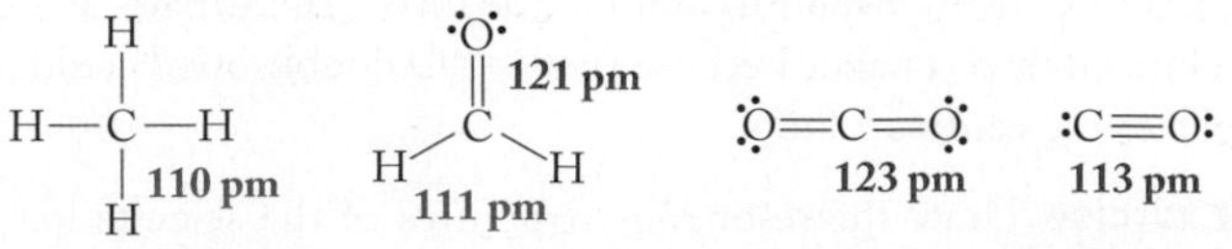

FIGURE 4.15 Bond lengths depend on the identity of the two atoms forming the bond and decrease with increasing bond order. The lengths of the C—H single bonds in CH$_2$O and CH$_4$ are nearly the same, as are the lengths of the C=O double bonds in CH$_2$O and CO$_2$. However, the C≡O triple bond in CO is much shorter than the C=O double bonds in CH$_2$O and CO$_2$.

TABLE 4.7 Average Lengths and Energies of Selected Covalent Bonds

Bond	Bond Length (pm)	Bond Energy (kJ/mol)	Bond	Bond Length (pm)	Bond Energy (kJ/mol)	Bond	Bond Length (pm)	Bond Energy (kJ/mol)
C—C	154	348	N—H	104	391	S—S	204	266
C=C	134	614	N—N	147	163	S—H	134	347
C≡C	120	839	N=N	124	418	H—H	74	436
C—N	147	293	N≡N	110	945	H—F	92	567
C=N	127	615	N—O	136	201	H—Cl	127	431
C≡N	116	891	N=O	122	607	H—Br	141	366
C—O	143	358	N≡O	106	678	H—I	161	299
C=O	123	743^a	O—O	148	146	F—F	143	155
C≡O	113	1072	O=O	121	498	Cl—Cl	200	243
C—H	110	413	O—H	96	463	Br—Br	228	193
C—F	133	485	S—O	151	265	I—I	266	151
C—Cl	177	328	S=O	143	523			

^{a}The bond energy of the C=O bond in CO_2 is 799 kJ/mol.

As the bond order between two atoms increases, bond length decreases. We see the trend in the lengths of the O—O bond in hydrogen peroxide, H_2O_2, and the O=O bond in O_2 (Figure 4.14b); in the C=O bonds in CO_2 and the C≡O bond in CO (Figure 4.15); and in the average lengths of several series of bonds listed in Table 4.7. It is logical, then, that bond length can be used to determine bond order, and *vice versa*, even when bond order is not a simple whole number. For example, the actual bond order in O_3 is neither 1 nor 2, but rather 1.5, with each pair of O atoms sharing 3 electrons instead of 2 or 4.

SAMPLE EXERCISE 4.17 **Determining Bond Order and Bond Length from Resonance Structures** **LO7**

Draw the resonance structures of the carbonate ion, CO_3^{2-}, and from them calculate the bond order of its carbon–oxygen bonds and estimate their length.

COLLECT AND ORGANIZE We are asked to draw the resonance structures of a polyatomic ion and then to determine the order and length of its bonds. A five-step procedure for drawing Lewis structures is described in Section 4.3. Table 4.7 contains average bond lengths, which vary inversely with bond order.

ANALYZE We drew the resonance structures of the nitrate ion, NO_3^-, in Sample Exercise 4.13. Carbon atoms (group 14) have one fewer valence electron than nitrogen atoms (group 15), but carbonate ions have one more negative charge than nitrate ions. Therefore, CO_3^{2-} and NO_3^- are isoelectronic and should have similar Lewis structures.

SOLVE

1. The number of valence electrons is

Element:	C	+	3 O
Valence electrons:	4	+	$(3 \times 6) = 22$
Additional electrons due to the 2− charge:			$\underline{2}$
Total valence electrons:			24

2. Connecting the carbon atom with single bonds to the three O atoms gives us this skeletal structure:

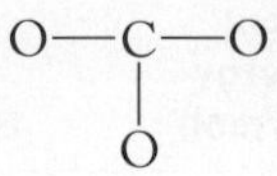

$$O-C-O$$
$$|$$
$$O$$

3. Each O atom needs three lone pairs of electrons to complete its octet:

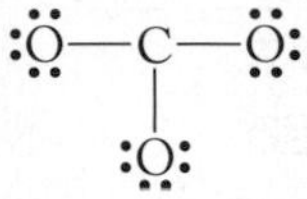

$$\ddot{O}-C-\ddot{O}:$$
$$|$$
$$\ddot{O}:$$

4. There are 24 valence electrons in the structure, which matches the number determined in step 1.
5. The central C atom has only 6 electrons. To provide it with the 2 more that it needs to complete its octet, we convert a lone pair on one of the oxygen atoms into a bonding pair:

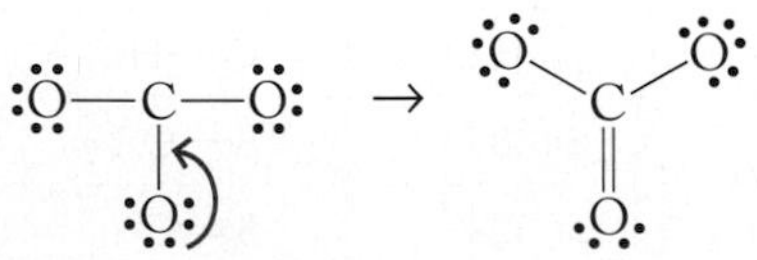

The carbon atom now has a complete octet. Adding brackets and the ionic charge, we have a complete Lewis structure:

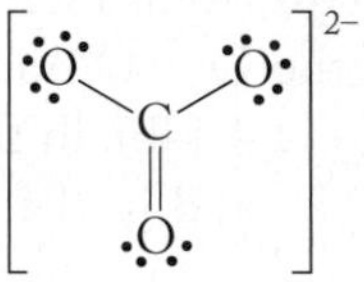

Using the O atom to the left or right of the central C atom to form the double bond creates two additional resonance forms, or three in all:

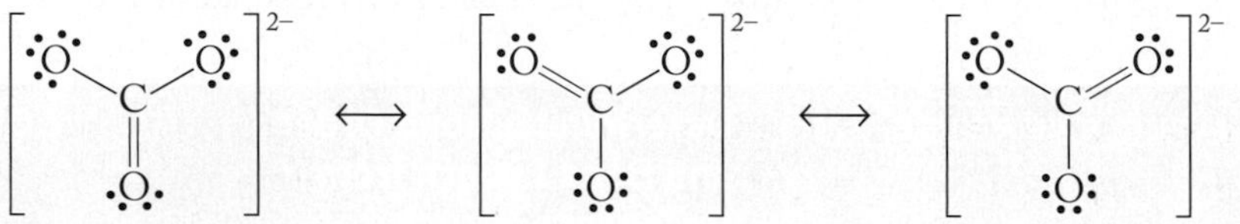

Each resonance structure contains two C—O bonds and one C=O bond, which means there are a total of four bonding pairs of electrons distributed evenly among three bonds. Therefore, each bond consists of 4/3 = 1.33 bonding pairs, giving it a bond order of 1.33.

According to Table 4.7, the average length of a C—O bond is 143 pm and the average length of a C=O bond is 123 pm. The length of the bonds in CO_3^{2-} ions (129 pm) lies in between these values.

THINK ABOUT IT In this exercise we used resonance structures to determine bond order and, in turn, to estimate bond length. In the text we used an experimentally determined length of the bonds in ozone to determine bond order and confirm the validity of the resonance structures we had drawn. The process seems to work well in both directions.

Practice Exercise Draw the resonance structures of nitrous acid, HNO_2, and from them estimate the length of the nitrogen–oxygen bonds in the acid. ✺

(Answers to Practice Exercises are in the back of the book.)

Bond Energies

The energy needed to break a H—H bond is represented by the depth of the "well" in Figure 4.3. That amount of energy is released when the bond between two H atoms is formed, and that same amount must be added to break the bond between them—to move them so far apart that they no longer interact. Bond energy (or *bond strength*) for any bond is usually expressed in terms of the energy needed to break 1 mole of bonds in the gas phase. Bond energies for some common covalent bonds, expressed in kilojoules per mole, are given in Table 4.7.

Like bond lengths, bond energies are average values because they vary depending on the structure of the rest of the molecule. For example, the bond energy of a C=O bond in carbon dioxide is 799 kJ/mol, but the C=O bond energy in formaldehyde is only 743 kJ/mol. Bond energies are always positive quantities because breaking bonds requires *the addition* of energy.

Another view of the variability in bond energy comes from breaking the C—H bonds in CH_4 in a step-by-step fashion:

Decomposition Step	Energy Needed (kJ/mol)
$CH_4 \rightarrow CH_3 + H$	435
$CH_3 \rightarrow CH_2 + H$	453
$CH_2 \rightarrow CH + H$	425
$CH \rightarrow C + H$	339
Total:	1652
Average:	413

The results tell us that the chemical environment of a bond can influence the energy required to break it: breaking the first C—H bond in methane is easier (requires less energy) than breaking the second but is more difficult than breaking the third or fourth. The total energy needed to break all four C—H bonds is 1652 kJ/mol, which is an average of 413 kJ/mol per bond.

The relationship between bond order and bond energy is also apparent in Table 4.7. Consider the bond energies of C—C, C=C, and C≡C bonds: they are 348, 614, and 839 kJ/mol, respectively. Note that the increase in bond energy is roughly proportional to the number of bonds between carbon atoms. The correlation between bond order and bond energy is true for other pairs of atoms: the higher the bond order, the greater the bond energy.

In this chapter we have explored the electrostatic potential energy between oppositely charged ions that leads to ionic bond formation. We have also explored the nature of the covalent bonds that hold together molecules and polyatomic ions, observing that the bonds owe their strength to the pairs of electrons shared between nuclei of atoms. Sharing does not necessarily mean equal sharing, and unequal sharing coupled with bond vibration accounts for the ability of some atmospheric gases to absorb and emit infrared radiation. In so doing, the molecules function as potent greenhouse gases.

Early in the chapter we noted that moderate concentrations of greenhouse gases are required for climate stability and to make our planet habitable. The escalating concern of many is that Earth's climate is currently being destabilized by too much of a good thing. Policies being made by the world's governments today will have a significant impact on the problem of global warming, one way or the other. As an informed member of the world community, you have the opportunity to influence how those policy decisions are made. We hope that you will make the most of that opportunity.

SAMPLE EXERCISE 4.18 **Integrating Concepts: Moth Balls**

A compound often referred to by the acronym PDB is the active ingredient in most moth balls. It is also used to control mold and mildew, as a deodorant, and as a disinfectant. Tablets containing it are often stuck under the lids of garbage cans or placed in the urinals in public restrooms, producing a distinctive aroma. Molecules of PDB have the following skeletal structure:

a. Draw the Lewis structure of PDB and note any nonzero formal charges.
b. Is the structure stabilized by resonance? If so, draw all resonance structures.
c. Which, if any, of the bonds in the structure you drew are polar?
d. Predict the average carbon–carbon bond length and bond strength in the structure you drew.

COLLECT AND ORGANIZE We are given the skeletal structure of a molecule and are asked to draw its Lewis structure, including all resonance structures, and to perform a formal charge analysis. We are also asked to identify any polar bonds in the structure and to predict the length and strength of the carbon–carbon bonds. Bond polarity depends on the difference in electronegativities of the bonded atoms, which are given in Figure 4.8. Table 4.7 contains average lengths and energies (strengths) of covalent bonds.

ANALYZE The five-step procedure used in Sample Exercises 4.9 through 4.11 to draw Lewis structures of other small molecules should be useful in drawing the Lewis structure of PDB. Resonance structures for PDB like those for benzene (Figure 4.12) should be possible if there are alternating single and double carbon–carbon bonds in PDB's six-carbon ring.

SOLVE a and b. The number of valence electrons is

Element: 6 C   + 2 Cl   + 4 H

Valence electrons: $(6 \times 4) + (2 \times 7) + (4 \times 1) = 42$

Completing the octets on the Cl atoms:

gives us a structure with 36 valence electrons (12 bonding pairs and 6 lone pairs). We need 6 more

electrons to reach 42. We can achieve that number if we create three more bonds between carbon atoms by turning single C—C bonds into double bonds. We have to distribute them evenly around the ring to avoid any C atoms with five bonds. Two equivalent resonance structures can be drawn to show the bonding pattern:

Resonance stabilizes the structure of PDB. Each C atom has four bonds and each H and Cl atom has one bond, so every atom has the number of bonds that matches its bond capacity. This means that all formal charges are zero.

c. The differences in electronegativities for the bonded pairs of atoms are

C—C	$\Delta\chi = 0$
Cl—C	$\Delta\chi = 3.0 - 2.5 = 0.5$
C—H	$\Delta\chi = 2.5 - 2.1 = 0.4$

Of the three pairs, only the Cl—C bond meets our polar bond guidelines ($0.4 < \Delta\chi < 2.0$).

d. The even distribution of a total of 9 bonding pairs of electrons among 6 C atoms means that, on average, each pair shares 1.5 pairs of bonding electrons. The corresponding bond length and bond strength should be about halfway between those of C—C single and C=C double bonds, given in Table 4.7:

Approximate bond length:

$$[(154 + 134)/2] \text{ pm} = 144 \text{ pm}$$

Approximate bond strength:

$$[(348 + 614)/2] \text{ kJ/mol} = 481 \text{ kJ/mol}$$

THINK ABOUT IT The resonance structures closely resemble those of benzene, which is reflected in the common name of PDB, *para-*di*chlorob*enzene. We will explore the rules for naming organic compounds like PDB in Chapter 19. For now, please note that the two polar C—Cl bonds in PDB are oriented *in opposite directions*. Thus, the unequal sharing of the bonding pair of electrons in the Cl—C bond on the left side of the molecule is offset by the unequal sharing of the bonding pair of electrons in the C—Cl bond on the right side. In Chapter 5 we will explore how offsetting bond polarities in symmetrical molecules like this one explain why substances such as PDB are, overall, nonpolar.

SUMMARY

Section 4.1 A chemical bond results from two ions being attracted to each other (an **ionic bond**) or from two atoms sharing electrons (a **covalent bond**). The atoms in metallic solids pool their electrons in forming **metallic bonds**.

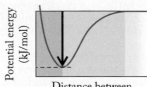

Distance between nuclei (pm)

Section 4.2 To name binary ionic compounds, first write the name of the cation's parent element, and then write the name of the anion's parent element. Change the ending of the name of the second element to *-ide*. Roman numerals in parentheses indicate the charges of transition metal cations. The names of **oxoanions** (**polyatomic ions** containing oxygen atoms) end in *-ate* or *-ite* and may have a *per-* or *hypo-* prefix to indicate the relative number of oxygen atoms per ion. To name binary molecular compounds, first write the name of the element that is to the left of, or, if the elements are in the same group, below the other one in the periodic table. Prefixes indicate the number of atoms of each element per molecule. The names of solutions of binary acids (general formula HX) begin with the prefix *hydro-* followed by the name of element X, but end in *-ic* followed by the word *acid*. The names of the **oxoacids** are similar to the names of their oxoanions, but their endings change from *-ate* to *-ic acid* and from *-ite* to *-ous acid*.

Section 4.3 **Lewis symbols** use dots to represent paired and unpaired electrons in the ground states of atoms. The number of unpaired electrons indicates the number of bonds the element is likely to form, that is, its **bonding capacity**. Chemical stability is achieved when atoms have 8 electrons in their valence *s* and *p* orbitals, following the **octet rule**. A **Lewis structure** shows the bonding pattern in molecules and polyatomic ions; pairs of dots represent **lone pairs** of electrons that do not contribute to bonding. A **single bond** consists of a single pair of electrons shared between two atoms; there are two shared pairs in a **double bond**, and three shared pairs in a **triple bond**.

Section 4.4 Unequal electron sharing between atoms of different elements results in **polar covalent bonds**. Bond polarity is a measure of how unequally the electrons in covalent bonds are shared. More polarity results from greater differ-

H——Cl

ences between the **electronegativities** of the bonded atoms. Electronegativity generally increases with increasing ionization energy.

Section 4.5 Covalent bonds behave more like flexible springs than rigid rods. They can undergo a variety of bond vibrations. The vibrations of polar bonds may create fluctuating electrical fields that allow molecules to absorb infrared electromagnetic radiation. When atmospheric gases absorb IR radiation they contribute to the greenhouse effect.

Section 4.6 Two or more equivalent Lewis structures—called **resonance structures**—can sometimes be drawn for one molecule or polyatomic ion. The actual bonding pattern in a molecule is an average of equivalent resonance structures.

Section 4.7 The preferred resonance structure of a molecule is one in which the **formal charges (FC)** on its atoms are zero or closest to zero and any negative formal charges are on the more electronegative atoms. The formal charge on an atom in a Lewis structure is the difference between the number of valence electrons in the free atom and the sum of the number of electrons in lone pairs and half the number of electrons in bonding pairs on the bonded atom.

Section 4.8 **Free radicals** include reactive molecules that have an odd number of valence electrons and contain atoms with incomplete octets. Atoms of elements in the third row of the periodic table with $Z > 12$ and beyond may expand their valence shells to accommodate more than an octet of electrons.

Section 4.9 **Bond order** is the number of bonding pairs in a covalent bond. *Bond energy* is the energy change that accompanies the breaking of 1 mole of a particular covalent bond in the gas phase. As the bond order between two atoms increases, the bond length decreases and the bond energy increases.

PROBLEM-SOLVING SUMMARY

TYPE OF PROBLEM	CONCEPTS AND EQUATIONS		SAMPLE EXERCISES
Calculating the electrostatic potential energy of ionic bonds	$$E_{el} = 2.31 \times 10^{-19}\,\text{J} \cdot \text{nm}\left(\frac{Q_1 \times Q_2}{d}\right)$$	(4.1)	4.1
Naming binary ionic compounds and writing their formulas	First write the name of the cation's parent element; if it is a transition metal that forms ions with different charges, use a Roman numeral to represent the charge. Then write the name of the anion's parent element with its ending changed to *-ide*.		4.2, 4.3
Naming compounds of polyatomic ions and writing their formulas	As with a binary compound, write the name of the cation followed by the name of the anion. Use Table 4.3 to find the names of oxoanions (which end in *-ate* or *-ite*) and other polyatomic ions.		4.4, 4.5

TYPE OF PROBLEM	CONCEPTS AND EQUATIONS	SAMPLE EXERCISES
Naming binary molecular compounds and writing their formulas	First write the name of the element that is to the left of, or, if the elements are in the same group, below the other one in the periodic table. Then write the name of the other element, changing its ending to *-ide*. Use the prefixes in Table 4.4 to indicate the number of atoms of each element.	4.6
Naming acids and writing their formulas	For a binary acid (HX), begin with the prefix *hydro-* followed by the name of element X, but change its ending to *-ic* followed by the word *acid*. For an oxoacid, change the name of its oxoanion (Table 4.3) from *-ate* to *-ic acid*, or from *-ite* to *-ous acid*.	4.7
Drawing Lewis structures	Connect the atoms with single covalent bonds, distributing the valence electrons to give each noncentral atom 8 valence electrons (except 2 for H); use multiple bonds where necessary to complete the central atom's octet.	4.8–4.11
Comparing bond polarities	Calculate the difference in electronegativity ($\Delta\chi$) between the two bonded atoms. If $\Delta\chi \geq 2.0$, the bond is considered ionic; if $0.4 < \Delta\chi < 2.0$, the bond is considered polar covalent; if $\Delta\chi \leq 0.4$, the bond is considered nonpolar covalent.	4.12
Drawing resonance structures	Include all possible arrangements of covalent bonds in the molecule if more than one equivalent structure can be drawn.	4.13
Selecting resonance structures based on formal charges	Calculate formal charge on each atom using $$FC = \left(\begin{array}{c}\text{number of}\\\text{valence e}^-\end{array}\right) - \left[\begin{array}{c}\text{number of}\\\text{unshared e}^-\end{array} + \frac{1}{2}\left(\begin{array}{c}\text{number of e}^-\\\text{in bonding pairs}\end{array}\right)\right] \quad (4.2)$$ Select structures with formal charges closest to zero and with negative formal charges on the most electronegative atoms.	4.14
Drawing Lewis structures of odd-electron molecules	Distribute the valence electrons in the Lewis structure to leave the most electronegative atom(s) with 8 valence electrons and the least electronegative atom with the odd number of electrons.	4.15
Drawing Lewis structures containing atoms with expanded valence shells	Distribute the valence electrons in the Lewis structure, allowing atoms of elements in period 3 and beyond to have more than 8 valence electrons if more than four bonds are needed, or if the structure with the expanded valence shell results in formal charges closer to zero.	4.16
Determining bond order and bond length from resonance structures	Draw resonance structures to determine the average bond order for the equivalent bonds. Relate bond order to bond length using Table 4.7.	4.17

VISUAL PROBLEMS ●●■

(Answers to boldface end-of-chapter questions and problems are in the back of the book.)

4.1. Which group highlighted in Figure P4.1 contains atoms that have the following? (a) 1 valence electron; (b) 4 valence electrons; (c) 6 valence electrons

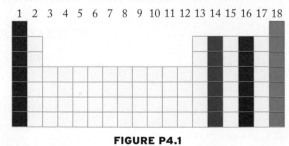

1 2 3 4 5 6 7 8 9 10 11 12 13 14 15 16 17 18

FIGURE P4.1

4.2. Which of the groups highlighted in Figure P4.2 contains atoms with the following? (a) 2 valence electrons; (b) 5 valence electrons; (c) 3 valence electrons

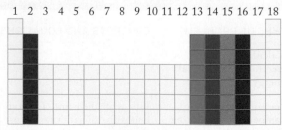

1 2 3 4 5 6 7 8 9 10 11 12 13 14 15 16 17 18

FIGURE P4.2

4.3. Which of the Lewis structures in Figure P4.3 correctly portrays the most stable ion of magnesium?

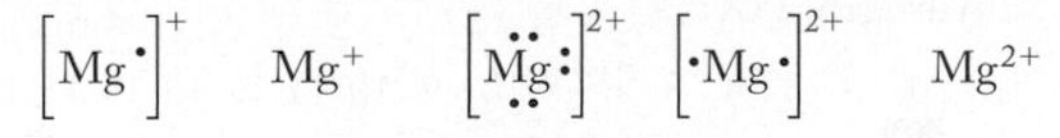

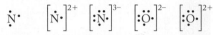

FIGURE P4.3

4.4. Which of the Lewis symbols in Figure P4.4 are correct?

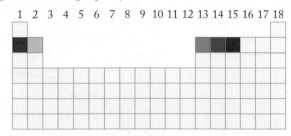

FIGURE P4.4

4.5. Which of the highlighted elements in Figure P4.5 has the greatest bonding capacity?

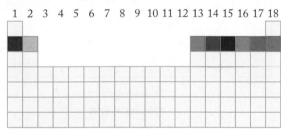

FIGURE P4.5

4.6. Which of the highlighted elements in Figure P4.6 has the greatest electronegativity?

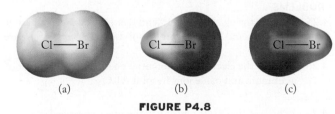

FIGURE P4.6

4.7. Which two of the highlighted elements in Figure P4.6 make up the pair that forms the bond with the most ionic character?

NOTE: The color scale used in Problems 4.8, 4.9, and 4.13 is the same as in Figure 4.7, where violet is a charge of 1+, red is a charge of 1−, and yellow-green is 0.

4.8. Which of the drawings in Figure P4.8 is the best description of the distribution of electrical charge in ClBr?

FIGURE P4.8

*4.9. Which of the drawings in Figure P4.9 best describes the distribution of electrical charge in LiF?

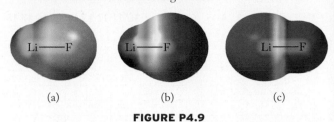

FIGURE P4.9

4.10. Are the three structures in Figure P4.10 resonance forms of the thiocyanate ion (SCN⁻)?

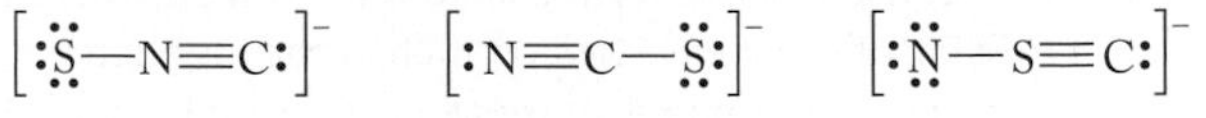

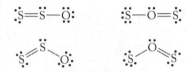

FIGURE P4.10

4.11. Why are the structures in Figure P4.11 not all resonance forms of the molecule S_2O?

FIGURE P4.11

4.12. Water in the atmosphere is a greenhouse gas, which means its molecules are transparent to visible light but may absorb photons of infrared radiation. Which of the three modes of bond vibration shown in Figure P4.12 are infrared active? Note that molecules of H_2O are not linear; the angle between the O—H bonds in the top image is 104.5°.

Asymmetric stretch · Symmetric stretch · Bend

FIGURE P4.12

*4.13. Which of the drawings in Figure P4.13 most accurately describes the distribution of electron density in SO_2? Explain your answer.

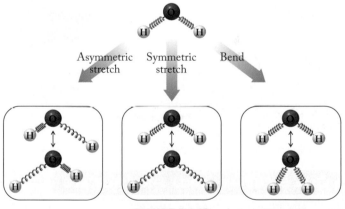

FIGURE P4.13

4.14. Which groups among main group elements in Figure P4.14 have an odd number of valence electrons?

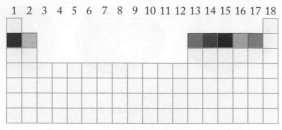

FIGURE P4.14

4.15. Krypton and xenon form compounds with only the most reactive of other elements. Which of the highlighted elements in Figure P4.15 is one of those highly reactive elements?

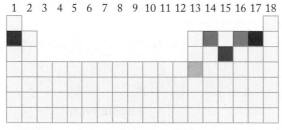

FIGURE P4.15

4.16. Which of the highlighted elements in Figure P4.16 expands its valence shell when bonding to a highly electronegative element?

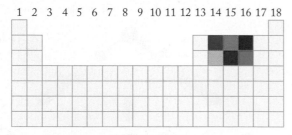

FIGURE P4.16

4.17. Which of the highlighted groups in Figure P4.17 have negative partial charges in binary compounds with hydrogen, HX?

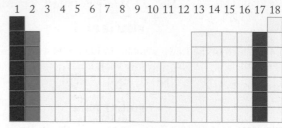

FIGURE P4.17

4.18. Figure P4.18 shows two graphs of electrostatic potential energy versus internuclear distance. One is for a pair of potassium and chloride ions, and the other is for a pair of potassium and fluoride ions. Which is which?

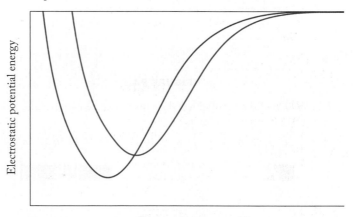

Distance between nuclei

FIGURE P4.18

QUESTIONS AND PROBLEMS

Chemical Bonds

CONCEPT REVIEW

4.19. Do all the elements in a group in the periodic table have the same number of valence electrons?

4.20. Which electrons in an atom are considered *valence* electrons?

4.21. Does the number of valence electrons in a neutral atom ever equal the atomic number?

4.22. Does the number of valence electrons in a neutral atom ever equal the group number?

4.23. Does the strength of an ion–ion attraction depend on the number of ions in the compound?

4.24. Describe the differences in bonding in *covalent* and *ionic* compounds.

PROBLEMS

4.25. What is the electrostatic potential energy between a pair of potassium and bromide ions in solid KBr? *Hint*: See Figure 3.35.

4.26. What is the electrostatic potential energy between a pair of aluminum and oxide ions in solid Al_2O_3?

4.27. Which of these substances has the most negative lattice energy? (a) KCl, (b) TiO_2, (c) $BaCl_2$, (d) KI

4.28. Rank the following ionic compounds, which have the same crystal structure, from least negative to most negative lattice energy: CsCl, CsBr, and CsI.

4.29. Rank the following ionic compounds in order of increasing electrostatic potential energy between their ions: KBr, SrBr$_2$, and CsBr.

4.30. Rank the following ionic compounds in order of increasing coulombic attraction between their ions: BaO, BaCl$_2$, and CaO.

Naming Compounds and Writing Formulas

CONCEPT REVIEW

4.31. What is the role of Roman numerals in the names of the compounds formed by transition metals?

4.32. Why does the name of a binary ionic compound in which the cation is from a group 1 or group 2 element not need a Roman numeral after the element's name?

4.33. Consider a mythical element X, which forms two oxoanions: XO$_2^{2-}$ and XO$_3^{2-}$. Which of the two has a name that ends in *-ite*?

4.34. Concerning the oxoanions in Problem 4.33, would the name of either of them require a prefix such as *hypo-* or *per-*? Explain why or why not.

PROBLEMS

4.35. What are the names of these compounds of nitrogen and oxygen? (a) NO$_3$; (b) N$_2$O$_5$; (c) N$_2$O$_4$; (d) NO$_2$; (e) N$_2$O$_3$; (f) NO; (g) N$_2$O; (h) N$_4$O

4.36. More than a dozen compounds containing sulfur and oxygen have been identified. What are the chemical formulas of the following six? (a) sulfur monoxide; (b) sulfur dioxide; (c) sulfur trioxide; (d) disulfur monoxide; (e) hexasulfur monoxide; (f) heptasulfur dioxide

4.37. Predict the formula and give the name of the ionic compound formed by these pairs of elements: (a) sodium and sulfur; (b) strontium and chlorine; (c) aluminum and oxygen; (d) lithium and hydrogen.

4.38. Predict the formula and give the name of the ionic compound formed by these pairs of elements: (a) potassium and bromine; (b) calcium and hydrogen; (c) lithium and nitrogen; (d) aluminum and chlorine.

4.39. What are the names of the cobalt oxides that have the following formulas? (a) CoO; (b) Co$_2$O$_3$; (c) CoO$_2$

4.40. What are the formulas of the following copper minerals?
a. cuprite, copper(I) oxide
b. chalcocite, copper(I) sulfide
c. covellite, copper(II) sulfide

4.41. Give the formula and charge of the oxoanion in each of the following compounds: (a) sodium hypobromite; (b) potassium sulfate; (c) lithium iodate; (d) magnesium nitrite.

*__4.42.__ Give the formula and charge of the oxoanion in each of the following compounds: (a) potassium tellurite; (b) sodium arsenate; (c) calcium selenite; (d) potassium chlorate.

4.43. What are the names of the following ionic compounds? (a) NiCO$_3$; (b) NaCN; (c) LiHCO$_3$; (d) Ca(ClO)$_2$

4.44. What are the names of the following ionic compounds? (a) Mg(ClO$_4$)$_2$; (b) NH$_4$NO$_3$; (c) Cu(CH$_3$COO)$_2$; (d) K$_2$SO$_3$

4.45. Give the name or chemical formula of each of the following acids: (a) HF; (b) HBrO$_3$; (c) phosphoric acid; (d) nitrous acid.

4.46. Give the name or chemical formula of each of the following acids: (a) HBr; (b) HIO$_4$; (c) selenous acid; (d) hydrocyanic acid.

4.47. What are the names of these compounds? (a) Na$_2$O; (b) Na$_2$S; (c) Na$_2$SO$_4$; (d) NaNO$_3$; (e) NaNO$_2$

4.48. What are the names of these compounds? (a) K$_3$PO$_4$; (b) K$_2$O; (c) K$_2$SO$_3$; (d) KNO$_3$; (e) KNO$_2$

4.49. Write the chemical formulas of these compounds: (a) potassium sulfide; (b) potassium selenide; (c) rubidium sulfate; (d) rubidium nitrite; (e) magnesium sulfate.

4.50. Write the chemical formulas of these compounds: (a) rubidium nitride; (b) potassium sulfite; (c) rubidium sulfite; (d) rubidium nitrate; (e) magnesium selenite.

4.51. What are the names of these compounds? (a) MnS; (b) V$_3$N$_2$; (c) Cr$_2$(SO$_4$)$_3$; (d) Co(NO$_3$)$_2$; (e) Fe$_2$O$_3$

4.52. What are the names of these compounds? (a) RuS; (b) PdCl$_2$; (c) Ag$_2$O; (d) WO$_3$; (e) PtO$_2$

4.53. Which is the formula of sodium sulfite? (a) Na$_2$S; (b) Na$_2$SO$_3$; (c) Na$_2$SO$_4$; (d) NaHS

4.54. Which is the formula of calcium nitrate? (a) Ca$_3$N$_2$; (b) Ca$_2$NO$_3$; (c) Ca$_2$(NO$_3$)$_2$; (d) Ca(NO$_3$)$_2$

Lewis Symbols and Lewis Structures

CONCEPT REVIEW

4.55. Some of his critics described G. N. Lewis's approach to explaining covalent bonding as an exercise in double counting and therefore invalid. Explain the basis for the criticism.

4.56. Does the octet rule mean that a diatomic molecule must have 16 valence electrons?

4.57. Why is the bonding pattern in water H—O—H and not H—H—O?

4.58. Does each atom in a pair that is covalently bonded always contribute the same number of valence electrons to form the bonds between them?

PROBLEMS

4.59. Draw Lewis symbols of cesium, barium, and aluminum.

4.60. Draw Lewis symbols of nitrogen, selenium, and chlorine.

4.61. Draw Lewis structures of Na$^+$, In$^+$, Ca^{2+}, and S^{2-}.

4.62. Draw Lewis structures of the most stable ions formed by lithium, magnesium, aluminum, and fluorine.

4.63. Which of the following ions have a complete valence-shell octet? B^{3+}, I$^-$, Ca^{2+}, or Pb^{2+}

4.64. How many valence electrons are in each of these atoms or ions? Xe, Sr^{2+}, Cl, and Cl$^-$

4.65. How many valence electrons does each of the following species contain? (a) BN; (b) HF; (c) OH$^-$; (d) CN$^-$

4.66. How many valence electrons does each of the following species contain? (a) N_2^+; (b) CS^+; (c) CN; (d) CO

4.67. Draw Lewis structures for the following diatomic molecules and ions: (a) CO; (b) O_2; (c) ClO^-; (d) CN^-.

4.68. Draw Lewis structures for the following diatomic molecules and ions: (a) F_2; (b) NO^+; (c) SO; (d) HI.

4.69. Draw Lewis structures for the following molecular compounds and ions: (a) CCl_4; (b) BH_3; (c) SiF_4; (d) BH_4^-; (e) PH_4^+.

4.70. Draw Lewis structures for the following molecular compounds and ions: (a) $AlCl_3$, (b) PH_3; (c) H_2Se; (d) NO_2^-; (e) AlH_4^-.

4.71. **Greenhouse Gases** Chlorofluorocarbons (CFCs) are compounds linked to depletion of stratospheric ozone. They are also greenhouse gases. Draw Lewis structures for the following CFCs:
   a. CCl_3F (Freon 11)
   b. CCl_2F_2 (Freon 12)
   c. $CClF_3$ (Freon 13)
   d. Cl_2FC—$CClF_2$ (Freon 113)
   e. ClF_2C—$CClF_2$ (Freon 114)

4.72. The replacement of a halogen atom in a CFC molecule with a hydrogen atom makes the compound more environmentally "friendly." Draw Lewis structures for the following such compounds:
   a. $CHCl_2F$ (Freon 21)
   b. CHF_2Cl (Freon 22)
   c. CH_2ClF (Freon 31)
   d. F_3C—$CHBrCl$ (Halon 2311)
   e. Cl_2FC—CH_3 (HCFC 141b)

4.73. Draw Lewis structures for the following oxoanions:
   (a) ClO_2^-; (b) SO_3^{2-}; (c) HCO_3^-.

4.74. Draw Lewis structures for the following oxoanions:
   (a) BrO_4^-; (b) SeO_4^{2-}; (c) HPO_4^{2-}.

4.75. **Skunks and Rotten Eggs** Many sulfur-containing organic compounds have characteristically foul odors: butanethiol ($CH_3CH_2CH_2CH_2SH$) is responsible for the odor of skunks, and rotten eggs smell the way they do because they produce tiny amounts of pungent hydrogen sulfide, H_2S. Draw the Lewis structures for $CH_3CH_2CH_2CH_2SH$ and H_2S.

4.76. **Acid in Ants** Formic acid, HCOOH, is the smallest organic acid and was originally isolated by distilling red ants. Draw its Lewis structure given the connectivity of the atoms as shown in Figure P4.76.

FIGURE P4.76

4.77. **Chlorine Bleach** Chlorine combines with oxygen in several proportions. Dichlorine monoxide (Cl_2O) is used in the manufacture of bleaching agents. Potassium chlorate ($KClO_3$) is used in oxygen generators aboard aircraft. Draw the Lewis structures for Cl_2O and ClO_3^-. Cl is the central atom in each case.

4.78. **Dangers of Mixing Cleansers** Labels on household cleansers caution against mixing bleach with ammonia (Figure P4.78) because they react with each other producing monochloramine (NH_2Cl) and hydrazine (N_2H_4), both of which are toxic. Draw the Lewis structures for monochloramine and hydrazine.

FIGURE P4.78

Electronegativity, Unequal Sharing, and Polar Bonds

CONCEPT REVIEW

4.79. How can we use electronegativity to predict whether a bond between two atoms is likely to be covalent or ionic?

4.80. How do the electronegativities of the elements change across a row and down a group in the periodic table?

4.81. How are trends in electronegativity related to trends in atomic size?

4.82. Is the element with the most valence electrons in a row of the periodic table also the most electronegative?

4.83. What is meant by the term *polar covalent bond*?

4.84. Why are the electrons in bonds between different elements not shared equally?

PROBLEMS

4.85. Which of the following bonds are polar? C—Se, C—O, Cl—Cl, O=O, N—H, C—H. In the bond or bonds that you selected, which atom has the greater electronegativity?

4.86. Which is the least polar bond? C—Se, C=O, Cl—Br, O=O, N—H, C—H

4.87. Which of the binary compounds formed by the following pairs of elements contain polar covalent bonds, and which are considered ionic compounds?
  a. C and S   c. Al and Cl
  b. C and O   d. Ca and O

4.88. Which of the beryllium halides, if any, are considered ionic compounds?

Vibrating Bonds and the Greenhouse Effect

CONCEPT REVIEW

4.89. Describe how atmospheric greenhouse gases act like the panes of glass in a greenhouse.

*4.90. Water vapor in the atmosphere contributes more to the greenhouse effect than carbon dioxide; yet, water vapor is not considered an important factor in climate change. Propose a reason why.

4.91. Increasing concentrations of nitrous oxide in the atmosphere may be contributing to climate change. Is the ability of N_2O to absorb IR radiation due to nitrogen–nitrogen bond stretching, nitrogen–oxygen bond stretching, or both? Explain your answer.

4.92. Is the ability of H_2O molecules to absorb photons of IR radiation due to symmetrical stretching or asymmetrical stretching of its O—H bonds, or both? Explain your answer. (*Hint*: The angle between the two O—H bonds in H_2O is 104.5°.)

4.93. Can molecules of carbon monoxide in the atmosphere absorb photons of IR radiation? Explain why or why not.

*4.94. How does the high-temperature conversion of limestone ($CaCO_3$) to lime (CaO) during the production of cement contribute to climate change?

4.95. Why does infrared radiation cause bonds to vibrate but not break (as UV radiation can)?

4.96. Argon is the third most abundant species in the atmosphere. Why isn't it a greenhouse gas?

*4.97. Would the energy required to cause the bond in CO to vibrate be more or less than that required by the carbon–oxygen bond in CO_2?

*4.98. Which compound absorbs IR radiation of a longer wavelength, NO or NO_2?

Resonance

CONCEPT REVIEW

4.99. Explain the concept of resonance.

4.100. How does resonance influence the stability of a molecule or an ion?

4.101. What factors determine whether or not a molecule or ion exhibits resonance?

4.102. What structural features do all the resonance forms of a molecule or ion have in common?

4.103. Explain why NO_2 is more likely to exhibit resonance than CO_2.

4.104. Are these two skeletal structures resonance forms: X—X—O and X—O—X?

PROBLEMS

4.105. Draw two Lewis structures showing the resonance that occurs in cyclobutadiene (C_4H_4), a cyclic molecule with a structure that includes a ring of four carbon atoms.

*4.106. Pyridine (C_5H_5N) and pyrazine ($C_4H_4N_2$) have structures similar to benzene's. Both compounds have structures with six atoms in a ring. Draw Lewis structures for pyridine and pyrazine showing all resonance forms. The N atoms in pyrazine are across the ring from each other.

---

*4.107. Oxygen and nitrogen combine to form a variety of nitrogen oxides, including the following two unstable compounds each with two nitrogen atoms per molecule: N_2O_2 and N_2O_3. Draw Lewis structures for the molecules showing all resonance forms.

*4.108. Oxygen and sulfur combine to form a variety of different sulfur oxides. Some are stable molecules and some, including S_2O_2 and S_2O_3, decompose when they are heated. Draw Lewis structures for the two compounds showing all resonance forms.

4.109. Draw Lewis structures for fulminic acid (HCNO), showing all resonance forms.

4.110. Draw Lewis structures for hydrazoic acid (HN_3), showing all resonance forms.

4.111. Draw Lewis structures showing the resonance that occurs in dinitrogen pentoxide. (*Hint*: N_2O_5 has an O atom at its center.)

4.112. **Bacteria Make Nitrites** Nitrogen-fixing bacteria convert urea [$H_2NC(O)NH_2$] into nitrite ions. Draw Lewis structures for the two species. Include all resonance forms. (*Hint*: There is a C=O bond in urea.)

Formal Charge: Choosing among Lewis Structures

CONCEPT REVIEW

4.113. Describe how formal charges are used to choose between possible molecular structures.

4.114. How do the electronegativities of elements influence the selection of which Lewis structure is favored?

4.115. In a molecule containing S and O atoms, is a structure with a negative formal charge on sulfur more likely to contribute to bonding than an alternative structure with a negative formal charge on oxygen?

4.116. In a cation containing N and O, why do Lewis structures with a positive formal charge on nitrogen contribute more to the actual bonding in the molecule than do those structures with a positive formal charge on oxygen?

PROBLEMS

4.117. Hydrogen isocyanide (HNC) has the same elemental composition as hydrogen cyanide (HCN), but the H in HNC is bonded to the nitrogen atom. Draw a Lewis structure for HNC, and assign formal charges to each atom. How do the formal charges on the atoms differ in the Lewis structures for HCN and HNC?

4.118. Molecules in Interstellar Space Hydrogen cyanide (HCN) and cyanoacetylene (HC_3N) have been detected in the interstellar regions of space. Draw Lewis structures for the molecules, and assign formal charges to each atom. The hydrogen atom is bonded to carbon in both cases.

4.119. Origins of Life The discovery of polyatomic organic molecules such as cyanamide (H_2NCN) in interstellar space has led some scientists to believe that the molecules from which life began on Earth may have come from space. Draw Lewis structures for cyanamide, and select the preferred structure on the basis of formal charges.

4.120. Complete the Lewis structures for and assign formal charges to the atoms in five of the resonance forms of thionitrosyl azide (SN_4). Indicate which of your structures should be most stable. The molecule is linear with S at one end.

*4.121. Nitrogen is the central atom in molecules of nitrous oxide (N_2O). Draw Lewis structures for another possible arrangement: N—O—N. Assign formal charges and suggest a reason why the structure is not likely to be stable.

4.122. More Molecules in Space Formamide ($HCONH_2$) and methyl formate (HCO_2CH_3) also have been detected in space. Draw the Lewis structures of the compounds, based on the skeletal structures in Figure P4.122, and assign formal charges:

$$
\begin{array}{cc}
\text{H—C—N} & \text{H—C—O—C—H} \\
\end{array}
$$

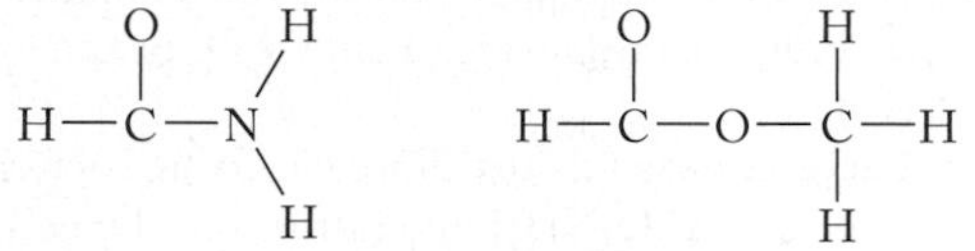

FIGURE P4.122

*4.123. Nitromethane (CH_3NO_2) reacts with hydrogen cyanide (HCN) to produce $CNNO_2$ and CH_4.

   a. Draw Lewis structures for CH_3NO_2, showing all resonance forms.

   b. Draw Lewis structures for $CNNO_2$, showing all resonance forms, based on the two possible skeletal structures for it in Figure P4.123. Assign formal charges, and predict which structure is more likely to exist.

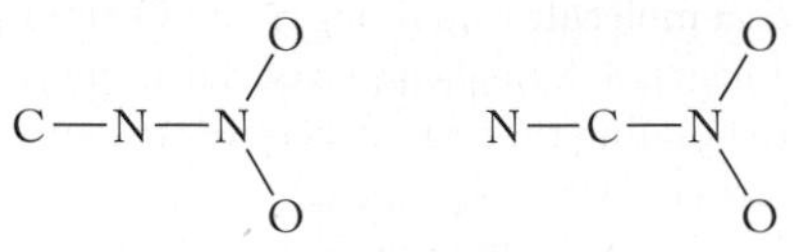

$$
\begin{array}{cc}
\text{C—N—N} & \text{N—C—N} \\
\end{array}
$$

FIGURE P4.123

   c. Are the two structures of $CNNO_2$ resonance forms of each other?

4.124. Use formal charges to determine which resonance form of each of the following ions is preferred: CNO^-, NCO^-, and CON^-.

Exceptions to the Octet Rule

CONCEPT REVIEW

4.125. Are all odd-electron molecules exceptions to the octet rule?

4.126. Describe the factors that contribute to the stability of structures in which the central atoms have more than 8 valence electrons.

4.127. Why do C, N, O, and F atoms in covalently bonded molecules and ions have no more than 8 valence electrons?

4.128. Do atoms in rows 3 and below always expand their valence shell? Explain your answer.

PROBLEMS

4.129. In which of the following molecules does the sulfur atom have an expanded valence shell? (a) SF_6; (b) SF_5; (c) SF_4; (d) SF_2

4.130. In which of the following molecules does the phosphorus atom have an expanded valence shell? (a) $POCl_3$; (b) PF_5; (c) PF_3; (d) P_2F_4 (which has a P—P bond)

4.131. How many electrons are there in the covalent bonds surrounding the sulfur atom in the following species? (a) SF_4O; (b) SOF_2; (c) SO_3; (d) SF_5^-

4.132. How many electrons are there in the covalent bonds surrounding the phosphorus atom in the following species? (a) $POCl_3$; (b) H_3PO_4; (c) H_3PO_3; (d) PF_6^-

*4.133. Draw the Lewis structures of NOF_3 and POF_3 in which the group 15 element is the central atom and the other atoms are bonded to it. What differences are there in the types of bonding in the molecules?

*4.134. The phosphate ion (PO_4^{3-}) is part of our DNA. The corresponding nitrogen-containing oxoanion, NO_4^{3-}, is not chemically stable. Draw Lewis structures showing any resonance forms of both oxoanions.

4.135. Dissolving NaF in selenium tetrafluoride (SeF_4) produces $NaSeF_5$. Draw the Lewis structures of SeF_4 and SeF_5^-. In which structure does Se have more than 8 valence electrons?

4.136. The reaction between NF_3, F_2, and SbF_3 at 200°C and 100 atm pressure produces the ionic compound NF_4SbF_6. Draw the Lewis structures of the ions in the product.

4.137. Ozone Depletion The compound Cl_2O_2 may play a role in ozone depletion in the stratosphere. Draw the Lewis structure of Cl_2O_2 based on the arrangement of atoms in Figure P4.137. Does either of the chlorine atoms in the structure have an expanded valence shell?

FIGURE P4.137

4.138. Cl_2O_2 decomposes to Cl_2 and ClO_2. Draw the Lewis structure of ClO_2.

4.139. Which of the following chlorine oxides are odd-electron molecules? (a) Cl_2O_7; (b) Cl_2O_6; (c) ClO_4; (d) ClO_3; (e) ClO_2

4.140. Which of the following nitrogen oxides are odd-electron molecules? (a) NO; (b) NO_2; (c) NO_3; (d) N_2O_4; (e) N_2O_5

4.141. In the following species, which atom is most likely to have an unpaired electron? (a) SO^+; (b) NO; (c) CN; (d) OH

4.142. In the following molecules, which atom is most likely to have an unpaired electron? (a) NO_2; (b) CNO; (c) ClO_2; (d) HO_2

4.143. Which of the Lewis structures in Figure P4.143 contributes most to the bonding in CNO?

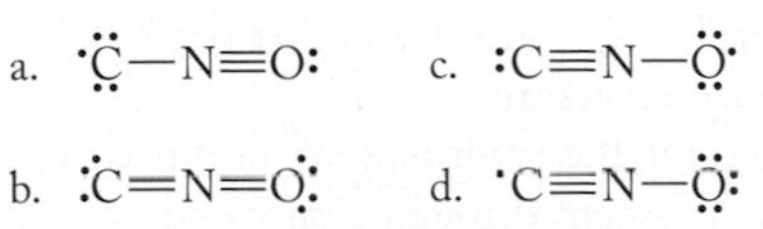

a. $\cdot\ddot{C}{-}N{\equiv}O{:}$ c. $:C{\equiv}N{-}\ddot{O}{:}$

b. $:C{=}N{=}\ddot{O}{:}$ d. $\cdot C{\equiv}N{-}\ddot{O}{:}$

FIGURE P4.143

4.144. Why is the Lewis structure in Figure P4.144 unlikely to contribute much to the bonding in NCO?

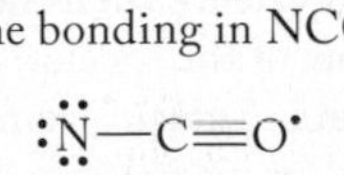

$:\ddot{N}{-}C{\equiv}O\cdot$

FIGURE P4.144

The Lengths and Strengths of Covalent Bonds

CONCEPT REVIEW

4.145. Do you expect the nitrogen–oxygen bond length in the nitrate ion to be the same as in the nitrite ion?

4.146. Why is the oxygen–oxygen bond length in O_3 not the same as in O_2?

4.147. Explain why the nitrogen–oxygen bond lengths in N_2O_4 (which has a nitrogen–nitrogen bond) and N_2O are nearly identical (118 and 119 pm, respectively).

4.148. Do you expect the sulfur–oxygen bond lengths in sulfite (SO_3^{2-}) and sulfate (SO_4^{2-}) ions to be about the same? Why?

4.149. Rank the following ions in order of increasing nitrogen–oxygen bond lengths: NO_2^-, NO^+, and NO_3^-.

4.150. Rank the following ions in order of increasing carbon–oxygen bond lengths: CO, CO_2, and CO_3^{2-}.

4.151. Rank the following ions in order of increasing nitrogen–oxygen bond energy: NO_2^-, NO^+, and NO_3^-.

4.152. Rank the following ions in order of increasing carbon–oxygen bond energy: CO, CO_2, and CO_3^{2-}.

Additional Problems

4.153. The unpaired dots in Lewis symbols of the elements represent valence electrons available for covalent bond formation. In Figure P4.153, which of the options for placing dots around the symbol for each element is preferred?

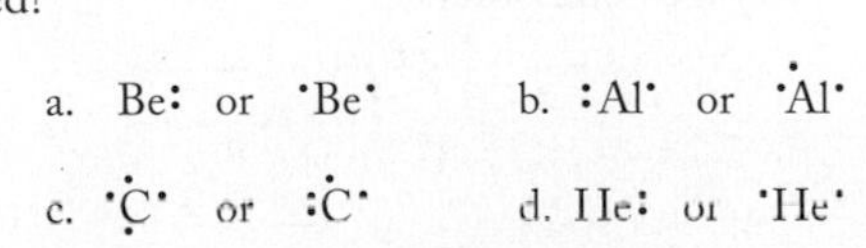

a. $Be{:}$ or $\cdot Be\cdot$ b. $:Al\cdot$ or $\cdot\dot{Al}\cdot$

c. $\cdot\dot{C}\cdot$ or $:\dot{C}\cdot$ d. $He{:}$ or $\cdot He\cdot$

FIGURE P4.153

4.154. Based on the Lewis symbols in Figure P4.154, predict to which group in the periodic table element X belongs.

a. $\cdot\dot{X}\cdot$ b. $\cdot\dot{X}{:}$ c. $\cdot\dot{X}{:}$ d. $:\ddot{X}{:}$

FIGURE P4.154

4.155. Use formal charges to predict whether the atoms in carbon disulfide are arranged CSS or SCS.

4.156. Use formal charges to predict whether the atoms in hypochlorous acid are arranged HOCl or HClO.

4.157. **Chemical Weapons** Draw the Lewis structure of phosgene, $COCl_2$, a poisonous gas used in chemical warfare during World War I.

4.158. The dinitramide anion $[N(NO_2)_2^-]$ was first isolated in 1996. The arrangement of atoms in $N(NO_2)_2^-$ is in Figure P4.158.

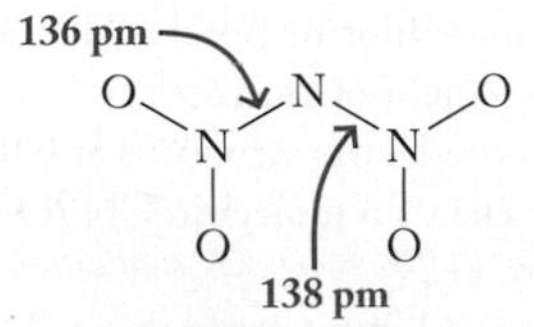

FIGURE P4.158

a. Complete the Lewis structure of $N(NO_2)_2^-$ including any resonance forms, and assign formal charges.

b. Explain why the nitrogen–oxygen bond lengths in $N(NO_2)_2^-$ and N_2O should (or should not) be similar.

c. $N(NO_2)_2^-$ was isolated as $[NH_4^+][N(NO_2)_2^-]$. Draw the Lewis structure of NH_4^+.

*4.159. Silver cyanate (AgOCN) is a source of the cyanate ion (OCN^-). Under certain conditions the species OCN is an anion with a charge of 1–; under others it is a neutral, odd-electron molecule, OCN.

a. Two molecules of OCN combine to form OCNNCO. Draw the Lewis structure of the molecule, including all resonance forms.

b. The OCN^- ion reacts with BrNO, forming the unstable molecule OCNNO. Draw the Lewis structures of BrNO and OCNNO, including all resonance forms.

c. The OCN^- ion reacts with Br_2 and NO_2 to produce N_2O, CO_2, BrNCO, and OCN(CO)NCO. Draw the resonance structures of OCN(CO)NCO, which has the arrangement of atoms shown in Figure P4.159.

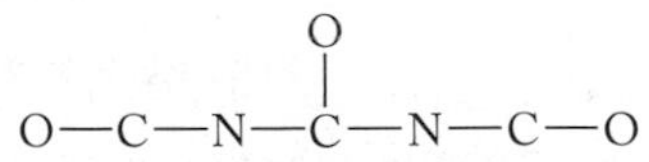

$$O{-}C{-}N{-}\overset{\overset{\displaystyle O}{|}}{C}{-}N{-}C{-}O$$

FIGURE P4.159

*4.160. During the reaction of the cyanate ion (OCN^-) with Br_2 and NO_2, a very unstable substance called an *intermediate* forms and then quickly falls apart. Its formula is O_2NNCO.

a. Draw three of the resonance forms for O_2NNCO, assign formal charges, and predict which of the three contributes the most to the bonding in O_2NNCO. Its skeletal structure is shown in Figure P4.160a.

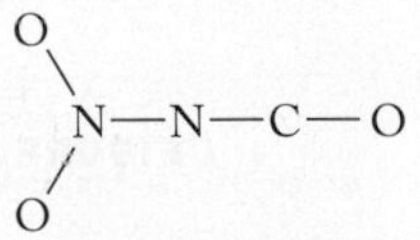

FIGURE P4.160a

b. Draw Lewis structures for a different arrangement of the N, C, and O atoms in O_2NNCO as shown in Figure P4.160b.

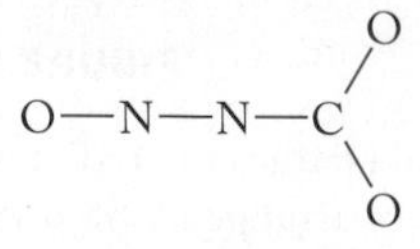

FIGURE P4.160b

4.161. A compound with the formula Cl_2O_6 decomposes to a mixture of ClO_2 and ClO_4. Draw two Lewis structures for Cl_2O_6: one with a chlorine–chlorine bond and one with a Cl—O—Cl arrangement of atoms.

*4.162. A compound consisting of chlorine and oxygen, Cl_2O_7, decomposes to ClO_4 and ClO_3.
   a. Draw two Lewis structures of Cl_2O_7: one with a chlorine–chlorine bond and one with a Cl—O—Cl arrangement of atoms.
   b. Draw the Lewis structure of ClO_3.

*4.163. The odd-electron molecule CN reacts with itself to form cyanogen, C_2N_2.
   a. Draw the Lewis structure of CN, and predict which arrangement for cyanogen is more likely: NCCN or CNNC.
   b. Cyanogen reacts slowly with water to produce oxalic acid ($H_2C_2O_4$) and ammonia; the Lewis structure of oxalic acid is shown in Figure P4.163. Compare the structure to your answer in part a. Do you still believe the structure you selected in part a is the better one?

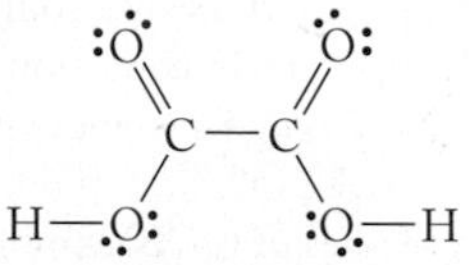

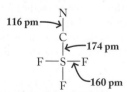

FIGURE P4.163

4.164. The odd-electron molecule SN forms S_2N_2, which has a cyclic structure (the atoms form a ring).
   a. Draw a Lewis structure of SN and complete the possible Lewis structures for S_2N_2 in Figure P4.164.
   b. Which of the two is the preferred structure for S_2N_2?

$$\begin{matrix} S—N & & S—N \\ | \quad | & & | \quad | \\ S—N & & N—S \end{matrix}$$

FIGURE P4.164

*4.165. The molecular structure of sulfur cyanide trifluoride (SF_3CN) has been shown to have the arrangement of atoms with the indicated bond lengths in Figure P4.165. Using the observed bond lengths as a guide, complete the Lewis structure of SF_3CN and assign formal charges.

$$\begin{matrix} & & N & \\ 116\text{ pm} \rightarrow & & \| & \\ & & C & \\ & & |\leftarrow & 174\text{ pm} \\ F— & S & —F & \\ & & | & \\ & & F & \nwarrow 160\text{ pm} \end{matrix}$$

FIGURE P4.165

4.166. Strike-Anywhere Matches Heating phosphorus with sulfur produces P_4S_3, a solid used in the heads of strike-anywhere matches. P_4S_3 has the skeletal structure shown in Figure P4.166. Draw its Lewis structure.

FIGURE P4.166

*4.167. The $TeOF_6{}^{2-}$ anion was first synthesized in 1993. Draw its Lewis structure.

*4.168. Sulfur in the Environment Sulfur is cycled in the environment through compounds such as dimethyl sulfide (CH_3SCH_3), hydrogen sulfide (H_2S), and sulfite and sulfate ions. Draw Lewis structures for the four species. Are expanded valence shells needed to minimize the formal charges for any of the species?

4.169. Antacid Tablets Antacids commonly contain calcium carbonate and/or magnesium hydroxide. Draw the Lewis structures for calcium carbonate and magnesium hydroxide.

4.170. How many pairs of electrons does xenon share in the following molecules and ions? (a) XeF_2; (b) $XeOF_2$; (c) XeF^+; (d) $XeF_5{}^+$; (e) XeO_4

*4.171. A short-lived allotrope of nitrogen, N_4, was reported in 2002.
   a. Draw the Lewis structures of all the resonance forms of linear N_4 (N—N—N—N).
   b. Assign formal charges, and determine which resonance structure is the best description of N_4.
   c. Draw a Lewis structure of a ring (cyclic) form of N_4, and assign formal charges.

*4.172. Scientists have predicted the existence of O_4 even though the molecule has never been observed. However, $O_4{}^{2-}$ has been detected. Draw the Lewis structures for O_4 and $O_4{}^{2-}$.

4.173. Which of the following molecules and ions contains an atom with an expanded valence shell? (a) Cl_2; (b) ClF_3; (c) ClI_3; (d) ClO^-

4.174. Which of the following molecules contains an atom with an expanded valence shell? (a) XeF_2; (b) $GaCl_3$; (c) ONF_3; (d) SeO_2F_2

*4.175. A linear nitrogen anion, $N_5{}^-$, was isolated for the first time in 1999.
   a. Draw the Lewis structures for four resonance forms of linear $N_5{}^-$.
   b. Assign formal charges to the atoms in the structures in part a, and identify the structures that contribute the most to the bonding in $N_5{}^-$.
   c. Compare the Lewis structures for $N_5{}^-$ and $N_3{}^-$. In which ion do the nitrogen–nitrogen bonds have the higher average bond order?

*4.176. Carbon tetroxide (CO_4) was discovered in 2003.
  a. Draw the Lewis structure of CO_4 based on the skeletal structure shown in Figure P4.176.

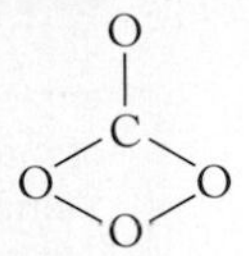

FIGURE P4.176

  b. Are there any resonance forms of the structure you drew that have zero formal charges on all atoms?
  c. Can you draw a structure in which all four oxygen atoms in CO_4 are bonded to carbon?

4.177. Plot the electronegativities of elements with $Z = 3$ to 9 (y-axis) versus their first ionization energy (x-axis). Is the plot linear? Use your graph to predict the electronegativity of neon, whose first ionization energy is 2081 kJ/mol.

*4.178. In the typical Lewis structure of BF_3 there are only 6 valence electrons on the boron atom and each B—F bond is a single bond. However, the length and strength of these bonds indicate that they have a small measure of double bond character, that is, their bond order is slightly greater than 1.
  a. Draw a Lewis structure, including all resonance structures, of BF_3 in which there is one B=F double bond.
  b. What is the formal charge on the B atom, and what is the average formal charge on each F atom?
  c. Based on formal charges alone, what should be the bond order of each B—F bond in BF_3?
  d. What factor might support a bond order slightly greater than 1?

4.179. The cation N_2F^+ is isoelectronic with N_2O.
  a. What does it mean to be isoelectronic?
  b. Draw the Lewis structure of N_2F^+. (*Hint*: The molecule contains a nitrogen–nitrogen bond.)
  c. Which atom has the +1 formal charge in the structure you drew in part b?
  d. Does N_2F^+ have resonance forms?
  e. Could the middle atom in the N_2F^+ ion be a fluorine atom? Explain your answer.

4.180. **Ozone Depletion** Methyl bromide (CH_3Br) is produced naturally by fungi. Methyl bromide has also been used in agriculture as a fumigant, but its use is being phased out because the compound has been linked to ozone depletion in the upper atmosphere.
  a. Draw the Lewis structure of CH_3Br.
  b. Which bond in CH_3Br is more polar, carbon–hydrogen or carbon–bromine?

5

Bonding Theories
Explaining Molecular Geometry

LEARNING OUTCOMES

LO1 Use VSEPR and the concept of steric number to predict the bond angles in molecules and the shapes of molecules with one central atom
Sample Exercises 5.1, 5.2, 5.3

LO2 Predict whether a substance is polar or nonpolar based on its molecular structure
Sample Exercise 5.4

LO3 Use valence bond theory to explain bond angles and molecular shape
Sample Exercise 5.5

LO4 Use atomic orbital hybridization to visualize molecular shape
Sample Exercise 5.6

LO5 Recognize chiral molecules
Sample Exercise 5.7

LO6 Draw molecular orbital (MO) diagrams of small molecules and use MO theory to predict bond order and explain magnetic properties and spectra
Sample Exercises 5.8, 5.9, 5.10

Biological Activity and Molecular Shape

Hold your hands out in front of you, palms up, fingers extended. Now rotate your wrists inward so that your thumbs point straight up. Your right hand looks the same as the image your left hand makes in a mirror. Does that mean that your two hands have the same shape? If you ever tried to put your right hand in a glove made for your left, you know that they do not have the same shape—very similar, but not the same. Many other objects in our world have a "handedness" about them, from scissors to golf clubs to the conch shells that wash up on tropical beaches.

This chapter focuses on the importance of shape at the molecular level. For example, the compound that produces the refreshing aroma of spearmint has the molecular formula $C_{10}H_{14}O$. The compound responsible for the musty aroma of caraway seeds has the same molecular formula *and* the same Lewis structure. To understand how two compounds could be so much alike and still have different properties, we have to consider their structures in three dimensions.

We perceive a difference in their aromas in part because each molecule has a unique site where it attaches to our nasal membranes. Just as a left hand only fits a left glove, the spearmint molecule fits only the spearmint-shaped site, and the caraway molecule fits only the caraway-shaped site. This phenomenon is called molecular recognition, and it enables biomolecular structures, such as nasal membranes, to recognize and react when a particular molecule with a specific shape fits into a part of the structure known as an *active site*. Many substances in the foods we eat and in the pharmaceuticals we take exert physiological effects because they are recognized by and bind to active sites in the molecules that make up our body.

What are the three-dimensional shapes of molecules, and what determines those shapes? Can we predict shapes if we know how atoms are bonded together in molecules? In this chapter, we examine theories of bonding that explain molecular shapes, and we begin to explore the impact of molecular shape on the physical, chemical, and biological properties of compounds.

Chiral Shells Conch shells are made mostly of $CaCO_3$ that grows outward in a clockwise spiral. Directional preference is common in nature and is linked to a phenomenon called chirality. ▶

5.1 Molecular Shape

The shape of a molecule can affect many properties of the substance, including its physical state at room temperature, its aroma, its biological activity, and its distribution in the environment. In Chapter 4 we drew Lewis structures to account for bonding in molecules, but Lewis structures are only two-dimensional representations of how atoms and the electron pairs that surround them are arranged in molecules. Lewis structures show how atoms are *connected* in molecules, but they don't show how the atoms are *oriented* in three dimensions, nor do they necessarily show the overall shape of the molecule.

To illustrate this point, let's consider the Lewis structures and ball-and-stick models (Figure 5.1) of two compounds: carbon dioxide and methane. The linear array of atoms and bonding electrons in the Lewis structure of CO_2 corresponds to the actual linear shape of the molecule as represented by the ball-and-stick model. The angle between the two C=O bonds is 180°, just as in the Lewis structure. On the other hand, the Lewis structure for methane does not convey the true orientation of the four C—H bonds in each molecule. The 90° angles between bonds in methane's Lewis structure do not match the actual H—C—H **bond angles** of 109.5°.

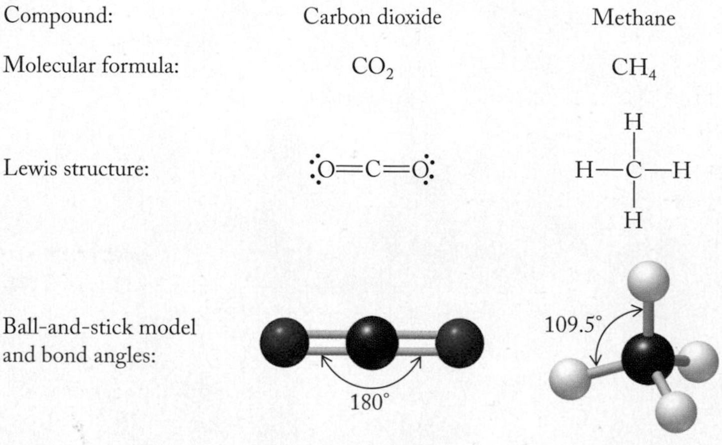

In this chapter, we explore theories of covalent bonding that account for and predict the shapes of molecules. We focus on small molecules with single central atoms, though the theories can be applied to molecules of any size.

Compound:	Carbon dioxide	Methane
Molecular formula:	CO_2	CH_4
Lewis structure:	:O=C=O:	H—C—H (with H above and below)
Ball-and-stick model and bond angles:	180°	109.5°

FIGURE 5.1 The Lewis structure of CO_2 matches its true molecular structure, but the Lewis structure of CH_4 does not because the C—H bonds in CH_4 actually extend in three dimensions.

▶❙❙ **CHEMTOUR** VSEPR Model

5.2 Valence-Shell Electron-Pair Repulsion Theory (VSEPR)

Let's start with a theory based on a fundamental chemical principle: electrons have negative charges and repel each other. **Valence-shell electron-pair repulsion theory (VSEPR)** applies this principle by assuming that pairs of valence electrons are arranged about central atoms in ways that minimize repulsions between the pairs. To predict molecular shape using VSEPR, we must consider two things: **electron-group geometry**, which defines the relative positions in three-dimensional space of all the bonding pairs and lone pairs of valence electrons on the central atom, and **molecular geometry**, which defines the relative positions of the atoms in a molecule. To accurately predict molecular geometry we first need to know electron-group geometry. If there are no lone pairs of electrons, then the process is simplified because the electron-group geometry *is* the molecular geometry. Let's begin with this simpler case and consider the shapes of molecules that have different numbers of atoms bonded to a central atom that has no lone pairs of electrons. To make our exploration even simpler, we will initially focus on molecules in which the same kind of atom is bonded to each central atom, so that all the bonds in each molecule are identical.

Central Atoms with No Lone Pairs

To determine the geometry of a molecule, we start by drawing its Lewis structure. From the Lewis structure, we determine the **steric number (SN)** of the central atom, which is the sum of the number of atoms bonded to that atom and the number of lone pairs on it:

$$SN = \left(\begin{array}{c}\text{number of atoms}\\\text{bonded to central atom}\end{array}\right) + \left(\begin{array}{c}\text{number of lone pairs}\\\text{on central atom}\end{array}\right) \qquad (5.1)$$

In molecules in which the central atom has no lone pairs, the *steric number equals the number of atoms bonded to the central atom*. In evaluating the shapes of the molecules, we will generate five common shapes that describe both electron-group geometries and molecular geometries.

The simplest molecular structure with a central atom is one that has only two other atoms bonded to the central atom. As long as there are no lone pairs of valence electrons on the central atom, its steric number is 2. How do the electron pairs in the two bonds arrange themselves to minimize their mutual repulsion? The answer is that they are as far from each other as possible on opposite sides of the central atom (Figure 5.2a). This gives a *linear* electron-group geometry and a **linear** molecular geometry. The three atoms in the molecule are arranged in a straight line, and the angle between the two bonds is 180°.

If the central atom is bonded to three other atoms and has no lone pairs of valence electrons, then SN = 3. The three bonding groups are as far apart as possible when they are located at the three corners of an equilateral triangle. The angle between each group of bonds is 120° (Figure 5.2b). The name of this electron-group and molecular geometry is **trigonal planar**.

If the central atom is bonded to four atoms and has no lone pairs, then SN = 4. The atoms bonded to the central atom occupy the four vertices of a *tetrahedron*, which is a four-sided pyramid (*tetra* is Greek, meaning "four"). The bonding pairs form bond angles of 109.5° with each other as shown in Figure 5.2(c). The electron-group geometry and molecular geometry are both **tetrahedral**.

When a central atom is bonded to five atoms and has no lone pairs, SN = 5 and the five other atoms occupy the five corners of two triangular pyramids that share the same base. The central atom of the molecule is at the common center of the two bases as shown in Figure 5.2(d). One bonding pair points to the top of the upper pyramid, one points to the bottom of the lower pyramid, and the other three point to the three vertices of the shared triangular base. It may be helpful to think of the three vertices as three points along a circle that is the equator of a sphere. In this model, the atoms that occupy the three sites and the bonds that connect them to the central atom are called *equatorial* atoms and bonds. The bond angles between the three equatorial bonds are 120° (just as in the triangle of the trigonal planar geometry in Figure 5.2b). The bond angle between an equatorial

valence-shell electron-pair repulsion theory (VSEPR) a model predicting the arrangement of valence electron pairs around a central atom that minimizes their mutual repulsion to produce the lowest energy orientations.

electron-group geometry the three-dimensional arrangement of bonding pairs and lone pairs of electrons about a central atom.

molecular geometry the three-dimensional arrangement of the atoms in a molecule.

steric number (SN) the sum of the number of atoms bonded to a central atom plus the number of lone pairs of electrons on the central atom.

linear molecular geometry about a central atom with a steric number of 2 and no lone pairs of electrons.

trigonal planar molecular geometry about a central atom with a steric number of 3 and no lone pairs of electrons.

tetrahedral molecular geometry about a central atom with a steric number of 4 and no lone pairs of electrons.

FIGURE 5.2 Electron-group geometries depend on the steric number (SN) of the central atom in a molecule. In these images there are no lone pairs of electrons on the central atoms (red dots), so the SN equals the number of atoms bonded to the central atom, and *electron-group geometry* is the same as *molecular geometry*. The black lines represent covalent bonds; the blue lines outline the geometric forms that give these shapes their names.

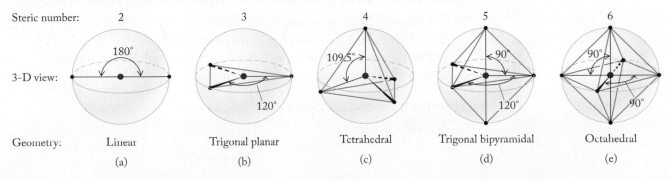

Steric number:	2	3	4	5	6
3-D view:	180°	120°	109.5°	90° / 120°	90° / 90°
Geometry:	Linear	Trigonal planar	Tetrahedral	Trigonal bipyramidal	Octahedral
	(a)	(b)	(c)	(d)	(e)

trigonal bipyramidal molecular geometry about a central atom with a steric number of 5 and no lone pairs of electrons, in which three atoms occupy equatorial sites and two other atoms occupy axial sites above and below the equatorial plane.

octahedral molecular geometry about a central atom with a steric number of 6 and no lone pairs of electrons, in which all six sites are equivalent.

bond and either vertical, or *axial*, bond is 90°, and the angle between the two axial bonds is 180°. A molecule in which the atoms are arranged this way is said to have a **trigonal bipyramidal** electron-group and molecular geometry.

For SN = 6, picture two pyramids that have a square base (Figure 5.2e). Put them together, base to base, and you form a shape in which all six positions are equivalent. We can think of the six bonding pairs of electrons as three sets of two electron pairs each. The two pairs in each set are oriented at 180° to each other and at 90° to the other two pairs, just like the axes of an *xyz* coordinate system. In our spherical model, there are four equatorial atoms at the four vertices of the common square base. The bonds to them are 90° apart. Another atom lies at the top of the upper pyramid, and a sixth lies at the bottom of the lower one. This arrangement defines **octahedral** electron-group and molecular geometries.

CONCEPT TEST ···

Assume that all of the bonds in molecules with the five shapes described above are single bonds and that none of the central atoms have lone pairs of electrons. Which SN values correspond to central atoms with less than an octet of valence electrons, which values correspond to central atoms with an expanded octet, and which value has a central atom with exactly 8 valence electrons?

(Answers to Concept Tests are in the back of the book.)

···

Now let's consider a real-world analogy to the above bond orientations. We start with a batch of fully inflated balloons. We tie together clusters of two, three, four, five, and six balloons (Figure 5.3) so tightly that they push against each other. If the tie points of the clusters represent the central atom in our balloon model, then the opposite ends represent the atoms that are bonded to the central atom. Note how our clusters of two, three, four, five, and six balloons produce the same orientations that resulted from selecting points that were as far apart as possible. The long axes of the balloons in Figure 5.3 provide an accurate representation of the bond directions in Figure 5.2.

FIGURE 5.3 Balloons tied tightly together push against each other and orient themselves as far away from each other as possible. In doing so they mimic the locations of different numbers of electron pairs about a central atom. Each balloon represents one electron pair. Note the similarities between the patterns of the balloons and the geometric shapes in Figure 5.2.

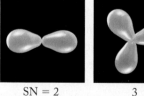   

SN = 2   3   4   5   6

Now let's look at five simple molecules that have no lone pairs about the central atom and apply VSEPR and the concept of steric number to predict their electron-group and molecular geometries. To do so, we follow these three steps:

1. Draw the Lewis structure for the molecule.
2. Determine the steric number of the central atom.
3. Use the steric number to predict the electron-group and molecular geometries using the images in Figure 5.2.

Example I: Carbon Dioxide, CO_2

1. Lewis structure:

$$\ddot{\text{O}}\!=\!\text{C}\!=\!\ddot{\text{O}}$$

2. The central carbon atom has two atoms bonded to it and no lone pairs, so the steric number is 2.
3. Because SN = 2, the O=C=O bond angle is 180° (see Figure 5.2a) and the electron-group and molecular geometries are both linear.

Example II: Boron Trifluoride, BF₃

1. The Lewis structure of BF₃ that results in zero formal charges on all atoms is

2. There are three fluorine atoms bonded to the central boron atom, and there are no lone pairs on boron. Therefore, SN = 3.
3. If SN = 3 then all four atoms lie in the same plane. The three F atoms form an equilateral triangle, and the F—B—F bond angles are 120°. The electron-group and molecular geometries are trigonal planar (Figure 5.4).

CONNECTION We learned in Chapter 4 that some molecules have central atoms with incomplete octets of electrons. The boron atom in BF₃ appears to fall into that category, though there are other bonding options in BF₃ (see Problem 5.128).

FIGURE 5.4 The ball-and-stick model shows the orientation of the atoms in boron trifluoride. All F—B—F bond angles are 120° in this trigonal planar molecular geometry.

Example III: Carbon Tetrachloride, CCl₄

1. Lewis structure:

2. There are four chlorine atoms bonded to the central carbon atom, which gives carbon a full octet. Therefore, SN = 4.
3. If SN = 4 then the chlorine atoms are located at the vertices of a tetrahedron and all four Cl—C—Cl bond angles are 109.5°, producing tetrahedral electron-group and molecular geometries (Figure 5.5).

Before we explore any more molecular geometries, some comments about the conventions used in drawing three-dimensional structures on two-dimensional paper are in order. To convey the structure of a molecule in three dimensions, we use a solid wedge (▬) to indicate a bond that comes out of the paper toward the viewer. The solid wedge in the CCl₄ structure in Figure 5.5, for example, means that the chlorine atom in that position points toward the viewer at a downward angle. A dashed wedge (⋯) indicates a bond that goes into the paper away from the viewer. Solid lines are used for bonds that lie in the plane of the paper.

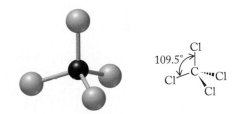

FIGURE 5.5 The ball-and-stick model shows the orientation of the atoms in carbon tetrachloride. All Cl—C—Cl bond angles are 109.5° in this tetrahedral molecular geometry.

Example IV: Phosphorus Pentafluoride, PF₅

1. Lewis structure:

2. There are five fluorine atoms bonded to the central phosphorus atom, which has no lone pairs. Therefore, SN = 5.

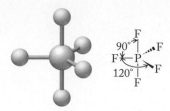

FIGURE 5.6 The ball-and-stick model shows the orientation of the atoms in phosphorus pentafluoride. The equatorial P—F bonds are 120° from each other. Each of them is 90° from the two bonds connecting the central phosphorus atom to the fluorine atoms in the axial positions. This molecular geometry is trigonal bipyramidal.

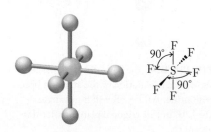

FIGURE 5.7 The ball-and-stick model shows how the six fluorine atoms are oriented in three dimensions about the central sulfur atom in this octahedral molecular geometry.

3. If SN = 5 then the fluorine atoms are located at the vertices of a trigonal bipyramid and the electron-group geometry and molecular geometry are trigonal bipyramidal (Figure 5.6).

Example V: Sulfur Hexafluoride, SF₆

1. Lewis structure:

2. There are six fluorine atoms bonded to the central sulfur atom, which has no lone pairs. Therefore, SN = 6.
3. If SN = 6 then the fluorine atoms are located at the vertices of an octahedron and the electron-group and molecular geometries are octahedral (Figure 5.7).

SAMPLE EXERCISE 5.1 **Using VSEPR to Predict Geometry I** **LO1**

Formaldehyde, CH₂O, is a gas at room temperature and is an ingredient in solutions used to preserve biological samples. Use VSEPR to predict the molecular geometry of formaldehyde.

COLLECT AND ORGANIZE We are given the molecular formula of formaldehyde. If there are no lone pairs of valence electrons on the central atom, the solution requires (1) drawing the Lewis structure, (2) determining the steric number, and (3) identifying the molecular geometry using Figure 5.2.

ANALYZE Carbon is the likely central atom of the molecule because it has a bonding capacity of 4 and is less electronegative than oxygen, whose bonding capacity is 2. If the three atoms bonded to the carbon atom are as far from each other as possible, the probable molecular geometry is trigonal planar.

SOLVE Following the procedure developed in Chapter 4 for drawing Lewis structures, we obtain this one for formaldehyde:

There are three atoms bonded to the central atom and no lone pairs on it, so SN = 3 and the molecular geometry is trigonal planar.

THINK ABOUT IT The key to predicting the correct molecular geometry of a molecule with no lone pairs on its central atom is to determine the steric number, which is simply a matter of counting the number of atoms bonded to the central atom.

Practice Exercise Use VSEPR to determine the molecular geometry of the chloroform molecule, CHCl₃, and draw the molecule using the solid-wedge, dashed-wedge convention.

(Answers to Practice Exercises are in the back of the book.)

---

angular or **bent** molecular geometry about a central atom with a steric number of 3 and one lone pair or a steric number 4 and two lone pairs.

Measurements of the bond angles in formaldehyde show that the H—C—H bond angle is slightly smaller than the 120° predicted for a trigonal planar geometry (Figure 5.2b), and the H—C—O bond angles are about 1° larger. The C=O double bond consists of two pairs of bonding electrons that exert greater repulsion than a single bonding pair would. This greater repulsion decreases the H—C—H bond angle. VSEPR does not enable us to predict the actual values of those bond angles, but it does allow us to correctly predict how bond angles deviate from the ideal values of an idealized trigonal planar molecule.

(a) Electron-group geometry = trigonal planar

(b) Molecular geometry = bent

FIGURE 5.8 (a) The electron-group geometry of O_3 is trigonal planar because the steric number of the central oxygen atom is 3 (it is bonded to two atoms and has one lone pair of electrons). (b) The molecular geometry is bent because there is no bonded atom, only a lone pair of electrons, on the central oxygen atom in the structure.

Central Atoms with Lone Pairs

We have explored the electron-group and molecular geometries of molecules whose central atoms have no lone pairs of electrons. Now let's see what happens when a central atom has one or more lone pairs. If SN = 2 and one of the electron groups on the central atom is a lone pair, then there is only one other atom in the molecule (Equation 5.1). Since two points define a straight line, two-atom molecules are all linear and they have no bond angles.

Moving on to SN = 3, let's revisit the angular shape of ozone molecules:

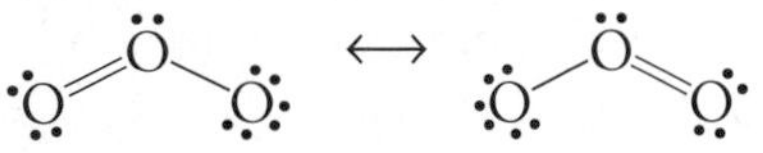

To calculate the steric number of the central O atom we need to add up the number of atoms and lone pairs of electrons that surround it. In both resonance structures, the central O atom is bonded to two other atoms and has one lone pair of electrons. Therefore its SN value in either resonance structure is 2 + 1 = 3. When SN = 3, the arrangement of atoms and lone pairs about the central atom is trigonal planar (Figure 5.2b). This means that the electron-group geometry, which takes into account both the atoms and the nonbonding pairs around the central atom, is trigonal planar (as shown in Figure 5.8a). However, the dashed lines in the structures in Figure 5.8(a) do not indicate a bond but rather point to a lone pair of electrons. The lone pair is located on the central O atom, and the attachment of two more oxygen atoms gives the **angular** (or **bent**) molecular geometry shown in Figure 5.8(b).

Experimental measurements establish that O_3 is indeed a bent molecule with a bond angle of 117°. The angle is smaller than the 120° we would expect in a symmetrical trigonal planar electron-group geometry. We can explain the smaller angle using VSEPR by comparing the amount of space occupied by bonding electrons with the amount of space occupied by electrons in a lone pair. Because the bonding electrons are attracted to two nuclei, they have a high probability of being located between the two atomic centers that share them. In contrast, the lone pair is not shared with a second atom and is spread out near the central O atom as shown in Figure 5.9. This puts the lone pair of electrons closer to the

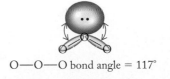

O—O—O bond angle = 117°

FIGURE 5.9 The lone pair of electrons on the central oxygen atom in O_3 occupies more space (larger purple region) than the electron pairs in the O—O bonds (smaller purple regions). The increased repulsion (represented by the double-headed arrows) resulting from the larger volume occupied by the lone pair forces the noncentral oxygen atoms closer together, making the bond angle 117° instead of 120°.

bonding pairs and produces greater repulsion. As a result, the lone pair pushes the bonding pairs closer together, thereby reducing the bond angle. In general,

➤ repulsion between lone pairs and bonding pairs is greater than repulsion between bonding pairs;

➤ repulsion caused by a lone pair is greater than repulsion caused by a double bond;

➤ repulsion caused by a double bond is greater than that caused by a single bond;

➤ two lone pairs of electrons on a central atom exert a greater repulsive force on the atom's bonding pairs than does one lone pair.

SAMPLE EXERCISE 5.2   **Predicting Relative Sizes of Bond Angles**   **LO1**

Rank NH_3, CH_4, and H_2O in order of decreasing bond angles in their molecular structures.

COLLECT AND ORGANIZE We are asked to predict the relative size of the bond angles in three molecules. We are given their molecular formulas. The information in Figure 5.2 links steric numbers to electron-group geometries and bond angles. Bond angles are also influenced by the presence of double bonds and lone pairs of electrons on the central atom.

ANALYZE We need to determine the steric numbers of the central atoms in the three molecules. To do that we first need to translate the molecular formulas into Lewis structures to determine how many lone pairs and/or double bonds they have.

SOLVE Using the method for drawing Lewis structures from Chapter 4 we obtain these results:

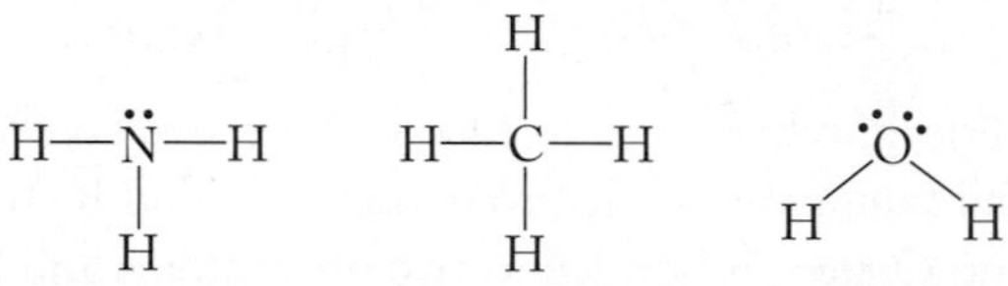

In each of the structures the central atom is surrounded by a total of four atoms or lone pairs of electrons, which means that each has a steric number of 4. There are no double bonds.

According to Figure 5.2, the electron-group geometry for SN = 4 is tetrahedral. In a molecule such as CH_4 in which all four tetrahedral electron pairs are equivalent bonding pairs, all bond angles are 109.5°. However, greater repulsion from the lone pairs of electrons in NH_3 and H_2O squeezes those molecules' bonds together, reducing the bond angles. The two lone pairs on the O atom in H_2O exert a greater repulsive force on its bonding pairs than the single lone pair on the N atom exerts on the bonding pairs in NH_3. Therefore, the bond angle in H_2O should be less than the bond angles in NH_3. Ranking the three molecules in order of decreasing bond angle we have:

$$CH_4 > NH_3 > H_2O$$

THINK ABOUT IT The logic used in answering this problem is supported by experimental evidence: the bond angles in molecules of CH_4, NH_3, and H_2O are 109.5°, 107.0°, and 104.5°, respectively.

Practice Exercise In which of the following species is the bond angle the largest and in which is it the smallest? NO_2, N_2O, and NO_2^- ⚙

(Answers to Practice Exercises are in the back of the book.)

trigonal pyramidal molecular geometry about a central atom with a steric number of 4 and one lone pair of electrons.

As we saw in the preceding Sample Exercise, three possible combinations of atoms and lone pairs are possible about a central atom with a SN of 4: four atoms and no lone pairs, three atoms and one lone pair, or two atoms and two lone pairs. The first case is illustrated by the molecular structure of methane, CH_4, in which a tetrahedral electron-group geometry translates into a tetrahedral molecular geometry. In a molecule of ammonia, NH_3, there are only three bonding pairs and one lone pair. This means that the Lewis structure in Figure 5.10(a) translates into a tetrahedral electron-group geometry in which one of the vertices is the lone pair on the N atom (Figure 5.10b). The resulting molecular geometry, which is essentially that of a tetrahedron that has been squashed a little, is called **trigonal pyramidal** (Figure 5.10c). As we discussed in Sample Exercise 5.2, the strong repulsion produced by the diffuse lone pair of electrons on the N atom pushes the N—H bonds closer together in NH_3 and reduces the angles between them as compared to the H—C—H bond angles in CH_4. Thus, the H—N—H bond angle in ammonia is only 107.0°.

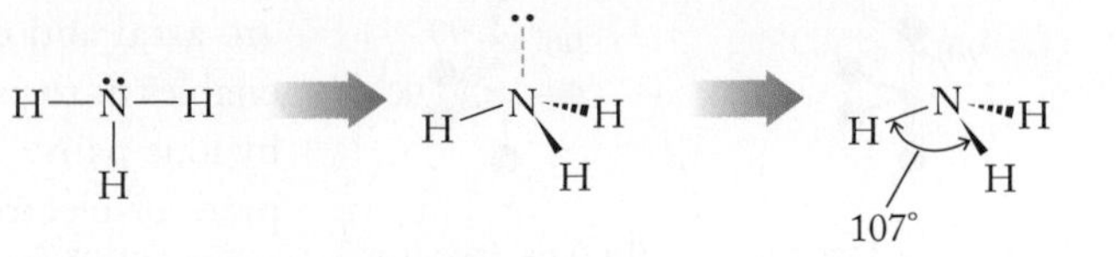

The central O atom in a molecule of H_2O has a SN of 4 because it is bonded to two atoms and has two lone pairs of electrons. The result is a tetrahedral electron-group geometry (Figure 5.11). The presence of two lone pairs of electrons means that two of the four tetrahedral vertices are missing, which leaves us with a bent (or angular) molecular geometry. As we also discussed in Sample Exercise 5.2, the H—O—H bond angle in water is smaller than 109.5° due to repulsion between the two lone pairs and each bonding pair. The actual bond angle is 104.5°.

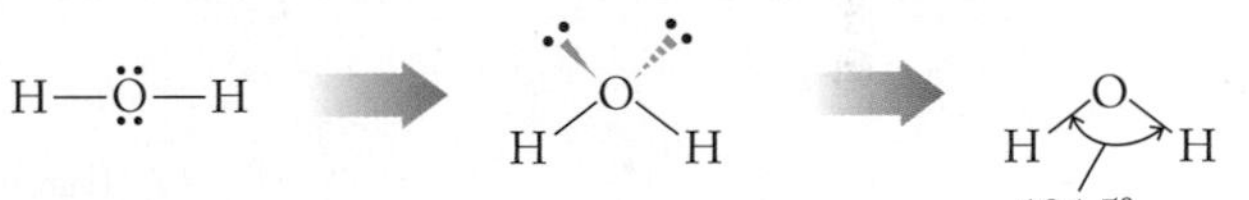

Molecules with trigonal bipyramidal electron-group geometry (SN = 5) have four possible molecular geometries depending on the number of lone pairs per molecule (see Table 5.1). These options are linked to the fact that there

(a) Lewis structure (b) Tetrahedral electron-group geometry (c) Trigonal pyramidal molecular geometry

FIGURE 5.10 The steric number of the N atom in NH_3 is 4 (it is bonded to three atoms and has one lone pair of electrons), so its electron-group geometry is tetrahedral. However, one of the vertices of the tetrahedron is occupied by a lone pair of electrons, not an atom, so the molecular geometry is trigonal pyramidal.

(a) Lewis structure (b) Tetrahedral electron-group geometry (c) Bent (angular) molecular geometry

FIGURE 5.11 The steric number of the O atom in H_2O is 4 (it is bonded to two atoms and has two lone pairs of electrons), so its electron-group geometry is tetrahedral. However, two of the vertices of the tetrahedron are occupied by lone pairs of electrons, meaning that only three atoms define the molecular geometry, which is bent.

TABLE 5.1 Electron-Group and Molecular Geometries

SN = 4	Electron-Group Geometry	No. of Bonded Atoms	No. of Lone Pairs	Molecular Geometry
	Tetrahedral	4	0	Tetrahedral
	Tetrahedral	3	1	Trigonal pyramidal
	Tetrahedral	2	2	Bent (angular)
SN = 5	**Electron-Group Geometry**	**No. of Bonded Atoms**	**No. of Lone Pairs**	**Molecular Geometry**
	Trigonal bipyramidal	5	0	Trigonal bipyramidal
	Trigonal bipyramidal	4	1	Seesaw
	Trigonal bipyramidal	3	2	T-shaped
	Trigonal bipyramidal	2	3	Linear
SN = 6	**Electron-Group Geometry**	**No. of Bonded Atoms**	**No. of Lone Pairs**	**Molecular Geometry**
	Octahedral	6	0	Octahedral
	Octahedral	5	1	Square pyramidal
	Octahedral	4	2	Square planar
	Octahedral	3	3	Although these geometries are possible, we will not encounter any molecules with them.
	Octahedral	2	4	

(a) Equatorial lone pair   (b) Axial lone pair

FIGURE 5.12 A lone pair of electrons in an equatorial position of a molecule with trigonal bipyramidal electron-group geometry interacts through 90° with two other electron pairs. A lone pair in an axial position interacts through 90° with *three* other electron pairs. Fewer 90° interactions reduce internal electron-pair repulsion and lead to greater stability.

FIGURE 5.13 A single lone pair of electrons in an equatorial position of a trigonal bipyramidal electron-group geometry produces a seesaw molecular geometry.

FIGURE 5.14 Two lone pairs of electrons in equatorial positions of a trigonal bipyramidal electron-group geometry produce a T-shaped molecular geometry.

(a) Trigonal bipyramidal electron-group geometry   (b) Linear molecular geometry

FIGURE 5.15 Three lone pairs of electrons in equatorial positions of a trigonal bipyramidal electron-group geometry produce a linear molecular geometry.

are axial and equatorial vertices in a trigonal bipyramid (Figure 5.2d). VSEPR enables us to predict which vertices are occupied by atoms and which are occupied by lone pairs. The key to these predictions is the fact that the repulsions between pairs of electrons decrease as the angle between them increases: two electron pairs at 90° experience a greater mutual repulsion than two at 120°, which have a greater repulsion than two at 180°. To minimize repulsions involving lone pairs, VSEPR predicts that they preferentially occupy equatorial rather than axial vertices. Why? Because an equatorial lone pair has *two* 90° repulsions with the two axial electron pairs as shown in Figure 5.12(a), but an axial lone pair has *three* 90° repulsions (with three equatorial electron pairs) as shown in Figure 5.12(b).

When we assign one, two, or three lone pairs of valence electrons to equatorial vertices, we get three of the molecular geometries for SN = 5 in Table 5.1. When a single lone pair occupies an equatorial site we get a molecular geometry called **seesaw** (Figure 5.13) because its shape, when rotated 90° clockwise, resembles a playground seesaw. (The formal name for this shape is *disphenoidal*.)

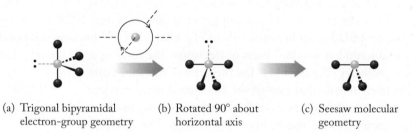

(a) Trigonal bipyramidal electron-group geometry   (b) Rotated 90° about horizontal axis   (c) Seesaw molecular geometry

When two lone pairs occupy equatorial sites, the molecular geometry that results is called **T-shaped**. This designation becomes more apparent when the structure in Figure 5.14(a) is rotated 90° counterclockwise. Lone-pair repulsions result in bond angles in T-shaped molecules that are slightly less than the 90° and 180° angles we would expect from a perfectly shaped "T" geometry.

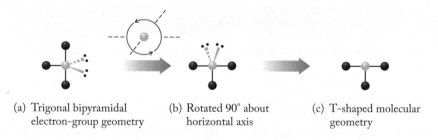

(a) Trigonal bipyramidal electron-group geometry   (b) Rotated 90° about horizontal axis   (c) T-shaped molecular geometry

Finally, a SN = 5 molecule with three lone pairs and two bonded atoms has a linear geometry because all three lone pairs occupy equatorial sites and the bonding pairs are in the two axial positions (Figure 5.15).

CONCEPT TEST •

The bond angles in a trigonal bipyramidal molecular structure are either 90°, 120°, or 180°. Are the corresponding bond angles in a seesaw structure likely to be larger than, the same as, or smaller than those values?

(Answers to Concept Tests are in the back of the book.)

• •

Molecules that have a central atom with a steric number of 6 have an octahedral electron-group geometry (Figure 5.2e) and the molecular geometries listed in Table 5.1. With one lone pair of electrons, there is only one possible molecular geometry because all the sites in an octahedron are equivalent (Figure 5.16). That one molecular geometry is called **square pyramidal**: a pyramid with a square base, four triangular sides, and the central atom "embedded" in the base.

Because of stronger repulsion between lone pairs and bonding pairs, we predict the bond angles in a square pyramidal molecule to be slightly less than the ideal angle of 90°. The molecule BrF_5, for example, has a square pyramidal molecular geometry, and the angles between its equatorial and axial bonds are 85°.

When two lone pairs are present at the central atom in a molecule with octahedral electron-group geometry, they occupy vertices on opposite sides of the octahedron to minimize the interactions between them. The resultant molecular geometry is called **square planar** because the molecule is shaped like a square and all five atoms reside in the same plane (Figure 5.17). Because the two lone pairs are on opposite sides of the bonding pairs, the presence of the lone pairs does not distort the bond angles: they are all 90°.

(a) Octahedral electron-group geometry

(b) Square pyramidal molecular geometry

FIGURE 5.16 A lone pair of electrons in an octahedral electron-group geometry produces a square pyramidal molecular geometry.

(a) Octahedral electron-group geometry

(b) Square planar molecular geometry

FIGURE 5.17 Two lone pairs of electrons on opposite sides of an octahedral electron-group geometry produce a square planar molecular geometry.

SAMPLE EXERCISE 5.3 — Using VSEPR to Predict Geometry II — LO1

The Lewis structure of sulfur tetrafluoride (SF_4) is

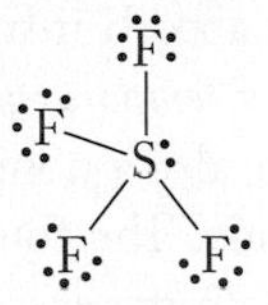

$$\ddot{F}-\ddot{S}\begin{matrix}:\ddot{F}:\\ \\ :\ddot{F}\quad\ddot{F}:\end{matrix}$$

What is its molecular geometry and what are the angles between the S—F bonds?

COLLECT AND ORGANIZE We are given the Lewis structure of SF_4 and can determine from it the steric number of the central atom in the molecule and its electron-group geometry.

ANALYZE Four atoms are bonded to the central S atom, which also has one lone pair of electrons. This means that SN = 5.

SOLVE With a steric number of 5 for its central atom, the electron-group geometry of SF_4 is trigonal bipyramidal. The presence of one lone pair of electrons on the S atom means that its molecular geometry is not the same as its electron-group geometry. Table 5.1 tells us that the molecular geometry that results when a lone pair occupies one of the three equatorial positions in a trigonal bipyramidal electron-group geometry is seesaw:

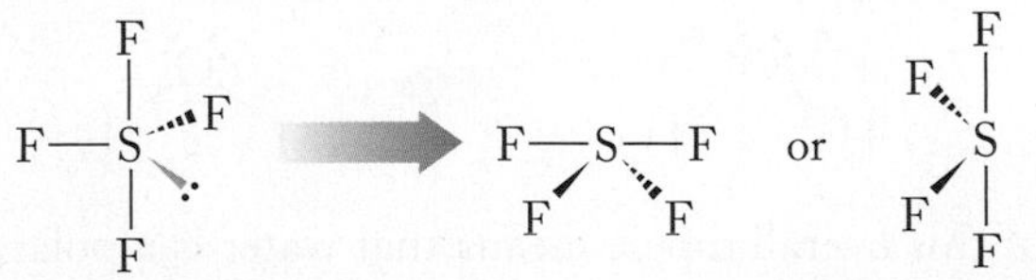

$$F-\underset{F}{\overset{F}{S}}\cdots F \quad \longrightarrow \quad F-\underset{F}{\overset{F}{S}}-F \quad \text{or} \quad \underset{F}{\overset{F}{S}}$$

The equatorial lone pair reduces the bond angles from their normal values of 90° between the axial and equatorial bonds, 120° between the two equatorial bonds, and 180° between the two axial bonds.

THINK ABOUT IT Any molecule with SN = 5 and one lone pair should have the same geometry as SF_4.

Practice Exercise Determine the molecular geometry and the bond angles of SO_2Cl_2. ⬡

(Answers to Practice Exercises are in the back of the book.)

---

seesaw molecular geometry about a central atom with a steric number of 5 and one lone pair of electrons in an equatorial position; the atoms occupy two axial sites and two equatorial sites.

T-shaped molecular geometry about a central atom with a steric number of 5 and two lone pairs of electrons that occupy equatorial positions; the atoms occupy two axial sites and one equatorial site.

square pyramidal molecular geometry about a central atom with a steric number of 6 and one lone pair of electrons; as typically drawn, the atoms occupy four equatorial sites and one axial site.

square planar molecular geometry about a central atom with a steric number of 6 and two lone pairs of electrons that occupy axial sites; the atoms occupy four equatorial positions.

5.3 Polar Bonds and Polar Molecules

We learned in Chapter 4 that an unequal distribution of bonding electrons between two atoms produces a partial negative charge on one end of the bond and a partial positive charge on the other. This charge separation, caused by the difference in electronegativity between the two atoms, makes a **bond dipole**. Now that we know something about the three-dimensional geometry of molecules, we can investigate how bond dipoles contribute to a molecule's *overall* polarity.

The polarity of a molecule can be determined by summing the polarities of all the bond dipoles in the molecule. This summing must take into account both the strengths of the individual bond dipoles and their orientations with respect to one another. In CO_2, for example, the two dipoles are equivalent in strength because they involve the same atoms and the same kind of bond ($C\!=\!O$). The linear shape of the molecule means that the direction of one $C\!=\!O$ bond dipole is opposite the direction of the other:

$$\overset{\longleftarrow}{\ddot{O}}\!=\!C\!=\!\overset{\longrightarrow}{\ddot{O}}$$

Each dipole exactly offsets the other so that, although the individual bonds are polar, CO_2 is a nonpolar molecule overall. Put another way, CO_2 has no overall **permanent dipole**. We use the adjective *permanent* here because, as we saw in Chapter 4, the vibration of the bonds in molecules like CO_2 can create fluctuating electric fields that allow CO_2 to absorb infrared electromagnetic radiation. The fluctuations create high-frequency *temporary* dipoles even in nonpolar molecules.

Carbon tetrachloride, CCl_4, is also a nonpolar substance even though its molecules contain polar C—Cl bonds. The four C—Cl bond dipoles all have the same strength because the same atoms are involved. The tetrahedral molecular geometry means that the bond dipoles offset one another so that, overall, the CCl_4 molecule is nonpolar:

Water molecules are bent, not linear like molecules of CO_2. Therefore, the dipoles of the two O—H bonds in a molecule of H_2O do not offset each other. Instead, the molecule has an overall permanent dipole with the negative end directed toward the O atom:

The presence of this overall dipole means that water is a polar molecule. Water's polarity leads to interactions between H and O atoms on adjacent H_2O molecules in liquid water or solid ice. Other polar molecules exhibit similar interactions. We will discuss those interactions and their consequences in Chapter 6.

The polarity of a molecule can be determined experimentally by measuring its permanent **dipole moment (μ)**. The value of μ expresses the extent of the overall separation of positive and negative charge in the molecule, and it is determined by measuring the degree to which the molecule aligns with a strong electric field, as shown in Figure 5.18. Note how the negative ends of the HF molecules in the figure are oriented toward the positive plate, and how their

---

CONNECTION In Chapter 4 we introduced electronegativity and the unequal distribution of electrons in polar covalent bonds. See Figure 4.6 to review the use of arrows to indicate bond polarity.

---

bond dipole separation of electrical charge created when atoms with different electronegativities form a covalent bond.

permanent dipole permanent separation of electrical charge in a molecule due to unequal distributions of bonding and/or lone pairs of electrons.

dipole moment (μ) a measure of the degree to which a molecule aligns itself in an applied electric field; a quantitative expression of the polarity of a molecule.

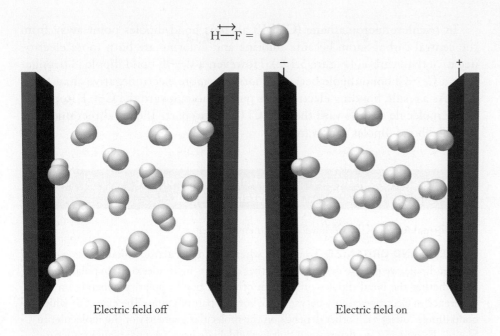

FIGURE 5.18 Gaseous HF molecules are oriented randomly in the absence of an electric field but align themselves when an electric field is applied to two metal plates. The negative (fluorine) end of each molecule is directed toward the positively charged plate; the positive (hydrogen) end is directed toward the negative plate.

positive ends are oriented toward the negative plate. The more polar a molecule is, the more strongly it aligns with such a field.

Dipole moments are usually expressed in units of *debyes* (D), where 1 D = 3.34×10^{-30} coulomb-meter. The dipole moments of several polar substances are shown in Table 5.2. Note the structures in the last two rows of the table. Chloroform ($CHCl_3$) has a relatively strong dipole moment because its bond dipoles differ both in the *degree* of polarity and in the *direction* of the polarity (Figure 5.19a). Chlorine is more electronegative than carbon, and the two electrons in each C—Cl bond are pulled away from C and toward Cl. Because hydrogen is less electronegative than carbon, the two electrons in the H—C bond are pulled away from H and toward C. Consequently the electron distribution is away from the top part of the molecule as drawn and toward the bottom.

TABLE 5.2	**Permanent Dipole Moments of Several Polar Molecules**		
Formula	**Structure with Bond Dipole(s)**	**Direction of Overall Dipole**	**Dipole Moment (debyes)**
HF	H→F	⟶	1.91
H_2O	H ·Ö· H	↕	1.85
NH_3	H—N—H H	↕	1.47
$CHCl_3$	H Cl—C—Cl Cl	↕	1.04
CCl_3F	F Cl—C—Cl Cl	↕	0.45

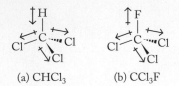

FIGURE 5.19 (a) The four bonds in chloroform ($CHCl_3$) are not equivalent, which causes an asymmetrical distribution of bonding electrons in the molecule, resulting in a permanent dipole directed toward the bottom of $CHCl_3$ as drawn. (b) The C—F bond dipole in CCl_3F is stronger than the C—Cl bond dipoles and is not completely offset by them, which makes CCl_3F a slightly polar substance in the direction of the fluorine atom.

In trichlorofluoromethane (CCl_3F) all four bond dipoles point away from the central carbon atom because fluorine and chlorine are both more electronegative than carbon (Figure 5.19b). However, a C—F bond dipole is stronger than a C—Cl bond dipole because fluorine is more electronegative than chlorine. As a result, bonding electrons are pulled more toward the C—F (top) part of the molecule than toward the C—Cl (bottom) part. The net direction of the summed bond dipoles is upward.

SAMPLE EXERCISE 5.4 **Predicting Whether or Not a Substance Is Polar**   **LO2**

Does formaldehyde (CH_2O) have a permanent dipole?

COLLECT AND ORGANIZE To predict whether a molecular compound has a permanent dipole, we need to determine whether or not its molecules contain polar bonds and whether the bond dipoles offset each other. A bond's polarity depends on the difference in electronegativity between the bonded pair of atoms. The extent to which bond dipoles offset each other depends on the molecular geometry of the molecule. In Sample Exercise 5.1 we determined that formaldehyde has a trigonal planar structure:

ANALYZE The electronegativities of the elements in the compound are H = 2.1, C = 2.5, and O = 3.5 (Figure 4.8). Given the differences in the electronegativities of the atoms bonded to the central atom, it is likely that formaldehyde has a permanent dipole.

SOLVE The hydrogen atoms are the least electronegative of the atoms in the molecule and oxygen is the most, so each of the bonds in the molecule has a bond dipole directed toward the oxygen atom:

Thus, the formaldehyde molecule has a permanent dipole.

THINK ABOUT IT When more than one kind of atom is bonded to a central atom, it is highly likely that the molecule will have a permanent dipole.

Practice Exercise Does carbon disulfide (CS_2), a gas present in small amounts in crude petroleum, have a dipole moment?

(Answers to Practice Exercises are in the back of the book.)

CONCEPT TEST

Water and hydrogen sulfide both have a bent molecular geometry with dipole moments of 1.85 D and 0.98 D, respectively. Why is the dipole moment of H_2S smaller than that of H_2O?

(Answers to Concept Tests are in the back of the book.)

5.4 Valence Bond Theory and Hybrid Orbitals

Drawing Lewis structures and applying VSEPR enable us to predict the geometry of many molecules reliably. However, we have avoided making a connection between molecular geometry and the picture of electrons in atoms developed in Chapter 3—namely, that electrons exist in atomic orbitals. The absence of a connection between the electronic structure of atoms and the electronic structure of molecules led to the development of bonding theories that use atomic orbitals to account for the molecular geometries predicted by VSEPR. We explore one of those theories now.

Valence bond theory evolved in the late 1920s when Linus Pauling merged quantum mechanics with Lewis's model of shared electron pairs to develop a new theory to explain molecular bonding. Valence bond theory assumes that a chemical bond forms when the atomic orbitals of two atoms **overlap**. The electrons in overlapping orbitals are attracted to the nuclei of both bonded atoms. This attraction leads to lower potential energy and greater chemical stability than if the atoms were completely free, as we discussed in our introduction to covalent bond formation in Chapter 4 (see Figure 4.3). Valence bond theory was an important advance over Lewis's model because it could be used to explain the molecular geometries we have been exploring in this chapter. Let's see how it works.

We begin by applying Pauling's valence bond theory to the simplest of diatomic molecules: H_2. A free H atom has a single electron in a $1s$ atomic orbital. According to valence bond theory, overlap of two half-filled $1s$ orbitals leads to increased electron density between the nuclei of the bonding atoms as the two nuclei share the two $1s$ electrons. As a result, a H—H covalent bond forms (Figure 5.20). The region of highest electron density lies along the bond axis between the two atoms, which makes the H—H bond an example of a **sigma (σ) bond**.

sp^3 Hybrid Orbitals

Now let's consider the bonding in the slightly larger molecule of methane, CH_4. We established in Chapter 3 that free carbon atoms have the electron configuration $[He]2s^2 2p^2$, that s orbitals are spherical, and that p orbitals are usually drawn with teardrop shapes (actually, they're shaped more like mushroom caps). The lobes of the three p orbitals in a subshell are oriented at 90° to one another as shown in Figure 5.21. Further, we learned in Chapter 4 that carbon atoms form four covalent bonds to complete their octets.

We saw in Section 5.2 that the central carbon atom in methane has a steric number of four (four bonds, no lone pairs) and a tetrahedral molecular geometry. How can a carbon atom with a filled, spherical $2s$ orbital and two half-filled $2p$ orbitals oriented 90° from each other form a tetrahedral molecule? Valence bond theory offers one explanation if the process of atomic orbital overlap and bond formation includes a reconfiguration of the $2s$ and $2p$ orbitals. This reconfiguration process is called **hybridization**.

Hybridization happens when, for example, 4 H atoms approach a single C atom in the gas phase. As the valence-shell atomic orbitals of the C atom begin to overlap with the $1s$ orbitals of the four H atoms, the $2s$, $2p_x$, $2p_y$, and $2p_z$ orbitals of the C atom no longer represent the lowest energy orientations of the valence orbitals. What happens next can be viewed as a mathematical mixing of the wave functions defining the atom's $2s$ and $2p$ orbitals. The product of mixing is a new

valence bond theory a quantum mechanics–based theory of bonding that assumes covalent bonds form when half-filled orbitals on different atoms overlap or occupy the same region in space.

overlap a term in valence bond theory describing bonds arising from two orbitals on different atoms that occupy the same region of space.

sigma (σ) bond a covalent bond in which the highest electron density lies between the two atoms along the bond axis.

hybridization in valence bond theory, the mixing of atomic orbitals to generate new sets of orbitals that then are available to form covalent bonds with other atoms.

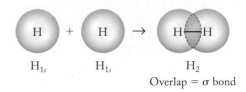

FIGURE 5.20 The overlap of the $1s$ orbitals on two hydrogen atoms produces a single σ bond that holds the two hydrogen atoms together in a H_2 molecule.

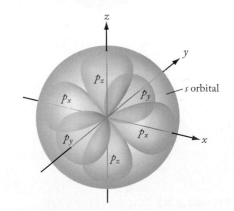

FIGURE 5.21 Relative orientation in space of one $2s$ orbital (spherical) and three $2p$ orbitals (teardrop-shaped) oriented at 90° to one another along the x-, y-, and z-axes of a coordinate system.

hybrid atomic orbital in valence bond theory, one of a set of equivalent orbitals about an atom created when specific atomic orbitals are mixed.

sp^3 hybrid orbitals a set of four hybrid orbitals with a tetrahedral orientation produced by mixing one s and three p atomic orbitals.

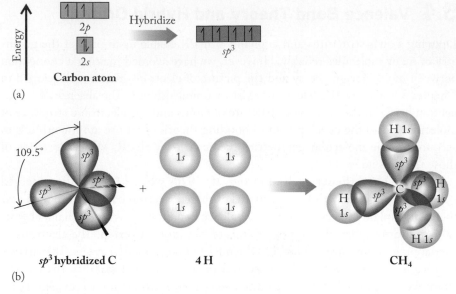

(a)

(b)

FIGURE 5.22 (a) When the $2s$ and $2p$ atomic orbitals of carbon are hybridized, the four orbitals are mixed to create four sp^3 hybrid orbitals, each containing one electron. (b) The hybrid orbitals are oriented 109.5° from one another, which is the orientation of the bonds in a molecule of methane and in other molecules with a tetrahedral molecular geometry.

◯◯ **CONNECTION** We noted in Chapter 3 that the shapes of the p orbitals we use in this book are longer and narrower than their true shapes to make it easier to see how they are oriented (see Figure 3.23).

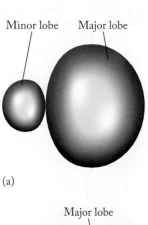

(a)

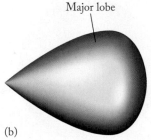

(b)

FIGURE 5.23 Two views of sp^3 hybrid orbitals. (a) Boundary surface of an sp^3 orbital derived from quantum mechanics, consisting of both a major and minor lobe. (b) To better show the orientation of bonds formed by hybrid orbitals, we elongate the major lobe and ignore the minor one (which is not involved in bond formation).

set of wave functions that define four equivalent **hybrid atomic orbitals** with shapes and orientations unlike those of $2s$ and $2p$ orbitals (Figure 5.22). Because a single s orbital and three p orbitals are involved in this mixing process, we call the result a set of **sp^3 hybrid orbitals** as shown in Figure 5.22(a). The lobes are all 109.5° from each other because the distance between them is maximized, which means the repulsion between bonding pairs of electrons in them is minimized. Overlap of the lobes with the $1s$ orbitals of four hydrogen atoms results in the formation of four σ bonds and one molecule of methane that has the tetrahedral molecular geometry shown in Figure 5.22(b).

These sp^3 hybrid orbitals appear to have one lobe each, but in fact every sp^3 hybrid orbital consists of a major lobe and a minor lobe, as shown in Figure 5.23. Because our current focus is bonding and the minor lobe is not involved in bonding, we ignore it in this chapter. Also, the shapes of the hybrid orbitals that we use in this book are more elongated than their boundary surfaces as calculated from quantum mechanics. As with our p orbital images, we stretch the lobes of hybrid orbitals a bit to better show their orientation.

According to valence bond theory, *any* atom can form a set of sp^3 hybrid orbitals as it takes center stage in a molecule with a tetrahedral electron-group geometry. This includes atoms in which one or more hybrid orbitals are filled before any bonding takes place. For example, sp^3 hybridization of the nitrogen atom, which has five valence electrons, produces three hybrid orbitals that are half-filled and one hybrid orbital that is completely filled, as in a molecule of ammonia (NH_3, Figure 5.24a). The three half-filled orbitals form the three σ bonds in ammonia by overlapping with the $1s$ orbitals of three hydrogen atoms. The hybrid orbital that was already full contains the N atom's lone pair of electrons in NH_3.

Similarly, the oxygen atom in water forms four sp^3 hybrid orbitals: two are filled before any bonding takes place, and two are half-filled and available for bond formation (Figure 5.24b). The latter two are the orbitals that overlap with $1s$ orbitals from hydrogen atoms and form two σ bonds in H_2O. Thus, a carbon atom

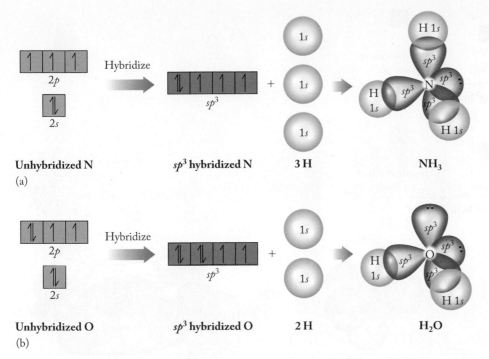

Unhybridized N
(a)

sp^3 **hybridized N**

3 H

NH_3

Unhybridized O
(b)

sp^3 **hybridized O**

2 H

H_2O

FIGURE 5.24 A set of four sp^3 hybrid orbitals point toward the vertices of a tetrahedron. (a) The nonbonding electron pair of the N in NH_3 occupies one of the four hybrid orbitals, giving the molecule a trigonal pyramidal molecular geometry. (b) The O atom in H_2O is also sp^3 hybridized. Only two of the four orbitals contain bonding pairs of electrons, giving H_2O a bent molecular geometry.

forming four σ bonds, a nitrogen atom with three σ bonds and one lone pair of electrons, and an oxygen atom with two σ bonds and two lone pairs of electrons all have the same steric number (4) and are all sp^3 hybridized atoms. This consistency is not accidental. As we will see, the SN of an atom with an octet of valence electrons is directly linked to both its hybridization and its molecular geometry.

sp^2 Hybrid Orbitals

We saw in Sample Exercise 5.1 that formaldehyde has a trigonal planar molecular geometry (Figure 5.25). According to the VSEPR model the SN of the central carbon atom is three, so three atomic orbitals must be mixed to make three equivalent hybrid orbitals. Instead of mixing all four atomic orbitals, as in sp^3 hybridization, we mix only three: the $2s$ orbital and two of the $2p$ orbitals. This leaves the third $2p$ orbital unhybridized (Figure 5.26a).

Mixing one s and two p orbitals generates three hybrid orbitals called **sp^2 hybrid orbitals**. The energy level of sp^2 orbitals is slightly lower than that of sp^3 orbitals because only two p orbitals are mixed with one s orbital in a set of sp^2 orbitals. The orbitals in an sp^2 hybridized atom all lie in the same plane and are 120° apart. The two lobes of the unhybridized p orbital lie above and below the plane of the triangle defined by the sp^2 hybrid orbitals. An sp^2 hybridized atom can form up to three σ bonds with its three hybridized orbitals, and the carbon atom in formaldehyde does so; it forms two σ bonds to the two H atoms and one σ bond to the O atom.

For its part, the O atom is also sp^2 hybridized (Figure 5.26b). One of its three sp^2 orbitals forms the σ bond to carbon, and the other two hold lone pairs of electrons. Its unhybridized p orbital overlaps sideways with the one on the carbon

σ bonds
π bond

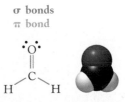

FIGURE 5.25 Formaldehyde has 3 σ bonds and 1 π bond.

sp^2 hybrid orbitals three hybrid orbitals in a trigonal planar orientation formed by mixing one s and two p orbitals.

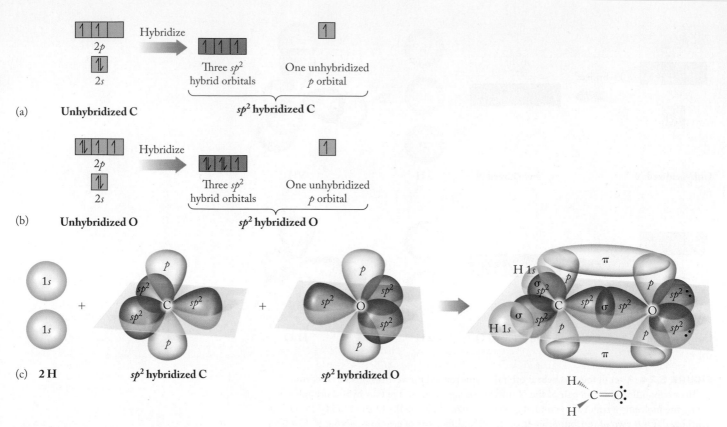

FIGURE 5.26 (a) Mixing the 2s orbital of a carbon atom and *only two* of its three 2p orbitals produces three sp^2 hybrid orbitals, leaving one unhybridized p orbital available for π bond formation. (b) The 2s and 2p orbitals of oxygen hybridize in the same way. (c) The carbon atom forms three σ bonds: two to the two H atoms and one to the O atom. It also forms one π bond to the O atom with its unhybridized p orbital, which is located above and below the plane defined by the sp^2 orbitals. An sp^2 hybridized oxygen atom forms one σ bond with one of its sp^2 hybrid orbitals and one π bond with its unhybridized p orbital. Its other two sp^2 orbitals hold the O atom's two lone pairs of electrons.

atom, producing a **pi (π) bond** in which electron densities are highest above and below the internuclear axis, as shown in the last image in Figure 5.26(c). Note that the π bond is one bond formed by the two atoms' p orbitals at their two separate points of contact—it is *not* two bonds. This π bond and the σ bond between C and O together make up the C═O double bond in CH_2O. Note how the trigonal planar geometries around the C and O atoms align with each other, giving an extended array of coplanar σ bonds, all about 120° from each other. Also, keep in mind that the widths of the p orbitals' lobes and the distances between atoms are not drawn to scale in Figure 5.26(c). The lobes are actually much wider than drawn (see Figure 3.23), and they really do overlap.

A nitrogen atom can also form sp^2 hybrid orbitals as shown in Figure 5.27. When it does, its lone pair occupies one of the three hybrid orbitals; the other two are half-filled and available to form two σ bonds, and the unhybridized half-filled p orbital is available to form one π bond. This combination of one lone pair plus the capacity to form two σ bonds to two other atoms gives N a steric number of 3, the same SN value as sp^2 hybridized C and O atoms. This link to SN − 3 means that central atoms with trigonal planar electron-group geometries are often sp^2 hybridized.

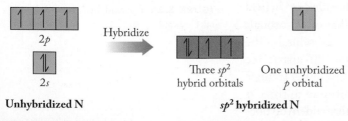

FIGURE 5.27 A nitrogen atom with sp^2 hybrid orbitals has the capacity to form two σ bonds and one π bond. One of its sp^2 orbitals contains a lone pair of electrons.

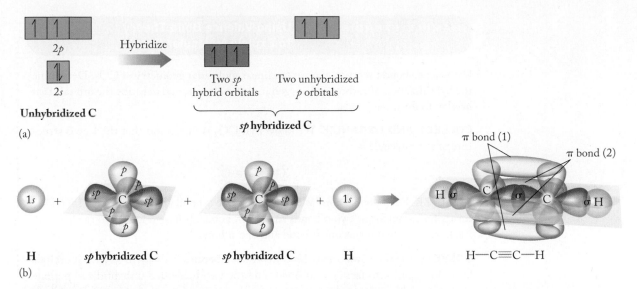

FIGURE 5.28 (a) Mixing one 2s orbital and one 2p orbital on a carbon atom creates two *sp* hybrid orbitals and leaves the carbon atom with two unhybridized *p* orbitals. (b) Two *sp* hybridized carbon atoms each bond to one hydrogen atom and to the other carbon atom via σ bonds to form the linear molecule C_2H_2. The two pairs of unhybridized *p* orbitals on the two carbon atoms overlap sideways, one pair above and below and the other in front of and behind the bonding axis, forming two π bonds.

sp Hybrid Orbitals

We have seen hybrid orbitals generated by mixing one *s* orbital with three *p* orbitals and by mixing one *s* orbital with two *p* orbitals. The remaining mixture, one *s* orbital with one *p* orbital, forms the final set of important hybrid orbitals made from *s* and *p* atomic orbitals.

In the linear molecule acetylene, C_2H_2, both carbon atoms have SN = 2 according to VSEPR. Because steric number always indicates the number of hybrid orbitals that must be created, we must mix two atomic orbitals to make two hybrid orbitals on each carbon atom. Mixing one 2s and one 2p orbital on a carbon atom and leaving the other two 2p orbitals unhybridized produces two **sp hybrid orbitals** with major lobes that are on opposite sides of the carbon atom as shown in Figure 5.28. The double lobes of the two unhybridized *p* orbitals are above and below and in front of and behind the axis of the *sp* hybrid orbitals.

The valence bond view of C_2H_2 shows a σ-bonding framework in which a hydrogen 1s orbital overlaps with one *sp* hybrid orbital on each carbon atom. The remaining *sp* hybrid orbitals, one on each carbon atom, overlap with each other (Figure 5.28b). This arrangement brings the two sets of unhybridized *p* orbitals on the two carbon atoms into parallel alignment with each other. They then overlap to form two π bonds between the two carbon atoms, so that the one σ bond and the two π bonds form the triple bond of HC≡CH. The steric number of each *sp* hybridized carbon atom in acetylene is 2 (no lone pairs and bonds to two atoms). The link between SN = 2 and *sp* hybridization also applies to nitrogen atoms in N_2. A lone pair occupies one of the two *sp* orbitals on each nitrogen atom, and the other *sp* hybrid orbital forms the σ bond as in Figure 5.29.

▶❚❚ **CHEMTOUR** Hybridization

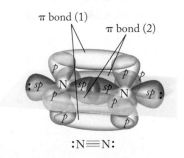

FIGURE 5.29 The triple bond between *sp* hybridized nitrogen atoms in N_2 consists of one σ bond and two π bonds.

pi (π) bond a covalent bond in which electron density is greatest above and below the bonding axis.

sp hybrid orbitals two hybrid orbitals on opposite sides of the hybridized atom formed by mixing one *s* and one *p* orbital.

Use valence bond theory to explain the linear molecular geometry of CO_2. Determine the hybridization of carbon and oxygen in this molecule, and describe the orbitals that overlap to form the bonds.

COLLECT AND ORGANIZE We know that CO_2 is linear and that the Lewis structure of the molecule is

$$\ddot{\text{O}}\!=\!\text{C}\!=\!\ddot{\text{O}}$$

ANALYZE Each double bond is composed of one σ bond and one π bond. Therefore, the carbon atom forms two σ bonds and two π bonds. It has no lone pairs, so SN = 2. Each oxygen atom forms one σ bond and one π bond.

SOLVE The carbon atom must be *sp* hybridized because half-filled *sp* hybrid orbitals have the capacity to form two σ bonds and the two half-filled unhybridized *p* orbitals are available to form the two π bonds. The oxygen atoms must be sp^2 hybridized because that hybridization leaves the O atoms with one unhybridized *p* orbital to form a π bond. The σ bonds form when each *sp* orbital on the carbon atom overlaps with one sp^2 orbital on an oxygen atom.

THINK ABOUT IT The *sp* hybridization of the carbon atom is consistent with the following: (1) its steric number (SN = 2); (2) its capacity to form two σ and two π bonds as shown in Figure 5.30; and (3) the linear geometry of the molecule. Notice that the two unhybridized *p* orbitals of carbon are oriented 90° to each other, and that the plane containing the three sp^2 orbitals of one oxygen atom is rotated 90° with respect to the plane containing the three sp^2 orbitals of the other oxygen atom. That orientation is necessary for the formation of the two π bonds because the two unhybridized *p* orbitals that overlap to form a π bond must be parallel. The lone pairs of electrons on each oxygen atom and its σ bond to carbon lie in a trigonal plane, which minimizes their repulsion of each other.

Practice Exercise In which of these molecules does the central atom have sp^3 hybrid orbitals? (a) CCl_4; (b) HCN; (c) SO_2; (d) PH_3

(Answers to Practice Exercises are in the back of the book.)

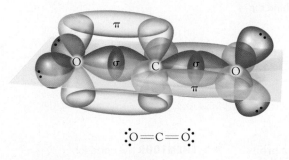

$$\ddot{\text{O}}\!=\!\text{C}\!=\!\ddot{\text{O}}$$

FIGURE 5.30 The bonding pattern in CO_2. The π bond to the left in the drawing is above and below the plane of the molecule; the π bond to the right is in front of and in back of the plane.

CONCEPT TEST

Why can't a carbon atom with sp^3 hybrid orbitals form π bonds?

(Answers to Concept Tests are in the back of the book.)

Hybrid Schemes for Expanded Octets

Valence bond theory can also account for the molecular geometries of molecules with central atoms that may have more than eight valence electrons. We learned in Chapter 4 that one way to explain the hypervalency of central atoms is to assume that valence-shell *d* orbitals are involved in bond formation. For example, to expand the octet of the central phosphorus atom in PF_5 to form a fifth P—F bond, we might include one of the phosphorus atom's $3d$ orbitals in bond formation. To account for the trigonal bipyramidal molecular shape of PF_5, we could hybridize five of the P atom's atomic orbitals—its $3s$, all three of its $3p$ orbitals, and one of its $3d$ orbitals—to produce a set of five **sp^3d hybrid orbitals** with lobes that point toward the vertices of a trigonal bipyramid (Figure 5.31a). To explain the presence of six S—F bonds and the octahedral shape of SF_6, we go one step further and incorporate *two* of the sulfur atom's d orbitals in a hybridization scheme that forms a set of six equivalent **sp^3d^2 hybrid orbitals** (Figure 5.31b). A summary of the shapes associated with all of the hybridization schemes we have discussed so far is given in Table 5.3.

Before ending our discussion of hybridization in hypervalent compounds, we need to acknowledge that many chemists don't support the use of d orbitals in the hybridization process. They note that the energies of d orbitals are significantly higher than the energies of the s and p orbitals in the same shell, which makes s–p–d orbital mixing problematic. Instead, other bonding models have been proposed to explain the bonding and geometries of hypervalent molecules—models that do not incorporate d orbitals. We will explore one of them, called molecular orbital (MO) theory, in detail in Section 5.7.

sp^3d hybrid orbitals five equivalent hybrid orbitals with lobes pointing toward the vertices of a trigonal bipyramid that form by mixing one s orbital, three p orbitals, and one d orbital from the same shell.

sp^3d^2 hybrid orbitals six equivalent hybrid orbitals pointing toward the vertices of an octahedron that form by mixing one s orbital, three p orbitals, and two d orbitals from the same shell.

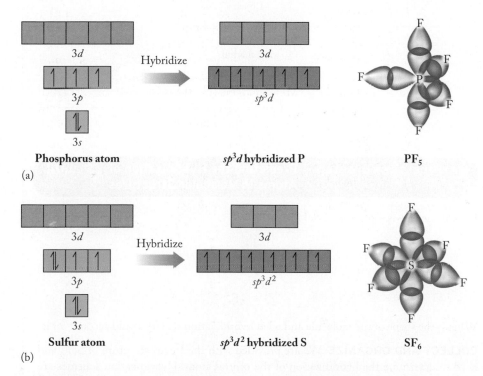

FIGURE 5.31 Hybrid orbitals can be generated by combining d orbitals with s and p orbitals in atoms that expand their octets. (a) One $3s$ orbital, three $3p$ orbitals, and one $3d$ orbital mix on the central phosphorus atom in PF_5 to form five sp^3d hybrid orbitals. (b) One $3s$, three $3p$, and two $3d$ orbitals mix on the central sulfur atom in SF_6 to form six sp^3d^2 hybrid orbitals. For simplicity, only the fluorine orbitals that overlap with the hybrid orbitals of P and S are shown.

TABLE 5.3 **Summary of Hybridization Schemes and Orbital Orientations**

Hybridization	Orientation of Hybrid Orbitals	Numbers of σ bonds	Molecular Geometries	Angles between Hybrid Orbitals
sp		2	Linear	180°
sp^2		3 2	Trigonal planar Bent	120°
sp^3		4 3 2	Tetrahedral Trigonal pyramidal Bent	109.5°
sp^3d		5 4 3 2	Trigonal bipyramidal Seesaw T-shaped Linear	90°, 120°, 180°
sp^3d^2		6 5 4	Octahedral Square pyramidal Square planar	90°, 180°

SAMPLE EXERCISE 5.6 **Predicting the Hybridization of Atoms in Molecular Structures** **LO4**

Here is the Lewis structure of SeF_4:

What is the shape of the molecule and what hybridization scheme could account for it?

COLLECT AND ORGANIZE We are provided with the Lewis structure of SeF_4 and need to determine the hybridization of the central atom. Hybridization schemes are used to explain molecular shapes, so we need to determine the shape of the molecule first.

ANALYZE The central atom has four single (σ) bonds and one lone pair, giving it SN = 5.

SOLVE The electron-group geometry associated with SN = 5 is trigonal bipyramidal (Figure 5.2). The presence of the lone pair means that the molecular geometry is not the same as the electron-group geometry. Instead, according to Table 5.1, the molecular shape is seesaw.

A steric number of 5 also suggests that the Se atom has an expanded octet of 10 valence electrons. A hybridization scheme that incorporates 10 electrons (5 orbitals) uses one *d* orbital with one *s* and three *p* orbitals, or sp^3d hybridization.

THINK ABOUT IT The information in Table 5.3 confirms that a seesaw molecular geometry is consistent with sp^3d hybridization.

Practice Exercise What is the hybridization of the iodine atom in IF_5 that is consistent with this Lewis structure?

(Answers to Practice Exercises are in the back of the book.)

> **molecular recognition** the process by which molecules interact with other molecules to produce a biological effect.

5.5 Molecules with Multiple "Central" Atoms

Living things respond to molecules that interact with regions in their tissues called *receptors* or *active sites*. The process by which molecules and sites interact is known as **molecular recognition**. Recognition does not usually involve covalent bond formation. Instead, interactions between molecules require that the biologically active molecules and the receptors that respond to them fit tightly together, which means that they must have complementary three-dimensional shapes (Figure 5.32). An example of a biological effect caused by molecular recognition is the process by which green tomatoes and other types of produce ripen. Tomatoes ripen faster when stored in a paper or plastic bag instead of sitting on a kitchen counter. Why? Because tomatoes give off ethylene gas as they ripen, and they also have receptors that respond to molecules of ethylene (C_2H_4) by accelerating the ripening process. A bag traps this gas.

Both carbon atoms in ethylene are bonded to three other atoms and have no lone pairs of electrons (Figure 5.33a). Therefore SN = 3. This means that the geometry around each carbon atom is trigonal planar and that the carbon atoms are both sp^2 hybridized (see Table 5.3). Two of the sp^2 orbitals form σ bonds with hydrogen 1*s* orbitals; the third forms the C=C σ bond as shown in Figure 5.33(b). The C=C π bond is formed by overlap of the unhybridized 2*p* orbitals on the carbon atoms. Taken together, the two trigonal planar carbon atoms produce an overall geometry for ethylene in which all six atoms lie in the same plane. The π bond in C_2H_4 molecules contributes to an overall distribution of electrons

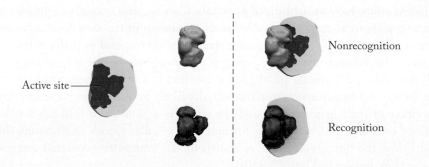

Active site

Nonrecognition

Recognition

FIGURE 5.32 Molecular recognition in biological systems. Only the blue molecule matches the shape of the active site and interacts with it.

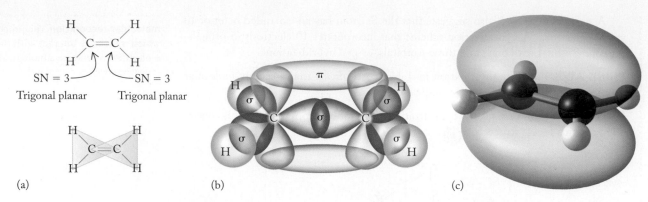

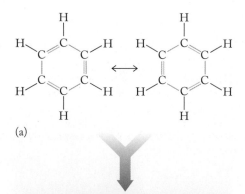

FIGURE 5.33 Ethylene molecules contain two carbon atoms that are (a) at the centers of two overlapping triangular planes that are also coplanar. This combination means that all of the atoms are in the same plane. (b) The carbon atoms are both sp^2 hybridized. (c) The π bond in ethylene results in electron density above and below the plane of the molecule.

with maximum densities above and below the plane, as shown by the boundary surfaces in Figure 5.33(c).

Now let's consider a biologically active molecule with three "central" atoms. Acrolein is one of the components of barbeque smoke that contributes to the distinctive odor of a cookout. It is also a possible cancer-causing compound. The Lewis structure of acrolein shows that the molecule contains three carbon atoms, each with a steric number of 3 and sp^2 hybridization:

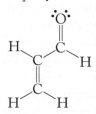

This network of three carbon atoms, each representing a trigonal planar "center," means that *all* of the atoms in a molecule of acrolein may be found in the same plane.

Another component of barbeque smoke, benzene, also consists of molecules in which all the atoms are in the same plane due to sp^2 hybridization and trigonal planar geometries on, in this case, six carbon atoms (Figure 5.34). As we saw in Chapter 4, a benzene molecule features a hexagon of carbon atoms, each bonded to two other carbon atoms and one hydrogen atom. This geometry is a perfect match for sp^2 hybridization because the 120° bond angles those orbitals produce is exactly the same as the 120° inside angles of a regular hexagon.

Bonding around the benzene ring consists of alternating single and double carbon–carbon bonds. This pattern is an example of *conjugation*—a term that applies to alternating single and multiple bonds in molecular compounds in which adjacent atoms have unhybridized p orbitals. Conjugation in benzene means there are two ways to arrange the double bonds, as shown in the two resonance structures in Figure 5.34(a). The resonance structures correspond to shifts in the locations of the π bonds formed by overlapping $2p$ orbitals on the six carbon atoms. All the carbon $2p$ orbitals are aligned above and below the plane of the ring and are the same distance apart, so all are equally likely to overlap with the $2p$ orbitals on either of their neighbors around the ring. As a result, the π bonds are delocalized over *all six* carbon atoms, forming two circular clouds of electrons above and below the hexagonal rings, as shown by the computer-generated images of π-electron density in Figure 5.34(b).

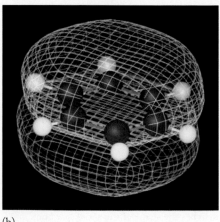

FIGURE 5.34 Two views of the resonance in benzene. (a) The Lewis resonance structures of benzene can be explained using valence bond theory by showing localized π bonds between adjacent carbon atoms. (b) Resonance leads to complete delocalization of the π bonds around the benzene ring.

In which of these three molecules and polyatomic ions are the π bonds delocalized?

(a)   (b)   (c)

(Answers to Concept Tests are in the back of the book.)

aromatic compound a cyclic, planar compound with delocalized π electrons above and below the plane of the molecule.

Delocalization of bonding pairs is an important phenomenon because it spreads out their electrons, thereby reducing their repulsion of one another and lowering the potential energy of a molecule or polyatomic ion, which makes it more stable. In general, the greater the degree of delocalization, the greater the stability. This trend is evident in a class of compounds called **aromatic compounds** that includes benzene and other molecules with planar ring structures. An important subclass of aromatic compounds, known as *polycyclic aromatic hydrocarbons* (PAHs), consists of molecules containing multiple six-carbon rings fused together as shown in Figure 5.35. PAHs are formed in fires fueled by coal, oil, or natural gas, and they are found in cigarette smoke. In 2004 they were discovered in interstellar space.

The chemical stability of PAHs means that they persist in the environment. Their planar shape contributes to their posing a particularly serious health hazard. If we inhale or ingest the compounds, they may bind to the DNA in our cells in a process called *intercalation*. Because PAHs are flat, they can slide into the double helix that DNA forms as shown in Figure 5.35. Once there, they may alter or prevent DNA replication and thereby damage or kill cells. Intercalation in DNA is one step in the process by which PAHs induce cancer.

CONNECTION Drawing molecular structures of benzene with a circle in the middle of the hexagon of carbon atoms (see Figure 4.12) is another way to represent the clouds of π electrons above and below the plane of the molecule.

FIGURE 5.35 The molecules of these polycyclic aromatic hydrocarbons consist of fused benzene rings whose π bonds are delocalized over all the rings in each molecule. Molecules with this planar shape can slip in between the strands of DNA and disrupt cell replication, which can lead to cell death or induce malignancy.

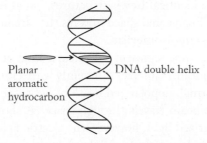

Intercalation of PAH in DNA

Naphthalene   Phenanthrene   Anthracene   Benzo[*a*]pyrene

5.6 Chirality and Molecular Recognition

Before ending our exploration of molecular shape, we need to address the subject of handedness introduced at the beginning of the chapter. There we described how two molecules with the same molecular formula and Lewis structure interact differently with receptors in our nasal membranes. As a result, one produces the smell of caraway seeds, and the other spearmint leaves. You may find the different odors surprising when you consider the molecular structures of the two compounds:

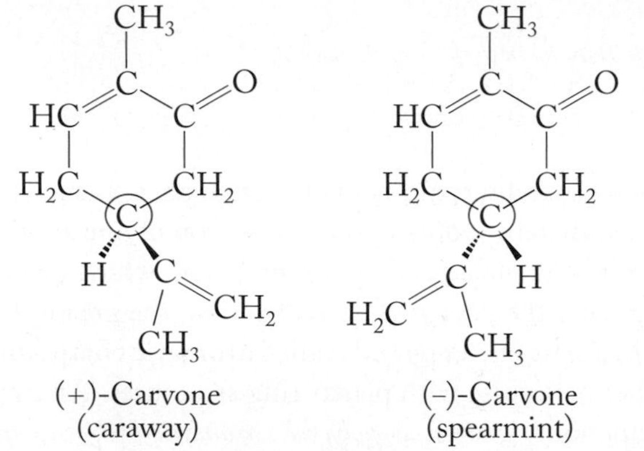

(+)-Carvone   (–)-Carvone
(caraway)   (spearmint)

At first glance they may seem identical. In fact, they have the same common chemical name, carvone. But look closely at the bonding pattern around the carbon atoms at the bottom of the rings highlighted with red circles. The hydrogen atom bonded directly to the bottom carbon atom is on the back side in the caraway compound, but it is on the front side of the ring in the spearmint compound. This minor difference in bond orientation creates a difference in molecular shape that is easily recognized by receptors in our noses.

Compounds that have the same chemical formula but different molecular structures are called **isomers**. The term is derived from *iso* (Greek for "same") and *mer* (Greek for "unit" or "part"). In this chapter and those that follow we will encounter several kinds of isomerism. The carvone molecules above represent one kind: they have structures with the same atoms bonded together in the same ways, that is, they have identical Lewis structures, but at least some of the bonds are oriented in three-dimensional space in ways that are not the same. That kind of isomerism is called **stereoisomerism**.

The two forms of carvone represent a particular kind of stereoisomerism—one in which the two stereoisomers interact differently with a kind of light called *plane-polarized* light. In a normal, unpolarized beam of light, the electric fields oscillate in all directions as the beam travels through space, as shown on the left in Figure 5.36. In plane-polarized light, however, the electric fields oscillate in only one plane. When a beam of plane-polarized light passes through a solution containing (+)-carvone, or any (+) stereoisomer, the plane rotates clockwise; if the beam passes through a solution of (–)-carvone, it rotates counterclockwise. Stereoisomers that can rotate plane-polarized electromagnetic radiation in this way are said to be *optically active* and are called *optical* isomers. They are also called **enantiomers**.

Observing how a compound interacts with plane-polarized light is one way to find out if it is optically active. Another way is to examine its molecular structure and, in particular, to determine whether or not it has any sp^3 carbon atoms that are bonded to four different atoms or groups of atoms. Such a carbon atom is called a **chiral** carbon atom or chiral center. Its presence makes the compound a chiral compound, which also means it is optically active. Many compounds of biological importance, including the proteins and carbohydrates that we consume

isomer one of a group of compounds having the same chemical formula but different molecular structures.

stereoisomerism isomerism created by differences in the orientations of the bonds between atoms in molecules.

enantiomer one of a pair of optical isomers of a compound.

chiral describes a molecule that is not superimposable on its mirror image.

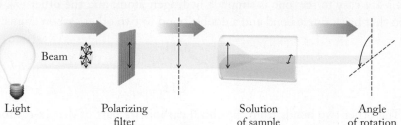

Light   Polarizing   Solution   Angle
                  filter   of sample   of rotation

FIGURE 5.36 A beam of plane-polarized light consists of electric field vectors that oscillate in only one direction. The plane of oscillation rotates if the beam passes through a solution of one enantiomer of an optically active compound. The (+) enantiomer causes the beam to rotate clockwise; the (−) enantiomer causes the beam to rotate counterclockwise.

each day, are chiral compounds. "Chirality" comes from the Greek word *cheir*, or "hand," and is quite correctly called "handedness." Although several features within molecules can lead to chirality, the most common one is the presence of a chiral carbon atom.

To see how chirality works, let's look at a compound that contains a central carbon atom bonded to four different atoms. The compound is bromochlorofluoromethane (CHBrClF) (Figure 5.37a). It is used in fire extinguishers on airplanes. We will compare its molecular structure to that of dibromochloromethane ($CHBr_2Cl$), a compound that may form during the purification of municipal water supplies with chlorine, and that has only three different atoms bonded to its central carbon atom (Figure 5.37b). In the first step of our comparison, we generate mirror images of both molecules. Then we rotate the mirror images 180° in an attempt to superimpose each mirror image on its original image. If the reflected, rotated image is superimposable on the original image, then the substance is *not* chiral. Scientists call it *achiral*. The molecular images of $CHBr_2Cl$ can be superimposed in this way, so it is not a chiral compound. However, the images of CHBrClF cannot be superimposed. For example, when we superimpose the F, C, and H atoms of the two structures in Figure 5.37(a), the Br and Cl atoms are not aligned. This means that CHBrClF *is* a chiral compound.

You may be wondering where the chiral carbon atom is in the molecules of the caraway- and spearmint-flavored compounds. If you guessed the circled carbon atoms, you were right. Two of the four different groups bonded to the circled

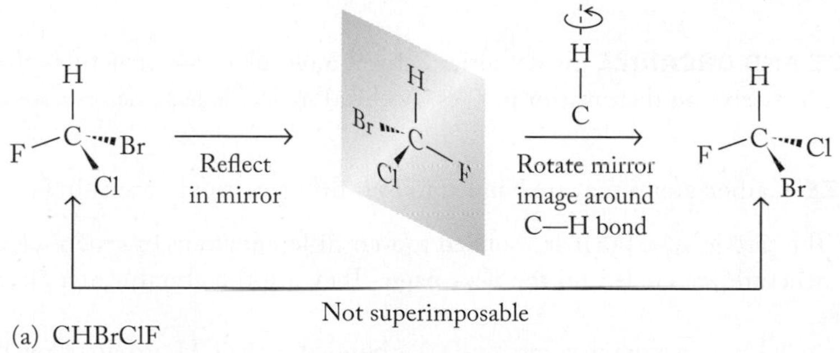

►II **CHEMTOUR** Chirality

►II **CHEMTOUR** Chiral Centers

FIGURE 5.37 (a) A molecule of a chiral compound, such as CHBrClF, is not superimposable on its mirror image. (b) A molecule of a compound that is not chiral, such as $CHBr_2Cl$, is superimposable on its mirror image.

H
|
F—C⋯Br   Reflect   Br⋯C—F   Rotate mirror   F—C⋯Cl
| in mirror | image around |
Cl   Cl   C—H bond   Br
                    H   H   H

Not superimposable

(a) CHBrClF

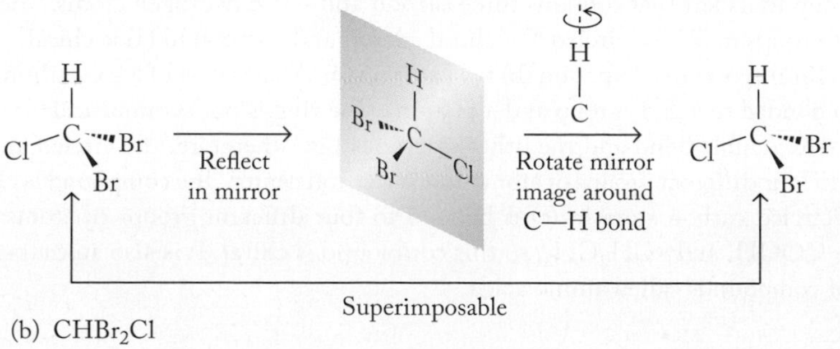

H
|
Cl—C⋯Br   Reflect   Br⋯C—Cl   Rotate mirror   Cl—C⋯Br
| in mirror | image around |
Br   Br   C—H bond   Br
                    H   H   H

Superimposable

(b) $CHBr_2Cl$

atoms are easy to see: one is simply a hydrogen atom, and the other is a carbon atom that has a single bond and a double bond to two other carbon atoms:

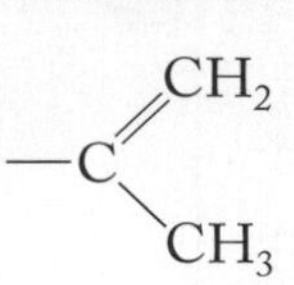

The other two bonds from the chiral carbon are part of the six-carbon ring. They both connect to $-CH_2-$ groups, so how are they different? The answer lies in other groups that the $-CH_2-$ groups are also bonded to. Note that the left side of the ring contains a $C{=}C$ double bond, and the right side contains a $C{=}O$ double bond. This asymmetry in the ring causes carvone to pass the not-superimposable-on-its-mirror-image test.

SAMPLE EXERCISE 5.7 **Recognizing Structures That Are Chiral** **LO5**

Identify which of these molecules are chiral, and circle their chiral centers. Some molecules may have more than one chiral center.

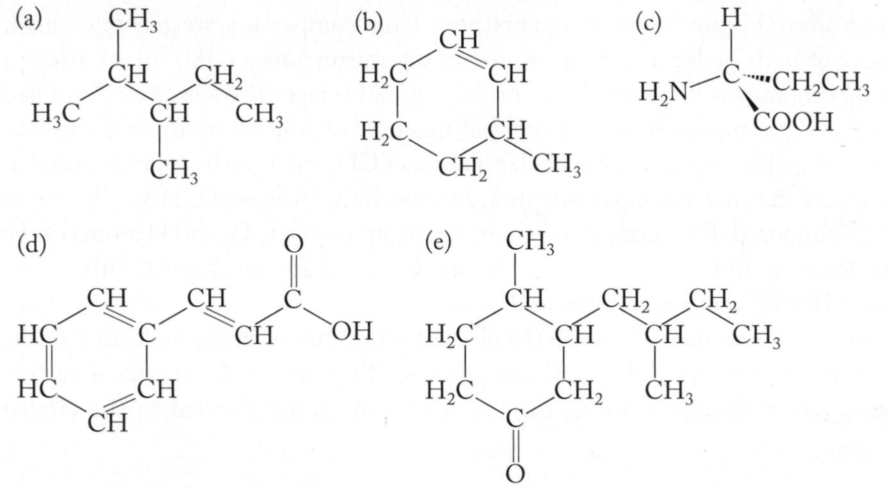

COLLECT AND ORGANIZE By definition, chiral molecules have structures that are not superimposable on their mirror images. A chiral molecule has one or more chiral centers.

ANALYZE Carbon atoms that are bonded to four different groups are chiral centers.

SOLVE The carbon atoms that are bonded to four different atoms or groups of atoms in each structure are circled on the next page. They are the chiral centers in these molecules.

The circled carbon atom in compound (a) is bonded to the CH_3 group below it, the H atom above it, a group to its right that contains two carbon and five hydrogen atoms, and a group to its left that contains three carbon and seven hydrogen atoms. The four different groups make the circled C a chiral center, and compound (a) is chiral.

The chiral center in compound (b) is a carbon atom that is part of a six-carbon ring. It is also bonded to a CH_3 group and a H atom. The ring is not symmetrical: one side has a $C{=}C$ double bond and the other side does not. Therefore, the circled atom is bonded to four different groups of atoms and is a chiral center. The compound is chiral.

The circled carbon atom in (c) is bonded to four different groups of atoms: $-H$, $-NH_2$, $-COOH$, and $-CH_2CH_3$, so this compound is chiral. It is also an example of a class of compounds called amino acids.

All of the carbon atoms in compound (d) are *sp²* hybridized, giving them planar molecular geometries. Planar molecules cannot be chiral because they have a super-imposable mirror image. Think of a mirror plane that contains all nine carbons in compound (d): the mirror image is exactly the same. Compound (d) is achiral.

Compound (e) has three chiral centers. Working from right to left, the first one is similar to the chiral center in compound (a). The other two are in the six-carbon ring. Each is bonded to different groups outside the ring, and the ring itself is not symmetrical: there is a CH_3 group bonded to the top carbon atom in the ring, and the bottom atom is double bonded to an oxygen atom. This asymmetry means that each of the two circled atoms is bonded to four different groups of atoms and is a chiral center. The presence of any one chiral center makes molecule (e) chiral.

THINK ABOUT IT To decide whether an *sp³* hybridized carbon atom is a chiral center, we often have to look beyond the atoms bonded directly to it. When the atom is in a ring of atoms, encountering different groups as we work our way around the ring can give it an asymmetry that ensures that the atom is bonded to at least two different groups. If its other two bonds also go to different groups, the atom is a chiral center.

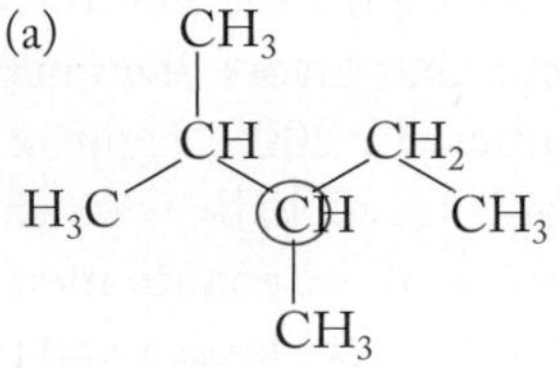

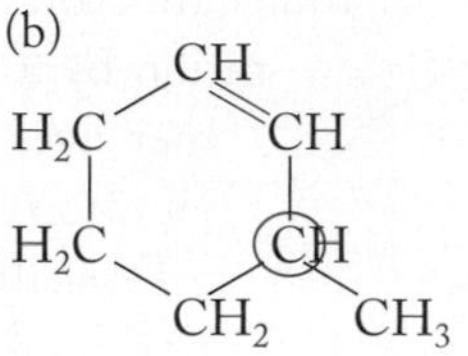

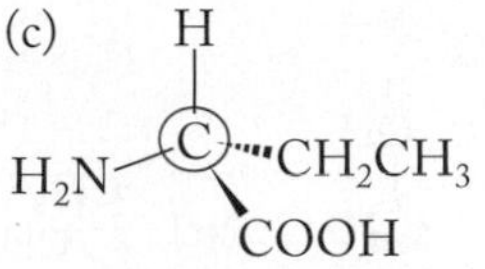

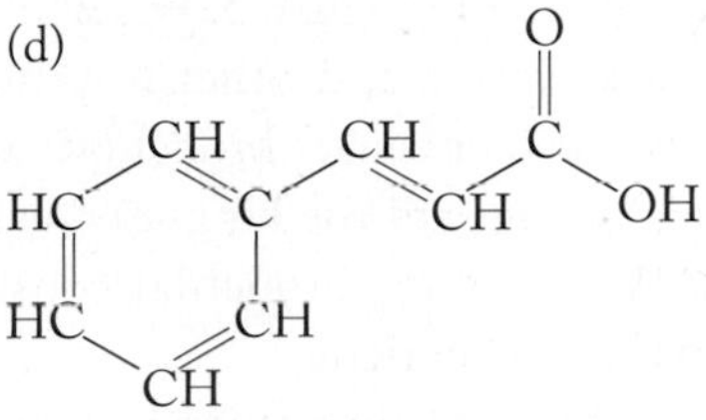

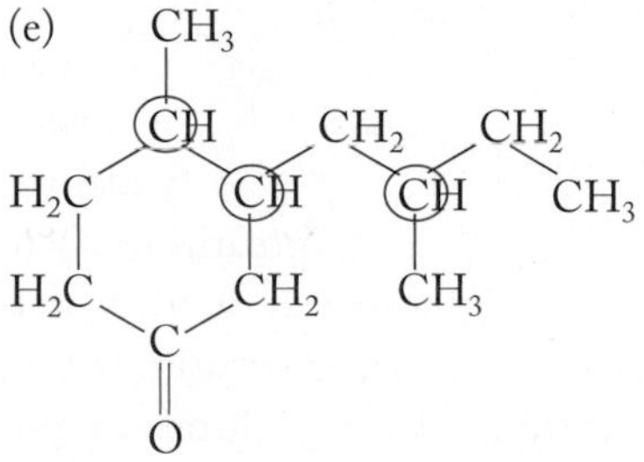

Practice Exercise Identify which of the molecules below are chiral. Circle the chiral centers in each structure.

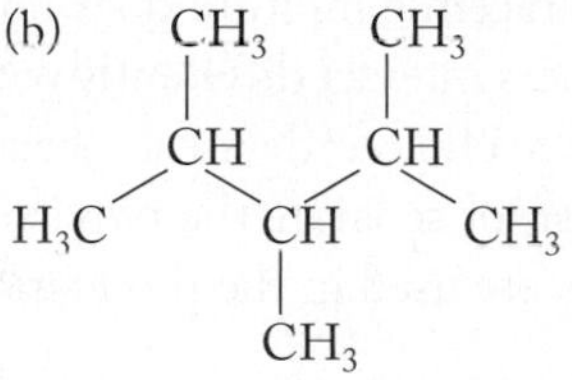

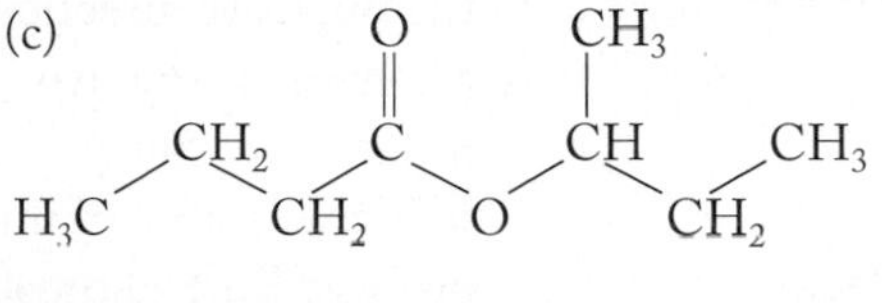

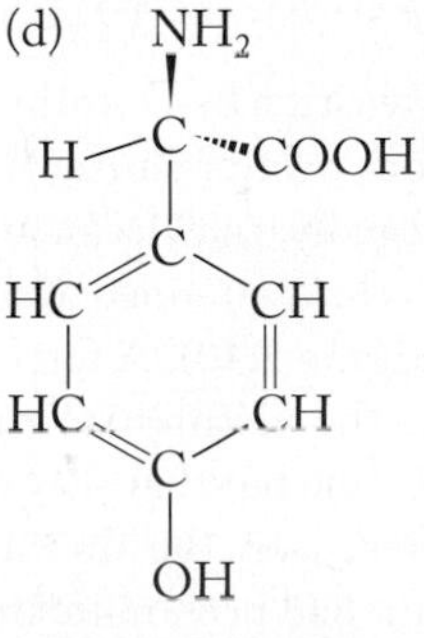

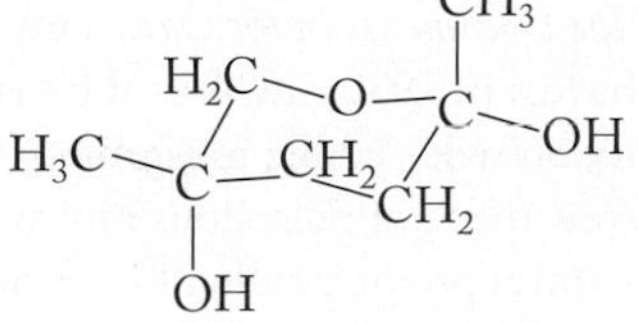

(Answers to Practice Exercises are in the back of the book.)

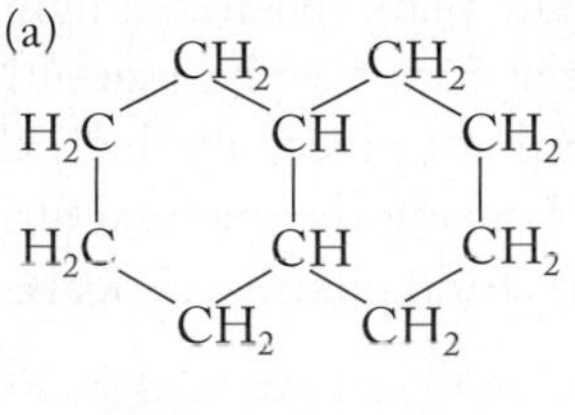

Chirality in Nature

Many of the compounds formed by living systems are chiral. For example, the proteins in our tissues and in the foods we consume are made up of amino acids that are nearly all chiral compounds. Most of them are like the amino acid alanine in that our bodies can make proteins from only one of the two alanine enantiomers—the one on the left in Figure 5.38. The origins of biological preference for one enantiomer over another are unknown, though the amino acids in meteorites striking Earth have slightly more of the enantiomers found in living things in Earth, which may indicate that the chiral preferences observed in Earth's biosphere are not limited to our planet.

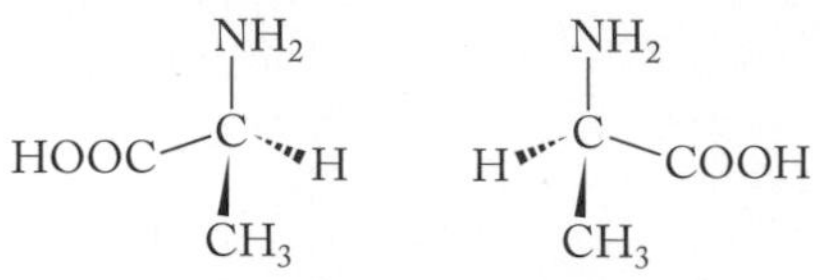

FIGURE 5.38 The two enantiomers of the amino acid alanine.

Whatever the origin of the preferences, processes requiring molecular recognition in living systems often depend on the selectivity conveyed by chirality. The human body is a chiral environment, so the handedness of molecules matters, including those we take into our bodies for therapeutic purposes. As many as half of the prescription drugs on the market today are chiral and owe their function to recognition by a receptor that favors one enantiomer over the other. In 2008, eight of the ten top-selling drugs globally were chiral. Typically only one of the enantiomers of a chiral drug is active. A classic example is the drug albuterol (Figure 5.39), which is used to treat asthma and other respiratory disorders. One isomer causes bronchodilation (widening the air passages of the lungs and easing breathing), while the other causes bronchial constriction and may actually be detrimental to the patient.

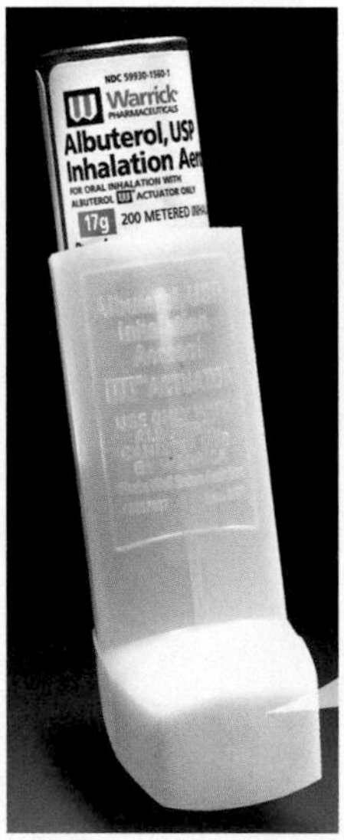

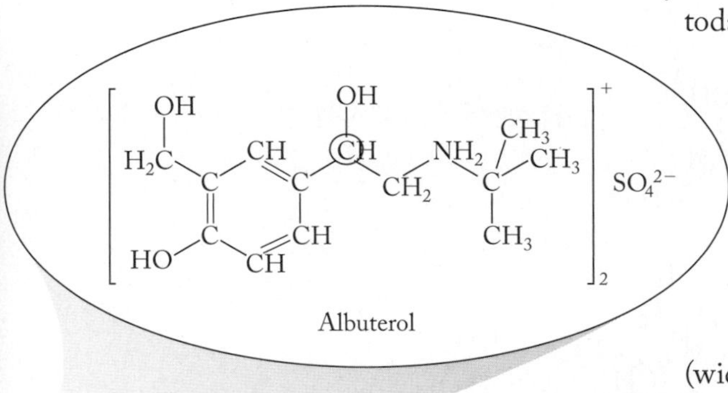

FIGURE 5.39 The antiasthma drug albuterol. The chiral center is circled.

As noted above, living organisms typically produce only one stereoisomer of a chiral compound. When a chiral compound such as albuterol is synthesized in a laboratory, however, both enantiomers are often produced in equal amounts. The resulting 50:50 mixture is called a **racemic mixture**. Because one isomer rotates the plane of polarized light in one direction, and the other to the same extent in the opposite direction, a racemic mixture does not rotate plane-polarized light at all. Because the two isomers interact differently with receptors, the pharmaceutical industry routinely faces two choices: devise a special synthetic procedure that yields only the isomer of interest, or separate the two isomers during the manufacturing process. Both approaches are used in the production of chiral pharmaceuticals.

CONCEPT TEST

The Documents in the Case, a mystery written by Dorothy L. Sayers in 1930, involves the suspicious death of an authority on wild, edible mushrooms. Allegedly, the victim ate a stew made of poisonous mushrooms that contained a toxic natural product called muscarine (Figure 5.40). A forensic specialist evaluated the contents of the victim's stomach and found that they contained muscarine but that they did not rotate plane-polarized light. Based on the findings, the coroner concluded that the victim had been murdered. How did the coroner reach that conclusion?

(Answers to Concept Tests are in the back of the book.)

FIGURE 5.40 Several varieties of mushrooms, including *Amanita muscaria*, contain the toxic substance muscarine.

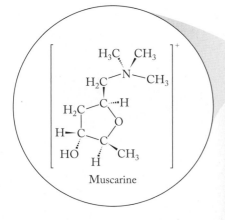

5.7 Molecular Orbital Theory

Lewis structures and valence bond theory help us understand the bonding capacities of individual atoms and the bonding patterns in molecules and molecular ions, while VSEPR and valence bond theory account for their molecular shapes. However, none of the models explain why O_2 is attracted to a magnetic field, or how molecules and molecular ions in Earth's upper atmosphere produce the shimmering, colorful display known as an *aurora* (Figure 5.41). To explain those phenomena we need another model that describes covalent bonding. We need **molecular orbital (MO) theory**.

MO theory is based on quantum mechanics. Like atomic orbitals, **molecular orbitals** are wave functions that represent discrete energy states inside molecules. As with atomic orbitals, the lowest energy MOs always fill first. Electrons in them move to higher energy MOs when molecules absorb quanta of electromagnetic radiation. When the electrons return to lower energy MOs, distinctive wavelengths of UV and visible radiation are emitted, producing, for example, some of the colors in an aurora. Note how these processes of molecular absorption and emission of electromagnetic radiation parallel the absorption and emission of radiation by free atoms, which we discussed in Chapter 3.

Unlike atomic orbitals, including hybrid atomic orbitals, MOs are not linked to single atoms but rather belong to all the atoms in a molecule. Thus, any electron in a molecule could be anywhere in the molecule. This delocalized view of covalent bonding is particularly effective at describing the bonding in molecules such as benzene, in which conjugation allows π bonds to spread out over more than one pair of atoms. The availability of electrons from other atoms also means that we do not have to involve a particular atom's *d* orbitals to explain expanded octets, as we sometimes need to in valence bond theory.

In MO theory, molecular orbitals are formed by combining atomic orbitals from each of the atoms in the molecule. In this book we limit our discussion of MO theory to simple molecules in which MOs are formed from the atomic orbitals of only a few atoms. Some MOs have lobes of high electron density that lie *between* bonded pairs of atoms; they are called **bonding orbitals**. The energies of bonding MOs are lower than the energies of the atomic orbitals that combined to form them, so populating them with electrons (each MO can hold two electrons, just like an atomic orbital) stabilizes the molecule and contributes to the strength of the bonds holding its atoms together.

There are other MOs with lobes of high electron density that are not located between the bonded atoms. They are **antibonding orbitals** and have energies that are higher than the atomic orbitals that combined to form them. When electrons are in antibonding orbitals they destabilize the molecule. As you might guess, if a molecule were to have the same number of electrons in its bonding and antibonding orbitals, then there would be no net energy holding the molecule together, and it would fall apart (actually, it would never have formed in the first place). Another key point is that the *total number of molecular orbitals must match the number of atomic orbitals involved in forming them*. For example, combining two atomic orbitals, one from each of two atoms, produces one low-energy bonding orbital and one high-energy antibonding orbital. To better understand how atomic orbitals combine to form molecular orbitals, let's apply MO theory to the bond that holds together the simplest of molecules, H_2.

racemic mixture a sample containing equal amounts of both enantiomers of a compound.

molecular orbital (MO) theory a bonding theory based on the mixing of atomic orbitals of similar shapes and energies to form molecular orbitals that belong to the molecule as a whole.

molecular orbital a region of characteristic shape and energy where electrons in a molecule are located.

bonding orbital term in MO theory describing regions of increased electron density between nuclear centers that serve to hold atoms together in molecules.

antibonding orbital term in MO theory describing regions of electron density in a molecule that destabilize the molecule because they do not increase the electron density between nuclear centers.

FIGURE 5.41 Auroras are spectacular displays of color produced by excited-state atoms, ions, and molecules in Earth's upper atmosphere.

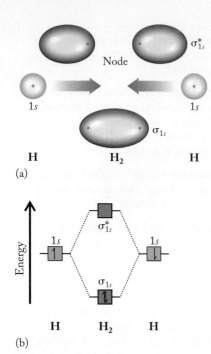

(a)

(b)

FIGURE 5.42 Mixing the 1s orbitals of two hydrogen atoms creates two molecular orbitals: a bonding σ_{1s} orbital containing two electrons and an empty antibonding σ_{1s}^* orbital. (a) The lower red oval is the bonding orbital. The two red ovals at the top together make up the antibonding orbital. *Note:* The two top ovals represent only *one* molecular orbital with a node of zero electron density in between. Dots show the locations of the two hydrogen nuclei. (b) A molecular orbital diagram shows the relative energies of the bonding and antibonding molecular orbitals and of the atomic orbitals that formed them.

◯◯ CONNECTION We used orbital diagrams in Chapter 3 to show electron transitions between energy levels as atoms absorb and emit electromagnetic radiation.

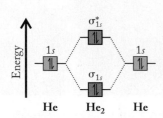

FIGURE 5.43 The molecular orbital diagram for nonexistent He_2 indicates that the same number of electrons occupy the antibonding orbital and the bonding orbital. Therefore, the bond order is 0; the molecule is not stable.

Molecular Orbitals of H₂

According to MO theory, a hydrogen molecule is formed when the 1s atomic orbitals on two free hydrogen atoms combine to form two molecular orbitals as shown in Figure 5.42(a). Molecular orbital theory stipulates that mixing two atomic orbitals creates two molecular orbitals. The lower energy, bonding molecular orbital is oval shaped and spans the two atomic centers. The high density of electrons between the two atoms makes this bonding MO a **sigma (σ) molecular orbital**. When two electrons occupy it, a single σ bond is formed. The σ bonding molecular orbital in H_2 is labeled σ_{1s} because it is formed by mixing two 1s atomic orbitals.

The high-energy, antibonding molecular orbital formed from two hydrogen atomic orbitals is designated σ_{1s}^* (pronounced "sigma star"). This antibonding orbital has two separate lobes of electron density oriented away from the internuclear axis and a region of zero electron density (a node) between the atoms, as shown in Figure 5.42(a).

Figure 5.42(b) is a **molecular orbital diagram**, analogous to the *atomic* orbital diagrams we discussed in Chapter 3. Note that the σ_{1s} MO is lower in energy and therefore more stable than the 1s atomic orbitals by about the same amount that the σ_{1s}^* MO is higher in energy than the 1s atomic orbitals. Therefore, the formation of the two MOs does not significantly change the total energy of the system. A hydrogen molecule has two valence electrons, one from each H atom, both residing in the lower energy σ_{1s} orbital. As in atomic orbitals, these two σ_{1s} electrons must have opposite spins. The electron configuration that corresponds to the molecular orbital diagram in Figure 5.42 is written $(\sigma_{1s})^2$ where the superscript indicates that there are two electrons in the σ_{1s} molecular orbital. Because the energy of the electrons in a $(\sigma_{1s})^2$ configuration is lower than the energy of the electrons in two isolated hydrogen atoms, H_2 molecules are lower in energy and therefore more stable than unbonded H atoms.

Hydrogen is a diatomic gas, but helium exists as free atoms and not as molecular He_2. MO theory explains why. Each He atom has two valence electrons in its 1s atomic orbital. Mixing two He 1s orbitals yields the same set of molecular orbitals (Figure 5.43) that H_2 has. Adding four electrons fills both of helium's MOs. The presence of two electrons in the σ_{1s}^* orbital cancels the stability gained from having two electrons in the σ_{1s} orbital. Because there is no net gain in stability, He_2, $(\sigma_{1s})^2(\sigma_{1s}^*)^2$, does not exist.

Another way to compare the bonding in H_2 and He_2 is to consider the *bond order* in each molecule. We previously defined bond order as the number of bonds between two atoms: a bond order of 1 for Cl—Cl, 2 for O=O, and 3 for N≡N. In MO theory we define bond order as follows:

$$\text{Bond order} = \tfrac{1}{2}\left[\left(\begin{array}{c}\text{number of}\\\text{bonding electrons}\end{array}\right) - \left(\begin{array}{c}\text{number of}\\\text{antibonding electrons}\end{array}\right)\right] \quad (5.2)$$

In a molecule of H_2, there are two electrons in the bonding MO and none in the antibonding MO, so

$$\text{Bond order in } H_2 = \tfrac{1}{2}(2 - 0) = 1$$

In nonexistent He_2, the bond order is 0 because an equal number of electrons reside in bonding and antibonding orbitals:

$$\text{Bond order in } He_2 = \tfrac{1}{2}(2 - 2) = 0$$

In general, the greater the bond order, the stronger the bond and the more stable the molecule.

SAMPLE EXERCISE 5.8   Using MO Theory to Predict   LO6
Bond Order I

Draw the MO diagram of the molecular ion H_2^-; determine the bond order of the ion, and predict whether or not it is stable.

COLLECT AND ORGANIZE We apply MO theory to draw an orbital diagram for the molecular ion H_2^-, and then determine bond order using Equation 5.2. If the bond order is greater than zero, the ion may be stable.

ANALYZE We should be able to base the MO diagram for H_2^- on the MO diagram for H_2 (Figure 5.42b) because the H_2^- ion has only one more electron than H_2 and the empty σ_{1s}^* orbital in H_2 can accommodate up to two more electrons.

SOLVE The σ_{1s} orbital is filled in H_2. The third electron of H_2^- goes into the σ_{1s}^* orbital, so the MO diagram is as shown in Figure 5.44. The notation for this electron configuration is $(\sigma_{1s})^2(\sigma_{1s}^*)^1$ (listing the molecular orbitals in order of increasing energy). The bond order is

$$\text{Bond order in } H_2^- = \tfrac{1}{2}(2-1) = \tfrac{1}{2}$$

The H_2^- ion is not as stable as H_2, but it is stable.

THINK ABOUT IT We encountered fractional bonds in Chapter 4 in the discussion of resonance, and we encounter them again here in the MO treatment of H_2^-. A bond order of $\tfrac{1}{2}$ in MO theory means that the bond between the two atoms in H_2^- is weaker than the single bond in H_2, making H_2^- a less stable species.

Practice Exercise Use MO theory to predict the bond order in a H_2^+ ion.

(Answers to Practice Exercises are in the back of the book.)

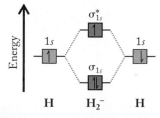

FIGURE 5.44 The molecular orbital diagram for H_2^-.

CONNECTION We first defined bond order in Chapter 4 when we related the lengths and strengths of bonds to the number of electron pairs shared by two atoms.

Molecular Orbitals of Other Homonuclear Diatomic Molecules

Molecular orbital diagrams for homonuclear (same-atom) diatomic molecules like N_2 and O_2 are more complex than the diagram for H_2 because of the greater number and variety of atomic orbitals in N_2 and O_2. Not all combinations of atomic orbitals result in effective bonding, but there are some general guidelines for constructing the molecular orbital diagram for any molecule:

1. The number of molecular orbitals equals the number of atomic orbitals used to create them.
2. Atomic orbitals with similar energy and orientation mix more effectively than do those that have different energies and orientations.
3. Better mixing leads to a larger energy difference between bonding and antibonding orbitals and thus greater stabilization of the bonding MOs.
4. A molecular orbital can accommodate a maximum of two electrons; two electrons in the same MO have opposite spins.
5. Electrons in ground-state molecules occupy the lowest energy molecular orbitals available, following the aufbau principle and Hund's rule.

In mixing atomic orbitals to create molecular orbitals, we consider *only the valence electrons* on the atoms because core electrons do not participate in bonding. Focusing on N_2 and O_2 as examples, we first mix their 2s orbitals. The mixing

CHEMTOUR Molecular Orbitals

sigma (σ) molecular orbital in MO theory, the orbital that results in the highest electron density between the two bonded atoms.

molecular orbital diagram in MO theory, an energy-level diagram showing the relative energies and electron occupancy of the molecular orbitals for a molecule.

pi (π) molecular orbital in MO theory, an orbital formed by the mixing of atomic orbitals oriented above and below, or in front of and behind, the bonding axis; electrons in π orbitals form π bonds.

process is analogous to the one we used for H_2, except that the resulting MOs are designated σ_{2s} and σ_{2s}^*.

Next we mix the three pairs of $2p$ orbitals, producing a total of six MOs. The different spatial orientations of the $2p_x$, $2p_y$, and $2p_z$ atomic orbitals result in different kinds of MOs (Figure 5.45). The $2p_z$ atomic orbitals point toward each other. When they mix, two molecular orbitals form: a σ_{2p} bonding orbital and a σ_{2p}^* antibonding orbital. The lobes of the $2p_x$ and $2p_y$ atomic orbitals are oriented at 90° to the bonding axis and also at 90° to each other. When the $2p_x$ orbitals mix together, and when the $2p_y$ orbitals mix together, they do so around the bonding axis instead of along it. This mixing produces **pi (π) molecular orbitals**. As with σ molecular orbitals, there are equal numbers of low-energy π and high-energy π* molecular orbitals. Electrons that occupy π orbitals contribute to the formation of π bonds; those that populate π* orbitals detract from π bond formation.

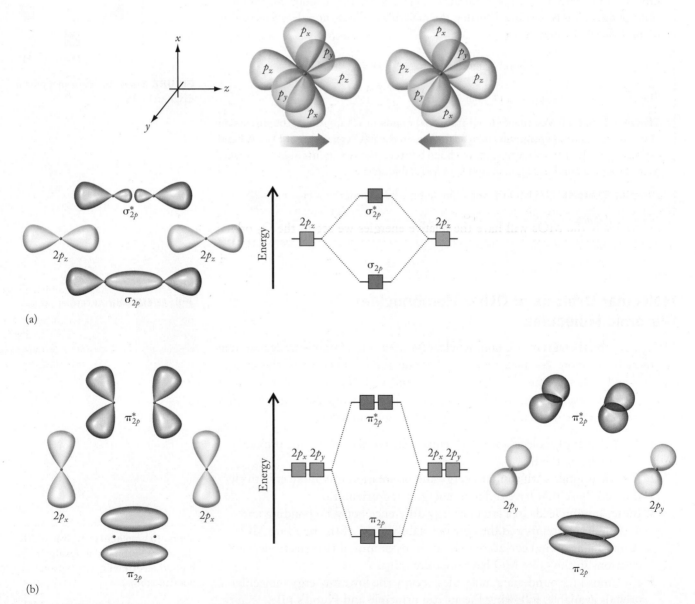

FIGURE 5.45 Two sets of three p atomic orbitals mix to form six molecular orbitals. (a) The $2p_z$ atomic orbitals create a σ_{2p} bonding orbital and a σ_{2p}^* antibonding orbital. (b) The $2p_x$ and $2p_y$ atomic orbitals mix to form two π_{2p} bonding molecular orbitals and two π_{2p}^* antibonding molecular orbitals.

The relative energies of σ and π molecular orbitals for N_2 and O_2 are shown in Figure 5.46. In each molecule, the energies of the σ_{2s} and σ_{2s}^* MOs are lower than the energy of the σ_{2p} MO for the same reason that a $2s$ atomic orbital is lower in energy than a $2p$ atomic orbital. However, the relative energies of the MOs formed by mixing the $2p$ orbitals in N_2 and O_2 are not the same. Let's begin with O_2 (Figure 5.46b) because its diagram is representative of most homonuclear diatomic molecules, including all the halogens. In order of increasing energy, the MOs are σ_{2p}, π_{2p}, π_{2p}^*, and σ_{2p}^*. Keep in mind that there are groups of two π_{2p} and two π_{2p}^* orbitals. This means that each group of two can hold up to 4 electrons. Adding 12 valence electrons (6 from each O atom in the O_2 molecule) to the MOs, starting with the lowest energy MO and working our way up, produces the following electron configuration:

$$O_2: \quad (\sigma_{2s})^2(\sigma_{2s}^*)^2(\sigma_{2p})^2(\pi_{2p})^4(\pi_{2p}^*)^2$$

The distribution of electrons in the MO diagram reflects this sequence. Note that the two π_{2p}^* orbitals are degenerate (equivalent in energy), so each contains a single electron, in accordance with Hund's rule. The MO diagram tells us that there are two unpaired electrons in a molecule of O_2. This is not the view obtained from the Lewis structure of O_2, or from VSEPR or valence bond theory, which all predict that the valence electrons are paired. We return to this point shortly, but for now let's consider the MO diagram for N_2.

When we compare the MO diagrams for N_2 and O_2 in Figure 5.46, we see a difference in the relative energies of two of the MOs. In N_2, the π_{2p} molecular orbitals are lower in energy than the σ_{2p} orbital. This difference in energy level stacking in N_2 compared to O_2 is linked to increasing energy differences between the $2s$ and $2p$ orbitals with increasing atomic number. The greater the difference, the more likely the MOs will have the relative energies we see in the O_2 orbital diagram. Lesser differences result in some mixing of the $2s$ and $2p$ orbitals during

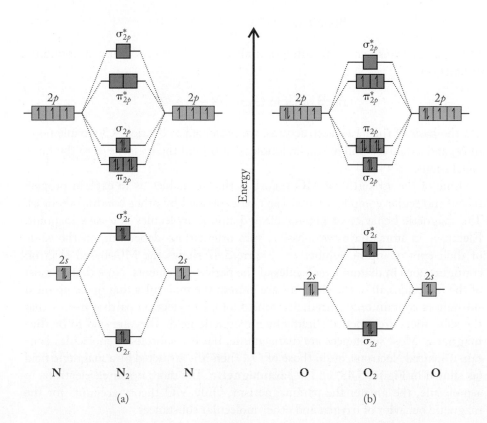

N N_2 N

(a)

O O_2 O

(b)

FIGURE 5.46 Molecular orbital diagrams of (a) N_2 and (b) O_2. The vertical sequence of orbitals in N_2 applies to homonuclear diatomic molecules of elements with $Z \le 7$. The O_2 sequence applies to homonuclear diatomic molecules of all elements beyond oxygen ($Z \ge 8$).

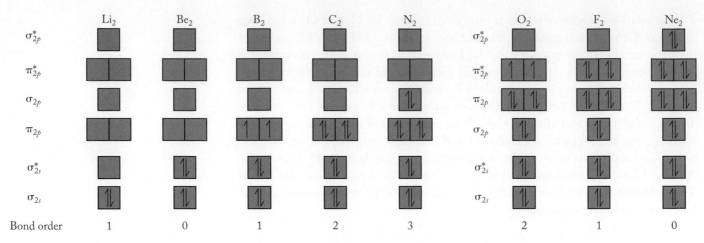

FIGURE 5.47 Valence-shell molecular orbital diagrams and bond orders of the homonuclear diatomic molecules of the second-row elements.

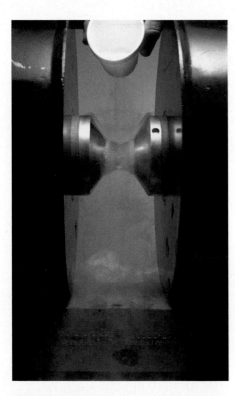

FIGURE 5.48 Liquid O_2 poured from a cup is suspended in the space between the poles of this magnet because the unpaired electrons in its molecules make O_2 paramagnetic. Paramagnetic substances are attracted to magnetic fields.

MO formation, which has the effect of lowering the energy of the σ_{2s} orbital and raising the energy of the σ_{2p} orbital—enough to put it above π_{2p} as in the N_2 orbital diagram. Adding 10 valence electrons to this stack of MOs in N_2, lowest energy first, produces this electron configuration:

$$N_2: \quad (\sigma_{2s})^2(\sigma_{2s}^*)^2(\pi_{2p})^4(\sigma_{2p})^2$$

The distribution of electrons in the MO diagram in Figure 5.46(a) reflects that electron configuration. There are a total of 8 electrons in bonding MOs and 2 in antibonding MOs. Using Equation 5.2 to calculate the bond order for N_2, we get

$$\text{Bond order in } N_2 = \tfrac{1}{2}(8-2) = 3$$

In O_2, 8 electrons occupy bonding orbitals and 4 electrons occupy antibonding orbitals, so

$$\text{Bond order in } O_2 = \tfrac{1}{2}(8-4) = 2$$

On the basis of their Lewis structures, we predicted in Section 4.3 a triple bond in N_2 and a double bond in O_2, and molecular orbital theory leads us to the same conclusions.

One of the strengths of MO theory is that it enables us to explain properties of molecular compounds that can't be explained by other bonding theories. The magnetic behavior of homonuclear diatomic molecules is a case in point. Electrons in atoms have two possible spin orientations depending on the value of their spin quantum number m_s. Figure 5.47 shows the MO-based electron configurations in diatomic molecules of the period 2 elements. Note that in most of the molecules all of the electrons are paired; the molecules that make up most substances contain only paired electrons. Complete electron pairing means that the substances are repelled slightly by a magnetic field. They are said to be **diamagnetic**. Most substances are diamagnetic, but if a substance's molecules contain unpaired electrons, as do those of O_2, then it is attracted by a magnetic field (as shown in Figure 5.48) and is **paramagnetic**. The more unpaired electrons in a molecule, the greater the paramagnetism. Only MO theory accounts for the magnetic behavior of oxygen and other molecular substances.

CONCEPT TEST

a. Which of the diatomic molecules in Figure 5.47 does not exist?
b. Which one is paramagnetic (besides O_2)?

(Answers to Concept Tests are in the back of the book.)

diamagnetic describes a substance with no unpaired electrons that is weakly repelled by a magnetic field.

paramagnetic describes a substance with unpaired electrons that is attracted to a magnetic field.

SAMPLE EXERCISE 5.9 Using MO Theory to Predict Bond Order II LO6

In which molecules in Figure 5.47 would there be an increase in bond order if they lost one electron each, forming singly charged diatomic cations (X_2^+)?

COLLECT AND ORGANIZE We must determine which molecules in Figure 5.47 acquire a higher bond order when one electron is removed. Equation 5.2 relates bond order to the difference in the numbers of electrons in bonding and antibonding orbitals.

ANALYZE Removing an electron from Li_2, Be_2, B_2, C_2, N_2, O_2, F_2, and Ne_2 will result in the molecular ions Li_2^+, Be_2^+, B_2^+, C_2^+, N_2^+, O_2^+, F_2^+, and Ne_2^+. We may assume that the molecular ions have MO diagrams with orbital energies in the same sequence as their parent molecules. Thus, the MO diagrams for the molecular ions as the same as those in Figure 5.47, but with one electron removed from the highest energy orbital. We expect an increase in bond order if the electron is removed from an antibonding orbital.

SOLVE Removing one electron from each MO diagram in Figure 5.47 gives us the results in the following table:

Ion	Electron Configuration	Bond Order
Li_2^+	$(\sigma_{2s})^1$	$\frac{1}{2}(1 - 0) = 0.5$
Be_2^+	$(\sigma_{2s})^2(\sigma_{2s}^*)^1$	$\frac{1}{2}(2 - 1) = 0.5$
B_2^+	$(\sigma_{2s})^2(\sigma_{2s}^*)^2(\pi_{2p})^1$	$\frac{1}{2}(3 - 2) = 0.5$
C_2^+	$(\sigma_{2s})^2(\sigma_{2s}^*)^2(\pi_{2p})^3$	$\frac{1}{2}(5 - 2) = 1.5$
N_2^+	$(\sigma_{2s})^2(\sigma_{2s}^*)^2(\pi_{2p})^4(\sigma_{2p})^1$	$\frac{1}{2}(7 - 2) = 2.5$
O_2^+	$(\sigma_{2s})^2(\sigma_{2s}^*)^2(\sigma_{2p})^2(\pi_{2p})^4(\pi_{2p}^*)^1$	$\frac{1}{2}(8 - 3) = 2.5$
F_2^+	$(\sigma_{2s})^2(\sigma_{2s}^*)^2(\sigma_{2p})^2(\pi_{2p})^4(\pi_{2p}^*)^3$	$\frac{1}{2}(8 - 5) = 1.5$
Ne_2^+	$(\sigma_{2s})^2(\sigma_{2s}^*)^2(\sigma_{2p})^2(\pi_{2p})^4(\pi_{2p}^*)^4(\sigma_{2p}^*)^1$	$\frac{1}{2}(8 - 7) = 0.5$

Comparing these values with the bond orders listed in Figure 5.47, we see that bond order increases for O_2^+ and F_2^+. Although Be_2 and Ne_2 are not stable (bond order = 0), both Be_2^+ and Ne_2^+ have a bond order of 0.5, an increase in bond order as well.

THINK ABOUT IT Removing an electron from Be_2, O_2, F_2, or Ne_2 reduces the number of electrons in antibonding molecular orbitals while leaving the number of electrons in bonding molecular orbitals unchanged. The result is an increase in bond order. In the other four homonuclear diatomic molecules, removing an electron reduces the number of electrons in bonding molecular orbitals while leaving the number of electrons in antibonding orbitals unchanged. The result is a reduction in bond order for Li_2^+, B_2^+, C_2^+, and N_2^+.

Practice Exercise Which molecules in Figure 5.47 show an increase in bond order when one electron is added to the molecule?

(Answers to Practice Exercises are in the back of the book.)

Molecular Orbitals of Heteronuclear Diatomic Molecules

Molecular orbital theory also enables us to account for the bonding in *heteronuclear* diatomic molecules, which are molecules containing two different atoms. The bonding in some of these molecules is difficult to explain using other bonding theories. For example, it is often difficult to draw a single Lewis structure for an odd-electron molecule like nitrogen monoxide, NO. In Chapter 4, we considered several arrangements of the valence electrons in NO, such as

$$:\!N\!=\!\ddot{O}\!: \qquad \cdot\ddot{N}\!=\!\ddot{O}\!:$$

We predicted that oxygen was more likely to have a complete octet of valence electrons because it is the more electronegative element. In addition, experimental evidence rules out structures with unpaired electrons on the oxygen atom. Therefore the preferred structure is the one shown in red. However, the length of the NO bond (115 pm) is considerably shorter than the value in Table 4.7 for an average N=O double bond (122 pm).

Molecular orbital theory helps explain both the bonding in NO and the shorter-than-expected bond length. Let's look at the bonding first. The MO diagram for nitrogen monoxide is different from the diagrams of homonuclear diatomic gases. Nitrogen and oxygen atoms have different numbers of protons and electrons, and the difference in effective nuclear charge between N and O atoms means that their atomic orbitals have different energies, as Figure 5.49 shows for the $2s$ and $2p$ orbitals.

In constructing the MO diagram for NO, the guidelines described previously still apply. The number of MOs formed must equal the number of atomic orbitals combined, and the energy and orientation of the atomic orbitals being mixed must be considered. One additional factor influences the energies of the MOs in heteronuclear diatomic molecules: *bonding* MOs tend to be closer in energy to the atomic orbitals of the more electronegative atom, and *antibonding* MOs tend to be closer in energy to the atomic orbitals of the less electronegative atom. The MO diagram for NO in Figure 5.49 illustrates this phenomenon. Note how the energy of the bonding σ_{2s} orbital is closer to that of the $2s$ orbital of the oxygen atom, and the energy of the antibonding σ_{2s}^* orbital is closer to that of the $2s$ orbital of the nitrogen atom. Similarly, the π_{2p} MOs in NO are closer in energy to the $2p$ orbitals of oxygen, and the π_{2p}^* MOs are closer in energy to the $2p$ orbitals of nitrogen. The proximity of the nitrogen $2p$ atomic orbitals to the π_{2p}^* MOs is consistent with the single electron in the π_{2p}^* MOs being on nitrogen rather than on oxygen. This prediction is consistent with our Lewis structure in which the odd electron in NO is on the nitrogen atom.

Molecular orbital theory also enables us to rationalize the relatively short bond length in NO. Equation 5.2 tells us that the bond order is $\frac{1}{2}(8-3) = 2.5$, halfway between the bond orders for N=O and N≡O and consistent with a bond length of 115 pm, halfway between the lengths of the N=O bond (122 pm) and the N≡O bond (106 pm).

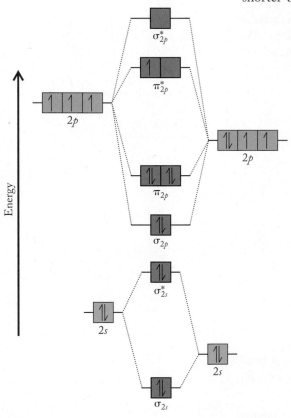

FIGURE 5.49 The molecular orbital diagram for NO shows that the unpaired electron occupies a π_{2p}^* antibonding orbital, which is closer in energy to the $2p$ atomic orbitals of nitrogen than to the $2p$ atomic orbitals of oxygen. As a result of this proximity, the electron has more nitrogen character and is more likely to be on the nitrogen atom than the oxygen atom.

SAMPLE EXERCISE 5.10   **Using MO Theory to Determine**   **LO6**
the Bond Order of Heteronuclear
Diatomic Molecules

Nitrogen monoxide reacts with many transition metals, including the iron in our blood. In these compounds, NO is sometimes considered to be NO^+ and at other times NO^-. Use Figure 5.49 to predict the bond order of NO^+ and NO^-.

COLLECT AND ORGANIZE We need to find the bond order of two diatomic ions based on the MO diagram of their parent molecule (Figure 5.49). Equation 5.2 relates bond order to the numbers of electrons in bonding and antibonding orbitals.

ANALYZE NO has 11 valence electrons. We can remove one electron from the MO diagram for NO to get the diagram for NO^+ and add one electron to get the diagram for NO^-.

SOLVE To generate NO^+, we remove the highest energy electron in NO, which is the one in the π_{2p}^* molecular orbital. That gives NO^+ this electron configuration:

$$NO^+: \quad (\sigma_{2s})^2(\sigma_{2s}^*)^2(\sigma_{2p})^2(\pi_{2p})^4$$

Adding an electron to the lowest energy MO available in NO (also π_{2p}^*) yields NO^- with this electron configuration:

$$NO^-: \quad (\sigma_{2s})^2(\sigma_{2s}^*)^2(\sigma_{2p})^2(\pi_{2p})^4(\pi_{2p}^*)^2$$

The bond orders of the two ions are

$$NO^+: \text{Bond order} = \tfrac{1}{2}(8-2) = 3$$
$$NO^-: \text{Bond order} = \tfrac{1}{2}(8-4) = 2$$

THINK ABOUT IT The bond orders in N_2 and O_2 are 3 and 2, respectively. The cation NO^+ is isoelectronic with N_2, so our calculated bond order for it makes sense. The anion NO^- is isoelectronic with O_2, and so our calculated bond order for it is also reasonable.

Practice Exercise Using Figure 5.49 as a guide, draw the MO diagram for carbon monoxide, and determine the bond order for the carbon–oxygen bond. ⚙

(Answers to Practice Exercises are in the back of the book.)

▶❚❚ **CHEMTOUR** Chemistry of the Upper Atmosphere

Molecular Orbitals of N_2^+ and the Colors of Auroras

In addition to predicting the magnetic properties of molecules, MO theory is particularly useful for predicting their spectroscopic properties. In Section 3.4, we noted that the light emitted by excited free atoms is quantized because it is linked to the movement of electrons between atomic orbitals with discrete energies. Broadly speaking, the same is true in molecules: electrons can move from one molecular orbital to another when molecules absorb or emit light.

We can use this information to look again at the phenomenon described at the opening of this section: how the colors of the auroras are produced. The principal chemical species involved are listed in Table 5.4. An asterisk indicates a molecule or molecular ion in an excited state. Excited N_2^+ ions produce blue-violet (391–470 nm) light, and excited N_2 molecules produce deep crimson red (650–680 nm) light. Comparing the MO diagrams of N_2^* and N_2 in Figure 5.50(a), we find that one of the two electrons originally in the σ_{2p} orbital in N_2 has been raised to a π_{2p}^*

TABLE 5.4 **Origins of Colors in the Aurora**

Wavelength (nm)	Color	Chemical Species
650–680	Deep red	N_2^*
630	Red	O^*
558	Green	O^*
391–470	Blue-violet	N_2^{+*}

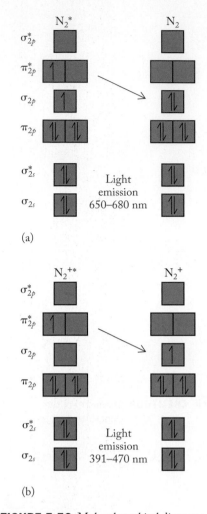

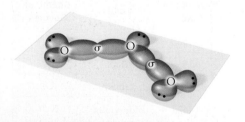

FIGURE 5.50 Molecular orbital diagrams for (a) N_2 and (b) N_2^+ show electronic transitions that result in the emission of visible light. High-energy collisions between particles in the upper atmosphere result in the promotion of electrons from the σ_{2p} orbital in N_2 molecules to the π_{2p}^* orbital and the formation of N_2^{+*} molecular ions that also have electrons in π_{2p}^* orbitals. When the electrons return to the ground state, red and blue-violet light is emitted.

FIGURE 5.51 Two sigma bonds and five lone pairs of electrons in a molecule of O_3.

orbital in N_2^*, leaving an unpaired σ_{2p} electron behind. Figure 5.50(b) shows us that N_2^{+*} also has one electron in a π_{2p}^* orbital, but its σ_{2p} orbital is empty because the other σ_{2p} electron originally in the N_2 molecule was lost when the molecule was ionized. As the π_{2p}^* electrons return from their antibonding orbital in the excited state to the bonding σ_{2p} orbital in the ground state, the distinctive celestial emissions of N_2^+ and N_2 are produced.

CONCEPT TEST ····································

Are the bond orders of the excited-state species in Figure 5.50 the same as the ground-state species?

(Answers to Concept Tests are in the back of the book.)

··

Using MO Theory to Explain Fractional Bond Orders and Resonance

In Chapter 4 our attempts to draw the Lewis structure of ozone led to the conclusion that molecules of O_3 are held together by two O—O single bonds plus a shared pair of electrons distributed equally across both bonds. The concept of resonance was used to describe delocalization of the second bonding pair and the equivalency of the bonds in O_3, which are shorter and stronger than O—O single bonds but not as short or strong as O=O double bonds. In this chapter we have learned how VSEPR explains the bent shape of O_3 molecules, and how valence bond theory and the formation of sp^2 hybrid orbitals help us visualize the trigonal planar orientation of bonding and lone pairs of electrons that account for ozone's molecular shape.

Now let's explore how MO theory provides another perspective on the delocalized bonding in O_3—one that does not require drawing resonance structures that, as we have admitted, do not accurately convey where all the bonding and lone pairs of electrons really are. MO theory is great at explaining bonding pair delocalization because MOs are not isolated between pairs of atoms but rather are spread across several atoms or even entire molecules. To simplify our exploration, we start with three O atoms that are connected by two single (σ) bonds and that have a total of five lone pairs of valence electrons (Figure 5.51). This trigonal planar pattern of electron pairs is what we would expect if each of the O atoms had sp^2 hybridized atomic orbitals. There are 14 valence electrons in Figure 5.51, leaving $(3 \times 6) - 14 = 4$ electrons unaccounted for. Let's distribute the 4 electrons among a set of molecular orbitals formed by mixing the three unhybridized p orbitals of the three O atoms.

MO theory says that mixing three p orbitals with lobes above and below the bonding plane produces three molecular orbitals. One of them is a low-energy π MO, and another is a high-energy π^* MO, but what about the third? It turns out that the third orbital is neither bonding nor antibonding. Instead, it is a *nonbonding* (n) molecular orbital. Electrons in nonbonding MOs neither lower the energy of a molecular structure, which would stabilize it, nor do they raise its energy and destabilize it. Rather, nonbonding MOs have the same energy as the atomic orbitals from which they formed, as shown in the partial MO diagram of ozone in Figure 5.52. Adding 4 electrons to these three molecular orbitals puts 2 electrons in the π orbital, 2 in the nonbonding orbital, and 0 in π^*. Therefore, there is $\frac{1}{2}(2 - 0) = 1$ π bond distributed across the entire molecule. The overall

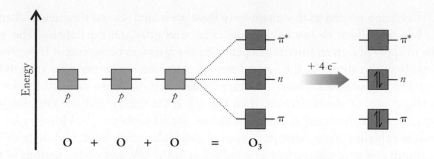

FIGURE 5.52 Forming molecular orbitals from unhybridized p orbitals in O_3.

bond order of 2 σ bonds + 1 π bond = 3 is divided equally between the two pairs of atoms, giving the familiar average bond order for ozone of 1.5. The pair of nonbonding electrons is also delocalized. Each of the two noncentral O atoms gets half ownership of this nonbonding pair for a total of 2.5 lone pairs of electrons each, which is the average of 2 lone pairs on one noncentral O atom and 3 on the other in the resonance structures of O_3:

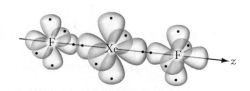

MO Theory for SN > 4

In Chapter 4 we used expanded octets to explain the bonding in compounds such as SF_6 in which the central atom has a steric number greater than 4. In this chapter we have used hybrid schemes such as sp^3d and sp^3d^2 to help visualize the bonding in those compounds and to account for their molecular shapes. Now we will see how MO theory provides an alternate explanation of bonding and molecular shape that accommodates steric numbers greater than 4 without incorporating d orbitals.

To see how the MO approach works, let's consider the bonding in XeF_2, which, in 1962, was one of the first noble gas compounds to be synthesized. Lewis theory predicts that there are two bonding pairs and three lone pairs of electrons around the central Xe atom. The molecule is linear, which makes sense if the lone pairs occupy the three equatorial positions of a trigonal bipyramidal electron-group array and the fluorine atoms are in the axial positions, as shown in Figure 5.53. This bonding pattern is consistent with a hybridization process involving all the 5s and 5p orbitals and one of the 5d orbitals of the central Xe atom, that is, sp^3d.

MO theory provides another explanation of the linear shape of XeF_2 and does not need d orbitals to do it. Let's start with a Xe atom and two F atoms aligned so that the lobes of their p_z orbitals are all in a single row (Figure 5.54). Alignment of their lobes allows the orbitals to mix together and form a set of bonding, nonbonding, and antibonding σ molecular orbitals as shown in Figure 5.55. There are a total of four electrons in these orbitals, and they fill the bonding and nonbonding MOs while leaving the antibonding orbital empty. This array produces a pair of bonding electrons and a bond order of 1 that is delocalized over the entire molecule. Therefore, the effective bond order of each of the two Xe—F bonds is only 0.5. You might think that a compound with bonds that are half the strength of single bonds might be unstable, and you would be correct. Xenon difluoride, though more stable than some other noble gas compounds, decomposes when it is exposed to sunlight and when it comes in contact with water.

FIGURE 5.53 Lewis structure of XeF_2.

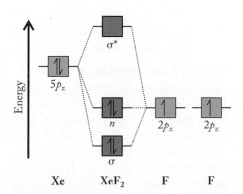

FIGURE 5.54 Alignment of valence-shell p_z orbitals of F and Xe prior to formation of bonding, nonbonding, and antibonding σ molecular orbitals in XeF_2.

FIGURE 5.55 Partial molecular orbital diagram of XeF_2. These are the only MOs that contribute to bonding.

In Chapter 4 and in this chapter we have presented several theories of chemical bonding. Each theory has its particular strengths and capabilities. The best one to apply in a given situation depends on the question being asked. If our focus is on how the atoms in the second row of elements in the periodic table bond together in molecules, Lewis structures usually suffice. If we are also interested in the shapes of the molecules, then valence bond theory and the visualization capability of hybridized atomic orbitals are useful tools to use. However, if we wish to explain the magnetic properties of molecular compounds and their ability to absorb and emit electromagnetic radiation in the UV and visible regions of the electromagnetic spectrum, then we must use molecular orbital theory.

SAMPLE EXERCISE 5.11   Integrating Concepts: Treatment for Alzheimer's Disease

In 2011 a compound with the molecular structure shown here was reported to be effective in treating neurodegenerative diseases, including Huntington's and Alzheimer's, in animals:

Note the two S atoms in the structure. One has two bonds, which is consistent with its Lewis bonding capacity, but the other has six. Answer these questions about this drug:

a. Many pharmaceuticals are chiral. Are any of the carbon atoms in this compound chiral centers?

b. What are the steric numbers and electron-group geometries of the two sulfur atoms?

c. What is the bond angle between the two $S=O$ double bonds?

d. There are two six-carbon rings in the structure. Are either of them aromatic rings? Which one(s)?

e. Does this compound have a permanent dipole?

COLLECT AND ORGANIZE We are given the molecular structure of a biologically active compound and are asked questions about it. A chiral carbon atom is one that is bonded to four different atoms or groups of atoms. The steric number of an atom is the sum of its lone pairs of valence electrons and the number of atoms bonded to it. Steric numbers allow us to predict electron-group geometries and bond angles, and, indirectly, molecular geometries. Aromatic rings have six sp^2 hybridized C atoms and are stabilized by the delocalization of the electrons in three π bonds around the ring. Molecules in which bond dipoles do not cancel have permanent dipoles, that is, molecules with asymmetrical distributions of valence electrons.

ANALYZE

a. Only carbon atoms with four single (σ) bonds can be chiral centers, which eliminates all those in rings that are double bonded. Also, the C atoms in $-CH_2-$ groups can't be chiral because two of the four atoms or groups bonded to them are the same (H atoms).

b. and c. Sulfur atoms complete their octets by forming two covalent bonds, giving them two bonding pairs and two lone pairs. The sulfur atom in the five-atom ring fits that description. The other sulfur has two single bonds and two double bonds for a total of six pairs of bonding electrons—clearly more than an octet. This bonding pattern looks a lot like the one for sulfur in sulfuric acid (see Section 4.8), which has no lone pairs in its Lewis structure.

d. The atoms in both six-carbon rings are bonded to each other with alternating single and double bonds, which allows the π electrons in the three double bonds to be completely delocalized around each ring.

e. The molecule consists mostly of C and H atoms, which form bonds that are not very polar, but it also contains C—O, N=O, N—H, and S=O bonds that are polar. There is also asymmetry in the overall structure of the molecule.

SOLVE

a. None of the sp^3 carbon atoms are bonded to four different atoms or groups of atoms, so none are chiral.

b. One of the two sulfur atoms has two atoms bonded to it and two lone pairs of valence electrons, which add up to SN = 4. The corresponding electron-group geometry is tetrahedral. The other S atom has six bonds to four atoms and no lone pairs, which means its SN is also 4 and it also has a tetrahedral electron-group geometry.

c. Given the tetrahedral electron-group geometry around S and lack of lone pairs, we anticipate tetrahedral molecular geometry and an $O=S=O$ bond angle of about 109°.

d. Due to the presence of delocalized π electrons, both six-carbon rings are aromatic.

e. Given its generally asymmetric structure and the presence of polar bonds, we can conclude that the molecule has a permanent dipole.

THINK ABOUT IT We established that the compound has no chiral carbon atoms, but does it have no chiral centers at all? Look carefully at the nitrogen atom in the middle of the structure. It is sp^3 hybridized and surrounded by (1) a H atom, (2) the right side of the molecule, which is different from (3) the left side, and (4) a lone pair of electrons. Do the four different groups (counting the lone pair as a "group") make this N atom a chiral center? They would, except for one thing: a process known as *inversion*. If we label the right side of the molecule R and the left side L, then the structure surrounding the N atom could have two enantiomers:

The double arrow symbol between them means that the bonds of one can reversibly flip around the N atom to form the other enantiomer. Because of this inversion of the bonds, we can't produce a sample of just one enantiomer.

SUMMARY

Section 5.1 The shape of a molecule reflects the arrangement of the atoms in three-dimensional space and is determined largely by characteristic **bond angles**.

Section 5.2 Minimizing repulsion between pairs of valence electrons (the **VSEPR** model) results in the lowest energy orientations of bonding and nonbonding electron pairs and accounts for the observed **molecular geometries** of molecules. The shape of a molecule can be determined by its **steric number** (the sum of the number of bonded atoms and lone pairs around a central atom) and the **electron-group geometry**, or arrangement of atoms and lone pairs. Molecules with SN = 2 and no lone pairs on the central atom have a **linear** electron-group geometry and linear molecular geometry, while the electron-group geometries of molecules with steric numbers 3 to 6 are **trigonal planar**, **tetrahedral**, **trigonal bipyramidal**, and **octahedral**, respectively. Molecular geometries called **angular (bent)**, **trigonal pyramidal**, **seesaw**, **T-shaped**, **square pyramidal**, and **square planar** result from different combinations of bonded atoms and lone pairs about a central atom. The observed bond angles in molecules deviate from the ideal values as a result of unequal repulsions between lone pairs and bonding pairs of electrons.

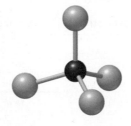

Section 5.3 Two covalently bonded atoms with different electronegativities have partial electrical charges of opposite sign, creating a **bond dipole**. If the individual bond dipoles in a molecule do not offset each other, the molecule is polar. If they do offset each other, the molecule is nonpolar. A polar molecule has a permanent **dipole moment (μ)**, which is a quantitative measure of the polarity of the molecule.

Section 5.4 According to **valence bond theory**, the **overlap** of half-filled orbitals results in covalent bonds between pairs of atoms in molecules. Molecular geometry is explained by the mixing, or **hybridization**, of atomic orbitals to create **hybrid atomic orbitals**. Mixing one s and three p orbitals forms four sp^3 **hybrid orbitals**. Overlap between sp^3 orbitals and other atomic or hybrid orbitals results in up to four **sigma (σ) bonds** and a tetrahedral orientation of valence electrons. Mixing one s and two p orbitals forms three sp^2 **hybrid orbitals**. Overlap between sp^2 orbitals and other atomic or hybrid orbitals results in up to three σ bonds and a trigonal planar orientation of valence electrons. Mixing one s and one p orbital forms two **sp hybrid orbitals**. Overlap between two sp hybrid orbitals results in up to two σ bonds oriented linearly to one another. Covalent bonds in which the electron density is greatest either above and below or in front of and behind the bonding axis are **pi (π) bonds**. Mixing one s orbital, three p orbitals, and one d orbital yields five equivalent sp^3d **hybrid orbitals** with lobes that point toward the vertices of a trigonal bipyramid. Overlap between sp^3d orbitals and other atomic or hybrid orbitals results in up to five σ bonds. Mixing one s orbital, three p orbitals, and two d orbitals gives six equivalent sp^3d^2 **hybrid orbitals** that point toward the vertices of an octahedron. Overlap between sp^3d^2 orbitals and other atomic or hybrid orbitals results in up to six σ bonds.

Section 5.5 The shape of a molecule with more than one central atom is a result of overlapping geometries around the atoms. Molecules with only sp^2 hybridized central atoms have extended planar geometries. The molecules of **aromatic compounds** contain planar rings of six sp^2 hybridized carbon atoms with alternating π bonds whose electrons are delocalized over the entire ring system.

Section 5.6 **Chiral** molecules exist in left- and right-handed forms that have different properties. Many contain an sp^3 hybridized carbon atom bonded to four different atoms or groups of atoms.

Section 5.7 Molecular orbital (MO) theory is based on the formation of **molecular orbitals**, which are orbitals belonging to an entire molecule. MO theory explains the bonding in molecules and the magnetic and spectroscopic properties of molecular compounds. Mixing two atomic orbitals creates one **bonding orbital** and one **antibonding orbital**. The region of highest electron density lies along the bond axis in a **sigma (σ) molecular orbital**. Electrons in σ molecular orbitals form σ bonds. The regions of highest electron density of **pi (π) molecular orbitals** are above and below or behind and in front of the bonding axis. Electrons occupying π orbitals form π bonds.

A **molecular orbital diagram** shows the relative energies of the molecular orbitals in a molecule. MO electron configurations use the designations σ, σ*, π, and π* to describe the type of molecular orbitals occupied by electrons; subscripts to identify the atomic orbitals that combined to form the MOs; and superscripts to indicate the number of electrons in each MO. Atoms, ions, and molecules with no unpaired electrons are **diamagnetic** and are slightly repelled by an applied magnetic field. Atoms, ions, and molecules containing at least one unpaired electron are **paramagnetic** and are attracted by an external magnetic field.

PROBLEM-SOLVING SUMMARY

TYPE OF PROBLEM	CONCEPTS AND EQUATIONS	SAMPLE EXERCISES
Predicting molecular geometry	Draw the Lewis structure of the molecule. Determine the steric number (SN) of the central atom, where $$SN = \left(\begin{array}{c}\text{number of atoms}\\\text{bonded to central atom}\end{array}\right) + \left(\begin{array}{c}\text{number of lone pairs}\\\text{on central atom}\end{array}\right) \quad (5.1)$$ Choose the electron-group geometry that corresponds to the SN. Choose a molecular geometry based on the electron-group geometry that accounts for the number of lone pairs of valence electrons on the central atom.	5.1, 5.3
Predicting bond angles	Repulsion from lone pairs on a central atom pushes bonding pairs closer together and decreases bond angles.	5.2
Predicting whether or not a substance is polar	A substance is polar if a molecule of it has an overall permanent dipole—that is, its bond dipoles and the locations of its lone pairs do not offset each other.	5.4
Using valence bond theory to explain molecular shape	Translate the observed electron-group geometry into the appropriate central-atom hybridization following these guidelines:	5.5, 5.6

Steric Number	Electron-Group Geometry	Hybridization
2	Linear	sp
3	Trigonal planar	sp^2
4	Tetrahedral	sp^3
5	Trigonal bipyramidal	sp^3d
6	Octahedral	sp^3d^2

TYPE OF PROBLEM	CONCEPTS AND EQUATIONS	SAMPLE EXERCISES
Recognizing chiral molecules	Look for an sp^3 central atom that is bonded to four different atoms or groups of atoms.	5.7
Using MO theory to predict bond order	Draw the molecular orbital diagram and count the numbers of electrons in bonding and antibonding orbitals. $$\text{Bond order} = \frac{1}{2}\left[\left(\begin{array}{c}\text{number of}\\\text{bonding electrons}\end{array}\right) - \left(\begin{array}{c}\text{number of}\\\text{antibonding electrons}\end{array}\right)\right] \quad (5.2)$$	5.8–5.10

VISUAL PROBLEMS

(Answers to boldface end-of-chapter questions and problems are in the back of the book.)

5.1. The two compounds with the molecular structures shown in Figure P5.1 have the same molecular formula: $C_2H_3F_3$. Which of the two molecules in Figure P5.1 has the greater dipole moment?

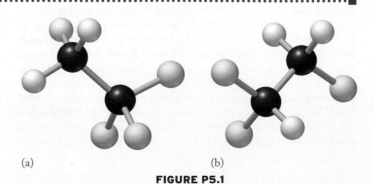

(a)   (b)

FIGURE P5.1

5.2. Could you distinguish between the two structures of N_2H_2 shown in Figure P5.2 by the magnitude of their dipole moments?

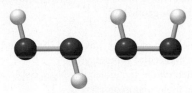

FIGURE P5.2

5.3. In which of the molecules shown in Figure P5.3 are all the atoms in a single plane? Are there delocalized π electrons in any of the molecules?

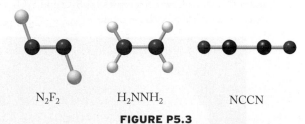

N₂F₂   H₂NNH₂   NCCN

FIGURE P5.3

5.4. Are all the atoms in C_4H_4 (Figure P5.4) in a single plane? Are there delocalized π electrons?

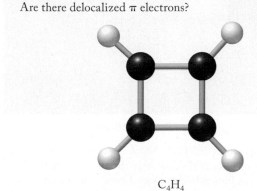

C_4H_4

FIGURE P5.4

5.5. Use the MO diagram in Figure P5.5 to predict whether O_2^+ has more or fewer electrons in antibonding molecular orbitals than O_2^{2+}.

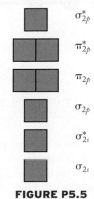

σ_{2p}^*

π_{2p}^*

π_{2p}

σ_{2p}

σ_{2s}^*

σ_{2s}

FIGURE P5.5

5.6. Under appropriate reaction conditions, diatomic molecules of iodine are ionized, forming I_2^+ cations. The corresponding

anion, I_2^-, is unknown. Use the molecular orbital diagram in Figure P5.6 to explain why I_2^+ is more stable than I_2^-.

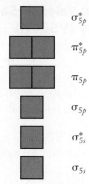

σ_{5p}^*

π_{5p}^*

π_{5p}

σ_{5p}

σ_{5s}^*

σ_{5s}

FIGURE P5.6

5.7. The molecular structure in Figure P5.7 is that of a constituent of pine oil that contributes to its characteristic aroma. Is the compound chiral? Explain your answer.

CH_3
H_2C — C — CH
H_2C — C — CH_2
HO
CH
H_3C — CH_3

FIGURE P5.7

*5.8. The molecular structure in Figure P5.8 is that of menthol, a chiral compound that gives mint leaves their characteristic aroma. Locate the chiral center(s) in the structure.

CH_3
CH
H_2C — CH_2
HO — CH — CH_2
CH
CH
H_3C — CH_3

FIGURE P5.8

5.9. The molecular geometry of ReF_7 is an uncommon structure called a pentagonal bipyramid, which is shown in Figure P5.9. What are the bond angles in a pentagonal bipyramid?

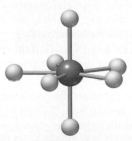

FIGURE P5.9

5.10. Suppose one of the bonding pairs of electrons in the structure in Figure P5.9 were actually a lone pair. Would the lone pair likely be in an axial or equatorial position? Why?

QUESTIONS AND PROBLEMS ······················ ■

Molecular Shape; Valence-Shell Electron-Pair Repulsion Theory (VSEPR)

CONCEPT REVIEW

5.11. Why is the shape of a molecule determined by repulsions between electron pairs and not by repulsions between nuclei?

5.12. In which molecular geometry do equatorial bonding pairs of electrons repel each other more: square pyramidal or trigonal bipyramidal?

5.13. Why do NO_3^- and NO_2^- ions have similar O–N–O bond angles even though they have different numbers of N–O bonds?

5.14. The O–N–O bond angle in NO_2 is slightly larger than it is in NO_2^-. Why?

5.15. In a molecule of ammonia, why is the repulsion between the lone pair and a bonding pair of electrons on nitrogen greater than the repulsion between two N—H bonding pairs?

5.16. Why do we need to draw the Lewis structure of a molecule before predicting its geometry?

5.17. Why does the seesaw structure have lower energy than a trigonal pyramidal structure when SN = 5?

*5.18. Do all resonance forms of a molecule have the same molecular geometry? Explain your answer.

PROBLEMS

5.19. Rank the following molecular geometries in order of increasing bond angles: (a) trigonal planar; (b) linear; (c) tetrahedral.

5.20. Rank the following molecules in order of increasing bond angles: (a) NH_3; (b) CH_4; (c) H_2O.

5.21. Which of the following electron-group geometries is not consistent with a linear molecular geometry, assuming three atoms per molecule? (a) tetrahedral; (b) octahedral; (c) trigonal planar.

5.22. How many lone pairs of electrons would there have to be on a SN = 6 central atom for it to have a linear molecular geometry?

5.23. Determine the molecular geometries of these molecules: (a) GeH_4; (b) PH_3; (c) H_2S; (d) $CHCl_3$.

5.24. Determine the molecular geometries of the following molecules and ions: (a) NO_3^-; (b) NO_4^{3-}; (c) NCN^{2-}; (d) NF_3.

5.25. Determine the molecular geometries of the following ions: (a) NH_4^+; (b) CO_3^{2-}; (c) NO_2^-; (d) XeF_5^+.

5.26. Determine the molecular geometries of the following ions: (a) SCN^-; (b) $CH_3PCl_3^+$ (P is the central atom and is bonded to the C atom of the methyl group); (c) ICl_2^-; (d) PO_3^{3-}.

5.27. Determine the molecular geometries of the following ions and molecules: (a) $S_2O_3^{2-}$; (b) PO_4^{3-}; (c) NO_3; (d) NCO.

5.28. Determine the molecular geometries of the following molecules: (a) ClO_2; (b) ClO_3; (c) IF_3; (d) SF_4.

5.29. Which two of the triatomic molecules O_3, SO_2, and CO_2 have the same molecular geometry?

5.30. Which two of the species N_3^-, O_3, and CO_2 have the same molecular geometry?

5.31. Which two of the ions SCN^-, CNO^-, and NO_2^- have the same molecular geometry?

5.32. Which two of the molecules N_2O, Se_2O, and CO_2 have the same molecular geometry?

5.33. **The Venusian Atmosphere** A number of sulfur oxides not found in Earth's atmosphere have been detected in the atmosphere of Venus (Figure P5.33), including S_2O and S_2O_2. Draw Lewis structures for S_2O and S_2O_2, and determine their molecular geometries.

FIGURE P5.33

5.34. The structures of NOCl, NO_2Cl, and NO_3Cl were determined in 1995. They have the skeletal structures shown in Figure P5.34. Draw Lewis structures for the three compounds and predict the electron-group geometry at each nitrogen atom.

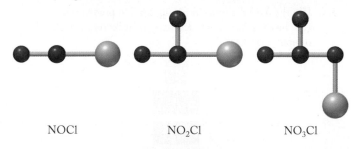

NOCl NO_2Cl NO_3Cl

FIGURE P5.34

*5.35. For many years, it was believed that the noble gases could not form covalently bonded compounds. However, xenon does react with fluorine. One of the products is the pentafluoroxenate anion, XeF_5^-. Draw the Lewis structure of XeF_5^- and predict its geometry.

*5.36. The first compound containing a xenon–sulfur bond was isolated in 1998. Draw a Lewis structure for HXeSH and determine its molecular geometry.

*5.37. **Chemical Terrorism** In 1995 a gang attacked the Tokyo subway system with the nerve gas Sarin and focused world attention on the dangers of chemical warfare agents. The structure in Figure P5.37 shows the connectivity of the atoms in the Sarin molecule. Complete the Lewis structure by adding bonds and lone pairs as necessary. Assign formal charges to the P and O atoms, and determine the molecular geometry around P.

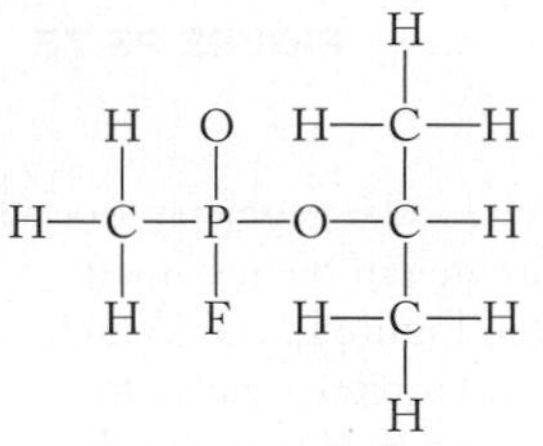

$$
\begin{array}{ccccc}
 & & \text{H} & & \\
\text{H} & \text{O} & \text{H—C—H} & & \\
| & \| & | & & \\
\text{H—C—P—O—C—H} & & & \\
| & | & | & & \\
\text{H} & \text{F} & \text{H—C—H} & & \\
 & & | & & \\
 & & \text{H} & &
\end{array}
$$

FIGURE P5.37

5.38. Determine the electron-group geometries around the nitrogen atoms in the following unstable nitrogen oxides: (a) N_2O_2; (b) N_2O_5; (c) N_2O_3. (N_2O_2 and N_2O_3 have N–N bonds; N_2O_5 does not.)

Polar Bonds and Polar Molecules

CONCEPT REVIEW

5.39. Explain the difference between a polar bond and a polar molecule.

5.40. Must a polar molecule contain polar covalent bonds? Why?

5.41. Can a nonpolar molecule contain polar covalent bonds?

5.42. What does a dipole moment measure?

PROBLEMS

5.43. The following molecules contain polar covalent bonds. Which of them are polar molecules and which have no permanent dipoles? (a) CCl_4; (b) $CHCl_3$; (c) CO_2; (d) H_2S; (e) SO_2

5.44. Which of the molecules has a permanent dipole? (a) C≡O; (b) N≡N; (c) H—C≡N; (d) H—C≡C—H

5.45. Compounds containing carbon, chlorine, and fluorine are known as chlorofluorocarbons (CFCs). Which of the following CFCs are polar and which are nonpolar? (a) $CFCl_3$; (b) CF_2Cl_2; (c) Cl_2FCCF_2Cl

5.46. Which of the following molecules has a permanent dipole? (a) C_4F_8 (cyclic structure); (b) $ClFCCF_2$; (c) $Cl_2HCCClF_2$

5.47. Predict which molecule in each of the following pairs is more polar: (a) $CClF_3$ or $CBrF_3$; (b) CF_2Cl_2 or CHF_2Cl

5.48. Which molecule in each of the following pairs is more polar? (a) NH_3 or PH_3; (b) CCl_2F_2 or CBr_2F_2

5.49. **Chemical Warfare Gas** A compound with the formula $COCl_2$ has been used as a chemical warfare agent. It and two similar compounds, $COBr_2$ and COI_2, are all eye irritants and cause skin to blister. The severity of the skin reactions is influenced by the polarity of the compounds. Rank the compounds in order of increasing polarity of their C—X bonds (where X is a halogen atom).

5.50. Among the diatomic molecules detected in interstellar space are CO, CS, SiO, SiS, SO, and NO. Arrange the molecules in order of increasing polarity of their bonds.

Valence Bond Theory and Hybrid Orbitals

CONCEPT REVIEW

5.51. What must atomic orbitals have in common to mix together and form hybrid orbitals?

5.52. Why do atomic orbitals form hybrid orbitals?

PROBLEMS

5.53. What is the hybridization of sulfur in each of the following molecules? (a) SO; (b) SO_2; (c) S_2O; (d) SO_3

5.54. What is the hybridization of nitrogen in each of the following ions and molecules? (a) NO_2^+; (b) NO_2^-; (c) N_2O; (d) N_2O_5; (e) N_2O_3

5.55. N_2F_2 has two possible structures as shown in Figure P5.55. Are the differences between the structures related to differences in the hybridization of nitrogen in N_2F_2? Identify the hybrid orbitals that account for the bonding in N_2F_2. Are they the same as those in acetylene, C_2H_2?

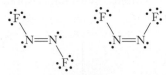

FIGURE P5.55

5.56. **Airbags** Azides such as sodium azide, NaN_3, are used in automobile airbags as a source of nitrogen gas. Another compound with three nitrogen atoms bonded together is N_3F. What differences are there between the arrangements of the electrons around the nitrogen atoms in the azide ion (N_3^-) and in N_3F? Is there a difference in the hybridization of the central nitrogen atom?

5.57. How does the hybridization of the central atom change in the series CO_2, NO_2, O_3, ClO_2?

5.58. How does the hybridization of the sulfur atom change in the series SF_2, SF_4, SF_6?

*5.59. **Perchlorate Ion and Human Health** Perchlorate compounds adversely affect human health by interfering with the uptake of iodine in the thyroid gland, but because of that behavior they are also used to treat hyperthyroidism, or overactive thyroid. Draw the Lewis structure(s) of the perchlorate ion, ClO_4^-, including all resonance forms, in which formal charges are closest to zero. What is the shape of the ion? Suggest a hybridization scheme for the central chlorine atom that accounts for this shape.

*5.60. **Bleaching Agents** Draw the Lewis structure of the chlorite ion, ClO_2^-, which is used as a bleaching agent. Include all resonance structures in which formal charges are closest to zero. What is the shape of the ion? Suggest a hybridization scheme for the central chlorine atom that accounts for the structures you have drawn.

5.61. Synthesis of the first compound of argon was reported in 2000. HArF was made by reacting Ar with HF. Draw the Lewis structure for HArF, and determine the hybridization of Ar in this molecule.

5.62. Draw a Lewis structure for Cl_3^+. Determine its molecular geometry and the hybridization of the central Cl atom.

5.63. Do all resonance forms of N_2O have the same hybridization at the central N atom?

5.64. The Lewis structure for N_4O, with the skeletal structure O–N–N–N–N, contains one N—N single bond, one N=N double bond, and one N≡N triple bond. Is the hybridization of all the nitrogen atoms the same?

*5.65. The trifluorosulfate anion, $SO_2F_3^-$, was isolated in 1999 as the tetramethylammonium salt, $[(CH_3)_4N]^+SO_2F_3^-$.
   a. What are the C–N–C bond angles in the cation?
   b. What is the hybridization of the N atom?
   c. The lengths of both S–O bonds in $SO_2F_3^-$ are 143 pm. Draw a Lewis structure of $SO_2F_3^-$ that is compatible with this bond length.
   d. What is the molecular geometry of the $SO_2F_3^-$ ion?

*5.66. Several resonance forms can be drawn for the anion $[C(CN)_3]^-$ including the two structures shown in Figure P5.66. Do they have the same geometry about the central carbon? What is the hybridization of each carbon atom?

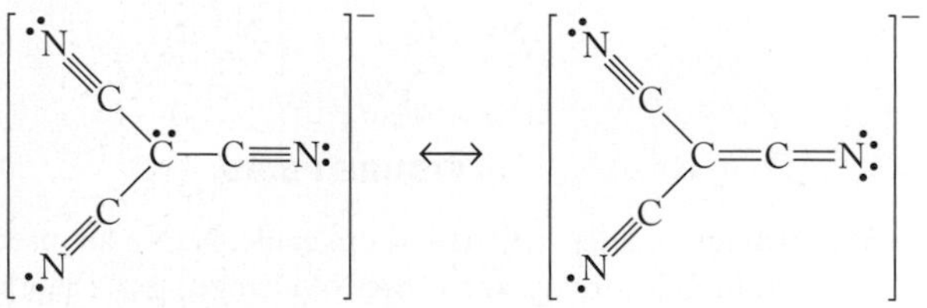

FIGURE P5.66

Molecules with Multiple "Central" Atoms

CONCEPT REVIEW

5.67. Can molecules with more than one central atom have resonance forms?

5.68. Why is it difficult to assign a single geometry to a molecule with more than one central atom?

*5.69. Are resonance structures examples of electron delocalization? Explain your answer.

*5.70. Can hybrid orbitals be associated with more than one atom?

PROBLEMS

5.71. The two nitrogen atoms in nitramide are connected with two oxygen atoms on one terminal nitrogen and two hydrogen atoms on the other (Figure P5.71). What is the molecular geometry of each nitrogen atom in nitramide? Is the hybridization of both nitrogen atoms the same?

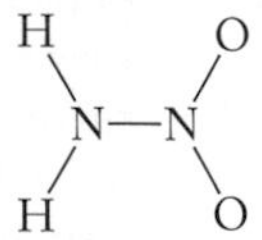

FIGURE P5.71

5.72. Cyclic structures exist for many compounds of carbon and hydrogen. Describe the molecular geometry and hybridization around each carbon atom in benzene (C_6H_6), cyclobutane (C_4H_8), and cyclobutene (C_4H_6) (Figure P5.72).

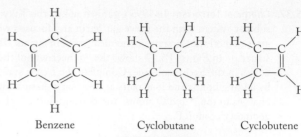

Benzene   Cyclobutane   Cyclobutene

FIGURE P5.72

5.73. What is the molecular geometry around sulfur and nitrogen in the sulfamate anion shown in Figure P5.73? Which atomic or hybrid orbitals overlap to form the S–O and S–N bonds in the sulfamate anion?

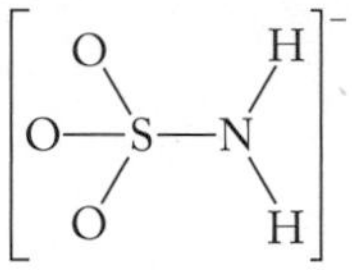

FIGURE P5.73

5.74. What is the geometry around each sulfur atom in the disulfate anion shown in Figure P5.74? What is the hybridization of the central oxygen atom?

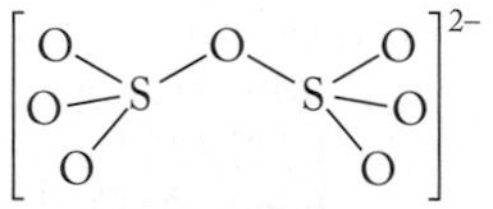

FIGURE P5.74

Chirality and Molecular Recognition

CONCEPT REVIEW

5.75. Which of the following objects are chiral? (a) a golf club; (b) a spoon; (c) a glove; (d) a shoe

5.76. Which of the following objects are chiral? (a) a key; (b) a screwdriver; (c) a fluorescent coil lightbulb; (d) a baseball

5.77. Could a *sp* hybridized carbon atom be a chiral center? Explain your answer.

5.78. Two compounds have the same Lewis structure and the same optical activity. Are they enantiomers or the same compound?

5.79. Are racemic mixtures homogeneous or heterogeneous?

5.80. Can a mixture of enantiomers rotate plane-polarized light? Explain your answer.

PROBLEMS

5.81. Which of the molecules in Figure P5.81 are chiral?

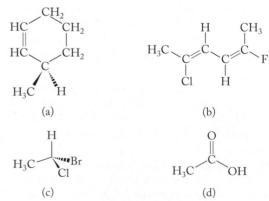

(a)   (b)

(c)   (d)

FIGURE P5.81

5.82. Which of the molecules in Figure P5.82 are chiral?

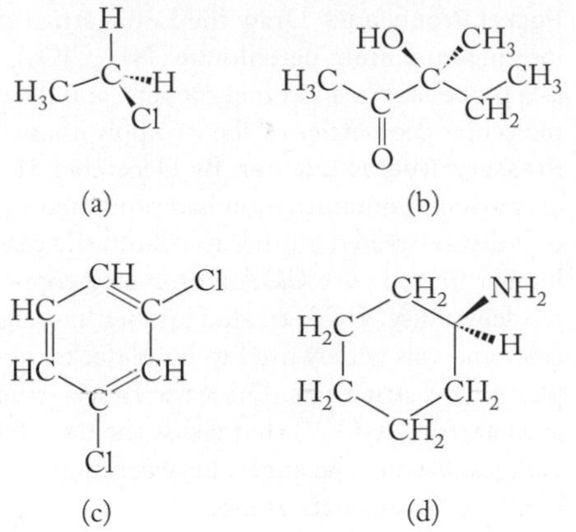

(a) (b)

(c) (d)

FIGURE P5.82

5.83. Which of the molecules in Figure P5.83 are chiral?

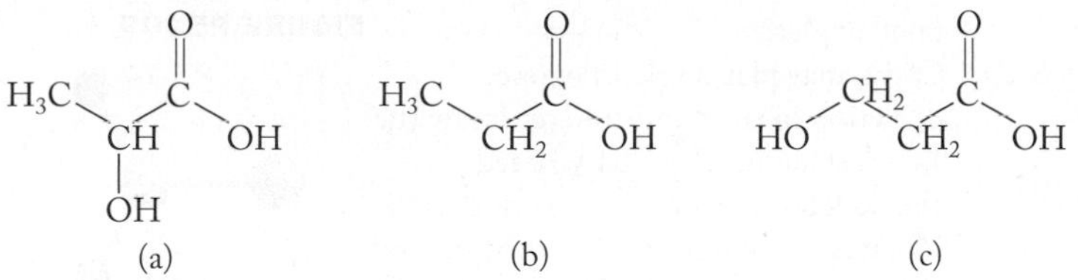

(a) (b) (c)

FIGURE P5.83

5.84. Which of the molecules in Figure P5.84 are chiral?

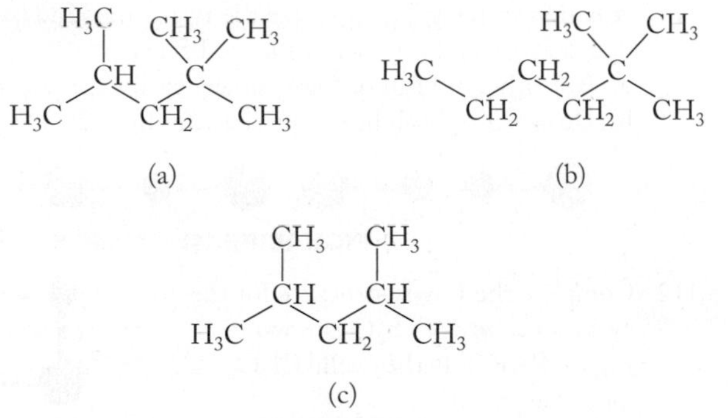

(a) (b)

(c)

FIGURE P5.84

5.85. Artificial Sweeteners Figure P5.85 shows three artificial sweeteners that have been used in food and beverages. Saccharin is the oldest, dating to 1879. Cyclamates were banned in the United States in 1969 but are still available in over 50 other countries. Aspartame may be more familiar to you by the brand name NutraSweet. Each sweetener contains between zero and two chiral carbon atoms. Circle the chiral centers in each compound.

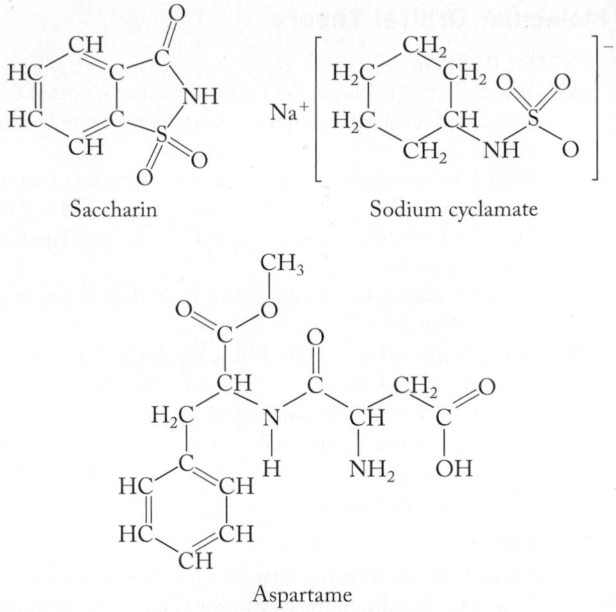

Saccharin Sodium cyclamate

Aspartame

FIGURE P5.85

5.86. Identify the chiral centers in each of the molecules in Figure P5.86.

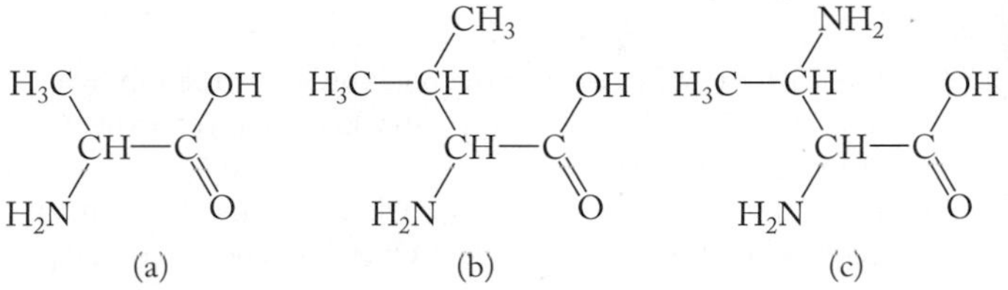

(a) (b) (c)

FIGURE P5.86

5.87. The Smell of Raspberries The compound with the structure in Figure P5.87 is a major contributor to the aroma of ripe raspberries. Identify any chiral center(s).

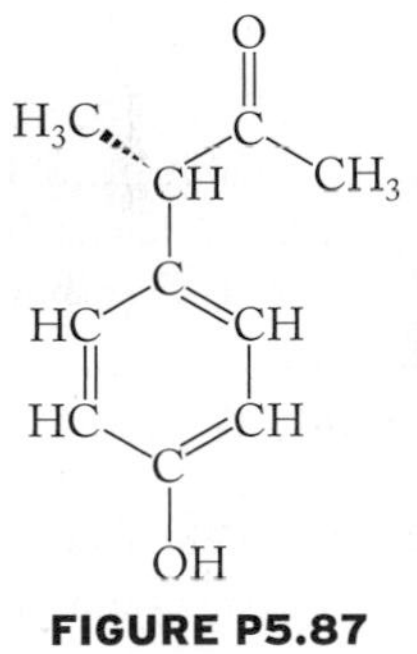

FIGURE P5.87

5.88. Antidepressants Figure P5.88 shows the structure of the drug bupropion, which is used to treat depression. Identify any chiral center(s).

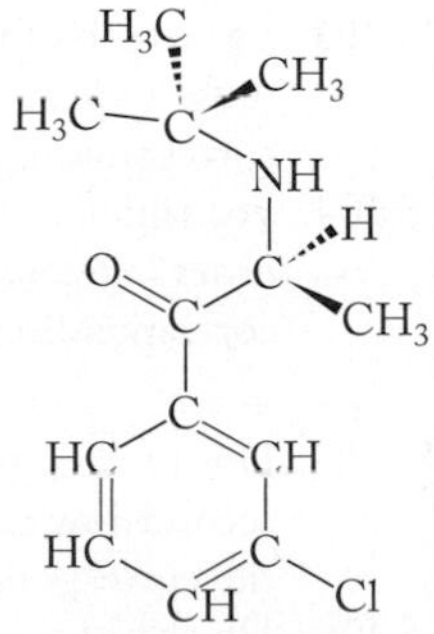

FIGURE P5.88

Molecular Orbital Theory

CONCEPT REVIEW

5.89. Which better explains the visible emission spectra of molecular substances: valence bond theory or molecular orbital theory?

5.90. Which better explains the magnetic properties of molecular substances: valence bond theory or molecular orbital theory?

5.91. Do all σ molecular orbitals result from the overlap of s atomic orbitals?

5.92. Do all π molecular orbitals result from the overlap of p atomic orbitals?

5.93. Are s atomic orbitals with different principal quantum numbers (n) as likely to overlap and form MOs as s atomic orbitals with the same value of n?

5.94. Are a $2p_x$ atomic orbital and a $2p_y$ atomic orbital on adjacent molecules as likely to form MOs as two $2p_z$ atomic orbitals on the same two atoms?

PROBLEMS

5.95. Make a sketch showing how two $1s$ orbitals overlap to form a σ_{1s} bonding molecular orbital and a σ_{1s}^* antibonding molecular orbital.

5.96. Make a sketch showing how two $2p_y$ orbitals overlap "sideways" to form a π_{2p} bonding molecular orbital and a π_{2p}^* antibonding molecular orbital.

5.97. Use MO theory to predict the bond orders of the following molecular ions: N_2^+, O_2^+, C_2^+, and Br_2^{2-}. Do you expect any of the species to exist?

5.98. Diatomic noble gas molecules, such as He_2 and Ne_2, do not exist. Would removing an electron create molecular ions, such as He_2^+ and Ne_2^+, that are more stable than He_2 and Ne_2?

5.99. Which of the following molecular ions are paramagnetic? (a) N_2^+; (b) O_2^+; (c) C_2^{2+}; (d) Br_2^{2-}

5.100. Which of the following molecular ions are diamagnetic? (a) O_2^-; (b) O_2^{2-}; (c) N_2^{2-}; (d) F_2^+

5.101. Which of the following molecular anions have electrons in π antibonding orbitals? (a) C_2^{2-}; (b) N_2^{2-}; (c) O_2^{2-}; (d) Br_2^{2-}

5.102. Which of the following molecular cations have electrons in π antibonding orbitals? (a) N_2^+; (b) O_2^+; (c) C_2^{2+}; (d) Br_2^{2+}

5.103. For which of the following diatomic molecules does the bond order increase with the gain of two electrons, forming the corresponding 2− anion? (a) B_2; (b) C_2; (c) N_2; (d) O_2

5.104. For which of the following diatomic molecules does the bond order increase with the loss of two electrons, forming the corresponding 2+ cation? (a) B_2; (b) C_2; (c) N_2; (d) O_2

5.105. Do the 1+ cations of homonuclear diatomic molecules of the second-row elements always have shorter bond lengths than the corresponding neutral molecules?

5.106. Do any of the anions of the homonuclear diatomic molecules formed by B, C, N, O, and F have shorter bond lengths than those of the corresponding neutral molecules? Consider only the 1− and 2− anions.

Additional Problems

5.107. **Rocket Propellants** Draw the Lewis structure for the two ions in ammonium perchlorate (NH_4ClO_4), which is used as a propellant in solid fuel rockets, and determine the molecular geometries of the two polyatomic ions.

5.108. **Pressure-Treated Lumber** By December 31, 2003, concerns over arsenic contamination had prompted the manufacturers of pressure-treated lumber to voluntarily cease producing lumber treated with CCA (chromated copper arsenate) for residential use. CCA-treated lumber has a light greenish color and was widely used to build decks, sand boxes, and playground structures. Draw the Lewis structure for the arsenate ion (AsO_4^{3-}) that yields the most favorable formal charges. Predict the angles between the arsenic–oxygen bonds in the arsenate anion.

5.109. Consider the molecular structure in Figure P5.109. What is the angle formed by the N—C—C bonds in the structure? What are the O=C—O and C—O—H bond angles?

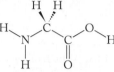

FIGURE P5.109

5.110. Cl_2O_2 may play a role in ozone depletion in the stratosphere. Draw the Lewis structure for Cl_2O_2 based on the skeletal structure in Figure P5.110. What is the geometry about the central chlorine atom?

FIGURE P5.110

5.111. Bombardment of Cl_2O_2 molecules with intense UV radiation is thought to produce the two compounds with the skeletal structures shown in Figure P5.111.
a. Are either or both of these molecules linear?
b. Do either or both have a permanent dipole?

FIGURE P5.111

*5.112. Complete the Lewis structure for the cyclic structure of Cl_2O_2 shown in Figure P5.112. Is the cyclic Cl_2O_2 molecule planar?

FIGURE P5.112

5.113. **Ozone Depletion** In 1999, the ClO^+ ion, a potential contributor to stratospheric ozone depletion, was isolated in the laboratory.
a. Draw the Lewis structure for ClO^+.
b. Using the molecular orbital diagram for ClO^+ in Figure P5.113, determine the order of the Cl–O bond in ClO^+.

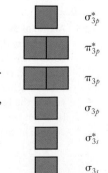

FIGURE P5.113

5.114. Early Earth Some scientists believe that an anion with the skeletal structure shown in Figure P5.114 may have played a role in the formation of nucleic acids before life existed on Earth.
 a. Complete the Lewis structure of this anion.
 b. Predict the C—P—O bond angle in the anion.

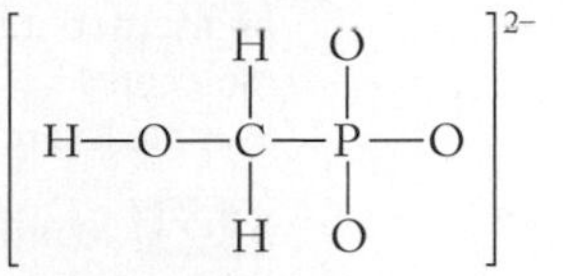

$$\left[\begin{array}{c} \text{H} \quad \text{O} \\ | \quad | \\ \text{H—O—C—P—O} \\ | \quad | \\ \text{H} \quad \text{O} \end{array}\right]^{2-}$$

FIGURE P5.114

5.115. Cola Beverages Phosphoric acid imparts a tart flavor to cola beverages. The structure of phosphoric acid is shown in Figure P5.115. Complete the Lewis structure for phosphoric acid in which formal charges are closest to zero. What is the molecular geometry around the phosphorus atom in your structure?

FIGURE P5.115

5.116. Fluoroaluminate anions AlF_4^- and AlF_6^{3-} have been known for over a century, but the structure of the pentafluoroaluminate ion, AlF_5^{2-}, was not determined until 2003. Draw the Lewis structures for AlF_3, AlF_4^-, AlF_5^{2-}, and AlF_6^{3-}. Determine the molecular geometry of each molecule or ion. Describe the bonding in AlF_3, AlF_4^-, AlF_5^{2-}, and AlF_6^{3-} using valence bond theory.

5.117. Two compounds formed by the reaction of boron with carbon monoxide have these skeletal structures: B–B–C–O and O–C–B–B–C–O.
 a. Draw the Lewis structures of both compounds that minimize formal charges.
 b. What are the B–B–C bond angles in the molecules?

*5.118. Boron reacts with NO, forming a compound with the formula BNO.
 a. Draw the Lewis structure for BNO including any resonance forms.
 b. Assign formal charges and predict which structure provides the best description of the bonding in this molecule.
 c. Predict the molecular geometry of BNO.

*5.119. Compounds That May Help Prevent Cancer Broccoli, cabbage, and kale contain compounds that break down in the human body to form isothiocyanates, whose presence may reduce the risk of certain types of cancer. The simplest isothiocyanate is methyl isothiocyanate, CH_3NCS.
 a. Draw the Lewis structure for methyl isothiocyanate, including all resonance forms.
 b. Assign formal charges and determine which structure is likely to contribute the most to bonding.
 c. Predict the molecular geometry of the molecule at both carbon atoms.

*5.120. Toxic to Insects and People Methyl thiocyanate (CH_3SCN) is used as an agricultural pesticide and fumigant. It is slightly water soluble and is readily absorbed through the skin, but it is highly toxic if ingested. Its toxicity stems in part from its metabolism to cyanide ion.
 a. Draw the Lewis structure for methyl thiocyanate, including all resonance forms.
 b. Assign formal charges and predict which structure contributes the most to bonding.
 c. Predict the molecular geometry of the molecule at both carbon atoms.

5.121. Borazine, $B_3N_3H_6$ (a cyclic compound with alternating B and N atoms in the ring), is isoelectronic with benzene (C_6H_6). Are there delocalized π electrons in borazine?

5.122. Draw a molecular orbital diagram for F_2. How many electrons are in antibonding molecular orbitals in F_2?

5.123. Some chemists think HArF consists of H^+ ions and ArF^- ions. Using an appropriate MO diagram, determine the bond order of the Ar–F bond in ArF^-.

*5.124. To model the bonding in SF_6 gas some chemists assume the existence of SF_4^{2+} cations surrounded by two F^- ions.
 a. Draw the Lewis structure of a SF_4^{2+} ion.
 b. What are the formal charges on S and F in the structure you drew?
 c. What is the shape of the ion?
 d. Does the S atom have an expanded octet?

*5.125. Which of the unstable nitrogen oxides N_2O_2, N_2O_5, and N_2O_3 are polar molecules? (N_2O_2 and N_2O_3 have N–N bonds; N_2O_5 does not.)

5.126. Hydrogen atoms have one electron. Does this mean that hydrogen gas is paramagnetic? Why or why not?

5.127. Use molecular orbital diagrams to determine the bond order of the peroxide (O_2^{2-}) and superoxide (O_2^-) ions. Are the bond order values consistent with those predicted from Lewis structures?

*5.128. Trimethylamine, $(CH_3)_3N$, has a trigonal *pyramid* structure while trisilylamine, $(SiH_3)_3N$, has a trigonal *planar* geometry. Draw Lewis structures for both compounds consistent with the observed geometries and explain your reasoning.

5.129. Elemental sulfur has several allotropic forms including cyclic S_8 molecules. What is the orbital hybridization of sulfur atoms in the S_8 allotrope? The bond angles are about 108°.

5.130. Using an appropriate molecular orbital diagram, show that the bond order in the disulfide anion S_2^{2-} is equal to 1. Is S_2^{2-} diamagnetic or paramagnetic?

*5.131. Ozone (O_3) has a small dipole moment (0.54 D). How can a molecule with only one kind of atom have a dipole moment?

*5.132. The bond angle in H_2O is 104.5°; however, the bond angles in H_2S, H_2Se, and H_2Te are all close to 90°. The small bond angles may lead you to wonder if the valence shell orbitals of S, Se, and Te are sp^3 hybridized in the compounds. Propose a reason why hybridization may not be so important for the large, many-electron atoms of S, Se, and Te.

6

Intermolecular Forces
Attractions between Particles

Ubiquitous, Essential, and Remarkable

Water is everywhere in our world. It covers over 70% of Earth's surface. Muscle tissue is about 75% (by mass) water, and our blood is about 95% water. Water's omnipresence and familiarity may lead us to take its remarkable properties for granted. For example, did you know that nearly all substances are denser as solids than as liquids, but water is not? That every other substance of comparable molar mass is a gas at room temperature and pressure, but not water? Objects that are denser than water can float on it because of a phenomenon called surface tension, and capillary action enables liquid water to overcome the force of gravity and climb to the top of the world's tallest trees. Then there is the essential role that water plays in sustaining all life-forms. Indeed, the search for life beyond Earth in our solar system and on planets orbiting distant stars begins with an analysis of whether or not water might be there.

In part, water sustains life through its ability to dissolve other substances. Water is sometimes called *nature's solvent* because ionic and molecular solids, other liquids, and atmospheric gases all dissolve in it, at least a little. An important substance in the last category is oxygen. For every million molecules of H_2O, only about 10 molecules of dissolved O_2 are required to sustain aquatic life. And even though we inhale atmospheric O_2 directly, it must dissolve in the moist tissues of our lungs before entering our bloodstream and being transported to our cells. The limited solubility of O_2 in water decreases when the temperature of water increases—a phenomenon that can be disastrous for aquatic life. Perhaps you have seen dead fish floating in water that has become unusually warm because of natural processes or human activity. It is usually not the heat that kills the fish, but rather the lack of O_2.

In this chapter we explore the interactions between atoms, ions, and molecules that produce phenomena such as those described above. This exploration will help explain the physical properties of substances, including their states at different temperatures and pressures, how much they dissolve in water and other solvents, and other phenomena that are key to understanding natural processes and those that we humans control—or try to.

Watering a Giant Sequoia Capillary action enables groundwater absorbed by the roots of giant sequoias, among the world's tallest trees, to flow up their trunks to their highest branches. This distance can approach 100 meters. A molecular view of capillary action shows that it is the result of adhesive forces between water molecules and —O— and —O—H groups in the tree's structure and cohesive forces between water molecules.

LEARNING OUTCOMES

LO1 Explain trends in boiling point, viscosity, and other physical properties of molecular compounds based on the nature and the strengths of interactions between their molecules
Sample Exercise 6.1

LO2 Identify the solute(s) and solvent in a solution
Sample Exercise 6.2

LO3 Identify the regions of a phase diagram and explain the effect of temperature and pressure on phase changes
Sample Exercise 6.3

LO4 Explain why ionic compounds and polar molecular compounds dissolve in polar solvents and nonpolar molecular compounds dissolve in nonpolar solvents

LO5 Describe the unusual properties of water and how they relate to the hydrogen bonds formed by water molecules

TABLE 6.1	Boiling Points of the Noble Gases		
Noble Gas	Atomic View	Z	Boiling Point (K)
He		2	4
Ne		10	27
Ar		18	87
Kr		36	120
Xe		54	165
Rn		86	211

∞ CONNECTION Polar bond formation and the color scales and symbols used to represent it are described in Chapter 4.

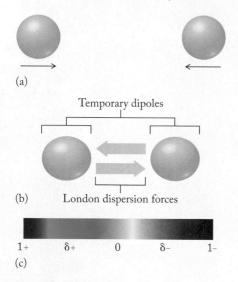

(a)

Temporary dipoles

(b) London dispersion forces

1+   δ+   0   δ−   1−
(c)

FIGURE 6.1 (a) Two atoms with a symmetrical distribution of electrons approach each other and (b) create two temporary dipoles as their nuclei and electron clouds interact. (c) The strengths of temporary dipoles are shown using the same color scale used in Chapter 4 to represent the strengths of permanent dipoles in molecules.

temporary dipole the separation of charge produced in an atom or molecule by a momentary uneven distribution of electrons; also called *induced dipole*.

London dispersion force an intermolecular force between nonpolar molecules caused by the presence of temporary dipoles in the molecules.

6.1 London Dispersion Forces: They're Everywhere

The macroscopic properties of substances, which we can readily observe and measure, arise from the microscopic interactions between the particles that make up those substances. Let's see how this happens by looking at a group of elements that exist as single atoms: the noble gases. Table 6.1 lists their atomic numbers and boiling points. Note the correlation between them: boiling points increase as atomic numbers increase. To understand why this correlation exists, let's think about what happens at the particle level when a liquid vaporizes. As we discussed in Chapter 1, the particles in liquids (and solids) are in direct contact with each other. However, when a liquid vaporizes, the contacts are broken: the gas-phase particles become essentially free and independent. Separating liquid-phase particles that are attracted to each other takes energy, and the stronger the particles' attraction for each other, the greater the energy needed to separate them. Greater energy requirements mean higher boiling points.

Why do atoms with greater atomic numbers interact more strongly? For that matter, *how* do atoms not bonded to each other interact? German-American physicist Fritz London (1900–1954) proposed one explanation for these interactions in 1930. It was based on the notion that when atoms approach each other (Figure 6.1a), they interact in ways that are similar to the electrostatic interactions involved in covalent bond formation. In other words, one atom's positive nucleus is attracted to the other atom's negative electrons, and vice versa, even as their electron clouds repel each other. These competing interactions can cause the electrons around each atom to be distributed unevenly, producing **temporary dipoles** of partial electrical charge (Figure 6.1b) that are attracted to regions of opposite partial charge on the adjacent atom. (This attraction is represented by the arrows in Figure 6.1b.) In Chapter 4 we used different colors to represent the partial electrical charges created by uneven sharing of bonding pairs of electrons. In this chapter we use the same colors (Figure 6.1c) to show partial electrical charges caused by uneven distributions of electrons in neutral atoms and over entire molecules.

The presence of temporary dipoles in molecules creates a way for them to interact with atoms and other molecules. Of course, in a molecule the partial charges in temporary dipoles are likely to be distributed over considerably greater distances than in atoms, as their atomic nuclei and the clouds of electrons shared by groups of atoms, or even the entire molecule, interact with the electrons and nuclei in neighboring atoms or molecules.

Interactions based on the presence of temporary dipoles are called **London dispersion forces** in honor of Fritz London's pioneering work. The strengths of the interactions increase as the numbers of electrons in atoms and molecules increase because the larger the cloud of electrons surrounding a nucleus or in a molecule, the more likely they are to be distributed unevenly, or *polarized*. Greater **polarizability** leads to stronger temporary dipoles and stronger intermolecular interactions, so London dispersion forces become stronger as atoms and molecules become larger. This trend in polarizability accounts for the correlation between the boiling points and atomic numbers of the noble gases.

It also explains similar trends in the boiling points of the halogens (Table 6.2). These elements exist as diatomic molecules in which equal sharing of the bonding pairs of electrons by identical atoms means that the molecules have no permanent dipoles. Comparing the boiling points of the halogens with their molar masses (as a measure of particle size) discloses a trend that mimics the one we saw with the noble gases: boiling point increases as particle size increases. London's explanation of this trend was also the same: larger clouds of increasing numbers of electrons per molecule are more polarizable. Greater polarizability means they are more likely to form temporary dipoles, which attract molecules to each other in the liquid phase and inhibit their vaporization.

A similar trend is also observed in the boiling points of a series of nonpolar hydrogen–carbon compounds, or **hydrocarbons** (Figure 6.2). In the hydrocarbons called **alkanes**, each carbon atom is bonded to four other atoms, which means all of the bonds in the compounds are single bonds. The carbon atoms in these particular alkane molecules are bonded to no more than two other carbon atoms, which gives the molecules a chainlike appearance when there are four or more carbon atoms per molecule. We call such hydrocarbons *normal* alkanes, or *n*-alkanes. We explore the names, structures, and properties of these compounds and others like them in Chapter 19. For now, be assured that there is nothing *abnormal* about alkanes whose carbon atoms do not form single chains.

> **CONCEPT TEST**

Explain why CF_4 is a gas at room temperature but CCl_4 is a liquid.

(Answers to Concept Tests are in the back of the book.)

The Importance of Shape

Molecular shape, as well as size, plays a role in determining the strength of London dispersion forces. Consider the molecular structures of the three hydrocarbons in Figure 6.3. These compounds all have the same formula, C_5H_{12}, and molar mass, 72 g/mol. However, the bonding patterns and shapes of the molecules are different. Compounds such as these, with the same molecular formula but different connections between the atoms in their molecules, are called **constitutional isomers** or *structural* isomers. Compounds that are constitutional isomers have different physical and chemical properties despite their shared molecular formulas. For example, all three of the C_5 (five carbon atoms per molecule) hydrocarbons are nonpolar, which means the only intermolecular forces they experience are London dispersion forces. The carbon atoms in a molecule of pentane form a single chain, which gives the molecule the overall shape of a stubby piece of chalk. This shape gives the molecule a larger surface area and more opportunity to interact with other pentane molecules than, for example, molecules of 2,2-dimethylpropane, which are shaped more like lumpy spheres, have a smaller surface-area-to-volume ratio, and interact less strongly with adjacent molecules. Weaker London dispersion forces explain why 2,2-dimethylpropane has a lower boiling point than pentane. The remaining molecule, 2-methylbutane, is neither as straight as pentane nor as spherical as 2,2-dimethylpropane. As you might guess, the compound boils at a temperature between the boiling points of the other two. In general, molecules with more branching in their structures have lower boiling points.

TABLE 6.2 Boiling Points of the Halogens

Halogen	Molecular View	Molar Mass (g/mol)	Boiling Point (K)
F_2		38	85
Cl_2		71	239
Br_2		160	332
I_2		254	457
At_2		420	610

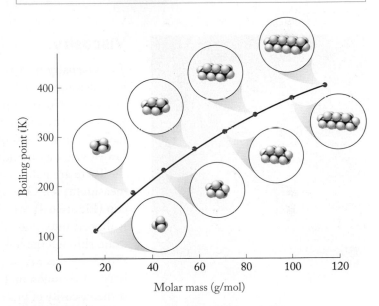

FIGURE 6.2 Boiling points of the C_1 to C_8 normal alkanes.

polarizability the relative ease with which the electron cloud in a molecule, ion, or atom can be distorted, inducing a temporary dipole.

hydrocarbon an organic compound whose molecules contain only carbon and hydrogen atoms.

alkane a hydrocarbon in which each carbon atom is bonded to four other atoms.

constitutional isomer one of a set of compounds with the same molecular formula but different connections between the atoms in their molecules; also called *structural isomer*.

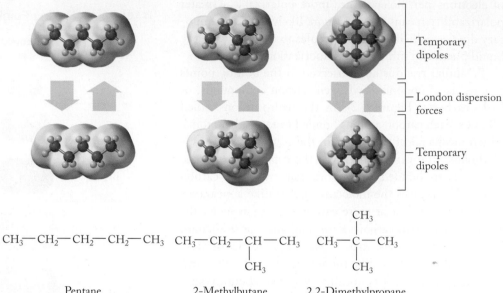

Temporary dipoles

London dispersion forces

Temporary dipoles

CH_3—CH_2—CH_2—CH_2—CH_3

Pentane
Boiling point 309 K

CH_3—CH_2—CH—CH_3
|
CH_3

2-Methylbutane
Boiling point 301 K

$\begin{array}{c} CH_3 \\ | \\ CH_3-C-CH_3 \\ | \\ CH_3 \end{array}$

2,2-Dimethylpropane
Boiling point 282 K

FIGURE 6.3 Molecular structures and boiling points of three C_5 isomers.

FIGURE 6.4 Determination of viscosity using a Zahn cup.

viscosity the measure of the resistance to flow of a fluid.

ion–dipole interaction an attractive force between an ion and a molecule that has a permanent dipole.

Viscosity

The **viscosity** of a fluid (a fluid is a liquid or a gas—anything that can flow) is a measure of its resistance to flow. Viscous or "thick" fluids such as honey, molasses, and the oil used to lubricate engines have high resistances to flow. Low-viscosity substances such as water and gasoline flow or pour easily. Gases have ultralow viscosities that are several orders of magnitude below those of the lowest viscosity liquids.

There are many ways to measure viscosity and to express the results of those measurements. One of the simplest measurements uses a device called a Zahn cup (Figure 6.4), which is a small metal cup with a hole in the bottom. The cup is dipped into a liquid sample and then lifted out. The time the liquid takes to drain through the hole is measured and converted into units of dynamic viscosity. (There are other types, but *dynamic* viscosity is the one most widely used in chemistry.) A common unit for expressing viscosity is the *centipoise* (cP), where 1.00 cP is the viscosity of pure water at 20°C.

Now let's consider the viscosities of a group of liquid *n*-alkanes with between 6 (hexane) and 16 (hexadecane) carbon atoms per molecule (Table 6.3). We would expect molecules of hexadecane to experience the strongest London dispersion forces in this group because they are the largest. Strong intermolecular attraction means that they do not slip past each other as easily as those experiencing weaker intermolecular attraction. More resistance to flow means higher viscosity. The data in Table 6.3 confirm this trend: viscosity increases with increasing molecular size.

6.2 Interactions Involving Polar Molecules

In Section 5.3 we saw how unequal distributions of bonding pairs of electrons result in partial negative charges on some bonded atoms and partial positive charges on others. When bond dipoles are arranged asymmetrically within a

TABLE 6.3	Viscosities of Some Liquid *n*-Alkanes		
Compound	**Molecular Structure**	**Molar Mass (g/mol)**	**Viscosity at 20°C (cP)**
Hexane		86	0.29
Octane		114	0.54
Decane		142	0.92
Dodecane		170	1.34
Hexadecane		226	3.34

molecule, the molecule itself has an overall permanent dipole. Permanent dipoles can interact with each other and with the ions in ionic compounds. The strengths of the interactions are much weaker than the strengths of ionic or covalent bonds, but they are strong enough to add another important mode of interaction to the London dispersion forces that all molecules experience. For example, interactions between polar liquids and ionic solids play a key role in salts dissolving in water, and interactions between the molecules of polar liquids are responsible for boiling points and viscosities that are higher than those of nonpolar substances of similar molar mass.

Ion–Dipole Interactions

One of the reasons why ionic compounds such as NaCl dissolve in water is that their ions interact with the permanent dipoles of water molecules. These **ion–dipole interactions** pull ions from the solid into solution (Figure 6.5). As an ion is removed from its solid-state neighbors, it is surrounded by a cluster of water molecules that form a **sphere of hydration** (Figure 6.6). The dissolved ions are said to be *hydrated*. The strengths of these multiple ion–dipole interactions (represented by the light green dashed lines in Figures 6.5 and 6.6) help overcome the lattice energy, that is, the energy of the ionic bonds that hold ions together in crystalline ionic solids. Within a sphere of hydration, the water molecules closest to the ion are oriented so that their oxygen atoms (negative poles) are directed toward cations and their hydrogen atoms (positive poles) are directed toward anions, as shown for Na^+ and Cl^- ions in Figure 6.6. The number of water molecules in the *inner sphere of hydration* depends on the size of the ion. Six water molecules hydrate most ions, but the number can be between four and nine.

When a substance dissolves in a liquid other than water, the cluster of host molecules is called a sphere of *solvation*. The dissolved particles are said to be *solvated*.

CONNECTION Bond energies of covalent bonds are discussed in Section 4.9.

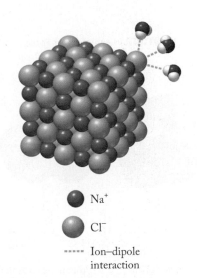

● Na^+

○ Cl^-

----- Ion–dipole interaction

FIGURE 6.5 Ion–dipole interactions. The hydrogen atoms (positive poles) of H_2O molecules are attracted to the Cl^- ions of NaCl. Note that the Na^+ cation is smaller than the Cl^- anion, as we saw in Figure 3.35.

sphere of hydration the cluster of water molecules surrounding an ion in an aqueous solution.

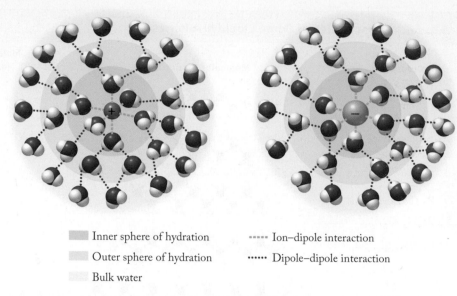

▬▬▬ Inner sphere of hydration

▬▬▬ Outer sphere of hydration

▬▬▬ Bulk water

····· Ion–dipole interaction

••••• Dipole–dipole interaction

FIGURE 6.6 Each hydrated Na^+ and Cl^- ion in a solution of NaCl is surrounded by six water molecules oriented toward the center ion as a result of ion–dipole interactions (light green dashed lines). Those water molecules make up an inner sphere of hydration. Water molecules in an outer sphere of hydration are oriented as a result of dipole–dipole interactions (dark blue dotted lines) with molecules in the inner sphere. All water molecules experience particularly strong dipole–dipole interactions.

Dipole–Dipole Interactions

▶❚❚ **CHEMTOUR** Intermolecular Forces

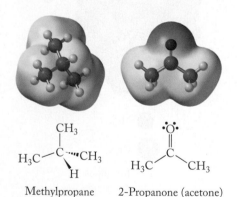

Methylpropane
Boiling point 261 K

2-Propanone (acetone)
Boiling point 329 K

FIGURE 6.7 Molecular structures and boiling points of methylpropane and acetone.

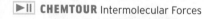

dipole–dipole interaction an attraction between regions of polar molecules that have partial charges of opposite sign.

The water molecules closest to the ions in Figure 6.6 are surrounded by other water molecules that form an *outer sphere of hydration*. They in turn are surrounded by, and interact with, many more water molecules that are not part of the sphere of hydration. The molecules in the outer sphere are oriented more randomly than those in the inner sphere, but not as randomly as those beyond. Molecules of H_2O in all of the regions experience another type of intermolecular force: **dipole–dipole interactions**, which occur between molecules that have permanent dipoles. In water molecules, partial negative charges on oxygen atoms and partial positive charges on hydrogen atoms cause the two hydrogen atoms of each molecule of H_2O to be attracted to the oxygen atoms of two other H_2O molecules. These interactions are represented by the dark blue dotted lines in Figure 6.6.

Dipole–dipole interactions are not as strong as ion–dipole interactions because dipole–dipole interactions involve atoms with only partial charges. In contrast, an ion involved in an ion–dipole interaction has completely lost or gained one or more electrons and has a charge of at least plus or minus 1. As you might expect, substances with large dipole moments are capable of stronger dipole–dipole interactions than substances with smaller dipole moments.

To appreciate how dipole–dipole forces add to London dispersion interactions, let's consider the properties of two compounds: methylpropane and 2-propanone (also known as acetone). The molecules of the compounds have the same molar mass (58 g/mol), and they have shapes that are not all that different despite the different hybridizations of their central carbon atoms (Figure 6.7).

We might expect the molecules to experience similar London dispersion forces, yet the boiling points of methylpropane and acetone are quite different:

261 K and 329 K, respectively. What explains the nearly 70-degree-higher boiling point of acetone? The answer is contained in the C=O bond in its molecular structure. The bond dipole between the highly electronegative (χ = 3.5) oxygen atom and the less electronegative (χ = 2.5) carbon atom gives acetone an overall dipole moment of 2.9 D and sets up dipole–dipole interactions between its molecules, which account for its higher boiling point. The C=O bond constitutes one of the most common functional groups in carbon-based molecules, or **organic compounds**. By functional groups, we mean particular combinations of atoms in molecular structures that impart characteristic properties to organic compounds. The C=O group in organic compounds is called a **carbonyl group**.

organic compounds compounds that contain carbon, hydrogen, and sometimes other elements including oxygen, nitrogen, sulfur, and a halogen.

carbonyl group a functional group that consists of a carbon atom with a double bond to an oxygen atom.

hydrogen bond the strongest dipole–dipole interaction, which occurs between a hydrogen atom bonded to a N, O, or F atom and another N, O, or F atom.

Hydrogen Bonds

Let's now consider the boiling points of the compounds whose molecules each contain one atom of a group 14, 15, 16, or 17 element bonded to enough hydrogen atoms to have a complete valence-shell octet (Figure 6.8). Most of the data follow a familiar trend: boiling points increase with increasing molar masses, as we saw with nonpolar hydrocarbons in Figure 6.2. Actually, *all* of the data points for the group 14 compounds fit this trend. However, the boiling points of the other three groups' lowest molecular mass compounds—NH_3, H_2O, and HF—are unusually high compared to the others in their series. To understand why, we need to focus on the polar bonds formed between H atoms and N, O, and F. The H atoms share only two electrons to begin with, and when they are drawn away from the H end of the bond to a N, O, or F atom, there is little electron density left, especially on the surface of the atom farthest from the bond. The lack of electron density means that electronegative N, O, and F atoms on neighboring molecules can get very close to the H atoms and interact strongly with them. The resulting superstrength dipole–dipole interactions are called **hydrogen bonds** and are represented by the dotted lines in Figure 6.9.

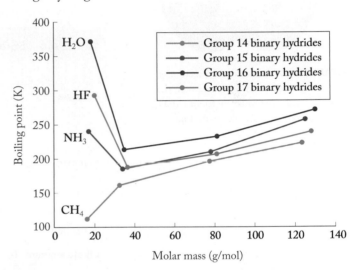

FIGURE 6.8 The boiling points of most of the group 14–17 binary hydrides increase with increasing molar mass, but not all. The boiling points of H_2O, HF, and NH_3 are much higher than one might expect. They are higher because of hydrogen bonding between their molecules.

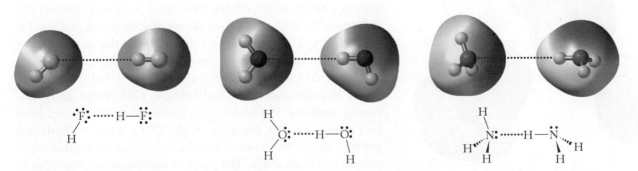

FIGURE 6.9 Hydrogen bonds (dotted lines) occur between hydrogen atoms bonded to F, O, or N atoms in one molecule and F, O, or N atoms in adjacent molecules.

∞ **CONNECTION** In Chapter 4 we learned that the strength of attraction between particles of opposite electrical charge increases as the distance between them decreases.

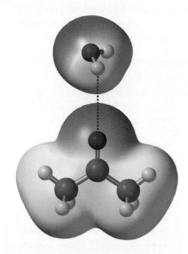

FIGURE 6.10 In solutions of acetone in water, hydrogen bonds form between the hydrogen atoms of water molecules and the oxygen atoms in molecules of acetone.

Hydrogen bonds can also form between molecules of different substances, even when one of them has no H atoms bonded to N, O, or F atoms. For example, when acetone dissolves in water, hydrogen bonds form between the H atoms of water molecules and the O atoms of acetone molecules as shown in Figure 6.10. These interactions happen even though the H atoms in acetone are bonded to C atoms and cannot form hydrogen bonds. The key to hydrogen bonding between acetone and water molecules is the negative partial charge of the O atoms in molecules of acetone, as indicated by the red color of the electrostatic potential map of acetone in Figure 6.10, and the blue color of the electron-deprived H atoms in molecules of H_2O.

To get another perspective on the importance of hydrogen bonding, let's consider the boiling points and intermolecular forces in three more compounds: ethane, formaldehyde, and methanol. Their molar masses are nearly the same, so they should experience similar London dispersion forces. Table 6.4 shows their molecular structures in electrostatic potential maps and lists some of their properties.

TABLE 6.4	Some Properties of Ethane, Formaldehyde, and Methanol		
	Ethane	**Formaldehyde**	**Methanol**
Formula	CH_3CH_3	CH_2O	CH_3OH
Structure			
ℳ (g/mol)	30.0	30.0	32.0
Dipole Moment (D)	0.00	2.33	1.69
Boiling Point (K)	184	254	338

We can explain the higher boiling point of formaldehyde compared to ethane (70 K higher) by the presence of the carbonyl group in formaldehyde, which gives it a dipole moment of 2.33 D. As a result, its molecules experience dipole–dipole interactions that nonpolar ethane molecules do not. However, the difference between the boiling points of ethane and methanol is even greater, 154 K, despite the fact that the dipole moment of methanol is *less than* that of formaldehyde. Why is the boiling point of methanol so high? The answer is found in the strength of the hydrogen bonds formed between OH groups on neighboring methanol molecules, as represented by the dotted lines in Figure 6.11. The presence of the OH groups, or **hydroxyl groups**, makes methanol a member of a class of organic compounds called **alcohols**. The physical and chemical properties of alcohols are linked closely to the hydroxyl groups in their structures and their capacity to form hydrogen bonds.

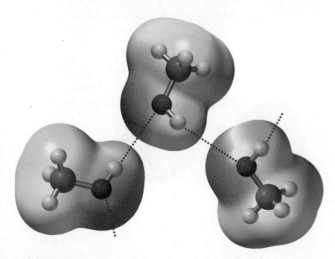

FIGURE 6.11 A methanol molecule can form hydrogen bonds with two other methanol molecules.

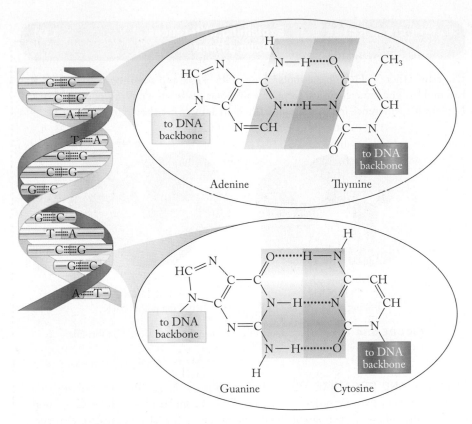

hydroxyl group a functional group that consists of an oxygen atom with a single bond to a hydrogen atom.

alcohol an organic compound whose molecular structure includes a hydroxyl group bonded to a carbon atom that is not bonded to any other functional group(s).

FIGURE 6.12 Hydrogen bonds (dotted lines) occur between hydrogen and nitrogen or oxygen atoms in adjacent strands of DNA, stabilizing its double-helix structure. The two detailed views show the four building blocks of DNA: guanine (G), cytosine (C), adenine (A), and thymine (T).

Hydrogen bonding also plays a role in defining the shapes of very large molecules, especially those of biological interest such as proteins and DNA. Proteins are so large that their long chains of atoms can fold back and wrap around themselves, allowing atoms to hydrogen bond with others in the same chain in much the way they interact with atoms in adjacent molecules. In addition, hydrogen bonding between atoms on adjacent chains can induce them to interact in ways that maximize the number of hydrogen bonds they form. The double strands of DNA form a three-dimensional shape called a double helix in which pairs of the molecular building blocks of DNA, called nucleotides, form hydrogen bonds that keep the strands linked together. Pairs of two nucleotides named guanine (G) and cytosine (C) on adjacent DNA strands form three hydrogen bonds, and pairs of adenine (A) and thymine (T) form two hydrogen bonds (Figure 6.12).

Hydrogen bond formation is a reversible process. The ones in DNA are broken as the double helices pull apart when living cells prepare to divide. Prior to cell division the DNA strands of the parent cell replicate, forming two new double helices as shown in Figure 6.13. During the replication process the new DNA strands are synthesized in such a way that the G nucleotides of the new strands are hydrogen bonded to C nucleotides on an original strand (and vice versa), and every A on a new strand is paired with a T on an original strand (and vice versa). We explore DNA replication in more detail in Chapter 20.

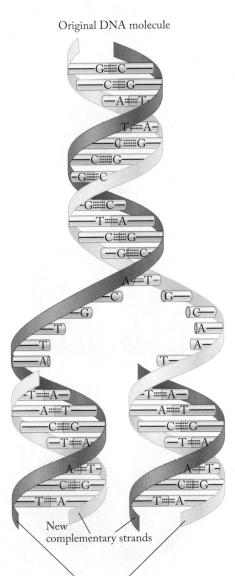

FIGURE 6.13 Hydrogen bonding plays a key role in DNA replication. Nucleotides in the new strands pair up with those in the parent strands: A and T are always paired, as are C and G.

SAMPLE EXERCISE 6.1 **Explaining Differences in Boiling Point** **LO1**

Dimethyl ether (C_2H_6O) has a molar mass of 46 g/mol and a boiling point of 248 K. Ethanol has the same chemical formula and molar mass but a boiling point of 351 K. Explain the difference in boiling point. Their structures are shown in Figure 6.14.

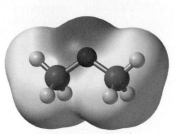
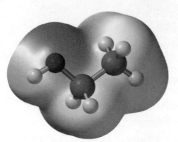

Dimethyl ether
CH_3OCH_3
Boiling point 248 K

Ethanol
CH_3CH_2OH
Boiling point 351 K

FIGURE 6.14 Dimethyl ether and ethanol have the same molecular formula but different boiling points.

COLLECT AND ORGANIZE We need to explain the large difference between the boiling points of two compounds that have the same molar mass. Molecules of both compounds contain oxygen atoms bonded to atoms of less electronegative elements. The electrostatic potential maps of the compounds indicate that both have overall dipole moments.

ANALYZE Because their molar masses are the same, the molecules of both compounds should experience similar London dispersion forces. Both also contain polar groups: the OH group in the ethanol molecule is polar because of unequal sharing of the bonding pair of electrons between O and H, and the C—O—C group in dimethyl ether is polar because of the bond dipoles between C and O, and because the bond angle between the three atoms should be a little less than 109° given the sp^3 hybridization. The resulting asymmetry in the molecule gives dimethyl ether a permanent dipole. In addition, hydrogen bonds can form between the OH groups on adjacent ethanol molecules.

SOLVE Hydrogen bonds are particularly strong dipole–dipole interactions, so the hydrogen bonds between molecules of ethanol should be stronger than the dipole–dipole interactions between molecules of dimethyl ether. More energy is required to overcome those stronger interactions, which is why ethanol has a higher boiling point.

THINK ABOUT IT Dimethyl ether and ethanol are constitutional isomers, yet their boiling points differ by more than 100 K—a difference directly linked to hydrogen bonding between the OH groups that molecules of ethanol have, and that molecules of dimethyl ether don't.

Practice Exercise Isopropanol is the compound commonly known as rubbing alcohol. Its boiling point is 355 K. Ethylene glycol, which is the principal ingredient in automotive antifreeze, has about the same molar mass and boils at 469 K. Why do the substances (Figure 6.15) have such different boiling points? ⚙

(Answers to Practice Exercises are in the back of the book.)

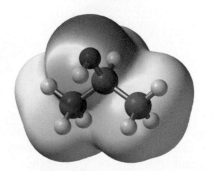

Isopropanol
$CH_3CH(OH)CH_3$
Boiling point 355 K

Ethylene glycol
$HOCH_2CH_2OH$
Boiling point 469 K

FIGURE 6.15 Isopropanol and ethylene glycol have similar molar masses but different boiling points.

Ethane and hydrazine (Figure 6.16) have similar molecular structures, yet the boiling point of hydrazine (387 K) is over 200 K greater than that of ethane (184 K). What intermolecular interactions account for this huge difference in boiling points?

Ethane	Hydrazine
CH_3CH_3	NH_2NH_2
Boiling point 184 K	Boiling point 387 K

FIGURE 6.16 Ethane and hydrazine have similar molecular structures but different boiling points.

6.3 Trends in Solubility

Boiling point and viscosity are not the only properties influenced by the strengths of interactions between particles. Another is the extent to which different substances dissolve in each other. Before we explore the interactions involved in forming solutions, we need to define a few key terms. In Chapter 1 we noted that a solution is a homogeneous mixture of two or more substances. The substance that is present in the greatest proportion (based on numbers of moles) is called the **solvent**. All the other substances in the solution are dissolved in the solvent and are called **solutes**. For example, the water in seawater functions as the solvent, and sodium chloride and the other sea salts dissolved in it are solutes. **Solubility** is a measure of how much solute can dissolve in a given volume of solution.

We explored the process by which an ionic salt dissolves in water in Section 6.2, noting the importance of ion–dipole interactions in overcoming the lattice energy of the salt. Dissolving a gas in a liquid or one liquid in another can also be explained in terms of intermolecular forces. For example, low-molar-mass alcohols such as methanol, isopropanol, and ethylene glycol dissolve in water because of the hydrogen bonds that form between the OH groups in the alcohols and in water. The strengths and numbers of the hydrogen bonds, as shown for ethylene glycol in Figure 6.17, allow these alcohols to dissolve in water and for water to dissolve them in all proportions. In describing their unlimited solubility in each other, chemists say that pairs of liquids are **miscible** with each other. On the other hand, a pair of liquids that have only limited solubility in each other are said to be **immiscible**. Gasoline, for example, is immiscible with water.

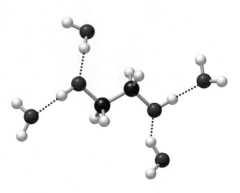

FIGURE 6.17 A single molecule of ethylene glycol can form multiple hydrogen bonds. The strength of these solute–solvent interactions makes ethylene glycol miscible with water.

SAMPLE EXERCISE 6.2   **Distinguishing Solute from Solvent**   **LO2**

The liquid in the cooling system of an automobile engine is a homogeneous mixture that is 30% by mass water and 70% by mass ethylene glycol. The freezing point of the mixture is −51°C (−60°F). Which ingredient is the solvent?

COLLECT AND ORGANIZE We are asked to identify which of the two components of a homogeneous mixture (a solution) is the solvent. We know the mass percentages of the components. The solvent is the component present in the greater number of moles.

ANALYZE We need to relate the mass proportions that are given to numbers of moles. If we assume that we have a 100 g sample of the coolant, then we can convert the percentage values into masses in grams. To convert masses into numbers of moles, we need to divide them by the molar masses of the two compounds. To calculate the molar masses we need to know their formulas. Water's is well known: H_2O. The formula of ethylene glycol is not given in the problem, but the molecular structure of ethylene glycol was shown in Figure 6.15.

---

solvent the component of a solution that is present in the largest number of moles.

solute any component in a solution other than the solvent. A solution may contain one or more solutes.

solubility the maximum quantity of a substance that can dissolve in a given volume of solution.

miscible liquids that are mutually soluble in any proportion.

SOLVE

1. Assume a 100 g sample to convert the percentage values into masses in grams:

$$30\% \text{ water} = 30 \text{ g water}$$

$$70\% \text{ ethylene glycol} = 70 \text{ g ethylene glycol}$$

2. Counting up the number of carbon, hydrogen, and oxygen atoms in the molecular structure of ethylene glycol, we get a molecular formula of $C_2H_6O_2$. The formula of water is, of course, H_2O.
3. The molar masses of the components are 62 g/mol for $C_2H_6O_2$ and 18 g/mol for H_2O.
4. The number of moles of the two components in the 100 g sample are

$$70 \text{ g } C_2H_6O_2 \times 1 \text{ mol} / 62 \text{ g} = 1.1 \text{ mol } C_2H_6O_2$$

$$30 \text{ g } H_2O \times 1 \text{ mol} / 18 \text{ g} = 1.7 \text{ mol } H_2O$$

5. There are more moles of water than ethylene glycol, so water is the solvent in the engine coolant.

THINK ABOUT IT Even though the coolant is 70% ethylene glycol and only 30% water, the much larger molar mass of ethylene glycol means that there are fewer moles of it, so water is the solvent.

Practice Exercise Most of the rum sold in liquor stores is about 30–40% ethanol, C_2H_5OH, by volume. However, a particular *overproof* variety used to prepare flaming drinks contains 75.5% ethanol by volume. Assuming the remaining 24.5% is mostly water, which of the two compounds is the solute and which is the solvent in overproof rum? The densities of ethanol and water are 0.789 g/mL and 0.998 g/mL, respectively, at 20°C.

To predict whether a given solute is likely to be highly soluble in a given solvent, or only slightly soluble, we need to consider the strengths and numbers of the interactions among solute particles and the strengths of the interactions among solvent particles. The stronger those interactions are, the harder it will be for the solute particles to separate from one another and dissolve in the solvent, and the harder it will be for the solvent particles to move away from each other to make space for particles of solute. On the other hand, the stronger the interactions between solute and solvent molecules are, the easier it will be for them to overcome solute–solute and solvent–solvent interactions and for the solute to dissolve.

Let's explore solute–solvent interactions first because they help drive the dissolution process. Polar solutes tend to dissolve in polar solvents when there are strong dipole–dipole interactions between molecules of solute and solvent, and especially when hydrogen bonds form between them (as when ethylene glycol or methanol dissolve in water). On the other hand, nonpolar solutes don't dissolve in polar solvents, or dissolve only a little, because the solute–solvent interactions that promote dissolution are much weaker than those that keep solute molecules together with other solute molecules and solvent molecules together with other solvent molecules. For example, *n*-octane does not dissolve in water because hydrogen bonding keeps the H_2O molecules together and London dispersion forces keep the octane molecules together, resulting in two separate liquid phases that neither mix with nor dissolve in each other when octane is added to water (Figure 6.18).

What little solubility octane or any nonpolar solute has in water or any polar solvent is promoted by **dipole–induced dipole interactions**. We saw in Section 6.1

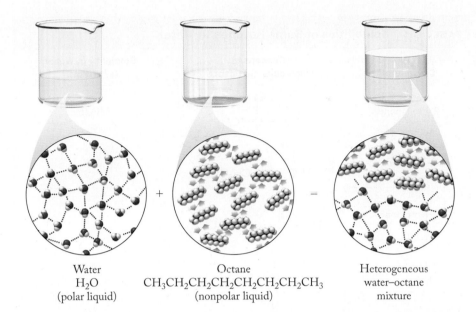

Water
H_2O
(polar liquid)

Octane
$CH_3CH_2CH_2CH_2CH_2CH_2CH_2CH_3$
(nonpolar liquid)

Heterogeneous
water–octane
mixture

FIGURE 6.18 The strength of London dispersion forces, represented by the blue arrows, holds nonpolar molecules of octane together in the liquid phase. Hydrogen bonding between water molecules keeps them together, too, and the lack of strong solute–solvent interactions needed to overcome these forces explains why octane and water are immiscible.

how temporary dipoles occur in nonpolar molecules through transient perturbations in the normally symmetrical distribution of electrons in molecules. An even stronger perturbation may occur when a polar molecule such as H_2O approaches a nonpolar molecule such as O_2 (Figure 6.19). If the O atom in H_2O approaches one end of an O_2 molecule, the partial negative charge on the O atom repels the electrons in O_2, pushing them toward the other end of that molecule and creating, or *inducing*, a dipole within the normally nonpolar O_2. The end of the O_2 molecule closest to the O atom of H_2O temporarily acquires a partial positive charge, and the opposite end has a temporary partial negative charge. When the two molecules move apart, the partial charges in the O_2 molecule disappear. The transient induced dipole is not as strong as the one that induced it, but it is strong enough to allow a measure of intermolecular interaction that leads to the slight solubility of O_2 in water—enough dissolved oxygen to sustain a multitude of aquatic life-forms.

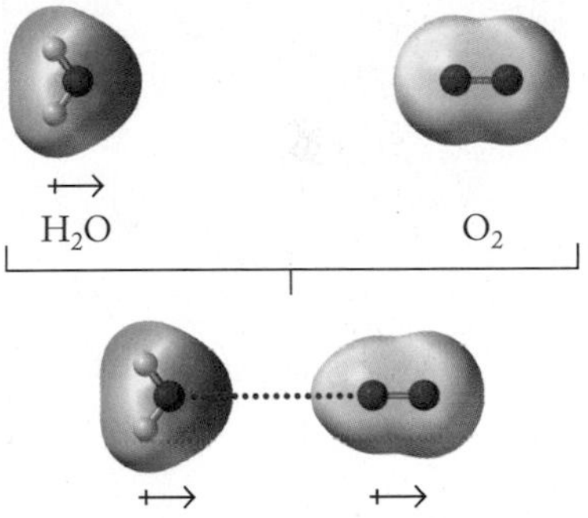

Competing Intermolecular Forces

If polar solutes readily dissolve in polar solvents, and nonpolar solutes do not, what is the solubility in a polar solvent, such as water, of solutes whose molecules contain both polar and nonpolar groups? To answer the question, let's consider the solubilities of the compounds in Table 6.5. All have C=O double bonds (carbonyl groups) bonded to nonpolar CH_2 and CH_3 groups. These carbonyl compounds are members of a group of organic compounds called **ketones**, in which the carbonyl carbon atom is bonded to two other carbon atoms.

Note how the solubility of the ketones in water decreases as the number of nonpolar CH_2 groups increases. The additional nonpolar groups strengthen the London dispersion forces between the ketone molecules but contribute little to the interactions between the ketone molecules and water molecules, which interact with each other mostly through hydrogen bonding. As we have seen, strong interactions between solute molecules that are not offset by strong solute–solvent interactions will tend to keep the solute molecules together in a single phase and inhibit their mixing with, and dissolving in, solvents.

Nonpolar interactions such as the London dispersion forces between the hydrocarbon chains in ketones are called **hydrophobic** (literally, "water-fearing")

H_2O O_2

FIGURE 6.19 Dipole–induced dipole interactions between molecules of H_2O and O_2. The permanent dipole of H_2O induces a temporary dipole in a normally nonpolar molecule of O_2.

dipole–induced dipole interaction an attraction between a polar molecule and the oppositely charged pole it induces in another molecule.

ketone an organic compound that contains a carbonyl group bonded to two other carbon atoms.

hydrophobic describes a "water-fearing" or repulsive interaction between a solute and water that diminishes water solubility.

hydrophilic describes a "water-loving" or attractive interaction between a solute and water that promotes water solubility.

TABLE 6.5   Solubilities of Some Ketones in Water

Compound	Condensed Molecular Structure	Solubility in Water (g/100 mL)
2-Propanone	$$H_3C\!-\!\overset{\overset{\textstyle O}{\|\|}}{C}\!-\!CH_3$$	Miscible
2-Butanone	$$H_3C\!-\!\overset{\overset{\textstyle O}{\|\|}}{C}\!-\!CH_2CH_3$$	25.6
2-Pentanone	$$H_3C\!-\!\overset{\overset{\textstyle O}{\|\|}}{C}\!-\!CH_2CH_2CH_3$$	4.3
2-Hexanone	$$H_3C\!-\!\overset{\overset{\textstyle O}{\|\|}}{C}\!-\!(CH_2)_3CH_3$$	1.4
2-Heptanone	$$H_3C\!-\!\overset{\overset{\textstyle O}{\|\|}}{C}\!-\!(CH_2)_4CH_3$$	0.4

interactions because they tend to keep the compounds from dissolving in water. On the other hand, dipole–dipole interactions, especially hydrogen bonding, promote solubility in water and are called **hydrophilic** ("water-loving") interactions. The solubility in water of compounds whose molecules contain both polar and nonpolar groups, as the ketones do, is the net result of offsetting interactions. Hydrophilic interactions between molecules of solute and water promote solubility, and hydrophobic interactions among solute molecules inhibit it. As the sizes of the nonpolar regions of solute molecules increase, the strengths of solute–solute hydrophobic interactions increase and their solubilities in water decrease.

CONCEPT TEST

Rank the normal alcohols with the molecular structures shown below from most soluble to least soluble in water:

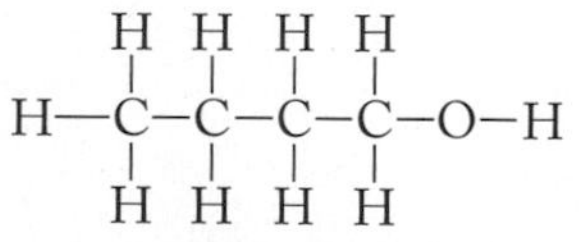

n-Butanol 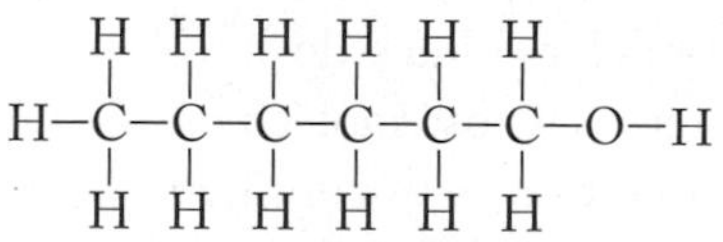   *n*-Hexanol 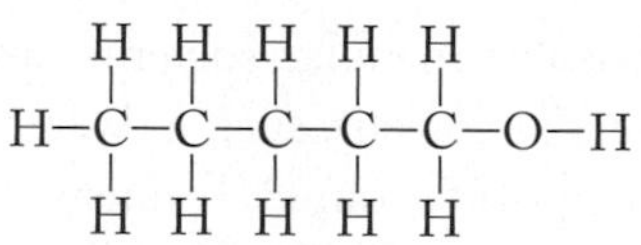   *n*-Pentanol 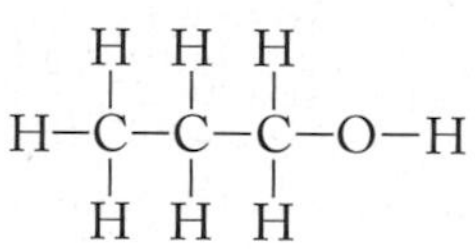   *n*-Propanol

FIGURE 6.20
Household cleaners such as this one are effective at removing grease stains and other nonpolar materials from surfaces and fabrics because the principal ingredient in them is a mixture of nonpolar hydrocarbons.

We have established that polar solutes tend to dissolve in polar solvents and that nonpolar solutes don't, at least not much. Given the solubility of polar solutes in polar solvents, isn't it reasonable that nonpolar solutes would tend to dissolve in nonpolar solvents? Indeed, they do, in accordance with a useful and general solubility guideline: *like dissolves like*. For example, the principal ingredient in some household cleaners (Figure 6.20) that remove grease, crayon wax, label adhesives, and other nonpolar materials from surfaces and fabrics is often labeled "petroleum distillate" or "petroleum naphtha." Those are common names for a mixture of hydrocarbons derived from crude oil, most of which have between 5 and 11 carbon atoms per molecule. They are effective in dissolving nonpolar substances because hydrocarbons are also nonpolar, and like dissolves like.

6.4 Phase Diagrams: Intermolecular Forces at Work

The strengths of the forces between particles control whether a substance is a solid, liquid, or gas at a given temperature: the stronger the forces, the better the chance that the substance will be a solid, or at least a liquid. Heating substances to higher temperatures provides the particles in them with more energy, which allows the particles to overcome the intermolecular forces that hold them rigidly in place in a solid or keep them close to each other in a liquid. As a result, the solid melts and the liquid vaporizes. However, it turns out that temperature and the strength of intermolecular forces are not the only factors that influence the physical state of a substance.

Pressure

Pressure (P), which is the ratio of force to surface area, also influences the physical state of substances. Why? Because the solid phases of most substances are the most dense and the vapor phases are the least dense. As a result, most substances (with the notable exception of H_2O) expand a little when they melt, and all substances expand a lot when they vaporize (or sublime). Increasing the pressure on a substance inhibits it from expanding, which raises the temperatures at which most substances melt and raises the temperatures at which *all* substances vaporize and sublime.

The pressure experienced by most substances of interest to us is *atmospheric* pressure, which is the weight of Earth's atmosphere pressing down on its surface divided by its surface area (Figure 6.21). We can calculate how strong this pressure is by multiplying the mass of all the gases in the atmosphere by the acceleration due to gravity experienced by all objects on Earth's surface, which is 9.807 m/s². The product of the multiplication is the force (F) that the mass of atmospheric gases exerts on Earth's surface due to gravity:

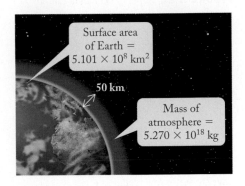

FIGURE 6.21 Atmospheric pressure results from the force exerted by the gases of Earth's atmosphere on Earth's surface.

$$F = (5.270 \times 10^{18} \text{ kg}) \times \left(9.807\frac{\text{m}}{\text{s}^2}\right) = 5.168 \times 10^{19}\,\frac{\text{kg m}}{\text{s}^2}$$

This combination of units, kg · m/s², is defined as a newton (N) in honor of Sir Isaac Newton. It is the SI unit of force. If we divide the force by the Earth's surface area (A), we obtain an average atmospheric pressure:

$$P = \frac{F}{A} \tag{6.1}$$

$$= \frac{(5.168 \times 10^{19} \text{ N})}{5.101 \times 10^8 \text{ km}^2 \left(\dfrac{1000 \text{ m}}{\text{km}}\right)^2}$$

$$= 1.013 \times 10^5\,\frac{\text{N}}{\text{m}^2}$$

One newton per square meter is defined as one pascal (Pa), the SI unit of pressure. The average pressure exerted by the atmosphere on Earth's surface (at sea level) is, therefore, 1.013×10^5 Pa. This average also defines a quantity of pressure called the **standard atmosphere (atm)**.

pressure (P) the ratio of a force to the surface area over which the force is applied.

standard atmosphere (atm) the average pressure at sea level on Earth.

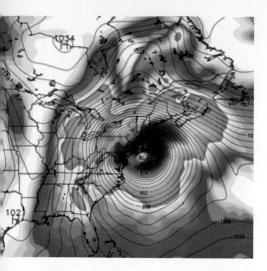

FIGURE 6.22 Superstorm Sandy caused significant damage to the eastern seaboard in 2012. The lines on this weather map mark different atmospheric pressures.

▶‖ **CHEMTOUR** Phase Diagrams

∞ **CONNECTION** Figure 1.2 in Chapter 1 shows the names of phase changes and the physical states involved.

phase diagram a graphical representation of the dependence of the stabilities of the physical states of a substance on temperature and pressure.

triple point the temperature and pressure where all three phases of a substance coexist. Freezing and melting, boiling and liquefaction, and sublimation and deposition all proceed at the same rate, so no net change takes place in the system.

critical point a specific temperature and pressure at which the liquid and gas phases of a substance have the same density and are indistinguishable from each other.

supercritical fluid the state of a substance that is above the temperature and pressure at the critical point, where the liquid and vapor phases are indistinguishable.

To avoid the need for exponents, atmospheric pressure is usually expressed in larger units such as the kilopascal (kPa), bar, or millibar (mb):

$$1 \text{ atm} = 101.3 \text{ kPa}$$
$$= 1.013 \text{ bar}$$
$$= 1013 \text{ mb}$$

The bar is not an official SI unit, but it is based on one: 1 bar = 10^5 Pa, and it is widely used in science. Meteorologists tend to use millibars to express atmospheric pressures, as shown in Figure 6.22. Doing so simplifies pressure values on weather maps by avoiding the need for decimal points.

Phase Diagrams

Scientists use **phase diagrams** to show which phases of a substance are the most stable at different combinations of temperature and pressure. Phase diagrams are graphs in which temperature is the *x* axis and pressure is the *y* axis. The scale of the *y* axis is usually logarithmic rather than linear to cover a wide range of pressures.

Phase diagrams like that of water (Figure 6.23) have at least three regions corresponding to the three states of matter. The lines separating the regions represent combinations of temperature and pressure at which the two phases on either side of each line coexist in equilibrium with each other. Thus, the blue line separating the solid and liquid regions in Figure 6.23 represents a series of melting (or freezing) points; the red line separating the liquid and gas regions represents a series of boiling (or condensation) points, and the green line separating the solid and gas regions represents a series of sublimation (or deposition) points.

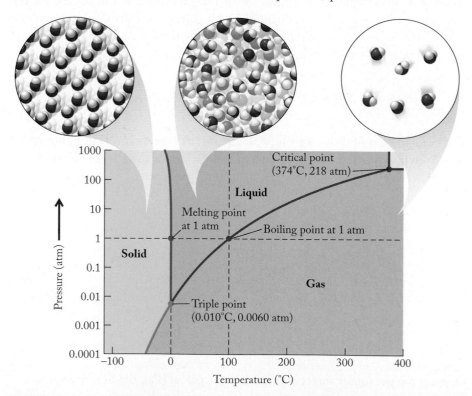

FIGURE 6.23 The phase diagram for water indicates in which phase (solid, liquid, or gas) water exists at various combinations of pressure and temperature. The yellow region indicates the region where water exists as a supercritical liquid.

Notice that the red line curves from lower left to upper right. This orientation makes sense because when the pressure above a liquid is increased, as in a pressure cooker (Figure 6.24), there is an increase in the energy required to overcome that pressure and convert molecules of liquid into molecules of gas where they take up much more space. Heating the liquid to an even higher temperature gives it that additional energy. The shape of the solid–gas curve also makes sense: higher pressures make it more difficult for molecules of ice to sublime into water vapor. In both cases, a phase transition from a dense, condensed (liquid or solid) phase to a much less dense vapor phase occurs at higher temperatures under increasing pressure.

On the other hand, the melting/freezing equilibrium line for water is nearly vertical at low to moderate pressures and then bends to the left at high pressures. Thus, the temperature at which ice melts *decreases* as pressure increases. This trend is opposite that observed for almost all other substances (note, for example, the curve of the blue line in the phase diagram of CO_2 in Figure 6.25). The reason for water's unusual melting/freezing behavior is that water *expands when it freezes*. As we have noted, the solid phases of nearly all substances are denser than in their liquid states, which means they contract when they freeze. Water expands because of hydrogen bonding. As it freezes, more hydrogen bonds form between its molecules as they cease to flow past each other and instead take up positions in a solid structure where they are surrounded by, and hydrogen bonded to, an extended array of other water molecules. These additional hydrogen bonds create a slightly more open structure in ice than in liquid water, which makes ice less dense. If we apply enough pressure to ice, we can force it to melt.

FIGURE 6.24 Steam is restricted from escaping a pressure cooker as the water inside it starts to boil. The higher resulting pressure causes the water to boil at a higher temperature, so the food inside cooks faster.

CONCEPT TEST

Figure 6.26 shows how a length of wire with heavy weights on each end can pass downward through a block of ice without cutting it in two. The temperature of the ice stays below its melting point the whole time, and the ice is still a single block after the wire has passed all the way through it. Explain how the wire can pass through a frozen block of ice without cutting it in two.

A point of special interest on a phase diagram is the point where all three phase transition lines meet. Known as the **triple point**, it identifies the temperature and pressure at which the liquid, solid, and vapor states of a substance coexist. The triple point of water is at a temperature of 0.010°C—just above its normal (1 standard atmosphere) freezing point of 0°C—but at a pressure of only 0.0060 atm.

Another point of interest is the place where the liquid/gas equilibrium line ends. At this **critical point**, the two states are indistinguishable from each other. This point is reached because thermal expansion at high temperature decreases the density of the liquid state while high pressure compresses the gas into a small volume, increasing its density. At the critical point the densities of the liquid and vapor states are equal. At temperature–pressure combinations above a substance's critical point, it exists as a **supercritical fluid**, as shown for CO_2 in Figure 6.25.

A supercritical fluid has the ultralow viscosity of a gas and can easily diffuse through many solid materials, yet it can dissolve substances in those materials as if it were a liquid. Supercritical CO_2 is used in the food processing industry to decaffeinate coffee beans and to remove fat from potato chips. It is also used

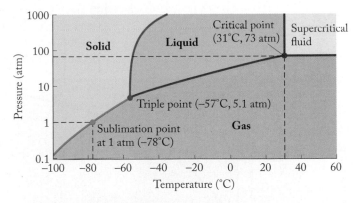

FIGURE 6.25 Phase diagram for carbon dioxide.

FIGURE 6.26 A weighted wire passes through a solid block of ice.

to extract essential oils from plants to make perfumes and to dry clean clothes, providing a "greener" alternative to the organic solvents that were once widely used. In these applications, supercritical CO_2 has the advantage of being non-toxic and nonflammable, and of forming at a relatively low temperature (31°C), which is an advantage in extracting pharmaceuticals and other compounds that may decompose at high temperature.

Another look at the phase diagram of CO_2 discloses that the blue region representing liquid CO_2 does not exist below a pressure of 5.1 atm. This means that solid CO_2 does not melt into a liquid at normal (atmospheric) pressures. Rather, it sublimes directly to CO_2 gas. This behavior is why solid CO_2 is commonly called *dry ice*; it is a cold solid that does not melt at normal pressures; it just disappears as a colorless gas.

SAMPLE EXERCISE 6.3   Interpreting Phase Diagrams   LO3

Describe the phase changes that take place when the pressure on a sample of water is increased from 0.0001 atm to 100 atm at a constant temperature of −25°C, and the sample is then warmed from −25°C to 350°C at a constant pressure of 100 atm.

COLLECT AND ORGANIZE We are asked to describe the phase changes a sample of water undergoes as its pressure increases at constant temperature and as its temperature increases at constant pressure. The phase diagram in Figure 6.23 shows which phases of water are stable at various combinations of temperature and pressure.

ANALYZE The change in pressure at constant temperature defines two points on the phase diagram of water with the coordinates (−25°C, 0.0001 atm) and (−25°C, 100 atm). Connecting those points will give us a vertical straight line. If the line crosses a phase boundary, there will be a change in physical state. The change in temperature at constant pressure defines a third point on the phase diagram, which will connect to the second point via a horizontal line.

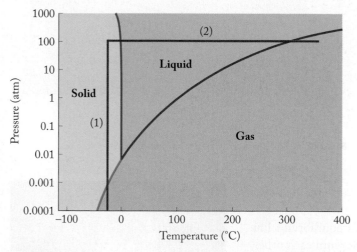

FIGURE 6.27 Phase diagram showing changes in pressure and temperature of a sample of water.

SOLVE Figure 6.27 is a plot of our sample's changes in pressure and temperature. At the bottom end of the vertical line 1, water is a gas (steam). As pressure increases along line 1, it crosses the boundary between gas and solid, which means that steam turns directly into solid ice, which is stable beyond 100 atm at −25°C. As the temperature of the ice increases along the horizontal line 2, it intersects the solid/liquid boundary and the ice melts. At even higher temperatures, just below 350°C, the line intersects the liquid/gas boundary and the liquid water vaporizes.

THINK ABOUT IT The solid-to-liquid and liquid-to-gas transitions with increasing temperature along line 2 are what we would expect when a solid substance is warmed to its melting point and then the liquid is heated to its boiling point at a given pressure. The transition along line 1 is less familiar because it is caused by increasing the pressure on a gas at a temperature that is below the triple point temperature, so the steam never condenses. Instead, it is deposited as a solid.

Practice Exercise Describe the phase changes that occur when the temperature of CO_2 is increased from −100°C to 50°C at a pressure of 25 atm and the pressure is then increased to 100 atm.

6.5 Some Remarkable Properties of Water

So far in this chapter we have examined the impact of intermolecular forces on the boiling points of substances, their viscosities, and their solubilities in each other. In this section we expand our list to include several macroscopic properties of water.

Surface Tension, Capillary Action, and the Density of Water

First we explore the phenomenon of **surface tension**, which is the ability of the surface of a liquid to resist an external force, such as the weight of an object. We usually think of surface tension as a property of liquids, such as the water on which a water strider literally walks (Figure 6.28), but it also applies to solids, where surface tension is called *surface energy* because it is expressed in units of energy per unit of surface area. It is also called *surface stress*, which is defined as the energy needed to stretch a surface so that its area increases by a unit amount.

To understand how surface tension depends on intermolecular forces, consider the microscopic view of a steel needle suspended on the surface of a dish of water (Figure 6.29). Water molecules in the interior of the water are surrounded by other water molecules and form hydrogen bonds to them in all directions. However, there are no molecules of liquid water above those at the surface, so those water molecules form fewer hydrogen bonds per molecule. Now consider the forces at work on the steel needle in Figure 6.29. Gravity pulls it downward, but to move downward the needle must penetrate the surface. This means pushing aside surface water molecules, increasing the area of the surface, and turning what were once interior water molecules into surface molecules. This transformation requires breaking some of the hydrogen bonds experienced by the interior molecules, which requires energy: 23 kJ of it per mole of hydrogen bonds between molecules of H_2O. The downward pressure of the needle on the surface is not enough to overcome that energy, so the needle floats on the surface.

Another phenomenon linked to the strength of intermolecular forces is the shape of a liquid's surface in a graduated cylinder or other small-diameter container made of glass, which is mostly SiO_2. As shown in Figure 6.30, the surface of water in a partially filled glass test tube is concave, but the surface of liquid mercury is convex. Either curved surface is called a **meniscus**. In both liquids, the meniscus is the result of two competing forces: *cohesive forces*, which are interactions between like particles (such as the hydrogen bonds between water molecules), and *adhesive forces*, which are interactions between materials made of different substances. In the water sample in Figure 6.30, the adhesive forces are dipole–dipole interactions between water molecules and polar Si—O—Si groups on the surface of the glass, and even hydrogen bonding between water molecules and Si—O—H groups on the surface. The adhesive forces are strong enough to cause the water to climb up the glass. Cohesive interactions with other water molecules pull them up to nearly the same height as the surface molecules. Water molecules farther from the inner wall of the tube are pulled up progressively less, creating the concave meniscus.

When mercury partially fills a similar glass test tube, it experiences strong cohesive forces in the form of metallic bonds between mercury atoms. The only

FIGURE 6.28 Surface tension allows a water strider to walk on water without sinking.

FIGURE 6.29 Intermolecular forces, including hydrogen bonding, are exerted equally in all directions in the interior of a liquid. However, there is no liquid water above a surface to exert attractive intermolecular forces. The resulting imbalance causes the surface water molecules to adhere tightly to one another, creating surface tension. This surface tension exceeds the downward force exerted on the surface water molecules by the needle, causing it to float even though it is much denser than water.

capillary action the rise of a liquid in a narrow tube as a result of adhesive forces between the liquid and the tube and cohesive forces within the liquid.

▶❚❚ **CHEMTOUR** Capillary Action

▶❚❚ **CHEMTOUR** Hydrogen Bonding in Water

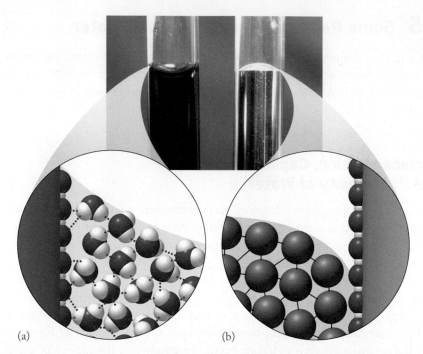

(a)   (b)

FIGURE 6.30 (a) A combination of adhesive and cohesive forces causes the water to form a concave meniscus in a test tube made of silica glass, which contains polar surface groups. (b) Atoms of liquid mercury do not adhere strongly to glass, so mercury forms a convex meniscus.

FIGURE 6.31 Because of capillary action, water (containing blue dye) rises in a stalk of celery.

adhesive forces are relatively weak interactions between induced dipoles in surface mercury atoms and the polar groups on the glass surface. In this case, mercury atoms are more strongly attracted to one another than to the glass because the dipole–induced dipole adhesive attractions are much weaker than the metallic bonds in liquid Hg. As a result, mercury pulls away from the surface, mounding up in a way that reduces its interaction with the test tube and increases cohesive interactions. The result is a convex meniscus.

The principle behind water's concave meniscus is taken to the extreme when water enters extremely narrow tubes called capillaries. The tiny inner diameter of capillaries means that all water molecules inside a capillary either experience adhesive intermolecular forces directly or are a small number of hydrogen bonds away from molecules that do. As a result, adhesion pulls up the water molecules along the wall of the capillary and cohesion pulls up all the others. The result of this combination of intermolecular forces is a phenomenon called **capillary action**, which is the ability of a liquid to flow against gravity, spontaneously rising in a narrow tube or in structures made up of narrow pores, such as celery stalks (Figure 6.31) or the trunks of trees, as in the photograph at the beginning of this chapter.

Capillary action is also responsible for wicking, the movement of a fluid away from its source through a porous material. Wicking is the process by which paper towels absorb spills, and by which some fabrics, such as those made of polyester microfibers, draw perspiration away from the skin. Undergarments made of such fabrics provide a base layer of clothing below other layers that provide thermal insulation. This combination keeps athletes warm while allowing them to practice and play in clothing that is not waterlogged with perspiration.

Is mercury spontaneously drawn up into a glass capillary tube? Why or why not?

Another unusual property of water is the way its density changes as its temperature drops to near its freezing point. Like nearly all substances, water's density increases as its temperature decreases. However, the density of water reaches a maximum at 4°C (Figure 6.32) and actually *decreases* as it cools toward its freezing point. When water begins to freeze, its density drops even more, to about 0.92 g/mL, as the oxygen atom in each water molecule becomes the center of an array of two covalent bonds (to the H atoms within the molecule) and two hydrogen bonds (to two other H_2O molecules), as shown in Figure 6.33(a). Note that the length of the hydrogen bonds is more than twice that of the covalent bonds in H_2O. However, the larger distance is not the reason ice is less dense than water; rather, it is the *directionality* of the hydrogen bonds in ice. Recall from Chapter 5 that the electron-group geometry of the valence electrons in the O atoms (SN = 4) is tetrahedral. However, ice crystals feature *hexagonal* arrays of water molecules (Figure 6.33b). This means that the angle between the H······O—H hydrogen bond and covalent bond at each corner is about 120°, which is larger than the tetrahedral ideal of 109.5°. The larger corner angles mean that ice's hexagonal arrangement creates more space between the molecules than liquid water's largely tetrahedral orientation of hydrogen-bonded molecules, which allows liquid water molecules to snuggle up closer to their nearest neighbors.

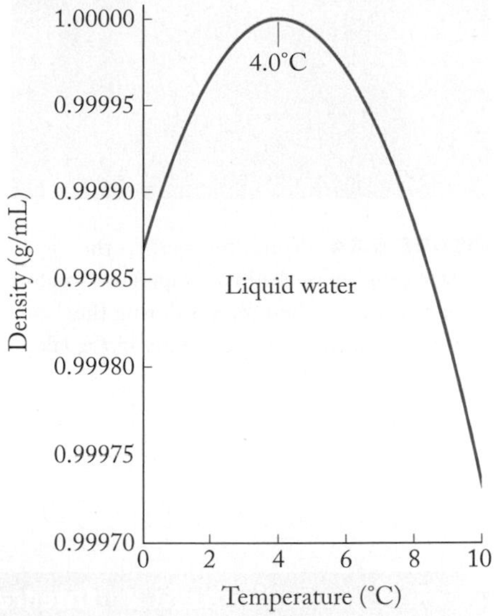

Water and Aquatic Life

The expanded microscopic structure of ice plays a crucial role in aquatic ecosystems in temperate and polar climates. The lower density of ice means that lakes, rivers, and polar oceans freeze from the top down, allowing fish and other aquatic life to survive in the denser layer of liquid water below the frozen surface.

FIGURE 6.32 As water is cooled, its density increases until the temperature has been lowered to 4°C. At that temperature the density has its maximum value, 1.000 g/mL. As the water cools from 4°C to its freezing point at 0°C, the density decreases.

(a) 197 pm 96 pm H H O

(b)

FIGURE 6.33 (a) The oxygen atoms in ice form two covalent bonds to H atoms and two hydrogen bonds to the H atoms in nearby molecules. (b) Ice has a lower density than water because of its three-dimensional molecular structure created by the covalent and hydrogen bonds within and between its molecules.

FIGURE 6.34 Wintering over. As the surface water of a lake in a temperate climate cools to 0°C and then freezes during the winter, the deepest, densest water in the lake remains a relatively warm 4°C.

Let's follow how the temperature and density of the water in a deep lake in a temperate climate changes over the course of a year. In early spring, rising air temperatures and more sunlight warm the surface waters of the lake. Continued heating creates an upper layer of warm, low-density water separated from the colder, denser water below by a *thermocline*, a sharp change in temperature between the two layers. Little mixing takes place between the two layers over the summer and into the fall. Then, declining air temperatures cool the surface waters until they are colder and denser than the water below. The cold surface water sinks, forcing warmer water to the surface where it also cools and sinks. The process continues until all of the water in the lake reaches 4°C and is uniformly mixed.

As the surface water cools further with the approach of winter, it becomes less dense and ice eventually forms. The layer of ice floats above and insulates the water below, including the 4°C water that fills the deepest parts of the lake because it is the densest (Figure 6.34). This water provides a liquid haven for aquatic life-forms and allows them to survive the subfreezing air temperatures of winter.

SAMPLE EXERCISE 6.4   **Integrating Concepts: Drug Efficacy and Partition Ratios**

The effectiveness of most drugs, particularly those taken orally, depends on how well they can reach their target organs and tissues and on their therapeutic strength and persistence once they get to their targets. These factors are linked both to how soluble a drug is in the aqueous environments of blood serum and cytosol (the liquid inside cells) and to how well the drug can penetrate nonaqueous, nonpolar parts of the body such as the epithelium surrounding the small intestine and the membranes surrounding individual cells. In other words, to be effective a drug must have at least some hydrophilic *and* hydrophobic character.

One way scientists can predict whether a molecular compound has a blend of hydrophilic and hydrophobic properties is to determine its relative solubility in water and in octanol, $CH_3(CH_2)_7OH$ (Figure 6.35a), which is immiscible in water. Typically, a quantity of the compound is added to a mixture of water and octanol. The mixture is shaken vigorously to allow the compound to dissolve in either liquid, the liquids are allowed to separate, and then the concentration of the compound in each liquid is measured. The ratio of the concentrations defines the compound's partition ratio:

$$\text{Partition ratio} = \frac{\text{concentration in octanol}}{\text{concentration in water}} \quad (6.2)$$

a. What types of intermolecular forces are experienced by the molecules in a sample of octanol?
b. Which of the forces cited in part a accounts for most of the interaction between molecules of octanol? (The immiscibility of octanol in water and the information in Figure 6.35 may help you decide.)
c. Are compounds with large partition ratios more or less hydrophobic than compounds with small partition ratios? Why?
d. What if equal volumes of octanol and water were used in one determination of a compound's partition ratio,

and then twice as much octanol was used in another determination? How would this difference in volumes influence the two experimental partition ratio values?

COLLECT AND ORGANIZE We know the molecular structure of octanol and that it is immiscible in water. We are asked what types of intermolecular forces octanol molecules experience, which force is the strongest, and how the partitioning of a compound between octanol and water relates to its hydrophobicity. We are also asked to predict how changing the relative volumes of octanol and water affects the ratio of the concentrations of a solute in the two liquids.

ANALYZE Octanol is an alcohol with one OH group at the end of a chain of 8 carbon atoms. The O and H atoms in OH groups can form hydrogen bonds to the H and O atoms on neighboring molecules. All molecules experience London dispersion forces that increase in strength with increasing molar mass. The immiscibility of octanol in water means that it is, on balance, hydrophobic and should be a good solvent for hydrophobic solutes. If the partition ratio of a compound is a constant, then the ratio of concentrations of the compound in octanol and water must be a constant that does not change with changes in the volumes of either or both of the solvents.

$CH_3(CH_2)_7OH$, $\mathcal{M} = 130$ g/mol
Boiling point 468 K

(a) Octanol

$CH_3(CH_2)_7CH_3$, $\mathcal{M} = 128$ g/mol
Boiling point 424 K

(b) Nonane

FIGURE 6.35 (a) Octanol. (b) Nonane.

SOLVE

a. Molecules of octanol interact with each other via London dispersion forces, as all molecules do, and via hydrogen bonding between the OH groups on adjacent molecules.

b. The hydrophobicity of octanol is an indication that its molecules experience strong London dispersion forces due to their long, nonpolar chains of carbon and hydrogen atoms. The importance of these interactions is reinforced by how relatively small the difference is between the boiling points of octanol and nonpolar nonane (Figure 6.35b)—a compound with nearly the same molar mass and molecular shape as octanol. Therefore, London dispersion forces are the principal intermolecular force among octanol molecules.

c. The long nonpolar chains in molecules of octanol make it an effective solvent for nonpolar, or hydrophobic, solutes. Those same solutes should have little solubility in water, which would give a relatively high concentration ratio in Equation 6.2 during a partition ratio determination. Thus, a compound with a large partition ratio is more hydrophobic than one with a smaller partition ratio.

d. Because the concentration ratio in Equation 6.2 is a constant, changing the relative volumes of octanol and water in a partition ratio determination should not alter the calculated value. On the other hand, doubling the volume of octanol would mean that twice as much of the total *quantity* of solute in the mixture would end up in the octanol phase.

THINK ABOUT IT Boiling points on the Kelvin scale provide a measure of the strength of intermolecular interaction because, theoretically, a substance that experiences no intermolecular interaction (admittedly, no such substance exists) would become a gas just as its temperature exceeded 0 K. Therefore, comparing the Kelvin-scale boiling points of compounds provides insights into the relative strengths of their intermolecular forces.

SUMMARY

Section 6.1 All atoms and molecules experience **London dispersion forces** due to their **polarizability** and the existence of **temporary dipoles** even in particles that have no permanent dipole. Polarizability increases with increasing particle size, so larger atoms and molecules experience stronger London dispersion forces than small ones. Stronger London dispersion forces lead to higher boiling points and greater **viscosities**. **Hydrocarbons** are compounds whose molecules contain only hydrogen and carbon atoms. **Alkanes** are hydrocarbons in which each carbon atom is bonded to four other atoms. **Constitutional isomers** have the same molecular formulas but different connections between the atoms in their molecules.

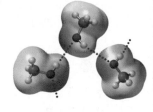

Section 6.2 Ions in aqueous solution interact with water molecules through **ion–dipole interactions**, forming a **sphere of hydration** around the ion. Molecules with permanent dipoles interact through a combination of London dispersion forces and **dipole–dipole interactions**. The strongest dipole–dipole interactions are **hydrogen bonds**, which form between H atoms bonded to N, O, and F atoms and other N, O, and F atoms. Polar **organic compounds** contain polar functional groups. Among them are **carbonyl groups**, which contain C=O bonds. When the carbonyl carbon atom is bonded to two other carbon atoms in a molecular structure, the compound is called a *ketone*. Other polar functional groups include the **hydroxyl group** (OH) in **alcohols**.

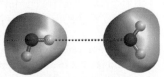

Section 6.3 Polar **solutes** dissolve in polar **solvents** when the dipole–dipole interactions between solute and solvent molecules offset the interactions that keep either solute molecules or solvent molecules together. The limited **solubility** of nonpolar solutes in polar solvents is a result of **dipole–induced dipole interactions**. **Hydrophilic** substances are more soluble in water than are **hydrophobic** substances, which are more soluble in nonpolar solvents. In general, polar solutes dissolve in polar solvents and nonpolar solutes dissolve in nonpolar solvents, that is, like dissolves like.

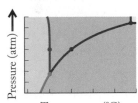

Section 6.4 The ratio of force to surface area is defined as **pressure (P)**, which is often expressed in **standard atmospheres (atm)**, a unit equal to Earth's average atmospheric pressure at sea level. The **phase diagram** of a substance indicates whether it exists as a solid, liquid, gas, or **supercritical fluid** at a particular combination of pressure and temperature. Two adjoining regions in a phase diagram are separated by a line representing pressures and temperatures at which the two phases can coexist. All three states (solid, liquid, and gas) exist in equilibrium at the **triple point**. Above the temperature and pressure of its **critical point**, a substance exists as a supercritical fluid with many of the physical properties of a gas but the ability to dissolve other substances as if it were a liquid.

Section 6.5 The remarkable behavior of water, including its unusually high **surface tension**, results from the strength of hydrogen bonding between its molecules. Those interactions also cause water to expand when it freezes and contribute to **capillary action**.

PROBLEM-SOLVING SUMMARY ■

TYPE OF PROBLEM	CONCEPTS AND EQUATIONS	SAMPLE EXERCISES
Explaining differences and trends in boiling points of liquids	Substances made of large molecules usually have higher boiling points than those with smaller molecules because of London dispersion forces. Polar compounds have higher boiling points than nonpolar compounds of similar molar mass because of dipole–dipole interactions. Compounds whose molecules form hydrogen bonds have even higher boiling points because H bonds are especially strong dipole–dipole interactions.	6.1
Distinguishing solute from solvent	Calculate the number of moles of each constituent of a solution. The one with the greatest number of moles is the solvent.	6.2
Interpreting phase diagrams	Locate the combination of temperature and pressure of interest on the phase diagram, and determine which physical state exists at that point. Changes in pressure at constant temperature are represented by vertical paths, and changes in temperature at constant pressure are represented by horizontal paths. If a path crosses a phase boundary line, there is a change in phase.	6.3

VISUAL PROBLEMS ■

(Answers to boldface end-of-chapter questions and problems are in the back of the book.)

6.1. Figure P6.1 contains molecular structures of four constitutional isomers of heptane. Which one has the highest boiling point?

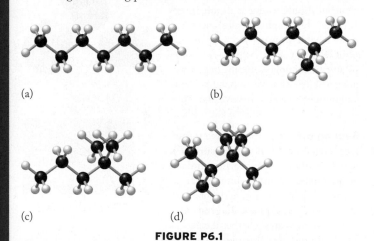

(a)   (b)

(c)   (d)

FIGURE P6.1

6.2. Which of the constitutional isomers of heptane in Figure P6.1 is the most viscous at 20°C?

6.3. The image in Figure P6.3 is the phase diagram of imaginary molecular compound X. If a sample of X is left outside in a sealed container on a summer day, will the X in the container be a solid, liquid, or a gas?

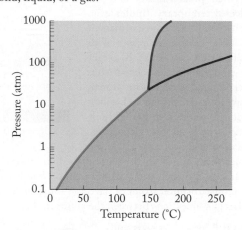

FIGURE P6.3

6.4. Suppose you bring the sample of X from Problem 6.3 inside and place it in an uncovered pot of boiling water on your kitchen stove. What phase changes, if any, will occur?

6.5. Another sample of X (Figure P6.3) is stored in a pressurized container (P = 50 atm) at 0°C. If the sample is then transferred to an oven and slowly warmed to 250°C, what phase changes, if any, will X undergo?

6.6. Does the green line in Figure P6.3 represent a series of (a) freezing points, (b) sublimation points, (c) boiling points, or (d) critical points?

6.7. Does the solid form of compound X (Figure P6.3) float on the liquid as the liquid begins to freeze at P = 300 atm?

6.8. The graph in Figure P6.8 is an expanded view of part of the phase diagram of water. In this version the pressure scale is linear instead of logarithmic. Which phases are represented by the blue and pink colors?

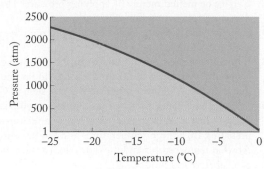

FIGURE P6.8

6.9. Referring to Figure P6.8, what phase change, if any, takes place if the temperature of a sample of water is increased from −25°C to −15°C while the pressure on it is decreased from 2500 atm to 1000 atm? Assume that the changes in temperature and pressure occur at constant rates over the same time interval.

*6.10. Suppose an ice skater weighing 50.0 kg and wearing newly sharpened ice skates stands on ice at a temperature of −10°C. If the surface area of the edges of the skates pressing down on the ice is 0.033 cm^2, will the ice under the skate edges melt? Figure P6.8 may help you in making your prediction.

QUESTIONS AND PROBLEMS ....■

London Dispersion Forces: They're Everywhere

CONCEPT REVIEW

6.11. Why does a branched alkane have a lower boiling point than a normal alkane of the same molar mass?

6.12. Why do the strengths of London dispersion forces increase with increasing molecular size?

PROBLEMS

6.13. Select the compound in each of the following pairs whose molecules experience stronger London dispersion forces. (a) CCl_4 or CF_4; (b) CH_4 or C_3H_8; (c) CS_2 or CO_2

6.14. The three most abundant gases in air are N_2, O_2, and Ar. Which of them has the highest boiling point, and which has the lowest boiling point?

6.15. **Fuels from Crude Oil** Petroleum (crude oil) is a complex mixture of mostly hydrocarbons that can be separated into useful fuels by distillation. Common petroleum-based fuels in order of increasing boiling point are gasoline, jet fuel, kerosene, fuel oil, and diesel oil. Which of the fuels contains hydrocarbons with the highest average molar mass?

6.16. Which of the fuels in Problem 6.15 is the most viscous at 20°C?

Interactions Involving Polar Molecules

CONCEPT REVIEW

6.17. How are water molecules oriented around anions in aqueous solutions?

6.18. How are water molecules oriented around cations in aqueous solutions?

6.19. Why are dipole–dipole interactions generally weaker than ion–dipole interactions?

6.20. Two liquids—one polar, one nonpolar—have the same molar mass. Which one is likely to have the higher boiling point?

6.21. Why are hydrogen bonds considered a special class of dipole–dipole interactions?

6.22. Can all polar hydrogen-containing molecules form hydrogen bonds?

PROBLEMS

6.23. Suggest two reasons why the boiling point of methyl fluoride, CH_3F, is higher than the boiling point of methane, CH_4.

6.24. Why is the boiling point of Br_2 lower than that of iodine monochloride, ICl, even though they have nearly the same molar mass?

_____

6.25. Why do molecules of methanol (CH_3OH) form hydrogen bonds, but molecules of methane (CH_4) do not?

6.26. The boiling point of PH_3 is lower than that of NH_3 even though PH_3 has twice the molar mass of NH_3. Why?

_____

6.27. In which of the following compounds do the molecules experience dipole–dipole interactions? (a) CF_4; (b) CF_2Cl_2; (c) CCl_4; (d) $CFCl_3$

6.28. Molecules of which of these compounds: CO_2, NO_2, SO_2, or H_2S, experience dipole–dipole interactions?

_____

(Before solving the following problems, you may find it useful to review Section 4.1 on the strengths of ionic bonds.)

6.29. In an aqueous solution containing chloride, bromide, and iodide salts, which anion would you expect to experience the strongest ion–dipole interactions with surrounding water molecules?

6.30. In an aqueous solution containing Na^+, Mg^{2+}, K^+, and Ca^{2+} salts, which cation would you expect to experience the strongest ion–dipole interactions?

Trends in Solubility

CONCEPT REVIEW

6.31. What is the difference, if there is any, between the terms *miscible* and *soluble*?

6.32. Which of these substances have little solubility in water? (a) benzene, C_6H_6; (b) KBr; (c) Ar

6.33. In what context do the terms *hydrophobic* and *hydrophilic* relate to the solubilities of substances in water?

6.34. A series of alcohols has the generic molecular formula $C_nH_{(2n+1)}OH$. How does the value of n affect the solubility of these alcohols in water?

PROBLEMS

6.35. In each of the following pairs of compounds, which compound is likely to be more soluble in water?
  a. CCl_4 or $CHCl_3$
  b. CH_3OH or $C_6H_{11}OH$
  c. NaF or MgO
  d. CaF_2 or BaF_2

6.36. In each of the following pairs of compounds, which compound is likely to be more soluble in CCl_4?
  a. Br_2 or NaBr
  b. CH_3CH_2OH or CH_3OCH_3
  c. CS_2 or KOH
  d. I_2 or CaF_2

6.37. Which of the following compounds is likely to be the most soluble in water? (a) NaCl; (b) KI; (c) $Ca(OH)_2$; (d) CaO

6.38. Which sulfur compound would you predict to be more soluble in nonpolar solvents: SO_2 or CS_2?

6.39. Which of these compounds is the most soluble in water?
  a. $CH_3(CH_2)_2O(CH_2)_2CH_3$
  b. $CH_3(CH_2)_3O(CH_2)_2CH_3$
  c. CH_3OCH_3
  d. $CH_3CH_2OCH_2CH_3$

6.40. Rank the ketones in Figure P6.40 from least soluble to most soluble in water.

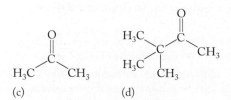

FIGURE P6.40

Phase Diagrams: Intermolecular Forces at Work

CONCEPT REVIEW

6.41. Explain the difference between sublimation and evaporation.

6.42. Can ice be melted merely by applying pressure? How about dry ice? Explain your answers.

6.43. Explain what is meant by the term *equilibrium line* as it is used in Section 6.4.

6.44. Explain how the solid–liquid line in the phase diagram of water differs in character from the solid–liquid line in the phase diagrams of most other substances, such as CO_2.

6.45. Which phase of a substance (gas, liquid, or solid) is most likely to be the stable phase: (a) at low temperatures and high pressures; (b) at high temperatures and low pressures?

6.46. At what temperatures and pressures does a substance behave as a supercritical fluid?

6.47. **Preserving Food** Freeze-drying is used to preserve food at low temperature with minimal loss of flavor. Freeze-drying works by freezing the food and then lowering the pressure with a vacuum pump to sublime the ice. Must the pressure be lower than the pressure at the triple point of H_2O?

6.48. Solid helium cannot be converted directly into the vapor phase. Does the phase diagram of He have a triple point?

PROBLEMS

(To solve Problems 6.49 through 6.56, you should consult Figures 6.23 and 6.25.)

6.49. What changes in temperature and pressure are required to bring a sample of water at 25°C and 1 atm of pressure to its triple point?

6.50. What is the freezing point of water at $P = 1000$ atm?

6.51. What phase changes, if any, does liquid water at 100°C undergo if the initial pressure of 5.0 atm is reduced to 0.5 atm at constant temperature?

6.52. What phase changes, if any, occur if a sample of CO_2 initially at −80°C and 5.0 atm is warmed to −25°C at 5.0 atm?

6.53. Below what temperature can solid CO_2 (dry ice) be converted into CO_2 gas simply by lowering the pressure?

6.54. What is the maximum pressure at which solid CO_2 (dry ice) can be converted into CO_2 gas without melting?

6.55. Predict the phase of water that exists under the following conditions:
  a. 2 atm of pressure and 110°C
  b. 0.5 atm of pressure and 80°C
  c. 7×10^{-3} atm of pressure and 3°C

6.56. Which phase or phases of water exist under the following conditions?
  a. 0.32 atm and 70°C
  b. 300 atm and 400°C
  c. 1 atm and 0°C

Some Remarkable Properties of Water

CONCEPT REVIEW

6.57. Explain why a needle floats on the surface of water but sinks in a container of methanol.

6.58. Explain why different liquids do not reach the same height in capillary tubes of the same diameter.

6.59. Explain why the plumbing in a house may burst if the temperature in the house drops below 0°C.

6.60. A hot needle sinks when placed on the surface of cold water, but a cold needle floats. Why?

6.61. The meniscus of mercury in a mercury thermometer is convex. Why?

6.62. The mercury level in a capillary tube inserted into a dish of mercury is below the surface of the mercury in the dish. Why?

6.63. Describe the origin of surface tension in terms of intermolecular interactions.

6.64. Would you expect water to rise to the same height in a tube made of a polyethylene plastic (essentially a very-long-chain hydrocarbon) as it does in a silica glass capillary tube of the same diameter?

Additional Problems

6.65. Methanol has a larger molar mass than water, but boils at a lower temperature. Suggest a reason why.

6.66. What kinds of intermolecular forces must be overcome as (a) solid CO_2 sublimes; (b) $CHCl_3$ boils; (c) ice melts?

*6.67. The dipole moment of CH_2F_2 (1.93 D) is larger than that of CH_2Cl_2 (1.60 D), yet the boiling point of CH_2Cl_2 (40°C) is much higher than that of CH_2F_2 (–52°C). Why?

6.68. How is it that the dipole moment of HCl (1.08 D) is larger than the dipole moment of HBr (0.82 D), yet HBr boils at a higher temperature?

6.69. Does the sublimation point of ice increase or decrease with increasing pressure?

6.70. Why is methanol miscible with water, but methane is not?

*6.71. The melting point of hydrogen is 15.0 K at 1.00 atm. The temperature of its triple point is 13.8 K. Does liquid H_2 expand or contract when it freezes?

6.72. Sketch a phase diagram for element Z, which has a triple point at (152 K, 0.371 atm), a boiling point of 166 K at a pressure of 1.00 bar, and a normal melting point of 161 K.

*6.73. Pick from among the compounds with the molecular structures shown in Figure P6.73 the ones that you think should be soluble in both water and octanol (see Sample Exercise 6.4).

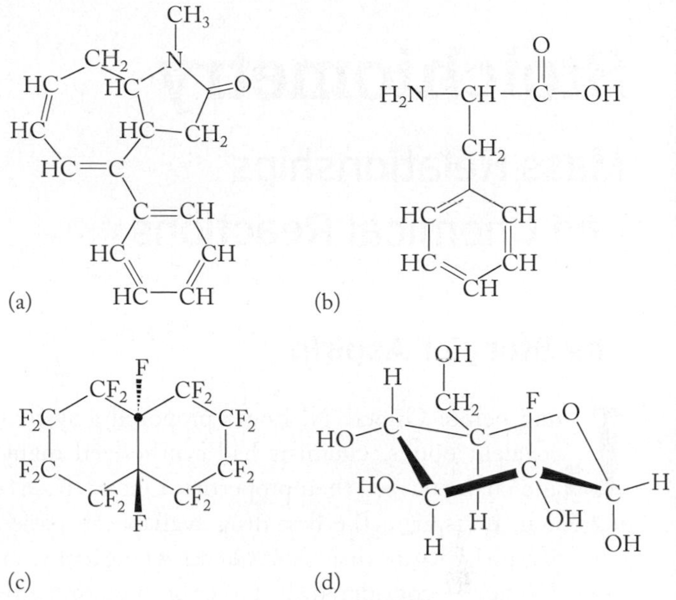

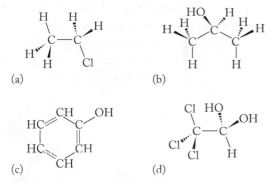

FIGURE P6.73

6.74. **First Aid for Bruises** Compounds with low boiling points may be sprayed on the skin as a topical anesthetic. They chill the skin as they evaporate and provide short-term relief from injuries. Predict which compound among those in Figure P6.74 has the lowest boiling point.

FIGURE P6.74

7

Stoichiometry

Mass Relationships and Chemical Reactions

The Story of Aspirin

Long before Gilbert N. Lewis proposed a systematic way of describing covalent bonds, chemists had synthesized many chemical compounds and characterized their properties. One of those compounds is the active ingredient in aspirin, the first drug available to patients in tablet form. The story of aspirin begins in ancient Greece where a physician named Hippocrates (460–370 BCE)—considered the father of modern medicine and memorialized in the Hippocratic oath physicians still take—recognized that something in willow tree bark was effective in treating fever, headaches, and other discomforts. The identity of that something eluded scientists and physicians for centuries. Then, in 1828, a professor of pharmacy at the University of Munich named Johann Buchner isolated a small quantity of a yellow, bitter-tasting substance from willow tree bark that had the analgesic properties described by Hippocrates. Buchner called this substance salicin.

A decade later, Italian chemist Raffaele Piria discovered that salicin consists of two bonded subunits: a sugar (the cyclic structure at lower left in the molecule below) and an aromatic unit (the structure in the upper right). The latter gave salicin its therapeutic power.

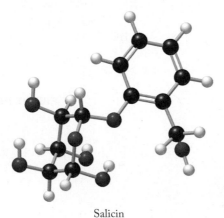

Salicin

LEARNING OUTCOMES

LO1 Write balanced chemical equations to describe chemical reactions
Sample Exercises 7.1, 7.2

LO2 Use balanced chemical equations to relate the masses of reactants and products
Sample Exercise 7.3

LO3 Interconvert the chemical formula and percent composition of a substance
Sample Exercises 7.4, 7.5, 7.6

LO4 Determine the molecular formula of a substance from its percent composition and molar mass
Sample Exercise 7.7

LO5 Use combustion analysis data to determine the empirical formula of a substance
Sample Exercise 7.8

LO6 Determine the limiting reactant in a reaction mixture
Sample Exercise 7.9

LO7 Calculate the theoretical and percent yields in a chemical reaction
Sample Exercise 7.10

Acetylene Artistry Creators of metal sculptures often use oxyacetylene torches ▶ to cut and weld pieces of metal. Acetylene reacts with oxygen, producing carbon dioxide and water.

Piria was able to separate the sugar from the therapeutic portion and then convert that portion to salicylic acid:

Salicylic acid

Salicylic acid proved to be as effective a pain reliever as salicin, but it frequently caused stomach discomfort. In 1853, the French chemist Charles Gerhardt solved the stomach-upset problem by preparing acetylsalicylic acid:

Acetylsalicylic acid

Gerhardt never marketed his new compound, but in 1899 Felix Hoffmann, a German chemist working for the Bayer pharmaceutical company, came across Gerhardt's work and prepared a sample of acetylsalicylic acid, allegedly to give to his father, who suffered from rheumatoid arthritis. The drug did indeed ease his father's pain and that of many other patients, leading Bayer to market it as Bayer *aspirin*. It is still sold today.

The theories of chemical bonding and molecular structure we explored in Chapter 5 were not developed until after 1900, but that did not stop Buchner, Piria, Gerhardt, and Hoffmann from using an extract of willow tree bark as the starting material for a series of chemical reactions that led to the synthesis of the most widely used pain reliever in the world. In this chapter we explore the fundamentals of chemical reactions, answering questions such as

➤ How can we describe the changes that occur in chemical reactions using balanced chemical equations?

➤ How can we use balanced chemical equations to relate the quantities of substances consumed and produced in chemical reactions?

➤ How can we determine a compound's elemental composition and chemical formula?

7.1 Chemical Reactions and the Conservation of Mass

John Dalton helped lay the foundation for modern chemistry in the early 1800s through his atomic theory of matter. As we noted in Chapter 1, Dalton's atomic theory explained Proust's *law of definite proportions*: compounds always contain the same proportions of their component elements. It also explained Dalton's own *law of multiple proportions*: elements combine in proportions that are ratios of small whole numbers.

Let's revisit the law of multiple proportions by considering two oxides of carbon: CO and CO_2. They are products of **combustion** reactions, which are reactions in which a fuel (carbon in this case) rapidly reacts with O_2 and in the process releases energy. During *complete* combustion, 12 grams of carbon combine with 32 grams of oxygen, forming 44 grams of CO_2. However, when there is an insufficient oxygen supply, the same 12 grams of carbon may combine with only 16 grams of oxygen, forming 28 grams of CO. The ratio of the two masses of oxygen is 32:16, or 2:1, which is indeed a ratio of two small whole numbers, in accordance with Dalton's law of multiple proportions.

The reaction between elemental carbon and oxygen that produces carbon dioxide can be represented by models of the particles involved. At the atomic level, it takes just one atom of carbon and one molecule (two atoms) of oxygen to form one molecule of carbon dioxide:

This reaction is an example of complete combustion because the CO_2 that is produced contains the highest proportion of oxygen to carbon possible. The reaction is also an example of a **combination reaction**, that is, a reaction in which two (or more) substances combine to form one **product**. In the reaction, C and O_2 are the **reactants** and CO_2 is the product.

The complete combustion of carbon can also be expressed in the form of a **chemical equation** in which we use the symbols of elements and compounds to represent the reactants and products:

$$C(s) + O_2(g) \rightarrow CO_2(g) \tag{7.1}$$

A reaction arrow links the reactants and product and shows the direction of the reaction. The symbols in parentheses after the chemical formulas indicate the physical states of the reactants and product: in this case (*s*) means solid and (*g*) means gas. In other equations we will encounter (ℓ) after liquids and (*aq*) for substances that are dissolved in water, which means they are in an *aqueous* solution.

An important feature of Equation 7.1, or any chemical equation, is that it is balanced: every atom that appears in the reactants to the left of the reaction arrow is also present in the products. Equation 7.1 has one C atom and two O atoms on the left side of the reaction arrow, and the same numbers of C and O atoms in the molecule of CO_2 on the right. The number of atoms must balance because atoms do not change their identities when reactants become products. Equality in the number of each type of atom means there is also a conservation of mass: the sum of the masses of the reactants always equals the sum of the masses of the products. This equality is known as the **law of conservation of mass**. It applies to all chemical reactions.

◉◉ **CONNECTION** Proust's law of definite proportions and Dalton's law of multiple proportions are described in Chapter 1.

combustion a rapid reaction between fuel and oxygen that produces energy.

combination reaction a reaction in which two or more substances form a single product.

chemical equation a description of the identities and proportions of **reactants** (substances consumed during a chemical reaction) and **products** (substances formed).

law of conservation of mass the principle that the sum of the masses of the reactants in a chemical reaction is equal to the sum of the masses of the products.

We can write a different chemical equation for the reaction of carbon with oxygen that produces carbon monoxide:

$$2\,C(s) + O_2(g) \rightarrow 2\,CO(g) \tag{7.2}$$

The "2" coefficients in front of C and CO indicate how many atoms of C and molecules of CO are needed to give us a balanced chemical equation. In Equation 7.2 there are 2 C atoms and 2 O atoms on the left and right sides of the reaction arrow. Balancing chemical equations is an exercise in using the appropriate coefficients to make the number of atoms of each element the same among the reactants and products. Keep in mind that we cannot balance a chemical equation by changing the subscripts in chemical formulas, which would change the identity of a reactant or product and the nature of the reaction itself.

Another example of a combination reaction takes place at the high temperatures inside a gasoline-fueled engine:

$$N_2(g) + O_2(g) \rightarrow 2\,NO(g) \tag{7.3}$$

When NO in automobile exhaust escapes into the atmosphere it may react with atmospheric O_2 in yet another combination reaction:

$$2\,NO(g) + O_2(g) \rightarrow 2\,NO_2(g) \tag{7.4}$$

Nitrogen monoxide is a colorless gas, but NO_2 is brown. If enough of it accumulates in the air over a high-traffic area, its presence can give a brown tinge to the air and sky (Figure 7.1). Under the right reaction conditions, NO_2 can combine with still more O_2 to form dinitrogen pentoxide:

$$4\,NO_2(g) + O_2(g) \rightarrow 2\,N_2O_5(g) \tag{7.5}$$

Let's take a closer look at Equations 7.3 through 7.5. Note that each is balanced because the numbers of N and O atoms on the left and right sides of the reaction arrows are the same. To achieve balance we need coefficients of 2 in front of NO and NO_2 in Equation 7.4, and we need coefficients of 4 and 2 in front of NO_2 and N_2O_5, respectively, in Equation 7.5.

FIGURE 7.1 The brown haze over a large city indicates the presence of NO_2 in the atmosphere.

Another combination reaction occurs when dinitrogen pentoxide dissolves in liquid water, forming an aqueous solution of nitric acid, HNO_3:

$$N_2O_5(g) \quad + \quad H_2O(\ell) \quad \rightarrow \quad 2\,HNO_3(aq) \qquad (7.6)$$

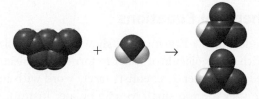

Rain that falls from an atmosphere polluted with nitrogen oxides contains trace concentrations of nitric acid and is an example of acid rain.

The atmosphere over eastern North America and much of Europe also contains sulfur oxides: SO_2 and SO_3. When SO_3 combines with water vapor in the atmosphere, the product is liquid sulfuric acid, H_2SO_4:

$$SO_3(g) \quad + \quad H_2O(g) \quad \rightarrow \quad H_2SO_4(\ell) \qquad (7.7)$$

Atmospheric H_2SO_4 may dissolve in rain to produce an aqueous solution of sulfuric acid, another form of acid rain. If this rain encounters calcium carbonate, $CaCO_3$, the following chemical reaction occurs:

$$CaCO_3(s) + H_2SO_4(aq) \rightarrow CaSO_4(aq) + H_2O(\ell) + CO_2(g) \qquad (7.8)$$

This is not a combination reaction like the previous ones because it has multiple products. One of them, calcium sulfate ($CaSO_4$), is soluble in water (unlike the reactant calcium carbonate). Thus, the reaction in Equation 7.8 can result in the loss of distinctive features of statues and other structures made of minerals such as limestone and marble, which are forms of $CaCO_3$ (Figure 7.2). The reaction is an example of *chemical weathering*, a process that is continually changing Earth's

1935

1994

FIGURE 7.2 Acidic precipitation reacts with limestone and marble statues such as this one of George Washington in New York City, converting solid calcium carbonate into soluble calcium sulfate, which is washed away by rain and melting snow.

stoichiometry the mole ratios among the reactants and products in a chemical reaction.

surface. Many reactions associated with chemical weathering involve acidic compounds in the atmosphere falling from the sky in rain or snow and dissolving minerals or, sadly, objects sculpted from those minerals.

Moles and Chemical Equations

The molecular models below Equation 7.7 show how one molecule of sulfur trioxide combines with one molecule of water to form one molecule of sulfuric acid. However, as noted in Chapter 2, chemists rarely work with individual atoms or molecules because they are too small to manipulate. Instead, chemists and students of chemistry usually work with macroscopic quantities of reactants and products, such as the mole, that they can see and measure. We can scale up Equation 7.7 using the concept of the mole and the molar masses of SO_3 (80.06 g/mol), H_2O (18.02 g/mol), and H_2SO_4 (98.08 g/mol) as follows:

1. 1 *mol* SO_3 reacts with 1 *mol* H_2O, producing 1 *mol* H_2SO_4.
2. 80.06 *g* SO_3 reacts with 18.02 *g* H_2O, producing 98.08 *g* H_2SO_4.

These microscopic and macroscopic interpretations of Equation 7.7 work for *any* number of moles (*x* mol) of SO_3 that react with *x* mol H_2O to produce *x* mol H_2SO_4, because the ratio of the coefficients of SO_3 to H_2O to H_2SO_4 in the chemical equation is 1:1:1. The mole ratio of reactants and products in a chemical equation is called the **stoichiometry** of the reaction. It allows us to calculate how much of one reactant is consumed by reacting with any quantity (*x* mol) of another reactant, or how much product can be made from a given quantity of reactant by multiplying *x* by the molar mass ($\mathcal{M}$) of the product. For the reaction in Equation 7.7, these calculations are summarized in Table 7.1. A similar table can be created based on the stoichiometry of any chemical reaction.

◉◉ CONNECTION In Chapter 2 we learned that the molecular mass of a compound is the sum of the average atomic masses of the atoms in one of its molecules, and that its molar mass ($\mathcal{M}$) has the same numerical value as its molecular mass, but expressed in grams per mole.

TABLE 7.1	Quantities of Reactants and Product in the Formation of H_2SO_4					
	$SO_3(g)$	+	$H_2O(g)$	→	$H_2SO_4(\ell)$	
Molecular Ratio	1 molecule	+	1 molecule	→	1 molecule	
Mole Ratio	1 mol	+	1 mol	→	1 mol	
Mass Ratio	80.06 g	+	18.02 g	→	98.08 g	
General Case (moles)	*x* mol	+	*x* mol	→	*x* mol	
General Case (masses)	*x*(80.06 g)	+	*x*(18.02 g)	→	*x*(98.08 g)	

CONCEPT TEST

The bright white patterns of light produced during fireworks displays are sometimes produced by the combustion of magnesium metal:

$$2\,Mg(s) + O_2(g) \rightarrow 2\,MgO(s) \tag{7.9}$$

Is the total mass of Mg and O_2 consumed equal to the mass of MgO produced? Is the mass of Mg consumed equal to the mass of O_2 consumed? Explain your two answers.

(Answers to Concept Tests are in the back of the book.)

7.2 Balancing Chemical Equations

The goal of this section is for you to hone your skills at balancing chemical equations as we explore some more chemical reactions linked to chemical weathering. Let's start with the combination reaction between SO_2 and O_2 that produces SO_3. We start by writing a preliminary chemical equation with the formulas of the reactants and product:

$$SO_2(g) + O_2(g) \rightarrow SO_3(g)$$

Taking an inventory of the atoms of S and O on each side of the reaction arrow, we discover that the S atoms are balanced (one on each side), but the O atoms are not (there are four on the left and only three on the right):

$$SO_2(g) + O_2(g) \rightarrow SO_3(g)$$

Atoms: $1\,S + 4\,O \rightarrow 1\,S + 3\,O$

The presence of an odd number of O atoms on the right is a problem because both sources of O atoms on the left contain even numbers of O atoms (2). One way to resolve the conflict is to make the number of O atoms on the right even by placing a coefficient of 2 in front of SO_3. Doing so gives us this atom inventory:

$$SO_2(g) + O_2(g) \rightarrow 2\,SO_3(g)$$

Atoms: $1\,S + 4\,O \rightarrow 2\,S + 6\,O$

Unfortunately, now both S and O are unbalanced. Since there is only one source of S atoms on the left and only one product, let's balance S first. One way to do that is by placing a coefficient of 2 in front of SO_2. Doing so gives us this atom inventory:

$$2\,SO_2(g) + O_2(g) \rightarrow 2\,SO_3(g) \qquad (7.10)$$

Atoms: $2\,S + 6\,O \rightarrow 2\,S + 6\,O$ ✓

The number of S atoms and the number of O atoms on both sides of the reaction arrow are equal, as we can confirm by counting up the atoms in the molecular models. This equality means that the equation has been successfully (✓) balanced.

Now let's solve a more challenging problem. When NO_2 dissolves in water, two water-soluble products form: HNO_3 and NO. To write a balanced chemical equation describing the reaction, let's start with a preliminary unbalanced expression consisting of one mole of each of the reactants and products:

$$NO_2(g) + H_2O(\ell) \rightarrow HNO_3(aq) + NO(aq)$$

We take an inventory of the atoms on each side of the reaction arrow:

$$1\,N + 2\,H + 3\,O \rightarrow 2\,N + 1\,H + 4\,O$$

None of the elements are balanced. To decide which element to balance first, it is often a good idea to pick the one that appears only once on each side of the reaction arrow. In this reaction there is only one reactant with H atoms and only one product with them, so let's balance H first. We can do that by placing a coefficient of 2 in front of HNO_3:

$$NO_2(g) + H_2O(\ell) \rightarrow 2\,HNO_3(aq) + NO(aq)$$

Atoms: $1\,N + 2\,H + 3\,O \rightarrow 3\,N + 2\,H + 7\,O$

Next we balance the N atoms because there is only one source of them on the left. There are 3 atoms of N on the right and only one on the left, so let's give NO_2 a coefficient of 3 and take another atom inventory:

$$3\,NO_2(g) + H_2O(\ell) \rightarrow 2\,HNO_3(aq) + NO(aq) \tag{7.11}$$

Atoms: $3\,N + 2\,H + 7\,O \rightarrow 3\,N + 2\,H + 7\,O$ ✓

▶❚❚ **CHEMTOUR** Balancing Equations

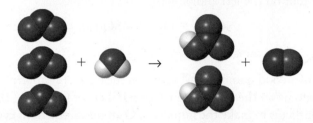

The atoms of all three elements are now balanced, and so is the equation.

There are other ways to balance chemical equations, but the approach we took with the last two reactions works for many of them. To summarize the steps we followed:

1. Write an expression using correct chemical formulas that includes one mole of the known reactants and products. Include symbols indicating physical states. Check whether the expression is already balanced. If so, your work is done.
2. If not, choose an element that appears in only one reactant and product to balance first. Insert the appropriate coefficient(s) to balance this element.
3. Choose the element that appears in the next fewest reactants and products and balance it. Repeat the process for additional elements if necessary.

SAMPLE EXERCISE 7.1   Balancing a Chemical Equation   LO1

Write a balanced chemical equation for the chemical weathering reaction between acid rain that contains sulfuric acid and the iron-containing mineral hematite (Fe_2O_3). One of the products of the reaction is water-soluble $Fe_2(SO_4)_3$; the other is water itself.

COLLECT AND ORGANIZE We are given the identities of the reactants and products. We also have information about their physical states: $Fe_2(SO_4)_3$ is soluble in water, so its symbol is (aq); H_2SO_4 is dissolved in rain, so its symbol is also (aq). A mineral is a solid substance of limited solubility in liquid water, so the symbol after Fe_2O_3 is (s).

ANALYZE To write a balanced chemical equation, we start with a preliminary expression that contains one molecule or formula unit of each reactant and product. Then we take inventories of the number of atoms of each of the elements in the reaction mixture and we balance them, starting with those that appear in only one reactant and product.

SOLVE

1. The preliminary expression relating reactants and products is:

$$Fe_2O_3(s) + H_2SO_4(aq) \rightarrow Fe_2(SO_4)_3(aq) + H_2O(\ell)$$

We take an inventory of the atoms on both sides:

Atoms: $2\,Fe + 2\,H + 1\,S + 7\,O \rightarrow 2\,Fe + 2\,H + 3\,S + 13\,O$

2. The iron atoms are already balanced, so let's focus on sulfur. There are three S atoms on the right but only one on the left. To balance sulfur, we place a coefficient of 3 in front of H_2SO_4 and recalculate the distribution of atoms on both sides:

$$Fe_2O_3(s) + 3\,H_2SO_4(aq) \rightarrow Fe_2(SO_4)_3(aq) + H_2O(\ell)$$

Atoms: $2\,Fe + 6\,H + 3\,S + 15\,O \rightarrow 2\,Fe + 2\,H + 3\,S + 13\,O$

3. Addressing the imbalance in hydrogen atoms next, there are six on the left and only two on the right. A coefficient of 3 in front of H_2O fixes this imbalance and balances the entire equation:

$$Fe_2O_3(s) + 3\,H_2SO_4(aq) \rightarrow Fe_2(SO_4)_3(aq) + 3\,H_2O(\ell) \qquad (7.12)$$

Atoms: $2\,Fe + 6\,H + 3\,S + 15\,O \rightarrow 2\,Fe + 6\,H + 3\,S + 15\,O$ ✓

THINK ABOUT IT Balancing the equation required balancing the numbers of atoms of four elements. Balancing the three that were part of only one reactant and one product automatically balanced the fourth (oxygen), which was part of all four reactants and products.

Practice Exercise Balance the chemical equation for the reaction between $P_4O_{10}(s)$ and liquid water that produces phosphoric acid, $H_3PO_4(\ell)$. ⚙

(Answers to Practice Exercises are in the back of the book.)

7.3 Combustion Reactions

Nearly one-fourth of the energy consumed in the United States comes from the combustion of natural gas, which is mostly methane, CH_4. This simplest of all alkanes produces more energy in the United States and in many other industrialized countries than any other single substance. The complete combustion of methane (or any hydrocarbon) rapidly consumes O_2 and produces CO_2 and H_2O. Let's write a balanced chemical equation describing the complete combustion of methane, starting with single molecules of each reactant and product:

$$CH_4(g) + O_2(g) \rightarrow CO_2(g) + H_2O$$

Atoms: $1\,C + 4\,H + 2\,O \rightarrow 1\,C + 2\,H + 3\,O$

Our inventory of the atoms on each side of the reaction arrow reveals that the number of C atoms is already balanced, but H and O are not. There are H atoms in only one reactant and product, so let's balance it next. There are four H atoms in the molecule of methane on the left and only two in the molecule of water on the right. To remedy this imbalance we place a coefficient of 2 in front of H_2O:

$$CH_4(g) + O_2(g) \rightarrow CO_2(g) + 2\,H_2O$$

Atoms: $1\,C + 4\,H + 2\,O \rightarrow 1\,C + 4\,H + 4\,O$

Now C and H are balanced but O is not: there are four atoms of O on the right but only two on the left. In this chemical equation, balancing C and H did not automatically balance O because there is a source of O atoms, O_2, that does not contain C or H atoms. The good news is that we can increase the number of O atoms on the left in multiples of two by increasing the number of O_2 molecules in the equation, and doing so does not change the numbers of C and H atoms. So, we place a coefficient of 2 in front of O_2 to get four atoms of O on both sides of the reaction arrow and a balanced chemical equation:

$$CH_4(g) + \mathbf{2}\,O_2(g) \rightarrow CO_2(g) + 2\,H_2O$$

$$\text{Atoms:} \quad 1\,C + 4\,H + 4\,O \rightarrow 1\,C + 4\,H + 4\,O \; \checkmark$$

Did you notice that we did not assign a physical state to H_2O in the balanced chemical equation? There is a reason for this omission: we did not specify temperature. It might have been reasonable to assume that, in the hot flame produced by the combustion of methane, the water would exist as water vapor. Under these conditions the complete version of the chemical equation would be

$$CH_4(g) + 2\,O_2(g) \rightarrow CO_2(g) + 2\,H_2O(g) \qquad (7.13)$$

However, the physical states of the products of combustion reactions are sometimes written assuming that the reaction mixture has cooled to room temperature. In this case, the complete reaction equation is

$$CH_4(g) + 2\,O_2(g) \rightarrow CO_2(g) + 2\,H_2O(\ell) \qquad (7.14)$$

This may seem like an unimportant detail. However, the physical states of reactants and products influence how much energy is released or absorbed in a chemical reaction. We will investigate this phenomenon in detail in Chapter 9. For now, be assured that expressing the right physical states in chemical equations is important.

If we could peer inside the blue flame (Figure 7.3) produced by complete combustion of methane and observe the molecules that are formed during the reaction, we would discover that the reactants are not transformed into final products all at once. Instead the reaction happens in stages. In an early stage, molecules of CH_4 split apart into molecular fragments that combine with oxygen, forming a mixture of carbon monoxide, hydrogen gas and water vapor:

$$CH_4(g) + O_2(g) \rightarrow CO(g) + H_2(g) + H_2O(g) \qquad (7.15)$$

Note that this reaction is balanced as written. Next, the hydrogen formed in the reaction combines with more oxygen to form more water vapor. Writing single molecules of these reactants and products in a preliminary reaction expression gives us

$$H_2(g) + O_2(g) \rightarrow H_2O(g)$$

$$\text{Atoms:} \quad 2\,H + 2\,O \rightarrow 2\,H + 1\,O$$

FIGURE 7.3 Complete combustion of natural gas (mostly CH_4) produces a characteristic blue flame due to emission from excited-state molecular fragments such as CH.

Our inventory discloses that the H atoms are in balance but the O atoms are not. We need at least two O atoms on the right side because their source is diatomic molecules of O_2 on the left. Placing a coefficient of 2 in front of H_2O resolves the O imbalance but creates a new imbalance in H atoms:

$$H_2(g) + O_2(g) \rightarrow 2\, H_2O(g)$$

Atoms: $2\,H + 2\,O \rightarrow 4\,H + 2\,O$

We can balance the number of H atoms by placing a coefficient of 2 in front of H_2, which gives us a balanced chemical equation:

$$2\,H_2(g) + O_2(g) \rightarrow 2\,H_2O(g) \tag{7.16}$$

Atoms: $4\,H + 2\,O \rightarrow 4\,H + 2\,O$ ✓

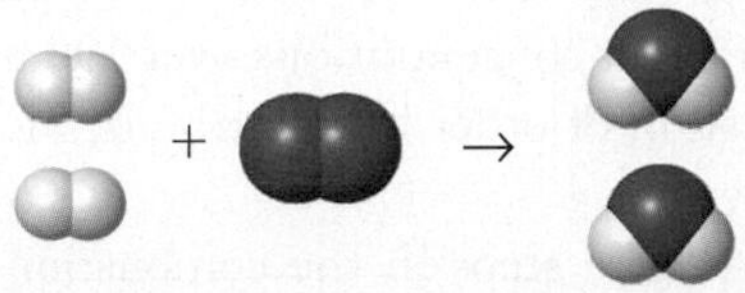

Next, the carbon monoxide formed in the first step of the process (Equation 7.15) combines with more O_2, forming CO_2. Writing single molecules of these reactants and product in a preliminary reaction expression gives us

$$CO(g) + O_2(g) \rightarrow CO_2(g)$$

Atoms: $1\,C + 3\,O \rightarrow 1\,C + 2\,O$

This inventory shows that the C atoms are in balance but the O atoms are not. Balancing the O atoms is complicated by the fact that the number on the right side must be an even number because there are two O atoms in each molecule of CO_2. Putting a coefficient of 2 in front of CO_2 in an attempt to increase the number of atoms on the right gives us one too many on the right and one too many C atoms as well:

$$CO(g) + O_2(g) \rightarrow 2\,CO_2(g)$$

Atoms: $1\,C + 3\,O \rightarrow 2\,C + 4\,O$

Giving CO_2 a coefficient of 2 may not seem like a smart move, but actually it is because we can now simultaneously bring both C and O into balance by adding a coefficient of 2 in front of CO:

$$2\,CO(g) + O_2(g) \rightarrow 2\,CO_2(g) \tag{7.17}$$

Atoms: $2\,C + 4\,O \rightarrow 2\,C + 4\,O$ ✓

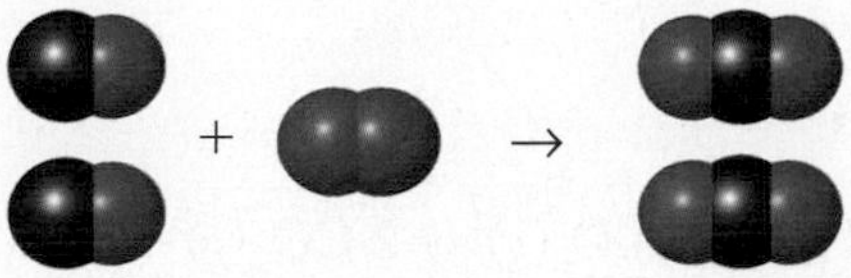

Writing Equation 7.17 illustrates an important feature of balancing chemical equations. The process is sometimes like playing a game of chess or solving a Rubik's Cube: you have to plan ahead, using one step to set up the next, even when it produces an expression that seems less balanced than the one that came before it.

We have just developed three balanced chemical equations that describe the stepwise combustion of methane:

Step 1 $\qquad CH_4(g) + O_2(g) \rightarrow CO(g) + H_2(g) + H_2O(g)$ $\qquad$ (7.15)

Step 2 $\qquad 2\,H_2(g) + O_2(g) \rightarrow 2\,H_2O(g)$ $\qquad$ (7.16)

Step 3 $\qquad 2\,CO(g) + O_2(g) \rightarrow 2\,CO_2(g)$ $\qquad$ (7.17)

All three reactions occur rapidly in a flame, though the rate of the last one, conversion of CO to CO_2, is not quite as rapid as the other two. Inevitably, some CO escapes from the reaction mixture (the flame) before the third reaction is complete. This is a hazardous situation because CO is toxic, producing nausea and headaches at concentrations between 10 and 100 parts per million (ppm) in the air, and CO can be fatal at concentrations over 100 ppm. Formation of CO is a concern in the combustion of other hydrocarbons, including gasoline in automobile engines.

We can combine the three steps in the combustion of methane (Equations 7.15–7.17) to produce the overall reaction (Equation 7.13). Doing so is an exercise in balancing the numbers of molecules of substances that are products of early steps and reactants in later steps so that they cancel out when we add the reaction equations. For instance, H_2 is a product of step 1 and a reactant in step 2, and CO is a product of step 1 and a reactant in step 3. Before we combine the three steps we need to balance the number of molecules of H_2 and CO in these steps. One way to do that is by multiplying all the coefficients in step 1 by two. This gives us two molecules of H_2 and CO on the product side, which balance the two molecules of H_2 and CO on the reactant sides of steps 2 and 3, respectively. So, multiplying step 1 by a factor of 2 and then adding it to steps 2 and 3 yields

$2 \times$ Step 1 $\qquad 2\,CH_4(g) + 2\,O_2(g) \rightarrow 2\,CO(g) + 2\,H_2(g) + 2\,H_2O(g)$

$+$ Step 2 $\qquad 2\,H_2(g) + O_2(g) \rightarrow 2\,H_2O(g)$

$+$ Step 3 $\qquad 2\,CO(g) + O_2(g) \rightarrow 2\,CO_2(g)$

$$2\,CH_4(g) + 2\,H_2(g) + 2\,CO(g) + 4\,O_2(g) \rightarrow$$
$$2\,CO(g) + 2\,H_2(g) + 2\,CO_2(g) + 4\,H_2O(g)$$

We cancel out the terms that appear on both sides of the reaction arrow:

$$2\,CH_4(g) + \cancel{2\,H_2(g)} + \cancel{2\,CO(g)} + 4\,O_2(g) \rightarrow$$
$$\cancel{2\,CO(g)} + \cancel{2\,H_2(g)} + 2\,CO_2(g) + 4\,H_2O(g)$$

or

$$2\,CH_4(g) + 4\,O_2(g) \rightarrow 2\,CO_2(g) + 4\,H_2O(g)$$

We can simplify this chemical equation by dividing all coefficients by 2:

$$CH_4(g) + 2\,O_2(g) \rightarrow CO_2(g) + 2\,H_2O(g)$$

to obtain Equation 7.13 describing the overall reaction.

SAMPLE EXERCISE 7.2 **Writing the Balanced Equation** **LO1**
for a Combustion Reaction

Methane is the principal ingredient in natural gas, but ethane, C_2H_6, is also present. Write a balanced chemical equation describing the complete combustion of C_2H_6.

COLLECT AND ORGANIZE We need to write a balanced chemical equation describing the complete combustion of ethane. Combustion reactions involve the rapid reaction of fuels (C_2H_6 in this case) with O_2. The products of the complete combustion of hydrocarbons such as ethane are CO_2 and H_2O.

ANALYZE The first step involves writing a reaction expression with single molecules of the reactants and products. Ethane and O_2 are the reactants and CO_2 and H_2O are the products. At the high temperature of a combustion flame all four ingredients in the reaction are gases. Next we balance the number of C and H atoms because they are each present in only one reactant and one product, and we finish by balancing the total number of O atoms on each side of the reaction arrow.

SOLVE The preliminary reaction expression and its atom inventory are

$$C_2H_6(g) + O_2(g) \rightarrow CO_2(g) + H_2O(g)$$

Atoms: $2\,C + 6\,H + 2\,O \rightarrow 1\,C + 2\,H + 3\,O$

We balance C atoms first by placing a coefficient of 2 on CO_2:

$$C_2H_6(g) + O_2(g) \rightarrow \mathbf{2}\,CO_2(g) + H_2O(g)$$

Atoms: $2\,C + 6\,H + 2\,O \rightarrow 2\,C + 2\,H + 5\,O$

and then balance H atoms by placing a coefficient of 3 on H_2O:

$$C_2H_6(g) + O_2(g) \rightarrow 2\,CO_2(g) + \mathbf{3}\,H_2O(g)$$

Atoms: $2\,C + 6\,H + 2\,O \rightarrow 2\,C + 6\,H + 7\,O$

This gives us an expression with an odd number of O atoms on the right side of the reaction arrow. The only source of O atoms is O_2, so we need an even number of O atoms on the right. One way to get an even number without disrupting the balance of H and O atoms is by multiplying all the coefficients in the expression by 2:

$$\mathbf{2}\,C_2H_6(g) + \mathbf{2}\,O_2(g) \rightarrow \mathbf{4}\,CO_2(g) + \mathbf{6}\,H_2O(g)$$

Atoms: $4\,C + 12\,H + 4\,O \rightarrow 4\,C + 12\,H + 14\,O$

Now we can balance the number of O atoms by replacing the coefficient of 2 in front of O_2 with 7, which gives us 5 more O_2 molecules, or 10 more O atoms, and a balanced equation:

$$2\,C_2H_6(g) + \mathbf{7}\,O_2(g) \rightarrow 4\,CO_2(g) + 6\,H_2O(g)$$

Atoms: $4\,C + 12\,H + 14\,O \rightarrow 4\,C + 12\,H + 14\,O$ ✓

THINK ABOUT IT Writing a balanced chemical equation for the combustion of C_2H_6 required one more step (to get an even number of O atoms on both sides of the reaction arrow) than writing the chemical equation describing the combustion of CH_4. This step was needed because we had an odd number of molecules of H_2O on the right side of the reaction arrow after balancing the number of H atoms. Can you see how this will happen every time the number of H atoms per molecule of a hydrocarbon fuel *is not divisible by 4*? For another illustration, try this Practice Exercise.

Practice Exercise Write the chemical equation describing the complete combustion of butane (C_4H_{10}), the fuel in disposable lighters. ⚙

7.4 Stoichiometric Calculations and the Carbon Cycle

The composition of Earth's atmosphere began to change dramatically about 2.5 billion years ago. Before then there was little atmospheric O_2, but organisms had evolved that could perform *photosynthesis*, harnessing the energy of sunlight to drive biochemical processes. There are several types of photosynthesis and many chemical steps in each type. The most familiar type—the one that occurs in green plants today—consumes two reactants, atmospheric CO_2 and liquid H_2O, and forms two products, glucose ($C_6H_{12}O_6$) and O_2 gas:

$$6\ CO_2(g) + 6\ H_2O(\ell) \rightarrow C_6H_{12}O_6(aq) + 6\ O_2(g) \tag{7.18}$$

The reverse reaction, *respiration*, is a principal source of energy for most living things on Earth:

$$C_6H_{12}O_6(aq) + 6\ O_2(g) \rightarrow 6\ CO_2(g) + 6\ H_2O(\ell) \tag{7.19}$$

Photosynthesis and respiration are key reactions in the *carbon cycle* (Figure 7.4). The two processes are nearly, but not exactly, in balance in Earth's biosphere. Each year about 0.01% of the decaying mass of dead plants and animals (called *detritus*) is incorporated into sediments and soils. Shielded from atmospheric O_2, this buried carbon is not converted back into CO_2, at least not right away.

FIGURE 7.4 The carbon cycle. ① Green plants and marine plants incorporate CO_2 into their biomass. ② Some of the plant biomass becomes the biomass of animals. ③ As plants and animals respire, they release CO_2 back into the environment. ④ When they die, the decay of their tissues releases most of their carbon content as CO_2, but about 0.01% is incorporated into ⑤ carbonate minerals and deposits of coal, petroleum, and natural gas (fossil fuels). ⑥ Mining and the combustion of fossil fuels for human use are shifting the natural equilibrium that has controlled the concentration of CO_2 in the atmosphere.

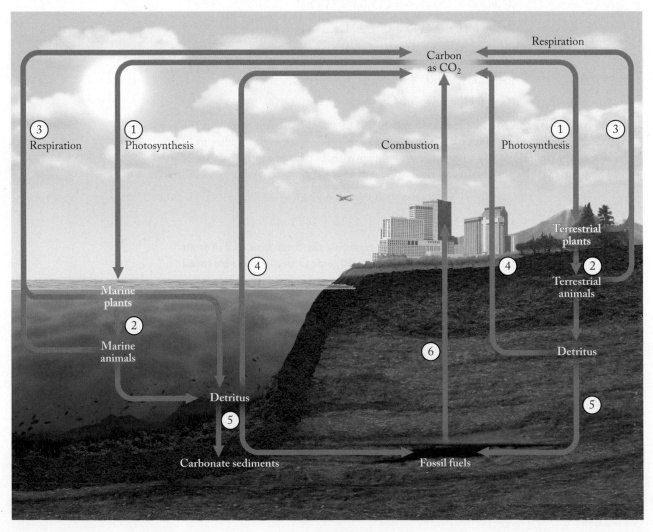

Although 0.01% may not seem like much, over hundreds of millions of years it has added up to the removal of about 10^{20} kilograms of carbon dioxide from the atmosphere. About 10^{15} kilograms of the buried carbon is in the form of fossil fuels: coal, petroleum, and natural gas.

Fossil fuels are now the primary source of energy in industrialized countries around the world. Each year approximately 8.2 trillion (8.2×10^{12}) kilograms of fossil carbon are added to the atmosphere as CO_2. As a result, the concentration of CO_2 in the atmosphere has increased sharply, especially over the past 50 years (Figure 7.5). Today it is approaching 400 ppm. To put this value in perspective, atmospheric CO_2 concentrations have oscillated between 190 and 280 ppm for the past 800,000 years and probably longer (Figure 7.6). The concentration of CO_2 in the atmosphere today is more than 40% higher than at any time between our evolution as a species and the beginning of the Industrial Revolution (~1750–1850).

CONNECTION The impact of increasing CO_2 concentration on global climate has been the subject of considerable debate, as we discussed in Chapter 4.

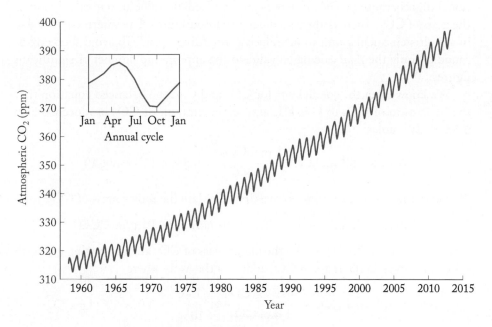

FIGURE 7.5 Increasing concentrations of CO_2 in the atmosphere. This graph is based on analyses done since 1958 at the Mauna Loa Observatory in Hawaii and is called the Keeling curve to honor Charles David Keeling (1928–2005), who began determining atmospheric CO_2 concentrations long before climate change was a widespread concern. The annual cycle in the data is the result of the different rates of photosynthesis by trees and other vegetation in the Northern Hemisphere during summer and winter months.

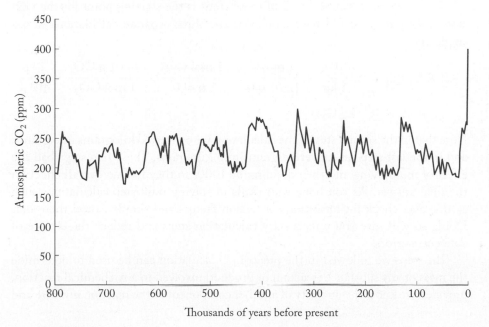

FIGURE 7.6 History of atmospheric concentration of CO_2. Analyses of bubbles of air trapped in Antarctic glaciers have enabled scientists to create a history of atmospheric CO_2 levels over the past 800,000 years. The data show that atmospheric concentrations of CO_2 oscillated between about 190 and 280 ppm. Low CO_2 concentrations coincide with ice ages in the Northern Hemisphere. High CO_2 concentrations correspond to relatively warm periods called *interglacials*, including the one we are now experiencing. The most recent data are based on the Keeling curve in Figure 7.5.

If the combustion of fossil fuels adds 8.2×10^{12} kilograms of carbon to the atmosphere each year as CO_2, what is the mass of added carbon dioxide? The CO_2 mass must be more than 8.2×10^{12} kg, because this amount represents only the mass due to carbon. The balanced chemical equation for the complete combustion of carbon is

$$C(s) + O_2(g) \rightarrow CO_2(g)$$

To use this equation to determine the mass of CO_2 released, we first convert mass of C into moles of C:

$$8.2 \times 10^{12} \text{ kg C} \times \frac{10^3 \text{ g}}{1 \text{ kg}} \times \frac{1 \text{ mol C}}{12.01 \text{ g C}} = 6.83 \times 10^{14} \text{ mol C}$$

Note that the "moles of C" value is expressed with three significant figures even though the mass of C is known to only two significant figures. We use an additional digit because our result here is an intermediate value in our calculation of the mass of CO_2. There is always a danger that rounding off an intermediate value in a multistep calculation can introduce a "rounding error." To avoid such errors, round off only the *final* calculated value to the appropriate number of significant figures.

We know from the coefficients for CO_2 and C in the balanced equation that the mole ratio of CO_2 to C is 1:1, and so the amount of CO_2 produced is also 6.83×10^{14} moles:

▶‖ **CHEMTOUR** Carbon Cycle

$$6.83 \times 10^{14} \text{ mol C} \times \frac{1 \text{ mol CO}_2}{1 \text{ mol C}} = 6.83 \times 10^{14} \text{ mol CO}_2$$

To convert moles of CO_2 to mass, we first calculate the molar mass of CO_2:

$$\mathcal{M}_{CO_2} = 12.01 \text{ g/mol} + 2(16.00 \text{ g/mol}) = 44.01 \text{ g/mol CO}_2$$

Multiplying the moles of CO_2 by the molar mass of CO_2 and converting grams to kilograms gives us the mass of CO_2 added to the atmosphere:

$$6.83 \times 10^{14} \text{ mol CO}_2 \times \frac{44.01 \text{ g CO}_2}{1 \text{ mol CO}_2} \times \frac{1 \text{ kg}}{10^3 \text{ g}} = 3.0 \times 10^{13} \text{ kg CO}_2$$

Note that the answer in each of these steps is the starting point for the next step. Therefore we could have combined the three separate calculations into a single one:

$$8.2 \times 10^{12} \text{ kg C} \times \frac{10^3 \text{ g}}{1 \text{ kg}} \times \frac{1 \text{ mol C}}{12.01 \text{ g C}} \times \frac{1 \text{ mol CO}_2}{1 \text{ mol C}} \times \frac{44.01 \text{ g CO}_2}{1 \text{ mol CO}_2} \times \frac{1 \text{ kg}}{10^3 \text{ g}}$$

$$= 3.0 \times 10^{13} \text{ kg CO}_2$$

Note that in the second step of the calculation we converted kilograms into grams, and in the last step we converted grams into kilograms. These two steps had the effect of multiplying and then dividing by 1000, so they had no overall effect on the final answer. As you hone your skills in solving multistep calculations such as this one, check for offsetting conversion factors and simply cancel them out. Doing so will save you unnecessary calculation steps and reduce the chance of data entry errors.

The steps we followed in the preceding calculation can be used to determine the mass of any substance (reactant or product) involved in any chemical reaction, provided we know the quantity of another component of the reaction mixture and the *stoichiometric relation* between the two substances—that is, their mole ratio in the balanced chemical equation describing the reaction. For example, the chemical reaction that produces aspirin from salicylic acid, which we discussed at the beginning of this chapter, is described by Equation 7.20:

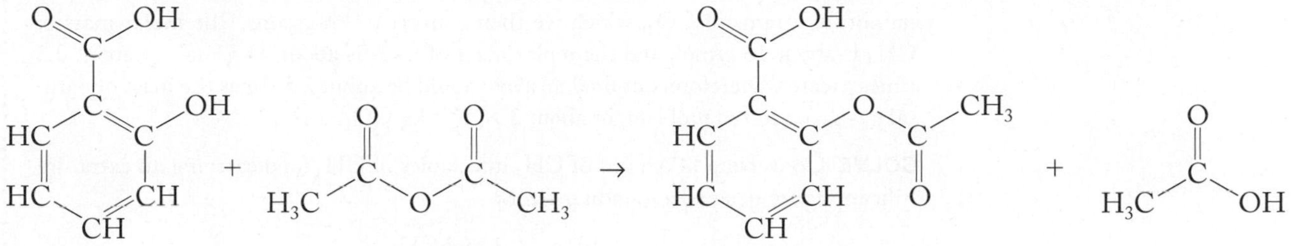

| Salicylic acid | Acetic anhydride | Acetylsalicylic acid | Acetic acid | (7.20) |

Suppose we wished to prepare 1.00 kilogram of acetylsalicylic acid ($C_9H_8O_4$, $\mathcal{M} = 180.16$ g/mol). How many grams of salicylic acid ($C_7H_6O_3$, $\mathcal{M} = 138.12$ g/mol) and how many grams of acetic anhydride ($C_4H_6O_3$, $\mathcal{M} = 102.09$ g/mol) would we need? To answer these questions we must convert 1.00 kg $C_9H_8O_4$ into an equivalent number of moles of $C_9H_8O_4$. This step is essential because the chemical equation tells us that one *mole* of salicylic acid reacts with one *mole* of acetic anhydride to produce one *mole* of acetylsalicylic acid. The 1:1:1 mole ratio means that once we convert 1.00 kg $C_9H_8O_4$ into moles of $C_9H_8O_4$, we will know how many moles of the two reactants we need. Multiplying this number of moles by the appropriate molar masses will give us the masses of the two reactants. The details of the unit conversions involved in these calculations are shown in the following two equations:

$$1.00 \text{ kg } C_9H_8O_4 \times \frac{10^3 \text{ g}}{1 \text{ kg}} \times \frac{1 \text{ mol } C_9H_8O_4}{180.16 \text{ g } C_9H_8O_4} \times \frac{1 \text{ mol } C_7H_6O_3}{1 \text{ mol } C_9H_8O_4}$$

$$\times \frac{138.12 \text{ g } C_7H_6O_3}{1 \text{ mol } C_7H_6O_3} = 767 \text{ g } C_7H_6O_3$$

$$1.00 \text{ kg } C_9H_8O_4 \times \frac{10^3 \text{ g}}{1 \text{ kg}} \times \frac{1 \text{ mol } C_9H_8O_4}{180.16 \text{ g } C_9H_8O_4} \times \frac{1 \text{ mol } C_4H_6O_3}{1 \text{ mol } C_9H_8O_4}$$

$$\times \frac{102.09 \text{ g } C_4H_6O_3}{1 \text{ mol } C_4H_6O_3} = 567 \text{ g } C_4H_6O_3$$

SAMPLE EXERCISE 7.3 **Calculating the Mass of a Product from the Mass of a Reactant** **LO2**

Each year, power plants in the United States consume about 1.1×10^{11} kilograms of natural gas. How many kilograms of CO_2 ($\mathcal{M} = 44.01$ g/mol) are released into the atmosphere from these power plants? Given that natural gas is mostly methane (CH_4, $\mathcal{M} = 16.04$ g/mol), base the calculation on its combustion reaction, Equation 7.13:

$$CH_4(g) + 2 \, O_2(g) \rightarrow CO_2(g) + 2 \, H_2O(g)$$

COLLECT AND ORGANIZE We are asked to calculate the mass of CO_2 released by the combustion of 1.1×10^{11} kg of CH_4 (our model compound for natural gas). We are given the molar masses of the two compounds and the balanced chemical equation relating the moles of CO_2 produced to the moles of CH_4 consumed.

ANALYZE The mole ratio of CO_2 to CH_4 in the chemical equation is 2:2, or 1:1. To use it, we must first convert the mass of CH_4 to moles of CH_4. That conversion involves dividing a mass in grams by the molar mass, so we will also need to convert kilograms to grams. The number of moles of CH_4 consumed is also the number of moles of CO_2 produced. When we multiply it by the molar mass of CO_2 we get the number of grams of CO_2, which we then convert to kilograms. The molar mass of CH_4 is about 16 g/mol, and the molar mass of CO_2 is about 44 g/mol, or about 2.5 times greater. Therefore, our final answer should be about 2.5 times the mass of natural gas given in the problem, or about 3×10^{11} kg CO_2.

SOLVE Converting kilograms of CH_4 into moles of CH_4 (and carrying an extra significant figure to avoid a rounding error):

$$1.1 \times 10^{11} \; \text{kg } CH_4 \times \frac{10^3 \; \text{g}}{1 \; \text{kg}} \times \frac{1 \; \text{mol } CH_4}{16.04 \; \text{g } CH_4} = 6.86 \times 10^{12} \; \text{mol } CH_4$$

Converting moles of CH_4 into moles of CO_2:

$$6.86 \times 10^{12} \; \text{mol } CH_4 \times \frac{1 \; \text{mol } CO_2}{1 \; \text{mol } CH_4} = 6.86 \times 10^{12} \; \text{mol } CO_2$$

Converting moles of CO_2 into kilograms of CO_2:

$$6.86 \times 10^{12} \; \text{mol } CO_2 \times \frac{44.01 \; \text{g } CO_2}{1 \; \text{mol } CO_2} \times \frac{1 \; \text{kg}}{10^3 \; \text{g}} = 3.0 \times 10^{11} \; \text{kg } CO_2$$

We can combine the three separate calculations into a single calculation, and in subsequent problems we will do this routinely and not show the individual steps:

$$1.1 \times 10^{11} \; \text{kg } CH_4 \times \frac{10^3 \; \text{g}}{1 \; \text{kg}} \times \frac{1 \; \text{mol } CH_4}{16.04 \; \text{g } CH_4} \times \frac{1 \; \text{mol } CO_2}{1 \; \text{mol } CH_4} \times \frac{44.01 \; \text{g } CO_2}{1 \; \text{mol } CO_2} \times \frac{1 \; \text{kg}}{10^3 \; \text{g}}$$

$$= 3.0 \times 10^{11} \; \text{kg } CO_2$$

THINK ABOUT IT The mass of CO_2 produced is close to our estimated value. It represents 1% of the fossil carbon that is estimated to be entering the atmosphere as CO_2 each year.

Practice Exercise How many grams of O_2 are consumed during the complete combustion of 1.00 g of butane, C_4H_{10}? 🔘

7.5 Percent Composition and Empirical Formulas

▶❚❚ **CHEMTOUR** Percent Composition

During the Industrial Revolution, access to geological deposits of iron oxides was a key to economic development. The more accessible an iron-containing mineral was and the higher its iron content, the more valuable it was. There are several ways to express the elemental content of substances in mixtures. A popular one is **percent composition**: the percentage by mass of the constituent elements of a compound with respect to its total mass. Let's consider two iron-containing minerals: wustite, FeO, and hematite, Fe_2O_3. Which has the higher iron content, and what is the iron content value expressed as a percent composition?

One way to answer the first question is to examine the mole ratios of Fe to O, which are 1:1 in FeO and 2:3 (or 1 : 1.5) in Fe_2O_3. There are 1.5 atoms of O for every atom of Fe in hematite, but only 1 atom of O per atom of Fe in wustite, which means there is a higher proportion of O, and hence a lower proportion of Fe, in Fe_2O_3. Therefore, FeO must have the higher iron content.

To find the percent composition of FeO, let's assume that we have 1 mole of it. If we do, then we have 1 mole of Fe and 1 mole of O. We can convert the

percent composition the composition of a compound expressed in terms of the percentage by mass of each element in the compound.

number of moles to equivalent masses using the molar masses of Fe (55.85 g/mol) and O (16.00 g/mol):

$$\left(1 \text{ mol Fe} \times \frac{55.85 \text{ g}}{1 \text{ mol}}\right) + \left(1 \text{ mol O} \times \frac{16.00 \text{ g}}{1 \text{ mol}}\right) = 71.85 \text{ g FeO}$$

Of this 71.85 g, Fe accounts for 55.85 g. Therefore, the Fe content of FeO is

$$\frac{\text{mass of Fe}}{\text{total mass}} = \frac{55.85 \text{ g Fe}}{71.85 \text{ g FeO}} = 0.7773 \quad \text{or} \quad 77.73\% \text{ Fe}$$

It follows that the oxygen content is

$$\frac{\text{mass of O}}{\text{total mass}} = \frac{16.00 \text{ g O}}{71.85 \text{ g FeO}} = 0.2227 \quad \text{or} \quad 22.27\%$$

Because FeO contains only Fe and O, we could also have determined the percentage of O by subtracting the percentage of Fe from 100%:

$$100.00\% - 77.73\% = 22.27\%$$

SAMPLE EXERCISE 7.4 **Calculating Percent Composition** **LO3**
from a Chemical Formula

What is the percent composition of the mineral forsterite, Mg_2SiO_4 (Figure 7.7)?

COLLECT AND ORGANIZE We are given the chemical formula of forsterite and asked to calculate its percent composition, which is its composition expressed in terms of the percentages by mass of its component elements. The chemical formula tells us the number of moles of each element in a mole of a compound.

ANALYZE A convenient way to start solving this problem is to assume that we have one mole of the mineral. Its formula tells us how many moles of each of its constituent elements we have. Multiplying each of these values by the element's molar mass gives us how many grams of each element there are in one mole of forsterite, and the sum is the molar mass of forsterite. Dividing each element's total mass by the molar mass of forsterite and expressing the results as percentages gives the percent composition of the mineral. We can predict that oxygen will have the highest percentage by mass because, even though O has the smallest molar mass of the three elements, there are four moles of O for every two moles of Mg and one mole of Si.

SOLVE The mass of each element in one mole of Mg_2SiO_4 and the molar mass of Mg_2SiO_4 are

$$2 \text{ mol Mg} \times 24.31 \text{ g/mol} = 48.62 \text{ g Mg}$$
$$+ \; 1 \text{ mol Si} \times 28.09 \text{ g/mol} = 28.09 \text{ g Si}$$
$$\underline{+ \; 4 \text{ mol O} \times 16.00 \text{ g/mol} = 64.00 \text{ g O}}$$
$$\text{Molar mass of } Mg_2SiO_4 = 140.71 \text{ g/mol}$$

Calculating percent composition:

$$\%Mg = \frac{48.62 \text{ g Mg}}{140.71 \text{ g}} \times 100\% = 34.55\% \text{ Mg}$$

$$\%Si = \frac{28.09 \text{ g Si}}{140.71 \text{ g}} \times 100\% = 19.96\% \text{ Si}$$

$$\%O = \frac{64.00 \text{ g O}}{140.71 \text{ g}} \times 100\% = 45.48\% \text{ O}$$

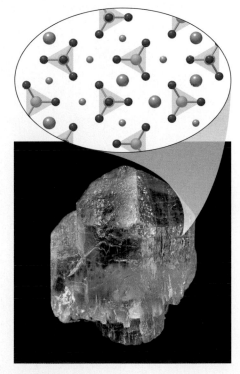

FIGURE 7.7 Crystals of pure Mg_2SiO_4 are colorless, but sometimes they contain impurities such as Fe^{2+} ions (in place of Mg^{2+} ions) that make the crystals green. The greater the Fe^{2+} content, the deeper the green color. The green crystals are made of three-dimensional arrays of silicate (SiO_4^{4-}) tetrahedra with Mg^{2+} (small, light gray spheres) and Fe^{2+} (larger gray spheres) ions in the spaces between the tetrahedra.

THINK ABOUT IT As we predicted, oxygen accounts for the highest percentage—nearly half the mass of forsterite. It also makes sense that the percentage of the mass of forsterite due to magnesium is nearly twice the percentage due to silicon, because the molar masses of Mg and Si are not that different and there are two moles of Mg for every one mole of Si.

Practice Exercise What is the percent composition of the antibiotic tetracycline, which has the molecular formula $C_{22}H_{24}N_2O_8$? ⚙

■ ....

Note that the three percentages in Sample Exercise 7.4 sum to 99.99%. Percent composition values should always sum to 100%, or very close to it, if we have accounted for all the elements in the compound and done the math correctly.

We typically determine the percent composition of a substance in the laboratory by measuring the amount of each element in a given mass of the substance. We can use these data to determine the substance's **empirical formula**, which is a formula based on the lowest whole-number ratio of its component elements. The formulas of nearly all ionic compounds *are* empirical formulas because they are based on formula units, which contain the fewest numbers of cations and anions that together have no overall charge. The word *empirical* is synonymous with *experimental*, meaning "derived from experimental data." Its use here is appropriate because, as noted above, empirical formulas are often derived from percent composition data that are obtained through chemical analyses.

To derive an empirical formula from percent composition data, we first assume that we have exactly 100 grams of the compound. This way, the percent composition values become the masses of each element expressed in grams. Suppose, for example, the results of an elemental analysis of a sample of an iron oxide show that it is 69.94% Fe and 30.06% O. In exactly 100 grams of the sample there would be 69.94 g Fe and 30.06 g O. Next, we convert these masses into equivalent numbers of moles by dividing by the appropriate molar masses:

$$69.94 \; \cancel{\text{g Fe}} \times \frac{1 \text{ mol Fe}}{55.85 \; \cancel{\text{g Fe}}} = 1.252 \text{ mol Fe}$$

and

$$30.06 \; \cancel{\text{g O}} \times \frac{1 \text{ mol O}}{16.00 \; \cancel{\text{g O}}} = 1.879 \text{ mol O}$$

To convert these values to a ratio of small whole numbers, we divide both by the smaller value:

$$\frac{1.879 \text{ mol O}}{1.252} = 1.501 \text{ mol O} \approx 3/2 \text{ mol O}$$

$$\frac{1.252 \text{ mol Fe}}{1.252} = 1.000 \text{ mol Fe}$$

To transform this mole ratio into whole numbers that will be the subscripts in an empirical formula, we need to recognize that 1.501 is really close to $1\frac{1}{2}$, or $\frac{3}{2}$. If we multiply each mole value by 2, we then have 2 moles of Fe and 3 moles of O, which translate into the empirical formula Fe_2O_3.

👓 **CONNECTION** We introduced the concept of formula units in Section 2.4. We used the term "empirical" to describe Balmer's and Rydberg's equations in Section 3.4.

empirical formula a formula showing the smallest whole-number ratio of elements in a compound.

Which pairs of the following compounds have the same empirical formula?

    a. Ethylene (C_2H_4), a gas used to ripen green tomatoes.
    b. Eicosene ($C_{20}H_{40}$), a compound used to trap Japanese beetles.
    c. Acetylene (C_2H_2), a gas used in torches for cutting steel and other metals.
    d. Benzene (C_6H_6), a carcinogenic compound found in petroleum.

SAMPLE EXERCISE 7.5   **Deriving an Empirical Formula**   **LO3**
                               from Percent Composition

The mineral magnetite (Figure 7.8) is an oxide of iron that is 72.36% iron by mass. What is its empirical formula?

COLLECT AND ORGANIZE We are given the iron content of magnetite and asked to determine its empirical formula, that is, the simplest ratio of iron ions to oxide ions in the mineral.

ANALYZE Because the only elements in the mineral are iron and oxygen, the percent composition by mass of the oxygen may be obtained by subtracting the percent Fe from 100%. We can use the steps described in the preceding discussion to:

    1. Convert percentage values into mass values in grams by assuming we have exactly 100 grams of magnetite.
    2. Convert mass values into mole values by dividing by the molar masses of Fe and O.
    3. Calculate a mole ratio by dividing both by the smaller of the two values.
    4. Convert the mole ratio from step 3 into small whole numbers, if necessary.

SOLVE

    1. In a 100 g sample there are 72.36 g Fe and there are

$$100.00 - 72.36 \text{ g Fe} = 27.64 \text{ g O}$$

    2. Converting grams into moles:

$$72.36 \text{ g Fe} \times \frac{1 \text{ mol Fe}}{55.85 \text{ g Fe}} = 1.296 \text{ mol Fe}$$

    and

$$27.64 \text{ g O} \times \frac{1 \text{ mol O}}{16.00 \text{ g O}} = 1.728 \text{ mol O}$$

    3. Calculating a mole ratio of O to Fe:

$$\frac{1.728 \text{ mol O}}{1.296 \text{ mol Fe}} = 1.333 \text{ mol O / mol Fe}$$

    4. To simplify this ratio, we need to recognize that 1.333 is the decimal equivalent of the improper fraction 4/3, which means the mole ratio of O to Fe is 4/3:1, or simply 4:3. The corresponding empirical formula is Fe_3O_4.

THINK ABOUT IT As we have seen again in this exercise, a key to deriving an empirical formula from percent composition data is translating mole ratios that are decimal values greater than 1 into equivalent improper fractions. Perhaps you are wondering how Fe_3O_4 can be a legitimate formula when, as we discussed in Chapter 2, iron occurs in nature as Fe^{2+} and Fe^{3+} ions. In magnetite, iron *appears* to have a fractional charge

FIGURE 7.8 A team of geologists and students from the University of British Columbia encountered this 8-metric ton (8000 kg) magnetite boulder on Vancouver Island during a 2005 field trip to an abandoned iron mine.

of $\frac{8}{3}$, or 2.67. To understand this fractional charge, we need to realize that the formula unit of Fe_3O_4 can contain two Fe^{3+} ions and one Fe^{2+} ion.

Practice Exercise For thousands of years, the mineral chalcocite (pronounced KAL-kuh-site; Figure 7.9) has been a valuable source of copper. Its chemical composition is 79.85% Cu and 20.15% S. What is its empirical formula?

FIGURE 7.9 Chalcocite is an important source of copper.

SAMPLE EXERCISE 7.6 **Deriving an Empirical Formula of a Compound Containing More than Two Elements** **LO3**

A sample of the carbonate mineral dolomite is 21.73% Ca, 13.18% Mg, 13.03% C, and 52.06% O. What is its empirical formula?

COLLECT AND ORGANIZE We are given the percent composition by mass of dolomite and asked to determine its empirical formula. The percent composition data should add up to 100% (which they do).

ANALYZE We follow the same steps in deriving the empirical formula of dolomite as we did in Sample Exercise 7.5, except that here we must calculate more than one mole ratio because there are four elements in the mineral.

SOLVE Converting percentage values into grams and then into moles:

$$21.73 \text{ g Ca} \times \frac{1 \text{ mol Ca}}{40.08 \text{ g Ca}} = 0.5422 \text{ mol Ca}$$

$$13.18 \text{ g Mg} \times \frac{1 \text{ mol Mg}}{24.31 \text{ g Mg}} = 0.5422 \text{ mol Mg}$$

$$13.03 \text{ g C} \times \frac{1 \text{ mol C}}{12.01 \text{ g C}} = 1.085 \text{ mol C}$$

$$52.06 \text{ g O} \times \frac{1 \text{ mol O}}{16.00 \text{ g O}} = 3.254 \text{ mol O}$$

Dividing each mole value by the smallest one (0.5422):

$$\frac{0.5422 \text{ mol Ca}}{0.5422} = 1.000 \text{ mol Ca} \qquad \frac{0.5422 \text{ mol Mg}}{0.5422} = 1.000 \text{ mol Mg}$$

$$\frac{1.085 \text{ mol C}}{0.5422} = 2.001 \text{ mol C} \qquad \frac{3.254 \text{ mol O}}{0.5422} = 6.001 \text{ mol O}$$

All of these results are either whole numbers or very close to whole numbers. Rounding them to single digits gives us the mole ratio: 1 Ca : 1 Mg : 2 C : 6 O, which corresponds to the empirical formula $CaMgC_2O_6$.

THINK ABOUT IT The problem states that dolomite is a carbonate mineral, which means that it contains CO_3^{2-} ions. This information can be used to rewrite the empirical formula in a way that conveys the fact that the mineral is actually a blend of two compounds, calcium carbonate and magnesium carbonate: $CaCO_3 \cdot MgCO_3$.

Practice Exercise Determine the empirical formula of a mineral that is 46.46% Cr, 24.95% Fe, and 28.59% O by mass.

7.6 Empirical and Molecular Formulas Compared

An empirical formula provides the simplest mole ratios of the elements in a compound. It represents one formula unit of an ionic compound, but it does not provide the number of atoms of each element in one molecule of a molecular compound. To illustrate the difference between empirical and molecular formulas, let's consider the molecular structure of glycolaldehyde in Figure 7.10. Counting up the atoms in the structure, we find that the molecular formula of glycolaldehyde is $C_2H_4O_2$. This molecular formula is not the simplest ratio of the moles of each element in the compound because it can be simplified by dividing each subscript by 2. Doing so produces the empirical formula CH_2O.

On the other hand, the molecular formula of a related compound—formaldehyde (Figure 7.11)—*is* CH_2O, which means that its molecular and empirical formulas are one and the same. This sameness is true for a few other compounds, but it is not true for most. For these others, we need additional information to determine their molecular formulas from percent composition data: we need to know the masses of their molecules.

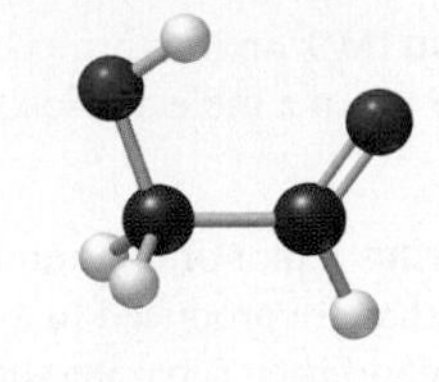

FIGURE 7.10 Molecular structure of glycolaldehyde, one of over 100 organic compounds detected in interstellar gases. It is also the smallest molecule classified as a sugar.

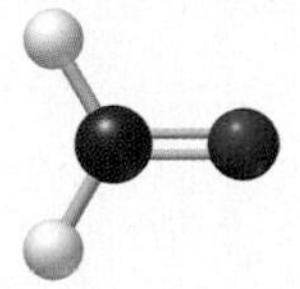

FIGURE 7.11 Molecular structure of formaldehyde, a compound whose molecular and empirical formulas are the same.

CONCEPT TEST ···

Which of the following compounds have empirical formulas that match their molecular formulas?

    a. ethylene glycol, $C_2H_6O_2$
    b. isopropanol, C_3H_8O
    c. glucose, $C_6H_{12}O_6$

··

To see how to use the empirical formula of a compound and its molecular mass to determine its molecular formula, let's reconsider glycolaldehyde. Its molecular formula ($C_2H_4O_2$) can be expressed in terms of its empirical formula (CH_2O) this way:

$$C_2H_4O_2 = (CH_2O)_2$$

Similarly, the molecular formula of glucose, or blood sugar ($C_6H_{12}O_6$), can be expressed in terms of its empirical formula, which is also CH_2O:

$$C_6H_{12}O_6 = (CH_2O)_6$$

Extending the last two equations to any molecular compound, we have the following general relation between molecular and empirical formulas:

$$(\text{Molecular formula}) = (\text{empirical formula})_n \qquad (7.21)$$

where n is a positive integer equal to or greater than 1.

The key to deriving a molecular formula from an empirical formula is to determine the value of n. That's where molecular mass comes in. Using glucose as an example, let's first calculate the mass of one of its molecules:

$$\left(6 \text{ atoms C} \times \frac{12.01 \text{ amu}}{1 \text{ atom C}}\right) + \left(12 \text{ atoms H} \times \frac{1.008 \text{ amu}}{1 \text{ atom H}}\right)$$

$$+ \left(6 \text{ atoms O} \times \frac{16.00 \text{ amu}}{1 \text{ atom O}}\right) = 180.16 \text{ amu}$$

molecular ion (M⁺) an ion formed in a mass spectrometer when a molecule loses an electron.

mass spectrum a plot of the number of ions (y axis) that are produced in a mass spectrometer and then separated based on their mass-to-charge (m/z) ratios (x axis).

Now we calculate the mass of its empirical formula unit (CH_2O):

$$\left(1 \; \text{atom C} \times \frac{12.01 \; \text{amu}}{1 \; \text{atom C}}\right) + \left(2 \; \text{atoms H} \times \frac{1.008 \; \text{amu}}{1 \; \text{atom H}}\right)$$

$$+ \left(1 \; \text{atom O} \times \frac{16.00 \; \text{amu}}{1 \; \text{atom O}}\right) = 30.03 \; \text{amu}$$

Pretend that we did not already know the molecular formula of glucose, but we did know its empirical formula and could calculate the mass of one empirical formula unit (30.03 amu). If we also knew that its molecular mass was 180.16 amu, then we could divide the molecular mass by the mass of one empirical formula unit to obtain the value of n to use in Equation 7.21:

$$n = \frac{\text{molecular mass}}{\text{mass of 1 empirical formula unit}} \qquad (7.22)$$

$$= \frac{180.16 \; \text{amu}}{30.03 \; \text{amu}} = 5.999 \approx 6$$

Inserting this value of n into Equation 7.21 yields the actual molecular formula of glucose:

$$(CH_2O)_6 = C_6H_{12}O_6$$

Molecular Mass and Mass Spectrometry

How do chemists who isolate unknown molecular compounds from reaction mixtures or biological systems determine the molecular masses of the compounds? They often use a powerful analytical technique called *mass spectrometry*.

Mass spectrometers ionize molecules and then separate the ions based on the ratio of their masses (m) to their electric charges (z). In many mass spectrometers, samples are vaporized and then bombarded with a beam of high-energy electrons (Figure 7.12). These electrons smash into gas-phase molecules with such force that they knock electrons off the molecules, forming positively charged **molecular ions (M⁺)**:

$$M + e^- \rightarrow M^+ + 2 \; e^-$$

Sometimes the collisions are so forceful that they break molecules into fragments and ionize the fragments. The charged fragments and molecular ions are ejected into a second region of the mass spectrometer where they are separated based on their m/z ratios. Then they pass into a detector and are counted. The resulting data are displayed as a **mass spectrum** in which the horizontal axis is m/z and the spectrum itself is a series of vertical bars at various m/z values. The heights of the bars indicate the number of ions reaching the detector with that particular m/z value. The bar with the highest m/z value in the spectrum often represents molecular ions, M⁺. If they are singly charged ions ($z = 1+$), then the m/z value of that bar is simply m, the molecular mass of the compound.

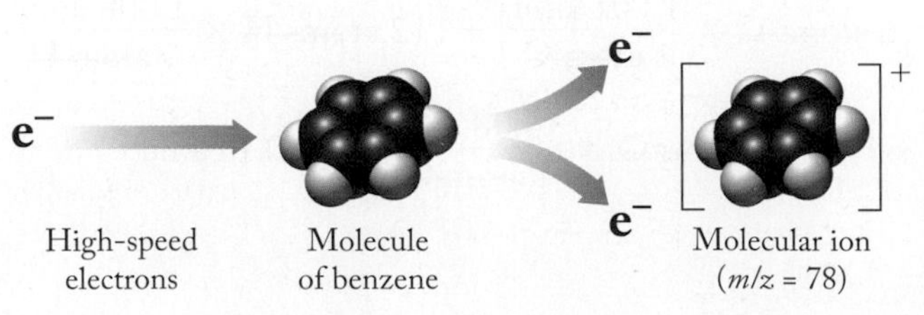

CONNECTION The positive-ray analyzer that Francis Aston built and used to separate two of the isotopes of neon (see Chapter 2) was the forerunner of today's mass spectrometers.

FIGURE 7.12 In many mass spectrometers, sample molecules are bombarded with beams of high-energy electrons, producing molecular ions with positive charges, as shown here for benzene.

High-speed electrons Molecule of benzene Molecular ion ($m/z = 78$)

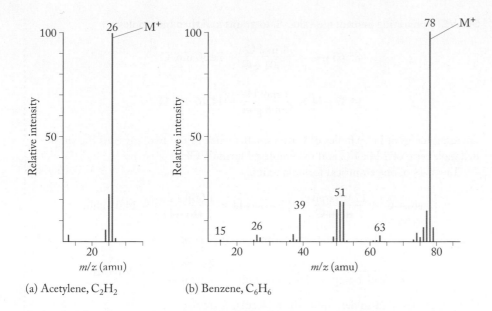

FIGURE 7.13 Mass spectra of (a) acetylene and (b) benzene. The molecular-ion bar in both spectra is labeled M^+. The bars at much smaller m/z values are produced when the ionizing beam also breaks apart the molecules into fragments. The molar masses of C_2H_2 and C_6H_6 are 26 and 78 g/mol, respectively.

Figure 7.13 shows the mass spectra of acetylene and benzene. In both spectra the tall bar with the greatest m/z value (at 26 amu in the acetylene spectrum and at 78 amu in benzene's) is the molecular ion. The bars at lower m/z values represent fragments. Distinctive fragmentation patterns help scientists confirm molecular structures.

CONCEPT TEST

Does the molecular ion in a mass spectrum correspond to the mass associated with the empirical formula or the molecular formula of a compound?

SAMPLE EXERCISE 7.7 | **Using Percent Composition and Molecular Mass to Derive a Molecular Formula** | **LO4**

Pheromones are chemical substances secreted by members of a species to stimulate a response in other individuals of the same species. The percent composition of eicosene, a compound similar to the Japanese beetle mating pheromone, is 85.63% C and 14.37% H. Its molecular mass, as determined by mass spectrometry, is 280 amu. What is the molecular formula of eicosene?

COLLECT AND ORGANIZE We are given the percent composition of eicosene and its molecular mass. The percent composition data can be used to derive the empirical formula. Equation 7.21 relates the empirical and molecular formulas of compounds. Equation 7.22 can be used to calculate the value of n from the mass of an empirical formula unit.

ANALYZE We can follow the procedure used in Sample Exercise 7.5 to determine the empirical formula of eicosene. Next we calculate the mass of one formula unit. Dividing that mass into the molecular mass (280 amu) given in the problem will give us the value of n for Equation 7.21 and allow us to convert the empirical formula into a molecular formula. Given the large molecular mass and low atomic masses of C (12.01 amu) and H (1.008 amu), n may be a large number.

SOLVE Converting percentage values into grams and then into moles:

$$85.63 \text{ g C} \times \frac{1 \text{ mol C}}{12.01 \text{ g C}} = 7.130 \text{ mol C}$$

$$14.37 \text{ g H} \times \frac{1 \text{ mol H}}{1.008 \text{ g H}} = 14.26 \text{ mol H}$$

Dividing moles of H by moles of C (the smaller value of the two) gives us the simplified mole ratio of 2 H : 1 C and the empirical formula CH_2.

The mass of one empirical formula unit is

$$\left(1 \text{ atom C} \times \frac{12.01 \text{ amu}}{1 \text{ atom C}}\right) + \left(2 \text{ atoms H} \times \frac{1.008 \text{ amu}}{1 \text{ atom H}}\right) = 14.03 \text{ amu}$$

Using Equation 7.22 to calculate the formula multiplier n:

$$n = \frac{\text{molecular mass}}{\text{mass of 1 empirical formula unit}} = \frac{280 \text{ amu}}{14.03 \text{ amu}} = 19.96 \approx 20$$

Using Equation 7.21 to determine the molecular formula:

$$\text{Molecular formula} = (CH_2)_n = (CH_2)_{20} = C_{20}H_{40}$$

THINK ABOUT IT The molecular formula of eicosene represents the number of C and H atoms in one molecule of eicosene. The compound is one of many hydrocarbons with the empirical formula CH_2.

Practice Exercise Determine the empirical and molecular formulas of a compound that contains 43.64% P and 56.36% O and has a molar mass of 284 g/mol. ⚙

7.7 Combustion Analysis

As noted in Section 7.3, combustion reactions involve burning substances in oxygen. In **combustion analysis**, the complete combustion of a compound followed by an analysis of the products enables chemists to determine the chemical composition of that compound. To ensure that combustion is complete, the process is carried out in an atmosphere of pure oxygen. This type of elemental analysis relies on complete combustion so that all of the carbon in the sample is converted to CO_2 and all of the hydrogen to H_2O. For a generic hydrocarbon C_aH_b, the overall reaction is

$$C_aH_b + \text{excess } O_2(g) \rightarrow a\, CO_2(g) + b/2\, H_2O(g) \qquad (7.23)$$

As shown in Figure 7.14, the products of combustion in an elemental analyzer flow through a tube packed with $Mg(ClO_4)_2(s)$, which selectively absorbs the $H_2O(g)$, and then through a tube containing $NaOH(s)$, which absorbs the $CO_2(g)$. The masses of the tubes are measured before and after combustion. The differences equal the masses of CO_2 and H_2O produced by the combustion of the sample. Suppose, for example, that complete combustion of a sample of a hydrocarbon results in a 1.320 g increase in the mass of the tube that traps $CO_2(g)$, and the mass of the tube that traps $H_2O(g)$ increases by 0.541 g. How can we use these results to determine the empirical formula of the hydrocarbon?

Empirical formulas express the mole ratios of the elements in a compound, so we start by converting the masses of CO_2 and H_2O caught in the traps into

combustion analysis a laboratory procedure in which a substance is burned completely in oxygen to produce known compounds whose masses are used to determine the composition of the original material.

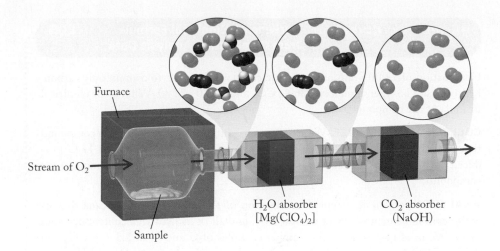

Furnace
Stream of O_2
Sample
H_2O absorber
[$Mg(ClO_4)_2$]
CO_2 absorber
(NaOH)

FIGURE 7.14 A carbon/hydrogen elemental analyzer relies on the complete combustion of organic compounds. The products are H_2O vapor and CO_2. Water vapor is absorbed by a cartridge packed with solid $Mg(ClO_4)_2$, and CO_2 is absorbed by solid NaOH. The empirical formula of the compound is calculated from the masses of H_2O and CO_2 absorbed.

moles of CO_2 and H_2O, and then to equivalent numbers of moles of C and H in the original sample:

$$1.320 \cancel{\text{g } CO_2} \times \frac{1 \cancel{\text{mol } CO_2}}{44.01 \cancel{\text{g } CO_2}} \times \frac{1 \text{ mol C}}{1 \cancel{\text{mol } CO_2}} = 0.02999 \text{ mol C} \approx 0.0300 \text{ mol C}$$

$$0.541 \cancel{\text{g } H_2O} \times \frac{1 \cancel{\text{mol } H_2O}}{18.02 \cancel{\text{g } H_2O}} \times \frac{2 \text{ mol H}}{1 \cancel{\text{mol } H_2O}} = 0.0600 \text{ mol H}$$

The mole ratio of H to C is (0.0600 mol/0.0300 mol) = 2, so the empirical formula is CH_2.

If we want to extend this method to determining a molecular formula, we need to know molecular mass. Suppose the mass spectrum of the hydrocarbon of interest has a molecular ion at 84 amu. The formula mass of CH_2 is 14 amu. Dividing the molecular mass by the formula mass, we get the multiplier, *n*:

$$n = \frac{84 \cancel{\text{amu}}}{14 \cancel{\text{amu}}} = 6$$

The molecular formula is therefore

$$(CH_2)_6 = C_6H_{12}$$

Note that in this problem we did not need to know the initial mass of the sample to determine its empirical formula. We only needed to know that the sample was a hydrocarbon and that it was completely converted into the stated amounts of CO_2 and H_2O.

What if we knew that our sample was *not* a hydrocarbon? What if we had isolated a pharmacologically promising compound from a tropical plant, and we knew that its molecules contained atoms of carbon, hydrogen, *and* oxygen? We would need to determine the proportion of oxygen in it, but there is no simple way to do that directly because the compound is burned in an atmosphere of O_2. However, the results of combustion analysis do allow us to calculate the number of moles of C and H in the original sample, which we can convert to grams of C and H. Subtracting their sum from the mass of the original sample provides a measure of the mass of O in the sample. That mass is converted to moles of O and used together with the moles of C and H to determine the empirical formula. The following Sample Exercise illustrates how all the calculations fit together.

SAMPLE EXERCISE 7.8 **Deriving an Empirical Formula** **LO5**
from Combustion Analysis Data

Combustion of 1.000 gram of an organic compound known to contain only carbon, hydrogen, and oxygen produces 2.360 g CO_2 and 0.640 g H_2O. What is the empirical formula of the compound?

COLLECT AND ORGANIZE We are given the initial mass of a sample of a compound that consists of C, H, and O, and we also know the masses of CO_2 and H_2O produced during its combustion. We are asked to determine the empirical formula of the compound.

ANALYZE We assume complete combustion of the sample, which means that its entire carbon content is converted into CO_2, and all of its hydrogen content becomes H_2O. We need to determine the number of moles of C and H in 2.360 g CO_2 and 0.640 g H_2O, respectively. We also need to calculate the masses of C and H because the difference between the sum of the masses of C and H and the mass of the original sample is equal to the O content of the original sample. We then determine the C:H:O mole ratio and convert it into a ratio of small whole numbers.

SOLVE The numbers of moles of C and H in the CO_2 and H_2O collected during combustion are

$$2.360 \text{ g CO}_2 \times \frac{1 \text{ mol CO}_2}{44.01 \text{ g CO}_2} \times \frac{1 \text{ mol C}}{1 \text{ mol CO}_2} = 0.05362 \text{ mol C}$$

$$0.640 \text{ g H}_2\text{O} \times \frac{1 \text{ mol H}_2\text{O}}{18.02 \text{ g H}_2\text{O}} \times \frac{2 \text{ mol H}}{1 \text{ mol H}_2\text{O}} = 0.07103 \text{ mol H}$$

The masses of C and H are

$$0.05362 \text{ mol C} \times \frac{12.01 \text{ g C}}{1 \text{ mol C}} = 0.6440 \text{ g C}$$

$$0.07103 \text{ mol H} \times \frac{1.008 \text{ g H}}{1 \text{ mol H}} = 0.07160 \text{ g H}$$

The difference between the original sample mass (1.000 g) and the sum of the masses of C and H (0.6440 g + 0.07160 g = 0.7156 g) is the mass of the oxygen in the sample:

$$\text{Mass of oxygen} = 1.000 \text{ g} - 0.7156 \text{ g} = 0.2844 \text{ g O}$$

The number of moles of O atoms in the sample is

$$0.2844 \text{ g O} \times \frac{1 \text{ mol O}}{16.00 \text{ g O}} = 0.01778 \text{ mol O}$$

Combining the above results, we get the mole ratio of the three elements in the sample:

$$0.05362 \text{ mol C} : 0.07103 \text{ mol H} : 0.01778 \text{ mol O}$$

To simplify the ratios we divide them by the smallest of their values (0.01778):

$$\frac{0.05362 \text{ mol C}}{0.01778} = 3.016 \approx 3 \text{ mol C}$$

$$\frac{0.07103 \text{ mol H}}{0.01778} = 3.995 \approx 4 \text{ mol H}$$

$$\frac{0.01778 \text{ mol O}}{0.01778} = 1 \text{ mol O}$$

The whole-number mole ratio that most closely matches the calculated values is 3:4:1, making the empirical formula of the sample C_3H_4O.

THINK ABOUT IT The results of our mole ratio calculations for C and H are very close to small whole numbers, which means that our calculations and the formula based on them are probably correct.

Practice Exercise Vanillin is the compound containing carbon, hydrogen, and oxygen that gives vanilla beans their distinctive flavor. The combustion of 30.4 mg of vanillin produces 70.4 mg of CO_2 and 14.4 mg of H_2O. The mass spectrum of vanillin shows a molecular-ion line at 152 amu. Use this information to determine the molecular formula of vanillin.

limiting reactant a reactant that is consumed completely in a chemical reaction. The amount of product formed depends on the amount of the limiting reactant available.

7.8 Limiting Reactants and Percent Yield

Let's return to the first reaction that we discussed in Section 7.3, the combustion of methane:

$$CH_4(g) + 2\,O_2(g) \rightarrow CO_2(g) + 2\,H_2O(\ell) \qquad (7.14)$$

The stoichiometry of the reaction tells us that one mole of CH_4 requires two moles of O_2 for complete combustion. In reality, the furnaces and hot water heaters fueled by natural gas are surrounded by an abundant supply of O_2, so the quantities of reactants consumed, products formed, and energy generated are limited only by the fuel supply. In this reaction mixture, CH_4 is said to be the **limiting reactant**, which means that the extent to which the reaction proceeds is limited only by the quantity of CH_4 available. Figure 7.15 provides another—perhaps more familiar—view of the concept of a limiting reactant.

CONCEPT TEST

Describe another chemical reaction in which at least one of the reactants is usually present in excess so that another is the limiting reactant.

(a)

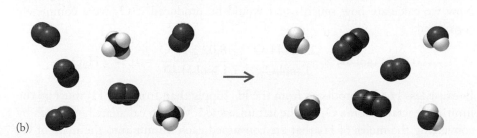

(b)

FIGURE 7.15 (a) Illustrating the principle of the limiting *ingredient*: suppose our goal is to prepare modest salami-and-cheese sandwiches (only one slice of salami and one slice of cheese per sandwich). If there are eight slices of bread, four slices of cheese, and three slices of salami, how many salami and cheese sandwiches can we make? Only three: salami is the limiting ingredient, and there are two slices of bread and one slice of cheese left over. (b) Suppose we have a combustion reaction mixture in which O_2 molecules outnumber CH_4 molecules 5 to 2. Complete combustion produces 2 molecules of CO_2 and 4 molecules of H_2O, leaving one molecule of O_2 left over. Methane is the limiting *reactant*.

Calculations Involving Limiting Reactants

Sometimes it is hard to say whether one reactant or another is the limiting reactant in a chemical reaction. We might know the masses of two reactants and be tempted to select the one with the smaller mass as the limiting reactant, but that is often not the case. Take the combustion of methane as an example. Suppose a sealed reaction vessel contains 16 g CH_4 and 48 g O_2. When the mixture is ignited, which reactant is the limiting reactant? You might select CH_4 because there is three times more O_2 (by mass) than CH_4 in the reaction mixture. However, the chemical equation for the combustion reaction tells us that we need *two moles* of O_2 for every one mole of CH_4, and the molar mass of O_2 (32.00 g/mol) is about *twice* that of CH_4 (16.04 g/mol). Twice as many moles of a reactant that has twice the molar mass means that we need *four* times as much of it by mass. However, the mass of O_2 in the reaction vessel is only *three* times the mass of CH_4, so O_2 is the limiting reactant.

▶❚❚ CHEMTOUR Limiting Reactant

Let's set up a procedure for identifying the limiting reactant for any reaction in which we know the quantities of two or more reactants. There are several ways to solve such a problem. We explore two of them here, starting with the generic chemical equation

$$A + B \rightarrow C$$

In our first method, we calculate how much product would be formed if reactant A were completely consumed in the reaction. Then we repeat the calculation for reactant B. Let's try this approach with the reaction that has produced electricity in spacecraft since the dawn of manned space flight in the 1960s. It makes use of fuel cells based on this chemical reaction:

$$2\,H_2(g) + O_2(g) \rightarrow 2\,H_2O(\ell) \tag{7.16}$$

Suppose a gas-delivery system pumps hydrogen and oxygen to a fuel cell at the rates of 0.055 g H_2 and 0.48 g O_2 per minute. Which gas is the limiting reactant?

Equation 7.16 tells us that two moles of H_2 combine with one mole of O_2, so let's convert the masses of the two gases delivered each minute into moles by dividing by their respective molar masses.

$$0.055 \,\cancel{g\,H_2} \times \frac{1\text{ mol }H_2}{2.016 \,\cancel{g\,H_2}} = 0.0273\text{ mol }H_2$$

$$0.48 \,\cancel{g\,O_2} \times \frac{1\text{ mol }O_2}{32.00 \,\cancel{g\,O_2}} = 0.0150\text{ mol }O_2$$

As is our standard practice, we carry an extra significant figure through these early stages of the calculation. Next we calculate how much water would be produced if H_2 were completely consumed:

$$0.0273 \,\cancel{\text{mol }H_2} \times \frac{2\,\cancel{\text{mol }H_2O}}{2\,\cancel{\text{mol }H_2}} \times \frac{18.02\text{ g }H_2O}{1\,\cancel{\text{mol }H_2O}} = 0.49\text{ g }H_2O$$

Now we calculate how much water would be produced if O_2 were completely consumed:

$$0.0150 \,\cancel{\text{mol }O_2} \times \frac{2\,\cancel{\text{mol }H_2O}}{1\,\cancel{\text{mol }O_2}} \times \frac{18.02\text{ g }H_2O}{1\,\cancel{\text{mol }H_2O}} = 0.54\text{ g }H_2O$$

Because less H_2O is produced from the H_2 supply than from O_2, H_2 must be the limiting reactant. Some O_2 will be left unreacted. We can calculate how much by converting the moles of H_2 that are consumed each minute into the mass of O_2

that is consumed. We use the 2:1 stoichiometric ratio between H_2 and O_2 and the molar mass of O_2:

$$0.0273 \text{ mol } H_2 \times \frac{1 \text{ mol } O_2}{2 \text{ mol } H_2} \times \frac{32.00 \text{ g } O_2}{1 \text{ mol } O_2} = 0.44 \text{ g } O_2$$

theoretical yield the maximum amount of product possible in a chemical reaction for given quantities of reactants; also called *stoichiometric yield*.

The difference between the mass of O_2 supplied (0.48 g) and the mass consumed (0.44 g) equals 0.04 g O_2 left over per minute of fuel cell operation.

Another way to identify limiting reactants focuses on their mole ratios in the reaction mixture, and in the balanced chemical equation for the reaction. To use this approach for the generic reaction $A + B \rightarrow C$, we:

1. Convert the masses of reactants A and B into moles.
2. Calculate the mole ratio of A to B.
3. Compare the calculated mole ratio with the stoichiometric mole ratio from the balanced chemical equation. If

$$\left(\frac{\text{mol } A}{\text{mol } B}\right)_{\text{given}} > \left(\frac{\text{mol } A}{\text{mol } B}\right)_{\text{stoichiometric}} \tag{7.24}$$

then there is an excess of A, which means B is the limiting reactant. Likewise, if the mole ratio for the given quantities is less than the stoichiometric mole ratio, then A is the limiting reactant. And if the two ratios are equal, then the two masses represent stoichiometric quantities of the two reactants and there is no limiting reactant—or, viewed another way, they both are.

Applying this approach to our fuel cell scenario, we know that the gas-delivery system provides 0.0273 mol H_2 and 0.0150 mol O_2 per minute to the fuel cell. Those values give us a H_2-to-O_2 mole ratio of

$$\frac{0.0273 \text{ mol } H_2}{0.0150 \text{ mol } O_2} = 1.82$$

This ratio is less than the stoichiometric ratio of 2 mol H_2 per mole O_2. Therefore H_2 is the limiting reactant.

Whether we calculate the quantity of a product twice or we use the mole ratio method, we always need to begin by converting masses of reactants into equivalent numbers of moles. The advantage of the first method is that it not only identifies the limiting reactant but also determines the **theoretical yield**, that is, the maximum quantity of product that could be obtained from the reaction mixture. The downside to the first approach is that more math is involved, particularly when our goal is only to identify the limiting reactant. When that is the case, comparing mole ratios is usually easier.

SAMPLE EXERCISE 7.9 **Identifying the Limiting Reactant in a Reaction Mixture** **LO6**

The flame in an oxyacetylene torch (shown in the photo at the beginning of this chapter) reaches temperatures as high as 3500°C as a result of the combustion of a mixture of acetylene (C_2H_2) and pure O_2. If these two gases flow from high-pressure tanks at the rates of 52.0 g C_2H_2 and 188 g O_2 per minute, which reactant is the limiting reactant, or is the mixture stoichiometric?

COLLECT AND ORGANIZE We are given the rates at which two reactants are introduced to a reaction system (the torch) and asked to determine whether one of them is limiting and, if so, which one. We are not given the chemical equation describing the combustion process, but we know that a stoichiometric ratio of the reactants results in complete combustion to CO_2 and H_2O. We are given the rates of reactant flow to the torch in grams per minute. These rates can be converted directly into masses in grams by simply assuming one minute of torch operation.

ANALYZE We need to write a balanced chemical equation describing the complete combustion of acetylene. We also need to convert the masses of the two reactants into moles so that we can compare their mole ratio to the stoichiometric mole ratio of the reactants in the balanced chemical equation.

SOLVE To write a balanced chemical equation, we start with one mole of each reactant and product and take an elemental inventory on both sides of the reaction arrow:

$$C_2H_2(g) + O_2(g) \rightarrow CO_2(g) + H_2O(\ell)$$

Atoms: $2\,C + 2\,H + 2\,O \rightarrow 1\,C + 3\,O + 2\,H$

To balance the C atoms we place a coefficient of 2 in front of CO_2:

$$C_2H_2(g) + O_2(g) \rightarrow \mathbf{2}\,CO_2(g) + H_2O(\ell)$$

Atoms: $2\,C + 2\,H + 2\,O \rightarrow 2\,C + 2\,H + 5\,O$

Balancing the O atoms requires multiplying all coefficients by 2 to get an even number of O atoms on the right side (without disrupting the previously balanced numbers of C and H atoms):

$$\mathbf{2}\,C_2H_2(g) + \mathbf{2}\,O_2(g) \rightarrow \mathbf{4}\,CO_2(g) + \mathbf{2}\,H_2O(\ell)$$

Atoms: $4\,C + 4\,H + 4\,O \rightarrow 4\,C + 4\,H + 10\,O$

Next we add 6 more atoms of O to the left side by changing the coefficient in front of O_2 from 2 to 5:

$$2\,C_2H_2(g) + \mathbf{5}\,O_2(g) \rightarrow 4\,CO_2(g) + 2\,H_2O(\ell)$$

Atoms: $4\,C + 4\,H + 10\,O \rightarrow 4\,C + 4\,H + 10\,O$ ✓

The stoichiometric ratio of O_2 to C_2H_2 is 5:2, or 2.5.

Converting the masses of the two reactants into numbers of moles:

$$188\ \text{g O}_2 \times \frac{1\ \text{mol O}_2}{(2 \times 16.00)\ \text{g O}_2} = 5.88\ \text{mol O}_2$$

$$52.0\ \text{g C}_2\text{H}_2 \times \frac{1\ \text{mol C}_2\text{H}_2}{(2 \times 12.01 + 2 \times 1.008)\ \text{g C}_2\text{H}_2} = 2.00\ \text{mol C}_2\text{H}_2$$

The mole ratio of O_2 to C_2H_2 is 5.88/2.00 = 2.94, which is greater than the stoichiometric ratio of 2.5 from the balanced chemical equation. Therefore, there is more O_2 in the reaction mixture than is needed to react completely with C_2H_2, and C_2H_2 is the limiting reactant.

THINK ABOUT IT We compared the mole ratios of reactants in the given reaction mixture and the balanced chemical equation to identify the limiting reactant. This approach involves fewer steps than calculating theoretical yields twice, so it is preferred when our goal is only to identify the limiting reactant and not to calculate the quantities of products.

Practice Exercise Any fuel–oxygen mixture that contains less fuel than oxygen needed to burn the fuel completely is called a *fuel-lean* (or simply *lean*) mixture, and a combustion mixture containing too little oxygen is called a *fuel-rich* mixture. A high-performance heater that burns propane, C_3H_8, is adjusted so that 100.0 g O_2 enters the system for every 100.0 g propane. Is the mixture fuel-rich or fuel-lean?

Percent Yield: Actual versus Theoretical

We have defined *theoretical* yield: the maximum quantity of product possible from given quantities of reactants. In nature, industry, or the laboratory, the *actual* yield is often less than the theoretical yield for several reasons. Sometimes reactants combine to produce products other than the ones desired. Other reactions are so slow that some reactants remain unreacted even after long reaction times. Still other reactions do not go to completion no matter how long they are allowed to run. That can happen when a reaction also runs in reverse, turning products back into reactants. When this happens, the result can be a mixture of reactants and products in which the composition does not change with time. These reactions have reached a state of *chemical equilibrium*, a concept we explore in considerable detail in Chapter 14. So, for various reasons, it is useful to distinguish between the theoretical and the actual yields of a chemical reaction and to calculate the **percent yield**:

$$\text{Percent yield} = \frac{\text{actual yield}}{\text{theoretical yield}} \times 100\% \qquad (7.25)$$

> **percent yield** the ratio, expressed as a percentage, of the actual yield of a chemical reaction to the theoretical yield.

SAMPLE EXERCISE 7.10   Calculating Percent Yield   LO7

The industrial process for making the ammonia used in fertilizer, explosives, and many other products is based on the reaction between nitrogen and hydrogen at high temperature and pressure:

$$N_2(g) + 3\,H_2(g) \rightarrow 2\,NH_3(g)$$

If 18.2 kg NH_3 ($\mathcal{M}$ = 17.03 g/mol) is produced by a reaction mixture that initially contains 6.00 kg H_2 ($\mathcal{M}$ = 2.016 g/mol) and an excess of N_2, what is the percent yield of the reaction?

COLLECT AND ORGANIZE We know that the actual yield of NH_3 is 18.2 kg. We also know that H_2 must be the limiting reactant because there is an excess of N_2, and that the reaction mixture initially contained 6.00 kg H_2. The mole ratio of NH_3 to H_2 in the chemical equation is 2:3.

ANALYZE One way to calculate percent yield involves these steps:

1. converting the mass of H_2 into moles of H_2;
2. converting the moles of H_2 consumed into moles of NH_3 produced using the 2:3 stoichiometric ratio in the balanced chemical equation;
3. converting moles of NH_3 into grams of NH_3 by multiplying by the molar mass of NH_3;
4. converting grams of NH_3 into kilograms of NH_3 to calculate the theoretical yield of the reaction;
5. dividing the actual yield (18.2 kg NH_3) by the theoretical yield from step 4 to obtain the percent yield.

SOLVE Calculating the theoretical yield (combining steps 1–4 into a single series of conversion factors):

$$6.00\ \cancel{\text{kg H}_2} \times \frac{10^3\ \cancel{\text{g}}}{1\ \cancel{\text{kg}}} \times \frac{1\ \cancel{\text{mol H}_2}}{2.016\ \cancel{\text{g H}_2}} \times \frac{2\ \cancel{\text{mol NH}_3}}{3\ \text{mol H}_2} \times \frac{17.03\ \cancel{\text{g}}\ NH_3}{1\ \cancel{\text{mol NH}_3}} \times \frac{1\ \text{kg}}{10^3\ \cancel{\text{g}}}$$

$$= 33.8\ \text{kg NH}_3$$

Dividing the actual yield by this theoretical yield (step 5):

$$\frac{18.2 \ \cancel{\text{kg NH}_3}}{33.8 \ \cancel{\text{kg NH}_3}} \times 100\% = 53.8\%$$

THINK ABOUT IT A yield of about 54% may seem low, but it is not unusual for a reaction such as this one, which reaches chemical equilibrium before the limiting reactant has been completely consumed. In Chapter 14 we will discuss ways that chemists and chemical engineers manipulate reaction conditions to improve the yields of reactions that reach equilibrium.

Practice Exercise The industrial synthesis of H_2 begins with the steam-reforming reaction, in which methane reacts with high-temperature steam:

$$CH_4(g) + H_2O(g) \rightarrow CO(g) + 3 \ H_2(g)$$

What is the percent yield when a reaction vessel that initially contains 64.0 kg CH_4 and excess steam yields 18.1 kg H_2?

SAMPLE EXERCISE 7.11   Integrating Concepts: Synthesizing Hydrogen Gas

Practice Exercise 7.10 describes the first stage in the *steam-reforming reaction* for synthesizing H_2 gas. The carbon monoxide that is a by-product of the reaction can be reacted with more steam in a second stage to produce more hydrogen and carbon dioxide.

a. Write a balanced chemical equation describing the reaction between CO and steam that produces H_2 and CO_2.
b. Combine your answer from part a with the chemical equation for the first stage of the steam-reforming reaction from Practice Exercise 7.10 to write an overall reaction in which methane and steam react to form hydrogen gas and carbon dioxide, CO_2.
c. Suppose a reaction vessel for the two-stage process initially contains 48.0 kg CH_4 and 118 kg $H_2O(g)$ and that the process yields 18 kg H_2. Is the initial reaction mixture stoichiometric? If not, which is the limiting reactant?
d. What is the percent yield of the reaction mixture in part c?

COLLECT AND ORGANIZE We need to write a balanced chemical equation describing the reaction between CO and H_2O that produces H_2 and CO_2, and then to combine it with the steam-reforming reaction to write an overall equation describing the formation of H_2 and CO_2 from CH_4 and H_2O. We must then determine whether or not a reaction mixture of CH_4 and H_2O is stoichiometric and calculate the percent yield of the reaction. We know the masses of CH_4 and H_2O in the reaction mixture and the mass of H_2 they produce. We also know the chemical equation of the steam-reforming reaction, which is the first step in the two-step process. From previous calculations in this chapter we know these molar masses: 18.02 g/mol H_2O, 16.04 g/mol CH_4, and 2.016 g/mol H_2.

ANALYZE To combine two chemical equations, we need to make sure that the coefficients of products in the first equation that are reactants in the second equation are the same, so that they cancel out when the equations are combined. One way to identify whether or not a reaction mixture is stoichiometric involves calculating the quantities of product each reactant could produce if it were completely consumed. Using this approach makes sense because we need to determine the percent yield of the reaction, which involves calculating the theoretical yield and comparing it to the actual yield.

SOLVE

a. First we write the chemical equation for the reaction between CO and steam that produces H_2 and CO_2. Starting with one molecule of each reactant and product and taking an inventory of all the atoms on both sides of the reaction arrow:

$$CO(g) + H_2O(g) \rightarrow H_2(g) + CO_2(g)$$

Atoms: $1\,C + 2\,H + 2\,O \rightarrow 1\,C + 2\,H + 2\,O$ ✓

we find that the equation is balanced already!

b. Comparing this equation to the one for the steam-reforming reaction:

$$CH_4(g) + H_2O(g) \rightarrow CO(g) + 3 \ H_2(g)$$

we find that CO is a product of the steam-reforming reaction and a reactant in the second-stage reaction. Its coefficient is 1 in both reactions, so the two can be combined without modification:

$$CH_4(g) + H_2O(g) \rightarrow \cancel{CO(g)} + 3 \ H_2(g)$$
$$\underline{+ \ \cancel{CO(g)} + H_2O(g) \rightarrow H_2(g) + CO_2(g)}$$
$$CH_4(g) + 2 \ H_2O(g) \rightarrow 4 \ H_2(g) + CO_2(g)$$

c. To determine whether the reaction mixture is stoichiometric, we calculate the theoretical yields of H_2 produced by each of the reactants:

$$48.0 \text{ kg CH}_4 \times \frac{10^3 \text{ g}}{1 \text{ kg}} \times \frac{1 \text{ mol CH}_4}{16.04 \text{ g CH}_4} \times \frac{4 \text{ mol H}_2}{1 \text{ mol CH}_4}$$

$$\times \frac{2.016 \text{ g H}_2}{1 \text{ mol H}_2} \times \frac{1 \text{ kg}}{10^3 \text{ g}} = 24.13 \text{ kg H}_2$$

$$118 \text{ kg H}_2\text{O} \times \frac{10^3 \text{ g}}{1 \text{ kg}} \times \frac{1 \text{ mol H}_2\text{O}}{18.02 \text{ g H}_2\text{O}} \times \frac{4 \text{ mol H}_2}{2 \text{ mol H}_2\text{O}}$$

$$\times \frac{2.016 \text{ g H}_2}{1 \text{ mol H}_2} \times \frac{1 \text{ kg}}{10^3 \text{ g}} = 26.40 \text{ kg H}_2$$

Methane produces less H_2, so it is the limiting reactant.

d. The percent yield calculation is based on the lesser theoretical yield (24.13 kg H_2):

$$\frac{18 \text{ kg H}_2}{24.13 \text{ kg H}_2} \times 100\% = 75\% \text{ yield}$$

THINK ABOUT IT The second-stage reaction in the synthesis of H_2 is called the *water–gas shift reaction*. Many important industrial processes rely on chemical reactions with yields that are less than 100%, and the production of hydrogen gas is one of them. As in other multistep calculations, we used extra significant figures for intermediate values and did not round off to two digits until the end. Another way to avoid rounding errors is to use values in calculator memory for each new step in a calculation rather than reentering the results of previous calculations.

SUMMARY

Section 7.1 In a **chemical equation**, the chemical formulas of **reactants** appear first, followed by a reaction arrow and the chemical formulas of the **products**. The proportions of the reactants and products are expressed by coefficients preceding their formulas. Symbols in parentheses are used to indicate physical states: (g) for gas, (ℓ) for liquid, (s) for solid, and (aq) for substances dissolved in water. The mole ratios of reactants and products in a chemical reaction are called the **stoichiometry** of the reaction. The **law of conservation of mass** states that the sum of the masses of the reactants in a chemical reaction is equal to the sum of the masses of the products.

Section 7.2 In a balanced chemical equation, the number of atoms of each element is the same on the two sides of the reaction arrow.

Section 7.3 *Combustion* reactions occur between oxygen and another element or compound. In the complete combustion of hydrocarbons and other compounds that contain hydrogen and carbon, the hydrogen combines with oxygen to form H_2O, and the carbon combines with oxygen to form CO_2.

Section 7.4 In the carbon cycle, CO_2 from the atmosphere is converted to glucose during photosynthesis in green plants. Most, but not all, of the carbon in green plants returns to the atmosphere as CO_2 during respiration. Combustion of fossil fuels adds about 3×10^{13} kg CO_2 to the atmosphere each year.

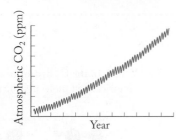

Section 7.5 The **percent composition** of a compound is the percentage by mass of each element in the compound. The **empirical formula** of a compound represents the lowest whole-number ratio of its elements.

Section 7.6 A compound's empirical formula may or may not be the same as its molecular formula, which indicates the number of each type of atom in one molecule. Converting the empirical formula of a compound into a molecular formula requires knowing the molecular mass of the compound, which can be determined by identifying the **molecular ion (M⁺)** in its **mass spectrum**.

Section 7.7 In **combustion analysis**, a known mass of an organic compound is burned in a stream of oxygen gas. The carbon in the sample is converted into CO_2 and the hydrogen is converted into H_2O. The masses of CO_2 and H_2O are measured and used to determine the masses of C and H in the organic compound and then the compound's empirical formula.

Section 7.8 The **limiting reactant** in a reaction mixture is the reactant that limits how much product can be made. The maximum amount of product that can form in a chemical reaction is the **theoretical yield**. The measured amount of product formed in a reaction is the actual yield, and the ratio of actual yield to theoretical yield expressed as a percentage is the **percent yield** for the reaction.

PROBLEM-SOLVING SUMMARY

TYPE OF PROBLEM	CONCEPTS AND EQUATIONS	SAMPLE EXERCISES
Balancing a chemical reaction	Adjust coefficients to balance number of atoms of each element on both sides of the reaction arrow. Start with elements that are in only one reactant and product on each side.	7.1

TYPE OF PROBLEM	CONCEPTS AND EQUATIONS	SAMPLE EXERCISES
Writing and balancing a chemical equation for a combustion reaction	C and H in organic compounds react with O_2 to form CO_2 and H_2O, for example: $$CH_4(g) + 2\,O_2(g) \rightarrow CO_2(g) + 2\,H_2O(g)$$ Balance the atoms of C first, then H, then O.	7.2
Calculating the mass of a product from the mass of a reactant	Convert mass of reactant into moles of reactant; use reaction stoichiometry to calculate moles of product, and then convert moles of product into mass of product.	7.3
Calculating percent composition from a chemical formula	Calculate the mass of each element in one mole of the compound. Divide each element's mass by the molar mass; express the results as percentages.	7.4
Deriving an empirical formula from percent composition	Assume a 100 g sample so that percentage values become masses in grams. Convert the mass of each element into moles by dividing by its molar mass. Divide these numbers of moles by the smallest of their values. If necessary, multiply by the appropriate factor to remove any fractions and obtain whole-number mole values, which are the subscripts in the empirical formula of the compound.	7.5, 7.6
Using percent composition and molecular mass to derive a molecular formula	Derive an empirical formula from percent composition data. Calculate the multiplier n by dividing the molecular mass by the mass of one empirical formula unit. Multiply subscripts in the empirical formula by n to obtain the molecular formula.	7.7
Deriving an empirical formula from combustion analysis data	For hydrocarbons, convert given masses of CO_2 and H_2O into moles of CO_2 and H_2O, and then to moles of C and H. For compounds also containing O, convert moles of C and H into masses of C and H and subtract the sum of these values from the sample mass to calculate the mass of O. Convert mass of O to moles of O. Simplify mole ratios of C to H to O.	7.8
Identifying the limiting reactant in a reaction mixture	Method 1: Calculate how much product each reactant could make; the reactant making the least amount of product is the limiting reactant. Method 2: Convert given masses of reactants into moles. Compare the mole ratio of reactants to the corresponding mole ratio in the stoichiometric equation. If $$\left(\frac{\text{mol A}}{\text{mol B}}\right)_{\text{given}} > \left(\frac{\text{mol A}}{\text{mol B}}\right)_{\text{stoichiometric}} \qquad (7.24)$$ then B is the limiting reactant. If not, then A is the limiting reactant or, if the ratios are equal, there is no single limiting reactant.	7.9
Calculating percent yield	Calculate the theoretical yield of product using the quantity of the limiting reactant. Divide actual yield (given) by theoretical yield: $$\text{Percent yield} = \frac{\text{actual yield}}{\text{theoretical yield}} \times 100\% \qquad (7.25)$$	7.10

VISUAL PROBLEMS

(Answers to boldface end-of-chapter questions and problems are in the back of the book.)

7.1. The molecular models in Figure P7.1 represent five oxides of nitrogen. Write the empirical and molecular formulas of the one(s) for which the two formulas are not the same.

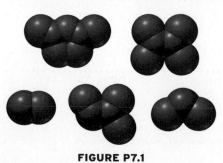

FIGURE P7.1

7.2. Which of the C_2 hydrocarbons in Figure P7.2 have different empirical and molecular formulas? What are those formulas?

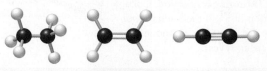

FIGURE P7.2

7.3. Each of the pairs of images in Figure P7.3 contains substances composed of two elements: X (red balls) and Y (blue balls). Write a balanced chemical equation describing the reaction taking place in each pair of images. Be sure to indicate the physical states of the reactants and products using the appropriate symbols in parentheses.

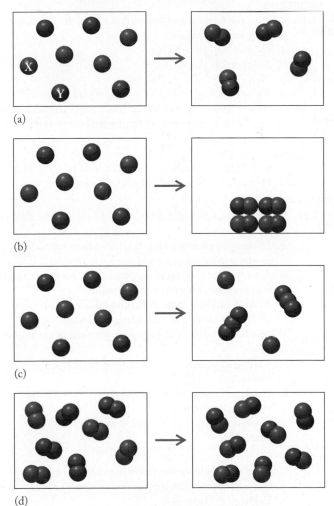

(a)

(b)

(c)

(d)

FIGURE P7.3

7.4. Identify the limiting reactant in each of the pairs of containers pictured in Figure P7.4. The red balls represent atoms of element X, and the blue balls are atoms of element Y. Each question mark means that there is unreacted reactant left over.

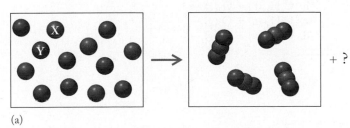

(a)

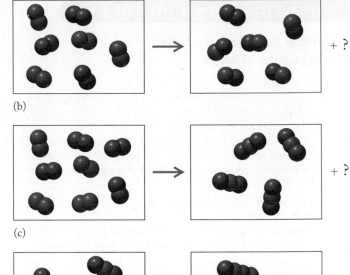

(b)

(c)

(d)

FIGURE P7.4

7.5. Which two of the hydrocarbons with molecular structures shown in Figure P7.5 have the same percent composition?

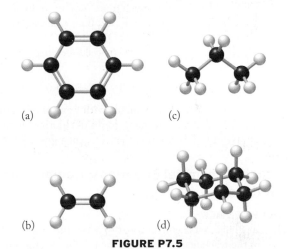

(a)

(b)

(c)

(d)

FIGURE P7.5

7.6. Suppose 1.000 g samples of the compounds with the molecular structures shown in Figure P7.6 undergo combustion analysis. Assuming complete combustion, which of the samples will produce the most CO_2?

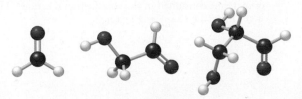

FIGURE P7.6

QUESTIONS AND PROBLEMS .... ■

Chemical Reactions and the Conservation of Mass

CONCEPT REVIEW

7.7. In the combination reaction A + 2 B → C, 1.00 gram of substance A and 4.00 grams of substance B are consumed. How many grams of substance C are formed?

7.8. In the reaction A + B → C + D, 3.00 grams of substance C and 3.00 grams of substance D are produced as 2.00 grams of substance A are consumed. How many grams of substance B are also consumed?

7.9. Among the four reactants and products in Question 7.8, which one has the largest molar mass?

*7.10. Among the three substances in Question 7.7 and the four substances in Question 7.8, which ones *could be* elements?

PROBLEMS

7.11. Depending on reaction conditions, O_2 may combine with N_2 to form NO or NO_2. If x grams of O_2 combine with y grams of N_2 to form NO, how many grams of O_2 combine with y grams of N_2 to form NO_2?

7.12. Combustion of sulfur may, depending on reaction conditions, produce SO_2 or SO_3. If x grams of O_2 combine with y grams of sulfur to form SO_2, how many grams of O_2 combine with y grams of sulfur to form SO_3?

7.13. Two of the more common oxides of iron have the formulas FeO and Fe_2O_3. How much more oxygen combines with a given mass of iron to form Fe_2O_3 than combines with the same mass of iron to form FeO?

7.14. Tin(IV) chloride is prepared by the following combination reaction:

$$Sn(s) + 2\ Cl_2(g) \rightarrow SnCl_4(\ell)$$

If x grams of chlorine combine with y grams of tin to form tin(IV) chloride, how many grams of chlorine are there in a sample of tin(II) chloride that contains y grams of tin?

Balancing Chemical Equations; Combustion Reactions

CONCEPT REVIEW

7.15. In a balanced chemical equation, does the number of atoms in the reactants always equal the number of atoms in the products?

7.16. In a balanced chemical equation describing the complete combustion of a hydrocarbon, what is the ratio of atoms of C in the hydrocarbon to molecules of CO_2 produced?

7.17. How many moles of water vapor are produced for every mole of methane consumed in the combustion reaction $CH_4(g)$ + 2 $O_2(g)$ → $CO_2(g)$ + 2 $H_2O(g)$?

7.18. Why is CO produced during the combustion of hydrocarbons even when enough O_2 is present for complete combustion?

PROBLEMS

7.19. Balance the following reactions describing the formation of nitrogen compounds:
   a. $N_2(g) + O_2(g) \rightarrow NO(g)$
   b. $N_2(g) + O_2(g) \rightarrow N_2O(g)$
   c. $NO(g) + NO_3(g) \rightarrow NO_2(g)$
   d. $NO(g) + O_2(g) + H_2O(\ell) \rightarrow HNO_2(\ell)$

7.20. **Chemistry of Geothermal Vents** Some scientists believe that life on Earth originated near geothermal vents. Balance the following reactions, which are among those taking place near such vents:
   a. $CH_3SH(aq) + CO(aq) \rightarrow CH_3COSCH_3(aq) + H_2S(aq)$
   b. $H_2S(aq) + CO(aq) \rightarrow CH_3CO_2H(aq) + S_8(s)$

*7.21. Write a balanced chemical equation for each of the following reactions:
   a. Dinitrogen pentoxide reacts with sodium metal to produce sodium nitrate and nitrogen dioxide.
   b. A mixture of nitric acid and nitrous acid is formed when water reacts with dinitrogen tetroxide.
   c. At high pressure, nitrogen monoxide decomposes to dinitrogen monoxide and nitrogen dioxide.

7.22. Write a balanced chemical equation for each of the following reactions:
   a. Carbon dioxide reacts with carbon to form carbon monoxide.
   b. Potassium reacts with water to give potassium hydroxide and the element hydrogen.
   c. Phosphorus (P_4) burns in air to give diphosphorus pentoxide.

7.23. Complete and balance the following chemical equations describing the complete combustion of several hydrocarbons.
   a. $C_3H_8(g) + O_2(g) \rightarrow$
   b. $C_4H_{10}(g) + O_2(g) \rightarrow$
   c. $C_6H_6(\ell) + O_2(g) \rightarrow$
   d. $C_8H_{18}(\ell) + O_2(g) \rightarrow$

7.24. Complete and balance the following chemical equations describing the complete combustion of several hydrocarbons.
   a. $C_5H_{10}(\ell) + O_2(g) \rightarrow$
   b. $C_6H_{14}(\ell) + O_2(g) \rightarrow$
   c. $C_8H_{10}(\ell) + O_2(g) \rightarrow$
   d. $C_9H_{12}(\ell) + O_2(g) \rightarrow$

7.25. Write balanced chemical equations describing the complete combustion of the gaseous hydrocarbons with the molecular structures in Figure P7.25.

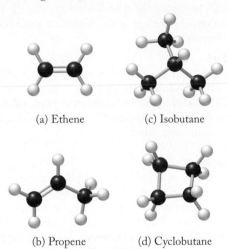

(a) Ethene   (c) Isobutane

(b) Propene   (d) Cyclobutane

FIGURE P7.25

7.26. Write balanced chemical equations describing the complete combustion of the liquid hydrocarbons with the molecular structures in Figure P7.26.

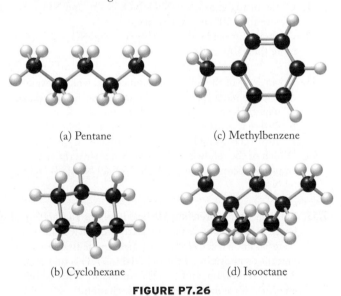

(a) Pentane   (c) Methylbenzene

(b) Cyclohexane   (d) Isooctane

FIGURE P7.26

7.27. Chemistry of Volcanic Gases Balance the following reactions that occur during volcanic eruptions:
a. $SO_2(g) + O_2(g) \rightarrow SO_3(g)$
b. $H_2S(g) + O_2(g) \rightarrow SO_2(g) + H_2O(g)$
*c. $H_2S(g) + SO_2(g) \rightarrow S_8(s) + H_2O(g)$

*7.28. Copper was one of the first metals used by humans because it can be recovered from several copper minerals including cuprite (Cu_2O), chalcocite (Cu_2S), and malachite [$Cu_2CO_3(OH)_2$]. Balance the following reactions for converting these minerals into copper metal:
a. $Cu_2O(s) + C(s) \rightarrow Cu(s) + CO_2(g)$
b. $Cu_2O(s) + Cu_2S(s) \rightarrow Cu(s) + SO_2(g)$
c. $Cu_2CO_3(OH)_2(s) + C(s) \rightarrow Cu(s) + CO_2(g) + H_2O(g)$

Stoichiometric Calculations and the Carbon Cycle

CONCEPT REVIEW

7.29. There are two ways to write the equation for the combustion of ethane:

$$C_2H_6(g) + 7/2\, O_2(g) \rightarrow 3\, H_2O(g) + 2\, CO_2(g)$$
$$2\, C_2H_6(g) + 7\, O_2(g) \rightarrow 6\, H_2O(g) + 4\, CO_2(g)$$

Do the different ways of writing the equation affect the calculation of how much CO_2 is produced from a known quantity of C_2H_6?

7.30. Suppose the same mass of the each of these components of natural gas was completely combusted: (a) CH_4, (b) C_2H_6, (c) C_3H_8, and (d) C_4H_{10}. Which one would produce the greatest mass of CO_2?

PROBLEMS

7.31. Land Management It has been estimated that better management of cropland, grazing land, and forests could reduce the amount of carbon dioxide in the atmosphere by 5.4×10^9 kilograms of carbon per year.
a. How many moles of carbon are present in 5.4×10^9 kilograms of carbon?
b. How many kilograms of carbon dioxide does this quantity of carbon represent?

7.32. Most of the CO_2 emitted by industrial sources comes from the combustion of fossil fuels, but some is also produced by cement manufacturing and the conversion of limestone ($CaCO_3$) to lime (CaO):

$$CaCO_3(s) \rightarrow CaO(s) + CO_2(g)$$

How many tons of CO_2 are produced per ton of lime?

7.33. When $NaHCO_3$ is heated above 270°C, it decomposes to $Na_2CO_3(s)$, $H_2O(g)$, and $CO_2(g)$.
a. Write a balanced chemical equation for the decomposition reaction.
b. Calculate the mass of CO_2 produced from the decomposition of 25.0 g $NaHCO_3$.

7.34. Egyptian Cosmetics PbCl(OH) is one of several lead compounds used in ancient Egyptian cosmetics (see Problem 7.50). It is prepared from PbO according to the following ancient recipe:

$$PbO(s) + NaCl(aq) + H_2O(\ell) \rightarrow PbCl(OH)(s) + NaOH(aq)$$

How many grams of PbO and how many grams of NaCl would be required to produce 10.0 g PbCl(OH)?

7.35. One step in the conversion of aluminum ore into aluminum metal involves the synthesis of cryolite (Na_3AlF_6) in the following reaction:

$$6\, HF(g) + 3\, NaAlO_2(s) \rightarrow Na_3AlF_6(s) + 3\, H_2O(\ell) + Al_2O_3(s)$$

How much $NaAlO_2$ (sodium aluminate) is required to produce 1.00 kg Na_3AlF_6?

7.36. Chromium metal can be produced from high-temperature reactions of chromium(III) oxide with silicon or aluminum:

$$Cr_2O_3(s) + 2\, Al(\ell) \rightarrow 2\, Cr(\ell) + Al_2O_3(s)$$
$$2\, Cr_2O_3(s) + 3\, Si(\ell) \rightarrow 4\, Cr(\ell) + 3\, SiO_2(s)$$

a. Calculate the mass of aluminum required to prepare 400.0 grams of chromium metal by the first reaction.

b. Calculate the mass of silicon required to prepare 400.0 grams of chromium metal by the second reaction.

7.37. Oxygen for First Responders In self-contained breathing devices used by first responders, potassium superoxide, KO_2, reacts with exhaled carbon dioxide producing potassium carbonate and oxygen:

$$4 \, KO_2(s) + 2 \, CO_2(g) \rightarrow 2 \, K_2CO_3(s) + 3 \, O_2(g)$$

How much O_2 could be produced from 85 g KO_2?

7.38. Charcoal (C) and propane (C_3H_8) are used as fuel in backyard grills.

a. Write balanced chemical equations for the complete combustion reactions of C and C_3H_8.

b. How many grams of carbon dioxide are produced from burning 500.0 grams of each of the two fuels?

*7.39. The uranium minerals found in nature must be refined and enriched in ^{235}U before the uranium can be used as a fuel in nuclear reactors. One procedure for enriching uranium relies on the reaction of UO_2 with HF to form UF_4, which is then converted into UF_6 by reaction with fluorine:

(1) $UO_2(g) + 4 \, HF(aq) \rightarrow UF_4(g) + 2 \, H_2O(\ell)$

(2) $UF_4(g) + F_2(g) \rightarrow UF_6(g)$

a. How many kilograms of HF are needed to completely react with 5.00 kg UO_2?

b. How much UF_6 can be produced from 850.0 g UO_2?

7.40. The mineral bauxite, which is mostly Al_2O_3, is the principal industrial source of aluminum metal. How much aluminum can be produced from 1.00 metric ton (1.00×10^3 kg) of Al_2O_3?

*7.41. Chalcopyrite ($CuFeS_2$) is an abundant copper mineral that can be converted into elemental copper. How much Cu is there in 1.00 kg $CuFeS_2$?

*7.42. Mining for Gold Unlike most metals, gold occurs in nature as the pure element. Miners in California in 1849 searched for gold nuggets and gold dust in streambeds, where the denser gold could be easily separated from sand and gravel. However, larger deposits of gold are found in veins of rock and can be separated chemically in a two-step process:

(1) $4 \, Au(s) + 8 \, NaCN(aq) + O_2(g) + 2 \, H_2O(\ell) \rightarrow$
$\qquad\qquad 4 \, NaAu(CN)_2(aq) + 4 \, NaOH(aq)$

(2) $2 \, NaAu(CN)_2(aq) + Zn(s) \rightarrow 2 \, Au(s) + Na_2[Zn(CN)_4](aq)$

If 23 kilograms of ore is 0.19% gold by mass, how much Zn is needed to react with the gold in the ore? Assume that reactions 1 and 2 are 100% efficient.

Percent Composition and Empirical and Molecular Formulas

CONCEPT REVIEW

7.43. What is the difference between an empirical formula and a molecular formula?

7.44. Do the empirical and molecular formulas of a compound have the same percent composition values?

7.45. Is the element with the largest atomic mass always the element present in the highest percentage by mass in a compound?

7.46. Sometimes the composition of a compound is expressed as a mole percentage or atom percentage. Are the values of these parameters likely to be the same for a given compound, or different?

PROBLEMS

7.47. Calculate the percent composition of (a) Na_2O, (b) NaOH, (c) $NaHCO_3$, and (d) Na_2CO_3.

7.48. Calculate the percent composition of (a) sodium sulfate, (b) dinitrogen tetroxide, (c) strontium nitrate, and (d) aluminum sulfide.

7.49. Organic Compounds in Space The following compounds have been detected in space. Which of them contains the greatest percentage of carbon by mass?
a. naphthalene, $C_{10}H_8$ c. pentacene, $C_{22}H_{14}$
b. chrysene, $C_{18}H_{12}$ d. pyrene, $C_{16}H_{10}$

7.50. Lead Compounds as Pigments Ancient Egyptians used lead compounds including PbS, $PbCO_3$, and $Pb_2Cl_2CO_3$ as pigments in cosmetics, and many people suffered from chronic lead poisoning as a result. Calculate the percentage of lead in each of the compounds.

7.51. Of the nitrogen oxides—N_2O, NO, N_2O_3, and NO_2—which is more than 50% oxygen by mass?

7.52. Of the sulfur oxides—S_2O, SO, SO_2, and SO_3—which is more than 50% oxygen by mass?

7.53. Do any two of the following compounds, which have been detected in outer space, have the same empirical formula?
a. naphthalene, $C_{10}H_8$ d. pyrene, $C_{16}H_{10}$
b. chrysene, $C_{18}H_{12}$ e. benzoperylene, $C_{22}H_{12}$
c. anthracene, $C_{14}H_{10}$ f. coronene, $C_{24}H_{12}$

7.54. Which of the following nitrogen oxides have the same empirical formulas? (a) N_2O; (b) NO; (c) NO_2; (d) N_2O_2; (e) N_2O_4

7.55. Surgical-Grade Titanium Medical implants and high-quality jewelry items for body piercings are frequently made of a material known as G23Ti or surgical-grade titanium. The percent composition of the material is 64.39% titanium, 24.19% aluminum, and 11.42% vanadium. What is the empirical formula of surgical-grade titanium?

7.56. A sample of an iron-containing compound is 22.0% iron, 50.2% oxygen, and 27.8% chlorine by mass. What is the empirical formula of this compound?

7.57. Phosphorus burns in pure oxygen with a brilliant white light. The product of combustion is 43.64% phosphorus and 56.36% oxygen.
a. What is the empirical formula of this compound?
b. The molar mass of the compound is 284 g/mol. What is its molecular formula?

7.58. Ferrophosphorus (Fe_2P) reacts with pyrite (FeS_2), producing iron(II) sulfide and a compound that is 27.87% P and 72.13% S by mass and has a molar mass of 444.56 g/mol.
a. Determine the empirical and molecular formulas of the compound.
b. Write a balanced chemical equation for the reaction.

7.59. Asbestos and Lung Disease Inhalation of asbestos fibers may lead to a lung disease known as asbestosis and to a form of lung cancer called mesothelioma. One form of asbestos, chrysotile, is 26.31% magnesium, 20.27% silicon, and 1.45% hydrogen by mass, with the remainder of the mass as oxygen. What is the empirical formula of chrysotile?

7.60. Chemistry of Soot A piece of glass held over a candle flame becomes coated with soot, which is the result of the incomplete combustion of candle wax. Elemental analysis of a compound extracted from a sample of soot gave these results: 7.74% H and 92.26% C by mass. Calculate the empirical formula of the compound.

7.61. What is the empirical formula of the compound that is 24.2% Cu, 27.0% Cl, and 48.8% O by mass?

7.62. A chlorine oxide used to kill anthrax spores in contaminated buildings is 52.6% Cl by mass. What is its empirical formula?

Combustion Analysis

CONCEPT REVIEW

7.63. Explain why it is important for combustion analysis to be carried out in an excess of oxygen.

7.64. Why is the quantity of CO_2 obtained in a combustion analysis not a direct measure of the oxygen content of the starting compound?

7.65. Can the results of a combustion analysis ever give the true molecular formula of a compound?

7.66. What additional information is needed to determine a molecular formula from the results of an elemental analysis of an organic compound?

PROBLEMS

7.67. A 0.100 g sample of a compound containing C, H, and O is burned in oxygen, producing 0.1783 g CO_2 and 0.0734 g H_2O. What is the empirical formula of the compound?

7.68. GRAS List for Food Additives The compound geraniol is on the U.S. Food and Drug Administration's GRAS (Generally Recognized as Safe) list and can be used in foods and personal care products. By itself, geraniol has a roselike odor, but it is frequently blended with other scents to produce the fruity fragrances of some personal care products. Complete combustion of 175 mg of geraniol produces 499 mg CO_2 and 184 mg H_2O. What is the empirical formula of geraniol?

7.69. Combustion of 135.0 mg of a hydrocarbon produces 440.0 mg CO_2 and 135.0 mg H_2O. The molar mass of the hydrocarbon is 270 g/mol. What are the empirical and molecular formulas of this compound?

7.70. The combustion of 40.5 mg of a compound extracted from the bark of the sassafras tree and known to contain C, H, and O produces 110.0 mg CO_2 and 22.5 mg H_2O. The molar mass of the compound is 162 g/mol. What are its empirical and molecular formulas?

Limiting Reactants and Percent Yield

CONCEPT REVIEW

7.71. If a reaction vessel contains equal masses of Fe and S, a mass of FeS corresponding to which of the following could theoretically be produced?

a. the sum of the masses of Fe and S
b. more than the sum of the masses of Fe and S
c. less than the sum of the masses of Fe and S

7.72. A reaction vessel contains equal masses of magnesium metal and oxygen gas. The mixture is ignited, forming MgO. After the reaction has gone to completion, the mass of the MgO is less than the mass of the reactants. Is this result a violation of the law of conservation of mass? Explain your answer.

7.73. Explain how the parameters of theoretical yield and percent yield differ.

7.74. Can the percent yield of a chemical reaction ever exceed 100%?

7.75. Give two reasons why the actual yield from a chemical reaction may be less than the theoretical yield.

7.76. A chemical reaction produces less than the expected amount of product. Is this result a violation of the law of conservation of mass?

PROBLEMS

7.77. Making Hollandaise Sauce A recipe for 1 cup of hollandaise sauce calls for $\frac{1}{2}$ cup of butter, $\frac{1}{4}$ cup of hot water, 4 egg yolks, and the juice of a medium-sized lemon. How many cups of sauce can be made from a pound (2 cups) of butter, a dozen eggs, 4 medium-sized lemons, and an unlimited supply of hot water?

7.78. A factory making toy wagons has 13,466 wheels, 3360 handles, and 2400 wagon beds in stock. What is the maximum number of wagons with four wheels the factory can make from these components?

*7.79. Suppose 75 metric tons of coal that is 3.0% sulfur by mass is burned at a power plant. During combustion the sulfur is converted into SO_2. Antipollution scrubbers installed in the smoke stacks of the power plant capture 3.9 metric tons of this SO_2. How efficient are the scrubbers in capturing SO_2? How many metric tons of SO_2 escaped?

7.80. One reaction in the production of sulfuric acid involves the conversion of sulfur dioxide to sulfur trioxide. In the presence of excess O_2, 88 kg SO_2 produces 106 kg SO_3. What is the percent yield?

7.81. Ammonia gas combines with hydrogen chloride gas, forming solid ammonium chloride.
a. Write a balanced chemical equation for the reaction.
b. In a reaction mixture of 3.0 g NH_3 and 5.0 g HCl, which is the limiting reactant?
c. How many grams of NH_4Cl could form from the reaction mixture in part b?
d. How much of which reactant is left over in the reaction mixture in part b?

7.82. A reaction vessel contains 10.0 g CO and 10.0 g O_2 which combine to form CO_2:

$$2\,CO(g) + O_2(g) \rightarrow 2\,CO_2(g)$$

a. Which reactant is the limiting reactant?
b. How many grams of CO_2 could be produced?
c. How many grams of the nonlimiting reactant are left over?

*7.83. Syngas An industrial process for producing hydrogen gas is based on the reaction between carbon heated to

incandescence and steam that produces a mixture of CO and H_2 called synthesis gas, or *syngas*.
a. Write a balanced chemical equation describing the production of syngas.
b. If a reaction vessel that initially contains 66 kilograms of incandescent carbon and excess steam produces 6.8 kg H_2, what is the percent yield?

*7.84. Baking soda ($NaHCO_3$) is produced on an industrial scale by the Solvay process. A key reaction in the process is

$$NaCl(aq) + NH_3(aq) + CO_2(aq) + H_2O(\ell) \rightarrow$$
$$NaHCO_3(s) + NH_4Cl(aq)$$

Suppose a reaction vessel initially contains 58.5 kg NaCl, 18.8 kg NH_3, and excess CO_2 and H_2O. If 66 kg $NaHCO_3$ is produced, what is the percent yield?

7.85. Chemistry of Fermentation Yeast converts glucose ($C_6H_{12}O_6$) in aqueous solution into ethanol (CH_3CH_2OH, $d = 0.789$ g/mL) in a process called fermentation. Carbon dioxide is also produced.
a. Write a balanced chemical equation for the fermentation reaction.
b. If 100.0 grams of glucose yields 50.0 mL of ethanol, what is the percent yield for the reaction?

*7.86. **Composition of Seawater** A 1-liter sample of seawater contains 19.4 g Cl^-, 10.8 g Na^+, and 1.29 g Mg^{2+}.
a. How many moles of each ion are present?
b. If the seawater were evaporated, would there be enough chloride ions present to form the chloride salts of all the sodium and magnesium ions present?

Additional Problems

*7.87. **Artificial Bones for Medical Implants** The material often used to make artificial bones is the same material that gives natural bones their strength. Its common name is hydroxyapatite, and its formula is $Ca_5(PO_4)_3OH$.
a. Propose a systematic name for this compound.
b. What is the mass percentage of calcium in it?
c. When treated with hydrogen fluoride, hydroxyapatite becomes fluorapatite [$Ca_5(PO_4)_3F$], an even stronger substance. Does the percent mass of Ca increase or decrease as a result of this substitution?

*7.88. As a solution of copper sulfate slowly evaporates, beautiful blue crystals form. Their chemical formula is $CuSO_4 \cdot 5H_2O$.
a. What is the percent water in this compound?
b. At high temperatures the water is driven off as steam. What fraction of the original sample's mass is lost as a result?

7.89. Aluminum is obtained from the mineral bauxite. Suppose that a sample of bauxite from Jamaica is 86% aluminum oxide. If 2.3 metric tons of aluminum metal is recovered from 5.1 metric tons of the bauxite ore, what is the percent yield of the recovery process?

7.90. **Chemistry of Copper Production** The mineral chalcopyrite ($CuFeS_2$) is an important source of copper metal, though recovering the metal requires several chemical reactions that transform $CuFeS_2$ into CuS, then Cu_2S, and finally Cu metal. The pennies minted in the United States between 1909 and 1982 weighed 3.11 g and were 95% by mass copper.

a. How much chalcopyrite had to be mined to produce one dollar's worth of these pennies?
b. How much chalcopyrite had to be mined to produce one dollar's worth of the pennies if the first reaction had a percent yield of 85% and the second and third reactions had percent yields of essentially 100%?
c. How much chalcopyrite had to be mined to produce one dollar's worth of the pennies if each of the reactions proceeded with 85% yield?

*7.91. **Mining for Gold** Gold can be extracted from the surrounding rock using a solution of sodium cyanide. While effective for isolating gold, toxic cyanide finds its way into watersheds, causing environmental damage and harming human health.

$$4\,Au(s) + 8\,NaCN(aq) + O_2(g) + 2\,H_2O(\ell) \rightarrow$$
$$4\,NaAu(CN)_2(aq) + 4\,NaOH(aq)$$

$$2\,NaAu(CN)_2(aq) + Zn(s) \rightarrow 2\,Au(s) + Na_2[Zn(CN)_4](aq)$$

a. If a sample of rock contains 0.009% gold by mass, how much NaCN is needed to extract the gold from 1 metric ton (10^3 kg) of rock as $NaAu(CN)_2$?
b. How much zinc is needed to convert the $NaAu(CN)_2$ from part a to metallic gold?
c. The gold recovered in part b is manufactured into a gold ingot in the shape of a cube. The density of gold is 19.3 g/cm^3. How big is the block of gold in cm^3?

*7.92. Phosgenite, a lead compound with the formula $Pb_2Cl_2CO_3$, is found in Egyptian cosmetics. Phosgenite was prepared by the reaction of PbO, NaCl, and CO_2. An unbalanced expression of the reactant mixture is

$$PbO(s) + NaCl(aq) + H_2O(\ell) + CO_2(g) \rightarrow$$
$$Pb_2Cl_2CO_3(s) + NaOH(aq)$$

a. Balance the equation.
b. How many grams of phosgenite can be obtained from 10.0 g PbO and 10.0 g NaCl in the presence of excess water and CO_2?
c. Phosgenite can be considered a mixture of two lead compounds. Which compounds appear to be combined to make phosgenite?

*7.93. Uranium oxides used in the preparation of fuel for nuclear reactors are separated from other metals in minerals by converting the uranium to $UO_x(NO_3)_y(H_2O)_z$, where uranium has a positive charge ranging from 3+ to 6+.
a. Roasting $UO_x(NO_3)_y(H_2O)_z$ at 400°C leads to loss of water and decomposition of the nitrate ion to nitrogen oxides, leaving behind a product with the formula U_aO_b that is 83.22% U by mass. What are the values of a and b? What is the charge on U in U_aO_b?
b. Higher temperatures produce a different uranium oxide, U_cO_d, with a higher uranium content, 84.8% U. What are the values of c and d? What is the charge on U in U_cO_d?
c. The values of x, y, and z in $UO_x(NO_3)_y(H_2O)_z$ are found by gently heating the compound to remove all of the water. In a laboratory experiment, 1.328 g $UO_x(NO_3)_y(H_2O)_z$ produced 1.042 g $UO_x(NO_3)_y$. Continued heating generated 0.742 g U_nO_m. Using the information in parts a and b, calculate x, y, and z.

*7.94. Corn farmers in the American Midwest typically use 5.0×10^3 kilograms of ammonium nitrate fertilizer per square kilometer of cornfield per year. Some of the fertilizer washes into the Mississippi River and eventually flows into the Gulf of Mexico, promoting the growth of algae and endangering other aquatic life.

a. Ammonium nitrate can be prepared by the following reaction:

$$NH_3(g) + HNO_3(aq) \rightarrow NH_4NO_3(aq)$$

How much nitric acid would be required to make the fertilizer needed for 1 km^2 of cornfield per year?

b. Ammonium ions dissolved in groundwater may be converted into NO_3^- ions by bacterial action:

$$NH_4^+(aq) + 2\,O_2(g) \rightarrow NO_3^-(aq) + H_2O(\ell) + 2\,H^+(aq)$$

If 10% of the ammonium component of 5.0×10^3 kilograms of fertilizer ends up as nitrate ions, how much oxygen would be consumed?

7.95. **Fiber in the Diet** Dietary fiber is a mixture of many compounds including xylose ($C_5H_{10}O_5$) and methyl galacturonate ($C_7H_{12}O_7$).

a. Do these compounds have the same empirical formula?

b. Write balanced chemical equations for the complete combustion of xylose and methyl galacturonate.

7.96. Some catalytic converters in automobiles contain two manganese oxides: Mn_2O_3 and MnO_2.

a. What are the names of these compounds?

b. What is the manganese content of each (expressed as a percent by mass)?

c. Explain how Mn_2O_3 and MnO_2 are consistent with the law of multiple proportions.

*7.97. A number of chemical reactions have been proposed for the formation of organic compounds from inorganic precursors. Here is one of them:

$$H_2S(g) + FeS(s) + CO_2(g) \rightarrow FeS_2(s) + HCO_2H(\ell)$$

a. Identify the ions in FeS and FeS_2.

b. What are the names of FeS and FeS_2?

c. How much HCO_2H is obtained by reacting 1.00 g FeS, 0.50 g H_2S, and 0.50 g CO_2 if the reaction results in a 50.0% yield?

*7.98. Organic compounds called *carbohydrates* may be formed in reactions between iron(II) sulfide and carbonic acid:

$$2\,FeS + H_2CO_3 \rightarrow 2\,FeO + \tfrac{1}{n}(CH_2O)_n + 2\,S$$

a. What is the empirical formula of these carbohydrates?

b. How much carbohydrate is produced from a reaction mixture that initially contains 211 g FeS and excess H_2CO_3 if the reaction results in a 78.5% yield?

c. If the carbohydrate product has a molecular mass of 300 amu, what is its molecular formula?

*7.99. **Marine Chemistry of Iron** On the seafloor, solid iron(II) oxide may react with water to form solid Fe_3O_4 and hydrogen gas.

a. Write a balanced chemical equation describing the reaction.

b. When CO_2 is also present, the product of the reaction is methane, not hydrogen. Write a balanced chemical equation describing this reaction.

7.100. Titanium dioxide and zinc oxide are common names of two of the active ingredients approved by the U.S. FDA for use in sunscreens.

a. What are the chemical formulas of these compounds?

b. What are the proper names of the compounds based on the rules for naming described in Chapter 4?

c. Which of the two contains the higher percentage of oxygen by mass?

*7.101. E-85 is an alternative fuel for automobiles and light trucks that consists of 85% (by volume) ethanol, CH_3CH_2OH, and 15% gasoline. The density of ethanol is 0.79 g/mL.

a. How many moles of ethanol are in a gallon of E-85?

b. How many moles of carbon dioxide are produced by the complete combustion of the ethanol in a gallon of E-85 fuel?

7.102. A sealed chamber contains 1.604 g CH_4 and 6.800 g O_2. The mixture is ignited. How many grams of CO_2 are produced?

*7.103. You are given a 0.6240 g sample of a substance with the generic formula $MCl_2 \cdot 2\,H_2O$. After completely drying the sample (which means removing the 2 mol of H_2O per mole of MCl_2), the sample has a mass of 0.5471 g. What is the identity of element M?

7.104. A compound found in crude oil is 93.71% C and 6.29% H by mass. The molar mass of the compound is 128 g/mol. What is its molecular formula?

7.105. A reaction vessel for synthesizing ammonia by reacting nitrogen and hydrogen is charged with 6.04 kg H_2 and excess N_2. If 28.0 kg NH_3 is produced, what is the percent yield of the reaction?

7.106. If a cube of table sugar, which is made of sucrose, $C_{12}H_{22}O_{11}$, is added to concentrated sulfuric acid, the acid "dehydrates" the sugar: it removes the hydrogen and oxygen from it, leaving behind a lump of carbon. What percentage of the initial mass of sugar is carbon?

7.107. **Reducing SO_2 Emissions** One way in which SO_2 is removed from the "stack" gases of coal-burning power plants is by spraying the gases with fine particles of solid calcium oxide suspended in O_2 gas. The product of the reaction of SO_2, CaO, and O_2 is calcium sulfate.

a. Write a balanced chemical equation describing the reaction.

b. How many metric tons of calcium sulfate would be produced from each ton of SO_2 that is trapped?

7.108. **Gas Grill Reaction** The burner in a gas grill mixes 24 volumes of air for every one volume of propane (C_3H_8) fuel. Like all gases, the volume that propane occupies is directly proportional to the number of moles of it at a given temperature and pressure. Air is 21% (by volume) O_2. Is the flame produced by the burner fuel-rich (excess propane in the reaction mixture), fuel-lean (not enough propane), or stoichiometric (just right)?

7.109. A common mineral in Earth's crust has the chemical composition 34.55% Mg, 19.96% Si, and 45.49% O. What is its empirical formula?

7.110. **Ozone Generators** Some indoor air-purification systems work by converting a little of the oxygen in the air to ozone, which kills mold and mildew spores and other biological air pollutants. The chemical equation for the ozone generation reaction is

$$3\,O_2(g) \rightarrow 2\,O_3(g)$$

It is claimed that one such system generates 4.0 g O_3 per hour from dry air passing through the purifier at a flow of 5.0 L/min. If 1 liter of indoor air contains 0.28 g O_2, what percentage of the O_2 is converted to O_3 by the air purifier?

8

Aqueous Solutions
Chemistry of the Hydrosphere

The Composition of Seawater: Steady as She Goes

Photographs taken from space make it clear why Earth is called the "water planet." Depressions in Earth's crust contain over 10^{21} L of freshwater and seawater—enough to cover 70% of the planet. All of this water contains dissolved ionic and molecular compounds. The concentrations of solutes vary considerably from one body of freshwater to the next, but the proportions of the major ions in seawater are essentially the same all over the world. Perhaps more remarkably, their concentrations appear to have changed little for over a billion years.

This constant oceanic composition may come as a surprise given the continuous influx of soluble and suspended material from the world's major rivers and from dust blowing off the continents. Each year, rivers deliver about 4×10^{12} kg of dissolved ionic compounds to the sea. This quantity may seem impressive, but the huge volume of freshwater flowing to the sea (4×10^{16} L per year) means that the total concentration of all ionic compounds in river water is typically less than 1/100 the salinity of seawater. In addition, the proportions of the ions in river water don't match those of seawater.

Given these differences in composition, how did the sea become so salty in the first place, and how has it stayed that way? Not all of the salt got there by physical erosion and chemical weathering of the land. Much of the Cl^- content, for example, probably came from HCl released by underwater volcanoes billions of years ago, and most of the Na^+ ions probably leached out of the ocean's floor as it first filled with water. Today, an elaborate system of physical and chemical processes operates within and above Earth's oceans to maintain their composition. About 90% of the water vapor in the atmosphere—the source of all that river water flowing into the sea—is evaporated seawater. Therefore, river water does not dilute the saltiness of seawater, because the evaporation process that created it actually made the sea a little saltier in the first place.

On the other hand, Ca^{2+} ions flowing into the sea don't remain in seawater very long. They are taken up by corals, shellfish, and other sea creatures to grow the hard parts of their bodies (made mostly of $CaCO_3$). Other ions such as Fe^{3+} and Mn^{2+} are soluble in freshwater—which is, on average, slightly acidic—but become insoluble when they reach the sea, which is slightly basic.

The Blue Planet Life exists on Earth's surface because liquid water exists here. Most of it is seawater—a solution of NaCl and other salts. The orange spheres represent Na^+, the green spheres are Cl^-, and Mg^{2+} ions are silver. ▶

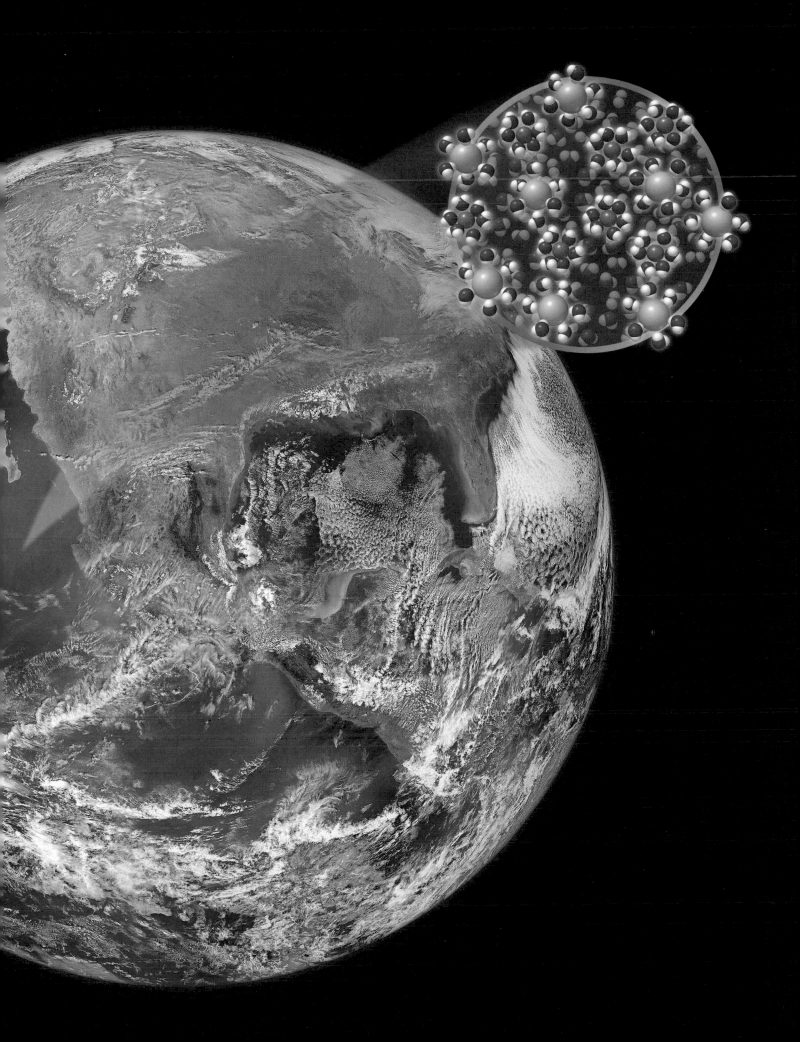

The composition of the sea and the survival of the creatures in it are linked to chemical and biochemical reactions that depend on acid/base balance, the formation of insoluble ionic compounds, and reactions in which elements are oxidized or reduced. In this chapter we explore each of these types of reactions. You will learn how to predict what reactions occur when solutions mix together, how to write chemical equations describing these reactions, and how to relate the quantities of reactants and products involved, including their concentrations in solution.

8.1 Solutions and Their Concentrations

As defined in Section 1.3, a solution is a homogeneous mixture of two or more substances. Though we usually think of solutions as liquids, homogeneous mixtures of solids and gases also exist. For example, familiar metals such as brass (a mixture of copper and zinc), bronze (copper and tin), and stainless steel (chromium and other metals mixed with iron) are solutions; so is filtered air. The substance present in the greatest proportion in a solution is called the solvent, and all the substances dissolved in it are solutes. Solutions in which water is the solvent are called *aqueous* solutions. They are the focus of this chapter.

The concentration of an aqueous solution is expressed as a ratio of the mass of the solute to the total volume or mass of the solution. Some concentration units are based on mass-to-mass ratios, such as milligrams of solute per kilogram of solution (mg/kg). Other concentration units are based on mass-to-volume ratios, such as milligrams of solute per liter of solution (mg/L). Actually, these two units are equivalent for dilute aqueous solutions because the densities of these solutions are nearly the same as that of water, which is 1.000 kg/L at 4°C. Though the density of water changes with temperature, it is within 0.003 units of 1.000 kg/L between 0 and 25°C. Another concentration unit that is synonymous with mg/kg is parts per million (ppm). These units are equivalent because $1 \text{ mg} = 10^{-3}$ g and $1 \text{ kg} = 10^3$ g. Therefore,

$$1 \text{ mg/kg} = 10^{-3} \text{ g}/10^3 \text{ g} = 1 \text{ g}/10^6 \text{ g} = 1 \text{ ppm}$$

One millionth of a gram per gram is the same as one gram per million grams, or simply one part per million. Even smaller concentrations may be expressed in micrograms per kilogram (μg/kg) or micrograms per liter (μg/L), which are also equivalent for dilute aqueous solutions and are the same as parts per billion (ppb).

Because we are interested in describing reactions in solutions, we will be working with quantities of dissolved reactants and products that are related by how many *moles* of each of them there are in balanced chemical equations. Consequently, we need to express their concentrations in units based on moles of solute per mass or volume of solution. One such unit is **molarity (M)**, which is the number of moles (n) of solute in a volume (V) of 1 liter of solution.[1] In equation form:

$$M = \frac{\text{moles of solute}}{\text{liter of solution}} = \frac{n}{V} \tag{8.1}$$

Often we know the volume and concentration of a solution and need to calculate the number of moles of solute in it. Rearranging Equation 8.1 to solve for n gives us an expression that meets this need:

$$n = V \times M = \cancel{L} \times \frac{\text{mol}}{\cancel{L}} \tag{8.2}$$

CONNECTION Solvents and solutes were introduced in Chapter 6, where we used intermolecular forces to predict the solubility of molecular substances.

▶❘❘ **CHEMTOUR** Molarity

molarity (M) the number of moles of solute per liter of solution: $M = n/V$; also called *molar concentration.*

[1]Note that molarity is symbolized by M while molar mass is $\mathcal{M}$ throughout this text.

To calculate the mass (m) in grams of the solute in a solution of a known volume and molar concentration, we can multiply the number of moles of solute obtained using Equation 8.2 by the molar mass ($\mathcal{M}$) of the solute:

$$m = (V \times M) \times \mathcal{M} = \left(\cancel{L} \times \frac{\cancel{mol}}{\cancel{L}}\right) \times \frac{g}{\cancel{mol}} \qquad (8.3)$$

Equation 8.3 is particularly useful when we need to calculate the mass of solute required to prepare a solution of a desired volume and concentration.

In many environmental and biological systems, solute concentrations are much less than 1.0 M. In Table 8.1, for instance, the average concentrations of most of the major ions in seawater and in human serum are more conveniently expressed in *milli*moles per liter (mmol/L), or millimolarity (mM), where 1 mM = 10^{-3} M. Even smaller concentrations of minor and trace elements in seawater and of many biologically active substances in blood, urine, and other biological liquids are expressed in micromolarity (μM, 1 μM = 10^{-6} M), nanomolarity (nM, 1 nM = 10^{-9} M), and even picomolarity (pM, 1 pM = 10^{-12} M).

The concentrations in the first column of numbers in Table 8.1 are expressed in millimoles of solute per kilogram of seawater. Environmental scientists and especially oceanographers prefer to use concentration units based on the masses of water samples rather than their volumes because the volume of a given mass of water varies with changing temperature and pressure, whereas its mass and the quantities of solutes dissolved in that mass remain constant.

CONCEPT TEST ···

Rank these solutions from most concentrated to least concentrated: (a) 0.0053 M NaCl; (b) 54 mM NaCl; (c) 550 μM NaCl; (d) 56,000 nM NaCl.

(Answers to Concept Tests are in the back of the book.)

···

TABLE 8.1 Average Concentrations of 11 Major Constituents of Seawater and Human Serum

Constituent	SEAWATER		HUMAN SERUM
	mmol/kg	mM	mM
Na$^+$	468.96	480.57	135–145
K$^+$	10.21	10.46	3.5–5.0
Mg^{2+}	52.83	54.14	0.08–0.12
Ca^{2+}	10.28	10.53	0.2–0.3
Sr^{2+}	0.0906	0.0928	$<3 \times 10^{-4}$
Cl$^-$	545.88	559.40	98–108
SO$_4^{2-}$	28.23	28.93	0.3
HCO$_3^-$	2.06	2.11	22–30
Br$^-$	0.844	0.865	0.04–0.06
B(OH)$_3$	0.416	0.426	$<8 \times 10^{-4}$
F$^-$	0.068	0.070	5–6

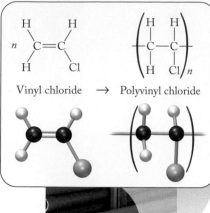

FIGURE 8.1 Vinyl chloride is the common name of a small molecule, called a monomer (Greek for "one unit"), from which very large molecules or polymers ("many units") called polyvinyl chloride (PVC) are made. PVC is found in many products including the pipes used for water mains and for drainage and sewer systems.

SAMPLE EXERCISE 8.1   **Converting Mass-per-Volume Concentrations into Molarity**   **LO1**

Vinyl chloride (Figure 8.1) is one of the most widely used industrial chemicals (about 3×10^{10} kg is produced each year, mostly for making polyvinyl chloride plastic). It is also one of the most toxic, and it's a known carcinogen. The maximum concentration of vinyl chloride allowed in drinking water in the United States is 0.002 mg/L. What is that concentration in moles per liter?

COLLECT AND ORGANIZE We are asked to convert a concentration from milligrams of solute per liter of solution to molarity (mol/L). The molar mass of a substance relates the mass and number of moles of a given quantity of the substance.

ANALYZE Because the solute mass is given in milligrams, we need to convert it to grams and then to moles. Given the number of C and Cl atoms in its molecular structure, the molar mass of vinyl chloride is likely to be between 50 and 100 g/mol, so 0.002 mg/L should translate into about 0.00002 mmol/L—or 2×10^{-8} M, or 0.02 μM.

SOLVE The molar mass of vinyl chloride (C_2H_3Cl) is

$$\mathcal{M} = 2(12.01 \text{ g/mol}) + 3(1.008 \text{ g/mol}) + 35.45 \text{ g/mol} = 62.49 \text{ g/mol}$$

The molarity equivalent of 0.002 mg/L of vinyl chloride is

$$\frac{0.002 \text{ mg}}{L} \times \frac{1 \text{ g}}{10^3 \text{ mg}} \times \frac{1 \text{ mol}}{62.49 \text{ g}} = \frac{3 \times 10^{-8} \text{ mol}}{L} = 3 \times 10^{-8} \text{ } M \quad \text{or} \quad 0.03 \text{ } \mu M$$

THINK ABOUT IT This concentration is close to the value we predicted. Expressing it using micromolarity units to one significant figure is appropriate because the initial concentration value was known to only one significant figure. Given the hazards vinyl chloride poses to human health, it is reasonable that the allowed concentration of it in drinking water would be very small.

Practice Exercise When a 1.00 L sample of water from the surface of the Dead Sea is evaporated, 179 g of $MgCl_2$ is recovered. What was the molarity of $MgCl_2$ in the water sample?

(Answers to Practice Exercises are in the back of the book.)

SAMPLE EXERCISE 8.2   **Converting Mass-per-Mass Concentrations into Molarity**   **LO1**

A water sample from the Great Salt Lake in Utah contains 83.6 mg Na^+ per gram of lake water. What is the molar concentration of Na^+ ions if the density of the lake water is 1.160 g/mL?

COLLECT AND ORGANIZE Our task is to convert concentration units from mg Na^+ per gram of solution to mol Na^+ per liter of solution. We know the lake water's density, which we will use to convert a concentration based on mass of solution to molarity, which is based on volume of solution. The molar mass of Na is 22.99 g/mol.

ANALYZE To convert a concentration based on mass of solute per mass of solution into molarity (moles of solute per volume of solution), we need to convert solute mass into moles and solution mass into a volume in liters. The solute conversion involves dividing by the molar mass of Na, and the solution conversion includes dividing by its density to obtain an equivalent volume value. The initial concentration value of 83.6 mg Na^+/g is the same as 83.6 g/kg. This mass of Na^+ corresponds to about 4 moles of Na^+, and the volume of a kilogram of lake water will be only a little less than a liter given a density of about 1.2 kg/L, so the answer should be close to 4 M.

SOLVE Calculating moles of Na^+:

$$83.6 \; \text{mg} \; Na^+ \times \frac{1 \; \text{g}}{1000 \; \text{mg}} \times \frac{1 \; \text{mol}}{22.99 \; \text{g}} = 3.64 \times 10^{-3} \; \text{mol} \; Na^+$$

The volume of exactly one gram of lake water is

$$1 \; \text{g} \times \frac{1 \; \text{mL}}{1.160 \; \text{g}} \times \frac{1 \; \text{L}}{1000 \; \text{mL}} = 8.621 \times 10^{-4} \; \text{L}$$

and the molarity of Na^+ ions in the lake is

$$M = \frac{3.64 \times 10^{-3} \; \text{mol} \; Na^+}{8.621 \times 10^{-4} \; \text{L}} = 4.22 \; M$$

THINK ABOUT IT According to Table 8.1, the average concentration of Na^+ in seawater is 480.57 mM or 0.48057 M. The concentration of Na^+ in the Great Salt Lake is nearly 10 times greater. A clue to the lake's salinity is provided by its density: 1.160 g/mL is 16% greater than the density of pure water.

Practice Exercise If the density of seawater at a depth of 10,000 m is 1.071 g/mL, and a 25.0 g sample of water from that depth contains 99.7 mg K^+, what is the molarity of potassium ions in the sample?

SAMPLE EXERCISE 8.3 **Calculating the Quantity of Solute Needed to Prepare a Solution** **LO2**

An aqueous solution called phosphate-buffered saline (PBS) is used in biology research to wash and store living cells. It contains ionic solutes including 10.0 mM $Na_2HPO_4 \cdot 2 \; H_2O$. How many grams of this solute would you need to prepare 10.0 L of PBS?

COLLECT AND ORGANIZE We are asked to calculate the mass of solute needed to prepare a solution of specified volume and concentration. We know the formula of the solute. Equation 8.3 relates the mass of a solute in a solution to the volume and concentration of the solution and the molar mass of the solute.

ANALYZE We need to use the formula of the solute to calculate its molar mass and then use that value and the volume and concentration of the solution in Equation 8.3 to calculate the mass we need. We need enough solute to make 10.0 liters of 10.0 mM (or 0.0100 M) solution, which translates into 10.0 L $\times$ 0.01 mol/L or 0.1 moles. Based on its formula, the molar mass of the solute should be around 200 g/mol, so 0.1 mol has a mass of about 20 g.

SOLVE Calculating the molar mass of $Na_2HPO_4 \cdot 2 \; H_2O$:

$$\mathcal{M} = 2(22.99 \; \text{g/mol}) + 5(1.008 \; \text{g/mol}) + 30.97 \; \text{g/mol} + 6(16.00 \; \text{g/mol})$$

$$= 177.99 \; \text{g/mol}$$

Inserting $\mathcal{M}$ and the given values of volume and molarity into Equation 8.3:

$$m = (V \times M) \times \mathcal{M}$$

$$= \left(10.0 \; \text{L} \times 10.0 \; \frac{\text{mmol}}{\text{L}} \times \frac{1 \; \text{mol}}{1000 \; \text{mmol}} \right) \times \frac{177.99 \; \text{g}}{1 \; \text{mol}} = 17.8 \; \text{g}$$

THINK ABOUT IT In solving the problem we converted "mmol/L" to "mol/L" so that the product ($V \times M$) yielded moles of solute, which we then converted into a mass in grams. The result of the calculation is reassuringly close to our estimate.

Practice Exercise An aqueous solution known as Ringer's lactate is administered intravenously to trauma victims suffering from blood loss or severe burns. The solution contains the chloride salts of sodium, potassium, and calcium and is 4.00 mM in sodium lactate NaCH$_3$CH(OH)CO$_2$. How many grams of sodium lactate are needed to prepare 10.0 L of Ringer's lactate? ⚙

8.2 Dilutions

▶❚❚ CHEMTOUR Dilution

Laboratory scientists and technicians who routinely analyze environmental, biological, or clinical samples often use commercially available **standard solutions**. Some of these solutions contain substances used in standardized tests required by regulatory agencies such as the U.S. EPA and Environment Canada. Other solutions are calibration standards: they contain known concentrations of the substances whose concentrations the lab workers need to determine. Calibration standards typically contain much higher concentrations of the substances of interest than occur in the sample being analyzed, so they need to be diluted.

Dilution is the process of reducing solute concentration by adding more solvent to a solution. This usually involves transferring a precisely measured volume of a concentrated solution to a volumetric flask and then filling the flask to its calibration mark with solvent. For example, suppose we wish to prepare 50.0 mL of an aqueous solution in which the concentration of Cu^{2+} ions is 20 ppm. We have available a commercial standard that is 1000 ppm in Cu^{2+} ions. What volume of the 1000 ppm solution should we transfer to a 50.0 mL volumetric flask to be diluted to its full volume with water? There are several ways to calculate the volume to be transferred. All of them are based on the same basic principle: the number of Cu^{2+} ions transferred from the 1000 ppm solution to the flask must be the same number of Cu^{2+} ions present in the entire volume of the 20 ppm solution. Adding water to the flask to prepare the dilute solution does not change the number of Cu^{2+} ions in it, only the volume they occupy.

One way to calculate the volume to be transferred starts with a comparison of the final and initial concentrations of Cu^{2+} ions: 1000 ppm and 20 ppm. The ratio of the two is 20/1000, or 1/50. If we are making 50 mL of a solution that has 1/50 the concentration of the solution transferred, then we need to transfer 1/50 of 50 mL, or 1 mL. To transfer a volume that small, we could use a volumetric pipet that is calibrated to deliver 1.00 mL.

Sometimes we need to use a standard solution to prepare a diluted solution with a particular *molar* concentration. For example, Figure 8.2 shows how to prepare 250.0 mL of 0.100 M CuSO$_4$ starting with a 1.00 M solution. To calculate the volume of the more concentrated solution to be transferred, we use a variation of Equation 8.2:

$$n = V_i \times M_i \tag{8.4}$$

where the "i" subscripts refer to quantities of the *initial* (concentrated) solution. The number of particles of solute transferred is unaffected by dilution, so a corresponding equality exists for the dilute solution:

$$n = V_f \times M_f \tag{8.5}$$

standard solution a solution of known concentration that is used in chemical analysis.

dilution the process of lowering the concentration of a solution by adding more solvent.

where the "f" subscripts refer to the volume and molarity of the *final* (dilute) solution. Logic tells us that two quantities equal to the same quantity (*n*) are equal to each other, so we can set as equal the right sides of Equations 8.4 and 8.5:

$$V_i \times M_i = V_f \times M_f \qquad (8.6)$$

Rearranging Equation 8.6 to solve for V_i, we obtain a handy equation for calculating the volume of a concentrated solution to be transferred to prepare a dilute solution:

$$V_i = \frac{V_f \times M_f}{M_i} \qquad (8.7)$$

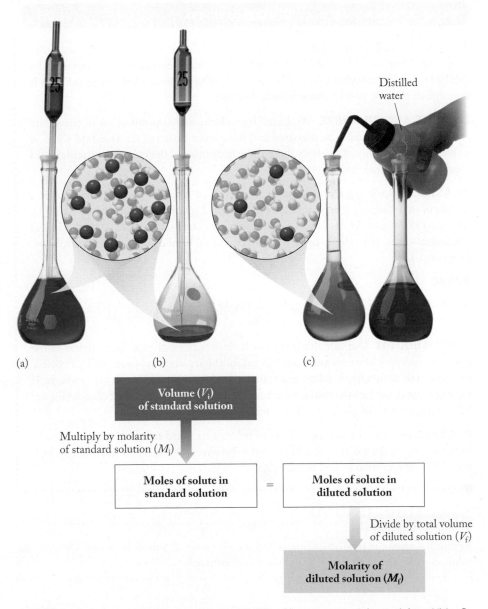

(a)   (b)   (c)

Distilled water

Volume (V_i) of standard solution

Multiply by molarity of standard solution (M_i)

Moles of solute in standard solution = **Moles of solute in diluted solution**

Divide by total volume of diluted solution (V_f)

Molarity of diluted solution (M_f)

FIGURE 8.2 To prepare 250.0 mL of 0.100 M CuSO$_4$, (a) a pipette is used to withdraw 25.0 mL of 1.00 M CuSO$_4$. (b) This volume of standard solution is transferred to a 250.0 mL volumetric flask. (c) Distilled water is added to bring the volume of the dilute solution to 250.0 mL. Note that the color of the dilute solution is lighter than the standard solution.

Inserting the values for our preparation of 0.100 M $CuSO_4$ yields

$$V_i = \frac{250.0 \text{ mL} \times 0.100 \text{ }\cancel{M}}{1.00 \text{ }\cancel{M}} = 25.0 \text{ mL}$$

Therefore, we would use a 25.0 mL volumetric pipette to transfer the concentrated solution, as shown in Figure 8.2.

Equation 8.7 can be used for *any* units of volume and concentration as long as the units used to express the initial and final volumes are the same, and the units used to express the concentration of the initial solution are the same as those used for the final one.

SAMPLE EXERCISE 8.4   Diluting Solutions   **L02**

The solution used in hospitals for intravenous infusion—called *physiological saline* or *normal saline*—is 0.155 M NaCl. It may be prepared by diluting a commercially available standard solution that is 1.76 M NaCl. What volume of standard solution is required to prepare 10.0 L of physiological saline?

COLLECT AND ORGANIZE We know the volume and concentration of the dilute solution (normal saline) to be prepared and the concentration of the standard solution to be diluted. Equation 8.7 relates these three quantities to the volume of the standard solution to be transferred.

ANALYZE Calculating the volume of the standard solution to be transferred is a matter of inserting the three quantities given into Equation 8.7, where V_f = 10.0 L, M_f = 0.155 M, and M_i = 1.76 M. The final concentration is about 1/10 the initial concentration, so the volume to be transferred should be about 1/10 the final volume, or about 1 L.

SOLVE

$$V_i = \frac{V_f \times M_f}{M_i} = \frac{10.0 \text{ L} \times 0.155 \text{ }\cancel{M}}{1.76 \text{ }\cancel{M}} = 0.881 \text{ L or } 881 \text{ mL}$$

THINK ABOUT IT Our calculated volume is close to our estimate. Its value raises the question: how would we measure out 881 mL of the standard solution with acceptable precision and accuracy? A 1-liter graduated cylinder might not be precise enough. If we knew the density of the initial solution, we could use an electronic balance (Figure 8.3) to dispense up to 10 kg of solution with a precision of 0.01 g.

Practice Exercise The concentration of Pb^{2+} in a standard solution is 1.000 mg/mL. What volume of the solution should be used to prepare 500.0 mL of a solution in which the Pb^{2+} concentration is 2.50 mg/L? ⚙

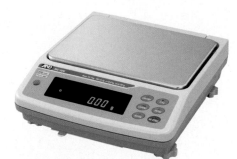

FIGURE 8.3 This electronic balance can be used to measure masses up to 10 kg to the nearest 0.01 g.

(a)   (b)   (c)   (d)   (e)

FIGURE 8.4 Aqueous solutions of cough syrup.

CONCEPT TEST

Figure 8.4 shows several solutions of red cough syrup dissolved in water. Order these solutions from the most dilute to the most concentrated.

8.3 Electrolytes and Nonelectrolytes

Consider the images in Figure 8.5. They show an apparatus in which a lightbulb and two graphite rods are connected in series to a source of electricity (not shown) so that the bulb lights up only when enough electric current flows between the rods. As you can see, this happens in some but not all of the images in Figure 8.5. It does not happen in Figure 8.5(a) when the two rods are immersed in distilled water. The lack of electric current means that distilled water is not a good conductor of electricity. The bulb does light when the rods are immersed in 0.50 M NaCl (Figure 8.5b), which means that this solution *is* a good conductor of electricity. How does adding NaCl make water a good conductor? Is there something inherently conductive about crystals of solid NaCl? The answer to that question is provided in Figure 8.5(c): the bulb does not light when the rods are immersed in a beaker of salt crystals.

The key to the conductivity of a solution of NaCl must be linked to the salt's dissolution in water. We know that NaCl is made of Na$^+$ and Cl$^-$ ions. When NaCl dissolves, its ions are liberated from their crystal lattice as shown in the molecular view of Figure 8.5(b), and they are able to migrate independently through the solvent. As they migrate, they take their electric charges with them. If they migrate in opposite directions—as they do when the Na$^+$ ions are attracted to and migrate toward the graphite rod, or **electrode**, connected to the battery's negative terminal, while the Cl$^-$ ions do the same toward the positive graphite electrode—then this ion migration allows electricity to flow through the solution.

Any medium (such as a solution of NaCl) that can conduct electricity because it contains free ions is called an **electrolyte**. Because NaCl dissociates completely into Na$^+$ and Cl$^-$ ions when it dissolves in water, it is called a **strong electrolyte** even though solid NaCl alone is not a good conductor of electricity. On the other hand, if NaCl is heated to a temperature above its melting point, the molten NaCl

electrode a solid electrical conductor that is used to make contact with a solution or other nonmetallic component of an electrical circuit.

electrolyte a material that conducts electricity because it contains free ions; ionic solutions and molten salts are examples of electrolytes.

strong electrolyte an ionic substance that dissociates completely when it dissolves in water.

▶II **CHEMTOUR** Migration of Ions in Solution

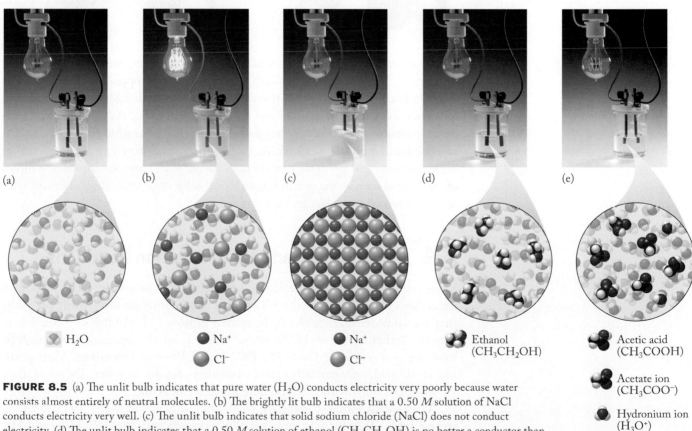

(a) (b) (c) (d) (e)

🔘 H$_2$O

⚫ Na$^+$
🔘 Cl$^-$

⚫ Na$^+$
🔘 Cl$^-$

Ethanol (CH$_3$CH$_2$OH)

Acetic acid (CH$_3$COOH)

Acetate ion (CH$_3$COO$^-$)

Hydronium ion (H$_3$O$^+$)

FIGURE 8.5 (a) The unlit bulb indicates that pure water (H$_2$O) conducts electricity very poorly because water consists almost entirely of neutral molecules. (b) The brightly lit bulb indicates that a 0.50 M solution of NaCl conducts electricity very well. (c) The unlit bulb indicates that solid sodium chloride (NaCl) does not conduct electricity. (d) The unlit bulb indicates that a 0.50 M solution of ethanol (CH$_3$CH$_2$OH) is no better a conductor than pure water. (e) The dimly lit bulb indicates that a 0.50 M solution of acetic acid (CH$_3$COOH) is a better conductor than pure water but not as good as 0.50 M NaCl.

nonelectrolyte a molecular substance that does not dissociate into ions when it dissolves in water.

hydronium ion (H_3O^+) an H^+ ion plus a water molecule, H_2O; the form in which the hydrogen ion is found in an aqueous solution.

weak electrolyte a substance that only partly dissociates into ions when it dissolves in water.

weak acid an acid that only partially dissociates in aqueous solutions.

strong acid an acid that completely dissociates into ions in aqueous solution.

that is produced *is* a good conductor of electricity because the ions in liquid NaCl are free to migrate.

Now consider the unlit bulb in Figure 8.5(d). Here the graphite electrodes are immersed in a 0.50 *M* solution of ethanol (CH_3CH_2OH) in water, which appears to conduct electricity no better than pure water. From this observation we may conclude that when molecules of CH_3CH_2OH dissolve in water, they stay intact as whole molecules and do not form ions. Solutes that do not form ions when they dissolve are called **nonelectrolytes**.

Finally, look closely at the lightbulb in Figure 8.5(e). It produces a faint glow, which tells us that at least some current is flowing through the 0.50 *M* solution of acetic acid in the beaker, though not as much as flows through 0.50 *M* NaCl. Acetic acid, like ethanol, is a molecular compound. Unlike ethanol, some of the acetic acid molecules ionize when they dissolve in water, donating H^+ ions to molecules of H_2O to form **hydronium ions, H_3O^+**, as shown in the molecular view of Equation 8.8. Actually, less than 1% of the acetic acid molecules in this solution ionize, but that is enough to produce a faint glow in the bulb and enough to qualify acetic acid as a **weak electrolyte**: a class of molecular compounds that partially ionize when they dissolve in water.

The following chemical equation represents the partial ionization of acetic acid:

$$CH_3COOH(aq) + H_2O(\ell) \rightleftharpoons CH_3COO^-(aq) + H_3O^+(aq) \quad (8.8)$$

The reactants and products in Equation 8.8 are not connected by the usual reaction arrow, but instead by two *half arrows* pointed in opposite directions. The lower half arrow represents the reverse reaction, in which hydronium and acetate ions recombine to form molecules of acetic acid and water. As we noted in Chapter 7, reversible reactions may not go to completion but rather may reach a state of chemical equilibrium, in which the reaction mixture contains quantities of both reactants and products that do not change with time. An aqueous solution of acetic acid such as the one in Figure 8.5(e) achieves chemical equilibrium after relatively few of its molecules have ionized.

8.4 Acids, Bases, and Neutralization Reactions

Many compounds besides acetic acid produce H_3O^+ ions when they dissolve in water. This behavior means that all of these compounds are *acids*. More precisely, they are considered *Arrhenius* acids, named in honor of Swedish chemist Svante August Arrhenius (1859–1927), whose research on the behavior of electrolytic solutions was recognized with the 1903 Nobel Prize in Chemistry. Most acids, like acetic acid, only partially ionize when they dissolve in water. Therefore, they are weak electrolytes and are also designated **weak acids**.

A few acids ionize completely when they dissolve in water and are called **strong acids**. They include three volatile hydrogen halides: HCl, HBr, and HI, which are completely transformed from polar molecules in the gas phase into H_3O^+ and halide ions in aqueous solution. For example, when a molecule of hydrogen chloride dissolves in water it completely ionizes, forming one chloride and one hydronium ion:

$$HCl(g) + H_2O(\ell) \rightarrow Cl^-(aq) + H_3O^+(aq) \qquad (8.9)$$

In this way hydrogen chloride gas in water forms a solution we call hydrochloric acid, in which every molecule of HCl reacts with a molecule of H_2O and produces one H_3O^+ ion (Figure 8.6a). Hydrogen bromide and hydrogen iodide react the same way, forming hydrobromic and hydroiodic acids, which are also strong acids.

However, hydrofluoric acid (formed from hydrogen fluoride) is a weak acid, meaning it produces far fewer H_3O^+ ions than the number of HF molecules dissolved in it (Figure 8.6b). Why? For one thing, the covalent bonds that hold HF molecules together are much stronger than the bonds in other hydrogen halides, and they don't break as easily. For another, even if the bonds do break and HF ionizes:

$$HF(g) + H_2O(\ell) \rightarrow F^-(aq) + H_3O^+(aq)$$

the ions that form may not move about freely in solution. Dissolved fluoride ions are much smaller than those of the other halogens, which means they can get closer, and are more strongly attracted, to hydronium ions. This electrostatic attraction is so strong the ions form stable ion pairs:

$$F^-(aq) + H_3O^+(aq) \rightleftharpoons [H_3O^+ \cdot F^-](aq) \qquad (8.10)$$

The reaction in Equation 8.10 is reversible and does not consume *all* the H_3O^+ ions in solution. The relatively few that are still free give hydrofluoric acid its weakly acidic character.

We should note that chemists often simplify the chemical equations describing the behavior of acids in aqueous solution by leaving out H_2O from among the reactants. Instead, they use *(aq)* symbols to indicate the physical state of all reactants, and they leave out the H_2O part of the hydronium ion formula, and instead write it $H^+(aq)$. Thus, the chemical equation describing the ionization of hydrochloric acid (Equation 8.9) becomes simply

$$HCl(aq) \rightarrow H^+(aq) + Cl^-(aq) \qquad (8.11)$$

We will encounter many other chemical equations in the paragraphs and chapters ahead that have been simplified by leaving out water molecules as individual molecules or parts of hydronium ions.

Table 8.2 contains the names and formulas of the most common strong acids and the gases that form them when those gases dissolve in water (or combine with water vapor). We have already discussed the acids formed by HCl, HBr, and HI, so

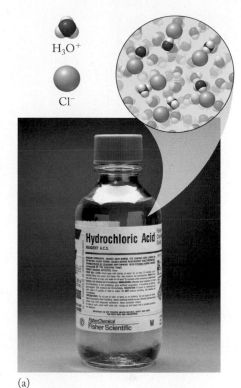

(a)

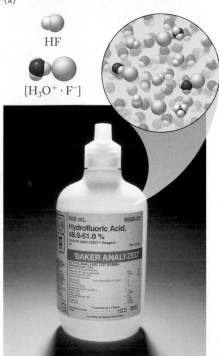

(b)

FIGURE 8.6 (a) Hydrochloric acid is a stronger acid than (b) hydrofluoric acid because molecules of HCl react with molecules of H_2O to form H_3O^+ and Cl^- ions. However, not all molecules of HF react with molecules of H_2O, and even when they do, they mostly form $[H_3O^+ \cdot F^-]$ ion pairs rather than free H_3O^+ ions.

TABLE 8.2	**Strong Acids and the Gases That Form Them**
Gases	**Acids**
Hydrogen chloride, HCl	Hydrochloric acid, $HCl(aq)$
Hydrogen bromide, HBr	Hydrobromic acid, $HBr(aq)$
Hydrogen iodide, HI	Hydroiodic acid, $HI(aq)$
Sulfur trioxide, SO_3	Sulfuric acid, $H_2SO_4(\ell)$
Nitrogen dioxide, NO_2	Nitric acid, $HNO_3(\ell)$
a	Perchloric acid, $HClO_4(aq)$

*a*Perchloric acid is prepared by reacting concentrated solutions of sodium perchlorate and hydrochloric acid: $NaClO_4(aq) + HCl(aq) \rightarrow HClO_4(aq) + NaCl(s)$.

let's focus on those formed by oxides of sulfur and nitrogen. The two strong acids they make, sulfuric and nitric, are among the most important industrial chemicals in the world. Sulfuric acid is made by combining sulfur trioxide and water vapor:

$$SO_3(g) + H_2O(g) \rightarrow H_2SO_4(\ell) \tag{8.12}$$

Nitric acid can be made from nitrogen dioxide gas and water vapor:

$$3\ NO_2(g) + H_2O(g) \rightarrow NO(g) + 2\ HNO_3(\ell) \tag{8.13}$$

Aqueous solutions of HNO_3 and H_2SO_4 are completely ionized:

$$HNO_3(aq) \rightarrow H^+(aq) + NO_3^-(aq) \tag{8.14}$$

$$H_2SO_4(aq) \rightarrow H^+(aq) + HSO_4^-(aq) \tag{8.15}$$

The hydrogen sulfate (HSO_4^-) ion formed in the second reaction has another H atom that is partially ionized:

$$HSO_4^-(aq) \rightleftharpoons H^+(aq) + SO_4^{2-}(aq) \tag{8.16}$$

Thus, H_2SO_4 is a strong acid, but HSO_4^- is a weak acid. The capacity of sulfuric acid to donate up to two H^+ ions per molecule makes it a *diprotic* acid (each molecule can donate *two protons*). Like most diprotic acids, the singly charged anion that forms when a molecule of the acid donates one H^+ ion is a weaker acid than the original one. This pattern holds for *triprotic* acids, too, which means, for example, that phosphoric acid (H_3PO_4) is a stronger acid than $H_2PO_4^-$, which is a stronger acid than HPO_4^{2-}.

> ⬤ **CONCEPT TEST** ··
>
> A molecule with the common name EDTA has four ionizable hydrogen atoms. To highlight this property, it is sometimes written as H_4EDTA. Rank the following species in order of increasing acid strength: H_4EDTA, H_3EDTA^-, H_2EDTA^{2-}, $HEDTA^{3-}$, $EDTA^{4-}$.
>
> ···

In 1923 chemists Johannes Brønsted (1879–1947) and Thomas Lowry (1874–1936) independently proposed a new definition of what it means to be an acid. In the **Brønsted–Lowry model**, an acid is any compound that donates H^+ ions. Arrhenius acids are included in this **Brønsted–Lowry acid** definition because producing H_3O^+ ions in water (as Arrhenius acids do) involves donating H^+ ions to H_2O molecules. However, the Brønsted–Lowry definition also includes compounds that donate H^+ ions to molecules other than H_2O. For example, when hydrogen chloride and ammonia gas are mixed together, they form solid NH_4Cl:

$$HCl(g) + NH_3(g) \rightarrow NH_4Cl(s) \tag{8.17}$$

Molecules of HCl ionize and donate H^+ ions to NH_3 molecules, forming NH_4^+ and Cl^- ions. In this reaction, HCl is a Brønsted–Lowry acid but not an Arrhenius acid.

If acids are H^+ ion donors, then what do we call the substances that accept the H^+ ions that acids donate? As suggested by the title of this section, they're called *bases*. In particular, they're called **Brønsted–Lowry bases**. Just as there are strong and weak acids, so too are there **strong bases** and **weak bases**. Among the strongest bases are hydroxide ions, OH^-, which combine with H^+ ions to form molecules of water:

$$H^+(aq) + OH^-(aq) \rightarrow H_2O(\ell) \tag{8.18}$$

Brønsted–Lowry model defines acids as H^+ ion donors and bases as H^+ ion acceptors.

Brønsted–Lowry acid a proton donor.

Brønsted–Lowry base a proton acceptor.

strong base a base that completely dissociates into ions in aqueous solution.

weak base a base that only partially dissociates in aqueous solutions.

$FeCl_3$	Na_3PO_4	HNO_3	~~NH_4Cl~~ NH_4Cl
Na_3PO_4 ✓	HNO_3	NH_4Cl	KOH
HNO_3 ✓	NH_4Cl	KOH	
~~NH_4Cl~~ ✓	KOH		
~~KOH~~			

Important sources of OH^- ions include the alkali metal hydroxides, which are solid ionic compounds that have the generic formula MOH (where M represents any alkali metal). They all dissociate completely into M^+ and OH^- ions when they dissolve in water. The hydroxides of Ca^{2+}, Sr^{2+}, and Ba^{2+} also dissociate completely when they dissolve in water, though they are not as soluble as the group 1 hydroxides. The limited solubility of the group 2 hydroxides means that the concentrations of dissolved OH^- ions in their solutions cannot match those in concentrated solutions of group 1 hydroxides.

A good example of a *weak* base is ammonia (NH_3) gas. When NH_3 dissolves in water, it produces a solution that conducts electricity weakly: if the apparatus in Figure 8.5 were immersed in 0.50 M NH_3, the bulb would light about as brightly as it does in 0.50 M acetic acid (Figure 8.5e). Thus, ammonia is a weak electrolyte as well as a weak base, just as acetic acid is both a weak electrolyte and weak acid. The partial ionization of ammonia in aqueous solution is represented by the following chemical equation:

$$NH_3(aq) + H_2O(\ell) \rightleftharpoons NH_4^+(aq) + OH^-(aq) \qquad (8.19)$$

At equilibrium, this solution contains mostly aqueous ammonia molecules with smaller concentrations of ammonium and hydroxide ions.

Some compounds are difficult to label as acids or bases because they can behave as both, accepting H^+ ions in the presence of acids and donating them in the presence of bases. One compound that exhibits this behavior is water. Consider the reactions described by Equations 8.9 and 8.19:

$$HCl(g) + H_2O(\ell) \rightarrow Cl^-(aq) + H_3O^+(aq) \qquad (8.9)$$

$$NH_3(aq) + H_2O(\ell) \rightleftharpoons NH_4^+(aq) + OH^-(aq) \qquad (8.19)$$

In a solution of hydrochloric acid, molecules of H_2O accept H^+ ions created by the ionization of molecules of HCl. In an aqueous solution of ammonia, molecules of H_2O ionize, donating H^+ ions to molecules of NH_3. This dual behavior earns water the label **amphiprotic**, which describes a substance that can function as either a Brønsted–Lowry base or a Brønsted–Lowry acid. Another common amphiprotic substance is sodium bicarbonate ($NaHCO_3$), the principal ingredient in baking soda. In a solution that also contains a strong acid, bicarbonate ions act like bases, accepting H^+ ions from the acid to form the weak acid H_2CO_3, which decomposes to water and CO_2.

$$H^+(aq) + HCO_3^-(aq) \rightarrow H_2CO_3(aq) \rightarrow H_2O(\ell) + CO_2(g) \qquad (8.20)$$

In a solution that also contains a strong base, bicarbonate acts like an acid, donating H^+ ions to the OH^- ions from the base:

$$OH^-(aq) + HCO_3^-(aq) \rightarrow H_2O(\ell) + CO_3^{2-}(aq) \qquad (8.21)$$

Neutralization Reactions and Net Ionic Equations

The reaction in which H^+ and OH^- ions combine to form H_2O (Equation 8.18) is at the heart of an important class of chemical reactions between acids and bases called **neutralization reactions**. It contains an important stoichiometric message: in any neutralization reaction, the number of moles of H^+ ions donated by the acid must equal the number of moles of H^+ ions accepted by the base. Put another way, *there must be one H^+ ion for every OH^- ion* in acid–base neutralization reactions in aqueous solution. For example, in the neutralization reaction between hydrochloric acid and sodium hydroxide:

$$HCl(aq) + NaOH(aq) \rightarrow NaCl(aq) + H_2O(\ell) \qquad (8.22)$$

amphiprotic describes a substance that can behave as either a proton acceptor or a proton donor.

neutralization reaction a reaction that takes place when an acid reacts with a base and produces a solution of a salt in water.

salt the product of a neutralization reaction; it is made up of the cation of the base in the reaction and the anion of the acid.

molecular equation a balanced equation that describes a reaction in solution in which the reactants are written as undissociated molecules.

spectator ion an ion that is unchanged by a chemical reaction.

net ionic equation a balanced equation that describes the actual reaction taking place in solution; it is obtained by eliminating the spectator ions from the total ionic equation.

we need one mole of NaOH for every mole of HCl, because each mole of HCl can donate one mole of H^+ ions and each mole of NaOH contains one mole of OH^- ions, that is, one mole of H^+ ion acceptors.

Equation 8.22 describes a classic neutralization reaction involving a strong acid and a strong base. It includes two products that are common to all these reactions: a salt and water. In fact, neutralization reactions provide one operational definition of a **salt**: the ionic compound produced in the neutralization reaction of an acid and a base.

Equation 8.22 is also an example of a traditional **molecular equation** for an aqueous reaction. This label means that each reactant and product is written as a neutral compound, which is the format we used for all the chemical equations in Chapter 7. To get a particle-level view of the reaction in Equation 8.22, let's rewrite it to show all of the ionic species in the aqueous reaction mixture:

$$\underbrace{HCl(aq)}_{H^+(aq) + Cl^-(aq)} + \underbrace{NaOH(aq)}_{Na^+(aq) + OH^-(aq)} \rightarrow$$

$$\underbrace{NaCl(aq)}_{Na^+(aq) + Cl^-(aq)} + H_2O(\ell) \tag{8.23}$$

Equation 8.23 is an example of a *total ionic equation*. To write such an equation we apply these rules:

1. Write all soluble ionic compounds as separate ions, with each of their symbols followed by (*aq*).
2. Write all soluble nonelectrolytes—and all weak electrolytes, such as acetic acid—using their molecular formulas.
3. Use the molecular formula $H_2O(\ell)$ for water when it is a reactant or product (as in a neutralization reaction).
4. Write insoluble solids and gases using their normal chemical formulas followed by (*s*) or (*g*) as appropriate.

Note that Na^+ and Cl^- ions appear on both sides of the reaction arrow in Equation 8.23. They are not changed by the reaction: they are free ions in solution before the reaction, and they are free ions in solution afterward. We call these unchanged ions **spectator ions** to indicate their lack of participation in the reaction. We can simplify the total ionic equation by canceling out the spectator ions:

$$H^+(aq) + \cancel{Cl^-(aq)} + \cancel{Na^+(aq)} + OH^-(aq) \rightarrow \cancel{Na^+(aq)} + \cancel{Cl^-(aq)} + H_2O(\ell)$$

which leaves us with

$$H^+(aq) + OH^-(aq) \rightarrow H_2O(\ell) \tag{8.18}$$

Equation 8.18 is the core acid–base neutralization reaction. It is also an example of a **net ionic equation**, which is a chemical equation of a reaction that contains only those species that are changed by the reaction.

SAMPLE EXERCISE 8.5 Writing the Net Ionic Equation for a Neutralization Reaction LO4

Write the net ionic equation describing the reaction that takes place when acid rain containing sulfuric acid reacts with the surface of a marble statue, as described in Chapter 7 (Figure 7.2) and by Equation 7.8:

$$CaCO_3(s) + H_2SO_4(aq) \rightarrow CaSO_4(aq) + H_2O(\ell) + CO_2(g) \tag{7.8}$$

COLLECT AND ORGANIZE We are given the balanced chemical equation describing the reaction that happens when an aqueous solution of H_2SO_4 comes in contact with a marble statue (solid $CaCO_3$). The products include soluble $CaSO_4$, CO_2 gas, and liquid H_2O. Our task is to write a net ionic equation describing the reaction. A net ionic equation is a total ionic equation without the spectator ions, which do not participate in the reaction.

ANALYZE To write the net ionic equation, we must first write the total ionic equation containing the formulas of all the individual particles in the reaction mixture. Sulfuric acid (H_2SO_4) is the H^+ ion donor in the reaction, making $CaCO_3$ the H^+ ion acceptor. Each mole of H_2SO_4 can donate 2 moles of H^+ ions. The products of the reaction include $CaSO_4$ and two others, H_2O and CO_2, which must have formed when one mole of CO_3^{2-} ions accepted the 2 moles of H^+ ions as the statue dissolved:

$$2\,H^+(aq) + CO_3^{2-}(aq) \rightarrow H_2O(\ell) + CO_2(g)$$

Assuming this equation is correct, carbonate ions are the species in solution acting as the base in this reaction.

Sulfuric acid is a strong acid and is completely ionized, forming H^+ and HSO_4^- ions (Equation 8.15). However, HSO_4^- is a weak acid and is only partially separated into H^+ and SO_4^{2-} ions (Equation 8.16), so it remains HSO_4^- in the total ionic equation. Calcium sulfate is a soluble ionic compound and is completely separated into Ca^{2+} and SO_4^{2-} ions in solution.

SOLVE Building a total ionic equation from the above analysis of the reaction in Equation 7.8:

$$CaCO_3(s) + \underbrace{H_2SO_4(aq)} \rightarrow \underbrace{CaSO_4(aq)} + H_2O(\ell) + CO_2(g)$$
$$CaCO_3(s) + H^+(aq) + HSO_4^-(aq) \rightarrow Ca^{2+}(aq) + SO_4^{2-}(aq) + H_2O(\ell) + CO_2(g)$$

To write a net ionic equation, we eliminate spectator ions from the total ionic equation. However, there aren't any in this case: none of the ions on the reactant side of the reaction arrow appear on the product side. Therefore, the total ionic equation *is* the net ionic equation:

$$CaCO_3(s) + H^+(aq) + HSO_4^-(aq) \rightarrow Ca^{2+}(aq) + SO_4^{2-}(aq) + H_2O(\ell) + CO_2(g)$$

THINK ABOUT IT Because one of the reactants was a solid and the products included a gas and liquid water, the number of possible spectator ions in the total ionic equation was not large to start with. That, coupled with the partial ionization of HSO_4^- ions, led to the complete absence of any spectator ions. We have seen in this section that adding enough acid to compounds that contain either HCO_3^- ions (Equation 8.20) or CO_3^{2-} ions (in this exercise) produces the same two products: liquid H_2O and CO_2 gas.

Practice Exercise Write balanced molecular, total ionic, and net ionic equations for the reaction between aqueous solutions of phosphoric acid (H_3PO_4) and sodium hydroxide. The products are sodium phosphate and water. ⚙

Neutralization reactions such as the one in Sample Exercise 8.5 highlight an important point about the behavior of acids and bases: even though weak acids and bases are only partially ionized in their aqueous solutions, they become completely ionized during neutralization reactions. For example, it takes two moles of a strong base such as NaOH to completely neutralize one mole of H_2SO_4 because there are two *ionizable* hydrogen atoms in one molecule of H_2SO_4. It does not matter that both H atoms are not actually ionized in a solution of sulfuric acid; they *do* ionize during the neutralization reaction. Similarly, it takes one mole of NaOH to completely neutralize one mole of acetic acid, even though less than 1% of the molecules of acetic acid in solution are ionized before the reaction begins.

The same principle applies to weak bases. For example, it takes one mole of hydrochloric acid to neutralize one mole of ammonia in solution, even though most of the molecules of NH_3 have not reacted with H_2O to form NH_4^+ and OH^- ions before HCl is added. As the neutralization reaction proceeds, more and more NH_3 reacts with H_2O to form NH_4^+ and OH^- ions, and the OH^- ions are consumed by combining with H^+ ions from the acid to form H_2O.

8.5 Precipitation Reactions

Some of the most abundant elements in Earth's crust, including silicon (Si), aluminum (Al), and iron (Fe), are not abundant in seawater for the simple reason that the minerals containing these elements have limited solubility in water. Many minerals contain ionic compounds. Why aren't they soluble? We explored some of the reasons why in Chapter 6, noting that compounds are not soluble when solute–solvent interactions (ion–dipole in this case) are not strong enough to off-set solute–solute interactions (ion–ion in this case) and solvent–solvent interactions such as hydrogen bonding between water molecules.

The strengths of ion–ion interactions increase with increasing charge and decrease with increasing ion size (see Equation 4.1). These trends are reflected in the solubility rules listed in Table 8.3 for some common ionic compounds. For example, all salts that contain alkali metal cations are soluble—as we would expect, given their 1+ charges. It also makes sense that all nitrate and acetate compounds are soluble because both ions are relatively large polyatomic ions and have charges of only 1−. On the other hand, nearly all ionic compounds in which the cations' charges are 3+ or 4+ and the anions' charges are 2− or 3− (many transition metal oxides and sulfides are in this category) have limited solubility, as we would expect given the strengths of the interactions between multiply charged ions. However, other factors do influence the solubilities of ionic compounds, complicating solubility trends based only on ionic charges. As a result, some compounds composed of 1+ and 2+ cations and 1− anions have limited solubility in water. These include the halides of silver(I), mercury(I), lead(II), and many transition metal hydroxides.

The solubility rules in Table 8.3 allow us to predict whether or not a **precipitate** (solid product) forms when two aqueous solutions are mixed together. For example, does a precipitate form when a solution of sodium iodide (NaI) is mixed

◉◉ CONNECTION The role of ion–dipole forces in the dissolution of ionic compounds was discussed in Chapter 6.

precipitate a solid product formed from a reaction in solution.

TABLE 8.3   Solubility Rules for Ionic Compounds

All compounds containing the following ions are soluble in water:
  • Cations: group 1 ions (alkali metals) and NH_4^+
  • Anions: NO_3^- and CH_3COO^- (acetate)

Compounds containing the following anions are soluble except as noted:
  • Group 17 ions (halides), except the halides of Ag^+, Cu^+, Hg_2^{2+}, and Pb^{2+}
  • SO_4^{2-}, except the sulfates of Ba^{2+}, Ca^{2+}, Hg_2^{2+}, Pb^{2+}, and Sr^{2+}

Compounds that are only slightly soluble include these:
  • All hydroxides except those of group 1 cations[a]
  • All sulfides except those of group 1 cations and NH_4^{+}[a]
  • All carbonates except those of group 1 cations and NH_4^+
  • All phosphates except those of group 1 cations and NH_4^+

[a]The solubilities of the group 2 hydroxides and sulfides increase with increasing atomic number. MgS decomposes in water, forming H_2S and $Mg(OH)_2$.

with a solution of lead(II) nitrate, $Pb(NO_3)_2$? Put another way, will either of the cations in the two solutions combine with the anion from the other solution to form a compound that has limited solubility? To answer this question we need to:

1. Identify the ions that are dissolved in the two solutions after the two ionic compounds dissolve and separate into their component ions. In this example, one solution contains Na^+ ions and I^- ions. The other contains Pb^{2+} ions and NO_3^- ions.

2. Determine whether either of the new anion/cation combinations produces a product of limited solubility. When the ions in the two solutions in this example swap partners, the new pairings are $NaNO_3$ and PbI_2. According to the Table 8.3 rules, sodium nitrate is soluble for two reasons: it consists of a group 1 cation (all of their compounds are soluble) and it's a nitrate compound (all of them are soluble, too). However, PbI_2 has limited solubility in water because it is a lead(II) halide, and it precipitates when the two solutions mix (Figure 8.7).

Let's write balanced molecular and ionic equations describing this precipitation reaction. We start with single formula units of reactants and products:

$$Pb(NO_3)_2(aq) + NaI(aq) \rightarrow PbI_2(s) + NaNO_3(aq)$$

To produce a balanced molecular equation we need to balance the number of iodide and nitrate ions on each side of the reaction arrow. We do this by placing coefficients of 2 in front of NaI and $NaNO_3$:

$$Pb(NO_3)_2(aq) + 2\,NaI(aq) \rightarrow PbI_2(s) + 2\,NaNO_3(aq)$$

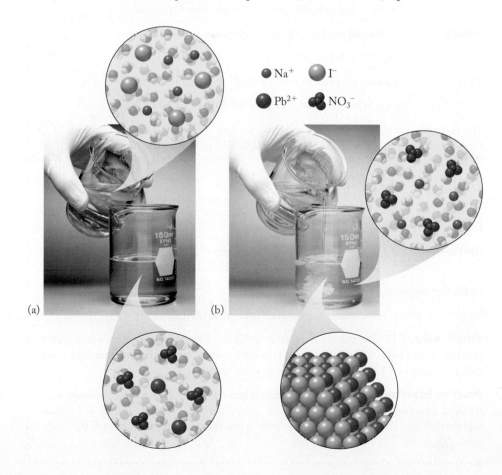

(a)   (b)

● Na^+ ○ I^-

● Pb^{2+} ❈ NO_3^-

FIGURE 8.7 (a) One beaker contains 0.1 M $Pb(NO_3)_2$, and the other contains 0.2 M NaI. Both solutions are colorless. (b) As the NaI solution is poured into the $Pb(NO_3)_2$ solution, a yellow precipitate of PbI_2 forms. Sodium ions and nitrate ions remain in solution.

Next we write a total ionic equation in which we separate all of the soluble salts into their component ions:

$$Pb^{2+}(aq) + 2\,NO_3^-(aq) + 2\,Na^+(aq) + 2\,I^-(aq) \rightarrow PbI_2(s) + 2\,Na^+(aq) + 2\,NO_3^-(aq)$$

Note how the Na^+ and NO_3^- ions are in solution on both sides of the reaction arrow. They are spectator ions in the reaction, so we delete them to write the net ionic equation:

$$Pb^{2+}(aq) + 2\,I^-(aq) \rightarrow PbI_2(s) \tag{8.24}$$

SAMPLE EXERCISE 8.6   **Writing the Net Ionic Equation**   **LO4**
for a Precipitation Reaction

A precipitate forms when aqueous solutions of ammonium sulfate and barium chloride are mixed. Write the net ionic equation for the reaction.

COLLECT AND ORGANIZE We know that mixing solutions of ammonium sulfate, $(NH_4)_2SO_4$, and barium chloride, $BaCl_2$, yields a product with limited solubility. We need to identify it.

ANALYZE Mixing the two solutions and allowing the salts in them to swap partners creates the opportunity for ammonium chloride or barium sulfate to precipitate. According to the Table 8.3 rules, ammonium salts are soluble as are most sulfates, but what about barium sulfate? Table 8.3 tells us that $BaSO_4$ is not soluble. To write a net ionic equation describing a reaction in which solid $BaSO_4$ is a product, we begin with a molecular equation and then write the corresponding total ionic equation, which includes all of the individual ions present in solutions of soluble reactants and products. Elimination of the spectator ions will give us the net ionic equation.

SOLVE The reactants and products of the reaction are

$$(NH_4)_2SO_4(aq) + BaCl_2(aq) \rightarrow BaSO_4(s) + NH_4Cl(aq)$$

To balance the equation we need to insert a coefficient of 2 in front of NH_4Cl:

$$(NH_4)_2SO_4(aq) + BaCl_2(aq) \rightarrow BaSO_4(s) + 2\,NH_4Cl(aq)$$

The reactants are soluble ionic compounds, so they are written as separate ions in a total ionic equation, as is soluble NH_4Cl. However, solid barium sulfate is written as $BaSO_4(s)$:

$$2\,NH_4^+(aq) + SO_4^{2-}(aq) + Ba^{2+}(aq) + 2\,Cl^-(aq) \rightarrow$$
$$BaSO_4(s) + 2\,NH_4^+(aq) + 2\,Cl^-(aq)$$

Eliminating the NH_4^+ and Cl^- spectator ions:

$$\cancel{2\,NH_4^+(aq)} + SO_4^{2-}(aq) + Ba^{2+}(aq) + \cancel{2\,Cl^-(aq)} \rightarrow$$
$$BaSO_4(s) + \cancel{2\,NH_4^+(aq)} + \cancel{2\,Cl^-(aq)}$$

yields the net ionic equation:

$$SO_4^{2-}(aq) + Ba^{2+}(aq) \rightarrow BaSO_4(s)$$

THINK ABOUT IT The solubility rules in Table 8.3 help us predict whether or not a precipitate will form when solutions of two ionic solutes are mixed together and their cations and anions have the opportunity to swap partners.

Practice Exercise Does a precipitate form when aqueous solutions (a) of sodium acetate and ammonium sulfate or (b) calcium chloride and mercury(I) nitrate are mixed together? If you answered yes in either case, write the net ionic equation for the reaction.

Lead(II) dichromate ($PbCr_2O_7$; the dichromate ion is $Cr_2O_7^{2-}$) is the solid pigment called school bus yellow that is used to paint lines on highways. Propose a synthesis of school bus yellow that uses a precipitation reaction.

Analytical chemists use precipitation reactions to determine the concentrations of ions in solution. For example, NaCl (in the form of rock salt) is used to melt ice and snow on roads during the winter. Some of this NaCl may find its way into nearby drinking water supplies. To determine if sources of drinking water have been contaminated with road salt, we can measure the concentration of chloride ion by reacting a sample of the water with a solution of silver nitrate, $AgNO_3$. We need to make sure that there is excess $AgNO_3$ in the reaction mixture so that any Cl^- ions in the water combine with Ag^+ ions, forming a precipitate of AgCl:

$$NaCl(aq) + AgNO_3(aq) \rightarrow AgCl(s) + NaNO_3(aq) \qquad (8.25)$$

The precipitate can be filtered and dried, and its mass determined. From this mass and the compound's molar mass, we can calculate the number of moles of AgCl on the filter, which is the same as the number of moles of Cl^- ions in the water sample. Precipitation reactions used in analytical chemistry sometimes include indicators that provide a visual signal (like a color change) when there is an excess of the precipitation agent to ensure that all of the ions of interest have precipitated. Put another way, Cl^- ion is the limiting reactant in this analysis based on the precipitation of AgCl.

CONNECTION The concept of a limiting reactant was introduced in Chapter 7.

SAMPLE EXERCISE 8.7 **Predicting the Mass of a Precipitate** **LO6**

Barium sulfate is used to enhance X-ray imaging of the upper and lower gastrointestinal tracts. In upper GI imaging, patients drink a suspension of solid $BaSO_4$ in water, which has the consistency of a dense, chalky milkshake and, even with flavoring added, tastes terrible. The compound is not toxic because of its low solubility. To make pure $BaSO_4$, a precipitation reaction is employed: aqueous solutions of soluble barium nitrate and sodium sulfate are mixed together, and solid $BaSO_4$ is separated from the reaction mixture by filtration. How many grams of $BaSO_4$ ($\mathcal{M}$ = 233.40 g/mol) will be produced if exactly one liter of 1.55 M $Ba(NO_3)_2$ is reacted with excess Na_2SO_4?

COLLECT AND ORGANIZE We know the soluble reactants [$Ba(NO_3)_2$ and Na_2SO_4] and the insoluble product ($BaSO_4$) of a precipitation reaction and the volume and molar concentration of a solution containing the limiting reactant. We are asked to calculate the mass of $BaSO_4$ that will be produced. Molarity means moles per liter.

ANALYZE This stoichiometry problem is like the ones we solved in Chapter 7, except that this one is based on quantities of dissolved reactants [$Ba(NO_3)_2$ and Na_2SO_4] and a solid product ($BaSO_4$):

$$Ba(NO_3)_2(aq) + Na_2SO_4(aq) \rightarrow BaSO_4(s) + 2\ NaNO_3(aq)$$

There are about 1.5 moles of $Ba(NO_3)_2$ in one liter of a 1.55 M solution, which translates into about 1.5 moles of Ba^{2+} ions and 1.5 moles of $BaSO_4$. Multiplying 1.5 by a molar mass of about 233 g/mol should give us an answer of about 350 grams.

SOLVE We first multiply the volume and concentration of the $Ba(NO_3)_2$ solution to obtain an equivalent number of moles of $Ba(NO_3)_2$. Two more conversion factors are then required to convert mol $Ba(NO_3)_2$ to mol $BaSO_4$, and finally into a mass of $BaSO_4$:

$$1 \, \cancel{L} \times \frac{1.55 \, \cancel{\text{mol Ba(NO}_3)_2}}{\cancel{L}} \times \frac{1 \, \cancel{\text{mol BaSO}_4}}{1 \, \cancel{\text{mol Ba(NO}_3)_2}} \times \frac{233.43 \text{ g BaSO}_4}{1 \, \cancel{\text{mol BaSO}_4}} = 362 \text{ g BaSO}_4$$

THINK ABOUT IT As in many calculations based on reactions in solution, using "mol/L" instead of "M" provides a guide to help us properly convert units and obtain a correct answer, which is close to our estimate.

Practice Exercise Vermilion, also known as Chinese red, is a very rare and expensive solid natural pigment used to print the artist's signature on works of art. It is mercury(II) sulfide, and it is insoluble in water. What mass of vermilion can be produced when 50.00 mL of 0.0150 M mercury(II) nitrate is mixed with a solution containing excess sodium sulfide? ⚙

SAMPLE EXERCISE 8.8   **Calculating Solute Concentration from Mass of a Precipitate**   **LO6**

To determine the concentration of chloride ion in a 100.0 mL sample of groundwater, a chemist adds a large enough volume of a solution of $AgNO_3$ to precipitate all the Cl^- ions as AgCl. The mass of the resulting precipitate is 71.7 mg. What is the Cl^- concentration in the sample in milligrams per liter?

COLLECT AND ORGANIZE We are given the sample volume and mass of AgCl formed. Our task is to determine the chloride concentration in milligrams of Cl^- per liter of groundwater.

ANALYZE We need to calculate the mass of the Cl^- ions in the weighed mass of AgCl and then divide that mass of Cl^- ions by the volume of the water sample to calculate a mg/L concentration. The molar mass of chlorine is about 1/3 that of silver and about 1/4 the formula mass of AgCl. So, 71.7 mg AgCl should contain a little less than 20 mg Cl^- ions, which were originally dissolved in a 100 mL sample. This mass-to-volume ratio is a little less than 200 mg/1000 mL, or 200 mg/L.

SOLVE The molar masses of Cl and Ag are 35.45 g/mol and 107.87 g/mol, respectively, so the molar mass of AgCl is 143.32 g/mol. The mass of Cl^- ions in 71.7 mg AgCl is

$$71.7 \, \cancel{\text{mg AgCl}} \times \frac{1 \, \cancel{\text{g}}}{1000 \, \cancel{\text{mg}}} \times \frac{1 \, \cancel{\text{mol AgCl}}}{143.32 \, \cancel{\text{g AgCl}}} \times \frac{1 \, \cancel{\text{mol Cl}^-}}{1 \, \cancel{\text{mol AgCl}}} \times \frac{35.45 \, \cancel{\text{g}} \, \text{Cl}^-}{1 \, \cancel{\text{mol Cl}^-}}$$

$$\times \frac{1000 \text{ mg}}{1 \, \cancel{\text{g}}} = 17.7 \text{ mg Cl}^-$$

This 17.7 mg mass of Cl^- ions was originally in a 100.0 mL sample. Converting this mass-to-volume ratio to mg/L yields

$$\frac{17.7 \text{ mg Cl}^-}{100.0 \, \cancel{\text{mL}}} \times \frac{1000 \, \cancel{\text{mL}}}{1 \text{ L}} = 177 \text{ mg Cl}^-/\text{L}$$

THINK ABOUT IT Did you notice that the point of the first step in the first calculation was to convert milligrams to grams (of AgCl), and that its last step converted grams to milligrams (of Cl^- ions)? Numerically, the two steps involved dividing by 1000 and then multiplying by 1000. Had we recognized that the initial and final quantities were both expressed in milligrams, we could have eliminated those two steps and carried the "milli-" prefix through all of the steps from first to last, thereby simplifying the calculation without affecting the result.

saturated solution a solution that contains the maximum concentration of a solute possible at a given temperature.

unsaturated solution a solution that contains less than the maximum quantity of solute predicted to be soluble in a given volume of solution at a given temperature.

supersaturated solution a solution that contains more than the maximum quantity of solute predicted to be soluble in a given volume of solution at a given temperature.

Practice Exercise The concentration of SO_4^{2-} ions in a 50.0 mL sample of coastal seawater is determined by adding a solution of $BaCl_2$ to the sample and precipitating the SO_4^{2-} as $BaSO_4$. The precipitate is removed from the sample by filtration and then dried and weighed. If the mass of $BaSO_4$ recovered from the sample is 0.311 g, what is the sulfate concentration of the sample expressed in mmol/L? ⚙

▶‖ **CHEMTOUR** Saturated Solutions

Saturated Solutions and Supersaturation

The aqueous solubilities of many ionic and molecular solids increase with increasing temperature. For example, more table sugar (or sucrose, $C_{12}H_{22}O_{11}$) dissolves in hot water than in cold—a phenomenon familiar to anyone who has attempted to sweeten a glass of iced tea. When as much sugar as possible has dissolved in hot water, the resulting solution is said to be **saturated** with sugar because it can't hold any more. If we continue to heat the solution without adding more sugar, it becomes **unsaturated**. If the solution is then cooled to a temperature at which less sugar can dissolve, the solution becomes **supersaturated**.

Supersaturated solutions can remain supersaturated for a long time if left undisturbed, but eventually mechanical shock or the addition of a *seed crystal* (a small crystal of the solute that provides a site for crystallization) causes the solute to rapidly precipitate (Figure 8.8).

FIGURE 8.9 Gypsum ($CaSO_4 \cdot 2\,H_2O$) crystals in this cave in Naica, Mexico, grew to enormous size over half a million years as gypsum slowly precipitated.

(a) (b) (c)

FIGURE 8.8 (a) A seed crystal is added to a supersaturated solution of sodium acetate. (b) The seed crystal becomes a site for rapid growth of sodium acetate crystals. (c) Crystal growth continues until the solution is no longer supersaturated but merely saturated with sodium acetate.

More dramatic examples of crystal growth from supersaturated solutions were discovered in 2000 by miners in Naica, Mexico (Figure 8.9). They happened upon a cave containing enormous crystals of the mineral gypsum (calcium sulfate), some nearly 12 m long and 2 m across. Apparently the crystals formed from Ca^{2+} and SO_4^{2-} ions dissolved in groundwater that had been geothermally heated to nearly 60°C. As this water cooled in the cave to about 54°C, it became slightly supersaturated, and crystals of gypsum began to grow very slowly for over half a million years. When the supersaturated gypsum solution was pumped out of the cave as part of current mining operations, the crystals were revealed.

Solutions may also become supersaturated when solvent evaporates. For example, groundwater leaching through porous limestone becomes saturated with $CaCO_3$. If it then seeps into a cave where it simultaneously drips from the ceiling and evaporates, $CaCO_3$ precipitates, forming deposits (Figure 8.10) called stalactites (attached to the ceiling) and stalagmites (built up from the floor).

FIGURE 8.10 In Carlsbad Caverns in New Mexico, stalagmites of limestone grow up from the cavern floor and stalactites grow downward from the ceiling.

8.6 Oxidation–Reduction Reactions

Oxygen is one of the most abundant elements on Earth; it makes up 50% of the crust and 89% of the water covering Earth's surface. Its molecules also make up 21% of all those in Earth's atmosphere. It is an essential element for nearly all creatures, even aquatic organisms, who must obtain their oxygen from the small concentrations of it that dissolve in water. Oxygen's name is applied to the chemical changes that are at the core of many chemical reactions that release energy and sustain life: *oxidation–reduction* reactions, or *redox* reactions.

Although this chapter focuses on reactions in solution, we begin our coverage of redox chemistry with oxygen reactions that occur in the air because many of them are both familiar and simple to describe with balanced chemical equations. Oxygen combines with nonmetals such as carbon, sulfur, and nitrogen to produce volatile oxides. It also combines with metals and semimetals, producing solid oxides. All of these reactions are examples of redox reactions. Indeed, *oxidation* was once defined as a reaction that increased the oxygen content of a substance. Under this definition, combustion of a hydrocarbon such as methane, which we discussed in Chapter 7, involves oxidation because the products, CO_2 and H_2O, contain more oxygen than CH_4:

$$CH_4(g) + 2\ O_2(g) \rightarrow CO_2(g) + 2\ H_2O(\ell) \tag{7.14}$$

Similarly, corrosion of metals involves their oxidation because the products of corrosion, such as rusted iron (Figure 8.11), have higher oxygen content than the original metal (which had none):

$$4\ Fe(s) + 3\ O_2(g) \rightarrow 2\ Fe_2O_3(s) \tag{8.26}$$

Reduction reactions were originally defined in a similar fashion—they were reactions in which the oxygen content of a substance was reduced, a classic example being the reduction of iron ores such as Fe_2O_3 to metallic Fe:

$$Fe_2O_3(s) + 3\ CO(g) \rightarrow 2\ Fe(s) + 3\ CO_2(g) \tag{8.27}$$

The reaction in Equation 8.27 describes the reduction of Fe_2O_3, but there is more to it than that. Oxidations and reductions don't happen in isolation; they happen together. This means that as Fe_2O_3 is reduced, an element in the other reactant must be oxidized. That other element is the carbon in CO, which is oxidized from CO to a compound with an even higher oxygen content, CO_2.

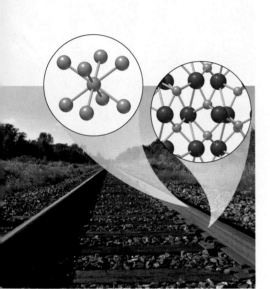

FIGURE 8.11 Corrosion (oxidation) leaves steel railroad tracks covered in a thin layer of orange Fe_2O_3. Any rust that forms on the top of the tracks is worn away by the wheels of trains passing over them, leaving shiny metallic surfaces of mostly iron.

CONCEPT TEST

The following equation describes the conversion of one iron mineral called magnetite (Fe_3O_4) into another called hematite (Fe_2O_3):

$$4\ Fe_3O_4(s) + O_2(g) \rightarrow 6\ Fe_2O_3(s)$$

Is iron oxidized or reduced in the reaction?

Oxidation Numbers

Today we use *redox* to describe other reactions in addition to those that increase or decrease the oxygen content of a substance. An element is said to undergo **oxidation** when it *loses* electrons and **reduction** when it *gains* them.[2] These definitions

---

[2]The mnemonic OIL RIG may be helpful for remembering these expanded definitions: "Oxidation Is Loss; Reduction Is Gain."

are broader than the older ones though still connected to them. To see how, let's reconsider the process of rust (Fe_2O_3) forming on an iron object.

$$4\,Fe(s) + 3\,O_2(g) \rightarrow 2\,Fe_2O_3(s) \qquad (8.26)$$

Clearly, Fe is oxidized according to the old definition—its oxygen content increases when it forms Fe_2O_3—but it does so by *losing* electrons. We may assume that the Fe atoms in elemental iron have their normal free-atom complement of 26 electrons each; however, the Fe^{3+} ions in Fe_2O_3 have only 23 electrons. This loss of electrons means that iron is oxidized under the new rules. At the same time, oxygen is reduced from its elemental state (where we again assume the O atoms have their normal complement of 8 electrons) to oxide (O^{2-}) ions with 10 electrons each. The simultaneous loss and gain of electrons are the key feature of all redox reactions: if one element loses electrons, another must gain them so that the losses and gains are in balance. In Equation 8.26 this balance is achieved because 4 Fe atoms each lose 3 electrons for a total loss of 12 electrons, while 6 O atoms gain 2 electrons each for a total gain of 12. Thus, Equation 8.26 is balanced because the numbers of Fe and O atoms on both sides of the reaction arrow are the same *and* because the number of electrons gained and lost is the same.

To help keep track of electron losses and gains in redox reactions, chemists follow changes in the oxidation states of the atoms in reactants and products. The **oxidation state**, or **oxidation number (O.N.)**, of an atom in a molecule or ion is a measure of the number of electrons it has compared to the number it would have if it were a free atom. The O.N. values of monatomic ions are the same as their electrical charges. In molecules and polyatomic ions, atoms have O.N. values that are related to the number of covalent bonds they form, but their O.N. values also depend on the elements they are bonded to and the O.N. values of those elements. The rules for assigning O.N. values are summarized in Table 8.4.

oxidation a chemical change in which an element loses electrons; the oxidation number of the element increases.

reduction a chemical change in which an element gains electrons; the oxidation number of the element decreases.

oxidation number (O.N.) or **oxidation state** a positive or negative number based on the number of electrons that an atom gains or loses when it forms an ion, or that it shares when it forms a covalent bond with an atom of another element.

TABLE 8.4   Oxidation Number (O.N.) Assignment Rules

1. O.N. = 0 for atoms in pure elements.
2. O.N. = the charge of monatomic ions.
3. O.N. of fluorine = −1 in all of its compounds.
4. O.N. of oxygen = −2 in *nearly* all its compounds. Exceptions occur when O is bonded to fluorine (for example, OF_2), where its O.N. = +2, or when O is bonded to another O atom in a peroxide (for example, H_2O_2, where its O.N. = −1), or in a superoxide (for example, KO_2, where its O.N. = $-\frac{1}{2}$).
5. O.N. of hydrogen = +1 in *nearly* all its compounds. The exception is hydrogen in metal hydrides (for example, LiH), where H is the more electronegative element and its O.N. = −1.

6. O.N. values of the atoms in a neutral molecule sum to zero.
7. O.N. values of the atoms in a polyatomic ion sum to the charge on the ion.

TABLE 8.5   Some Oxidation States of Chlorine

Chlorine Compound	Chlorine O.N. Value
HCl	−1
HClO	+1
$HClO_2$	+3
ClO_2	+4
$HClO_3$	+5
$HClO_4$	+7

There is a hierarchy in assigning O.N. values to the atoms in molecules that is based on electronegativity. Fluorine is always in a −1 oxidation state in molecular compounds because it is the most electronegative element and it forms one bond per atom. The other halogens also have O.N. values of −1 in their compounds, but not when they are bonded to the more electronegative fluorine or oxygen. As illustrated in Table 8.5, chlorine (and the higher Z halogens) can be in oxidation states up to +7 depending on the number of O atoms they are bonded to.

Oxygen nearly always has an O.N. of −2 in its compounds because it is the next most electronegative element after fluorine and tends to form two bonds per atom. However, its O.N. is +2 in OF_2 because of rule 3 in Table 8.4 (fluorine's O.N. is always −1 in its compounds) and rule 6 (the O.N. values of the elements in a compound must sum to zero). The oxidation state of O is also different when it bonds to itself in compounds called peroxides. These compounds have —O—O— functional groups in which oxygen's O.N. equals −1.

SAMPLE EXERCISE 8.9   **Determining Oxidation Numbers**   **LO7**

What are the oxidation states of sulfur in (a) S_8, (b) SO_2, (c) Na_2S, and (d) $CaSO_4$?

COLLECT AND ORGANIZE We are asked to assign oxidation numbers to sulfur in S_8 and in three sulfur compounds. The rules for assigning oxidation numbers are given in Table 8.4.

ANALYZE
   a. S_8 is a molecular form of the element sulfur.
   b. In SO_2, sulfur is bonded to oxygen, which is a more electronegative element and has an O.N. of −2 (rule 4).
   c. Na_2S is a binary ionic compound; the charge of the sulfide ion is 2−.
   d. Sulfur in $CaSO_4$ is part of the sulfate ion, which means its oxidation number added to those of the four O atoms must add up to the charge of the ion, which is 2− (rule 7).

SOLVE
   a. The oxidation state of sulfur in elemental sulfur (S_8) is 0 (rule 1).
   b. The sum of the oxidation numbers for S and the two O atoms in this neutral molecule must be zero (rule 6). Assigning an O.N. of −2 to each oxygen atom, and letting the oxidation state of sulfur be x, we can set up this equation:

$$x + 2(-2) = 0$$

$$x = +4$$

   So, the oxidation state of sulfur in SO_2 is +4.
   c. The oxidation state of the atom in a monatomic ion is the same as the charge on the ion (rule 2). Therefore, the oxidation state of sulfur in a sulfide ion is −2.
   d. The charge on a SO_4^{2-} ion is 2−. Each O atom has an O.N. of −2. Letting x be the O.N. of sulfur and applying rule 7:

$$x + 4(-2) = -2$$

$$x = +6$$

   So, the oxidation state of S in $CaSO_4$ is +6.

THINK ABOUT IT We found in this exercise that sulfur, like chlorine in the compounds in Table 8.5, can exist in oxidation states that range from negative to quite positive values. The most positive values occur when single sulfur atoms are bonded to multiple oxygen atoms: the more O atoms, the higher the oxidation state.

Practice Exercise Determine the oxidation state of nitrogen in (a) NO_2; (b) N_2O; (c) HNO_3.

Electron Transfer in Redox Reactions

Redox reactions involve the transfer of electrons between atoms, and these transfers result in changes in the oxidation states of the atoms. We know from rules 1 and 2 in Table 8.4 that when an atom acquires an electron, forming a 1− anion, its oxidation number is reduced from 0 to −1 (ΔO.N. = −1). On the other hand, when an atom loses an electron and forms a 1+ cation, its oxidation number increases from 0 to +1 (ΔO.N. = +1). We can generalize this pattern to describe the gain or loss of any number (n) of electrons:

| Oxidation | Lose n electrons | ΔO.N. = $+n$ |
| Reduction | Gain n electrons | ΔO.N. = $-n$ |

This pattern can help us track the flow of electrons from the elements losing them (being oxidized) to the elements gaining them (being reduced) in a reaction. The tracking system involves calculating the changes in oxidation number of the elements that are oxidized and reduced.

To see how the system works, let's examine the redox properties of two neighbors in the periodic table: copper ($Z = 29$) and zinc ($Z = 30$). Elemental Zn has considerable electron-donating power, and Cu^{2+} ions tend to accept electrons. These complementary properties are illustrated in Figure 8.12.

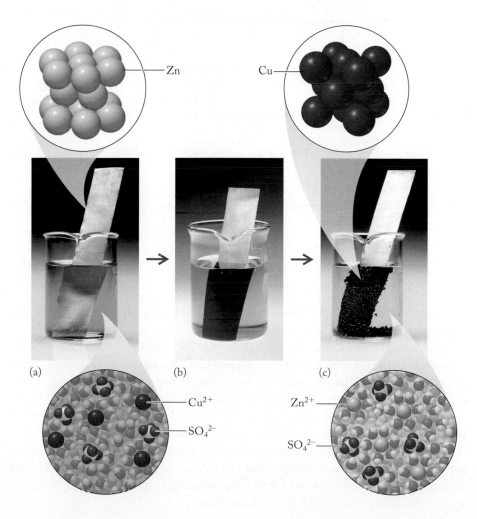

(a) (b) (c)

FIGURE 8.12 Redox reaction of Zn metal with Cu^{2+} ions. (a) A strip of Zn is placed in a solution of $CuSO_4$. (b) Over time, the surface of the Zn strip darkens as it is coated by a layer of Cu atoms. (c) Eventually the blue color of Cu^{2+} ions in solution fades as Cu atoms form. They accumulate on the Zn strip and then fall to the bottom of the beaker.

When a strip of Zn metal is placed in a solution of $CuSO_4$, that is, a solution of Cu^{2+} ions and SO_4^{2-} ions, Zn atoms spontaneously donate electrons to Cu^{2+} ions, forming Zn^{2+} ions and Cu atoms. We can see this redox reaction happen: the shiny zinc surface turns dark brown as a textured layer of copper metal accumulates on it, and the distinctive blue color of $Cu^{2+}(aq)$ ions fades as they acquire electrons and become atoms of copper metal. The chemical equation describing this reaction is

$$Zn(s) + CuSO_4(aq) \rightarrow ZnSO_4(aq) + Cu(s) \qquad (8.28)$$

and the total ionic equation including all ions in solution is

$$Zn(s) + Cu^{2+}(aq) + SO_4^{2-}(aq) \rightarrow Cu(s) + Zn^{2+}(aq) + SO_4^{2-}(aq)$$

Eliminating SO_4^{2-} ions, which are unchanged in the reaction, produces this net ionic equation:

$$Zn(s) + Cu^{2+}(aq) \rightarrow Cu(s) + Zn^{2+}(aq) \qquad (8.29)$$

Now let's assign oxidation numbers to the Cu and Zn atoms and ions in Equation 8.29. According to rules 1 and 2 in Table 8.4, the oxidation states of the atoms in Zn and Cu metal are 0, and the oxidation states of the atoms in both of their ions are +2. Therefore, the changes in oxidation number (ΔO.N.) are +2 for Zn atoms and −2 for Cu^{2+} ions:

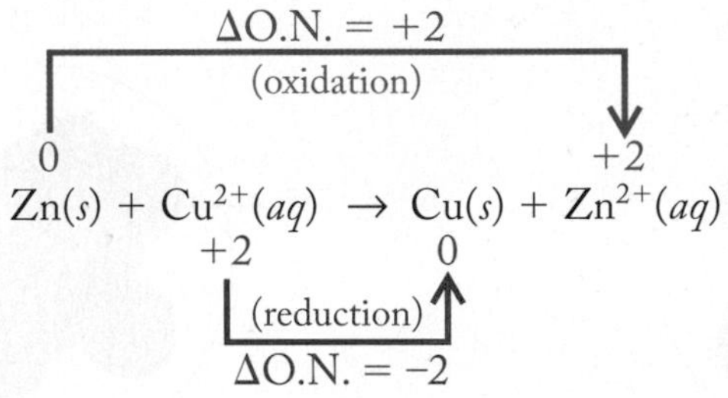

These changes mean that each Zn atom *loses* two electrons and each Cu^{2+} ion *gains* two electrons.

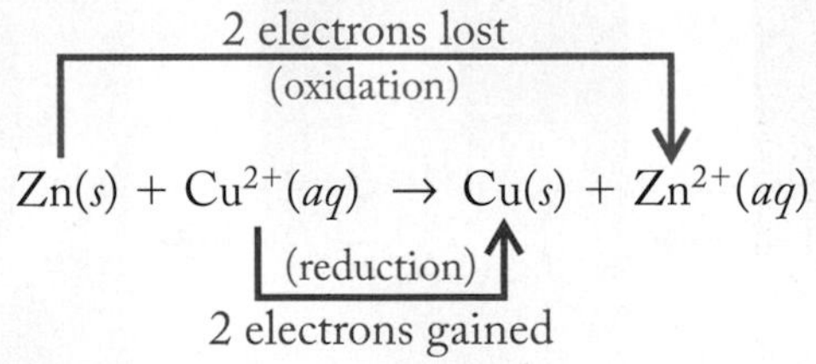

The gains and losses of electrons are the same, as they must be in any balanced chemical equation describing a redox reaction.

One way to think about this reaction is that Zn atoms give electrons to Cu^{2+} ions, causing the Cu^{2+} ions to be *reduced*, and Cu^{2+} ions take electrons away from Zn atoms, causing the Zn atoms to be *oxidized*. From this perspective, Cu^{2+} ions are forcing Zn atoms to be oxidized and are called **oxidizing agents**. Similarly, Zn atoms are causing Cu^{2+} ions to be reduced, so Zn atoms are called the **reducing agents** in the reaction. These terms are widely used in describing redox reactions, but they can take some getting used to because the *oxidizing* agent is actually *reduced* and the *reducing* agent is *oxidized*. Every redox reaction has an oxidizing agent and a reducing agent. The reducing agent is the electron donor, and the oxidizing agent is the electron acceptor.

oxidizing agent a reactant that accepts electrons from another in a redox reaction, thereby oxidizing the other reactant; the oxidizing agent is reduced in the reaction.

reducing agent a reactant that donates electrons to another in a redox reaction, thereby reducing the other reactant; the reducing agent is oxidized in the reaction.

Which of the following reactions are redox reactions? For those that are, identify the oxidizing agent and the reducing agent:

a. $Sn^{2+}(aq) + Br_2(aq) \rightarrow Sn^{4+}(aq) + 2\,Br^-(aq)$

b. $2\,F_2(g) + 2\,H_2O(\ell) \rightarrow 4\,HF(aq) + O_2(g)$

c. $NaHCO_3(aq) + HCl(aq) \rightarrow NaCl(aq) + CO_2(g) + H_2O(\ell)$

SAMPLE EXERCISE 8.10   **Identifying Oxidizing and Reducing Agents and Determining Number of Electrons Transferred**   **LO7**

Energy released by the reaction of hydrazine and dinitrogen tetroxide:

$$2\,N_2H_4(\ell) + N_2O_4(g) \rightarrow 3\,N_2(g) + 4\,H_2O(g)$$

is used to orient and maneuver spacecraft and to propel rockets into space (Figure 8.13). Identify the elements that are oxidized and reduced, the oxidizing agent, and the reducing agent, and determine the number of electrons transferred in the balanced chemical equation.

COLLECT AND ORGANIZE The reactants are hydrazine, N_2H_4, and dinitrogen tetroxide, N_2O_4, and the products are N_2 and H_2O. According to the rules in Table 8.4, the oxidation state of nitrogen in molecules of N_2 (a pure element) is 0, and the O.N. values of O and H in the three reactant and product compounds are −2 and +1, respectively.

ANALYZE Because the O.N. values of O and H are −2 and +1, respectively, on both sides of the reaction arrow, the only element that is oxidized and reduced in the reaction is nitrogen. Rule 6 in Table 8.4 can be used to calculate the oxidation states of nitrogen in N_2H_4 and N_2O_4.

SOLVE The oxidation numbers of N and H in N_2H_4 must sum to zero (rule 6). The O.N. value of hydrogen is +1, so the O.N. of nitrogen is

$$(x)(2) + (+1)(4) = 0$$

$$x = -2$$

Applying rule 6 to N_2O_4, we find that the O.N. of nitrogen in this molecule is:

$$(x)(2) + (-2)(4) = 0$$

$$x = +4$$

FIGURE 8.13 The NASA Sky Crane, which was used to land the Mars rover *Curiosity* in August 2012, is powered by the redox reaction between hydrazine (N_2H_4) and dinitrogen tetroxide (N_2O_4).

The overall changes in oxidation state are shown in the chemical equation below. The O.N. of each N atom in N_2H_4 changes from -2 to zero for a gain of $+2$. There are two of these N atoms per molecule, and there are two N_2H_4 molecules in the chemical equation, so the overall ΔO.N. is $2 \times 2 \times (+2) = +8$. This change corresponds to a loss of 8 electrons. The O.N. of each N atom in N_2O_4 changes from $+4$ to zero for a ΔO.N. of -4. There are two N atoms in the chemical equation's single N_2O_4 molecule, so the overall ΔO.N. is $2 \times (-4) = -8$, which corresponds to an overall gain of 8 electrons.

$$\Delta\text{O.N.} = 2 \times 2 \times (+2) = +8$$
$$\text{(oxidation)}$$
$$\underset{-2}{2\,N_2H_4(g)} + \underset{+4}{N_2O_4(g)} \rightarrow \underset{0}{3\,N_2(g)} + 4\,H_2O(g)$$
$$\text{(reduction)}$$
$$\Delta\text{O.N.} = 2 \times (-4) = -8$$

In this reaction N_2H_4 is the reducing agent because the N atoms in it are oxidized (lose electrons), and N_2O_4 is the oxidizing agent because the N atoms in it are reduced (gain electrons).

THINK ABOUT IT We tend to identify the specific *atoms* that are oxidized or reduced, while whole *molecules or ions* are identified as the oxidizing agents or reducing agents. In the equation in this exercise, nitrogen is oxidized but N_2H_4 is the reducing agent.

Practice Exercise In the reaction between $O_2(g)$ and $SO_2(g)$ to make $SO_3(g)$, identify the species oxidized, the species reduced, the oxidizing agent, the reducing agent, and the number of electrons transferred. ⚙

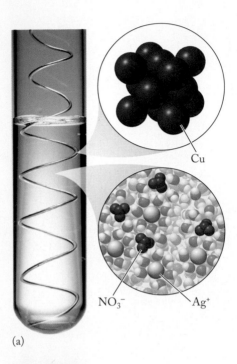

(a)

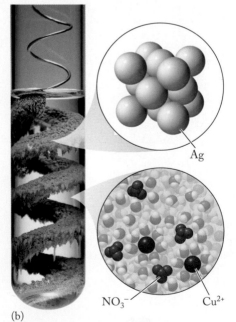

(b)

FIGURE 8.14 (a) When a Cu wire is immersed in a solution of $AgNO_3$, Cu metal oxidizes to Cu^{2+} ions as Ag^+ ions are reduced to Ag metal. (b) A day later, the solution has the blue color of Cu^{2+} ions in solution, and the wire is coated with Ag metal.

Balancing Redox Reaction Equations

We have noted that the losses and gains of electrons must balance in any redox reaction. To apply this principle, let's examine the reaction that occurs when a piece of copper wire [elemental copper, $Cu(s)$] is placed in a colorless solution of silver nitrate (containing Ag^+ and NO_3^- ions). The solution gradually takes on the blue color of a solution of Cu^{2+} ions, and branchlike structures of solid Ag form on the copper wire (Figure 8.14). In this chemical reaction, Cu metal is oxidized to Cu^{2+} ions and Ag^+ ions are reduced to Ag metal. If we ignore the nitrate ions, which are only spectator ions in the reaction, the reactants and products that should be in the net ionic equation are

$$Ag^+(aq) + Cu(s) \rightarrow Ag(s) + Cu^{2+}(aq)$$

However, this expression is *not* the net ionic equation because the electrical charges are not balanced: the sum of the charges is $1+$ for the reactants but $2+$ for the products. The charges are not balanced because the loss and gain of electrons by Cu metal and Ag^+ ions are not balanced, which is reflected in the changes in oxidation state: -1 for Ag^+ ions becoming Ag atoms, and $+2$ for Cu atoms becoming Cu^{2+} ions.

$$\overset{\Delta O.N. = +2 \downarrow}{\underset{+1 \qquad\qquad\quad 0}{\overset{0 \qquad\qquad\quad +2}{Ag^+(aq) + Cu(s) \rightarrow Ag(s) + Cu^{2+}(aq)}}}$$

$$\Delta O.N. = -1 \uparrow$$

To bring these changes in oxidation state and the loss and gain of electrons into balance, we need twice as many Ag^+ ions, which each consume one electron, as Cu atoms, which each donate two electrons. So we place a coefficient of 2 in front of Ag^+ on the reactant side—and also in front of Ag on the product side to keep the number of Ag atoms in balance. These additions give us a balanced net ionic equation:

$$2\,Ag^+(aq) + Cu(s) \rightarrow 2\,Ag(s) + Cu^{2+}(aq)$$

Now let's write a balanced net ionic equation for a slightly more complicated redox reaction. In this one, Mn^{2+} ions react with O_2 in a slightly acidic solution, forming solid MnO_2 (Figure 8.15). We start with a preliminary expression relating the known reactants and product:

$$Mn^{2+}(aq) + O_2(g) \rightarrow MnO_2(s)$$

At first glance this expression looks balanced, but closer inspection reveals that there is an ionic charge of 2+ on the reactant side and no charge on the product side. Oxygen is present as a free element (O.N. = 0) on the reactant side and in a compound (O.N. = −2) on the product side, so O_2 is reduced in the reaction. This means that the reaction is a redox reaction, which means that Mn^{2+} must be oxidized. An O.N. analysis confirms this: the oxidation state of Mn increases from +2 in Mn^{2+} to +4 in MnO_2:

$$\overset{\Delta O.N. = +2 \downarrow}{\underset{0 \qquad\qquad\quad -2}{\overset{+2 \qquad\qquad\quad +4}{Mn^{2+}(aq) + O_2(g) \rightarrow MnO_2(s)}}}$$

$$\Delta O.N. = 2 \times (-2) = -4$$

Inspection of ΔO.N. values reveals that each Mn atom increases by +2 and each O atom decreases by −2. However, the O atoms come in pairs—as O_2 molecules—and the total change in oxidation number per molecule is $2 \times (-2) = -4$. Therefore, we need twice as many Mn^{2+} ions as O_2 molecules to balance the changes in O.N. and the losses and gains of electrons. So we place a coefficient of 2 in front of Mn^{2+}, and we do the same in front of MnO_2 to keep the number of Mn atoms in balance:

$$\overset{\Delta O.N. = 2 \times (+2) = +4}{\underset{0 \qquad\qquad\quad -2}{\overset{+2 \qquad\qquad\quad +4}{2\,Mn^{2+}(aq) + O_2(g) \rightarrow 2\,MnO_2(s)}}}$$

$$\Delta O.N. = 2 \times (-2) = -4$$

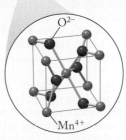

FIGURE 8.15 Oxidation of manganese(II) compounds on the surfaces of rocks in arid regions of the American Southwest builds up a layer of black MnO_2 that is commonly called desert varnish. Native Americans have etched petroglyphs such as these in desert varnish for thousands of years.

Now the number of atoms of Mn is balanced, but not much else is. We have an ionic charge of 4+ on the reactant side but still no charge on the product side, and the number of O atoms is no longer balanced. Our first priority is to balance the charge; the question is how to do that. A clue is provided by the fact that the reaction takes place in a slightly acidic solution, which means that there are H^+ ions available. If we add four of them to the product side, we have taken care of the charge imbalance:

$$2\,Mn^{2+}(aq) + O_2(g) \rightarrow 2\,MnO_2(s) + 4\,H^+(aq)$$

Now we have an expression with four more H atoms and two more O atoms on the product side than the reactant side. At this point, the key is to recognize that this is the same number of H and O atoms as in two molecules of H_2O. There are plenty of them in an aqueous solution, so let's add two to the reactant side:

$$2\,Mn^{2+}(aq) + 2\,H_2O(\ell) + O_2(g) \rightarrow 2\,MnO_2(s) + 4\,H^+(aq)$$

A final check reveals that we have a balanced net ionic equation.

The procedure for writing a net ionic equation that we just employed can be used to balance other redox reaction equations that happen in acidic or neutral solutions. What about those in basic solutions? The difference comes in the charge-balancing step. Instead of adding H^+ ions, we add OH^- ions to either side of the reaction equation in basic solutions. Doing so will probably produce an imbalance in the number of H and O atoms that we can correct by adding molecules of H_2O to either side. Table 8.6 summarizes the steps to follow in writing balanced chemical equations for redox reactions in aqueous solution.

TABLE 8.6 Steps in Balancing Chemical Equations for Redox Reactions in Aqueous Solution

1. Write a preliminary reaction expression using one mole of each of the known reactants and products.
2. Calculate the $\Delta O.N.$ values of the elements that are reduced and oxidized.
3. Insert coefficients as appropriate to balance the $\Delta O.N.$ values (and the losses and gains of electrons). This step includes balancing the numbers of atoms of the elements that are oxidized and reduced, but skip balancing O atoms at this step.
4. Balance ionic charges by adding H^+ ions in acidic or neutral solutions, or by adding OH^- ions in basic solutions.
5. Add H_2O molecules to balance the numbers of H and O atoms if necessary.

SAMPLE EXERCISE 8.11 Balancing Redox Reaction Equations I: Acidic Solutions LO8

The concentration of N_2O in the atmosphere has been increasing in recent years (Figure 8.16). Environmental chemists believe that one reason for the increase is heavy agricultural use of fertilizers, which can result in elevated levels of nitrates in soils and groundwater. Microorganisms called denitrifying bacteria that grow in waterlogged soil convert NO_3^- ions into N_2O gas as they feed on dead plant tissue (empirical formula: CH_2O), converting it into CO_2 and H_2O. Write a balanced net ionic equation describing this conversion of dissolved nitrates to N_2O gas. Assume the reaction occurs in slightly acidic water.

COLLECT AND ORGANIZE We need to write a balanced net ionic equation for a reaction in which the reactants include dissolved nitrate ions, $NO_3^-(aq)$, and dead plant tissue, $CH_2O(s)$, and the products include N_2O gas, CO_2 gas, and liquid H_2O.

ANALYZE Table 8.6 outlines the steps to follow in writing a balanced net ionic equation for a redox reaction. We know that the oxidation numbers of O and H have their usual values in these compounds: -2 and $+1$, respectively. That leaves N and C as the elements to be oxidized and reduced. The problem states that the water is slightly acidic, which means that we should add H^+ ions to balance ionic charges in step 4 of Table 8.6.

SOLVE

1. We write a preliminary expression relating the known reactants and products:

$$NO_3^-(aq) + CH_2O(s) \rightarrow N_2O(g) + CO_2(g) + H_2O(\ell)$$

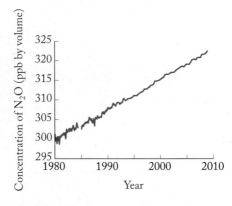

FIGURE 8.16 Atmospheric concentrations of N_2O have been steadily increasing in recent years, due in part to increasing use of nitrogen-containing fertilizers in agriculture.

To begin balancing this equation, we need a coefficient of 2 in front of the NO_3^- ions on the left side to provide the two N atoms needed to form one molecule of N_2O:

$$2\ NO_3^-(aq) + CH_2O(s) \rightarrow N_2O(g) + CO_2(g) + H_2O(\ell)$$

2. We perform a ΔO.N. analysis for nitrogen and carbon:

$$\Delta\text{O.N.} = +4$$

$$\begin{array}{cccc} 0 & & +4 \\ 2\ NO_3^-(aq) + CH_2O(s) & \rightarrow & N_2O(g) + CO_2(g) + H_2O(\ell) \\ +5 & & +1 \end{array}$$

$$\Delta\text{O.N.} = 2 \times (-4) = -8$$

3. To balance these ΔO.N. values (and the number of electrons gained and lost), we need a coefficient of 2 in front of CO_2. This also requires a 2 in front of CH_2O to balance the number of C atoms:

$$2\ NO_3^-(aq) + 2\ CH_2O(s) \rightarrow N_2O(g) + 2\ CO_2(g) + H_2O(\ell)$$

4. To balance the ionic charges (2– on the reactant side, 0 on the product side), we add two H^+ ions to the left side because the reaction is happening in slightly acidic water:

$$2\ NO_3^-(aq) + 2\ CH_2O(s) + 2\ H^+(aq) \rightarrow N_2O(g) + 2\ CO_2(g) + H_2O(\ell)$$

5. Taking an inventory of the atoms on both sides of the reaction arrow, we find that the left side has four more H atoms and two more O atoms than the right side. We correct this imbalance by adding two more molecules of H_2O to the right side for a total of three:

$$2\ NO_3^-(aq) + 2\ CH_2O(s) + 2\ H^+(aq) \rightarrow N_2O(g) + 2\ CO_2(g) + 3\ H_2O(\ell)$$

Now the equation is balanced.

THINK ABOUT IT Many organisms, including humans, get energy through respiration processes in which digested food (empirical formula: CH_2O) is oxidized to CO_2 and H_2O using molecular oxygen, O_2. Denitrifying bacteria are able to extract oxygen from nitrate ions to convert similar food (chemically) into CO_2 and H_2O and energy.

Practice Exercise A nail made of $Fe(s)$ is placed in an aqueous solution of a palladium(II) salt. The nail disappears and the solution turns a yellow color indicative of the presence of dissolved Fe^{3+} ions. Particles of solid palladium metal also form. Write a balanced equation that describes this reaction. ⚙

Redox processes play a major role in determining the character of Earth's rocks and soils. For example, the orange-red spires called hoodoos in Bryce Canyon National Park in Utah owe their distinctive color to an iron(III) oxide mineral called hematite (Figure 8.17). These rocks are also relatively fragile and susceptible to physical and chemical weathering, which has resulted in the fascinating shapes of the hoodoos for which Bryce is famous. Iron(III) oxide is also the form of iron known as rust, the crumbly, red-brown solid that forms on iron objects exposed to water and oxygen.

FIGURE 8.17 Red-orange rock formations called hoodoos in Bryce Canyon, Utah, owe their color to high concentrations of iron(III) oxide.

Fe^{3+} Fe^{2+}

FIGURE 8.18 (a) Wetland soil has a blue-gray color because of the presence of iron(II) compounds. (b) Red-orange mottling indicates the presence of iron(III) oxides formed as a result of O_2 permeation through channels made by plant roots.

Soil color is influenced by mineral content and by the availability of molecular oxygen. Wetlands are areas that are either saturated with water year-round or flooded during much of the year. Even though wetland soils have considerable iron content, they do not have the distinctive red-orange color of iron(III) compounds, because the waterlogged soils have so little O_2 that the dominant oxidation state of Fe is +2 instead of +3. In fact, soil color is one of the ways environmental scientists can tell whether an area is a wetland (Figure 8.18).

We can observe the characteristic colors associated with soils rich in iron(II) or iron(III) compounds in the laboratory. Figure 8.19(a) shows a solution of iron(II) ammonium sulfate, $(NH_4)_2Fe(SO_4)_2$, as a solution of NaOH is added to it. Addition of base causes $Fe(OH)_2$ to precipitate as a blue-gray solid, similar in color to the wetland soil shown in Figure 8.18(a). When this mixture is filtered, the iron(II) hydroxide residue immediately starts to darken (Figure 8.19b). After about 20 minutes of exposure to oxygen in the air, the precipitate turns the orange-red color of iron(III) hydroxide (Figure 8.19c). The moist iron(II) hydroxide has been oxidized to iron(III) hydroxide—in much the way the iron(II) compounds in a wetland soil are oxidized when exposed to oxygen. The following Sample Exercise illustrates how to write the net ionic equation for a redox reaction such as this one that is carried out in a basic solution.

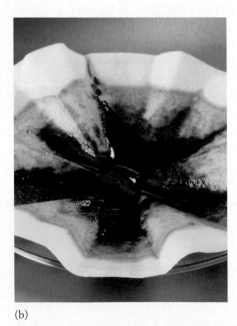

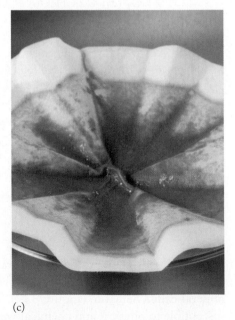

FIGURE 8.19 (a) A solution of iron(II) ammonium sulfate forms a blue-green precipitate of iron(II) hydroxide when NaOH is added. (b) When the precipitate is isolated on a paper filter, it rapidly turns a darker color. (c) After about 20 minutes of exposure to air, the precipitate has changed to the red-orange color of iron(III) oxide.

SAMPLE EXERCISE 8.12 Balancing Redox Reaction Equations II: Basic Solutions LO8

The brown residues that collect along the inner walls and bottoms of toilet tanks are made mostly of solid $Fe(OH)_3$ that forms when Fe^{2+} ions in water are oxidized by dissolved O_2. Write a balanced net ionic equation for this reaction in a slightly basic solution.

COLLECT AND ORGANIZE We know that the reactants include dissolved O_2 and Fe^{2+} ions, and that one of the products is solid $Fe(OH)_3$. We need to write a net ionic equation describing the reaction, that is, an equation that represents all of the species involved in the reaction less those that are not changed by it.

ANALYZE Table 8.6 outlines the steps to follow in writing a balanced net ionic equation for a redox reaction. The reaction occurs in basic solution, so we balance charges by adding OH^- ions if necessary.

SOLVE

1. Write a preliminary expression relating the known reactants and products:

$$Fe^{2+}(aq) + O_2(aq) \rightarrow Fe(OH)_3(s)$$

2. Perform a ΔO.N. analysis for iron and oxygen:

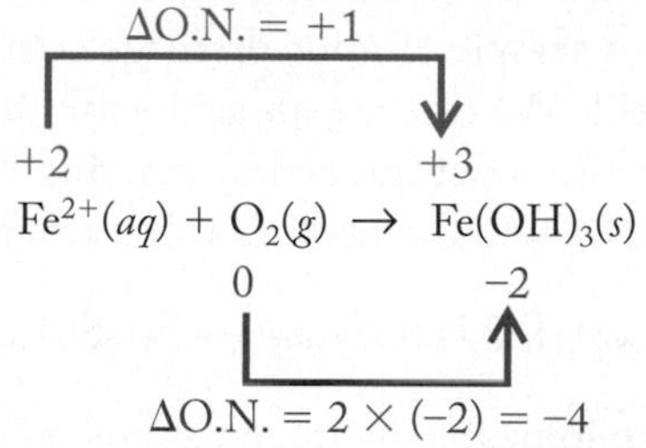

3. To balance these ΔO.N. values (and the number of electrons gained and lost), we need a coefficient of 4 in front of Fe^{2+}. This also requires a 4 in front of $Fe(OH)_3$ to balance the number of Fe atoms:

$$4\ Fe^{2+}(aq) + O_2(g) \rightarrow 4\ Fe(OH)_3(s)$$

4. To balance the electrical charges (8+ on the left, 0 on the right) we add 8 OH^- ions to the reactant side:

$$4\ Fe^{2+}(aq) + O_2(g) + 8\ OH^-(aq) \rightarrow 4\ Fe(OH)_3(s)$$

5. Taking an inventory of the atoms on both sides of the reaction arrow, we find that the right side has four more H atoms and two more O atoms than the left side. We correct this imbalance by adding two molecules of H_2O to the left side:

$$4\ Fe^{2+}(aq) + O_2(g) + 8\ OH^-(aq) + 2\ H_2O(\ell) \rightarrow 4\ Fe(OH)_3(s)$$

Now the equation is balanced.

THINK ABOUT IT It may have seemed that we were just lucky to need twice as many H atoms as O atoms in step 5: an imbalance easily resolved by adding the necessary number of H_2O molecules. Actually, 2:1 imbalances in H and O atoms are frequently encountered in the final step of writing a net ionic equation describing a redox reaction and are often an indicator that the first four steps were done correctly.

Practice Exercise Hydroperoxide ions, HO_2^-, react with permanganate ions, MnO_4^-, producing MnO_2 and O_2 gas. Write a balanced net ionic equation for this reaction in a basic solution.

8.7 Titrations

Aqueous-phase reactions can be used to determine the concentrations of dissolved substances with excellent precision and accuracy. The analytical technique that is used is called **titration**. It is classified as a *volumetric* method of analysis because it is based on measuring the volumes of both a standard solution and a sample solution. The standard solution, called the **titrant**, contains a known concentration of a substance that reacts with a substance in the sample. The goal is to precisely determine the concentration of this second substance, called the **analyte**. The chemistries of titration reactions are often based on acid–base neutralization reactions, though precipitation and redox reactions are also widely used.

To get a feel for how titrations work, let's consider an application from environmental science: analysis of acidic water draining from abandoned coal mines. Sulfide-containing minerals called pyrites are often present in and around seams of coal. Mining the coal exposes the pyrite to chemical and biochemical weathering. Those processes oxidize the sulfur in the sulfides to sulfuric acid, which dissolves in groundwater seeping through abandoned mines. To assess and control the environmental impact of this acidic water, scientists and engineers need to know how much sulfuric acid is dissolved in it. That's where titrations come in.

Suppose we have a sample of mine drainage containing an unknown concentration of sulfuric acid. We can use an acid–base titration in which the sulfuric acid in the water sample is neutralized by reacting it with a standard solution of strong base, such as NaOH. The neutralization reaction is

$$H_2SO_4(aq) + 2\,NaOH(aq) \rightarrow Na_2SO_4(aq) + 2\,H_2O(\ell)$$

Note that both of the ionizable hydrogen atoms on each molecule of H_2SO_4 are neutralized in the reaction. Neutralizing both takes two moles of NaOH for every one mole of H_2SO_4.

To determine the concentration of analyte (sulfuric acid) in the sample, we need to measure the volume of titrant (standard solution of NaOH) needed to neutralize it. We start by carefully transferring a known volume of the sample into a reaction vessel such as an Erlenmeyer flask. A few drops of a color indicator are added to detect when just enough titrant has been added to neutralize the sulfuric acid (Figure 8.20). The titrant is added slowly using a buret, a narrow glass cylinder with volume markings and fitted with a stopcock to quickly stop the flow when just enough titrant has been added. This stage of a titration, when the reaction is just complete, is called the **equivalence point** because a quantity of titrant has been added that is *equivalent* to the quantity of the analyte. It is as if both the analyte and the titrant were limiting reactants in the titration reaction.

In the spirit of full disclosure, we should note that the volume of titrant required to make the indicator change color may not be *exactly* the same as the volume of titrant needed to completely consume all of the analyte in the sample. The volume needed to make the indicator change color is actually the **end point** of the titration. It is the volume we record and use in calculating how much analyte was in the sample. With the right indicator and proper technique, the difference in volume between the equivalence point and the end point in acid–base titrations is insignificant.

Suppose it takes 22.40 mL of 0.00100 *M* NaOH to react completely with the H_2SO_4 in a 100.0 mL sample of acidic mine drainage. We know the volume and molarity of the NaOH standard solution, so we can adapt Equation 8.2 to calcu-

titration an analytical method for determining the concentration of a solute in a sample by reacting the solute with a solution of known concentration.

titrant the standard solution added to the sample in a titration.

analyte the substance whose concentration is to be determined in a chemical analysis.

equivalence point the point in a titration when just enough titrant has been added to react with all of the analyte in the sample.

end point the point in a titration when a color change or other signal indicates that enough titrant has been added to react with all of the analyte in the sample.

late the number of moles of NaOH consumed after first converting the volume into liters:

$$n_{NaOH} = V_{NaOH} \times M_{NaOH}$$

$$= 0.02240 \text{ L} \times 1.00 \times 10^{-3} \frac{\text{mol NaOH}}{\text{L}} = 2.240 \times 10^{-5} \text{ mol NaOH}$$

We also know from the stoichiometry of the reaction that 2 moles of NaOH are required to neutralize 1 mole of H_2SO_4, so the number of moles of H_2SO_4 in the 100.0 mL sample must have been

$$n_{H_2SO_4} = 2.240 \times 10^{-5} \text{ mol NaOH} \times \frac{1 \text{ mol } H_2SO_4}{2 \text{ mol NaOH}} = 1.120 \times 10^{-5} \text{ mol } H_2SO_4$$

Dividing this number of moles of H_2SO_4 by the volume of the sample (V_S) in liters gives us the molar concentration of H_2SO_4:

$$M_{H_2SO_4} = \frac{n_{H_2SO_4}}{V_S}$$

$$= \frac{1.120 \times 10^{-5} \text{ mol } H_2SO_4}{0.1000 \text{ L}} = 1.120 \times 10^{-4} \frac{\text{mol } H_2SO_4}{\text{L}}$$

$$= 1.120 \times 10^{-4} \text{ } M \text{ } H_2SO_4$$

To detect the end point in a sulfuric acid–sodium hydroxide titration, we might use the indicator *phenolphthalein*, which is colorless in acidic solutions but pink in basic solutions (Figure 8.20). The part of the titration that requires the most skill is adding just enough NaOH solution to reach the end point, the point at which a pink color first persists in the solution being titrated. To catch the end point precisely, the standard solution must be added no faster than one drop at a time, with thorough mixing between drops.

(a)

(b)

FIGURE 8.20 Determining a sulfuric acid concentration. (a) A known volume of a sample containing H_2SO_4 is placed in the flask. The buret is filled with the titrant, which is an aqueous NaOH solution of known concentration. A few drops of phenolphthalein indicator solution are added to the flask. (b) Titrant is carefully added to the flask until the indicator changes from colorless to pink, signaling that the acid has been neutralized.

SAMPLE EXERCISE 8.13 **Determining Solute Concentration from Titration Data** **LO9**

Apple cider vinegar is an aqueous solution of acetic acid (CH_3COOH, $M = 60.05$ g/mol) made from fermented apple juice. Commercial vinegar must contain at least 4 grams of acetic acid per 100 mL of vinegar. Suppose the titration of a 25.00 mL sample of vinegar requires 12.15 mL of 1.885 M NaOH. What is the molarity of acetic acid in the vinegar sample? Does it meet the 4 g/100 mL standard?

COLLECT AND ORGANIZE We have the results of a titration of acetic acid in vinegar with a strongly basic titrant. We know the volume and concentration of the titrant and the volume of the sample and need to calculate the concentration of acetic acid in the sample.

ANALYZE We can use Equation 8.2 to calculate the number of moles of basic titrant consumed from its volume and concentration. We need a balanced chemical equation to relate the number of moles of titrant (base) to the number of moles of acid in the sample during the titration reaction. There is one ionizable H atom in a molecule of CH_3COOH (highlighted in red), and there is one OH^- ion per formula unit of NaOH. Therefore, the mole ratio of CH_3COOH to NaOH in the balanced chemical equation describing the titration reaction is 1:1. The volume of titrant consumed is about half the sample volume, which means the concentration of acetic acid in the sample should be about half the concentration of NaOH in the titrant, or about 1 M.

SOLVE The balanced chemical equation for the neutralization reaction is

$$CH_3COOH(aq) + NaOH(aq) \rightarrow H_2O(\ell) + CH_3COONa(aq)$$

The number of moles of NaOH required to reach the end point is:

$$n_{NaOH} = V_{NaOH} \times M_{NaOH}$$

$$= \left(12.15 \text{ mL} \times \frac{1 \text{ L}}{1000 \text{ mL}}\right) \times \frac{1.885 \text{ mol}}{L} = 0.02290 \text{ mol NaOH}$$

Converting moles of NaOH into moles of CH_3COOH:

$$n_{CH_3COOH} = 0.02290 \text{ mol NaOH} \times \frac{1 \text{ mol } CH_3COOH}{1 \text{ mol NaOH}} = 0.02290 \text{ mol } CH_3COOH$$

Dividing moles of CH_3COOH by the sample volume (in liters), we obtain the molar concentration of CH_3COOH in the sample:

$$M_{CH_3COOH} = \frac{n_{CH_3COOH}}{V_S}$$

$$= \frac{0.02290 \text{ mol } CH_3COOH}{25.00 \text{ mL}} \times \frac{1000 \text{ mL}}{1 \text{ L}} = 0.9160 \text{ mol/L} = 0.9160 \ M$$

Converting this molar concentration into grams per 100 mL:

$$\frac{0.9160 \text{ mol}}{L} \times \frac{60.05 \text{ g}}{\text{mol}} \times \frac{1 \text{ L}}{1000 \text{ mL}} = 0.05501 \text{ g/mL} \quad \text{or} \quad 5.501 \text{ g/100 mL}$$

The sample meets the 4 g/100 mL standard for apple cider vinegar.

THINK ABOUT IT The calculated molar concentration of acetic acid in the vinegar sample was close to our 1 M estimate, and the equivalent concentration expressed in grams per 100 mL is just above the standard for apple cider vinegar, so our results are entirely reasonable.

Practice Exercise Citric acid ($\mathcal{M}$ = 192.12 g/mol) is a triprotic acid (it has three carboxylic acid groups, as shown in Figure 8.21) that occurs in lemon juice and other citrus products, and it is widely used as a flavoring in beverages. What is the molarity of citric acid in lemon juice if 24.26 mL of 0.904 M NaOH is required in a titration to neutralize 25.00 mL of juice? How many grams of citric acid are in 100 mL of juice?

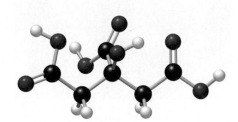

FIGURE 8.21 Citric acid, $C_6H_8O_7$.

8.8 Ion Exchange

In Section 8.5 we saw how dissolved ions can exchange partners, forming insoluble precipitates. Actually, ions exchanging partners in solution is at the heart of most aqueous-phase chemistry. It is also the basis for a method of water purification called **ion exchange**. Water containing certain metal ions—principally Ca^{2+} and Mg^{2+}—is called *hard water*, and it causes problems in industrial processes and in homes. Because hard water combines with soap to form a gray scum, clothes washed in hard water appear gray and dingy. Hard water forms scale (an incrustation) in boilers, pipes, and kettles, diminishing their ability to conduct heat and carry water; and hard water sometimes has an unpleasant taste.

To *soften* hard water (remove the ions responsible for its hardness), it is pumped through tanks filled with small plastic beads called resin, labeled "R" in Figure 8.22. In the tanks, Ca^{2+} and Mg^{2+} ions in the water are exchanged for Na^+ ions that are loosely bound to ion-exchange sites on the resin. In many resins these sites consist of carboxylate ($R—COO^-$) groups. As hard water flows through the

ion exchange a process in which one ion is displaced by another.

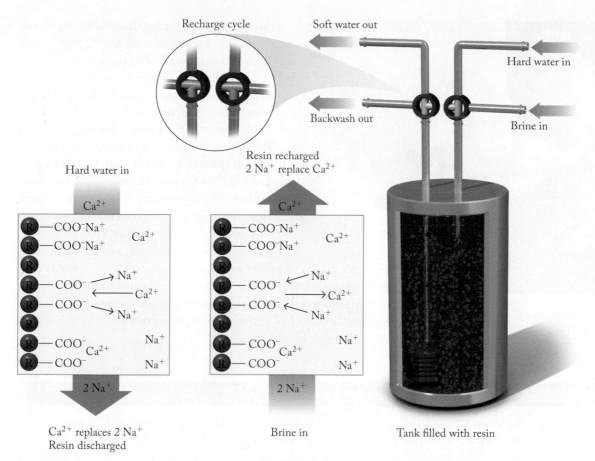

FIGURE 8.22 Residential water softeners use ion exchange to remove 2+ ions (such as Ca^{2+}) that make water hard. The ion-exchange resin contains cation-exchange sites that are initially occupied by Na^+ ions. These ions are replaced by 2+ "hardness" ions as water flows through the resin. Eventually most of the ion-exchange sites are occupied by 2+ ions, and the system loses its water-softening ability. The resin is then backwashed with a saturated solution of NaCl (*brine*), which displaces the hardness ions, restoring the resin to its Na^+ form.

resin, Ca^{2+} and Mg^{2+} ions exchange places with Na^+ ions on the resin because cations with 2+ and 3+ charges bind more strongly to the R—COO$^-$ groups than Na^+ ions. The following chemical equation illustrates this ion-exchange process:

$$2\,(R\!-\!COO^-)Na^+(s) + Ca^{2+}(aq) \rightarrow (R\!-\!COO^-)_2Ca^{2+}(s) + 2\,Na^+(aq)$$

Hard water that has been softened in this way contains increased concentrations of sodium ion. Although this may not be a problem for healthy children and adults, people suffering from high blood pressure often must limit their intake of Na^+ and so should not drink water softened by this kind of ion-exchange reaction.

Naturally occurring minerals called **zeolites** are also capable of ion-exchange reactions. Zeolites are formed in nature as a result of chemical reactions between salt water and aluminum and silicon oxides in molten lava. They can also be synthesized by precipitating compounds called aluminosilicates from supersaturated solutions of aluminum and silicon oxides. On an atomic scale, natural zeolites consist of three-dimensional arrays of interconnecting tunnels and cages (Figure 8.23). The inner surfaces of the tunnels have negative charges that are balanced by cations electrostatically bound to them. They function as ion-exchange sites in much the same way that plastic resins do in water softeners: as water flows

zeolites natural crystalline minerals or synthetic materials consisting of three-dimensional networks of channels that contain sodium or other 1+ cations.

through tunnels (pores) lined with Na^+ ions, these ions exchange with Ca^{2+}, Mg^{2+}, and other cations dissolved in the water.

Zeolites are used in a variety of industrial processes and commercial products, from water softeners and purifiers to livestock feed additives and odor suppressants. They have replaced environmentally harmful phosphates in detergents to bind Ca^{2+} and Mg^{2+} and soften the water used to wash clothes. They are used in municipal water filtration plants to treat drinking water, to remove toxic metals such as Pb^{2+} from contaminated waste streams, and in swimming pools to improve water clarity. Zeolites may be poured directly on wounds to stop bleeding by absorbing water from the blood, thereby concentrating clotting factors to promote coagulation. If the ion-exchange sites in zeolites contain Ag^+ ions, the resulting product also has antimicrobial properties and not only stems bleeding but reduces the risk of infection.

FIGURE 8.23 A sample of zeolite, which contains microscopic cages and tunnels that include cation-exchange sites, shown here in their Na^+ form.

SAMPLE EXERCISE 8.14 **Integrating Concepts: Selecting an Antacid**

A 5.0 mL dose of the liquid antacid in Figure 8.24(a) contains 200 mg $Mg(OH)_2$ and 200 mg $Al(OH)_3$. A tablet of the antacid in Figure 8.24(b) contains 500 mg $CaCO_3$. According to their labels, the bottle of liquid antacid holds 12 fluid ounces (355 mL), and the other bottle contains 150 tablets. In many drugstores, the two bottles sell for about the same price.

FIGURE 8.24 Over-the-counter antacids: (a) liquid Mylanta containing suspended $Mg(OH)_2$ and $Al(OH)_3$, and (b) Tums tablets composed mostly of $CaCO_3$.

(a)   (b)

a. If the solubility of $Mg(OH)_2$ is $1.2 \times 10^{-4}\ M$, what fraction of the $Mg(OH)_2$ in the liquid antacid is dissolved and not solid particles suspended in the viscous liquid?

b. Write net ionic equations for the neutralization of excess stomach acid, HCl(*aq*), by the three active ingredients in the two antacids. Assume that all three are solids.

c. How many moles of HCl could be neutralized by one 5.0 mL dose of the liquid antacid?

d. How many moles of HCl could be neutralized by one 500 mg tablet of the solid antacid?

e. Which of the two antacids is a better buy in terms of acid-neutralizing capacity per bottle?

COLLECT AND ORGANIZE We know the quantities and identities of the active ingredients in the doses of two antacids: 200 mg $Mg(OH)_2$ and 200 mg $Al(OH)_3$ in 5.0 mL of the liquid, and 500 mg $CaCO_3$ per tablet of the solid. There are 150 tablets in the bottle of solid antacid and 355/5.0 = 71 doses in one bottle of the liquid. Our tasks include calculating how much of the $Mg(OH)_2$ is dissolved in the liquid and not just suspended in it, writing balanced net ionic equations describing the neutralization reactions, calculating the quantity of stomach acid one dose of each antacid can neutralize, and determining which is the better buy.

ANALYZE To answer part a, we could use the molar solubility of $Mg(OH)_2$ to calculate the mass of $Mg(OH)_2$ that dissolves in 5.0 mL and compare this quantity to 200 mg. Because molar masses are given in grams per mole and molarity is the same as moles per liter, it might be simpler to scale up all the mass and volume quantities by 1000 and calculate the mass of $Mg(OH)_2$ in grams that dissolves in 5.0 L and then compare this value to 200 g. All of the active ingredients in the antacids are solids, so they should appear as whole formula units and not individual ions in the net ionic equations in part b. The balanced equations in b and the calculations in part c will be based on the following stoichiometric considerations: (1) one mole of $Mg(OH)_2$ contains 2 moles of OH^- ions, so it reacts with 2 moles of HCl; (2) one mole of $Al(OH)_3$ contains 3 moles of OH^- ions, so it can neutralize 3 moles of HCl; (3) the one mole of CO_3^{2-} ions in a mole of $CaCO_3$ can accept 2 moles of H^+ ions, which means it can neutralize 2 moles of HCl. The molar masses of the three reactants are 100.09 g $CaCO_3$/mol, 58.32 g $Mg(OH)_2$/mol, and 78.00 g $Al(OH)_3$/mol.

SOLVE

a. If there is 200 mg of $Mg(OH)_2$ in 5.0 mL of the liquid antacid, then there is 200 g of $Mg(OH)_2$ in 5.0 L of it. Therefore, we convert the solubility of $Mg(OH)_2$ from molarity to the number of grams of it that will dissolve in 5.0 liters:

$$1.2 \times 10^{-4} \; \frac{\text{mol Mg(OH)}_2}{\text{L}} \times \frac{58.32 \text{ g Mg(OH)}_2}{1 \text{ mol Mg(OH)}_2} \times 5.0 \text{ L}$$

$$= 0.0350 \text{ g Mg(OH)}_2$$

We divide this result by 200 g, expressing the quotient as the percentage of $Mg(OH)_2$ that is dissolved (the rest is tiny particles in suspension):

$$\frac{0.0350 \text{ g}}{200 \text{ g}} \times 100\% = 0.017\% \text{ dissolved}$$

b. Based on the stoichiometric ratios developed in the Analyze section, the molecular equations for the three neutralization reactions with the appropriate salts and moles of water are

$$Mg(OH)_2(s) + 2 \, HCl(aq) \rightarrow MgCl_2(aq) + 2 \, H_2O(\ell)$$

$$Al(OH)_3(s) + 3 \, HCl(aq) \rightarrow AlCl_3(aq) + 3 \, H_2O(\ell)$$

$$CaCO_3(s) + 2 \, HCl(aq) \rightarrow CaCl_2(aq) + H_2O(\ell) + CO_2(g)$$

Separating the soluble ionic compounds into their free ions, we obtain total ionic equations:

$$Mg(OH)_2(s) + 2 \, H^+(aq) + 2 \, Cl^-(aq) \rightarrow$$
$$Mg^{2+}(aq) + 2 \, Cl^-(aq) + 2 \, H_2O(\ell)$$

$$Al(OH)_3(s) + 3 \, H^+(aq) + 3 \, Cl^-(aq) \rightarrow$$
$$Al^{3+}(aq) + 3 \, Cl^-(aq) + 3 \, H_2O(\ell)$$

$$CaCO_3(s) + 2 \, H^+(aq) + 2 \, Cl^-(aq) \rightarrow$$
$$Ca^{2+}(aq) + 2 \, Cl^-(aq) + H_2O(\ell) + CO_2(g)$$

In each of the above equations, Cl^- ions are the only spectator ions. Deleting them yields these net ionic equations:

$$Mg(OH)_2(s) + 2 \, H^+(aq) \rightarrow Mg^{2+}(aq) + 2 \, H_2O(\ell)$$

$$Al(OH)_3(s) + 3 \, H^+(aq) \rightarrow Al^{3+}(aq) + 3 \, H_2O(\ell)$$

$$CaCO_3(s) + 2 \, H^+(aq) \rightarrow Ca^{2+}(aq) + H_2O(\ell) + CO_2(g)$$

c. Calculating the number of moles of HCl neutralized by one dose (5.0 mL) of the liquid antacid (containing 200 mg of each active ingredient):

$$200 \text{ mg Mg(OH)}_2 \times \frac{1 \text{ g}}{1000 \text{ mg}} \times \frac{1 \text{ mol Mg(OH)}_2}{58.32 \text{ g Mg(OH)}_2}$$

$$\times \frac{2 \text{ mol HCl}}{1 \text{ mol Mg(OH)}_2} = 0.00686 \text{ mol HCl}$$

$$+ \; 200 \text{ mg Al(OH)}_3 \times \frac{1 \text{ g}}{1000 \text{ mg}} \times \frac{1 \text{ mol Al(OH)}_3}{78.00 \text{ g Al(OH)}_3}$$

$$\times \frac{3 \text{ mol HCl}}{1 \text{ mol Al(OH)}_3} = 0.00769 \text{ mol HCl}$$

$$= 0.01455 \text{ mol HCl}$$

d. Calculating the number of moles of HCl neutralized by one tablet (500 mg) of the solid antacid:

$$500 \text{ mg CaCO}_3 \times \frac{1 \text{ g}}{1000 \text{ mg}} \times \frac{1 \text{ mol CaCO}_3}{100.09 \text{ g CaCO}_3}$$

$$\times \frac{2 \text{ mol HCl}}{1 \text{ mol CaCO}_3} = 0.00999 \text{ mol HCl}$$

The 5.0 mL dose of the liquid contains $0.01455/0.00999 \approx 1.5$ times more acid-neutralizing power than one tablet of the solid antacid.

e. There are 71 doses in one bottle of the liquid antacid, which have the capacity to neutralize (71 doses × 0.01455 mol HCl/dose) = 1.043 mol HCl. There are 150 tablets in a bottle of the solid antacid, which have the capacity to neutralize (150 tablets × 0.00999 mol HCl/tablet) = 1.50 mol HCl. Therefore, there is more neutralizing capacity in a bottle of the solid antacid, and it is the better buy.

THINK ABOUT IT All three active ingredients are solids with limited solubility in water. However, they react with—and dissolve in—acid solutions such as stomach acid. The solid antacid provides about half again as much acid-neutralizing capacity as the liquid for the same (or at least a comparable) price. In fairness to this particular liquid product, it also contains an ingredient that relieves gas.

SUMMARY

Section 8.1 The concentration of solute in a solution can be expressed as mass of solute per mass of solution, such as milligrams of solute per kilogram of solution, which is the same as parts per million (ppm). Concentrations can also be expressed as mass of solute per volume of solution, such as moles of solute per liter of solution (mol/L), which is called **molarity (M)**.

Section 8.2 During **dilution** the quantity of solute in a sample of a concentrated solution does not change, but adding solvent increases the volume of the sample and decreases the concentration of solute.

Section 8.3 A solute that dissociates into ions in aqueous solution is called an **electrolyte**. Mobility of these ions makes the solution a better electrical conductor than pure water. **Strong electrolytes** dissociate completely in water, **weak electrolytes** dissociate partially, and **nonelectrolytes** do not dissociate at all.

Section 8.4 A **Brønsted–Lowry acid** is a proton (H^+) donor, and a **Brønsted–Lowry base** is a proton acceptor. In an aqueous-phase **neutralization reaction**, H^+ ions from the acid combine with OH^- ions from the base, forming H_2O and a **salt**. The **net ionic equation** of a reaction includes only the species that change during the reaction and omits the **spectator ions**.

Section 8.5 In a precipitation reaction, mixing dissolved reactants produces an insoluble **precipitate** as cations and anions switch partners. A **saturated solution** contains the maximum amount of solute possible at a given temperature. A **supersaturated solution** contains more than the maximum concentration of a solute.

Section 8.6 In a redox reaction, substances either gain electrons and thereby undergo **reduction**, or they lose electrons and undergo **oxidation**. The two processes are complementary: the electrons lost by the substance being oxidized are gained by the substance being reduced. A reaction is a redox reaction if the **oxidation numbers (O.N.)**, or **oxidation states**, of elements in the reactants change during the reaction.

Section 8.7 **Titrations** are volumetric methods of chemical analysis in which the concentrations of solutes called **analytes** are determined by reacting them with known volumes and concentrations of standard solutions called **titrants**.

Section 8.8 In an **ion-exchange** reaction, ions in solution displace ions held at ion-exchange sites on the surface of a solid. In a water softener, Ca^{2+} and Mg^{2+} ions dissolved in hard water displace Na^+ ions from an ion-exchange resin.

PROBLEM-SOLVING SUMMARY

TYPE OF PROBLEM	CONCEPTS AND EQUATIONS		SAMPLE EXERCISES
Converting concentration units	Use the solute/solvent ratios in concentration units as conversion factors. Use mol/L to represent molarity.		8.1, 8.2
Calculating mass of solute to prepare a solution	$$m = (V \times M) \times \mathcal{M} = \left(L \times \frac{mol}{L} \right) \times \frac{g}{mol}$$	(8.3)	8.3
Preparing dilute solutions	$$V_i \times M_i = V_f \times M_f$$	(8.6)	8.4
Writing neutralization reaction equations	Balance the total ionic equation, using separate ions to represent strong electrolytes. Delete spectator ions to write the net ionic equation. Balance the number of H^+ ions donated by the acid and accepted by the base.		8.5
Writing a net ionic equation for a precipitation reaction	Balance the total ionic equation, writing the precipitates as solids. Delete spectator ions to write the net ionic equation.		8.6
Predicting the mass of a precipitate	Identify the limiting reactant if necessary. Use the stoichiometry of the net ionic equation to calculate moles of precipitate, and then convert moles to mass.		8.7
Calculating solute concentration from mass of a precipitate	Convert precipitate mass into moles by dividing by its molar mass. Convert moles of precipitate into moles of solute. Calculate molarity of the solute in the sample by dividing by the volume of the sample in liters.		8.8
Determining oxidation numbers (O.N.)	Follow the rules in Table 8.4. In general, the O.N. of a pure element is 0, the O.N. of a monatomic ion is the charge on the ion, and the O.N. values of O and H in most compounds are −2 and +1, respectively.		8.9
Identifying oxidizing and reducing agents and determining number of electrons transferred	The oxidizing agent contains an atom whose O.N. decreases during the reaction; the reducing agent contains an atom whose O.N. increases; the changes in O.N. determine the number of electrons transferred.		8.10
Balancing redox reaction equations	Follow the steps in Table 8.6. In general, use coefficients to balance ΔO.N. values (the gains and losses of electrons). In aqueous solutions, balance ionic charges by adding H^+ ions (in acidic solutions) or OH^- ions (in basic solutions). Balance H and O atoms by adding H_2O.		8.11, 8.12
Determining solute concentration from titration data	Calculate moles of titrant from its concentration and the volume needed to reach the end point. Use the stoichiometry of the titration reaction to calculate moles of analyte and then divide by sample volume to calculate analyte concentration.		8.13

VISUAL PROBLEMS

(Answers to boldface end-of-chapter questions and problems are in the back of the book.)

8.1. Figure P8.1 shows a solution containing three binary (HX) acids. One of them is a weak acid and the other two are strong acids. Which color sphere is formed by ionization of the weak acid?

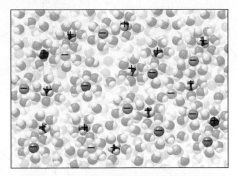

FIGURE P8.1

8.2. Figure P8.2 represents the products of the reaction that occurs when solutions of NaCl and $AgNO_3$ are mixed together. The silver spheres represent Ag^+ ions. Which spheres represent (a) Na^+; (b) Cl^-; (c) NO_3^- ions?

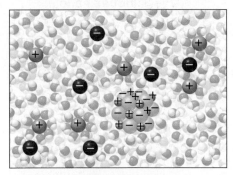

FIGURE P8.2

8.3. Figure P8.3 represents the reaction mixture at the equivalence point in a titration of a sample of battery acid (H_2SO_4) using a standard solution of NaOH as the titrant. Which ion is represented by the green spheres and which by the blue spheres?

FIGURE P8.3

8.4. Suppose that two antacid tablets each containing 0.50 g $CaCO_3$ ($\mathcal{M} = 100$ g/mol) react with 0.02 mol HCl. Which of the images in Figure P8.4, in which green spheres represent

cations and blue spheres represent anions, accurately reflects the population of ions in solution after the reaction is over?

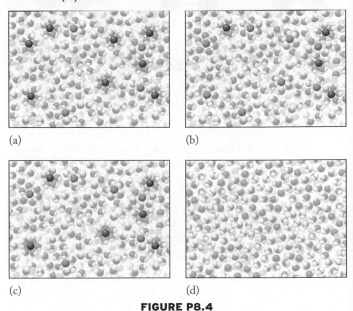

(a)   (b)

(c)   (d)

FIGURE P8.4

8.5. What is the change in oxidation state of nitrogen in the reaction described by the molecular models in Figure P8.5? (Blue spheres are nitrogen, red spheres are oxygen.)

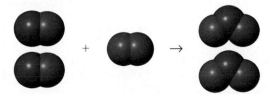

FIGURE P8.5

8.6. Identify the oxidizing and reducing agents in the reaction described by the molecular models in Figure P8.5.

8.7. Rank the nitrogen oxides depicted by the molecular models in Figure P8.7 in order of decreasing oxidation state of the nitrogen (blue spheres) in them. (The red spheres are oxygen atoms.)

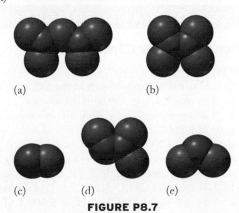

(a)   (b)

(c)   (d)   (e)

FIGURE P8.7

8.8. Rank the hydrocarbons depicted by the molecular models in Figure P8.8 based on decreasing oxidation state of the carbon (black spheres) in them.

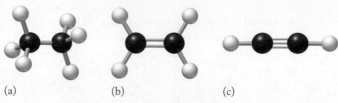

(a)   (b)   (c)

FIGURE P8.8

*8.9. One way to follow the progress of a titration and detect its equivalence point is by monitoring the conductivity of the titration reaction mixture. For example, consider the way the conductivity of a sample of sulfuric acid changes as it is titrated with a standard solution of barium hydroxide before and then after the equivalence point.
a. Write the total ionic equation for the titration reaction.
b. Which of the four graphs in Figure P8.9 comes closest to representing the changes in conductivity during the titration? (The zero point on the *y*-axis of these graphs represents the conductivity of pure water; the break points on the *x*-axis represent the equivalence point in the titration.)

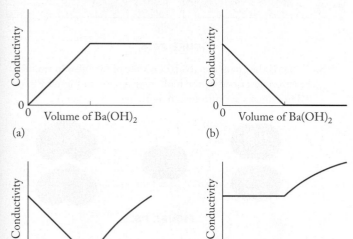

(a)   (b)
(c)   (d)

FIGURE P8.9

*8.10. Which of the graphs in Figure P8.10 best represents the changes in conductivity that occur before and after the equivalence point in each of the following titrations?
a. Sample, Cl^- (*aq*); titrant, $AgNO_3$(*aq*)
b. Sample, HCl(*aq*); titrant, LiOH(*aq*)
c. Sample, CH_3COOH(*aq*); titrant, NaOH(*aq*)

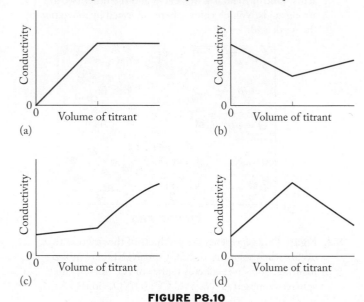

(a)   (b)
(c)   (d)

FIGURE P8.10

QUESTIONS AND PROBLEMS

Solutions and Their Concentrations

PROBLEMS

8.11. Calculate the molarity of each of the following solutions:
a. 0.56 mol $BaCl_2$ in 100.0 mL of solution
b. 0.200 mol Na_2CO_3 in 200.0 mL of solution
c. 0.325 mol $C_6H_{12}O_6$ in 250.0 mL of solution
d. 1.48 mol KNO_3 in 250.0 mL of solution

8.12. Calculate the molarity of each of the following solutions:
a. 0.150 mol urea $[(NH_2)_2CO]$ in 250.0 mL of solution
b. 1.46 mol $NH_4CH_3CO_2$ in 1.000 L of solution
c. 1.94 mol methanol (CH_3OH) in 5.000 L of solution
d. 0.045 mol sucrose $(C_{12}H_{22}O_{11})$ in 50.0 mL of solution

8.13. Calculate the molarity of Na^+ ions in each of the following solutions:
a. 0.29 *M* $NaNO_3$
b. 0.33 g NaCl in 25 mL of solution
c. 0.88 *M* Na_2SO_4
d. 0.46 g Na_3PO_4 in 100 mL of solution

8.14. Calculate the molarity of each of the following:
a. 64.7 g LiCl in 250.0 mL of solution
b. 29.3 g $NiSO_4$ in 200.0 mL of solution
c. 50.0 g KCN in 500.0 mL of solution
d. 0.155 g $AgNO_3$ in 100.0 mL of solution

8.15. How many grams of solute are needed to prepare each of the following solutions?
  a. 1.000 L of 0.200 M NaCl
  b. 250.0 mL of 0.125 M $CuSO_4$
  c. 500.0 mL of 0.400 M CH_3OH

8.16. How many grams of solute are needed to prepare each of the following solutions?
  a. 500.0 mL of 0.250 M KBr
  b. 25.0 mL of 0.200 M $NaNO_3$
  c. 100.0 mL of 0.375 M CH_3OH

8.17. River Water The Mackenzie River in northern Canada contains, on average, 0.820 mM Ca^{2+}, 0.430 mM Mg^{2+}, 0.300 mM Na^+, 0.0200 M K^+, 0.250 mM Cl^-, 0.380 mM SO_4^{2-}, and 1.82 mM HCO_3^-. What, on average, is the total mass of these ions in 2.75 L of Mackenzie River water?

8.18. Zinc, copper, lead, and mercury ions are toxic to Atlantic salmon at concentrations of 6.42×10^{-2} mM, 7.16×10^{-3} mM, 0.965 mM, and 5.00×10^{-2} mM, respectively. What are the corresponding concentrations in milligrams per liter?

8.19. Calculate the number of moles of solute contained in the following volumes of aqueous solutions of four pesticides:
  a. 0.400 L of 0.024 M lindane
  b. 1.65 L of 0.473 mM dieldrin
  c. 25.8 L of 3.4 mM DDT
  d. 154 L of 27.4 mM aldrin

8.20. A sample of crude oil contains 3.13 mM naphthalene, 12.0 mM methylnaphthalene, 23.8 mM dimethylnaphthalene, and 14.1 mM trimethylnaphthalene. What is the total number of moles of all the naphthalene compounds in 100.0 mL of the oil?

8.21. The pesticide DDT ($C_{14}H_9Cl_5$) was banned in the United States in 1972 because of dire environmental impacts. Just prior to its being banned, analyses of groundwater samples in Pennsylvania between 1969 and 1971 yielded the following results:

Location	Sample Size	Mass of DDT
Orchard	250.0 mL	0.030 mg
Residential	1.750 L	0.035 mg
Residential after a storm	50.0 mL	0.57 mg

Express these data in units of millimoles of DDT per liter.

8.22. Pesticide concentrations in the Rhine River between Germany and France between 1969 and 1975 averaged 0.55 mg/L of hexachlorobenzene (C_6Cl_6), 0.06 mg/L of dieldrin ($C_{12}H_8Cl_6O$), and 1.02 mg/L of hexachlorocyclohexane ($C_6H_6Cl_6$). Express these concentrations in millimoles per liter.

8.23. Effluent from municipal sewers often contains high concentrations of zinc. A sewer pipe discharges effluent that contains 10 mg Zn^{2+}/L. What is the molarity of Zn^{2+} in the effluent?

*__8.24.__ The concentration of copper(II) sulfate in one brand of soluble plant fertilizer is 0.07% by mass. If a 20 g sample of the fertilizer is dissolved in 2.0 L of solution, what is the molarity of Cu^{2+}?

*__8.25.__ For which of the following compounds is it possible to make a 1.0 M solution at 0°C?
  a. $CuSO_4 \cdot 5\,H_2O$, solubility = 23.1 g/100 mL
  b. $AgNO_3$, solubility = 122 g/100 mL
  c. $Fe(NO_3)_2 \cdot 6\,H_2O$, solubility = 113 g/100 mL
  d. $Ca(OH)_2$, solubility = 0.185 g/100 mL

8.26. Gold in the Ocean About 6×10^9 g of gold is thought to be dissolved in the oceans of the world. If the total volume of the oceans is 1.5×10^{21} L, what is the average molarity of gold in seawater?

8.27. The concentration of Mg^{2+} in a sample of coastal seawater is 1.09 g/kg. What is the molarity of Mg^{2+} in this seawater with a density of 1.02 g/mL?

8.28. Hemoglobin in Blood A typical adult body contains 6.0 liters of blood. The hemoglobin content of blood is about 15.5 g/100.0 mL of blood. The approximate molar mass of hemoglobin is 64,500 g/mol. How many moles of hemoglobin are present in a typical adult?

Dilutions

PROBLEMS

8.29. Calculate the final concentrations of the following aqueous solutions after each has been diluted to a final volume of 25.0 mL:
  a. 1.00 mL of 0.452 M Na^+
  b. 2.00 mL of 3.4 mM LiCl
  c. 5.00 mL of 6.42×10^{-2} mM Zn^{2+}

8.30. Chemists who analyze samples for dissolved trace elements may buy standard solutions that contain 1.000 g/L concentrations of the elements. If a chemist wishes to prepare 0.500 L of a working standard that has a concentration of 5.00 mg/L, what volume of the 1.000 g/L standard is needed?

*__8.31.__ A puddle of coastal seawater, caught in a depression formed by some coastal rocks at high tide, begins to evaporate on a hot summer day as the tide goes out. If the volume of the puddle decreases to 23% of its initial volume, what is the concentration of Na^+ after evaporation if initially it was 0.449 M?

8.32. What volume of 2.5 M $SrCl_2$ is needed to prepare 500.0 mL of 5.0 mM solution?

*__8.33. Dilution of Adult-Strength Cough Syrup__ A standard dose of an over-the-counter cough suppressant for adults is 20.0 mL. A portion this size contains 35 mg of the active pharmaceutical ingredient (API). A pediatrician prescribes the drug for a 6-year-old child, but the child may take only 10.0 mL at a time and receive a maximum of 4.00 mg of the API. What is the concentration in mg/mL of the adult-strength medication, and how many milliliters of it should be diluted to make 100.0 mL of child-strength cough syrup?

***8.34. Mixing Fertilizer** The label on a bottle of "organic" liquid fertilizer concentrate states that it contains 8 grams of phosphate per 100.0 mL and that 16 fluid ounces should be diluted with water to make 32 gallons of fertilizer to be applied to growing plants. What is the phosphate concentration in grams per liter in the diluted fertilizer? (1 gallon = 128 fluid ounces.)

Electrolytes and Nonelectrolytes

CONCEPT REVIEW

8.35. A solution of table salt is a good conductor of electricity, but a solution containing an equal molar concentration of table sugar is not. Why?

8.36. Why can scientists use conductivity to study the mixing of fresh water and seawater in estuaries?

8.37. Explain why liquid methanol, CH_3OH, cannot conduct electricity whereas molten NaOH can.

8.38. Fuel Cells The electrolyte in an electricity-generating device called *a fuel cell* consists of a mixture of Li_2CO_3 and K_2CO_3 heated to 650°C. At this temperature these ionic solids melt. Explain how the mixture of molten carbonates can conduct electricity.

***8.39.** Rank the following solutions on the basis of their ability to conduct electricity, starting with the most conductive: (a) 1.0 M NaCl; (b) 1.2 M KCl; (c) 1.0 M Na_2SO_4; (d) 0.75 M LiCl.

8.40. Rank the conductivities of 1 M aqueous solutions of each of the following solutes, starting with the most conductive: (a) acetic acid; (b) methanol; (c) sucrose (table sugar); (d) hydrochloric acid.

PROBLEMS

8.41. What is the molarity of Na^+ ions in a 0.025 M aqueous solution of (a) NaBr; (b) Na_2SO_4; (c) Na_3PO_4?

8.42. What is the molarity of each ion in a 0.025 M aqueous solution of (a) KCl; (b) $CuSO_4$; (c) $CaCl_2$?

Acids, Bases, and Neutralization Reactions

CONCEPT REVIEW

8.43. What chemical property of an acid makes it an acid?

8.44. What is the difference between a strong acid and a weak acid?

8.45. Give the formulas of two strong acids and two weak acids.

8.46. Why is $HSO_4^-(aq)$ a weaker acid than H_2SO_4?

8.47. What chemical property of a base makes it a base?

8.48. What is the difference between a strong base and a weak base?

8.49. Give the formulas of two strong bases and two weak bases.

8.50. Write the net ionic equation for the neutralization of a strong acid by a strong base.

PROBLEMS

8.51. For each of the following acid–base reactions, identify the acid and the base and then write the net ionic equation:
a. $H_2SO_4(aq) + Ca(OH)_2(s) \rightarrow CaSO_4(aq) + 2\,H_2O(\ell)$
b. $PbCO_3(s) + H_2SO_4(aq) \rightarrow$
$$PbSO_4(s) + CO_2(g) + H_2O(\ell)$$
c. $Ca(OH)_2(s) + 2\,CH_3COOH(aq) \rightarrow$
$$Ca(CH_3COO)_2(aq) + 2\,H_2O(\ell)$$

8.52. Complete and balance each of the following neutralization reactions, name the products, and write the net ionic equations.
a. $HBr(aq) + KOH(aq) \rightarrow$
b. $H_3PO_4(aq) + Ba(OH)_2(aq) \rightarrow$
c. $Al(OH)_3(s) + HCl(aq) \rightarrow$
d. $CH_3COOH(aq) + Sr(OH)_2(aq) \rightarrow$

8.53. Write a balanced molecular equation and a net ionic equation for the following reactions:
a. Solid magnesium hydroxide reacts with a solution of sulfuric acid.
b. Solid magnesium carbonate reacts with a solution of hydrochloric acid.
c. Ammonia gas reacts with hydrogen chloride gas.

8.54. Write a balanced molecular equation and a net ionic equation for the following reactions:
a. Solid aluminum hydroxide reacts with a solution of hydrobromic acid.
b. A solution of sulfuric acid reacts with solid sodium carbonate.
c. A solution of calcium hydroxide reacts with a solution of nitric acid.

8.55. Toxicity of Lead Pigments The use of lead(II) carbonate and lead(II) hydroxide as white pigments in paint was discontinued in the United States in 1978 because these compounds dissolved in the stomachs of young children who ingested paint chips. The Pb^{2+} ions released when the compounds dissolve interfere with neurotransmissions in the brain, causing neurological disorders. Using net ionic equations, show why lead(II) carbonate and lead(II) hydroxide dissolve in acidic solutions.

8.56. Lawn Care Many homeowners treat their lawns with $CaCO_3(s)$ to reduce the acidity of the soil. Write a net ionic equation for the reaction of $CaCO_3(s)$ with a strong acid.

Precipitation Reactions

CONCEPT REVIEW

8.57. What is the difference between a saturated solution and a supersaturated solution?

8.58. What are common solubility units?

8.59. What is a precipitation reaction?

8.60. A precipitate may appear when two completely clear aqueous solutions are mixed. What circumstances are responsible for this event?

8.61. Is a saturated solution always a concentrated solution? Explain.

8.62. Honey is a concentrated solution of sugar molecules in water. Clear, viscous honey becomes cloudy after being stored for long periods. Explain how this transition illustrates supersaturation.

PROBLEMS

8.63. According to the solubility rules in Table 8.3, which of the following compounds have limited solubility in water? (a) barium sulfate; (b) barium hydroxide; (c) lanthanum nitrate; (d) sodium acetate; (e) lead hydroxide; (f) calcium phosphate

8.64. Ocean Vents The black "smoke" that flows out of deep ocean hydrothermal vents (Figure P8.64) is made of insoluble metal sulfides suspended in seawater. Of the following cations that are present in the water flowing up through these vents, which ones could contribute to the formation of the black smoke? Na^+, Li^+, Mn^{2+}, Fe^{2+}, Ca^{2+}, Mg^{2+}, Zn^{2+}, Pb^{2+}, Cu^{2+}

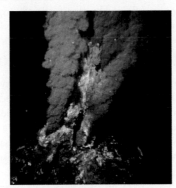

FIGURE P8.64

8.65. Complete and balance the chemical equations for the precipitation reactions, if any, between the following pairs of reactants, and write the net ionic equations:
a. $Pb(NO_3)_2(aq) + Na_2SO_4(aq) \rightarrow$
b. $NiCl_2(aq) + NH_4NO_3(aq) \rightarrow$
c. $FeCl_2(aq) + Na_2S(aq) \rightarrow$
d. $MgSO_4(aq) + BaCl_2(aq) \rightarrow$

*8.66. Wastewater Treatment Show with appropriate net ionic reactions how Cr^{3+} and Cd^{2+} can be removed from wastewater by treatment with solutions of sodium hydroxide.

*8.67. An aqueous solution containing Ca^{2+}, Cl^-, CO_3^{2-}, and NO_3^- is allowed to evaporate. Which compound will precipitate first?

8.68. Ten milliliters of a 5×10^{-3} M solution of Cl^- ions is reacted with a 0.500 M solution of $AgNO_3$. What is the maximum mass of $AgCl$ that precipitates?

8.69. Calculate the mass of $MgCO_3$ precipitated by mixing 10.0 mL of a 0.200 M Na_2CO_3 solution with 5.00 mL of 0.0500 M $Mg(NO_3)_2$ solution.

8.70. Toxic chromate can be precipitated from an aqueous solution by bubbling SO_2 through the solution. How many grams of SO_2 are required to treat 3.0×10^8 L of 0.050 mM Cr^{6+}?

$$2\,CrO_4^{2-}(aq) + 3\,SO_2(g) + 4\,H^+(aq) \rightarrow$$
$$Cr_2(SO_4)_3(s) + 2\,H_2O(\ell)$$

8.71. Iron(II) can be precipitated from a slightly basic aqueous solution by bubbling oxygen through the solution, which converts Fe^{2+} to insoluble Fe^{3+}:

$$4\,Fe(OH)^+(aq) + 4\,OH^-(aq) + O_2(g) + 2\,H_2O(\ell) \rightarrow$$
$$4\,Fe(OH)_3(s)$$

How many grams of O_2 are consumed to precipitate all of the iron in 75 mL of 0.090 M Fe^{2+}?

8.72. Given the following equation, how many grams of $PbCO_3$ will dissolve when 1.00 L of 1.00 M H^+ is added to 5.00 g of $PbCO_3$?

$$PbCO_3(s) + 2\,H^+(aq) \rightarrow Pb^{2+}(aq) + H_2O(\ell) + CO_2(g)$$

*8.73. Drinking Water Purification Phosphate can be removed from drinking-water supplies by treating the water with $Ca(OH)_2$:

$$5\,Ca(OH)_2(aq) + 3\,PO_4^{3-}(aq) \rightarrow$$
$$Ca_5(OH)(PO_4)_3(s) + 9\,OH^-(aq)$$

How much $Ca(OH)_2$ is required to remove 90% of the PO_4^{3-} from 4.5×10^6 L of drinking water containing 25 mg/L of PO_4^{3-}?

8.74. Toxic cyanide ions can be removed from wastewater by adding hypochlorite:

$$2\,CN^-(aq) + 5\,OCl^-(aq) + H_2O(\ell) \rightarrow$$
$$N_2(g) + 2\,HCO_3^-(aq) + 5\,Cl^-(aq)$$

How many liters of 0.125 M OCl^- are required to remove the CN^- in 3.4×10^6 L of wastewater in which the CN^- concentration is 0.58 mg/L?

Oxidation–Reduction Reactions

CONCEPT REVIEW

8.75. How are the gains or losses of electrons related to changes in oxidation numbers?

8.76. What is the sum of the oxidation numbers of the atoms in a molecule?

8.77. What is the sum of the oxidation numbers of all the atoms in each of the following polyatomic ions? (a) OH^-; (b) NH_4^+; (c) SO_4^{2-}; (d) PO_4^{3-}

8.78. Gold does not dissolve in concentrated H_2SO_4 but readily dissolves in H_2SeO_4 (selenic acid). Which acid is the stronger oxidizing agent?

8.79. Silver dissolves in sulfuric acid to form silver sulfate and H_2, but gold does not dissolve in sulfuric acid to form gold sulfate. Which of the two metals is the better reducing agent?

8.80. Rank the following oxoanions in order of decreasing oxidation number on chlorine: (a) ClO^-; (b) ClO_2^-; (c) ClO_3^-; (d) ClO_4^-.

*8.81. The generic formula of normal alkanes is C_nH_{2n+2}. How does the oxidation state of carbon in these compounds change with increasing n?

8.82. Why is the oxidizing agent in a redox reaction reduced and the reducing agent oxidized?

PROBLEMS

8.83. What is the oxidation number of chlorine in each of the following oxoacids? (a) hypochlorous acid ($HClO$); (b) chloric acid ($HClO_3$); (c) perchloric acid ($HClO_4$)

8.84. What is the oxidation number of nitrogen in each of the following species? (a) elemental nitrogen (N_2); (b) hydrazine (N_2H_4); (c) ammonium ion (NH_4^+)

8.85. How many moles of electrons are transferred during the dehydration of one mole of sucrose?

$$C_{12}H_{22}O_{11}(s) \rightarrow 12\ C(s) + 11\ H_2O(\ell)$$

8.86. Nitrogen Cycle In one stage of the nitrogen cycle, *Nitrosomonas* bacteria convert ammonia and oxygen into nitrite ions.
  a. What is the change in oxidation state of nitrogen during the reaction?
  b. Write a balanced net ionic equation describing the reaction in acidic groundwater.

8.87. Natural Weathering of Ores Iron is oxidized in a number of chemical weathering processes. How many moles of O_2 are consumed in converting one mole of magnetite (Fe_3O_4) to hematite (Fe_2O_3)?

***8.88.** The mineral rhodochrosite [manganese(II) carbonate, $MnCO_3$] is a commercially important source of manganese. How many moles of O_2 are consumed in converting one mole $MnCO_3$ to one mole of MnO_2?

8.89. Earth's Crust The following chemical reactions have helped to shape Earth's crust. Determine the oxidation numbers of all the elements in the reactants and products, and identify which elements are oxidized and which are reduced:
  a. $3\ SiO_2(s) + 2\ Fe_3O_4(s) \rightarrow 3\ Fe_2SiO_4(s) + O_2(g)$
  b. $SiO_2(s) + 2\ Fe(s) + O_2(g) \rightarrow Fe_2SiO_4(s)$
  c. $4\ FeO(s) + O_2(g) + 6\ H_2O(\ell) \rightarrow 4\ Fe(OH)_3(s)$

8.90. Determine the oxidation numbers of each of the elements in the following reactions, and identify which of them are oxidized or reduced, if any:
  a. $SiO_2(s) + 2\ H_2O(\ell) \rightarrow H_4SiO_4(aq)$
  b. $2\ MnCO_3(s) + O_2(g) \rightarrow 2\ MnO_2(s) + 2\ CO_2(g)$
  c. $3\ NO_2(g) + H_2O(\ell) \rightarrow 2\ NO_3^-(aq) + NO(g) + 2\ H^+(aq)$

8.91. How many moles of O_2 are consumed in the conversion of one mole of $FeCO_3$ to each of the following compounds? (a) Fe_2O_3; (b) Fe_3O_4

8.92. Uranium is found in Earth's crust as UO_2 and in an assortment of compounds containing UO_2^{n+} cations. How many moles of electrons are transferred in the conversion of one mole of UO_2 to each of the following species? In which of the conversions is uranium oxidized? (a) $UO_2(CO_3)_3^{4-}(aq)$; (b) $UO_2(HPO_4)_2^{2-}(aq)$

8.93. Nitrogen in the hydrosphere is found primarily as ammonium ions and nitrate ions. Complete and balance the following chemical equation describing the oxidation of ammonium ions to nitrate ions in acid solution:

$$NH_4^+(aq) + O_2(g) \rightarrow NO_3^-(aq)$$

8.94. In sediments and waterlogged soil, dissolved O_2 concentrations are so low that the microorganisms living there must rely on other sources of oxygen for respiration. Some bacteria can extract the oxygen from sulfate ions, reducing the sulfur in them to hydrogen sulfide gas and giving the sediments or soil a distinctive rotten-egg odor.

  a. What is the change in oxidation state of sulfur as a result of this reaction?
  b. Write the balanced net ionic equation for the reaction under acidic conditions that releases O_2 from sulfate and forms hydrogen sulfide gas.

8.95. The solubilities of Fe and Mn compounds in freshwater streams are affected by changes in the oxidation states of these metals. Complete and balance the following redox reaction in which soluble Mn^{2+} becomes solid MnO_2:

$$Fe(OH)_2^+(aq) + Mn^{2+}(aq) \rightarrow MnO_2(s) + Fe^{2+}(aq)$$

8.96. A method for determining the quantity of dissolved oxygen in natural waters requires a series of redox reactions. Balance the following chemical equations in that series under the conditions indicated:
  a. $Mn^{2+}(aq) + O_2(g) \rightarrow MnO_2(s)$ (basic solution)
  b. $MnO_2(s) + I^-(aq) \rightarrow Mn^{2+}(aq) + I_2(s)$ (acidic solution)
  c. $I_2(s) + S_2O_3^{2-}(aq) \rightarrow$
  $$I^-(aq) + S_4O_6^{2-}(aq)\quad\text{(neutral solution)}$$

8.97. Silver can be extracted from rocks using cyanide ion. Complete and balance the following reaction for this process:

$$Ag(s) + CN^-(aq) + O_2(g) \rightarrow$$
$$Ag(CN)_2^-(aq)\quad\text{(basic solution)}$$

8.98. Permanganate ion (MnO_4^-) is used in water purification to remove oxidizable substances. Complete and balance the following reactions for the removal of sulfide, cyanide, and sulfite. Assume that reaction conditions are basic:
  a. $MnO_4^-(aq) + S^{2-}(aq) \rightarrow MnS(s) + S_8(s)$
  b. $MnO_4^-(aq) + CN^-(aq) \rightarrow MnO_2(s) + CNO^-(aq)$
  c. $MnO_4^-(aq) + SO_3^{2-}(aq) \rightarrow MnO_2(s) + SO_4^{2-}(aq)$

8.99. Biocide Chemistry The water-soluble gas ClO_2 is known as an oxidative biocide. It destroys bacteria by oxidizing their cell walls and viruses by attacking their viral envelopes. ClO_2 may be prepared for use as a decontaminating agent from several different starting materials in slightly acidic solutions. Complete and balance the following chemical reactions for the synthesis of ClO_2.
  a. $ClO_3^-(aq) + SO_2(g) \rightarrow ClO_2(g) + SO_4^{2-}(aq)$
  b. $ClO_3^-(aq) + Cl^-(aq) \rightarrow ClO_2(g) + Cl_2(g)$
  c. $ClO_3^-(aq) + Cl_2(g) \rightarrow ClO_2(g) + O_2(g)$

8.100. Soluble hypochlorite (OCl^-) compounds are widely used to treat wastewater. In this environment, a reaction occurs between hypochlorite and nitrite ions that produces nitrate and chloride ions. Write a net ionic equation describing this reaction in slightly basic wastewater.

Titrations

PROBLEMS

8.101. How many milliliters of 0.100 M NaOH are required to neutralize the following solutions?
  a. 10.0 mL of 0.0500 M HCl
  b. 25.0 mL of 0.126 M HNO$_3$
  c. 50.0 mL of 0.215 M H$_2$SO$_4$

8.102. How many milliliters of 0.100 M HNO_3 are needed to neutralize the following solutions?
    a. 45.0 mL of 0.667 M KOH
    b. 58.5 mL of 0.0100 M $Al(OH)_3$
    c. 34.7 mL of 0.775 M NaOH

*8.103. The solubility of slaked lime, $Ca(OH)_2$, in water at 20°C is 0.185 g/100.0 mL. What volume of 0.00100 M HCl is needed to neutralize 10.0 mL of a saturated $Ca(OH)_2$ solution?

8.104. The solubility of magnesium hydroxide, $Mg(OH)_2$, in water is 9.0×10^{-4} g/100.0 mL. What volume of 0.00100 M HNO_3 is required to neutralize 1.00 L of a saturated $Mg(OH)_2$ solution?

8.105. **Chlorinity of Seawater** Scientists can precisely determine the chloride ion concentration, or *chlorinity*, of seawater samples using a titration called the Mohr method. The titrant is a solution of $AgNO_3$. The indicator is a few drops of K_2CrO_4 solution, which imparts a light yellow color to the titration mixture before the equivalence point (Figure P8.105a). However, after all the Cl^- ions have been precipitated as AgCl, the first excess of Ag^+ ions combines with CrO_4^{2-} ions to form Ag_2CrO_4, which makes the milky suspension of AgCl look pink (Figure P8.105b). If it takes 27.80 mL of 0.5000 M $AgNO_3$ to titrate a 25.00 mL sample of seawater, what is the concentration of Cl^- in the sample? Express your answer in mM and in g/kg (the density of the sample is 1.025 g/mL).

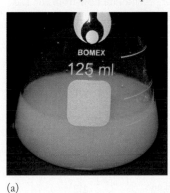 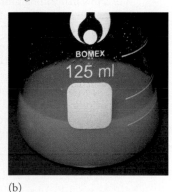

(a)   (b)

FIGURE P8.105

8.106. **Exercise Physiology** Lactic acid accumulates in muscle when glucose is metabolized under conditions where oxygen is the limiting reagent. Lactic acid has one carboxylic acid functional group per molecule (Figure P8.106). To determine the concentration of a solution of lactic acid, a chemist titrates a 20.00 mL sample of it with 0.1010 M NaOH, and finds that 12.77 mL of titrant is required to reach the equivalence point. What is the molarity of the lactic acid solution?

FIGURE P8.106

Ion Exchange

CONCEPT REVIEW

8.107. Explain how a mixture of anion and cation exchangers can be used to deionize water.

8.108. Describe the process by which the ion exchanger in a home water softener is regenerated for further use.

8.109. Why is there no advantage to using a K^+ resin rather than a Na^+ resin to deionize water?

8.110. A piece of Zn metal is placed in a solution containing Cu^{2+} ions. At the surface of the Zn metal, Cu^{2+} ions react with Zn atoms, forming Cu atoms and Zn^{2+} ions. Is this reaction an example of ion exchange? Explain why or why not.

Additional Problems

8.111. To determine the concentration of SO_4^{2-} ion in a sample of groundwater, 100.0 mL of the sample is titrated with 0.0250 M $Ba(NO_3)_2$, forming insoluble $BaSO_4$. If 3.19 mL of the $Ba(NO_3)_2$ solution is required to reach the end point of the titration, what is the molarity of the SO_4^{2-}?

8.112. **Antifreeze** Ethylene glycol is the common name for the liquid used to keep the coolant in automobile cooling systems from freezing. It is 38.7% carbon, 9.7% hydrogen, and 51.6% oxygen by mass. Its molar mass is 62.07 g/mol, and its density is 1.106 g/mL at 20°C.
    a. What is the empirical formula of ethylene glycol?
    b. What is the molecular formula of ethylene glycol?
    c. In a solution prepared by mixing equal volumes of water and ethylene glycol, which ingredient is the solute and which is the solvent?

*8.113. According to the label on a bottle of concentrated hydrochloric acid, the contents are 36.0% HCl by mass and have a density of 1.18 g/mL.
    a. What is the molarity of this concentrated HCl?
    b. What volume of it would you need to prepare 0.250 L of 2.00 M HCl?
    c. What mass of sodium hydrogen carbonate would be needed to neutralize the spill if a bottle containing 1.75 L of this concentrated HCl dropped on a lab floor and broke open?

8.114. **Synthesis and Toxicity of Chlorine** Chlorine was first prepared in 1774 by heating a mixture of NaCl and MnO_2 in sulfuric acid:

$$NaCl(aq) + H_2SO_4(aq) + MnO_2(s) \rightarrow$$
$$Na_2SO_4(aq) + MnCl_2(aq) + H_2O(\ell) + Cl_2(g)$$

a. Assign oxidation numbers to the elements in each compound, and balance the redox reaction in acid solution.
b. Write a net ionic equation describing the reaction for formation of chlorine.
c. If chlorine gas is inhaled, it causes pulmonary edema (fluid in the lungs) because it reacts with water in the alveolar sacs of the lungs to produce the strong acid HCl and the weaker acid HOCl. Balance the equation for the conversion of Cl_2 to HCl and HOCl.

*8.115. When a solution of dithionate ions ($S_2O_4^{2-}$) is added to a solution of chromate ions (CrO_4^{2-}), the products of the ensuing chemical reaction that occurs under basic conditions include soluble sulfite ions and solid chromium(III) hydroxide. This reaction is used to remove Cr^{6+} from wastewater generated by factories that make chrome-plated metals.
  a. Write the net ionic equation for this redox reaction.
  b. Which element is oxidized and which is reduced?
  c. Identify the oxidizing and reducing agents in the reaction.
  d. How many grams of sodium dithionate would be needed to remove the Cr^{6+} in 100.0 L of wastewater that contains 0.00148 M chromate ion?

*8.116. A prototype battery based on iron compounds with large, positive oxidation numbers was developed in 1999. In the following reactions, assign oxidation numbers to the elements in each compound and balance the redox reactions in basic solution:
  a. $FeO_4^{2-}(aq) + H_2O(\ell) \rightarrow FeOOH(s) + O_2(g) + OH^-(aq)$
  b. $FeO_4^{2-}(aq) + H_2O(\ell) \rightarrow Fe_2O_3(s) + O_2(g) + OH^-(aq)$

*8.117. **Polishing Silver** Silver tarnish is the result of silver metal reacting with sulfur compounds, such as H_2S, and O_2 in the air. The tarnish on silverware (Ag_2S) can be removed by soaking the silverware in a slightly basic solution of $NaHCO_3$ (baking soda) in a basin lined with aluminum foil.
  a. Write a balanced chemical equation for the tarnish formation reaction.
  b. Write a balanced net ionic equation for the tarnish removal process in which Ag_2S reacts with Al metal, forming $Al(OH)_3(s)$, Ag metal, and HS^- ions.

8.118. Give the formula and name of the acids formed in the following chemical reactions of chlorine oxides.
  a. $ClO + H_2O \rightarrow ? + ?$
  b. $Cl_2O + H_2O \rightarrow HCl + ?$
  c. $Cl_2O_6 + H_2O \rightarrow ? + ?$

8.119. Many nonmetal oxides react with water to form acidic solutions. Give the formula and name for the acids produced from the following reactions:
  a. $P_4O_{10} + 6\,H_2O \rightarrow ?$
  b. $SeO_2 + H_2O \rightarrow ?$
  c. $B_2O_3 + 3\,H_2O \rightarrow ?$

8.120. Write net ionic equations for the reactions that occur when
  a. a sample of acetic acid is titrated with a solution of KOH.
  b. a solution of sodium carbonate is mixed with a solution of calcium chloride.
  c. calcium oxide dissolves in water.

8.121. One way to determine the concentration of hypochlorite ions (ClO^-) in solution is by first reacting them with I^- ions. Under acidic conditions the products of the reaction are I_2 and Cl^- ions. Then the I_2 produced in the first reaction is titrated with a solution of thiosulfate ions ($S_2O_3^{2-}$). The products of the titration reaction are $S_4O_6^{2-}$ and I^- ions. Write net ionic equations for the two reactions.

*8.122. **Fluoride Ion in Drinking Water** Sodium fluoride is added to drinking water in many municipalities to protect teeth against cavities. The target of the fluoridation is hydroxyapatite, $Ca_{10}(PO_4)_6(OH)_2$, a compound in tooth enamel. There is concern, however, that fluoride ions in water may contribute to skeletal fluorosis, an arthritis-like disease.
  a. Write a net ionic equation for the reaction between hydroxyapatite and sodium fluoride that produces fluorapatite, $Ca_{10}(PO_4)_6F_2$.
  b. The U.S. EPA currently restricts the concentration of F^- in drinking water to 4 mg/L. Express this concentration of F^- in molarity.
  c. One study of skeletal fluorosis suggests that drinking water with a fluoride concentration of 4 mg/L for 20 years raises the fluoride content in bone to 6 mg/g, a level at which a patient may experience stiff joints and other symptoms. How much fluoride (in milligrams) is present in a 100 mg sample of bone with this fluoride concentration?

*8.123. **Rocket Fuel in Drinking Water** Near Las Vegas, improper disposal of perchlorates used to manufacture rocket fuel contaminated a stream flowing into Lake Mead, the largest artificial lake in the United States and a major supply of drinking and irrigation water for the American Southwest. The U.S. EPA has proposed an advisory range for perchlorate concentrations in drinking water of 4 to 18 µg/L. The perchlorate concentration in the stream averages 700.0 µg/L, and the stream flows at an average rate of 161 million gallons per day (1 gal = 3.785 L).
  a. What are the formulas of sodium perchlorate and ammonium perchlorate?
  b. How many kilograms of perchlorate flow from the Las Vegas stream into Lake Mead each day?
  c. What volume of perchlorate-free lake water would have to mix with the stream water each day to dilute the stream's perchlorate concentration from 700.0 to 4 µg/L?
  d. Since 2003, the states of Maryland, Massachusetts, and New Mexico have limited perchlorate concentrations in drinking water to 0.1 µg/L. Five replicate samples were analyzed for perchlorates by laboratories in each state, and the following data (µg/L) were collected:

MD	MA	NM
1.1	0.90	1.2
1.1	0.95	1.2
1.4	0.92	1.3
1.3	0.90	1.4
0.9	0.93	1.1

Which of the labs produced the most precise analytical results?

*8.124. **Mine Drainage** Water draining from abandoned mines on Iron Mountain in California is extremely acidic and leaches iron, zinc, and other metals from the underlying rock (Figure P8.124). One liter of drainage contains as much as 80.0 g of dissolved iron and 6 g of zinc.
  a. Calculate the molarity of iron and of zinc in the drainage.

b. One source of the dissolved iron is iron(II) oxide in soil, which reacts with drainage containing H_2SO_4 to form soluble iron(II) sulfate. Write a net ionic equation for the reaction.

c. Sources of dissolved zinc include the mineral smithsonite, $ZnCO_3$. Write a balanced net ionic equation for the reaction between smithsonite and H_2SO_4 that produces soluble zinc sulfate.

d. One member of a class of minerals called ferrites contains a mixture of zinc(II), iron(II), and iron(III) oxides. The generic formula for the mineral is $Zn_xFe_{1-x}O \cdot Fe_2O_3$. If acidic mine waste flowing through a deposit of franklinite (a ferrite mineral) contains 80 g of Fe and 6 g of Zn as a result of dissolution of the mineral, what is the value of x in the formula of the mineral in the deposit?

FIGURE P8.124

*8.125. **Making Apple Cider Vinegar** Some people who prefer natural foods make their own apple cider vinegar. They start with freshly squeezed apple juice that contains about 6% natural sugars. These sugars, which all have nearly the same empirical formula, CH_2O, are fermented with yeast in a chemical reaction that produces equal numbers of moles of ethanol, C_2H_5OH, and carbon dioxide. The product of fermentation, called hard cider, undergoes an acid fermentation step in which ethanol and dissolved oxygen gas react together to form acetic acid, CH_3COOH, and water. This acetic acid is the principal solute in vinegar.

a. Write a balanced chemical equation describing the fermentation of natural sugars to ethanol and carbon dioxide. You may use in the equation the empirical formula given in the above paragraph.

b. Write a balanced chemical equation describing the acid fermentation of ethanol to acetic acid.

c. What are the oxidation states of carbon in the reactants and products of the two fermentation reactions?

d. If a sample of apple juice contains 1.00×10^2 g of natural sugar, what is the maximum quantity of acetic acid that could be produced by the two fermentation reactions?

8.126. A food chemist determines the concentration of acetic acid in a sample of apple cider vinegar (see Problem 8.125) by acid–base titration. What is the concentration of acetic acid

in the vinegar if the density of the sample is 1.01 g/mL, the titrant is 1.002 M NaOH, and the average volume of titrant required to titrate 25.00 mL subsamples of the vinegar is 20.78 mL? Express your answer the way a food chemist probably would: as percent by mass.

8.127. **Cave Formations** The stalactites and stalagmites in most caves are made of calcium carbonate (see Figure 8.10). However, in the Lower Kane Cave in Wyoming they are made of gypsum (calcium sulfate). The presence of $CaSO_4$ is explained by the following sequence of reactions:

$$H_2S(aq) + 2\,O_2(g) \rightarrow H_2SO_4(aq)$$

$$H_2SO_4(aq) + CaCO_3(s) \rightarrow CaSO_4(s) + H_2O(\ell) + CO_2(g)$$

a. Which (if either) of these reactions is a redox reaction? How many electrons are transferred?

b. Write a net ionic equation for the reaction of H_2SO_4 with $CaCO_3$.

c. How would the net ionic equation be different if the reaction were written as follows?

$$H_2SO_4(aq) + CaCO_3(s) \rightarrow CaSO_4(s) + H_2CO_3(aq)$$

8.128. **Gardening Chemistry** Dolomite is a mixed carbonate mineral (Figure P8.128) with the formula $MgCa(CO_3)_2$. Gardeners add dolomite granules to soil and potting mixes to reduce acidity and provide a source of Mg^{2+} ions, which plants need to grow. Would a 50-pound bag of dolomite granules neutralize more acid than a 50-pound bag of limestone ($CaCO_3$) granules? How much more? Express your answer as a percentage.

FIGURE P8.128

8.129. Which of the following reactions of calcium compounds is or are redox reactions?

a. $CaCO_3(s) \rightarrow CaO(s) + CO_2(g)$
b. $CaO(s) + SO_2(g) \rightarrow CaSO_3(s)$
c. $CaCl_2(s) \rightarrow Ca(s) + Cl_2(g)$
d. $3\,Ca(s) + N_2(g) \rightarrow Ca_3N_2(s)$

8.130. HF is prepared by reacting CaF_2 with H_2SO_4:

$$CaF_2(s) + H_2SO_4(\ell) \rightarrow 2\,HF(g) + CaSO_4(s)$$

HF can in turn be electrolyzed when dissolved in molten KF to produce fluorine gas:

$$2\,HF(\ell) \rightarrow F_2(g) + H_2(g)$$

Fluorine is extremely reactive, so it is typically sold as a 5% mixture by volume in an inert gas such as helium. How much CaF_2 is required to produce 500.0 L of 5% F_2 in helium? Assume the density of F_2 gas is 1.70 g/L.

9

Thermochemistry

Energy Changes in Chemical Reactions

Sunlight Unwinding

Energy—to power an automobile, heat a home, or support life—may seem to be an abstract idea. Energy has no mass and it has no volume. However, we see energy interacting with matter all the time—electricity flowing through light fixtures illuminates your room at night, solar energy warms the Earth, a gas flame heats a pot of water and converts some of it into steam. The energy stored in gasoline propels millions of cars and their passengers, unless of course, they are stuck in traffic.

Combustion reactions produce most of the energy we consume, but where does this energy come from? Combustion of coal (which is mostly carbon) generates much of the electricity in China, India, and the United States; combustion of natural gas (mostly CH_4) heats water, warms homes, and is also used to generate electricity; and then there is the combustion of gasoline, which many of us rely on to get us where we need to go. All of the fuels in these combustion reactions are fossil fuels, the decomposed remains of plants and animals that lived millions of years ago. To trace the origin of the energy in fossil fuels, we need to analyze the food chains that supplied nutrition to those plants and animals. Nearly all of those food chains started with green plants that harvested the energy of the sun—that is, with photosynthesis.

The connection to the sun is perhaps more obvious when we consider renewable sources of energy. In many parts of the world, wood is the principal fuel for heating homes and cooking food. Dry wood is mostly cellulose, the most common organic compound in the world, consisting of long chains of glucose molecules. As we discussed in Chapter 7, green plants harness the energy of the sun to synthesize glucose from carbon dioxide and water via photosynthesis. Trees require hundreds to thousands of molecules of glucose to synthesize one molecule of cellulose.

Burning wood in a stove (or food inside us) turns this process around, converting cellulose (or glucose) and O_2 back into CO_2 and H_2O and liberating the energy of the sun that was originally harvested by green plants. R. Buckminster Fuller (1895–1983), an architect, inventor, and futurist, once described the release of energy that happens when wood burns as "all that sunlight unwinding."

Burning Fuels Burning wood for recreational and other purposes involves the combustion of cellulose and other large molecules rich in carbon and hydrogen. The products are CO_2, H_2O, and a considerable quantity of energy. ▶

chemical energy potential energy stored in chemical bonds.

internal energy (*E*) the sum of all the kinetic and potential energies of all of the components of a system.

thermochemistry the study of the changes in energy that accompany chemical reactions.

thermodynamics the study of energy and its transformations.

first law of thermodynamics the principle that the energy gained or lost by a system must equal the energy lost or gained by the surroundings.

system the part of the universe that is the focus of a thermochemical study.

FIGURE 9.1 Water about to go over a waterfall has more potential energy than water at the bottom of the waterfall. Waterwheels were developed by the ancient Greeks and widely used in medieval Europe and colonial America to convert this potential energy into mechanical energy to grind grain into flour, weave yarn into cloth, mill lumber, and shape metals.

Nearly all chemical reactions, including those we discussed in Chapters 7 and 8, either consume or release energy. In this chapter we explore how these changes in energy are linked to changes in the molecular structure of reactants as they become products. In this exploration we will find out why hydrogen provides more energy per gram than any other fuel, why natural gas yields more energy per gram than any other fossil fuel, and why adding oxygenated fuels like ethanol to gasoline, though it may help reduce air pollution from automobile exhaust, also reduces the number of miles you can drive on a tank of gasoline.

9.1 Energy as a Reactant or Product

We learned in Chapter 1 that there are two broad categories of energy: *kinetic energy*, which is the energy of an object in motion, and *potential energy*, which can be the energy that an object has because of its position, as illustrated in Figure 9.1. Another form of potential energy is the energy that a substance has because of its composition, that is, the way in which the atoms inside it are bonded together. This kind of potential energy is called **chemical energy**. The sum of all the kinetic and potential (including chemical) energies of an object, or a collection of many objects, is the **internal energy (*E*)** of the object or collection. When the collective internal energies of the reactants in a chemical reaction are different from those of the products—and they nearly always are—the difference constitutes the *change in energy (ΔE)* that accompanies the reaction.

As a practical matter, the *absolute* value of a system's internal energy is extremely difficult to determine, but *changes* in energy are fairly easy to measure because they correspond to changes in the system's physical state or temperature. The branch of chemistry that explores these changes in energy is called **thermochemistry**, our primary focus in this chapter. Thermochemistry is related to a more expansive scientific domain called **thermodynamics**, which is the study of energy and its transformations.

In our exploration of the energy changes that accompany chemical reactions, we will often encounter reactions in which the collective internal energy of the reactants is greater than that of the products, such as the combustion of hydrogen (Figure 9.2). The result is a release of energy, which has to go somewhere. It can't just disappear, because that would violate the principle articulated by the law of conservation of energy and the **first law of thermodynamics**: energy cannot be created nor destroyed, though it can change from one form to another. Expressed another way, the total energy of the universe is a constant. The impact of this principle may be clearer if we consider the universe to consist of only two parts: that part we are interested in, which we call the **system**, and everything else,

FIGURE 9.2 Energy is released during chemical reactions when high-energy reactants, such as H_2 and O_2, form lower energy products, such as H_2O.

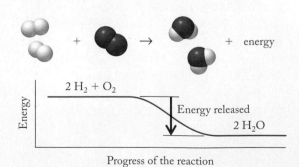

which we call the **surroundings**. We can express these definitions in the form of an equation:

$$\text{Universe} = \text{system} + \text{surroundings} \tag{9.1}$$

Let's use Equation 9.1 to connect changes in the internal energy of a system and its surroundings to the principle of energy conservation. If energy cannot be created or destroyed, then it follows that the energy of the universe (E_{univ}) is a constant, which means that ΔE_{univ} during any process is always equal to zero. If so, then:

$$\Delta E_{univ} = \Delta E_{sys} + \Delta E_{surr} = 0$$

and

$$\Delta E_{sys} = -\Delta E_{surr} \tag{9.2}$$

When a thermodynamic system releases energy, as when molecules of H_2 and O_2 combine to form molecules of H_2O, the energy flows into its surroundings. Whatever energy the system loses ($\Delta E_{sys} < 0$), its surroundings gain ($\Delta E_{surr} > 0$).

What effect does this increase in energy have on the surroundings? To get one answer to this question, let's consider the combustion of hydrogen that powers the engines of a Delta IV rocket (Figure 9.3). If the combustion gases (the reactants and products) are the system, then everything else in the universe—including the rocket, its launch pad, and the air through which the rocket travels on its way into space—make up the surroundings. The most significant energy transfer from the system to the surroundings occurs when the rocket and its cargo are launched into space. This involves exerting a large[1] force (F) through a considerable distance (d), which meets our definition from Chapter 1 of doing work (w):

$$w = F \times d \tag{1.1}$$

In addition to doing work, the hot gases of the system heat their surroundings. These two processes, doing work and transferring heat, represent the two ways in which energy moves between a system and its surroundings. In general, if heat flows *out from* a system, or if work is done *by* the system on its surroundings, the internal energy E of the system decreases. On the other hand, if heat flows *into* a system or if work is done *on* a system, the internal energy of the system increases. We can express this last relationship in an equation, using the symbol q to represent heat:

$$\Delta E = q + w \tag{9.3}$$

Adding heat to a system means that q has a positive value. Doing work on a system means that w has a positive value. If both q and w are positive, then so is ΔE, that is, $\Delta E > 0$. When a system does work *on* its surroundings, like the reaction mixture in the rocket engines in Figure 9.3, then w is negative, and if heat flows out from the system (see again Figure 9.3), then q is also negative. Taken together, these changes produce a decrease in the internal energy of the system ($\Delta E < 0$). The various combinations of heat flow and work and their impact on ΔE are summarized in Table 9.1 and in Figure 9.4.

An important point about changing the internal energy of a system is that no matter how it happens—whether only one of the processes in Table 9.1 is

surroundings everything in a thermochemical study that is not part of the system.

▶‖ **CHEMTOUR** Internal Energy

◉◉ **CONNECTION** In Chapter 1, we defined *energy* as the ability to do work. We presented the *law of conservation of energy* and the concept that energy cannot be created or destroyed, but can be changed from one form of energy to another.

FIGURE 9.3 Up to three hydrogen-fueled engines make up the first stage of a Delta IV rocket such as this one, which can send a satellite weighing 13 metric tons into orbit around the Earth.

◉◉ **CONNECTION** In Chapter 1, we defined *heat* as the spontaneous transfer of energy from warmer objects to cooler ones.

---

[1] At full throttle, one of the Delta IV engines produces up to 3.2×10^6 newtons of thrust (force), which is about 15 times that of a Boeing 747 engine at takeoff.

TABLE 9.1	Flows of Heat and Work and Their Impact on E_{sys}	
Processes That Increase E_{sys}		**Result**
Surroundings hotter than the system, so heat flows into the system		$q > 0$
Surroundings do work on the system		$w > 0$
Processes That Decrease E_{sys}		**Result**
System hotter than its surroundings, so heat flows into surroundings		$q < 0$
System does work on its surroundings		$w < 0$

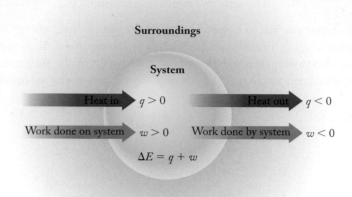

FIGURE 9.4 The change in internal energy (ΔE) of a system is positive when heat flows into it (q is positive) or work is done on it (w is positive). A system loses internal energy (ΔE is negative) when heat flows out from it (q is negative) or the system does work on its surroundings (w is negative).

▶‖ **CHEMTOUR** State Functions and Path Functions

state function a property of an entity based solely on its chemical or physical state or both, but not on how it achieved that state.

thermal energy the portion of the total internal energy of a system that is proportional to its absolute temperature.

involved or some combination of them is—the final internal energy of the system is independent of how it was acquired. It makes no difference whether the system lost or gained heat, nor whether it did work or had work done on it to achieve a particular internal energy. In other words, internal energy is a **state function**: it depends only on how much potential and kinetic energy the system has and not at all on the pathway or sequence of events that produced those levels of energy. Think of skiers ascending to the top of a mountain for their next run: whether they ride straight up on a ski lift or hike up a winding path, they arrive at the same height in the end (Figure 9.5). Their altitude is a state function, as is the change in altitude they experienced from the base of the mountain to its summit: both are independent of the path the skiers take. Similarly, any change in the internal energy of a system is independent of the processes causing the change. Only the initial and final energy levels of the system define ΔE:

$$\Delta E = E_{\text{final}} - E_{\text{initial}}$$

We will encounter other examples of system properties that are state functions in later sections of this chapter and in later chapters of this book.

At this point we need to be clear on what we mean by *heat*. Heat is energy that is in the process of being transferred from a higher temperature object to a lower temperature one. An operating engine of a Delta IV rocket gets very hot, which is reflected in the high temperature of the material that lines the combustion chamber. However, the quantity of **thermal energy** in this hot material is determined not only by its temperature but also by the composition of the material and how massive it is (a concept called *heat capacity* that we explore in Section 9.4). By *thermal* energy, we mean that part of the total internal energy of a system that is linked to how hot it is.

At the atomic level, the thermal energy of a system is the sum of the kinetic energies of the particles that make up the system. We learned in Chapter 1 that

FIGURE 9.5 Whether skiers at the base of a mountain ① ride a ski lift or ② hike up a mountain trail, their increase in altitude is the same.

the kinetic energy of any object in motion, be it a rocket or an atom, is a function of its mass (m) and its speed (u):

$$\text{KE} = \tfrac{1}{2}mu^2 \qquad (1.2)$$

The kinetic energies of atom-size particles such as molecules of H_2, O_2, or H_2O depend only on their temperature. In fact, the average kinetic energy of a population of gaseous atoms and molecules *is directly proportional* to their *absolute* temperature (that is, their temperature in kelvin): if T doubles, so does KE. Higher average kinetic energies mean higher average speeds, but because the speed term is squared in Equation 1.2, particle speed is not a linear function of temperature. For example, when the absolute temperature of a gas doubles, the average speed of the particles in it increases by only $\sqrt{2}$, or 1.414, times.

The motion of particles of gases as they bounce around inside the space they occupy is called *translational* kinetic energy (Figure 9.6a). Molecules also have non-translational, or *internal*, forms of kinetic energy. These include *rotational* kinetic energy (Figure 9.6b), and *vibrational* kinetic energy (Figure 9.6c). Bond vibrations also provide additional forms of potential energy in molecules in much the way that stretching, compressing, or bending a spring increases its potential energy.

9.2 Transferring Heat and Doing Work

Systems come in different sizes. They can be as large as galaxies or as small as single living cells. Most of the systems we discuss in this chapter fit on a laboratory bench, and we usually limit our view of their surroundings to that part of the universe that can exchange energy and matter with the system. In evaluating how much energy is gained or lost in a chemical reaction, the system may be just the particles involved in the reaction, but it may also include the vessel that contains them.

Isolated, Closed, and Open Systems

As we explore a system's energy production and consumption, we need to be clear on how the system interacts with its surroundings. For example, can energy be transferred from the system to its surroundings? Can matter be exchanged between

◉◉ CONNECTION We noted in Chapter 4 that there are several kinds of bond vibrations in molecules with more than two atoms, including symmetric and asymmetric stretching and bending as the angles between bonds increase and decrease.

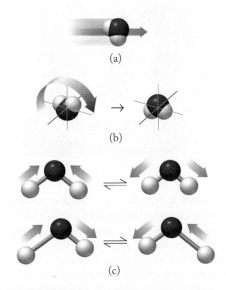

(a)

(b)

(c)

FIGURE 9.6 Some of the types of molecular motion that contribute to the internal energy of a system: (a) translational motion, motion from place to place along a path; (b) rotational motion, motion about a fixed axis; and (c) vibrational motion, movement back and forth from some central position.

(a) **Isolated system:** A thermos bottle containing hot soup with the lid screwed on tightly

(b) **Closed system:** A cup of hot soup with a lid

(c) **Open system:** An open cup of hot soup

FIGURE 9.7 Illustrations of isolated, closed, and open systems: (a) Hot soup in a tightly sealed thermos bottle approximates an isolated system: no vapor escapes, no matter is added or removed, and no energy is transferred to the surroundings. (b) Hot soup in a cup with a lid is a closed system; the soup transfers heat to the surroundings as it cools; however, no matter escapes and none is added. (c) Hot soup in a cup with no lid is an open system; it transfers both matter (steam) and energy (heat) to the surroundings as it cools. Matter in the form of pepper, grated cheese, or other flavor enhancers might also be added to the system.

the system and its surroundings? The answers to these questions define the type of system we have. If both matter and energy can be transferred between a system and its surroundings, we have an *open* system. If only energy can flow between them, we have a *closed* system. And if a system is completely cut off from its surroundings so that neither matter nor energy can be transferred, we have an *isolated* system.

The containers of hot soup in Figure 9.7 illustrate the three types of systems. Figure 9.7(a) shows a (theoretically) isolated system: a serving of soup inside the ultimate thermos bottle. Let's pretend that the bottle has a perfect seal and that it's perfectly insulated, allowing neither heat nor matter to escape or enter. The mass of the system never changes, nor does its temperature, nor does its energy content. Of course, no such thermos bottle actually exists. Even the best ones can't keep soup hot forever. For short times, however, substances in sealed, well-insulated containers approximate isolated systems.

Hot soup in a cup with a lid on it is an example of a *closed* thermodynamic system (Figure 9.7b). The lid keeps the soup and any vapor it produces from escaping, and it keeps any matter from getting into the soup. Only energy is exchanged between the soup and its surroundings as heat is transferred from the soup through the cup and into the air and the tabletop. Gradually the soup cools to room temperature. Many real thermodynamic systems are closed systems.

Soup in an open cup is an example of an *open* thermodynamic system that can exchange both energy and matter with its surroundings (Figure 9.7c). Heat flowing from the soup into its surroundings and matter—in the form of water vapor, for example—is free to leave the system. Matter can also enter the system—perhaps a little seasoning to enhance the flavor.

CONCEPT TEST •••••••••••••••••••••••••••••••••••

Suppose two identical pots of water are heated on a stove until the water inside them begins to boil. Both pots are then removed. One of the two is covered with a tight lid; the other is not, and both are allowed to cool.

a. What category of thermodynamic system—open, closed, or isolated—describes each of the cooling pots?

b. Which pot cools faster? Why?

(Answers to Concept Tests are in the back of the book.)

•••

Exothermic and Endothermic Processes

Chemists classify reactions based on whether they give off or absorb energy. A chemical reaction that results in the transfer of energy—usually in the form of heat—from the reaction mixture (the system) to its surroundings is an **exothermic** reaction. This energy flow can usually be detected by an increase in the temperature of the surroundings. Combustion reactions are good examples of exothermic reactions. Chemical reactions that absorb energy from their surroundings are **endothermic** reactions. The same terms are used to describe nonchemical processes that give out or take in heat. For example, your body cools itself during strenuous exercise by sweating. Cooling happens as the sweat on your skin evaporates because conversion of a liquid to a gas is an endothermic process, requiring the flow of heat into the system (sweat) from its surroundings (your body), as shown in Figure 9.8. On the other hand, the reverse process, water vapor condensing, as on the outside of a bottle of cold water on a humid summer day, is exothermic. Heat is transferred from the system (condensing water vapor) to its surroundings, including the water in the bottle in Figure 9.8.

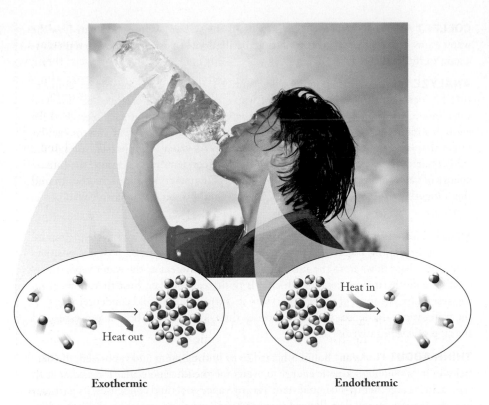

one in which energy—usually in the form of heat—flows from a system into its surroundings.

endothermic process one in which energy—usually in the form of heat—flows from the surroundings into the system.

Exothermic

Endothermic

FIGURE 9.8 A process that is exothermic in one direction, such as the condensation of atmospheric water vapor on the outside of a cold water bottle, is endothermic in the reverse direction, as when perspiration evaporates from an athlete's skin and clothing.

SAMPLE EXERCISE 9.1 **Identifying Exothermic and Endothermic Processes** **LO1**

Describe the flow of thermal energy (heat) during the purification of water by distillation (Figure 9.9), identify the steps as either endothermic or exothermic, and give the sign of q associated with each step. Consider the water being purified to be the system.

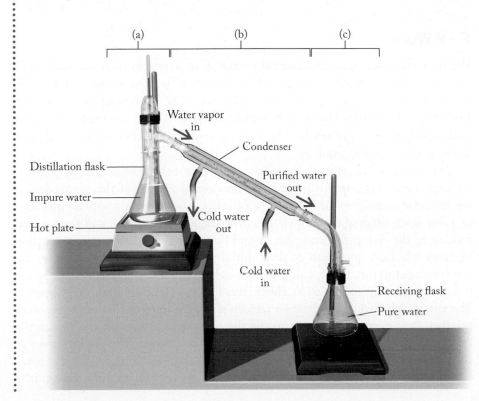

Water vapor in

Condenser

Distillation flask

Purified water out

Impure water

Hot plate

Cold water out

Cold water in

Receiving flask

Pure water

FIGURE 9.9 A laboratory setup for distilling water. (a) Impure water is heated to the boiling point in the distillation flask. (b) Water vapor rises and enters the condenser, where it is liquefied. (c) The purified liquid is collected in the receiving flask.

COLLECT AND ORGANIZE Since the water is the system, we must evaluate how the water gains or loses thermal energy during distillation. A positive value of q represents a gain in thermal energy.

ANALYZE During distillation, heat flows into the impure water, raising its temperature by increasing the kinetic energy of water molecules and particles of dissolved substances. Additional heating vaporizes the water by providing its molecules with enough kinetic energy to overcome the hydrogen bonds and other intermolecular forces that keep them together in the liquid phase. As water vapor passes through the cold condenser, heat flows from the vapor into the condenser. As a result, the kinetic energy of vapor molecules drops, which allows their attractions for each other to pull them together. These molecules form a layer of liquid water on the inside wall of the condenser.

SOLVE Heat flows from the surroundings (hot plate) to heat the system (impure water) to its boiling point. That process is endothermic and the sign of q is positive. Additional heat flows from the surroundings into the system as the water vaporizes—another endothermic process for which q is positive. However, heat then flows from the system (in the form of water vapor) into its surroundings (cold condenser walls) as the water vapor condenses and further cools: these are both exothermic processes and the sign of q is negative for them.

THINK ABOUT IT As any liquid is heated to its boiling point and vaporized, the particles in it gain enough kinetic energy to overcome the intermolecular forces that keep the particles together in the liquid state. As any vapor cools and condenses, its particles lose the kinetic energy that allowed them to overcome their attraction for each other, and they come together in the liquid state.

Practice Exercise What is the sign of q as (a) a match burns, (b) drops of molten candle wax solidify, and (c) a piece of dry ice disappears at room temperature? In each case, define the system and indicate whether the process is endothermic or exothermic. ✸

(Answers to Practice Exercises are in the back of the book.)

P–V Work

We have discussed how the internal energy E of a system increases ($\Delta E > 0$) when heat flows into it ($q > 0$) or work is done on it by its surroundings ($w > 0$), and how the internal energy of a system decreases ($\Delta E < 0$) when heat flows out from it ($q < 0$) or it does work on its surroundings ($w < 0$). Let's explore a system that rapidly cycles between having its surroundings do work on it and then it doing work on its surroundings.

The system is the gases in the combustion chamber of one of the cylinders in a diesel engine (Figure 9.10). First, the downward motion of the piston during the intake stroke allows air to enter the cylinder, where it mixes with an injection of a few microliters of diesel fuel. Then the intake valve closes, and the upward motion of the piston squeezes the air and fuel vapors (the system) in the cylinder into less than one-tenth of their initial volume. The compression heats up the reaction mixture (a clear sign its internal energy has increased) until, at the point of maximum compression, the mixture gets so hot it spontaneously ignites. The energy released by combustion rapidly raises the temperature of the reaction mixture even further, which increases the pressure it exerts on the top of the cylinder. This pressure drives the piston downward during the power stroke as the volume of the combustion mixture rapidly expands. Compressing the reaction mixture increased the mixture's internal energy because its surroundings (the

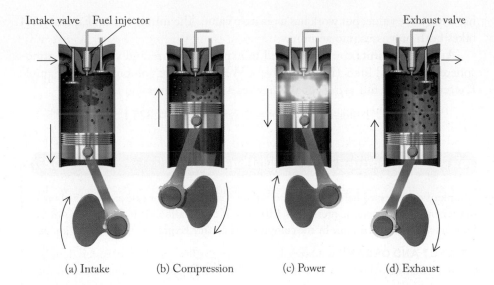

Intake valve   Fuel injector   Exhaust valve

(a) Intake   (b) Compression   (c) Power   (d) Exhaust

FIGURE 9.10 Diesel engines are based on a four-stroke cycle in which (a) fuel is injected and air is drawn into the combustion cylinder as the piston goes down, (b) the mixture is compressed as the piston moves up, (c) compression heating causes the mixture to ignite spontaneously and release energy that pushes the piston down, and (d) the products of combustion are pushed out as the piston moves up.

piston) did work on the system. Then, expansion of the hot products of combustion did work on their surroundings (the piston again), releasing energy they had gained through compression and combustion by pushing the piston downward.

The work done *on* the gases in an engine cylinder as they are compressed and the work done *by* the gases as they expand are two examples of **pressure–volume (P–V) work**. The second example is the work that any gaseous system does when its volume increases by an amount ΔV (where $\Delta V = V_f - V_i$) against an opposing pressure P. The amount of work done is the product $P \times \Delta V$. This pressure–volume work is equivalent to force $\times$ distance ($F \times d$) work. To understand why, keep in mind that pressure is force per unit of surface area:

$$P = \frac{F}{A} \tag{6.1}$$

Area has dimensions of distance squared (d^2), and volume is often expressed in units of distance cubed (d^3). If we substitute d^2 and d^3 for area and volume change, then:

$$P \times \Delta V = \frac{F}{d^2} \times d^3 = F \times d$$

Having established the equivalency of pressure–volume work and force–distance work, let's address how to convert the units used to express P–V work into units of energy, for example, joules. Doing so enables us to combine the quantities of heat transferred and P–V work done to calculate changes in the internal energy (ΔE) of a gaseous system. We use a modified version of Equation 9.3 in which w is replaced with $-P\Delta V$:

$$\Delta E = q - P\Delta V \tag{9.4}$$

Perhaps you are wondering about the minus sign in Equation 9.4. It is there because an increase in the volume of a system ($\Delta V > 0$) means that the system is expanding, which means that it is pushing aside a portion of its surroundings, which involves doing work *on* its surroundings. This kind of work has a negative value from the perspective of the system and results in a decrease in its internal energy (see Figure 9.4). On the other hand, when the surroundings do work on the system (for example, when a gaseous system is compressed), volume change

⊙⊙ **CONNECTION** In Chapter 6, we defined pressure and its units, including the pascal (Pa), the standard atmosphere (atm), and the bar.

▶‖ **CHEMTOUR** Pressure–Volume Work

pressure–volume (P–V) work the work associated with the expansion or compression of a gas.

has a negative value, but work has a positive value. The minus sign in Equation 9.4 takes both processes into account.

Among the units commonly used to express pressure and volume are atmospheres (atm) and liters (L), respectively. When they are combined, as in Sample Exercise 9.2, the following equality makes a handy conversion factor:

$$1 \text{ liter-atmosphere} = 1 \text{ L} \cdot \text{atm} = 101.325 \text{ J}$$

SAMPLE EXERCISE 9.2 **Calculating *P–V* Work** **LO2**

A tank of compressed helium is used to inflate 100 balloons for sale at a carnival on a day when the atmospheric pressure is 1.01 atm. If each balloon is inflated with 4.8 L, how much *P–V* work is done by the compressed helium? Express your answer in joules.

COLLECT AND ORGANIZE Our task is to determine how much *P–V* work is done by the 100 inflating balloons, each with 4.8 L of helium expanding against an opposing pressure of 1.01 atm.

ANALYZE We focus on the work done on the atmosphere (the surroundings) by the helium (the system) as the volume of the balloons increases. Inflating each of 100 balloons with about 5 L of helium against a pressure of about 1 atm means that *P*Δ*V* will be about 500 liter-atmospheres and about 100 times that number of joules, or 50,000 J.

SOLVE The work (*w*) done by the compressed helium is

$$w = -P\Delta V$$

$$= -1.01 \text{ atm} \times \frac{4.8 \text{ L}}{1 \text{ balloon}} \times 100 \text{ balloons} \times \frac{101.325 \text{ J}}{1 \text{ L} \cdot \text{atm}}$$

$$= -4.9 \times 10^4 \text{ J or } -49 \text{ kJ}$$

THINK ABOUT IT The internal energy of the compressed helium (the system) decreases when some of it expands to inflate the balloons because work is done *by* the system on its surroundings. Lower internal energy in this case corresponds to the lower kinetic energy (and lower temperature) of both the helium atoms in the balloons and those remaining in the tank.

Practice Exercise The balloon *Spirit of Freedom* (Figure 9.11), flown around the world by American aviator Steve Fossett (1944–2007) in 2002, contained 550,000 cubic feet of helium. How much *P–V* work was done to inflate the balloon, assuming the atmospheric pressure was 1.00 atm? (1 m³ = 1000 L = 35.3 ft³.)

FIGURE 9.11 The balloon *Spirit of Freedom* was flown around the world in 2002.

SAMPLE EXERCISE 9.3 **Relating Δ*E*, *q*, and *w*** **LO2**

The racing cars in Figure 9.12 are powered by V8 engines in which the motion of each piston in its cylinder displaces a volume of 0.733 L. If combustion of the mixture of gasoline vapor and air in one cylinder releases 1.68 kJ of energy, and if 33% of the energy does *P–V* work, how much pressure, on average, does the combustion reaction mixture exert on each piston? How much heat flows from the reaction mixture to its surroundings?

COLLECT AND ORGANIZE We know the quantity of energy (1.68 kJ) released by a combustion reaction. This energy flows from the reaction mixture to its surroundings in the form of heat and *P–V* work done on the surroundings as described by Equation 9.4:

$$\Delta E = q - P\Delta V$$

The value of ΔV is 0.733 L, and $P\Delta V$ is 33% of the energy released. We need to calculate the values of P and q.

ANALYZE The gases in each engine cylinder react exothermically, releasing 1.68 kJ. This loss of energy means that ΔE of the system is −1.68 kJ, 33% of which does P–V work on the surroundings. This corresponds to slightly more than 0.5 kJ (500 J) of energy for ∼3/4 L, or between 6.5 and 7.5 atm after converting J/L to atm. The other 67% must be transferred as heat (q). The calculated value of q should be about 2/3 of −1.68 kJ, or about −1 kJ.

SOLVE Average pressure:

$$-P\Delta V = 0.33 \times \Delta E = 0.33 \times (-1.68 \text{ kJ}) = -0.554 \text{ kJ} = -554 \text{ J}$$

$$P\Delta V = 554 \text{ J}$$

$$P = \frac{554 \text{ J}}{\Delta V} = \frac{554 \text{ J}}{0.733 \text{ L}} \times \frac{1 \text{ L} \cdot \text{atm}}{101.325 \text{ J}} = 7.5 \text{ atm}$$

Heat transferred:

$$q = \Delta E + P\Delta V = -1.68 \text{ kJ} + 0.554 \text{ kJ} = -1.13 \text{ kJ}$$

THINK ABOUT IT We used a negative sign in front of the $P\Delta V$ term because $P\Delta V$ contains positive values of pressure and volume change, but represents work done *by* the system and is part of the energy lost by the system (−1.68 kJ). Similarly, the heat transferred has a negative sign because it is thermal energy lost by the system. Its value is close to the one we estimated.

Practice Exercise The air compressor used with a paint sprayer does 64 J of work pumping air into a tank. Warmed by the process, air in the tank gives off 32 J of energy to its surroundings. What is the change in the internal energy of the air in the tank as a result of the two processes? ⚙

FIGURE 9.12 These racing cars are powered by internal combustion engines in which hot gases do P–V work that is harnessed by the engine and drive train and propels the cars to speeds over 200 miles per hour (320 km/hr).

9.3 Enthalpy and Enthalpy Changes

Many physical and chemical changes take place at constant pressure. These include chemical and biochemical reactions that occur in organisms living on Earth's surface, and they include the chemical reactions done in open beakers and flasks on laboratory benches. We use a thermodynamic parameter called *enthalpy* to track the changes in energy and the flow of heat into or out from these and other constant-pressure systems. **Enthalpy (H)** is a measure of the total energy of a system. By *total* we mean the system's internal energy (E) *and* the energy expended to push aside its surroundings to make space for the system at a given pressure. In equation form this definition is

$$H = E + PV \qquad (9.5)$$

A thermodynamic system's enthalpy is hard to quantify. However, the *change* in enthalpy that accompanies a chemical reaction or physical change is a readily accessible quantity because **enthalpy change (ΔH)** is equal to the transfer of heat into or out from a system at constant pressure, that is, q_P.

To see why ΔH and q_P are equivalent, let's turn Equation 9.5 into a description of the changes in H, E, and V that accompany a process occurring at constant pressure:

$$\Delta H = \Delta E + P\Delta V \qquad (9.6)$$

enthalpy (H) the sum of the internal energy and the pressure–volume product of a system; $H = E + PV$.

enthalpy change (ΔH) the energy absorbed by the reactants (endothermic reaction) or the energy given off by the products (exothermic reaction) for a reaction carried out at constant pressure.

Now let's revisit Equation 9.4 and apply it to a process occurring at constant pressure so that q is q_P:

$$\Delta E = q_P - P\Delta V$$

If we rearrange the terms to isolate q_P on the left side, we get an equation with the same right side as Equation 9.6:

$$q_P = \Delta E + P\Delta V$$

In other words:

$$\Delta H = q_P \tag{9.7}$$

Like ΔE, ΔH is also a state function. The value of ΔH depends only on the difference in enthalpy between the initial and final states, not on the pathway followed in going from one to the other.

Recall that we use the labels *exothermic* and *endothermic* to describe processes that release or absorb energy, usually as thermal energy or heat. These labels also describe processes in which the changes in enthalpy of the system are less than or greater than zero:

Description	Enthalpy Change
Exothermic	$\Delta H < 0$
Endothermic	$\Delta H > 0$

When an exothermic process runs in reverse, it becomes an endothermic process, and vice versa. For example, 6.01 kJ of thermal energy must be added to melt 1 mole of ice at 0°C, an endothermic process, but 6.01 kJ of thermal energy must be removed from a mole of liquid water at 0°C to freeze it, an exothermic process. This enthalpy change of 6.01 kJ/mol is called the **enthalpy of fusion (ΔH_{fus})** of water, often referred to as *heat of fusion*, where the subscript "fus" indicates melting (or *fusion*) of the solid. The enthalpy changes associated with other paired phase changes—vaporization and condensation, and sublimation and deposition—also have complementary values. For example, it takes 40.7 kJ of thermal energy to vaporize a mole of liquid water at 100°C, an endothermic process, and the same 40.7 kJ of thermal energy is released when one mole of water vapor condenses, an exothermic process. Thus, the **enthalpy of vaporization (ΔH_{vap})** of water, often referred to as *heat of vaporization*, is 40.7 kJ/mol.

Table 9.2 contains ΔH_{fus} and ΔH_{vap} values for several common compounds. Note how the values for the alkanes containing one to four carbons increase with increasing molar mass. This pattern makes sense if we recall that these molecules interact via London dispersion forces that increase with increasing molecular size. It is logical, then, that the thermal energy required to convert them from solids into liquids and from liquids into gases also increases with increasing molecular size.

Most of the heat of vaporization values in Table 9.2 are based on the heat required to vaporize substances at their normal boiling points. However, the ΔH_{vap} values for water include additional values at temperatures below the normal boiling point of 100°C. As you can see, ΔH_{vap} values decrease as temperature increases. Why? Because the kinetic energy of the H_2O molecules in liquid water increases as temperature increases. More kinetic energy means greater molecular motion, which also means less interaction between molecules: the average number of hydrogen bonds per water molecule decreases from an estimated 3.6 at 25°C to only 3.2 at 100°C. This decrease in interaction between molecules contributes to the decrease in energy required to separate them into independent, gas-phase particles.

enthalpy of fusion (ΔH_{fus}) the energy required to convert 1 mole of a solid substance at its melting point into the liquid state; also called *heat of fusion*.

enthalpy of vaporization (ΔH_{vap}) the energy required to convert 1 mole of a liquid substance at its boiling point into the vapor state; also called *heat of vaporization*.

Using intermolecular forces, explain why water has the highest ΔH_{vap} value of all the compounds in Table 9.2 (and the highest boiling point, too).

TABLE 9.2	Enthalpy of Fusion and Vaporization			
Compound and Molecular Structure		$\mathcal{M}$ (g/mol)	ΔH_{fus} (kJ/mol)	ΔH_{vap} (kJ/mol)
Methane		16.04	0.94	8.2
Ethane		30.07	2.86	14.7
Propane		44.10	3.53	15.7
Butane	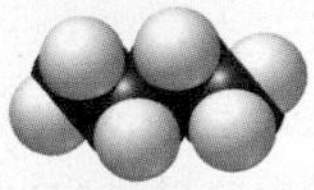	58.12	4.66	21.0
Methanol		32.04	3.18	35.3
Ethanol		46.07	5.02	38.6
Acetone		46.07	5.69	31.3
Ammonia		17.03	5.65	23.4
Water		18.02	6.01	44.0 at 25°C 43.5 at 40°C 42.5 at 60°C 41.6 at 80°C 40.7 at 100°C

9.4 Heating Curves and Heat Capacity

Hikers use portable stoves fueled by propane or butane to prepare hot meals. For winter hikers, the only source of water may be ice or snow. Let's consider the changes in temperature and physical state that water undergoes as the heat from a portable stove turns a saucepan filled with snow initially at −18.0°C into boiling water. We'll track the changes using a type of graph called a heating curve (Figure 9.13), which plots the increasing temperature of a system while it is heated at a constant rate.

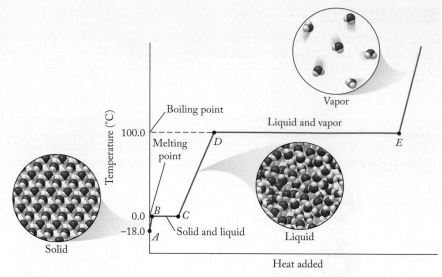

FIGURE 9.13 The energy required to melt snow and boil the resultant water is illustrated by the four line segments on the heating curve of water: heating snow to its melting point ($\overline{AB}$); melting the snow to form liquid water ($\overline{BC}$); heating the water to its boiling point ($\overline{CD}$); and boiling the water to convert it to vapor ($\overline{DE}$).

At first, the heat raises the temperature of the snow as shown by the $\overline{AB}$ segment of the heating curve in Figure 9.13. The temperature stops rising when the snow reaches its melting point at 0.0°C, and it stays there (horizontal segment $\overline{BC}$) until all the snow has melted. Continued heating causes the temperature of the pot's contents, now melted snow, to rise again until the liquid reaches its boiling point at 100.0°C (segment $\overline{CD}$). At 100.0°C, the temperature of the water remains steady again as liquid water becomes water vapor (horizontal segment $\overline{DE}$). Only if all of the liquid water in the pan vaporized (which would be too bad if the goal was to make a hot meal) could the temperature of the water vapor in the pan rise above 100.0°C, as indicated by the final line segment in Figure 9.13.

The difference between the x-axis coordinates at the beginning and end of each line segment in Figure 9.13 indicates how much energy is required during each step of the overall process. In the first step (line segment $\overline{AB}$), the energy required to raise the temperature of the snow from −18.0°C to 0.0°C is related to the snow's **heat capacity (C_P)**, a factor representing how much thermal energy the snow can store. Mathematically, C_P is the quantity of heat that must flow into an object at constant pressure (hence the "P" subscript) to raise its temperature by 1°C. Thus, the units of C_P are J/°C. If we know the value of C_P we can calculate how much heat (q) must be transferred to an object to raise its temperature by an amount ΔT:

$$q = C_P \Delta T \tag{9.8}$$

When an object is a pure substance (like the snow in the hiker's pot), its heat capacity can then be calculated using that substance's *specific heat capacity,* or **specific heat (c_P)**, which is the quantity of heat required to raise the temperature of one gram of a substance by 1°C at constant pressure. Thus, the units of c_P are J/(g · °C). The product of the mass (m) in grams of an object composed of a single substance times the specific heat of the substance (c_P) is the same as the heat capacity of the object:

$$C_P \; (\text{J/°C}) = m \; (\text{g}) \times c_P \; [\text{J/(g} \cdot \text{°C)}]$$

▶‖ **CHEMTOUR** Heating Curves

heat capacity (C_P) the energy required to raise the temperature of an object by 1°C at constant pressure.

specific heat (c_P) the energy required to raise the temperature of 1 g of a substance by 1°C at constant pressure.

Substituting (mc_P) for C_P in Equation 9.8 gives us an equation that relates heat flow to the change in temperature of a mass m of a pure substance:

$$q = mc_P\Delta T \qquad (9.9)$$

Heat capacity is an extensive property that depends on how massive an object is as well as what it's made of, whereas specific heat is an intensive property that is characteristic of a particular substance in a particular physical state (see Table 9.3). For example, the c_P of liquid water, 4.18 J/(g · °C), is about twice that of ice. Why the difference? It has to do with the interactions between water molecules and their freedom of motion. Water molecules in solid ice have little freedom of motion, though they can vibrate in place. Some of these vibrations stretch and compress the O—H bonds in the molecules or change the angle between the bonds. Still others stretch and compress the hydrogen bonds between water molecules. These various modes of vibration provide ways for ice to disperse any heat flowing into it. The more ways there are, the greater the specific heat. The molecules in liquid water enjoy even more freedom of motion and even more ways to disperse absorbed thermal energy. This is why liquid water has a higher specific heat than ice.

Another intensive property related to specific heat is **molar heat capacity** ($c_{P,n}$), which is the quantity of energy required to raise the temperature of 1 mole

molar heat capacity ($c_{P,n}$) the energy required to raise the temperature of 1 mole of a substance by 1°C at constant pressure.

TABLE 9.3 Specific Heat and Molar Heat Capacity Values

SUBSTANCE	PHASE	c_P [J/(g · °C)]	$c_{P,n}$ [J/(mol · °C)]
Elements			
Aluminum	(s)	0.897	24.2
Carbon (graphite)	(s)	0.71	8.5
Chromium	(s)	0.449	23.3
Copper	(s)	0.385	24.5
Gold	(s)	0.129	25.4
Iron	(s)	0.45	25.1
Lead	(s)	0.129	26.7
Silver	(s)	0.233	25.1
Tin	(s)	0.227	26.9
Titanium	(s)	0.523	25.0
Zinc	(s)	0.387	25.3
Compounds			
Silicon dioxide	(s)	0.703	42.2
Water (−10°C)	(s)	2.11	38.0
Water (25°C)	(ℓ)	4.18	75.3
Water (102°C)	(g)	1.89	34.1
Ammonia	(ℓ)	4.75	80.9
Mixture			
Air[a]	(g)	1.003	29.1

[a]Dry air at 0°C and 1 atmosphere of pressure.

of a substance by 1°C at constant pressure. If we know the number of moles (n) of a substance in a sample and also the temperature change in degrees Celsius (ΔT) that the substance experiences, we can calculate q for the process:

$$q = nc_{P,n}\Delta T \qquad (9.10)$$

Hot Soup on a Cold Day

Suppose the winter hikers decide to make soup starting with 275 g of snow. They put the snow, which has an initial temperature (T_i) of −18.0°C, into the pan and start to heat it up. At first the heat is consumed to warm the snow to its melting point, which makes the final temperature of the snow (T_f) 0.0°C. We can use Equation 9.9 to calculate how many joules of thermal energy must be consumed to do this. The value of m is 275 g, c_P is the specific heat of ice [2.11 J/(g · °C)], and the change in temperature is the difference between the final and initial temperatures, or

$$\Delta T = T_f - T_i = [0.0 - (-18.0)]°C = 18.0°C$$

We insert the values in Equation 9.9:

$$q = mc_P\Delta T$$

$$= 275 \; \cancel{g} \times \frac{2.11 \; J}{\cancel{g} \cdot \cancel{°C}} \times 18.0 \; \cancel{°C} = 1.04 \times 10^4 \; J = 10.4 \; kJ$$

Notice that this value is positive, which means that the system (the snow) gains energy as it warms up.

The thermal energy absorbed in the next step is that required to melt (or *fuse*) the snow, which is the process that occurs along line segment $\overline{BC}$ in Figure 9.13. The value of q for this step can be calculated using the heat of fusion (ΔH_{fus}) of water and the number of moles (n) of water in 275 g of snow:

$$q = n\Delta H_{fus} \qquad (9.11)$$

$$= 275 \; \cancel{g} \times \frac{1 \; \cancel{mol}}{18.02 \; \cancel{g}} \times \frac{6.01 \; kJ}{\cancel{mol}} = 91.7 \; kJ$$

This value is also positive because heat flows into the system even though its temperature doesn't change. Thermal energy breaks some of the hydrogen bonds that hold molecules of water in rigid three-dimensional structures in the solid water. Breaking the bonds increases the *potential* energy of the water molecules because they are no longer confined to a single location—they have gained the freedom to flow past their nearest molecular neighbors. However, this newly acquired ability does not result in actual motion of the H_2O molecules, and a corresponding increase in their average *kinetic* energies or temperature, until all of the ice has melted.

As heat continues to flow into the pot of 0.0°C water, its temperature rises. If enough heat is added, the temperature of the water reaches 100.0°C. The relation between this increase in temperature (ΔT) of 100.0°C and the quantity of heat that was absorbed by the water can again be calculated using Equation 9.9, but this time we use the specific heat of liquid water:

$$q = mc_P\Delta T$$

$$= 275 \; \cancel{g} \times \frac{4.18 \; J}{\cancel{g} \cdot \cancel{°C}} \times 100.0 \; \cancel{°C} = 1.15 \times 10^5 \; J = 115 \; kJ$$

At this point in our story we assume that our hiker–chefs accidentally leave the boiling water unattended and it vaporizes completely. As line segment $\overline{DE}$ in Figure 9.13 shows, the temperature of the water remains at 100.0°C until all of it is vaporized. The enthalpy change and the heat transferred is the product of the number of moles of liquid water and water's heat of vaporization (ΔH_{vap}):

$$q = n\Delta H_{vap} \tag{9.12}$$

$$= 275 \text{ g} \times \frac{1 \text{ mol}}{18.02 \text{ g}} \times \frac{40.7 \text{ kJ}}{\text{mol}} = 621 \text{ kJ}$$

As happened when the snow sample melted, the heat absorbed by molecules of liquid H_2O during vaporization is converted into greater *potential* energy of the molecules, as essentially all of the hydrogen bonds between them are broken. However, their average kinetic energy (and temperature) does not increase. Only after all the liquid water has completely vaporized does the temperature of the water vapor increase above 100.0°C.

Notice in Figure 9.13 that line segment $\overline{DE}$, representing the phase change from liquid water to water vapor, is much longer than line segment $\overline{BC}$, which represents the change from solid snow to liquid water. The relative lengths of the lines reflect the fact that the heat of vaporization of water (40.67 kJ/mol) is much larger than the heat of fusion of ice (6.01 kJ/mol). Why does it take more energy to boil 1 mole of water than to melt 1 mole of ice? Only a fraction of the hydrogen bonds in ice are broken upon melting, whereas essentially all of the hydrogen bonds in liquid water are broken when it vaporizes.

SAMPLE EXERCISE 9.4 Calculating Heat Transfer   LO3

Between periods of a hockey game, an ice-resurfacing machine (Figure 9.14) spreads 3.00×10^2 liters of hot (40.0°C) water across a skating rink. How much heat must the water lose as it cools to its freezing point, freezes, and further cools to −10.0°C? Assume that the water is the system, and that its density is 0.992 g/mL at 40.0°C.

COLLECT AND ORGANIZE We know the temperature, volume, and density of water used to resurface a skating rink and are asked to calculate how much heat the water loses as: (1) it cools from 40.0°C to its freezing point (0.0°C); (2) it freezes; and (3) the ice it forms cools from 0.0 to −10.0°C. Table 9.3 lists specific heat and molar heat capacity values for liquid water and solid ice. The heat of fusion (ΔH_{fus}) of water is 6.01 kJ/mol.

ANALYZE Dividing the volume of 40.0°C water applied to the ice by its density will give us the mass of water in grams. Dividing that number by the molar mass of H_2O (18.02 g/mol) will give us the number of moles, which is a value we can use with the molar heat capacities of liquid water and solid ice to calculate the losses of heat during stages (1) and (3), and which, when multiplied by the heat of fusion, gives us how much heat must be lost to freeze the water in stage (2). Given the large number of moles of H_2O in 300 L (~300 kg) of water and the high molar heat capacity of liquid water, we expect a large, negative value for heat.

SOLVE First, we convert 3.00×10^2 L of water into moles:

$$3.00 \times 10^2 \text{ L} \times \frac{1000 \text{ mL}}{1 \text{ L}} \times \frac{0.992 \text{ g}}{1 \text{ mL}} \times \frac{1 \text{ mol}}{18.02 \text{ g}} = 1.651 \times 10^4 \text{ mol}$$

Then we calculate the heat lost in each stage of the cooling/freezing process.

FIGURE 9.14 An ice-resurfacing machine uses hot water to melt the ice, which then rapidly refreezes to create a smooth rink for skating.

1. Cooling 40.0°C water to its freezing point:

$$q_1 = nc_{P,n}\Delta T$$

$$= 1.651 \times 10^4 \, \text{mol} \times \frac{75.3 \, \text{J}}{\text{mol} \cdot °\text{C}} \times (0.0 - 40.0)°\text{C}$$

$$= -4.973 \times 10^7 \, \text{J} = -4.973 \times 10^4 \, \text{kJ}$$

2. Freezing the water:

$$q_2 = n(-\Delta H_{fus})$$

$$= 1.651 \times 10^4 \, \text{mol} \times \frac{-6.01 \, \text{kJ}}{\text{mol}} = -9.923 \times 10^4 \, \text{kJ}$$

3. Cooling the ice:

$$q_3 = nc_{P,n}\Delta T$$

$$= 1.651 \times 10^4 \, \text{mol} \times \frac{36.0 \, \text{J}}{\text{mol} \cdot °\text{C}} \times (-10.0 - 0.0)°\text{C}$$

$$= -5.944 \times 10^6 \, \text{J} = -5.944 \times 10^3 \, \text{kJ}$$

Summing these three q values and rounding off the total to three significant figures:

$$q_1 + q_2 + q_3 = -1.55 \times 10^5 \, \text{kJ}$$

THINK ABOUT IT The large negative total value for q is reasonable given the large volume of water involved in resurfacing a hockey rink and the very high molar heat capacity of water, and it is consistent with our prediction.

Practice Exercise The flame in a torch used to cut metal is produced by burning acetylene (C_2H_2) in pure oxygen. Assuming the combustion of 1 mole of acetylene releases 1251 kJ of energy, what mass of acetylene is needed to cut through a piece of steel if the process requires 5.42×10^4 kJ of energy? ⚙

Water is an extraordinary substance for many reasons, and its high specific heat value (the second highest one in Table 9.3) is one of those reasons. The ability of water to absorb large quantities of thermal energy is one reason it is used as a *heat sink* both in automobile radiators and in our bodies. The term "heat sink" is often used to identify matter that can absorb heat without changing phase or significantly changing its temperature. Weather and climate changes are largely driven and regulated by cycles involving retention of energy by our planet's oceans, which serve as enormous heat sinks.

Why does liquid water have such an unusually high specific heat? One of the principal reasons is hydrogen bonding. As we discussed in Chapter 6, molecules of H_2O can form up to four hydrogen bonds each. These strong intermolecular interactions contribute to a lower potential energy. As water absorbs thermal energy, its internal energy increases. Part of the increase goes toward increasing the kinetic energies of the water molecules—and raising the temperature of the water. However, a significant part goes toward increasing the potential energy of the water by breaking some of the hydrogen bonds between its molecules via increased molecular motion. As we have seen in this section, increasing potential energy produces no increase in temperature. As a result, even more heat must be absorbed to produce a temperature change of, say, 1°C. In this way hydrogen bonding contributes to the rather high molar heat capacity of water (compared to the other compounds in Table 9.3). When we factor in its small molar mass

∞ **CONNECTION** As shown in Figure 6.33, a molecule of H_2O in ice or liquid water can form up to four hydrogen bonds: each of its H atoms can hydrogen-bond to the O atom of a nearby water molecule, and its O atom, with its two lone pairs of electrons, can form two hydrogen bonds to nearby H atoms.

compared to most other compounds, the extraordinarily high specific heat of liquid water begins to make sense.

Cold Drinks on a Hot Day

Let's consider another application of heat transfer. Suppose we throw a party and plan to chill three cases (72 aluminum cans, each containing 355 mL) of beverages by placing the cans, which are initially at a temperature of 25.0°C, in an insulated cooler and covering them with ice cubes that are initially at −18.0°C. If we assume that (1) the ice is sold in 10-pound bags; (2) the mass of aluminum in each can is 12.5 grams; (3) the cans contain mostly water, which has a density of 1.00 g/mL; and (4) the other ingredients are present in such small concentrations that they will not affect our heat transfer calculation, then how many bags of ice do we need to chill the cans and their contents to 0.0°C (as in "ice cold")?

You may already have an idea that more than 1 bag, but probably fewer than 10, will be needed. We can use the heat transfer relationships we have defined to more precisely predict how much ice is required. In doing so, we assume that whatever heat is absorbed by the ice is lost by the cans and the beverages in them. In equation form this assumption is

$$q_{lost} = -q_{gained} \tag{9.13}$$

and it applies to heat transfer processes that occur between warm and cold objects in any isolated system.

Let's first consider the energy lost by the cans and their contents. By substance, their masses are

$$72 \text{ cans} \times \frac{12.5 \text{ g Al}}{\text{can}} = 9.00 \times 10^2 \text{ g Al}$$

plus

$$72 \text{ cans} \times \frac{355 \text{ mL H}_2\text{O}}{\text{can}} \times \frac{1.00 \text{ g}}{\text{mL}} = 2.56 \times 10^4 \text{ g H}_2\text{O}$$

We can use Equation 9.9 and the specific heat values for aluminum and liquid water in Table 9.3 to calculate the quantity of heat that must be transferred to lower the temperatures of the beverages by −25.0°C:

$$q = mc_P\Delta T$$

$$= 900 \text{ g Al} \times \frac{0.897 \text{ J}}{\text{g Al} \cdot {}^\circ\text{C}} \times (-25.0{}^\circ\text{C})$$

$$+ 2.56 \times 10^4 \text{ g H}_2\text{O} \times \frac{4.18 \text{ J}}{\text{g H}_2\text{O} \cdot {}^\circ\text{C}} \times (-25.0{}^\circ\text{C})$$

$$= -2.02 \times 10^4 \text{ J} + (-2.68 \times 10^6 \text{ J}) = -2.70 \times 10^6 \text{ J} = -2.70 \times 10^3 \text{ kJ}$$

If the beverages transfer −2.70 × 10³ kJ, then according to Equation 9.13 (and the law of conservation of energy), the ice gains +2.70 × 10³ kJ of heat. This heat serves two purposes: warming ice cubes to 0.0°C and then melting them. To calculate how many bags of ice melt, we'll need to solve for n, the number of moles of ice that we need. We focus on moles rather than grams because the heat of fusion is expressed in kJ/mol, and we can calculate the heat required to warm the ice using its molar heat capacity [$c_{P,n} = 38.0$ J/(mol · °C)] from Table 9.3 and Equation 9.10.

To keep all energy units the same, let's convert $c_{P,n}$ to 0.0380 kJ/(mol · °C) before combining the heat absorbed by (1) warming and (2) melting the ice:

$$q_1 + q_2 = nc_{P,n}\Delta T + n\Delta H_{fus} = 2.70 \times 10^3 \text{ kJ}$$

$$\left(n \times \frac{0.0380 \text{ kJ}}{\text{mol} \cdot {}^{\circ}\text{C}} \times 18.0 {}^{\circ}\text{C}\right) + \left(n \times \frac{6.01 \text{ kJ}}{\text{mol}}\right) = 2.70 \times 10^3 \text{ kJ}$$

$$n \times \left(\frac{0.684 \text{ kJ}}{\text{mol}} + \frac{6.01 \text{ kJ}}{\text{mol}}\right) = 2.70 \times 10^3 \text{ kJ}$$

$$n = 403 \text{ mol}$$

Converting moles into grams, then pounds, and then 10-pound bags:

$$403 \text{ mol} \times \frac{18.02 \text{ g}}{\text{mol}} \times \frac{1 \text{ lb}}{453.6 \text{ g}} \times \frac{1 \text{ bag}}{10 \text{ lbs}} = 1.6 \text{ bags}$$

We can't buy a fraction of a bag, so we'll need to buy two bags of ice for our party.

SAMPLE EXERCISE 9.5 **Calculating a Final Temperature from Heat Gain and Loss**   **LO3**

Suppose you wish to make a glass of freshly brewed iced tea. You start with exactly 1 cup (237 g) of hot (100.0°C) brewed tea in an insulated mug and add 2.50×10^2 g of ice initially at −18.0°C. All of the ice melts. What is the final temperature of the tea? Assume that tea has the same thermal properties as water.

COLLECT AND ORGANIZE We know the mass of tea, its initial temperature, and its specific heat (the c_P value for liquid water). We know the mass of ice and its initial temperature. From Tables 9.2 and 9.3 we know the specific heats and molar heat capacities of ice and liquid water and the heat of fusion of ice. Our task is to find the final temperature of the tea.

ANALYZE The heat lost by the tea has the same absolute value as, but the opposite sign of, the heat gained by the ice cubes. In equation form:

$$q_{ice} = -q_{tea}$$

Based on the way the problem is presented, we have to assume that once all the ice has melted, the 250 grams of 0.0°C water produced will mix with and be warmed by the still warmer 237 grams of tea. Together they reach the same final temperature. Therefore, there are three parts to q_{ice}: the heat gained when the ice warms from −18.0°C to 0.0°C, the heat gained when the ice melts, and the heat gained when the temperature of the melted ice increases from 0.0°C to the final temperature (which we'll call x°C). We have to calculate the number of moles of H_2O in the ice to use ΔH_{fus}, so we'll use the molar heat capacities of ice and water to calculate q_{ice}. The tea undergoes no phase change, so its only loss of heat is due to its temperature change from 100.0°C to x.

SOLVE

1. Converting grams of ice to moles:

$$2.50 \times 10^2 \text{ g} \times \frac{1 \text{ mol}}{18.02 \text{ g}} = 13.87 \text{ mol}$$

2. The ice gains heat as it:
   a. Warms to 0.0°C:

$$q = nc_{P,n}\Delta T = 13.87 \text{ mol} \times \frac{38.0 \text{ J}}{\text{mol} \cdot {}^{\circ}\text{C}} \times (18.0 {}^{\circ}\text{C}) \times \frac{1 \text{ kJ}}{1000 \text{ J}} = 9.49 \text{ kJ}$$

b. Melts:

$$q = n\Delta H_{fus} = 13.87 \text{ mol} \times \frac{6.01 \text{ kJ}}{\text{mol}} = 83.4 \text{ kJ}$$

c. Warms to final temperature x:

$$q = nc_{P,n}\Delta T = 13.87 \text{ mol} \times \frac{75.3 \text{ J}}{\text{mol} \cdot {}^\circ\text{C}} \times \frac{1 \text{ kJ}}{1000 \text{ J}} \times (x - 0.0){}^\circ\text{C} = 1.044(x) \text{ kJ}$$

3. The heat lost by the tea (q_{tea}) as it cools to $x{}^\circ$C is

$$q_{tea} = mc_P\Delta T = 237 \text{ g} \times \frac{4.18 \text{ J}}{\text{g} \cdot {}^\circ\text{C}} \times (x - 100.0){}^\circ\text{C} \times \frac{1 \text{ kJ}}{1000 \text{ J}} = 0.991(x - 100.0) \text{ kJ}$$

4. Balancing the loss and gain of heat:

$$q_{ice} = {}^-q_{tea}$$

$$(9.49 + 83.4 + 1.044x) \text{ kJ} = -(0.991x - 99.1) \text{ kJ}$$

$$x = 3.1{}^\circ\text{C}$$

THINK ABOUT IT The answer is consistent with the assumption that after all of the ice melted, the temperature of the liquid water thus produced would increase a little to reach the same temperature as the tea. There were no temperature units in the final values used to calculate x, but remember that x was assumed to have units of °C when we inserted it into the ΔT terms in the first steps of the calculation.

Practice Exercise Calculate the final temperature of a mixture of 350 g of ice initially at −18°C and 237 g of water initially at 100.0°C. ⚙

Determining Specific Heat

Suppose your laboratory instructor gives you some metal pellets and asks you to determine whether the metal is aluminum, titanium, zinc, tin, or lead. You are not allowed to chemically alter the metal in your determination of its identity. In addition to the glassware in your lab drawer, you have access to a hot plate, a balance, and Styrofoam coffee cups. How can you decide which metal it is? (Perhaps the title of this subsection and the data in Table 9.3 provide a hint.)

Table 9.3 tells us that the five possible metals have quite different specific heat values, ranging from 0.129 to 0.897 J/(g · °C). How can we determine the specific heat of the unknown metal? We need to link its mass (we have a balance) and a change in its temperature (easily measured with a thermometer) to the quantity of heat transferred during a temperature change. That is the challenge. One way to meet it is to couple the heat gained or lost by the metal to the heat lost or gained by a known substance. A strong candidate for this is water.

We can use an experimental setup like the one in Figure 9.15. It includes a hot plate that we can use to heat a beaker of water to its boiling point: 100.0°C. Then we measure the mass of the pellets—let's say it is 25.0 g—and place them in a test tube (to keep them dry). The test tube is immersed in the boiling water. After a while, the temperature of the metal should also be 100.0°C. Meanwhile we add 50.0 grams of water to two nested Styrofoam coffee cups. We use the coffee cups because they have good insulating capability and negligible mass compared to the water.

The critical part of the experiment comes when we measure the temperature of the water in the nested cups—it's 19.3°C—and then quickly remove the test tube from the boiling water bath and pour the metal out of it and into the water

100.0°C

25.0 g metal

Heat

(a)

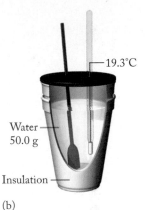

19.3°C

Water
50.0 g

Insulation

(b)

27.1°C

Insulation

(c)

FIGURE 9.15 Experimental setup to determine the specific heat of a metal. (a) Heat 25.0 grams of the metal to 100.0°C in boiling water. (b) Add 50.0 grams of water at 19.3°C to nested Styrofoam cups. (c) Transfer the hot metal to the water, and record the peak temperature of the metal/water mixture, which reaches 27.1°C in this experiment.

in the cups. Then we measure the temperature of that water until it reaches a maximum value, which is 27.1°C. At this temperature we may assume that the heat gained by the water and lost by the metal are in balance:

$$q_{\text{water}} = -q_{\text{metal}}$$

The left side of the equation is a quantity we can calculate using Equation 9.9:

$$q_{\text{water}} = mc_{\text{P}}\Delta T$$

$$= 50.0 \text{ g} \times \frac{4.18 \text{ J}}{\text{g} \cdot {}^\circ\text{C}} \times (27.1 - 19.3){}^\circ\text{C} = 1.63 \times 10^3 \text{ J}$$

We apply this value to what we know about the right side of the equation:

$$1.63 \times 10^3 \text{ J} = -q_{\text{metal}} = -mc_{\text{P}}\Delta T$$

$$= -25.0 \text{ g} \times c_{\text{P}} \times (27.1 - 100.0){}^\circ\text{C} = (1823 \text{ g} \cdot {}^\circ\text{C}) \, c_{\text{P}}$$

$$c_{\text{P}} = 0.89 \text{ J/(g} \cdot {}^\circ\text{C})$$

A check of the c_{P} values in Table 9.3 indicates that the mystery metal is aluminum.

CONCEPT TEST ··

In the preceding determination of the specific heat of a metal, how would the following factors have affected the experimental c_{P} value?

    a. Slow transfer of the metal from the hot water bath to the Styrofoam cups so that the metal's temperature was less than 100.0°C when it hit the water.
    b. Drops of hot water adhering to the metal when it was transferred to the Styrofoam cups.
    c. Ignoring the heat capacity of the thermometer.
    d. Heat transfer through the Styrofoam cups or the open top. Assume the lab temperature was 22°C.

9.5 Heats of Reaction and Calorimetry

▶‖ **CHEMTOUR** Calorimetry

calorimetry the experimental determination of the quantity of energy transferred during a physical change or chemical process.

calorimeter a device used to measure the absorption or release of energy by a physical change or chemical process.

enthalpy of reaction (ΔH_{rxn}) the enthalpy change that accompanies a chemical reaction; also called *heat of reaction.*

In previous sections we have used specific heat, molar heat capacity, and heats of fusion and vaporization to track flows of thermal energy as substances were warmed or cooled or changed physical state, but we have not explained how we know the values of those parameters. One of the ways we know them is through experiments based on calorimetry. **Calorimetry** is an analytical technique in which changes in the temperature of a device with a known heat capacity, called a **calorimeter**, are used to determine the heat released or absorbed by a process occurring inside the calorimeter. The process may be a physical exchange of thermal energy, or it can be a chemical reaction in which heat flows either from or into a reaction mixture.

The magnitude of the heat absorbed or released during a chemical reaction is proportional to the quantity of the reactants consumed and to a property of the reaction called its **enthalpy of reaction (ΔH_{rxn})**, often referred to as *heat of reaction*. The subscript may be changed to reflect a specific type of reaction. For example, ΔH_{comb} is sometimes used to represent the enthalpy change that accompanies a combustion reaction.

Logically, the value of ΔH_{rxn} of a particular reaction depends on the difference in enthalpy between its reactants and products. It also depends on how we choose to represent the reaction in a balanced chemical equation. For example, suppose we describe the complete combustion of butane, the fuel in disposable lighters, in this way:

$$2\ C_4H_{10}(g) + 13\ O_2(g) \rightarrow 8\ CO_2(g) + 10\ H_2O(\ell)$$

We can expand this chemical equation, turning it into a **thermochemical equation,** by adding the enthalpy change that accompanies the combustion of 2 moles of C_4H_{10} by 13 moles of O_2 to form 8 moles of CO_2 and 10 moles of liquid H_2O:

$$2\ C_4H_{10}(g) + 13\ O_2(g) \rightarrow 8\ CO_2(g) + 10\ H_2O(\ell) \qquad \Delta H_{rxn} = -5754\ \text{kJ}$$

However, we might wish to express the enthalpy change that accompanies the combustion of only 1 mole of butane. As you might expect, burning half the fuel releases half the heat and is accompanied by only half the change in enthalpy, as represented by the following thermochemical equation:

$$C_4H_{10}(g) + 13/2\ O_2(g) \rightarrow 4\ CO_2(g) + 5\ H_2O(\ell) \qquad \Delta H_{rxn} = -2877\ \text{kJ}$$

> **thermochemical equation** the chemical equation of a reaction that includes the change in enthalpy that accompanies the reaction.

CONCEPT TEST •

The enthalpy change accompanying the combustion of one mole of hydrogen gas, producing one mole of liquid water, is 286 kJ. What is the value of ΔH_{rxn} for the reaction below?

$$2\ H_2(g) + O_2(g) \rightarrow 2\ H_2O(\ell) \qquad \Delta H_{rxn} = \ ?$$

As noted in Section 9.3, enthalpy values are difficult to determine, but we can determine *changes* in enthalpy experimentally. The apparatus in Figure 9.15 is a kind of calorimeter called a *coffee-cup calorimeter* that is particularly useful for determining ΔH_{rxn} for reactions in aqueous solutions. Because the reactions happen at constant (ambient) pressure, the heat absorbed or released by the reaction (q_{rxn}) is equal to the enthalpy change that accompanies the reaction (see Equation 9.7). If all of this heat is transferred to or from the contents of the calorimeter, then:

$$q_{rxn} = -q_{calorimeter}$$

The value of $q_{calorimeter}$ can be calculated from the measured change in temperature (ΔT) and the heat capacity of the calorimeter ($C_{calorimeter}$):

$$q_{calorimeter} = C_{calorimeter}\Delta T$$

In reactions involving dilute aqueous solutions, nearly all of the mass (m) of the calorimeter comes from the mass of the water in it. This fact and the very high specific heat of water [$c_{P,H_2O} = 4.18\ \text{J/(g} \cdot {}^\circ\text{C)}$] allow us to assume that the heat capacity of the calorimeter is essentially the same as the heat capacity of the water in it:

$$q_{calorimeter} = 4.18\ \text{J/(g} \cdot {}^\circ\text{C)}m\Delta T \qquad (9.14)$$

SAMPLE EXERCISE 9.6 Calculating ΔH_{rxn} from Calorimetry Data   **LO4**

When 0.200 L of 0.200 M HCl is mixed with 0.200 L of 0.200 M NaOH in a coffee-cup calorimeter, the temperature of the mixture increases from 22.15°C to 23.48°C. If the densities of the two solutions are 1.00 g/mL, what is the value of ΔH_{rxn} for the following reaction?

$$HCl(aq) + NaOH(aq) \rightarrow NaCl(aq) + H_2O(\ell) \qquad \Delta H_{rxn} = ?$$

COLLECT AND ORGANIZE We know the (a) concentrations, (b) volumes, (c) densities, and (d) initial and final temperatures of two solutions: a strong acid and a strong base. We are asked to calculate the value of ΔH_{rxn} for the reaction in which they neutralize each other.

ANALYZE The densities of both aqueous solutions are nearly the same as that of water (1.00 g/mL), which confirms that they are dilute solutions with heat capacities that are essentially the same as that of water. Therefore, Equation 9.14 can be used to calculate $q_{calorimeter}$. Their common density also means that the total mass of the solutions in grams is the same as their total volume in milliliters: 400. Using this value in Equation 9.14 and a ΔT value of about 1.3 gives a $q_{calorimeter}$ of about $4 \times 400 \times 1.3 \approx 2000$ J, or 2 kJ. The heat gained by the solution represents about −2 kJ released by the exothermic neutralization reaction (q_{rxn}). To calculate ΔH_{rxn} for the reaction as written (1 mole of each reactant), we will need to calculate the number of moles of each reactant in the reaction mixture. The volumes of each solution are 200 mL = 1/5 of a liter and their concentrations are 1/5 molar, so there is only $1/5 \times 1/5 = 1/25$ of a mole of each reactant. Therefore, the value of ΔH_{rxn} for a reaction in which 1 mole of each reactant is consumed should be about −2 kJ × 25 = −50 kJ.

SOLVE Calculating the value of $q_{calorimeter}$ ($-q_{rxn}$):

$$q_{calorimeter} = 4.18 \text{ J/(g} \cdot °\text{C)} m\Delta T$$

$$= 4.18 \frac{\text{J}}{\text{g} \cdot °\text{C}} \times 400 \text{ g} \times (23.48 - 22.15)°\text{C} = 2224 \text{ J} = 2.224 \text{ kJ}$$

Therefore, $q_{rxn} = -2.224$ kJ. Calculating the value of ΔH_{rxn}:

$$\Delta H_{rxn} = \frac{-2.224 \text{ kJ}}{0.200 \text{ L} \times \dfrac{0.200 \text{ mol}}{\text{L}}} = -55.6 \text{ kJ/mol reactant, or } -55.6 \text{ kJ}$$

THINK ABOUT IT In the second step we converted the heat released by the reaction mixture (q_{rxn}) into the heat that would have been released had there been one mole of each reactant, which is the quantity of each of them in the balanced chemical equation.

Practice Exercise Addition of 1.31 g of zinc metal to 100.0 mL of 0.200 M HCl in a coffee-cup calorimeter causes the temperature to decrease from 15.21 to 11.96°C. If the density of the HCl solution is 1.00 g/mL, what is the value of ΔH_{rxn} for the following reaction?

$$Zn(s) + 2 \text{ HCl}(aq) \rightarrow ZnCl_2(aq) + H_2(g) \; \text{⚙}$$

Bomb Calorimetry

bomb calorimeter a constant-volume device used to measure the energy released during a combustion reaction.

calorimeter constant ($C_{calorimeter}$) the heat capacity of a calorimeter.

One of the most important categories of chemical reactions in terms of their transfer of thermal energy is combustion reactions. The quantities of energy they produce can be determined with devices called **bomb calorimeters** (Figure 9.16). To use these instruments, a combustible sample is placed in a sealed vessel (called a *bomb*) capable of withstanding high pressures and is submerged in a large volume of water in a heavily insulated container. Oxygen is introduced into the bomb, and the mixture is ignited with an electric spark. As combustion occurs, thermal energy generated by the reaction flows into the walls of the bomb and then into the water surrounding it.

A good bomb calorimeter keeps the system (the chemical reaction) contained within the bomb and ensures that all heat released by the reaction stays inside the calorimeter. Specifically, the calorimeter consists of the bomb, the water, the insulated container, and minor components (stirrer, thermometer, and any other materials). The energy produced by the reaction is determined by measuring the temperature of the water before and after the reaction. The water is at the same temperature as all the other insulated parts, so the temperature change of the water tracks the temperature change of the entire calorimeter.

Measuring the change in temperature of the water is not the whole story. We also need to know the heat capacity of it and all the other insulated components of the calorimeter, that is, its **calorimeter constant**, $C_{calorimeter}$. Why? Unlike in our coffee-cup calorimeter, where essentially all the heat warms a mass of water, the heat in a bomb calorimeter is absorbed by its many components, each of which has a heat capacity of its own. If we know the value of $C_{calorimeter}$ and if we can measure the change in water temperature, then we can calculate the quantity of energy that flows from a reaction mixture into a calorimeter.

How is $C_{calorimeter}$ determined? One way is to burn a known quantity of a material with a known heat of combustion. The quantity of heat released by the reaction is then known, and the value of $C_{calorimeter}$ can be calculated from the change in temperature of the calorimeter produced by that quantity of heat. Benzoic acid ($C_7H_6O_2$) is often used for this purpose because very pure samples of it can be obtained. Complete combustion of exactly one gram of it is known to release 26.38 kJ of thermal energy. Once $C_{calorimeter}$ has been determined, the calorimeter can be used to determine the quantities of energy produced by other combustion reactions on a per-gram or per-mole basis.

Because there is no change in the volume of the reaction mixture in a bomb calorimeter, this technique is referred to as *constant-volume calorimetry*. No $P-V$ work is done, so Equation 9.4:

$$\Delta E = q - P\Delta V$$

simplifies to

$$q = \Delta E$$

The heat lost by the reaction mixture in a bomb calorimeter is the heat gained by the calorimeter, so an increase in the temperature of a bomb calorimeter, which provides a measure of the heat flowing into it, also provides a measure of the energy released by the reaction happening inside it.

How is the ΔE measured in a bomb calorimeter related to ΔH? Unlike the coffee-cup calorimeter in Section 9.4 (constant pressure), reactions in a bomb calorimeter are carried out at constant volume. For most reactions we consider in this chapter, ΔE and ΔH are more or less equal, so we do not need to account for the small differences between ΔE and ΔH.

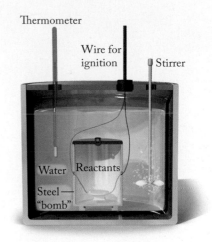

FIGURE 9.16 A bomb calorimeter.

SAMPLE EXERCISE 9.7 **Determining the Heat Capacity** **LO4**
of a Calorimeter

Before we can determine the energy change of a reaction run in a calorimeter, we must determine the heat capacity of the calorimeter, $C_{calorimeter}$. What is the value of $C_{calorimeter}$

if burning 1.000 g of benzoic acid in a calorimeter increases its temperature by 7.248°C? Combustion of benzoic acid releases 26.38 kJ of energy per gram of benzoic acid.

COLLECT AND ORGANIZE We have been asked to find the calorimeter constant of a calorimeter, which is the heat required to increase the temperature of the calorimeter by 1°C. We know that a reaction that releases 26.38 kJ of energy increases the temperature of the calorimeter by 7.248°C.

ANALYZE We know how much heat (26.38 kJ) produces a ΔT of 7.248°C. The heat needed to produce a ΔT of 1°C should be about 1/7 of 26.38 kJ, or about 4 kJ.

SOLVE Applying Equation 9.8 and solving for $C_{\text{calorimeter}}$:

$$q_{\text{calorimeter}} = C_{\text{calorimeter}}\Delta T$$

$$C_{\text{calorimeter}} = \frac{q_{\text{calorimeter}}}{\Delta T}$$

$$= \frac{26.38 \text{ kJ}}{7.248°C} = 3.640 \text{ kJ/°C}$$

THINK ABOUT IT The calorimeter constant is determined for a specific calorimeter. Once $C_{\text{calorimeter}}$ is known, the calorimeter can be used to determine the enthalpy of combustion of any combustible material. If any of the internal parts of the calorimeter change or are replaced, a new constant must be determined. Notice how large $C_{\text{calorimeter}}$ is compared to most of the substances in Table 9.3.

Practice Exercise When a 0.500 g mixture of hydrocarbons is burned in the bomb calorimeter from Sample Exercise 9.7, its temperature rises by 6.76°C. How much energy (in kilojoules) is released during combustion? How much energy is released with the combustion of 1.000 g of the same mixture? ⚙

9.6 Hess's Law and Standard Heats of Reaction

As noted in the previous section, calorimetry can be used to determine the energy and enthalpy changes that accompany chemical reactions. However, there may be times when determining ΔH_{rxn} directly is not possible. For example, CO_2 is the principal product of the combustion of carbon in the form of charcoal:

$$(1) \quad C(s) + O_2(g) \rightarrow CO_2(g)$$

When the oxygen supply is limited, however, the products may include carbon monoxide:

$$(2) \quad 2\,C(s) + O_2(g) \rightarrow 2\,CO(g)$$

It is difficult to directly determine the enthalpy change that accompanies reaction (2) because as long as any oxygen is present, some of the CO formed may combine with O_2 to form CO_2, yielding a mixture of CO and CO_2. However, we can calculate ΔH for reaction (2) *indirectly* by starting with ΔH_{rxn} values that we *can* determine. For example, we can determine the enthalpy changes that accompany both the reaction in equation (1) and the combustion of a sample of pure CO gas:

$$(3) \quad 2\,CO(g) + O_2(g) \rightarrow 2\,CO_2(g)$$

Look closely at the reactants and products of chemical equations (1), (2), and (3). Equation (1) represents the complete combustion of carbon to CO_2. Equations (2) and (3) represent the stepwise combustion of carbon: first to CO and then to CO_2. Given this relationship, it is reasonable that the enthalpy change that accompanies the overall reaction (1) is related to the sum of the enthalpy changes associated with the stepwise reactions (2) and (3). Why? Because, as noted in Section 9.3, enthalpy change is a state function. This means that the ΔH_{rxn} value of an overall reaction that may occur in two or more steps, such as reaction (1), is the sum of the ΔH_{rxn} values of those steps. This is so because the various steps consume the same reactants and eventually form the same products as the overall reaction.

Combining heats of reaction in this way is in accordance with **Hess's law**, also known as *Hess's law of constant heat of summation*, which states that the change in enthalpy that accompanies a process that occurs in more than one step is the sum of the enthalpy changes that occur in each of those steps. Let's put Hess's law to work by deriving an expression for enthalpy change that accompanies the incomplete combustion of C to CO as described in reaction (2) using the measurable ΔH_{rxn} values of reactions (1) and (3).

Combining both the equations that describe chemical reactions and their ΔH_{rxn} values is an exercise in pattern recognition. The key is to look for the reactants and products of the reaction whose ΔH_{rxn} we wish to find in chemical equations describing the reactions whose ΔH_{rxn} values we do know. In this example, we need to combine the chemical equations describing reactions (1) and (3) in a way that gives us the equation for reaction (2). Carbon and O_2 are reactants in equation (2), and CO is the only product. Inspecting the other two equations, we find that CO is a reactant in (3). Because CO is on the product side of equation (2), we flip equation (3) so that the reaction runs in reverse. We noted in Section 9.3 that reversing a reaction *changes the sign* of its ΔH value. Applying this principle to the reaction in equation (3), we get:

$$(4) \qquad 2\,CO_2(g) \rightarrow 2\,CO(g) + O_2(g) \qquad \Delta H_4 = -\Delta H_3$$

Flipping (3) puts O_2 on the product side of equation (4), which is not where we want it. O_2 is on the reactant side of equation (1), so combining (1) and (4) may result in canceling out O_2 from the product side. However, we can't combine (1) and (4) just yet because there are two molecules of CO_2 on the reactant side of equation (4) but only one on the product side of equation (1). We'd prefer that these values be the same so that they cancel out when we combine equations (1) and (4), because there are no CO_2 terms at all in equation (2). To get two molecules of CO_2 on the product side of equation (1) we multiply all of the terms in the equation, *including* ΔH_1, by 2:

$$(5) \qquad 2\,C(s) + 2\,O_2(g) \rightarrow 2\,CO_2(g) \qquad \Delta H_5 = 2\Delta H_1$$

Now we combine equations (4) and (5), *which includes summing their ΔH_{rxn} values*:

$$(4) \qquad 2\,CO_2(g) \rightarrow 2\,CO(g) + O_2(g) \qquad \Delta H_4 = -\Delta H_3$$
$$+\,(5) \qquad 2\,C(s) + 2\,O_2(g) \rightarrow 2\,CO_2(g) \qquad \Delta H_5 = 2\Delta H_1$$
$$\overline{\;\cancel{2\,CO_2(g)} + 2\,C(s) + \cancel{2}\,O_2(g) \rightarrow 2\,CO(g) + \cancel{O_2(g)} + \cancel{2\,CO_2(g)}\;}$$

or

$$(2) \qquad 2\,C(s) + O_2(g) \rightarrow 2\,CO(g) \qquad \Delta H_2 = 2\Delta H_1 - \Delta H_3$$

Perhaps you are wondering why we multiplied the enthalpy change that accompanies reaction (1) by 2 when we multiplied the coefficients of each of the reactants and product by 2. After all, why should changing the way we write the chemical equation impact the enthalpy change that accompanies the formation of the same product from the same reactants? Recall that the coefficients in chemical equations can represent the numbers of moles of reactants and products (as well as

Hess's law the principle that the heat of reaction ΔH_{rxn} for a process that is the sum of two or more reactions is equal to the sum of the ΔH_{rxn} values of the constituent reactions; also called *Hess's law of constant heat of summation*.

▶❙❙ **CHEMTOUR** Hess's Law

the numbers of atoms and molecules). Thus, reaction (1) describes the incomplete combustion of *1 mole* of carbon, but reaction (5) describes the incomplete combustion of *2 moles* of carbon. It is only reasonable that burning twice as much fuel should generate twice as much heat—and be accompanied by twice the change in enthalpy.

Standard Heat of Reaction (ΔH°_{rxn})

Now let's put some values on the enthalpy changes that accompany chemical reactions. Our focus will be the reactions that make up the industrial process for producing hydrogen gas (see Sample Exercise 7.11). In the first step of hydrogen production, methane reacts with a limited supply of high-temperature steam to produce carbon monoxide and hydrogen gas:

$$CH_4(g) + H_2O(g) \rightarrow CO(g) + 3\,H_2(g)$$

This reaction is called the steam-reforming reaction and is quite endothermic, requiring a flow of thermal energy into the reaction mixture to turn reactants into products. To describe *how* endothermic the reaction is, we use a thermodynamic property called **standard enthalpy of reaction (ΔH°_{rxn})**, which is often called *standard heat of reaction*. The adjective *standard*, indicated by the symbol °, describes the enthalpy change that accompanies a reaction under **standard conditions**, which means at a constant pressure of 1 atm. There is no universal standard temperature, though many tables of thermodynamic data, including those in the appendix of this book, apply to processes occurring at 25°C.

Implied in our notion of standard conditions is the assumption that parameters such as ΔH change with temperature and pressure. That assumption is correct, as we saw with heats of vaporization in Section 9.3, but the changes are so small that we ignore them in the calculations in this chapter and those that follow. We also use the term **standard state** to describe the most stable physical state of a substance under standard conditions. For example, at $P = 1$ atm and $T = 25°C$, most metals and metalloids are solids, mercury and bromine are liquids, and H_2, N_2, O_2, F_2, Cl_2, and the group 18 elements are gases. These are the standard states of these elements.

Returning to the steam-reforming reaction and including its standard heat of reaction value yields this thermochemical equation:

$$(1) \quad CH_4(g) + H_2O(g) \rightarrow CO(g) + 3\,H_2(g) \qquad \Delta H^\circ_1 = 206 \text{ kJ}$$

In the second step in hydrogen production, CO from the first step reacts with more steam to produce CO_2 and more H_2 gas in a reaction called the *water–gas shift reaction*:

$$(2) \quad CO(g) + H_2O(g) \rightarrow CO_2(g) + H_2(g) \qquad \Delta H^\circ_2 = -41 \text{ kJ}$$

In accordance with Hess's law, we can write an overall thermochemical equation for the process by adding reactions 1 and 2 and their ΔH° values:

standard enthalpy of reaction (ΔH°_{rxn}) the enthalpy change associated with a reaction that takes place under standard conditions; also called *standard heat of reaction*.

standard conditions in thermodynamics: a pressure of 1 atm (~1 bar) and some specified temperature, assumed to be 25°C unless otherwise stated; for solutions, a concentration of 1 M is specified.

standard state the most stable form of a substance under 1 atm pressure and some specified temperature (usually 25.0°C).

$$
\begin{array}{lll}
\text{(1)} & \text{CH}_4(g) + \text{H}_2\text{O}(g) \rightarrow \cancel{\text{CO}(g)} + 3\,\text{H}_2(g) & \Delta H_1^\circ = 206\ \text{kJ} \\
+\ \text{(2)} & \cancel{\text{CO}(g)} + \text{H}_2\text{O}(g) \rightarrow \text{CO}_2(g) + \text{H}_2(g) & \Delta H_2^\circ = -41\ \text{kJ} \\
\hline
\text{(3)} & \text{CH}_4(g) + 2\,\text{H}_2\text{O}(g) \rightarrow \text{CO}_2(g) + 4\,\text{H}_2(g) & \Delta H_3^\circ = 165\ \text{kJ}
\end{array}
$$

This result is illustrated graphically in Figure 9.17.

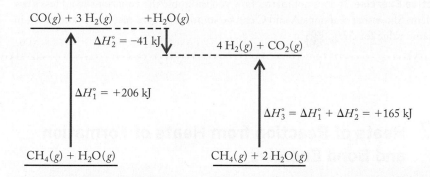

FIGURE 9.17 Hess's law predicts that the enthalpy change (ΔH_3°) for the reaction in which 1 mole of CH_4 and 2 moles of H_2O vapor produce 4 moles of H_2 and 1 mole of CO_2 is the sum of the enthalpy changes that accompany each of the two steps in the overall reaction: (1) 1 mole of CH_4 reacts with 1 mole of H_2O vapor to produce 1 mole of CO and 3 moles of H_2 (ΔH_1°), and (2) the reaction of the CO produced in the first step with another mole of H_2O vapor to form a fourth mole of H_2 and 1 mole of CO_2 (ΔH_2°).

SAMPLE EXERCISE 9.8 **Calculating ΔH_{rxn}° Using Hess's Law** **LO5**

One reason furnaces and hot-water heaters fueled by natural gas need to be vented is that incomplete combustion can produce toxic carbon monoxide:

Equation A: $\quad 2\,\text{CH}_4(g) + 3\,\text{O}_2(g) \rightarrow 2\,\text{CO}(g) + 4\,\text{H}_2\text{O}(g) \qquad \Delta H_\text{A}^\circ = ?$

Use these thermochemical equations to calculate ΔH_A°:

Equation B: $\quad \text{CH}_4(g) + 2\,\text{O}_2(g) \rightarrow \text{CO}_2(g) + 2\,\text{H}_2\text{O}(g) \qquad \Delta H_\text{B}^\circ = -802\ \text{kJ}$

Equation C: $\quad 2\,\text{CO}(g) + \text{O}_2(g) \rightarrow 2\,\text{CO}_2(g) \qquad\qquad\qquad \Delta H_\text{C}^\circ = -566\ \text{kJ}$

COLLECT AND ORGANIZE We are given two equations (B and C) with thermochemical data and a third (A) for which we are asked to find ΔH°. All the reactants and products in equation A are present in B and/or C.

ANALYZE We can manipulate equations B and C so that they sum to give the equation for which ΔH° is unknown. Then we can calculate the unknown value by applying Hess's law. Methane is a reactant in A and B, so we will use B in the direction written. CO is a product in A but a reactant in C, so we have to reverse C to get CO on the product side. Reversing C means that we must change the sign of ΔH_C°. If the coefficients in B and the reverse of C do not allow us to sum the two equations to obtain equation A, we will need to multiply one or both by appropriate factors.

SOLVE Comparing equation B as written and the reverse of C:

(B) $\quad \text{CH}_4(g) + 2\,\text{O}_2(g) \rightarrow \text{CO}_2(g) + 2\,\text{H}_2\text{O}(g) \qquad \Delta H_\text{B}^\circ = -802\ \text{kJ}$

(C, reversed) $\quad 2\,\text{CO}_2(g) \rightarrow 2\,\text{CO}(g) + \text{O}_2(g) \qquad\qquad -\Delta H_\text{C}^\circ = +566\ \text{kJ}$

with Equation A, we find that the coefficient of CH_4 is 2 in A but only 1 in B, so we need to multiply all the terms in B by 2, including ΔH_B°:

(2B) $\quad 2\,\text{CH}_4(g) + 4\,\text{O}_2(g) \rightarrow 2\,\text{CO}_2(g) + 4\,\text{H}_2\text{O}(g) \qquad 2\,\Delta H_\text{B}^\circ = -1604\ \text{kJ}$

When we sum C (reversed) and 2B, the CO_2 terms cancel out and we obtain equation A:

$$
\begin{array}{lll}
\text{(C, reversed)} & \cancel{2\,\text{CO}_2(g)} \rightarrow 2\,\text{CO}(g) + \cancel{\text{O}_2(g)} & -\Delta H_\text{C}^\circ = +566\ \text{kJ} \\
 & \qquad\qquad\quad {}^{3} & \\
+\ \text{(2B)} & 2\,\text{CH}_4(g) + \overset{3}{\cancel{4}}\,\text{O}_2(g) \rightarrow \cancel{2\,\text{CO}_2(g)} + 4\,\text{H}_2\text{O}(g) & 2\,\Delta H_\text{B}^\circ = -1604\ \text{kJ} \\
\hline
\text{(A)} & 2\,\text{CH}_4(g) + 3\,\text{O}_2(g) \rightarrow 2\,\text{CO}(g) + 4\,\text{H}_2\text{O}(g) & \Delta H_\text{A}^\circ = -1038\ \text{kJ}
\end{array}
$$

THINK ABOUT IT Our calculation shows that incomplete combustion of 2 moles of methane is less exothermic ($\Delta H^\circ_A = -1038$ kJ) than their complete combustion ($2 \Delta H^\circ_B = -1604$ kJ), which makes sense because the CO produced in incomplete combustion reacts exothermically with more O_2 to form CO_2. In fact, the value of ΔH°_C for the reaction $2 CO(g) + O_2(g) \rightarrow 2 CO_2(g)$ is the difference between -1604 kJ and -1038 kJ.

Practice Exercise It does not matter how you assemble the equations in a Hess's law problem. Show that reactions A and C can be summed to give reaction B and result in the same value for ΔH°_B. ⚙

TABLE 9.4	Standard Heats of Formation of Selected Substances at 25°C	
Substance	**ΔH°_f (kJ/mol)**
$O_2(g)$	0
$H_2(g)$	0
$H_2O(g)$	−241.8
$H_2O(\ell)$	−285.8
$C(s, graphite)$	0
$CH_4(g)$, methane	−74.8
$C_2H_2(g)$, acetylene	226.7
$C_2H_4(g)$, ethylene	52.4
$C_2H_6(g)$, ethane	−84.67
$C_3H_8(g)$, propane	−103.8
$C_4H_{10}(g)$, butane	−125.6
$CO_2(g)$	−393.5
$CO(g)$	−110.5
$N_2(g)$	0
$NH_3(g)$, ammonia	−46.1
$N_2H_4(g)$, hydrazine	95.35
$N_2H_4(\ell)$	50.63
$NO(g)$	90.3
$Br_2(\ell)$	0
$CH_3OH(\ell)$, methanol	−238.7
$CH_3CH_2OH(\ell)$, ethanol	−277.7
$CH_3COOH(\ell)$, acetic acid	−484.5

9.7 Heats of Reaction from Heats of Formation and Bond Energies

As noted in Sections 9.1 and 9.3, it is impossible to measure the *absolute* value of the internal energy of a substance, and the same is true for the enthalpy of a substance. However, we can establish *relative* enthalpy values that are referenced to a convenient standard. This approach is similar to using the freezing point of water as the zero point on the Celsius temperature scale, or sea level as the zero point for expressing altitude. The enthalpy value referenced to this zero point is a substance's **standard enthalpy of formation (ΔH°_f)**, or *standard heat of formation*, defined as the enthalpy change that takes place at a constant pressure of 1 atm when 1 mole of a substance is formed from its constituent elements in their standard states. A reaction that fits this description is known as a **formation reaction**.

The standard heat of formation of any pure element in its standard state is, by definition, zero. This is the zero point of all other enthalpy values. Standard heat of formation values for several compounds are given in Table 9.4, and a more complete list can be found in Appendix 4. Because the definition of a formation reaction specifies 1 mole of product, writing balanced equations for formation reactions may require the use of something we have tried to avoid until now: fractional coefficients in the final form of our balanced equations. For example, the balanced chemical equation describing the formation of nitrogen monoxide is usually written

$$N_2(g) + O_2(g) \rightarrow 2 NO(g) \qquad \Delta H^\circ_{rxn} = 180.6 \text{ kJ}$$

Although all reactants in the equation are in their standard states, it is not a formation reaction because 2 moles of product are formed. Therefore, ΔH°_{rxn} does not equal ΔH°_f. In fact, it is twice ΔH°_f because the equation as written describes the formation of *two* moles of NO. To write an equation describing the formation of one mole we divide each coefficient (and ΔH°_{rxn}) in the above equation by 2:

$$\tfrac{1}{2} N_2(g) + \tfrac{1}{2} O_2(g) \rightarrow NO(g) \qquad \Delta H^\circ_f = 90.3 \text{ kJ}$$

SAMPLE EXERCISE 9.9 Recognizing Formation Reactions **LO5**

Consider the ΔH°_{rxn} values in the following thermochemical equations. Which of them are ΔH°_f values, assuming the reaction takes place at 25°C and a constant pressure of 1 atm? For those that are not formation reactions, explain why not.

a. $H_2(g) + \frac{1}{2} O_2(g) \rightarrow H_2O(g)$ $\qquad$ $\Delta H°_{rxn} = -241.4$ kJ

b. $C_{graphite}(s) + 2 H_2(g) + \frac{1}{2} O_2(g) \rightarrow CH_3OH(\ell)$ $\qquad$ $\Delta H°_{rxn} = -238.7$ kJ

c. $CH_4(g) + 2 O_2(g) \rightarrow CO_2(g) + 2 H_2O(g)$ $\qquad$ $\Delta H°_{rxn} = -802.3$ kJ

d. $P_4(s, white) + 6 Cl_2(g) \rightarrow 4 PCl_3(\ell)$ $\qquad$ $\Delta H°_{rxn} = -1278$ kJ

COLLECT AND ORGANIZE We are given four balanced thermochemical equations, and we need to determine which of the $\Delta H°_{rxn}$ values are also $\Delta H°_f$ values. The standard heat of formation of a substance is the enthalpy of a reaction in which one mole of the substance is formed from its constituent elements each in their standard state.

ANALYZE Each reaction must be evaluated for the quantity of the product compound (whether there is only one mole of it) and for the state of each reactant.

SOLVE

a. One mole of water vapor is formed from its constituent elements in their standard states. Therefore, this is the formation reaction for $H_2O(g)$, and its heat of reaction is $\Delta H°_f$.

b. One mole of liquid methanol is formed from its constituent elements in their standard states. Therefore the equation describes a formation reaction, and the heat of reaction is $\Delta H°_f$.

c. The reactants are not elements in their standard states and more than one product is formed, so the heat of reaction is not $\Delta H°_f$.

d. A check of Table A4.3 in Appendix 4 discloses that the most stable form of phosphorus is white phosphorus, which has the molecular formula P_4, so both reactants are elements in their standard states. However, the reaction produces *four* moles of PCl_3, so the heat of reaction is not the $\Delta H°_f$ of PCl_3. Actually, it is four times $\Delta H°_f$.

THINK ABOUT IT Just because we can write formation reactions for substances like methanol does not mean that anyone would ever use that reaction to make methanol. Formation reactions are defined to provide a standard so that the flows of thermal energy in reactions can be evaluated.

Practice Exercise Write formation reactions for (a) $CaCO_3(s)$; (b) $CH_3COOH(\ell)$ (acetic acid); (c) $KMnO_4(s)$. ⚙

Standard heats of formation can be used to predict standard heats of reaction. To do so, we break down each reaction into a series of formation reactions for the reactants and products. Then we apply Hess's law to combine the $\Delta H°_f$ values of the reactants and products into an overall $\Delta H°_{rxn}$ value. Let's use the combustion of methane to illustrate how this is done. As we saw in Chapter 7, the overall combustion reaction can be written

$$CH_4(g) + 2 O_2(g) \rightarrow CO_2(g) + 2 H_2O(g) \qquad (7.13)$$

The formation reactions for three of the four reactants and products (O_2 is a pure element in its standard state, so it doesn't have a formation reaction) are:

(A) $\qquad C(s, graphite) + 2 H_2(g) \rightarrow CH_4(g) \qquad \Delta H°_f = -74.8$ kJ

(B) $\qquad C(s, graphite) + O_2(g) \rightarrow CO_2(g) \qquad \Delta H°_f = -393.5$ kJ

(C) $\qquad H_2(g) + \frac{1}{2} O_2(g) \rightarrow H_2O(g) \qquad \Delta H°_f = -241.8$ kJ

To combine equations A, B, and C to end up with Equation 7.13, we need to first reverse equation A to put CH_4 on the reactant side. We also need to multiply equation C by 2 because we need a coefficient of 2 in front of H_2O in Equation 7.13, and also because we need a coefficient of 2 in front of H_2 to cancel out the

standard enthalpy of formation ($\Delta H°_f$) the enthalpy change of a formation reaction; also called *standard heat of formation.*

formation reaction a reaction in which 1 mole of a substance is formed from its component elements in their standard states.

H_2 term in the reverse of equation A. Making these changes and summing the three equations that result:

(A, reversed) $\qquad$ $CH_4(g) \rightarrow C(s, \text{graphite}) + 2\,H_2(g)$ $\quad \Delta H_f^\circ = 74.8 \text{ kJ}$

$+$ (B) $\qquad$ $C(s, \text{graphite}) + O_2(g) \rightarrow CO_2(g)$ $\qquad\qquad \Delta H_f^\circ = -393.5 \text{ kJ}$

$+$ (2C) $\qquad$ $2\,H_2(g) + O_2(g) \rightarrow 2\,H_2O(g)$ $\qquad\qquad \Delta H_f^\circ = -483.6 \text{ kJ}$

$CH_4(g) + \cancel{C(s, \text{graphite})} + 2\,O_2(g) + \cancel{2\,H_2(g)} \rightarrow$

$\qquad\qquad\qquad\qquad \cancel{C(s, \text{graphite})} + \cancel{2\,H_2(g)} + CO_2(g) + 2\,H_2O(g)$

or

$$CH_4(g) + 2\,O_2(g) \rightarrow CO_2(g) + 2\,H_2O(g) \qquad \Delta H_{rxn}^\circ = -802.3 \text{ kJ}$$

where $\Delta H_{rxn}^\circ = (74.8 - 393.5 - 483.6) \text{ kJ} = -802.3 \text{ kJ}$.

Let's analyze the math in this application of Hess's law. By leaving B alone while doubling C and reversing A, we combined ΔH_f° of CO_2 with $2 \times \Delta H_f^\circ$ of H_2O and subtracted ΔH_f° of CH_4. We can generalize these operations to fit any chemical equation by taking the following approach: sum the ΔH_f° values of the products, each multiplied by its coefficient in the chemical equation, and then subtract the sum of the ΔH_f° values of the reactants, each multiplied by its coefficient. Expressing this in equation form:

$$\Delta H_{rxn}^\circ = \sum n_{products}\,\Delta H_{f,products}^\circ - \sum n_{reactants}\,\Delta H_{f,reactants}^\circ \qquad (9.15)$$

where $n_{products}$ is the number of moles of each product in the balanced equation and $n_{reactants}$ is the number of moles of each reactant. Applying Equation 9.15 to the combustion of methane:

$$\Delta H_{rxn}^\circ = [(1 \text{ mol } CO_2)(-393.5 \text{ kJ/mol}) + (2 \text{ mol } H_2O)(-241.8 \text{ kJ/mol})]$$

$$- [(1 \text{ mol } CH_4)(-74.8 \text{ kJ/mol}) + (2 \text{ mol } O_2)(0.0 \text{ kJ/mol})]$$

$$= [(-393.5 \text{ kJ}) + (-483.6 \text{ kJ})] - [(-74.8 \text{ kJ}) + (0.0 \text{ kJ})]$$

$$= -802.3 \text{ kJ}$$

SAMPLE EXERCISE 9.10 **Calculating Standard Heats of Reaction from Standard Heats of Formation** **LO5**

Use the appropriate standard heat of formation values to calculate ΔH_{rxn}° for the complete combustion of propane: $C_3H_8(g) + 5\,O_2(g) \rightarrow 3\,CO_2(g) + 4\,H_2O(g)$.

COLLECT AND ORGANIZE The standard heats of formation of all the reactants and products are listed in Table 9.4. Our task is to use the balanced reaction and the ΔH_f° data from Table 9.4 to calculate the heat of combustion.

ANALYZE Equation 9.15 defines the relation between standard heats of formation of reactants and products and the standard heat of a reaction. Because O_2 gas is a pure element in its standard state, its ΔH_f° value is zero. We expect the reaction to be very exothermic ($\Delta H_{rxn}^\circ \ll 0$) because it represents combustion.

SOLVE Inserting ΔH_f° values for the products and reactants and their coefficients from the balanced chemical equation into Equation 9.15:

$$\Delta H_{rxn}^\circ = \left[(3 \text{ mol } CO_2)\left(-393.5\,\frac{\text{kJ}}{\text{mol}}\right) + (4 \text{ mol } H_2O)\left(-241.8\,\frac{\text{kJ}}{\text{mol}}\right) \right]$$

$$- \left[(1 \text{ mol } C_3H_8)\left(-103.8\,\frac{\text{kJ}}{\text{mol}}\right) + (5 \text{ mol } O_2)\left(0.0\,\frac{\text{kJ}}{\text{mol}}\right) \right]$$

$$= -2043.9 \text{ kJ}$$

THINK ABOUT IT The large negative value is expected for combustion of a hydrocarbon fuel. Propane is used in many backyard grills.

Practice Exercise Use standard heat of reaction values to calculate ΔH°_{rxn} for the water–gas shift reaction:

$$CO(g) + H_2O(g) \rightarrow CO_2(g) + H_2(g)$$

Heats of Reaction and Bond Energies

The energy changes associated with chemical reactions depend on how much energy is required to break the bonds in the reactants and how much is released as their atoms recombine to form products. For example, in the methane combustion reaction [$CH_4(g) + 2\,O_2(g) \rightarrow CO_2(g) + 2\,H_2O(g)$], the C—H bonds in CH_4 and the O=O bonds in O_2 must be broken before the C=O bonds in CO_2 and the O—H bonds in H_2O can form. Breaking bonds is endothermic (the "Energy in" arrow in Figure 9.18), and forming bonds is exothermic (the "Energy out" arrows). If a chemical reaction is exothermic, as methane combustion is, more energy is released in forming the bonds in molecules of products than is consumed in breaking the bonds in molecules of reactants.

Bond energy (or bond strength) is the enthalpy change (ΔH) that occurs when 1 mole of bonds in the gas phase is broken. The quantity of energy needed to break a particular bond is equal in magnitude but opposite in sign to the quantity of energy released when that same bond forms from free atoms. In other words, breaking a bond requires an investment of energy and is a highly endothermic process ($\Delta H > 0$), whereas energy is released when free atoms come together to make a covalent bond (and a more stable molecular structure). Thus, bond formation is a highly exothermic process ($\Delta H < 0$).

During the combustion of 1 mole of CH_4 (Figure 9.18), 4 moles of C—H bonds and 2 moles of O=O bonds must be broken. The formation of 1 mole of

CONNECTION We first encountered bond energy in Chapter 4. Bond energies for some common covalent bonds are listed in Table 4.7, and a more complete list is found in Appendix 4.

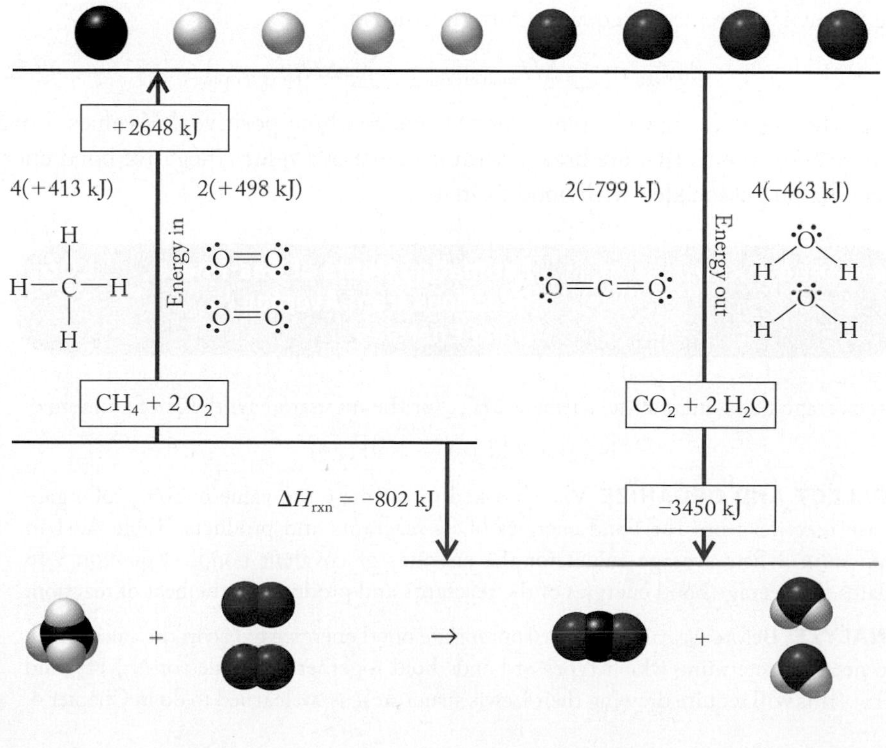

FIGURE 9.18 The complete combustion of methane to CO_2 and water vapor requires that 4 moles of C—H bonds and 2 moles of O=O bonds be broken. Breaking bonds requires energy and is accompanied by an increase in enthalpy. Formation of 2 moles of C=O bonds and 4 moles of O—H bonds is accompanied by an even greater decrease in enthalpy, so the overall reaction is accompanied by a decrease in enthalpy and is exothermic.

CO_2 and 2 moles of H_2O requires the formation of 2 moles of $C{=}O$ bonds and 4 moles of $O{-}H$ bonds. The net change in energy resulting from breaking reactant bonds and forming the bonds in the products can be estimated from the average bond energies. We start by taking an inventory of the bond energies involved:

	Bond	Number of Bonds (mol)	Bond Energy (kJ/mol)	ΔH
Bonds Broken	$C{-}H$	4	413	$4 \text{ mol} \times 413 \text{ kJ/mol}$
	$O{=}O$	2	498	$2 \text{ mol} \times 498 \text{ kJ/mol}$
Bonds Formed	$O{-}H$	4	463	$-(4 \text{ mol} \times 463 \text{ kJ/mol})$
	$C{=}O$	2	799	$-(2 \text{ mol} \times 799 \text{ kJ/mol})$

Next we sum the positive ΔH values associated with breaking bonds and the negative ΔH values associated with forming them to obtain an estimate of the enthalpy change of the overall reaction:

$$\Delta H_{rxn} = (4 \text{ mol} \times 413 \text{ kJ/mol})$$
$$+ (2 \text{ mol} \times 498 \text{ kJ/mol})$$
$$- (4 \text{ mol} \times 463 \text{ kJ/mol})$$
$$- (2 \text{ mol} \times 799 \text{ kJ/mol})$$
$$= -802 \text{ kJ}$$

This estimate is essentially the same as the ΔH°_{rxn} value, -802.3 kJ, which we derived from the standard heats of formation of the reactants and products. We need to keep in mind that the energy of a particular type of bond, such as a $C{-}H$ bond, can vary from one molecule to another or even within a molecule. However, bond energies are usually close enough to their average values that the averages can be used to calculate reliable estimates of the ΔH_{rxn} values of gas-phase reactions (including the two in the following Sample and Practice Exercises). The procedure we follow is expressed in equation form in this way:

$$\Delta H_{rxn} = \sum \Delta H_{\text{bond breaking}} - \sum \Delta H_{\text{bond forming}} \qquad (9.16)$$

▶❚❚ **CHEMTOUR** Estimating Enthalpy Changes

Here the negative sign is a reminder that we combine positive ΔH values (bond energies) for bonds that are broken with *negative* ΔH values (negative bond energies) that are associated with bond formation.

SAMPLE EXERCISE 9.11 **Estimating Heats of Gas-Phase Reactions from Average Bond Energies** **LO6**

Use average bond energies to estimate ΔH_{rxn} for the industrial synthesis of ammonia:

$$N_2(g) + 3 H_2(g) \rightarrow 2 NH_3(g)$$

COLLECT AND ORGANIZE We are asked to estimate the value of ΔH_{rxn} of a gas-phase reaction from the bond energies of its reactants and products. Table A4.1 in Appendix 4 lists average values for the energies of covalent bonds. Equation 9.16 relates the average bond energies of the reactants and products to the heat of reaction.

ANALYZE Before we can select the appropriate bond energies to use in the calculation, we need to determine which types of bonds hold together molecules of N_2, H_2, and NH_3. This will require drawing their Lewis structures, as we learned to do in Chapter 4.

Once we have the structures we can take an inventory of the number of each type of bond that is broken or formed, look up the average bond energies for the bonds, and combine them using Equation 9.16. The reaction involves the formation of 2 moles of a compound from its component elements in their standard states. Therefore the result of this calculation should be close to 2 times the standard heat of formation (ΔH_f°) of NH_3.

SOLVE Let's insert Lewis structures of the reactants and products into the balanced chemical equation to help get a count of all the bonds involved:

$$:N\equiv N: \; + 3\,H—H \; \rightarrow \; 2\; H—\overset{\cdot\cdot}{N}—N$$
$$\underset{\displaystyle H}{|}$$

We then construct a table using the appropriate bond energies from Appendix 4:

	Bond	Number of Bonds (mol)	Bond Energy (kJ/mol)	ΔH
Bonds Broken	$N\equiv N$	1	945	1 mol × 945 kJ/mol
	H—H	3	436	3 mol × 436 kJ/mol
Bonds Formed	N—H	6	391	−(6 mol × 391 kJ/mol)

Combining the values in the right hand column yields ΔH_{rxn}:

$$\Delta H_{rxn} = (1\text{ mol} \times 945\text{ kJ/mol}) + (3\text{ mol} \times 436\text{ kJ/mol}) - (6\text{ mol} \times 391\text{ kJ/mol})$$

$$= -93\text{ kJ}$$

THINK ABOUT IT An enthalpy change of −93 kJ means that the reaction is exothermic. The value for ΔH_f° of NH_3 (see Table 9.4, or Appendix 4) is −46.1 kJ/mol. Two moles of NH_3 are produced in the reaction so multiplying by 2, we get −92.2 kJ, which is very close to the calculated value.

Practice Exercise Use average bond energies to estimate ΔH_{rxn} for the water–gas shift reaction, which is part of the industrial process for making hydrogen gas:

$$CO(g) + H_2O(g) \rightarrow CO_2(g) + H_2(g)$$

9.8 More Applications of Thermochemistry

In the previous section we calculated the standard heats of reaction for the combustion of one mole of methane (−802.3 kJ) and one mole of propane (−2043.9 kJ). Does the much more negative (exothermic) value for propane make it an inherently better (higher energy) fuel? Not necessarily. Expressing ΔH_{rxn}° values on a per-mole basis is the only way to ensure that we are talking about the same number of molecules. However, we do not purchase fuels, or anything else for that matter, in units of moles. Depending on the fuel, we are more likely to buy it by mass or by volume (gasoline and diesel fuel, for example).

To better compare methane and propane as fuels, let's calculate the enthalpy change that takes place when 1 g of each burns in air to produce CO_2 and water vapor. We divide the value of ΔH_{rxn}° (in kilojoules per mole) for each reaction by the molar mass of the hydrocarbon to determine the number of kilojoules of energy released per gram of substance:

$$CH_4: \quad \frac{-802.3\text{ kJ}}{\text{mol}} \times \frac{1\ \cancel{\text{mol}}}{16.04\text{ g}} = -50.02\text{ kJ/g}$$

$$C_3H_8: \quad \frac{-2043.9\text{ kJ}}{\text{mol}} \times \frac{1\ \cancel{\text{mol}}}{44.10\text{ g}} = -46.35\text{ kJ/g}$$

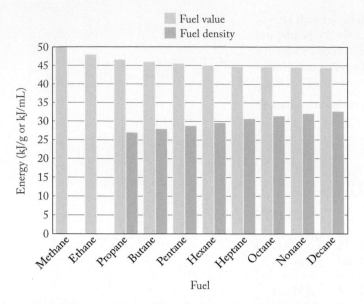

FIGURE 9.19 Fuel values and fuel densities of the C_1 to C_{10} alkanes based on their complete combustion to CO_2 and H_2O vapor.

TABLE 9.5	Fuel Values of Some Common Fuels
Fuel	**Fuel Value[a] (kJ/g)**
Hydrogen	141.8
Natural gas	54.0
Propane	50.3
Butane	49.5
Gasoline	47.4
Diesel	44.8
Ethanol	29.8
Coal (anthracite)	27
Wood	17

[a]Based on complete combustion to CO_2 and liquid H_2O.

From the perspective of the surroundings that absorb this energy, such as the air in a gas furnace, the water in a hot-water heater, or the food on a stove or backyard grill, equal masses of the two fuels provide comparable quantities of energy. Energies like these, calculated on a per-gram basis, are called **fuel values**.

Fuel values of the C_1 to C_{10} alkanes are shown in Figure 9.19. Note the trend of decreasing fuel values with increasing carbon number (and molar mass). What is different about these compounds (besides the sizes of their molecules) that could explain this trend? One difference is the hydrogen-to-carbon ratios. As the number of carbon atoms per molecule increases, the hydrogen-to-carbon ratio decreases: from 4 atoms of H per atom of C in methane to about 2 atoms of H per atom of C in higher-molar-mass alkanes. To understand why this ratio is important, remember that it takes only 2.0 g of hydrogen atoms to make one mole of water vapor and release 241.8 kJ of thermal energy. However, it takes 12.0 g of carbon to form one mole of CO_2 and release 393.5 kJ. Thus, it takes six times as much carbon (by mass) to produce only about half again more thermal energy than hydrogen. This analysis explains why pure hydrogen has the highest fuel value of all fuels.

Fuel values of the major fossil and renewable fuels are listed in Table 9.5. These data are based on combustion reactions that produce CO_2 and *liquid* H_2O, and they are somewhat higher than those plotted in Figure 9.19 for the same compounds (propane, for example) based on their combustion to CO_2 and H_2O *vapor*. Why the difference? Because it takes 44.0 kJ of thermal energy (water's heat of vaporization) to vaporize a mole of liquid water. Therefore, liquid water is a lower energy product than water vapor, and a combustion process that produces liquid water, such as the combustion of methane (Figure 9.20), is accompanied by a greater decrease in enthalpy than if water vapor had been produced.

Another term used to compare the energy content of liquid fuels is **fuel density**. Fuel density is the quantity of energy released per unit volume of a liquid fuel. Many liquid fuels are priced based on volume (gallons in the United States,

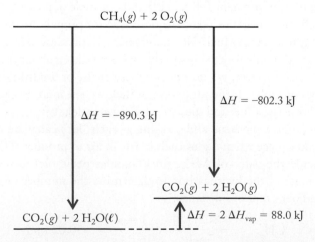

FIGURE 9.20 Methane's heats of combustion have different values depending on whether water vapor or liquid water is produced. The difference is due to the heat of vaporization of water.

liters in the rest of the world), so fuel density can be more useful than fuel value in comparing them as sources of thermal energy, as we do in Sample Exercise 9.12.

fuel value the quantity of energy released during the complete combustion of 1 g of a substance.

fuel density the quantity of energy released during the complete combustion of a particular volume of a liquid fuel.

CONCEPT TEST

The data in Figure 9.19 show that the fuel densities of the C_3–C_{10} alkanes increase with increasing carbon number (and molar mass) even though their fuel values decrease. Why?

SAMPLE EXERCISE 9.12 **Comparing Fuel Values and Fuel Densities** **LO7**

Most automobiles run on either gasoline or diesel fuel. At most gas stations in the United States and Canada the price of diesel fuel is higher than regular-grade gasoline. Suppose a particular station sells diesel fuel (density = 0.83 g/mL) for exactly 10% more per unit volume than regular-grade gasoline (density = 0.75 g/mL). Which fuel contains more thermal energy per dollar?

COLLECT AND ORGANIZE We know the densities and relative prices of gasoline and diesel fuel. Table 9.5 lists their fuel values: 47.4 kJ/g for gasoline and 44.8 kJ/g for diesel fuel.

ANALYZE We can use the given density values to convert fuel values (in kJ/g) into fuel densities (in kJ/mL) and then compare the relative volumes the same amount of money can buy. We are not given the individual prices of the fuels, so to make the units convenient let's make x the price of 1 mL of gasoline. That makes the price of 1 mL of diesel fuel $1.10x$. Gasoline has a slightly higher fuel value but significantly lower density, so it is likely diesel fuel will have the greater fuel density. However, diesel costs more per unit of volume, so the difference in energy content per x dollars (or cents) may be small.

SOLVE

Gasoline: $\dfrac{47.4 \text{ kJ}}{\text{g}} \times \dfrac{0.75 \text{ g}}{\text{mL}} \times \dfrac{1 \text{ mL}}{x} = 36 \text{ kJ}/x$

Diesel fuel: $\dfrac{44.8 \text{ kJ}}{\text{g}} \times \dfrac{0.83 \text{ g}}{\text{mL}} \times \dfrac{1 \text{ mL}}{1.10 \ x} = 34 \text{ kJ}/x$

Therefore, gasoline provides slightly more energy content (about 6%) for the same amount of money x.

THINK ABOUT IT Comparable energy content for the money is only part of the energy story in selecting fuels (or buying cars). Another part is the efficiencies with which internal combustion (gasoline) engines and diesel engines convert the chemical energy in their fuels to mechanical work. On this score diesel engines are considerably better, which translates into much greater distances traveled on the same fuel budget.

Practice Exercise In November 2011, an Atlas V rocket launched NASA's Mars Science Laboratory, including the Mars rover *Curiosity*, into space (Figure 9.21). The first-stage engine of the Atlas V is fueled by kerosene, a hydrocarbon mixture with a density of 0.79 g/mL and a fuel value of 46 kJ/g. At full throttle the Atlas V engine generates 20 gigawatts (2.0×10^{10} J/s) of power. If the engine is 50% efficient at converting chemical energy into mechanical energy (thrust lifting the rocket into space), how rapidly (in liters per second) must kerosene be supplied to and completely burned in the engine? ⚙

FIGURE 9.21 Launch of the Mars Science Laboratory.

Energy from Food

Food is the fuel of living systems. The overall biochemical processes that convert foods into energy resemble combustion reactions, though each process involves many more steps to convert the carbon and hydrogen content of foods, such as carbohydrates, into carbon dioxide and water. The energy content of food can be determined using the same instruments used to evaluate the energy released by other combustible substances. For example, we can burn dry or dehydrated foods in bomb calorimeters to determine the energy they would provide if we ate them.

As an illustration, let's determine the fuel value of peanuts. Suppose our sample is a single peanut with a mass of 1.89 g. We put it in a calorimeter for which $C_{\text{calorimeter}} = 19.31$ kJ/°C and burn it completely in excess oxygen. The resulting temperature increase of the calorimeter is 2.334°C. We determine the fuel value of the peanut much as we would determine the heat of combustion of any combustible material: by first calculating the quantity of heat required to raise the temperature of the calorimeter by 2.334°C:

$$q_{\text{calorimeter}} = C_{\text{calorimeter}} \, \Delta T$$

$$= (19.31 \text{ kJ/°C})(2.334°C)$$

$$= 45.07 \text{ kJ}$$

This quantity of thermal energy is generated by the combustion of 1.89 g, so the fuel value of the peanut is

$$\frac{45.07 \text{ kJ}}{1.89 \text{ g}} = 23.8 \text{ kJ/g}$$

The energy content of food is often expressed in Calories (Cal), where 1 Calorie = 10^3 cal = 4.184 kJ. The fuel value of the peanut in Calories is

$$23.8 \text{ kJ/g} \times \frac{1 \text{ Cal}}{4.184 \text{ kJ}} = 5.69 \text{ Cal/g}$$

SAMPLE EXERCISE 9.13   Calculating the Fuel Value of Food   LO7

Glucose ($C_6H_{12}O_6$) is a simple sugar formed by photosynthesis in plants. The complete combustion of 0.5763 g of glucose in a calorimeter ($C_{\text{calorimeter}} = 6.20$ kJ/°C) raises the temperature of the calorimeter by 1.45°C. What is the fuel value of glucose in Calories per gram?

COLLECT AND ORGANIZE We are asked to determine the fuel value of glucose, which means the energy given off when 1 g is completely converted to CO_2 and H_2O liquid. We know the mass of glucose sample that is completely combusted, and the temperature change produced by the heat released during combustion in a calorimeter with a known calorimeter constant.

ANALYZE The value of the heat released during combustion of the sample and gained by the calorimeter is the product of the calorimeter constant and the change in temperature. We can convert this quantity into Calories by using the conversion factor 1 Cal = 4.184 kJ. For this small sample of glucose, we anticipate a fuel value similar to that of our peanut.

SOLVE

$$q_{\text{calorimeter}} = C_{\text{calorimeter}} \Delta T = (6.20 \text{ kJ/°C})(1.45°C) = 8.99 \text{ kJ}$$

To convert this quantity of energy to a fuel value, we divide by the sample mass:

$$\frac{8.99 \text{ kJ}}{0.5763 \text{ g}} = 15.6 \text{ kJ/g}$$

and then convert into Calories:

$$(15.6 \text{ kJ/g})\left(\frac{1 \text{ Cal}}{4.184 \text{ kJ}}\right) = 3.73 \text{ Cal/g}$$

THINK ABOUT IT Food and nutrition scientists persist in using Calories to describe the energy content of foods even though the SI unit of energy, which is widely used in the physical sciences, is the joule. To add to unit confusion, there are calories (lowercase "c") and there are Calories (uppercase "C"). They are not the same: 1 Cal = 1000 cal = 1 kcal. All three units are widely used in the health sciences and medicine.

Practice Exercise Sucrose (table sugar) has the formula $C_{12}H_{22}O_{11}$ ($\mathcal{M} = 342.30$ g/mol) and a fuel value of 16.4 kJ/g. Determine the calorimeter constant of the calorimeter in which the combustion of 1.337 g of sucrose raises the temperature by 1.96°C. ☼

CONCEPT TEST

Combustion of one gram of glucose ($C_6H_{12}O_6$) produces less energy than combustion of one gram of sucrose ($C_{12}H_{22}O_{11}$). Suggest a reason why.

Recycling Aluminum

Analyses of the energy required to carry out industrial procedures are frequently done to assess costs of operations and support new approaches to producing materials. The following evaluation of the energy requirements associated with the production and recycling of aluminum illustrates a practical application of thermochemical concepts.

Over the last century, aluminum, both alone and in combination with other metals, replaced steel for building structures in which high strength-to-weight ratios and corrosion resistance are paramount. These structures include major components of airplanes, motor vehicles, and the facades of buildings.

The industrial process for converting aluminum ore (Al_2O_3) into aluminum metal was developed by two 23-year-old chemists, Charles Hall and Paul Louis-Toussaint Héroult (both 1863–1914), working independently in the United States (Hall) and France (Héroult) (Figure 9.22). Their process is based on passing an electric current through a solution of aluminum oxide dissolved in molten cryolite (Na_3AlF_6). As electricity passes through the solution, Al^{3+} ions are reduced to aluminum metal while a positively charged carbon electrode is oxidized to carbon dioxide. The process is described by the following reaction:

$$2 \text{ Al}_2\text{O}_3(\text{in molten Na}_3\text{AlF}_6) + 3 \text{ C}(s) \rightarrow 4 \text{ Al}(\ell) + 3 \text{ CO}_2(g) \qquad (9.17)$$

FIGURE 9.22 Charles M. Hall (top) and Paul Louis-Toussaint Héroult (middle) independently developed the same electrolytic process for producing aluminum metal from aluminum ore. Hall's sister Julia (bottom), also a chemistry major at Oberlin College, assisted her brother in the lab, and her business skills made their aluminum production company a financial success. The company became the Aluminum Company of America, shortened to Alcoa.

The principal energy cost of the Hall–Héroult process is the electricity needed to reduce Al(III) to Al. The major cost in recycling aluminum is the energy required to melt aluminum metal so that it can be reshaped into new products. We can use thermochemical principles to estimate the energy required for the Hall–Héroult process and compare it to the energy needed for recycling.

We can obtain an estimate of the energy involved in producing aluminum metal from aluminum ore from the standard heats of formation of the reactants and products in Equation 9.17.

$$\Delta H^\circ_{rxn} = [3(\Delta H^\circ_{f,CO_2} + 4(\Delta H^\circ_{f,Al(\ell)})] - [2(\Delta H^\circ_{f,Al_2O_3}) + 3(\Delta H^\circ_{f,C(s)})]$$

$$= \left[(3 \text{ mol } CO_2)\left(\frac{-393.5 \text{ kJ}}{1 \text{ mol } CO_2}\right) + (4 \text{ mol Al})\left(\frac{10.79 \text{ kJ}}{1 \text{ mol Al}}\right) \right]$$

$$- \left[(2 \text{ mol } Al_2O_3)\left(\frac{-1675.7 \text{ kJ}}{1 \text{ mol } Al_2O_3}\right) + (3 \text{ mol C})\left(\frac{0.0 \text{ kJ}}{1 \text{ mol C}}\right) \right]$$

$$= +2214.1 \text{ kJ}$$

Dividing this value by the 4 moles of aluminum produced in the reaction as written, we get

$$\frac{2214.1 \text{ kJ}}{4 \text{ mol Al}} = 553.52 \frac{\text{kJ}}{\text{mol Al}}$$

In this calculation we used a nonzero standard heat of formation value for aluminum metal. This may seem odd because ΔH°_f values of pure elements are normally zero. However, the zero value applies only to an element in its standard state, which for aluminum is solid, not liquid (molten aluminum is produced in the Hall–Héroult process). Therefore we used the heat of fusion of aluminum (10.79 kJ/mol) as the standard heat of formation for liquid aluminum.

Our estimate of the energy requirements to reduce aluminum ore to aluminum metal did not include the energy needed to melt the cryolite in which the ore dissolves (its melting point is 1012°C), nor did we consider the inefficiency of using electricity to drive a chemical reaction. These and other factors make the energy requirements for producing one mole of Al from Al(III) ore in a modern refinery closer to 1000 kJ.

We can estimate the energy required to recycle one mole of aluminum by calculating the heat required to raise its temperature from 25°C to its melting point (660°C) and then melt it. The molar heat capacity of Al is 24.2 J/(mol · °C), so the heat needed to warm it to its melting point is

$$q = nc_{P,n}\Delta T$$

$$= 1.00 \text{ mol} \times 24.2 \frac{J}{\text{mol} \cdot {}^\circ C} \times (660 - 25)^\circ C \times \frac{1 \text{ kJ}}{1000 \text{ J}}$$

$$= 15.4 \text{ kJ}$$

Once aluminum reaches its melting point, the energy required to melt it ($\Delta H_{fus} = 10.79$ kJ/mol) is

$$10.79 \frac{\text{kJ}}{\text{mol}} \times 1.00 \text{ mol} = 10.8 \text{ kJ}$$

The estimated total energy to heat and melt 1.00 mole of aluminum is then

$$15.4 \text{ kJ} + 10.8 \text{ kJ} = 26.2 \text{ kJ}$$

This value represents only a small percentage of the energy needed to produce 1 mole of aluminum from its ore.

High energy costs make recycling aluminum economically attractive in addition to environmentally sound. The production and recycling of aluminum both incur energy costs, but detailed analyses of both processes have shown that recycling saves aluminum manufacturers about 95% of the energy required to produce the metal from ore. This energy saving has inspired the rapid growth of a global aluminum recycling industry. In the United States alone, aluminum recycling is a $1 billion per year business. In most industrialized countries nearly all of the aluminum in motor vehicles and building materials is recycled, as is about half the aluminum in food and beverage containers.

In this chapter we have explored how the internal energies of systems change as heat flows into, or out from, them and when work is done on, or by, them. Flows of heat or work produce (or accompany) changes in temperature, physical state, or chemical composition. The quantity of energy involved in physical changes can be calculated based on characteristic intensive properties such as specific heat (or molar heat capacity), heat of fusion, and heat of vaporization. Energy is also a product or a reactant in virtually all chemical reactions and as such behaves stoichiometrically, which means that predictable quantities of energy are consumed or released during chemical reactions depending on the quantities of the reactants consumed and the difference in chemical energy between them and the products. Combustion reactions supply most of the world's energy needs as chemical energy stored in different fuels is released when their carbon and hydrogen content is converted to CO_2 and H_2O. The fuel value of the foods that supply the energy needs of living organisms is similarly linked to a series of biochemical reactions that convert C and H atoms to molecules of CO_2 and H_2O.

SAMPLE EXERCISE 9.14 **Integrating Concepts: Selecting a Heating System**

Suppose some friends who happen to be avid skiers decide to pool their resources and buy a vacation home near their favorite ski resort. They find one that meets their needs, but then discover that it needs a new furnace, though it does have an operational wood stove. They collect information about replacement furnace options, focusing on two fuels: home heating oil, which is derived from crude oil and has a hydrocarbon distribution similar to that of diesel fuel, and propane. Their research on the two fuels yields the results summarized in the table below.

Fuel	Price of Fuel ($/U.S. gallon)	Fuel Value (kJ/g)	Density (g/mL)	Furnace Efficiency
Heating oil	3.50	44.9	0.83	83%
Propane	3.35	50.3	0.62	95%

a. Based on these data, which of the two fuels is the better buy?

b. The brochure for the propane furnace they are considering describes it as a "condensing" furnace, noting that it is 10% more efficient than a noncondensing model. What does "condensing" mean in this context, and how much does it contribute to the higher efficiency of the furnace?

c. The most frugal of the friends suggests that they use the wood stove to save fuel costs. A local firewood dealer sells one cord of seasoned hardwood, which weighs about one metric ton, for $250. How much money would the friends save burning a cord of wood instead of the better fuel identified in part a? Assume the fuel value of the wood is 17 kJ/g and that the wood stove is 50% efficient.

d. Which of the fuels under consideration produces the most "fossil" CO_2 per kJ of heat delivered to the home's living space?

COLLECT AND ORGANIZE We know the price per U.S. gallon of two fossil fuels and the price of a renewable alternative, wood. We also know the densities of the fossil fuels and the mass of a cord of wood, and we know the fuel values of all three fuels. We need to select which of the two fossil fuels is the better energy buy and then compare its cost-effectiveness with that of wood. The efficiencies with which stoves and furnaces convert the chemical energy of fuels into heat for the living space are also a factor, and we have to explain why a condensing furnace for burning propane is so efficient. Among the unit conversion factors in the back of the book is 1 U.S. gallon = 3.785 L.

ANALYZE One way to compare the costs of the three fuels is to start with the fuel values and convert them from kJ/g into kJ/$. The prices are based on volumes, so we will need to use the densities of the fuels as one of the needed conversion factors. Fuel values such as those given in Table 9.5 and in this problem are based on combustion reactions that convert the carbon and hydrogen content of the fuel into CO_2 and liquid H_2O. To compare fuel costs we need to convert price per volume values into price per mass and then price per unit of energy produced, and then use the efficiencies of the furnace to calculate the price per unit of heat delivered to the vacation home's living space.

The CO_2 produced by the combustion of propane and heating oil is considered fossil CO_2 because the fuels are derived from natural gas and crude oil, making them fossil fuels. The fossil fuel with the higher carbon-to-hydrogen ratio will be the one that produces the most fossil CO_2 per kilojoule of energy released.

SOLVE

a. Converting the fuel value of heating oil into kilojoules of heat for the home:

$$\frac{44.9 \text{ kJ}}{\text{g}} \times \frac{0.83 \text{ g}}{\text{mL}} \times \frac{1000 \text{ mL}}{\text{L}} \times \frac{3.785 \text{ L}}{\text{gallon}}$$

$$\times \frac{1 \text{ gallon}}{\$3.50} \times \frac{83\%}{100\%} = 3.35 \times 10^4 \text{ kJ/\$}$$

Doing the same conversion for propane:

$$\frac{50.3 \text{ kJ}}{\text{g}} \times \frac{0.62 \text{ g}}{\text{mL}} \times \frac{1000 \text{ mL}}{\text{L}} \times \frac{3.785 \text{ L}}{\text{gallon}}$$

$$\times \frac{1 \text{ gallon}}{\$3.35} \times \frac{95\%}{100\%} = 3.35 \times 10^4 \text{ kJ/\$}$$

Expressing the kJ/\$ values to three significant figures (which is one more than the two we are entitled to), we find that the two fuels deliver the same energy per dollar.

b. The higher efficiency of the *condensing* propane furnace is the result of extracting additional heat from the reaction mixture as it cools and the water vapor in it *condenses*, giving this overall combustion reaction:

$$C_3H_8(g) + 5 \text{ O}_2(g) \rightarrow 3 \text{ CO}_2(g) + 4 \text{ H}_2\text{O}(\ell)$$

instead of this one (see Sample Exercise 9.10):

$$C_3H_8(g) + 5 \text{ O}_2(g) \rightarrow 3 \text{ CO}_2(g) + 4 \text{ H}_2\text{O}(g)$$
$$\Delta H^\circ_{\text{rxn}} = -2043.9 \text{ kJ}$$

The difference between the $\Delta H^\circ_{\text{rxn}}$ values of these two reactions is four times the heat of vaporization of one mole of water at 298 K:

$$H_2\text{O}(\ell) \rightarrow H_2\text{O}(g) \qquad \Delta H_{\text{vap}} = 44.0 \text{ kJ}$$

According to Hess's law, we can reverse this vaporization equation and multiply it by four to match the four moles of H_2O in the combustion reactions. We add it to the thermochemical equation from Sample Exercise 9.10 to get our desired equation:

$$C_3H_8(g) + 5 \text{ O}_2(g) \rightarrow 3 \text{ CO}_2(g) + 4 \text{ H}_2\text{O}(g)$$
$$\Delta H^\circ_{\text{rxn}} = -2043.9 \text{ kJ}$$

$$+ 4 \text{ H}_2\text{O}(g) \rightarrow 4 \text{ H}_2\text{O}(\ell)$$
$$4 \times (-\Delta H_{\text{vap}}) = -176.0 \text{ kJ}$$

$$\overline{C_3H_8(g) + 5 \text{ O}_2(g) \rightarrow 3 \text{ CO}_2(g) + 4 \text{ H}_2\text{O}(\ell)}$$
$$\Delta H^\circ_{\text{rxn}} = -2219.9 \text{ kJ}$$

The ratio of $\Delta H^\circ_{\text{rxn}}$ values is

$$\frac{-2219.9 \text{ kJ}}{-2043.9 \text{ kJ}} = 1.0861$$

This nearly 9% increase in $\Delta H^\circ_{\text{rxn}}$ accounts for most of the greater efficiency of the condensing propane furnace as compared to the noncondensing model.

c. Calculating the heat obtained by burning a cord of hardwood in the wood stove:

$$\frac{17 \text{ kJ}}{\text{g}} \times \frac{1000 \text{ g}}{\text{kg}} \times \frac{1000 \text{ kg}}{\text{ton}} \times \frac{1 \text{ ton}}{\text{cord}} \times 1 \text{ cord} \times \frac{50\%}{100\%}$$
$$= 8.5 \times 10^6 \text{ kJ}$$

The cost of the heating oil needed to produce this much heat is

$$8.5 \times 10^6 \text{ kJ} \times \frac{\$1}{3.35 \times 10^4 \text{ kJ}} = \$250$$

This cost happens to match that of the cost of a cord of hardwood, so burning wood would achieve no cost savings over heating oil.

d. We know from its formula that the ratio of C to H atoms in propane is 3:8, but we are not given this information for the hydrocarbons in heating oil. However, we do know that heating oil has a lower fuel value than propane, which means that it is composed of hydrocarbons with higher molar masses that have higher carbon-to-hydrogen ratios than propane. Therefore, more of heating oil's fuel value comes from the conversion of C to CO_2 than from the conversion of H to H_2O. This means it produces more CO_2 per kilojoule of heat released than propane. On top of that, the heating oil furnace is less efficient than the propane furnace, so heating oil produces even more fossil CO_2 to heat the home than propane. Wood is a renewable biofuel and produces no fossil CO_2 during combustion.

THINK ABOUT IT These calculations of energy costs considered the price and fuel values of three fuels and the efficiencies with which their chemical energies could be used to heat a living space. All three factors are important, though there are others we did not consider, such as the cost of installing a new furnace and connecting it to the home's existing heating system. Of course, environmental concerns surrounding climate change might lead the skiing friends to use the wood stove to heat the home on ski weekends.

SUMMARY

Section 9.1 Thermodynamics is the study of energy and its transformations, and the study of the changes in energy that accompany chemical reactions is known as **thermochemistry**. A **system** is the part of the universe under study, and everything not part of the system is the **surroundings**. The **first**

Surroundings

System
$q > 0$
$w > 0$

law of thermodynamics states that energy is neither created nor destroyed, so the energy gained or lost by a system equals the energy lost or gained by the surroundings. Heating a system or doing work on it increases the **internal energy (E)** of a system, which includes the kinetic energies of the particles in it and the **chemical energy** related to its composition. Internal energy is a **state function**, which means that its value does not depend on the path the system took to achieve that energy. The system's **thermal energy** is proportional to its absolute temperature.

Section 9.2 *Isolated* thermodynamic systems exchange neither energy nor matter with their surroundings, *closed* systems exchange only energy, and *open* systems exchange energy and matter. We use the symbol q to represent the *quantity* of energy transferred between a system and its surroundings. In an **exothermic process** the system loses energy by heating its surroundings ($q < 0$); in an **endothermic process** the system absorbs energy from its warmer surroundings ($q > 0$). The internal energy of a system is increased ($\Delta E = E_{\text{final}} - E_{\text{initial}}$ is positive) when it is heated ($q > 0$) or if work is done on it ($w > 0$).

Section 9.3 The **enthalpy (H)** of a system is given by $H = E + PV$. The **enthalpy change (ΔH)** of a system is equal to the energy (q_P) added to or removed from the system at constant pressure: $\Delta H > 0$ for endothermic reactions and $\Delta H < 0$ for exothermic reactions. The **enthalpy of fusion (ΔH_{fus})** is the change in enthalpy when 1 mole of the solid substance melts at its melting point. The **enthalpy of vaporization (ΔH_{vap})** is the enthalpy change that occurs when 1 mole of the liquid substance is vaporized at its boiling point.

Section 9.4 The **heat capacity (C_P)** of an object at constant pressure is the amount of energy required to increase the temperature of the object by 1°C at constant pressure. The **specific heat (c_P)** of a substance is the amount of energy required to increase the temperature of 1 g of the substance by 1°C. The **molar heat capacity ($c_{P,n}$)** of a substance is the amount of energy required to increase the temperature of 1 mole of the substance by 1°C.

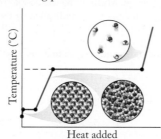

Section 9.5 A **calorimeter** is a device for determining how much heat is transferred from a system (often a reaction mixture) to its surroundings. A coffee-cup calorimeter operates at constant (atmospheric) pressure and is useful for determining the enthalpy

changes or **enthalpies of reaction (ΔH_{rxn})** of aqueous-phase chemical reactions. A **bomb calorimeter** is a device used to determine the energy released during combustion reactions. A **thermochemical equation** includes the change in enthalpy that accompanies a chemical reaction.

Section 9.6 A **standard enthalpy of reaction ($\Delta H^{\circ}_{\text{rxn}}$)** is the enthalpy change associated with a reaction occurring under **standard conditions**: a constant pressure of 1 atm and a specified temperature. A substance's **standard state** is its most stable physical state under standard conditions. **Hess's law** states that the heat of a reaction that occurs in more than one step is the sum of the heats of reaction of the steps. Hess's law can be used to calculate enthalpy changes in reactions that are hard or impossible to measure directly.

$$\Delta H^{\circ}_2 = -41 \text{ kJ}$$

$$\Delta H^{\circ}_1 = +206 \text{ kJ} \qquad \Delta H^{\circ}_3 = \Delta H^{\circ}_1 + \Delta H^{\circ}_2 = +165 \text{ kJ}$$

Section 9.7 The **standard enthalpy of formation (ΔH°_f)** of a substance is the enthalpy change that accompanies a **formation reaction**, in which 1 mole of the substance is made from its constituent elements in their standard states. The ΔH°_f of an element in its standard state is zero. Heats of reaction can be calculated from the heats of formation of reactants and products.

Section 9.8 **Fuel value** is the amount of energy released during complete combustion of 1 g of a fuel. An alternative measure of a fuel's energy content, **fuel density**, is the amount of energy released per unit volume of a liquid fuel. Heats of combustion can be used to evaluate the fuel values of dried foods because the enthalpy change that accompanies respiration in organisms is the same as that of combustion because the products are the same: CO_2 and H_2O. Nutritionists often express the energy content of foods in Calories rather than in the SI unit of joules.

PROBLEM-SOLVING SUMMARY

TYPE OF PROBLEM	CONCEPTS AND EQUATIONS		SAMPLE EXERCISES
Identifying endothermic and exothermic processes	During an endothermic process, heat flows into the system from its surroundings. During an exothermic process, heat flows out from the system into its surroundings.		9.1
Calculating P–V work	$w = -P\Delta V$		9.2
Relating ΔE, q, and w	$\Delta E = q + w = q - P\Delta V$	(9.3, 9.4)	9.3

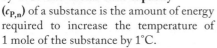

TYPE OF PROBLEM	CONCEPTS AND EQUATIONS		SAMPLE EXERCISES
Calculating heat transfer (q) associated with a change of temperature or state of a substance	Heating either an object: $$q = C_p\Delta T \qquad (9.8)$$ or a mass (m) of a pure substance: $$q = mc_p\Delta T \qquad (9.9)$$ or a quantity of a pure substance in moles (n): $$q = nc_{P,n}\Delta T \qquad (9.10)$$ Melting a solid at its melting point: $$q = n\Delta H_{fus} \qquad (9.11)$$ Vaporizing a liquid at its boiling point: $$q = n\Delta H_{vap} \qquad (9.12)$$		9.4, 9.5
Calculating $C_{calorimeter}$ and ΔH_{rxn} from calorimetry data	$$q_{rxn} = -q_{calorimeter} = -C_{calorimeter}\Delta T$$		9.6, 9.7
Calculating ΔH_{rxn} using Hess's law	Reorganize the information so that the reactions add together as desired. Reversing a reaction changes the sign of the reaction's ΔH_{rxn} value. Multiplying the coefficients in a reaction by a factor means the reaction's ΔH_{rxn} value has to be multiplied by the same factor.		9.8
Recognizing formation reactions	The reactants must be elements in their standard states and the product must be 1 mole of a single compound.		9.9
Calculating $\Delta H°_{rxn}$ from standard heats of formation	$$\Delta H°_{rxn} = \sum n_{products}\,\Delta H°_{f,products} - \sum n_{reactants}\,\Delta H°_{f,reactants} \qquad (9.15)$$		9.10
Estimating heats of gas-phase reactions from average bond energies	$$\Delta H_{rxn} = \sum \Delta H_{bond\ breaking} - \sum \Delta H_{bond\ forming} \qquad (9.16)$$		9.11
Calculating fuel values and fuel densities	The fuel value of a substance is the energy released by the complete combustion of 1 g of the substance. Fuel density is the energy released during combustion of a unit volume (such as 1 mL) of a substance.		9.12, 9.13

VISUAL PROBLEMS ●●●■

(Answers to boldface end-of-chapter questions and problems are in the back of the book.)

9.1. Figure P9.1 shows the compression stroke of a diesel engine. How does upward motion of the piston alter the internal energy of the gases trapped in the cylinder?

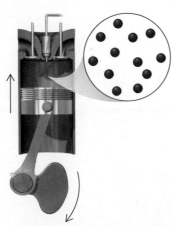

Compression

FIGURE P9.1

9.2. Figure P9.2 shows the power stroke of a diesel engine as energy released by the rapid combustion of the air and fuel vapor inside the cylinder pushes the cylinder downward. If the gases inside the cylinder are a thermodynamic system, how does the internal energy of the system change as a result of the combustion reaction and downward motion of the piston? In your description indicate the signs on ΔE, q, and w.

Power

FIGURE P9.2

9.3. Based on their molecular structures, predict which of the four hydrocarbons in Figure P9.3 has the highest fuel value and which has the lowest.

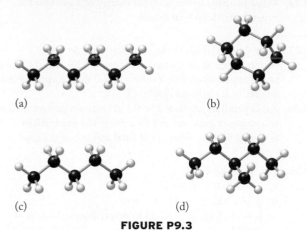

(a)   (b)

(c)   (d)

FIGURE P9.3

9.4. The diagram in Figure P9.4 shows how the volume of a reaction mixture at constant pressure and temperature changes as N_2 and H_2 combine, forming NH_3.
   a. In this reaction, does the reaction mixture do work on the surroundings, or vice versa?
   b. Use data from Appendix 4 to calculate $\Delta H°_{rxn}$ for the formation of 1 mole of product.
   c. To achieve a final temperature that is the same as the initial one, does heat flow out from, or into, the reaction mixture?

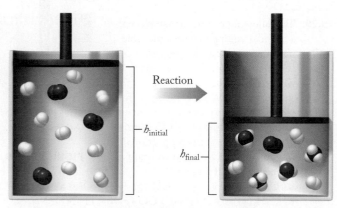

FIGURE P9.4

9.5. Assuming the reaction mixture in Figure P9.4 is a thermodynamic system, what are the signs on q, w, and ΔE? If each molecule in the figure represents one mole of reactant or product, what is the percent yield of the reaction?

9.6. The diagram in Figure P9.6 is based on the standard heats of formation ($\Delta H°_f$) values for four compounds made from the elements listed on the "zero" line of the vertical axis.
   a. Why are the elements all on the same horizontal line?
   b. Why is $C_2H_2(g)$ sometimes called an "endothermic" compound?

c. Based on these $\Delta H°_f$ values, predict which of these two reactions has the more negative enthalpy change:

$$2\,C_2H_2(g) + 3\,O_2(g) \rightarrow 4\,CO(g) + 2\,H_2O(g)$$

or

$$2\,C_2H_2(g) + 5\,O_2(g) \rightarrow 4\,CO_2(g) + 2\,H_2O(g)$$

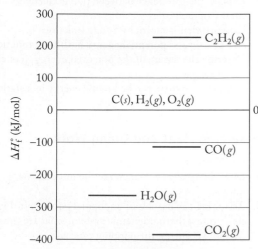

FIGURE P9.6

9.7. Figure P9.7 represents a chemical reaction taking place at constant temperature and pressure.
   a. Write a balanced chemical equation for the reaction.
   b. Use data from Appendix 4 to calculate $\Delta H°_{rxn}$ for the formation of 1 mole of product.
   c. To achieve a final temperature that is the same as the initial one, does heat flow out from, or into, the reaction mixture?

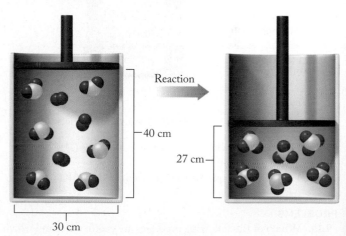

FIGURE P9.7

9.8. If the reaction mixture in Figure P9.7 is a thermodynamic system and the pressure on the system is a constant 1.00 atm, what is the value of w? (The volume of a cylinder is $V = \pi r^2 h$.)

QUESTIONS AND PROBLEMS ···■

Energy as a Reactant or Product

CONCEPT REVIEW

9.9. How are energy and work related?

9.10. Explain the difference between potential energy and kinetic energy in molecules.

9.11. Explain what is meant by a state function.

9.12. Are kinetic and potential energy both state functions?

9.13. Describe the nature of the potential energy in one mole of acetylene gas.

9.14. Explain how there can be kinetic energy in a stationary ice cube.

Transferring Heat and Doing Work

CONCEPT REVIEW

9.15. Describe two ways to increase the internal energy of a gas sample.

9.16. Assuming the kernels of unpopped popcorn in Figure P9.16 constitute a thermodynamic system, what are the signs of q, w, and ΔE during the popping process?

FIGURE P9.16

9.17. How can the product of pressure and volume (P–V work) have energy units?

9.18. Why is there a negative sign in front of the $P\Delta V$ term in $\Delta E = q - P\Delta V$?

PROBLEMS

9.19. Which of the following processes are exothermic, and which are endothermic?
  a. Molten aluminum solidifies.
  b. Rubbing alcohol evaporates from the skin.
  c. Fog forms over San Francisco Bay.

9.20. Which of the following processes are exothermic, and which are endothermic?
  a. Ice cubes solidify in the freezer.
  b. Ice cubes in a frost-free freezer slowly lose mass.
  c. Dew forms on a lawn overnight.

9.21. What happens to the internal energy of a liquid at its boiling point when it vaporizes?

9.22. What happens to the internal energy of a gas when it expands (with no heat flow)?

9.23. How much P–V work does a gas system do on its surroundings at a constant pressure of 1.00 atm if the volume of gas triples from 250.0 mL to 750.0 mL? Express your answer in L · atm and joules (J).

9.24. An expanding gas does 150.0 J of work on its surroundings at a constant pressure of 1.01 atm. If the gas initially occupied 68 mL, what is the final volume of the gas?

9.25. Calculate ΔE when
  a. $q = 100.0$ J; $w = -50.0$ J
  b. $q = 6.2$ kJ; $w = 0.70$ L · atm
  c. $q = -615$ kJ; $w = -3.25$ kilowatt-hours (1 kWh = 3600 kJ)

9.26. Calculate ΔE for a system that absorbs 726 kJ of heat from its surroundings and does 526 kJ of work on its surroundings.

9.27. Calculate ΔE for the combustion of a gas that releases 210.0 kJ of heat to its surroundings and does 65.5 kJ of work on its surroundings.

9.28. Calculate ΔE for a chemical reaction that releases 90.7 kJ of heat to its surroundings but does no work on them.

*9.29. The following reactions take place in a cylinder equipped with a movable piston at atmospheric pressure (Figure P9.29). Which reactions will result in work being done on the surroundings? What is the sign of w? Assume the system returns to an initial temperature of 110°C. *Hint*: The volume of a gas is proportional to number of moles (n) at constant temperature and pressure.
  a. $CH_4(g) + 2\,O_2(g) \rightarrow CO_2(g) + 2\,H_2O(g)$
  b. $C_3H_8(g) + 5\,O_2(g) \rightarrow 3\,CO_2(g) + 4\,H_2O(g)$
  c. $N_2(g) + 2\,O_2(g) \rightarrow 2\,NO_2(g)$

FIGURE P9.29

*9.30. In which direction will the piston described in Problem 9.29 move when the following reactions are carried out at atmospheric pressure inside the cylinder and after the system has returned to its initial temperature of 110°C? What is the sign on w? *Hint*: The volume of a gas is proportional to the number of moles (n) at constant temperature and pressure.
  a. $N_2(g) + 3\,H_2(g) \rightarrow 2\,NH_3(g)$
  b. $C(s) + O_2(g) \rightarrow CO_2(g)$
  c. $CH_3CH_2OH(g) + 3\,O_2(g) \rightarrow 2\,CO_2(g) + 3\,H_2O(g)$

Enthalpy and Enthalpy Changes

CONCEPT REVIEW

9.31. What is meant by an *enthalpy change*?

9.32. Describe the difference between an internal energy change (ΔE) and an enthalpy change (ΔH).

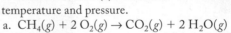

9.33. Why is the sign of ΔH negative for an exothermic process?

9.34. What happens to the magnitude and sign of the enthalpy change when a process is reversed?

PROBLEMS

9.35. Adding Drano to a clogged sink causes the drainpipe to get warm. What is the sign of ΔH when Drano dissolves in water?

9.36. Instant Cold Pack Chemistry Breaking the small pouch of water inside a chemical cold pack containing ammonium nitrate activates the pack, which is used by sports trainers for injured athletes. What is the sign of ΔH for the process taking place in the cold pack?

9.37. The stable form of oxygen at room temperature and pressure is the diatomic molecule O_2. What is the sign of ΔH for the following process?

$$O_2(g) \rightarrow 2\ O(g)$$

9.38. Plaster of Paris Gypsum is the common name of calcium sulfate dihydrate ($CaSO_4 \cdot 2\ H_2O$). When gypsum is heated to 150°C, it loses most of the water in its formula and forms plaster of Paris ($CaSO_4 \cdot 0.5\ H_2O$):

$$2\ CaSO_4 \cdot 2\ H_2O(s) \rightarrow 2\ CaSO_4 \cdot 0.5\ H_2O(s) + 3\ H_2O(g)$$

What is the sign of ΔH for making plaster of Paris from gypsum?

9.39. A solid with metallic properties is formed when hydrogen gas is compressed under extremely high pressures. What is the sign of ΔH for the deposition process: $H_2(g) \rightarrow H_2(s)$?

9.40. In a simple "kitchen chemistry" experiment, some vinegar is poured into an empty soda bottle. A deflated balloon containing baking soda is stretched over the mouth of the bottle. Holding up the balloon and shaking it allows the baking soda to fall into the vinegar, which starts the following reaction and inflates the balloon:

$$NaHCO_3(aq) + CH_3COOH(aq) \rightarrow$$
$$CH_3COONa(aq) + CO_2(g) + H_2O(\ell)$$

If the contents of the bottle are the system, is work being done on the surroundings or on the system?

Heating Curves and Heat Capacity

CONCEPT REVIEW

9.41. What is the difference between *specific heat* and *heat capacity*?

9.42. Which has more heat capacity: one liter of water or one cubic meter of water? Which has more molar heat capacity?

9.43. Why is the heat of vaporization of water so much greater than its heat of fusion?

9.44. The same quantity of heat warms equal masses of metals A and B. Does the metal with the larger specific heat reach the higher temperature?

*9.45. Cooling an Automobile Engine Most automobile engines are cooled by water circulating through them and a radiator. However, the original Volkswagen Beetle had an air-cooled engine. Why might car designers choose water cooling over air cooling?

*9.46. Nuclear Reactor Coolants The reactor-core cooling systems in some nuclear power plants use liquid sodium as the coolant. Sodium has a thermal conductivity of 1.42 J/(cm · s · K), which is quite high compared with that of water [6.1×10^{-3} J/(cm · s · K)]. The respective molar heat capacities are 28.3 J/(mol · K) and 75.3 kJ/(mol · K). What is the advantage of using liquid sodium over water in this application?

PROBLEMS

9.47. How much heat must be absorbed by 100.0 grams of water to raise its temperature from 30.0°C to 100.0°C?

9.48. At an elevation where the boiling point of water is 93°C, 1.33 kg of water at 30°C absorbs 290.0 kJ from a mountain climber's stove. Is this amount of thermal energy sufficient to heat the water to its boiling point?

9.49. Use the following data to sketch a heating curve for 1 mole of methanol. Start the curve at −100°C and end it at 100°C.

Boiling point	65°C
Melting point	−94°C
Heat of vaporization (ΔH_{vap})	35.3 kJ/mol
Heat of fusion (ΔH_{fus})	3.18 kJ/mol
Molar heat capacity (ℓ)	81.1 J/(mol · °C)
Molar heat capacity (g)	43.9 J/(mol · °C)
Molar heat capacity (s)	48.7 J/(mol · °C)

9.50. Use the following data to sketch a heating curve for 1 mole of octane. Start the curve at −57°C and end it at 150°C.

Boiling point	125.7°C
Melting point	−56.8°C
Heat of vaporization (ΔH_{vap})	41.5 kJ/mol
Heat of fusion (ΔH_{fus})	20.7 kJ/mol
Molar heat capacity (ℓ)	254.6 J/(mol · °C)
Molar heat capacity (g)	316.9 J/(mol · °C)

9.51. Keeping an Athlete Cool During a strenuous workout, an athlete generates 233 kJ of thermal energy. What mass of water would have to evaporate from the athlete's skin to dissipate this energy?

9.52. The same quantity of thermal energy is added to equal masses of four metals: titanium, iron, silver, and lead. All of them are initially at the same temperature. Which metal has the highest final temperature?

9.53. Exactly 10.0 mL of water at 25.0°C is added to a hot iron skillet. All of the water is converted into steam at 100.0°C. The mass of the pan is 1.20 kg. What is the change in temperature of the skillet?

*9.54. A 20.0 g piece of iron and a 20.0 g piece of gold at 100.0°C are dropped into 1.00 L of water at 20.0°C. What is the final temperature of the water and pieces of metal?

Heats of Reaction and Calorimetry

CONCEPT REVIEW

9.55. Why is it necessary to know the heat capacity of a calorimeter?

9.56. Could an endothermic reaction be used to determine the heat capacity of a calorimeter?

9.57. If we replace the water in a bomb calorimeter with another liquid, do we need to redetermine the heat capacity of the calorimeter?

9.58. When measuring the heat of combustion of a very small amount of material, would you prefer to use a calorimeter having a heat capacity that is small or large?

PROBLEMS

9.59. Calculate the heat capacity of a calorimeter if the combustion of 5.000 g of benzoic acid produces a temperature increase of 16.397°C.

9.60. Calculate the heat capacity of a calorimeter if the combustion of 4.663 g of benzoic acid produces an increase in temperature of 7.149°C.

9.61. The complete combustion of 1.200 g of cinnamaldehyde (C_9H_8O, one of the compounds in cinnamon) in a bomb calorimeter ($C_{calorimeter}$ = 3.640 kJ/°C) produced an increase in temperature of 12.79°C. How much thermal energy is produced during the complete combustion of one mole of cinnamaldehyde?

9.62. **Spices** The aromatic hydrocarbon cymene ($C_{10}H_{14}$) is found in nearly 100 spices and fragrances including coriander, anise, and thyme. The complete combustion of 1.608 g of cymene in a bomb calorimeter ($C_{calorimeter}$ = 3.640 kJ/°C) produced an increase in temperature of 19.35°C. How much thermal energy is produced during the complete combustion of one mole of cymene?

9.63. **Hormone Mimics** Phthalates that are used to make plastics flexible are among the most abundant industrial contaminants in the environment. Several have been shown to act as hormone mimics in humans by activating the receptors for estrogen, a female sex hormone. Combustion of one mole of one of these compounds, dimethyl phthalate ($C_{10}H_{10}O_4$), produces 4685 kJ of thermal energy. If 1.00 gram of dimethyl phthalate is combusted in a bomb calorimeter whose heat capacity ($C_{calorimeter}$) is 7.854 kJ/°C, what is the change in temperature of the calorimeter?

9.64. **Flavorings** The flavor of anise is due to anethole, a compound with the molecular formula $C_{10}H_{12}O$. Combustion of one mole of anethole produces 5541 kJ of thermal energy. If 0.950 g of anethole is combusted in a bomb calorimeter whose heat capacity ($C_{calorimeter}$) is 7.854 kJ/°C, what is the change in temperature of the calorimeter?

9.65. A coffee-cup calorimeter contains 100.0 mL of 1.00 M HCl at 22.4°C. When 0.243 g of Mg metal is added to the acid, the ensuing reaction:

$$Mg(s) + 2\ HCl(aq) \rightarrow MgCl_2(aq) + H_2(g) \qquad \Delta H_{rxn} = ?$$

causes the temperature of the solution to increase to 33.4°C. What is the value of ΔH_{rxn} of the reaction? Assume the density of the solution is 1.01 g/mL and that its specific heat is 4.18 J/(g · °C).

9.66. **Chemical Cold Pack** When 4.00 g NH_4NO_3 (M = 80.04 g/mol)—the active ingredient in some chemical cold packs—is dissolved in 96.0 g H_2O, the temperature of the resulting solution is 3.07°C colder than the water and ammonium nitrate were before they were mixed together. What is the value of ΔH for the following dissolution process?

$$NH_4NO_3(s) \rightarrow NH_4NO_3(aq) \qquad \Delta H = ?$$

Hess's Law and Standard Heats of Reaction

CONCEPT REVIEW

9.67. How is Hess's law consistent with the law of conservation of energy?

9.68. Would Hess's law be valid if enthalpy were not a state function? Why or why not?

PROBLEMS

9.69. Use the ΔH°_{rxn} values of the following reactions:

$$2\ SO_2(g) + O_2(g) \rightarrow 2\ SO_3(g) \qquad \Delta H^\circ_{rxn} = -196\ kJ$$
$$\tfrac{1}{4} S_8(s) + 3\ O_2(g) \rightarrow 2\ SO_3(g) \qquad \Delta H^\circ_{rxn} = -790\ kJ$$

to calculate the ΔH°_{rxn} value of this reaction:

$$\tfrac{1}{8} S_8(s) + O_2(g) \rightarrow SO_2(g) \qquad \Delta H^\circ_{rxn} = ?$$

9.70. **Ozone Layer** The destruction of the ozone layer by chlorofluorocarbons (CFCs) can be described by the following reactions:

$$ClO(g) + O_3(g) \rightarrow Cl(g) + 2\ O_2(g) \qquad \Delta H^\circ_{rxn} = -29.90\ kJ$$
$$2\ O_3(g) \rightarrow 3\ O_2(g) \qquad \Delta H^\circ_{rxn} = +24.18\ kJ$$

Use the above ΔH°_{rxn} values to determine the value of the standard heat of reaction for this reaction:

$$Cl(g) + O_3(g) \rightarrow ClO(g) + O_2(g) \qquad \Delta H^\circ_{rxn} = ?$$

9.71. The mineral spodumene ($LiAlSi_2O_6$) exists in two crystalline forms called α and β. Use Hess's law and the following information to calculate ΔH°_{rxn} for the conversion of α-spodumene into β-spodumene:

$$Li_2O(s) + 2\ Al(s) + 4\ SiO_2(s) + \tfrac{3}{2} O_2(g) \rightarrow 2\ \alpha\text{-}LiAlSi_2O_6(s)$$
$$\Delta H^\circ_{rxn} = -1870.6\ kJ$$

$$Li_2O(s) + 2\ Al(s) + 4\ SiO_2(s) + \tfrac{3}{2} O_2(g) \rightarrow 2\ \beta\text{-}LiAlSi_2O_6(s)$$
$$\Delta H^\circ_{rxn} = -1814.6\ kJ$$

9.72. Use the following data to determine whether the conversion of diamond into graphite is exothermic or endothermic:

$$C(s, diamond) + O_2(g) \rightarrow CO_2(g) \qquad \Delta H^\circ_{rxn} = -395.4\ kJ$$
$$2\ CO_2(g) \rightarrow 2\ CO(g) + O_2(g) \qquad \Delta H^\circ_{rxn} = +566.0\ kJ$$
$$2\ CO(g) \rightarrow C(s, graphite) + CO_2(g) \qquad \Delta H^\circ_{rxn} = -172.5\ kJ$$
$$C(s, diamond) \rightarrow C(s, graphite) \qquad \Delta H^\circ_{rxn} = ?$$

9.73. You are given the following thermochemical data:

$$\tfrac{1}{2} N_2(g) + \tfrac{1}{2} O_2(g) \rightarrow \tfrac{1}{2} NO(g) \qquad \Delta H^\circ_{rxn} = +90.3\ kJ$$
$$NO(g) + \tfrac{1}{2} Cl_2(g) \rightarrow NOCl(g) \qquad \Delta H^\circ_{rxn} = -38.6\ kJ$$

What is the value of ΔH°_{rxn} for the decomposition of NOCl?

$$2 \, NOCl(g) \rightarrow N_2(g) + O_2(g) + Cl_2(g) \qquad \Delta H^\circ_{rxn} = ?$$

9.74. Use the information in the following two thermochemical equations:

$$NO_2Cl(g) \rightarrow NO_2(g) + \tfrac{1}{2} Cl_2(g) \qquad \Delta H^\circ_{rxn} = -114 \, kJ$$

$$\tfrac{1}{2} N_2(g) + O_2(g) \rightarrow NO_2(g) \qquad \Delta H^\circ_{rxn} = +33.2 \, kJ$$

to calculate the value of ΔH°_{rxn} for this reaction:

$$\tfrac{1}{2} N_2(g) + O_2(g) + \tfrac{1}{2} Cl_2(g) \rightarrow NO_2Cl(g) \qquad \Delta H^\circ_{rxn} = ?$$

Heats of Reaction from Heats of Formation and Bond Energies

CONCEPT REVIEW

9.75. Why is the standard heat of formation of $CO(g)$ difficult to measure experimentally?

9.76. Explain how the use of ΔH°_f to calculate ΔH°_{rxn} is an example of Hess's law.

9.77. Oxygen and ozone are both forms of elemental oxygen. Are the standard heats of formation of oxygen and ozone the same? Why or why not?

9.78. Why are the standard heats of formation of elements in their standard states assigned a value of zero?

9.79. Why must the stoichiometry of a reaction be known in order to estimate the enthalpy change from bond energies?

9.80. Why must the structures of the reactants and products be known in order to estimate the enthalpy change of a reaction from bond energies?

*9.81. When calculating the enthalpy change for a chemical reaction using bond energies, why is it important that the reactants and products all be gases?

*9.82. If the energy needed to break 2 moles of $C=O$ bonds is greater than the sum of the energies needed to break the $O=O$ bonds in 1 mole of O_2 and vaporize 1 mole of carbon, why does the combustion of pure carbon release heat?

PROBLEMS

9.83. For which of the following reactions does ΔH°_{rxn} represent a heat of formation?
   a. $C(s) + O_2(g) \rightarrow CO_2(g)$
   b. $CO_2(g) + C(s) \rightarrow 2 \, CO(g)$
   c. $CO_2(g) + H_2(g) \rightarrow H_2O(g) + CO(g)$
   d. $2 \, H_2(g) + C(s) \rightarrow CH_4(g)$

9.84. For which of the following reactions does ΔH°_{rxn} also represent a heat of formation?
   a. $2 \, N_2(g) + 3 \, O_2(g) \rightarrow 2 \, NO_2(g) + 2 \, NO(g)$
   b. $N_2(g) + O_2(g) \rightarrow 2 \, NO(g)$
   c. $2 \, NO_2(g) \rightarrow N_2O_4(g)$
   d. $N_2(g) + 2 \, O_2(g) \rightarrow 2 \, NO_2(g)$

9.85. **Methanogenesis** Use standard heats of formation from Appendix 4 to calculate the standard heat of reaction for the following methane-generating reaction of methanogenic bacteria:

$$4 \, H_2(g) + CO_2(g) \rightarrow CH_4(g) + 2 \, H_2O(\ell)$$

9.86. Use standard heats of formation from Appendix 4 to calculate the standard heat of reaction for the following methane-generating reaction of methanogenic bacteria, given ΔH°_f of $CH_3NH_2(g) = -22.97 \, kJ/mol$:

$$4 \, CH_3NH_2(g) + 2 \, H_2O(\ell) \rightarrow 3 \, CH_4(g) + CO_2(g) + 4 \, NH_3(g)$$

9.87. Ammonium nitrate decomposes to N_2O and water vapor at temperatures between 250°C and 300°C. Write a balanced chemical reaction describing the decomposition of ammonium nitrate, and calculate its ΔH°_{rxn} using the appropriate ΔH°_f values from Appendix 4.

*9.88. **Chemistry of TNT** Trinitrotoluene (TNT) is a highly explosive compound. The thermal decomposition of TNT is described by the following chemical equation:

$$2 \, C_7H_5N_3O_6(s) \rightarrow 12 \, CO(g) + 5 \, H_2(g) + 3 \, N_2(g) + 2 \, C(s)$$

If ΔH°_{rxn} for the reaction is $-10,153 \, kJ/mol$, how much TNT is needed to equal the explosive power of 1 mole of ammonium nitrate (see Problem 9.87)?

9.89. **Explosives** Fertilizer (ammonium nitrate) and fuel oil (a mixture of long chain hydrocarbons similar to decane, $C_{10}H_{22}$) can form an explosive mixture. Determine the enthalpy change of the following explosive reaction by using the appropriate enthalpies of formation ($\Delta H^\circ_{f,C_{10}H_{22}} = 249.7 \, kJ/mol$):

$$3 \, NH_4NO_3(s) + C_{10}H_{22}(\ell) + 14 \, O_2(g) \rightarrow 3 \, N_2(g) + 17 \, H_2O(g) + 10 \, CO_2(g)$$

9.90. **Military Explosives** Explosives called amatols are mixtures of ammonium nitrate and trinitrotoluene (TNT) introduced during World War I when TNT was in short supply. The mixtures can provide 30% more explosive power than TNT alone. Above 300°C, ammonium nitrate decomposes to N_2, O_2, and H_2O. Write a balanced chemical reaction describing the decomposition of ammonium nitrate, and calculate its ΔH°_{rxn} using the appropriate ΔH°_f values from Appendix 4.

NOTE: Use the average bond energy values in Table A4.1 of Appendix 4 to answer Problems 9.91–9.96.

9.91. Use average bond energies to estimate the enthalpy changes of the following reactions:
   a. $N_2(g) + 3 \, H_2(g) \rightarrow 2 \, NH_3(g)$
   b. $N_2(g) + 2 \, H_2(g) \rightarrow H_2NNH_2(g)$
   c. $2 \, N_2(g) + O_2(g) \rightarrow 2 \, N_2O(g)$

9.92. Use average bond energies to estimate the enthalpy changes of the following reactions:
   a. $CO_2(g) + H_2(g) \rightarrow H_2O(g) + CO(g)$
   b. $N_2(g) + O_2(g) \rightarrow 2 \, NO(g)$
   *c. $C(s) + CO_2(g) \rightarrow 2 \, CO(g)$
   [*Hint:* The heat of sublimation of graphite, $C(s)$, is 719 kJ/mol.]

9.93. Estimate how much less energy is released during the incomplete combustion of 1 mole of ethane to carbon monoxide and water vapor than is released in the complete combustion of ethane to carbon dioxide and water vapor.

9.94. Estimate how much more energy is released by the reaction

$$C(s) + O_2(g) \rightarrow CO_2(g)$$

than by the reaction

$$C(s) + \tfrac{1}{2}O_2(g) \rightarrow CO(g)$$

*9.95. Estimate $\Delta H°_{rxn}$ for the following reaction:

$$4\,NH_3(g) + 7\,O_2(g) \rightarrow 4\,NO_2(g) + 6\,H_2O(g)$$

*9.96. The value of ΔH_{rxn} for the reaction

$$2\,H_2S(g) + 3\,O_2(g) \rightarrow 2\,SO_2(g) + 2\,H_2O(g)$$

is −1036 kJ. Estimate the energy of the bonds in SO_2.

More Applications of Thermochemistry

CONCEPT REVIEW

9.97. What is meant by *fuel value*?

9.98. What are the units of fuel values?

9.99. Without doing any calculations, predict which compound in each pair releases more energy during combustion:
 a. 1 mole of CH_4 or 1 mole of H_2
 b. 1 g of CH_4 or 1 g of H_2

9.100. Is fuel value or fuel density a more useful measure of energy content of liquid fuels? Explain your answer.

PROBLEMS

9.101. If all the energy obtained from burning 1.00 pound of propane is used to heat water, how many kilograms of water can be heated from 20.0°C to 45.0°C?

9.102. An article in *Discover* magazine on world-class sprinters contained the following statement: "In one race, a field of eight runners releases enough energy to boil a gallon jug of ice at 0.0°C in ten seconds!" How much "energy" do the runners release in 10 seconds? Assume that the ice has a mass of 128 ounces.

9.103. Lightweight camping stoves typically use *white gas*, a mixture of C_5 and C_6 hydrocarbons.
 a. Calculate the fuel value of C_5H_{12}, given that $\Delta H°_{comb} = -3535$ kJ/mol.
 b. How much heat is released during the combustion of 1.00 kg of C_5H_{12}?
 c. How many grams of C_5H_{12} must be burned to heat 1.00 kg of water from 20.0°C to 90.0°C? Assume that all the heat released during combustion is used to heat the water.

9.104. The heavier (more dense) hydrocarbons in camp stove fuel are hexanes (C_6H_{14}).
 a. Calculate the fuel value of C_6H_{14}, given that $\Delta H°_{comb} = -4163$ kJ/mol.
 b. How much heat is released during the combustion of 1.00 kg of C_6H_{14}?
 c. How many grams of C_6H_{14} are needed to heat 1.00 kg of water from 25.0°C to 85.0°C? Assume that all of the heat released during combustion is used to heat the water.
 d. Assume white gas is 25% C_5 hydrocarbons (see Problem 9.103) and 75% C_6 hydrocarbons; how many grams of white gas are needed to heat 1.00 kg of water from 25.0°C to 85.0°C?

Additional Problems

*9.105. **Hung Out to Dry** Laundry left to dry on a clothesline in freezing weather dries by sublimation instead of vaporization. The increase in internal energy of water produced by sublimation is less than the amount of thermal energy absorbed. Explain why.

9.106. Chlorofluorocarbons (CFCs) such as CF_2Cl_2 are refrigerants whose use has been phased out because of their destructive effect on Earth's ozone layer. The standard heat of evaporation of CF_2Cl_2 is 17.4 kJ/mol. How many grams of liquid CF_2Cl_2 must vaporize in a refrigeration system to cool 200.0 g of water from 50.0 to 40.0°C?

9.107. A 100.0 mL sample of 1.0 M NaOH is mixed with 50.0 mL of 1.0 M H_2SO_4 in a large Styrofoam coffee cup; a thermometer is mounted in the lid of the cup to measure the temperature of the contents. The temperature of each solution before mixing is 22.3°C. After mixing, their temperature reaches 31.4°C. Assume: (1) the density of the mixed solutions is 1.00 g/mL, (2) the specific heat of the mixed solutions is 4.18 J/(g · °C), and (3) no heat is lost to the surroundings.
 a. Write a balanced chemical equation for the reaction that takes place in the cup.
 b. Is any NaOH or H_2SO_4 left in the cup when the reaction is over?
 c. Calculate the enthalpy change per mole of H_2O produced in the reaction.

9.108. With reference to Problem 9.107, what if 65.0 mL of 1.0 M H_2SO_4 is mixed with 100.0 mL of 1.0 M NaOH? Will the increase in temperature be less than, more than, or the same as that measured in Problem 9.107?

*9.109. An insulated container holds 50.0 g of water at 25.0°C. A 7.25 g sample of copper that had been heated to 100.1°C is dropped into the water. What is the final, shared temperature of the copper and the water?

9.110. Magnetite (Fe_3O_4) is magnetic, whereas iron(II) oxide is not.
 a. Write and balance the chemical equation for the formation of magnetite from iron(II) oxide and oxygen.
 b. Given that 318 kJ of heat is released for each mole of Fe_3O_4 formed, what is the enthalpy change of the balanced reaction of formation of Fe_3O_4 from iron(II) oxide and oxygen?

9.111. Endothermic compounds have positive standard heats of formation. An example is acetylene, C_2H_2 ($\Delta H°_f = 226.7$ kJ/mol). Combustion of acetylene in pure oxygen produces a flame hot enough to cut and weld steel.
 a. What is the standard heat of combustion of acetylene?
 b. What is the fuel value of acetylene, assuming the products are CO_2 and H_2O vapor?

9.112. Balance the following chemical equation, name the reactants and products, and calculate the enthalpy change under standard conditions.

$$FeO(s) + O_2(g) \rightarrow Fe_2O_3(s) \qquad \Delta H°_{rxn} = ?$$

*9.113. **Polymer Chemistry** Use appropriate bond energies from Table A4.1 of Appendix 4 to predict whether the reaction in which ethylene forms polyethylene plastic is exothermic, endothermic, or involves no change in enthalpy. The reaction can be written:

$$n \, CH_2{=}CH_2 \rightarrow \text{---}\!\!\left[CH_2\text{---}CH_2\right]\!\!\text{---}_n$$

where the structure in the brackets is the *repeating unit* of polyethylene and the value of n is typically in the thousands.

9.114. **Metabolism of Methanol** In December, 2011, over 100 people died in West Bengal, India, after drinking bootleg (illegal) liquor spiked with methanol. Methanol is toxic because it is metabolized in a two-step process that produces formic acid ($HCOOH$):

$$O_2(g) + 2 \, CH_3OH(aq) \rightarrow 2 \, HCOOH(aq) + 2 \, H_2O(\ell)$$
$$\Delta H^\circ_{rxn} = -1019.6 \text{ kJ}$$

   a. Is the reaction endothermic or exothermic?
   b. What change in enthalpy accompanies the metabolism of 60.0 g of methanol?
   c. In the first step of the process, methanol is converted into formaldehyde (CH_2O), which is then converted into formic acid. Would you expect ΔH°_{rxn} for the metabolism of one mole of methanol to formaldehyde to be larger or smaller than the conversion of one mole of methanol to formic acid?

9.115. In a high-temperature gas-phase reaction, methanol (CH_3OH) reacts with N_2 to produce HCN and NH_3. The reaction is endothermic, requiring 164 kJ of thermal energy per mole of methanol under standard conditions.
   a. Write a balanced chemical equation describing this reaction.
   b. Is energy a reactant or a product?
   c. What is the change in enthalpy under standard conditions if 60.0 grams of $CH_3OH(g)$ reacts with excess $N_2(g)$, HCN(g), and $NH_3(g)$?

9.116. Calculate ΔH°_{rxn} for the reaction

$$2 \, Ni(s) + \tfrac{1}{4} \, S_8(s) + 3 \, O_2(g) \rightarrow 2 \, NiSO_3(s) \qquad \Delta H^\circ_{rxn} = ?$$

   from the following information:

   (1) $NiSO_3(s) \rightarrow NiO(s) + SO_2(g) \qquad \Delta H^\circ_{rxn} = 156 \text{ kJ}$

   (2) $\tfrac{1}{8} \, S_8(s) + O_2(g) \rightarrow SO_2(g) \qquad \Delta H^\circ_{rxn} = -297 \text{ kJ}$

   (3) $Ni(s) + \tfrac{1}{2} \, O_2(g) \rightarrow NiO(s) \qquad \Delta H^\circ_{rxn} = -241 \text{ kJ}$

9.117. Use the information in thermochemical equations (1) through (3) to calculate the value of ΔH°_{rxn} for the reaction in equation (4).

   (1) $Pb(s) + \tfrac{1}{2} \, O_2(g) \rightarrow PbO(s) \qquad \Delta H^\circ_{rxn} = -219 \text{ kJ}$

   (2) $C(s) + O_2(g) \rightarrow CO_2(g) \qquad \Delta H^\circ_{rxn} = -394 \text{ kJ}$

   (3) $PbCO_3(s) \rightarrow PbO(s) + CO_2(g) \qquad \Delta H^\circ_{rxn} = 86 \text{ kJ}$

   (4) $2 \, Pb(s) + 2 \, C(s) + 3 \, O_2(g) \rightarrow 2 \, PbCO_3(s) \qquad \Delta H^\circ_{rxn} = ?$

9.118. **Ethanol as Automobile Fuel** Most of the new cars sold in Brazil are *flex-fuel* vehicles, which means they can run on blends of ethanol and gasoline or on pure ethanol alone. Ethanol is a popular liquid fuel in Brazil because it can be efficiently produced by fermenting sugar extracted from sugarcane. Use appropriate standard heats of formation

to calculate the enthalpy change that accompanies the combustion of one mole of liquid ethanol in which the products are CO_2 and H_2O vapor.

9.119. **Baking soda** ($NaHCO_3$) thermally decomposes to soda ash (Na_2CO_3), CO_2, and H_2O:

$$2 \, NaHCO_3(s) \rightarrow Na_2CO_3(s) + CO_2(g) + H_2O(g)$$

   Use appropriate standard heats of formation to calculate the enthalpy change that accompanies the thermal decomposition of one mole of $NaHCO_3$.

*9.120. **Specific Heats of Metals** In 1819, Pierre Dulong and Alexis Petit reported that the product of the atomic mass of a metal times its specific heat is approximately constant, an observation called the *law of Dulong and Petit*.
   a. Explain how the specific heat and molar heat capacity data for the metals in Table 9.3 support this law.
   b. Use these data to predict the specific heat values of nickel and platinum.

9.121. **Odor of Urine** Urine odor gets worse with time because it contains the metabolic product urea, $CO(NH_2)_2$, a compound that is slowly converted to ammonia, which has a sharp, unpleasant odor, and carbon dioxide:

$$CO(NH_2)_2(aq) + H_2O(\ell) \rightarrow CO_2(aq) + 2 \, NH_3(aq)$$

   This reaction is much too slow for ΔH°_{rxn} to be determined experimentally by measuring a change in temperature, but it can be calculated from the appropriate ΔH°_f values. Calculate ΔH°_{rxn}. [The value of ΔH°_f of $CO(NH_2)_2(aq)$ is −319.2 kJ/mol.]

9.122. Propane is a gas at 1 bar of pressure above −42°C, though it is transported as a liquid under pressure. How does the physical state of propane affect its fuel value and its fuel density?

9.123. **Rocket Fuels** The payload of a rocket includes a fuel and oxygen for combustion of the fuel. Reactions 1 and 2 describe the combustion of dimethylhydrazine and hydrogen, respectively. Pound for pound, which is the better rocket fuel?

   (1) $(CH_3)_2NNH_2(\ell) + 4 \, O_2(g) \rightarrow$
$$N_2(g) + 4 \, H_2O(g) + 2 \, CO_2(g) \qquad \Delta H^\circ_{rxn} = -1694 \text{ kJ}$$

   (2) $H_2(g) + \tfrac{1}{2} \, O_2(g) \rightarrow H_2O(g) \qquad \Delta H^\circ_{rxn} = -241.8 \text{ kJ}$

9.124. At high temperatures, such as those in the combustion chambers of automobile engines, nitrogen and oxygen form nitrogen monoxide:

$$N_2(g) + O_2(g) \rightarrow 2 \, NO(g) \qquad \Delta H^\circ_{comb} = +180 \text{ kJ}$$

   Any NO released into the environment may be oxidized to NO_2:

$$2 \, NO(g) + O_2(g) \rightarrow 2 \, NO_2(g) \qquad \Delta H^\circ_{comb} = -112 \text{ kJ}$$

   Is the overall reaction

$$N_2(g) + 2 \, O_2(g) \rightarrow 2 \, NO_2(g)$$

   exothermic or endothermic? What is ΔH°_{comb} for this reaction?

10

Properties of Gases

The Air We Breathe

An Invisible Necessity

How often do we think about air? We are reminded how important it is when we dive into a pool of water or hike to the top of a tall mountain. Otherwise we don't give much thought to breathing or to the mixture of gases that makes up the air we breathe. This may be because the oxygen we must inhale to stay alive is colorless, tasteless, odorless—and free. Despite our lack of attention to the process, the exchange of gases between our lungs and the surrounding atmosphere is crucial to our lives; air is an invisible necessity.

In a hospital operating room, having the right mixture of gases in our bodies is critical during surgery, which is why anesthesiologists constantly monitor blood levels of oxygen and carbon dioxide. Managing the delicate balance of gases entering and leaving a patient can mean the difference between normal recovery and irreversible coma.

Gases are intimately involved in chemical reactions in living systems as well as in the material world. Most life in our biosphere requires oxygen. Fish, insects, birds, mammals, and even plants must bring O_2 gas into their systems to metabolize nutrients. All the fuels we burn to provide heat and light for our buildings and power our vehicles rely on atmospheric O_2 to support combustion. The products of the complete metabolism of food and the complete combustion of most fuels are CO_2, H_2O, and the energy we need to stay alive and make life easier.

Gases are also important and useful because of their physical properties. We rely on their capacity to expand to fill the space available or to be compressed when subjected to greater pressure. Our ability to understand how the chemical and physical properties of gases fit into our lives is based on our knowledge of how gases behave in response to changing volume, temperature, and pressure. Why, for example, do helium-filled party balloons, or much larger ones filled with hot air, rise? Why must an undersea explorer breathe a custom blend of gases instead of air? What makes the pressure in automobile tires change as we drive or when the seasons change? In this chapter we answer these questions and others as we explore the properties of gases on both the macroscopic and molecular levels.

LEARNING OUTCOMES

LO1 Use kinetic molecular theory to explain the properties of gases

LO2 Use Graham's law to calculate the relative rates of effusion of gases
Sample Exercise 10.1

LO3 Describe how a manometer works and interpret the results obtained with one
Sample Exercise 10.2

LO4 Calculate changes in the volume, temperature, pressure, and number of moles of an ideal gas using gas laws including the ideal gas law
Sample Exercises 10.3, 10.4, 10.5, 10.6, 10.7

LO5 Use the ideal gas law to calculate the density and molar mass of an ideal gas
Sample Exercises 10.8, 10.9

LO6 Use balanced chemical equations to relate the volumes of gas-phase reactants and products using the stoichiometry of the reaction and the ideal gas law
Sample Exercise 10.10

LO7 Relate the mole fraction, partial pressure, and quantity of a gas in a mixture
Sample Exercises 10.11, 10.12

LO8 Calculate the solubility of a gas using Henry's law
Sample Exercise 10.13

LO9 Calculate the root-mean-square speed of a gas from its mass and absolute temperature
Sample Exercise 10.14

LO10 Use the van der Waals equation to correct for nonideal behavior of gases
Sample Exercise 10.15

Scuba Diving The air pressure inside a "full" scuba tank is about 10 times that surrounding a diver 10 meters below the surface. This means the same quantity of gas inside the tank takes up 1/10 the volume it does outside. ▶

FIGURE 10.1 Some behaviors of gases. (a) The volume of a weather balloon expands as it ascends into the sky and experiences decreasing atmospheric pressure with increasing altitude. It eventually bursts, and its instrument package and data record fall back to Earth. (b) The internal pressure of a scuba tank is a direct measure of the amount of air inside it.

10.1 The Properties of Gases

Gases have properties that set them apart from liquids and solids. If we observe how gases behave at temperatures and pressures such as those found in nature or on a laboratory bench, we find that:

1. The volume a gas occupies is inversely proportional to its pressure ($V \propto 1/P$). This is why the volume of a weather balloon increases as it ascends and the pressure of the air surrounding it decreases (Figure 10.1a).

2. The pressure of a quantity of gas in a rigid container is directly proportional to its absolute temperature ($P \propto T$). This property is one reason why cold winter temperatures may set off low-tire-pressure sensors in automobiles: lower tire temperatures mean less air pressure inside, even when no air has leaked out.

3. The gas pressure inside a rigid container at constant temperature is directly proportional to the quantity (number of moles) of gas in the container ($P \propto n$). For example, the gauge on a scuba tank (Figure 10.1b) measures the internal pressure of the tank but also provides a diver with a measure of how much air is left in the tank.

4. All gases are miscible with all other gases; that is, all gas mixtures are homogeneous. This means that when the scuba diver takes a breath of air from his or her tank, the proportion of O_2 in the air is always the same, whether it's the first breath taken at the beginning of a dive or the last one before reaching the surface.

5. At a given temperature and pressure, the densities of gaseous substances are directly proportional to their molar masses and are much lower than their densities as solids or liquids. For example, the density of liquid butane (the fuel in disposable lighters) at its boiling point of 0°C is 223 times greater than the density of butane vapor at 0°C.

6. Gases expand to occupy the entire volume of their containers. This is why the shape of a party balloon is pretty much the same whether it is partially or fully inflated: the gas in the balloon occupies all of the space available to it uniformly, pushing outward with the same pressure in all directions.

7. Gases in inflated party balloons escape at rates that are inversely proportional to their molar masses. Those with the smallest molar masses, such as helium, escape the most rapidly.

In this chapter we investigate these and other properties of gases in more detail, and we explain why gases behave the way they do. Our explanations are based on a theory of how the atoms and molecules of gases move and interact with each other and with the walls of their containers. As we will see, this theory provides microscale explanations of all the macroscale properties listed above.

CONCEPT TEST

Nitrogen dioxide, NO_2, is a red-brown gas that contributes to photochemical smog formation. Which of the images in Figure 10.2 best depicts a sealed container of NO_2?

(Answers to Concept Tests are in the back of the book.)

(a) (b) (c)

FIGURE 10.2

10.2 Effusion and the Kinetic Molecular Theory of Gases

Let's begin our exploration of how and why gases behave the way they do by examining the last of the seven properties of gases listed in Section 10.1. Why is it that helium-filled party balloons, which are made of latex rubber, deflate and are no longer buoyant after a day or two, whereas the same balloons filled with air remain inflated for many days? The answer to this question has two parts: one based on the structure of party balloons and the other based on the motion of the atoms and molecules inside them.

There are tiny imperfections in latex rubber balloons—microscopic holes, sometimes called pinholes—that allow particles of gas to escape (Figure 10.3). Gas leaking through pin holes from a higher pressure space (inside an inflated balloon) to a lower pressure space (outside the balloon) is called **effusion**. This phenomenon was investigated in the early 19th century by Scottish chemist Thomas Graham (1805–1869), who discovered that the rate of effusion of different gases was inversely proportional to the square root of their densities. As noted in property 5 in Section 10.1, the densities of gases are proportional to their molar masses, so the modern interpretation of **Graham's law of effusion** states that the effusion rate of a gas is inversely proportional to the square root of its molar mass:

$$\text{Effusion rate} \propto \sqrt{\frac{1}{\text{density}}} \propto \sqrt{\frac{1}{\text{molar mass}}}$$

Why do gases with lower molar masses effuse faster? This behavior can be explained using the **kinetic molecular theory (KMT)** of gases. KMT is based on a set of assumptions about the particulate nature of gases:

➤ Gas molecules have tiny volumes compared with the collective volume they occupy. Their individual volumes are so small as to be considered negligible, allowing particles in a gas to be treated as *point masses*—masses with essentially no volume. Gas molecules are separated by large distances; hence a gas is mostly empty space. This assumption is consistent with the low density of gases compared to liquids and solids. For example, because liquid butane contains 223 times as many molecules as the same volume of butane vapor at 0°C (as noted earlier in property 5), butane molecules must be spaced much farther apart in the vapor phase.

➤ Because gas particles are so spread out, they don't interact with each other as much as the particles in liquids and solids do. In fact, KMT assumes they don't interact *at all*.

➤ Gas molecules move constantly and randomly throughout the space they occupy, continually colliding with one another and with their container walls (Figure 10.4). These collisions are *elastic*. To get a mental picture of what an elastic collision is like, imagine a cue ball colliding with one or more stationary balls on a pool table (Figure 10.5). Most of the balls exit the collision in different directions at different speeds. Some of the kinetic energy of the cue ball has transferred to the others, but the *average* kinetic energy of all of the balls is the same immediately after the collision as it was before.

➤ The average kinetic energy of the molecules in a gas is proportional to the absolute temperature of the gas. All populations of gas molecules at the same temperature have the same average kinetic energy.

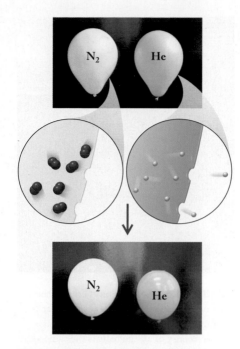

FIGURE 10.3 Two balloons at the same temperature and pressure, one filled with nitrogen gas and the other filled with an equal volume of helium gas. Over time, the volume of the helium balloon decreases much faster.

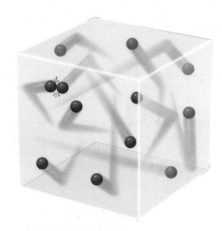

FIGURE 10.4 Gas molecules continually collide with one another and with their container walls.

effusion the process by which a gas escapes from its container through a tiny hole into a region of lower pressure.

Graham's law of effusion the rate of effusion of a gas is inversely proportional to the square root of its molar mass.

kinetic molecular theory (KMT) a model that explains the behavior of gases based on the motion of the particles that make them up.

FIGURE 10.5 The collisions between the balls on a pool table are a good model for elastic collisions, which means the average kinetic energy of all the balls is the same after a collision as it was before.

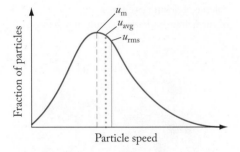

FIGURE 10.6 At any given temperature, the speeds of the particles that make up a gaseous substance cover a range of values. The dashed line represents the most probable particle speed (u_m); the dotted line is the arithmetic average speed (u_{avg}); and the solid line is the root-mean-square speed (u_{rms}), which is directly proportional to the average kinetic energy of the particles.

▶❙❙ **CHEMTOUR** Molecular Motion

Recall from Chapter 1 that the kinetic energy of any object in motion—including a gas-phase atom or molecule—is proportional to its mass (m) and the square of its speed (u):

$$KE = \tfrac{1}{2}mu^2 \qquad (1.2)$$

We use the term *average* kinetic energy in describing the behavior of molecules of a gas because the speeds of particles with the same mass and at the same temperature are not all the same. Instead they are distributed over a range of values, as shown by the graph in Figure 10.6. The *x*-axis of the graph represents particle speed, and the *y*-axis records the relative number of particles that have that speed. The peak in the speed distribution curve represents the *most probable speed* (u_m). It is the speed that the largest number of particles have. Because the distribution of speeds is not symmetrical, the *average speed* (u_{avg}), which is simply the arithmetic average of the individual speeds of all the particles, is a little higher than the most probable speed. A third kind of average speed is called the **root-mean-square speed (u_{rms})**. Its name comes from its theoretical definition: the square root of the average of the squared speeds of all the particles of a gas. It is important to us because u_{rms} is the speed of a particle whose kinetic energy is exactly the same as the *average* kinetic energy of all the particles in a gas. The equation form of this definition is

$$KE_{avg} = \tfrac{1}{2}mu_{rms}^2 \qquad (10.1)$$

We can use Equation 10.1 to compare the u_{rms} speeds of different gases at the same temperature. For example, suppose we have a mixture of O_2 ($\mathcal{M}$ = 32.00 g/mol), N_2 ($\mathcal{M}$ = 28.02 g/mol), He ($\mathcal{M}$ = 4.00 g/mol), and H_2 ($\mathcal{M}$ = 2.02 g/mol). Because all four gases have the same temperature, their particles all have the same average kinetic energies. If KE_{avg} is the same for all four gases, then for each of them the product $\tfrac{1}{2}mu_{rms}^2$ is also the same. However, the masses of their particles are not the same, which means their u_{rms} speeds cannot be the same, either: the least massive particles (molecules of H_2) should have the highest u_{rms} speeds, and the most massive particles (molecules of O_2) should have the slowest u_{rms} speeds. This prediction is supported by the speed distribution curves for these four elements in Figure 10.7. Note how particle speeds are inversely related to particle masses and that the least massive particles (atoms of He and molecules of H_2)

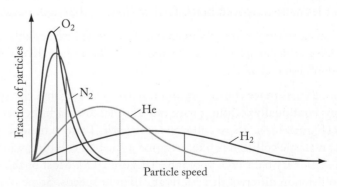

FIGURE 10.7 The speeds of gas particles at a given temperature are inversely related to their masses: the most massive (molecules of O_2 in this case) move, on average, the slowest, and the least massive (molecules of H_2) move the fastest and have the most widely distributed speeds.

have not only the highest average speeds but also the most widely distributed speed profiles.

Now let's use Equation 10.1 to make a more quantitative link between particle speed and particle mass. We start with a mixture of a heavier gas (N_2) and a lighter gas (He) from Figure 10.7. The mixture has a single temperature, so the two gases have the same average kinetic energies:

root-mean-square speed (u_{rms}) the square root of the average of the squared speeds of all the particles in a population of gas particles; a particle possessing the average kinetic energy moves at this speed.

$$KE_{avg,He} = KE_{avg,N_2} \qquad (10.2)$$

Substituting the right side of Equation 10.1 into both sides of Equation 10.2:

$$\tfrac{1}{2}m_{He}u_{rms,He}^2 = \tfrac{1}{2}m_{N_2}u_{rms,N_2}^2 \qquad (10.3)$$

Rearranging the terms in Equation 10.3 so that the masses are on one side of the equation and u_{rms} speeds are on the other:

$$\frac{u_{rms,He}^2}{u_{rms,N_2}^2} = \frac{m_{N_2}}{m_{He}}$$

Taking the square root of both sides:

$$\frac{u_{rms,He}}{u_{rms,N_2}} = \sqrt{\frac{m_{N_2}}{m_{He}}} \qquad (10.4)$$

Now let's scale up the mass values on the right side of Equation 10.4 by multiplying both of them by Avogadro's number. Doing so converts molecular masses into molar masses without changing the value of their ratio:

$$\frac{u_{rms,He}}{u_{rms,N_2}} = \sqrt{\frac{\mathcal{M}_{N_2}}{\mathcal{M}_{He}}} = \sqrt{\frac{28.02 \ \cancel{g/mol}}{4.00 \ \cancel{g/mol}}} = 2.65 \qquad (10.5)$$

Equation 10.5 tells us that the root-mean-square speeds of the two gases are inversely proportional to the square roots of their molar masses. Therefore, atoms of He collide with the walls of their container (such as the green party balloon in Figure 10.3) 2.65 times more frequently than molecules of N_2 at the same pressure and temperature. This means He atoms encounter and pass through atomic-scale escape routes in their balloon 2.65 times more frequently than N_2 molecules do. Therefore, it is only reasonable that He should effuse 2.65 times faster than N_2, and that a helium-filled balloon should deflate much faster than one filled with air (mostly N_2).

Equation 10.5 can be used with any pair of gases that behave as predicted by KMT. We can write a generic form of it for any two gases A and B:

$$\frac{u_{rms,A}}{u_{rms,B}} = \sqrt{\frac{\mathcal{M}_B}{\mathcal{M}_A}} \qquad (10.6)$$

If particles of gas A have higher speeds than particles of gas B, it is only logical that gas A particles collide more frequently with the walls of their container, which increases the frequency with which they encounter and pass through microscopic holes in the wall and undergo effusion. Expressing this logic in equation form:

$$\frac{\text{effusion rate}_A}{\text{effusion rate}_B} = \sqrt{\frac{\mathcal{M}_B}{\mathcal{M}_A}} \qquad (10.7)$$

Equation 10.7 is a mathematical representation of Graham's law that can be used to calculate the relative rates of effusion of pairs of gases, as illustrated in the following Sample Exercise.

TABLE 10.1	Composition of Dry Air[a]
Compound	**% (by volume)**
Nitrogen	78.08
Oxygen	20.95
Argon	0.934
Carbon dioxide	0.0395[b]
Neon	0.0018
Helium	0.00052
Methane	0.00018
Krypton	0.00011

[a]Includes major and minor gases (with concentrations >1 ppm by volume).
[b]Value as of January 2013. Atmospheric CO_2 is increasing by about 2 ppm each year.

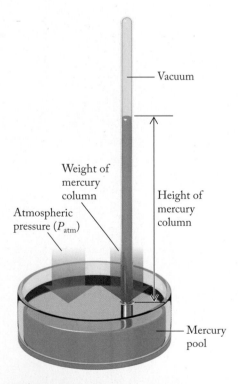

FIGURE 10.8 The height of the mercury column in this simple barometer designed by Evangelista Torricelli is proportional to atmospheric pressure.

Labels in figure: Vacuum; Weight of mercury column; Height of mercury column; Atmospheric pressure (P_{atm}); Mercury pool

SAMPLE EXERCISE 10.1 Calculating Relative Rates of Effusion **LO2**

An odorous gas emitted by a hot spring was found to effuse at 0.342 times the rate at which helium effuses. What is the molar mass of the emitted gas?

COLLECT AND ORGANIZE We are asked to determine the molar mass of an unknown gas based on its rate of effusion relative to the rate for helium ($\mathcal{M}_{He} = 4.003$ g/mol).

ANALYZE Equation 10.7 provides the mathematical relationship between the effusion rates and molar masses of a pair of gases. Because the unknown gas effuses about 1/3 as fast as He does, the molar mass should be nearly 3^2 or 9 times the mass of He, or about $9 \times 4 = 36$ g/mol.

SOLVE Using the symbol X for the unknown gas in Equation 10.7:

$$\frac{\text{effusion rate}_X}{\text{effusion rate}_{He}} = \sqrt{\frac{\mathcal{M}_{He}}{\mathcal{M}_X}} = 0.342$$

Squaring both sides of the rightmost equality yields

$$\frac{\mathcal{M}_{He}}{\mathcal{M}_X} = (0.342)^2$$

$$\mathcal{M}_X = \frac{4.003 \text{ g/mol}}{(0.342)^2} = 34.2 \text{ g/mol}$$

THINK ABOUT IT The molar mass of the unidentified gas is 34.2 g/mol, which is consistent with our prediction. One possibility for the identity of this gas is $H_2S(g)$, $\mathcal{M} = 34.1$ g/mol, a toxic gas with a foul smell associated with rotten eggs (or low tide on a seaweed-covered shoreline).

Practice Exercise Helium effuses 3.16 times as fast as which other noble gas?

(Answers to Practice Exercises are in the back of the book.)

10.3 Atmospheric Pressure

In this section we continue our exploration of the behavior of gases by focusing on the ones we take into our lungs with each breath. Earth is surrounded by an atmosphere that is about 50 km thick and composed primarily of nitrogen (78%), oxygen (21%), and lesser proportions of other gases (Table 10.1). To put the thickness of the atmosphere in perspective, if Earth were the size of an apple, the atmosphere would be about as thick as the apple's skin.

As we discussed in Chapter 6, Earth's atmosphere is pulled downward by gravity and exerts a force that is spread across the entire surface of the planet. The ratio of force, F, to surface area, A, defines atmospheric pressure P:

$$P = \frac{F}{A} \tag{6.1}$$

Atmospheric pressure is measured with an instrument called a **barometer**. A simple but effective barometer design consists of a narrow tube about 1 meter long, sealed at one end and filled with mercury. The tube is inverted, and its open end is placed into a pool of mercury that is open to the atmosphere (Figure 10.8). Gravity pulls downward on the mercury in the tube, creating a vacuum at the top of the tube. At the same time, atmospheric pressure pushes downward on the mercury in the pool, which has the effect of pushing it up into the tube. The opposing forces on the mercury column in the tube create a stable column height that provides a measure of atmospheric pressure.

Atmospheric pressure varies from place to place and with changing weather conditions. As we noted in Chapter 6, the average atmospheric pressure at sea level defines the standard atmosphere (1 atm) of pressure. This pressure is capable of supporting a column of mercury 760 mm high, which is the basis for another non-SI unit of pressure: millimeters of mercury (mmHg). Pressure in millimeters of mercury is also expressed in a unit called the torr in honor of Evangelista Torricelli (1608–1647), the Italian mathematician and physicist who invented the barometer. Thus,

$$1 \text{ atm} = 760 \text{ mmHg} = 760 \text{ torr}$$

Other non-SI units for expressing the pressure of gases are derived from masses and areas in the U.S. Customary System, such as pounds per square inch (psi) and bars. Relationships between different units of pressure are summarized in Table 10.2.

The SI unit of pressure is the pascal (Pa), named in honor of French mathematician and physicist Blaise Pascal (1623–1662), who was the first to propose that atmospheric pressure decreases with increasing altitude. We can explain this phenomenon by noting that the atmospheric pressure at any given location on Earth's surface is related to the mass of the column of air *above* that location. As altitude increases, the mass of the column of air *above* that altitude decreases. Less mass means a smaller force exerted downward by the air at higher altitude, which means less pressure (Figure 10.9).

barometer an instrument that measures atmospheric pressure.

CONNECTION In Chapter 6 we discussed how the average value of Earth's atmospheric pressure is the ratio of the force exerted by the atmosphere to the surface area of the planet.

TABLE 10.2 **Units for Expressing Pressure**

Unit	Value
Standard atmosphere (atm)	1 atm
Pascal (Pa)	1 atm = 1.01325×10^5 Pa
Kilopascal (kPa)	1 atm = 101.325 kPa
Millimeter of mercury (mmHg)	1 atm = 760 mmHg
Torr	1 atm = 760 torr
Bar	1 atm = 1.01325 bar
Millibar (mbar or mb)	1 atm = 1013.25 mbar
Pounds per square inch (psi)	1 atm = 14.7 psi
Inches of mercury	1 atm = 29.92 inches of Hg

FIGURE 10.9 Atmospheric pressure decreases with increasing altitude because the mass of the column of air above a given area decreases with increasing altitude.

manometer an instrument for measuring the pressure exerted by a gas.

Scientists conducting experiments with gases usually need to know the pressures of gases in closed systems. A **manometer** is an instrument that has long been used to measure these pressures. The two common types of manometers, called *closed end* and *open end*, are illustrated in Figure 10.10. Each contains a U-shaped tube filled with mercury (or another dense liquid). One end of the tube is connected via a valve to a sample vessel containing the gas of interest. The difference between the two manometers is whether the other end of the tube is closed or open to the atmosphere. Closed-end manometers are particularly useful for measuring pressures that are less than atmospheric pressure, as shown in Figure 10.10(a) and (b). Before the manometer is connected to the sample,

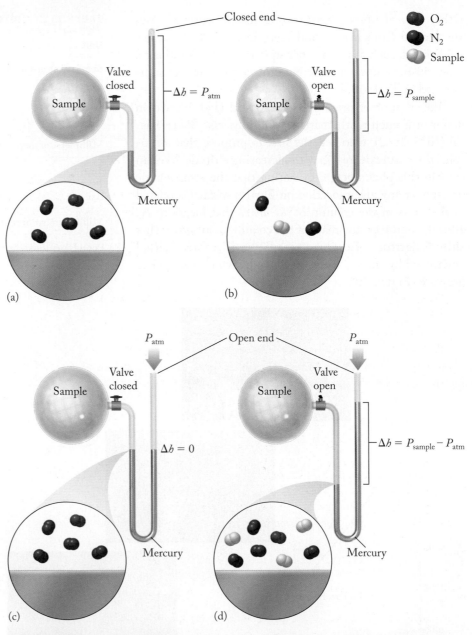

FIGURE 10.10 (a and b) A closed-end manometer is designed for measuring sample pressures that are less than atmospheric pressure. (c and d) A manometer open to the atmosphere is designed for measuring sample pressures that are greater than atmospheric pressure.

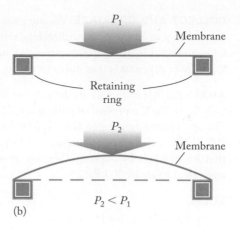

P_1

Membrane

Retaining ring

P_2

Membrane

$P_2 < P_1$

(a)   (b)

FIGURE 10.11 (a) The pressure sensor in this barograph is a partially evacuated corrugated metal can. (b) As atmospheric pressure decreases, the lid of the can distorts outward. This motion is amplified by a series of levers and transmitted via the horizontal arm to a pen tip that records the pressure on the graph paper as the drum slowly turns. A week's worth of barometric data can be recorded in this way.

the difference in the height of the mercury in the two arms of the tube (Δh) is a direct measure of atmospheric pressure. In this way the manometer initially acts like a barometer. The manometer is then connected to the sample while the valve between them is kept closed (Figure 10.10a). When the valve is opened, the lower pressure of the sample gas sucks air from the manometer up into the flask, which raises the level of mercury in the left-hand side of the manometer, lowers it on the right side, and produces a smaller Δh, which is a direct measure of the pressure in the sample vessel (Figure 10.10b).

Open-end manometers are particularly useful for determining gas pressures greater than atmospheric pressure. Initially both sides of the tube are open to the atmosphere and there is no difference in their mercury levels. After the manometer is connected to the sample vessel with its valve closed (Figure 10.10c), the valve is opened and the higher pressure of the sample gas pushes down on the mercury in its side of the tube. This change creates a Δh that is a measure of the difference between the sample pressure and atmospheric pressure (Figure 10.10d).

Today manometers have been largely replaced by electronic pressure sensors based on flexible metallic or ceramic diaphragms. As the pressure on one side of the diaphragm increases, it distorts away from that side. This is the same mechanism used to sense changes in atmospheric pressure in most barometers, including the recording barometer, or *barograph*, shown in Figure 10.11.

SAMPLE EXERCISE 10.2 **Measuring Gas Pressure with a Manometer** **LO3**

For centuries, kilns such as the one in Figure 10.12 have been used to convert limestone ($CaCO_3$) into quicklime (CaO), which is used to "sweeten" (reduce the acidity of) soil in agriculture and to make mortar and concrete. High temperatures inside the kilns drive the endothermic decomposition of $CaCO_3$:

$$CaCO_3(s) \rightarrow CaO(s) + CO_2(g) \qquad \Delta H° = 178.1 \text{ kJ/mol}$$

Suppose a chemist studying the decomposition reaction places a quantity of $CaCO_3$ in a flask, pumps the air out of the flask with a vacuum pump, and then connects the flask to a closed-end manometer as shown in Figure 10.13(a). Then the flask is heated to a temperature at which the $CaCO_3$ decomposes. After the source of heat is removed and the flask has returned to its initial temperature, the difference in levels of mercury (Δh) in the arms of the manometer is 144 mm (Figure 10.13b). Calculate the pressure of the CO_2 in (a) torr, (b) atmospheres, and (c) kilopascals.

FIGURE 10.12 Historic lime kiln.

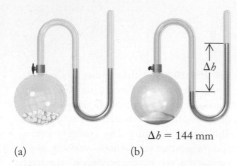

(a)   (b)

$\Delta h = 144$ mm

FIGURE 10.13 Apparatus for monitoring the thermal decomposition of $CaCO_3$ using a closed-end manometer. (a) An evacuated flask containing chunks of $CaCO_3$. (b) The same setup after a decomposition reaction that releases CO_2 into the flask.

FIGURE 10.14 Apparatus for monitoring the thermal decomposition of $CaCO_3$ using an open-end manometer.

COLLECT AND ORGANIZE We are given the difference in height of the mercury column in a manometer in millimeters, which we need to convert to the pressure of CO_2 produced by the decomposition reaction. We are asked to express this pressure using three different sets of units: torr, atm, and kPa.

ANALYZE The difference in the mercury levels is a direct measure of the pressure of CO_2 in the flask expressed in millimeters of mercury, which has the same value in torr. The appropriate conversion factors, found in Table 10.2, are 1 atm = 760 mmHg = 101.325 kPa. The measured pressure of 144 mm is about 1/5 of 760, so we can estimate that it will be about 0.2 atm. There are about 100 kPa in 1 atmosphere, so 0.2 atm is equivalent to about 20 kPa.

SOLVE

a. Converting mmHg to torr:

$$144 \; \text{mmHg} \times \frac{1 \; \text{torr}}{1 \; \text{mmHg}} = 144 \; \text{torr}$$

b. Converting torr to atmospheres:

$$144 \; \text{torr} \times \frac{1 \; \text{atm}}{760 \; \text{torr}} = 0.189 \; \text{atm}$$

c. Converting atmospheres to kilopascals:

$$0.189 \; \text{atm} \times \frac{101.325 \; \text{kPa}}{1 \; \text{atm}} = 19.2 \; \text{kPa}$$

THINK ABOUT IT The calculated values are close to our estimated values. All are expressed to three significant figures because there were three in the initial manometer measurement of 144 mmHg.

Practice Exercise Suppose the closed-end manometer in Figure 10.13 is replaced with an open-end manometer (Figure 10.14), and the same quantity of $CaCO_3$ is roasted as in Sample Exercise 10.2. If atmospheric pressure and the pressure inside the flask are 760 torr at the start of the experiment, what will be the value of Δh in millimeters of mercury at the end?

10.4 Relating *P*, *T*, and *V*: The Gas Laws

In Section 10.1, we summarized some of the properties of gases and described the effect of pressure and temperature on volume, mostly in qualitative terms. Our knowledge of the quantitative relationships among *P*, *T*, and *V* goes back more than three centuries, to a time before the field of chemistry as we know it even existed. Some of the experiments that led to our understanding of how gases behave were inspired by people's fascination with manned flight and the development of the first hot-air balloons that were big enough (and safe enough) to carry passengers in the 18th century (Figure 10.15).

Boyle's Law: Relating Pressure and Volume

Gases are compressible, which means that large quantities of gases can be transported in relatively small cylinders under high pressure. The inverse relation between the pressure and volume of a fixed quantity of gas (as in number of

Boyle's law the volume of a gas at constant temperature is inversely proportional to its pressure.

moles, *n*) at constant temperature was discovered by British chemist Robert Boyle (1627–1691) and is known as **Boyle's law**:

$$P \propto \frac{1}{V} \qquad \text{(at constant } n \text{ and } T) \qquad (10.8)$$

In other words, as the pressure on a quantity of gas at a constant temperature increases, the volume of the gas decreases; conversely, as the pressure on the gas decreases, its volume increases. In his experiments, Boyle used a J-shaped tube similar in construction to a closed-end manometer. He could increase the pressure on the gas trapped in the closed end by pouring mercury into the open end. With no mercury in the tube (Figure 10.16a), the pressure on the air inside was simply atmospheric pressure (let's assume it was 760 mmHg). When just enough mercury was added to fill up the bottom of the J-tube (Figure 10.16b), the pressure on the trapped air was still 760 mmHg. However, adding more mercury increased the pressure on the trapped air by an amount equal to the difference in height (Δ*h*) of the mercury in the two sides of the tube (Figure 10.16c). The total pressure on the air was 760 mmHg plus Δ*h*. As the total pressure increased, the volume of the trapped air decreased as predicted by Equation 10.8.

We can turn Equation 10.8 from a proportionality into a true equation by multiplying its right side by a constant:

$$P = (\text{constant})\frac{1}{V}$$

or

$$PV = \text{constant} \qquad (10.9)$$

where the value of the constant depends on how much air is trapped (*n*) and its temperature (*T*). If *PV* is a constant, then the product of the pressure and volume of a given quantity of gas under one set of conditions, that is, $P_1 \times V_1$, has the same value under any other set of conditions, say, $P_2 \times V_2$, as long as temperature remains the same. Putting this equality in equation form gives us a handy mathematical expression of Boyle's law:

$$P_1V_1 = P_2V_2 \qquad (10.10)$$

FIGURE 10.15 Hot-air balloons then and now. Fascination with hot-air ballooning in the late 18th and early 19th centuries led to important discoveries about the properties of gases.

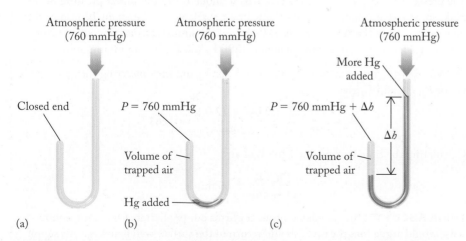

(a) (b) (c)

FIGURE 10.16 Boyle used a J-shaped tube for his experiments on the relation between pressure and volume. The volume of the air trapped in the closed end of the tube decreased as more mercury was added and the difference in the height of mercury in the two sides of the tube—a measure of the pressure on the trapped gas—increased.

CONCEPT TEST

Which of the graphs in Figure 10.17 correctly describes the relationship between the product of pressure and volume (PV) as a function of pressure (P) for a given quantity of gas at constant temperature?

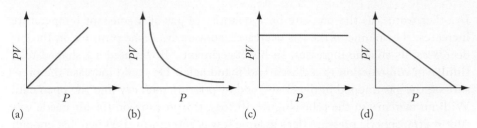

(a) (b) (c) (d)

FIGURE 10.17

SAMPLE EXERCISE 10.3   Applying Boyle's Law   LO4

A popular scuba tank for sport diving has an internal volume of 12.0 liters and can be filled with air up to a pressure of 232 bar. Suppose a diver consumes air at the rate of 21 L/min while diving on a coral reef where the sum of atmospheric pressure (1.0 bar) and water pressure averages 2.2 bar. How long will it take the diver to use up a full tank of air? The temperature of the air and water on the reef are the same: 28°C.

COLLECT AND ORGANIZE We know that the diver's scuba tank contains 12.0 L of compressed air at a pressure of 232 bar. The diver inhales air from the tank at a rate of 21 L/min and an ambient pressure of 2.2 bar. We are asked to calculate how long it will take the diver to use up the air in the tank. The temperature of the gas does not change.

ANALYZE We know initial values of the pressure (232 bar) and volume (12.0 L) of the air in a scuba tank, that is, we know P_1 and V_1. The air is consumed at an ambient pressure (P_2) of 2.2 bar. We can use Boyle's law to calculate the volume of air available to the diver at the lower pressure, that is, V_2. Dividing this volume by the rate at which it is consumed will provide us with the time it takes for the air in the tank to be consumed. The pressure the diver experiences (2.2 bar) is about 1/100 the initial pressure in the tank, so we can estimate that the volume of air available will be about 100 times the internal volume of the tank, or about 1200 L. It is consumed at a rate close to 20 L/min, which translates into a dive time of about (1200 L)/(20 L/min), or 60 minutes.

SOLVE Rearranging Equation 10.10 to solve for V_2 and then inserting the given values of P_1, P_2, and V_1 gives us

$$V_2 = \frac{P_1 V_1}{P_2} = \frac{232 \text{ bar} \times 12.0 \text{ L}}{2.2 \text{ bar}} = 1265 \text{ L}$$

Calculating the time to consume 1265 L of air:

$$\frac{1265 \text{ L}}{21 \text{ L/min}} = 60 \text{ min}$$

THINK ABOUT IT The calculated result confirms our prediction. Of course, common sense would argue that the diver should return to the surface well before the air supply is completely consumed. Note that the question specified the temperature of the air and water (28°C), but the value was not used in the calculation. The importance of the single value is that the temperature of the air in the tank stayed the same during the dive, which allowed us to use Boyle's law and Equation 10.10 to solve the problem.

Charles's law the volume of a gas at constant pressure is directly proportional to its absolute temperature.

absolute temperature temperature expressed on the Kelvin scale whose zero value is absolute zero, the lowest possible temperature.

Practice Exercise As the scuba diver in Sample Exercise 10.3 exhales a lungful of air (see the photo at the beginning of this chapter), by what percentage does the volume of exhaled air increase as it bubbles up to the surface of the ocean? Assume the temperature of the air bubbles does not change on their way to the surface.

How can Boyle's law be explained using kinetic molecular theory? To answer this question, consider the effect of compressing a collection of gas particles into a smaller space—as happens, for example, when the piston in Figure 10.18 moves up in the cylinder. Kinetic molecular theory asserts that particles of gas are in constant random motion, meaning that they are constantly colliding with one another and with the interior surface of their container. Their collisions with the interior surface are responsible for the pressure exerted by the gas. If the particles are squeezed into a smaller volume of space, then the density of the particles is greater and the frequency with which they collide with the walls of their container is greater. The more frequent the collisions, the greater the force exerted by the particles per unit of interior surface area, and the greater the pressure. The converse is also true: as the volume of a container of gas increases, collision frequency decreases, and pressure drops.

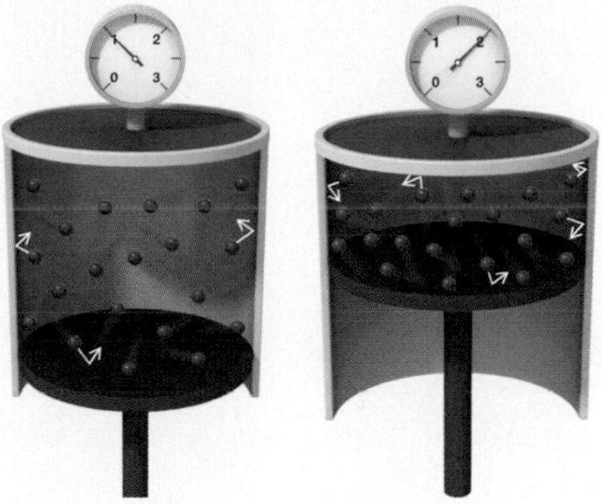

FIGURE 10.18 Gas particles are in constant motion and exert pressure through collisions with the interior surface of their container. When a quantity of gas is squeezed into half its original volume, the particles collide more frequently with their container, causing pressure to double.

Charles's Law: Relating Volume and Temperature

Nearly a century after Boyle's discovery of the inverse relation between the pressure exerted by a gas and the volume of the gas, French scientist Jacques Charles (1746–1823) documented the linear relation between the volume and temperature of a constant quantity of gas at constant pressure. Now known as **Charles's law**, the relation states that, when the pressure exerted on a gas is held constant, the volume of a fixed quantity of gas is directly proportional to the **absolute temperature**, the temperature expressed in kelvin, of the gas:

CONNECTION In Chapter 1, we defined a temperature on the Kelvin scale as equal to the temperature in degrees Celsius plus 273.15: $T(\text{K}) = T(°\text{C}) + 273.15$.

$$V \propto T \qquad \text{(at constant } P \text{ and } n\text{)} \tag{10.11}$$

The effects of Charles's law can be seen in Figure 10.19, in which a balloon has been attached to a flask, trapping a fixed amount of gas in the apparatus. Heating the flask causes the gas to expand, inflating the balloon. As with Boyle's law, an equals sign can replace the proportionality symbol in Equation 10.11 if we include a constant:

$$V = (\text{constant}) \times T$$

or,

$$\frac{V}{T} = \text{constant} \tag{10.12}$$

The value of the constant depends on the number of moles of gas in the sample (n) and on the pressure of the gas (P). When these two parameters are held constant, the ratio V/T does not change, and any two combinations of volume and temperature are related as follows:

$$\frac{V_1}{T_1} = \frac{V_2}{T_2} \tag{10.13}$$

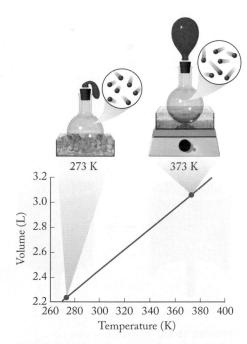

FIGURE 10.19 A balloon attached to a flask inflates as the temperature of the gas inside the flask increases from 273 K to 373 K at constant pressure. This behavior is described by Charles's law.

We can rearrange the terms in Equation 10.13 to obtain another expression of the proportionality between the volume of a gas and its absolute temperature:

$$\frac{V_1}{V_2} = \frac{T_1}{T_2} \tag{10.14}$$

SAMPLE EXERCISE 10.4 **Applying Charles's Law** **LO4**

Several students at a northern New England campus are hosting a party celebrating the mid-January start of "spring" semester classes. They decide to decorate the front door of their apartment building with party balloons. The air in the inflated balloons is initially 70°F. After an hour outside, the temperature of the balloons is –12°F. Assuming no air leaks from the balloons and that the pressure in them does not change significantly, how much does their volume change? Express your answer as a percentage of initial volume.

COLLECT AND ORGANIZE We are asked to calculate the change in volume that accompanies a temperature change from 70°F to –12°F. According to Charles's law, the volume of a fixed quantity of a gas at constant pressure is proportional to the *absolute* temperature of the gas. The inside back cover of the book contains conversion factors for different temperature scales: $T(°\text{F}) = \frac{9}{5}T(°\text{C}) + 32$; $T(\text{K}) = T(°\text{C}) + 273.15$. After rearranging and combining, we can get an equation to convert Fahrenheit temperatures into kelvin: $T(\text{K}) = \frac{5}{9}[T(°\text{F}) - 32] + 273.15$.

ANALYZE There is more than one way to express Charles's law mathematically. Because we are given temperature values and need to calculate the corresponding ratio of volumes, Equation 10.14 is particularly well suited to our calculation:

$$\frac{V_1}{V_2} = \frac{T_1}{T_2}$$

Before we can use Equation 10.14, we will have to convert Fahrenheit temperatures to degrees Celsius and kelvin. We can estimate that the given temperatures are approximately 20° above and 20° below zero on the Celsius scale, or about 293 and 253 K. The ratio of the smaller to the larger temperatures (253/293) is a little more than 5/6, or about 85%. Therefore, the change in volume should be about –15%.

SOLVE Converting 70°F and –12°F into kelvin:

$$\frac{5}{9}(70°\text{F} - 32) + 273.15 = 294.3 \text{ K}$$

$$\frac{5}{9}(-12°\text{F} - 32) + 273.15 = 248.7 \text{ K}$$

Letting T_1 be the lower temperature, the value of the right side of Equation 10.14 is

$$\frac{T_1}{T_2} = \frac{248.7 \text{ K}}{294.3 \text{ K}} = 0.8451 = 84.5\%$$

This is the same value as the ratio of the volumes (V_1/V_2) on the left side, so the answer we seek is the difference between 100% and 85%, or –15%.

THINK ABOUT IT A substantial decrease in temperature (from a human's creature comfort perspective) translates into a relatively small decrease in volume, as we predicted. The small change makes sense because the difference in temperature on the Kelvin scale is the difference between two large values close to one another in value, which means their ratio is not very far from one.

Practice Exercise Hot, expanding gases can be used to perform useful work in a cylinder fitted with a movable piston, as in Figure 10.20. If the temperature of a gas confined to such a cylinder is raised from 245°C to 560°C, what is the ratio of the initial volume to the final volume if the pressure inside the cylinder remains constant? ⚙

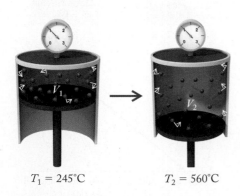

$T_1 = 245°\text{C}$ $T_2 = 560°\text{C}$

FIGURE 10.20 A molecular view of Charles's law. The volume of a gas increases when its temperature increases and pressure stays the same.

How does kinetic molecular theory explain Charles's law? Temperature is directly related to molecular motion. The average speed at which gas-phase particles move increases with increasing temperature. For a given number of particles, increasing temperature increases their motion and therefore increases the frequency and average kinetic energy with which they collide with the walls of their container. More frequent, higher energy collisions would produce an increase in pressure if volume stayed the same. However if pressure is to remain constant—as required by Charles's law—then the volume must expand so that the faster moving particles have to travel farther to collide with the walls of their container. Thus, a higher temperature means more energetic particle–wall collisions, but a larger volume means that these collisions happen less frequently so that, overall, pressure does not change.

Charles's law can also be viewed from a thermodynamic perspective. Suppose a gaseous system is heated, which raises the average kinetic energy of the particles in the system and the internal energy (*E*) of the system itself. To maintain constant pressure, the system expands by an amount Δ*V* against a constant opposing pressure *P*, thereby doing *P*–*V* work, which lowers the internal energy of the system.

Charles's law made possible the experimental determination that absolute zero (0 K) is equal to –273.15°C. Consider what happens when we plot the volumes of fixed quantities of gases at constant pressures as a function of temperature, as in Figure 10.21. All three curves show a linear relationship between volume and temperature, but we are limited by how cold we can make the gases before we reach their boiling points, that is, the temperatures at which the gases condense. However, we can *extrapolate* from our experimental data points to the temperature at which the volume of a sample of gas would reach zero if condensation did not occur. The dashed lines in Figure 10.21 all cross the temperature axis (*V* = 0) at –273.15°C, which is 0.00 K, or absolute zero. Thus, the volume of any gas that obeys Charles's law is directly proportional to its absolute temperature at a given pressure.

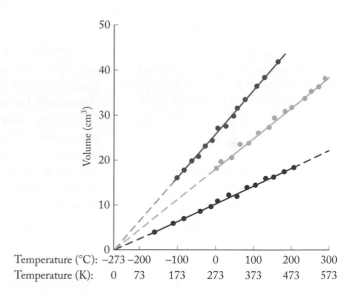

FIGURE 10.21 A plot of the volumes of three gases at constant pressure versus temperature. When the curves are extrapolated to *V* = 0, the corresponding temperatures are all –273.15°C on the Celsius scale, or 0 K (absolute zero).

Avogadro's Law: Relating Volume and Quantity of Gas

Boyle's and Charles's laws apply to isolated systems that contain unchanging quantities of gases. However, many physical and chemical systems are open systems in which the quantities of gases *do* change. An example of such a system is the air in a scuba tank. As we noted in Section 10.1, the pressure of a scuba tank during a dive is a reliable indicator of how much air is left in the tank because when temperature and volume are constant, the pressure (*P*) of a gas, or of a mixture of gases, is proportional to the quantity (moles, *n*) of gas in its container:

$$P \propto n \quad \text{(at constant } V \text{ and } T)$$

or

$$\frac{P}{n} = \text{constant} \tag{10.15}$$

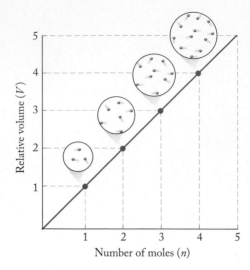

FIGURE 10.22 The volume of a gas at constant temperature and pressure is directly related to the quantity (number of moles) of the gas, a relationship known as Avogadro's law.

⦿⦿ **CONNECTION** The number of particles in a mole is called Avogadro's number (also called Avogadro's constant) in recognition of his early work with gases that led to the determination of atomic masses, as described in Chapter 2.

Kinetic molecular theory can be used to explain this proportionality between P and n: as the number of particles of a gas increases, the frequency of particle collisions with the walls of the container increases, assuming volume and temperature are constant. For example, if the number of particles doubles, the number of collisions doubles. More collisions mean higher pressure.

Some containers (balloons, for example) are not rigid. When gas is added to them, their volumes increase. If temperature and pressure remain constant, then the increase in volume (V) is directly proportional to the quantity (moles, n) of gas added. Expressing this relationship mathematically:

$$V \propto n \qquad \text{(at constant } P \text{ and } T)$$

or

$$\frac{V}{n} = \text{constant} \qquad (10.16)$$

Amedeo Avogadro (1776–1856), whom we know from Avogadro's number (N_A), first recognized that the volume of a gas is proportional to the number of particles in it at constant temperature and pressure, and this relationship is called **Avogadro's law** in his honor. Avogadro's law is also explained by the kinetic molecular theory. As we discussed above, adding particles of gas to a container increases the number of them per unit volume, which results in an increase in the frequency of particle collisions with the walls of the container, which increases pressure—*unless* the volume of the container increases. If the volume increases enough to maintain the same density of particles and the same particle collision frequency, then pressure will remain constant. Therefore, pressure remains constant as long as the volume of the system is proportional to the number of particles of gas in it, as shown in Figure 10.22.

CONCEPT TEST

Which graph in Figure 10.23 correctly describes the relationship between n and the value of V/n (at constant P and T)?

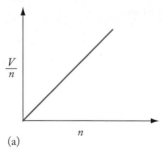

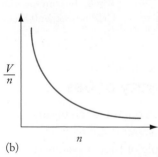

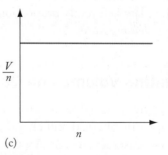

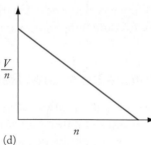

(a)   (b)   (c)   (d)

FIGURE 10.23

Amontons's Law: Relating Pressure and Temperature

Earlier in this section we applied kinetic molecular theory to explain Charles's law: that the volume of a gas is proportional to its temperature at constant pressure. We noted that an increase in temperature increases the speed of gas particles, which in turn increases both the frequency and force of their collisions with the walls of their container. These increases produce an increase in pressure if volume remains constant. However, an increase in volume relieves the increase in pressure, which is the essence of Charles's law. On the other hand, if the gas is confined to a rigid container, then volume remains constant and pressure increases. In general, the pressure of any quantity of gas is directly proportional to its absolute temperature if its volume does not change:

$$P \propto T \quad \text{or} \quad \frac{P}{T} = \text{constant} \quad (V \text{ and } n \text{ fixed}) \qquad (10.17)$$

This relationship between *P* and *T* is called **Amontons's law** in honor of French physicist Guillaume Amontons (1663–1705), a contemporary of Robert Boyle's who constructed a thermometer based on the observation that the pressure of a gas is directly proportional to its temperature.

Where do we see evidence of Amontons's law? Consider a bicycle tire inflated to a prescribed pressure of 100 psi (6.8 atm) on an afternoon in late autumn when the temperature is 25°C. The next morning, following an early freeze in which the temperature drops to 0°C, the pressure in the tire drops to about 91 psi (6.2 atm). Did the tire leak? Probably not, because the decrease in pressure can also be explained by Amontons's law. We can express the law in the form of an equation relating the two pressures and temperatures:

$$\frac{P_1}{T_1} = \frac{P_2}{T_2} \qquad (10.18)$$

Solving for P_2 (the pressure at the lower temperature), we get the expected result:

$$P_2 = \frac{P_1 T_2}{T_1} = \frac{(6.8 \text{ atm})(273 \text{ K})}{298 \text{ K}} = 6.2 \text{ atm}$$

> **Avogadro's law** the volume of a gas at constant temperature and pressure is proportional to the quantity (number of moles) of the gas.
>
> **Amontons's law** the pressure of a gas is proportional to its absolute temperature if its volume does not change.

SAMPLE EXERCISE 10.5 **Applying Amontons's Law** **LO4**

Labels on aerosol cans caution against their incineration because the cans may explode when the pressure inside them exceeds 3.00 atm. At what temperature in degrees Celsius might an aerosol can burst if its internal pressure is 2.00 atm at 25°C?

COLLECT AND ORGANIZE We are given the temperature ($T_1 = 25$°C) and pressure ($P_1 = 2.00$ atm) of a gas and asked to determine the temperature (T_2) at which the pressure (P_2) reaches 3.00 atm.

ANALYZE Because the gas is isolated in a rigid aerosol can, we know that the quantity of gas and its volume are constant. Amontons's law (Equation 10.18) relates the pressures of a confined quantity of gas at two different temperatures. To estimate our answer, we note that the pressure in the can must increase by 50% to reach 3.00 atm. Pressure is directly proportional to absolute temperature, so it, too, must increase by 50%. The initial temperature of 25°C is nearly 300 K, so a 50% increase in absolute temperature corresponds to a final temperature near 450 K, or about 175°C.

SOLVE Rearranging Equation 10.18 to solve for T_2:

$$T_2 = \frac{T_1 P_2}{P_1}$$

and inserting the given *T* and *P* values:

$$T_2 = \frac{[(25 + 273) \text{ K}](3.00 \text{ atm})}{2.00 \text{ atm}} = 447 \text{ K}$$

Converting T_2 to degrees Celsius:

$$T_2 = 447 \text{ K} - 273 = 174°C$$

THINK ABOUT IT This temperature is close to our estimated value. It is also well below the temperatures that solid waste experiences in the fires of an operating incinerator, which makes the warning label on the can all the more important.

Practice Exercise Air pressure in each of the tires of an automobile is adjusted to 34 psi at a gas station in San Diego, California, where the air temperature is 68°F. After a 3-hour drive along Interstate Highway 8, the car and driver are in Yuma, Arizona, where the temperature is 110°F. What is the pressure in the tires? ⚙

10.5 The Combined Gas Law

In the previous section we discussed Boyle's discovery of the inverse relationship between the pressure and volume of a gas at a constant temperature—a relationship that can be expressed in equation form like this:

$$P_1 V_1 = P_2 V_2 \tag{10.10}$$

We also discussed Guillaume Amontons's discovery that the pressure of a gas is directly proportional to its temperature, which corresponds to this equation for relating the pressures of a gas at two temperatures:

$$\frac{P_1}{T_1} = \frac{P_2}{T_2} \tag{10.18}$$

Now let's combine Equations 10.10 and 10.18 to obtain a single equation that relates the pressure, volume, and temperature of a quantity of gas changing from one set of conditions (P_1, V_1, T_1) to another (P_2, V_2, T_2):

$$\frac{P_1 V_1}{T_1} = \frac{P_2 V_2}{T_2} \tag{10.19}$$

Equation 10.19 is known as the **combined gas law**. It is extremely useful for calculating the impact of changing temperature and pressure on the volume of a gaseous system, as we do in Sample Exercise 10.6.

SAMPLE EXERCISE 10.6 Applying the Combined Gas Law   LO4

The pressure inside a weather balloon as it is released is 798 mmHg. If the volume and temperature of the balloon are 131 L and 20°C, what is the volume of the balloon when it reaches an altitude where its internal pressure is 235 mmHg and $T = -52°C$?

COLLECT AND ORGANIZE We are given the initial temperature, pressure, and volume of a gas, and we are asked to determine the final volume after the pressure and temperature have both decreased.

ANALYZE The quantity of gas is a constant, so PV/T is also a constant according to the combined gas law (Equation 10.19) as long as we convert the Celsius temperatures to kelvin. We expect the balloon's volume to decrease as its temperature decreases, but to increase as its pressure decreases. The final pressure is only about 1/3 of the initial pressure, which would result in a tripling of the balloon's volume were it not for the simultaneous decrease in temperature. However, the relative change in absolute temperature is not as great as the decrease in pressure, so we can predict that the volume of the balloon should increase by almost, but not quite, a factor of 3.

combined gas law the ratio PV/T for a given quantity of gas is a constant.

SOLVE First we convert the given Celsius temperatures to kelvin:

$$T_1 = 20°C + 273 = 293 \text{ K}$$

$$T_2 = -52°C + 273 = 221 \text{ K}$$

We then solve Equation 10.19 for V_2:

$$\frac{P_1 V_1}{T_1} = \frac{P_2 V_2}{T_2}$$

$$V_2 = V_1 \times \frac{P_1}{P_2} \times \frac{T_2}{T_1}$$

$$V_2 = 131 \text{ L} \times \frac{798 \text{ mmHg}}{235 \text{ mmHg}} \times \frac{221 \text{ K}}{293 \text{ K}} = 336 \text{ L}$$

THINK ABOUT IT The volume increases by a factor of 336 L/131 L, or 2.56 times, which is in good agreement with our estimate of nearly a threefold increase.

Practice Exercise The balloon in Sample Exercise 10.6 is designed to continue its ascent to an altitude of 30 km, where it bursts, releasing a package of meteorological instruments that parachute back to Earth. If the pressure inside the balloon at 30 km is 33 mmHg and the temperature is −45°C, what is the volume of the balloon when it bursts? ⚙

10.6 Ideal Gases and the Ideal Gas Law

Gases that behave in accordance with the combined gas law are called **ideal gases**. Most gases exhibit ideal behavior at the pressures and temperatures we typically encounter in nature (and while doing general chemistry lab experiments). Under these conditions, gases behave ideally because the assumptions about their composition that we described in Section 10.1 are valid. This means that the volumes of the individual gas particles are insignificant compared to the overall volume occupied by the gas, and the particles do not interact with one another. Instead they move independently with speeds that are related to their masses and to the temperature of the gas.

▶️⏸ **CHEMTOUR** The Ideal Gas Law

In Section 10.4 we explored how the pressure of an ideal gas is

1. Proportional to the moles of particles in it: $P \propto n$

2. Proportional to its absolute temperature: $P \propto T$

3. Inversely proportional to its volume: $P \propto \dfrac{1}{V}$

Combining these three expressions into one, we obtain

$$P \propto n \times T \times \frac{1}{V}$$

Now let's use the symbol R to represent the constant of proportionality that links pressure to the variables on the right side of the expression:

$$P = R \times n \times T \times \frac{1}{V}$$

ideal gas a gas whose behavior is predicted by the linear relations defined by the combined gas law.

ideal gas equation (also called **ideal gas law**) the pressure, volume, number of moles, and temperature of an ideal gas are related by the equation $PV = nRT$, where R is the universal gas constant.

universal gas constant the constant R in the ideal gas equation; its value and units depend on the units used for the variables in the equation.

standard temperature and pressure (STP) 0°C and 1 bar as defined by IUPAC; 0°C and 1 atm are commonly used in the United States.

TABLE 10.3 **Values of the Universal Gas Constant (R)**

Value of R	Units
0.08206	$L \cdot atm/(mol \cdot K)$
8.314	$kg \cdot m^2/(s^2 \cdot mol \cdot K)$
8.314	$J/(mol \cdot K)$
8.314	$m^3 \cdot Pa/(mol \cdot K)$
62.37	$L \cdot torr/(mol \cdot K)$

◯◯ **CONNECTION** In Chapter 1 we defined energy as the ability to do work or transfer heat.

◯◯ **CONNECTION** In Chapter 9 we defined standard conditions of temperature and pressure as they apply to thermochemistry. Note that *STP* and *standard conditions* are not the same. STP applies strictly to calculations involving the gas laws, while standard conditions apply to thermochemical data.

Rearranging the terms and simplifying a bit, we get

$$PV = nRT \tag{10.20}$$

Equation 10.20 is probably the most important one in this chapter. It is called the **ideal gas equation**—a mathematical expression of the **ideal gas law**. The constant represented by R is called the **universal gas constant**. Its value depends on the units used for pressure and volume. (We always express the quantity of gas in moles, and we use the Kelvin scale for temperature.) If we use SI units for volume (cubic meters) and pressure (pascals), then

$$R = 8.314 \frac{m^3 \cdot Pa}{mol \cdot K}$$

As a practical matter, it is often more convenient to express volumes using units smaller than cubic meters, such as liters ($1 \text{ L} = 10^{-3} \text{ m}^3$), and to express pressures in larger units such as kilopascals ($1 \text{ kPa} = 10^3 \text{ Pa}$). Because these units are 1/1000 and 1000 times the size of the first set of units, the two factors cancel out and R has the same value:

$$R = 8.314 \frac{L \cdot kPa}{mol \cdot K}$$

In many of the calculations in this book we express pressure in standard atmospheres (atm), so we will also use an R value based on the equality 1 atm = 101.325 kPa:

$$R = 8.314 \frac{L \cdot \cancel{kPa}}{mol \cdot K} \times \frac{1 \text{ atm}}{101.325 \cancel{kPa}}$$

$$= 0.08206 \frac{L \cdot atm}{mol \cdot K}$$

Note how the numerators of all these R values represent volume × pressure. In Chapter 9 we discussed how the product of volume and pressure is the kind of work, called *P–V* work, that a gas does when it expands against an opposing pressure. Energy is the ability to do work, so the numerator of R can also have energy units. The SI unit of energy is the joule, where 1 J is equivalent to $1 \text{ m}^3 \cdot Pa$, which gives another set of units for a familiar R value:

$$R = 8.314 \frac{J}{mol \cdot K}$$

These R values are summarized in Table 10.3.

Before closing this section, we should note the existence of some useful reference conditions in the properties of gases. One of them is **standard temperature and pressure (STP)**. Since 1982 the International Union of Pure and Applied Chemistry (IUPAC) has defined STP as 0°C and 1 bar of pressure. Before that, STP was defined as 0°C and 1 standard atmosphere of pressure, and that definition has lived on, especially among some scientists and science teachers in the United States. One reason for the persistence of the old definition is that the standard atmosphere is equal to 1.01325 bar, so the difference between the two pressures is relatively small.

One value that depends on how we define STP is *molar volume*, which is the volume that one mole of an ideal gas occupies at STP (Figure 10.24). If we use the old definition of STP pressure (1 atm), the volume of one mole of an ideal gas at 0°C (273 K) is

$$V = \frac{nRT}{P}$$

$$= \frac{(1 \ \text{mol})\left(0.08206 \ \dfrac{\text{L} \cdot \text{atm}}{\text{mol} \cdot \text{K}}\right)(273 \ \text{K})}{1 \ \text{atm}} = 22.4 \ \text{L}$$

Molar volume based on the modern definition of STP ($P = 1$ bar or 100 kPa) is slightly larger:

$$V = \frac{(1 \ \text{mol})\left(8.314 \ \dfrac{\text{L} \cdot \text{kPa}}{\text{mol} \cdot \text{K}}\right)(273 \ \text{K})}{100 \ \text{kPa}} = 22.7 \ \text{L}$$

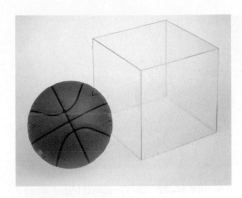

FIGURE 10.24 This box represents the volume of 1 mole of gas at STP. A basketball would fit loosely inside the box.

Many chemical and biochemical processes (including those that sustain us) take place at pressures near 1 atm (or 1 bar) and at temperatures between 0°C and 40°C. Under these conditions, the volume of one mole of a gas is no more than about 15% greater than the molar volume. Therefore, the volume of a gas can be estimated if the number of moles of the gas is known. An important feature of the molar volume of a gas at ambient temperatures and pressures is that it applies to essentially all gases independent of their chemical composition, because they all behave like ideal gases.

SAMPLE EXERCISE 10.7   **Applying the Ideal Gas Law**   **LO4**

Bottles of compressed O_2 carried by climbers ascending Mt. Everest are designed to hold one kilogram of the gas. What volume of O_2 can one bottle deliver to a climber at an altitude where the temperature is −38°C and the atmospheric pressure is 0.35 atm? Assume that each bottle contains 1.00 kg O_2.

COLLECT AND ORGANIZE We are given the pressure, mass, and temperature of a gas and asked to determine its volume.

ANALYZE The ideal gas equation enables us to relate P, V, T, and n, the number of moles of O_2. We are not given n, but we do know the mass of O_2, and we can use its molar mass ($16.00 \times 2 = 32.00$ g/mol) to calculate n and then use the ideal gas equation to calculate V. To estimate our answer, we note that a kilogram of O_2 contains 1000/32, or about 30 moles of the gas, which would occupy 30×22.4, or about 660 liters at STP. Of course, conditions on the top of Mt. Everest are a far cry from STP. The temperature on an absolute scale is somewhat lower, which would tend to reduce the volume of gas, but the pressure is only about 1/3 of what it is at STP, which means the volume should be nearly 3 times 660 L, or nearly 2000 L.

SOLVE Let's start by converting the mass of O_2 into moles:

$$1.00 \ \text{kg} \ O_2 \times \frac{1000 \ \text{g}}{1 \ \text{kg}} \times \frac{1 \ \text{mol} \ O_2}{32.00 \ \text{g} \ O_2} = 31.25 \ \text{mol} \ O_2$$

Solving the ideal gas equation for V and inserting the values known:

$$PV = nRT$$

$$V = \frac{nRT}{P}$$

$$= \frac{(31.25 \ \text{mol})\left(0.08206 \ \dfrac{\text{L} \cdot \text{atm}}{\text{mol} \cdot \text{K}}\right)(273 - 38) \ \text{K}}{0.35 \ \text{atm}} = 1.7 \times 10^3 \ \text{L}$$

FIGURE 10.25 The Goodyear blimp *Spirit of Innovation*.

THINK ABOUT IT Our answer is certainly reasonable based on our estimate. Most climbers require several of these bottles to climb Mt. Everest and return. We used the ideal gas law to solve this problem because all four gas law variables: P, V, T, and n, factored into the calculation.

Practice Exercise Buoyancy for the Goodyear blimp *Spirit of Innovation* comes from 2.03×10^5 cubic feet of helium (Figure 10.25).

a. What is the mass of this much helium at 25°C and 1.00 atm of pressure?
b. What is the buoyancy of the balloon—that is, the difference between the mass of the He and the mass of the same volume of dry air at the same temperature and pressure?

10.7 Densities of Gases

Carbon dioxide is a relatively minor component of our atmosphere, but it has received considerable attention in recent years due to its increasing concentration in the atmosphere and the resulting impact on global climate. While this increase has been linked to human activity, many natural sources such as volcanoes also add CO_2 to the atmosphere. The quantities of CO_2 involved in volcanic eruptions are not normally a threat to human health, but when this gas collects in the waters of a deep crater lake and is then released all at once, the results can be catastrophic. On August 12, 1986, volcanic Lake Nyos in Cameroon, Africa, suddenly released about a cubic kilometer of CO_2 into the air. The gas emerged from the lake in a frothy mist an estimated 100 meters high, which flowed out of the crater and into surrounding valleys, killing over 1800 people and their farm animals (Figure 10.26). Because CO_2 is denser than air, the deadly mist formed a ground-hugging layer that displaced the air and the oxygen necessary for life.

The density of a gas at STP can be calculated by dividing its molar mass by the molar volume. Carbon dioxide, for example, has a molar mass of 44.0 g/mol. Therefore, the density of CO_2 at 0°C and 1 atm is

$$\frac{44.0 \text{ g/mol}}{22.4 \text{ L/mol}} = 1.96 \text{ g/L}$$

The density of air, which is mostly N_2 ($\mathcal{M}$ = 28.0 g/mol) and O_2 ($\mathcal{M}$ = 32.0 g/mol), is only 1.29 g/L at STP, so it is not surprising that 1 km³ of CO_2, a colorless, odorless gas, could displace enough air to become a silent killer. The greater density of CO_2 and the fact that it does not support combustion also makes it effective in fighting small fires. The gas from a CO_2 fire extinguisher effectively blankets burning fuel, depriving it of O_2 and extinguishing the fire.

We can use the ideal gas equation to calculate the density of a gas at any temperature and pressure. It's a matter of dividing the mass of one mole of the gas, that is, its molar mass ($\mathcal{M}$), by the volume (V) that one mole of it occupies at a given temperature and pressure—a value we can calculate by solving the ideal gas equation for V when $n = 1$:

(a)

(b)

FIGURE 10.26 As many as 1800 people may have been asphyxiated by a dense cloud of CO_2 that was released by Lake Nyos in northwestern Cameroon on August 21, 1986. (a) The lake normally has a deep blue color. (b) The CO_2 event stirred up sediment from the lake floor, turning the water brown.

$$V = \frac{nRT}{P} = \frac{RT}{P}$$

Combining this equation with the definition of density and assuming one mole of gas (so that $m = \mathcal{M}$) give us an equation for calculating the density of a known gas at any temperature and pressure:

$$d = \frac{m}{V} = \mathcal{M} \frac{1}{V}$$

$$d = \mathcal{M} \frac{P}{RT} \tag{10.21}$$

Alternatively, we may be able to determine the density of a gas experimentally using an apparatus such as the one shown in Figure 10.27. In this apparatus, a glass bulb of known volume is attached to a vacuum pump and the air is removed. The mass of the bulb is determined twice: first while it is evacuated, and again after the bulb is filled with the test gas. The difference in the masses divided by the volume of the bulb is the density of the gas.

If we also knew the temperature and pressure of the gas in the bulb in Figure 10.27, we could use the values of V, T, and P to calculate the number of moles of gas in the bulb:

$$n = \frac{PV}{RT}$$

Dividing the mass (m) of the gas in the bulb (in grams) by the number of moles of gas (n) gives the molar mass of the gas:

$$\mathcal{M} = \frac{m}{n}$$

(a)

(b)

(c)

FIGURE 10.27 Apparatus for determining the density of a gas. (a) A gas collection tube with an internal volume of 235 mL is evacuated by connecting it to a vacuum pump. (b) The mass of the empty tube is measured. (c) The tube is then opened to the atmosphere to be refilled with air. The difference in mass between the filled and evacuated tube (0.273 g, or 273 mg) is the mass of the air inside it. The density of the air sample is 273 mg/235 mL, or 1.16 mg/mL, or 1.16 g/L.

CONCEPT TEST •

Which graph in Figure 10.28 best approximates the following for an ideal gas?

   a. The relationship between density and pressure (n and T constant)

   b. The relationship between density and temperature (n and P constant)

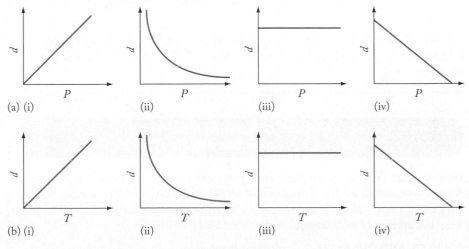

(a) (i)   (ii)   (iii)   (iv)

(b) (i)   (ii)   (iii)   (iv)

FIGURE 10.28

SAMPLE EXERCISE 10.8 Calculating the Density of a Gas   LO5

According to the U.S. National Weather Service, the air temperature in Phoenix, Arizona, reached 78°F on January 1, 2012, when atmospheric pressure was 1024 millibars. What was the density of the air? Assume the average molar mass of air is 28.8 g/mol, which is the weighted average of the molar masses of the various gases in dry air (see Table 10.1).

COLLECT AND ORGANIZE We are provided with the average molar mass, temperature, and atmospheric pressure of an air mass and asked to calculate its density, where $d = m/V$.

ANALYZE We may assume that the air behaves like an ideal gas and obeys the ideal gas law. One way to calculate the density of air (or any ideal gas) is to divide the mass (m) of one mole of it (28.8 g in this case) by the volume (V) that one mole occupies. This volume can be calculated using the ideal gas equation and the temperature and pressure values provided. We will need to convert T and P to units that are compatible with one of our R values. Because the density of air at STP is about 1.3 g/L, we can estimate that the air density at Phoenix's warm temperature is about 10% lower than at STP, or about 1.2 g/L.

SOLVE First we need to convert the Fahrenheit temperature to kelvin:

$$T(K) = \tfrac{5}{9}(78°F - 32) + 273 = 299 \text{ K}$$

There are several options for converting millibars to another pressure unit. Let's use the equality 1 atm = 1013.25 millibar (mb) and calculate the volume of one mole of air, starting with the ideal gas equation: $PV = nRT$. Solving for V:

$$V = \frac{nRT}{P}$$

$$= \frac{(1 \text{ mol})\left(0.08206 \dfrac{\text{L} \cdot \text{atm}}{\text{mol} \cdot \text{K}}\right)(299 \text{ K})}{(1024 \text{ mb})\left(\dfrac{1 \text{ atm}}{1013.25 \text{ mb}}\right)} = 24.28 \text{ L}$$

Calculating density:

$$d = \frac{m}{V} = \frac{28.8 \text{ g}}{24.28 \text{ L}} = 1.19 \text{ g/L}$$

THINK ABOUT IT The calculated result is consistent with our estimate and with the general trend that gases are about 10% less dense at comfortable ambient temperatures than at STP (0°C).

Practice Exercise Air is a mixture of mostly nitrogen and oxygen. A balloon is filled with pure oxygen and released in a room full of air. Will it sink to the floor or float to the ceiling?

**SAMPLE EXERCISE 10.9 Calculating Molar Mass   LO5
                        from Density**

Vent pipes at solid-waste landfills emit gases released by decomposing material. A sample of such a gas has a density of 0.65 g/L at 25.0°C and 757 mmHg. What is its molar mass? (Note: If the sample is a mixture of gases, the answer will be the weighted average of molar masses of the individual gases.)

COLLECT AND ORGANIZE We are given the density, temperature, and pressure (in mmHg) of a gaseous sample and asked to calculate the molar mass of the gas.

ANALYZE The density of the sample tells us the mass of one liter of it: 0.65 g. The ideal gas equation enables us to use the temperature and pressure of the gas to calculate the number of moles in one liter. Dividing the mass (m) of one liter by the number of moles (n) in one liter will give us the mass of one mole, which is molar mass ($\mathcal{M}$). Before we can use the ideal gas equation, we need to convert the Celsius temperature to a Kelvin temperature, and we need to convert mmHg to a more convenient pressure unit, such as atm (1 atm = 760 mmHg). We can use the concept of molar volume to estimate the molar mass of the gas. We saw in the previous Sample Exercise that, at 25°C and a pressure near 1 atm, the volume of one mole of an ideal gas is about 24 L. If the mass of one liter is 0.65 g, or about 2/3 of a gram, then the mass of 24 L is about 2/3 of 24, or 16 g/mol.

SOLVE Solving $PV = nRT$ for n:

$$n = \frac{PV}{RT}$$

The number of moles of the gas in a 1-liter sample is

$$n = \frac{(757 \ \cancel{mmHg})\left(\dfrac{1 \ \cancel{atm}}{760 \ \cancel{mmHg}}\right)(1 \ \cancel{L})}{\left(0.08206 \ \dfrac{\cancel{L} \cdot \cancel{atm}}{mol \cdot \cancel{K}}\right)(25 + 273) \ \cancel{K}} = 0.0407 \ mol$$

Molar mass is the ratio of the mass (0.65 g) to the number of moles (0.0407 mol) in 1 liter of the gas:

$$\mathcal{M} = \frac{m}{n} = \frac{0.65 \ g}{0.0407 \ mol} = 16 \ g/mol$$

THINK ABOUT IT There is another (maybe easier) way to solve this problem, starting with Equation 10.21: $d = \mathcal{M}P/RT$. Rearranging the terms to solve for $\mathcal{M}$ gives us

$$\mathcal{M} = \frac{dRT}{P}$$

We know all of the terms on the right side of the equation and inserting them yields the molar mass of the gas in one step:

$$\mathcal{M} = \frac{dRT}{P} = \frac{0.65 \ \dfrac{g}{\cancel{L}} \times 0.08206 \ \dfrac{\cancel{L} \cdot \cancel{atm}}{mol \cdot \cancel{K}} \times (25 + 273) \ \cancel{K}}{(757 \ \cancel{mmHg})\left(\dfrac{1 \ \cancel{atm}}{760 \ \cancel{mmHg}}\right)} = 16 \ g/mol$$

The only common gas with this $\mathcal{M}$ value is methane, CH_4, a principal component of the gases emitted by decomposing solid waste. Methane itself is odorless and the smell of waste comes from small quantities of other gases produced by decomposing matter.

Practice Exercise When HCl(aq) and $NaHCO_3$(aq) are mixed together, a chemical reaction takes place in which a gas is one of the products. A sample of the gas has a density of 1.81 g/L at 1.00 atm and 23.0°C. What is the molar mass of the gas? Can you identify the gas? ⚙

10.8 Gases in Chemical Reactions

Many important chemical reactions, including those that supply us with most of our energy needs, involve gaseous reactants or products. Even when the fuel is a solid or a liquid, the actual combustion reaction takes place in the gas phase, as when gasoline burns in the combustion chamber of a car engine, or the flames of a charcoal fire cook dinner at a backyard barbecue.

In a chemical reaction involving a gas as either reactant or product, the quantity of it (in moles) in the reaction mixture can be determined using the ideal gas law if we know the volume of the gas and the temperature and pressure at which the reaction proceeds. Once we find the value of n, we can calculate quantities of other reactants or products, including the quantity of energy released or consumed, from the stoichiometry of the reaction. Alternatively, we may know the quantity of another reactant and need to calculate the volume of a gaseous reactant that is needed for a complete reaction. The following Sample Exercise provides an example of such a calculation from the world of backyard barbecues.

FIGURE 10.29 A gas grill.

SAMPLE EXERCISE 10.10 **Combining Stoichiometry and the Ideal Gas Law** **LO6**

The maximum flow rate of propane fuel through one of the burners in the gas grill in Figure 10.29 is 5.5 g C_3H_8 per minute. What volume of air must mix with the propane ($\mathcal{M}$ = 44.1 g/mol) each minute to provide enough O_2 for complete combustion? Assume P = 1.00 atm and T = 25°C.

COLLECT AND ORGANIZE We know that each minute 5.5 g C_3H_8 is burned, and we are asked to calculate the volume of air required for complete combustion. According to the data in Table 10.1, air is 20.95% O_2. The volume of an ideal gas is related to its amount, temperature, and pressure by the ideal gas law: $PV = nRT$.

ANALYZE The quantities of two reactants in any chemical reaction are related by the stoichiometry of the balanced chemical equation describing the reaction, so we need to write that equation. We also need to convert 5.5 g C_3H_8 to moles C_3H_8 and then to moles O_2 using the stoichiometry of the reaction. Moles of O_2 can be converted to a volume of O_2 using the ideal gas law, and to a corresponding volume of air. To estimate our answer, we can figure that the mass of propane per minute represents 5.5/44, or a little more than 1/10, of a mole. We will probably need about five times that many moles of O_2 to convert the 3 C atoms per molecule to CO_2 and the 8 H atoms to H_2O. Assuming we need about half a mole of O_2, the volume of O_2 will be a little more than half the molar volume, or about 12 liters. Air is about 1/5 O_2, so we will need about 12 × 5, or 60, liters of air.

SOLVE To write the balanced chemical equation for the reaction, we balance the number of C, then H, and finally the O atoms (as in Chapter 7), which yields

$$C_3H_8(g) + 5\ O_2(g) \rightarrow 3\ CO_2(g) + 4\ H_2O(g)$$

The number of moles of C_3H_8 is

$$5.5\ \cancel{g\ C_3H_8} \times \frac{1\ mol\ C_3H_8}{44.1\ \cancel{g\ C_3H_8}} = 0.125\ mol\ C_3H_8$$

Applying the 5:1 stoichiometric ratio of O_2 to C_3H_8, we obtain

$$0.125\ \cancel{mol\ C_3H_8} \times \frac{5\ mol\ O_2}{1\ \cancel{mol\ C_3H_8}} = 0.625\ mol\ O_2$$

The pressure is given as 1.00 atm, so we use the R value with atmospheres in the numerator in the ideal gas law to convert moles O_2 to liters of O_2:

$$V = \frac{(0.625\ \cancel{mol\ O_2})\left(0.08206\ \dfrac{L \cdot \cancel{atm}}{\cancel{mol} \cdot \cancel{K}}\right)(25 + 273)\ \cancel{K}}{1.00\ \cancel{atm}} = 15.3\ L\ O_2$$

Air is 20.95% O_2, so the volume of air that contains 15.3 L of O_2 is

$$V_{air} = 15.3\ L\ \cancel{O_2} \times \frac{100.00\ \%\ air}{20.95\ \cancel{\%\ O_2}} = 73\ L\ air$$

THINK ABOUT IT Our calculated result is not too far from the predicted value, and it is reasonable given the size of a molar volume and the fact that a molar volume of O_2 corresponds to five molar volumes of air. In working through our estimate, we assumed that half a mole of O_2 would occupy a little more than half a molar volume, but we used the STP molar volume of 22.4 L, which assumes an absolute temperature of 273 K. The actual temperature in the exercise is 298 K. Multiplying 22.4 L by the ratio 298/273, as allowed by Charles's law, gives us an adjusted molar volume of 24.5 L/mol at 298 K at 1 atm, which explains why our estimate was low.

Practice Exercise Automobile air bags (Figure 10.30) inflate during a crash or sudden stop by the rapid generation of nitrogen gas from sodium azide. The first step of the air bag reaction may be written

$$2 \, NaN_3(s) \rightarrow 2 \, Na(\ell) + 3 \, N_2(g)$$

How many grams of sodium azide are needed to provide sufficient nitrogen gas to fill a $45 \times 45 \times 25$ cm bag to a pressure of 1.20 atm at 15°C?

FIGURE 10.30 An automobile air bag inflates when solid NaN_3 rapidly decomposes, producing N_2 gas.

10.9 Mixtures of Gases

As noted in Table 10.1, dry air is mostly N_2 and O_2 plus much smaller concentrations of other gases. Each gas in air, or in any mixture of gases, exerts its own pressure, called its **partial pressure**. Atmospheric pressure is the sum of the partial pressures of all of the gases in the air:

$$P_{atm} = P_{N_2} + P_{O_2} + P_{Ar} + P_{CO_2} + \cdots$$

A similar expression can be written for any mixture that contains "i" gases where $i = 1, 2, 3, \ldots$:

$$P_{total} = P_1 + P_2 + P_3 + \cdots \tag{10.22}$$

Equation 10.22 is a mathematical expression of **Dalton's law of partial pressures**: the total pressure of any mixture of gases is the sum of the partial pressures of the gases in the mixture.

You might guess that the most abundant gases in a mixture (such as N_2 and O_2 in air) have the greatest partial pressures and contribute the most to the total pressure of the mixture. The mathematical term used to express the abundance of component i in a mixture of gases is its **mole fraction (x_i)**,

$$x_i = \frac{n_i}{n_{total}} \tag{10.23}$$

where n_i is the number of moles of component i and n_{total} is the total number of moles of gas in the mixture. The partial pressure of each component is the product of its mole fraction times the total pressure of the mixture:

$$P_i = x_i P_{total} \tag{10.24}$$

Let's apply Equation 10.24 to the major gases in air. If the atmospheric pressure of a particularly hot, dry air mass is 998 mbar, the partial pressures of N_2 and O_2 are

$$P_{N_2} = x_{N_2} P_{total} = (0.7808)(998 \text{ mbar}) = 779 \text{ mbar}$$

$$P_{O_2} = x_{O_2} P_{total} = (0.2095)(998 \text{ mbar}) = 209 \text{ mbar}$$

CONNECTION Dalton's law of partial pressures was observed by English chemist John Dalton, who we discussed in Chapters 1 and 2.

partial pressure the contribution to the total pressure made by a component in a mixture of gases.

Dalton's law of partial pressures the total pressure of a mixture of gases is the sum of the partial pressures of all the gases in the mixture.

mole fraction (x_i) the ratio of the number of moles of a particular component i in a mixture to the total number of moles in the mixture.

▶II **CHEMTOUR** Dalton's Law

Note that the partial pressure of each of these gases depends only on total pressure and its mole fraction; it does not depend on the identity of the gas or of the other gases in the mixture. Another point: mole fractions have no units. This means that they can be used when describing the behavior of any kind of homogeneous mixture (solid, liquid, or gas). Finally, the sum of the mole fractions of all the components in a mixture should always add up to 1 (or, because of rounding, very close to 1).

The kinetic molecular theory of gases explains the additive nature of partial pressures and the dependence of partial pressure on mole fraction. According to the theory, the particles of a pure gas, or of a mixture of gases, collide with the walls of the mixture's container. The force of these collisions, when averaged over the surface area of the container, produces a total pressure. The more particles there are, the more frequent the collisions are and the greater the pressure is. The theory assumes that all particles have, on average, the same kinetic energy, which means that the frequency and force of their collisions, and the contributions these collisions make to overall pressure, depend only on how many of the particles are present. The theory further assumes that the particles do not interact with each other, so their identities are irrelevant—indeed, the particles of all of the components of a mixture are equal-opportunity colliders. The particles of each gas contribute to the total pressure of the mixture in exact proportion to how many of them are in the container as compared to the total number of particles.

SAMPLE EXERCISE 10.11 **Calculating Mole Fractions and Partial Pressures** **LO7**

Scuba divers who dive to depths below 50 meters may breathe a gas mixture called trimix during the deepest parts of their dives. One formulation of the mixture, called *trimix 10/70*, is 10% oxygen, 70% helium, and 20% nitrogen by volume. What is the mole fraction of each gas in this mixture, and what is the partial pressure of oxygen in the lungs of a diver at a depth of 60 meters (where the ambient pressure is 7.0 atm)?

COLLECT AND ORGANIZE We are given the composition of a scuba diver's breathing mixture expressed in percent by volume of its three components: O_2, He, and N_2. We are asked to determine the mole fractions of the gases and the partial pressure of O_2 in the lungs of a diver breathing the mixture at a depth where the pressure is 7.0 atm.

ANALYZE According to Avogadro's law, equal volumes of ideal gases at the same temperature and pressure contain equal numbers of moles of gas. Therefore, the mole fraction of each gas in a mixture is equal to the volume of that gas divided by the sum of the volumes of all the gases that were mixed together. Thus, the mole fractions of the gases are the same as their percent by volume values. The partial pressure of a component of a gas mixture may be calculated using Equation 10.24:

$$P_i = x_i P_{\text{total}}$$

SOLVE Expressing the percent by volume values as mole fractions:

Gas	% by Volume	Mole Fraction
O_2	10	0.10
He	70	0.70
N_2	20	0.20

Calculating P_{O_2}:

$$P_{O_2} = x_{O_2} P_{\text{total}} = 0.10 \times 7.0 \text{ atm} = 0.70 \text{ atm}$$

THINK ABOUT IT This exercise introduces a useful equality: the concentrations of the gases in a mixture, such as those listed in Table 10.1 for dry air, are equivalent to the mole fractions of the gases, though some moving of decimal points may be required: for example, two places to the left to convert percent values into mole fractions, or six places to the left to convert part per million values.

Practice Exercise What is the partial pressure, in atmospheres, of O_2 in the air outside an airliner cruising at an altitude of 10 km, where the atmospheric pressure is 190 mmHg? How much must the outside air be compressed to produce a cabin pressure in which $P_{O_2} = 0.200$ atm? ⚙

CONCEPT TEST

Do each of the gases in an equimolar mixture of four gases have the same mole fraction?

Sample Exercise 10.11 and its Practice Exercise illustrate how our health and well-being rely on access to the appropriate breathing gases when we venture a few kilometers above sea level, or a few tens of meters below it (underwater). A scuba diver can't breathe compressed air at depths of 70 meters or more because the P_{O_2} in a diver's lungs would exceed 1.6 atm, which is enough to induce oxygen toxicity (damage to the central nervous system). At even shallower depths, high P_{N_2} can lead to a condition called nitrogen narcosis, or "rapture of the deep," in which high concentrations of nitrogen in the blood lead to hallucinations. Alpine climbers on the tallest peaks on Earth experience the opposite problem. For them the problem is a lack of O_2 and a condition called *anoxia*, characterized by fatigue and an inability to think clearly because of lack of dissolved oxygen in muscle tissues and the brain. The phrase "death zone" is used to describe altitudes above 8 km (26,000 feet), where P_{O_2} is less than 0.07 atm. People breathing such thin air, even those who are acclimated to high altitudes, don't have enough oxygen in their blood to stay alive.

At more dweller-friendly altitudes, such as those of most chemistry laboratories, Dalton's law of partial pressures is used to determine the quantities of gaseous products in chemical reactions. For example, heating potassium chlorate ($KClO_3$) in the presence of MnO_2 causes it to decompose into $KCl(s)$ and $O_2(g)$. The oxygen gas produced by this reaction can be collected by bubbling the gas into an inverted bottle that is initially filled with water (Figure 10.31). As the reaction proceeds, $O_2(g)$ displaces the water in the bottle. When the reaction is complete, the volume of water displaced provides a measure of the volume of O_2 produced. If the temperature of the water and atmospheric pressure are known, the ideal gas law can be used to determine the number of moles of O_2 produced.

Water displacement can be used to collect and measure the volume of any gas that neither reacts with nor dissolves appreciably in water. However, one additional step is needed to apply the ideal gas law to calculating the number of moles of gas produced. At room temperature (nominally 20–25°C), any enclosed space above a pool of liquid water contains water vapor in addition to the gas produced by the reaction. Therefore, the gas collected in the $KClO_3$ reaction in Figure 10.31 is a mixture of O_2 and H_2O vapor. Each gas exerts its own partial pressure, and Dalton's law of partial pressures tells us the total pressure of the mixture when the water level inside the collection vessel matches the water level outside the vessel:

$$P_{total} = P_{atm} = P_{O_2} + P_{H_2O} \tag{10.25}$$

FIGURE 10.31 Collecting O_2 gas by water displacement. Oxygen is generated by the thermal decomposition of $KClO_3$. The delivery tube is removed when the height of the water in the collection jar matches the level of water around it, ensuring that the pressure in the jar is equal to atmospheric pressure.

TABLE 10.4	Partial Pressure of Water Vapor at Selected Temperatures
Temperature (°C)	**Pressure (mmHg)**
5	6.5
10	9.2
15	12.8
20	17.5
25	23.8
30	31.8
35	42.2
40	55.3
45	71.9
50	92.5

To calculate the quantity of oxygen produced using the ideal gas law, we need to know P_{O_2}, which we get by subtracting P_{H_2O} (Table 10.4) from P_{atm}. If we know the values of P_{O_2}, T, and V, we can calculate the number of moles or number of grams of oxygen produced.

SAMPLE EXERCISE 10.12 **Calculating the Quantity of a Gas Collected by Water Displacement** **LO7**

During the decomposition of $KClO_3$, 92.0 mL of gas is collected by the displacement of water at 25.0°C. If atmospheric pressure is 756 mmHg, what mass of O_2 is collected?

COLLECT AND ORGANIZE We are given values for the total pressure (756 mmHg), volume (92.0 mL), and temperature (25.0°C) of the gas collected by water displacement during the decomposition of $KClO_3$, and we are asked to calculate the mass of O_2 in the gas. Table 10.4 contains values for the partial pressure of water vapor at several temperatures. The ideal gas law relates the number of moles of an ideal gas to its pressure, volume, and absolute temperature.

ANALYZE Oxygen is collected over water, so the collection vessel contains both O_2 and water vapor. To calculate P_{O_2} we subtract the partial pressure of water vapor at 25°C (23.8 mmHg, according to Table 10.4) from atmospheric (total) pressure. To calculate the number of moles of O_2 using the ideal gas equation, we need to convert the pressure value to atm and the temperature to kelvin. Multiplying that number of moles by the molar mass of O_2 will give us the mass of O_2 collected. To estimate the result, let's round off the volume of the gas to 100 mL, or 0.1 L. At 25°C the molar volume of an ideal gas is about 24 L, so the volume collected corresponds to 0.1/24, or about 0.004 moles of gas. The partial pressure of water vapor is small compared to atmospheric pressure, so most of the gas collected is O_2. The mass of this quantity of O_2 is 0.004 mol × 32 g/mol, or about 0.13 g.

SOLVE The value of P_{O_2} is

$$P_{O_2} = P_{atm} - P_{H_2O} = (756 - 23.8) \text{ mmHg} = 732 \text{ mmHg}$$

The number of moles of O_2 is

$$n = \frac{\left(732 \text{ mmHg} \times \frac{1 \text{ atm}}{760 \text{ mmHg}}\right)\left(92.0 \text{ mL} \times \frac{1 \text{ L}}{1000 \text{ mL}}\right)}{\left(0.08206 \frac{\text{L} \cdot \text{atm}}{\text{mol} \cdot \text{K}}\right)(25 + 273) \text{ K}} = 0.00362 \text{ mol}$$

and the mass of O_2 is

$$m = 0.00362 \text{ mol} \left(\frac{32.00 \text{ g}}{1 \text{ mol}}\right) = 0.116 \text{ g}$$

THINK ABOUT IT The calculated mass of O_2 is about what was expected. The value of P_{O_2} is (732/756) or 96.8% of P_{atm}—close enough to justify our ignoring the contribution of P_{H_2O} in our estimate, but not so close that it could be ignored in the actual calculation.

Practice Exercise Electrical energy can be used to separate water into O_2 and H_2. In one demonstration of this reaction, 27 mL of H_2 is collected over water at 25°C. Atmospheric pressure is 761 mmHg. How many grams of H_2 are collected? ⚙

Henry's law the concentration of a sparingly soluble gas in a liquid is proportional to the partial pressure of the gas.

10.10 Solubilities of Gases and Henry's Law

Oxygen is essential to most life-forms, including those that live underwater. In Chapter 6 we observed that relatively weak dipole–induced dipole forces between polar water molecules and nonpolar oxygen molecules lead to a concentration of dissolved oxygen sufficient to sustain aquatic plants and fish. The solubility of O_2 (or any sparingly soluble gas) in a liquid such as water is directly proportional to the partial pressure of the gas above the surface of the liquid. This relationship is known as **Henry's law**, in honor of William Henry (1775–1836), a British physician who first proposed the relationship. In equation form the relationship is

$$C_{gas} = k_H P_{gas} \tag{10.26}$$

where C_{gas} represents the maximum concentration (solubility) of a gas in a particular solvent, k_H is the Henry's law constant for the gas in that solvent, and P_{gas} is the partial pressure of the gas in the environment surrounding the solvent. When C_{gas} is expressed in molarity, the units of the Henry's law constant are moles per liter-atmosphere, mol/(L · atm). Table 10.5 lists k_H values for several common gases in water.

According to Henry's law, the concentration of dissolved oxygen in blood is proportional to the partial pressure of oxygen in the air we inhale and thus proportional to atmospheric pressure. This is an accurate statement of Henry's law, but it does not tell the whole story. It does not mean, for example, that residents of Denver, Colorado (average P_{O_2} = 0.178 atm) live with less blood oxygen than residents of New York City or anywhere near sea level (where P_{O_2} averages 0.209 atm).

The concentration of O_2 in the blood also depends on the concentration of hemoglobin, an oxygen-transporting protein that attaches to molecules of O_2 as blood passes through the lungs. The hemoglobin binding sites in most people can be saturated with O_2 even when the P_{O_2} in their lungs is as low as 0.110 atm. If P_{O_2} decreases to about 0.066 atm (as it does on high mountains), only about 80% of the hemoglobin molecules pick up molecules of O_2 passing through the lungs. Over several weeks, the body responds to low P_{O_2} by producing more hemoglobin so that the same amount of O_2 is delivered to tissues. Some endurance athletes choose to train at high altitudes so that, when they compete at lower altitudes, their bodies transport O_2 more efficiently.

The solubility of gases in water and other liquids also depends on their temperature: the warmer the liquid, the less soluble are gases in it. You may have seen the impact of this trend when opening a can of warm soda: the release of CO_2 can border on the explosive (Figure 10.32) because less CO_2 dissolves in warm water than in cold water (or soda). This same trend shows up in the way the solubility of O_2 in water decreases from 0°C to 50°C (Figure 10.33) and in the way the k_H value for O_2 decreases as water temperature increases (Table 10.6).

| SAMPLE EXERCISE 10.13 | Calculating Gas Solubility Using Henry's Law | LO8 |

Lake Titicaca (Figure 10.34) is located high in the Andes Mountains between Peru and Bolivia. Its surface is 3811 m above sea level, where the average atmospheric pressure is 0.636 atm. During the summer, the average temperature of the water's surface rarely exceeds 15°C. What is the solubility of oxygen in Lake Titicaca at that temperature? Express your answer in molarity and in mg/L.

TABLE 10.5 Henry's Law Constants for Several Gases in Water at 20°C

Gas	k_H [mol/(L · atm)]
He	3.5×10^{-4}
O_2	1.3×10^{-3}
N_2	6.7×10^{-4}
CO_2	3.5×10^{-2}

▶ll CHEMTOUR Henry's Law

FIGURE 10.32 Opening a warm can of soda can be risky because less of the CO_2 gas inside dissolves in warm soda (or water) than in cold soda, which increases the pressure of CO_2 gas in the can.

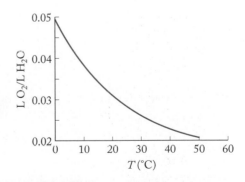

FIGURE 10.33 The solubility of O_2 in water decreases with increasing temperature (data are based on P_{O_2} = 1.00 atm).

TABLE 10.6	Henry's Law Constants for O_2 Gas in Water
Temperature (°C)	k_H [mol/(L · atm)]
0	2.18×10^{-3}
5	1.90×10^{-3}
10	1.68×10^{-3}
15	1.50×10^{-3}
20	1.35×10^{-3}
25	1.23×10^{-3}
30	1.12×10^{-3}
35	1.04×10^{-3}
40	0.96×10^{-3}
45	0.90×10^{-3}
50	0.84×10^{-3}

FIGURE 10.34 Lake Titicaca is the highest altitude navigable lake in the world.

COLLECT AND ORGANIZE We are asked to determine the solubility (the maximum concentration) of oxygen gas in water at an atmospheric pressure of 0.636 atm and 15°C. According to Table 10.1, dry air is 20.95% O_2. Henry's law (Equation 10.26) relates the solubility of gases to their partial pressures. The Henry's law constants for O_2 in water at different temperatures are listed in Table 10.6.

ANALYZE We need the partial pressure of oxygen to calculate its solubility using Equation 10.26. Partial pressure is the product of the mole fraction of O_2 in air (20.95% expressed as a decimal value) and atmospheric pressure (0.636 atm). The product of partial pressure and the Henry's law constant for O_2 in water at 15°C is the solubility value in molarity that we seek. To express solubility in mg/L will require converting moles to grams by multiplying by the molar mass of O_2. To estimate our answer, we note that all the Henry's law constants for O_2 in water are about 10^{-3} mol/(L · atm), and the value of P_{O_2} is about 0.2×0.6, or 0.12 atm, so the product of $k_H \times P_{O_2}$ (which is solubility according to Equation 10.26) will be around 10^{-4} mol/L. Multiplying this value by the molar mass of O_2 (32 g/mol) should give us a solubility near 3×10^{-3} g/L, or 3 mg/L.

SOLVE The partial pressure of oxygen is calculated using Equation 10.24:

$$P_{O_2} = x_{O_2}P_{total} = (0.2095)(0.636 \text{ atm}) = 0.133 \text{ atm}$$

We use this value for P_{O_2} in Equation 10.26 and the k_H for O_2 in water at 15°C from Table 10.6 to calculate solubility (C_{O_2}):

$$C_{O_2} = k_H P_{O_2} = \left(1.50 \times 10^{-3} \frac{\text{mol}}{\text{L} \cdot \text{atm}}\right)(0.133 \text{ atm}) = 2.00 \times 10^{-4} \text{ mol/L}$$

$$(2.00 \times 10^{-4} \text{ mol/L})(32.00 \text{ g/mol}) = 6.40 \times 10^{-3} \text{ g/L} = 6.40 \text{ mg/L}$$

THINK ABOUT IT The calculated solubility values in molarity and in mg/L are in the same ballpark as those we predicted and illustrate the limited solubility of O_2 in water. It makes sense that little O_2 dissolves because it is a nonpolar solute and water is a polar solvent, so the solute–solvent intermolecular forces that promote solubility are weak in this case.

Practice Exercise If the pressure of CO_2 inside a 1-liter bottle of seltzer is 44 psi at 20°C, how much CO_2 is dissolved in the seltzer? Express your answer in grams. ⚙

10.11 Gas Diffusion: Molecules Moving Rapidly

In Section 10.2 we learned that the root-mean-square speed of the atoms or molecules of a gas is inversely proportional to the square root of the molar mass of the gas. We used this relationship to calculate the ratios of the root-mean-square speeds of the particles of different gases at the same temperature:

$$\frac{u_{rms,A}}{u_{rms,B}} = \sqrt{\frac{\mathcal{M}_B}{\mathcal{M}_A}} \tag{10.6}$$

Equation 10.6 allows us to compare u_{rms} values for any two gases, but it does not provide actual u_{rms} values. A different equation, which is derived from kinetic molecular theory, does:

$$u_{rms} = \sqrt{\frac{3RT}{\mathcal{M}}} \tag{10.27}$$

The value of R in this equation is equivalent to the others we have used in this chapter, though the units are different:

$$R = 8.314 \frac{\text{kg} \cdot \text{m}^2}{\text{s}^2 \cdot \text{mol} \cdot \text{K}}$$

This version of R has the same numerical value as the one with joules in the numerator:

$$R = 8.314 \, \frac{J}{mol \cdot K}$$

and for good reason: $1 \, J = 1 \, kg \cdot (m/s)^2$. In using Equation 10.27, we need to be careful to express molar masses in kilograms (not grams) per mole to be consistent with the units on R. As always, temperature must be expressed in kelvin. The units on u_{rms} are meters per second. The following Sample Exercise illustrates how to calculate the speeds of gas-phase atoms and molecules.

▶❙❙ **CHEMTOUR** Molecular Speed

SAMPLE EXERCISE 10.14 **Calculating Root-Mean-Square Speeds** **LO9**

Calculate the root-mean-square speed of nitrogen molecules at 25°C.

COLLECT AND ORGANIZE We are asked to calculate the root-mean-square speed (u_{rms}) of N_2 molecules at 25°C (298 K). Equation 10.27 relates the u_{rms} speed of the particles in a gas to its absolute temperature and molar mass.

ANALYZE Nitrogen gas (N_2) has a molar mass of 28.02 g/mol. We can estimate the value of u_{rms} from the approximate values of R, T, and $\mathcal{M}$ in the right side of Equation 10.27: R is about 10 and T is about 300 K, so the value of the numerator is approximately $3 \times 10 \times 300$, or nearly 10,000. Expressing the molar mass of N_2 in kg/mol makes the denominator about 0.03, and the fraction inside the square root sign is about 10,000/0.03 or 333,000, the square root of which is between 500 and 600 m/s.

SOLVE

$$u_{rms} = \sqrt{\frac{3RT}{\mathcal{M}}}$$

$$= \sqrt{\frac{(3)\left(8.314 \, \dfrac{kg \cdot m^2}{s^2 \cdot mol \cdot K}\right)(273 + 25) \, K}{\left(\dfrac{28.02 \, g}{1 \, mol}\right)\left(\dfrac{1 \, kg}{1000 \, g}\right)}} = 515 \text{ m/s}$$

THINK ABOUT IT The calculated result is within our estimated range, and it suggests that the molecules in the air surrounding us are moving *very* rapidly. In U.S. Customary units, 515 m/s corresponds to 1150 miles per hour. An airplane flying that fast would be supersonic.

Practice Exercise Calculate the root-mean-square speed of helium at 25°C in meters per second, and compare your result with the root-mean-square speed of nitrogen calculated in Sample Exercise 10.14. ⚙

Equation 10.27 and the foundations of kinetic molecular theory tell us that the root-mean-square speed of the particles of a gas is proportional to the square root of the absolute temperature of the gas. This increase in u_{rms} with increasing T is illustrated in Figure 10.35. Note that increasing u_{rms} values are accompanied by wider distributions in particle speeds with increasing temperature. The profiles in Figure 10.35 also show that no matter how hot a gas is, there is a finite possibility that at least some of the particles in it are hardly moving at all.

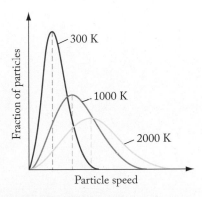

FIGURE 10.35 The most probable speeds (dashed lines) of the particles of a gas increase with increasing temperature. Notice that the distributions of particle speeds also broaden as temperature increases.

diffusion the spread of one substance (usually a gas or liquid) through another.

mean free path the average distance that a particle can travel through air or any gas before colliding with another particle.

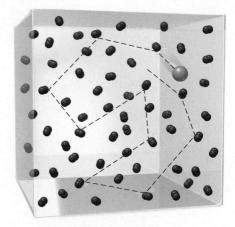

FIGURE 10.36 Diffusion rates of gases are slowed by frequent collisions between gas-phase particles. The dashed line shows the path followed by one particle as it collides with others. The average distance between collisions is called the mean free path.

In Section 10.2 we linked the rate at which a gas effuses, that is, leaks through pinholes, to the root-mean-square speed of its particles. Effusion of gases is closely related to the **diffusion** of gases, which is defined as the spread of one substance through another. Gas-phase diffusion occurs when there are differences in composition of a mixture of gases within the space they occupy. Diffusion is one of the processes by which odors spread from their source through the air, such as the smell of perfume from the person wearing it or the smell of baking bread from the kitchen. Given the high speeds at which gas particles move, you might expect diffusion in air to be very rapid. It's not. If the air is still, it may take several minutes for the scent of an open bottle of perfume to spread throughout a space the size of a bedroom. It takes this long because the molecules of scent keep colliding with molecules of N_2, O_2, and other components of air (Figure 10.36). Even though ambient air is mostly empty space between these molecules, they still take up enough space to limit how far any molecule can travel before it collides with one of them.

The average distance that a particle can travel through air or any gas before colliding with another one is called the **mean free path** of the particle. As you might expect, mean free paths depend on how densely the air particles are packed, and that depends on their pressure. At 1 atm, the mean free path is about 6.8×10^{-8} meter. For particles traveling at hundreds of meters per second, such a tiny distance translates into frequent collisions: about 10^{10} of them per second. A high collision frequency translates into a lot of bouncing around with little forward progress—and a slow rate of diffusion.

Recall that we also described Thomas Graham's effusion experiments and formulated *Graham's law of effusion*:

$$\frac{\text{effusion rate}_A}{\text{effusion rate}_B} = \sqrt{\frac{\mathcal{M}_B}{\mathcal{M}_A}} \qquad (10.7)$$

Graham conducted many gas *diffusion* experiments as well. A favorite experimental design of his involved measuring the time it took different gases to diffuse through porous plugs made of plaster of Paris. The results of the experiments were essentially the same as those for gas effusion: gases diffuse at rates that are inversely proportional to the square root of their densities (and molar masses). Thus, Graham's law of effusion and *Graham's law of diffusion* are interchangeable.

CONCEPT TEST

Which graph in Figure 10.37 best describes the ratio of the rates of diffusion of two gases (X and Y) as a function of temperature?

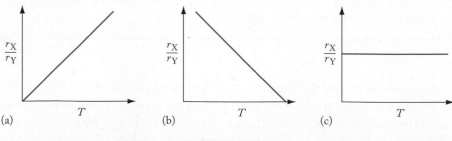

FIGURE 10.37

10.12 Real Gases

Up to now all of our calculations of P, V, T, and n have assumed ideal gas behavior, and this is acceptable because, under typical atmospheric pressures and temperatures, most gases *do* behave ideally. We have assumed, according to the kinetic molecular theory, that the volume occupied by individual gas particles is negligible compared with the total volume occupied by the gas. In addition, we have assumed that no interactions other than random elastic collisions occur between particles. These assumptions are not valid, however, when gases are subjected to extremely high pressures or to temperatures so low that they approach the temperature at which the gases condense.

Deviations from Ideality

Let's start by considering the behavior of one mole of a gas as we increase the pressure on the gas. From the ideal gas law, we know that $PV/RT = n$, so for one mole of gas, PV/RT should remain equal to 1 regardless of how we change the pressure. This relationship between PV/RT and P for an ideal gas is shown by the horizontal purple line in Figure 10.38. However, the curves for CH_4, H_2, and CO_2 at pressures above 10 atm are not horizontal straight lines. Not only do their curves diverge from the ideal, but the shapes of the curves also differ for each gas, indicating that deviations from ideal behavior depend on the identity of the gas.

Why don't these real gases behave like ideal gases at high pressure? One reason is that particles of ideal gases do not interact, but real atoms and molecules *do* interact with one another in various ways, as we discussed in Chapter 6. There are more opportunities for intermolecular interactions when extremely high pressures push gas-phase particles closer together, or when lower temperatures force the particles to move more slowly and they take longer to pass by one another. Both conditions—high pressure and low temperature—favor intermolecular interactions (represented by the double arrows in the right-side molecular view in Figure 10.39) that the ideal gas law does not take into account. The more often that particles interact with each other, the less frequently and forcefully they collide with the walls of their container, and the lower the P value in PV/RT.

Another reason for nonideal behavior is that atoms and molecules of gases do have some volume. Under extremely high pressure, the volume of the particles themselves becomes a significant fraction of the total volume of the gas as the particles are squeezed more tightly together. However, it is only the empty space between particles that makes gases compressible. The particles themselves are not compressible: they have finite dimensions and do not shrink with increasing pressure. Therefore, when the pressure doubles, the volume of a real gas does not shrink by 50% (as it does in an ideal gas) because

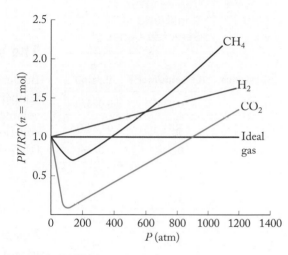

FIGURE 10.38 The effect of very high pressure on the behavior of 1 mole of real and ideal gases. The curves for real gases diverge from ideal behavior in a manner unique to each gas and due to two competing factors: interactions between particles (which lower the value of PV/RT) and the incompressibility of the particles themselves (which raises the value of PV/RT).

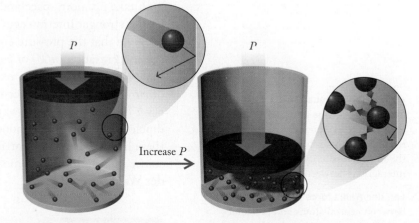

FIGURE 10.39 At very high pressures there are more interactions between particles (represented by the broad red arrows in the expanded view). These interactions reduce the frequency and force of particle collisions with the walls of the container (red arrow), thereby reducing the pressure.

CONNECTION Intermolecular forces including dipole–dipole interactions, dipole-induced dipole interactions, and London forces were discussed in Chapter 6.

TABLE 10.7	Van der Waals Constants of Selected Gases	
Substance	a ($L^2 \cdot atm/mol^2$)	b (L/mol)
He	0.0341	0.02370
Ar	1.34	0.0322
H_2	0.244	0.0266
N_2	1.39	0.0391
O_2	1.36	0.0318
CH_4	2.25	0.0428
CO_2	3.59	0.0427
CO	1.45	0.0395
H_2O	5.46	0.0305
NO	1.34	0.02789
NO_2	5.28	0.04424
HCl	3.67	0.04081
SO_2	6.71	0.05636

van der Waals equation an equation describing how the pressure, volume, and temperature of a quantity of a real gas are related; it includes terms that account for the incompressibility of gas particles and interactions between them.

van der Waals force any interaction between neutral atoms and molecules including hydrogen bonds, other dipole–dipole interactions, and London dispersion forces; the term does not apply to interactions involving ions.

a significant part of that volume can't shrink at all. If the volume shrinks by less than 50% when pressure doubles, then the product $P \times V$ in the numerator of PV/RT is no longer a constant. Instead, $P \times V$ and the PV/RT ratio increase as P increases. Because all atoms and molecules have a finite volume, this positive deviation in PV/RT is observed for all gases, including the three in Figure 10.38, if P is high enough.

So two competing factors, namely, intermolecular interactions and the volumes of gas-phase particles, can cause PV/RT to decrease or increase with increasing external pressure. At extremely high pressures, the size of the particles and the space they occupy becomes the dominant factor and PV/RT increases with increasing P. At less extreme pressures, intermolecular interactions may offset the incompressibility of particles, and the value of PV/RT decreases with increasing P for some gases, including CH_4 and CO_2 as shown in Figure 10.38.

The van der Waals Equation for Real Gases

Because the ideal gas equation does not hold at extremely high pressures and low temperatures, we need to modify it if we are to relate the quantities, pressures, volumes, and temperatures of real gases under these nonideal conditions. An equation developed in 1873 by Dutch scientist Johannes Diderik van der Waals (1837–1923) incorporates the needed modifications:

$$\left[P + a \left(\frac{n}{V} \right)^2 \right] (V - nb) = nRT \qquad (10.28)$$

Equation 10.28 is called the **van der Waals equation** and includes two terms to correct for (1) intermolecular interactions that lower the number of independent particles and the pressure they create by colliding with container walls [the $a(n/V)^2$ term] and (2) the volume taken up by the particles of a gas, which are not compressible (the nb term). The values of a and b in these terms, called *van der Waals constants*, have been determined experimentally for many gases (Table 10.7). The values of both constants increase with increasing molar mass and with the number of atoms per molecule, which makes sense because larger molecules both take up more space and experience stronger London dispersion forces that lead to stronger intermolecular actions.

Note that the pressure correction term, $a(n/V)^2$, increases with the square of the concentration (n/V) of particles. This dependency makes sense if we consider that the chance that two particles will get close enough to interact with each other depends on the concentration of both particles, that is $(n/V)^2$. In Chapter 6 we explored the many ways that molecules can interact: London dispersion, dipole–dipole, hydrogen bonding, and dipole–induced dipole interactions. All of these interactions may contribute to the value of the a constant of a gaseous substance, depending on its molecular structure, so they are collectively called **van der Waals forces**.

CONCEPT TEST

The van der Waals constant a for SO_2, 6.71 $L^2 \cdot atm/mol^2$, is nearly twice the value of a for CO_2, 3.59 $L^2 \cdot atm/mol^2$. Suggest a reason for this difference.

SAMPLE EXERCISE 10.15 **Using the van der Waals Equation** **LO10**

Patients suffering from chronic lung disease often rely on portable tanks of compressed O_2 when they are out and about (Figure 10.40). The tanks have an internal volume of 2.24 liters and typically contain 0.500 kg O_2 when filled to their recommended pressure. What is that pressure at 20°C? Calculate your answer using both the ideal gas equation and the van der Waals equation.

COLLECT AND ORGANIZE We are given the mass, volume, and temperature of a quantity of O_2, and we are asked to calculate its pressure using both the van der Waals equation and the ideal gas equation. The values of van der Waals constants a and b for O_2 are listed in Table 10.7.

ANALYZE The van der Waals constants for O_2 in Table 10.7 are $a = 1.36\ L^2 \cdot atm/mol^2$ and $b = 0.0318\ L/mol$. These values are a little more than half the a and b values for CH_4, which exhibits nonideal behavior as shown in Figure 10.38: intermolecular interactions produce a decrease in PV/RT below 400 atm. O_2 molecules should experience weaker, but still significant, intermolecular interactions that lead to pressures below those predicted by the ideal gas law.

SOLVE Let's start by converting the mass of O_2 into the corresponding number of moles because n appears several times in both the ideal gas and van der Waals equations.

$$0.500\ \text{kg O}_2 \left(\frac{1000\ \text{g}}{1\ \text{kg}} \right)\left(\frac{1\ \text{mol O}_2}{32.00\ \text{g O}_2} \right) = 15.62\ \text{mol O}_2$$

Assuming ideal behavior and solving the ideal gas equation for P yields

$$P = \frac{nRT}{V} = \frac{(15.62\ \text{mol})\left(0.08206\ \frac{\text{L} \cdot \text{atm}}{\text{mol} \cdot \text{K}} \right)(273 + 20)\ \text{K}}{(2.24\ \text{L})} = 168\ \text{atm}$$

Repeating this calculation using the van der Waals equation, we obtain

$$P = \frac{nRT}{V - nb} - a\left(\frac{n}{V} \right)^2$$

$$= \frac{(15.62\ \text{mol})\left(0.08206\ \frac{\text{L} \cdot \text{atm}}{\text{mol} \cdot \text{K}} \right)(273 + 20)\ \text{K}}{2.24\ \text{L} - (15.62\ \text{mol})\left(0.0318\ \frac{\text{L}}{\text{mol}} \right)} - \left(1.36\ \frac{\text{L}^2 \cdot \text{atm}}{\text{mol}^2} \right)\left(\frac{15.62\ \text{mol}}{2.24\ \text{L}} \right)^2$$

$$= \qquad 215.4\ \text{atm} \qquad\quad - \qquad 66.1\ \text{atm}$$

$$= 149\ \text{atm}$$

THINK ABOUT IT As predicted, the "real" pressure is less than the "ideal" pressure due to intermolecular interactions between O_2 molecules. The last step in the calculation shows that the pressure in the tank would have been much higher (215 atm) than in an ideal gas (168 atm) because of the volume of the gas occupied by O_2 molecules—the nb correction term. However, this increase was more than offset by the decrease in pressure (−66 atm) due to interactions between molecules—the $a(n/V)^2$ correction term.

Practice Exercise Thousands of buses in the United States and other countries run on compressed natural gas (CNG) to reduce pollution and save fuel costs (Figure 10.41). CNG, which is mostly methane, is stored on the roofs of the buses in large tanks in which the density of CH_4 is as high as 165 g/L. Use the van der Waals equation and the ideal gas equation to calculate the pressure in one of these tanks at 25°C. ⚙

FIGURE 10.40 A patient breathing supplemental oxygen from a portable tank.

FIGURE 10.41 A bus fueled by compressed natural gas.

SAMPLE EXERCISE 10.16   Integrating Concepts: Air for a Jet Engine

The Boeing 767 (Figure 10.42) is one of the most popular wide-body commercial airliners (over 1000 have been built), and it's also one of the most fuel efficient. While cruising at 851 km/hr (530 mph) at an altitude of 11,000 m (36,000 ft), a 767-200ER (the extended-range version of the plane) consumes about 1720 U.S. gallons of jet fuel per hour.

We are given the following facts: (1) the density of jet fuel is 0.80 g/mL; (2) at an altitude of 11,000 m, P_{atm} = 210 mmHg and T = –56°C; (3) 1 U.S. gallon = 3.785 L; and (4) dodecane, $C_{12}H_{26}$, is considered an appropriate model hydrocarbon for jet fuel.

    a. What volume of air, in liters, does a cruising 767 need so that it can completely burn an hour's worth of jet fuel?
    b. Fuel efficiency is often based on number of passengers times distance traveled per volume of fuel consumed. In the United States, this value is typically expressed in units of passenger-miles per gallon. However, in much of the rest of the world, efficiency units are inverted and are typically expressed in liters per 100 km per passenger. Express the fuel efficiency of a full 767-200ER (which holds 224 passengers) in both sets of units.

COLLECT AND ORGANIZE We know the quantity of dodecane to be combusted, and we are asked to calculate the volume of air needed for complete combustion—that is, to convert its C and H content into CO_2 and H_2O. We know the pressure and temperature of the air. According to Table 10.1, dry air is 20.95% O_2.

ANALYZE This exercise involves a chemical reaction, so writing a balanced chemical equation describing it is a good place to start. Next we need to convert the volume of fuel into an equivalent number of moles of fuel, and then to use the stoichiometry of the reaction to convert that value to moles of O_2. We will then calculate the equivalent volume of O_2 using the ideal gas equation. (The pressure is below 1 atm, so there should be no need to correct for nonideal behavior.) Finally we will convert the volume of O_2 to the corresponding volume of air. The volume of air needed each hour by the engines of a 767 should be enormous.

SOLVE
    a. Calculating volume of air needed in 1 hour:
       1. Write the balanced chemical equation describing the combustion reaction. The reactants and products are

$$C_{12}H_{26}(\ell) + O_2(g) \rightarrow CO_2(g) + H_2O(g)$$

We first balance the numbers of C and H atoms:

$$C_{12}H_{26}(\ell) + O_2(g) \rightarrow 12\ CO_2(g) + 13\ H_2O(g)$$

This leaves us with an odd number of O atoms on the right, requiring that we multiply all the terms by 2 and then balance the number of O atoms:

$$2\ C_{12}H_{26}(\ell) + 37\ O_2(g) \rightarrow 24\ CO_2(g) + 26\ H_2O(g)$$

FIGURE 10.42 A Boeing 767-200ER.

    2. Converting the volume of fuel consumed in 1 hour into an equivalent number of moles of $C_{12}H_{26}$:

$$1720\ \text{gal} \left(\frac{3.785\ \text{L}}{1\ \text{gal}}\right)\left(\frac{1000\ \text{mL}}{1\ \text{L}}\right)\left(\frac{0.80\ \text{g}}{\text{mL}}\right)\left(\frac{1\ \text{mol}}{170.33\ \text{g}}\right)$$

$$= 3.06 \times 10^4\ \text{mol}\ C_{12}H_{26}$$

    3. Converting moles of $C_{12}H_{26}$ into moles of O_2:

$$3.06 \times 10^4\ \text{mol}\ C_{12}H_{26}\left(\frac{37\ \text{mol}\ O_2}{2\ \text{mol}\ C_{12}H_{26}}\right) = 5.66 \times 10^5\ \text{mol}\ O_2$$

    4. In converting moles of O_2 into a volume of air using the ideal gas law, we need to convert pressure units from mmHg to atm and temperature to the Kelvin scale:

$$V = \frac{nRT}{P} = \frac{(5.66 \times 10^5\ \text{mol})\left(0.08206\ \dfrac{\text{L} \cdot \text{atm}}{\text{mol} \cdot \text{K}}\right)(273 - 56)\ \text{K}}{210\ \text{mmHg}\left(\dfrac{1\ \text{atm}}{760\ \text{mmHg}}\right)}$$

$$= 3.65 \times 10^7\ \text{L}\ O_2$$

    5. Air is 20.95% O_2 by volume, so the volume of air the engines must take in each hour is

$$3.65 \times 10^7\ \text{L}\ O_2\left(\frac{100\ \text{L air}}{20.95\ \text{L}\ O_2}\right) = 1.74 \times 10^8\ \text{L air}$$

    b. Calculating fuel efficiency in U.S. Customary units based on the distance traveled and fuel consumed in 1 hour:

$$\frac{224\ \text{passengers} \times 530\ \text{miles}}{1720\ \text{gal}} = 69.0\ \text{passenger-miles/gallon}$$

The corresponding efficiency value in liters per 100 km per passenger is

$$\frac{1720\ \text{gal} \times \dfrac{3.785\ \text{L}}{1\ \text{gal}}}{224\ \text{passengers} \times 851\ \text{km}} \times 100 = 3.41\ \text{L/100 km-passenger}$$

THINK ABOUT IT The calculated volume of air is 174 million liters. That is a lot of air, but then 1720 U.S. gallons of jet fuel is a lot of fuel. Perhaps a more interesting value is the 69.0 passenger-miles per gallon: better than that of most automobiles with two occupants, and you reach your destination much faster.

SUMMARY

Section 10.1 All gases occupy the entire volume of their container. The volume occupied by a gas changes significantly with pressure and temperature. Gases are miscible, mixing in any proportion, and they are much less dense than liquids or solids.

Section 10.2 The behavior of gases is explained by **kinetic molecular theory (KMT)**, which assumes that gases are composed of particles in constant random motion, that the volume of the particles is insignificant compared to the

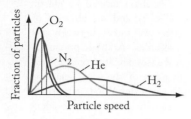

volume occupied by the gas, and that particles' collisions are elastic. The particles are moving at **root-mean-square speeds (u_{rms})** that are inversely proportional to the square root of their molar masses, which explains **Graham's law of effusion**: the rate of **effusion** (escape through a pinhole) of a gas at a fixed temperature is inversely proportional to the square root of its molar mass.

Section 10.3 The mass of the gases in Earth's atmosphere combined with the force of gravity result in an atmospheric pressure on the surface of the planet. Pressure is defined as the ratio of force to surface area and is measured with **barometers** and **manometers**.

Section 10.4 The laws describing how pressure, volume, and temperature of a gas are related bear the names of the scientists who discovered them: **Boyle's law, Charles's law, Avogadro's law, Amontons's law** (Figure 10.43).

Section 10.5 The **combined gas law** relates the pressure, volume, and temperature of a fixed quantity of gas.

Section 10.6 The **ideal gas law** and the **ideal gas equation** ($PV = nRT$), where R is the **universal gas constant**) describe the behavior of gases under normal conditions. At **standard temperature and pressure (STP)**, which in the United States is traditionally defined as 0°C and 1 atm, the *molar volume* of an **ideal gas** is 22.4 L.

Section 10.7 The density of a gas is proportional to its molar mass and pressure, and inversely proportional to its absolute temperature.

Section 10.8 The ideal gas equation and the stoichiometry of a chemical reaction can be used to calculate the volumes of gases required or produced in the reaction.

(a) **Boyle's law:** volume inversely proportional to pressure; *n* and *T* fixed

$$PV = \text{constant}$$

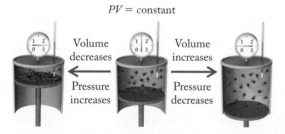

(b) **Charles's law:** volume directly proportional to temperature; *n* and *P* fixed

$$\frac{V}{T} = \text{constant}$$

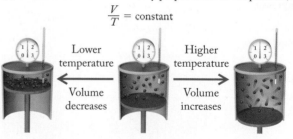

(c) **Avogadro's law:** volume directly proportional to number of moles; *T* and *P* fixed

$$\frac{V}{n} = \text{constant}$$

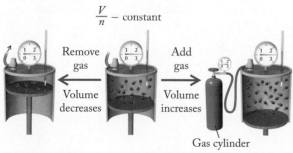

(d) **Amontons's law:** pressure directly proportional to temperature; *n* and *V* fixed

$$\frac{P}{T} = \text{constant}$$

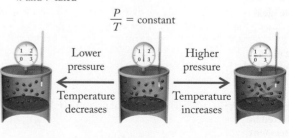

FIGURE 10.43 Summary of ideal gas behavior. The relations between: (a) pressure and volume (Boyle's law); (b) volume and temperature (Charles's law); (c) volume and quantity (Avogadro's law), and (d) pressure and temperature (Amontons's law).

Section 10.9 In a contribution each component gas makes to the total gas pressure is called the **partial pressure** of that gas. **Dalton's law of partial pressures** states that the total pressure of a gas mixture is the sum of the partial pressures of its components. Dalton's law allows us to calculate the partial pressure (P_i) of any constituent gas i in a gas mixture if we know its **mole fraction (x_i)** and the total pressure. The total pressure of a gas sample collected over water is the sum of the partial pressure of water vapor and the partial pressure of the gas.

Section 10.10 According to **Henry's law** the solubilities of gases increase with increasing partial pressure. Solubilities decrease with increasing temperature.

Section 10.11 Gas **diffusion** is the spread of one gas through another. Graham's law of effusion describes diffusion as well: at a fixed temperature, the rate at which a gas spreads is inversely proportional to the square root of its molar mass.

Section 10.12 Ideal gas behavior is observed at moderate temperatures and low pressures, where $PV = nRT$ holds. Ideal gas behavior is characterized by elastic collisions and an absence of attractive forces between gas molecules. At high pressures, real gases deviate from the predictions of the ideal gas law. The **van der Waals equation**, a modified form of the ideal gas equation, accounts for the behavior of real gases.

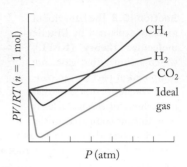

PROBLEM-SOLVING SUMMARY

TYPE OF PROBLEM	CONCEPTS AND EQUATIONS		SAMPLE EXERCISES
Calculating relative rates of effusion	$\dfrac{\text{effusion rate}_A}{\text{effusion rate}_B} = \sqrt{\dfrac{\mathcal{M}_B}{\mathcal{M}_A}}$	(10.7)	10.1
Measuring gas pressure with a manometer; converting pressure units	See the conversion factors inside the back cover and Table 10.2.		10.2
Applying Boyle's law	$P_1V_1 = P_2V_2$	(10.10)	10.3
Applying Charles's law	$\dfrac{V_1}{V_2} = \dfrac{T_1}{T_2}$	(10.14)	10.4
Applying Amontons's law	$\dfrac{P_1}{T_1} = \dfrac{P_2}{T_2}$	(10.18)	10.5
Applying the combined gas law	$\dfrac{P_1V_1}{T_1} = \dfrac{P_2V_2}{T_2}$	(10.19)	10.6
Applying the ideal gas law	$PV = nRT$	(10.20)	10.7–10.10
Calculating mole fractions and partial pressures	$x_i = \dfrac{n_i}{n_{\text{total}}}$ $P_i = x_i P_{\text{total}}$	(10.23) (10.24)	10.11
Calculating the quantity of a gas collected by water displacement	Calculate the partial pressure (P_i) of the collected gas using the equation $P_i = P_{\text{atm}} - P_{H_2O}$		10.12
Calculating gas solubility using Henry's law	$C_{\text{gas}} = k_H P_{\text{gas}}$	(10.26)	10.13
Calculating root-mean-square speeds	$u_{\text{rms}} = \sqrt{\dfrac{3RT}{\mathcal{M}}}$	(10.27)	10.14
Using the van der Waals equation	$\left[P + a\left(\dfrac{n}{V}\right)^2\right](V - nb) = nRT$	(10.28)	10.15

VISUAL PROBLEMS ●● ■

(Answers to boldface end-of-chapter questions and problems are in the back of the book.)

10.1. Which of the drawings in Figure P10.1 most accurately reflects the distribution of gas molecules in the balloon?

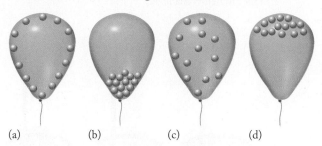

(a) (b) (c) (d)

FIGURE P10.1

10.2. How does the downward motion of the piston in Figure P10.2(a) affect the pressure of the gas in the cylinder? Assume that the temperature of the gas does not change significantly, but the volume is reduced by half.

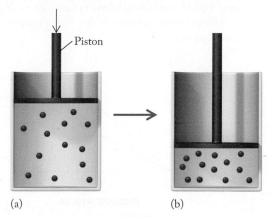

Piston

(a) (b)

FIGURE P10.2

10.3. How does the motion of the piston in Problem 10.2 affect the root-mean-square speed of the gas particles in the cylinder?

10.4. Suppose the temperature of the gas in Figure P10.2(b) increases from 200 K to 400 K but that the pressure on the piston remains constant. Does the position of the piston change? If so, by how much does the volume of the gas change?

10.5. Enough gas is added isothermally to the cylinder in Figure P10.5(a) to double the number of particles inside the cylinder.
  a. Assuming the position of the piston does not change, by how much does pressure inside the cylinder change?
  b. By how much does the frequency of the collisions between gas particles and the inner walls of the cylinder change?
  c. By how much does the most probable speed of the gas particles change?

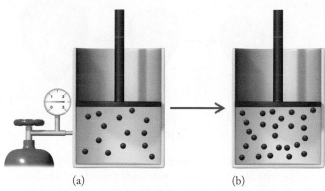

(a) (b)

FIGURE P10.5

10.6. With reference to Problem 10.5, suppose the same quantity of gas was added to the cylinder in Figure P10.5 at constant temperature *and* constant pressure.
  a. In which direction would the piston move?
  b. By how much would the volume of the gas in the cylinder change?

10.7. Consider the two gas mixtures in Figure P10.7. Assume their volumes and temperatures are the same. Molecules of gas A are represented by red spheres.
  a. In which mixture is the mole fraction of A greater?
  b. In which mixture is the partial pressure of A greater?
  c. In which mixture is the total pressure greater?

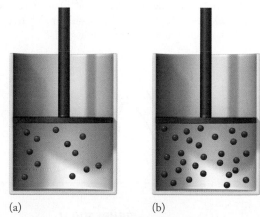

(a) (b)

FIGURE P10.7

10.8. Suppose the piston in Figure P10.7(a) was pushed downward so that the volume of the gas mixture was about 2/3 its initial value. Which of the following values would increase, which would decrease, and which would remain the same assuming no change in temperature or in the numbers of red and blue spheres occurs?
  a. the mole fraction of A (whose molecules are represented by red spheres)
  b. the partial pressure of A
  c. the total pressure of the mixture
  d. the most probable speed of the molecules of A

10.9. Which of the two outcomes diagrammed in Figure P10.9 more accurately illustrates the effusion of helium from a balloon at constant atmospheric pressure?

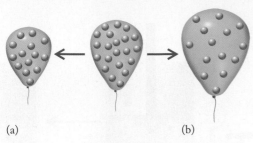

(a)   (b)

FIGURE P10.9

10.10. Which of the two outcomes shown in Figure P10.10 more accurately illustrates the effusion of gases from a balloon at constant atmospheric pressure if the red spheres have a greater root-mean-square speed than the blue spheres?

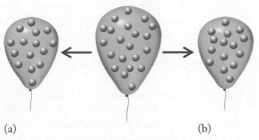

(a)   (b)

FIGURE P10.10

10.11. Figure P10.11 shows the distribution of molecular speeds of CO_2 and SO_2 molecules at 25°C. Which curve is the profile for SO_2? Which of the profiles should match that of propane (C_3H_8), a common fuel in portable grills?

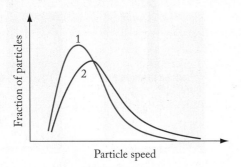

FIGURE P10.11

10.12. How would a graph showing the distribution of molecular speeds of CO_2 at –100°C differ from the curve for CO_2 shown in Figure P10.11?

10.13. Consider the three Torricelli barometers in Figure P10.13. Which one is most likely sensing atmospheric pressure at each of these locations?
1. San Diego, California (sea level)
2. The summit of Mount Everest (8.8 km above sea level)
3. The bottom of the Tau Tona gold mine in South Africa (3.9 km below sea level)

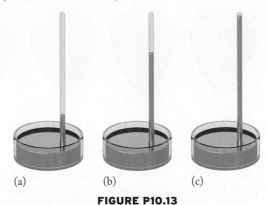

(a)   (b)   (c)

FIGURE P10.13

10.14. The two plots of volume versus reciprocal pressure in Figure P10.14 correspond to the same quantity of gas at two different temperatures. Which line corresponds to the higher temperature?

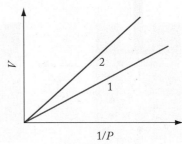

FIGURE P10.14

10.15. Suppose the two plots of volume versus reciprocal pressure in Figure P10.14 were for two different quantities of the same gas at the same pressure. Which line corresponds to the greater quantity?

10.16. In Figure P10.16, which of the two plots of volume versus pressure at constant temperature is consistent with the ideal gas law?

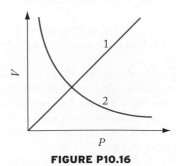

FIGURE P10.16

10.17. In Figure P10.17, which of the two plots of volume versus temperature at constant pressure is not consistent with the ideal gas law?

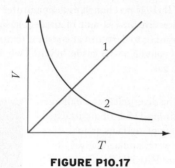

FIGURE P10.17

10.18. Figure P10.18 contains plots of the densities of propane and butane gas versus pressure at the same temperature. Which plot is based on propane densities?

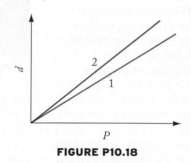

FIGURE P10.18

10.19. Which of the images in Figure P10.19 best describes the effect of pressure on the solubility of a gas?

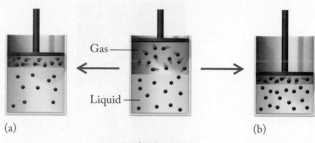

FIGURE P10.19

10.20. Which of the images in Figure P10.20 represents the gas with the greatest Henry's law constant, k_H?

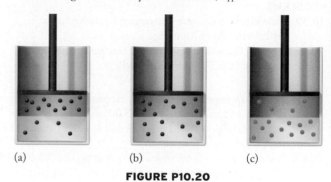

FIGURE P10.20

QUESTIONS AND PROBLEMS

Effusion and the Kinetic Molecular Theory of Gases

CONCEPT REVIEW

10.21. What is meant by the *root-mean-square* speed of gas particles?

***10.22.** Why is the root-mean-square speed of gas particles greater than the simple average of their speeds?

10.23. Does pressure affect the root-mean-square speed of the particles in a gas? Why or why not?

10.24. How is the rate of effusion of a gas related to its: (a) molar mass; (b) root-mean-square speed; (c) temperature?

PROBLEMS

10.25. Rank the gases NO, NO_2, N_2O_4, and N_2O_5 in order of increasing root-mean-square speed at 0°C.

10.26. In a mixture of the major components of natural gas: CH_4, C_2H_6, C_3H_8, and C_4H_{10}, which gas effuses the slowest and which the fastest?

10.27. Molecular hydrogen effuses 4 times as fast as gas X at the same temperature. What is the molar mass of gas X?

10.28. Which noble gas effuses 3.2 times faster than argon at 0°C?

10.29. If an unknown gas has one-third the root-mean-square speed of H_2 at 300 K, what is its molar mass?

10.30. How much faster does F_2 effuse than Cl_2 at 25°C?

10.31. In the 1930s hydrogen gas was used to inflate an earlier generation of today's helium-filled blimps that fly over championship football games. Which gas would effuse more rapidly from the same balloon?

10.32. Compounds sensitive to oxygen are often manipulated in glove boxes that may contain an atmosphere of pure nitrogen or pure argon. A rubber balloon filled with carbon monoxide was placed in such a glove box. After 24 hours, the volume of the balloon was unchanged. Did the glove box contain N_2 or Ar?

Atmospheric Pressure

CONCEPT REVIEW

10.33. Describe the difference between force and pressure.

10.34. Why does atmospheric pressure decrease with increasing elevation?

10.35. Three barometers based on Torricelli's design are constructed using water (density $d = 1.00$ g/mL), ethanol ($d = 0.789$ g/mL), and mercury ($d = 13.546$ g/mL). Which barometer contains the tallest column of liquid?

10.36. In constructing a barometer based on Torricelli's design, what advantage is there in choosing a dense liquid?

10.37. Why does an ice skater exert more pressure on ice when wearing newly sharpened skates than when wearing skates with dull blades?

10.38. Why is it easier to travel over deep snow when wearing boots and snowshoes (Figure P10.38) rather than just boots?

FIGURE P10.38

PROBLEMS

10.39. Convert the following pressures into atmospheres:
(a) 2.0 kPa; (b) 562 mmHg.

10.40. Convert the following pressures into millimeters of mercury: (a) 0.541 atm; (b) 2.8 kPa.

10.41. Use the appropriate datum from Appendix 3 to calculate the mass of a cube of tin that is 5.00 cm on a side and to calculate the downward pressure due to gravity exerted by the bottom face of the cube.

10.42. A gold brick is 4.0 cm wide, 12.0 cm long, and 3.0 cm high. Use the appropriate datum from Appendix 3 to calculate:
a. The mass of the brick
b. The downward pressure due to gravity exerted by the bottom face of the brick
c. The downward pressure due to gravity exerted by the brick when it is stood up on its end

10.43. **Record High Atmospheric Pressure** The highest atmospheric pressure ever recorded on Earth was 108.6 kPa at Tosontsengel, Mongolia, on December 19, 2001. Express this pressure in (a) millimeters of mercury, (b) atmospheres, and (c) millibars.

10.44. **Record Low Atmospheric Pressure** Despite the destruction from Hurricane Katrina in August, 2005, the lowest pressure for a hurricane in the Atlantic Ocean was measured several weeks after Katrina. Hurricane Wilma registered an atmospheric pressure of 88.2 kPa on October 19, 2005, about 2 kPa lower than Hurricane Katrina. What was the *difference* in pressure between the two hurricanes in (a) millimeters of mercury, (b) atmospheres, and (c) millibars?

Relating *P*, *T*, and *V*: The Gas Laws

CONCEPT REVIEW

10.45. How does kinetic molecular theory explain why pressure is directly proportional to temperature at fixed volume (Amontons's law)?

10.46. Explain Boyle's law using kinetic molecular theory.

10.47. A hot-air balloonist is rising too fast for her taste. Should she increase the temperature of the gas in the balloon, or decrease it?

10.48. Could the pilot of the balloon in Problem 10.47 reduce her rate of ascent by allowing some gas to leak out of the balloon? Explain your answer.

PROBLEMS

10.49. The volume of a quantity of gas at 1.00 atm is compressed from 3.25 liters to 2.24 L. What is the final pressure of the gas if there is no change in temperature?

10.50. The pressure on a sample of an ideal gas is increased from 715 mmHg to 3.55 atm at constant temperature. If the initial volume of the gas is 485 mL, what is the final volume of the gas?

10.51. A scuba diver releases a balloon containing 115 L of air attached to a tray of artifacts at an underwater archaeological site (Figure P10.51). When the balloon reaches the surface, it has expanded to a volume of 352 L. The pressure at the surface is 1.00 atm.
a. What is the pressure at the underwater site? Assume that water temperature is constant.
b. If pressure increases by 1.0 atm for every 10 m of depth, at what depth was the diver working?

FIGURE P10.51

10.52. **Breath-Hold Diving** The world record for diving without supplemental air tanks ("breath-hold diving") is about 125 m, a depth at which water pressure is about 12.5 atm. If a diver's lungs have a volume of 6 L at the surface of the water, what is their volume at a depth of 125 m?

10.53. A 4.66 L sample of gas is warmed from 273 K to a final temperature of 398 K. Assuming no change in pressure, what is the final volume of the gas?

10.54. A 22.5 L sample of gas is cooled from 145°C to a temperature at which its volume is 18.3 L. What is the new temperature? Assume no change in pressure of the gas.

10.55. Balloons for a New Year's Eve party in Potsdam, New York, are filled to a volume of 5.0 L at a temperature of 20°C and then hung outside where the temperature is −25°C. What is the volume of the balloons after they have cooled to the outside temperature? Assume that atmospheric pressure inside and outside the house is the same.

10.56. The air inside a balloon is heated to 45°C and then cools to 25°C. By what percentage does the volume of the balloon change during cooling?

10.57. Which of the following actions would produce the greatest increase in the volume of a gas sample?
a. Lowering the pressure from 760 mmHg to 700 mmHg at constant temperature
b. Raising the temperature from 10°C to 35°C at constant pressure

10.58. Which of the following actions would produce the greatest increase in the volume of a gas sample?
a. Doubling the amount of gas in the sample at constant temperature and pressure
b. Raising the temperature from 244°C to 1100°C

10.59. What happens to the volume of gas in a cylinder with a movable piston under the following conditions?
  a. Both the absolute temperature and the external pressure on the piston double.
  b. The absolute temperature is halved, and the external pressure on the piston doubles.
  c. The absolute temperature increases by 75%, and the external pressure on the piston increases by 50%.

10.60. What happens to the pressure of a gas under the following conditions?
  a. The absolute temperature is halved and the volume doubles.
  b. Both the absolute temperature and the volume double.
  c. The absolute temperature increases by 75%, and the volume decreases by 50%.

10.61. A 150.0 L weather balloon contains 6.1 moles of helium but loses it at a rate of 10 mmol/h. What is the volume of the balloon after 24 h?

10.62. Which has the greater effect on the volume of a gas at constant temperature: doubling the number of moles of gas or reducing the pressure by half?

10.63. **Temperature Effects on Bicycle Tires** A bicycle racer inflates his tires to 7.1 atm on a warm autumn afternoon when temperatures reached 27°C. By morning the temperature has dropped to 5.0°C. What is the pressure in the tires if we assume that the volume of the tire does not change significantly?

10.64. The volume of a weather balloon is 200.0 L and its internal pressure is 1.17 atm when it is launched at 20°C. The balloon rises to an altitude in the stratosphere where its internal pressure is 63 mmHg and the temperature is 210 K. What is the volume of the balloon then?

Ideal Gases and the Ideal Gas Law

CONCEPT REVIEW

10.65. What is meant by standard temperature and pressure (STP)? What is the volume of 1 mole of an ideal gas at STP?

10.66. Which of the following are not characteristics of an ideal gas?
  a. The molecules of gas have little volume compared with the volume that they occupy.
  b. Its volume is independent of temperature.
  c. The density of all ideal gases is the same.
  d. Gas atoms or molecules do not interact with one another.

PROBLEMS

10.67. How many moles of air must there be in a racing bicycle tire with a volume of 2.36 L if it has an internal pressure of 6.8 atm at 17.0°C?

10.68. At what temperature will 1.00 mol of an ideal gas in a 1.00 L container exert a pressure of 1.00 atm?

10.69. **Hyperbaric Oxygen Therapy** Hyperbaric oxygen chambers are used to treat divers suffering from decompression sickness (the "bends") with pure oxygen at greater than atmospheric pressure. Other clinical uses include treatment

of patients with thermal burns and CO poisoning. What is the pressure in a chamber with a volume of 4.85×10^3 L that contains 5.00 kg of $O_2(g)$ at a temperature of 298 K?

10.70. What is the volume of 8.80 g of CO_2 vapor at 37°C and 0.98 atm?

*10.71. Suppose atmospheric temperature and pressure at the top of a ski run are −5°C and 713 mmHg. At the bottom of the run, the temperature and pressure are 0°C and 734 mmHg. How many more moles of oxygen does a skier take in with a lungful of air at the bottom of the run than at the top? Express your answer as a percentage.

*10.72. A balloon vendor at a street fair is using a tank of helium to fill her balloons. The tank has an internal volume of 45.0 L and a pressure of 195 atm at 22°C. After a while she notices that the valve has not been closed properly and the pressure has dropped to 115 atm. How many moles of He have been lost?

10.73. In some cities public buses are fueled by compressed natural gas (mostly methane, CH_4). How many grams of CH_4 gas are in a 250 L fuel tank at a pressure of 255 bar at 20°C?

10.74. **Liquid Nitrogen-Powered Car** Students at the University of North Texas and the University of Washington built a car propelled by compressed nitrogen gas. The gas was obtained by boiling liquid nitrogen stored in a 182 L tank. What volume of N_2 is released at 0.927 atm of pressure and 25°C from a tank full of liquid N_2 ($d = 0.808$ g/mL)?

10.75. The volume of a 4.0-g sample of a gaseous substance was measured at the temperatures listed below. The pressure of the gas was 1.00 atm at all times.
  a. How many moles of a gas were in the sample?
  b. What was the probable identity of the gas?

V (L)	T (K)
7.88	96
3.94	48
1.97	24
0.79	9.6
0.39	4.8

10.76. The volume of 0.50 mol of a noble gas was measured at the temperatures listed below. The pressure of the gas was 1.00 atm at all times. Use the data in the table to calculate the value of the ideal gas law constant.

V (L)	T (K)
3.94	96
1.97	48
0.79	24
0.39	9.6
0.20	4.8

Densities of Gases

CONCEPT REVIEW

10.77. Do all gases at the same pressure and temperature have the same density? Explain your answer.

10.78. Birds and sailplanes take advantage of thermals (rising columns of warm air) to gain altitude with less effort than usual. Why does warm air rise?

10.79. How does the density of a gas sample change when (a) its pressure is increased, and (b) its temperature is decreased?

10.80. How would you measure the density of a gas sample of known molar mass?

PROBLEMS

10.81. Biological Effects of Radon Exposure Radon is a naturally occurring radioactive gas found in the ground and in building materials. It is easily inhaled and emits α particles when it decays. Cumulative radon exposure is a significant risk factor for lung cancer.
  a. Calculate the density of radon at 298 K and 1 atm of pressure.
  b. Are radon concentrations likely to be greater in the basement or on the top floor of a building?

*10.82. Four empty balloons, each with a mass of 10.0 g, are inflated to a volume of 20.0 L. The first balloon contains He, the second Ne, the third CO_2, and the fourth CO. If the density of air at 25°C and 1.00 atm is 1.17 g/L, how many of the balloons float in it?

10.83. A 30.0 mL flask contains 0.078 g of a volatile oxide of sulfur. The pressure in the flask is 750 mmHg, and the temperature is 22°C. Is the gas SO_2 or SO_3?

10.84. A 100.0 mL flask contains 0.193 g of a volatile oxide of nitrogen. The pressure in the flask is 760 mmHg at 17°C. Is the gas NO, NO_2, or N_2O_5?

10.85. The density of an unknown gas is 1.107 g/L at 300 K and 740 mmHg. Could this gas be CO or CO_2?

10.86. A gas containing chlorine and oxygen has a density of 2.875 g/L at 756 mmHg and 11°C. What is the most likely molecular formula of the gas?

Gases in Chemical Reactions

PROBLEMS

10.87. Miners' Lamps Before the development of reliable batteries, miners' lamps burned acetylene produced by the reaction of calcium carbide with water:

$$CaC_2(s) + H_2O(\ell) \rightarrow C_2H_2(g) + CaO(s)$$

Suppose a lamp uses 4.8 liters of acetylene per hour at 1.02 atm pressure and 25°C.
  a. How many moles of C_2H_2 are used per hour?
  b. How many grams of calcium carbide are consumed for a 4-hour shift?

10.88. Acid precipitation dripping on limestone produces carbon dioxide by the following reaction:

$$CaCO_3(s) + 2 H^+(aq) \rightarrow Ca^{2+}(aq) + CO_2(g) + H_2O(\ell)$$

If 8.6 mL of CO_2 were produced at 15°C and 760 mmHg, then
  a. How many moles of CO_2 were produced?
  b. How many milligrams of $CaCO_3$ were consumed?

10.89. Oxygen is generated by the thermal decomposition of potassium chlorate:

$$2 KClO_3(s) \rightarrow 2 KCl(s) + 3 O_2(g)$$

How much $KClO_3$ is needed to generate 200.0 L of oxygen at 0.85 atm and 273 K?

10.90. Calculate the volume of carbon dioxide at 20°C and 1.00 atm produced from the complete combustion of 1.00 kg of methane. Compare your result with the volume of CO_2 produced from the complete combustion of 1.00 kg of propane (C_3H_8).

10.91. Healthy Air for Submariners The CO_2 that builds up in the air of a submerged submarine can be removed by reacting it with sodium peroxide:

$$2 Na_2O_2(s) + 2 CO_2(g) \rightarrow 2 Na_2CO_3(s) + O_2(g)$$

If a sailor exhales 125 mL of CO_2 per minute at 23°C and 1.02 atm, how much sodium peroxide is needed per sailor in a 24-hour period?

10.92. **Rescue Breathing Devices** Self-contained self-rescue breathing devices, like the one shown in Figure P10.92, convert CO_2 into O_2 according to the following reaction:

$$4 KO_2(s) + 2 CO_2(g) \rightarrow 2 K_2CO_3(s) + 3 O_2(g)$$

How many grams of KO_2 are needed to produce 100.0 L of O_2 at 20°C and 1.00 atm?

FIGURE P10.92

Mixtures of Gases

CONCEPT REVIEW

10.93. What is meant by the *partial pressure* of a gas?

10.94. Can a barometer be used to measure just the partial pressure of oxygen in the atmosphere? Why or why not?

PROBLEMS

10.95. A gas mixture contains 0.70 mol N_2, 0.20 mol H_2, and 0.10 mol CH_4. What is the mole fraction of H_2 in the mixture?

10.96. A gas mixture contains 7.0 g N_2, 2.0 g H_2, and 16.0 g CH_4. What is the mole fraction of H_2 in the mixture?

10.97. Calculate the pressure of the gas mixture and the partial pressure of each constituent gas in Problem 10.95 if the mixture is in a 0.75 L vessel at 10°C.

10.98. Calculate the pressure of the gas mixture and the partial pressure of each constituent gas in Problem 10.96 if the mixture is in a 5.0 L vessel at 20°C.

10.99. A sample of oxygen is collected over water at 25°C and 1.00 atm. If the total sample volume is 0.480 L, how many moles of O_2 are collected?

10.100. Water vapor is removed from the O_2 sample in Problem 10.99. What is the volume of the dry O_2 at 25°C and 1.00 atm?

10.101. The following reactions are carried out in sealed containers. Will the total pressure after each reaction is complete be greater than, less than, or equal to the total pressure before the reaction? Assume all reactants and products are gases at the same temperature.
   a. $N_2O_5(g) + NO_2(g) \rightarrow 3\ NO(g) + 2\ O_2(g)$
   b. $2\ SO_2(g) + O_2(g) \rightarrow 2\ SO_3(g)$
   c. $C_3H_8(g) + 5\ O_2(g) \rightarrow 3\ CO_2(g) + 4\ H_2O(g)$

10.102. In each of the following gas-phase reactions, determine whether the total pressure at the end of the reaction (carried out in a sealed, rigid vessel) will be greater than, less than, or equal to the total pressure at the beginning. Assume all reactants and products are gases at the same temperature.
   a. $H_2(g) + Cl_2(g) \rightarrow 2\ HCl(g)$
   b. $4\ NH_3(g) + 5\ O_2(g) \rightarrow 4\ NO(g) + 6\ H_2O(g)$
   c. $2\ NO(g) + O_2(g) \rightarrow 2\ NO_2(g)$

*10.103. **High-Altitude Mountaineering** Most alpine climbers breathe pure oxygen near the summits of the world's highest mountains. How much more O_2 is there in a lungful of pure O_2 at an elevation where atmospheric pressure is 266 mmHg than in a lungful of air at sea level? Express your answer as a percentage.

10.104. **Scuba Diving** A scuba diver is at a depth of 50 m, where the pressure is 5.0 atm. What should be the mole fraction of O_2 in the gas mixture the diver breathes to achieve the same P_{O_2} as at sea level?

10.105. Carbon monoxide at a pressure of 650 mmHg reacts completely with O_2 at a pressure of 325 mmHg in a sealed vessel to produce CO_2. What is the final pressure in the flask?

10.106. Ozone reacts completely with NO, producing NO_2 and O_2. A 2.50 L vessel is filled with 0.150 mol NO and 0.150 mol O_3 at 125°C. What is the partial pressure of each

product and the total pressure in the flask at the end of the reaction?

*10.107. Ammonia is produced industrially from the reaction of hydrogen with nitrogen under pressure in a sealed reactor. What is the percent decrease in total pressure of a sealed reaction vessel during the reaction between H_2 at a partial pressure of 2.4 atm and N_2 at a partial pressure of 3.6 atm if half of the H_2 is consumed? No other gases are present initially in the reactor.

*10.108. A mixture of 0.156 mol C is reacted with 0.117 mol O_2 in a sealed, 10.0 L vessel at 500 K, producing a mixture of CO and CO_2. The total pressure is 0.640 atm. What is the partial pressure of CO?

Solubilities of Gases and Henry's Law

CONCEPT REVIEW

10.109. Why is the Henry's law constant for CO_2 so much larger than those for N_2 and O_2 at the same temperature? *Hint*: Does CO_2 react with water?

10.110. As water in a beaker is heated, bubbles form inside the beaker at temperatures well below the boiling point of water. What gas is in the bubbles?

10.111. What type of intermolecular interaction accounts for the limited solubility of methane in water?

*10.112. Air is primarily a mixture of nitrogen and oxygen. Is the Henry's law constant for the solubility of air in water the sum of k_H for N_2 and k_H for O_2? Explain why or why not.

PROBLEMS

*10.113. **Arterial Blood** Arterial blood contains about 0.25 g of oxygen per liter at 37°C and standard atmospheric pressure. What is the Henry's law constant, in mol/(L · atm), for O_2 dissolution in blood?

10.114. The solubility of O_2 in water is 6.5 mg/L at an atmospheric pressure of 1 atm and a temperature of 40°C. Calculate the Henry's law constant of O_2 at 40°C.

*10.115. **Oxygen for Climbers and Divers** Use the Henry's law constant for O_2 dissolved in arterial blood from Problem 10.113 to calculate the solubility of O_2 in the blood of (a) a climber on Mt. Everest ($P_{atm} = 0.35$ atm) and (b) a scuba diver breathing air at a depth of 20 meters ($P \approx 3.0$ atm).

*10.116. The solubility of air in water is approximately 7.9 × 10^{-4} M at 20°C and 1.0 atm. Calculate the Henry's law constant for air.

Gas Diffusion: Molecules Moving Rapidly

PROBLEMS

10.117. What is the root-mean-square speed of CH_4 at 298 K?

10.118. Air is approximately 21% O_2 and 78% N_2 by mass. Calculate the root-mean-square speed of each gas at 273 K.

10.119. Calculate the root-mean-square speed of Ar atoms at the temperature at which their average kinetic energy is 5.18 kJ/mol.

10.120. Determine the root-mean-square speed of CO_2 molecules that have an average kinetic energy of 3.2×10^{-21} J per molecule.

10.121. A flask of ammonia is connected to a flask of an unknown acid HX by a 1.00 m glass tube. As the two gases diffuse down the tube, a white ring of NH_4X forms 68.5 cm from the ammonia flask. Identify element X.

10.122. **Enriching Uranium** The two isotopes of uranium, ^{238}U and ^{235}U, can be separated by diffusion of the corresponding UF_6 gases. What is the ratio of the root-mean-square speed of $^{238}UF_6$ to that of $^{235}UF_6$ at constant temperature?

Real Gases

CONCEPT REVIEW

10.123. Why do real gases behave nonideally at very low temperatures and very high pressures?

10.124. Under what conditions is the pressure exerted by a real gas *less* than that predicted for an ideal gas?

10.125. Why do the values of the van der Waals constant *b* of the noble gas elements increase with atomic number?

10.126. Why does the value of the constant *a* in the van der Waals equation generally increase with the molar mass of the gas?

PROBLEMS

10.127. Explain why the van der Waals constant *a* for Ar is greater than *a* for He.

10.128. The van der Waals constant *a* for CO_2 is 3.59 $L^2 \cdot$ atm/mol^2. Would you expect the value of *a* for CS_2 to be larger or smaller than 3.59 $L^2 \cdot$ atm/mol^2?

10.129. The graphs of PV/RT versus P (see Figure 10.38) for 1 mole of CH_4 and 1 mole of H_2 differ in how they deviate from ideal behavior. For which gas is the effect of the volume occupied by the gas molecules more important than the attractive forces between molecules at $P = 200$ atm?

10.130. Which noble gas is expected to deviate the most from ideal behavior in a graph of PV/RT versus P?

10.131. At high pressures, real gases do not behave ideally.
  a. Use the van der Waals equation and data in the text to calculate the pressure exerted by 50.0 g of H_2 at 20°C in a 1.00 L container.
  b. Repeat the calculation assuming that the gas behaves like an ideal gas.

10.132. Calculate the pressure exerted by 5.00 mol of CO_2 in a 1.00 L vessel at 300 K (a) assuming the gas behaves ideally and (b) using the van der Waals equation.

Additional Problems

10.133. Each of the cylinders in the engine of the sports car in Figure P10.133 contains 633 mL of air and gasoline vapor before the motion of a piston increases the pressure inside the cylinder by 9.0 times.
  a. What is the volume of the air and gasoline vapor mixture after the compression, assuming the temperature of the mixture does not change significantly?
  b. Is the assumption of no temperature change realistic? Why or why not?

FIGURE P10.133

10.134. **Planetary Atmospheres** Saturn's largest moon, Titan, has a surface atmospheric pressure of 1220 torr. The atmosphere consists of 82% N_2, 12% Ar, and 6% CH_4 by volume. Calculate the partial pressure of each gas in Titan's atmosphere. Chilly temperatures aside, could life as we know it exist on Titan?

10.135. Scientists have used laser light to slow atoms to speeds corresponding to temperatures below 0.00010 K. At this temperature, what is the root-mean-square speed of argon atoms?

10.136. **Blood Pressure** A typical blood pressure in a resting adult is "120 over 80," meaning 120 mmHg with each beat of the heart and 80 mmHg of pressure between heartbeats. Express these pressures in the following units: (a) torr; (b) atm; (c) bar; (d) kPa.

*10.137. A popular style of scuba tank is called the "aluminum 80" because it can deliver 80 cubic feet of air at "normal" temperature (72°F) and pressure (1.00 atm) when filled with air at a pressure of 200 atm. A particular aluminum 80 tank has a mass of 15 kg empty. What is its mass when filled with air at 200 atm?

10.138. The flame produced by the burner of a gas (propane) grill is a pale blue color when enough air mixes with the propane (C_3H_8) to burn it completely. For every gram of propane that flows through the burner, what volume of air is needed to burn it completely? Assume that the temperature of the burner is 200°C, the pressure is 1.00 atm, and the mole fraction of O_2 in air is 0.21.

10.139. **Anesthesia** Halothane is a common anesthesia gas with the structure shown in Figure P10.139. Liquid halothane boils at 50.2°C and 1.00 atm. Assume that halothane behaves as an ideal gas.
  a. What volume would 6.0 mL of liquid halothane ($d = 1.87$ g/mL) occupy at 37°C and 1.00 atm of pressure?
  b. What would be its density?

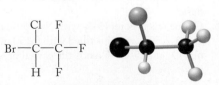

FIGURE P10.139

10.140. A cotton ball soaked in ammonia and another soaked in hydrochloric acid were placed at opposite ends of a 1.00 m glass tube (Figure P10.140). The vapors diffused toward the middle of the tube and formed a white ring of ammonium chloride where they met.
  a. Write the chemical equation for this reaction.
  b. Should the ammonium chloride ring be closer to the end of the tube with ammonia or the end with hydrochloric acid? Explain your answer.
  c. Calculate the distance from the ammonia end to the position of the ammonium chloride ring.

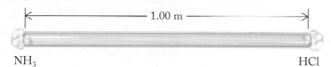

NH_3 HCl

FIGURE P10.140

*10.141. The same apparatus described in Problem 10.140 was used in another series of experiments. A cotton ball soaked in either hydrochloric acid (HCl) or acetic acid (CH_3COOH) was placed at one end. Another cotton ball soaked in one of three amines (a class of organic compounds)—CH_3NH_2, $(CH_3)_2NH$, or $(CH_3)_3N$—was placed in the other end.
  a. In one combination of acid and amine, a white ring was observed almost exactly halfway between the two ends. Which acid and which amine were used?
  b. Which combination of acid and amine would produce a ring closest to the amine end of the tube?
  c. Would any two of the six combinations result in the formation of product at the same position in the ring? Assume measurements can be made to the nearest centimeter.

10.142. The pressure in an aerosol can is 1.2 atm at 27°C. The can will withstand a pressure of 3.0 atm. Will it burst if heated in a campfire to 450°C?

10.143. Uranus has a total atmospheric pressure of 130 kPa and consists of the following gases: 83% H_2, 15% He, and 2% CH_4 by volume. Calculate the partial pressure of each gas in Uranus's atmosphere.

10.144. Derive an equation that expresses the ratio of the densities (d_1 and d_2) of a gas under two different combinations of temperature and pressure: (T_1, P_1) and (T_2, P_2).

10.145. **Denitrification in the Environment** In some aquatic ecosystems, nitrate (NO_3^-) is converted to nitrite (NO_2^-), which then decomposes to nitrogen and water. As an example of this second reaction, consider the decomposition of ammonium nitrite:

$$NH_4NO_2(aq) \rightarrow N_2(g) + 2 H_2O(\ell)$$

What is the change in pressure in a sealed 10.0 L vessel due to the formation of N_2 gas when the ammonium nitrite in 1.00 L of 1.0 M NH_4NO_2 decomposes at 25°C?

*10.146. When sulfur dioxide bubbles through a solution containing nitrite, chemical reactions that produce gaseous N_2O and NO may occur.
  a. How much faster on average are NO molecules moving than N_2O molecules in such a reaction mixture?
  b. If these two nitrogen oxides are separated based on differences in their rates of effusion, will unreacted SO_2 interfere with the separation? Explain your answer.

*10.147. **Using Wetlands to Treat Agricultural Waste** Wetlands can play a significant role in removing fertilizer residues from rain runoff and groundwater. One way they do this is through denitrification, which converts nitrate ions to nitrogen gas:

$$2 NO_3^-(aq) + 5 CO(g) + 2 H^+(aq) \rightarrow N_2(g) + H_2O(\ell) + 5 CO_2(g)$$

Suppose 200.0 g of NO_3^- flows into a swamp each day.
  a. What volume of N_2 would be produced at 17°C and 1.00 atm if the denitrification process were complete?
  b. What volume of CO_2 would be produced?
  c. Suppose the gas mixture produced by the decomposition reaction is trapped in a container at 17°C; what is the density of the mixture assuming $P_{total} = 1.00$ atm?

10.148. Ammonium nitrate decomposes on heating. The products depend on the reaction temperature:

$$NH_4NO_3(s) \xrightarrow{>300°C} N_2(g) + \tfrac{1}{2} O_2(g) + H_2O(g)$$

$$\xrightarrow{200-260°C} N_2O(g) + 2 H_2O(g)$$

A sample of NH_4NO_3 decomposes at an unspecified temperature, and the resulting gases are collected over water at 20°C.
  a. Without completing a calculation, predict whether the volume of gases collected can be used to distinguish between the two reaction pathways. Explain your answer.
  b. The gas produced during the thermal decomposition of 0.256 g of NH_4NO_3 displaces 79 mL of water at 20°C and 760 mmHg of atmospheric pressure. Is the gas N_2O or a mixture of N_2 and O_2?

10.149. In Problem 10.74, nitrogen gas obtained by boiling liquid nitrogen ($d = 0.808$ g/mL) from a 182 L tank was used to power a car. How much hydrazine (in grams) is needed to produce an equivalent amount of N_2 gas by the following reaction with hydrogen peroxide?

$$2 H_2O_2(\ell) + N_2H_4(\ell) \rightarrow N_2(g) + 4 H_2O(g)$$

10.150. **Air Bag Chemistry** Use this chemical equation describing the overall reaction in an automobile air bag:

$$20 NaN_3(s) + 6 SiO_2(s) + 4 KNO_3(s) \rightarrow$$
$$32 N_2(g) + 5 Na_4SiO_4(s) + K_4SiO_4(s)$$

to calculate how many grams of sodium azide (NaN_3) are needed to inflate a 40 × 40 × 20 cm bag to a pressure of 1.25 atm at a temperature of 20°C. How much more sodium azide is needed if the air bag must produce the same pressure at 10°C?

11

Properties of Solutions

Their Concentrations and Colligative Properties

Water, Water, Everywhere

In his "Rime of the Ancient Mariner," British poet Samuel Taylor Coleridge (1772–1834) wrote: "Water, water, everywhere, and all the boards did shrink. Water, water, everywhere, nor any drop to drink." The water in Coleridge's verse is seawater, which is mostly an aqueous solution of NaCl and other salts. Even though all life-forms need water, people who drink seawater suffer from dehydration, and drinking enough of it can prove fatal.

Why is drinking seawater a really bad idea? The problem is the difference in salinity (salt concentration) between seawater and the water in most living cells (even those in the tissues of marine organisms). If individual cells are immersed in seawater, they dehydrate, shrivel up, and die. The reason why is that there are simply too many ions—mostly Na^+ and Cl^-—present in a given volume of seawater. Too many dissolved ions mean too few individual molecules of H_2O in the same volume. In effect, the concentration of *water* in the fluids in most living cells is so much greater than it is in seawater that when these cells come in contact with seawater the water inside them literally oozes out, and they die.

However, cells do contain *some* dissolved salt, which is why fluids administered by intravenous (IV) injection in hospitals to deliver medications or to rehydrate patients contain the equivalent of about 0.155 *M* NaCl. This value matches the concentration of the electrolytes in most of our cells. Just as too high a concentration of salt can be dangerous, too little can be, too. An IV injection without the needed electrolytes could cause a patient to retain too much water, which can have life-threatening consequences.

Other properties of water besides its potability (drinkability) are influenced by the presence of solutes. Aqueous solutions freeze at lower temperatures, which is why antifreeze is added to the water in the cooling systems of automobile engines. In this chapter we explore how and why the properties of solutions depend on the concentration of solutes in them as we probe the science behind Coleridge's poetry. We also explore a process by which modern mariners make the seawater they sail on fit to drink.

Sea Spray Ocean waves crashing on a shoreline produce a spray that carries the ions dissolved in the water into the air. ▶

LEARNING OUTCOMES

LO1 Use the Born–Haber cycle to calculate enthalpy changes as substances dissolve
Sample Exercise 11.1

LO2 Use the Clausius–Clapeyron equation to calculate vapor pressure
Sample Exercise 11.2

LO3 Calculate vapor pressures of solutions using Raoult's law
Sample Exercises 11.3, 11.4

LO4 Express concentrations in molality
Sample Exercise 11.5

LO5 Calculate the freezing points and boiling points of solutions
Sample Exercises 11.6, 11.7

LO6 Use the van 't Hoff factor to account for solution properties
Sample Exercises 11.8, 11.9

LO7 Predict the direction of solvent flow in osmosis and calculate osmotic pressure
Sample Exercises 11.10, 11.11

LO8 Use colligative properties to determine molar masses
Sample Exercises 11.12, 11.13

11.1 Energy Changes when Substances Dissolve

On many college campuses, winter brings subfreezing temperatures and snow—sometimes a lot of it. To keep walkways clear on these campuses after plowing or shoveling, pellets of calcium chloride are often spread to melt packed snow and ice and to prevent new ice from forming. Meanwhile, road crews apply mixtures of sand and sodium chloride to keep streets and highways navigable (Figure 11.1).

Later in this chapter we will explore why $CaCl_2$ and $NaCl$ are effective in melting ice. (*Hint*: It has to do with their capacity to lower the freezing point of water.) For now, the question to be addressed is why $CaCl_2$, which costs much more than $NaCl$, is so widely used to keep sidewalks ice-free. There are several reasons, but we focus here on just one: when $NaCl$ dissolves in melting ice, there is little change in the temperature of the water, but when $CaCl_2$ dissolves, the water heats up, which helps melt more ice. In this section we investigate why the two salts produce such different temperature changes when they dissolve—that is, why dissolving $CaCl_2$ is a decidedly exothermic process, but dissolving $NaCl$ is slightly endothermic (Figure 11.2). To understand why, we need to analyze, step by step, how salts dissolve in water, and we need to track the changes in enthalpy that accompany each of these steps.

As we have discussed in prior chapters, ionic compounds are held together by strong ionic bonds that have bond energies of many hundreds of kilojoules per mole. For their part, molecules of H_2O interact with each other principally through hydrogen bonding. When an ionic solid dissolves in water, the ionic bonds that hold its ions together must be broken. Similarly, many hydrogen bonds must be broken as molecules of H_2O must make space for and cluster around the dissolved ions. These endothermic processes require investments of thermal energy, which we can calculate by summing the enthalpy change that accompanies each process:

$$\text{Thermal energy invested} = \Delta H_{\text{ionic bonds}} + \Delta H_{H_2O-H_2O}$$

FIGURE 11.1 Sodium chloride helps keep streets and highways clear of snow and ice during the winter.

FIGURE 11.2 The enthalpy of solution ($\Delta H_{\text{solution}}$) of NaCl is the sum of the enthalpy changes that accompany breaking the hydrogen bonds between molecules of water (to make space for Na⁺ and Cl⁻ ions), separating the Na⁺ and Cl⁻ ions in solid NaCl, and forming ion–dipole interactions in solution.

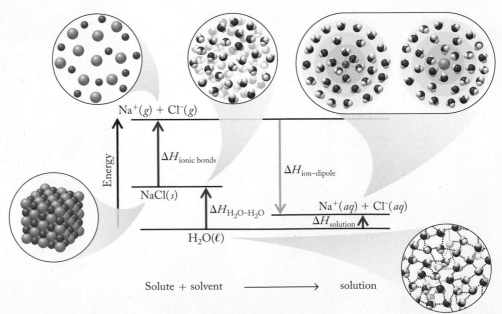

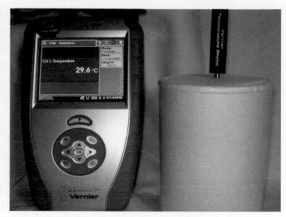

Initial water temperature = 21.2°C   Adding CaCl₂ to water   Final temperature of CaCl₂ solution = 29.6°C

FIGURE 11.3 Determining the enthalpy of solution. The temperature of 100 mL of water increases from 21.2°C to 29.6°C when 5.0 g of $CaCl_2$ dissolve in it. This temperature change and the heat capacity of water can be used to calculate the heat of solution of $CaCl_2$.

When ionic compounds dissolve in water, their ions interact with the permanent dipoles of water molecules that form spheres of hydration around the ions. We assign the change in enthalpy that accompanies this exothermic process the symbol $\Delta H_{ion-dipole}$. The overall change in enthalpy that accompanies the dissolution process, called the **enthalpy of solution ($\Delta H_{solution}$)** or *heat of solution*, is the sum of the thermal energies absorbed and released (see Figure 11.2):

$$\Delta H_{solution} = \underbrace{[\Delta H_{ionic\ bonds} + \Delta H_{H_2O-H_2O}]}_{\text{Endothermic}} + \underbrace{\Delta H_{ion-dipole}}_{\text{Exothermic}} \qquad (11.1)$$

Keep in mind that the energy required to break the ionic bonds in one mole of an ionic compound is equal in magnitude but opposite in sign to the lattice energy (U) of the compound:

$$\Delta H_{ionic\ bonds} = -U \qquad (11.2)$$

We can determine the value of $\Delta H_{solution}$ experimentally using calorimetric methods such as the one shown in Figure 11.3. The heat of solution can be positive or negative, that is, the dissolution process can be endothermic—as it is for NaCl (Figure 11.2)—or exothermic, as it is for $CaCl_2$. Table 11.1 lists $\Delta H_{solution}$

CONNECTION The use of calorimetry to determine enthalpy changes in chemical reactions and physical processes was described in Chapter 9.

TABLE 11.1 Enthalpies of Solution of Some Common Ionic Compounds in Water

ENDOTHERMIC		EXOTHERMIC	
Compound	$\Delta H_{solution}$ (kJ/mol)	Compound	$\Delta H_{solution}$ (kJ/mol)
NH_4NO_3	25.7	$CaCl_2$	82.2
KCl	17.2	KOH	−57.6
NH_4Cl	15.2	NaOH	−44.5
NaCl	4.0	LiCl	−37.1

enthalpy of solution ($\Delta H_{solution}$) the overall change in enthalpy that occurs when a solute is dissolved in a solvent; also called *heat of solution*.

∞ CONNECTION Lattice energy is the energy released when gas-phase ions come together to form one mole of an ionic compound (see Chapter 4). This quantity of energy is consumed in separating a mole of the compound into individual gas-phase ions.

∞ CONNECTION Hess's law is used to calculate the enthalpy change that accompanies a multistep reaction by summing the enthalpy changes that occur in each of its steps, as we discussed in Chapter 9.

values for several common ionic compounds. The compounds in the left column of the table (including NaCl) all have positive heats of solution, meaning that the temperature of the water in which they dissolve decreases as they dissolve. The compounds on the right (including $CaCl_2$) have negative heats of solution, which means that they warm the water in which they dissolve.

Calculating Lattice Energies Using the Born–Haber Cycle

Lattice energies are difficult to determine directly, but the lattice energy of a binary ionic compound can be calculated from its standard heat of formation and a judicious application of Hess's law. To illustrate how this works, let's calculate the lattice energy of NaCl. We start with its standard heat of formation:

$$Na(s) + \tfrac{1}{2} Cl_2(g) \rightarrow NaCl(s) \qquad \Delta H_f^\circ = -411 \text{ kJ}$$

The value of ΔH_f° can be determined experimentally with a calorimeter containing known quantities of elemental Na and Cl_2. Think of the formation reaction as consisting of five steps, as shown in Figure 11.4 and described below:

1. Sublimation of 1 mole of Na metal, producing 1 mole of gas-phase Na atoms. The enthalpy change, $\Delta H_{sub,Na}$, that accompanies this step is the heat of sublimation of sodium. Sublimation is an endothermic process.
2. Breaking the covalent bonds in 0.5 mole of Cl_2 molecules, producing 1 mole of gas-phase Cl atoms. The enthalpy change here is half the bond energy (BE) of 1 mole of Cl—Cl bonds, or $\tfrac{1}{2} BE_{Cl_2}$ (also endothermic).
3. Ionization of 1 mole of Na atoms to 1 mole of Na^+ ions in the gas phase, which requires adding a quantity of thermal energy equal to the first ionization energy of sodium, $IE_{1,Na}$ (also endothermic).

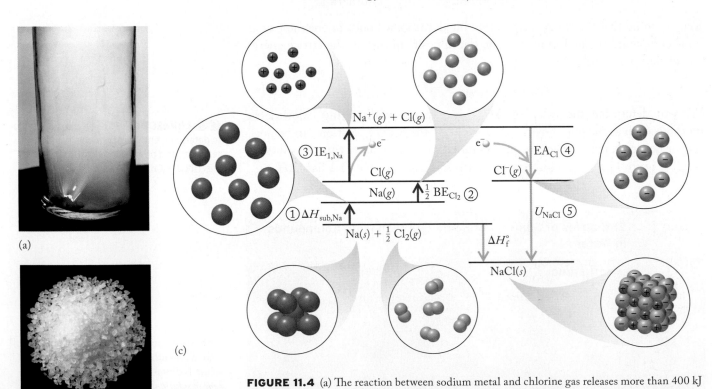

(a)

(b)

(c)

FIGURE 11.4 (a) The reaction between sodium metal and chlorine gas releases more than 400 kJ of energy per mole of NaCl produced. (b) Solid sodium chloride. (c) The Born–Haber cycle shows that the principal reason for this violently exothermic reaction is the energy released in step 5 when free sodium ions $Na^+(g)$ and free chloride ions $Cl^-(g)$ combine to form NaCl(s).

TABLE 11.2 Born–Haber Cycle for Formation of NaCl(s)

Step	PROCESS		Enthalpy Change (kJ)
	Description	**Chemical Equation**	
1	Sublime 1 mol Na(s)	$Na(s) \rightarrow Na(g)$	$\Delta H_{sub,Na} = 108$
2	Break $\frac{1}{2}$ mol Cl—Cl bonds	$\frac{1}{2} Cl_2(g) \rightarrow Cl(g)$	$\frac{1}{2} BE_{Cl_2} = \frac{1}{2}(243) = 121.5$
3	Ionize 1 mol Na(g) atoms forming 1 mol Na$^+$ ions	$Na(g) \rightarrow Na^+(g) + e^-$	$IE_{1,Na} = 495$
4	1 mol Cl atoms acquires 1 mol electrons forming 1 mol Cl$^-$ ions	$Cl(g) + e^- \rightarrow Cl^-(g)$	$EA_{Cl} = -349$
5	1 mol Na$^+$ ions combine with 1 mol Cl$^-$ ions forming 1 mol NaCl(s)	$Na^+(g) + Cl^-(g) \rightarrow NaCl(s)$	$U_{NaCl} = ?$

4. Combining 1 mole of Cl atoms with 1 mole of electrons to form 1 mole of Cl$^-$ ions, which is accompanied by an enthalpy change equal to the electron affinity of chlorine, EA_{Cl} (an exothermic process for Cl and many other elements).

5. Formation of 1 mole of solid NaCl from 1 mole each of gas-phase Na$^+$ and Cl$^-$ ions. The enthalpy change that accompanies this step is the quantity we seek: the lattice energy (U_{NaCl}) of NaCl.

This sequence of steps is diagrammed in Figure 11.4 and summarized in Table 11.2. It is called the **Born–Haber cycle**. The enthalpy changes that accompany the five steps add up to the standard heat of formation of NaCl, which is −411 kJ/mol:

$$\Delta H^\circ_{f,NaCl} = \Delta H_{sub,Na} + \tfrac{1}{2} BE_{Cl_2} + IE_{1,Na} + EA_{Cl} + U_{NaCl}$$

We insert the values from Table 11.2 and solve for U:

$$-411 \text{ kJ} = (108 \text{ kJ}) + \tfrac{1}{2}(243 \text{ kJ}) + (495 \text{ kJ}) + (-349 \text{ kJ}) + U_{NaCl}$$

$$U_{NaCl} = (-411 \text{ kJ}) - (108 \text{ kJ}) - (121.5 \text{ kJ}) - (495 \text{ kJ}) - (-349 \text{ kJ}) = -786 \text{ kJ}$$

The Born–Haber cycle can also be used to calculate enthalpy changes in other steps in the cycle that are difficult to determine experimentally, such as electron affinities. Of course, to do so we need to know the enthalpy changes of all the other steps and the value of ΔH°_f.

SAMPLE EXERCISE 11.1   Calculating Lattice Energy   LO1

Calcium fluoride occurs in nature as the mineral fluorite, which is the principal source of the world's supply of fluorine. (Highly reactive fluorine gas was first detected in nature in a German fluorite mine in 2012.) Use the following data to calculate the lattice energy of CaF_2.

$\Delta H_{sub,Ca} = 168$ kJ/mol

$BE_{F_2} = 155$ kJ/mol

$EA_F = -328$ kJ/mol

$IE_{1,Ca} = 590$ kJ/mol

$IE_{2,Ca} = 1145$ kJ/mol

Born–Haber cycle a series of steps with corresponding enthalpy changes that describes the formation of an ionic solid from its constituent elements.

COLLECT AND ORGANIZE We are asked to calculate the lattice energy of CaF_2 using enthalpy changes that accompany steps in the Born–Haber cycle. Table A4.3 in Appendix 4 contains the standard heat of formation of solid CaF_2: -1228.0 kJ/mol.

ANALYZE The standard enthalpy of formation of CaF_2 is the enthalpy change for the overall process:

$$Ca(s) + F_2(g) \rightarrow CaF_2(s) \qquad \Delta H_f^\circ = -1228 \text{ kJ/mol}$$

When we break down this reaction into the steps of the Born–Haber cycle, we need to include both the first and second ionization energies of Ca because its atoms lose two electrons when forming Ca^{2+} ions. Two moles of fluorine atoms are needed to react with one mole of calcium atoms, so we need to break the F—F bonds in one mole of F_2, and we need to multiply the electron affinity of F by 2 to calculate the enthalpy change accompanying the formation of *two* moles of F^- ions. Figure 11.5 summarizes the Born–Haber cycle for calculating the lattice energy of CaF_2.

To estimate what a reasonable lattice energy value is, we may refer to Table 4.1, which contains the lattice energy of another halide of an alkaline earth metal: $MgCl_2$ ($U_{MgCl_2} = -2540$ kJ/mol). The charges on the ions in CaF_2 and $MgCl_2$ are the same, and even though Ca^{2+} ions are larger than Mg^{2+} ions, F^- ions are smaller than Cl^- ions, so the overall differences in ionic radii may offset, which would lead to comparable d values in Coulomb's law expressions (Equation 4.1) for the two compounds. Therefore, a U_{CaF_2} value near -2540 kJ/mol would be reasonable.

SOLVE The Born–Haber cycle for forming CaF_2 is

$$\Delta H_f^\circ = \Delta H_{sub,Ca} + BE_{F_2} + IE_{1,Ca} + IE_{2,Ca} + 2\,EA_F + U_{CaF_2}$$

Substituting in the values:

$$-1228 \text{ kJ} = 168 \text{ kJ} + 155 \text{ kJ} + 590 \text{ kJ} + 1145 \text{ kJ} + 2(-328 \text{ kJ}) + U_{CaF_2}$$

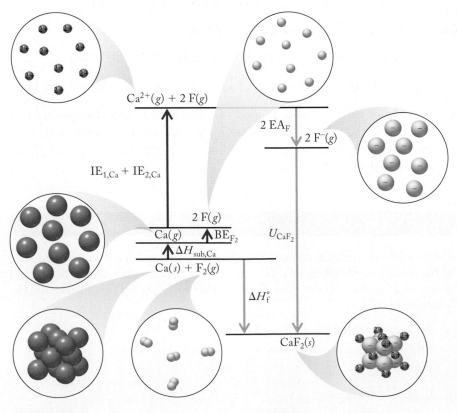

FIGURE 11.5 Born–Haber cycle for the formation of CaF_2.

Solving for lattice energy:

$$U_{CaF_2} = -2630 \text{ kJ/mol } CaF_2$$

THINK ABOUT IT The lattice energies of CaF_2 and $MgCl_2$ are similar in strength, as we thought they might be.

Practice Exercise Burning magnesium metal in air produces MgO and a very bright white light, making the reaction popular in fireworks and signaling devices:

$$Mg(s) + \tfrac{1}{2}O_2(g) \rightarrow MgO(s) \qquad \Delta H_f^\circ = -602 \text{ kJ}$$

Calculate the lattice energy of MgO from the following endothermic changes in enthalpy:

Process	Enthalpy Change (kJ/mol)	Process	Enthalpy Change (kJ/mol)
$Mg(s) \rightarrow Mg(g)$	148	$Mg(g) \rightarrow Mg^{2+}(g) + 2\,e^-$	2188
$O_2(g) \rightarrow 2\,O(g)$	498	$O(g) + 2\,e^- \rightarrow O^{2-}(g)$	605

(Answers to Practice Exercises are in the back of the book.)

Because the lattice energy of an ionic compound must be overcome if the compound is to dissolve, it's logical that ionic compounds with very large lattice energies often have limited solubility in water. This expectation is aligned with the solubility rules described in Chapter 8: ionic compounds formed by the group 1 cations (all 1+) and the ammonium ion (also 1+) are all soluble in water, as are all nitrates (1−) and acetates (also 1−). We would expect compounds containing these ions to require less energy to separate into their component ions because of their small charges, which result in relatively small values in the numerator of Equation 4.1, which describes coulombic attraction:

$$E_{el} = 2.31 \times 10^{-19}\,\text{J} \cdot \text{nm}\left(\frac{Q_1 \times Q_2}{d}\right) \tag{4.1}$$

Moreover, NH_4^+, NO_3^-, and CH_3COO^- have relatively large ionic radii, which lead directly to relatively long distances between anion and cation centers and large d values in the denominator. We might predict, for example, that NaF ($U = -910$ kJ/mol) is more soluble in water than MgO ($U = -3791$ kJ/mol) due to the greater attraction of the 2+ and 2− charges of the ions in MgO, as compared to the 1+ and 1− charges of the ions of comparable size in NaF. We would be correct in our prediction: the solubility of NaF in water is 12 g/100 mL at 25°C, and the solubility of MgO is only 0.000062 g/100 mL.

Molecular Solutes

Until now our focus has been on enthalpy changes that accompany the dissolution of ionic solutes in water. However, many polar molecular compounds, especially those that can form hydrogen bonds with water molecules, are also soluble in water. In Chapter 6 we learned that low-molar-mass alcohols such as methanol (CH_3OH), ethanol (CH_3CH_2OH), and ethylene glycol ($HOCH_2CH_2OH$) are *miscible* with water, that is, they dissolve in water—and water dissolves in them—in all proportions. The heats of solution of several molecular compounds are listed in Table 11.3. Like all $\Delta H_{solution}$ values, the ones in Table 11.3 represent

TABLE 11.3	Enthalpies of Solution of Some Common Molecular Compounds in Water	
Compound	**$\Delta H_{\text{solution}}$ (kJ/mol)**	
HCl	−74.8	
NH$_3$	−30.5	
CH$_3$CH$_2$OH	−10.6	
CH$_3$OH	−3.0	
CH$_3$COOH	−1.5	

∞ **CONNECTION** The lower solubility in water of alcohols with greater molecular masses (and more nonpolar −CH$_2$− groups per molecule) was discussed in Chapter 6.

the net sum of the energy investments needed to overcome solvent–solvent and solute–solute interactions and the energy released when particles of solute and solvent interact. Equation 11.3 presents this relationship for the case of methanol dissolving in water.

$$\Delta H_{\text{solution}} = \Delta H_{\text{H}_2\text{O–H}_2\text{O}} + \Delta H_{\text{CH}_3\text{OH–CH}_3\text{OH}} + \Delta H_{\text{CH}_3\text{OH–H}_2\text{O}} \qquad (11.3)$$

As we have seen for ionic solutes, the value of $\Delta H_{\text{solution}}$ can be determined using calorimetric methods, but determining the three terms on the right side of Equation 11.3 independently is difficult. When heats of solution have values near zero, as in the case of methanol, at least we know that the energies required to disrupt solvent–solvent and solute–solute interactions are nearly balanced by the energy released when new interactions form between molecules of solute and solvent. In the case of methanol, there is a balance between the hydrogen bonds between water molecules and between methanol molecules that are broken, and the hydrogen bonds that form between molecules of water and methanol.

The negative $\Delta H_{\text{solution}}$ values in Table 11.3 mean that the solute–solvent intermolecular interactions are stronger than solute–solute and solvent–solvent intermolecular interactions. However, not all molecular solutes have exothermic heats of solution in water. Among those that don't are alcohols with four or more carbon atoms per molecule. Stronger London dispersion forces between the nonpolar hydrocarbon tails of these molecules lead to stronger solute–solute interactions and contribute to endothermic heats of reaction. The increasing nonpolar (hydrophobic) nature of these alcohols also causes them to be less soluble in water.

11.2 Vapor Pressure

The level of water in a glass left on a countertop slowly drops as molecules on the surface escape from it and enter the gas phase. They do so because the water molecules at the surface of the liquid have a distribution of kinetic energies and speeds (as we saw with gas-phase molecules in Chapter 10). At least some molecules have sufficient energy to escape the intermolecular forces holding them on the surface. Once gone, they have taken their kinetic energy with them; the remaining molecules redistribute their energy, drawing heat from their surroundings, and the evaporation process continues. Eventually the glass will go dry without ever being above room temperature. The rate at which molecules make this jump to the vapor phase depends on temperature, on the surface area of the liquid, and on the strength of the intermolecular forces that hold the molecules together in the liquid:

1. The higher the temperature, the larger the number of molecules with sufficient kinetic energy to break the attractive forces that hold them together in the liquid and enable them to enter the gas phase.
2. The greater the surface area of the liquid, the larger the number of molecules on the surface in a position to enter the gas phase.
3. The stronger the intermolecular forces, the greater the kinetic energy needed for a molecule to escape the surface, and the smaller the number of molecules in the population that have this energy.

If an identical glass of water at the same temperature is covered (Figure 11.6), water molecules still evaporate at the same rate, but they are then confined to the headspace above the water. Some molecules of vapor may condense on the cover or walls of the glass or, more significantly, at the liquid surface where they

↑ Evaporation ↓ Condensation

FIGURE 11.6 A covered glass of water achieves a dynamic equilibrium where the rate at which liquid water is lost to evaporation equals the rate at which liquid water is gained by condensation.

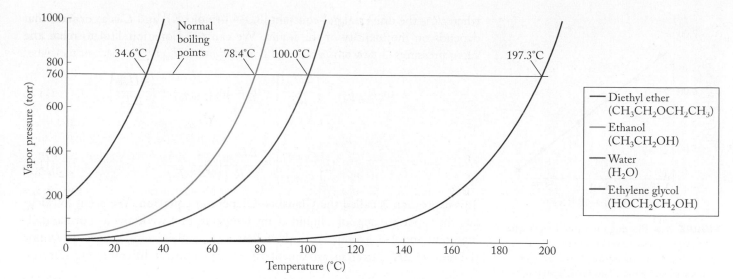

FIGURE 11.7 A graph of vapor pressure versus temperature for four liquids shows that vapor pressure increases with increasing temperature. The temperature at which the vapor pressure equals 1 atm is the normal boiling point of the liquid.

return to the liquid phase. In a short time the rates of evaporation and condensation equalize, as water molecules in the liquid and gas phases achieve a state of dynamic equilibrium. Their concentration in the gas phase remains constant as long as the temperature does not change and the cover stays on.

Like any component in a mixture of gases, the water vapor in the headspace of the glass exerts a partial pressure that is proportional to both the ambient atmospheric pressure (P_{atm}) and the mole fraction of the particles in the headspace that are molecules of water (x_{H_2O}). In other words, water vapor obeys Dalton's law of partial pressures. The partial pressure exerted by a gas in equilibrium with its liquid state is called the **vapor pressure** of the liquid. A liquid is described as **volatile** when enough of its molecules vaporize to produce a significant vapor pressure at a given temperature: the higher its vapor pressure, the more volatile it is.

As the temperature of a volatile liquid increases, the fraction of its molecules with enough energy to break away from its surface and enter the gas phase increases, and so does its vapor pressure. This trend is shown in Figure 11.7. The less volatile liquids in the figure, such as ethylene glycol, require higher temperatures to exert significant vapor pressures; the most volatile liquid, diethyl ether, vaporizes at much lower temperatures. If the temperature of a liquid is high enough, its vapor pressure reaches ambient atmospheric pressure (the horizontal line in Figure 11.7). Then the liquid vaporizes readily because it has reached its boiling point. When this ambient pressure is a standard atmosphere (1 atm), liquids are said to boil at their **normal boiling points**, such as 100.0°C for water.

The Clausius–Clapeyron Equation

We have seen that the vapor pressures of volatile liquids increase with increasing temperature. As Figure 11.7 shows, this increase is not linear. However, if we graph the natural logarithm of the vapor pressure versus $1/T$, where T is the absolute temperature, we do get a straight line (Figure 11.8). The slope of this line depends on the heat of vaporization (ΔH_{vap}) of the liquid, as reflected in the generic equation for such graphs:

$$\ln(P_{vap}) = -\frac{\Delta H_{vap}}{R}\left(\frac{1}{T}\right) + C \qquad (11.4)$$

⊙⊙ **CONNECTION** According to Dalton's law of partial pressures, each gas in a mixture contributes a partial pressure equal to the product of the mole fraction of each gas times the total pressure of the mixture (Chapter 10).

vapor pressure the pressure exerted by a gas in equilibrium with its liquid phase at a given temperature.

volatile having a significant vapor pressure at a given temperature.

normal boiling point the temperature at which the vapor pressure of a liquid equals 1 atm (760 torr).

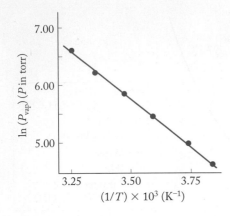

FIGURE 11.8 Plotting the natural logarithm of the vapor pressure versus the reciprocal of the absolute temperature gives a straight line described by the Clausius–Clapeyron equation. The graph shows the plot for pentane.

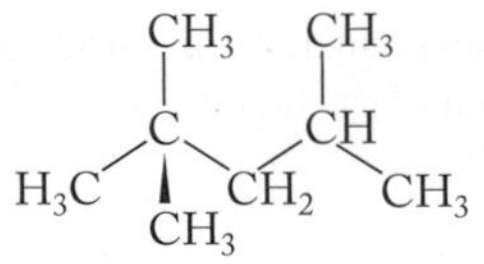

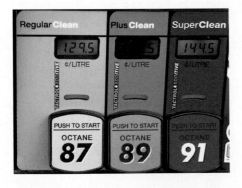

FIGURE 11.9 Structural formula of isooctane.

FIGURE 11.10 Octane ratings for gasoline.

Clausius–Clapeyron equation relates the vapor pressure of a substance at different temperatures to its heat of vaporization.

where R is the universal gas constant [8.314 J/(mol · K)] and C is a constant that depends on the identity of the liquid. We can use Equation 11.4 to relate the vapor pressures at two temperatures (T_1 and T_2) to ΔH_{vap}:

$$\ln\left(P_{vap,T_1}\right) + \left(\frac{\Delta H_{vap}}{RT_1}\right) = C = \ln\left(P_{vap,T_2}\right) + \left(\frac{\Delta H_{vap}}{RT_2}\right)$$

or

$$\ln\left(\frac{P_{vap,T_1}}{P_{vap,T_2}}\right) = \frac{\Delta H_{vap}}{R}\left(\frac{1}{T_2} - \frac{1}{T_1}\right) \tag{11.5}$$

This expression is called the **Clausius–Clapeyron equation**. We use it to calculate the vapor pressure of a liquid at any temperature if we know its normal boiling point and ΔH_{vap} value, as in the following Sample Exercise, or to determine the value of ΔH_{vap} from measurements of vapor pressure at different temperatures.

SAMPLE EXERCISE 11.2 Calculating Vapor Pressure Using ΔH_{vap} LO2

The compound with the common name *isooctane* (its official name is 2,2,4-trimethyl-pentane) has the structure shown in Figure 11.9. It defines the "100" value on the octane rating scale used to grade gasoline (Figure 11.10). Its normal boiling point is 99°C, and its heat of vaporization is 35.2 kJ/mol. What is the vapor pressure of isooctane at 25°C in torr?

COLLECT AND ORGANIZE We are given the normal boiling point and heat of vaporization of isooctane. We are asked to calculate its vapor pressure at 25°C. The Clausius–Clapeyron equation relates the vapor pressure values of a liquid at two temperatures to its ΔH_{vap} value.

ANALYZE To use the Clausius–Clapeyron equation, we must express temperatures on the Kelvin scale and convert ΔH_{vap} to joules per mole to be compatible with the units on R, J/(mol · K). Vapor pressure decreases sharply (Figure 11.7) as temperatures decrease below the normal boiling point of a liquid, so the vapor pressure of isooctane at 25°C should be only a fraction of 760 torr.

SOLVE We express the two temperatures in kelvin:

$$T_1 = 99°C + 273 = 372\ K \quad \text{and} \quad T_2 = 25°C + 273 = 298\ K$$

ΔH_{vap} in joules per mole is

$$35.2\ \frac{kJ}{mol} \times \frac{1000\ J}{kJ} = 35{,}200\ \frac{J}{mol}$$

Entering these values in Equation 11.5 and solving for P_{vap,T_2} yields

$$\ln\left(\frac{P_{vap,T_1}}{P_{vap,T_2}}\right) = \frac{\Delta H_{vap}}{R}\left(\frac{1}{T_2} - \frac{1}{T_1}\right)$$

$$\ln\left(\frac{760\ torr}{P_{vap,T_2}}\right) = \frac{\left(35{,}200\ \dfrac{J}{mol}\right)}{\left(8.314\ \dfrac{J}{mol \cdot K}\right)}\left(\frac{1}{298\ K} - \frac{1}{372\ K}\right) = 2.826$$

$$P_{vap,T_2} = 45.0\ torr$$

THINK ABOUT IT We expected isooctane to have a relatively low vapor pressure at a temperature well below its normal boiling point, so this number makes sense. Its value and that of other hydrocarbons in gasoline at summerlike temperatures are well above the vapor pressure of water (23.8 torr), which means that their evaporation rates are higher—high enough to impact the quality of the air near gas stations, especially in urban areas.

Practice Exercise Pentane (C_5H_{12}) gas is used to blow the bubbles in molten polystyrene that turn it into Styrofoam, which is used in coffee cups and other products that have good thermal insulation properties. The normal boiling point of pentane is 36°C; its vapor pressure at 25°C is 505 torr. What is the heat of vaporization of pentane? ⚙

fractional distillation a method of separating a mixture of compounds on the basis of their different boiling points.

CONCEPT TEST

Diesel fuel is made of hydrocarbons with an average of 13 carbon atoms per molecule, and gasoline is made of hydrocarbons with an average of 7 carbon atoms per molecule. Which fuel has the higher vapor pressure at room temperature?

(Answers to Concept Tests are in the back of the book.)

11.3 Mixtures of Volatile Substances

Small differences in the vapor pressures of volatile liquids can be used to separate them from each other. Two familiar examples of such mixtures are gasoline and the material from which it is derived, crude oil. Both are complex mixtures of hydrocarbons, molecular compounds composed entirely of atoms of carbon and hydrogen bonded together. Crude oil is the source of several important classes of fuels in addition to gasoline, including diesel oil, jet fuel, kerosene, and heating oil. The process by which crude oil is separated into these fuels is called **fractional distillation** (Figure 11.11). Like simple distillation, fractional distillation separates the components of mixtures based on differences in their volatilities through selective evaporation and condensation. Unlike simple distillation, fractional distillation employs many evaporation/condensation cycles to separate substances that have only slightly different volatilities.

The separation depicted in Figure 11.11 begins with the injection of crude oil at the bottom of a fractionating tower, where the oil is heated to a temperature at which nearly all of its components vaporize. As the vapors percolate upward in the tower, they encounter cooler surfaces where the least volatile of them condense, collecting in trays. The more volatile vapors do not condense until they reach higher, cooler regions of the tower. As heat continues to flow upward in the tower, the substances collected in each tray reach temperatures at which they revaporize. However, the vapor produced by the liquid in each tray has a composition that is different from the liquid that condensed into the tray: the vapor-phase mixture is richer in the more volatile components of the liquid. When this enriched vapor condenses in another,

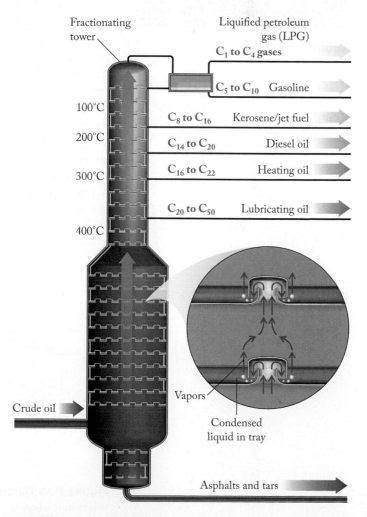

FIGURE 11.11 Fractional distillation separates crude oil into products that are used as fuels, lubricants, and building materials.

∞ CONNECTION Simple distillation, such as the desalination of seawater as described in Chapter 1, is used to separate nonvolatile solutes from volatile solvents.

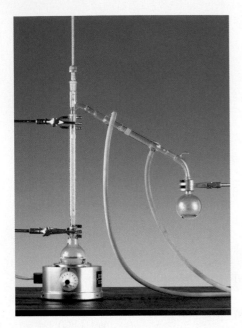

FIGURE 11.12 Fractional distillation apparatus. Vapors rise through a fractionating column, where they repeatedly condense and vaporize. The most volatile component distills first and is first to pass through the condenser and into the collecting flask. Increasingly less volatile, higher boiling components are distilled in turn. The progress of the distillation process is monitored using the thermometer at the top of the fractionating column.

▶‖ **CHEMTOUR** Fractional Distillation

cooler tray and then revaporizes, the vapor produced this time is even richer in the more volatile components. In a fractional distillation apparatus, these three steps of vaporization, condensation, and revaporization happen over and over again, which allows components with only slightly different vapor pressures to be separated from one another.

To explore how fractional distillation works in more detail, let's employ it to separate a 50:50 (by moles) mixture of two volatile hydrocarbons present in gasoline: heptane (C_7H_{16}, boiling point 98°C) and octane (C_8H_{18}, boiling point 126°C). It makes sense that octane has the higher boiling point because its molecules are larger and experience stronger London dispersion forces. These forces keep octane molecules together in the liquid state and inhibit their vaporization until they acquire greater kinetic energies at higher temperatures.

Suppose a sample of our 50:50 mixture is heated in a conventional distillation apparatus. The mixture starts to boil at a temperature at which the sum of their vapor pressures reaches ambient pressure (760 torr). This temperature turns out to be 108°C, which is between the normal boiling points of the two components of the mixture. At this temperature, heptane has the greater vapor pressure because it is more volatile, but octane has a significant vapor pressure, too, so the vapors that condense in the distillation apparatus are enriched in heptane but still have a significant octane component. This codistillation of the two components means that a one-step distillation does not separate them completely.

To obtain a more complete separation, we use an apparatus that includes a fractionating column above the distillation flask (Figure 11.12). Like the towers used to refine crude oil, fractionating columns used in labs contain surfaces (sometimes they are packed with glass beads to increase surface area) where hot vapors rising up from the boiling flask condense. As heptane-enriched vapors from a boiling 50:50 mixture reach the bottom of such a column, they condense, forming a liquid that is also enriched in heptane. The process is shown graphically in Figure 11.13. The blue curve represents temperatures at which different liquid mixtures of the two components boil. The temperature at point 1 on the curve confirms that the initial boiling point of a 50:50 solution is about 108°C. The red line on the graph shows the composition of the vapor that is formed as these

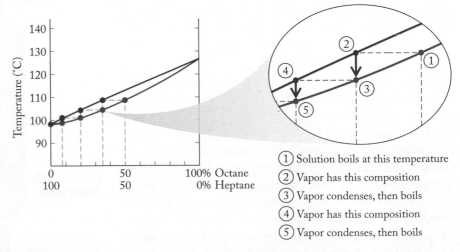

① Solution boils at this temperature
② Vapor has this composition
③ Vapor condenses, then boils
④ Vapor has this composition
⑤ Vapor condenses, then boils

FIGURE 11.13 Distillation of a 50:50 mixture of heptane and octane. The blue line shows the temperature where a mixture of a given composition boils. The red line shows the composition of the vapor arising from those solutions. The stair-step line illustrates what happens in a fractionating column.

solutions boil. To find the composition of the vapor produced by a 50:50 mixture of the two liquids boiling at 108°C, we move horizontally to the left along the dashed line between points 1 and 2. The *x*-coordinate of point 2 tells us that the composition of the vapor produced by the liquid boiling at point 1 is about 65% heptane and only 35% octane.

This 65:35 vapor rises up in the distillation column, cools, and condenses as a 65:35 liquid in a process represented by the red arrow from point 2 to point 3. Continued heating of the column warms this liquid, and it vaporizes at about 104°C (the *y*-coordinate of point 3). To find the composition of the vapor above this boiling liquid, we again move left on the temperature axis until we intersect the red curve (point 4). Reading down from point 4 to the concentration axis, we see that the vapor concentration is now about 80% heptane and only 20% octane. This vapor with 80:20 composition rises up, where it cools and condenses, and the distillation cycle is repeated.

If we continue this process of redistilling mixtures with increasing concentrations of heptane and then cooling and condensing the vapors, we eventually obtain a condensate that is pure heptane. If we monitor the temperature at which vapors condense at the very top of our distillation column, we will see a profile of temperature versus volume of distillate produced that looks like Figure 11.14. The first liquid to be produced is nearly pure heptane, which has a boiling point of 98°C. Ideally, all the heptane in the original sample is recovered before octane is collected as the temperature at the top rises to 126°C.

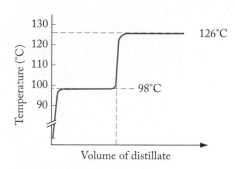

FIGURE 11.14 Fractional distillation of a 50:50 mixture of heptane and octane produces two plateaus at the boiling points of the two components. The first distillate to be collected is nearly pure heptane and the second is nearly pure octane.

CONCEPT TEST

The Lewis structures and dipole moments of dimethyl ether and acetone are shown in Figure 11.15. Which of the two compounds would you expect to be present in higher concentration in the vapors of a boiling 50:50 (mol/mol) mixture?

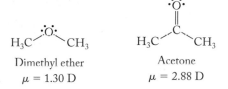

Dimethyl ether
$\mu = 1.30$ D

Acetone
$\mu = 2.88$ D

FIGURE 11.15 Lewis structures and dipole moments of dimethyl ether and acetone.

The fractional distillation process discussed above illustrates how most mixtures of volatile substances behave. This behavior was extensively studied by French chemist François Marie Raoult (1830–1901), who published the following succinct description of it in 1882: *The total vapor pressure of an ideal solution depends on the vapor pressure of each component in the solution and its mole fraction in the liquid mixture.* In equation form, this description is known as **Raoult's law**:

$$P_{total} = x_1 P_1 + x_2 P_2 + x_3 P_3 + \cdots \tag{11.6}$$

where x_i is the mole fraction of each volatile component of a solution and P_i is the vapor pressure of the pure component. Each term in the series represents the contribution each volatile component of a solution makes to the total vapor pressure of the solution. Thus, Raoult's law is analogous to Dalton's law of partial pressures, but for volatile liquids.

▶❚❚ **CHEMTOUR** Raoult's Law

| SAMPLE EXERCISE 11.3 | Calculating the Vapor Pressure of a Solution | LO3 |

What is the vapor pressure of a solution prepared by dissolving 13 g of heptane (C_7H_{16}) in 87 g of octane (C_8H_{18}) at 25°C? How much higher is the mole ratio of heptane to octane in the vapor above the solution than in the solution itself? The vapor pressures of heptane and octane at 25°C are 31 torr and 11 torr, respectively.

Raoult's law the vapor pressure of a solution is the sum of the vapor pressures of the volatile components of the solution, which are each the product of the vapor pressure of the pure component and its mole fraction in the solution.

COLLECT AND ORGANIZE We are asked to calculate the vapor pressure of a solution of a volatile solute (heptane) dissolved in a volatile solvent (octane) and to predict the composition of the vapor produced by the solution at 25°C. We know the vapor pressures of the pure solute and pure solvent at this temperature. Raoult's law (Equation 11.6) relates the vapor pressure of a mixture of volatile substances to the individual vapor pressures of the components of the mixture.

ANALYZE To calculate the vapor pressures of the volatile components of a mixture (the $x_i P_i$ terms in Equation 11.6), we need to first calculate their mole fractions in the mixture. The relative concentrations of the vapors trapped above a solution of volatile components should be proportional to the ratios of their vapor pressures—assuming the vapors behave like ideal gases. If they do, then they obey the ideal gas law, $PV = nRT$, and the partial pressure of each is proportional to its concentration (moles per liter) in the gas phase:

$$P \propto \frac{n}{V} \quad \text{at constant } T$$

The vapor pressure of the mixture should be between that of pure heptane and pure octane—and closer to the octane value because there is more of it. The vapor pressure of heptane is nearly three times that of octane, so the vapor should be enriched by about a factor of three in heptane compared to the composition of the liquid.

SOLVE The number of moles of each component is

$$87 \text{ g C}_8\text{H}_{18} \times \frac{1 \text{ mol C}_8\text{H}_{18}}{114.23 \text{ g C}_8\text{H}_{18}} = 0.762 \text{ mol C}_8\text{H}_{18}$$

$$13 \text{ g C}_7\text{H}_{16} \times \frac{1 \text{ mol C}_7\text{H}_{16}}{100.20 \text{ g C}_7\text{H}_{16}} = 0.130 \text{ mol C}_7\text{H}_{16}$$

The mole fraction of each component in the mixture is

$$x_{\text{octane}} = \frac{0.762 \text{ mol}}{(0.762 + 0.130) \text{ mol}} = 0.854$$

$$x_{\text{heptane}} = 1 - x_{\text{octane}} = 0.146$$

Using these mole fraction values and the vapor pressures of the two hydrocarbons in Equation 11.6, we have

$$P_{\text{total}} = x_{\text{heptane}} P_{\text{heptane}} + x_{\text{octane}} P_{\text{octane}}$$

$$= 0.146(31 \text{ torr}) + 0.854(11 \text{ torr})$$

$$= 4.5 \text{ torr} + 9.4 \text{ torr} = 13.9 \text{ torr}$$

As discussed above, the heptane/octane concentration ratio in the gas phase should be the same as the ratio of their vapor pressures:

$$\frac{4.5 \text{ torr}}{9.4 \text{ torr}} = 0.48$$

The mole ratio of heptane to octane in the liquid mixture is

$$\frac{0.13 \text{ mol}}{0.76 \text{ mol}} = 0.17$$

Therefore, the vapor phase is enriched in heptane by a factor of

$$\frac{0.48}{0.17} = 2.8$$

THINK ABOUT IT As expected, the vapor pressure of the mixture is between the vapor pressures of the separate components, and the vapors are enriched in the more volatile component, heptane, by about a factor of three compared to the liquid.

Practice Exercise Benzene (C_6H_6) is a trace component of gasoline. What is the mole ratio of benzene to octane in the vapor above a solution of 10% benzene and 90% octane by mass at 25°C? The vapor pressures of octane and benzene at 25°C are 11 torr and 95 torr, respectively. ⚙

ideal solution one that obeys Raoult's law.

You may have noticed that Raoult's law refers to **ideal solutions**. Let's be clear on what that term means. Solutions such as the hydrocarbons in gasoline and crude oil obey Raoult's law when the intermolecular interactions between all the molecules in the mixture have comparable strengths. In the case of a binary mixture in which the more abundant component (in terms of number of moles) is designated the *solvent* and the other component is the *solute*, an ideal solution is one in which the strengths of solvent–solvent, solute–solute, and solute–solvent interactions are much the same.

However, solute–solvent interactions are sometimes stronger than solvent–solvent or solute–solute interactions. When this happens, the *adhesive* forces between solute and solvent molecules are greater than the *cohesive* forces between solute molecules and between solvent molecules. As a result, vaporization of both the solute and solvent is inhibited by their strong attraction for each other in the liquid phase, which produces *negative* deviations from the vapor pressures predicted by Raoult's law, as shown in Figure 11.16(a). Solutions of chloroform and acetone (Figure 11.17) exhibit this behavior. It has been proposed that strong dipole–dipole interactions between acetone and chloroform molecules result in the deviation from Raoult's law in Figure 11.16(a).

Other solute/solvent pairs exhibit *positive* deviations from Raoult's law (Figure 11.16b). Their molecules experience *weaker* adhesive interactions than the cohesive forces between solute molecules and between solvent molecules. Chloroform and ethanol are one such pair of compounds. In their solutions, ethanol molecules hydrogen-bond more strongly with each other than with molecules of $CHCl_3$, which reduces the likelihood of strong solute–solvent interaction.

CONNECTION In Chapter 6 we learned that strong *adhesive* forces between the molecules of water and the molecular structure of a capillary coupled with strong *cohesive* forces between molecules of water produce capillary action.

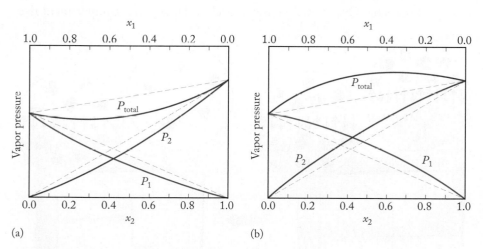

(a)   (b)

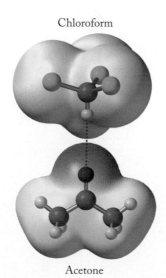

Chloroform

Acetone

FIGURE 11.16 In a mixture of two volatile substances, the vapor pressures P_1 and P_2 may deviate from the ideal behavior predicted by Raoult's law and described by the dashed lines. (a) If solute–solvent interactions are stronger than solvent–solvent or solute–solute interactions, the deviations from Raoult's law are negative. (b) If solute–solvent interactions are weaker than solvent–solvent or solute–solute interactions, the deviations are positive.

FIGURE 11.17 In a solution of chloroform and acetone, solute–solvent interactions are stronger than solute–solute or solvent–solvent interactions.

Which of the following solutions is least likely to follow Raoult's law: (a) acetone and ethanol; (b) pentane and hexane; (c) pentanol and water? The structures of the compounds are given in Figure 11.18.

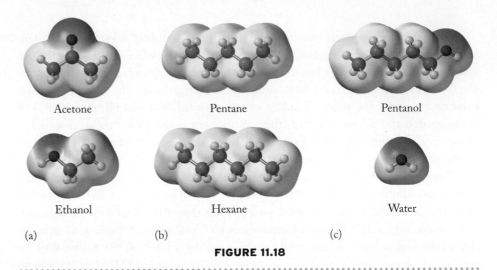

Acetone   Pentane   Pentanol

Ethanol   Hexane   Water

(a)   (b)   (c)

FIGURE 11.18

11.4 Colligative Properties of Solutions

We have seen how Raoult's law lets us predict the vapor pressure of solutions of volatile compounds. In this section we examine the impact of nonvolatile solutes on the vapor pressure (and other properties) of volatile solvents. We start with the solution that covers most of Earth's surface: seawater. The salts in seawater have no vapor pressure of their own, but they can alter the volatility of the water they are dissolved in. Consider the images in Figure 11.19(a). The left-side compartment in the figure contains pure water, and the right side initially contains the same volume of seawater. Over time, the water level in the seawater compartment rises

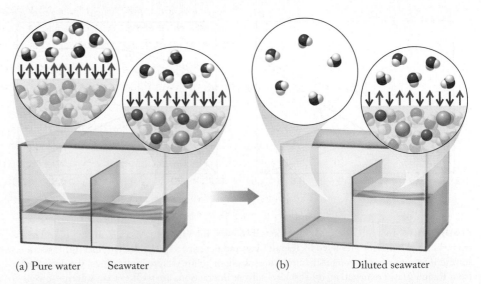

(a) Pure water   Seawater   (b)   Diluted seawater

FIGURE 11.19 (a) Adjoining compartments are partially filled with pure water and seawater. (b) The slightly higher vapor pressure of the pure water leads to a net transfer of water from the pure-water compartment to the seawater compartment.

as the level in the pure-water compartment drops at a matching rate (assuming the system is well sealed). Eventually, nearly all the water ends up in the seawater compartment (Figure 11.19b).

Let's explore how and why water moves from one compartment to the other. Water in both compartments evaporates, producing water vapor in the headspace above them. Eventually the concentration of water vapor stabilizes, exerting a partial pressure that matches the vapor pressure of water at the temperature inside the chamber. At this point, the overall rate of evaporation from the compartments matches the rate of condensation so that P_{H_2O} remains constant.

Because both compartments share a common headspace, the concentration of water vapor is the same throughout it. It is logical, then, that the rate of condensation into both compartments should be the same. If the rates of evaporation were also the same, the two liquid levels would not change over time. However, the levels clearly do change, which means the rates of evaporation are not the same. The rate of evaporation of H_2O must be greater from pure water than from seawater to explain the transport of water from the pure-water compartment to the other. We may conclude that the presence of solute particles *lowers the vapor pressure* and, therefore, the evaporation rate of the water in seawater.

Raoult's Law Revisited

The lower vapor pressure of seawater compared to pure water makes sense in the context of Raoult's law. In this case the vapor pressure of a volatile solvent, $P_{solution}$, is related to its mole fraction times the vapor pressure of the pure solvent, $P_{solvent}$, according to this variation of Raoult's law:

$$P_{solution} = x_{solvent}P_{solvent} \tag{11.7}$$

Equation 11.7 tells us that the more nonvolatile solute particles there are in a solution, the lower the mole fraction of volatile solvent particles, and the lower the vapor pressure of the solution. In addition, the value of $P_{solution}$ depends only on the number of solute particles, and not on their nature. They could be ions, or they could be molecules of a nonvolatile molecular compound. They could be cations with 1+ or 2+ charges, or anions with 1− or 2− charges; it makes no difference as long as the sum of all their charges adds up to zero—and the solutions behave *ideally*. Properties of solutions that depend only on the concentration of particles and not on their identity are called **colligative properties**.

Let's apply Raoult's law to the liquids in the two compartments in Figure 11.19. If the temperature of the liquids is 20°C, the vapor pressure in the pure-water compartment is 0.0231 atm. To calculate the vapor pressure of the seawater, which contains 53.6 moles of water and 1.120 moles of ions per kilogram, we first calculate the mole fraction of water in seawater:

$$x_{H_2O} = \frac{53.6 \ \text{mol} \ H_2O}{(53.6 + 1.120) \ \text{mol}} = 0.980$$

Inserting this value in Equation 11.7, we find that the vapor pressure for seawater at 20°C is

$$P_{seawater} = (0.980)(0.0231 \ \text{atm}) = 0.0226 \ \text{atm}$$

This slightly lower vapor pressure of seawater as compared to pure water results in a greater evaporation rate of pure water, causing the migration of its molecules to the seawater compartment.

colligative properties characteristics of solutions that depend on the concentration and not the identity of particles dissolved in the solvent.

CONCEPT TEST

Is the vapor pressure of a pure solvent an intensive or extensive property? Is the vapor pressure of a solution an intensive or extensive property?

SAMPLE EXERCISE 11.4 **Calculating the Vapor Pressure of a Solution of One or More Nonvolatile Solutes** **LO3**

Most of the world's supply of maple syrup is produced in Quebec, Canada, where the sap from maple trees is evaporated until the concentration of sugar (mostly sucrose) in the sap reaches at least 66% by weight. What is the vapor pressure in atm of a 66% aqueous solution of sucrose ($\mathcal{M}$ = 342 g/mol) at 100°C? Assume that the solution obeys Raoult's law.

COLLECT AND ORGANIZE We are asked to calculate the vapor pressure of a 66% solution by weight of sucrose. The vapor pressure of the solution is a colligative property that depends on the mole fraction of solvent (x_{H_2O}) in it:

$$P_{solution} = x_{H_2O}P_{H_2O}$$

The vapor pressure of pure water (P_{H_2O}) at 100°C (its normal boiling point) is 1.00 atm. Its molar mass is 18.02 g/mol.

ANALYZE A concentration of 66% means that a 100-gram sample of the solution contains 66 grams of sucrose and 34 grams of water. To calculate the mole fraction of water in a 100 g sample, we need to convert these two masses to moles of sucrose and moles of water and then calculate the mole fraction of water. A mass of 66 grams of sucrose ($\mathcal{M}$ = 342 g/mol) represents about 0.2 mole. There are about 2 moles of water in 34 grams of water ($\mathcal{M} \approx$ 18 g/mol), so the mole fraction of water is about 2/2.2, or about 0.9, and the vapor pressure should be about 0.9 atm.

SOLVE In 100 grams of syrup there are

$$66 \text{ g sucrose} \times \frac{1 \text{ mol sucrose}}{342 \text{ g sucrose}} = 0.193 \text{ mol sucrose}$$

$$34 \text{ g water} \times \frac{1 \text{ mol water}}{18.02 \text{ g water}} = 1.89 \text{ mol water}$$

The mole fraction of water is

$$x_{H_2O} = \frac{1.89 \text{ mol}}{(1.89 + 0.193) \text{ mol}} = 0.91$$

Vapor pressure of the solution:

$$P_{solution} = x_{H_2O}P_{H_2O} = 0.91 \times 1.00 \text{ atm} = 0.91 \text{ atm}$$

THINK ABOUT IT The calculated value is close to the estimated one and less than the vapor pressure of pure water, as we expect in a concentrated solution. Because its vapor pressure is less than atmospheric pressure, maple syrup does not boil at 100°C. Its boiling point at P = 1.00 atm, like those of other aqueous solutions, is higher than 100°C.

Practice Exercise The liquid used in automobile cooling systems is prepared by dissolving ethylene glycol ($HOCH_2CH_2OH$, $\mathcal{M}$ = 62.07 g/mol) in water. What is the vapor pressure at 50°C of a solution prepared by mixing equal volumes of ethylene glycol (density 1.114 g/mL) and water (density 1.000 g/mL)? Express your answer to the nearest torr. The vapor pressure of pure water at 50°C is 92 torr. The vapor pressure of ethylene glycol is less than 1 torr at 50°C. ⚙

The impact of solute particles on the properties of a concentrated aqueous solution can be seen in the phase diagram in Figure 11.20. The orange line represents the boiling points of pure water at different pressures and temperatures. Keep in mind that the pressure values along this line, which represent a series of boiling points, are actually vapor pressure values at different temperatures. This is because water boils when its vapor pressure reaches ambient pressure. The more intense red line in Figure 11.20 represents vapor pressures of a concentrated aqueous solution, such as antifreeze, over the same range of temperatures. Notice that the vapor pressure of the solution is lower than that of pure water at all temperatures, and that the solution reaches a vapor pressure of 1 atm at a temperature well above 100°C. This phenomenon is called *boiling point elevation*. The difference between the normal boiling point and the boiling point of the solution is labeled ΔT_b and is proportional to the total concentration of all the solute particles in the solution—it is, in fact, a colligative property.

At its lower end, the orange line intersects a faint blue line separating solid ice and pure liquid water (representing melting points) and a faint green line separating ice and steam (representing sublimation points). This point of intersection is the triple point of water. There is an analogous triple point for the solution, and there are analogous lines for melting point (blue) and sublimation point (green). The separation between the blue lines reflects the fact that solutions freeze at lower temperatures than their pure solvents—a phenomenon called *freezing point depression*. The symbol ΔT_f represents the difference in freezing points. As with boiling point elevation, the size of ΔT_f is a colligative property; it is proportional to the total concentration of all the solute particles in solution. Before we use this proportionality to calculate the values of ΔT_b and ΔT_f, or to measure these temperature differences and use them to determine the concentrations of solutions, we need to become acquainted with the concentration unit known as *molality*.

CONNECTION We introduced phase diagrams in Chapter 6 to show how the physical states of substances change with changing temperatures and pressures.

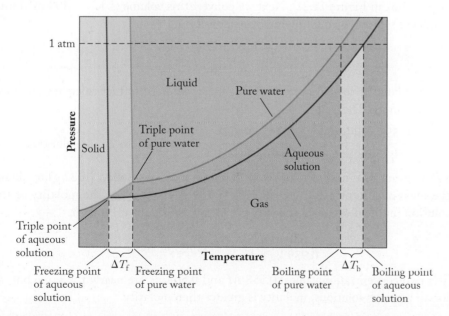

FIGURE 11.20 Combined phase diagram for pure water and a solution of a nonvolatile solute in water. Notice that the solution's boiling point is higher than the boiling point of the water, and that the solution's freezing point is lower than the freezing point of the water.

molality (*m*) concentration expressed as the number of moles of solute per kilogram of solvent.

Molality

Molarity, or moles of solute per liter of solution, is probably the most widely used unit of concentration in chemistry. Molality is closely related to molarity. It, too, is based on the number of moles of solute in solution, but instead of moles per liter of solution, **molality (*m*)** is based on moles of solute *per kilogram of solvent*:

$$m = \frac{n_{solute}}{\text{kg solvent}} \tag{11.8}$$

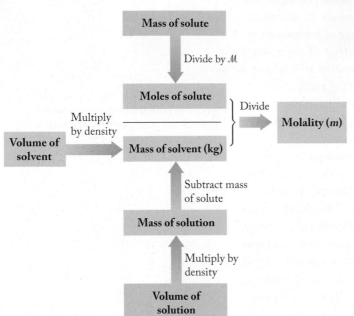

FIGURE 11.21 Flow diagram for calculating molality.

Molality is used to express concentrations of solutes when calculating boiling point elevation or freezing point depression. Why not use molarity? The reason is because liquids, like all forms of matter, tend to expand as their temperatures increase. An increase in volume means, for example, that exactly 1 kilogram of water has a volume of 1.0000 L at 4.0°C, but it expands to occupy 1.0434 L when heated to 100°C. This increase in volume means that the molarity of an aqueous solution is about 4% lower near its boiling point than it is at 4.0°C. However, the mole fractions of solute and solvent (which define shifts in vapor pressure, boiling point, and freezing point) remain unchanged. To avoid the problem of shifting molarities due to thermal expansion and contraction, we use molality in calculations of ΔT_b and ΔT_f.

Figure 11.21 summarizes pathways that can be followed in calculating molality concentrations. All lead to a central ratio of the number of moles of solute to mass of solvent. Let's follow the pathway that starts at the bottom of Figure 11.21 to calculate the molality of an aqueous solution of sodium chloride that is 0.558 *M* in NaCl (approximately the NaCl concentration in seawater) and that has a density of 1.022 g/mL. We start by assuming a solution volume of one liter (the bottom box in Figure 11.21). Next we convert this volume (1 L = 1000 mL) into a mass of solution by multiplying by its density:

$$1000 \text{ mL} \times \frac{1.022 \text{ g}}{\text{mL}} = 1022 \text{ g}$$

To calculate the solvent portion of this mass, we need to first calculate the mass of NaCl in it:

$$\frac{0.558 \text{ mol NaCl}}{1 \text{ L solution}} \times \frac{58.44 \text{ g NaCl}}{1 \text{ mol NaCl}} = 32.6 \text{ g NaCl in 1 L of solution}$$

We subtract this mass of solute from the total mass of solution (1022 g) to obtain the mass of solvent: (1022 g − 32.6 g) = 989 g = 0.989 kg. The molality of the solution (number of moles of solute per kilogram of solvent) is then:

$$\frac{0.558 \text{ mol NaCl}}{0.989 \text{ kg solvent}} = 0.564 \ m \text{ NaCl}$$

The two concentration values, 0.558 *M* and 0.564 *m*, are nearly the same but, as for all aqueous solutions, molality is greater than molarity.

CONCEPT TEST ••

The difference between the molar concentration and molal concentration of any dilute aqueous solution is small. Why?

SAMPLE EXERCISE 11.5 **Calculating the Molality** **LO4**
of a Solution

A popular recipe[1] for preparing corned beef (a St. Patrick's Day favorite in many places) calls for preparing a seasoned brine (salt solution) that contains 0.50 pound of kosher salt (NaCl) dissolved in 3.0 quarts of water. Assuming the density of water is 1.00 g/mL, what is the molality of NaCl in this solution for "corning" beef?

COLLECT AND ORGANIZE We are asked to calculate the molality of a salt solution prepared from a known mass of salt and volume of water. Molality is moles of solute per kilogram of solvent. We know the density of water: 1.00 g/mL, which is the same as 1.00 kg/L; the latter value should be more useful in this calculation given the volume of the brine. The table of conversion factors at the back of the book includes 1 pound = 453.59 grams and 1 gallon = 3.785 L. There are four quarts in one gallon.

ANALYZE We need to convert 3.0 quarts of water into an equivalent mass in kilograms. We also need to calculate the number of moles of NaCl in 0.50 pound. Its molar mass is not given, but we have used it on many previous occasions; it's 58.44 g/mol. Half a pound of NaCl contains (453.59 g/2)/(58.44 g) moles, or about 4 mol. Three quarts of water is a little less than three liters, so its mass is a little less than 3 kg. Therefore, the results of our molarity calculation should be about 4 mol/3 kg, or 1.3 m.

SOLVE The mass of water in the brine is

$$3.0 \text{ qt} \times \frac{1 \text{ gal}}{4 \text{ qt}} \times \frac{3.785 \text{ L}}{1 \text{ gal}} \times \frac{1.00 \text{ kg}}{1 \text{ L}} = 2.84 \text{ kg}$$

The number of moles of NaCl is

$$0.50 \text{ lb NaCl} \times \frac{453.59 \text{ g}}{1 \text{ lb}} \times \frac{1 \text{ mol NaCl}}{58.44 \text{ g NaCl}} = 3.88 \text{ mol NaCl}$$

The molality of the brine is

$$\frac{3.88 \text{ mol NaCl}}{2.84 \text{ kg}} = 1.4 \; m \text{ NaCl}$$

THINK ABOUT IT The calculated molality is consistent with our prediction. It is rounded off to only two significant figures, which may seem stingy, but it's appropriate given the low number of significant figures in the amounts of ingredients in the recipe. Actually, the quantities in this and most recipes are typically expressed with only one significant figure.

Practice Exercise What is the molality of a solution prepared by dissolving 78.2 g of ethylene glycol, $HOCH_2CH_2OH$, in 1.50 L of water? Assume the density of water is 1.00 g/mL. ⚙

Calculations Involving Boiling Point Elevation

As noted earlier, boiling point elevation is a colligative property that does not depend on the identity of the solute particles. However, each solvent has its own sensitivity to the presence of solute particles among its own molecules, so each solvent has its own *boiling-point-elevation constant*, K_b. This constant is used to relate boiling point elevation (ΔT_b) to the concentration of solute particles expressed in molality (m):

▶❚❚ **CHEMTOUR** Boiling and Freezing Points

$$\Delta T_b = K_b m \qquad (11.9)$$

[1]A. Brown, *Good Eats 3*, 2011, p. 32.

Boiling points are typically expressed in Celsius degrees, so the units of K_b are °C/m. The K_b of water is 0.52°C/m, which means that dissolving one mole of particles in 1 kg of water increases the boiling point of the water by 0.52°C:

$$\Delta T_b = K_b m = \frac{0.52°C}{m} \times 1 \; m = 0.52°C$$

SAMPLE EXERCISE 11.6   **Calculating the Boiling Point   LO5 Elevation of an Aqueous Solution**

What is the boiling point of seawater? Assume that the average concentration of sea salt in seawater is equivalent to 0.560 mol NaCl/kg, and that the average density of seawater is 1.025 g/mL.

COLLECT AND ORGANIZE We are asked to calculate the boiling point of 0.560 mol/kg NaCl. The K_b of water is 0.52°C/m, and its normal boiling point is 100.0°C. Boiling point elevation can be calculated using Equation 11.9: $\Delta T_b = K_b m$.

ANALYZE To use Equation 11.9 we need to express the salt concentration in moles of dissolved particles per kilogram of *solvent*. The value we are given is in moles of NaCl per kilogram of *solution* (seawater). Therefore, we need to calculate the mass of H_2O in one kilogram of seawater by subtracting the mass of NaCl, and then use the mass of H_2O to calculate molality. Sodium chloride is a binary ionic compound that dissociates into Na^+ and Cl^- ions when it dissolves. Therefore, one mole of NaCl should in theory produce *two* moles of dissolved particles. Assuming the molal concentration we calculate is similar to the mol/kg concentration we started with, then the value of m that we use in Equation 11.9 will be about 0.56 m × 2 = 1.12 m. This concentration will elevate the boiling point of water by (1.12 × 0.52)°C, or about 0.6°C.

SOLVE Mass of H_2O in one kilogram of seawater:

$$\left(1.000 \text{ kg solution} \times \frac{1000 \text{ g}}{1 \text{ kg}}\right) - \left(0.560 \text{ mol NaCl} \times \frac{58.44 \text{ g NaCl}}{1 \text{ mol NaCl}}\right) = 967 \text{ g} = 0.967 \text{ kg}$$

Concentration of NaCl:

$$\frac{0.560 \text{ mol NaCl}}{0.967 \text{ kg } H_2O} = 0.579 \; m \text{ NaCl}$$

The total concentration of Na^+ and Cl^- ions in solution:

$$0.579 \; m \times 2 = 1.158 \; m$$

Boiling point elevation:

$$\Delta T_b = K_b m = 0.52 \frac{°C}{m} \times 1.158 \; m = 0.60°C$$

Boiling point of seawater:

$$(100.0 + 0.60)°C = 100.6°C$$

THINK ABOUT IT As predicted, the boiling point of seawater is 0.6°C higher than the boiling point of pure water. The change seems small given the saltiness of seawater, but consider that the value of K_b for water is itself small.

Practice Exercise Crude oil pumped out of the ground may be accompanied by saline *formation water*. If the boiling point of a sample of formation water is 2.3°C above the normal boiling point of pure water, what is the molality of dissolved particles in the sample? 🔩

Freezing Point Depression

As shown in Figure 11.20, solutions have lower freezing points as well as higher boiling points than pure solvents. *Freezing point depression* is another colligative property that is directly proportional to the molal concentration of dissolved solute:

$$\Delta T_f = K_f m \qquad (11.10)$$

Here, ΔT_f is the change in the freezing temperature of the solvent, K_f is the *freezing-point-depression constant* of the solvent, and m is the molality of the solution.

SAMPLE EXERCISE 11.7   **Calculating the Freezing**   **LO5**
                            Point of a Solution

What is the freezing point of automobile radiator fluid prepared by mixing equal volumes of ethylene glycol ($\mathcal{M}$ = 62.07 g/mol) and water at a temperature where the density of ethylene glycol is 1.114 g/mL and the density of water is 1.000 g/mL? The freezing-point-depression constant of water, K_f, is 1.86°C/m.

COLLECT AND ORGANIZE We are asked to determine the freezing point of a 50:50 (by volume) solution of ethylene glycol in water. We are given the densities of the two liquids and the value of K_f, which relates the molality of ethylene glycol to freezing point depression (Equation 11.10).

ANALYZE To use Equation 11.10, we need to convert the 50:50 volume ratio into a solute concentration expressed in moles of ethylene glycol per kilogram of water. To simplify the calculation, let's assume we mix 1.000 liter of each of the two liquids. This is a handy volume because one liter of water has a mass of one kilogram (because a density of 1.000 g/mL is the same as 1.000 kg/L). The mass of one liter of ethylene glycol is 1114 g. It contains 1114/62.07 or about 1200/60 = 20 moles of ethylene glycol. The freezing point depression of a 20 m aqueous solution is (20 × 1.86)°C, or nearly 40°C. Therefore, the freezing point of the radiator fluid should be a bit above −40°C.

SOLVE The solvent mass is

$$1.000\ \cancel{L}\ H_2O \times \frac{1000\ \cancel{mL}}{1\ \cancel{L}} \times \frac{1.000\ \cancel{g}}{\cancel{mL}} \times \frac{0.001\ kg}{1\ \cancel{g}} = 1.000\ kg\ H_2O$$

Moles of ethylene glycol in 1.000 L:

$$1.000\ \cancel{L\ ethylene\ glycol} \times \frac{1000\ \cancel{mL}}{1\ \cancel{L}} \times \frac{1.114\ \cancel{g}}{1\ \cancel{mL}} \times \frac{1\ mol\ ethylene\ glycol}{62.07\ \cancel{g\ ethylene\ glycol}}$$

$$= 17.95\ mol\ ethylene\ glycol$$

Concentration of ethylene glycol:

$$m = \frac{17.95\ mol\ ethylene\ glycol}{1.000\ kg\ H_2O} = 17.95\ m$$

Using Equation 11.10 gives us

$$\Delta T_f = K_f m$$

$$= 1.86\ \frac{°C}{\cancel{m}} \times 17.95\ \cancel{m} = 33.4°C$$

Subtracting this temperature change from the normal freezing point of water, 0.0°C, we have

$$Freezing\ point\ of\ radiator\ fluid = 0.0°C - 33.4°C = -33.4°C$$

THINK ABOUT IT The answer is about what we expected. It makes sense because the freezing point of radiator fluid should be below the coldest expected temperatures, and −33°C (−27°F) is colder than most places in the United States ever get. In much of Alaska and Canada, however, 60–70% by volume solutions of ethylene glycol would be advisable to prevent engines from freezing up.

Practice Exercise What is the boiling point of the automobile radiator fluid in the preceding Sample Exercise? The K_b of water is 0.52°C/m. ⚙

The van 't Hoff Factor

The molality (m) terms in Equations 11.9 and 11.10 refer to *moles of solute particles per kg* of solvent. We saw in Sample Exercise 11.6 that this does not always mean simply *moles of solute per kg*. In calculating the boiling point of a solution of NaCl, we multiplied the concentration of NaCl by 2 because the dissolution of one mole of NaCl in water produces two moles of ions—one mole each of $Na^+(aq)$ and $Cl^-(aq)$—and each ion counts as a separate particle. This multiplicity of particles applies to the solutions of all ionic compounds.

In the late 19th century, Dutch chemist Jacobus van 't Hoff (1852–1911) showed that the physical properties of solutions, such as their boiling and freezing points, were linked to the concentrations of dissolved particles. In his studies van 't Hoff defined the term *i*, now called the **van 't Hoff factor**, which is a ratio of the experimentally measured value of a colligative property to the value expected if the solute did not dissociate into ions when it dissolved. As you might guess, the value of *i* for NaCl should be 2 because it produces two moles of particles per mole of NaCl. This means that one mole of NaCl raises boiling points and lowers freezing points twice as much as one mole of a nonelectrolyte such as an aqueous solution of ethanol or ethylene glycol. Following this logic, the *i* value of $MgSO_4$ is also 2, but it is 3 for $CaCl_2$ and 4 for Na_3PO_4—at least theoretically.

In reality, *i* values based on the formulas of ionic compounds—the moles of ions per mole of compound—accurately predict the colligative properties of *dilute* solutions. However, when the concentrations of these compounds are above 0.01 m, observed values of freezing point depression and boiling point elevation tend to be less than the predicted values. This nonideal behavior happens when cations and anions do not behave as completely independent particles in solution. Instead, as their concentrations increase, they cluster together. The simplest and most common cluster is an **ion pair**: one cation and one anion acting as a single particle. In 0.010 m NaCl, for example, there is little ion pairing and the properties of the solution are consistent with the *theoretical* van 't Hoff factor of 2.0 (Figure 11.22a). However, ion pairing reduces the concentration of Na^+ and Cl^- ions in 0.10 m NaCl (Figure 11.22b), and its freezing point corresponds to a concentration of only 0.19 m particles instead of 0.20 m. Thus, the *experimental* van 't Hoff factor for 0.10 m NaCl is 1.9 and not 2.0. Ion pairing and deviations from ideal behavior are observed for other electrolytes, as shown by the data in Figure 11.23. Note how the difference between theoretical and experimental values, reflecting greater ion pairing, increases with increasing ionic charge. This behavior is consistent with Coulomb's law and the greater attraction between cations and anions of higher charge.

We incorporate experimentally verified *i* factors in colligative property calculations using the following modified versions of Equations 11.9 and 11.10:

$$\Delta T_b = iK_b m \tag{11.11}$$

$$\Delta T_f = iK_f m \tag{11.12}$$

van 't Hoff factor (*i*) the ratio of the concentration of solute particles in a solution to the concentration of particles that would be there if the solute did not dissociate.

ion pair a cluster formed when a cation and an anion associate with each other in solution.

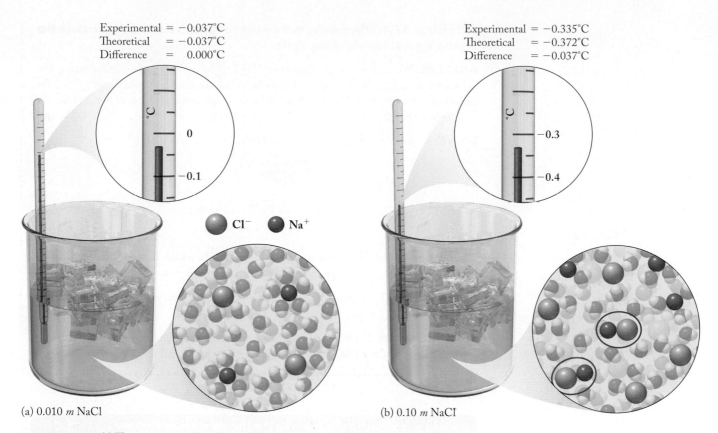

Experimental = −0.037°C
Theoretical = −0.037°C
Difference   =   0.000°C

Experimental = −0.335°C
Theoretical = −0.372°C
Difference   = −0.037°C

Cl⁻ Na⁺

(a) 0.010 m NaCl

(b) 0.10 m NaCl

FIGURE 11.22 (a) The experimentally measured freezing point of a 0.010 m solution of NaCl is the same as the theoretical value obtained with Equation 11.12. This agreement means that the solution behaves ideally and little or no ion pairing takes place. (b) The experimentally measured freezing point of a 0.10 m solution is about 0.04°C higher than the theoretical value because some of the Na⁺ and Cl⁻ ions form ion pairs, as shown inside the red ovals. The formation of ion pairs causes the concentration of solute particles to be less than the theoretical number; as a result, the van 't Hoff factor for the solution is less than the theoretical value of 2, and the decrease in freezing point is less than expected.

The m term is simply the concentration of solute—be it molecular or ionic. If the solute is molecular (such as ethanol or ethylene glycol), then the value of i is usually 1 because each mole of solute produces 1 mole of dissolved particles. There is an important exception to this rule: weak electrolytes such as organic acids, which partially ionize in water, have i values slightly greater than 1 depending on the degree to which they ionize.

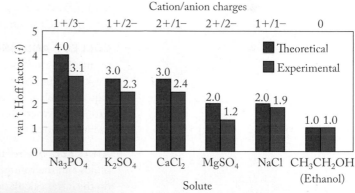

FIGURE 11.23 Theoretical and experimentally measured values for the van 't Hoff factors for 0.1 m solutions of several electrolytes and the nonelectrolyte ethanol. The higher the charge on the ions, the greater the difference between theoretical and experimentally measured values.

SAMPLE EXERCISE 11.8 **Assessing Particle Interactions in Solution** **LO6**

The experimentally measured freezing point of a 1.90 m aqueous solution of NaCl is −6.57°C. What is the value of the van 't Hoff factor for this solution? Is the solution behaving ideally, or is there evidence that solute particles are interacting with one another? The freezing-point-depression constant of water is $K_f = 1.86$°C/m. Assume the freezing point of pure water is 0.00°C.

COLLECT AND ORGANIZE We are asked to calculate the i value of a solution of known molality. We are given the measured freezing point and the K_f value. Equation

11.12 relates ΔT_f, K_f, the molality of the solution, and the value of i. We are also asked whether the solution is behaving ideally.

ANALYZE We can rearrange Equation 11.12 to solve for i before substituting the values for ΔT_f, K_f, and m. If the solution behaved ideally, we would expect $i = 2$ for the strong electrolyte NaCl. However, given the high concentration of NaCl (1.90 m), a value less than 2 is likely.

SOLVE Rearranging Equation 11.12 and solving for i:

$$i = \frac{\Delta T_{f,\text{measured}}}{K_f m}$$

$$= \frac{6.57°\text{C}}{\left(\dfrac{1.86°\text{C}}{m}\right)1.90 \, m} = 1.86$$

The value of i is not 2 (the theoretical value for NaCl), which tells us that the solution is not behaving ideally. Ion pairs must be forming in solution.

THINK ABOUT IT As predicted, the value for i for this solution is less than the theoretical value of 2.

Practice Exercise Is the van 't Hoff factor for a 0.050 m aqueous solution of magnesium sulfate likely to be: (a) greater than 2; (b) between 1 and 2; (c) less than 1?

SAMPLE EXERCISE 11.9 **Using the van 't Hoff Factor** **LO6**

Some Thanksgiving dinner chefs "brine" their turkeys prior to roasting them to help the birds retain moisture and to season the meat. Brining involves completely immersing a turkey for about 6 hours in a brine (salt solution) prepared by dissolving 2.0 pounds of salt (NaCl, $\mathcal{M} = 58.44$ g/mol) in 2.0 gallons of water (a recipe from an American cookbook). Suppose a turkey soaking in such a solution is left for 6 hours on an unheated porch on a day when the air temperature is 18°F (−8°C). Are the brine (and turkey) in danger of freezing? The K_f of water is 1.86°C/m; assume that $i = 1.85$ for this NaCl solution.

COLLECT AND ORGANIZE We are asked, indirectly, to determine the freezing point of a salt solution. We know the mass of the solute and volume of solvent, and we know the van 't Hoff factor for the solute at its concentration in the solution. Equation 11.12 relates freezing point depression to solute concentration and the van 't Hoff factor of the solution. The conversion factors at the back of the book include 1 pound = 453.59 g and 1 gallon = 3.785 L.

ANALYZE Before using Equation 11.12, we need to convert the mass of NaCl to moles and the volume of water to a mass in kilograms. We need the density of water to do the latter calculation, which is not given, but we may assume that it is 1.00 g/mL or 1.00 kg/L. To estimate the expected result, let's start by assuming that 2.0 pounds of NaCl is close to 1 kg or 1000 g, which means the solution contains (1000 g)/(~60 mol/g), or about 17 mol NaCl. The mass of H_2O is (2.0 gal) × (~4 L/gal) × (1 kg/L), or about 8 kg. The concentration of NaCl is about 17 mol/8 kg or 2 m. The value of K_f is close to 2, as is the value of i, so ΔT_f should be about (2 × 2°C/m × 2 m) or 8°C, which gives us a freezing point close to the temperature on the porch (−8°C).

SOLVE Moles of solute (NaCl):

$$2.0 \, \text{lb NaCl} \times \frac{453.59 \, \text{g}}{1 \, \text{lb}} \times \frac{1 \, \text{mol NaCl}}{58.44 \, \text{g NaCl}} = 15.5 \, \text{mol NaCl}$$

Mass of solvent (H_2O):

$$2.0 \text{ gal } H_2O \times \frac{3.785 \text{ L}}{1 \text{ gal}} \times \frac{1 \text{ kg}}{\text{L}} = 7.57 \text{ kg } H_2O$$

NaCl concentration:

$$\frac{15.5 \text{ mol NaCl}}{7.57 \text{ kg } H_2O} = 2.05 \; m \text{ NaCl}$$

Freezing point depression:

$$\Delta T_f = iK_f m = 1.85 \times 1.86 \frac{°C}{m} \times 2.05 \; m = 7.0°C$$

The freezing point of the brine is $(0.0 - 7.0)°C = -7.0°C$. The temperature of the air on the porch is one degree below the freezing point, which indicates that the brine solution could freeze if it sat outside long enough.

THINK ABOUT IT The calculated freezing-point-depression value is close to our estimate. As for whether the brine and bird would actually freeze in 6 hours, it is not likely that either would freeze, though a layer of slushy ice might start to form on the surface of the brine.

Practice Exercise One molal solutions of the following compounds are prepared. Rank them in order of decreasing freezing point: (a) NaCl, (b) CH_3COOH, (c) $Ca(NO_3)_2$, (d) methanol, (e) Na_3PO_4. ⚙

> **osmosis** the flow of a fluid through a semipermeable membrane to balance the concentration of solutes in solutions on the two sides of the membrane. The flow of solvent molecules proceeds from the more dilute solution into the more concentrated one.

11.5 Osmosis and Osmotic Pressure

Let's explore one more colligative property of solutions: their *osmotic pressure*. To understand what this phrase means, consider what happens when two liquids, one a solution and the other a pure solvent, are separated by a *semipermeable* membrane (Figure 11.24). By semipermeable, we mean that it allows molecules of solvent to pass through it but not particles of solute. Initially, the volumes and liquid levels in the two compartments are the same. However, over time the volume of pure solvent decreases and the volume of the solution increases as molecules of solvent migrate through the membrane from the pure solvent side to the solution side. This movement of solvent is called **osmosis**.

Osmosis happens because the concentration of solvent (water) molecules on the pure solvent side of the membrane is higher than it is on the solution side. This higher concentration means that they pass through the semipermeable membrane at a faster rate than those on the solution side pass through in the opposite direction, as symbolized by the greater number of blue arrows than red ones in Figure 11.24(a). This migration of water molecules is analogous to the effusion that occurs in a helium-filled party balloon. Helium atoms

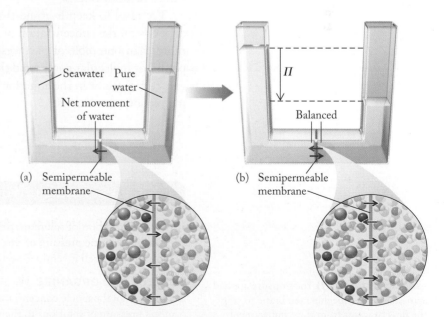

FIGURE 11.24 (a) When equal volumes of seawater and pure water are separated by a semipermeable membrane, water moves by osmosis from the pure water to the seawater. (b) Net osmotic flow stops when the difference between the heights of the two liquids creates a sufficiently large opposing osmotic pressure (Π).

⊙⊙ CONNECTION In Chapter 10 we learned that gas effusion takes place when particles of gas on the higher pressure side of a permeable membrane pass through it.

effuse out of a rubber balloon because the concentration of He atoms (in terms of moles of atoms per unit volume) is much higher inside the balloon than outside. Therefore, He atoms that are inside the balloon pass through pinholes in the balloon membrane much more rapidly than the outside He atoms effuse into the balloon through the same holes. The net outward flow of He atoms continues until the balloon has deflated.

The osmotic flow of water shown in Figure 11.24 continues until the rate of concentration-driven movement of particles toward the solution compartment is balanced by an opposing pressure created by the difference in the levels of the liquids in the two compartments. This opposing pressure, which increases the flow of water molecules in the direction of the red arrows in Figure 11.24, is called the **osmotic pressure (Π)** of the solution. The Greek capital letter pi (Π) is used as the symbol for osmotic pressure to distinguish it from the pressure (P) exerted by gases. However, the two kinds of pressure are analogous, as described in the preceding paragraph, and as we can see if we solve the ideal gas law $PV = nRT$ for pressure:

$$P = \left(\frac{n}{V}\right)RT$$

This arrangement groups the ratio n/V because it is often expressed in moles per liter, which is synonymous with molarity. Thus, the pressure of a gas is proportional to the product of the molar concentration of the particles that make it up (M), the gas law constant (R), and absolute temperature (T). An analogous equation can be written that links the osmotic pressure (Π) of a solution to the molarity (M) of its solute(s), R, and T:

$$\Pi = MRT$$

When we use an R value of 0.08206 L · atm/(mol · K), we get osmotic pressure values in atmospheres.

We need to keep in mind that osmotic pressure is a colligative property that depends on the concentration of dissolved solute particles. If the solute produces more than one mole of particles per mole of solute (because it is an ionic compound or molecular compound that ionizes in water), then we need to include the van 't Hoff factor in the equation above. Thus, the complete equation for calculating osmotic pressure is

$$\Pi = iMRT \tag{11.13}$$

SAMPLE EXERCISE 11.10 Calculating Osmotic Pressure LO7

If the concentration of solute particles in the fluid inside a red blood cell is 0.310 M, what is the osmotic pressure of this fluid at body temperature (37°C)? The membrane surrounding the cell is semipermeable.

COLLECT AND ORGANIZE We are asked to calculate the osmotic pressure of a solution with a total particle concentration of 0.310 M at 37°C. Equation 11.13 relates the osmotic pressure of solutions to their solute concentrations and absolute temperatures.

ANALYZE We know the concentration of solute particles, which is equivalent to the product of the i and M terms in Equation 11.13. The absolute temperature is (37°C + 273) = 310 K.

osmotic pressure (Π) the pressure applied across a semipermeable membrane to stop the flow of water from the compartment containing pure solvent or a less concentrated solution to the compartment containing a more concentrated solution.

SOLVE

$$\Pi = iMRT = 0.310\ \frac{\cancel{mol}}{\cancel{L}} \times \frac{0.08206\ \cancel{L} \cdot atm}{\cancel{mol} \cdot \cancel{K}} \times 310\ \cancel{K} = 7.89\ atm$$

THINK ABOUT IT The calculated osmotic pressure is nearly 8 atm. That is a lot of pressure—comparable to that in high-performance bicycle tires. In the absence of this opposing pressure, water would migrate across the semipermeable membrane of the red blood cell if it were immersed in distilled water, causing the cell to swell up (Figure 11.25a) and eventually burst.

Practice Exercise Calculate the osmotic pressure across a semipermeable membrane separating pure water from seawater at 25°C. The total concentration of all the ions in seawater is 1.09 *M*. ⚙

▶❙❙ **CHEMTOUR** Osmotic Pressure

What if a semipermeable membrane separates two solutions that have different concentrations of solutes? For example, what if we immerse red blood cells in seawater, as shown in Figure 11.25(c)? As noted in Sample Exercise 11.10, the membranes surrounding red blood cells are semipermeable, allowing water molecules to leave the cell (where solute concentration is lower) and enter the seawater (where solute concentration is higher). It makes sense that the osmotic pressure needed to halt this migration of H_2O molecules would be related to *the difference in solute concentrations* on opposite sides of the membrane. In the case of seawater surrounding a red blood cell, this difference is (1.09 − 0.31) = 0.78 *M*. Inserting this value into Equation 11.13 and assuming a human body temperature (37°C),

$$\Pi = iMRT = 0.78\ \frac{\cancel{mol}}{\cancel{L}} \times \frac{0.08206\ \cancel{L} \cdot atm}{\cancel{mol} \cdot \cancel{K}} \times (37 + 273)\ \cancel{K} = 19.8\ atm$$

In the absence of this opposing pressure, the water inside a red blood cell immersed in seawater spontaneously flows out through the cell membrane into the seawater, causing the cell to collapse.

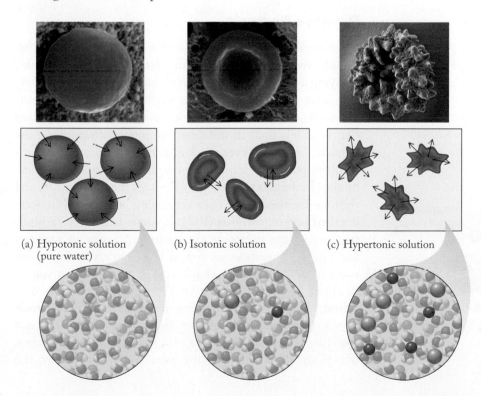

(a) Hypotonic solution (pure water) (b) Isotonic solution (c) Hypertonic solution

FIGURE 11.25 The membrane of a red blood cell is semipermeable, which means that water easily flows by osmosis into and out of the cell to equalize the solute concentrations on the two sides of the membrane. (a) When the cell is bathed in pure water, the solute concentration is lower outside the cell than inside (hypotonic conditions), so water flows by osmosis from the region of zero solute concentration to the region of high solute concentration—into the cell—and the cell expands. (b) When a cell is bathed in a solution in which the solute concentration equals the solute concentration inside the cell (isotonic conditions), the flow of water into the cell is exactly balanced by the flow of water out of the cell, and the cell size does not change. (c) When the cell is bathed in a solution in which the solute concentration is higher than inside the cell (hypertonic conditions), water flows by osmosis from the region of low solute concentration to the region of high solute concentration—out of the cell—and the cell shrinks.

CONCEPT TEST ··

If a 1.0 *M* glucose solution is inside a semipermeable membrane and 1.0 *M* KCl is on the outside, in which direction do water molecules migrate through the membrane?

··

The preceding osmotic pressure calculations show that relatively small differences in solute concentration can produce large osmotic pressures that can distort and even destroy living cells. Drinking seawater may prove fatal because the ingested seawater pulls water out of the organs of the digestive system, causing them to dehydrate and ultimately fail. Similarly, fluids administered intravenously to hospital patients must have nearly the same concentration of solute particles as the patients' blood and blood cells. Such fluids are said to be *isotonic*. As we saw in Figure 11.25, solutions with lower (*hypotonic*) or higher (*hypertonic*) solute concentrations cause cells to swell or collapse as water molecules either enter or leave the cells.

Two common IV solutions are physiological saline, which contains 9.0 g/L NaCl, and D5W, which is 50 g/L dextrose ($\mathcal{M} = 180.16$ g/mol). Let's convert these concentrations to molarity values:

$$\text{Physiological saline:} \quad \frac{9.0 \text{ g NaCl}}{1 \text{ L}} \times \frac{1 \text{ mol NaCl}}{58.44 \text{ g NaCl}} = 0.15 \text{ } M \text{ NaCl}$$

$$\text{D5W:} \quad \frac{50 \text{ g dextrose}}{1 \text{ L}} \times \frac{1 \text{ mol dextrose}}{180.16 \text{ g dextrose}} = 0.28 \text{ } M \text{ dextrose}$$

While the *molarity* of D5W is nearly twice that of physiological saline, the van 't Hoff factor of a 0.15 *M* solution of NaCl is nearly 2. Dextrose, on the other hand, is a molecular compound that does not ionize in water and has a van 't Hoff factor of 1. Thus, the concentrations of dissolved particles in the two solutions are essentially the same.

Reverse Osmosis

We have developed a molecular-level explanation based on osmosis for why people shouldn't drink seawater. However, applying a variation on osmosis can make seawater drinkable. The process involves removing most of the ions from seawater, or *desalination*. Distillation is a desalination method that is widely used in the desert countries surrounding the Persian Gulf; however, a less energy-intensive approach involves pumping seawater through semipermeable membranes. Molecules of water pass through the membranes, and the ions dissolved in the seawater are left behind. This process is called **reverse osmosis (RO)**. It works because very-high-pressure pumps overcome the osmotic pressure of seawater and force the spontaneous flow of H_2O molecules from pure water into seawater to reverse itself, flowing the other way.

The RO system shown in Figure 11.26 consists of a cylindrical chamber containing many narrow tubes made of a semipermeable membrane. Seawater is forced through the chamber so that it washes over the exterior of the tubes, which are filled with flowing pure water. Ordinarily, the

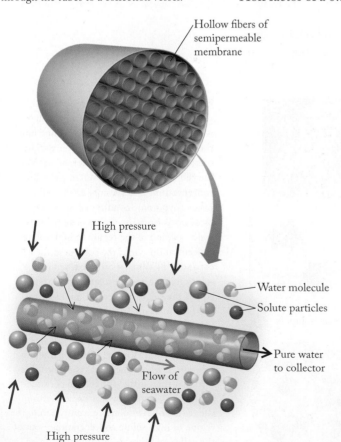

FIGURE 11.26 Seawater can be desalinated by reverse osmosis. In this apparatus, seawater flows at a pressure greater than its osmotic pressure around bundles of tubes with semipermeable walls. Water molecules pass from the seawater into the tubes and flow through the tubes to a collection vessel.

Hollow fibers of semipermeable membrane

High pressure

Water molecule

Solute particles

Pure water to collector

Flow of seawater

High pressure

direction of osmosis would be from the inner tubes into the surrounding seawater. However, when a pressure greater than the osmotic pressure of seawater is exerted on the seawater in the outer chamber, water flows in the other direction: from the chamber, through the membranes, into the tubes, and out of the apparatus to a pure water collector.

Some municipal water-supply systems use reverse osmosis to make slightly saline (brackish) water fit to drink. Some industries use this method to purify conventional tap water, and it provides a renewable supply of drinking water for sailors on ocean-crossing yachts. Strong pumps and fittings and rugged membranes are required because reverse osmosis systems must operate at very high pressures (illustrated in Sample Exercise 11.11). We should also note that the semipermeable membranes are not 100% efficient, so RO systems cannot turn seawater into completely deionized water. Some solvated Na^+ and Cl^- ions inevitably pass through the membrane, but their concentrations in the product water are reduced to levels that make the water drinkable.

reverse osmosis (RO) a purification process in which solvent is forced through semipermeable membranes, leaving dissolved impurities behind.

**SAMPLE EXERCISE 11.11   Calculating Pressure   LO7
                         for Reverse Osmosis**

What is the reverse osmotic pressure required at 20°C to purify brackish well water that is equivalent to 0.177 M NaCl if the product water is to meet the drinking water standard that the concentration of Na^+ ions be no greater than 20 mg/L? Assume the i value of the well water electrolytes is 1.85.

COLLECT AND ORGANIZE We are asked to calculate the pressure needed to purify water by reverse osmosis. We are given the temperature, the concentration of NaCl in the water to be purified, and the target concentration of Na^+ ions in the water to be produced.

ANALYZE We can adapt Equation 11.13 to calculate the difference in osmotic pressure between the source and the product water, but first we need to convert the concentration of Na^+ ions in the purified water to molarity. We may assume that this value also represents the concentration of Cl^- ions in the drinking water, which must be there to balance ionic charges. This concentration value will probably be small compared to the salt concentration of the well water, so we can estimate the osmotic pressure that must be overcome based on the fact that the NaCl concentration in the well water is close to that of physiological saline, which has an osmotic pressure of about 8 atm (see Sample Exercise 11.10).

SOLVE Convert 20 mg/L Na^+ to molarity:

$$\frac{20 \text{ mg Na}^+}{L} \times \frac{1 \text{ g}}{1000 \text{ mg}} \times \frac{1 \text{ mol Na}^+}{22.99 \text{ g Na}^+} = \frac{8.7 \times 10^{-4} \text{ mol Na}^+}{L}$$

The difference in NaCl concentration between the well water and target drinking water is

$$(0.177 - 0.00087) = 0.176 \ M \text{ NaCl}$$

The corresponding osmotic pressure is

$$\Pi = iMRT$$

$$= 1.85 \times 0.176 \ \frac{\text{mol}}{L} \times 0.08206 \ \frac{L \cdot \text{atm}}{\text{mol} \cdot K} \times (20 + 273) \ K$$

$$= 7.8 \text{ atm}$$

THINK ABOUT IT In the absence of any external pressure, solvent will flow from the product water to the well water. However, an external pressure greater than 7.8 atm will force the spontaneous flow to reverse, forming drinkable product water from brackish well water.

Practice Exercise Calculate the minimum external pressure that must be applied in a reverse osmosis system to reduce the salinity of seawater from 3.5% NaCl ($i = 1.84$) to 174 mg NaCl per liter at 25°C.

11.6 Using Colligative Properties to Determine Molar Mass

CONNECTION In Chapter 7 we used molar masses determined by mass spectrometry to convert empirical formulas derived from elemental analyses to molecular formulas.

CONNECTION In Chapter 10 we used measurements of density and the ideal gas law to calculate molar masses of gases.

In principle, the molar mass of any solute can be determined by dissolving a known quantity of the solute in a known quantity of solvent and then measuring the effect the dissolved solute has on a colligative property of the solvent. In practice, this method works only for nonelectrolytes, which have a van 't Hoff factor of 1. Freezing-point-depression measurements, for example, can be used to find the molar mass of a molecular compound if it is sufficiently soluble in a solvent whose K_f value is known, as we illustrate in Sample Exercise 11.12.

SAMPLE EXERCISE 11.12 **Using Freezing Point Depression to Determine Molar Mass** **LO8**

Eicosene is a hydrocarbon used as an additive in the high-pressure fracturing of rock layers to enhance recovery of petroleum and natural gas, a process sometimes called *fracking*. The freezing point of a solution prepared by dissolving 100 mg of eicosene in 1.00 g of benzene is 1.75°C lower than the freezing point of pure benzene. What is the molar mass of eicosene? (K_f for benzene is 4.90°C/m.)

COLLECT AND ORGANIZE We are asked to determine the molar mass of a hydrocarbon, which is a molecular compound. We are given the mass of it that lowers the freezing point of a solvent by a known amount, and we know the K_f of the solvent. Equation 11.12 relates the solute concentration to the change in freezing point.

ANALYZE Eicosene is a nonelectrolyte and has a van 't Hoff factor of 1, so we can use freezing point depression to determine its molal concentration in solution. We already know the concentration of such a solution in milligrams of eicosene per gram of solvent. Combining the two concentration values should let us determine the number of moles of eicosene in the mass of it in the sample. To get a quick estimate of molar mass, we scale up the quantities of solute and solvent by a factor of 1000 and get a concentration of 100 g/kg. Dividing the value of ΔT_f (1.75°C) by the K_f value of 4.90°C/m gives a molality of about 0.3 m (mol/kg). Dividing 100 g/kg by 0.3 mol/kg gives us a molar mass of about 300 g/mol.

SOLVE The molality of the eicosene solution is

$$\Delta T_f = iK_f m = K_f m$$

$$m = \frac{\Delta T_f}{K_f} = \frac{1.75°\text{C}}{4.90°\text{C}/m} = 0.357\ m = 0.357\ \text{mol/kg}$$

The sample contains 100 mg eicosene/g. If we multiply the top and bottom of this ratio by 1000, we get an equivalent concentration of 100 g/1000 g, or 100 g/kg. The last ratio is a handy one because the quantity of solvent, 1 kg, matches that used in molality. If we divide this concentration by the molality value, we obtain molar mass directly:

$$\frac{100 \text{ g/kg}}{0.357 \text{ mol/kg}} = 280 \text{ g/mol}$$

THINK ABOUT IT The molar mass of eicosene, 280 g/mol, is close to our estimate. We saved several conversion steps in the solution by scaling up the concentration units of the sample to match the quantity of solvent used in the definition of molality, namely, one kilogram.

Practice Exercise A solution prepared by dissolving 360 mg of a common sweetener in 1.00 g of water freezes at −3.72°C. What is the molar mass of this sweetener? The K_f value of water is 1.86°C/m. ⚙

Measuring the osmotic pressure of the solution of a nonelectrolyte can also be used to determine molar mass. Osmotic pressure measurements have the advantage of producing large, easily measured pressures in comparison to the small differences in, for example, the freezing points of aqueous solutions. Consider that a nonelectrolyte concentration of 0.0100 M produces an osmotic pressure at 25°C equal to

$$\Pi = iMRT$$

$$= 1 \times 0.0100 \; \frac{\text{mol}}{\text{L}} \times 0.08206 \; \frac{\text{L} \cdot \text{atm}}{\text{mol} \cdot \text{K}} \times (25 + 273) \text{ K}$$

$$= 0.245 \text{ atm or } 186 \text{ torr}$$

On the other hand, the freezing point depression of a 0.0100 m nonelectrolyte solution is

$$\Delta T_f = iK_f m$$

$$= 1 \times 1.86°C/m \times 0.0100 \; m = 0.0186°C$$

Moreover, biologically important substances such as proteins and carbohydrates are often available only in small quantities, and they have large molar masses. These two factors further reduce solution concentration and the size of ΔT_f or ΔT_b.

SAMPLE EXERCISE 11.13 **Using Osmotic Pressure to Determine Molar Mass** **LO8**

A molecular compound that is a nonelectrolyte is isolated from a South African tree. A 47 mg sample is dissolved in water to make 2.50 mL of solution at 25°C. The osmotic pressure of the solution is 0.489 atm. Calculate the molar mass of the compound.

COLLECT AND ORGANIZE We know the concentration of a solution of an unknown material of a substance, and we know its osmotic pressure at 25°C. Its van 't Hoff factor is 1 because it is a nonelectrolyte. We need to find its molar mass.

ANALYZE Given our known quantities i and T, we can use Equation 11.13 to calculate the molarity of the solution. We can then compare that moles-per-volume concentration value to the known mass-per-volume concentration of the sample to calculate molar mass.

SOLVE Rearranging Equation 11.13 to solve for M and substituting the given values,

$$M = \Pi/iRT = \frac{0.489 \text{ atm}}{1 \times 0.08206 \frac{\text{L} \cdot \text{atm}}{\text{mol} \cdot \text{K}} \times 298 \text{ K}}$$

$$= 2.00 \times 10^{-2} \text{ mol/L}$$

The sample concentration is

$$\frac{47 \text{ mg}}{2.50 \text{ mL}} = 18.8 \text{ mg/mL} = 18.8 \text{ g/L}$$

To calculate molar mass, we take the ratio of the g/L concentration to the mol/L concentration, which gives us g/mol:

$$\frac{18.8 \text{ g/L}}{2.00 \times 10^{-2} \text{ mol/L}} = 9.4 \times 10^2 \text{ g/mol}$$

THINK ABOUT IT One of the advantages of determining molar mass by osmotic pressure is that only a small quantity of sample is required. Only a 47 mg sample of a substance with a large molar mass produced an osmotic pressure sufficiently large (0.489 atm) to allow for an accurate determination of molar mass.

Practice Exercise A solution is made by dissolving 5.00 mg of a polysaccharide in water to give a final volume of 1.00 mL. The osmotic pressure of this nonelectrolyte solution is 1.91×10^{-3} atm at 25°C. What is the molar mass of the polysaccharide? ⚙

SAMPLE EXERCISE 11.14 Integrating Concepts: Antifreeze in Car Batteries

The aqueous electrolyte inside a fully charged lead–acid battery (the kind used to start automobile engines) is 35% (by volume) sulfuric acid, H_2SO_4. Pure sulfuric acid is a dense (1.84 g/mL), nonvolatile liquid (its vapor pressure at 25°C is <0.001 torr). It is also a strong acid that is completely ionized in concentrated aqueous solutions:

$$H_2SO_4(aq) \rightarrow H^+(aq) + HSO_4^-(aq)$$

The power to start the engine comes from the following chemical reaction as the battery is discharged:

$$Pb(s) + PbO_2(s) + 2\,H_2SO_4(aq) \rightarrow 2\,PbSO_4(s) + 2\,H_2O(\ell)$$

   a. What is the vapor pressure of the solution at 25°C?
   b. What is the freezing point of the solution?
   c. How do vapor pressure and freezing point change as the battery is discharged?

COLLECT AND ORGANIZE We know the concentration of a solution expressed as the volume ratio of a nonvolatile liquid solute (H_2SO_4, $\mathcal{M} = 98.08$ g/mol) to solvent (H_2O). We are asked to calculate the freezing point of the solution and its vapor pressure at 25°C. We also need to describe how these freezing point and vapor pressure values change during the course of a reaction in which H_2SO_4 is a reactant and H_2O is a product. The vapor pressure of such a solution can be calculated using Equation 11.7: $P_{\text{solution}} = x_{\text{solvent}} P_{\text{solvent}}$. Its freezing point can be calculated using Equation 11.12: $\Delta T_f = iK_f m$. The vapor pressure of pure water at 25°C is 23.8 torr, and its K_f value is 1.86°C/m.

ANALYZE Solutions of nonvolatile solutes in volatile solvents have lower vapor pressures and freezing points than their pure solvents. To find out how much lower our solution's values are, we must first calculate the mole fraction of the solvent in solution (needed in Equation 11.7) and the molal concentration of solute (needed in Equation 11.12). As the battery is discharged, the concentration of H_2SO_4 decreases. Assume the molecules of H_2SO_4 in the electrolyte form H^+ and HSO_4^- ions, so the van 't Hoff factor of sulfuric acid is 2.

SOLVE
   a. If we assume a sample volume of 100 mL, then 35% H_2SO_4 translates into 35 mL H_2SO_4 and (100 − 35) = 65 mL H_2O. Converting these volumes into masses and then moles,

$$35 \text{ mL } H_2SO_4 \times 1.84 \frac{\text{g}}{\text{mL}} \times \frac{1 \text{ mol } H_2SO_4}{98.08 \text{ g } H_2SO_4} \times \frac{2 \text{ mol ions}}{1 \text{ mol } H_2SO_4}$$

$$= 1.31 \text{ mol ions}$$

$$65 \text{ mL } H_2O \times 1.00 \frac{\text{g}}{\text{mL}} \times \frac{1 \text{ mol } H_2O}{18.02 \text{ g } H_2O}$$

$$= 3.61 \text{ mol } H_2O$$

The mole fraction of H_2O is then

$$3.61 \text{ mol } H_2O/(3.61 + 1.31) \text{ total mol} = 0.734$$

And the vapor pressure of the battery acid (via Equation 11.7) is

$$P_{solution} = x_{solvent}P_{solvent} = 0.734 \times 23.8 \text{ torr} = 17.5 \text{ torr}$$

b. To calculate freezing point depression, we first need to calculate the molality of solute particles:

$$\frac{1.31 \text{ mol ions}}{65 \text{ mL } H_2O \times 1.00 \text{ g/mL}} \times \frac{1000 \text{ g}}{\text{kg}} = 20.2 \text{ mol/kg} = 20.2 \text{ } m$$

Calculating freezing point depression (ΔT_f):

$$\Delta T_f = K_f m = 1.86°C/m \times 20.2 \text{ } m = 38°C$$

Therefore, the freezing point of battery acid is 38°C below the freezing point of pure water (0.0°C), or −38°C.

c. The chemical reaction that provides the battery's energy consumes solute, converting it into more solvent. The result is a less concentrated solution of sulfuric acid and a greater mole fraction of H_2O. Such a solution has a higher freezing point (closer to that of pure water) and a greater vapor pressure (again, closer to that of pure water).

THINK ABOUT IT We did not include the van 't Hoff factor of 2 when using Equation 11.12 in part b because it had already been used to calculate the number of moles of solute particles in part a. Thus, the m value we used in part b was actually the product of i and m.

SUMMARY

Section 11.1 The **enthalpy of solution ($\Delta H_{solution}$)** of an ionic compound in water is the sum of the enthalpy changes that accompany breaking the bonds between ions in the solid solute and breaking some of the hydrogen bonds between molecules of H_2O, which are endothermic processes, and the enthalpy change that accompanies the formation of ion–dipole interactions between dissolved ions and water molecules that surround them. The lattice energies of ionic compounds can be calculated using the **Born–Haber cycle**.

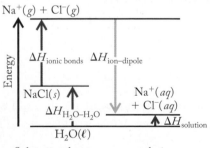

Solute + solvent → solution

Section 11.2 Molecules at the surface of a liquid evaporate by breaking intermolecular interactions with neighboring molecules and entering the gas phase. The **vapor pressure** of a liquid is proportional to the fraction of its molecules that enter the gas phase. If the liquid produces a significant vapor pressure at a given temperature, it is considered **volatile**. A greater proportion of liquid molecules enter the gas phase as the temperature increases, leading to higher vapor pressures at higher temperatures. The vapor pressure of a pure substance as a function of temperature is determined by the **Clausius–Clapeyron equation**.

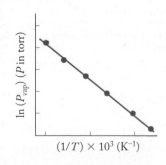

Section 11.3 Crude oil is composed primarily of hydrocarbons that can be separated by **fractional distillation** into gasoline and other useful products. The vapor pressure of an **ideal solution** of volatile compounds follows **Raoult's law**.

Section 11.4 The vapor pressure of a solution is a **colligative property**, which is defined as any property of a solution that depends only on the concentration of solutes, not on their identity. Other colligative properties include boiling point elevation and freezing point depression. The concentration units used for colligative property measurements include molarity and **molality (m)**, which is the number of moles of solute per kilogram of solvent. Solutes in solution elevate the solvent's boiling point and depress its freezing point. The **van 't Hoff factor (i)** accounts for the colligative properties of electrolytes and the formation of solute **ion pairs** in concentrated solutions.

Section 11.5 In **osmosis**, solvent flows through a semipermeable membrane from a solution of lower solute concentration into a solution of higher solute concentration. **Osmotic pressure (Π)**, a colligative property, is defined as the pressure required to halt the net flow of solvent through a semipermeable membrane. **Reverse osmosis (RO)** is used to purify water.

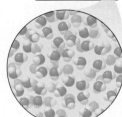

Section 11.6 The molar mass of a nonelectrolyte can be determined by measuring the freezing point depression, boiling point elevation, or osmotic pressure of a solution of the compound.

PROBLEM-SOLVING SUMMARY ••• ■

TYPE OF PROBLEM	CONCEPTS AND EQUATIONS	SAMPLE EXERCISES
Calculating lattice energy	The lattice energy of an ionic compound can be calculated from the heat of formation of the compound by applying Hess's law to the steps involved in converting free elements into an ionic solid, that is, the Born–Haber cycle.	11.1
Calculating vapor pressure or ΔH_{vap} using the Clausius–Clapeyron equation	$$\ln\left(\frac{P_{vap,T_1}}{P_{vap,T_2}}\right) = \frac{\Delta H_{vap}}{R}\left(\frac{1}{T_2}-\frac{1}{T_1}\right) \qquad (11.5)$$	11.2
Calculating the vapor pressure of a solution of volatile compounds	Obtain the vapor pressures, P_i, of the components of a solution as pure compounds and determine their mole fractions (x_i) in the solution. Use Raoult's law to calculate the vapor pressure of the solution: $$P_{total} = x_1 P_1 + x_2 P_2 + x_3 P_3 + \cdots \qquad (11.6)$$	11.3
Calculating the vapor pressure of a solution of one or more nonvolatile solutes	Use the form of Raoult's law for nonvolatile solutes: $$P_{solution} = x_{solvent} P_{solvent} \qquad (11.7)$$ where $P_{solution}$ is the vapor pressure of the solution, $x_{solvent}$ is the mole fraction of the solvent in the solution, and $P_{solvent}$ is the vapor pressure of the pure solvent.	11.4
Calculating the molality of a solution	$$m = \frac{n_{solute}}{\text{kg solvent}} \qquad (11.8)$$	11.5
Calculating boiling points and freezing points of solutions	For nonelectrolyte solutes, use $$\Delta T_b = K_b m \qquad (11.9)$$ where ΔT_b is the elevation in the boiling point of the solvent and $$\Delta T_f = K_f m \qquad (11.10)$$ where ΔT_f is the depression in the freezing point of the solvent. For electrolytes, use $$\Delta T_b = iK_b m \qquad (11.11)$$ and $$\Delta T_f = iK_f m \qquad (11.12)$$ where i is the van 't Hoff factor.	11.6, 11.7, 11.9
Assessing interactions among particles in solution by comparing the theoretical value of the van 't Hoff factor i with the experimentally measured value	Use measured freezing point depression or boiling point elevation values and Equations 11.11 or 11.12 to determine the value of the van 't Hoff factor.	11.8
Calculating osmotic pressure and reverse osmotic pressure	$$\Pi = iMRT \qquad (11.13)$$ where Π is the osmotic pressure, i is the van 't Hoff factor, M is the molar concentration of the solution, and T is the absolute temperature of the solution.	11.10, 11.11
Determining molar mass from boiling point elevation, freezing point depression, or osmotic pressure	Use $\Delta T_b = iK_b m$, $\Delta T_f = iK_f m$, or $\Pi = iMRT$ to calculate the molality m or molarity M of a solution containing a known mass of solute.	11.12, 11.13

VISUAL PROBLEMS ••• ■

(Answers to boldface end-of-chapter questions and problems are in the back of the book.)

11.1. Figure P11.1 provides a particle-level view of a sealed container partially filled with a solution that has two components: X (blue spheres) and Y (red spheres). Which of the following statements about substances X and Y are true?

a. X is the solvent in this solution.
b. Pure Y is a volatile liquid.
c. If Y were not present, there would be fewer X particles in the gas above the liquid solution.
d. The presence of Y increases the vapor pressure of X.

FIGURE P11.1

11.2. Figure P11.2 provides a particle-level view of a sealed container partially filled with a solution of two miscible liquids: X (blue spheres) and Y (red spheres). Which of the following statements about substances X and Y are true?

a. Y is the solvent in this solution.
b. Pure Y has a higher vapor pressure than pure X.
c. The presence of Y in the solution lowers the vapor pressure of X.
d. If Y were not present, there would be fewer total particles in the gas above the liquid solution.

FIGURE P11.2

11.3. Figure P11.3 provides particle-level views of 0.001 *M* aqueous solutions of the following four solutes: $C_6H_{12}O_6$, NaCl, $MgCl_2$, and K_3PO_4. The blue spheres represent particles of solute. Which compounds are represented in images (a)–(d)?

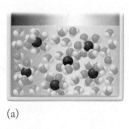

(a)

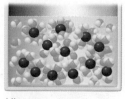

(b)

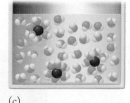

(c)

(d)

FIGURE P11.3

11.4. Which of the four solutions in Figure P11.3 has the highest: (a) vapor pressure; (b) boiling point; (c) freezing point; (d) osmotic pressure?

11.5. Use the graph in Figure P11.5 to estimate the normal boiling points of substances X and Y. Molecules of which substance experience the stronger intermolecular forces?

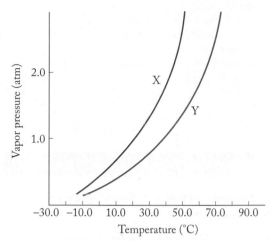

FIGURE P11.5

11.6. The graph in Figure P11.6 shows the decrease in the freezing point of water ΔT_f for solutions of two molecular solutes, A (triangles) and B (circles), in water.
a. Explain why they overlap.
b. How would the curve of an ionic solute differ from the one for substances A and B?

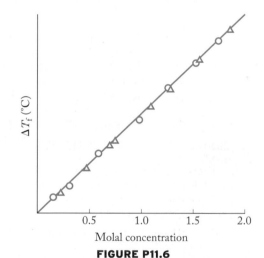

FIGURE P11.6

11.7. The arrow in Figure P11.7 indicates the direction of solvent flow through a semipermeable membrane in equipment designed to measure osmotic pressure. Which solution, A or B, is the more concentrated? Explain your answer.

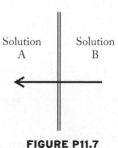

FIGURE P11.7

*11.8. **Kidney Dialysis** Semipermeable membranes of the sort used in kidney dialysis do not allow large molecules and cells to pass but do allow small ions and water to pass. Figure P11.8 shows such a membrane separating fluids of various compositions.

   a. In which direction does the water flow in each solution?

   b. In which direction do the sodium ions flow in each solution?

   c. In which direction do the potassium ions flow in each solution?

Pure H$_2$O	0.2 *M* KCl in H$_2$O
Membrane	Membrane
Proteins 0.5 *M* NaCl 0.1 *M* KCl in H$_2$O	Proteins 0.5 *M* NaCl 0.1 *M* KCl in H$_2$O
(i)	(ii)

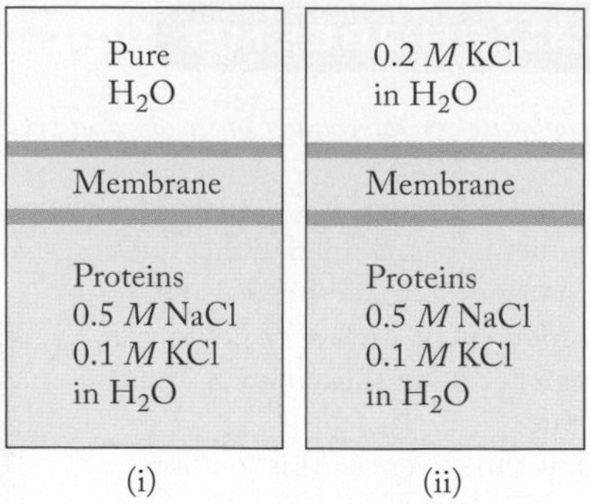

FIGURE P11.8

QUESTIONS AND PROBLEMS ·····································■

Energy Changes when Substances Dissolve; Vapor Pressure

CONCEPT REVIEW

11.9. Use kinetic molecular theory to explain why the vapor pressure of a liquid increases with increasing temperature.

11.10. How do ion–dipole interactions influence whether an ionic compound's heat of solution is exothermic or endothermic?

11.11. Generally speaking, how is the vapor pressure of a liquid affected by the strength of intermolecular forces?

11.12. Is vapor pressure an intensive or extensive property of a volatile substance?

PROBLEMS

11.13. Use a Born–Haber cycle to calculate the lattice energy of potassium chloride (KCl) from the following data:

   Ionization energy of K(g) = 419 kJ/mol
   Electron affinity of Cl(g) = −349 kJ/mol
   Energy to sublime K(s) = 89 kJ/mol
   Bond energy of Cl$_2$(g) = 243 kJ/mol
   Standard heat of formation of KCl = −436.5 kJ/mol

11.14. Calculate the lattice energy of sodium oxide (Na$_2$O) from the following data:

   Ionization energy of Na(g) = 495 kJ/mol
   Electron affinity of O(g) for 2 electrons = 603 kJ/mol
   Energy to sublime Na(s) = 109 kJ/mol
   Bond energy of O$_2$(g) = 498 kJ/mol

$$\Delta H_{rxn} \text{ for } 2\,Na(s) + \tfrac{1}{2}\,O_2(g) \rightarrow Na_2O(s) = -416 \text{ kJ/mol}$$

11.15. Rank the following compounds in order of increasing vapor pressure at 298 K: (a) CH$_3$CH$_2$OH, (b) CH$_3$OCH$_3$, and (c) CH$_3$CH$_2$CH$_3$.

11.16. Rank the compounds in Figure P11.16 in order of increasing vapor pressure at 298 K.

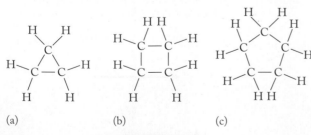

   (a)   (b)   (c)

FIGURE P11.16

11.17. **Pine Oil** The smell of fresh cut pine is due in part to the cyclic alkene pinene, whose carbon-skeleton structure is shown in Figure P11.17. Use the data in the table to calculate the heat of vaporization, ΔH_{vap}, of pinene.

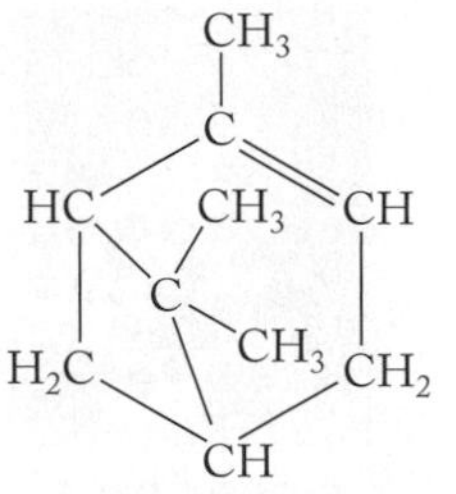

Pinene

FIGURE P11.17

Vapor Pressure (torr)	Temperature (K)
760	429
515	415
340	401
218	387
135	373

11.18. **Almonds and Cherries** Almonds and almond extracts are common ingredients in baked goods. Almonds contain the compound benzaldehyde (shown in Figure P11.18), which accounts for the odor of the nut. Benzaldehyde is also partly responsible for the aroma of cherries. Use the data in the table to calculate the heat of vaporization, ΔH_{vap}, of benzaldehyde.

Vapor Pressure (torr)	Temperature (K)
50	373
111	393
230	413
442	433
805	453

Benzaldehyde

FIGURE P11.18

11.19. High-Octane Gasoline Gasoline is a complex mixture of hydrocarbons. Gasoline is sold with a variety of octane ratings that are based on the comparison of the gasoline with the combustion properties of isooctane, a compound with the molecular formula C_8H_{18}. The Lewis structures of isooctane and another compound with molecular formula C_8H_{18} are shown in Figure P11.19, along with their normal boiling points and heats of vaporization. Determine the vapor pressure of each isomer on a day when the temperature is 38°C.

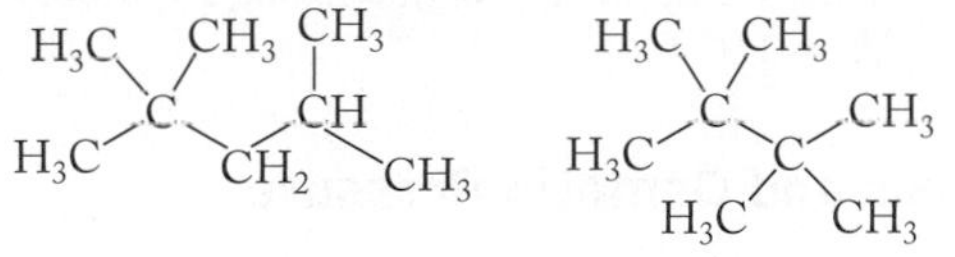

Isooctane
bp = 98.2°C
ΔH_{vap} = 35.8 kJ/mol

Tetramethylbutane
bp = 106.5°C
ΔH_{vap} = 43.3 kJ/mol

FIGURE P11.19

11.20. Camping Fuel Portable lanterns and stoves used for camping and backpacking often use a mixture of C_5 and C_6 hydrocarbons known as "white gas." Figure P11.20 shows the carbon-skeleton structure of pentane, C_5H_{12}, along with its normal boiling point and heat of vaporization. Determine the vapor pressure of pentane on a morning when the temperature is 5°C.

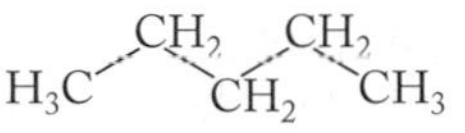

Pentane
bp = 36.0°C
ΔH_{vap} = 27.6 kJ/mol

FIGURE P11.20

Mixtures of Volatile Substances

CONCEPT REVIEW

11.21. What physical property of the components of crude oil is used to separate them?

11.22. What is the difference between distillation and fractional distillation?

11.23. In an equimolar mixture of C_5H_{12} and C_7H_{16}, which compound is present in higher concentration in the vapor above the solution?

11.24. Why does the boiling point of a mixture of volatile hydrocarbons increase over time during a distillation?

PROBLEMS

11.25. At 20°C, the vapor pressure of ethanol (CH_3CH_2OH) is 45 torr and the vapor pressure of methanol (CH_3OH) is 92 torr. What is the vapor pressure at 20°C of a solution prepared by mixing 25 g of methanol and 75 g of ethanol?

11.26. A bottle is half-filled with a 50:50 (by moles) mixture of heptane (C_7H_{16}) and octane (C_8H_{18}) at 25°C. What is the mole ratio of heptane vapor to octane vapor in the air space above the liquid in the bottle? The vapor pressures of heptane and octane at 25°C are 31 torr and 11 torr, respectively.

Colligative Properties of Solutions

CONCEPT REVIEW

11.27. Explain how nearly all the water ends up in one compartment in Figure 11.19.

11.28. A dam separates the freshwater of the Charles River from the seawater of Boston Harbor. If the two bodies of water are at the same temperature, which one evaporates faster on a hot summer day?

11.29. What is the difference between molarity and molality?

11.30. As a solution of NaCl becomes more concentrated, does the difference between its molarity and its molality increase or decrease?

11.31. Explain why seawater has a lower freezing point than freshwater.

11.32. The thermostat in a refrigerator filled with cans of soft drinks malfunctions and the temperature of the refrigerator drops below 0°C. The contents of the cans of diet soft drinks freeze, rupturing many of the cans and causing an awful mess. However, none of the cans containing regular, nondiet soft drinks rupture. Why?

11.33. Why is it important to know if a substance is a strong electrolyte before predicting its effect on the boiling and freezing points of a solvent?

11.34. Why are the van 't Hoff factors of electrolytes greater than those of nonelectrolytes?

11.35. Why are theoretical and experimental van 't Hoff factors of electrolytes sometimes different?

11.36. Can an experimentally measured value of a van 't Hoff factor be greater than the theoretical value? Why or why not?

PROBLEMS

11.37. A solution contains 3.5 mol of water and 1.5 mol of glucose ($C_6H_{12}O_6$). What is the mole fraction of water in the solution? What is the vapor pressure of the solution at 25°C, given that the vapor pressure of pure water at 25°C is 23.8 torr?

11.38. A solution contains 4.5 mol of water, 0.3 mol of sucrose ($C_{12}H_{22}O_{11}$), and 0.2 mol of glucose. Sucrose and glucose are nonvolatile. What is the mole fraction of water in the solution? What is the vapor pressure of the solution at 35°C, given that the vapor pressure of pure water at 35°C is 42.2 torr?

11.39. Calculate the molality of each of the following solutions:
  a. 0.875 mol of glucose ($C_6H_{12}O_6$) in 1.5 kg of water
  b. 11.5 mmol of acetic acid (CH_3COOH) in 65 g of water
  c. 0.325 mol of baking soda ($NaHCO_3$) in 290.0 g of water

*11.40. Table 8.1 lists molar concentrations of major ions in seawater. Using a density of 1.025 g/mL for seawater, convert the concentrations into molalities.

11.41. What mass of the following solutions contains 0.100 mol of solute? (a) 0.334 m NH_4NO_3; (b) 1.24 m ethylene glycol, $HOCH_2CH_2OH$; (c) 5.65 m $CaCl_2$

11.42. How many moles of solute are there in the following solutions?
  a. 0.150 m glucose solution made by dissolving the glucose in 100.0 kg of water
  b. 0.028 m Na_2CrO_4 solution made by dissolving the Na_2CrO_4 in 1000.0 g of water
  c. 0.100 m urea solution made by dissolving the urea in 500.0 g of water

11.43. Fish Kills High concentrations of ammonia (NH_3), nitrite ion, and nitrate ion in water can kill fish. Lethal concentrations of these species for rainbow trout are 1.1 mg/L, 0.40 mg/L, and 1361 mg/L, respectively. Express these concentrations in molality units, assuming a solution density of 1.00 g/mL.

11.44. The concentrations of six important elements in a sample of river water are 0.050 mg/kg of Al^{3+}, 0.040 mg/kg of Fe^{3+}, 13.4 mg/kg of Ca^{2+}, 5.2 mg/kg of Na^+, 1.3 mg/kg of K^+, and 3.4 mg/kg of Mg^{2+}. Express each of these concentrations in molality units.

11.45. Cinnamon Cinnamon owes its flavor and odor to cinnamaldehyde (C_9H_8O). Determine the freezing point of a solution of 75 mg of cinnamaldehyde dissolved in 1.00 g of benzene (K_f = 4.3°C/m; normal freezing point = 5.5°C).

11.46. Spearmint Determine the boiling point elevation of a solution of 125 mg of carvone ($C_{10}H_{14}O$, oil of spearmint) dissolved in 1.50 g of carbon disulfide (K_b = 2.34°C/m).

11.47. What molality of a nonvolatile, nonelectrolyte solute is needed to lower the melting point of camphor by 1.000°C (K_f = 39.7°C/m)?

11.48. What molality of a nonvolatile, nonelectrolyte solute is needed to raise the boiling point of water by 7.60°C (K_b = 0.52°C/m)?

11.49. Saccharin Determine the melting point of an aqueous solution made by adding 186 mg of saccharin ($C_7H_5O_3NS$) to 1.00 mL of water (density = 1.00 g/mL, K_f = 1.86°C/m).

11.50. Determine the boiling point of an aqueous solution that is 2.50 m ethylene glycol ($HOCH_2CH_2OH$); K_b for water is 0.52°C/m. Assume that the boiling point of pure water is 100.00°C.

11.51. Which aqueous solution has the lowest freezing point: 0.5 m glucose, 0.5 m NaCl, or 0.5 m $CaCl_2$?

11.52. Which aqueous solution has the highest boiling point: 0.5 m glucose, 0.5 m NaCl, or 0.5 m $CaCl_2$?

11.53. Which one of the following aqueous solutions should have the highest boiling point: 0.0200 m CH_3OH, 0.0125 m KCl, or 0.0100 m $Ca(NO_3)_2$?

11.54. Which one of the following aqueous solutions should have the lowest freezing point: 0.0500 m $C_6H_{12}O_6$, 0.0300 m KBr, or 0.0150 m Na_2SO_4?

11.55. Arrange the following aqueous solutions in order of increasing boiling point:
  a. 0.06 m $FeCl_3$ (i = 3.4)
  b. 0.10 m $MgCl_2$ (i = 2.7)
  c. 0.20 m KCl (i = 1.9)

11.56. Arrange the following solutions in order of increasing freezing point depression:
  a. 0.10 m $MgCl_2$ in water, i = 2.7, K_f = 1.86°C/m
  b. 0.20 m toluene in diethyl ether, i = 1.00, K_f = 1.79°C/m
  c. 0.20 m ethylene glycol in ethanol, i = 1.00, K_f = 1.99°C/m

Osmosis and Osmotic Pressure

CONCEPT REVIEW

11.57. What is a semipermeable membrane?

11.58. A pure solvent is separated from a solution containing the same solvent by a semipermeable membrane. In which direction does the solvent flow across the membrane, and why?

11.59. A dilute solution is separated from a more concentrated solution containing the same solvent by a semipermeable membrane. In which direction does the solvent tend to flow across the membrane, and why?

11.60. How is the osmotic pressure of a solution related to its molar concentration and its temperature?

11.61. What is reverse osmosis? List the basic components of equipment used to purify seawater by reverse osmosis.

11.62. Explain how the minimum pressure for purification of seawater by reverse osmosis can be estimated from its composition.

PROBLEMS

11.63. The following pairs of aqueous solutions are separated by a semipermeable membrane. In which direction will the solvent flow?
  a. A = 1.25 M NaCl; B = 1.50 M KCl
  b. A = 3.45 M $CaCl_2$; B = 3.45 M NaBr
  c. A = 4.68 M glucose; B = 3.00 M NaCl

11.64. The following pairs of aqueous solutions are separated by a semipermeable membrane. In which direction will the solvent flow?
  a. A = 0.48 M NaCl; B = 55.85 g of NaCl dissolved in 1.00 L of solution
  b. A = 100 mL of 0.982 M $CaCl_2$; B = 16 g of NaCl in 100 mL of solution
  c. A = 100 mL of 6.56 mM $MgSO_4$; B = 5.24 g of $MgCl_2$ in 250 mL of solution

11.65. Calculate the osmotic pressure of each of the following aqueous solutions at 20°C:
  a. 2.39 M methanol (CH_3OH)
  b. 9.45 mM $MgCl_2$
  c. 40.0 mL of glycerol ($C_3H_8O_3$) in 250.0 mL of aqueous solution (density of glycerol = 1.265 g/mL)
  d. 25 g of $CaCl_2$ in 350 mL of solution

11.66. Calculate the osmotic pressure of each of the following aqueous solutions at 27°C:
  a. 10.0 g of NaCl in 1.50 L of solution
  b. 10.0 mg/L of $LiNO_3$
  c. 0.222 M glucose
  d. 0.00764 M K_2SO_4

11.67. Determine the molarity of each of the following solutions from its osmotic pressure at 25°C. Include the van 't Hoff factor for the solution when the factor is given.
  a. Π = 0.674 atm for a solution of ethanol (CH_3CH_2OH)
  b. Π = 0.0271 atm for a solution of aspirin ($C_9H_8O_4$)
  c. Π = 0.605 atm for a solution of $CaCl_2$, i = 2.47

11.68. Determine the molarity of each of the following solutions from its osmotic pressure at 25°C. Include the van 't Hoff factor for the solution when the factor is given.
  a. Π = 0.0259 atm for a solution of urea [$H_2NC(O)NH_2$]
  b. Π = 1.56 atm for a solution of sucrose ($C_{12}H_{22}O_{11}$)
  c. Π = 0.697 atm for a solution of KI, i = 1.90

11.69. Is the following statement true or false? For solutions of the same reverse osmotic pressure at the same temperature, the molarity of a solution of NaCl will always be less than the molarity of a solution of $CaCl_2$. Explain your answer.

11.70. Suppose you have 1.00 M aqueous solutions of each of the following solutes: glucose ($C_6H_{12}O_6$), NaCl, and acetic acid (CH_3COOH). Which solution has the highest pressure requirement for reverse osmosis?

Using Colligative Properties to Determine Molar Mass

CONCEPT REVIEW

11.71. What effect does dissolving a solute have on the following properties of a solvent? (a) osmotic pressure; (b) freezing point; (c) boiling point

11.72. How can measurements of osmotic pressure, freezing point depression, and boiling point elevation be used to find the molar mass of a solute? Why is it important to know whether the solute is an electrolyte or a nonelectrolyte?

PROBLEMS

11.73. Throat Lozenges A 188 mg sample of a nonelectrolyte isolated from throat lozenges is dissolved in enough water to make 10.0 mL of solution at 25°C. The osmotic pressure of the resulting solution is 4.89 atm. Calculate the molar mass of the compound.

*11.74. An unknown compound (152 mg) is dissolved in water to make 75.0 mL of solution. The solution does not conduct electricity and has an osmotic pressure of 0.328 atm at 27°C. Elemental analysis reveals the substance to be 78.90% C, 10.59% H, and 10.51% O. Determine the molecular formula of this compound.

---

*11.75. **Cloves** Eugenol is one of the compounds responsible for the flavor of cloves. A 111 mg sample of eugenol is dissolved in 1.00 g of chloroform (K_b = 3.63°C/m), increasing the boiling point of chloroform by 2.45°C. Calculate eugenol's molar mass. Eugenol is 73.17% C, 7.32% H, and 19.51% O by mass. What is the molecular formula of eugenol?

*11.76. **Caffeine** The freezing point of a solution prepared by dissolving 150 mg of caffeine in 10.0 g of camphor is lower by 3.07°C than that of pure camphor (K_f = 39.7°C/m). What is the molar mass of caffeine? Elemental analysis of caffeine yields the following results: 49.49% C, 5.15% H, 28.87% N, and the remainder is O. What is the molecular formula of caffeine?

Additional Problems

*11.77. **Melting Ice** $CaCl_2$ is used to melt ice on sidewalks. Could $CaCl_2$ melt ice at −20°C? Assume that the solubility of $CaCl_2$ at this temperature is 70.1 g $CaCl_2$/100.0 g of H_2O and that the van 't Hoff factor for a saturated solution of $CaCl_2$ is 2.5.

*11.78. **Making Ice Cream** A mixture of table salt and ice is used to chill the contents of hand-operated ice-cream makers. What is the melting point of a mixture of 2.00 lb of NaCl and 12.00 lb of ice if exactly half of the ice melts? Assume that all the NaCl dissolves in the melted ice and that the van 't Hoff factor for the resulting solution is 1.44.

11.79. The freezing points of 0.0935 m ammonium chloride and 0.0378 m ammonium sulfate in water were found to be −0.322°C and −0.173°C, respectively. What are the values of the van 't Hoff factors for these salts?

11.80. The following data were collected for three compounds in aqueous solution. Determine the value of the van 't Hoff factor for each salt (K_f for water = 1.86°C/m).

Compound	Concentration (g/kg)	Experimentally Measured ΔT_f (°C)
LiCl	5.0	0.410
HCl	5.0	0.486
NaCl	5.0	0.299

11.81. Physiological Saline After 100.0 mL of a solution of physiological saline (0.90% NaCl by mass) is diluted by the addition of 250.0 mL of water, what is the osmotic pressure of the final solution at 37°C? Assume that NaCl dissociates completely into $Na^+(aq)$ and $Cl^-(aq)$.

11.82. Suppose 100.0 mL of 2.50 mM NaCl is mixed with 80.0 mL of 3.60 mM $MgCl_2$ at 20°C. Calculate the osmotic pressure of each starting solution and that of the mixture, assuming that the volumes are additive and that both salts dissociate completely into their component ions.

11.83. A solution of 7.50 mg of a small protein in 5.00 mL aqueous solution has an osmotic pressure of 6.50 torr at 23.1°C. What is the molar mass of the protein?

11.84. Kidney Dialysis Hemodialysis, a method of removing waste products from the blood if the kidneys have failed, uses a tube made of a cellulose membrane that is immersed in a large volume of aqueous solution. Blood is pumped through the tube and is then returned to the patient's vein. The membrane does not allow passage of large protein molecules and cells but does allow small ions, urea, and water to pass through it. Assume that a physician wants to decrease the concentration of sodium ion and urea in a patient's blood while maintaining the concentration of potassium ion and chloride ion in the blood. What materials must be dissolved in the aqueous solution in which the dialysis tube is immersed? How must the concentrations of ions in the immersion fluid compare to those in blood?

12

Thermodynamics
Why Chemical Reactions Happen

The Game of Energy Conversion

Some processes are so familiar that we tend not to consider why they happen. If a car tire is punctured, the air inside rushes out and the tire goes flat; on the other hand, air does not rush back into a punctured tire and reinflate it. Objects made of iron left on the ground or underwater for months or years turn into clumps of rust; they don't turn back into shiny metal all by themselves. A tray of ice cubes accidentally left on a kitchen counter turns into a tray of liquid water; as long as the kitchen stays warm there is no way the water in the tray will re-form as ice cubes.

The processes described above have something in common: they are all spontaneous. This designation means that they all happen without any ongoing intervention. In each case the reverse process is nonspontaneous—it cannot happen on its own. So, why are some processes spontaneous and others not? We find one explanation in the second law of thermodynamics and a thermodynamic parameter called entropy.

Recall that the first law of thermodynamics tells us that energy cannot be created or destroyed. This means that when we play the game of energy conversion, we can't win. We cannot do more work than allowed by the available energy. The second law of thermodynamics says, in effect, that not all of the energy released by a spontaneous reaction, such as burning gasoline in a car engine, is available to do useful work. In other words, in the game of energy conversion, not only can we not win, but we can't even break even.

Energy cannot be destroyed, so what happens to the energy that is unavailable to do useful work? The answer is that this energy spreads out, becoming less concentrated over time. This dispersion of energy is called entropy. In this chapter we explore the meaning and some of the impacts of entropy and the second law of thermodynamics on familiar processes. We explore that component of the energy, called *free energy*, that is available to do work and how it is related to reaction spontaneity. We will also see how spontaneous reactions can be coupled with nonspontaneous reactions to make multistep chemical and biochemical processes happen. This coupling sustains the molecular processes that keep us alive.

Corrosion at Sea Most metal objects in contact with seawater, including the remains of this shipwreck, corrode as a result of spontaneous chemical reactions. Here, iron metal has turned into iron(III) oxide. ▶

LEARNING OUTCOMES

LO1 Predict the signs of entropy changes for chemical reactions and physical processes
Sample Exercise 12.1

LO2 Predict the relative entropies of substances based on their molecular structures
Sample Exercise 12.2

LO3 Calculate entropy changes in chemical reactions using standard molar entropies
Sample Exercise 12.3

LO4 Calculate free-energy changes in chemical reactions from the changes in enthalpy and entropy for the reactions
Sample Exercise 12.4

LO5 Calculate standard free-energy changes in chemical reactions using the appropriate standard free energy of formation values
Sample Exercise 12.5

LO6 Predict the spontaneity of a chemical reaction as a function of temperature
Sample Exercise 12.6

LO7 Calculate the net change in free energy of coupled spontaneous and nonspontaneous reactions
Sample Exercise 12.7

spontaneous process a process that proceeds without outside intervention.

12.1 Spontaneous Processes

Before Thomas Edison invented the lightbulb, city streets were often illuminated at night by gas-fueled lamps. Once lit, they burned through the night until their fuel was cut off as dawn approached. Chemical reactions like the combustion of gas-lamp fuel are examples of **spontaneous** reactions: once started they proceed without outside assistance as long as there are sufficient quantities of reactants available. The reverse reaction—in this case, converting carbon dioxide and water into fuel and oxygen—is *nonspontaneous*: it cannot happen on its own.

The word *spontaneous* can be a bit misleading because many spontaneous reactions don't start all by themselves: they need a little energy boost, such as a spark or external flame that ignites the flame in a gas lamp or the gasoline/air mixture in an automobile engine. However, once initiated, a spontaneous reaction continues on its own. It is the self-sustaining nature of these reactions that earns them the label *spontaneous*.

In addition, spontaneous does not necessarily mean fast. Though combustion reactions certainly proceed rapidly, other spontaneous reactions, such as the formation of a layer of rust on an object made of iron, can take a very long time depending on temperature and the rate at which O_2 reaches the iron surface:

$$4 \, Fe(s) + 3 \, O_2(g) \rightarrow 2 \, Fe_2O_3(s) \qquad \Delta H^\circ_{rxn} = -1648 \text{ kJ}$$

Still, iron turns into rust without intervention, which means that this reaction is spontaneous.

Spontaneous is also *not* a synonym for exothermic, though many scientists once thought so. In the mid-19th century, many of them believed that exothermic reactions in which high-enthalpy reactants formed low-enthalpy (more stable) products should always be spontaneous, and it is true that many exothermic reactions, including combustion reactions and metal corrosion, *are* spontaneous. Later in this chapter we explore why so many are. However, some exothermic reactions are not spontaneous, and some endothermic chemical reactions are spontaneous, depending on reaction conditions. Consider, for example, the reaction that occurs when baking soda (sodium bicarbonate) is added to room-temperature vinegar (dilute acetic acid) as shown in Figure 12.1. The foaming reaction mixture tells us that a chemical change is taking place and a decrease in the temperature of the reaction mixture tells us that the reaction is endothermic ($\Delta H_{rxn} > 0$). A similar drop in temperature occurs when solutions containing bicarbonate and calcium ions are mixed together:

$$2 \, HCO_3^-(aq) + Ca^{2+}(aq) \rightarrow CaCO_3(s) + H_2O(\ell) + CO_2(g) \qquad \Delta H^\circ_{rxn} = 40.6 \text{ kJ}$$

CONNECTION We learned in Chapter 9 that enthalpy change (ΔH) is a measure of how much heat flows into or out from a system during a process occurring at constant pressure.

$$NaHCO_3(s) + CH_3COOH(aq) \longrightarrow Na^+(aq) + CH_3COO^-(aq) + H_2O + CO_2(g)$$

FIGURE 12.1 Adding baking soda to vinegar produces sodium ions, acetate ions, water, and carbon dioxide in an endothermic ($\Delta H^\circ_{rxn} = 48.5$ kJ/mol) yet spontaneous reaction.

Why are these two endothermic reactions spontaneous? The answer to this question lies in something they have in common: the particles that make up their products are more spread out and/or have more freedom of motion than the particles in their reactants. For example, the reactants in both reactions are liquids and solids, but the products of both reactions include a gas (in both cases here: CO_2). The particles that make up these three states of matter have quite different freedoms of motion, as illustrated for the molecules of water in ice, liquid water, and steam in Figure 12.2. The particles that make up ice or any crystalline solid occupy fixed positions, and their motion is limited to vibrating in place without going anywhere (unless the whole crystal moves). The particles in liquids are more mobile: they are able to flow past and tumble over each other, which means they have some *translational* and *rotational* freedom of motion. Particles in the gas phase are much more spread out—which is why gases are compressible whereas liquids and solids are not—and they have much more translational and rotational freedom of motion. Therefore, when either solid or liquid reactants form one or more products that are gases, there is a substantial increase in particulate freedom of motion *and* the gas-phase particles are much more spread out.

Now let's consider the process that makes instant cold packs cold (Figure 12.3). There are two compartments in an instant cold pack. One is filled with water, and the other contains a water-soluble compound such as ammonium nitrate that has a positive enthalpy of solution ($\Delta H_{solution} > 0$). When the membrane separating the two compartments is ruptured, the solid compound mixes with and dissolves in water. The resulting solution gets very cold, becoming an effective anti-inflammatory treatment for bruises and

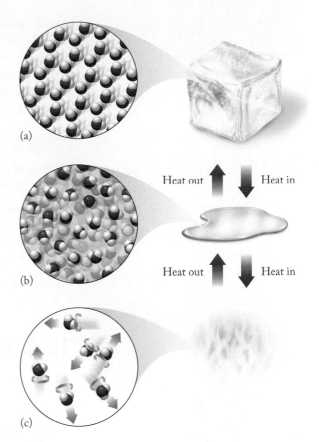

FIGURE 12.2 Molecular motion in the three phases of water. (a) Molecules in ice can only vibrate in place. (b) Molecules in water can flow past and tumble over each other. (c) Molecules of water vapor have much more translational and rotational motion.

FIGURE 12.3 An instant cold pack gets cold when a membrane inside it is ruptured, allowing a water-soluble compound such as ammonium nitrate to dissolve. The NH_4^+ and NO_3^- ions in solid NH_4NO_3 experience increased freedom of motion as they form hydrated NH_4^+ and NO_3^- ions in solution. The temperature of the solution drops because the dissolution process is endothermic.

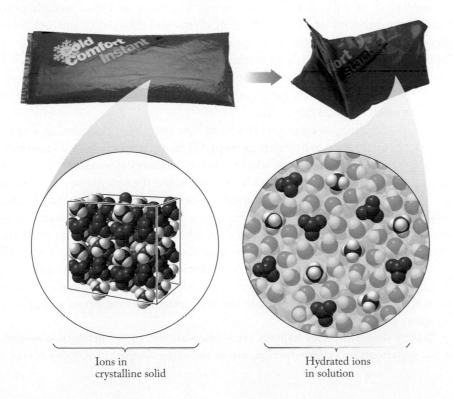

Ions in crystalline solid

Hydrated ions in solution

▶❚❚ **CHEMTOUR** Dissolution of Ammonium Nitrate

muscle sprains. We can link the spontaneity of this process to the production of particles (dissolved NH_4^+ and NO_3^- ions) that are more spread out and have more freedom of motion than in the solid phase. This pattern applies to the dissolution of all solids in liquid solvents. It also applies to melting solids because their particles are usually more spread out and always have more freedom of motion in the liquid state, as shown for ice and water in Figure 12.2.

> **CONCEPT TEST** ●●
>
> In which of the following processes do particles become more spread out and gain more freedom of motion?
>
> a. Perspiration evaporates.
> b. Dew forms overnight on blades of grass.
> c. A spoonful of sugar dissolves in a cup of coffee.
> d. A log of wood burns in a fireplace.
>
> *(Answers to Concept Tests are in the back of the book.)*

12.2 Entropy

Whenever an isolated thermodynamic system undergoes a change in which the particles in the system disperse into a larger volume and/or gain freedom of motion, the process is *always* spontaneous. The system is said to experience an increase in entropy. The **second law of thermodynamics** states that the entropy of an isolated thermodynamic system *always increases* during a spontaneous process. What exactly is entropy? In this section we explore two complementary definitions of entropy, but before we do let's be clear about what entropy *is not*.

In many science courses entropy is defined as disorder or randomness. Actually, these descriptors were used by the 19th-century scientists who introduced the concept of entropy, but they used words like these as visual metaphors to describe the behavior of *particles*, such as atoms and molecules, and not macroscopic objects. Thus, disorder as applied to entropy does not mean messy or untidy, as in articles of clothing scattered around a student's bedroom. Such a mess does not come into existence spontaneously from a neat and tidy room; rather, the occupant expends energy creating the mess, and must expend more to make the room neat and tidy again.

▶❚❚ **CHEMTOUR** Entropy

The real meaning of entropy is linked to how energy is distributed in a thermodynamic system. In particular, **entropy (S)** is a *measure of how dispersed the energy of a system is*. To illustrate what we mean by energy dispersion, let's revisit the experiment in Chapter 9 in which we determined the specific heat of a metal (see Figure 9.15). Recall how the metal pellets are first heated to 100°C in a bath of boiling water and then transferred to a small volume of cool water in an insulated container. There is an increase in the temperature of the water in the container that can be used to calculate how much energy flows from the metal into the water as the metal and water (constituting a closed thermodynamic system) come to thermal equilibrium. Energy initially confined to the metal flows *spontaneously* from the hot metal into the cool water and is *dispersed* throughout the metal/water system.

Energy dispersion can happen even when all the components of a system, or the system and its surroundings, are at the same temperature. Let's revisit a

second law of thermodynamics the principle that the total entropy of the universe increases in any spontaneous process.

entropy (S) a measure of how dispersed the energy in a system is at a specific temperature.

spontaneous, isothermal process from Chapter 10: the diffusion of ideal gases. The apparatus in Figure 12.4 contains two equal-volume chambers: one contains neon, and the other holds the same number of atoms of argon. We assume that the gases behave as ideal gases, which means their atoms are essentially tiny points of mass that do not interact with each other (except for frequent elastic collisions). We also assume the temperatures of both chambers are the same, which means that their pressures are also the same.

When the partition between the two chambers is removed, atoms of each gas spontaneously diffuse into the space originally occupied by the other gas. This behavior is explained by kinetic molecular theory and the random motion of the particles that make up ideal gases. Graham's law tells us that the less massive neon atoms will diffuse faster into the space occupied by the more massive argon atoms, but, given enough time, the atoms of both gases will end up evenly distributed throughout the combined volume.

Though kinetic molecular theory is a powerful tool for predicting that gases mix spontaneously, there is another way to account for the spontaneity of mixing—one based on a statistical view of energy dispersion and entropy. To illustrate how it works, let's start with a vastly simplified model of neon and argon mixing (Figure 12.5). We start with 2 red marbles representing atoms of one of the gases and 2 blue marbles representing atoms of the other. Initially they are separated into two compartments as shown in Figure 12.5(a). Then the partition separating them is removed and the marbles are able to occupy any of the marble-holding sites in either compartment. There are six ways to arrange the 2 red and 2 blue marbles among the 4 sites as shown in Figure 12.5(b). In only two of these (numbered 1 and 6) are the same color marbles in the same compartment. In the other four they are mixed. If there is an equal chance any marble could be in any one of the positions, then it is twice as likely that they will be mixed together (arrangements 2–5) rather than separate (arrangements 1 and 6).

Now consider what happens when we double the number of marbles and marble-holding sites as shown in Figure 12.6(a). If we allow all 8 of the marbles to be randomly arranged among the 8 sites available to them, there are again only two possible arrangements in which each compartment contains marbles of only one color (Figure 12.6b). There are several mixing options: either 1 or 2 or 3 of the red marbles could end up on the side that originally contained

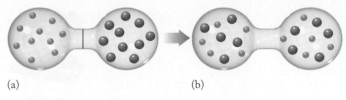

FIGURE 12.4 Entropy of mixing. (a) When the partition separating two containers with different noble gases is removed, (b) the two gases spontaneously diffuse into each other's compartment, resulting in a uniform mixture of randomly distributed atoms of the two gases.

∞ CONNECTION In Chapter 10 we learned how kinetic molecular theory explains many of the physical properties of gases including the rates at which they effuse and diffuse, as described by Graham's laws.

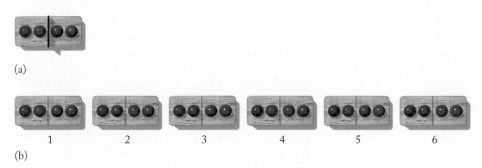

FIGURE 12.5 A simple model of the entropy of mixing. (a) Initially pairs of red and blue marbles are in separate compartments. (b) When allowed to mix, the marbles can be in six different patterns. In two of the arrangements (1 and 6), the marbles are still in separate compartments, but in four of them (2–5), they are mixed together: one of each color marble in both compartments. Thus, they are twice as likely to be mixed as to remain separate.

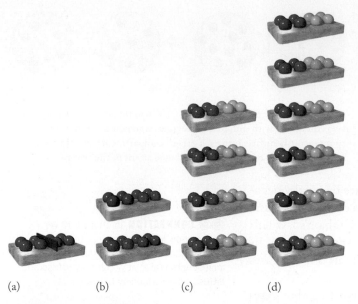

(a)   (b)   (c)   (d)

FIGURE 12.6 A larger model of the entropy of mixing. (a) Initially 4 red and 4 blue marbles are in separate compartments. (b) When allowed to mix, there are two arrangements of the marbles in which the colors are still separate. (c) There are four possible arrangements in each compartment when it exchanges 1 marble with the other compartment. Therefore, there are 4 × 4 = 16 different arrangements for both compartments, and another 16 (not shown) when the compartments exchange 3 marbles. (d) The greatest number of arrangements: six for each compartment, or 6 × 6 = 36 for both, occurs when each compartment contains 2 red and 2 blue marbles. Therefore, the most probable arrangement is even mixing, the next most probable is uneven mixing, and by far the least probable is no mixing at all.

only blue marbles, and vice versa. The exchange of 1 marble produces four options for how to arrange the blue one and the remaining 3 red ones in the originally all-red compartment, as shown in Figure 12.6(c). For each of these four arrangements there are four ways to arrange the 1 red marble and 3 blue ones in the other compartment. This means that there are 4 × 4 = 16 different ways to arrange the marbles if 1 red and 1 blue marble are exchanged between compartments. Can you see how there are also 16 different ways to arrange the marbles if 3 red and 3 blue ones are exchanged between compartments? The third option is equal mixing so that each compartment has 2 red and 2 blue marbles. There are six ways to arrange these marbles in each compartment (Figure 12.6d) for a total of 6 × 6 = 36 arrangements. If all arrangements are equally probable, then out of a total of 2 + 16 + 16 + 36 = 70 arrangements, only 2 (or 2/70 = 0.029 = 2.9%) of them produce no mixing of the marbles. The remaining 68 arrangements (97.1%) produce at least some mixing, and the most probable mixing pattern (36 out of 70) produces a uniform distribution of marbles between the two compartments.

If we extend the message contained in the visual images in Figures 12.5 and 12.6 and our statistical analyses of them to extremely large numbers (think Avogadro's number) of particles, we come to the following conclusions:

1. Substances (such as gases) that are miscible in each other mix with each other spontaneously because there is a higher probability of a mixed distribution.
2. The most probable mixing pattern produces a uniform distribution of particles throughout the volume they occupy.

Now let's connect this probabilistic view of particle behavior to the entropy of thermodynamic systems. This connection was made in the 1870s by German physicist Ludwig Boltzmann (1844–1906) following his pioneering work on the kinetic molecular theory of gases. Boltzmann proposed that the entropy of a system was related to the number of different ways the particles in it could be arranged that was consistent with the system's total internal energy. This number of energy-equivalent arrangements of particles in which each of them is in a particular position and has a particular momentum (the product of its mass and speed) defined a variable Boltzmann called *Wahrscheinlichkeit* (W)—the German word for probability. He then linked the entropy (S) of the system to the number of probable arrangements (W) of its particles and particle energies

$$S = k_B \ln W \qquad (12.1)$$

where k_B is the Boltzmann constant, which is equal to the gas constant divided by Avogadro's number:

accessible microstate a unique arrangement of the positions and momenta of the particles in a thermodynamic system.

$$k_B = \frac{R}{N_A} = \frac{8.314 \ \dfrac{J}{\text{mol} \cdot K}}{6.0221 \times 10^{23}/\text{mol}} = 1.381 \times 10^{-23} \ J/K$$

The probable number of arrangements of the particles in a system at a given temperature has come to be called the number of **accessible microstates**. To get a feel for what this phrase means, let's revisit Figure 12.5 and the arrangements numbered 2–5. For now, think of the marbles in these arrangements as stationary atoms in an isolated thermodynamic system. Each of the four arrangements represents an accessible microstate—an equivalent arrangement of the *macro*state in which two atoms of two elements are mixed so that each compartment has one atom of each element. Images 1 and 6 are equivalent microstates of the macrostate in which the four atoms are not mixed. The fact that the mixed macrostate has more microstates means that it has greater entropy. This means that the mixing process is accompanied by an increase in the number of accessible microstates and an increase in entropy, and, according to the second law, is spontaneous.

Boltzmann's probabilistic W also incorporates particle motion as well as position. In addressing particle motion we need to think of all the ways that particles can move. Put another way, we need to consider the kinds of *motional* energy they have. As we discussed in Chapter 9 (see Figure 9.6) molecules (and atoms) can have translational energy, rotational energy, and (in molecules) bond vibrational energy. Recall from Chapter 3 that energy on the atomic scale is not continuous; rather, it is quantized. Thus, the translational and rotational energies of atoms and molecules and the vibrational energies of the bonds in molecules are all quantized. The differences between translational and rotational energy states of gas-phase atoms and molecules are so small that they have access to enormous numbers of energy states at room temperature. To illustrate how many, let's use Equation 12.1 to calculate the number of microstates that are accessible to a single neon atom at 25°C. To calculate the number of accessible microstates, we rearrange the terms in Equation 12.1 to solve for W and then insert the entropy value of 1 mole of Ne under standard conditions at 25°C (146.3 J/mol · K, see Section 12.3) and the quantity of Ne (1 atom):

$$W = e^{S/k_B} = e^{\dfrac{\left(146.3 \,\frac{J}{mol \cdot K}\right)\left(\dfrac{1 \, atom}{6.0221 \times 10^{23} \,\frac{atoms}{mol}}\right)}{1.381 \times 10^{-23} \,\frac{J}{K}}}$$

$$= e^{17.59} = 4.4 \times 10^7$$

This value illustrates the enormous number of ways that the energy of a single gas-phase atom can be dispersed. Of course, the motion of an atom in a solid or a liquid is much more restricted, which means it has access to many fewer microstates. Consider, for example, an atom of carbon in the rigid crystalline structure of a diamond. It is restricted to vibrating within a very confined space. There is little uncertainty in the position of the atom, and it has little motional energy. These limitations are reflected in the small entropy value of diamond, 2.4 J/(mol · K), and in the number of microstates accessible to one of its carbon atoms:

$$W = e^{S/k_B} = e^{\dfrac{\left(2.4 \,\frac{J}{mol \cdot K}\right)\left(\dfrac{1 \, atom}{6.0221 \times 10^{23} \,\frac{atoms}{mol}}\right)}{1.381 \times 10^{-23} \,\frac{J}{K}}}$$

$$= e^{0.29} = 1.3$$

This result means that there is on average little more than one microstate available to each atom in a diamond at 25°C and 1 bar, but that value may be misleading.

third law of thermodynamics the principle that the entropy of a perfect crystal is zero at absolute zero.

standard molar entropy (S°) the absolute entropy of 1 mole of a substance in its standard state.

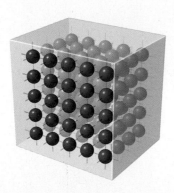

FIGURE 12.7 A perfect crystal at 0 K has zero entropy because all particles are locked in place and have zero freedom of motion and no kinetic energy.

When we attempt to calculate the number of accessible microstates in one *mole* of diamond at 25°C and 1 bar, the result is so enormous most calculators can't process it:

$$W = e^{S/k_B} = e^{\dfrac{\left(2.4 \frac{J}{mol \cdot K}\right) \times 1\, mol}{1.381 \times 10^{-23} \frac{J}{K}}}$$

$$= e^{1.7 \times 10^{23}} = ?$$

Clearly the number of microstates accessible to a mole of particles is enormous, even when they are locked in place in a crystalline solid.

The statistical definition of entropy based on accessible microstates is conceptually compatible with the thermodynamic, macroscopic view based on energy dispersion. After all, each accessible microstate represents one of the multitude of ways that energy can be dispersed in a thermodynamic system. Actually, some facets of entropy are easier to understand using the microstate model. One of them is the concept of absolute entropy, and the notion embedded in the **third law of thermodynamics**, which states that *a perfect crystalline solid has zero entropy at absolute zero.* If the particles of a crystalline solid are perfectly aligned (Figure 12.7) and in their lowest possible energy states, the crystal has only one microstate, and, according to Equation 12.1, its entropy is zero:

$$S = k_B \ln W = k_B \ln 1 = k_B \times 0 = 0$$

CONCEPT TEST

Predict which one of the following values comes closest to the number of microstates accessible to one molecule of liquid water at 25°C: (a) 1, (b) 10^2, (c) 10^6, (d) 10^{12}, (e) 10^{18}. Use the appropriate $S°$ value from Appendix 4 and Equation 12.1 to determine whether your selection was the best one.

TABLE 12.1 Standard States of Pure Substances and Solutions

Physical State	Standard State	Pressure[a]	Temperature[b]
Solid	Pure solid, most stable allotrope of an element	1 bar	25°C
Liquid	Pure liquid	1 bar	25°C
Gas	Pure gas	1 bar	25°C
Solution	1 M	1 bar	25°C

[a]Since 1982 1 bar has been the standard pressure for tabulating all thermodynamic data. Prior to 1982 standard pressure was 1 atmosphere (atm) = 1.01325 bar.
[b]The thermodynamic data in Appendix 4 and used elsewhere in this book are based on a temperature of 25°C (298 K). Note that this temperature is not the same as the STP value we use for ideal gases, which is 273 K.

12.3 Absolute Entropy and Molecular Structure

The third law of thermodynamics provides a vital reference point for the absolute entropy scale: a perfect crystal at 0 K has zero entropy. From this starting point and by carefully measuring the rise in temperature of substances as known quantities of thermal energy are added to them, scientists have been able to establish absolute entropy values for many substances at many temperatures. The most common quantity for expressing these values is **standard molar entropy (S°)**, which is the entropy of one mole of a substance in its standard state (see Table 12.1) at a pressure of 1 bar (~1 atm). Some sample $S°$ values at 298 K are listed in Table 12.2.

The $S°$ values for liquid water and water vapor in Table 12.2 illustrate a difference between the entropies of liquids and gases that we have discussed before in this chapter: the molecules in a gas under standard conditions are much more widely dispersed than the molecules in a liquid, and the entropies of the different phases of a substance at a given temperature follow the order $S_{solid} < S_{liquid} < S_{gas}$.

The entropy changes that occur as one mole of ice at 0 K is heated are shown in Figure 12.8. Note the jump in entropy as the ice melts and the even bigger jump as the liquid water vaporizes. Also note that the lines between the phase changes are curved. The change in entropy, ΔS, with temperature is not linear because heating a substance at a higher temperature produces a smaller entropy increase than adding the same quantity of heat to the same substance at a lower temperature (see Equation 12.3 in the next section).

Let's summarize the factors that affect entropy change:

1. Entropy increases when temperature increases.
2. Entropy increases when volume increases.
3. Entropy increases when the number of independent particles increases.

TABLE 12.2   Selected Standard Molar Entropy Values^a

Formula	$S°$ [J/(mol · K)]	Formula	Name	$S°$ [J/(mol · K)]
$Br_2(g)$	245.5	$CH_4(g)$	Methane	186.2
$Br_2(\ell)$	152.2	$CH_3CH_3(g)$	Ethane	229.5
$C_{diamond}(s)$	2.4	$CH_3CH_2CH_3(g)$	Propane	269.9
$C_{graphite}(s)$	5.7	$CH_3(CH_2)_2CH_3(g)$	Butane	310.0
$CO(g)$	197.7	$CH_3(CH_2)_2CH_3(\ell)$		231.0
$CO_2(g)$	213.8	$CH_3OH(g)$	Methanol	239.9
$H_2(g)$	130.6	$CH_3OH(\ell)$		126.8
$N_2(g)$	191.5	$CH_3CH_2OH(g)$	Ethanol	282.6
$O_2(g)$	205.0	$CH_3CH_2OH(\ell)$		160.7
$H_2O(g)$	188.8	$C_6H_6(g)$	Benzene	269.2
$H_2O(\ell)$	69.9	$C_6H_6(\ell)$		172.9
$NH_3(g)$	192.5	$C_{12}H_{22}O_{11}(s)$	Sucrose	360.2

^{a}Values for additional substances are given in Appendix 4.

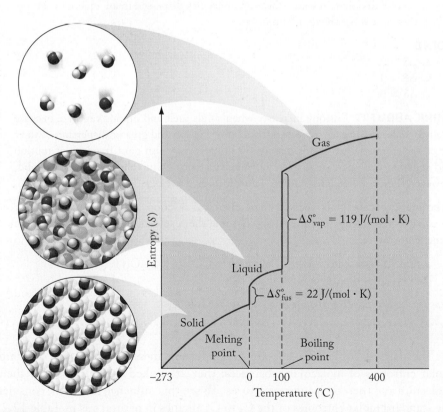

FIGURE 12.8 Abrupt increases in entropy accompany changes of state, as shown for water, with the greater increase occurring during the transition from liquid to gas.

In all three cases, entropy increases because each change increases the dispersion of the energy of a system's particles. In other words, the particles have access to more microstates. We can often make qualitative predictions about entropy changes that accompany chemical reactions based on these three factors, even if we have no thermodynamic data about the reactants and products.

SAMPLE EXERCISE 12.1 **Predicting the Sign of an Entropy Change** **LO1**

Predict whether an increase or decrease in entropy accompanies each of these processes when they occur at constant temperature:

a. $H_2O(\ell) \rightarrow H_2O(g)$

b. $NH_3(g) + HCl(g) \rightarrow NH_4Cl(s)$

c. $C_{12}H_{22}O_{11}(s) \xrightarrow{H_2O} C_{12}H_{22}O_{11}(aq)$

COLLECT AND ORGANIZE We are given equations describing three processes and are asked to predict the signs of the accompanying entropy changes.

ANALYZE

a. Molecules of liquid water become molecules of (gaseous) water vapor, which means the molecules and their kinetic energies are more dispersed.

b. Two gaseous compounds form a single solid compound, which means the particles of product occupy much less space and have much less freedom of motion.

c. A solid dissolves in water, forming molecules dispersed in an aqueous solution that have more freedom of motion.

SOLVE

a. $\Delta S > 0$.

b. $\Delta S < 0$.

c. $\Delta S > 0$.

THINK ABOUT IT Entropy increases when solids melt and liquids vaporize because their particles experience increased freedom of motion and greater dispersion of their kinetic energies. When solids dissolve in liquids, they also gain freedom of motion and experience an increase in entropy. When gases combine to form a liquid or solid, however, they lose freedom of motion and entropy decreases.

Practice Exercise Does each of these chemical reactions result in an increase or decrease in the entropy of the system? Assume the reactants and products are at the same temperature.

a. $CaCO_3(s) + 2\,HCl(aq) \rightarrow CaCl_2(aq) + CO_2(g) + H_2O(\ell)$

b. $NH_3(g) + BF_3(g) \rightarrow NH_3BF_3(s)$ ⚙

(Answers to Practice Exercises are in the back of the book.)

The data in Table 12.2 contain an important message about the standard molar entropies of molecular substances: they are linked to the masses of their molecules and their molecular structures. To see this influence in action, consider the standard molar entropies of the C_1 to C_4 alkanes in natural gas in Table 12.2. Note how $S°$ values increase with increasing molecular size. This trend is due to the different ways that these molecules move in addition to simple translational motion in three-dimensional space. For example, we have noted that molecules in the gas phase are free to rotate as they move through the space they occupy, and the enormous number of quantized rotational energy states that they can occupy adds to the number of accessible microstates. Even liquid molecules tumble over each other as they flow past their nearest neighbors, adding to the ways that energy can be dispersed among them.

The quantities of energy involved in rotational motion depend on how massive the molecules are *and* on how their masses are distributed. Consider, for example, the $S°$ values of octane and isooctane in Figure 12.9. The isomers have the same molar masses but different molecular shapes. The longer, narrower shape of octane means that more of its mass is farther from the center of the molecule and farther from the axis of rotation when the molecule tumbles end over end. When the mass of a molecule is more spread out, the spacing between its rotational energy states is smaller and more are accessible. Therefore, molecules of octane have greater entropy than the same number of molecules of isooctane at the same temperature.

Another structural feature that influences entropy is rigidity. The two most common forms of carbon, namely, diamond and graphite, are both polymeric network solids. However, diamond is an extremely hard substance because it has a rigid three-dimensional atomic structure. Its atoms have access to fewer microstates and it has less entropy [$S° = 2.4$ J/(mol · K)] than soft, easily deformed graphite [$S° = 5.7$ J/(mol · K)], which consists of two-dimensional sheets that may slide past each other when a modest shear force is applied.

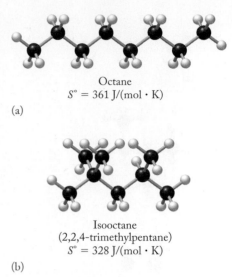

(a)

Octane
$S° = 361$ J/(mol · K)

(b)

Isooctane
(2,2,4-trimethylpentane)
$S° = 328$ J/(mol · K)

FIGURE 12.9 Ball-and-stick models and $S°$ values of (a) octane and (b) isooctane (2,2,4-trimethylpentane).

SAMPLE EXERCISE 12.2 **Comparing Standard Molar Entropy Values** **LO2**

Without consulting any reference sources, select the component in each of the following pairs that has the greater standard molar entropy at 298 K. Assume there is one mole of each component in its standard state (the pressure of each gas is 1 bar and the concentration of each solution is 1 M).

a. HCl(g), HCl(aq)
b. $CH_3OH(\ell)$, $CH_3CH_2OH(\ell)$

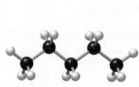 

c. $CH_3(CH_2)_3CH_3(\ell)$ $CH_3C(CH_3)_2CH_3(\ell)$

COLLECT AND ORGANIZE We are given chemical formulas or molecular models of pairs of substances, and we are asked to select which component of each pair has the greater entropy per mole under standard conditions—that is, the greater standard molar entropy ($S°$) at 298 K.

ANALYZE Particles in the vapor state of a substance are more dispersed and have more entropy than they do in the liquid state at the same temperature, which in turn have more entropy than particles in the solid state. Substances with larger molar masses tend to have greater $S°$ values than those in the same physical state that are

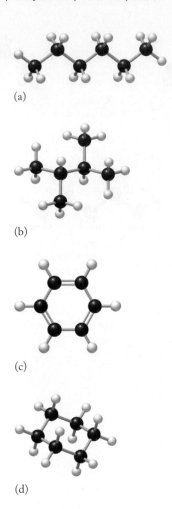

(a)

(b)

(c)

(d)

FIGURE 12.10

CONNECTION In Chapter 9 we defined an isolated thermodynamic system as one that exchanges neither energy nor matter with its surroundings. A closed system exchanges energy but not matter, and an open system exchanges both.

composed of less massive particles. Substances composed of compact molecules have lower standard molar entropies than those composed of elongated molecules of the same mass. The volume of one mole of an ideal gas at $P = 1$ bar and $T = 298$ K is about 24.8 liters; the volume of one mole of solute in a 1 M solution is only 1 liter.

SOLVE

a. One mole of HCl dissolved in one liter of water occupies much less volume and its particles have much less freedom of motion than one mole of HCl gas. Therefore, HCl gas has a greater $S°$ value than HCl in solution.

b. Both compounds are liquid alcohols, but molecules of ethanol (CH_3CH_2OH) are more massive than those of methanol (CH_3OH). Therefore, ethanol has a greater $S°$ value than methanol.

c. The two compounds are isomers with the same molar mass. However, molecules of $CH_3(CH_2)_3CH_3$ are more elongated and have more opportunities than $CH_3C(CH_3)_2CH_3$ to disperse their kinetic energies. Therefore, $CH_3(CH_2)_3CH_3$ has a greater $S°$ value.

THINK ABOUT IT The comparison in part a is complicated by the fact that HCl is a strong acid, which means that each molecule produces two ions in solution. However, they are liquid-phase ions and have less total entropy than half as many gas-phase HCl molecules. [Compare the $S°$ values of HCl(g), H$^+$(aq), and Cl$^-$(aq) in Appendix 4 to see for yourself.]

Practice Exercise Figure 12.10 shows ball-and-stick models of four hydrocarbons that each contain six carbon atoms per molecule. Rank these compounds in order of decreasing $S°$ values.

12.4 Applications of the Second Law

We saw in Section 12.2 how the second law of thermodynamics accounts for the spontaneity of processes occurring in isolated systems. It can also be applied to processes occurring in systems that are not isolated (which is good, because most are not—they are either open or closed). Recall from Chapter 9 that in thermodynamics we divide up the universe into two parts: the part we are interested in (the system) and everything else (the surroundings). Mathematically,

$$\text{Universe} = \text{system} + \text{surroundings}$$

It is logical, then, that the overall change in the entropy of the universe as a result of a physical or chemical change is the sum of the entropy changes experienced by the system and its surroundings:

$$\Delta S_{univ} = \Delta S_{sys} + \Delta S_{surr} \tag{12.2}$$

When a spontaneous process occurs in an isolated system, ΔS_{sys} is greater than zero. Because the system is isolated, the process has no impact on its surroundings, so $\Delta S_{surr} = 0$. Therefore, according to Equation 12.2, ΔS_{univ} must be greater than zero because ΔS_{sys} is greater than zero. The positive value of ΔS_{univ} is the basis for another way of expressing the second law of thermodynamics that applies to all systems: *a spontaneous process produces an increase in the entropy of the universe.*

The latter version of the second law is built on the assumption that a physical or chemical change in a closed or open thermodynamic system can alter the entropies of both the system and its surroundings. The second law says that a process is spontaneous when ΔS_{univ} is greater than zero. The second law also provides a thermodynamic requirement for *non*spontaneity: a process that produces a *decrease* in the entropy of the universe will not occur on its own. Therefore, the reverse of any spontaneous process has to be nonspontaneous because it can be spontaneous in only one direction—the one for which $\Delta S_{univ} > 0$. Reversing a spontaneous process reverses the sign of ΔS_{univ}, making it less than zero, which means the process is nonspontaneous.

➤ If $\Delta S_{univ} > 0$, then a process is spontaneous.

➤ If $\Delta S_{univ} < 0$, then a process is nonspontaneous.

To see how a process affects the entropy of its surroundings, let's focus on a familiar exothermic reaction: the combustion of natural gas (methane):

$$CH_4(g) + 2\,O_2(g) \rightarrow CO_2(g) + 2\,H_2O(\ell) \qquad \Delta H° = -890 \text{ kJ}$$

As written, the reaction consumes 3 moles of gases, and it produces 2 moles of a liquid product and 1 mole of CO_2 gas. Note that there are fewer moles of gases on the product side of the reaction equation. Given the much greater freedom of motion of particles in the gas phase, we can accurately predict that there will be a decrease in entropy of the reaction mixture, which is our thermodynamic system:

$$\Delta S_{sys} < 0$$

However, we know that the reaction is spontaneous, which means

$$\Delta S_{univ} > 0$$

How do we reconcile these opposing inequalities? Equation 12.2 supplies an explanation. If ΔS_{univ} is greater than zero, then the sum of ΔS_{sys} and ΔS_{surr} must also be greater than zero. The fact that ΔS_{sys} is less than zero simply means that ΔS_{surr} is not only greater than zero, it must have a large enough positive value to more than offset the negative value of ΔS_{sys}. Expressing this relationship in terms of the absolute values of ΔS_{surr} and ΔS_{sys}:

$$|\Delta S_{surr}| > |\Delta S_{sys}|$$

Is the combustion of methane likely to produce a large, positive ΔS_{surr}? Absolutely it will—because the reaction is highly exothermic: combustion of only one mole (16 grams) of methane releases 890 kJ of thermal energy. As energy flows from the system into its surroundings, dispersion of this energy produces a positive ΔS_{surr} that more than compensates for the unfavorable (negative) value of ΔS_{sys}.

The combinations of ΔS_{sys} and ΔS_{surr} that produce a positive value for ΔS_{univ} are shown in Figure 12.11. Note how processes in which ΔS_{sys} and ΔS_{surr} are both greater than zero (Figure 12.11a) inevitably result in an increase in ΔS_{univ}, which means they are always spontaneous. On the other hand, processes in which ΔS_{sys} and ΔS_{surr} are both less than zero (Figure 12.11d) always produce a decrease in ΔS_{univ}, which means they are always nonspontaneous. The other four cases in Figure 12.11 represent pairs of ΔS_{sys} and ΔS_{surr} values with opposite signs and which may or may not combine to produce positive ΔS_{univ} values. It all depends on the absolute values of ΔS_{sys} and ΔS_{surr}. If the larger of the two has the positive value, then ΔS_{univ} increases, and the process is spontaneous.

All exothermic reactions have the capacity to raise the entropy of their surroundings. The more energy that flows into the surroundings, or into any collection of particles, the greater the dispersion of energy among the particles and

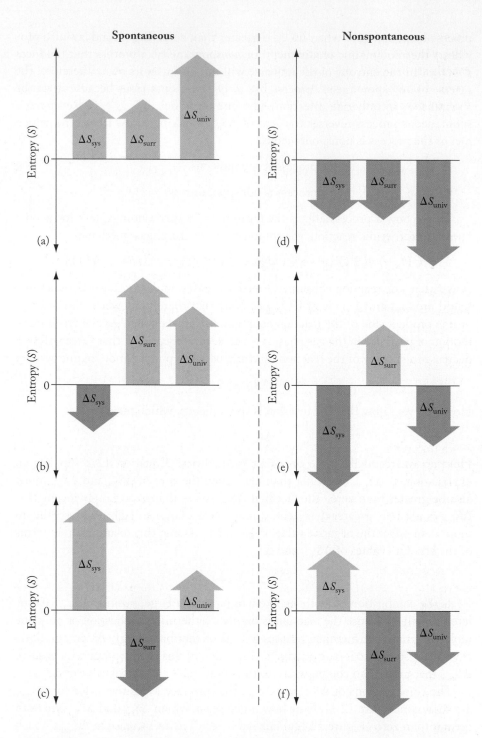

Spontaneous

Nonspontaneous

FIGURE 12.11 The relationships between ΔS_{sys}, ΔS_{surr}, and ΔS_{univ} in spontaneous and nonspontaneous processes. A process is spontaneous whenever the entropy of the universe increases: $\Delta S_{univ} > 0$. A process is nonspontaneous whenever the entropy of the universe decreases: $\Delta S_{univ} < 0$.

the greater the increase in their entropy (ΔS). However, heating particles that are already hot produces a smaller gain in entropy than heating the same particles at a lower temperature. This inverse relationship between entropy gain and temperature is reflected in Equation 12.3:

$$\Delta S = \frac{q_{rev}}{T} \tag{12.3}$$

where q_{rev} represents the *reversible* energy flow caused by the difference in temperature between the system and its surroundings.

Theoretically, a **reversible process** is one that happens so slowly that after an incremental change in the system has occurred, the process can be reversed by a tiny change that restores the original state of the system with no net flow of energy into or out from it. In other words, the system and its surroundings finish exactly as they were before the process began. At best, real chemical reactions and physical changes are only approximately reversible, but many are close enough that we can adapt Equation 12.3 to calculate ΔS_{sys}:

reversible process a process that happens so slowly that an incremental change can be reversed by another tiny change, restoring the original state of the system with no net flow of energy between the system and its surroundings.

$$\Delta S_{sys} = \frac{q_{sys}}{T} \qquad (12.4)$$

Let's apply Equation 12.4 to a familiar change of physical state: ice melting. It takes 6.01 kJ (or 6.01×10^3 J) of energy to melt 1.00 mol of ice at 0°C. Assuming the process occurs reversibly, then

$$\Delta S_{sys} = \frac{q_{sys}}{T} = \frac{(1.00 \ \text{mol})(6.01 \times 10^3 \ \text{J/mol})}{273 \ \text{K}} = 22.0 \ \text{J/K}$$

Because energy flows into the ice, q_{sys} is positive, which also makes ΔS_{sys} positive, as we would expect given the greater freedom of motion of particles in the liquid phase. Note that the units of entropy in this calculation are joules per kelvin. We will continue to use these units in all entropy calculations in this chapter.

Now suppose that the melting process occurs as heat flows into a 1-mole cube of ice from room-temperature (22°C or 295 K) surroundings. The change in entropy of the surroundings can be calculated using another adaptation of Equation 12.3:

$$\Delta S_{surr} = \frac{q_{surr}}{T} = \frac{(1.00 \ \text{mol})(-6.01 \times 10^3 \ \text{J/mol})}{295 \ \text{K}} = -20.4 \ \text{J/K}$$

Note that the sign of q_{surr} is negative because energy flows from the surroundings into the system (ice cube). Also note that (1) the value of ΔS_{surr} is less than zero, and (2) the magnitude of the decrease in ΔS_{surr} is less than the increase in ΔS_{sys}, because the same absolute value of q is divided by a higher temperature to calculate ΔS_{surr}. Therefore, when we sum ΔS_{sys} and ΔS_{surr}, we get a positive value for ΔS_{univ}:

$$\Delta S_{univ} = \Delta S_{sys} + \Delta S_{surr} = (22.0 - 20.4) \ \text{J/K} = 1.6 \ \text{J/K}$$

The result of this calculation is one example of a general truth about the flow of thermal energy that comes as no surprise: it flows spontaneously into a system when the system is cooler than its surroundings—or, even more generally, energy flows spontaneously from a warm object to an adjacent cooler object.

You may wonder why we did not consider the change in temperature of the surroundings as heat flowed from it into the melting ice. The answer lies in the sheer size of the surroundings (the universe minus a small cube of ice). The temperature of such a gigantic thermal mass does not change significantly.

Now let's consider how entropy changes when the temperature of the surroundings is *lower* than the temperature of the system. Suppose 1.00 mol of liquid water at 0°C is placed in a freezer at −10°C. Because the temperature of the surroundings (the freezer) is lower than the temperature of the liquid water, energy spontaneously flows from the water into its surroundings. The net entropy change for this process is

$$\Delta S_{univ} = \Delta S_{sys} + \Delta S_{surr}$$

$$= \frac{(1.00 \ \text{mol})(-6.01 \times 10^3 \ \text{J/mol})}{273 \ \text{K}} + \frac{(1.00 \ \text{mol})(+6.01 \times 10^3 \ \text{J/mol})}{263 \ \text{K}}$$

$$= (-22.0 \ \text{J/K}) + (+22.9 \ \text{J/K})$$

$$= 0.9 \ \text{J/K}$$

FIGURE 12.12 Water vapor exhaled by this Inuit hunter in the Northwest Territories of Canada was spontaneously deposited on his facial hair as frost—a testament to how cold it was when this photo was taken.

Once again there is an increase in the entropy of the universe as energy flows spontaneously from the warmer object (liquid water at 0°C) into the colder surroundings (the freezer at −10°C).

CONCEPT TEST

Is ΔS_{univ} greater than, less than, or equal to zero when water vapor exhaled by the Inuit hunter in Figure 12.12 is deposited as crystals of ice on his beard?

Entropy-change calculations based on Equation 12.3 assume process reversibility, which, as we have discussed, is an idealized, theoretical concept. In reality, the ΔS values calculated in this way are *minimum* ΔS values. When processes take place in the real world, the accompanying changes in entropy are inevitably greater than those calculated using Equation 12.3.

12.5 Calculating Entropy Changes

The entropy of a system (like its enthalpy and internal energy) is a state function, which means that the change in entropy that accompanies a process depends only on the initial and final states of the system, not on the pathway of the process. Therefore the change in entropy experienced by the system is simply the difference between its initial and final absolute entropy levels:

CONNECTION State functions were defined in Chapter 9.

$$\Delta S_{\text{sys}} = S_{\text{final}} - S_{\text{initial}} \tag{12.5}$$

We can adapt Equation 12.5 to calculate the change in entropy that accompanies a chemical reaction under standard conditions, $\Delta S_{\text{rxn}}^{\circ}$, from the difference between the standard molar entropies of $n_{\text{reactants}}$ moles of reactants (the equivalent of S_{initial} in Equation 12.5) and the standard molar entropies of n_{products} moles of products (that is, S_{final}):

$$\Delta S_{\text{rxn}}^{\circ} = \sum n_{\text{products}} S_{\text{products}}^{\circ} - \sum n_{\text{reactants}} S_{\text{reactants}}^{\circ} \tag{12.6}$$

Each individual $S°$ value for a product or reactant is multiplied by the appropriate number of moles from the balanced chemical equation. In other words, $\Delta S_{\text{rxn}}^{\circ}$, like $\Delta H_{\text{rxn}}^{\circ}$, is an extensive thermodynamic property that depends on the amounts of substances consumed or produced in a reaction. Standard molar entropies of selected substances are listed in Appendix 4.

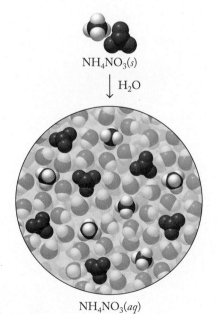

$NH_4NO_3(s)$

↓ H_2O

$NH_4NO_3(aq)$

FIGURE 12.13 Dissolution of solid ammonium nitrate into NH_4^+ and NO_3^- ions is endothermic but spontaneous under standard conditions because $\Delta S°$ is greater than zero, which contributes to $\Delta S_{\text{univ}} > 0$.

SAMPLE EXERCISE 12.3 Calculating $\Delta S°$ Values · · · **LO3**

Given the following standard molar entropy values at 298 K, what is the value of $\Delta S°$ for the dissolution of ammonium nitrate (Figure 12.13) under standard conditions?

$$NH_4NO_3(s) \rightarrow NH_4^+(aq) + NO_3^-(aq)$$

	$NH_4NO_3(s)$	$NH_4^+(aq)$	$NO_3^-(aq)$
$S°$ [J/(mol · K)]	151.1	113.4	146.4

COLLECT AND ORGANIZE We are given the standard molar entropy values of a solid ionic compound and its ions in aqueous solution, and we are asked to calculate the change in entropy that occurs when it dissolves under standard conditions. Entropy changes associated with chemical reactions depend on the entropies of the reactants and products as described in Equation 12.6.

ANALYZE It is debatable whether or not the dissolution of an ionic solid in water constitutes a chemical reaction, but Equation 12.6 still allows us to calculate $\Delta S°$ of the dissolution process from the difference in standard molar entropies of the solid solute and its ions in solution. Dissolving an ionic solid in water increases the freedom of motion of the solute ions, so $\Delta S°$ should be greater than zero.

SOLVE

$$\Delta S° = \sum n_{\text{products}} S°_{\text{products}} - \sum n_{\text{reactants}} S°_{\text{reactants}}$$

$$= \left(1 \text{ mol} \times 113.4 \frac{J}{\text{mol} \cdot K} + 1 \text{ mol} \times 146.4 \frac{J}{\text{mol} \cdot K}\right) - 1 \text{ mol} \times 151.1 \frac{J}{\text{mol} \cdot K}$$

$$= 108.7 \text{ J/K}$$

THINK ABOUT IT As predicted, $\Delta S°$ is greater than zero because the freedom of motion and the dispersion of the energy of solute particles increase when the solid solute dissolves.

Practice Exercise Calculate the standard molar entropy change for the combustion of methane gas using $S°$ values from Table 12.2. Before carrying out the calculation, predict whether the entropy of the system increases or decreases. Assume that liquid water is one of the products. ⚙

12.6 Free Energy

In Sample Exercise 12.3 we used standard molar entropy values to calculate $\Delta S°$. The solute, solvent, and resulting solution together constituted a closed thermodynamic system, which meant that heat could flow into it as its temperature dropped but matter was not exchanged with its surroundings. Therefore the $\Delta S°$ value that was calculated applied only to the system. There was no evaluation of the change in the entropy of the system's surroundings accompanying the dissolution process, though we might predict that heat flowing from the surroundings into the system would produce a decrease in S_{surr}. (Had the dissolution process been exothermic, heat would have flowed from the system into its surroundings and ΔS_{surr} would have been greater than zero.)

Thus, the entropy change experienced by the surroundings of any chemical thermodynamic system depends on whether the process occurring in the system is exothermic or endothermic. When energy flows from an exothermic process occurring at constant pressure into a system's surroundings, the quantity of energy is equal in magnitude but opposite in sign to the enthalpy change of the system:

$$q_{\text{surr}} = -\Delta H_{\text{sys}} \tag{12.7}$$

When a system undergoes an endothermic process, as in Sample Exercise 12.3, the direction of energy flow is reversed, but Equation 12.7 still applies. Assuming these transfers of heat occur reversibly, we can calculate the value of ΔS_{surr} using this modification of Equation 12.3:

$$\Delta S_{\text{surr}} = \frac{q_{\text{surr}}}{T}$$

which we combine with Equation 12.7:

$$\Delta S_{\text{surr}} = -\frac{\Delta H_{\text{sys}}}{T}$$

CONNECTION In Chapter 9 we defined a change in enthalpy (ΔH) as the energy gained or lost as heat during a process taking place at constant pressure.

▶❙❙ **CHEMTOUR** Gibbs Free Energy

and substitute this expression into Equation 12.2 ($\Delta S_{univ} = \Delta S_{sys} + \Delta S_{surr}$):

$$\Delta S_{univ} = \Delta S_{sys} - \frac{\Delta H_{sys}}{T} \qquad (12.8)$$

The beauty of Equation 12.8 is that it allows us to predict whether or not a process is spontaneous at a particular temperature once we calculate the enthalpy and entropy changes accompanying the process. The downside of Equation 12.8 is that spontaneity relies on the value of a parameter (ΔS_{univ}) that is impossible to determine directly and that has little physical meaning. It would be great if we could substitute a single thermodynamic parameter for ΔS_{univ} that is based only on the system and not the entire universe. Such a parameter exists. It is called the change in the system's **free energy**. Note that *free* in this context does not mean "at no cost"; it means energy that is *freed* during a process to do useful work.

In chemistry we focus on a particular kind of free energy, which is called **Gibbs free energy (G)** in honor of American scientist J. Willard Gibbs (1839–1903). Gibbs free energy is the energy available to do useful work in processes happening at constant temperature and pressure or once the temperatures and pressures of reaction mixtures have returned to their initial values. These conditions are met in the chemical reactions and physical changes we discuss in the following sections, so the changes in free energy that accompany these processes will be changes in Gibbs free energy, and we will use the symbol ΔG to represent these changes.

Like many of the thermodynamic properties we have examined, absolute free energy values of substances are often of less interest than the *changes* in free energy that accompany chemical reactions and other processes. Gibbs proposed that the change in free energy (ΔG_{sys}) of a process occurring at constant temperature and pressure is linked directly to that temperature and ΔS_{univ}:

$$\Delta G_{sys} = -T\,\Delta S_{univ}$$

Because of the $(-T)$ multiplier, *negative* values of ΔG_{sys} correspond to *positive* values of ΔS_{univ}. Therefore

➤ If $\Delta G_{sys} < 0$, then $\Delta S_{univ} > 0$ and the reaction is spontaneous.

➤ If $\Delta G_{sys} > 0$, then $\Delta S_{univ} < 0$ and the reaction is nonspontaneous. Instead, the reaction running in reverse of the process is spontaneous.

➤ If $\Delta G_{sys} = 0$, then $\Delta S_{univ} = 0$ and the composition of the reaction mixture does not change with time. In other words, the reaction has reached chemical equilibrium.

Let's combine Gibbs's equation with Equation 12.8. It's easy once we multiply all of the terms in Equation 12.8 by $(-T)$:

$$-T\,\Delta S_{univ} = -T\,\Delta S_{sys} + \Delta H_{sys} \qquad (12.9)$$

Note that the left side of Equation 12.9 is equal to ΔG_{sys}. Making that substitution and rearranging the terms on the right side,

$$\Delta G_{sys} = \Delta H_{sys} - T\,\Delta S_{sys}$$

Since all of the parameters in this equation apply to the system, we typically simplify the equation by eliminating the subscripts:

$$\Delta G = \Delta H - T\,\Delta S \qquad (12.10)$$

CONCEPT TEST ••

    a. Given the thermodynamic data in Table A4.3, is the conversion of diamond to graphite spontaneous? Explain your answer.

    b. If yes, does knowing that the conversion is spontaneous tell you how rapid the conversion is?

Equation 12.10 highlights the two thermodynamic factors that contribute to a decrease in free energy and to making a process spontaneous:

    1. The system experiences an increase in entropy ($\Delta S > 0$).
    2. The process is exothermic ($\Delta H < 0$).

Table 12.3 summarizes how the signs of ΔH and ΔS impact the sign of ΔG and how they define the conditions under which a process is spontaneous.

The change in free energy of a process can be calculated using Equation 12.10 if we first calculate the values of ΔH and ΔS. In the case of a chemical reaction occurring under standard conditions, the *standard* change in free energy ΔG°_{rxn} can be calculated using a modified version of Equation 12.10:

$$\Delta G^\circ_{rxn} = \Delta H^\circ_{rxn} - T\Delta S^\circ_{rxn} \qquad (12.11)$$

In Chapter 9 we calculated ΔH°_{rxn} values from the difference in the standard heats of formation (ΔH°_f) values of products and reactants:

$$\Delta H^\circ_{rxn} = \sum n_{products} \Delta H^\circ_{f,products} - \sum n_{reactants} \Delta H^\circ_{f,reactants} \qquad (9.15)$$

The value of ΔS°_{rxn} can be calculated using Equation 12.6:

$$\Delta S^\circ_{rxn} = \sum n_{products} S^\circ_{products} - \sum n_{reactants} S^\circ_{reactants} \qquad (12.6)$$

The results of these two calculations are combined using Equation 12.11 to calculate ΔG°_{rxn}, as illustrated in Sample Exercise 12.4.

TABLE 12.3		Effects of ΔH, ΔS, and T on ΔG and Spontaneity	
ΔH	ΔS	ΔG	Spontaneity
−	+	Always < 0	Always spontaneous
−	−	< 0 at lower temperature	Spontaneous at lower temperature
+	+	< 0 at higher temperature	Spontaneous at higher temperature
+	−	Always > 0	Never spontaneous

SAMPLE EXERCISE 12.4   **Predicting Reaction Spontaneity**   **LO4**
                                    under Standard Conditions

Consider the reaction of nitrogen gas and hydrogen gas (Figure 12.14) at 298 K to make ammonia at the same temperature:

$$N_2(g) + 3\,H_2(g) \rightarrow 2\,NH_3(g)$$

    a. Before doing any calculations, predict the sign of ΔS°_{rxn}.
    b. What is the actual value of ΔS°_{rxn}?
    c. What is the value of ΔH°_{rxn}?
    d. What is the value of ΔG°_{rxn} at 298 K?
    e. Is the reaction spontaneous at 298 K and 1 bar of pressure?

COLLECT AND ORGANIZE We are asked to predict the sign and calculate the value of the standard entropy change of a reaction. We are also asked to calculate the enthalpy change of the reaction and its change in free energy, and then to predict its spontaneity under standard conditions. Standard molar entropies and standard heats of formation of the reactants and product are in Table 12.2 and Appendix 4.

ANALYZE Figure 12.14 reinforces the point that there are more moles of gaseous reactants than products in the reaction. Equation 12.6 can be used to calculate entropy changes under standard conditions:

$$\Delta S^\circ_{rxn} = \sum n_{products} S^\circ_{products} - \sum n_{reactants} S^\circ_{reactants} \qquad (12.6)$$

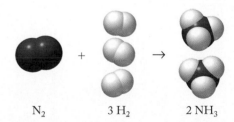

$$N_2 \qquad\qquad 3\,H_2 \qquad\qquad 2\,NH_3$$

FIGURE 12.14 In the synthesis of ammonia, one molecule of nitrogen reacts with three molecules of hydrogen to yield two molecules of ammonia. All substances are gases.

We learned in Chapter 9 how to calculate ΔH°_{rxn} from standard heats of formation values (in Appendix 4) using this equation:

$$\Delta H^\circ_{rxn} = \sum n_{products}\, \Delta H^\circ_{f,products} - \sum n_{reactants}\, \Delta H^\circ_{f,reactants} \qquad (9.15)$$

The calculated values of ΔS°_{rxn} and ΔH°_{rxn} can be combined to calculate ΔG°_{rxn}:

$$\Delta G^\circ_{rxn} = \Delta H^\circ_{rxn} - T\,\Delta S^\circ_{rxn} \qquad (12.11)$$

The sign of ΔG°_{rxn} will allow us to predict whether or not the reaction is spontaneous under standard conditions (and $T = 298$ K). It is difficult to predict the sign of ΔG°_{rxn} and whether this reaction is spontaneous without knowing the magnitudes of ΔH°_{rxn} and $T\,\Delta S^\circ_{rxn}$.

SOLVE

a. The number of gas-phase molecules decreases as the reaction proceeds, so the sign of ΔS°_{rxn} is probably negative.

b. Using data from Table 12.2 in Equation 12.6 to calculate ΔS°_{rxn},

$$\Delta S^\circ_{rxn} = \sum n_{products} S^\circ_{products} - \sum n_{reactants} S^\circ_{reactants}$$

$$= \left[2\ \text{mol} \times \left(192.5\ \frac{\text{J}}{\text{mol} \cdot \text{K}}\right)\right] - \left[1\ \text{mol} \times \left(191.5\ \frac{\text{J}}{\text{mol} \cdot \text{K}}\right) + 3\ \text{mol} \times \left(130.6\ \frac{\text{J}}{\text{mol} \cdot \text{K}}\right)\right]$$

$$= -198.3\ \text{J/K}$$

The entropy change is negative, as predicted.

c. The change in enthalpy that accompanies the reaction under standard conditions is

$$\Delta H^\circ_{rxn} = \sum n_{products}\, \Delta H^\circ_{f,products} - \sum n_{reactants}\, \Delta H^\circ_{f,reactants}$$

$$= \left[2\ \text{mol} \times \left(-46.1\ \frac{\text{kJ}}{\text{mol}}\right)\right] - \left[1\ \text{mol} \times \left(0.0\ \frac{\text{kJ}}{\text{mol}}\right) + 3\ \text{mol} \times \left(0.0\ \frac{\text{kJ}}{\text{mol}}\right)\right]$$

$$= -92.2\ \text{kJ}$$

d. Using the values of ΔS°_{rxn} and ΔH°_{rxn} calculated in parts b and c in Equation 12.11,

$$\Delta G^\circ_{rxn} = \Delta H^\circ_{rxn} - T\,\Delta S^\circ_{rxn}$$

$$= -92.2\ \text{kJ} - (298\ \text{K})\left(-198.3\ \frac{\text{J}}{\text{K}}\right)\left(\frac{1\ \text{kJ}}{1000\ \text{J}}\right)$$

$$= -33.1\ \text{kJ}$$

e. The decrease in free energy tells us that the reaction is spontaneous under standard conditions and $T = 298$ K.

THINK ABOUT IT We were unable to make a prediction about reaction spontaneity because we didn't have a feel for the magnitudes of ΔH°_{rxn} and ΔS°_{rxn}. In this case, an unfavorable entropy change ($\Delta S^\circ_{rxn} < 0$) is more than offset by a favorable enthalpy change ($\Delta H^\circ_{rxn} < 0$) so that, overall, $\Delta G^\circ_{rxn} < 0$.

Practice Exercise For the reaction $2\ H_2(g) + O_2(g) \rightarrow 2\ H_2O(\ell)$,

a. Predict the sign of the entropy change for the reaction.
b. What is the value of ΔS°_{rxn}?
c. What is the value of ΔH°_{rxn}?
d. Is the reaction spontaneous at 298 K and 1 bar pressure?

CONCEPT TEST

The preparation of ammonia from nitrogen and hydrogen in Sample Exercise 12.4 is predicted to be spontaneous under standard conditions, yet if we mix the two gases at 298 K, no reaction is observed. Suggest a reason why.

standard free energy of formation (ΔG_f°) the change in free energy associated with the formation of 1 mole of a compound in its standard state from its elements in their standard states.

There is another way to calculate the change in free energy of a reaction under standard conditions. It is based on another thermodynamic property of substances listed in Appendix 4: **standard free energy of formation (ΔG_f°)**. A compound's ΔG_f° value is the change in free energy associated with the formation of 1 mole of it in its standard state from its elements in their standard states.

In much the way we calculated standard heats of reactions (ΔH_{rxn}°) from the difference in the standard heats of formation (ΔH_f°) of their products and reactants in Chapter 9, so, too, can we calculate the change in standard free energy of a reaction under standard conditions from the difference in the standard free energies of formation of its products and reactants. As with standard heat of formation, the standard free energy of formation of the most stable forms of elements in their standard states is zero. The similarities in the two calculations can be seen from the similar formats of the equations used to calculate ΔG_{rxn}°

$$\Delta G_{rxn}^\circ = \sum n_{products}\,\Delta G_{f,products}^\circ - \sum n_{reactants}\,\Delta G_{f,reactants}^\circ \tag{12.12}$$

and ΔH_{rxn}°:

$$\Delta H_{rxn}^\circ = \sum n_{products}\,\Delta H_{f,products}^\circ - \sum n_{reactants}\,\Delta H_{f,reactants}^\circ \tag{9.15}$$

Sample Exercise 12.5 illustrates just how similar the two calculations are.

(a) Octane
$\Delta G_f^\circ = 16.3$ kJ/mol

(b) 2-Methylheptane
$\Delta G_f^\circ = 11.7$ kJ/mol

(c) 3,3-Dimethylhexane
$\Delta G_f^\circ = 12.6$ kJ/mol

FIGURE 12.15 Molecular structures and ΔG_f° values of three C_8H_{18} isomers.

CONCEPT TEST

Molecular models and standard free energies of formation for three structural isomers with the molecular formula C_8H_{18} are shown in Figure 12.15. All three isomers burn in air, as described by the same chemical equation:

$$2\ C_8H_{18}(\ell) + 25\ O_2(g) \rightarrow 16\ CO_2(g) + 18\ H_2O(g)$$

Are the ΔG_{rxn}° values for the three combustion reactions also the same? Why or why not?

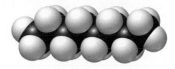

SAMPLE EXERCISE 12.5 Calculating ΔG_{rxn}° Using Appropriate ΔG_f° Values LO5

Use the appropriate standard free energy of formation values in Appendix 4 to calculate the change in free energy as ethanol burns under standard conditions. Assume the reaction proceeds as described by the following chemical equation:

$$CH_3CH_2OH(\ell) + 3\ O_2(g) \rightarrow 2\ CO_2(g) + 3\ H_2O(\ell)$$

COLLECT AND ORGANIZE We are asked to calculate the value of ΔG_{rxn}° for the combustion of ethanol using the "appropriate" ΔG_f° values, which are, according to Equation 12.12, the ΔG_f° values of the reactants and products in the combustion reaction:

Substance	$CH_3CH_2OH(\ell)$	$O_2(g)$	$CO_2(g)$	$H_2O(\ell)$
ΔG_f° (kJ/mol)	−174.9	0	−394.4	−237.2

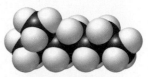

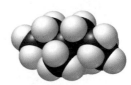

ANALYZE The reaction consumes 1 mole of liquid ethanol and 3 moles of oxygen and produces 2 moles of CO_2 gas and 3 moles of liquid H_2O. Multiplying the above ΔG_f° values by the appropriate numbers of moles and subtracting the sum of the reactant values from the sum of the product values, as described in Equation 12.12, will yield the value of ΔG_{rxn}°. The combustion of ethanol, a common additive in gasoline in the United States, is spontaneous, so ΔG_{rxn}° should be less than zero.

SOLVE Inserting the appropriate numbers of moles and ΔG_f° values in Equation 12.12 and doing the math,

$$\Delta G_{rxn}^\circ = \sum n_{products} \Delta G_{f,products}^\circ - \sum n_{reactants} \Delta G_{f,reactants}^\circ$$

$$= [2 \text{ mol } CO_2 \times (-394.4 \text{ kJ/mol } CO_2) + 3 \text{ mol } H_2O \times (-237.2 \text{ kJ/mol } H_2O)]$$

$$- 1 \text{ mol } CH_3CH_2OH \times (-174.9 \text{ kJ/mol } CH_3CH_2OH)$$

$$= -1325.5 \text{ kJ}$$

THINK ABOUT IT The calculated value represents that part of the total energy released by the combustion of one mole of ethanol under standard conditions that is available to do useful work.

Practice Exercise Use the appropriate standard free energy of formation values in Appendix 4 to calculate the value of ΔG_{rxn}° for the steam–methane reforming reaction used to produce H_2 gas:

$$CH_4(g) + H_2O(g) \rightarrow CO(g) + 3\,H_2(g)$$

The Meaning of *Free* Energy

What exactly does "energy available to do useful work" mean? Let's attempt to answer this question by considering the internal combustion (gasoline) engines used to power most automobiles. A combustion reaction is a thermodynamic system that experiences a decrease in internal energy (ΔE) as heat (q) flows from it into its surroundings and as it does work (w) on its surroundings. These three variables are related by Equation 9.3:

$$\Delta E = q + w \tag{9.3}$$

and all three are less than zero from the perspective of the system. Internal combustion engines have cooling systems to manage the dissipation of heat, which is wasted energy that does nothing to power the car. The energy that does move the car is derived from the rapid expansion of the gaseous products of combustion in the cylinders of its engine. As Figure 12.16 shows, this expansion pushes down on the piston of a cylinder, increasing the volume of the reaction mixture. The product of the pressure exerted by the reacting gases and their products on the piston and the resulting volume change is $P–V$ work (discussed in Section 9.2), which propels the car. At constant pressure:

$$w = -P\,\Delta V$$

Free energy is a measure of the *maximum* amount of work that can be done by the energy released during combustion. To see how this theoretical quantity of work compares with the total energy released, let's rearrange Equation 12.10 by isolating the ΔH term:

$$\Delta H = \Delta G + T\,\Delta S \tag{12.13}$$

FIGURE 12.16 Thermal expansion of the gases in a cylinder in a car engine pushes down on a piston with a pressure P represented by the blue arrow. The product of P and the change in volume of the gases ΔV is the work done by the expanding gases.

Equation 12.13 tells us that the enthalpy change that accompanies making and breaking chemical bonds during a chemical reaction may be divided into two parts. One part, ΔG, is the energy that can theoretically be converted into motion and other useful work (like generating electricity for the car's electrical system). The other part, $T\Delta S$, is not usable: it is the portion of energy that spreads out when, for example, hot gases flow out the end of an automobile exhaust pipe. This part of ΔH is wasted. Consequently, conversion of chemical energy into useful mechanical energy (ΔG) is never 100% efficient. In addition, some portion of ΔG is also wasted because the combustion and energy conversion happen quickly and, therefore, irreversibly. Maximum efficiency comes with very slow, reversible reactions and energy conversion rates, but that is not how automobile engines operate. It turns out that gasoline engines convert only about 30% of the energy produced during combustion into useful work.

12.7 Temperature and Spontaneity

Let's revisit the process of ice melting (Figure 12.2), this time focusing on how the values of the three terms in Equation 12.10, namely, ΔH, $T\Delta S$, and ΔG, change as the temperature of a mixture of ice and water increases from −10°C to +10°C (Figure 12.17). As temperatures rise over this range there is little impact on the heat of fusion, ΔH, as shown by the nearly flat green line in Figure 12.17. However, it is only reasonable that increasing the value of T increases the value $T\Delta S$ (as shown by the upward slope of the purple line). After all, the ΔS of ice (or any solid) melting has a positive value, so $T\Delta S$ must increase as T increases. The green and purple lines intersect at 0°C, which means ΔH is equal to $T\Delta S$ at that temperature. Put another way, the difference between ΔH and $T\Delta S$ at 0°C is zero—and so is ΔG (remember: $\Delta G = \Delta H - T\Delta S$). When $\Delta G = 0$, a process is at equilibrium and ice and water coexist and are in *equilibrium* with each other at 0°C.

We also know that ice melts spontaneously above 0°C, which means that ΔG for the melting process must be less than zero. The graph in Figure 12.17 shows that indeed it is. Above 0°C the value of $T\Delta S$ is greater than ΔH. Therefore the difference between them ($\Delta H - T\Delta S$) is less than zero and becomes more negative with increasing temperature, as illustrated by the distance from the purple line to the green line in Figure 12.17.

Below 0°C, ice does not melt spontaneously, which means ΔG is greater than zero. The graph shows why this is true: at $T < 273$ K (0°C), $T\Delta S$ (purple line) values are less than ΔH (green line) values, which means that $\Delta H - T\Delta S$ (and ΔG) is greater than zero. Positive ΔG values mean that the process is nonspontaneous below 273 K and ice does not melt below its freezing point. However, the opposite process—liquid water freezing—*is* spontaneous because reversing a process keeps the absolute values but switches the signs of ΔH, ΔS, and ΔG. Therefore, if the melting process is nonspontaneous, then the exothermic freezing process *is* spontaneous at low temperatures.

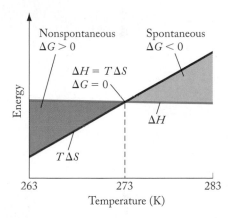

FIGURE 12.17 Changes in the values of ΔH, the quantity $T\Delta S$, and ΔG for ice melting with changing temperature from −10°C (263 K) to +10°C (283 K).

∞ **CONNECTION** In Chapter 6 we discussed the equilibrium between phases of a substance and used phase diagrams to illustrate which physical states are stable at various combinations of temperature and pressure.

CONCEPT TEST •

A process is spontaneous at lower temperature and nonspontaneous at higher temperature. Sketch a graph similar to Figure 12.17 for this process.

• •

The temperature at the point in Figure 12.17 where $\Delta G = 0$ defines the melting (or freezing) point of water, which is the temperature at which solid ice and liquid water coexist in a state of thermodynamic equilibrium. Similarly, at 100°C and 1 atm, liquid water and water vapor coexist because the free energy changes that accompany vaporization and condensation both equal zero.

The temperature at which $\Delta G = 0$ for a process can be calculated from Equation 12.10 if we know the values of ΔH and ΔS. These values for the fusion of ice are

$$H_2O(s) \rightarrow H_2O(\ell) \qquad \Delta H° = 6.01 \times 10^3 \text{ J/mol}; \Delta S° = 22.0 \text{ J/(mol} \cdot \text{K)}$$

Assuming $\Delta H°$ and $\Delta S°$ do not change significantly with small changes in temperature, we can use these values in calculations at temperatures other than 298 K. This turns out to be a reasonable assumption so that

$$\Delta G = \Delta H - T \Delta S \approx \Delta H° - T \Delta S°$$

Inserting the values of $\Delta H°$ and $\Delta S°$ and using the fact that $\Delta G = 0$ for a process at equilibrium, we get

$$\Delta G = (6.01 \times 10^3 \text{ J/mol}) - T[22.0 \text{ J/(mol} \cdot \text{K)}] = 0$$

$$T = \frac{6.01 \times 10^3 \text{ J/mol}}{22.0 \text{ J/(mol} \cdot \text{K)}} = 273 \text{ K} = 0°\text{C}$$

which is the familiar value for the melting point of ice.

Like physical processes, chemical reactions can have zero change in free energy. As a spontaneous reaction ($\Delta G_{rxn} < 0$) proceeds, reactants become products and ΔG_{rxn} becomes less negative, as we will see in the next section. In fact, ΔG_{rxn} may reach zero before all the reactants are consumed. In this case, there is no further increase in the quantities of products and no further loss in the quantities of reactants in the reaction mixture. Rather, reactants and products coexist in equilibrium with each other. Reactants still react, and products are still formed, but the reverse reaction, in which products become reactants, proceeds at the same rate as the forward reaction. This is the state we know as chemical equilibrium. No net change in the amounts of reactants and products occurs once equilibrium is reached.

SAMPLE EXERCISE 12.6 **Relating Reaction Spontaneity** **LO6**
to ΔH_{rxn} and ΔS_{rxn}

A certain chemical reaction is spontaneous at low temperatures but not at high temperatures. What are the signs of $\Delta H°_{rxn}$ and $\Delta S°_{rxn}$ for this reaction?

COLLECT AND ORGANIZE We are asked to determine the signs of $\Delta H°_{rxn}$ and $\Delta S°_{rxn}$ that are consistent with a reaction being spontaneous only at low temperatures. The following approximation of Equation 12.11 relates these variables to ΔG_{rxn}:

$$\Delta G_{rxn} \approx \Delta H°_{rxn} - T \Delta S°_{rxn}$$

ANALYZE There are two possible combinations for $\Delta H°_{rxn}$ and $\Delta S°_{rxn}$ that might lead to $\Delta G_{rxn} < 0$ depending on temperature: $\Delta H°_{rxn} > 0$ and $\Delta S°_{rxn} > 0$; $\Delta H°_{rxn} < 0$ and $\Delta S°_{rxn} < 0$. We need to identify which of these combinations will satisfy the condition that $\Delta G_{rxn} < 0$ only at low temperatures.

SOLVE The importance of $\Delta S°_{rxn}$ increases with increasing temperature because the product $T \Delta S°_{rxn}$ appears in the approximation of Equation 12.11 for nonstandard conditions. The reaction is nonspontaneous at higher temperatures, where the magnitude of $T\,\Delta S^\circ_{rxn}$ is more likely to be larger than the magnitude of ΔH°_{rxn}. The reaction is spontaneous at low temperatures, however, where the impact of a negative ΔS°_{rxn} value is more than offset by a decrease in enthalpy, a change that favors the reaction. The reaction must have negative ΔS°_{rxn} and negative ΔH°_{rxn} values.

THINK ABOUT IT Table 12.3 confirms our prediction that a process that is spontaneous only at low temperatures is one in which there is a decrease in both entropy and enthalpy.

Practice Exercise At high temperatures, ammonia decomposes to nitrogen and hydrogen gases:

$$2\,NH_3(g) \rightarrow N_2(g) + 3\,H_2(g)$$

ΔH°_{rxn} for the reaction is positive and ΔS°_{rxn} is positive. Predict whether the reaction is spontaneous at all temperatures, at only high temperatures, or at only low temperatures. ⚙

12.8 Driving the Human Engine: Coupled Reactions

The laws of thermodynamics that govern chemical reactions in the laboratory also govern all the chemical reactions that take place in living systems (Figure 12.18). Organisms carry out reactions that release the energy contained in the chemical bonds of food molecules and then use that energy to do work and to sustain an array of other essential biological functions. Just like mechanical engines, however, humans and other life-forms are far from 100% efficient, which means that life requires a continuous input of energy. Thus we humans must constantly absorb energy in the form of the caloric content of the food we eat and then release heat and waste products into our surroundings. Young women have an average daily nutritional need of 2100 Cal; for young men the figure is 2900 Cal. This level of caloric intake provides the energy we need in order to function at all levels, from thinking to getting out of bed in the morning.

CONNECTION In Chapter 9 we defined 1 Calorie, the "calorie" used in discussing food, as equivalent to 1 kcal.

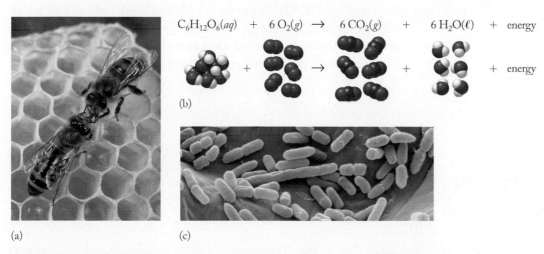

$$C_6H_{12}O_6(aq) \;+\; 6\,O_2(g) \;\rightarrow\; 6\,CO_2(g) \;+\; 6\,H_2O(\ell) \;+\; energy$$

(b)

(a)

(c)

FIGURE 12.18 The rules of thermodynamics apply to all living systems. (a) Honeybees extract energy from nutrients to support life. As bees collect and process plant nectars, they store the energy as honey, which is then consumed by the bees, by humans, or by other animals. (b) When the sugar in honey is digested, carbon dioxide, water, and energy are released, increasing the entropy of the universe. (c) The spontaneous life processes of all organisms—including the *E. coli* bacteria that live in our gastrointestinal tracts—increase the entropy of the universe.

glycolysis a series of reactions that converts glucose into pyruvate; a major anaerobic (no oxygen required) pathway for the metabolism of glucose in the cells of almost all living organisms.

phosphorylation a reaction resulting in the addition of a phosphate group to an organic molecule.

CONNECTION Photosynthesis and the carbon cycle were introduced in Chapter 7.

Glucose

↓

Glycolysis

↓

Pyruvate ion

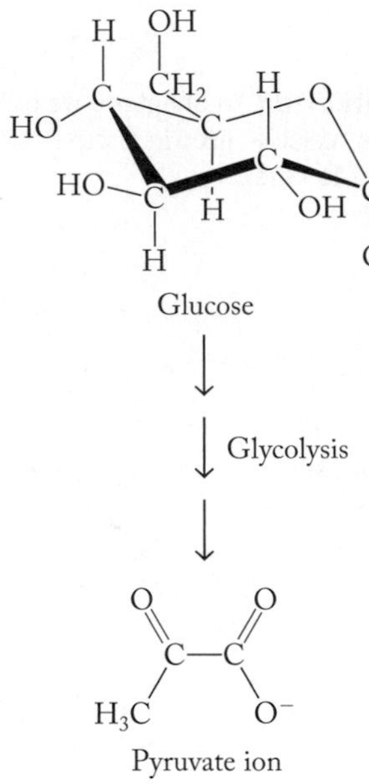

FIGURE 12.19 In glycolysis, molecules of glucose are converted into twice their number of pyruvate ions.

In living systems, spontaneous reactions ($\Delta G_{rxn} < 0$) typically involve breaking food down, while nonspontaneous reactions ($\Delta G_{rxn} > 0$) involve building molecules needed by the body. Living systems use the energy from spontaneous reactions to run nonspontaneous reactions; we say that the spontaneous reactions are *coupled* to the nonspontaneous reactions. Part of the study of biochemistry involves deciphering the molecular mechanisms that enable reaction coupling. This topic is covered in Chapter 20, but for now it is sufficient to know that elegant molecular processes have evolved to enable living systems to couple chemical reactions so that the energy obtained from spontaneous reactions can be used to drive the nonspontaneous reactions that maintain life. The metabolic chemical reactions we look at here are *not* presented for you to memorize. Rather, the intent is to aid your understanding of how changes in free energy allow spontaneous reactions to drive nonspontaneous reactions and to illustrate operationally what the phrase *coupled reactions* actually means.

As noted in the essay at the beginning of Chapter 9, the energy contained in the food we eat comes indirectly from solar energy. Green plants store energy from sunlight in their tissues in the chemical bonds of molecules such as glucose ($C_6H_{12}O_6$), which they produce from CO_2 and H_2O during photosynthesis.

Production of glucose by green plants is a nonspontaneous process, which is why the plants require the energy of sunlight to carry out this reaction. Animals that consume plants use the energy stored in glucose and other molecules, and they release CO_2 and H_2O back into the environment:

$$C_6H_{12}O_6(s) + 6\ O_2(g) \rightarrow 6\ CO_2(g) + 6\ H_2O(\ell)$$

This reaction also releases energy, an event that increases the entropy of the universe, and is, therefore, spontaneous. However, the reaction takes place in living organisms in many steps, some of which are nonspontaneous. Figure 12.19 illustrates one portion of glucose metabolism—**glycolysis**—that involves a series of spontaneous and nonspontaneous reactions.

In glycolysis, each mole of glucose is converted into 2 moles of pyruvate ion (CH_3COCOO^-). An early step is conversion of glucose into glucose 6-phosphate (Figure 12.20), an example of a **phosphorylation** reaction. Glucose reacts with hydrogen phosphate ion (HPO_4^{2-}), producing glucose 6-phosphate and water. This reaction is not spontaneous ($\Delta G^\circ_{rxn} = +13.8$ kJ/mol). The energy needed to make it happen comes from an ion called adenosine triphosphate (ATP^{4-}), which functions in our cells both as a storehouse of energy and as an energy-transfer

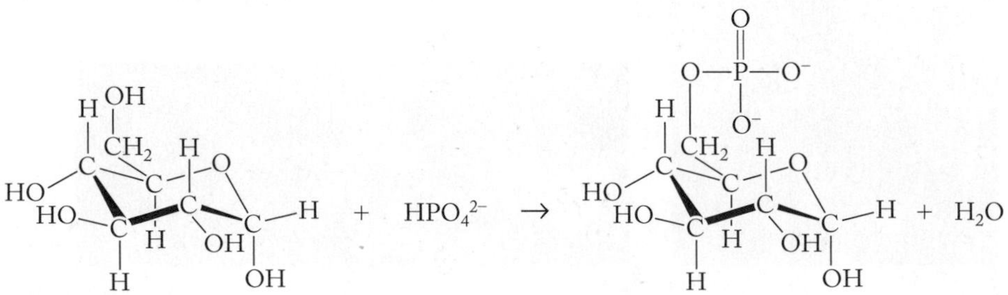

Glucose(*aq*) + $HPO_4^{2-}(aq)$ → Glucose 6-phosphate(*aq*) + $H_2O(\ell)$

FIGURE 12.20 The conversion of glucose into glucose 6-phosphate is an early step in glycolysis. This reaction is nonspontaneous, which means energy must be added to make the reaction go: $\Delta G^\circ_{rxn} = 13.8$ kJ/mol.

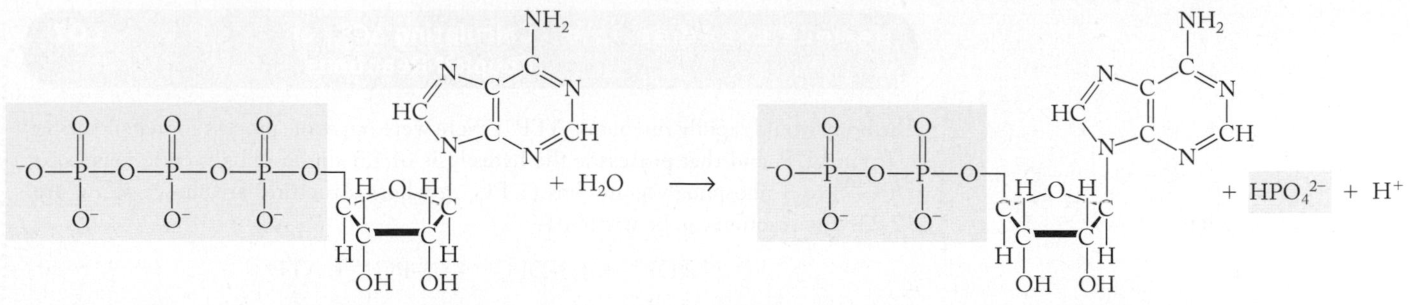

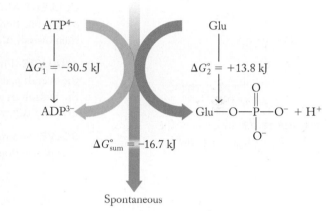

Adenosine triphosphate (ATP^{4-})   Adenosine diphosphate (ADP^{3-})

FIGURE 12.21 The hydrolysis of ATP to ADP is a spontaneous reaction: $\Delta G° = -30.5$ kJ/mol. The body couples this reaction to nonspontaneous reactions so that the energy released can drive them (Figure 12.19).

agent: it hydrolyzes to the adenosine diphosphate ion (ADP^{3-}) in a reaction (Figure 12.21) that produces a hydrogen phosphate ion and energy ($\Delta G°_{rxn} = -30.5$ kJ/mol).

In a living system, spontaneous hydrolysis of ATP^{4-} consumes water and produces HPO$_4{}^{2-}$ and H$^+$, whereas the nonspontaneous phosphorylation of glucose consumes HPO$_4{}^{2-}$ and produces water. The two reactions are coupled: the spontaneous ATP$^{4-} \rightarrow$ ADP^{3-} reaction supplies the energy that drives the nonspontaneous formation of glucose 6-phosphate (Figure 12.22).

This example illustrates another important general point about reactions and $\Delta G°_{rxn}$ values: those for coupled reactions (and also for sequential reactions) are additive. This is true for any set of reactions, not just those occurring in living systems.

For the first two steps in glycolysis:

(1) $\text{ATP}^{4-}(aq) + \text{H}_2\text{O}(\ell) \rightarrow \text{ADP}^{3-}(aq) + \text{HPO}_4{}^{2-}(aq) + \text{H}^+(aq)$

$$\Delta G°_{rxn} = -30.5 \text{ kJ}$$

(2) $\text{C}_6\text{H}_{12}\text{O}_6(aq) + \text{HPO}_4{}^{2-}(aq) \rightarrow \text{C}_6\text{H}_{11}\text{O}_6\text{PO}_3{}^{2-}(aq) + \text{H}_2\text{O}(\ell)$

  Glucose   Hydrogen   Glucose
            phosphate   6-phosphate $\Delta G°_{rxn} = 13.8 \text{ kJ}$

If we add these reactions and their free energies, we get

$\text{ATP}^{4-}(aq) + \cancel{\text{H}_2\text{O}(\ell)} + \text{C}_6\text{H}_{12}\text{O}_6(aq) + \cancel{\text{HPO}_4{}^{2-}(aq)} \rightarrow$
$\quad \text{ADP}^{3-}(aq) + \cancel{\text{HPO}_4{}^{2-}(aq)} + \text{H}^+(aq) + \text{C}_6\text{H}_{11}\text{O}_6\text{PO}_3{}^{2-}(aq) + \cancel{\text{H}_2\text{O}(\ell)}$

$$\Delta G°_{rxn} = (-30.5 + 13.8) \text{ kJ} = -16.7 \text{ kJ}$$

Because equal quantities of H$_2$O and HPO$_4{}^{2-}$ appear on both sides of the combined equation, they cancel out, leaving the net reaction:

$\text{C}_6\text{H}_{12}\text{O}_6(aq) + \text{ATP}^{4-}(aq) \rightarrow \text{ADP}^{3-}(aq) + \text{C}_6\text{H}_{11}\text{O}_6\text{PO}_3{}^{2-}(aq) + \text{H}^+(aq)$

$$\Delta G°_{rxn} = -16.7 \text{ kJ}$$

Since $\Delta G°_{rxn}$ for the net reaction is negative, the reaction is spontaneous.

FIGURE 12.22 The spontaneous hydrolysis of ATP is coupled to the nonspontaneous phosphorylation of glucose (Glu). The overall reaction—the sum of the two individual reactions—is spontaneous.

∞ **CONNECTION** In Chapter 9 we used Hess's law (the enthalpy change of a reaction that is the sum of two or more reactions equals the sum of the enthalpy changes of the constituent reactions) to calculate $\Delta H°_{rxn}$. We apply a similar principle here when adding $\Delta G°_{rxn}$ values of coupled reactions.

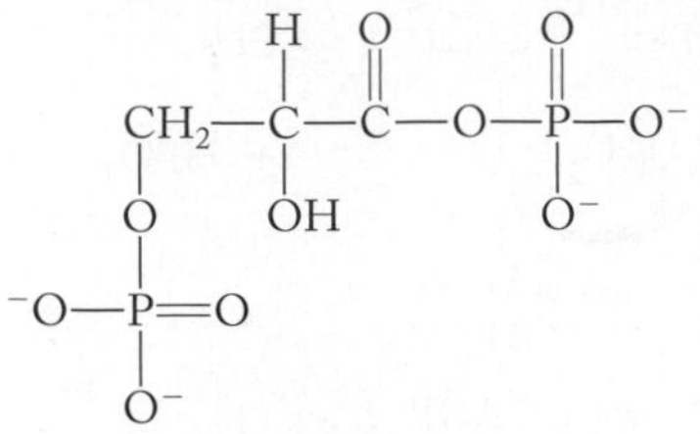

1,3-Diphosphoglycerate
(1,3-DPG^{4-})

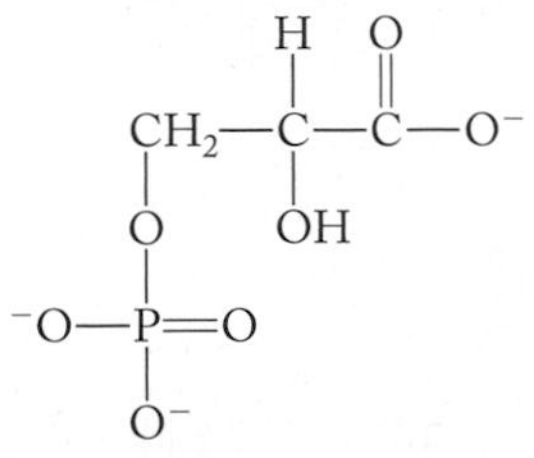

3-Phosphoglycerate
(3-PG^{3-})

FIGURE 12.23 Structures of 1,3-diphosphoglycerate and 3-phosphoglycerate.

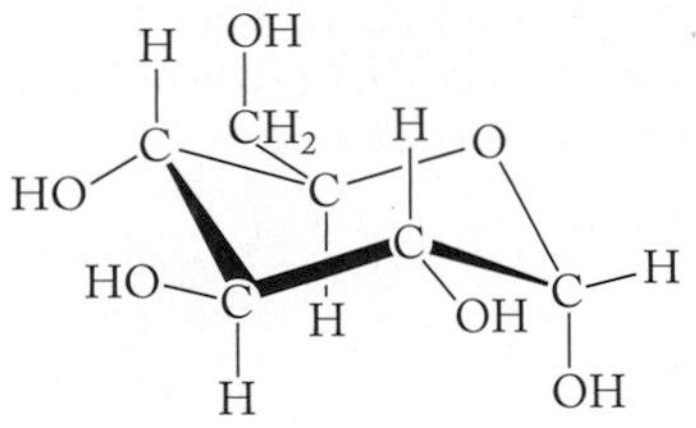

Glucose

Lactic acid

FIGURE 12.24 Structures of glucose and lactic acid.

SAMPLE EXERCISE 12.7 **Calculating ΔG°_{rxn} of Coupled Reactions** **LO7**

The body would rapidly run out of ATP if there were not some process for regenerating it from ADP, and that process is the hydrolysis of 1,3-diphosphoglycerate ions (1,3-DPG^{4-}) to 3-phosphoglycerate ions (3-PG^{3-}), whose structures are shown in Figure 12.23. The reaction can be written as

$$ADP^{3-} + 1,3\text{-DPG}^{4-} \rightarrow 3\text{-PG}^{3-} + ATP^{4-}$$

This hydrolysis is spontaneous. Calculate its ΔG° value from these values:

(1) $\quad 1,3\text{-DPG}^{4-}(aq) + H_2O(\ell) \rightarrow 3\text{-PG}^{3-}(aq) + HPO_4^{2-}(aq) + H^+(aq)$
$$\Delta G^\circ_{rxn} = -49.0 \text{ kJ}$$

(2) $\quad ADP^{3-}(aq) + HPO_4^{2-}(aq) + H^+(aq) \rightarrow ATP^{4-}(aq) + H_2O(\ell)$
$$\Delta G^\circ_{rxn} = 30.5 \text{ kJ}$$

COLLECT AND ORGANIZE We need to calculate ΔG°_{rxn} for a reaction that is the sum of two reactions. If the reactions in equations 1 and 2 add up to the overall reaction, then overall ΔG°_{rxn} is the sum of the ΔG°_{rxn} values for the individual reactions.

ANALYZE First, we add the reactions described by equations 1 and 2. Assuming that the overall reaction between ADP and 1,3-DPG is the sum of the reactions describing the hydrolysis of 1,3-DPG and the phosphorylation of ADP, we know that the sum of ΔG°_1 and ΔG°_2 will be less than zero because the overall reaction is spontaneous.

SOLVE Summing the reactions in equations 1 and 2 confirms that they equal the overall reaction:

(1) $\quad 1,3\text{-DPG}^{4-}(aq) + H_2O(\ell) \rightarrow 3\text{-PG}^{3-}(aq) + HPO_4^{2-}(aq) + H^+(aq)$

(2) $\quad ADP^{3-}(aq) + HPO_4^{2-}(aq) + H^+(aq) \rightarrow ATP^{4-}(aq) + H_2O(\ell)$
_____

$1,3\text{-DPG}^{4-}(aq) + \cancel{H_2O(\ell)} + ADP^{3-}(aq) + \cancel{HPO_4^{2-}(aq)} + \cancel{H^+(aq)} \rightarrow$
$\quad 3\text{-PG}^{3-}(aq) + \cancel{HPO_4^{2-}(aq)} + \cancel{H^+(aq)} + ATP^{4-}(aq) + \cancel{H_2O(\ell)}$

We sum the ΔG°_{rxn} values for steps 1 and 2 to determine ΔG°_{rxn} for the overall reaction:

$$\Delta G^\circ_{overall} = \Delta G^\circ_1 + \Delta G^\circ_2 = [(-49.0) + (30.5)] \text{ kJ} = -18.5 \text{ kJ}$$

THINK ABOUT IT The hydrolysis of 1,3-DPG provides more than sufficient energy for the conversion of ADP into ATP.

Practice Exercise The conversion of glucose into lactic acid (Figure 12.24) drives the phosphorylation of 2 moles of ADP to ATP:

$C_6H_{12}O_6(aq) + 2\ HPO_4^{2-}(aq) + 2\ ADP^{3-}(aq) + 2\ H^+(aq) \rightarrow$
Glucose

$\quad 2\ CH_3CH(OH)COOH(aq) + 2\ ATP^{4-}(aq) + 2\ H_2O(\ell)$
Lactic acid

$$\Delta G^\circ_{rxn} = -135 \text{ kJ/mol}$$

What is ΔG°_{rxn} for the conversion of glucose into lactic acid?

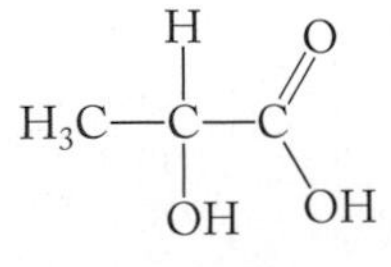

$$C_6H_{12}O_6(aq) \rightarrow 2\ CH_3CH(OH)COOH(aq)$$

The ATP produced from the breakdown of glucose (as, for example, via the first reaction in the previous Practice Exercise) is used to drive nonspontaneous reactions in cells. The metabolism of fats and proteins also relies on a series of chemical cycles, all of which involve coupled reactions.

In this chapter we addressed the reasons why some reactions and processes are spontaneous while others are not. We have learned that the free-energy change (ΔG) determines the spontaneity of a chemical reaction or process. Together, enthalpy and entropy changes for chemical reactions allow us to determine ΔG, which also represents the maximum amount of work that can be done by the energy associated with a change. For changes carried out in the real world, the maximum amount of work is always actually less than ΔG, because—to paraphrase the second law of thermodynamics once more—in the game of energy, not only can you not win, you can't even break even.

SAMPLE EXERCISE 12.8   **Integrating Concepts: Trouton's Rule**

In 1883–1884, while still an undergraduate student at Trinity College, Dublin, Ireland, Frederick Trouton (1863–1922) published two short papers describing what we now call Trouton's rule. Trouton's rule states that the ratio of the enthalpy of vaporization (ΔH_{vap}) of a liquid to its normal boiling point (in kelvin, K) is approximately constant:

$$\frac{\Delta H_{vap}}{T_b} \approx 88 \; \frac{J}{mol \cdot K}$$

This ratio is also known as the entropy of vaporization (ΔS_{vap}) and may be used to estimate ΔH_{vap} values of liquids whose boiling points are known.

  a. Suggest why $\Delta H_{vap}/T$ should be approximately constant.

  b. Calculate the values of ΔH_{vap} for the substances given in the table below (whose structures are shown in Figure 12.25) using Trouton's rule, and compare them to the experimentally determined values. Also calculate ΔS_{vap} for each substance by dividing the experimentally determined ΔH_{vap} values by the normal boiling points of the liquids. Identify those compounds whose actual ΔH_{vap} values deviate by more than 10% from those estimated using Trouton's rule.

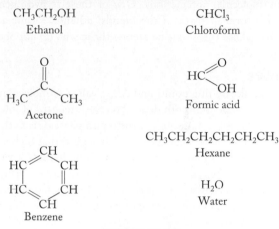

FIGURE 12.25

  c. Based on the results in part b and the molecular structures of the seven compounds, explain why some of them deviate from Trouton's rule.

  d. Predict which of the substances listed below (whose structures are shown in Figure 12.26), obey Trouton's rule: (i) chloromethane; (ii) toluene; (iii) methanol; (iv) diethyl ether; (v) acetic acid; (vi) methylamine.

  e. What is the sign of the change in standard free energy of vaporization (ΔG°_{vap}) of each of the seven liquids in the table at 298 K?

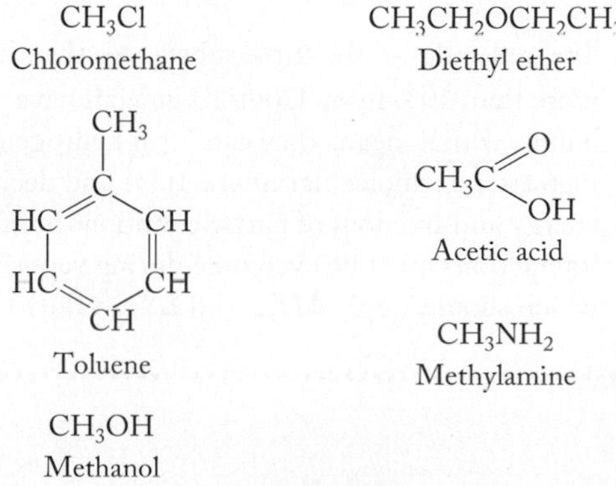

Substance	Boiling Point T_b (°C)	Calculated ΔH_{vap} (J/mol)	Experimentally Determined ΔH_{vap} (J/mol)	Actual Value of ΔS_{vap} [J/(mol · K)]
Ethanol	78.3		38,600	
Acetone	56.0		29,100	
Benzene	80.0		30,700	
Chloroform	61.1		29,200	
Formic acid	100.8		22,700	
Hexane	68.7		28,900	
Water	100.0		40,660	

FIGURE 12.26

COLLECT AND ORGANIZE We are given the boiling points of seven liquids. Trouton's rule relates the boiling points of liquids to their ΔH_{vap} values. In part a we are asked to explain why this relationship exists. In part b we use Trouton's rule to estimate the ΔH_{vap} values of seven liquids, calculate actual ΔS_{vap} values, and compare these results to their actual ΔH_{vap} values, explaining any deviations in part c. In part d we predict how well Trouton's rule will apply to six other liquids, and in part e we are asked to predict the sign of ΔG°_{vap} of the seven liquids at 298 K.

ANALYZE The molecular structures of compounds influence how they interact with each other, and the strengths of intermolecular interactions help define physical properties such as boiling points and heats of vaporization. Our task is to look for structural features that are associated with particularly strong intermolecular interactions. One of these is hydrogen bonds, which form in three of the liquids: ethanol, formic acid, and water. These three may be among those who do not obey Trouton's rule.

SOLVE

a. The boiling points and ΔH_{vap} values of molecular compounds both depend on the strengths of interactions between molecules in the liquid phase. Given this mutual dependence, it is only logical that higher boiling points would be associated with higher ΔH_{vap} values so that the ratio between the two is (nearly) constant.

b. The calculated values of ΔH_{vap} for ethanol, formic acid, and water all vary by more than 10% from the measured values.

Substance	Boiling Point T_b (°C)	Calculated ΔH_{vap} (J/mol)	Experimentally Determined ΔH_{vap} (J/mol)	Actual Value of ΔS_{vap} [J/(mol · K)]
Ethanol	78.3	30,900	38,600	110
Acetone	56.0	29,000	29,100	88
Benzene	80.0	31,100	30,700	87
Chloroform	61.1	29,400	29,200	87
Formic acid	100.8	32,900	22,700	61
Hexane	68.7	30,100	28,900	85
Water	100.0	32,800	40,660	109

c. The molecules of the three substances that deviate more than 10% from Trouton's rule all have —O—H bonds, which means they can form hydrogen bonds that increase intermolecular interactions and decrease the energy and freedom of particle motions (entropy). These interactions must be overcome during vaporization, which should imply ΔH_{vap} and ΔS_{vap} values that are

larger than predicted by Trouton's rule. Both ethanol and water fit this pattern, but formic acid does not. In fact, the calculated ΔS_{vap} value of formic acid is *lower* than 88 J/(mol · K). Something else must be happening to formic acid in the gas phase that decreases its entropy. There is evidence that formic acid forms hydrogen-bonded dimers in both the liquid phase *and* the gas phase:

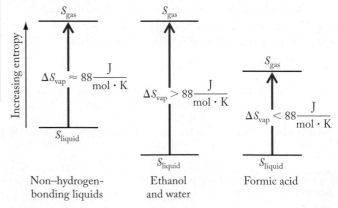

Their presence in both phases lowers the entropy of both, but the decrease in the gas phase is greater because the absolute entropy of the gas phase is much greater. Figure 12.27 illustrates the relative relationships between the entropies of the liquid and gas phases of these compounds.

d. The ΔH_{vap} values of (i) chloromethane, (ii) toluene, and (iv) diethyl ether that are calculated using Trouton's rule should be close to the experimental ΔH_{vap} values because these compounds do not form hydrogen bonds.

e. The boiling points of all seven liquids are above 25°C (298 K). Therefore, vaporization of their liquids under standard conditions and at $T = 298$ K is nonspontaneous, and the ΔG°_{vap} values of all seven should be greater than zero.

THINK ABOUT IT Trouton's rule was developed empirically, but our answer to part a provides an explanation of why it should be true. This exercise also shows that compounds experiencing intermolecular forces such as hydrogen bonds that add significantly to London dispersion forces in the liquid phase have ΔH_{vap} values that are significantly greater than those predicted by Trouton's rule.

FIGURE 12.27

SUMMARY ∙∙∙∎

Section 12.1 Spontaneous processes happen on their own without continuing intervention. Nonspontaneous processes, which are spontaneous processes in reverse, do not happen on their own. Spontaneous processes may be exothermic or endothermic and are often accompanied by an increase in the freedom of motion of the particles involved in the process. Spontaneous reactions are not necessarily rapid.

Section 12.2 Entropy (S) is a thermodynamic property that provides a measure of how dispersed the energy is in a system at a given temperature. According to the **second law of thermodynamics**, a spontaneous process is accompanied by an increase in the entropy of an isolated system. The entropy of a system increases as the number of probable arrangements of its particles, called **accessible microstates**, increases. According to the **third law of thermodynamics**, a perfect crystal of a pure substance has zero entropy at absolute zero. All substances have positive entropies at temperatures above absolute zero.

Section 12.3 Standard molar entropies (S°) are entropy values for substances in their standard states. The entropy of a system increases with increasing molecular complexity and with increasing temperature.

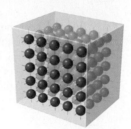

Section 12.4 Any process is spontaneous if it produces an increase in the entropy of the universe. A **reversible process** takes place in very small steps and very slowly so that the system can be restored to its initial state with no net flow of energy to or from its surroundings.

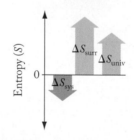

Section 12.5 The entropy change in a reaction under standard conditions can be calculated from the standard entropies of the products and reactants and their coefficients in the balanced chemical equation.

Section 12.6 Gibbs free energy (G) is defined as the energy available to do useful work for a process occurring at constant temperature and pressure. When a process results in a decrease in **free energy** of a system ($\Delta G < 0$), the process is spontaneous; when $\Delta G > 0$, the process is nonspontaneous. The change in free energy of a reaction under standard conditions can be calculated either from the **standard free energies of formation ($\Delta G_f°$)** of the products and reactants, or from the values of $\Delta H_{rxn}°$ and $\Delta S_{rxn}°$.

Section 12.7 The temperature range over which a process is spontaneous depends on the relative magnitudes of ΔH and ΔS.

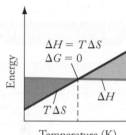

Section 12.8 Many important biochemical processes, including **glycolysis** and **phosphorylation**, are made possible by coupled spontaneous and nonspontaneous reactions. The free energy released in the spontaneous processes going on in the body is used to drive nonspontaneous processes.

PROBLEM-SOLVING SUMMARY ∙∙∙∎

TYPE OF PROBLEM	CONCEPTS AND EQUATIONS	SAMPLE EXERCISES
Predicting the sign of an entropy change	Look for fewer moles of gas as products than as reactants or precipitation of a solute from solution ($\Delta S < 0$ for both). For the reverse processes, $\Delta S > 0$.	12.1
Comparing standard molar entropy values	Substances composed of larger, less rigid molecules have more entropy. Among substances with similar molar masses, gases have more entropy than liquids, which have more than solids.	12.2
Calculating the entropy change of a chemical reaction	Use $$\Delta S_{rxn}° = \sum n_{products} S_{products}° - \sum n_{reactants} S_{reactants}° \qquad (12.6)$$ where $S_{products}°$ and $S_{reactants}°$ are the standard molar entropies, and $n_{products}$ and $n_{reactants}$ are the stoichiometric coefficients for the process.	12.3
Predicting reaction spontaneity under standard conditions	A reaction is spontaneous if $\Delta G_{rxn}° < 0$, where $$\Delta G_{rxn}° = \Delta H_{rxn}° - T\Delta S_{rxn}° \qquad (12.11)$$	12.4

TYPE OF PROBLEM	CONCEPTS AND EQUATIONS	SAMPLE EXERCISES
Calculating ΔG°_{rxn} using appropriate ΔG°_f values	Use $$\Delta G^\circ_{rxn} = \sum n_{products}\, \Delta G^\circ_{f,products} - \sum n_{reactants}\, \Delta G^\circ_{f,reactants} \qquad (12.12)$$ where $\Delta G^\circ_{f,products}$ and $\Delta G^\circ_{f,reactants}$ are the standard molar free energies of formation, and $n_{products}$ and $n_{reactants}$ are the stoichiometric coefficients for the process.	12.5
Relating reaction spontaneity to ΔH_{rxn} and ΔS_{rxn}	Use $$\Delta G_{rxn} = \Delta H_{rxn} - T\,\Delta S_{rxn}$$	12.6
Calculating ΔG°_{rxn} of coupled reactions	Free-energy changes of coupled reactions are additive.	12.7

VISUAL PROBLEMS ••• ■

(Answers to boldface end-of-chapter questions and problems are in the back of the book.)

12.1. Two tires shown in cross-section in Figure P12.1 are inflated at the same temperature to the same volume, though more air is used to inflate the tire on the right. In which tire is the gas under greater internal pressure and in which does the gas have greater entropy?

FIGURE P12.1

12.2. Two cubic containers (Figure P12.2) contain the same quantity of gas at the same temperature.
   a. Which cube contains the gas with more entropy?
   b. If the sample in cube b is left unchanged but the sample in cube a is cooled so that it condenses, which sample has the higher entropy?

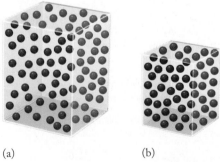

(a)   (b)

FIGURE P12.2

12.3. Figure P12.3 shows two connected bulbs that have just been filled with a mixture of two ideal gases: A (red spheres) and B (blue spheres). If the molar mass of A is twice that of B, will the atoms of A eventually fill the bottom bulb and the atoms of B fill the top bulb? Why or why not?

FIGURE P12.3

12.4. The box on the left in Figure P12.4 represents a mixture of two diatomic gases: A_2 (red spheres) and B_2 (blue spheres). As a result of the process depicted by the arrow, how do the entropies of A_2 and B_2 change?

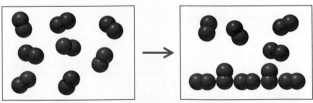

FIGURE P12.4

12.5. Is the process in Figure P12.4 more likely to be spontaneous at high temperature or low temperature, or is it unaffected by changing temperature?

12.6. Figure P12.6 shows the plots of ΔH and $T\,\Delta S$ for a phase change as a function of temperature.
  a. What is the status of the process at the point where the two lines intersect?
  b. Over what temperature range is the process spontaneous?

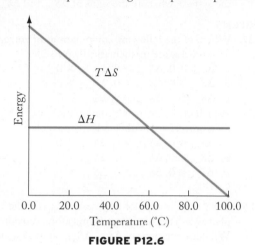

FIGURE P12.6

12.7. Of the six phase changes—melting, vaporization, condensation, freezing, sublimation, and deposition—which ones have thermodynamic profiles that fit the pattern in Figure P12.6?

12.8. Figure P12.8 presents the ΔG_f° values of several elements and compounds selected from Appendix 4. Which of the following conversions are spontaneous? (a) $C_6H_6(\ell)$ to $CO_2(g)$ and $H_2O(\ell)$; (b) $CO_2(g)$ to $C_2H_2(g)$; (c) $H_2(g)$ and $O_2(g)$ to $H_2O(\ell)$. Explain your reasoning.

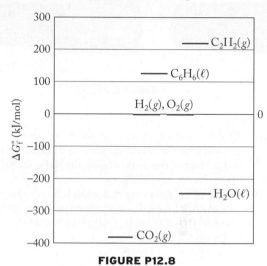

FIGURE P12.8

QUESTIONS AND PROBLEMS

Spontaneous Processes; Entropy

CONCEPT REVIEW

12.9. How is the entropy change that accompanies a reaction related to the entropy change that happens when the reaction runs in reverse?

12.10. Identify the following processes as spontaneous or nonspontaneous, and explain your choice.
  a. A photovoltaic cell in a solar panel produces electricity.
  b. Helium gas escapes from a latex party balloon.
  c. A sample of pitchblende (uranium ore) emits alpha particles.

12.11. You flip three coins, assigning the values +1 for heads and −1 for tails. Each outcome of the three flips constitutes a microstate. How many different microstates are possible from flipping the three coins? Which value or values for the sums in the microstates are most likely? *Hint*: The sequence HHT (+1 +1 −1) is one possible outcome, or microstate. Note, however, that this outcome differs from THH (−1 +1 +1), even though the two sequences sum to the same value.

12.12. Imagine you have four identical chairs to arrange on four steps leading up to a stage, one chair on each step. The chairs have numbers on their backs: 1, 2, 3, and 4. How many different microstates for the chairs are possible? (Notice that when viewed from the front, all the microstates look the same. Viewed from the back, you can identify the different microstates because you can distinguish the chairs by their numbers.)

PROBLEMS

12.13. Use the appropriate standard molar entropy value in Appendix 4 to calculate how many microstates are available to a single molecule of liquid H_2O at 298 K.

12.14. Use the appropriate standard molar entropy value in Appendix 4 to calculate how many microstates are available to a single molecule of N_2 at 298 K.

12.15. The three identical glass spheres in Figure P12.15 contain the same number of particles at the same temperature. Rank the containers in order of increasing number of microstates accessible to the particles inside them.

(a)　　　　　(b)　　　　　(c)

FIGURE P12.15

12.16. Figure P12.16(a) shows a cylinder within a cylinder that contains a population of gaseous molecules. The volume occupied by the molecules can be increased by pulling the inside cylinder out (Figure P12.16b), much like a telescope. The number of molecules within the cylinder remains the same during this operation. Compare the number of microstates available to the molecules in Figure P12.16(a) and (b).

(a)

(b)

FIGURE P12.16

12.17. Which of the following ionic solutes experiences the greatest increase in entropy when 0.0100 mol of it dissolves in 1.00 liter of water? (a) $CaCl_2$, (b) NaBr, (c) KCl, (d) $Cr(NO_3)_3$, (e) LiOH

12.18. Which of the following molecular solutes experience an increase in entropy when dissolved in water? (a) $CO_2(g)$, (b) $HF(g)$, (c) $CH_3OH(\ell)$, (d) $CH_3COOH(\ell)$, (e) $C_{12}H_{22}O_{11}(s)$

Absolute Entropy and Molecular Structure

CONCEPT REVIEW

12.19. Which component in each of the following pairs has the greater entropy?
a. 1 mole of $S_2(g)$ or 1 mole of $S_8(g)$
b. 1 mole of $S_2(g)$ or 1 mole of $S_8(s)$
c. 1 mole of $O_2(g)$ or 1 mole of $O_3(g)$
d. 1 gram of $O_2(g)$ or 1 gram of $O_3(g)$

12.20. **Digestion** During digestion, complex carbohydrates decompose into simple sugars. Do the carbohydrates experience an increase or decrease in entropy?

*12.21. Diamond and the fullerenes are two allotropes of carbon. On the basis of their different structures and properties, predict which has the higher standard molar entropy.

12.22. **Superfluids** The 1996 Nobel Prize in Physics was awarded to Douglas Osheroff, Robert Richardson, and David Lee for discovering *superfluidity* (apparently frictionless flow) in ^3He. When ^3He is cooled to 2.7 mK, the liquid settles into an *ordered* superfluid state. Predict the sign of the entropy change for the conversion of liquid ^3He into its superfluid state.

PROBLEMS

12.23. Rank the compounds in each of the following groups in order of increasing standard molar entropy ($S°$):
a. $CH_4(g)$, $CF_4(g)$, and $CCl_4(g)$
b. $CH_2O(g)$, $CH_3CHO(g)$, and $CH_3CH_2CHO(g)$
c. $HF(g)$, $H_2O(g)$, and $NH_3(g)$

12.24. Rank the compounds in each of the following groups in order of increasing standard molar entropy ($S°$):
a. $CH_4(g)$, $CH_3CH_3(g)$, and $CH_3CH_2CH_3(g)$
b. $CCl_4(\ell)$, $CHCl_3(\ell)$, and $CH_2Cl_2(\ell)$
c. $CO_2(\ell)$, $CO_2(g)$, and $CS_2(g)$

Applications of the Second Law

CONCEPT REVIEW

12.25. Ice cubes melt in a glass of lemonade, cooling the lemonade from 10.0°C to 0.0°C. If the ice cubes are the system, what are the signs of ΔS_{sys} and ΔS_{surr}?

12.26. Adding sidewalk deicer (calcium chloride) to water causes the temperature of the water to increase. If solid $CaCl_2$ is the system, what are the signs of ΔS_{sys} and ΔS_{surr}?

PROBLEMS

12.27. Which of the following combinations of entropy changes for a process are mathematically possible?
a. $\Delta S_{sys} > 0$, $\Delta S_{surr} > 0$, $\Delta S_{univ} > 0$
b. $\Delta S_{sys} > 0$, $\Delta S_{surr} < 0$, $\Delta S_{univ} > 0$
c. $\Delta S_{sys} > 0$, $\Delta S_{surr} > 0$, $\Delta S_{univ} < 0$

12.28. Which of the following combinations of entropy changes for a process are mathematically possible?
a. $\Delta S_{sys} < 0$, $\Delta S_{surr} > 0$, $\Delta S_{univ} > 0$
b. $\Delta S_{sys} < 0$, $\Delta S_{surr} < 0$, $\Delta S_{univ} > 0$
c. $\Delta S_{sys} < 0$, $\Delta S_{surr} > 0$, $\Delta S_{univ} < 0$

12.29. What are the signs of ΔS_{sys} and ΔS_{univ} for the photosynthesis of glucose from carbon dioxide and water?

12.30. What are the signs of ΔS_{sys}, ΔS_{surr}, and ΔS_{univ} for the complete combustion of propane in which the products include water vapor and carbon dioxide?

12.31. The nonspontaneous reaction A + B → C decreases the system entropy by 66.0 J/K. What is the minimum value of the entropy change of the surroundings?

12.32. The spontaneous reaction D → E + F increases the system entropy by 72.0 J/K. What is the maximum value of the entropy change of the surroundings?

Calculating Entropy Changes

CONCEPT REVIEW

12.33. Under standard conditions the products of a reaction have, overall, greater entropy than the reactants. What is the sign of $\Delta S°_{rxn}$?

12.34. Do decomposition reactions tend to have $\Delta S°_{rxn}$ values that are greater than zero or less than zero? Why?

12.35. Do precipitation reactions tend to have $\Delta S°_{rxn}$ values that are greater than zero or less than zero? Why?

*12.36. Suppose compound $A(s)$ decomposes to substances $B(\ell)$ and $C(\ell)$ at moderately high temperatures. How would running the decomposition reaction at even higher temperatures (above the melting point of A, but still below the boiling points of B and C) affect the value of $\Delta S°_{rxn}$?

PROBLEMS

12.37. **Smog** Use the standard molar entropies in Appendix 4 to calculate $\Delta S°$ values for each of the following atmospheric reactions that contribute to the formation of photochemical smog.
a. $N_2(g) + O_2(g) \rightarrow 2\,NO(g)$
b. $2\,NO(g) + O_2(g) \rightarrow 2\,NO_2(g)$
c. $NO(g) + \frac{1}{2}O_2(g) \rightarrow NO_2(g)$
d. $2\,NO_2(g) \rightarrow N_2O_4(g)$

12.38. Use the standard molar entropies in Appendix 4 to calculate the $\Delta S°$ value for each of the following reactions of sulfur compounds.
  a. $H_2S(g) + \frac{3}{2} O_2(g) \rightarrow H_2O(g) + SO_2(g)$
  b. $2 SO_2(g) + O_2(g) \rightarrow 2 SO_3(g)$
  c. $SO_3(g) + H_2O(\ell) \rightarrow H_2SO_4(aq)$
  d. $S(g) + O_2(g) \rightarrow SO_2(g)$

12.39. **Ozone Layer** The following reaction plays a key role in the destruction of ozone in the atmosphere:

$$Cl(g) + O_3(g) \rightarrow ClO(g) + O_2(g)$$

The standard entropy change ($\Delta S°_{rxn}$) is 19.9 J/(mol · K). Use the standard molar entropies ($S°$) in Appendix 4 to calculate the $S°$ value of $ClO(g)$.

12.40. Calculate the $\Delta S°$ value for the conversion of ozone to oxygen

$$2 O_3(g) \rightarrow 3 O_2(g)$$

in the absence of Cl atoms, and compare it with the $\Delta S°$ value in Problem 12.39.

Free Energy

CONCEPT REVIEW

12.41. What does the sign of ΔG tell you about the spontaneity of a process?

12.42. What does the sign of ΔG tell you about the rate of a reaction?

*12.43. Many 19th-century scientists believed that all exothermic reactions were spontaneous. Why did so many of them share this belief?

12.44. In which direction does a reaction proceed when: (a) $\Delta G_{rxn} < 0$; (b) $\Delta G_{rxn} = 0$; (c) $\Delta G_{rxn} > 0$?

12.45. What are the signs of ΔS, ΔH, and ΔG for the sublimation of dry ice (solid CO_2) at 25°C?

12.46. What are the signs of ΔS, ΔH, and ΔG for the formation of dew on a cool night?

12.47. Which of the following processes is/are spontaneous?
  a. A tornado forms.
  b. A broken cell phone fixes itself.
  c. You get an A in this course.
  d. Hot soup gets cold before it is served.

12.48. Which of the following processes is/are spontaneous?
  a. Wood burns in air.
  b. Water vapor condenses on the sides of a glass of ice tea.
  c. Salt dissolves in water.
  d. Photosynthesis.

PROBLEMS

12.49. Calculate the free-energy change for the dissolution in water of 1 mole of NaBr and 1 mole of NaI at 298 K from the values in the table below.

	$\Delta H°_{solution}$ (kJ/mol)	$\Delta S°_{solution}$ [J/(mol · K)]
NaBr	−1	57
NaI	−7	74

12.50. The values of $\Delta H°_{rxn}$ and $\Delta S°_{rxn}$ for the reaction

$$2 NO(g) + O_2(g) \rightarrow 2 NO_2(g)$$

are −12 kJ and −146 J/K.
  a. Use these values to calculate $\Delta G°_{rxn}$ at 298 K.
  b. Explain why the value of $\Delta S°_{rxn}$ is negative.

12.51. A mixture of $CO(g)$ and $H_2(g)$ is produced by passing steam over hot charcoal:

$$H_2O(g) + C(s) \rightarrow H_2(g) + CO(g)$$

Calculate the $\Delta G°_{rxn}$ value for the reaction from the appropriate $\Delta G°_f$ data in Appendix 4.

12.52. Use the appropriate $\Delta G°_f$ data in Appendix 4 to calculate $\Delta G°_{rxn}$ for the complete combustion of methanol:

$$2 CH_3OH(g) + 3 O_2(g) \rightarrow 2 CO_2(g) + 4 H_2O(g)$$

12.53. **Photochemical Smog** Use the appropriate $\Delta G°_f$ data in Appendix 4 to calculate $\Delta G°_{rxn}$ for the oxidation of NO to NO_2—a key reaction in the formation of photochemical smog:

$$NO(g) + \frac{1}{2} O_2(g) \rightarrow NO_2(g)$$

12.54. Use the free energies of formation from Appendix 4 to calculate the standard free-energy change for the decomposition of ammonia in the following reaction:

$$2 NH_3(g) \rightarrow N_2(g) + 3 H_2(g)$$

Is the reaction spontaneous under standard conditions?

12.55. **Acid Precipitation** Aerosols (fine droplets) of sulfuric acid form in the atmosphere as a result of the reaction below. Use the appropriate $\Delta G°_f$ data in Appendix 4 to calculate $\Delta G°_{rxn}$ for the combination reaction

$$SO_3(g) + H_2O(g) \rightarrow H_2SO_4(\ell)$$

12.56. One source of sulfuric acid aerosols in the atmosphere is combustion of high-sulfur fuels, which releases SO_2 gas that then is further oxidized to SO_3:

$$2 SO_2(g) + O_2(g) \rightarrow 2 SO_3(g)$$

Use the appropriate $\Delta G°_f$ data in Appendix 4 to calculate $\Delta G°_{rxn}$ for this combination reaction. Is it spontaneous under standard conditions?

Temperature and Spontaneity

CONCEPT REVIEW

12.57. Are exothermic reactions spontaneous only at low temperature? Explain your answer.

12.58. Are endothermic reactions never spontaneous at low temperature? Explain your answer.

PROBLEMS

12.59. What is the lowest temperature at which the following reaction (see Problem 12.51) is spontaneous?

$$H_2O(g) + C(s) \rightarrow H_2(g) + CO(g)$$

12.60. Above what temperature does nitrogen monoxide form from nitrogen and oxygen?

$$N_2(g) + O_2(g) \rightarrow 2\ NO(g)$$

Assume that the values of ΔH°_{rxn} and ΔS°_{rxn} do not change appreciably with temperature.

12.61. Use the data in Appendix 4 to calculate ΔH° and ΔS° for the vaporization of hydrogen peroxide:

$$H_2O_2(\ell) \rightarrow H_2O_2(g)$$

Assuming that the calculated values are independent of temperature, what is the boiling point of hydrogen peroxide at $P = 1.00$ atm?

12.62. **Volcanoes** Deposits of elemental sulfur are often seen near active volcanoes. Their presence there may be due to the following reaction of SO_2 with H_2S:

$$SO_2(g) + 2\ H_2S(g) \rightarrow \tfrac{3}{8} S_8(s) + 2\ H_2O(g)$$

Assuming the values of ΔH°_{rxn} and ΔS°_{rxn} do not change appreciably with temperature, over what temperature range is the reaction spontaneous?

12.63. Which of the following reactions is spontaneous: (i) only at low temperatures; (ii) only at high temperatures; (iii) at all temperatures?
a. $2\ NO(g) + O_2(g) \rightarrow 2\ NO_2(g)$
b. $2\ NH_3(g) + 2\ O_2(g) \rightarrow N_2O(g) + 3\ H_2O(g)$
c. $NH_4NO_3(s) \rightarrow 2\ H_2O(g) + N_2O(g)$

12.64. Which of the following reactions is spontaneous: (i) only at low temperatures; (ii) only at high temperatures; (iii) at all temperatures?
a. $2\ H_2S(g) + 3\ O_2(g) \rightarrow 2\ H_2O(g) + 2\ SO_2(g)$
b. $SO_2(g) + H_2O_2(\ell) \rightarrow H_2SO_4(\ell)$
c. $S(g) + O_2(g) \rightarrow SO_2(g)$

Driving the Human Engine: Coupled Reactions

CONCEPT REVIEW

12.65. Describe the ways in which two chemical reactions must complement each other so that the decrease in free energy of the spontaneous one can drive the nonspontaneous reaction.

12.66. How do you calculate the value of ΔG° for a reaction that is the result of coupling a spontaneous reaction ($\Delta G^{\circ}_{spon} < 0$) and a nonspontaneous reaction ($\Delta G^{\circ}_{nonspon} > 0$)?

12.67. Why is it important that at least some of the spontaneous steps in glycolysis convert ADP to ATP?

12.68. The second step in glycolysis converts glucose 6-phosphate into fructose 6-phosphate (Figure P12.68). Suggest a reason why ΔG° for this reaction is close to zero.

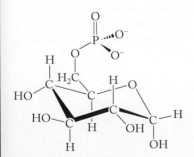
Glucose 6-phosphate

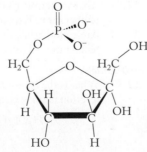
Fructose 6-phosphate

FIGURE P12.68

PROBLEMS

12.69. The methane in natural gas is an important starting material, or feedstock, for producing industrial chemicals, including H_2 gas.
a. Use the appropriate ΔG°_f value(s) from Appendix 4 to calculate ΔG°_{rxn} for the reaction known as *steam–methane reforming*:

$$CH_4(g) + H_2O(g) \rightarrow CO(g) + 3\ H_2(g)$$

b. To drive this nonspontaneous reaction the CO that is produced can be oxidized to CO_2 using more steam:

$$CO(g) + H_2O(g) \rightarrow CO_2(g) + H_2(g)$$

Use the appropriate ΔG°_f value(s) from Appendix 4 to calculate ΔG°_{rxn} for this reaction, which is known as the *water–gas shift reaction*.
c. Combine these two reactions and write the chemical equation of the overall reaction in which methane and steam combine to produce hydrogen gas and carbon dioxide.
d. Calculate the ΔG°_{rxn} value of the overall reaction. Is it spontaneous under standard conditions?

12.70. In addition to the reactions described in Problem 12.69, methane can, in theory, be used to produce hydrogen gas by a process in which it decomposes into elemental carbon and hydrogen:

$$(1)\quad CH_4(g) \rightarrow C(s) + 2\ H_2(g)$$

and the carbon produced in the first step is then oxidized to CO_2:

$$(2)\quad C(s) + O_2(g) \rightarrow CO_2(g)$$

a. Calculate the ΔG°_{rxn} values of the above two reactions.
b. Write a balanced chemical equation describing the overall reaction obtained by coupling the above two and calculate its ΔG°_{rxn} value. Is the coupled reaction spontaneous under standard conditions?

Additional Problems

12.71. Chlorofluorocarbons (CFCs) are no longer used as refrigerants because they catalyze the decomposition of stratospheric ozone. Trichlorofluoromethane (CCl_3F) boils at 23.8°C and its molar heat of vaporization is 24.8 kJ/mol. What is the molar entropy of evaporation of $CCl_3F(\ell)$?

12.72. **Methane-Producing Bacteria** Methanogenic bacteria convert liquid acetic acid (CH_3COOH) into $CO_2(g)$ and $CH_4(g)$.
a. Is this process endothermic or exothermic under standard conditions?
b. Is the reaction spontaneous under standard conditions?

*12.73. At what temperature is the free-energy change for the following reaction equal to zero?

$$NH_4Cl(s) \rightarrow NH_3(g) + HCl(g)$$

12.74. Consider the precipitation reactions described by the following net ionic equations:

$$Mg^{2+}(aq) + 2\ OH^-(aq) \rightarrow Mg(OH)_2(s)$$
$$Ag^+(aq) + Cl^-(aq) \rightarrow AgCl(s)$$

a. Predict the sign of ΔS°_{rxn} for the reactions.
b. Using the values for S° from Appendix 4, calculate ΔS° for these reactions.
c. Do your calculations support your prediction?

*12.75. Calculate the standard free-energy change of the following reaction. Is it spontaneous?

$$2 \, NO(g) + 2 \, H_2(g) \rightarrow N_2(g) + 2 \, H_2O(g)$$

12.76. Which of these processes result in a decrease in entropy of the system?
a. Diluting hydrochloric acid with water
b. Boiling water
c. $2 \, NO(g) + O_2(g) \rightarrow 2 \, NO_2(g)$
d. Making ice cubes in the freezer

*12.77. Show that hydrogen cyanide (HCN) is a gas at 25°C by estimating its normal boiling point from the following data:

	ΔH_f° (kJ/mol)	S° [J/(mol · K)]
HCN(ℓ)	108.9	113
HCN(g)	135.1	202

*12.78. Estimate the free-energy change of the following reaction at 225°C:

$$C_2H_4(g) + 3 \, O_2(g) \rightarrow 2 \, CO_2(g) + 2 \, H_2O(g)$$

12.79. Lightbulb Filaments Tungsten (W) is the favored metal for lightbulb filaments, in part because of its high melting point of 3422°C. The enthalpy of fusion of tungsten is 35.4 kJ/mol. What is its entropy of fusion?

12.80. Making Methanol The element hydrogen (H_2) is not abundant in nature, but it is a useful reagent in, for example, the potential synthesis of the liquid fuel methanol from gaseous carbon monoxide:

$$2 \, H_2(g) + CO(g) \rightarrow CH_3OH(\ell)$$

Under what temperature conditions is this reaction spontaneous?

12.81. Two allotropes (A and B) of sulfur interconvert at 369 K and 1 atm pressure:

$$S_8(s, A) \rightarrow S_8(s, B)$$

The enthalpy change in this transition is 297 J/mol. What is the entropy change?

*12.82. Over what temperature range is the reduction of tungsten(VI) oxide by hydrogen to give metallic tungsten and water spontaneous? The standard heat of formation of $WO_3(s)$ is −843 kJ/mol, and its standard molar entropy is 76 J/(mol · K).

*12.83. Lime Enormous amounts of lime (CaO) are used in steel industry blast furnaces to remove impurities from iron. Lime is made by heating limestone and other solid forms of $CaCO_3(s)$. Why is the standard molar entropy of $CaCO_3(s)$ higher than that of $CaO(s)$? At what temperature is the pressure of $CO_2(g)$ over $CaCO_3(s)$ equal to 1.0 atm?

	ΔH_f° (kJ/mol)	S° [J/(mol · K)]
CaCO₃(s)	−1207	93
CaO(s)	−636	40
CO₂(g)	−394	214

*12.84. Copper forms two oxides, Cu_2O and CuO.
a. Name these oxides.
b. Predict over what temperature range this reaction is spontaneous using the following thermodynamic data:

$$Cu_2O(s) \rightarrow CuO(s) + Cu(s)$$

	ΔH_f° (kJ/mol)	S° [J/(mol · K)]
Cu₂O(s)	−170.7	92.4
CuO(s)	−156.1	42.6

c. Why is the standard molar entropy of $Cu_2O(s)$ larger than that of $CuO(s)$?

*12.85. Melting Organic Compounds When dicarboxylic acids (compounds with two −COOH groups in their structures) melt, they frequently decompose to produce 1 mole of CO_2 gas for every 1 mole of dicarboxylic acid melted (shown in Figure P12.85).
a. What are the signs of ΔH and ΔS for the process as written?
b. Do you think the dicarboxylic acid will re-form when the melted material cools? Why or why not?

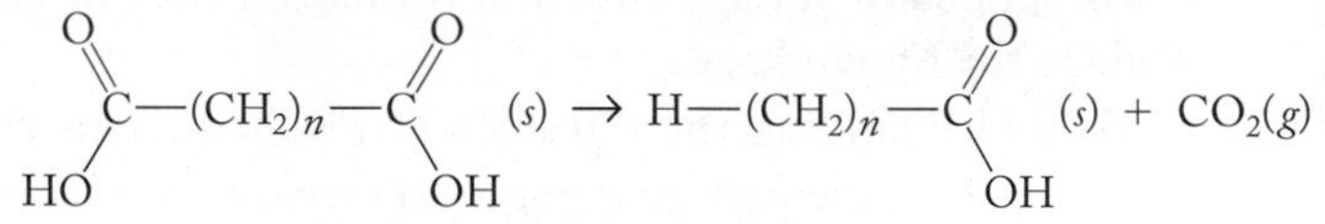

FIGURE P12.85

*12.86. Melting DNA When a solution of DNA in water is heated, the DNA double helix separates into two single strands.
a. What is the sign of ΔS for the separation process?
b. The DNA double helix re-forms as the system cools. What is the sign of ΔS for the process by which two single strands re-form the double helix?
c. The melting point of DNA is defined as the temperature at which $\Delta G = 0$. At that temperature, the melting reaction produces two single strands as fast as two single strands recombine to form the double helix. Write an equation that defines the melting temperature (T) of DNA in terms of ΔH and ΔS.

13

Chemical Kinetics

Clearing the Air

Reaction Rates and Smog Formation

Take a look at the photographs at the beginning of this chapter. Note the yellow-brown haze that colors the left side of the picture and obscures the mountains in the distance. The murky color is the result of *smog*, so called because it combines the effects of smoke and fog, reducing visibility and causing respiratory difficulties. *Photochemical smog* forms when vehicle emissions and other air pollutants collect over urban centers. The interaction of sunlight with these pollutants initiates a host of chemical reactions that produce the brown haze.

By understanding the rates of chemical reactions that produce pollutants in car engines, scientists and engineers have been able to design internal combustion engines that burn fuel more efficiently than ever before. To further reduce pollutant concentrations, engine exhaust passes through a device called a catalytic converter. Since they were developed in the early 1970s, catalytic converters have removed billions of tons of pollution from urban air in the United States and around the world.

Designing catalytic converters required extensive study of the by-products of the combustion reactions in automobile engines and the rates of the reactions that convert these substances into less noxious ones. The target compounds include hydrocarbons, carbon monoxide, and various oxides of nitrogen. Engineers and scientists designed catalytic converters that sped up the oxidation of hydrocarbons and CO to CO_2 and H_2O and the reduction of nitrogen oxides to N_2 and O_2. Their goal was to make these reactions happen during the fraction of a second that molecules spend in contact with catalytic surfaces. The devices they have designed do not remove the target pollutants completely, but they almost do.

Removing pollutants from automobile exhaust is just one example of how altering the rates of chemical reactions can significantly impact the composition of the atmosphere—and human health. Another takes place high in the stratosphere, where molecules of pollutants called chlorofluorocarbons encounter intense ultraviolet (UV) radiation from the sun and break up into their component atoms. One of them, chlorine, enhances the rate at which

Urban Smog A light snowfall and change in wind direction dramatically affect visibility and air quality over Salt Lake City, Utah, clearing the photochemical smog from the air. These photos were taken on January 28, 2007, at 4:15 PM (left) and January 31, 2007, at 3:53 PM (right).

LEARNING OUTCOMES

LO1 Use the stoichiometry of a reaction to relate the rates at which the concentrations of its reactants and products change as the reaction proceeds
Sample Exercise 13.1

LO2 Determine average and instantaneous reaction rates
Sample Exercise 13.2

LO3 Determine the rate law and overall order for a chemical reaction using initial rate data
Sample Exercise 13.3

LO4 Use integrated rate laws to identify the orders of reactions and determine their rate constants
Sample Exercises 13.4, 13.6, 13.8

LO5 Calculate the half-lives of chemical reactions
Sample Exercises 13.5, 13.7

LO6 Explain the effect of temperature on the rates of reactions and calculate their activation energy values
Sample Exercise 13.9

LO7 Use rate laws to assess the validity of a reaction mechanism
Sample Exercises 13.10, 13.11

LO8 Identify catalysts and describe their impact on reaction rates and mechanisms
Sample Exercise 13.12

photochemical smog a mixture of gases formed in the lower atmosphere when sunlight interacts with compounds produced in internal combustion engines and other pollutants.

ozone, O_3, decomposes into O_2. This has damaged the layer of stratospheric ozone that covers the Earth and protects its inhabitants from the full impact of the sun's UV rays.

As we begin this chapter on rates of reactions, consider the following questions: How are the rates of smog formation and other chemical reactions measured? How does knowing these rates allow us to better understand how chemical reactions occur on a molecular and atomic level? How can we manipulate the rates of chemical reactions to improve the quality of our environment and make our lives better?

13.1 Cars, Trucks, and Air Quality

The left half of the photograph at the beginning of this chapter provides a view of an all too common air quality problem: **photochemical smog**. Photochemical smog is created by the interaction of sunlight with nitrogen oxides from vehicle exhaust and other pollutants from anthropogenic and natural sources.

A key first step in smog formation takes place inside internal combustion engines, where N_2 and O_2 combine to form nitrogen monoxide:

$$N_2(g) + O_2(g) \rightarrow 2\,NO(g) \qquad \Delta H^\circ_{rxn} = 180.6 \text{ kJ} \qquad (13.1)$$

CONCEPT TEST

The reaction shown in Equation 13.1 is endothermic ($\Delta H^\circ_{rxn} > 0$), but it is spontaneous ($\Delta G_{rxn} < 0$) at very high temperatures.

a. What is the sign of ΔS°_{rxn}?
b. Why is the sign of ΔS°_{rxn} difficult to predict based only on the information in Equation 13.1?

(Answers to Concept Tests are in the back of the book.)

Once NO enters the atmosphere, it reacts with more oxygen, producing brown nitrogen dioxide gas (Figure 13.1):

$$2\,NO(g) + O_2(g) \rightarrow 2\,NO_2(g) \qquad \Delta H^\circ_{rxn} = -114.2 \text{ kJ} \qquad (13.2)$$

Sunlight provides sufficient energy to break the bonds in NO_2, forming NO and very reactive oxygen atoms:

$$NO_2(g) \xrightarrow{\text{sunlight}} NO(g) + O(g) \qquad (13.3)$$

This photochemically generated atomic oxygen combines with molecular oxygen, producing ozone,

$$O_2(g) + O(g) \rightarrow O_3(g) \qquad (13.4)$$

and with water vapor to produce hydroxyl radicals:

$$O(g) + H_2O(g) \rightarrow 2\,OH(g) \qquad (13.5)$$

In the atmosphere, O_3 and OH radicals react with volatile organic compounds (VOCs) to form a variety of noxious compounds. In one such reaction, acetaldehyde (a widely used industrial VOC) is transformed into peroxyacetyl nitrate, which is highly irritating to one's eyes, nose, and throat and can cause respiratory distress:

FIGURE 13.1 The orange-brown color of the air in this photo of Denver, Colorado, is due mostly to the presence of NO_2, one of many noxious compounds in photochemical smog.

CONNECTION In Chapter 4 we learned that odd-electron species such as OH belong to a highly reactive group of substances called free radicals.

$$N_2(g) + 3 O_2(g) + OH(g) + CH_3CHO(g) \rightarrow$$

Acetaldehyde

$$CH_3C(O)O_2NO_2(g) + H_2O(g) + NO_2(g) \qquad (13.6)$$

Peroxyacetyl nitrate

Look carefully at the above chemical equations, and note how many of the substances in photochemical smog are produced in some of the reactions and consumed in others. The rates of these and other atmospheric reactions influence when smog events happen, how intense they are, and how long they persist. Figure 13.2 shows that the maximum NO concentration occurs during the morning rush hour. Later in the morning, the concentration of NO_2 reaches a maximum. This sequence makes sense because NO is a precursor of NO_2. The highest ozone concentrations are reached in the middle of the afternoon, when the reactions shown in Equations 13.2 and 13.3 are in full swing, producing a supply of free O atoms for the formation of ozone (Equation 13.4), hydroxyl radicals (Equation 13.5), and peroxyacetyl nitrate (Equation 13.6).

Ozone formed in the reaction in Equation 13.4 also reacts with NO to form NO_2 and O_2:

$$O_3(g) + NO(g) \rightarrow O_2(g) + NO_2(g) \qquad (13.7)$$

However, NO_2 concentrations drop in the afternoon on smoggy days because NO_2, ozone, and hydrocarbons react to form an array of other (mostly unpleasant) compounds.

The catalytic converter, now a standard feature on most automobiles, helps combat the production of photochemical smog by removing NO and unburned or partially oxidized hydrocarbons from exhaust gases. Knowledge of **chemical kinetics**, the study of the rates at which reactant and product concentrations change during a chemical reaction, has provided the understanding required to develop catalytic converters and other devices aimed at improving our air quality. At least 75% of the NO produced by vehicles equipped with catalytic converters is converted back to N_2 and O_2 before being emitted into the air.

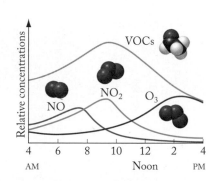

FIGURE 13.2 In photochemical smog NO from engine exhaust builds up in the early morning and then decreases as the NO reacts with atmospheric O_2, forming NO_2, the concentration of which is highest in late morning. Photodecomposition of NO_2 leads to the formation of high levels of O_3 in the afternoon.

13.2 Reaction Rates

The concentration of NO in the exhaust from an automobile engine depends on how rapidly the reaction in Equation 13.1 proceeds. We can express **reaction rate** as the change in the concentration of a product, such as NO, or a reactant, such as N_2 or O_2, that occurs over some interval of time. The rate of change in the concentration of a *reactant* (Δ[reactant]/Δt) has a *negative* value because reactant concentrations *decrease* as a reaction proceeds. However, the measured rate of any reaction is defined as a positive quantity because it describes the rate at which

chemical kinetics the study of the rates of change of concentrations of substances involved in chemical reactions.

reaction rate how rapidly a reaction occurs; it is related to rates of change in the concentrations of reactants and products over time.

▶❙❙ **CHEMTOUR** Reaction Rate

reactants form products, so a minus sign is used with $\Delta[\text{reactant}]/\Delta t$ values to obtain an overall positive value for reaction rate. Thus, the rate at which N_2 and O_2 combine to form NO can be expressed this way:

$$\text{Reaction rate} = (-)\ \text{rate of } N_2 \text{ consumption} = -\frac{\Delta[N_2]}{\Delta t} \qquad (13.8)$$

We use brackets around N_2 to represent its concentration in moles per liter—a practice we will follow in the remaining chapters of this book. Reaction rates are usually expressed in units of concentration per time, as in the fraction on the right in Equation 13.8. A commonly encountered example is molarity per second (M/s).

Nitrogen and oxygen gas are consumed at the same rate in the reaction in Equation 13.1 because their coefficients are both 1. However, 2 moles of NO form for every 1 mole of N_2 or O_2 consumed. Therefore, the rate of increase in the concentration of NO is *twice* the rate of decrease in the concentrations of N_2 and O_2. Let's express these relative rates of change using an equation:

$$\frac{\Delta[NO]}{\Delta t} = -2\frac{\Delta[N_2]}{\Delta t} = -2\frac{\Delta[O_2]}{\Delta t} \qquad (13.9)$$

The negative signs are needed to convert the rates at which $[N_2]$ and $[O_2]$ are *decreasing* into positive values. When these values are multiplied by 2, they equal the *increasing* rate of change in $[NO]$. Now let's divide all of the terms in Equation 13.9 by 2 and rearrange them so that they match the sequence of reactants and products in Equation 13.1:

$$-\frac{\Delta[N_2]}{\Delta t} = -\frac{\Delta[O_2]}{\Delta t} = \frac{1}{2}\frac{\Delta[NO]}{\Delta t} \qquad (13.10)$$

Equation 13.10 means that the concentrations of N_2 and O_2 decrease at half the rate at which the concentration of NO increases. Note how the coefficient of 2 in front of NO in the chemical equation has become its reciprocal, $\frac{1}{2}$, in Equation 13.10. This pattern holds for all reactions: the coefficients in an equation expressing the relative rates of change of reactants and products are the *reciprocals* of their coefficients in a balanced chemical equation describing the reaction. To give all of the terms positive values, the rates representing decreasing reactant concentrations receive minus signs. Sample Exercise 13.1 provides practice in applying this approach to the changing concentrations of reactants and product in an important industrial process: the synthesis of ammonia.

SAMPLE EXERCISE 13.1   **Predicting Relative Rates of Concentration Change in a Chemical Reaction**   **LO1**

Each year about 130 million metric tons (1.3×10^{11} kg) of ammonia is produced worldwide, mostly for use as fertilizer in agriculture. The process for making ammonia is based on the reaction

$$N_2(g) + 3\,H_2(g) \rightarrow 2\,NH_3(g)$$

How is the rate of formation of NH_3 related to the rates of consumption of N_2 and H_2?

COLLECT AND ORGANIZE The rate of formation of NH_3 is related to the rates of consumption of N_2 and H_2 by the balanced chemical equation. Rates are expressed as $-\Delta[\text{reactant}]/\Delta t$ or $\Delta[\text{product}]/\Delta t$.

ANALYZE The coefficients in an equation relating the rates of change in the concentrations of N_2, H_2, and NH_3 are the reciprocals of their coefficients (1, 3, and 2, respectively) in the balanced chemical equation. Nitrogen and hydrogen are consumed in the reaction, so their rate terms are preceded by minus signs.

SOLVE Inserting the reciprocals of 1, 3, and 2 in front of the rates of change in the concentrations of N_2, H_2, and NH_3, and placing minus signs before the two reactant terms, give us the relative rates:

$$-\frac{\Delta[N_2]}{\Delta t} = -\frac{1}{3}\frac{\Delta[H_2]}{\Delta t} = \frac{1}{2}\frac{\Delta[NH_3]}{\Delta t}$$

THINK ABOUT IT The balanced equation indicates that 2 moles of NH_3 are formed for every 1 mole of N_2 consumed. That means the rate of consumption of N_2 is half the rate of formation of ammonia. Similarly, we know that 3 moles of H_2 are consumed for every 1 mole of N_2 consumed. Therefore, the rate at which N_2 is consumed is one-third the rate at which H_2 is consumed.

Practice Exercise In the oxidation of carbon monoxide to carbon dioxide,

$$2\,CO(g) + O_2(g) \rightarrow 2\,CO_2(g)$$

which reactant is consumed at the higher rate? How is the rate of change in the concentration of CO_2 related to the rate of change in the concentration of O_2?

(Answers to Practice Exercises are in the back of the book.)

Reaction Rate Values

Up to this point our focus has been on the relative rates at which reactants are consumed and products are formed in chemical reactions. We have seen how these rates differ depending on the values of the coefficients in balanced chemical equations, that is, depending on reaction stoichiometry. The questions to answer now are these:

> ➤ How do we know the actual rates at which concentrations change in a chemical reaction?

> ➤ Which of these rates do we use to express the rate of the reaction?

The answer to the first question is that rates of change are determined experimentally. Suppose, for example, that analysis of the gas mixture in the ammonia synthesis in Sample Exercise 13.1 reveals that, over a 10-second time interval, the concentration of ammonia increases from 0.133 M to 0.605 M. These data mean that the average rate of change in ammonia concentration over the interval is

$$\frac{\Delta[NH_3]}{\Delta t} = \frac{(0.605 - 0.133)\,M}{10\text{ s}} = 0.0472\,M\text{/s}$$

To calculate the rate of change in the concentration of N_2, we use the equation we derived in the exercise:

$$-\frac{\Delta[N_2]}{\Delta t} = \frac{1}{2}\frac{\Delta[NH_3]}{\Delta t} = \frac{1}{2}(0.0472\,M\text{/s})$$

$$\frac{\Delta[N_2]}{\Delta t} = -0.0236\,M\text{/s}$$

Which of these two values should we use to express the rate of the reaction? Both values work as long as we have referenced the compound in the chemical equation whose concentration was monitored as a function of time. However, it is conventional to use the rate of change of the reactant or product with a coefficient of "1" in the chemical equation—in this case, N_2—as the basis for expressing the rate of the overall reaction. Therefore, the rate of the reaction (which is always a positive value) is

$$\text{Rate} = -\frac{\Delta[N_2]}{\Delta t} = 0.0236 \; M/\text{s}$$

Average and Instantaneous Rates of Formation of NO

Suppose we run an experiment to determine the rate of formation of NO in an automobile engine. In the laboratory, we use a reaction vessel as hot as the combustion chambers in the engine and we monitor the concentrations of reactants and products over time, obtaining the data in Table 13.1, which are plotted in Figure 13.3. These data can be used to calculate a reaction rate based on the change in the concentration of any product or reactant over a particular time interval, as we did in the preceding paragraphs for the ammonia reaction. Using Table 13.1, let's calculate the average rate of formation of NO between 5.0 and 10.0 μs:

$$\frac{\Delta[NO]}{\Delta t} = \frac{[NO]_{10.0\,\mu s} - [NO]_{5.0\,\mu s}}{t_{10.0\,\mu s} - t_{5.0\,\mu s}} = \frac{(14.8 - 7.8)\;\mu M}{(10.0 - 5.0)\;\mu s}$$

$$= 1.4 \; M/\text{s}$$

During the same interval, the average rate of change in the concentration of N_2 (and O_2) is

$$\frac{\Delta[N_2]}{\Delta t} = \frac{[N_2]_{10.0\,\mu s} - [N_2]_{5.0\,\mu s}}{t_{10.0\,\mu s} - t_{5.0\,\mu s}} = \frac{(9.6 - 13.1)\;\mu M}{(10.0 - 5.0)\;\mu s}$$

$$= -0.70 \; M/\text{s}$$

These results give us two values on which to base the average rate of the reaction. In this example we select the rate of change in $[N_2]$ or $[O_2]$ because they each have a coefficient of 1 in the balanced chemical equation. Therefore, the average rate of this reaction between 5.0 and 10.0 μs is

$$\text{Rate} = -\frac{\Delta[N_2]}{\Delta t} = -(-0.70 \; M/\text{s}) = 0.70 \; M/\text{s}$$

The curvature of the lines in Figure 13.3 tells us that this average rate value applies only to that particular 5.0 μs time interval. Any other time interval has a different average rate. For instance, from $t = 25.0$ μs to $t = 30.0$ μs, the rate of the reaction is

$$-\frac{\Delta[N_2]}{\Delta t} = -\frac{[N_2]_{30.0\,\mu s} - [N_2]_{25.0\,\mu s}}{t_{30.0\,\mu s} - t_{25.0\,\mu s}}$$

$$= -\frac{(3.6 - 4.5)\;\mu M}{(30.0 - 25.0)\;\mu s} = 0.18 \; M/\text{s}$$

We can also determine the *instantaneous* rate of a reaction, that is, the rate of the reaction at a particular time after it began. The difference between average and instantaneous reaction rates is analogous to the difference between the average and instantaneous speeds of a runner. If a competitor in a marathon runs from mile 10 to mile 20 in 1 hr, her average speed over that distance was 10 mi/hr. At a given instant during the run, however, her instantaneous speed could have been 12 mi/hr while going downhill, and 8 mi/hr while going uphill.

TABLE 13.1 **Changing Concentrations of Reactants and Products for the Reaction $N_2(g) + O_2(g) \rightarrow 2\,NO(g)$**

Time (μs)	$[N_2]$, $[O_2]$ (μM)	[NO] (μM)
0	17.0	0.0
5.0	13.1	7.8
10.0	9.6	14.8
15.0	7.6	18.6
20.0	5.8	22.2
25.0	4.5	24.8
30.0	3.6	26.7

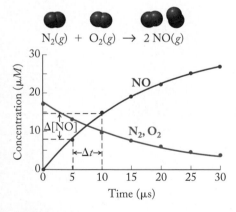

FIGURE 13.3 Concentrations of N_2, O_2, and NO over 30.0 μs for the reaction $N_2(g) + O_2(g) \rightarrow 2\,NO(g)$, plotted from data in Table 13.1.

Let's again consider the conversion of NO in engine exhaust into NO_2 (Equation 13.2), the gas responsible for much of the brown color in photochemical smog. The conversion takes place when NO reacts with oxygen in the air:

$$2\,NO(g) + O_2(g) \rightarrow 2\,NO_2(g)$$

The rate of this reaction is described by the data in Table 13.2, specifically the rate of consumption of O_2, the reactant with a coefficient of 1. The data are plotted in Figure 13.4, where we calculate the instantaneous rate by drawing a tangent to a particular point on the curve.

TABLE 13.2 Changing Concentrations of Reactants and Products for the Reaction $2\,NO(g) + O_2(g) \rightarrow 2\,NO_2(g)$ at 25°C

Time (s)	[NO] (M)	[O₂] (M)	[NO₂] (M)
0	0.0100	0.0100	0.0000
285	0.0090	0.0095	0.0010
660	0.0080	0.0090	0.0020
1175	0.0070	0.0085	0.0030
1895	0.0060	0.0080	0.0040
2975	0.0050	0.0075	0.0050
4700	0.0040	0.0070	0.0060
7800	0.0030	0.0065	0.0070

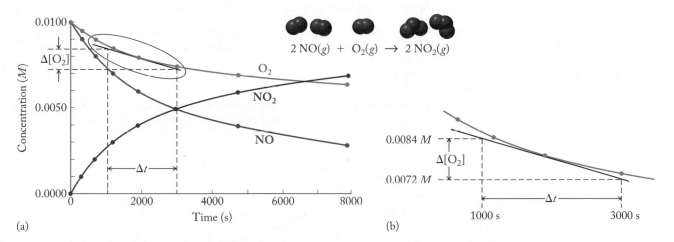

FIGURE 13.4 (a) The instantaneous rate of change in $[O_2]$ in the reaction $2\,NO(g) + O_2(g) \rightarrow 2\,NO_2(g)$ is equal to the slope of a tangent to the curve of $[O_2]$ versus time. (b) An expanded view of the instantaneous rate of change in $[O_2]$ at $t = 2000$ s.

CONCEPT TEST •

Which of the following statements is true about the instantaneous rate for the chemical reaction $A \rightarrow B$ as the reaction progresses?

a. $-\Delta[A]/\Delta t$ increases, $\Delta[B]/\Delta t$ decreases
b. $-\Delta[A]/\Delta t$ decreases, $\Delta[B]/\Delta t$ increases
c. $-\Delta[A]/\Delta t$ and $\Delta[B]/\Delta t$ both increase
d. $-\Delta[A]/\Delta t$ and $\Delta[B]/\Delta t$ both decrease

Any instantaneous reaction rate can be determined from a graph of the concentration-versus-time data for the participants in the reaction. If, for the NO_2 reaction, we wish to determine the instantaneous rate of change of $[O_2]$ at $t = 2000$ s, we draw a tangent to the point on the green curve in Figure 13.4 corresponding to $t = 2000$ s and then select two convenient points—for example, $t = 1000$ s and $t = 3000$ s—along that tangent. From the difference in the y-axis coordinates for $[O_2]_{3000\ s}$ and $[O_2]_{1000\ s}$ at these two points and the difference in the x-axis coordinates, 3000 s and 1000 s, we calculate the slope of the line, which is a measure of the instantaneous rate of change of $[O_2]$, the negative value of which is the instantaneous rate of the reaction at $t = 2000$ s:

$$\text{Slope} = \frac{\Delta[O_2]}{\Delta t} = \frac{(0.0072 - 0.0084)\ M}{(3000 - 1000)\ s} = -6.0 \times 10^{-7}\ M\,s^{-1}$$

$$\text{Rate} = -\frac{\Delta[O_2]}{\Delta t} = -(-6.0 \times 10^{-7}\ M\,s^{-1}) = 6.0 \times 10^{-7}\ M\,s^{-1}$$

Note that we substitute the concentrations corresponding to time points along the tangent line, not along the curve drawn from the data points.[1]

SAMPLE EXERCISE 13.2 **Determining an Instantaneous Reaction Rate** **LO2**

a. What is the instantaneous rate of change of $[NO]$ at $t = 2000$ s in the experiment that produced the data in Table 13.2?
b. What is the rate of the reaction based on your result in part a?

COLLECT AND ORGANIZE We are asked to determine the instantaneous rate of change of $[NO]$ and the corresponding reaction rate at $t = 2000$ s. The coefficient of NO is 2 in the balanced chemical equation: $2\ NO(g) + O_2(g) \rightarrow 2\ NO_2(g)$.

ANALYZE The instantaneous rate of change in $[NO]$ at the stated time can be determined from a graph of the data. The corresponding reaction rate will be $(-1/2)$ this value.

SOLVE
a. First we plot $[NO]$ versus time and draw a tangent to the curve at the point $t = 2000$ s (Figure 13.5). We then choose two points along the tangent,

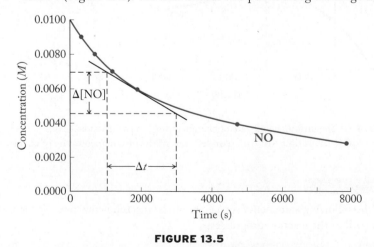

FIGURE 13.5

---

[1] If you have studied calculus, you may realize that the average rate of a reaction approaches the instantaneous rate as Δt approaches zero. Using calculus, we would say $\lim_{\Delta t \to 0} \Delta[O_2]/\Delta t = d[O_2]/dt$. The slope of a tangent to a curve is the derivative of the curve at a given point and can be calculated using many scientific calculators.

$t = 1000$ and 3000 s, and we determine the concentrations corresponding to those times along the vertical axis. Using those values, we calculate the slope of the line:

$$\frac{\Delta[\text{NO}]}{\Delta t} = \frac{(0.0046 - 0.0070)\ M}{(3000 - 1000)\ \text{s}} = -1.2 \times 10^{-6}\ M/\text{s}$$

b. The corresponding reaction rate is

$$\text{Rate} = -\frac{1}{2}\frac{\Delta[\text{NO}]}{\Delta t} = -\frac{1}{2}(-1.2 \times 10^{-6}\ M/\text{s}) = 6.0 \times 10^{-7}\ M/\text{s}$$

THINK ABOUT IT The sign of $\Delta[\text{NO}]/\Delta t$ is negative because NO is a reactant whose concentration decreases with time. However, the rates of chemical reactions have positive values. Therefore, we need a minus sign in front of the $\Delta[\text{NO}]/\Delta t$ term in the solution to part b. Note that the instantaneous reaction rate calculated in this Sample Exercise is the same as the rate we calculated earlier based on the rate of change of $[\text{O}_2]$.

Practice Exercise What is the instantaneous rate of change in $[\text{NO}_2]$ at $t = 2000$ s in the experiment that produced the data in Table 13.2? ⚙

13.3 Effect of Concentration on Reaction Rate

Figure 13.6 shows a typical result for a reactant concentration plotted as a function of time. Tangents are drawn to the line at three points: (a) at the instant the reaction begins, (b) at an instant when the reaction is about halfway to completion, and (c) when the reaction is nearly over. Point (a) defines the *initial* rate of the reaction, which is the rate that occurs at the instant the reactants are mixed at $t = 0$.

The key observation from Figure 13.6 is that the slopes of the tangents approach zero as the reaction proceeds. When the reaction is over, no more change occurs in the concentration of any remaining reactant or in the concentration of any product, and in this region of the curve the slope of any tangent is zero. This behavior is typical: the most rapid changes in reactant and product concentrations take place early in most reactions.

Kinetic molecular theory and our picture of molecules in the gas phase provide us with a way to explain this trend. If we assume that reactions take place as a result of collisions between reactant molecules, then the more reactant molecules there are in a given space, such as a flask, the more collisions per unit time and the more opportunities for reactants to turn into products. As reactant concentrations decrease, fewer reactant molecules occupy the space in the flask, so the frequency of collisions between molecules decreases and the rate of conversion of reactants to products slows down.

What happens to the rate of some of the reactions we have discussed when we change the initial concentrations of the reactants? Doubling the concentration of O_2 in Equation 13.2 doubles the rate of the reaction. However, doubling the concentration of NO *quadruples* the rate. Why does the rate of reaction have different dependencies on the concentrations of the two reactants?

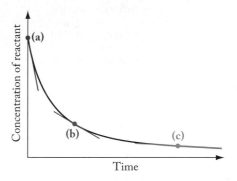

FIGURE 13.6 Typical plot of reactant concentration as a function of time: (a) tangent at $t = 0$; (b) tangent at the midpoint of the reaction; (c) tangent close to the end of the reaction.

CONNECTION We discussed kinetic molecular theory, a model that describes the behavior of gases, in Chapter 10.

Reaction Order and Rate Constants

Experimental observations and theoretical considerations tell us that reaction rates depend on reactant concentrations. However, they do not tell us *to what extent* rates depend on reactant concentrations. For example, if the concentration

▶❙❙ **CHEMTOUR** Reaction Order

TABLE 13.3 Effect of Reactant Concentrations on Initial Reaction Rates for the Reaction 2 NO(g) + O₂(g) → 2 NO₂(g) at 25°C

Experiment	[NO]$_0$ (M)	[O$_2$]$_0$ (M)	Initial Reaction Rate (M/s)
1	0.0100	0.0100	1.0×10^{-6}
2	0.0100	0.0050	0.5×10^{-6}
3	0.0050	0.0100	2.5×10^{-7}

of a reactant doubles, does the reaction rate double? The answer is expressed by the **reaction order**, a parameter derived from experiments that tells us how reaction rate depends on reactant concentrations. Knowing the order of a reaction provides insights into *how* the reaction takes place—which molecules collide with which other molecules as bonds break, new bonds form, and reactants are transformed into products.

Let's look at one way in which reaction order is determined by revisiting the reaction between nitrogen monoxide and oxygen:

$$2 \text{ NO}(g) + \text{O}_2(g) \rightarrow 2 \text{ NO}_2(g)$$

In experiments to evaluate the kinetics of this reaction, different initial concentrations of NO and O_2 are introduced into a reaction vessel at 25°C, and the initial reaction rate is determined in each case. Table 13.3 shows three of these initial reaction rates. Remember that the initial rate is determined from the slope of a line tangent to the curve of concentration versus time. Why do we choose to calculate the initial rate (instantaneous rate at $t = 0$)? In the previous section we observed that reaction rates depend on collisions between reactants. Choosing $t = 0$ as our point on the concentration-versus-time curve means that the concentration of products is essentially zero. This choice is important because reactions can also run in reverse. If the product concentration is essentially zero, then no collisions occur between product molecules and the rate of the reverse reaction can be largely ignored. If we chose to work with rates at a point where $t \neq 0$, then we would have to account for the different rates of the forward and reverse reactions.

To interpret the data in Table 13.3, we select pairs of experiments in which the concentrations of one reactant differ but the concentrations of the other are the same. For example, [NO]$_0$ is the same in experiments 1 and 2, but [O$_2$]$_0$ in experiment 1 is twice [O$_2$]$_0$ in experiment 2. (The zero subscripts indicate that the concentrations of NO and O_2 are the values at the start of the experiments, when $t = 0$.) The initial reaction rate in experiment 1 is twice that in experiment 2, allowing us to state that doubling the concentration of O_2 while holding the NO concentration constant doubles the initial reaction rate. We conclude that the initial reaction rate is proportional to O_2 concentration:

$$\text{Rate} \propto [\text{O}_2]$$

In experiments 1 and 3, [O$_2$]$_0$ is the same but [NO]$_0$ in experiment 1 is twice [NO]$_0$ in experiment 3. Comparing the initial reaction rates for these two experiments, we find that the rate in experiment 1 is four times the rate in experiment 3. Thus, the reaction rate is proportional to [NO] squared:

$$\text{Rate} \propto [\text{NO}]^2$$

We combine these two rate expressions to get an overall rate expression for the reaction by multiplying the right sides of the rate expressions for the two reactants:

$$\text{Rate} \propto [\text{NO}]^2[\text{O}_2]$$

To understand why we multiply the concentration terms together, let's consider another reaction between nitrogen monoxide and oxygen. This one involves NO and ozone (O_3) and produces NO_2 and O_2:

$$\text{NO}(g) + \text{O}_3(g) \rightarrow \text{NO}_2(g) + \text{O}_2(g)$$

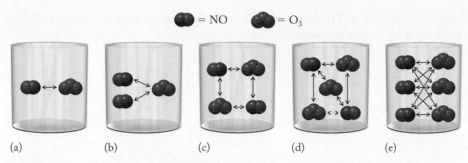

FIGURE 13.7 Increasing the concentration increases the number of possible collisions (double-headed arrows) and therefore the number of potential reaction events. Reaction rate depends on the number of collisions, which are shown for the reaction between NO and ozone (O_3) that produces NO_2 and O_2. (a) With only one NO molecule and one O_3 molecule, each molecule can collide only with the other, giving a relative reaction rate of $1 \times 1 = 1$. (e) Three molecules of NO can collide with three molecules of O_3 for a relative reaction rate of $3 \times 3 = 9$ times the relative rate in a.

Figure 13.7 shows how different numbers of NO and O_3 molecules in a reaction vessel might collide together and react with each other. Note how increasing the numbers of molecules in the containers in Figure 13.7(a)–(e) produces increasing numbers of collisions that are proportional to the *product* of the number of molecules of each reactant. Increasing the number of molecules of each type in the vessels is equivalent to increasing the concentrations of the two gases. Therefore, the rate of the reaction should (and does) depend on the product of the concentrations of NO and O_3:

$$\text{Rate} \propto [\text{NO}][\text{O}_3] \tag{13.11}$$

Similar patterns occur with all other chemical reactions whose rate depends on the concentration of more than one reactant. We can modify Equation 13.11 to obtain a **rate law**, an equation that defines the relation between reactant concentrations and reaction rate. We convert the proportionality to an equation by inserting a proportionality constant k, called the **rate constant** for the reaction:

$$\text{Rate} = k[\text{NO}][\text{O}_3] \tag{13.12}$$

Now let's consider a generic chemical reaction with two reactants, A and B:

$$\text{A} + \text{B} \rightarrow \text{C}$$

The rate law expression for this may be written

$$\text{Rate} = k[\text{A}]^m[\text{B}]^n \tag{13.13}$$

where m is the reaction order with respect to A and n is the reaction order with respect to B. We determine reaction order by comparing differences in reaction rates to differences in reactant concentrations. Sometimes these comparisons are easy to make, as they were when we examined the relative rates of the experiments in Table 13.3. Other times, however, the comparisons are not so easy to make. On these occasions, we can use a more mathematical approach. For example, suppose we run three experiments to solve for the values of m and n in Equation 13.13. In experiments 1 and 2 the value of [A] is the same, but the values of [B] are different. In experiments 2 and 3 we keep [B] constant and vary [A]. The ratio of the reaction rates in experiments 1 (Rate_1) and 2 (Rate_2) is related to the ratio of the concentrations of B: $[\text{B}]_1/[\text{B}]_2$, and to the dependence of reaction rate on [B]; that

rate law an equation that defines the experimentally determined relation between the concentration of reactants in a chemical reaction and the rate of that reaction.

rate constant the proportionality constant that relates the rate of a reaction to the concentrations of reactants.

is, on the value of n. We can express this relationship in equation form and then solve for n by taking the logarithm of both sides:

$$\frac{Rate_1}{Rate_2} = \left(\frac{[B]_1}{[B]_2}\right)^n \qquad \log\left(\frac{Rate_1}{Rate_2}\right) = n \log\left(\frac{[B]_1}{[B]_2}\right)$$

Rearranging the terms,

$$n = \frac{\log\left(\dfrac{Rate_1}{Rate_2}\right)}{\log\left(\dfrac{[B]_1}{[B]_2}\right)}$$

A corresponding equation expresses the value of m with respect to [A] for the results of experiments 2 and 3. In general, we can use such an equation to calculate the order n of any reaction with a reactant X whose initial concentration is the sole variable in a pair of reaction rate experiments:

$$n = \frac{\log\left(\dfrac{Rate_1}{Rate_2}\right)}{\log\left(\dfrac{[X]_1}{[X]_2}\right)} \tag{13.14}$$

CONCEPT TEST •••

In the reaction A → B, the rate of the reaction triples when [A] is tripled.

    a. What is the correct value of m in the rate law: Rate = $k[A]^m$?
    b. What is the value of m if the rate is unchanged when [A] is tripled?

•••

The power to which a concentration term is raised is the order of the reaction in terms of that reactant. Thus in our example from Table 13.3, the reaction of NO and O_2 is *second order* in NO and *first order* in O_2. The **overall reaction order** for a reaction is the sum of the powers in the rate equation, so the reaction of NO and O_2 is *third order* overall.

An exponent in a rate law may be a fraction, zero, or, in rare cases, negative. It is important to remember that rate laws and reaction orders are different from relative rates and must be determined experimentally. They cannot always be predicted from the coefficients in a balanced chemical equation. The significance of reaction order in describing how a reaction takes place is addressed in Section 13.5.

The value of the rate constant k is unique to each particular reaction at a given temperature, and right now we may simply consider k as a proportionality constant that does not change with concentration. In other words, *reaction rate depends on the concentration of the reactants, but the rate constant does not*. The rate constant changes with changing temperature or in the presence of a catalyst.

We can calculate k for a reaction run at some specified temperature from initial reaction rate data. Let's do so for the reaction of NO and O_2 by selecting the results of one experiment in Table 13.3. Which experiment we use doesn't matter. As long as the temperature is the same, the value of k is the same. To determine k, let's use Equation 13.13 and insert the data from experiment 1:

----

overall reaction order the sum of the exponents of the concentration terms in the rate law.

$$\text{Rate} = k[\text{NO}]^2[\text{O}_2]$$

$$1.0 \times 10^{-6} \ M/s = k(0.0100 \ M)^2(0.0100 \ M)$$

$$k = \frac{1.0 \times 10^{-6} \ \cancel{M}/s}{(0.0100 \ M)^2(0.0100 \ \cancel{M})} = 1.0 \ M^{-2} \ s^{-1} \qquad (13.15)$$

Keep in mind that the value calculated from experimental data for a rate constant is valid only at the temperature at which the experiments were carried out.

The units of k differ from one reaction to another. Remember that reaction rates always have units of concentration per time interval. Therefore, for a first-order reaction in which concentration is expressed in molarity and time in seconds, the units of k must be per second (s^{-1}):

$$\text{Rate} = k[\text{X}]$$

$$k = \frac{\text{Rate}}{[\text{X}]} = \frac{\cancel{M}/s}{\cancel{M}} = \frac{1}{s} = s^{-1}$$

The units of the rate constant for a reaction that is second order overall can be derived in a similar way. If the reaction is first order in reactants X and Y, we have

$$\text{Rate} = k[\text{X}][\text{Y}]$$

$$k = \frac{\text{Rate}}{[\text{X}][\text{Y}]} = \frac{\cancel{M}/s}{M^{\cancel{2}}} = M^{-1} \ s^{-1}$$

The units of k for a reaction that is third order overall are $M^{-2} \ s^{-1}$, as we determined in Equation 13.15.

SAMPLE EXERCISE 13.3 **Deriving a Rate Law from Initial Reaction Rate Data** **LO3**

Write the rate law for the reaction of N_2 with O_2

$$N_2(g) + O_2(g) \rightarrow 2 \ NO(g)$$

using the data in Table 13.4. Determine the overall reaction order and the value of the rate constant.

COLLECT AND ORGANIZE We are asked to determine the rate law and rate constant for a reaction. We are given the initial reaction rate (note the subscript zero on the concentration terms in Table 13.4) for each of three sets of initial concentrations.

ANALYZE The general form of the rate law for any reaction between reactants A and B is

$$\text{Rate} = k[\text{A}]^m[\text{B}]^n$$

We can use the experimental data given to find the values of k, m, and n for the reaction in which A = N_2 and B = O_2. The overall order of the reaction is the sum of the orders of the individual reactants. Once we have established the rate law, the rate constant is calculated using concentrations of reactants from any single row in Table 13.4. The rate constant for the reaction must have units that express the reaction rate in $M \ s^{-1}$.

SOLVE To determine the reaction order m with respect to N_2, we use the data from experiments 2 and 3, because in these two experiments the values of $[N_2]$ are different but the $[O_2]$ values are the same. When the concentration of N_2 is increased by a

TABLE 13.4 **Initial Reaction Rates for the Formation of NO from the Reaction $N_2(g) + O_2(g) \rightarrow 2 \ NO(g)$ at Constant Temperature**

Experiment	$[N_2]_0$ (M)	$[O_2]_0$ (M)	Initial Reaction Rate (M/s)
1	0.040	0.020	707
2	0.040	0.010	500
3	0.010	0.010	125

factor of 4, the rate increases by a factor of 4. Thus, the reaction rate is proportional to N_2 concentration, which means that $m = 1$. To calculate n, we note that Experiments 1 and 2 have different $[O_2]$ values, but $[N_2]$ is the same. The ratio of the reaction rates (707/500) is not a whole number, so let's apply Equation 13.14:

$$n = \frac{\log\left(\dfrac{Rate_1}{Rate_2}\right)}{\log\left(\dfrac{[O_2]_1}{[O_2]_2}\right)} = \frac{\log\left(\dfrac{707\ \cancel{M/s}}{500\ \cancel{M/s}}\right)}{\log\left(\dfrac{0.020\ \cancel{M}}{0.010\ \cancel{M}}\right)} = \frac{\log 1.41}{\log 2} = 0.50$$

Thus the reaction is one-half order with respect to O_2, first order with respect to N_2, and $(\frac{1}{2} + 1) = \frac{3}{2}$ order overall:

$$Rate = k\,[N_2][O_2]^{1/2}$$

We can use the data from any experiment to obtain the value of k. Let's use experiment 1:

$$707\ M/s = k(0.040\ M)(0.020\ M)^{1/2} = k(0.0057)$$

$$k = 1.2 \times 10^5\ M^{-1/2}\ s^{-1}$$

THINK ABOUT IT An order of $\frac{1}{2}$ for oxygen in this reaction is determined from experimental data, not from the balanced chemical equation. The units on the rate constant, $M^{-1/2}\ s^{-1}$, are appropriate because when we substitute the units for each term into the rate law, we end up with the correct units for a rate: $(\cancel{M}^{-1/2}\ s^{-1})(M)(\cancel{M^{1/2}}) = M\,s^{-1}$.

Practice Exercise Nitric oxide reacts rapidly with unstable nitrogen trioxide (NO_3) to form NO_2:

$$NO(g) + NO_3(g) \rightarrow 2\,NO_2(g)$$

Determine the rate law for the reaction and calculate the rate constant from the data in Table 13.5.

TABLE 13.5	Initial Reaction Rates for the Formation of NO_2 from the Reaction $NO(g) + NO_3(g) \rightarrow 2\,NO_2(g)$ at 25°C		
Experiment	$[NO]_0$ (M)	$[NO_3]_0$ (M)	Initial Reaction Rate (M/s)
1	1.25×10^{-3}	1.25×10^{-3}	2.45×10^4
2	2.50×10^{-3}	1.25×10^{-3}	4.90×10^4
3	2.50×10^{-3}	2.50×10^{-3}	9.80×10^4

Integrated Rate Laws: First-Order Reactions

Determining a rate law using initial reaction rate data has two distinct disadvantages. The method requires several experiments with different concentrations of reactants, varied in a systematic fashion. We also must accurately determine the reaction rate at the instant the reaction begins. It would be much easier if we could determine the rate law and calculate the rate constant for a reaction from the plot of concentration versus time alone. In fact, this can be accomplished for reactions in which the reaction rate depends on the concentration of only one substance. One such reaction is the photochemical decomposition of ozone:

$$O_3(g) \xrightarrow{\text{sunlight}} O_2(g) + O(g)$$

This reaction can be studied in the laboratory using high-intensity ultraviolet light to simulate sunlight; one such study yielded the results listed in Table 13.6 and plotted in Figure 13.8(a). Because ozone is the only reactant, the rate law for the reaction should depend only on the ozone concentration. In Table 13.6, we start with the assumption that the reaction is first order in O_3, which means that the rate law can be written

$$\text{Rate} = k[O_3]$$

Because the coefficient of O_3 in the balanced equation is 1, the O_3 consumption rate $-\Delta[O_3]/\Delta t$ is equal to the reaction rate, and we can write

$$\text{Rate} = -\frac{\Delta[O_3]}{\Delta t} = k[O_3]$$

This rate law can be transformed using calculus into an expression that relates the concentration of ozone $[O_3]$ at any instant during the reaction to the initial concentration $[O_3]_0$:

$$\ln\frac{[O_3]}{[O_3]_0} = -kt \qquad (13.16)$$

This version of the rate law is called an **integrated rate law** because integral calculus is used to derive it, and it describes the change in reactant concentration with time. The general integrated rate law for any reaction that is first order in reactant X is

$$\ln\frac{[X]}{[X]_0} = -kt \qquad (13.17)$$

Using the identity $\ln(a/b) = \ln a - \ln b$, we can rearrange this equation to

$$\ln[X] = -kt + \ln[X]_0 \qquad (13.18)$$

which is the equation of a straight line of the form

$$y = mx + b$$

where $\ln[X]$ is the y variable and t is the x variable. The slope of the line (m) is $-k$, and the y-intercept (b) is $\ln[X]_0$. Rearranging Equation 13.16 to fit the format of Equation 13.18 gives

$$\ln[O_3] = -kt + \ln[O_3]_0 \qquad (13.19)$$

integrated rate law a mathematical expression that describes the change in concentration of a reactant in a chemical reaction with time.

TABLE 13.6 Rate of Photochemical Decomposition of Ozone

Time (s)	$[O_3]$ (M)	$\ln[O_3]$
0	1.000×10^{-4}	−9.2103
100	0.896×10^{-4}	−9.320
200	0.803×10^{-4}	−9.430
300	0.719×10^{-4}	−9.540
400	0.644×10^{-4}	−9.650
500	0.577×10^{-4}	−9.760
600	0.517×10^{-4}	−9.870

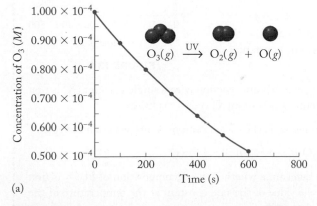

(a)

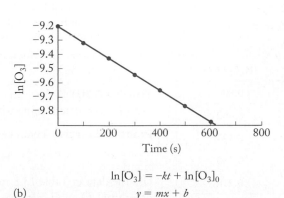

$$\ln[O_3] = -kt + \ln[O_3]_0$$
$$y = mx + b$$

(b)

FIGURE 13.8 The decomposition of O_3 is plotted as (a) $[O_3]$ versus time and (b) $\ln[O_3]$ versus time. The line in b is straight, indicating that the decomposition reaction is first order in O_3.

Note in Figure 13.8(b) that a graph plotting the natural logarithm of $[O_3]$ versus time is indeed a straight line. This linearity means that our assumption was correct and the reaction is first order in O_3. Calculating the slope of the line, $\Delta(\ln[O_3])/\Delta t$, from the two points at $t = 100$ s and $t = 400$ s in Figure 13.8(b) yields the value of the rate constant:

$$\text{Slope} = \frac{-9.650 - (-9.320)}{400 \text{ s} - 100 \text{ s}} = -1.10 \times 10^{-3} \text{ s}^{-1}$$

$$k = 1.10 \times 10^{-3} \text{ s}^{-1}$$

TABLE 13.7(a) Concentration of N_2O_5 as a Function of Time	
Time (s)	$[N_2O_5]$ (M)
0	0.1000
50	0.0707
100	0.0500
200	0.0250
300	0.0125
400	0.00625

TABLE 13.7(b) Concentration and Natural Logarithm of Concentration of N_2O_5 as a Function of Time		
Time (s)	$[N_2O_5]$ (M)	$\ln[N_2O_5]$
0	0.1000	−2.303
50	0.0707	−2.649
100	0.0500	−2.996
200	0.0250	−3.689
300	0.0125	−4.382
400	0.00625	−5.075

TABLE 13.8 Concentration of Hydrogen Peroxide as a Function of Time	
Time (s)	$[H_2O_2]$ (M)
0	0.500
100	0.460
200	0.424
500	0.330
1000	0.218
1500	0.144

SAMPLE EXERCISE 13.4 **Using Integrated Rate Laws to Determine the Rate Law Expression and Rate Constant of a Reaction** **LO4**

One of the less abundant nitrogen oxides in the atmosphere is dinitrogen pentoxide. One reason concentrations of this oxide are low is that the molecule is unstable and rapidly decomposes to N_2O_4 and O_2:

$$2 N_2O_5(g) \rightarrow 2 N_2O_4(g) + O_2(g)$$

A kinetic study of the decomposition of N_2O_5 at a particular temperature yields the data in Table 13.7(a). Assume that the decomposition of N_2O_5 is first order in N_2O_5.

 a. Test the validity of your assumption.
 b. Determine the value of the rate constant.

COLLECT AND ORGANIZE We are given experimental data showing the concentration of a single reactant as a function of time and told to assume a first order reaction. We can verify this assumption using the integrated rate law for a first order reaction (Equation 13.18) and then calculate the rate constant.

ANALYZE Equation 13.18 has the form $y = mx + b$, which means that a plot of $\ln[N_2O_5]$ (y) versus time (x) should be linear if the decomposition of N_2O_5 is first order. The slope (m) of the graph corresponds to $-k$, the negative of the rate constant.

SOLVE Our first step is to determine $\ln[N_2O_5]$ values (Table 13.7b).

 a. The plot of $\ln[N_2O_5]$ versus t is shown in Figure 13.9. The fact that the curve in Figure 13.9 is a straight line indicates that the reaction is first order in N_2O_5.
 b. The line that best fits the data points in Figure 13.9 has a slope of -0.00693 s^{-1}. The slope of the line equals $-k$. Therefore, the rate constant $k = 0.00693$ s^{-1}.

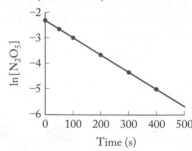

FIGURE 13.9

THINK ABOUT IT We can test whether any reaction with a single reactant (X) is first order in X by determining whether a plot of $\ln[X]$ versus t is linear.

Practice Exercise Hydrogen peroxide (H_2O_2) decomposes into water and oxygen:

$$H_2O_2(\ell) \rightarrow H_2O(\ell) + \tfrac{1}{2} O_2(g)$$

Use the data in Table 13.8 to determine whether the decomposition of H_2O_2 is first order in H_2O_2, and calculate the value of the rate constant at the temperature of the experiment that produced the data.

Reaction Half-Lives

A parameter frequently cited in kinetic studies is the **half-life ($t_{1/2}$)** of a reaction, which is the interval during which the concentration of a reactant decreases by half, as shown for the decomposition of N_2O (laughing gas) in Figure 13.10:

$$N_2O(g) \rightarrow N_2(g) + \tfrac{1}{2} O_2(g)$$

Over each half-life interval, one-half of the quantity of reactant present at the beginning of the interval is consumed; the other half remains.

Half-life is inversely related to the rate constant of a reaction: the higher the reaction rate, the shorter the half-life. For first-order reactions, such as the decomposition of N_2O, we can derive a mathematical relation between half-life $t_{1/2}$ and rate constant k by starting with Equation 13.17:

$$\ln \frac{[X]}{[X]_0} = -kt$$

After one half-life has passed ($t = t_{1/2}$), the concentration of X is half its original value: $[X] = \tfrac{1}{2}[X]_0$. Inserting these values for $[X]$ and t into the equation yields

$$\ln \frac{\tfrac{1}{2}\cancel{[X]_0}}{\cancel{[X]_0}} = -kt_{1/2}$$

$$\ln \tfrac{1}{2} = -kt_{1/2}$$

The natural log of $\tfrac{1}{2}$ is -0.693, so

$$-0.693 = -kt_{1/2}$$

$$t_{1/2} = \frac{0.693}{k} \tag{13.20}$$

Thus, the half-life of a first-order reaction is inversely proportional to the rate constant, as noted at the beginning of this discussion. The absence of any concentration term in Equation 13.20 means that no matter what the initial concentration of reactant is, half of it is consumed in one half-life.

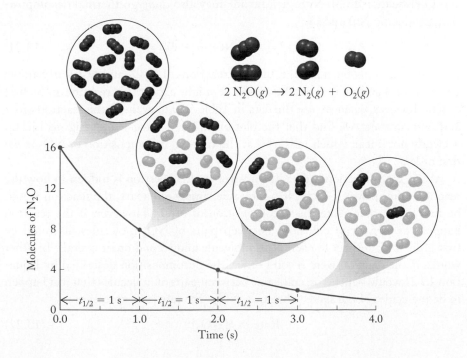

$$2\,N_2O(g) \rightarrow 2\,N_2(g) + O_2(g)$$

FIGURE 13.10 The decomposition of $N_2O(g)$ is first order in N_2O. At a particular temperature the half-life of the reaction is 1.0 s, which means that, on average, half of a population of 16 N_2O molecules decomposes in 1.0 s, half of the remaining 8 molecules decompose in the next 1.0 s, and so on.

SAMPLE EXERCISE 13.5 Calculating the Half-Life of a First-Order Reaction **LO5**

The rate constant for the decomposition of N_2O_5 at a particular temperature is 7.8×10^{-3} s^{-1}. What is the half-life of N_2O_5 at that temperature?

COLLECT AND ORGANIZE The half-life is the time required for the concentration of a reactant to decrease by half. We are asked to determine the half-life of N_2O_5 for its decomposition reaction. We know from Sample Exercise 13.4 that the decomposition of N_2O_5 is a first-order process.

ANALYZE Equation 13.20 relates k for a first-order reaction to the reaction half-life. We are given no information about concentrations of reactants or products, but we know from Equation 13.20 that the half-life of a first-order reaction is independent of concentration. Because half-life is inversely proportional to k, and the value of k is on the order of 10^{-2} s^{-1}, we predict that the half-life will be on the order of 10^2 s.

SOLVE

$$t_{1/2} = \frac{0.693}{k} = \frac{0.693}{7.8 \times 10^{-3} \text{ s}^{-1}} = 89 \text{ s}$$

THINK ABOUT IT The calculated value of $t_{1/2}$ is a little less than 100, the predicted value. Remember that Equation 13.20 is valid only for first-order reactions.

Practice Exercise Environmental scientists calculating half-lives of pollutants often define a *transport rate constant* that is analogous to a reaction rate constant and describes how a pollutant moves out of an ecosystem. In a study of the gasoline additive MTBE in Donner Lake, California, scientists from the University of California, Davis, found that in the summer the half-life of MTBE in the lake was 28 days. Assume that the transport process is first order. What was the transport rate constant of MTBE out of Donner Lake during the study? Express your answer in reciprocal days (day^{-1}).

Integrated Rate Laws: Second-Order Reactions

In Section 13.1 we described how NO_2 exposed to sunlight decomposes to NO and O (Equation 13.3). Nitrogen dioxide may also undergo thermal decomposition, producing NO and O_2:

$$2\, NO_2(g) \rightarrow 2\, NO(g) + O_2(g) \tag{13.21}$$

Because it has only one reactant, like the reactions of N_2O_5 and H_2O_2 in Sample Exercise 13.4 and its Practice Exercise, we might expect this reaction to be first order. However, when we use the data in Table 13.9 to evaluate the reaction order and rate constant, we find that the plot of $\ln[NO_2]$ versus time (Figure 13.11a) is clearly not linear, which tells us that the thermal decomposition of NO_2 is *not* first order.

What is the reaction order? The answer to this question is hidden in how the reaction takes place. If each NO_2 molecule simply fell apart, the reaction would be first order, much like the decomposition of N_2O_5. However, if the reaction happens as a result of collisions between pairs of NO_2 molecules, it would be first order with respect to each NO_2 molecule and second order overall. In other words, if the reaction were second order, the decomposition described by Equation 13.21 would depend on collisions between pairs of molecules that just happen to be molecules of the same substance, NO_2. The rate law expression is:

$$\text{Rate} = k[NO_2]^2 \tag{13.22}$$

TABLE 13.9 Rate of Decomposition of NO_2 to NO and O_2

Time (s)	$[NO_2]$ (M)	$\ln[NO_2]$	$1/[NO_2]$ (1/M)
0	1.00×10^{-2}	−4.605	100
100	6.48×10^{-3}	−5.039	154
200	4.79×10^{-3}	−5.341	209
300	3.80×10^{-3}	−5.573	263
400	3.15×10^{-3}	−5.760	317
500	2.69×10^{-3}	−5.918	372
600	2.35×10^{-3}	−6.057	426

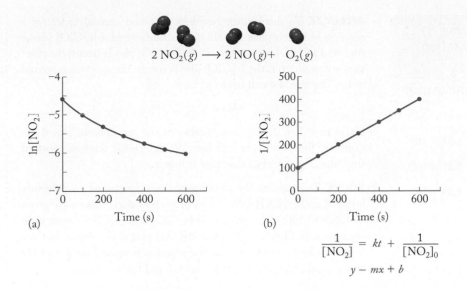

$$2\,NO_2(g) \longrightarrow 2\,NO(g) + O_2(g)$$

(a)

(b)

$$\frac{1}{[NO_2]} = kt + \frac{1}{[NO_2]_0}$$

$$y - mx + b$$

FIGURE 13.11 At high temperatures, NO_2 slowly decomposes into NO and O_2. (a) The plot of $\ln[NO_2]$ versus time is not linear, indicating that the reaction is not first order. (b) The plot of $1/[NO_2]$ versus time is linear, indicating that the reaction is second order in NO_2. The slope of the line in this graph equals the rate constant.

How can we determine whether this decomposition is really second order? One way is to assume that it is and then test that assumption. The test entails transforming the rate law in Equation 13.22 into the integrated rate law for a second-order reaction, again using calculus. The result of the transformation is

$$\frac{1}{[NO_2]} = kt + \frac{1}{[NO_2]_0} \tag{13.23}$$

which, like Equation 13.19, has the form $y = mx + b$ and is the equation of a straight line, this time with $1/[NO_2]$ as the y variable and t as the x variable.

The graph obtained using data from columns 1 and 4 of Table 13.9 is shown in Figure 13.11(b). That the curve is linear tells us that this NO_2 decomposition reaction is second order overall. The slope of the line provides a direct measure of k, which is $0.544\ M^{-1}\ s^{-1}$.

A general form of Equation 13.23 that applies to any reaction that is second order in a single reactant (X) is

$$\frac{1}{[X]} = kt + \frac{1}{[X]_0} \tag{13.24}$$

SAMPLE EXERCISE 13.6 **Distinguishing between First- and Second-Order Reactions**   **LO4**

Chlorine monoxide accumulates in the stratosphere above Antarctica each winter and plays a key role in the formation of the ozone hole above the South Pole each spring. Eventually, ClO decomposes according to the equation

$$2\,ClO(g) \rightarrow Cl_2(g) + O_2(g)$$

The kinetics of this reaction were studied in a laboratory experiment at 298 K, and the data are shown in Table 13.10(a). Determine the order of the reaction, the rate law, and the value of k at 298 K.

COLLECT AND ORGANIZE We are given experimental data describing the variation of concentration of ClO with time at 298 K and we are asked to determine the order of the decomposition reaction of ClO. Two of the choices we have are first and second order.

TABLE 13.10(a) **Concentration of Chlorine Monoxide as a Function of Time**

Time (ms)	[ClO] (M)
0	1.50×10^{-8}
10	7.19×10^{-9}
20	4.74×10^{-9}
30	3.52×10^{-9}
40	2.81×10^{-9}
100	1.27×10^{-9}
200	0.66×10^{-9}

TABLE 13.10(b) **Concentration of Chlorine Monoxide, ln[ClO], and 1/[ClO] as a Function of Time**

Time (ms)	[ClO] (M)	ln[ClO]	1/[ClO] (1/M)
0	1.50×10^{-8}	−18.015	6.67×10^{7}
10	7.19×10^{-9}	−18.751	1.39×10^{8}
20	4.74×10^{-9}	−19.167	2.11×10^{8}
30	3.52×10^{-9}	−19.465	2.84×10^{8}
40	2.81×10^{-9}	−19.690	3.56×10^{8}
100	1.27×10^{-9}	−20.484	7.89×10^{8}
200	0.66×10^{-9}	−21.139	1.51×10^{9}

ANALYZE To distinguish between first and second order for a reaction in which there is a single reactant, we plot ln[ClO] versus time and 1/[ClO] versus time. If the ln[ClO] plot is linear, the reaction is first order; if the 1/[ClO] plot is linear, the reaction is second order. The rate law will have the form

$$\text{Rate} = k[\text{ClO}]^{m}$$

where $m = 1$ or 2. The overall order of the reaction will equal the exponent for ClO in the rate law. We determine the rate constant from the slope of whichever plot is linear.

SOLVE To evaluate the two possibilities, we need to calculate ln[ClO] and 1/[ClO] values; sets of both of these values are given in Table 13.10(b). The graphs for ln[ClO] and 1/[ClO] versus time are shown in Figure 13.12. The ln[ClO] plot is not linear, but the 1/[ClO] plot is, meaning that the reaction is second order in ClO and second order overall. Thus, $m = 2$ and the rate law is

$$\text{Rate} = k[\text{ClO}]^{2}$$

Substituting [ClO] into the generic integrated rate law (Equation 13.24) gives

$$\frac{1}{[\text{ClO}]} = kt + \frac{1}{[\text{ClO}]_{0}}$$

From the general equation $y = mx + b$, we know that k is the slope of the graph of 1/[ClO] versus time. The line that best fits the data has a slope of $7.22 \times 10^{9}\ M^{-1}\ s^{-1}$, which is the value of k.

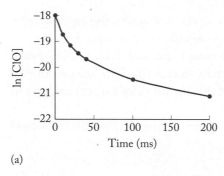
(a)

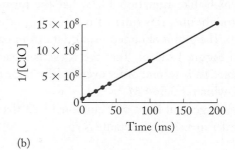
(b)

FIGURE 13.12

THINK ABOUT IT Sample Exercise 13.4 and this one show we can use integrated rate laws to distinguish between first- and second-order reactions in a single reactant.

Practice Exercise Experimental evidence shows that in the reaction below, the reaction rate depends only on the concentration of NO_2:

$$NO_2(g) + CO(g) \rightarrow NO(g) + CO_2(g)$$

Determine whether the reaction is first or second order in NO_2, and calculate the rate constant from the data in Table 13.11, which were obtained at 488 K. ⚙

TABLE 13.11 **Concentration of NO_2 as a Function of Time**

Time (hr)	[NO_2] (M)
0.00	0.250
1.39	0.198
3.06	0.159
4.72	0.132
6.39	0.114
8.06	0.099
9.72	0.088
11.39	0.080

The concept of half-life can also be applied to second-order reactions. The relation between rate constant and half-life for the decomposition of NO_2 can be derived from Equation 13.24 if we first rearrange the terms to solve for kt:

$$kt = \frac{1}{[\text{X}]} - \frac{1}{[\text{X}]_{0}}$$

After one half-life has elapsed ($t = t_{1/2}$), [X] has decreased to half its initial concentration. Substituting this information into the preceding equation, we have

$$kt_{1/2} = \frac{1}{\frac{1}{2}[X]_0} - \frac{1}{[X]_0}$$

$$= \frac{2}{[X]_0} - \frac{1}{[X]_0} = \frac{1}{[X]_0}$$

or

$$t_{1/2} = \frac{1}{k[X]_0} \qquad (13.25)$$

Note that the value of $t_{1/2}$ is inversely proportional to the initial concentration of X. This dependence on concentration is unlike the $t_{1/2}$ values of first-order reactions, which are independent of concentration.

SAMPLE EXERCISE 13.7 **Calculating the Half-Lives** **LO5**
of Second-Order Reactions

Calculate the half-life of the second-order decomposition of NO_2 (Equation 13.21) if the rate constant is $0.544\ M^{-1}\ s^{-1}$ at a particular temperature and the initial concentration of NO_2 is $0.0100\ M$.

COLLECT AND ORGANIZE We are asked to find the reaction's half-life, the time required for the concentration of NO_2 to decrease by one-half. We are given the rate constant and initial concentration of NO_2 and told the reaction is second order. The half-life of a second-order reaction in which there is only one reactant is given by Equation 13.25.

ANALYZE The half-life is inversely proportional to the rate constant k and to the initial concentration of NO_2. The value of k is about $0.5\ M^{-1}\ s^{-1}$ at an initial concentration of $10^{-2}\ M$, so $t_{1/2}$ is inversely proportional to a quantity of about $5 \times 10^{-3}\ s^{-1}$. Thus we predict a half-life of about 200 s.

SOLVE

$$t_{1/2} = \frac{1}{k[NO_2]_0} = \frac{1}{(0.544\ \cancel{M}^{-1}\ s^{-1})(1.00 \times 10^{-2}\ \cancel{M})} = 184\ s$$

THINK ABOUT IT The calculated value of $t_{1/2}$ is very close to the predicted value of 200 s. We needed to know $[NO_2]_0$ to calculate the value of $t_{1/2}$ for this second-order reaction in NO_2.

Practice Exercise The rate constant for the decomposition of ClO (Sample Exercise 13.6) is $7.22 \times 10^9\ M^{-1}\ s^{-1}$ at 298 K. Determine the half-life of ClO when its initial concentration is $1.50 \times 10^{-8}\ M$. ⚙

Pseudo-First-Order Reactions

The integrated rate law in Equation 13.24 applies only to reactions that are second order in a single reactant. It does not apply to reactions that are second order overall but are first order in two reactants, such as

$$NO(g) + O_3(g) \rightarrow NO_2(g) + O_2(g)$$

or

$$NO_2(g) + O_3(g) \rightarrow NO_3(g) + O_2(g)$$

pseudo-first-order a reaction in which all the reactants but one are present at such high concentrations that they do not decrease significantly during the course of the reaction, so that the reaction rate is controlled by the concentration of the limiting reactant.

Because the integrated rate law for a reaction that is first order in two reactants is complicated, kineticists frequently adjust reaction conditions so that a simple rate law can be used. One approach is to have one of the reactants present at a much higher concentration than the other. This condition is common for many components in the urban atmosphere where, for example, ozone concentrations are often hundreds to thousands of times greater than NO concentrations. With such a large excess, the ozone concentration remains virtually constant over the course of the reaction

$$NO(g) + O_3(g) \rightarrow NO_2(g) + O_2(g)$$

Thus, this reaction's rate law

$$\text{Rate} = k[NO][O_3]$$

may be simplified to

$$\text{Rate} = k'[NO] \tag{13.26}$$

where

$$k' = k[O_3]_0 \tag{13.27}$$

and where $[O_3]_0$ is the initial concentration of ozone, which remains virtually constant throughout the reaction.

Equation 13.26 looks like the rate law for a first-order reaction (rate $= k[X]$). It is considered a **pseudo-first-order** rate law because it *appears* to obey first-order kinetics. A pseudo-first-order reaction has the same integrated rate law as a first-order reaction, but the rate of the pseudo-first-order reaction depends on the concentration of more than one reactant.

SAMPLE EXERCISE 13.8 **Deriving Pseudo-First-Order Rate Laws** **LO4**

The data in Table 13.12(a) were obtained in a study of the oxidation of trace levels of NO in the presence of a large excess of ozone at 298 K:

$$NO(g) + O_3(g) \rightarrow NO_2(g) + O_2(g)$$

(Note the NO concentration units in Table 13.12a. When the molar concentrations of reactants are very low, as in the case of many atmospheric pollutants, it is easier to express them in units of molecules/cm^3.)

a. Verify that the reaction is pseudo-first-order and determine the pseudo-first-order rate constant k'.
b. If the ozone concentration is 100 times the NO concentration at $t = 0$, what is the second-order rate constant k? Express this k in $M^{-1}\,s^{-1}$.

COLLECT AND ORGANIZE Under pseudo-first-order conditions, a large excess of one reactant is maintained so that we can use the integrated rate law for a first-order reaction. We are asked to verify that the reaction is pseudo-first-order and to determine the pseudo-first-order rate constant. We are given experimental data showing the change of concentration of NO with time. We are also given the initial concentration of O_3 and asked to calculate the second-order rate constant for the reaction.

ANALYZE We verify our assumption by treating the data as we would for a first-order reaction and plotting ln[NO] versus time. The plot should be a straight line with a slope equal to $-k'$. We can solve for the second-order rate constant k using Equation 13.27. We expect the second-order rate constant to be rather large since the data (concentration changes) were collected on a microsecond scale.

TABLE 13.12(a) **Concentration of NO as a Function of Time**

Time (μs)	Concentration of NO (molecules/cm³)
0	1.00×10^9
100	8.36×10^8
200	6.98×10^8
300	5.83×10^8
400	4.87×10^8
500	4.07×10^8
1000	1.65×10^8

SOLVE

a. The plot of ln[NO] values (Table 13.12b) versus time is linear as shown in Figure 13.13, which indicates that the reaction is indeed pseudo-first-order. The slope of the line that best fits the data is -1.81×10^{-3} μs^{-1}. Changing the sign of the slope and converting the units from μs^{-1} to s^{-1} gives us the value of the pseudo-first-order rate constant k', 1.81×10^3 s^{-1}.

b. We calculate the second-order rate constant k using Equation 13.27, $k' = k[O_3]_0$, and

$$[O_3]_0 = 100[NO]_0 = 100(1.00 \times 10^9 \text{ molecules/cm}^3)$$

Solving for k,

$$k = \frac{k'}{[O_3]_0} = \frac{1.81 \times 10^3 \text{ s}^{-1}}{1.00 \times 10^{11} \text{ molecules/cm}^3} = \frac{1.81 \times 10^{-8} \text{ cm}^3}{\text{molecules} \cdot \text{s}}$$

To convert the units to M^{-1} s^{-1}, we need to convert the reciprocal concentration units of cm³/molecule into M^{-1}:

$$k = \frac{1.81 \times 10^{-8} \cancel{\text{cm}^3} \text{ s}^{-1}}{\cancel{\text{molecules}}} \times \frac{6.022 \times 10^{23} \cancel{\text{molecules}}}{1 \text{ mol}} \times \frac{1 \text{ L}}{1000 \cancel{\text{cm}^3}}$$

$$= 1.09 \times 10^{13} \frac{\text{L} \cdot \text{s}^{-1}}{\text{mol}} = 1.09 \times 10^{13} \text{ } M^{-1} \text{ s}^{-1}$$

THINK ABOUT IT As we predicted, the second-order rate constant for the reaction is large. The pseudo-first-order (k') and second-order (k) rate constants are very different because k' contains a term for the initial concentration of ozone, which is very small when expressed in moles per liter. The value of k' will be different for every initial concentration of O_3, but the value of k will be independent of $[O_3]_0$.

Practice Exercise The reaction

$$Cl(g) + O_3(g) \rightarrow ClO(g) + O_2(g)$$

is first order in both reactants. Determine the pseudo-first-order and second-order rate constants for the reaction from the data in Table 13.13 if the initial ozone concentration is 8.5×10^{-11} M. ⚙

CONCEPT TEST

A student measured the pseudo-first-order rate constant for the reaction of NO with O_3 in Sample Exercise 13.8 at four initial concentrations of ozone. Assuming that all four $[O_3]_0$ values were much greater than [NO], how could the student determine the second-order rate constant graphically?

Zero-Order Reactions

In the Practice Exercise accompanying Sample Exercise 13.6, we were introduced to the reaction

$$NO_2(g) + CO(g) \rightarrow NO(g) + CO_2(g)$$

and the solution revealed that the rate law for the reaction is

$$\text{Rate} = k[NO_2]^2 \tag{13.28}$$

The rate of the reaction does not change with changing [CO], even when the concentrations of CO and NO_2 are comparable. This situation is not the same as in

TABLE 13.12(b) Concentration of NO and ln[NO] as a Function of Time		
Time (μs)	[NO] (molecules/cm³)	ln[NO]
0	1.00×10^9	20.723
100	8.36×10^8	20.544
200	6.98×10^8	20.364
300	5.83×10^8	20.184
400	4.87×10^8	20.004
500	4.07×10^8	19.824
1000	1.65×10^8	18.915

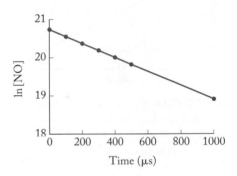

FIGURE 13.13

TABLE 13.13 Concentration of Chlorine Atoms as a Function of Time	
Time (μs)	[Cl] (M)
0	5.60×10^{-14}
100	5.27×10^{-14}
600	3.89×10^{-14}
1200	2.69×10^{-14}
1850	1.81×10^{-14}

pseudo-first-order reactions, where the rate does not depend on the concentration of one reactant because that reactant is present in large excess.

One interpretation of Equation 13.28 is that it contains a [CO] term to the zeroth power, making the reaction *zero order* in that reactant. Because any value raised to the zeroth power is 1, we have

$$\text{Rate} = k[NO_2]^2[CO]^0 = k[NO_2]^2(1) = k[NO_2]^2$$

Reactions with a true zero-order rate law are rare, but let's consider a generic reaction involving a single reactant X, with the reaction zero order in reactant X and zero order overall. The rate law is

$$\text{Rate} = -\Delta[X]/\Delta t = k[X]^0 = k$$

and the integrated rate law is

$$[X] = -kt + [X]_0$$

The slope of a plot of reactant concentration versus time (Figure 13.14) equals the negative of the zero-order rate constant k.

We can calculate the half-life of a zero-order reaction by substituting $t = t_{1/2}$ and $[X] = [X]_0/2$ into the integrated rate law:

$$[X]_0/2 = -kt_{1/2} + [X]_0$$

$$kt_{1/2} = [X]_0 - [X]_0/2 = [X]_0/2$$

$$t_{1/2} = [X]_0/2k$$

For now we leave the discussion of zero-order reactions with this purely mathematical treatment. We return to them and examine their meaning at the molecular level in Sections 13.5 and 13.6.

Figure 13.15 summarizes how to determine whether a reaction in which a reactant X forms one or more products is zero, first, or second order. The figure

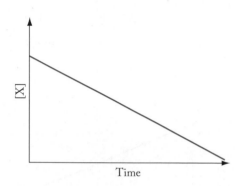

FIGURE 13.14 The change in concentration of the reactant X in the zero-order reaction X → Y is constant over time.

FIGURE 13.15 Summary of how to distinguish between zero-order, first-order, and second-order kinetics for reactions involving a single reactant (X).

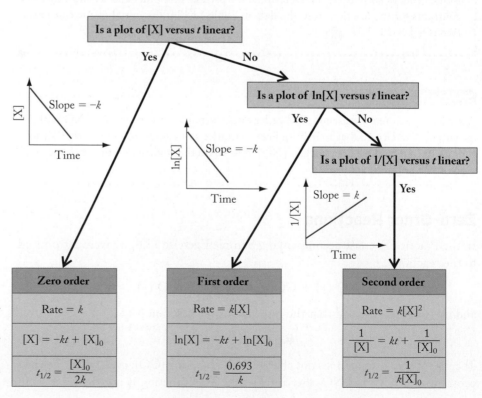

	Zero order	First order	Second order
Rate law	Rate = k	Rate = $k[X]$	Rate = $k[X]^2$
Integrated rate law	$[X] = -kt + [X]_0$	$\ln[X] = -kt + \ln[X]_0$	$\dfrac{1}{[X]} = kt + \dfrac{1}{[X]_0}$
Half-life expression	$t_{1/2} = \dfrac{[X]_0}{2k}$	$t_{1/2} = \dfrac{0.693}{k}$	$t_{1/2} = \dfrac{1}{k[X]_0}$

also lists the rate laws and integrated rate laws for these reactions and the equations used to calculate the half-lives of X. The decision-making process shown in Figure 13.15 can also be used to determine the kinetics of reactions involving two reactants (X and Y) if the reaction is zero order in Y. In such cases we focus on how the concentration of X changes with time. Finally, if a reaction is first order in both X and Y, which means second order overall, and Y is present in the reaction mixture at a much higher concentration than X, we can calculate the pseudo-first-order rate constant k' from a plot of $\ln[X]$ versus t, and then calculate the second-order rate constant by dividing k' by $[Y]$.

activation energy (E_a) the minimum energy molecules need to react when they collide.

13.4 Reaction Rates, Temperature, and the Arrhenius Equation

Why do rate constants for different reactions have different values? Why do they have the particular numerical values that we measure? In Section 13.3, we introduced the idea that chemical reactions take place in the gas phase when molecules collide with sufficient energy to break bonds in reactants and allow bonds in products to form. The minimum amount of energy that enables this to happen is called the **activation energy (E_a)**, and every chemical reaction has a characteristic activation energy, usually expressed in kilojoules per mole. Activation energy is an energy barrier that must be overcome if a reaction is to proceed—like the hill that must be climbed to hike the trail shown in Figure 13.16. Generally, the greater the activation energy, the slower the reaction.

According to kinetic molecular theory, molecules move at higher speeds and have more kinetic energy at higher temperatures. This means that the fraction of molecules with kinetic energies greater than a given activation energy increases with increasing temperature, as shown in Figure 13.17(a). Therefore, the rates of

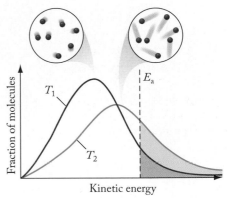

Fraction of molecules in sample with sufficient energy to react at T_1

Increase in number of molecules in sample with sufficient energy to react at T_2; $T_2 > T_1$

(a)

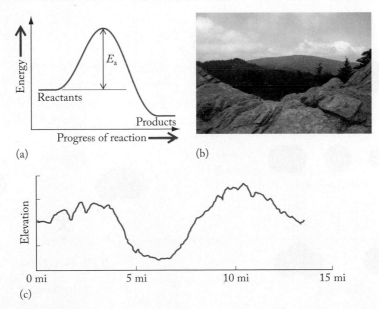

FIGURE 13.16 (a) The energy profile of a reaction includes an activation energy barrier E_a that must be overcome before the reaction can proceed. (b) A real-world analogy confronts a hiker climbing a series of hills. The Beartree Gap Trail forms part of the Iron Mountain Loop in the Mount Rogers National Recreation Area in southwestern Virginia. (c) From the starting point at mile zero, hikers must climb over two "energy barriers": Straight Mountain (at 3 mi) and Iron Mountain (at 10 mi) before returning to their starting point after a 13.5-mile (22 km) hike.

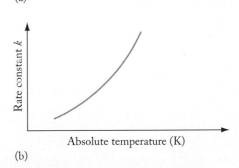

(b)

FIGURE 13.17 (a) According to kinetic molecular theory, a fraction of reactant molecules have kinetic energies equal to or greater than the activation energy (E_a) of the reaction. As temperature increases from T_1 to T_2, the number of molecules with energies exceeding E_a increases, leading to an increase in reaction rate. (b) The rate constant for any reaction increases with increasing temperature.

chemical reactions should increase with increasing temperature, and indeed they do (Figure 13.17b).

In the late 19th century, experiments carried out in the laboratories of Jacobus van 't Hoff and Svante Arrhenius led to a fundamental advance in understanding how temperature affects rates of chemical reactions. The mathematical connection between temperature, the rate constant k for a reaction, and its activation energy is given by the **Arrhenius equation**:

$$k = A\, e^{-E_a/RT} \tag{13.29}$$

where R is the gas constant in J/(mol · K) and T is the reaction temperature in kelvin. The factor A, called the **frequency factor**, is the product of collision frequency and a term that corrects for the fact that not every collision results in a chemical reaction. Because the exponential part of the equation is unitless, the frequency factor A must have the same units as the rate constant k.

Some collisions do not lead to products because the colliding molecules are not oriented relative to each other in the right way. To examine the importance of molecular orientation during collisions, let's revisit the reaction (Equation 13.7) between O_3 and NO:

$$O_3(g) + NO(g) \rightarrow O_2(g) + NO_2(g)$$

Two ways that ozone and nitric oxide molecules might approach each other are shown in Figure 13.18. Only one of these orientations, the one in which an O_3 molecule approaches the nitrogen atom of NO, leads to a chemical reaction between the two molecules.

A collision between O_3 and NO molecules with the correct orientation and enough kinetic energy may result in the formation of the **activated complex** shown in Figure 13.18(a). In this species, one of the O—O bonds in the O_3 molecule has started to break and the new N—O bond is beginning to form. Activated complexes represent midway points in chemical reactions. They have extremely brief

CONNECTION The van 't Hoff equation (Chapter 11) is named after the same Jacobus van 't Hoff who studied the temperature dependence of reaction rates. Arrhenius acids (Chapter 8) are named for the same Svante Arrhenius who proposed what we now call the Arrhenius equation.

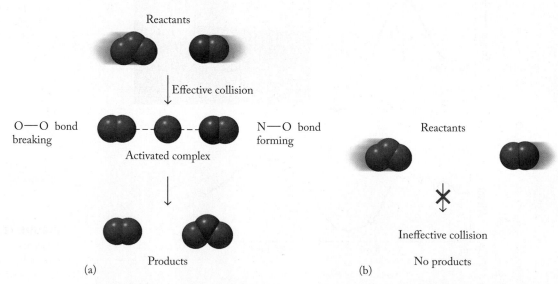

FIGURE 13.18 The effect of molecular orientation on reaction rate. (a) When an O_3 and a NO molecule are oriented such that the collision is between an O_3 oxygen and the NO nitrogen, the collision is effective and an activated complex forms, yielding the two product molecules NO_2 and O_2. (b) When the reactant molecules are oriented such that the collision is between an O_3 oxygen and the NO oxygen, no activated complex forms and no reaction occurs.

lifetimes and fall apart rapidly, either forming products or re-forming reactants. Activated complexes are formed by reacting species that have acquired enough potential energy to react with each other. The internal energy of an activated complex represents a high-energy **transition state** of the reaction. In fact, the energy of an activated complex for a reaction defines the height of the activation energy barrier for the reaction. The magnitudes of activation energies can vary from a few kilojoules to hundreds of kilojoules per mole.

We can draw an **energy profile** for a chemical reaction that shows the changes in potential energy for the reaction as a function of the *progress of the reaction* from reactants to products. We have already seen one energy profile in Figure 13.16. Now consider the energy profile for the reaction between nitric oxide and ozone, shown in Figure 13.19(a). The *x*-axis represents the progress of the reaction and the *y*-axis represents chemical energy. The activation energy is equivalent to the difference in energy between the transition state and either the reactants or the products. The size of the activation energy barrier depends on the direction from which it is approached. In the forward direction ($NO + O_3 \rightarrow NO_2 + O_2$, Figure 13.19a), E_a is smaller than in the reverse direction ($NO_2 + O_2 \rightarrow NO + O_3$, Figure 13.19b). A smaller activation energy barrier means that the forward reaction proceeds at a higher rate than the reverse reaction, assuming equal concentrations of reactants and products.

One of the many uses of the Arrhenius equation is to calculate the value of E_a for a chemical reaction. When we take the natural logarithm of both sides of Equation 13.29:

$$\ln k = -\frac{E_a}{R}\left(\frac{1}{T}\right) + \ln A \qquad (13.30)$$

the result fits the general equation of a straight line ($y = mx + b$) if we make ($\ln k$) the *y* variable and ($1/T$) the *x* variable. We can calculate E_a by determining the

▶II **CHEMTOUR** Arrhenius Equation

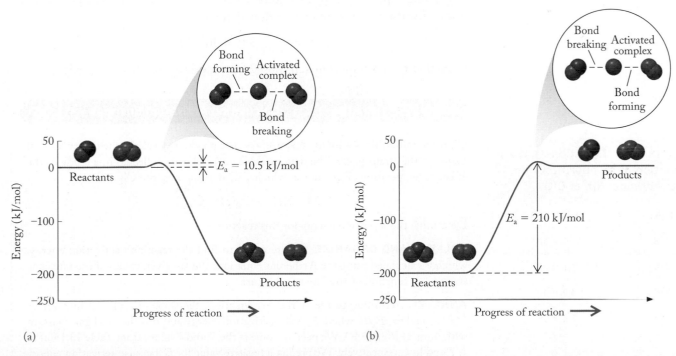

(a) (b)

FIGURE 13.19 (a) The energy profile for the reaction $NO(g) + O_3(g) \rightarrow NO_2(g) + O_2(g)$ includes an activation energy barrier of 10.5 kJ/mol. (b) The reverse reaction has a much larger activation energy of 210 kJ/mol.

TABLE 13.14 **Temperature Dependence of the Rate of the Reaction** $NO(g) + O_3(g) \rightarrow NO_2(g) + O_2(g)$

T (K)	k (M⁻¹ s⁻¹)	ln k	1/T (K⁻¹)
300	1.21×10^{10}	23.216	3.33×10^{-3}
325	1.67×10^{10}	23.539	3.08×10^{-3}
350	2.20×10^{10}	23.814	2.86×10^{-3}
375	2.79×10^{10}	24.052	2.67×10^{-3}
400	3.45×10^{10}	24.264	2.50×10^{-3}
425	4.15×10^{10}	24.449	2.35×10^{-3}

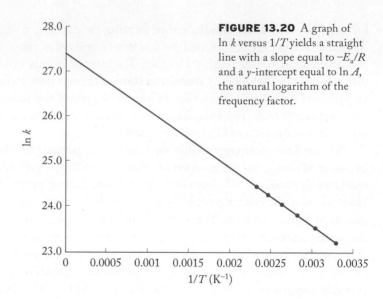

FIGURE 13.20 A graph of ln k versus $1/T$ yields a straight line with a slope equal to $-E_a/R$ and a y-intercept equal to ln A, the natural logarithm of the frequency factor.

▶❚❚ **CHEMTOUR** Collision Theory

rate constant k for the reaction at several temperatures. Plotting ln k versus $1/T$ should give a straight line, the slope of which is $-E_a/R$. Table 13.14 and Figure 13.20 show data for the reaction between NO and O$_3$ at six temperatures. The straight line that best fits the points in the figure has a slope of -1.26×10^3 K, and the activation energy of the reaction is

$$E_a = -\text{slope} \times R$$

$$= -(-1.26 \times 10^3 \text{ K}) \times \left(8.314 \frac{\text{J}}{\text{mol} \cdot \text{K}}\right) = 1.05 \times 10^4 \text{ J/mol}$$

$$= 10.5 \text{ kJ/mol}$$

The y-intercept ($1/T = 0$) in Figure 13.20 is 27.41. From Equation 13.30, we know that this value represents ln A, which means

$$A = e^{27.41} = 8.0 \times 10^{11} \ M^{-1} \ s^{-1}$$

Note that the units on A match those of the k values used to calculate it.

SAMPLE EXERCISE 13.9 **Calculating Activation Energies** **LO6**

TABLE 13.15(a) **Rate Constant as a Function of Temperature for the Decomposition of ClO**

T (K)	k (M⁻¹ s⁻¹)
238	1.9×10^9
258	3.1×10^9
278	4.9×10^9
298	7.2×10^9

Chlorine monoxide is a highly reactive gas that plays a key role in the destruction of ozone in the stratosphere. The data in Table 13.15(a) were collected in a study of the effect of temperature on the rate at which ClO decomposes into Cl$_2$ and O$_2$:

$$2 \ ClO(g) \rightarrow Cl_2(g) + O_2(g)$$

Determine the activation energy for this reaction.

COLLECT AND ORGANIZE We are asked to find the reaction's activation energy, which is calculated using the Arrhenius equation. We are given values of the rate constant k as a function of absolute temperature.

ANALYZE According to the Arrhenius equation, the slope of a plot of ln k versus $1/T$ is equal to $-E_a/R$, where E_a is the activation energy and R is the ideal gas constant with units of J/(mol · K). We need to convert the T and k values from Table 13.15(a) to $1/T$ and ln k, respectively. We predict a positive value for E_a because activation energy represents a barrier that costs energy to overcome. The rate constant for the reaction is fairly large, about $10^9 \ M^{-1} \ s^{-1}$, so we predict that the activation energy barrier will be relatively low.

TABLE 13.15(b) **Summary of T, 1/T, k, and ln k for the Decomposition of ClO**

T (K)	1/T (K⁻¹)	k (M⁻¹ s⁻¹)	ln k
238	4.20×10^{-3}	1.9×10^{9}	21.365
258	3.88×10^{-3}	3.1×10^{9}	21.855
278	3.60×10^{-3}	4.9×10^{9}	22.313
298	3.36×10^{-3}	7.2×10^{9}	22.697

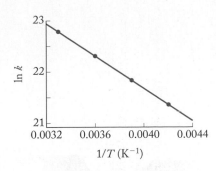

FIGURE 13.21

SOLVE Expanding the data table to include columns of ln k and $1/T$ yields Table 13.15(b). A plot of ln k versus $1/T$ gives us a straight line, the slope of which is -1590 K (Figure 13.21). We use the values of the slope and R [8.314 J/(mol · K)] to calculate the value of E_a:

$$E_a = -\text{slope} \times R$$

$$= -(-1590\ \cancel{K}) \times 8.314\ \frac{J}{mol \cdot \cancel{K}} = 1.32 \times 10^4\ J/mol$$

$$= 13.2\ kJ/mol$$

THINK ABOUT IT The activation energy is the height of a barrier that has to be overcome before a reaction can proceed. As predicted, the activation energy for ClO decomposition has a small positive value.

Practice Exercise The rate constant for the reaction

$$Br(g) + O_3(g) \rightarrow BrO(g) + O_2(g)$$

was determined at the four temperatures shown in Table 13.16. Calculate the activation energy for this reaction.

TABLE 13.16 **Rate Constants as a Function of Temperature for the Reaction of Br with O₃**

T (K)	k [cm³/(molecule · s)]
238	5.9×10^{-13}
258	7.7×10^{-13}
278	9.6×10^{-13}
298	1.2×10^{-12}

Calculation of activation energies using the graphical method generally requires measurement of the rate constant at a minimum of three different temperatures to verify that the plot of ln k versus $1/T$ is a straight line. Once we know the value of E_a, we can use it and the value of the rate constant (k_1) of a reaction at one temperature (T_1) to calculate the value of the rate constant (k_2) at another temperature (T_2). We start by substituting k_1, k_2, T_1, and T_2 into Equation 13.30:

$$\ln k_1 = -\frac{E_a}{R}\left(\frac{1}{T_1}\right) + \ln A \qquad \ln k_2 = -\frac{E_a}{R}\left(\frac{1}{T_2}\right) + \ln A$$

Subtracting these two expressions, $\ln k_1 - \ln k_2$, gives us

$$\ln k_1 - \ln k_2 = \left[-\frac{E_a}{R}\left(\frac{1}{T_1}\right) + \ln A\right] - \left[-\frac{E_a}{R}\left(\frac{1}{T_2}\right) + \ln A\right]$$

Using the mathematical properties of logarithms, we can rearrange this equation:

$$\ln \frac{k_1}{k_2} = \frac{E_a}{R}\left(\frac{1}{T_2}\right) - \frac{E_a}{R}\left(\frac{1}{T_1}\right)$$

$$= \frac{E_a}{R}\left(\frac{1}{T_2} - \frac{1}{T_1}\right) \qquad\qquad (13.31)$$

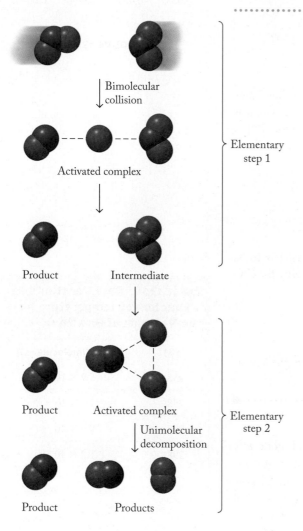

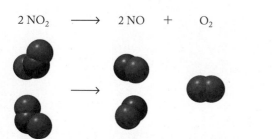

FIGURE 13.22 The decomposition of NO_2 begins when two NO_2 molecules collide, producing NO and NO_3 (elementary step 1). The NO_3 then rapidly decomposes into NO and O_2 (elementary step 2).

13.5 Reaction Mechanisms

We ended the previous section with an equation that lets us calculate the rate constant of a reaction at any temperature from the value of k at a particular temperature and the value of E_a. There is an assumption built into doing such a calculation: that the reaction proceeds in the same way no matter what the temperature is.[2] In other words, Equation 13.31 holds as long as the reaction happens via the same set of collisions and involves the same sequence of bonds breaking in reactants and forming in products at both T_1 and T_2. These molecular steps define its **reaction mechanism**. In this section we explore how kinetics data enable chemists to acquire this molecular view of how chemical reactions happen.

Elementary Steps

Let's start by revisiting the thermal decomposition of NO_2:

$$2 NO_2(g) \rightarrow 2 NO(g) + O_2(g)$$

We noted in Section 13.3 that this reaction is second order in NO_2 because it takes place as a result of the collisions of *pairs* of NO_2 molecules. How do the atoms in two colliding NO_2 molecules rearrange themselves to form two NO molecules and one O_2 molecule? One answer to this question is contained in the two-step reaction pathway shown in Figure 13.22. In the first step, a collision between two NO_2 molecules produces a molecule of NO and a molecule of NO_3. In a second step, the NO_3 decomposes to NO and O_2. Both steps in the mechanism involve transient activated complexes. In the activated complex of the first step, two molecules share an oxygen atom. The bonds in the activated complex of the second step rearrange so that two oxygen atoms become bonded together, forming a molecule of O_2 and leaving behind a molecule of NO. The molecule NO_3 is an **intermediate** in this mechanism because it is produced in one step and consumed in the next. Intermediates are not considered reactants or products and do not appear in the equation describing a reaction. In some cases, intermediates in chemical reactions are sufficiently long-lived to be

---

[2]This assumption is not always valid, especially when T_1 and T_2 are far apart. Under these circumstances Equation 13.31 may not hold.

isolated. In contrast, activated complexes have never been isolated, although they have been detected.

This reaction mechanism is a combination of two **elementary steps**. An elementary step that involves a single molecule is called **unimolecular**, and one that involves a collision between two molecules is **bimolecular**. Bimolecular elementary steps are much more common than **termolecular** (three-molecule) elementary steps because the chance of three molecules colliding at exactly the same time is much smaller. The terms *uni-*, *bi-*, and *termolecular* are used by chemists to describe the **molecularity** of an elementary step, which refers to the number of reacting atoms, ions, or molecules involved in that step. The molecularity of an elementary step is the same as its reaction order.

A valid reaction mechanism must be consistent with the stoichiometry of the reaction. In other words, the sum of the elementary steps in Figure 13.22 must be consistent with the observed proportions of reactants and products as defined in the balanced chemical equation. In this case, the sum matches the stoichiometry:

Elementary step 1 $2 NO_2(g) \rightarrow NO(g) + NO_3(g)$

Elementary step 2 $NO_3(g) \rightarrow NO(g) + O_2(g)$

Summing the two elementary steps and canceling out the intermediate (NO_3) terms yields the net reaction:

$$2 NO_2(g) + \cancel{NO_3(g)} \rightarrow 2 NO(g) + \cancel{NO_3(g)} + O_2(g)$$

$$2 NO_2(g) \rightarrow 2 NO(g) + O_2(g)$$

Before ending this discussion, let's consider how activation energy applies to a two-step reaction such as this one. The two elementary steps produce an energy profile with two maxima. In elementary step 1, collisions between pairs of NO_2 molecules result in the formation of an activated complex associated with the first transition state in Figure 13.23. As this activated complex transforms into NO and NO_3, the energy of the system drops to the bottom of the trough between the two maxima. In elementary step 2, NO_3 forms the activated complex associated with the second transition state. As this complex transforms into the final products, NO and O_2, the energy of the system drops to its final level.

Figure 13.23 shows that the energy barrier for elementary step 1 is much greater than that for elementary step 2. This difference is consistent with the relative rates of the two steps: step 1 is slower than step 2. If the reaction were to proceed in the reverse direction (as NO and O_2 react, forming NO_2), the first energy barrier would be the smaller of the two, and the first elementary step would be the more rapid one. Experimental evidence supports these expectations.

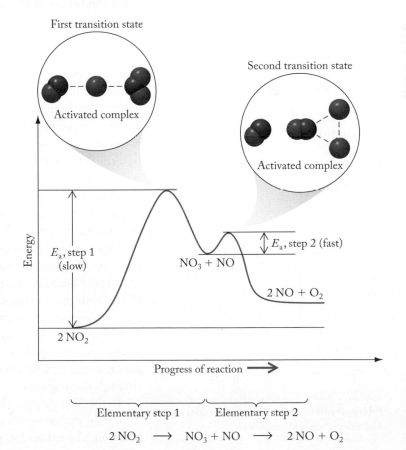

FIGURE 13.23 The energy profile for the decomposition of NO_2 to NO and O_2 shows activation energy barriers for both elementary steps. The activation energy of the first step is larger than that of the second step, so the first step is the slower of the two.

► II **CHEMTOUR** Reaction Mechanisms

For a reaction mechanism with n elementary steps, how many activation energies does the overall reaction have?

Rate Laws and Reaction Mechanisms

Any mechanism proposed for a reaction must be consistent with the rate law derived from experimental data. Is the mechanism proposed in Figure 13.22 consistent with the reaction's second-order rate law? It is if the molecularity of one of the elementary steps in the mechanism is consistent with the reaction order expressed in the rate law.

As noted in Section 13.3, the experimentally determined rate law for NO_2 decomposition is second order in NO_2:

$$\text{Rate} = k[NO_2]^2 \qquad (13.32)$$

Remember that the balanced chemical equation for a reaction does not give us enough information to write the overall rate law for the reaction. However, for any *elementary step* in a reaction mechanism, we can use the balanced chemical equation to write a rate law for that step. For the decomposition of NO_2, the rate law for the first elementary step, $2\,NO_2(g) \rightarrow NO_3(g) + NO(g)$, is

$$\text{Rate}_1 = k_1[NO_2]^2 \qquad (13.33)$$

which we obtain by writing the concentration for each reactant raised to a power equal to the coefficient of that reactant in the balanced equation for that step. For the second elementary step in the NO_2 decomposition, $NO_3(g) \rightarrow NO(g) + O_2(g)$, the sole reactant, NO_3, has a coefficient of 1, so the rate law is

$$\text{Rate}_2 = k_2[NO_3] \qquad (13.34)$$

How do we use the rate laws in Equations 13.33 and 13.34 to determine the validity of the mechanism, that is, to show that they conform to the observed rate law for the decomposition of NO_2 in Equation 13.32?

The two steps in the mechanism proceed at different rates, with different activation energies. One of the steps is slower than the other, and this is the **rate-determining step** in the reaction. The rate-determining step is the slowest elementary step in a chemical reaction, and the rate of this step controls the overall reaction rate. We can use the rate laws for the two steps to identify the rate-determining step. Because the rate law for step 1 matches the experimentally determined rate law, we may assume that step 1 defines how rapidly the reaction proceeds.

One way of visualizing the concept of a rate-determining step is to analyze the flow of people through a busy airport. Many travelers arrive at the airport with their boarding passes in hand or print them from convenient kiosks that allow them to avoid lines at ticket counters and move quickly. The next step in the process is passing through security. Typically, the number of people in the line outside the security point greatly exceeds the number of available security gates, so the time required to make it to a flight depends mostly on the time needed to pass through security. Security screening is the rate-determining step on the way through the airport.

If the first step in the mechanism in Figure 13.22 is the rate-determining step, then the value of k in Equation 13.32 is equal to k_1 from Equation 13.33. In addition, the value of k_1 must be smaller than the value of k_2. NO_3 is sufficiently

rate-determining step the slowest step in a multistep chemical reaction.

stable that we can make small amounts of it and test whether $k_1 < k_2$. Experiments run at 300 K starting with NO_2 or NO_3 yield the vastly different values $k_1 \approx 1 \times 10^{-10}\ M^{-1}\ s^{-1}$ and $k_2 \approx 6.3 \times 10^4\ s^{-1}$. Therefore, as soon as any NO_3 forms in step 1, it rapidly falls apart to NO and O_2 in step 2.

Now let's consider the reaction that is the reverse of NO_2 decomposition, namely, the formation of NO_2 from NO and O_2:

$$\text{Overall reaction} \quad 2\,NO(g) + O_2(g) \rightarrow 2\,NO_2(g)$$

We determined in Section 13.3 that this reaction is second order in NO and first order in O_2:

$$\text{Rate} = k[NO]^2[O_2] \tag{13.35}$$

One proposed mechanism is shown in Figure 13.24. It has two elementary steps:

Step 1 $\quad NO(g) + O_2(g) \rightleftharpoons NO_3(g) \quad$ Rate $= k_1[NO][O_2] \quad$ (13.36)

Step 2 $\quad NO_3(g) + NO(g) \rightarrow 2\,NO_2(g) \quad$ Rate $= k_2[NO_3][NO] \quad$ (13.37)

The rate laws in Equations 13.36 and 13.37 were obtained by writing the concentration for each reactant raised to a power equal to the reactant's coefficient in the balanced equation. If step 1 were the rate-determining step, the reaction would be first order in NO and O_2, but that is not what the experimentally determined rate law indicates. If step 2 were the rate-determining step, the reaction would be first order in NO and NO_3, but that is not consistent with the rate law either. How can we account for the experimental rate law?

Consider what happens if step 2 is slow while step 1 is fast *and reversible*, which means NO_3 forms rapidly from NO and O_2 but decomposes just as rapidly back into NO and O_2. We indicate reversibility with a double arrow. Expressing this situation in equation form:

$$\text{Rate of forward reaction} = k_f[NO][O_2] = \text{fast}$$

$$\text{Rate of reverse reaction} = k_r[NO_3] = \text{equally fast}$$

The subscripts "f" and "r" refer to the forward and reverse reactions, respectively.

Combining these two expressions, we have

$$k_f[NO][O_2] = k_r[NO_3]$$

$$[NO_3] = \frac{k_f}{k_r}[NO][O_2] \tag{13.38}$$

Now, if we replace the $[NO_3]$ term in the rate law of Equation 13.37 (which we select because step 2 is the rate-determining step) with the right side of Equation 13.38, we get

$$\text{Rate} = k_2\frac{k_f}{k_r}[NO]^2[O_2]$$

The three rate constants can be combined,

$$k_{\text{overall}} = k_2\frac{k_f}{k_r}$$

and the rate law for the overall reaction becomes

$$\text{Rate} = k_{\text{overall}}[NO]^2[O_2]$$

This expression matches the overall rate law in Equation 13.35, so the proposed mechanism, a fast and reversible step 1 followed by a slow step 2, may be valid.

(1) Formation of intermediate: elementary step 1

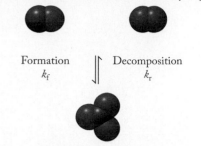

Formation k_f Decomposition k_r

(2) Reaction of intermediate: elementary step 2

k_2

Overall reaction:

$2\,NO \quad + \quad O_2 \quad \longrightarrow \quad 2\,NO_2$

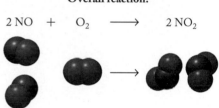

FIGURE 13.24 A mechanism for the formation of NO_2 from NO and O_2 has two elementary steps: (1) a fast, reversible bimolecular reaction in which NO and O_2 form NO_3 and (2) a slower, rate-determining bimolecular reaction in which NO_3 reacts with a molecule of NO to form two molecules of NO_2.

As a final point regarding reaction mechanisms, even though the proposed reaction mechanism is consistent with the overall stoichiometry of the reaction and with the experimentally derived rate law, these consistencies do not *prove* that the proposed mechanism is correct. We need more experimental evidence, such as detecting the presence of an intermediate in the reaction mixture. On the other hand, *not* finding a reactive (and transient) intermediate would not necessarily disprove a reaction mechanism.

SAMPLE EXERCISE 13.10   **Linking Reaction Mechanisms**   **LO7**
                                                       to Experimental Rate Laws

The experimentally determined rate law for the reduction reaction

$$2\,H_2(g) + 2\,NO(g) \rightarrow N_2(g) + 2\,H_2O(g)$$

which occurs at high temperatures is

$$Rate = k[NO]^2[H_2]$$

A proposed mechanism for the reaction is

Elementary step 1 $H_2(g) + 2\,NO(g) \rightarrow N_2O(g) + H_2O(g)$

Elementary step 2 $H_2(g) + N_2O(g) \rightarrow N_2(g) + H_2O(g)$

Is this reaction mechanism consistent with the stoichiometry of the reaction and with the rate law? If so, which is the rate-determining step?

COLLECT AND ORGANIZE We are first asked if the proposed mechanism is consistent with the reaction stoichiometry. We can use the two elementary steps to answer that question. We are then asked if the mechanism is consistent with the experimentally determined rate law. A rate-determining step has the smallest rate constant of all steps in the mechanism.

ANALYZE To answer the questions posed in this exercise we need to do the following:

1. Determine whether the chemical equations of the elementary steps add up to the overall reaction equation.
2. Write rate laws for each elementary step.
3. Compare the rate laws of the elementary steps to the experimental rate law of the overall reaction to assess the validity of the mechanism.
4. Determine which step is rate determining by matching its rate law to the observed rate law.

SOLVE Let's test whether the elementary steps add up to the overall reaction:

(1) $H_2(g) + 2\,NO(g) \rightarrow \cancel{N_2O}(g) + H_2O(g)$

(2) $\underline{H_2(g) + \cancel{N_2O}(g) \rightarrow N_2(g) + H_2O(g)}$

       $2\,H_2(g) + 2\,NO(g) \rightarrow N_2(g) + 2\,H_2O(g)$

This is indeed the equation of the overall reaction. The elementary steps are consistent with the stoichiometry of the overall reaction.

Next we need to focus on the reaction mechanism. Elementary step 1 involves two molecules of NO colliding with one molecule of H_2 in a rare termolecular reaction. Elementary step 2 is a bimolecular reaction between the N_2O produced in step 1 and another molecule of H_2. We apply the fact that the rate law for any elementary step can be written directly from the balanced equations, using the equation coefficients as exponents in the rate law:

Step 1 $H_2(g) + 2\,NO(g) \rightarrow N_2O(g) + H_2O(g)$ $Rate = k_1[H_2][NO]^2$

Step 2 $H_2(g) + N_2O(g) \rightarrow N_2(g) + H_2O(g)$ $Rate = k_2[H_2][N_2O]$

The rate law of step 1 matches the observed rate law of the overall reaction. Therefore, the proposed two-step mechanism is consistent with the experimental rate law, and step 1 is the rate-determining step.

THINK ABOUT IT We do not have direct proof that this is the correct mechanism, though it is consistent with the available data. A plausible mechanism for a reaction must yield a balanced equation whose stoichiometry matches the overall reaction and must be consistent with the rate law for the overall process. Both of these conditions are met in this Sample Exercise.

Practice Exercise The following mechanism is proposed for a reaction between compounds A and B:

Step 1	$2\,A(g) + B(g) \rightleftharpoons C(g)$	fast and reversible
Step 2	$B(g) + C(g) \rightarrow D(g)$	slow
Overall	$2\,A(g) + 2\,B(g) \rightarrow D(g)$	

What is the rate law for the overall reaction based on the proposed mechanism?

SAMPLE EXERCISE 13.11 **Testing Proposed Reaction Mechanisms** **LO7**

One proposed mechanism for the decomposition of N_2O_5 to NO_2, shown below, involves three elementary steps:

Step 1	$2\,N_2O_5(g) \rightleftharpoons N_4O_{10}(g)$	fast and reversible
Step 2	$N_4O_{10}(g) \rightarrow N_2O_3(g) + 2\,NO_2(g) + O_3(g)$	slow
Step 3	$N_2O_3(g) + O_3(g) \rightarrow 2\,NO_2(g) + O_2(g)$	fast
Overall:	$2\,N_2O_5(g) \rightarrow 4\,NO_2(g) + O_2(g)$	

What is the rate law for the overall reaction based on the proposed mechanism?

COLLECT AND ORGANIZE We are asked to find the rate law for the overall reaction, which will reflect the rate laws for the elementary reactions preceding and including the rate-determining step. We are given three elementary steps and their relative rates.

ANALYZE The rate law for an elementary step can be written directly from the balanced chemical equation for that step. We are told that step 1 is fast *and reversible*, which means that as the reaction proceeds and $[N_4O_{10}]$ increases, the rate of step 1 in the forward direction is eventually matched by the rate of step 1 in the reverse direction. Only elementary steps 1 and 2 will contribute to the rate law for the overall reaction because step 2, the only slow step, must be the rate-determining step.

SOLVE The rate laws for step 1 in the forward and reverse directions are

$$\text{Rate of forward step 1} = k_f[N_2O_5]^2$$

$$\text{Rate of reverse step 1} = k_r[N_4O_{10}]$$

The rate law for step 2 is

$$\text{Rate of step 2} = k_2[N_4O_{10}]$$

To find the overall rate law, we need to express $[N_4O_{10}]$ in terms of $[N_2O_5]$ by setting the step 1 rates equal to each other and rearranging the terms:

$$[N_4O_{10}] = \frac{k_f}{k_r}[N_2O_5]^2$$

We then substitute this expression for $[N_4O_{10}]$ into the rate law for step 2:

$$\text{Rate} = \frac{k_f k_2}{k_r}[N_2O_5]^2$$

THINK ABOUT IT The rate law for the overall reaction must depend on the concentration of N_2O_5. The order of the reaction in N_2O_5 depends on the proposed mechanism for the reaction. Notice that the rate of step 3 has no impact on the rate law for this mechanism because it follows the slowest step. In addition, note that the rate law for the mechanism in this Sample Exercise does not match the experimentally determined rate law from Sample Exercise 13.4: Rate $= k[N_2O_5]$. Therefore, the mechanism that starts with the dimerization of N_2O_5 cannot be correct.

Practice Exercise Here is another proposed mechanism for the reaction of NO with H_2 (Sample Exercise 13.10):

Elementary step 1 $H_2(g) + NO(g) \rightarrow N(g) + H_2O(g)$

Elementary step 2 $N(g) + NO(g) \rightarrow N_2(g) + O(g)$

Elementary step 3 $H_2(g) + O(g) \rightarrow H_2O(g)$

Is this a valid mechanism?

Mechanisms and One Meaning of Zero Order

Before we leave reaction mechanisms, let's revisit the reaction between NO_2 and CO, which has an experimentally determined rate law that is zero order in CO, second order in NO_2, and second order overall:

$$NO_2(g) + CO(g) \rightarrow NO(g) + CO_2(g) \qquad \text{Rate} = k[NO_2]^2$$

What does this overall rate law tell us about the reaction? Remember that the overall rate law depends on the concentration of reactants in the rate-determining step, which means that CO is not a reactant in the rate-determining step. Carbon monoxide is clearly involved in the reaction—it is converted into CO_2—but whatever step involves CO must not be the rate-determining step. This leads us to conclude that the reaction must have at least two elementary steps, one rate-determining and one not.

It has been proposed that the reaction mechanism is

(1) $2\,NO_2(g) \rightarrow NO_3(g) + NO(g)$ Rate $= k_1[NO_2]^2$

(2) $NO_3(g) + CO(g) \rightarrow NO_2(g) + CO_2\,(g)$ Rate $= k_2[NO_3][CO]$

The experimentally determined overall rate law matches the rate law for the first step, which must be the slower, rate-determining step. The overall reaction is zero order in CO because CO is not a reactant in that step.

CONCEPT TEST

The rate law for the reaction $XO_2(g) + M(g) \rightarrow XO(g) + MO(g)$, where X and M represent metallic elements, is second order in $[XO_2]$ and independent of $[M]$ for a wide variety of compounds XO_2 and M. Why can we conclude that these reactions likely proceed by the same mechanism?

13.6 Catalysts

We noted in Section 13.4 that reactions may be slow if they have a high activation energy. Suppose we wanted to increase the rate of such a reaction. How could we do it? One way is to increase the temperature of the reaction mixture. However, in some chemical reactions, elevated temperatures can lead to undesired products or to lower yields. Another way is to add a **catalyst**, a substance that increases the rate of a reaction but is not consumed in the process.

catalyst a substance added to a reaction that increases the rate of the reaction but is not consumed in the process.

Catalysts and the Ozone Layer

As we discussed in Chapter 4, the way we think about ozone depends on where the ozone is located. Ozone in the stratosphere between 10 and 40 km above Earth's surface is necessary to protect us from UV radiation, but ozone at ground level is hazardous to our health. In this section, we discuss the role of catalysts in the breakdown of stratospheric ozone that has led to the annual formation of ozone holes over Antarctica during early spring in the Southern Hemisphere (September and October).

The natural photodecomposition of ozone in the stratosphere

$$2\, O_3(g) \xrightarrow{\text{sunlight}} 3\, O_2(g) \tag{13.39}$$

begins with the absorption of UV radiation from the sun and the generation of atomic oxygen:

$$(1) \qquad O_3(g) \rightarrow O_2(g) + O(g)$$

The oxygen atom can react with an ozone molecule to form two more molecules of oxygen:

$$(2) \qquad O_3(g) + O(g) \rightarrow 2\, O_2(g)$$

The rate of the second elementary step is low because its activation energy is relatively high: 17.7 kJ/mol.

In 1974, two American scientists, Sherwood Rowland (1927–2012) and Mario Molina (b. 1943), predicted significant depletion of stratospheric ozone because of the release of a class of volatile compounds called chlorofluorocarbons (CFCs) into the atmosphere at ground level that ultimately enter the stratosphere. This prediction was later supported by experimental evidence of a thinning of the ozone layer and formation of annual ozone holes over Antarctica. In September 2000 stratospheric ozone concentrations over Antarctica were the lowest ever observed—less than half of what they were in 1980—and the ozone hole covered nearly all of Antarctica and the tip of South America (Figure 13.25).

In 1987 an international agreement known as the Montreal Protocol called for an end to the production of ozone-depleting CFCs. The Montreal Protocol has had a dramatic effect on

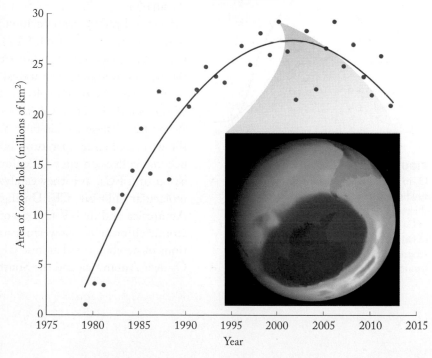

FIGURE 13.25 A 30-year trend in stratospheric ozone depletion over the South Pole. The data points represent the maximum area of the ozone hole observed during each spring in the Southern Hemisphere (usually in September or October). The satellite image shows the hole at its maximum in 2000. The violet color over Antarctica represents ozone concentrations that are less than half their normal values.

CFC production and emission into the atmosphere, and as the trend line in Figure 13.25 indicates, the ozone layer over Antarctica may be slowly recovering. In October 2012 the area of the ozone hole reached 21.2 million km², down from the 2000 maximum of 29.4 million km².

How do CFCs contribute to the destruction of ozone? Three of the more widely used CFCs were CCl_2F_2, CCl_3F, and $CClF_3$. In the stratosphere, they encounter UV radiation with enough energy to break C—Cl bonds, releasing chlorine atoms:

$$CCl_3F(g) \rightarrow CCl_2F(g) + Cl(g) \qquad (13.40)$$

Free chlorine atoms react with ozone, forming chlorine monoxide:

$$Cl(g) + O_3(g) \rightarrow ClO(g) + O_2(g) \qquad (13.41)$$

Chlorine monoxide then reacts with ozone, producing oxygen and regenerating atomic chlorine:

$$ClO(g) + O_3(g) \rightarrow Cl(g) + 2\,O_2(g) \qquad (13.42)$$

If we add Equations 13.41 and 13.42 and cancel species as needed, we get

$$2\,O_3(g) \rightarrow 3\,O_2(g)$$

The overall reaction is exactly the same as the natural photodecomposition of ozone. The difference is the presence of chlorine atoms in Equations 13.41 and 13.42. Chlorine atoms act as a catalyst for the destruction of ozone because they speed up the reaction but are not consumed by it. Rather, they are consumed in one elementary step but then regenerated in a later elementary step. Chlorine is a catalyst, not an intermediate, because a catalyst is consumed in an early step of a reaction mechanism and then regenerated in a later step, whereas an intermediate is produced before it is consumed. A single chlorine atom can catalyze the destruction of hundreds to thousands of stratospheric O_3 molecules before it combines with other atoms and forms a less reactive molecule.

The activation energy for the Cl-catalyzed reaction is only 2.2 kJ/mol, whereas the activation energy for the uncatalyzed reaction is 17.7 kJ/mol (Figure 13.26). Its smaller E_a value means that the catalyzed destruction of ozone is faster than the natural photodecomposition process. In the reaction describing the destruction of ozone, the catalyst, $Cl(g)$, and reactant, $O_3(g)$, exist in the same physical phase. When a catalyst and the reacting species are in the same phase, we call the catalyst a **homogeneous catalyst**.

The rate of ozone depletion accelerates over the South Pole toward the end of the Southern Hemisphere winter because of the clouds that cover much of Antarctica. These clouds form as crystals of ice and tiny drops of nitric acid in the winter when temperatures in the lower stratosphere dip to −80°C. The clouds become collection sites for chlorine compounds formed by the photodecomposition of CFCs, and they catalyze the transformation of these compounds into molecular chlorine, Cl_2. During late winter and early spring the sun rises over Antarctica and its UV rays decompose Cl_2 into free Cl atoms. As we have seen, atomic chlorine is a powerful catalyst for ozone destruction, and high concentrations of newly formed atomic Cl mean rapid and widespread loss of stratospheric O_3 over Antarctica and the Southern Ocean (Figure 13.25).

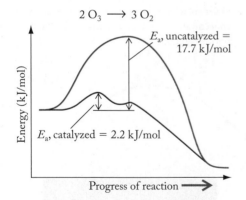

FIGURE 13.26 The decomposition of O_3 in the presence of chlorine atoms has a smaller activation energy (2.2 kJ/mol) than the naturally occurring photodecomposition of O_3 to O_2 (17.7 kJ/mol). The catalytic effect of chlorine is a key factor in the depletion of stratospheric ozone and the formation of an ozone hole over the South Pole.

SAMPLE EXERCISE 13.12 **Identifying Catalysts in Reaction Mechanisms** **LO8**

A reaction mechanism proposed for the decomposition of ozone in the presence of NO at high temperatures consists of three elementary steps:

$$(1) \quad O_3(g) + NO(g) \rightarrow O_2(g) + NO_2(g)$$

$$(2) \quad NO_2(g) \rightarrow NO(g) + O(g)$$

$$(3) \quad O(g) + O_3(g) \rightarrow 2\,O_2(g)$$

If the rate of the overall reaction is higher than the rate of the uncatalyzed decomposition of ozone to oxygen (Equation 13.39), is NO a catalyst in the reaction?

COLLECT AND ORGANIZE We are asked to determine whether NO is a catalyst in a reaction. A catalyst increases the rate of a reaction and is not consumed by the overall reaction. We are given the elementary steps of the reaction and are told that the reaction is more rapid in the presence of NO.

ANALYZE We can sum the reactions to determine the overall reaction. If NO is consumed in an early step before it is regenerated in a later one and it is not consumed in the overall process, it is a catalyst.

SOLVE Summing the three elementary steps gives the overall reaction:

$$O_3(g) + \cancel{NO(g)} + \cancel{NO_2(g)} + \cancel{O(g)} + O_3(g) \rightarrow$$
$$O_2(g) + \cancel{NO_2(g)} + \cancel{NO(g)} + \cancel{O(g)} + 2\,O_2(g)$$

$$2\,O_3(g) \rightarrow 3\,O_2(g)$$

This equation does not include NO, and the rate of the reaction is higher when NO is present. Thus NO fulfills both requirements for being a catalyst.

THINK ABOUT IT NO behaves much like the Cl atoms in Equations 13.41 and 13.42. NO is not an intermediate because it is consumed before it is produced.

Practice Exercise The combustion of fossil fuels results in the release of SO_2 into the atmosphere, where it reacts with oxygen to form SO_3:

$$2\,SO_2(g) + O_2(g) \rightarrow 2\,SO_3(g)$$

In the atmosphere, SO_2 may react with NO_2, forming SO_3 and NO:

$$NO_2(g) + SO_2(g) \rightarrow NO(g) + SO_3(g)$$

The rate of reaction of SO_2 with NO_2 is faster than the rate of reaction of SO_2 with oxygen. If the NO produced in the reaction of NO_2 and SO_2 is then oxidized to NO_2:

$$2\,NO(g) + O_2(g) \rightarrow 2\,NO_2(g)$$

is NO_2 a catalyst in the reaction of SO_2 with O_2?

Catalytic Converters

We started this chapter with a discussion of air pollution caused by vehicles and of the technology that has been developed to clean the air. Figure 13.27 shows a catalytic converter in a car's exhaust system and how it provides a surface on which NO in engine exhaust decomposes to N_2 and O_2. A catalytic converter is an example of a **heterogeneous catalyst**, where reactants and catalyst have different phases. To maximize the surface area inside a catalytic converter, it is filled with a fine honeycomb mesh coated with transition metals such as palladium,

heterogeneous catalyst a catalyst in a different phase than the reactants.

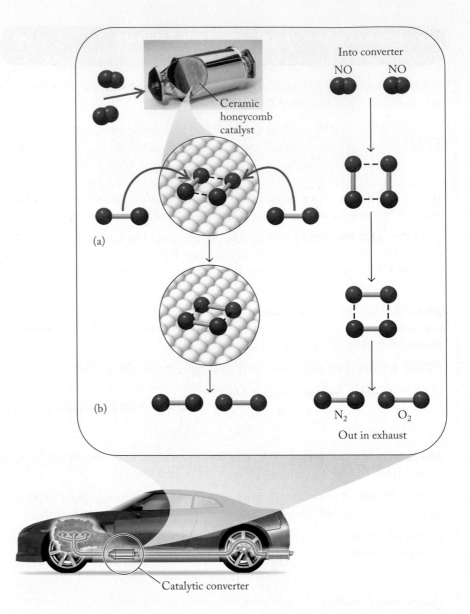

FIGURE 13.27 Catalytic converters in automobiles reduce emissions of NO by lowering the activation energy of its decomposition into N_2 and O_2. Metal catalysts are supported on a porous ceramic honeycomb. At the molecular level, (a) NO molecules are adsorbed onto the surface of metal clusters where their NO bonds are broken, and (b) pairs of O atoms and N atoms form O_2 and N_2. The O_2 and N_2 desorb from the surface and are released to the atmosphere.

platinum, and rhodium. The metals that make up catalytic coatings are first dissolved as metal salts and dispersed on the mesh, where they are reduced to metallic clusters 2 to 10 nm in diameter. The small size of these clusters and the large number of them add to the total surface area available to promote the removal of pollutants. Some of these clusters specifically speed up the decomposition of NO, while others promote the oxidation of carbon monoxide to CO_2 and unburned hydrocarbons to CO_2 and water vapor.

Many gas-phase reactions that occur on solid heterogeneous catalysts exhibit zero-order reaction kinetics. This happens when there are more reactant molecules in the system than there are catalytic surface sites available to promote their decomposition. Under these conditions, adding more reactant has no effect on the catalyzed reaction rate, and the reaction appears to be zero order.

In the reactions responsible for the production of photochemical smog, we saw that determinations of the rates of chemical reactions are crucial for us to comprehend processes on a molecular level. The mechanisms of the reactions of volatile oxides produced during combustion and by other natural events must be thoroughly understood so that problems arising because of the presence of these substances in the environment can be effectively managed. The continued development of catalytic converters for vehicles to diminish the problems caused by burning fossil fuels and to clean our air requires a thorough knowledge of the kinetics and mechanisms of many of the reactions discussed here.

SAMPLE EXERCISE 13.13   Integrating Concepts: Chocolate-Covered Cherries

Chocolate-covered cherries consist of a solid chocolate shell enclosing a sweet liquid that surrounds a cherry. To make chocolate-covered cherries, the fruits are initially covered with fondant, a doughlike mixture of powdered sugar (sucrose), butter, cherry juice, and the enzyme invertase. They are then dipped in chocolate, and they must sit for a period of time before they are eaten.

The reason for these instructions involves the reactions of sucrose (table sugar) and products derived from it. Sucrose hydrolyzes to produce two isomeric sugars, glucose and fructose, as shown in Figure 13.28. Invertase acts as a catalyst for this reaction. The mixture of glucose and fructose is called invert sugar, and it is a liquid. Table 13.17(a) contains data on the concentration of sucrose in the fondant as a function of time at 5°C, the temperature of a chilled case where the coated cherries are allowed to rest until ready. If the cherries will be ready to eat when 95% of the sucrose has been converted to glucose and fructose, how long do you have to wait before eating the cherries? What would happen if we left the invertase catalyst out of the recipe (assume the reaction is spontaneous)?

COLLECT AND ORGANIZE We are given data on changing concentration with time. We are asked how long it will take for a specific amount of the reactant to have been consumed, and we need to think about the role of the catalyst in this reaction.

ANALYZE We can use the available data to derive a rate law for the reaction and to estimate the time we have to wait for the cherries to be ready. The reaction takes place in an aqueous solution, so we may assume that the concentration of water is very large and does not change during the course of the reaction. Therefore, any change in the rate of the reaction is due only to consumption of sucrose. Inspection of the data reveals that half of the initial concentration of sucrose is consumed between 48 and 96 hours of reaction time. For 95% of the sugar to be gone, a time equivalent to between 4 and 5 half-lives will be required, so we estimate that we will have to wait a long while (more than 1 week) before we can eat the cherries.

SOLVE Let's start by testing whether the reaction is first order in sucrose. If it is, a plot of ln[sucrose] versus time (Table 13.17b) will be linear.

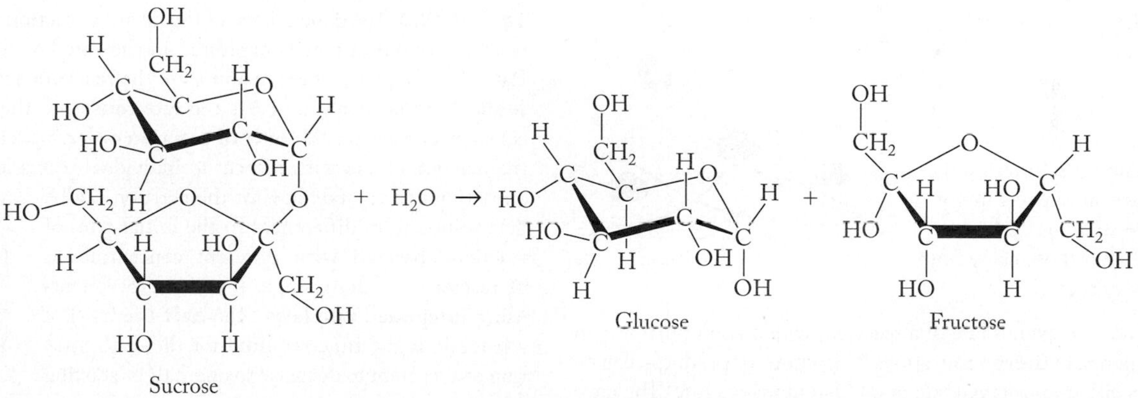

FIGURE 13.28

TABLE 13.17(a) Concentration of Sucrose as a Function of Time	
Time (hr)	[Sucrose] in Fondant (g/cm³)
0	0.844
24	0.681
48	0.550
96	0.357
192	0.153

TABLE 13.17(b) Concentration of Sucrose and ln[Sucrose] as a Function of Time		
Time (hr)	[Sucrose] in Fondant (g/cm³)	ln[Sucrose]
0	0.844	−0.170
24	0.681	−0.384
48	0.550	−0.598
96	0.357	−1.030
192	0.153	−1.877

Plotting ln[sucrose] versus t does indeed yield a straight line (Figure 13.29), which means the reaction is first order in sucrose. The slope of the line that best fits the data points is -0.00889 hr^{-1}, which equals $-k$, so the rate expression is

$$\text{Rate} = (0.00889 \text{ hr}^{-1})[\text{sucrose}]$$

Using the integrated rate law for a first-order reaction (Equation 13.17), we can write

$$\ln \frac{[\text{sucrose}]}{[\text{sucrose}]_0} = -kt$$

and use this expression to calculate the time t necessary for 95% of the sucrose to react. If 95% of the sugar has reacted, that means the amount of sugar remaining is 5% of the original amount, or

$$(0.05)(0.844 \text{ g/cm}^3) = 0.0422 \text{ g/cm}^3$$

We substitute this value into the integrated rate law and solve for t:

$$\ln \frac{0.0422 \text{ g/cm}^3}{0.844 \text{ g/cm}^3} = -(0.00889 \text{ hr}^{-1})t$$

$$t = 337 \text{ hr, or about 14 days}$$

If we left the catalyst out, the reaction would still be spontaneous, but a lot slower. A catalyst does not change the value of ΔG for a process; it only speeds up the process.

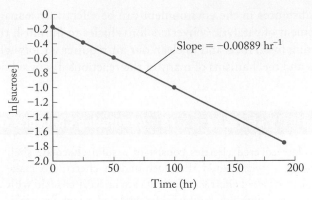

FIGURE 13.29

THINK ABOUT IT Based on our estimation considering half-lives, our answer seems reasonable. Enzymes are catalysts in biological systems, and invertase speeds up this reaction in the fondant, which would otherwise be even slower. We could probably make the reaction go a bit faster by keeping the cherries at a higher temperature, but then the chocolate would soften and melt, so the candy is usually stored at a temperature lower than room temperature while the liquid centers are being formed.

SUMMARY

Section 13.1 The reactions that follow the formation of NO in vehicle engines can produce **photochemical smog**, a major air pollutant in urban areas. Familiarity with **chemical kinetics**, the study of reaction rates, is important in understanding how natural processes and those caused by human activity occur.

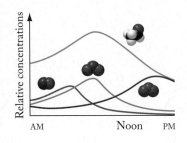

Section 13.2 The overall rate of a reaction, which is proportional to the rate of change in the concentrations of reactants or products, can be expressed as either an average rate or an instantaneous rate. The rates of consumption of reactants and formation of products are related by the stoichiometry of the reaction and are typically expressed in molarity per unit time. **Reaction rates** are typically determined from experimental measurements. In most reactions, the reaction rate decreases as the reaction proceeds.

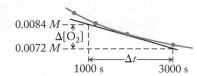

Section 13.3 The dependence of the rate of a reaction $X + Y \rightarrow Z$ on reactant concentrations is expressed in the **rate law** for the reaction: Rate $= k[X]^m[Y]^n$, where m and n are the **reaction order** in reactants X and Y, respectively, and k is the **rate constant**. The units of a rate constant depend on the **overall reaction order**, which is the sum of the reaction orders with respect to individual reactants. The order of a reaction and the rate law for the reaction can be determined from differences in the initial rates of reaction observed with different concentrations of reactants or from single kinetics experiments using **integrated rate laws**. The **half-life ($t_{1/2}$)** of a reaction is the time required for the concentration of a reactant to decrease to one-half its starting concentration.

Section 13.4 Increasing the temperature of a chemical reaction increases its rate. The **activation energy (E_a)** of a reaction is a barrier that separates the sum of the internal energies of the reactants from the energies of the products. The top of the energy barrier is the **transition state** related to the internal energy of a short-lived **activated complex**. Reactions with large activation energies are usually slow. Measuring the rate constant of a reaction at different temperatures allows the calculation of activation energies using the **Arrhenius equation**.

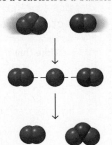

Section 13.5 Rate studies give insight into **reaction mechanisms,** which describe what is happening at a molecular level. A reaction mechanism consists of one or more **elementary steps** that describe how the reaction takes place. The overall reaction is the sum of these elementary steps. Elementary steps that involve one, two, or three molecules are said to be **unimolecular, bimolecular,** and **termolecular,** respectively. The rate law for a reaction applies to the slowest elementary step, which is called the **rate-determining step.** The proposed mechanism for any reaction must be consistent with the observed rate law and with the stoichiometry of the overall reaction.

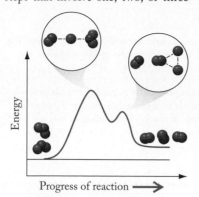

Progress of reaction ⟶

Section 13.6 Catalysts increase reaction rates by changing the mechanism of a reaction and by decreasing the activation energy. A **homogeneous catalyst** is one that is in the same phase as the reactants in the reaction being catalyzed. A **heterogeneous catalyst** is one that is in a phase different from the phase of the reactants.

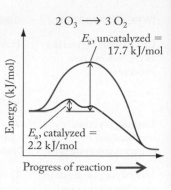

PROBLEM-SOLVING SUMMARY ▪

TYPE OF PROBLEM	CONCEPTS AND EQUATIONS	SAMPLE EXERCISES
Predicting relative reaction rates	Determine relative rates from the stoichiometry of the balanced chemical equation.	13.1
Determining an instantaneous reaction rate	Determine the slope of a line tangent to a point on the plot of concentration versus time.	13.2
Deriving a rate law from initial reaction rate data	Compare the change in rate when the concentration of one reactant is changed (while the concentrations of other reactants are kept constant) to determine the reaction order (usually whole numbers) with respect to that reactant. $$n = \frac{\log\left(\dfrac{\text{Rate}_1}{\text{Rate}_2}\right)}{\log\left(\dfrac{[X]_1}{[X]_2}\right)} \qquad (13.14)$$	13.3
Using integrated rate laws	A linear plot of the natural logarithm of concentration versus time indicates a first-order reaction with a slope of $-k$.	13.4
Calculating the half-life of a first-order reaction	In a first-order reaction, $$t_{1/2} = 0.693/k \qquad (13.20)$$	13.5
Distinguishing between first- and second-order reactions	A linear plot of the natural logarithm of reactant concentration ($\ln[X]$) versus time indicates a first-order reaction, whereas a linear plot of the reciprocal of reactant concentration ($1/[X]$) versus time indicates a second-order reaction.	13.6
Calculating the half-life of a second-order reaction	In a second-order reaction, $$t_{1/2} = 1/k[X]_0 \qquad (13.25)$$	13.7
Deriving pseudo-first-order rate laws	A plot of the natural logarithm of concentration of the limiting reactant versus time is linear. The slope of the plot gives $k[\text{excess reactant}] = k'$.	13.8

TYPE OF PROBLEM	CONCEPTS AND EQUATIONS	SAMPLE EXERCISES
Calculating activation energies from rate constants	Using the logarithmic form of the Arrhenius equation, $$\ln k = -\frac{E_a}{R}\left(\frac{1}{T}\right) + \ln A \qquad (13.30)$$ plot $\ln k$ versus $1/T$. The slope is $-E_a/R$.	13.9
Linking reaction mechanisms to experimental rate laws	The order of each reactant in an elementary step equals its coefficient in that step. The rate law for the mechanism must be the same as the observed rate law and does not include intermediates.	13.10, 13.11
Identifying catalysts in reaction mechanisms	Determine whether or not a potential catalyst is present by summing the elementary steps in the reaction to get the overall reaction. If that procedure reveals a potential catalyst, determine whether it increases the rate of reaction and whether it is initially consumed and then regenerated in the process.	13.12

VISUAL PROBLEMS

(Answers to boldface end-of-chapter questions and problems are in the back of the book.)

13.1. Nitrous oxide decomposes to nitrogen and oxygen in the following reaction:

$$2\,N_2O(g) \rightarrow 2\,N_2(g) + O_2(g)$$

In Figure P13.1, which curve represents $[N_2O]$ and which curve represents $[O_2]$?

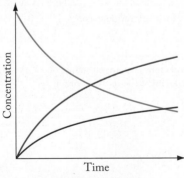

FIGURE P13.1

13.2. Sulfur trioxide is formed in the reaction

$$SO_2(g) + \tfrac{1}{2}\,O_2(g) \rightarrow SO_3(g)$$

In Figure P13.2, which curve represents $[SO_2]$ and which curve represents $[O_2]$? All three gases are present initially.

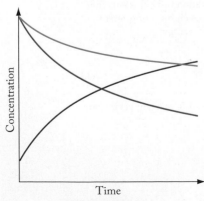

FIGURE P13.2

13.3. The rate law for the reaction $2\,A \rightarrow B$ is second order in A. Figure P13.3 represents samples with different concentrations of A; the red spheres represent molecules of A. In which sample will the reaction $A \rightarrow B$ proceed most rapidly?

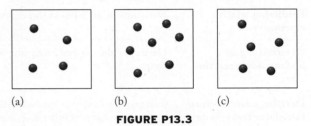

(a)   (b)   (c)

FIGURE P13.3

13.4. The rate law for the reaction $A + B \rightarrow C$ is first order in both A and B. Figure P13.4 represents samples with different concentrations of A (red spheres) and B (blue spheres). In which sample will the reaction $A + B \rightarrow C$ proceed most rapidly?

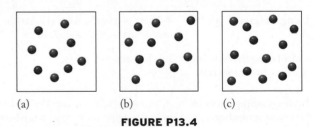

(a)   (b)   (c)

FIGURE P13.4

13.5. Using the numbers 1–5 on the reaction profile (Figure P13.5), identify the following:
   a. The energy of the reactants
   b. The energy of the products
   c. The activation energy of the forward reaction
   d. The activation energy of the reverse reaction
   e. The energy change of the reaction

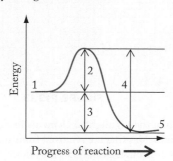

FIGURE P13.5

13.6. Select the diagram from those given in Figure P13.6 that best corresponds to the following:
   a. A highly exothermic reaction with a large activation energy
   b. A highly endothermic reaction with a small activation energy
   c. A reaction involving a stable intermediate

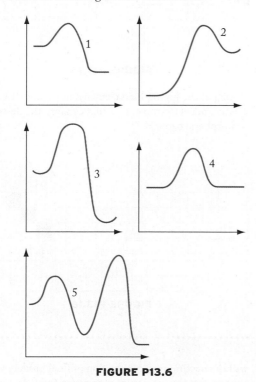

FIGURE P13.6

13.7. Which of the reaction profiles in Figure P13.7 represents the reaction with the smallest rate constant at constant T?

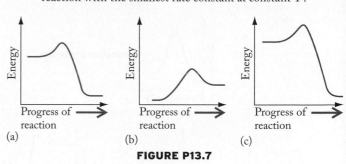

FIGURE P13.7

13.8. Which of the reaction profiles in Figure P13.8 represents the reaction with the largest rate constant at constant T?

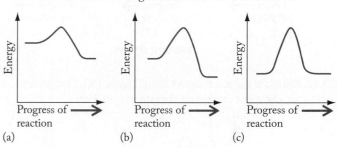

FIGURE P13.8

13.9. Which of the following mechanisms is consistent with the reaction profile shown in Figure P13.9?
   a. $2\,A \xrightarrow{\text{slow}} B$
      $B \xrightarrow{\text{fast}} C$
   b. $A + B \rightarrow C$
   c. $2\,A \underset{}{\overset{\text{fast}}{\rightleftharpoons}} B$
      $B \xrightarrow{\text{slow}} C$

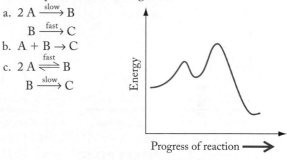

FIGURE P13.9

13.10. Which of the following mechanisms is consistent with the reaction profile shown in Figure P13.10?
   a. $A + B \xrightarrow{\text{slow}} C$
      $C \xrightarrow{\text{fast}} D$
   b. $A + B \rightarrow C$
   c. $2\,A \xrightarrow{\text{fast}} B$
      $B + C \xrightarrow{\text{slow}} D$

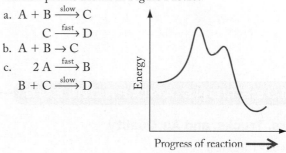

FIGURE P13.10

13.11. Which of the reaction profiles in Figure P13.11 represents the effect of a catalyst on the rate of a reaction?

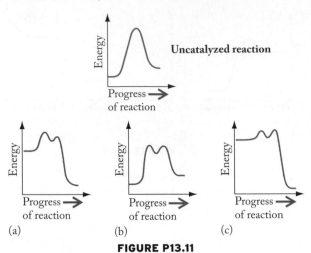

FIGURE P13.11

*13.12. Which of the reaction profiles in Figure P13.12 represents the effect of a catalyst on the rate of a reaction?

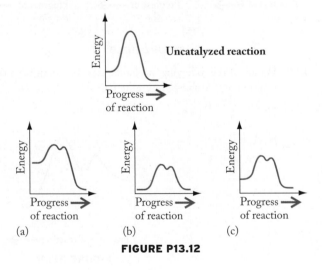

FIGURE P13.12

13.13. Which of the highlighted elements in Figure P13.13 forms volatile oxides associated with photochemical smog formation?

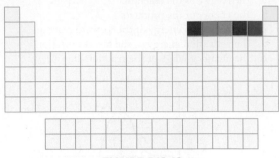

FIGURE P13.13

13.14. Which of the highlighted elements in Figure P13.13 forms noxious oxides that are removed from automobile exhaust as it passes through a catalytic converter?

13.15. Which of the highlighted elements in Figure P13.15 are widely used as heterogeneous catalysts?

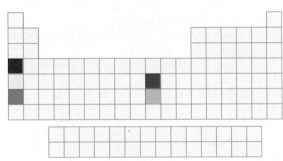

FIGURE P13.15

13.16. Which of the highlighted elements in Figure P13.16 forms volatile, odd-electron oxides that catalyze the destruction of stratospheric ozone?

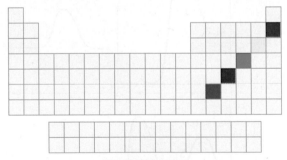

FIGURE P13.16

QUESTIONS AND PROBLEMS

Cars, Trucks, and Air Quality

CONCEPT REVIEW

13.17. Why does the maximum concentration of ozone in Figure 13.2 occur much later in the day than the maximum concentration of NO and NO_2?

13.18. If we plot the concentration of reactants and products as a function of time for any sequence of two spontaneous chemical reactions, such as

$$A \rightarrow B \rightarrow C$$

will the maximum concentration of final product C always appear after the maximum concentration of B?

13.19. Why isn't there an increase in NO concentration after the evening rush hour?

13.20. If ozone can react with NO to form NO_2, why does the ozone concentration reach a maximum in the early afternoon?

PROBLEMS

13.21. Using data in Appendix 4, calculate $\Delta H°$ for the reaction

$$2 NO(g) + O_2(g) \rightarrow 2 NO_2(g)$$

13.22. Using data in Appendix 4, calculate $\Delta H°$ for the reaction

$$O_3(g) + NO(g) \rightarrow O_2(g) + NO_2(g)$$

13.23. Nitrogen and oxygen can combine to form different nitrogen oxides that play a minor role in the chemistry of smog. Write balanced chemical equations for the reactions of N_2 and O_2 that produce (a) N_2O and (b) N_2O_5.

13.24. Nitrogen oxides such as N_2O and N_2O_5 are present in the air in low concentrations, in part because of their reactivity. Write balanced chemical equations for these reactions:
   a. The conversion of N_2O to NO_2 in the presence of oxygen
   b. The decomposition of N_2O_5 to NO_2 and O_2

Reaction Rates

CONCEPT REVIEW

13.25. Explain the difference between the average rate and the instantaneous rate of a chemical reaction.

13.26. Can the average rate and instantaneous rate of a chemical reaction ever be the same?

13.27. Why do the average rates of most reactions change with time?

13.28. Does the instantaneous rate of a chemical reaction change with time?

PROBLEMS

13.29. Bacterial Degradation of Ammonia *Nitrosomonas* bacteria convert ammonia into nitrite in the presence of oxygen by the following reaction:

$$2\,NH_3(aq) + 3\,O_2(g) \rightarrow 2\,H^+(aq) + 2\,NO_2^-(aq) + 2\,H_2O(\ell)$$

   a. How are the rates of formation of H^+ and NO_2^- related to the rate of consumption of NH_3?
   b. How is the rate of formation of NO_2^- related to the rate of consumption of O_2?
   c. How is the rate of consumption of NH_3 related to the rate of consumption of O_2?

13.30. Catalytic Converters and Combustion Catalytic converters in automobiles combat air pollution by converting NO and CO into N_2 and CO_2:

$$2\,CO(g) + 2\,NO(g) \rightarrow N_2(g) + 2\,CO_2(g)$$

   a. How is the rate of formation of N_2 related to the rate of consumption of CO?
   b. How is the rate of formation of CO_2 related to the rate of consumption of NO?
   c. How is the rate of consumption of CO related to the rate of consumption of NO?

13.31. Write expressions for the rate of formation of products and the rate of consumption of reactants in each of the following reactions:
   a. $H_2O_2(g) \rightarrow 2\,OH(g)$
   b. $ClO(g) + O_2(g) \rightarrow ClO_3(g)$
   c. $N_2O_5(g) + H_2O(g) \rightarrow 2\,HNO_3(g)$

13.32. Write expressions for the rate of formation of products and the rate of consumption of reactants in each of the following reactions:
   a. $Cl_2O_2(g) \rightarrow 2\,ClO(g)$
   b. $N_2O_5(g) \rightarrow NO_2(g) + NO_3(g)$
   c. $2\,INO(g) \rightarrow I_2(g) + 2\,NO(g)$

13.33. Power-Plant Emissions Sulfur dioxide emissions in power-plant stack gases may react with carbon monoxide as follows:

$$SO_2(g) + 3\,CO(g) \rightarrow 2\,CO_2(g) + COS(g)$$

Write an equation relating each of the following pairs of rates:
   a. The rate of formation of CO_2 to the rate of consumption of CO
   b. The rate of formation of COS to the rate of consumption of SO_2
   c. The rate of consumption of CO to the rate of consumption of SO_2

13.34. Reducing Nitric Oxide Emissions from Power Plants Nitric oxide (NO) can be removed from gas-fired power-plant emissions by reaction with methane as follows:

$$CH_4(g) + 4\,NO(g) \rightarrow 2\,N_2(g) + CO_2(g) + 2\,H_2O(g)$$

Write an equation relating each of the following pairs of rates:
   a. The rate of formation of N_2 to the rate of formation of CO_2
   b. The rate of formation of CO_2 to the rate of consumption of NO
   c. The rate of consumption of CH_4 to the rate of formation of H_2O

13.35. Stratospheric Ozone Depletion Chlorine monoxide (ClO) plays a major role in the creation of the ozone holes in the stratosphere over Earth's polar regions.
   a. If $\Delta[ClO]/\Delta t$ at 298 K is -2.3×10^7 M/s, what is the rate of change in $[Cl_2]$ and $[O_2]$ in the following reaction?

$$2\,ClO(g) \rightarrow Cl_2(g) + O_2(g)$$

   b. If $\Delta[ClO]/\Delta t$ is -2.9×10^4 M/s, what is the rate of formation of oxygen and ClO_2 in the following reaction?

$$ClO(g) + O_3(g) \rightarrow O_2(g) + ClO_2(g)$$

13.36. The chemistry of smog formation includes NO_3 as an intermediate in several reactions.
   a. If $\Delta[NO_3]/\Delta t$ is -2.2×10^5 mM/min in the following reaction, what is the rate of formation of NO_2?

$$NO_3(g) + NO(g) \rightarrow 2\,NO_2(g)$$

   b. What is the rate of change of $[NO_2]$ in the following reaction if $\Delta[NO_3]/\Delta t$ is -2.3 mM/min?

$$2\,NO_3(g) \rightarrow 2\,NO_2(g) + O_2(g)$$

13.37. Nitrite ion reacts with ozone in aqueous solution, producing nitrate ion and oxygen:

$$NO_2^-(aq) + O_3(g) \rightarrow NO_3^-(aq) + O_2(g)$$

The following data were collected for this reaction at 298 K. Calculate the average reaction rate between 0 and 100 μs (microseconds) and between 200 and 300 μs.

Time (μs)	$[O_3]$ (M)
0	1.13×10^{-2}
100	9.93×10^{-3}
200	8.70×10^{-3}
300	8.15×10^{-3}

13.38. Dinitrogen pentoxide (N_2O_5) decomposes as follows to nitrogen dioxide and nitrogen trioxide:

$$N_2O_5(g) \rightarrow NO_2(g) + NO_3(g)$$

Calculate the average rate of this reaction between consecutive measurement times listed in the following table.

Time (s)	$[N_2O_5]$ (molecules/cm^3)
0	1.500×10^{12}
1.45	1.357×10^{12}
2.90	1.228×10^{12}
4.35	1.111×10^{12}
5.80	1.005×10^{12}

13.39. The following data were collected for the dimerization of ClO to Cl_2O_2 at 298 K.

Time (s)	[ClO] (molecules/cm^3)
0	2.60×10^{11}
1	1.08×10^{11}
2	6.83×10^{10}
3	4.99×10^{10}
4	3.93×10^{10}
5	3.24×10^{10}
6	2.76×10^{10}

Plot [ClO] and $[Cl_2O_2]$ as a function of time and determine the instantaneous rates of change in both at 1 s.

13.40. **Tropospheric Ozone** Tropospheric (lower atmosphere) ozone is rapidly consumed in many reactions, including

$$O_3(g) + NO(g) \rightarrow NO_2(g) + O_2(g)$$

Use the following data to calculate the instantaneous rate of the reaction at $t = 0.000$ s and $t = 0.052$ s.

Time (s)	[NO] (M)
0.000	2.0×10^{-8}
0.011	1.8×10^{-8}
0.027	1.6×10^{-8}
0.052	1.4×10^{-8}
0.102	1.2×10^{-8}

Effect of Concentration on Reaction Rate

CONCEPT REVIEW

13.41. Can two different chemical reactions have the same rate law?

13.42. Why are the units of the rate constants different for reactions of different order?

13.43. Does the half-life of a second-order reaction have the same units as the half-life for a first-order reaction?

13.44. Does the half-life of a first-order reaction depend on the concentration of the reactants?

13.45. What effect does doubling the initial concentration of a reactant have on the half-life of a reaction that is second order in the reactant?

13.46. Two first-order decomposition reactions of the form $A \rightarrow B + C$ have the same rate constant at a given temperature. Do the reactants in the two reactions have the same half-lives at this temperature?

PROBLEMS

13.47. For each of the following rate laws, determine the order with respect to each reactant and the overall reaction order.
   a. Rate = $k[A][B]$
   b. Rate = $k[A]^2[B]$
   c. Rate = $k[A][B]^3$

13.48. Determine the overall order of the following rate laws and the order with respect to each reactant.
   a. Rate = $k[A]^2[B]^{1/2}$
   b. Rate = $k[A]^2[B][C]$
   c. Rate = $k[A][B]^3[C]^{1/2}$

13.49. Write rate laws and determine the units of the rate constant (by using the units M for concentration and s for time) for the following reactions:
   a. The reaction of oxygen atoms with NO_2 is first order in both reactants.
   b. The reaction between NO and Cl_2 is second order in NO and first order in Cl_2.
   c. The reaction between Cl_2 and chloroform ($CHCl_3$) is first order in $CHCl_3$ and one-half order in Cl_2.
   *d. The decomposition of ozone (O_3) to O_2 is second order in O_3 and an order of -1 in O atoms.

13.50. Compounds A and B react to give a single product, C. Write the rate law for each of the following cases and determine the units of the rate constant by using the units M for concentration and s for time:
   a. The reaction is first order in A and second order in B.
   b. The reaction is first order in A and second order overall.
   c. The reaction is independent of the concentration of A and second order overall.
   d. The reaction is second order in both A and B.

13.51. Predict the rate law for the reaction $2 BrO(g) \rightarrow Br_2(g) + O_2(g)$ under each of the following conditions:
   a. The rate doubles when [BrO] doubles.
   b. The rate quadruples when [BrO] doubles.
   c. The rate is halved when [BrO] is halved.
   d. The rate is unchanged when [BrO] is doubled.

13.52. Predict the rate law for the reaction $NO(g) + Br_2(g) \rightarrow NOBr_2(g)$ under each of the following conditions:
   a. The rate doubles when [NO] is doubled and [Br_2] remains constant.
   b. The rate doubles when [Br_2] is doubled and [NO] remains constant.
   c. The rate increases by 1.56 times when [NO] is increased 1.25 times and [Br_2] remains constant.
   d. The rate is halved when [NO] is doubled and [Br_2] remains constant.

13.53. In the reaction of NO with ClO,

$$NO(g) + ClO(g) \rightarrow NO_2(g) + Cl(g)$$

the initial rate of reaction quadruples when the concentrations of both reactants are doubled. What additional information do we need to determine whether the reaction is first order in each reactant?

13.54. The reaction between chlorine monoxide and nitrogen dioxide

$$ClO(g) + NO_2(g) + M(g) \rightarrow ClONO_2(g) + M(g)$$

produces chlorine nitrate ($ClONO_2$). A third molecule (M) takes part in the reaction but is unchanged by it. The reaction is first order in NO_2 and in ClO.
a. Write the rate law for this reaction.
b. What is the reaction order with respect to M?

13.55. Rate Laws for Destruction of Tropospheric Ozone The reaction of NO_2 with ozone produces NO_3 in a second-order reaction overall:

$$NO_2(g) + O_3(g) \rightarrow NO_3(g) + O_2(g)$$

a. Write the rate law for the reaction if the reaction is first order in each reactant.
b. The rate constant for the reaction is $1.93 \times 10^4 \ M^{-1} \ s^{-1}$ at 298 K. What is the rate of the reaction when $[NO_2] = 1.8 \times 10^{-8} \ M$ and $[O_3] = 1.4 \times 10^{-7} \ M$?
c. What is the rate of formation of NO_3 under these conditions?
d. What happens to the rate of the reaction if the concentration of $O_3(g)$ is doubled?

13.56. Sources of Nitric Acid in the Atmosphere The reaction between N_2O_5 and water is a source of nitric acid in the atmosphere:

$$N_2O_5(g) + H_2O(g) \rightarrow 2 HNO_3(g)$$

a. The reaction is first order in each reactant. Write the rate law for the reaction.
b. When $[N_2O_5]$ is 0.132 mM and $[H_2O]$ is 230 mM, the rate of the reaction is 4.55×10^{-4} mM/min^{-1}. What is the rate constant for the reaction?

13.57. Each of the following reactions is first order in the reactants and second order overall. Which reaction is fastest if the initial concentrations of the reactants are the same? All reactions are at 298 K.
a. $ClO_2(g) + O_3(g) \rightarrow ClO_3(g) + O_2(g)$
$$k = 3.0 \times 10^{-19} \ cm^3/(molecule \cdot s)$$
b. $ClO_2(g) + NO(g) \rightarrow NO_2(g) + ClO(g)$
$$k = 3.4 \times 10^{-13} \ cm^3/(molecule \cdot s)$$
c. $ClO(g) + NO(g) \rightarrow Cl(g) + NO_2(g)$
$$k = 1.7 \times 10^{-11} \ cm^3/(molecule \cdot s)$$
d. $ClO(g) + O_3(g) \rightarrow ClO_2(g) + O_2(g)$
$$k = 1.5 \times 10^{-17} \ cm^3/(molecule \cdot s)$$

13.58. Two reactions in which there is a single reactant have nearly the same magnitude rate constant. One is first order; the other is second order.
a. If the initial concentrations of the reactants are both 1.0 mM, which reaction will proceed at the higher rate?
b. If the initial concentrations of the reactants are both 2.0 M, which reaction will proceed at the higher rate?

13.59. In the presence of water, the species NO and NO_2 react to form nitrous acid (HNO_2) by the following reaction:

$$NO(g) + NO_2(g) + H_2O(\ell) \rightarrow 2 HNO_2(aq)$$

When the concentration of NO or NO_2 is doubled, the initial rate of reaction doubles. If the rate of the reaction does not depend on $[H_2O]$, what is the rate law for this reaction?

13.60. Hydroperoxyl Radicals in the Atmosphere During a smog event, trace amounts of many highly reactive substances are present in the atmosphere. One of these is the hydroperoxyl radical, HO_2, which reacts with sulfur trioxide, SO_3. The rate constant for the reaction

$$2 HO_2(g) + SO_3(g) \rightarrow H_2SO_3(g) + 2 O_2(g)$$

at 298 K is $2.6 \times 10^{11} \ M^{-1} \ s^{-1}$. The initial rate of the reaction doubles when the concentration of SO_3 or HO_2 is doubled. What is the rate law for the reaction?

13.61. Disinfecting Municipal Water Supplies Chlorine dioxide (ClO_2) is a disinfectant used in municipal water-treatment plants (Figure P13.61). It dissolves in basic solution, producing ClO_3^- and ClO_2^-:

$$2 ClO_2(g) + 2 OH^-(aq) \rightarrow ClO_3^-(aq) + ClO_2^-(aq) + H_2O(\ell)$$

FIGURE P13.61

The following kinetic data were obtained at 298 K for the reaction:

Experiment	$[ClO_2]_0$ (M)	$[OH^-]_0$ (M)	Initial Rate (M/s)
1	0.060	0.030	0.0248
2	0.020	0.030	0.00827
3	0.020	0.090	0.0247

Determine the rate law and the rate constant for this reaction at 298 K.

13.62. The following kinetic data were collected at 298 K for the reaction of ozone with nitrite ion, producing nitrate and oxygen:

$$NO_2^-(aq) + O_3(g) \rightarrow NO_3^-(aq) + O_2(g)$$

Experiment	$[NO_2^-]_0$ (M)	$[O_3]_0$ (M)	Initial Rate (M/s)
1	0.0100	0.0050	25
2	0.0150	0.0050	37.5
3	0.0200	0.0050	50.0
4	0.0200	0.0200	200.0

Determine the rate law and the rate constant for this reaction at 298 K.

13.63. Hydrogen gas reduces NO to N_2 in the following reaction:

$$2 H_2(g) + 2 NO(g) \rightarrow 2 H_2O(g) + N_2(g)$$

The initial reaction rates of four mixtures of H_2 and NO were measured at 900°C with the following results:

Experiment	$[H_2]_0$ (M)	$[NO]_0$ (M)	Initial Rate (M/s)
1	0.212	0.136	0.0248
2	0.212	0.272	0.0991
3	0.424	0.544	0.793
4	0.848	0.544	1.59

Determine the rate law and the rate constant for the reaction at 900°C.

13.64. The rate of the reaction

$$NO_2(g) + CO(g) \rightarrow NO(g) + CO_2(g)$$

was determined in three experiments at 225°C. The results are given in the following table:

Experiment	$[NO_2]_0$ (M)	$[CO]_0$ (M)	Initial Rate $-\Delta[NO_2]/\Delta t$ (M/s)
1	0.263	0.826	1.44×10^{-5}
2	0.263	0.413	1.44×10^{-5}
3	0.526	0.413	5.76×10^{-5}

a. Determine the rate law for the reaction.
b. Calculate the value of the rate constant at 225°C.
c. Calculate the rate of appearance of CO_2 when $[NO_2] = [CO] = 0.500$ M.

13.65. Nitrogen trioxide decomposes to NO_2 and O_2 in the following reaction:

$$2 NO_3(g) \rightarrow 2 NO_2(g) + O_2(g)$$

The following data were collected at 298 K:

Time (min)	$[NO_3]$ (μM)
0	1.470×10^{-3}
10	1.463×10^{-3}
100	1.404×10^{-3}
200	1.344×10^{-3}
300	1.288×10^{-3}
400	1.237×10^{-3}
500	1.190×10^{-3}

Calculate the value of the second-order rate constant at 298 K.

13.66. Two structural isomers of ClO_2 are shown in Figure P13.66.

FIGURE P13.66

The isomer with the Cl—O—O skeletal arrangement is unstable and rapidly decomposes according to the reaction $2 ClOO(g) \rightarrow Cl_2(g) + 2 O_2(g)$. The following data were collected for the decomposition of ClOO at 298 K:

Time (μs)	[ClOO] (M)
0.0	1.76×10^{-6}
0.7	2.36×10^{-7}
1.3	3.56×10^{-8}
2.1	3.23×10^{-9}
2.8	3.96×10^{-10}

Determine the rate law and the rate constant for the reaction at 298 K.

13.67. At high temperatures, ammonia spontaneously decomposes into N_2 and H_2. The following data were collected at one such temperature:

Time (s)	$[NH_3]$ (M)
0	2.56×10^{-2}
12	2.47×10^{-2}
56	2.16×10^{-2}
224	1.31×10^{-2}
532	5.19×10^{-3}
746	2.73×10^{-3}

Determine the rate law for the decomposition of ammonia and the value of the rate constant at the temperature of the experiment.

13.68. Atmospheric Chemistry of Hydroperoxyl Radicals
Atmospheric chemistry involves highly reactive, odd-electron molecules such as the hydroperoxyl radical HO_2, which decomposes into H_2O_2 and O_2. Determine the rate law for the reaction and the value of the rate constant at 298 K by using the following data obtained at 298 K.

Time (μs)	[HO_2] (μM)
0.0	8.5
0.6	5.1
1.0	3.6
1.4	2.6
1.8	1.8
2.4	1.1

13.69. In addition to being studied in the gas phase, the decomposition of N_2O_5 has been evaluated in solution. In carbon tetrachloride (CCl_4) at 45°C,

$$2\,N_2O_5 \rightarrow 4\,NO_2 + O_2$$

is a first-order reaction and $k = 6.32 \times 10^{-4}$ s^{-1}. How much N_2O_5 remains in solution after 1 hr if the initial concentration of N_2O_5 was 0.50 mol/L? What percent of the N_2O_5 has reacted at that point?

13.70. Because the units of concentration in the term $\ln[X]/[X]_0$ cancel out in the integrated rate law for first-order reactions (Equation 13.17), molar concentration can be replaced by any concentration term that is proportional to mol/L. With gases, for example, partial pressures may be used. The decomposition of phosphine gas (PH_3) at 600°C is first order in PH_3, and $k = 0.023$ s^{-1}.

$$4\,PH_3(g) \rightarrow P_4(g) + 6\,H_2(g)$$

If the initial partial pressure of PH_3 is 375 torr, what percent of PH_3 reacts in 1 min?

13.71. Laughing Gas Nitrous oxide (N_2O) is used as an anesthetic (laughing gas) and in aerosol cans to produce whipped cream. It is a potent greenhouse gas and decomposes slowly to N_2 and O_2:

$$2\,N_2O(g) \rightarrow 2\,N_2(g) + O_2(g)$$

a. If the plot of $\ln[N_2O]$ as a function of time is linear, what is the rate law for the reaction?
b. How many half-lives will it take for the concentration of N_2O to reach 6.25% of its original concentration? [*Hint*: The amount of reactant remaining after time t (A_t) is related to the amount initially present (A_0) by the equation $A_t/A_0 = (0.5)^n$, where n is the number of half-lives in time t.]

13.72. The unsaturated hydrocarbon butadiene (C_4H_6) dimerizes to 4-vinylcyclohexene (C_8H_{12}). When data collected in studies of the kinetics of this reaction were plotted against reaction time, plots of [C_4H_6] or $\ln[C_4H_6]$ produced curved lines, but the plot of $1/[C_4H_6]$ was linear.
a. What is the rate law for the reaction?
b. How many half-lives will it take for the [C_4H_6] to decrease to 3.1% of its original concentration?

13.73. Tracing Phosphorus in Organisms Radioactive isotopes such as ^{32}P are used to follow biological processes. The following radioactivity data (in relative radioactivity values) were collected for a sample containing ^{32}P:

Time (days)	Relative Radioactivity
0	10.00
1	9.53
2	9.08
5	7.85
10	6.16
20	3.79

a. Write the rate law for the decay of ^{32}P.
b. Determine the value of the rate constant.
c. Determine the half-life of ^{32}P.

13.74. Nitrous acid slowly decomposes to NO, NO_2, and water in the following second-order reaction:

$$2\,HNO_2(aq) \rightarrow NO(g) + NO_2(g) + H_2O(\ell)$$

a. Use the data below to determine the rate constant for this reaction at 298 K:

Time (min)	[HNO_2] (μM)
0	0.1560
1000	0.1466
1500	0.1424
2000	0.1383
2500	0.1345
3000	0.1309

b. Determine the half-life for the decomposition of HNO_2.

13.75. The dimerization of ClO,

$$2\,ClO(g) \rightarrow Cl_2O_2(g)$$

is second order in ClO. Use the following data to determine the value of k at 298 K.

Time (s)	[ClO] (molecules/cm^3)
0	2.60×10^{11}
1	1.08×10^{11}
2	6.83×10^{10}
3	4.99×10^{10}
4	3.93×10^{10}

Determine the half-life for the dimerization of ClO.

13.76. Kinetic data for the reaction $Cl_2O_2(g) \rightarrow 2\,ClO(g)$ are summarized in the following table. Determine the value of the rate constant.

Time (μs)	$[Cl_2O_2]$ (M)
0	6.60×10^{-8}
172	5.68×10^{-8}
345	4.89×10^{-8}
517	4.21×10^{-8}
690	3.62×10^{-8}
862	3.12×10^{-8}

Determine the half-life for the decomposition of Cl_2O_2.

13.77. Kinetics of Sucrose Hydrolysis The metabolism of table sugar (sucrose, $C_{12}H_{22}O_{11}$) begins with the hydrolysis of the disaccharide to glucose and fructose (both $C_6H_{12}O_6$):

$$C_{12}H_{22}O_{11}(aq) + H_2O(\ell) \rightarrow 2\,C_6H_{12}O_6(aq)$$

The kinetics of the reaction were studied at 24°C in a reaction system with a large excess of water, so the reaction was pseudo-first-order in sucrose. Determine the rate law and the pseudo-first-order rate constant for the reaction from the following data:

Time (s)	$[C_{12}H_{22}O_{11}]$ (M)
0	0.562
612	0.541
1600	0.509
2420	0.484
3160	0.462
4800	0.442

13.78. Hydroperoxyl radicals react rapidly with ozone to produce oxygen and OH radicals:

$$HO_2(g) + O_3(g) \rightarrow OH(g) + 2\,O_2(g)$$

The rate of this reaction was studied in the presence of a large excess of ozone. Determine the pseudo-first-order rate constant and the second-order rate constant for the reaction from the following data:

Time (ms)	$[HO_2]$ (M)	$[O_3]$ (M)
0	3.2×10^{-6}	1.0×10^{-3}
10	2.9×10^{-6}	1.0×10^{-3}
20	2.6×10^{-6}	1.0×10^{-3}
30	2.4×10^{-6}	1.0×10^{-3}
80	1.4×10^{-6}	1.0×10^{-3}

Reaction Rates, Temperature, and the Arrhenius Equation

CONCEPT REVIEW

13.79. How does the size of a reaction's activation energy influence the rate of a reaction?

13.80. Do all spontaneous reactions happen instantaneously at room temperature?

13.81. Under what circumstances is the activation energy of a reaction proceeding in the forward direction less than the activation energy of it happening in reverse?

13.82. Under what circumstances is the activation energy of a reaction proceeding in the forward direction greater than the activation energy of it happening in reverse?

*13.83. The order of a reaction is independent of temperature, but the value of the rate constant varies with temperature. Why?

13.84. Does reducing the activation energy of a reaction by $\frac{1}{2}$ increase its rate constant by a factor of 2?

*13.85. Two first-order reactions have activation energies of 15 and 150 kJ/mol. Which reaction will show the larger increase in rate as temperature is increased?

13.86. According to the Arrhenius equation, does the activation energy of a chemical reaction depend on temperature? Explain your answer.

PROBLEMS

13.87. The rate constant for the reaction of ozone with oxygen atoms was determined at four temperatures. Calculate the activation energy and frequency factor A for the reaction

$$O(g) + O_3(g) \rightarrow 2\,O_2(g)$$

given the following data:

T (K)	k [cm³/(molecule · s)]
250	2.64×10^{-4}
275	5.58×10^{-4}
300	1.04×10^{-3}
325	1.77×10^{-3}

13.88. The rate constant for the reaction

$$NO_2(g) + O_3(g) \rightarrow NO_3(g) + O_2(g)$$

was determined over a temperature range of 40 K, with the following results:

T (K)	k (M⁻¹ s⁻¹)
203	4.14×10^5
213	7.30×10^5
223	1.22×10^5
233	1.96×10^6
243	3.02×10^6

a. Determine the activation energy for the reaction.
b. Calculate the rate constant of the reaction at 300 K.

13.89. Activation Energy of Smog-Forming Reactions The initial step in the formation of smog is the reaction between nitrogen and oxygen. The activation energy of the reaction can be determined from the temperature dependence of the rate constants. At the temperatures indicated, values of the rate constant of the reaction

$$N_2(g) + O_2(g) \rightarrow 2\,NO(g)$$

are as follows:

T (K)	k ($M^{-1/2}\,s^{-1}$)
2000	318
2100	782
2200	1770
2300	3733
2400	7396

a. Calculate the activation energy of the reaction.
b. Calculate the frequency factor for the reaction.
c. Calculate the value of the rate constant at ambient temperature, $T = 300$ K.

13.90. Values of the rate constant for the decomposition of N_2O_5 gas at four different temperatures are as follows:

T (K)	k (s^{-1})
658	2.14×10^5
673	3.23×10^5
688	4.81×10^5
703	7.03×10^5

a. Determine the activation energy of the decomposition reaction.
b. Calculate the value of the rate constant at 300 K.

13.91. Activation Energy of Stratospheric Ozone Destruction Reactions The kinetics of the reaction between chlorine dioxide and ozone are relevant to the study of atmospheric ozone destruction. The activation energy of the reaction can be determined from the temperature dependence of the rate constant. The value of the rate constant for the reaction between chlorine dioxide and ozone was measured at four temperatures between 193 and 208 K. The results were as follows:

T (K)	k ($M^{-1}\,s^{-1}$)
193	34.0
198	62.8
203	112.8
208	196.7

Calculate the values of the activation energy and the frequency factor for the reaction.

13.92. Chlorine atoms react with methane, forming HCl and CH_3. The rate constant for the reaction is $6.0 \times 10^7\,M^{-1}\,s^{-1}$ at 298 K. When the experiment was repeated at three other temperatures, the following data were collected:

T (K)	k ($M^{-1}\,s^{-1}$)
303	6.5×10^7
308	7.0×10^7
313	7.5×10^7

Calculate the values of the activation energy and the frequency factor for the reaction.

13.93. A rule of thumb states that a reaction rate doubles for every 10°C rise in temperature. This is not always the case, but it is a convenient concept to apply when making estimations. Azomethane (CH_3NNCH_3) is a heat-sensitive explosive that decomposes to yield nitrogen gas and methyl radicals:

$$CH_3NNCH_3(g) \rightarrow N_2(g) + 2\,CH_3(g)$$

At 600°C, the reaction has an activation energy of 2.14×10^4 J/mol and $k = 2.00 \times 10^8\,s^{-1}$. Does the rule of thumb hold for this reaction?

13.94. The compound 1,1-difluoroethane decomposes at elevated temperatures to give fluoroethylene and hydrogen fluoride:

$$CH_3CHF_2(g) \rightarrow CH_2CHF(g) + HF(g)$$

At 460°C, $k = 5.8 \times 10^{-6}\,s^{-1}$ and $E_a = 265$ kJ/mol. To what temperature would you have to raise the reaction to make it go four times as fast?

Reaction Mechanisms

CONCEPT REVIEW

13.95. The rate law for the reaction between NO and H_2 is second order in NO and third order overall, whereas the reaction of NO with Cl_2 is first order in each reactant and second order overall. Do these reactions proceed by similar mechanisms?

13.96. The rate law for the reaction of NO with Cl_2 (Rate = $k[NO][Cl_2]$) is the same as that for the reaction of NO_2 with F_2 (Rate = $k[NO_2][F_2]$). Is it possible that these reactions have similar mechanisms?

*13.97. Under what reaction conditions does a bimolecular reaction obey pseudo-first-order reaction kinetics?

*13.98. If a reaction is zero order in a reactant, does that mean the reactant is never involved in collisions with other reactants? Explain your answer.

PROBLEMS

13.99. The hypothetical reaction A → B has an activation energy of 50.0 kJ/mol. Draw a reaction profile for each of the following mechanisms:
a. A single elementary step.
b. A two-step reaction in which the activation energy of the second step is 15 kJ/mol.
c. A two-step reaction in which the activation energy of the second step is the rate-determining barrier.

13.100. For the spontaneous reaction $A + B \rightarrow C \rightarrow D + E$, draw three reaction profiles, one for each of the following mechanisms:
   a. C is an activated complex.
   b. The reaction has two elementary steps; the first step is rate determining and C is an intermediate.
   c. The reaction has two elementary steps; the second step is rate determining and C is an intermediate.

13.101. Write the rate laws for the following elementary steps and identify them as uni-, bi-, or termolecular steps:
   a. $SO_2Cl_2(g) \rightarrow SO_2(g) + Cl_2(g)$
   b. $NO_2(g) + CO(g) \rightarrow NO(g) + CO_2(g)$
   c. $2\,NO_2(g) \rightarrow NO_3(g) + NO(g)$

13.102. Write the rate laws for the following elementary steps and identify them as uni-, bi-, or termolecular steps:
   a. $Cl(g) + O_3(g) \rightarrow ClO(g) + O_2(g)$
   b. $2\,NO_2(g) \rightarrow N_2O_4(g)$
   *c. $^{14}_{6}C \rightarrow {}^{14}_{7}N + {}^{0}_{-1}\beta$

13.103. Write the overall reaction that consists of the following elementary steps:

$$N_2O_5(g) \rightarrow NO_3(g) + NO_2(g)$$
$$NO_3(g) \rightarrow NO_2(g) + O(g)$$
$$2\,O(g) \rightarrow O_2(g)$$

13.104. What overall reaction consists of the following elementary steps?

$$ClO^-(aq) + H_2O(\ell) \rightarrow HClO(aq) + OH^-(aq)$$
$$I^-(aq) + HClO(aq) \rightarrow HIO(aq) + Cl^-(aq)$$
$$OH^-(aq) + HIO(aq) \rightarrow H_2O(\ell) + IO^-(aq)$$

*__13.105.__ In the following mechanism for NO formation, oxygen atoms are produced by breaking $O{=}O$ bonds at high temperature in a fast reversible reaction. If $\Delta[NO]/\Delta t = k[N_2][O_2]^{1/2}$, which step in the mechanism is the rate-determining step?

   (1) $O_2(g) \rightleftharpoons 2\,O(g)$

   (2) $O(g) + N_2(g) \rightarrow NO(g) + N(g)$

   (3) $N(g) + O(g) \rightarrow NO(g)$

   Overall: $N_2(g) + O_2(g) \rightarrow 2\,NO(g)$

13.106. A proposed mechanism for the decomposition of hydrogen peroxide consists of three elementary steps:

$$H_2O_2(g) \rightarrow 2\,OH(g)$$
$$H_2O_2(g) + OH(g) \rightarrow H_2O(g) + HO_2(g)$$
$$HO_2(g) + OH(g) \rightarrow H_2O(g) + O_2(g)$$

If the rate law for the reaction is first order in H_2O_2, which step in the mechanism is the rate-determining step?

13.107. At a given temperature, the rate of the reaction between NO and Cl_2 is proportional to the product of the concentrations of the two gases: $[NO][Cl_2]$. The following two-step mechanism has been proposed for the reaction:

   (1) $NO(g) + Cl_2(g) \rightarrow NOCl_2(g)$

   (2) $NOCl_2(g) + NO(g) \rightarrow 2\,NOCl(g)$

   Overall: $2\,NO(g) + Cl_2(g) \rightarrow NOCl_2(g)$

Which step must be the rate-determining step if this mechanism is correct?

13.108. Mechanism of Ozone Destruction Ozone decomposes thermally to oxygen in the following reaction:

$$2\,O_3(g) \rightarrow 3\,O_2(g)$$

The following mechanism has been proposed:

$$O_3(g) \rightarrow O(g) + O_2(g)$$
$$O(g) + O_3(g) \rightarrow 2\,O_2(g)$$

The reaction is second order in ozone. What properties of the two elementary steps (specifically, relative rate and reversibility) are consistent with this mechanism?

13.109. Mechanism of NO_2 Destruction The rate laws for the thermal and photochemical decomposition of NO_2 are different. Which of the following mechanisms are possible for the thermal decomposition of NO_2, and which are possible for the photochemical decomposition of NO_2? For thermal decomposition, Rate $= k[NO_2]^2$, and for photochemical decomposition, Rate $= k[NO_2]$.

   a. $NO_2(g) \xrightarrow{\text{slow}} NO(g) + O(g)$
           $O(g) + NO_2(g) \xrightarrow{\text{fast}} NO(g) + O_2(g)$
   b. $NO_2(g) + NO_2(g) \xrightarrow{\text{fast}} N_2O_4(g)$
           $N_2O_4(g) \xrightarrow{\text{slow}} NO(g) + NO_3(g)$
           $NO_3(g) \xrightarrow{\text{fast}} NO(g) + O_2(g)$
   c. $NO_2(g) + NO_2(g) \xrightarrow{\text{slow}} NO(g) + NO_3(g)$
           $NO_3(g) \xrightarrow{\text{fast}} NO(g) + O_2(g)$

13.110. The rate laws for the thermal and photochemical decomposition of NO_2 are different. Which of the following mechanisms are possible for the thermal decomposition of NO_2, and which are possible for the photochemical decomposition of NO_2? For thermal decomposition, Rate $= k[NO_2]^2$, and for photochemical decomposition, Rate $= k[NO_2]$.

   a. $NO_2(g) + NO_2(g) \xrightarrow{\text{slow}} N_2O_4(g)$
           $N_2O_4(g) \xrightarrow{\text{fast}} N_2O_3(g) + O(g)$
       $N_2O_3(g) + O(g) \xrightarrow{\text{fast}} N_2O_2(g) + O_2(g)$
           $N_2O_2(g) \xrightarrow{\text{fast}} 2\,NO(g)$
   b. $NO_2(g) + NO_2(g) \xrightarrow{\text{slow}} NO(g) + NO_3(g)$
           $NO_3(g) \xrightarrow{\text{fast}} NO(g) + O_2(g)$
   c. $NO_2(g) \xrightarrow{\text{slow}} N(g) + O_2(g)$
         $N(g) + NO_2(g) \xrightarrow{\text{fast}} N_2O_2(g)$
           $N_2O_2(g) \xrightarrow{\text{slow}} 2\,NO(g)$

Catalysts

CONCEPT REVIEW

13.111. Does a catalyst affect both the rate and the rate constant of a reaction?

13.112. Is the rate law for a catalyzed reaction the same as that for the uncatalyzed reaction?

13.113. Does a substance that increases the rate of a reaction also increase the rate of the reverse reaction?

13.114. The rate of the reaction between NO_2 and CO is independent of $[CO]$. Does this mean that CO is a catalyst for the reaction?

13.115. Why doesn't the concentration of a homogeneous catalyst appear in the rate law for the reaction it catalyzes?

*13.116. The rate of a chemical reaction is too slow to measure at room temperature. We could either raise the temperature or add a catalyst. Which would be a better solution for making an accurate determination of the rate constant?

PROBLEMS

13.117. Is NO a catalyst for the decomposition of N_2O in the following two-step reaction mechanism, or is N_2O a catalyst for the conversion of NO to NO_2?

$$(1) \qquad NO(g) + N_2O(g) \rightarrow N_2(g) + NO_2(g)$$

$$(2) \qquad 2\,NO_2(g) \rightarrow 2\,NO(g) + O_2(g)$$

13.118. **NO as a Catalyst for Ozone Destruction** Explain why NO is a catalyst in the following two-step process that results in the depletion of ozone in the stratosphere:

$$(1) \qquad NO(g) + O_3(g) \rightarrow NO_2(g) + O_2(g)$$

$$(2) \qquad O(g) + NO_2(g) \rightarrow NO(g) + O_2(g)$$

$$\text{Overall:} \qquad O(g) + O_3(g) \rightarrow 2\,O_2(g)$$

13.119. On the basis of the frequency factors and activation energy values of the following two reactions, determine which one will have the larger rate constant at room temperature (298 K).

$$O_3(g) + O(g) \rightarrow O_2(g) + O_2(g)$$

$A = 8.0 \times 10^{-12}$ cm³/(molecules · s) $\qquad E_a = 17.1$ kJ/mol

$$O_3(g) + Cl(g) \rightarrow ClO(g) + O_2(g)$$

$A = 2.9 \times 10^{-11}$ cm³/(molecules · s) $\qquad E_a = 2.16$ kJ/mol

13.120. On the basis of the frequency factors and activation energy values of the following two reactions, determine which one will have the larger rate constant at room temperature (298 K).

$$O_3(g) + Cl(g) \rightarrow ClO(g) + O_2(g)$$

$A = 2.9 \times 10^{-11}$ cm³/(molecules · s) $\qquad E_a = 2.16$ kJ/mol

$$O_3(g) + NO(g) \rightarrow NO_2(g) + O_2(g)$$

$A = 2.0 \times 10^{-12}$ cm³/(molecules · s) $\qquad E_a = 11.6$ kJ/mol

Additional Problems

13.121. A student inserts a glowing wood splint into a test tube filled with O_2. The splint quickly catches on fire (Figure P13.121). Why does the splint burn so much faster in pure O_2 than in air?

FIGURE P13.121

*13.122. A backyard chef turns on the propane gas to a barbecue grill. Even though the reaction between propane and oxygen is spontaneous, the gas does not begin to burn until the chef pushes an igniter button to produce a spark. Why is the spark needed?

13.123. On average, someone who falls through the ice covering a frozen lake is less likely to experience anoxia (lack of oxygen) than someone who falls into a warm pool and is underwater for the same length of time. Why?

*13.124. Why doesn't a quadrupling of the rate correspond to a reaction order of 4, for example, Rate $\propto [NO]^4$?

13.125. If the rate of the reverse reaction is much slower than the rate of the forward reaction, does the method used to determine a rate law from initial concentrations and initial rates also work at some other time t? What concentrations would we use in the case where we use the rate when $t \neq 0$?

13.126. Identify what is wrong with the following statement: The reaction rate and the rate constant for a reaction both depend on the number of collisions and on the concentrations of the reactants.

13.127. How do we find k if we plot $1/[X] - 1/[X]_0$ as a function of t?

13.128. Many reactions are first order, fewer are second order in a single reactant, and third-order reactions in a single reactant are practically nonexistent. Can you suggest why?

13.129. Why can't an elementary step in a mechanism have a rate law that is zero order in a reactant?

13.130. During the decomposition of dinitrogen pentoxide,

$$2\,N_2O_5(g) \rightarrow 4\,NO_2(g) + O_2(g)$$

how is the rate of consumption of N_2O_5 related to the rate of formation of NO_2 and O_2?

13.131. In the reaction between nitrogen dioxide and ozone,

$$2\,NO_2(g) + O_3(g) \rightarrow N_2O_5(g) + O_2(g)$$

how are the rates of change in the concentrations of the reactants and products related?

13.132. Determine the order of the decomposition reaction of N_2O_5, by using the initial rate data from the following table:

Experiment	$[N_2O_5]_0$ (M)	Initial Rate (M/s)
1	0.050	1.8×10^{-5}
2	0.100	3.6×10^{-5}

13.133. At the temperature at which the experiments were carried out in the previous problem, what is the rate constant for the decomposition of N_2O_5? Write the complete rate law for the decomposition reaction.

13.134. The table below contains kinetics data for the reaction

$$2\,NO(g) + Cl_2(g) \rightarrow 2\,NOCl(g)$$

Experiment	$[NO]_0$ (M)	$[Cl_2]_0$ (M)	Initial Rate (M/s)
1	0.20	0.10	0.63
2	0.20	0.30	5.70
3	0.80	0.10	2.58
4	0.40	0.20	?

Predict the initial rate of reaction in experiment 4.

13.135. An important reaction in the formation of photochemical smog is the reaction between ozone and NO:

$$NO(g) + O_3(g) \rightarrow NO_2(g) + O_2(g)$$

The reaction is first order in NO and O_3. The rate constant of the reaction is 80 M^{-1} s^{-1} at 25°C and 3000 M^{-1} s^{-1} at 75°C.
   a. If this reaction were to occur in a single step, would the rate law be consistent with the observed order of the reaction for NO and O_3?
   b. What is the value of the activation energy of the reaction?
   c. What is the rate of the reaction at 25°C when $[NO] = 3 \times 10^{-6}\,M$ and $[O_3] = 5 \times 10^{-9}\,M$?
   d. Predict the values of the rate constant at 10°C and 35°C.

13.136. Ammonia reacts with nitrous acid to form an intermediate, ammonium nitrite (NH_4NO_2), which decomposes to N_2 and H_2O:

$$NH_3(g) + HNO_2(aq) \rightarrow NH_4NO_2(aq) \rightarrow N_2(g) + 2\,H_2O(\ell)$$

   a. The reaction is first order in ammonia and second order in nitrous acid. What is the rate law for the reaction? What are the units on the rate constant if concentrations are expressed in molarity and time in seconds?

   b. The rate law for the reaction has also been written as

$$Rate = k[NH_4^+][NO_2^-][HNO_2]$$

   Is this expression equivalent to the one you wrote in part a?
   c. With the data in Appendix 4, calculate the value of ΔH°_{rxn} of the overall reaction ($\Delta H^\circ_{f,HNO_2} = -43.1$ kJ/mol).
   d. Draw a reaction-energy profile for the process with the assumption that E_a of the first step is lower than E_a of the second step.

*13.137. When ionic compounds such as NaCl dissolve in water, the sodium ions are surrounded by six water molecules. The bound water molecules exchange with those in bulk solution as described by the reaction involving ^{18}O-enriched water:

$$Na(H_2O)_6^+(aq) + H_2^{18}O(\ell) \rightarrow Na(H_2O)_5(H_2^{18}O)(aq) + H_2O(\ell)$$

   a. The following reaction mechanism has been proposed:

(1) $$Na(H_2O)_6^+(aq) \rightarrow Na(H_2O)_5^+(aq) + H_2O(\ell)$$

(2) $$Na(H_2O)_5^+(aq) + H_2^{18}O(\ell) \rightarrow Na(H_2O)_5(H_2^{18}O)^+(aq)$$

   What is the rate law if the first step is the rate-determining step?
   b. If you were to sketch a reaction-energy profile, which would you draw with the higher energy, the reactants or the products?

13.138. Lachrymators in Smog The combination of ozone, volatile hydrocarbons, nitrogen oxide, and sunlight in urban environments produces peroxyacetyl nitrate (PAN), a potent lachrymator (a substance that causes eyes to tear). PAN decomposes to acetyl radicals and nitrogen dioxide in a process that is second order in PAN, as shown in Figure P13.138:

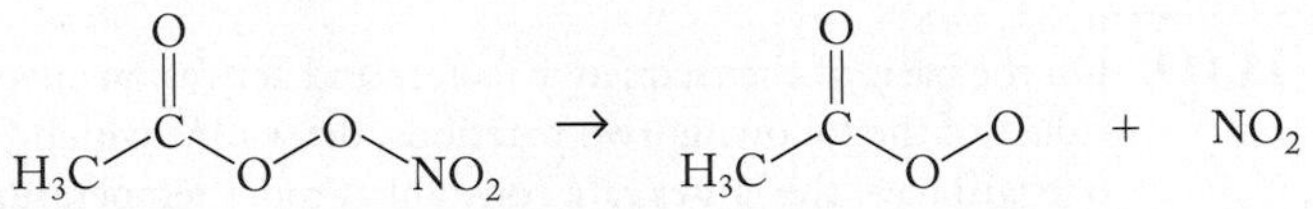

FIGURE P13.138

   a. The half-life of the reaction, at 23°C and $P_{CH_3CO_3NO_2} = 10.5$ torr, is 100 hr. Calculate the rate constant for the reaction.
   b. Determine the rate of the reaction at 23°C and $P_{CH_3CO_3NO_2} = 10.5$ torr.
   c. Draw a graph showing P_{PAN} as a function of time from 0 to 200 hr starting with $P_{CH_3CO_3NO_2} = 10.5$ torr.

13.139. Nitric Oxide in the Human Body Nitric oxide (NO) is a gaseous free radical that plays many biological roles including regulating neurotransmission and the human immune system. One of its many reactions involves the peroxynitrite ion ($ONOO^-$):

$$NO(g) + ONOO^-(aq) \rightarrow NO_2(g) + NO_2^-(aq)$$

a. Use the following data to determine the rate law and rate constant of the reaction at the experimental temperature at which these data were generated.

Experiment	$[NO]_0$ (M)	$[ONOO^-]_0$ (M)	Rate (M/s)
1	1.25×10^{-4}	1.25×10^{-4}	2.03×10^{-11}
2	1.25×10^{-4}	0.625×10^{-4}	1.02×10^{-11}
3	0.625×10^{-4}	2.50×10^{-4}	2.03×10^{-11}
4	0.625×10^{-4}	3.75×10^{-4}	3.05×10^{-11}

b. Draw the Lewis structure of peroxynitrite ion (including all resonance forms) and assign formal charges. Note which form is preferred.

c. Use the average bond energies in Appendix Table A4.1 to estimate the value of ΔH°_{rxn} using the preferred structure from part b.

13.140. Kinetics of Protein Chemistry In the presence of O_2, NO reacts with sulfur-containing proteins to form S-nitrosothiols, such as $C_6H_{13}SNO$. This compound decomposes to form a disulfide and NO:

$$2\ C_6H_{13}SNO(aq) \rightarrow 2\ NO(g) + C_{12}H_{26}S_2(aq)$$

The following data were collected for the decomposition reaction at 69°C.

Time (min)	$[C_6H_{13}SNO]$ (M)
0	1.05×10^{-3}
10	9.84×10^{-4}
20	9.22×10^{-4}
30	8.64×10^{-4}
60	7.11×10^{-4}

Calculate the value of the first-order rate constant for the reaction.

13.141. Solutions of nitrous acid, HNO_2, in ^{18}O-labeled water undergo isotope exchange:

$$HNO_2(aq) + H_2{}^{18}O(\ell) \rightarrow HN^{18}O_2(aq) + H_2O(\ell)$$

a. Use the following data at 24°C to determine the dependence of the reaction rate on the concentration of HNO_2.

Time (min)	$[HN^{18}O_2]$
0	5.4×10^{-2}
20	1.5×10^{-3}
40	7.7×10^{-4}
60	5.2×10^{-4}

b. Does the reaction rate depend on the concentration of $H_2{}^{18}O$?

13.142. Ethylene (C_2H_4) reacts with ozone to form 2 mol of formaldehyde (a probable human carcinogen) per mole of ethylene as shown in Figure P13.142. The following kinetic data were collected at 298 K.

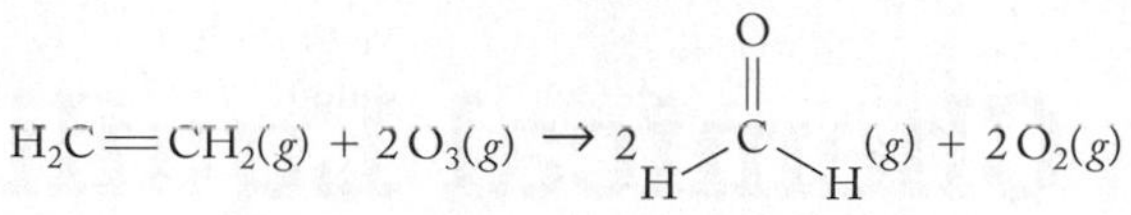

$$H_2C=CH_2(g) + 2\,O_3(g) \rightarrow 2\ \underset{H}{\overset{O}{\underset{}{\overset{\|}{C}}}}\!\!\diagdown_H(g) + 2\,O_2(g)$$

FIGURE P13.142

Experiment	$[O_3]_0$ (M)	$[C_2H_4]_0$ (M)	Rate (M/s)
1	0.86×10^{-2}	1.00×10^{-2}	0.0877
2	0.43×10^{-2}	1.00×10^{-2}	0.0439
3	0.22×10^{-2}	0.50×10^{-2}	0.0110

a. Determine the rate law and the value of the rate constant of the reaction at 298 K.

b. The rate constant was determined at several additional temperatures. Calculate the activation energy of the reaction from the following data.

T (K)	k ($M^{-1}\,s^{-1}$)
263	3.28×10^2
273	4.73×10^2
283	6.65×10^2
293	9.13×10^2

13.143. Reducing NO Emissions Adding NH_3 to the stack gases at an electric power generating plant can reduce NO_x emissions. This selective noncatalytic reduction (SNR) process depends on the reaction between NH_2 (an odd-electron compound) and NO:

$$NH_2(g) + NO(g) \rightarrow N_2(g) + H_2O(g)$$

The following kinetic data were collected at 1200 K.

Experiment	$[NH_2]_0$ (M)	$[NO]_0$ (M)	Rate (M/s)
1	1.00×10^{-5}	1.00×10^{-5}	0.12
2	2.00×10^{-5}	1.00×10^{-5}	0.24
3	2.00×10^{-5}	1.50×10^{-5}	0.36
4	2.50×10^{-5}	1.50×10^{-5}	0.45

a. What is the rate law for the reaction?

b. What is the value of the rate constant at 1200 K?

14

Chemical Equilibrium
Equal but Opposite Reaction Rates

Food for Thought

For centuries farmers have applied fertilizer to their fields, providing their crops with nitrogen, potassium, and phosphorus compounds that are essential to plant growth. In the past farmers used mostly biological fertilizers: manure and pulverized bones from farm animals and the droppings of wild ones, including guano collected from bat caves. Today, large-scale production of corn, wheat, and other grains has created a demand for fertilizer that is too large to be met by traditional biological sources. Instead, most of the potassium and phosphorus comes from mined minerals, and most of the nitrogen comes from a synthetic chemistry process in which atmospheric N_2 is combined with H_2 to form ammonia, NH_3. Each year about 200 million tons of ammonia and nitrogen compounds derived from it are used to enhance the world's food supply.

The industrial process for synthesizing ammonia was developed in Germany just before the outbreak of World War I, and much of the ammonia first produced with it was not used to grow food, but rather to manufacture explosives that sustained the German war effort. It is ironic that a process that sustains human life in many lands today contributed to the deaths of millions of people between 1914 and 1918.

Ammonia synthesis is based on this chemical reaction:

$$N_2(g) + 3\,H_2(g) \rightleftharpoons 2\,NH_3(g)$$

The reaction is exothermic ($\Delta H^\circ_{rxn} < 0$) and spontaneous ($\Delta G^\circ_{rxn} < 0$) under standard conditions but it's also very slow at 25°C. Catalysts and elevated temperatures increase the rate of the reaction, but lower its yield. In other words, less ammonia is synthesized at higher temperatures but what is made is made faster. Why does yield decrease as temperatures rise? A clue is contained in the fact that it takes 4 moles of reactant gases to make 2 moles of ammonia gas. This decrease in the number of gas-phase molecules means that the reaction mixture experiences a decrease in entropy ($\Delta S_{rxn} < 0$), which means that the value of ΔG_{rxn} in this key equation from Chapter 12:

$$\Delta G_{rxn} = \Delta H_{rxn} - T\,\Delta S_{rxn}$$

Chemistry and the Food Supply Most crops require nitrogen fertilizers and the source of most of this nitrogen is ammonia, which is synthesized by the reaction $N_2(g) + 3\,H_2(g) \rightleftharpoons 2\,NH_3(g)$. ▶

LEARNING OUTCOMES

LO1 Describe the dynamic nature of chemical equilibria

LO2 Write mass action or equilibrium constant expressions for reversible reactions including those involving heterogeneous equilibria
Sample Exercises 14.1, 14.8

LO3 Use equilibrium concentration or partial pressure data to calculate the value of K_c or K_p
Sample Exercises 14.2, 14.3

LO4 Interconvert K_c and K_p values of gas-phase reactions
Sample Exercise 14.4

LO5 Calculate K values of related chemical reactions
Sample Exercises 14.5, 14.6

LO6 Calculate the value of a reaction quotient and use it to predict the direction of a reversible chemical reaction
Sample Exercise 14.7

LO7 Predict how a reaction at equilibrium responds to changes in reaction conditions
Sample Exercises 14.9, 14.10, 14.11

LO8 Calculate the concentrations or partial pressures of reactants and products in a reaction mixture at equilibrium from their starting values and the value of K
Sample Exercises 14.12, 14.13

LO9 Relate the standard free-energy change of a reversible reaction to its equilibrium constant
Sample Exercise 14.14

LO10 Predict the value of K at any temperature from thermodynamic data
Sample Exercise 14.15

chemical equilibrium a dynamic process in which the concentrations of reactants and products remain constant over time and the rate of a reaction in the forward direction matches its rate in the reverse direction.

▶❚❚ **CHEMTOUR** Equilibrium

increases as T increases and eventually becomes greater than zero. In Chapter 12 we associated positive values of ΔG_{rxn} with nonspontaneous reactions. In this chapter we discover that reaction spontaneity is not a yes or no proposition, but rather a measure of how far a reaction proceeds until it reaches *chemical equilibrium* and there is no further increase in product concentration. The more positive the value of ΔG_{rxn}, the less product forms before equilibrium is achieved.

How, then, does the chemical industry produce millions of tons of ammonia each year? Scientists and engineers have developed ways to manipulate the equilibrium state in ammonia synthesis reactors. For example, they pump the equilibrium reaction mixture through condensers that liquefy and remove NH_3 but leave N_2 and H_2 in the gas phase where they combine to replace the lost ammonia. In this chapter we explore why this technique is so effective at tricking the reaction mixture into making more ammonia and providing the fertilizer that helps feed billions of people.

14.1 The Dynamics of Chemical Equilibrium

Chemical equilibrium is a *dynamic* process because reactants form products and products form reactants at equal but opposite rates. Equilibrium is represented by a pair of single-headed arrows pointing in opposite directions:

$$\text{Reactants} \rightleftharpoons \text{products}$$

and by the following equality in reaction rates at equilibrium:

$$\text{Rate}_{forward} = \text{rate}_{reverse}$$

Some reactions reach equilibrium only after nearly all the reactants have formed products. We say that these equilibria *lie far to the right* (the direction of the forward reaction arrow). In other reactions little product is formed. These equilibria are said to favor reactants and *lie far to the left* (the direction of the reverse reaction arrow). Keep in mind that nothing can be inferred from the position of the equilibrium regarding how much time the reaction takes to reach equilibrium. Studies of equilibrium reveal only the extent to which a reaction proceeds, not how rapidly it proceeds.

To explore the dynamics of chemical equilibrium, let's look at a two-step industrial process for making H_2 gas. The reactants are methane and steam. In the first step, methane reacts with steam in the presence of a nickel or iron oxide catalyst at temperatures near 1000°C:

$$CH_4(g) + H_2O(g) \rightleftharpoons CO(g) + 3\,H_2(g) \tag{14.1}$$

This reaction, called the steam-reforming reaction, is followed by another, called the water–gas shift reaction, in which the carbon monoxide formed in the first step is reacted with more steam in the presence of a Cu/ZnO catalyst at around 200°C:

$$H_2O(g) + CO(g) \rightleftharpoons H_2(g) + CO_2(g) \tag{14.2}$$

Let's explore the dynamics of the second reaction. If we put an equal number of moles of water vapor and carbon monoxide in a closed chamber and allow them to react, the concentrations of CO and H_2O initially fall as the concentrations of H_2 and CO_2 increase, as shown in Figure 14.1. Because H_2O and CO react in a 1:1 stoichiometric ratio, their concentrations decrease at the same rate and are

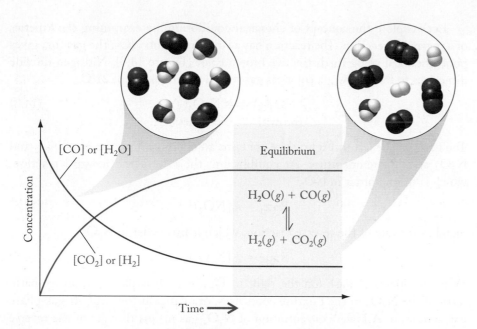

FIGURE 14.1 Concentrations of reactants and products in the water–gas shift reaction change over time until equilibrium is reached. At equilibrium in the water–gas shift reaction, H_2 and CO_2 are being formed at the same rate at which they are reacting to re-form H_2O and CO.

always equivalent in the reaction chamber. Similarly, H_2 and CO_2 are formed in a 1:1 ratio, and their concentrations increase at the same rate and are always equal.

The water–gas shift reaction is reversible. Reversibility means that if we intervene in the course of the reaction by, for example, changing the reaction temperature or removing or adding a reactant or product, we might be able to force the reaction to run in reverse, re-forming reactants from products. One requirement for reversibility is that the products must remain in contact with each other. If one or more of the products is a gas, then we need to run the reaction in a sealed chamber.

The graph in Figure 14.1 shows that eventually the concentrations of reactants and products in the water–gas shift reaction no longer change with time, at which point the reaction has reached chemical equilibrium. Note, however, that the concentrations of the reactants (CO and H_2O) do not go to zero. The presence of these reactants in the reaction mixture when the reaction has reached equilibrium means that the yield of the reaction never reaches 100%.

Now let's think about the changes in the *rates* of the forward and reverse reactions during the course of the water–gas shift reaction. When the CO and H_2O are initially mixed, their concentrations are at their maximum and the rate of the forward reaction, as indicated by the slope of the [CO] and $[H_2O]$ curve at $t = 0$ in Figure 14.1, is also at its maximum. As the reaction proceeds, reactant concentrations decrease, and the rate of the forward reaction also decreases because the likelihood of collisions between reactant molecules decreases.

The concentrations of products are initially zero and so is the rate of the reverse reaction. As product molecules form, the likelihood of their colliding with one another to re-form reactant molecules increases and so does the rate of the reverse reaction. Ultimately, the rate of the reverse reaction equals the rate of the forward reaction, as shown in Figure 14.2. At this point equilibrium has been achieved.

CONNECTION The method for calculating the percent yield of a reaction was described in Section 7.8.

CONNECTION In Chapter 13 we discussed how reaction rate depends on molecular collisions; the higher the concentrations of reactants, the more frequently molecules collide, and the faster a reaction proceeds.

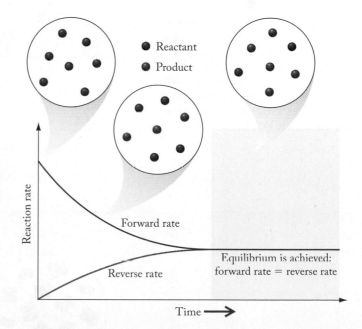

FIGURE 14.2 The rates of forward and reverse reactions are the same when equilibrium is achieved.

Let's explore the concept of chemical equilibrium by examining the kinetics of a reversible reaction. The reaction has a single reactant: NO_2, the gas that gives photochemical smog its distinctive brown color (Figure 14.3). Nitrogen dioxide dimerizes to form N_2O_4, a colorless gas with a boiling point of 21°C:

$$2\ NO_2(g) \rightleftharpoons N_2O_4(g) \tag{14.3}$$
$$\text{(brown)} \qquad \text{(colorless)}$$

The flask on the left in Figure 14.4 contains an equilibrium mixture of NO_2 and N_2O_4 at room temperature. At equilibrium, the rate of the forward reaction, which is second order in NO_2:

$$\text{Rate}_f = k_f[NO_2]^2 \tag{14.4}$$

matches the rate of the reverse reaction, which is first order in N_2O_4:

$$\text{Rate}_r = k_r[N_2O_4] \tag{14.5}$$

When an identical flask (on the right in Figure 14.4) is placed in an ice bath, most of the N_2O_4 in the mixture condenses, dramatically lowering its gas-phase concentration. A lower concentration of N_2O_4 gas means the rate of the reverse reaction in Equation 14.5 decreases. As a result, the forward reaction dominates, consuming NO_2 gas and causing the brown color of the reaction mixture to fade. As the concentration of NO_2 gas decreases, the rate of the forward reaction decreases until it again matches the rate of the reverse reaction:

$$\text{Rate}_f = \text{rate}_r \tag{14.6}$$

and equilibrium is reestablished.

Replacing the terms in Equation 14.6 with the right sides of Equations 14.4 and 14.5,

$$k_f[NO_2]^2 = k_r[N_2O_4]$$

FIGURE 14.3 Air quality in Los Angeles has improved since this photograph of "brown LA haze" was taken. The color was caused by high concentrations of NO_2.

∞ CONNECTION We discussed the rates of reactions of nitrogen oxides in the atmosphere in Chapter 13 in our discussion of the chemistry of photochemical smog.

FIGURE 14.4 Shifting equilibrium. The flask on the left contains an equilibrium reaction mixture of brown NO_2 gas and colorless N_2O_4 gas at room temperature. When an identical flask is placed in an ice bath, N_2O_4 condenses, which shifts the equilibrium to produce more of it, lowering the concentration of NO_2 gas and causing the brown color of the gas mixture to fade.

which we can rearrange to

$$\frac{k_f}{k_r} = \frac{[N_2O_4]}{[NO_2]^2} \qquad (14.7)$$

The ratio k_f/k_r is the ratio of two constants, which is simply another constant. This constant has a special name: it is called an **equilibrium constant (K)**. Equating it to the ratio of product concentrations to reactant concentrations on the right of Equation 14.7 gives us the **equilibrium constant expression** for this reaction:

$$K = \frac{[N_2O_4]}{[NO_2]^2} \qquad (14.8)$$

We will shortly explore other ways to formulate equilibrium constant expressions that do not require knowledge of the rate laws.

equilibrium constant (K) the value of the ratio of concentration (or partial pressure) terms in the equilibrium constant expression at a specific temperature.

equilibrium constant expression the ratio of the equilibrium concentrations or partial pressures of products to reactants, each term raised to a power equal to the coefficient of that substance in the balanced chemical equation for the reaction.

CONCEPT TEST

In the reversible reaction A $\rightleftharpoons$ B, the rate constant of the forward reaction at a particular temperature is 3.0 times the value of the rate constant of the reverse reaction. What is the value of the equilibrium constant at that temperature?

(Answers to Concept Tests are in the back of the book.)

Equilibrium constants can have values that range from almost too large to determine to almost too small to determine. If K is very large ($K \gg 1$), the concentrations of products in a reaction mixture at equilibrium are much larger than the concentrations of reactants. If K is very small ($K \ll 1$), the concentrations of products at equilibrium are much smaller than the concentrations of reactants. Always keep in mind that the equilibrium constant says nothing about how fast a reaction reaches equilibrium; it only tells us the extent of the reaction once equilibrium is reached.

▶Ⅱ **CHEMTOUR** Equilibrium in the Gas Phase

14.2 Writing Equilibrium Constant Expressions

Let's revisit the water–gas shift reaction:

$$H_2O(g) + CO(g) \rightleftharpoons H_2(g) + CO_2(g) \qquad (14.2)$$

The data in Table 14.1 describe the results of four experiments in which different quantities of H_2O vapor, CO, H_2, and CO_2 are injected into a sealed reaction vessel heated to 500 K. The four gases are allowed to react and the final concentrations of all four are determined when the reaction has reached equilibrium.

TABLE 14.1 Initial and Equilibrium Concentrations of the Reactants and Products in the Water–Gas Shift Reaction [$H_2O(g) + CO(g) \rightleftharpoons H_2(g) + CO_2(g)$] at 500 K

Experiment	INITIAL CONCENTRATION (*M*)				EQUILIBRIUM CONCENTRATION (*M*)			
	[H₂O]	[CO]	[H₂]	[CO₂]	[H₂O]	[CO]	[H₂]	[CO₂]
1	0.0200	0.0200	0	0	0.0034	0.0034	0.0166	0.0166
2	0	0	0.0200	0.0200	0.0034	0.0034	0.0166	0.0166
3	0.0100	0.0200	0.0300	0.0400	0.0046	0.0146	0.0354	0.0454
4	0.0200	0.0100	0.0200	0.0100	0.0118	0.0018	0.0282	0.0182

law of mass action the ratio of the concentrations or partial pressures of products to reactants at equilibrium has a characteristic value at a given temperature when each term is raised to a power equal to the coefficient of that substance in the balanced chemical equation for the reaction.

mass action expression equivalent to the equilibrium constant expression, but applied to reaction mixtures that may or may not be at equilibrium.

In Experiment 1 the reaction vessel initially contains equimolar concentrations of H_2O and CO but no H_2 or CO_2. In Experiment 2 the vessel initially contains equimolar concentrations of H_2 and CO_2 but no H_2O or CO. The data in Table 14.1 indicate that when the reaction mixtures in Experiments 1 and 2 achieve chemical equilibrium, the concentrations of H_2O and CO are the same (0.0034 M) in both experiments and so too are the concentrations of H_2 and CO_2 (0.0166 M). (Note that these statements about equilibrium concentrations are true in this case because the stoichiometric coefficients are all "1.") These results indicate that the composition of a reaction mixture at equilibrium is independent of the direction in which a particular reaction ran to achieve equilibrium.

Additional data in Table 14.1 show that the forward reaction takes place when the concentrations of the products are initially the same as the concentrations of the reactants (Experiment 4), or even higher (Experiment 3). The significance of these observations, taken with those from Experiments 1 and 2, can be appreciated if we do the following math: multiply the equilibrium concentrations of the products (H_2 and CO_2) together and divide that product by the product of the equilibrium concentrations of the reactants (H_2O and CO):

$$\text{Experiments 1 and 2} \qquad \frac{[H_2][CO_2]}{[H_2O][CO]} = \frac{(0.0166)(0.0166)}{(0.0034)(0.0034)} = 24$$

$$\text{Experiment 3} \qquad \frac{[H_2][CO_2]}{[H_2O][CO]} = \frac{(0.0354)(0.0454)}{(0.0046)(0.0146)} = 24$$

$$\text{Experiment 4} \qquad \frac{[H_2][CO_2]}{[H_2O][CO]} = \frac{(0.0282)(0.0182)}{(0.0118)(0.0018)} = 24$$

In every experiment this calculation yields the same result.

As you might guess, we would get the same ratio of product to reactant concentrations at equilibrium from *any* combination of initial concentrations of these four gases at 500 K. This constancy applies to other reaction mixtures and has been known since the mid-19th century, when Norwegian chemists Cato Guldberg (1836–1902) and Peter Waage (1833–1900) discovered that any reversible reaction eventually reaches a state in which the ratio of the concentrations of products to reactants, with each value raised to a power corresponding to the coefficient for that substance in the balanced chemical equation for the reaction, has a characteristic value at a given temperature. They called this phenomenon the **law of mass action**. This ratio of concentration terms is the equilibrium constant expression for the reaction. It is also called the **mass action expression**, but we will limit the use of that term to those reactions that are not at equilibrium. For the water–gas shift reaction at 500 K, the equilibrium constant expression is

$$K = \frac{[CO_2][H_2]}{[CO][H_2O]} = 24$$

Remember that the exponents in the equilibrium constant expression must be the same as the coefficients in the balanced equation. This is in contrast to rate law expressions, where the exponents are frequently not the same as the coefficients in the balanced equation because they reflect the stoichiometry of the rate-determining step, not necessarily the overall reaction.

CONCEPT TEST

Refer to the water–gas shift reaction, Equation 14.2. Equal numbers of moles of water vapor, carbon dioxide, carbon monoxide, and hydrogen gas are injected into a rigid, sealed reaction vessel and heated to 500 K. Which expression about the composition of the equilibrium reaction mixture is true?

a. $[CO] = [H_2] = [CO_2] = [H_2O]$
b. $[CO_2] = [H_2] > [CO] = [H_2O]$
c. $[CO_2] = [H_2] < [CO] = [H_2O]$
d. $[H_2O] = [H_2] > [CO_2] = [CO]$
e. $[H_2O] = [H_2] < [CO_2] = [CO]$

In the preceding paragraphs we used equilibrium constant expressions based on the molar concentrations of products and reactants. For the generic reaction in which a moles of reactant A react with b moles of B to form c moles of substance C and d moles of D:

$$a\,A + b\,B \rightleftharpoons c\,C + d\,D$$

the equilibrium constant expression is

$$K_c = \frac{[C]^c[D]^d}{[A]^a[B]^b} \qquad (14.9)$$

where the subscript "c" of the equilibrium constant represents *concentration*. If substances A, B, C, and D are gases, then the equilibrium constant may also be expressed in terms of their *partial pressures*:

$$K_p = \frac{(P_C)^c(P_D)^d}{(P_A)^a(P_B)^b} \qquad (14.10)$$

As we'll see later in this chapter, the values of K_c and K_p for a given reaction and temperature may or may not be the same. It depends on whether the number of moles of gaseous reactants is the same as the number of moles of gaseous products.

Also, *equilibrium constants never have units*. This is true even when there are different numbers of moles of reactants and products. Why? It is because the concentrations and partial pressures in K_c or K_p expressions are (theoretically) *ratios* of concentrations or partial pressures to an *ideal standard* concentration (1.000 *M*) or partial pressure (1.000 atm). These ratios take into account the nonideal behavior of substances. Because the terms in equilibrium constant expressions are ratios of values having the same units, their units cancel out, leaving unitless equilibrium constants. In reality, we use concentration and partial pressure values directly in equilibrium calculations. When we do we are assuming that all the reactants and products are behaving ideally.

∞ CONNECTION We introduced the concept of nonideal behavior in Section 10.12 when we described the reasons why real gases may not behave exactly like ideal gases.

SAMPLE EXERCISE 14.1 **Writing Equilibrium Constant Expressions** **LO2**

A key reaction in the formation of acid rain involves the reversible combination of SO_2 and O_2 in the atmosphere, producing SO_3:

$$2\,SO_2(g) + O_2(g) \rightleftharpoons 2\,SO_3(g)$$

Write the K_c and K_p expressions for this reaction.

COLLECT AND ORGANIZE We are given the balanced chemical equation for a reaction and asked to write K_c and K_p expressions for the reaction.

ANALYZE Equilibrium constant expressions are ratios of the concentrations (K_c) or partial pressures (K_p) of products to reactants, with each term raised to the power equal to its coefficient in the balanced chemical equation of the reaction. In the reaction of interest, the coefficients of SO_2 and SO_3 are both 2, so the SO_2 and SO_3 terms in the K_c and K_p expressions will be squared.

SOLVE

$$K_c = \frac{[SO_3]^2}{[SO_2]^2[O_2]}$$

$$K_p = \frac{(P_{SO_3})^2}{(P_{SO_2})^2(P_{O_2})}$$

THINK ABOUT IT The K_c and K_p expressions have the same form: their numerators and denominators contain terms for the same products and reactants, each raised to the same power. The difference between them is the nature of the terms: molar concentrations in the K_c expression and pressures in the K_p expression.

Practice Exercise Write the equilibrium constant expressions K_c and K_p for this reaction:

$$CH_4(g) + H_2O(g) \rightleftharpoons CO(g) + 3\,H_2(g)$$ ⚙

(Answers to Practice Exercises are in the back of the book.)

■ •

SAMPLE EXERCISE 14.2   Calculating the Value of K_c   **LO3**

Table 14.2 contains data from four experiments on the dimerization of NO_2:

$$2\,NO_2(g) \rightleftharpoons N_2O_4(g)$$

The experiments were run at 100°C in a rigid, closed container. Use the data from each experiment to calculate a value of the equilibrium constant K_c for the dimerization reaction.

TABLE 14.2   Data for the Reaction $2\,NO_2(g) \rightleftharpoons N_2O_4(g)$ at 100°C

Experiment	INITIAL CONCENTRATION (*M*)		EQUILIBRIUM CONCENTRATION (*M*)	
	[NO₂]	[N₂O₄]	[NO₂]	[N₂O₄]
1	0.0200	0.0000	0.0172	0.00139
2	0.0300	0.0000	0.0244	0.00280
3	0.0400	0.0000	0.0310	0.00452
4	0.0000	0.0200	0.0310	0.00452

COLLECT AND ORGANIZE We are given four sets of data that contain initial and equilibrium concentrations of a reactant and product. We are asked to determine the value of the equilibrium constant K_c in each experiment. We know the equilibrium constant expression (Equation 14.8) for this reaction:

$$K_c = \frac{[N_2O_4]}{[NO_2]^2}$$

ANALYZE In each of the four experiments, the concentration of NO_2 is nearly 10 times the concentration of N_2O_4 at equilibrium. However, both values in each experiment are much less than 1, and the [NO_2] term is squared. These two factors taken together mean that the values of the numerators in the equilibrium constant expressions will probably be greater than the denominators, so the calculated values of K_c will be greater than 1.

SOLVE

$$\text{Experiment 1:} \quad K_c = \frac{[N_2O_4]}{[NO_2]^2} = \frac{0.00139}{(0.0172)^2} = 4.70$$

$$\text{Experiment 2:} \quad K_c = \frac{0.00280}{(0.0244)^2} = 4.70$$

$$\text{Experiment 3:} \quad K_c = \frac{0.00452}{(0.0310)^2} = 4.70$$

$$\text{Experiment 4:} \quad K_c = \frac{0.00452}{(0.0310)^2} = 4.70$$

THINK ABOUT IT The values calculated for K_c are the same, as they should be for the same reaction at the same temperature, and, as predicted, are greater than 1.

Practice Exercise A mixture of gaseous CO and H_2, called *synthesis gas*, is used commercially to prepare methanol (CH_3OH), a compound considered an alternative fuel to gasoline. Under equilibrium conditions at 700 K, $[H_2] = 0.074\ M$, $[CO] = 0.025\ M$, and $[CH_3OH] = 0.040\ M$. What is the value of K_c for this reaction at 700 K?

SAMPLE EXERCISE 14.3   Calculating the Value of K_p LO3

A sealed chamber contains an equilibrium mixture of NO_2 and N_2O_4 at 300°C and partial pressures $P_{NO_2} = 0.101$ atm and $P_{N_2O_4} = 0.074$ atm. What is the value of K_p for the following reaction under these conditions?

$$2\ NO_2(g) \rightleftharpoons N_2O_4(g)$$

COLLECT AND ORGANIZE We are asked to determine the value of the equilibrium constant K_p for the dimerization reaction of NO_2 to form N_2O_4. We are given the partial pressure values of both gases at equilibrium. The equilibrium constant expression written in terms of concentrations for the reaction (Equation 14.8) is

$$K_c = \frac{[N_2O_4]}{[NO_2]^2}$$

ANALYZE The form of the K_p expression for the reaction is the same as the K_c expression. The only difference is that the concentration terms in the K_p expression are replaced with partial pressure terms:

$$K_p = \frac{P_{N_2O_4}}{(P_{NO_2})^2}$$

The partial pressure of NO_2 is slightly greater than the partial pressure of N_2O_4 at equilibrium, but both values are less than 1. Moreover, the P_{NO_2} term is squared. These two factors taken together mean that the value of the numerator in the K_p expression will probably be greater than that of the denominator, so the calculated value of K_p will be greater than 1.

SOLVE

$$K_p = \frac{0.074}{(0.101)^2} = 7.3$$

THINK ABOUT IT As we predicted, the value of K_p is greater than 1, even though the equilibrium partial pressure of the product ($P_{N_2O_4} = 0.074$ atm) is actually less than that of the reactant ($P_{NO_2} = 0.101$ atm). Relatively little N_2O_4 forms in this case because P_{NO_2} is so low, and two moles of NO_2 are required to make one mole of N_2O_4.

As we will see in Section 14.7, low pressures favor the side of a gas-phase reaction that has more moles of gas. To illustrate this point, let's calculate the value of $P_{N_2O_4}$ that would be in equilibrium with 1 atm of NO_2 at 300°C:

$$K_p = \frac{P_{N_2O_4}}{(P_{NO_2})^2} = 7.3 = \frac{P_{N_2O_4}}{1^2}$$

$$P_{N_2O_4} = 7.3 \text{ atm}$$

The value of $P_{N_2O_4}$ is 7.3 times the P_{NO_2} value at these higher overall pressures, whereas $P_{N_2O_4}$ was lower than P_{NO_2} at the lower pressures in this exercise.

Practice Exercise A reaction vessel contains an equilibrium mixture of SO_2, O_2, and SO_3. Given the partial pressures $P_{SO_2} = 0.0018$ atm, $P_{O_2} = 0.0032$ atm, and $P_{SO_3} = 0.0166$ atm, calculate the value of K_p for the reaction:

$$2 \, SO_2(g) + O_2(g) \rightleftharpoons 2 \, SO_3(g)$$

The value of K indicates how far a reaction proceeds at a given temperature. The values we have seen thus far—$K_c = 24$ for the water–gas shift reaction at 500 K, and $K_p = 7.3$, $K_c = 4.7$ for the dimerization of NO_2 at 300°C and 100°C, respectively—are considered intermediate values. Because the range of K values is so large ($0 < K < \infty$), all three of these values are considered *close* to 1, which means that comparable concentrations of reactants and products are likely to be present at equilibrium.

In contrast, the reaction between H_2 and O_2 to form water proceeds until one of the reactants is almost completely consumed. This observation is consistent with a large value of K at 25°C:

$$2 \, H_2(g) + O_2(g) \rightleftharpoons 2 \, H_2O(g) \qquad K_c = 3 \times 10^{81}$$

On the other hand, the decomposition of CO_2 to CO and O_2 at 25°C proceeds hardly at all and is consistent with a very small value of K and virtually no product being formed:

$$2 \, CO_2(g) \rightleftharpoons 2 \, CO(g) + O_2(g) \qquad K_c = 3 \times 10^{-92}$$

14.3 Relationships between K_c and K_p Values

As we noted in Section 14.2, the values of K_c and K_p for a given reaction and temperature may or may not be the same, depending on the numbers of moles of gaseous reactants and products. To better understand this relationship, we begin with the ideal gas law:

$$PV = nRT$$

If we solve for P and express volume in liters, then n/V has units of moles per liter, which is the same as molarity (M):

$$P = \frac{n}{V}RT$$

$$P = MRT \qquad\qquad (14.11)$$

Let's apply Equation 14.11 to the gases in the NO_2/N_2O_4 equilibrium from Sample Exercise 14.3:

$$P_{NO_2} = \frac{n_{NO_2}}{V}RT = [NO_2]RT$$

$$P_{N_2O_4} = \frac{n_{N_2O_4}}{V}RT = [N_2O_4]RT$$

Substituting these values into the expression for K_p from Sample Exercise 14.3, we get

$$K_p = \frac{P_{N_2O_4}}{(P_{NO_2})^2} = \frac{[N_2O_4]RT}{([NO_2]RT)^2} = \frac{[N_2O_4]\cancel{RT}}{[NO_2]^2(RT)^{\cancel{2}}}$$

The ratio of concentration terms in the expression on the right, $[N_2O_4]/[NO_2]^2$, is the same as the K_c expression for this reaction. Substituting K_c for those terms and simplifying the RT terms:

$$K_p = K_c\frac{1}{RT}$$

This last equation defines the specific relationship between K_c and K_p for this reaction. A more general expression can be derived for the generic reaction of gases A and B forming gases C and D:

$$a\,A + b\,B \rightleftharpoons c\,C + d\,D$$

As noted in the previous section, the K_p expression for this reaction is

$$K_p = \frac{(P_C)^c(P_D)^d}{(P_A)^a(P_B)^b} \tag{14.10}$$

Replacing each partial pressure term in Equation 14.10 with the corresponding molar concentration term $\times RT$ (from Equation 14.11) gives us the general expression

$$K_p = \frac{([C]RT)^c([D]RT)^d}{([A]RT)^a([B]RT)^b} \tag{14.12}$$

Combining the RT terms, we get

$$K_p = \frac{[C]^c[D]^d}{[A]^a[B]^b} \times RT^{[(c+d)-(a+b)]} \tag{14.13}$$

The concentration ratio on the right side of this equation matches the K_c expression for this generic reaction (see Equation 14.9). Substituting this equality into Equation 14.13 gives us

$$K_p = K_c(RT)^{[(c+d)-(a+b)]} \tag{14.14}$$

To simplify Equation 14.14, consider this: $(c + d)$ represents the sum of the coefficients of the gaseous products in the reaction—the sum of the number of moles of gases produced. Similarly, $(a + b)$ represents the sum of the number of moles of gaseous reactants consumed. The difference between the two sums,

$(c + d) - (a + b)$, represents the *change in the number of moles of gases* between the product and reactant sides of the balanced chemical equation. We use the symbol Δn to represent this change. Substituting Δn for $(c + d) - (a + b)$ in Equation 14.14 gives us

$$K_p = K_c(RT)^{\Delta n} \tag{14.15}$$

Equation 14.15 provides a quantitative interpretation of the opening statement of this section: the relationship between the K_p and K_c values of a chemical reaction involving gases depends on the number of moles of gaseous reactants and products. In reactions such as the water–gas shift reaction:

$$H_2O(g) + CO(g) \rightleftharpoons H_2(g) + CO_2(g) \tag{14.2}$$

in which the number of moles of gas on both sides of the reaction arrow is the same, $\Delta n = 0$ and $K_p = K_c$. However, in the steam-reforming reaction:

$$CH_4(g) + H_2O(g) \rightleftharpoons CO(g) + 3\,H_2(g) \tag{14.1}$$

2 moles of gaseous reactants form 4 moles of gaseous products. Therefore:

$$\Delta n = 4\text{ mol} - 2\text{ mol} = 2\text{ mol}$$

Inserting this value for Δn in Equation 14.15 gives us the relationship between K_p and K_c for this reaction:

$$K_p = K_c(RT)^2$$

A final point about the relative sizes of K_p and K_c values: One mole of an ideal gas at 0°C and 1 atm of pressure (the combination of temperature and pressure referred to as STP) occupies a volume of 22.4 liters. Therefore, its molar concentration is 1 mol/22.4 L = 0.0446 M. Thus, the pressure of a pure gas at STP is 22.4 times its molar concentration. The RT term in Equation 14.15 is essentially a conversion factor for changing molar concentrations into partial pressures. The value of R that we use depends on the units used to express concentration, pressure, and temperature. If we stick with molarity, atmospheres, and T in kelvin, the value of R is 0.08206 (L · atm)/(mol · K).

SAMPLE EXERCISE 14.4 **Calculating K_c from K_p**   **LO4**

In Sample Exercise 14.3 we calculated the value of K_p (7.3) for the dimerization of NO_2 to N_2O_4 at 300°C. What is the value of K_c for this reaction at 300°C?

COLLECT AND ORGANIZE We are given a K_p value and asked to calculate the corresponding K_c value for the same reaction at the same temperature. Equation 14.15 relates K_p and K_c values:

$$K_p = K_c(RT)^{\Delta n}$$

where Δn represents the change in the number of moles of gas when going from reactant to product.

ANALYZE We need a balanced chemical equation to determine the value of Δn. From Sample Exercise 14.2 we know that that equation is

$$2\,NO_2(g) \rightleftharpoons N_2O_4(g)$$

Because the number of moles of gas is not the same on both sides of the reaction arrow, we predict that the values of K_c and K_p will not be the same.

SOLVE According to the balanced equation, 2 moles of gaseous reactants yield 1 mole of gaseous product. Therefore:

$$\Delta n = 1 \text{ mol} - 2 \text{ mol} = -1 \text{ mol}$$

Inserting this value, the known value of K_p, and the temperature into Equation 14.15,

$$K_p = K_c(RT)^{\Delta n}$$

$$7.3 = K_c[0.08206 \times (273 + 300)]^{-1}$$

$$K_c = (7.3)(0.08206)(573) = 3.4 \times 10^2$$

THINK ABOUT IT The value of K_c differs from the value of K_p because two moles of gaseous reactants produce only one mole of gaseous product. Actually, the value of K_c is nearly 50 times larger than that of K_p. The K_c/K_p ratio is over twice the STP molar volume factor (22.4 L/mol) because the absolute temperature of the reaction is more than twice the temperature at STP.

Practice Exercise An important industrial process for synthesizing the ammonia used in agricultural fertilizers involves the combination of N_2 and H_2:

$$N_2(g) + 3 \text{ H}_2(g) \rightleftharpoons 2 \text{ NH}_3(g) \qquad K_c = 2.8 \times 10^{-9} \text{ at } 30°C$$

What is the value of K_p of this reaction at 30°C?

14.4 Manipulating Equilibrium Constant Expressions

Suppose the following hypothetical combination reaction has a K_c value of 100:

$$A + B \rightleftharpoons 2 \text{ C} \qquad K_c = 100$$

What would be the K_c value for the reverse reaction:

$$2 \text{ C} \rightleftharpoons A + B \qquad K_{c,reverse} = ?$$

and what would be the K_c value for the reaction in which only one mole of C is made, which means that all coefficients are half what they were in the original equation:

$$\tfrac{1}{2} A + \tfrac{1}{2} B \rightleftharpoons C \qquad K_{c,1/2} = ?$$

To answer questions like these we need to think about how the equilibrium constant expressions for these reactions are related. We explore these relationships in this section.

K for Reverse Reactions

Let's write the K_c expression for our combination reaction $A + B \rightleftharpoons 2 \text{ C}$:

$$K_c = \frac{[C]^2}{[A][B]} = 100$$

The K_c expression for the reverse reaction, $2 \text{ C} \rightleftharpoons A + B$, is

$$K_{c,reverse} = \frac{[A][B]}{[C]^2}$$

Comparing the K_c expressions for the forward and reverse reactions, we see that $K_{c,reverse}$ *is the reciprocal of* K_c, which means its value must be $1/100 = 0.0100$:

$$2\,C \rightleftharpoons A + B \qquad K_{c,reverse} = 0.0100$$

Expressing this result for any chemical reaction, we have

$$K_{forward} = \frac{1}{K_{reverse}} \tag{14.16}$$

This inverse relationship between $K_{forward}$ and $K_{reverse}$ makes sense when we apply it to our hypothetical reaction. The relatively large value of $K_{forward}$ (100) means that the concentration of product C in a reaction mixture at equilibrium is likely to be much higher than the concentrations of reactants A and B. The corresponding small value of $K_{reverse}$ (0.0100) means the same thing in terms of the relative values of [A], [B], and [C]. The difference is that the equilibrium mixture in the reverse reaction contains mostly reactant (C) and little of the products (A and B), as we would expect given the small value of $K_{reverse}$.

K for an Equation Multiplied by a Number

Now let's write the K_c expression for our combination reaction $\frac{1}{2}A + \frac{1}{2}B \rightleftharpoons C$, representing the preparation of only 1 mole of C:

$$K_{c,1/2} = \frac{[C]}{[A]^{1/2}[B]^{1/2}}$$

Comparing this expression with the K_c expression for the forward reaction:

$$K_c = \frac{[C]^2}{[A][B]}$$

we see that the $K_{c,1/2}$ expression is equal to the square root of the reaction's K_c expression: the C term is squared in K_c but is only raised to the first power in $K_{c,1/2}$, and the A and B terms are raised to the first power in K_c but to the $\frac{1}{2}$ power in $K_{c,1/2}$. Therefore, we may conclude that the value of $K_{c,1/2}$ is the square root of the value of K_c, or $\sqrt{100} = 10.0$. We can extend this pattern to all chemical equilibria. If the balanced chemical equation of a reaction is multiplied by some factor n, then the value of K is raised to the nth power (K^n).

SAMPLE EXERCISE 14.5 **Calculating the *K* Values** **LO5**
of Related Chemical Reactions

We learned in Chapter 13 that a key reaction in the formation of photochemical smog is the one between NO and atmospheric O_2 that forms NO_2:

$$(1) \qquad 2\,NO(g) + O_2(g) \rightleftharpoons 2\,NO_2(g)$$

The K_p value of this reaction is 2.4×10^{12} at 25°C.

   a. What is the K_p value of the decomposition of $NO_2(g)$?

$$(2) \qquad 2\,NO_2(g) \rightleftharpoons 2\,NO(g) + O_2(g)$$

   b. What is the K_p value of the reaction in which NO and O_2 combine to form only one mole of NO_2?

$$(3) \qquad NO(g) + \tfrac{1}{2}O_2(g) \rightleftharpoons NO_2(g)$$

COLLECT AND ORGANIZE Three gases (NO, O_2, and NO_2) are involved in a reaction for which we know the K_p value and two reactions for which we don't know the K_p value.

ANALYZE Reaction 2 is the reverse of reaction 1. Therefore, the K_p value of reaction 2 is the reciprocal of the K_p value of reaction 1. Reaction 3 has the same reactants and product as reaction 1, but the coefficients of the terms in its chemical equation are one-half of the values used in the chemical equation for reaction 1. When the coefficients of a chemical equation are multiplied by a factor n ($\frac{1}{2}$ in this case), the value of the equilibrium constant is raised to the nth power, which in this case means that the K_p value of reaction 3 is the square root of the K_p value of reaction 1. The value of $K_{p,1}$ is huge—a bit more than 10^{12}—which means the value of $K_{p,2}$ for the reverse reaction should be very small—somewhat less than 10^{-12}. The value of $K_{p,3}$ is equal to $\sqrt{K_{p,1}}$ or about 10^6.

SOLVE

a. $K_{p,2} = \dfrac{1}{K_{p,1}} = \dfrac{1}{2.4 \times 10^{12}} = 4.2 \times 10^{-13}$

b. $K_{p,3} = \sqrt{K_{p,1}} = \sqrt{2.4 \times 10^{12}} = 1.5 \times 10^6$

THINK ABOUT IT As predicted, the value of $K_{p,2}$ is very small and the value of $K_{p,3}$ is quite large. Both values carry the same message: that NO should (eventually) be oxidized to NO_2 in the atmosphere. How long the reaction takes, as we saw in Chapter 13, is another matter.

Practice Exercise The equilibrium constant K_c for the synthesis of ammonia,

$$N_2(g) + 3\,H_2(g) \rightleftharpoons 2\,NH_3(g)$$

is 9.6 at 300°C. What is the value of K_c for the decomposition of ammonia,

$$NH_3(g) \rightleftharpoons \tfrac{1}{2}\,N_2(g) + \tfrac{3}{2}\,H_2(g)$$

at 300°C?

Given two equally legitimate equilibrium constant expressions for the same reaction, you may wonder how the same reaction with the same reactants and products can have two or more equilibrium constant values. Surely the same ingredients should be present in the same proportions at equilibrium no matter how we choose to write a balanced equation describing their reaction. In fact, they are. The difference in K values is not chemical; it is only mathematical. It does not affect the composition of a reaction mixture at equilibrium—only the arithmetic we use to predict it.

Combining *K* Values

In Chapter 9, we applied Hess's law to calculate the enthalpies of combined reactions. We carry out a similar process here to determine the overall K for a reaction that is the sum of two or more other reactions.

Consider two reactions from Chapter 13 involved in the formation of photochemical smog, wherein the NO produced in a car's engine at high temperatures is oxidized to NO_2 in the atmosphere:

(1)	$N_2(g) + O_2(g) \rightleftharpoons 2\,\cancel{NO(g)}$
(2)	$\cancel{2\,NO(g)} + O_2(g) \rightleftharpoons 2\,NO_2(g)$
Overall:	$N_2(g) + 2\,O_2(g) \rightleftharpoons 2\,NO_2(g)$

The equilibrium constant expression for the overall reaction is

$$K_c = \frac{[NO_2]^2}{[N_2][O_2]^2}$$

We can derive this expression from the equilibrium constant expressions for reactions 1 and 2,

$$K_1 = \frac{[NO]^2}{[N_2][O_2]} \quad \text{and} \quad K_2 = \frac{[NO_2]^2}{[NO]^2[O_2]}$$

if we multiply K_1 by K_2:

$$K_1 \times K_2 = \frac{\cancel{[NO]^2}}{[N_2][O_2]} \times \frac{[NO_2]^2}{\cancel{[NO]^2}[O_2]} = \frac{[NO_2]^2}{[N_2][O_2]^2} = K_{overall}$$

This approach works for all series of reactions, and as a general rule:

$$K_{overall} = K_1 \times K_2 \times K_3 \times \cdots \times K_n \tag{14.17}$$

The overall equilibrium constant for a sum of two or more reactions is the product of the equilibrium constants of the individual reactions. Thus, the value of K_c for the overall reaction for the formation of NO_2 from N_2 and O_2 at 1000 K is the product of the equilibrium constants for reactions 1 and 2:

$$K_1 = \frac{[NO]^2}{[N_2][O_2]} = 7.2 \times 10^{-9}$$

$$K_2 = \frac{[NO_2]^2}{[NO]^2[O_2]} = 0.020$$

$$K_{overall} = K_1 \times K_2 = 7.2 \times 10^{-9} \times 0.020 = 1.4 \times 10^{-10}$$

Remember that the equilibrium constant expression for the overall reaction must contain the appropriate terms for the products and reactants of that reaction. Just as with Hess's law in thermochemical calculations, we may need to reverse an equation or multiply an equation by a factor when we combine it with another to create the equation of interest. If we reverse a reaction, we must take the reciprocal of its K. If we multiply a reaction by a constant, we must raise its K to that power.

SAMPLE EXERCISE 14.6   **Calculating Overall K Values**   **LO5**
of Combined Reactions

At 1000 K, the K_c values of these reactions are as follows:

(1) $N_2O_4(g) \rightleftharpoons 2\,NO_2(g)$ $K_c = 1.5 \times 10^6$

(2) $N_2(g) + 2\,O_2(g) \rightleftharpoons 2\,NO_2(g)$ $K_c = 1.4 \times 10^{-10}$

What is the K_c value at 1000 K of the reaction?

$$N_2(g) + 2\,O_2(g) \rightleftharpoons N_2O_4(g)$$

COLLECT AND ORGANIZE We are given two reactions and their K_c values. We need to combine the two reactions in such a way that N_2 and O_2 are on the reactant side of the overall equation and N_2O_4 is on the product side and then calculate the K_c value of the overall reaction.

ANALYZE The overall reaction is the sum of the reverse of reaction 1 and reaction 2 as written:

Reaction 1 reversed	$2\,NO_2(g) \rightleftharpoons N_2O_4(g)$
Reaction 2	$N_2(g) + 2\,O_2(g) \rightleftharpoons 2\,NO_2(g)$
Overall	$N_2(g) + 2\,O_2(g) \rightleftharpoons N_2O_4(g)$

Reversing a chemical reaction requires taking the reciprocal of its K_c value, and when two reactions are added, the value of K_c of the overall reaction is the product of the K_c values of the two reactions. The K_c value of reaction 1 is large ($>10^6$), which means its reciprocal is small ($<10^{-6}$). The product of the latter value and the even smaller K_c value of reaction 2 ($\sim 10^{-10}$) should be a very small value—about 10^{-16}.

SOLVE

$$K_{overall} = \frac{1}{K_1} \times K_2 = \frac{1}{1.5 \times 10^6} \times 1.4 \times 10^{-10} = 9.3 \times 10^{-17}$$

THINK ABOUT IT The result of the calculation is close to our estimated value. The overall equilibrium of the combined reactions lies far to the left; little N_2O_4 forms from N_2 and O_2 at 1000 K.

Practice Exercise Calculate the value of K_c for the hypothetical reaction

$$Q(g) + X(g) \rightleftharpoons M(g)$$

from the following information:

$$2\,M(g) \rightleftharpoons Z(g) \qquad\qquad K_c = 6.2 \times 10^{-4}$$

$$Z(g) \rightleftharpoons 2\,Q(g) + 2\,X(g) \qquad K_c = 5.6 \times 10^{-2}$$

To summarize the key points for manipulating equilibrium constants:

➤ The K value of a reaction running in reverse is the reciprocal of the K of the forward reaction.

➤ If the original chemical equation describing an equilibrium is multiplied by a factor n, the value of K of the new equilibrium constant expression is the value of the original K raised to the nth power.

➤ If an overall chemical reaction is the sum of two or more other reactions, the overall value of K is the product of the K values of the other reactions.

14.5 Equilibrium Constants and Reaction Quotients

In Sections 14.1 and 14.2 we introduced two key terms: *equilibrium constant expression* and *mass action expression*. Until now we have used the first term almost exclusively because we have been dealing with chemical reactions that have achieved equilibrium. Now it is time to reintroduce the concept of a mass action expression because we can apply it not only to the concentrations (or partial pressures) of products and reactants in reaction mixtures that have reached equilibrium, but also to reaction mixtures that are on their way to equilibrium but are not there yet.

reaction quotient (Q) the numerical value of the mass action expression for *any values* of the concentrations (or partial pressures) of reactants and products; at equilibrium, $Q = K$.

Even if a reversible chemical reaction has not reached equilibrium, we can still insert reactant and product concentrations (or partial pressures) into its mass action expression. The mathematical result is not a K value because the reaction is not yet at equilibrium. Instead it is a Q value, where Q stands for **reaction quotient**.

The value of Q provides us with a kind of status report on how a reaction is proceeding. To see how, let's revisit the water–gas shift reaction

$$H_2O(g) + CO(g) \rightleftharpoons H_2(g) + CO_2(g) \qquad K_c = 24 \text{ at } 500 \text{ K}$$

and the data from Experiment 3 in Table 14.1. In Experiment 3, the initial concentrations of reactants and products are as follows:

[H₂O] (M)	[CO] (M)	[H₂] (M)	[CO₂] (M)
0.0100	0.0200	0.0300	0.0400

Inserting these values into the mass action expression for the reaction yields a value for Q based on concentration, that is, Q_c:

$$Q_c = \frac{[H_2][CO_2]}{[H_2O][CO]} = \frac{(0.0300)(0.0400)}{(0.0100)(0.0200)} = 6.00$$

Now we compare this value of Q_c to the reaction's K_c value, which is 24. Clearly Q_c is less than K_c, which means there are proportionally smaller concentrations of products and larger concentrations of reactants in the initial reaction mixture than there will be at equilibrium. To achieve equilibrium, some of the reactants must form products, increasing the value of Q_c until it matches the value of K_c and the following equilibrium concentrations of reactants and products are present in the reaction vessel:

[H₂O] (M)	[CO] (M)	[H₂] (M)	[CO₂] (M)
0.0046	0.0146	0.0354	0.0454

To put the results from the data in Table 14.1 in context, let's consider the curves in Figure 14.5. Starting at the left end of the graph (zone a), we have the initial conditions of Experiment 1 from Table 14.1: equal concentrations of reactants are present, but no products. Over time, reactant concentrations (the red curve) decrease as product concentrations (the blue curve) increase. At the right end of the graph (zone c), we have the initial conditions of Experiment 2: products are present, but no reactants. Over time, reactant concentrations increase as product concentrations decrease. In the middle of the graph (zone b), no net change occurs in the composition because the reaction is at equilibrium.

We can also characterize the three zones on the basis of the value of Q compared to K. In zone (a), Q values are less than K and there is a net conversion of reactants into products as the forward reaction dominates. In zone (c), Q values are greater than K and a net conversion of products into reactants takes place as the reverse reaction dominates. In the middle, zone (b), $Q = K$ and no change in the composition of the reaction mixture occurs over time. The relative values of Q and K and their consequences are summarized in Table 14.3.

CONCEPT TEST ●

In which of the three zones of Figure 14.5 do the initial reaction conditions in Experiments 3 and 4 in Table 14.1 fall?

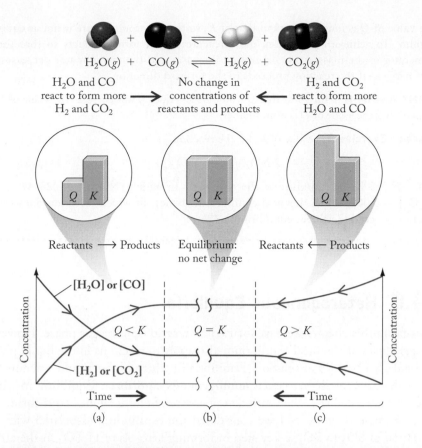

$$H_2O(g) + CO(g) \rightleftharpoons H_2(g) + CO_2(g)$$

H_2O and CO react to form more H_2 and CO_2 → No change in concentrations of reactants and products ← H_2 and CO_2 react to form more H_2O and CO

Reactants $\longrightarrow$ Products   Equilibrium: no net change   Reactants $\longleftarrow$ Products

[H_2O] or [CO]

$Q < K$ $Q = K$ $Q > K$

[H_2] or [CO_2]

Time $\longrightarrow$ $\longleftarrow$ Time

(a)   (b)   (c)

FIGURE 14.5 The value of the reaction quotient Q relative to the equilibrium constant K for the water–gas shift reaction. (a) Reactant concentrations (red) are higher than they are once equilibrium is reached, and product concentrations (blue) are lower than at equilibrium; $Q < K$, and the reactants form more products. (b) Equilibrium concentrations are achieved; $Q = K$, and no net change in concentrations takes place. (c) Product concentrations are higher than they are at equilibrium, and reactant concentrations are lower than at equilibrium; $Q > K$, and products form more reactants as the reaction runs in reverse.

TABLE 14.3	Comparison of Q and K Values	
Value of Q	**What It Means**
$Q < K$	Reaction as written proceeds in forward direction ($\rightarrow$)
$Q = K$	Reaction is at equilibrium ($\rightleftharpoons$)
$Q > K$	Reaction as written proceeds in reverse direction ($\leftarrow$)

**SAMPLE EXERCISE 14.7   Using Q and K Values to Predict   LO6
the Direction of a Reaction**

At 2300 K the value of K of the following reaction is 1.5×10^{-3}:

$$N_2(g) + O_2(g) \rightleftharpoons 2\,NO(g)$$

At the instant when a reaction vessel at 2300 K contains 0.50 M N_2, 0.25 M O_2, and 0.0042 M NO, is the reaction mixture at equilibrium? If not, in which direction will the reaction proceed to reach equilibrium?

COLLECT AND ORGANIZE We are asked whether or not a reaction mixture is at equilibrium. We are given the value of K and the concentrations of reactants and product.

ANALYZE The mass action expression for this reaction based on concentrations is

$$Q = \frac{[NO]^2}{[N_2][O_2]}$$

If we insert the given concentration values into this expression, we can determine the value of Q. Comparing Q with K enables us to determine (1) whether the reaction is at equilibrium and (2) if it is not at equilibrium, in which direction the reaction proceeds. The value of [NO] is about 10^{-2} times the concentrations of the reactants, and the [NO] term is squared in the mass action expression. These two factors together make the value of the numerator about 10^{-4} that of the denominator. Therefore, the value of Q may be less than the value of K.

SOLVE

$$Q = \frac{(0.0042)^2}{(0.50)(0.25)} = 1.4 \times 10^{-4}$$

homogeneous equilibrium involves reactants and products in the same phase.

heterogeneous equilibrium involves reactants and products in more than one phase.

The value of Q is indeed less than that of K, so the reaction mixture is not at equilibrium. To achieve equilibrium, more reactants must form products so that the numerator of the mass action expression increases and the denominator decreases. This happens if the reaction proceeds in the forward direction.

THINK ABOUT IT Our guess that Q would be less than K was correct. The value of K is small, but the value of Q is even smaller.

Practice Exercise The value of K_c for the reaction

$$2\,NO_2(g) \rightleftharpoons N_2O_4(g)$$

is 4.7 at 373 K. Is a mixture of the two gases in which $[NO_2] = 0.025\ M$ and $[N_2O_4] = 0.0014\ M$ in chemical equilibrium? If not, in which direction does the reaction proceed to achieve equilibrium?

14.6 Heterogeneous Equilibria

Thus far in this chapter we have focused on reactions in the gas phase. However, the principles of chemical equilibrium also apply to reactions in the liquid phase, particularly reactions in solution. Equilibria in which products and reactants are all in the same phase are called **homogeneous equilibria**. Equilibria in which reactants and products are in different phases are **heterogeneous equilibria**.

In Sample Exercise 14.1 we considered the equilibrium associated with the oxidation of SO_2 to SO_3, a key step in forming aerosols of H_2SO_4 in the atmosphere. One way to prevent this reaction from happening is to "scrub" SO_2 from the exhaust gases emitted at factories where sulfur-containing fuels are burned. Solid lime, CaO, is a widely used scrubbing agent. Sprayed into the exhaust gases, it combines with SO_2 to form calcium sulfite:

$$CaO(s) + SO_2(g) \rightleftharpoons CaSO_3(s)$$

The large quantities of lime needed for this reaction and for many other industrial and agricultural uses come from heating pulverized limestone, which is mostly $CaCO_3$, in kilns (Figure 14.6) operated at temperatures near 1000°C. At these temperatures, $CaCO_3$ decomposes into lime (CaO) and CO_2 gas:

$$CaCO_3(s) \rightleftharpoons CaO(s) + CO_2(g) \qquad \Delta H^\circ_{rxn} = 178.1\ kJ$$

We might write the concentration-based equilibrium constant expression for this reaction as follows:

$$K_c = \frac{[CaO][CO_2]}{[CaCO_3]}$$

This expression contains concentration terms for two solids: CaO and $CaCO_3$. But what do we mean by the concentration of a solid? Any pure solid has a constant concentration because its mass (and number of moles) per unit volume is always the same. As long as there is *any* CaO or $CaCO_3$ present, the effective "concentration" of both is 1, and there is no need for [CaO] and [$CaCO_3$] terms in the equilibrium constant expression. This leaves us with

$$K_c = [CO_2]$$

This expression means that, as long as some CaO or $CaCO_3$ is present, the equilibrium concentration of CO_2 gas does not vary at a given temperature, as shown

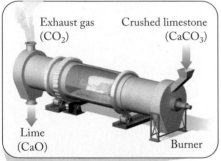

FIGURE 14.6 Rotary kilns operating at 1000°C are used to convert limestone into lime. Rotation of these large cylindrical kilns ensures that crushed limestone ($CaCO_3$) is uniformly heated and has thermally decomposed to lime (CaO) and CO_2 gas by the time it passes through the kiln.

Exhaust gas (CO₂) Crushed limestone (CaCO₃) Lime (CaO) Burner

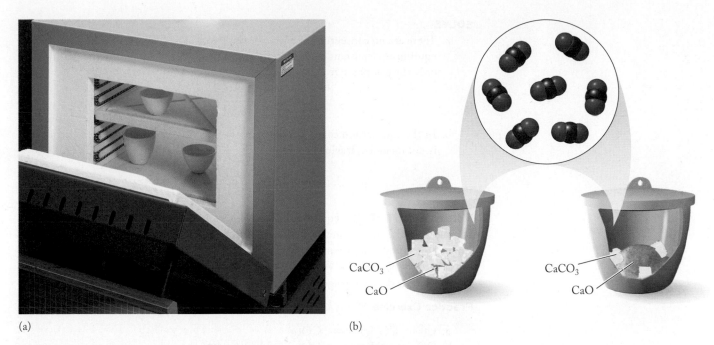

FIGURE 14.7 The position of a heterogeneous equilibrium between $CaCO_3$, CaO, and CO_2 at constant temperature depends only on the concentration of CO_2 gas present. As long as some solid is in the system, the equilibrium concentration of CO_2 remains the same. (a) Muffle furnace for heating crucibles containing $CaCO_3$. (b) Two crucibles containing different amounts of solid material at the same temperature have the same concentration of CO_2 gas.

in Figure 14.7. Instead, the concentration of CO_2 is the same as the value of K_c at that temperature.

The same concept of constant concentration applies to pure liquids that are involved in reversible chemical reactions. As long as the liquid is present, its "concentration" is considered constant during the course of the reaction and does not appear in the equilibrium constant expression. Similarly, K_c expressions for most reactions in aqueous solutions do not include a term for $[H_2O]$ even when water is a reactant or product because its concentration does not change significantly. In writing equilibrium constant expressions for heterogeneous equilibria, we follow the rules we learned earlier, with the additional rule that pure liquids and solids do not appear in the expression.

SAMPLE EXERCISE 14.8   **Writing Equilibrium Constant Expressions for Heterogeneous Equilibria**   **LO2**

Write K_c expressions for these reactions:

    a. $CaO(s) + SO_2(g) \rightleftharpoons CaSO_3(s)$
    b. $CO_2(g) + H_2O(\ell) \rightleftharpoons H_2CO_3(aq)$

COLLECT AND ORGANIZE We are given reversible reactions involving reactants and products in more than one phase, and we are asked to write equilibrium constant expressions for their equilibria. We know that pure liquids and solids are considered to have constant concentrations throughout such reactions.

ANALYZE We need to identify the pure liquids and solids involved in the equilibria and to exclude terms for them in the equilibrium constant expressions. The first equilibrium involves two solids: CaO and $CaSO_3$. The second involves liquid H_2O.

SOLVE

a. There are no concentration terms for the two solids, CaO and $CaSO_3$, in the equilibrium constant expression for reaction a, which leaves us with a term for only the gas-phase reactant, SO_2, in the denominator:

$$K_c = \frac{1}{[SO_2]}$$

b. In the equilibrium constant expression for reaction b, $[H_2O]$ is a constant and is not included, leaving

$$K_c = \frac{[H_2CO_3]}{[CO_2]}$$

THINK ABOUT IT The equilibrium constant expressions we have written in this Sample Exercise and in previous exercises included terms for reactants and products whose concentrations or partial pressures were likely to change significantly during the course of the reaction. We exclude terms for pure solids and liquids because their concentrations do not change.

Practice Exercise Write K_p expressions for the reactions

a. $C(s) + CO_2(g) \rightleftharpoons 2\,CO(g)$
b. $CO_2(g) + H_2(g) \rightleftharpoons CO(g) + H_2O(\ell)$ ⚙

CONCEPT TEST

Explain why $K_c = [H_2O(g)]$ is the mass action expression for the equilibrium $H_2O(\ell) \rightleftharpoons H_2O(g)$.

14.7 Le Châtelier's Principle

▶❚❚ **CHEMTOUR** Le Châtelier's Principle

We can perturb chemical reactions at equilibrium in several ways—for example, by changing the concentration or partial pressure of a reactant or product, or by changing the temperature of the system. Adding or removing an ingredient in the system alters the value of the reaction quotient Q so that it is no longer equal to the value of K. On the other hand, changing the temperature of a system changes the value of K. Either way, the perturbed system is not at equilibrium and the composition of the system must change to restore equilibrium.

One of the first scientists to study and then successfully predict how chemical equilibria respond to such perturbations was French chemist Henri Louis Le Châtelier (1850–1936). He articulated what is now known as **Le Châtelier's principle**, which states that, if a system at equilibrium is perturbed (or *stressed*), the position of the equilibrium shifts in the direction that relieves the stress. Through the years, chemists have used Le Châtelier's principle to increase the yields of chemical reactions that would otherwise have produced very little of a desired compound.

Effects of Adding or Removing Reactants or Products

Le Châtelier's principle a system at equilibrium responds to a stress in such a way that it relieves that stress.

When a reactant or product is added or removed, a system at chemical equilibrium is perturbed. Following Le Châtelier's principle, the system responds in such a way as to restore equilibrium (Figure 14.8).

To explore how industrial chemists exploit Le Châtelier's principle, let's revisit the water–gas shift reaction for making hydrogen:

$$H_2O(g) + CO(g) \rightleftharpoons H_2(g) + CO_2(g) \tag{14.2}$$

To shift the equilibrium toward the production of more H_2, chemists pass the reaction mixture through a scrubber containing a concentrated aqueous solution of K_2CO_3. Doing this removes CO_2 from the gaseous mixture as a result of the following reaction:

$$CO_2(g) + H_2O(\ell) + K_2CO_3(aq) \rightleftharpoons 2\,KHCO_3(s)$$

Removing CO_2 means fewer molecules of it are available to collide with molecules of H_2 to drive the reverse reaction in Equation 14.2. As a result, the rate of the reverse reaction becomes slower than the rate of the forward reaction. This means the system is no longer in equilibrium. To return to equilibrium, the reaction proceeds in the forward direction (we say that the reaction *shifts to the right*), making more product to restore some of what was removed until a new equilibrium is achieved. The new equilibrium is like the old one in that the value of the mass action expression:

$$\frac{[H_2][CO_2]}{[H_2O][CO]}$$

is equal to the value of K. This is true even though the concentrations of the individual reactants and products have changed; the *overall* ratio in the K expression is restored.

Another way to shift a reaction mixture at equilibrium is to add more reactant or product. If the goal is to form more products, then adding more reactants increases their concentration in the reaction mixture, which increases the rate of the forward reaction. Some of the added reactants are converted into additional products. This approach works even when only one reactant is added in a multi-reactant reaction as long as there is still some of the other reactant(s) available. The reason why can be gleaned from the impact of increasing just one of the reactant concentration terms in the equilibrium constant expression. The result is a reaction quotient Q that is less than K, which means the reaction will proceed in the forward direction, forming more products, until equilibrium is restored.

> **CONCEPT TEST** ··
>
> The reaction mixture in the water–gas shift reaction is at equilibrium and some CO_2 is rapidly removed.
>
> a. How does this removal affect the value of the reaction quotient Q?
> b. When equilibrium is restored, which of the four compounds in the system is present at a higher concentration, which is present at a lower concentration, and which, if any, has the same concentration as in the original equilibrium mixture?
>
> ···

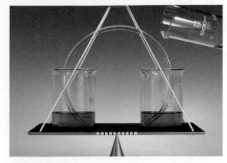

(a)

(b)

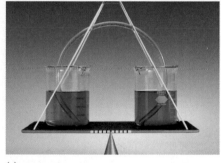

(c)

FIGURE 14.8 Chemical equilibrium is a bit like two beakers of water connected by a siphon. (a) The system is at equilibrium when the levels of water in both beakers are the same. (b) If we add water to one beaker, the added water drives a flow of water through the siphon into the opposite beaker until (c) equilibrium is restored. The opposite response would occur if we removed some water from one of the beakers; the flow through the siphon would be toward that beaker.

SAMPLE EXERCISE 14.9   **Adding or Removing Reactants or Products to Stress an Equilibrium**   **LO7**

Suggest three ways the production of ammonia via the reaction

$$N_2(g) + 3\,H_2(g) \rightleftharpoons 2\,NH_3(g)$$

could be increased without changing the reaction temperature.

COLLECT AND ORGANIZE We are given the balanced chemical equation of a reversible reaction and are asked to suggest three ways to increase its yield, that is, to shift its equilibrium to the right.

ANALYZE This reaction's equilibrium can be shifted to the right by removing NH_3, which reduces the value of the numerator of the mass action expression:

$$Q_p = \frac{(P_{NH_3})^2}{(P_{H_2})^3(P_{N_2})}$$

The equilibrium can also be shifted to the right by increasing the partial pressure of N_2 or H_2, which increases the value of the denominator. Either approach will produce a smaller reaction quotient Q_p. If $Q_p < K_p$, the reaction system will respond by consuming N_2 and H_2 and forming more NH_3.

SOLVE If we (1) increase the partial pressure of N_2, (2) increase the partial pressure of H_2, or (3) remove NH_3 from the system, the equilibrium will shift to the right.

THINK ABOUT IT The industrial synthesis of ammonia relies on shifting the equilibrium to the right by running the reaction at high partial pressures of the reactants, and also by removing the product NH_3 by passing the reaction mixture through chilled condensers. Ammonia can be removed because it condenses at a higher temperature than N_2 or H_2.

Practice Exercise Describe the changes that occur under the following conditions in a gas-phase equilibrium based on the following reaction:

$$2\,H_2S(g) + 3\,O_2(g) \rightleftharpoons 2\,SO_2(g) + 2\,H_2O(g)$$

a. The mixture is cooled and water vapor condenses.
b. SO_2 gas dissolves in liquid water as it condenses.
c. More O_2 is added.

We can express the effects of stressing equilibria in general terms. First, increasing the partial pressure (or concentration) of a reactant or product shifts the equilibrium so that more of that substance is consumed in the reaction. Second, decreasing the partial pressure (or concentration) of a reactant or product shifts the equilibrium toward the production of more of that substance.

Effects of Changes in Pressure and Volume

A reaction involving gaseous reactants or products may be perturbed by altering the partial pressures of the reactants and products, and a simple way to do that is by changing the volume of the reaction mixture while keeping temperature constant. To see how such a change in volume perturbs a gas-phase equilibrium, let's revisit the equilibrium between NO_2 and its dimer, N_2O_4:

$$2\,NO_2(g) \rightleftharpoons N_2O_4(g) \tag{14.3}$$

Suppose a reaction mixture of the two gases is at equilibrium. Then the mixture is compressed at constant temperature into a volume half its original size. The resulting increases in the partial pressures of the gases stress the equilibrium and change the value of its reaction quotient. To see how, let's assume that the

CONNECTION According to Boyle's law (Chapter 10) the pressure of an ideal gas is inversely proportional to its volume at constant temperature.

equilibrium partial pressures of the two gases are $P_{NO_2} = X$ and $P_{N_2O_4} = Y$. This means that the value of K_p is

$$K_p = \frac{P_{N_2O_4}}{(P_{NO_2})^2} = \frac{Y}{X^2}$$

Now we rapidly compress the reaction mixture into half its equilibrium volume, keeping T constant. In accordance with Boyle's law, the partial pressure of each gas doubles: $P_{NO_2} = 2X$ and $P_{N_2O_4} = 2Y$. Inserting these new values into the mass action expression gives us a value of the reaction quotient that is half the value of K_p:

$$Q_p = \frac{2Y}{(2X)^2} = \frac{2}{4}\frac{Y}{X^2} = \frac{1}{2}K_p$$

When the value of Q_p for any reaction is less than the value of K_p, the reaction proceeds in the forward direction, consuming reactants and forming products. In this case, NO_2 is consumed and N_2O_4 forms.

Did you notice that converting NO_2 into N_2O_4 reduced the total number of moles of gas in the reaction mixture (because two moles of NO_2 are consumed for every one mole of N_2O_4 produced)? There is an important message in this observation: any time a reaction mixture at equilibrium is compressed—and the partial pressures of its gas-phase reactants and products increase—the equilibrium shifts toward the side of the reaction equation *with fewer moles of gases*. In doing so the reaction mixture relieves some of the stress induced by the increase in pressure due to compression. On the other hand, allowing a reaction mixture to expand at constant temperature lowers the partial pressures of all the gaseous reactants and products and shifts the equilibrium toward the side of the reaction equation *with more moles of gases*. These shifts occur in all gas-phase reactions in which chemical equilibrium is perturbed by isothermal compression or expansion.

SAMPLE EXERCISE 14.10 **Assessing the Effect of Compression on Gas-Phase Equilibria** **LO7**

In which of the following reactions would isothermal compression of a reaction mixture at equilibrium promote the formation of more product(s)?

    a. $N_2(g) + O_2(g) \rightleftharpoons 2\,NO(g)$
    b. $2\,NO(g) + O_2(g) \rightleftharpoons 2\,NO_2(g)$
    c. $N_2O_4(g) \rightleftharpoons 2\,NO_2(g)$
    d. $H_2O(\ell) + CO_2(g) \rightleftharpoons H_2CO_3(aq)$
    e. $CaCO_3(s) \rightleftharpoons CaO(s) + CO_2(g)$

COLLECT AND ORGANIZE We are asked to identify the reactions for which an increase in pressure causes an increase in product formation. We know that increasing pressure shifts a chemical equilibrium involving gases toward the side of the reaction with fewer moles of gas.

ANALYZE We need to identify those reactions in which there are fewer moles of gaseous products than there are gaseous reactants.

SOLVE We summarize the number of moles of gas on the reactant side and product side in the accompanying table. The only two reactions with fewer moles of gaseous products than reactants are reactions b and d. Therefore, they are the only two in which an increase in the total pressure of the reacting gases increases product formation.

Reaction	Moles of Gaseous Reactants	Moles of Gaseous Products
a	2	2
b	3	2
c	1	2
d	1	0
e	0	1

THINK ABOUT IT In the case of reaction a, the number of moles of gas on the reactant side is the same as the number of moles on the product side, so changing pressure does not cause the equilibrium to shift. In reactions c and e, the number of moles on the product side is greater than on the reactant side, so increasing pressure favors the reverse of the reaction as written and decreases product formation.

Practice Exercise How does isothermal compression impact the composition of an equilibrium reaction mixture in the synthesis of ammonia?

$$N_2(g) + 3\,H_2(g) \rightleftharpoons 2\,NH_3(g)\ \text{⚙}$$

Effect of Temperature Changes

In Chapter 9 we explored the flow of energy that accompanies many chemical reactions. Now we explore the stresses produced on chemical equilibria by adding or removing energy by increasing or decreasing temperature. Let's start with an exothermic reaction, the synthesis of ammonia:

$$N_2(g) + 3\,H_2(g) \rightleftharpoons 2\,NH_3(g) + \text{energy}$$

If we think of energy as a product in the forward reaction, then raising the temperature of the reaction mixture favors the reverse reaction, whereas lowering the temperature favors the forward reaction.

There is one major difference, however, between applying Le Châtelier's principle to concentration or pressure changes and applying it to temperature changes: changing temperature does not perturb an equilibrium by changing the ratio of product to reactant concentrations or partial pressures, that is, by changing the value of Q. Instead, changing temperature *changes the value of K*. As a result, Q and K are no longer equal and the reaction proceeds in the direction that makes them equal again.

Increasing temperature reduces the yield of ammonia because increasing temperature reduces the value of K. In general, the value of K decreases as temperature increases for exothermic reactions. We will look more closely at the influence of temperature on K values in Section 14.10, but for now this general analysis enables us to predict the direction of a shift in equilibrium with changing temperature.

SAMPLE EXERCISE 14.11 **Predicting How Temperature Changes Impact Chemical Equilibria** **LO7**

The color of an aqueous acidic solution of cobalt(II) chloride depends on the temperature (Figure 14.9). In aqueous HCl, the solution is pink at 0°C, magenta at 25°C, and dark blue at 75°C. Is the reaction producing the pink-to-blue color change exothermic or endothermic?

COLLECT AND ORGANIZE We are given a reversible reaction and asked to determine whether it is exothermic or endothermic. Asked another way, is energy a product or a reactant in this reaction?

ANALYZE If the reaction is exothermic, then increasing the temperature is the equivalent of adding a product, causing a shift toward the pink reactant side. If the reaction is endothermic, then increasing the temperature is the equivalent of adding a reactant, causing a shift toward the side of the blue product.

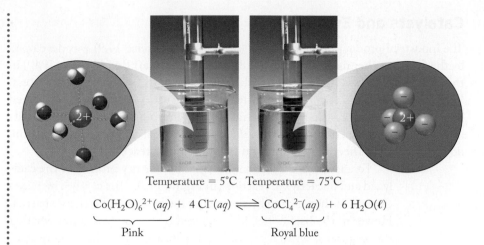

Temperature = 5°C   Temperature = 75°C

$$Co(H_2O)_6^{2+}(aq) + 4\,Cl^-(aq) \rightleftharpoons CoCl_4^{2-}(aq) + 6\,H_2O(\ell)$$

Pink   Royal blue

FIGURE 14.9 Two forms of cobalt, one pink and one blue, are in equilibrium in aqueous hydrochloric acid solution. The position of equilibrium shifts to the right as the temperature changes, causing the color of the solution to change as more and more of the blue $CoCl_4^{2-}$ ion forms.

SOLVE When energy is added to the system, its color changes from pink to blue. This means energy is a reactant, and the reaction as written must be endothermic:

$$Energy + pink \rightleftharpoons blue$$

THINK ABOUT IT This Sample Exercise illustrates how determining the effect of changing the temperature of an equilibrium reaction mixture can tell us whether the reaction is exothermic or endothermic. The fact that the reaction mixture in this exercise is magenta at room temperature tells us that both the pink and blue forms are present. This must mean that the value of K at room temperature is close to 1.

Practice Exercise Predict how the value of the equilibrium constant of the reaction

$$N_2(g) + O_2(g) \rightleftharpoons 2\,NO(g) \qquad \Delta H^\circ_{rxn} = 181\ kJ$$

changes with increasing temperature.

Table 14.4 summarizes how an exothermic system at equilibrium responds to various stresses.

TABLE 14.4 Responses of an Exothermic Reaction [2 A(g) ⇌ B(g)] at Equilibrium to Different Kinds of Stress

Kind of Stress	How Stress Is Relieved	Direction of Shift
Add A	Consume A	To the right
Remove A	Produce A	To the left
Add B	Consume B	To the left
Remove B	Produce B	To the right
Increase pressure by compressing the reaction mixture	Consume A to relieve pressure increase	To the right
Decrease pressure by expanding volume	Produce A to maintain equilibrium pressure	To the left
Increase temperature by adding energy	Consume some of the energy	To the left
Decrease temperature by removing energy	Produce energy	To the right

Catalysts and Equilibrium

The industrial production of ammonia (see Sample Exercise 14.9) was developed by the German chemists Fritz Haber (1868–1934) and Carl Bosch (1874–1940) in the early 20th century and is still widely referred to as the Haber–Bosch process. What makes the process commercially feasible is the use of catalysts. As discussed in Chapter 13, a catalyst increases the rate of a chemical reaction by lowering its activation energy. The question is this: If a catalyst increases the rate of a reaction, does that catalyst affect the equilibrium constant of the reaction?

To answer this question, consider the energy profiles of the catalyzed and uncatalyzed reactions in Figure 14.10. The catalyst increases the rate of the reaction by decreasing the height of the energy barrier. However, the barrier height is reduced by the same amount whether the reaction as written proceeds in the forward direction or in reverse. As a result, the increase in reaction rate produced by the catalyst is the same in both directions. Therefore a catalyst has no effect on the equilibrium constant of a reaction or on the composition of an equilibrium reaction mixture. A catalyst does, however, decrease the amount of time needed for a reaction to reach equilibrium.

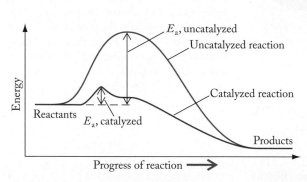

FIGURE 14.10 The effect of a catalyst on a reaction. A catalyst lowers the activation energy barrier, and as a result the rate of the reaction increases. However, because both the forward reaction and the reverse reaction occur more rapidly, the position of equilibrium (that is, the value of K) does not change. The system comes to equilibrium more rapidly, but the relative amounts of product and reactant present at equilibrium do not change.

14.8 Calculations Based on K

Reference books and the tables in Appendix 5 of this book contain lists of equilibrium constants for chemical reactions. These values are used in several kinds of calculations, including those in which:

1. We want to determine whether a reaction mixture has reached equilibrium (Sample Exercise 14.7).
2. We know the value of K and the starting concentrations or partial pressures of reactants and/or products, and we want to calculate their equilibrium concentrations or pressures.

In this section we focus on the second type of calculation and introduce a useful way of handling such problems: a table of reactant and product concentration (or partial pressure) values called a *RICE table*. The acronym RICE means that the table starts with the balanced chemical equation describing the **R**eaction followed by rows that contain **I**nitial concentration values, **C**hanges in those initial values as the reaction proceeds toward equilibrium, and **E**quilibrium values.

In our first example, we calculate how much nitrogen monoxide forms in a sample of air heated to a temperature at which the K_p of the following reaction is 1.00×10^{-5}:

$$N_2(g) + O_2(g) \rightleftharpoons 2\,NO(g)$$

The initial partial pressures are $P_{N_2} = 0.79$ atm and $P_{O_2} = 0.21$ atm, and we assume no NO is present.

▶❙❙ **CHEMTOUR** Solving Equilibrium Problems

We start by writing the given (initial) information in our RICE table:

Reaction	$N_2(g)$ +	$O_2(g)$ ⇌	2 NO(g)
	P_{N_2} (atm)	P_{O_2} (atm)	P_{NO} (atm)
Initial	0.79	0.21	0
Change			
Equilibrium			

We know the reaction will proceed in the forward direction because there is no product initially present, so $Q_p = 0 < K_p$.

We need to use algebra to fill in rows C and E. We don't know how much N_2 or O_2 will be consumed or how much NO will be formed. We can define the change in partial pressure of N_2 as $-x$ because N_2 is consumed during the reaction. Because the mole ratio of N_2 to O_2 in the reaction is 1:1, the change in O_2 is also $-x$. Two moles of NO are produced from each mole of N_2 and O_2, so the change in P_{NO} is $+2x$. Inserting these values in the C row, we have

Reaction	$N_2(g)$ +	$O_2(g)$ ⇌	2 NO(g)
	P_{N_2} (atm)	P_{O_2} (atm)	P_{NO} (atm)
Initial	0.79	0.21	0
Change	$-x$	$-x$	$+2x$
Equilibrium			

Combining the I and C rows, we obtain expressions for the three partial pressures at equilibrium:

Reaction	$N_2(g)$ +	$O_2(g)$ ⇌	2 NO(g)
	P_{N_2} (atm)	P_{O_2} (atm)	P_{NO} (atm)
Initial	0.79	0.21	0
Change	$-x$	$-x$	$+2x$
Equilibrium	$0.79 - x$	$0.21 - x$	$2x$

The next step is to substitute the terms from the E row into the K_p expression for the reaction:

$$K_p = \frac{(P_{NO})^2}{(P_{N_2})(P_{O_2})}$$

$$= \frac{(2x)^2}{(0.79 - x)(0.21 - x)} \qquad (14.18)$$

Multiplying the terms in the denominator of Equation 14.18 gives

$$K_p = \frac{4x^2}{0.1659 - 1.00x + x^2} = 1.00 \times 10^{-5}$$

Cross-multiplying, we get

$$1.659 \times 10^{-6} - (1.00 \times 10^{-5})x + (1.00 \times 10^{-5})x^2 = 4x^2$$

Combining the x^2 terms and rearranging, we have

$$3.99999\,x^2 + (1.00 \times 10^{-5})x - 1.659 \times 10^{-6} = 0$$

You may recognize this equation as one that fits the general form of a quadratic equation:

$$ax^2 + bx + c = 0$$

which can be solved for x either with a scientific calculator or using the quadratic formula:

$$x = \frac{-b \pm \sqrt{b^2 - 4ac}}{2a}$$

Two values are possible for x, but only one is positive: 6.428×10^{-4} atm. We focus on this value because a gas cannot have a negative partial pressure. Calculating equilibrium partial pressures to two significant figures:

$$P_{O_2} = 0.21 - x$$

$$= 0.21 - (6.428 \times 10^{-4}) = 0.21 \text{ atm}$$

$$P_{N_2} = 0.79 - x$$

$$= 0.79 - (6.428 \times 10^{-4}) = 0.79 \text{ atm}$$

$$P_{NO} = 2x$$

$$= 2(6.428 \times 10^{-4}) = 1.2855 \times 10^{-3} = 0.0013 \text{ atm}$$

The small quantity of NO produced by the reaction means that there is no significant change in the partial pressures of N_2 or O_2.

Because the x terms in the denominator of Equation 14.18 are very much smaller than the initial partial pressures, a simpler approach to calculating P_{NO} is possible: let's ignore the x terms in the denominator and use the initial values for P_{N_2} and P_{O_2} instead:

$$K_p = 1.0 \times 10^{-5} = \frac{(P_{NO})^2}{(P_{N_2})(P_{O_2})}$$

$$= \frac{4x^2}{(0.79 - x)(0.21 - x)} \approx \frac{4x^2}{(0.79)(0.21)}$$

$$4x^2 = (0.79)(0.21)(1.0 \times 10^{-5})$$

$$= 1.659 \times 10^{-6}$$

$$x^2 = 4.148 \times 10^{-7}$$

$$x = 6.440 \times 10^{-4} \text{ atm}$$

This value of x does not differ significantly from that obtained by solving the quadratic equation (6.428×10^{-4} atm), given that we know the initial partial pressures to only two significant figures. Generally speaking, we can ignore the x component of a reactant term if the value of K is less than 3×10^{-5} and the initial concentration is greater than or equal to 0.01 M.

The calculations associated with the terms in the RICE table can also be simplified when the initial concentrations of the reactants are the same. For example, suppose we have a vessel containing 0.100 $M\,N_2$ and 0.100 $M\,O_2$ at a temperature where K_c for the NO formation reaction is 0.100. What is the equilibrium concentration of NO? The RICE table in this case is

Reaction	$N_2(g)$ +	$O_2(g)$ ⇌	2 NO(g)
	$[N_2]$ (M)	$[O_2]$ (M)	$[NO]$ (M)
Initial	0.100	0.100	0
Change	$-x$	$-x$	$+2x$
Equilibrium	$0.100 - x$	$0.100 - x$	$2x$

Inserting the values from the E row into the expression for K_c gives

$$K_c = \frac{(2x)^2}{(0.100 - x)(0.100 - x)} = 0.100$$

Taking the square root of each side:

$$\frac{2x}{0.100 - x} = 0.316$$

Solving for x, we get $x = 0.0136\ M$, and the equilibrium concentrations are $[N_2] = [O_2] = 0.100\ M - 0.0136\ M = 0.086\ M$ and $[NO] = 2x = 0.0272\ M$.

It's a good idea to check that these concentrations are consistent with the known value of K_c. Substituting into the K_c expression, we get

$$K_c = \frac{(0.0272)^2}{(0.086)^2} = 0.100$$

which is the value of K given for the reaction in question.

SAMPLE EXERCISE 14.12 **Calculating an Equilibrium** **LO8**
Partial Pressure I

Much of the H_2 used in the Haber–Bosch process is produced by the water–gas shift reaction:

$$CO(g) + H_2O(g) \rightleftharpoons CO_2(g) + H_2(g) \qquad (14.2)$$

If a reaction vessel at 400°C is filled with an equimolar mixture of CO and steam such that $P_{CO} = P_{H_2O} = 2.00$ atm, what is the partial pressure of H_2 at equilibrium? The equilibrium constant $K_p = 10$ at 400°C.

COLLECT AND ORGANIZE We are asked to find the partial pressure of a gaseous product in an equilibrium mixture given the initial partial pressures of reactants and the value of K_p. One way to approach this problem involves (1) setting up a RICE table, (2) using the partial pressures from the E row in the equilibrium constant expression, and (3) solving for P_{H_2}.

ANALYZE The system initially contains no product. This means that the reaction quotient Q_p is equal to zero and thus less than K. Therefore, the reaction proceeds in the forward direction, decreasing the partial pressures of the reactants while increasing those of the products. A K_p value of 10 means that most of the reactants should be converted into products, but there should be significant partial pressures of both reactants and products at equilibrium.

SOLVE Let x be the increase in partial pressure of H_2 as a result of the reaction. The stoichiometry of the reaction tells us that the change in P_{CO_2} is also x and that the changes in both P_{CO} and P_{H_2} are $-x$:

Reaction	CO(g) +	H₂O(g) ⇌	CO₂(g) +	H₂(g)
	P_{CO} (atm)	P_{H_2O} (atm)	P_{CO_2} (atm)	P_{H_2} (atm)
Initial	2.00	2.00	0.00	0.00
Change	$-x$	$-x$	$+x$	$+x$
Equilibrium	$2.00 - x$	$2.00 - x$	x	x

Inserting these equilibrium terms into the equilibrium constant expression for the reaction gives

$$K_p = \frac{(P_{CO_2})(P_{H_2})}{(P_{CO})(P_{H_2O})} = \frac{(x)(x)}{(2.00 - x)(2.00 - x)} = 10$$

The latter equation can be simplified by taking the square root of both sides:

$$\frac{x}{2.00 - x} = \sqrt{10} = 3.16$$

Solving for x gives $x = 1.52$ atm, which is the equilibrium partial pressure of H_2 (and CO_2).

THINK ABOUT IT Our prediction that most, but far from all, of the reactants would form products was correct. It is a good idea to substitute the results into the equilibrium constant expression as a check on the validity of the solution. The calculated partial pressures are $P_{H_2} = P_{CO_2} = 1.52$ atm and $P_{H_2O} = P_{CO} = 2.00 - 1.52 = 0.48$ atm. Inserting these values in the equilibrium constant expression gives $K_p = (1.52)^2/(0.48)^2 = 10.03$, which is not significantly different from the given value of 10 and which confirms that our calculation is correct.

Practice Exercise The chemical equation for the formation of hydrogen iodide from H_2 and I_2 is

$$H_2(g) + I_2(g) \rightleftharpoons 2\,HI(g)$$

The value of K_p for the reaction is 50 at 450°C. What is the partial pressure of HI in a sealed reaction vessel at 450°C if the initial partial pressures of H_2 and I_2 are both 0.100 atm and initially there is no HI present? ⚙

**SAMPLE EXERCISE 14.13   Calculating an Equilibrium   LO8
Partial Pressure II**

Suppose that in a reaction vessel running the water–gas shift reaction,

$$CO(g) + H_2O(g) \rightleftharpoons CO_2(g) + H_2(g)$$

at 400°C the initial partial pressures are $P_{CO} = 2.00$ atm, $P_{H_2O} = 2.00$ atm, $P_{H_2} = 0.15$ atm, and $P_{CO_2} = 0.00$ atm. What is the partial pressure of H_2 at equilibrium, given $K_p = 10$ at 400°C?

COLLECT AND ORGANIZE We are asked to calculate the partial pressure of a product at equilibrium. We know the initial partial pressures of all reactants and of both products as well as the value of K_p. The difference between this problem and the preceding Sample Exercise is that here we have product present before the reaction starts.

ANALYZE Comparing the reaction quotient Q with the value of K lets us know in which direction the reaction proceeds to attain equilibrium. There is no CO_2 initially present, so $Q = 0$. Therefore, the reaction as written proceeds in the forward direction. Our strategy is to solve the problem by setting up a RICE table.

SOLVE Let x be the increase in P_{H_2}. The change in P_{CO_2} is also x, and the changes in P_{CO} and P_{H_2O} are $-x$:

Reaction	CO(g) +	H₂O(g) ⇌	CO₂(g) +	H₂(g)
	P_{CO} (atm)	P_{H_2O} (atm)	P_{CO_2} (atm)	P_{H_2} (atm)
Initial	2.00	2.00	0.00	0.15
Change	$-x$	$-x$	$+x$	$+x$
Equilibrium	$2.00 - x$	$2.00 - x$	x	$0.15 + x$

$$K_p = \frac{(P_{CO_2})(P_{H_2})}{(P_{CO})(P_{H_2O})}$$

$$= \frac{(x)(0.15 + x)}{(2.00 - x)(2.00 - x)} = 10$$

Solving for x gives two values, $x = 1.50$ and 2.96. The value $x = 2.96$ is not possible because using it in the equilibrium terms results in negative partial pressures of CO and H_2O, a physical impossibility. Therefore, $x = 1.50$, and at equilibrium $P_{H_2} = 1.50 + 0.15 = 1.65$ atm.

THINK ABOUT IT Using the value of x in the terms in the E row of the RICE table, we find that the equilibrium partial pressures of CO, H_2O, and CO_2 are 0.50, 0.50, and 1.50, respectively. These values result in a K_p value of $(1.65)(1.50)/(0.5)(0.5) = 9.9$, which is acceptably close to the given value of 10. Comparing the results of this Sample Exercise to the previous one, we find that having an initial P_{H_2} of 0.15 atm results in a higher final P_{H_2} value (1.65 versus 1.52 atm); however, the presence of H_2 at the start of the reaction results in slightly less conversion of reactants to products than when no H_2 is present initially. This result makes sense because the presence of some product before the reaction starts means that less product has to form before $Q_p = K_p$.

Practice Exercise The value of K_c for the reaction

$$N_2O_4(g) \rightleftharpoons 2\,NO_2(g)$$

is 0.21 at 373 K. If a reaction vessel at that temperature initially contains 0.030 M NO_2 and 0.030 M N_2O_4, what are the concentrations of the two gases at equilibrium? ⚙

14.9 Equilibrium and Thermodynamics

In Section 12.5 we explored how the change in free energy, ΔG, of a chemical reaction provides us with an indication of whether or not it will proceed at a particular temperature and pressure. If ΔG is negative, a reaction is spontaneous as written and proceeds in the forward direction. If ΔG is positive, the reaction as written is nonspontaneous; the reverse reaction *is* spontaneous, and the reaction as written proceeds in the reverse direction. As a spontaneous reaction under constant temperature and pressure proceeds, the concentrations of reactants and products change, and the free energy of the system changes as well. Eventually ΔG reaches zero. When it does, no free energy is left to do useful work. The reaction has achieved chemical equilibrium.

The magnitude of ΔG—how far it is from zero in either a negative or positive direction—indicates how far a system is from its equilibrium position. Similarly, when Q is much larger or smaller than K, we know that a chemical system is far from equilibrium. It is reasonable, then, to think that the separation between the values of Q and K and the sign and magnitude of ΔG are somehow related. Indeed they are, and their mathematical relationship is one of the most important connections in chemistry because it enables us to relate the thermodynamics of a chemical reaction to the composition of a reaction mixture at equilibrium.

The thermodynamic view of equilibrium and the relationship between ΔG and Q are described by the equation

$$\Delta G = \Delta G^\circ + RT \ln Q \qquad (14.19)$$

where ΔG° is the change in free energy under standard conditions. Let's explore what this means by looking at a reaction we have examined several times in this chapter, the decomposition of N_2O_4:

$$N_2O_4(g) \rightleftharpoons 2\,NO_2(g)$$

CONNECTION In Chapter 12 we defined the change in free energy of a reaction, ΔG_{rxn}, as the energy available to do useful work at a particular temperature and pressure.

▶❚❚ **CHEMTOUR** Equilibrium and Thermodynamics

We can calculate the change in standard free energy for the reaction (ΔG°_{rxn}) as we did in Chapter 12, using standard free energy of formation (ΔG°_f) values from Table A4.3 in Appendix 4 and the formula

$$\Delta G^\circ_{rxn} = \sum n_{products} \, \Delta G^\circ_{f,products} - \sum n_{reactants} \, \Delta G^\circ_{f,reactants} \qquad (12.12)$$

$$= 2 \text{ mol } (51.3 \text{ kJ/mol}) - 1 \text{ mol } (97.8 \text{ kJ/mol}) = +4.8 \text{ kJ}$$

The positive value of ΔG°_{rxn} indicates that the reaction as written is not spontaneous in the forward direction at 298 K. However, this value of ΔG°_{rxn} applies only under standard conditions when $P_{NO_2} = P_{N_2O_4} = 1$ atm. That is not the case when the partial pressures vary from 1 atm.

Let's explore what happens in a reaction vessel at 298 K that initially contains just 1 mole of pure N_2O_4 at a partial pressure of 1 atm and no (or hardly any) NO_2. Under these conditions the value of the reaction quotient (Q_p) is zero:

$$Q_p = \frac{(P_{NO_2})^2}{P_{N_2O_4}} = \frac{0}{1} = 0$$

Although we do not know the value of K, it must be greater than zero. Therefore, $Q_p < K$, and the system responds by spontaneously forming NO_2 from N_2O_4.

Because the reaction in the forward direction is spontaneous, the change in free energy of the reaction (ΔG) must be less than zero, even though the value of ΔG° is positive (4.8 kJ/mol). Furthermore, we know that the value of ΔG is negative because the ($RT \ln Q$) term in Equation 14.19 approaches $-\infty$ as Q approaches zero. Any reversible reaction, regardless of its ΔG° value, has a negative ΔG value and is spontaneous when there is only reactant and no product (or practically none) in the system.

The opposite situation occurs if we have 2 moles of NO_2 in our reaction vessel and essentially no N_2O_4. Under these conditions, the denominator of the reaction quotient is nearly zero, making the value of Q_p enormous. Likewise, the ($RT \ln Q_p$) term in Equation 14.19 has a large positive value, guaranteeing that $\Delta G > 0$. This means that the forward reaction is not spontaneous, but the reverse reaction is. We also know that NO_2 in the reaction vessel will combine to form N_2O_4 because the enormous Q_p must be greater than K_p. Therefore, the reaction will run in reverse until $Q_p = K_p$.

Figure 14.11(a) shows how free energy changes as the quantities of N_2O_4 and NO_2 in the reaction mixture change. The minimum of the curve (point 3) corresponds to a free energy that is lower than that of either pure N_2O_4 (point 1) or pure NO_2 (point 2). The vertical dashed line, marking the minimum free energy, intersects the top and bottom axes at values that tell us the reaction mixture's composition at equilibrium: about 0.83 mol N_2O_4 and about 0.34 mol NO_2.

FIGURE 14.11 The change in standard free energy ΔG°_{rxn} is a constant for a given reaction. (a) Point 3, the point of minimum free energy, defines the composition of a system at equilibrium. At point 1, the system consists of reactants only; at point 2, the system consists of products only. (b) ΔG is the "distance" from equilibrium in terms of free energy. As a spontaneous reaction proceeds, the composition of the system changes and ΔG approaches 0 (point 3), at which point no further change in composition occurs.

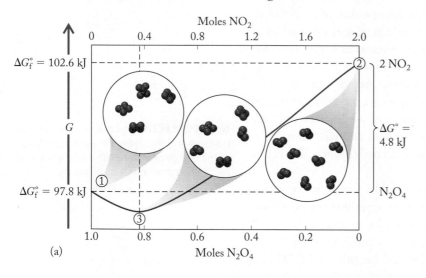

(a)

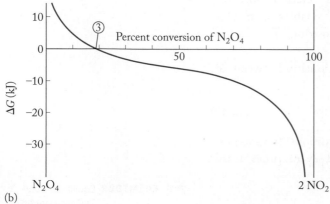

(b)

The curve in Figure 14.11(b) delivers the same message as in part a. In part b, the y-axis represents the distance from equilibrium of N_2O_4/NO_2 reaction mixtures of different composition. Clearly, mixtures that are either pure reactant or product are very far away from equilibrium, as we just discussed. The reaction curve crosses the x-axis ($\Delta G = 0$) at point 3, which defines the composition of the reaction mixture at equilibrium. This value, expressed as the percentage of N_2O_4 that has dissociated into NO_2, corresponds to the same N_2O_4/NO_2 ratio as the minimum in the curve in Figure 14.11(a).

CONCEPT TEST ··

Suppose the ΔG°_{rxn} value of the hypothetical chemical reaction $A \rightleftharpoons B$ is -3.0 kJ/mol. Which of the following statements about an equilibrium mixture of A and B at 298 K is true?

    a. There is only A present.
    b. There is only B present.
    c. There is an equimolar mixture of A and B present.
    d. There is more A than B present.
    e. There is more B than A present.

Once a reaction has reached equilibrium, $Q = K$, $\Delta G = 0$, and Equation 14.19 becomes

$$\Delta G = \Delta G^\circ + RT \ln K = 0$$

or

$$\Delta G^\circ = -RT \ln K \tag{14.20}$$

Rearranging Equation 14.20 allows us to calculate the K value for a reaction from its change in standard free energy and absolute temperature. First, we rearrange the terms:

$$\ln K = \frac{-\Delta G^\circ}{RT} \tag{14.21}$$

and then take the antilogarithm of both sides:

$$K = e^{-\Delta G^\circ/RT} \tag{14.22}$$

Equation 14.22 provides the following interpretation of reaction spontaneity under standard conditions. Whenever ΔG° is negative, the exponent $-\Delta G^\circ/RT$ in Equation 14.22 is positive, and $e^{-\Delta G^\circ/RT} > 1$, making $K > 1$. Therefore, any reversible reaction with an equilibrium constant greater than 1 is spontaneous under standard conditions, as shown in Figure 14.12(a). This spontaneity has its limits. As reactants are consumed and products are formed, the value of the reaction quotient increases, making the value of ΔG less negative. When it reaches zero, there is no further change in the composition of the reaction mixture because chemical equilibrium has been achieved.

It follows that a reversible reaction with a less negative value of ΔG° (Figure 14.12b) is still spontaneous but has a smaller equilibrium constant, so less reactant is consumed and less product is formed before the value of ΔG reaches zero. Finally, a reaction that has a positive value of ΔG° (Figure 14.12c) has an equilibrium constant that is less than 1, and it is not spontaneous under standard conditions. Instead, the reverse of the reaction is spontaneous.

Let's apply this concept to the equilibrium between N_2O_4 and NO_2:

$$N_2O_4(g) \rightleftharpoons 2\,NO_2(g) \qquad \Delta G^\circ_{rxn} = 4.8 \text{ kJ/mol}$$

and focus on the composition of the reaction mixture of these two gases at equilibrium in Figure 14.11. We start by calculating the value of the exponent in Equation 14.22:

$$-\frac{\Delta G^{\circ}_{rxn}}{RT} = -\frac{\left(\dfrac{4.8 \ \cancel{kJ}}{\cancel{mol}}\right)\left(\dfrac{1000 \ J}{1 \ \cancel{kJ}}\right)}{\left(\dfrac{8.314 \ J}{\cancel{mol} \cdot \cancel{K}}\right)(298 \ \cancel{K})} = -1.94$$

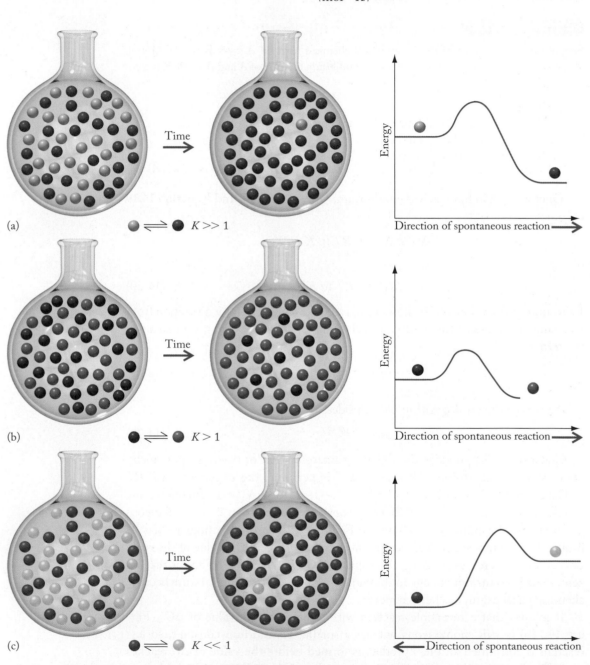

FIGURE 14.12 The three reaction vessels on the left each contain equimolar mixtures of reactant and product gases. Their initial partial pressures are each 1 atm. (a) The equilibrium constant for the formation of "red" gas from "green" gas is very large, which means very little "green" gas is present at equilibrium. (b) The equilibrium constant for the formation of "orange" gas from "blue" gas is greater than 1 but not as large as in part a, which means there is more "blue" gas present at equilibrium than there was "green" gas in part a. (c) The equilibrium constant for the formation of "yellow" gas from "purple" gas is much less than 1, so the reaction runs in reverse, forming purple gas from yellow gas, until equilibrium is achieved.

Inserting this value into Equation 14.22 gives

$$K = e^{-\Delta G°_{rxn}/RT} = e^{-1.94} = 0.144$$

This result is consistent with the composition of the reaction mixture at equilibrium in Figure 14.11, where $P_{N_2O_4} \approx 0.83$ atm and $P_{NO_2} \approx 0.34$ atm. Inserting these values into the equilibrium constant expression for the reaction gives us an approximate value of K_p that is close to the one we calculated from $\Delta G°_{rxn}$:

$$K_p = \frac{(P_{NO_2})^2}{P_{N_2O_4}} \approx \frac{(0.34)^2}{0.83} = 0.14$$

In this example, a positive $\Delta G°_{rxn}$ value of only few kilojoules per mole corresponds to an equilibrium constant value that is less than 1 but still greater than 0.1. As a result, the equilibrium reaction mixture contains more reactant than product, but less than an order of magnitude more. Similarly, an equilibrium reaction mixture produced by a reaction with a negative $\Delta G°_{rxn}$ value of only a few kilojoules per mole is likely to contain more products than reactants, but with significant quantities of both.

SAMPLE EXERCISE 14.14 **Relating K and $\Delta G°_{rxn}$** **LO9**

Use $\Delta G°_f$ values from Table A4.3 to calculate $\Delta G°_{rxn}$ and the value of K for the formation of NO_2 from NO and O_2 at 298 K:

$$NO(g) + \tfrac{1}{2} O_2(g) \rightleftharpoons NO_2(g)$$

COLLECT AND ORGANIZE We are asked to calculate the values of $\Delta G°_{rxn}$ and K for a reaction starting with $\Delta G°_f$ values from Table A4.3: 51.3 kJ/mol for NO_2 and 86.6 kJ/mol for NO. Because O_2 gas is the most stable form of the element at STP, its $\Delta G°_f$ value is 0.0 kJ/mol.

ANALYZE We can use Equation 12.12 to calculate $\Delta G°_{rxn}$ and then Equation 14.22 to calculate the value of K. The $\Delta G°_f$ value of NO_2 is less than that of NO, as $\Delta G°_{rxn}$ is negative and the value of the exponent in Equation 14.22 is positive. That translates into a K value greater than 1.

SOLVE

$$\Delta G°_{rxn} = [\Delta G°_f(NO_2)] - [\Delta G°_f(NO) + \tfrac{1}{2}\Delta G°_f(O_2)]$$

$$= [1 \text{ mol } (51.3 \text{ kJ/mol})] - [1 \text{ mol } (86.6 \text{ kJ/mol}) + \tfrac{1}{2} \text{ mol } (0.0 \text{ kJ/mol})]$$

$$= -35.3 \text{ kJ or } -35,300 \text{ J per mole of } NO_2 \text{ produced}$$

The exponent in Equation 14.22 is

$$-\frac{\Delta G°_{rxn}}{RT} = -\frac{\left(\dfrac{-35,300 \text{ J}}{\text{mol}}\right)}{\left(\dfrac{8.314 \text{ J}}{\text{mol} \cdot \text{K}}\right)(298 \text{ K})} = 14.2$$

The corresponding value of K_p is

$$K_p = e^{-\Delta G°_{rxn}/RT} = e^{14.2} = 1.5 \times 10^6$$

THINK ABOUT IT The exponential relationship between $\Delta G°_f$ and K_p means that a moderately negative free-energy change of -35.3 kJ/mol corresponds to a very large value of K_p: in this case, greater than 10^6.

Practice Exercise The standard free energy of formation of ammonia at 298 K is -16.5 kJ/mol. What is the value of K for the reaction at 298 K?

$$N_2(g) + 3 H_2(g) \rightleftharpoons 2 NH_3(g)$$

We have yet to explain which kind of equilibrium constant, K_c or K_p, is related to $\Delta G°$ by Equation 14.22. The symbol $\Delta G°$ represents a change in free energy under standard conditions. The standard state for a gaseous reactant or product is one in which its *partial pressure* is 1 bar ($\approx$ 1 atm). Thus, the $\Delta G°$ of a reaction *in the gas phase* is linked by Equation 14.22 to its K_p value. However, standard conditions for reactions in solution (the focus of Chapter 15) mean that all dissolved reactants and products are present at a *concentration* of 1.00 *M*. Thus, the $\Delta G°$ of a reaction *in solution* is related by Equation 14.22 to its K_c value.

The values of $\Delta G°$ and K are also linked by their dependence on how we choose to write the chemical equation describing a reaction. For example, multiplying the coefficients in an equation by n means multiplying its $\Delta G°$ value by n. However, in this situation K is not multiplied by n but rather is raised to the nth power: K becomes K^n. This difference makes sense given the logarithmic relationship between $\Delta G°$ and K, and the fact that $\ln (X^n) = n \ln X$. This logarithmic relationship also explains why reversing a reaction changes the sign of $\Delta G°$ but produces a K that is the reciprocal of the original: $\ln (1/X) = -\ln X$.

14.10 Changing *K* with Changing Temperature

We have noted often in this chapter that the value of K changes with changing temperature. In this section we look more closely at that relationship, developing several important equations that link K and T.

Temperature, *K*, and ΔG°

Let's begin by combining a key equation from Chapter 12:

$$\Delta G° = \Delta H° - T \Delta S° \tag{12.11}$$

with one from this chapter:

$$\ln K = \frac{-\Delta G°}{RT} \tag{14.21}$$

to derive an equation that relates K to $\Delta H°$ and $\Delta S°$:

$$\ln K = \frac{-\Delta G°}{RT} = -\frac{\Delta H°}{RT} + \frac{T \Delta S°}{RT}$$

$$= -\frac{\Delta H°}{RT} + \frac{\Delta S°}{R} \tag{14.23}$$

Note how a negative value of $\Delta H°$ or a positive value of $\Delta S°$ contributes to a large value of K. These dependencies make sense because negative values of $\Delta H°$ and positive values of $\Delta S°$ are the two factors that contribute to making reactions spontaneous.

Because we are discussing the influence of temperature on K, let's identify the factors affected by changes in T in Equation 14.23. First of all, the values of $\Delta H°$ and $\Delta S°$ such as those in Appendix 4 for $T = 298$ K change very little with changing temperature, so we will treat $\Delta H°$ and $\Delta S°$ as constants in this discussion. However, the value of the first term on the right side of Equation 14.23 is inversely proportional to temperature, which means *the influence* of $\Delta H°$ on K decreases as temperature increases. Thus, a favorable (negative) $\Delta H°$ contributes less to increasing the value of K as temperature increases. This temperature dependence makes sense if we invoke Le Châtelier's principle and the notion that energy is a product of exothermic ($\Delta H° < 0$) reactions and a reactant in endothermic reactions. Increasing temperature shifts a reaction toward the side opposite the energy term: it promotes endothermic reactions and inhibits exothermic reactions.

If $\Delta H°$ and $\Delta S°$ do not vary much with temperature, then Equation 14.23 predicts that ln K will be a linear function of $1/T$. Furthermore, we expect a graph of ln K versus $1/T$ to have a positive slope for an exothermic process and a negative slope for an endothermic process. We can determine $\Delta H°$ from the slope of the line and $\Delta S°$ from its y-intercept. Thus, we can calculate fundamental thermodynamic values of a reaction at equilibrium by determining its equilibrium constant at different temperatures. For example, let's determine the values of $\Delta H°$ and $\Delta S°$ for the reaction

$$2\ CO_2(g) \rightleftharpoons 2\ CO(g) + O_2(g)$$

starting with its K_p values of 2.57×10^{-11} at 1500 K, 1.42×10^{-3} at 2500 K, and 0.112 at 3000 K. The fact that K_p increases as temperature increases tells us that energy is a reactant and the reaction is endothermic. We can use these data to determine the values of $\Delta H°$ and $\Delta S°$ by plotting ln K_p values versus $1/T$. To see how this graphical method works, let's rewrite Equation 14.23 so that it fits the form of the equation for a straight line ($y = mx + b$):

$$\ln K = -\frac{\Delta H°}{R}\left(\frac{1}{T}\right) + \frac{\Delta S°}{R} \tag{14.24}$$

The graph of ln K_p versus $1/T$ (Figure 14.13) is indeed a straight line. The slope of this line is −66,662 K, which is equal to $-\Delta H°_{rxn}/R$. The corresponding value of $\Delta H°_{rxn}$ is

$$\Delta H°_{rxn} = -\text{slope} \times R$$

$$= -(-66,662\ \cancel{K})\left(\frac{8.314\ J}{mol \cdot \cancel{K}}\right)$$

$$= 554{,}228\ J/mol = 554.2\ kJ/mol$$

The y-intercept ($\Delta S°_{rxn}/R$) of the graph is 20.1. The corresponding value of $\Delta S°_{rxn}$ is

$$\Delta S°_{rxn} = (20.1)\left(\frac{8.314\ J}{mol \cdot K}\right) = 167\ \frac{J}{mol \cdot K}$$

This positive entropy change is logical because the forward reaction converts 2 moles of gaseous CO_2 into 3 moles of gaseous products.

We can use Equation 14.24 to compare the K values of a reaction at two different temperatures. In the two-temperature version of Equation 14.24, the $\Delta S°/R$ terms drop out and we have

$$\ln\left(\frac{K_2}{K_1}\right) = -\frac{\Delta H°}{R}\left(\frac{1}{T_2} - \frac{1}{T_1}\right) \tag{14.25}$$

This equation looks very much like the *Clausius–Clapeyron equation* we developed in Chapter 11, which described the relationship between the vapor pressures of a liquid at two different temperatures:

$$\ln\left(\frac{P_{vap,T_1}}{P_{vap,T_2}}\right) = \frac{\Delta H_{vap}}{R}\left(\frac{1}{T_2} - \frac{1}{T_1}\right) \tag{11.5}$$

CONNECTION In Chapter 9 we defined exothermic reactions as those giving off energy as heat. They have a negative enthalpy change ($\Delta H < 0$), while endothermic reactions absorb energy ($\Delta H > 0$).

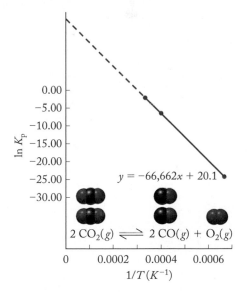

FIGURE 14.13 The graph of ln K_p versus $1/T$ is a straight line. We calculate $\Delta H°_{rxn}$ from the slope and extrapolate the line to the y-intercept, from which we calculate $\Delta S°_{rxn}$.

Indeed, the Clausius–Clapeyron equation is just a special case of Equation 14.25 applied to a change in physical state (liquid → gas) instead of a change in chemical composition. The pressures in Equation 11.5 are actually equilibrium vapor pressures, and the enthalpy of reaction is the enthalpy of vaporization (ΔH_{vap}). Equation 14.25 is called the *van 't Hoff equation* because it was first derived by Jacobus van 't Hoff. It is particularly useful for calculating the value of K at a very high or low temperature if we already know K at a standard reference temperature, such as 298 K. We use such a calculation in the following exercise.

SAMPLE EXERCISE 14.15 **Calculating an Equilibrium** **LO10**
Constant Value at a
Specific Temperature

Use data from Appendix 4 to calculate the equilibrium constant K_p for the exothermic reaction

$$N_2(g) + 3\,H_2(g) \rightleftharpoons 2\,NH_3(g)$$

at 298 K and at 773 K, a typical temperature used in the Haber–Bosch process for synthesizing ammonia.

COLLECT AND ORGANIZE We are asked to calculate the K_p value for a reaction at two temperatures. One of the temperatures is 298 K, which is the reference temperature for the standard thermodynamic data in Appendix 4. The ΔG_f° value of NH_3 is −16.5 kJ/mol.

ANALYZE We can use Equation 14.22 to calculate the value of K_p at 298 K from the value of ΔG_{rxn}°. Then we can use Equation 14.25 to calculate the value of K_p at 773 K from the value of K_p at 298 K and the value of ΔH_{rxn}°. The ΔH_{rxn}° value can be calculated from the enthalpy of formation of NH_3 gas (−46.1 kJ/mol). The negative value of ΔH_f° means that ΔH_{rxn}° is also negative, which means that the reaction is exothermic. Because energy is a product of the reaction, we can predict that the value of K_p will be smaller at 773 K than at 298 K.

SOLVE The value of ΔG_{rxn}° for the reaction is

$$\Delta G_{rxn}^\circ = \frac{2\;\text{mol NH}_3}{1\;\text{mol N}_2} \times \frac{-16.5\;\text{kJ}}{1\;\text{mol NH}_3} \times \frac{1000\;\text{J}}{1\;\text{kJ}} = -33{,}000\;\frac{\text{J}}{\text{mol N}_2}$$

Using this value in Equation 14.22 to calculate the value of K_p,

$$K_p = e^{-\Delta G_{rxn}^\circ/RT}$$

$$= \exp\left|\frac{-\left(-33{,}000\;\dfrac{\text{J}}{\text{mol}}\right)}{8.314\;\dfrac{\text{J}}{\text{mol}\cdot\text{K}} \times 298\;\text{K}}\right|$$

$$= 6.1 \times 10^5$$

Once we know the value of K_p at 298 K, we can calculate K_p at 773 K using Equation 14.25. We must first calculate the value of ΔH_{rxn}°, which is twice the ΔH_f° of NH_3:

$$\Delta H_{rxn}^\circ = \frac{2\;\text{mol NH}_3}{1\;\text{mol N}_2} \times \frac{-46.1\;\text{kJ}}{1\;\text{mol NH}_3} \times \frac{1000\;\text{J}}{1\;\text{kJ}}$$

$$= -92.2\;\text{kJ, or} -92{,}200\;\frac{\text{J}}{\text{mol N}_2}$$

After substituting $K_1 = 6.1 \times 10^5$, $T_1 = 298$ K, and $T_2 = 773$ K into the equation, we solve for K_2:

$$\ln\left(\frac{K_2}{6.1 \times 10^5}\right) = -\frac{-\left(-92{,}200 \, \frac{\text{J}}{\text{mol}}\right)}{8.314 \frac{\text{J}}{\text{mol} \cdot \text{K}}}\left(\frac{1}{773 \, \text{K}} - \frac{1}{298 \, \text{K}}\right)$$

$$\ln\left(\frac{K_2}{6.1 \times 10^5}\right) = -22.87$$

$$\frac{K_2}{6.1 \times 10^5} = e^{-22.87} = 1.2 \times 10^{-10}$$

$$K_2 = 7.3 \times 10^{-5} \text{ at 773 K}$$

THINK ABOUT IT The equilibrium constant decreases markedly when the temperature of the reaction is raised from 298 K to 773 K. This decrease fits our prediction for this exothermic reaction. Note that the standard thermodynamic data for the reaction ($\Delta H^\circ_{\text{rxn}}$ and $\Delta G^\circ_{\text{rxn}}$) are expressed per mole of the reactant with a coefficient of 1 in the balanced chemical equation for N_2.

Practice Exercise Use data from Appendix 4 to calculate the value of K_p for the following reaction at 298 K and 2000 K.

$$2\,N_2(g) + O_2(g) \rightleftharpoons 2\,N_2O(g) \quad \text{\textcircled{\tiny❂}}$$

The equilibrium constant determined in Sample Exercise 14.15 for the ammonia synthesis reaction is greater than 10^5 at room temperature but about 10^{-5} at 773 K. That's ten orders of magnitude smaller. We now have a more complete picture of this reaction, which was the basis for Sample Exercise 14.9. Ammonia is usually synthesized from N_2 and H_2 at temperatures near 400°C. These high temperatures *increase the rate* of this exothermic reaction even though they significantly *decrease* K_p and decrease the amount of ammonia produced at equilibrium. In this case, the practical benefit of a more favorable reaction rate outweighs the less favorable thermodynamics of running the reaction at high temperature. In practice, the ammonia produced in the reaction is continuously removed from the reaction vessel by condensation, shifting the equilibrium toward the formation of more product as predicted by Le Châtelier's principle.

We have seen throughout this chapter that every reversible chemical reaction has a unique equilibrium constant at a specific temperature and that the value of that constant provides us with a measure of how much product can form from given quantities of reactants. We have also seen how the value of K is linked to the value of $\Delta G^\circ_{\text{rxn}}$ and how this linkage provides us with a more sophisticated interpretation of what a negative value of $\Delta G^\circ_{\text{rxn}}$ (and reaction spontaneity) means. Our new interpretation of $\Delta G^\circ_{\text{rxn}} < 0$ is that the reaction proceeds in the forward direction, producing reaction mixtures with an abundance of products and smaller quantities of reactants left over. The more negative the value of $\Delta G^\circ_{\text{rxn}}$, the greater the yield of products before the reaction reaches chemical equilibrium and there is no further increase in the ratio of products to reactants.

On the other hand, a positive value of $\Delta G^\circ_{\text{rxn}}$ means more than that a reaction is simply *nonspontaneous*. It really means that a mixture of reactants may form *some* products (just not very much) before chemical equilibrium is reached and there is no further increase in the ratio of products to reactants. The more positive the value of $\Delta G^\circ_{\text{rxn}}$, the smaller the yield of products.

Dry chemical fire extinguishers are the most common fire extinguishers purchased for use in the home. They are also used to fight fires at airport crash rescue sites. A dry chemical extinguisher sprays a very fine powder of sodium bicarbonate ($NaHCO_3$, baking soda) or potassium bicarbonate ($KHCO_3$). Both substances decompose at the elevated temperatures of a fire and release carbon dioxide to extinguish it. The baking soda decomposition reaction

$$2\, NaHCO_3(s) \rightleftharpoons Na_2CO_3(s) + CO_2(g) + H_2O(g)$$

has a K_p value of 0.25 at 125°C.

a. Neither $NaHCO_3$ nor $KHCO_3$ decomposes at room temperature, but both decompose rapidly at high temperatures. Are their decomposition reactions exothermic or endothermic? What does this mean about the stability of these compounds in fire extinguishers?

b. $KHCO_3$ is marketed as being about two times as effective on fires involving flammable liquids or gases (class B fires) as $NaHCO_3$. Use thermochemical data from Appendix 4 to assess this claim by comparing the amounts of CO_2 generated when each material decomposes. Assume that the temperature of the fire is 1200°C.

COLLECT AND ORGANIZE We are asked how long fire extinguishers containing $NaHCO_3$ or $KHCO_3$ can be kept at room temperature without losing their efficacy and whether manufacturers' claims about fighting fires with these compounds are valid. We are given an equilibrium constant for one reaction at 125°C, the fact that the compounds do not decompose at room temperature but do at elevated temperatures, and tables in Appendix 4 of thermochemical data on both compounds as well as the reaction products. To assess the claim of greater efficacy of $KHCO_3$, we need to determine K_p values for reactions of both substances at the fire temperature. We can then see if one produces more CO_2 than the other.

ANALYZE We can use ΔG_f° values for the reactants and products to calculate ΔG_{rxn}° at 298 K for the decomposition of $KHCO_3$, and then use that value to calculate K_p for the reaction at 298 K. Then we can use ΔH_f° values for the reactants and products to calculate ΔH_{rxn}° values for both reactions. These values and their K_p values at 298 K can be used with Equation 14.25 to calculate their K_p values at 1200°C. Comparing these two equilibrium constant values will help us determine whether one or the other bicarbonate compound releases more CO_2 at flame temperatures.

SOLVE

a. The fact that $NaHCO_3$ and $KHCO_3$ do not decompose at room temperature but do at flame temperatures allows us to predict that their decomposition reactions

are endothermic. Energy is essentially a reactant in endothermic reactions and adding heat means adding an essential reactant to the reaction mixture. These compounds should be stable in fire extinguishers stored at room temperature.

b. Let's first calculate the value of K_p for the decomposition of $NaHCO_3$ at 1200°C using Equation 14.25:

$$\ln\left(\frac{K_2}{K_1}\right) = -\frac{\Delta H^\circ}{R}\left(\frac{1}{T_2} - \frac{1}{T_1}\right)$$

We need ΔH° for the reaction of $NaHCO_3$; we can calculate that from data given in Appendix 4 and methods we learned in Chapter 9.

$$\Delta H^\circ = \sum n_{products}\, \Delta H_{f,products}^\circ - \sum n_{reactants}\, \Delta H_{f,reactants}^\circ \quad (9.15)$$
$$= (\Delta H_{f,Na_2CO_3}^\circ + \Delta H_{f,H_2O}^\circ + \Delta H_{f,CO_2}^\circ) - 2(\Delta H_{f,NaHCO_3}^\circ)$$
$$= [(-1130.7) + (-241.8) + (-393.5) - (2)(-950.8)]\, kJ/mol = 135.6\, kJ/mol$$
$$= 135,600\, J/mol$$

Then

$$\ln\frac{K_2}{0.25} = -\frac{135,600\, \text{J/mol}}{8.314\, \dfrac{\text{J}}{\text{mol}\cdot\text{K}}}\left(\frac{1}{1473\, \text{K}} - \frac{1}{398\, \text{K}}\right)$$
$$\ln K_2 - \ln(0.25) = 29.91$$
$$\ln K_2 = 28.52$$
$$K_2 = 2.43 \times 10^{12} = K_{p,NaHCO_3}\ \text{at 1200°C}$$

Now we need to calculate K_p for $KHCO_3$ at 1200°C using Equation 14.22:

$$K = e^{-\Delta G^\circ/RT}$$

We can use ΔG° values from the Appendix to calculate ΔG_{rxn}° as we learned in Chapter 12 and from that determine K_p at 25°C. Then we can use Equation 14.25 and ΔH° values from the Appendix to estimate $K_{p,KHCO_3}$ at 1200°C. The reaction of interest is

$$2\, KHCO_3(s) \rightleftharpoons K_2CO_3(s) + CO_2(g) + H_2O(g)$$
$$\Delta G^\circ = \sum n_{products}\, \Delta G_{f,products}^\circ - \sum n_{reactants}\, \Delta G_{f,reactants}^\circ \quad (12.12)$$
$$= (\Delta G_{f,K_2CO_3}^\circ + \Delta G_{f,H_2O}^\circ + \Delta G_{f,CO_2}^\circ) - 2(\Delta G_{f,KHCO_3}^\circ)$$
$$= [(-1063.5) + (-228.6) + (-394.4) - (2)(-863.6)]\, kJ/mol$$
$$= 40.7\, kJ/mol = 40,700\, J/mol$$

Then

$$K = e^{-40,700/(8.314\times298)} = e^{-16.43} = 7.34 \times 10^{-8}$$

We can now use $\Delta H°$ values from the Appendix to find $\Delta H°_{rxn}$ and apply Equation 14.25 again to calculate the value of $K_{p,KHCO_3}$ at 1200°C:

$$\Delta H° = \sum n_{products}\, \Delta H°_{f,products} - \sum n_{reactants}\, \Delta H°_{f,reactants}$$

$$= (\Delta H°_{f,K_2CO_3} + \Delta H°_{f,H_2O} + \Delta H°_{f,CO_2})$$
$$\quad - 2(\Delta H°_{f,KHCO_3})$$

$$= [(-1151.0) + (-241.8) + (-393.5)$$
$$\quad - (2)(-963.2)]\ \text{kJ/mol} = 140.1\ \text{kJ/mol}$$

$$= 140{,}100\ \text{J/mol}$$

Then

$$\ln \frac{K_2}{7.34 \times 10^{-8}} = -\frac{140{,}100\ \cancel{\text{J/mol}}}{8.314\ \dfrac{\cancel{\text{J}}}{\text{mol} \cdot \cancel{\text{K}}}}\left(\frac{1}{1473\ \cancel{\text{K}}} - \frac{1}{298\ \cancel{\text{K}}}\right)$$

$$\ln K_2 = 28.70 \quad \text{and} \quad K_{p,KHCO_3} = 2.91 \times 10^{12}$$

The expressions for both equilibrium constants are the same:

$$K_{p,NaHCO_3} = [P_{CO_2}][P_{H_2O}]$$

$$K_{p,KHCO_3} = [P_{CO_2}][P_{H_2O}]$$

because the other ingredients are solids. We may assume the initial partial pressures of CO_2 and water vapor are insignificant compared to their equilibrium partial pressures. Their 1:1 mole ratio in the two reactions allows us to represent the equilibrium partial pressure of each with the same symbol. Let's use x in the $NaHCO_3$ reaction and y in the $KHCO_3$ reaction:

$$K_{p,NaHCO_3} = (P_{CO_2})(P_{H_2O}) = (x)(x) = 2.43 \times 10^{12}$$

$$K_{p,KHCO_3} = (P_{CO_2})(P_{H_2O}) = (y)(y) = 2.91 \times 10^{12}$$

Solving for x and y, we find that the partial pressure of CO_2 (P_{CO_2}) in a fire at 1200°C is 1.56×10^6 atm from the decomposition of $NaHCO_3$ and 1.71×10^6 atm from the decomposition of $KHCO_3$. These values are nearly the same, so the advantage of using $KHCO_3$ is apparently not based on its ability to produce more CO_2.

THINK ABOUT IT The $\Delta H°_{rxn}$ values for both compounds are positive, which agrees with our prediction that the reactions are endothermic. The similar P_{CO_2} values mean that $KHCO_3$ must do something else chemically that makes it a better fire extinguishing agent than $NaHCO_3$. In fact, it does do something different: $KHCO_3$ better inhibits reactions involving free radical intermediates formed in a fire.

SUMMARY

Section 14.1 Chemical equilibrium can be approached from either reaction direction and is achieved when the forward and reverse reaction rates are the same. At equilibrium, a reaction vessel may contain comparable amounts of reactants and products, may contain mostly reactants (the equilibrium *lies to the left*), or may contain mostly products (the equilibrium *lies to the right*). The value of the reaction's **equilibrium constant (K)** tells us whether products or reactants are favored at equilibrium.

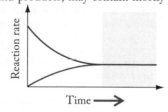

Section 14.2 According to the **law of mass action**, the *equilibrium constant expression* (or **mass action expression**) for K_c is the ratio of the equilibrium molar concentrations of the products divided by the equilibrium molar concentrations of the reactants, each raised to the respective stoichiometric coefficient in the balanced equation. If the substances involved in an equilibrium are gases, then an equilibrium constant K_p may also be written based on the partial pressures of the gases. Whether expressed as K_c or K_p, equilibrium constants have no units.

Section 14.3 The relationship between K_c and K_p depends on the number of moles of gaseous reactants and products in the balanced chemical equation.

Section 14.4 The reverse of a reaction has an equilibrium constant that is the reciprocal of K for the forward reaction. If the balanced equation for a reaction is multiplied by some factor n, the value of K for that reaction is raised to the nth power. If reactions are summed to give an overall reaction, their equilibrium constants are multiplied together to obtain an overall equilibrium constant.

Section 14.5 The **reaction quotient (Q)** is the value of the mass action expression at any instant during a reaction.

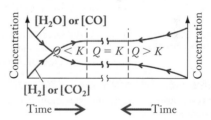

Section 14.6 Heterogeneous equilibria involve more than one phase. The concentrations of pure liquids and solids do not change during a reaction and so are omitted from equilibrium constant expressions.

Section 14.7 According to **Le Châtelier's principle**, chemical reactions at equilibrium respond to stress by shifting position to relieve the stress. Adding or removing a reactant or product, changing the partial pressures of reacting gases, and changing the temperature all create stresses that shift the equilibrium position. A catalyst decreases the time it takes a system to achieve equilibrium but does not change the value of the equilibrium constant.

Section 14.8 Equilibrium concentrations or partial pressures of reactants and products can be calculated from initial concentrations or pressures, the reaction stoichiometry, and the value of the equilibrium constant.

Section 14.9 A negative value of $\Delta G°$ corresponds to $K > 1$ (products favored), and a positive value of $\Delta G°$ corresponds to $K < 1$ (reactants favored). Concentrations of reactants and products at equilibrium are similar if the value of $\Delta G°$ is near 0 (K is near 1).

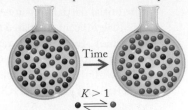

Time

$K > 1$

Section 14.10 Higher reaction temperatures increase the equilibrium constant of an endothermic reaction but decrease the equilibrium constant of an exothermic reaction. The slope of a plot of $\ln K$ versus $1/T$ for an equilibrium system is used to determine $\Delta H°_{rxn}$ and the y-intercept of the plot is used to determine the value of $\Delta S°_{rxn}$.

PROBLEM-SOLVING SUMMARY

TYPE OF PROBLEM	CONCEPTS AND EQUATIONS	SAMPLE EXERCISES
Writing equilibrium constant expressions	For the reaction $$a\,A + b\,B \rightleftharpoons c\,C + d\,D$$ $$K_c = \frac{[C]^c[D]^d}{[A]^a[B]^b} \qquad (14.9)$$ and $$K_p = \frac{(P_C)^c(P_D)^d}{(P_A)^a(P_B)^b} \qquad (14.10)$$	14.1
Calculating K_c or K_p	Insert equilibrium molar concentrations or partial pressures into the equilibrium constant expression.	14.2, 14.3
Interconverting K_c and K_p	Use the relation $$K_p = K_c(RT)^{\Delta n} \qquad (14.15)$$ where Δn is the number of moles of product gas minus the number of moles of reactant gas in the balanced chemical equation.	14.4
Calculating K values of related reactions	To calculate K for the reverse of a reaction, take the reciprocal of K of the forward reaction. If all the coefficients in a chemical equation are multiplied by n, the value of K increases by the power of n. If reactions are summed to give an overall reaction, their equilibrium constants are multiplied together to obtain an overall K.	14.5, 14.6
Using Q and K values to predict the direction of a reaction	If $Q < K$, the reaction proceeds in the forward direction to make more products; if $Q = K$, the reaction is at equilibrium; if $Q > K$, the reaction proceeds in the reverse direction to make more reactants.	14.7
Writing equilibrium constant expressions for heterogeneous equilibria	Molar concentrations of pure liquids and pure solids are omitted from equilibrium constant expressions because such concentrations are constant.	14.8
Adding or removing reactants or products to stress an equilibrium	Decreasing the concentration of a substance involved in an equilibrium shifts the equilibrium toward the production of more of that substance. Increasing the concentration of a substance shifts the equilibrium so that some of that substance is consumed in the reaction.	14.9
Predicting the effect of compression or expansion on gas-phase equilibria	Equilibria involving different numbers of gaseous reactants and products shift in response to an increase (or decrease) in volume toward the side with more (or fewer) moles of gases.	14.10
Predicting how temperature changes impact chemical equilibrium	The value of K for an endothermic reaction increases with increasing temperature; the value of K for an exothermic reaction decreases with increasing temperature.	14.11
Calculating concentrations or partial pressures of reactants and products at equilibrium	Use a RICE table to develop algebraic terms for each reactant's and product's partial pressure or concentration at equilibrium. Let x be the change in partial pressure or concentration of one component of the reaction. Express the changes in the other components in terms of x. Substitute these terms into the expression for K and solve for x.	14.12, 14.13

TYPE OF PROBLEM	CONCEPTS AND EQUATIONS		SAMPLE EXERCISES
Relating K and ΔG°_{rxn}	Calculate ΔG°_{rxn} from		14.14
	$$\Delta G^{\circ}_{rxn} = \sum n_{products}\,\Delta G^{\circ}_{f,products} - \sum n_{reactants}\,\Delta G^{\circ}_{f,reactants}$$	(12.12)	
	Then use		
	$$K = e^{-\Delta G^{\circ}/RT}$$	(14.22)	
	Be sure to convert ΔG° to joules per mole and use $R = 8.314$ J/(mol · K).		
Calculating equilibrium constant values at specific temperatures	Use		14.15
	$$\ln\left(\frac{K_2}{K_1}\right) = -\frac{\Delta H^{\circ}}{R}\left(\frac{1}{T_2} - \frac{1}{T_1}\right)$$	(14.25)	
	Convert ΔH° from kilojoules per mole to joules per mole to match the units of R.		

VISUAL PROBLEMS ●●■

(Answers to boldface end-of-chapter questions and problems are in the back of the book.)

14.1. Figure P14.1 shows the energy profiles of reactions A $\rightleftharpoons$ B and C $\rightleftharpoons$ D, respectively. Which reaction has the larger forward rate constant? Which reaction has the smaller reverse rate constant? Which reaction has the larger equilibrium constant K_c?

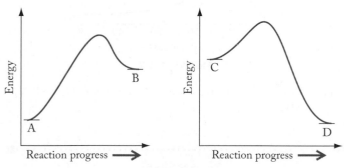

FIGURE P14.1

14.2. The progress with time of a reaction system is depicted in Figure P14.2. Red spheres represent the molar concentration of substance A and blue spheres represent the molar concentration of substance B.
  a. Does the system reach equilibrium?
  b. In which direction (A → B or B → A) is equilibrium attained?
  c. What is the value of the equilibrium constant K_c?

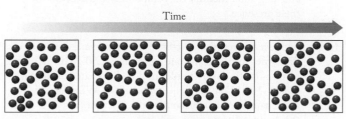

FIGURE P14.2

14.3. In Figure P14.3 the red spheres represent reactant A and the blue spheres represent product B in equilibrium with A.
  a. Write a chemical equation that describes the equilibrium.
  b. What is the value of the equilibrium constant K_c?

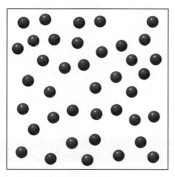

FIGURE P14.3

14.4. The equilibrium constant K_c for the reaction

A (red spheres) + B (blue spheres) $\rightleftharpoons$ AB

is 3.0 at 300.0 K. Does the situation depicted in Figure P14.4 correspond to equilibrium? If not, in what direction (to the left or to the right) will the system shift to attain equilibrium?

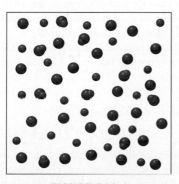

FIGURE P14.4

14.5. The left and right diagrams in Figure P14.5 represent equilibrium states of the reaction

$$A \text{ (red spheres)} + B \text{ (blue spheres)} \rightleftharpoons AB$$

at 300 K and 400 K, respectively. Is this reaction endothermic or exothermic? Explain.

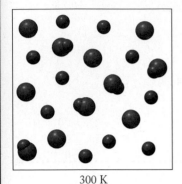 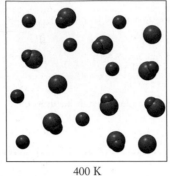

| 300 K | 400 K |

FIGURE P14.5

14.6. Figure P14.6 shows a plot of ln K_c versus $1/T$ for the reaction $A + 2B \rightleftharpoons AB_2$. Is the reaction endothermic or exothermic?

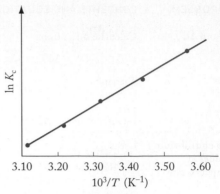

FIGURE P14.6

14.7. Does the reaction $A \rightarrow 2B$ represented in Figure P14.7 reach equilibrium in 20 μs? Explain your answer.

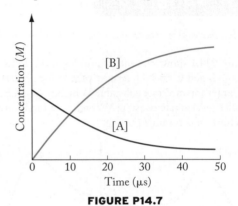

FIGURE P14.7

14.8. How are forward and reverse reaction rates in Figure P14.7 related at 30 μs?

QUESTIONS AND PROBLEMS ••• ■

The Dynamics of Chemical Equilibrium

CONCEPT REVIEW

14.9. Describe an example of a dynamic equilibrium that you experienced today.

14.10. At equilibrium, is the sum of the concentrations of all the reactants always equal to the sum of the concentrations of the products? Explain.

14.11. Suppose the forward rate constant of the reaction $A \rightleftharpoons B$ is greater than the rate constant of the reverse reaction at a given temperature. Is the value of the equilibrium constant less than, greater than, or equal to 1?

14.12. Explain how it is possible for a reaction to have a large equilibrium constant but small forward and reverse rate constants.

PROBLEMS

14.13. In a study of the reaction

$$2 N_2O(g) \rightleftharpoons 2 N_2(g) + O_2(g)$$

quantities of all three gases were injected into a reaction vessel. The N_2O consisted entirely of isotopically labeled $^{15}N_2O$. Analysis of the reaction mixture after 1 day

revealed the presence of compounds with molar masses 28, 29, 30, 32, 44, 45, and 46 g/mol. Identify the compounds and account for their presence.

14.14. A mixture of ^{13}CO, $^{12}CO_2$, and O_2 in a sealed reaction vessel was used to follow the reaction

$$2 CO(g) + O_2(g) \rightleftharpoons 2 CO_2(g)$$

Analysis of the reaction mixture after 1 day revealed the presence of compounds with molar masses 28, 29, 32, 44, and 45 g/mol. Identify the compounds and account for their presence.

14.15. Suppose the reaction $A \rightleftharpoons B$ in the forward direction is first order in A and the rate constant is 1.50×10^{-2} s^{-1}. The reverse reaction is first order in B and the rate constant is 4.50×10^{-2} s^{-1} at the same temperature. What is the value of the equilibrium constant for the reaction $A \rightleftharpoons B$ at this temperature?

14.16. At 700 K the equilibrium constant K_c for the gas-phase reaction between NO and O_2 forming NO_2 is 8.7×10^6. The rate constant for the reverse reaction at this temperature is $0.54\ M^{-1}$ s^{-1}. What is the value of the rate constant for the forward reaction at 700 K?

Writing Equilibrium Constant Expressions; Relationships between K_c and K_p Values

CONCEPT REVIEW

14.17. Under what conditions are the numerical values of K_c and K_p equal?

14.18. At what temperature are the K_c and K_p values of the following reaction equal?

$$N_2(g) + O_2(g) \rightleftharpoons 2\, NO(g)$$

14.19. Nitrogen oxides play important roles in air pollution. Write expressions for K_c and K_p for the following reactions involving nitrogen oxides.
a. $N_2(g) + 2\, O_2(g) \rightleftharpoons N_2O_4(g)$
b. $3\, NO(g) \rightleftharpoons NO_2(g) + N_2O(g)$
c. $2\, N_2O(g) \rightleftharpoons 2\, N_2(g) + O_2(g)$

14.20. Write expressions for K_c and K_p for the following reactions, which contribute to the destruction of stratospheric ozone.
a. $Cl(g) + O_3(g) \rightleftharpoons ClO(g) + O_2(g)$
b. $2\, ClO(g) \rightleftharpoons 2\, Cl(g) + O_2(g)$
c. $2\, O_3(g) \rightleftharpoons 3\, O_2(g)$

PROBLEMS

14.21. Use the graph in Figure P14.21 to estimate the value of the equilibrium constant K_c for the reaction

$$N_2O(g) \rightleftharpoons N_2(g) + \tfrac{1}{2}\, O_2(g)$$

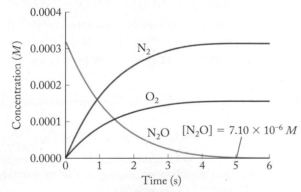

FIGURE P14.21

14.22. Estimate the value of the equilibrium constant K_c for the reaction

$$2\, NO(g) + O_2(g) \rightleftharpoons 2\, NO_2(g)$$

from the data in Figure P14.22.

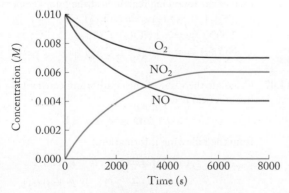

FIGURE P14.22

14.23. At 1200 K the partial pressures of an equilibrium mixture of H_2S, H_2, and S are 0.020, 0.045, and 0.030 atm, respectively. Calculate the value of the equilibrium constant K_p at 1200 K.

$$H_2S(g) \rightleftharpoons H_2(g) + S(g)$$

14.24. At 1045 K the partial pressures of an equilibrium mixture of H_2O, H_2, and O_2 are 0.040, 0.0045, and 0.0030 atm, respectively. Calculate the value of the equilibrium constant K_p at 1045 K.

$$2\, H_2O(g) \rightleftharpoons 2\, H_2(g) + O_2(g)$$

14.25. At equilibrium, the concentrations of gaseous N_2, O_2, and NO in a sealed reaction vessel are $[N_2] = 3.3 \times 10^{-3}\, M$, $[O_2] = 5.8 \times 10^{-3}\, M$, and $[NO] = 3.1 \times 10^{-3}\, M$. What is the value of K_c for the following reaction at the temperature of the mixture?

$$N_2(g) + O_2(g) \rightleftharpoons 2\, NO(g)$$

14.26. Analyses of an equilibrium mixture of gaseous N_2O_4 and NO_2 gave the following results: $[NO_2] = 4.2 \times 10^{-3}\, M$ and $[N_2O_4] = 2.9 \times 10^{-3}\, M$. What is the value of the equilibrium constant K_c for the following reaction at the temperature of the mixture?

$$2\, NO_2(g) \rightleftharpoons N_2O_4(g)$$

14.27. A sealed reaction vessel initially contains 1.50×10^{-2} mol water vapor and 1.50×10^{-2} mol CO. After the reaction

$$H_2O(g) + CO(g) \rightleftharpoons H_2(g) + CO_2(g)$$

has come to equilibrium, the vessel contains 8.3×10^{-3} mol CO_2. What is the value of the equilibrium constant K_c of the reaction at the temperature of the vessel?

*14.28. A 100 mL reaction vessel initially contains 2.60×10^{-2} mol NO and 1.30×10^{-2} mol H_2. At equilibrium, the concentration of NO in the vessel is $0.161\, M$. At equilibrium the vessel also contains N_2, H_2O, and H_2. What is the value of the equilibrium constant K_c for the following reaction?

$$2\, H_2(g) + 2\, NO(g) \rightleftharpoons 2\, H_2O(g) + N_2(g)$$

14.29. The equilibrium constant K_p for the following equilibrium is 32 at 298 K. What is the value of K_c for this same equilibrium at 298 K?

$$A(g) + B(g) \rightleftharpoons AB(g)$$

14.30. The equilibrium constant K_c for the following equilibrium is 6.0×10^4 at 500 K. What is the value of K_p for this same equilibrium at 500 K?

$$CD(g) \rightleftharpoons C(g) + D(g)$$

14.31. At 500°C, the equilibrium constant K_p for the synthesis of ammonia,

$$N_2(g) + 3\, H_2(g) \rightleftharpoons 2\, NH_3(g)$$

is 1.45×10^{-5}. What is the value of K_c at 500°C?

14.32. If the value of the equilibrium constant, K_c, for the following reaction is 5×10^5 at 298 K, what is the value of K_p at 298 K?

$$2 CO(g) + O_2(g) \rightleftharpoons 2 CO_2(g)$$

14.33. For which of the following reactions are the values of K_c and K_p equal?
a. $2 SO_2(g) + O_2(g) \rightleftharpoons 2 SO_3(g)$
b. $Fe(s) + CO_2(g) \rightleftharpoons FeO(s) + CO(g)$
c. $H_2O(g) + CO(g) \rightleftharpoons H_2(g) + CO_2(g)$

14.34. For which of the following reactions are the values of K_c and K_p different?
a. $SO_2Cl_2(g) \rightleftharpoons SO_2(g) + Cl_2(g)$
b. $2 NO(g) + O_2(g) \rightleftharpoons 2 NO_2(g)$
c. $2 O_3(g) \rightleftharpoons 3 O_2(g)$

14.35. Bulletproof Glass Phosgene ($COCl_2$) is used in the manufacture of foam rubber and bulletproof glass. It is formed from carbon monoxide and chlorine in the following reaction:

$$Cl_2(g) + CO(g) \rightleftharpoons COCl_2(g)$$

The value of K_c for the reaction is 5.0 at 327°C. What is the value of K_p at 327°C?

14.36. If the value of K_p for the following reaction

$$SO_2(g) + NO_2(g) \rightleftharpoons NO(g) + SO_3(g)$$

is 3.45 at 298 K, what is the value of K_c for the reverse reaction?

Manipulating Equilibrium Constant Expressions

CONCEPT REVIEW

14.37. How is the value of the equilibrium constant affected by scaling up or down the coefficients of the reactants and products in the chemical equation describing the reaction?

14.38. Why must a written form of a reaction and the temperature be given when reporting the value of an equilibrium constant?

PROBLEMS

14.39. The equilibrium constant for the reaction

$$I_2(g) + Br_2(g) \rightleftharpoons 2 IBr(g)$$

is 120 at 425 K. What is the value of K_c for the equilibrium

$$\tfrac{1}{2} I_2(g) + \tfrac{1}{2} Br_2(g) \rightleftharpoons IBr(g)$$

at 425 K?

14.40. The equilibrium constant K_p for the synthesis of ammonia,

$$N_2(g) + 3 H_2(g) \rightleftharpoons 2 NH_3(g)$$

is 4.3×10^{-3} at 300°C. What is the value of K_p for the equilibrium

$$\tfrac{1}{2} N_2(g) + \tfrac{3}{2} H_2(g) \rightleftharpoons NH_3(g)$$

at 300°C?

14.41. The following reaction is one of the elementary steps in the oxidation of NO:

$$NO(g) + NO_3(g) \rightleftharpoons 2 NO_2(g)$$

Write an expression for the equilibrium constant K_c for this reaction and for the reverse reaction:

$$2 NO_2(g) \rightleftharpoons NO(g) + NO_3(g)$$

How are the two K_c expressions related?

14.42. Making Ammonia The value of the equilibrium constant K_p for the formation of ammonia,

$$N_2(g) + 3 H_2(g) \rightleftharpoons 2 NH_3(g)$$

is 4.5×10^{-5} at 450°C. What is the value of K_p for the following reaction at 450°C?

$$2 NH_3(g) \rightleftharpoons N_2(g) + 3 H_2(g)$$

14.43. Air Pollutants Sulfur oxides are major air pollutants. The reaction between sulfur dioxide and oxygen can be written in two ways:

$$SO_2(g) + \tfrac{1}{2} O_2(g) \rightleftharpoons SO_3(g)$$

and

$$2 SO_2(g) + O_2(g) \rightleftharpoons 2 SO_3(g)$$

Write expressions for the equilibrium constants for both reactions. How are they related?

14.44. At a given temperature, the equilibrium constant K_c for the reaction

$$2 NO(g) + 2 H_2(g) \rightleftharpoons N_2(g) + 2 H_2O(g)$$

is 0.11. What is the equilibrium constant for the following reaction?

$$NO(g) + H_2(g) \rightleftharpoons \tfrac{1}{2} N_2(g) + H_2O(g)$$

14.45. At a given temperature, the equilibrium constant K_c for the reaction

$$2 SO_2(g) + O_2(g) \rightleftharpoons 2 SO_3(g)$$

is 2.4×10^{-3}. What is the value of the equilibrium constant for each of the following reactions at that temperature?
a. $SO_2(g) + \tfrac{1}{2} O_2(g) \rightleftharpoons SO_3(g)$
b. $2 SO_3(g) \rightleftharpoons 2 SO_2(g) + O_2(g)$
c. $SO_3(g) \rightleftharpoons SO_2(g) + \tfrac{1}{2} O_2(g)$

14.46. If the equilibrium constant K_c for the reaction

$$2 NO(g) + O_2(g) \rightleftharpoons 2 NO_2(g)$$

is 5×10^{12}, what is the value of the equilibrium constant of each of the following reactions at the same temperature?
a. $NO(g) + \tfrac{1}{2} O_2(g) \rightleftharpoons NO_2(g)$
b. $2 NO_2(g) \rightleftharpoons 2 NO(g) + O_2(g)$
c. $NO_2(g) \rightleftharpoons NO(g) + \tfrac{1}{2} O_2(g)$

14.47. Calculate the value of the equilibrium constant K for the hypothetical reaction

$$2 D \rightleftharpoons A + 2 B$$

from the following information:

$$A + 2 B \rightleftharpoons C \qquad K_c = 3.3$$
$$C \rightleftharpoons 2 D \qquad K_c = 0.041$$

14.48. Calculate the value of the equilibrium constant K for the hypothetical reaction

$$E + F \rightleftharpoons G$$

from the following information:

$2\,G \rightleftharpoons H$	$K_c = 3.1 \times 10^{-4}$
$H \rightleftharpoons 2\,E + 2\,F$	$K_c = 2.8 \times 10^{22}$

Equilibrium Constants and Reaction Quotients

CONCEPT REVIEW

14.49. What is a reaction quotient?

14.50. How is an equilibrium constant different from a reaction quotient?

14.51. What does it mean when the reaction quotient Q is numerically equal to the equilibrium constant K?

14.52. Explain how knowing Q and K for an equilibrium system enables you to say whether it is at equilibrium or whether it will shift in one direction or another.

PROBLEMS

14.53. If the equilibrium constant K_c for the hypothetical reaction $A(g) \rightleftharpoons B(g)$ is 22 at a given temperature, and if $[A] = 0.10\ M$ and $[B] = 2.0\ M$ in a reaction mixture at that temperature, is the reaction at chemical equilibrium? If not, in which direction will the reaction proceed to reach equilibrium?

14.54. The equilibrium constant K_c for the hypothetical reaction $2\,C \rightleftharpoons D + E$ is 3×10^{-3}. At a particular time, the composition of the reaction mixture is $[C] = [D] = [E] = 5 \times 10^{-4}\ M$. In which direction will the reaction proceed to reach equilibrium?

14.55. Suppose the value of the equilibrium constant K_p of the following hypothetical reaction

$$A(g) + B(g) \rightleftharpoons C(g)$$

is 1.00 at 300 K. Are either of the following reaction mixtures at chemical equilibrium at 300 K?
a. $P_A = P_B = P_C = 1.0$ atm
b. $[A] = [B] = [C] = 1.0\ M$

14.56. In which direction will the following hypothetical reaction proceed to reach equilibrium under the conditions given?

$$A(g) + B(g) \rightleftharpoons C(g) \qquad K_p = 1.00 \text{ at } 300 \text{ K}$$

a. $P_A = P_C = 1.0$ atm, $P_B = 0.50$ atm
b. $[A] = [B] = [C] = 1.0\ M$

14.57. If the equilibrium constant K_c for the reaction

$$N_2(g) + O_2(g) \rightleftharpoons 2\,NO(g)$$

is 1.5×10^{-3}, in which direction will the reaction proceed if the partial pressures of the three gases are all 1.00×10^{-3} atm?

14.58. At 650 K, the value of the equilibrium constant K_p for the ammonia synthesis reaction

$$N_2(g) + 3\,H_2(g) \rightleftharpoons 2\,NH_3(g)$$

is 4.3×10^{-4}. If a vessel contains a reaction mixture in which $[N_2] = 0.010\ M$, $[H_2] = 0.030\ M$, and $[NH_3] = 0.00020\ M$, will more ammonia form?

14.59. The hypothetical equilibrium $X + Y \rightleftharpoons Z$ has $K_c = 1.00$ at 350 K. If the initial molar concentrations of X, Y, and Z in a solution are all 0.2 M, in which direction will the reaction shift to reach equilibrium?
a. To the left, making more X and Y
b. To the right, making more Z
c. The system is at equilibrium and the concentrations will not change

14.60. In Problem 14.59, when the equilibrium shifts, does the concentration of X increase or decrease?

Heterogeneous Equilibria

CONCEPT REVIEW

14.61. Write the K_c expression for the following reaction:

$$CuS(s) \rightleftharpoons Cu^{2+}(aq) + S^{2-}(aq)$$

14.62. Write the K_c expression for the following reaction:

$$Al_2O_3(s) + 3\,H_2O(\ell) \rightleftharpoons 2\,Al^{3+}(aq) + 6\,OH^-(aq)$$

14.63. Why does the K_c expression for the reaction

$$CaCO_3(s) \rightleftharpoons CaO(s) + CO_2(g)$$

not contain terms for the concentrations of $CaCO_3$ and CaO?

14.64. How many partial pressure terms are there in the K_p expression for the thermal decomposition of sodium bicarbonate?

$$2\,NaHCO_3(s) \rightleftharpoons Na_2CO_3(s) + CO_2(g) + H_2O(g)$$

Le Châtelier's Principle

CONCEPT REVIEW

14.65. Does adding reactants to a system at equilibrium increase the value of the equilibrium constant?

14.66. Increasing the concentration of a reactant shifts the position of chemical equilibrium toward formation of more products. What effect does adding a reactant have on the rates of the forward and reverse reactions?

14.67. **Carbon Monoxide Poisoning** Patients suffering from carbon monoxide poisoning are treated with pure oxygen to remove CO from the hemoglobin (Hb) in their blood. The two relevant equilibria are

$$Hb + 4\,CO(g) \rightleftharpoons Hb(CO)_4$$
$$Hb + 4\,O_2(g) \rightleftharpoons Hb(O_2)_4$$

The value of the equilibrium constant for CO binding to Hb is greater than that for O_2. How, then, does this treatment work?

14.68. Is the equilibrium constant K_p for the reaction

$$2\,NO_2(g) \rightleftharpoons N_2O_4(g)$$

in air the same in Los Angeles as in Denver if the atmospheric pressure in Denver is lower but the temperature is the same?

14.69. Henry's law (Chapter 10) predicts that the solubility of a gas in a liquid increases with its partial pressure. Explain Henry's law in relation to Le Châtelier's principle.

*__14.70.__ For the reaction

$$2\,CO(g) + O_2(g) \rightleftharpoons 2\,CO_2(g)$$

why does adding an inert gas such as argon to an equilibrium mixture of CO, O_2, and CO_2 in a sealed vessel increase the total pressure of the system but not affect the position of the equilibrium?

PROBLEMS

14.71. Which of the following equilibria will shift toward formation of more products if an equilibrium mixture is compressed into half its volume?
   a. $2\,N_2O(g) \rightleftharpoons 2\,N_2(g) + O_2(g)$
   b. $2\,CO(g) + O_2(g) \rightleftharpoons 2\,CO_2(g)$
   c. $N_2(g) + O_2(g) \rightleftharpoons 2\,NO(g)$
   d. $2\,NO(g) + O_2(g) \rightleftharpoons 2\,NO_2(g)$

14.72. Which of the following equilibria will shift toward formation of more products if the volume of a reaction mixture at equilibrium increases by a factor of 2?
   a. $2\,SO_2(g) + O_2(g) \rightleftharpoons 2\,SO_3(g)$
   b. $NO(g) + O_3(g) \rightleftharpoons NO_2(g) + O_2(g)$
   c. $2\,N_2O_5(g) \rightleftharpoons 4\,NO_2(g) + O_2(g)$
   d. $N_2O_4(g) \rightleftharpoons 2\,NO_2(g)$

14.73. What would be the effect of the changes listed on the equilibrium concentrations of reactants and products in the following reaction?

$$2\,O_3(g) \rightleftharpoons 3\,O_2(g)$$

   a. O_3 is added to the system.
   b. O_2 is added to the system.
   c. The mixture is compressed to one-tenth its initial volume.

14.74. How will the changes listed affect the position of the following equilibrium?

$$2\,NO_2(g) \rightleftharpoons NO(g) + NO_3(g)$$

   a. The concentration of NO is increased.
   b. The concentration of NO_2 is increased.
   c. The volume of the system is allowed to expand to 5 times its initial value.

14.75. How would reducing the partial pressure of $O_2(g)$ affect the position of the equilibrium in the following reaction?

$$2\,SO_2(g) + O_2(g) \rightleftharpoons 2\,SO_3(g)$$

*__14.76.__ Ammonia is added to a gaseous reaction mixture containing H_2, Cl_2, and HCl that is at chemical equilibrium. How will the addition of ammonia affect the relative concentrations of H_2, Cl_2, and HCl if the equilibrium constant of reaction 2 is much greater than the equilibrium constant of reaction 1?

   (1) $H_2(g) + Cl_2(g) \rightleftharpoons 2\,HCl(g)$
   (2) $HCl(g) + NH_3(g) \rightleftharpoons NH_4Cl(s)$

14.77. In which of the following hypothetical equilibria does the product yield increase with increasing temperature?
   a. $A + 2\,B \rightleftharpoons C$ $\Delta H > 0$
   b. $A + 2\,B \rightleftharpoons C$ $\Delta H = 0$
   c. $A + 2\,B \rightleftharpoons C$ $\Delta H < 0$

14.78. In which of the following hypothetical equilibria does the product yield decrease with increasing temperature?
   a. $2\,X + Y \rightleftharpoons Z$ $\Delta H > 0$
   b. $2\,X + Y \rightleftharpoons Z$ $\Delta H = 0$
   c. $2\,X + Y \rightleftharpoons Z$ $\Delta H < 0$

Calculations Based on *K*

CONCEPT REVIEW

14.79. Why are calculations of how much product is formed in the reaction

$$X + Y \rightleftharpoons Z$$

often simpler when there is no Z initially present and the value of K is very small ($<10^{-6}$)?

14.80. Could the quadratic equation be used to solve for the equilibrium concentration of NO_2 in the following reaction?

$$2\,NO(g) + O_2(g) \rightleftharpoons 2\,NO_2(g)$$

PROBLEMS

14.81. Consider this reaction:

$$PCl_5(g) \rightleftharpoons PCl_3(g) + Cl_2(g) \qquad K_p = 23.6 \text{ at } 500 \text{ K}$$

   a. Calculate the equilibrium partial pressures of the reactants and products if the initial pressures are $P_{PCl_5} = 0.560$ atm and $P_{PCl_3} = 0.500$ atm.
   b. If more chlorine is added after equilibrium is reached, how will the concentrations of PCl_5 and PCl_3 change?

14.82. Enough NO_2 gas is injected into a cylindrical vessel to produce a partial pressure, P_{NO_2}, of 0.900 atm at 298 K. Calculate the equilibrium partial pressures of NO_2 and N_2O_4, given

$$2\,NO_2(g) \rightleftharpoons N_2O_4(g) \qquad K_p = 4 \text{ at } 298 \text{ K}$$

14.83. The value of K_c for the reaction between water vapor and dichlorine monoxide

$$H_2O(g) + Cl_2O(g) \rightleftharpoons 2\,HOCl(g)$$

is 0.0900 at 25°C. Determine the equilibrium concentrations of all three compounds if the starting concentrations of both reactants are 0.00432 M and no HOCl is present.

14.84. The value of K_p for the reaction

$$3\,H_2(g) + N_2(g) \rightleftharpoons 2\,NH_3(g)$$

is 4.3×10^{-4} at 648 K. Determine the equilibrium partial pressure of NH_3 in a reaction vessel that initially contained 0.900 atm N_2 and 0.500 atm H_2 at 648 K.

14.85. The value of K_p for the reaction

$$NO(g) + \tfrac{1}{2} O_2(g) \rightleftharpoons NO_2(g)$$

is 2×10^6 at 25°C. At equilibrium, what is the ratio of P_{NO_2} to P_{NO} in air at 25°C? Assume that $P_{O_2} = 0.21$ atm and does not change.

*14.86. Making Hydrogen Gas Passing steam over hot carbon produces a mixture of carbon monoxide and hydrogen:

$$H_2O(g) + C(s) \rightleftharpoons CO(g) + H_2(g)$$

The value of K_c for the reaction at 1000°C is 3.0×10^{-2}.
a. Calculate the equilibrium partial pressures of the products and reactants if $P_{H_2O} = 0.442$ atm and $P_{CO} = 5.0$ atm at the start of the reaction. Assume that the carbon is in excess.
b. Determine the equilibrium partial pressures of the reactants and products after sufficient CO and H_2 are added to the equilibrium mixture in part a to initially increase the partial pressures of both gases by 0.075 atm.

14.87. The value of K_p for the reaction

$$CO_2(g) + C(s) \rightleftharpoons 2\, CO(g)$$

is 1.5 at 700°C. Calculate the equilibrium partial pressures of CO and CO_2 if initially $P_{CO_2} = 5.0$ atm and $P_{CO} = 0.0$. Pure graphite is present both initially and when equilibrium is achieved.

14.88. Jupiter's Atmosphere Ammonium hydrogen sulfide (NH$_4$SH) has been detected in the atmosphere of Jupiter, where it probably exists in equilibrium with ammonia and hydrogen sulfide:

$$NH_4SH(s) \rightleftharpoons NH_3(g) + H_2S(g)$$

The value of K_p for the reaction at 24°C is 0.126. Suppose a sealed flask contains an equilibrium mixture of NH$_4$SH, NH$_3$, and H$_2$S. At equilibrium, the partial pressure of H$_2$S is 0.355 atm. What is the partial pressure of NH$_3$?

*14.89. A flask containing pure NO$_2$ is heated to 1000 K, a temperature at which the value of K_p for the decomposition of NO$_2$ is 158.

$$2\, NO_2(g) \rightleftharpoons 2\, NO(g) + O_2(g)$$

The partial pressure of O$_2$ at equilibrium is 0.136 atm.
a. Calculate the partial pressures of NO and NO$_2$.
b. Calculate the total pressure in the flask at equilibrium.

14.90. The equilibrium constant K_p of the reaction

$$2\, SO_3(g) \rightleftharpoons 2\, SO_2(g) + O_2(g)$$

is 7.69 at 830°C. If a vessel at this temperature initially contains pure SO$_3$ and if the partial pressure of SO$_3$ at equilibrium is 0.100 atm, what is the partial pressure of O$_2$ in the flask at equilibrium?

*14.91. NO$_x$ Pollution In a study of the formation of NO$_x$ air pollution, a chamber heated to 2200°C was filled with air (0.79 atm N$_2$, 0.21 atm O$_2$). What are the equilibrium

partial pressures of N$_2$, O$_2$, and NO if $K_p = 0.050$ for the following reaction at 2200°C?

$$N_2(g) + O_2(g) \rightleftharpoons 2\, NO(g)$$

*14.92. The equilibrium constant K_p for the thermal decomposition of NO$_2$

$$2\, NO_2(g) \rightleftharpoons 2\, NO(g) + O_2(g)$$

is 6.5×10^{-6} at 450°C. If a reaction vessel at this temperature initially contains 0.500 atm NO$_2$, what will be the partial pressures of NO$_2$, NO, and O$_2$ in the vessel when equilibrium has been attained?

14.93. The value of K_c for the thermal decomposition of hydrogen sulfide

$$2\, H_2S(g) \rightleftharpoons 2\, H_2(g) + S_2(g)$$

is 2.2×10^{-4} at 1400 K. A sample of gas in which [H$_2$S] = 6.00 M is heated to 1400 K in a sealed high-pressure vessel. After chemical equilibrium has been achieved, what is the value of [H$_2$S]? Assume that no H$_2$ or S$_2$ was present in the original sample.

14.94. Urban Air On a very smoggy day, the equilibrium concentration of NO$_2$ in the air over an urban area reaches 2.2×10^{-7} M. If the temperature of the air is 25°C, what is the concentration of the dimer N$_2$O$_4$ in the air? Given:

$$N_2O_4(g) \rightleftharpoons 2\, NO_2(g) \qquad K_c = 6.1 \times 10^{-3}$$

*14.95. Chemical Weapon Phosgene, COCl$_2$, gained notoriety as a chemical weapon in World War I. Phosgene is produced by the reaction of carbon monoxide with chlorine:

$$CO(g) + Cl_2(g) \rightleftharpoons COCl_2(g)$$

The value of K_c for this reaction is 5.0 at 600 K. What are the equilibrium partial pressures of the three gases if a reaction vessel initially contains a mixture of the reactants in which $P_{CO} = P_{Cl_2} = 0.265$ atm and $P_{COCl_2} = 0.000$ atm?

*14.96. At 2000°C, the value of K_c for the reaction

$$2\, CO(g) + O_2(g) \rightleftharpoons 2\, CO_2(g)$$

is 1.0. What is the ratio of [CO] to [CO$_2$] in an atmosphere in which [O$_2$] = 0.0045 M?

*14.97. The water–gas shift reaction is an important source of hydrogen. The value of K_c for the reaction

$$CO(g) + H_2O(g) \rightleftharpoons CO_2(g) + H_2(g)$$

at 700 K is 5.1. Calculate the equilibrium concentrations of the four gases if the initial concentration of each of them is 0.050 M.

*14.98. Sulfur dioxide reacts with NO$_2$, forming SO$_3$ and NO:

$$SO_2(g) + NO_2(g) \rightleftharpoons SO_3(g) + NO(g)$$

If the value of K_c for the reaction is 2.50, what are the equilibrium concentrations of the products if the reaction mixture was initially 0.50 M SO$_2$, 0.50 M NO$_2$, 0.0050 M SO$_3$, and 0.0050 M NO?

14.99. Nitrogen dioxide and nitric oxide react to form dinitrogen trioxide. The K_p for this reaction is 0.535 at 25°C:

$$NO(g) + NO_2(g) \rightleftharpoons N_2O_3(g)$$

If a 4.0 L flask contains 15 g NO and 69 g NO_2, what are the concentrations of all species at equilibrium?

14.100. For the decomposition of $HI(g)$ into $H_2(g)$ and $I_2(g)$ at 400°C, $K_c = 0.0183$:

$$2 HI(g) \rightleftharpoons H_2(g) + I_2(g)$$

If 80.0 g of $HI(g)$ is placed in a 2.5 L chamber at 400°C, what are the concentrations of all species when the system comes to equilibrium?

Equilibrium and Thermodynamics

CONCEPT REVIEW

14.101. Do all reactions with equilibrium constants < 1 have values of $\Delta G° > 0$?

*__14.102.__ The equation $\Delta G° = -RT \ln K$ relates the value of K_p, not K_c, to the change in standard free energy for a reaction in the gas phase. Explain why.

14.103. Starting with pure reactants, in which direction will an equilibrium shift if $\Delta G° < 0$?

14.104. Starting with pure products, in which direction will an equilibrium shift if $\Delta G° < 0$?

PROBLEMS

14.105. Which of the following reactions has the largest equilibrium constant at 25°C?
a. $Cl_2(g) + F_2(g) \rightleftharpoons 2 ClF(g)$ $\Delta G° = 115.4$ kJ
b. $Cl_2(g) + Br_2(g) \rightleftharpoons 2 ClBr(g)$ $\Delta G° = -2.0$ kJ
c. $Cl_2(g) + I_2(g) \rightleftharpoons 2 ICl(g)$ $\Delta G° = -27.9$ kJ

14.106. The value of $\Delta G°$ for the reaction

$$N_2O(g) + \tfrac{1}{2} O_2(g) \rightleftharpoons 2 NO(g)$$

is 68.9 kJ. What is the value of the equilibrium constant for this reaction at 298 K?

*__14.107.__ At a temperature of 1000 K, $SO_2(g)$ combines with oxygen to make $SO_3(g)$:

$$2 SO_2(g) + O_2(g) \rightleftharpoons 2 SO_3(g)$$

Under these conditions, $K_p = 3.4$.
a. Use the appropriate thermodynamic data in Appendix 4 to calculate the value of $\Delta H°_{rxn}$ for this reaction.
b. What is the value of K_p for this reaction at 298 K?
c. Use the answer from part b to calculate the value of $\Delta G°_{rxn}$ at 298 K, and compare it to the value you obtained using the $\Delta G°_f$ values in Appendix 4.

*__14.108.__ The value of the equilibrium constant K_p for the reaction

$$H_2(g) + CO_2(g) \rightleftharpoons H_2O(g) + CO(g)$$

is 0.534 at 700°C.
a. What is the value of K_p for this reaction at 298 K? (*Hint*: To perform this K_p calculation, you will need to calculate the value of $\Delta H°_{rxn}$ using the appropriate data from Appendix 4.)

b. Calculate the value of $\Delta G°_{rxn}$ at 298 K using your answer from part a. Then compare it to the $\Delta G°_{rxn}$ value you obtained using the $\Delta G°_f$ values in Appendix 4.

14.109. Use the following data to calculate the value of K_p at 298 K for the reaction:

$$N_2(g) + 2 O_2(g) \rightleftharpoons 2 NO_2(g)$$

Given:

$$N_2(g) + O_2(g) \rightleftharpoons 2 NO(g) \qquad \Delta G° = 173.2 \text{ kJ}$$
$$2 NO(g) + O_2(g) \rightleftharpoons 2 NO_2(g) \qquad \Delta G° = -69.7 \text{ kJ}$$

14.110. Under the appropriate conditions, NO forms N_2O and NO_2:

$$3 NO(g) \rightleftharpoons N_2O(g) + NO_2(g)$$

Use the values for $\Delta G°$ for the following reactions to calculate the value of K_p for the above reaction at 500°C.

$$2 NO(g) + O_2(g) \rightleftharpoons 2 NO_2(g) \qquad \Delta G° = -69.7 \text{ kJ}$$
$$2 N_2O(g) \rightleftharpoons 2 NO(g) + N_2(g) \qquad \Delta G° = -33.8 \text{ kJ}$$
$$N_2(g) + O_2(g) \rightleftharpoons 2 NO(g) \qquad \Delta G° = 173.2 \text{ kJ}$$

Changing K with Changing Temperature

CONCEPT REVIEW

14.111. The value of the equilibrium constant of a reaction decreases with increasing temperature. Is this reaction endothermic or exothermic?

14.112. The reaction

$$2 CO(g) + O_2(g) \rightleftharpoons 2 CO_2(g)$$

is exothermic. Does the value of K_p increase or decrease with increasing temperature?

14.113. The value of K_p for the water–gas shift reaction

$$CO(g) + H_2O(g) \rightleftharpoons H_2(g) + CO_2(g)$$

increases as the temperature decreases. Is the reaction exothermic or endothermic?

14.114. Does the value of K_p for the reaction

$$CH_4(g) + H_2O(g) \rightleftharpoons 3 H_2(g) + CO(g) \qquad \Delta H° = 206 \text{ kJ}$$

increase, decrease, or remain unchanged as temperature increases?

PROBLEMS

14.115. **Air Pollution** Automobiles and trucks pollute the air with NO. At 2000°C, K_c for the reaction

$$N_2(g) + O_2(g) \rightleftharpoons 2 NO(g)$$

is 4.10×10^{-4}, and $\Delta H° = 180.6$ kJ. What is the value of K_c at 25°C?

14.116. At 400 K the value of K_p for the reaction

$$N_2(g) + 3 H_2(g) \rightleftharpoons 2 NH_3(g)$$

is 41, and $\Delta H° = -92.2$ kJ. What is the value of K_p at 700 K?

14.117. The equilibrium constant for the reaction

$$2 NO(g) + O_2(g) \rightleftharpoons 2 NO_2(g)$$

decreases from 1.5×10^5 at 430°C to 23 at 1000°C. From these data, calculate the value of $\Delta H°$ for the reaction.

14.118. The value of K_c for the reaction A $\rightleftharpoons$ B is 0.455 at 50°C and 0.655 at 100°C. Calculate $\Delta H°$ for the reaction.

Additional Problems

*14.119. CO as a Fuel Is carbon dioxide a viable source of the fuel CO? Pure carbon dioxide ($P_{CO_2} = 1$ atm) decomposes at high temperatures. For the system

$$2 CO_2(g) \rightleftharpoons 2 CO(g) + O_2(g)$$

the percentage of decomposition of $CO_2(g)$ changes with temperature as follows:

Temperature (K)	Decomposition (%)
1500	0.048
2500	17.6
3000	54.8

Is the reaction endothermic? Calculate the value of K_p at each temperature and discuss the results. Is the decomposition of CO_2 an antidote for global warming?

14.120. Ammonia decomposes at high temperatures. In an experiment to explore this behavior, 2.000 mol of gaseous NH_3 is sealed in a rigid 1-liter vessel. The vessel is heated to 800 K and some of the NH_3 decomposes in the following reaction:

$$2 NH_3(g) \rightleftharpoons N_2(g) + 3 H_2(g)$$

The system eventually reaches equilibrium and is found to contain 0.0040 mol of NH_3. What are the values of K_p and K_c for the decomposition reaction at 800 K?

*14.121. Elements of group 16 form hydrides with the generic formula H_2X. When gaseous H_2X is bubbled through a solution containing 0.3 M hydrochloric acid, the solution becomes saturated and $[H_2X] = 0.1$ M. The following equilibria exist in this solution:

$$H_2X(aq) + H_2O(\ell) \rightleftharpoons HX^-(aq) + H_3O^+(aq) \qquad K_1 = 8.3 \times 10^{-8}$$

$$HX^-(aq) + H_2O(\ell) \rightleftharpoons X^{2-}(aq) + H_3O^+(aq) \qquad K_2 = 1 \times 10^{-14}$$

Calculate the concentration of X^{2-} in the solution.

*14.122. Nitrogen dioxide reacts with SO_2 to form SO_3 and NO:

$$NO_2(g) + SO_2(g) \rightleftharpoons NO(g) + SO_3(g)$$

An equilibrium mixture is analyzed at a certain temperature and found to contain $[NO_2] = 0.100$ M, $[SO_2] = 0.300$ M, $[NO] = 2.00$ M, and $[SO_3] = 0.600$ M. At the same temperature, extra $SO_2(g)$ is added to make $[SO_2] = 0.800$ M. Calculate the composition of the mixture when equilibrium has been reestablished.

*14.123. Carbon disulfide is a foul-smelling solvent that dissolves sulfur and other nonpolar substances. It can be made by heating sulfur in an atmosphere of methane:

$$4 CH_4(g) + S_8(s) \rightleftharpoons 4 CS_2(g) + 8 H_2(g)$$

Starting with the appropriate data in Appendix 4, calculate the values of K_p for the reaction at 25°C and 500°C.

*14.124. Making Hydrogen Debate continues on the practicality of using H_2 gas as a fuel for cars. The equilibrium constant K_c for the reaction

$$CO(g) + H_2O(g) \rightleftharpoons CO_2(g) + H_2(g)$$

is 1.0×10^5 at 25°C. Starting with this value, calculate the value of $\Delta G°_{rxn}$ at 25°C, and, without doing any calculations, guess the sign of $\Delta H°_{rxn}$.

*14.125. Air Pollution Control Calcium oxide is used to remove the pollutant SO_2 from smokestack gases. The $\Delta G°$ of the overall reaction

$$CaO(s) + SO_2(g) + \tfrac{1}{2} O_2(g) \rightleftharpoons CaSO_4(s)$$

is −418.6 kJ. What is P_{SO_2} in equilibrium with air ($P_{O_2} = 0.21$ atm) and solid CaO?

*14.126. Volcanic Eruptions During volcanic eruptions, gases as hot as 700°C and rich in SO_2 are released into the atmosphere. As air mixes with these gases, the following reaction converts some of this SO_2 into SO_3:

$$2 SO_2(g) + O_2(g) \rightleftharpoons 2 SO_3(g)$$

Calculate the value of K_p for this reaction at 700°C. What is the ratio of P_{SO_2} to P_{SO_3} in equilibrium with $P_{O_2} = 0.21$ atm?

15

Aqueous Equilibria
Chemistry of the Water World

A Balancing Act

For centuries, farmers have added ground limestone (a mix of $CaCO_3$ and $MgCO_3$) to their fields to "sweeten" the soil (to make it less acidic). Adding limestone neutralizes the organic acids that are produced when biological matter, such as the plowed-under remnants of last year's crop, decays. In recent times, farmers and gardeners have added limestone to soil to neutralize acidic precipitation seeping into it.

Normal rainwater is slightly acidic because atmospheric CO_2 dissolves in it, forming a dilute solution of weak carbonic acid. *Acid rain*, however, contains dissolved oxides of nitrogen and sulfur that enter the air mostly as by-products of fossil fuel combustion. These oxides form both weak and strong acids when they dissolve in water. As a result, acid rain may be from 10 to 1000 times more acidic than uncontaminated rainwater. Though some plants, such as blueberries, actually thrive in acidic soil, most crops do not. At the very least their growth is stunted. Sometimes they don't grow at all.

Farmers seek to achieve a balanced soil chemistry that is *neutral*—neither acidic nor basic—or nearly so. Most plants grow best in neutral soil because the nutrients they need to grow are more available to them under such conditions. Sometimes this availability depends on microorganisms that thrive under near-neutral conditions and transform the nutrients into species that are more readily absorbed by plant roots.

The bioavailability of nutrients and nonessential elements may also be a matter of their simply being more soluble under certain conditions and thus more easily absorbed by plants. Consider, for example, the chemical basis for the variation in the color of hydrangea blossoms, which is controlled by the availability of Al^{3+} ions in the soil. In acidic soil, hydrangeas can absorb Al^{3+} ions because the ions are soluble in acidic solutions. As a result the plants form blue blossoms. In neutral to slightly basic soil, aluminum ions precipitate as aluminum hydroxide and are not available to plants. The blossoms of hydrangeas grown in these soils lack the blue pigment and are pink.

Other biological systems—including the human body—are highly sensitive to even small changes in acid–base balance. For example, we produce CO_2 during metabolism, and we eliminate it every time we exhale. Our breathing maintains an acid–base balance by controlling the concentration of dissolved

Shades of Purple, Pink, and Blue Flowers The color of hydrangea flowers ▶ results from different dye molecules produced by the plant in response to the acidity of the soil in which they are grown.

LEARNING OUTCOMES

LO1 Identify the Brønsted–Lowry acids and bases and their conjugate bases and acids in chemical reactions
Sample Exercise 15.1

LO2 Interconvert [H$^+$], [OH$^-$], pH, and pOH values
Sample Exercises 15.2, 15.3

LO3 Relate the strengths of acids and bases to their K_a, K_b, pK_a, and pK_b values and to their percent ionization or dissociation in water
Sample Exercise 15.4

LO4 Calculate the pH values of solutions of weak acids and bases and the salts of their conjugate bases and acids
Sample Exercises 15.5, 15.6, 15.8

LO5 Predict whether a salt is acidic, basic, or neutral
Sample Exercise 15.7

LO6 Explain how pH buffers control pH, and calculate the pH of a conjugate acid–base pair
Sample Exercises 15.9, 15.10

LO7 Interpret the results of an acid–base titration
Sample Exercise 15.11

LO8 Relate the solubility of an ionic compound to its solubility product and solution pH
Sample Exercises 15.12, 15.13, 15.14, 15.15

CO$_2$ and the carbonic acid that it forms in our blood. When lung function is impaired by a chronic disease, such as emphysema, CO$_2$ concentrations in the blood and tissue fluids increase, making them more acidic, which can lead to coma and even death.

In this chapter we explore processes that influence the acid–base balance in the world around us and within our bodies. We address questions such as why these changes happen, how extensive they are, and what they mean to our own health and that of our planet.

▶‖ **CHEMTOUR** Acid Rain

15.1 Acids and Bases: The Brønsted–Lowry Model

In Chapter 8 we defined acids as substances that donate H$^+$ ions and bases as substances that accept H$^+$ ions. These descriptions were developed independently by Danish chemist Johannes Brønsted and English chemist Thomas Lowry and constitute the *Brønsted–Lowry model* of acids and bases.

∞ **CONNECTION** We introduced acids, bases, and neutralization reactions in Chapter 8, where we defined Brønsted–Lowry acids as proton donors and bases as proton acceptors. Neutralization reactions between acids and bases produce water and a salt.

Strong and Weak Acids

As we discussed in Section 8.4, acids in aqueous solution are classified as strong or weak depending on the extent to which they ionize, donating H$^+$ ions to molecules of water and forming hydronium (H$_3$O$^+$) ions. Strong acids are completely ionized in water; weak acids are only partially ionized. Figure 15.1 compares the ionization of two acids with similar chemical structures: nitric acid (HNO$_3$), which is strong, and nitrous acid (HNO$_2$), which is weak. As shown in the molecular view in Figure 15.1(a), HNO$_3$ in a 0.010 *M* solution is completely ionized: there are no intact molecules of HNO$_3$ in the solution. The only solute particles are nitrate (NO$_3^-$) ions and H$^+$ ions that have combined with H$_2$O molecules to form H$_3$O$^+$ ions as a result of the following reaction:

0.010 *M* HNO$_3$
(a)

0.010 *M* HNO$_2$
(b)

NO$_3^-$ H$_3$O$^+$ NO$_2^-$ HNO$_2$

FIGURE 15.1 Difference in degree of ionization for strong and weak acids. (a) The strong acid HNO$_3$ ionizes completely to H$_3$O$^+$ and NO$_3^-$ ions. (b) The weak acid HNO$_2$ ionizes very little.

$$\textbf{HNO}_3\textbf{(aq)} \quad + \quad \textbf{H}_2\textbf{O}(\ell) \quad \rightarrow \quad NO_3^-(aq) + H_3O^+(aq) \qquad (15.1)$$
$$(\text{H}^+\ \text{donor}) \qquad (\text{H}^+\ \text{acceptor})$$

In this equation we have highlighted (in red) nitric acid as the H$^+$ ion donor (Brønsted–Lowry acid) and water (in blue) as the H$^+$ ion acceptor (Brønsted–Lowry base). Essentially all of the molecules of HNO$_3$ in an aqueous solution of the acid are ionized; chemists say that the reaction in Equation 15.1 *goes to completion*. The single reaction arrow in Equation 15.1 conveys the message that HNO$_3$ is completely ionized. In contrast, only a small percentage of the nitrous acid molecules in 0.010 *M* HNO$_2$ ionize before chemical equilibrium is achieved, which is why HNO$_2$ is considered a weak acid (Figure 15.1b):

$$HNO_2(aq) + H_2O(\ell) \rightleftharpoons NO_2^-(aq) + H_3O^+(aq) \qquad (15.2)$$

Two types of strong acids are listed in Table 15.1. First are three binary acids with the generic formula HX, where X is Cl, Br, or I. Next are three oxoacids with the generic formula H$_m$XO$_n$, where X is a nonmetal. Molecules of these acids are particularly good at donating H$^+$ ions to molecules of water because of the stability of the anions they form when they lose H$^+$ ions. In all strong acids, ionizable H atoms are bonded to atoms of one of these highly electronegative elements: O, Cl, Br, or I. When one of these bonds breaks and its H atom is released as an H$^+$ ion, the bonding pair of electrons (and 1− electrical charge) remains on the electronegative atom, as shown for HCl in Figure 15.2. In the case of a

TABLE 15.1 **The Molecular Structures of Strong Acids and Their Ionization Reactions in Water**

Name (. . . acid)	Molecular Structure	Reaction in Water
Hydrochloric	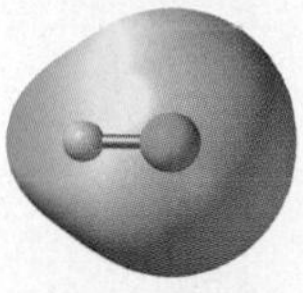	$HCl(aq) + H_2O(\ell) \rightarrow Cl^-(aq) + H_3O^+(aq)$
Hydrobromic	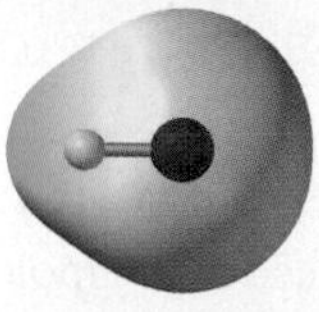	$HBr(aq) + H_2O(\ell) \rightarrow Br^-(aq) + H_3O^+(aq)$
Hydroiodic	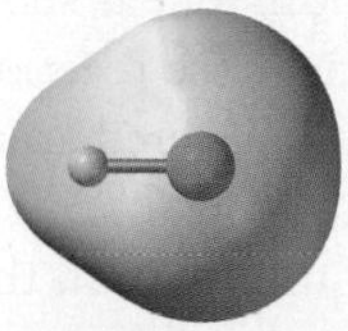	$HI(aq) + H_2O(\ell) \rightarrow I^-(aq) + H_3O^+(aq)$
Nitric	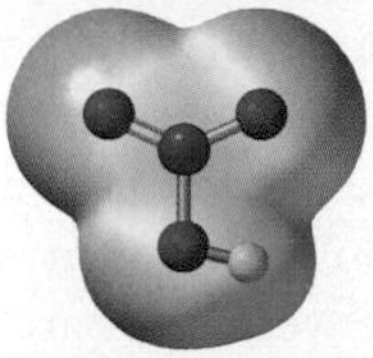	$HNO_3(aq) + H_2O(\ell) \rightarrow NO_3^-(aq) + H_3O^+(aq)$
Perchloric	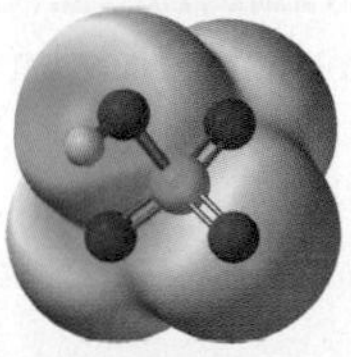	$HClO_4(aq) + H_2O(\ell) \rightarrow ClO_4^-(aq) + H_3O^+(aq)$
Sulfuric	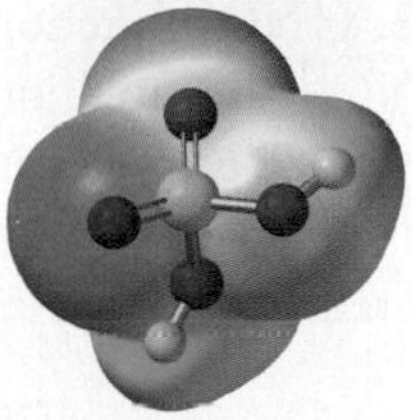	$H_2SO_4(aq) + H_2O(\ell) \rightarrow HSO_4^-(aq) + H_3O^+(aq)$ $HSO_4^-(aq) + H_2O(\ell) \rightleftharpoons SO_4^{2-}(aq) + H_3O^+(aq)$

▶❚❚ **CHEMTOUR** Acid−Base Ionization

H_mXO_n acid, the negative charge is distributed over several atoms of electronegative elements. The anions that are formed in these ionization reactions are among the most common ones found in nature and are chemically stable. As we have seen in previous chapters, reactions that produce stable (low energy) products are favored over those that don't.

Ionization of a strong acid (actually *any* acid) in water is also facilitated by strong dipole–dipole interactions between its ionizable H atoms and the O atoms of H_2O molecules. Lone pairs of electrons on these O atoms become bonding pairs as polar covalent O—H bonds form between H^+ ions and O atoms and molecules of H_2O are transformed into H_3O^+ ions (Figure 15.2).

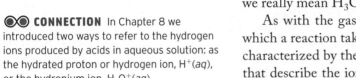

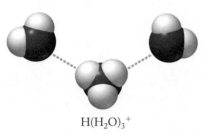

FIGURE 15.2 When hydrogen chloride dissolves in water, HCl ionizes and forms Cl^- and H_3O^+ ions.

The positive charge of each H_3O^+ ion attracts the partial negative charges of the oxygen atoms of nearby H_2O molecules. These ion–dipole interactions create clusters of water molecules around each hydronium ion, as shown in Figure 15.3. As we describe concentrations of H^+ ions in acid–base reactions, remember that we really mean H_3O^+ ions with water molecules clustered around them.

⊙⊙ **CONNECTION** In Chapter 8 we introduced two ways to refer to the hydrogen ions produced by acids in aqueous solution: as the hydrated proton or hydrogen ion, $H^+(aq)$, or the hydronium ion, $H_3O^+(aq)$.

As with the gas-phase equilibria we discussed in Chapter 14, the extent to which a reaction takes place before it reaches equilibrium in the aqueous phase is characterized by the value of its equilibrium constant. The equilibrium constants that describe the ionization of acids in water are given the symbol K_a. They are concentration-based equilibrium constants, but the subscript "c" is replaced with "a" to indicate that the equilibrium involves the ionization of an *acid*. Table 15.2 lists some common weak acids, their ionization reactions in water, and their K_a values at 25°C. Note that the K_a values of these weak acids are all much less than 1. These small values contrast with the large K_a values of the strong acids in Table 15.1, which are often expressed "$K_a \gg 1$" to indicate that ionization of the acid is essentially complete in aqueous solutions.

Conjugate Acid–Base Pairs

In the chemical equilibrium in Equation 15.2, HNO_2 functions as a Brønsted–Lowry acid in the forward reaction and the NO_2^- ion functions as a Brønsted–Lowry base in the reverse reaction. The difference between these two species is the H^+ ion that the molecule of HNO_2 gives away when it forms an NO_2^- ion. Chemists call an acid and a base that are related in this way a **conjugate acid–**

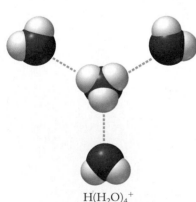

$H(H_2O)_3^+$

$H(H_2O)_4^+$

FIGURE 15.3 Water molecules cluster around hydronium ions, forming species with the general formula $H(H_2O)_n^+$.

TABLE 15.2	Some Common Weak Acids and Their Ionization Reactions in Water	
Weak Acid	**Reaction in Water**	K_a
Acetic	$CH_3COOH(aq) + H_2O(\ell) \rightleftharpoons CH_3COO^-(aq) + H_3O^+(aq)$	1.8×10^{-5}
Formic	$HCOOH(aq) + H_2O(\ell) \rightleftharpoons HCOO^-(aq) + H_3O^+(aq)$	1.8×10^{-4}
Hydrofluoric	$HF(aq) + H_2O(\ell) \rightleftharpoons F^-(aq) + H_3O^+(aq)$	6.8×10^{-4}
Hypochlorous	$HClO(aq) + H_2O(\ell) \rightleftharpoons ClO^-(aq) + H_3O^+(aq)$	2.9×10^{-8}
Nitrous	$HNO_2(aq) + H_2O(\ell) \rightleftharpoons NO_2^-(aq) + H_3O^+(aq)$	4.0×10^{-4}

base pair. An acid forms its **conjugate base** when it loses an H^+ ion, and every base (like NH_3) forms its **conjugate acid** (in this case, NH_4^+) when it gains an H^+ ion. This relationship is expressed in these simple chemical equations:

$$HNO_2 + H_2O \rightleftharpoons NO_2^- + H_3O^+ \qquad NH_3 + H_2O \rightleftharpoons NH_4^+ + OH^-$$

Acid · · · · · · · · · · Conjugate · · · · · · · · · · · Base · · · · · · · · · · · · · Conjugate
· · · · · · · · · · · · · · · · base · acid

The corresponding general equations for equilibria in aqueous solutions are

$$\text{Acid}(aq) + H_2O(\ell) \rightleftharpoons \text{conjugate base}(aq) + H_3O^+(aq) \qquad (15.3)$$

$$\text{Base}(aq) + H_2O(\ell) \rightleftharpoons \text{conjugate acid}(aq) + OH^-(aq) \qquad (15.4)$$

Water and the hydronium ion form another conjugate acid−base pair: H_2O is the base and H_3O^+ is its conjugate acid in the reaction with nitrous acid; H_2O is the acid and OH^- is its conjugate base in the reaction with ammonia.

> **conjugate acid−base pair** a Brønsted−Lowry acid and base differing from each other only by the presence or absence of a H^+ ion: acid $\rightleftharpoons$ conjugate base + H^+.
>
> **conjugate base** formed when a Brønsted−Lowry acid donates a H^+ ion.
>
> **conjugate acid** formed when a Brønsted−Lowry base accepts a H^+ ion.

SAMPLE EXERCISE 15.1 **Identifying Conjugate Acid−Base Pairs** **LO1**

Identify the conjugate acid−base pairs in the reactions that result when perchloric acid, $HClO_4$, and formic acid, $HCOOH$, dissolve in water.

COLLECT AND ORGANIZE We are asked to identify the conjugate acid−base pairs that form in aqueous solutions of $HClO_4$ and $HCOOH$. We know from Table 15.1 that perchloric acid is a strong acid and from Table 15.2 that formic acid is a weak acid.

ANALYZE Acids in aqueous solutions form their conjugate bases by donating H^+ ions to molecules of H_2O as described by Equation 15.3. Therefore, the formulas of their conjugate bases are the formulas of the original acids minus a H^+ ion. The formula of perchloric acid has only one H atom in it, which must be the one that it loses as a H^+ ion. The formula of formic acid has two H atoms. The ionizable one is bonded to an O atom in the carboxylic acid group (H atoms bonded to C atoms are not acidic).

SOLVE Rewriting Equation 15.3 for aqueous solutions of the two acids, we have

Perchloric acid: $\qquad HClO_4(aq) + H_2O(\ell) \rightarrow ClO_4^-(aq) + H_3O^+(aq)$
$\qquad\qquad\qquad\qquad\qquad$ Acid $\qquad\qquad\qquad$ Conjugate base

Formic acid: $\qquad HCOOH(aq) + H_2O(\ell) \rightleftharpoons HCOO^-(aq) + H_3O^+(aq)$
$\qquad\qquad\qquad\qquad\quad$ Acid $\qquad\qquad\qquad$ Conjugate base

In both reactions H_3O^+ and H_2O are also a conjugate acid−base pair; H_2O is the base, and H_3O^+ is its conjugate acid.

THINK ABOUT IT A single arrow is used in the $HClO_4$ equation because perchloric acid is a strong acid that ionizes completely in water. Equilibrium arrows are used for $HCOOH$ because formic acid is a weak acid and a significant concentration of $HCOOH$ is usually present in solution at equilibrium.

Practice Exercise Identify the conjugate acid−base pairs in the reaction that takes place when the weak organic acid acetic acid (CH_3COOH) dissolves in water. ⚙

(Answers to Practice Exercises are in the back of the book.)

TABLE 15.3 Strong Bases and Their Ionization Reactions in Water

Strong Base	Reaction in Water
Lithium hydroxide	$LiOH(aq) \rightarrow Li^+(aq) + OH^-(aq)$
Sodium hydroxide	$NaOH(aq) \rightarrow Na^+(aq) + OH^-(aq)$
Potassium hydroxide	$KOH(aq) \rightarrow K^+(aq) + OH^-(aq)$
Calcium hydroxide	$Ca(OH)_2(aq) \rightarrow Ca^{2+}(aq) + 2\,OH^-(aq)$
Barium hydroxide	$Ba(OH)_2(aq) \rightarrow Ba^{2+}(aq) + 2\,OH^-(aq)$
Strontium hydroxide	$Sr(OH)_2(aq) \rightarrow Sr^{2+}(aq) + 2\,OH^-(aq)$

∞ CONNECTION Magnesium hydroxide is considered a weak base because of its limited solubility in water (see Table 8.3 in Chapter 8).

Strong and Weak Bases

The most common strong bases are the soluble hydroxides of group 1 and 2 metals. Table 15.3 shows how these ionic compounds dissociate when they dissolve in water. Their corresponding equilibrium constants all have values much greater than one ($K_b \gg 1$, where the "b" subscript indicates that the reactant functions as a *base*). The hydroxide ions produced when these bases dissolve in water are very effective H^+ acceptors and so these compounds are strong Brønsted–Lowry bases.

The Brønsted–Lowry model can also be used to explain what happens when a weak base (Table 15.4) dissolves in water. Let's use ammonia as an example. In aqueous solution, NH_3 molecules accept H^+ ions from molecules of water to form NH_4^+ and OH^- ions:

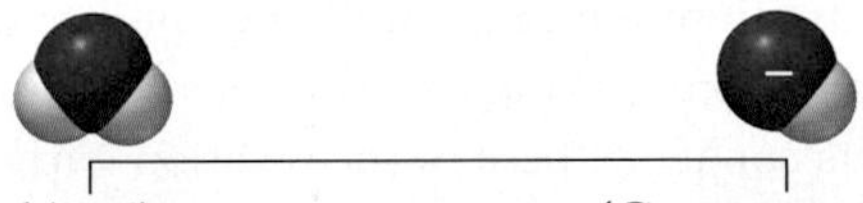

(Acid)   (Conjugate base)

$$NH_3(aq) + H_2O(\ell) \rightleftharpoons NH_4^+(aq) + OH^-(aq) \qquad (15.5)$$

(Base)   (Conjugate acid)

Note how the NH_4^+ ion is the conjugate acid of NH_3, a relationship that may be clearer if we consider the reaction that occurs when an ammonium salt dissolves in water:

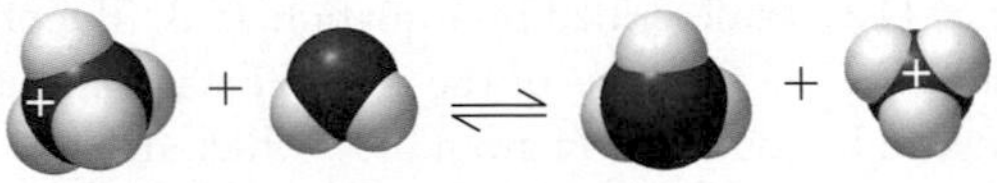

$$NH_4^+(aq) + H_2O(\ell) \rightleftharpoons NH_3(aq) + H_3O^+(aq) \qquad (15.6)$$

Here NH_4^+, acting like an acid, donates a H^+ ion and thereby forms NH_3, its conjugate base. Also note that water molecules are proton donors in Equation 15.5, making H_2O a Brønsted–Lowry acid in that reaction. However, water molecules are proton acceptors in Equation 15.6, making water a Brønsted–Lowry base. We revisit the acid–base duality of water throughout this chapter.

TABLE 15.4 Some Common Weak Bases and Their Ionization Reactions in Water

Weak Base	Reaction in Water	K_b
Ammonia	$NH_3(aq) + H_2O(\ell) \rightleftharpoons NH_4^+(aq) + OH^-(aq)$	1.8×10^{-5}
Aniline	$C_6H_5NH_2(aq) + H_2O(\ell) \rightleftharpoons C_6H_5NH_3^+(aq) + OH^-(aq)$	4.0×10^{-10}
Dimethylamine	$(CH_3)_2NH(aq) + H_2O(\ell) \rightleftharpoons (CH_3)_2NH_2^+(aq) + OH^-(aq)$	5.9×10^{-4}
Methylamine	$CH_3NH_2(aq) + H_2O(\ell) \rightleftharpoons CH_3NH_3^+(aq) + OH^-(aq)$	4.4×10^{-4}
Pyridine	$C_5H_5N(aq) + H_2O(\ell) \rightleftharpoons C_5H_5NH^+(aq) + OH^-(aq)$	1.7×10^{-9}

Relative Strengths of Acids and Bases

We can use the concepts we have developed for describing equilibria to evaluate the relative strengths of acids and bases. To do so, let's examine the ionization of HCl:

$$HCl(aq) + H_2O(\ell) \rightarrow Cl^-(aq) + H_3O^+(aq)$$

Because HCl is a strong acid, this reaction goes to completion, which means that a Cl^- ion (the conjugate base of HCl) must be a very weak base with little tendency to accept a proton from H_3O^+. The contrast in relative strengths applies to all conjugate pairs: strong acids have very weak conjugate bases and strong bases have very weak conjugate acids (Figure 15.4).

Between these extremes are many substances of intermediate strength: weak acids with weak conjugate bases. For example, HNO_2 is a weak acid ($K_a = 4.0 \times 10^{-4}$), which means that its conjugate base, NO_2^-, is a weak base.

All the strong acids in Figure 15.4 ionize completely in water. The H_2O molecules in their solutions readily accept H^+ ions from 100% of the acid molecules. In this context, water is said to *level* the strengths of these acids; they all are equally strong acids because they cannot be more than 100% ionized. This **leveling effect** means that the conjugate acid of H_2O, H_3O^+, is the strongest H^+ donor that can exist in water. An even stronger acid, no matter how much stronger it is, simply donates all its ionizable H atoms to water molecules, forming H_3O^+ ions.

On the other hand, weak acids are differentiated by their ability to donate their ionizable H atoms to water molecules. The weak acids higher on the list in Figure 15.4 form more acidic aqueous solutions than acids lower on the list.

A similar pattern is evident in the strengths of bases. The strongest base that can exist in water is the conjugate base of H_2O, which is the OH^- ion. Any base that is stronger than OH^- hydrolyzes in water to produce OH^- ions. The oxide ion (O^{2-}), for example, is a very strong base, and reacts with water to produce pairs of OH^- ions:

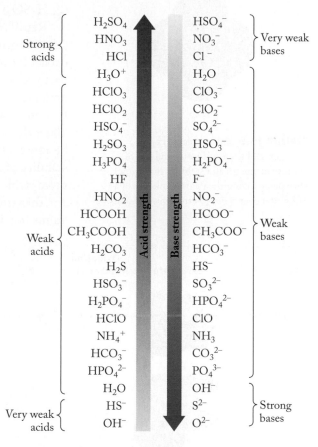

FIGURE 15.4 Opposing trends characterize the relative strengths of acids and their conjugate bases: the stronger the acid, the weaker its conjugate base. The same is true for bases: the stronger the base, the weaker its conjugate acid.

$$O^{2-}(aq) + H_2O(\ell) \longrightarrow 2\,OH^-(aq)$$

The strengths of bases weaker than OH^- ions can be differentiated by the fraction of their molecules that accept H^+ ions from water molecules in aqueous solutions. Weaker bases are higher on the list in Figure 15.4; stronger bases are lower on the list.

leveling effect the observation that strong acids all have the same strength in water and are completely converted into solutions of H_3O^+ ions; strong bases are likewise leveled in water and are completely converted into solutions of OH^- ions.

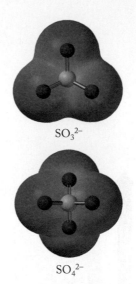

SO_3^{2-}

SO_4^{2-}

FIGURE 15.5 Sulfuric acid (H_2SO_4) is a stronger acid than sulfurous acid (H_2SO_3) because of the greater stability that comes with delocalizing the negative charge of a SO_4^{2-} ion over more electronegative atoms.

15.2 Acid Strength and Molecular Structure

Why are some acids and bases strong, but most are weak? What, for example, makes sulfuric acid (H_2SO_4) a strong acid, but sulfurous acid (H_2SO_3) a weak one? The answer can be found in subtle differences in molecular structure (Figure 15.5). The ionizable hydrogen atoms in both molecules are bonded to oxygen atoms that are also bonded to the central sulfur atoms. The difference between the two is that the central sulfur atom is also bonded to either one (in H_2SO_3) or two (in H_2SO_4) additional oxygen atoms.

Recall from Chapter 4 that oxygen is the second most electronegative element (after fluorine). This means that oxygen atoms bonded to the central atom of an oxoacid attract electron density toward themselves. The more electron density that is drawn away from the hydrogen ends of O—H bonds, the more polar the bonds. As a result, the hydrogen atom is more readily released as a H^+ ion to the O atom of a water molecule. In addition, more O atoms bonded to a central atom allow the negative charge on the anion that is formed when a H^+ ion leaves to be more spread out or *delocalized*. Charge delocalization increases the chemical stability of ions and makes their formation more favorable. Thus, SO_4^{2-} ions are more stable than SO_3^{2-} ions, which makes H_2SO_4 a stronger acid than H_2SO_3.

This trend of increasing acid strength with increasing numbers of oxygen atoms bonded to the central atom (that is, with increasing oxidation number of the central atom) is true for all oxoacids with the same central atom. The trend is illustrated by the strong acidity of HNO_3 and the weak acidity of HNO_2, and by the strengths of the oxoacids of chlorine, shown in Figure 15.6.

The strength of an oxoacid is also related to the electron-withdrawing power of the central atom. Consider, for example, the relative strengths of the three hypohalous acids in Figure 15.7. The most electronegative of the three halogen atoms (Cl) has the

Acid	Structure	Oxidation Number of Cl	K_a
Hypochlorous HClO	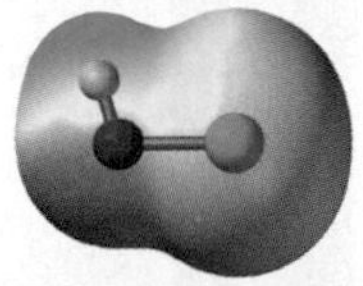	+1	2.9×10^{-8}
Chlorous HClO$_2$	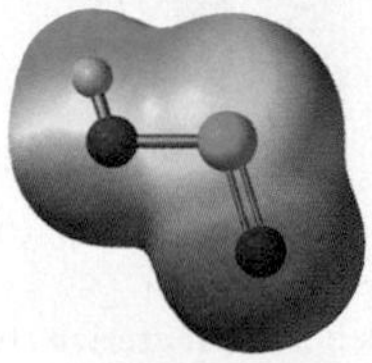	+3	1.1×10^{-2}
Chloric HClO$_3$	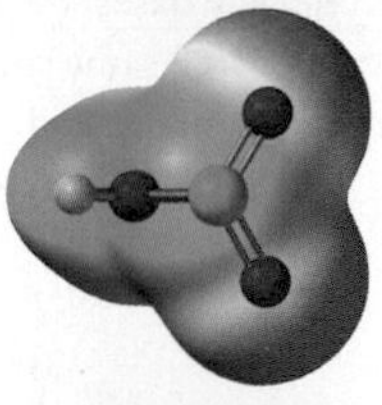	+5	~1
Perchloric HClO$_4$	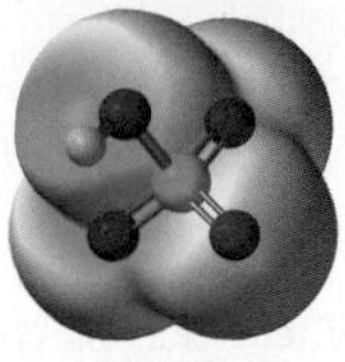	+7	Strong acid

FIGURE 15.6 In the oxoacids of chlorine, acid strength increases with increasing Cl oxidation number. The higher the oxidation number, the greater the number of O atoms bonded to the Cl. The greater the number of O atoms bonded to Cl, the greater the ability to delocalize the negative charge on the anion created when each acid loses its H.

Acid	Structure	Electronegativity of Halogen Atom	K_a
Hypochlorous HClO	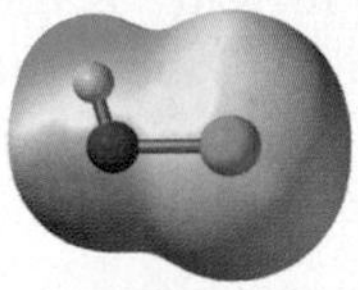	3.0	2.9×10^{-8}
Hypobromous HBrO	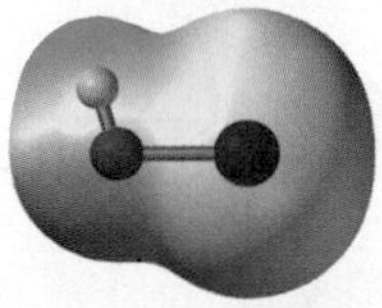	2.8	2.3×10^{-9}
Hypoiodous HIO	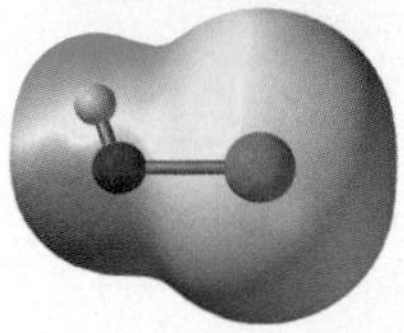	2.5	2.3×10^{-11}

FIGURE 15.7 The strengths of these three hypohalous acids are related to the electronegativities of their halogen atoms. The more electronegative the halogen atom, the more it pulls electron density away from the hydrogen end of the molecule. The result is a more stable anion and a stronger acid.

greatest attraction for the pair of electrons it shares with oxygen. This attraction draws electron density away from hydrogen toward chlorine and toward the oxygen end of the already polar O—H bond. Thus, HClO(*aq*) is the strongest of the three acids, followed by hypobromous acid [HBrO(*aq*)] and hypoiodous acid [HIO(*aq*)].

CONNECTION The electronegativities of the elements are given in Figure 4.8 in Chapter 4.

CONCEPT TEST

Rank the following compounds in order of decreasing acid strength: H_3AsO_4, H_3BiO_4, H_3PO_4, and H_3SbO_4.

▶❙❙ CHEMTOUR Acid Strength and Molecular Structure

(Answers to Concept Tests are in the back of the book.)

15.3 pH and the Autoionization of Water

We have seen that acidity is directly related to the concentration of H_3O^+ ions in a particular solution. In this section we examine another way to express acidity. To understand this alternative we need to first understand the **autoionization** of water, which is the process in which one water molecule, acting as an acid, donates a hydrogen ion to another, which acts as a base:

▶❙❙ CHEMTOUR Autoionization of Water

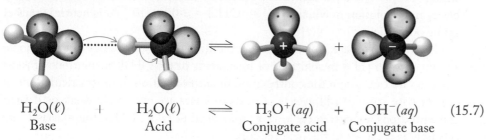

$$H_2O(\ell) \quad + \quad H_2O(\ell) \quad \rightleftharpoons \quad H_3O^+(aq) \quad + \quad OH^-(aq) \qquad (15.7)$$

Base   Acid   Conjugate acid   Conjugate base

The donor H_2O forms its conjugate base (OH^-), and the acceptor H_2O forms its conjugate acid (H_3O^+). We have already encountered this dual nature of water: molecules of H_2O act as H^+ ion acceptors in solutions of acidic solutes, and as H^+ ion donors in solutions of basic solutes. As we discussed in Chapter 8, any substance that can act as either an acid or a base is said to be amphiprotic. The autoionization of water is an example of amphiprotic behavior.

CONNECTION We defined *amphiprotic* compounds in Chapter 8 as having both acidic and basic properties.

The equilibrium constant for the autoionization of water described in Equation 15.7 is:

$$K_c = \frac{[H_3O^+][OH^-]}{[H_2O][H_2O]} \qquad (15.8)$$

We can simplify Equation 15.8 by first recalling an important concept from Chapter 14: mass action expressions do not contain terms for pure solids and liquids. This allows us to simplify Equation 15.8:

CONNECTION In Chapter 14, we learned that terms for pure solids or liquids in equilibrium constant expressions are equal to 1 and do not appear in K_c and K_p expressions.

$$K_c = [H_3O^+][OH^-]$$

Frequently this expression is simplified even more by substituting H^+ for H_3O^+. This simplification is done with the understanding that H^+ ions in aqueous solutions do not exist as lone protons; rather, they are bonded to molecules of water. The result of this simplification is a K_c expression so important to aqueous equilibria that it merits its own symbol, K_w:

autoionization the process that produces equal and very small concentrations of H_3O^+ and OH^- ions in pure water.

$$K_w = [H^+][OH^-] \qquad (15.9)$$

In pure water at 25°C, $[H^+] = [OH^-] = 1.00 \times 10^{-7}\ M$. Inserting these values into Equation 15.9 gives

$$K_w = [H^+][OH^-] = (1.00 \times 10^{-7})(1.00 \times 10^{-7}) = 1.00 \times 10^{-14} \quad (15.10)$$

Such a tiny value of K_w confirms that a very tiny fraction of water molecules undergoes autoionization. The reverse of autoionization—the reaction between $[H^+]$ and $[OH^-]$ to produce H_2O—has an equilibrium constant of $1/K_w = 1.00 \times 10^{14}$ and essentially goes to completion:

$$H^+(aq) + OH^-(aq) \rightleftharpoons H_2O(\ell) \qquad K = 1/K_w = 1.00 \times 10^{14}$$

The value $K_w = 1.00 \times 10^{-14}$ applies to all aqueous solutions at 25°C, and we will assume this temperature for all aqueous solutions we explore in this chapter. We will not include $[H_2O]$ terms in K_c expressions for the reason we did not include them in the K_w expression, and because water in these aqueous solutions has *nearly* the concentration of pure water and does not change significantly as a result of interacting with the solutes dissolved in it.

Equation 15.10 contains an important message about the values of $[H^+]$ and $[OH^-]$ in any aqueous sample: as the value of one increases, the value of the other must decrease so that the product of the two is always 1.00×10^{-14}. A solution in which $[H^+] > [OH^-]$ is considered acidic, a solution in which $[H^+] < [OH^-]$ is basic, and a solution in which $[H^+] = [OH^-] = 1.00 \times 10^{-7}\ M$ is neutral (neither acidic nor basic).

The tiny value of K_w means that autoionization of water does not contribute significantly to $[H^+]$ in solutions of most acids or to $[OH^-]$ in solutions of most bases, so we can ignore the contribution of autoionization in most calculations of acid or base strength. However, if acids or bases are extremely weak or if their concentrations are extremely low, we may need to take H_2O autoionization into account, as we'll see in the next section.

The pH Scale

In the early 1900s, scientists developed a device called the *hydrogen electrode* to determine the $[H^+]$ of solutions. The electrical voltage, or *potential*, produced by the hydrogen electrode is a linear function of the logarithm of $[H^+]$. This relation led Danish biochemist Søren Sørensen (1868–1939) to propose a scale for expressing acidity and basicity based on what he termed "the *potential* of the hydrogen ion," abbreviated **pH**. Mathematically, we define pH as the negative logarithm of $[H^+]$:

$$pH = -\log[H^+] \qquad (15.11)$$

For example, the pH of a solution in which $[H^+] = 5.0 \times 10^{-3}\ M$ is

$$pH = -\log(5.0 \times 10^{-3}) = -(-2.30) = 2.30$$

The negative sign in front of the logarithmic term means that most pH values, except for concentrated solutions of strong acids or bases, are positive numbers between 0 and 14. It also means that *large pH values* correspond to *small values of $[H^+]$*. Acidic solutions have pH values less than 7.00 ($[H^+] > 1.00 \times 10^{-7}\ M$), and basic solutions have pH values greater than 7.00 ($[H^+] < 1.00 \times 10^{-7}\ M$). A solution with a pH of exactly 7 is neutral. The pH values for some common aqueous solutions are shown in Figure 15.8.

The pH scale has several attractive features. Because it is logarithmic, there are no exponents, as are commonly encountered in values of $[H^+]$. The logarithmic

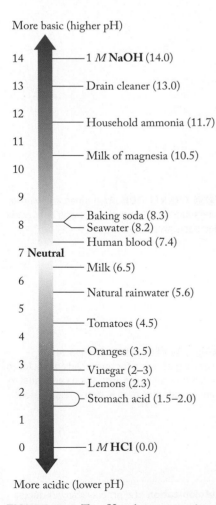

More basic (higher pH)

14	1 M **NaOH** (14.0)
13	Drain cleaner (13.0)
12	Household ammonia (11.7)
11	
10	Milk of magnesia (10.5)
9	
8	Baking soda (8.3) Seawater (8.2) Human blood (7.4)
7 Neutral	
6	Milk (6.5)
5	Natural rainwater (5.6)
4	Tomatoes (4.5)
3	Oranges (3.5) Vinegar (2–3)
2	Lemons (2.3) Stomach acid (1.5–2.0)
1	
0	1 M **HCl** (0.0)

More acidic (lower pH)

FIGURE 15.8 The pH scale is a convenient way to express the range of acidic or basic properties of some common materials.

scale also means that a change of one pH unit corresponds to a 10-fold change in $[H^+]$, so that a solution with a pH of 5.0 has 10 times the $[H^+]$ of a solution with a pH of 6.0 and is 10 times as acidic. Similarly, a solution with a pH of 10.0 has 1/100 the $[H^+]$, or 100 times the $[OH^-]$, as a solution with a pH of 8.0 and is 100 times more basic.

pH the negative logarithm of the hydrogen ion concentration in an aqueous solution.

CONCEPT TEST

Suppose solution A has a pH of 6.0 and solution B has a pH of 7.0. Which of the following statements about the two solutions is/are true?

a. Solution A is 10 times more acidic than solution B.
b. Solution B is neither acidic nor basic.
c. The concentration of OH^- ions in solution B is 10 times their concentration in solution A.
d. $[OH^-] = [H^+]$ in solution B.
e. The value of $[H^+]$ in solution A is 10 times that in solution B.

▶❚❚ **CHEMTOUR** pH Scale

A note about expressing pH values to the appropriate number of significant figures is necessary. Because any pH value is the negative logarithm of the hydrogen ion concentration, the first number in the value defines the location of the decimal point in the concentration term. As such, it is not considered when determining the number of significant figures. For example, a hydrogen ion concentration of $2.7 \times 10^{-4}\ M$ has two significant figures in the coefficient of the power of 10. The corresponding pH value with two significant figures is 3.57. The "3" in pH 3.57 is not considered a significant figure because it just means that the $[H^+]$ is between $1.0 \times 10^{-3}\ M$ and $1.0 \times 10^{-4}\ M$. It is worth noting that pH values are often reported to two places beyond the decimal point, which reflects the precision of many pH meters.

SAMPLE EXERCISE 15.2   Interconverting pH and [H⁺]   LO2

Is solution A with a pH of 9.58 more or less acidic than solution B in which $[H^+] = 4.3 \times 10^{-10}\ M$?

COLLECT AND ORGANIZE We are asked to compare the acidities of two solutions. We know the pH of one and the $[H^+]$ of the other. These two parameters are related by Equation 15.11:

$$pH = -\log[H^+]$$

ANALYZE We can either convert the given pH value into the corresponding $[H^+]$ value, or convert the given $[H^+]$ value into pH. Doing both will provide us with a check on our calculations. The given $[H^+]$ value is between 10^{-9} and $10^{-10}\ M$, which means that the corresponding pH value will have a 9 in front of the decimal place.

SOLVE pH of solution B:

$$pH = -\log(4.3 \times 10^{-10})$$

$$= -(-9.37) = 9.37$$

The pH of solution B (9.37) is lower than the pH of solution A (9.58). Therefore, solution A is less acidic. Alternatively, $[H^+]$ of solution A:

$$9.58 = -\log[H^+]$$

$$[H^+] = 10^{-9.58} = 2.6 \times 10^{-10}\ M$$

The [H$^+$] of solution B (4.3×10^{-10} M) is greater than that of solution A (2.6×10^{-10} M). Therefore, solution A is less acidic.

THINK ABOUT IT Our two results agree. Solution A is slightly less acidic than solution B in both calculations. Actually, both solutions have pH values greater than 7 and are weakly basic. Therefore, a better way of expressing their relative acid–base balance would be to say that solution A is slightly more basic than solution B.

Practice Exercise What is the pH of 6.92×10^{-3} M HCl?

pOH

The letter p as used in pH is also used with other symbols to mean *the negative logarithm* of the variable that follows it. For example, just as every aqueous solution has a pH value, it also has a **pOH** value, defined as

$$\text{pOH} = -\log[\text{OH}^-] \tag{15.12}$$

We can use Equation 15.10 to relate pOH to pH. We start by taking the negative logarithm of both sides of the equation:

$$K_\text{w} = [\text{H}^+][\text{OH}^-] = 1.00 \times 10^{-14}$$

$$-\log K_\text{w} = -\log([\text{H}^+][\text{OH}^-]) = -\log(1.00 \times 10^{-14})$$

$$\text{p}K_\text{w} = -(\log[\text{H}^+] + \log[\text{OH}^-]) = -(-14.00)$$

$$\text{p}K_\text{w} = \text{pH} + \text{pOH} = 14.00 \tag{15.13}$$

Many tables of equilibrium constants list pK values rather than K values at 25°C because doing so does not require the use of exponential notation and is more convenient. For example, the K_a of acetic acid is 1.8×10^{-5}; its pK_a is 4.75. The tables in Appendix 5 of this book contain both sets of values. Use them whenever you need a K or pK value that is not provided in a problem.

SAMPLE EXERCISE 15.3 **Relating [H$^+$], [OH$^-$], pH, and pOH** **LO2**

Carbonic acid that forms when atmospheric CO_2 dissolves in rainwater gives rain a normal pH of about 5.6. The pH of acid rain (Figure 15.9), however, can be one or more pH units lower than 5.6. Calculate the values of [H$^+$], pOH, and [OH$^-$] in rainwater of pH 5.6 and in a sample of acid rain with a pH of 4.3.

COLLECT AND ORGANIZE We are given pH values of two samples of rainwater and asked to determine the corresponding [H$^+$], pOH, and [OH$^-$] values. These variables are related by the following equations:

$$K_\text{w} = [\text{H}^+][\text{OH}^-] = 1.00 \times 10^{-14} \tag{15.10}$$

$$\text{pH} = -\log[\text{H}^+] \tag{15.11}$$

$$\text{pOH} = -\log[\text{OH}^-] \tag{15.12}$$

$$\text{p}K_\text{w} = \text{pH} + \text{pOH} = 14.00 \tag{15.13}$$

ANALYZE We can start with Equation 15.11 to convert pH into [H$^+$], use Equation 15.13 to convert pH into pOH, and then use Equation 15.12 to convert pOH into [OH$^-$]. Both samples are weakly acidic with pH values below 7. This means, according to Equation 15.13, that the corresponding pOH values must be greater than 7. The acid rain sample has a lower pH than natural rainwater and should have a higher [H$^+$] value, which means a smaller [OH$^-$] value and a higher pOH.

pOH the negative logarithm of the hydroxide ion concentration in an aqueous solution.

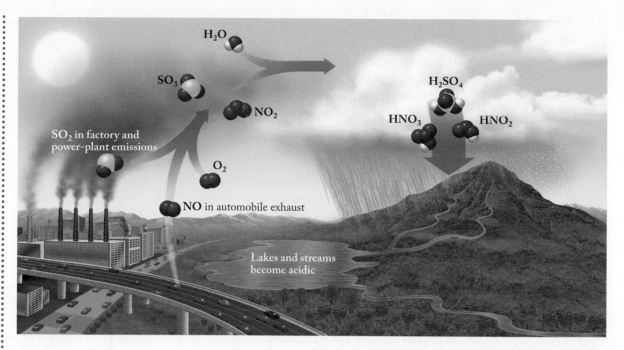

FIGURE 15.9 Acid rain forms when volatile nonmetal oxides such as NO and SO_2 are further oxidized in the atmosphere and dissolve in rain to form nitric (HNO_3), nitrous (HNO_2), and sulfuric (H_2SO_4) acids. These acids ionize, forming NO_3^-, NO_2^-, and SO_4^{2-} ions, respectively, and the hydronium (H_3O^+) ions that make rain acidic.

SOLVE For $[H^+]$,

Normal rain: $\quad pH = 5.6 = -\log[H^+]$

$\quad\quad\quad\quad\quad [H^+] = 10^{-5.6} = 3 \times 10^{-6}\ M$

Acid rain: $\quad pH = 4.3 = -\log[H^+]$

$\quad\quad\quad\quad\quad [H^+] = 10^{-4.3} = 5 \times 10^{-5}\ M$

For pOH,

$$pK_w = pH + pOH = 14.00$$

$$pOH = 14.00 - pH$$

Normal rain: $\quad pOH = 14.00 - 5.6 = 8.4$

Acid rain: $\quad pOH = 14.00 - 4.3 = 9.7$

For $[OH^-]$,

$$pOH = -\log[OH^-]$$

$$[OH^-] = 10^{-pOH}$$

Normal rain: $\quad [OH^-] = 10^{-8.4} = 4 \times 10^{-9}\ M$

Acid rain: $\quad [OH^-] = 10^{-9.7} = 2 \times 10^{-10}\ M$

THINK ABOUT IT The $[H^+]$ in the acid rain sample is nearly 20 times higher than its concentration in normal rain, and its pH value differs by 1.3 units. These differences make sense because (1) the pH scale is logarithmic, so one unit difference in pH means a 10-fold difference in $[H^+]$ or $[OH^-]$, and (2) pH values decrease as $[H^+]$ increases. The differences in pOH values make sense for the same reasons. All calculated concentration values are rounded off to only one significant figure because each starting pH value had only one significant figure—the single digit after the decimal point.

Practice Exercise What are the values of $[H^+]$ and $[OH^-]$ in household ammonia, an aqueous solution of NH_3 that has a pH of 11.7?

15.4 Calculations Involving pH, K_a, and K_b

If we know the pH and concentration of a solution of a weakly acidic or basic substance, we can calculate the acid or base equilibrium constant K_a or K_b for that substance. Of more practical importance, if we know the value of K_a or K_b we can calculate the pH of an aqueous solution of a weak acid or weak base.

Weak Acids

The vast majority of the acids on our planet are weak acids. Among them, as we saw in Section 15.1, is nitrous acid:

$$HNO_2(aq) + H_2O(\ell) \rightleftharpoons NO_2^-(aq) + H_3O^+(aq) \tag{15.2}$$

Let's begin our quantitative analysis of this equilibrium by writing the equilibrium constant expression for the reaction:

$$K_a = \frac{[NO_2^-][H^+]}{[HNO_2]}$$

A generic form of this K_a expression that applies to the ionization reaction of any acid (HA),

$$HA(aq) \rightleftharpoons A^-(aq) + H^+(aq)$$

is written

$$K_a = \frac{[A^-][H^+]}{[HA]} \tag{15.14}$$

We can use Equation 15.14 to calculate the K_a value of an unknown weak acid if we know the pH of a solution of the acid and the value of [HA]. Suppose we measure the pH of a 0.100 M solution and find that it is 2.20. To calculate K_a we then first convert the pH value to [H$^+$]:

$$pH = -\log[H^+] = 2.20$$

$$[H^+] = 10^{-2.20} = 6.3 \times 10^{-3}\ M$$

Assuming the only source of H$^+$ ions is ionization of HA, then [A$^-$] must also be $6.3 \times 10^{-3}\ M$ because the stoichiometry of the reaction is that 1 mole of HA ionizes to form 1 mole of H$^+$ and 1 mole of A$^-$. If the ionization reaction yields a [H$^+$] of $6.3 \times 10^{-3}\ M$, then [HA] must have decreased by the same amount. Therefore, at equilibrium,

$$[HA] = (0.100 - 6.3 \times 10^{-3})\ M = 0.094\ M$$

Using the three calculated equilibrium concentrations in the K_a expression,

$$K_a = \frac{[A^-][H^+]}{[HA]} = \frac{(6.3 \times 10^{-3})(6.3 \times 10^{-3})}{(0.094)} = 4.2 \times 10^{-4}$$

The small value of K_a confirms that HA is a weak acid.

The ratio of the concentration of H$^+$ ions at equilibrium to the initial concentration of HA represents the **degree of ionization** of HA, which is usually expressed as a percentage of the initial acid concentration. It is also called **percent ionization**. In equation form this relationship is

$$\text{Percent ionization} = \frac{[H^+]_{\text{equilibrium}}}{[HA]_{\text{initial}}} \times 100\% \tag{15.15}$$

degree of ionization the ratio of the quantity of a substance that is ionized to the concentration of the substance before ionization; when expressed as a percentage, also called **percent ionization**.

Inserting the data from the previous calculation into Equation 15.15 gives

$$\text{Percent ionization} = \frac{6.3 \times 10^{-3} \, M}{0.100 \, M} \times 100\% = 6.3\%$$

A plot of percent ionization as a function of initial concentration of nitrous acid is shown in Figure 15.10. This same pattern is observed for all weak acids: the degree to which they ionize increases as their concentration decreases. The reason why is linked to the fact that the acid ionization process produces two ions for every one molecule of acid consumed—and to Le Châtelier's principle. We learned in Chapter 14 that decreasing the pressure of an equilibrium mixture of gaseous reactants and products shifts the equilibrium toward the side of the reaction with more moles of gases. Similarly, diluting an equilibrium mixture of aqueous reactants and products shifts the equilibrium toward the side of the reaction with more dissolved particles. In this case, that means shifting the ionization process toward the production of more H^+ and A^- ions.

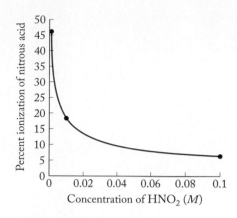

FIGURE 15.10 The degree of ionization of a weak acid increases with decreasing acid concentration. Here the degree of ionization of nitrous acid increases from about 6% in a 0.100 M solution to 18% in a 0.010 M solution to 46% in a 0.001 M solution.

SAMPLE EXERCISE 15.4 **Relating pH, K_a, and Percent Ionization of a Weak Acid** **LO3**

Formic acid (HCOOH) is a weak organic acid with this molecular structure:

$$\underset{H}{}\overset{\displaystyle O}{\underset{\displaystyle \parallel}{C}}\!\!\underset{O}{}\!\!H$$

The pH of 1.00 M HCOOH is 1.88.

a. What is the percent ionization of 1.00 M HCOOH?
b. What is the K_a value of the acid?
c. What is the percent ionization of 0.0100 M HCOOH?

COLLECT AND ORGANIZE We are asked to determine the K_a value of formic acid and its percent ionization in two solutions of known concentration. We know the pH of one of the solutions. Equation 15.11,

$$pH = -\log[H^+]$$

relates pH to $[H^+]$. Equation 15.14,

$$K_a = \frac{[A^-][H^+]}{[HA]}$$

is the generic equilibrium constant expression for a weak acid, and Equation 15.15,

$$\text{Percent ionization} = \frac{[H^+]_{\text{equilibrium}}}{[HA]_{\text{initial}}} \times 100\%$$

is the formula for calculating percent ionization.

ANALYZE We can use Equation 15.11 to convert pH to $[H^+]$ and then use Equation 15.15 to calculate percent ionization of the 1.00 M solution. The 1:1:1 stoichiometry of the ionization reaction,

$$HCOOH(aq) \rightleftharpoons HCOO^-(aq) + H^+(aq)$$

tells us that in a solution of HCOOH at equilibrium, $[HCOO^-] = [H^+]$, and $[HCOOH]$ is equal to the initial concentration of the acid minus the portion of it that ionized, or

$$[HCOOH]_{\text{equilibrium}} = [HCOOH]_{\text{initial}} - [H^+]_{\text{equilibrium}}$$

Inserting these concentration values into the equilibrium constant expression for formic acid based on Equation 15.14:

$$K_a = \frac{[HCOO^-][H^+]}{[HCOOH]}$$

enables us to calculate the value of K_a. Once we know K_a, we can use the equilibrium constant expression to calculate $[H^+]$ in any solution of formic acid, and, from $[H^+]$, the percent ionization of the acid in that solution.

The pH value of the 1.00 M solution is close to 2, which corresponds to $[H^+] = 10^{-2}\ M$. Therefore the percent ionization of formic acid in this solution should be about 1%. The more dilute solution should have a higher percent ionization.

SOLVE

a. The pH of the 1.00 M solution is 1.88. The corresponding $[H^+]$ is

$$[H^+] = 10^{-1.88} = 1.32 \times 10^{-2}\ M$$

Inserting this value and the initial concentration of HCOOH in Equation 15.15:

$$\text{Percent ionization} = \frac{[H^+]_{equilibrium}}{[HCOOH]_{initial}} \times 100\%$$

$$= \frac{1.32 \times 10^{-2}\ M}{1.00\ M} \times 100\% = 1.32\%$$

b. At equilibrium, $[HCOO^-] = [H^+] = 1.32 \times 10^{-2}\ M$, and the equilibrium concentration of HCOOH is

$$(1.00 - 1.32 \times 10^{-2})\ M = 0.99\ M$$

Inserting these values in Equation 15.14,

$$K_a = \frac{[HCOO^-][H^+]}{[HCOOH]} = \frac{(1.32 \times 10^{-2})(1.32 \times 10^{-2})}{(0.99)} = 1.76 \times 10^{-4}$$

c. To calculate $[H^+]$ in 0.0100 M HCOOH, we use a RICE table (as in the equilibrium calculations in Chapter 14) to solve for $[H^+]$ at equilibrium. Because $[H^+]$ is the unknown in the calculation, we give it the symbol x. Filling in the other cells in the RICE table:

Reaction	HCOOH(aq)	⇌	HCOO⁻(aq)	+	H⁺(aq)
	[HCOOH] (M)		**[HCOO⁻] (M)**		**[H⁺] (M)**
Initial	0.0100		0.0000		0.0000
Change	$-x$		$+x$		$+x$
Equilibrium	$0.0100 - x$		x		x

Inserting the equilibrium terms into the K_a expression and using the value of K_a from part b:

$$K_a = \frac{(x)(x)}{(0.0100 - x)} = 1.76 \times 10^{-4}$$

Solving for x using the quadratic equation or an equation-solving program:

$$x = 1.24 \times 10^{-3}\ M$$

The value of x is equal to $[H^+]$ at equilibrium. Therefore, the percent ionization of formic acid in a 0.0100 M solution is

$$\text{Percent ionization} = \frac{[H^+]_{equilibrium}}{[HCOOH]_{initial}} \times 100\% = \frac{1.24 \times 10^{-3}}{0.0100} \times 100\% = 12.4\%$$

THINK ABOUT IT The solutions to parts a and c agree with our estimates: the percent ionization of HCOOH in the 1.00 M solution (1.32%) is indeed near 1%, and the percent ionization in the 0.0100 M solution is significantly higher: 12.4%. However, we need to keep in mind that 12.4% of 0.0100 M is a lot less than 1.32% of 1.00 M, so the more dilute solution is still much less acidic. The calculated K_a value also agrees with the tabulated value for formic acid (Appendix 5).

Practice Exercise The value of [H$^+$] in a 0.050 M solution of an organic acid is 5.9×10^{-3} M. What are the pH of the solution, the percent ionization of the acid, and its K_a value?

CONCEPT TEST

The K_a values of three weak acids are given in the table to the right. Which of the three acids is the most extensively ionized in a 0.100 M solution of the acid? Which of the three acids has the lowest percent ionization in a 1.00 M solution of the acid?

Acid	K_a
A	3.6×10^{-5}
B	4.9×10^{-4}
C	9.2×10^{-4}

Weak Bases

Now let's consider what happens when a weakly basic compound dissolves in water. As noted in Section 15.1, the weak base ammonia accepts hydrogen ions from water as described by the following chemical equation:

$$NH_3(aq) + H_2O(\ell) \rightleftharpoons NH_4^+(aq) + OH^-(aq) \tag{15.5}$$

This reaction is the result of strong intermolecular forces that lead to covalent bonds breaking and new bonds forming, as shown in Figure 15.11. Not all the NH_3 molecules in an ammonia solution accept hydrogen ions; in fact most do not. Instead, the reaction reaches an equilibrium in which ammonia is present mostly as dissolved NH_3 molecules rather than NH_4^+ ions. The limited strength of ammonia as a base is reflected in its small K_b value at 25°C:

$$K_b = \frac{[NH_4^+][OH^-]}{[NH_3]} = 1.8 \times 10^{-5} \tag{15.16}$$

(As with K_a expressions, we leave out the [H$_2$O] term in this and other K_b expressions.)

We can calculate K_b if we know the initial concentration of base and the pH of the solution, or we can work in the other direction and calculate pH from a known K_b value. One additional step is necessary: the unknown in this equilibrium calculation is [OH$^-$] instead of [H$^+$], but once we know [OH$^-$] we can take its negative logarithm to calculate pOH and then use that value to calculate pH using Equation 15.13.

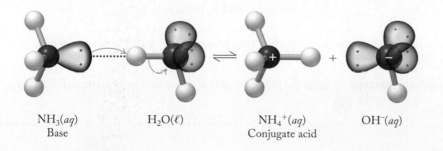

NH$_3$(aq)   H$_2$O(ℓ)   NH$_4^+$(aq)   OH$^-$(aq)
Base   Conjugate acid

FIGURE 15.11 Ammonia is considered a weak base because only a fraction of its molecules react with water, producing ammonium ions and hydroxide ions in solution. The lone pair of electrons on the nitrogen of NH_3 is shared with a transferred H$^+$ ion to make the fourth N—H bond in NH$_4^+$.

SAMPLE EXERCISE 15.5   **Calculating the pH of a Solution of a Weak Base**   **LO4**

The concentration of NH_3 in household ammonia ranges between 50 and 100 g/L, or from about 3 M to almost 6 M. What is the pH of a 3.0 M solution of NH_3?

COLLECT AND ORGANIZE We are asked to determine the pH of a 3.0 M solution of ammonia. Equation 15.5 describes the behavior of ammonia as a base in aqueous solutions, and Equation 15.16 provides the equilibrium constant expression and K_b value.

ANALYZE We can calculate the equilibrium concentration of OH^- ions using the K_b value expression and then convert $[OH^-]$ to pOH and finally to pH. Given the 1:1:1 stoichiometry of NH_3, NH_4^+, and OH^- in Equation 15.16, $[NH_4^+] = [OH^-]$ at equilibrium, and if that value is x, then the change in $[NH_3]$ during the course of the reaction is $-x$. The pH value of a fairly concentrated solution of a base with a K_b value near 10^{-5} should be well above 7 but below 14.

SOLVE We begin by setting up a RICE table based on Equations 15.5 and 15.16 and letting $[NH_4^+] = [OH^-] = x$ at equilibrium:

Reaction	$NH_3(aq) + H_2O(\ell) \rightleftharpoons$	$NH_4^+(aq)$ +	$OH^-(aq)$
	$[NH_3]$ (M)	$[NH_4^+]$ (M)	$[OH^-]$ (M)
Initial	3.0	0.0	0.0
Change	$-x$	$+x$	$+x$
Equilibrium	$3.0 - x$	x	x

Because K_b is small (1.8×10^{-5}) relative to the initial concentration of base (3.0 M), we can make the simplifying assumption that x is small compared with 3.0 M, and so $3.0 - x \approx 3.0$. With this assumption, our equilibrium constant expression is

$$K_b = \frac{[NH_4^+][OH^-]}{[NH_3]} = \frac{(x)(x)}{(3.0)} = 1.8 \times 10^{-5}$$

Solving for x gives us

$$x = [OH^-] = \sqrt{5.4 \times 10^{-5}} = 7.3 \times 10^{-3} \, M$$

Taking the negative logarithm of $[OH^-]$ to calculate pOH:

$$pOH = -\log[OH^-] = -\log(7.3 \times 10^{-3} \, M) = 2.14$$

Then we subtract this value from 14.00 to obtain the pH:

$$pH = 14.00 - pOH = 14.00 - 2.14 = 11.86$$

THINK ABOUT IT The calculated pH value falls in the range we predicted given the small K_b value and relatively high initial concentration of ammonia. To check our simplifying assumption, let's compare the value of x to $[NH_3]_{initial}$ (3.0 M):

$$\frac{7.3 \times 10^{-3}}{3.0} = 0.0024 \times 100\% = 0.24\%$$

This small percentage is acceptable, which means our simplifying assumption was justified.

Practice Exercise What is the pH of a 0.200 M solution of methylamine (CH_3NH_2, $K_b = 4.4 \times 10^{-4}$)?

pH of Very Dilute Solutions

In the pH calculations examined thus far, we have not had to consider how much the autoionization of water contributes to the concentration of H^+ or OH^-. Let's now look at a case where we do have to take this into account. Suppose we want to calculate the pH of 1.00×10^{-8} M HCl. The acid is completely ionized so that $[H^+] = 1.00 \times 10^{-8}$ and pH (from Equation 15.11) is

$$pH = -\log[H^+] = -\log(1.00 \times 10^{-8}) = 8.00$$

This answer is not reasonable—how could a solution of a strong acid, no matter how dilute, have a weakly basic pH? We expect the solution to be at least slightly acidic (pH < 7).

To calculate the pH of a solution this dilute, we must consider two sources of H^+ ions: ionization of the acid (1.00×10^{-8} M) and the autoionization of water. Let's use x to represent $[H^+]$ and $[OH^-]$ resulting from autoionization. The $[H^+]$ term in the K_w expression is the sum of x and 1.00×10^{-8} M:

$$K_w = [H^+][OH^-]$$

$$1.00 \times 10^{-14} = (x + 1.00 \times 10^{-8})(x)$$

Rearranging this equation to solve for x gives

$$x^2 + (1.00 \times 10^{-8})x - (1.00 \times 10^{-14}) = 0$$

$$x = 9.5 \times 10^{-8} \ M$$

The concentration of hydrogen ion in the solution is therefore

$$[H^+] = (1.00 \times 10^{-8} \ M) + (9.5 \times 10^{-8} \ M) = 10.5 \times 10^{-8} \ M = 1.05 \times 10^{-7} \ M$$

and the pH is

$$pH = -\log(1.05 \times 10^{-7} \ M) = 6.98$$

This value agrees with our prediction that the solution should be slightly acidic.

CONCEPT TEST ....

In the pH calculations in this chapter we routinely ignore the concentrations of H_3O^+ and OH^- ions produced by the autoionization of water. Suppose calculations of the pH of six different salt solutions that did not take autoionization of water into account produced the results shown in the table below.

Solution	A	B	C	D	E	F
pH	2.66	4.12	6.39	7.27	9.10	12.88

Which, if any, of the calculations should have taken into account the autoionization of water to obtain an accurate result?

....

15.5 Polyprotic Acids

Up to this point we have dealt with **monoprotic acids**, which have only one ionizable hydrogen atom per molecule. Acids that contain more than one ionizable hydrogen—such as sulfuric acid (H_2SO_4) and phosphoric acid (H_3PO_4)—are called **polyprotic acids**. For molecules with two and three ionizable hydrogen

monoprotic acid has one ionizable hydrogen atom per molecule.

polyprotic acid has two or more ionizable hydrogen atoms per molecule.

atoms, we use the more specific terms *diprotic acids* and *triprotic acids*, respectively. Let's consider the acidic properties of an important strong diprotic acid: sulfuric acid. Sulfuric acid is a strong acid (Table 15.1) because ionization of the first proton:

$$H_2SO_4(aq) \rightarrow HSO_4^-(aq) + H^+(aq) \tag{15.17}$$

is complete ($K_{a_1} \gg 1$). However, the second ionization is not; it has $K_{a_2} < 1$:

$$HSO_4^-(aq) \rightleftharpoons SO_4^{2-}(aq) + H^+(aq) \qquad K_{a_2} = 1.2 \times 10^{-2} \tag{15.18}$$

Note that these equilibrium constant symbols have an additional subscript, 1 or 2, corresponding to the loss of a first and then a second H^+ ion per molecule.

The combination of one complete and one incomplete ionization means that many solutions of H_2SO_4 contain more than 1 mole but less than 2 moles of $H^+(aq)$ for every mole of H_2SO_4 dissolved. Let's be more quantitative about this observation and determine the pH of a 0.100 M solution of H_2SO_4. The starting point in this calculation is a solution in which all the H_2SO_4 has ionized as described in Equation 15.17. Therefore, as the second step begins, $[HSO_4^-] = [H^+] = 0.100$ M. Ionization of HSO_4^- (Equation 15.18) produces additional H^+ ions. To analyze the effect of the second ionization, we set up a RICE table in which $+x$ is the change in $[H^+]$ produced by the second ionization. Given the 1:1:1 stoichiometry of the balanced equation for this step, the change in $[SO_4^{2-}]$ is also $+x$, and the change in $[HSO_4^-]$ is $-x$. Inserting these values in the RICE table and completing the third row,

Reaction	$HSO_4^-(aq)$	$\rightleftharpoons$	$SO_4^{2-}(aq)$	$+$	$H^+(aq)$
	$[HSO_4^-]$ (M)		$[SO_4^{2-}]$ (M)		$[H^+]$ (M)
Initial	0.100		0.000		0.100
Change	$-x$		$+x$		$+x$
Equilibrium	$0.100 - x$		x		$0.100 + x$

Inserting the equilibrium concentrations in the equilibrium constant expression for K_{a_2},

$$K_a = \frac{[SO_4^{2-}][H^+]}{[HSO_4^-]} = \frac{(x)(0.100 + x)}{(0.100 - x)} = 1.2 \times 10^{-2}$$

Solving for x using the quadratic equation or a solver program, we get

$$x = +0.010 \quad \text{or} \quad -0.12$$

The negative value for x has no physical meaning because it gives us a negative $[SO_4^{2-}]$ value. Therefore, we use only the positive x value. At equilibrium,

$$[H^+] = (0.100 + x)\ M = (0.100 + 0.010)\ M = 0.110\ M$$

The corresponding pH is

$$pH = -\log[H^+] = -\log(0.110\ M) = 0.96$$

As predicted, the value of $[H^+]$, 0.110 M, is between one and two times the initial concentration of H_2SO_4 (0.100 M). The degree of ionization of HSO_4^- is

$$\frac{[SO_4^{2-}]_{\text{equilibrium}}}{[HSO_4^-]_{\text{initial}}} = \frac{0.010\ \cancel{M}}{0.100\ \cancel{M}} \times 100\% = 10\%$$

so we were correct in our decision not to use the simplifying assumption to avoid solving a quadratic equation.

TABLE 15.5 **Ionization Equilibria for Two Diprotic Acids**

Acid	Ionization Equilibria		K_a
Carbonic acid	Step 1:	$H_2CO_3(aq) \rightleftharpoons HCO_3^-(aq) + H^+(aq)$	$K_{a_1} = 4.3 \times 10^{-7}$
	Step 2:	$HCO_3^-(aq) \rightleftharpoons CO_3^{2-}(aq) + H^+(aq)$	$K_{a_2} = 4.7 \times 10^{-11}$
Sulfurous acid	Step 1:	$H_2SO_3(aq) \rightleftharpoons HSO_3^-(aq) + H^+(aq)$	$K_{a_1} = 1.7 \times 10^{-2}$
	Step 2:	$HSO_3^-(aq) \rightleftharpoons SO_3^{2-}(aq) + H^+(aq)$	$K_{a_2} = 6.2 \times 10^{-8}$

Calculating the pH of a solution of a weak diprotic acid, such as carbonic or sulfurous acid, is actually easier than the above calculation for sulfuric acid. To see why, look closely at the K_a values in Table 15.5. In each of the two pairs of values, K_{a_2} is much smaller than K_{a_1}. We can rationalize the difference on the basis of electrostatic attractions between oppositely charged ions. The first ionization produces a negatively charged oxoanion—HCO_3^- or HSO_3^-. The second ionization requires that a positive ion (H^+) dissociate from a negative ion to produce an even more negative oxoanion. Separating oppositely charged ions that are naturally attracted to each other is not a process that we would expect to be favored, and the smaller values for K_{a_2} confirm our expectations. In general, the K_{a_2} of any diprotic acid is less, and often much less, than K_{a_1}. The consequence of these large differences is that essentially all of the limited strength of weak polyprotic acids is due to the first ionization reaction.

To see how this separation of ionization steps plays out, let's focus on carbonic acid (H_2CO_3), which is present in every drop of rain that falls from the sky. It gets there because the atmosphere is about 0.039% (by volume) CO_2. Carbon dioxide is slightly soluble in water, and when it dissolves, some of it forms carbonic acid:

$$CO_2(aq) + H_2O(\ell) \rightleftharpoons H_2CO_3(aq) \qquad K_c = 1.7 \times 10^{-3}$$

The small value of K_c tells us that most of the CO_2 that dissolves in water remains as $CO_2(aq)$; however, to simplify equilibrium calculations involving solutions of carbon dioxide, we routinely represent the total concentration of $CO_2(aq)$ and $H_2CO_3(aq)$ as $[H_2CO_3]$ even though carbonic acid is not the principal species in solution. Thus, the acidic properties of dissolved CO_2 are represented by the following acid ionization reaction:

$$H_2CO_3(aq) \rightleftharpoons HCO_3^-(aq) + H^+(aq)$$

In the following Sample Exercise we will use this equilibrium to calculate the natural pH of rainwater.

SAMPLE EXERCISE 15.6 **Calculating the pH of a Solution of a Weak Diprotic Acid** **LO4**

What is the pH of rainwater at 25°C in which atmospheric CO_2 has dissolved, producing a constant $[H_2CO_3]$ of $1.3 \times 10^{-5}\ M$?

COLLECT AND ORGANIZE We are asked to determine the pH of a dilute solution of H_2CO_3. There are two ionizable H atoms in H_2CO_3. The K_{a_1} and K_{a_2} values are given in Table 15.5. Any H_2CO_3 consumed by the reaction is replaced by the dissolution of more CO_2 so that $[H_2CO_3]$ remains a constant $1.3 \times 10^{-5}\ M$.

ANALYZE The large difference between the K_{a_1} and K_{a_2} values indicates that the pH of the solution is controlled by the first ionization equilibrium:

$$H_2CO_3(aq) \rightleftharpoons HCO_3^-(aq) + H^+(aq) \qquad K_{a_1} = 4.3 \times 10^{-7}$$

Because of the small value of K_{a_1} and the small concentration of H_2CO_3, we expect to obtain a pH value that is less than 7, but closer to 7 than to 0.

SOLVE First we set up a RICE table in which $x = [H^+] = [HCO_3^-]$ at equilibrium and the value of $[H_2CO_3]$ at equilibrium is 1.3×10^{-5} M.

Reaction	$H_2CO_3(aq)$	$\rightleftharpoons$	$HCO_3^-(aq)$	+	$H^+(aq)$
	$[H_2CO_3]$ (M)		$[HCO_3^-]$ (M)		$[H^+]$ (M)
Initial	1.3×10^{-5}		0		0
Change	0		$+x$		$+x$
Equilibrium	1.3×10^{-5}		x		x

$$K_{a_1} = \frac{[HCO_3^-][H^+]}{[H_2CO_3]} = \frac{(x)(x)}{1.3 \times 10^{-5}} = 4.3 \times 10^{-7}$$

$$x = [H^+] = 2.36 \times 10^{-6}\ M$$

Taking the negative logarithm of $[H^+]$ to calculate pH,

$$pH = -\log[H^+] = -\log(2.36 \times 10^{-6}\ M) = 5.63$$

THINK ABOUT IT Carbonic acid is a weak acid, and its concentration here is small, so obtaining a pH value that is only about 1.4 units below neutral pH (7.00) is reasonable.

Practice Exercise The pH value in Sample Exercise 15.6 is not far from 7.00, and it raises the question of whether the autoionization of water that produces a $[H^+]$ of 1.00×10^{-7} M contributes significantly to $[H^+]$ in the rainwater sample. Recalculate the pH of the rainwater sample taking into account the autoionization of water.

Some acids have three ionizable H atoms per molecule. Two important ones are phosphoric acid, H_3PO_4, and citric acid, the acid responsible for the tart flavor of citrus fruits. Note in Table 15.6 how $K_{a_1} > K_{a_2} > K_{a_3}$ for both acids. This pattern is much like that for the $K_{a_1} > K_{a_2}$ values of diprotic acids and for the same reason: it is more difficult to remove a second H^+ ion from the negatively charged ion formed after the first H^+ ion is removed, and it is even more difficult to remove a third H^+ ion from an ion with a 2− charge.

CONCEPT TEST

Do you expect the second and third acid ionization steps in phosphoric acid and citric acid to influence the pH of 0.100 M solutions of either acid?

15.6 pH of Salt Solutions

Seawater and the freshwater in many rivers and lakes have pH values that range from weakly basic to weakly acidic. How can these waters be more basic than the acidic rainwater (pH ≤ 5.6) that serves, directly or indirectly, as their water supply? The answer is that, when rain soaks into the ground, its pH changes as it flows through soils that contain basic materials. To understand the chemical processes

TABLE 15.6 Ionization Equilibria for Two Triprotic Acids

Phosphoric Acid

(1)

$K_{a_1} = 7.11 \times 10^{-3}$

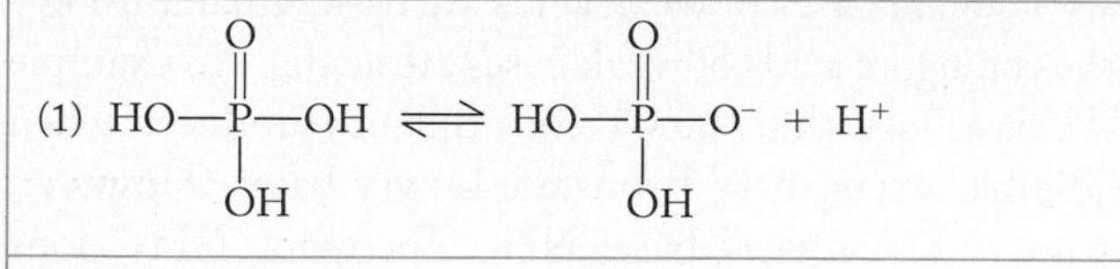

(2)

$K_{a_2} = 6.32 \times 10^{-8}$

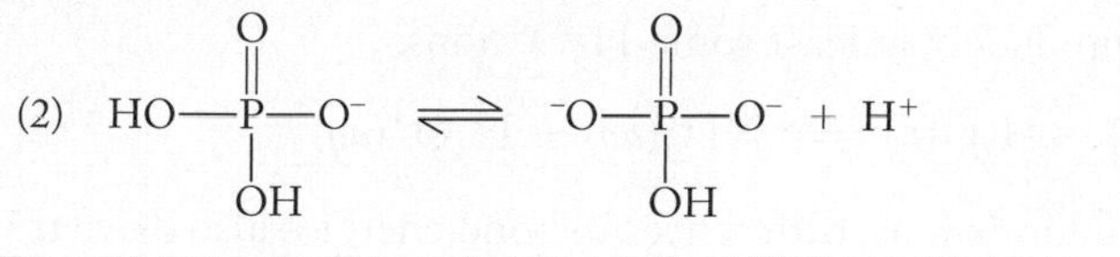

(3)

$K_{a_3} = 4.5 \times 10^{-13}$

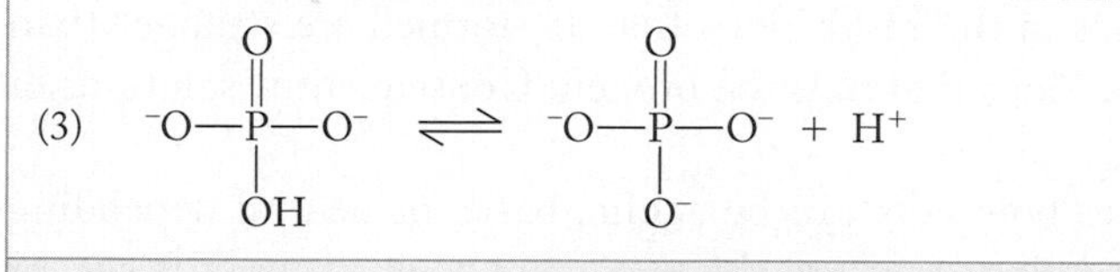

Citric Acid

(1)

$K_{a_1} = 7.44 \times 10^{-4}$

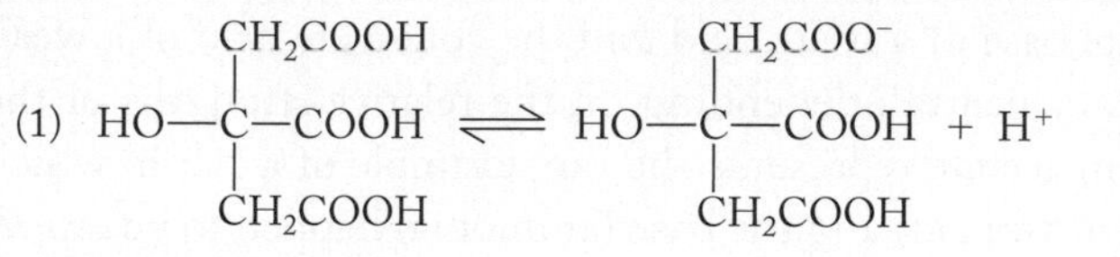

(2)

$K_{a_2} = 1.73 \times 10^{-5}$

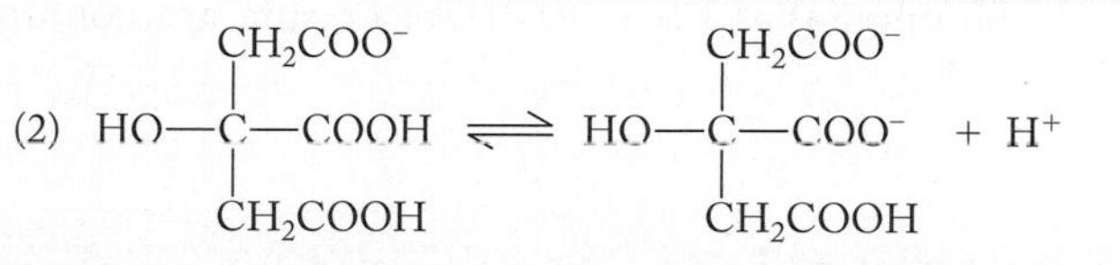

(3)

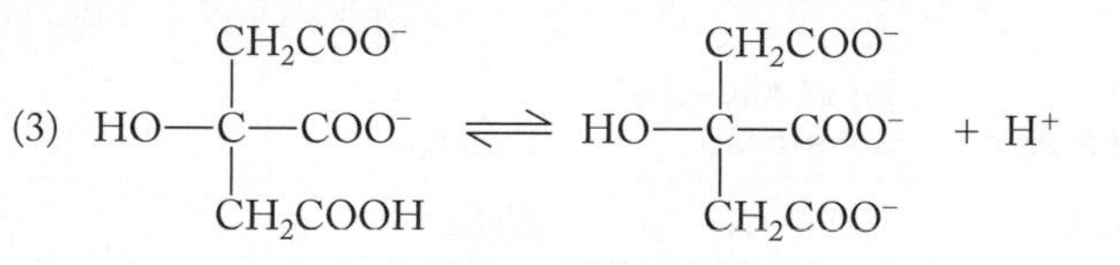

$K_{a_3} = 4.02 \times 10^{-7}$

that produce neutral or slightly basic groundwater, we first need to examine the acid–base properties of some common ionic compounds present in these waters.

As we discussed in Chapter 8, soluble ionic compounds separate into their component ions when they dissolve in water. For example, a 0.01 M solution of NaCl contains 0.01 M Na$^+$ ions and 0.01 M Cl$^-$ ions. It is also a neutral solution. Neither Na$^+$ ions nor Cl$^-$ ions react with water to form either H$_3$O$^+$ or OH$^-$ ions. Similarly, none of the other group 1 ions nor Br$^-$ or I$^-$ ions exhibit acidic or basic properties in aqueous solutions. This is because the Cl$^-$ ion is the conjugate base of a strong acid (HCl) and, therefore, must be a very weak base (see Figure 15.4).

Now let's consider another sodium salt, NaF. When it dissolves in water it produces F$^-$ ions, which, unlike the other halide ions, are the conjugate base of a *weak* acid, HF. Therefore, F$^-$ ions are weakly basic, producing at least some OH$^-$ ions when they dissolve in water:

$$F^-(aq) + H_2O(\ell) \rightleftharpoons HF(aq) + OH^-(aq)$$

A factor favoring this reaction is that it results in the formation of H—F bonds, though it requires breaking O—H bonds. As the data in Appendix 4 show, H—F bonds are stronger than O—H bonds. On the other hand, the H—X bonds of the other halogens are all weaker than the O—H bonds in H_2O, which helps explain why the other halide ions do not react with water. Solutions of NaF are weakly basic.

If salts that contain the conjugate bases of weak acids are basic, then it is logical that salts that contain the conjugate acids of weak bases are acidic. An example of such a salt is NH_4Cl. We have seen that the Cl^- ions that are produced when NH_4Cl dissolves have negligible strength as Brønsted–Lowry bases. However, NH_4^+ ions are the conjugate acid of a weak base, NH_3. Therefore, NH_4^+ ions should be weakly acidic, producing at least some H_3O^+ ions:

$$NH_4^+(aq) + H_2O(\ell) \rightleftharpoons NH_3(aq) + H_3O^+(aq)$$

As with the hydrolysis of fluoride ions, differences in bond energies also drive this reaction: the O—H bonds in the H_3O^+ ions that are formed are stronger than the N—H bonds in NH_4^+ ions that must be broken. Consequently, solutions of NH_4Cl are weakly acidic.

Table 15.7 summarizes how salts can be acidic, basic, or neutral depending on whether they include cations that are the conjugate acids of weak bases, or anions that are the conjugate bases of weak acids, or both. Note that salts that contain both the conjugate base of a weak acid *and* the conjugate acid of a weak base may be acidic, basic, or neutral, depending on the relative strengths of the acid and base. Ammonium acetate represents the rare example of a salt in which the strengths of the acid (acetic acid) and the base (ammonia) happen to be *exactly the same* [K_a (acetic acid) = K_b (ammonia) = 1.8×10^{-5}]. As a result, ammonium acetate is a neutral salt.

TABLE 15.7 Acid–Base Properties of Some Common Salts

Anion Is Derived from a	Cation Is Derived from a	pH of Aqueous Solutions	Example
Strong acid	Strong base	7	NaCl
Strong acid	Weak base	≤7	NH_4Cl
Weak acid	Strong base	≥7	NaF
Weak acid	Weak base	Depends on relative values of K_a and K_b	$K_a > K_b$, acidic; NH_4F $K_b > K_a$, basic; NH_4HCO_3 $K_a = K_b$, neutral; NH_4CH_3COO

SAMPLE EXERCISE 15.7 Distinguishing Acidic, Basic, and Neutral Salts **LO5**

Is an aqueous solution of NaClO acidic, basic, or neutral?

COLLECT AND ORGANIZE We are asked whether a solution of NaClO is acidic, basic, or neutral. When this salt dissolves in water, it dissociates into Na^+ ions and ClO^- ions.

ANALYZE Sodium ions do not hydrolyze and hence do not produce acidic solutions in water. However, ClO^- ions are the conjugate base of HClO, which is a weak acid (Table 15.2). Therefore, ClO^- ions should partially hydrolyze in water, forming OH^- ions:

$$ClO^-(aq) + H_2O(\ell) \rightleftharpoons HClO(aq) + OH^-(aq)$$

SOLVE Because hydrolysis of ClO^- ions produces OH^- ions, solutions of NaClO are weakly basic.

THINK ABOUT IT Any sodium salt containing an anion that is the conjugate base of a weak acid produces weakly basic aqueous solutions.

Practice Exercise Write a chemical equation for the hydrolysis reaction that explains why an aqueous solution of K_2SO_4 is basic.

Having established that salts can be acidic, basic, or neutral, let's now explore a strategy for calculating the pH values of their aqueous solutions. Our strategy is much like the approach we have taken to calculate the pH values of solutions of weak acids and bases. Let's start with a 0.100 M solution of sodium carbonate, Na_2CO_3. We start with a carbonate salt because the pH of many natural waters as well as biological systems is controlled by the presence of CO_3^{2-} ions and their conjugate acid, HCO_3^- ions. The $[CO_3^{2-}]$ and $[HCO_3^-]$ values are linked by the second ionization reaction of H_2CO_3 (see Table 15.5):

$$HCO_3^-(aq) + H_2O(\ell) \rightleftharpoons CO_3^{2-}(aq) + H_3O^+(aq) \quad K_{a_2} = 4.7 \times 10^{-11} \quad (15.19)$$

Carbonate and bicarbonate are also linked by the chemical reaction in which the carbonate ion acts like a Brønsted–Lowry base:

$$CO_3^{2-}(aq) + H_2O(\ell) \rightleftharpoons HCO_3^-(aq) + OH^-(aq) \quad (15.20)$$

None of the tables in this chapter or Appendix 5 contains the K_b value of the reaction in Equation 15.20, so we have to calculate it. We start with the equilibrium constant expression for the reaction:

$$K_{b_1} = \frac{[HCO_3^-][OH^-]}{[CO_3^{2-}]}$$

We have labeled this constant K_{b_1} because it describes the first of two possible reactions that produce OH^- ions. In the second, the bicarbonate ion produced in the first reaction also acts like a Brønsted–Lowry base:

$$HCO_3^-(aq) + H_2O(\ell) \rightleftharpoons H_2CO_3(aq) + OH^-(aq)$$

We give the equilibrium constant for this reaction the symbol K_{b_2}:

$$K_{b_2} = \frac{[H_2CO_3][OH^-]}{[HCO_3^-]}$$

To calculate the values of K_{b_1} and K_{b_2}, we begin with the K_{a_1} and K_{a_2} equilibrium constant expressions for H_2CO_3:

$$K_{a_1} = \frac{[HCO_3^-][H^+]}{[H_2CO_3]} = 4.3 \times 10^{-7}$$

$$K_{a_2} = \frac{[CO_3^{2-}][H^+]}{[HCO_3^-]} = 4.7 \times 10^{-11}$$

Now compare the K_{b_1} expression for the carbonate ion and the K_{a_2} expression for carbonic acid:

$$K_{b_1} = \frac{[HCO_3^-][OH^-]}{[CO_3^{2-}]} \qquad K_{a_2} = \frac{[CO_3^{2-}][H^+]}{[HCO_3^-]} = 4.7 \times 10^{-11}$$

CONNECTION We noted in Chapter 4 that *bicarbonate* is a more common name for the HCO_3^- ion than *hydrogen carbonate*.

Their similarity becomes more apparent when we write the reciprocal of the K_{a_2} expression:

$$K_{b_1} = \frac{[HCO_3^-][OH^-]}{[CO_3^{2-}]} \qquad \frac{1}{K_{a_2}} = \frac{[HCO_3^-]}{[CO_3^{2-}][H^+]} = \frac{1}{4.7 \times 10^{-11}}$$

The only differences between the two are the $[OH^-]$ term in the K_{b_1} expression and the $[H^+]$ term in the $1/K_{a_2}$ expression. As we have seen, $[H^+]$ and $[OH^-]$ are linked by K_w (Equation 15.10). Consider what happens when we multiply $1/K_{a_2}$ by K_w:

$$\left(\frac{1}{K_{a_2}}\right) \times K_w = \left(\frac{[HCO_3^-]}{[\cancel{H^+}][CO_3^{2-}]}\right)([\cancel{H^+}][OH^-]) = \left(\frac{[HCO_3^-][OH^-]}{[CO_3^{2-}]}\right)$$

The resulting expression is the K_{b_1} expression for the carbonate ion. We know the values of K_w and K_{a_2}, so we can calculate K_{b_1}:

$$K_{b_1} = \frac{K_w}{K_{a_2}} = \frac{1.0 \times 10^{-14}}{4.7 \times 10^{-11}} = 2.1 \times 10^{-4}$$

The inverse relation between the K_{a_2} of H_2CO_3 and the K_{b_1} of CO_3^{2-} also holds for the K_{a_1} of H_2CO_3 and the K_{b_2} of CO_3^{2-}. The connection between the acidic strength of H_2CO_3 and the basic strength of its conjugate base, HCO_3^-, is given by

$$K_{b_2} = \frac{K_w}{K_{a_1}} = \frac{1.0 \times 10^{-14}}{4.3 \times 10^{-7}} = 2.3 \times 10^{-8}$$

The connection between K_w, K_a, and K_b values for carbonic acid equilibria applies to any conjugate acid–base pair:

$$K_b = \frac{K_w}{K_a} \quad \text{or} \quad K_w = K_a \times K_b \tag{15.21}$$

Equation 15.21 reinforces the complementary nature of an acid and its conjugate base: as the strength (K_a) of the acid increases, the strength (K_b) of its conjugate base decreases, and vice versa, as we saw in Figure 15.4. In the carbonate and carbonic acid examples, the CO_3^{2-} ion is a stronger base ($K_{b_1} = 2.1 \times 10^{-4}$) than its conjugate acid, the HCO_3^- ion, is an acid ($K_{a_2} = 4.7 \times 10^{-11}$), but the HCO_3^- ion is a weaker base ($K_{b_2} = 2.3 \times 10^{-8}$) than its conjugate acid, H_2CO_3, is an acid ($K_{a_1} = 4.3 \times 10^{-7}$).

Now that we have a value for $K_{b_1} = 2.1 \times 10^{-4}$ for the carbonate ion, we can calculate the pH of the 0.100 M solution of Na_2CO_3. In setting up a RICE table, we assume that the only important source of OH^- is hydrolysis of the carbonate ion and that the autoionization of water does not contribute significantly to $[OH^-]$ at equilibrium. Let x be the equilibrium value of $[OH^-]$. Then $[HCO_3^-]$ is also x. The changes in the two must both be $+x$, and the change in $[CO_3^{2-}]$ must be $-x$. Completing the RICE table, we have

Reaction	$CO_3^{2-}(aq) + H_2O(\ell)$ $\rightleftharpoons$	$HCO_3^-(aq)$ +	$OH^-(aq)$
	$[CO_3^{2-}]$ (M)	$[HCO_3^-]$ (M)	$[OH^-]$ (M)
Initial	0.100	0	0
Change	$-x$	$+x$	$+x$
Equilibrium	$0.100 - x$	x	x

Solving for x,

$$K_{b_1} = 2.1 \times 10^{-4} = \frac{[HCO_3^-][OH^-]}{[CO_3^{2-}]} = \frac{(x)(x)}{(0.100 - x)}$$

$$x = 4.5 \times 10^{-3}\ M = [OH^-]$$

The calculated [OH$^-$] is much greater than [OH$^-$] in pure water ($1.0 \times 10^{-7}\ M$), so the assumption that water autoionization is unimportant in this calculation is valid. To calculate pH from [OH$^-$], we first calculate pOH:

$$pOH = -\log[OH^-] = -\log(4.5 \times 10^{-3}) = 2.35$$

and then use Equation 15.13 to calculate pH:

$$pH = pK_w - pOH = 14.00 - 2.35 = 11.65$$

A pH of 11.65 is quite basic. If you swam in a pool of water at that pH, you would experience skin irritation and painful burning in your eyes. Sodium carbonate *is* often used to adjust the pH of pools, but to make sure just the right amount is used, the pH of the water is tested to ensure that the pool is not too basic. We will see how this is done later in this chapter.

SAMPLE EXERCISE 15.8 **Calculating the pH of a Solution of an Acidic Salt** **LO4**

What is the pH of a 0.25 M solution of NH$_4$Cl?

COLLECT AND ORGANIZE We are asked to calculate the pH of a solution of NH$_4$Cl. When NH$_4$Cl dissolves in water, NH$_4^+$ and Cl$^-$ ions are released into solution. The NH$_4^+$ ion is the conjugate acid of NH$_3$; the Cl$^-$ ion is the conjugate base of HCl.

ANALYZE We have seen that the Cl$^-$ ion has negligible strength as a Brønsted–Lowry base, so it does not contribute to the acid–base properties of NH$_4$Cl. Ammonia is a weak base, which means that its conjugate acid, NH$_4^+$, is a weak acid. Therefore some of the ammonium ions in solution donate H$^+$ ions to molecules of water:

$$NH_4^+(aq) + H_2O(\ell) \rightleftharpoons NH_3(aq) + H_3O^+(aq)$$

The K_a value of NH$_4^+$ is not given in the problem and is not listed in Appendix 5. However, the K_b value of its conjugate base, NH$_3$, is in Appendix 5 (Table A5.3): 1.8×10^{-5}. The value of K_a can be calculated from K_b using Equation 15.21.

The K_b value of ammonia is close to 10^{-5}. The product of $K_a \times K_b$ of a conjugate acid–base pair is 10^{-14}; therefore, the K_a value of the ammonium ion will be close to 10^{-9}. Because the ammonium ion is a very weak acid, we can anticipate a pH value that is less than 7 but probably closer to 7 than to 0.

SOLVE The simplified K_a expression for the NH$_4^+$ ion is

$$K_a = \frac{[NH_3][H^+]}{[NH_4^+]}$$

Rearranging Equation 15.21 to solve for K_a:

$$K_a = \frac{K_w}{K_b} = \frac{1.00 \times 10^{-14}}{1.8 \times 10^{-5}} = 5.56 \times 10^{-10} = \frac{[NH_3][H^+]}{[NH_4^+]}$$

We set up a RICE table in which we make the usual assumptions that the reaction is the only significant source of H$^+$ and that $x = [H^+] = [NH_3]$ at equilibrium:

Reaction	$NH_4^+(aq) + H_2O(\ell) \rightleftharpoons$	$NH_3(aq) +$	$H_3O^+(aq)$
	$[NH_4^+]$ (M)	$[NH_3]$ (M)	$[H^+]$ (M)
Initial	0.25	0	0
Change	$-x$	$+x$	$+x$
Equilibrium	$0.25 - x$	x	x

$$K_a = 5.56 \times 10^{-10} = \frac{[NH_3][H^+]}{[NH_4^+]} = \frac{(x)(x)}{0.25 - (x)}$$

Given the very small value of K_a, we can make the simplifying assumption that $0.25\ M - x \approx 0.25\ M$, which gives us

$$\frac{x^2}{0.25} = 5.56 \times 10^{-10}$$

$$x^2 = 1.39 \times 10^{-10}$$

$$x = 1.18 \times 10^{-5} = [H^+]$$

$$pH = -\log[H^+] = -\log(1.18 \times 10^{-5}) = 4.93$$

THINK ABOUT IT This result matches our prediction: the pH of the solution is less than 7, but closer to 7 than to 0. The calculated $[H^+]$ is less than 5% of the initial concentration of NH_4^+, so our simplifying assumption was valid.

Practice Exercise What is the pH of a 0.25 M solution of sodium acetate? (*Hint*: The acetate ion is the conjugate base of acetic acid.)

The reactions of carbonate minerals in soils and rocks with acidic groundwater are critically important in mitigating the effects of acidic precipitation. When, for example, rain containing dilute sulfuric acid soaks into soil containing $CaCO_3$ in the form of limestone, marble, or shellfish shells, the acid is converted either into environmentally more benign carbonic acid,

$$CaCO_3(s) + H_2SO_4(aq) \rightleftharpoons CaSO_4(s) + H_2CO_3(aq)$$

or, if enough $CaCO_3$ is available, into calcium sulfate and soluble calcium bicarbonate:

$$2\ CaCO_3(s) + H_2SO_4(aq) \rightleftharpoons CaSO_4(s) + Ca(HCO_3)_2(aq)$$

As long as carbonates and other basic substances are present in soils and in the sediments of rivers and lakes, nature has the capacity to neutralize the acid in acid rain and maintain pH in a range that supports aquatic life. Unfortunately these minerals are scarce in the soils and sediments of some watersheds, and severe acidification and loss of aquatic life has occurred as a result.

CONNECTION The reaction of acidic groundwater with calcium carbonate and its connection to the formation of limestone caves are described in Chapter 8.

15.7 The Common-Ion Effect

Thus far in this chapter we have worked with reactions consisting of a single acidic or basic reactant. Natural systems are often more complicated than that: they typically have multiple reactants that can influence pH or resist a pH change. Suppose, for example, that a sample of river water contains $1.3 \times 10^{-5}\ M\ H_2CO_3$ as a result of atmospheric carbon dioxide dissolving in the water. Suppose also that river sediment suspended in the water contains tiny particles of solid calcium carbonate ($CaCO_3$). We have seen how $CaCO_3$ can neutralize strong acids; it can also neutralize weak acids, as shown in the following reaction that produces soluble calcium bicarbonate:

$$CaCO_3(s) + H_2CO_3(aq) \rightleftharpoons Ca(HCO_3)_2(aq)$$

Other acidic substances in the river water could also be neutralized by calcium carbonate, producing additional soluble bicarbonate salts. Suppose that the total concentration of bicarbonate ions in the river water from these reactions is 1.0×10^{-4} M. How would the presence of this much HCO_3^- affect the pH of the water, assuming that the water also contains 1.3×10^{-5} M H_2CO_3? To answer this question, we need to keep in mind that H_2CO_3 and HCO_3^- represent a conjugate acid–base pair. They are related by the first ionization of carbonic acid,

$$H_2CO_3(aq) \rightleftharpoons HCO_3^-(aq) + H^+(aq) \tag{15.22}$$

and by its equilibrium constant expression:

$$K_{a_1} = \frac{[HCO_3^-][H^+]}{[H_2CO_3]} = 4.3 \times 10^{-7}$$

Let's insert the given values of $[H_2CO_3]$ and $[HCO_3^-]$ into the K_{a_1} expression, letting $x = [H^+]$:

$$K_{a_1} = \frac{[HCO_3^-][H^+]}{[H_2CO_3]} = \frac{(1.0 \times 10^{-4})(x)}{(1.3 \times 10^{-5})} = 4.3 \times 10^{-7}$$

Solving for x:

$$x = 4.3 \times 10^{-7} \times \frac{1.3 \times 10^{-5}}{1.0 \times 10^{-4}} = 5.6 \times 10^{-8} \ M = [H^+]$$

Taking the negative logarithm to obtain pH:

$$pH = -\log[H^+] = -\log(5.6 \times 10^{-8} \ M) = 7.25$$

In Sample Exercise 15.6 we calculated that the pH of 1.3×10^{-5} M H_2CO_3 is 5.63. When H_2CO_3 is the only solute, ionization of H_2CO_3 is the only source of HCO_3^-. Dissolution of carbonate minerals provides a second source. The additional HCO_3^- causes the pH of the carbonic acid solution to increase by nearly 2 units. This increase in pH corresponds to a *decrease in [H$^+$]* of nearly 2 orders of magnitude.

Is that the sort of change we should have expected? It does make sense because $[HCO_3^-]$ in equilibrium with 1.3×10^{-5} M H_2CO_3 alone is only $10^{-5.63}$ = 2.3×10^{-6} M. By increasing $[HCO_3^-]$ to 1.0×10^{-4} M, we drive the equilibrium in Equation 15.22 to the left, as predicted by Le Châtelier's principle. The shift to the left lowers $[H^+]$ and raises the pH.

This phenomenon illustrates a principle known as the **common-ion effect**: in any equilibrium involving ions, the reaction that produces an ion is suppressed when another source of the same ion is added to the system. In the sample of river water, ionization of H_2CO_3 (Equation 15.22) is suppressed when HCO_3^- from carbonate minerals is added.

We can apply the common-ion effect to any equilibrium involving a weak acid and its conjugate base,

$$Acid(aq) \rightleftharpoons H^+(aq) + base(aq)$$

common-ion effect the shift in the position of an equilibrium caused by the addition of an ion taking part in the reaction.

which has the following equilibrium constant expression:

$$K_a = \frac{[H^+][\text{base}]}{[\text{acid}]}$$

If we take the negative logarithm of both sides of this expression, we transform $[H^+]$ into pH and K_a into pK_a:

$$pK_a = pH - \log\frac{[\text{base}]}{[\text{acid}]}$$

$$pH = pK_a + \log\frac{[\text{base}]}{[\text{acid}]} \qquad (15.23)$$

Equation 15.23 is particularly useful in calculating the pH of a solution in which there are independent sources of both an acid and its conjugate base (or a base and its conjugate acid). This equation is called the **Henderson–Hasselbalch equation**.

Consider what happens to the logarithmic term in the Henderson–Hasselbalch equation when the concentrations of the acid and base are the same. Then the numerator and denominator in the log term are equal, and the value of the fraction is 1. The log of 1 is 0, and $pH = pK_a$. This equality serves as a handy reference point in an acid–conjugate base system. If the concentration of the basic component is greater than that of the acid, the logarithmic term is greater than zero and $pH > pK_a$. If the concentration of the basic component is less than that of the acid, the logarithmic term is less than zero, and $pH < pK_a$.

Consider the case in which the concentration of the base is 10 times the concentration of the acid, that is, $[\text{base}] = 10[\text{acid}]$. Substituting this equality into Equation 15.23, we have

$$pH = pK_a + \log\frac{10[\text{acid}]}{[\text{acid}]}$$

$$= pK_a + \log 10 = pK_a + 1$$

A 10-fold higher concentration of base produces a pH one unit above the pK_a value. Similarly, if the concentration of the acid component is 10 times that of the base, then $pH = pK_a - 1$.

To demonstrate how the Henderson–Hasselbalch equation simplifies pH calculations when we know the concentrations of a weak acid and its conjugate base, let's use the equation to recalculate the pH of our sample of river water. Our starting information includes the concentrations of a weak acid, $[H_2CO_3] = 1.3 \times 10^{-5}$ M, and its conjugate base, $[HCO_3^-] = 1.0 \times 10^{-4}$ M. They are linked by the equilibrium

$$H_2CO_3(aq) \rightleftharpoons HCO_3^-(aq) + H^+(aq) \qquad K_{a_1} = 4.3 \times 10^{-7}$$

First we take the negative log of K_{a_1},

$$pK_{a_1} = -\log K_{a_1} = -\log(4.3 \times 10^{-7}) = 6.37$$

and then insert that value and the concentrations into the Henderson–Hasselbalch equation:

$$pH = pK_a + \log\frac{[\text{base}]}{[\text{acid}]}$$

$$= 6.37 + \log\frac{1.0 \times 10^{-4}}{1.3 \times 10^{-5}} = 7.26$$

$$= 7.26$$

Henderson–Hasselbalch equation used to calculate the pH of a solution in which the concentrations of an acid and conjugate base are known.

This result is within 0.01 pH unit of the value we calculated earlier.

The Henderson–Hasselbalch equation can also be used to calculate the pH of a solution of a weak base and its conjugate acid. An extra step may be involved in that calculation because we may know the value of the K_b of the base but not of the K_a of its conjugate acid, and only K_a can be used in the Henderson–Hasselbalch equation. Equation 15.21 allows us to interconvert K_a and K_b values using K_w. Interconverting pK_a and pK_b values is even easier. All we have to do is a logarithmic transformation of Equation 15.21:

$$-\log K_b = -\log(1.00 \times 10^{-14}) - (-\log K_a)$$

$$pK_b = 14.00 - pK_a$$

$$pK_b + pK_a = 14.00 \qquad (15.24)$$

Thus, converting a pK_b value into the pK_a of its conjugate acid is simply a matter of subtracting the pK_b value from 14.00.

SAMPLE EXERCISE 15.9   **Calculating the pH of a Solution of a Weak Base and Its Conjugate Acid**   **LO6**

Calculate the pH of a solution that is 0.200 M in NH_3 and 0.300 M in NH_4Cl.

COLLECT AND ORGANIZE We are asked to calculate the pH of a solution containing known concentrations of a weak base (NH_3) and a salt of its conjugate acid (NH_4^+). The Henderson–Hasselbalch equation may be used to calculate the pH of such a solution from the concentrations of the two components and the pK_a of the acid. Table A5.3 in Appendix 5 contains K_b values of common bases. The pK_a and pK_b values of a conjugate acid–base pair are related by Equation 15.24:

$$pK_b + pK_a = 14.00$$

ANALYZE Our approach involves converting the pK_b value of ammonia from Appendix 5 (4.75) into pK_a. Addition of ammonium ion to a solution of ammonia should produce a solution that is still basic, but not as basic as a solution of only ammonia.

SOLVE Inserting the value of pK_b in Equation 15.24 and solving for pK_a,

$$pK_a = 14.00 - pK_b = 14.00 - 4.75 = 9.25$$

Using this value and the given concentrations of NH_3 and NH_4^+ in Equation 15.23:

$$pH = pK_a + \log\frac{[\text{base}]}{[\text{acid}]}$$

$$= 9.25 + \log\frac{0.200}{0.300} = 9.07$$

THINK ABOUT IT We predicted a result that would be a basic pH, but not as basic as a solution of ammonia alone. To confirm this prediction, we can calculate the pH of 0.200 M NH_3 using the approach in Sample Exercise 15.5. The result is a pH of 11.27—over 2 pH units higher and more than 100 times more basic than the mixture of ammonia and ammonium chloride in this exercise.

Practice Exercise Calculate the pH of a solution that is 0.150 M in benzoic acid and 0.100 M in sodium benzoate.

15.8 pH Buffers

The presence of both carbonic acid and bicarbonate ion in river water, as discussed in the preceding section, gives the water a capacity to resist pH change when acid rain falls into it. Consider, for example, what happens when a quantity of strong acid is added to our model river water in which $[H_2CO_3] = 1.3 \times 10^{-5}$ M and $[HCO_3^-] = 1.0 \times 10^{-4}$ M. For reference purposes we will also evaluate the impact of adding the same quantity of acid to pure water (pH = 7.00). We start with 1.00 L each of river water and pure water and add 10.0 mL of 1.0×10^{-3} M HNO_3 to both.

Adding acid to pure water is an exercise in dilution, a concept introduced in Chapter 8. In this case, an initial volume of 10.0 mL (V_i) of 1.0×10^{-3} M HNO_3 (M_i) is diluted to a final volume (V_f) of 1.01 L. To calculate the final concentration (M_f) of HNO_3 (and H^+ ions), we use Equation 8.6:

▶‖ **CHEMTOUR** Buffers

$$V_i \times M_i = V_f \times M_f$$

Solving for M_f,

$$M_f = [H^+] = \frac{V_i \times M_i}{V_f} = \frac{0.0100 \text{ L} \times (1.0 \times 10^{-3} \text{ } M)}{1.01 \text{ L}} = 9.9 \times 10^{-6} \text{ } M$$

The corresponding pH is $-\log(9.9 \times 10^{-6} \text{ } M) = 5.00$. Thus, the addition of acid dropped the pH of the water by 2.00 pH units.

To calculate the change in pH when the same quantity of acid is added to 1.00 L of pH 7.25 river water, we need to focus on the carbonic acid–bicarbonate equilibrium and the pK_{a_1} of H_2CO_3:

$$H_2CO_3(aq) \rightleftharpoons HCO_3^-(aq) + H^+(aq) \qquad pK_{a_1} = 6.37 \qquad (15.25)$$

Any acid added to an equilibrium mixture of H_2CO_3 and HCO_3^- reacts with HCO_3^-, producing H_2CO_3 as the reaction runs in reverse. The number of moles of H^+ added is

$$(0.010 \text{ L}) \times \left(1.0 \times 10^{-3} \frac{\text{mol H}^+}{\text{L}} \right) = 1.0 \times 10^{-5} \text{ mol H}^+$$

This quantity of HCO_3^- in the river water sample is consumed as the reaction in Equation 15.25 runs in the reverse direction. The number of moles of bicarbonate present initially is

$$\left(1.0 \times 10^{-4} \frac{\text{mol}}{\text{L}} \right) \times (1.00 \text{ L}) = 1.0 \times 10^{-4} \text{ mol}$$

After subtracting the number of moles of HCO_3^- consumed, we have 9×10^{-5} moles left in a final volume of 1.01 L. As we have done in previous calculations, we assume that the value of $[H_2CO_3]$ is controlled by the solubility of CO_2 in water and does not change from the initial value of 1.3×10^{-5}. With this information we can calculate the pH of the river water using the Henderson–Hasselbalch equation:

$$pH = pK_a + \log\frac{[\text{base}]}{[\text{acid}]} = 6.37 + \log\left(\frac{\dfrac{9 \times 10^{-5} \text{ mol}}{1.01 \text{ L}}}{1.3 \times 10^{-5} \dfrac{\text{mol}}{\text{L}}} \right) = 7.21$$

Equal concentrations of
weak acid (⬤) and conjugate base (⬤)

FIGURE 15.12 How buffers control pH. A pH 6.50 buffer (the middle beaker) contains an equal number of moles of a weak acid and its conjugate base. After an acidic substance is added that reacts with 25% of the base, converting it into the weak acid, the pH of the buffer drops, but only by 0.22 pH units. Addition of base that consumes 25% of the weak acid raises the pH of the buffer, but again only by 0.22 units.

The original pH was 7.25. Therefore, the addition of 10.0 mL of strong acid lowered the pH by only 0.04 pH units in the river water sample—in contrast to a decrease of 2.00 pH units in pure water. River water acts as a **pH buffer**, defined as a solution that has the capacity to resist pH change by neutralizing small additions of acid or base. A buffer is typically a solution of a weak acid and its conjugate base, like the carbonic acid and bicarbonate ion in river water. Small additions of basic substances are neutralized by the weak acid and small additions of acidic substances are neutralized by its conjugate base so that, in both cases, there is little change in pH, as shown in Figure 15.12.

SAMPLE EXERCISE 15.10 **Calculating Buffer Response to Addition of Acid or Base** **LO6**

Calculate the change in pH when 1.0 mL of 1.00 M HCl is added to 100 mL of a solution that is 0.100 M sodium acetate and 0.100 M acetic acid.

COLLECT AND ORGANIZE We need to determine how much the pH of a solution of acetic acid and sodium acetate changes when a quantity of strong acid is added to it. This solution functions as a buffer. We are given the volume and composition of the buffer and of the strong acid added to it. The pH of a solution with known concentrations of a conjugate acid–base pair can be calculated using the Henderson–Hasselbalch equation:

$$pH = pK_a + \log \frac{[\text{base}]}{[\text{acid}]}$$

The pK_a value of acetic acid is 4.75 (Appendix 5).

pH buffer a solution that resists changes in pH when acids or bases are added to it; typically a solution of a weak acid and its conjugate base.

ANALYZE The pH of this buffer is controlled by the ionization equilibrium of acetic acid:

$$CH_3COOH(aq) \rightleftharpoons CH_3COO^-(aq) + H^+(aq)$$

When a strong acid is added to a solution containing acetic acid and an acetate salt, some of the CH_3COO^- ions react with the added H^+ ions, forming more CH_3COOH as the ionization reaction runs in reverse. Initially $[CH_3COOH] = [CH_3COO^-]$, which means the log term in the Henderson–Hasselbalch equation is zero and $pH = pK_a = 4.75$. Addition of a quantity of strong acid that does not consume all the acetate ion should result in a pH that is slightly lower than 4.75.

SOLVE The initial quantities of CH_3COO^- and CH_3COOH in the buffer are both

$$100 \text{ mL} \times \frac{0.100 \text{ mol}}{L} \times \frac{1 \text{ L}}{1000 \text{ mL}} = 0.0100 \text{ mol}$$

The quantity of H^+ added represents the quantity of CH_3COO^- converted into CH_3COOH:

$$1.0 \text{ mL} \times \frac{1.00 \text{ mol}}{L} \times \frac{1 \text{ L}}{1000 \text{ mL}} = 0.00100 \text{ mol}$$

The impact of this conversion is summarized in the following table:

	CH₃COOH (mol)	CH₃COO⁻ (mol)
Initial	0.0100	0.0100
Change	+0.00100	−0.00100
Final	0.0110	0.0090

Both the final quantities are in the same total volume, 101 mL. Therefore we can use the ratio of quantities in lieu of concentration values in Equation 15.23 to calculate pH. Inserting these quantities and the pK_a value of acetic acid in Equation 15.23 gives

$$pH = pK_a + \log\frac{[CH_3COO^-]}{[CH_3COOH]} = 4.75 + \log\frac{0.0090}{0.0110} = 4.66$$

The change in the pH of the solution after adding the acid is $4.75 - 4.66 = 0.09$ pH units.

THINK ABOUT IT The result is reasonable because adding strong acid produces a buffer solution with more acid and less conjugate base, so it has a slightly lower final pH (4.66) than it had initially (4.75).

Practice Exercise Calculate the change in pH when 10.0 mL of a 0.100 M solution of NaOH is added to 1.00 L of a solution that is 1.00 M in sodium acetate and 1.00 M in acetic acid. ⚙

A Physiological Buffer

Buffers are vital components of living systems because most biochemical reactions involved in life-sustaining processes, like metabolism, respiration, and transmission of nerve impulses, take place only within a narrow pH range. The pH of blood, for instance, needs to be buffered against perturbations caused by the ingestion or internal production of acidic or basic substances. One of the buffer systems the human body relies on to control pH is the same one we have discussed in the context of environmental waters: the carbonic acid–bicarbonate system. This system is intimately tied to respiration, and a key feature of pH control in this system is the role of breathing to maintain pH balance by regulating the concentration of dissolved CO_2.

Carbon dioxide is a product of cell metabolism, and one of the jobs of the blood is to carry CO_2 from cells to the lungs, where it is eliminated as we exhale. If lung function is impaired through disease or injury, the concentration of CO_2 dissolved in the blood may increase, which increases the concentration of carbonic acid, which in turn raises [H$^+$] and lowers pH. This sequence of events is summarized by shifts from left to right in the following equilibria:

$$CO_2(aq) + H_2O(\ell) \rightleftharpoons H_2CO_3(aq) \rightleftharpoons H^+(aq) + HCO_3^-(aq)$$

The resulting drop in pH below 7.35 produces a condition called *respiratory acidosis*. On the other hand, hyperventilation—fast, overly deep breathing—may cause too much CO_2 to be exhaled, shifting the above equilibria to the left and causing a decrease in [H$^+$] and a rise in pH above 7.45. The result is a condition called *respiratory alkalosis*.

Buffer Range and Capacity

We have seen how buffers resist pH change when either acid or base is added. When scientists prepare buffers to control pH in particular experiments or processes, they need to answer two questions:

1. What is the desired pH range to be maintained? The appropriate buffer is one whose weak acid has a pK_a that is within one pH unit of the desired pH. The Henderson–Hasselbalch equation tells us that over this pH range the ratio [base]/[acid] varies from 10:1 to 1:10. Expressed another way, if this condition is met:

$$0.1 < \frac{[\text{base}]}{[\text{acid}]} < 10$$

then both components are available to neutralize additions of either acid or base and to maintain the desired pH.

2. How much acid or base can the system consume without a large change in pH? The answer to this question defines the **buffer capacity**. A buffer is best able to resist changes in pH when the initial concentrations of acid and conjugate base are comparable to each other and greater than the concentration of acid or base that might be added. The greater the concentration of the buffer components, the greater the buffer capacity (Figure 15.13).

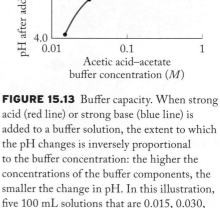

FIGURE 15.13 Buffer capacity. When strong acid (red line) or strong base (blue line) is added to a buffer solution, the extent to which the pH changes is inversely proportional to the buffer concentration: the higher the concentrations of the buffer components, the smaller the change in pH. In this illustration, five 100 mL solutions that are 0.015, 0.030, 0.100, 0.300, and 1.000 *M* acetic acid and sodium acetate all have an initial pH of 4.75. The graph shows the pH values of these solutions after 1.00 mL of 1.00 *M* HCl or 1.00 mL of 1.00 *M* NaOH has been added.

CONCEPT TEST

Select from the list in Appendix 5 a weak acid that, when mixed with the sodium salt of its conjugate base in roughly equimolar proportions, produces a buffer with an arterial blood pH of 7.41. Indicate whether the buffer will contain *exactly* the same concentrations of acid and conjugate base, or slightly more acid or base.

15.9 pH Indicators and Acid–Base Titrations

In Section 15.6 we noted that swimming pool operators routinely check the pH of the pool. They often use a test kit that includes a **pH indicator** (Figure 15.14), a substance that changes color as pH changes. One such substance is phenol red. It is a weak acid ($pK_a = 7.6$) that is yellow in its un-ionized form (which, for convenience, we assign the generic formula HIn, where "In" stands for indicator) and violet in its ionized (In^-) form. At a pH one unit above the pK_a—at pH 8.6—the ratio $[In^-]/[HIn]$ is 10:1 according to the Henderson–Hasselbalch equation, and a phenol red solution is violet. At a pH less than 6.6, phenol red is largely un-ionized, so a solution of the indicator is yellow. In the pH range from about 6.8 to 8.8, the color of a phenol red solution changes from yellow to orange to red to violet with increasing pH (Figure 15.15).

As with buffers, different pH indicators are useful in different pH ranges. For example, phenol red is the best choice if the pH values being monitored fall between pH 6.6 and 8.6; one of the other indicators shown in Figure 15.15 would be a better choice in another pH range. Every indicator has a useful pH range defined by its $pK_a \pm 1.0$ pH unit. In addition to their role in determining pH values, indicators are also used to detect the large changes in pH that occur in acid–base titrations.

Acid–Base Titrations

We discussed titration methods in Section 8.7, and we summarize here how they are carried out. Figure 15.16 shows a typical titration apparatus. There are four steps in its use:

1. Accurately transfer a known volume of sample to a flask or beaker.
2. Either add a few drops of an indicator solution to the sample, or insert the probe of a pH meter.
3. Fill a buret with a solution, the *titrant*, of known concentration of a substance that will react with a solute, the *analyte*, in the sample.
4. Slowly add titrant to the sample, and monitor the change in pH. The volume of titrant needed to completely consume the analyte is indicated by either a change in indicator color or a rapid change in the reading on the pH meter.

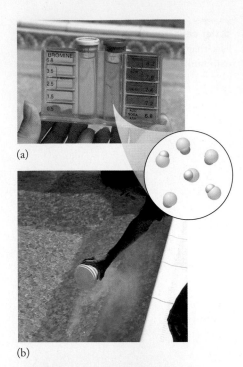

(a)

(b)

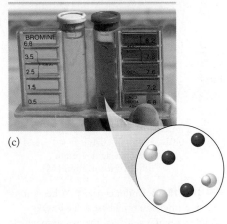

(c)

FIGURE 15.14 Many pool test kits include the pH indicator phenol red. A few drops are added to a sample of pool water collected in the tube with the red cap. (a) After a rainstorm, the pH of the pool water is 6.8 (or less), as indicated by the yellow color, which is that of the acid form of the phenol red. (b) Sodium carbonate is added to the pool to raise the pH. (c) A follow-up test produces a red-orange color, which is the color produced by a mixture of acid (yellow) and basic (magenta) forms of phenol red. This color indicates that the pH of the pool has been properly adjusted to a value near the pK_a of the indicator.

pH indicator a water-soluble weak organic acid that changes color as pH changes.

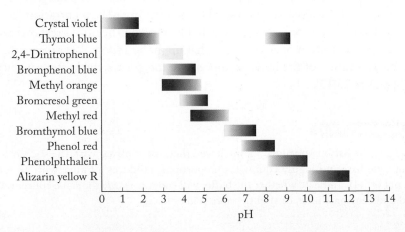

FIGURE 15.15 A pH indicator is useful within a range of 1 pH unit above and below the pK_a value of the indicator. This array of indicators could be used to determine pH values from 0 to 12.

We use the recorded titrant volume to calculate the concentration of analyte in the sample.

The neutralization titrations in Chapter 8 involved titrating strong acids with strong bases and vice versa. Here we look at another type, called an *alkalinity titration*. The term *alkalinity* is used to indicate the buffer capacity of a solution or sample of natural water against additions of acid. Although the buffers found in natural waters have more species involved than we included in our model of river water, the familiar carbonic acid–bicarbonate system is usually the most important buffer.

An alkalinity titration is more complex than a strong acid/strong base titration because it may have *two* equivalence points as we titrate first carbonate ions and then bicarbonate ions. Let's work our way up to dealing with the complexity of an alkalinity titration by starting with titrations of artificial samples containing known concentrations of weak and strong monoprotic acids and *monobasic* bases. (A monobasic base accepts one hydrogen ion per molecule.) In these examples we will monitor the change in pH with addition of titrant using a pH electrode, though the appropriate color indicators would work just as well.

In the first example, let's compare the titration curves of two 20.0 mL samples. One contains 0.100 M NaOH; the other contains 0.100 M NH$_3$. Both are titrated with 0.100 M HCl. The two titration curves are shown in Figure 15.17. The initial pH of the NaOH solution is higher than that of the NH$_3$ solution because NaOH is a strong base.

In the titration of the strong base NaOH with the strong acid HCl, the pH of the sample as acid is added does not change much until it is close to the equivalence point. As the equivalence point is approached, the principal ions present in the reaction mixture are Na$^+$(aq), Cl$^-$(aq), and OH$^-$(aq). The pH of the solution is determined only by the [OH$^-$] still present. When enough acid has been added to completely consume all the OH$^-$ ions in the sample, the equivalence point is reached. The solution now consists of water and NaCl. The ions in solution (Na$^+$ and Cl$^-$) do not hydrolyze and do not influence pH; therefore, at the equivalence point, pH = 7.00.

The pH of the sample containing the weak base NH$_3$ changes abruptly with the first few drops of added acid, but then the changes become smaller and the titration curve levels out. In this nearly flat region, additions of acidic titrant are consumed as NH$_3$ combines with H$^+$ ions to form NH$_4$$^+$ ions, and the solution acts like a pH buffer (a solution of a weak base and its conjugate acid). As long as there are significant concentrations of NH$_3$ and NH$_4$$^+$ in the sample, the changes in pH caused by added titrant are small.

When enough titrant has been added to consume all of the base in either of the two samples—in other words, at the equivalence point of each titration—pH drops sharply. At the equivalence point, the same volume of acid has been added to both samples, because the volumes of the two samples and their concentrations of base are the same. Whether the base is strong or weak does not affect the volume of titrant needed to reach the equivalence point: only the number of moles of base present determines the number of moles of acid required to neutralize it. Beyond the equivalence point, the curves are identical because the pH is determined only by the amount of HCl added and the total volume.

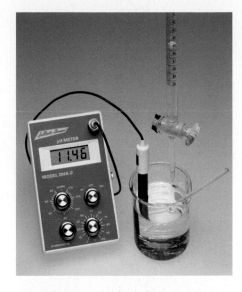

FIGURE 15.16 A digital pH meter is used to measure pH during a titration.

▶❚❚ **CHEMTOUR** Acid/Base Titrations

◯◯ **CONNECTION** As noted in Chapter 8, the equivalence point is that point in a titration when just enough titrant has been added to completely react with all the analyte.

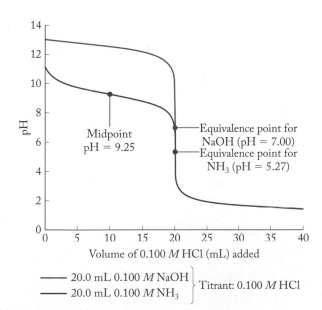

— 20.0 mL 0.100 M NaOH
— 20.0 mL 0.100 M NH$_3$ } Titrant: 0.100 M HCl

FIGURE 15.17 The red curve describes a titration of 20.0 mL of 0.100 M NH$_3$ with 0.100 M HCl titrant. The blue curve describes a titration of 20.0 mL of 0.100 M NaOH with 0.100 M HCl titrant. Before the equivalence point, the pH of the weak base (NH$_3$) solution is always lower than the pH of the strong base (NaOH) solution. Once the equivalence point is reached, the two solutions have the same pH as additional titrant is added.

The pH values at the equivalence points of the two curves, however, are different: 7.00 (neutral) in the strong acid–strong base titration, but 5.27 (slightly acidic) in the strong acid–weak base titration. The neutralization of NH_3 with HCl does not produce a neutral solution because the product of the titration reaction is NH_4Cl:

$$HCl(aq) + NH_3(aq) \rightarrow NH_4^+(aq) + Cl^-(aq)$$

And, as we saw in Sample Exercise 15.8, ammonium ions are weakly acidic:

$$NH_4^+(aq) + H_2O(\ell) \rightleftharpoons NH_3(aq) + H_3O^+(aq)$$

As a general rule, the pH at the equivalence point in a titration of a weak base with a strong acid is less than 7 because the cation of the salt produced in the neutralization reaction hydrolyzes to produce an acidic solution.

One other important point in the NH_3 titration curve lies halfway to the equivalence point. At this *midpoint*, half of the NH_3 initially in the sample has been converted into NH_4^+ ions. Therefore, $[NH_3] = [NH_4^+]$. Because we know the relative concentrations of the acid and its conjugate base, we can use the Henderson–Hasselbalch equation to determine its pH. Because $[NH_3] = [NH_4^+]$, the logarithmic term is zero, and pH = pK_a. The K_a of NH_4^+ is 5.68×10^{-10} (see Sample Exercise 15.8), so

$$pH = -\log(5.68 \times 10^{-10}) = 9.25$$

Just as the concentration of basic compounds in aqueous solution can be determined by titration with known quantities of a strong acid, the concentration of acidic compounds in aqueous solution can be determined by titration with known quantities of a strong base. Figure 15.18 illustrates two such titrations: 20.0 mL of 0.100 M HCl and 20.0 mL of 0.100 M acetic acid (CH_3COOH), each titrated with 0.100 M NaOH. Until the equivalence points are reached, the two titration curves differ, showing consistently higher pH values for CH_3COOH, the weak acid. The curves overlap beyond the equivalence points, where pH is controlled only by the increasing concentration of NaOH titrant.

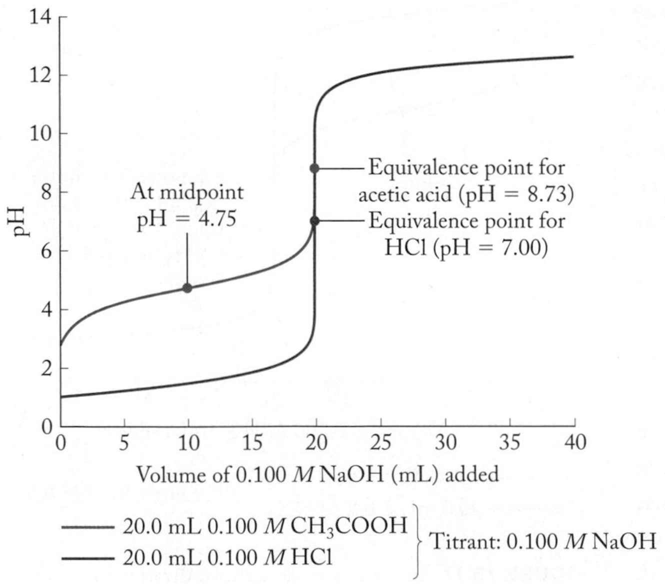

The titration curve for acetic acid about halfway to its equivalence point is nearly flat for the same reason the ammonia curve is nearly flat in the corresponding region of Figure 15.17: the partially titrated sample acts like a pH buffer. In this case the buffer consists of a weak acid—the acetic acid in the original sample—and its conjugate base, the acetate ions produced as a result of the titration reaction. The pH *exactly* halfway to the equivalence point is 4.75, which is the pK_a of acetic acid. At this point, half of the acetic acid in the sample has been neutralized, that is, turned into acetate ions, its conjugate base. Therefore, the concentrations of the acid and its conjugate base are the same, and according to the Henderson–Hasselbalch equation, pH = pK_a.

The pH at the equivalence point in the titration of the acetic acid titration is 8.73. This value is well above 7 even though it is the point in the titration at which just enough NaOH has been added to exactly neutralize all the acetic acid in the sample. The product of the titration reaction,

$$CH_3COOH(aq) + NaOH(aq) \rightarrow CH_3COONa(aq) + H_2O(\ell)$$

▶❙❙ **CHEMTOUR** Titrations of Weak Acids

At midpoint
pH = 4.75

Equivalence point for acetic acid (pH = 8.73)
Equivalence point for HCl (pH = 7.00)

Volume of 0.100 M NaOH (mL) added

—— 20.0 mL 0.100 M CH_3COOH
—— 20.0 mL 0.100 M HCl } Titrant: 0.100 M NaOH

FIGURE 15.18 The blue curve describes a titration of 20.0 mL of 0.100 M acetic acid (CH_3COOH) with 0.100 M NaOH titrant. The red curve describes a titration of 20.0 mL of 0.100 M HCl with 0.100 M NaOH titrant. Before the equivalence point, the pH of the weak acid (acetic acid) solution is always higher than the pH of the strong acid (HCl) solution. After the equivalence point is reached, the two curves overlap.

is a solution of sodium acetate (CH_3COONa). Aqueous solutions of sodium acetate are basic because the acetate ion hydrolyzes:

$$CH_3COO^-(aq) + H_2O(\ell) \rightleftharpoons CH_3COOH(aq) + OH^-(aq)$$

As a general rule, the pH at the equivalence point in the titration of a weak acid with a strong base is greater than 7 because hydrolysis of the anion of the salt produced in the neutralization reaction produces a basic solution.

> **CONCEPT TEST** ··
>
> The titration of a weak acid with a weak base is usually not done. Why?
> ···

To review the acid–base titration calculations described in Chapter 8, suppose we titrate 10.00 mL of vinegar with 0.1050 M NaOH and find that it requires 16.24 mL of 0.1050 M NaOH solution to reach the equivalence point. To determine the concentration of acetic acid in the vinegar, we start with the balanced equation for the reaction:

$$CH_3COOH(aq) + NaOH(aq) \rightarrow CH_3COONa(aq) + H_2O(\ell)$$

One mole of CH_3COOH is consumed per mole of NaOH added. Because we are working with volumes in milliliters, a more useful quantity is the millimole (mmol), which is 10^{-3} mole. The definition of molarity (M) is moles of solute per liter of solution, and we can convert moles to millimoles and liters to milliliters by dividing both by 1000:

$$M = \frac{\text{mol}}{\text{L}} = \frac{\text{mol}/1000}{\text{L}/1000} = \frac{\text{mmol}}{\text{mL}}$$

Moles per liter (mol/L) is therefore equivalent to millimoles per milliliter (mmol/mL). The number of millimoles of a solute in solution is equal to the volume of the solution in milliliters times the molarity of the solute. Thus we can write

$$(V_{CH_3COOH})(M_{CH_3COOH}) = (V_{NaOH})(M_{NaOH}) \tag{15.26}$$

$$M_{CH_3COOH} = \frac{(V_{NaOH})(M_{NaOH})}{(V_{CH_3COOH})}$$

and solve for the concentration of acetic acid by inserting the experimental data:

$$M_{CH_3COOH} = \frac{(16.24 \text{ mL})(0.1050 \text{ } M)}{(10.00 \text{ mL})} = 0.1705 \text{ } M$$

We can write an equation for calculating the concentration of any acidic or basic solute from the results of a titration by using a general version of Equation 15.26:

$$V_A M_A = \frac{n_A}{n_B} V_B M_B \tag{15.27}$$

in which V_A and M_A represent the volume and molarity of the acid, V_B and M_B represent the volume and molarity of the base, and n_A/n_B is the ratio of the moles of acid to moles of base in the balanced chemical equation describing the reaction. For example, in a titration of a solution of sulfuric acid with NaOH,

$$H_2SO_4(aq) + 2\,NaOH(aq) \rightarrow Na_2SO_4(aq) + 2\,H_2O(\ell)$$

the number of moles of NaOH consumed is twice the number of moles of H_2SO_4 consumed. Therefore, $n_A/n_B = 1/2$ and Equation 15.27 becomes

$$(V_{H_2SO_4})(M_{H_2SO_4}) = \tfrac{1}{2}(V_{NaOH})(M_{NaOH})$$

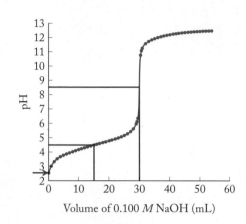

FIGURE 15.19

SAMPLE EXERCISE 15.11 Interpreting the Results of a Titration LO7

Suppose a 20.0 mL sample of an unknown acidic substance is titrated with 0.1000 M NaOH and the titration curve shown in Figure 15.19 is obtained.

  a Is the substance a weak or strong acid?
  b. What is the concentration of the acid in the sample?
  c. What is the pK_a of the acid?

COLLECT AND ORGANIZE We are provided a titration curve of a sample of an acidic substance and asked to determine if it's a weak or strong acid, its pK_a value, and its concentration in the sample. Equation 15.27 can be used to calculate the concentration of a reactant in solution if the stoichiometry of the titration reaction can be determined.

ANALYZE The initial pH of the sample is above 2.5, as noted in Figure 15.19, and it increases gradually between pH 4 and 5 over much of the titration prior to the equivalence point. The pH at the equivalence point, which appears to align exactly with a titrant volume of 30 mL, is 8.5. There is only one equivalence point in the titration curve, which means there is only one ionizable hydrogen atom per molecule of the unknown substance. Therefore, we can assign it the formula HX and use the following equation to describe the titration reaction:

$$HX(aq) + NaOH(aq) \rightarrow NaX(aq) + H_2O(\ell)$$

The 1:1 stoichiometry of acid to base in this equation means that Equation 15.27 simplifies to

$$V_A M_A = V_B M_B$$

SOLVE
  a. Given the weakly acidic pH of the sample and the basic pH at the equivalence point, we may conclude that the substance is a weak acid.
  b. According to the titration curve in Figure 15.19, the titrant (0.1000 M NaOH) volume at the equivalence point is very close to 30 mL. Therefore, the concentration of the acid (M_A) in the sample is

$$M_A = \frac{V_B M_B}{V_A} = \frac{30 \text{ mL} \times 0.1000 \text{ } M}{20.0 \text{ mL}} = 0.15 \text{ } M$$

  c. The pH of a titration of a weak acid exactly halfway to the equivalence point is equal to the pK_a of the acid. The pH that corresponds to a titrant volume of one-half of 30 mL, or 15 mL, on the titration curve in Figure 15.19 appears to be 4.5, which is also the pK_a of the acid.

THINK ABOUT IT In actual pH titrations, pH values are usually recorded to the nearest 0.01 pH unit and the volume of titrant added at the equivalence point is usually known to at least the nearest 0.1 mL, and often to the nearest 0.01 mL. Therefore, the determinations of analyte concentration are usually expressed with at least 3 and often 4 significant figures and determinations of pK_a values are usually expressed to the nearest 0.01.

Practice Exercise Your job is to determine the concentration of a basic active ingredient in a commercial window cleaner. Titration of a 25.0 mL sample of commercial window cleaner produces the titration curve shown in Figure 15.20.

    a. What is the concentration of the active ingredient in the sample?
    b. Could the active ingredient be ammonia?

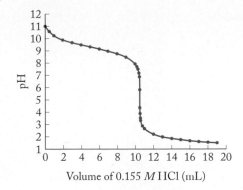

FIGURE 15.20

Alkalinity Titrations

Let's return to the application we discussed early in this section: determination of the alkalinity of a sample of natural water—by which we mean the capacity of the water to neutralize additions of acid because it contains CO_3^{2-} and HCO_3^- ions, the conjugate bases of weak acids. We start with a known volume of the sample and slowly add a strongly acidic titrant from a buret, monitoring the change in pH produced by each drop. If both CO_3^{2-} and HCO_3^- ions are present in the sample, the first additions of titrant react with CO_3^{2-} ions first because they are the stronger of the bases, converting them into HCO_3^- ions:

$$CO_3^{2-}(aq) + H^+(aq) \rightarrow HCO_3^-(aq)$$

This reaction is the first stage of the alkalinity titration. In the second stage, the bicarbonate formed in the first stage plus any bicarbonate present in the original sample reacts with additional titrant to form carbonic acid:

$$HCO_3^-(aq) + H^+(aq) \rightarrow H_2CO_3(aq)$$

If more carbonic acid is produced than is soluble (remember, carbonic acid is a solution of CO_2 in water), then the carbonic acid leaves the solution in the form of bubbles of carbon dioxide:

$$H_2CO_3(aq) \rightarrow H_2O(\ell) + CO_2(g)$$

As shown in Figure 15.21, additions of acidic titrant produce gradual changes in pH until the first equivalence point. Gradual change is the result of HCO_3^- and CO_3^{2-} ions functioning together as a weak acid–conjugate base pH buffer. When the CO_3^{2-} in the sample is completely consumed, the pH drops sharply, producing the first equivalence point. During the second stage of the titration the conversion of HCO_3^- ions to H_2CO_3 produces a second pH buffer and another plateau of gradual pH change before the HCO_3^- ions are completely consumed and pH drops sharply for a second time, creating a second equivalence point.

Note that the initial pH of the sample in Figure 15.21 is slightly above 10, which is quite basic and above the pH range tolerated by many species of aquatic life. Such highly basic water may be found in arid regions such as the U.S. Southwest, where rocks containing $CaCO_3$ and other basic compounds are in contact with water. The pH of seawater is about 8.2, lower than the first equivalence point on the alkalinity titration curve in Figure 15.21. This means that the dominant carbonate species in seawater is actually bicarbonate. For this reason, alkalinity titration curves for seawater (and most freshwater samples) have only one equivalence point, coinciding with the pH of the second equivalence point in Figure 15.21.

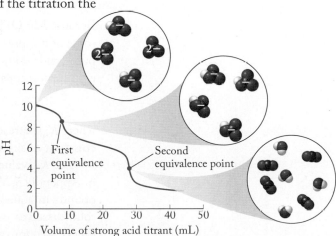

FIGURE 15.21 The titration curve for an alkalinity titration has two equivalence points. The first marks the complete conversion of carbonate into bicarbonate, and the second marks the conversion of bicarbonate into carbonic acid.

In the alkalinity titration of a sample that initially contains both CO_3^{2-} and HCO_3^-, the volume of titrant required to reach the first equivalence point is less than that required to titrate from the first equivalence point to the second. Why?

We can detect both equivalence points in an alkalinity titration with appropriate color indicators. Phenol red would not be a good choice because it changes color between pH 6.8 and 8.4. That range is just below the pH of the first equivalence point and well above the pH of the second equivalence point. To detect the first equivalence point, we need an indicator with a pK_a near the pH of the solution at the first equivalence point (8.5). Looking at Figure 15.15, we see that one candidate is phenolphthalein ($pK_a = 8.8$), which is pink in its basic form and colorless at low pH.

To detect the second equivalence point, we could add bromcresol green ($pK_a = 4.6$) after the first equivalence point has been reached. We would not add it before the first equivalence point because its blue-green color in basic solutions would obscure the pink-to-colorless transition of phenolphthalein. We do not need to be concerned about the phenolphthalein obscuring the bromcresol green color change because phenolphthalein is colorless in acidic solutions.

15.10 Solubility Equilibria

CONNECTION The solubilities of ionic compounds are summarized in Table 8.3.

In Section 15.1 we noted that the common strong bases are all hydroxides of alkali or alkaline earth elements. However, aqueous solutions of one of these compounds, magnesium hydroxide, are only weakly basic because the concentrations of hydroxide ions in them are determined by the limited solubility of $Mg(OH)_2$ in water. Magnesium hydroxide is the active ingredient in a product found in many medicine cabinets: the antacid called *milk of magnesia*. This liquid appears "milky" because it is an aqueous *suspension* (not solution) of solid, white $Mg(OH)_2$. We can express the limited solubility of solid $Mg(OH)_2$ by the following equation:

$$Mg(OH)_2(s) \rightleftharpoons Mg^{2+}(aq) + 2\,OH^-(aq)$$

Because $Mg(OH)_2$ is a solid, its effective concentration does not change as long as some of it is present in the system. As was the case for pure liquids (Section 15.3), pure solids do not appear in the equilibrium constant expression. Therefore, the equilibrium constant for the dissolution of $Mg(OH)_2$ is

$$K_{sp} = [Mg^{2+}][OH^-]^2$$

where K_{sp} represents an equilibrium constant called either the **solubility-product constant** or simply the **solubility product**.

The K_{sp} values of $Mg(OH)_2$ and other slightly soluble compounds are listed in Appendix 5. We can use these values to calculate concentrations of these compounds in aqueous solutions. Two terms are used in discussing how much of a solid dissolves in a solvent: *solubility*, which is expressed in grams of solute per liter of solution, and *molar solubility*, which is expressed in moles of solute per liter of solution. Let's use the K_{sp} of $Mg(OH)_2$ (5.6×10^{-12}) to calculate the molar solubility of $Mg(OH)_2$ at 25°C and how many grams of it dissolve in 50.0 mL at 25°C. We start with the K_{sp} expression:

$$K_{sp} = [Mg^{2+}][OH^-]^2 = 5.6 \times 10^{-12}$$

solubility-product constant (also called **solubility product, K_{sp}**) an equilibrium constant that describes the formation of a saturated solution of a slightly soluble salt.

We let x be the number of moles of $Mg(OH)_2$ that dissolve in 1 L of solution, producing a solution in which $[Mg^{2+}] = x$ and $[OH^-] = 2x$. Inserting these values in the K_{sp} expression,

$$K_{sp} = (x)(2x)^2 = x(4x^2) = 4x^3 = 5.6 \times 10^{-12}$$

$$x = 1.1 \times 10^{-4} \, M$$

Note that the entire algebraic expression for $[OH^-]$, $2x$, is squared in this calculation. Forgetting to square the coefficient is a common mistake. Also note that the molar solubility of $Mg(OH)_2$ is much greater than its solubility product. This difference is true for all sparingly soluble ionic compounds because each K_{sp} is the product of small values of concentration multiplied together, producing an even smaller overall K_{sp} value.

Using the molar solubility of $Mg(OH)_2$ that we just calculated (1.1×10^{-4} mol/L), we can determine how many grams of $Mg(OH)_2$ dissolve in 50.0 mL of a saturated solution. We convert molar solubility into an equivalent number of grams of $Mg(OH)_2$ per liter using the molar mass of $Mg(OH)_2$, 58.32 g/mol, and then into the mass that dissolves in 50.0 mL of solution:

$$\frac{1.1 \times 10^{-4} \, \cancel{mol}}{1 \, \cancel{L}} \times \frac{58.32 \, g}{1 \, \cancel{mol}} \times \frac{1 \, \cancel{L}}{1000 \, \cancel{mL}} \times 50.0 \, \cancel{mL} = 3.2 \times 10^{-4} \, g$$

We can derive a general equation relating the molar solubility of any ionic compound M_mX_x to its K_{sp} value. We start with the dissolution equilibrium,

$$M_mX_x(s) \rightleftharpoons m \, M^{n+}(aq) + x \, X^{y-}(aq)$$

If S represents the molar solubility of M_mX_x, then

$$[M^{n+}] = (m \times S) \quad \text{and} \quad [X^{y-}] = (x \times S)$$

Inserting these concentrations into the solubility product expression for M_mX_x,

$$K_{sp} = [M^{n+}]^m[X^{y-}]^x = (m \times S)^m(x \times S)^x$$

$$K_{sp} = (m^m x^x) \, S^{(m+x)} \tag{15.28}$$

Before we get lost in all of these letters, let's apply Equation 15.28 to calculate the molar solubility of calcium phosphate, $Ca_3(PO_4)_2$:

$$K_{sp} = (m^m x^x) \, S^{(m+x)} = (3^3 2^2) \, S^{(3+2)} = 108 \, S^5$$

Note that there are really only two terms in Equation 15.28. The first contains each subscript raised to the same power as its value: every "2" is squared, every "3" is cubed, and so on. The second term is simply molar solubility (S) raised to a power equal to the sum of the subscripts. One warning about subscripts and polyatomic ions: use only the subscript outside the parentheses, not the subscripts that are part of the formula of the ion.

Now let's insert the K_{sp} value listed for $Ca_3(PO_4)_2$ in Appendix 5 and solve for S.

$$K_{sp} = 2.1 \times 10^{-33} = 108 \, S^5$$

$$S = \sqrt[5]{\frac{2.1 \times 10^{-33}}{108}} = 1.1 \times 10^{-7} \, M$$

SAMPLE EXERCISE 15.12   **Calculating Molar Solubility**   **LO8**
from K_{sp}

The mineral barite is mostly barium sulfate ($BaSO_4$) and is widely used in industry and in medical imaging of the digestive system. Calculate the molar solubility at 25°C of $BaSO_4$ in (a) pure water and (b) seawater in which the concentration of sulfate ions is 2.8 g/L.

COLLECT AND ORGANIZE We are asked to calculate the molar solubility of $BaSO_4$ in both pure water and in seawater that already contains sulfate ions. The K_{sp} value of $BaSO_4$ given in Appendix 5 is 9.1×10^{-11}.

ANALYZE The dissolution reaction,

$$BaSO_4(s) \rightleftharpoons Ba^{2+}(aq) + SO_4^{2-}(aq)$$

indicates that 1 mole of Ba^{2+} ions and 1 mole of SO_4^{2-} ions form from each mole of $BaSO_4$ that dissolves in pure water. If S mol/L $BaSO_4$ dissolves in pure water, then $[Ba^{2+}] = [SO_4^{2-}] = S$. However, seawater has a background concentration of SO_4^{2-}. According to Le Châtelier's principle and the common-ion effect, the sulfate ion already in seawater should shift the dissolution equilibrium to the left, which means that less $BaSO_4$ should dissolve in seawater than in pure water.

SOLVE

a. In pure water,

$$K_{sp} = [Ba^{2+}][SO_4^{2-}] = (S)(S) = 9.1 \times 10^{-11}$$

$$S^2 = 9.1 \times 10^{-11}$$

$$S = 9.5 \times 10^{-6}\ M$$

b. In seawater, we first need to calculate the value of $[SO_4^{2-}]$ before any $BaSO_4$ dissolves:

$$\text{Initial } [SO_4^{2-}] = \frac{2.8\ \cancel{g}}{L} \times \frac{1\ mol}{96.06\ \cancel{g}} = \frac{0.029\ mol}{L}$$

The value of SO_4^{2-} at equilibrium is the sum of the background concentration (0.029 M) and the additional SO_4^{2-} ions from the dissolution of $BaSO_4$ (S). Incorporating this value into the $[SO_4^{2-}]$ term in the K_{sp} expression gives

$$K_{sp} = [Ba^{2+}][SO_4^{2-}] = (S)(0.029 + S) = 9.1 \times 10^{-11}$$

Solving for S is simplified if we assume that the K_{sp} of $BaSO_4$ is so small that we can ignore its contribution to the total SO_4^{2-} concentration. This assumption is reasonable because S is likely to be much less than 0.029 M given the limited solubility of $BaSO_4$ in pure water. Therefore,

$$(S)(0.029 + S) \approx (S)(0.029) = 9.1 \times 10^{-11}$$

$$S = 3.1 \times 10^{-9}\ M$$

THINK ABOUT IT The calculated molar solubility of $BaSO_4$ in seawater is much less than the initial $[SO_4^{2-}]$ value, so our simplifying assumption was justified. The much lower solubility of $BaSO_4$ in seawater is another illustration of the common-ion effect: the dissolution of $BaSO_4$ is suppressed by the SO_4^{2-} ions already present in seawater.

Practice Exercise What is the molar solubility of $MgCO_3$, a component of the mineral dolomite, at 25°C?

In the preceding Sample Exercises we saw how the common-ion effect can suppress the solubility of an ionic compound. Other perturbations to solubility equilibria can actually promote solubility, as happens when the anion of the compound is the conjugate base of a weak acid. The molar solubilities of such compounds increase when strong acid is added and pH is lowered. We will see why as we solve the following problem.

SAMPLE EXERCISE 15.13 **Calculating the Effect of pH** **LO8**
on Solubility

What is the molar solubility of CaF_2 at 25°C in (a) pure water and (b) an acidic buffer in which $[H^+]$ is a constant 0.050 M?

COLLECT AND ORGANIZE We are asked to calculate the solubility of CaF_2 in both pure water and in an acidic buffer. The dissolution process is described by the equilibrium:

$$(1) \qquad CaF_2(s) \rightleftharpoons Ca^{2+}(aq) + 2\,F^-(aq)$$

The equilibrium constant expression for the process and the K_{sp} value from Appendix 5 are

$$K_{sp} = [Ca^{2+}][F^-]^2 = 3.9 \times 10^{-11}$$

Calcium fluoride is a basic salt because the F^- ion is the conjugate base of the weak acid HF ($K_a = 6.8 \times 10^{-4}$).

ANALYZE To account for the effect of acid on the solubility of a fluoride salt, we need to consider the chemical equilibrium in which the fluoride ion acts as a Brønsted–Lowry base (H^+ ion acceptor):

$$(2) \qquad F^-(aq) + H^+(aq) \rightleftharpoons HF(aq)$$

This reaction is the reverse of the acid ionization reaction:

$$HF(aq) \rightleftharpoons F^-(aq) + H^+(aq)$$

Therefore, the simplified equilibrium constant for reaction 2 is the reciprocal of the K_a of HF:

$$K_2 = \frac{[HF]}{[H^+][F^-]} = \frac{1}{6.8 \times 10^{-4}} = 1.47 \times 10^3$$

As reaction 2 proceeds, F^- ions are consumed, which shifts the equilibrium in reaction 1 to the right and so increases CaF_2 solubility. Therefore we anticipate an increase in the solubility of CaF_2 when acid is present.

SOLVE

a. Let S be the molar solubility of CaF_2 in pure water. According to the stoichiometry of reaction 1, $[Ca^{2+}] = S$ mol/L and $[F^-] = 2S$ mol/L. Inserting these symbols in the K_{sp} expression,

$$K_{sp} = [Ca^{2+}][F^-]^2 = (S)(2S)^2 = 4S^3$$

$$4S^3 = 3.9 \times 10^{-11}$$

$$S = 2.1 \times 10^{-4} \, M$$

b. In the acidic buffer, $[H^+] = 0.050\ M$. The F^- and H^+ ions combine as in reaction 2. Assuming the buffer pH is constant, $[H^+] = 0.050\ M$ and the equilibrium constant expression for reaction 2 is

$$K_2 = \frac{[HF]}{[0.050][F^-]} = 1.47 \times 10^3$$

$$\frac{[HF]}{[F^-]} = 73.5$$

(3) $[HF] = 73.5[F^-]$

This calculation tells us that most of the F^- produced when calcium fluoride dissolves is converted into HF. The tiny fraction of all the F^- ions produced by CaF_2 that remains as free F^- ions is defined by this ratio:

(4) $$\frac{[F^-]}{[F^-] + [HF]}$$

where the numerator is the concentration of free F^- ions at equilibrium and the denominator is the concentration of all the F^- ions produced when CaF_2 dissolved. Combining expressions 3 and 4,

$$\frac{[F^-]}{[F^-] + [HF]} = \frac{[F^-]}{[F^-] + 73.5[F^-]} = \frac{[F^-]}{74.5[F^-]} = 0.0134$$

If S is the molar solubility of CaF_2 in the acid, S mol/L Ca^{2+} and $2S$ mol/L F^- are produced. However, most of the fluoride ions produced as CaF_2 dissolves are converted into HF, and, as we just calculated, the free F^- ion concentration is only 0.0134 of that produced when CaF_2 dissolved. Therefore, the $[F^-]$ term in the K_{sp} expression is not $2S$, but only $[0.0134 \times (2S)] = 0.0268\,S$. Inserting this value in the K_{sp} expression:

$$K_{sp} = [Ca^{2+}][F^-]^2 = (S)(0.0268\,S)^2 = 7.18 \times 10^{-4}\,S^3$$

$$7.18 \times 10^{-4}\,S^3 = 3.9 \times 10^{-11}$$

$$S = 3.8 \times 10^{-3}\,M$$

THINK ABOUT IT A comparison of the results from parts a and b reveals that the molar solubility of CaF_2 is about 10 times higher in the acidic buffer, as we predicted, because most of the F^- ions produced when CaF_2 dissolves in the buffer form HF. This conversion of F^- ions to HF means a product (free F^- ion) is being removed from the dissolution equilibrium. Le Châtelier's principle tells us that the result will be a shift in the position of the equilibrium to the right, in favor of forming product. In this case greater product formation means greater solubility.

Practice Exercise What is the molar solubility of $Fe(OH)_2$ at 25°C in one liter of water with a pH of 6.00? ⚙

CONCEPT TEST

In part b of Sample Exercise 15.13, we assumed that $[H^+]$ did not change significantly as a result of the reaction $F^-(aq) + H^+(aq) \rightleftharpoons HF(aq)$. If $[H^+]$ had dropped significantly, how would the solubility of CaF_2 have been affected?

K_{sp} and Q

In the preceding Sample Exercises we used calculations involving the K_{sp} of slightly soluble salts in cases where solubility of the salt was either diminished or enhanced. We can also use K_{sp} values to determine if a precipitate will form when two solutions are mixed. In answering such questions, we can use the concept of the reaction quotient Q that we developed in Chapter 14. When applied to the equilibrium governing a slightly soluble salt, Q is sometimes called the *ion product*, a name that describes exactly what Q is in this case: the product of the concentrations of the ions in the precipitate raised to powers equal to their coefficients in the balanced equation. If the calculated Q value is greater than the K_{sp} of a salt ($Q > K_{sp}$), then that salt will precipitate when the solutions are mixed. If $Q < K_{sp}$, no precipitate will form.

SAMPLE EXERCISE 15.14 **Determining if a Precipitate** **LO8**
Forms when Solutions Are Mixed

Lead(II) chloride ($K_{sp} = 1.60 \times 10^{-5}$) is a white pigment used in 15th-century European sculpture. Will $PbCl_2$ precipitate when 275 mL of a 0.134 M solution of $Pb(NO_3)_2$ is added to 125 mL of a 0.0339 M solution of NaCl?

COLLECT AND ORGANIZE We are asked if $PbCl_2$ will precipitate when solutions containing Pb^{2+} and Cl^- ions are mixed. The process is described by this equilibrium:

$$Pb^{2+}(aq) + 2\ Cl^-(aq) \rightleftharpoons PbCl_2(s)$$

ANALYZE To determine if a precipitate of solid $PbCl_2$ forms, we need to first calculate the concentrations of Pb^{2+} and Cl^- ions after the two solutions are mixed together. Then we use those concentration values in the K_{sp} expression for $PbCl_2$ to calculate the value of the reaction quotient (Q) that these concentrations represent:

$$Q = [Pb^{2+}][Cl^-]^2$$

If the value of Q is greater than the K_{sp} value (1.60×10^{-5}), $PbCl_2$ will precipitate.

SOLVE The values of $[Pb^{2+}]$ and $[Cl^-]$ immediately after mixing are

$$[Pb^{2+}] = \frac{0.134\ M \times 0.275\ \text{L}}{0.400\ \text{L}} = 0.0921\ M$$

$$[Cl^-] = \frac{0.0339\ M \times 0.125\ \text{L}}{0.400\ \text{L}} = 0.0106\ M$$

The value of Q is

$$Q = [Pb^{2+}][Cl^-]^2 = (0.0921)(0.0106)^2 = 1.03 \times 10^{-5}$$

The value of Q is less than 1.60×10^{-5}, so no precipitate forms.

THINK ABOUT IT To predict if a precipitate forms when solutions of known ion concentration are combined, we compare the Q value based on the concentrations of the ions immediately after the solutions are mixed to the K_{sp} value of the salt. If Q equals K_{sp}, then the solution is saturated and contains the maximum quantity of solute.

Practice Exercise Will calcium fluoride ($K_{sp} = 3.9 \times 10^{-11}$) precipitate when 175 mL of a 4.78×10^{-3} M solution of $Ca(NO_3)_2$ is added to 135 mL of a 7.35×10^{-3} M solution of KF?

We can use differences in the solubilities of ionic compounds to separate mixtures of ions by selectively removing them from solution. For example, suppose a solution contains 0.10 M Ca^{2+} ion and 0.020 M Mg^{2+} ion. The hydroxide salts of both ions are slightly soluble. Is it possible to remove the Mg^{2+} ions from solution while leaving the Ca^{2+} ions in solution? Consider the following solubility equilibria and K_{sp} values:

$$Ca(OH)_2(s) \rightleftharpoons Ca^{2+}(aq) + 2\ OH^-(aq) \qquad K_{sp} = [Ca^{2+}][OH^-]^2 = 4.7 \times 10^{-6}$$

$$Mg(OH)_2(s) \rightleftharpoons Mg^{2+}(aq) + 2\ OH^-(aq) \qquad K_{sp} = [Mg^{2+}][OH^-]^2 = 5.6 \times 10^{-12}$$

Note that $Ca(OH)_2$ has a larger K_{sp} than $Mg(OH)_2$. We can therefore ask: What is the maximum concentration (x) of OH^- ions that will *not* cause the calcium ion to precipitate? We can calculate that directly from the K_{sp} of calcium hydroxide:

$$K_{sp} = 4.7 \times 10^{-6} = [Ca^{2+}][OH^-]^2 = (0.10)(x)^2$$

$$x = \sqrt{\frac{4.7 \times 10^{-6}}{0.10}} = 6.9 \times 10^{-3}\ M$$

How much of the Mg^{2+} will precipitate if $[OH^-] = 6.9 \times 10^{-3}\ M$? The concentration x of magnesium ion in the presence of $6.9 \times 10^{-3}\ M$ hydroxide ion may be calculated from its K_{sp}:

$$K_{sp} = 5.6 \times 10^{-12} = [Mg^{2+}][OH^-]^2 = (x)(6.9 \times 10^{-3})^2$$

$$x = \frac{5.6 \times 10^{-12}}{(6.9 \times 10^{-3})^2} = 1.2 \times 10^{-7}\ M$$

Because the original solution was $0.020\ M$ in Mg^{2+} ions, a concentration of $1.2 \times 10^{-7}\ M\ Mg^{2+}$ means that only $[(1.2 \times 10^{-7})/0.020] \times 100\%$, or only 0.00060% of the original amount of Mg^{2+} remains in solution. This corresponds to virtually complete precipitation of Mg^{2+} and successful separation of the solution's calcium ions from its magnesium ions. Indeed, magnesium can be separated from the less abundant calcium ions in seawater by this technique.

SAMPLE EXERCISE 15.15 **Separating Ions in Solution** **LO8**

Both lead(II) chloride and lead(II) fluoride are slightly soluble salts. A solution of lead(II) nitrate is added to a solution that is $0.275\ M$ in both $Cl^-(aq)$ and $F^-(aq)$. Can we use this method to separate the two halide ions? If "complete precipitation" is defined as there being less than 0.10% of a particular ion left in solution, is the precipitation of the first salt complete before the second salt begins to precipitate? (For $PbCl_2$, $K_{sp} = 1.6 \times 10^{-5}$; for PbF_2, $K_{sp} = 3.2 \times 10^{-8}$.)

COLLECT AND ORGANIZE We are given a solution that contains two ions that form slightly soluble lead(II) salts and are asked if one ion can be completely removed before the second one starts to precipitate when lead(II) ion is added to the solution. We have the K_{sp} values for both salts and the initial concentration of both ions in the solution.

ANALYZE The equilibrium constant expressions for both ions are

$$PbCl_2(s) \rightleftharpoons Pb^{2+}(aq) + 2\,Cl^-(aq) \qquad K_{sp} = [Pb^{2+}][Cl^-]^2 = 1.6 \times 10^{-5}$$

$$PbF_2(s) \rightleftharpoons Pb^{2+}(aq) + 2\,F^-(aq) \qquad K_{sp} = [Pb^{2+}][F^-]^2 = 3.2 \times 10^{-8}$$

In both equilibrium expressions, the concentration of lead ion is raised to the first power and the concentration of the halide ion is raised to the second power, so we can compare the influence of the ion concentrations on the K_{sp} values directly. The K_{sp} of $PbCl_2(s)$ is almost 10^3 times larger than the K_{sp} of $PbF_2(s)$, so we can assume the $PbF_2(s)$ will precipitate first. We can therefore ask, what is the maximum Pb^{2+} concentration in the solution that will not cause the $PbCl_2(s)$ to precipitate? When we determine that value, we can calculate how much $F^-(aq)$ is in solution under those conditions and determine if the precipitation of F^- is complete.

SOLVE The maximum amount of Pb^{2+} in the solution that will not cause the chloride ion to precipitate is

$$K_{sp} = 1.6 \times 10^{-5} = [Pb^{2+}][Cl^-]^2 = (x)(0.275)^2$$

$$x = \frac{1.6 \times 10^{-5}}{(0.275)^2} = 2.12 \times 10^{-4}\ M$$

The concentration of $F^-(aq)$ in the solution at this concentration of lead(II) ion is

$$K_{sp} = 3.2 \times 10^{-8} = (2.12 \times 10^{-4})(x)^2$$

$$x = \sqrt{\frac{3.2 \times 10^{-8}}{2.12 \times 10^{-4}}} = 0.0123 \; M$$

The original solution was 0.275 M in $F^-(aq)$, and a concentration of 0.0123 M means that $(0.0123/0.275) \times 100\% = 4.5\%$ of the original amount of $F^-(aq)$ remains in solution, so the precipitation of $PbF_2(s)$ is not complete and we could not use this method to separate the two ions.

THINK ABOUT IT Determining which salt precipitates first was easy in this example because the concentrations of both ions were the same and both were raised to the same power in the K_{sp} expressions. If the concentrations differ, or if the concentrations of the ions in the K_{sp} expressions are raised to different powers, the species that precipitates first may have to be determined by calculation rather than simple inspection.

Practice Exercise An aqueous solution of sodium fluoride is slowly added to a water sample that contains barium ion (0.0375 M) and calcium ion (0.0667 M). Consult Appendix 5 to help you answer these questions:

a. Are both barium fluoride and calcium fluoride slightly soluble salts?
b. Applying the definition of complete precipitation given in Sample Exercise 15.15, determine if Ba^{2+} ions and Ca^{2+} ions in solution can be completely separated by selective precipitation with F^- ions. ⚙

SAMPLE EXERCISE 15.16 — Integrating Concepts: More about Fire Extinguishers

The Sample Exercise at the end of Chapter 14 dealt with dry chemical fire extinguishers that contain either sodium or potassium bicarbonate, which are *class BC* dry chemical agents. Another material, ammonium dihydrogen phosphate ($NH_4H_2PO_4$), is a *class ABC* or *multipurpose* agent used in dry extinguishers. This chemical extinguishes fires by melting and flowing over burning fuel, effectively stopping its reaction with oxygen in the air. The following comments appear in the product literature and contrast the behavior of $NH_4H_2PO_4$ to that of the bicarbonate agents:

a. Any residue from the use of sodium and potassium bicarbonate extinguishers can be removed with a solution of vinegar and water, unlike the residue from using a dihydrogen phosphate extinguisher.
b. Purple-K agents (potassium bicarbonate) should never be mixed with phosphate-based fire suppression agents, especially in the presence of water, because the ensuing chemical reaction will destroy their efficacy.
c. The dihydrogen phosphate agent is more corrosive than other dry chemical agents. It is recommended that the multipurpose agent not be used on aircraft because it is highly corrosive to aluminum if moisture is present, and it cannot be washed away like BC agents.

Explain each of these recommendations using appropriate chemical equations.

COLLECT AND ORGANIZE We are given three chemical agents that are used in fire extinguishers—$NaHCO_3$, $KHCO_3$, and $NH_4H_2PO_4$—and asked to write equations that explain the recommendations on how they are to be handled. Their differences involve their reactions with a weak acid (vinegar–water solution), with each other, and with aluminum. Appendix 5 has any K values we need.

ANALYZE

a. The two ions involved, HCO_3^- and $H_2PO_4^-$, have the capacity to donate or accept H^+ ions, but they would have to accept H^+ ions to react with acetic acid. Apparently HCO_3^- does, but $H_2PO_4^-$ does not. This difference must be linked to the K_c values of these two reactions:

[1] $HCO_3^-(aq) + H^+(aq) \rightleftharpoons H_2CO_3(aq)$

[2] $H_2PO_4^-(aq) + H^+(aq) \rightleftharpoons H_3PO_4(aq)$

Reactions 1 and 2 are the reverse of the first ionization reactions of H_2CO_3 and H_3PO_4, respectively, which means their K_c values are the reciprocals of the K_{a_1} values of H_2CO_3 and H_3PO_4 (Appendix 5):

$$K_{c[1]} = \frac{1}{K_{a_1,H_2CO_3}} = \frac{1}{4.3 \times 10^{-7}} = 2.3 \times 10^6$$

$$K_{c[2]} = \frac{1}{K_{a_1,H_3PO_4}} = \frac{1}{7.11 \times 10^{-3}} = 141$$

The greater value of $K_{c[1]}$ tells us that HCO_3^- is a better H^+ ion acceptor (and stronger base) than $H_2PO_4^-$.

b. Dissolving compounds that contain $H_2PO_4^-$ and HCO_3^- ions gives these ions the opportunity to react with each other. Based on the K_a values and our analysis in part 1, we know that the conjugate acid of $H_2PO_4^-$ (H_3PO_4) is a much stronger acid than the conjugate acid of HCO_3^- (H_2CO_3). Therefore, $H_2PO_4^-$ may be able to donate H^+ ions to HCO_3^- ions:

[3] $\quad HCO_3^-(aq) + H_2PO_4^-(aq) \overset{?}{\rightarrow}$
$$H_2CO_3(aq) + HPO_4^{2-}(aq)$$

If this reaction happens, the carbonic acid that is produced will decompose into water and CO_2 gas.

c. The chemistry of aluminum is based on its tendency to form Al^{3+} ions, so the product of aluminum corrosion that cannot be washed away must be an Al(III) compound with limited solubility in water or weakly acidic solutions.

SOLVE

a. To quantitatively evaluate the extent to which reactions 1 and 2 occur when solids containing HCO_3^- and $H_2PO_4^-$ ions are treated with acetic acid, we need to combine reactions 1 and 2 and their K_c values with the acid ionization reaction of acetic acid and its K_a value. (Remember: Adding the reactions requires that we multiply their K values.)

[1] $\quad HCO_3^-(aq) + \cancel{H^+(aq)} \rightleftharpoons H_2CO_3(aq)$
$$K_{c[1]} = 2.3 \times 10^6$$

$\quad CH_3COOH(aq) \rightleftharpoons CH_3COO^-(aq) + \cancel{H^+(aq)}$
$$K_a = 1.8 \times 10^{-5}$$

$\overline{HCO_3^-(aq) + CH_3COOH(aq) \rightleftharpoons}$
$$H_2CO_3(aq) + CH_3COO^-(aq)$$
$$K_c = 41$$

[2] $\quad H_2PO_4^-(aq) + \cancel{H^+(aq)} \rightleftharpoons H_3PO_4(aq)$
$$K_{c[2]} = 141$$

$\quad CH_3COOH(aq) \rightleftharpoons CH_3COO^-(aq) + \cancel{H^+(aq)}$
$$K_a = 1.8 \times 10^{-5}$$

$\overline{H_2PO_4^-(aq) + CH_3COOH(aq) \rightleftharpoons}$
$$H_3PO_4(aq) + CH_3COO^-(aq)$$
$$K_c = 2.5 \times 10^{-3}$$

The K_c value for the reaction between HCO_3^- and vinegar is greater than 1 and the K_c value for the reaction between $H_2PO_4^-$ and vinegar is less than 1. Therefore, acetic acid has the strength to neutralize bicarbonate compounds, turning them into H_2CO_3, which will spontaneously decompose into CO_2 gas. However, acetic acid is not a strong enough acid to

react with compounds containing $H_2PO_4^-$ ions, so it will not be effective in cleaning up the debris of fires treated with phosphate compounds.

b. To quantitatively evaluate the extent to which reaction 3 occurs, we can combine the K_c values of the reaction in which $H_2PO_4^-$ ions donate H^+ ions (the second ionization reaction of H_3PO_4) and the reaction in which HCO_3^- ions accept H^+ ions (the reverse of the first acid ionization reaction of H_2CO_3):

$$H_2PO_4^-(aq) \rightleftharpoons HPO_4^{2-}(aq) + H^+(aq)$$
$$K_{a_2} = 6.32 \times 10^{-8}$$

$$HCO_3^-(aq) + H^+(aq) \rightleftharpoons H_2CO_3(aq)$$
$$K_{c[1]} = 2.3 \times 10^6$$

$\overline{HCO_3^-(aq) + H_2PO_4^-(aq) \rightleftharpoons}$
$$H_2CO_3(aq) + HPO_4^{2-}(aq)$$
$$K = 0.14$$

This is not a very large K value; however, production and spontaneous decomposition of H_2CO_3 mean that the reaction releases CO_2 gas, removing a product and shifting the equilibrium in favor of making more of it. Decomposition of the bicarbonate component of the mixture reduces its capacity to extinguish fires.

c. A check of the solubility product values in Table A5.4 of Appendix 5 discloses this equilibrium:

$$AlPO_4(s) \rightleftharpoons Al^{3+}(aq) + PO_4^{3-}(aq)$$
$$K_{sp} = 9.8 \times 10^{-21}$$

This tiny K_{sp} value means the concentration of free PO_4^{3-} ions in the water used to wash away fire residues must be extremely small. For example, if $[Al^{3+}]$ were 1×10^{-3} M, then $[PO_4^{3-}]$ could be no more than

$$[PO_4^{3-}] = \frac{K_{sp}}{[Al^{3+}]} = \frac{9.8 \times 10^{-21}}{1 \times 10^{-3}} = 1 \times 10^{-17} M$$

Stepwise ionization of $H_2PO_4^-$ to HPO_4^{2-} and then PO_4^{3-} might account for the presence of such a small concentration of phosphate ions. The overall reaction would be

$$H_2PO_4^-(aq) \rightleftharpoons 2 H^+(aq) + PO_4^{3-}(aq)$$

The value of K_c for this reaction is the product of the K_{a_2} and K_{a_3} values of phosphoric acid:

$$K_c = K_{a_2} \times K_{a_3} = (6.32 \times 10^{-8})(4.5 \times 10^{-13}) = 2.8 \times 10^{-20}$$

We can use this K_c value to determine how large a concentration of $H_2PO_4^-$ ions is needed to produce a $[PO_4^{3-}]$ value of 1×10^{-17} M, but first we need to pick a value for $[H^+]$. Assuming solutions of ammonium dihydrogen phosphate are weakly acidic because both its cation (NH_4^+) and anion ($H_2PO_4^-$) are weak acids, let's let $[H^+] = 1 \times 10^{-4}$ and see what happens.

$$K_c = 2.84 \times 10^{-20} = \frac{[H^+]^2[PO_4^{3-}]}{[H_2PO_4^-]} = \frac{(1 \times 10^{-4})^2(1 \times 10^{-17})}{[H_2PO_4^-]}$$

and $[H_2PO_4^-] = 4 \times 10^{-6}$. This value represents the minimum $[H_2PO_4^-]$ that would be needed to precipitate aluminum phosphate. In reality, the concentration of $H_2PO_4^-$ ions should actually be orders of magnitude greater than this, so the probable aluminum(III) compound that resists clean-up is $AlPO_4$.

THINK ABOUT IT This exercise demonstrates how instructions for using different kinds of dry chemical fire extinguishers may be understood by applying the principles of acid–base and solubility equilibria. In doing so we considered some of the factors that influence the performance of chemical fire extinguishers and how they interact with materials. There are certainly many others.

SUMMARY

Section 15.1 The Brønsted–Lowry model of acids and bases defines acids as H^+ ion donors and bases as H^+ ion acceptors. Strong acids include binary acids HX, where X is any group 17 element other than fluorine, and some oxoacids H_mXO_n, where X is a nonmetal. All strong 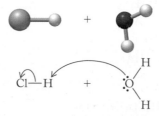 acids are completely ionized in water. The H^+ ions they release combine with water molecules to form hydronium (H_3O^+) ions. Most acids are weak acids that ionize only partially in water. When weak acid HA ionizes, it forms its **conjugate base**, A^-. When base B acquires a H^+ ion, it forms its **conjugate acid**, HB^+. Strong bases are those group 1 and 2 hydroxides that are soluble in water and dissociate completely when they dissolve. The strongest acid in water is the H_3O^+ ion; the strongest base is the OH^- ion.

Section 15.2 The strength of an oxoacid is related to the stability of the anion formed when the acid ionizes. This stability increases as the electronegativity of the central atom increases and as the number of oxygen atoms bonded to the central atom increases.

Section 15.3 Water is an amphiprotic substance in that it is capable of behaving both as an acid and as a base. This behavior is evident in the **autoionization** of water in which strong hydrogen bonding between water molecules results in formation of a hydronium ion and a hydroxide ion:

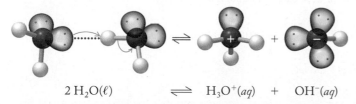

$$2\,H_2O(\ell) \quad \rightleftharpoons \quad H_3O^+(aq) \quad + \quad OH^-(aq)$$

In a neutral solution, $[H_3O^+]$ (or $[H^+(aq)]$) $= [OH^-] = 1.00 \times 10^{-7}$. The **pH** scale is a logarithmic scale for expressing the acidic or basic strength of solutions. Acidic solutions have pH values less than 7.00; basic solutions have pH values greater than 7.00. Because pH is the negative logarithm of H^+ concentration, the higher the pH, the lower the H^+ concentration. An increase in one pH unit represents a decrease in $[H^+]$ to 1/10 its initial value.

Section 15.4 The **degree of ionization** of a weak acid HA is the ratio of $[H^+]$ to the initial (total) acid concentration and is usually expressed as a percentage.

Section 15.5 Weak **polyprotic acids** can undergo more than one acid ionization reaction, but the first one is usually the one that controls pH.

Section 15.6 A salt solution is acidic if the cation in the salt is the conjugate acid of a weak base and the anion is the conjugate base of a strong acid. A salt solution is basic if the anion in the salt is the conjugate base of a weak acid and the cation is the conjugate acid of a strong base.

Section 15.7 Adding the salt of a weak acid HA to a solution of the acid provides a second source of the conjugate base A^-. As predicted by Le Châtelier's principle, the added A^- inhibits the acid ionization reaction, causing pH to rise. Adding a salt of a weak base B to a solution of the base provides a second source of its conjugate acid HB^+, which lowers pH. These shifts are examples of the **common-ion effect**.

Section 15.8 A **pH buffer** is a solution that contains either a weak acid and a salt of its conjugate base or a weak base and a salt of its conjugate acid. Buffer solutions resist pH change when acids or bases are added. Additions of acid are neutralized by the basic component of a buffer; additions of base are neutralized by the acidic component. The result is a change in the concentration ratio in the *Henderson–Hasselbalch equation*, but the changes in the logarithm of the ratio, and the impact on pH, tend to be small as long as neither component of the buffer is completely consumed.

Section 15.9 A **pH indicator** is a weak acid or base that has a color which is different from that of its conjugate base or acid. Indicators can be used to determine the pH of a solution that is within 1.0 pH unit of the pK_a or pK_b of the indicator. These indicators are also used to detect the equivalence points in pH titrations, which are used to determine the concentration of an acid or a base in an aqueous sample.

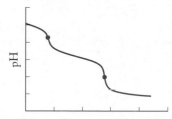

Volume of strong acid titrant (mL)

Section 15.10 The solubility of slightly soluble ionic compounds is described by their K_{sp} or **solubility product**, which is the value of the equilibrium constant for their dissolution.

PROBLEM-SOLVING SUMMARY

TYPE OF PROBLEM	CONCEPTS AND EQUATIONS	SAMPLE EXERCISES
Identifying conjugate acid–base pairs	The formula of the base in a conjugate pair is the formula of the acid less one H^+ ion.	15.1
Interconverting $[H^+]$, $[OH^-]$, pH, and pOH	Use the following: $$pH = -\log[H^+] \qquad (15.11)$$ $$pOH = -\log[OH^-] \qquad (15.12)$$ $$pK_w = pH + pOH = 14.00 \qquad (15.13)$$	15.2, 15.3
Relating pH, K_a, and percent ionization of a weak acid HA	Use the following: $$[H^+] = 10^{-pH}$$ $$K_a = \frac{[A^-][H^+]}{[HA]} \qquad (15.14)$$ $$\text{Percent ionization} = \frac{[H^+]_{equilibrium}}{[HA]_{initial}} \times 100\% \qquad (15.15)$$	15.4
Calculating the pH of a solution of a weak base B	Set up a RICE table based on the equilibrium $$B(aq) + H_2O(\ell) \rightleftharpoons HB^+(aq) + OH^-(aq)$$ Let $x = [OH^-] = [HB^+]$ at equilibrium. Calculate x using $$K_b = \frac{x^2}{[B] - x}$$ Then use $$pOH = -\log[OH^-]$$ $$pH = 14.00 - pOH$$	15.5
Calculating the pH of a solution of a weak diprotic acid	Set up a RICE table based on the K_{a_1} equilibrium $$H_2A\,(aq) \rightleftharpoons H^+(aq) + HA^-(aq)$$ Let $x = [H^+] = [HA^-]$ at equilibrium. Calculate x using $$K_{a_1} = \frac{x^2}{[H_2A] - x}$$ Then calculate $pH = -\log[H^+]$.	15.6
Distinguishing acidic, basic, and neutral salts	The cations in acidic salts are the conjugate acids of weak bases. The anions in basic salts are the conjugate bases of weak acids.	15.7
Calculating the pH of a solution of an acidic salt	Assume the salt completely dissociates into BH^+ and X^-. Set up a RICE table for the equilibrium $$BH^+(aq) \rightleftharpoons B(aq) + H^+(aq)$$ Let $x = [H^+] = [B]$ at equilibrium. Calculate x using $$K_a = \frac{K_w}{K_b} = \frac{x^2}{[BH^+] - x}$$ Then calculate pH.	15.8

TYPE OF PROBLEM	CONCEPTS AND EQUATIONS		SAMPLE EXERCISES
Calculating the pH of a solution of a weak base and its conjugate acid (or weak acid and its conjugate base)	Use the Henderson–Hasselbalch equation: $$pH = pK_a + \log \frac{[base]}{[acid]}$$	(15.23)	15.9, 15.10
Interpreting results of an alkalinity titration	Use the relation $$V_A M_A = \frac{n_A}{n_B} V_B M_B$$ where V_A and M_A are the volume and molarity of the acid, V_B and M_B are the volume and molarity of the base, and n_A and n_B are the coefficients of the acid and base, respectively, in the balanced equation.	(15.27)	15.11
Calculating molar solubility (S) of M_mX_x from K_{sp}	Use the relation $$K_{sp} = (m^m x^x)\, S^{(m+x)}$$	(15.28)	15.12
Calculating the effect of pH on the molar solubility (S) of MZ	If Z^- is the conjugate base of a weak acid, calculate the fraction of Z^- that remains as the free ion. Use this fraction as the coefficient of S in the K_{sp} expression.		15.13
Determining if a precipitate forms when solutions are mixed and which precipitate forms if more than one is possible	Compare the ion product Q to K_{sp} to determine if a precipitate will form; use K_{sp} expressions to calculate maximum concentrations of one ion in solution that will not cause another ion to precipitate.		15.14, 15.15

VISUAL PROBLEMS ....■

(Answers to boldface end-of-chapter questions and problems are in the back of the book.)

15.1. Which of the lines in Figure P15.1 best represents the dependence of the degree of ionization of acetic acid on its concentration in aqueous solution?

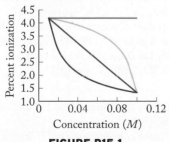

FIGURE P15.1

15.2. The bar graph in Figure P15.2 shows the degree of ionization of 1×10^{-3} M solutions of three hypohalous acids: HClO, HBrO, and HIO. Which bar is the one for HIO?

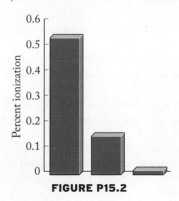

FIGURE P15.2

15.3. The graph in Figure P15.3 shows the titration curves of a 1 M solution of a weak acid with a strong base and a 1 M solution of a strong acid with the same base. Which curve is which?

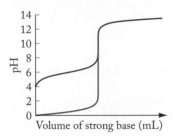

FIGURE P15.3

15.4. Estimate to within one pH unit the pH of a 0.5 M solution of the sodium salt of the weak acid in Problem 15.3.

15.5. Suppose you have four color indicators to choose from for detecting the equivalence point of the titration reaction represented by the red curve (upper curve on the left side of the plot) in Figure P15.3. The pK_a values of the four indicators are 3.3, 5.0, 7.0, and 9.0. Which indicator would be the best one to choose?

15.6. What is the pK_a value of the weak acid in Figure P15.3?

15.7. One of the titration curves in Figure P15.7 represents the titration of an aqueous sample of Na_2CO_3 with strong acid; the other represents the titration of an aqueous sample of $NaHCO_3$ with the same acid. Which curve is which?

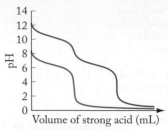

FIGURE P15.7

15.8. Consider the three beakers in Figure P15.8. Each contains a few drops of the color indicator bromthymol blue, which is yellow in acidic solutions and blue in basic solutions. One beaker contains a solution of ammonium chloride, one contains ammonium acetate, and the third contains sodium acetate. Which beaker contains which salt?

FIGURE P15.8

QUESTIONS AND PROBLEMS

Acids and Bases: The Brønsted–Lowry Model

CONCEPT REVIEW

15.9. In an aqueous solution of HBr, which compound acts as a Brønsted–Lowry acid and which is the Brønsted–Lowry base?

15.10. In an aqueous solution of HNO_3, which compound acts as a Brønsted–Lowry acid and which is the Brønsted–Lowry base?

15.11. In an aqueous solution of NaOH, which species acts as a Brønsted–Lowry acid and which is the Brønsted–Lowry base?

15.12. Both NaOH and $Ca(OH)_2$ are strong bases. Does this mean that solutions of the two compounds with the same molarity have the same capacity to neutralize strong acids? Why or why not?

15.13. Identify the acids and bases in the following reactions:
 a. $HNO_3(aq) + NaOH(aq) \rightarrow NaNO_3(aq) + H_2O(\ell)$
 b. $CaCO_3(s) + 2\ HCl(aq) \rightarrow CaCl_2(aq) + CO_2(g) + H_2O(\ell)$
 c. $NH_3(aq) + HCN(aq) \rightarrow NH_4CN(aq)$

15.14. Identify the acids and bases in the following reactions:
 a. $NH_2^-(aq) + H_2O(\ell) \rightleftharpoons NH_3(aq) + OH^-(aq)$
 b. $HClO_4(aq) + H_2O(\ell) \rightleftharpoons ClO_4^-(aq) + H_3O^+(aq)$
 c. $HSO_4^-(aq) + CO_3^{2-}(aq) \rightleftharpoons SO_4^{2-}(aq) + HCO_3^-(aq)$

15.15. Identify the conjugate base of each of the following compounds: HNO_2, $HClO$, H_3PO_4, and NH_3.

15.16. Identify the conjugate acid of each of the following species: NH_3, ClO_2^-, SO_4^{2-}, and OH^-.

PROBLEMS

15.17. What is the concentration of H^+ ions in a 1.50 M solution of HNO_3?

15.18. What is the concentration of H^+ ions in a solution of hydrochloric acid that was prepared by diluting 20.0 mL of concentrated (11.6 M) HCl to a final volume of 500 mL?

15.19. What is the value of $[OH^-]$ in a 0.0800 M solution of $Sr(OH)_2$?

15.20. A particular drain cleaner contains NaOH. What is the value of $[OH^-]$ in a solution produced when 5.0 g of NaOH dissolves in enough water to make 250 mL of solution?

15.21. Describe how you would prepare 2.50 L of a NaOH solution in which $[OH^-] = 0.70\ M$, starting with solid NaOH.

15.22. How many milliliters of a $1.00\ M$ solution of NaOH do you need to prepare 250 mL of a solution in which $[OH^-] = 0.0200\ M$?

Acid Strength and Molecular Structure

CONCEPT REVIEW

15.23. Explain why the K_{a_1} of H_2SO_4 is much greater than the K_{a_1} of H_2SeO_4.

15.24. Explain why the K_{a_1} of H_2SO_4 is much greater than the K_{a_1} of H_2SO_3.

15.25. Predict which acid in the following pairs of acids is the stronger acid: (a) H_2SO_3 or H_2SeO_3; (b) H_2SeO_4 or H_2SeO_3.

15.26. Predict which acid in the following pairs of acids is the stronger acid: (a) $HBrO$ or $HBrO_2$; (b) $HClO$ or $HBrO$.

pH and the Autoionization of Water

CONCEPT REVIEW

15.27. Explain why pH values decrease as acidity increases.

15.28. Solution A is 100 times more acidic than solution B. What is the difference in the pH values of solution A and solution B?

15.29. Under what conditions is the pH of a solution negative?

***15.30.** In principle, ethanol (CH_3CH_2OH) can undergo autoionization. Propose an explanation for why the value of the equilibrium constant for the autoionization of ethanol is much less than that of water.

PROBLEMS

15.31. Calculate the pH and pOH of the solutions with the following hydrogen ion or hydroxide ion concentrations. Indicate which solutions are acidic, basic, or neutral.
   a. $[H^+] = 3.45 \times 10^{-8}\ M$
   b. $[H^+] = 2.0 \times 10^{-5}\ M$
   c. $[H^+] = 7.0 \times 10^{-8}\ M$
   d. $[OH^-] - 8.56 \times 10^{-4}\ M$

15.32. Calculate the pH and pOH of the solutions with the following hydrogen ion or hydroxide ion concentrations. Indicate which solutions are acidic, basic, or neutral.
   a. $[OH^-] = 7.69 \times 10^{-3}\ M$
   b. $[OH^-] = 2.18 \times 10^{-9}\ M$
   c. $[H^+] = 4.0 \times 10^{-8}\ M$
   d. $[H^+] = 3.56 \times 10^{-4}\ M$

15.33. What is the pH of stomach acid in which $[HCl] = 0.155\ M$?

15.34. What is the pH of $0.00500\ M\ HNO_3$?

15.35. What are the pH and pOH of $0.0450\ M$ NaOH?

15.36. What are the pH and pOH of $0.160\ M$ KOH?

15.37. What is the pH of $2.3 \times 10^{-8}\ M\ HNO_3$?

***15.38.** What is the pH of $6.9 \times 10^{-8}\ M$ LiOH?

15.39. Suppose the pH of a volume of dilute hydrochloric acid is 2.0. What is the pH after an equal volume of deionized water is mixed with the dilute acid?

15.40. **pH of Blood** The pH of venous blood, which carries CO_2 back to the lungs, is 7.35, while that of arterial blood is 7.45. How many times more concentrated in H^+ is venous blood compared with arterial blood?

Calculations Involving pH, K_a, and K_b

CONCEPT REVIEW

15.41. One-molar solutions of the following acids are prepared: CH_3COOH, HNO_2, $HClO$, HCN. Rank them in order of decreasing $[H^+]$.

15.42. On the basis of the following degree-of-ionization data for $0.100\ M$ solutions, select which acid has the smallest K_a.

Acid	Degree of Ionization (%)
C_6H_5COOH	2.5
HF	8.5
HN_3	1.4
CH_3COOH	1.3

15.43. A $1.0\ M$ aqueous solution of $NaNO_2$ is a much better conductor of electricity than is a $1.0\ M$ solution of HNO_2. Explain why.

15.44. Hydrogen chloride and water are molecular compounds, yet a solution of HCl dissolved in H_2O is an excellent conductor of electricity. Explain why.

15.45. Hydrofluoric acid is a weak acid. Write the mass action expression for its acid ionization reaction.

15.46. In the formula of formic acid, HCOOH, one H atom is ionizable. Write the mass action expression for the acid ionization equilibrium of formic acid.

***15.47.** The K_a values of weak acids depend on the solvent in which they dissolve. For example, the K_a of alanine in aqueous ethanol is less than its K_a in water.
   a. In which solvent does alanine ionize to the largest degree?
   b. Which is the stronger Brønsted–Lowry base: water or ethanol?

***15.48.** The K_a of proline is 2.5×10^{-11} in water, 2.8×10^{-11} in an aqueous solution that is 28% ethanol, and 1.66×10^{-8} in 37% aqueous formaldehyde at 25°C.
   a. In which solvent is proline the strongest acid?
   b. Rank these compounds on the basis of their strengths as Brønsted–Lowry bases: water, ethanol, and formaldehyde.

15.49. When methylamine, CH_3NH_2, dissolves in water, the resulting solution is slightly basic. Which compound is the Brønsted–Lowry acid and which is the base?

***15.50.** When 1,2-diaminoethane, $H_2NCH_2CH_2NH_2$, dissolves in water, the resulting solution is basic. Write the formula of the ionic compound that is formed when hydrochloric acid is added to a solution of 1,2-diaminoethane.

PROBLEMS

15.51. Sore Muscles Muscles *burn* during strenuous exercise because of the buildup of lactic acid and other metabolites in muscle tissues. In a 1.00 *M* aqueous solution, 2.94% of lactic acid is ionized. What is the value of its K_a?

15.52. Rancid Butter The odor of spoiled butter is due in part to butanoic acid, which results from the chemical breakdown of butter fat. A 0.100 *M* solution of butanoic acid is 1.23% ionized. Calculate the value of K_a for butanoic acid.

15.53. At equilibrium, the value of $[H^+]$ in a 0.250 *M* solution of an unknown acid is 4.07×10^{-3} *M*. Determine the degree of ionization and the K_a of the acid.

15.54. Nitric acid (HNO_3) is a strong acid that is completely ionized in aqueous solutions of concentrations ranging from 1% to 10% (1.5 *M*). However, in more concentrated solutions, part of the nitric acid is present as un-ionized molecules of HNO_3. For example, in a 50% solution (7.5 *M*) at 25°C, only 33% of the molecules of HNO_3 dissociate into H^+ and NO_3^-. What is the value of K_a of HNO_3?

15.55. Ant Bites The venom of biting ants contains formic acid, HCOOH, $K_a = 1.8 \times 10^{-4}$ at 25°C. Calculate the pH of a 0.060 *M* solution of formic acid.

15.56. Gout Uric acid can collect in joints, giving rise to a medical condition known as gout. If the pK_a of uric acid is 3.89, what is the pH of a 0.0150 *M* solution of uric acid?

15.57. Acid Rain I A weather system moving through the American Midwest produced rain with an average pH of 5.02. By the time the system reached New England, the rain it produced had an average pH of 4.66. How much more acidic was the rain falling in New England?

15.58. Acid Rain II A newspaper reported that the "level of acidity" in a sample taken from an extensively studied watershed in New Hampshire in February 1998 was "an astounding 200 times lower than the worst measurement" taken in the preceding 23 years. What is this difference expressed in units of pH?

15.59. The K_b of dimethylamine $[(CH_3)_2NH]$ is 5.9×10^{-4} at 25°C. Calculate the pH of a 1.20×10^{-3} *M* solution of dimethylamine.

15.60. Painkiller I Morphine is an effective painkiller but is also highly addictive. Calculate the pH of a 0.115 *M* solution of morphine if its $pK_b = 5.79$.

15.61. Painkiller II Codeine is a popular prescription painkiller because it is much less addictive than morphine. Codeine contains a basic nitrogen atom that can be protonated to give the conjugate acid of codeine. Calculate the pH of a 3.42×10^{-4} *M* solution of codeine if the pK_a of the conjugate acid is 8.21.

15.62. Pyridine (C_5H_5N) is a particularly foul-smelling substance used in manufacturing pesticides and plastic resins. Calculate the pH of a 0.125 *M* solution of pyridine.

*15.63. Aspirin Aspirin (Figure P15.63) is an acid substance whose $pK_a = 3.50$ at body temperature (37°C). One standard adult aspirin contains 325 mg of the drug. If you take 1 aspirin on a full stomach (assume a volume of 1 L and a pH of 2.0), what percent of the aspirin in your stomach is present as the anion?

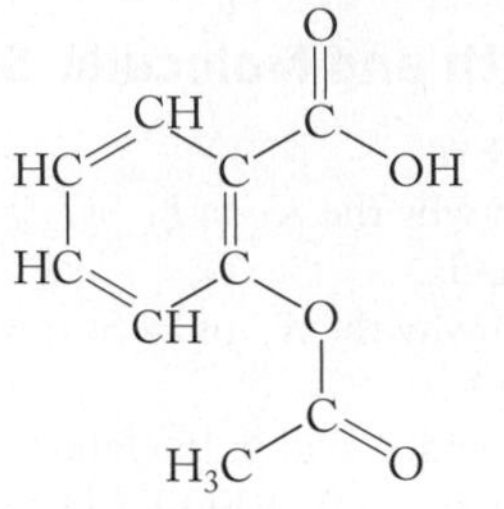

FIGURE P15.63

*15.64. Cough Syrup A typical cough syrup contains 30.0 mg of pseudoephedrine hydrochloride (Figure P15.64) in 1 tsp (5 mL) of medicine. Calculate the pH of cough syrup based on the presence of this substance if the pK_a of pseudoephedrine hydrochloride is 9.22.

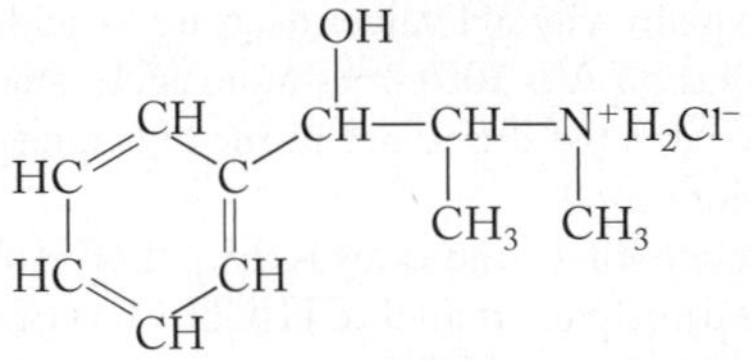

FIGURE P15.64

Polyprotic Acids

CONCEPT REVIEW

15.65. Why is the K_{a_2} value of phosphoric acid less than its K_{a_1} value but greater than its K_{a_3} value?

15.66. In calculating the pH of a 1.0 *M* solution of sulfurous acid, we can ignore the H^+ ions produced by the ionization of the bisulfite ion; however, in calculating the pH of a 1.0 *M* solution of sulfuric acid, we cannot ignore the H^+ ions produced by the ionization of the bisulfate ion. Why?

PROBLEMS

15.67. What is the pH of a 0.300 *M* solution of H_2SO_4?

15.68. What is the pH of a 0.150 *M* solution of sulfurous acid?

15.69. Ascorbic acid (vitamin C) is a triprotic acid. What is the pH of a 0.250 *M* solution of ascorbic acid?

15.70. Rhubarb Pie The leaves of the rhubarb plant contain high concentrations of diprotic oxalic acid (HOOCCOOH) and must be removed before the stems are used to make rhubarb pie. What is the pH of a 0.0288 *M* solution of oxalic acid?

15.71. Addiction to Tobacco Nicotine is responsible for the addictive properties of tobacco. What is the pH of a 1.00×10^{-3} *M* solution of nicotine?

15.72. 1,2-Diaminoethane, $H_2NCH_2CH_2NH_2$, is used extensively in the synthesis of compounds containing transition metals in water. If $pK_{b_1} = 3.29$ and $pK_{b_2} = 6.44$, what is the pH of a 2.50×10^{-4} M solution of 1,2-diaminoethane?

15.73. **Malaria Treatment** Quinine occurs naturally in the bark of the cinchona tree. For centuries it was the only treatment for malaria. Calculate the pH of a 0.01050 M solution of quinine in water.

15.74. Dozens of pharmaceuticals ranging from Cyclizine for motion sickness to Viagra for impotence are derived from the organic compound piperazine, whose structure is shown in Figure P15.74.

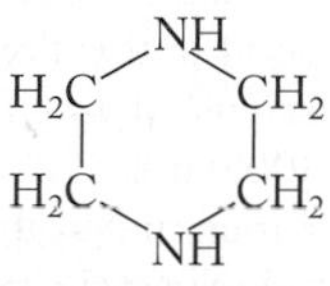

FIGURE P15.74

a. Solutions of piperazine are basic ($K_{b_1} = 5.38 \times 10^{-5}$; $K_{b_2} = 2.15 \times 10^{-9}$). What is the pH of a 0.0133 M solution of piperazine?

*b. Draw the structure of the ionic form of piperazine that would be present in stomach acid (about 0.15 M HCl).

pH of Salt Solutions

CONCEPT REVIEW

*15.75. The pK_a values of the ions shown in Figure P15.75 increase as more methyl ($-CH_3$) groups are added. Do more methyl groups increase or decrease the strength of the neutral conjugate bases of these ions?

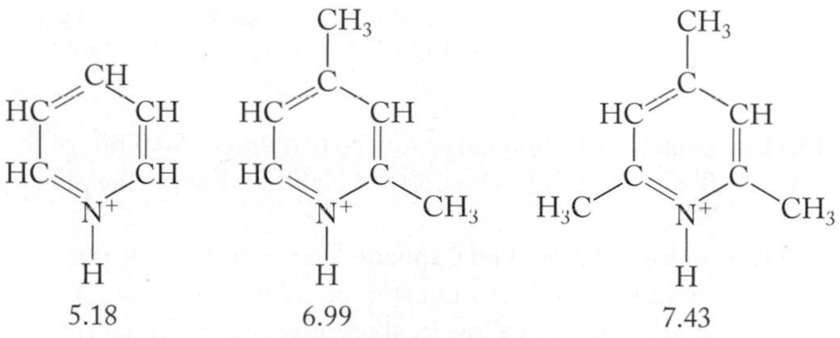

FIGURE P15.75

15.76. Why is it unnecessary to publish tables of K_b values of the conjugate bases of weak acids whose K_a values are known?

15.77. Which of the following salts produces an acidic solution in water: ammonium acetate, ammonium nitrate, or sodium formate?

15.78. Which of the following salts produces a basic solution in water: NaF, KCl, or NH_4Cl?

15.79. **Neutralizing the Smell of Fish** Trimethylamine, $(CH_3)_3N$, $K_b = 6.5 \times 10^{-5}$ at 25°C, is a contributor to the "fishy" odor of not-so-fresh seafood. Some people squeeze fresh lemon juice (which contains a high concentration of citric acid) on cooked fish to reduce the fishy odor. Why is this practice effective?

15.80. **Nutritional Value of Beets** Beets contain high concentrations of the calcium salt of a dicarboxylic acid with the common name malonic acid and the formula $HOOCCH_2COOH$. Could the presence of the calcium salt of malonic acid affect the pH balance of beets? If so, in which direction? Explain why.

PROBLEMS

15.81. If the K_a of the conjugate acid of the artificial sweetener saccharin is 2.1×10^{-11}, what is the pK_b for saccharin?

15.82. If the K_{a_1} value of chromic acid (H_2CrO_4) is 0.16 and its K_{a_2} value is 3.2×10^{-7}, what are the values of K_{b_1} and K_{b_2} of the CrO_4^{2-} anion?

15.83. **Dental Health** Sodium fluoride is added to many municipal water supplies to reduce tooth decay. Calculate the pH of a 0.00339 M solution of NaF at 25°C.

15.84. Calculate the pH of a 1.25×10^{-2} M solution of the decongestant ephedrine hydrochloride if the pK_b of ephedrine (its conjugate base) is 3.86.

The Common-Ion Effect and pH Buffers

CONCEPT REVIEW

15.85. Why is a solution of sodium acetate and acetic acid a much better pH buffer than is a solution of sodium chloride and hydrochloric acid?

15.86. Why does a solution of a weak base and its conjugate acid act as a better buffer than does a solution of the weak base alone?

PROBLEMS

15.87. A buffer contains 0.244 M acetic acid and 0.122 M sodium acetate.
   a. What is its pH at 25°C?
   b. What is its pH at 0°C (K_a of acetic acid $= 1.64 \times 10^{-5}$)?

15.88. What is the pH of a buffer that is 0.100 M pyridine and 0.275 M pyridinium chloride at 25°C?

15.89. Calculate the pH and pOH of 500.0 mL of a phosphate buffer that is 0.225 M HPO_4^{2-} and 0.225 M PO_4^{3-} at 25°C.

15.90. Determine the pH and pOH of 0.250 L of a buffer that is 0.0200 M boric acid and 0.0250 M sodium borate at 25°C.

15.91. What is the ratio of acetate ion to acetic acid in a buffer containing these compounds at pH $= 3.56$?

15.92. What is the ratio of lactic acid to lactate in a solution with pH $= 4.00$?

15.93. What is the pH at 25°C of a solution that results from mixing together equal volumes of a 0.05 M solution of ammonia and a 0.025 M solution of hydrochloric acid?

15.94. What is the pH at 25°C of a solution that results from mixing together equal volumes of a 0.05 M solution of acetic acid and a 0.025 M solution of sodium hydroxide?

*15.95. How much 10.0 M HNO_3 must be added to 1.00 L of a buffer that is 0.010 M acetic acid and 0.10 M sodium acetate to reduce the pH to 5.00 at 25°C?

*15.96. How much 6.0 M NaOH must be added to 0.500 L of a buffer that is 0.0200 M acetic acid and 0.0250 M sodium acetate to raise the pH to 5.75 at 25°C?

*15.97. Calculate the pH at 25°C of 1.00 L of a buffer that is 0.120 M HNO_2 and 0.150 M $NaNO_2$ before and after the addition of 1.00 mL of 12.0 M HCl.

*15.98. Calculate the pH at 25°C of 100.0 mL of a buffer that is 0.100 M NH_4Cl and 0.100 M NH_3 before and after the addition of 1.0 mL of 6 M HNO_3.

pH Indicators and Acid–Base Titrations

CONCEPT REVIEW

15.99. What are the differences between the titration curve of a strong acid titrated with a strong base and that of a weak acid titrated with a strong base?

15.100. Do all titrations of a strong base with a strong acid have the same pH at the equivalence point?

15.101. Do all titrations of a weak acid with a strong base have the same pH at the equivalence point?

15.102. What properties must a compound have to serve as an acid–base indicator?

PROBLEMS

15.103. A 25.0 mL sample of 0.100 M acetic acid is titrated with 0.125 M NaOH. Calculate the pH at 25°C of the titration mixture after 10.0, 20.0, and 30.0 mL of the base have been added.

15.104. A 25.0 mL sample of a 0.100 M solution of aqueous trimethylamine is titrated with a 0.125 M solution of HCl. Calculate the pH of the solution after 10.0, 20.0, and 30.0 mL of acid have been added; pK_b of $(CH_3)_3N = 4.19$ at 25°C.

15.105. What is the concentration of ammonia in a solution if 22.35 mL of 0.1145 M HCl is needed to titrate a 100.0 mL sample of the solution?

15.106. **Alkaline Hot Springs** In an alkalinity titration of a 100.0 mL sample of water from a hot spring, 2.56 mL of a 0.0355 M solution of HCl is needed to reach the first equivalence point (pH = 8.3) and another 10.42 mL is needed to reach the second equivalence point (pH = 4.0). If the alkalinity of the spring water is due only to the presence of carbonate and bicarbonate, what are the concentrations of each?

15.107. What volume of 0.100 M HCl is required to titrate 250 mL of 0.0100 M Na_2CO_3 to the first equivalence point? How much more 0.100 M HCl is needed to reach the second equivalence point?

15.108. What volumes of 0.0100 M HCl are required to titrate 250 mL of 0.0100 M Na_2CO_3 and to titrate 250 mL of 0.0100 M HCO_3^-?

15.109. In the titration of a solution of a weak monoprotic acid with a 0.1025 M solution of NaOH, the pH halfway to the equivalence point was 4.44. In the titration of a second solution of the same acid, exactly twice as much of a 0.1025 M solution of NaOH was needed to reach the equivalence point. What was the pH halfway to the equivalence point in this titration?

15.110. A 125.0 mg sample of an unknown monoprotic acid was dissolved in 100.0 mL of distilled water and titrated with a 0.050 M solution of NaOH. The pH of the solution was monitored throughout the titration, and the following data were collected. Determine the K_a of the acid.

Volume of OH⁻ Added (mL)	pH	Volume of OH⁻ Added (mL)	pH
0	3.09	22.0	5.93
5	3.65	22.2	6.24
10	4.10	22.6	9.91
15	4.50	22.8	10.21
17	4.55	23.0	10.38
18	4.71	24.0	10.81
19	4.94	25.0	11.02
20	5.11	30.0	11.49
21	5.37	40.0	11.85

15.111. Sketch a titration curve for the titration of 50.0 mL of 0.250 M HNO_2 with 1.00 M NaOH. What is the pH at the equivalence point?

15.112. **Indicators from Red Cabbage** Red cabbage juice is a sensitive acid–base indicator; its colors range from red at acidic pH to yellow in alkaline solutions. What color would red cabbage juice have when 25 mL of a 0.10 M solution of acetic acid is titrated with 0.10 M NaOH to its equivalence point?

15.113. Sketch a titration curve for the titration of the malaria drug quinine if 40.0 mL of a 0.100 M solution of quinine is titrated with a 0.100 M solution of HCl.

15.114. Sketch a titration curve for the titration of 100 mL of 1.25×10^{-2} M ascorbic acid with 1.00×10^{-2} M NaOH. How many equivalence points should the curve have, and what color indicator(s) could be used?

Solubility Equilibria

CONCEPT REVIEW

15.115. What is the difference between *molar solubility* and *solubility product*?

15.116. Give an example of how the common-ion effect limits the dissolution of a sparingly soluble ionic compound.

15.117. Which of the following cations will precipitate first as a carbonate mineral from an equimolar solution of Mg^{2+}, Ca^{2+}, and Sr^{2+}?

15.118. If the solubility of a compound increases with increasing temperature, does K_{sp} increase or decrease?

15.119. The K_{sp} of strontium sulfate increases from 2.8×10^{-7} at 37°C to 3.8×10^{-7} at 77°C. Is the dissolution of strontium sulfate endothermic or exothermic?

15.120. How will adding concentrated $NaOH(aq)$ affect the solubility of an Al(III) salt?

15.121. Chemistry of Tooth Decay Tooth enamel is composed of a mineral known as hydroxyapatite with the formula $Ca_5(PO_4)_3(OH)$. Explain why tooth enamel can be eroded by acidic substances released by bacteria growing in the mouth.

15.122. **Fluoride and Dental Hygiene** Fluoride ions in drinking water and toothpaste convert hydroxyapatite in tooth enamel into fluorapatite:

$$Ca_5(PO_4)_3(OH)(s) + F^-(aq) \rightleftharpoons Ca_5(PO_4)_3F(s) + OH^-(aq)$$

Why is fluorapatite less susceptible than hydroxyapatite to erosion by acids?

PROBLEMS

15.123. At a particular temperature the value of $[Ba^{2+}]$ in a saturated solution of barium sulfate is 1.04×10^{-5} *M*. Starting with this information, calculate the K_{sp} value of barium sulfate at this temperature.

15.124. Suppose a saturated solution of barium fluoride contains 1.5×10^{-2} *M* F^-. What is the K_{sp} value of BaF_2?

15.125. What are the equilibrium concentrations of Cu^+ and Cl^- in a saturated solution of copper(I) chloride if $K_{sp} = 1.02 \times 10^{-6}$?

15.126. What are the equilibrium concentrations of Pb^{2+} and F^- in a saturated solution of lead fluoride if the K_{sp} value of PbF_2 is 3.2×10^{-8}?

15.127. What is the solubility of calcite ($CaCO_3$) in grams per milliliter at a temperature at which its $K_{sp} = 9.9 \times 10^{-9}$?

15.128. What is the solubility of silver iodide in grams per milliliter at a temperature at which its $K_{sp} = 1.50 \times 10^{-16}$?

15.129. What is the pH at 25°C of a saturated solution of silver hydroxide?

15.130. **pH of Milk of Magnesia** What is the pH at 25°C of a saturated solution of magnesium hydroxide (the active ingredient in the antacid milk of magnesia)?

15.131. Suppose you have 100 mL of each of the following solutions. In which will the most $CaCO_3$ dissolve? (a) 0.1 *M* NaCl; (b) 0.1 *M* Na_2CO_3; (c) 0.1 *M* NaOH; (d) 0.1 *M* HCl

15.132. In which of the following solutions will CaF_2 be most soluble? (a) 0.010 *M* $Ca(NO_3)_2$; (b) 0.01 *M* NaF; (c) 0.001 *M* NaF; (d) 0.10 *M* $Ca(NO_3)_2$

15.133. Composition of Seawater The average concentration of sulfate in surface seawater is about 0.028 *M*. The average concentration of Sr^{2+} is 9×10^{-5} *M*. If the K_{sp} value of strontium sulfate is 3.4×10^{-7}, is the concentration of strontium in the sea probably controlled by the insolubility of its sulfate salt?

15.134. **Fertilizing the Sea to Combat Global Warming** Some scientists have proposed adding Fe(III) compounds to large expanses of the open ocean to promote the growth of phytoplankton that would in turn remove CO_2 from the atmosphere through photosynthesis. The average pH of open ocean water is 8.1. What is the maximum value of $[Fe^{3+}]$ in seawater if the K_{sp} value of $Fe(OH)_3$ is 1.1×10^{-36}?

15.135. Will calcium fluoride precipitate when 125 mL of 0.375 *M* $Ca(NO_3)_2$ is added to 245 mL of 0.255 *M* NaF at 25°C?

15.136. Will strontium sulfate precipitate at 25°C when 345 mL of a solution that is 0.0100 *M* in $Sr^{2+}(aq)$ is added to 75 mL of 0.175 *M* K_2SO_4?

15.137. A solution is 0.010 *M* in both Br^- and SO_4^{2-}. A 0.250 *M* solution of lead(II) nitrate is slowly added to it.
   a. Which anion will precipitate first?
   b. What is the concentration in the solution of the first ion when the second one starts to precipitate at 25°C?

*15.138. Solution A is 0.0250 *M* in Ag^+ ion and Pb^{2+} ion. You have access to two other solutions: (B) 0.500 *M* NaCl and (C) 0.500 *M* NaBr.
   a. Which would be the better solution to add to solution A to separate lead from silver by precipitation? (The better solution is the one that has less silver remaining in solution when the lead begins to precipitate.)
   b. Using the solution you selected in part a, is the separation of the two ions complete? ("Complete" is defined as the point when less than 0.10% of the silver ion is left in the solution when the lead ion begins to precipitate.)

Additional Problems

15.139. Describe the intermolecular forces and changes in bonding that lead to the formation of a basic solution when methylamine (CH_3NH_2) dissolves in water.

15.140. Describe the chemical reactions of sulfur that begin with the burning of high-sulfur fossil fuel and that end with the reaction between acid rain and building exteriors made of marble ($CaCO_3$).

15.141. Suggest acid/conjugate base pairs that could be used to prepare buffers at each of these pH values: 3.00, 5.00, 7.00, and 12.00.

15.142. The value of K_{a_1} of phosphorous acid, H_3PO_3, is nearly the same as the K_{a_1} of phosphoric acid, H_3PO_4. Identify the ionizable hydrogen atoms in Figure P15.142.

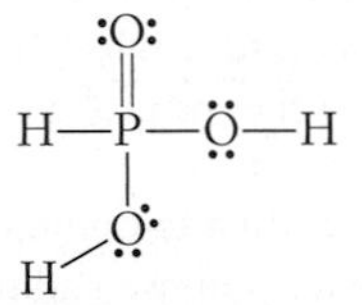

FIGURE P15.142

*15.143. pH of Baking Soda A cook dissolves a teaspoon of baking soda ($NaHCO_3$) in a cup of water, then discovers that the recipe calls for a tablespoon, not a teaspoon. So, the cook adds two more teaspoons of baking soda to make up the difference. Does the additional baking soda change the pH of the solution? Explain why or why not.

*15.144. Antacid Tablets Antacids contain a variety of bases such as $NaHCO_3$, $MgCO_3$, $CaCO_3$, and $Mg(OH)_2$. Only $NaHCO_3$ has appreciable solubility in water.
  a. Write a net ionic equation for the reaction of each base with aqueous HCl.
  b. Explain how insoluble substances can act as effective antacids.

*15.145. pH of Natural Waters I In a 1985 study of Little Rock Lake in Wisconsin, 400 gallons of 18 M sulfuric acid were added to the lake over 6 years. The initial pH of the lake was 6.1 and the final pH was 4.7. If none of the acid was consumed in chemical reactions, estimate the volume of the lake.

15.146. pH of Natural Waters II Between 1993 and 1995, sodium phosphate was added to Seathwaite Tarn in the English Lake District to increase its pH. Explain why addition of this compound increased pH.

15.147. Acid–Base Properties of Pharmaceuticals I Zoloft is a common prescription drug for the treatment of depression. It is sold as a salt of HCl.

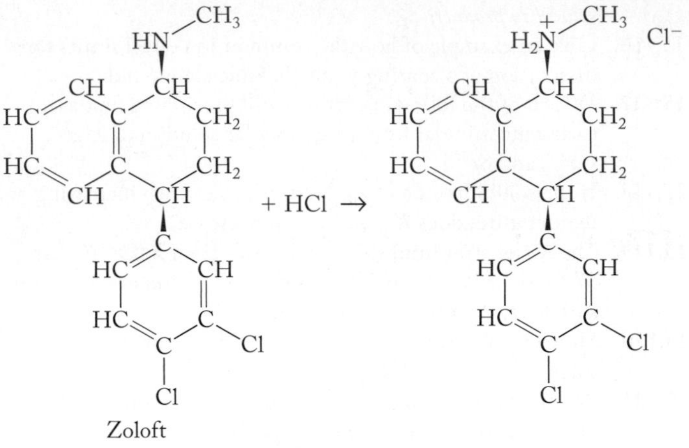

FIGURE P15.147

  a. In the reaction shown in Figure P15.147, which structure is that of the acid salt?
  b. When Zoloft dissolves in water, will the resulting solution be acidic or basic?

15.148. Acid–Base Properties of Pharmaceuticals II Prozac is another popular antidepressant drug. Its structure is given in Figure P15.148.

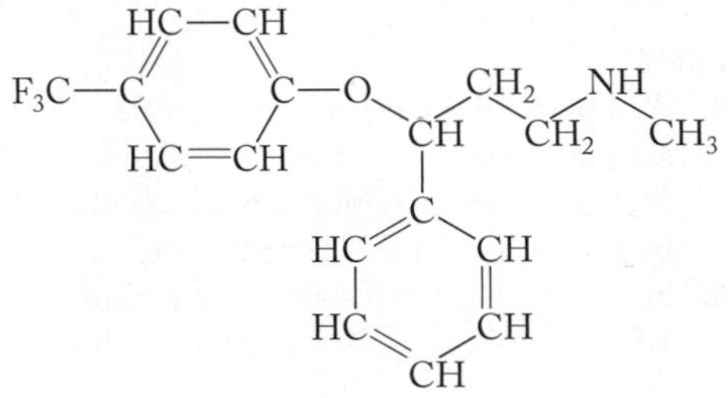

FIGURE P15.148

  a. Is a solution of Prozac in water likely to be slightly basic or slightly acidic? Explain your answer.
  b. Prozac is also sold as a salt of HCl. Which atom, N or O, is more likely to react with HCl?
  c. Prozac is sold as a salt of HCl because the solubility of the salt in water is higher than Prozac itself. Why is the salt more soluble?

15.149. Hydrogen fluoride (HF) behaves as a weak acid in aqueous solution. Two equilibria influence which fluorine-containing species are present in solution:

$$HF(aq) + H_2O(\ell) \rightleftharpoons H_3O^+(aq) + F^-(aq) \qquad K_a = 1.1 \times 10^{-3}$$

$$F^-(aq) + HF(aq) \rightleftharpoons HF_2^-(aq) \qquad\qquad K = 2.6 \times 10^{-1}$$

a. What is the equilibrium constant for this equilibrium?

$$2\,HF(aq) + H_2O(\ell) \rightleftharpoons H_3O^+(aq) + HF_2^-(aq)$$

b. What is the equilibrium concentration of HF_2^- in a 0.150 M solution of HF?

*15.150. Pentafluorocyclopentadiene, which has the structure shown in Figure P15.150, is a strong acid.

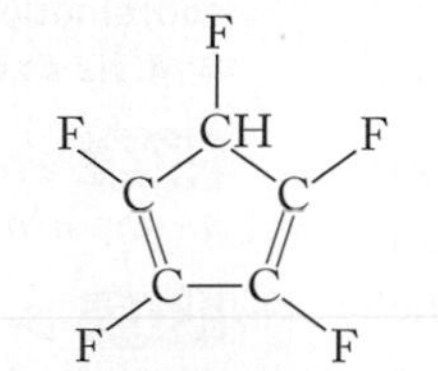

FIGURE P15.150

a. Draw the conjugate base of C_5F_5H.
b. Why is the compound so acidic when most organic acids are weak?

15.151. **Anti-inflammatories** Naproxen (also known as Aleve) is an anti-inflammatory drug used to reduce pain, fever, inflammation, and stiffness caused by conditions such as osteoarthritis and rheumatoid arthritis. Naproxen is an organic acid; its structure is shown in Figure P15.151. Naproxen has limited solubility in water, so it is sold as its sodium salt.

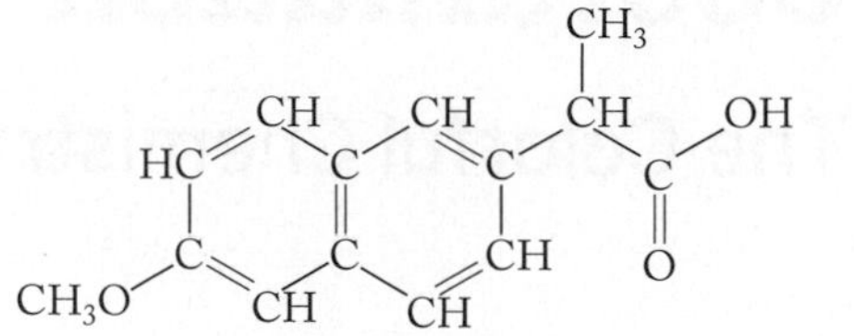

FIGURE P15.151

a. Draw the molecular structure of the sodium salt.
b. Should a solution of the salt be acidic or basic? Explain why.
c. Explain why the salt is more soluble than naproxen itself.

*15.152. **Greenhouse Gases and Ocean pH** Some climate models predict a decrease in the pH of the oceans of 0.77 pH units because of increases in atmospheric carbon dioxide.

a. Explain, by using the appropriate chemical reactions and equilibria, how an increase in atmospheric CO_2 could produce a decrease in oceanic pH.
b. How much more acidic (in terms of $[H^+]$) would the oceans be if their pH dropped this much?
c. Oceanographers are concerned about the impact of a drop in oceanic pH on the survival of coral reefs. Why?

16

Coordination Compounds
The Colorful Chemistry of Metals

The Company They Keep

Many of the metallic elements in the periodic table are essential to good health. For example, copper, zinc, and cobalt play key roles in protein function; iron is needed to transport oxygen from our lungs to all the cells of our body; and nearly everyone knows the importance of calcium in building strong teeth and bones.

These and other essential metallic elements should be present either in our diets or in the supplements many of us rely on for balanced nutrition. However, the mere presence of these elements is not sufficient—they must be in a form that our cells can use. Swallowing an 18 mg steel pellet as if it were an aspirin tablet would not be a good way for a young woman to get her recommended daily allowance of iron. Chewing a gram of calcium metal would be even less pleasant. If we are to benefit from consuming essential metals in food and nutritional supplements, the metals need to be in compounds, not free elements, and the compounds must be absorbable by the body.

All the metallic elements essential to human health occur in nature in ionic compounds, but not all ionic forms are absorbed equally well. For example, most of the iron in fish, poultry, and red meat is readily absorbed because it is present in a form called *heme iron*. However, the iron in plants is mostly nonheme and is not as readily absorbed. Eating a meal that includes both meat and vegetables improves the absorption of the nonheme iron in the vegetables, as does consuming foods high in vitamin C. All these dietary factors work together at the molecular level to provide us with the nutrients we need to survive.

Metal ions are essential to good health, but the nonmetal ions and molecules that accompany them are equally important because they make the metals chemically reactive and biologically available. Interactions between metals and these ions and molecules influence the solubility of the metals, which is a key factor that determines their activity in plants and animals. These interactions also influence other properties, including the wavelengths of visible light the metals absorb and therefore the colors of their compounds and solutions. In this chapter we explore how the chemical environment of metal ions in solids and in solutions affects their physical, chemical, and biological properties. We answer such questions as why many, but not all, metal compounds have distinctive colors, and how, through the formation of complex ions with biomolecules, metals play key roles in many biological processes.

Green Leaves The pigments in green plants include chlorophyll *a*, whose molecules ▶
each contain a ring of nitrogen atoms surrounding a Mg^{2+} ion.

LEARNING OUTCOMES

LO1 Evaluate the capacity of Lewis acids and bases to accept and donate pairs of electrons
Sample Exercises 16.1, 16.4

LO2 Recognize complex ions and their counter ions in chemical formulas
Sample Exercise 16.2

LO3 Interconvert the names and formulas of complex ions and coordination compounds
Sample Exercise 16.3

LO4 Explain the chelate effect and its importance

LO5 Describe the factors that lead to high-spin or low-spin magnetic states of complex ions
Sample Exercise 16.5

LO6 Identify stereoisomers of coordination compounds
Sample Exercise 16.6

LO7 Use formation constants to calculate the concentrations of free and complexed metal ions in solution
Sample Exercise 16.7

Lewis base a substance that *donates* a lone pair of electrons in a chemical reaction.

Lewis acid a substance that *accepts* a lone pair of electrons in a chemical reaction.

16.1 Lewis Acids and Bases

In this chapter we focus on the interactions between metal ions and the other ions and molecules that surround them in solids and solutions. To understand these interactions, we need to reconsider the definitions of acids and bases we used in Chapter 15. Let's begin by revisiting what happens when ammonia gas dissolves in water:

$$NH_3(g) + H_2O(\ell) \rightleftharpoons NH_4^+(aq) + OH^-(aq)$$

Figure 16.1(a) shows a Brønsted–Lowry interpretation of this reaction: in donating H^+ ions to ammonia, H_2O acts as a Brønsted–Lowry acid, and in accepting the protons NH_3 acts as a Brønsted–Lowry base.

Another way to view this reaction is illustrated in Figure 16.1(b). Instead of focusing on the transfer of hydrogen ions, consider the two reactants as a donor and an acceptor of *electron pairs*. In this view, the N atom in NH_3 shares its lone pair of electrons with one of the H atoms in H_2O. In the process, one of the H—O bonds in H_2O is broken in such a way that the bonding pair of electrons remains with the O atom. The lone pair from the N atom forms a fourth N—H covalent bond. The result is the same as in the Brønsted–Lowry model: a molecule of NH_3 bonds to a H^+ ion, forming an NH_4^+ ion. The molecule of H_2O that lost the H^+ ion becomes a OH^- ion.

Viewing this process as the donation and acceptance of an electron pair by two atoms, resulting in the formation of a covalent bond between them, provides the following basis for defining acids and bases:

➤ A **Lewis base** is a substance that *donates* a lone pair of electrons in a chemical reaction.

➤ A **Lewis acid** is a substance that *accepts* a lone pair of electrons in a chemical reaction.

FIGURE 16.1 (a) Brønsted–Lowry view of the reaction between H_2O (proton donor) and NH_3 (proton acceptor). (b) Lewis view of the reaction: H_2O acts as a Lewis acid (electron-pair acceptor) and NH_3 acts as a Lewis base (electron-pair donor).

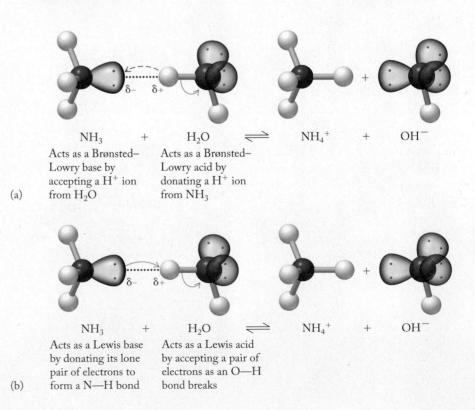

$$NH_3 \qquad + \qquad H_2O \qquad \rightleftharpoons \qquad NH_4^+ \qquad + \qquad OH^-$$

Acts as a Brønsted–Lowry base by accepting a H^+ ion from H_2O | Acts as a Brønsted–Lowry acid by donating a H^+ ion from NH_3

(a)

$$NH_3 \qquad + \qquad H_2O \qquad \rightleftharpoons \qquad NH_4^+ \qquad + \qquad OH^-$$

Acts as a Lewis base by donating its lone pair of electrons to form a N—H bond | Acts as a Lewis acid by accepting a pair of electrons as an O—H bond breaks

(b)

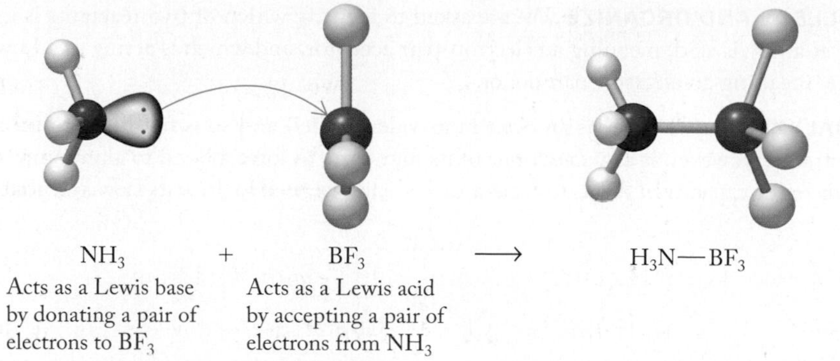

$$NH_3 \quad + \quad BF_3 \quad \longrightarrow \quad H_3N\text{—}BF_3$$

Acts as a Lewis base by donating a pair of electrons to BF_3

Acts as a Lewis acid by accepting a pair of electrons from NH_3

FIGURE 16.2 In the reaction between NH_3 and BF_3, NH_3 acts as a Lewis base and BF_3 acts as a Lewis acid.

These definitions are named after their developer, Gilbert N. Lewis, who pioneered research into the nature of chemical bonds. The Lewis definition of a base is consistent with the Brønsted–Lowry model because a substance must be able to donate a pair of electrons if it is to bond with a H^+ ion. However, the same parallelism is not true for acids. The Brønsted–Lowry model defines an acid as a hydrogen-ion donor, but the Lewis definition encompasses species that have no hydrogen ions to donate but that can still accept electrons. One such compound is boron trifluoride, BF_3.

With only six valence electrons and an empty valence-shell orbital, the boron atom in BF_3 can accept another pair of electrons to complete its octet. NH_3 is a suitable electron-pair donor, as shown in Figure 16.2. There is no transfer of H^+ ions in this reaction, so it is not an acid–base reaction according to the Brønsted–Lowry model. However, NH_3 donates a lone pair of electrons and BF_3 accepts them, so it is an acid–base reaction according to the broader Lewis model.

Many important Lewis bases are anions, including the halide ions, OH^-, and O^{2-}. To see how O^{2-} functions as a Lewis base, let's revisit the reaction described in Section 14.6 between SO_2 and CaO that is used to reduce emissions from power stations and smelters:

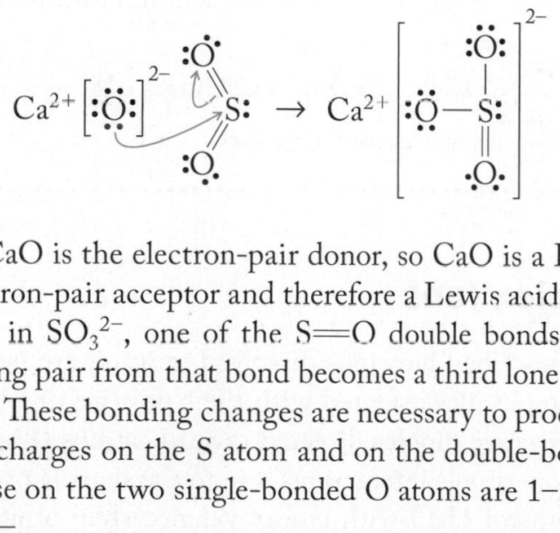

The oxide ion in CaO is the electron-pair donor, so CaO is a Lewis base. Sulfur dioxide is the electron-pair acceptor and therefore a Lewis acid. To accommodate the third O atom in SO_3^{2-}, one of the $S{=}O$ double bonds becomes a single bond, and a bonding pair from that bond becomes a third lone pair on one of the O atoms in SO_3^{2-}. These bonding changes are necessary to produce a structure in which the formal charges on the S atom and on the double-bonded O atom are both zero and those on the two single-bonded O atoms are 1–, giving the ion an overall charge of 2–.

CONNECTION Lewis's pioneering theories of the nature of covalent bonding are described in Section 4.3.

CONNECTION The concept of formal charge and its calculation are described in Section 4.7.

SAMPLE EXERCISE 16.1 **Identifying Lewis Acids and Bases** **LO1**

Which species is a Lewis acid and which is a Lewis base in this reaction?

$$AlCl_3 + Cl^- \rightarrow AlCl_4^-$$

coordinate bond formed when one anion or molecule donates a pair of electrons to another ion or molecule to form a covalent bond.

ligand a Lewis base bonded to the central metal ion of a complex ion.

complex ion an ionic species consisting of a metal ion bonded to one or more Lewis bases.

inner coordination sphere the ligands that are bound directly to a metal via coordinate bonds.

COLLECT AND ORGANIZE We are asked to identify which of two reactants is acting as a Lewis acid, meaning an electron-pair acceptor, and which is acting as a Lewis base, meaning an electron-pair donor.

ANALYZE The Cl^- ion has an octet in its valence shell and so is not likely to accept electrons. However, it can *donate* one of its four pairs to form a bond to aluminum. To analyze the capacity of $AlCl_3$ to act as a Lewis acid, we need to draw its Lewis structure:

This structure accounts for all the valence electrons (3×7 from 3 Cl atoms + 3 from Al = 24) with three Al—Cl single bonds and *no lone pairs* on the Al atom. This leaves Al with only 6 valence electrons and thus the capacity to accept one more pair—it can act as a Lewis acid.

SOLVE In this reaction, $AlCl_3$ is a Lewis acid and the Cl^- ion is a Lewis base. We can represent the reaction using Lewis structures:

THINK ABOUT IT Drawing the Lewis structure of $AlCl_3$ is the key to solving the problem. With its incomplete octet, $AlCl_3$ has the capacity to accept an additional pair of electrons, much like BF_3, so it may act as a Lewis acid.

These Lewis structures assume that $AlCl_3$ is a molecular compound and not, like most other metal chlorides, an ionic compound. This assumption is supported by the physical properties of $AlCl_3$ and particularly by the fact that it sublimes at 178°C and 1 atm. Most metal halides do not even melt until heated to temperatures many hundreds of degrees higher than that.

Practice Exercise Which reactant is the Lewis acid and which is the Lewis base in this reaction?

$$CO_2(g) + CaO(s) \rightarrow CaCO_3(s)$$

(Answers to Practice Exercises are in the back of the book.)

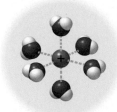

FIGURE 16.3 In Chapter 6 we explored how ions in aqueous solutions are surrounded by water molecules oriented with their positive dipoles (H atoms) directed toward anions and their negative dipoles (O atoms) directed toward cations. When these O atoms donate a lone pair of electrons to an empty orbital of a cation, a coordinate bond is formed.

16.2 Complex Ions

In Chapter 6, we described how ions dissolved in water are *hydrated*, that is, surrounded by water molecules oriented with their positive dipoles directed toward anions and their negative dipoles directed toward cations (Figure 16.3). In some hydrated cations, ion–dipole interactions lead to the sharing of lone-pair electrons on the oxygen atoms of H_2O with empty valence-shell orbitals on the cations. These shared electron pairs meet our definition of covalent bonds, but these particular bonds are called *coordinate covalent bonds*, or simply **coordinate bonds**. Such bonds form when either a molecule or an anion donates a lone pair of electrons to an empty valence-shell orbital of an atom, molecule, or cation. Once formed, a coordinate bond is indistinguishable from any other kind of covalent bond. Much of the chemistry of the transition metals is associated with their ability to form coordinate bonds with molecules or anions.

•••••••••••••••••••••••••••••••••••••

Is the N—B bond that forms between NH_3 and BF_3 (as shown in Figure 16.2) a coordinate bond? Explain your answer.

(Answers to Concept Tests are in the back of the book.)

•••

Molecules or anions (Figure 16.4) that function as Lewis bases and form coordinate bonds with metal cations are called **ligands**. The resulting species, which are composed of central metal ions and the surrounding ligands, are called **complex ions** or simply *complexes*. Direct bonding to a central cation means that the ligands in a complex occupy the **inner coordination sphere** of the cation. Recall Sample Exercise 14.11, in which we discussed the equilibrium between two forms of cobalt(II), one pink and one blue, in an aqueous solution of HCl:

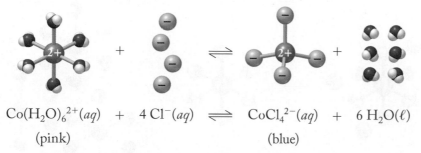

$$Co(H_2O)_6^{2+}(aq) + 4\,Cl^-(aq) \rightleftharpoons CoCl_4^{2-}(aq) + 6\,H_2O(\ell)$$
(pink) (blue)

Both the pink and blue forms are complexes; the pink cobalt(II) species has 6 water ligands in its inner coordination sphere, and the blue species has 4 chloride ions. The charge on each complex ion is the sum of the charges of the metal ion and the ligands: 2+ for $Co(H_2O)_6^{2+}$ because the charge of the cobalt ion is 2+ and water molecules are neutral, and 2− for $CoCl_4^{2-}$ because the sum of the 2+ charge on the cobalt ion and four 1− charges on the chloride ions is 2−.

The foundation of our understanding of the bonding and structure of complex ions comes from the pioneering research of Swiss chemist Alfred Werner (1866–1919), for which he was awarded the Nobel Prize in Chemistry in 1913. Some of Werner's research addressed the unusual behavior of different compounds formed by dissolving cobalt(II) chloride in aqueous ammonia and oxidizing it to cobalt(III) by bubbling air through the solution. One of the redox reactions produces an orange compound (Figure 16.5a) that contains 3 moles of Cl^- ions for every one mole of Co^{3+} ions and 6 moles of ammonia. Another reaction produces a purple compound (Figure 16.5b) that has the same proportions of Cl^- and Co^{3+} ions, but with only 5 moles of ammonia per mole of cobalt(III). Both compounds are water-soluble solids that react with aqueous solutions of $AgNO_3$, forming solid AgCl. However, one mole of the orange compound produces 3 moles of solid AgCl, whereas one mole of the purple one produces only 2 moles of AgCl. Results like these inspired Werner to study the electrical conductivity of aqueous solutions of the two compounds. He found that the orange compound was the better conductor, indicating that it produced more ions in solution than the purple one.

Werner concluded that the differences in the two compounds' composition, in their capacities to react with Ag^+ ions, and in their electrolytic properties are all caused by the presence of two different types of Co—Cl bonds inside them. He proposed that these bonds are all ionic in the orange compound, but that only two-thirds of them are ionic in the purple compound. The remaining one-third are not ionic, but rather coordinate covalent.

This capacity to form two different kinds of bonds means that Co^{3+} and other transition metal ions have two different kinds of bonding capacity, or *valence*. The

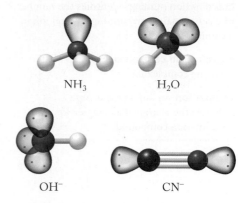

NH₃ H₂O

OH⁻ CN⁻

FIGURE 16.4 Anions or molecules with lone pairs of electrons have the capacity to donate those electrons to metal cations. When these compounds do this, they are called ligands. All these ligands except NH_3 have more than one lone pair, but only one pair at a time can be directed toward a single metal cation.

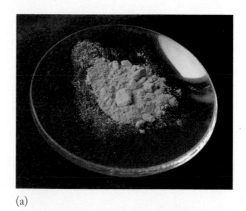

(a)

(b)

FIGURE 16.5 Two compounds of cobalt(III) chloride and ammonia. (a) $[Co(NH_3)_6]Cl_3$, (b) $[Co(NH_3)_5Cl]Cl_2$.

CONNECTION We discussed in Chapter 8 how the electrical conductivity of aqueous solutions depends on the concentrations of dissolved ions in the solutions.

coordination number identifies the number of electron pairs surrounding a metal ion in a complex.

coordination compound made up of at least one complex ion.

counter ion an ion whose charge balances the charge of a complex ion in a coordination compound.

first kind involves ionic bonds and is based on the number of electrons a metal atom loses when it forms an ion. This valence is equivalent to its oxidation number, which is +3 for cobalt in both the orange and purple compounds. The second kind of valence is based on the capacity of metal ions to form coordinate bonds. This property corresponds to an ion's **coordination number**, the number of sites around a central metal ion where bonds to ligands form.

The formulas of the reactants in the two chemical equations below fit the chemical composition and explain the properties of the orange and purple cobalt(III) compounds. Note how brackets in the formulas of these **coordination compounds** set off the complex ions from the ionically bonded chloride **counter ions**, which balance the charges on the complex ions: 3+ for the orange one but only 2+ for the purple one because it contains a negatively charged Cl^- ion *within its inner coordination sphere*.

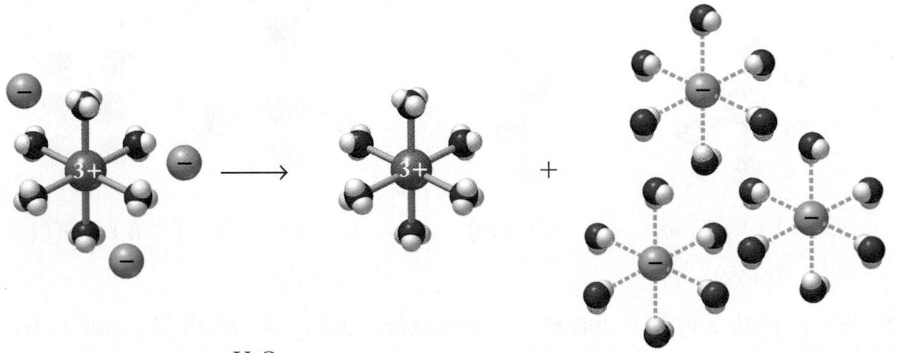

$$[Co(NH_3)_6]Cl_3(s) \xrightarrow{H_2O} Co(NH_3)_6^{3+}(aq) + 3\ Cl^-(aq)$$
(orange)

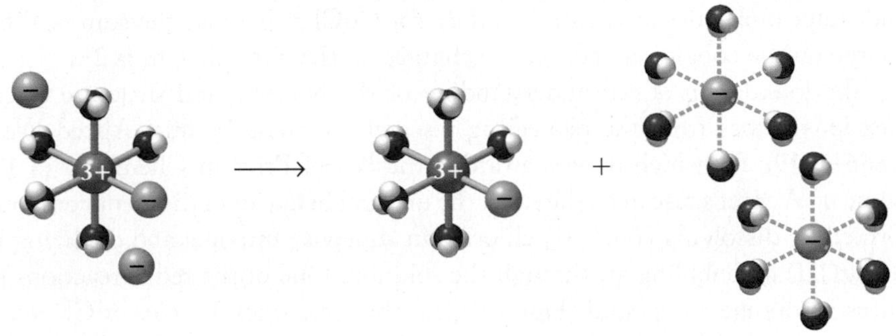

$$[Co(NH_3)_5Cl]Cl_2(s) \xrightarrow{H_2O} Co(NH_3)_5Cl^{2+}(aq) + 2\ Cl^-(aq)$$
(purple)

The release of 3 moles of chloride ions from the orange compound, but only 2 from the purple compound, explains the compounds' chemical and electrolytic properties.

Werner proposed, correctly, that the six ligands in the inner coordination sphere of cobalt(III) ions are arranged in an octahedral geometry, as shown in the structures above. This geometry is consistent with the formation of six bonds to a central atom (or central ion in this case), as we learned in Chapter 5. Many other transition metal ions form six coordinate bonds and octahedral complex ions. Some are listed in Table 16.1, as are complex ions that have only two or four ligands. Those with two have linear structures, whereas those with four may be tetrahedral or square planar, with only four of the pairs of electrons involved in bond formation. Both geometries are seen in cobalt(II) ions: $CoCl_4^{2-}$ is tetrahedral while $Co(CN)_4^{2-}$ is square planar.

CONNECTION We learned in Chapter 5 how the shapes of molecules depend on the number of bonding pairs and lone pairs of valence electrons on the central atom. The same rules apply to complex ions.

TABLE 16.1 Common Coordination Numbers and Shapes of Complex Ions

Coordination Number	Steric Number	Shape	Structure	Examples
6	6	Octahedral		$Fe(H_2O)_6{}^{3+}$ $Ni(H_2O)_6{}^{2+}$ $Co(H_2O)_6{}^{3+}$
4	6	Square planar		$Pt(NH_3)_4{}^{2+}$
4	4	Tetrahedral		$Zn(H_2O)_4{}^{2+}$
2	2	Linear		$Ag(NH_3)_2{}^{+}$

CONCEPT TEST

In the coordination compound $Na_3[Fe(CN)_6]$, which ions occupy the inner coordination sphere of the Fe^{3+} ion, and which ions are counter ions? Of the four cobalt complexes described in this section, which would have conductivity in aqueous solution similar to that of $Na_3[Fe(CN)_6]$?

SAMPLE EXERCISE 16.2 Writing Formulas of Coordination Compounds LO2

The first modern synthetic color, Prussian blue, was made in 1704 in Berlin, Germany. It was heavily used in the 19th-century Japanese woodblock print *The Great Wave off Kanagawa* (Figure 16.6), and it continues to be popular with artists worldwide. The formula for the insoluble pigment is $Fe_4[Fe(CN)_6]_3$.

a. What is the formula of the complex ion in this compound and what is its charge if the counter ion is Fe^{3+}?

b. What is the oxidation state of iron in the complex ion?

COLLECT AND ORGANIZE We have the formula of a coordination compound and are asked to identify the complex ion and the counter ion.

ANALYZE The complex ion is contained in brackets, the counter ion is outside the brackets, and the compound must be electrically neutral.

SOLVE

a. The counter ion is given as Fe^{3+}. The complex ion is $[Fe(CN)_6]^{n-}$. If four Fe^{3+} $(4 \times 3+) = 12+$, then to balance the charge on three complex ions, n must equal $3 \times n = 12-$ and n must equal 4.

$$[Fe(CN)_6]^{4-}$$

b. Six cyanide ligands, CN^-, occupy the inner coordination sphere for a total charge of 6−. Therefore the charge on Fe must be 2+ for the charge on the complex ion to equal 4−.

FIGURE 16.6 The source of the blue color in *The Great Wave off Kanagawa*, a famous woodblock print by the Japanese artist Hokusai, is the pigment Prussian blue.

THINK ABOUT IT Complex ions may have positive or negative charges. Correspondingly, counter ions will have charges opposite to that of the complex to ensure electrical neutrality.

Practice Exercise A coordination compound of ruthenium, $[Ru(NH_3)_4Cl_2]Cl$, has shown some activity against leukemia in animal studies. Identify the complex ion, determine the oxidation number of the metal, and identify the counter ion. ⚙

16.3 Naming Complex Ions and Coordination Compounds

The names of complex ions and coordination compounds tell us the identity and oxidation state of the central ion, the names and numbers of ligands, the charge in the case of complex ions, and the identity of counter ions. To convey all this information, we need to follow some naming rules.

Complex Ions with a Positive Charge

1. Start with the identities of the ligand(s). Names of common ligands appear in Table 16.2. If there is more than one kind of ligand, list the names alphabetically.
2. Use the usual prefix(es) in front of the name(s) written in step 1 to indicate the number of each type of ligand (Table 16.3).
3. Write the name of the metal ion with a Roman numeral indicating its oxidation state.

Examples:

Formula	Name	Structure
$Ni(H_2O)_6{}^{2+}$	Hexaaquanickel(II)	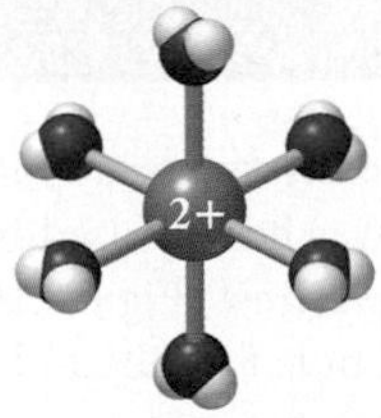
$Co(NH_3)_6{}^{3+}$	Hexaamminecobalt(III)	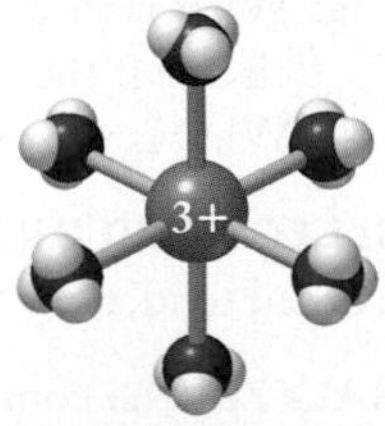
$Cu(NH_3)_4(H_2O)_2{}^{2+}$	Tetraamminediaquacopper(II)	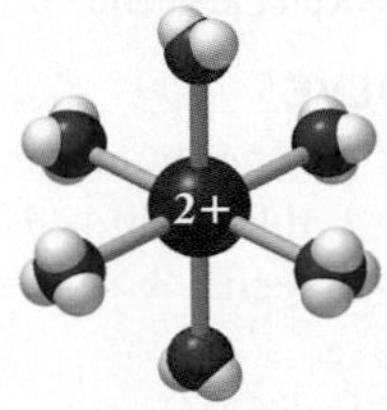

TABLE 16.2 **Names and Structures of Common Ligands**

Ligand	Name within Complex Ion	Structure	Charge	Number of Donor Groups
Iodide	Iodo	I$^-$	1–	1
Bromide	Bromo	Br$^-$	1–	1
Chloride	Chloro	Cl$^-$	1–	1
Fluoride	Fluoro	F$^-$	1–	1
Nitrite	Nitro	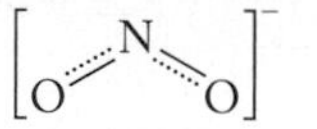 $\left[O\overset{N}{\underset{}{}}O\right]^-$	1–	1
Hydroxide	Hydroxo	[O—H]$^-$	1–	1
Water	Aqua	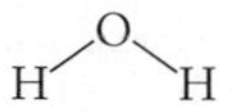 H—O—H	0	1
Pyridine (py)	Pyridyl	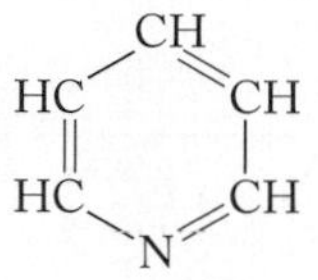 (pyridine ring structure)	0	1
Ammonia	Ammine	NH_3	0	1
Ethylenediamine (en)	(same)[a]	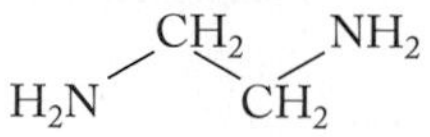 $H_2N—CH_2—CH_2—NH_2$	0	2
2,2'-Bipyridine (bipy)	Bipyridyl	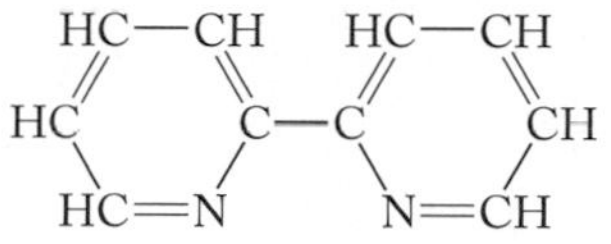 (bipyridine structure)	0	2
1,10-Phenanthroline (phen)	(same)[a]	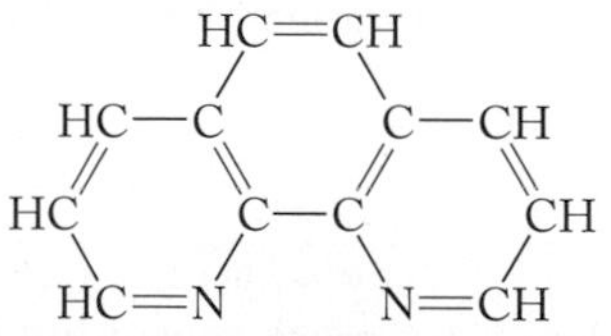 (phenanthroline structure)	0	2
Cyanide[b]	Cyano	[C≡N]$^-$	1–	1
Carbon monoxide[b]	Carbonyl	C≡O	0	1

[a]The names of some electrically neutral ligands in complexes are the same as the names of the molecules.
[b]Carbon atoms are the lone pair donors in these ligands.

TABLE 16.3 **Common Prefixes Used in the Names of Complex Ions**

Number of Ligands	Prefix
2	di-
3	tri-
4	tetra-
5	penta-
6	hexa-

It may seem strange having two *a*'s together in these names, but it is permitted under current naming rules. Prefixes are ignored in determining alphabetical order, which is why *ammine* comes before *aqua* rather than *di* before *tetra* in tetraamminediaquacopper(II).

In these three examples the ligands are all electrically neutral. This makes determining the oxidation state of the central metal ion a simple task because the charge on the complex ion is the same as the charge on the metal ion, which is the oxidation state of the metal. When the ligands include anions, determining the oxidation state of the central metal ion requires us to account for these charges.

Complex Ions with a Negative Charge

1. Follow the steps for naming positively charged complexes.
2. Add *-ate* to the name of the central metal ion to indicate that the complex ion carries a negative charge (just as we use *-ate* to end the names of oxoanions). For some metals, the base name changes, too. The two most common examples are iron, which becomes *ferrate*, and copper, which becomes *cuprate*.

Examples:

Formula	Name	Structure
$Fe(CN)_6^{3-}$	Hexacyanoferrate(III)	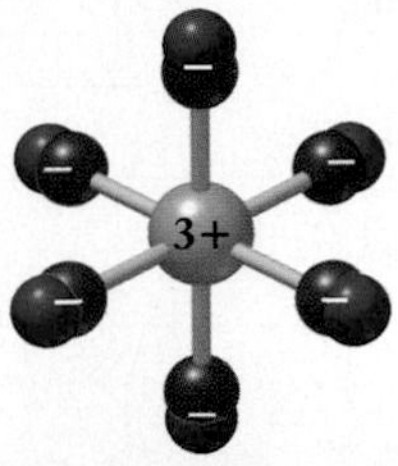
$[Fe(H_2O)(CN)_5]^{3-}$	Aquapentacyanoferrate(II)	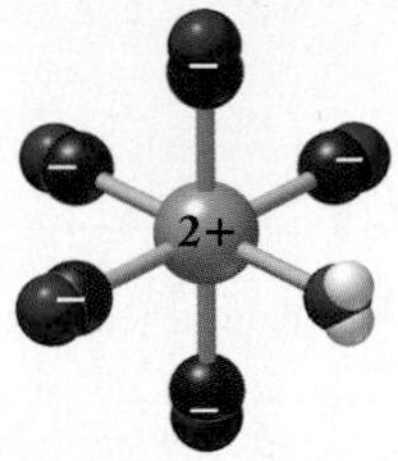
$[Al(H_2O)_2(OH)_4]^-$	Diaquatetrahydroxoaluminate	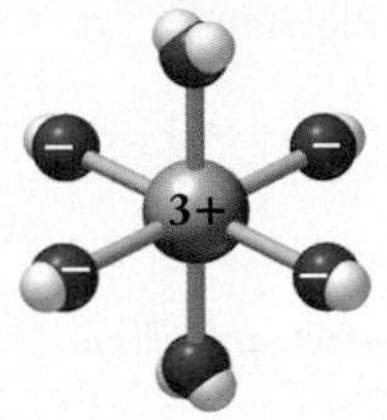

In the first two examples we must determine the oxidation state of Fe. We start with the charge on the complex ion and then take into account the charges on the ligand anions to calculate the charge on the metal ion. For example, the overall charge of the aquapentacyanoferrate(II) ion is 3−. It contains five CN^- ions. To reduce the combined charge of 5− from these cyanide ions to an overall charge of 3−, the charge on Fe must be 2+.

CONCEPT TEST

What is the name of the complex anion with the formula $PtCl_4^{2-}$?

Coordination Compounds

1. If the counter ion of the complex ion is a cation, the cation name goes first, followed by the name of the anionic complex ion.
2. If the counter ion of the complex ion is an anion, the name of the cationic complex ion goes first, followed by the name of the anion.

Examples:

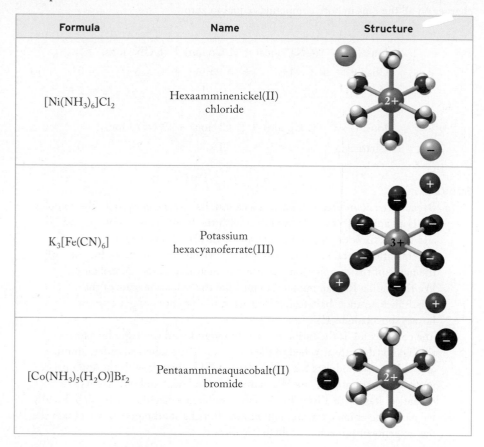

Formula	Name	Structure
[Ni(NH$_3$)$_6$]Cl$_2$	Hexaamminenickel(II) chloride	
K$_3$[Fe(CN)$_6$]	Potassium hexacyanoferrate(III)	
[Co(NH$_3$)$_5$(H$_2$O)]Br$_2$	Pentaammineaquacobalt(II) bromide	

A key to naming coordination compounds is to recognize from their formulas that they *are* coordination compounds. For help with this, look for formulas that have the atomic symbols of a metallic element and one or more ligands, all in brackets, either followed by the atomic symbol of an anion, as in [Co(NH$_3$)$_5$(H$_2$O)]Br$_2$, or preceded by the symbol of a cation, as in K$_3$[Fe(CN)$_6$].

SAMPLE EXERCISE 16.3 Naming Coordination Compounds LO3

Name the coordination compounds (a) Na$_4$[Co(CN)$_6$] and (b) [Co(NH$_3$)$_5$Cl](NO$_3$)$_2$.

COLLECT AND ORGANIZE We are asked to write a name for each compound that unambiguously identifies its composition. The formulas of the complex ions appear in brackets in both compounds. Because cobalt, the central metal ion in both, is a transition metal, we express its oxidation state using Roman numerals. The names of common ligands are given in Table 16.2.

ANALYZE It is useful to take an inventory of the ligands and counter ions:

Compound	Counter Ion	LIGAND			
		Formula	Name	Number	Prefix
Na$_4$[Co(CN)$_6$]	Na$^+$	CN$^-$	Cyano	6	Hexa-
[Co(NH$_3$)$_5$Cl](NO$_3$)$_2$	NO$_3^-$	NH$_3$	Ammine	5	Penta-
		Cl$^-$	Chloro	1	—

The oxidation state of each cobalt ion can be calculated by setting the sum of the charges on all the ions in both compounds equal to zero:

a.

Ions:	(4 Na$^+$ ions) + (1 Co ion) + (6 CN$^-$ ions)
Charges:	4+ + x + 6– = 0

$$x = 2+$$

b.

Ions:	(1 Co ion) + (1 Cl$^-$ ion) + (2 NO$_3^-$ ions)
Charges:	x + 1– + 2– = 0

$$x = 3+$$

SOLVE

a. Because the counter ion, *sodium*, is a cation, its name comes first. The complex ion is an anion. To name it, we begin with the ligand *cyano*, to which we add the prefix *hexa–* and write *hexacyano*. This is followed by the name of the transition metal ion: hexacyano*cobalt*. We add *-ate* to the ending of the name of the complex ion because it is an anion: hexacyanocobalt*ate*. We then add a Roman numeral to indicate the oxidation state of the cobalt: hexacyanocobaltate(*II*). Putting it all together, we get sodium hexacyanocobaltate(II).

b. The complex ion is the cation in this compound, and we begin by naming the ligands directly attached to the metal ion in alphabetical order: ammine and chloro. We indicate the number (5) of NH_3 ligands with the appropriate prefix: *penta*amminechloro. We name the metal next and indicate its oxidation state with a Roman numeral: pentaamminechloro*cobalt(III)*. Finally we name the anionic counter ion: *nitrate*. Putting it all together, we obtain the name: pentaamminechlorocobalt(III) nitrate.

THINK ABOUT IT Naming coordination compounds requires us to (1) distinguish between ligands and counter ions and (2) recall which ligands are electrically neutral and which are anions. The structures of the complex ions in the named coordination compounds are shown in Figure 16.7.

Practice Exercise Identify the ligands and counter ions in (a) [Zn(NH$_3$)$_4$]Cl$_2$ and (b) [Co(NH$_3$)$_4$(H$_2$O)$_2$](NO$_2$)$_2$, and name each compound. ⚙

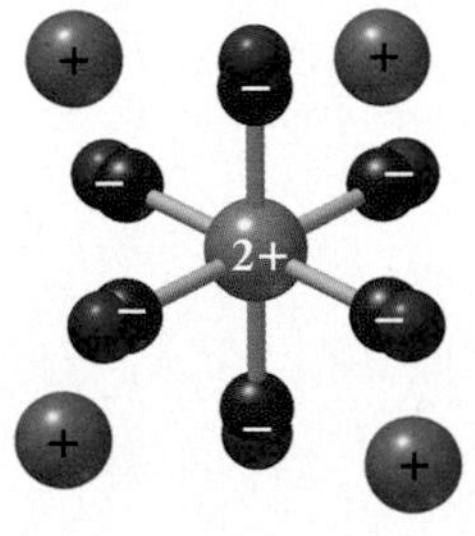

(a)

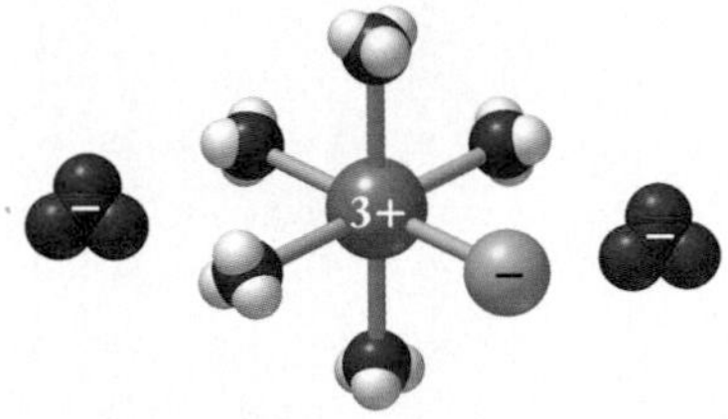

(b)

FIGURE 16.7 The structures of
(a) sodium hexacyanocobaltate(II) and
(b) pentaamminechlorocobalt(III) nitrate.

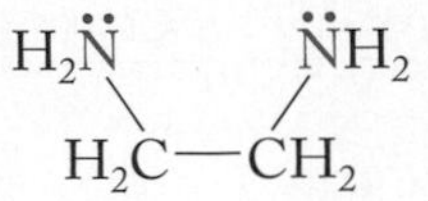

monodentate ligand a species that forms only a single coordinate bond to a metal ion in a complex.

polydentate ligand a species that can form more than one coordinate bond per molecule.

chelation the interaction of a metal with a polydentate ligand (chelating agent); pairs of electrons on one molecule of the ligand occupy two or more coordination sites on the central metal.

16.4 Polydentate Ligands

The ligands in Figure 16.4 and many of those in Table 16.2 can donate only one pair of electrons to a single metal ion. Even those with more than one lone pair can donate only one pair at a time to a given metal ion because the other lone pair or pairs are oriented away from the metal ion. Because these ligands have effectively only one donor group, they are called **monodentate ligands**, which literally means "single-toothed."

Molecules larger than those in Figure 16.4 may be able to donate more than one lone pair of electrons and therefore form more than one coordinate bond to a central metal ion. Ligands in this category are called **polydentate ligands**, or more specifically *bidentate*, *tridentate*, and so on. One group of polydentate ligands is the polyamines, which includes the compound ethylenediamine, a bidentate ligand that has the structure

$$H_2\ddot{N} \qquad \ddot{N}H_2$$
$$H_2C—CH_2$$

The lone pairs on the two $-NH_2$ groups are separated from each other by two $-CH_2-$ groups. This combination means that a molecule of ethylenediamine can partially encircle a metal ion so that both lone pairs can bond to the same metal ion.

Formation of the ethylenediamine complex of $Ni^{2+}(aq)$ is shown in Figure 16.8(a). The two bonding orbitals that accept lone pairs of electrons from a molecule of ethylenediamine must be on the same side of the Ni^{2+} ion. However, two more ethylenediamine molecules can bond to other pairs of bonding sites, displacing additional pairs of water molecules and forming a complex in which the Ni^{2+} ion is surrounded by three bidentate ethylenediamine molecules, as shown in Figure 16.8(b).

Note that each ethylenediamine forms a five-atom ring with the metal ion. If the ring were a regular pentagon (meaning all bond lengths and bond angles were exactly the same), each of its bond angles would be 108°. These pentagons are not perfect, but each ring's preferred octahedral bond angles of 90° for the N—Ni—N bond and of 107° to 109° for all the other bonds are accommodated with only a little strain on the ideal bond angles.

An even larger ligand, diethylenetriamine ($H_2NCH_2CH_2NHCH_2CH_2NH_2$), is shown in Figure 16.9(a). The lone pairs of electrons on its three nitrogen atoms give diethylenetriamine the capacity to form three coordinate bonds to a metal ion, meaning this is a *tridentate* ligand (Figure 16.9b).

As you may imagine, larger molecules may have even more atoms per molecule that can bond to a single metal ion. The interaction of a metal ion with a ligand having multiple donor atoms is called **chelation** (pronounced *key-LAY-shun*). The word comes from the Greek *chele*, meaning "claw." The polydentate ligands that take part in these interactions are called *chelating agents*.

Many chelating agents have more than one kind of Lewis base. *Aminocarboxylic acids* represent one family of such compounds. The most important of them is ethylenediaminetetraacetic acid, EDTA, the molecular structure of which is shown in Figure 16.10(a). Note that one molecule of EDTA contains two amine (nitrogen-containing) groups and four carboxylic acid (–COOH) groups. When the acid groups release their H^+ ions, they form four carboxylate anions, $-COO^-$, in which either of the O atoms can donate a pair of electrons to a central metal ion. When O

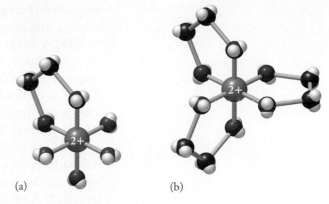

FIGURE 16.8 (a) The bidentate ligand ethylenediamine has two N atoms that can each donate a pair of electrons to empty orbitals of adjacent octahedral bonding sites on the same $Ni^{2+}(aq)$ ion (gold sphere), displacing two molecules of water. (b) Three ethylenediamine molecules occupy all six octahedral coordination sites of a Ni^{2+} ion.

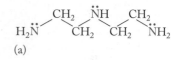

(a)

(b)

FIGURE 16.9 Tridentate chelation. (a) The three amine groups in the tridentate ligand diethylenetriamine are all potential electron-pair donor groups. (b) When these groups donate their lone pairs of electrons to a $Ni^{2+}(aq)$ ion (gold sphere), they occupy three of the six coordination sites on the ion.

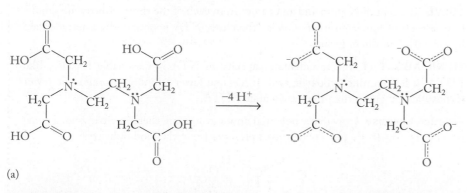

(a)

(b)

FIGURE 16.10 (a) In the hexadentate ligand EDTA, the six donor groups are the two amine groups and the four carboxylic acid groups. The acid groups ionize to form carboxylate anions. (b) All six Lewis base groups in ionized EDTA can form a coordinate bond with the same metal ion, such as Co^{3+} (the gold sphere) shown here. In the process they form four 5-membered rings.

atoms on all four groups do so and the two amine groups do as well, six octahedral bonding sites around the metal ion can be occupied, as shown in Figure 16.10(b).

EDTA forms very stable complex ions and is used as a metal ion *sequestering agent*, that is, a chelating agent that binds metal ions so tightly that they are "sequestered" and prevented from reacting with other substances. For example, EDTA is used as a preservative in many beverages and prepared foods because it sequesters iron, copper, zinc, manganese, and other transition metal ions often present in these foods that can catalyze the degradation of ingredients in the foods. Many foods are fortified with ascorbic acid (vitamin C), which is particularly vulnerable to metal-catalyzed degradation because it is also a polydentate ligand and is more likely to be oxidized when chelated to one of the above metal ions. EDTA effectively shields vitamin C from these ions. We explore the preferential binding of metal ions to different ligands in the next section.

SAMPLE EXERCISE 16.4 **Identifying the Potential Electron-Pair-Donor Groups in a Molecule** **LO1**

How many donor groups does this polydentate ligand, nitrilotriacetic acid (NTA), have?

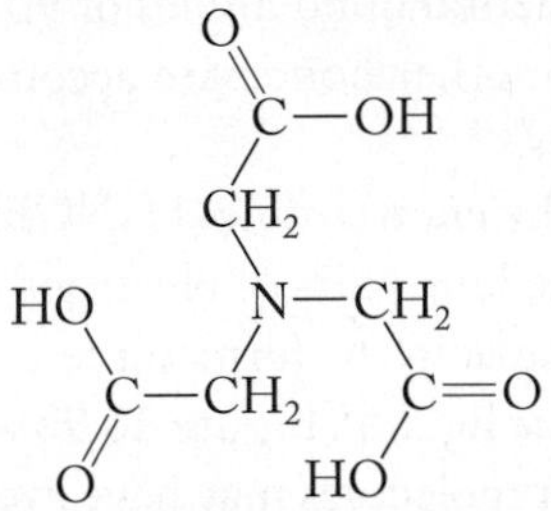

COLLECT AND ORGANIZE We need to examine this molecular structure to find electron pairs that can be donated. The molecule has a nitrogen atom with three single bonds and three carboxylic acid groups.

ANALYZE The three single bonds around the N atom mean its valence shell must also contain a lone pair of electrons. When all three carboxylic acid groups are ionized, there are three carboxylate groups in the molecule. Each carboxylate group can donate one nonbonding pair of electrons from one of its two oxygen atoms to a metal atom. (The other O atom would be oriented away from the metal ion and unable to also bond with it.) When an O atom from a carboxylate group and the center N atom simultaneously form coordinate bonds with a metal ion, a five-atom ring is produced, as in EDTA complexes.

SOLVE The central N atom and an O atom from each of the three carboxylate groups form a total of four coordinate bonds. Therefore, NTA is potentially a tetradentate ligand with four donor groups.

THINK ABOUT IT The tetradentate capacity of NTA is reasonable because, like EDTA, it is an aminocarboxylic acid. It has one fewer amino group and one fewer carboxylic acid group than the hexadentate EDTA.

Practice Exercise How many potential donor groups are there in citric acid, a component of citrus fruits and a widely used preservative in the food industry?

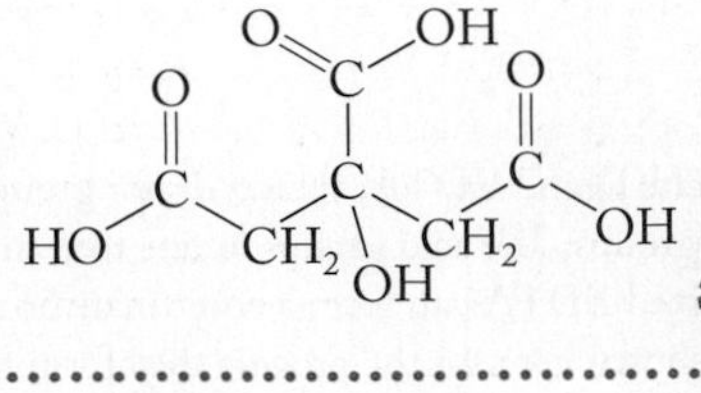

16.5 Ligand Strength and the Chelate Effect

Let's begin our discussion of the relative strengths of ligands as Lewis bases by exploring the relative affinity of $Ni^{2+}(aq)$ ions for two common monodentate ligands, H_2O and NH_3. We start by dissolving crystals of nickel(II) chloride hexahydrate in water. The formula of this compound is often written $NiCl_2 \cdot 6\,H_2O$. The dot connecting the two halves of the formula and the prefix *hexa* tell us that crystals of nickel(II) chloride contain Ni^{2+} ions that are surrounded by six water molecules. Brilliant green crystals of $NiCl_2 \cdot 6\,H_2O$ form green aqueous solutions (Figure 16.11). The fact that the solid and its solution are both green tells us that the same clustering of water molecules around Ni^{2+} ions occurs in both solid $NiCl_2 \cdot 6\,H_2O$ and aqueous solutions of Ni^{2+} ions. Thus, the Ni^{2+} ion in a solution of nickel(II) chloride is $Ni(H_2O)_6{}^{2+}$.

Now let's bubble colorless ammonia gas through a green solution of $Ni(H_2O)_6{}^{2+}$ ions. As shown in the middle test tube in Figure 16.11(b), the green solution turns blue. The color change means that different ligands are bonded to the Ni^{2+} ions. We may conclude that NH_3 molecules have displaced at least some H_2O molecules around the Ni^{2+} ions. If all the molecules of H_2O are displaced, the complex $Ni(NH_3)_6{}^{2+}$ is formed. The following chemical equation describes this change:

$$Ni(H_2O)_6{}^{2+}(aq) + 6\,NH_3(g) \rightarrow Ni(NH_3)_6{}^{2+}(aq) + 6\,H_2O(\ell)$$

Keep in mind that the hydrated ion $Ni(H_2O)_6{}^{2+}$ is often expressed as $Ni^{2+}(aq)$.

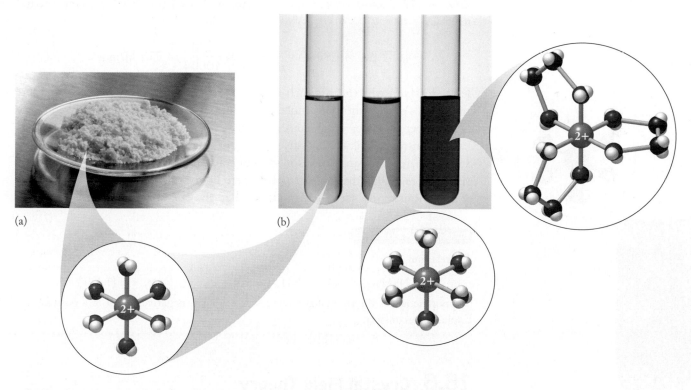

(a) (b)

FIGURE 16.11 (a) Solid nickel(II) chloride hexahydrate is green. (b) When it dissolves in water, the resulting solution has the same green color, telling us that each Ni^{2+} ion (gold sphere) must be surrounded by H_2O molecules both in the solid and in the solution. When ammonia gas is bubbled through a solution of $Ni(H_2O)_6{}^{2+}$, the color changes to blue as NH_3 replaces H_2O in the Ni^{2+} ion's inner coordination sphere. When ethylenediamine is added to a solution of $Ni(NH_3)_6{}^{2+}$, the color turns from blue to purple as the ethylenediamine displaces the ammonia ligands and the $Ni(en)_3{}^{2+}$ complex forms.

chelate effect the greater affinity of metal ions for polydentate ligands than for monodentate ligands.

This *ligand displacement* reaction tells us that Ni^{2+} ions have a greater affinity for molecules of NH_3 than for molecules of H_2O. Many other transition metal ions also have a greater affinity for ammonia than for water. We may conclude that ammonia is inherently a better electron-pair donor and hence a stronger Lewis base than water. This conclusion is reasonable because we saw in Chapter 15 that ammonia was also a stronger Brønsted–Lowry base than H_2O.

Next we add ethylenediamine to the blue solution of $Ni(NH_3)_6^{2+}$ ions. The solution changes color again: from blue to purple (Figure 16.11b), indicating yet another change in the ligands surrounding the Ni^{2+} ions. Molecules of ethylenediamine (en) displace ammonia molecules from the inner coordination sphere of Ni^{2+} ions as described in the following chemical equation:

$$Ni(NH_3)_6^{2+}(aq) + 3\ en(aq) \rightarrow Ni(en)_3^{2+}(aq) + 6\ NH_3(aq) \qquad (16.1)$$

Why should the affinity of Ni^{2+} ions for ethylenediamine be so much greater than their affinity for ammonia? After all, in both ligands the coordinate bonds are formed by lone pairs of electrons on N atoms. The color change tells us that this reaction as written is spontaneous. As we discussed in Chapter 12, spontaneous reactions are those in which free energy decreases ($\Delta G < 0$). Furthermore, under standard conditions the change in free energy ($\Delta G°$) is related to the changes in enthalpy and entropy that accompany the reaction:

$$\Delta G° = \Delta H° - T\,\Delta S°$$

The displacement of NH_3 by ethylenediamine is exothermic, but only slightly ($\Delta H° = -12$ kJ/mol). More important, $\Delta S° = +185$ J/(mol·K). This means that at 25°C,

$$T\,\Delta S° = 298\ \cancel{K} \times \frac{185\ \cancel{J}}{mol \cdot \cancel{K}} \times \frac{1\ kJ}{1000\ \cancel{J}} = 55.1\ kJ/mol$$

To understand why there is such a large increase in entropy, consider that there are 4 moles of reactants but 7 moles of products in Equation 16.1. Nearly doubling the number of moles of aqueous products over reactants translates into a large gain in entropy. It is this positive $\Delta S°$ more than the negative $\Delta H°$ value that drives the reaction and makes it spontaneous. Entropy gains drive many complexation reactions that involve polydentate ligands. The entropy-driven affinity of metal ions for polydentate ligands is called the **chelate effect**.

CONCEPT TEST

The reaction

$$Ni(H_2O)_6^{2+}(aq) + 6\ NH_3(aq) \rightleftharpoons Ni(NH_3)_6^{2+}(aq) + 6\ H_2O(\ell)$$

is spontaneous. Is the spontaneity due principally to a negative ΔH or to a positive ΔS? Explain your answer.

16.6 Crystal Field Theory

We have seen that formation of complex ions can change the color of solutions of transition metals. Why is this? The colors of transition metal compounds and ions in solution are due to transitions of d-orbital electrons. Let's explore these transitions using Cr^{3+} as our model transition metal ion (Figure 16.12).

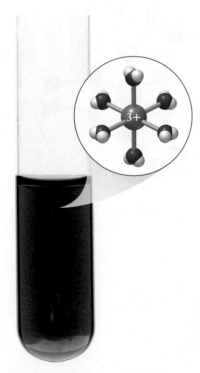

FIGURE 16.12 When chromium(III) nitrate dissolves in water, the resulting solution has a distinctive violet color due to the presence of $Cr(H_2O)_6^{3+}$ ions.

A Cr^{3+} ion has the electron configuration $[Ar]3d^3$. When a Cr^{3+} ion (or any atom or ion) is in the gas phase, all the orbitals in a given subshell have the same energy (Figure 16.13a). However, when a Cr^{3+} ion is in an aqueous solution and surrounded by an octahedral array of water molecules in $Cr(H_2O)_6^{3+}$, the energies of its $3d$ orbitals are no longer all the same. As the six oxygen atoms of the water molecules approach the Cr^{3+} ion during coordinate bond formation, the $3d$ electrons of Cr^{3+} and the lone pairs of electrons on the ligands repel one another. These repulsions raise the energies of all the d orbitals, but to different extents. The $3d_{xy}$, $3d_{yz}$, and $3d_{xz}$ orbitals experience some increase in energy, but the energies of the $3d_{x^2-y^2}$ and $3d_{z^2}$ orbitals increase even more (Figure 16.13b) because the lobes of the $3d_{x^2-y^2}$ and $3d_{z^2}$ orbitals point directly toward the H_2O molecules' oxygen atoms at the corners of the octahedron formed by the ligands, and are repelled by the electrons on those O atoms (Figure 16.13c). The energies of the $3d_{xy}$, $3d_{yz}$, and $3d_{xz}$ orbitals are not raised as much because the lobes of these three orbitals do not point directly toward the corners of the octahedron, so the electron repulsion they experience is weaker.

This process of changing *degenerate* (equal-energy) orbitals into orbitals of different energies is known as **crystal field splitting**, and the difference in energy created by crystal field splitting is called **crystal field splitting energy (Δ)**. The name was originally used to describe splitting of d-orbital energies in ionic crystals, but the theory is also routinely applied to species in aqueous solutions.

In a $Cr(H_2O)_6^{3+}$ ion, three electrons are distributed among five $3d$ orbitals (Figure 16.13b). According to Hund's rule, each of the three electrons should occupy one of the three lower energy orbitals, leaving the two higher energy

▶❚❚ **CHEMTOUR** Crystal Field Splitting

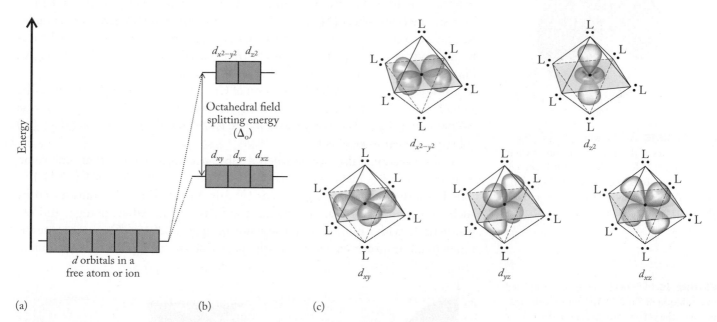

FIGURE 16.13 Octahedral crystal field splitting. (a) In an atom or ion in the gas phase, all orbitals in a subshell are degenerate, as shown here for the five $3d$ orbitals. (b) When an ion is part of a complex ion in a compound or solution, repulsions between electrons in the ion's d orbitals and ligand electrons raise the energy of the orbitals, as shown here for an octahedral field. (c) The greatest repulsion is experienced by electrons in the $d_{x^2-y^2}$ and d_{z^2} orbitals because the lobes of these orbitals are directed toward the corners of the octahedron and so are closest to the lone pairs on the ligands (L). The lobes of the lower energy d_{xy}, d_{yz}, and d_{xz} orbitals are directed toward points that lie between the corners of the octahedron, so electrons in them experience less repulsion.

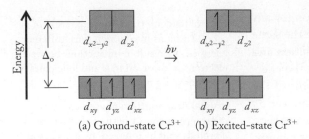

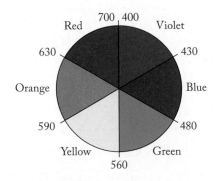

FIGURE 16.14 A Cr^{3+} ion, $[Ar]3d^3$, in an octahedral field can absorb a photon of light that has energy ($h\nu$) equal to Δ_o. This energy raises a $3d$ electron from (a) one of the lower energy d orbitals to (b) one of the higher energy d orbitals.

CONNECTION In Chapter 3, we first used the equation $E = h\nu = hc/\lambda$ in discussing the energy of light in the electromagnetic spectrum.

FIGURE 16.15 A color wheel. Colors on opposite sides of the wheel are complementary to each other. When we look at a solution or object that absorbs light corresponding to a given color, we see the complementary color. Wavelengths are in nanometers.

orbitals unoccupied, as shown in Figure 16.14(a). The energy difference between the two subsets of orbitals is symbolized by Δ_o, where the subscript indicates that the energy split was caused by an *octa*hedral array of electron repulsions.

What if a Cr^{3+} ion absorbs a photon whose energy is exactly equal to Δ_o? As the photon is absorbed, a $3d$ electron moves from a lower energy orbital to a higher energy orbital (Figure 16.14b). The wavelength λ of the absorbed photon is related to the energy difference between the two groups of orbitals—in other words, to the crystal field splitting energy—by Equation 16.2:

$$E = \frac{hc}{\lambda} = \Delta_o \tag{16.2}$$

As we discussed in Chapter 3, the energy and wavelength of a photon are inversely proportional to each other. Therefore, the larger the crystal field splitting in a complex ion, the shorter the wavelength of the photons the ion absorbs.

The size of the energy gap between split d orbitals often corresponds to radiation in the visible region of the electromagnetic spectrum. This means that the colors of solutions of metal complexes depend on the strengths of metal–ligand interactions. When white light (which contains all colors of visible light) passes through a solution containing complex ions, the ions may absorb energy corresponding to a particular color. The light leaving the solution and reaching our eyes is missing that color.

The color we perceive for any transparent object is not the color it absorbs but rather the color(s) that it transmits. To relate the color of a solution to the wavelengths of light it absorbs, we need to consider complementary colors as defined by a simple color wheel (Figure 16.15). For example, red and green are complementary colors; therefore, a solution that absorbs green light appears red to us because our eyes and brains process the transmitted colors—red, orange, yellow, blue, and violet—as the average of those colors, which is red.

Now we can explain why the solution of $Cr^{3+}(aq)$ in Figure 16.12 is violet. The radiation absorbed by the electron transition shown in Figure 16.14 happens to be yellow-orange light. Because yellow-orange is the complement of violet, $Cr^{3+}(aq)$ solutions appear to be violet.

Color-averaging also occurs when a substance absorbs more than one color, as many transition metal solutions do. For example, a solution of $Cu(NH_3)_4^{2+}$ ions has a distinctive deep blue color (Figure 16.16). The light transmitted by such a solution features an absorption band that spans yellow, orange, and red wavelengths with a minimum transmission of light at 620 nm. Our eyes sense the range that is transmitted, which mostly spans violet to blue-green. Then our brain

FIGURE 16.16 The visible light transmitted by a solution of $Cu(NH_3)_4^{2+}$ ions is missing much of the yellow, orange, and red portions of the visible spectrum because of a broad absorption band centered at 620 nm. Our eyes and brain perceive the transmitted colors as navy blue.

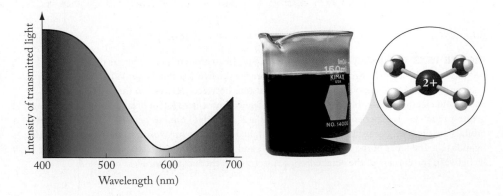

processes this band of transmitted colors and signals to us the average of these colors, a deep navy blue.

The nickel(II) and chromium(III) complexes we have examined up to this point are all octahedral and involve six ligands. However, in the solution of $Cu(NH_3)_4^{2+}$ the complex ions have only four ligands. This means that the deep blue color of this complex ion is caused by a different type of ligand–metal interaction.

Four ligands around a central metal have either a tetrahedral arrangement or a square planar arrangement (Table 16.1). Square planar geometries tend to be limited to the transition metal ions with nearly filled valence-shell d orbitals, particularly those with d^8 or d^9 electron configurations. One such ion is Cu^{2+}, which has the electron configuration $[Ar]3d^9$. The $Cu(NH_3)_4^{2+}$ complex (Figure 16.16) is square planar, which means that the strongest interactions occur between the $3d$ orbitals on the central ion and the nitrogen atom lone pairs at the four corners of the equatorial plane of the octahedron, as shown in Figure 16.17. The $3d$ orbital with the strongest interactions and so the highest energy is the $d_{x^2-y^2}$ orbital because its lobes are oriented directly at the four corners of the plane. The d_{xy} orbital has slightly less energy because its lobes, though in the xy plane, are directed 45° away from the corners. Electrons in the three d orbitals with most of their electron density out of the xy plane interact even less with the lone pairs of the ligand and thus have even lower energies.

Finally, let's consider the d orbital crystal field splitting that occurs in a tetrahedral complex (Figure 16.18a). In this geometry, the greatest electron–electron repulsions are experienced in the d_{xy}, d_{yz}, and d_{xz} orbitals because the lobes of these orbitals are oriented most directly to the corners of the tetrahedron, which are occupied by ligand electron pairs (Figure 16.18b). The two other d orbitals are less affected because their lobes do not point toward the corners. The difference in energy between the two subsets of d orbitals resulting from the tetrahedral interactions is labeled Δ_t.

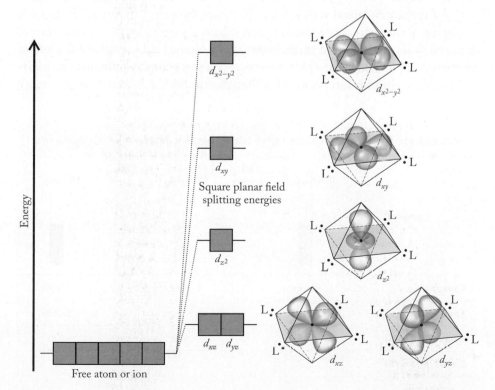

FIGURE 16.17 Square planar crystal field splitting. The d orbitals of a transition metal ion in a square planar field are split into several energy levels depending on how close the orbital lobes are to the ligand electrons at the four corners of the square. The $d_{x^2-y^2}$ orbital has the highest energy because its lobes are directed right at the four corners of the square plane.

Energy

Square planar field splitting energies

$d_{x^2-y^2}$

d_{xy}

d_{z^2}

d_{xz} d_{yz}

Free atom or ion

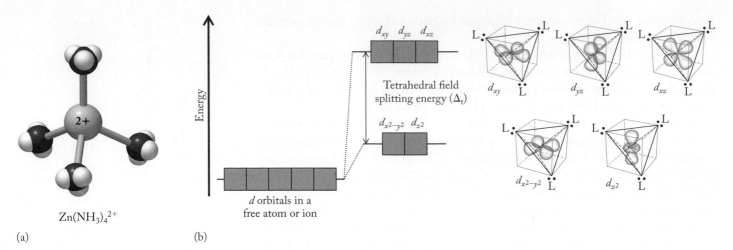

(a)

Zn(NH₃)₄²⁺

(b)

FIGURE 16.18 (a) In a tetrahedral complex ion, such as $Zn(NH_3)_4^{2+}$, the d orbitals of the metal ion undergo tetrahedral crystal field splitting. (b) The lobes of the higher energy orbitals—d_{xy}, d_{yz}, and d_{xz}—are closer to the ligands at the four corners of the tetrahedron than the lobes of the lower energy orbitals are. (One of the four corners of the tetrahedron is hidden in these drawings.)

Before ending this discussion on the colors of transition metal ions, let's revisit the color changes we saw in Figure 16.11, when first ammonia and then ethylenediamine were added to a solution of Ni^{2+} ions. Let's think in terms of the colors these solutions *absorb*. A solution of $Ni^{2+}(aq)$ ions is green because it absorbs colors at the red end of the spectrum. Similarly, a solution of $Ni(NH_3)_6^{2+}$ is blue because it absorbs colors opposite blue, which are centered on orange, and a solution of $Ni(en)_3^{2+}$ is violet because it absorbs colors centered on the complement of violet, which is yellow. Note how the colors these three solutions absorb are in the sequence red, orange, and yellow. This sequence runs from longest wavelength to shortest, from about 730 nm for red to about 570 nm for yellow. Radiant energy is inversely proportional to wavelength (Equation 16.2); therefore, the ability of these ligands to split the energies of the d orbitals of Ni^{2+} ions is en > NH_3 > H_2O. Table 16.4 summarizes the observed colors of these Ni^{2+} ion complexes and the meaning we can infer from their colors.

Chemists use the parameter *field strength* to describe the relative abilities of ligands to split the energies of d orbitals in metal ions, ranking ligands in what is called a **spectrochemical series**. Table 16.5 contains one such series. The ligands at the top of the chart create the strongest repulsions with electrons on the central ion, and those at the bottom create the weakest. As the field strength of the ligand increases, the crystal field splitting energy (Δ) increases. Consequently, high-field-strength ligands form complexes that absorb short-wavelength, high-energy

spectrochemical series a list of ligands rank-ordered by their ability to split the energies of the d orbitals of transition metal ions.

TABLE 16.4	**Light Transmitted and Absorbed by Three Ni²⁺ Complexes**				
Complex	$Ni(H_2O)_6^{2+}$	$\xrightarrow{NH_3}$	$Ni(NH_3)_6^{2+}$	$\xrightarrow{en}$	$Ni(en)_3^{2+}$
Appearance	Green		Blue		Violet
Absorbs	Red		Orange		Yellow
Absorbed λ (nm)	730	>	600	>	570
$E = hc/\lambda$ (J × 10¹⁹)	2.7	<	3.3	<	3.5

light, whereas complexes of low-field-strength ligands absorb long-wavelength, low-energy light.

CONCEPT TEST

Figure 16.19 is the absorption spectrum of a complex found in nature. What color is it?

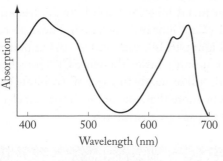

FIGURE 16.19

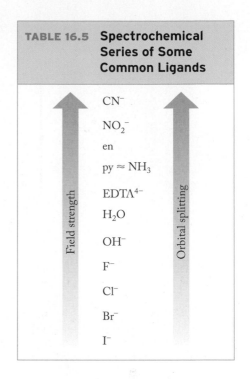

TABLE 16.5 Spectrochemical Series of Some Common Ligands

Field strength →

CN^-

NO_2^-

en

py ≈ NH_3

$EDTA^{4-}$

H_2O

OH^-

F^-

Cl^-

Br^-

I^-

← Orbital splitting

16.7 Magnetism and Spin States

In addition to determining the color of transition metal ions, crystal field splitting influences their magnetic properties because these properties depend on the number of unpaired electrons in the valence shell d orbitals. The larger this number, the more paramagnetic the ion. For example, an Fe^{3+} ion has five $3d$ electrons (Figure 16.20a). In an octahedral field there are two ways to distribute the five $3d$ electrons among these orbitals. One way conforms to Hund's rule and has a single electron in each orbital, leaving them all unpaired (Figure 16.20b). However, when Δ_o is large, as shown in Figure 16.20(c), all five electrons occupy the three lower energy orbitals. This pattern of electron distribution occurs when the energy of repulsion between two electrons in the same orbital is less than the energy needed to promote an electron to a higher energy orbital. In this configuration, only one electron is unpaired.

∞ **CONNECTION** We introduced the magnetic behavior of matter in Chapter 5 in our discussion of molecular orbital theory. Substances made of atoms, ions, or molecules that contain unpaired electrons are paramagnetic.

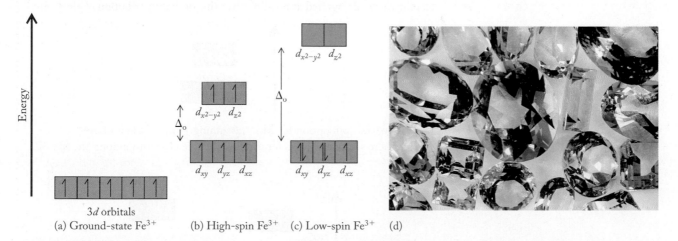

FIGURE 16.20 Low-spin and high-spin complexes. (a) The ground state of a free Fe^{3+} ion has a degenerate, half-filled set of $3d$ orbitals. (b) A weak octahedral field (Δ_o < electron-pairing energy) produces the high-spin state: five unpaired electrons, each in its own orbital. (c) In a strong octahedral field (Δ_o > electron-pairing energy), the energies of the $3d$ orbitals are split enough to produce the low-spin state: two sets of paired electrons, one unpaired electron, and two empty, higher energy orbitals. (d) The Fe^{3+} ions in crystals of aquamarine are high spin.

The configuration with all five electrons unpaired is called the *high-spin state* because the spin on all five electrons is in the same direction, resulting in the maximum magnetic field produced by the spins. The configuration with only one electron unpaired is called the *low-spin state*. Both configurations are paramagnetic because both have at least one unpaired electron. However, the high-spin state is much more paramagnetic.

Not all transition metal ions can have both high-spin and low-spin states. Consider, for example, Cr^{3+} ions in an octahedral field. Because each ion has only three $3d$ electrons (Figure 16.14), each electron is unpaired whether the orbital energies are split a lot or only a little. Therefore Cr^{3+} ions have only one spin state. Other metal ions with full or nearly full sets of d orbitals have only one spin state because there cannot be more than one or two unpaired electrons in these orbitals no matter how the electrons are distributed.

SAMPLE EXERCISE 16.5 **Predicting Spin States** **LO5**

Determine which of these ions can have high-spin and low-spin configurations when part of an octahedral complex: (a) Mn^{4+}; (b) Mn^{2+}; (c) Cu^{2+}.

COLLECT AND ORGANIZE Mn and Cu are in groups 7 and 11 of the periodic table, so their atoms have 7 and 11 valence electrons, respectively. In an octahedral field, a set of five d orbitals splits into a low-energy subset of three orbitals and a high-energy subset of two orbitals.

ANALYZE To determine whether high-spin and low-spin states are possible, we need to determine the number of d electrons in each ion. Then we need to distribute them among sets of d orbitals split by an octahedral field to see if it is possible for the ions to have different spin states. The electron configurations of the atoms of the three elements are $[Ar]3d^54s^2$ for Mn and $[Ar]3d^{10}4s^1$ for Cu. When they form cations, the atoms of these transition metals lose their $4s$ electrons first and then their $3d$ electrons. Therefore, the numbers of d electrons in the ions are 3 in Mn^{4+}, 5 in Mn^{2+}, and 9 in Cu^{2+}. It is likely that the ion with the fewest d electrons (Mn^{4+}) and the one with nearly the most d electrons possible (Cu^{2+}) will each have only one spin state.

SOLVE

a. Mn^{4+}: Putting three electrons into the lowest energy $3d$ orbitals available and keeping them as unpaired as possible gives this orbital distribution of electrons:

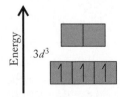

There is no low-spin option for Mn^{4+} assuming Hund's rule is obeyed.

b. Mn^{2+}: There are two options for distributing five electrons among the five $3d$ orbitals:

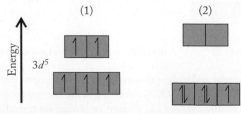

Thus Mn^{2+} can have a high-spin (on the left) or a low-spin (on the right) configuration in an octahedral field.

c. Cu^{2+}: The nine $3d$ electrons completely fill the lower energy orbitals and nearly fill the higher energy ones. No arrangement is possible other than the one shown, so Cu^{2+} has only one spin state:

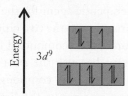

THINK ABOUT IT In an octahedral field, metal ions with 4, 5, 6, or 7 d electrons can exist in high-spin and low-spin states. Those ions with 3 or fewer d electrons have only one spin state, in which all the electrons are unpaired and in the lower energy set of orbitals. Ions with 8 or more d electrons have only one spin state because their lower energy set of orbitals is completely filled. The magnitude of the crystal field splitting energy, Δ_o, determines which spin state an ion with 4, 5, 6, or 7 d electrons occupies.

Practice Exercise Which of these ions can have high-spin and low-spin configurations when part of an octahedral complex: (a) V^{4+}; (b) Cr^{3+}; (c) Ni^{3+}? Are any of the possible spin configurations diamagnetic? ⚙

As noted earlier, whether a transition metal ion is in a high-spin state or a low-spin state depends on whether less energy is needed to promote an electron to a higher energy orbital or to overcome the repulsion experienced by two electrons sharing the same lower energy orbital. Several factors affect the size of Δ_o. We have already discussed a major one in the context of the spectrochemical series shown in Table 16.5: the different field strengths of different ligands. Because molecules containing nitrogen atoms with lone pairs of electrons are stronger field-splitting ligands than H_2O molecules, hydrated metal ions (small Δ_o) are more likely to be in high-spin states, and metals surrounded by nitrogen-containing ligands (large Δ_o) are more likely to be in low-spin states. Another factor affecting spin state is the oxidation state of the metal ion. The higher the oxidation number (and ionic charge), the stronger the attraction of the electron pairs on the ligands for the ion. Greater attraction leads to more ligand–d orbital interaction and therefore to a larger Δ_o.

Complexes of transition metals in the fifth and sixth rows tend to be low spin because their $4d$ and $5d$ orbitals extend farther from the nucleus than do $3d$ orbitals. These larger d orbitals overlap more and interact more strongly with the lone pairs of electrons on the ligands, leading to greater crystal field splitting.

Our discussion of high-spin and low-spin states has focused entirely on d orbitals split by octahedral fields. What about spin states in tetrahedral fields? Most tetrahedral complexes are high spin because tetrahedral fields, which are created by only four ligands, are weaker than octahedral fields created by six. Weaker field strength means less d-orbital splitting—not enough to offset the energies associated with pairing two electrons in the same orbitals. Therefore, Hund's rule is obeyed.

CONCEPT TEST

Explain the following:

a. $Mn(pyridine)_6^{2+}$ is a high-spin complex ion, but $Mn(CN)_6^{4-}$ is low spin.
b. $Fe(NH_3)_6^{2+}$ is high spin, but $Ru(NH_3)_6^{2+}$ is low spin.

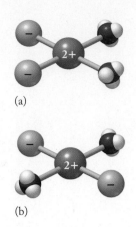

(a)

(b)

FIGURE 16.21 Two ways to orient the Cl⁻ ions and NH₃ molecules around a Pt²⁺ ion (gold sphere) in the square planar coordination compound Pt(NH₃)₂Cl₂. (a) The two members of each pair of ligands are on the same side of the square in *cis*-diamminedichloroplatinum(II). (b) The two members of each pair are at opposite corners in *trans*-diamminedichloroplatinum(II).

◉◉ CONNECTION We introduced the concept of isomers in Chapter 5 when we discussed chirality. Enantiomers, also called optical isomers, are one type of stereoisomer; they are molecules that are not superimposable on their mirror image and that rotate the plane of polarized light.

16.8 Isomerism in Coordination Compounds

The coordination compound $Pt(NH_3)_2Cl_2$ was first described in 1849 and is now widely used as a chemotherapeutic agent for the treatment of several types of cancer. Both molecules of ammonia and the two chloride ions are coordinately bonded to the Pt^{2+} ion, so the name of the compound is diamminedichloroplatinum(II). No counter ions are present because the sum of the charges on the ligands is zero and the complex is neutral. The compound is square planar, and when we draw it, there are two ways to arrange the ligands about the central metal ion: the two chloride ions and the two ammonia ligands can be at adjacent corners of the square (Figure 16.21a) or at opposite corners (Figure 16.21b). Molecules like these two that have the same composition and the same connections between parts but differ in the three-dimensional arrangement of those parts are called **stereoisomers**.

The two molecules in Figure 16.21 have different physical and chemical properties, and we must find a way to distinguish between them when we name them. Isomer a, with two members of each pair of ligands at adjacent corners, is called *cis*-diamminedichloroplatinum(II). Isomer b, with pairs of ligands at opposite corners, is *trans*-diamminedichloroplatinum(II). The prefixes *cis*- and *trans*- are derived from Latin: *cis* means "on the same side," and *trans* means "across." These types of stereoisomers are called cis–trans isomers.

To illustrate the importance of stereoisomerism, consider this: *cis*-diamminedichloroplatinum(II) is a widely used anticancer drug with the common name *cisplatin*, but the trans isomer is ineffective in fighting cancer. The therapeutic power of cisplatin comes from its structurally specific reactions with DNA. During these reactions the two Cl atoms of $Pt(NH_3)_2Cl_2$ are replaced by two nitrogen-containing bases on a strand of DNA in the nucleus of a cancerous cell, as shown in Figure 16.22. The ability of cisplatin to cross-link these bases distorts the molecular shape of the DNA (Figure 16.22d) and disrupts its normal function. Most importantly, this kind of cell DNA damage inhibits the ability of cancerous cells to grow and replicate. The trans isomer forms complexes with other intracellular compounds more readily than with DNA so it is ineffective as an anticancer agent.

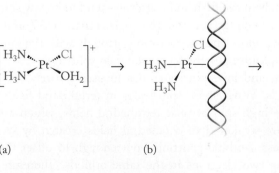

(a)   (b)   (c)

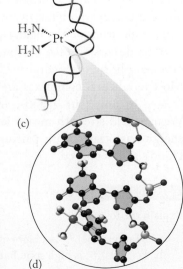

(d)

FIGURE 16.22 The attack of *cis*-diamminedichloroplatinum(II) on DNA. (a) After the drug is administered, one of its chloride ions is displaced by a molecule of water, turning each molecule into a 1+ ion. (b) The ion is attracted to a nitrogen-containing base on a strand of DNA, which displaces the water molecule and forms a coordinate bond to Pt. (c) A nearby base forms a bond to Pt by displacing the chloride ion. Forming bonds to the two DNA bases distorts the DNA molecule so much that it cannot function properly. (d) A magnified view of the bonds that form between platinum(II) and nitrogen atoms in DNA. We explore the structure and function of DNA in Chapter 20.

(a) *cis*-Tetraamminedichlorocobalt(III) chloride

(b) *trans*-Tetraamminedichlorocobalt(III) chloride

FIGURE 16.23 The two stereoisomers of the coordination compound with the formula $[Co(NH_3)_4Cl_2]Cl$: (a) *cis*-tetraamminedichlorocobalt(III) chloride and (b) *trans*-tetraamminedichlorocobalt(III) chloride.

Cis–trans isomers are also possible in octahedral complexes containing more than one type of ligand. For example, there are two possible stereoisomers of $[Co(NH_3)_4Cl_2]Cl$ (Figure 16.23). The two chloro ligands in $[Co(NH_3)_4Cl_2]^+$ are either on the same side of the complex with a 90° Cl—Co—Cl bond angle (the cis isomer), or across from each other so that the Cl—Co—Cl angle is 180° (the trans isomer). *cis*-Tetraamminedichlorocobalt(III) chloride is violet, and *trans*-tetraamminedichlorocobalt(III) chloride is green.

SAMPLE EXERCISE 16.6 **Identifying Stereoisomers of Coordination Compounds** **LO6**

Sketch the structures and name the stereoisomers of $Ni(NH_3)_4Cl_2$.

COLLECT AND ORGANIZE We are given the formula and asked to name and draw the structures of the stereoisomers of a coordination compound. The Ni^{2+} complexes we have seen so far in the chapter have all been octahedral. Ammonia molecules and Cl^- ions are both monodentate ligands.

ANALYZE The formula contains no brackets, so the Cl^- ions are not counter ions; they must be covalently bonded to the Ni ion. There are two Cl^- ions, so the charge on Ni must be 2+. There are a total of six ligands, which confirms that the compound is octahedral.

SOLVE There are two ways to orient the chloride ions: opposite each other with a Cl—Ni—Cl bond angle of 180° or on the same side of the octahedron with a Cl—Ni—Cl bond angle of 90°:

> **stereoisomers** molecules with the same formulas and the same connectivities between their atoms, but with different spatial arrangements of their atoms.

The first isomer is *trans*-tetraamminedichloronickel(II); the second is *cis*-tetraammine-dichloronickel(II).

THINK ABOUT IT We can draw other tetraamminedichloronickel(II) structures that do not look exactly like these two structures. If we flip or rotate them, however, they will match one of the two structures shown.

Practice Exercise Sketch the stereoisomers of $[CoBr_2(en)(NH_3)_2]^+$ and name them. ⚙

Enantiomers and Linkage Isomers

Consider the octahedral Co(III) complex ion containing two ethylenediamine molecules and two chloride ions. There are two ways to arrange the chloride ions: on adjacent bonding sites in a cis isomer or on opposite sides of the octahedron in a trans isomer. These two molecules are cis–trans isomers, but the cis isomer (Figure 16.24) illustrates another kind of stereoisomerism that is possible in complex ions and coordination compounds.

Figure 16.24 shows that *cis*-Co(en)$_2$Cl$_2^+$ is chiral: it has a mirror image that is not identical to the original. The difference is demonstrated by the fact that there is no way to rotate the mirror image so that its atoms align exactly with those in the original. In other words, the two structures are not *superimposable*. We encountered this phenomenon in Chapter 5 and noted that such nonsuperimposable stereoisomers are called *enantiomers*.

Naming this cis isomer also requires an additional rule. The complex ion is named *cis*-dichlorobis(ethylenediamine)cobalt(III). Note the prefix *bis-* just before *(ethylenediamine)*. In naming complex ions containing a polydentate ligand, we use *bis-* instead of *di-* to indicate that two molecules of the ligand are present and to avoid the use of two *di-* prefixes in the same ligand name. Other prefixes used in this fashion are *tris-* for three polydentate ligands and *tetrakis-* for four.

FIGURE 16.24 The complex ion *cis*-dichlorobis(ethylenediamine)cobalt(III) is chiral, which means that its mirror image is not superimposable on the original complex. To illustrate this point, we rotate the mirror image 180° about its vertical axis so that it looks as much like the original as possible. However, note that the top ethylenediamine ligand is located behind the plane of the page in the original but in front of the plane of the page in the rotated mirror image. Thus, the mirror images are not superimposable.

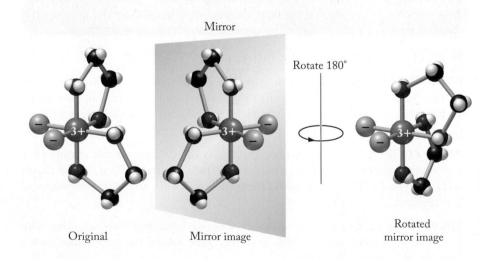

Mirror

Rotate 180°

Original   Mirror image   Rotated mirror image

CONCEPT TEST

Would four different ligands arranged in a square planar geometry produce a chiral complex ion? What about four different ligands in a tetrahedral geometry?

A third kind of isomerism called *linkage isomerism* occurs in coordination compounds when a ligand can bind to a central metal ion using either of two possible electron-pair-donating atoms. The two complexes of Co^{3+} shown in Figure 16.25 are a pair of linkage isomers. One of the ligands in each of two complexes contains this sequence of atoms: S–C–N. It is called a thiocyanate (SCN^-) ion when the S atom forms the covalent bond to the metal (Figure 16.25a) and an isothiocyanate (NCS^-) ion when the N atom forms the bond (Figure 16.25b). Another ligand that forms linkage isomers with metals is NO_2^-, called nitro when the N atom donates an electron pair and nitrito when an O atom is the donor. Note that because these pairs of molecules have different connectivities, linkage isomers are *not* stereoisomers. Instead, linkage isomers are a type of constitutional isomer.

16.9 Complex Ions in Biomolecules

At the beginning of the chapter, we noted that metals essential to human health must be present in foods in forms the body can absorb. In this section we explore some biological polydentate ligands that help metal ions participate in processes that are essential to nutrition and good health.

Let's begin with photosynthesis, a chemical process at the base of our food chain. Green plants can harness solar energy because they contain large biomolecules we collectively call *chlorophyll*. All molecules of chlorophyll contain ring-shaped tetradentate ligands called *chlorins*. The skeletal structures of chlorins (Figure 16.26a) are similar to those of **porphyrins**, another class of tetradentate ligands found in biological systems (Figure 16.26b). Chlorins and porphyrins are members of a larger category of compounds known as **macrocyclic ligands**. (*Macrocycle* means, literally, "big ring.")

When either ring forms coordinate bonds with a metal ion M^{n+} (Figure 16.26c), the two hydrogen atoms bonded to nitrogen atoms ionize, giving the ring a charge of 2– and the complex ion an overall charge of $(n - 2)$. The lone pairs of electrons on the N atoms in the ionized structure are all oriented toward the center of the ring. These lone pairs can occupy either the four equatorial coordination sites in an octahedral complex ion or all four coordination sites in a square planar complex ion. In octahedral complex ions, each central metal ion still has its two axial sites available for bonding to other ligands. Depending on the charge of the central ion, the complex ion may be either ionic or electrically neutral.

CONNECTION We introduced constitutional isomers, which have the same molecular formula but different connections between their atoms, in Chapter 6.

(a) $[Co(CN)_5SCN]^{3-}$

(b) $[Co(CN)_5NCS]^{3-}$

FIGURE 16.25 The SCN^- ligand is bound to the Co^{3+} ion through the S atom in (a) the pentacyanothiocyanatocobaltate(III) ion and through the N atom in (b) the pentacyanoisothiocyanatocobaltate(III) ion.

Chlorin ring system
(a)

Porphyrin ring system
(b)

$+ M^{n+} \longrightarrow$

Metal–porphyrin complex
(c)

$+ 2H^+$

FIGURE 16.26 Skeletal structures of (a) chlorin and (b) porphyrin. (In these structures the bonds formed by carbon atoms are shown, but not the atoms themselves nor their bonds to hydrogen atoms). The principal difference between the structures is a C=C in the porphyrin structure (shown in red) that is a single bond in chlorin rings. The innermost atoms in each ring are four nitrogen atoms with lone pairs of electrons. All four N atoms form coordinate bonds with a metal ion, as shown with the porphyrin ring in (c).

porphyrin a type of tetradentate macrocyclic ligand.

macrocyclic ligand a ring containing multiple electron-pair donors that bind to a metal ion.

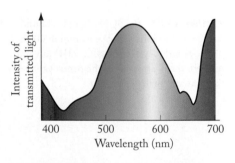

Chlorophyll *a*

FIGURE 16.27 Chlorophyll absorbs sunlight and initiates a series of reactions that convert sunlight, carbon dioxide, and water into chemical energy stored in the bonds of carbohydrates. All forms of chlorophyll, including chlorophyll *a* shown here, have a Mg^{2+} ion in a chlorin ring as part of their structure.

FIGURE 16.28 Combined transmission spectrum of chlorophyll and other pigments in a typical green leaf. The pigments absorb most of the visible radiation emitted by the sun except yellow-green.

Porphyrins and chlorins are widespread in nature and play many biochemical roles. Their chemical and physical properties depend on the following:

1. The identity of the central metal ion.
2. The species occupying the axial coordination sites of octahedral complexes.
3. The number and identity of organic groups attached to the outside of the ring.

The chlorin ring in chlorophyll *a* has a Mg^{2+} ion at its center (Figure 16.27). Delocalized *p* electrons in the conjugated double bonds in and around the ring stabilize it and give it and other plant pigments the ability to absorb wavelengths of red and blue-violet light. Because plant leaves absorb these colors, most of them are green and yellow-green—the colors that are not absorbed (Figure 16.28). In temperate climates, the leaves of trees such as maples and oaks lose chlorophyll at the end of each growing season, revealing the colors of other pigments in the beautiful leaves of autumn (Figure 16.29).

An important porphyrin complex called the *heme* group has a central Fe^{2+} ion (Figure 16.30). The heme group enables the proteins hemoglobin and myoglobin to transport O_2 in the blood and to store O_2 in muscle tissues, respectively. The four nitrogen atoms in the porphyrin ring of a heme group occupy equatorial positions in an octahedral complex of an Fe^{2+} ion. Below the ring a fifth bond is formed between the Fe^{2+} ion and a lone pair of electrons on another nitrogen atom in the protein. The sixth ligand, located above the porphyrin ring, is typically a molecule of O_2, as in the oxygenated forms of hemoglobin in blood leaving the lungs.

Each O_2 molecule can act as a Lewis base and donate one of its lone pairs of electrons to the iron in heme. This coordinate covalent bond is strong enough to carry oxygen from the lungs to the cells in our bodies but weak enough to break easily when the oxygen reaches a cell. Other ligands of similar size can bind to the Fe^{2+} ion in heme, and problems arise when some of them do. Carbon monoxide is such a ligand. Similar in size to O_2, a CO molecule easily fits into the sixth binding site. Unfortunately, it binds about 200 times more strongly than O_2. If a person breathes air containing carbon monoxide, CO prevents O_2 from being

FIGURE 16.29 The colors of fall in southern Vermont are characterized by the reds, yellows, and golds of leaves that have lost their green pigments and are about to fall.

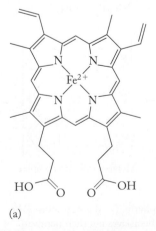

(a)

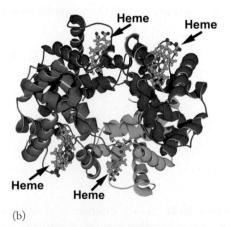

(b)

FIGURE 16.30 (a) In the porphyrin known as heme, the four nitrogen atoms occupy equatorial positions in an octahedral complex in which the central metal ion is Fe^{2+}. (b) Hemoglobin contains four heme groups, each of which is bound to a protein chain symbolized by colored ribbons. The four protein chains are held together by intermolecular forces.

taken up by blood flowing through the lungs, causing the symptoms of CO poisoning and sometimes death.

Proteins called *cytochromes* also contain heme groups (Figure 16.31). Cytochromes mediate oxidation and reduction processes connected with energy production in cells as iron ions flip back and forth between their +2 and +3 oxidation states, reversibly consuming or releasing electrons needed in other redox reactions. There are many kinds of cytochrome proteins with different substituents on the porphyrin rings and different axial ligands, each of which influences the function of the complex. This last point has been repeated several times in this chapter: the chemical properties and biological functions of transition metals that are essential to living organisms are linked to their molecular environments and to the formation of stable complex ions with ligands that are strong electron donors, that is, strong Lewis bases.

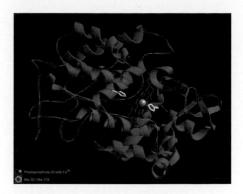

FIGURE 16.31 The structure of cytochrome proteins, such as cytochrome *c* shown here, includes heme complexes (shown in red) that mediate energy production and redox reactions in living cells. Different cytochromes have different ligands (shown in yellow) occupying the fifth and sixth octahedral coordination sites, and different groups in the protein may be attached to the porphyrin ring.

16.10 Complex-Ion Equilibria

Now let's take a quantitative look at the formation of complex ions using mathematical tools we developed in Chapters 14 and 15. These tools are appropriate because complex formation processes are reversible and usually reach chemical equilibrium rapidly. We start this investigation with two aqueous solutions, one containing copper(II) sulfate ($CuSO_4$), the other NH_3 (Figure 16.32). The $CuSO_4$ solution is robin's-egg blue, the color characteristic of $Cu^{2+}(aq)$ ions, and the ammonia solution is colorless.

When the solutions are mixed, the robin's-egg blue turns a dark navy blue, as shown on the right in Figure 16.32. This is the color of tetraamminecopper(II),

FIGURE 16.32 The beaker on the left contains a solution of $Cu^{2+}(aq)$, which is a characteristic robin's-egg blue. As a colorless solution of ammonia is added (from the bottle in the middle), the mixture of the two solutions turns dark blue (beaker on the right), which is the color of the $Cu(NH_3)_4^{2+}$ complex ion.

formation constant (K_f) equilibrium constant describing the formation of a metal complex from a free metal ion and its ligands.

$Cu(NH_3)_4^{2+}$. The change in color provides visual evidence that the following equilibrium:

$$Cu^{2+}(aq) + 4\,NH_3(aq) \rightleftharpoons Cu(NH_3)_4^{2+}(aq)$$

lies far to the right, favoring complex formation. This conclusion is supported by the large equilibrium constant for the reaction:

$$K_f = \frac{[Cu(NH_3)_4^{2+}]}{[Cu^{2+}][NH_3]^4} = 5.0 \times 10^{13}$$

This equilibrium constant K_f is called a **formation constant** because it describes the formation of a complex ion. For the generic complex formation reaction in which metal ion M combines with n moles of ligand X,

$$M(aq) + n\,X(aq) \rightleftharpoons MX_n(aq)$$

the formation constant expression is

$$K_f = \frac{[MX_n]}{[M][X]^n}$$

Formation constants can be used to calculate the concentration of complex ions in solution, or to calculate the concentration of free, uncomplexed metal ions in equilibrium with a given (usually larger) concentration of ligand, as in the following Sample Exercise.

SAMPLE EXERCISE 16.7   **Calculating the Concentration**   **LO7**
of a Free Metal Ion in
Equilibrium with a Complex

Ammonia gas is dissolved in a 1.00×10^{-4} M solution of $CuSO_4$, so that initially $[NH_3] = 2.00 \times 10^{-3}$ M. Calculate the concentration of $Cu^{2+}(aq)$ ions in the solution after the reaction mixture has come to chemical equilibrium.

COLLECT AND ORGANIZE The concentration of $CuSO_4$ means that $[Cu^{2+}]$ before complex formation is 1.00×10^{-4} M. The initial concentration of the ligand (NH_3) is 2.00×10^{-3} M. We know from the text preceding this exercise that the reaction involves the formation of the tetraamminecopper(II) complex ion,

$$Cu^{2+}(aq) + 4\,NH_3(aq) \rightleftharpoons Cu(NH_3)_4^{2+}(aq)$$

and that the values of $[NH_3]$, $[Cu^{2+}]$, and $[Cu(NH_3)_4^{2+}]$ are related by the formation constant expression:

$$K_f = \frac{[Cu(NH_3)_4^{2+}]}{[Cu^{2+}][NH_3]^4} = 5.0 \times 10^{13}$$

ANALYZE The stoichiometric ratio of NH_3 to Cu^{2+} is 4:1, but initially $[NH_3]/[Cu^{2+}] = 20$, so there is more than enough NH_3 to convert all the Cu^{2+} into $Cu(NH_3)_4^{2+}$. Because K_f is large, we can assume that nearly all the Cu^{2+} ions are converted and that only a tiny concentration of free Cu^{2+} ions (x) remains uncomplexed at equilibrium. Therefore, at equilibrium:

$$[Cu^{2+}] = x$$

and

$$[Cu(NH_3)_4^{2+}] = (1.00 \times 10^{-4}) - x$$

The balanced equation tells us that 4 moles of NH_3 are consumed for every mole of $Cu(NH_3)_4^{2+}$ produced. This means that if the change in $[Cu(NH_3)_4^{2+}]$ is $(1.00 \times 10^{-4} - x)$, the change in $[NH_3]$ is a *decrease* equal to four times that value: $4(1.00 \times 10^{-4} - x)$.

Given the large value of K_f, we should obtain a $[Cu^{2+}]$ value at equilibrium that is much less than 1.00×10^{-4} M.

SOLVE Completing the RICE table for the equilibrium:

Reaction	$Cu^{2+}(aq)$	+	$4\ NH_3(aq)$	$\rightleftharpoons$	$Cu(NH_3)_4^{2+}(aq)$
	$[Cu^{2+}]$ (M)		$[NH_3]$ (M)		$[Cu(NH_3)_4^{2+}]$ (M)
Initial	1.00×10^{-4}		2.00×10^{-3}		0
Change	$-(1.00 \times 10^{-4} - x)$		$-4(1.00 \times 10^{-4} - x)$		$+(1.00 \times 10^{-4} - x)$
Equilibrium	x		$(1.60 \times 10^{-3}) + 4x$		$1.00 \times 10^{-4} - x$

Now we make the simplifying assumption that x is much smaller than 1.00×10^{-4} M. If it is, then $4x$ must be much smaller than 1.60×10^{-3} M. Therefore, we can ignore the x terms in the equilibrium values of $[NH_3]$ and $[Cu(NH_3)_4^{2+}]$ and write

$$K_f = \frac{[Cu(NH_3)_4^{2+}]}{[Cu^{2+}][NH_3]^4} = \frac{1.00 \times 10^{-4}}{(x)(1.60 \times 10^{-3})^4} = 5.0 \times 10^{13}$$

$$x = \frac{1.00 \times 10^{-4}}{(1.60 \times 10^{-3})^4 (5.0 \times 10^{13})}$$

$$= 3.1 \times 10^{-7}\ M = [Cu^{2+}]$$

THINK ABOUT IT This result confirms our simplifying assumption and also our prediction that $[Cu^{2+}]$ at equilibrium is much less than $[Cu^{2+}]$ initially. In fact, more than 99% of the Cu(II) in the solution is present as $Cu(NH_3)_4^{2+}$.

Practice Exercise Calculate the equilibrium concentration of $Ag^+(aq)$ in a solution that is initially 0.100 M $AgNO_3$ and 0.800 M NH_3 after this reaction takes place:

$$Ag^+(aq) + 2\ NH_3(aq) \rightleftharpoons Ag(NH_3)_2^+(aq) \qquad K_f = 1.7 \times 10^7$$

16.11 Hydrated Metal Ions as Acids

Consider what happens when $FeCl_3$ dissolves in water. The compound dissociates as it dissolves, forming $Fe(H_2O)_6^{3+}$ and Cl^- ions. As we saw in Chapter 15, Cl^- ions are very weak bases and do not influence the pH of the solution. However, a check of the pH of the $FeCl_3$ solution discloses that it is quite acidic. We must conclude that $Fe(H_2O)_6^{3+}$ ions can donate H^+ ions in aqueous solutions, but where does this acid strength come from?

To answer this question, let's revisit the discussion of acid strength and molecular structure in Chapter 15. In particular, recall that the strength of an acid increases as the number of highly electronegative atoms near an –OH group in the molecule increases. The ability of these electronegative elements to draw electron density and electrical charge toward their part of the molecule and away from the O—H bond stabilizes the anion, releasing a H^+ ion.

A similar distortion of electron density and dispersion of the negative charge left behind when a H^+ ion leaves can happen in $Fe(H_2O)_6^{3+}$ ions, too. The electrons in the O—H bonds of the water molecules in the inner coordination sphere are attracted to the positively charged central metal ion and away from the H ends of the bonds. This electron distribution allows a water molecule in the inner

CONNECTION In Chapter 15 we learned that the solutions of some salts are acidic or basic because of the reaction of their ions with water.

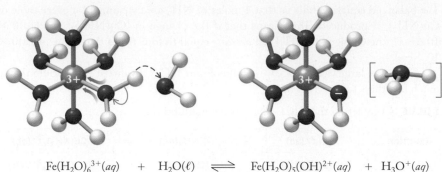

FIGURE 16.33 A hydrated Fe^{3+} cation draws electron density away from the water molecules of its inner coordination sphere, which makes it possible for one or more of these molecules to donate a H^+ ion to a water molecule outside the sphere. The hydrated Fe^{3+} ion is left with one fewer H_2O molecule and one OH^- ion.

$$Fe(H_2O)_6{}^{3+}(aq) \ + \ H_2O(\ell) \ \rightleftharpoons \ Fe(H_2O)_5(OH)^{2+}(aq) \ + \ H_3O^+(aq)$$

Ion	K_a
$Fe^{3+}(aq)$	3×10^{-3}
$Cr^{3+}(aq)$	1×10^{-4}
$Al^{3+}(aq)$	1×10^{-5}
$Cu^{2+}(aq)$	3×10^{-8}
$Pb^{2+}(aq)$	3×10^{-8}
$Zn^{2+}(aq)$	1×10^{-9}
$Co^{2+}(aq)$	2×10^{-10}
$Ni^{2+}(aq)$	1×10^{-10}

TABLE 16.6 K_a Values of Hydrated Metal Ions

Acid strength

coordination sphere to act as a Brønsted–Lowry acid and donate a H^+ ion to a water molecule outside that sphere, as shown in Figure 16.33 and as described by the following chemical equilibrium:

$$M(H_2O)_6{}^{n+}(aq) + H_2O(\ell) \rightleftharpoons M(H_2O)_5(OH)^{(n-1)+}(aq) + H_3O^+(aq)$$

Many hydrated metal ions, particularly those with charges of 2+ or more, are Brønsted–Lowry acids. The acidic properties of these ions are reflected in the K_a values in Table 16.6. Note that in the product in Figure 16.33, one of the complex's original six water molecules has been converted into a hydroxide ion, reducing the charge on the complex from 3+ to 2+. The $Fe(H_2O)_5(OH)^{2+}$ ion can also act as a Brønsted–Lowry acid (H^+ donor) in a second acid ionization:

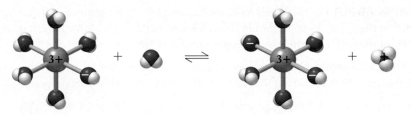

$$Fe(H_2O)_5(OH)^{2+}(aq) \ + \ H_2O(\ell) \ \rightleftharpoons \ Fe(H_2O)_4(OH)_2{}^+(aq) \ + \ H_3O^+(aq)$$

A third acid ionization is possible, producing solid iron(III) hydroxide:

$$Fe(H_2O)_4(OH)_2{}^+(aq) \ + \ H_2O(\ell) \ \rightleftharpoons \ Fe(H_2O)_3(OH)_3(s) \ + \ H_3O^+(aq)$$

For simplicity, we usually write the formula of iron(III) hydroxide as $Fe(OH)_3(s)$ even though the neutral material contains some water of hydration.

Other 3+ cations, including Cr^{3+} and Al^{3+}, display similar behavior. However, these two trivalent ions form electrically neutral hydroxide compounds that have an unusual solubility pattern. Whereas $Fe(OH)_3(s)$ and most other transition metal hydroxides have very limited solubility in basic solutions, $Cr(OH)_3$ and $Al(OH)_3$ are more soluble in strongly basic solutions (pH > 11) than in weakly basic solutions (pH ≈ 8) because solid $Cr(OH)_3$ and $Al(OH)_3$ may react with OH^- ions at high pH, forming soluble anionic complex ions:

$$Cr(H_2O)_3(OH)_3(s) + OH^-(aq) \rightleftharpoons Cr(H_2O)_2(OH)_4^-(aq) + H_2O(\ell)$$

$$Al(H_2O)_3(OH)_3(s) + OH^-(aq) \rightleftharpoons Al(H_2O)_2(OH)_4^-(aq) + H_2O(\ell)$$

Formation of complex ions explains the solubility of Cr^{3+} and Al^{3+} ions at low and high pH. These ions are soluble in strongly acidic solutions (pH $\approx$ 2), where they exist as $Cr(H_2O)_6^{3+}(aq)$ and $Al(H_2O)_6^{3+}(aq)$. In less acidic solutions they exist as positively charged complex ions with generic formulas such as $M(H_2O)_5(OH)^{2+}(aq)$ and $M(H_2O)_4(OH)_2^+(aq)$, and in strongly basic solutions they exist as $M(H_2O)_2(OH)_4^-(aq)$. Zinc hydroxide, $Zn(OH)_2$, is the only other transition metal hydroxide that is soluble at high pH.

Nearly all transition metals share one common characteristic: They exist as $M^{n+}(aq)$ ions only in strongly acidic solutions. They occur as complex ions in which at least one of their ligands is an OH^- ion in aqueous solutions that range from slightly acidic to slightly basic (3 < pH < 9). This pH range includes most environmental waters and biological fluids.

CONCEPT TEST

Can $Al(OH)_3$ act as both a Brønsted–Lowry acid and a Brønsted–Lowry base? Use balanced chemical equations to support your answer.

SAMPLE EXERCISE 16.8 **Integrating Concepts: Analysis of an Alloy**

The presence of iron, cobalt, and nickel in homogeneous mixtures of metals called alloys can be detected by a series of chemical tests. The alloy sample is prepared by dissolving it in acid to produce an aqueous solution of Fe^{2+}, Co^{2+}, and Ni^{2+} ions. After pH adjustment, H_2S gas is bubbled through the solution, producing sulfide precipitates of the three ions.

1. The precipitated mass is collected and treated with hydrochloric acid to dissolve only the most soluble of the three metal sulfides. This solution is retained as solution 1.
2. The precipitate from step 1, containing the two less soluble metal sulfides, is dissolved in nitric acid. This solution is retained as solution 2.
3. Solution 1 is treated with HNO_3, which oxidizes the metal ions in solution from their +2 to +3 oxidation states. In the process, HNO_3 is converted into NO.
4. A few crystals of NH_4SCN are added to the solution from step 3, producing a blood-red solution indicating the presence of $Fe(NCS)(H_2O)_5^{2+}$ ions.
5. $NH_3(aq)$ is added to solution 2 to form ammine complexes of the ions present. The resultant solution is divided into two portions: 5A and 5B.
6. The bidentate ligand dimethylglyoxime (molecular DMG: $C_4H_8N_2O_2$) is added to solution 5A, producing a red precipitate, which is analyzed and determined to have the following percent composition:

Ni: 20.31%; C: 33.25%; H: 4.89%; N: 19.40%.

7. Hydrochloric acid, which converts ammine complexes back to aqua complexes, is added to solution 5B. A few crystals of NH_4SCN are also added, and a blue solution is produced indicating the presence of $Co(SCN)_4^{2-}$, which is a square-planar complex.

Identify the metals involved in each reaction, and write net ionic equations for the reactions involved in these steps. Draw structures of the complex ions that form and identify any possible isomers. The K_{sp} values for the three sulfides are 6.3×10^{-18} for FeS, 4×10^{-21} for CoS, and 1×10^{-24} for NiS.

COLLECT AND ORGANIZE A metal sample is dissolved in strong acid, producing a solution containing Fe^{2+}, Co^{2+}, and Ni^{2+} ions. These ions are precipitated as their sulfides and then selectively redissolved with more acids, including nitric acid, which oxidizes one of the metals to its +3 oxidation state as HNO_3 is reduced to NO. The presence of the three metals is confirmed by adding ligands that form complexes whose colors indicate the presence of a particular metal ion. We are asked to write net ionic equations describing the reactions involved in the tests and to draw the structures of the complex ions that are formed.

ANALYZE The K_{sp} values for the binary sulfides of the three metals provide a direct measure of which of them is the most soluble and therefore dissolves in step 1. The reactions of the ion of the most soluble sulfide are described in steps 1, 3, and 4. Writing a balanced net ionic equation for the redox reaction in step 3 will require an analysis of the changes in oxidation states of the reactants and products. The reaction takes place in an acidic solution, so the final steps involving H and O atoms will be done by adding H_2O and H^+ ions.

The chemistry of the other two ions is described in steps 2, 5, 6, and 7. In step 6 we are provided elemental composition information and will need to use it to calculate the empirical formula of the coordination compound nickel forms with DMG. From that formula, we can calculate the ratio of DMG ligands per nickel ion. The elemental composition data do not add up to 100%, but note that a value is missing for oxygen, which is present in the molecular formula of DMG. Oxygen's proportion can be calculated by adding up the other percent composition values and subtracting from 100%.

SOLVE Of the metal sulfides in step 1, FeS has the largest K_{sp}, so it is the one that dissolves when the sulfide precipitate is treated with hydrochloric acid, producing soluble iron(II) chloride and hydrogen sulfide:

$$FeS(s) + 2\,H^+(aq) \rightarrow Fe^{2+}(aq) + H_2S(g)$$

The collected precipitate from step 1 contains $CoS(s)$ and $NiS(s)$. These solids are dissolved in nitric acid in step 2, producing solutions of cobalt(II) and nickel(II) nitrates and hydrogen sulfide gas:

$$CoS(s) + 2\,H^+(aq) \rightarrow Co^{2+}(aq) + H_2S(g)$$

$$NiS(s) + 2\,H^+(aq) \rightarrow Ni^{2+}(aq) + H_2S(g)$$

Nitric acid oxidizes $Fe^{2+}(aq)$ to $Fe^{3+}(aq)$ in step 3 and forms $NO(g)$ in the process. To write a balanced net ionic equation describing this reaction we start with the known reactants and products:

Unbalanced: $\quad Fe^{2+}(aq) + NO_3^-(aq) \rightarrow Fe^{3+}(aq) + NO(g)$

Analyzing the changes in oxidation state:

and balancing the changes in oxidation state by multiplying the iron terms by 3, we get

$$3\,Fe^{2+}(aq) + NO_3^-(aq) \rightarrow 3\,Fe^{3+}(aq) + NO(g)$$

Balancing the numbers of O and then H atoms on both sides by adding H_2O and H^+ ions as needed,

$$3\,Fe^{2+}(aq) + NO_3^-(aq) \rightarrow 3\,Fe^{3+}(aq) + NO(g) + \mathbf{2\,H_2O}(\ell)$$

$$3\,Fe^{2+}(aq) + NO_3^-(aq) + \mathbf{4\,H^+}(aq) \rightarrow$$
$$3\,Fe^{3+}(aq) + NO(g) + 2\,H_2O(\ell)$$

A final check of the charges (9+ on both sides of the reaction arrow) confirms that we have a balanced net ionic equation describing the redox reaction.

In step 4, hydrated Fe^{3+} ions combine with dissolved thiocyanate ions, forming a red solution of $Fe(NCS)(H_2O)_5^{2+}$ complex ions. The formula of the complex tells us that the nitrogen end of one SCN^- ion is bonded to the Fe^{3+} ion (see Figure 16.25). The net ionic equation describing the complex formation reaction is

$$Fe(H_2O)_6^{3+}(aq) + SCN^-(aq) \rightarrow$$
$$Fe(NCS)(H_2O)_5^{2+}(aq) + H_2O(\ell)$$

and the structure of the complex ion may be drawn

The added ammonia in step 5 displaces water from the inner coordination sphere and produces the hexaammine complexes of the $Co^{2+}(aq)$ and $Ni^{2+}(aq)$ ions:

$$Co(H_2O)_6^{2+}(aq) + 6\,NH_3(aq) \rightarrow Co(NH_3)_6^{2+}(aq) + 6\,H_2O(\ell)$$

$$Ni(H_2O)_6^{2+}(aq) + 6\,NH_3(aq) \rightarrow Ni(NH_3)_6^{2+}(aq) + 6\,H_2O(\ell)$$

The DMG ligand in step 6 must have a negative charge to form a precipitate (neutral compound) with Ni^{2+}, but we do not know what charge it has or how many DMG ligands bind to one Ni^{2+} ion. We can determine the complex's empirical formula after calculating the percent oxygen by summing the other values and subtracting the sum from 100.00%. That value is 22.15%.

Ni:	20.31 g/(58.69 g/mol) = 0.3461 mol	0.3461/0.3461 ≈ 1
C:	33.25 g/(12.01 g/mol) = 2.769 mol	2.769/0.3461 ≈ 8
H:	4.89 g/(1.008 g/mol) = 4.85 mol	4.85/0.3461 ≈ 14
N:	19.40 g/(14.01 g/mol) = 1.385 mol	1.385/0.3461 ≈ 4
O:	22.15 g/(16.00 g/mol) = 1.384 mol	1.384/0.3461 ≈ 4

The empirical formula of nickel dimethylglyoxime is $NiC_8H_{14}N_4O_4$, which indicates that there are 2 DMG ligands bonded to each Ni^{2+} ion. However, if the ligands were neutral, 2 DMG would be $2(C_4H_8N_2O_2)$ or $C_8H_{16}N_4O_4$, and not $C_8H_{14}N_4O_4$. The difference of 2 H atoms indicates that each ligand must have lost one H^+ ion, giving it a 1− charge. Therefore the molecular formula of the neutral (insoluble) coordination compound is $Ni(C_4H_7N_2O_2)_2$, and the balanced net ionic equation describing its formation in this experiment is

$$Ni(NH_3)_6^{2+}(aq) + 2\,C_4H_8N_2O_2(aq) \rightarrow$$
$$Ni(C_4H_7N_2O_2)_2(s) + 2\,NH_4^+(aq) + 4\,NH_3(aq)$$

DMG is a bidentate ligand, so four sites around the nickel ion are occupied by electron pairs from DMG. A coordination number of 4 could yield either tetrahedral or square planar geometry. However, analyses of the structure of nickel dimethylglyoxime using a technique called X-ray diffraction, which we discuss in Chapter 18, have shown that it is square planar:

In the solution 5B reaction in step 7, strong acid is added to the hexaamminecobalt(II) complex, which results in regeneration of the hexaaquacobalt(II) complex:

$$Co(NH_3)_6{}^{2+}(aq) + 6\,H^+(aq) + 6\,H_2O(\ell) \rightarrow$$
$$Co(H_2O)_6{}^{2+}(aq) + 6\,NH_4{}^+(aq)$$

Adding SCN^- results in formation of $Co(SCN)_4{}^{2-}$:

$$Co(H_2O)_6{}^{2+}(aq) + 4\,SCN^-(aq) \rightarrow Co(SCN)_4{}^{2-}(aq) + 6\,H_2O(\ell)$$

which has a square planar structure:

THINK ABOUT IT The analysis of the alloy involves many of the reactions we have seen involving metal ions—precipitation, redox, and complexation—in addition to acid–base chemistry. Steps 4 and 6 produce vivid colors and illustrate reactions that are used in qualitative tests for the presence of Fe^{3+} and Ni^{2+} ions in aqueous solution. The formulas of the two thiocyanate complexes indicate that the thiocyanate ligand coordinates to Fe^{3+} ions through the N atom, but it coordinates to Co^{2+} ions through the S atom. When it was first discovered, the nickel–DMG complex was thought to be tetrahedral; it is now known to be among the many 4-coordinate complexes of nickel that are square planar.

SUMMARY

Section 16.1 A **Lewis base** is a substance that donates a lone pair of electrons to a **Lewis acid**, defined as an electron-pair acceptor. The donated electron pair forms a covalent bond. In some Lewis acid–Lewis base reactions, other bonds must break to accommodate the new one.

Section 16.2 Transition metal ions form **complex ions** when **ligands** donate pairs of electrons to empty valence-shell orbitals on the metal ion, forming **coordinate bonds**. The number of coordinate bonds in a complex defines the **coordination number** of the metal ion. The metal ion acts as a Lewis acid, and the ligands act as Lewis bases. Compounds that contain complex ions are called **coordination compounds**; the net charges on complex ions are balanced by charges from **counter ions**. Ligands occupy binding sites in the **inner coordination sphere** of a metal ion.

Section 16.3 The names of complex ions and coordinate compounds provide information about the identities and numbers of ligands, the identity and oxidation state of the central metal ion, and the identity of counter ions.

Section 16.4 A **monodentate ligand** donates one pair of electrons in a complex ion; a **polydentate ligand** donates more than one in a process called **chelation**. EDTA is a particularly effective sequestering agent, which is a chelating agent that prevents metal ions in solution from reacting with other substances.

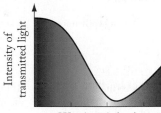

Section 16.5 Polydentate ligands are particularly effective at forming complex ions. This phenomenon is called the **chelate effect** and can be explained by the increase in entropy that accompanies chelation.

Section 16.6 The colors of transition metals can be explained by the interactions between electrons in different *d* orbitals and the lone pairs of electrons on surrounding ligands. These interactions create **crystal field splitting** of the energies of the *d* orbitals. A **spectrochemical series** ranks ligands on the basis of their *field strength* and the wavelengths of electromagnetic radiation absorbed by their complex ions. The stronger the field the ligand produces, the greater the **crystal field splitting energy (Δ)** is, and the shorter the wavelength of radiation the complex absorbs. The color of a complex ion in solution or in a crystalline solid is the complement of the color(s) it absorbs.

Section 16.7 Strong repulsions and large values of crystal field splitting energy can lead to electron pairing in lower energy orbitals and an electron configuration called a low-spin state. Metals and their ions are less paramagnetic in low-spin states than when their *d* electrons are evenly distributed across all the *d* orbitals in the valence shell—a configuration called a high-spin state.

Section 16.8 Complex metal ions containing more than one type of ligand may form **stereoisomers**. When one type occupies two adjacent corners of a square planar or octahedral complex (forming a 90° ligand–metal–ligand bond angle), the complex is a *cis* isomer; when the same ligand occupies opposite corners (180° bond angle), it is a *trans* isomer. Enantiomers and linkage isomers are also possible in coordination compounds.

Section 16.9 Complex ions play key roles in many biochemical processes. Photosynthesis is mediated by chlorophyll, a molecule that contains tetradentate chlorin rings coordinately bonded to central Mg^{2+} ions. Oxygen transport in the body is based on the reversible bonding of O_2 molecules to heme groups of Fe^{2+} ions in **porphyrin** rings. Energy production in cells is mediated by cytochromes, proteins that contain heme groups and metals in different oxidation states.

Section 16.10 The stability of any complex ion is expressed mathematically by its **formation constant (K_f)**, which can be used to calculate the equilibrium concentration of free metal ions in a solution of complex ions.

Section 16.11 Hydrated metal ions with charges of 2+ or greater can act as Brønsted–Lowry acids, which is why solutions of their soluble salts are acidic. Most transition metal ions have limited solubility in concentrated basic solutions. Metals that are exceptions to this rule form anionic complexes, such as $Cr(OH)_4^-$. The solubility of an ionic compound may be enhanced if the cation forms stable complex ions in solution.

PROBLEM-SOLVING SUMMARY

TYPE OF PROBLEM	CONCEPTS AND EQUATIONS	SAMPLE EXERCISES
Identifying Lewis acids and bases	Determine which reactant donates a pair of electrons (the Lewis base) and which one accepts them (the Lewis acid).	16.1
Writing formulas of coordination compounds	Write formulas that distinguish ligands in the inner and outer coordination spheres. Use square brackets to contain a complex ion.	16.2
Naming coordination compounds	Follow the naming rules in Section 16.3.	16.3
Identifying the potential electron-pair-donor groups in a molecule	Examine the molecular structure of a compound and find lone pairs that can be donated to a metal.	16.4
Predicting spin states	Sketch a *d*-orbital diagram based on crystal field splitting. Fill the lowest energy orbitals with the valence electrons. If the number of electron pairs in the diagram is greater than the number of pairs you would have if you distributed the electrons evenly over all five *d* orbitals, multiple spin states are possible.	16.5
Identifying stereoisomers of coordination compounds	If ligands of one type are all on the same side of the complex ion, it is a *cis* isomer. If ligands of one type are on opposite sides, it is a *trans* isomer.	16.6
Calculating the concentration of a free metal ion in equilibrium with a complex	Set up a RICE table based on formation of the complex. Let *x* represent the concentration of metal ion that *does not* form the complex. If the value of K_f is large (it usually is), assume *x* is much less than the other concentrations.	16.7

VISUAL PROBLEMS

(Answers to boldface end-of-chapter questions and problems are in the back of the book.)

16.1. The chlorides of two of the four highlighted elements in Figure P16.1 are colored. Which ones?

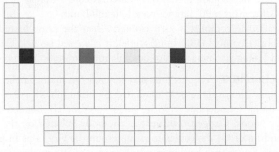

FIGURE P16.1

16.2. Which of the highlighted transition metals in Figure P16.2 form M^{2+} cations that cannot have high-spin and low-spin states?

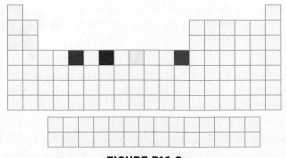

FIGURE P16.2

16.3. Which of the highlighted transition metals in Figure P16.3 have M^{2+} cations that form colorless tetrahedral complex ions?

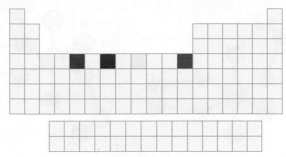

FIGURE P16.3

16.4. Smoky quartz has distinctive lavender and purple colors due to the presence of manganese impurities in crystals of silicon dioxide. Which of the orbital diagrams in Figure P16.4 best describes the Mn^{2+} ion in a tetrahedral field?

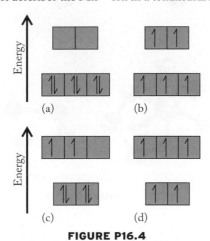

FIGURE P16.4

16.5. Chelation Therapy I The compound with the structure shown in Figure P16.5 is widely used in chelation therapy to remove excessive lead or mercury in patients exposed to these metals. How many electron-pair-donor groups ("teeth") does the sequestering agent have when the carboxylic acid groups are ionized?

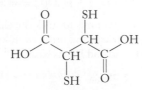

FIGURE P16.5

16.6. Chelation Therapy II The compound with the structure shown in Figure P16.6 has been used to treat people exposed to plutonium, americium, and other actinide metal ions. How many donor groups does the sequestering agent have when the carboxylic acid groups are ionized?

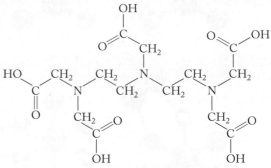

FIGURE P16.6

16.7. The three beakers in Figure P16.7 contain solutions of $[CoF_6]^{3-}$, $[Co(NH_3)_6]^{3+}$, and $[Co(CN)_6]^{3-}$. Based on the colors of the three solutions, which compound is present in each of the beakers?

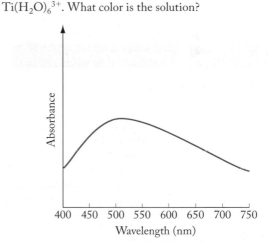

FIGURE P16.7

16.8. Figure P16.8 shows the absorption spectrum of a solution of $Ti(H_2O)_6^{3+}$. What color is the solution?

FIGURE P16.8

16.9. For each pair of complexes in Figure P16.9, indicate whether the complexes are (i) identical, (ii) isomers, or (iii) neither identical nor isomers.

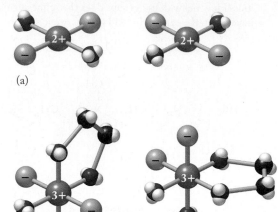

(a)

(b)

(c)

FIGURE P16.9

16.10. For each pair of complexes in Figure P16.10, indicate whether the complexes are (i) identical, (ii) isomers, or (iii) neither identical nor isomers.

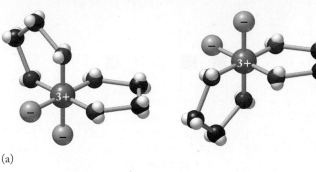

(a)

(b)

(c)

FIGURE P16.10

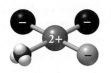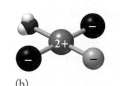

```
QUESTIONS AND PROBLEMS
```
■

Lewis Acids and Bases

CONCEPT REVIEW

16.11. Can a substance be a Lewis base and not be a Brønsted–Lowry base? Explain why or why not.

16.12. Can a substance be a Brønsted–Lowry acid and not be a Lewis acid? Explain why or why not.

16.13. Why is BF_3 a Lewis acid but not a Brønsted–Lowry acid?

16.14. Would you expect NH_3 or H_2O to be a stronger Lewis base? Explain your selection.

PROBLEMS

16.15. Use Lewis structures to show how electron pairs move and bonds form in the following reaction, and identify the Lewis acid and Lewis base.

$$BF_3(g) + F^-(aq) \rightarrow BF_4^-(aq)$$

16.16. Use Lewis structures to show how electron pairs move and bonds form and break in this reaction, and identify the Lewis acid and Lewis base.

$$MgO(s) + CO_2(g) \rightarrow MgCO_3(s)$$

16.17. Use Lewis structures to track the electron pairs and show which bonds form and break in this reaction, and identify the Lewis acid and Lewis base.

$$CO_2(g) + H_2O(\ell) \rightleftharpoons H_2CO_3(aq)$$

16.18. Use Lewis structures to track the electron pairs and show which bonds form and break in this reaction, and identify the Lewis acid and Lewis base.

$$SO_3(g) + H_2O(\ell) \rightarrow H_2SO_4(aq)$$

16.19. Use Lewis structures to track the electron pairs and show which bonds form and break in this reaction, and identify the Lewis acid and Lewis base.

$$B(OH)_3(aq) + H_2O(\ell) \rightleftharpoons B(OH)_4^-(aq) + H^+(aq)$$

***16.20.** Use Lewis structures to track the electron pairs and show which bonds form and break in this reaction, and identify the Lewis acid and Lewis base. *Hint*: $HSbF_6$ is an ionic compound and one of the strongest Brønsted–Lowry acids known.

$$SbF_5(s) + HF(g) \rightleftharpoons HSbF_6(s)$$

Complex Ions

CONCEPT REVIEW

16.21. When NaCl dissolves in water, which molecules or ions occupy the inner coordination sphere around the Na^+ ions?

16.22. When $Cr(NO_3)_3$ dissolves in water, which of the following species are nearest the Cr^{3+} ions?
  a. Other Cr^{3+} ions
  b. NO_3^- ions
  c. H_2O molecules with the O atoms closest to the Cr^{3+}
  d. H_2O molecules with the H atoms closest to the Cr^{3+}

16.23. When $Ni(NO_3)_2$ dissolves in water, what molecules or ions occupy the inner coordination sphere around the Ni^{2+} ions?

16.24. When $[Ni(NH_3)_6]Cl_2$ dissolves in water, which molecules or ions occupy the inner coordination sphere around the Ni^{2+} ions?

16.25. Which ion is the counter ion in the coordination compound $Na_2[Zn(CN)_4]$?

16.26. Which ion is the counter ion in the coordination compound $[Co(NH_3)_4Cl_2]NO_3$?

16.27. The table below contains data from reactions of solutions of a series of octahedral platinum(IV) complexes with $AgNO_3(aq)$. The compounds have the general formula $Pt(NH_3)_xCl_4$, where x is 6, 5, 4, 3, or 2. Write formulas for each compound.

Composition of Complex	Number of Moles of AgCl Produced per Mole of Ag^+ Added
$Pt(NH_3)_6Cl_4$	4
$Pt(NH_3)_5Cl_4$	3
$Pt(NH_3)_4Cl_4$	2
$Pt(NH_3)_3Cl_4$	1
$Pt(NH_3)_2Cl_4$	0

16.28. The compositions of three compounds of chromium(III) and chloride ion are known: $Cr(H_2O)_6Cl_3$, $Cr(H_2O)_5Cl_3$, and $Cr(H_2O)_4Cl_3$. The table below summarizes their properties.

Compound	Color of Aqueous Solution	Number of Moles of AgCl Produced per Mole of Ag^+ Added
A	Dark green	1
B	Gray-blue	3
C	Light green	2

Write the correct formulas for compounds A, B, and C.

Naming Complex Ions and Coordination Compounds

PROBLEMS

16.29. Name the complex ions of platinum(IV) in Problem 16.27.

16.30. Name the complex ions of chromium(III) in Problem 16.28.

16.31. What are the names of the following complex ions?
  a. $Cr(NH_3)_6^{3+}$ b. $Co(H_2O)_6^{3+}$ c. $[Fe(NH_3)_5Cl]^{2+}$

16.32. What are the names of the following complex ions?
  a. $Cu(NH_3)_2^+$
  b. $Ti(H_2O)_4(OH)_2^{2+}$
  c. $Ni(NH_3)_4(H_2O)_2^{2+}$

16.33. What are the names of the following complex ions?
  a. $CoBr_4^{2-}$ b. $Zn(H_2O)(OH)_3^-$ c. $Ni(CN)_5^{3-}$

16.34. What are the names of the following complex ions?
  a. CoI_4^{2-} b. $CuCl_4^{2-}$ c. $[Cr(en)(OH)_4]^-$

16.35. What are the names of the following coordination compounds?
  a. $[Zn(en)]SO_4$
  b. $[Ni(NH_3)_5(H_2O)]Cl_2$
  c. $K_4Fe(CN)_6$

16.36. What are the names of the following coordination compounds?
  a. $(NH_4)_3[Co(CN)_6]$
  b. $[Co(en)_2Cl](NO_3)_2$
  c. $[Fe(H_2O)_4(OH)_2]Cl$

Polydentate Ligands; Ligand Strength and the Chelate Effect

CONCEPT REVIEW

16.37. What is meant by the term *sequestering agent*? What properties make a substance an effective sequestering agent?

16.38. The structures of two compounds that each contain two $-NH_2$ groups are shown in Figure P16.38. The one on the left is ethylenediamine, a bidentate ligand. Does the molecule on the right have the same ability to donate two pairs of electrons to a metal ion? Explain why you think it does or does not.

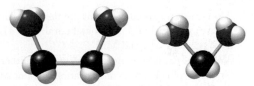

FIGURE P16.38

16.39. How does the chelating ability of an *aminocarboxylate* vary with changing pH?

*16.40. **Food Preservative** The EDTA that is widely used as a food preservative is added to food, not as the undissociated acid, but rather as the calcium disodium salt: $Na_2[CaEDTA]$. This salt is actually a coordination compound with a Ca^{2+} ion at the center of a complex ion. Draw the structure of this compound.

Crystal Field Theory

CONCEPT REVIEW

16.41. Explain why the compounds of most of the first-row transition metals are colored.

16.42. Unlike the compounds of most transition metal ions, those of Ti^{4+} are colorless. Why?

16.43. Why is the d_{xy} orbital higher in energy than the d_{xz} and d_{yz} orbitals in a square planar crystal field?

16.44. On average, the d orbitals of a transition metal ion in an octahedral field are higher in energy than they are when the ion is in the gas phase. Why?

PROBLEMS

16.45. Aqueous solutions of one of the following complex ions of Cr(III) are violet; solutions of the other are yellow. Which is which? (a) $Cr(H_2O)_6^{3+}$; (b) $Cr(NH_3)_6^{3+}$

16.46. Which of the following complex ions should absorb the shortest wavelengths of electromagnetic radiation? (a) $CuCl_4^{2-}$; (b) CuF_4^{2-}; (c) CuI_4^{2-}; (d) $CuBr_4^{2-}$

16.47. The octahedral crystal field splitting energy Δ_o of $Co(phen)_3^{3+}$ is 5.21×10^{-19} J/ion. What is the color of a solution of this complex ion?

16.48. The octahedral crystal field splitting energy Δ_o of $Co(CN)_6^{3-}$ is 6.74×10^{-19} J/ion. What is the color of a solution of this complex ion?

16.49. Solutions of $NiCl_4^{2-}$ and $NiBr_4^{2-}$ absorb light at 702 nm and 756 nm, respectively. In which ion is the split of d-orbital energies greater?

16.50. Chromium(III) chloride forms six-coordinate complexes with bipyridine, including *cis*-$[Cr(bipy)_2Cl_2]^+$, which reacts slowly with water to produce two products, *cis*-$[Cr(bipy)_2(H_2O)Cl]^{2+}$ and *cis*-$[Cr(bipy)_2(H_2O)_2]^{3+}$. In which of these complexes should Δ_o be the largest?

Magnetism and Spin States

CONCEPT REVIEW

16.51. What determines whether a transition metal ion is in a *high-spin* configuration or a *low-spin* configuration?

16.52. Would you expect a solution of a high-spin complex of a transition metal ion to be the same color as a solution of a low-spin complex? Why?

PROBLEMS

16.53. How many unpaired electrons are there in the following transition-metal ions in an octahedral field? High-spin Fe^{2+}, Cu^{2+}, Co^{2+}, and Mn^{3+}

16.54. Which of the following cations can have either a high-spin or a low-spin electron configuration in an octahedral field? Fe^{2+}, Co^{3+}, Mn^{2+}, and Cr^{3+}

16.55. Which of the following cations can, in principle, have either a high-spin or a low-spin electron configuration in a tetrahedral field? Co^{2+}, Cr^{3+}, Ni^{2+}, and Zn^{2+}

16.56. How many unpaired electrons are in the following transition metal ions in an octahedral crystal field? High-spin Fe^{3+}, Rh^+, V^{3+}, and low-spin Mn^{3+}

16.57. The manganese minerals pyrolusite, MnO_2, and hausmannite, Mn_3O_4, contain manganese ions surrounded by oxide ions.
  a. What are the charges of the Mn ions in each mineral?
  b. In which of these compounds could there be high-spin and low-spin Mn ions?

16.58. Dietary Supplement Chromium picolinate is an over-the-counter diet aid sold in many pharmacies. The Cr^{3+} ions in this coordination compound are in an octahedral field. Is the compound paramagnetic or diamagnetic?

*16.59. Refining Cobalt One method for refining cobalt involves the formation of the complex ion $CoCl_4^{2-}$. This anion is tetrahedral. Is this complex paramagnetic or diamagnetic?

*16.60. Why is it that $Ni(CN)_4^{2-}$ is diamagnetic, but $NiCl_4^{2-}$ is paramagnetic?

Isomerism in Coordination Compounds; Complex Ions in Biomolecules

CONCEPT REVIEW

16.61. What do the prefixes *cis*- and *trans*- mean in the context of an octahedral complex ion?

16.62. What do the prefixes *cis*- and *trans*- mean in the context of a square planar complex?

16.63. How many different types of donor groups are required in order to have stereoisomers of a square planar complex?

16.64. With respect to your answer to the previous question, do all square planar complexes with this many different types of donor groups have stereoisomers?

PROBLEMS

16.65. Does $Co(en)(H_2O)_2Cl_2$ have stereoisomers?

16.66. Does the complex ion $Fe(en)_3^{3+}$ have stereoisomers?

*16.67. Sketch the stereoisomers of the square planar complex ion $CuCl_2Br_2^{2-}$. Are any of these isomers chiral?

*16.68. Sketch the stereoisomers of the complex ion $Ni(en)$ $Cl_2(CN)_2^{2-}$. Are any of these isomers chiral?

Complex-Ion Equilibria

CONCEPT REVIEW

16.69. Trace Metal Toxicity Dissolved concentrations of Cu^{2+} as low as 10^{-6} M are toxic to phytoplankton (microscopic algae). However, a solution that is 10^{-3} M in $Cu(NO_3)_2$ and 10^{-3} M in EDTA is not toxic. Why?

16.70. When a strong base is added to a solution of $CuSO_4$, which is pale blue, a precipitate forms and the solution above the precipitate is colorless. When ammonia is added, the precipitate dissolves and the solution turns a deep navy blue. Use appropriate chemical equations to explain why the observed changes occur.

16.71. Cleaning Labware A lab technician cleaning glassware that contains residues of AgCl washes the glassware with an aqueous solution of ammonia. The AgCl, which is insoluble in water, rapidly dissolves in the ammonia solution. Why?

16.72. The procedure used in the previous question dissolves AgCl but not AgI. Why?

PROBLEMS

NOTE: Appendix 5 contains formation constant (K_f) values that may be useful in solving the following problems.

16.73. One millimole of $Ni(NO_3)_2$ dissolves in 250 mL of a solution that is 0.500 M in ammonia.
  a. What is the concentration of $Ni(NO_3)_2$ in the solution?
  b. What is the concentration of $Ni^{2+}(aq)$ in the solution?

16.74. A 1.00 L solution contains 3.00×10^{-4} M $Cu(NO_3)_2$ and 1.00×10^{-3} M ethylenediamine. What is the concentration of $Cu^{2+}(aq)$ in the solution?

16.75. Suppose a 500 mL solution contains 1.00 mmol of $Co(NO_3)_2$, 100 mmol of NH_3, and 100 mmol of ethylenediamine. What is the concentration of $Co^{2+}(aq)$ in the solution?

*16.76. To a 250 mL volumetric flask are added 1.00 mL volumes of three solutions: 0.0100 M $AgNO_3$, 0.100 M $NaBr$, and 0.100 M $NaCN$. The mixture is diluted with deionized water to the mark and shaken vigorously. Are the contents of the flask cloudy or clear? Support your answer with the appropriate calculations. (*Hint*: The K_{sp} of AgBr is 5.4×10^{-13}.)

16.77. What is the maximum concentration of Fe^{3+} in a solution that contains 0.010 M $K_3[Fe(C_2O_4)_3]$ and 0.010 M $C_2O_4^{2-}$?

*16.78. Rhubarb People are generally advised not to eat rhubarb leaves, which contain up to 1.3% by weight oxalic acid ($H_2C_2O_4$), which forms a compound with calcium ions and forms complexes with many transition metals. Calcium oxalate and compounds containing the iron(III) oxalate anion are frequently found in kidney stones. The K_{sp} of CaC_2O_4 is 2.0×10^{-8}. Consider the addition of small portions of an aqueous solution that is 0.010 M in oxalate ion to equimolar aqueous solutions of Ca^{2+}, Co^{2+}, Cu^{2+}, and Fe^{3+}.
  a. In which solution would a precipitate appear first?
  b. Suggest a reason why Fe^{3+} and Ca^{2+} are found more abundantly in kidney stones than are complexes of the other metals.

Hydrated Metal Ions as Acids

CONCEPT REVIEW

16.79. Which, if any, aqueous solutions of the following compounds are acidic? (a) $CaCl_2$; (b) $CrCl_3$; (c) $NaCl$; (d) $FeCl_2$

16.80. If 0.1 M aqueous solutions of each of these compounds were prepared, which one has the lowest pH? (a) $BaCl_2$; (b) $NiCl_2$; (c) KCl; (d) $TiCl_4$

16.81. When ozone is bubbled through a solution of iron(II) nitrate, dissolved Fe(II) ions are oxidized to Fe(III) ions. How does the oxidation process affect the pH of the solution?

16.82. As an aqueous solution of KOH is slowly added to a stirred solution of $AlCl_3$, the mixture becomes cloudy, but then clears when more KOH is added.
  a. Explain the chemical changes responsible for the changes in the appearance of the mixture.
  b. Would you expect to observe the same changes if KOH were added to a solution of $FeCl_3$? Explain why or why not.

16.83. Chromium(III) hydroxide is amphiprotic. Write chemical equations showing how an aqueous suspension of this compound reacts to the addition of a strong acid and a strong base.

16.84. Zinc hydroxide is amphiprotic. Write chemical equations showing how an aqueous suspension of this compound reacts to the addition of a strong acid and a strong base.

16.85. Refining Aluminum To remove impurities such as calcium and magnesium carbonates and Fe(III) oxides from aluminum ore (which is mostly Al_2O_3), the ore is treated with a strongly basic solution. In this treatment Al(III) dissolves but the other metal ions do not. Why?

*16.86. The same quantity of $FeCl_3$ dissolves in 1 M aqueous solutions of each of these acids: HNO_3, HNO_2, H_2SO_3, and CH_3COOH. Is the concentration of $Fe^{3+}(aq)$ the same in all four solutions? Explain why or why not.

PROBLEMS

16.87. What is the pH of 0.50 M $Al(NO_3)_3$?

16.88. What is the pH of 0.25 M $CrCl_3$?

16.89. What is the pH of 0.100 M $Fe(NO_3)_3$?

16.90. What is the pH of 1.00 M $Cu(NO_3)_2$?

16.91. Sketch the titration curve (pH versus volume of 0.50 M NaOH) for a 25 mL sample of 0.5 M $FeCl_3$.

16.92. Sketch the titration curve that results from the addition of 0.50 M NaOH to a sample containing 0.5 M $KFe(SO_4)_2$.

Additional Problems

16.93. Photographic Film Processing During the processing of black-and-white photographic film, excess silver(I) halides are removed by washing the film in a bath containing sodium thiosulfate. This treatment is based on the following complexation reaction:

$$Ag^+(aq) + 2\,S_2O_3^{2-}(aq) \rightleftharpoons Ag(S_2O_3)_2^{3-}(aq) \qquad K_f = 5 \times 10^{13}$$

What is the ratio of $[Ag^+]$ to $[Ag(S_2O_3)_2^{3-}]$ in a bath in which $[S_2O_3^{2-}] = 0.233\ M$?

16.94. Lead Poisoning Children used to be treated for lead poisoning with intravenous injections of EDTA. If the concentration of EDTA in the blood of a patient is 2.5×10^{-8} M and the formation constant for the complex $Pb(EDTA)^{2-}$ is 2.0×10^{18}, what is the concentration ratio of the free (and potentially toxic) $Pb^{2+}(aq)$ in the blood to the much less toxic Pb^{2+}–EDTA complex?

16.95. Dissolving cobalt(II) nitrate in water gives a beautiful purple solution. There are three unpaired electrons in this cobalt(II) complex. When cobalt(II) nitrate is dissolved in aqueous ammonia and oxidized with air, the resulting yellow complex has no unpaired electrons. Which cobalt complex has the larger crystal field splitting energy Δ_o?

*16.96. A solid compound containing Fe(II) in an octahedral crystal field has four unpaired electrons at 298 K. When the compound is cooled to 80 K, the same sample appears to have no unpaired electrons. How do you explain this change in the compound's properties?

16.97. When Ag_2O reacts with peroxodisulfate ($S_2O_8^{2-}$) ion (a powerful oxidizing agent), AgO is produced. Crystallographic and magnetic analyses of AgO suggest that it is not simply Ag(II) oxide, but rather a blend of Ag(I) and Ag(III) in a square planar environment. The Ag^{2+} ion is paramagnetic but, like AgO, Ag^+ and Ag^{3+} are diamagnetic. Explain why.

16.98. The iron(II) compound $Fe(bipy)_2(SCN)_2$ is paramagnetic, but the corresponding cyanide compound $Fe(bipy)_2(CN)_2$ is diamagnetic. Why do these two compounds have different magnetic properties?

16.99. Aqueous solutions of copper(II)–ammonia complexes are dark blue. Will the color of the series of complexes $Cu(H_2O)_{(6-x)}(NH_3)_x{}^+$ shift toward shorter or longer wavelengths as the value of x increases from 0 to 6?

17

Electrochemistry

The Quest for Clean Energy

Running on Electricity

The first decade of the 21st century saw remarkable swings in the price of gasoline. These swings, coupled with growing concerns over climate change, invigorated development of innovative propulsion systems for vehicles that consume less fossil fuel. Some of these vehicles, called hybrids, are propelled by combinations of electric motors powered by rechargeable batteries and small gasoline engines. Others are completely electric, powered by banks of high-performance batteries or fuel cells.

To give all-electric cars the driving range and performance many motorists want, scientists and engineers continue to develop lighter, higher capacity, and more reliable batteries to power these vehicles. Batteries convert chemical energy into electrical energy; motors then convert that electrical energy into mechanical energy. These two processes together are more efficient than the conversion of chemical energy directly to mechanical energy in gasoline engines. Moreover, electric motors consume no energy when the vehicles they power are stopped in traffic. A vehicle's electric motor can even help recharge its batteries when the brakes are applied. As the car slows, the motor becomes an electric *generator*, turning the vehicle's kinetic energy into electrical energy.

Ultimately, the driving range of a battery-powered vehicle is limited by the capacity of its battery to store and deliver electrical energy. The sports sedan in the photo can make trips as long as 400 km (240 mi) on a single charge, though this range has a cost: its battery pack, which contains 8000 individual cells each about the size of a AA battery, accounts for nearly one-third the weight of the car. The smaller battery packs in vehicles known as plug-in hybrids (their batteries can be recharged using electricity from the power grid) give them driving ranges of about 50 km (30 mi) on battery power alone.

In this chapter, we examine the chemistry of modern batteries and of another source of mobile electrical power, fuel cells. Batteries and fuel cells harness the chemical energy released by spontaneous redox reactions, so we begin the chapter by reviewing what we learned about redox chemistry in Chapter 8. Then we explore how the electrochemical energy released inside batteries can be harnessed to pump electrons through devices like electric motors, and how a battery's chemical energy can be restored by forcing its spontaneous cell reaction to run in reverse. We also investigate how the energy released in fuel cells when H_2 and O_2 combine to form H_2O is used to do electric work and propel automobiles more efficiently and with less environmental impact than the cars we drive today.

High Performance (Batteries Included) This all-electric Tesla sports sedan runs on batteries in which lithium ions (the blue spheres) flow between electrodes made of carbon or silicon to others made of transition metal oxides. ▶

LEARNING OUTCOMES

LO1 Combine the appropriate half-reactions to write net ionic equations describing redox reactions
Sample Exercises 17.1, 17.2

LO2 Draw cell diagrams and describe the components of electrochemical cells and their roles in interconverting chemical and electrical energy
Sample Exercise 17.3

LO3 Calculate standard cell potentials from standard reduction potentials
Sample Exercise 17.4

LO4 Interconvert a cell's standard potential and the change in standard free energy of the cell reaction
Sample Exercise 17.5

LO5 Use the Nernst equation to calculate nonstandard cell potentials
Sample Exercise 17.6

LO6 Interconvert a cell's standard potential with the value of the equilibrium constant of the cell reaction under standard conditions
Sample Exercise 17.7

LO7 Interconvert masses of reactants in cell reactions with quantities of electrical charge
Sample Exercises 17.8, 17.9

17.1 Redox Chemistry Revisited

In Chapters 13 through 15 we examined some of the environmental problems associated with the combustion of fossil fuels and the operation of internal combustion engines. In this chapter, we explore alternative propulsion technologies that are more efficient than internal combustion engines at converting chemical energy into mechanical energy and that have the potential to dramatically reduce air pollution. These technologies are based on **electrochemistry**, the branch of chemistry that links chemical reactions to the production or consumption of electrical energy. At the heart of electrochemistry are chemical reactions in which electrons are gained and lost at electrode surfaces. In other words, electrochemistry is based on *red*uction and *ox*idation, or *redox*, chemistry.

The principles of redox reactions were introduced in Section 8.6. Let's briefly review them here:

➤ A redox reaction consists of two complementary processes: the *reduction* of a substance that gains electrons, and the *oxidation* of a substance that loses electrons.

➤ Reduction and oxidation happen simultaneously so that the number of electrons gained during reduction exactly matches the number lost during oxidation.

➤ A substance that is easily oxidized is one that readily gives up electrons. This electron-donating power makes the substance an effective reducing agent. In any redox reaction the *reducing agent is always oxidized*.

➤ A substance that readily accepts electrons and is thereby reduced is an effective oxidizing agent. In any redox reaction the *oxidizing agent is always reduced*.

CONNECTION Redox reactions were discussed in detail in Section 8.6.

In Chapter 8 we introduced redox chemistry by exploring the reaction between zinc atoms and Cu^{2+} ions described in the following net ionic equation:

$$Zn(s) + Cu^{2+}(aq) \rightarrow Cu(s) + Zn^{2+}(aq) \tag{8.29}$$

We can think of the Zn/Cu^{2+} reaction, or any redox reaction, as consisting of two **half-reactions**: one oxidation reaction and one reduction reaction. Here, each Zn atom that is oxidized *loses* 2 electrons in an oxidation half-reaction:

$$Zn(s) \rightarrow Zn^{2+}(aq) + 2 \ e^-$$

and each Cu^{2+} ion that is reduced *accepts* 2 electrons in a reduction half-reaction:

$$Cu^{2+}(aq) + 2 \ e^- \rightarrow Cu(s)$$

Because the number of electrons lost and gained is the same in the two half-reactions, writing a net ionic equation to describe the overall redox reaction is simply a matter of adding the two half-reactions together:

$$Zn(s) \rightarrow Zn^{2+}(aq) + 2 \ e^-$$
$$\underline{Cu^{2+}(aq) + 2 \ e^- \rightarrow Cu(s)}$$
$$Zn(s) + Cu^{2+}(aq) + \cancel{2 \ e^-} \rightarrow Cu(s) + Zn^{2+}(aq) + \cancel{2 \ e^-}$$

Canceling out the equal numbers of electrons gained and lost, we get Equation 8.29.

electrochemistry the branch of chemistry that examines the transformations between chemical and electrical energy.

half-reaction one of the two halves of a redox reaction; one half-reaction is the oxidation component, and the other is the reduction component.

Combining half-reactions can be a useful way to write the net ionic equation describing any redox reaction. Generally, we can accomplish this by following these five steps:

1. Write separate expressions linking the known reactants and products in the two half-reactions.
2. Balance the number of atoms of each element in each half-reaction. For some reactions in solution this step may involve adding molecules of H_2O and either H^+ or OH^- ions to balance the numbers of H and O atoms.
3. Balance the charge in each half-reaction by adding electrons to the appropriate side.
4. Multiply each half-reaction by the appropriate whole number so that the numbers of electrons match.
5. Add the two half-reactions together, and cancel any terms that appear on both sides.

This procedure is especially convenient when balanced equations for the two half-reactions are available, which allows us to skip steps 1–3 and start with step 4. One handy source for such equations is Table A6.1 in Appendix 6, which contains dozens of common half-reactions. Note that they are all written as reduction half-reactions. There is no need for a separate table of *oxidation* half-reactions because any reduction half-reaction can always be reversed to obtain the corresponding oxidation half-reaction. Therefore, you will find the half-reaction describing the reduction of Cu^{2+} ions to Cu metal in the table, but you will not find a half-reaction describing the oxidation of Zn metal to Zn^{2+} ions. However, you will find a half-reaction reduction of Zn^{2+} ions to Zn metal in the table. So, writing a net ionic equation for the redox reaction depicted in Figure 17.1 involves (1) scanning the list of *reactants* in Table A6.1 for the *product* of the oxidation half-reaction (Zn^{2+} ions), and (2) making sure that the *product* of the reduction half-reaction in the table matches the *reactant* of our oxidation half-reaction (Zn metal).

Many of the half-reactions in Table A6.1 are written as if they occur in acidic solutions. In fact, H^+ ions are used to balance the number of H atoms in most of those in which water is a reactant or product. For example, the reduction of O_2 to H_2O is written

$$O_2(g) + 4\,H^+(aq) + 4\,e^- \rightarrow 2\,H_2O(\ell)$$

However, a few half-reactions in Table A6.1 contain OH^- ions, which means they apply to reactions in basic solutions. One of them involves the reduction of O_2:

$$O_2(g) + 2\,H_2O(\ell) + 4\,e^- \rightarrow 4\,OH^-$$

In the following exercises we need to select the half-reaction that matches the pH conditions of the overall redox reaction.

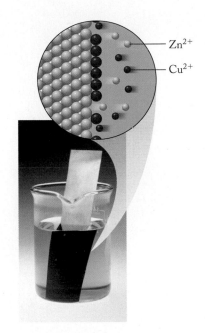

FIGURE 17.1 A strip of zinc is immersed in an aqueous solution of blue copper(II) sulfate. It becomes encrusted with a dark layer of copper as Cu^{2+} ions are reduced to Cu atoms and Zn atoms are oxidized to colorless Zn^{2+} ions. The solution becomes a paler blue as Cu^{2+} ions are removed.

∞∞ CONNECTION Keep in mind that hydrogen ions in aqueous solutions are really hydronium ions, $H_3O^+(aq)$, as described in Chapters 8 and 15.

**SAMPLE EXERCISE 17.1   Writing Redox Reaction Equations   LO1
by Combining Half-Reactions I**

Write a balanced net ionic equation describing the oxidation of $Fe^{2+}(aq)$ by O_2 gas in an acidic solution. *Hint:* Water is a component of the O_2 reduction half-reaction.

COLLECT AND ORGANIZE We find these half-reactions for O_2 reduction in Table A6.1:

$$O_2(g) + 4\,H^+(aq) + 4\,e^- \rightarrow 2\,H_2O(\ell)$$
$$O_2(g) + 2\,H_2O(\ell) + 4\,e^- \rightarrow 4\,OH^-(aq)$$

All of the half-reactions in Table A6.1 are reductions, so there is no half-reaction for the oxidation of Fe^{2+} ions. However, Fe^{2+} ions appear as a product in the reduction of Fe^{3+}:

$$Fe^{3+}(aq) + e^- \rightarrow Fe^{2+}(aq)$$

We can reverse this reduction half-reaction to make Fe^{2+} the reactant in an oxidation half-reaction:

$$Fe^{2+}(aq) \rightarrow Fe^{3+}(aq) + e^-$$

ANALYZE The problem specifies acidic conditions, so we use the O_2 half-reaction with H^+ rather than OH^- ions in it. In any redox reaction the number of electrons gained by the substance that is reduced must equal the number of electrons lost by the substance that is oxidized. There is a gain of 4 moles of electrons in the O_2 half-reaction and a loss of 1 mole of electrons in the Fe^{2+} half-reaction. Therefore, we multiply the Fe^{2+} half-reaction by 4 to obtain a balanced overall redox reaction equation.

SOLVE Multiplying the Fe^{2+} half-reaction by 4 and adding it to the O_2 half-reaction containing H^+ ions, we get

$$4\,Fe^{2+}(aq) \rightarrow 4\,Fe^{3+}(aq) + 4\,e^-$$

$$O_2(g) + 4\,H^+(aq) + 4\,e^- \rightarrow 2\,H_2O(\ell)$$
_____
$$4\,Fe^{2+}(aq) + O_2(g) + 4\,H^+(aq) + \cancel{4\,e^-} \rightarrow 4\,Fe^{3+}(aq) + 2\,H_2O(\ell) + \cancel{4\,e^-}$$

Simplifying gives

$$4\,Fe^{2+}(aq) + O_2(g) + 4\,H^+(aq) \rightarrow 4\,Fe^{3+}(aq) + 2\,H_2O(\ell)$$

THINK ABOUT IT We can verify that this is a balanced net ionic equation by confirming that the numbers of Fe, O, and H atoms are the same on the two sides of the reaction arrow, and that the total electrical charge is the same on the two sides (both are 12+). Note that the result of combining these two half-reactions is not a molecular equation but a net ionic equation.

Practice Exercise Write a net ionic equation describing the oxidation of NO_2^- to NO_3^- by O_2 in a basic solution. ⚙

(Answers to Practice Exercises are in the back of the book.)

In the preceding Sample Exercise we combined half-reactions from Appendix 6 to write the net ionic equation for the oxidation of Fe^{2+} by O_2 under acidic conditions. What if we needed to write a balanced chemical equation describing the oxidation of iron under basic conditions, in which solid $Fe(OH)_2$ is oxidized to $Fe(OH)_3$? There is no half-reaction in Appendix 6 with these iron hydroxides, so we need to write our own half-reaction describing the oxidation of $Fe(OH)_2$ to $Fe(OH)_3$. We can then combine it with the half-reaction for the reduction of O_2 under basic conditions. Let's work through steps 1–5 from the previous page:

1. From Table A6.1, as seen in Sample Exercise 17.1, we already have a balanced half-reaction for the reduction of O_2:

$$O_2(g) + 2\,H_2O(\ell) + 4\,e^- \rightarrow 4\,OH^-(aq)$$

To write a half-reaction for the oxidation of Fe^{2+}, we begin with the formulas of the reduced and oxidized forms of the element of interest:

$$Fe(OH)_2(s) \rightarrow Fe(OH)_3(s)$$

2. We need to balance the number of atoms of each element in the oxidation half-reaction. In the expression above, Fe is already balanced, but the right side has one more O and H atom than the left. The simplest solution to this imbalance is to add a OH^- ion (which would be present under basic reaction conditions) to the left side:

$$Fe(OH)_2(s) + OH^-(aq) \rightarrow Fe(OH)_3(s)$$

3. We add one electron to the right side to balance the charges, obtaining a balanced oxidation half-reaction:

$$Fe(OH)_2(s) + OH^-(aq) \rightarrow Fe(OH)_3(s) + e^-$$

4. This oxidation half-reaction has one electron, but the O_2 reduction half-reaction consumes four electrons, so we need to multiply the oxidation half-reaction by 4:

$$4\,Fe(OH)_2(s) + 4\,OH^-(aq) \rightarrow 4\,Fe(OH)_3(s) + 4\,e^-$$

5. We can now add the two half-reactions together and cancel the terms common to both sides:

$$4\,Fe(OH)_2(s) + 4\,OH^-(aq) \rightarrow 4\,Fe(OH)_3(s) + 4\,e^-$$

$$O_2(g) + 2\,H_2O(\ell) + 4\,e^- \rightarrow 4\,OH^-(aq)$$

$$4\,Fe(OH)_2(s) + 4\,\cancel{OH^-(aq)} + O_2(g) + 2\,H_2O(\ell) + \cancel{4\,e^-} \rightarrow$$
$$4\,Fe(OH)_3(s) + \cancel{4\,e^-} + 4\,\cancel{OH^-(aq)}$$

which yields a balanced equation describing the redox reaction:

$$4\,Fe(OH)_2(s) + O_2(g) + 2\,H_2O(\ell) \rightarrow 4\,Fe(OH)_3(s)$$

SAMPLE EXERCISE 17.2 **Writing Redox Reaction Equations   LO1**
by Combining Half-Reactions II

One of the ways to determine the concentration of ethanol (CH_3CH_2OH) in blood involves a titration in which ethanol is reacted with a standard solution of dichromate ($Cr_2O_7^{2-}$) ions under acidic conditions. The products of the reaction are $Cr^{3+}(aq)$, $CO_2(g)$, and $H_2O(\ell)$. Write a balanced net ionic equation describing the titration reaction.

COLLECT AND ORGANIZE We need to write the net ionic equation describing a reaction in which $CH_3CH_2OH(aq)$ and $Cr_2O_7^{2-}(aq)$ are reactants and $Cr^{3+}(aq)$, $CO_2(g)$, and $H_2O(\ell)$ are products. A check of Appendix 6 discloses the following reduction half-reaction containing both forms of chromium:

$$Cr_2O_7^{2-}(aq) + 14\,H^+(aq) + 6\,e^- \rightarrow 2\,Cr^{3+}(aq) + 7\,H_2O(\ell)$$

There is no half-reaction involving the oxidation or reduction of ethanol in Appendix 6.

ANALYZE Dichromate is reduced to chromium(III) in the reduction half-reaction above, which means that ethanol is oxidized to carbon dioxide and water in the titration reaction. To write a net ionic equation describing the reaction, we need to first develop a half-reaction for the oxidation of ethanol following steps 1–3 from the previous pages. Then, following steps 4 and 5, we will need to balance the losses and gains of electrons and combine the chromium and ethanol half-reactions.

SOLVE

1. Starting with the formulas of the reactant and products in the ethanol half-reaction,

$$CH_3CH_2OH(aq) \rightarrow CO_2(g) + H_2O(\ell)$$

2. To balance the numbers of atoms of each element, let's start with carbon, placing a coefficient of 2 in front of CO_2:

$$CH_3CH_2OH(aq) \rightarrow \mathbf{2}\,CO_2(g) + H_2O(\ell)$$

There is now 1 O atom on the left and 5 on the right. To balance them, we add 4 molecules of H_2O to the left side:

$$CH_3CH_2OH(aq) + \mathbf{4\,H_2O}(\ell) \rightarrow 2\,CO_2(g) + H_2O(\ell)$$

which simplifies to

$$CH_3CH_2OH(aq) + \mathbf{3}\,H_2O(\ell) \rightarrow 2\,CO_2(g)$$

Now we have 12 H atoms on the left and none on the right, so we add 12 H^+ ions to the right side:

$$CH_3CH_2OH(aq) + 3\,H_2O(\ell) \rightarrow 2\,CO_2(g) + \mathbf{12\,H^+}(aq)$$

3. Finally, we add 12 electrons to the right side to balance the charges of the 12 H^+ ions, yielding a balanced oxidation half-reaction:

$$CH_3CH_2OH(aq) + 3\,H_2O(\ell) \rightarrow 2\,CO_2(g) + 12\,H^+(aq) + \mathbf{12\,e^-}$$

4. We now have two balanced half-reactions. To balance the gain and loss of electrons, we multiply the reduction half-reaction, which requires six electrons, by 2:

$$\mathbf{2}\,Cr_2O_7^{2-}(aq) + \mathbf{28}\,H^+(aq) + \mathbf{12}\,e^- \rightarrow \mathbf{4}\,Cr^{3+}(aq) + \mathbf{14}\,H_2O(\ell)$$

5. Adding both half-reactions and then simplifying the combined equation,

$$2\,Cr_2O_7^{2-}(aq) + 28\,H^+(aq) + 12\,e^- \rightarrow 4\,Cr^{3+}(aq) + 14\,H_2O(\ell)$$
$$CH_3CH_2OH(aq) + 3\,H_2O(\ell) \rightarrow 2\,CO_2(g) + 12\,H^+(aq) + 12\,e^-$$

$$\overline{2\,Cr_2O_7^{2-}(aq) + 16\ \cancel{28}\,H^+(aq) + \cancel{12\,e^-} + CH_3CH_2OH(aq) + \cancel{3\,H_2O(\ell)} \rightarrow}$$
$$4\,Cr^{3+}(aq) + 11\ \cancel{14}\,H_2O(\ell) + 2\,CO_2(g) + \cancel{12\,H^+(aq)} + \cancel{12\,e^-}$$

or,

$$2\,Cr_2O_7^{2-}(aq) + 16\,H^+(aq) + CH_3CH_2OH(aq) \rightarrow$$
$$4\,Cr^{3+}(aq) + 11\,H_2O(\ell) + 2\,CO_2(g)$$

This equation is now complete and balanced.

THINK ABOUT IT We could use the chromium half-reaction from Appendix 6 in this exercise because it contains the right chemical forms of the elements and because it is written for acidic reaction conditions. Acidic conditions are also the reason why H^+ ions are added to balance the H atoms in writing the ethanol half-reaction. The balanced net ionic equation contains a key piece of information for the analyst using this reaction to determine blood alcohol concentrations: dichromate and alcohol react in a mole ratio of 2:1.

Practice Exercise An acidic solution of permanganate (MnO_4^-) ions can oxidize nitrous acid to nitrate ions. The products include dissolved Mn^{2+} ions. Write chemical equations describing the oxidation of nitrous acid and the reduction of permanganate, and combine the two half-reactions to write a net ionic equation describing the overall redox reaction. ⚙

electrochemical cell an apparatus that converts chemical energy into electrical work or electrical work into chemical energy.

anode an electrode at which an oxidation half-reaction (loss of electrons) takes place.

cathode an electrode at which a reduction half-reaction (gain of electrons) takes place.

17.2 Electrochemical Cells

Having shown how the redox reaction between Zn metal and Cu^{2+} ions is the net result of two separate half-reactions, one involving the oxidation of Zn metal and the other the reduction of Cu^{2+} ions, let's now physically separate the two half-reactions using a device called an **electrochemical cell**. Doing so forces the electrons lost by the Zn atoms to flow through an external circuit before they can be acquired by Cu^{2+} ions. Figure 17.2 provides a view of an electrochemical cell based on the Zn/Cu^{2+} reaction. One compartment in the Zn/Cu^{2+} cell contains a strip of zinc metal immersed in 1.00 M $ZnSO_4$; the other contains a strip of copper metal immersed in 1.00 M $CuSO_4$. Sulfate ions are spectator ions in this reaction and are not included in the net ionic equations shown in the figure. The two metal strips serve as the *electrodes* in the cell, providing pathways along which the electrons produced and consumed in the two half-reactions flow through an external circuit.

As the cell reaction proceeds, oxidation of Zn atoms produces electrons, which travel from the Zn electrode through the external circuit to the surface of the Cu electrode where they combine with Cu^{2+} ions, forming Cu metal. In an electrochemical cell, the electrode at which the *oxidation* half-reaction takes place (the zinc electrode in this case) is always called the **anode**, and the electrode at which the *reduction* half-reaction takes place is always called the **cathode**.

You might think that production of Zn^{2+} ions in the left compartment in Figure 17.2 would result in a buildup of positive charge on that side of the cell, and that conversion of Cu^{2+} ions to Cu metal would create an excess of SO_4^{2-} ions and thus a negative charge in the Cu compartment. If that happened, the cell reaction would shut down. However, no such charge buildup occurs because the two compartments are connected by a permeable bridge and because the solutions surrounding both electrodes contain a *background electrolyte* made of ions that are not involved in either half-reaction. In the Zn/Cu^{2+} cell, Na_2SO_4 makes an ideal background electrolyte. Migration of Na^+ ions toward the Cu compartment and SO_4^{2-} ions toward the Zn compartment through the permeable bridge as shown in the middle molecular view in Figure 17.2 balances the flow of electrons in the external circuit and eliminates any accumulation of ionic charge in either compartment.

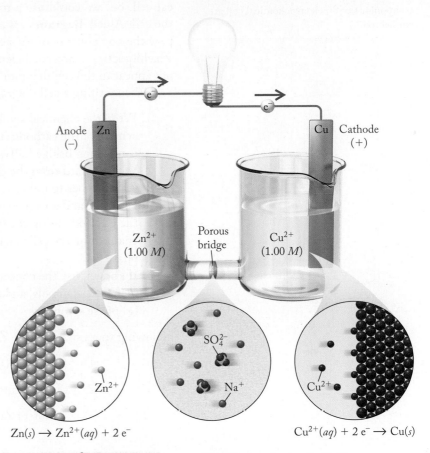

$$Zn(s) \rightarrow Zn^{2+}(aq) + 2\,e^-$$

$$Cu^{2+}(aq) + 2\,e^- \rightarrow Cu(s)$$

FIGURE 17.2 In this Zn/Cu^{2+} electrochemical cell, zinc metal is oxidized to Zn^{2+} ions at the anode in the left-hand compartment while Cu^{2+} ions are reduced to Cu metal in the right-hand compartment. A porous bridge provides an electrical connection through which ions of a background electrolyte (Na_2SO_4) migrate from one compartment to the other.

▶❙❙ **CHEMTOUR** Zinc–Copper Cell

CONCEPT TEST

As the cell reaction in Figure 17.2 proceeds, is the increase in mass of the copper strip the same as the decrease in mass of the zinc strip?

(Answers to Concept Tests are in the back of the book.)

cell diagram symbols that show how the components of an electrochemical cell are connected.

Figure 17.2 provides a view of the physical reality of a Zn/Cu^{2+} electrochemical cell, but we could use a more compact way to represent the components of the cell. A **cell diagram** uses a string of chemical formulas and symbols to show how the components of the cell are connected. A cell diagram does not convey stoichiometry, so any coefficients in the balanced equation for the cell reaction do not appear in the cell diagram.

When writing a cell diagram,

1. Write the chemical symbol of the anode at the far left of the diagram, the symbol of the cathode at the far right, and double vertical lines for the connecting bridge halfway between them.
2. Work inward from the electrodes toward the connecting bridge, using vertical lines to indicate phase changes (like that between a solid metal electrode and an aqueous solution). Represent the electrolytes surrounding the electrode using the symbols of the ions or compounds that are changed by the cell reaction. Use commas to separate species in the same phase.
3. If known, use the concentrations of the dissolved species in place of the (*aq*) phase symbols, and add the partial pressures of any gases within their (*g*) phase symbols.

Following these steps for the Zn/Cu^{2+} electrochemical cell in Figure 17.2,

$$(1) \qquad Zn(s) \dots\dots\dots \| \dots\dots\dots Cu(s)$$

$$(2) \qquad Zn(s) \mid Zn^{2+}(aq) \| Cu^{2+}(aq) \mid Cu(s)$$

$$(3) \qquad Zn(s) \mid Zn^{2+}(1.00\ M) \| Cu^{2+}(1.00\ M) \mid Cu(s)$$

SAMPLE EXERCISE 17.3 **Diagramming an Electrochemical Cell** **LO2**

Figure 17.3 depicts an electrochemical cell in which a copper electrode immersed in a 1.00 *M* solution of Cu^{2+} ions is connected to a silver electrode immersed in a 1.00 *M* solution of Ag^+ ions. Write a balanced chemical equation for this cell reaction and write a cell diagram for this cell.

COLLECT AND ORGANIZE We have a cell reaction in which electrons spontaneously flow from a copper electrode in contact with a solution of Cu^{2+} ions through an external circuit to a silver electrode in contact with a solution of Ag^+ ions. The half-reactions in Table A6.1 involving these metals and ions are

$$Cu^{2+}(aq) + 2\ e^- \rightarrow Cu(s)$$

$$Ag^+(aq) + e^- \rightarrow Ag(s)$$

In a cell diagram, the anode and the species involved in the oxidation half-reaction are on the left and the cathode and the species involved in the reduction half-reaction are on the right. We use single lines to separate phases and a double line to represent the porous bridge separating the two compartments of the cell.

ANALYZE In an electrochemical cell, electrons flow from the anode through an external circuit to the cathode. Therefore, in the cell in Figure 17.3, copper is the anode and silver is the cathode. This means that the Cu reduction half-reaction must run in reverse as an oxidation half-reaction:

$$Cu(s) \rightarrow Cu^{2+}(aq) + 2\ e^-$$

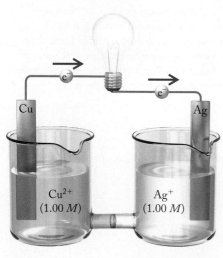

FIGURE 17.3 A Cu/Ag^+ electrochemical cell.

Two moles of electrons are produced in the anode half-reaction, but only 1 mole of electrons is consumed in the cathode half-reaction at the Ag electrode. We therefore need to multiply the silver half-reaction by 2 before combining the two equations.

SOLVE Multiplying the Ag^+ half-reaction by 2 and adding it to the Cu half-reaction, we get

$$2\,Ag^+(aq) + 2\,e^- \rightarrow 2\,Ag(s)$$

$$Cu(s) \rightarrow Cu^{2+}(aq) + 2\,e^-$$

$$\overline{2\,Ag^+(aq) + \cancel{2\,e^-} + Cu(s) \rightarrow 2\,Ag(s) + Cu^{2+}(aq) + \cancel{2\,e^-}}$$

$$2\,Ag^+(aq) + Cu(s) \rightarrow 2\,Ag(s) + Cu^{2+}(aq)$$

The equation is balanced, and we are finished with the first part of the exercise.

Applying the rules for writing a cell diagram:

1. Anode on the left, cathode on the right, bridge in the middle:

$$Cu(s) \dotsb \| \dotsb Ag(s)$$

2. Adding electrode/solution boundaries and the formulas of the ions produced and consumed in the cell reaction completes the cell diagram:

$$Cu(s) \mid Cu^{2+}(aq) \| Ag^+(aq) \mid Ag(s)$$

3. Adding concentration terms,

$$Cu(s) \mid Cu^{2+}(1.00\ M) \| Ag^+(1.00\ M) \mid Ag(s)$$

THINK ABOUT IT To test the validity of the cell diagram, let's translate it into a sentence: *Cu atoms at a copper anode are oxidized to aqueous Cu^{2+} ions, which are separated by a permeable bridge from an aqueous solution of Ag^+ ions that are reduced to Ag atoms at a silver cathode.* This description matches the net ionic equation for the reaction and is consistent with the layout of, and flow of electrons in, Figure 17.3.

Practice Exercise Write a balanced chemical equation and the cell diagram for an electrochemical cell that has a copper cathode immersed in a solution of Cu^{2+} ions and an aluminum anode immersed in a solution of Al^{3+} ions. ⚙

17.3 Standard Potentials

Table A6.1 lists half-reactions in order of a parameter called **standard reduction potential ($E°$)**, expressed in volts (V). The more positive the value of $E°$, the greater the probability that the reduction half-reaction will couple with an oxidation half-reaction to produce a spontaneous redox reaction. The most positive $E°$ value in Table A6.1 is that for the reduction of fluorine:

$$F_2(g) + 2\,e^- \rightarrow 2\,F^-(aq) \qquad E° = 2.866\ V$$

Fluorine's top position means that it is the most easily reduced reactant in Table A6.1. It also means that F_2 is the strongest oxidizing agent in the table. It is capable of oxidizing any of the substances on the product side of the half-reactions lower in the table. We have seen this affinity of fluorine for electrons before. We saw in Chapter 3 that the electron affinity of fluorine atoms is −328 kJ/mol (the second most negative EA value—meaning the second greatest affinity for electrons—of all the elements). And, as we saw in Chapter 4, fluorine is the most electronegative of all the elements.

The reactants with very negative $E°$ values at the bottom of Table A6.1 include the major cations found in biological systems: Na^+, K^+, Mg^{2+}, and Ca^{2+}.

standard reduction potential ($E°$) the potential of a reduction half-reaction in which all reactants and products are in their standard states at 25°C.

standard cell potential (E°_{cell}) a measure of how forcefully an electrochemical cell, in which all reactants and products are in their standard states, can pump electrons through an external circuit.

electromotive force (emf) also called voltage, the force pushing electrons through an electrical circuit.

voltaic cell or **galvanic cell** an electrochemical cell in which chemical energy is transformed into electrical work by a spontaneous redox reaction.

cell potential (E_{cell}) the electromotive force with which an electrochemical cell can pump electrons through an external circuit.

FIGURE 17.4 Alessandro Volta is credited with building the first battery in 1798. It consisted of a stack of alternating layers of zinc, blotter paper soaked in salt water, and silver.

Their negative E° values tell us that these ions are not easily reduced to their free metals. This message makes sense because none of these elements occur in nature in the atomic state, but rather as the listed cations. It is also consistent with the fact that they have the lowest ionization energies of all the elements.

Now let's consider what happens to E° when the half-reactions in Table A6.1 run in reverse. This means that the products of reduction half-reactions become the reactants of oxidation half-reactions. It also means that the half-reaction at the very bottom of the table with the most negative E° value:

$$Li^+(aq) + e^- \rightarrow Li(s) \qquad E^\circ = -3.05 \text{ V}$$

is likely to run in reverse as an oxidation half-reaction:

$$Li(s) \rightarrow Li^+(aq) + e^-$$

Thus, Li^+ ions in aqueous solution have little ability to oxidize anything, but Li metal is a very powerful reducing agent that can reduce any of the substances on the reactant side of the half-reactions above it in Table A6.1.

> **CONCEPT TEST** ···
>
> Given the positions of the reactants and products in Table A6.1, predict which of the following reactions are spontaneous under standard conditions:
>
> a. $Cu(s) + 2 Fe^{3+}(aq) \rightarrow Cu^{2+}(aq) + 2 Fe^{2+}(aq)$
> b. $2 Ag(s) + Zn^{2+}(aq) \rightarrow 2 Ag^+(aq) + Zn(s)$
> c. $Hg(\ell) + 2 H^+(aq) \rightarrow Hg^{2+}(aq) + H_2(g)$

We can use the standard reduction potentials in Table A6.1 to calculate the **standard cell potentials (E°_{cell})** of electrochemical cells. Standard cell potentials are measures of the **electromotive force (emf)** generated by cell reactions, which reflects how forcefully cells pump electrons out from their anodes through external circuits and into their cathodes. In the process of pumping electrons, these electrochemical cells convert the chemical energy of a spontaneous cell reaction into electrical work. Cells that do this are called **voltaic cells** in honor of Italian physicist Alessandro Volta (1745–1827, Figure 17.4); they can also be called **galvanic cells** after the physicist Luigi Galvani (1737–1798), a contemporary of Volta. We can combine the standard reduction potentials of the cathode and anode of a voltaic cell to calculate the force of its electron pump under standard conditions; this is the value of E°_{cell}. The value of E°_{cell} is the difference between the standard reduction potentials of its cathode and anode:

$$E^\circ_{cell} = E^\circ_{cathode} - E^\circ_{anode} \qquad (17.1)$$

To understand why we subtract E°_{anode} from $E^\circ_{cathode}$, keep in mind that E°_{anode}, like *all electrode potentials*, is a standard *reduction* potential. However, the anode is the electrode in a cell where *oxidation* takes place. The half-reaction happening at the anode is essentially a reduction half-reaction running in reverse. As we saw in Chapters 9 and 12, when we reverse a reaction we change the signs of its thermodynamic properties, such as ΔH°, ΔS°, and ΔG°. Standard cell potential is another thermodynamic property, which is, as we'll see in Section 17.4, directly linked to the change in standard free energy, ΔG°_{cell}. Therefore, it is reasonable that the sign of E°_{anode} changes when the direction of its half-reaction is reversed. That change in sign is the reason for the minus sign in Equation 17.1.

The superscript (°) in Equation 17.1 has its usual thermodynamic meaning—all reactants and products are in their standard states, that is, the concentrations of all dissolved substances are 1 M and the partial pressures of all gases are 1 atm. If we want to denote a generalized cell potential in which reactants and products are not necessarily in their standard states, we simply refer to the **cell potential (E_{cell})**.

Let's use Equation 17.1 to calculate $E°_{cell}$ for the Zn/Cu^{2+} voltaic cell in Figure 17.2. The standard reduction potential of the cathode half-reaction is

$$Cu^{2+} + 2\,e^- \rightarrow Cu \qquad E° = 0.342\ V$$

To obtain the standard potential for the oxidation half-reaction at the zinc anode, we start with the standard reduction potential of Zn^{2+} ions in Table A6.1:

$$Zn^{2+} + 2\,e^- \rightarrow Zn \qquad E° = -0.762\ V$$

Now we use Equation 17.1 to calculate $E°_{cell}$:

$$E°_{cell} = E°_{cathode} - E°_{anode}$$
$$= 0.342 - (-0.762) = 1.104\ V$$

This is the cell potential we measure if we connect a device called a voltmeter across the two electrodes, as shown in Figure 17.5.

To use Equation 17.1 we need to know which half-reaction occurs at the cathode (reduction) and which occurs at the anode (oxidation). In other words, we need to know which component of the spontaneous cell reaction is oxidized and which is reduced. As we have seen, this decision can be made based on the data in Table A6.1. In our Zn/Cu^{2+} cell, for instance, the value of $E°$ for the reduction of Cu^{2+} ions to Cu metal is 0.342 V, which is greater than the value of $E°$ for reducing Zn^{2+} ions to Zn metal (−0.762 V). Therefore, in the Zn/Cu^{2+} voltaic cell, Cu^{2+} ions are reduced and Zn metal is oxidized. We can generalize this observation to the cell reaction of any voltaic cell: the half-reaction with the more positive value of $E°$ runs as a reduction and the other one runs in reverse as an oxidation. This way we are assured that $E°_{cell}$ is a positive value.

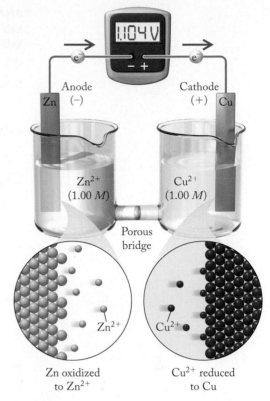

FIGURE 17.5 A voltmeter displays a cell potential of 1.104 V between a Zn electrode immersed in a 1.00 M solution of Zn^{2+} ions and a Cu electrode immersed in a 1.00 M solution of Cu^{2+} ions.

▶❙❙ **CHEMTOUR** Cell Potential

SAMPLE EXERCISE 17.4 | **Identifying Anode and Cathode Half-Reactions and Calculating the Value of $E°_{cell}$** | **LO3**

The standard reduction potentials of the half-reactions in single-use alkaline batteries are

$$ZnO(s) + H_2O(\ell) + 2\,e^- \rightarrow Zn(s) + 2\,OH^-(aq) \qquad E° = -1.25\ V$$
$$2\,MnO_2(s) + H_2O(\ell) + 2\,e^- \rightarrow Mn_2O_3(s) + 2\,OH^-(aq) \qquad E° = 0.15\ V$$

What is the net ionic equation for the cell reaction and the value of $E°_{cell}$?

COLLECT AND ORGANIZE We can calculate $E°_{cell}$ using Equation 17.1:

$$E°_{cell} = E°_{cathode} - E°_{anode}$$

but first we need to decide which half-reaction occurs at the cathode and which at the anode. The equation for a cell reaction is written by combining half-reactions once the loss and gain of electrons in the two half-reactions are balanced.

▶❙❙ **CHEMTOUR** Alkaline Battery

ANALYZE The MnO_2 half-reaction has the more positive $E°$, making it our reduction half-reaction. We must reverse the ZnO half-reaction, turning it into an oxidation half-reaction. Both half-reactions involve the transfer of 2 electrons, so they may be combined by simply adding them together.

SOLVE The oxidation half-reaction at the anode is

$$Zn(s) + 2\,OH^-(aq) \rightarrow ZnO(s) + H_2O(\ell) + 2\,e^-$$

and the reduction half-reaction at the cathode is

$$2\,MnO_2(s) + H_2O(\ell) + 2\,e^- \rightarrow Mn_2O_3(s) + 2\,OH^-(aq)$$

Combining these half-reactions to obtain the overall cell reaction, we get:

$$2\,MnO_2(s) + \cancel{H_2O(\ell)} + Zn(s) + \cancel{2\,OH^-(aq)} + \cancel{2\,e^-} \rightarrow$$
$$Mn_2O_3(s) + \cancel{2\,OH^-(aq)} + ZnO(s) + \cancel{H_2O(\ell)} + \cancel{2\,e^-}$$

Simplifying gives us the net ionic equation for the cell reaction:

$$2\,MnO_2(s) + Zn(s) \rightarrow Mn_2O_3(s) + ZnO(s)$$

The overall $E°_{cell}$ for this reaction is obtained using Equation 17.1:

$$E°_{cell} = E°_{cathode} - E°_{anode}$$
$$= 0.15\,V - (-1.25\,V) = 1.40\,V$$

THINK ABOUT IT The $E°_{cell}$ value is reasonable because the potential of most alkaline batteries is nominally 1.5 V. In this particular cell reaction the net ionic equation is also the complete molecular equation.

Practice Exercise The half-reactions in nicad (nickel–cadmium) batteries are

$$Cd(OH)_2(s) + 2\,e^- \rightarrow Cd(s) + 2\,OH^-(aq) \qquad E° = -0.403\,V$$
$$2\,NiO(OH)(s) + 2\,H_2O(\ell) + 2\,e^- \rightarrow 2\,Ni(OH)_2(s) + 2\,OH^-(aq) \qquad E° = 1.32\,V$$

Write the net ionic equation for the cell reaction and calculate the value of $E°_{cell}$. ⚙

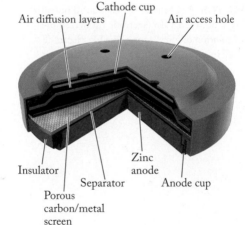

Cathode cup

Air diffusion layers   Air access hole

Insulator

Zinc anode

Separator   Anode cup

Porous carbon/metal screen

FIGURE 17.6 Most of the internal volume of a zinc–air battery is occupied by the anode: a paste of Zn particles in an aqueous solution of KOH surrounded by a metal cup that serves as the negative terminal of the battery. Oxygen from the air is the reactant at the cathode. Air enters through holes in an inverted metal cup that serves as the positive terminal of the battery. Once inside the battery, air diffuses through layers of gas-permeable plastic film that let air in but keep electrolyte from leaking out. Oxygen in the air is reduced at the porous carbon/metal cathode to OH^- ions that migrate toward the anode, where they are consumed in the Zn oxidation half-reaction.

Before closing this discussion of standard cell potentials, let's combine two half-reactions in which different numbers of electrons are gained and lost. This combination occurs in a type of battery that has a limitless supply of one of its reactants. It is called the zinc–air battery (Figure 17.6), and it powers devices in which small battery size and low mass are high priorities, such as hearing aids. Most of the internal volume of one of these batteries is occupied by an anode consisting of a paste of zinc particles packed in an aqueous solution of KOH. As in alkaline batteries (Sample Exercise 17.4), the anode half-reaction is

$$Zn(s) + 2\,OH^-(aq) \rightarrow ZnO(s) + H_2O(\ell) + 2\,e^-$$

which is the reverse of the reaction in Table A6.1:

$$ZnO(s) + H_2O(\ell) + 2\,e^- \rightarrow Zn(s) + 2\,OH^-(aq) \qquad E° = -1.25\,V$$

The cathode consists of porous carbon supported by a metal screen. Air diffuses through small holes in the battery and across a layer of plastic film that lets gases pass through but keeps electrolyte from leaking out. As air passes through the cathode, oxygen is reduced to hydroxide ions:

$$O_2(g) + 2\,H_2O(\ell) + 4\,e^- \rightarrow 4\,OH^-(aq) \qquad E°_{cathode} = 0.401\,V$$

To write the overall cell reaction, we need to multiply the oxidation half-reaction by 2 before combining it with the reduction half-reaction:

$$2[Zn(s) + 2\,OH^-(aq) \rightarrow ZnO(s) + H_2O(\ell) + 2\,e^-]$$

$$O_2(g) + 2\,H_2O(\ell) + 4\,e^- \rightarrow 4\,OH^-(aq)$$

$$2\,Zn(s) + 4\,\cancel{OH^-(aq)} + O_2(g) + 2\,\cancel{H_2O(\ell)} + 4\,\cancel{e^-} \rightarrow$$
$$2\,ZnO(s) + 2\,\cancel{H_2O(\ell)} + 4\,\cancel{OH^-(aq)} + 4\,\cancel{e^-}$$

or

$$2\,Zn(s) + O_2(g) \rightarrow 2\,ZnO(s)$$

$$E°_{cell} = E°_{cathode} - E°_{anode} = 0.401\,V - (-1.25\,V) = 1.65\,V$$

Note that when we multiplied the anode half-reaction by 2 and added it to the cathode half-reaction, *we did not multiply the $E°$* of the anode half-reaction by 2. The reason we did not is that $E°$ is an *intensive* property of a half-reaction or a complete cell reaction. It does not change when the quantities of reactants and products change. Thus, a zinc–air battery the size of a pea has the same $E°_{cell}$ as one the size of a book (like those being developed for electric vehicles). On the other hand, the amount of electrical work a zinc–air battery can do *does* depend on how much zinc is inside it because, as we are about to see, the electrical work that a voltaic cell can do depends on both cell potential *and* the quantity of charge it can deliver at that potential.

17.4 Chemical Energy and Electrical Work

When we connect the Zn and Cu electrodes from Figure 17.2 to a digital voltmeter as in Figure 17.5—the Zn electrode to the negative terminal of the meter and the Cu electrode to the positive terminal—the meter reads 1.104 V. These connections tell us that the battery is pumping electrons from the Zn electrode through the external circuit to the Cu electrode. The reading tells us how much electromotive force (emf) is pushing the electrons through the circuit. Under standard conditions this emf is the same as the standard cell potential of the cell ($E°_{cell}$); under any other conditions it is simply E_{cell}.

    When a voltaic cell pumps electrons through an external circuit, those moving electrons are doing electrical work, like lighting a lightbulb or turning an electric motor. The change in chemical free energy (ΔG_{cell}) that accompanies a spontaneous cell reaction is a measure of the electrical work (w_{elec}) that may result:

$$\Delta G_{cell} = w_{elec} \qquad (17.2)$$

The sign of ΔG_{cell} is negative because the reaction in a voltaic cell is spontaneous, which means there is a decrease in the free energy of the system (cell reaction mixture) as reactants become products. The sign of w_{elec} is also negative because it represents work done *by* the system *on* its surroundings, which has a negative value according to the sign convention summarized in Table 9.1 and Figure 9.4.

    The work done by a voltaic cell on its surroundings is the product of the quantity of electric charge (Q) the cell pushes through an external circuit times the force (emf) pushing that charge. That force is the same as the cell potential, so the connection between E_{cell} and w_{elec} is

$$w_{elec} = -QE_{cell} \qquad (17.3)$$

▶❚❚ **CHEMTOUR** Free Energy

◉◉ **CONNECTION** The sign conventions used for work done *on* a thermodynamic system (+) and the work done *by* the system (−) were explained in Section 9.1.

Faraday constant (F) the magnitude of electric charge in 1 mole of electrons. Its value to three significant figures is 9.65×10^4 C/mol.

where the negative sign reflects the fact that work done *by* a voltaic cell *on* its surroundings (the external circuit) corresponds to energy lost by the cell.

The quantity of charge is proportional to the number of electrons flowing through the circuit. As noted in Chapter 2, the magnitude of the charge on a single electron is 1.602×10^{-19} coulomb (C). The magnitude of electric charge on 1 mole of electrons is

$$\frac{1.602 \times 10^{-19} \text{ C}}{e^-} \times \frac{6.0221 \times 10^{23} \text{ } e^-}{1 \text{ mol } e^-} = \frac{9.65 \times 10^4 \text{ C}}{\text{mol } e^-}$$

This quantity of charge, 9.65×10^4 C/mol, is called the **Faraday constant (F)** after Michael Faraday (1791–1867), the English chemist and physicist who discovered that redox reactions take place when electrons are transferred from one species to another. The quantity of charge Q flowing through an electrical circuit is the product of the number of moles of electrons n times the Faraday constant:

$$Q = nF \tag{17.4}$$

Combining Equations 17.3 and 17.4 gives us an equation relating w_{elec} and E_{cell}:

$$w_{elec} = -nFE_{cell} \tag{17.5}$$

Equations 17.2 and 17.5 connect the quantity of electrical work a voltaic cell does on its surroundings to the change in free energy in the cell. Under standard conditions this connection is

$$\Delta G^{\circ}_{cell} = -nFE^{\circ}_{cell} \tag{17.6}$$

Perhaps you are wondering how the product on the right side of Equation 17.6 is the equivalent of energy. It works out that way because the units are

$$\text{moles} \times \frac{\text{coulombs}}{\text{mole}} \times \text{volts} = \text{coulombs-volts}$$

A coulomb-volt is the same quantity of energy as a joule:

$$1 \text{ coulomb-volt} = 1 \text{ joule}$$

$$1 \text{ C} \cdot \text{V} = 1 \text{ J}$$

The negative sign on the right side of Equation 17.6 tells us that the E_{cell} of any voltaic cell must have a positive value because the sign of ΔG_{cell} for the spontaneous chemical reaction inside the cell must be negative.

Let's use Equation 17.6 to calculate the change in standard free energy of the Zn/Cu^{2+} cell reaction. Let's start with the standard cell potential we calculated in Section 17.3:

$$E^{\circ}_{cell(Zn/Cu^{2+})} = 1.104 \text{ V}$$

We can convert this standard cell potential into a change in standard free energy $(\Delta G^{\circ}_{cell})$ using Equation 17.6 under standard conditions so that $\Delta G = \Delta G^{\circ}$ and $\Delta E = \Delta E^{\circ}$. We need to use $n = 2$ because 2 moles of electrons are transferred in the half-reactions for this cell:

$$\Delta G^{\circ}_{cell} = -nFE^{\circ}_{cell}$$

$$= -\left(2 \text{ mol} \times \frac{9.65 \times 10^4 \text{ C}}{\text{mol}} \times 1.104 \text{ V}\right)$$

$$= -2.13 \times 10^5 \text{ C} \cdot \text{V} = -2.13 \times 10^5 \text{ J} = -213 \text{ kJ}$$

To put this value in perspective, on a mole-for-mole basis, the Zn/Cu^{2+} reaction produces nearly as much useful energy as the combustion of hydrogen gas:

$$H_2(g) + \tfrac{1}{2} O_2(g) \rightarrow H_2O(g) \qquad \Delta G° = -228.6 \text{ kJ}$$

CONCEPT TEST ···

When a rechargeable battery, such as the one used to start a car's engine, is recharged, an external source of electrical power forces the voltaic cell reaction to run in reverse. What are the signs of ΔG_{cell} and E_{cell} during the recharging process?

···

SAMPLE EXERCISE 17.5 **Relating $\Delta G°_{cell}$ and $E°_{cell}$** **LO4**

Many of the "button" batteries used in electric watches consist of a Zn anode and Ag$_2$O cathode separated by a membrane soaked in a concentrated solution of KOH (Figure 17.7). At the cathode, Ag$_2$O is reduced to Ag metal, and at the anode Zn is oxidized to solid Zn(OH)$_2$. Write the net ionic equation for the reaction, and, using the appropriate standard reduction potentials from Table A6.1, calculate the values of $E°_{cell}$ and $\Delta G°_{cell}$.

COLLECT AND ORGANIZE We know the reactants and products of the anode and cathode reactions and that the reactions occur in a basic solution. The following equations should be useful in calculating $E°_{cell}$ and $\Delta G°_{cell}$ from the appropriate standard potentials:

$$E°_{cell} = E°_{cathode} - E°_{anode}$$

$$\Delta G°_{cell} = -nFE°_{cell}$$

ANALYZE The half-reaction at the cathode is based on the reduction of Ag$_2$O to Ag. The appropriate half-reaction in Table A6.1 is

$$Ag_2O(s) + H_2O(\ell) + 2\,e^- \rightarrow 2\,Ag(s) + 2\,OH^-(aq) \qquad E°_{cathode} = 0.342 \text{ V}$$

We must reverse the anode's oxidation half-reaction to find an entry in Table A6.1 in which Zn(OH)$_2$ is the reactant and Zn is the product:

$$Zn(OH)_2(s) + 2\,e^- \rightarrow Zn(s) + 2\,OH^-(aq) \qquad E°_{anode} = -1.249 \text{ V}$$

However, in order to write the net ionic equation, we need to use the actual oxidation half-reaction at the anode—the reverse of this half-reaction—and combine it with the cathode's reduction half-reaction. The two half-reactions involve the transfer of the same number of electrons, so combining them simply means adding them together. The value of $E°_{cell}$ will be about [0.35 − (−1.25)] or 1.60 V. This value is half again as large as the $E°_{cell}$ of the Zn/Cu^{2+} cell. Therefore, the magnitude of its $\Delta G°_{cell}$ value should be half again as large as −213 kJ/mol, or about −300 kJ/mol.

SOLVE Reversing the Zn(OH)$_2$ half-reaction and adding it to the Ag$_2$O half-reaction, we get:

$$Zn(s) + 2\,OH^-(aq) \rightarrow Zn(OH)_2(s) + 2\,e^-$$

$$Ag_2O(s) + H_2O(\ell) + 2\,e^- \rightarrow 2\,Ag(s) + 2\,OH^-(aq)$$

$$\overline{\rule{0pt}{1em}}$$

$$Ag_2O(s) + H_2O(\ell) + Zn(s) + \cancel{2\,OH^-(aq)} + \cancel{2\,e^-} \rightarrow$$
$$2\,Ag(s) + \cancel{2\,OH^-(aq)} + Zn(OH)_2(s) + \cancel{2\,e^-}$$

or

$$Ag_2O(s) + H_2O(\ell) + Zn(s) \rightarrow 2\,Ag(s) + Zn(OH)_2(s)$$

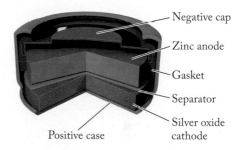

FIGURE 17.7 Many of the button batteries that power small electronic devices incorporate a Zn anode and a Ag$_2$O cathode separated by a membrane containing KOH electrolyte.

Negative cap
Zinc anode
Gasket
Separator
Silver oxide cathode
Positive case

Calculating $E°_{cell}$,

$$E°_{cell} = E°_{cathode} - E°_{anode}$$
$$= 0.342\ V - (-1.249\ V)$$
$$= 1.591\ V$$

Calculating $\Delta G°_{cell}$,

$$\Delta G°_{cell} = -nFE°_{cell}$$
$$= -\left(2\ \text{mol} \times \frac{9.65 \times 10^4\ C}{\text{mol}} \times 1.591\ V\right)$$
$$= -3.07 \times 10^5\ C \cdot V = -3.07 \times 10^5\ J = -307\ kJ$$

THINK ABOUT IT The positive value of $E°_{cell}$ and negative value of $\Delta G°_{cell}$ are expected because voltaic cell reactions are spontaneous. The calculated values are close to those we estimated.

Practice Exercise The alkaline batteries used in flashlights (Sample Exercise 17.4) produce a cell potential of 1.50 V. What is the value of ΔG_{cell}? ⚙

Some final thoughts about the $\Delta G°_{cell}$ value calculated in Sample Exercise 17.5 are in order. First, it is based on the reaction of 1 mole of Ag_2O and 1 mole of Zn, which corresponds to 232 g of Ag_2O and 65 g of Zn. The energy stored in a button battery (Figure 17.7), which has a mass of only 1 or 2 g, would be a tiny fraction of the calculated value. Also, note that no ions appear in the net ionic equation. Because all the reactants and products in the silver oxide battery reaction are solids, its net ionic equation and its molecular equation are identical.

17.5 A Reference Point: The Standard Hydrogen Electrode

We can measure the value of E_{cell} using a voltmeter, but how do we measure the individual electrode potentials of the cathode and anode? The answer to this question is that we arbitrarily assign a value of zero volts to the standard potential for the reduction of hydrogen ions to hydrogen gas:

$$2\ H^+(aq) + 2\ e^- \rightarrow H_2(g) \qquad E° = 0.000\ V \qquad (17.7)$$

An electrode that generates this reference potential, called the **standard hydrogen electrode (SHE)**, consists of a platinum electrode in contact with a solution of a strong acid ($[H^+] = 1.00\ M$) and hydrogen gas at a pressure of 1.00 atm (Figure 17.8). The platinum is unchanged—neither oxidized nor reduced—by the electrode reaction. Rather, it serves as a chemically inert conveyor of electrons. Electrons may be consumed at the electrode surface if H^+ ions are reduced to hydrogen gas, or they may flow away from the surface into the Pt electrode if hydrogen gas is oxidized to H^+ ions. The potential of the SHE, which is the same for both half-reactions, is defined to be 0.000 V.

To write the cell diagram for a cell in which the SHE serves as the anode, we represent the SHE half of the cell as follows:

$$Pt\ |\ H_2(g,\ 1.00\ atm)\ |\ H^+(1.00\ M)\ ||$$

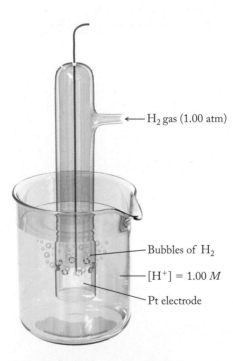

← H_2 gas (1.00 atm)

— Bubbles of H_2

— $[H^+] = 1.00\ M$

— Pt electrode

FIGURE 17.8 The standard hydrogen electrode consists of a platinum electrode immersed in a 1.00 M solution of $H^+(aq)$ and bathed in a stream of pure H_2 gas at a pressure of 1.00 atm. Its potential is the same (0.000 V) whether $H^+(aq)$ ions are reduced or H_2 gas is oxidized.

to indicate that the anode half-reaction involves the oxidation of H_2 gas to H^+ ions. If the SHE is the cathode, then we diagram its half of the cell this way:

$$|| H^+(1.00 \ M) \ | \ H_2(g, 1.00 \ atm) \ | \ Pt$$

standard hydrogen electrode (SHE) a reference electrode based on the half-reaction $2 \ H^+(aq) + 2 \ e^- \rightarrow H_2(g)$ that produces a standard electrode potential defined to be 0.000 V.

to indicate that the cathode half-reaction involves the reduction of H^+ ions to H_2 gas.

Because the standard reduction potential of the SHE is 0.000 V, the measured E_{cell} of any voltaic cell in which a SHE is one of the two electrodes—either cathode or anode—is the potential produced by the other electrode. This means that if we attach a voltmeter to the cell, the meter reading is the electrode potential of the other electrode. Suppose, for example, that a voltaic cell consists of a strip of zinc metal immersed in a 1.00 M solution of Zn^{2+} ions in one compartment and a SHE in the other (Figure 17.9a). Also suppose that a voltmeter is connected to the cell so that it measures the potential at which the cell pumps electrons from the zinc electrode to the SHE. This direction of electron flow means that the zinc electrode is the cell's anode and the SHE is the cathode of the cell. At 25°C the meter reads 0.762 V. We know that the value of $E°_{cathode}$ is that of the SHE (0.000 V), and that $E°_{anode}$ is $E°_{Zn}$. Inserting these values and symbols into Equation 17.1:

$$E°_{cell} = E°_{cathode} - E°_{anode}$$

$$E°_{cell} = E°_{SHE} - E°_{Zn}$$

$$0.762 \ V = 0.000 \ V - E°_{Zn}$$

$$E°_{Zn} = -0.762 \ V$$

This value is equal to the standard reduction potential of Zn^{2+} in Table A6.1:

$$Zn^{2+}(aq) + 2 \ e^- \rightarrow Zn(s) \qquad E° = -0.762 \ V$$

In Figure 17.9(b), the SHE is coupled to a copper electrode immersed in a 1.00 M solution of Cu^{2+} ions. In this cell, electrons flow from the SHE through an external circuit to the copper electrode at a cell potential of 0.342 V at 25°C.

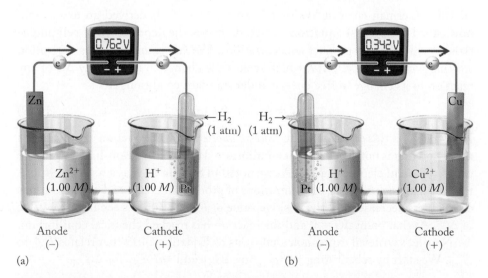

FIGURE 17.9 The standard hydrogen electrode allows us to determine the standard potential of any half reaction. (a) When coupled to a Zn electrode under standard conditions, the SHE is the cathode and the Zn electrode is the anode. When the SHE is connected to the positive terminal of a voltmeter and the Zn electrode to the negative terminal, the meter measures a cell potential of 0.762 V. (b) When coupled to a Cu electrode under standard conditions, the SHE is the anode, the Cu electrode is the cathode, and the meter measures a cell potential of 0.342 V.

The direction of current flow means that the electrons are consumed at the copper electrode, making it the cathode. The value of $E°$ for the copper half-reaction is calculated as follows:

$$E°_{cell} = E°_{cathode} - E°_{anode}$$

$$= E°_{Cu} - E°_{SHE}$$

$$0.342 \text{ V} = E°_{Cu} - 0.000 \text{ V}$$

$$E°_{Cu} = 0.342 \text{ V}$$

This half-reaction potential matches the value of $E°$ for the reduction of Cu^{2+} to Cu metal.

FIGURE 17.10

CONCEPT TEST

A cell consists of a SHE in one compartment and a Ni electrode immersed in a 1.00 M solution of Ni^{2+} ions in the other. If a voltmeter is connected to the electrode as shown in Figure 17.10, what will be the value in the voltmeter's display?

17.6 The Effect of Concentration on E_{cell}

Reactions stop when one of the reactants is completely consumed. This concept was the basis for our discussion of limiting reactants in Chapter 7. However, a commercial battery usually stops operating at its rated cell potential—1.5 V for a flashlight battery—before its reactants are completely consumed. This happens because the cell potential of a voltaic cell is determined by the concentrations of the reactants and products.

The Nernst Equation

In 1889, German chemist Walther Nernst (1864–1941) derived an expression, now called the **Nernst equation**, which describes the dependence of cell potentials on reactant and product concentrations. We can reconstruct his derivation starting with Equation 14.19, which relates the change in free energy ΔG of any reaction to its change in free energy under standard conditions $\Delta G°$:

$$\Delta G = \Delta G° + RT \ln Q \qquad (14.19)$$

Note that Q in this equation is the *reaction quotient*, the mass action expression for a reaction that is not necessarily at equilibrium. (It has no relationship to the Q we use to represent electric charge.) As we noted in Section 14.9, as a reversible spontaneous reaction proceeds, concentrations of products increase and concentrations of reactants decrease until the positive value of $RT \ln Q$ offsets the negative value of $\Delta G°$. At that point $\Delta G = 0$ and the reaction has reached chemical equilibrium.

Now let's write an expression analogous to Equation 14.19 that relates E_{cell} to $E°_{cell}$. We start by substituting $-nFE_{cell}$ for ΔG_{cell} and $-nFE°_{cell}$ for $\Delta G°_{cell}$:

$$-nFE_{cell} = -nFE°_{cell} + RT \ln Q$$

Dividing all terms by $-nF$ gives

$$E_{cell} = E°_{cell} - \frac{RT \ln Q}{nF} \qquad (17.8)$$

CONNECTION In Chapter 14 we discussed the relationship between free energy changes and the reaction quotient Q.

Nernst equation an equation relating the potential of a cell (or half-cell) reaction to its standard potential ($E°$) and to the concentrations of its reactants and products.

This is the equation Walther Nernst developed in 1889. We can obtain a very useful form of it if we insert values for R [8.314 J/(mol · K)] and F (9.65 × 10^4 C/mol), make T = 298 K, and convert the natural logarithm to a base-10 logarithm: ln Q = 2.303 log Q. With these changes the Nernst equation becomes

$$E_{cell} = E°_{cell} - \frac{0.0592}{n} \log Q \qquad (17.9)$$

Equation 17.9 enables us to predict how the potential of a voltaic cell changes as the cell reaction proceeds and the concentrations of products inside the cell increase and the concentrations of reactants decrease. As they do, Q increases and so does the value of $0.0592/n \times \log Q$. The negative sign in front of this term in Equation 17.9 means that the value of E_{cell} decreases as reactants are converted into products. Eventually E_{cell} approaches zero. When it reaches zero, the cell reaction has achieved chemical equilibrium. The cell can no longer pump electrons through an external circuit. In other words, it's dead.

Batteries are voltaic cells, so their cell potentials should drop with use. Let's consider how much they drop by focusing on the *lead–acid* battery used to start most car engines (Figure 17.11). These batteries each contain six electrochemical cells. Their anodes are made of lead (Pb) and their cathodes are made of lead(IV) oxide (PbO$_2$). Both electrodes are immersed in 4.5 M H$_2$SO$_4$. The value of a fully charged cell is about 2.0 V. The six cells are connected in series so that the operating potential of the battery is the sum of the six cell potentials, or 12.0 V.

As the battery discharges, PbO$_2$(s) is reduced to PbSO$_4$(s) at the cathodes:

$$PbO_2(s) + 3\,H^+(aq) + HSO_4^-(aq) + 2\,e^- \rightarrow PbSO_4(s) + 2\,H_2O(\ell)$$

$$E° = 1.685\ V$$

and Pb(s) is oxidized to PbSO$_4$(s) at the anodes:

$$Pb(s) + HSO_4^-(aq) \rightarrow PbSO_4(s) + H^+(aq) + 2\,e^- \qquad E° = -0.356\ V$$

The reduction half-reaction consumes 2 moles of electrons, and the oxidation half-reaction involves the loss of 2 moles of electrons for each mole of Pb.

The net ionic equation for the overall cell reaction is the sum of the two half-reactions:

$$PbO_2(s) + Pb(s) + 2\,H^+(aq) + 2\,HSO_4^-(aq) \rightarrow 2\,PbSO_4(s) + 2\,H_2O(\ell)$$

The value of $E°_{cell}$ is

$$E°_{cell} = E°_{cathode} - E°_{anode} = 1.685\ V - (-0.356\ V) = 2.041\ V$$

Most of the reactants and products are pure solids or liquids, and the Q expression contains only terms for H$^+$ and HSO$_4^-$. As the battery discharges, [H$^+$] and [HSO$_4^-$] decrease, and so does the value of E_{cell} calculated from the Nernst equation:

$$E_{cell} = 2.041\ V - \frac{0.0592}{2} \log \frac{1}{[H^+]^2[HSO_4^-]^2}$$

However, the decrease in E_{cell} is very gradual, not falling below 2.0 V until the battery is about 97% discharged, as shown in Figure 17.12. The gradual decrease makes sense because of the logarithmic relationship between Q and E_{cell}. If [H$_2$SO$_4$] decreased by an order of magnitude—for example, from 1.00 M to 0.100 M—the value of E_{cell} would decrease from 2.041 V by less than 6%:

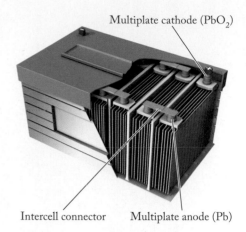

FIGURE 17.11 The lead–acid battery that provides power to start most motor vehicles contains six cells. Each has an anode made of lead and a cathode made of PbO$_2$ immersed in a background electrolyte of 4.5 M H$_2$SO$_4$. The electrodes are formed into plates and held in place by grids made of a lead alloy. The grids connect the cells together in series so that the operating potential of the battery (12.0 V) is the sum of the six cell potentials (each 2.0 V).

Multiplate cathode (PbO$_2$)

Intercell connector   Multiplate anode (Pb)

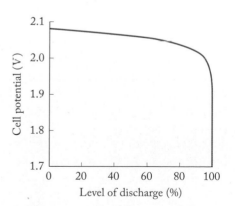

FIGURE 17.12 The potential of a cell in a lead–acid battery decreases as reactants are converted into products, but the change in potential is small until the battery is nearly completely discharged.

$$E_{cell} = 2.041 \text{ V} - \frac{0.0592}{2} \log \frac{1}{0.100^4} = 1.923 \text{ V}$$

The logarithmic relationship between Q and E_{cell} means that most batteries can deliver current at a cell potential close to their fully charged potential until they are nearly completely discharged.

SAMPLE EXERCISE 17.6   **Calculating E_{cell} from $E°_{cell}$ and the Concentrations of Reactants and Products**   **LO5**

The standard potential ($E°_{cell}$) of a voltaic cell based on the Zn/Cu^{2+} ion reaction:

$$Zn(s) + Cu^{2+}(aq) \rightarrow Zn^{2+}(aq) + Cu(s)$$

is 1.104 V. What is the value of E_{cell} at 25°C when $[Cu^{2+}] = 0.100 \, M$ and $[Zn^{2+}] = 1.90 \, M$?

COLLECT AND ORGANIZE We are given the standard cell potential and are asked to determine the value of E_{cell} when $[Cu^{2+}] = 0.100 \, M$ and $[Zn^{2+}] = 1.90 \, M$. The Nernst equation (Equation 17.9) enables us to calculate E_{cell} values for different concentrations of reactants and products. This equation requires us to work with the reaction quotient Q, which we know from Section 14.5 to be the mass action expression for the reaction. Solid copper and zinc are also part of the reaction system, but no terms for pure solids appear in reaction quotients.

ANALYZE The only term in the numerator of the Q expression for this cell reaction is $[Zn^{2+}]$ and the only one in the denominator is $[Cu^{2+}]$. Each Cu^{2+} ion acquires 2 electrons, and each Zn atom donates 2, so the value of n in the Nernst equation is 2. The value of $[Zn^{2+}]$ is greater than $[Cu^{2+}]$, which makes $Q > 1$. The negative sign in front of the $0.0592/n \times \log Q$ term in Equation 17.9 means that the calculated value of E_{cell} should be less than the value of $E°_{cell}$.

SOLVE Substituting the values of $[Zn^{2+}]$ and $[Cu^{2+}]$ in the Nernst equation gives

$$E_{cell} = E°_{cell} - \frac{0.0592}{n} \log Q = E°_{cell} - \frac{0.0592}{n} \log \frac{[Zn^{2+}]}{[Cu^{2+}]} = 1.104 \text{ V} - \frac{0.0592}{2} \log \frac{1.90}{0.100}$$

$$E_{cell} = 1.104 \text{ V} - \frac{0.0592}{2}(1.28) = 1.066 \text{ V}$$

THINK ABOUT IT The calculated E_{cell} value is only 0.038 V less than $E°_{cell}$ because the logarithmic dependence of cell potential on reactant and product concentrations minimizes the impact of changing concentrations.

Practice Exercise The standard cell potential of the zinc–air battery (Figure 17.6) is 1.65 V. If at 25°C the partial pressure of oxygen in the air diffusing through its cathode is 0.21 atm, what is the cell potential? Assume the cell reaction is

$$2 \, Zn(s) + O_2(g) \rightarrow 2 \, ZnO(s)$$

$E°$ and K

When the cell reaction of a voltaic cell reaches chemical equilibrium, $\Delta G_{cell} = E_{cell} = 0$ and $Q = K$. Equation 17.9 then becomes

$$0 = E°_{cell} - \frac{0.0592}{n} \log K$$

which we can rearrange to

$$\log K = \frac{nE^\circ_{cell}}{0.0592} \qquad (17.10)$$

We can use Equation 17.10 to calculate the equilibrium constant for any redox reaction at 25°C, not just those in electrochemical cells. For the more general case, we substitute E°_{rxn} for E°_{cell}:

$$\log K = \frac{nE^\circ_{rxn}}{0.0592} \qquad (17.11)$$

SAMPLE EXERCISE 17.7   **Calculating K for a Redox**   **LO6**
Reaction from the Standard
Potentials of Its Half-Reactions

Many procedures for determining mercury levels in environmental samples begin by oxidizing the mercury to Hg^{2+} and then reducing the Hg^{2+} to elemental Hg with Sn^{2+}. Use the appropriate standard reduction potentials from Table A6.1 to calculate the equilibrium constant at 25°C for this reaction:

$$Sn^{2+}(aq) + Hg^{2+}(aq) \rightarrow Sn^{4+}(aq) + Hg(\ell)$$

COLLECT AND ORGANIZE We can calculate the equilibrium constant for a redox reaction from the standard potentials of its half-reactions. Equation 17.11 relates the equilibrium constant for any redox reaction to the standard potential E°_{rxn}. To calculate E°_{rxn}, we need to combine the appropriate standard potentials. Table A6.1 lists two reduction half-reactions involving our reactants and products:

$$Hg^{2+}(aq) + 2\,e^- \rightarrow Hg(\ell) \qquad E^\circ = 0.851\ V$$

$$Sn^{4+}(aq) + 2\,e^- \rightarrow Sn^{2+}(aq) \qquad E^\circ = 0.154\ V$$

ANALYZE The problem states that Hg^{2+} is reduced by Sn^{2+}, so Sn^{2+} is the reducing agent in the reaction, which means that it must be oxidized. Therefore, the second reaction runs in reverse, as an oxidation, and we must subtract its standard potential from that of the mercury half-reaction. The difference between the two half-reaction potentials is about +0.7 V. Therefore, the right side of Equation 17.11 will be about $(2 \times 0.7)/0.06 \approx 23$, and the value of K should be about 10^{23}.

SOLVE We obtain the standard potential for the reaction from a modified version of Equation 17.1 where E°_{rxn} equals E°_{cell} for any redox reaction:

$$E^\circ_{rxn} = E^\circ_{Hg} - E^\circ_{Sn} = 0.851\ V - 0.154\ V = 0.697\ V$$

Using this value for E°_{rxn} in Equation 17.11 and a value of 2 for n, we have

$$\log K = \frac{nE^\circ_{rxn}}{0.0592} = \frac{2(0.697\ V)}{0.0592} = 23.5$$

$$K = 10^{23.5} = 3 \times 10^{23}$$

THINK ABOUT IT The calculated value is quite close to what we estimated. Note how a relatively small (<1 V) positive value of E°_{rxn} corresponds to a huge equilibrium constant, indicating that the reaction essentially goes to completion. That the reaction goes to completion is one reason it can be reliably used to determine the concentrations of mercury in aqueous samples containing Hg^{2+} ions.

Practice Exercise Use the appropriate standard reduction potentials from Table A6.1 to calculate the value of K at 25°C for the following reaction:

$$5\ Fe^{2+}(aq) + MnO_4^-(aq) + 8\ H^+(aq) \rightarrow 5\ Fe^{3+}(aq) + Mn^{2+}(aq) + 4\ H_2O(\ell)$$

| | **TABLE 17.1** | | Relations between K, E°_{cell}, and $\Delta G^{\circ}_{\text{cell}}$ Values of Electrochemical Reactions | |

K	E°_{cell}	$\Delta G^{\circ}_{\text{cell}}$	Favors Formation of
<1	<0	>0	Reactants
>1	>0	<0	Products
1	0	0	Neither

Before ending our discussion of how to derive equilibrium constant values at 25°C from E°_{cell} values, we should note that measuring the potential of an electrochemical reaction allows us to calculate equilibrium constant values that may be too large or too small to determine from the values of equilibrium concentrations of reactants and products. A value of K as large as that calculated in Sample Exercise 17.7 could not be obtained by analyzing the composition of an equilibrium reaction mixture because the concentrations of the reactants would be too small to be determined accurately. Similarly, a cell potential of about −1 V would correspond to a tiny K value and concentrations of products that are too small to be determined quantitatively.

Table 17.1 summarizes how the values of K and E°_{cell} are related to each other and to the change in free energy ($\Delta G^{\circ}_{\text{cell}}$) of a cell reaction under standard conditions. Note that spontaneous electrochemical reactions have E°_{cell} values greater than zero and K values greater than one. The connection between positive cell potential and reaction spontaneity applies even under nonstandard conditions. Also keep in mind that small positive values of E°_{cell} (only a fraction of a volt, for example) correspond to very large K values and to cell reactions that go nearly to completion.

17.7 Relating Battery Capacity to Quantities of Reactants

An important performance characteristic of a battery is its *capacity* to do electrical work, that is, to deliver electric charge at the labeled cell potential. This capacity—the amount of electrical work done—is defined by Equation 17.5,

$$w_{\text{elec}} = -nFE_{\text{cell}}$$

where nF is the quantity of electric charge delivered, in coulombs (C).

Another important unit in electricity is the *ampere* (A), which is the SI base unit of electric current. An ampere is defined as a current of 1 coulomb per second:

$$1 \text{ ampere} = 1 \text{ coulomb/second}$$

which we can rearrange to

$$1 \text{ coulomb} = 1 \text{ ampere-second}$$

Multiplying both sides of the latter equation by volts, and recalling that 1 joule of electrical energy is equivalent to 1 coulomb-volt of electrical work, we get

$$1 \text{ (coulomb)(volt)} = 1 \text{ joule} = 1 \text{ (ampere-second)(volt)} \qquad (17.12)$$

Joules are small energy units, however, so battery capacities are usually expressed in energy units with time intervals longer than seconds. For example, the *energy ratings* of rechargeable AA batteries (Figure 17.13) are often expressed in ampere-hours at the rated cell potential.

We need even bigger units to express the capacities of the large battery packs used in hybrid vehicles. They are the *watt* (W), the SI unit of power, and the *kilowatt-hour*, a unit of energy equal to over 3 million joules, as shown in the following unit conversions:

$$1 \text{ watt} = 1 \text{ joule/second}$$
$$1 \text{ kilowatt} = 1000 \text{ W} = 1000 \text{ J/s}$$
$$1 \text{ kilowatt} \cdot \text{hour} = (1000 \text{ W})(1 \text{ hr})$$
$$= (1000 \text{ J/s} \times 60 \text{ s/min} \times 60 \text{ min/hr})(1 \text{ hr})$$
$$= 3.6 \times 10^6 \text{ J}$$

FIGURE 17.13 The electrical energy rating of these rechargeable nickel–metal hydride AA batteries is 1.6 A · hr at 1.2 V.

Which energy unit—watt-second (joule), watt-minute, watt-hour, or kilowatt-hour—would be most appropriate for expressing (a) the capacity of a cell phone battery, (b) the annual electrical energy consumed by an Energy Star dishwasher, and (c) the monthly electricity bill for a student apartment?

Nickel–Metal Hydride Batteries

Hybrid vehicles such as the Toyota Prius (Figure 17.14) are powered by combinations of small gasoline engines and electric motors. The motors are powered by battery packs made of dozens of nickel–metal hydride (NiMH) cells (Figure 17.15). At the cathodes in these cells, NiO(OH) is reduced to Ni(OH)$_2$; at the anodes, which are made of one or more transition metals, hydrogen atoms are oxidized to H$^+$ ions, which combine with OH$^-$ to form water. The electrodes are separated by aqueous KOH.

The cathode half-reaction is

$$\text{NiO(OH)}(s) + \text{H}_2\text{O}(\ell) + \text{e}^- \rightarrow \text{Ni(OH)}_2(s) + \text{OH}^-(aq) \qquad E° = 1.32 \text{ V}$$

At the anode, hydrogen is present as a *metal hydride*. To write the anode half-reaction, we use the generic formula MH, where M stands for a transition metal or metal alloy that forms a hydride. In a basic background electrolyte, the anode oxidation half-reaction is

$$\text{MH}(s) + \text{OH}^-(aq) \rightarrow \text{M}(s) + \text{H}_2\text{O}(\ell) + \text{e}^-$$

The standard potential of this half-reaction depends on the chemical properties of MH, but generally the value is near that of the SHE, or about 0.0 V.

The overall cell reaction from these two half-reactions is

$$\text{MH}(s) + \text{NiO(OH)}(s) \rightarrow \text{M}(s) + \text{Ni(OH)}_2(s)$$

The value of $E°_{\text{cell}}$ for the NiMH battery cannot be calculated precisely because we have only an approximate value of $E°_{\text{anode}}$. Most NiMH cells are rated at about 1.2 V.

FIGURE 17.14 The 2012 Toyota Prius is powered by a combination of a gasoline engine and an electric motor. Electricity for the motor comes from a battery pack. The battery pack is recharged when the engine is running or when the brakes are applied, as the car's electric motor acts like a generator to convert the car's kinetic energy into electrical energy.

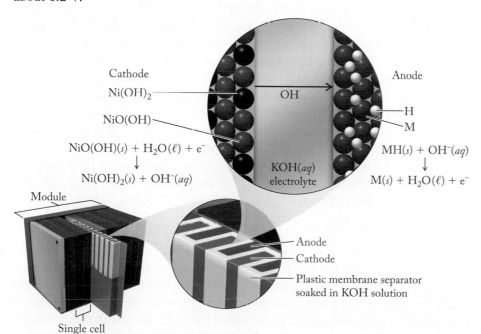

FIGURE 17.15 In the cells of a nickel–metal hydride battery pack, H atoms are oxidized to H$^+$ ions at the anodes (blue plates), and NiO(OH) is reduced to Ni(OH)$_2$ at the cathodes (red plates). The OH$^-$ ions produced by the cathode half-reaction migrate across a KOH-soaked porous membrane and are consumed in the anode half-reaction. The H$^+$ ions combine with OH$^-$ to form water. The anodes are connected to the case of the battery pack, which serves as the (−) terminal, and the cathodes are connected to the cap, which is the (+) terminal.

Now let's relate the electrical energy stored in a battery (in other words, its energy rating) to the quantities of reactants needed to produce that energy. Consider a rechargeable AA NiMH battery rated to deliver 2.5 ampere-hours of electrical charge at 1.2 V. How much NiO(OH) has to be converted to Ni(OH)$_2$ to deliver this much charge? To answer the question, we need to relate the quantity of charge to a number of moles of electrons and then convert that to an equivalent number of moles of reactant and finally to a mass of reactant. Let's begin by recalling that an ampere is defined as a coulomb per second, which means the quantity of electrical charge delivered is

$$2.5 \ \text{A} \cdot \text{hr} \times \frac{1 \ \text{C}}{1 \ \text{A} \cdot \text{s}} \times \frac{60 \ \text{min}}{1 \ \text{hr}} \times \frac{60 \ \text{s}}{1 \ \text{min}} = 9.0 \times 10^3 \ \text{C}$$

The Faraday constant tells us that 1 mole of charge is equivalent to 9.65×10^4 C, so the number of moles of charge, which is equal to the number of moles of electrons that flow from the battery, is

$$9.0 \times 10^3 \ \text{C} \left(\frac{1 \ \text{mol e}^-}{9.65 \times 10^4 \ \text{C}} \right) = 0.0933 \ \text{mol e}^-$$

The stoichiometry of the cathode half-reaction tells us that the mole ratio of NiO(OH) to electrons is 1:1. Therefore, the mass of NiO(OH) consumed is

$$0.0933 \ \text{mol e}^- \left(\frac{1 \ \text{mol NiO(OH)}}{1 \ \text{mol e}^-} \right) \left(\frac{91.70 \ \text{g NiO(OH)}}{1 \ \text{mol NiO(OH)}} \right) = 8.6 \ \text{g NiO(OH)}$$

The mass of a AA battery is 30 g, so this mass for the NiO(OH) is reasonable if we allow for the mass of the anode, background electrolyte, and exterior shell.

Lithium–Ion Batteries

The NiMH batteries used in hybrid vehicles do not have the capacity to power the vehicles at highway speeds nor for extended distances. Nor do these batteries have the energy capacity to power all-electric vehicles such as the Tesla sports sedan on the opening page of this chapter or plug-in hybrids such as the Chevrolet Volt (Figure 17.16). The electrical power demands of these vehicles require batteries with much greater ratios of energy capacity to battery size. The technology of choice in these applications is the lithium–ion battery (Figure 17.17), the same kind of battery that powers laptop computers, cell phones, and digital cameras.

In a lithium–ion battery, Li$^+$ ions are typically stored in a graphite anode, though other anode materials such as nanowires made of silicon have been recently developed. During discharge, these ions migrate through a nonaqueous electrolyte to a porous cathode. These cathodes are made of transition metal oxides or phosphates that can form stable complexes with Li$^+$ ions. One popular cathode material is cobalt(IV) oxide. Lithium–ion batteries with these cathodes and graphite anodes have cell potentials of about 3.6 V (three times that of a NiMH battery). The cell reaction is

$$\text{Li}_{1-x}\text{CoO}_2(s) + \text{Li}_x\text{C}_6(s) \rightarrow 6 \ \text{C}(s) + \text{LiCoO}_2(s) \tag{17.13}$$

In a fully charged cell, $x = 1$, which makes the cathode lithium-free CoO$_2$. As the cell discharges and Li$^+$ ions migrate from the carbon anode to the cobalt oxide cathode, the value of x falls toward zero. To balance this flow of positive charges inside the cell, electrons flow from the anode to the cathode through an external circuit. When fully discharged, the cathode is LiCoO$_2$, and the oxidation number

FIGURE 17.16 The 2013 Chevrolet Volt has a lithium–ion battery pack that can store 16 kW · hr of electrical energy, giving the Volt a driving range of about 60 km (~37 mi). The battery pack is recharged either by plugging it into an electrical outlet or by running an onboard gasoline-powered generator. The generator gives the Volt an overall driving range of about 600 km (~370 mi).

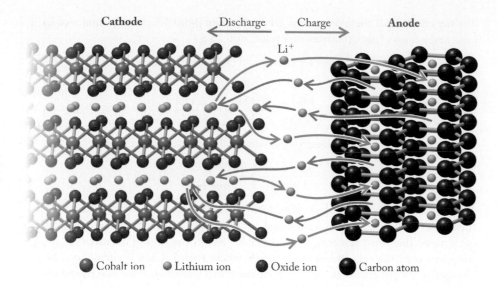

Cathode ← Discharge | Charge → **Anode**

Li⁺

● Cobalt ion   ● Lithium ion   ● Oxide ion   ● Carbon atom

FIGURE 17.17 In a discharging lithium–ion battery, Li^+ ions stored in graphite layers of the anode travel to the cathode, which is made of CoO_2 in this example. During recharging, the direction of ion migration reverses.

of Co is reduced to +3. The electrodes in a lithium–ion battery may react with oxygen and water, so the electrolytes (for example, $LiPF_6$) are dissolved in polar organic solvents, such as tetrahydrofuran, ethylene carbonate, or propylene carbonate (Figure 17.18).

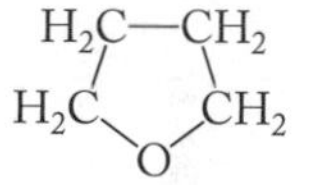

Tetrahydrofuran

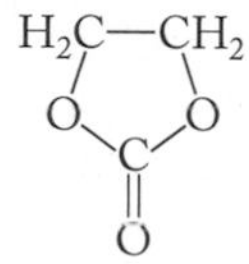

Ethylene carbonate

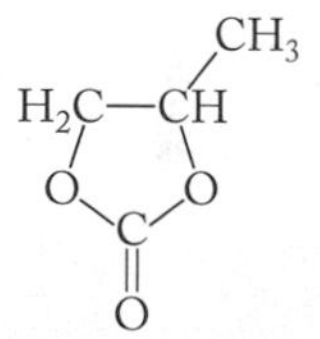

Propylene carbonate

FIGURE 17.18 Polar organic compounds such as these molecules are the solvents for the electrolytes in a lithium–ion battery.

SAMPLE EXERCISE 17.8 **Relating the Mass of a Reactant**   **LO7**
in a Cell Reaction to a Quantity
of Electrical Charge

The capacity of the lithium–ion battery in a digital camera is 3.4 W · hr at 3.6 V. How many grams of Li^+ ions must migrate from anode to cathode to produce this much electrical energy?

COLLECT AND ORGANIZE We are asked to relate the electrical energy generated by an electrochemical cell to the mass of the ions involved in generating that energy. We know the cell potential and its capacity in the energy unit of watt-hours. Given these starting points and our need to calculate moles and then grams of Li^+ ions, the following equivalencies may be useful:

$$1 \text{ watt (W)} = 1 \text{ ampere-volt (A · V)}$$

$$1 \text{ coulomb (C)} = 1 \text{ ampere-second (A · s)}$$

We may also need to use the Faraday constant, 9.65×10^4 C/mol.

ANALYZE The energy capacity of the battery is the product of the charge (electrons) it can deliver times the cell potential pumping that charge. This exercise focuses on the quantity of charge, so we need to separate the contribution of the cell potential to the energy rating from the contribution of the charge. To do that we need to divide the energy rating (3.4 W · hr) by the battery's cell potential (3.6 V), which translates into about 1 ampere-hour, and then follow that division with these unit conversions to get to moles of charge:

$$\frac{\text{watt-hours}}{\text{volts}} \times \frac{1 \text{ ampere-volt}}{1 \text{ watt}} \times \frac{1 \text{ coulomb}}{1 \text{ ampere-second}} \times \frac{3600 \text{ seconds}}{1 \text{ hour}}$$

$$\times \frac{1 \text{ mol } e^-}{9.65 \times 10^4 \text{ coulombs}}$$

The net effect of all these conversion factors is about $3600/10^5$ or 0.036 moles of Li^+ ions, and a mass of about 7 times that value, or 0.25 g.

SOLVE Using the given values for energy and cell potential in the above unit conversion series, we get

$$\frac{3.4 \; W \cdot hr}{3.6 \; V} \times \frac{1 \; A \cdot V}{1 \; W} \times \frac{1 \; C}{1 \; A \cdot s} \times \frac{3600 \; s}{1 \; hr} \times \frac{1 \; mol \; e^-}{9.65 \times 10^4 \; C} = 0.0352 \; mol \; e^-$$

Converting 0.0352 moles of electrons into an equivalent mass of Li^+ ions,

$$0.0352 \; mol \; e^- \times \frac{1 \; mol \; Li^+}{1 \; mol \; e^-} \times \frac{6.941 \; g \; Li^+}{1 \; mol \; Li^+} = 0.24 \; g \; Li^+$$

THINK ABOUT IT The result is quite close to our estimated value. The battery that is the subject of this exercise has a mass of about 22 g, so Li^+ ions make up only about 1% of its mass. This small percentage is reasonable given the masses of the other required components of the cell, including an anode where each Li^+ ion is surrounded by a hexagon of six carbon atoms and a cathode made of a transition metal compound with a molar mass many times that of Li metal.

Practice Exercise Magnesium metal is produced by passing an electric current through molten $MgCl_2$. The reaction at the cathode is

$$Mg^{2+}(\ell) + 2 \; e^- \rightarrow Mg(\ell)$$

How many grams of magnesium metal are produced if an average current of 63.7 A flows for 4.50 hr? Assume all of the current is consumed by the half-reaction shown. ⚙

17.8 Electrolytic Cells and Rechargeable Batteries

Lead–acid, NiMH, and lithium–ion batteries are rechargeable, which means that their spontaneous ($\Delta G < 0$) cell reactions that convert chemical energy into electrical energy can be forced to run in reverse. Recharging happens when external sources of electrical energy are applied to the batteries. This electrical energy is converted into chemical energy as it drives nonspontaneous ($\Delta G > 0$) reverse cell reactions, re-forming reactants from products. To make this possible, the products of the original cell reactions must be substances that either adhere to or are embedded in the electrodes, and thus are available to react with the electrons supplied to the cathodes and drawn away from the anodes by the external power supply.

Recharging a battery is an example of **electrolysis**, which is defined as any chemical reaction driven by electricity. During recharging, a battery is transformed from a voltaic cell into an **electrolytic cell** (Figure 17.19). To explore this transformation, let's revisit the lead–acid battery used to start car engines. Its discharge and recharge cycles are shown in Figure 17.20. Note that electrons flow in one direction when the battery discharges—out of the negative terminal and into the positive terminal—but in the opposite direction when the battery is recharging. Thus, the Pb electrodes connected to the negative terminal in Figure 17.20 serve as anodes during discharge but as cathodes during recharge. Any $PbSO_4$ that forms on these Pb electrodes during discharge is reduced back to Pb metal during recharge:

$$PbSO_4(s) + H^+(aq) + 2 \; e^- \rightarrow Pb(s) + HSO_4^-(aq)$$

electrolysis a process in which electrical energy is used to drive a nonspontaneous chemical reaction.

electrolytic cell a device in which an external source of electrical energy does work on a chemical system, turning reactant(s) into higher energy product(s).

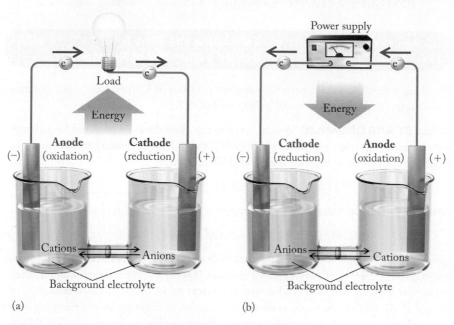

| **Voltaic Cell**
Spontaneous cell reaction
does work on its surroundings | **Electrolytic Cell**
External source of electrical power
does work on system |

(a) (b)

FIGURE 17.19 Voltaic versus electrolytic cells. (a) In a voltaic cell, a spontaneous reaction produces electrical energy and does electrical work on its surroundings, such as lighting a lightbulb. (b) In an electrolytic cell, an external supply of electrical energy does work on the chemical system in the cell, driving a nonspontaneous reaction.

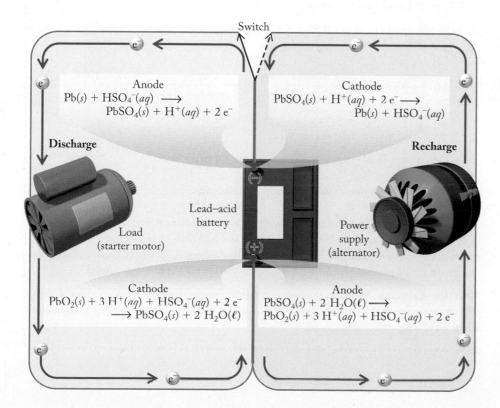

FIGURE 17.20 The lead–acid battery used in many vehicles is based on oxidation of Pb and reduction of PbO_2. As the battery discharges (left circuit), Pb is oxidized to $PbSO_4$ and PbO_2 is reduced to $PbSO_4$. When the engine is running (right circuit), a device called an alternator generates electrical energy that flows into the battery, recharging it as both electrode reactions are reversed: $PbSO_4$ is oxidized to PbO_2, and $PbSO_4$ is reduced to Pb.

Similarly, the PbO_2 electrodes connected to the positive terminal serve as cathodes during discharge but as anodes during recharge. Any $PbSO_4$ that forms on these PbO_2 electrodes during discharge is oxidized back to PbO_2 during recharge:

$$PbSO_4(s) + 2\ H_2O(\ell) \rightarrow PbO_2(s) + 3\ H^+(aq) + HSO_4^-(aq) + 2\ e^-$$

SAMPLE EXERCISE 17.9 **Calculating the Time to Oxidize a Quantity of Reactant** **LO7**

If a charger for AA NiMH batteries supplies a current of 1.00 A, how many minutes does it take to oxidize 0.649 g of $Ni(OH)_2$ to $NiO(OH)$?

COLLECT AND ORGANIZE We are asked to calculate the time required for a charging current to oxidize a given mass of $Ni(OH)_2$ to $NiO(OH)$. During the discharge of a NiMH battery the spontaneous cathode half-reaction is

$$NiO(OH)(s) + H_2O(\ell) + e^- \rightarrow Ni(OH)_2(s) + OH^-(aq) \qquad E° = 1.32\ V$$

ANALYZE During recharging the spontaneous cathode half-reaction runs in reverse:

$$Ni(OH)_2(s) + OH^-(aq) \rightarrow NiO(OH)(s) + H_2O(\ell) + e^-$$

One mole of electrons is produced for each mole of $Ni(OH)_2$ consumed. Our first steps are to convert 0.649 g of $Ni(OH)_2$ into moles of $Ni(OH)_2$ and then into moles of electrons. The Faraday constant can be used to convert moles of electrons to coulombs of charge. A coulomb is the same as an ampere-second, so dividing by the charging current will give us seconds of charging current. Assuming the molar mass of $Ni(OH)_2$ is about 100, the unit conversions involved are

$$g\ Ni(OH)_2 \times \frac{1\ mol\ Ni(OH)_2}{\sim 100\ g\ Ni(OH)_2} \times \frac{1\ mol\ e^-}{1\ mol\ Ni(OH)_2} \times \frac{9.65 \times 10^4\ coulombs}{1\ mol\ e^-}$$

$$\times \frac{1\ A \cdot s}{1\ coulomb} \times \frac{1}{1.00\ A}$$

so the result should be about $0.65\ g \times 1/(10^2\ g) \times 10^5$ s, or 650 s.

SOLVE

$$0.649\ g\ Ni(OH)_2 \times \frac{1\ mol\ Ni(OH)_2}{92.71\ g\ Ni(OH)_2} \times \frac{1\ mol\ e^-}{1\ mol\ Ni(OH)_2} \times \frac{9.65 \times 10^4\ C}{1\ mol\ e^-}$$

$$\times \frac{1\ A \cdot s}{1\ C} \times \frac{1}{1.00\ A} = 676\ s$$

Thus, the charger must deliver 1.00 A of current for 676 s, or:

$$676\ s \times \frac{1\ min}{60\ s} = 11.3\ min$$

THINK ABOUT IT The calculated result agrees well with our estimate. A charging time of 11.3 min may seem short, but the quantity of the $Ni(OH)_2$ to be oxidized (0.649 g) is much less than the quantity of $NiO(OH)$ in a fully charged AA NiMH battery (8.6 g, as calculated in the previous section), so this battery was only slightly discharged.

Practice Exercise Suppose that a car's starter motor draws 230 A of current for 6.0 s to start the car. What mass of Pb is oxidized in the battery to supply this much electricity?

Electrolysis is used in many processes other than recharging batteries. Electrolytic cells are used to electroplate thin layers of silver, gold, and other metals onto objects, giving these objects the appearance, resistance to corrosion, and other properties of the electroplated metal, but at a fraction of the cost of fabricating the entire object out of the metal (Figure 17.21).

In the chemical industry, the electrolysis of molten salts is used to produce highly reactive substances such as sodium, chlorine, and fluorine, as well as alkali and alkaline earth metals and aluminum. When NaCl, for instance, is heated to just above its melting point (above 800°C), it becomes an ionic liquid that can conduct electricity. If a sufficiently large potential is applied to carbon electrodes immersed in the molten NaCl, the sodium ions are attracted to the negative electrode and are reduced to sodium metal while the chloride ions are attracted to the positive electrode and oxidized to Cl_2 gas:

$$2\,Na^+(\ell) + 2\,Cl^-(\ell) \rightarrow 2\,Na(\ell) + Cl_2(g)$$

A final note about anode and cathode polarity is in order. The reactions in voltaic cells are spontaneous. These cells pump electric current through external circuits and electrical devices with a force equal to their cell potentials. The anode in a voltaic cell is negative because an oxidation half-reaction supplies negatively charged electrons to the device powered by the cell. Electrons flow from the device into the positive battery terminal which is connected to the cathode, where these electrons are consumed in a reduction half-reaction.

The reactions in electrolytic cells are nonspontaneous. They require electrical energy from an external power supply. When the negative terminal of such a power supply is connected to the cathode of the battery, the power supply pumps electrons into the cathode, where they are consumed in reduction half-reactions. Electrons are pumped away from the anode, where they must have been generated in an oxidation half-reaction, toward the positive terminal of the power supply. Thus, as we can see in Figure 17.19, the cathode of an electrolytic cell is the negative electrode, but the cathode of a voltaic cell is the positive electrode. Similarly, the anode of an electrolytic cell is the positive electrode, but the anode of a voltaic cell is the negative electrode. These "pole reversals" make sense if we keep in mind the fundamental definitions:

➤ Anodes are electrodes where oxidation takes place.

➤ Cathodes are electrodes where reduction takes place.

To verify your understanding of the pole reversal process, study the flow of electrons to and from the battery terminals in Figure 17.20 when the battery is being discharged (and working like a voltaic cell) and when it is being recharged (and operating as an electrolytic cell).

FIGURE 17.21 The cutlery known as silver plate has only a thin coating of silver, applied over a base metal core by a type of electrolysis known as electroplating. The positive terminal of an electric power supply is attached to a piece of pure silver, and the negative terminal is connected to the piece to be electroplated, such as the spoon shown here. The oxidation half-reaction at the silver electrode (anode), $Ag \rightarrow Ag^+ + e^-$, runs in reverse at the cathode (spoon) as Ag^+ ion is reduced to the free metal on the spoon surface.

CONCEPT TEST

The electrolysis of molten NaCl produces liquid Na metal at the cathode and Cl_2 gas at the anode. However, the electrolysis of an aqueous solution of NaCl produces gases at both cathode and anode. On the basis of the standard potentials in Table A6.1, predict the gas formed at each electrode.

fuel cell a voltaic cell based on the oxidation of a continuously supplied fuel. The reaction is the equivalent of combustion, but chemical energy is converted directly into electrical energy.

17.9 Fuel Cells

Fuel cells are promising energy sources for many applications, from powering office buildings to cruise ships to electric vehicles. Fuel cells are voltaic cells, but they are different from batteries in that their supplies of reactants are constantly renewed. Therefore, they do not "discharge"; they never run down, and they don't die unless their fuel supply is cut off. From a thermodynamic perspective, batteries are closed systems and fuel cells are open systems.

In a typical fuel cell, electrons are supplied to an external circuit by the oxidation of H_2 at the anode. Electrons are consumed by the reduction of O_2 at the cathode, so the fuel cell reaction is

$$2\,H_2(g) + O_2(g) \rightarrow 2\,H_2O(\ell)$$

The fuel cells used to power electric vehicles consist of metallic or graphite electrodes separated by a hydrated polymeric material called a *proton-exchange membrane* (PEM). The PEM serves as both an electrolyte and a barrier that prevents crossover and mixing of the fuel and oxidant (Figure 17.22). The surfaces of the electrodes are coated with transition metal catalysts to speed up the electrode half-reactions. Platinum catalysts promote H—H bond breaking during the oxidation of H_2 gas to H^+ ions at the anode:

$$H_2(g) \rightarrow 2\,H^+(aq) + 2\,e^- \qquad E° = 0.000\ \text{V}$$

and a platinum–nickel alloy with the formula Pt_3Ni is particularly effective in catalyzing the formation of free O atoms from O_2 molecules, which is part of the reduction half-reaction at the cathode:

$$O_2(g) + 4\,H^+(aq) + 4\,e^- \rightarrow 2\,H_2O(\ell) \qquad E° = 1.229\ \text{V}$$

Hydrogen ions that form at the anode migrate through the PEM to the cathode, where they combine with O_2. This migration of positive charges inside the fuel cell drives the flow of electrons in the electrical device attached to it.

▶❚❚ **CHEMTOUR** Fuel Cell

FIGURE 17.22 Most fuel cells used in vehicles have a proton-exchange membrane between the two halves of the cell. Hydrogen gas diffuses to the anode, and oxygen gas diffuses to the cathode. These electrodes are made of porous material, such as graphite, that have a relatively high surface area for a given mass of material. Catalysts on the electrode surfaces also increase the rate of the half-reactions at the anode ($H_2 \rightarrow 2\,H^+ + 2\,e^-$) and the cathode ($O_2 + 4\,H^+ + 4\,e^- \rightarrow 2\,H_2O$).

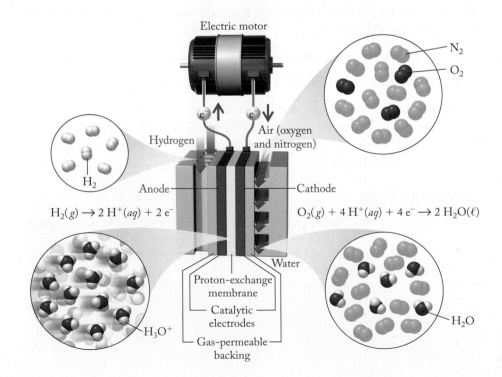

Electric motor

N₂

O₂

Hydrogen

Air (oxygen and nitrogen)

H₂

Anode

Cathode

$$H_2(g) \rightarrow 2\,H^+(aq) + 2\,e^-$$

$$O_2(g) + 4\,H^+(aq) + 4\,e^- \rightarrow 2\,H_2O(\ell)$$

Water

Proton-exchange membrane

Catalytic electrodes

H₃O⁺

Gas-permeable backing

H₂O

A single PEM fuel cell typically has a cell potential of about 1.0 V. When hundreds of these cells are assembled into fuel cell *stacks*, they are capable of producing 100 kW of electrical power. That is enough to give a midsize car such as the one in Figure 17.23 a top speed over 160 km/hr (99 mi/hr).

PEM fuel cells are well suited for use in vehicles because they are compact, lightweight, and operate at fairly low temperatures of 60 to 80°C. The performance of nearly all fuel cells is better at above-ambient temperatures because the higher rates of their half-reactions at these temperatures mean that they generate more electrical power.

Other fuel cells use basic electrolytes such as concentrated KOH. Pure O_2 is supplied to a cathode made of porous graphite containing a nickel catalyst, and H_2 gas is supplied to a graphite anode containing nickel(II) oxide. Hydroxide ions formed during O_2 reduction at the cathode,

$$O_2(g) + 2\,H_2O(\ell) + 4\,e^- \rightarrow 4\,OH^-(aq) \qquad E° = 0.401\ V$$

migrate through the cell to the anode, where they combine with H_2 as it is oxidized to water:

$$H_2(g) + 2\,OH^-(aq) \rightarrow 2\,H_2O(\ell) + 2\,e^-$$

This oxidation half-reaction is the reverse of the following reduction half-reaction from Table A6.1:

$$2\,H_2O(\ell) + 2\,e^- \rightarrow H_2(g) + 2\,OH^-(aq) \qquad E° = -0.828\ V$$

Note that this basic pair of standard electrode potentials yields the same $E°_{cell}$ value:

$$E°_{cell} = 0.401\ V - (-0.828\ V) = 1.229\ V$$

as the acidic pair described earlier:

$$E°_{cell} = 1.229\ V - 0.000\ V = 1.229\ V$$

This equality is logical because $\Delta G°$, the energy released under standard conditions by the oxidation of hydrogen gas to form liquid water,

$$2\,H_2(g) + O_2(g) \rightarrow 2\,H_2O(\ell)$$

should have only one value, which means that $E°_{cell}$ should have only one value independent of the pH of the electrolyte.

CONCEPT TEST ....

During the operation of molten alkali metal carbonate fuel cells, carbonate ions are generated at one electrode, migrate across the cell, and are consumed at the other electrode. Do the carbonate ions migrate toward the cathode or the anode?

....

The same chemical energy that is released in fuel cells could also be obtained by burning hydrogen gas in an internal combustion engine. However, only about 20–25% of the chemical energy in the fuel burned in such an engine is typically converted into mechanical energy; most is lost to the surroundings as heat. In contrast, fuel cell technologies can convert up to about 80% of the energy released in a fuel cell's redox reaction into electrical energy. The electric motors they power are also about 80% efficient at converting electrical energy into mechanical energy. Thus, the overall conversion efficiency of a fuel cell–powered car is theoretically as high as (80% × 80%) or 64%. Actually, the measured efficiency of the propulsion

FIGURE 17.23 The Honda FCX Clarity is powered by a 100 kW fuel cell stack and 100 kW electric motor that give the car a top speed of 160 km/hr (99 mi/hr).

system of the car in Figure 17.23 is about 60%, which is still more than twice that of an internal combustion engine. In addition, H_2-fueled vehicles emit only water vapor; they produce no oxides of nitrogen, no carbon monoxide, and no CO_2.

As fuel cells become even more efficient and less expensive, the principal limit on their use in passenger cars will be the availability and cost of hydrogen fuel. Some people worry about the safety of storing hydrogen in a high-pressure tank in a car. In fact, H_2 has a higher ignition temperature than gasoline and spreads through the air more quickly, reducing the risk of fire. Still, hydrogen in air burns over a much wider range of concentrations than gasoline, and its flame is almost invisible.

Because of the lack of an extensive hydrogen distribution network, it is likely that most fuel cell–powered vehicles will be buses and fleet vehicles operating from a central location where hydrogen gas is available. With their very large fuel tanks, buses powered by fuel cells have an operating range of 400 km (~248 mi). Longer ranges would be possible if better methods for storing hydrogen gas were available. Currently under development are mobile chemical-processing plants that can extract hydrogen from gasoline, methane, or methanol. Onboard production of hydrogen from fossil fuels also generates carbon dioxide as a by-product, so these fuels still produce greenhouse gases. However, the efficiencies of fuel cells and the electric motors they power mean that much less of these fuels will be needed and less CO_2 will be emitted.

◉◉ CONNECTION Processes for producing hydrogen gas by reacting methane and other fuels with steam were discussed in Chapter 14.

SAMPLE EXERCISE 17.10 **Integrating Concepts: Electrolysis of Salt**

The electrolysis of sodium chloride is used industrially to produce several commercially important chemicals. One process uses a Downs cell (Figure 17.24), which has a carbon anode and an iron cathode, and which uses molten NaCl as the electrolyte and source of the reactants in the cell reaction. Typically, calcium chloride (melting point 782°C) is added to the sodium chloride (melting point 801°C; density 2.16 g/cm³), and the process is carried out at temperatures around 600°C. The products of the electrolysis are liquid sodium (melting point 97.8°C; density 0.93 g/cm³) and chlorine gas. An iron screen in the cell separates the electrodes so that the two products do not come into contact with each other.

a. Write chemical equations for the half-reactions at the anode and cathode, and for the overall cell reaction.

b. Use the appropriate standard reduction potentials in Appendix 6 to estimate the minimum potential difference between anode and cathode that must be applied to electrolyze molten NaCl.

c. Migration of which ions carries electric charge between the anodes and cathodes in a Downs cell?

d. Calcium chloride is added to the NaCl to depress its melting point. However, calcium metal is not produced in this cell along with the liquid sodium. Why?

e. Why is it important to keep the two products separated?

f. In a Downs cell the electrodes are usually positioned at the bottom of the molten NaCl bath, but the products are collected at the top. Why?

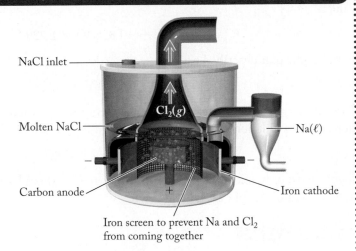

FIGURE 17.24 The Downs cell is the primary unit used to produce sodium commercially. It is also a minor source of industrially produced chlorine. An iron screen prevents contact between the two products.

COLLECT AND ORGANIZE We are given a description of a Downs cell, the conditions under which it is run, and the physical properties of the reactants and products of the electrolytic process taking place in it. We are asked to write the cell half-reactions and overall reaction and to estimate the voltage needed to electrolyze molten NaCl. We are also asked how electricity flows through the cell, why Ca^{2+} ions are not also reduced at the cathode, why elemental sodium and chlorine need to be kept separated, and why they are harvested at the top of the cell even though the electrodes at which they are generated are near the bottom.

ANALYZE Sodium ion is reduced at the cathode to elemental sodium, and chloride ion is oxidized at the anode to chlorine gas. Creating these unstable elements from a stable compound like NaCl will require a significant investment in electrical energy and, very likely, the application of a cell potential of several volts. Contact between the products would allow them to recombine back into NaCl. The spontaneity of this reaction is indicated by the ΔG_f° of NaCl in Appendix 4 (Table A4.3): −384.2 kJ/mol. Sodium ions will be preferentially reduced over Ca^{2+} ions if the standard reduction potential of Na^+ ions is less negative than that of Ca^{2+} ions. The manner of how the products are recovered from the electrolysis cell depends on their physical properties, especially their densities relative to the reactant, molten NaCl.

SOLVE
a. The anode half-reaction is

$$2\,Cl^-(\ell) \rightarrow Cl_2(g) + 2\,e^-$$

The cathode half-reaction is

$$Na^+(\ell) + e^- \rightarrow 2\,Na(\ell)$$

The overall reaction is the sum of the two reactions. Summing them requires first multiplying the Na reaction by 2 to balance the loss and gain of electrons:

$$2\,Na^+(\ell) + 2\,Cl^-(\ell) \rightarrow 2\,Na(\ell) + Cl_2(g)$$

b. Calculating E_{cell}° requires the appropriate $E_{cathode}^\circ$ and E_{anode}° values. A search of Appendix 6 uncovers these half-reactions and E° values:

$$Cl_2(g) + 2\,e^- \rightarrow 2\,Cl^-(aq) \quad E^\circ = 1.3583\,V$$
$$Na^+(aq) + e^- \rightarrow Na(s) \quad E^\circ = -2.71\,V$$

Combining the standard reduction potentials,

$$E_{cell}^\circ = E_{cathode}^\circ - E_{anode}^\circ = -2.71 - 1.3583 = -4.07\,V$$

The negative E_{cell}° value is a measure of how much voltage must be applied to drive the nonspontaneous cell reaction, that is, 4.07 V. However, the actual minimum voltage will not be exactly 4.07 V. For one thing, the sodium and chloride ions are in a nonaqueous environment of molten NaCl, and their concentrations are probably not 1 M. However, we have seen that cell potentials change relatively little as the concentrations of reactants and products change, so 4 volts is a reasonable estimate of the voltage needed to drive the reaction in a Downs cell.

c. Migration of Cl^- ions toward the positively charged anode, where they are oxidized to Cl_2 gas, carries part of the electrolysis current through the cell. The rest is carried by the migration of Na^+ ions toward the negatively charged cathode, where they are reduced to Na metal.

d. The standard reduction potential of Ca^{2+} ions (Appendix 6) is −2.868 V, which is more negative than the standard reduction potential of Na^+ ions (−2.71 V). Therefore, Na^+ ions are preferentially reduced in the presence of Ca^{2+} ions.

e. The ΔG_f° of solid NaCl is −384.2 kJ/mol, which tells us that elemental sodium and chlorine spontaneously combine to form NaCl. Therefore these two products of electrolysis must be kept away from each other.

f. Liquid sodium is less dense than molten sodium chloride, and as it forms in the cell it floats on top of the molten salt. Chlorine is a gas and bubbles out of molten NaCl and is collected in hoods above the liquid.

THINK ABOUT IT The Downs cell is an electrolytic cell that requires a large input of electrical energy to drive the nonspontaneous cell reaction. Sodium and chlorine are both so reactive that special care and extreme caution are required to collect and transport them.

SUMMARY

Section 17.1 Electrochemistry is the branch of chemistry that links redox reactions to the production or consumption of electrical energy. Any redox reaction can be broken down into oxidation and reduction **half-reactions**.

Section 17.2 In an **electrochemical cell** the oxidation half-reaction occurs at the **anode** and the reduction half-reaction occurs at the **cathode**. Migration of the ions in the cell's electrolyte allows electric charges to flow between the cathode and anode compartments as electrons flow through an external electrical circuit. A **cell diagram** shows how the components of the cathode and anode compartments of the cell are connected.

Section 17.3 The difference between the **standard reduction potentials (E°)** of a cell's cathode and anode half-reactions is equal to the **standard cell potential (E_{cell}°)**. The value of E_{cell}° is a measure of the **electromotive force (emf)** with which a **voltaic cell** can pump electrons through an external circuit under standard conditions. If reactants and products are not in their standard states, this value is simply called the **cell potential (E_{cell})**.

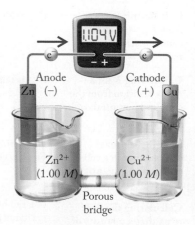

Section 17.4 A voltaic cell has a positive cell potential ($E_{cell} > 0$) and its cell reaction has a negative change in free energy ($\Delta G_{cell}^{\circ} < 0$). This decrease in free energy in a voltaic cell is available to do work in an external electrical circuit. The **Faraday constant (F)** relates the quantity of electric charge to the number of moles of electrons and indirectly to the number of moles of reactants.

Section 17.5 All standard cell potentials are referenced to the cell potential of the **standard hydrogen electrode (SHE)**: $E_{SHE}^{\circ} = 0.000$ V.

Section 17.6 The potential of a voltaic cell decreases as reactants turn into products. The **Nernst equation** describes how cell potential changes with concentration changes. The potential of a voltaic cell approaches zero as the cell reaction approaches chemical equilibrium, at which point $E_{cell} = 0$ and $Q = K$.

Section 17.7 The quantities of reactants consumed in a voltaic cell reaction are directly proportional to the coulombs of electric charge delivered by the cell.

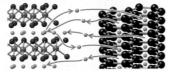

Nickel–metal hydride batteries supply electricity when H atoms are oxidized to H^+ ions at the anodes and NiO(OH) is reduced to $Ni(OH)_2$ at the cathodes. In lithium–ion batteries, electricity is produced when Li^+ ions stored in graphite anodes migrate toward and are incorporated into transition metal oxide or phosphate cathodes.

Section 17.8 A spontaneous cell reaction can be reversed by applying an opposing potential greater than E_{cell} using an external power supply. Using electrical power to force a nonspontaneous cell reaction is called **electrolysis**, and this process turns the cell into an **electrolytic cell**. The polarities of cathodes and anodes in electrolytic cells are opposite the polarities of these electrodes in voltaic cells.

Section 17.9 **Fuel cells** directly convert the chemical free energy released during the reaction $2\,H_2 + O_2 \rightarrow 2\,H_2O$ into electrical energy. The electrodes in fuel cells often incorporate catalysts to speed up the half-reactions involving H_2 and O_2 gas. Fuel cells with proton-exchange membranes have been developed as power supplies for electric vehicles.

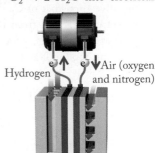

Hydrogen

Air (oxygen and nitrogen)

Water

PROBLEM-SOLVING SUMMARY

TYPE OF PROBLEM	CONCEPTS AND EQUATIONS	SAMPLE EXERCISES
Writing redox reaction equations by combining half-reactions	Combine the reduction and oxidation half-reactions after balancing the gain and loss of electrons.	17.1, 17.2
Diagramming an electrochemical cell	Start with the symbol for the anode on the left; use double vertical lines to separate it from the symbol for the cathode. Using single lines to separate phases, list species in anode compartments, then those in the cathode compartment; separate species in the same phase with commas. Insert concentrations if known.	17.3
Identifying anode and cathode half-reactions and calculating the value of E_{cell}°	The half-reaction with the more positive standard reduction potential is the cathode half-reaction. $$E_{cell}^{\circ} = E_{cathode}^{\circ} - E_{anode}^{\circ} \qquad (17.1)$$	17.4
Relating ΔG_{cell}° and E_{cell}°	$$\Delta G_{cell}^{\circ} = -nFE_{cell}^{\circ} \qquad (17.6)$$ where n is the number of moles of electrons transferred in the cell reaction and F is the Faraday constant, 9.65×10^4 C/mol.	17.5
Calculating E_{cell} from E_{cell}° and the concentrations of reactants and products	E_{cell} at 25°C is related to E_{cell}° and the cell reaction quotient by the Nernst equation: $$E_{cell} = E_{cell}^{\circ} - \frac{0.0592}{n} \log Q \qquad (17.9)$$	17.6
Calculating K for a redox reaction from the standard potentials of its half-reactions	E_{rxn}° is related to K at 298 K by $$\log K = \frac{nE_{rxn}^{\circ}}{0.0592} \qquad (17.11)$$	17.7
Relating the mass of a reactant in a cell reaction to a quantity of electrical charge	Determine the ratio of moles of reactants to moles of electrons transferred; use the Faraday constant to relate coulombs of charge to moles of electrons.	17.8
Calculating the time to oxidize a quantity of reactant	Use the Faraday constant and the relation between moles of electrons and moles of reactants to describe an electrolytic process.	17.9

VISUAL PROBLEMS ·· ■

(Answers to boldface end-of-chapter questions and problems are in the back of the book.)

17.1. In the voltaic cell shown in Figure P17.1, the greater density of a concentrated solution of $CuSO_4$ allows a less concentrated solution of $ZnSO_4$ solution to be (carefully) layered on top of it. Why is a porous separator not needed in this cell?

FIGURE P17.1

17.2. In the voltaic cell shown in Figure P17.2, the concentrations of Cu^{2+} and Cd^{2+} are 1.00 *M*. On the basis of the standard potentials in Appendix 6, identify which electrode is the anode and which is the cathode. Indicate the direction of electron flow.

FIGURE P17.2

17.3. In the voltaic cell shown in Figure P17.3, $[Ag^+] = [H^+] = 1.00$ *M*. On the basis of the standard potentials in Appendix 6, identify which electrode is the anode and which is the cathode. Indicate the direction of electron flow.

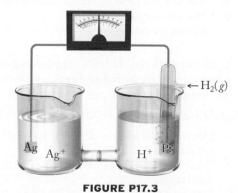

FIGURE P17.3

*17.4. In many electrochemical cells the electrodes are metals that carry electrons to and from the cell but are not chemically changed by the cell reaction. Each of the highlighted clusters in the periodic table in Figure P17.4 consists of three metals. Which of the highlighted clusters is best suited to form inert electrodes?

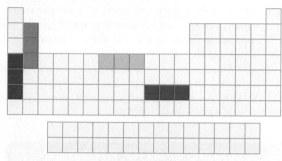

FIGURE P17.4

17.5. Which of the four curves in Figure P17.5 best represents the dependence of the potential of a lead–acid battery on the concentration of sulfuric acid? Note that the scale of the *x*-axis is logarithmic.

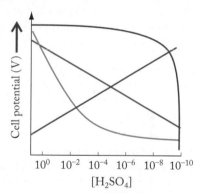

FIGURE P17.5

17.6. Consider the four types of batteries in Figure P17.6. From top to bottom the sizes are AAA, AA, C, and D. The performance of batteries like these is often expressed in units such as (a) volts, (b) watt-hours, or (c) milliampere-hours. Which of the values differ significantly between the four batteries?

FIGURE P17.6

17.7. The apparatus in Figure P17.7 is used for the electrolysis of water. Hydrogen and oxygen gas are collected in the two inverted burets. An inert electrode at the bottom of the left buret is connected to the negative terminal of a 6-volt battery; the electrode in the buret on the right is connected to the positive terminal. A small quantity of sulfuric acid is added to speed up the electrolytic reaction.

  a. What are the half-reactions at the left and right electrodes and their standard potentials?

  b. Why does sulfuric acid make the electrolysis reaction go more rapidly?

17.8. An electrolytic apparatus identical to the one shown in Figure P17.7 is used to electrolyze water, but the reaction is speeded up by the addition of sodium carbonate instead of sulfuric acid.

  a. What are the half-reactions and the standard potentials for the electrodes on the left and right?

  b. Why does sodium carbonate make the electrolysis reaction go more rapidly?

Overall cell reaction
$$H_2O(\ell) \rightarrow H_2(g) + \tfrac{1}{2} O_2(g)$$

FIGURE P17.7

QUESTIONS AND PROBLEMS

Redox Chemistry Revisited and Electrochemical Cells

CONCEPT REVIEW

17.9. What is meant by a half-reaction?

17.10. The Zn/Cu^{2+} reactions in Figures 17.1 and 17.2 are the same. However, the reaction in the cell in Figure 17.2 generates electricity; the reaction in the beaker in Figure 17.1 does not. Why?

17.11. Why can't a wire perform the same function as a porous separator in an electrochemical cell?

17.12. In a voltaic cell, why is the cathode labeled the positive terminal and the anode the negative terminal?

PROBLEMS

17.13. Balance the following half-reactions by adding the appropriate number of electrons. Identify the oxidation half-reactions and the reduction half-reactions.

  a. $Br_2(\ell) \rightarrow 2\ Br^-(aq)$

  b. $Pb(s) + 2\ Cl^-(aq) \rightarrow PbCl_2(s)$

  c. $O_3(g) + 2\ H^+(aq) \rightarrow O_2(g) + H_2O(\ell)$

  d. $H_2S(g) \rightarrow S(s) + 2\ H^+(aq)$

17.14. Balance the following half-reactions by adding the appropriate number of electrons. Which are oxidation half-reactions and which are reduction half-reactions?

  a. $Fe^{2+}(aq) \rightarrow Fe^{3+}(aq)$

  b. $AgI(s) \rightarrow Ag(s) + I^-(aq)$

  c. $VO_2^+(aq) + 2\ H^+(aq) \rightarrow VO^{2+}(aq) + H_2O(\ell)$

  d. $I_2(s) + 6\ H_2O(\ell) \rightarrow 2\ IO_3^-(aq) + 12\ H^+(aq)$

***17.15. Groundwater Chemistry** Write a half-reaction for the oxidation of magnetite (Fe_3O_4) to hematite (Fe_2O_3) in acidic groundwater. Add H_2O, H^+, and electrons as needed to balance the half-reaction.

***17.16.** Write a half-reaction for the oxidation of the manganese in $MnCO_3$ to MnO_2 in neutral groundwater where the principal carbonate species is HCO_3^-. Add H_2O, H^+, and electrons as needed to balance the half-reaction.

17.17. A voltaic cell with an aqueous electrolyte is based on the reaction between $Pb^{2+}(aq)$ and $Zn(s)$, producing $Pb(s)$ and $Zn^{2+}(aq)$.

  a. Write half-reactions for the anode and cathode.

  b. Write a balanced cell reaction.

  c. Diagram the cell.

17.18. A voltaic cell is based on the reaction between $Ag^+(aq)$ and $Ni(s)$, producing $Ag(s)$ and $Ni^{2+}(aq)$.

  a. Write the anode and cathode half-reactions.

  b. Write a balanced cell reaction.

  c. Diagram the cell.

17.19. A voltaic cell with a basic aqueous background electrolyte is based on the oxidation of $Cd(s)$ to $Cd(OH)_2(s)$ and the reduction of $MnO_4^-(aq)$ to $MnO_2(s)$.

  a. Write half-reactions for the cell's anode and cathode.

  b. Write a balanced cell reaction.

  c. Diagram the cell.

17.20. A voltaic cell is based on the reduction of $Ag^+(aq)$ to $Ag(s)$ and the oxidation of $Sn(s)$ to $Sn^{2+}(aq)$.

  a. Write half-reactions for the cell's anode and cathode.

  b. Write a balanced cell reaction.

  c. Diagram the cell.

17.21. Super Iron Batteries In 1999, scientists in Israel developed a battery based on the following cell reaction with iron(VI), nicknamed "super iron":

$$2\ K_2FeO_4(aq) + 3\ Zn(s) \rightarrow Fe_2O_3(s) + ZnO(s) + 2\ K_2ZnO_2(aq)$$

  a. Determine the number of electrons transferred in the cell reaction.

  b. What are the oxidation states of the transition metals in the reaction?

  c. Diagram the cell.

17.22. Aluminum–Air Batteries In recent years engineers have been working on an aluminum–air battery as an alternative energy source for electric vehicles. The battery consists of an aluminum anode, which is oxidized to solid aluminum hydroxide, immersed in an electrolyte of aqueous KOH. At the cathode, oxygen from the air is reduced to hydroxide ions on an inert metal surface. Write the two half-reactions for the battery and diagram the cell. Use the generic $M(s)$ symbol for the metallic cathode material.

Standard Potentials

CONCEPT REVIEW

17.23. In some textbooks the formula used to calculate standard cell potentials from the standard reduction potentials of the half-reactions occurring at the cathode and anode is given as

$$E°_{cell} = E°_{reduction}(cathode) + E°_{oxidation}(anode)$$

Show how this equation is equivalent to Equation 17.1.

*17.24. Why is O_2 a stronger oxidizing agent in acid than in base? Use standard reduction potentials from Appendix 6 to support your answer.

PROBLEMS

17.25. Using standard reduction potentials from Appendix 6, and the following equations for half-reactions:

$$Cu^{2+}(aq) + 2\ e^- \rightarrow Cu(s)$$
$$Co^{2+}(aq) + 2\ e^- \rightarrow Co(s)$$
$$Hg^{2+}(aq) + 2\ e^- \rightarrow Hg(s)$$

a. Identify the combination of half-reactions that would lead to the largest value of $E°_{cell}$.
b. Identify the combination of half-reactions that would lead to the smallest positive value of $E°_{cell}$.

17.26. From the table of standard reduction potentials in Appendix 6,
a. Select an oxidizing agent that will oxidize $Cr(s)$ to $Cr^{3+}(aq)$ but not $Cd(s)$ to $Cd^{2+}(aq)$.
b. Select a reducing agent that will reduce $Br_2(\ell)$ to $Br^-(aq)$ but not $I_2(s)$ to $I^-(aq)$.

17.27. If a piece of silver is placed in a solution in which $[Ag^+] = [Cu^{2+}] = 1.00\ M$, will the following reaction proceed spontaneously?

$$2\ Ag(s) + Cu^{2+}(aq) \rightarrow 2\ Ag^+(aq) + Cu(s)$$

17.28. A piece of cadmium is placed in a solution in which $[Cd^{2+}] = [Sn^{2+}] = 1.00\ M$. Will the following reaction proceed spontaneously?

$$Cd(s) + Sn^{2+}(aq) \rightarrow Cd^{2+}(aq) + Sn(s)$$

17.29. In a voltaic cell similar to the Zn/Cu^{2+} cell in Figure 17.2, the Cu electrode is replaced with one made of Ni immersed in a solution of $NiSO_4$. Will the standard potential of this cell be greater than, the same as, or less than 1.10 V?

17.30. Suppose the copper half of the Zn/Cu^{2+} cell in Figure 17.2 was replaced with a silver wire in contact with $1\ M\ Ag^+(aq)$.
a. What would be the value of $E°_{cell}$?
b. Which electrode would be the anode?

17.31. Voltaic cells based on the following pairs of half-reactions are prepared so that all reactants and products are in their standard states. For each pair, write a balanced equation

for the cell reaction, and identify which half-reaction takes place at the anode and which at the cathode.
a. $Hg^{2+}(aq) + 2\ e^- \rightarrow Hg(\ell)$
   $Zn^{2+}(aq) + 2\ e^- \rightarrow Zn(s)$
b. $ZnO(s) + H_2O(\ell) + 2\ e^- \rightarrow Zn(s) + 2\ OH^-(aq)$
   $Ag_2O(s) + H_2O(\ell) + 2\ e^- \rightarrow 2\ Ag(s) + 2\ OH^-(aq)$
c. $Ni(OH)_2(s) + 2\ e^- \rightarrow Ni(s) + 2\ OH^-(aq)$
   $O_2(g) + 2\ H_2O(\ell) + 4\ e^- \rightarrow 4\ OH^-(aq)$

17.32. Voltaic cells based on the following pairs of half-reactions are constructed. For each pair, write a balanced equation for the cell reaction, and identify which half-reaction takes place at each anode and cathode.
a. $Cd^{2+}(aq) + 2\ e^- \rightarrow Cd(s)$
   $Ag^+(aq) + e^- \rightarrow Ag(s)$
b. $AgBr(s) + e^- \rightarrow Ag(s) + Br^-(aq)$
   $MnO_2(s) + 4\ H^+(aq) + 2\ e^- \rightarrow Mn^{2+}(aq) + 2\ H_2O(\ell)$
c. $PtCl_4^{2-}(aq) + 2\ e^- \rightarrow Pt(s) + 4\ Cl^-(aq)$
   $AgCl(s) + e^- \rightarrow Ag(s) + Cl^-(aq)$

Chemical Energy and Electrical Work

CONCEPT REVIEW

17.33. The negative sign in Equation 17.3 ($w_{elec} = -QE_{cell}$) seems to indicate that a voltaic cell with a *positive* cell potential does *negative* electrical work. How is this possible?

*17.34. Mechanical work (w) is done by exerting a force (F) to move an object through a distance (d) according to the equation $w = F \times d$. Explain how this definition of work relates to electrical work.

PROBLEMS

17.35. Starting with the appropriate standard free energies of formation from Appendix 4, calculate the value of $\Delta G°$ and $E°_{cell}$ of the following reactions:
a. $2\ Cu^+(aq) \rightarrow Cu^{2+}(aq) + Cu(s)$
b. $Ag(s) + Fe^{3+}(aq) \rightarrow Ag^+(aq) + Fe^{2+}(aq)$

17.36. Starting with the appropriate standard free energies of formation from Appendix 4, calculate the values of $\Delta G°$ and $E°_{cell}$ of the following reactions:
a. $FeO(s) + H_2(g) \rightarrow Fe(s) + H_2O(\ell)$
b. $2\ Pb(s) + O_2(g) + 2\ H_2SO_4(aq) \rightarrow$
   $$2\ PbSO_4(s) + 2\ H_2O(\ell)$$

17.37. Flashlights For many years the 1.50 V batteries used to power flashlights were based on the following cell reaction:

$$Zn(s) + 2\ NH_4Cl(s) + 2\ MnO_2(s) \rightarrow$$
$$Zn(NH_3)_2Cl_2(s) + Mn_2O_3(s) + H_2O(\ell)$$

What is the value of ΔG_{cell}?

17.38. Laptops The first generation of laptop computers was powered by nickel–cadmium (nicad) batteries, which generated 1.20 V based on the following cell reaction:

$$Cd(s) + 2\ NiO(OH)(s) + 2\ H_2O(\ell) \rightarrow$$
$$Cd(OH)_2(s) + 2\ Ni(OH)_2(s)$$

What is the value of ΔG_{cell}?

17.39. The cells in the nickel–metal hydride battery packs used in many hybrid vehicles produce 1.20 V based on the following cell reaction:

$$MH(s) + NiO(OH)(s) \rightarrow M(s) + Ni(OH)_2(s)$$

What is the value of ΔG_{cell}?

17.40. A cell in a lead–acid battery delivers exactly 2.00 V of cell potential based on the following cell reaction:

$$Pb(s) + PbO_2(s) + 2 H_2SO_4(aq) \rightarrow 2 PbSO_4(s) + 2 H_2O(\ell)$$

What is the value of ΔG_{cell}?

17.41. Starting with standard potentials listed in Appendix 6, calculate the values of $E°_{cell}$ and $\Delta°G$ of the following reactions.
a. $Cu(s) + Sn^{2+}(aq) \rightarrow Cu^{2+}(aq) + Sn(s)$
b. $Zn(s) + Ni^{2+}(aq) \rightarrow Zn^{2+}(aq) + Ni(s)$

17.42. Starting with the standard potentials listed in Appendix 6, calculate the values of $E°_{cell}$ and $\Delta G°$ of the following reactions.
a. $Fe(s) + Cu^{2+}(aq) \rightarrow Fe^{2+}(aq) + Cu(s)$
b. $Ag(s) + Fe^{3+}(aq) \rightarrow Ag^+(aq) + Fe^{2+}(aq)$

A Reference Point: The Standard Hydrogen Electrode

CONCEPT REVIEW

17.43. What is the function of platinum in the standard hydrogen electrode?

17.44. Is it possible to build a battery in which the anode chemistry is based on a half-reaction in which none of the species is a solid conductor? For example,

$$Fe^{2+}(aq) \rightarrow Fe^{3+}(aq) + e^-$$

The Effect of Concentration on E_{cell}

CONCEPT REVIEW

17.45. Why does the operating cell potential of most batteries change little until the battery is nearly discharged?

17.46. The standard potential of the Zn/Cu^{2+} cell reaction

$$Zn(s) + Cu^{2+}(aq) \rightarrow Zn^{2+}(aq) + Cu(s)$$

is 1.10 V. Would the potential of the cell differ from 1.10 V if the concentrations of both Cu^{2+} and Zn^{2+} were 0.25 M?

PROBLEMS

17.47. Calculate the E_{cell} value at 298 K for the cell based on the reaction

$$Fe^{3+}(aq) + Cr^{2+}(aq) \rightarrow Fe^{2+}(aq) + Cr^{3+}(aq)$$

when $[Fe^{3+}] = [Cr^{2+}] = 1.50 \times 10^{-3} \, M$ and $[Fe^{2+}] = [Cr^{3+}] = 2.5 \times 10^{-4} \, M$.

17.48. Calculate the E_{cell} value at 298 K for the cell based on the reaction

$$Cu(s) + 2 Ag^+(aq) \rightarrow Cu^{2+}(aq) + 2 Ag(s)$$

when $[Ag^+] = 2.56 \times 10^{-3} \, M$ and $[Cu^{2+}] = 8.25 \times 10^{-4} \, M$.

17.49. Using the appropriate standard potentials from Appendix 6, determine the equilibrium constant for the following reaction at 298 K:

$$Fe^{3+}(aq) + Cr^{2+}(aq) \rightarrow Fe^{2+}(aq) + Cr^{3+}(aq)$$

17.50. Using the appropriate standard potentials from Appendix 6, determine the equilibrium constant at 298 K for the following reaction between MnO$_2$ and Fe^{2+} in acid solution:

$$4 H^+(aq) + MnO_2(s) + 2 Fe^{2+}(aq) \rightarrow$$
$$Mn^{2+}(aq) + 2 Fe^{3+}(aq) + 2 H_2O(\ell)$$

17.51. If the potential of a hydrogen electrode based on the half-reaction

$$2 H^+(aq) + 2 e^- \rightarrow H_2(g)$$

is 0.000 V at pH = 0.00, what is the potential of the same electrode at pH = 7.00?

17.52. Glucose Metabolism The standard potentials for the reduction of nicotinamide adenine dinucleotide (NAD$^+$) and oxaloacetate (reactants in the multistep metabolism of glucose) are as follows:

$$NAD^+(aq) + 2 H^+(aq) + 2 e^- \rightarrow NADH(aq) + H^+(aq)$$
$$E° = -0.320 \text{ V}$$

$$Oxaloacetate(aq) + 2 H^+(aq) + 2 e^- \rightarrow malate(aq)$$
$$E° = -0.166 \text{ V}$$

a. Calculate the standard potential for the following reaction:

$$Oxaloacetate(aq) + NADH(aq) + H^+(aq) \rightarrow$$
$$malate(aq) + NAD^+(aq)$$

b. Calculate the equilibrium constant for the reaction at 298 K.

17.53. Permanganate ion can oxidize sulfite to sulfate in basic solution as follows:

$$2 MnO_4^-(aq) + 3 SO_3^{2-}(aq) + H_2O(\ell) \rightarrow$$
$$2 MnO_2(s) + 3 SO_4^{2-}(aq) + 2 OH^-(aq)$$

Determine the potential for the reaction (E_{rxn}) at 298 K when the concentrations of the reactants and products are as follows: $[MnO_4^-] = 0.150 \, M$, $[SO_3^{2-}] = 0.256 \, M$, $[SO_4^{2-}] = 0.178 \, M$, and $[OH^-] = 0.0100 \, M$. Will the value of E_{rxn} increase or decrease as the reaction proceeds?

*****17.54.** Manganese dioxide is reduced by iodide ion in acid solution as follows:

$$MnO_2(s) + 2 I^-(aq) + 4 H^+(aq) \rightarrow$$
$$Mn^{2+}(aq) + I_2(aq) + 2 H_2O(\ell)$$

Determine the electrical potential of the reaction at 298 K when the initial concentrations of the components are as follows: $[I^-] = 0.225 \, M$, $[H^+] = 0.900 \, M$, $[Mn^{2+}] = 0.100 \, M$, and $[I_2] = 0.00114 \, M$. If the solubility of iodine in water is approximately 0.114 M, will the value of E_{rxn} increase or decrease as the reaction proceeds?

17.55. A copper penny dropped into a solution of nitric acid produces a mixture of nitrogen oxides. The following reaction describes the formation of NO, one of the products:

$$3 Cu(s) + 8 H^+(aq) + 2 NO_3^-(aq) \rightarrow$$
$$2 NO(g) + 3 Cu^{2+}(aq) + 4 H_2O(\ell)$$

a. Starting with the appropriate standard potentials from Appendix 6, calculate $E°_{rxn}$ for this reaction.
b. Calculate E_{rxn} at 298 K when $[H^+] = 0.100 \, M$, $[NO_3^-] = 0.0250 \, M$, $[Cu^{2+}] = 0.0375 \, M$, and the partial pressure of NO = 0.00150 atm.

17.56. Chlorine dioxide (ClO$_2$) is produced by the following reaction of chlorate (ClO$_3^-$) with Cl$^-$ in acid solution:

$$2 ClO_3^-(aq) + 2 Cl^-(aq) + 4 H^+(aq) \rightarrow$$
$$2 ClO_2(g) + Cl_2(g) + 2 H_2O(\ell)$$

a. Determine $E°$ for the reaction.

b. The reaction produces a mixture of gases in the reaction vessel in which P_{ClO_2} = 2.0 atm; P_{Cl_2} = 1.00 atm. Calculate $[ClO_3^-]$ if, at equilibrium (T = 298 K), $[H^+]$ = $[Cl^-]$ = 10.0 M.

---

*17.57. The oxidation of NH_4^+ to NO_3^- in acid solution is described by the following equation:

$$NH_4^+(aq) + 2\,O_2(g) \rightarrow NO_3^-(aq) + 2\,H^+(aq) + H_2O(\ell)$$

a. Calculate $E°$ for the overall reaction.

b. If the reaction is in equilibrium with air (P_{O_2} = 0.21 atm) at pH 5.60, what is the ratio of $[NO_3^-]$ to $[NH_4^+]$ at 298 K?

17.58. What is the value of $E°$ for the following reaction?

$$2\,AgCl(s) + H_2(g) \rightarrow 2\,Ag(s) + 2\,HCl(aq)$$

Relating Battery Capacity to Quantities of Reactants

CONCEPT REVIEW

17.59. One 12-volt lead–acid battery has a higher ampere-hour rating than another. Which of the following parameters are likely to be different for the two batteries?
a. Individual cell potentials
b. Anode half-reactions
c. Total masses of electrode materials
d. Number of cells
e. Electrolyte composition
f. Combined surface areas of their electrodes

17.60. In a voltaic cell based on the Zn/Cu^{2+} cell reaction,

$$Zn(s) + Cu^{2+}(aq) \rightarrow Cu(s) + Zn^{2+}(aq)$$

there is exactly 1 mole of each reactant and product. A second cell based on this cell reaction:

$$Cd(s) + Cu^{2+}(aq) \rightarrow Cu(s) + Cd^{2+}(aq)$$

also has exactly 1 mole of each reactant and product. Which of the following statements about these two cells is true?
a. Their cell potentials are the same.
b. The masses of their electrodes are the same.
c. The quantities of electric charge that they can produce are the same.
d. The quantities of electric energy that they can produce are the same.

PROBLEMS

17.61. Which of the following voltaic cells will produce the greater quantity of electric charge per gram of anode material?

$$Cd(s) + 2\,NiO(OH)(s) + 2\,H_2O(\ell) \rightarrow$$
$$2\,Ni(OH)_2(s) + Cd(OH)_2(s)$$

or

$$4\,Al(s) + 3\,O_2(g) + 6\,H_2O(\ell) + 4\,OH^-(aq) \rightarrow$$
$$4\,Al(OH)_4^-(aq)$$

17.62. Which of the following voltaic cells will produce the greater quantity of electric charge per gram of anode material?

$$Zn(s) + MnO_2(s) + H_2O(\ell) \rightarrow ZnO(s) + Mn(OH)_2(s)$$

or

$$Li(s) + MnO_2(s) \rightarrow LiMnO_2(s)$$

*17.63. Which of the following voltaic cell reactions delivers more electrical energy per gram of anode material at 298 K?

$$Zn(s) + 2\,NiO(OH)(s) + 2\,H_2O(\ell) \rightarrow$$
$$2\,Ni(OH)_2(s) + Zn(OH)_2(s) \qquad E°_{cell} = 1.20\ V$$

or

$$Li(s) + MnO_2(s) \rightarrow LiMnO_2(s) \qquad E°_{cell} = 3.15\ V$$

*17.64. Which of the following voltaic cell reactions delivers more electrical energy per gram of anode material at 298 K?

$$Zn(s) + Ni(OH)_2(s) \rightarrow Zn(OH)_2(s) + Ni(s)$$
$$E°_{cell} = 1.50\ V$$

or

$$2\,Zn(s) + O_2(g) \rightarrow 2\,ZnO(s) \qquad E°_{cell} = 2.08\ V$$

Electrolytic Cells and Rechargeable Batteries

CONCEPT REVIEW

17.65. The positive terminal of a voltaic cell is the cathode. However, the cathode of an electrolytic cell is connected to the negative terminal of a power supply. Explain this difference in polarity.

17.66. The anode in an electrochemical cell is defined as the electrode where oxidation takes place. Why is the anode in an electrolytic cell connected to the positive (+) terminal of an external supply, whereas the anode in a voltaic cell battery is connected to the negative (−) terminal?

17.67. The salts obtained from the evaporation of seawater can act as a source of halogens, principally Cl_2 and Br_2, through the electrolysis of the molten alkali metal halides. As the potential of the anode in an electrolytic cell is increased, which of these two halogens forms first?

17.68. In the electrolysis described in Problem 17.67, why is it necessary to use molten salts rather than seawater itself?

17.69. Quantitative Analysis Electrolysis can be used to determine the concentration of Cu^{2+} in a given volume of solution by electrolyzing the solution in a cell equipped with a platinum cathode. If all of the Cu^{2+} is reduced to Cu metal at the cathode, the increase in mass of the electrode provides a measure of the concentration of Cu^{2+} in the original solution. To ensure the complete (99.99%) removal of the Cu^{2+} from a solution in which $[Cu^{2+}]$ is initially about 1.0 M, will the potential of the cathode (versus SHE) have to be more negative or less negative than 0.34 V (the standard potential for $Cu^{2+} + 2\,e^- \rightarrow Cu$)?

17.70. A high school chemistry student wishes to demonstrate how water can be separated into hydrogen and oxygen by electrolysis. She knows that the reaction will proceed more rapidly if an electrolyte is added to the water. She has access to 2.00 M solutions of these compounds: H_2SO_4, HBr, NaI, NaCl, Na_2SO_4, and Na_2CO_3. Which one(s) should she use? Explain your selection(s).

PROBLEMS

17.71. Suppose the current from a battery is used to electroplate an object with silver. Calculate the mass of silver that would be deposited by a battery that delivers 1.7 A · hr of charge.

17.72. A battery charger used to recharge the NiMH batteries in a digital camera can deliver as much as 0.50 A of current to each battery. If it takes 100 min to recharge one battery, how much $Ni(OH)_2$ (in grams) is oxidized to NiO(OH)?

17.73. A quantity of electric charge deposits 0.732 g of Ag(s) from an aqueous solution of silver nitrate. When that same quantity of charge is passed through a solution of a gold salt, 0.446 g of Au(s) is formed. What is the oxidation state of the gold ion in the salt?

17.74. What amount of current is required to deposit 0.750 g of Pt(s) from a solution containing the $[PtCl_4]^{2-}$ ion within a time of 2.5 hours?

17.75. A NiMH battery containing 4.10 g of NiO(OH) is 50% discharged when it is connected to a charger with an output of 2.00 A at 1.3 V. How long does it take to recharge the battery?

*17.76. How long does it take to deposit a coating of gold 1.00 μm thick on a disk-shaped medallion 4.0 cm in diameter and 2.0 mm thick at a constant current of 85 A? The density for the electroplating process is 19.3 g/cm³. The electroplating solution contains gold(III).

*17.77. Oxygen Supply in Submarines Nuclear submarines can stay under water nearly indefinitely because they can produce their own oxygen by the electrolysis of water.
  a. How many liters of O_2 at 298 K and 1.00 bar are produced in 1 hr in an electrolytic cell operating at a current of 0.025 A?
  b. Could seawater be used as the source of oxygen in this electrolysis? Explain why or why not.

17.78. In the electrolysis of water, how long will it take to produce 125 liters of H_2 at 20°C and a pressure of 750 torr using an electrolytic cell through which the current is 52 mA?

17.79. Calculate the minimum (least negative) cathode potential (versus SHE) needed to begin electroplating nickel from 0.35 M Ni^{2+} onto a piece of iron.

*17.80. What is the minimum (least negative) cathode potential (versus SHE) needed to electroplate silver onto cutlery in a solution of Ag^+ and NH_3 in which most of the silver ions are present as the complex $Ag(NH_3)_2^+$ and the concentration of $Ag^+(aq)$ is only 3.50×10^{-5} M?

Fuel Cells

CONCEPT REVIEW

17.81. Describe two advantages of hybrid (gasoline engine–electric motor) power systems over all-electric systems based on fuel cells. Describe two disadvantages.

17.82. Describe three factors limiting the more widespread use of cars and other vehicles powered by fuel cells.

17.83. Methane can serve as the fuel for electric cars powered by fuel cells. Carbon dioxide is a product of the fuel cell reaction. All cars powered by internal combustion engines burning natural gas (mostly methane) produce CO_2. Why are electric vehicles powered by fuel cells likely to produce less CO_2 per mile?

17.84. To make the refueling of fuel cells easier, several manufacturers are developing converters that turn readily available fuels—such as natural gas, propane, and methanol—into H_2 and CO_2. Although vehicles with such power systems are not truly "zero emission," they still offer significant environmental benefits over vehicles powered by internal combustion engines. Describe a few of them.

PROBLEMS

17.85. Fuel cells with molten alkali metal carbonates as electrolytes can use methane as a fuel. The methane is first converted into hydrogen in a two-step process:

$$CH_4(g) + H_2O(g) \rightarrow CO(g) + 3\,H_2(g)$$
$$CO(g) + H_2O(g) \rightarrow H_2(g) + CO_2(g)$$

  a. Assign oxidation numbers to carbon and hydrogen in the reactants and products.
  b. Using the standard free energy of formation values from Appendix 4, calculate the standard free-energy changes in the two reactions and the overall $\Delta G°$ for the formation of $H_2 + CO_2$ from methane and steam.

*17.86. Molten carbonate fuel cells fueled with H_2 convert as much as 60% of the free energy released by the formation of water from H_2 and O_2 into electrical energy. Determine the quantity of electrical energy obtained from converting 1 mole of H_2 into $H_2O(\ell)$ in such a fuel cell.

Additional Problems

17.87. Suppose there were a scale for expressing electrode potentials in which the standard potential for the reduction of water in base:

$$2\,H_2O(\ell) + 2\,e^- \rightarrow H_2(g) + 2\,OH^-(aq)$$

was assigned an $E°$ value of 0.000 V. How would the standard potential values on this new scale differ from those in Appendix 6?

*17.88. To inhibit corrosion of steel structures in contact with seawater, pieces of other metals (often zinc) are attached to the structures to serve as "sacrificial anodes." Explain how these attached pieces of metal might protect the structures, and describe which properties of zinc make it a good selection.

17.89. Sometimes the anode half-reaction in the zinc–air battery (Figure 17.6) is written with zincate ion, $Zn(OH)_4^{2-}$, as the product. Write a balanced equation for the cell reaction based on this product.

*17.90. Solar-Powered Lamps Rechargeable nickel–cadmium (nicad) batteries are used to store energy in solar-powered landscape lamps (Figure P17.90). The batteries contain Cd anodes and cathodes made of NiO(OH). The products of the cell reaction include cadmium(II) and nickel(II) hydroxide. Write a net ionic equation describing the cell reaction.

FIGURE P17.90

17.91. The half-reactions and standard potentials for a nickel–metal hydride battery with a titanium–zirconium anode are as follows:

Cathode: $NiO(OH)(s) + H_2O(\ell) + e^- \rightarrow Ni(OH)_2(s) + OH^-(aq)$
$$E° = 1.32 \text{ V}$$

Anode: $TiZr_2H(s) + OH^-(aq) \rightarrow TiZr_2(s) + H_2O(\ell) + e^-$
$$E° = 0.00 \text{ V}$$

a. Write the overall cell reaction for this battery.

b. Calculate the standard cell potential.

*17.92. **Lithium–Ion Batteries** Scientists at the University of Texas, Austin, and at MIT developed a cathode material for lithium–ion batteries based on $LiFePO_4$, which is the composition of the cathode when the battery is fully discharged. Batteries with this cathode are more powerful than those of the same mass with $LiCoO_2$ cathodes. They are also more stable at high temperatures.

a. What is the formula of the $LiFePO_4$ cathode when the battery is fully charged?

b. Is Fe oxidized or reduced as the battery discharges?

c. Is the cell potential of a lithium–ion battery with an iron phosphate cathode likely to differ from one with a cobalt oxide cathode? Explain your answer.

17.93. Describe in your own words the meaning of this cell diagram:

$$Sn(s) \mid Sn^{2+}(aq) \; (0.025 \; M) \parallel Ag^+(aq) \; (0.010 \; M) \mid Ag(s)$$

Start your description with "The cell consists of a tin anode, which is. . . ."

17.94. In a NiMH battery, what are the oxidation states of (a) Ni in NiO(OH), (b) H in MH, (c) M in MH, and (d) H in H_2O?

17.95. A magnesium battery can be constructed from an anode of magnesium metal and a cathode of molybdenum sulfide, Mo_3S_4. The half-reactions are

Anode: $Mg(s) \rightarrow Mg^{2+}(aq) + 2 \; e^-$ $\quad E°_{anode} = 2.37 \; V$

Cathode: $Mg^{2+}(aq) + Mo_3S_4(s) + 2 \; e^- \rightarrow MgMo_3S_4(s)$
$$E°_{cathode} = ?$$

a. If the standard cell potential for the battery is 1.50 V, what is the value of $E°$ for the reduction of Mo_3S_4?

b. What are the apparent oxidation states and electron configurations of Mo in Mo_3S_4 and in $MgMo_3S_4$?

*c. The electrolyte in the battery contains a complex magnesium salt, $Mg(AlCl_3CH_3)_2$. Why is it necessary to include Mg^{2+} ions in the electrolyte?

17.96. **Clinical Chemistry** The concentration of Na^+ ions in red blood cells (11 mM) and in the surrounding plasma (140 mM) are quite different. Calculate the electrochemical potential (emf) across the cell membrane at 37°C as a result of this concentration gradient.

*17.97. The element fluorine, F_2, was first produced in 1886 by electrolysis of HF. Chemical syntheses of F_2 did not happen until 1986 when Karl O. Christe successfully prepared F_2 by this reaction:

$$K_2MnF_6(s) + 2 \; SbF_5(\ell) \rightarrow 2 \; KSbF_6(s) + MnF_3(s) + \tfrac{1}{2} F_2(g)$$

a. Assign oxidation numbers to each compound and determine the number of electrons involved in the process.

b. Using the following $\Delta H°_f$ values, calculate $\Delta H°$ for the reaction.

$\Delta H°_{f,SbF_5(\ell)} = -1324 \; kJ/mol \quad \Delta H°_{f,K_2MnF_6(s)} = -2435 \; kJ/mol$

$\Delta H°_{f,MnF_3(s)} = -1579 \; kJ/mol \quad \Delta H°_{f,KSbF_6(s)} = -2080 \; kJ/mol$

c. If we assume that ΔS is relatively small, such that $\Delta G \approx \Delta H$, estimate $E°$ for this reaction.

d. If ΔS for the reaction is greater than zero, is our value for $E°$ in part c too high or too low?

e. The electrochemical synthesis of F_2 is described by the electrolytic cell reaction

$$2 \; KHF_2(\ell) \rightarrow 2 \; KF(\ell) + H_2(g) + F_2(g)$$

Assign oxidation numbers and determine the number of electrons involved in this process.

17.98. **Corrosion of Copper Pipes** The copper pipes frequently used in household plumbing may corrode and eventually leak. The corrosion reaction is believed to involve the formation of copper(I) chloride:

$$2 \; Cu(s) + Cl_2(aq) \rightarrow 2 \; CuCl(s)$$

a. Write balanced equations for the half-reactions in this redox reaction.

b. Calculate $E°_{rxn}$ and $\Delta G°_{rxn}$ for the reaction.

17.99. Elemental uranium may be produced from uranium dioxide by the following two-step process:

$$UO_2(s) + 4 \; HF(g) \rightarrow UF_4(s) + 2 \; H_2O(\ell)$$

$$UF_4(s) + 2 \; Mg(s) \rightarrow U(s) + 2 \; MgF_2(aq)$$

a. Identify the reducing agent.

b. Identify the element that is reduced.

c. Using data from the table of standard reduction potentials in Appendix 6, find the maximum $E°$ value for the reduction of UF_4 for the second reaction.

d. Will 1.00 g of Mg(s) be sufficient to produce 1.00 g of uranium?

17.100. Which element is oxidized and which is reduced in the lithium–ion cell reaction described by Equation 17.13?

*17.101. **Electrolysis of Seawater** Magnesium metal is obtained by the electrolysis of molten Mg^{2+} salts obtained from evaporated seawater.

a. Does elemental Mg form at the cathode or anode?

b. Do you think the principal ingredient in sea salt (NaCl) needs to be separated from the Mg^{2+} salts before electrolysis? Explain your answer.

c. Would electrolysis of an aqueous solution of $MgCl_2$ also produce elemental Mg?

d. If your answer to part c was no, what would the products of electrolysis be?

*17.102. **Silverware Tarnish** Low concentrations of hydrogen sulfide in air react with silver to form Ag_2S, more familiar to us as tarnish. Silver polish contains aluminum metal powder in a basic suspension.

a. Write a balanced net ionic equation for the redox reaction between Ag_2S and Al metal that produces Ag metal and $Al(OH)_3$.

b. Calculate $E°$ for the reaction. *Hint*: Derive $E°$ values for the half-reactions in which Ag_2S is reduced to Ag metal and $Al(OH)_3$ is reduced to Al metal. Then, replace the $[Ag^+]$ and $[Al^{3+}]$ terms in the Nernst equations for these two half-reactions with terms based on the K_{sp} values of Ag_2S and $Al(OH)_3$ and the concentrations of sulfide and hydroxide ions (both of which are equal to one molar under standard conditions).

18

The Solid State

A Particulate View

Stronger, Tougher, Harder

Did you know that most gold rings and other gold jewelry are not made of pure gold? Because gold is one of the softer metals, it is very easy to bend jewelry made of pure gold. In contrast, jewelry made of gold blended with other metals such as palladium, silver, or copper is more resistant to physical damage and can last a lifetime. Any mixture of two or more metals in the solid state is called an *alloy*, and there are many thousands of them. Some are solid solutions, which means they are homogeneous at the atomic level just as liquid solutions are, and others are heterogeneous mixtures. The properties of alloys can be controlled by varying the proportions of their constituent metals.

Brass is an alloy of copper and zinc that is both harder and stronger than either element alone. It is commonly used in doorknobs and hinges, but over time brass fixtures tarnish, looking dull, even dirty. Because stainless steel doorknobs look clean and sleek, they are frequently used in public places such as hospitals and schools, where the appearance of cleanliness is important. However, brass may be the better choice in these settings because it kills bacteria that come in contact with it; stainless steel does not. The bactericidal action of brass is due to copper, a metal that has long been known for its antimicrobial properties.

Aluminum alloys are ideal for making aircraft because they are strong and lightweight. The wing of a 747-400 airplane is about 1.8 m longer than the wing of a regular 747 but about 2 metric tons lighter because it is made of an alloy of 95% aluminum blended with copper, manganese, and iron. Even sports equipment benefits from the development of new alloys: a patented five-metal alloy of nickel, zirconium, titanium, copper, and beryllium is advertised as possibly changing the game of golf because the alloy produces a stronger, lighter, more resilient club that enables a golfer to transfer more energy from the swing into the ball.

In this chapter we explore the links between the physical properties of solids at the macroscopic level and the structures of these solids at the atomic level. We start with metals and their alloys and conclude with an equally important class of nonmetallic compounds called ceramics, seeking answers to

Strong Enough to Reach the Sky One World Trade Center, the tallest building in the Western Hemisphere when completed in 2013, is supported by structural steel, an alloy of iron and carbon. ▶

LEARNING OUTCOMES

LO1 Use band theory to explain the conductivity of metals and semiconductors

LO2 Relate unit cell dimensions to the radii of the particles that make up the cell
Sample Exercises 18.1, 18.4

LO3 Relate the densities of solids to the masses of the particles in their unit cells and the dimensions of the cells
Sample Exercises 18.2, 18.5

LO4 Distinguish between substitutional and interstitial alloys
Sample Exercise 18.3

LO5 Calculate the interlayer spacing in crystalline solids using the Bragg equation
Sample Exercise 18.6

CONNECTION In Chapter 5 we discussed the valence bond theory of chemical bond formation as well as combining atomic orbitals into molecular orbitals.

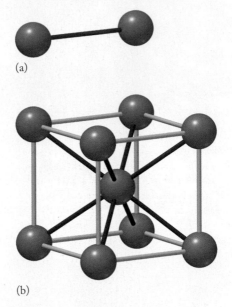

(a)

(b)

FIGURE 18.1 Covalent bonds differ from metallic bonds. (a) This ball-and-stick model of the molecule Na_2 is based on the assumption that the two atoms share their $3s$ electrons to form a covalent bond. (b) The atoms in solid sodium metal are actually arranged in a cubic structure in which each atom is bonded to 8 others. As a result, the bonds in sodium and other metals are much more diffuse than the covalent bonds in small molecules.

FIGURE 18.2 Solid sodium is a soft metal that can be cut with a knife because the metallic bonds between sodium atoms are weak.

band theory an extension of molecular orbital theory that describes bonding in solids.

such questions as these: Why do metals bend? Why are they such good conductors of electricity compared to ceramics or salts? Why are metal alloys so much stronger, tougher, and harder than the pure metals from which they are made?

18.1 Metallic Bonds and Conduction Bands

Most elements are metals, which means they are typically hard, shiny, malleable (easily shaped), ductile (easily drawn out), and able to conduct electricity. In this section we explore *why* they have these properties by examining metals at the atomic level and by exploring models describing the bonds that hold metal atoms together.

According to valence bond theory, a covalent bond forms between two atoms when partially filled atomic orbitals—one from each atom—overlap. Our focus in Chapter 5 was on covalent bonding in gas-phase molecules. In this section we explore the bonds that form between the densely packed atoms in metallic solids. Dense packing means that the valence orbitals of atoms overlap with orbitals of many nearby atoms. This large number of interactions makes metals strong. At the same time, sharing a limited number of valence electrons with many bonding partners makes the bond linking any two metal atoms relatively weak.

To understand this point, consider the bond that would form if two sodium atoms bonded together in a molecule of Na_2. As we learned in Chapter 3, sodium has the electron configuration $[Ne]3s^1$. When the partially filled $3s$ orbitals of the two Na atoms overlap, the result should be a diatomic molecule held together by a single covalent bond (Figure 18.1a). However, the Na atoms in solid sodium are surrounded by and bonded to 8 other Na atoms, and each of them is bonded to 8 of its neighbors (Figure 18.1b). This means each Na atom must share its $3s$ electron with 8 other atoms, not just one. Inevitably, this dispersion of bonding electrons weakens the Na—Na bond between each pair of atoms. The weakness of these bonds contributes to the unusual softness of sodium metal: you can, literally, cut it with a knife (Figure 18.2).

In Chapter 4 we described metal atoms "floating" in seas of mobile bonding electrons where the electrons are shared by all of the nuclei in the sample. The diffuse nature of metallic bonding described in the preceding paragraph certainly fits the sea-of-electrons model, but a more sophisticated approach, called **band theory**, better explains the bonding in metals and other solids. Let's apply band theory, which is an extension of molecular orbital theory, to explain the bonding between atoms in sodium metal. When the $3s$ atomic orbitals on two $Na(g)$ atoms overlap to form Na_2, the atomic orbitals combine to form two molecular orbitals with different energies equally spaced above and below the initial value (Figure 18.3). This is analogous to the formation of lower-energy bonding and higher-energy antibonding molecular orbitals (Section 5.7). If another two Na atoms join the first two to form a molecule of Na_4, the $3s$ atomic orbitals of four Na atoms combine to form four molecular orbitals. If we add another four atoms to make Na_8, a total of eight sodium $3s$ atomic orbitals combine to form eight molecular orbitals. In all these molecules the lower energy orbitals are filled with the available $3s$ electrons and the upper orbitals are empty, as shown in Figure 18.3. If we apply this model to the enormous number of atoms in a piece of solid Na, an equally enormous number of molecular orbitals are created. The lower energy half of them are occupied by electrons; the higher energy half are empty. There are so many of these orbitals that they form a continuous *band* of energies with no gap between the occupied lower half and the empty upper half. Because this band of MOs was formed by combining valence-shell orbitals, it is called a **valence band**.

Band theory explains the electrical conductivity of many metals by assuming that there is essentially no gap between the energy of the occupied lower portion of the valence band and the empty upper portion. This proximity means valence electrons can move easily from the filled lower portion to the empty upper portion, where they are free to move from one empty orbital to the next and flow throughout the solid.

The model of a partially filled valence shell explains the conductivity of many metals, though not all. Consider, for example, the excellent conductivity of zinc. Its electron configuration, $[Ar]3d^{10}4s^2$, tells us that all of its valence-shell electrons reside in filled orbitals, which means the valence band in solid zinc should be filled (Figure 18.4). With no empty space in the valence band to accommodate additional electrons, it might seem that the valence-shell electrons in Zn would be immobile. However, they are actually quite mobile, and band theory explains why. The theory assumes that *all* atomic orbitals of comparable shape and energy, including the empty $4p$ orbitals in zinc, combine to form additional energy bands. The energy band produced by combining empty $4p$ orbitals, called a **conduction band**, is also empty and is broad enough to overlap the valence band. This overlap means that electrons from the valence band can move to the conduction band, where they are free to migrate from atom to atom in solid zinc, thereby conducting electricity.

Before ending this discussion, we need to disclose that the two views of valence bands provided in Figures 18.3 and 18.4 are not mutually exclusive. The overlap of filled valence and empty conduction bands in Zn that is shown in Figure 18.4 can also be viewed as a partially filled valence band (as in Figure 18.3) if we assume that molecular orbitals can form from the mixing of the filled *s* and empty *p* orbitals of many Zn atoms. The important message is that either model leads us to the same conclusion: the valence electrons in zinc metal are quite mobile, making it a good conductor.

> **CONCEPT TEST**
>
> Use band theory to explain the electrical conductivity of magnesium metal.

(Answers to Concept Tests are in the back of the book.)

18.2 Semiconductors

To the right of the metals in the periodic table there is a "staircase" of elements that tend to have the physical properties of metals and the chemical properties of nonmetals. These semimetals (or metalloids) are not as good at conducting electricity as metals, but they are better at it than nonmetals. We can use band theory

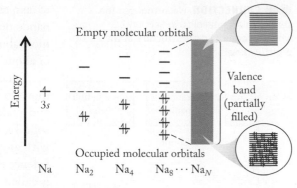

FIGURE 18.3 The half-filled $3s$ atomic orbitals of increasing numbers of sodium atoms combine to form molecular orbitals. As more MOs form, their energies get closer together until a continuous energy band is formed—a valence band that is half-filled with electrons. The electrons can move from the filled half (purple) to the slightly higher energy upper half (orange), where they are free to migrate through delocalized empty orbitals throughout the entire solid.

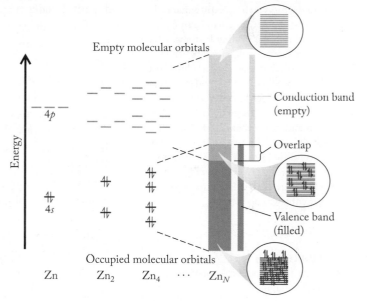

FIGURE 18.4 As the filled $4s$ atomic orbitals of an increasing number of Zn atoms overlap, they form a filled valence band (purple). An empty conduction band (gray) is produced by combining the empty $4p$ orbitals. The valence and conduction bands overlap each other, and electrons move easily from the filled valence band to the empty conduction band.

valence band a band of orbitals that are filled or partially filled by valence electrons.

conduction band an unoccupied band higher in energy than a valence band in which electrons are free to migrate.

⊙⊙ **CONNECTION** We introduced the semimetals (metalloids) in Section 2.3 when we described the structure of the periodic table.

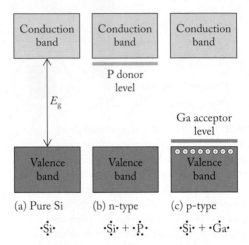

FIGURE 18.5 The electrical conductivity of semiconductors can be greatly enhanced by doping. (a) Pure Si has a band gap E_g of about 100 kJ/mol. (b) Adding P atoms, which have one more valence electron each than Si atoms, creates a narrow P donor level just below the Si conduction band and an n-type semiconductor. (c) Adding atoms of Ga, which each have one fewer valence electron than Si atoms, creates a narrow Ga acceptor level just above the Si valence electrons and a p-type semiconductor. Valence electrons move into the acceptor level, leaving behind positively charged holes (⊕) in the valence band that enhance conductivity.

⊙⊙ **CONNECTION** Emission spectra obtained from gas-discharge tubes were discussed in Chapter 3.

....

band gap (E_g) the energy gap between the valence and conduction bands.

semiconductor a semimetal (metalloid) with electrical conductivity between that of metals and insulators that can be chemically altered to increase its electrical conductivity.

n-type semiconductor a semiconductor containing an electron-rich dopant.

p-type semiconductor a semiconductor containing an electron-poor dopant.

to explain this intermediate behavior. In semimetals, conduction and valence bands do not overlap but instead are separated by an energy gap. In silicon, the most abundant semimetal, band theory predicts an energy gap, or **band gap (E_g)**, of about 100 kJ/mol (Figure 18.5a).

Generally, only a few valence-band electrons in Si have sufficient energy to move to the conduction band, which limits silicon's ability to conduct electricity and makes it a **semiconductor**. However, we can enhance the conductivity of solid Si, or of any other elemental semimetal, by replacing some of the Si atoms with atoms of an element of similar atomic radius but with a different number of valence electrons. The replacement process is called *doping*, and the added element is called a *dopant*.

Suppose the dopant is a group 15 element such as phosphorus. Each P atom has one more valence electron than the atom of silicon it replaces (compare the Lewis symbols of Si and P in Figure 18.5). The energy of these additional electrons is different from the energy of the silicon electrons. They populate a narrow band labeled the P *donor level* (Figure 18.5b) that is only about 4 kJ/mol below the Si conduction band. This small energy difference means that the donor electrons can easily reach the Si conduction band, resulting in enhanced conductivity. Phosphorus-doped silicon is an example of an **n-type semiconductor** because the dopant donates **n**egative charges (electrons) to the structure of the host element.

The conductivity of solid silicon can also be enhanced by replacing some Si atoms with atoms of a group 13 element, such as gallium (Figure 18.5c). Because Ga atoms have one fewer valence electron than Si atoms, substituting them in the Si structure means fewer valence electrons in the solid. The result is the creation of a narrow Ga *acceptor level* located about 7 kJ/mol above the Si valence band. Si valence electrons can move from the valence band into the acceptor level, leaving behind positively charged "holes" represented by the ⊕ symbols in Figure 18.5(c). The presence of the **p**ositive holes and the migration of electrons between them, filling one hole and creating another, enhances conductivity and makes Si doped with Ga a **p-type semiconductor**. The semiconductors used in solid-state electronics are combinations of n-type and p-type.

Doping is not the only way to change the conductivity of semimetals. Compounds prepared from combinations of group 13 and group 15 elements, such as gallium arsenide, may also behave as semiconductors. Semiconductors made of GaAs have the same *average* number of valence electrons per atom as silicon: $(3 + 5)/2 = 4$. But their band gaps are larger and they function better at higher frequencies, which makes them better suited to applications like satellite communications and in cell phones. Gallium arsenide is also capable of emitting infrared radiation ($\lambda = 874$ nm) when connected to an electrical circuit. This emission is used in many devices including remote control units for televisions and DVD players. When electrical energy is applied to the material, electrons are raised to the conduction band. When they fall back to the valence band, they emit radiation. If aluminum is substituted for gallium in GaAs, the band gap increases, and predictably the wavelength of emitted light decreases. For example, a material with the empirical formula AlGaAs$_2$ emits orange-red light ($\lambda = 620$ nm). Many of the multicolored indicator lights in electronic devices use AlGaAs$_2$ semiconductors.

CONCEPT TEST ••

Gallium arsenide (GaAs) can be made an n-type or a p-type semiconductor by replacing some of the As atoms with another element. Which element—Se or Sn—would form an n-type semiconductor with GaAs?

••

18.3 Structures of Metals

In Section 18.1 we learned that the atoms in a piece of sodium metal are arranged in such a way that each atom touches 8 others (Figure 18.1b). In this section we take a more detailed look at how the atoms in sodium and other metals are stacked together.

Stacking Patterns

When a metal is heated above its melting point and then allowed to slowly cool, it usually solidifies into a **crystalline solid**, that is, a solid in which atoms are arranged in an ordered three-dimensional array called a *crystal lattice*. Think of a crystal lattice as stacked layers (designated *a*, *b*, *c*, . . .) of metal atoms that are packed together as tightly as possible to maximize the strengths of the metallic bonds that hold them together, thereby minimizing the energy of the structure and enhancing its stability. In the most tightly packed arrangements, each atom touches 6 others in its layer (denoted layer *a*) as shown in Figure 18.6(a). The atoms in a second layer (*b*) nestle into some of the spaces created by the first layer (Figure 18.6b), just like oranges in a fruit-stand display or a stack of cannonballs (Figure 18.7). Similarly, the atoms in a third layer (*c*) nestle among those in the *b* layer. However, two different alignments are possible for the atoms in the *c* layer. They can sit directly above the atoms in *a* (Figure 18.6c), or they can nestle into the atoms of *b* in such a way that they are not aligned directly above the atoms in layer *a* (Figure 18.6d). When this happens and a fourth layer (*d*) of atoms is then nestled into the spaces of *c*, the *d* layer atoms lie directly above the *a* atoms. Therefore, in Figure 18.6(c) we have a stacking pattern *ababab* . . . throughout the crystal, and in Figure 18.6(d) we show the stacking pattern *abcabc* . . . .

crystalline solid a solid made of an ordered array of atoms, ions, or molecules.

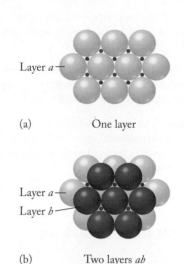

(a)   One layer

(b)   Two layers *ab*

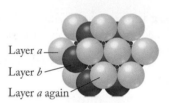

(c)   Three layers *ababab* . . .

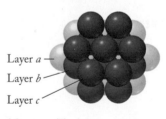

(d)   Three layers *abcabc* . . .

FIGURE 18.6 The *ababab* . . . and *abcabc* . . . stacking patterns represent two equally efficient ways to stack layers of atoms (or any particles of equal size).

(a)   (b)

FIGURE 18.7 (a) Stacks of oranges in a grocery store and (b) cannonballs at El Morro in San Juan, Puerto Rico, illustrate closest-packed arrays of spherical objects.

Stacking Spheres and Unit Cells

Whether the atoms in a metal are stacked in an *ababab* or an *abcabc* pattern determines the shape of the crystals the metal forms when it slowly cools and solidifies from the molten state. To see how crystal structures are linked to stacking patterns, let's take a closer look at a cluster of atoms in the *ababab* . . . stacking pattern (Figure 18.8). This cluster forms a *hexagonal* (six-sided) prism of closely packed atoms. In fact, they are as tightly packed as they can be, so the crystal structure

hexagonal closest-packed (hcp) a crystal lattice in which the layers of atoms or ions have an *ababab*... stacking pattern.

unit cell the basic repeating unit of the arrangement of atoms, ions, or molecules in a crystalline solid.

hexagonal unit cell an array of 9 closest-packed particles that are the repeating unit in a hexagonal closest-packed crystal.

crystal structure an ordered arrangement in three-dimensional space of the particles (atoms, ions, or molecules) that make up a crystalline solid.

▶❙❙ **CHEMTOUR** Crystal Packing

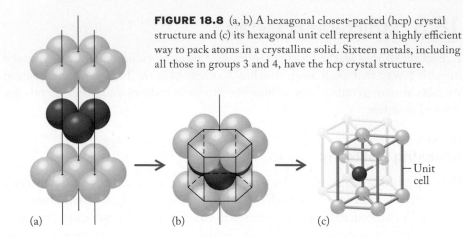

FIGURE 18.8 (a, b) A hexagonal closest-packed (hcp) crystal structure and (c) its hexagonal unit cell represent a highly efficient way to pack atoms in a crystalline solid. Sixteen metals, including all those in groups 3 and 4, have the hcp crystal structure.

is called **hexagonal closest-packed (hcp)**. As Figure 18.9 shows, the atoms of 16 elements have an hcp crystal lattice. In these lattices the 9-atom cluster highlighted in Figure 18.8(c) serves as an atomic-scale building block—a pattern of atoms repeated over and over again in all three dimensions in the elements.

FIGURE 18.9 Crystal structures of some metals and semimetals. The five elements designated "Other" have unit cells more complicated than can easily be described in this book.

▶❙❙ **CHEMTOUR** Unit Cell

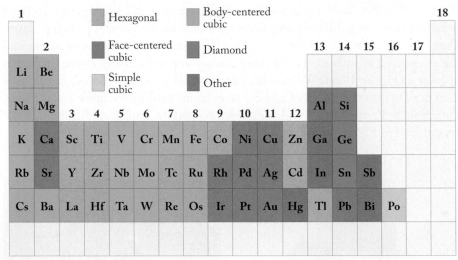

We call each of these building blocks a **unit cell**, and in the example in Figure 18.8 we have a **hexagonal unit cell**. A unit cell represents the minimum repeating pattern that describes the array of atoms forming the crystal lattice of any crystalline solid, including metals. Think of unit cells as three-dimensional microscopic analogs of the two-dimensional repeating pattern in fabrics, wrapping paper, or even a checkerboard. Look carefully at Figure 18.10 to confirm that the outlined portion represents the minimum repeating pattern in the checkerboard and in the paper. A unit cell has the same role in the **crystal structure** of a solid. The crystal structure of an element or compound describes the location of its atoms in three-dimensional space.

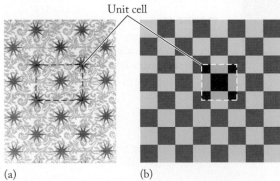

FIGURE 18.10 Two-dimensional repeating patterns. (a) The repeating pattern outlined with the dashed lines represents a "unit cell" for this wrapping-paper design. (b) The portions of the squares inside the white-dashed lines are one way to represent the repeating pattern in this checkerboard. Can you sketch an alternative way?

Atoms in other metals adopt the *abcabc . . .* stacking pattern when they solidify. What is the unit cell in this stacking pattern? Consider what happens when we take the 14-atom cluster in Figure 18.11(a), rotate it, and tip it 45° to obtain the orientation shown in Figure 18.11(b). The black outline in Figure 18.11(b) shows that the atoms form a cube: one atom at each of the eight corners of the cube and one at the center of each of the six faces. Because the atoms are stacked together as closely as possible, this crystal lattice is called **cubic closest-packed (ccp)**, and the corresponding unit cell is called a **face-centered cubic (fcc) unit cell**. Because this unit cell is a cube, the edges are all of equal length, and the angle between any two edges is 90°. Note that in Figure 18.11(b), the corner atoms do not touch one another, but adjacent atoms along the face diagonals do touch each other.

So far we have introduced two *closest-packed* crystal lattices—hexagonal and cubic—along with their associated unit cells (hexagonal and face-centered cubic). The hcp and ccp crystal lattices represent the most efficient ways of arranging solid spheres of equal radius. We can express the **packing efficiency** as the percentage of the total volume of the unit cell occupied by the spheres:

$$\text{Packing efficiency (\%)} = \frac{\text{volume occupied by spheres}}{\text{volume of unit cell}} \times 100\% \quad (18.1)$$

For both hcp and ccp crystal lattices, the packing efficiency is approximately 74%. We will come back to this calculation at the end of the next subsection, but first, let's look at some other packing arrangements.

Stacking patterns also exist in which the atoms are arranged close together but not as efficiently as hcp and ccp lattices. Two such patterns are shown in Figures 18.12 and 18.13. We can arrange the atoms in an *a* layer so that each atom touches four adjacent atoms, an arrangement called *square packing* (Figure 18.12a). If we add a second layer of spheres directly above the first, we create the *aaa . . .* stacking pattern shown in Figure 18.12(b) that is called *cubic packing*. The three-dimensional repeating pattern of this arrangement is called a **simple cubic (sc) unit cell**. It is the least efficiently packed of the cubic unit cells and is quite rare among metals: only radioactive polonium (Po) forms a simple cubic unit cell.

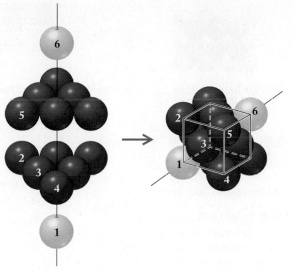

(a) *abcabc...* layering   (b) Face-centered cubic unit cell

FIGURE 18.11 (a) The stacking pattern *abcabc . . .* has a face-centered cubic (fcc) unit cell. (b) The shape of the unit cell is more easily seen when the layers are tipped 45° and rotated. Note that atoms at adjacent corners do not touch each other, but the three atoms along the diagonal of any face of the cube—atoms 2, 3, and 4 here—do touch each other.

cubic closest-packed (ccp) a crystal structure composed of face-centered cubic unit cells and layers of particles having an *abcabc . . .* stacking pattern.

face-centered cubic (fcc) unit cell an array of closest-packed particles that has eight of the particles at the corners of a cube and six of them at the centers of each face of the cube.

packing efficiency percentage of the total volume of a unit cell occupied by the spheres.

simple cubic (sc) unit cell a cell with atoms only at the eight corners of a cube.

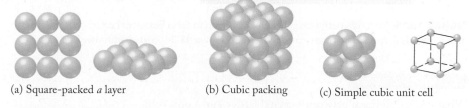

(a) Square-packed *a* layer   (b) Cubic packing   (c) Simple cubic unit cell

FIGURE 18.12 (a) In a square-packed *a* layer, each atom (like the one in the middle) touches four others. (b) If the atoms in all other layers are directly above those in the *a* layer, the stacking pattern is called *cubic packing*. (c) The repeating unit of this pattern is called a *simple cubic* (sc) unit cell with eight atoms at the eight corners of a cube.

(a) Two square-packed layers in an *ab* pattern

(b) Body-centered cubic unit cell

FIGURE 18.13 (a) The atoms represented by purple spheres in the *b* layer nestle into the spaces between the square-packed atoms (yellow spheres) in the *a* layer. (b) Atoms in the third layer are directly above those in the first, producing an *abab* . . . stacking pattern and a *body-centered cubic* (bcc) unit cell.

TABLE 18.1	Summary of Unit Cells, Stacking Patterns, and Packing Efficiencies for Solid Spheres				
Lattice Name	**Unit Cell**	**Type of Packing**	**Stacking Pattern**	**Number of Nearest Neighbors**	**Packing Efficiency**
Hexagonal closest-packed (hcp)	Hexagonal	Close packing	*ababab* . . .	12	74%
Cubic closest-packed (ccp)	Face-centered cubic (fcc)	Close packing	*abcabc* . . .	12	74%
Body-centered cubic packing	Body-centered cubic (bcc)	Square packing	*ababab* . . .	8	68%
Cubic packing	Simple cubic (sc)	Square packing	*aaaa* . . .	6	52%

If each atom in a second layer is nestled in the space created by four atoms in a square-packed *a* layer (Figure 18.13a), we have two layers in an *ab* stacking pattern. If the atoms in the third layer are directly above those in the first, then we have an *ababab* . . . stacking pattern based on layers of square-packed atoms. The simplest three-dimensional repeating unit of this pattern is called a **body-centered cubic (bcc) unit cell**. It consists of nine atoms, one at each of the eight corners of a cube and one in the middle of the cube (Figure 18.13b). All the group 1 metals and many transition metals have bcc unit cells. Table 18.1 summarizes the different stacking patterns, packing efficiencies, and unit cells described in this section.

Solids with cubic unit cells form crystals that often contain cubes or four-sided prisms (Figure 18.14). Crystalline solids with hexagonal unit cells tend to form hexagonal crystals (Figure 18.15).

FIGURE 18.14 Palladium (and other metals) with fcc unit cells may form crystals that contain four-sided prisms such as those in this photomicrograph.

FIGURE 18.15 Materials with hexagonal unit cells tend to form hexagonal crystals. There are two such materials in this photo: quartz crystals (see also Figure 18.36) coated with titanium metal.

CONCEPT TEST

What is the difference between a crystal lattice and a unit cell?

body-centered cubic (bcc) unit cell a cell with atoms at the eight corners of a cube and at the center of the cell.

Unit Cell Dimensions

Figure 18.16 shows whole-atom and cutaway views of sc, fcc, and bcc unit cells. These views also provide us with a way to determine how many equivalent atoms are in each type of cubic unit cell.

Let's start with the simple cubic unit cell (Figure 18.16a). Note how only a fraction of each corner atom is inside the unit cell boundary. In a crystal lattice with this unit cell, each atom is a corner atom in eight unit cells (Figure 18.17a). Thus each atom contributes the equivalent of one-eighth of an atom to the unit cell. There are eight corners in a cube, so there is a total of

$$\tfrac{1}{8} \text{ corner atom/\sout{corner}} \times 8 \text{ \sout{corners}/unit cell} = 1 \text{ atom/unit cell}$$

This calculation applies to the corner atoms in any type of cubic unit cell. Note that the two corner atoms along each edge in Figure 18.16(a) touch each other. Therefore the edge length ℓ in the simple cubic unit cell is equal to twice the atomic radius:

$$\ell = 2r$$

In an fcc unit cell (Figure 18.16b) there are eight corner atoms and one atom in the center of each of the six faces (Figure 18.17b). Each face atom is shared by the two unit cells that abut each other at that face. Therefore each unit cell "owns" half of each face atom, making a total of

$$\tfrac{1}{2} \text{ face atom/\sout{face}} \times 6 \text{ \sout{faces}/fcc unit cell} = 3 \text{ face atoms/fcc unit cell}$$

As just noted, every cubic unit cell owns the equivalent of one corner atom. Therefore, an fcc unit cell consists of

$$1 \text{ corner atom} + 3 \text{ face atoms} = 4 \text{ atoms per fcc unit cell}$$

To relate the size of these atoms to the dimensions of the fcc unit cell, note in the cutaway view of Figure 18.16(b) that the corner atoms do not touch one another but adjacent atoms along the face diagonal do touch each other. A face diagonal spans the radius r of two corner atoms and the diameter (2 radii = $2r$) of a face atom. Therefore the length of a face diagonal is $1 + 2 + 1 = 4$ atomic radii = $4r$. A face diagonal connects the ends of two edges and forms a right

(a) Simple cubic:
Atoms touch along edge

(b) Face-centered cubic:
Atoms touch along face diagonal

(c) Body-centered cubic:
Atoms touch along body diagonal

FIGURE 18.16 Whole-atom and cutaway views of cubic unit cells. (a) In a simple cubic unit cell, each corner atom of the unit cell is part of eight unit cells. Atoms along each edge touch. (b) In a face-centered cubic unit cell, the face atoms are part of two unit cells. Atoms along the face diagonal touch. (c) In a body-centered cubic unit cell, one atom in the center lies entirely in one unit cell. The atoms along the body diagonal touch.

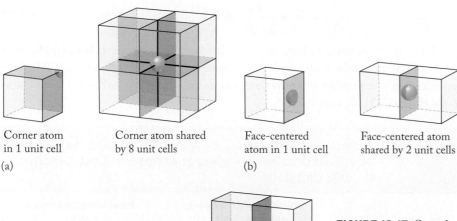

Corner atom in 1 unit cell   Corner atom shared by 8 unit cells   Face-centered atom in 1 unit cell   Face-centered atom shared by 2 unit cells

(a)   (b)

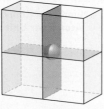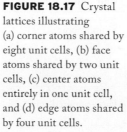

Body-centered atom in 1 unit cell   Edge atom in 1 unit cell   Edge atom shared by 4 unit cells

(c)   (d)

FIGURE 18.17 Crystal lattices illustrating (a) corner atoms shared by eight unit cells, (b) face atoms shared by two unit cells, (c) center atoms entirely in one unit cell, and (d) edge atoms shared by four unit cells.

(a) fcc unit cell dimensions

(b) bcc unit cell dimensions

FIGURE 18.18 (a) In a face-centered cubic unit cell, the face diagonal forms a right triangle with two adjoining edges each of length ℓ. Applying the Pythagorean theorem to this triangle yields $r = 0.3536\ell$. (b) In a body-centered cubic unit cell, the atoms touch along a body diagonal. Applying the Pythagorean theorem to the right triangle formed by an edge, a face diagonal, and a body diagonal yields $r = 0.4330\ell$.

triangle with those two edges, each of length ℓ (Figure 18.18a). Therefore, according to the Pythagorean theorem, the length of the face diagonal is

$$\text{Face diagonal} = 4r = \sqrt{\ell^2 + \ell^2} = \sqrt{2\ell^2} = \ell\sqrt{2}$$

$$r = \frac{\ell\sqrt{2}}{4} = 0.3536\ell \tag{18.2}$$

Now let's focus on the bcc unit cell in Figure 18.16(c). In addition to the one atom from one-eighth of an atom at each of the eight corners, there is also one atom in the center of the cell that is entirely within the cell (Figure 18.17c). This means a bcc unit cell consists of

$$1 \text{ corner atom} + 1 \text{ center atom} = 2 \text{ atoms per bcc unit cell}$$

Relating unit cell edge length to atomic radius in a bcc cell is complicated by the fact that, in addition to not touching along the edges, adjacent atoms along any face diagonal do not touch each other. However, each corner atom does touch the atom in the center of the cell, which means that the atoms touch along a *body diagonal*, which runs between opposite corners through the center of the cube. In the cutaway view in Figure 18.18(b), the diagonal runs from the bottom left corner of the front face to the top right corner of the rear face. It spans (1) the radius of the front-face bottom left corner atom, (2) the diameter (2 radii) of the central atom, and (3) the radius of the rear-face top right atom, making the length of the body diagonal equivalent to $4r$.

We can again use the Pythagorean theorem to determine the relationship between ℓ and r. A right triangle is formed by an edge, a face diagonal, and a body diagonal serving as the hypotenuse of the triangle (Figure 18.18b). Using the face-diagonal value $\ell\sqrt{2}$, we get

$$\text{Body diagonal} = 4r = \sqrt{(\text{edge length})^2 + (\text{face diagonal})^2}$$

$$= \sqrt{\ell^2 + (\ell\sqrt{2})^2} = \sqrt{\ell^2 + \ell^2(2)} = \sqrt{3\ell^2} = \ell\sqrt{3}$$

so

$$r = \frac{\ell\sqrt{3}}{4} = 0.4330\ell \tag{18.3}$$

Table 18.2 summarizes how atoms in different locations in sc, bcc, and fcc unit cells contribute to the total number of atoms in each unit cell, while Table 18.3 summarizes the number of equivalent atoms and the relationship between r and ℓ for the three cubic unit cells.

TABLE 18.2	Contributions of Atoms to Cubic Unit Cells	
Atom Position	**Contribution to Unit Cell**	**Unit Cell**
Center	1 atom	bcc
Face	$\frac{1}{2}$ atom	fcc
Corner	$\frac{1}{8}$ atom	bcc, fcc, sc

TABLE 18.3	Summary of Unit Cells, Equivalent Atoms, and the Relationship between Radius of Atoms and Edge Length of Cubic Unit Cells	
Unit Cell	**Number of Equivalent Atoms per Unit Cell**	**Relationship between r and ℓ**
Simple cubic	1	$r = \dfrac{\ell}{2} = 0.50\ell$
Body-centered cubic	2	$r = \dfrac{\ell\sqrt{3}}{4} = 0.4330\ell$
Face-centered cubic	4	$r = \dfrac{\ell\sqrt{2}}{4} = 0.3536\ell$

SAMPLE EXERCISE 18.1   **Calculating Atomic Radius   LO2
from Unit Cell Dimensions**

The most stable form of iron at room temperature, called *ferrite*, is shown in Figure 18.19. Its unit cell has an edge length of 287 pm. Calculate the radius in picometers of the iron atoms in it. Check your answer against the data in Appendix 3.

COLLECT AND ORGANIZE The unit cell of ferrite consists of eight corner atoms and one atom in the center, a bcc unit cell. We know the edge length of the cell: 287 pm.

ANALYZE The iron atoms do not touch along the unit cell edges or along any face diagonal but do touch along the body diagonals. Equation 18.3 gives the relationship between r and ℓ for a bcc unit cell. We know that the radius is a little less than half the edge length ($\ell/2$), so we expect a value less than 144 pm for r.

SOLVE

$$r = 0.4330 \times 287 \text{ pm} = 124 \text{ pm}$$

THINK ABOUT IT The value for r, 124 pm, is indeed less than half the value of the edge length ($\ell/2 = 144$ pm). The average atomic radius of iron atoms is 126 pm (see Table A3.1 in Appendix 3), so the result of this calculation is reasonable.

Practice Exercise At 1070°C, the most stable form of iron is *austenite* (Figure 18.20). The edge length of its fcc unit cell is 361 pm. What is the atomic radius of iron in austenite?

(Answers to Practice Exercises are in the back of the book.)

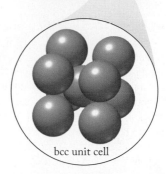

bcc unit cell

FIGURE 18.19 The pattern in this newly cut surface of the meteorite that caused the Odessa crater in Texas about 50,000 years ago is due to crystals of ferrite, which grew when the meteor formed from molten iron 4.5 billion years ago. The microview shows the bcc unit cell of ferrite.

FIGURE 18.20 The fcc unit cell of austenite.

SAMPLE EXERCISE 18.2   **Using Unit Cell Dimensions   LO3
to Calculate Density**

Calculate the density of iron (ferrite) in grams per cubic centimeter at 25°C, given that its bcc unit cell has an edge length of 287 pm.

COLLECT AND ORGANIZE We are asked to calculate the density of iron at 25°C knowing its unit cell geometry and edge length. The fact that ferrite has a bcc unit cell means that there are two Fe atoms per cell. The molar mass of Fe is 55.845 g/mol, and the formula for the volume of a cube of edge length ℓ is $V = \ell^3$.

ANALYZE We assume that the density of the unit cell is the same as the density of solid Fe. The density of the Fe bcc unit cell is the mass of two Fe atoms divided by the volume of the cell. We need to calculate the mass of two Fe atoms starting with the molar mass. The conversion includes dividing by Avogadro's number to calculate the mass of each Fe atom in grams. A piece of iron sinks when immersed in water, so we predict that the density of iron should be greater than 1.0 g/cm³, the density of liquid water.

SOLVE First we calculate the mass m of two Fe atoms:

$$m = \frac{55.845 \text{ g Fe}}{1 \text{ mol Fe}} \times \frac{1 \text{ mol Fe}}{6.0221 \times 10^{23} \text{ atoms Fe}} \times 2 \text{ atoms Fe} = 1.8547 \times 10^{-22} \text{ g Fe}$$

and then the volume of the cell in cubic centimeters:

$$V = \ell^3 = (287 \text{ pm})^3 \times \frac{(10^{-10} \text{ cm})^3}{(1 \text{ pm})^3} = 2.364 \times 10^{-23} \text{ cm}^3$$

The density is

$$d = \frac{m}{V} = \frac{1.8547 \times 10^{-22}\,\text{g}}{2.364 \times 10^{-23}\,\text{cm}^3} = 7.85\,\text{g/cm}^3$$

THINK ABOUT IT According to the data in Table A3.2 of Appendix 3, the density of iron is 7.874 g/mL, which is nearly the same as the result of our calculation. (Note that 1 mL = 1 cm³.) The density of the unit cell must equal the density of a bulk sample because the latter is composed of a crystal lattice of bcc unit cells.

Practice Exercise Silver and gold both crystallize in face-centered cubic unit cells with edge lengths of 407.7 and 407.0 pm, respectively. Calculate the density of each metal and compare your answers with the densities listed in Appendix 3. ⚙

Earlier in this section we mentioned that ccp and hcp are the most efficient packing schemes for spheres. Now that we know more about the fcc unit cell, let's take another look at the calculation of packing efficiency using Equation 18.1. The unit cell for a ccp arrangement of solid spheres of radius r is an fcc unit cell that contains four equivalent spheres. The volume occupied by these spheres is

$$V_{\text{spheres}} = 4\left(\tfrac{4}{3}\pi r^3\right) = \tfrac{16}{3}\pi r^3$$

From Table 18.3, $r = (\ell\sqrt{2})/4$, and the unit cell edge (ℓ) is

$$\ell = \frac{4r}{\sqrt{2}} = 2.828r$$

This equation enables us to express the volume of the unit cell in terms of r:

$$V_{\text{unit cell}} = \ell^3 = (2.828r)^3 = 22.62r^3$$

Substituting V_{spheres} and $V_{\text{unit cell}}$ into Equation 18.1:

$$\text{Packing efficiency (\%)} = \frac{V_{\text{spheres}}}{V_{\text{unit cell}}} \times 100\% = \frac{\tfrac{16}{3}\pi r^3}{22.62 r^3} \times 100\% = 74.1\%$$

This means that the most efficient packing of spheres results in about 26% of the total volume being empty.

18.4 Alloys

People have been using metals for tools, weapons, currency, and jewelry for tens of thousands of years. For most of that time, only three elements—copper, silver, and gold—were used because these are the only ones found as free metals in Earth's crust. Even so, most silver and copper occur not as pure metals but in ores such as argentite (Ag_2S), chalcopyrite ($CuFeS_2$), and chalcocite (Cu_2S). **Ores** are naturally occurring compounds or mixtures of compounds from which elements can be extracted.

Around 6000 years ago, metal technology took a giant leap forward when artisans in areas that are now Iraq and Pakistan discovered how to convert copper ore, principally $CuFeS_2$, to copper metal. The process involved pulverizing the ore and then baking it in ovens. Baking initiated a chemical reaction with O_2 (from air), which converted the Cu in $CuFeS_2$ to CuO. In the second step in the process, CuO was reacted with carbon monoxide, which was produced by burning wood or charcoal (mostly carbon) in a furnace with an insufficient supply of air:

$$CuO(s) + CO(g) \rightarrow Cu(s) + CO_2(g) \qquad (18.4)$$

ore a mineral that contains one or more metals valuable enough to be mined.

alloy a blend of a host metal and one or more other elements, which may or may not be metals, that are added to change the properties of the host metal.

substitutional alloy an alloy in which atoms of the nonhost element replace host atoms in the crystal lattice.

One disadvantage of copper tools and weapons is that the metal is very malleable, which means that copper objects are easily bent and damaged. We can explain the malleability of Cu (and other metals) in terms of the metallic bonds between its atoms and its cubic closest-packed crystal structure. We have noted that copper atoms are weakly bonded to their nearest neighbors. This arrangement makes it possible for the atoms in one layer, under stress, to slip past atoms in an adjacent layer (Figure 18.21). When the stress is relieved and the atoms stop slipping, many have different atoms as their nearest neighbors but the overall crystal structure is still cubic closest-packed. The ease with which copper atoms slip past each other made it easy for prehistoric metalworkers to hammer copper metal into spear points and shields, but it also meant that those objects could easily be damaged in battle.

Substitutional Alloys

About 5500 years ago, people living around the Aegean Sea discovered that mixing molten tin and copper produced bronze, a material that was much stronger than either tin or copper alone. Its discovery ushered in the Bronze Age. Bronze is an **alloy**, a metallic material made when a host metal is blended with one or more other elements, which may or may not be metals, thereby changing the properties of the host metal.

Like the mixtures discussed in Chapter 1, alloys can be classified according to their composition as homogeneous or heterogeneous mixtures. Bronze is a *homogeneous alloy*, a solid solution in which the atoms of the added element(s) (in this case tin) are randomly but uniformly distributed among the atoms of the host (copper in this example). *Heterogeneous alloys* consist of matrices of atoms of host metals interspersed with small "islands" made up of individual atoms of other elements. In both cases, the compositions may vary over a limited range. In contrast, *intermetallic compounds* have a reproducible stoichiometry and constant composition (just like chemical compounds) but are still commonly referred to as alloys and are considered to be a subgroup within homogeneous alloys. An example of an intermetallic compound is Ag_3Sn, an alloy of silver and tin used in dental fillings. It is a homogeneous mixture of silver and tin atoms in exactly a 3:1 ratio.

Now let's address how the atoms are arranged in the crystal lattices and unit cells of alloys. These arrangements depend on the type of alloy in question. A **substitutional alloy** is one in which atoms of the nonhost element replace host atoms in the crystal lattice. Bronze is a *homogeneous, substitutional alloy* in which the tin concentration can be as high as 30% by mass. Substitutional alloys may form between metals that have the same crystal lattice and have atomic radii that are within about 15% of each other.

Figure 18.22 illustrates one layer of the crystal lattice of bronze. The radii of copper and tin atoms are similar—128 pm and 140 pm, respectively. Inserting the

FIGURE 18.21 Copper and other metals are malleable because their atoms are stacked in layers that can slip past each other under stress. Slippage is possible because of the diffuse nature of metallic bonds and the relatively weak interactions between pairs of atoms in adjoining layers.

CONNECTION The classification of homogeneous and heterogeneous mixtures and the difference between compounds and mixtures were described in Chapter 1.

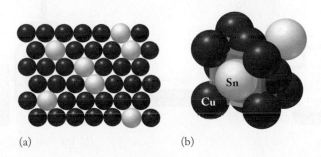

(a) (b)

FIGURE 18.22 Two atomic-scale views of one type of bronze, a substitutional alloy. (a) A layer of close-packed copper (Cu) atoms interspersed with a few atoms of tin (Sn). (b) One possible unit cell for bronze. In this case tin atoms have replaced one corner Cu atom and one face Cu atom.

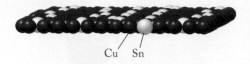

FIGURE 18.23 The larger Sn atoms in bronze disturb the Cu crystal lattice, producing atomic-scale bumps in the slip plane between layers of Cu atoms. These bumps make it more difficult for Cu atoms to slide by each other when an external force is applied.

slightly larger Sn atoms in the cubic closest-packed Cu crystal lattice disturbs the structure a little, making the planes of copper atoms "bumpy" instead of uniform (Figure 18.23). This atomic-scale roughness makes it more difficult for the copper atoms to slip past one another. Less slippage makes bronze less malleable than copper, but being less malleable also means that bronze is harder and stronger.

Many other substitutional alloys are known. Some are also copper based, including brass (zinc alloyed with copper) and pewter (tin alloyed with copper and antimony). However, most substitutional alloys in the modern world are *ferrous alloys*, so called because the host metal is iron. An important class of ferrous alloys is that of the rust-resistant *stainless steels*, which contain about 10% nickel and up to 20% chromium. When atoms of Cr on the surface of a piece of stainless steel combine with oxygen, they form a layer of Cr_2O_3 that bonds tightly to the surface and protects the metallic material beneath from further oxidation. This resistance to surface discoloration due to corrosion means that these alloys "stain less" than pure iron.

Interstitial Alloys

The Bronze Age began to wane about 3000 years ago with the discovery that iron oxides could be reduced to iron metal by limiting the air supplied to a wood or charcoal fire. As in the conversion of CuO to Cu (Equation 18.4), the reducing agent for iron oxide is carbon monoxide:

$$Fe_2O_3(s) + 3\ CO(g) \rightarrow 2\ Fe(s) + 3\ CO_2(g) \tag{18.5}$$

Iron quickly replaced bronze as the metallic material of choice for fabricating tools and weapons both because iron ore is much more abundant in Earth's crust than the ores of copper and tin and because tools and weapons made of iron or ferrous alloys are much stronger than those made of bronze.

Today, the reduction of iron ore is done in blast furnaces, enormous reaction vessels that operate at about 1600°C (Figure 18.24). Iron ore, hot carbon (*coke*), and limestone are added to the top of the vessel. The solid by-products (*slag*) float on top of the molten iron, which is harvested from the bottom. Blast furnaces get their name from blasts of hot air (up to 1000°C) that are injected through nozzles near the bottom of the furnace and that suspend the reactants until iron reduction

FIGURE 18.24 (a) Blast furnaces operate continuously at temperatures near 1600°C to convert iron ore into iron. Blasts of hot air inject O_2 into the furnace, which converts C to CO. Limestone is added to react with Si and P impurities. The products of these reactions become part of the slag layer. (b) Molten iron from a blast furnace is further purified in a second furnace, where $O_2(g)$ is injected instead of air.

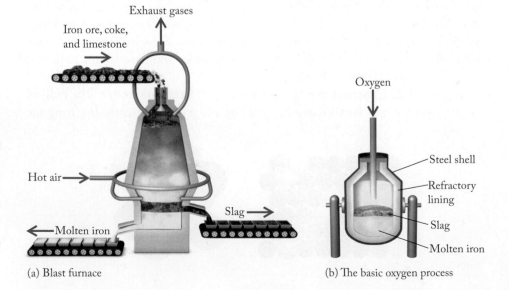

(a) Blast furnace

(b) The basic oxygen process

is complete. It may take as long as 8 hours for a batch of reactants to fall to the bottom of a blast furnace. On their way down, O_2 in the hot-air blasts partially oxidizes the coke to carbon monoxide, and the CO reduces the iron in iron ore as described in Equation 18.5. The limestone ($CaCO_3$) added along with the iron ore decomposes to calcium oxide (CaO, also called lime):

$$CaCO_3(s) \rightarrow CaO(s) + CO_2(g) \qquad (18.6)$$

The lime then reacts with silica impurities in the ore to form calcium silicate:

$$CaO(s) + SiO_2(s) \rightarrow CaSiO_3(\ell) \qquad (18.7)$$

Calcium silicate becomes part of the slag that floats on the denser molten iron at the bottom of the furnace.

Molten iron produced in a blast furnace may contain up to 5% carbon. To reduce the carbon content, the molten iron is transferred to a second furnace. When hot O_2 and additional CaO are injected, some of the carbon is oxidized to CO_2 and any remaining silicon impurities form more $CaSiO_3$ slag.

When the molten iron coming out of this second furnace cools to its melting point of 1538°C, it crystallizes in a body-centered cubic structure before undergoing a phase transition at around 1390°C to austenite, a form of solid iron made up of face-centered cubic unit cells. The spaces, or *holes*, between iron atoms in austenite can accommodate carbon atoms, forming an **interstitial alloy**, so named because the carbon atoms occupy spaces, or *interstices*, between the iron atoms (Figure 18.25).

All interstices in a crystal lattice are not equivalent. Indeed, holes of two different sizes occur between the atoms in any closest-packed crystal lattice (Figure 18.26). The larger holes are surrounded by clusters of six host atoms in the shape of an octahedron and are called *octahedral holes*. The smaller holes are located between clusters of four host atoms and are called *tetrahedral holes*. The data in Table 18.4 show which holes are more likely to be occupied based on the relative sizes of nonhost and host atoms. According to Appendix 3, the atomic radii of C and Fe are 77 and 126 pm, respectively. According to Table 18.4, the ratio 77/126 = 0.61 means that C atoms should fit in the octahedral holes of austenite, as shown in Figure 18.25, but not in the tetrahedral holes.

As austenite with its fcc unit cell cools to room temperature, it converts into the crystalline solid form of iron called ferrite, which is body-centered cubic. The holes in ferrite are smaller than those in austenite, so many fewer carbon atoms can be accommodated in a ferrite structure than in an austenite structure. The carbon that cannot be

interstitial alloy an alloy in which the nonhost atoms occupy spaces between atoms of the host.

FIGURE 18.25 Carbon steel is an interstitial alloy of carbon in iron.

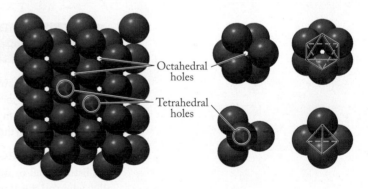

FIGURE 18.26 Close-packed atoms in adjacent layers of a crystal lattice produce octahedral holes surrounded by six host atoms and tetrahedral holes surrounded by four host atoms. Octahedral holes are larger than tetrahedral holes and can accommodate larger nonhost atoms in interstitial alloys. Note that all the atoms are identical here; the colors are only to distinguish one atom from another.

TABLE 18.4	**Atomic Radius Ratios and Locations of Nonhost Atoms in Unit Cells of Interstitial Alloys**	
Lattice Name	**Hole Type**	**Atomic Radius Ratio $r_{nonhost}/r_{host}$[a]**
hcp or ccp	Tetrahedral	0.22–0.41
hcp or ccp	Octahedral	0.41–0.73
Cubic packing	Cubic	0.73–1.00

[a]Radius ratios as predictors of the crystal lattice in crystalline solids are of limited value because atoms are not truly solid spheres with a constant radius; because the radius of an atom may differ in different compounds, these ranges are approximate.

TABLE 18.5	Effect of Carbon Content on the Properties of Steel		
Carbon Content (%)	Designation	Properties	Used to Make
0.05–0.19	Low carbon	Malleable, ductile	Nails, cables
0.20–0.49	Medium carbon	High strength	Construction girders
0.5–3.0	High carbon	Hard but brittle	Cutting tools

accommodated precipitates as clusters of carbon atoms, or it reacts with iron to form iron carbide, Fe_3C. The clusters of carbon and Fe_3C disrupt ferrite's crystalline lattice and inhibit the host iron atoms from slipping past each other when a stress is applied. This resistance to slippage, which is much like that experienced by the copper atoms in bronze (Figure 18.23), makes iron–carbon alloys, known as *carbon steel*, much harder and stronger than pure iron. In general, the higher the carbon concentration, the stronger the steel. However, there is a trade-off in this relationship. As Table 18.5 notes, increased strength and hardness come at the cost of increased brittleness.

CONCEPT TEST

In Chapter 3 we learned that atomic radius increases down a group and decreases across a period. Which would you predict to have the larger radius ratio, a lithium–aluminum or a lithium–magnesium alloy?

SAMPLE EXERCISE 18.3 Predicting the Crystal Structure   LO4
of a Two-Element Alloy

Sterling silver, which is 93% Ag and 7% Cu by mass, is widely used in jewelry. The presence of Cu inhibits tarnishing and strengthens the alloy. Is this copper–silver alloy a substitutional or an interstitial alloy? Silver has a cubic closest-packed crystal lattice with face-centered cubic unit cells.

COLLECT AND ORGANIZE We are asked whether the Cu–Ag alloy in sterling silver is substitutional or interstitial. Atoms of two or more elements form a substitutional alloy when all the atoms are of similar size. We are told that the atoms in solid Ag form fcc unit cells and have a cubic closest-packed crystal lattice.

ANALYZE The highest $r_{nonhost}/r_{host}$ ratio in Table 18.4 for a ccp lattice, 0.73, means that an interstitial alloy can form only when the radius of the nonhost atoms is less than 73% of the radius of the host atoms. The atomic radii found in Appendix 3 are 144 pm for Ag and 128 pm for Cu. The ratio of the atomic radii of Cu to Ag is 128 pm/144 pm = 0.89. Elements with similar radii are more likely to form substitutional alloys, as we saw with bronze.

SOLVE The $r_{nonhost}/r_{host}$ ratio of 0.89 indicates that the copper atoms are too big to fit into interstices in the lattice of silver atoms, regardless of whether the holes are tetrahedral or octahedral. Thus an interstitial alloy is impossible. Atoms of Cu can substitute for atoms of Ag in the Ag ccp lattice, however, with some room to spare. Thus, sterling silver is a substitutional alloy.

THINK ABOUT IT The guidelines for two metals forming a substitutional alloy are that they have the same type of crystal lattice and that their atomic radii are within

15% of each other. The radii of Ag and Cu differ by only 11%, so we expect copper and silver to form a substitutional alloy.

Practice Exercise Would you expect gold (atomic radius 144 pm) to form a substitutional alloy with silver (atomic radius 144 pm)? With copper (atomic radius 128 pm)?

Aluminum forms alloys with many other elements. Alloys with Mg, Si, Cu, and Zn are the most common and are widely used in aircraft construction. The tabs on beverage cans are manufactured from an aluminum alloy containing Mg and Mn. Because Li has the smallest molar mass (6.941 g/mol) and lowest density (0.534 g/cm^3) of all the metallic elements, aluminum alloys containing Li are attractive for applications where low mass is essential.

Aluminum and aluminum alloys are widely used as building materials and in fuselages of airplanes because they are corrosion resistant. This resistance is noteworthy because aluminum is actually a very reactive metal. However, when the surface of a piece of aluminum starts to oxidize in air, a thin layer of Al_2O_3 is produced that adheres strongly to the aluminum metal below it and acts as a protective shield against further oxidation.

18.5 Structures of Some Crystalline Nonmetals

In Figure 18.9, the color key identifies the unit cells of silicon, germanium, and tin as "diamond." Carbon, of course, is a nonmetal, but the specific arrangement of carbon atoms in diamond (Figure 18.27a) is also found in some metallic materials. In addition, some diamonds may include metal impurities in the crystal lattice that give rise to distinctive colors.

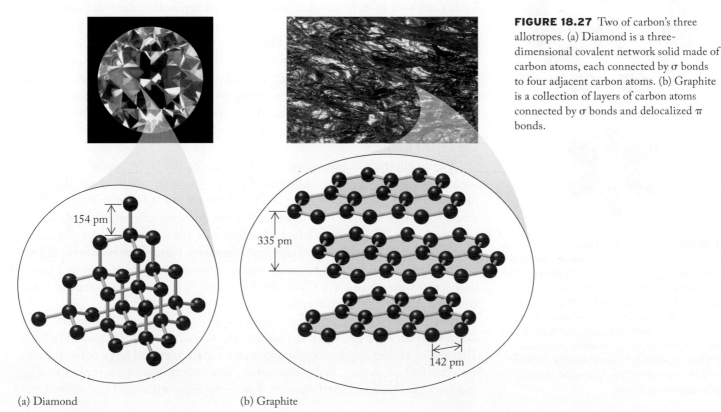

FIGURE 18.27 Two of carbon's three allotropes. (a) Diamond is a three-dimensional covalent network solid made of carbon atoms, each connected by σ bonds to four adjacent carbon atoms. (b) Graphite is a collection of layers of carbon atoms connected by σ bonds and delocalized π bonds.

(a) Diamond (b) Graphite

CONNECTION Allotropes are structurally different forms of the same physical state of an element, as explained in Chapter 4.

CONNECTION We introduced London dispersion forces between molecules in Chapter 6.

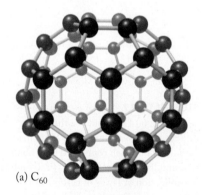

(a) C_{60}

(b) B_{12}

FIGURE 18.28 Some solids are described as clusters, a category between covalent network solids and molecular solids. (a) Fullerenes, the third allotrope of carbon, are clusters. One of them, buckminsterfullerene (C_{60}), is made up of 60 sp^2-hybridized carbon atoms. Both five- and six-membered rings are required to construct the nearly spherical molecule. (b) One crystalline form of boron is a cluster of 12 boron atoms. The clusters are arranged in a close-packed array.

Diamond is one of three allotropes of carbon, the other two being graphite and fullerenes (Figures 18.27b and 18.28a). Diamond is classified as a crystalline **covalent network solid** because it consists of atoms held together in an extended three-dimensional network of covalent bonds. Each carbon atom in diamond forms bonds by overlapping one of its sp^3 orbitals with an sp^3 orbital in each of four neighboring carbon atoms, creating a network of carbon tetrahedra. The atoms in these tetrahedra are connected by localized σ bonds, making diamond a poor electrical conductor. The sigma-bond network is extremely rigid, however, making diamond the hardest natural material known. The atoms of other group 14 elements also form covalent network solids based on the diamond crystal lattice.

Natural diamond forms from graphite under intense heat (>1700 K) and pressure (>50,000 atm) deep in Earth. Industrial diamonds are synthesized at high temperatures and pressures from graphite or any other source rich in carbon, and are used as abrasives and for coating the tips and edges of cutting tools. Diamond has the highest thermal conductivity of any natural substance (five times higher than copper or silver, the most thermally conductive metals), so tools made from diamond do not become overheated. However, industrial diamonds generally lack the size and optical clarity of gemstones, and laboratories also find it difficult to duplicate the naturally occurring impurities that produce rare and valuable colored diamonds.

By far the most abundant allotrope of carbon is graphite, another covalent network solid. Graphite is a principal ingredient in soot and smoke and is used to make pencils, lubricants, and gunpowder. Graphite contains sheets of carbon atoms connected by overlapping sp^2 orbitals to three neighboring carbon atoms in a two-dimensional covalent network of six-membered rings (Figure 18.27b). Each carbon–carbon σ bond is 142 pm, which is shorter than the C—C σ bond in diamond (154 pm). Overlapping unhybridized p orbitals on the carbon atoms form a network of π bonds that are delocalized across the plane defined by the rings. The mobility of these delocalized electrons makes graphite a good conductor of electricity in this plane. However, graphite is a poor conductor in the direction perpendicular to this plane because the layers of fused rings are 335 pm apart (Figure 18.27b). This distance is much too long to be a covalent bonding distance, so the sheets are held together only by London dispersion forces and not by shared electrons. On the other hand, the relatively weak interactions between adjacent sheets allow them to slide past each other, making graphite soft, flexible, and a good lubricant.

A third allotrope of carbon was discovered in the 1980s. Networks of five- and six-atom carbon rings form molecules of 60, 70, or more carbon atoms that look like miniature soccer balls (Figure 18.28a). They are called *fullerenes* because their shape resembles the geodesic domes designed by American architect R. Buckminster Fuller (1895–1983). Many chemists call them *buckyballs* for the same reason. When fullerenes were discovered, they were believed to be a form of carbon rarely found in nature. In recent years, however, analyses of soot and emission spectra from giant stars have disclosed that fullerenes are present in trace amounts throughout the universe.

Based on their size and properties, buckyballs are too small to be classified as covalent network solids but too large to be molecular solids (discussed below). They fall in an ambiguous zone between small molecules and large networks and are classified as *clusters*. Some nonmetals, like boron, have structures like C_{60} and may also be considered clusters. For example, one form of boron contains closest-packed arrays of 12-vertex, 20-sided *icosahedra* (singular *icosahedron*; Figure 18.28b) composed of 12 boron atoms.

Another form of carbon is both the simplest and perhaps the most challenging to produce and study: a single sheet of hexagonally bonded carbon atoms. Material with this structure is called *graphene*. Isolating a single layer of graphitic carbon atoms is not easy, nor is preserving it as a purely two-dimensional material. It tends to curl and buckle, forming three-dimensional structures with less surface area. One of the first successful approaches to isolating a single sheet of graphene was elegant in its simplicity: single layers of C atoms were pulled away from graphite using pieces of adhesive tape. Graphene is both the thinnest material known and the strongest—hundreds of times stronger than steel. It conducts heat better than any other known material and conducts electricity better than any metal. It is also nearly transparent, even though its carbon atoms and bonding electrons are so densely packed that atoms of helium gas cannot pass through it.

Crystalline **molecular solids** consist of molecules held together by intermolecular forces. Ice, CO_2, glucose, and most organic molecules crystallize as molecular solids. One of the two most common allotropes of phosphorus, white phosphorus, is a molecular solid consisting of P_4 tetrahedra arranged in a cubic array (Figure 18.29a). White phosphorus is a waxy material that oxidizes rapidly under standard conditions. It gives off a yellow-green light in a phenomenon called *phosphorescence.*

The other common phosphorus allotrope, red phosphorus, is not a molecular solid but rather a covalent network solid made of chains of P_4 tetrahedra connected by covalent phosphorus–phosphorus bonds (Figure 18.29b). Both red and white phosphorus melt to give the same liquid consisting of symmetrical P_4 tetrahedral molecules.

Sulfur has more allotropic forms than any other element. Most are molecular solids. This variety of forms arises because sulfur atoms form cyclic (ring) compounds of different sizes, which means that different crystalline arrangements of the molecules are possible. The most common allotropes of sulfur consist of

covalent network solid a solid consisting of atoms held together by extended arrays of covalent bonds.

molecular solid a solid formed by neutral, covalently bonded molecules held together by intermolecular attractive forces.

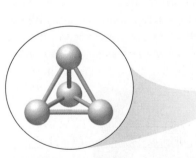

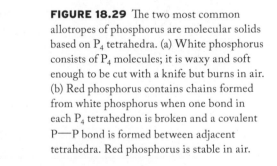

(a) White phosphorus

FIGURE 18.29 The two most common allotropes of phosphorus are molecular solids based on P_4 tetrahedra. (a) White phosphorus consists of P_4 molecules; it is waxy and soft enough to be cut with a knife but burns in air. (b) Red phosphorus contains chains formed from white phosphorus when one bond in each P_4 tetrahedron is broken and a covalent P—P bond is formed between adjacent tetrahedra. Red phosphorus is stable in air.

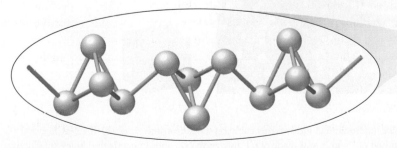

(b) Red phosphorus

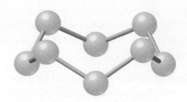

FIGURE 18.30 One form of sulfur is a molecular solid based on puckered S_8 rings.

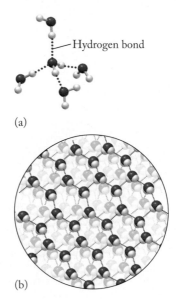

(a)

(b)

FIGURE 18.31 The crystal lattice of ice.

puckered rings (Figure 18.30) containing eight covalently bonded sulfur atoms. London dispersion forces hold one ring to another in solid sulfur in staggered stacking patterns. The weakness of these interactions is the reason elemental sulfur is soft and melts at only 115.21°C.

> **CONCEPT TEST** ··
>
> The crystal structure of ice is shown in Figure 18.31. Is ice best described as a molecular solid or a network solid? You may wish to refer to Section 6.5.
>
> ···

18.6 Salt Crystals: Ionic Solids

Most of Earth's crust is composed of **ionic solids** consisting of monatomic or polyatomic ions held together by ionic bonds. Most of these solids are crystalline. The simplest crystal structures are those of binary salts, such as NaCl (Figure 18.32a). The cubic shape of large NaCl crystals is a reflection of the cubic shape of the NaCl unit cell.

The unit cell of NaCl (Figure 18.32b) is a face-centered cubic arrangement of Cl^- ions at the corners and in the center of each face with the smaller Na^+ ions occupying the 12 octahedral holes along the edges of the unit cell and the single octahedral hole in the middle of the cell. The Na^+ ions fit into the octahedral holes because the radius ratio of Na^+ to Cl^- is

$$\frac{r_+}{r_-} = \frac{102 \text{ pm}}{181 \text{ pm}} = 0.564$$

This ratio is too large for Na^+ to occupy a tetrahedral hole (see Table 18.4) but well within the range for occupying an octahedral hole.

Let's take an inventory of the ions in a unit cell of NaCl. Like the metal atoms in the fcc unit cell in Figure 18.16(b), an isolated unit cell contains portions of fourteen Cl^- ions: one at each corner and one in each of the six faces of the cube. As in the analysis of Figure 18.16, accounting for partial contributions gives us

ionic solid a solid consisting of monatomic or polyatomic ions held together by ionic bonds.

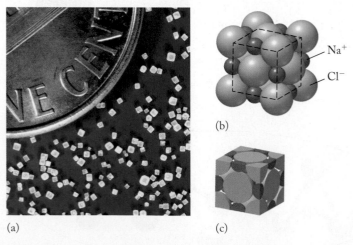

(a)

(b)

Na⁺

Cl⁻

(c)

FIGURE 18.32 (a) Sodium chloride forms cubic crystals of various sizes. (b) The NaCl crystal lattice is composed of fcc unit cells made of Cl^- ions with Na^+ ions in octahedral holes. (c) Cutaway view of a unit cell.

a total of four Cl^- ions in the unit cell. To count the Na^+ ions we note that one Na^+ ion occupies the central octahedral hole, and one Na^+ ion fits into each of the twelve octahedral holes along the edges of the cell. Because each Na^+ ion along an edge is shared by four unit cells, only one-fourth of each Na^+ ion on an edge is in each cell (see Figure 18.17d). Only the Na^+ in the center belongs completely to the unit cell. Therefore, the total number of Na^+ ions in the unit cell is

$$\left(12 \times \tfrac{1}{4}\right) + 1 = 4\ Na^+ \text{ ions}$$

The ratio of Na^+ to Cl^- ions in the unit cell is therefore 4:4, consistent with the chemical formula NaCl. Because the four Na^+ ions occupy all the octahedral holes in the unit cell, this calculation also tells us that each fcc unit cell contains the equivalent of four octahedral holes.

Note in Figure 18.32 that adjacent Cl^- ions along any face diagonal do not touch each other the way adjacent face-diagonal metal atoms do in Figure 18.16(b), because the Cl^- ions have to spread out a little to accommodate the Na^+ ions in the octahedral holes. Sodium ions and chloride ions touch along each edge of the unit cell, however, which means that each Na^+ ion touches six Cl^- ions and each Cl^- ion touches six Na^+ ions. This arrangement of positive and negative ions is common enough among binary ionic compounds to be assigned its own name: the *rock salt structure*.

In other binary ionic solids, the smaller ion is small enough to fit into the tetrahedral holes formed by the larger ions. For example, in the unit cell of the mineral sphalerite (zinc sulfide) the S^{2-} anions (ionic radius 184 pm) are arranged in an fcc unit cell (Figure 18.33), and half of the eight tetrahedral holes inside the cell are occupied by Zn^{2+} cations (74 pm). Therefore the unit cell contains four Zn^{2+} ions that balance the charges on the four S^{2-} ions. This pattern of half-filled tetrahedral holes in an fcc unit cell is called the *sphalerite structure*.

The crystal structure of the mineral fluorite (CaF_2) is based on an fcc unit cell of smaller Ca^{2+} ions at the cube's eight corners and six face centers, with all eight tetrahedral holes filled by larger F^- ions. Because there are a total of four Ca^{2+} ions in the unit cell and the eight F^- ions are all completely inside the cell, this arrangement satisfies the 1:2 mole ratio of Ca^{2+} ions to F^- ions. This structure is so common that it too has its own name: the *fluorite structure* (Figure 18.34). Other compounds having this structure are SrF_2, $BaCl_2$, and PbF_2.

Some compounds in which the cation-to-anion mole ratio is 2:1 have an *antifluorite structure*. In the crystal lattices of these compounds, which include Li_2O and K_2S, the smaller cations occupy the tetrahedral holes in an fcc unit cell formed by cubic closest-packing of the larger anions.

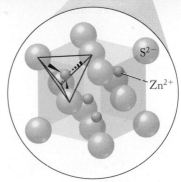

FIGURE 18.33 Many crystals of the mineral sphalerite (ZnS), like the largest one in this photograph, have a tetrahedral shape. The crystal lattice of sphalerite is based on an fcc unit cell of S^{2-} ions with Zn^{2+} ions in four of the eight tetrahedral holes. In the expanded view of the sphalerite unit cell each Zn^{2+} ion is in a tetrahedral hole such as the one outlined in red that is formed by one corner S^{2-} ion and three face-centered S^{2-} ions.

∞ **CONNECTION** Periodic trends in atomic radii were discussed in Chapter 3.

FIGURE 18.34 The mineral fluorite (CaF_2) forms cubic crystals. The crystal lattice of CaF_2 is based on an fcc array of Ca^{2+} ions, with F^- ions occupying all eight tetrahedral holes. Because they are bigger than Ca^{2+} ions, the F^- ions do not fit in the tetrahedral holes of a cubic closest-packed array of Ca^{2+} ions. Instead, the Ca^{2+} ions, while maintaining an fcc unit cell arrangement, spread out to accommodate the larger F^- ions. Note how adjacent Ca^{2+} ions along any face diagonal do not touch each other the way they do in the ideal fcc unit cell in Figure 18.16(b).

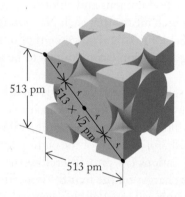

FIGURE 18.35

The unit cell of lithium chloride (LiCl) contains an fcc arrangement of Cl^- ions (Figure 18.35). In LiCl the Li^+ cations (radius 76 pm) are small enough to allow adjacent Cl^- ions to touch along any face diagonal.

   a. If the edge length of the LiCl fcc cell is 513 pm, what is the radius of the Cl^- ion?

   b. Use that value to predict the type of hole the Li^+ ion occupies.

COLLECT AND ORGANIZE We are given the edge length of a LiCl unit cell, the radius of the Li^+ cation, and a picture of the unit cell. The unit cell is an fcc array of Cl^- ions that touch along the face diagonal.

ANALYZE The face diagonal of an fcc unit cell is equal to four times the radius of the spheres (chloride ions) that make up the cell, and it's also equal to $\sqrt{2}$ times the cell length (513 pm):

The radius ratio of Li^+ to Cl^- ions and Table 18.4 allow us to predict which type of hole Li^+ occupies. We expect Li^+ ions to be smaller than Cl^- ions, but they may not be small enough to fit in tetrahedral holes.

SOLVE

   a. Substituting the edge length into Equation 18.2, we calculate the radius of Cl^-:

$$r = (513 \times \sqrt{2})/4 \text{ pm} = 181 \text{ pm}$$

   b. The ratio of the radius of a Li^+ ion to the radius of a Cl^- ion is

$$76 \text{ pm}/181 \text{ pm} = 0.42$$

According to Table 18.4, the Li^+ cations occupy octahedral holes in the lattice formed by the larger Cl^- anions.

THINK ABOUT IT The average ionic radius value in Figure 3.35 for Cl^- ions is 181 pm, so our calculation is correct. Figure 18.35 shows the structure of LiCl to be similar to the rock salt structure of NaCl in Figure 18.32(b).

Practice Exercise Assuming the Cl^- radius in NaCl is also 181 pm, what is the radius of the Na^+ ion in NaCl if the edge length of the NaCl unit cell is 564 pm but the anions do not touch along any face diagonal? ⚙

SAMPLE EXERCISE 18.5 **Calculating the Density of a Salt** **LO3**
from Its Unit Cell Dimensions

What is the density of LiCl at 25°C if the edge length of its fcc unit cell is 513 pm?

COLLECT AND ORGANIZE We are given the edge length of an fcc unit cell of LiCl and are asked to calculate its density. Density is the ratio of mass to volume, and the volume of a cubic cell is the cube of its edge length: $V = \ell^3$.

ANALYZE As in Sample Exercise 18.2, we assume that the density of the unit cell is the same as the density of the crystalline solid. The fact that LiCl has an fcc unit cell (like NaCl) means that there are four Cl ions and four Li$^+$ ions in the cell. The density of the unit cell is the sum of the masses of these eight ions divided by the volume of the cell, (513 pm)^3. As in Sample Exercise 18.2, we need to divide the molar masses of Li and Cl by Avogadro's number to obtain the masses of individual atoms (or monatomic ions) of Li and Cl in grams.

SOLVE Calculating the mass of four Cl$^-$ ions,

$$m = \frac{35.45 \text{ g Cl}^-}{1 \text{ mol Cl}^-} \times \frac{1 \text{ mol Cl}^-}{6.0221 \times 10^{23} \text{ ions Cl}^-} \times 4 \text{ ions Cl}^- = 2.355 \times 10^{-22} \text{ g Cl}^-$$

and the mass of four Li$^+$ ions,

$$m = \frac{6.941 \text{ g Li}^+}{1 \text{ mol Li}^\pm} \times \frac{1 \text{ mol Li}^\pm}{6.0221 \times 10^{23} \text{ ions Li}^\pm} \times 4 \text{ ions Li}^\pm = 0.4610 \times 10^{-22} \text{ g Li}^+$$

Combining the masses of the two kinds of ions in the unit cell,

$$2.355 \times 10^{-22} \text{ g} + 0.4610 \times 10^{-22} \text{ g} = 2.816 \times 10^{-22} \text{ g}$$

The volume of the cell in cubic centimeters is

$$V = \ell^3 = (513 \text{ pm})^3 \times \frac{(10^{-10} \text{ cm})^3}{(1 \text{ pm})^3} = 1.350 \times 10^{-22} \text{ cm}^3$$

Taking the ratio of mass to volume, we have

$$d = \frac{m}{V} = \frac{2.816 \times 10^{-22} \text{ g}}{1.350 \times 10^{-22} \text{ cm}^3} = 2.09 \text{ g/cm}^3$$

THINK ABOUT IT The result is reasonable in that most minerals are more dense than water but less dense than common metals.

Practice Exercise What is the density of NaCl at 25°C if the edge length of its fcc unit cell is 564 pm? ⚙

18.7 Ceramics: Insulators to Superconductors

A **ceramic** is a solid inorganic compound or mixture of compounds that has been heated to transform it into a harder and more heat-resistant material. The use of ceramic materials preceded metal technology by many thousands of years. The first ceramics were probably made of *clay*, the fine-grained soil produced by the physical and chemical weathering of igneous rocks (rocks of volcanic origin).

ceramic a solid inorganic compound or mixture that has been transformed into a harder, more heat resistant material by heating.

Moist clay is easily molded into a desired shape and then hardened over fires or in wood-burning kilns, an ancient process still in use today.

In this section we examine some of the physical properties of both primitive earthenware (ceramics fired at low temperatures) and modern ceramic materials (typically fired at high temperatures) and relate those properties to the chemical composition of these materials and to the chemical changes that occur when they are heated.

Polymorphs of Silica

One of the most abundant families of minerals found in igneous rocks has the chemical composition SiO_2. The correct chemical name is silicon dioxide, but the more common name is *silica*. Silica is a covalent network solid in which each silicon atom is covalently bonded to four oxygen atoms, forming a tetrahedron with an oxygen atom at each corner and the silicon atom at the center (Figure 18.36). Each corner oxygen atom is covalently bonded to two silicon atoms, thereby linking the tetrahedra into an extended three-dimensional network. Because each oxygen atom is bonded to two silicon atoms, each silicon atom gets only half "ownership" of the four oxygen atoms to which it is bonded—hence the formula SiO_2.

At least eight different minerals have the empirical formula SiO_2. The members of a family of substances with the same empirical formula but different crystal structures and properties are called *polymorphs*. The most abundant silica polymorph is quartz, a type of SiO_2 that can form impressively large, nearly transparent crystals (Figure 18.36). Note how the hexagonal ordering of the SiO_2 tetrahedra translates into hexagonal crystals.

Most, but not all, silica polymorphs are crystalline. When lava containing molten SiO_2 flows from a volcano into a sea or lake, it cools so quickly that the Si and O atoms may not have enough time to achieve an ordered crystal lattice as the lava solidifies. The solid formed in this way is an *amorphous* (disordered, non-crystalline) polymorph of silica known as either volcanic glass or obsidian (Figure 18.37). *Glass* is a term scientists and engineers use to describe any solid that has either no crystalline structure or only very tiny crystals surrounded by disordered arrays of atoms. This definition applies to laboratory glassware and the drinking glasses we use at home, as well as to more esoteric solids such as obsidian.

Ionic Silicates

In addition to covalent silica, igneous rocks also contain ionic minerals made of silicon and oxygen. These minerals have some of the tetrahedral crystal structure of silica, but not all the oxygen corner atoms are bonded to two Si atoms. Instead, some of the O atoms have an extra electron. The result is a *silicate* anion. One of the most common ionic silicates is chrysotile, a type of asbestos that consists of sheets of linked silicon–oxygen tetrahedra that form hexagonal clusters of six tetrahedra each (Figure 18.38). Each tetrahedron has three O atoms that it shares with other tetrahedra and one O atom—the one with the extra electron—that it does not share. Thus the basic tetrahedral unit consists of one Si atom and $[1 + 3(\frac{1}{2})] = 2.5$ oxygen atoms as well as a negative charge. This gives the sheet the empirical formula $SiO_{2.5}{}^-$. We generally use whole-number subscripts in chemical formulas when possible. In this case, multiplying $SiO_{2.5}{}^-$ by 2 gives

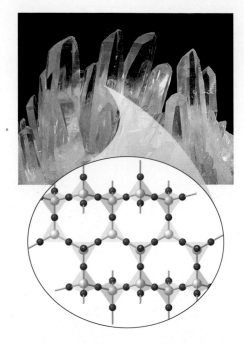

FIGURE 18.36 Quartz crystals are hexagonal. Their crystal structure features hexagonal arrays of silicon–oxygen tetrahedra that form an extended three-dimensional network. (All atoms are drawn undersized relative to the volume they actually occupy to make the structure easier to see.)

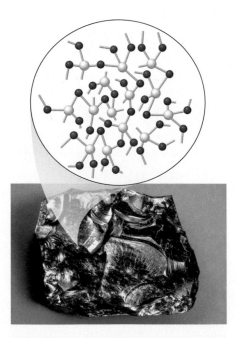

FIGURE 18.37 Obsidian (volcanic glass) is an unusual form of silica in that it is not crystalline. Obsidian contains mostly amorphous silica with random arrangements of silicon and oxygen atoms.

us the empirical formula $Si_2O_5^{2-}$ for this silicate anion. The subscript n in the formula in Figure 18.38 indicates that there are many empirical formula units in a single crystal.

Silicate minerals are neutral materials, so they must contain cations to balance the negative charges on the $Si_2O_5^{2-}$ layers. When this cation is Al^{3+}, the minerals are called *aluminosilicates*. One of the most common aluminosilicates is the clay mineral kaolinite (Figure 18.39). At least a little kaolinite is found in practically every soil, but rich deposits of nearly pure, brilliantly white kaolinite are found in highly weathered soils. For centuries these deposits have been mined for the clay used to make fine china and white porcelain. Today the greatest demand for kaolinite is in the production of glossy white paper, which is used in most magazines and books (including this one).

Common metal ions found in igneous rocks—including Na^+, K^+, Ca^{2+}, Mg^{2+}, and Fe^{3+}—are largely absent in kaolinite. Their absence indicates that kaolinite deposits form under acidic weathering conditions, during which H^+ ions displace other cations from ion-exchange sites. For example, $-O^-Na^+$ sites exchange H^+ for Na^+, leaving behind $-OH$ groups such as those shown in Figure 18.39.

The strong ionic interactions and hydrogen bonds in kaolinite make it hard to separate its layers, and water molecules cannot penetrate between them. For thousands of years this property has made kaolinite pots handy vessels for carrying water. For the same reason, kaolinite does not expand when water is added, nor does it shrink as much as most other clays when dehydrated at high temperatures, which makes it a desirable starting material for ceramics. Finally, moist kaolinite is *plastic*, meaning that it can be molded into a shape, and it keeps that shape during heating and cooling.

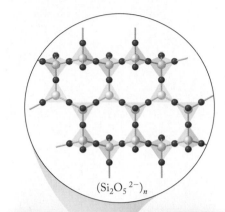

CONNECTION Ion-exchange reactions were discussed in Chapter 8.

$(Si_2O_5{}^{2-})_n$

FIGURE 18.38 Chrysotile, one of the two principal forms of asbestos formerly used in building construction as thermal insulation, is an ionic silicate compound. The ease with which thin fibers of chrysotile can flake off is related to its layered crystal lattice and the relatively weak intermolecular interactions between layers. The O atoms that are bonded to only one Si atom have an extra electron and a negative charge.

FIGURE 18.39 An edge-on view of the structure of kaolinite shows a top layer of OH^- ions bonded to a middle layer of Al^{3+} ions followed by a bottom layer of silicate ions ($Si_2O_5^{2-}$). The structure is repeated in subsequent layers, and the empirical formula of this crystal lattice is $Al_2(Si_2O_5)(OH)_4$. This enormous kaolinite mine in Bulgaria contributes to a worldwide production of about 40 trillion metric tons of the mineral per year.

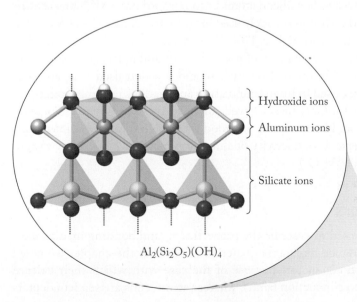

Hydroxide ions

Aluminum ions

Silicate ions

$Al_2(Si_2O_5)(OH)_4$

Magnesium ion, Mg^{2+}, can substitute for Al^{3+} in kaolinite. What is the formula of the mineral obtained—that is, what is the value of x (and y) in $Mg_xAl_y(Si_2O_5)(OH)_4$ if $x = y$?

From Clay to Ceramic

Creating ceramic objects from kaolinite and other clays takes several steps. First, moist clay is formed into pots, bricks, and other objects on a potter's wheel or in molds or presses. Drying at just above 100°C removes much of the water that made the clay plastic. Further heating to about 450°C removes water that was adsorbed onto the surfaces of the clay particles or between the layers of non-kaolinite clays. Above 450°C, kaolinite begins to decompose, releasing molecules of H_2O as some of the aluminum ions that were bonded to octahedra of six oxide ions in kaolinite (Figure 18.39) end up bonded to tetrahedra of only four oxide ions in a substance called metakaolin ($Al_2Si_2O_7$) that has an amorphous structure. The overall decomposition reaction, which is complete at temperatures near 900°C, is described by the equation

$$Al_2Si_2O_5(OH)_4(s) \xrightarrow{450-900°C} Al_2Si_2O_7(s) + 2\,H_2O(g)$$

The next compositional change occurs just below 1000°C when $Al_2Si_2O_7$ decomposes into a mixture of yet another disordered aluminosilicate, $Al_4Si_3O_{12}$, and $SiO_2(s)$:

$$2\,Al_2Si_2O_7(s) \xrightarrow{\sim950°C} Al_4Si_3O_{12}(s) + SiO_2(s)$$

At even higher temperatures, $Al_4Si_3O_{12}$ continues forming mixtures of compounds with increasing aluminum content and SiO_2:

$$Al_4Si_3O_{12}(s) \rightarrow Al_4Si_2O_{10}(s) + SiO_2(s)$$

$Al_4Si_2O_{10}$ is a member of a family of ceramic materials called mullite. These materials are actually solid solutions containing different proportions of aluminum and silicon oxides. At the atomic level their structures are a mixture of layers that contain octahedra of six-coordinate Al^{3+} ions each bonded to six oxide ions, or tetrahedra with either silicon atoms or four-coordinate Al^{3+} ions at their centers (Figure 18.40). Mullite is widely used in the manufacture of products that must tolerate temperatures as high as 1700°C: furnaces, boilers, ladles, and kilns. These products are used as containers of molten metals and in the glass, chemical, and cement industries. Mullite is also very hard and is widely used as an abrasive.

Other ceramics are used in high-temperature applications ranging from cookware to fireplace bricks. Ceramics are well suited to these uses because of their high melting points and because they are good thermal and electrical insulators. For example, the thermal conductivity of aluminum metal at 100°C is over eight times that of alumina (Al_2O_3).

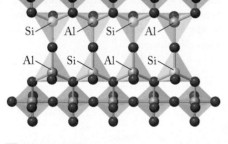

▨ SiO_4 and AlO_4 tetrahedra

▨ AlO_6 octahedra

FIGURE 18.40 One of many possible crystal structures of a family of ceramic materials called mullite.

Superconductors

Band theory can be used to describe the properties of, and bonding in, all solids, including the compounds in ceramic materials. Earlier in the chapter we noted that metals are good conductors because of the ease with which their valence electrons can move to conduction bands, whereas semimetals are semiconductors because they have significant energy gaps between their valence and conduction bands (although doping can shrink the gaps). Similarly, the insulating properties of ceramics can be explained by the large energy gaps between their valence and conduction bands. Because of the large gaps, their valence electrons have very limited mobility.

In conventional metallic conductors, the vibration of the metal atoms in the crystal lattice can interfere with the flow of free electrons: the higher the temperature, the greater the atomic vibration and the greater the resistance to electron flow. However, at temperatures approaching absolute zero, these vibrations become very small. In the early 20th century, scientists discovered that the lattice vibrations in mercury and some metal alloys are so small at temperatures below about 20 K, called the **critical temperature (T_c)** for each substance, that electrons can pass freely through these materials and they become **superconductors**. Today superconducting alloys, such as Nb_3Sn, are widely used in devices that require very high electrical currents, such as the electromagnets of magnetic resonance imaging (MRI) instruments.

Unfortunately, it is difficult and expensive to chill materials to 20 K. In 1986 it was discovered that $YBa_2Cu_3O_7$ and several other ceramic materials become superconductors when cooled to temperatures just above the boiling point of liquid nitrogen (77 K). This represented an important economical advance because it is much easier and cheaper to cool a material with liquid nitrogen than with liquid helium (boiling point = 4 K), which must be used to achieve superconductivity in Hg and metal alloys. Scientists have since produced superconducting ceramic materials with critical temperatures as high as 135 K.

Let's take a closer look at the crystal structure of $YBa_2Cu_3O_7$. The unit cell (Figure 18.41) is a stack of three cubes with a Y^{3+} ion in the center of the middle cube, Ba^{2+} ions in the center of the top and bottom cubes, sixteen Cu^{2+} ions at the corners of each cube, and twenty O^{2-} ions along the edges of the three stacked cubes.

Applying our usual practice of assigning fractions of atoms and ions to unit cells, we find that the stack has 7 oxide ions ($\frac{1}{4}$ of the 12 that occupy edges of the top and bottom cubes in the stacked array plus $\frac{1}{2}$ of the remaining 8 ions that occupy faces—7 in all) and 3 copper ions ($\frac{1}{8}$ of the 8 ions on the top and bottom surfaces plus $\frac{1}{4}$ of the 8 ions around the middle—3 in all). The two Ba^{2+} ions and the Y^{3+} ion lie entirely within the unit cell. Thus, the material with the crystal structure shown in Figure 18.41 has the chemical formula $YBa_2Cu_3O_7$.

Yttrium–barium–copper (YBC) oxides and related materials are superconductors because of the formation of electron pairs called *Cooper pairs* after American physicist Leon Cooper (1930–). In an undisturbed crystal lattice the ions are spaced uniformly, as shown in Figure 18.42(a). It is possible to add one or more electrons to the lattice from an external power supply. When an electron is introduced into the lattice of a YBC or related oxide (Figure 18.42b), the attractive electrostatic force between the electron and the cations in the oxide causes cations near the electron to move slightly away from their original positions as they are drawn toward the electron. Note that the central cluster in Figure 18.42(b) has one electron surrounded by four cations, which means the overall charge in that region is 3+. When a second electron is added, it is attracted to this

critical temperature (T_c) the temperature below which a material becomes a superconductor.

superconductor a material that has zero resistance to the flow of electric current.

▶‖ **CHEMTOUR** Superconductors

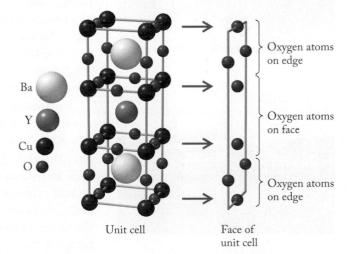

FIGURE 18.41 The unit cell of $YBa_2Cu_3O_7$, a high-temperature superconductor, contains a central barium ion in the top and bottom cubes, a central yttrium ion in the middle cube, and oxide ions on the edges of the unit cell.

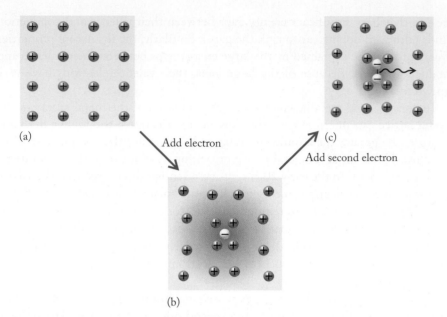

FIGURE 18.42 (a) In an undisturbed crystal lattice of an ionic solid, the cations are spaced uniformly. (b) An electron inserted into the lattice distorts the uniform spacing of the cations because the cations are attracted to the negative electrical charge of the electron. (c) A second electron in the lattice is attracted to the area of localized positive charge. It and the first electron create a Cooper pair.

FIGURE 18.43 (a) The Meissner effect: A cylindrical magnet floats above a superconductor chilled to below its critical temperature. (b) The Shanghai Maglev Train: The fastest passenger train in the world.

3+ region and comes into close proximity to the first electron, resulting in the formation of a Cooper pair of electrons (Figure 18.42c). The movement of Cooper pairs rather than individual electrons through the material accounts for its superconductivity. Because the superconductor in Figure 18.42 is part of a circuit connected to a source of electricity, additional Cooper pairs form and migrate through the solid.

To understand the superconductivity of a Cooper pair, consider the difference between two single horses and a pair of horses harnessed together running through a field of boulders. The harnessed horses must travel forward together, never separating from each other, and they reach the far end of the field together. Single animals, however, can choose completely different paths across the same field; they can be far apart from each other and scatter over a wide area, and they can exit the field at different locations. Similarly, single electrons are easily deflected in different directions by the atoms or ions in a solid, but electrons scatter much less when harnessed in a Cooper pair. Less scattering lowers electrical resistance.

Superconductors are of technological interest because, in principle, large amounts of electric charge can flow through them with no resistance. This property is attractive for the design of extremely fast computers. Superconductors also have the ability to exclude magnetic fields—a property known as the *Meissner effect* after German physicist Walther Meissner (1882–1974). In Figure 18.43(a) a small cylindrical magnet floats above a superconducting material that has been cooled to below its critical temperature. The magnetic lines of force surrounding the magnet cannot penetrate the superconductor, resulting in so much repulsion between them that the magnet floats above the superconductor.

This phenomenon, also called magnetic levitation, or maglev, has been used to build trains with superconducting undercarriages that float above electromagnetic tracks and guidance systems. The first high-speed maglev train went into commercial service near Shanghai, China in 2004 (Figure 18.43b). Its operational top speed is 431 km/hr (268 mph), making it the fastest passenger train in the world.

18.8 X-ray Diffraction: How We Know Crystal Structures

X-ray diffraction (XRD) a technique for determining the arrangement of atoms or ions in a crystal by analyzing the pattern that results when X-rays are scattered after bombarding the crystal.

Unit cell dimensions are determined using **X-ray diffraction (XRD)**. X-rays are well suited for the task of crystal structure determination because the wavelengths of X-rays (10^2 to 10^3 pm) are similar to the radii of atoms and ions and, therefore, the distances between them in crystals. To see how XRD works, we will look at atoms in a crystalline metal, but keep in mind that our description also applies to the ions in any crystalline ionic solid.

A narrow beam of X-rays is directed at some angle of incidence θ at the surface of the metal, as shown in Figure 18.44(a). Some of the X-rays collide with atoms in the surface layer, and some of those that do so are reradiated by atoms at an angle equal to the angle of incidence θ. For these X-rays the total change in their direction is the sum of the two angles, or 2θ, which is called their *angle of diffraction*.

Some of the incident X-rays pass through the surface layer and hit atoms in a second layer a distance d below the surface. Some of these X-rays are also diffracted through an angle 2θ. The two sets of X-rays can interfere with each other either constructively or destructively. They interfere constructively when they are in phase, as in Figure 18.44(b). To be in phase, the extra distance traveled by the X-ray diffracting off the second layer must be some whole-number multiple of the wavelength of the X-rays. This extra distance traveled is the sum of the lengths of line segments $\overline{XY}$ and $\overline{YZ}$ in Figure 18.44(b). The two right triangles

▶II **CHEMTOUR** X-ray Diffraction

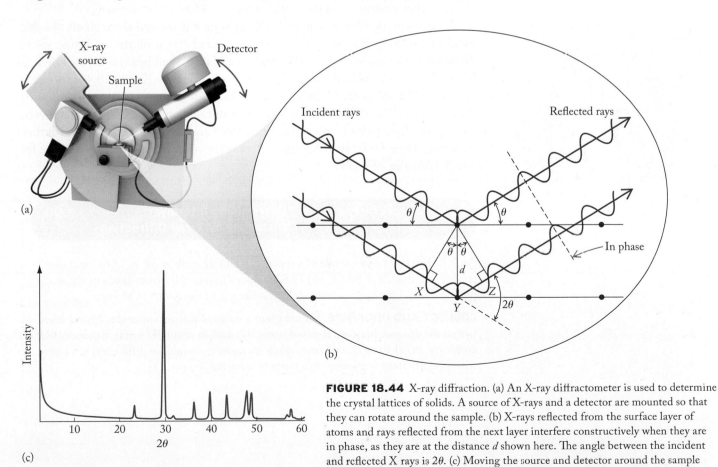

(a)

(b)

(c)

FIGURE 18.44 X-ray diffraction. (a) An X-ray diffractometer is used to determine the crystal lattices of solids. A source of X-rays and a detector are mounted so that they can rotate around the sample. (b) X-rays reflected from the surface layer of atoms and rays reflected from the next layer interfere constructively when they are in phase, as they are at the distance d shown here. The angle between the incident and reflected X rays is 2θ. (c) Moving the source and detector around the sample produces a scan such as this one for quartz. The peaks at different values of 2θ represent reflections from different planes of atoms in the solid.

Bragg equation relates the angle of diffraction (2θ) of X-rays to the spacing (d) between the layers of ions or atoms in a crystal: $n\lambda = 2d \sin \theta$.

incorporating these line segments share a hypotenuse d. Geometry tells us that the angles opposite $\overline{XY}$ and $\overline{YZ}$ are both equal to θ. According to trigonometry, the ratio of either $\overline{XY}$ or $\overline{YZ}$ to d is

$$\frac{\overline{XY}}{d} = \sin \theta \qquad \frac{\overline{YZ}}{d} = \sin \theta$$

which means

$$d \sin \theta = \overline{XY} \qquad d \sin \theta = \overline{YZ} \qquad (18.8)$$

As noted above, we are interested in $\overline{XY} + \overline{YZ}$, the extra distance traveled by the second ray. We therefore add Equations 18.8 to get

$$\overline{XY} + \overline{YZ} = d \sin \theta + d \sin \theta = 2d \sin \theta$$

When this extra distance $\overline{XY} + \overline{YZ}$ equals a whole-number multiple (n) of the wavelength (λ) of the X-rays, we have

$$\overline{XY} + \overline{YZ} = n\lambda$$

or

$$n\lambda = 2d \sin \theta \qquad (18.9)$$

Equation 18.9 is called the **Bragg equation** after William Henry Bragg (1862–1942) and his son William Lawrence Bragg (1890–1971), who discovered the relationship between wavelength and the spacings between layers in crystalline solids. Whenever $2d \sin \theta$ equals $n\lambda$, the crests and troughs of the two X-rays are in phase as they emerge from the metal and so interfere constructively.

To measure this interference, the X-ray source is rotated through all possible incident angles θ, and the intensity of the reflected rays is plotted as a function of 2θ (because the angle between the incident and diffracted beam is 2θ, as shown in Figure 18.44). Peaks in the intensity of scattered X-rays (Figure 18.44c) occur at angles of diffraction that satisfy Equation 18.9. From these angles, and knowing the wavelength (λ) of the X-rays, scientists can calculate the distance (d) between the layers of particles in a crystalline sample and determine its type of lattice structure. In most X-ray diffraction patterns the most intense peaks are those for which the value of n is 1.

SAMPLE EXERCISE 18.6 **Determining Interlayer** **LO5**
 Distances by X-ray Diffraction

An XRD analysis of a sample of copper has a major peak at $2\theta = 24.64°$ and much smaller peaks at $2\theta = 50.54°$ and $79.62°$. What distance d between layers of Cu atoms produces this diffraction pattern if the wavelength of the X-rays is 154 pm?

COLLECT AND ORGANIZE We are given a series of diffraction angles 2θ and asked to find the distance between layers of atoms in a sample of copper metal. Equation 18.9 enables us to calculate this distance when we know the angles of diffraction associated with constructive interference in a beam of reflected X-rays.

ANALYZE We must rearrange Equation 18.9 to solve for the parameter we are after: d.

$$d = \frac{n\lambda}{2 \sin \theta}$$

We are given the value of λ and can get values of θ from the given values of 2θ. The radius of a copper atom is on the order of 10^2 pm, so the distance between the layers should also be tens or hundreds of picometers.

SOLVE The key to determining n is to look for a pattern in the θ values. Notice that the higher values of θ are approximately two and three times the lowest value:

$$24.64°/2 = 12.32° \qquad 50.54°/2 = 25.27° \qquad 79.62°/2 = 39.81°$$

$$\frac{25.27°}{12.32°} = 2.051 \qquad \frac{39.81°}{12.32°} = 3.231$$

This pattern and the intensities of the peaks suggest that the values of n for this set of data are 1, 2, and 3, so let's use these combinations of n and θ to see whether they all give the same value of d:

$$d = \frac{(1)(154 \text{ pm})}{2 \sin 12.32°} = \frac{154 \text{ pm}}{(2)(0.2134)} = 361 \text{ pm}$$

$$d = \frac{(2)(154 \text{ pm})}{2 \sin 25.27°} = \frac{308 \text{ pm}}{(2)(0.4269)} = 361 \text{ pm}$$

$$d = \frac{(3)(154 \text{ pm})}{2 \sin 39.81°} = \frac{462 \text{ pm}}{(2)(0.6402)} = 361 \text{ pm}$$

We do indeed get the same value of d, so $d = 361$ pm is the distance between Cu atoms that produced the three peaks.

THINK ABOUT IT These consistent results mean that our assumption about the values of n for the three values of θ was correct. Also, the value of d is in the range of the edge lengths of unit cells we used in several Sample Exercises earlier in the chapter and is therefore reasonable.

Practice Exercise An X-ray diffraction analysis of crystalline CsCl using X-rays of wavelength 142.4 pm has a prominent peak at $2\theta = 19.9°$. What is the spacing between ion layers that produced the peak? ⚙

As we have seen throughout this text, the arrangement of atoms, ions, and molecules in a sample of matter is responsible for many of the properties of materials. XRD is a powerful tool for analyzing the atomic-level structure of a wide range of materials, including metals, minerals, polymers, plastics, ceramics, pharmaceuticals, and semiconductors. The technique is indispensable for scientific research and industrial production. The link between structure and properties is robust, and understanding the ordered arrangements of atoms and ions that characterize pure metals, alloys, semiconductors, ionic compounds, and ceramics enables us to design stronger, tougher, harder, and more heat-resistant solid materials.

SAMPLE EXERCISE 18.7 Integrating Concepts: Imperfect Crystals and Film Photography

Most of the crystals we have discussed and illustrated in this chapter have been perfect, in the sense that all sites in the crystal structure that should be occupied by an atom or ion *are* occupied. However, many crystals we find in the real world are imperfect, and several types of defects have been recognized as important. Two scientists, Walter Schottky (1886–1976) and Yakov Frenkel (1894–1952), studied the defects known as *point defects* because they occur at specific points (locations) within a crystal structure. A Schottky defect forms when positively and negatively charged ions both leave their normal sites in the structure, creating vacancies at those sites. A Frenkel defect forms when an atom or ion leaves its location in the structure and occupies a site not usually occupied by a particle. Crystals of silver bromide (AgBr), which has the rock salt structure like NaCl, may display both kinds of defects.

Silver bromide is used in black-and-white film photography. When light falls on a grain of AgBr embedded in film, a bromide ion loses an electron, which combines with a silver ion in a Frenkel defect and produces a silver atom. Later the film is bathed in a developer that reduces more silver ions to intensify the image. After the image is developed, the film is *fixed*, which involves dissolving any remaining silver bromide in an aqueous solution (about 0.2 M) of sodium thiosulfate, $Na_2S_2O_3$.

a. If a Schottky defect occurs in a unit cell of silver bromide (Figure 18.45a), does the overall stoichiometry of the unit cell change? If the defect occurs consistently throughout the structure, does the density of the silver bromide change?

b. If a Ag^+ ion that normally occupies an octahedral hole in the cubic closest-packed array of Br^- ions forms a Frenkel defect by moving to a tetrahedral hole, does the overall composition of the unit cell change? If the defect occurs consistently throughout a sample of AgBr, does its formula and its density change?

c. Why is AgBr, which is insoluble in water, soluble in a solution of sodium thiosulfate?

d. Silver ions are recovered from photographic processing solutions by passing them through columns containing steel wool (Fe). The reaction between silver(I) ions and Fe produces silver metal. Write a balanced net ionic equation describing this reaction and suggest why it is spontaneous.

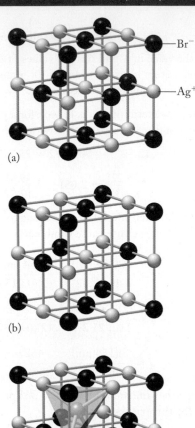

(a)

(b)

(c)

FIGURE 18.45 (a) A perfect crystal structure of AgBr, without defects. (b) One Ag^+ ion and one Br^- ion are missing from the front face, resulting in a Schottky defect. (c) One Ag^+ ion has moved from its normal location at the center to the highlighted tetrahedral hole, resulting in a Frenkel defect.

COLLECT AND ORGANIZE We are asked if Schottky and Frenkel defects in AgBr cause changes in the stoichiometry of the unit cell and the density of the bulk material. We are then asked to explain the solubility of AgBr in a solution that contains thiosulfate ions and the chemistry behind the recovery of silver as the metal by reacting silver(I) compounds with iron metal.

ANALYZE To determine the impact of Schottky and Frenkel defects on AgBr, we start with a unit cell of AgBr (Figure 18.45a) and then incorporate the changes described. By determining the ratio of the ions in defective unit cells and the number of ions per unit cell, we can determine whether there will be changes in the overall formula and density of a sample. To explain the solubility of AgBr in thiosulfate solutions, we need to consider how Ag^+ and Br^- ions interact with thiosulfate ($S_2O_3^{2-}$) ions. According to Table A5.5 in Appendix 5, Ag^+ ions combine with $S_2O_3^{2-}$ ions to produce a complex ion, $Ag(S_2O_3)_2^{3-}$, that has a formation constant of 4.7×10^{13}. The K_{sp} of AgBr (Table A5.4) is 5.4×10^{-13}. Our solution to this problem will need to link these two equilbria. Finally, conversion of silver(I) to Ag metal with iron must be a redox process. Checking the standard reduction potentials in Table A6.1 in Appendix 6, we find the following relevant half-reactions:

$$Ag^+(aq) + e^- \rightarrow Ag(s) \qquad E° = 0.7996 \text{ V}$$

$$Fe^{2+}(aq) + 2\,e^- \rightarrow Fe(s) \qquad E° = -0.447 \text{ V}$$

The positive $E°$ value of the silver half-reaction and the negative $E°$ value of the iron half-reaction indicate that Fe may be a reducing agent strong enough to convert the silver(I) to free silver metal.

SOLVE

a. A Schottky defect in AgBr occurs when a pair of Ag^+ and Br^- ions are lost, perhaps producing a unit cell like the one shown in Figure 18.45(b). Though the composition of this particular unit cell has changed because the missing ions occupied different positions that contributed differently to the total ion count in the cell, the overall ratio of silver ions to bromide ions in the solid should still be 1:1. However, if enough ion pairs are lost without changing the cell dimensions, the density of the solid will decrease.

b. If a silver ion moves from the center (octahedral) site to an unoccupied tetrahedral site (Figure 18.45c), the composition of the cell does not change, so neither the formula of the solid nor its density changes in the presence of Frenkel defects.

c. The K_{sp} of AgBr can be used to calculate the concentration of Ag^+ ions in equilibrium with solid AgBr as a result of the reaction $AgBr(s) \rightleftharpoons Ag^+(aq) + Br^-(aq)$:

$$K_{sp} = 5.4 \times 10^{-13} = [Ag^+][Br^-] = x \cdot x = x^2$$

$$x = [Ag^+] = 7.3 \times 10^{-7} \, M$$

Let's use this concentration to determine whether adding $0.2\ M$ thiosulfate ions will result in the formation of $Ag(S_2O_3)_2^{3-}$ complex ions. Inserting the known concentrations into the formation constant expression and solving for $[Ag(S_2O_3)_2^{3-}]$,

$$K_f = \frac{[Ag(S_2O_3)_2^{3-}]}{[Ag^+][S_2O_3^{2-}]^2} = \frac{x}{(7.3 \times 10^{-7})(0.2)^2} = 4.7 \times 10^{13}$$

$$x = [Ag(S_2O_3)_2^{3-}] = 1.4 \times 10^6 \, M$$

This very large concentration, as compared to $[Ag^+]$, means that any Ag^+ ions that are produced when AgBr dissolves will form complex ions, effectively removing free Ag^+ ions from solution. According to Le Châtelier's principle, removing one of the product ions shifts the K_{sp} equilibrium described above toward making more dissolved ions, which keeps on happening until all of the AgBr has dissolved (assuming there is an adequate supply of thiosulfate ions).

d. The two half-reactions in the Analyze section can be combined to write a net ionic equation describing the reaction between silver(I) and iron if we (1) reverse the iron half-reaction and (2) multiply the silver half-reaction by 2 to balance the loss and gain of electrons:

$$2\,Ag^+(aq) + 2\,e^- \rightarrow 2\,Ag(s) \qquad\qquad E° = 0.7996 \text{ V}$$

$$\underline{Fe(s) \rightarrow Fe^{2+}(aq) + 2\,e^- \qquad\qquad E° = 0.447 \text{ V}}$$

$$2\,Ag^+(aq) + Fe(s) \rightarrow 2\,Ag(s) + Fe^{2+}(aq) \quad E°_{rxn} = 1.247 \text{ V}$$

The positive reaction potential indicates that the reaction is spontaneous *under standard conditions*, that is, when $[Ag^+] = [Fe^{2+}] = 1\ M$. As noted above, the concentration of Ag^+ ions is much less than $1\ M$. However, we learned in Chapter 17 that the potentials of half-reactions change relatively little even as the concentrations of reactants and products change a lot. Therefore, we can confidently predict that the reaction will have a positive reaction potential and be spontaneous.

THINK ABOUT IT Physical and chemical properties of crystalline materials are closely related to their structure, and structural defects contribute to these properties as well. Other ionic substances with large size differences between the anion and the cation may also demonstrate Frenkel defects. AgCl, with a smaller anion (181 pm for Cl^- versus 195 pm for Br^-), is also photoreactive but not to the same extent as AgBr because the Ag^+ ion (126 pm) is too big to fit into tetrahedral holes in an array of Cl^- ions.

SUMMARY ·······································■

Section 18.1 Most metals are malleable and ductile. The electrical conductivity of metals can be explained by **band theory** as the easy movement of electrons from the **valence band** to the **conduction band**.

Section 18.2 Semimetals are **semiconductors**, intermediate in electrical conducting ability between metals and nonmetals. In semiconductors, the conduction band and the valence band are separated by a **band gap** (E_g). Substituting electron-rich atoms into a semiconductor results in **n-type semiconductors**. Substituting electron-poor atoms results in **p-type semiconductors**. Both types of substitution increase the conductivity of the semiconductor by decreasing its band gap.

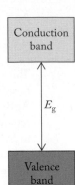

Section 18.3 Many metallic crystals are based on crystal lattices of the **cubic closest-packed (ccp)** and **hexagonal closest-packed (hcp)** types, which are the two most efficient ways of packing atoms in a solid. **Crystalline solids** contain repeating **unit cells**, which can be **simple cubic (sc)**, **body-centered cubic (bcc)**, or **face-centered cubic (fcc)**. The dimensions of the unit cell in a crystalline solid can be used to determine the radius of the atoms or ions and to predict density.

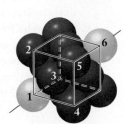

Section 18.4 Alloys are blends of a host metal and one or more other elements (which may or may not be metals) added

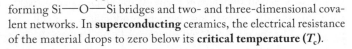

to enhance the properties of the host, including strength, hardness, and corrosion resistance. In **substitutional alloys**, atoms of the added elements replace atoms of the host metal in the crystal lattice. In **interstitial alloys**, atoms of added elements are located in the tetrahedral and/or octahedral holes between atoms of the host metal. Aluminum and aluminum alloys are highly desirable for applications requiring corrosion resistance and low mass.

Section 18.5 Two allotropes of carbon are the **covalent network solids** graphite and diamond. Many nonmetals form **molecular solids**, including sulfur, which forms puckered rings of eight sulfur atoms, and phosphorus, which forms P_4 tetrahedra.

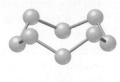

Section 18.6 Many **ionic solids** consist of crystals with some number of either cations or anions forming the unit cell and the opposite ion occupying octahedral or tetrahedral holes in the unit cell. The edge lengths of unit cells of ionic solids can be used to calculate ionic radii and to predict densities.

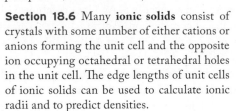

Section 18.7 Heating selected solid inorganic compounds (such as clays, which are aluminosilicate minerals) to high temperature alters their chemical composition and makes them harder, denser, and stronger. The resulting heat- and chemical-resistant materials (**ceramics**) are electrical insulators due to the large energy gap between their filled valence and empty conduction bands. The polymorphs of silica consist of tetrahedra made up of four O at the corners surrounding a central Si. Each tetrahedron can share some or all its oxygen atoms with other tetrahedra, forming Si—O—Si bridges and two- and three-dimensional covalent networks. In **superconducting** ceramics, the electrical resistance of the material drops to zero below its **critical temperature** (T_c).

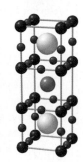

Section 18.8 X-ray diffraction (XRD) is an analytical method that records constructive interference of X-rays reflecting off different layers of atoms or ions in a crystalline solid. The distances between layers are calculated using the **Bragg equation**. X-ray diffraction makes it possible to determine the crystal structure of crystalline solids.

PROBLEM-SOLVING SUMMARY ··■

TYPE OF PROBLEM	CONCEPTS AND EQUATIONS	SAMPLE EXERCISES
Calculating atomic or ionic radii from unit cell dimensions	Determine the length ℓ of an edge, face diagonal, or body diagonal along which adjacent atoms touch in a unit cell. Use the relationship between ℓ and the atomic/ionic radius r to calculate the value of r: $r = \ell/2$ for sc unit cells, $r = 0.3536\ell$ for fcc unit cells, and $r = 0.4330\ell$ for bcc unit cells.	18.1, 18.4
Using unit cells to calculate the density of a solid	Determine the mass of the atoms in the unit cell from their molar mass and the volume of the unit cell from the edge length, then calculate the density: $$d = \frac{m}{V}$$	18.2
Predicting the crystal structure of two-element alloys	Compare the radii of the alloying elements. Similarly sized radii (within 15%) predict a substitutional alloy. When the radius of the smaller atom is <73% of the radius of the larger atom, an interstitial alloy forms.	18.3

TYPE OF PROBLEM	CONCEPTS AND EQUATIONS	SAMPLE EXERCISES
Calculating the density of a salt from its unit cell dimensions	Determine the masses of the ions in the unit cell from their molar masses and the volume of the unit cell from the edge length. Then apply $$d = \frac{m}{V}$$	18.5
Determining interlayer distances by X-ray diffraction	Apply the Bragg equation, $$n\lambda = 2d\sin\theta \qquad (18.9)$$	18.6

VISUAL PROBLEMS ···■

(Answers to boldface end-of-chapter questions and problems are in the back of the book.)

18.1. In Figure P18.1, which drawings are analogous to a crystalline solid, and which are analogous to an amorphous solid?

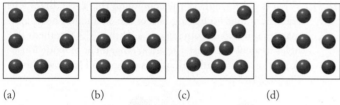

(a) (b) (c) (d)

FIGURE P18.1

18.2. The unit cells in Figure P18.2 continue infinitely in two dimensions. Draw a box around the unit cell in each pattern. How many light squares and how many dark squares are in each unit cell?

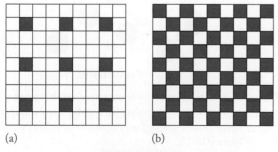

(a) (b)

FIGURE P18.2

18.3. The pattern in Figure P18.3 continues indefinitely in three dimensions. Draw a box around the unit cell. If the red circles represent element A and the blue circles represent element B, what is the chemical formula of the compound?

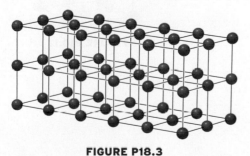

FIGURE P18.3

18.4. How many cations (A) and anions (B) are there in the unit cell in Figure P18.4?

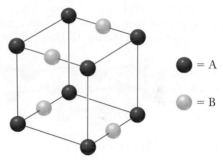

= A

= B

FIGURE P18.4

18.5. How many equivalent atoms of elements A and B are there in the portion of a unit cell in Figure P18.5?

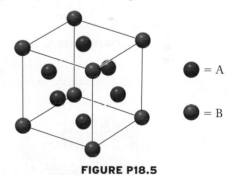

= A

= B

FIGURE P18.5

18.6. What is the chemical formula of the compound a portion of whose unit cell is shown in Figure P18.6?

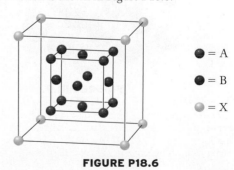

= A

= B

= X

FIGURE P18.6

18.7. What is the chemical formula of the ionic compound a portion of whose unit cell is shown in Figure P18.7? (A and B are cations, X is an anion.)

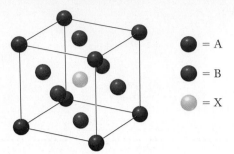

= A

= B

= X

FIGURE P18.7

18.8. When amorphous red phosphorus is heated at high pressure, it is transformed into the allotrope black phosphorus, which can exist in one of several forms. One form consists of six-membered rings of phosphorus atoms (Figure P18.8a). Why are the six-atom rings in black phosphorus puckered, whereas the six-atom rings in graphite (Figure P18.8b) are planar?

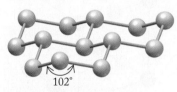

102°

(a) Black phosphorus

(b) Graphite

FIGURE P18.8

18.9. The distance between atoms in a cubic form of phosphorus is 238 pm (Figure P18.9). Calculate the density of this form of phosphorus.

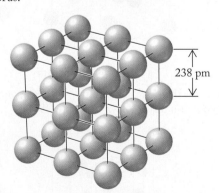

238 pm

FIGURE P18.9

18.10. In the unit cell in Figure P18.10 the small spheres are in tetrahedral holes formed by the large spheres.

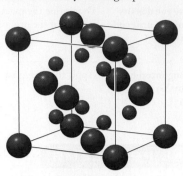

FIGURE P18.10

a. How many of the large spheres and how many of the small ones are assignable to the unit cell?
b. If the radius of the large spheres is 140 pm, what (theoretically) is the maximum size of the small spheres?

18.11. What is the formula of the compound that crystallizes with aluminum ions occupying half of the octahedral holes and magnesium ions occupying one-eighth of the tetrahedral holes in a cubic closest-packed arrangement of oxide ions? A portion of the unit cell of this compound is shown in Figure P18.11.

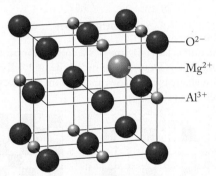

O²⁻

Mg²⁺

Al³⁺

FIGURE P18.11

18.12. What is the formula of the compound that crystallizes with metal (M) ions occupying half of the tetrahedral holes in a cubic closest-packed arrangement of sulfide ions? See Figure P18.12.

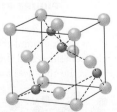

FIGURE P18.12

18.13. What is the formula of a transition metal sulfide that has the unit cell shown in Figure P18.13? Use "M" as the symbol for the metal ion.

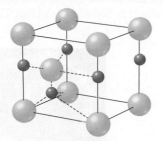

FIGURE P18.13

18.14. Figure P18.14 shows the unit cell of CsCl. From the information given and the radius of the chloride (corner) ions of 181 pm, calculate the radius of Cs^+ ions.

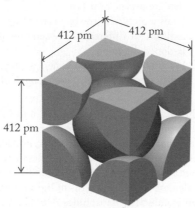

FIGURE P18.14

18.15. A number of metal chlorides adopt the rock salt crystal structure, in which the metal ions occupy all the octahedral holes in a face-centered cubic array of chloride ions. For at least one (maybe more) of the metallic elements highlighted in Figure P18.15, this crystal structure is not possible. Which one(s) cannot form a rock salt structure with Cl^- ions?

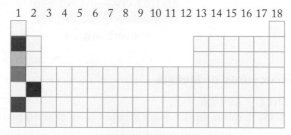

FIGURE P18.15

18.16. A number of metal fluorides adopt the fluorite crystal structure, in which the fluoride ions occupy all the tetrahedral holes in a face-centered cubic array of metal ions. For at least one (maybe more) of the highlighted

metallic elements in Figure P18.16, this crystal structure is not possible. Which one(s)?

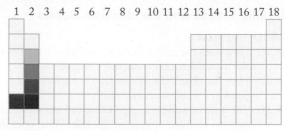

FIGURE P18.16

*18.17. **Superconducting Materials I** In 2000, magnesium boride was observed to behave as a superconductor. Its unit cell is shown in Figure P18.17. What is the formula of magnesium boride? A boron atom is in the center of the unit cell (on the left), which is part of the hexagonal closest-packed crystal structure (right).

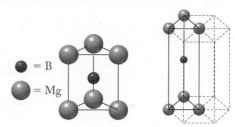

FIGURE P18.17

*18.18. **Superconducting Materials II** The 1987 Nobel Prize in physics was awarded to J. G. Bednorz and K. A. Müller for their discovery of superconducting ceramic materials such as $YBa_2Cu_3O_7$. Figure P18.18 shows the unit cell of another yttrium–barium–copper oxide.

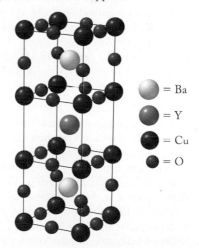

FIGURE P18.18

a. What is the chemical formula of this compound?
b. Eight oxygen atoms must be removed from the unit cell shown here to produce the unit cell of $YBa_2Cu_3O_7$. Does it make a difference which oxygen atoms are removed?

QUESTIONS AND PROBLEMS

Metallic Bonds and Conduction Bands

CONCEPT REVIEW

18.19. How does the sea-of-electrons model (Chapter 4) explain the high electrical conductivity of gold?

*18.20. How does band theory explain the high electrical conductivity of mercury?

18.21. The melting and boiling points of sodium metal are much lower than those of sodium chloride. What does this difference reveal about the relative strengths of metallic bonds and ionic bonds?

*18.22. Which metal do you expect to have the higher melting point—Al or Na? Explain your answer.

18.23. Some scientists believe that the solid hydrogen that forms at very low temperatures and high pressures may conduct electricity. Is this hypothesis supported by band theory?

18.24. Would you expect solid helium to conduct electricity?

Semiconductors

CONCEPT REVIEW

18.25. Which groups in the periodic table contain metals with filled valence bands?

18.26. Insulators are materials that do not conduct electricity; conductors are substances that allow electricity to flow through them easily. Rank the following in order of increasing band gap: semiconductor, insulator, conductor.

18.27. Why is it important to keep phosphorus out of silicon chips during their manufacture?

18.28. How might doping of silicon with germanium affect the conductivity of silicon?

*18.29. Antimony (Sb) combines with sulfur to form the semiconductor compound Sb_2S_3. In which group of the periodic table might you find elements for doping Sb_2S_3 to form a p-type semiconductor?

*18.30. In which group of the periodic table might you find elements for doping Sb_2S_3 to form an n-type semiconductor?

PROBLEMS

18.31. Thin films of doped diamond hold promise as semiconductor materials. Trace amounts of nitrogen impart a yellow color to otherwise colorless pure diamonds.
  a. Are nitrogen-doped diamonds examples of semiconductors that are p-type or n-type?
  b. Draw a picture of the band structure of diamond to indicate the difference between pure diamond and N-doped (nitrogen-doped) diamond.
  *c. N-doped diamonds absorb violet light at about 425 nm. What is the magnitude of E_g that corresponds to this wavelength?

18.32. **Hope Diamond** Trace amounts of boron give diamonds (including the Smithsonian's Hope Diamond) a blue color (Figure P18.32).

FIGURE P18.32

  a. Are boron-doped diamonds examples of semiconductors that are p-type or n-type?
  b. Draw a picture of the band structure of diamond to indicate the difference between pure diamond and B-doped diamond.
  *c. What is the band gap in energy if blue diamonds absorb red-orange light with a wavelength of 675 nm?

*18.33. The nitride ceramics AlN, GaN, and InN are all semiconductors used in the microelectronics industry. Their band gaps are 580.6, 322.1, and 192.9 kJ/mol, respectively. Which, if any, of these energies correspond to radiation in the visible region of the spectrum?

*18.34. Calculate the wavelengths of light emitted by the semiconducting phosphides AlP, GaP, and InP, which have band gaps of 241.1, 216.0, and 122.5 kJ/mol, respectively, and are used in the type of light source shown in Figure P18.34.

FIGURE P18.34

Structures of Metals

CONCEPT REVIEW

18.35. Explain the difference between cubic closest-packed and hexagonal closest-packed arrangements of identical spheres.

*18.36. Is it possible to have a closest-packed crystal lattice with four different repeating layers, *abcdabcd . . .* ?

18.37. Which unit cell has the greater packing efficiency, simple cubic or body-centered cubic?

18.38. Consult Figure 18.16 to predict which unit cell has the greater packing efficiency: body-centered cubic or face-centered cubic.

18.39. The unit cell in iron metal is either fcc or bcc, depending on temperature (see Sample Exercise 18.1). Are the fcc form of iron and the bcc form allotropes? Explain your answer.

*18.40. At low temperatures, the unit cell of calcium metal is found to be fcc, a closest-packed crystal lattice. At higher temperatures, the unit cell of calcium metal is found to be bcc, a crystal lattice that is not a closest-packed structure. What might be a reason for this difference?

PROBLEMS

18.41. Derive the edge length in bcc and fcc unit cells in terms of the radius (r) of the atoms in the cells in Figure 18.16.

*18.42. Derive the length of the body diagonal in simple cubic and fcc unit cells in terms of the radius (r) of the atoms in the unit cells in Figure 18.16.

———————

18.43. **Fluorescent Lighting** Europium is a lanthanide element that is used in the phosphors in fluorescent lamps. Metallic Eu forms bcc unit cells with an edge length of 240.6 pm. Use this information to calculate the radius of a europium atom.

18.44. Nickel has an fcc unit cell with an edge length of 350.7 pm. Use this information to calculate the radius of a nickel atom.

———————

18.45. What is the length of an edge of the unit cell when barium (atomic radius 222 pm) crystallizes in a crystal lattice of bcc unit cells?

18.46. What is the length of an edge of the unit cell when aluminum (atomic radius 143 pm) crystallizes in a crystal lattice of fcc unit cells?

———————

18.47. A crystalline form of copper has a density of 8.95 g/cm³. If the radius of copper atoms is 127.8 pm, is the copper unit cell (a) simple cubic; (b) body-centered cubic; or (c) face-centered cubic?

18.48. A crystalline form of molybdenum has a density of 10.28 g/cm³ at a temperature at which the radius of a molybdenum atom is 139 pm. Which unit cell is consistent with these data: (a) simple cubic, (b) body-centered cubic, or (c) face-centered cubic?

Alloys

CONCEPT REVIEW

18.49. Is there a difference between a solid solution and a homogeneous alloy?

18.50. **White Gold** White gold was originally developed to give the appearance of platinum. One formulation of white gold contains 25% nickel and 75% gold. Which is more malleable, white gold or pure gold?

*18.51. Explain why an alloy that is 28% Cu and 72% Ag melts at a lower temperature than the melting points of either Cu or Ag.

18.52. Is it possible for an alloy to be both substitutional and interstitial?

18.53. The interstitial alloy tungsten carbide (WC) is one of the hardest materials known. It is used on the tips of cutting tools. Without consulting a table of radii, decide which element you think is the host and which occupies the holes.

18.54. Why are the alloys that second-row nonmetals—such as B, C, and N—form with transition metals more likely to be interstitial than substitutional?

PROBLEMS

18.55. The unit cell of a substitutional alloy consists of a face-centered cube that has an atom of element X at each corner and an atom of element Y at the center of each face.
  a. What is the formula of the alloy?
  b. What would the formula of the alloy be if the positions of the two elements were reversed in the unit cell?

18.56. The bcc unit cell of a substitutional alloy has atoms of element A at the corners of the unit cell and an atom of element B at the center of the unit cell.
  a. What is the formula of the alloy?
  b. What would the formula of the alloy be if the positions of the two elements were reversed in the unit cell?

———————

18.57. Vanadium reacts with carbon to form vanadium carbide, an interstitial alloy. Given the atomic radii of V (135 pm) and C (77 pm), which holes in a cubic closest-packed array of vanadium atoms do you think the carbon atoms are more likely to occupy—octahedral or tetrahedral?

18.58. What is the minimum atomic radius required for a cubic closest-packed metal to accommodate boron atoms (radius 88 pm) in its octahedral holes?

———————

18.59. **Dental Fillings** Dental fillings are mixtures of several alloys including one with the formula Ag_3Sn. Silver (radius 144 pm) and tin (140 pm) both crystallize in an fcc unit cell. Is this alloy likely to be a substitutional alloy or an interstitial alloy?

18.60. An alloy used in dental fillings has the formula Sn_3Hg. The radii of tin and mercury atoms are 140 pm and 151 pm, respectively. Which alloy has a smaller mismatch (percent difference in atomic radii), Sn_3Hg or bronze (Cu/Sn alloys)?

———————

*18.61. **Hardening Metal Surfaces** Plasma nitriding is a process for embedding nitrogen atoms in the surfaces of metals that hardens the surfaces and makes them more corrosion resistant. Do the nitrogen atoms in the nitrided surface of a sample of cubic closest-packed iron fit in the octahedral holes of the crystal lattice? (Assume that the atomic radii of N and Fe are 75 and 126 pm, respectively.)

*18.62. **Hydrogen Storage** A number of crystalline transition metals (including titanium, zirconium, and hafnium) can store hydrogen as metal hydrides for use as fuel in a hydrogen-powered vehicle. Which metal or metals are most likely to accommodate H atoms (radius 37 pm) with a radius ratio closest to the ideal value in Table 18.4, given that their atomic radii are 147, 160, and 159 pm, respectively?

18.63. An interstitial alloy is prepared from metals A and B where B has the smaller atomic radius. The unit cell of metal A is fcc. What is the formula of the alloy if B occupies (a) all of the octahedral holes; (b) half of the octahedral holes; (c) half of the tetrahedral holes?

18.64. An interstitial alloy was prepared from two metals. Metal A with the larger atomic radius has a hexagonal closest-packed crystal lattice. What is the formula of the alloy if atoms of metal B occupy (a) all of the tetrahedral holes; (b) half of the tetrahedral holes; (c) half of the octahedral holes?

18.65. An interstitial alloy with an fcc unit cell contains one atom of B for every five atoms of host element A. What fraction of the octahedral holes is occupied in this alloy?

18.66. If the B atoms in the alloy described in Problem 18.65 occupy tetrahedral holes in A, what percentage of the holes would they occupy?

18.67. If the unit cell of a substitutional alloy of copper and tin has the same edge length as the unit cell of copper, will the alloy have a greater density than copper?

18.68. If the unit cell of an interstitial alloy of vanadium and carbon has the same edge length as the unit cell of vanadium, will the alloy have a greater density than vanadium?

Structures of Some Crystalline Nonmetals

CONCEPT REVIEW

18.69. S_8 is not a flat octagon—why?

*18.70. Selenium exists either as Se_8 rings or in a structure with helical chains of Se atoms. Are these two structures of selenium allotropes? Explain your answer.

*18.71. If the carbon atoms in graphite are replaced by alternating B and N atoms, would the resulting structure contain puckered rings like black phosphorus or flat ones like graphite (see Figure P18.8)?

*18.72. Cyclic allotropes of sulfur containing up to 20 sulfur atoms have been isolated and characterized. Propose a reason why the bond angles in S_n (where $n = 10, 12, 18,$ and 20) are all close to 106°.

PROBLEMS

18.73. Ice is a network solid. However, theory predicts that, under high pressure, ice (solid H_2O) becomes an ionic compound composed of H^+ and O^{2-} ions. The proposed unit cell for ice under these conditions is a bcc unit cell of oxygen ions with hydrogen ions in holes.
   a. How many H^+ and O^{2-} ions are in each unit cell?
   b. Draw a Lewis structure for "ionic" ice.

18.74. **Ice under Pressure** Kurt Vonnegut's novel *Cat's Cradle* describes an imaginary, high-pressure form of ice called "ice nine." With the assumption that ice nine has a cubic closest-packed arrangement of oxygen atoms with hydrogen atoms in the appropriate holes, what type of hole will accommodate the H atoms?

18.75. A chemical reaction between H_2S_4 and S_2Cl_2 produces cyclic S_6. What are the bond angles in S_6?

18.76. Reaction between S_8 and six equivalents of AsF_5 yields $[S_4^{2+}][AsF_6^-]_2$ by the reaction

$$S_8 + 6\,AsF_5 \rightarrow 2\,[S_4^{2+}][AsF_6^-]_2 + 2\,AsF_3$$

The S_4^{2+} ion has a cyclic structure. Are all four sulfur atoms in one plane?

Salt Crystals: Ionic Solids

CONCEPT REVIEW

18.77. Crystals of both LiCl and KCl have the rock salt structure. In the unit cell of LiCl adjacent Cl^- ions touch each other. In KCl they don't. Why?

18.78. Can $CaCl_2$ have the rock salt structure?

*18.79. In some books the unit cell of CsCl is described as being body-centered cubic (Figure P18.79); in others, as simple cubic (see Figure 18.16a). Explain how CsCl crystals might be described by either unit cell type.

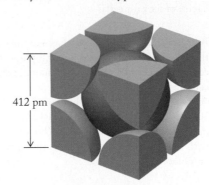

412 pm

FIGURE P18.79

18.80. In the crystals of ionic compounds, how do the relative sizes of the ions influence the location of the smaller ions?

*18.81. Instead of describing the unit cell of NaCl as an fcc array of Cl^- ions with Na^+ ions in octahedral holes, might we describe it as an fcc array of Na^+ ions with Cl^- ions in octahedral holes? Explain why or why not.

18.82. Why isn't crystalline sodium chloride considered a network solid?

*18.83. As the cation–anion radius ratio increases for an ionic compound with the rock salt crystal structure, is the calculated density more likely to be greater than or less than the measured value?

18.84. As the cation–anion radius ratio increases for an ionic compound with the rock salt crystal structure, is the length of the edge of the unit cell calculated from ionic radii likely to be greater than or less than the observed edge length?

PROBLEMS

18.85. What is the formula of the oxide that crystallizes with Fe^{3+} ions in one-fourth of the octahedral holes, Fe^{3+} ions in one-eighth of the tetrahedral holes, and Mg^{2+} in one-fourth of the octahedral holes of a cubic closest-packed arrangement of oxide ions (O^{2-})?

18.86. What is the chemical formula of the compound that crystallizes a simple cubic arrangement of fluoride ions with Ba^{2+} ions occupying half of the cubic holes?

18.87. A compound of uranium and oxygen consists of a cubic close-packed arrangement of uranium ions with oxide ions in all the tetrahedral holes. What is the formula of this compound?

18.88. A mixture of gallium and arsenic is a widely used semiconductor. The arsenide ions are in a cubic close-packed arrangement and half the tetrahedral holes are occupied by the gallium ions. What is the formula of this compound?

18.89. **The Vinland Map** At Yale University there is a map, believed to date from the 1400s, of a landmass labeled "Vinland" (Figure P18.89). The map is thought to be evidence of early Viking exploration of North America. Debate over the map's authenticity centers on yellow stains on the map paralleling the black ink lines. One analysis suggests the yellow color is from the mineral anatase, a form of TiO_2 that was not used in 15th-century inks.

FIGURE P18.89

a. The crystal structure of anatase is approximated by a ccp arrangement of oxide ions with titanium(IV) ions in holes. Which type of hole are Ti^{4+} ions likely to occupy?

b. What fraction of these holes is likely to be occupied? (The radius of Ti^{4+} is 60.5 pm.)

*18.90. The crystal structure of olivine—M_2SiO_4 (M = Mg, Fe)—can be viewed as a ccp arrangement of oxide ions with silicon(IV) ions in tetrahedral holes and metal ions in octahedral holes.
a. What fraction of each type of hole is occupied?
b. The unit cells of Mg_2SiO_4 and Fe_2SiO_4 have volumes of 2.91×10^{-26} cm³ and 3.08×10^{-26} cm³. Why is the volume of Fe_2SiO_4 larger?

*18.91. In nature, cadmium(II) sulfide (CdS) exists as two minerals. One of them, called *hawleyite*, has a sphalerite structure, and its density at 25°C is 4.83 g/cm³. A hypothetical form of CdS with the rock salt structure would have a density of 5.72 g/cm³. Why should the rock salt structure of CdS be denser? The ionic radii of Cd^{2+} and S^{2-} are 95 pm and 184 pm, respectively.

18.92. There are two crystalline forms of manganese(II) sulfide (MnS): the α form has a rock salt structure; the β form has a sphalerite structure.
a. Describe the differences between the two structures of MnS.
b. The ionic radii of Mn^{2+} and S^{2-} are 67 and 184 pm. Which type of hole in a ccp lattice of sulfide ions could theoretically accommodate a Mn^{2+} ion?

18.93. The unit cell of rhenium trioxide (ReO_3) consists of a cube with rhenium atoms at the corners and an oxygen atom on each of the 12 edges. The atoms touch along the edge of the unit cell. The radii of Re and O atoms in ReO_3 are 137 and 73 pm. Calculate the density of ReO_3.

18.94. With reference to Figure P18.79, calculate the density of simple cubic CsCl.

18.95. Magnesium oxide crystallizes in the rock salt structure. Its density is 3.60 g/cm³. What is the edge length of the fcc unit cell of MgO?

18.96. Crystalline potassium bromide (KBr) has a rock salt structure and a density of 2.75 g/cm³. Calculate the edge length of its unit cell.

Ceramics: Insulators to Superconductors

CONCEPT REVIEW

18.97. Which of the following properties are associated with ceramics and which are associated with metals: (a) ductile; (b) thermal insulator; (c) electrically conductive; (d) malleable?

18.98. Many ceramics such as TiO_2 are electrical insulators. What differences are there in the band structure of TiO_2 compared with Ti metal that account for the different electrical properties?

18.99. Replacement of Al^{3+} ions in kaolinite [$Al_2(Si_2O_5)(OH)_4$] with Mg^{2+} ions yields the mineral antigorite. What is its formula?

18.100. What is the formula of the silicate mineral talc, obtained by the replacement of Al^{3+} ions in pyrophyllite $[Al_4Si_8O_{20}(OH)_4]$ with Mg^{2+} ions?

PROBLEMS

18.101. Kaolinite $[Al_2(Si_2O_5)(OH)_4]$ is formed by weathering of the mineral $KAlSi_3O_8$ in the presence of carbon dioxide and water, as described by the following unbalanced reaction:

$$KAlSi_3O_8(s) + H_2O(\ell) + CO_2(aq) \rightarrow$$
$$Al_2(Si_2O_5)(OH)_4(s) + SiO_2(s) + K_2CO_3(aq)$$

Balance the reaction and determine whether or not it is a redox reaction.

18.102. Albite, a feldspar mineral with an ideal composition of $NaAlSi_3O_8$, can be converted to jadeite $(NaAlSi_2O_6)$ and quartz. Write a balanced chemical equation describing this transformation.

18.103. Under the high pressures in Earth's crust, the mineral anorthite $(CaAl_2Si_2O_8)$ is converted to a mixture of three minerals: grossular $[Ca_3Al_2(SiO_4)_3]$, kyanite (Al_2SiO_5), and quartz (SiO_2). (a) Write a balanced chemical equation describing this transformation. (b) Determine the charges and formulas of the silicate anions in anorthite, grossular, and kyanite.

*18.104. The calcium silicate mineral grossular is also formed under pressure in a reaction between anorthite $(CaAl_2Si_2O_8)$, gehlenite $(Ca_2Al_2SiO_7)$, and wollastonite $(CaSiO_3)$:

$$CaAl_2Si_2O_8 + Ca_2Al_2SiO_7 + CaSiO_3 \rightarrow Ca_3Al_2(SiO_4)_3$$

Anorthite   Gehlenite   Wollastonite   Grossular

   a. Balance this chemical equation.
   b. Express the composition of gehlenite the way mineralogists often do: as the percentage of the metal and semimetal oxides in it, that is, %CaO, %Al_2O_3, and %SiO_2.

18.105. The ceramic material barium titanate $(BaTiO_3)$ is used in devices that measure pressure. The radii of Ba^{2+}, Ti^{4+}, and O^{2-} are 135, 60.5, and 140 pm, respectively. If the O^{2-} ions are in a closest-packed structure, which hole(s) can accommodate the metal cations?

18.106. The mixed metal oxide $LiMnTiO_4$ has a structure with cubic closest-packed oxide ions and metal ions in both octahedral and tetrahedral holes. Which metal ion is most likely to be found in the tetrahedral holes? The ionic radii of Li^+, Mn^{3+}, Ti^{4+}, and O^{2-} are 76, 67, 60.5, and 140 pm, respectively.

X-ray Diffraction: How We Know Crystal Structures

CONCEPT REVIEW

18.107. Why does an amorphous solid not produce an XRD scan with sharp peaks?

18.108. X-ray diffraction cannot be used to determine the structures of compounds in solution—why?

18.109. Why are X-rays rather than microwaves chosen for diffraction studies of crystalline solids?

18.110. The radiation sources used in X-ray diffraction can be changed. Figure P18.110 shows a diffraction pattern made by a short-wavelength source. How would changing to a longer wavelength source affect the pattern?

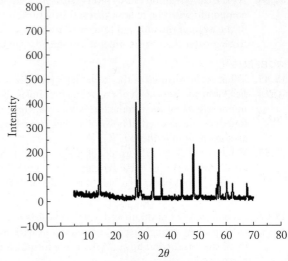

FIGURE P18.110

*18.111. Why might a crystallographer (a scientist who studies crystal structures) use different X-ray wavelengths to determine a crystal structure? *Hint*: Consider what mechanical limits are inherent in the design of the instrument depicted in Figure 18.44, and how those limits impact the 2θ scanning range.

*18.112. Where in earlier chapters have we seen diffraction used to acquire structural information?

PROBLEMS

18.113. The spacing between the layers of ions in sylvite (the mineral form of KCl) is larger than in halite (NaCl). Which crystal will diffract X-rays of a given wavelength through larger 2θ values?

18.114. Silver halides are used in black-and-white photography. In which compound would you expect to see a larger distance between ion layers, AgCl or AgBr? Which compound would you expect to diffract X-rays through larger values of 2θ if the same wavelength of X-ray were used?

18.115. Galena, Illinois, is named for the rich deposits of lead(II) sulfide (PbS) found nearby. When PbS is exposed to X-rays with $\lambda = 71.2$ pm, strong reflections from a single crystal of PbS are observed at 13.98° and 21.25°. Determine the values of n to which these reflections correspond, and calculate the spacing between the crystal layers.

18.116. **Pigments in Ceramics** Cobalt(II) oxide is used as a pigment in ceramics. It has the same type of crystal structure as NaCl. When cobalt(II) oxide is exposed to X-rays with $\lambda = 154$ pm, reflections are observed at 42.38°, 65.68°, and 92.60°. Determine the values of n to which these reflections correspond, and calculate the spacing between the crystal layers.

18.117. Pyrophyllite [$Al_2Si_4O_{10}(OH)_2$] is a silicate mineral with a layered structure. The distance between the layers is 1855 pm. What is the smallest angle of diffraction of X-rays with $\lambda = 154$ pm from this solid?

18.118. Minnesotaite [$Fe_3Si_4O_{10}(OH)_2$] is a silicate mineral with a layered structure similar to that of kaolinite. The distance between the layers in minnesotaite is 1940 ± 10 pm. What is the smallest angle of diffraction of X-rays with $\lambda = 154$ pm from this solid?

Additional Problems

18.119. A unit cell consists of a cube that has an ion of element X at each corner, an ion of element Y at the center of the cube, and an ion of element Z at the center of each face. What is the formula of the compound?

18.120. An element crystallizes in the cubic close-packed structure. The length of an edge of the unit cell is 408 pm. The density of the element is 19.27 g/cm³. Identify the element.

*18.121. The packing efficiency for a unit cell can be calculated using Equation 18.1 (repeated here):

$$\text{Packing efficiency (\%)} = \frac{\text{volume occupied by spheres}}{\text{volume of unit cell}} \times 100\%$$

What is the packing efficiency of the Si atoms in pure Si if the radius of one Si atom is 117 pm? The density of pure silicon is 2.33 g/mL.

*18.122. **The Composition of Light-Emitting Diodes** The colored lights on many electronic devices are light-emitting diodes (LEDs). One of the compounds used to make them is aluminum phosphide (AlP), which crystallizes in a sphalerite crystal structure.
   a. If AlP were an ionic compound, would the ionic radii of Al^{3+} and P^{3-} be consistent with the size requirements of the ions in a sphalerite crystal structure?
   b. If AlP were a covalent compound, would the atomic radii of Al and P be consistent with the size requirements of atoms in a sphalerite crystal structure?

18.123. Under the appropriate reaction conditions, small cubes of molybdenum, 4.8 nm on a side, can be deposited on carbon surfaces. These "nanocubes" are made of bcc arrays of Mo atoms.
   a. If the edge of each nanocube corresponds to 15 unit cell lengths, what is the effective radius of a molybdenum atom in these structures?
   b. What is the density of each molybdenum nanocube?
   c. How many Mo atoms are in each nanocube?

18.124. In the fullerene known as buckminsterfullerene, C_{60}, molecules of C_{60} form a cubic closest-packed array of spheres with a unit cell edge length of 1410 pm.
   a. What is the density of crystalline C_{60}?
   b. If we treat each C_{60} molecule as a sphere of 60 carbon atoms, what is the radius of the C_{60} molecule?

   c. C_{60} reacts with alkali metals to form M_3C_{60} (where M = Na or K). The crystal structure of M_3C_{60} contains cubic closest-packed spheres of C_{60} with metal ions in holes. If the radius of a K^+ ion is 138 pm, which type of hole is a K^+ ion likely to occupy? What fraction of the holes will be occupied?
   d. Under certain conditions, a different substance, M_6C_{60}, can be formed in which the C_{60} molecules have a bcc unit cell. Calculate the density of a crystal of M_6C_{60}.

18.125. **Earth's Core** The center of Earth is composed of a solid iron core within a molten iron outer core. When molten iron cools, it crystallizes in different ways depending on pressure—in a bcc unit cell at low pressure and in a hexagonal unit cell at high pressures like those at Earth's center.
   a. Calculate the density of bcc iron given that the radius of an iron atom is 126 pm.
   b. Calculate the density of hexagonal iron given a unit cell volume of 5.414×10^{-23} cm³.
   *c. Seismic studies suggest that the density of Earth's solid core is only about 90% of that of hexagonal Fe. Laboratory studies have shown that up to 4% by mass of Si can be substituted for Fe without changing the hcp crystal structure built on hexagonal unit cells. Calculate the density of such a crystal.

18.126. The unit cell of an alloy with a 1:1 ratio of magnesium and strontium is identical to the unit cell of CsCl. The unit cell edge length of MgSr is 390 pm. What is the density of MgSr?

18.127. Gold and silver can be separately alloyed with zinc to form AuZn (unit cell edge = 319 pm) and AgZn (unit cell edge = 316 pm). The two alloys have the same unit cell. Which alloy is more dense?

*18.128. Removing two electrons from cyclo-S_8 yields the dication cyclo-S_8^{2+}. Will all of the sulfur atoms be in one plane in the S_8^{2+} cation?

18.129. Use Equation 18.1 to calculate the packing efficiency in a simple cubic unit cell.

18.130. Use Equation 18.1 to calculate the packing efficiency in a body-centered cubic unit cell.

18.131. Manganese steels are a mixture of iron, manganese, and carbon. Is the manganese likely to occupy holes in the austenite lattice or are manganese steels substitutional alloys?

18.132. Aluminum forms alloys with lithium (LiAl), gold ($AuAl_2$), and titanium (Al_3Ti). Based on their crystal lattices, each of these alloys is considered to be a substitutional alloy.
   a. Do these alloys fit the general size requirements for substitutional alloys? The atomic radii for Li, Al, Au, and Ti are 152, 143, 144, and 147 pm, respectively.
   b. If the unit cell of LiAl is bcc, what is the density of LiAl?

*18.133. The aluminum alloy Cu_3Al crystallizes in a bcc unit cell. Propose a way that the Cu and Al atoms could be allocated between bcc unit cells that is consistent with the formula of the alloy.

19

Organic Chemistry

Fuels, Pharmaceuticals, and Modern Materials

The Stuff of Daily Life

Think for a moment about the stuff of everyday life. We drive cars powered by gasoline or diesel fuel. We cook chicken on a grill that burns propane. Most of us live in homes heated by natural gas, fuel oil, or electricity, and much of the electricity we use is produced by burning coal. All of these fuels are organic compounds—which means they are carbon-based compounds—and they are used as sources of heat because their combustion reactions are highly exothermic. The same is true of the food we eat. Most of the medicines we use are also organic compounds, whether they are natural products derived from living plants or animals or are synthesized by reacting other organic compounds together abiotically.

Your favorite soft T-shirt is made of cotton, an organic natural fiber. Because of the composition, shape, and orientation of the large molecules that form cotton fibers, the material is soft and absorbent. In contrast, the soles of running shoes are made of a synthetic rubber, a polymer manufactured from compounds of carbon and hydrogen derived from crude oil. Because of the size, shape, and orientation of the polymer molecules in synthetic rubber, the material is springy, nonabsorbent, and resistant to wear.

Fans at a hockey game sit on molded plastic seats and munch on hot dogs wrapped in plastic wrap while watching players who wear helmets of impact-resistant polymeric materials. Plastics, cling wrap, and hockey helmets are all made from compounds derived from crude oil—in other words, they are organic compounds. The plastic seat is strong, lightweight, and capable of bearing considerable mass without breaking. The sandwich wrap must be flexible yet prevent oxygen from reaching the hot dog and bun before you open it to eat it, and the players' helmets must be lightweight but still able to withstand the impact of a hockey puck. All these physical properties can be designed into the materials at the molecular level because they are directly linked to the structure and composition of the molecules.

In short, compounds of carbon are everywhere, and they are so varied in size, shape, and properties that an entire field of chemistry is devoted to their study. This chapter is a brief introduction to that field: organic chemistry. As you read it, think about questions such as these: Why do we burn some organic compounds as fuel? What makes Kevlar strong but allows nylon to stretch? What kinds of molecules do we find in common headache and fever remedies?

Olympic Polymers The elastic clothing worn by Olympic athletes is made of ▶ synthetic organic polymers.

LEARNING OUTCOMES

LO1 Distinguish between the molecular structures and properties of the different classes of hydrocarbons
Sample Exercise 19.1

LO2 Draw and name organic compounds
Sample Exercises 19.2, 19.3

LO3 Identify constitutional isomers and stereoisomers
Sample Exercises 19.4, 19.5

LO4 Identify the monomers in addition and condensation polymers
Sample Exercises 19.6, 19.9, 19.11

LO5 Compare the fuel values of hydrocarbons and oxygenated fuels
Sample Exercise 19.7

LO6 Relate the properties of polymers to their molecular structures
Sample Exercises 19.8, 19.10

LO7 Explain the differences between alcohols, ethers, amines, ketones, and aldehydes

LO8 Explain the differences between carboxylic acids, esters, and amides

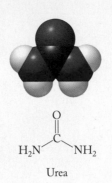

Urea

FIGURE 19.1 Urea was the first naturally occurring organic compound synthesized in the laboratory. It was prepared in 1828 by Friedrich Wöhler.

⊙⊙ **CONNECTION** In Chapter 6, we learned that compounds with the same molecular formula but different arrangements of atoms differ in physical properties such as melting point and boiling point.

⊙⊙ **CONNECTION** We first used the term *functional group* in Chapter 4 to describe the groups of atoms in molecules that are most responsible for the substance's physical and chemical properties.

....

organic chemistry the study of compounds containing C—C and/or C—H bonds.

R symbol in a general formula standing for an organic group that has one available bond; it is used to indicate the variable part of a molecule so that the focus is placed on the functional group.

monomer a small molecule that bonds with others to form polymers.

polymer a very large molecule with high molar mass; the root word *meros* is Greek for "part" or "unit," so *polymer* literally means "many units."

....

19.1 Carbon: The Scope of Organic Chemistry

The designation *organic* for carbon-containing compounds was once limited to substances produced by living organisms, but that definition has been broadened for two reasons. First, since 1828, when Friedrich Wöhler (1800–1882) discovered how to prepare urea (Figure 19.1) in the laboratory "without the intervention of a kidney," scientists have learned to synthesize many materials previously thought to be the products only of living systems. Second, chemists have learned how to synthesize many carbon-based materials that have never been produced by living systems. Chemical Abstracts Service (CAS), an organization that tracks and collects all chemical information published worldwide, maintains a registry of all the known organic and inorganic substances. There are over 70 million of them; most of them are organic, and updated information on about 15,000 of them is added to the registry *each day*. These statistics help explain why the study of **organic chemistry** occupies a special place within the discipline.

Families Based on Functional Groups

Much of the variety in the chemistry of carbon is derived from a carbon atom's ability to form four covalent bonds with a variety of other elements as well as other carbon atoms. The resulting compounds may contain a few carbon atoms or many thousands. Additional structural variety comes from the many patterns that large numbers of carbon atoms can adopt: from long chains to heavily branched structures to planar and puckered rings.

Managing the wealth of information about the millions of organic compounds that exist requires some organizing concepts. Chemists group organic compounds into families on the basis of *functional groups*: subunits of only a few atoms that confer particular chemical and physical properties to molecules.

We have already discussed several functional groups in prior chapters, including the –COOH group in carboxylic acids and the hydroxyl (–OH) group in alcohols. In this chapter we examine these and the other functional groups summarized in Table 19.1. When discussing functional groups, the convention is to use **R** to represent the entire molecule except the functional group. (Do not confuse it with the italic *R* representing the universal gas constant.)

Monomers and Polymers

A second organizing principle for organic molecules is based on size. Those having masses less than 1000 amu are typically considered small molecules. Very large ones, with masses up to and exceeding 1,000,000 amu, are called macromolecules. Many of these very large molecules consist of long chains made up of subunits called **monomers** that are chemically bonded together; the overall macromolecule is called a **polymer**. These mass boundaries are somewhat arbitrary, and medium-sized organic molecules called *oligomers* inhabit the realm between small molecules and polymers.

Organic compounds made of small molecules have constant composition and well-defined properties such as melting points and boiling points. In contrast, many synthetic polymers do not have constant composition or well-defined physical properties because they are mixtures of molecules that are similar but not identical: they usually differ in the number and arrangement of monomers in their molecular structures.

TABLE 19.1 Functional Groups of Organic Compounds

Name	Structural Formula of Group	Example and Name	
Alkane	R—H	$CH_3CH_2CH_3$	Propane
Alkene	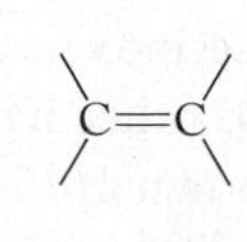	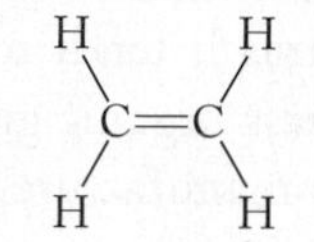	Ethylene (ethene)
Alkyne	—C≡C—	H—C≡C—H	Acetylene (ethyne)
Aromatic	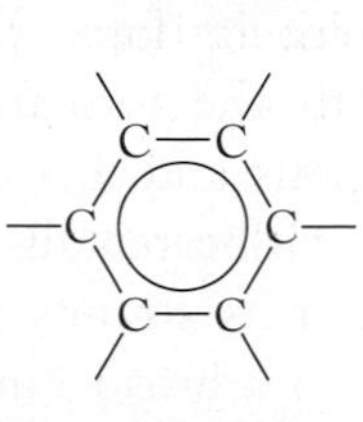	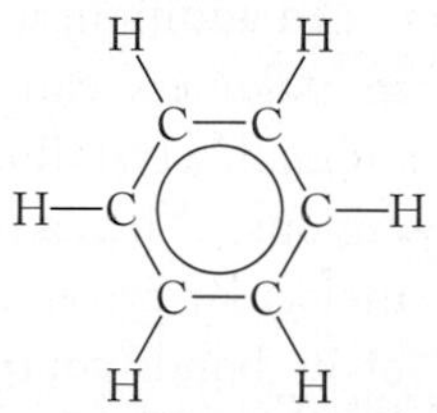	Benzene
Amine	R—NH$_2$ R—NHR R—NR$_2$	H_3C—NH_2	Methylamine
Alcohol	R—OH	CH_3CH_2OH	Ethanol
Ether	R—O—R	$CH_3CH_2OCH_2CH_3$	Diethyl ether
Aldehyde	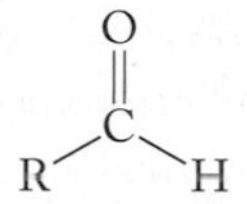	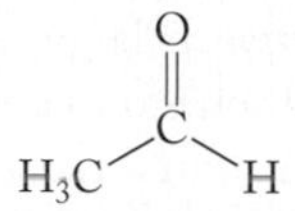	Acetaldehyde
Ketone	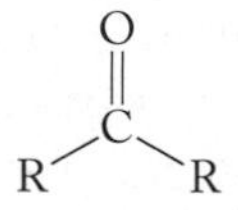	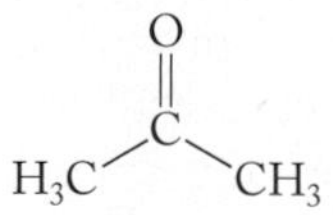	Acetone
Carboxylic acid	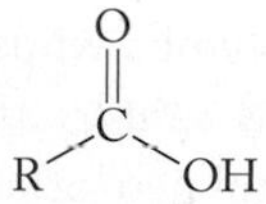	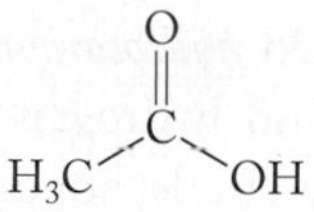	Acetic acid
Ester	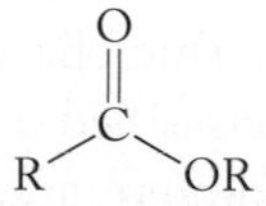	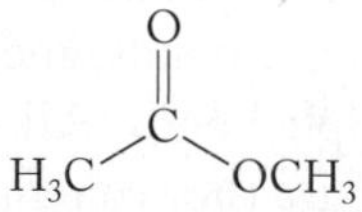	Methyl acetate
Amide	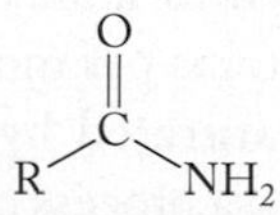	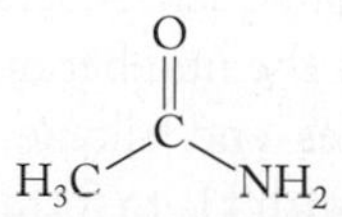	Acetamide

Many polymer molecules are so large that they cannot acquire sufficient kinetic energy to enter the gas phase. Synthetic organic polymers tend to soften and become more malleable as they are heated. Those that eventually melt usually form very viscous liquids; others remain solids up to temperatures at which their covalent bonds break and they undergo thermal decomposition. For example, plastic grocery bags are made of the polymer polyethylene. The monomer from which it is made is commonly called ethylene (H_2C=CH_2), though its

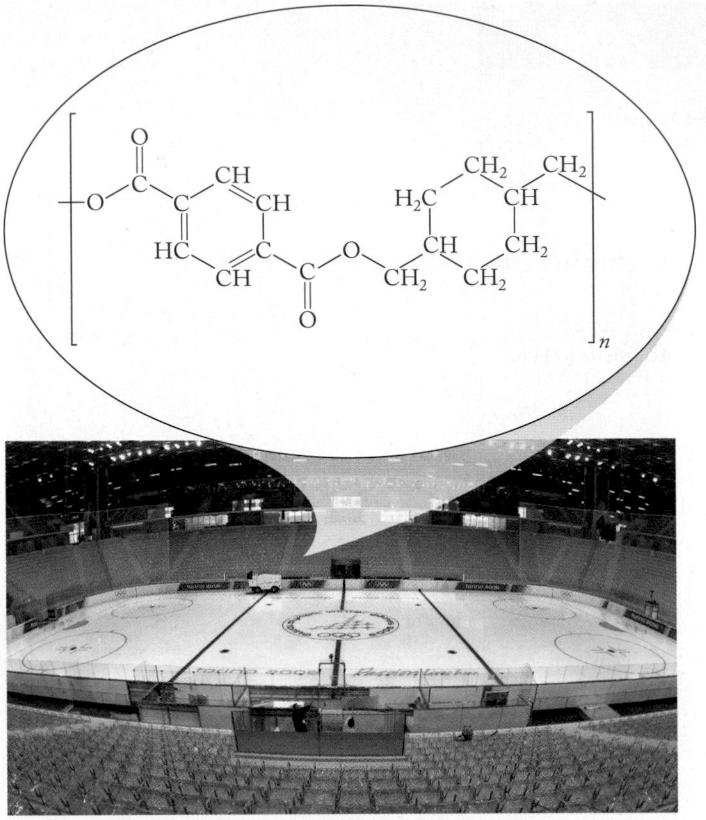

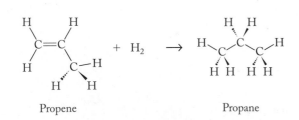

FIGURE 19.2 Seats in this Olympic ice hockey arena were made to look like ice using a special polymer called DuraStar.

official IUPAC (International Union of Pure and Applied Chemistry) name is ethene. The melting point of ethylene is −169°C; its boiling point is −104°C. The polyethylene from which grocery bags are made is a tough and flexible solid at room temperature. It softens gradually over a range of temperatures from 85°C to 110°C, and at higher temperatures it tends to decompose into molecules of low molar-mass organic gases. On the other hand, the polymer used to manufacture the seats shown in Figure 19.2 is a liquid at high temperatures. It is injected into molds and then cooled to make one-piece solid objects.

In addition, intermolecular forces between long chains in polymers can lead to the formation of both highly ordered, crystalline regions and less ordered, amorphous regions. When a polymer is heated, disruption of the inter-molecular forces between its strands and rearrangements of its bonds contribute to a broad range of melting temperatures. For now, we conclude that matter composed of polymers behaves very differently from matter composed of small organic molecules.

CONCEPT TEST

Which types of intermolecular forces are most likely to dominate between large molecules (polymers) containing long chains of carbon atoms bound to hydrogen atoms?

(Answers to Concept Tests are in the back of the book.)

19.2 Alkanes

In Chapter 6, we defined *hydrocarbons* as compounds whose molecules are composed of only carbon and hydrogen atoms, and *alkanes* as hydrocarbons whose molecules contain only single bonds. There are two other major categories of hydrocarbons: **alkenes**, whose molecules have one or more carbon–carbon double bonds, and **alkynes**, which have one or more carbon–carbon triple bonds. Alkanes are classified as **saturated hydrocarbons** because they contain the maximum ratio of hydrogen atoms to carbon atoms. The general molecular formula of alkanes is C_nH_{2n+2}, where n is the number of carbon atoms per molecule.

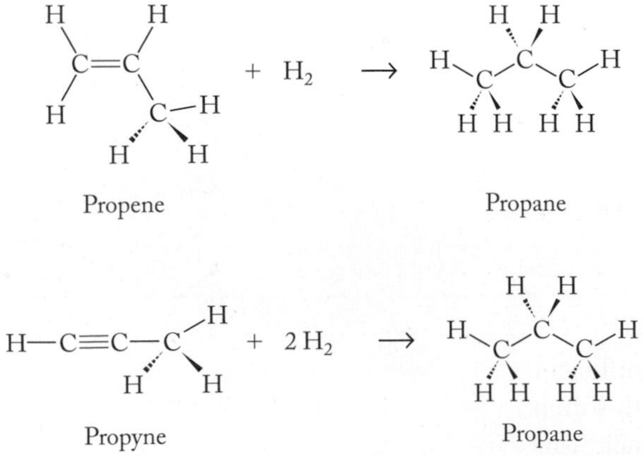

Alkenes and alkynes are **unsaturated hydrocarbons** that can combine with H_2 to form alkanes in a process called **hydrogenation**. A molecule with one C=C bond is described as having one *degree of unsaturation*. It combines with one molecule of H_2 to form one molecule of alkane (Figure 19.3). A molecule with one C≡C bond has two degrees of unsaturation and combines with two molecules of H_2, and a molecule with one double bond and one triple bond has three degrees of unsaturation. Determining the degree of unsaturation of an unknown compound using a hydrogenation reaction can help to identify its structure.

Propene + H₂ → Propane

Propyne + 2 H₂ → Propane

FIGURE 19.3 Hydrogenation of propene and propyne requires 1 mole of H_2 per mole of propene and 2 moles of H_2 per mole of propyne.

SAMPLE EXERCISE 19.1 **Distinguishing among Alkanes, Alkenes, and Alkynes** **LO1**

Three hydrocarbons, each containing four carbon atoms, are stored in separate, unlabeled flasks. If one is an alkane, one is an alkene with one double bond, and one is an alkyne with one triple bond, design an experiment based on their reactivity with H_2 gas to determine which is the alkane, which is the alkene, and which is the alkyne, given that each flask contains one mole of compound.

COLLECT AND ORGANIZE We are reacting three compounds with hydrogen. The compounds contain the same number of carbon atoms, but one compound contains a carbon–carbon double bond and another contains a carbon–carbon triple bond.

ANALYZE The alkane is a saturated hydrocarbon, which means it already contains the maximum number of hydrogen atoms possible and does not react with hydrogen. However, the alkene and alkyne are unsaturated, so they can both react with H_2 gas. One mole of the alkene with one double bond (or one degree of unsaturation) reacts with 1 mole of H_2. One mole of the alkyne with one triple bond (or two degrees of unsaturation) reacts with 2 moles of H_2. The flask containing the alkyne will require the most H_2 to react completely, while the flask containing the alkane will not react at all with H_2.

SOLVE If we allow 1 mole of each hydrocarbon to react with 2 moles of hydrogen and measure how much hydrogen is consumed, we can identify the compounds. We pick 2 moles of H_2 because that is the maximum amount that any of the three samples requires for complete reaction:

$$1 \text{ mol alkane} + 2 \text{ mol } H_2 \xrightarrow{\text{no reaction}} 2 \text{ mol } H_2 \text{ left}$$

$$1 \text{ mol alkene} + 2 \text{ mol } H_2 \rightarrow 1 \text{ mol alkane} + 1 \text{ mol } H_2 \text{ left}$$

$$1 \text{ mol alkyne} + 2 \text{ mol } H_2 \rightarrow 1 \text{ mol alkane} + 0 \text{ mol } H_2 \text{ left}$$

THINK ABOUT IT Measuring the amount of hydrogen that reacts with a hydrocarbon is a way to distinguish saturated from unsaturated hydrocarbons. As predicted, the alkyne, with two degrees of unsaturation, reacted with the most hydrogen. Unsaturated hydrocarbons may be distinguished if their degree of unsaturation differs.

Practice Exercise The labels have fallen off two containers in the lab. One label has structure A printed on it, and the other has structure B printed on it as shown below:

$$CH_3{-}CH_2{-}CH{=}CH{-}CH{=}CH{-}CH_3$$
Compound A

$$CH_3{-}CH{=}CH{-}CH{=}CH{-}CH{=}CH_2$$
Compound B

Can you use hydrogenation reactions to decide which label belongs on which container? What is the structure of the product in each case?

(Answers to Practice Exercises are in the back of the book.)

alkene hydrocarbon containing one or more carbon–carbon double bonds.

alkyne hydrocarbon containing one or more carbon–carbon triple bonds.

saturated hydrocarbon an alkane.

unsaturated hydrocarbon an alkene or alkyne.

hydrogenation the reaction of an unsaturated hydrocarbon with hydrogen.

CONCEPT TEST

Could we use information from hydrogenation reactions to distinguish a hydrocarbon with one triple bond from a hydrocarbon with two double bonds?

In the remainder of this section we look at the properties of the alkanes, deferring until Section 19.3 our coverage of the alkenes and alkynes.

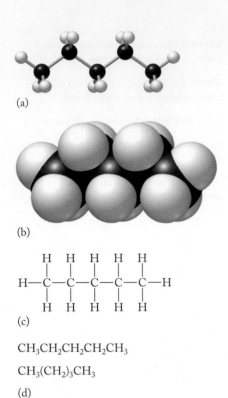

FIGURE 19.4 Several representations of the molecular structure of pentane: (a) ball-and-stick model; (b) space-filling model; (c) Lewis or Kekulé structure; (d) two condensed structures; (e) carbon-skeleton structure.

CONNECTION Kekulé structures are the same as the structural formulas we have been drawing throughout the text since Chapter 1.

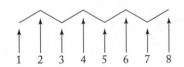

(a) Condensed structure

(b) Carbon-skeleton structure

FIGURE 19.5 When converting from (a) a condensed structure to (b) a carbon-skeleton structure, the carbon atoms are represented by junctions between lines, and each junction is assumed to have a sufficient number of H atoms to give that carbon atom four bonds.

Drawing Organic Molecules

Figure 19.4 shows some of the ways we represent the molecular structure of organic compounds—in this case, the five-carbon alkane pentane. In its Lewis structure, all of the bonds in the molecule are shown along with any lone pairs on the atoms. Alkanes do not have lone pairs on any of their atoms, but lone pairs are common in other organic molecules. When we draw structures of organic molecules showing all the bonds with lines but leaving off any lone pairs, the structural formulas are called **Kekulé structures** after August Kekulé (1829–1896), the German chemist who first used this method of illustrating molecules.

As you might imagine, writing Lewis or Kekulé structures for large organic molecules can be tedious. For this reason, chemists use various shorter notations to convey structures in organic chemistry, such as the *condensed structures* shown in Figure 19.4(d). These structures are condensed in that they do not show the individual bonds between atoms the way a Lewis structure does.

Some condensed structures use subscripts to indicate the number of times a particular subgroup is repeated. For example, the condensed structure of pentane can be written as $CH_3(CH_2)_3CH_3$ (Figure 19.4d). The numerical subscript "3" after the parenthetical methylene group means that three of these groups connect the two terminal methyl groups in this compound. (Note that the subscript indicating the number of methylene groups comes *after* the closing parenthesis. The subscript "2" inside the parentheses is for the two H atoms on the C atom of each methylene group.)

The most minimal notation, shown in Figure 19.4(e), is the *carbon-skeleton structure*, where no alphabetic symbols are used for carbon and hydrogen atoms. (Atoms other than C and H are shown in carbon-skeleton structures, as we see in Section 19.5.) Figure 19.5 shows how a carbon-skeleton structure is created. Short line segments are drawn at angles to one another, and each line segment symbolizes one carbon–carbon bond in the molecule. The angles represent the bond angle between the two carbon atoms (109.5° in the case of sp^3 hybridized carbon atoms in alkanes). Each end of the zigzag line is a –CH$_3$ group, and every intersection of two line segments is a –CH$_2$– group. In other words, it is understood that each carbon atom has the appropriate number of hydrogen atoms to give it a steric number of four. Hydrogen atoms are not shown in a carbon-skeleton structure because all carbon atoms are known to make four bonds, and any bonds not shown are understood to be C—H bonds.

Physical Properties and Structures of Alkanes

Alkanes are also known as *paraffins*, a name derived from Latin meaning "little affinity." This is a perfect description of the alkanes, which tend to be much less reactive than the other hydrocarbon families. Despite their lack of reactivity, alkanes are important because they are widely used as fuels, oils, and lubricants. *Unreactive* may not seem like the correct term to apply to compounds that are fuels, but alkanes do not react readily, even with oxygen. As evidence of this, consider that most fuels need some source of energy, such as a spark from an engine's spark plug, to initiate combustion.

In the language of organic chemistry, a **homologous series** is defined as a series of compounds in which members can be described by a general formula and similar chemical properties. The **normal (straight-chain) hydrocarbons** in the alkane family are a homologous series with the general formula C_nH_{2n+2}. Those whose molecules have three or more carbon atoms have the generic condensed structure $CH_3(CH_2)_nCH_3$. Each of these molecules differs from the next by one $-CH_2-$ unit, called a **methylene group**. The terminal $-CH_3$ groups are **methyl groups**. As we noted in Chapter 6, the physical properties of normal alkanes, including their viscosities, melting points, and boiling points, increase with increasing molar mass (see Figure 6.2). This trend is typical of all homologous series: as molecules increase in size, they experience stronger London dispersion forces, which means, for example, that higher temperatures are required to melt them if they are solids or to vaporize them if they are liquids.

CONNECTION We learned in Chapter 6 that the larger clouds of electrons in larger molecules are more polarizable, which leads to their experiencing stronger London dispersion forces.

SAMPLE EXERCISE 19.2   **Drawing Structures of Alkanes**   **LO2**

Write the condensed structure and the carbon-skeleton structure for the following normal alkanes: (a) the 3-carbon propane and (b) the 12-carbon dodecane.

COLLECT AND ORGANIZE Figure 19.4 summarizes the differences between condensed and carbon-skeleton structures, which we need to apply to normal hydrocarbons with 3 and 12 carbon atoms.

ANALYZE Condensed structures show the symbol for each element in a molecule and subscripts indicating the numbers of atoms of each element, but they do not show C—H or single C—C bonds. Each propane and dodecane chain has methyl groups ($-CH_3$) at both ends and one or more methylene groups ($-CH_2-$) in between. We can group methylene groups that are bonded together in a condensed structure into one term by using parentheses followed by a subscript showing the number of $-CH_2-$ groups. Carbon-skeleton structures use short line segments to represent the carbon–carbon bonds in molecules. The segments form a zigzag line because the intersections of these segments represent carbon atoms, which in alkanes form bonds at angles of about 109.5°.

SOLVE

a. Propane has three carbon atoms and the condensed structure $CH_3CH_2CH_3$. The carbon-skeleton structure consists of two segments representing the two carbon–carbon bonds:

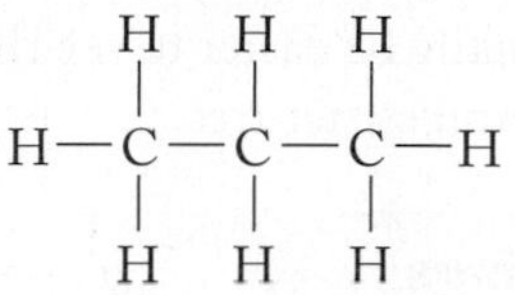

Lewis structure   Condensed structure   Carbon-skeleton structure

b. Dodecane is a continuous chain of 12 carbon atoms that has two methyl groups at the ends and ten methylene groups in between. The condensed structure is therefore $CH_3CH_2CH_2CH_2CH_2CH_2CH_2CH_2CH_2CH_2CH_2CH_3$ or, gathering the ten $-CH_2-$ groups: $CH_3(CH_2)_{10}CH_3$. The carbon-skeleton structure must contain as many ends and intersections as there are carbon atoms in the molecule:

THINK ABOUT IT Condensed and carbon-skeleton structures must both contain the same number of atoms and the same number of bonds, even though the bonds are not shown in the former and not all of the atoms or bonds are shown in the latter.

Kekulé structure a structure using lines to show all of the bonds in a covalently bonded molecule, but not showing lone pairs on the atoms.

homologous series a set of related organic compounds that differ from one another by the number of common subgroups, such as $-CH_2-$, in their molecular structures.

normal (straight-chain) hydrocarbon a hydrocarbon in which the carbon atoms are bonded together in one continuous line. Linear alkane chains have a methyl group at each end with methylene groups connecting them.

methylene group ($-CH_2-$), a structural unit that can make two bonds.

methyl group ($-CH_3$), a structural unit that can make only one bond.

Practice Exercise Draw the carbon-skeleton structure of hexane, $CH_3(CH_2)_4CH_3$, and a condensed structure of heptane:

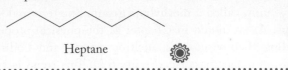

Heptane

Constitutional (Structural) Isomers

The alkane family would be huge even if it consisted of only straight-chain molecules, but another structural possibility arises that makes the family even larger. We first saw this possibility in Chapter 6 when we compared the boiling points of compounds having the same molecular formula, and hence the same molar mass, but different shapes (see Figure 6.3). For example, the four-carbon alkane has the molecular formula C_4H_{10}. However, the straight-chain molecule shown on the left in Table 19.2 is not the only shape possible, because with four or more carbons, a *branched* structure also exists, as shown on the right side of the table. A *branch* is a side chain attached to the main carbon skeleton.

The molecular formula of both structures in Table 19.2 is C_4H_{10}, but they represent different compounds with different properties and names (butane and 2-methylpropane). At this point it is not important to be able to name them but to recognize that they are *constitutional isomers* (also called *structural isomers*). As we learned in Chapter 6, such compounds have the same molecular formula but have their atoms connected in different bonding patterns. This difference makes constitutional isomers chemically distinct from one another.

2-Methylpropane is an example of a **branched-chain hydrocarbon**, which means that its molecular structure contains a main chain (the longest carbon chain in the molecule) and at least one side chain. We can illustrate constitutional isomerism with either condensed or carbon-skeleton structures as shown in Table 19.2. There are two ways to show the location of the methyl group. One way is to use condensed structures and to put the CH_3 group in parentheses next to the carbon to which it is bonded: $CH_3CH(CH_3)CH_3$. We can also show the carbon–carbon bond between the CH_3 and the center carbon of the three-carbon main chain. In this instance, it may actually be easier to see the constitutional isomers of C_4H_{10} using the carbon-skeleton structures.

TABLE 19.2	**Comparing Constitutional Isomers**	
	Butane	**2-Methylpropane**
Condensed Structure	$CH_3CH_2CH_2CH_3$	$CH_3CH(CH_3)CH_3$ or CH_3CHCH_3 | CH_3
Carbon-Skeleton Structure		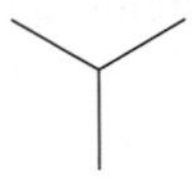
Melting Point (°C)	−138	−160
Normal Boiling Point (°C)	0	−12
Density of Gas (g/L)	0.5788	0.5934

branched-chain hydrocarbon an organic molecule in which the chain of carbon atoms is not linear.

SAMPLE EXERCISE 19.3   **Identifying the Longest Chain in Organic Molecules**   **LO2**

Determine the number of carbon atoms in the longest chain in the following two branched, saturated hydrocarbons:

a. $CH_3CH_2CH_2CH_2CHCH_3$
    |
    CH_2CH_3

$$\qquad\qquad CH_3$$
b. $CH_3CHCHCH_2CH_2CH_3$
    |
    CH_2CH_3

COLLECT AND ORGANIZE A branched, saturated hydrocarbon can be drawn in a variety of ways. We are asked to identify the longest carbon chain or *main chain* in two compounds. Both compounds are represented by condensed structures.

ANALYZE We assume that there is a bond between each carbon atom in the condensed structures. Carbon bonds to additional chains are indicated by a single line. We need to determine which of these branches belongs to the *main chain* and which belong to the *side chains*.

SOLVE We start at one end of any branch and assign numbers to each carbon atom. If the chain branches, we must choose one branch to follow. Later, we may wish to return to the compound and number the carbons starting from a different end of the molecule or taking a different branch.

a. Starting with the carbon atom furthest to the left and numbering consecutively from left to right, we reach a branch at carbon atom 5. If we continue to the right, we find that the chain contains six carbon atoms. If we take the branch at carbon atom 5, however, we find that the chain contains seven carbon atoms. This makes the longest chain in the compound seven carbon atoms long.

$$\overset{1}{C}H_3\overset{2}{C}H_2\overset{3}{C}H_2\overset{4}{C}H_2\overset{5}{C}H\overset{6}{C}H_3$$
|
CH_2CH_3

$$\overset{1}{C}H_3\overset{2}{C}H_2\overset{3}{C}H_2\overset{4}{C}H_2\overset{5}{C}H CH_3$$
|
$\underset{6\ \ 7}{CH_2CH_3}$

If we start at the right end of the molecule, we do not find a longer chain; the only other chain in the molecule is four carbon atoms long.

b. There are four places to start counting carbon atoms in structure b. Some possible ways to number the molecule are as follows:

$$CH_3$$
$$\overset{1}{C}H_3\overset{2}{C}H\overset{|3}{C}H\overset{4}{C}H_2\overset{5}{C}H_2\overset{6}{C}H_3$$
|
CH_2CH_3

$$CH_3$$
$$\overset{5}{C}H_3\overset{4}{C}H\overset{|3}{C}H\overset{2}{C}H_2\overset{1}{C}H_2 CH_3$$

The longest chain in this molecule (the main chain) also has seven carbon atoms.

THINK ABOUT IT The main chain in an organic compound may not always be the one running horizontally across the page.

Practice Exercise Determine the number of carbon atoms in the longest chain in the following two branched, saturated hydrocarbons:

a. 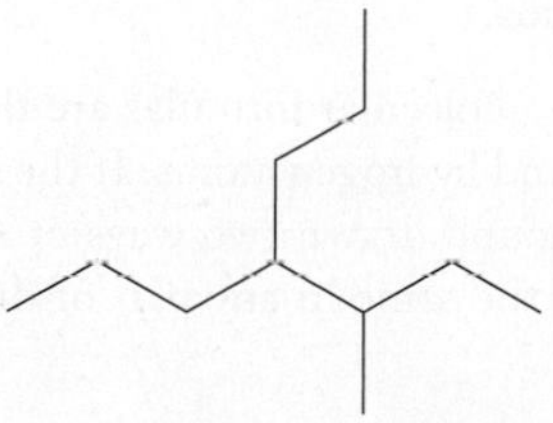   b.

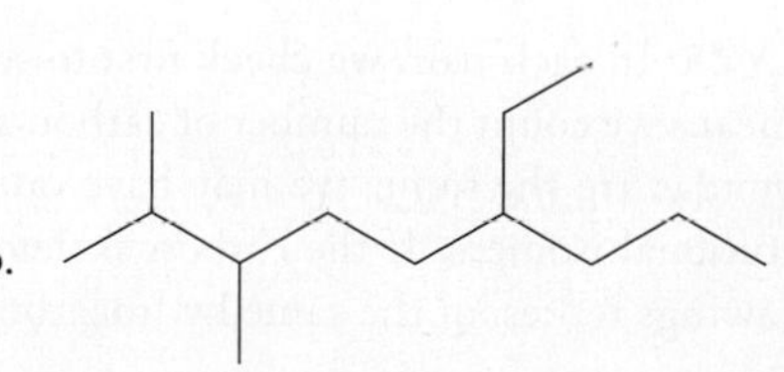

To determine whether two condensed structures or carbon-skeleton structures represent two different compounds, a single compound, or two constitutional isomers, our first step is to translate each structure into a molecular formula. If the molecular formulas are different, the structures represent two different compounds. If the molecular formulas are the same, we must compare the way the atoms are connected in the two structures. In doing so, we may have to reverse or rotate one of the structures. For example, structures 1 and 2 below may seem to be different, but rotating structure 2 by 180° shows that it is the same as structure 1. The two represent the same compound: a four-carbon main chain with a one-carbon side chain connected to the carbon atom next to a terminal carbon.

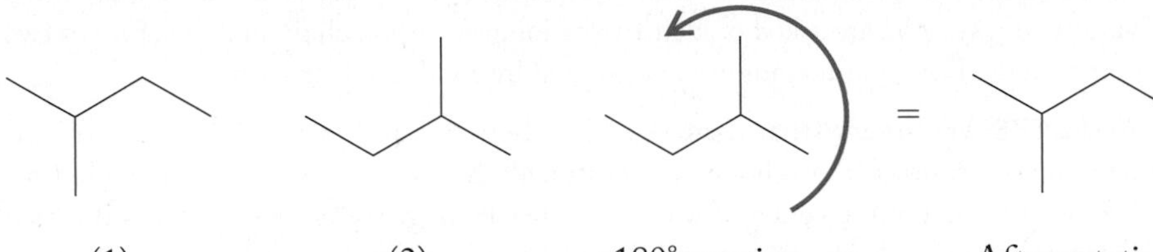

| (1) | (2) | 180° rotation | After rotation |

To determine whether two structures are identical or not, draw the structures with the longest chain horizontal, and then check to see whether the same side chains are attached at the same positions along the longest chain.

**SAMPLE EXERCISE 19.4 Recognizing Constitutional   LO3
Isomers**

Do the two structures in each set describe the same compound, constitutional isomers, or compounds with different molecular formulas?

a. $(CH_3)_2CHCH_2CH(CH_3)_2$

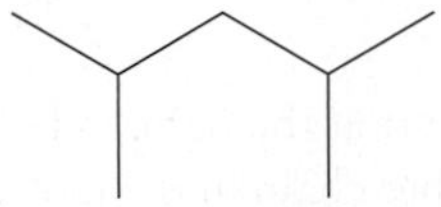

b.

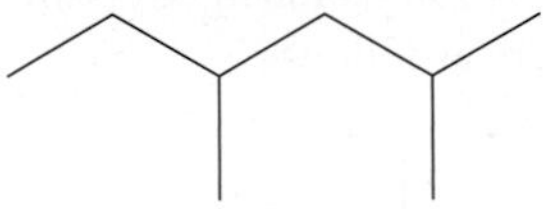

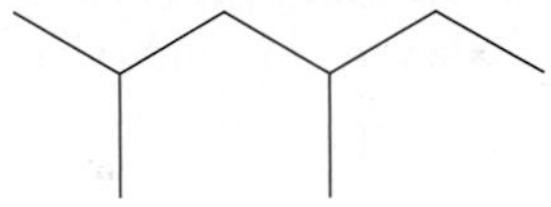

c.

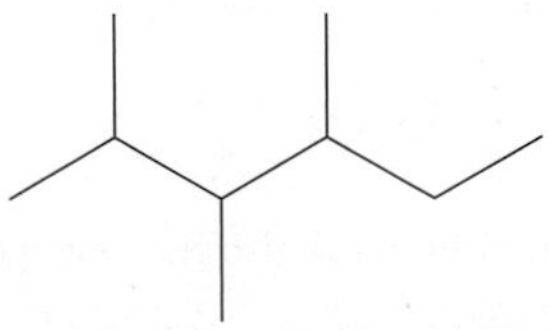

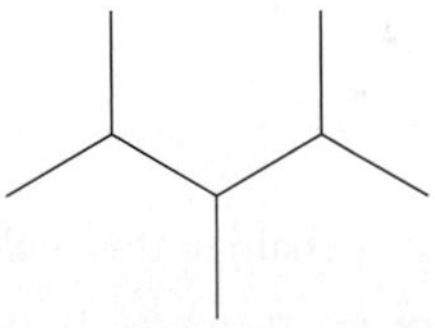

COLLECT AND ORGANIZE Identical compounds have the same molecular formula and the same connectivity and spatial arrangement of the atoms. Constitutional isomers have the same molecular formula but different connectivity. Compounds with different molecular formulas are different compounds.

ANALYZE In each pair, we check first to see if the molecular formulas are the same. That means we count the number of carbon atoms and hydrogen atoms. If the molecular formulas are the same, we may have one compound drawn two ways or a pair of constitutional isomers. If the carbon skeletons are the same in any pair of drawings, the drawings represent the same hydrocarbon.

SOLVE

a. The molecular formulas are the same: C_7H_{16}. Converting the condensed structure to a carbon-skeleton structure gives us

$$(CH_3)_2CHCH_2CH(CH_3)_2 \quad \longrightarrow$$

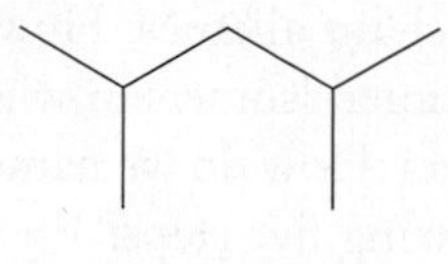

This structure is identical to the carbon-skeleton structure in this set. Therefore the condensed structure and carbon-skeleton structure represent the same molecule.

b. Both molecules in this set have the molecular formula C_8H_{18}. Both have six carbon atoms in the longest chain. Both have a methyl group attached to the second carbon atom from one end of the chain and another methyl group attached to the third carbon atom from the other end. Therefore the two structures represent the same compound:

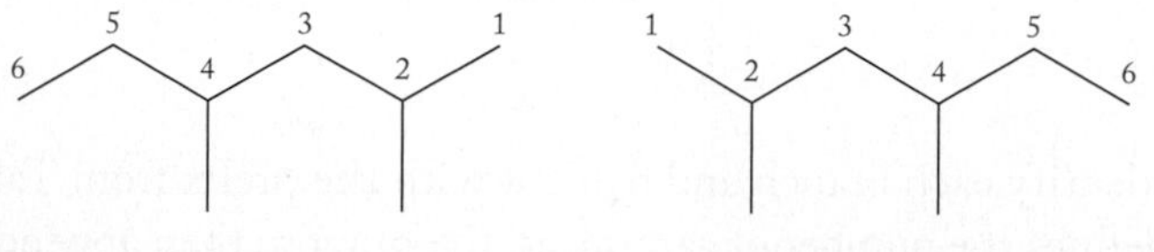

c. The left structure contains nine carbon atoms, but the right structure contains only eight. Therefore the two structures represent different compounds.

THINK ABOUT IT Even if two compounds have the same molecular formula, they are not necessarily identical. We must determine whether they are constitutional isomers.

Practice Exercise Do the two structures in each set describe the same compound, two different compounds, or a pair of constitutional isomers?

a.

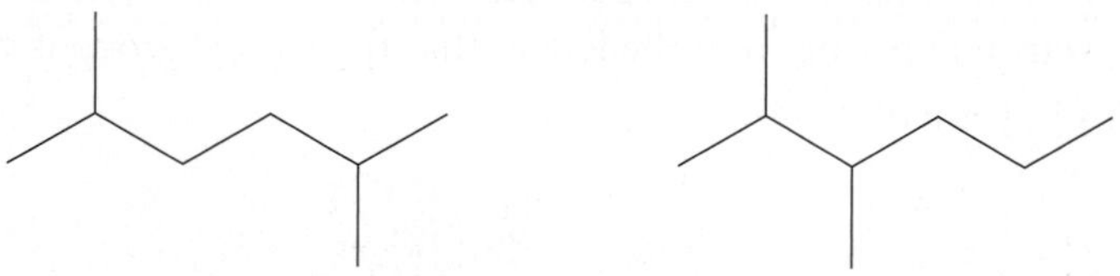

b.

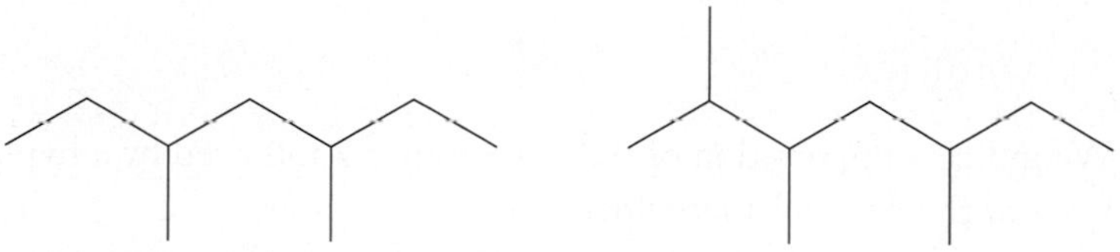

c.

Naming Alkanes

Now that we know how to draw alkanes, let's look at a few simple rules for naming them. The nomenclature system follows the same pattern for all families of organic compounds, so we begin with rules for naming alkanes and will add a few more for alkenes and alkynes. These rules are called IUPAC rules after the organization that defined them. Appendix 7 contains a more extensive treatment of nomenclature for organic compounds.

TABLE 19.3	Prefixes for Naming Alkanes and Other Hydrocarbons	
Prefix	Condensed Structure of Alkane	Name of Alkane
Meth-	CH_4	Methane
Eth-	CH_3CH_3	Ethane
Prop-	$CH_3CH_2CH_3$	Propane
But-	$CH_3(CH_2)_2CH_3$	Butane
Pent-	$CH_3(CH_2)_3CH_3$	Pentane
Hex-	$CH_3(CH_2)_4CH_3$	Hexane
Hept-	$CH_3(CH_2)_5CH_3$	Heptane
Oct-	$CH_3(CH_2)_6CH_3$	Octane
Non-	$CH_3(CH_2)_7CH_3$	Nonane
Dec-	$CH_3(CH_2)_8CH_3$	Decane

The names of the C_1 through C_{10} alkanes, in which all carbon atoms are bonded to no more than two other carbon atoms, start with the prefixes in Table 19.3 followed by -ane. Those molecules with four or more carbon atoms are called straight-chain alkanes. However, we have seen that straight-chain alkanes have constitutional isomers in which at least one carbon atom is bonded to more than two others. How do we name these branched alkanes? The answer is that we take the following five steps:

1. Select the longest chain of carbon atoms and use the prefix in Table 19.3 to name this as the *parent structure*. For the structure below, the parent name is pentane because the longest chain is 5 carbon atoms long:

$$CH_3CH(CH_3)CH_2CH_2CH_3$$

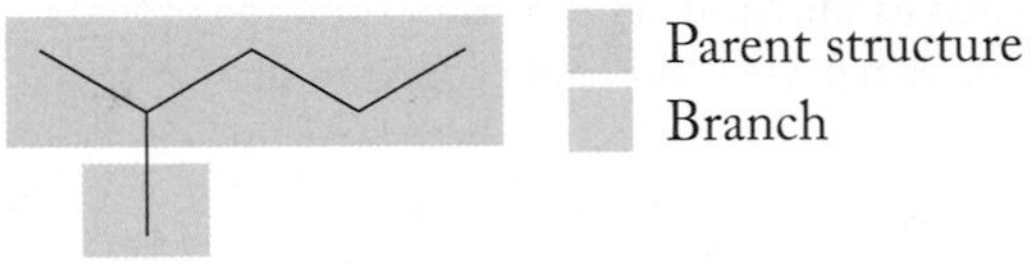

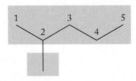

   Parent structure
   Branch

2. Identify each branch and name it with the prefix from Table 19.3 that defines the number of carbons in the branch; then append the suffix -yl to the prefix. A $-CH_3$ group is methyl, a $-CH_2CH_3$ group is ethyl, and so on. The structure from step 1 has a methyl group branch. The name of the branch comes before the name of the parent structure, and the two are written together as one word: methylpentane.

3. To indicate the point where the branch is attached, we number the carbon atoms in the parent chain so that the branch (or branches, if there are more than one) has the lowest possible number. In the molecule from step 1, we start numbering from the left so that the methyl group is on carbon atom 2 in the parent chain:

   We indicate the position of the branch by a 2 followed by a hyphen in front of the name: 2-methylpentane.

4. If the same group is attached more than once to the parent structure, we use the prefixes di-, tri-, tetra-, and so on to indicate the number of groups present. The position of each group is indicated by the appropriate number before the group name, with the numbers separated by commas. Thus the name of the following structure is 2,4-dimethylpentane:

$$CH_3CH(CH_3)CH_2CH(CH_3)CH_3$$
$$1 \quad 2 \quad\quad 3 \quad 4 \quad\quad 5$$

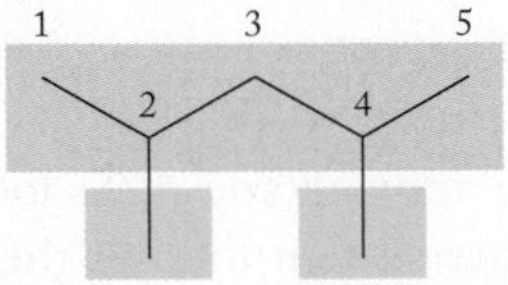

5. If different groups are attached to a parent chain, they are named in alphabetical order, as in 3-ethyl-4-methylheptane for the following structure:

$$\underset{1}{\text{CH}_3}\underset{2}{\text{CH}_2}\underset{3}{\text{CH}}(\text{CH}_2\text{CH}_3)\underset{4}{\text{CH}}(\text{CH}_3)\underset{5}{\text{CH}_2}\underset{6}{\text{CH}_2}\underset{7}{\text{CH}_3}$$

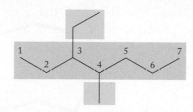

Note that we begin numbering the parent chain in this structure at the carbon on the far left, as described in step 3, so that the branch locations have the lowest possible numbers. For example, the compound shown here is 5-ethyl-2-methyloctane, not 4-ethyl-7-methyloctane:

$$\underset{1}{\text{CH}_3}\underset{2}{\text{CH}}\underset{3}{\text{CH}_2}\underset{4}{\text{CH}_2}\underset{5}{\text{CH}}\underset{6}{\text{CH}_2}\underset{7}{\text{CH}_2}\underset{8}{\text{CH}_3}$$

with $\underset{}{\overset{\text{CH}_3}{|}}$ at position 2 and $\underset{}{\overset{\text{CH}_2\text{CH}_3}{|}}$ at position 5

Correct numbering

$$\underset{8}{\text{CH}_3}\underset{7}{\text{CH}}\underset{6}{\text{CH}_2}\underset{5}{\text{CH}_2}\underset{4}{\text{CH}}\underset{3}{\text{CH}_2}\underset{2}{\text{CH}_2}\underset{1}{\text{CH}_3}$$

with $\underset{}{\overset{\text{CH}_3}{|}}$ and $\underset{}{\overset{\text{CH}_2\text{CH}_3}{|}}$

Incorrect numbering

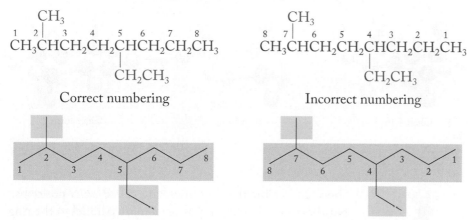

Cycloalkanes

Alkanes can form ring structures (Figure 19.6). These **cycloalkanes** have the general formula C_nH_{2n}, which is different from the general formula for straight-chain alkanes (C_nH_{2n+2}) because a cycloalkane has one more carbon–carbon bond and two fewer hydrogen atoms per molecule than a linear or branched alkane with the same number of carbon atoms. Although cycloalkanes have no C=C or C≡C bonds, 1 mole of a cycloalkane can, at least in theory, react with 1 mole of hydrogen to make 1 mole of a straight-chain alkane. In practice, however, only cycloalkanes with three-carbon and four-carbon rings react with hydrogen.

The left column of Figure 19.6(a) shows the condensed structure and carbon-skeleton structure of cyclohexane drawn in two dimensions, where the C—C—C bond angles appear to be 120°. Because the carbon atoms have sp^3 hybrid orbitals, we expect the angles to be 109.5°, however, and indeed they are in this molecule. A more accurate representation of the ring is shown in the adjacent structures, in which the ring is puckered instead of flat and the bond angles are 109.5°.

There are two possible conformations of cyclohexane that have the greatest structural stability. They are called *chair* forms, and a molecule of cyclohexane spends over 99% of its time in one chair form or the other. It can also flip back and forth between the two chair conformations as shown in Figure 19.6(b). However, to get from one chair conformation to the other the molecule must pass through higher energy transition-state conformations. One of these high-energy states is called the *boat* conformation in which repulsion between the two hydrogen atoms across the ring from each other adds to internal energy and reduces stability. Note how the H atoms in the left chair structure that are highlighted in light blue are

▶‖ **CHEMTOUR** Cyclohexane in 3-D

▶‖ **CHEMTOUR** Structure of Cyclohexane

cycloalkane a ring-containing alkane with the general formula C_nH_{2n}.

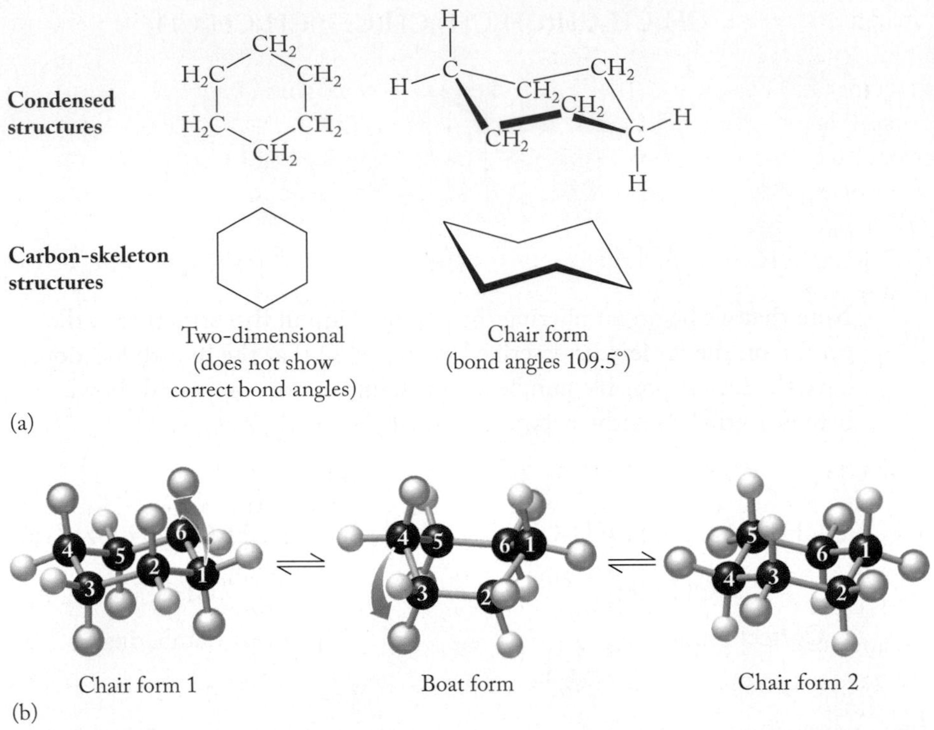

Condensed structures

Carbon-skeleton structures

Two-dimensional
(does not show
correct bond angles)

Chair form
(bond angles 109.5°)

(a)

(b)

Chair form 1   Boat form   Chair form 2

FIGURE 19.6 (a) The six-membered ring of cyclohexane may be drawn either flat or in styles that show its actual three-dimensional puckered conformations. (b) The most stable are two chair-shaped conformations shown in the left and right structures. A molecule can flip back and forth between them, but it has to temporarily adopt less stable shapes, such as the boat conformation shown in the middle structure, to do so.

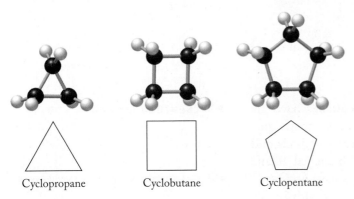

Cyclopropane   Cyclobutane   Cyclopentane

FIGURE 19.7 Cyclic alkanes have the general formula C_nH_{2n} and are considered to be unsaturated hydrocarbons. Shown here are cyclopropane ($n = 3$), cyclobutane ($n = 4$), and cyclopentane ($n = 5$).

above and below the ring (these are called *axial* positions), but these H atoms are in positions more parallel to the ring (called *equatorial* positions) when the molecule flips to the chair structure on the right.

Cycloalkanes with fewer than six carbon atoms are possible, but they are not as stable as cyclohexane. One of the important features that determine stability is the C—C—C bond angle. For example, cyclopropane (a C_3 ring) exists but is a highly reactive species because the interior ring angles are 60°, far from the ideal bond angle of 109.5° for a sp^3 hybridized carbon atom (Figure 19.7). No puckering is possible in a three-membered ring to relieve the strain in this system, so cyclopropane tends to react in a fashion that opens up the ring and relieves the strain. The C—C—C bond angles in cyclobutane and cyclopentane are larger, so there is less ring strain. Rings of six sp^3 hybridized carbons and beyond are essentially the same as straight-chain alkanes in terms of bond angle and have no ring strain. Six-membered rings are the most favored, because other thermodynamic factors have an impact on the formation of rings with seven or more carbons.

Sources and Uses of Alkanes

The principal source of liquid alkanes on Earth is crude oil. Natural gas is the major source for the simplest alkane, methane, as well as smaller quantities of low-molar-mass alkanes including ethane, propane, and perhaps some butanes. Methane is often associated with oil deposits but is also produced during bacterial decomposition of vegetable matter in the absence of air, a condition that frequently arises in swamps. Hence, methane's common name is swamp gas or marsh gas (Figure 19.8). Frequently, in the reductive environment of a marsh, a molecule known as phosphine (PH_3) also forms. Phosphine and methane together spontaneously ignite; this produces a ghostly flame known as will-o'-the-wisp that features prominently in some legends and gothic mysteries. Methane can also be produced in coal mines, where it can be especially dangerous because it forms explosive mixtures with humid air, giving rise to another common name for the gas: firedamp.

By far the most common use of alkanes in our lives is as fuels. Combustion reactions between alkanes and oxygen provide heat to power vehicles, generate electricity, warm our homes, and prepare our meals. Gasoline, kerosene, and diesel fuels are mostly mixtures of alkanes. Alkanes with higher boiling points are viscous liquids used as lubricating oils. Low-melting solid alkanes (C_{20}–C_{40}) are used in candles and in manufacturing matches. Very heavy hydrocarbon gums and solid residues (C_{36} and up) are used as asphalt for paving roads.

The combustion of fossil fuels is a timely topic for several reasons. The production of CO_2, a greenhouse gas, leads to the potential for climate change. The supplies of crude oil are also limited, and with increasing world demand for fossil fuels, society faces choices in how best to allocate finite supplies of oil. The alkanes distilled from crude oil serve not only as fuels but also as the starting material for building more complex molecules. They are the major source of carbon for the chemical manufacturing industry.

Some alkanes have medicinal use. Mineral oil is a mixture of C_{15}–C_{24} alkanes used as a skin ointment ("baby oil") to treat diaper rash and to alleviate some forms of eczema. It is used in many cosmetics, creams, and ointments. Taken orally, mineral oil acts as a laxative. As we continue our study of organic compounds, we will return to these dual themes of fuels and pharmaceuticals with the functional groups that we encounter.

FIGURE 19.8 Methane bubbles trapped in a frozen pond. Methane is produced by rotting organic matter at the bottom of the pond.

◎◎ CONNECTION In Chapter 11 we discussed the distillation of crude oil to produce gasoline, kerosene, and other hydrocarbon mixtures useful as fuels and as feedstocks for the chemical industry.

◎◎ CONNECTION In Chapter 4 we discussed methane's role as a greenhouse gas.

19.3 Alkenes and Alkynes

In Section 19.2 we introduced alkenes, compounds having one or more carbon–carbon double bonds (two or more sp^2 hybridized carbons), and alkynes, compounds having one or more carbon–carbon triple bonds (two or more sp hybridized carbons). Both these families represent unsaturated hydrocarbons. Alkenes are minor components of crude oil and are prevalent in many plants including pine needles, celery, and ginger (Figure 19.9).

FIGURE 19.9 Some naturally occurring alkenes.

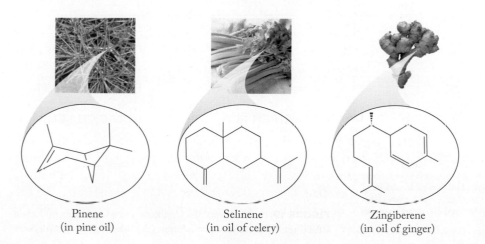

Pinene
(in pine oil)

Selinene
(in oil of celery)

Zingiberene
(in oil of ginger)

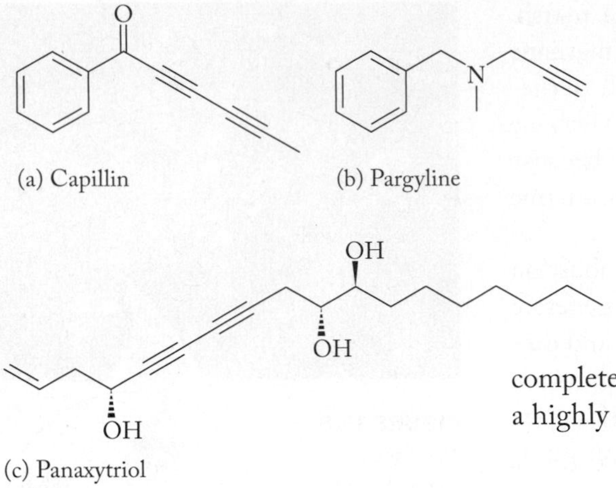

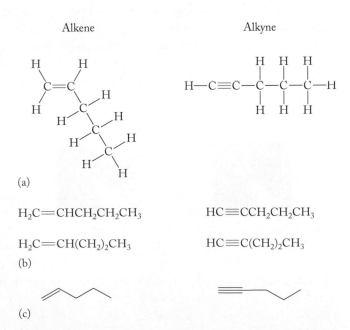

(a) Capillin   (b) Pargyline

(c) Panaxytriol

FIGURE 19.10 (a) Capillin, an antifungal drug, (b) paragyline, used to treat hypertension, and (c) panaxytriol, a potent antitumor drug, all contain the alkyne functional group.

Alkynes are found in crude oil as well, but they are generally more difficult to find in nature. Nevertheless, carbon–carbon triple bonds are found in some drugs along with other functional groups. Some examples are shown in Figure 19.10: capillin, an antifungal drug; pargyline, used to treat hypertension; and panaxytriol, a potent antitumor compound isolated from ginseng. The simplest alkyne, C_2H_2, has the official name ethyne but is better known by its common name, acetylene.

Acetylene and other alkynes are manufactured by the controlled oxidation of alkanes. When the simplest alkane, methane, is oxidized completely, the products are CO_2 and H_2O. But if the oxidation is carried out in a highly controlled process, acetylene is one of the products:

$$6\ CH_4(g) + O_2(g) \rightarrow 2\ HC\equiv CH(g) + 2\ CO(g) + 10\ H_2(g) \quad (19.1)$$

This reaction illustrates an important difference between alkanes and the unsaturated hydrocarbons: alkanes are the most reduced form of carbon. The sp^2 hybridized carbons in alkenes are in a higher oxidation state than the sp^3 hybridized carbons of alkanes, and the sp hybridized carbons in alkynes are in an even higher oxidation state. Their capacity to be reduced makes alkenes and alkynes more reactive than alkanes, and the carbon–carbon double bond in alkenes is one of the most versatile functional groups in organic chemistry. Structures of a C_5 alkene and a C_5 alkyne are shown in Figure 19.11.

Alkenes and alkynes share many features of alkanes. The melting and boiling points of homologous series of these compounds (Table 19.4) vary with molar mass, just as with the alkanes.

Molecules may contain more than one alkene or alkyne group, as shown in Figure 19.10. In particular, many molecules found in crude oil or produced by living systems have several double bonds. We discuss molecules

TABLE 19.4 Melting Points and Normal Boiling Points of Homologous Series of Alkenes and Alkynes

Condensed Structure: Alkene	Melting Point (°C)[a]	Normal Boiling Point (°C)
$H_2C\!=\!CHCH_3$	−185	−47
$H_2C\!=\!CHCH_2CH_3$	−185	−6
$H_2C\!=\!CH(CH_2)_2CH_3$	−138	30
$H_2C\!=\!CH(CH_2)_3CH_3$	−140	63
$H_2C\!=\!CH(CH_2)_4CH_3$	−119	94
$H_2C\!=\!CH(CH_2)_5CH_3$	−104	123
$H_2C\!=\!CH(CH_2)_6CH_3$	−81	146
$H_2C\!=\!CH(CH_2)_7CH_3$	−87	171
Condensed Structure: Alkyne		
$HC\equiv CCH_3$	−102	−23
$HC\equiv CCH_2CH_3$	−126	8
$HC\equiv C(CH_2)_2CH_3$	−90	40
$HC\equiv C(CH_2)_3CH_3$	−132	71
$HC\equiv C(CH_2)_4CH_3$	−81	100

[a]Melting points increase with molar mass but also depend on how molecules fit into crystal lattices. Melting points of alkenes with even numbers of carbon atoms form one series that follows this trend; alkenes with odd numbers of carbon atoms form another series.

Alkene   Alkyne

(a)

$H_2C\!=\!CHCH_2CH_2CH_3$ $HC\equiv CCH_2CH_2CH_3$

$H_2C\!=\!CH(CH_2)_2CH_3$ $HC\equiv C(CH_2)_2CH_3$

(b)

(c)

FIGURE 19.11 Structures of a C_5 alkene and a C_5 alkyne. (a) Lewis structures. (b) Condensed structures. (c) Carbon-skeleton structures.

with multiple double bonds in Chapter 20, but for now we concentrate on the properties associated with small molecules containing only one or a small number of double or triple bonds.

addition reaction a reaction in which two molecules couple together and form one product.

Chemical Reactivities of Alkenes and Alkynes

Figure 19.12 shows the electron distributions in the π bonds in an alkene and an alkyne. Electron density in these bonds is greatest around the bonding axis, which makes π electrons more accessible to reactants than the electrons in the σ bonds between carbon atoms. For this reason, unsaturated hydrocarbons are more reactive than saturated ones.

As an illustration of this difference in reactivity, consider how alkanes and alkenes react with the hydrogen halides HCl, HBr, and HI. With an alkane, there is no reaction:

$$HX(g) + H_3C\!-\!CH_3(g) \rightarrow \text{no reaction} \qquad (19.2)$$

With an alkene, however, the hydrogen halide reacts with the double bond to make an alkyl halide (an alkane in which a halogen has been substituted for one of the hydrogen atoms):

$$\underset{\substack{\text{Hydrogen} \\ \text{halide}}}{HX(g)} + \underset{\text{Alkene}}{H_2C\!=\!CH_2(g)} \rightarrow \underset{\substack{\text{Alkyl} \\ \text{halide}}}{CH_3CH_2X(\ell)} \qquad (19.3)$$

This reaction is called an **addition reaction** because two reactants combine (that is, they add together) to form one product.

Alkynes also react with hydrogen halides. They differ, however, in that two molecules of a hydrogen halide react with the triple bond to produce an alkane bearing two halogen atoms:

$$2\,HBr(g) + HC\!\equiv\!CH(g) \rightarrow CH_3CHBr_2(\ell) \qquad (19.4)$$

Because both double and triple bonds react with many reagents in addition to hydrogen halides, alkenes and alkynes are useful substances in the industrial production of other compounds.

Isomers of Alkenes and Alkynes

Molecules that contain alkene and alkyne functional groups can have straight or branched chains, with the same types of constitutional isomers we saw with alkanes. One additional facet of constitutional isomerization involves the location of the double or triple bond in a molecule. For example, consider the straight-chain isomers of pentene, the alkene that contains five carbon atoms and one double bond (Figure 19.13). Applying the test used in Section 19.2 for alkanes, we see that structures a and d in Figure 19.13 are the same. This is easier to see if we look at the carbon-skeleton structures in the figure. In both cases the double bond is between carbon atoms 1 and 2, that is, between C1 and C2. Remember, with branched-chain alkanes, we learned to number the carbons from whichever end gives the carbon attached to the branch the lowest number. The same rule holds for functional groups like the double bond shown here: the carbons in structure d must be numbered from right to left.

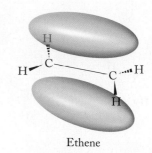
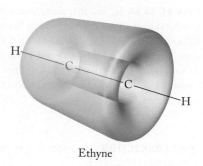

FIGURE 19.12 Shapes of the π bonds in ethene and ethyne.

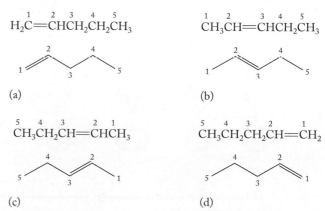

FIGURE 19.13 Four possible constitutional isomers of pentene, C_5H_{10}. Notice that structures a and d are identical, as are b and c.

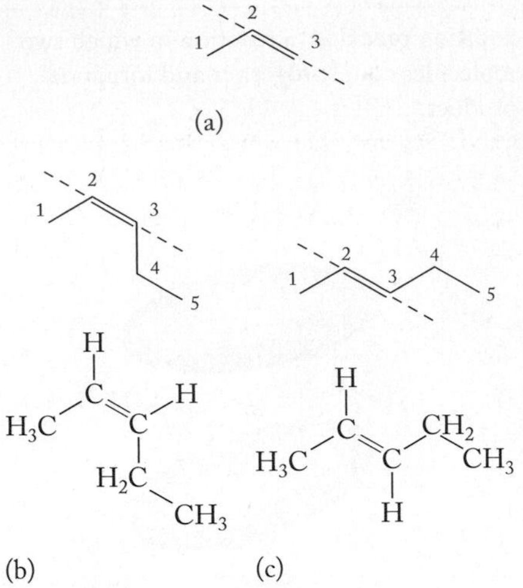

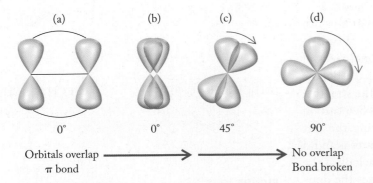

(a)

(b)   (c)

FIGURE 19.14 (a) The first three atoms of a carbon chain with a double bond between C2 and C3. (b) The chain continues on the same side of the double bond as C1 in the cis isomer. (c) The chain continues on the opposite side of the double bond as C1 in the trans isomer.

∞ CONNECTION Stereoisomerism was introduced in Chapter 5.

∞ CONNECTION The use of cis and trans in the names of simple organic molecules with C=C bonds parallels the use of those terms to indicate positions of ligands about a metal center in a complex ion (Chapter 16).

∞ CONNECTION In Chapter 9 we introduced rotations about bonds as one of the types of motion molecules experience as part of their overall kinetic energy.

Structures b and c are constitutional isomers of a and d because they have the same chemical formula but have the double bond in a different location. Drawing the carbon-skeleton structures of b and c, however, presents us with a new situation. After we draw the first three atoms of structure b, as in Figure 19.14(a), and draw a straight dashed line through the double bond, we see that we have two options for how to orient the rest of the molecule relative to the double bond. We can place the bond between C3 and C4 on the same side as the methyl group at C1 (Figure 19.14b) or on the opposite side (Figure 19.14c).

These two molecules are isomers of each other because they have the same molecular formula but different structures and therefore different properties. The isomer in Figure 19.14(b) is called either the **Z isomer** (Z stands for the German word *zusammen* or "together") or the **cis isomer** (cis is Latin for "on this side"), which in this case translates to "the methyl group and the chain after the double bond are both *together* or *on this side* of the structure." The isomer in Figure 19.14(c) is called either the **E isomer** (E for *entgegen* or "opposite") or **trans isomer** (trans is Latin for "across"). Because these isomers are characterized by differences in the spatial arrangement of their atoms, and not by how those atoms are connected to each other, they are *stereoisomers*. The cis and trans prefixes are widely used in the names of alkenes with simple structures; the *E/Z* system is needed for more complex molecules in which more than two different substituents are attached to a double bond.

Cis/trans isomers exist because there is no free rotation about the double bond. Recall from Chapter 5 that a double bond is formed from the overlap of two unhybridized *p* orbitals on adjacent carbon atoms. As Figure 19.15 shows, a carbon atom joined in a double bond cannot rotate freely about the bond axis without eliminating orbital overlap and breaking the bond. Breaking a π bond in 1 mole of an alkene costs about 290 kJ of energy, and that much energy is not available to the molecules at room temperature. This situation gives rise to restricted rotation about a carbon–carbon double bond and to the existence of stereoisomers.

Now let's examine the stereoisomers of structure c from Figure 19.13. In Figure 19.16(a), the two hydrogen atoms are on the same side of the double bond; this is the cis isomer. In Figure 19.16(b), the two hydrogen atoms are on opposite sides of the double bond; this is the trans isomer. The two molecules are stereoisomers.

A comparison of the stereoisomers in Figures 19.14 and 19.16 shows that the two cis structures are identical and the two trans structures are identical. There-

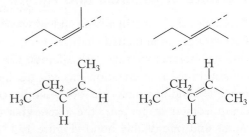

(a) The hydrogen atoms on the double bond are cis

(b) The hydrogen atoms on the double bond are trans

FIGURE 19.16 The positions of the H atoms in alkenes can be used to distinguish between (a) cis and (b) trans isomers.

(a) 0°   (b) 0°   (c) 45°   (d) 90°

Orbitals overlap ⟶ ⟶ No overlap
π bond   Bond broken

FIGURE 19.15 (a) To form a π bond, p_z orbitals overlap to establish a region of shared electron density above and below the plane of the carbon–carbon bond. (b) If you look down the bond axis, the orbitals line up. (c) If you rotate one carbon while keeping the other fixed, the orbitals are no longer parallel and do not overlap. (d) If you rotate one of the two bonded C atoms far enough, the π bond breaks.

fore, the straight-chain alkenes with five carbon atoms exist as three isomers: structure a from Figure 19.13, plus the cis and trans isomers of structure b. All three isomers are chemically distinct molecules.

CONCEPT TEST

Which of the following alkenes has cis and trans isomers?

a. CH_2=$CHCH_2CH_3$ b. CH=CH with CH_2 c. $(CH_3)_2C$=$C(CH_3)_2$

d. $(CH_3)_2C$=CH_2 e. CH_3CH=$CHCH_3$

cis isomer (also called **Z isomer**) molecule with two like groups (such as two R groups or two hydrogen atoms) on the same side of the molecule.

trans isomer (also called **E isomer**) molecule with two like groups (such as two R groups or two hydrogen atoms) on opposite sides of the molecule.

Naming Alkenes and Alkynes

To name straight-chain alkenes and alkynes, the prefixes in Table 19.3 are used to identify the length of the chain. The suffix *-ene* is appended if the compound is an alkene and *-yne* if the substance is an alkyne. The carbon atoms in the chain are numbered so that the first carbon atom in the double or triple bond has the lowest number possible, and that number precedes the name, followed by a dash. Stereoisomers are identified by writing *cis-* or *trans-* before the number. Thus the compounds in Figure 19.16(a) and (b) are *cis*-2-pentene and *trans*-2-pentene, respectively.

SAMPLE EXERCISE 19.5   Identifying and Naming   LO3 Stereoisomers and Constitutional Isomers

Draw the condensed structures and carbon-skeleton structures of the five isomers of the six-carbon straight-chain alkene containing one double bond, and name each isomer.

COLLECT AND ORGANIZE We are asked to draw and name five isomers of molecules that each have six carbon atoms in a single chain and one C=C double bond. Figures 19.11, 19.13, 19.14. and 19.16 show examples of condensed and carbon-skeleton structures of alkenes.

ANALYZE We must consider two kinds of isomers: constitutional isomers, which depend on where the double bond is located in the chain, and stereoisomers, which depend on the orientation of groups about the double bond (cis/trans isomers). A straight-chain six-carbon alkene has a maximum of five places where a C=C can be placed; however, some locations may result in identical molecules. Not all alkenes have cis and trans isomers.

SOLVE Let's start with the constitutional isomers, which have the double bond at different locations, and then draw the stereoisomers (cis/trans isomers) where possible. If all six carbons are in one straight chain, then a double bond can be located between C1 and C2, C2 and C3, or C3 and C4. These isomers are named 1-hexene, 2-hexene, and 3-hexene, respectively, where the number indicates the location of the double bond along the chain. Chains with a double bond between C4 and C5 or C5 and C6 are identical to the isomers with a double bond between C2 and C3 or C1 and C2, respectively.

1-Hexene does not have stereoisomers because the carbon atoms that form the double bond have three H atoms and only one nonhydrogen atom attached. Only

2-hexene and 3-hexene can have stereoisomers. *cis*-2-Hexene and *cis*-3-hexene have their two R groups on the same side of the double bond. *trans*-2-Hexene and *trans*-3-hexene have R groups on opposite sides of the double bond.

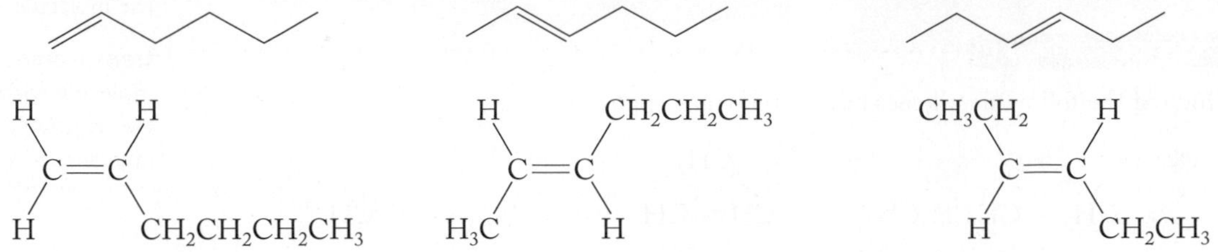

1-Hexene

trans-2-Hexene

trans-3-Hexene

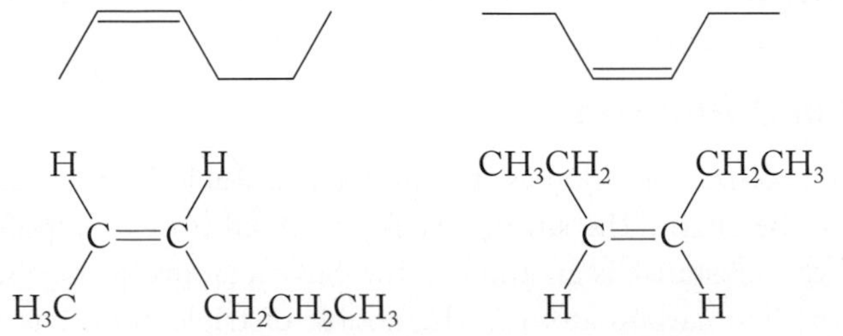

cis-2-Hexene

cis-3-Hexene

THINK ABOUT IT The number of isomers depends on the number of independent locations for a double bond in an alkene. Not all alkenes have stereoisomers.

Practice Exercise Draw and name the five isomers of the molecule with this carbon skeleton and one carbon–carbon double bond:

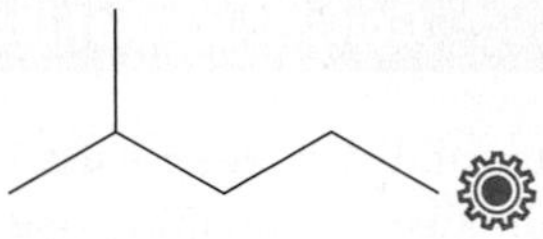

Polymers of Alkenes

Some widely used polymeric alkanes are produced industrially from small alkenes. The alkane polymer with the simplest structure is linear polyethylene (PE), produced from ethylene ($CH_2{=}CH_2$) at high temperature and pressure:

$$n\, H_2C{=}CH_2 \rightarrow -\!\!\left[CH_2{-}CH_2\right]\!\!_n \tag{19.5}$$

PE is a **homopolymer**, which means it is composed of only one type of monomer. The condensed structure could be written $CH_3(CH_2)_nCH_3$, but there are so many more methylene groups than methyl groups that the repeating unit is frequently written as shown in Equation 19.5 to highlight the structure and composition of the monomer. Polyethylene is also an example of an **addition polymer**, a polymer constructed by adding many molecules together to form the polymer chain.

In most products made of polyethylene, *n* is a very large number ranging from 1000 to almost 1 million. Polyethylene has a wide range of properties that depend on the value of *n* and on whether the polymer chains are straight or branched. In low-density PE (LDPE), a stretchable, soft plastic used in films and wrappers, *n* ranges from 350 to 3500 and the chains are branched (Figure 19.17). When the bagger at the grocery store asks, "Paper or plastic?" the plastic in question is LDPE.

homopolymer a polymer composed of only one kind of monomer unit.

addition polymer macromolecule prepared by adding monomers to a growing polymer chain.

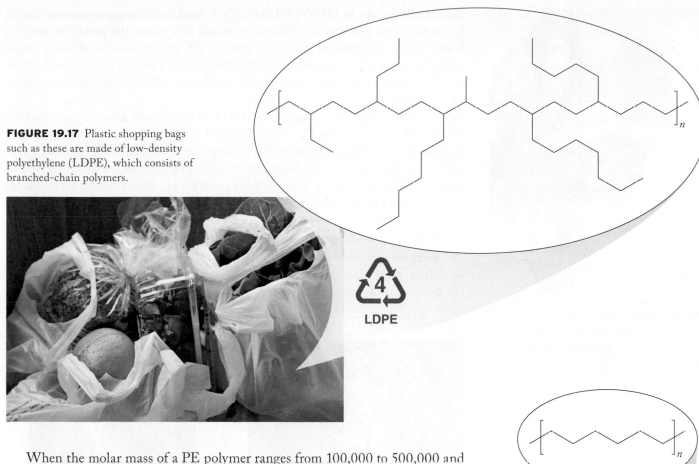

FIGURE 19.17 Plastic shopping bags such as these are made of low-density polyethylene (LDPE), which consists of branched-chain polymers.

When the molar mass of a PE polymer ranges from 100,000 to 500,000 and the chains are straight, the polymer has physical properties different from those of grocery bags. This straight-chain polymer—a rigid, translucent solid called high-density polyethylene (HDPE)—is used in beverage containers (Figure 19.18), electrical insulation, and toys.

Why are the properties of straight-chain HDPE (rigid, tough) so different from those of branched-chain LDPE (stretchable, soft)? Think of the branched polymer as a tree branch with lots of smaller branches attached to it. The polymer can have branches that come out of the plane of the paper, so it has three dimensions. In contrast, the straight-chain polymer is like a long, straight pole. Suppose you had a pile of 100 tree branches and a pile of 100 poles, and your task was to stack each pile into the smallest possible volume to fit into a truck. You can certainly pile the branches on top of one another, but they will not fit together neatly, and probably the best you can do is to make the pile a bit more compact. In contrast, you can stack the poles into a very compact pile.

The same situation arises with the branched and linear molecules of polyethylene. The density of LDPE is low compared to HDPE, primarily because the branched molecules of LDPE stack less efficiently. Lower density coupled with weaker intermolecular forces between polymer strands that are farther apart makes LDPE more deformable and softer. HDPE is more rigid and even has some regions that are crystalline because the packing is so uniform and the polymers are more rigidly held in place. These different structures mean that objects made of HDPE and LDPE must be separated before they can be processed for recycling, as indicated by their different recycling symbols.

Ultrahigh-molecular-weight PE (UHMWPE; $n > 100,000$) is an even tougher material because its molecules are both straight and significantly larger

FIGURE 19.18 Many translucent plastic bottles are made of high-density polyethylene (HDPE), which consists of straight polymers that stack together efficiently, making the bottles rigid and tough.

FIGURE 19.19 Devices used in knee-replacement surgery often incorporate a pad made of durable ultrahigh-molecular-weight polyethylene (UHMWPE) to cushion the joint.

than the molecules in HDPE. UHMWPE is used as a coating on some artificial ball-and-socket joints for hip replacements and to separate the prosthetic ends of bones in knee replacement surgery (Figure 19.19). The material is extremely resistant to abrasion and makes the joints last longer. The different forms of polyethylene illustrate how the size and shape of its molecules affect the physical properties of a material.

Of course, chemical composition also plays a role in determining properties. If the hydrogen atoms in PE are all replaced with fluorine atoms, the resultant polymer is chemically very unreactive, is capable of withstanding high temperatures, and has a very low coefficient of friction, which means other things do not stick to it. This polymer is Teflon, $-(CF_2-CF_2)_n$, most familiar for its use as a nonstick surface in cookware. However, Teflon tubing is also used in the grafts inserted into small-diameter blood vessels during vascular surgery on limbs (Figure 19.20). The analogous material cannot be formed with chlorine, but a polymer does exist in which every other $-CH_2-$ in the polymer chain is $-CCl_2-$. Its repeating monomer unit is $-(CH_2-CCl_2)_n$ (Figure 19.21), and the polymer is the familiar thin, flexible plastic used as Saran wrap.

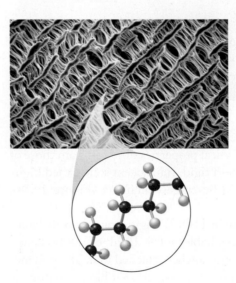

FIGURE 19.20 Fabrics made of woven strands of poly(tetrafluoroethylene) (Teflon) are used during surgical procedures to promote tissue growth. The strong, chemically inert mesh supports tissue that grows into its pores.

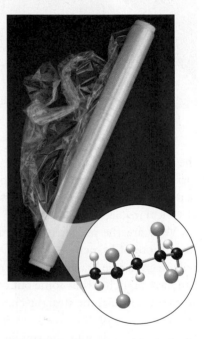

FIGURE 19.21 Saran wrap is made of poly(1,1-dichloroethylene).

Hydrocarbons containing two or more double bonds are frequently used to manufacture polymers. For example, polymerization of butadiene yields a stretchy, synthetic rubber useful in rubber bands (Figure 19.22a). Polyisoprene, prepared from 2-methyl-1,3-butadiene, is used in surgical gloves (Figure 19.22b).

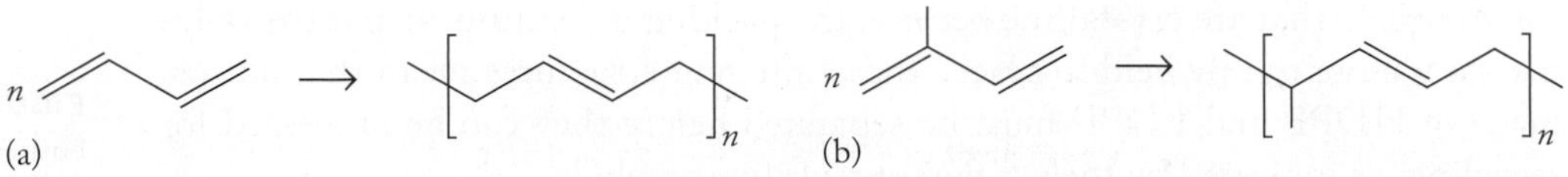

FIGURE 19.22 Monomer units and polymer structures for (a) butadiene and (b) a methylated butadiene known as isoprene.

SAMPLE EXERCISE 19.6 **Identifying Monomers**   **LO4**

Polypropylene, $-[CH_2CH(CH_3)]_n$, is an addition polymer used in the manufacture of fabrics, ropes, and other materials (Figure 19.23). Draw condensed and carbon-skeleton structures of the monomer used to prepare polypropylene, and name the monomer.

COLLECT AND ORGANIZE We are given a condensed structure of a polymer and asked to identify the monomer used in its preparation. We know that polypropylene is an addition polymer, so the monomer must be an alkene.

ANALYZE To understand the relationship between the polymer and the monomer from which it is made, let's look at Equation 19.5 in the reverse direction:

$$-[CH_2-CH_2]_n \rightarrow n\,CH_2{=}CH_2 \qquad (19.6)$$

Breaking the blue bonds in Equation 19.6 and making the red bond a double bond illustrates the relationship between polyethylene and its monomer, ethylene. We need to apply a similar analysis to polypropylene.

SOLVE The relationship between polypropylene and its monomer is illustrated by a corresponding equation:

$$-[CH_2-CH(CH_3)]_n \rightarrow n\,CH_2{=}CH(CH_3) \qquad (19.7)$$

Breaking the two blue bonds in Equation 19.7 and making the red C–C bond a double bond yields the alkene shown in Figure 19.24. The official name of this three-carbon alkene is propene (though it is also known as propylene). There is no need to precede the name of the monomer with a number to indicate the position of the C=C bond, because it has to be between the C1 and C2 carbon atoms.

THINK ABOUT IT The constitutional difference between propylene and ethylene is the presence of a –CH_3 group bonded to one of the two sp^2 carbon atoms in propylene instead of the H atom found in ethylene.

Practice Exercise Draw the condensed and carbon-skeleton structures of the monomer used to make poly(methyl methacrylate), PMMA, a polymer used in shatterproof transparent plastic that can take the place of glass:

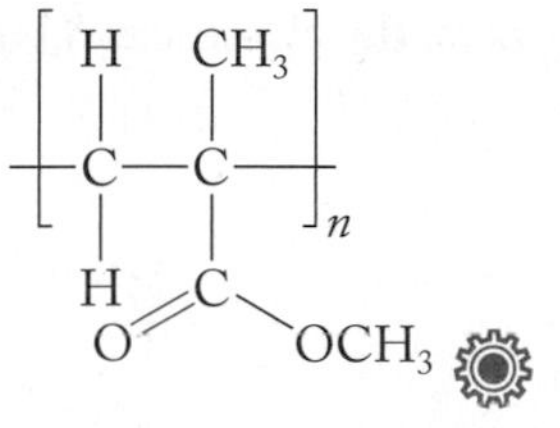

$$\left[\begin{array}{cc} H & CH_3 \\ | & | \\ {-}C{-}C{-} \\ | & | \\ H & C \\ & \diagup\diagdown \\ O & OCH_3 \end{array}\right]_n$$

FIGURE 19.23 Polypropylene is a common material for furniture, containers, clothing, lighting fixtures, and even objects of art. In addition to being moldable into many shapes, polypropylene is a good thermal insulator and does not absorb water easily.

$$CH_3CH{=}CH_2 \qquad \diagup\diagdown\diagup$$

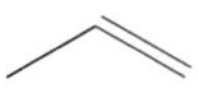

FIGURE 19.24 The monomer propene is polymerized to make polypropylene.

All addition polymers based on addition reactions of monosubstituted ethylene are called **vinyl polymers** because the $CH_2{=}CH{-}$ subunit is called the **vinyl group**, a name derived from *vinum* (Latin for wine). The name was given to the group by 18th-century chemists who prepared ethylene ($CH_2{=}CH_2$) from ethanol (CH_3CH_2OH), the alcohol in wine and other liquors. Polyvinyl chloride, PVC, is widely used in commercial articles ranging from plastic pipes for plumbing to computer cases. Classic vinyl phonograph records are also made from PVC. The condensed structures for the monomer and the polymer are

$$n\,CH_2{=}CHCl \rightarrow -[CH_2CHCl]_n \qquad (19.8)$$

Vinyl polymers are the world's second largest selling plastics, and polymers in this category are extraordinarily versatile. You probably encounter five to ten

vinyl polymer one of the family of polymers formed from monomers containing the subgroup $CH_2{=}CH{-}$.

vinyl group the subgroup $CH_2{=}CH{-}$.

vinyl polymers before you leave your room in the morning: vinyl shower curtains, vinyl drain pipes, vinyl flooring, vinyl insulation around electrical conduits. As we explore more organic functional groups, the basic concepts developed for the vinyl polymers will apply to polymers in other categories: the features of functional group, size, and shape determine the chemical and physical properties of these extraordinarily useful materials.

19.4 Aromatic Compounds

Among the components of gasoline that play an important role in increasing its octane rating (a measure of the ignition temperature of the fuel and its ability to resist engine "knock") is the class of compounds called *aromatic* hydrocarbons. Benzene is one of them: a cyclic, planar molecule with delocalized π electrons above and below the plane containing its six carbon and six hydrogen atoms.

As their class name implies, aromatic hydrocarbons have distinctive odors such as the smell of mothballs and recently cleaned public restrooms. To a chemist, *aromaticity* means more than odor; it means cyclic, planar molecules with sp^2 hybridized carbon atoms joined by a combination of σ and π bonds (Figure 19.25). Aromatic compounds are relatives of alkenes because they contain carbon–carbon double bonds. However, because their chemical and physical properties are unique and distinct from those of alkenes, they merit designation as a separate family.

The most common aromatic compound is benzene, C_6H_6. The different ways we view the bonding in benzene using Lewis theory and valence bond theory are summarized in Figure 19.25. As noted in Section 4.6, the delocalized electrons in benzene lead to considerable resonance stability in this molecule, and this is true of all other aromatic molecules as well.

The stability of aromatic systems has an impact on their chemical reactivity. In Section 19.3 we saw that alkenes react rapidly with hydrogen halides to make halogenated alkanes. In contrast, benzene does not react at all if HBr gas is bubbled through it. Because alkenes and aromatic compounds differ with respect to this reaction and many others, the classification of aromatic systems as a unique family is justified.

▶‖ **CHEMTOUR** Structure of Benzene

∞ **CONNECTION** In Chapters 4 and 5 we described bonding in the benzene molecule using Lewis theory and valence bond theory.

FIGURE 19.25 Different views of the bonding in benzene. (a) Carbon-skeleton structures showing resonance and double-bond delocalization. (b) Hexagonal array of σ bonds. (c) The unhybridized p_z orbitals on the sp^2 hybridized carbon atoms. (d) Delocalized π electrons above and below the ring.

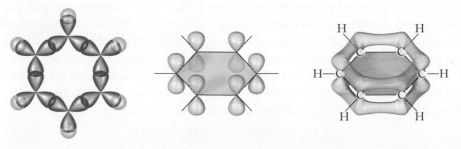

(a) Carbon-skeleton structures showing resonance forms of benzene

Skeletal symbol of benzene ring

(b) Sigma bonds in benzene

(c) Unhybridized p orbitals of carbon atoms

(d) Delocalized π cloud of electrons above and below plane of ring

Toluene

Constitutional Isomers of Aromatic Compounds

Many compounds can be formed by replacing the hydrogen atoms in an aromatic ring with other substituents. For example, when one methyl group replaces a hydrogen atom in benzene, we get methylbenzene, also known by its common name, toluene. All the positions around the benzene ring are equivalent, so it does not matter which carbon atom is bonded to the methyl group. This is why we do not have to write a number in the name methylbenzene.

There are three options for attaching two methyl groups to a benzene ring, so there are three constitutional isomers of dimethylbenzene, also known as xylene. We distinguish between the three constitutional isomers by numbering the carbon atoms to give the substituents the lowest possible numbers. From left to right below, the three dimethylbenzenes are called 1,2-dimethylbenzene; 1,3-dimethylbenzene; and 1,4-dimethylbenzene. Toluene and xylenes are used in inks, glues, and disinfectants.

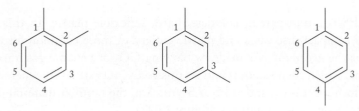

1,2-Dimethylbenzene   1,3-Dimethylbenzene   1,4-Dimethylbenzene

Benzene rings can share one or more of their hexagonal sides and thereby form polycyclic aromatic molecules. Three such compounds are naphthalene, anthracene, and phenanthrene:

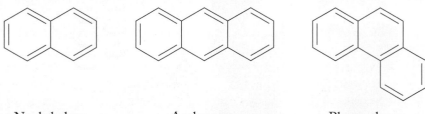

Naphthalene   Anthracene   Phenanthrene

Extensive delocalization of the π electrons over all the rings makes these structures particularly stable. In addition to being found in fossil fuels, they may be formed during the incomplete combustion of hydrocarbons and are present in particularly high concentrations in the soot from incinerators and diesel engines. They have also been identified in interstellar dust clouds and in blackened portions of grilled meat. When introduced into the environment, these compounds persist and are among the most long-lived of hydrocarbons.

CONCEPT TEST ●

Why isn't hexatriene considered an aromatic compound?

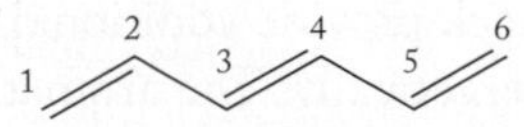

1,3,5-Hexatriene

FIGURE 19.26 The aromatic rings on neighboring chains in polystyrene stack together and provide strength to the material.

Polymers Containing Aromatic Rings

Individual aromatic rings, as well as fused rings, are flat molecules. They tend to stack neatly (Figure 19.26), and this tendency gives rise to useful properties in materials that incorporate aromatic systems. Replacing one hydrogen atom in benzene with a vinyl group gives the monomer styrene, and the polymer made from this monomer is polystyrene (PS):

Styrene   Polystyrene

Solid PS is a transparent, colorless, hard, inflexible plastic. In this form it is used for compact disc cases and plastic cutlery. A more common form of PS, however, is the *expanded solid* made by blowing CO_2 or pentane gas into molten polystyrene, which then expands and retains voids in its structure when it solidifies. One form of this expanded PS is Styrofoam, the familiar material of coffee cups and take-out food containers (Figure 19.27).

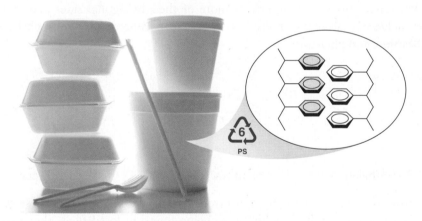

FIGURE 19.27 In its nonexpanded form, polystyrene is rigid and strong, suitable for making plastic utensils. In its expanded form, it is Styrofoam, used in carry-out food containers, packing materials, and thermal insulation in buildings.

FIGURE 19.28 When the Styrofoam coffee cup on the left is placed under pressure, some of the air between the polystyrene chains is forced out. The cup shrinks to the size on the right but retains its overall shape.

The difference in properties between transparent, colorless, inflexible nonexpanded PS and opaque, white, pliable Styrofoam can be explained by considering the role of the aromatic ring in aligning the polymer chains. Branches in the chains have the same effect that we saw with polyethylene, but the aromatic rings and their tendency to stack provide additional interactions that make chain alignment more favorable energetically. The aromatic rings along two neighboring chains can stack, and this stacking makes nonexpanded PS rigid. When the chains are blown apart by a gas, the stacking is disrupted and the chains open to make cavities that fill with air, making expanded PS a good thermal insulator and packing material. The presence of air in Styrofoam is illustrated in Figure 19.28.

19.5 Amines

Nitrogen atoms are the defining components of functional groups in another important family of organic molecules called **amines**. In organic compounds, nitrogen atoms—and any atoms other than carbon, hydrogen, or metals—are called **heteroatoms** and are shown in Kekulé structures (without showing the lone pairs), condensed structures, and carbon-skeleton structures. For example, notice the heteroatom N at the junction of three lines in the carbon-skeleton structures of Benadryl and pargyline in Figure 19.29.

Amines containing an alkyl or aromatic group are thought of as being derived from ammonia, NH_3. If one hydrogen atom in ammonia is replaced by an R group, the compound is called a *primary amine*. If two hydrogen atoms are replaced by R groups, the compound is a *secondary amine*; if all three hydrogen atoms are replaced by R groups, it is a *tertiary amine*:

$$RNH_2 \qquad R_2NH \qquad R_3N$$

Primary amine   Secondary amine   Tertiary amine

The R groups in a secondary or tertiary amine may be the same or different organic subunits. You may be familiar with the odor of trimethylamine, $(CH_3)_3N$, the compound responsible for the smell of decaying fish.

The amine functional group is found in many natural products and drugs. Some relatively simple examples include amphetamine (Figure 19.29), a stimulant that also contains an aromatic group, and Benadryl, an antihistamine used to treat the symptoms associated with allergies. Another amine, adrenaline, is produced by our bodies in glands near the kidneys and plays an important role in our nervous system.

Amines are organic bases, and their basicity is their defining chemical characteristic. They all react with water to some extent to produce hydroxide ions and protonated cations, just like ammonia:

$$NH_3(aq) + H_2O(\ell) \rightleftharpoons NH_4^+(aq) + OH^-(aq) \tag{19.9}$$

$$RNH_2(aq) + H_2O(\ell) \rightleftharpoons RNH_3^+(aq) + OH^-(aq) \tag{19.10}$$

Amines also react readily with acids like hydrochloric acid to form water-soluble salts:

$$RNH_2(aq) + HCl(aq) \rightarrow RNH_3^+(aq) + Cl^-(aq) \tag{19.11}$$

Many pharmaceuticals (like Benadryl) that contain the amine functional group are sold as hydrochloride salts to improve their solubility in water.

Amine groups are polar and can form hydrogen bonds with water, which also promotes their solubility in water. On the other hand, the portions of amine molecules composed of carbon and hydrogen are nonpolar and detract from aqueous solubility because the larger the nonpolar portion, the stronger the London dispersion forces between amine (solute) molecules that inhibit their solubility in water.

Bacteria of the genus *Methanosarcina* convert primary, secondary, and tertiary methylamines to methane, carbon dioxide, and ammonia:

$$4\ CH_3NH_2(aq) + 2\ H_2O(\ell) \rightarrow 3\ CH_4(g) + CO_2(g) + 4\ NH_3(aq) \tag{19.12}$$

$$2\ (CH_3)_2NH(aq) + 2\ H_2O(\ell) \rightarrow 3\ CH_4(g) + CO_2(g) + 2\ NH_3(aq) \tag{19.13}$$

$$4\ (CH_3)_3N(aq) + 6\ H_2O(\ell) \rightarrow 9\ CH_4(g) + 3\ CO_2(g) + 4\ NH_3(aq) \tag{19.14}$$

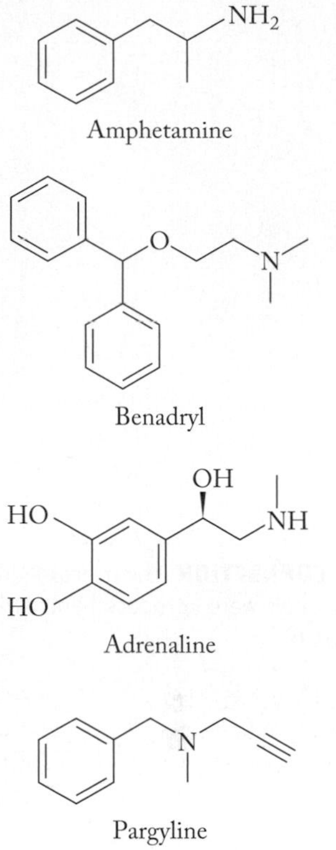

Amphetamine

Benadryl

Adrenaline

Pargyline

FIGURE 19.29 Amphetamine, a drug known to produce increased wakefulness and focus; Benadryl, an antihistamine; adrenaline, a hormone produced in our bodies and involved in the fight or flight response; and pargyline, a drug used to treat hypertension. Each of these physiologically active compounds contains the amine functional group.

amine organic compound that contains a group with the general formula RNH_2, R_2NH, or R_3N, where R is any organic subgroup.

heteroatom any atom other than a carbon, hydrogen, or metal atom in an organic compound.

These reactions describe pathways by which methane can be produced from the decay of biomass, serving as potential sources of fuel for heating and transportation. Amines represent only a small fraction of the material in plants, however, and the industrial development of fuel production from amines has been slow, mainly because fossil fuels are still plentiful enough and cheap enough to make processing amines cost-ineffective. This economic imbalance may change as fossil fuels are depleted and become more expensive.

CONCEPT TEST ....

Are amphetamine, Benadryl, adrenaline, and pargyline (Figure 19.29) primary, secondary, or tertiary amines?

....

19.6 Alcohols, Ethers, and Reformulated Gasoline

CONNECTION The hydroxyl group and alcohols were introduced and defined in Chapter 6.

In 1995, air-quality regulations went into effect in many U.S. cities mandating reductions in atmospheric pollutants from gasoline-fueled engines. The regulations led to the widespread use of "reformulated" gasoline containing additives to promote complete combustion and boost octane ratings. These additives are often organic compounds that contain oxygen in addition to hydrogen and carbon. Oxygen atoms in an organic compound are also heteroatoms and are components of functional groups in two important families of organic molecules: alcohols and ethers.

Alcohols: Methanol and Ethanol

Alcohols have the general formula R—OH, where R is any alkyl group. The R group can be a straight chain, branched chain, or ring. Like the N atoms in amines, the O atoms in alcohols are shown in carbon-skeleton structures. The chemical and physical properties of alcohols can be understood if we recognize that an alcohol looks like a combination of an alkane and water:

$$R\text{—}H \qquad H\text{—}OH \qquad R\text{—}OH$$
Alkane $\qquad$ Water $\qquad$ Alcohol

As we saw in Section 6.3, if the R group in the molecule is small, the alcohol behaves like water; as the R group gets larger, the alcohol behaves more like a hydrocarbon. As an illustration of this behavior, Table 19.5 shows the water solubilities of a homologous series of alcohols. The polar –OH group makes one end of these compounds "water-like." However, as the number of carbon atoms increases, a greater proportion of the alcohol molecule is "oil-like." Therefore, these alcohols' solubilities in water decrease until about C_8, beyond which their solubilities are comparable to those of the corresponding hydrocarbons.

As the names of the two simplest alcohols—methanol and ethanol—indicate, the chemical names of alcohols end in *-ol*; this suffix identifies the compound as an alcoh*ol*. Methanol (CH_3OH) is also known as methyl alcohol or wood alcohol. The latter name comes from one former source of this alcohol: it was made by collecting the vapors given off when wood is heated to the point of decomposition in the absence of oxygen. Methanol is a widely used industrial organic chemical. It is the starting material in the preparation of several organic compounds used

TABLE 19.5	Solubilities of a Homologous Series of Alcohols in Water at 20°C	
Condensed Structure	**Water Solubility (g/100 mL)**
CH_3OH	Miscible
CH_3CH_2OH	Miscible
$CH_3(CH_2)_2OH$	Miscible
$CH_3(CH_2)_3OH$	7.9
$CH_3(CH_2)_4OH$	2.3
$CH_3(CH_2)_5OH$	0.6
$CH_3(CH_2)_6OH$	0.2
$CH_3(CH_2)_7OH$	0.05

to make polymers. Its industrial synthesis is based on reducing carbon monoxide with hydrogen:

$$CO(g) + 2\,H_2(g) \rightarrow CH_3OH(\ell)$$

The CO and H_2 used to make methanol come from the steam–reforming reaction of methane and water that we discussed in Chapter 9:

$$CH_4(g) + H_2O(g) \rightarrow CO(g) + 3\,H_2(g)$$

Methanol burns according to the thermochemical equation

$$2\,CH_3OH(\ell) + 3\,O_2(g) \rightarrow 2\,CO_2(g) + 4\,H_2O(\ell) \qquad \Delta H^\circ_{comb} = -1454\ kJ$$

If we divide the absolute value of ΔH°_{comb} by twice the molar mass of methanol (because the reaction consumes 2 moles of methanol), we get a fuel value for methanol of

$$\frac{1454\ kJ}{\left(\dfrac{32.04\ g}{mol}\right)(2\ mol)} = 22.69\ kJ/g$$

Let's compare the fuel value of methanol with that of octane:

$$2\,C_8H_{18}(\ell) + 25\,O_2(g) \rightarrow 16\,CO_2(g) + 18\,H_2O(\ell) \quad \Delta H^\circ_{comb} = -1.091 \times 10^4\ kJ$$

$$\frac{1.091 \times 10^4\ kJ}{\left(\dfrac{114.22\ g}{mol}\right)(2\ mol)} = 47.76\ kJ/g$$

The fuel value of methanol is less than half that of octane (and most of the other hydrocarbons in gasoline). Why is this the case? The answer involves the molecular structure of methanol. The amount of energy released during combustion depends on the number of carbon atoms available for forming $C=O$ bonds in CO_2 and the number of hydrogen atoms available for forming $O-H$ bonds in H_2O. The presence of oxygen in CH_3OH adds significantly to its mass (methanol is 50% oxygen by mass) but adds nothing to its fuel value. The oxygen content of a combustible substance essentially dilutes its energy content: the more oxygen a fuel contains, the lower its fuel value is.

As noted earlier, ethanol (CH_3CH_2OH), also known as ethyl alcohol, is the alcohol in alcoholic beverages. It is formed by the fermentation of sugar from an amazing variety of vegetable sources. Indeed, any plant matter containing sufficient sugar may be used to produce ethanol. Grains are commonly used, from which ethanol derives its trivial name *grain alcohol*. Ethanol may be the oldest organic chemical used by humans, and it is still one of the most important. For industrial purposes, ethanol is prepared by the reaction of water and ethylene.

⊙⊙ CONNECTION We introduced fuel values in Chapter 9 as the amount of heat given off when one gram of fuel is burned.

SAMPLE EXERCISE 19.7 **Comparing Fuels for Camping Stoves** **LO5**

Some of the small, single-burner stoves that campers and hikers use to prepare meals and boil water use ethanol as fuel; others use a petroleum distillate called white gas, which is a mixture of C_6 through C_{12} hydrocarbons. Assuming nonane ($\mathcal{M} = 128.26$ g/mol; $\Delta H^\circ_{comb} = -6160$ kJ/mol) is a representative hydrocarbon in white gas, how much more thermal energy is available in 1 liter of white gas than in 1 liter of ethanol ($\mathcal{M} = 46.07$ g/mol; $\Delta H^\circ_{comb} = -1367$ kJ/mol)? The densities of nonane and ethanol are 0.718 and 0.789 g/mL, respectively.

COLLECT AND ORGANIZE We are given the densities, molar masses, and enthalpies of combustion of nonane and ethanol and are asked to determine how much more thermal energy is produced by the combustion of 1 liter of nonane than by burning the same volume of ethanol.

ANALYZE The heats of combustion are expressed in kilojoules per mole, so the comparison of the two fuels will require calculating how many moles of each are in 1 liter. The standard enthalpy of combustion of nonane is between 4 and 5 times that of ethanol, but the molar mass of nonane is about 3 times greater. Therefore, on balance, combustion of nonane should produce about 4.5/3 = 1.5 times more thermal energy than burning the same mass of ethanol. However, ethanol is about 10% more dense, so nonane should have about 1.4 times the energy content as the same volume of ethanol.

SOLVE Calculating the enthalpy change that accompanies the combustion of 1 liter of each fuel under standard conditions,

$$1 \text{ L nonane} \times \frac{1000 \text{ mL}}{1 \text{ L}} \times \frac{0.718 \text{ g}}{\text{mL}} \times \frac{1 \text{ mol}}{128.26 \text{ g}} \times \frac{-6160 \text{ kJ}}{\text{mol}} = -3.448 \times 10^4 \text{ kJ}$$

$$1 \text{ L ethanol} \times \frac{1000 \text{ mL}}{1 \text{ L}} \times \frac{0.789 \text{ g}}{\text{mL}} \times \frac{1 \text{ mol}}{46.07 \text{ g}} \times \frac{-1367 \text{ kJ}}{\text{mol}} = -2.341 \times 10^4 \text{ kJ}$$

Taking the ratio of the two values,

$$\frac{-3.448 \times 10^4 \text{ kJ}}{-2.341 \times 10^4 \text{ kJ}} = 1.47$$

Therefore, nonane (white gas) has 1.47 times the capacity to heat food and water as the same volume of ethanol.

THINK ABOUT IT The presence of oxygen accounts for 16 of ethanol's 46 grams of mass per mole, or 35%. It is reasonable, then, that an equal volume of a hydrocarbon of similar density would have significantly higher energy content. On the other hand, ethanol burns cleaner and, in principle, is a renewable fuel.

Practice Exercise Bottled propane and butane are also used as fuels in camping stoves. Which one has the greater fuel density? By how much?

	Propane	Butane
ΔH°_{comb} (kJ/mol)	−2200	−2877
Density (g/mL)	0.493	0.573

Most of the gasoline sold in the United States currently contains 10% ethanol, and there are efforts underway to increase this value to 15%. Most of this ethanol is produced by fermentation of sugar derived from corn, with annual production in 2011 reaching nearly 14 billion U.S. gallons. There are, however, several challenges limiting the greater use of ethanol as an automobile fuel. The enthalpy of combustion is significantly less negative than that of the hydrocarbons in gasoline. The change in free energy that accompanies the combustion of ethanol is also less negative than the combustion of the same volume of the hydrocarbons. Therefore, there is less energy available to do useful work, such as propelling cars and trucks down the road.

Furthermore, considerable energy, fertilizer, irrigation water, and valuable farmland are needed to grow and harvest corn and to convert corn starch into sugar and sugar into ethanol, and to separate the ethanol from the fermentation mixture. It is estimated that more than two-thirds of the energy released in the combustion of ethanol derived from corn is consumed in its production. Ethanol produced in this way is more expensive than gasoline, even at today's high prices for fossil fuels.

Alcohols are also components of many natural products and consumer goods derived from them. The distinctive odor of mint leaves comes from menthol, which is an alcohol, as is terpineol (oil of turpentine), an oil distilled from the resin of pine trees (Figure 19.30). Notice the similarity in the carbon-skeleton structures of these two compounds: both contain a six-carbon ring and an –OH group. Only the location of the OH group and the presence of a C=C double bond distinguish these two compounds.

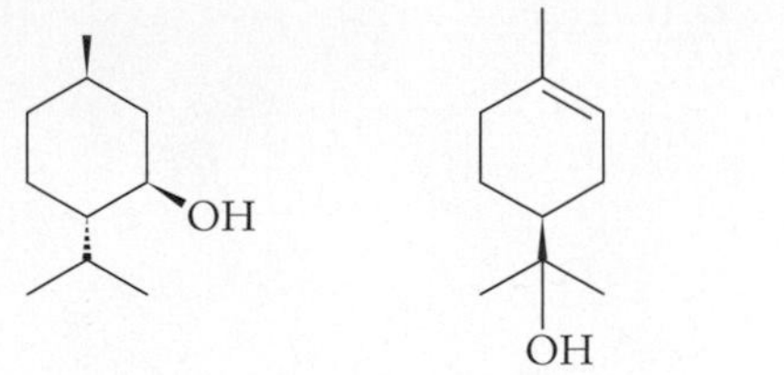

(a) Menthol (oil of mint) (b) Terpineol (oil of turpentine)

FIGURE 19.30 (a) Menthol and (b) terpineol are two naturally occurring alcohols present in mint leaves and pine needles, respectively.

Ethers: Diethyl Ether

Ethers have the general formula R—O—R, where R is any alkyl group or an aromatic ring. Just as with alcohols, we can think of ethers as water molecules in which the two H atoms have been replaced by two organic groups (R and R′, which may be the same or different):

$$R—H \qquad H—O—H \qquad H—R' \qquad R—O—R'$$
Alkane   Water   Alkane   Ether

Because the C—O—C bond angle is close to the tetrahedral bond angle of 109.5°, the bond dipoles of the two C—O bonds in an ether do not cancel, which means that ethers are polar molecules. This structural feature gives rise to the properties of typical ethers: their water solubilities are comparable to those of alcohols of similar molar mass, but their boiling points are about the same as alkanes of comparable molar mass (Table 19.6).

The most important ether industrially is diethyl ether, $CH_3CH_2OCH_2CH_3$. You may have heard of this as the substance simply called "ether" that has had wide use in medicine as an anesthetic since 1842. Although exactly how an anesthetic dulls nerves and puts patients to sleep is still unknown, certain properties of diethyl ether play a role in determining its behavior as a medicinal agent. Because

TABLE 19.6 Functional Groups Affect Physical Properties

	Molar Mass (g/mol)	Normal Boiling Point (°C)	Solubility in Water (g/100 mL at 20°C)
CH_3CH_2—O—CH_2CH_3 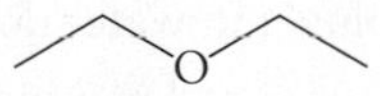 Diethyl ether	74	35	6.9
$CH_3CH_2CH_2CH_2CH_3$  Pentane	72	36	0.0038
$CH_3CH_2CH_2CH_2OH$  Butanol	74	117	7.9

ether organic compound with the general formula R—O—R′, where R and R′ are any alkyl group or aromatic ring; the R and R′ groups may be the same.

diethyl ether has a low boiling point, 35°C, it vaporizes easily, and a patient can inhale it. Because diethyl ether has a significant solubility in water, it is soluble in blood, which means that once inhaled, it can be easily transported throughout the body. Its low polarity and short saturated hydrocarbon chains combine to make it soluble in cell membranes, where it blocks stimuli coming into nerves. Ether has the unfortunate side effect of inducing nausea and headaches, and it has been replaced by newer anesthetics in modern hospitals. For many years, however, ether was the anesthetic of choice for surgical procedures.

A second common use of diethyl ether stems from another property that caused difficulty in the clinical setting—it is extremely flammable. Flammability is a liability in an operating room, but this property is used to our advantage when we spray ether in diesel engines to start them in cold weather when it is too cold for diesel fuel to ignite.

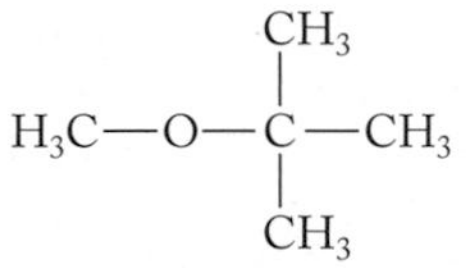

One ether widely used as a gasoline additive in the 1990s was methyl *tert*-butyl ether (MTBE, Figure 19.31), added to promote complete combustion (*tert*- is an abbreviation for *tertiary*, referring to a carbon atom bonded to three other carbon atoms). Unlike the nonpolar hydrocarbons in gasoline, MTBE is soluble in water. Consequently, gasoline spills, leakage from storage tanks, and release from watercraft can produce extensive MTBE contamination of groundwater and drinking water. After toxicity tests showed MTBE to be a possible carcinogen (cancer-causing agent), several states—including California, where more than 25% of the world's production of MTBE was used in gasoline—banned the use of MTBE as a gasoline additive. Most oil companies stopped adding MTBE to their gasolines in 2006. These changes raised the question of which additives would replace MTBE. The leading candidate to date has been ethanol.

FIGURE 19.31 MTBE, methyl *tert*-butyl ether, was used in the early 1990s as a fuel additive, but its use was curtailed when it was found to be an environmental pollutant.

CONCEPT TEST

Rank the following compounds in order of decreasing fuel value: diethyl ether, MTBE, methanol, and ethanol.

Polymers of Alcohols and Ethers

Over one million tons of the addition polymer poly(vinyl alcohol) (PVAL) are used worldwide each year to make adhesives, emulsions, and materials known as sizing, which change the surface properties of textiles and paper to make them less porous, less able to absorb liquids, and smoother. PVAL is the material of choice for laboratory gloves that are resistant to organic solvents. Because its chains are studded with –OH groups (Figure 19.32a), its surface is very polar and very water-like, and hydrocarbon solvents that are not soluble in water do not penetrate PVAL barriers.

PVAL is also impenetrable to carbon dioxide, and this property has led to its use in soda bottles, in which it is blended with the polymer poly(ethylene terephthalate) (PETE, Figure 19.32b). The two polymers do not mix but separate into layers (Figure 19.32c). The PETE makes the bottle strong enough to bear pressure changes due to temperature changes and survive the impact of falling off tables. However, CO_2, the dissolved gas that makes soda fizz, passes readily through PETE but not through the PVAL layers, with the result that the soda does not go flat. Polymers with different properties are frequently combined to create new materials with desired properties.

copolymer a macromolecule formed from the chemical combination of two different monomers.

heteropolymer a polymer made of three or more different monomer units.

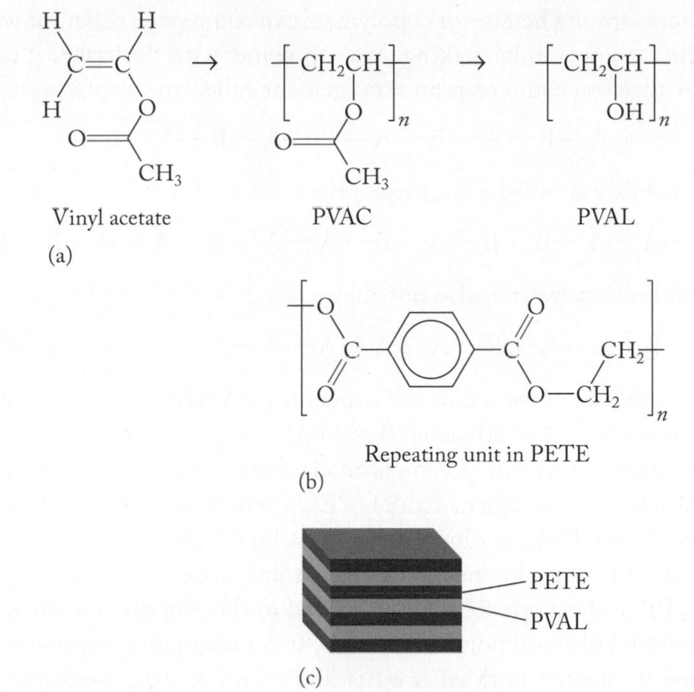

Vinyl acetate   PVAC   PVAL

(a)

Repeating unit in PETE

(b)

PETE
PVAL

(c)

FIGURE 19.32 (a) Poly(vinyl alcohol), or PVAL, is synthesized from vinyl acetate. The intermediate polymer, poly(vinyl acetate), or PVAC, is reacted with water to produce PVAL. (b) The repeating unit in PETE. (c) Layers of the polymers PVAL and PETE are used to make soda bottles. PETE makes the bottle strong, and PVAL keeps the carbon dioxide in.

The monomer from which poly(vinyl alcohol) is made is not the one you might expect based on our discussion of how addition polymers are synthesized. The monomer "vinyl alcohol" does not exist:

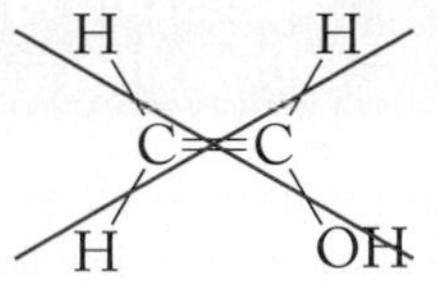

Instead, the monomer vinyl acetate is polymerized to make poly(vinyl acetate) (PVAC), which is then reacted with water to replace the acetate group with an –OH group. This replacement reaction turns PVAC into PVAL as shown in Figure 19.32(a).

The blend of PVAL and PETE in early soda bottles was a physical mixture of the two polymers. New materials can also be made by combining different monomer units in one polymer molecule. This type of molecule is called a **copolymer** when two different monomers are combined and a **heteropolymer** when three or more different monomers are combined. One example of an addition copolymer is a material called EVAL, made from ethylene and vinyl acetate (Figure 19.33). EVAL is used in food wrappings when preservation of aroma and flavor are required. Food usually deteriorates in the presence of oxygen, and packages made of EVAL provide an excellent barrier to the entry of oxygen while retaining the flavor and fragrance of the packaged food.

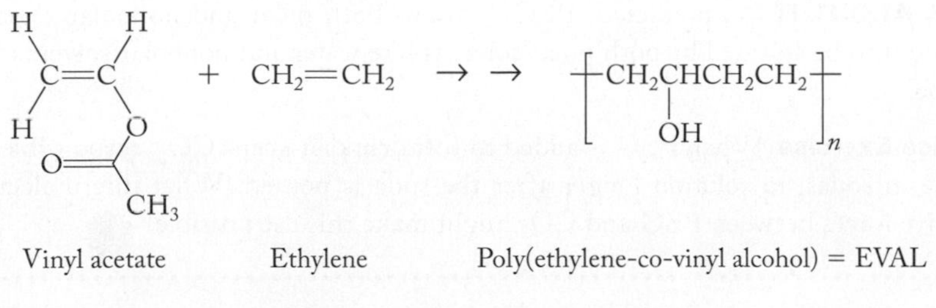

Vinyl acetate   Ethylene   Poly(ethylene-co-vinyl alcohol) = EVAL

FIGURE 19.33 EVAL is a copolymer of ethylene and vinyl acetate.

Monomers forming hetero- or copolymers can combine in different ways. If we represent the monomer units making up a copolymer with the letters A and B, one possible way they can combine is an arrangement called an *alternating copolymer*:

$$\text{---[A---B---A---B---A---B---A---B---A---B]---}$$

Another possibility is called a *block copolymer*:

$$\text{---[A---A---A---A---B---B---B---B---A---A---A---A---B---B---B---B]---}$$

Finally, a *random copolymer* is also possible:

$$\text{---[B---A---B---B---A---B---A---A---A---A---B---B---A---B---B---B]---}$$

This is the arrangement we see in the copolymer EVAL—it is a random copolymer of the monomers A = ethylene, B = vinyl acetate.

Commercially important polymers made from ethers include poly(ethylene glycol) (PEG) and poly(ethylene oxide) (PEO), which are made of the same subunit (Figure 19.34). PEG is a low-molar-mass liquid oligomer made from ethylene glycol, and PEO is a higher-molar-mass solid made from ethylene oxide. As a polyether, PEG has properties closely related to those of diethyl ether, in that it is soluble in both polar and nonpolar liquids. It is a common component in toothpaste because it interacts both with water and with the water-insoluble materials in the paste and keeps the toothpaste uniform both in the tube and during use. PEGs of many lengths are finding increasing use as attachments to pharmaceutical agents to improve their solubility and biodistribution in the body.

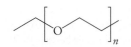

Repeating unit in PEG and PEO

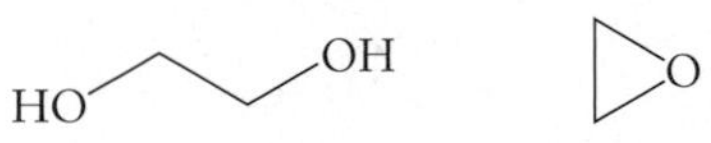

Ethylene glycol   Ethylene oxide

Monomers

FIGURE 19.34 Poly(ethylene glycol) (PEG) and poly(ethylene oxide) (PEO) have the same repeating unit. The two polymers differ only in their molar masses. PEG is typically made from ethylene glycol, and ethylene oxide is the monomer of choice for making PEO.

SAMPLE EXERCISE 19.8 **Assessing Properties of Polymers** **LO6**

The polymer PEG (Figure 19.34) is used to blend materials that are not soluble in each other. It is soluble both in water and in benzene, a nonpolar solvent. Describe the structural features of PEG that make it soluble in these two liquids of very different polarities.

COLLECT AND ORGANIZE The relationship between structure and solubility of compounds was discussed in Chapter 6, where we learned that "like dissolves like."

ANALYZE The statement "like dissolves like" refers to the polarity of the molecules and to the attractive and repulsive forces between molecules. We need to describe the intermolecular forces in PEG and see if one part is water-like and another benzene-like. Water is a polar molecule and benzene is a nonpolar molecule, so we predict that PEG contains both polar and nonpolar regions.

SOLVE The structure of PEG consists of $-CH_2CH_2-$ groups connected by oxygen atoms. The oxygen atoms are capable of hydrogen bonding with water molecules, so the attractive force between PEG and water is due to hydrogen bonding. Benzene is nonpolar and is attracted to the $-CH_2CH_2-$ groups in the polymer. Nonpolar groups interact via London dispersion forces, so those forces must be responsible for the solubility of PEG in nonpolar benzene.

THINK ABOUT IT As predicted, PEG contains both polar and nonpolar regions, allowing it to be solvated by both polar solvents like water and nonpolar solvents like benzene.

Practice Exercise When PEG is added to soft drinks it keeps CO_2, responsible for the fizz in sodas, in solution longer after the soda is poured. What intermolecular attractive forces between PEG and CO_2 might make this use possible? ⚙

19.7 Aldehydes, Ketones, Carboxylic Acids, Esters, and Amides

Five functional groups—aldehydes, ketones, carboxylic acids, esters, and amides—all contain a carbon atom double-bonded to an oxygen atom (Figure 19.35). Aldehydes and ketones are collectively referred to as *carbonyl compounds* because the carbonyl groups

in their molecules are bonded only to R groups or H atoms. Carboxylic acids, as their name implies, are acidic, and their chemistry is determined by the –COOH subunit, referred to as a *carboxylic acid group*. Esters and amides can be made from carboxylic acids by reacting them with alcohols and amines.

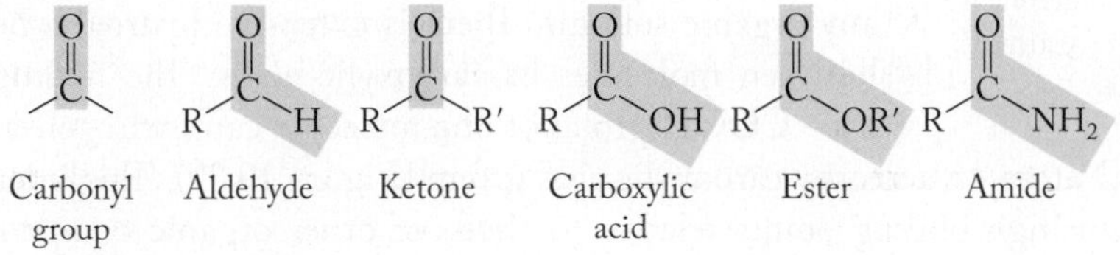

FIGURE 19.35 The carbonyl group is found in five important functional groups: aldehydes, ketones, carboxylic acids, esters, and amides. The R and R′ groups may be any organic group.

Aldehydes and Ketones

Aldehydes and ketones closely resemble each other in both chemical and physical properties. An **aldehyde** contains a carbonyl group bound to one R group and one hydrogen atom; its general formula is RCHO or R(C=O)H. A *ketone* contains a carbonyl group bound to two R groups; its general formula is RCOR or R(C=O)R. The R groups may be the same as in acetone, $CH_3C(O)CH_3$, or different as in 2-heptanone (Figure 19.36), which is found in cloves, blue cheese, and many fruits and dairy products.

The double bond in the carbonyl group accounts for the reactivity of aldehydes and ketones. It is different from the double bond in an alkene, however, because it is polar (Figure 19.37). The electronegative oxygen pulls electron density toward itself, and the chemistry of aldehydes and ketones is linked to the polarity of the C=O bond. Other polar species tend to react with carbonyls when electron-rich regions of their molecules approach the δ+ carbon atom of the carbonyl group.

Aldehydes and ketones are polar, and they tend to parallel the ethers with respect to water solubility. They cannot hydrogen-bond with other aldehyde and ketone molecules because they contain only carbon-bonded hydrogen atoms, so they have lower boiling points than alcohols of comparable molar mass.

Because of the C=O bond, aldehydes and ketones are in a higher oxidation state than alcohols, and indeed many of the smaller aldehydes and ketones are made by oxidizing alcohols of the same carbon number. Aldehydes and ketones do not polymerize through their carbonyl groups. Many polymers have carbonyl functional groups as part of their structure, but these groups themselves do not react to form long chains.

We have already seen several examples of aldehydes and ketones in earlier chapters. In Chapter 5, we were introduced to formaldehyde and acrolein, two

aldehyde organic compound containing a carbonyl group bonded to one R group and one hydrogen; its general formula is RCHO.

CONNECTION We first introduced carboxylic acids in our discussion of acids in Chapter 4. We defined ketones in Chapter 6.

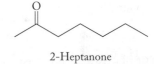

2-Heptanone

FIGURE 19.36 The ketone 2-heptanone is found in many plants and dairy products.

FIGURE 19.37 The electron distribution in a carbonyl group is skewed toward the oxygen end of the bond because oxygen is more electronegative than carbon.

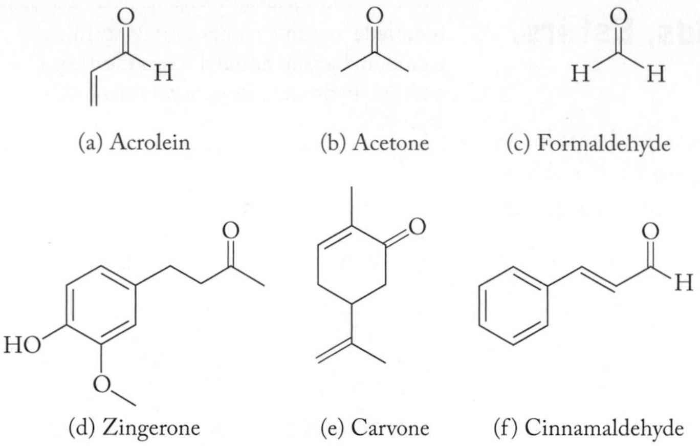

FIGURE 19.38 The ketone and aldehyde functional groups are common among organic compounds: (a) acrolein is found in barbeque smoke, (b) acetone is used in nail polish remover, (c) aqueous solutions of formaldehyde are used to preserve biological specimens, (d) zingerone is found in the spice ginger, (e) carvone is found in the leaves of spearmint, and (f) cinnamon owes its flavor and odor to cinnamaldehyde.

(a) Acrolein
(b) Acetone
(c) Formaldehyde
(d) Zingerone
(e) Carvone
(f) Cinnamaldehyde

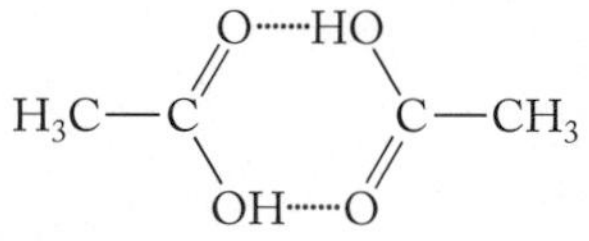

FIGURE 19.39 The high boiling points of carboxylic acids are the result of strong hydrogen bonds between neighboring molecules.

CONNECTION We discussed quantitative aspects of equilibria involving weak acids and weak bases in Chapter 15.

aldehydes shown in Figure 19.38. Acetone, $CH_3C(O)CH_3$, is a widely used solvent found in nail polish remover. The flavors and aromas of ginger (zingerone), spearmint (carvone), and cinnamon (cinnamaldehyde) all come from compounds that contain a carbonyl group.

CONCEPT TEST

What other functional groups are present in zingerone, carvone, and cinnamaldehyde besides the carbonyl group?

Carboxylic Acids

Carboxylic acids are organic compounds that are proton donors, which means they are Brønsted–Lowry acids (Sections 8.4 and 15.1). The R group in RCOOH may be any organic subunit. There is extensive hydrogen bonding between molecules of carboxylic acids. The hydrogen on the –COOH group of one molecule can hydrogen-bond to either O atom on a nearby carboxylic acid group (Figure 19.39). This interaction results in high boiling points relative to those of other organic compounds of comparable molar mass.

Donating a proton leaves the carboxylic acid with a negatively charged oxygen whose electron density is delocalized over the entire carbonyl group. This delocalization contributes to the stability of the carboxylate anion. The common carboxylic acids are weak acids, which means that they are present in aqueous solutions as mostly neutral molecules, a small fraction of which are ionized, donating H^+ ions to molecules of water (Figure 19.40).

Vinegar is a dilute aqueous solution of the carboxylic acid acetic acid. Large quantities of vinegar are produced commercially by the air oxidation of ethanol in the presence of enzymes from *Acetobacter* bacteria. Bacteria can also convert acetic acid and other constituents in biomass to methane. For example, the digestive systems of cows introduce significant amounts of methane to the atmosphere, about 100–200 liters per day per animal. Translating this process to an industrial scale is an attractive future source of hydrocarbons, provided the complexities of bacterial action can be adapted for large-scale production. Recall that we described the timescale for the formation of fossil fuels in Chapter 7. Converting organic matter into hydrocarbon fuel may be possible without waiting millennia for the anaerobic processes deep within Earth to do so.

The production of methane from plant residues that are mostly cellulose (a carbohydrate) requires the sequential action of several types of bacteria. In the first stages, selected bacteria break up cellulose into mixtures of small molecules. Depending on the bacterial strain, these small-molecule products include H_2 and CO_2, acetic acid, formic acid, or methanol or some other small alcohol. All these products then undergo reactions promoted by the metabolism of **methanogenic**

FIGURE 19.40 Carboxylic acids such as acetic acid (vinegar) are weak acids in water.

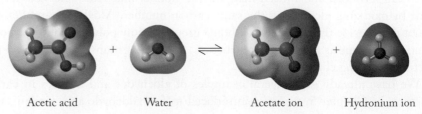

Acetic acid   Water   Acetate ion   Hydronium ion

(methane-producing) **bacteria**, which consume hydrogen and simple organic compounds for energy and produce methane gas in the process:

$$4\,H_2(g) + CO_2(g) \rightarrow CH_4(g) + 2\,H_2O(\ell)$$

$$\underset{\text{Acetic acid}}{CH_3COOH(aq)} \rightarrow CH_4(g) + CO_2(g)$$

$$\underset{\text{Formic acid}}{4\,HCOOH(aq)} \rightarrow CH_4(g) + 3\,CO_2(g) + 2\,H_2O(\ell)$$

$$\underset{\text{Methanol}}{4\,CH_3OH(aq)} \rightarrow 3\,CH_4(g) + CO_2(g) + 2\,H_2O(\ell)$$

Methanogenic bacteria have a measurable effect on Earth's atmosphere and climate because methane is a potent greenhouse gas, trapping about 20 times more heat per molecule than carbon dioxide.

methanogenic bacteria bacteria using simple organic compounds and hydrogen for energy; their respiration produces methane, carbon dioxide, and water, depending on the compounds they consume.

ester organic compound in which the −OH of a carboxylic acid group is replaced by −OR, where R can be any organic group.

condensation reaction two molecules combining to form a larger molecule and a small molecule (typically water).

Esters and Amides

A number of chemical families are closely related to the carboxylic acids. We consider only two of them here: esters and amides. In **esters**, the −COOH group of a carboxylic acid becomes a −COOR group, where R can be any organic group. The presence of these groups makes esters polar, and their boiling points are comparable to those of aldehydes and ketones of similar size. An ester is prepared by the *esterification* of an acid with an alcohol (Figure 19.41a). Esterification is a **condensation reaction**: two molecules combine ("condense") to create a larger molecule while a small molecule (typically water) is also formed.

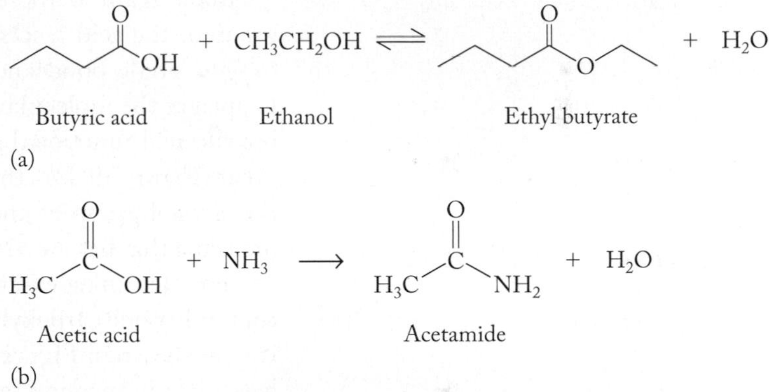

Esters frequently have very pleasant fragrances that are much different from the acids from which they are derived. For example, the carboxylic acid butyric acid, with a straight chain of 4 carbon atoms, is responsible for the odor of rancid butter. The ethyl ester of this carboxylic acid (ethyl butyrate) is responsible for the aroma of ripe pineapple. Esters are widely used in the personal products industry to provide pleasant scents for products like shampoos and soaps.

Many medications consist of molecules that contain carboxylic acid and ester groups. Figure 19.42 illustrates the carbon-skeleton structures of three common pain relievers—aspirin, ibuprofen, and naproxen—each of which contains a carboxylic acid group. Aspirin also contains an ester group.

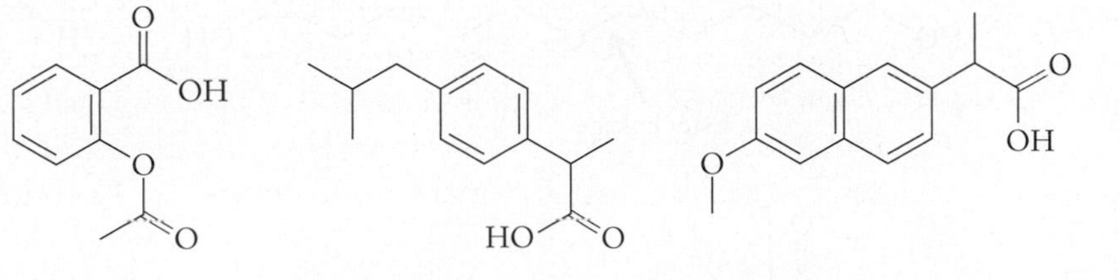

(a)

(b)

FIGURE 19.41 (a) Condensation reactions between carboxylic acids and alcohols produce esters. Here butyric acid reacts with ethanol, forming ethyl butyrate. (b) Condensation reactions between carboxylic acids and ammonia (or amines) produce amides. Here acetic acid reacts with ammonia, forming acetamide.

(a) Aspirin (b) Ibuprofen (c) Naproxen

FIGURE 19.42 Three pain medications, all of which contain carboxylic acid functional groups. (a) Aspirin was the first medication to be available in tablet form. (b) Ibuprofen has fewer side effects than aspirin. (c) Naproxen belongs to a class of compounds called nonsteroidal anti-inflammatory drugs (NSAIDs) and is used for the management of mild to moderate pain.

amide organic compound in which the –OH of a carboxylic acid group is replaced by –NH$_2$, –NHR, or –NR$_2$, where R can be any organic group.

condensation polymer macromolecule formed by the reaction of monomers yielding a polymer and water or another small molecule as products of the reaction.

▶❚❚ **CHEMTOUR** Polymers

Amides are made in condensation reactions between carboxylic acids and either ammonia or a primary or secondary amine (Figure 19.41b). Amides are polar, and hydrogen atoms bonded to nitrogen can hydrogen-bond with the oxygen atom of an adjacent amide group. These hydrogen bonds cause the boiling points of amides to be higher than those of esters of comparable molar mass.

Polyesters and Polyamides

Prior to this point, many of the compounds we have examined have been monofunctional, which means they have only one functional group that identifies their family. With the polymers of carboxylic acids and their derivatives, we enter the world of *difunctional* molecules, which are molecules with two functional groups. The key point to remember is that the functional groups for the most part still retain their individual chemical reactivity, even if they are in a molecule with another functional group. Also remember that the same features we enumerated for all other polymers still apply here: for polymers, function is determined by composition, structure, and size.

Look again at the esterification reaction in Figure 19.41(a). The –COOH group of the acid reacts with the –OH group of the alcohol to form a carbon–oxygen single bond and release a molecule of water. Think about what could happen at the molecular level if we had a single compound that contained a carboxylic acid functional group at one end and an alcohol functional group at the other (Figure 19.43). The carboxylic acid group of one molecule could react with the alcohol group of another molecule in a condensation reaction to generate a molecule that has a carboxylic acid group at one end, an alcohol at the other end, and an ester linkage in between. If this reaction happens repeatedly, a monomer containing one carboxylic acid and one hydroxyl group (a hydroxy acid) polymerizes, as shown in Figure 19.43, to form a polyester, a **condensation polymer**. We have already encountered one condensation polymer, PETE, in Section 19.6. In general, condensation polymers are formed by the reaction of monomers, yielding a polymer and water or another small molecule as products of the reaction. In addition to its use in plastic soda bottles, PETE is used extensively in medicine. Artificial heart valves and grafts for arteries are made from PETE.

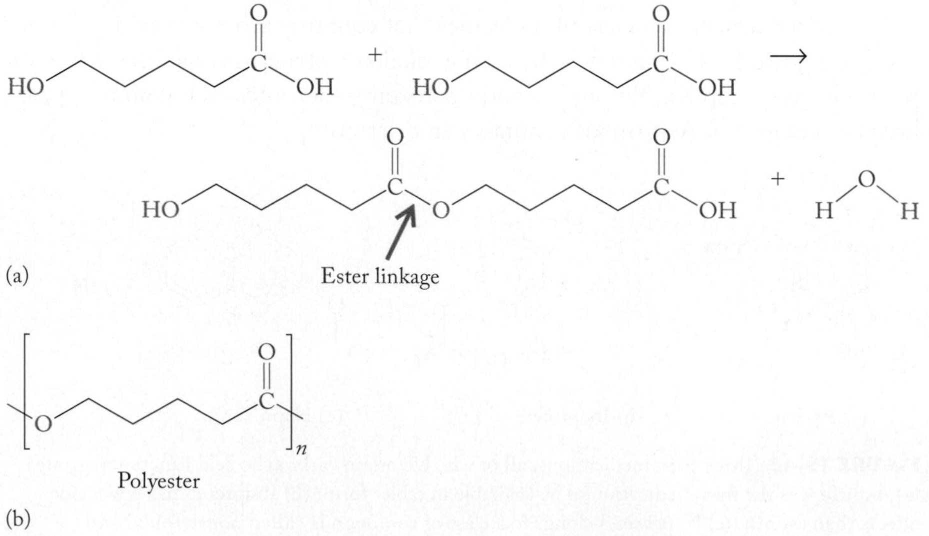

FIGURE 19.43 (a) Synthesis of an ester from a condensation reaction between two identical difunctional molecules, each one containing an alcohol group and a carboxylic acid group. The diester can then react with additional difunctional molecules at its –OH and –COOH ends, forming a triester, and the reaction repeats over and over, forming (b) the polyester made up of the repeating monomer unit shown.

(a)   Ester linkage

Polyester

(b)

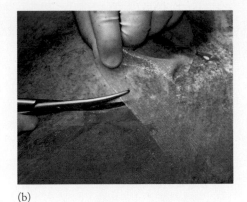

(a)

Glycolic acid   Lactic acid   A polyester

FIGURE 19.44 (a) The condensation polymer prepared from glycolic acid and lactic acid is used to make sutures that dissolve and artificial skins that protect against infection while promoting the regrowth of skin cells. (b) Synthetic skin being applied to a burn patient.

(b)

A copolymer formed by reaction of glycolic acid and lactic acid (Figure 19.44) is used to support the growth of skin cells used in grafts for burn victims. This condensation polymer is also used in making dissolving sutures. Esterification reactions used to make polyesters can be reversed by the addition of water, breaking the ester linkage and forming alcohol and acid functional groups. We examine this process in greater detail in Chapter 20.

SAMPLE EXERCISE 19.9 Making a Polyester   LO4

Show how a polyester can be synthesized from the difunctional alcohol $HO(CH_2)_3OH$ and the difunctional carboxylic acid $HOOC(CH_2)_3COOH$.

COLLECT AND ORGANIZE An ester is the product of a reaction between a carboxylic acid and an alcohol. A polyester is a polymer with a repeating unit containing an ester functional group. We are given an alcohol and a carboxylic acid to react to make the ester monomer.

ANALYZE Because the starting materials are difunctional, the alcohol can react with two molecules of carboxylic acid, and the acid can react with two molecules of alcohol.

SOLVE The reaction is

$$HOCH_2CH_2CH_2OH + HO\overset{O}{\underset{}{C}}CH_2CH_2CH_2\overset{O}{\underset{}{C}}OH \rightarrow$$

$$HOCH_2CH_2CH_2O\overset{O}{\underset{}{C}}CH_2CH_2CH_2\overset{O}{\underset{}{C}}OH + H_2O$$

The two OH groups shown in blue react to form an ester at one end of the carboxylic acid. The product molecule has an alcohol group on one end (shown in red) that can react with another molecule of carboxylic acid and a carboxylic acid group (green) on the other end that can react with another molecule of alcohol. Continuing these condensation reactions results in the formation of a polymer whose repeating unit is

$$\left[CH_2CH_2CH_2O\overset{O}{\underset{}{C}}CH_2CH_2CH_2\overset{O}{\underset{}{C}}O\right]_n$$

THINK ABOUT IT The repeating unit in the polyester contains one section that came from the alcohol and a second section that came from the carboxylic acid, which makes sense because esters are formed from alcohols and acids.

Practice Exercise A difunctional molecule may contain two different functional groups, such as this one with an alcohol group and a carboxylic acid group:

$$HO—CH_2CH_2CH_2C \overset{\displaystyle O}{\underset{\displaystyle OH}{\diagdown}}$$

Draw the repeating unit of the polyester made from this molecule.

SAMPLE EXERCISE 19.10 **Comparing Properties of Polymers** **LO6**

Clothes made from the polyester fabric known as Dacron may be less comfortable in hot weather than clothes made of cotton (also a polymer) because Dacron does not absorb perspiration as effectively as the cotton. Based on the repeating units of these two polymers (Figure 19.45), suggest a structural reason why cotton absorbs perspiration (water) better than Dacron.

COLLECT AND ORGANIZE Cotton and Dacron are polymers that differ in the functional groups in their repeating units. We need to identify the different functional groups and the interactions between these functional groups and water, a polar molecule.

ANALYZE The absorption of water by a polymer depends on the intermolecular forces present. Water is a polar molecule and is attracted to polar groups. We need to compare the groups in each monomer to see which has the greatest number of polar functional groups or atoms that can form hydrogen bonds. This will be the polymer more likely to absorb more water.

SOLVE Each monomer unit in cotton has three –OH groups attached to it, all polar and capable of hydrogen bond formation, which means they are likely to interact with the water in perspiration and thereby draw it away from the body. The monomer unit in Dacron has oxygen atoms in it, but no –OH groups. Although regions in the Dacron monomer are polar, they are not nearly as polar as the –OH groups in the cotton monomer unit. Cotton will absorb more perspiration than Dacron.

THINK ABOUT IT The principle of like interacting with like works for polymers just as it does for small molecules.

Practice Exercise Gloves made of a woven blend of cotton and polyester fibers protect the hands from exposure to oil and grease but are comfortable to wear because they "breathe"—they allow perspiration to evaporate and pass through them, thereby cooling the skin. Suggest how these gloves work at the molecular level.

Combining difunctional molecules in a condensation reaction can be used to make *polyamides*, another class of very useful synthetic polymers. The functional groups are a carboxylic acid and an amine. The difunctional monomer units can be identical (Figure 19.46a), each containing one carboxylic acid group and one amine group, or they can be different (Figure 19.46b), with one monomer containing two carboxylic acid groups (a dicarboxylic acid) and the other containing two amine groups (a diamine).

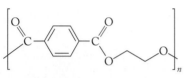

Repeating unit in Dacron

(a)

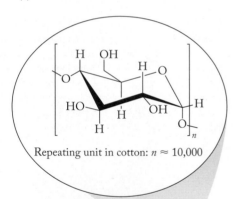

Repeating unit in cotton: $n \approx 10,000$

(b)

FIGURE 19.45 Based on the molecular structures of (a) Dacron and (b) cotton, why is cotton the better material for making tee shirts worn during strenuous exercise?

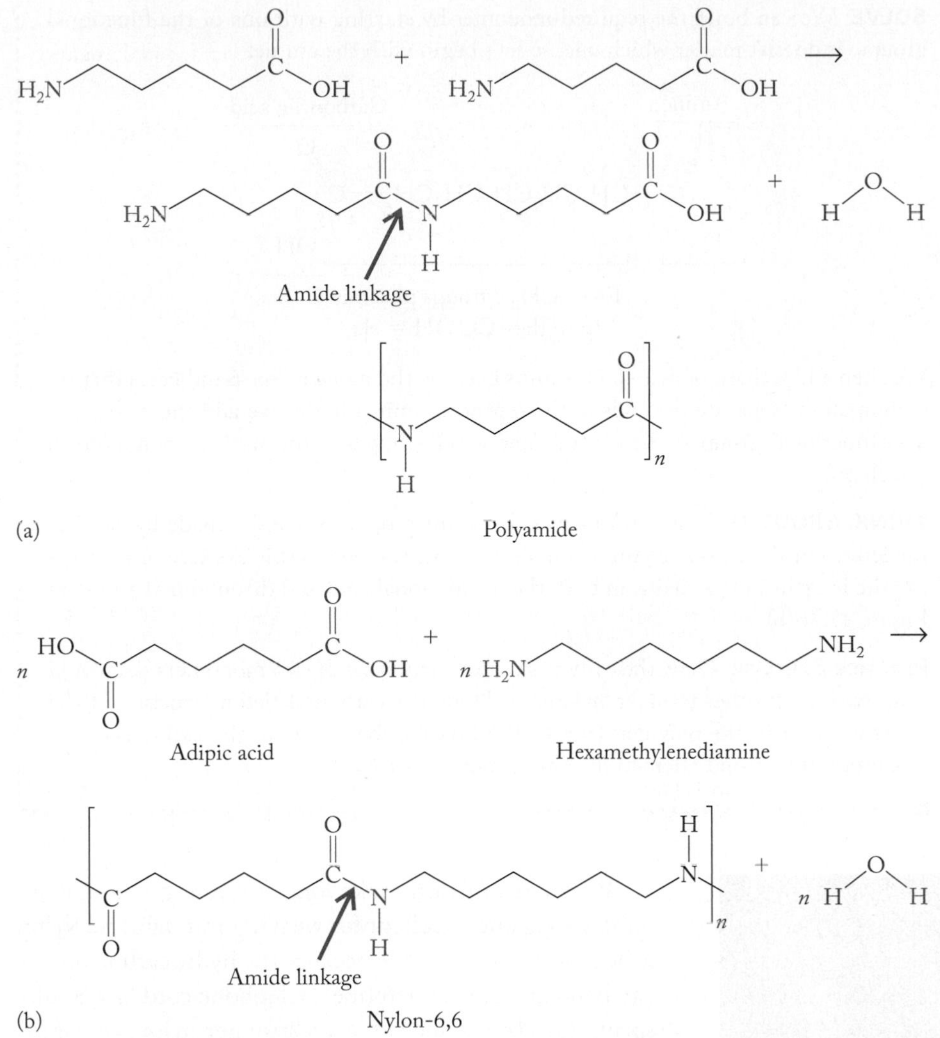

(a)   Polyamide

Adipic acid   Hexamethylenediamine

(b)   Nylon-6,6

FIGURE 19.46 (a) Synthesis of a polyamide from two identical monomers, each containing a carboxylic acid functional group and an amine functional group. (b) Synthesis of the polyamide nylon-6,6 from nonidentical monomers: adipic acid (a dicarboxylic acid) and hexamethylenediamine (a diamine).

Probably the most famous polyamide is nylon-6,6, made from the monomers shown in Figure 19.46(b). Each monomer contains six carbon atoms, which is what the digits in the name represent.

SAMPLE EXERCISE 19.11   Identifying Monomers   LO4

Another form of nylon is nylon-6, with the single 6 indicating that the polymer is made from the reaction of a series of identical six-carbon monomers. By analogy with the polyester in Sample Exercise 19.9 and the accompanying Practice Exercise, draw the condensed structure of a six-carbon molecule that could polymerize to make nylon-6.

COLLECT AND ORGANIZE Amides form from carboxylic acids and amines. We are asked to draw the condensed structure for a monomer that contains both functional groups and that could react with identical monomers to form a polyamide with a repeating unit six carbon atoms long. The exercise refers us to an example with polyesters to use as an analogy.

ANALYZE By analogy to Sample Exercise 19.9 and its accompanying Practice Exercise, we should suggest a difunctional molecule that has a carboxylic acid on one end and an amine on the other because these are the two functional groups that react to form the amide linkage in a polyamide.

Benzene-1,4-dicarboxylic acid
(terephthalic acid)

1,4-Diaminobenzene

Repeating unit in Kevlar

FIGURE 19.47 The monomers used to make Kevlar are a dicarboxylic acid and a diamine. The amide bond in the repeating unit is highlighted.

SOLVE We can build the required monomer by starting with one of the functional groups. It doesn't matter which one, so let's begin with the amine:

Amine   Carboxylic acid

$$ \underset{\text{H}}{\overset{\text{H}}{\text{N}}}-CH_2CH_2CH_2CH_2CH_2-\overset{O}{\underset{OH}{C}} $$

Five $-CH_2-$ groups plus one C
from the $-COOH$ = six C

We then add a chain of five $-CH_2-$ units because the name nylon-6 indicates that six carbon atoms separate the ends of the repeating unit. Finally, we add the carboxylic acid functional group as the second functional group and the sixth carbon atom in the chain.

THINK ABOUT IT Many nylons with different properties can be made by varying the length of the carbon chain in a monomer like the one in this exercise or by varying the lengths of the chains in both the difunctional acid and difunctional amine in Figure 19.46(b).

Practice Exercise Draw the carbon-skeleton structures of two monomers that could react with each other to make nylon-5,4. Draw the carbon-skeleton structure of the repeating unit in the polymer. (NOTE: The first number refers to the carboxylic acid monomer; the second refers to the amine monomer.)

Polymers of nylon make long, straight fibers that are quite strong and excellent for weaving into fabrics. Nylon is flexible and stretchable because the hydrocarbon chains can bend and curl, much like a telephone cord or a Slinky spring toy. To produce an even stronger nylon, researchers recognized they had to find some way to reduce the ability of the chains to form coils. They discovered this could be done using monomer units containing functional groups that made it difficult for the chains to bend. One product of this work was a polyamide called Kevlar, invented by Stephanie Kwolek of DuPont in 1965. Kevlar is formed from a dicarboxylic acid of benzene and a diamine of benzene (Figure 19.47). When these two monomers polymerize, the flat, rigid aromatic rings keep the chains straight.

Two additional intermolecular interactions orient the chains and hold them tightly together (Figure 19.48). First, the $-NH$ hydrogen atoms form hydrogen bonds with the oxygen atoms of carbonyl groups on adjacent chains. Second, the rings stack on top of one another (as in polystyrene) and provide additional interactions that hold the chains together in parallel arrays. The result is

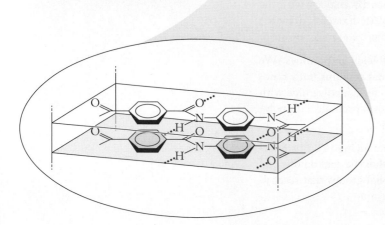

FIGURE 19.48 A bullet fired point-blank does not puncture fabric made of Kevlar. The strength is due in part to the strong interactions between the functional groups in the molecular structure of Kevlar.

TABLE 19.7 Summary of Common Polymers and Their Uses

NAME	ABBREVIATION	FUNCTIONAL GROUP	USE
Addition Polymers			
Polyethylene	PE	Alkane	Plastic bags and films
Poly(tetrafluoroethylene)	Teflon	Fluoroalkane	Nonstick coatings
Poly(1,1-dichloroethylene)	Saran	Chloroalkane	Plastic wrap
Polyvinyl chloride	PVC	Chloroalkane	Drain pipes
Poly(methyl methacrylate)	PMMA	Alkane and ester	Shatter-resistant glass, such as Plexiglas, Lucite
Polystyrene	PS	Aromatic hydrocarbon	Dishes, insulation
Poly(vinyl alcohol)	PVAL	Alcohol	Gloves, bottles
Condensation Polymers			
Poly(ethylene glycol)	PEG	Ether	Pharmaceuticals, consumer products
Poly(ethylene terephthalate)	PETE	Ester	Plastic bottles
Nylon		Amide	Clothing
Kevlar		Amide	Protective equipment
Dacron		Ester	Clothing

a fiber that is very strong but still flexible. Fabrics and helmets made of Kevlar resist puncture, even by bullets and hockey pucks fired at them, and they are also resistant to flames and reactive chemicals.

We have seen how the macroscopic properties of Kevlar and other polymers are influenced by their microscopic structure. Table 19.7 reinforces this message with descriptions of the molecular structures of several addition and condensation polymers and examples of familiar materials made from them.

19.8 A Brief Survey of Isomers

The existence of isomers is partly responsible for the enormous number and wide variety of organic compounds in the world. We have already discussed isomers in Chapters 5, 6, and 16, and in several sections of this chapter. This brief survey collects all the information we have presented so far, and Figure 19.49 summarizes the relationships between the types of isomers we have mentioned. Remember that isomers are molecules that have the same number and same kinds of atoms but differ in how those atoms are arranged. Because the molecules' structures are not the same, their physical and chemical properties differ as well.

As Figure 19.49 shows, isomers fall into two general categories: constitutional isomers and stereoisomers. Constitutional isomers, which we introduced in Chapter 6 and revisited with alkanes in Section 19.2, differ in the way in which the atoms comprising them are connected. Stereoisomers have their atoms connected in the same way, but the three-dimensional arrangement of their atoms differs. Let's review constitutional isomers first.

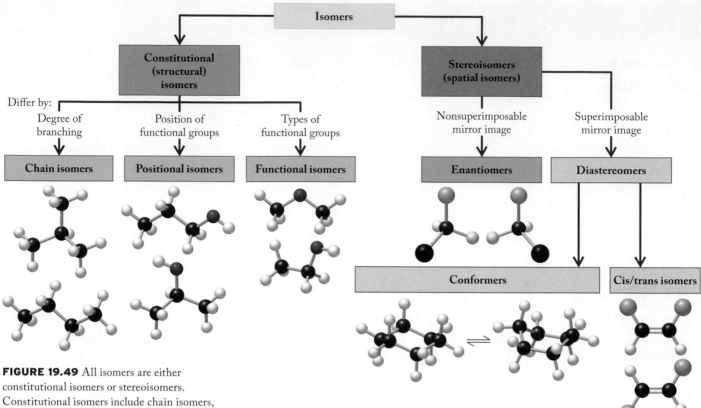

FIGURE 19.49 All isomers are either constitutional isomers or stereoisomers. Constitutional isomers include chain isomers, positional isomers, and functional isomers. Stereoisomers are further categorized as enantiomers (optically active compounds) or diastereomers.

In Chapter 6 we saw three kinds of constitutional isomers, although we did not name them at the time. Now that we have a better understanding of organic chemistry, we can categorize them more clearly as chain isomers, positional isomers, and functional isomers. *Chain isomers* are molecules having different arrangements of their carbon skeletons, like butane and 2-methylpropane. *Positional isomers* like propanol and isopropanol have the same functional group (in this case, the –OH group) bonded to different carbon atoms. *Functional isomers* like ethanol and dimethyl ether have different functional groups because of the arrangement of their atoms. The differing shapes and polarities of constitutional isomers give rise to differences in their physical properties such as melting points and boiling points, as we saw in Figure 6.3.

We introduced stereoisomers in Chapter 5 in our discussion of chirality. We saw them again in Chapter 16 when we learned about metal complexes, and then again in this chapter in Section 19.3 on alkenes. Stereoisomers are different from constitutional isomers because their atoms have the same connectivity: the same atoms are connected to each other in the same way. What makes stereoisomers special is that the arrangement of the atoms in three-dimensional space differs. That's why the prefix "stereo-" is used in the name; it means we need to consider what the molecule looks like in three dimensions.

Stereoisomers fall into two categories: *enantiomers* and *diastereomers*. Enantiomers are mirror-image isomers; diastereomers are stereoisomers that are not enantiomers. Let's first review enantiomers.

Several structural features can give rise to enantiomers, but the most important one for our purposes is the presence of a carbon atom that has four different atoms or groups of atoms attached to it. This arrangement about a carbon atom makes a molecule nonsuperimposable on its mirror image, much like your right and left hands. Such a molecule is chiral. This molecule and its mirror image rotate beams of plane-polarized light in opposite directions, as we saw in Figure 5.36. This behavior gives rise to another name for enantiomers: *optically active compounds* or *optical isomers*.

The second category of stereoisomers is the diastereomers, which include the cis and trans isomers that can occur because of the C=C double bonds in alkenes. Cis/trans isomers are also possible in metal complexes, as we learned in Chapter 16. *Conformers*, such as the chair and boat conformations of cyclohexane, are also diastereomers. Other structural features may give rise to enantiomers and diastereomers as well, but they are beyond the scope of this book.

In this chapter, we introduced the major functional groups in organic chemistry, and we discussed how the structure of a molecule and the functional groups it contains determine its chemical and physical properties. We examined a few reactions that produce several types of polymeric materials common in the modern world, ranging from carpets and insulation to bulletproof vests. Other organic compounds are used as fuels, as pharmaceuticals, and as building materials. Despite the range of materials we have discussed, this treatment can give only a brief taste of the importance of organic chemistry.

SAMPLE EXERCISE 19.12   Integrating Concepts: Taxol

Taxol, known generically as paclitaxel (Figure 19.50a; $C_{47}H_{51}NO_{14}$; molar mass 853.9 g/mol), is an important drug in the treatment of several types of cancer. The compound was originally harvested from the bark of the very slow-growing Pacific yew tree (*Taxus brevifolia*). Unfortunately, the concentration of taxol in the bark is only 0.01%. To conduct the clinical trials to test the effectiveness of taxol, *9000* mature trees were harvested. Using Pacific yew trees as the sole source of taxol would have led to their extinction, so other sources were needed. A number of research groups have reported the synthesis of taxol using either organic chemistry or biosynthetic processes using cultures of *Taxus* cells. The organic chemistry approaches start with other, simpler compounds found in nature. One of them uses a starting material called verbenone (Figure 19.50b), which is a commercially available derivative of pinene, a component of the resin exuded by pine trees. The synthesis of taxol from verbenone requires 36 steps and results in an overall yield of only 0.1% (moles of taxol/mole of verbenone).

a. How many aromatic rings are in taxol?
b. How many (i) ester, (ii) ether, (ii) alcohol, (iv) amide, (v) ketone, and (vi) alkene functional groups and (vii) chiral centers are in the taxol molecule?
c. Suppose 10.0 kg of taxol is required for a clinical trial of the substance in cancer patients. How many kilograms of the starting material verbenone are needed to prepare that amount?

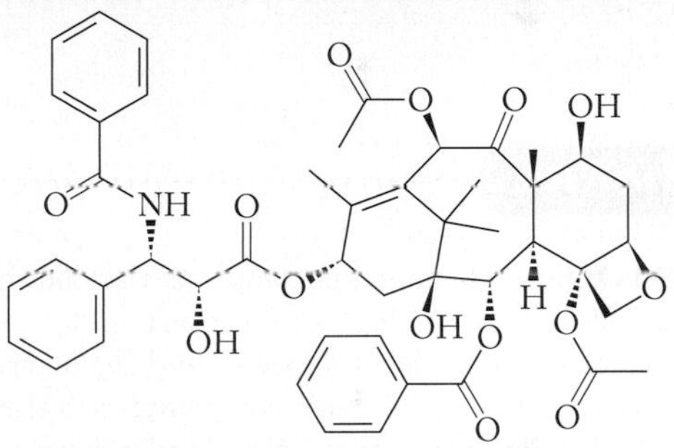

COLLECT AND ORGANIZE We have the structural formula of taxol, which consists of a variety of functional groups and chiral carbon centers. We also know the amount of taxol we need for a study and the overall yield of the 36-step reaction used to synthesize it. We are given the structure of the starting material used in step 1 of the process.

(a) Taxol

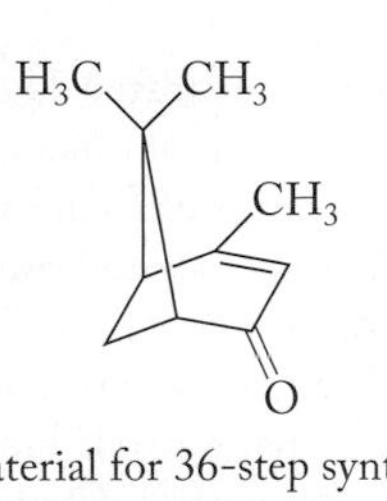

(b) Starting material for 36-step synthesis of taxol

FIGURE 19.50 One of the laboratory syntheses of (a) taxol begins with (b) the starting material verbenone.

ANALYZE We can use Table 19.1 to help us identify the types of functional groups in the molecule. The overall yield of the synthesis is 0.1%, so we can estimate that we need 1000 times as much starting material to set up the reaction to prepare 10.0 kg of the final product. We can calculate the mass of verbenone we need for the first step of the reaction, but we need to know the number of moles of verbenone required. Verbenone is a much smaller molecule than taxol; its formula is $C_{10}H_{14}O$ and its molecular weight is 150.24 g, making its molar mass about 1/5 that of taxol. We estimate that we will need around 2000 kg of it for the reaction.

SOLVE Figure 19.51 shows the highlighted functional groups in the molecule:

a. There are 3 aromatic rings (gray).
b. (i) There are 4 esters (yellow); (ii) 1 ether (green); (iii) 3 alcohols (orange); (iv) 1 amide (blue); (v) 1 ketone (red); (vi) 1 alkene (pink); and (vii) 11 chiral centers (carbon atoms with asterisks).
c. The overall yield of the multistep process is 0.1%, or 0.001 mol taxol/mol verbenone, so the mass (kg) of verbenone needed to produce 10.0 kg (or 1.00×10^4 g) of taxol is

$$1.00 \times 10^4 \; \cancel{\text{g taxol}} \times \frac{1 \; \cancel{\text{mol taxol}}}{853.9 \; \cancel{\text{g taxol}}} \times \frac{1 \; \cancel{\text{mol verbenone}}}{0.001 \; \cancel{\text{mol taxol}}}$$

$$\times \frac{150.24 \; \text{g verbenone}}{1 \; \cancel{\text{mol verbenone}}} \times \frac{1 \; \text{kg}}{10^3 \; \text{g}} = 2 \times 10^3 \; \text{kg verbenone}$$

Thus, we need 2000 kilograms of verbenone to produce just 10 kilograms of taxol.

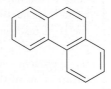

FIGURE 19.51

THINK ABOUT IT The answer matches our initial estimate (to one significant figure, which is all we are entitled to, given the 0.1% yield of the process). Like many natural substances with complex structures, taxol is a challenging substance to prepare synthetically. The overall yield of a series of reactions is equal to the product of the yields of 36 individual reactions. If the average yield of the reactions is n, then $n^{36} = 0.001$. Solving for n, we find that the average yield of each reaction is about 83%, which is pretty good for most synthetic work.

SUMMARY

Section 19.1 Organic chemistry encompasses the study of all carbon compounds, classified on the basis of functional groups—subunits of structure that confer on molecules specific and characteristic chemical and physical properties. Organic compounds are also differentiated based on size. **Polymers**, or macromolecules, have molar masses from several thousand to over 1,000,000 g/mol and are composed of repeating **monomer** units.

Section 19.2 Alkanes are **saturated hydrocarbons** because their molecules contain the maximum number of hydrogen atoms per carbon atom. A **homologous series** of alkanes is generated by sequential addition of –CH₂– units (**methylene groups**) into the chain, which has a –CH₃ (**methyl group**) at each end. Alkanes with more than 3 carbon atoms per molecule may have constitutional isomers, molecules with the same molecular formula but different arrangements of C–C bonds and physical properties. **Cycloalkanes** are alkanes containing rings of carbon atoms.

Section 19.3 *Alkenes* and *alkynes* are *unsaturated hydrocarbons* because the double bonds in alkenes and triple bonds in alkynes can be *hydrogenated* to incorporate more hydrogen into their molecular structures. In addition to having constitutional isomers, alkenes also have stereoisomers: *E*, or *trans*, **isomers** and *Z*, or *cis*, **isomers**, depending on the arrangement of the groups around the double bond. Alkenes undergo **addition reactions** with hydrogen and hydrogen halides. Alkenes can also be polymer-

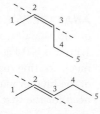

ized to **homopolymers** used in construction, in fabrics, as wrapping and packaging material, and in medical devices. Most of these polymers are **addition polymers**.

Section 19.4 Aromatic hydrocarbons are characterized by planar rings in which sp^2 hybridized carbon atoms are joined by a combination of σ and π bonds. The π bond electrons are delocalized over all the carbon atoms in the rings, making the structures particularly stable.

Section 19.5 Amines are organic compounds with the general formula RNH₂, R₂NH, or R₃N, where R is any organic subgroup.

Section 19.6 The alcohol (R—OH) and **ether** (R—O—R′) functional groups (where R and R′ are alkyl groups or aromatic rings) represent two ways of incorporating the *heteroatom* oxygen into organic compounds. Different monomer units derived from alcohols or ethers can be chemically combined to make **heteropolymers** or **copolymers** whose properties depend on the arrangement of the monomer units in the polymer.

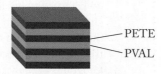

PETE
PVAL

Section 19.7 The carbonyl group, C═O, is found in **aldehydes** (RCHO), ketones (RCOR′), carboxylic acids (RCOOH), **esters** (RCOOR′), and **amides** (RCONH₂, RCONHR′, or RCONR′₂). The chemical reactivity of aldehydes and ketones centers on the C═O bond. In carboxylic acids, the COOH group imparts acidic properties to the molecules. Carboxylic acids react with alcohols via **condensation reactions** to form esters and with ammonia or amines to form amides. Condensation reactions are used to prepare **condensation polymers** such as polyesters and polyamides from difunctional compounds for use in fabrics under the familiar names Dacron and nylon.

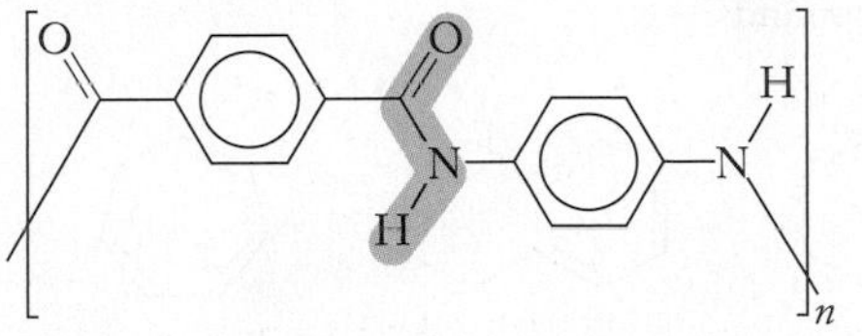

Section 19.8 Isomers are either constitutional isomers (chain isomers, positional isomers, and functional isomers) or stereoisomers. Stereoisomers are further divided into enantiomers (optical isomers) and diastereomers (cis/trans isomers).

PROBLEM-SOLVING SUMMARY

TYPE OF PROBLEM	CONCEPTS AND EQUATIONS	SAMPLE EXERCISES
Distinguishing among alkanes, alkenes, and alkynes	Alkanes contain only C—C single bonds; they are saturated hydrocarbons and do not react with H_2. Alkenes contain at least one C=C double bond that can combine with a molecule of H_2. Alkynes contain at least one C≡C triple bond that can combine with two molecules of H_2.	19.1
Drawing alkane structures	Use the Lewis structure to create a condensed structure, then gather all methylene groups inside parentheses to create the final condensed structure. To create a carbon-skeleton structure, use lines to depict single covalent bonds between carbon atoms. A sufficient number of H atoms to complete the valence of the carbon atoms are assumed.	19.2
Identifying the longest chain in organic molecules	Start at one end of any branch and assign numbers to each carbon atom. If the chain branches, choose one branch to follow. Repeat the process to explore other side chains and identify the longest chain.	19.3
Recognizing constitutional isomers	Establish that the compounds have the same molecular formula, and if they do, look for different arrangements of C—C bonds.	19.4
Identifying and naming stereoisomers and constitutional isomers	If molecules have the same formula, look for different arrangements of their bonds. Molecules with two groups on the same side of a C=C bond are cis isomers; those with groups on opposite sides are trans isomers. 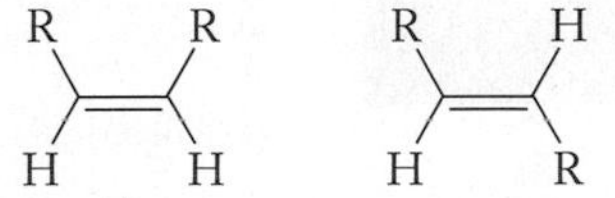	19.5
Identifying monomers in polymers	Find the smallest portion of the polymer that is repeated.	19.6, 19.11
Comparing the energy content of fuel mixtures	Determine the number of moles of each component, then multiply that number by the corresponding value of $\Delta H°_{comb}$.	19.7
Assessing properties of polymers	Evaluate the polarity of the functional groups in the polymer, and assess the relative importance of all types of intermolecular forces possible in the molecules.	19.8, 19.10
Making a polyester	Combine monomers with alcohol functional groups (ROH) and carboxylic acid functional groups (RCOOH) to form water (H_2O) and ester groups (RCOOR').	19.9

VISUAL PROBLEMS

(Answers to boldface end-of-chapter questions and problems are in the back of the book.)

19.1. How many degrees of unsaturation does each of the hydrocarbons shown in Figure P19.1 have?

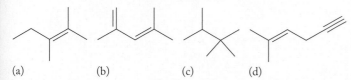

(a)   (b)   (c)   (d)

FIGURE P19.1

19.2. Which of the hydrocarbons in Figure P19.2 are constitutional isomers of each other?

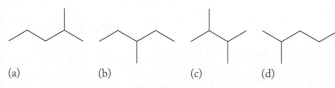

(a)   (b)   (c)   (d)

FIGURE P19.2

19.3. Figure P19.3 shows the carbon-skeleton structures of four organic compounds found in nature as fragrant oils. Which are alkenes?

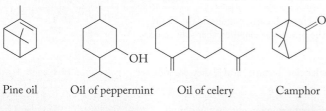

Pine oil   Oil of peppermint   Oil of celery   Camphor

FIGURE P19.3

19.4. Figure P19.4 shows three molecules: acrylonitrile (found in barbeque smoke), capillin (an antifungal drug), and pargyline (an antihypertensive drug). Which of these molecules does not contain the alkyne functional group?

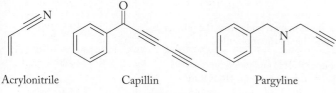

Acrylonitrile   Capillin   Pargyline

FIGURE P19.4

19.5. Which molecules in Figure P19.5 are considered aromatic compounds?

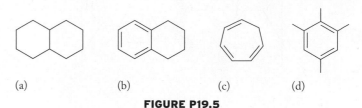

(a)   (b)   (c)   (d)

FIGURE P19.5

19.6. Benzyl acetate, carvone, and cinnamaldehyde are all naturally occurring oils. Their carbon-skeleton structures are shown in Figure P19.6. Which ones contain an aromatic ring?

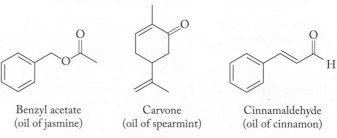

Benzyl acetate (oil of jasmine)   Carvone (oil of spearmint)   Cinnamaldehyde (oil of cinnamon)

FIGURE P19.6

19.7. In addition to the aromatic ring, what other functional groups can you identify in the molecules in Problem 19.6?

*19.8.** The three polymers shown in Figure P19.8 are widely used in the plastics industry. In which of them are the intermolecular forces per mole of monomer the strongest?

(a) Polyethylene   (b) Poly(vinyl chloride)   (c) Poly(1,1-dichloroethylene)

FIGURE P19.8

*19.9.** **Silly Putty** Silly Putty is a condensation polymer of dihydroxydimethylsilane (Figure P19.9). Draw the condensed structure of the repeating monomer unit in Silly Putty.

$$HO-\underset{\underset{CH_3}{|}}{\overset{\overset{CH_3}{|}}{Si}}-OH$$

Dihydroxydimethylsilane

FIGURE P19.9

19.10. Orlon and Acrilon Fibers Figure P19.10 shows the carbon-skeleton structure of polyacrylonitrile, which is marketed as Orlon and Acrilon. Draw the Kekulé structure of the monomeric reactant that produces this polymer.

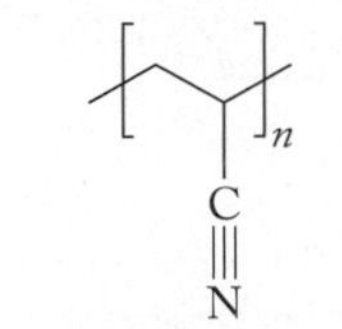

Polyacrylonitrile

FIGURE P19.10

19.11. Rubber is a polymer of isoprene. It is sometimes called polyisoprene. There are two forms of polyisoprene (Figure P19.11): *cis*-polyisoprene is the soft, flexible material we associate with the term "rubber"; gutta-percha, or *trans*-polyisoprene, is a much harder material. Draw the monomeric units of *cis*- and *trans*-polyisoprene.

cis-Polyisoprene   *trans*-Polyisoprene

FIGURE P19.11

19.12. Drugs and Enantiomeric Purity Thalidomide was marketed in the late 1950s as a drug to relieve morning sickness. Unfortunately, one isomer caused birth defects. Circle the chiral carbon atom(s) in the thalidomide molecule (Figure P19.12).

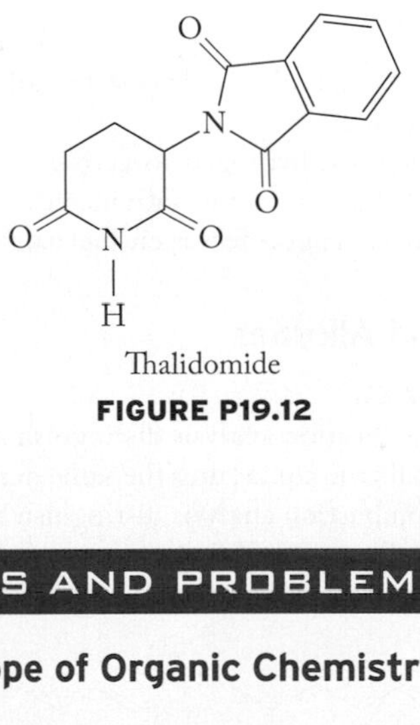

Thalidomide

FIGURE P19.12

19.13. Cholesterol-Lowering Drugs High serum cholesterol levels often correlate with increased risk of heart attacks. The drug sold under the trade name Mevacor has proven to be effective in lowering serum cholesterol. How many chiral carbon atoms are there in Mevacor (Figure P19.13)?

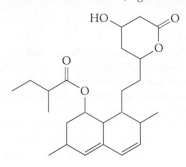

Mevacor

FIGURE P19.13

*19.14. The two compounds shown in Figure P19.14 are both considered to be amino acids. Identify the structural difference between them and explain why they are both amino acids.

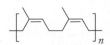

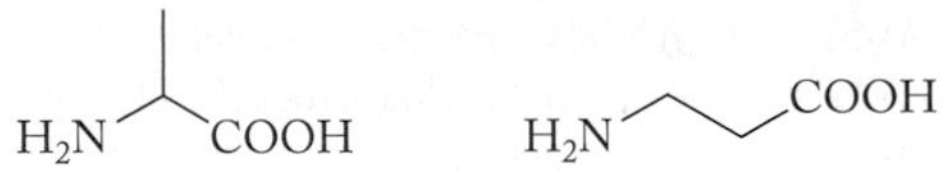

FIGURE P19.14

QUESTIONS AND PROBLEMS ● ■

Carbon: The Scope of Organic Chemistry

CONCEPT REVIEW

19.15. Describe three ways in which carbon atoms can form bonds to other carbon atoms using hybrid orbitals (valence bond theory).

19.16. What is the principal difference between an oligomer and a polymer formed from the same monomer?

*19.17. Is the interstitial alloy tungsten carbide (WC) considered to be an organic compound?

*19.18. Calcium carbide, CaC_2, was used in miner's lamps. Reaction of CaC_2 with water yields acetylene which, when ignited, gives light. Is calcium carbide considered an organic compound?

19.19. Which of the functional groups in Table 19.1 are polar?

19.20. Molecules of which of the compounds in Table 19.1 form hydrogen bonds with each other?

19.21. If the average molar mass of a polyethylene sample (A) is twice that of sample B, which sample begins to soften at a higher temperature? Explain why.

19.22. Which of the following properties of polyethylene increases as the number of monomer units per molecule of the polymer increases? (a) melting point; (b) viscosity; (c) density; (d) C:H ratio; (e) fuel value

Alkanes

CONCEPT REVIEW

19.23. Do linear and branched alkanes with the same number of carbon atoms all have the same empirical formula?

19.24. If an alkane and a cycloalkane have equal numbers of carbon atoms per molecule, do they have the same number of hydrogen atoms?

19.25. What is the hybridization of carbon in alkanes?

19.26. Figure P19.26 shows the carbon-skeleton structures of hexane and cyclohexane. Are hexane and cyclohexane constitutional isomers?

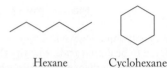

Hexane   Cyclohexane

FIGURE P19.26

19.27. Why isn't cyclohexane a planar molecule?

19.28. Which of the simple cycloalkanes (C_nH_{2n}, $n = 3$–8) has a nearly planar geometry?

19.29. Are cycloalkanes saturated hydrocarbons?

19.30. Do constitutional isomers always have the same molecular formula?

19.31. Do constitutional isomers always have the same chemical properties?

19.32. Are constitutional isomers members of a homologous series?

PROBLEMS

19.33. Draw and name all the constitutional isomers of C_5H_{12}.

19.34. Draw and name all the constitutional isomers of C_6H_{14}.

19.35. Which of the molecules in Figure P19.35 are constitutional isomers of octane (C_8H_{18})? Name these molecules.

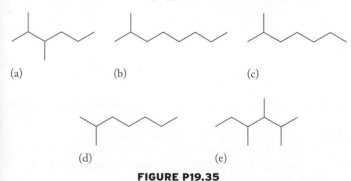

(a)   (b)   (c)

(d)   (e)

FIGURE P19.35

19.36. Which of the molecules in Figure P19.36 are constitutional isomers of heptane (C_7H_{16})? Name these molecules.

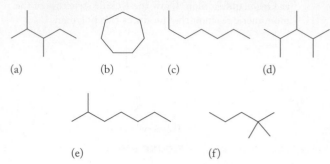

(a)   (b)   (c)   (d)

(e)   (f)

FIGURE P19.36

19.37. Convert the carbon-skeleton structures in Problem 19.35 to molecular formulas.

19.38. Convert the carbon-skeleton structures in Problem 19.36 to molecular formulas.

19.39. Place the following molecules in order of increasing boiling point: C_3H_8, $C_{14}H_{30}$, cyclooctane (C_8H_{16}).

***19.40.** Rank the molecules in Figure P19.40 in order of decreasing van der Waals forces.

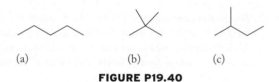

(a)   (b)   (c)

FIGURE P19.40

19.41. Which has a higher hydrogen-to-carbon ratio: hexane or cyclohexane?

***19.42.** How does the hydrogen-to-carbon ratio of a series of normal alkanes change with increasing molar mass? How does this change affect their fuel values?

Alkenes and Alkynes

CONCEPT REVIEW

19.43. Can combustion analysis distinguish between an alkene and a cycloalkane containing the same number of carbon atoms?

19.44. Can combustion analysis distinguish between an alkyne and a cycloalkene containing the same number of carbon atoms?

19.45. Why don't the alkenes in Figure P19.45 have cis and trans isomers?

FIGURE P19.45

19.46. Why don't alkynes have cis and trans isomers?

*19.47. Figure P19.47 shows the carbon-skeleton structure of carvone, which is found in oil of spearmint. Why doesn't the molecule carvone have cis and trans isomers?

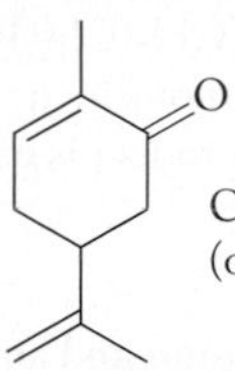

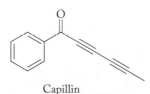

Carvone
(oil of spearmint)

FIGURE P19.47

*19.48. Figure P19.48 shows the carbon-skeleton structure of the antifungal compound capillin. Are the π electrons in capillin delocalized?

Capillin

FIGURE P19.48

19.49. Ethylene reacts quickly with HBr at room temperature, but polyethylene is chemically unreactive toward HBr. Explain why these related substances have such different properties.

*19.50. Polymerization of butadiene (CH$_2$=CHCH=CH$_2$) does not yield the same polymer as polymerization of ethylene (CH$_2$=CH$_2$). How could we convert poly(butadiene) into poly(ethylene)?

PROBLEMS

19.51. Using the average bond strengths given in Appendix 4, estimate the molar heat of hydrogenation, $\Delta H_{hydrogenation}$, for the conversion of C$_2$H$_4$ to C$_2$H$_6$.

$$CH_2=CH_2(g) + H_2(g) \rightarrow CH_3CH_3(g)$$

19.52. Using the average bond strengths given in Appendix 4, estimate the molar heat of hydrogenation, $\Delta H_{hydrogenation}$, for the conversion of C$_2$H$_2$ to C$_2$H$_6$.

$$CH\equiv CH(g) + 2 H_2(g) \rightarrow CH_3CH_3(g)$$

19.53. **Cinnamon** Label the isomers of cinnamaldehyde (oil of cinnamon) in Figure P19.53 as cis or trans and *E* or *Z*.

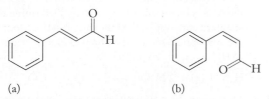

(a)   (b)

FIGURE P19.53

19.54. Prostaglandins, naturally occurring compounds in our bodies that cause inflammation and other physiological responses, are formed from arachidonic acid, an unsaturated hydrocarbon containing four C=C bonds and a carboxylic acid functional group. The stereoisomer containing all

cis double bonds is shown in Figure P19.54. How many stereoisomers other than this one are possible? Draw the isomer containing all trans double bonds.

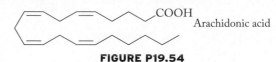

Arachidonic acid

FIGURE P19.54

19.55. Using data in Appendix 4, calculate ΔH_{rxn} for the production of acetylene from the controlled combustion of methane:

$$6 CH_4(g) + O_2(g) \rightarrow 2 C_2H_2(g) + 2 CO(g) + 10 H_2(g)$$

Is this an endothermic or an exothermic reaction?

19.56. Using data in Appendix 4, calculate ΔH_{rxn} for the production of acetylene from the reaction between calcium carbide and water, given that the ΔH_f° of CaC$_2$ is −59.8 kJ/mol:

$$CaC_2(s) + 2 H_2O(\ell) \rightarrow C_2H_2(g) + Ca(OH)_2(s)$$

Is this an endothermic or an exothermic reaction?

*19.57. Given the following two reactions and thermodynamic data from Appendix 4, estimate ΔH_{rxn} for the hydrogenation of acetylene (C$_2$H$_2$) with 1 mole of hydrogen gas to make ethylene (C$_2$H$_4$).

$$HC\equiv CH(g) + 2 H_2(g) \rightarrow H_3C-CH_3(g)$$
Acetylene   Ethane

$$H_2C=CH_2(g) + H_2(g) \rightarrow H_3C-CH_3(g)$$
Ethylene   Ethane

*19.58. The heat of hydrogenation of *cis*-2-butene is −119.7 kJ/mol; that of the trans isomer is −115.5 kJ/mol. Draw both isomers and the product of the hydrogenation reaction, and locate them on the graph of relative energy given in Figure P19.58. The condensed structural formula of 2-butene is CH$_3$CH=CHCH$_3$.

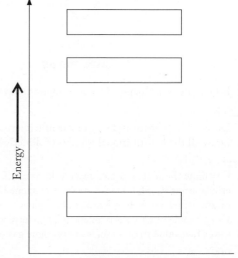

FIGURE P19.58

19.59. Making Glue Wood glue or "carpenter's glue" is made of poly(vinyl acetate). Draw the carbon-skeleton structure of this polymer. The monomer is shown in Figure P19.59.

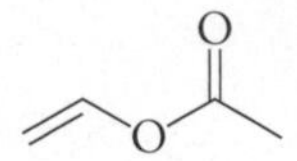

Vinyl acetate

FIGURE P19.59

19.60. The 2000 Nobel Prize in Chemistry was awarded for research on the electrically conductive polymer polyacetylene.
  a. Draw the carbon-skeleton structure of three monomeric units of the addition polymer that results from polymerization of acetylene, HC≡CH.
  *b. There are two possible stereoisomers of polyacetylene. Describe the two isomeric forms.

Aromatic Compounds

CONCEPT REVIEW

19.61. Why is benzene a planar molecule?
19.62. Why are aromatic molecules stable?
19.63. Do tetramethylbenzene and pentamethylbenzene have constitutional isomers?
19.64. Why aren't butadiene, C_4H_6, and 1,3-cyclohexadiene, C_6H_8 (Figure P19.64), considered aromatic molecules?

Butadiene   1,3-Cyclohexadiene

FIGURE P19.64

***19.65.** Pyridine (Figure P19.65) has the molecular formula C_6H_5N. Is pyridine an aromatic molecule?

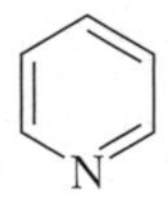

Pyridine

FIGURE P19.65

***19.66.** Is graphite (see Chapter 18) an aromatic compound?

PROBLEMS

19.67. Draw all the constitutional isomers of trimethylbenzene.
19.68. Draw all the constitutional isomers of dimethylnaphthalene.

19.69. Calculate the fuel values of gaseous benzene (C_6H_6) and ethylene gas (C_2H_4). Does 1 mole of benzene have a higher or lower fuel value than 3 moles of ethylene?
19.70. Does 1 mole of gaseous benzene (C_6H_6) have a higher or lower fuel value than 3 moles of acetylene gas (C_2H_2)?

Amines

CONCEPT REVIEW

19.71. Explain why methylamine (CH_3NH_2) is more soluble in water than butylamine [$CH_3(CH_2)_3NH_2$].
19.72. Combustion of hydrocarbons in air yields carbon dioxide and water. What other product is expected in the combustion of amines?

PROBLEMS

19.73. Are You Hungry? Serotonin and amphetamine both contain the amine functional group (Figure P19.73). Serotonin is responsible, in part, for signaling that we have had enough to eat. Amphetamine, an addictive drug, can be used as an appetite suppressant. Identify the primary and secondary amine functional groups in these molecules.

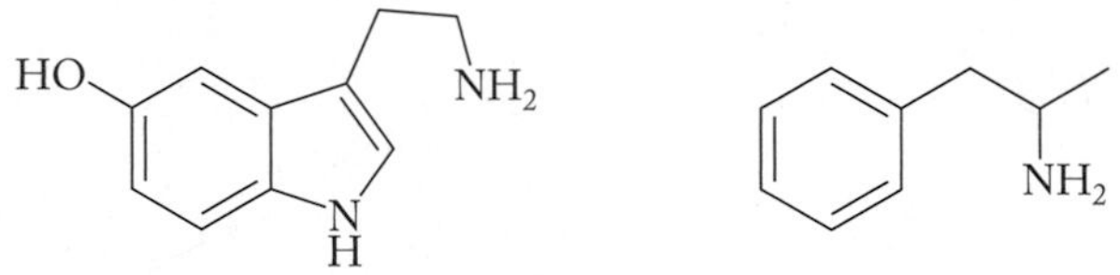

Serotonin   Amphetamine

FIGURE P19.73

19.74. Coffee Caffeine, the active ingredient in coffee, contains four nitrogen atoms per molecule. Aspartame is an artificial sweetener containing two nitrogen atoms. Structures of caffeine and aspartame are shown in Figure P19.74. Which nitrogen atoms represent secondary amines and which ones represent tertiary amines?

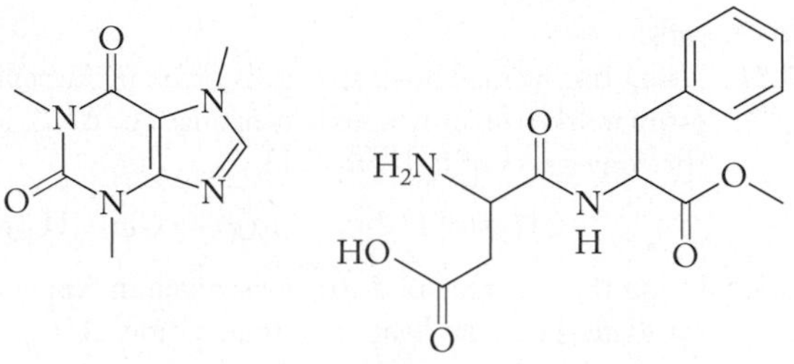

Caffeine   Aspartame

FIGURE P19.74

19.75. Renewable Energy Bacteria of the genus *Methanosarcina* convert amines to methane. Their action helps make methane a renewable energy source. Determine the standard enthalpy of the following reaction from the appropriate standard enthalpies of formation ($\Delta H^\circ_{f,CH_3NH_2} = -23.0$ kJ/mol):

$$4\,CH_3NH_2(g) + 2\,H_2O(\ell) \rightarrow 3\,CH_4(g) + CO_2(g) + 4\,NH_3(g)$$

19.76. Determine the ΔH°_{rxn} values of these combustion reactions of methylamine. ($\Delta H^\circ_{f,CH_3NH_2} = -23.0$ kJ/mol.)

$$4\,CH_3NH_2(g) + 13\,O_2(g) \rightarrow 4\,CO_2(g) + 4\,NO_2(g) + 10\,H_2O(\ell)$$

$$4\,CH_3NH_2(g) + 6\,O_2(g) \rightarrow 4\,CO_2(g) + 4\,NH_3(g) + 4\,H_2O(\ell)$$

*19.77. Methylamine is a weak base.
  a. Use information in Appendix 5 to sketch the titration curve for the titration of 125 mL of a 0.015 *M* solution of methylamine with 0.100 *M* HCl.
  b. Label the curve with the pH of the analyte solution and with the pH and titrant volumes halfway to the equivalence point and at the equivalence point.
  c. Draw the structures of the species present in the solution at the equivalence point.

*19.78. The pK_b values in Appendix 5 tell us that methylamine is a stronger base than ammonia and that dimethylamine is even stronger. Use the differences in their molecular structures to explain this trend in the strengths of these three bases.

Alcohols, Ethers, and Reformulated Gasoline

CONCEPT REVIEW

19.79. Why are the fuel values of ethanol and dimethyl ether (Figure P19.79) lower than that of ethane?

Dimethyl ether   Ethanol

FIGURE P19.79

19.80. Would you expect the fuel value of alcohols to increase or decrease as the number of carbon atoms in the alcohol increases?

19.81. Why do ethers typically boil at lower temperatures than alcohols with the same molecular formula?

19.82. Which do you expect to be more soluble in water, MTBE or 2,2-dimethylbutane (Figure P19.82)? Explain your answer.

MTBE   2,2-Dimethylbutane

FIGURE P19.82

*19.83. Clean Skin Disposable wipes used to clean the skin prior to a getting an immunization shot contain ethanol. After wiping your arm, your skin feels cold. Why?

*19.84. Dry Gas In the winter months in cold climates, water condensing in a vehicle's gas tank reduces engine performance. An auto mechanic recommends adding "dry gas" to the tank during your next fill-up. Dry gas is typically an alcohol that dissolves in gasoline and absorbs water. Based on the structures shown in Figure P19.84, which product would you predict would do a better job—methanol or 2-propanol?

Methanol   2-Propanol

FIGURE P19.84

PROBLEMS

19.85. Which of the compounds in Figure P19.85 are alcohols and which ones are ethers? Place them in order of increasing boiling point.

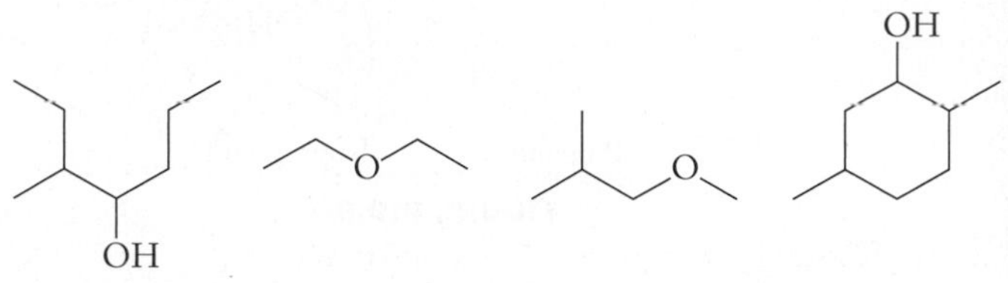

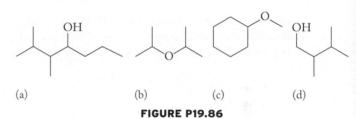

(a)   (b)   (c)   (d)

FIGURE P19.85

19.86. Which of the compounds in Figure P19.86 are alcohols and which ones are ethers? Place them in order of increasing vapor pressure at 25°C.

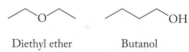

(a)   (b)   (c)   (d)

FIGURE P19.86

Consult tables of thermochemical data in Appendix 4 for any values you may need to solve Problems 19.87 through 19.90.

19.87. Calculate the fuel values of butanol and diethyl ether (Figure P19.87). Which has the higher fuel value?

Diethyl ether   Butanol

FIGURE P19.87

19.88. Calculate the fuel value of liquid diethyl ether and methyl propyl ether (Figure P19.88). Which has the higher fuel value? ($\Delta H^{\circ}_{\text{f,methyl propyl ether}} = -266.0$ kJ/mol.)

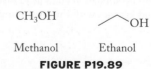

Diethyl ether   Methyl propyl ether

FIGURE P19.88

19.89. Problem 19.80 asked you to predict whether the fuel value of alcohols increases or decreases with the number of carbon atoms in the alcohol. Calculate the fuel values of liquid methanol and ethanol (Figure P19.89). Does your answer support the prediction you made in Problem 19.80?

CH$_3$OH   ⌿OH

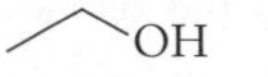

Methanol   Ethanol

FIGURE P19.89

19.90. Calculate the fuel values of liquid propanol and isopropanol (Figure P19.90). Which has the higher fuel value? ($\Delta H^{\circ}_{\text{f,propanol},\ell} = -302.6$ kJ/mol and $\Delta H^{\circ}_{\text{f,isopropanol},\ell} = -318.1$ kJ/mol.)

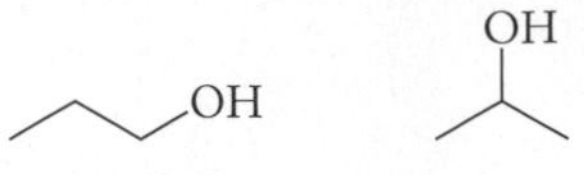

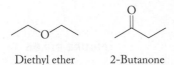

Propanol   Isopropanol

FIGURE P19.90

Aldehydes, Ketones, Carboxylic Acids, Esters, and Amides

CONCEPT REVIEW

19.91. Explain why carboxylic acids tend to be more soluble in water than aldehydes with the same number of carbon atoms.

19.92. In reference books, diethyl ether (Figure P19.92) is usually listed as "slightly soluble" in water, but 2-butanone is listed as "very soluble." Suggest why 2-butanone is more soluble.

Diethyl ether   2-Butanone

FIGURE P19.92

19.93. Are butanal and 2-butanone (Figure P19.93) constitutional isomers?

Butanal   2-Butanone

FIGURE P19.93

19.94. **Apples** The two esters shown in Figure P19.94 are both found in apples and contribute to the flavor and aroma of the fruit. Are the two compounds identical, constitutional isomers, or stereoisomers, or do they have different molecular formulas?

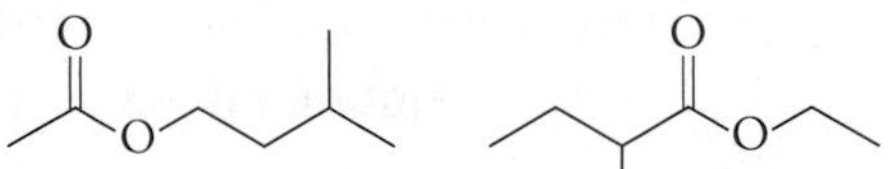

FIGURE P19.94

19.95. Can we use combustion analysis to distinguish between ketones and aldehydes with the same number of carbon atoms?

19.96. Can we use combustion analysis to distinguish between ethers and ketones with the same number of carbon atoms?

19.97. Resonance forms for acetic acid are shown in Figure P19.97. Which one contributes more to bonding? Explain your choice.

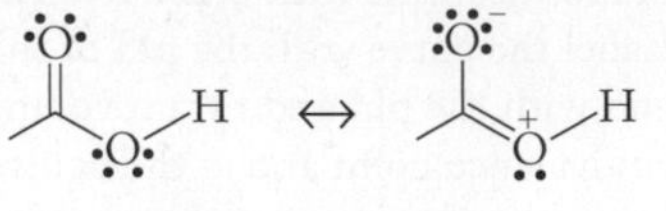

(a)   (b)

FIGURE P19.97

19.98. Figure P19.98 shows resonance forms for acetamide and acetic acid. Does resonance form (a) contribute more to the bonding in acetamide than does resonance form (b) to the bonding in acetic acid? Explain your answer.

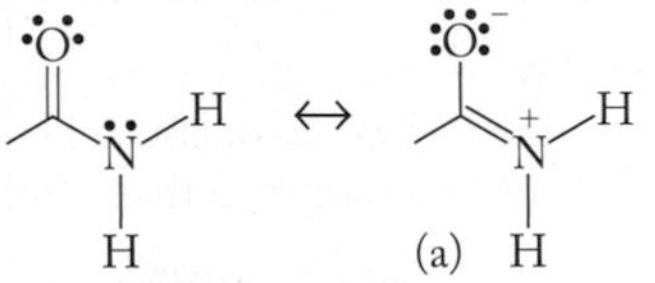

(a)

Acetamide

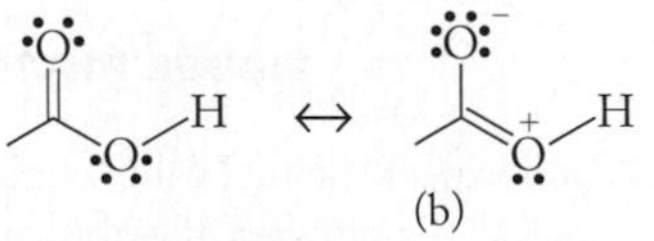

(b)

Acetic acid

FIGURE P19.98

19.99. What distinguishes an amine from an amide?

*19.100. Why can't we use tertiary amines to prepare amides?

PROBLEMS

19.101. Which of the compounds in Figure P19.101 are constitutional isomers of the aldehyde $C_5H_{10}O$?

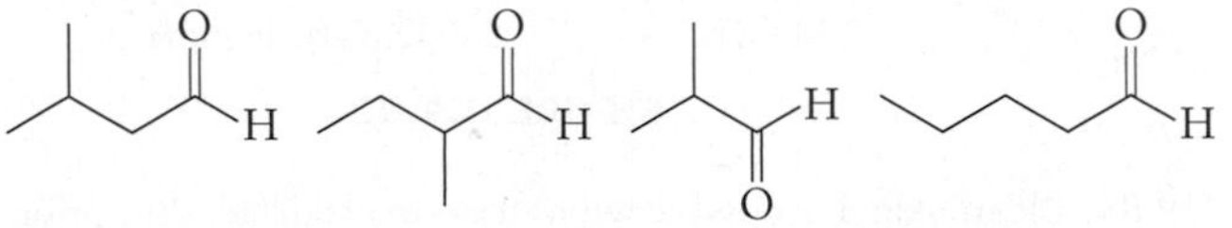

(a)   (b)   (c)   (d)

FIGURE P19.101

19.102. Each of the natural products in Figure P19.102 contains more than one functional group. Which of the compounds is an aldehyde?

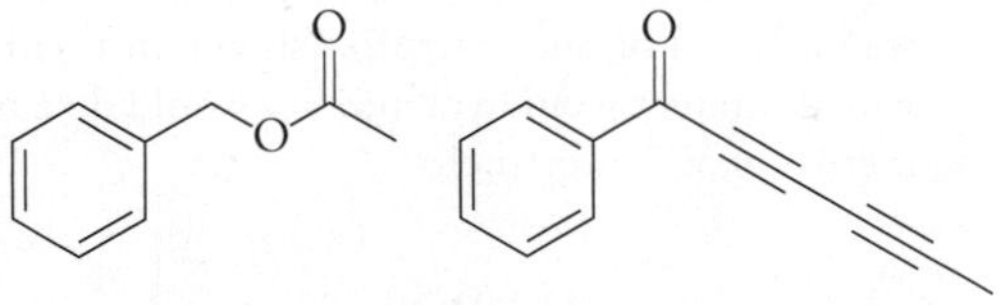

(a)   (b)

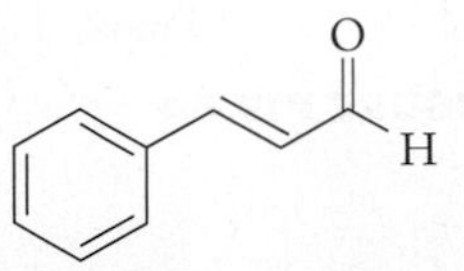

(c)

FIGURE P19.102

19.103. Which of the compounds in Figure P19.103 is a ketone?

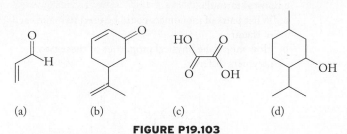

(a)   (b)   (c)   (d)

FIGURE P19.103

19.104. Propanal and acetone (2-propanone) have the same molecular formula, C_3H_6O, but different structures (Figure P19.104). Which compound is a ketone?

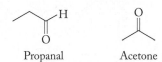

Propanal   Acetone

FIGURE P19.104

19.105. Plot the carbon-to-hydrogen ratio in aldehydes with one to six carbons as a function of the number of carbon atoms. Does this graph correlate better with the plot of C:H ratios for alkanes or for alkenes?

19.106. Plot the carbon-to-hydrogen ratio in carboxylic acids with one to six carbons as a function of the number of carbon atoms. Does this graph correlate better with the plot of C:H ratios for alkanes or for alkenes?

19.107. Esters are responsible for the odors of fruits including apples, bananas, and pineapples. Figure P19.107 shows three esters from these fruits. Identify the alcohol and carboxylic acid that react to form these compounds.

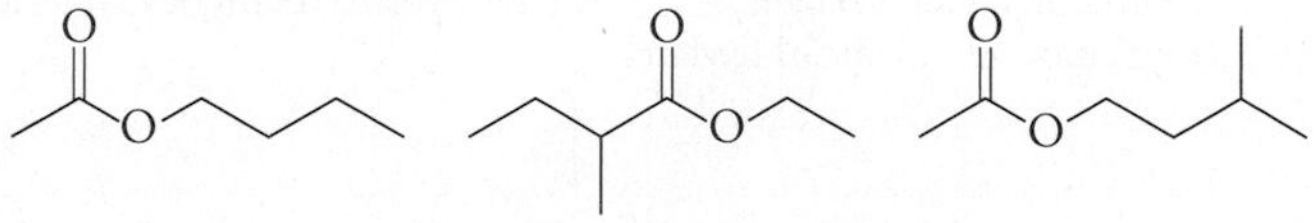

(a) Pineapples   (b) Bananas   (c) Apples

FIGURE P19.107

19.108. **Beeswax** The waxy material found in beehives (Figure P19.108) is an ester composed of an alcohol and a carboxylic acid with a long hydrocarbon chain. Identify the alcohol and acid in beeswax.

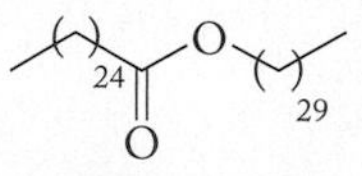

FIGURE P19.108

19.109. **Stimulants and Tranquilizers** Nicotine is a stimulant found in tobacco. Valium is a tranquilizer. Both molecules contain two nitrogen atoms in addition to other functional groups. In Figure P19.109, identify the nitrogen atoms shown in blue as belonging to an amine or an amide.

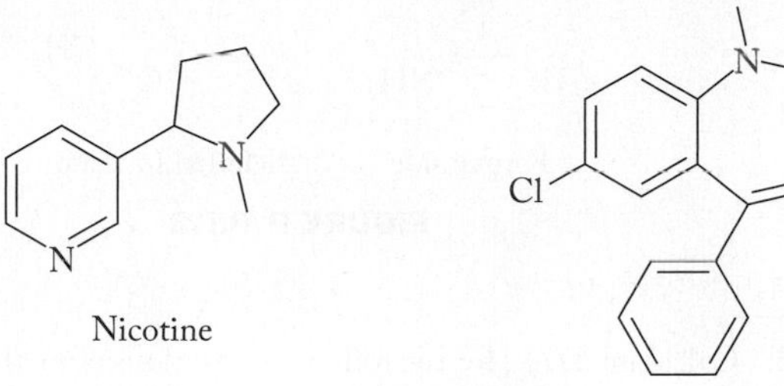

Nicotine   Valium

FIGURE P19.109

19.110. **Hot Peppers** Piperine and capsaicin are ingredients of peppers that give "heat" to spicy foods. In Figure P19.110, identify the nitrogen atoms shown in blue as belonging to an amine or an amide.

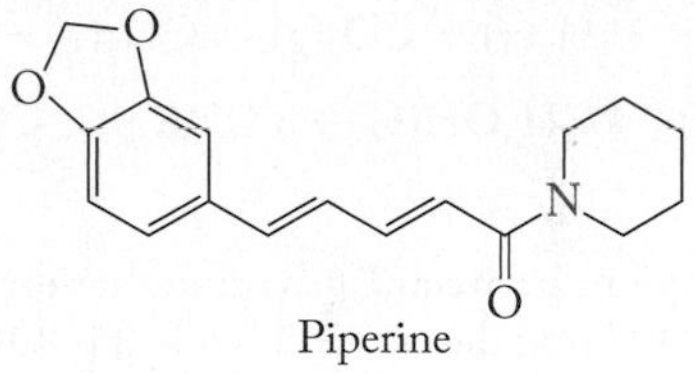

Piperine

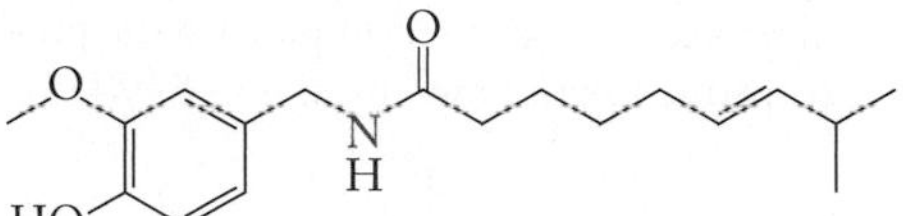

HO

Capsaicin

FIGURE P19.110

Consult tables of thermochemical data in Appendix 4 for any values you may need to solve Problems 19.111 through 19.114.

19.111. Calculate the fuel values of formaldehyde gas and liquid formic acid (Figure P19.111). The ΔH_f° of formaldehyde gas is −108.6 kJ/mol and the ΔH_f° of liquid formic acid is −425.0 kJ/mol. Which has the higher fuel value?

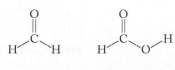

Formaldehyde   Formic acid

FIGURE P19.111

19.112. Calculate the fuel values of formamide and methyl formate (Figure P19.112), which have ΔH_f° values of −251 and −391 kJ/mol, respectively. Assume $NO_2(g)$ is a product of formamide combustion.

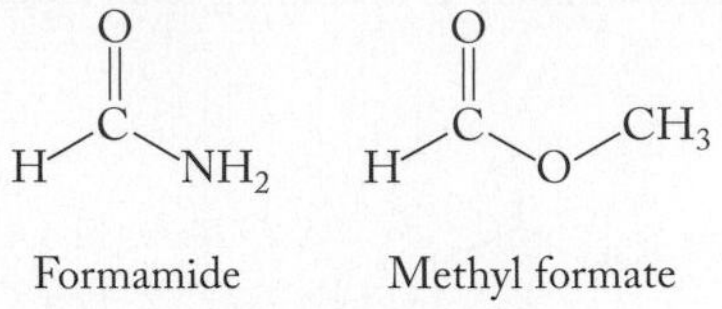

Formamide   Methyl formate

FIGURE P19.112

19.113. Calculate ΔH_{rxn}° for the following reactions of methanogenic bacteria, given $\Delta H_{f,HCOOH}^\circ = -425.0$ kJ/mol:

(1) $CH_3COOH(\ell) \rightarrow CH_4(g) + CO_2(g)$

(2) $4\,HCOOH(\ell) \rightarrow CH_4(g) + 3\,CO_2(g) + 2\,H_2O(\ell)$

19.114. Calculate ΔH_{rxn}° for the following reactions of methanogenic bacteria:

(1) $4\,H_2(g) + CO_2(g) \rightarrow CH_4(g) + 2\,H_2O(\ell)$

(2) $4\,CH_3OH(\ell) \rightarrow 3\,CH_4(g) + CO_2(g) + 2\,H_2O(\ell)$

19.115. Reactions between 1,6-diaminohexane, $H_2N(CH_2)_6NH_2$, and different dicarboxylic acids, $HOOC(CH_2)_nCOOH$, are used to prepare polymers that have a structure similar to that of nylon. How many carbon atoms were in the dicarboxylic acids used to prepare the polymers with the repeating units shown in Figure P19.115?

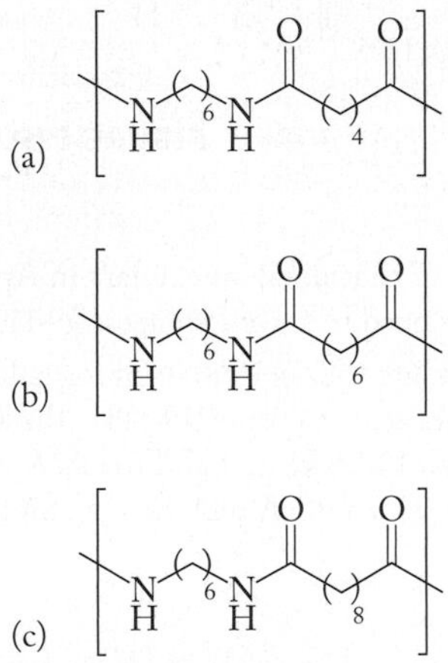

FIGURE P19.115

19.116. The two polymers in Figure P19.116 have the same empirical formula.
  a. What pairs of monomers could be used to make each of them?
  b. How might the physical properties of these two polymers differ?

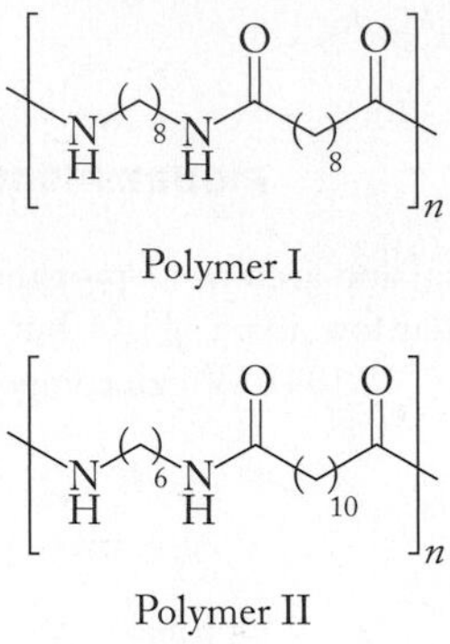

Polymer I

Polymer II

FIGURE P19.116

19.117. The polyester called Kodel is made with polymeric strands prepared by the reaction of dimethyl terephthalate with 1,4-di(hydroxymethyl)cyclohexane (Figure P19.117).
  a. Is Kodel a condensation polymer or an addition polymer? What is the other product of the reaction?
  *b. Dacron (Figure 19.45) is made from dimethyl terephthalate and ethylene glycol. What properties of Kodel fibers might make them better than Dacron as a clothing material?

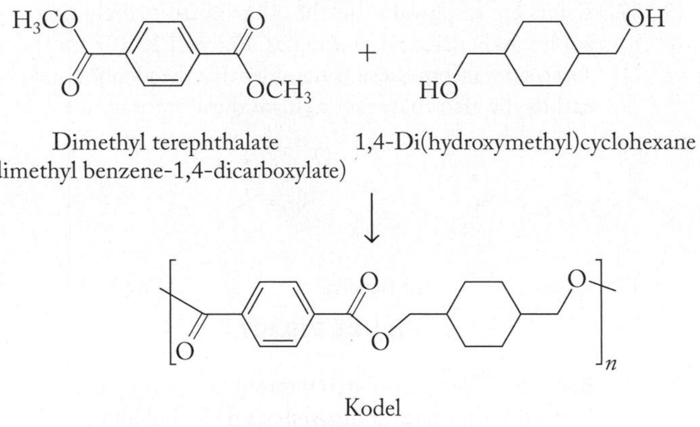

Dimethyl terephthalate   1,4-Di(hydroxymethyl)cyclohexane
(dimethyl benzene-1,4-dicarboxylate)

Kodel

FIGURE P19.117

19.118. Lexan is a polymer belonging to the class of materials called polycarbonates. Figure P19.118 shows the polymerization reaction for Lexan.
  a. What other compound is formed in the polymerization reaction?
  *b. Why is Lexan called a "polycarbonate"?

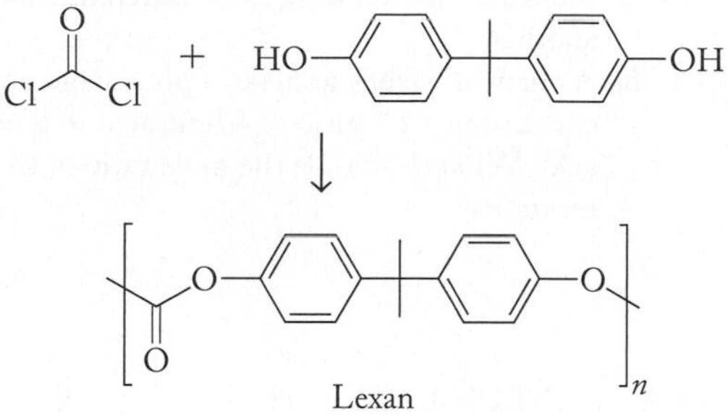

FIGURE P19.118

A Brief Survey of Isomers

CONCEPT REVIEW

19.119. Can all of the terms *enantiomer*, *achiral*, and *optically active* be used to describe a single compound? Explain.

*19.120. Could a racemic mixture be distinguished from an achiral compound based on optical activity? Explain your answer.

19.121. Can stereoisomers of molecules, such as cis and trans RCH=CHR, also have optical isomers? (R may be any of the functional groups we have encountered in this textbook.) Explain your answer.

*19.122. Could an oxygen atom in an alcohol, ketone, or ether ever be a chiral center in the molecule?

PROBLEMS

19.123. The scent associated with pine trees is derived from the molecule terpineol. One enantiomer of terpineol is shown in Figure P19.123. Draw the other optical isomer.

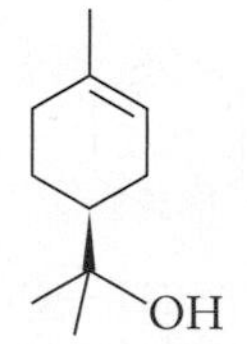

Terpineol

FIGURE P19.123

*19.124. The condensed structure of a molecule is CH$_3$CH=C(CH$_3$)CH(OH)CH$_3$CH$_3$. Draw all the possible stereoisomers of this compound.

19.125. Which of the molecules in Figure P19.125 do not have any possible constitutional isomers?

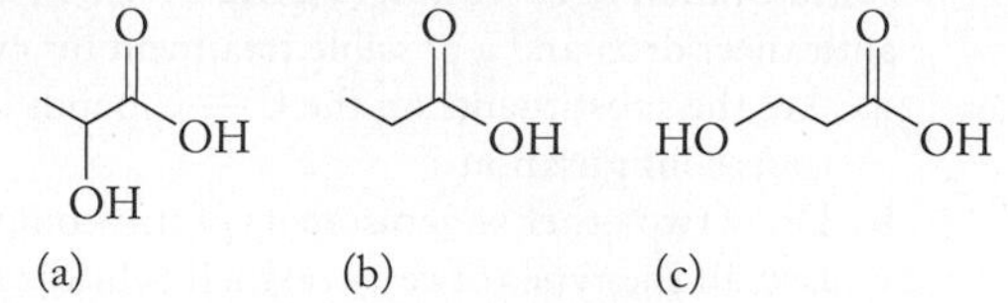

(a)   (b)   (c)

FIGURE P19.125

19.126. Which of the molecules in Figure P19.126 do not have any possible stereoisomers?

(a)   (b)   (c)

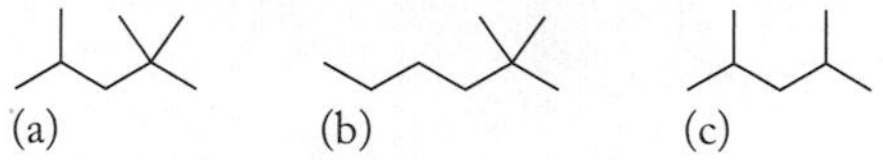

FIGURE P19.126

Additional Problems

19.127. How many grams of liquid methanol must be combusted to raise the temperature of 454 g of water from 20.0°C to 50.0°C? Assume that the transfer of heat to the water is 100% efficient. How many grams of carbon dioxide are produced in this combustion reaction?

19.128. Two compounds, both with molar masses of 74.12 g/mol, were combusted in a bomb calorimeter with $C_{calorimeter}$ = 3.640 kJ/°C. Combustion of 0.9842 g of compound A led to an increase in temperature of 10.33°C, while combustion of 1.110 g of compound B caused the temperature to rise 11.03°C. Which compound is butanol and which is diethyl ether?

19.129. Why should methane be more soluble in decane (C$_{10}$H$_{22}$) than in water?

19.130. **Salsa** Salsa made with cilantro has antibacterial properties because cilantro contains the aldehyde dodecenal (Figure P19.130).
  a. How many carbon atoms are in dodecenal?
  b. What functional groups are present in dodecenal?
  c. What types of isomerism are possible in dodecenal?

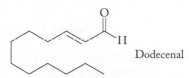

Dodecenal

FIGURE P19.130

19.131. Turmeric Turmeric is commonly used as a spice in Indian and Southeast Asian dishes. Turmeric contains a high concentration of curcumin (Figure P19.131), a potential anticancer drug and a possible treatment for cystic fibrosis.
a. Are the substituents on the C=C bonds in cis or trans configurations?
b. Draw two other stereoisomers of this compound.
c. List all the types of valence-shell hybridization of the carbon atoms in curcumin.

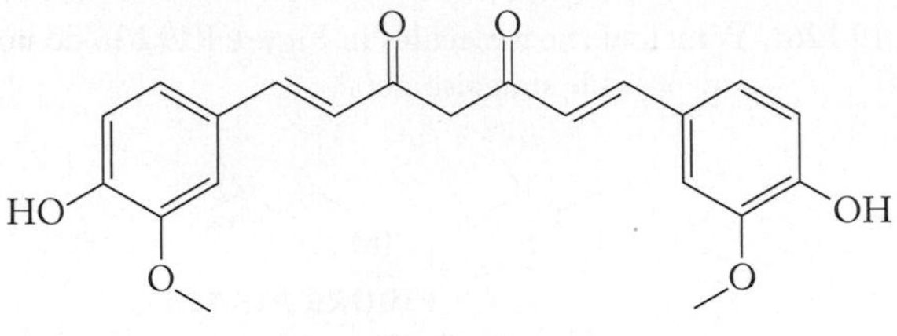

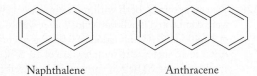

Curcumin

FIGURE P19.131

19.132. Polycyclic aromatic hydrocarbons are potent carcinogens. They are produced during combustion of fossil fuels and have also been found in meteorites. Can we use combustion analysis to distinguish between naphthalene and anthracene (Figure P19.132)?

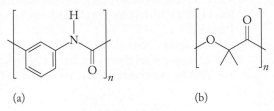

Naphthalene   Anthracene

FIGURE P19.132

19.133. Identify the reactants in the polymerization reactions that produce the polymers shown in Figure P19.133.

(a)   (b)

FIGURE P19.133

*19.134. **Raincoats** "Waterproof" nylon garments have a coating to prevent water from penetrating the hydrophilic fibers. Which functional groups in the nylon molecule make it hydrophilic?

19.135. Draw the carbon-skeleton structure of the condensation polymer of $H_2N(CH_2)_6COOH$. How does this polymer compare with nylon-6?

*19.136. Putrescine, $H_2N(CH_2)_4NH_2$, is one of the compounds that form in rotting meat.
a. Draw the carbon-skeleton structures of all the trimers (a molecule formed from three monomers) that can be formed from putrescine, adipic acid, and terephthalic acid (Figure P19.136). The three monomers forming the trimer do not have to be different from one another.
b. A chemist wishes to make a putrescine polymer containing a 1:1 ratio of adipic acid to terephthalic acid. What should be the mole ratio of the three reactants?

$$HOOCCH_2CH_2CH_2CH_2COOH$$

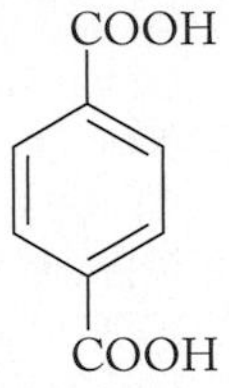

COOH

COOH

Adipic acid   Terephthalic acid
                     (benzene-1,4-dicarboxylic acid)

FIGURE P19.136

*19.137. Polymer chemists can modify the physical properties of polystyrene by copolymerizing divinylbenzene with styrene (Figure P19.137). The resulting polymer has strands of polystyrene cross-linked with divinylbenzene. Predict how the physical properties of the copolymer might differ from those of 100% polystyrene.

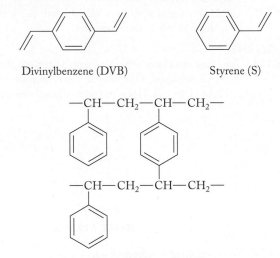

Divinylbenzene (DVB)   Styrene (S)

S cross-linked with DVB

FIGURE P19.137

19.138. Maleic anhydride and styrene (Figure P19.138) form a polymer with alternating units of each monomer.
  a. Draw two repeating monomer units of the polymer.
  b. Based on the structure of the copolymer, predict how its physical properties might differ from those of polystyrene.

Maleic anhydride   Styrene

FIGURE P19.138

19.139. **Superglue** The active ingredient in Superglue is methyl 2-cyanoacrylate (Figure P19.139). The liquid glue hardens rapidly when methyl 2-cyanoacrylate polymerizes. This happens when it contacts a surface containing traces of water or other compounds containing –OH or –NH– groups. Draw the carbon-skeleton structure of two repeating monomer units of poly(methyl 2-cyanoacrylate).

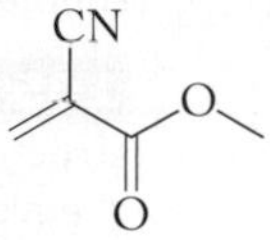

Methyl 2-cyanoacrylate

FIGURE P19.139

*19.140. Silicones are polymeric materials with the formula $[R_2SiO]_n$ (Figure P19.140). They are prepared by reaction of R_2SiCl_2 with water to yield the polymer and aqueous HCl. Consider this reaction as taking place in two steps: (1) water reacts with 1 mole of R_2SiCl_2 to produce a new monomer and 1 mole of HCl(aq); (2) one new monomer molecule reacts with another new monomer molecule to eliminate one molecule of HCl and make a dimer with a Si—O—Si bond.
  a. Suggest two balanced equations describing these reactions that occur over and over again to produce a silicone polymer.
  b. Why are silicones water repellent?

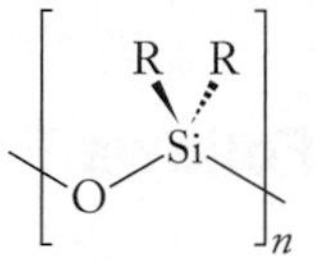

Silicone

FIGURE P19.140

19.141. Molecules of piperine and capsaicin (see Figure P19.110) contain amide functional groups.
  a. Draw the amine and the carboxylic acid that could react to form these two compounds.
  b. Are the double bonds in these molecules cis or trans?
  c. Name the functional groups that contain the oxygen atoms in these compounds.

19.142. The heats of combustion of two constitutional isomers are the same when estimated from average bond energies, but they are different when determined experimentally. Why?

Biochemistry

The Compounds of Life

Function Follows Form

For all the stunning diversity of the biosphere, from single-cell organisms to elephants, whales, and giant redwoods, all life-forms consist of substances made from only about 40 or 50 different small molecules. Huge variations among life-forms are possible when unique sequences of these few starting materials link together to form larger molecules and biopolymers. Unfortunately, subtle differences in these sequences can have devastating effects on health and even survival.

It is estimated, for example, that 70% of inherited diseases in humans are caused by the production of proteins that have the wrong composition, which leads to the wrong molecular shape. For example, the malformation or the total absence of the protein dystrophin causes a number of debilitating diseases referred to as muscular dystrophy (MD). In mild forms of MD, misshapen molecules of dystrophin are unable to build muscle fibers of sufficient strength to function normally. As a result, muscle tissue wastes away and the patient becomes physically disabled. In a severe form of the disease, Duchenne MD (DMD), functional dystrophin is absent, and the disability often results in early death. Knowledge of the mechanisms that produce dystrophin has led to promising therapies, some involving grafting of healthy muscle cells that can synthesize normal dystrophin into the tissues of MD patients. Some treatments involve injection of stem cells that fuse with and genetically complement dystrophic muscle in DMD patients. Similar approaches have resulted in the development of new drugs and cell therapies for other inherited and contagious diseases.

In this chapter we explore the composition, structure, and function of proteins and three other major classes of biomolecular compounds: carbohydrates, lipids, and nucleic acids. Each class performs similar functions in all the life-forms in which they occur. Many of these molecules are large and complex, but the knowledge we have developed about the chemical behavior of small molecules will serve as a useful framework on which to build an understanding of the behavior of the larger assemblies of molecules that form cells. As we explore the relationships between the structures and functions of these classes of compounds, we can begin to address such questions as these: How do nucleic

Genetically Engineered Hawaiian Papaya The U.S. FDA has approved more ▶ than 25 genetically engineered foods, including these papaya that resist a virus that threatened the Hawaiian papaya crop.

protein a biological polymer made of amino acids.

biomolecule an organic molecule present naturally in a living system.

amino acid a molecule that contains at least one amine group and one carboxylic acid group; in an *α-amino acid*, the two groups are attached to the same (α) carbon atom.

essential amino acid any of the 8 amino acids that make up peptides and proteins but are not synthesized in the human body and must be obtained through the food we eat.

$$H_2N-\underset{\underset{H}{|}}{\overset{\overset{R}{|}}{C}}-COOH$$

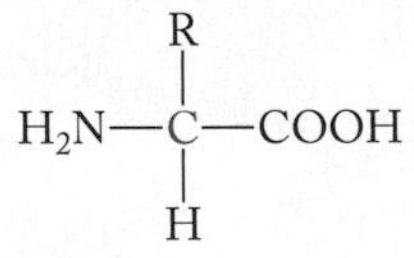

FIGURE 20.1 General structure of an α-amino acid.

∞ CONNECTION As in Chapter 19, we use the generic symbol R to represent any group of covalently bonded atoms that are connected to the main structure of a molecule via a C—C bond.

∞ CONNECTION The weakly basic properties of amines were described in Section 19.5.

acids convey genetic information from one generation to the next? How do some proteins provide structure to living organisms while others mediate biochemical processes? How do carbohydrates and lipids provide the energy we need and allow us to store the energy we don't need?

20.1 The Composition of Proteins

We begin this chapter with a discussion of **proteins**, the most abundant class of **biomolecules** in all animals, including us. Proteins account for about half of the mass of the human body that is not water. They are the major component in skin, muscles, cartilage, hair, and nails. Most of the enzymes that catalyze biochemical reactions are proteins, as are the molecules that transport oxygen to our cells and many of the hormones that regulate cell function and growth. Most proteins are large, with molar masses of 10^5 g/mol or more. They are examples of *biopolymers*.

Amino Acids

The molecular structure and biological function of big biomolecules depend on the identities of their small-molecule building blocks and the sequence in which those small molecules occur. The molecular building blocks of proteins are **amino acids**, so named because each of them contains at least one amine ($-NH_2$) group and at least one carboxylic acid ($-COOH$) group. The amino acids in proteins are *α-amino acids* because each of their molecules contains an α-carbon atom (highlighted in blue in Figure 20.1) that is bonded to both an $-NH_2$ and a $-COOH$ group.

> **CONCEPT TEST**
>
> According to valence bond theory, what is the hybridization of the α-carbon atom in the amino acid in Figure 20.1, and what are the approximate angles between the four bonds surrounding it?
>
> *(Answers to Concept Tests are in the back of the book.)*

In addition to its single bonds to the $-NH_2$ and $-COOH$ groups, the α-carbon atom in each of the amino acids that make up human proteins is bonded to a hydrogen atom and to one of 20 different *R groups*. The structures of these R groups, often called *side-chain* groups, are highlighted in red in the amino acid structures in Table 20.1. The amino acids are arranged in three categories based on their R groups. In the first category, containing 9 amino acids, the R groups contain mostly carbon and hydrogen atoms, and are nonpolar. The R groups of the remaining 11 amino acids contain at least one heteroatom (O, N, or S) bonded to a hydrogen atom and are polar. Of these amino acids, 2 (aspartic acid and glutamic acid) have R groups that contain carboxylic acid functional groups, and 3 (histidine, lysine, and arginine) contain groups with nitrogen atoms that are weakly basic.

Our bodies can synthesize 12 of the 20 amino acids in Table 20.1, but the other 8 must be present in the food we eat. These 8 are marked with a superscript *a* and are referred to as the **essential amino acids**. Most proteins from animal sources, including those in meats, eggs, and dairy products, contain all the essential amino acids needed by the human body, and in close to the correct

TABLE 20.1 **Structures and Abbreviations of the 20 Common Amino Acids**

Nonpolar R Groups

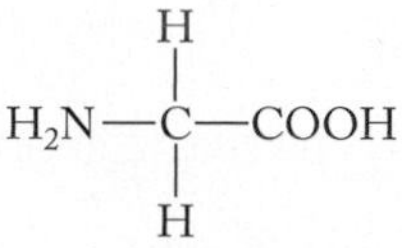

Glycine
(Gly)

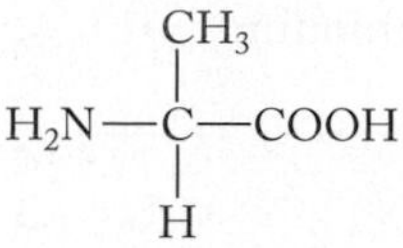

Alanine
(Ala)

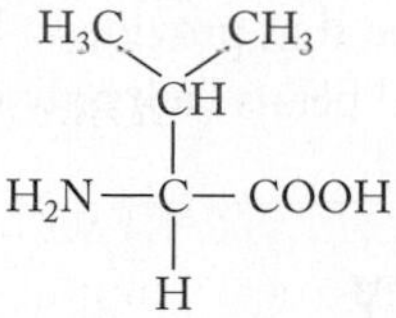

Valine[a]
(Val)

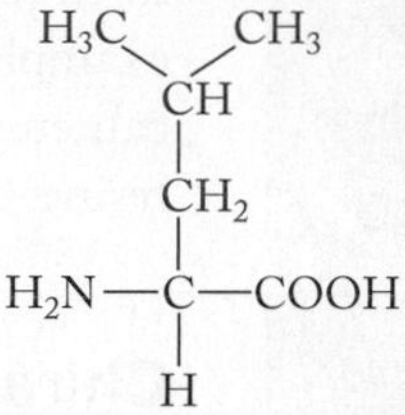

Leucine[a]
(Leu)

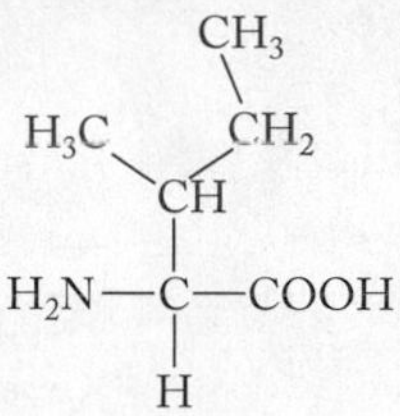

Isoleucine[a]
(Ile)

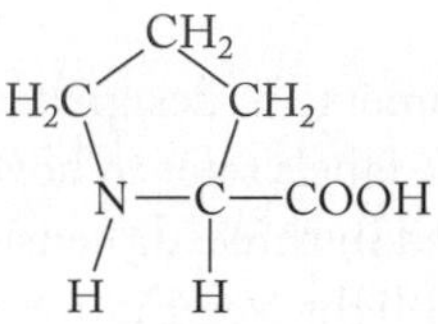

Proline
(Pro)

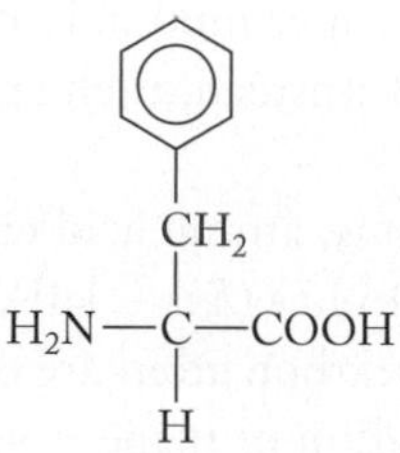

Phenylalanine[a]
(Phe)

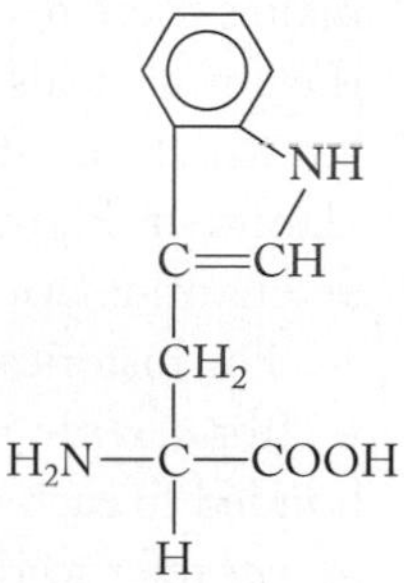

Tryptophan
(Trp)

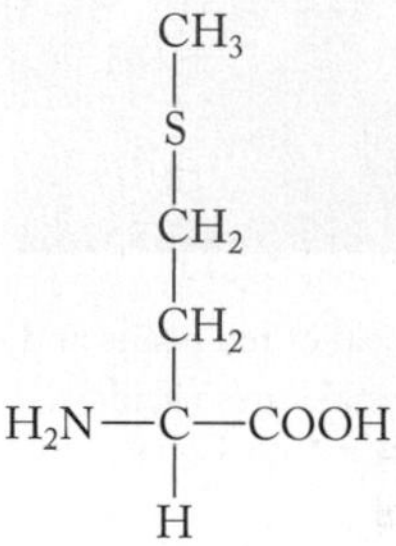

Methionine[a]
(Met)

Polar R Groups

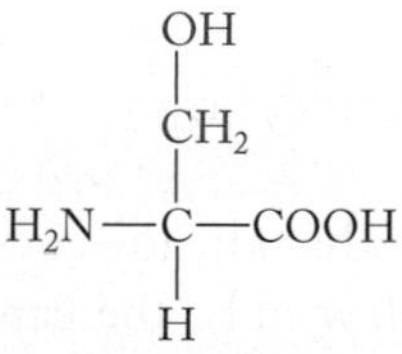

Serine
(Ser)

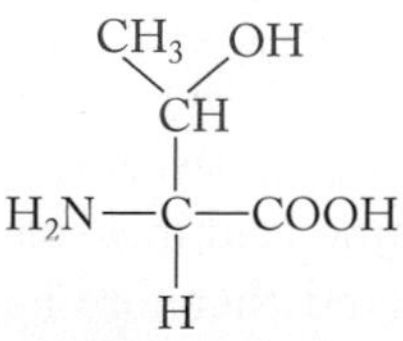

Threonine[a]
(Thr)

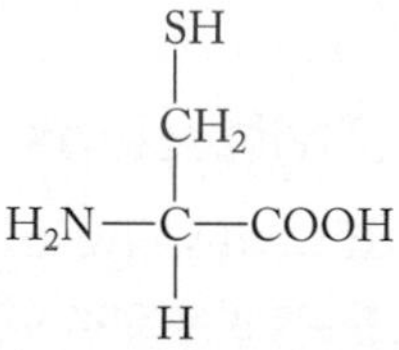

Cysteine
(Cys)

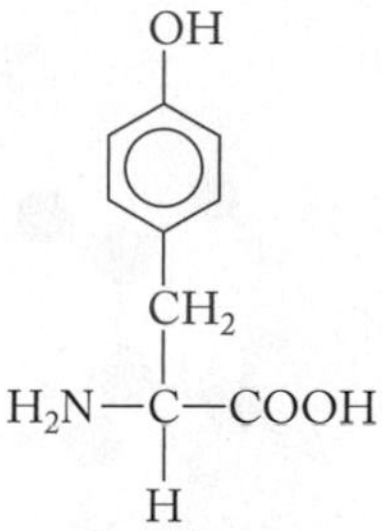

Tyrosine[a]
(Tyr)

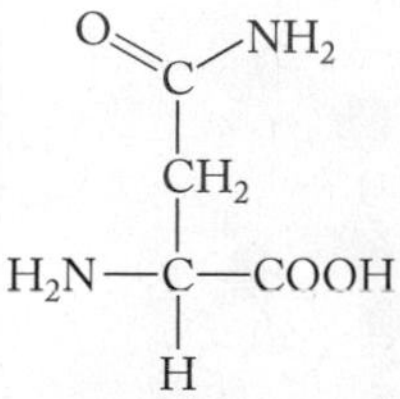

Asparagine
(Asn)

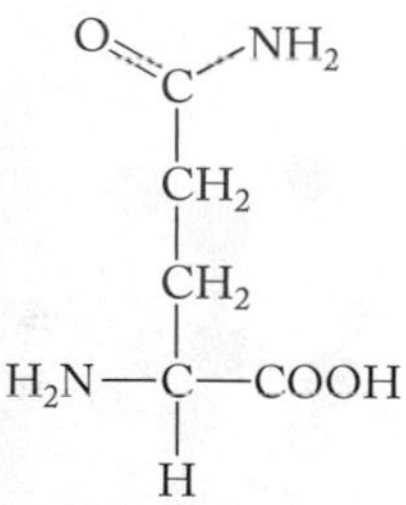

Glutamine
(Gln)

Acid R Groups

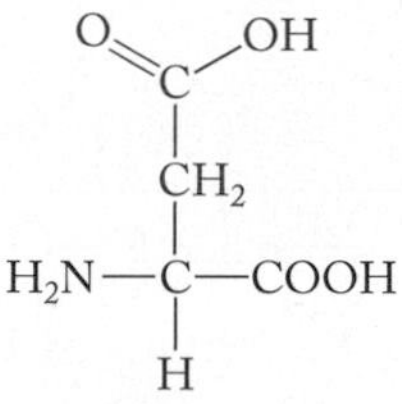

Aspartic acid
(Asp)

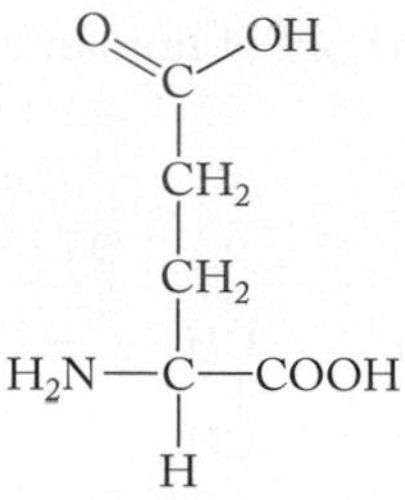

Glutamic acid
(Glu)

Basic R Groups

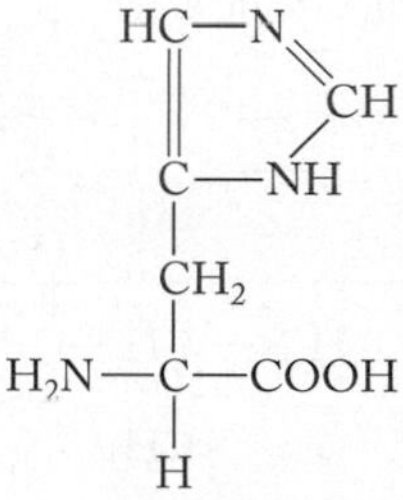

Histidine
(His)

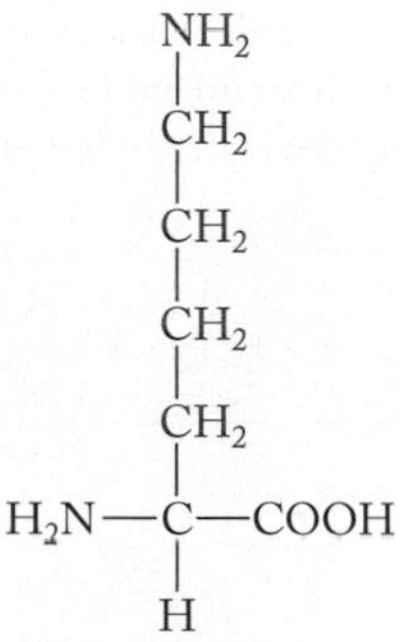

Lysine[a]
(Lys)

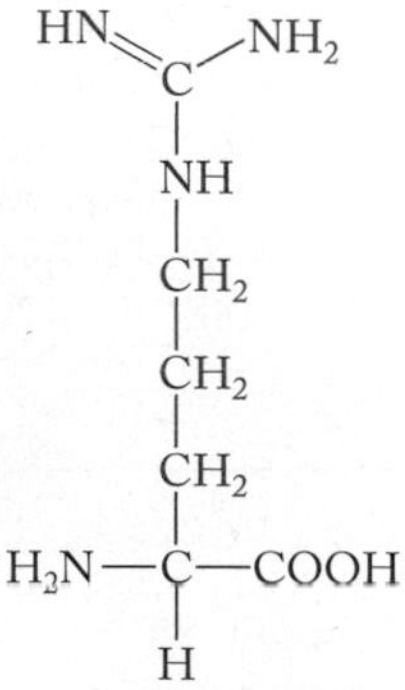

Arginine
(Arg)

[a]The eight essential amino acids for adults (histidine is essential for children).

FIGURE 20.2 A meal of red beans and rice provides all the essential amino acids.

∞ **CONNECTION** Enantiomers and chirality were discussed in Section 5.6.

proportions. These foods are sometimes referred to as *perfect foods* or, more precisely, *complete proteins*. In contrast, most plant proteins from foods such as legumes and vegetables are not complete proteins, so vegetarians must be careful to eat a combination of foods that provide all the essential amino acids. A good example is red beans and rice (Figure 20.2), a traditional dish in Latin American cuisine that provides a balance of these amino acids: rice has all of them but lysine, and beans lack only methionine.

Chirality

Look carefully at the generic structure of an amino acid in Figure 20.1. Note how the α-carbon atom is bonded to four different groups. This bonding pattern makes these α-carbon atoms *chiral centers* (see Section 5.6). It also means that all the amino acids in Table 20.1 except glycine are chiral compounds: their molecular structures are not superimposable on their mirror images, as illustrated for alanine in Figure 20.3. Instead, each amino acid and its mirror image constitute an enantiomeric pair.

For historical reasons, amino acid enantiomers are designated by the prefixes D- (for *dextro-*, right) and L- (*levo-*, left). These labels refer to how the four groups bonded to each chiral carbon atom are oriented in three-dimensional space. They *do not* refer to the rotation of plane-polarized light (see Figure 5.36) as it passes through a solution of the amino acid. In other words, there is no connection between the D- and L- prefixes in the names of amino acids and the *dextrorotary* and *levorotary* enantiomers that are designated with (+) and (−) signs based on their optical properties. All the chiral amino acids in the proteins in our bodies are L-enantiomers, even though 9 of the 19 are actually dextrorotary.

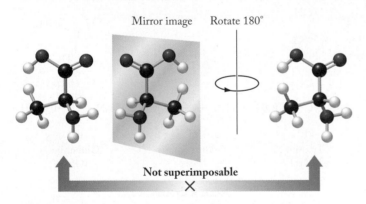

Mirror image   Rotate 180°

Not superimposable

FIGURE 20.3 Like most α-amino acids, alanine is chiral, which means that its molecular structure is not superimposable on its mirror image. Rotating the mirror image 180° so that the −CH₃ and −COOH groups overlap does not produce a structure in which −NH₂ and −H overlap.

Zwitterions

If we dissolve an amino acid in a solution that already contains a strong acid (and therefore has a low pH), the carboxylic acid group on each molecule does not ionize. Each amine group, however, being a weak base, accepts a H^+ ion to form a *protonated* $-NH_3^+$ group. The result is a molecular ion with an overall positive charge, as illustrated for alanine in Figure 20.4(a).

Now suppose we add a strong base to this solution to neutralize the strong acid and to raise the pH to about 7.4, close to the pH of human blood and of the fluids in most of

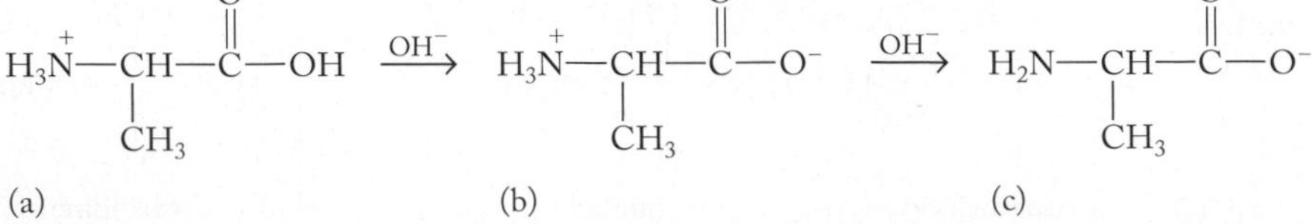

(a)

(b)

(c)

FIGURE 20.4 (a) At low pH, alanine (like many other amino acids) exists as a 1+ ion. (b) At physiological pH (7.4), the −COOH group is ionized and the molecule becomes a zwitterion with an overall charge of zero. (c) In basic solutions the charge decreases to 1− as the $-NH_3^+$ group deprotonates.

our tissues. At this *physiological pH* the –COOH groups in amino acids are mostly ionized, but nearly all the amine groups are still in the protonated –NH$_3^+$ form because –NH$_3^+$ is the conjugate acid of a weak base and therefore a *very* weak acid. As a result, most of the protonated amine groups still have positive charges and most of the ionized carboxylic acid groups have negative charges. A molecule with this distribution of charges is called a **zwitterion** (literally, a *hybrid ion*) because it has both a positive and a negative functional group (Figure 20.4b). Its net charge is zero.

If we add more base, alanine loses a H$^+$ from its NH$_3^+$ group (we say that it *deprotonates*), and we have a molecular ion with an overall charge of 1– (Figure 20.4c). The titration curve for alanine in Figure 20.5 shows this two-step neutralization process: the carboxylic acid is neutralized first, the protonated amine second.

CONNECTION The inverse relationship between the strengths of acid–base conjugate pairs was discussed in Chapter 15 and illustrated in Figure 15.4.

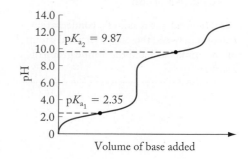

FIGURE 20.5 The titration curve of alanine resembles that of a weak diprotic acid. The carboxylic acid group (pK_a = 2.35) is neutralized first. The protonated amine group (pK_a = 9.87) is neutralized in the second step of the titration.

SAMPLE EXERCISE 20.1 Interpreting Acid–Base Titration Curves of Amino Acids   LO1

The –NH$_2$ groups in α-amino acids are weak bases (pK_b ≈ 4–5), so their conjugate acids (–NH$_3^+$) are even weaker acids (pK_a ≈ 9–10). Figure 20.6 shows that the titration curve for aspartic acid has three equivalence points. Draw the molecular structures of the principal form of aspartic acid that is in solution at the start of the titration and at each equivalence point.

COLLECT AND ORGANIZE We are asked to draw the molecular structures of aspartic acid at three different characteristic pH values. The molecular structure of the unionized form of aspartic acid is in Table 20.1.

ANALYZE There are two carboxylic acid groups and one amine group in aspartic acid. At low pH the carboxylic acid groups are not ionized and the amine group is protonated. As pH is raised, the –COOH groups ionize and then, under basic conditions, the protonated amine group releases its proton. The stronger of the two acid groups will ionize first. The –COOH group bonded to the α-carbon atom is only one carbon atom away from a protonated NH$_3^+$ group, which pulls electron density away from the –COOH group. The side-chain –COOH group is two carbon atoms away, so the carboxylate ion is further away, and less affected by the NH$_3^+$ group. Therefore, the –COOH group bonded to the α-carbon atom should ionize first.

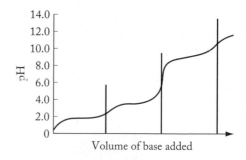

FIGURE 20.6 Titration curve of aspartic acid. The vertical lines identify equivalence points in the titration.

SOLVE At the beginning of the titration, the fully protonated 1+ ion is present:

At the first equivalence point, the –COOH group bonded to the α-carbon atom is ionized:

---

zwitterion a molecule that has both positively and negatively charged groups in its structure.

peptide a compound of two or more amino acids joined by peptide bonds. Small peptides containing up to 20 amino acids are *oligopeptides*; the term *polypeptide* is used for chains longer than 20 amino acids but shorter than proteins.

peptide bond the result of a condensation reaction between the carboxylic acid group of one amino acid and the amine group of another.

At the second equivalence point, the side-chain –COOH group is ionized:

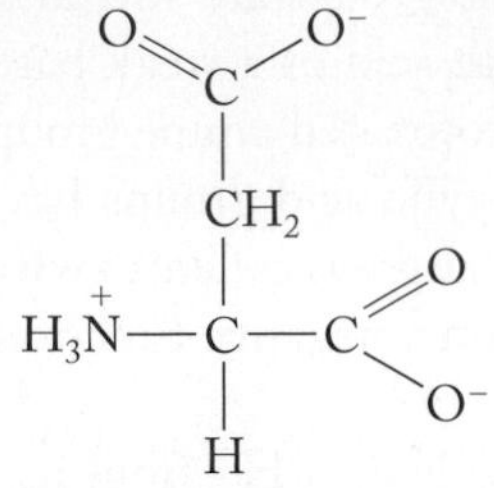

And at the third equivalence point in the titration, the amine group is deprotonated:

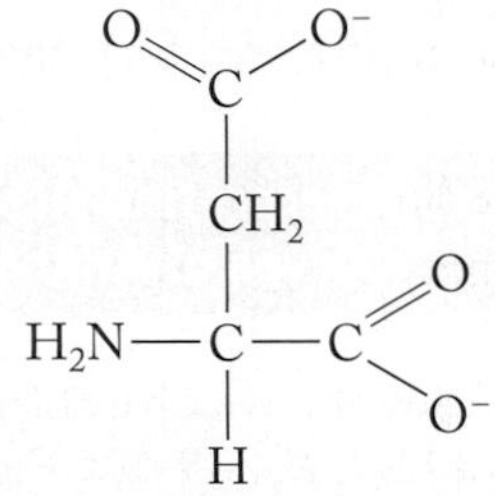

THINK ABOUT IT We can estimate the pK_a values of the two carboxylic acid groups from the midpoint pH values in the first two steps of the titration: about 1.9 and 3.7. They differ by nearly 2 pH units, which means that the α-COOH group is almost 10^2 times stronger an acid than the side-chain –COOH group, due to the presence of the α-NH_3^+ group.

Practice Exercise Sketch the titration curve for lysine starting at pH = 1.0, and draw the molecular structures of the principal form of lysine that is in solution at each equivalence point. ⚙

(Answers to Practice Exercises are in the back of the book.)

Peptides

Amino acids bond together to form chainlike molecules with a wide range of sizes. Amino acid *residues* (the name we give to amino acids that are part of a larger molecule) make up each link in the chain. The longest chains, some with molar masses as high as 3 million g/mol, are called proteins, but the shortest chains, called **peptides**, are only a few links long. The smallest peptides contain only two or three amino acid residues and are called *dipeptides* and *tripeptides*. Peptides up to 20 residues long are called *oligopeptides*. Those made of more than 20 are called *polypeptides*. The size at which a polypeptide becomes a protein is arbitrary but is typically set around 50–75 amino acid residues.

The bond linking the amino acids in peptides and proteins is called a **peptide bond** or *peptide linkage*. For example, the peptide bond that links molecules of valine (Val) and serine (Ser) is highlighted in blue in Figure 20.7. It forms when the α-carboxylic acid group of one amino acid (Val in this case) reacts with the α-amine group of another. The products are a peptide bond linking the two amino acids together and a molecule of water (which makes the reaction a condensation reaction). Note that peptide bonds have the same structure as the amide bonds that hold together the monomeric units of synthetic polyamides such as nylon.

The convention for drawing the structures of peptides begins by placing the amino acid that has a free α-amine group at the left end of the peptide chain and the amino acid with a free α-carboxylic acid at the right end. The left end is called

CONNECTION The formation of amide bonds in reactions between carboxylic acids and amines was described in Chapter 19 and illustrated in Figure 19.46.

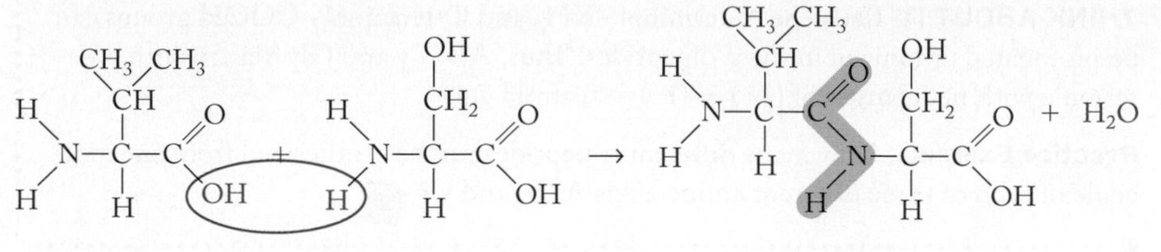

FIGURE 20.7 When the carboxylic acid group of valine forms a peptide bond with the amine group of serine, the products are the dipeptide valylserine and water.

▶❙❙ **CHEMTOUR** Condensation of Biological Polymers

the *amine* (or *N-*) *terminus* of the peptide and the right end is called the *carboxylic acid* (or *C-*) *terminus*. The name of a peptide is based on the names of its amino acids starting with the one at the N-terminus, except that the ending of all but the C-terminus amino acid is changed to *-yl*. Thus, the dipeptide formed by linking valine and serine in Figure 20.7 is called valylserine (abbreviated ValSer).

The artificial sweetener aspartame is the methyl ester of the dipeptide aspartyl-phenylalanine (Figure 20.8). At pH 7.4, aspartame exists as a zwitterion because the aspartic acid amine group is protonated and the carboxylic acid in its R group is ionized. In contrast, its parent dipeptide has a net charge of 1− because both −COOH groups are ionized at that pH. In an amino acid that has an amine group in its R group, such as lysine or arginine, that amine group is probably protonated at physiological pH, giving the amino acid a net charge of 1+. To sum up, the overall charge on a peptide at physiological pH is the sum of the positive charges on protonated amine groups and the negative charges of ionized carboxylic acid groups.

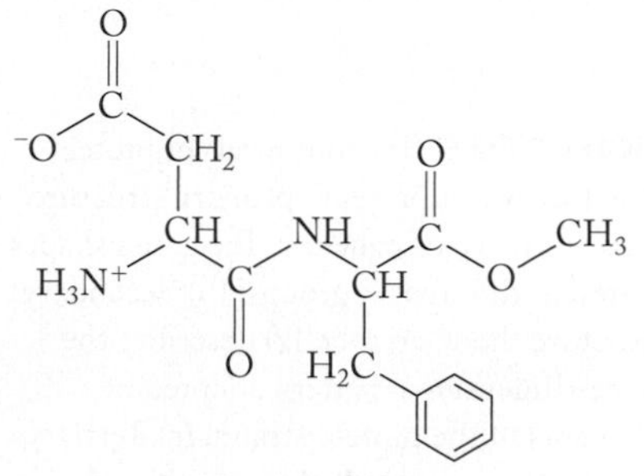

SAMPLE EXERCISE 20.2   Drawing and Naming Peptides   LO2

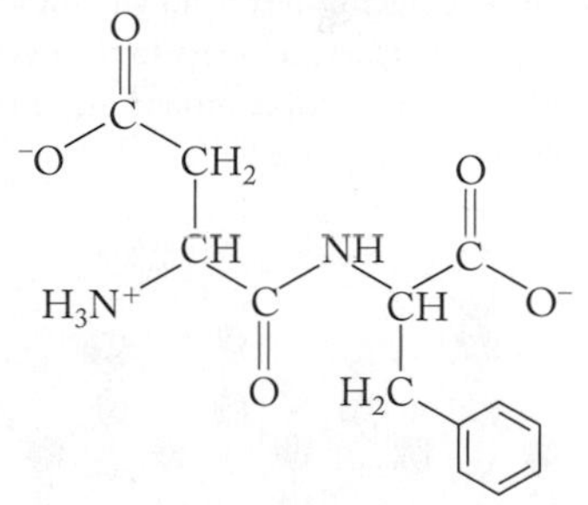

a. Name all the dipeptides that can be made by reacting alanine with glycine.
b. Draw their molecular structures in solution at pH 7.4.

COLLECT AND ORGANIZE We are asked to draw and name the dipeptides that are produced when alanine and glycine are linked by a peptide bond. Amino acid structures are in Table 20.1.

ANALYZE Two different peptides can be made from two amino acids by changing their sequence from the N-terminus to the C-terminus. For the two dipeptides, the sequences are AlaGly and GlyAla. The R groups in alanine (−CH₃) and glycine (−H) have no acidic or basic properties. At pH 7.4, the carboxylic acid terminus −COOH groups should be ionized and the amine terminus −NH₂ groups should be protonated.

SOLVE

a. The names of the two peptides with the amino acid sequence GlyAla and AlaGly are glycylalanine and alanylglycine, respectively.
b. At pH 7.4, the −NH₂ groups of amino acids are protonated and the −COOH groups are ionized, so the principal forms of these dipeptides are

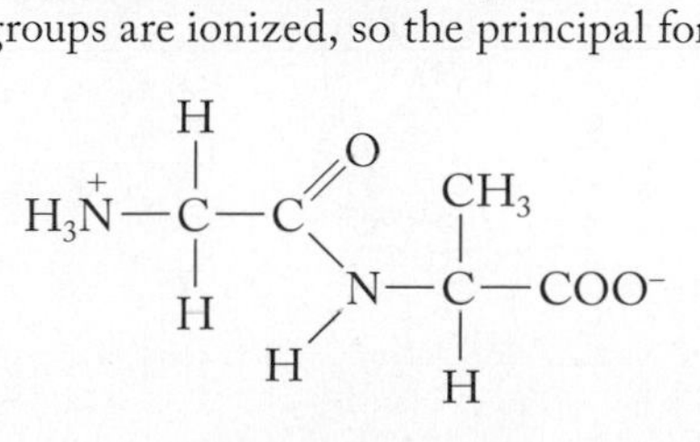

Glycylalanine
(GlyAla)

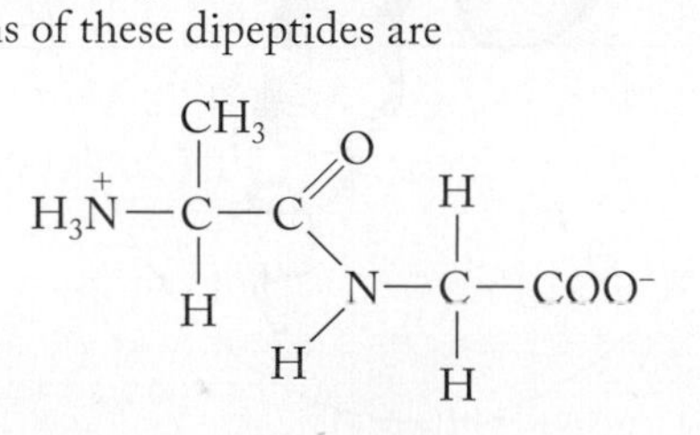

Alanylglycine
(AlaGly)

Aspartame

Aspartylphenylalanine

FIGURE 20.8 Aspartame is the methyl ester of the dipeptide aspartylphenylalanine and has one fewer ionized −COOH group than its parent dipeptide. Therefore, the overall charge of a molecule of aspartame at pH 7.4 is zero, whereas the charge on aspartylphenylalanine is 1−.

THINK ABOUT IT Only the N-terminal –NH₂ and C-terminal –COOH groups can be protonated or ionized in these dipeptides. Thus, AlaGly and GlyAla are both zwitterionic with net charges of $(1+) + (1-) = 0$ at pH 7.4.

Practice Exercise How many different tripeptides can be synthesized from one molecule of each of three different amino acids A, B, and C? ⚙

20.2 Protein Structure and Function

The structure of a protein is crucial to its function. In the essay at the beginning of this chapter, we mentioned the catastrophic impact of a malformed muscle protein. The functions of nonstructural proteins, such as the ability of enzymes to catalyze biochemical reactions, are also closely linked to their structures. These large biomolecules must assume particular three-dimensional conformations to interact with other molecules and to fulfill their biological mission.

Primary Structure

The **primary (1°) structure** of a protein is the sequence of the amino acids in it, starting with the N-terminus (Figure 20.9a). If two proteins are made up of the same number and type of amino acids but have different amino acid sequences, they are different proteins.

FIGURE 20.9 The four levels of protein structure. (a) A protein's primary structure is its amino acid sequence. The green shapes represent different R groups. (b) Secondary structure (here, an α helix) describes the three-dimensional pattern adopted by segments of the protein strand. (c) Tertiary structure is the overall shape of the molecule as segments of it bend and fold. (d) Quaternary structure refers to the overall shape adopted by multiple protein strands that assemble into a single unit.

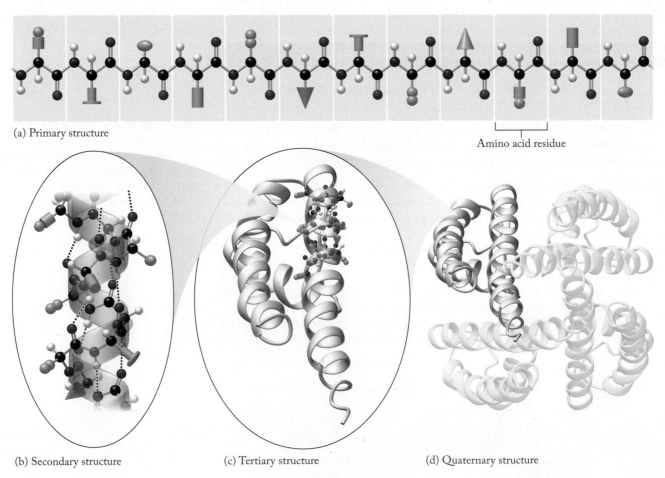

(a) Primary structure

Amino acid residue

(b) Secondary structure

(c) Tertiary structure

(d) Quaternary structure

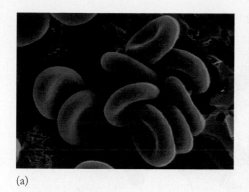

(a)   (b)

FIGURE 20.10 (a) Normal red blood cells are plump discs, whereas (b) those in patients suffering from sickle-cell anemia are distorted.

Changing only one amino acid can dramatically alter a protein's function. For example, in hemoglobin the sixth amino acid from the N-terminus of a protein strand that is 146 amino acids long is normally glutamic acid. In some people, valine substitutes for glutamic acid at this position. This one substitution alters the solubility of the protein and causes the red blood cell to take on a sickle shape instead of the normal, plump disc shape (Figure 20.10). These sickled cells do not pass through capillaries easily and may impede blood circulation. They do not last as long as normal blood cells. These factors diminish the capacity of the blood to carry oxygen, which is a characteristic of the disease called sickle-cell anemia.

Why does switching valine for glutamic acid affect the solubility of hemoglobin? The answer lies in the R groups of these two amino acids (Figure 20.11). The side-chain –COOH group of glutamic acid is ionized at physiological pH, and strong ion–dipole interactions with water molecules enhance the protein's solubility. However, if valine with its nonpolar isopropyl R group replaces glutamic acid, that strong intermolecular interaction with water is lost. The valine creates a hydrophobic region on the surface of the protein, which causes deoxygenated hemoglobins to stick to each other, deforming the cells.

Sickle-cell anemia is a debilitating disease, but it provides a survival advantage in regions where malaria is endemic. Having sickle-cell anemia does not protect people from contracting malaria or make them invulnerable to the parasite that causes it. However, children infected with malaria are more likely to survive the illness if they have sickle-cell anemia. Why sickle-cell anemia has this effect is not completely understood.

> **CONCEPT TEST** ●
>
> Which other amino acids besides valine might diminish hemoglobin solubility when substituted for glutamic acid?

● ●

Secondary Structure

In order for proteins to do what they are supposed to do, they must fold into a particular three-dimensional conformation. The first stage in the folding process is called a protein's **secondary (2°) structure**. Proteins acquire particular 2° structures as a result of a combination of intermolecular forces dominated by hydrogen bonds between the peptide linkages, dipole–dipole interactions and

∞ CONNECTION We discussed in Chapter 6 how ion–dipole and dipole–dipole interactions are the key to understanding the solubility of solutes in water.

Primary structure

Normal protein:
  Val - His - Leu - Thr - Pro - Glu - Lys - . . .

Abnormal protein:
  Val - His - Leu - Thr - Pro - Val - Lys - . . .

(a)

O=C–OH
  |
  CH$_2$
  |
  CH$_2$ H$_3$C   CH$_3$
  | \ CH /
H$_2$N–C–COOH   H$_2$N–C–COOH
  | |
  H   H

Glutamic acid   Valine

(b)

FIGURE 20.11 (a) The primary structure of the amine end of a protein in normal hemoglobin and in the abnormal hemoglobin responsible for sickle-cell anemia. (b) The replacement of glutamic acid with its hydrophilic R group by valine with its hydrophobic group is responsible for the disease.

....

primary (1°) structure the sequence in which amino acid monomers occur in a protein chain.

secondary (2°) structure the pattern of arrangement of segments of a protein chain.
....

α helix a coil in a protein chain's secondary structure.

β-pleated sheet a puckered two-dimensional array of protein strands held together by hydrogen bonds.

random coil an irregular or rapidly changing part of the secondary structure of a protein.

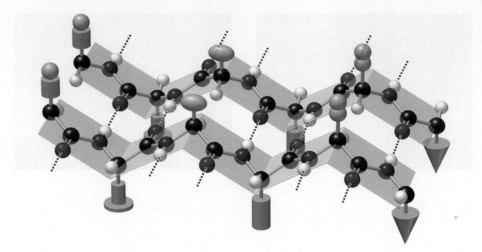

FIGURE 20.12 Each amino acid chain in a β-pleated sheet is folded in a zigzag pattern. Adjacent chains in a sheet are held together by hydrogen bonds (blue dotted lines). The R groups (green structures) extend above and below the sheet, linking it to adjacent sheets via intermolecular forces.

▶❙❙ **CHEMTOUR** Fiber Strength and Elasticity

$$\left|-CH_2CH_2COO^- \cdots H_3\overset{+}{N}-(CH_2)_4-\right|$$

(a)

$$\left|-CH_2-\overset{..}{\underset{H}{O}}\cdots H-O\overset{CH-}{\underset{CH_3}{}}\right|$$

(b)

$$\left|-CH\overset{CH_3}{\underset{CH_3}{}}\Rightarrow CH_3\atop CH_3-\right|$$

(c)

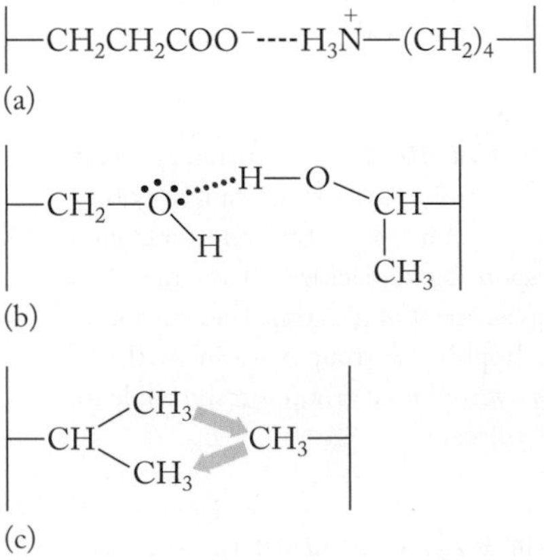

FIGURE 20.13 Interactions that influence the secondary and tertiary structures of proteins include (a) ion–ion interactions between acidic and basic R groups, (b) hydrogen bonding, and (c) hydrophobic interactions between nonpolar side chains.

hydrogen bonds between polar side groups, and attractions between nonpolar side groups based on London dispersion forces. Among the most common geometric patterns that segments of amino acid chains make (Figure 20.9b) is the **α helix**, a coiled arrangement with the R groups pointing outward. The helical structure is maintained by hydrogen bonds between –NH groups on one part of the chain and C=O groups on nearby amino acids. The α helix looks very much like a spring. Muscle tissue that stretches and contracts is made largely of α-helical proteins, as are keratins, the proteins in hair and fingernails.

Another common 2° structure is the **β-pleated sheet**. These sheets are assemblies of multiple amino acid chains aligned side by side. The pleats are caused by the tetrahedral molecular geometries of the atoms along the chains (Figure 20.12). Adjacent chains are linked together by hydrogen bonds, and the collection of side-by-side zigzag chains forms a continuous sheet. R groups extend above and below the pleats. β-pleated sheets may stack on top of one another like two pieces of corrugated roofing. Stacked sheets are held together by the same interactions that hold all proteins together, including ion–ion interactions, hydrogen bonding, and London dispersion forces, depending on the pairs of R groups involved (Figure 20.13).

The proteins that make up strands of silk form thin, planar crystals of β-pleated sheets that are only a few nanometers on a side. Enormous numbers of these crystals form long arrays of sheets stacked together like nanoscale pancakes. Hydrogen bonds hold the stacks together and reinforce adjacent sheets. As a result of these interactions, strands of silk are stronger than strands of steel with the same mass; indeed, silk is one of the toughest known materials—natural or synthetic.

Some single-strand proteins exist as α helices on their own but form β-pleated sheets when they clump together in multi-strand aggregates. One consequence of this clumping is the formation of insoluble protein deposits called plaque. Abnormal accumulation of plaque can be a serious health risk: plaque formed by a protein called amyloid β has been linked to the onset of Alzheimer's disease.

If part of a protein is characterized by an irregular or rapidly changing structure, it is said to have a **random coil** 2° structure. The amino acid chain may fold back on itself and around itself, but it has no regular features the way an α helix or a β-pleated sheet does. When proteins *denature*, losing their secondary structure because of heat or change in pH, they may become random coils.

Large protein molecules may contain all three types of 2° structure. In describing a protein, scientists may indicate the percentage of amino acids involved in each type—for example, 50% α-helical, 30% β-pleated sheet, and 20% random coil. Figure 20.14 shows a model of a protein called carbonic anhydrase, which illustrates all three types.

Tertiary and Quaternary Structure

Large proteins have structure beyond the 1° and 2° levels, folding back on themselves as a result of interactions between R groups on amino acids that are considerable distances apart along the protein chain. These interactions may be ion–ion, ion–dipole, or van der Waals forces. They may even involve the formation of covalent bonds. For example, the –SH groups on two cysteine residues may undergo an oxidation half-reaction that results in the formation of a disulfide linkage that holds two parts of the protein strand together via an intrastrand –S—S– covalent bond (Figure 20.15). Interactions and reactions such as these determine **tertiary (3°) structure**, the overall three-dimensional shape of the protein that is key to its biological activity (Figure 20.9c). Because the proteins in living systems exist in an aqueous environment, hydrophobic R groups tend to reside in the interiors of their 3° structures, whereas hydrophilic groups (as we saw in normal hemoglobin) are oriented toward the outside, where they interact with nearby molecules of water. Hydrophobic interactions are the primary cause of protein folding and compaction, but all the other modes of interaction help a large protein find its unique 3° structure.

Hemoglobin and some other proteins exhibit an even higher order of structure. One hemoglobin unit (Figure 20.16a) contains four protein strands, each of which enfolds a porphyrin ring containing one Fe^{2+} ion. The combination of four protein strands to make one hemoglobin assembly is an example of **quaternary (4°) structure** (Figure 20.9d). In the 4° structures of keratins (Figure 20.16b),

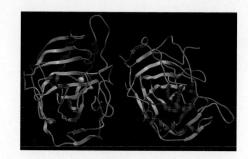

FIGURE 20.14 Two views of carbonic anhydrase. The structure of the protein has α-helical regions (red), β-pleated sheet regions (green), and random coil (blue). The light blue sphere in the center is a zinc ion.

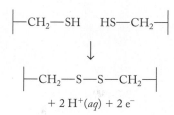

FIGURE 20.15 The tertiary structure of some proteins is stabilized by intrastrand –S—S– bonds that form between the –SH groups on the side chains of cysteine residues.

∞ CONNECTION All the types of intermolecular forces described in Chapter 6 are involved in the *intra*molecular interactions that give proteins their unique 2° and 3° structures.

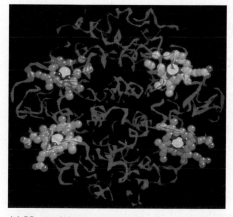

(a) Hemoglobin

(b) Keratin

FIGURE 20.16 Quaternary structure of proteins. (a) Four protein chains form a single unit in the quaternary structure of hemoglobin. The iron-containing porphyrins are bright green. (b) Pairs of α-helical chains (blue) wound together and linked by interstrand –S—S– bonds (yellow) stabilize the quaternary structure of the keratin in hair and fingernails.

tertiary (3°) structure the three-dimensional, biologically active structure of the protein that arises because of interactions between R groups on amino acids.

quaternary (4°) structure the larger structure functioning as a single unit that results when two or more proteins associate.

protein strands with α-helical 2° structures coil around each other to make even larger coils. When the protein strands are held together mostly by van der Waals forces, as they are in the keratin in skin tissue, the structures are flexible and elastic. If they are also restrained by many covalent bonds as in Figure 20.16(b), they produce tissues that are hard and less flexible, like fingernails and the beaks of birds.

Enzymes: Proteins as Catalysts

The chemical reactions involved with metabolism—both *catabolism* (breaking down of molecules) and *anabolism* (synthesis of complex materials from simple feedstocks)—are mediated in large part by proteins called **enzymes**. Enzymes are biological catalysts. Both catabolism and anabolism are organized in sequences of reactions called *metabolic pathways*, and each step in a metabolic pathway is catalyzed by a specific enzyme. For example, carbonic anhydrase (Figure 20.14) is an enzyme that speeds up the hydrolysis of CO_2 and its ionization to bicarbonate and hydrogen ions:

$$CO_2(aq) + H_2O(\ell) \rightleftharpoons HCO_3^-(aq) + H^+(aq) \tag{20.1}$$

In the presence of carbonic anhydrase, this reaction proceeds about 10 million times faster than in its absence. Without carbonic anhydrase, we would not be able to expel carbon dioxide fast enough to survive (as the reaction in Equation 20.1 runs in reverse). One molecule of carbonic anhydrase can catalyze the hydrolysis and ionization of more than a million molecules of CO_2 in 1 s. This value is called the *turnover number* for the enzyme; in general, the higher the turnover number, the faster the enzyme-catalyzed reaction proceeds. Turnover numbers for enzymes typically range from 10^3 to 10^7. The higher the turnover number, the lower the activation energy of the catalyzed reaction. As we learned in Chapter 13, molecules must collide with the proper orientation for reactions to proceed. Carbonic anhydrase is essentially a perfect enzyme because it catalyzes the hydrolysis reaction nearly every time it collides with CO_2.

Enzymes are highly selective: each catalyzes a particular reaction involving a particular reactant. For example, an enzyme called lactase catalyzes only the reaction by which lactose, the sugar in milk, is broken down during digestion. People who lack this enzyme cannot metabolize the sugar; they are said to be *lactose intolerant*. If they consume dairy products, unmetabolized lactose passes into their large intestines where bacteria ferment it, and unpleasant and painful abdominal disturbances result.

Even in the simplest organism, hundreds of enzyme-catalyzed chemical reactions are constantly underway. Most of these enzymes are effective only under limited reaction conditions: in aqueous media, at temperatures between 4 and 37°C, and over a narrow pH range. These hundreds of catalyzed reactions require hundreds of different enzymes, many operating with exquisite efficiency—sometimes on only one member of a pair of enantiomers, for example, producing only one enantiomerically pure product.

Enzyme selectivity is the main reason natural products tend to be optically pure materials while the same products synthesized in the laboratory tend to be racemic mixtures. In the pharmaceutical industry, research is currently focused on using enzymes outside the living system to produce enantiomerically pure materials needed as drugs. This research area, called **biocatalysis**, uses enzymes to catalyze chemical reactions run in industrial-sized reactors.

CONNECTION In Section 13.6 we discussed CFCs, inorganic homogeneous catalysts that contribute to the destruction of ozone. Enzymes are bio-organic homogeneous catalysts that selectively speed up biochemical reactions.

CONNECTION A racemic mixture contains equal proportions of the two enantiomers of a chiral compound, as we discussed in Section 5.6.

Synthetic reaction pathways catalyzed by enzymes can significantly reduce the production cost of, for example, biological pharmaceuticals. However, biocatalytic reactions often run best in very dilute solutions, which limits production. A key issue driving interest in such processes is that an enantiomerically pure pharmaceutical is likely to be more potent and produce fewer side effects than racemic mixtures of the same product. As an example, the drug thalidomide is a racemic mixture of two enantiomeric forms: one is very effective in treating morning sickness, but the second is a teratogen (an agent that disturbs the development of a fetus). In perhaps the worst medical tragedy in modern times, more than 10,000 children were born in the late 1950s and early 1960s with serious, frequently fatal, deformities as a result of their mothers having taken the racemic drug.

The molecular structure of enzymes contains a region called an **active site** that binds the reactant molecule, called the **substrate**. The action of enzymes was originally explained by a lock-and-key analogy in which the substrate is the key and the active site is the lock (Figure 20.17). The substrate is held in the active site by the same kinds of intermolecular interactions that hold other biomolecules together. Some enzymes become covalently bound to intermediates in the catalytic process. Once in the active site, the substrate is converted into product via a reaction having a lower energy transition state than it would without the enzyme. The reaction of a substrate S with an enzyme E produces an *enzyme–substrate* (E–S) *complex* that decomposes, forming a product P and regenerating the enzyme:

$$E + S \rightleftharpoons E\text{–}S \rightarrow E + P$$

The lock-and-key analogy does not fully account for enzyme behavior. A more accurate view is provided by the *induced-fit model*, which assumes that the substrate does more than just fit into the existing shape of an active site. This model assumes that as the E–S complex forms, the binding site undergoes subtle changes in its shape to more precisely fit the three-dimensional structure of the transition state. A simplified illustration of such an interaction is shown in Figure 20.18(a).

The induced-fit model helps explain the behavior of compounds called **inhibitors** that can diminish or destroy the effectiveness of enzymes. An inhibitor may bind to an active site and block it from interacting with the substrate (Figure 20.18b), or it may disable the enzyme by preventing it from assuming its active shape (Figure 20.18c).

active site the location on an enzyme where a reactive substance binds.

substrate the reactant that binds to the active site in an enzyme-catalyzed reaction.

inhibitor a compound that diminishes or destroys the ability of an enzyme to catalyze a reaction.

CONNECTION The phenomenon of molecular recognition in biological systems based on complementary molecular shapes was introduced in Chapter 5.

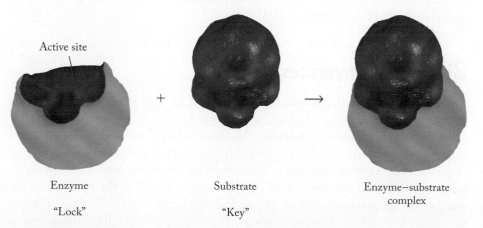

Enzyme "Lock" + Substrate "Key" → Enzyme–substrate complex

FIGURE 20.17 In the lock-and-key model of enzyme activity, the substrate fits exactly into the active site of the enzyme that catalyzes a chemical reaction involving the substrate.

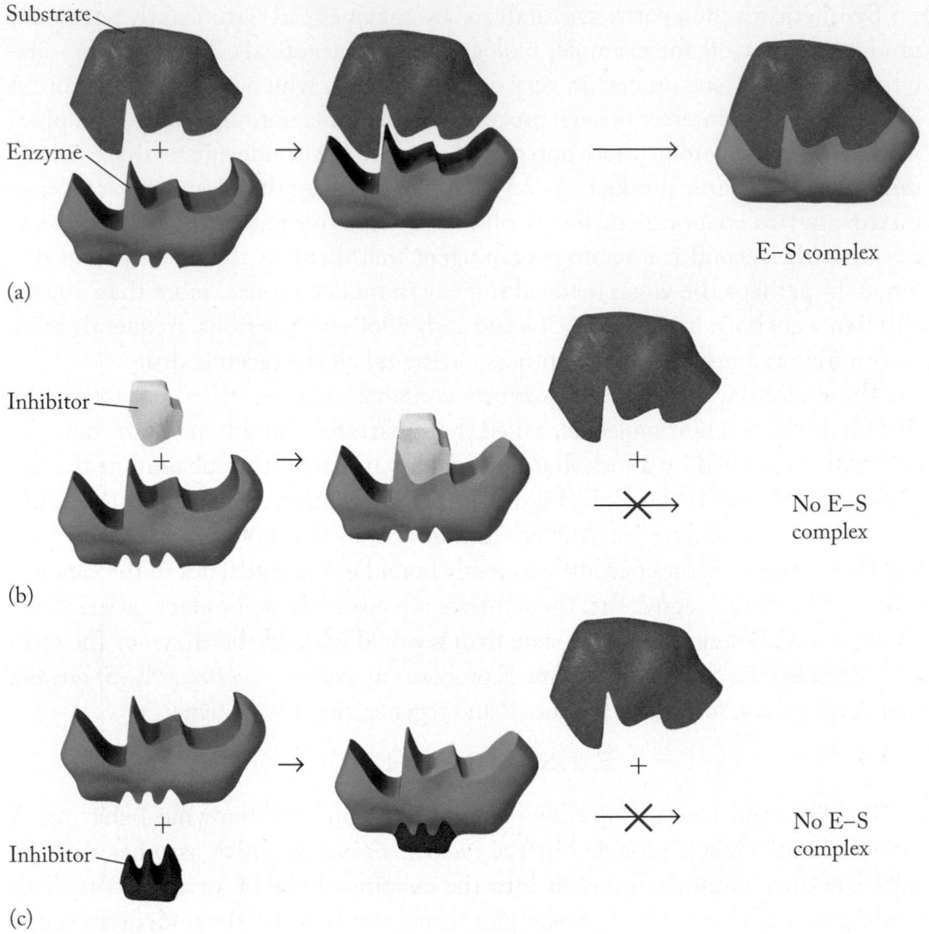

FIGURE 20.18 (a) The induced-fit model assumes that the shape of the enzyme changes to accommodate the substrate and form the enzyme–substrate (E–S) complex. Inhibitors (yellow and red molecules) may (b) block the enzyme's binding site or (c) cause a change elsewhere in the enzyme's structure that prevents the active site from attaining the shape it needs to form the E–S complex.

Natural enzyme inhibitors play important roles in regulating the rates of reactions that are catalyzed by enzymes. For example, in a multistep reaction pathway, the product of a later step may inhibit an enzyme that catalyzes an earlier reaction. This kind of negative feedback keeps the sequence of reactions from running too quickly and perhaps jeopardizing the health of the organism due to the accumulation of undesirable products or intermediates. Enzyme inhibitors may also be used to fight disease. Powerful drugs have been developed that inhibit the enzymes that allow viruses to break down proteins. Several such drugs have been particularly effective in treating HIV.

20.3 Carbohydrates

Carbohydrates have the generic formula $C_x(H_2O)_y$. This formula gives us a clue about where the name *carbohydrate*, or *hydrate of carbon*, comes from. The smallest carbohydrates are **monosaccharides** (the name means "one sugar"), which bond together to form more complex carbohydrates called **polysaccharides**. Many organisms use monosaccharides as their main energy source but convert them to polysaccharides for the purpose of energy storage. Starch is the most abundant energy-storage polysaccharide in plants. Polysaccharides in the form of cellulose also provide structural support in plants. Plants produce over 100 billion tons of cellulose each year—the woody parts of trees are over 50% cellulose, and cotton is 99% cellulose.

carbohydrate an organic molecule with the generic formula $C_x(H_2O)_y$.

monosaccharide a single-sugar unit and the simplest carbohydrate.

polysaccharide a polymer of monosaccharides.

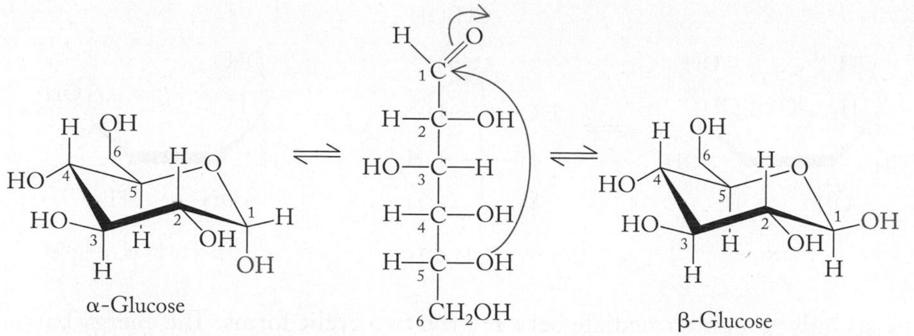

FIGURE 20.19 An equilibrium exists between the linear structure of glucose and the two cyclic forms α-glucose and β-glucose. The difference between the two cyclic structures is the orientation of the –OH group on C1 (highlighted in blue).

The principal building block of both starch and cellulose is the monosaccharide glucose (Figure 20.19). The different properties and functions of these polysaccharides come from the subtle differences in the molecular geometries of the glucose monomers in their structures and how the monomers are bonded together. Molecules of most monosaccharides, including glucose, contain several chiral centers, so multiple isomers exist. This complexity gives rise to the third major function of carbohydrates: molecular recognition. For example, combinations of carbohydrates and proteins, called *glycoproteins*, on the surfaces of blood cells determine the blood type of an individual.

Molecular Structures of Glucose and Fructose

Glucose is the most abundant monosaccharide in nature and in the human body. It is also called *dextrose*—a kind of abbreviation for *dextro*-glucose. Fructose is the principal monosaccharide in many fruits and root vegetables. Given the importance and abundance of these sugars, let's examine their structures and properties more closely.

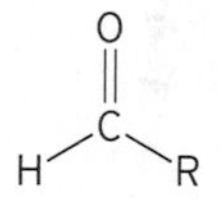

Three molecular views of glucose are provided by the structures in Figure 20.19. All three have the same chemical formula ($C_6H_{12}O_6$), and they are all structural isomers of one another. Note that the middle structure contains an aldehyde (–CH=O) group and all three structures contain multiple alcohol (–C—OH) groups. The polarities of these groups and their capacities to form hydrogen bonds give glucose its high molar solubility in water (5.0 M at 25°C).

The cyclic structures of glucose form when the carbon backbone of the linear form curls around so that the –OH group bonded to the carbon atom labeled number 5 (C5) comes close to the aldehyde group on C1 (as shown by the red arrow in Figure 20.19). The aldehyde and alcohol groups react to form a six-membered ring made up of five carbon atoms (C1 to C5) and the oxygen atom of the C5 alcohol. Note a small but significant difference in the two cyclic structures. In the molecule on the left, called α-glucose, the –OH group on C1 points down. In the molecule on the right, β-glucose, the C1 –OH group points up. These two orientations are possible because there are two ways that the carbon chain in the middle structure can form a cyclic structure: the –OH group on C5 can approach C1 from either of the two sides of the plane defined by the C1 carbon atom and the C and H atoms bonded to it. When the C5 –OH group approaches C1 as shown in Figure 20.20, the α isomer is produced. Approaching C1 from the other side yields the β isomer. Notice that in the linear form, C1 is an aldehyde and is not a chiral center. The cyclization creates a new chiral center in the cyclic molecule.

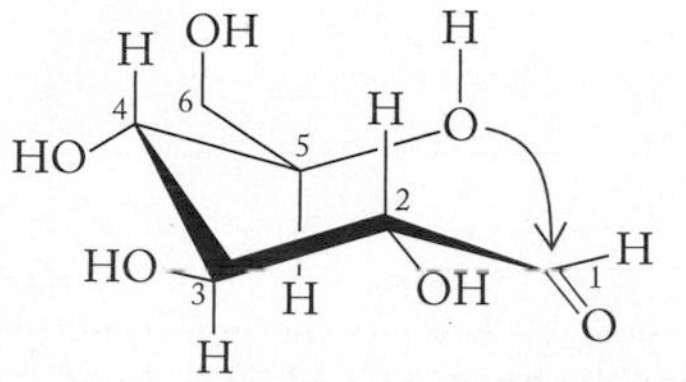

The β form is slightly more stable than the α form and accounts for 64% of glucose molecules in aqueous solution; the α form accounts for the remaining 36%. Both cyclic forms are more stable than the straight-chain form, which

CONNECTION An aldehyde group (Chapter 19) includes these atoms and bonds:

FIGURE 20.20 The –OH group on C5 may approach the aldehyde on C1 from either side of the plane defined by the C, H, and O atoms in the aldehyde group. For the approach shown here the product will be the α isomer.

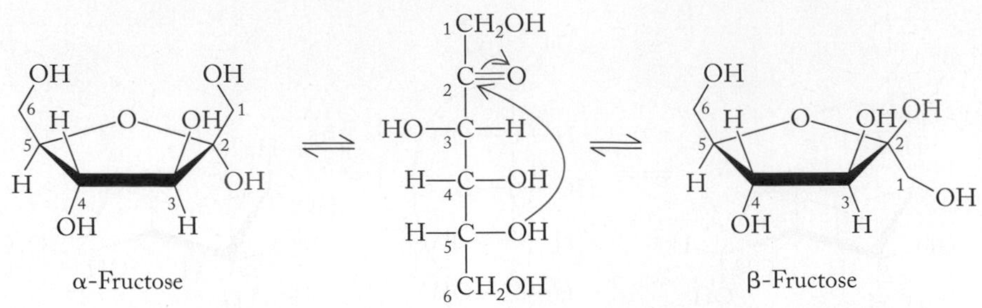

FIGURE 20.21 The cyclization of fructose proceeds via the –OH group on C5 and the ketone on C2.

◉◉ **CONNECTION** A ketone group (Chapter 19) contains these atoms and bonds:

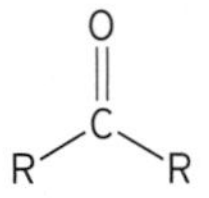

▶‖ **CHEMTOUR** Formation of Sucrose

exists only as an intermediate between the two cyclic forms. The energy barrier to flipping conformations is relatively small, so glucose molecules in solution are constantly opening and closing in a dynamic structural equilibrium.

Figure 20.21 shows the structures of the linear and cyclic forms of fructose, which also has the formula $C_6H_{12}O_6$. Note that the C=O group in the linear form of fructose is not on the terminal carbon atom. In other words, fructose is a ketone rather than an aldehyde. It forms a five-membered ring, not a six-membered ring, when the –OH group at C5 reacts with the carbonyl carbon atom at the C2 position. The product is a ring with two –CH$_2$OH groups that are either on the same side of the ring (α-fructose) or on opposite sides (β-fructose).

Disaccharides and Polysaccharides

Sucrose, or ordinary table sugar, is a *disaccharide* ("two sugars") that consists of one molecule of α-glucose bonded to one molecule of β-fructose (Figure 20.22). The bond between them forms when the –OH group on the C1 carbon atom of glucose reacts with the C2 carbon atom of fructose, producing a C—O—C **glycosidic bond** or *glycosidic linkage*. Water is also produced, making this reaction another example of a condensation reaction. The glycosidic bond in sucrose is called an α,β-1,2 linkage because of the orientations (α and β) of the two –OH groups involved and their positions (C1 and C2) in the cyclic structures of the two monosaccharides.

Another important glycosidic bond involves the –OH groups on the C1 and C4 carbon atoms of glucose molecules. When glucose molecules link at these positions they can form long-chain polysaccharides. If the starting monomer is

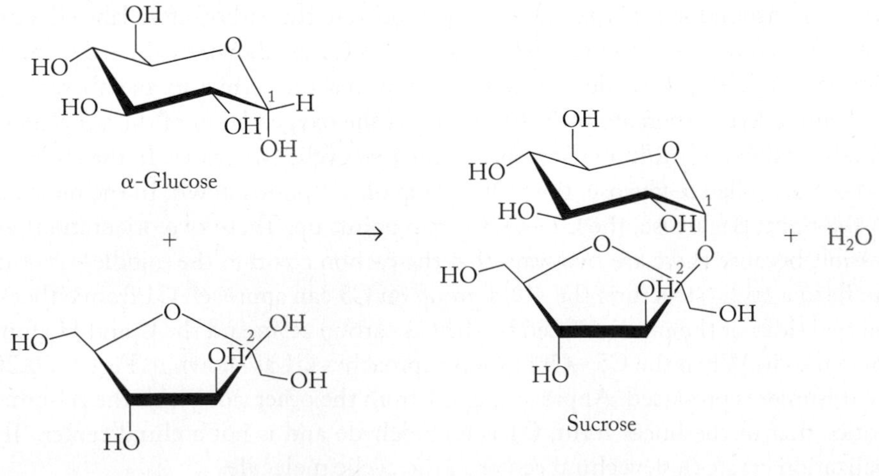

glycosidic bond a C—O—C bond between sugar molecules.

biomass the sum total of the mass of organic matter in any given ecological system.

FIGURE 20.22 α-Glucose and β-fructose combine to form a molecule of sucrose, ordinary table sugar.

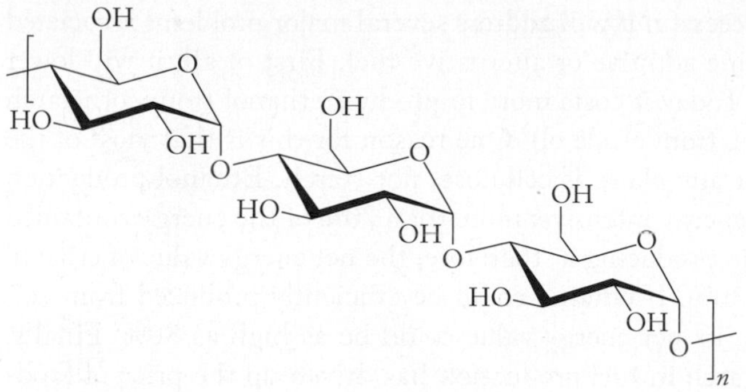

FIGURE 20.23 (a) Starch is a polysaccharide of α-glucose molecules joined by α-1,4 glycosidic bonds. (b) Cellulose is a polysaccharide of β-glucose molecules joined by β-1,4 glycosidic bonds.

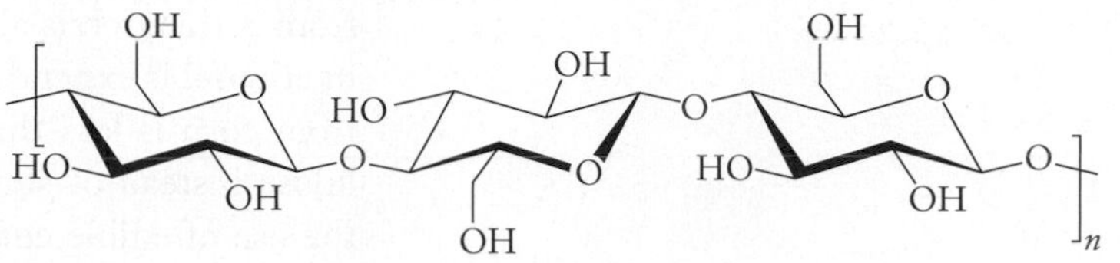

(a) Starch

(b) Cellulose

α-glucose, the bonds between them are α-1,4 glycosidic linkages, and the product of their formation is starch (Figure 20.23a). The conversion of α-glucose into starch is an effective way for plants to store energy because formation of the α-1,4 linkage is reversible. With the aid of digestive enzymes, α-1,4 bonds can be hydrolyzed and starch converted back into glucose by plants that make the starch, or by animals that eat the plants.

The cellulose that plants synthesize to build stems and other organs has a structure (Figure 20.23b) slightly different from that of starch because the building blocks of cellulose are β-glucose instead of α-glucose, so the monomers are linked by β-1,4 glycosidic bonds. This structural difference between α- and β-glycosidic bonds is important because it enables starch to coil and make granules for efficient storage while cellulose forms structural fibers.

Carbohydrates in plants are a major part of the total organic matter in any given ecological system—that is, they make up much of the system's **biomass**. People have used various forms of biomass, including wood and animal dung, as fuel for thousands of years. More recently, we have begun to convert biomass into a liquid *biofuel*, ethanol, for use as a gasoline additive or substitute. Because ethanol contains oxygen, it improves the combustion of a hydrocarbon fuel like gasoline and reduces carbon monoxide emissions. An important industrial application of starch hydrolysis is the conversion of cornstarch into glucose and then, through fermentation, into ethanol:

$$C_6H_{12}O_6(aq) \rightarrow 2\ CH_3CH_2OH(aq) + 2\ CO_2(g) \qquad (20.2)$$

This exothermic reaction provides energy to the yeast cells whose biological processes drive fermentation.

Unlike grazing animals, humans cannot digest cellulose because our digestive tracts do not contain microorganisms that have enzymes called cellulases, which catalyze hydrolysis of β-glycosidic bonds. The challenge of reproducing what cellulose-eating bacteria do in a laboratory or on an industrial scale is the focus of an enormous research effort as scientists try to develop efficient procedures for converting cellulose to glucose and then to ethanol. This research has focused on more efficient, less energy-intensive ways to break apart cellulose fibers. In addition, scientists are genetically engineering microorganisms like those in cattle stomachs to increase the supply of cellulases.

CONCEPT TEST

Cellobiose is a disaccharide made from the degradation of cellulose. We cannot digest cellobiose. Which of the two structures in Figure 20.24 represents a molecule of cellobiose?

(a)

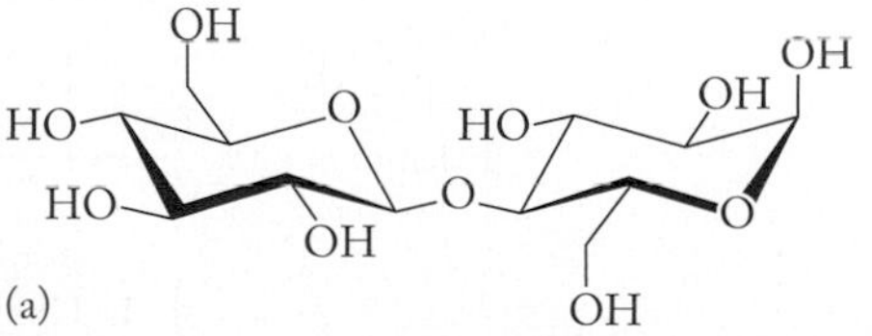

(b)

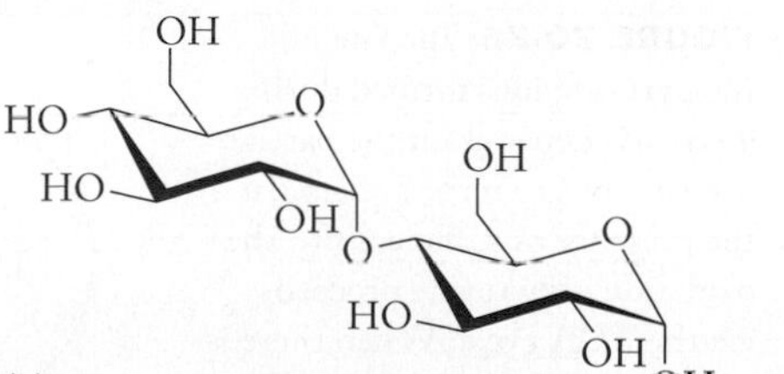

FIGURE 20.24

If this research is successful it will address several major problems associated with ethanol as a gasoline additive or alternative fuel. First of all, it will lower the cost of production. Today it costs more to produce ethanol from cornstarch than to produce gasoline from crude oil. One reason for this is that most of the mass of a corn plant, or any plant, is cellulose, not starch. Ethanol production from cornstarch is also energy intensive: more than 70% of the energy contained in ethanol is expended in producing it; therefore, the net energy value of ethanol from corn is less than 30%. If ethanol could be efficiently produced from cellulose instead of starch, its net energy value could be as high as 80%. Finally, the use of edible cornstarch in fuel production has driven up the price of foods derived from corn. The impacts of this inflation have been felt worldwide and have been particularly painful in developing countries. The use of agricultural land and consumption of increasingly scarce water resources for ethanol production raise further concerns.

CONNECTION In Chapter 12 we introduced glycolysis in a discussion of the thermodynamics of coupled reactions.

Glycolysis Revisited

Most cells, including those in our bodies, use glucose as a fuel in glycolysis (Figure 20.25), a series of reactions that oxidize glucose to pyruvate ion, the conjugate base of pyruvic acid. Pyruvate sits at a metabolic crossroads and can be converted into different products depending on the type of cell in which it is generated, the enzymes that are present, and the availability of oxygen (Figure 20.26). In yeast cells growing under low-oxygen conditions, pyruvate is converted into ethanol and CO_2. Another series of reactions, called the **tricarboxylic acid (TCA) cycle**, occurs in the presence of sufficient dissolved O_2 and is fundamental to the conversion of glucose to energy in humans and other animals. A key step prior to the TCA cycle occurs when pyruvate loses CO_2 and forms an acetyl group, which then combines with coenzyme A. The resulting product, acetyl-coenzyme A, is a reactant in many biosynthetic pathways, including the production and breakdown of cholesterol.

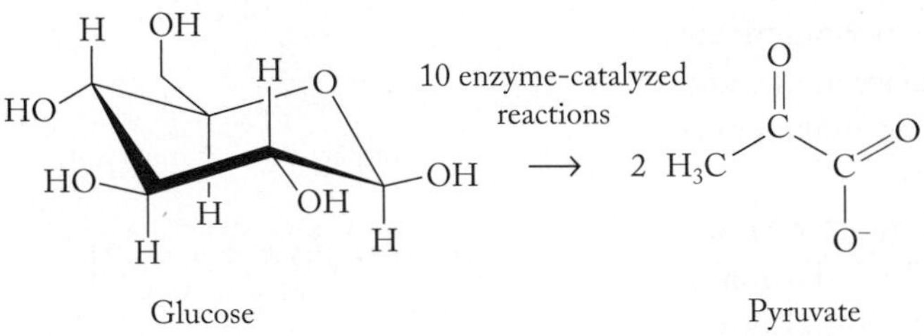

Cholesterol is a key component in the structure of cell membranes. It is also a precursor of the bile acids that aid in digestion and of steroid hormones, which regulate the development of the sex organs and secondary sexual traits, stimulate the biosynthesis of proteins, and regulate the balance of electrolytes in the kidneys. In a healthy person, synthesis and use of cholesterol are tightly regulated to prevent overaccumulation and consequent deposition of cholesterol in coronary arteries. We clearly need cholesterol, but persistent deposition in the arteries can lead to serious coronary disease (Figure 20.27).

FIGURE 20.25 In glycolysis, molecules of glucose are broken down to pyruvate ions.

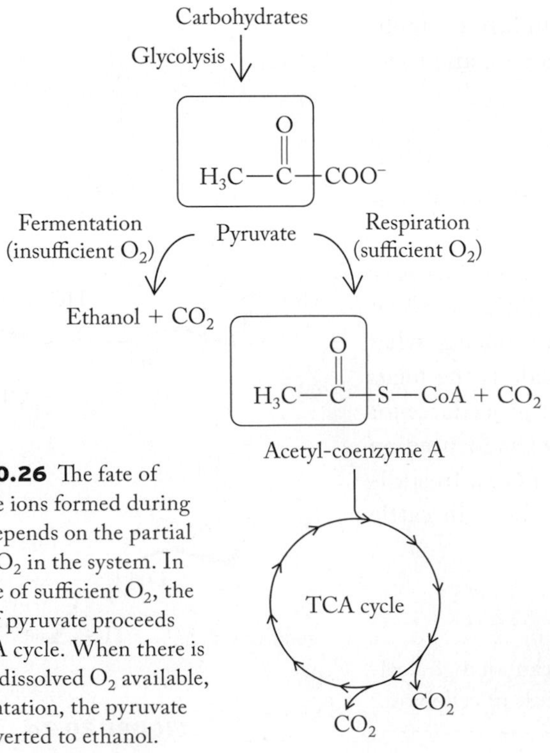

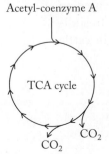

FIGURE 20.26 The fate of the pyruvate ions formed during glycolysis depends on the partial pressure of O_2 in the system. In the presence of sufficient O_2, the oxidation of pyruvate proceeds via the TCA cycle. When there is insufficient dissolved O_2 available, as in fermentation, the pyruvate may be converted to ethanol.

20.4 Lipids

Lipids differ from carbohydrates and proteins in that they are not biopolymers. Lipids are best described by their physical properties rather than any common structural subunit: lipids do not dissolve in water but are soluble in nonpolar solvents, and they are oily to the touch. Because they are insoluble in water, they are ideal components of cell membranes, which separate the aqueous solutions within cells from the aqueous environments outside them. An important class of lipids called **glycerides** are esters formed between glycerol and long-chain fatty acids (Figure 20.28). The three –OH groups on glycerol allow for mono-, di-, and triglycerides, with the latter being the most abundant. Glycerides account for over 98% of the lipids in the fatty tissues of mammals.

Table 20.2 lists some common fatty acids. The most abundant ones have an even number of carbon atoms because the fatty acids are built from two-carbon units. Their biosynthesis begins, as does the biosynthesis of cholesterol, with the conversion of pyruvate to acetyl-coenzyme A. Most fatty acids contain between 14 and 22 carbon atoms.

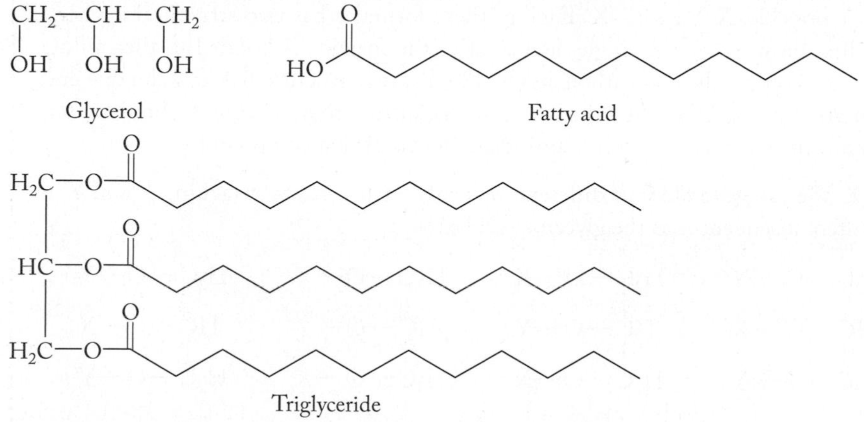

Glycerol

Fatty acid

Triglyceride

FIGURE 20.28 Glycerides are esters that form when glycerol combines with fatty acids. When all three –OH groups on glycerol react to form ester bonds, the product is a *tri*glyceride.

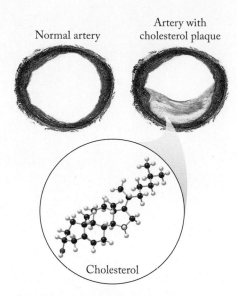

FIGURE 20.27 Cholesterol deposits are responsible for restricted blood flow, which results in a variety of sometimes catastrophic cardiovascular problems.

TABLE 20.2	Names, Formulas, and Structures of Common Fatty Acids
COMMON NAME (CHEMICAL NAME) (SOURCE)	**FORMULA**
Saturated Fatty Acids	
Lauric acid (dodecanoic acid) (coconut oil)	$CH_3(CH_2)_{10}COOH$
Myristic acid (tetradecanoic acid) (nutmeg butter)	$CH_3(CH_2)_{12}COOH$
Palmitic acid (hexadecanoic acid) (animal and vegetable fats)	$CH_3(CH_2)_{14}COOH$
Stearic acid (octadecanoic acid) (animal and vegetable fats)	$CH_3(CH_2)_{16}COOH$
Unsaturated Fatty Acids	
Oleic acid (*cis*-9-octadecenoic acid) (animal and vegetable fats)	$CH_3(CH_2)_7CH{=}CH(CH_2)_7COOH$
Linoleic acid (*cis,cis*-9,12-octadecadienoic acid) (linseed oil, cottonseed oil)	$CH_3(CH_2)_4CH{=}CHCH_2CH{=}CH(CH_2)_7COOH$
Linolenic acid (*cis,cis,cis*-9,12,15-octadecatrienoic acid) (linseed oil)	$CH_3CH_2CH{=}CHCH_2CH{=}CHCH_2CH{=}CH(CH_2)_7COOH$

tricarboxylic acid (TCA) cycle a series of reactions that continue the oxidation of pyruvate formed in glycolysis.

lipid a class of water-insoluble, oily organic compounds that are common structural materials in cells.

glyceride lipid consisting of esters formed between fatty acids and the alcohol glycerol.

How many different triglycerides (including structural isomers and stereoisomers) can be made from glycerol combining with two different fatty acids (X and Y) if each molecule of triglyceride contains at least one molecule of each fatty acid?

COLLECT AND ORGANIZE We are asked to determine how many different triglycerides can be made from glycerol and two different fatty acids. A triglyceride contains three fatty acid units.

ANALYZE Each fatty acid may bond to one of three –OH groups in glycerol. Each triglyceride has at least one X and one Y residue, which makes two combinations of X and Y possible: X_2Y and Y_2X. Each of these formulas has two structural isomers, depending on whether the single fatty acid in the formula is bonded to the middle carbon or to one of the end carbon atoms. Finally, if a structure has an X on one end carbon atom and a Y on the other, then the middle carbon atom is a chiral center, which means that there are two enantiomeric forms of that compound.

SOLVE We can generate four different molecular structures by attaching X and Y in four different sequences to the glycerol –OH groups:

$$
\begin{array}{cccc}
H_2C-O-X & H_2C-O-X & H_2C-O-Y & H_2C-O-Y \\
HC-O-X & HC-O-Y & HC-O-Y & HC-O-X \\
H_2C-O-Y & H_2C-O-X & H_2C-O-X & H_2C-O-Y \\
(1) & (2) & (3) & (4)
\end{array}
$$

Structures 1 and 3 have chiral centers, so each of these isomers has two enantiomeric forms. Therefore there are a total of six different triglycerides possible.

THINK ABOUT IT The central carbon atoms in structures 1 and 3 are chiral because they are each bonded to a H atom, an O atom, and to two C atoms that are themselves bonded to different fatty acids: X and Y.

Practice Exercise How many different triglycerides can be made from glycerol and one molecule each of three different fatty acids A, B, and C? ⚙

Function and Metabolism of Lipids

Lipids are an important energy source in our diets, providing more energy per gram than carbohydrates or proteins. As indicated in Table 20.2, fatty acids are either saturated, containing no carbon–carbon double bonds, or unsaturated, containing one or more carbon–carbon double bonds. **Fats** are glycerides composed primarily of saturated fatty acids. They are solids at room temperature. **Oils** are glycerides composed predominantly of unsaturated fatty acids and are liquids at room temperature. Oils can be converted into solid, saturated glycerides by hydrogenation. For example, in the hydrogenation of corn oil, which is a mixture of mostly two unsaturated fatty acids (oleic and linoleic acids), hydrogen is added to convert some or all of the –CH=CH– subunits into –CH_2—CH_2– subunits. The resulting solid can be whipped with skim milk, coloring agents, and vitamins to produce the food spread we know as margarine.

The consumption of too much saturated fat is associated with coronary heart disease. One of the purported advantages of the so-called Mediterranean diet is that olive oil, a liquid composed of glycerides containing over 80% oleic acid, is

fat solid triglyceride containing primarily saturated fatty acids.

oil liquid triglyceride containing primarily unsaturated fatty acids.

used in cooking rather than fats like butter and lard. In addition, this diet tends to be richer in fish and vegetables, both of which contain unsaturated fats. Most animal fats, like the marbling in beef that enhances its flavor, are saturated.

Another problem with hydrogenating vegetable oil arises when the oils are only partially hydrogenated. Partial hydrogenation alters the molecular structure around the remaining C=C bonds, changing them from their natural cis isomers into trans isomers (Figure 20.29). Unsaturated trans fats like elaidic acid tend to be solids at room temperature because their molecules pack together more uniformly than do molecules of cis unsaturated fatty acids. Consumption of trans fatty acids (a.k.a. *trans fats*) is associated with increased levels of cholesterol in the blood and other health risks.

The lipids in some prepared foods have been modified to reduce their caloric content but still provide the taste, aroma, and "mouth feel" we associate with lipid-rich foods. The active sites of enzymes that break down natural lipids accommodate triglycerides formed from glycerol and fatty acids. However, chemically modified esters made from the same fatty acids but attached to an alcohol other than glycerol cannot be metabolized by these enzymes. Such molecules, if they have the appropriate physical properties and are nontoxic, can be incorporated into foods without adding any calories because they are not metabolized.

Olestra is one such product (Figure 20.30). It is an ester made from long-chain fatty acids and the carbohydrate sucrose. (Remember, sugars have –OH groups and technically are alcohols.) Each of the eight –OH groups in a sucrose molecule reacts with a molecule of fatty acid to make the ester in olestra. The resultant material is used to deep-fry potato chips. Any olestra that remains on the chip does not add calories because it cannot be processed by enzymes that recognize only fatty acid esters on a glycerol scaffold.

CONCEPT TEST ··

Olestra may be "calorie-free" as a food subject to metabolism in the human body, but how would it compare with a common triglyceride in terms of kilojoules of heat released per gram in a calorimeter experiment?

··

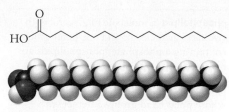

(a) Stearic acid: a saturated fatty acid

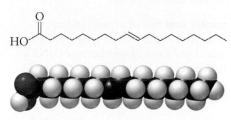

(b) Elaidic acid: a trans unsaturated fatty acid

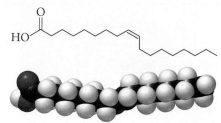

(c) Oleic acid: a cis unsaturated fatty acid

FIGURE 20.29 Types of fatty acids. (a) Stearic acid, a saturated C_{18} fatty acid. (b) Elaidic acid, an unsaturated C_{18} fatty acid (trans isomer). (c) Oleic acid, an unsaturated C_{18} fatty acid (cis isomer).

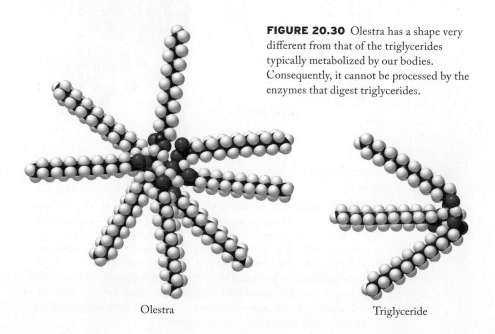

FIGURE 20.30 Olestra has a shape very different from that of the triglycerides typically metabolized by our bodies. Consequently, it cannot be processed by the enzymes that digest triglycerides.

Olestra

Triglyceride

phospholipid a molecule of glycerol with two fatty acid chains and one polar group containing a phosphate; phospholipids are major constituents of cell membranes.

lipid bilayer a double layer of molecules whose polar head groups interact with water molecules and whose nonpolar tails interact with each other.

nucleic acid one of a family of large molecules, which includes deoxyribonucleic acid (DNA) and ribonucleic acid (RNA), that stores the genetic blueprint of an organism and controls the production of proteins.

nucleotide a monomer unit from which nucleic acids are made.

The enzymes that metabolize triglycerides hydrolyze the esters and release glycerol and the fatty acids bonded to it. Glycerol enters the metabolic pathway for glucose. The fatty acids are oxidized in a series of reactions known as β-oxidation, a process that removes two carbon atoms at a time. For example, stearic acid (the saturated C_{18} fatty acid) is transformed into a C_2 fragment and the C_{16} acid, palmitic acid. Palmitic acid yields another C_2 fragment and myristic acid, and so forth until the fatty acid is completely degraded. Electrons released from this oxidative process are eventually donated to O_2. The energy released by this process powers metabolism.

Other Types of Lipids

Cells contain other types of lipids in addition to triglycerides. One type, **phospholipids** (Figure 20.31a), plays a key role in cell structure. A phospholipid molecule consists of a glycerol molecule bonded to two fatty acid chains and to one phosphate group that is also bonded to a polar substituent. The presence of nonpolar fatty acid chains and a polar region in the same molecule makes phospholipids ideal for forming cell membranes. In an aqueous medium, phospholipids form a **lipid bilayer**, a double layer enclosing each cell and isolating its interior from the outside environment. The phospholipid molecules of the bilayer align so that the nonpolar groups interact with each other within the membrane while the polar groups interact with water molecules outside the membrane (Figure 20.31b). Membranes exist both to isolate the contents of cells and to serve as the locus of communication between intracellular and extracellular processes.

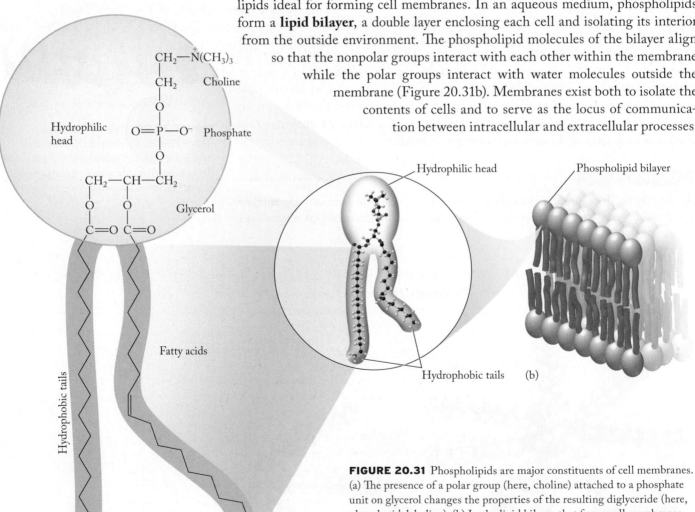

FIGURE 20.31 Phospholipids are major constituents of cell membranes. (a) The presence of a polar group (here, choline) attached to a phosphate unit on glycerol changes the properties of the resulting diglyceride (here, phosphatidylcholine). (b) In the lipid bilayer that forms cell membranes, phospholipids orient themselves so that the polar groups in one half of the bilayer face the aqueous environment outside the cell while the polar groups in the other half of the bilayer face the aqueous environment of the cell interior. This arrangement leaves the nonpolar fatty acid part of the phospholipids in the interior of the bilayer.

20.5 Nucleotides and Nucleic Acids

Nucleic acids are our fourth class of biomolecules and third class of biopolymers. We focus on two types: deoxyribonucleic acid (DNA) and ribonucleic acid (RNA). Even though nucleic acids make up only about 1% of an organism's mass, they code for the enzymes that control the metabolic activity of all its cells. DNA carries the genetic blueprint of an organism, and a variety of RNAs use that DNA blueprint to guide the production of proteins.

A nucleic acid is a polymer of monomeric units called **nucleotides**. Each nucleotide unit is in turn composed of three subunits: a five-carbon sugar, a phosphate group, and a nitrogen-containing base (Figure 20.32). The phosphate group in each nucleotide is attached to the carbon in the side chain of the sugar called the 5′ carbon atom. (The numbers 1′ through 5′ refer to the position of carbon atoms in the sugar molecule.) The nitrogen-containing base is attached to the sugar's 1′ carbon atom. The sugar in Figure 20.32 is called *ribose*, which makes this a nucleotide in a strand of *ribo*nucleic acid, or RNA. If the sugar were *deoxyribose* instead, there would be a H atom in place of the –OH group on the 2′ carbon atom, and the nucleotide would be a building block of *deoxyribo*nucleic acid, or DNA.

The nitrogen-containing base in Figure 20.32 is called adenine (A). Structural formulas of adenine and the other four bases in nucleic acids—cytosine (C), guanine (G), thymine (T), and uracil (U)—are shown in Figure 20.33. The point of attachment of the sugar–phosphate groups on each base is indicated by –R. In addition to the difference in their sugars, RNA and DNA also differ in one of the bases in their nucleotides: RNA contains A, C, G, and U, but DNA contains A, C, G, and T.

In a polymeric strand of nucleic acid, each phosphate is also linked to the 3′ carbon atom in the sugar of the monomer that precedes it in the chain, as shown for a strand of DNA in Figure 20.34. Both DNA and RNA strands are synthesized in the cell from the 5′ to the 3′ direction (downward in Figure 20.34). The structures of DNA and RNA are frequently written using only the single-letter labels of their bases, beginning with the free phosphate group on the 5′ end of the chain and reading toward the free 3′ hydroxyl group at the other terminus.

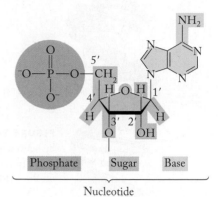

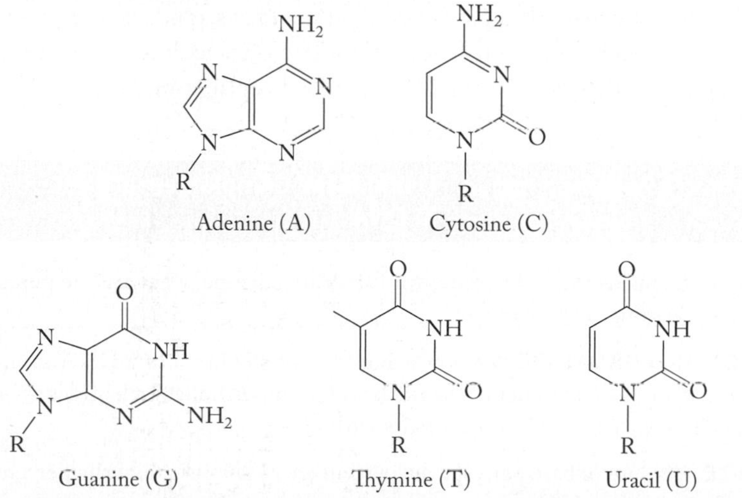

FIGURE 20.32 A nucleotide consists of a phosphate group and a nitrogen-containing base that are both bonded to a 5-carbon sugar.

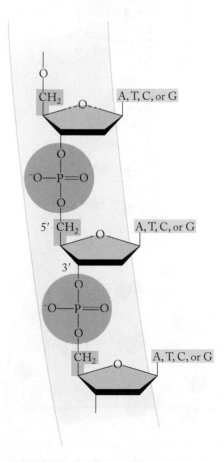

FIGURE 20.34 The backbone of the polymer chain in DNA consists of alternating sugar units (yellow) and phosphate units (pink). The bases (blue) are attached to the backbone through the 1′ carbon atom of the sugar unit.

Adenine (A)

Cytosine (C)

Guanine (G)

Thymine (T)

Uracil (U)

FIGURE 20.33 Structural formulas of the five nitrogen-containing bases in nucleotides. The nucleotides in DNA contain A, C, G, and T; those in RNA contain A, C, G, and U. The R-group identifies the point of attachment of the sugar residue.

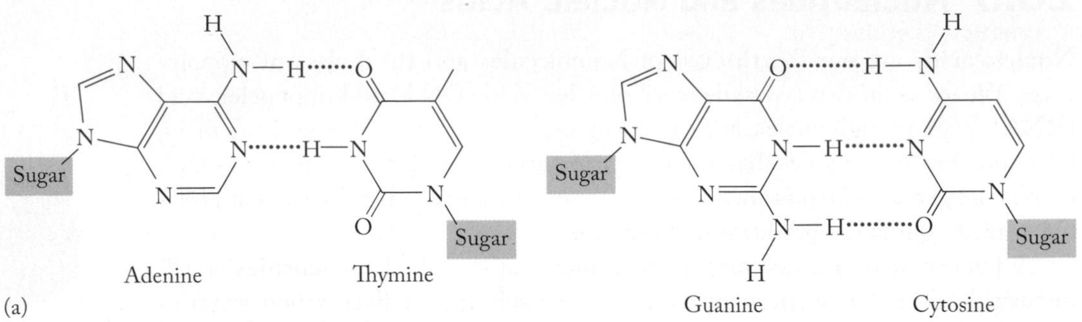

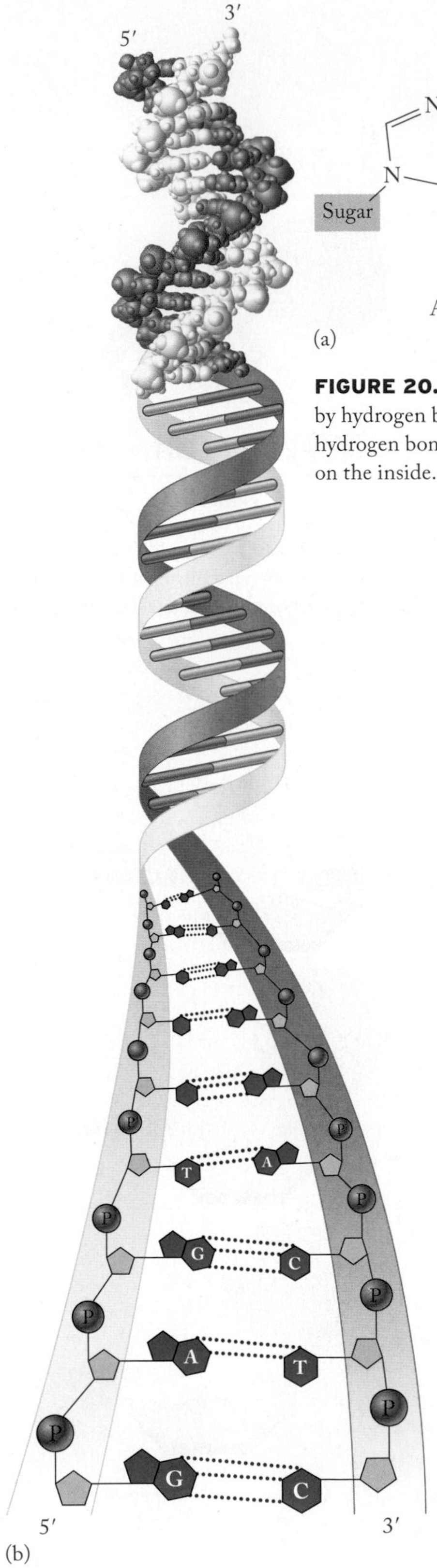

FIGURE 20.35 The nitrogen-containing bases on one strand of DNA pair with the bases on a second strand by hydrogen bonding. (a) Adenine and thymine pair via two hydrogen bonds; guanine and cytosine pair via three hydrogen bonds. (b) DNA as a double helix with the sugar–phosphate backbone on the outside and the base pairs on the inside.

When scientists first isolated DNA and began to analyze its composition, they made a pivotal observation about the abundance of the nitrogen-containing bases. A typical molecule of DNA consists of thousands of nucleotides, and the percentages of the four bases in different samples of DNA can vary over a wide range. However, the percentage of A in a sample always matches the percentage of T. Likewise, the percentage of C always matches that of G. This result makes sense if the bases are paired because a molecule of A can form two hydrogen bonds to a molecule of T, whereas a molecule of C can form three hydrogen bonds with a molecule of G. Therefore, A–T and G–C pairings maximize the number of hydrogen bonds possible (Figure 20.35a). More importantly, the base pairs assembled this way all have the same width and fit together in a regular structure to make a uniform double helix.

The normal structure of DNA features two strands of nucleotides wrapped around each other in a form that is called a *double helix* (Figure 20.35b). The nucleotide backbone is on the outside of the spiraling strands, with hydrogen bonds between the complementary bases keeping the two strands together. The base pairs are parallel to each other and perpendicular to the helical axis. The fidelity of this base pairing—A always with T, and C always with G—gives DNA the ability to copy itself. If a pair of complementary strands is unzipped into two single strands, each strand provides a template on which a new complementary strand can be synthesized via the process called **replication** (Figure 20.36).

SAMPLE EXERCISE 20.4 | **Using Base Complementarity in DNA** | **LO6**

If 31.6% of the nucleotides in a sample of DNA are adenine, what are the percentages of cytosine, guanine, and thymine?

COLLECT AND ORGANIZE We know how much adenine is in a DNA sample and need to calculate the remainder of the nucleotide composition. Nucleotides are paired: A always pairs with T, and C always pairs with G.

ANALYZE Because of base pairing, the percentage of T must equal the percentage of A, and the percentage of C must equal the percentage of G.

SOLVE If A = 31.6%, then T = 31.6%. This leaves $(100 - 2 \times 31.6)\% = 36.8\%$ left to be equally distributed between G and C. Therefore G = C = 18.4%.

replication the process by which one double-stranded DNA forms two new DNA molecules, each one containing one strand from the original molecule and one new strand.

THINK ABOUT IT The percentages should total 100%, and they do.

Practice Exercise Indicate the sequence of the complementary strand on the double helix formed by each of these sequences of nucleotides: (a) CGGTATCCGAT; (b) TTAAGCCGCTAG. ⚙

Original DNA molecule

During replication, the two strands are separated, forming a structure called the *replication fork*. The fork advances through the DNA as the process proceeds. One of the two resulting strands, the leading strand, is unzipped in the 3′–5′ direction, which allows the new complementary strand to be continuously synthesized in the 5′–3′ direction. The replication of the second strand, called the lagging strand, is more complicated. It proceeds in short fragments, which are then assembled into a continuous strand later in the process.

DNA's double-stranded structure is also the key to its ability to preserve genetic information. The two strands carry the same information, much like an old-fashioned photograph and its negative. Genetic information is duplicated every time a DNA molecule is replicated, a process that is essential whenever a cell divides into two new cells.

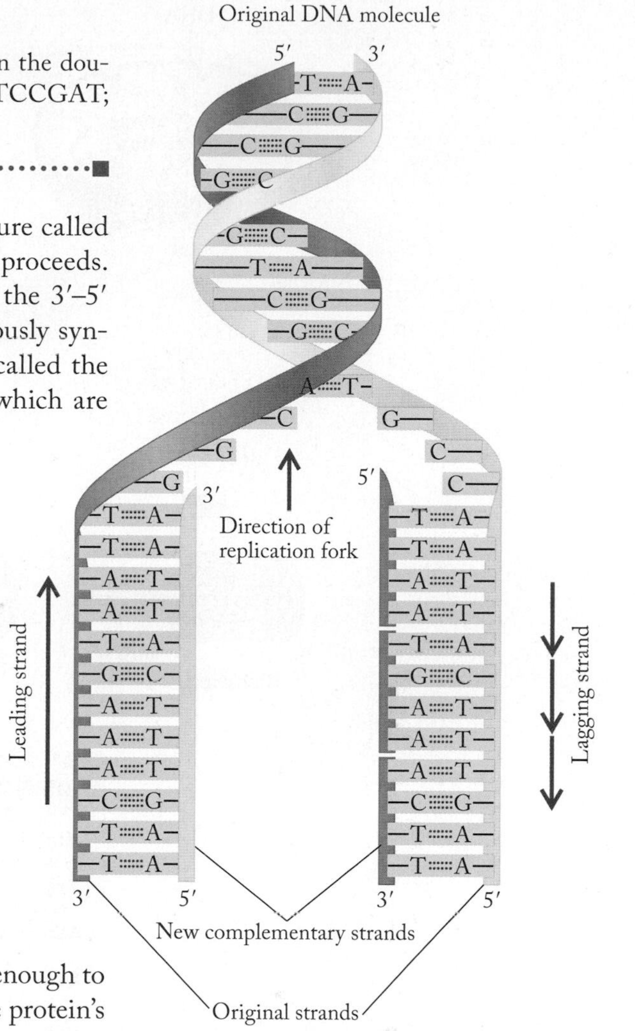

From DNA to New Proteins

Proteins are formed from amino acids in accordance with the *genetic code* contained in the base sequences of DNA strands. The bases A, T, G, and C are the alphabet in this code, and the "words" in the code are three-letter combinations of these four letters, with each word representing a particular amino acid or a signal to begin or end the protein synthesis process. Using four letters to write three-letter words means there are $4^3 = 64$ combinations possible, more than enough to encode the 20 amino acids found in cells. The genetic code specifies the protein's primary structure—the sequence of amino acids in proteins.

Protein synthesis begins with a process called **transcription** (Figure 20.37a), in which double-stranded DNA unwinds and its genetic information guides the synthesis of a single strand of a molecule called **messenger RNA (mRNA)**. This strand of mRNA has the complementary base sequence of the original DNA. It carries the 3-letter words of that DNA, in the form of three-base sequences called **codons** (Table 20.3), from the nucleus of the cell into the cytoplasm, where the mRNA binds with a cellular structure called a ribosome. Keep in mind that a sequence of A, C, G, and T in the original DNA is transcribed into an mRNA sequence containing A, C, G, and U:

DNA: ...ACGTTGAGC...   mRNA: ...UGCAACUCG...

At the ribosome the genetic information in the mRNA directs the synthesis of particular proteins in a process called **translation**. Another type of RNA, called **transfer RNA (tRNA)**, plays a key role in translation. There are 20 different forms of tRNA in the cell, one for each amino acid. To see how tRNA works, let's look at Figure 20.37(b). The first codon in this piece of an mRNA strand is AUG, which codes for the amino acid methionine (see Table 20.3). In the cytoplasm surrounding the ribosome, molecules of tRNA are reversibly bonded to molecules of every amino acid. The particular tRNA molecules that are bonded to methionine also contain the *anticodon* UAC, the complement of AUG, at a site that allows the tRNA to interact with mRNA. As Figure 20.37(b) shows, the

FIGURE 20.36 When DNA replicates, the two strands of a short portion of the double helix are unzipped, and each strand joins with a new complementary strand to produce two new DNA molecules.

transcription the process of copying the information in DNA to RNA.

messenger RNA (mRNA) the form of RNA that carries the code for synthesizing proteins from DNA to the site of protein synthesis in a cell.

codon a three-nucleotide sequence that codes for a specific amino acid.

translation the process of assembling proteins from the information encoded in RNA.

transfer RNA (tRNA) the form of RNA that delivers amino acids, one at a time, to polypeptide chains being assembled by the ribosome–mRNA complex.

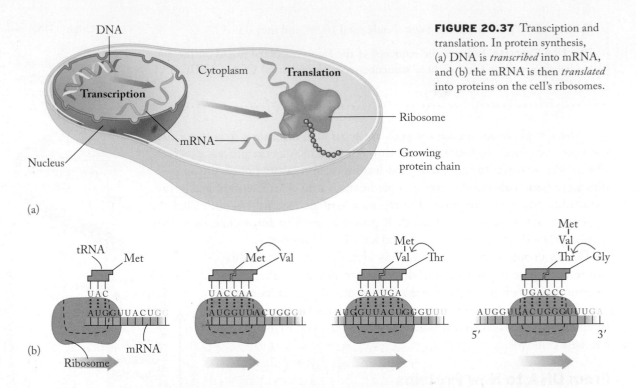

FIGURE 20.37 Transciption and translation. In protein synthesis, (a) DNA is *transcribed* into mRNA, and (b) the mRNA is then *translated* into proteins on the cell's ribosomes.

TABLE 20.3	**mRNA Codons**		
Ala	GCU, GCC, GCA, GCG	**Leu**	UUA, UUG, CUU, CUC, CUA, CUG
Arg	CGU, CGC, CGA, CGG, AGA, AGG	**Lys**	AAA, AAG
Asn	AAU, AAC	**Met**	AUG
Asp	GAU, GAC	**Phe**	UUU, UUC
Cys	UGU, UGC	**Pro**	CCU, CCC, CCA, CCG
Gln	CAA, CAG	**Ser**	UCU, UCC, UCA, UCG, AGU, AGC
Glu	GAA, GAG	**Thr**	ACU, ACC, ACA, ACG
Gly	GGU, GGC, GGA, GGG	**Trp**	UGG
His	CAU, CAC	**Tyr**	UAU, UAC
Ile	AUU, AUC, AUA	**Val**	GUU, GUC, GUA, GUG
Start	AUG	**Stop**	UAG, UGA, UAA

segment of mRNA with the AUG codon links with the complementary anticodon on the tRNA molecule bonded to methionine. In doing so, the methionine is put into a position to unlink from the tRNA and to be the first amino acid residue in the protein being synthesized.

The sequence of events described in the preceding paragraph is repeated many times. In this illustration the next codon, GUU, links up with a molecule of tRNA that has a CAA binding site and a molecule of valine in tow. In this way valine moves into position to become the next amino acid residue in the protein and to form a peptide bond with the N-terminal methionine. Valine is followed by threonine, which is followed by glycine, and so on, until a "stop" codon signals the end of the translation process.

If a GUU codon attracts valine to the translation site, does a UUG codon do the same thing? Explain your answer.

20.6 From Biomolecules to Living Cells

We end this chapter by addressing two fundamental questions about the major classes of biomolecules and their roles in sustaining life: how were they first formed on prebiotic Earth, and how did they assemble into living cells? Experiments conducted in the 1950s at the University of Chicago by Professor Harold Urey (1893–1981) and his student Stanley Miller (1930–2007) showed that amino acids could form from H_2O, CH_4, NH_3, and H_2 (Figure 20.38). Although the reactants the two scientists chose are now thought to be different from those present on the early Earth, the result still stands: inorganic molecules can react to produce the organic molecules found in living systems.

There is also evidence that some biomolecules may have reached Earth from extraterrestrial origins. In 2006 the NASA spacecraft Stardust returned to Earth with samples collected from the tail of a comet that is believed to have formed at about the same time as the solar system (Figure 20.39). Subsequent analyses disclosed the presence of glycine in the comet dust. This was not the first experiment to detect amino acids in space. A class of meteorites called carbonaceous chondrites contain isovaline and other amino acids. Interestingly, some of these amino acids are not racemic mixtures, but instead are more than 50% L-amino acids, the enantiomeric form that dominates our biosphere. These observations have led some to suggest that L-amino acids are somehow "favored" and that life on Earth was "seeded" from elsewhere.

As we have seen in this chapter, RNA is needed to guide the assembly of amino acids into proteins. Since the 1990s, groups of scientists have explored the possibility that strands of RNA may have formed spontaneously from solutions of nucleotides in contact with mineral surfaces that guided the self-assembly of the nucleotides into long strands of RNA. Oligonucleotides made in this way might have been able to catalyze their own replication. This ability to self-replicate is crucial to life, and the observation that molecules are capable of speeding up their own replication on a clay surface suggests that processes essential to the formation of living cells could have happened spontaneously. Much controversy still exists about these ideas, but the capacity of RNA to both store information like DNA *and* act as a catalyst like an enzyme has motivated the *RNA world hypothesis*. This hypothesis proposes that a world filled with cellular or precellular life based on RNA alone predates the current world of life based on DNA and proteins.

Current research is also testing the theory that life on Earth may have evolved near deep-ocean hydrothermal vents. Entire ecosystems have been discovered at these locations since they were first explored in the 1970s. They are sustained by geothermal and chemical energy rather than energy from the sun. It may be that life actually began in such environments, with hydrothermal energy driving reactions in which inorganic compounds like carbon dioxide and hydrogen sulfide formed small organic compounds. As with the reactions on the surfaces of clay minerals, the synthesis reactions at hydrothermal vents may have been catalyzed and guided by reactants adsorbed on solid compounds such as FeS

FIGURE 20.38 The apparatus used by Miller and Urey to simulate the synthesis of amino acids in the atmosphere of early (prebiotic) Earth. Discharges between the tungsten electrodes were meant to provide the sort of energy that might have come from lightning.

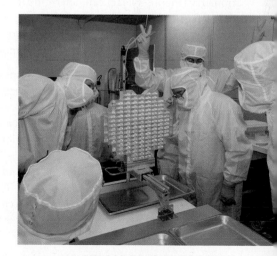

FIGURE 20.39 NASA scientists who analyzed the samples collected by the Stardust spacecraft were careful to avoid contaminating it with biological material from Earth. This meant isolating themselves from the sample as they prepared it for analysis.

FIGURE 20.40 Black clouds of transition metal oxides and sulfides flow into the sea through chimneys like this one at a deep-ocean hydrothermal vent. Some scientists believe that these particles may have guided and catalyzed the formation of the first self-replicating molecules on Earth.

and MnO_2, which pour into the sea in dense black clouds near some vents (Figure 20.40). Among the known products of these reactions are acetate ions (CH_3COO^-). Acetate is a key intermediate in many biosynthetic pathways in living organisms. In modern bacteria, the systems that make acetate depend on a catalyst made of iron, nickel, and sulfur that has a structure much like that of particles produced by "black smokers" on the ocean floor.

To take the next step toward forming living cells, large biomolecules must have assembled themselves into even larger structures, such as membranes, that allow cells and structures within them to collect materials and retain them at concentrations different from those in the surrounding medium. Molecules in these assemblies are not necessarily connected by covalent bonds, but rather are held together by the intermolecular interactions we have discussed in this chapter.

SAMPLE EXERCISE 20.5 Integrating Concepts: PKU Screening in Infants

Phenylketonuria (PKU) is a rare genetic disorder in which a baby is born without the ability to properly break down the amino acid phenylalanine (Phe). The gene for the liver enzyme phenylalanine hydroxylase (PAH), which is necessary for the metabolism of Phe to the amino acid tyrosine (Tyr) (Figure 20.41a), is mutated, and PAH is absent or significantly decreased in activity. As a result, Phe accumulates, which hinders brain development, causing mental retardation. Instead of being converted into Tyr, Phe is converted into phenylpyruvate (Figure 20.41b), which may be detected in the urine. A blood test may also be done to determine the amount of Phe present; normal levels are less than 2.0 mg/dL of blood. A different test using mass spectrometry (Figure 20.41c) can determine the ratio of Phe to Tyr; data suggest that in a healthy infant, the ratio should be less than 2.5 ([Phe]/[Tyr] < 2.5). Previous work has shown that IQ is impacted when Phe levels exceed 800 μmol/L.

a. Describe the changes in Phe in terms of functional groups and charge when it is converted to Tyr and phenylpyruvate.
b. What should be the concentration of Tyr (in mg/dL) in a healthy infant if the concentration of Phe reaches 2 mg/dL?
c. Express the concentrations of Tyr and Phe from part b in mol/L.
d. Assuming the level of Tyr does not change from the value you calculated in part c, what is the [Phe]/[Tyr] ratio when [Phe] = 800 μmol/L?

COLLECT AND ORGANIZE We are given the structural formulas for Phe, Tyr, and phenylpyruvate. We also know normal Phe levels in mg/dL, the normal [Phe]/[Tyr] ratio, and the level at which the amount of Phe damages mental abilities in μmol/L. We need to convert the numbers to one set of units to enable the comparisons requested.

ANALYZE From the structural formulas, we can determine the changes caused by metabolism. We need to calculate the molecular masses of Phe and Tyr to convert between mg/dL and mol/L. The value of [Tyr] in part b will be 1/2.5 × 2.0 mg/dL or a little less than 1 mg/dL. Based on their molecular structures we can estimate that the molar masses of Tyr and Phe will be between 100 and 200 g/mol. Therefore the molar concentrations in part c that are equivalent to 1 or 2 mg/dL (or 10–20 mg/L = 0.01–0.02 g/L) will be about 0.01 g/L / 100 g/mol, or ~10^{-4} mol/L.

SOLVE

a. Phenylalanine is a neutral amino acid with a nonpolar R group. It contains an aromatic ring in addition to its amine and carboxylic acid functional groups. PAH converts Phe into Tyr, which has an –OH group on the aromatic ring and hence a polar R group. Tyr is also neutral. In the absence of PAH, Phe is converted into phenylpyruvate, which is not an amino acid. Phenylpyruvate contains a ketone group and a carboxylic acid group that is ionized at physiological pH, so it is negatively charged.

b. If the concentration of Phe reaches 2 mg/dL, and the Phe/Tyr ratio is to be <2.5, then the *minimum* concentration of Tyr (*x*) in a healthy infant would be

$$\frac{2.0 \text{ mg/dL}}{x} = 2.5, \qquad x = 0.80 \text{ mg/dL}$$

c. Using the structural formulas given, Phe is $C_9H_{11}NO_2$ with a molecular mass of 165.19 g/mol; Tyr is $C_9H_{11}NO_3$ with a molecular mass of 181.19 g/mol. In units of mol/L, the concentrations of Phe and Tyr from part b are

$$[\text{Phe}] = \frac{2.0 \text{ mg}}{\text{dL}} \times \frac{1 \text{ g}}{1000 \text{ mg}} \times \frac{1 \text{ mol}}{165.19 \text{ g}} \times \frac{10 \text{ dL}}{1 \text{ L}}$$

$$= 1.2 \times 10^{-4} \, M$$

$$[Tyr] = \frac{0.80 \; \cancel{mg}}{\cancel{dL}} \times \frac{1 \; \cancel{g}}{1000 \; \cancel{mg}} \times \frac{1 \; mol}{181.19 \; \cancel{g}} \times \frac{10 \; \cancel{dL}}{1 \; L}$$

$$= 4.4 \times 10^{-5} \; M$$

d. If [Phe] = 800 μmol/L, we can express that in terms of M as

$$800 \; \frac{\cancel{\mu mol}}{L} \times \frac{1 \; mol}{10^6 \; \cancel{\mu mol}} = 8.0 \times 10^{-4} \; M$$

If we assume [Tyr] is 4.4×10^{-5} M in this situation, then the [Phe]/[Tyr] ratio is

$$\frac{8.0 \times 10^{-4} \; \cancel{M}}{4.4 \times 10^{-5} \; \cancel{M}} = 18$$

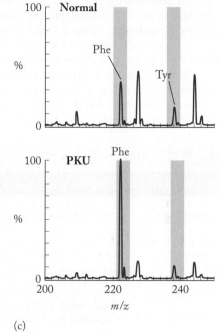

THINK ABOUT IT Mandatory screening programs identify infants with PKU at birth. Early diagnosis is essential for the successful management of this disorder. Infants diagnosed with PKU require a lifelong diet that is extremely low in Phe. This diet is challenging to maintain because Phe occurs in significant amounts in many common foods, including milk and eggs.

(a)

Phenylalanine → Tyrosine

(b)

Phenylalanine → Phenylpyruvate

(c)

FIGURE 20.41 (a) The amino acid phenylalanine is normally converted into the amino acid tyrosine. (b) In PKU, phenylalanine is instead converted into phenylpyruvate. (c) Amino acid profiles from a mass spectrometer of normal blood (top) and blood from a newborn infant with PKU (bottom). Adapted from S. A. Banta and R. D. Steiner in *Journal of Perinatal and Neonatal Nursing*, 18(1), 41 (2004).

SUMMARY

Section 20.1 The **proteins** and **peptides** in the human body are composed of 20 α-**amino acids** reversibly linked together by **peptide bonds**. At physiological pH (7.4), the amino acids exist as **zwitterions**.

Section 20.2 Protein molecules have four levels of structure. **Primary (1°) structure** is the amino acid sequence. **Secondary (2°) structure** is the shape the amino acid chain takes (**α helix**, **β-pleated sheet**, or **random coil**). **Tertiary (3°) structure** is the three-dimensional shape that results from attractive forces between amino acids located in various parts of the chain. **Quaternary (4°) structure** results when two or more proteins associate with each other to make larger functional units. Proteins called **enzymes** mediate the chemical reactions involved in metabolism. The structure of an enzyme contains a distinct region called an **active site** that binds to a **substrate**.

The *induced-fit model* suggests that the binding of a substrate to an enzyme changes the shape of the enzyme so that the reaction can take place. **Biocatalysis** seeks to replicate the selectivity of enzymes in industrial settings.

Section 20.3 Carbohydrates are produced via photosynthesis from CO_2 and H_2O. **Monosaccharides** are joined into **polysaccharides** through **glycosidic bonds**. The isomers α-glucose and β-glucose form the polysaccharides starch and cellulose, respectively. Organisms derive energy from glycolysis and the **tricarboxylic acid (TCA) cycle**, the reaction pathways by which glucose is oxidized to CO_2 and H_2O.

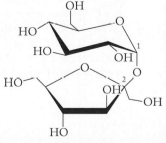

Section 20.4 Lipids are a major source of energy in our diet. The lipids known as **glycerides** are esters of the alcohol glycerol and up to three long-chain fatty acids. **Fats** are solid glycerides composed primarily of saturated fatty acids; **oils** are mostly unsaturated fatty acids and are liquids at room temperature. **Phospholipids** form **lipid bilayer** cell membranes.

Section 20.5 The **nucleic acids** DNA and RNA contain an organism's genetic information and control protein synthesis through **transcription** and **translation**. Genetic information is transmitted to new cells during cell division in a process called **replication**. DNA and RNA are made of chains of **nucleotides**, molecules composed of a five-carbon sugar, a phosphate group, and one of five nitrogen-containing organic bases. DNA consists of two nucleotide chains coiled into a double helix and connected via hydrogen bonds between complementary bases—A with T, and C with G. RNA,

in which the base pairings are A with U and C with G, is a single-stranded nucleic acid involved in protein synthesis through **messenger RNA (mRNA)** and **transfer RNA (tRNA)**.

Section 20.6 Living cells may have formed as a result of chemical reactions in which inorganic molecules combined to form small organic molecules such as amino acids and nucleotides. The latter may have self-assembled, perhaps in polymerization reactions catalyzed by adsorption onto the surface of clay minerals or transition metal sulfides or oxides, into self-replicating biopolymers.

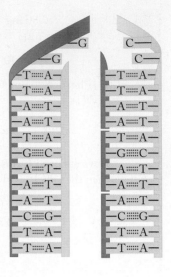

PROBLEM-SOLVING SUMMARY

TYPE OF PROBLEM	CONCEPTS AND EQUATIONS	SAMPLE EXERCISES
Interpreting acid–base titration curves of amino acids	At low pH all amino acids have at least two ionizable H atoms, one each from $-COOH$ and $-NH_3^+$. Side-chain carboxylic acid and amine groups may also impart acidic and basic strength to amino acids.	20.1
Drawing and naming peptides	Connect the α-amine of one amino acid to the α-carboxylic acid of another with a peptide bond. Starting with the free amine (N-) terminus on the left, name each amino acid residue by changing the ending of the name of the parent amino acid to -*yl* in all but the last (C-terminal) amino acid.	20.2
Identifying triglycerides	The $-COOH$ groups of fatty acids react with the $-OH$ groups in glycerol to form triglycerides and water.	20.3
Using base complementarity in DNA	Identify the base pairs: A pairs with T; G pairs with C. The percentage of T should equal the percentage of A, and the percentage of C should equal the percentage of G.	20.4

VISUAL PROBLEMS

(Answers to boldface end-of-chapter questions and problems are in the back of the book.)

20.1. Figure P20.1 shows how a molecule of glucose changes from an α-glucose to a β-glucose conformation. Draw a graph that shows how the energy of the molecule (or a mole of them) changes during the transformation from α-glucose to a β-glucose.

FIGURE P20.1

20.2. The nucleotides in DNA contain the bases with the structures shown in Figure P20.2. How many primary and secondary amine groups are in each structure?

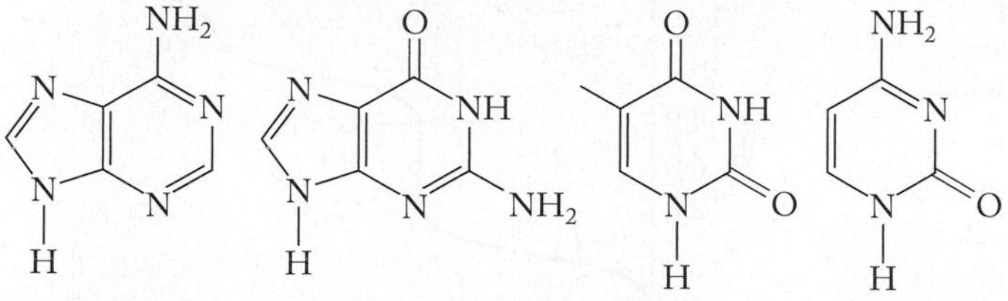

Adenine   Guanine   Thymine   Cytosine

FIGURE P20.2

20.3. Olive Oil Olive oil contains triglycerides such as those shown in Figure P20.3.

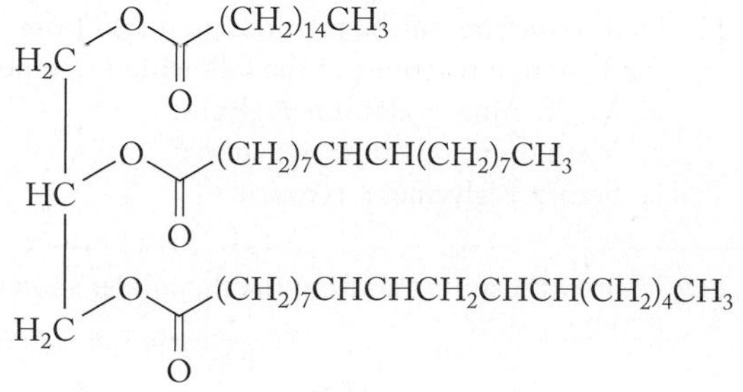

(a)

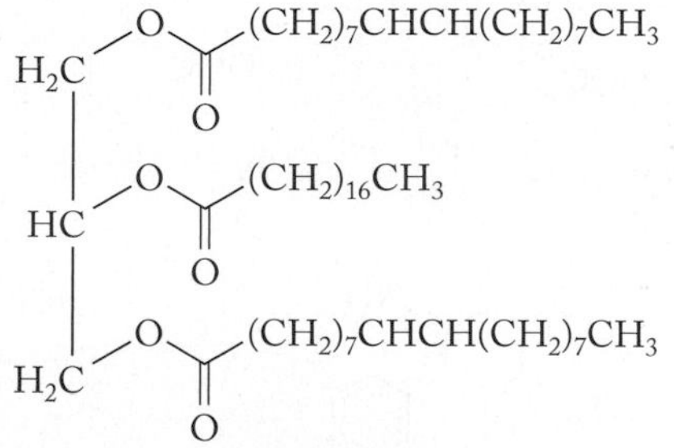

(b)

FIGURE P20.3

a. Which of the fatty acids in these triglycerides is/are saturated?
b. Are the unsaturated fatty acids likely to be cis or trans isomers?

20.4. The two compounds shown in Figure P20.4 are both amino acids. Which of them is an α-amino acid? Explain your choice.

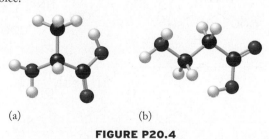

(a)   (b)

FIGURE P20.4

20.5. Natural Painkillers The human brain produces polypeptides called *endorphins* that help in controlling pain. The pentapeptide in Figure P20.5 is called enkephalin. Identify the five amino acids that make up enkephalin.

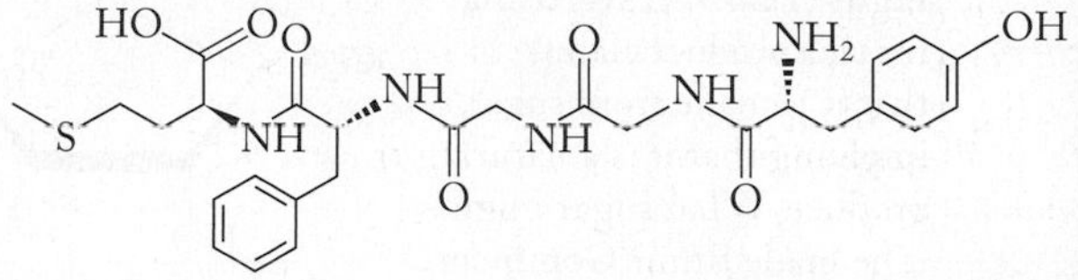

Enkephalin

FIGURE P20.5

20.6. Regulating Blood Pressure Angiotensin II is a polypeptide that regulates blood pressure. The structure of angiotensin II is shown in Figure P20.6. Which amino acids are in the structure?

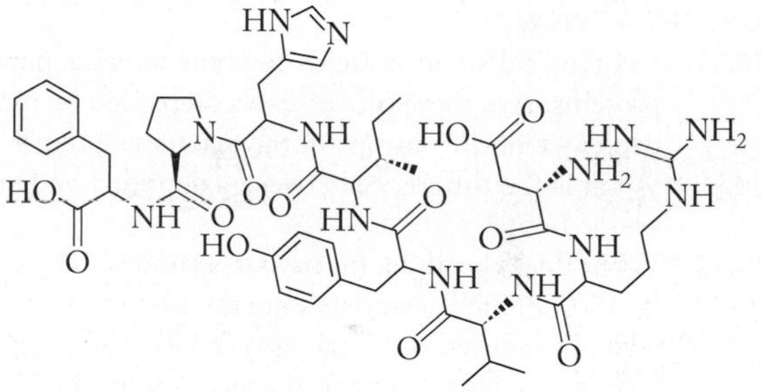

Angiotensin II

FIGURE P20.6

20.7. Trans Fats The role of "trans fats" in human health has been extensively debated both in the scientific community and in the popular press. What type of isomerism does the word "trans fat" refer to? Which of the molecules in Figure P20.7 are considered trans fats?

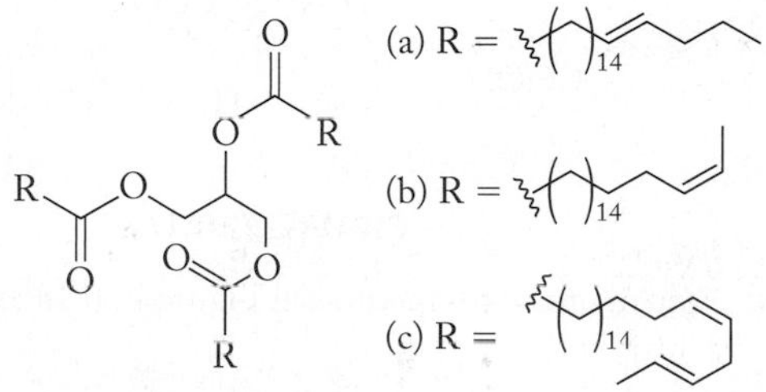

(a) R =

(b) R =

(c) R =

FIGURE P20.7

20.8. Cocoa Butter Cocoa butter (Figure P20.8) is a key ingredient in chocolate. Cocoa butter is a triglyceride that results from esterification of glycerol with three different fatty acids. Identify the fatty acids produced by hydrolysis of cocoa butter.

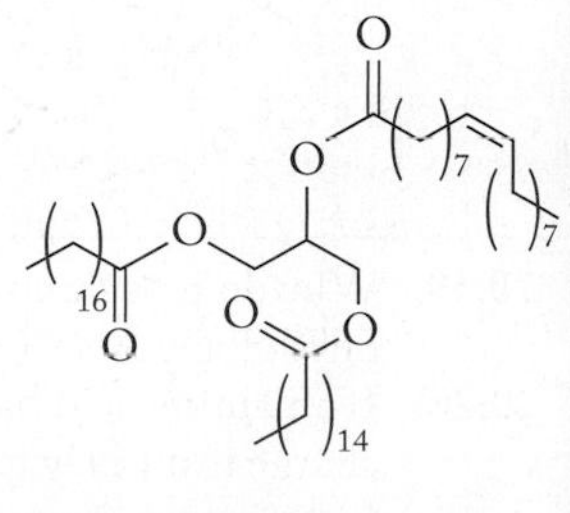

Cocoa butter

FIGURE P20.8

20.9. Sucralose The molecular structure of the artificial sweetener sucralose (trade name Splenda) is shown in Figure P20.9. Advertising for this product claims that it is made from sugar, implying that it is a natural product. What sugar might it be made from? Comment on the implication that it is a "natural" product.

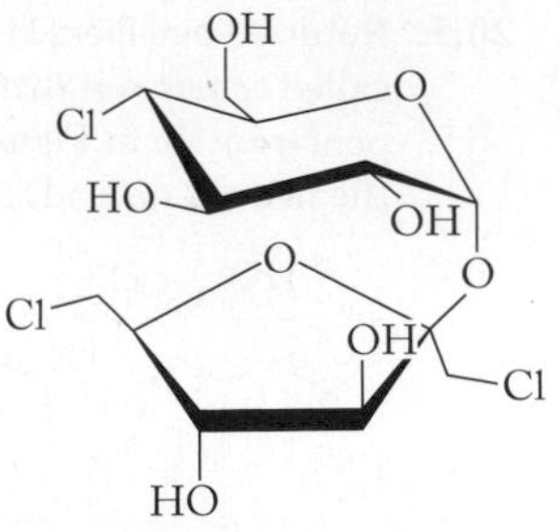

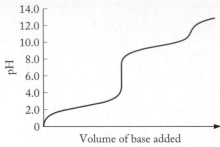

Sucralose

FIGURE P20.9

20.10. Figure P20.10 contains the titration curve of which of these amino acids: leucine, histidine, or lysine?

FIGURE P20.10

QUESTIONS AND PROBLEMS

The Composition of Proteins

CONCEPT REVIEW

20.11. In living cells, amino acids combine to make peptides and proteins. Are these processes accompanied by increases or decreases in the entropy of the reaction system?

20.12. What is the difference between a peptide bond and an amide bond?

20.13. What does the alpha mean in α-amino acid?

20.14. In 1806 French scientists were the first to isolate an amino acid. The source was asparagus shoots. The compound had a negative charge in neutral aqueous solutions. Can you identify it?

20.15. Meteorites contain more L-amino acids, which are the forms that make up the proteins in our bodies, than D-amino acids. What do the prefixes L- and D- mean?

20.16. Do any of the amino acids in Table 20.1 have more than one chiral carbon atom per molecule?

20.17. Which of the compounds in Figure P20.17 is not an α-amino acid?

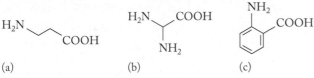

(a)   (b)   (c)

FIGURE P20.17

20.18. Which of the compounds in Figure P20.18 are α-amino acids?

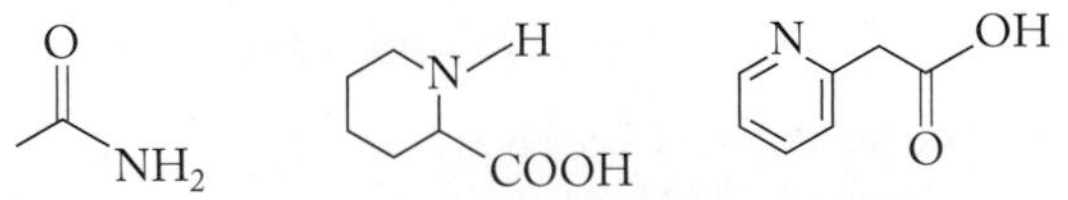

(a)   (b)   (c)

FIGURE P20.18

20.19. Why do most amino acids exist in the zwitterionic form at physiological pH (pH ≈ 7.4)?

20.20. If an amino acid has no net charge at pH = 7.4, what charge is it likely to have at pH = 2.3 and at pH = 10.3? Explain your answers.

PROBLEMS

20.21. Draw structures of the peptides produced from condensation reactions of the following L-amino acids:
a. Alanine + serine
b. Alanine + phenylalanine
c. Alanine + valine

20.22. Draw structures of the peptides produced from condensation reactions of the following L-amino acids:
a. Methionine + alanine + glycine
b. Methionine + valine + alanine
c. Serine + glycine + tyrosine

20.23. Identify the amino acids in the dipeptides shown in Figure P20.23.

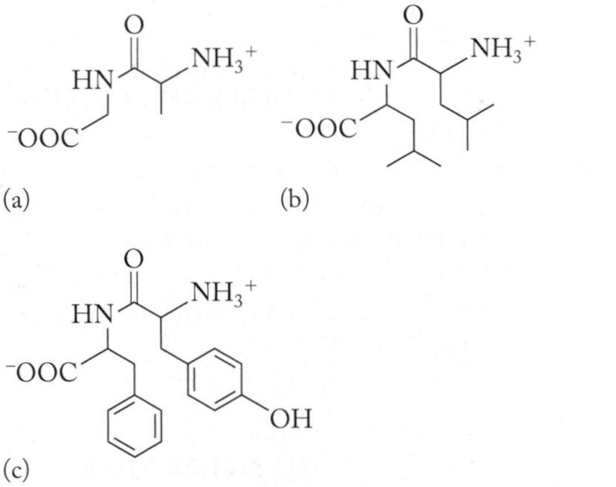

(a)   (b)

(c)

FIGURE P20.23

20.24. Identify the amino acids in the tripeptides shown in Figure P20.24.

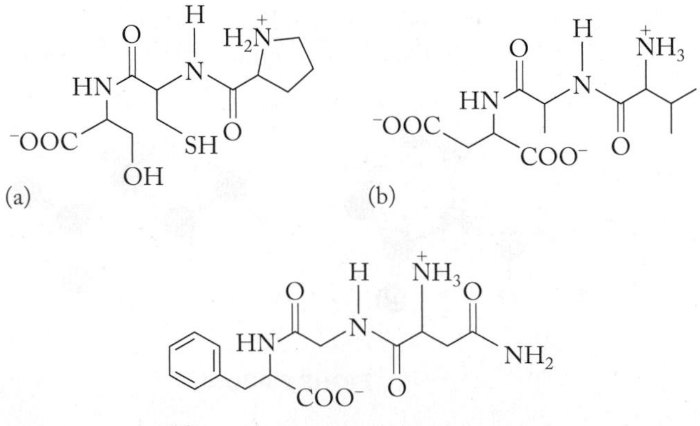

(a)   (b)

(c)

FIGURE P20.24

20.25. Identify the missing product in the metabolic reaction shown in Figure P20.25.

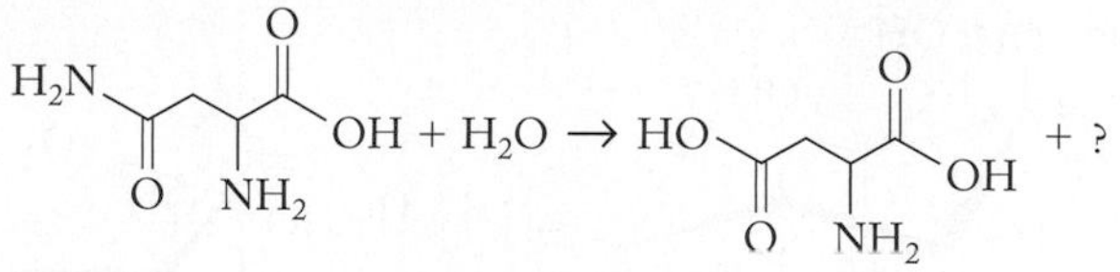

FIGURE P20.25

20.26. Identify the missing product in the metabolic reaction shown in Figure P20.26.

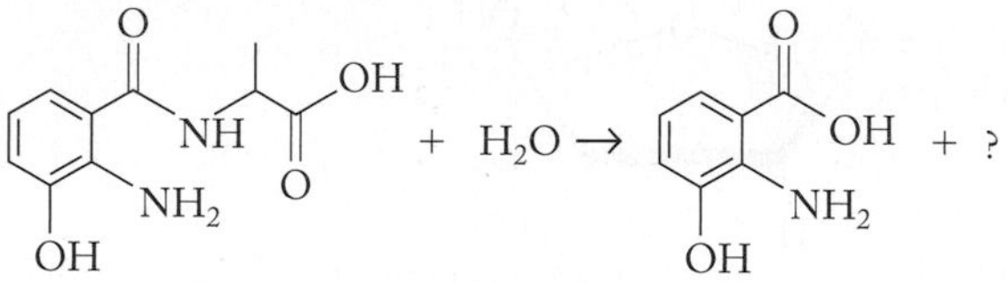

FIGURE P20.26

Protein Structure and Function

CONCEPT REVIEW

20.27. Which of the four levels of protein structure is most closely associated with the sequence of amino acids in a protein?

20.28. Which type of intermolecular interaction plays the dominant role in holding strands of proteins together in β-pleated sheets and stabilizing α-helices?

20.29. Which level of protein structure is associated with ion–ion interactions and disulfide bond formation?

20.30. Protein structure and function can be influenced by hydrophobic interactions between R groups. Which amino acids contribute to these interactions?

20.31. Describe the *induced-fit* theory of enzyme activity.

20.32. Describe two ways in which an inhibitor can reduce the catalytic properties of an enzyme.

20.33. Describe the role that molecular structure plays in the specificity of enzyme activity.

20.34. The rates of most reactions increase with increasing temperature, but above a critical temperature the rate of an enzyme-mediated reaction decreases with increasing temperature. Explain why.

20.35. When protein strands fold back on themselves in forming stable tertiary structures, lysine residues are often paired up with glutamic acid residues. Why?

20.36. Ion–ion interactions are particularly effective at stabilizing tertiary structures of proteins. Suggest a pair of amino acid residues that would be attracted to each other via ion–ion interactions at pH 7.4.

Carbohydrates

CONCEPT REVIEW

20.37. What are the structural differences between starch and cellulose?

20.38. Why is the discovery of enzymes that catalyze cellulose hydrolysis a worthwhile objective?

20.39. Is the fuel value of glucose in the linear form the same as that in the cyclic form?

*20.40.** Without doing the actual calculation, estimate the fuel values of glucose and starch by considering average bond energies. Do you predict the fuel values of the two substances to be the same or different?

20.41. The second step in glycolysis converts glucose 6-phosphate into fructose 6-phosphate. Can you think of a reason why $\Delta G°$ for this reaction is close to zero?

*20.42.** Why is it important that at least some of the steps in glycolysis convert ADP to ATP?

20.43. How do we calculate the overall free-energy change of a process consisting of two steps?

20.44. During glycolysis a monosaccharide is converted to pyruvate. Do you think this process produces an increase or decrease in the entropy of the system? Explain your answer.

PROBLEMS

20.45. Draw a diagram that shows how isomers of the sugar shown as a flat Kekulé structure in Figure P20.45 result when the linear molecule forms a 6-atom ring.

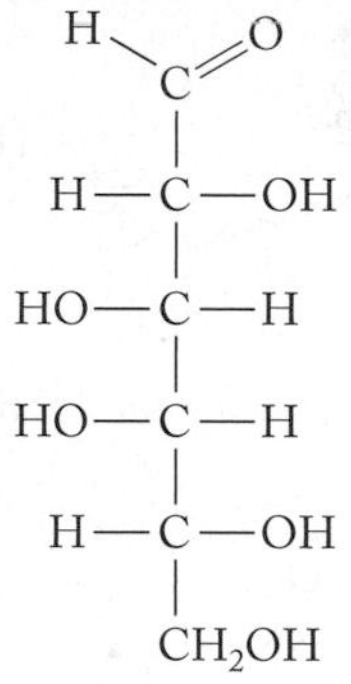

Galactose

FIGURE P20.45

20.46. Draw a diagram that shows how isomers of the sugar shown as a flat Kekulé structure in Figure P20.46 result when the linear molecule forms a 5-atom ring.

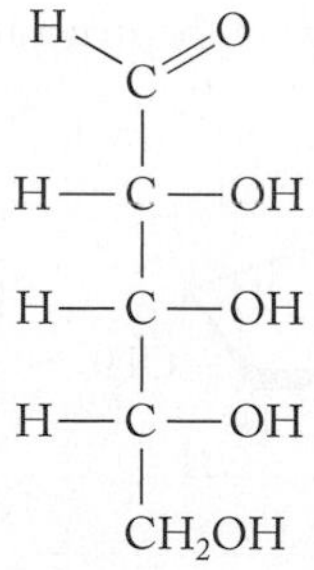

Ribose

FIGURE P20.46

20.47. Which, if any, of the structures in Figure P20.47 are β isomers of a monosaccharide?

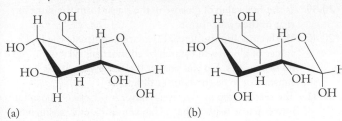

(a)   (b)

(c)

FIGURE P20.47

20.48. Which, if any, of the structures in Figure P20.48 are β isomers?

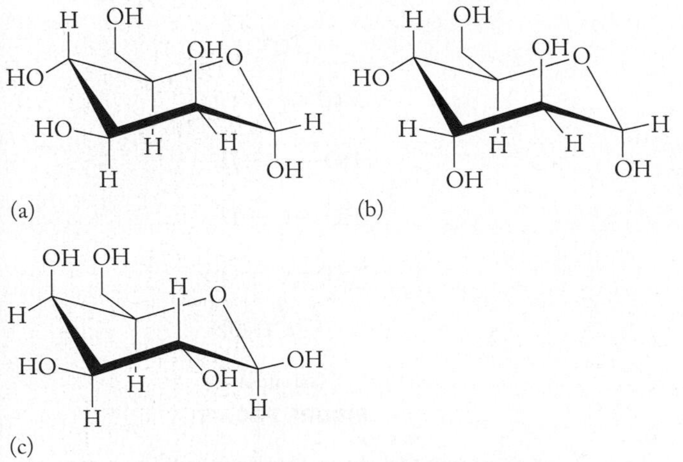

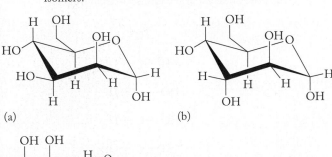

(a)   (b)

(c)

FIGURE P20.48

20.49. Which, if any, of the structures in Figure P20.49 are α isomers?

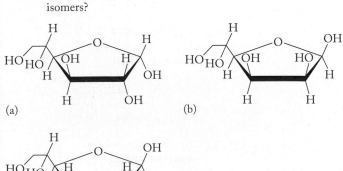

(a)   (b)

(c)

FIGURE P20.49

20.50. Which, if any, of the structures in Figure P20.50 are α isomers?

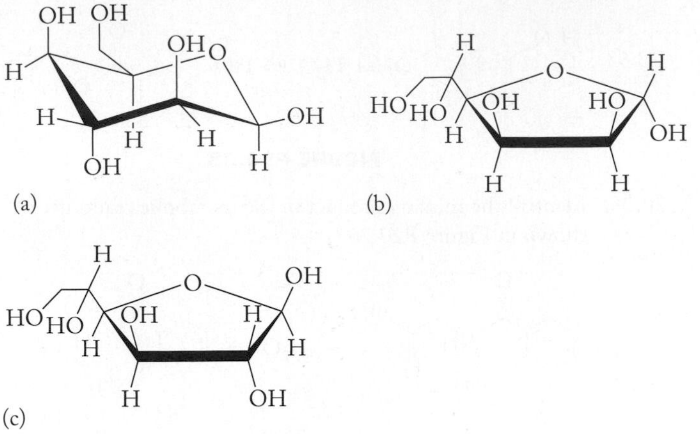

(a)   (b)

(c)

FIGURE P20.50

20.51. Which of the saccharides in Figure P20.51 is digestible by humans?

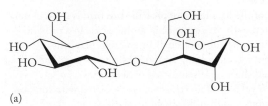

(a)

(b)

(c)

FIGURE P20.51

*20.52. For any of the disaccharides in Problem 20.51 that are not digestible by humans, draw an isomer that would be digested.

20.53. The structure of the disaccharide maltose appears in Figure P20.53. Hydrolysis of 1 mol of maltose ($\Delta G_f^\circ = -2246.6$ kJ/mol) produces 2 mol of glucose ($\Delta G_f^\circ = -1274.4$ kJ/mol):

$$\text{Maltose} + H_2O \rightarrow 2 \text{ glucose}$$

If the value of ΔG_f° of water is -285.8 kJ/mol, what is the change in free energy of the hydrolysis reaction?

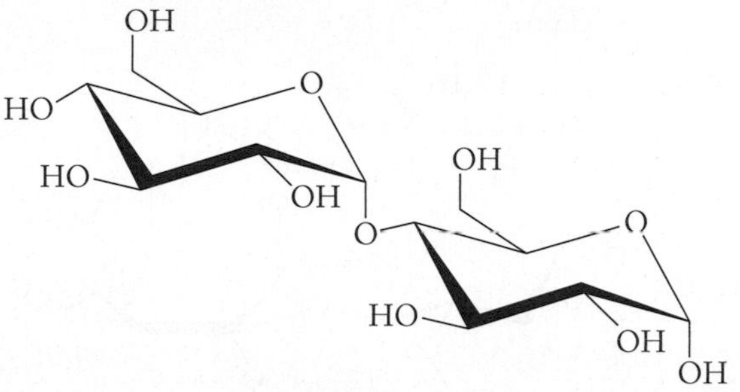

Maltose

FIGURE P20.53

20.54. If the maltose in Problem 20.53 were replaced by another disaccharide, would you expect the free-energy change for the hydrolysis to be exactly the same or just similar in value? Explain your answer.

Lipids

CONCEPT REVIEW

20.55. What is the difference between a saturated and an unsaturated fatty acid?

20.56. Why are the average fuel values of fats higher than those of carbohydrates and proteins?

20.57. **Polar Exploration** Some Arctic explorers have eaten sticks of butter on their explorations. Give a nutritional reason for this unusual cuisine.

20.58. **Salad Dressing** Salad dressings containing oil and vinegar quickly separate on standing. Explain the observed separation of layers based on the structure and properties of aqueous vinegar and oil.

20.59. Do triglycerides have a chiral center? Explain your answer.

*20.60. Using your knowledge of molecular geometry and intermolecular forces, why might polyunsaturated triglycerides be more likely to be liquid than saturated triglycerides?

PROBLEMS

20.61. Which of the triglycerides in Figure P20.61 are unsaturated fats?

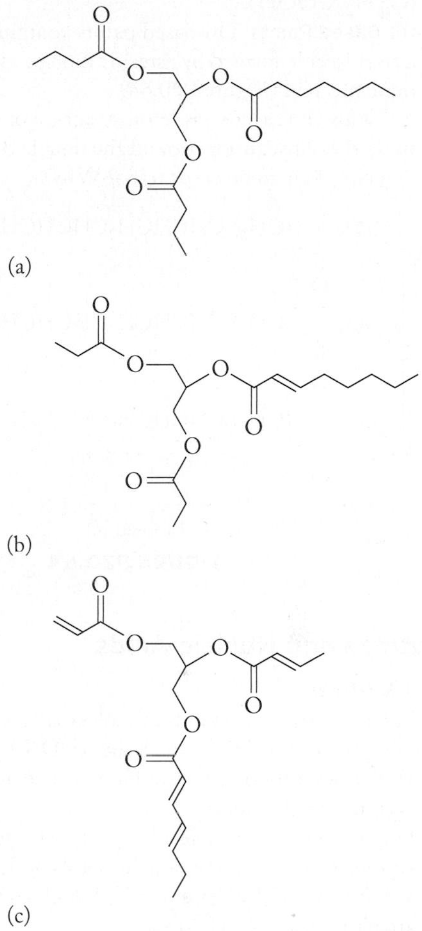

(a)

(b)

(c)

FIGURE P20.61

20.62. For each of the pairs of fatty acids in Figure P20.62, indicate whether they are structural isomers, stereoisomers, or unrelated compounds.

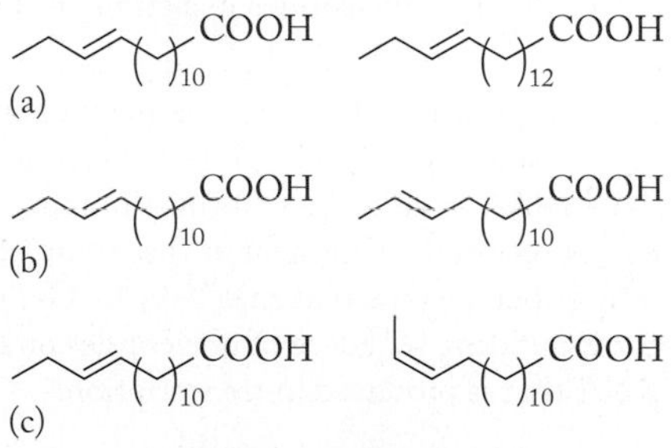

(a)

(b)

(c)

FIGURE P20.62

20.63. Draw the structures of the three fats formed by reaction of glycerol with (a) octanoic acid ($C_7H_{15}COOH$), (b) decanoic acid ($C_9H_{19}COOH$), and (c) dodecanoic acid ($C_{11}H_{23}COOH$).

20.64. Oil-Based Paints Oil-based paints contain linseed oil, a triglyceride formed by esterification of glycerol with linolenic acid (Figure P20.64).
   a. Draw the carbon-skeleton structure of linolenic acid.
   *b. Is the substitution around the double bonds in linolenic acid likely to be cis or trans? Why?

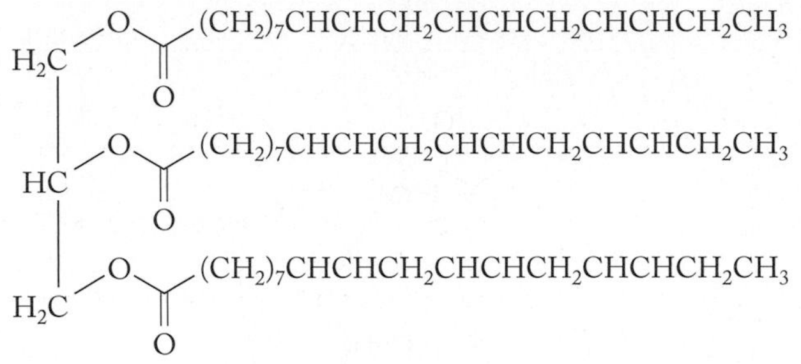

Linseed oil
FIGURE P20.64

Nucleotides and Nucleic Acids

CONCEPT REVIEW

20.65. What are the three kinds of molecular subunits in DNA? Which two form the "backbone" of DNA strands?

20.66. Why does a codon consist of a sequence of three, and not two, ribonucleotides?

20.67. What kind of intermolecular force holds together the strands of DNA in the double-helix configuration?

20.68. What is meant by "base pairing" in DNA? Which bases are paired?

PROBLEMS

20.69. Draw the structure of adenosine 5′-monophosphate, one of the four ribonucleotides in a strand of RNA.

20.70. Draw the structure of deoxythymidine 5′-monophosphate, one of the four nucleotides in a strand of DNA.

20.71. In the replication of DNA, a segment of an original strand has the sequence T-C-G-G-T-A. What is the sequence of the double-stranded helix formed in replication?

20.72. In transcription, a segment of the strand of DNA that is transcribed has the sequence T-C-G-G-T-A. What is the corresponding sequence of nucleotides on the messenger RNA that is produced in transcription?

Additional Problems

20.73. Olestra Olestra is a calorie-free fat substitute. The core of the olestra molecule (Figure P20.73) is a disaccharide that has reacted with a carboxylic acid; this results in the conversion of hydroxyl groups on the disaccharide into the depicted structure.
   a. What is the name of the disaccharide core of the olestra molecule?
   b. What functional group has replaced the hydroxyl groups on the disaccharide?
   c. What is the formula of the carboxylic acid used to make olestra?

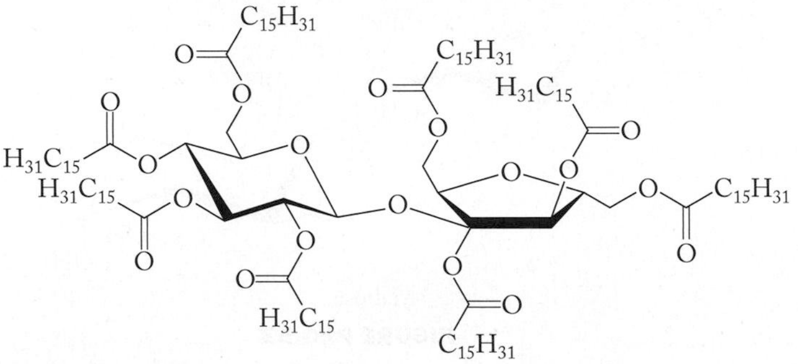

Olestra
FIGURE P20.73

20.74. When scientists at UC Santa Cruz directed UV radiation at an ice crystal containing methanol, ammonia, and hydrogen cyanide, three amino acids (glycine, alanine, and serine) were detected among the products of photochemical reactions. The formation of these amino acids suggests that they may also be synthesized in comets approaching the sun (and Earth). Determine the standard free-energy change of the hypothetical formation of glycine in comets, using standard free energies of formation of the reactants and products in this reaction [ΔG_f° for HCN(g) is +125 kJ/mol; ΔG_f° for solid glycine is −368.4 kJ/mol; other ΔG_f° values are in Appendix 4].

$$CH_3OH(\ell) + HCN(g) + H_2O(\ell) \rightarrow$$
$$H_2NCH_2COOH(s) + H_2(g)$$

20.75. Homocysteine (Figure P20.75) is formed during the metabolism of amino acids. A mutation in some people's genes leads to high concentrations of homocysteine in the blood and a consequent increase in their risk of heart disease and incidence of bone fractures in old age.

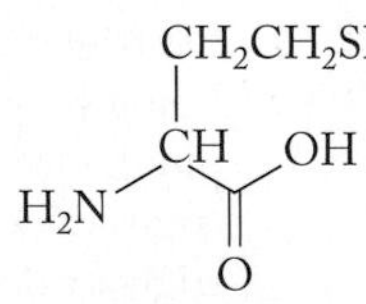

Homocysteine
FIGURE P20.75

   a. What is the structural difference between homocysteine and cysteine?
   b. Cysteine is a chiral compound. Is homocysteine chiral?

20.76. Molecules in Meteorites Some scientists believe life on Earth can be traced to amino acids and other molecules brought to Earth by comets and meteorites. In 2004, a new class of amino acids called diamino acids (Figure P20.76) were found in the Murchison meteorite.
a. Which of these diamino acids is not an α-amino acid?
b. Which of these amino acids is chiral?

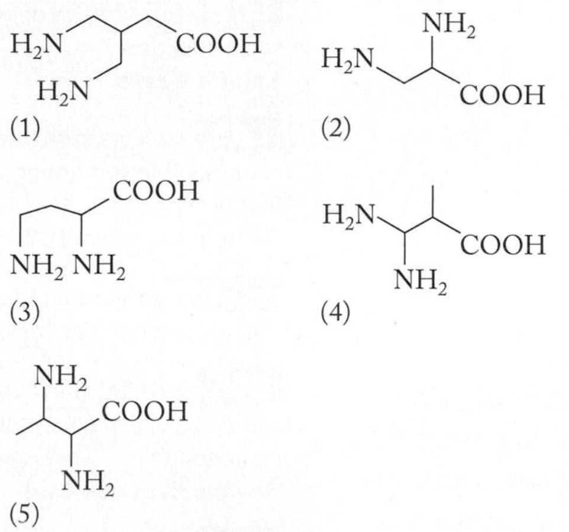

(1)
(2)
(3)
(4)
(5)

FIGURE P20.76

20.77. Ackee Ackee, the national fruit of Jamaica, is a staple in many Jamaican diets. Unfortunately, a potentially fatal sickness known as Jamaican vomiting disease is caused by the consumption of unripe ackee fruit, which contains the amino acid hypoglycin (Figure P20.77). Is hypoglycin an α-amino acid?

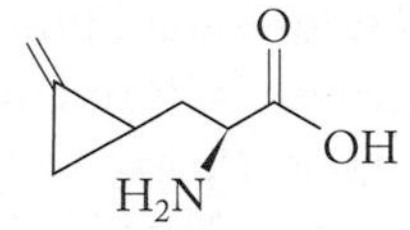

Hypoglycin

FIGURE P20.77

20.78. In late 2003, researchers at the Scripps Research Institute reported the development of genetically modified *E. coli* that could incorporate five new amino acids into proteins. These five amino acids, shown in Figure P20.78, are not among the 20 naturally occurring amino acids. Which naturally occurring amino acids are these compounds most similar to?

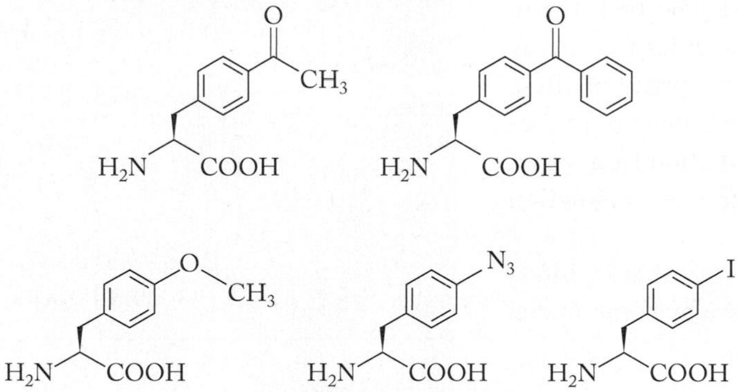

FIGURE P20.78

20.79. Creatine Creatine (Figure P20.79) is an amino acid produced by the human body. Body builders sometimes take creatine supplements to help gain muscle strength. A 2003 study reported that creatine may boost memory and cognitive thinking.
a. Is creatine an α-amino acid?
b. Draw the two dipeptides that can be formed from glycine and creatine.

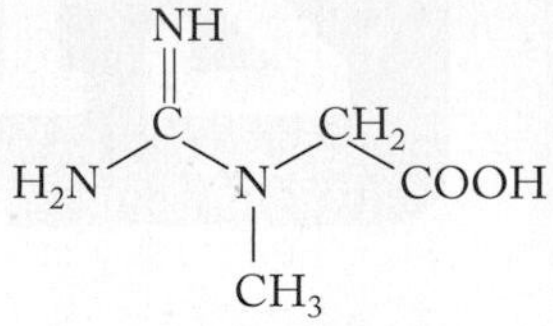

Creatine

FIGURE P20.79

20.80. In response to specific neural messages, the human hypothalamus may secrete a number of polypeptides including the tripeptide shown in Figure P20.80.
a. Sketch the structures of the three amino acids that combine to make this tripeptide.
b. Which, if any, of the constituent amino acids are among the 20 α-amino acids in proteins?

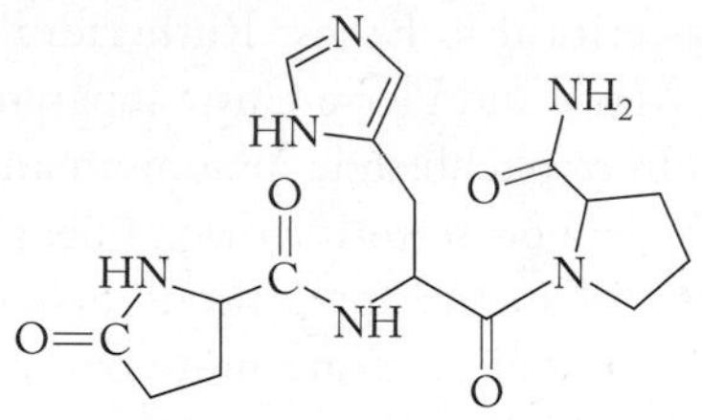

Thyrotropin-releasing factor

FIGURE P20.80

20.81. Glutathione (Figure P20.81) is an essential molecule in the human body. It acts as an activator for enzymes and protects lipids from oxidation. Which three amino acids combine to make glutathione?

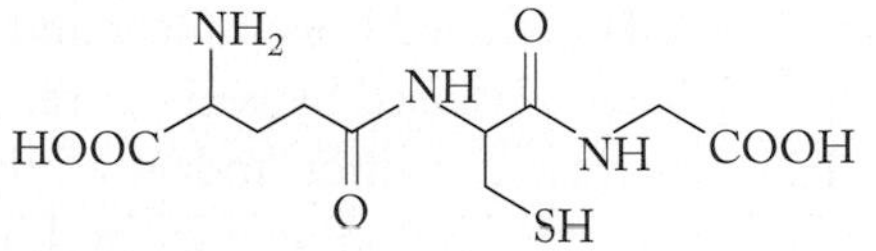

Glutathione

FIGURE P20.81

*20.82. Polyunsaturated fats have a tendency to polymerize. What kind of polymerization reaction might account for this observation?

*20.83. Without doing the actual calculation, estimate the fuel values of leucine and isoleucine by considering average bond energies. Should the fuel values of the two amino acids be the same? Actual calorimetric measurements show that isoleucine has a lower fuel value than leucine. Explain why.

20.84. Sucralose (see Figure P20.9) is about 600 times sweeter than sucrose (see Figure 20.22). All substances that taste sweet have functional groups that form hydrogen bonds with "sweetness" receptor sites on the tongue. What does the difference in sweetness between sucralose and sucrose tell you about additional intermolecular interactions between sweet compounds and receptor sites that contribute to their sweet taste?

21

Nuclear Chemistry

The Risks and Benefits

The Age of Radioactivity

In Section 2.1 we briefly described how Henri Becquerel's 1896 discovery of the radioactivity emitted by pitchblende, a uranium-containing mineral, was critical to Ernest Rutherford's efforts to unravel atomic structure. In 1898, Marie and Pierre Curie separated and purified two new radioactive elements from pitchblende, polonium and radium. These experiments and others launched a new scientific discipline: radiochemistry.

Interest in the applications of radium grew rapidly. Radium-containing paint was used to create luminous dials for watches and gauges. Ointments containing radium were prescribed as treatments for skin lesions. People were encouraged to visit health spas such as those in Saratoga Springs, New York, where the waters contained low concentrations of dissolved radium salts. Proliferation of radium-containing products also led to the discovery that nuclear radiation is dangerous. Young women hired to paint the dials of watches during World War I were instructed to lick the tips of their brushes, inadvertently introducing radium into their teeth and bones. Within a few years, many of these women developed bone cancer and died. Marie Curie herself succumbed to aplastic anemia caused by years of research with radioactive materials.

In recent years, a better understanding of the biological impacts of radiation has led to the development of methods for diagnosing and treating disease that minimize the hazards of radiation to patients and those who treat them. Nuclear reactors provide doctors and scientists with radionuclides that decay by predictable pathways within minutes or hours. Selective uptake of these nuclides by different organs in the body allows doctors to evaluate organ function and prescribe treatment when function is impaired. Radiation from other nuclides that concentrate in cancerous tissues can be used, often in conjunction with other therapies, to destroy malignant tumors.

In this chapter we examine the origins of nuclear reactions; that is, interactions that take place in the nuclei of atoms. We address why some nuclei

Positron Emission Tomography (PET) Cancer cells have abnormally high rates of glucose metabolism. In this photograph, a patient has been given glucose labeled with a positron emitter that is incorporated into tissues with a high metabolic demand for glucose. The colored regions reveal both normal brain function as well as abnormalities in the patient's cranial mediastinal lymph nodes. ▶

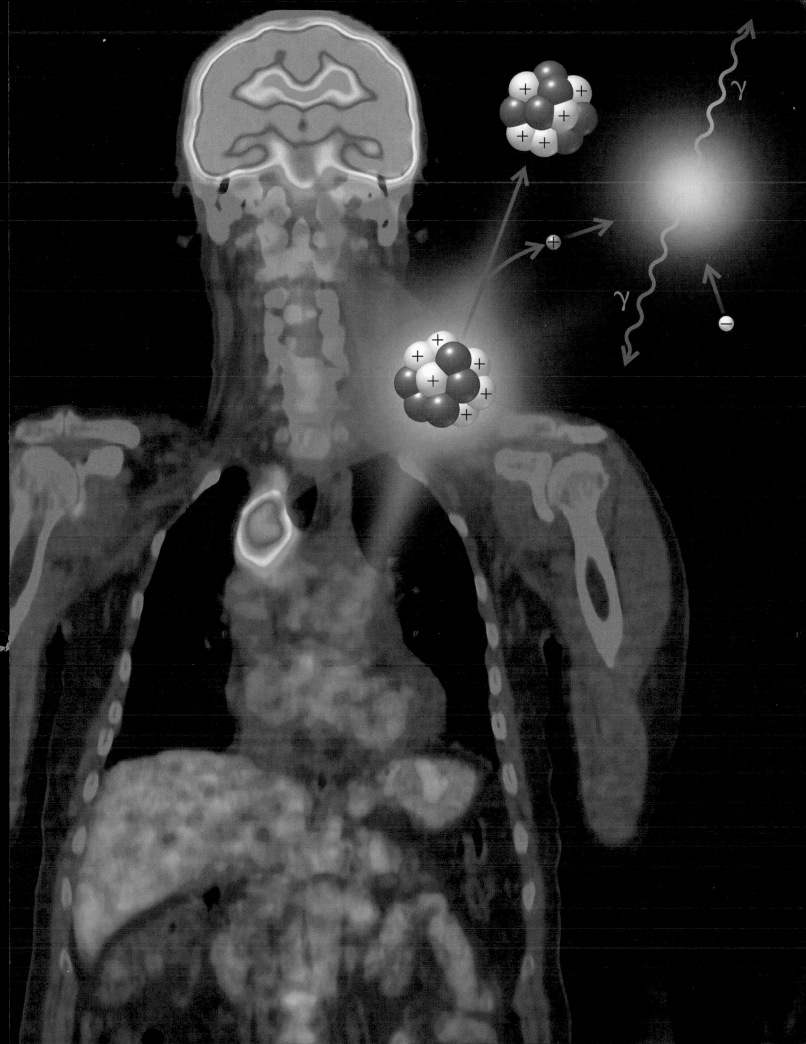

nuclear chemistry the study of reactions that involve changes in the nuclei of atoms.

nuclear fusion a nuclear reaction in which subatomic particles or atomic nuclei collide with each other at very high speeds and fuse together, forming more massive nuclei and releasing energy.

are stable and others are not, the kinds of spontaneous nuclear reactions that unstable nuclei undergo, and how the products of these reactions can be used to generate electrical power and to diagnose and treat disease. We look at how harnessing nuclear reactions similar to those that fuel the sun might someday meet the world's energy needs safely and sustainably. We also address some of the dangers radioactive substances pose to human health and how we can shield ourselves from them.

21.1 Changing the Identities of Atoms

Throughout this book we have seen that the identities of atoms remain unchanged in chemical reactions, in keeping with the law of conservation of mass. Now we turn to *nuclear* reactions, in which atoms' identities *do* change—because their nuclei change. The field of chemistry that studies such reactions is called **nuclear chemistry**.

We introduced some of the basic principles of nuclear chemistry in Chapter 2 as we discussed the rapid transformation of energy into matter following the Big Bang. Such transformations are described by Einstein's famous equation,

$$E = mc^2 \tag{2.5}$$

where E is the amount of energy transformed into matter (or vice versa), m is the mass of that matter, and c is the speed of light. The magnitude of the speed of light (squared) means that nuclear reactions that result in minor changes in mass are accompanied by enormous changes in energy, much greater than those that accompany chemical reactions.

We learned in Chapter 2 that primordial nucleosynthesis created an early universe mainly composed of two elements: hydrogen and helium. More massive nuclei containing up to 26 protons in each nucleus—that is, the nuclei of iron atoms—formed later in the cores of the giant stars that populated the first generation of galaxies. Even more massive nuclei formed during the enormous releases of energy that mark the deaths of giant stars, events we know as supernovae and have witnessed, from a safe distance and with a considerable time delay, here on Earth (see Figure 2.24). We begin this chapter with an investigation of a class of nuclear reactions that has played a central role in both primordial and stellar nucleosynthesis and the formation of the universe around us today: nuclear fusion.

CONNECTION We explored primordial and stellar nucleosynthesis in Section 2.6.

21.2 Fusion and the Quest for Clean Energy

The energy of our sun is derived from the high-speed collision and fusion of hydrogen nuclei to form helium. This **nuclear fusion** process involves more steps than the process that probably took place during primordial nucleosynthesis when protons (hydrogen nuclei) and neutrons fused together to form deuterons (nuclei of the hydrogen isotope deuterium, D). We can describe this process with the following nuclear equation:

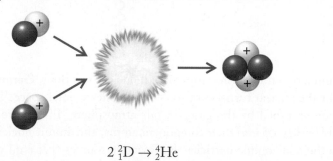

$$\,^1_1p + \,^1_0n \rightarrow \,^2_1D \tag{2.3}$$

Once deuterons formed, they also collided with each other and fused together, forming alpha particles, which are the nuclei of helium-4 atoms:

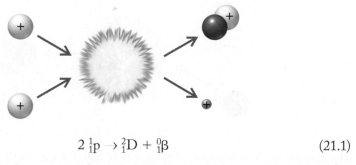

$$2\,^2_1D \rightarrow \,^4_2He \tag{2.4}$$

Recall from Chapter 2 that each subscript in a nuclear equation represents the electrical charge of a particle and each superscript represents the particle's mass number. When a particle is the nucleus of an atom, its electrical charge is equal to the number of protons in it, that is, its atomic number.

Hydrogen fusion in our sun follows a different path because free neutron concentrations are far lower there than they were in the primordial universe. In the sun, colliding protons may fuse together to form a deuteron and a second particle with hardly any mass and a charge of 1+:

CONNECTION We learned how to write and balance nuclear equations in Section 2.6.

CHEMTOUR Balancing Nuclear Equations

$$2\,^1_1p \rightarrow \,^2_1D + \,^0_1\beta \tag{21.1}$$

We use the electron (beta particle) symbol "β" for this tiny positive particle, but the positive charge on it means that it is actually a "positive electron" or **positron**. Positrons belong to a group of subatomic particles that have the opposite charge of particles typically found in atoms. In addition to these electrons with positive charges, there are protons with negative charges, called *antiprotons*, $\,^1_{-1}p$. These charge opposites are particles of **antimatter**.

Particles of matter and their antimatter opposites are like mortal enemies. If they collide, they instantly annihilate each other. In their mutual destruction, they cease to exist as matter, and all of their mass is released as energy in an amount predicted by Einstein's equation. The positrons produced from hydrogen

positron a particle with the mass of an electron but with a positive charge.

antimatter particles that are the charge opposites of normal subatomic particles.

CHEMTOUR Fusion of Hydrogen

fusion are rapidly annihilated in reactions with electrons. The sole product of the reaction is energy in the form of two or more gamma (γ) rays:

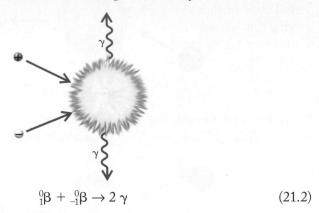

CONNECTION Gamma rays are the highest energy form of electromagnetic radiation (see Figure 3.1).

$$^{0}_{1}\beta + ^{\ 0}_{-1}\beta \rightarrow 2\ \gamma \qquad (21.2)$$

Gamma ray emission accompanies all nuclear reactions. Gamma rays are generated by the nuclear furnaces of stars and permeate outer space. Those that reach Earth are absorbed by the gases in our atmosphere. In the process, molecular gases are broken up into their component atoms, and atomic nuclei may be broken up into their subatomic particles. Radioactive sources that emit gamma rays are also used by oncologists to break up the molecules in, and kill, cancer cells.

CONCEPT TEST

What particle is formed when a proton fuses with an electron?

(Answers to Concept Tests are in the back of the book.)

In the second stage of solar fusion, protons fuse with deuterons to form helium-3 nuclei:

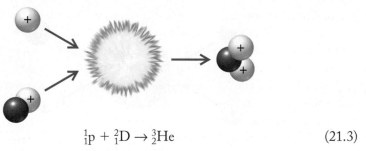

$$^{1}_{1}p + ^{2}_{1}D \rightarrow ^{3}_{2}He \qquad (21.3)$$

The superscript 3 indicates that the particle has 3 nucleons (2 protons and 1 neutron). Recall from Chapter 2 that any atom with 2 protons in its nucleus is by definition a helium atom, and that atoms of the same element with different numbers of nucleons are called isotopes. Finally, fusion of two helium-3 nuclei produces a helium-4 nucleus and 2 protons:

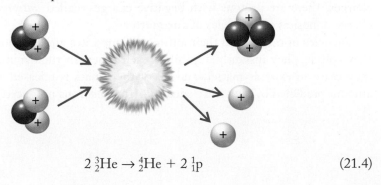

$$2\ ^{3}_{2}He \rightarrow ^{4}_{2}He + 2\ ^{1}_{1}p \qquad (21.4)$$

Deuterium and ^{3}He nuclei are intermediates in the hydrogen-fusion process because they are made in one step but then consumed in another. To write an overall equation for solar fusion, we combine Equations 21.1, 21.3, and 21.4, multiplying Equations 21.1 and 21.3 by 2 to balance the production and consumption of the intermediate particles:

$$2\,[2\,{}^1_1p \rightarrow {}^2_1D + {}^0_1\beta]$$
$$+\ 2\,[{}^1_1p + {}^2_1D \rightarrow {}^3_2He]$$
$$+\ \ \ \ \ 2\,{}^3_2He \rightarrow {}^4_2He + 2\,{}^1_1p$$

$$\overline{4\ 6\,{}^1_1p + \cancel{2\,{}^2_1D} + \cancel{2\,{}^3_2He} \rightarrow \cancel{2\,{}^2_1D} + \cancel{2\,{}^3_2He} + {}^4_2He + \cancel{2\,{}^1_1p} + 2\,{}^0_1\beta}$$

which reduces to

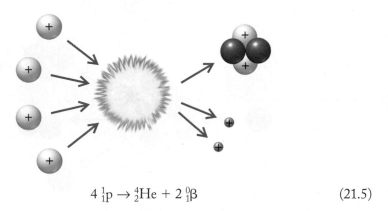

$$4\,{}^1_1p \rightarrow {}^4_2He + 2\,{}^0_1\beta \qquad (21.5)$$

Annihilation reactions between the positrons produced in Equation 21.5 and electrons in the matter surrounding the reactants release considerable energy, but most of the energy from hydrogen fusion comes from the loss in mass as four protons are transformed into an alpha particle (that is, a helium-4 nucleus) and two positrons (Table 21.1). In Sample Exercise 21.1 we will use Einstein's equation, $E = mc^2$, to calculate how much energy this is.

For decades scientists and engineers have sought to harness the enormous energy released during hydrogen fusion for peaceful purposes. In 2009 construction began on ITER (originally an acronym for International Thermonuclear Experimental Reactor), a project to build the world's largest nuclear fusion reactor. Located at the Cadarache facility in southern France, the project is funded and run by the European Union, India, Japan, China, Russia, South Korea, and the United States. When it is operational, ITER will use a device called a *tokamak* (Figure 21.1) to heat a mixture of deuterium ($^2_1$H) and tritium ($^3_1$H) to temperatures near 1.5×10^8 K. At such temperatures all these atoms are ionized, forming an incandescent plasma that is confined by the tokamak's powerful magnets to the center of a donut-shaped tunnel. High-speed collisions between deuterium and tritium nuclei in the plasma produce nuclei of helium-4:

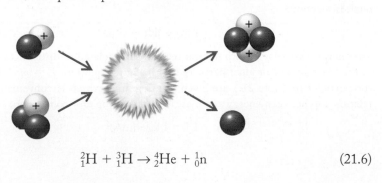

$${}^2_1H + {}^3_1H \rightarrow {}^4_2He + {}^1_0n \qquad (21.6)$$

∞● CONNECTION In Section 2.6 we used Einstein's equation to calculate the binding energy of the helium nucleus based on the mass defect (Δm) between it and its constituent nucleons.

TABLE 21.1	**Symbols and Masses of Subatomic Particles and Small Nuclei**	
Particle	**Symbol**	**Mass (kg)**
Neutron	${}^1_0n$	1.67493×10^{-27}
Proton	${}^1_1p$ or ${}^1_1H$	1.67262×10^{-27}
Electron (β particle)	${}^0_{-1}\beta$ or ${}^0_{-1}e$	9.10938×10^{-31}
Deuteron	${}^2_1D$ or ${}^2_1H$	3.34370×10^{-27}
α particle	${}^4_2\alpha$ or ${}^4_2He$	6.64465×10^{-27}
Positron	${}^0_1\beta$	9.10938×10^{-31}

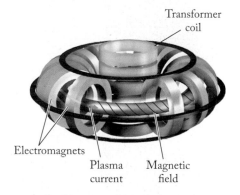

FIGURE 21.1 A tokamak transmits electrical energy into a toroidal (donut-shaped) chamber containing deuterium and tritium, causing these isotopes of hydrogen to ionize and form a plasma of nuclei and free electrons with a temperature above 10^8 K. Combinations of electromagnets confine the plasma to the interior of the torus, where collisions between ^{2}H and ^{3}H nuclei result in their fusing together, forming ^{4}He nuclei and free neutrons. The neutrons then collide with Li atoms in the walls of the chamber, initiating additional nuclear reactions that produce more tritium fuel.

The neutrons produced in the reaction collide with the nuclei of Li atoms in "breeder" blankets surrounding the hydrogen plasma. Two nuclear reactions are initiated by these collisions depending on which isotope of Li is involved:

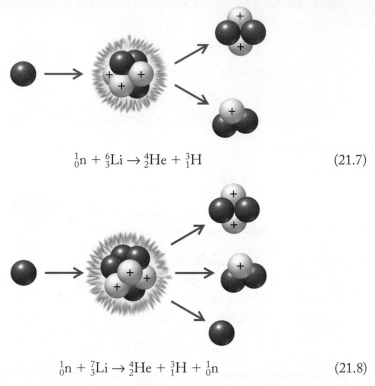

$$\,^1_0n + \,^6_3Li \rightarrow \,^4_2He + \,^3_1H \qquad (21.7)$$

$$\,^1_0n + \,^7_3Li \rightarrow \,^4_2He + \,^3_1H + \,^1_0n \qquad (21.8)$$

Note that the reactions in Equations 21.7 and 21.8 produce tritium nuclei. Thus, these reactions supply more fuel for the primary fusion reaction. The world's supply of deuterium, the other fuel, is enormous (seawater contains about 15 mg of deuterium per kilogram). On the other hand, tritium is not abundant in nature because it is radioactive with a half-life of only 12.3 years. One disadvantage of these reactions is their reliance on lithium during a time when expanding production of lithium–ion batteries (see Section 17.7) is increasing our demand for the element.

**SAMPLE EXERCISE 21.1   Calculating the Energy Released   LO1
in a Nuclear Reaction**

How much energy in joules is released by the overall fusion process in which four protons undergo nuclear fusion, producing an α particle and two positrons (Equation 21.5)?

COLLECT AND ORGANIZE We are asked to calculate the energy released in this nuclear reaction:

$$4\,^1_1p \rightarrow \,^4_2He + 2\,^0_1\beta \qquad (21.5)$$

The energies associated with nuclear reactions are related to differences in the masses of the reactant and product particles by Einstein's equation, $E = mc^2$. The masses of the particles in Table 21.1 are given in kilograms, which is convenient because the relationship between energy and mass is linked to the unit conversion:

$$1\,J = 1\,kg \cdot (m/s)^2$$

ANALYZE Given the value of the masses in Table 21.1, the difference in mass will probably be less than 10^{-27} kg. When multiplied by the square of the speed of light $(2.998 \times 10^8$ m/s$)^2 \approx 10^{17}$, the calculated value of E should be less than 10^{-10} J.

SOLVE First we calculate the change in mass:

$$\Delta m = (m_{\alpha\ particle} + 2\ m_{positron}) - 4m_{proton}$$

$$= (6.64465 \times 10^{-27} + 2 \times 9.10938 \times 10^{-31})\ kg - 4 \times 1.67262 \times 10^{-27}\ kg$$

$$= -4.40081 \times 10^{-29}\ kg$$

The energy corresponding to this loss in mass is calculated using Einstein's equation where $m = -4.40081 \times 10^{-29}$ kg:

$$E = mc^2$$

$$= -4.40081 \times 10^{-29}\ kg \times (2.998 \times 10^8\ m/s)^2$$

$$= -3.955 \times 10^{-12}\ kg \cdot (m/s)^2 = -3.955 \times 10^{-12}\ J$$

THINK ABOUT IT The decrease in mass translates into energy lost by the reaction system to its surroundings. As we predicted, the absolute value of this energy is less (actually much less) than 10^{-10} J, which seems like an awfully small value compared to the world's energy needs. However, this value applies to the formation of a single α particle. If we multiply it by Avogadro's number and convert to a value in kilojoules per mole, a unit we typically use in thermochemistry, we get

$$\frac{-3.955 \times 10^{-12}\ \cancel{J}}{\cancel{particle}} \times \frac{6.0221 \times 10^{23}\ \cancel{particles}}{1\ mol} \times \frac{1\ kJ}{1000\ \cancel{J}} = -2.382 \times 10^9\ kJ/mol$$

To put this value in perspective, it is about 10^7 times the change in free energy from the combustion of one mole of hydrogen gas.

Practice Exercise How much energy is released in the nuclear reaction described by Equation 21.6? Express your answer in kJ/mol. NOTE: The mass of a tritium nucleus is 5.00827×10^{-27} kg.

(Answers to Practice Exercises are in the back of the book.)

21.3 The Belt of Stability

The values of the atomic masses and mass numbers of the elements in the periodic table tell us about the ratios of neutrons to protons in the nuclei of their stable isotopes. The lighter elements have atomic masses that are about twice their atomic numbers and have neutron-to-proton ratios close to unity. For example, ^{12}C has 6 neutrons and 6 protons, and most oxygen atoms have 8 neutrons and 8 protons.

However, with increasing values of Z, the ratios of neutrons to protons increase. This trend is illustrated in Figure 21.2, where the green dots represent combinations of neutrons and protons that form stable nuclides. The band of green dots runs diagonally through the graph, defining the **belt of stability**. Note how the belt curves upward away from the purple (n = p) straight line. This curvature shows how the neutron-to-proton ratios increase from about 1:1 for the lightest stable nuclides to about 1.5:1 for the most massive ones.

The nuclides represented by orange dots in Figure 21.2 are *radionuclides*. They are not stable but instead undergo **radioactive decay**, the spontaneous disintegration of radioactive nuclei accompanied by the release of nuclear radiation. Their

belt of stability the region on a graph of number of neutrons versus number of protons that includes all stable nuclei.

radioactive decay the spontaneous disintegration of unstable particles accompanied by the release of radiation.

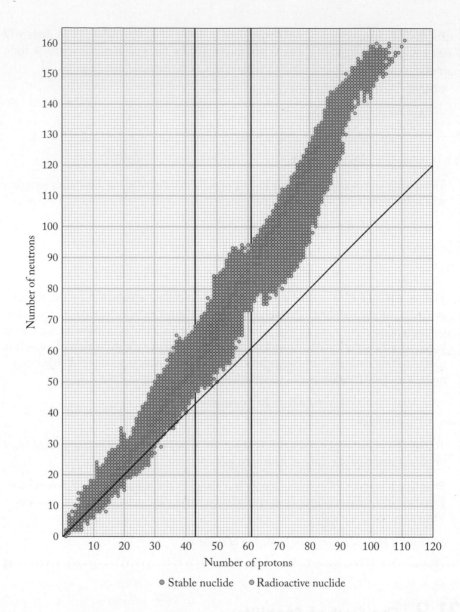

FIGURE 21.2 The belt of stability. Green dots represent stable combinations of protons and neutrons. Orange dots represent known radioactive (unstable) nuclides. Nuclides that fall along the purple line have equal numbers of neutrons and protons. Note that there are no stable nuclides (no green dots) for $Z = 43$ (technetium) and $Z = 61$ (promethium), as indicated by the two vertical red lines. These elements are the only two among the first 83 that are not found in nature.

• Stable nuclide   • Radioactive nuclide

mode of radioactive decay depends on whether they are above or below the belt of stability. Those radionuclides above the belt of stability, such as carbon-14, are *neutron rich* and tend to undergo decay reactions that reduce their neutron-to-proton ratio—specifically, β decay. For example, when ^{14}C undergoes β decay, its atomic number increases by one while its mass number stays the same. A neutron in its nucleus becomes a proton as a result of this nuclear reaction, and a high-energy electron is ejected from the nucleus:

CONNECTION We defined β decay when we explored the formation of elements more massive than iron in the cores of collapsing stars (Section 2.6).

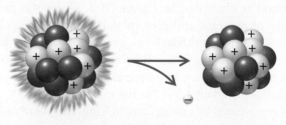

$$^{14}_{6}C \rightarrow {}^{14}_{7}N + {}^{0}_{-1}\beta$$

Nuclides below the belt of stability are *neutron poor* and undergo positron emission and electron capture, decay processes that *increase* their neutron-to-proton ratio. As its name implies, **positron emission** involves the release of a positron from a nucleus. As a result, the nucleus that is produced contains one less proton and one more neutron. Carbon-11 is a nuclide that undergoes positron emission:

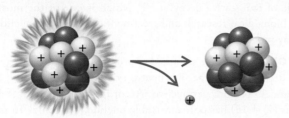

$$^{11}_{6}C \rightarrow ^{11}_{5}B + ^{0}_{1}\beta$$

The boron-11 produced in this reaction is a stable isotope. In fact, 80.2% of all boron atoms in nature are boron-11.

A carbon-11 atom can also increase its neutron-to-proton ratio by drawing one of its six electrons into its nucleus. When it does, the negatively charged electron combines with a positively charged proton. The product of this combination reaction is a nucleon with zero charge, that is, a neutron. The effect of this **electron capture** process is the same as positron emission: the number of protons *decreases* by one and the number of neutrons *increases* by one, so that again the product is boron-11.

$$^{11}_{6}C + ^{0}_{-1}e \rightarrow ^{11}_{5}B$$

Table 21.2 illustrates the impact of being neutron rich or neutron poor on some isotopes of carbon. Note that carbon has two stable isotopes: ^{12}C and ^{13}C. The isotopes with mass numbers greater than 13 are neutron rich and undergo beta decay. Those with mass numbers less than 12 are neutron poor and undergo either positron emission or electron capture.

positron emission the spontaneous emission of a positron from a neutron-poor nucleus.

electron capture a nuclear reaction in which a neutron-poor nucleus draws in one of its surrounding electrons, which transforms a proton in the nucleus into a neutron.

▶❙❙ **CHEMTOUR** Radioactive Decay Modes

TABLE 21.2	**Isotopes of Carbon and Their Radioactive Decay Products**			
Name	**Symbol**	**Mode(s) of Decay**	**Half-Life**	**Natural Abundance (%)**
Carbon-10	$^{10}_{6}C$	Positron emission	19.45 s	—
Carbon-11	$^{11}_{6}C$	Positron emission, electron capture	20.3 min	—
Carbon-12	$^{12}_{6}C$	—	(Stable)	98.89
Carbon-13	$^{13}_{6}C$	—	(Stable)	1.11
Carbon-14	$^{14}_{6}C$	β decay	5730 yr	—
Carbon-15	$^{15}_{6}C$	β decay	2.4 s	—
Carbon-16	$^{16}_{6}C$	β decay	0.74 s	—

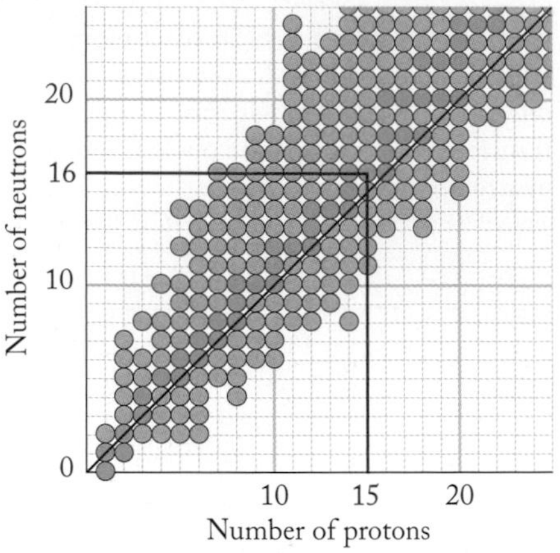

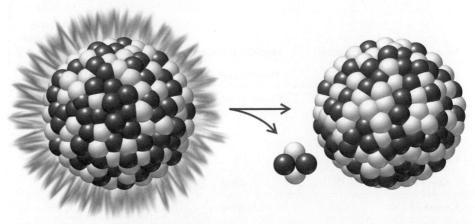

FIGURE 21.3

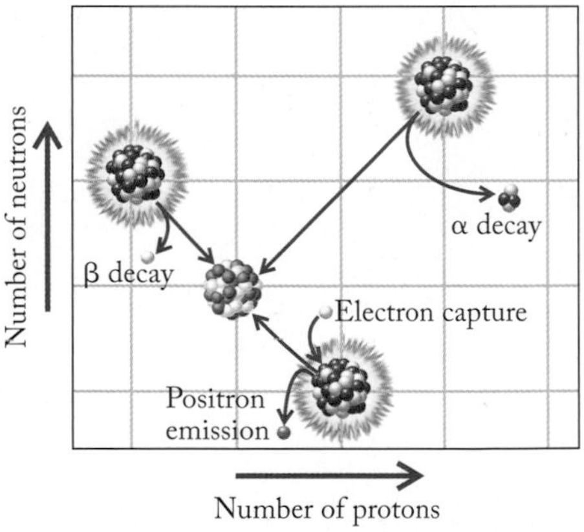

FIGURE 21.4 Radioactive decay results in predictable changes in the number of protons and neutrons in a nucleus. In alpha decay the nucleus loses 2 neutrons and 2 protons, so there is a decrease of 2 in atomic number and 4 in mass number. Beta decay leads to an increase of 1 proton at the expense of 1 neutron, so the atomic number increases by 1 but the mass number is unchanged. In positron emission and electron capture, the number of protons decreases by 1 and the number of neutrons increases by 1, so the atomic number decreases by 1 but the mass number remains the same.

SAMPLE EXERCISE 21.2 **Predicting the Modes and Products of Radioactive Decay** **LO2**

Predict the mode of radioactive decay of ^{32}P, which is one of the most widely used radionuclides in biomedical research and treatment. Identify the nuclide that is produced in the decay process.

COLLECT AND ORGANIZE We are asked to predict the mode of decay of a radionuclide, which will depend on whether it is neutron rich or neutron poor. Phosphorus-32 has 17 neutrons and 15 protons per nucleus. According to the belt of stability (Figure 21.3), phosphorus ($Z = 15$) has only one stable nuclide, which has 16 neutrons and a mass number of 31.

ANALYZE Phosphorus-32 is represented by the orange dot directly above the ^{31}P green dot, which means that it is (1) radioactive and (2) neutron rich. Neutron-rich radioisotopes of lighter elements undergo beta decay, so the decay of ^{32}P will produce a β particle. We learned how to balance nuclear equations in Section 2.7: the sums of the superscripts on the left and right sides must be equal, as must the sums of the subscripts.

SOLVE Letting a beta particle be one product of the decay reaction gives the incomplete nuclear equation:

$$^{32}_{15}P \rightarrow ? + \,^{0}_{-1}\beta$$

The missing product must have an atomic number of 16 (so that the subscripts on the right side add up to 15), which makes it an isotope of S. Its mass number must be 32, so the product is sulfur-32:

$$^{32}_{15}P \rightarrow \,^{32}_{16}S + \,^{0}_{-1}\beta$$

THINK ABOUT IT By emitting a beta particle, the nucleus increased its number of protons by 1 and decreased its number of neutrons by 1, thereby reducing its neutron "richness" and forming a stable isotope of sulfur.

Practice Exercise What is the mode of radioactive decay of ^{28}Al? Identify the nuclide produced by the decay process. ⚙

All known nuclides with more than 83 protons are radioactive. Because there is no stable reference point in the pattern of green dots in Figure 21.2, it is hard to say whether any given $Z > 83$ nuclide is neutron rich or neutron poor. We can make one general statement, though: these most massive nuclides tend to undergo either β decay or **α decay**. In the latter process they produce a nuclide with two fewer protons and two fewer neutrons, as in the case of uranium-238:

$$^{238}_{92}U \rightarrow \,^{234}_{90}Th + \,^{4}_{2}\alpha$$

Figure 21.4 summarizes the changes in atomic number and mass number caused by the various modes of decay.

For the most massive radioactive isotopes, one radioactive decay process often leads to another in what is referred to as a *radioactive decay series*. Consider, for example, the decay series that begins with the α decay of ^{238}U to ^{234}Th (Figure 21.5). Thorium has no stable isotopes and undergoes two β decay steps to produce ^{234}U. In a series of subsequent alpha decay steps, ^{234}U turns into thorium-230, radium-226, radon-222, polonium-218, and finally lead-214. Although some isotopes of lead ($Z = 82$) are stable, ^{214}Pb is not one of them. Therefore, the radioactive decay series continues as shown at the bottom left of Figure 21.5 and does not end until the stable nuclide ^{206}Pb is produced.

alpha (α) decay a nuclear reaction in which an unstable nuclide spontaneously emits an alpha particle.

CONCEPT TEST

In the ^{238}U radioactive decay series, five α decay steps in a row transform ^{234}U into ^{214}Pb. Given the shape of the belt of stability, why does it make sense that the product of these α decay steps would be a neutron-rich nuclide that undergoes β decay?

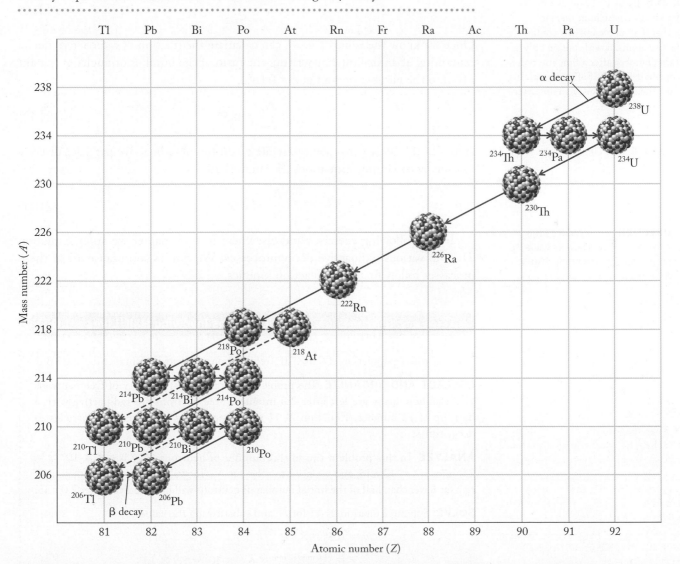

FIGURE 21.5 Uranium-238 radioactive decay series. The long diagonal arrows represent α decay events; the short horizontal ones represent β decay events. The dashed arrows are alternative pathways representing less than 1% of the decay events in this series. Note that, whether decay proceeds by the solid-line or dashed-line pathway, the end product is always stable lead-206. (Also note that the vertical axis in this figure is the mass number, not the number of neutrons as in the previous figures.)

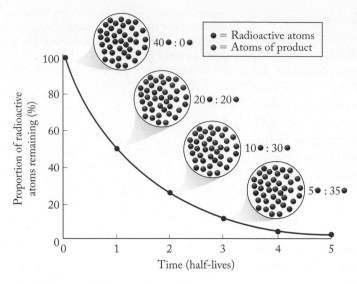

FIGURE 21.6 Radioactive decay follows first-order kinetics, which means, for example, that if a sample initially contains 40 radioactive atoms, it will contain only one-half that number after a time interval equal to one half-life. Half of the remaining half, or 10 radioactive atoms, remains after two half-lives, and so on.

▶❙❙ **CHEMTOUR** Half-Life

∞ **CONNECTION** We introduced the concept of *half-life*, $t_{1/2}$, and how its value is inversely proportional to the rate constant, k, of a first-order reaction in Chapter 13.

21.4 Rates of Radioactive Decay

In the preceding section we explored *how* radionuclides undergo radioactive decay; in this section we focus on *how rapidly* they decay. We defined radioactive decay as the *spontaneous* disintegration of unstable nuclei, but, as with chemical reactions, spontaneous does not necessarily mean rapid. All radioactive decay processes follow first-order kinetics (Section 13.3), and each has a characteristic *half-life* ($t_{1/2}$), the time interval during which the quantity of radioactive particles decreases by one-half (Figure 21.6). The faster the decay process, the shorter the half-life.

To calculate what fraction of radionuclide remains after decay time t, we can convert this time into an equivalent number of half-lives n:

$$n = \frac{t}{t_{1/2}} \tag{21.9}$$

Once we know the value of n, we can calculate the fraction of radioactive nuclei remaining at an instant t by writing the ratio of the number of nuclei at instant t (N_t) to the number present at $t = 0$ (N_0):

$$\frac{N_t}{N_0} = 0.5^n \tag{21.10}$$

Actually, it is more convenient to relate quantities of radioactive particles directly to time by combining Equations 21.9 and 21.10:

$$\frac{N_t}{N_0} = 0.5^{t/t_{1/2}} \tag{21.11}$$

In the following exercises and elsewhere in this chapter we apply Equation 21.11 to various radioactive decay processes. We can do so because all of these processes follow first-order reaction kinetics.

SAMPLE EXERCISE 21.3 **Calculations Involving Half-Lives** **LO3**

Starting with a population of 6.6×10^5 free neutrons, how many remain after 2.0 min?

COLLECT AND ORGANIZE The problem gives an initial quantity of free neutrons and asks how many are left after 2.0 minutes. Free neutrons are radioactive with a half-life of 12 minutes. Equation 21.11 relates quantities of radioactive particles to decay times.

ANALYZE In this problem the initial number of neutrons (N_0) is 6.6×10^5, $t = 2.0$ min, $t_{1/2} = 12$ min, and we need to solve for N_t. The value of t is only a fraction of $t_{1/2}$; far fewer than half of the initial number of neutrons will have decayed after 2.0 min.

SOLVE Solving Equation 21.11 for N_t and substituting the values,

$$N_t = 0.5^{t/t_{1/2}} \times N_0$$

$$= 0.5^{2.0 \text{ min}/12 \text{ min}} \times 6.6 \times 10^5 = 5.9 \times 10^5$$

THINK ABOUT IT The value of N_t is reasonable because, as we predicted, only a small fraction of the initial quantity of free neutrons decayed in 2.0 minutes.

Practice Exercise Cesium-131 is a short-lived radionuclide ($t_{1/2}$ = 9.7 d) used to treat prostate cancer. How much therapeutic strength does a cesium-131 source lose over 60 days? Express your answer as a percentage of the strength the source had at the beginning of the first day. ⚙

nuclear fission a nuclear reaction in which the nucleus of an element splits into two lighter nuclei. The process is usually accompanied by the release of one or more neutrons and energy.

chain reaction a self-sustaining series of fission reactions in which the neutrons released when nuclei split apart initiate additional fission events and sustain the reaction.

critical mass the minimum quantity of fissionable material needed to sustain a chain reaction.

21.5 Nuclear Fission

When an atom of uranium-235 captures a neutron, the nucleus of the unstable product, uranium-236, splits into two lighter nuclei in a process called **nuclear fission**. There are several uranium-235 fission reactions, including these three:

$$^{235}_{92}\text{U} + ^{1}_{0}\text{n} \rightarrow ^{141}_{56}\text{Ba} + ^{92}_{36}\text{Kr} + 3\,^{1}_{0}\text{n}$$

$$^{235}_{92}\text{U} + ^{1}_{0}\text{n} \rightarrow ^{137}_{52}\text{Te} + ^{97}_{40}\text{Zr} + 2\,^{1}_{0}\text{n}$$

$$^{235}_{92}\text{U} + ^{1}_{0}\text{n} \rightarrow ^{138}_{55}\text{Cs} + ^{96}_{37}\text{Rb} + 2\,^{1}_{0}\text{n}$$

In all these reactions, the sums of the masses of the products are slightly less than the sums of the masses of the reactants. As we observed for hydrogen fusion in Sample Exercise 21.1, this decrease in mass is released as energy in accordance with Einstein's equation ($E = mc^2$).

These reactions also produce additional neutrons, which can smash into other uranium-235 nuclei and initiate more fission events in a **chain reaction** (Figure 21.7). The reaction proceeds as long as there are enough uranium-235 nuclei present to absorb the neutrons being produced. On average, at least one neutron from each fission event must cause another nucleus to split apart for the chain reaction to be self-sustaining. The quantity of fissionable material needed to assure that every fission event produces another is called the **critical mass**. For uranium-235, the critical mass is about 1 kg of the pure isotope.

Uranium-235 is the most abundant fissionable isotope, but it makes up only 0.72% of the uranium in the principal uranium ore, pitchblende (Figure 21.8a). The uranium in nuclear reactors must be at least 3% to 4% uranium-235, and enrichment to about 85% is needed for nuclear weapons. The most common method for enriching uranium ore involves extracting the uranium in a process that yields a material called yellowcake, which is mostly U_3O_8 (Figure 21.8b). This oxide is then converted to UF_6, which, despite a molar mass of over 300 g/mol, is a volatile solid that sublimes at 56°C. The volatility of this nonpolar molecular compound can be explained by the relatively weak London dispersion forces experienced by its compact, symmetrical molecules (Figure 21.9). Fissionable $^{235}UF_6$ is separated from $^{238}UF_6$ based on their slightly different densities. Elaborate centrifuge systems are used to exploit this difference and speed up the separation (Figure 21.8c).

Harnessing the energy released by nuclear fission to generate electricity began in the middle of the 20th century. In a typical nuclear power plant (Figure 21.10), fuel rods containing 3% to 4% uranium-235 are interspersed with

FIGURE 21.7 Each fission event in the chain reaction of a uranium-235 nucleus begins when the nucleus captures a neutron, forming an unstable uranium-236 nucleus that then splits apart (fissions) in one of several ways. In the first process shown here, the uranium-236 nucleus splits into krypton-92, barium-141, and three neutrons. If, on average, at least one of the three neutrons from each fission event causes the fission of another uranium-235 nucleus, then the process is sustained in a chain reaction.

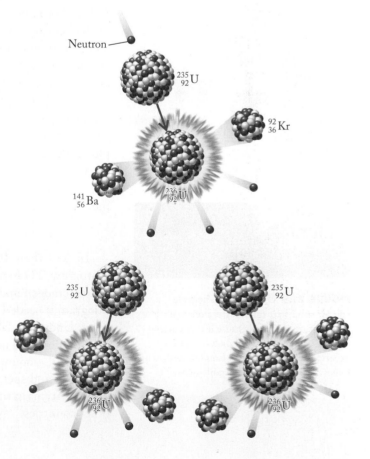

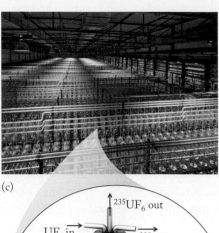

(c)

(a)   (b)

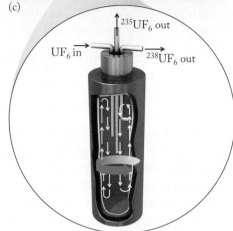

FIGURE 21.8 Preparing uranium fuel. (a) A piece of pitchblende, source of the uranium fuel for nuclear reactors. (b) Pitchblende ore is ground up and extracted with strong acid. The uranium compounds (mostly U_3O_8) obtained from the extract are called yellowcake. (c) Uranium oxides are converted to volatile UF_6, which is centrifuged at very high speed to separate $^{235}UF_6$ from $^{238}UF_6$. The less dense and less abundant $^{235}UF_6$ is enriched near the center of the centrifuge cylinder and separated from the heavier $^{238}UF_6$.

FIGURE 21.9 Uranium hexafluoride is a volatile solid that sublimes at a relatively low temperature of 56°C because it is composed of compact, symmetrical molecules that experience relatively weak London dispersion forces despite their considerable mass.

rods of boron or cadmium that control the rate of the fission by absorbing some of the neutrons produced during fission. Pressurized water flows around the fuel and control rods, removing the heat created during fission and transferring it to a steam generator. The water also acts as a moderator, slowing down the neutrons and thereby allowing for their more efficient capture by ^{235}U atoms.

In 1952 the first **breeder reactor** was built, so called because in addition to producing energy to make electricity, the reactor makes ("breeds") its own fuel. The reactor starts out with a mixture of plutonium-239 and uranium-238. As the plutonium fissions and the energy from those reactions is collected to produce electricity, some of the neutrons that are produced sustain the fission chain reaction just as in the reactor of Figure 21.10, while others convert the uranium into more plutonium fuel:

$$^{238}_{92}U + ^{1}_{0}n \rightarrow ^{239}_{92}U + \gamma \rightarrow ^{239}_{94}Pu + 2\,^{0}_{-1}\beta$$

In less than 10 years of operation, a breeder reactor can make enough plutonium-239 to refuel itself *and* another reactor. Unfortunately, plutonium-239 is a carcinogen and one of the most toxic substances known. Only about half a kilogram is needed to make an atomic bomb, and it has a long half-life: 2.4×10^4 years. Understandably, extreme caution and tight security surround the handling of plutonium fuel and the transportation and storage of nuclear wastes containing even small amounts of plutonium. Health and safety matters related to reactor operation and spent-fuel disposal are the principal reasons there are no breeder power stations in the United States, although they have been built in at least seven other countries.

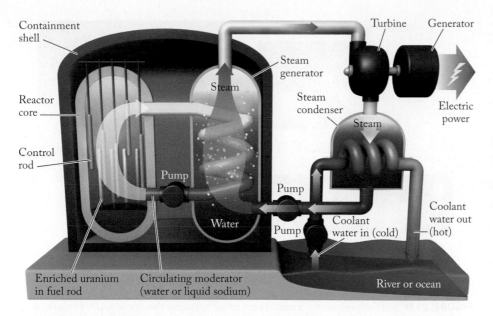

FIGURE 21.10 A pressurized, water-cooled nuclear power plant uses fuel rods containing uranium enriched to about 4% uranium-235. The fission chain reaction is regulated with control rods and a moderator that is either water or liquid sodium. The moderator slows down the neutrons released by fission so that they can be more efficiently captured by other uranium-235 nuclei. It also transfers the heat produced by the fission reaction to a steam generator. The steam generated by this heat drives a turbine that generates electricity.

CONCEPT TEST

Nuclear reactors powered by the energy released by the fission of uranium-235 have been operating since the 1950s, but a reactor powered by the energy released by the fusion of hydrogen has yet to be built. Why is it taking so long to build a fusion reactor?

21.6 Measuring Radioactivity

French scientist Henri Becquerel discovered radioactivity in 1896 when he observed that uranium and other substances produce radiation that fogs photographic film. Photographic film is still used to detect radioactivity, as in the film dosimeter badges worn by people working with radioactive materials to record their exposure to radiation. Detectors called **scintillation counters** use materials called *phosphors* to absorb energy released during radioactive decay. The phosphors then release the absorbed energy as visible light, the intensity of which is a measure of the amount of radiation initially emitted.

Radioactivity also can be measured with a **Geiger counter**, which detects the common products of radioactivity—α particles, β particles, and γ rays—on the basis of their abilities to ionize atoms (Figure 21.11). A sealed metal cylinder filled with gas (usually argon) and a positively charged electrode has a window that allows α particles, β particles, and γ rays to enter. Once inside the cylinder, these particles ionize argon atoms into Ar^+ ions and free electrons. If an electrical potential difference is applied between the cylinder shell and the central electrode, free electrons migrate toward the positive electrode and argon ions migrate toward the negatively charged shell. This ion migration produces a pulse of electrical current whenever radiation enters the cylinder. The current is amplified and read out to a meter and a microphone that makes a clicking sound.

One measure of radioactivity in a sample is the number of decay events per unit time. This parameter is called the radioactivity (*A*) of the sample. The SI unit of radioactivity is the **becquerel (Bq),** named in honor of Henri Becquerel and

breeder reactor a nuclear reactor in which fissionable material is produced during normal reactor operation.

scintillation counter an instrument that determines the level of radioactivity in samples by measuring the intensity of light emitted by phosphors in contact with the samples.

Geiger counter a portable device for determining nuclear radiation levels by measuring how much the radiation ionizes the gas in a sealed detector.

becquerel (Bq) the SI unit of radioactivity; one becquerel equals one decay event per second.

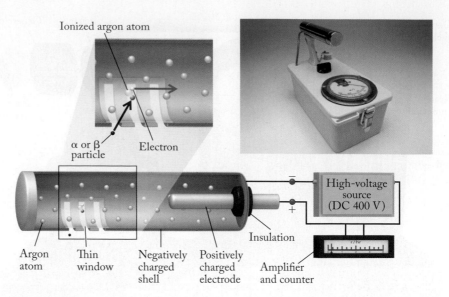

FIGURE 21.11 In a Geiger counter, a particle produced by radioactive decay passes through a thin window usually made of beryllium or a plastic film. Inside the tube, the particle collides with atoms of argon gas and ionizes them. The resulting argon cations migrate toward the negatively charged tube housing, and the electrons migrate toward a positive electrode, creating a pulse of current through the tube. The current pulses are amplified and recorded via a meter and a speaker that produces an audible "click" for each pulse.

equal to one decay event per second. An older radioactivity unit is the **curie (Ci)**, named in honor of Marie and Pierre Curie, where

$$1 \text{ Ci} = 3.70 \times 10^{10} \text{ Bq} = 3.70 \times 10^{10} \text{ decay events/s}$$

Both the becquerel and the curie quantify the *rate* at which a radioactive substance decays, which provides a measure of how much of it is in a sample. As noted in Section 21.4, all decay processes follow first-order kinetics. Recall from Section 13.3 that for a first-order reaction, the rate constant k is related to the concentration of a reactant R according to the equation

$$\text{Rate} = k[R]$$

The same mathematical relationships apply to radioactive decay processes except that we refer to radioactivity instead of rate of decay, and to the number of atoms (N) of a radionuclide in a sample instead of its concentration:

$$A = kN \tag{21.12}$$

Because radioactivity is the number of decay events per second, the units of the *decay rate constant k* are decay events per atom per second.

The value of k is inversely proportional to the half-life of a radionuclide:

$$t_{1/2} = \frac{0.693}{k} \tag{13.20}$$

Scientists usually express the quantity of a radionuclide in a sample not in terms of its mass but rather in terms of its radioactivity, because the latter value is much more important in determining how the substance is used and how we handle it safely. This substitution is valid because the quantity of a radionuclide is directly proportional to its radioactivity.

curie (Ci) non-SI unit of radioactivity; 1 Ci = 3.70×10^{10} decay events per second.

SAMPLE EXERCISE 21.4 **Calculating the Radioactivity** LO5
of a Sample

Radium-223 undergoes β decay with a half-life of 11.4 days. What is the radioactivity of a sample that contains 1.00 μg of ^{223}Ra? Express your answer in becquerels and in curies.

COLLECT AND ORGANIZE We are given the half-life and quantity of a radioactive substance and are asked to determine its radioactivity, that is, its rate of decay. All radioactive decay processes are first order, so the decay rate constant k is related to half-life by Equation 13.20:

$$t_{1/2} = \frac{0.693}{k}$$

and the radioactivity is the product of the rate constant and the number of atoms of radionuclide in the sample (Equation 21.12, $A = kN$).

ANALYZE Before using Equation 13.20, we must convert the half-life into seconds because both of the radioactivity units we need to calculate are based on decay events per second. To calculate radioactivity, we determine the number of atoms in 1.00 μg of radium. This will likely be a very large number, which should translate into a large number of decay events per second given the relatively short half-life of ^{223}Ra.

SOLVE The half-life is

$$11.4 \text{ d} \times \frac{24 \text{ hr}}{1 \text{ d}} \times \frac{60 \text{ min}}{1 \text{ hr}} \times \frac{60 \text{ s}}{1 \text{ min}} = 9.850 \times 10^5 \text{ s}$$

Using this value in Equation 13.20 and solving for k,

$$k = \frac{0.693}{9.850 \times 10^5 \text{ s}}$$

$$= 7.036 \times 10^{-7} \text{ s}^{-1} = 7.036 \times 10^{-7} \text{ decay events/(atom} \cdot \text{s)}$$

The number of atoms (N) of ^{223}Ra is

$$N = 1.00 \text{ μg} \times \frac{10^{-6} \text{ g}}{\text{μg}} \times \frac{1 \text{ mol}}{223 \text{ g}} \times \frac{6.0221 \times 10^{23} \text{ atoms}}{1 \text{ mol}} = 2.700 \times 10^{15} \text{ atoms}$$

Inserting these values of k and N into Equation 21.12,

$$A = kN$$

$$= 7.036 \times 10^{-7} \frac{\text{decay events}}{\text{atom} \cdot \text{s}} \times 2.700 \times 10^{15} \text{ atoms}$$

$$= 1.90 \times 10^9 \text{ decay events/s}$$

Because 1 Bq = 1 decay event/s, the radioactivity of the sample is 1.90×10^9 Bq. Expressing radioactivity in curies,

$$1.90 \times 10^9 \text{ decay events/s} \times \frac{1 \text{ Ci}}{3.70 \times 10^{10} \text{ decay event/s}} = 0.0514 \text{ Ci}$$

THINK ABOUT IT The large number of decay events per second that we calculated meets our expectation of a high level of radioactivity given the large number of radioactive atoms in the sample.

Practice Exercise Determine the radioactivity in 1.00 mg of radium-226 ($t_{1/2} = 1.6 \times 10^3$ yr) in becquerels and in millicuries. ⚙

21.7 Biological Effects of Radioactivity

The γ rays and many of the α and β particles produced by nuclear reactions have more than enough energy to tear chemical bonds apart, producing odd-electron radicals or free electrons and cations. Consequently, these rays and particles are classified as **ionizing radiation**. Other examples are X-rays and short-wavelength ultraviolet rays. The ionization of atoms and molecules in living tissue can lead to radiation sickness, cancer, birth defects, and death. The scientists who first worked with radioactive materials were not aware of these hazards, and some of them suffered for it. Marie Curie died of aplastic anemia caused by her many years of radiation exposure, and leukemia claimed her daughter Irène Joliot-Curie, who continued the research program started by her parents.

In medicine, the term *ionizing radiation* is limited to photons and particles that have sufficient energy to remove an electron from water:

$$H_2O(\ell) \xrightarrow{\text{1216 kJ/mol}} H_2O^+(aq) + e^-$$

The logic behind this definition is that the human body is composed largely of water. Therefore water molecules are the most abundant ionizable targets in an organism exposed to nuclear radiation. The cation H_2O^+ reacts with another water molecule in the body to form a hydronium ion and a hydroxyl free radical:

$$H_2O^+(aq) + H_2O(\ell) \rightarrow H_3O^+(aq) + \cdot OH(aq)$$

The rapid reactions of free radicals with biomolecules can threaten the integrity of cells and the health of the entire organism.

Radiation-induced alterations to the biochemical machinery that controls cell growth are most likely to occur in tissues in which cells grow and divide rapidly. One such tissue is bone marrow, where billions of white blood cells are produced each day to fortify the body's immune system. Molecular damage to bone marrow can lead to leukemia, an uncontrolled production of nonfunctioning white blood cells that spread throughout the body, crowding out healthy cells. Ionizing radiation can also cause molecular alterations in the genes and chromosomes of sperm and egg cells, increasing the chances of birth defects in offspring.

Radiation Dosage

The biological impact of ionizing radiation depends on how much of it is absorbed by an organism. If the radiation is coming from one radioactive source, then the amount absorbed depends on the radioactivity of the source and the energy of the radiation that is produced per decay event. Tables of radioactive isotopes often include information about their modes of decay and the energies of the particles and gamma rays they emit.

Absorbed dose is the quantity of ionizing radiation absorbed by a unit mass of living tissue. The SI unit of absorbed dose is the **gray (Gy)**. One gray is equal to the absorption of 1 J of radiation energy per kilogram of body mass:

$$1 \text{ Gy} = 1 \text{ J/kg}$$

Grays express dosage, but they do not indicate the amount of *tissue damage* caused by that dosage. Different products of nuclear reactions affect living tissue differently. Exposure to 1 Gy of γ rays produces about the same amount of tissue damage as exposure to 1 Gy of β particles. However, 1 Gy of α particles, which

ionizing radiation high-energy products of radioactive decay that can ionize molecules.

gray (Gy) the SI unit of absorbed radiation; 1 Gy = 1 J/kg of tissue.

relative biological effectiveness (RBE) a factor that accounts for the differences in physical damage caused by different types of radiation.

sievert (Sv) SI unit used to express the amount of biological damage caused by ionizing radiation.

move about 10 times slower than β particles but have nearly 10^4 times the mass, causes as much as 20 times as much damage as 1 Gy of γ rays. Neutrons cause 3 to 5 times as much damage. To account for these differences, values of **relative biological effectiveness (RBE)** have been established for the various forms of ionizing radiation (Table 21.3). When absorbed dose in grays is multiplied by an RBE factor, the product is called *effective* dose, a measure of tissue damage. The SI unit of effective dose is the **sievert (Sv)**.

Table 21.4 summarizes the various units used to express quantities of radiation and their biological impact. Two non-SI units are listed that predate their SI counterparts but are still often used. They are radiation absorbed dose, or rad, which is equivalent to 0.01 Gy, and the rem for tissue damage, which is an acronym for *roentgen equivalent man*. One rem is the product of one rad of ionization times the appropriate RBE factor. There are 100 rems in 1 Sv.

TABLE 21.3	RBE Values of Nuclear Radiation
Radiation	**RBE**
γ rays	1.0
β particles	1.0–1.5
Neutrons	3–5
Protons	10
α particles	20

TABLE 21.4 Units for Expressing Quantities of Ionizing Radiation

Parameter	SI Unit	Description	Alternative Common Unit	Description
Radioactivity	becquerel (Bq)	1 decay event/s	curie (Ci)	3.70×10^{10} decay events/s
Ionizing energy absorbed	gray (Gy)	1 J/kg of tissue	rad	0.01 J/kg of tissue
Amount of tissue damage	sievert (Sv)	1 Gy × RBE[a]	rem	1 rad × RBE[a]

[a]RBE, relative biological effectiveness.

The RBE of 20 for α particles may lead you to believe that these particles pose the greatest health threat from radioactivity. Not exactly. Alpha particles are so big that they have little penetrating power; they are stopped by a sheet of paper, clothing, or even a layer of dead skin (Figure 21.12). However, if you ingest or inhale an α emitter, tissue damage can be severe because the relatively massive α particles do not have to travel far to cause cell damage. Gamma rays are considered the most dangerous form of radiation emanating from a source outside the body because they have the greatest penetrating power.

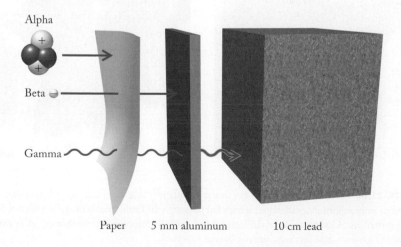

Alpha

Beta

Gamma

Paper   5 mm aluminum   10 cm lead

FIGURE 21.12 The tissue damage caused by α particles, β particles, and γ rays depends on their ability to penetrate materials shielding you from their source. Alpha particles are stopped by paper or clothing but are extremely dangerous if formed inside the body because their low penetrating power traps them inside. Stopping gamma rays requires a thick layer of lead or several meters of concrete or soil.

TABLE 21.5 Acute Effects of Single Whole-Body Effective Doses of Ionizing Radiation

Effective Dose (Sv)	Toxic Effect
0.05–0.25	No acute effect, possible carcinogenic or mutagenic damage to DNA
0.25–1.0	Temporary reduction in white blood cell count
1.0–2.0	Radiation sickness: fatigue, vomiting, diarrhea, impaired immune system
2.0–4.0	Severe radiation sickness: intestinal bleeding, bone marrow destruction
4.0–10.0	Death, usually through infection, within weeks
>10.0	Death within hours

FIGURE 21.13 The ruins of the nuclear reactor at Chernobyl, Ukraine, that exploded in 1986.

The effects of exposure to different single effective doses of radiation are summarized in Table 21.5. To put these data in perspective, the effective dose from a typical dental X-ray is about 25 μSv, or about 2000 times smaller than the lowest exposure level cited in the table.

Widespread exposure to very high levels of radiation occurred after the 1986 explosion at the Chernobyl nuclear reactor in what is now Ukraine (Figure 21.13). Many plant workers and first responders were exposed to more than 1.0 Sv of radiation. At least 30 of them died in the weeks after the accident. Many of the more than 300,000 workers who cleaned up the area around the reactor exhibited symptoms of radiation sickness, and at least 5 million people in Ukraine, Belarus, and Russia were exposed to fallout in the days following the accident. Studies conducted in the early 1990s uncovered high incidences of thyroid cancer in children in southern Belarus due to ^{131}I released in the Chernobyl accident, and children born in the region nearly a decade after the accident had unusually high rates of mutations in their DNA because of their parents' exposure to ionizing radiation. Genetic damage was also widespread among plants and animals living in the region (Figure 21.14).

Radiation exposure was not confined to Ukraine and Belarus. After the accident a cloud of radioactive material spread rapidly across northern Europe,

(a)   (b)

FIGURE 21.14 Wildlife surrounding the destroyed nuclear reactor at Chernobyl, Ukraine, was exposed to intense ionizing radiation, which led to deaths and sublethal biological effects such as genetic mutations. One example of the latter is the partially albino barn swallow (a). A normal swallow (b) has no white feathers directly beneath its beak.

and within 2 weeks increased levels of radioactivity were detected throughout the Northern Hemisphere (Figure 21.15). The accident produced a global increase in human exposure to ionizing radiation estimated to be equivalent to 0.05 mSv per year.

Evaluating the Risks of Radiation

To put global radiation exposure from Chernobyl in perspective, we need to consider typical annual exposure levels. For many people the principal source of radiation is radon gas in indoor air and in well water (Figure 21.16). Like all noble gases, radon is chemically inert. Unlike the others, all of its isotopes are radioactive. The most common isotope, radon-222, is produced when uranium-238 in rocks and soil decays to lead-206 (Figure 21.5). The radon gas formed in this decay series percolates upward and can enter a building through cracks and pores in its foundation.

If you breathe radon-contaminated air and then exhale before it decays, no harm is done. However, if radon-222 decays inside the lungs, it emits an α particle that can damage lung tissue. The nuclide produced by the α decay of ^{222}Rn is radioactive polonium-218 that may become attached to tissue in the respiratory system and undergo a second α decay, forming lead-214:

$$^{222}_{86}\text{Rn} \rightarrow {}^{218}_{84}\text{Po} + {}^{4}_{2}\alpha \qquad t_{1/2} = 3.8 \text{ d}$$

$$^{218}_{84}\text{Po} \rightarrow {}^{214}_{82}\text{Pb} + {}^{4}_{2}\alpha \qquad t_{1/2} = 3.1 \text{ min}$$

As we have seen, α particles are the most damaging product of nuclear decay when formed *inside the body*. How big a threat does radon pose to human health? Concentrations of indoor radon depend on local geology (Figure 21.17) and on how gastight building foundations are. The air in many buildings contains concentrations of radon in the range of 1 pCi per liter of air. How hazardous are such tiny concentrations? There appears to be no simple answer. The U.S. Environmental Protection Agency has established 4 pCi/L as an "action level," meaning that people occupying houses with higher concentrations should take measures to minimize their exposure.

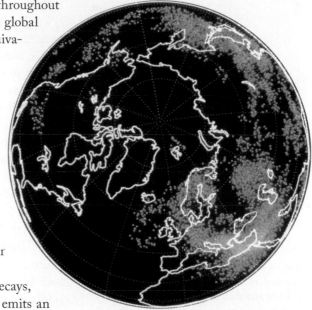

FIGURE 21.15 Radioactive fallout (shown in pink) from the Chernobyl accident in 1986 was detected throughout the Northern Hemisphere.

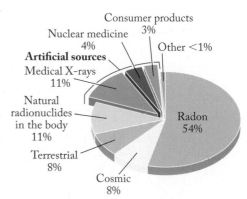

FIGURE 21.16 Sources of radiation exposure of the U.S. population. On average, a person living in the United States is exposed to 0.0036 Sv of radiation each year. More than 80% of this exposure comes from natural sources, mainly radon in the air and water. Artificial sources account for about 18% of the total exposure.

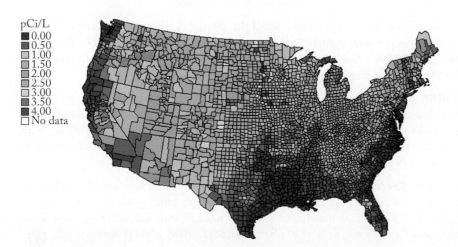

FIGURE 21.17 Levels of radon gas in soils and rocks across the United States.

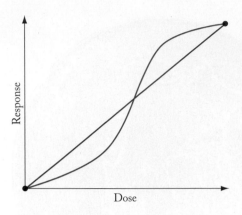

FIGURE 21.18 The risk of death from radiation-induced cancer may follow one of two models. In the linear response model (red line), risk is directly proportional to the radiation exposure. In the S-shaped model (blue line), risk remains low below a critical threshold and then increases rapidly as the exposure increases. In the S-shaped model, the risk is less than for the linear model at low doses but is higher at higher doses.

This action level is based on studies of the incidence of lung cancer in workers in uranium mines. These workers are exposed to radon concentrations (and concentrations of other radionuclides) that are much higher than the concentrations in homes and other buildings. However, many scientists believe that people exposed to very low levels of radon for many years are as much at risk as miners exposed to high levels of radiation for shorter periods. Some researchers use a model that assumes a linear relation between radon exposure and incidence of lung cancer. This model is represented by the red line in Figure 21.18, which graphs cancer deaths as a function of radiation absorbed. On the basis of this dose–response model, an estimated 15,000 Americans die of lung cancer each year because of exposure to indoor radon. This number comprises 10% of all lung-cancer fatalities and 30% of those among nonsmokers.

Is this linear model valid? Perhaps—but some scientists believe that there may be a threshold of exposure below which radon poses no significant threat to public health. They advocate an S-shaped dose–response curve, shown by the blue line in Figure 21.18. Notice that the risk of death from cancer in the S-shaped curve is much lower than in the linear response model at low radiation exposure but rises rapidly above a critical value.

SAMPLE EXERCISE 21.5 **Calculating Effective Dose** **LO6**

It has been estimated that a person living in a home where the air radon concentration is 4.0 pCi/L receives an annual absorbed dose of ionizing radiation equivalent to 0.40 mGy. What is the person's annual effective dose from this radon in millisieverts? Use information from Figure 21.16 to compare this annual effective dose of radon with the average annual effective dose of radon estimated for persons living in the United States.

COLLECT AND ORGANIZE We are given an absorbed radiation dose of 0.40 mGy. Radon isotopes emit alpha particles, which we know from Table 21.3 have a relative biological effectiveness of 20.

ANALYZE The effective dose caused by an absorbed dose of ionizing radiation is the absorbed dose multiplied by the RBE of the radiation.

SOLVE

$$0.40 \text{ mGy} \times 20 = 8.0 \text{ mSv}$$

The caption to Figure 21.16 tells us that the average American is exposed to 3.6 mSv of radiation per year, with 54% of that amount, or 1.9 mSv, from radon. The person living in the home described in the problem has an effective dose slightly more than four times the average value.

THINK ABOUT IT The calculated value is more than twice the average annual effective dose of 3.6 mSv from all sources of radiation. The U.S. National Research Council has estimated that a nonsmoker living in air contaminated with 4.0 pCi/L of radon has a 1% chance of dying from lung cancer due to this exposure. The cancer risk for a smoker is close to 5%.

Practice Exercise A dental X-ray for imaging impacted wisdom teeth produces an effective dose of 10 μSv. If a dental X-ray machine emits X-rays with an energy of 6.0×10^{-17} J each, how many of these X-rays must be absorbed per kilogram of tissue to produce an effective dose of 10 μSv? Assume the RBE of these X-rays is 1.2. ⚙

21.8 Medical Applications of Radionuclides

Radionuclides are used in both the detection and the treatment of diseases, and they are key agents in the respective medical fields of *diagnostic radiology* and *therapeutic radiology*. In diagnostic radiology, radionuclides are used alongside magnetic resonance imaging (MRI) and other imaging systems that involve only nonionizing radiation. Therapeutic radiology, however, is based almost entirely on the ionizing radiation that comes from nuclear processes.

Therapeutic Radiology

Because ionizing radiation causes the most damage to cells that grow and divide rapidly, it is a powerful tool in the fight *against* cancer. Radiation therapy consists of exposing cancerous tissue to γ radiation. Often the radiation source is external to the patient, but sometimes it is encased in a platinum capsule and surgically implanted in a cancerous tumor. The platinum provides a chemically inert outer layer and acts as a filter, absorbing α and β particles emitted by the radionuclide but allowing γ rays to pass into the tumor.

A nuclide's chemical properties can be exploited to direct it to a tumor site. For example, most iodine in the body is concentrated in the thyroid gland, so an effective therapy against thyroid cancer starts with ingestion of potassium iodide containing radioactive iodine-123. Other radionuclides used in cancer therapy are listed in Table 21.6.

Surgically inaccessible tumors can be treated with beams of γ rays from a radiation source outside the body. Unfortunately, γ radiation destroys both cancer cells and healthy ones. Thus, patients receiving radiation therapy frequently suffer symptoms of radiation sickness, including nausea and vomiting (the tissues that make up intestinal walls are especially susceptible to radiation-induced damage), fatigue, weakened immune response, and hair loss. To reduce the severity of these side effects, radiologists must carefully control the dosage a patient receives.

Diagnostic Radiology

The movement of radionuclides in the body and their accumulation in certain organs provide ways to assess organ function. A tiny quantity of a radioactive isotope is used, together with a much larger amount of a stable isotope of the same element. The radioactive isotope is called a *tracer*, and the stable isotope is the *carrier*. For example, the circulatory system can be imaged by injecting into the blood a solution of sodium chloride containing a trace amount of $^{24}NaCl$. Circulation is monitored by measuring the γ rays emitted by ^{24}Na as it decays.

The ideal tracer for medical imaging is one that has a half-life about equal to the length of time required to perform the imaging measurements. It should emit moderate-energy γ rays but no α particles or β particles that might cause tissue damage. Sodium-24 (a gamma emitter with a half-life of 15 hours) meets both these criteria. Table 21.7 lists several other radionuclides that are used in medical imaging.

TABLE 21.6 Some Radionuclides Used in Radiation Therapy

Nuclide	Radiation	Half-Life	Treatment
^{32}P	β	14.3 d	Leukemia therapy
^{60}Co	β, γ	5.3 yr	Cancer therapy
^{131}I	β	8.1 d	Thyroid therapy
^{131}Cs	γ	9.7 d	Prostate cancer therapy
^{192}Ir	β, γ	74 d	Coronary disease

TABLE 21.7 Selected Radionuclides Used for Medical Imaging

Nuclide	Radiation	Half-Life (hr)	Use
^{99}Tc	γ	6.0	Bones, circulatory system, various organs
^{67}Ga	γ	78	Tumors in the brain and other organs
^{201}Tl	γ	73	Coronary arteries, heart muscle
^{123}I	γ	13.3	Thyroxine production in thyroid gland

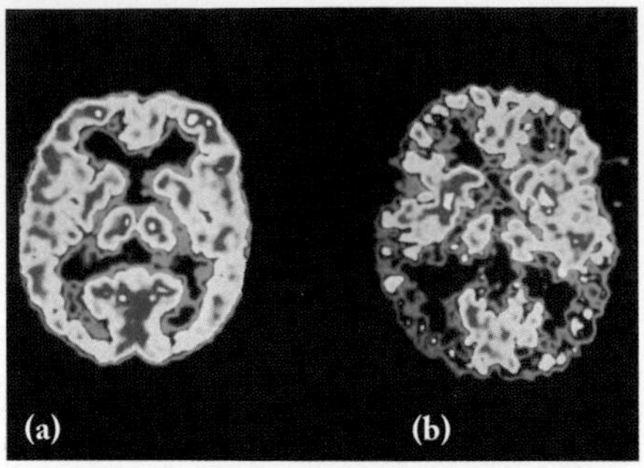

FIGURE 21.19 Positron emission tomography (PET) is used to monitor cell activity in organs such as the brain. (a) Brain function in a healthy person. The red and yellow regions indicate high brain activity; blue and black indicate low activity. (b) Brain function in a patient suffering from Alzheimer's disease.

CONNECTION In Section 2.4 we saw that mass spectrometry can be used to determine the abundances of the isotopes of elements in a sample.

....

radiometric dating a method for determining the age of an object based on the quantity of a radioactive nuclide and/or the products of its decay that the object contains.

....

Positron emission tomography (PET) is a powerful tool for diagnosing organ and cell function. PET uses short-lived, neutron-poor, positron-emitting radionuclides such as carbon-11, oxygen-15, and fluorine-18. A patient might be administered a solution of glucose in which some of the sugar molecules contain atoms of ^{11}C, ^{15}O, or ^{18}F. The rate at which glucose is metabolized in various regions of the brain is monitored by detecting the γ rays produced by positron–electron annihilations (Equation 21.2). Unusual patterns in PET images of brains (Figure 21.19) can indicate damage from strokes, schizophrenia, manic depression, Alzheimer's disease, and even nicotine addiction in tobacco smokers.

21.9 Radiometric Dating

Radiometric dating is a term used to describe methods for determining the age of objects based on their naturally occurring radionuclides and/or the nuclides' decay products. The concept was invented in the early 1900s by Ernest Rutherford, who had already recognized in his pioneering studies on radioactivity that radioactive decay processes have characteristic half-lives. Rutherford proposed to use this concept to determine the age of rocks and even the age of Earth itself. The basis for his initial attempt was the emission of α particles from uranium ore. He correctly suspected that alpha particles were part of helium atoms, and so he proposed to determine the age of uranium ore samples by determining the concentration of helium gas trapped inside them.

Rutherford's helium method did not yield very accurate results, but it did inspire a young American chemist, Bertram Boltwood (1870–1927), who had determined that the decay of radioactive uranium involves a series of decay events ending with the formation of stable lead (illustrated for ^{238}U in Figure 21.5). In 1907, Boltwood published the results of dating 43 samples of uranium-containing minerals based on the ratio of lead to uranium in them. The ages he reported spanned hundreds of millions to over a billion years and probably represent the first successful attempt at radiometric dating.

In recent years the development of the mass spectrometer for accurately determining the abundances of individual isotopes of elements, coupled with more accurate half-life values for decay events such as those in Figure 21.5, has allowed scientists to use the ratio of ^{206}Pb to ^{238}U in geological samples to determine their ages with a precision of about $\pm 1\%$. Other methods, including one based on the decay of ^{235}U to ^{207}Pb ($t_{1/2} = 7.0 \times 10^6$ yr), may be used to analyze the same samples, providing multiple independent determinations that mutually assure more accurate results. These analyses have shown that the oldest rocks on Earth are over 4.0 billion years old and that meteorites that formed as the solar system formed are 4.5 billion years old.

The radiometric methods described above yield reliable results only when the sample is a *closed system*, which means that the only loss of the radionuclide is via radioactive decay, and that all of the nuclides produced by the decay processes remain in the sample. In addition, those decay processes must be the *only* source of the product nuclides. For these reasons, scientists must exercise care in selecting the types of samples they subject to radiometric dating analysis. For example, the presence of the mineral zircon ($ZrSiO_4$) in a geological sample is good news for scientists interested in dating it because U^{4+} ions readily substitute for Zr^{4+} ions as crystals of $ZrSiO_4$ solidify from the molten state, but Pb^{2+} ions do not. Therefore, the only source of ^{206}Pb and ^{207}Pb in a zircon sample should be the decay of ^{238}U and ^{235}U, respectively.

In 1947, American chemist Willard Libby (1908–1980) developed a radiometric dating technique, called **radiocarbon dating**, for determining the age of artifacts from prehistory and early civilizations. The method is based on determining the carbon-14 content of samples derived from plants or the animals that consumed them. Carbon-14 originates in the upper atmosphere where cosmic rays break apart the nuclei of atoms, forming free protons and neutrons. When one of these neutrons collides with a nitrogen-14 atom, they form an atom of radioactive carbon-14 and a proton:

$$^{14}_{7}N + ^{1}_{0}n \rightarrow ^{14}_{6}C + ^{1}_{1}p$$

Atmospheric carbon-14 combines with oxygen, forming $^{14}CO_2$. The atmospheric concentration of $^{14}CO_2$ amounts to only about 10^{-12} of all the molecules of CO_2 in the air. These traces of radioactive CO_2, along with the stable forms $^{12}CO_2$ and $^{13}CO_2$, are incorporated into the structures of green plants during photosynthesis. The tiny fraction of the plant's mass that is ^{14}C gets even tinier after a plant dies, or after a part of it stops growing and photosynthesizing, because ^{14}C undergoes β decay as we described in Section 21.3:

$$^{14}_{6}C \rightarrow ^{14}_{7}N + ^{0}_{-1}\beta$$

The half-life of the decay process is 5730 years.

If we can determine the ^{14}C content (N_t) of an object of historical interest, such as a piece of wood from an ancient building, charcoal from a prehistoric campfire, or papyrus from an early Egyptian scroll, and if we know (or can predict) its ^{14}C content when the material in it was alive (N_0), then we can apply Equation 21.11:

$$\frac{N_t}{N_0} = 0.5^{t/t_{1/2}}$$

to determine its age. Predicting the value of N_0 is usually done by analyzing samples from growing plants—that is, samples for which the ^{14}C decay time is zero.

To facilitate radiocarbon dating calculations, let's solve Equation 21.11 for t by first taking the natural log of both sides (keeping in mind that ln 0.5 = −0.693):

$$\ln\frac{N_t}{N_0} = -0.693\frac{t}{t_{1/2}}$$

Rearranging the terms to solve for t gives us a useful equation for determining the radiocarbon age t:

$$t = -\frac{t_{1/2}}{0.693}\ln\frac{N_t}{N_0} \tag{21.13}$$

radiocarbon dating a method for establishing the age of a carbon-containing object by measuring the amount of radioactive carbon-14 remaining in the object.

SAMPLE EXERCISE 21.6 Radiocarbon Dating   LO7

The ^{14}C content of a wooden harpoon handle found in the remains of an Inuit encampment in western Alaska is 61.9% of the ^{14}C content of the same type of wood from a recently cut tree. How old is the harpoon?

COLLECT AND ORGANIZE We are asked to calculate the age of a sample that contains 61.9% of the ^{14}C in a modern sample of the same material. The half-life of carbon-14 is 5730 years. Equation 21.13 provides the age t of the artifact if we know the ratio of the ^{14}C in it today to its initial ^{14}C content.

ANALYZE The ^{14}C content of the modern sample can be used as a surrogate for the initial ^{14}C content of the artifact. Therefore 61.9% (or 0.619) represents the ratio N_t/N_0. This value is greater than 0.5, which means that the age of the sample is less than one half-life (5730 yr).

SOLVE

$$t = -\frac{t_{1/2}}{0.693} \ln\frac{N_t}{N_0}$$

$$= -\frac{5730 \text{ yr}}{0.693} \ln(0.619)$$

$$= 3966 \text{ yr} = 3.97 \times 10^3 \text{ yr}$$

THINK ABOUT IT The resulting age is less than one half-life, which is reasonable because it contained more than half the original carbon-14 content. The result is expressed with three significant figures to match that of the starting composition (61.9%).

Practice Exercise The carbon-14 radioactivity in papyrus growing along the Nile River today is 231 Bq per kilogram of carbon. If a papyrus scroll found near the Great Pyramid at Cairo has a carbon-14 radioactivity of 127 Bq per kilogram of carbon, how old is the scroll?

FIGURE 21.20 Radiocarbon dating relies on knowing the atmospheric concentration of carbon-14 over time. Ancient living trees, such as the bristlecone pines in the American Southwest, act as a check of the atmospheric carbon-14 levels over thousands of years. The ages of the rings can be determined by counting them, and their carbon-14 content can be determined by mass spectrometry.

The accuracy of radiocarbon dating can be checked by determining the ^{14}C content of the annual growth rings of very old trees, such as the bristlecone pines that grow in the American Southwest (Figure 21.20). When scientists plot the radiocarbon ages of these rings against their actual ages obtained by counting rings starting from the outer growth layer of the tree (representing $t = 0$), they find that the two sets of ages do not agree exactly, as shown in Figure 21.21. There are several reasons for this lack of agreement, including variability in the rates of ^{14}C production due to changing intensity of the cosmic rays striking Earth's upper atmosphere. To assure accurate ^{14}C results, scientists must correct for these and other variations, and they must be careful to avoid contaminating ancient

FIGURE 21.21 Calibration curves for radiocarbon dating allow scientists to accurately calculate the ages of archaeological objects. If the rate of ^{14}C production in the upper atmosphere were constant, then the age of objects based on their ^{14}C content would match their actual age—a condition represented by the red dashed line. However, analyses of tree rings, corals, and lake sediments tell us that the rate of ^{14}C production in the upper atmosphere is variable, so a real plot of ^{14}C age versus actual age produces the jagged blue line. This plot allows scientists to convert ^{14}C ages into actual ages.

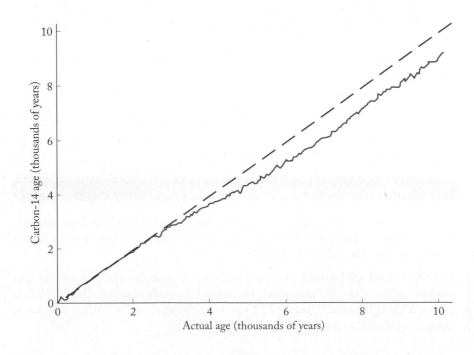

samples with modern carbonaceous material. With proper analytical technique, radiocarbon dating results are generally accurate to within ±40 yr for samples that are 500–50,000 years old.

CONCEPT TEST

How might the increased consumption of fossil fuels over the last century affect the ^{14}C content of growing plant tissues?

SAMPLE EXERCISE 21.7   Integrating Concepts: Radium Girls and Safety in the Workplace

Radium was discovered by Pierre and Marie Curie in 1898. By 1902 the new element had its first practical use: radium compounds were mixed with zinc sulfide (ZnS) to make paint that glowed in the dark as alpha particles emitted by the decay of ^{226}Ra ($t_{1/2}$ = 1600 years) caused ZnS crystals to emit a greenish fluorescence (Figure 21.22). The paint was used to make dials for watches, clocks, and instruments used on naval vessels and, a few years later, in military and civilian airplanes.

By 1914, U.S. companies were making radium-painted dials and employing young women in their late teens and early 20s as dial painters. Soon after their employment many of the women became very sick, suffering from anemia and other symptoms we now associate with overexposure to nuclear radiation. Some developed bone cancer and over one hundred of them, who became known around the world as the Radium Girls, died. Their deaths were linked to the practice of "tipping" the fine paint brushes they used, which meant using their lips to make fine points on the brushes to help them paint the tiny numerals and hands on watch faces (Figure 21.23). In the process they ingested some of the radioactive paint. Tests later determined that about 20% of the radium ingested was incorporated into their bones, where it attacked the bone marrow and caused malignancies known as osteosarcomas.

a. Suggest a reason why radium was concentrated in the victims' bones.
b. Studies of radiation levels and incidence of cancer in over 1000 female dial painters yielded the results in the table below. Which of the two dose–response curves in Figure 21.18 best fits these results?

Radium Exposure (μg ingested)	Occurrence of Malignancy (% of workers exposed)
1	0
3	0
10	0
30	0
100	5
300	53
1000	85

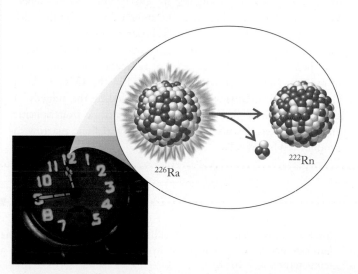

FIGURE 21.22 During the 20th century, many millions of watches and clocks had dials that glowed in the dark as high-energy α particles emitted by ^{226}Ra caused crystals of ZnS to fluoresce.

FIGURE 21.23 This editorial cartoon appeared in Sunday newspapers on February 28, 1926. It portrayed the deadly consequences of young women "tipping" their brushes with their lips as they painted the dials of watches with paint that contained radioactive radium.

c. The green luminescence of radium watch dials began to fade after a few years. Was this loss in luminosity due to decreased radioactivity in the paint? Explain why or why not.

d. In one study, the levels of radioactivity in pocket watches with radium-painted dials were found to be between 0.6 and 1.39 µCi per watch. How many micrograms of ^{226}Ra produce 1.39 µCi of radioactivity?

COLLECT AND ORGANIZE We are asked: (a) why ingested ^{226}Ra concentrates in bones; (b) whether malignancy in dial painters was proportional to their exposure to ^{226}Ra radiation or followed an S-shaped dose–response curve; (c) why the luminosity of radium-activated paint fades after a few years, and (d) how many micrograms of ^{226}Ra are needed to produce 1.39 µCi of radiation. One curie (Ci) is equal to 3.70×10^{10} decay events/s. The half-life ($t_{1/2}$) of ^{226}Ra is 1600 years and is related to the first-order rate constant (k) of the decay reaction by the equation

$$t_{1/2} = 0.693/k$$

The level of radioactivity (A) in a sample of radium is equal to the product of the rate constant and the number (N) of ^{226}Ra atoms: $A = kN$.

ANALYZE Radium is a group 2 element and should have chemical and biochemical properties that are similar to those of the other elements in that group, including its association with biological tissues. High concentrations of another group 2 element, calcium, occur in teeth and bones. Worker exposure levels in the table on the previous page cover a wide range, but exposure to up to nearly 100 µg ^{226}Ra caused few malignancies, whereas concentrations above 100 µg caused many. The half-life of ^{226}Ra is 1600 years, so the radioactivity of a sample decreases little over a few years or even over many decades. Relating a half-life expressed in years to a level of radioactivity expressed in a multiple of decay events per second will require converting units of time and then quantities of radioactive atoms to moles and then micrograms.

SOLVE

a. Radium likely accumulates in bones because its chemistry is similar to that of calcium, which means that ^{226}Ra^{2+} ions are likely to take the place of Ca^{2+} ions in bone tissue.

b. Malignancies did not occur among the dial painters who ingested less than 100 µg of ^{226}Ra; however, the percentage of the women who suffered from them increased sharply with exposure between 100 and 1000 µg. This pattern is described by the S-shaped (blue) curve in Figure 21.18.

c. Given the 1600-year half-life of ^{226}Ra, the loss of watch dial luminescence was not the result of depleted radioactivity. Rather, it must have been due to less efficient conversion of the energy of radioactive decay into visible light by ZnS crystals.

d. Let's first convert the half-life of ^{226}Ra into a decay rate constant in units of s^{-1}:

$$k = \frac{0.693}{t_{1/2}} = \frac{0.693}{1600 \text{ yr}} \times \frac{1 \text{ yr}}{365.25 \text{ d}} \times \frac{1 \text{ d}}{24 \text{ hr}} \times \frac{1 \text{ hr}}{3600 \text{ s}}$$

$$= 1.372 \times 10^{-11} \text{ s}^{-1}$$

Next, we solve the equation $A = kN$ for N, and we use the above rate constant and the radioactivity of the watch to calculate the number of ^{226}Ra atoms in the dial:

$$N = \frac{A}{k} = \frac{1.39 \text{ µCi}}{1.372 \times 10^{-11} \text{ s}^{-1}} \times \frac{1 \text{ Ci}}{1 \times 10^6 \text{ µCi}}$$

$$\times \frac{3.70 \times 10^{10} \text{ s}^{-1}}{\text{Ci}} = 3.747 \times 10^{15} \text{ atoms}$$

The corresponding mass in micrograms is

$$3.747 \times 10^{15} \text{ atoms} \times \frac{1 \text{ mol}}{6.0221 \times 10^{23} \text{ atoms}}$$

$$\times \frac{226 \text{ g}}{1 \text{ mol}} \times \frac{1 \times 10^6 \text{ µg}}{1 \text{ g}} = 1.41 \text{ µg} \ ^{226}\text{Ra}$$

THINK ABOUT IT Did you notice the similarity between the level of radioactivity (1.39 µCi) in the watch dial and the mass of radium (1.41 µg) producing it? This is not a coincidence. When the curie was adopted as the standard unit of radioactivity in the early 20th century, it was chosen to honor the pioneering work of Marie and Pierre Curie, and it was based on what was then believed to be the level of radioactivity in one gram of radium. Newspaper articles published in the 1920s make it clear that Marie Curie was deeply troubled by the tragedy of the Radium Girls. Sadly, in 1934 she would die from aplastic anemia—a disease caused by the inability of bone marrow to produce red blood cells.

SUMMARY

Section 21.1 Nuclear chemistry is the study and application of reactions that involve changes in atomic nuclei.

Section 21.2 Nuclear fusion occurs when subatomic particles or atomic nuclei collide with each other and fuse together, forming clusters of particles or more massive nuclei. When a particle of matter encounters a particle of **antimatter**, both are converted into energy (they annihilate one another), yielding gamma rays. Facilities that harness the enormous energy of hydrogen fusion must operate at temperatures over 10^8 K.

Section 21.3 Stable nuclei have neutron-to-proton ratios that fall within a range of values called the **belt of stability**. Unstable nuclides undergo **radioactive decay**. Neutron-rich nuclides (mass number greater than the average atomic mass) undergo β decay; neutron-poor nuclides undergo **positron emission** or **electron capture**. Very large nuclides ($Z > 83$) may undergo β decay or **α decay**.

Section 21.4 Radioactive decay follows first-order kinetics, so the half-life ($t_{1/2}$) of a radionuclide is a characteristic value of the decay process.

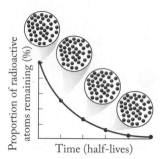

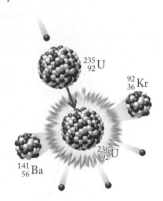

Section 21.5 Neutron absorption by uranium-235 and a few other massive isotopes may lead to **nuclear fission** into lighter nuclei accompanied by the release of energy that can be harnessed to generate electricity. A **chain reaction** happens when the neutrons released during fission collide with other fissionable nuclei. They require a **critical mass** of a fissionable isotope. A **breeder reactor** is used to make plutonium-239 from uranium-238 while also producing energy to make electricity.

Section 21.6 **Scintillation counters** and **Geiger counters** are used to measure levels of nuclear radiation. Radioactivity is the number of decay events per unit time. Common units are the **becquerel (Bq)** (1 decay event/s) and the **curie (Ci)** (1 Ci = 3.70×10^{10} Bq).

Section 21.7 Alpha particles, β particles, and γ rays have enough energy to break up molecules into electrons and cations and are examples of **ionizing radiation**, radiation that can damage body tissue and DNA. The quantity of ionizing radiation energy absorbed per kilogram of body mass is called the *absorbed dose* and is expressed in **grays (Gy)**: 1 Gy = 1.00 J/kg. The effective dose of any type of ionizing radiation is the product of the absorbed dose in grays and the **relative biological effectiveness (RBE)** of the radiation; the unit of effective dose is the **sievert (Sv)**. Alpha particles have a larger RBE than β particles and γ rays but have the least penetrating power of these three types of ionizing radiation.

Section 21.8 Selected radioactive isotopes are useful as tracers in the human body to map biological activity and diagnose diseases. Other radioactive isotopes are used to treat cancers.

Section 21.9 **Radiometric dating** is used to determine the age of an object based on its content of a radionuclide and/or its decay product. **Radiocarbon dating** involves measuring the amount of radioactive carbon-14 that remains in an object derived from plant or animal tissue to calculate the age of the object. To improve the accuracy of the technique, scientists calibrate the results of radiometric analysis of samples of known age, such as growth rings in ancient trees.

PROBLEM-SOLVING SUMMARY

TYPE OF PROBLEM	CONCEPTS AND EQUATIONS		SAMPLE EXERCISES
Calculating the energy released in a nuclear reaction	$E = mc^2$, where m is loss in mass as reactants form products. When the units of m are kg and the units of c are m/s, E is in joules because 1 J = 1 kg $\cdot$ (m/s)^2.		21.1
Predicting the modes and products of radioactive decay	Neutron-rich nuclides tend to undergo β decay; neutron-poor nuclides undergo positron emission or electron capture.		21.2
Calculations involving half-lives	$$\frac{N_t}{N_0} = 0.5^{t/t_{1/2}}$$ where N_t/N_0 is the ratio of the quantity of radionuclide present in a sample at time t (N_t) to the quantity at $t = 0$ (N_0).	(21.11)	21.3
Calculating the radioactivity of a sample	$$A = kN$$ where $k = 0.693/t_{1/2}$.	(21.12)	21.4
Calculating effective dose	Effective dose = absorbed dose $\times$ RBE		21.5
Radiometric dating	$$t = -\frac{t_{1/2}}{0.693} \ln \frac{N_t}{N_0}$$	(21.13)	21.6

VISUAL PROBLEMS

(Answers to boldface end-of-chapter questions and problems are in the back of the book.)

21.1. Which of the highlighted elements in Figure P21.1 currently plays a key role in the controlled fusion of hydrogen?

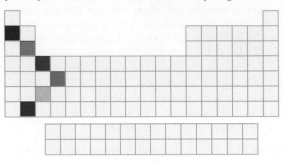

FIGURE P21.1

21.2. Exposure to which of the highlighted elements in Figure P21.1 could cause anemia and bone disease?

21.3. Which of the highlighted elements in Figure P21.1 are products of the decay of uranium-238?

21.4. Which of the graphs in Figure P21.4 illustrates α decay? Which decay pathway does the other graph illustrate?

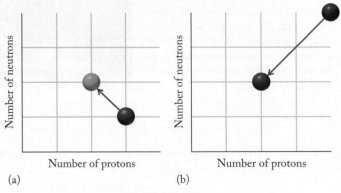

(a)   (b)

FIGURE P21.4

21.5. Which of the graphs in Figure P21.5 illustrates β decay?

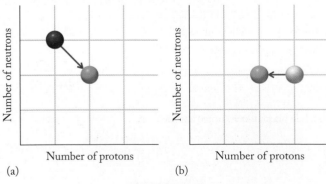

(a)   (b)

FIGURE P21.5

21.6. Which of the graphs in Figure P21.6 illustrates the overall effect of neutron capture followed by β decay?

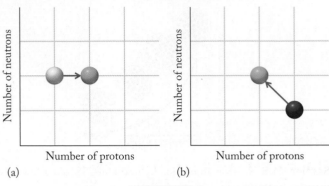

(a)   (b)

FIGURE P21.6

21.7. Which of the curves in Figure P21.7 represents the decay of an isotope that has a half-life of 2.0 days?

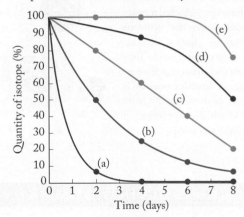

FIGURE P21.7

21.8. Which of the curves in Figure P21.7 do not represent a radioactive decay curve?

21.9. Which of the models in Figure P21.9 represents fission and which represents fusion?

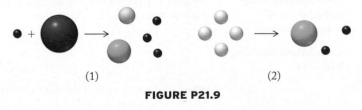

(1)   (2)

FIGURE P21.9

21.10. Isotopes in a nuclear decay series emit particles with a positive charge and particles with a negative charge. The two kinds of particles penetrate a column of water as shown in Figure P21.10. Is the "X" particle the positive or the negative one?

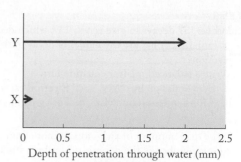

Depth of penetration through water (mm)

FIGURE P21.10

<hr>

QUESTIONS AND PROBLEMS

Fusion and the Quest for Clean Energy

CONCEPT REVIEW

21.11. Arrange the following particles in order of increasing mass: electron, β particle, positron, proton, neutron, α particle, deuteron.

21.12. Electromagnetic radiation is emitted when a neutron and proton fuse to make a deuteron. In which region of the electromagnetic spectrum is the radiation?

21.13. Scientists at the Fermi National Accelerator Laboratory in Illinois announced in the fall of 1996 that they had created "antihydrogen." How does antihydrogen differ from hydrogen?

21.14. Describe an antiproton.

21.15. In what ways are the fusion reactions that formed alpha particles during primordial nucleosynthesis different from those that fuel our sun today?

21.16. How are the fusion reactions that are the basis for power production in the tokamak described in Section 21.2 different from those that power our sun?

PROBLEMS

21.17. Calculate the energy and wavelength of the two gamma rays released by the annihilation of a proton and an antiproton.

21.18. Calculate the energy released and the wavelength of the two photons emitted in the annihilation of an electron and a positron.

<hr>

21.19. All of the following fusion reactions produce ^{28}Si. Calculate the energy released in each reaction from the masses of the isotopes: ^{2}H (2.0146 amu), ^{4}He (4.00260 amu), ^{10}B (10.0129 amu), ^{12}C (12.000 amu), ^{14}N (14.00307 amu), ^{16}O (15.99491 amu), ^{24}Mg (23.98504 amu), ^{28}Si (27.97693 amu).
 a. ^{14}N + ^{14}N → ^{28}Si
 b. ^{10}B + ^{16}O + ^{2}H → ^{28}Si
 c. ^{16}O + ^{12}C → ^{28}Si
 d. ^{24}Mg + ^{4}He → ^{28}Si

21.20. All of the following fusion reactions produce ^{32}S. Calculate the energy released in each reaction from the masses of the isotopes: ^{4}He (4.00260 amu), ^{6}Li (6.01512 amu), ^{12}C (12.000 amu), ^{14}N (14.00307 amu), ^{16}O (15.99491 amu), ^{24}Mg (23.98504 amu), ^{28}Si (27.97693 amu), ^{32}S (31.97207 amu).
 a. ^{16}O + ^{16}O → ^{32}S
 b. ^{28}Si + ^{4}He → ^{32}S
 c. ^{14}N + ^{12}C + ^{6}Li → ^{32}S
 d. ^{24}Mg + 2 ^{4}He → ^{32}S

<hr>

21.21. Tokamak Radiochemistry How much energy is released per nucleus of tritium produced during the following reactions? The mass of $^{3}_{1}$H is 5.00827×10^{-27} kg.
 a. $^{1}_{0}$n + $^{6}_{3}$Li → $^{4}_{2}$He + $^{3}_{1}$H
 b. $^{1}_{0}$n + $^{7}_{3}$Li → $^{4}_{2}$He + $^{3}_{1}$H + $^{1}_{0}$n

21.22. It has been proposed that electrical power production in the future might be based on the fusion of deuterium to helium-4.
 a. Write a radiochemical equation describing the reaction (assume that ^{4}He is the only product).
 b. Calculate how much energy is released during the formation of one mole of ^{4}He. The masses of the two nuclides are ^{2}D = 2.014102 amu and ^{4}He = 4.002603 amu.

The Belt of Stability

CONCEPT REVIEW

21.23. How can the belt of stability be used to predict the probable decay mode of an unstable nuclide?

21.24. Compare and contrast positron-emission and electron-capture processes.

21.25. The ratio of neutrons to protons in stable nuclei increases with increasing atomic number. Use this trend to explain why multiple α decay steps in the ^{238}U decay series are often followed by β decay.

21.26. If a ^{10}B nucleus absorbs a proton, the radionuclide that is produced may undergo α decay. What nuclide is produced in the decay process? Is it stable or radioactive?

PROBLEMS

21.27. Iodine-137 decays to give xenon-137, which decays to give cesium-137. What are the modes of decay in these two reactions?

21.28. Write a balanced nuclear equation for:
  a. Beta emission by ^{28}Mg c. Electron capture by ^{129}Cs
  b. Alpha emission by ^{255}Lr d. Positron emission by ^{25}Al

21.29. If the mass number of an isotope is more than twice the atomic number, is the neutron-to-proton ratio less than, greater than, or equal to 1?

21.30. In each of the following pairs of isotopes, which isotope has more protons and which has more neutrons? (a) ^{127}I or ^{131}I; (b) ^{188}Re or ^{188}W; (c) ^{14}N or ^{14}C

21.31. Aluminum atoms in Earth's crust are all ^{27}Al. However, ^{26}Al is formed in stars. It decays to ^{26}Mg with a half-life of 7.4×10^5 years. Write an equation describing the decay of ^{26}Al to ^{26}Mg.

21.32. Which nuclide is produced by the β decay of ^{131}I?

21.33. Predict the modes of decay for the following radioactive isotopes: (a) ^{10}C; (b) ^{19}Ne; (c) ^{50}Ti.

21.34. Predict the decay pathways of the following radioactive isotopes: (a) ^{56}Ni; (b) ^{90}Sr; (c) ^{116}Sb.

21.35. Elements in a Supernova The isotopes ^{56}Co and ^{44}Ti were detected in supernova SN 1987A. Predict the decay pathway for these radioactive isotopes.

21.36. Nine isotopes of sulfur have mass numbers ranging from 30 to 38. Five of the nine are radioactive: ^{30}S, ^{31}S, ^{35}S, ^{37}S, and ^{38}S. Which of these isotopes do you expect to decay by β decay?

Rates of Radioactive Decay

PROBLEMS

21.37. What percentage of a sample's original radioactivity remains after two half-lives?

21.38. What percentage of a sample's original radioactivity remains after five half-lives?

21.39. What is the half-life of a radionuclide if 87.5% of it decays in 6.6 days?

21.40. What is the half-life of a radionuclide if only 6.25% of it remains after 8 hours and 20 minutes?

21.41. Explosions at a disabled nuclear power station in Fukushima, Japan, in 2011 may have released more cesium-137 ($t_{1/2} = 30.2$ years) into the ocean than any other single event. How long will it take the radioactivity of this radionuclide to decay to 5.0% of the level released in 2011?

21.42. Spent fuel removed from nuclear power stations contains plutonium-239 ($t_{1/2} = 2.41 \times 10^4$ years). How long will it take a sample of this radionuclide to reach a level of radioactivity that is 2.5% of the level it had when it was removed from a reactor?

Nuclear Fission

CONCEPT REVIEW

21.43. How is the rate of energy release controlled in a nuclear reactor?

21.44. How does a breeder reactor create fuel and energy at the same time?

*21.45. Why are neutrons always by-products of the fission of most massive nuclides? (*Hint:* Look closely at the neutron-to-proton ratios shown in Figure 21.2.)

21.46. Seaborgium (Sg, element 106) is prepared by the bombardment of curium-248 with neon-22, which produces two isotopes, ^{265}Sg and ^{266}Sg. Write balanced nuclear reactions for the formation of both isotopes. Are these reactions better described as fusion or fission processes?

PROBLEMS

21.47. The fission of uranium produces dozens of isotopes. For each of the following fission reactions, determine the identity of the unknown nuclide:
  a. $^{235}U + {}^1n \rightarrow {}^{96}Zr + ? + 2\,{}^1n$
  b. $^{235}U + {}^1n \rightarrow {}^{99}Nb + ? + 4\,{}^1n$
  c. $^{235}U + {}^1n \rightarrow {}^{90}Rb + ? + 3\,{}^1n$

21.48. For each of the following fission reactions, determine the identity of the unknown nuclide:
  a. $^{235}U + {}^1n \rightarrow {}^{137}I + ? + 2\,{}^1n$
  b. $^{235}U + {}^1n \rightarrow {}^{137}Cs + ? + 3\,{}^1n$
  c. $^{235}U + {}^1n \rightarrow {}^{141}Ce + ? + 2\,{}^1n$

21.49. For each of the following fission reactions, determine the identity of the unknown nuclide:
  a. $^{235}U + {}^1n \rightarrow {}^{131}I + ? + 2\,{}^1n$
  b. $^{235}U + {}^1n \rightarrow {}^{103}Ru + ? + 3\,{}^1n$
  c. $^{235}U + {}^1n \rightarrow {}^{95}Zr + ? + 3\,{}^1n$

21.50. For each of the following fission reactions, determine the identity of the unknown nuclide:
  a. $^{235}U + {}^1n \rightarrow {}^{147}Pm + ? + 2\,{}^1n$
  b. $^{235}U + {}^1n \rightarrow {}^{94}Kr + ? + 2\,{}^1n$
  c. $^{235}U + {}^1n \rightarrow {}^{95}Sr + ? + 3\,{}^1n$

Measuring Radioactivity; Biological Effects of Radioactivity

CONCEPT REVIEW

21.51. What is the difference between a *level* of radioactivity and a *dose* of radioactivity?

21.52. What are some of the molecular effects of exposure to radioactivity?

21.53. Describe the dangers of exposure to radon-222.

21.54. Food Safety Periodic outbreaks of food poisoning from *E. coli*–contaminated meat have renewed the debate about irradiation as an effective treatment of food. In one newspaper article on the subject, the following statement appeared: "Irradiating food destroys bacteria by breaking apart their molecular structure." How would you improve or expand on this explanation?

PROBLEMS

21.55. Radiation Exposure from Dental X-rays Dental X-rays expose patients to about 5 μSv of radiation. Given an RBE of 1 for X-rays, how many grays of radiation does 5 μSv represent? For a 50 kg person, how much energy does 5 μSv correspond to?

*21.56. **Radiation Exposure at Chernobyl** Some workers responding to the explosion at the Chernobyl nuclear power plant were exposed to 5 Sv of radiation, resulting in death for many of them. If the exposure was primarily in the form of γ rays with an energy of 3.3×10^{-14} J and an RBE of 1, how many γ rays did an 80 kg person absorb?

*21.57. **Strontium-90 in Milk** In the years immediately following the explosion at the Chernobyl nuclear power plant, the concentration of ^{90}Sr in cow's milk in southern Europe was slightly elevated. Some samples contained as much as 1.25 Bq/L of ^{90}Sr radioactivity. The half-life of strontium-90 is 28.8 years.
   a. Write a balanced nuclear equation describing the decay of ^{90}Sr.
   b. How many atoms of ^{90}Sr are in a 200 mL glass of milk with 1.25 Bq/L of ^{90}Sr radioactivity?
   c. Why would strontium-90 be more concentrated in milk than other foods, such as grains, fruits, or vegetables?

*21.58. **Radium Watch Dials** If exactly 1.00 μg of ^{226}Ra was used to paint the glow-in-the-dark dial of a wristwatch made in 1914, how radioactive is the watch today? Express your answer in microcuries and becquerels. The half-life of ^{226}Ra is 1.60×10^3 years.

21.59. In 1999, the U.S. Environmental Protection Agency set a maximum radon level for drinking water at 4.0 pCi per milliliter.
   a. How many decay events occur per second in a milliliter of water for this level of radon radioactivity?
   b. If the above radioactivity were due to decay of ^{222}Rn ($t_{1/2} = 3.8$ days), how many ^{222}Rn atoms would there be in 1.0 mL of water?

21.60. A former Russian spy died from radiation sickness in 2006 after dining at a London restaurant where he apparently ingested polonium-210. The other people at his table did not suffer from radiation sickness, even though they were very near the radioactive food the victim ate. Why were they not affected?

Medical Applications of Radionuclides

CONCEPT REVIEW

21.61. How does the selection of an isotope for radiotherapy relate to (a) its half-life, (b) its mode of decay, and (c) the properties of the products of decay?

21.62. Are the same radioactive isotopes likely to be used for both imaging and cancer treatment? Why or why not?

PROBLEMS

21.63. Predict the most likely mode of decay for the following isotopes used as imaging agents in nuclear medicine: (a) ^{197}Hg (kidney); (b) ^{75}Se (parathyroid gland); (c) ^{18}F (bone).

21.64. Predict the most likely mode of decay for the following isotopes used as imaging agents in nuclear medicine: (a) ^{133}Xe (cerebral blood flow); (b) ^{57}Co (tumor detection); (c) ^{51}Cr (red blood cell mass); (d) ^{67}Ga (tumor detection).

21.65. A 1.00 mg sample of ^{192}Ir was inserted into the artery of a heart patient. After 30 days, 0.756 mg remained. What is the half-life of ^{192}Ir?

21.66. In a treatment that decreases pain and reduces inflammation of the lining of the knee joint, a sample of dysprosium-165 with an radioactivity of 1100 counts per second was injected into the knee of a patient suffering from rheumatoid arthritis. After 24 hr, the radioactivity had dropped to 1.14 counts per second. Calculate the half-life of ^{165}Dy.

21.67. **Treatment of Tourette's Syndrome** Tourette's syndrome is a condition whose symptoms include sudden movements and vocalizations. Iodine isotopes are used in brain imaging of people suffering from Tourette's syndrome. To study the uptake and distribution of iodine in cells, mammalian brain cells in culture were treated with a solution containing ^{131}I with an initial radioactivity of 108 counts per minute. The cells were removed after 30 days, and the remaining solution was found to have a radioactivity of 4.1 counts per minute. Did the brain cells absorb any ^{131}I ($t_{1/2} = 8.1$ days)?

21.68. A patient is administered mercury-197 to evaluate kidney function. Mercury-197 has a half-life of 65 hr. What fraction of an initial dose of mercury-197 remains after 6 days?

21.69. Carbon-11 is an isotope used in positron-emission tomography and has a half-life of 20.4 min. How long will it take for 99% of the ^{11}C injected into a patient to decay?

21.70. Sodium-24 is used to treat leukemia and has a half-life of 15 hr. A patient was injected with a salt solution containing sodium-24. What percentage of the ^{24}Na remained after 48 hr?

*21.71. **Boron Neutron-Capture Therapy** In boron neutron-capture therapy (BNCT), a patient is given a compound containing ^{10}B that accumulates inside cancer tumors. Then the tumors are irradiated with neutrons, which are absorbed by ^{10}B nuclei. The product of neutron capture is an unstable form of ^{11}B that undergoes α decay to ^{7}Li.
   a. Write a balanced nuclear equation for the neutron absorption and α decay process.
   b. Calculate the energy released by each nucleus of boron-10 that captures a neutron and undergoes α decay, given the following masses of the particles in the process: ^{10}B (10.0129 amu), ^{7}Li (7.01600 amu), ^{4}He (4.00260 amu), and ^{1}n (1.00866 amu).
   c. Why is the formation of a nuclide that undergoes α decay a particularly effective cancer therapy?

21.72. Balloon Angioplasty and Arteriosclerosis Balloon angioplasty is a common procedure for unclogging arteries in patients suffering from arteriosclerosis. Iridium-192 therapy is being tested as a treatment to prevent reclogging of the arteries. In the procedure, a thin ribbon containing pellets of ^{192}Ir is threaded into the artery. The half-life of ^{192}Ir is 74 days. How long will it take for 99% of the radioactivity from 1.00 mg of ^{192}Ir to disappear?

Radiometric Dating

CONCEPT REVIEW

21.73. Explain why radiocarbon dating is reliable only for artifacts and fossils younger than about 50,000 years.

21.74. Which of the following statements about ^{14}C dating are true?
   a. The amount of ^{14}C in all objects is the same.
   b. Carbon-14 is unstable and is readily lost from the atmosphere.
   c. The ratio of ^{14}C to ^{12}C in the atmosphere is a constant.
   d. Living tissue will absorb ^{12}C but not ^{14}C.

21.75. Why is ^{40}K dating ($t_{1/2} = 1.28 \times 10^9$ years) useful only for rocks older than 300,000 years?

21.76. Where does the ^{14}C found in plants come from?

PROBLEMS

21.77. First Humans in South America Archeologists continue to debate the origins and dates of arrival of the first humans in the Western Hemisphere. Radiocarbon dating of charcoal from a cave in Chile was used to establish the earliest date of human habitation in South America as 8700 years ago. What fraction of the ^{14}C present initially remained in the charcoal after 8700 years?

21.78. For thousands of years native Americans living along the north coast of Peru used knotted cotton strands called *quipu* (Figure P21.78) to record financial transactions and governmental actions. A particular quipu sample is 4800 years old. Compared with the fibers of cotton plants growing today, what is the ratio of carbon-14 to carbon-12 in the sample?

FIGURE P21.78

*21.79. Figure P21.79 is a close-up of the center of a giant sequoia tree cut down in 1891 in what is now Kings Canyon National Park. It contained 1342 annual growth rings. If samples of the tree were removed for radiocarbon dating today, what would be the difference in ^{14}C/^{12}C ratio in the innermost (oldest) ring compared with that ratio in the youngest ring?

FIGURE P21.79

*21.80. Geologists who study volcanoes can develop historical profiles of previous eruptions by determining the ^{14}C/^{12}C ratios of charred plant remains trapped in old magma and ash flows. If the uncertainty in determining these ratios is 0.1%, could radiocarbon dating distinguish between debris from the eruptions of Mt. Vesuvius that occurred in the years 472 and 512? (*Hint*: Calculate the ^{14}C/^{12}C ratios for samples from the two dates.)

21.81. Figure P21.81 shows a carved mammoth tusk that was uncovered at an ancient camp site in the Ural Mountains in 2001. The ^{14}C/^{12}C ratio in the tusk was only 1.19% of that in modern elephant tusks. How old is the mammoth tusk?

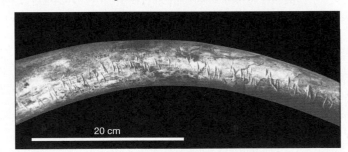

20 cm

FIGURE P21.81

21.82. The Destruction of Jericho The Bible describes the Exodus as a period of 40 years that began with plagues in Egypt and ended with the destruction of Jericho. Archeologists seeking to establish the exact dates of these events have proposed that the plagues coincided with a huge eruption of the volcano Thera in the Aegean Sea.
   a. Radiocarbon dating suggests that the eruption occurred around 1360 BCE, though other records place the eruption of Thera in the year 1628 BCE. What is the percent difference in the ^{14}C decay rate in biological samples from these two dates?
   b. Radiocarbon dating of blackened grains from the site of ancient Jericho provides a date of 1315 BCE ±13 years for the fall of the city. What is the ^{14}C/^{12}C ratio in the blackened grains compared with that of grain harvested last year?

Additional Problems

21.83. Thirty years before the creation of antihydrogen, television producer Gene Roddenberry (1921–1991) proposed to use this form of antimatter to fuel the powerful "warp" engines of the fictional starship *Enterprise*.
   a. Why would antihydrogen have been a particularly suitable fuel?
   b. Describe the challenges of storing such a fuel on a starship.

21.84. Tiny concentrations of radioactive tritium ($^3_1$H) occur naturally in rain and groundwater. The half-life of $^3_1$H is 12 years. Assuming that tiny concentrations of tritium can be determined accurately, could the isotope be used to determine whether a bottle of wine with the year 1969 on its label actually contained wine made from grapes that were grown in 1969? Explain your answer.

21.85. The energy released during the fission of ^{235}U is about 3.2×10^{-11} J per atom of the isotope. Compare this quantity of energy with that released by the fusion of four hydrogen atoms to make an atom of helium-4:

$$4\,^1_1\text{H} \rightarrow\ ^4_2\text{He} + 2\,^0_1\beta$$

Assume that the positrons are annihilated in collisions with electrons so that the masses of the positrons are converted into energy. In your comparison, express the energies released by the fission and fusion processes in joules per nucleon for ^{235}U and ^{4}He, respectively.

21.86. How much energy is required to remove a neutron from the nucleus of an atom of carbon-13 (mass = 13.00335 amu)? (*Hint*: The mass of an atom of carbon-12 is exactly 12.00000 amu.)

21.87. **Smoke Detectors** Americium-241 ($t_{1/2}$ = 433 yr) is used in smoke detectors. The α particles from this isotope ionize nitrogen and oxygen in the air, creating an electric current. When smoke is present, the current decreases, setting off the alarm.
   a. Does a smoke detector bear a closer resemblance to a Geiger counter or to a scintillation counter?
   b. How long will it take for the radioactivity of a sample of ^{241}Am to drop to 1% of its original radioactivity?
   c. Why are smoke detectors containing ^{241}Am safe to handle without protective equipment?

21.88. **Colorectal Cancer Treatment** Cancer therapy with radioactive rhenium-188 shows promise in patients suffering from colorectal cancer.
   a. Write the symbol for rhenium-188 and determine the number of neutrons, protons, and electrons.
   b. Are most rhenium isotopes likely to have fewer neutrons than rhenium-188?
   c. The half-life of rhenium-188 is 17 hours. If it takes 30 minutes to bind the isotope to an antibody that delivers the rhenium to the tumor, what percentage of the rhenium remains after binding to the antibody?
   d. The effectiveness of rhenium-188 is thought to result from penetration of β particles as deep as 8 mm into the tumor. Why wouldn't an α emitter be more effective?

   e. Using an appropriate reference text, such as the *CRC Handbook of Chemistry and Physics*, pick out the two most abundant isotopes of rhenium. List their natural abundances and explain why the one that is radioactive decays by the pathway that it does.

21.89. In 2006 an international team of scientists confirmed the synthesis of a total of three atoms of $^{294}_{118}$Uuo in experiments run in 2002 and 2005. They bombarded a ^{249}Cf target with ^{48}Ca nuclei.
   a. Write a balanced nuclear equation describing the synthesis of $^{294}_{118}$Uuo.
   b. The synthesized isotope of Uuo undergoes α decay ($t_{1/2}$ = 0.9 ms). What nuclide is produced by the decay process?
   c. The nuclide produced in part b also undergoes α decay ($t_{1/2}$ = 10 ms). What nuclide is produced by this decay process?
   d. The nuclide produced in part c also undergoes α decay ($t_{1/2}$ = 0.16 s). What nuclide is produced by this decay process?
   e. If you had to select an element that occurs in nature and that has physical and chemical properties similar to Uuo, which element would it be?

*21.90. Consider the following decay series:

$$\text{A}\ (t_{1/2} = 4.5\ \text{s}) \rightarrow \text{B}\ (t_{1/2} = 15.0\ \text{days}) \rightarrow \text{C}$$

If we start with 10^6 atoms of A, how many atoms of A, B, and C are there after 30 days?

21.91. Which element in the following series will be present in the greatest amount after 1 year?

$$^{214}_{83}\text{Bi} \xrightarrow{\alpha}\ ^{210}_{81}\text{Tl} \xrightarrow{\beta}\ ^{210}_{82}\text{Pb} \xrightarrow{\beta}\ ^{210}_{83}\text{Bi} \longrightarrow$$
$$t_{1/2} =\quad 20\ \text{min}\quad 1.3\ \text{min}\quad 20\ \text{yr}\quad 5\ \text{d}$$

*21.92. **Dating Cave Paintings** Cave paintings in Gua Saleh Cave in Borneo have been dated by measuring the amount of ^{14}C in calcium carbonate that formed over the pigments used in the paint. The source of the carbonate ion was atmospheric CO_2.
   a. What is the ratio of the ^{14}C radioactivity in calcium carbonate that formed 9900 years ago to that in calcium carbonate formed today?
   b. The archeologists also used a second method, uranium–thorium dating, to confirm the age of the paintings by measuring trace quantities of these elements present as contaminants in the calcium carbonate. Shown below are two candidates for the U–Th dating method. Which isotope of uranium do you suppose was chosen? Explain your answer.

$$^{235}_{92}\text{U} \rightarrow\ ^{231}_{90}\text{Th} \rightarrow\ ^{231}_{91}\text{Pa} \rightarrow$$
$$t_{1/2} =\quad 7.04 \times 10^8\ \text{yr}\quad 25.6\ \text{hr}\quad 3.25 \times 10^4\ \text{yr}$$

$$^{234}_{92}\text{U} \rightarrow\ ^{230}_{90}\text{Th} \rightarrow\ ^{226}_{88}\text{Pa} \rightarrow$$
$$t_{1/2} =\quad 2.44 \times 10^5\ \text{yr}\quad 7.7 \times 10^4\ \text{hr}\quad 1600\ \text{yr}$$

21.93. The synthesis of new elements and specific isotopes of known elements in linear accelerators involves the fusion of smaller nuclei.
  a. An isotope of platinum can be prepared from nickel-64 and tin-124. Write a balanced equation for this nuclear reaction. (You may assume that no neutrons are ejected in the fusion reaction.)
  b. Substitution of tin-132 for tin-124 increases the rate of the fusion reaction 10 times. Which isotope of Pt is formed in this reaction?

21.94. A sample of drinking water collected from a suburban Boston municipal water system in 2002 contained 0.5 pCi/L of radon. Assume that this level of radioactivity was due to the decay of ^{222}Rn ($t_{1/2} = 3.8$ days).
  a. What was the level of radioactivity (Bq/L) of this nuclide in the sample?
  b. How many decay events per hour would occur in 2.5 L of the water?

21.95. Stone Age Skeletons The discovery of six skeletons in an Italian cave at the beginning of the 20th century was considered a significant find in Stone Age archaeology. The age of these bones has been debated. The first attempt at radiocarbon dating indicated an age of 15,000 years. Redetermination of the age in 2004 indicated an older age for two bones, between 23,300 and 26,400 years. What is the ratio of ^{14}C in a sample 15,000 years old to one 25,000 years old?

21.96. There was once a plan to store radioactive waste that contained plutonium-239 in the reefs of the Marshall Islands. The planners claimed that the plutonium would be "reasonably safe" after 240,000 years. If the half-life is 24,400 years, what percentage of the ^{239}Pu would remain after 240,000 years?

21.97. Dating Prehistoric Bones In 1997 anthropologists uncovered three partial skulls of prehistoric humans in the Ethiopian village of Herto. Based on the amount of ^{40}Ar in the volcanic ash in which the remains were buried, their age was estimated at between 154,000 and 160,000 years old.
  a. ^{40}Ar is produced by the decay of ^{40}K ($t_{1/2} = 1.28 \times 10^9$ yr). Propose a decay mechanism for ^{40}K to ^{40}Ar.
  b. Why did the researchers choose ^{40}Ar rather than ^{14}C as the isotope for dating these remains?

*21.98. **Biblical Archeology** The Old Testament describes the construction of the Siloam Tunnel, used to carry water into Jerusalem under the reign of King Hezekiah (727–698 BCE). An inscription on the tunnel has been interpreted as evidence that the tunnel was not built until 200–100 BCE. ^{14}C dating (in 2003) indicated a date close to 700 BCE. What is the ratio of ^{14}C in a wooden object made in 100 BCE to one made from the same kind of wood in 700 BCE?

21.99. Thorium-232 slowly decays to bismuth-212 ($t_{1/2} = 1.4 \times 10^{10}$ yr). Bismuth-212 decays to lead-208 by two pathways: first β and then α decay, or α and then β decay. The intermediate nuclide in the second pathway is thallium-208. The thallium-208 can be separated from a sample of thorium nitrate by passing a solution of the sample through a filter pad containing ammonium phosphomolybdate. The radioactivity of ^{208}Tl trapped on the filter is measured as a function of time. In one such experiment, the following data were collected:

Time (s)	Counts/min
60	62
120	40
180	35
240	22
300	16
360	10

Use the data in the table to determine the half-life of ^{208}Tl.

Mathematical Procedures

Working with Scientific Notation

Quantities that scientists work with are sometimes very large, such as Earth's mass, and other times very small, such as the mass of an electron. It is easier to work with these values when they are expressed in scientific notation.

The general form of standard scientific notation is a value between 1 and 10 multiplied by 10 raised to an integral power. According to this definition, 598×10^{22} kg (Earth's mass) is not in standard scientific notation, but 5.98×10^{24} kg is. It is good practice to use and report values in standard scientific notation.

1. **To convert an "ordinary" number to standard scientific notation,** move the decimal point to the left for a large number, or to the right for a small one, so that the decimal point is located after the first nonzero digit.

   A. For example, to express Earth's average density (5517 kg/m^3) in scientific notation requires moving the decimal point three places to the left. Doing so is the same as dividing the number by 1000, or 10^3. To keep the value the same we multiply it by 10^3. So, Earth's density in standard scientific notation is 5.517×10^3 kg/m^3.

   B. If we move the decimal point of a value less than 1 to the right to express it in scientific notation, then the exponent is a negative integer equal to the number of places we moved the decimal point to the right. For example, the value of R used in solving ideal gas law problems is 0.08206 L $\cdot$ atm/(mol $\cdot$ K). Moving the decimal point two places to the right converts the value of R to scientific notation: 8.206×10^{-2} L $\cdot$ atm/(mol $\cdot$ K).

   C. Another value of R, 8.314 J/(mol $\cdot$ K), does not need an exponent, though it could be written 8.314×10^0 J/(mol $\cdot$ K) because any value raised to the zero power is equal to 1.

2. **For calculations with numbers in scientific notation,** most calculators have a function key for entering the exponents of values expressed in scientific notation. In many calculators it is labeled "E" or "EE" or "Exp." To enter, for example, the speed of light in meters per second, 2.998×10^8, we enter the value before the exponent, 2.998, followed by the exponent key and then the value of the exponent (8). To enter a value with a negative exponent, use the sign-change key. Sometimes it is labeled "(−)" or "+/−" or "±."

Working with Logarithms

A logarithm to the base 10 has the following form:

$$\log_{10} x = \log x = p, \text{ where } x = 10^p$$

We usually abbreviate the logarithm function "log" if the logarithm is to the base 10, which means the scale in which log 10 = 1.

A logarithm to the base e, called a *natural logarithm*, has the following form:

$$\log_e x = \ln x = q, \text{ where } x = e^q$$

Scientific calculators have (LOG) and (LN) keys, so it is easy to convert a number into its log or ln form. The directions below apply to most calculators.

Sample Exercise 1 Find the logarithm to the base 10 of 2.247 (log 2.247).

Solution In some calculators you enter 2.247 first and then press the (LOG) key. In others, such as the TI 84/89 series, you press the (LOG) key first followed by 2.247 and then the (ENTER) key. Either way, the answer should be 0.3516 (to four significant figures).

Sample Exercise 2 Find the natural logarithm of 2.247 (ln 2.247).

Solution Follow the same procedure as in Sample Exercise 1 except use the $\boxed{LN}$ key. The answer should be 0.8096. This answer is (0.8096/0.3516) = 2.303 times larger than log value. That is,

$$\ln x = 2.303 \log x$$

Sample Exercise 3 Find the log of 6.0221×10^{23}.

Solution Following the procedures described above for entering a value with an exponent into your calculator and then taking its log to the base 10, you should obtain the value 23.77974796. We know the original value to five significant figures; the log value should have the same precision, but how do we express it? You might think the log value should be 23.780; however, the 23 to the left of the decimal point reflects the value of the exponent in the original value, and exponents don't count in determining the number of significant figures in a value. Therefore, the value of the logarithm to five significant figures is 23.77975. To understand better why this is so, calculate the log of 6.0221. You should get 0.77975 to five significant figures. Note how the difference between the two log values (23.77975 and 0.77975) is simply the value of the exponent in the first value.

Combining Logs The following equations summarize how logarithms of the products or quotients of two or more values are related to the individual logs of those values:

$$\text{logarithm } ab = \text{logarithm } a + \text{logarithm } b$$

and

$$\text{logarithm } a/b = \text{logarithm } a - \text{logarithm } b$$

Converting Logarithms into Numbers

If we know the value of $\log x$, what is the value of x? This question is frequently asked when working with pH, which is the negative log of the concentration of hydrogen ions, $[H^+]$, in solution:

$$pH = -\log[H^+]$$

Suppose the pH of a solution of a weak acid is 2.50. The concentration of H^+ is related to this pH value as follows:

$$2.50 = -\log[H^+]$$

or

$$-2.50 = \log[H^+]$$

To find the value of $[H^+]$, we enter 2.5 into the calculator and press the appropriate key to change its sign to -2.5. The next step depends on the type of calculator. If yours has a $\boxed{10^x}$ key (often accessible using a second function key), use it to find the value of $10^{-2.5}$, which is the value we are looking for. The corresponding keystrokes with many graphing calculators are $\boxed{10^x}$, $\boxed{(-)}$, 2.5, $\boxed{\text{ENTER}}$. Some calculators, including the virtual one in many Windows operating systems, have an $\boxed{x^y}$ key. We can use it for this problem by entering 10 and pushing the $\boxed{x^y}$ key, then entering 2.5 and pushing the $\boxed{+/-}$ key followed by the $\boxed{=}$ key. All of these approaches do the same calculation, taking 10 to the -2.50 power, and give the same answer, $[H^+] = 3.2 \times 10^{-3}$ to two significant figures. (Remember, pH is a log value, so the digit before the decimal point is not significant.)

Solving Quadratic Equations

If the terms in an equation can be rearranged so that they take the form

$$ax^2 + bx + c = 0$$

they have the form of a quadratic equation. The value(s) of x can be determined from the values of the coefficients a, b, and c by using the equation

$$x = \frac{-b \pm \sqrt{b^2 - 4ac}}{2a}$$

For example, if the solution to a problem yields the following expression where x is the concentration of a solute:

$$x^2 + 0.112x - 1.2 \times 10^{-3} = 0$$

Then the value of x can be determined as follows:

$$x = \frac{-b \pm \sqrt{b^2 - 4ac}}{2a}$$

$$= \frac{-0.112 \pm \sqrt{(0.112)^2 - 4(1)(-1.2 \times 10^{-3})}}{2(1)}$$

$$= \frac{-0.112 \pm \sqrt{0.01254 + 0.0048}}{2}$$

$$= \frac{-0.112 \pm 0.132}{2} = +0.010 \text{ or } -0.122$$

In this example, the negative value for x satisfies the equation, but it has no meaning because we cannot have negative concentration values; therefore we use only the $+0.010$ value.

Expressing Data in Graphical Form

Fitting curves to plots of experimental data is a powerful tool in determining the relationships between variables. Many natural phenomena obey exponential functions. For example, the rate constant (k) of a chemical reaction increases exponentially with increasing absolute temperature (T). This relationship is described by the Arrhenius equation:

$$k = A\,e^{-E_a/RT}$$

where A is a constant for a particular reaction (called the frequency factor), E_a is the activation energy of the reaction, and R is the ideal gas constant. Taking the natural logarithms of both sides of the Arrhenius equation gives

$$\ln k = \ln A - \left(\frac{E_a}{RT}\right)$$

This equation fits the general equation of a straight line ($y = mx + b$) if ($\ln k$) is the y-variable and ($1/T$) is the x-variable. Plotting ($\ln k$) versus ($1/T$) should give a straight line with a slope equal to $-E_a/R$. The slopes of these plots are negative because the activation energies, E_a, of chemical reactions are positive. The data for a reaction given in columns 2 and 4 of Table A1.1 are plotted in Figure A1.1. The program that generated the graph also gives us the equation of the straight line that best fits the data. The slope (-1281 K) of this line is used to calculate the value of E_a:

$$-1281 \text{ K} = -\frac{E_a}{R}$$

$$E_a = -(-1281 \text{ K})[8.314 \text{ J/(mol} \cdot \text{K)}]$$

$$= 10{,}650 \text{ J/mol} = 10.65 \text{ kJ/mol}$$

Temperature T (K)	$1/T$ (K^{-1})	Rate Constant k	ln k
500	0.0020	0.030	−3.51
550	0.0018	0.38	−0.97
600	0.0017	2.9	1.06
650	0.0015	17	2.83
700	0.0014	75	4.32

TABLE A1.1 **Rate Constant k as a Function of Temperature T**

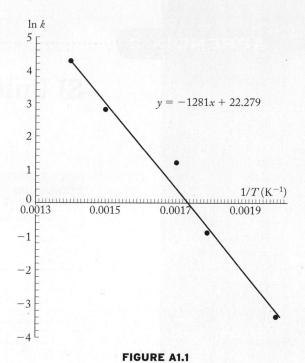

$$y = -1281x + 22.279$$

FIGURE A1.1

SI Units and Conversion Factors

TABLE A2.1 Six SI Base Units

SI Base Quantity	Unit	Symbol
length	meter	m
mass	kilogram	kg
time	second	s
amount of substance	mole	mol
temperature	kelvin	K
electric current	ampere	A

TABLE A2.2 Some SI–Derived Units

SI–Derived Quantity	Unit	Symbol	Dimensions
electric charge	coulomb	C	$A \cdot s$
electric potential	volt	V	J/C
force	newton	N	$kg \cdot m/s^2$
frequency	hertz	Hz	s^{-1}
momentum	newton-second	$N \cdot s$	$kg \cdot m/s$
power	watt	W	J/s
pressure	pascal	Pa	N/m^2
radioactivity	becquerel	Bq	s^{-1}
speed or velocity	meter per second	m/s	m/s
energy	joule (newton-meter)	$J (N \cdot m)$	$kg \cdot m^2/s^2$

TABLE A2.3 SI Prefixes

Prefix	Symbol	Multiplier	Prefix	Symbol	Multiplier
deci	d	10^{-1}	deka	da	10^1
centi	c	10^{-2}	hecto	h	10^2
milli	m	10^{-3}	kilo	k	10^3
micro	μ	10^{-6}	mega	M	10^6
nano	n	10^{-9}	giga	G	10^9
pico	p	10^{-12}	tera	T	10^{12}
femto	f	10^{-15}	peta	P	10^{15}
atto	a	10^{-18}	exa	E	10^{18}
zepto	z	10^{-21}	zetta	Z	10^{21}

TABLE A2.4 Special Units and Conversion Factors

Quantity	Unit	Symbol	Conversion[a]
energy[a]	electron-volt	eV	$1\ \text{eV} = 1.6022 \times 10^{-19}\ \text{J}$
energy[a]	kilowatt-hour	kWh	$1\ \text{kWh} = 3600\ \text{kJ}$
energy	calorie	cal	$1\ \text{cal} = 4.184\ \text{J}$
mass	pound	lb	$1\ \text{lb} = 453.59\ \text{g}$
mass[a]	atomic mass unit	amu	$1\ \text{amu} = 1.66054 \times 10^{-27}\ \text{kg}$
length	angstrom	Å	$1\ \text{Å} = 10^{-8}\ \text{cm} = 10^{-10}\ \text{m}$
length	inch	in	$1\ \text{in} = 2.54\ \text{cm}$
length	mile	mi	$1\ \text{mi} = 5280\ \text{ft} = 1.6093\ \text{km}$
pressure	atmosphere	atm	$1\ \text{atm} = 1.01325 \times 10^5\ \text{Pa}$
pressure	torr	torr	$1\ \text{torr} = 1/760\ \text{atm}$
temperature	Celsius scale	°C	$T(°\text{C}) = T(\text{K}) - 273.15$
temperature	Fahrenheit scale	°F	$T(°\text{F}) = \frac{9}{5} T(°\text{C}) + 32$
volume	liter	L	$1\ \text{L} = 1\ \text{dm}^3 = 10^{-3}\ \text{m}^3$
volume	cubic centimeter	cm^3, cc	$1\ \text{cm}^3 = 1\ \text{mL} = 10^{-3}\ \text{L}$
volume	cubic foot	ft^3	$1\ \text{ft}^3 = 7.4805\ \text{gal}$
volume	gallon (U.S.)	gal	$1\ \text{gal} = 3.785\ \text{L}$

[a]From http://physics.nist.gov/cuu/Constants/index.html.

TABLE A2.5 Physical Constants[a]

Quantity	Symbol	Value
acceleration due to gravity (Earth)	g	$9.807\ \text{m/s}^2$
Avogadro's number	N_A	$6.0221 \times 10^{23}\ \text{mol}^{-1}$
Bohr radius	a_0	$5.29 \times 10^{-11}\ \text{m}$
Boltzmann constant	k_B	$1.3806 \times 10^{-23}\ \text{J/K}$
electron charge-to-mass ratio	$-e/m_e$	$1.7588 \times 10^{11}\ \text{C/kg}$
elementary charge	e	$1.602 \times 10^{-19}\ \text{C}$
Faraday constant	F	$9.65 \times 10^4\ \text{C/mol}$
mass of an electron	m_e	$9.10938 \times 10^{-31}\ \text{kg}$
mass of a neutron	m_n	$1.67493 \times 10^{-27}\ \text{kg}$
mass of a proton	m_p	$1.67262 \times 10^{-27}\ \text{kg}$
molar volume of ideal gas at 0°C and 1 atm	V_m	$22.4\ \text{L/mol}$
Planck constant	h	$6.626 \times 10^{-34}\ \text{J} \cdot \text{s}$
speed of light in vacuum	c	$2.998 \times 10^8\ \text{m/s}$
universal gas constant	R	$8.314\ \text{J/(mol} \cdot \text{K)}$ $0.08206\ \text{L} \cdot \text{atm/(mol} \cdot \text{K)}$

[a]From http://physics.nist.gov/cuu/Constants/index.html.

The Elements and Their Properties

TABLE A3.1	Ground-State Electron Configurations, Atomic Radii, and First Ionization Energies of the Elements				
Element	Symbol	Atomic Number Z	Ground-State Configuration	Atomic Radius (pm)	First Ionization Energy (kJ/mol)
hydrogen	H	1	$1s^1$	37	1312.0
helium	He	2	$1s^2$	32	2372.3
lithium	Li	3	$[He]2s^1$	152	520.2
beryllium	Be	4	$[He]2s^2$	112	899.5
boron	B	5	$[He]2s^22p^1$	88	800.6
carbon	C	6	$[He]2s^22p^2$	77	1086.5
nitrogen	N	7	$[He]2s^22p^3$	75	1402.3
oxygen	O	8	$[He]2s^22p^4$	73	1313.9
fluorine	F	9	$[He]2s^22p^5$	71	1681.0
neon	Ne	10	$[He]2s^22p^6$	69	2080.7
sodium	Na	11	$[Ne]3s^1$	186	495.3
magnesium	Mg	12	$[Ne]3s^2$	160	737.7
aluminum	Al	13	$[Ne]3s^23p^1$	143	577.5
silicon	Si	14	$[Ne]3s^23p^2$	117	786.5
phosphorus	P	15	$[Ne]3s^23p^3$	110	1011.8
sulfur	S	16	$[Ne]3s^23p^4$	103	999.6
chlorine	Cl	17	$[Ne]3s^23p^5$	99	1251.2
argon	Ar	18	$[Ne]3s^23p^6$	97	1520.6
potassium	K	19	$[Ar]4s^1$	227	418.8
calcium	Ca	20	$[Ar]4s^2$	197	589.8
scandium	Sc	21	$[Ar]4s^23d^1$	162	633.1
titanium	Ti	22	$[Ar]4s^23d^2$	147	658.8
vanadium	V	23	$[Ar]4s^23d^3$	135	650.9
chromium	Cr	24	$[Ar]4s^13d^5$	128	652.9
manganese	Mn	25	$[Ar]4s^23d^5$	127	717.3
iron	Fe	26	$[Ar]4s^23d^6$	126	762.5
cobalt	Co	27	$[Ar]4s^23d^7$	125	760.4
nickel	Ni	28	$[Ar]4s^23d^8$	124	737.1
copper	Cu	29	$[Ar]4s^13d^{10}$	128	745.5
zinc	Zn	30	$[Ar]4s^23d^{10}$	134	906.4
gallium	Ga	31	$[Ar]4s^23d^{10}4p^1$	135	578.8
germanium	Ge	32	$[Ar]4s^23d^{10}4p^2$	122	762.2
arsenic	As	33	$[Ar]4s^23d^{10}4p^3$	121	947.0
selenium	Se	34	$[Ar]4s^23d^{10}4p^4$	119	941.0

TABLE A3.1 **Ground-State Electron Configurations, Atomic Radii, and First Ionization Energies of the Elements (Continued)**

Element	Symbol	Atomic Number Z	Ground-State Configuration	Atomic Radius (pm)	First Ionization Energy (kJ/mol)
bromine	Br	35	$[\text{Ar}]4s^23d^{10}4p^5$	114	1139.9
krypton	Kr	36	$[\text{Ar}]4s^23d^{10}4p^6$	110	1350.8
rubidium	Rb	37	$[\text{Kr}]5s^1$	247	403.0
strontium	Sr	38	$[\text{Kr}]5s^2$	215	549.5
yttrium	Y	39	$[\text{Kr}]5s^24d^1$	180	599.8
zirconium	Zr	40	$[\text{Kr}]5s^24d^2$	160	640.1
niobium	Nb	41	$[\text{Kr}]5s^14d^4$	146	652.1
molybdenum	Mo	42	$[\text{Kr}]5s^14d^5$	139	684.3
technetium	Tc	43	$[\text{Kr}]5s^24d^5$	136	702.4
ruthenium	Ru	44	$[\text{Kr}]5s^14d^7$	134	710.2
rhodium	Rh	45	$[\text{Kr}]5s^14d^8$	134	719.7
palladium	Pd	46	$[\text{Kr}]4d^{10}$	137	804.4
silver	Ag	47	$[\text{Kr}]5s^14d^{10}$	144	731.0
cadmium	Cd	48	$[\text{Kr}]5s^24d^{10}$	151	867.8
indium	In	49	$[\text{Kr}]5s^24d^{10}5p^1$	167	558.3
tin	Sn	50	$[\text{Kr}]5s^24d^{10}5p^2$	140	708.6
antimony	Sb	51	$[\text{Kr}]5s^24d^{10}5p^3$	141	833.6
tellurium	Te	52	$[\text{Kr}]5s^24d^{10}5p^4$	143	869.3
iodine	I	53	$[\text{Kr}]5s^24d^{10}5p^5$	133	1008.4
xenon	Xe	54	$[\text{Kr}]5s^24d^{10}5p^6$	130	1170.4
cesium	Cs	55	$[\text{Xe}]6s^1$	265	375.7
barium	Ba	56	$[\text{Xe}]6s^2$	222	502.9
lanthanum	La	57	$[\text{Xe}]6s^25d^1$	187	538.1
cerium	Ce	58	$[\text{Xe}]6s^24f^15d^1$	182	534.4
praseodymium	Pr	59	$[\text{Xe}]6s^24f^3$	182	527.2
neodymium	Nd	60	$[\text{Xe}]6s^24f^4$	181	533.1
promethium	Pm	61	$[\text{Xe}]6s^24f^5$	183	535.5
samarium	Sm	62	$[\text{Xe}]6s^24f^6$	180	544.5
europium	Eu	63	$[\text{Xe}]6s^24f^7$	208	547.1
gadolinium	Gd	64	$[\text{Xe}]6s^24f^75d^1$	180	593.4
terbium	Tb	65	$[\text{Xe}]6s^24f^9$	177	565.8
dysprosium	Dy	66	$[\text{Xe}]6s^24f^{10}$	178	573.0
holmium	Ho	67	$[\text{Xe}]6s^24f^{11}$	176	581.0
erbium	Er	68	$[\text{Xe}]6s^24f^{12}$	176	589.3
thulium	Tm	69	$[\text{Xe}]6s^24f^{13}$	176	596.7
ytterbium	Yb	70	$[\text{Xe}]6s^24f^{14}$	193	603.4
lutetium	Lu	71	$[\text{Xe}]6s^24f^{14}5d^1$	174	523.5
hafnium	Hf	72	$[\text{Xe}]6s^24f^{14}5d^2$	159	658.5
tantalum	Ta	73	$[\text{Xe}]6s^24f^{14}5d^3$	146	761.3
tungsten	W	74	$[\text{Xe}]6s^24f^{14}5d^4$	139	770.0
rhenium	Re	75	$[\text{Xe}]6s^24f^{14}5d^5$	137	760.3
osmium	Os	76	$[\text{Xe}]6s^24f^{14}5d^6$	135	839.4
iridium	Ir	77	$[\text{Xe}]6s^24f^{14}5d^7$	136	878.0

Continued on next page

TABLE A3.1	**Ground-State Electron Configurations, Atomic Radii, and First Ionization Energies of the Elements** *(Continued)*				
Element	Symbol	Atomic Number Z	Ground-State Configuration	Atomic Radius (pm)	First Ionization Energy (kJ/mol)
platinum	Pt	78	$[\text{Xe}]6s^14f^{14}5d^9$	139	868.4
gold	Au	79	$[\text{Xe}]6s^14f^{14}5d^{10}$	144	890.1
mercury	Hg	80	$[\text{Xe}]6s^24f^{14}5d^{10}$	151	1007.1
thallium	Tl	81	$[\text{Xe}]6s^24f^{14}5d^{10}6p^1$	170	589.4
lead	Pb	82	$[\text{Xe}]6s^24f^{14}5d^{10}6p^2$	154	715.6
bismuth	Bi	83	$[\text{Xe}]6s^24f^{14}5d^{10}6p^3$	150	703.3
polonium	Po	84	$[\text{Xe}]6s^24f^{14}5d^{10}6p^4$	167	812.1
astatine	At	85	$[\text{Xe}]6s^24f^{14}5d^{10}6p^5$	140	924.6
radon	Rn	86	$[\text{Xe}]6s^24f^{14}5d^{10}6p^6$	145	1037.1
francium	Fr	87	$[\text{Rn}]7s^1$	242	380
radium	Ra	88	$[\text{Rn}]7s^2$	211	509.3
actinium	Ac	89	$[\text{Rn}]7s^26d^1$	188	499
thorium	Th	90	$[\text{Rn}]7s^26d^2$	179	587
protactinium	Pa	91	$[\text{Rn}]7s^25f^26d^1$	163	568
uranium	U	92	$[\text{Rn}]7s^25f^36d^1$	156	587
neptunium	Np	93	$[\text{Rn}]7s^25f^46d^1$	155	597
plutonium	Pu	94	$[\text{Rn}]7s^25f^6$	159	585
americium	Am	95	$[\text{Rn}]7s^25f^7$	173	578
curium	Cm	96	$[\text{Rn}]7s^25f^76d^1$	174	581
berkelium	Bk	97	$[\text{Rn}]7s^25f^9$	170	601
californium	Cf	98	$[\text{Rn}]7s^25f^{10}$	186	608
einsteinium	Es	99	$[\text{Rn}]7s^25f^{11}$	186	619
fermium	Fm	100	$[\text{Rn}]7s^25f^{12}$	167	627
mendelevium	Md	101	$[\text{Rn}]7s^25f^{13}$	173	635
nobelium	No	102	$[\text{Rn}]7s^25f^{14}$	176	642
lawrencium	Lr	103	$[\text{Rn}]7s^25f^{14}6d^1$	161	—
rutherfordium	Rf	104	$[\text{Rn}]7s^25f^{14}6d^2$	157	—
dubnium	Db	105	$[\text{Rn}]7s^25f^{14}6d^3$	149	—
seaborgium	Sg	106	$[\text{Rn}]7s^25f^{14}6d^4$	143	—
bohrium	Bh	107	$[\text{Rn}]7s^25f^{14}6d^5$	141	—
hassium	Hs	108	$[\text{Rn}]7s^25f^{14}6d^6$	134	—
meitnerium	Mt	109	$[\text{Rn}]7s^25f^{14}6d^7$	129	—
darmstadtium	Ds	110	$[\text{Rn}]7s^25f^{14}6d^8$	128	—
roentgenium	Rg	111	$[\text{Rn}]7s^25f^{14}6d^9$	121	—
copernicium	Cn	112	$[\text{Rn}]7s^25f^{14}6d^{10}$	122	—
ununtrium	Uut	113	$[\text{Rn}]7s^25f^{14}6d^{10}7p^1$	136	—
flerovium	Fl	114	$[\text{Rn}]7s^25f^{14}6d^{10}7p^2$	143	—
ununpentium	Uup	115	$[\text{Rn}]7s^25f^{14}6d^{10}7p^3$	162	—
livermorium	Lv	116	$[\text{Rn}]7s^25f^{14}6d^{10}7p^4$	175	—
ununseptium	Uus	117	$[\text{Rn}]7s^25f^{14}6d^{10}7p^5$	165	—
ununoctium	Uuo	118	$[\text{Rn}]7s^25f^{14}6d^{10}7p^6$	157	—

TABLE A3.2 Miscellaneous Physical Properties of the Elements[a]

Element	Symbol	Atomic Number	Physical State[b,c]	Density[d] (g/cm^3)	Melting Point (°C)	Boiling Point (°C)
hydrogen	H	1	gas	0.000090	−259.14	−252.87
helium	He	2	gas	0.000179	<−272.2	−268.93
lithium	Li	3	solid	0.534	180.5	1347
beryllium	Be	4	solid	1.848	1283	2484
boron	B	5	solid	2.34	2300	3650
carbon	C	6	solid (gr)	1.9–2.3	~3350	sublimes
nitrogen	N	7	gas	0.00125	−210.00	−195.8
oxygen	O	8	gas	0.00143	−218.8	−182.95
fluorine	F	9	gas	0.00170	−219.62	−188.12
neon	Ne	10	gas	0.00090	−248.59	−246.08
sodium	Na	11	solid	0.971	97.72	883
magnesium	Mg	12	solid	1.738	650	1090
aluminum	Al	13	solid	2.6989	660.32	2467
silicon	Si	14	solid	2.33	1414	2355
phosphorus	P	15	solid (wh)	1.82	44.15	280
sulfur	S	16	solid	2.07	115.21	444.60
chlorine	Cl	17	gas	0.00321	−101.5	−34.04
argon	Ar	18	gas	0.00178	−189.3	−185.9
potassium	K	19	solid	0.862	63.28	759
calcium	Ca	20	solid	1.55	842	1484
scandium	Sc	21	solid	2.989	1541	2380
titanium	Ti	22	solid	4.54	1668	3287
vanadium	V	23	solid	6.11	1910	3407
chromium	Cr	24	solid	7.19	1857	2671
manganese	Mn	25	solid	7.3	1246	1962
iron	Fe	26	solid	7.874	1538	2750
cobalt	Co	27	solid	8.9	1495	2870
nickel	Ni	28	solid	8.902	1455	2730
copper	Cu	29	solid	8.96	1084.6	2562
zinc	Zn	30	solid	7.133	419.53	907
gallium	Ga	31	solid	5.904	29.76	2403
germanium	Ge	32	solid	5.323	938.25	2833
arsenic	As	33	solid (gy)	5.727	614	sublimes
selenium	Se	34	solid (gy)	4.79	221	685
bromine	Br	35	liquid	3.12	−7.2	58.78
krypton	Kr	36	gas	0.00373	−157.36	−153.22
rubidium	Rb	37	solid	1.532	39.31	688
strontium	Sr	38	solid	2.64	777	1382
yttrium	Y	39	solid	4.469	1526	3336
zirconium	Zr	40	solid	6.506	1855	4409
niobium	Nb	41	solid	8.57	2477	4744

Continued on next page

TABLE A3.2 **Miscellaneous Physical Properties of the Elements[a] (Continued)**

Element	Symbol	Atomic Number	Physical State[b,c]	Density[d] (g/cm³)	Melting Point (°C)	Boiling Point (°C)
molybdenum	Mo	42	solid	10.22	2623	4639
technetium	Tc	43	solid	11.50	2157	4538
ruthenium	Ru	44	solid	12.41	2334	3900
rhodium	Rh	45	solid	12.41	1964	3695
palladium	Pd	46	solid	12.02	1555	2963
silver	Ag	47	solid	10.50	961.78	2212
cadmium	Cd	48	solid	8.65	321.07	767
indium	In	49	solid	7.31	156.60	2072
tin	Sn	50	solid (wh)	7.31	231.9	2270
antimony	Sb	51	solid	6.691	630.63	1750
tellurium	Te	52	solid	6.24	449.5	998
iodine	I	53	solid	4.93	113.7	184.4
xenon	Xe	54	gas	0.00589	−111.75	−108.0
cesium	Cs	55	solid	1.873	28.44	671
barium	Ba	56	solid	3.5	727	1640
lanthanum	La	57	solid	6.145	920	3455
cerium	Ce	58	solid	6.770	799	3424
praseodymium	Pr	59	solid	6.773	931	3510
neodymium	Nd	60	solid	7.008	1016	3066
promethium	Pm	61	solid	7.264	1042	~3000
samarium	Sm	62	solid	7.520	1072	1790
europium	Eu	63	solid	5.244	822	1596
gadolinium	Gd	64	solid	7.901	1314	3264
terbium	Tb	65	solid	8.230	1359	3221
dysprosium	Dy	66	solid	8.551	1411	2561
holmium	Ho	67	solid	8.795	1472	2694
erbium	Er	68	solid	9.066	1529	2862
thulium	Tm	69	solid	9.321	1545	1946
ytterbium	Yb	70	solid	6.966	824	1194
lutetium	Lu	71	solid	9.841	1663	3393
hafnium	Hf	72	solid	13.31	2233	4603
tantalum	Ta	73	solid	16.654	3017	5458
tungsten	W	74	solid	19.3	3422	5660
rhenium	Re	75	solid	21.02	3186	5596
osmium	Os	76	solid	22.57	3033	5012
iridium	Ir	77	solid	22.42	2446	4130
platinum	Pt	78	solid	21.45	1768.4	3825
gold	Au	79	solid	19.3	1064.18	2856
mercury	Hg	80	liquid	13.546	−38.83	356.73
thallium	Tl	81	solid	11.85	304	1473
lead	Pb	82	solid	11.35	327.46	1749
bismuth	Bi	83	solid	9.747	271.4	1564

TABLE A3.2	Miscellaneous Physical Properties of the Elements[a] (Continued)					
Element	**Symbol**	**Atomic Number**	**Physical State**[b,c]	**Density**[d] **(g/cm³)**	**Melting Point (°C)**	**Boiling Point (°C)**
polonium	Po	84	solid	9.32	254	962
astatine	At	85	solid	unknown	302	337
radon	Rn	86	gas	0.00973	−71	−61.7
francium	Fr	87	solid	unknown	27	677
radium	Ra	88	solid	5	700	1737
actinium	Ac	89	solid	10.07	1051	~3200
thorium	Th	90	solid	11.72	1750	4788
protactinium	Pa	91	solid	15.37	1572	unknown
uranium	U	92	solid	19.05	1132	3818

[a]For relative atomic masses and alphabetical listing of the elements, see the flyleaf at the front of this volume.
[b]Normal state at 25°C and 1 atm.
[c]Allotropes: gr = graphite, gy = gray, wh = white.
[d]Liquids and solids at 25°C and 1 atm; gases at 0°C and 1 atm (STP).

TABLE A3.3	A Selection of Stable Isotopes[a]					
Isotope ^{A}**X**	**Natural Abundance (%)**	**Atomic Number** Z	**Neutron Number** N	**Mass Number** A	**Atomic Mass (amu)**	**Binding Energy per Nucleon (MeV)**[b]
^{1}H	99.985	1	0	1	1.007825	—
^{2}H	0.015	1	1	2	2.014000	1.160
^{3}He	0.000137	2	1	3	3.016030	2.572
^{4}He	99.999863	2	2	4	4.002603	7.075
^{6}Li	7.5	3	3	6	6.015121	5.333
^{7}Li	92.5	3	4	7	7.016003	5.606
^{9}Be	100.0	4	5	9	9.012182	6.463
^{10}B	19.9	5	5	10	10.012937	6.475
^{11}B	80.1	5	6	11	11.009305	6.928
^{12}C	98.90	6	6	12	12.000000	7.680
^{13}C	1.10	6	7	13	13.003355	7.470
^{14}N	99.634	7	7	14	14.003074	7.476
^{15}N	0.366	7	8	15	15.000108	7.699
^{16}O	99.762	8	8	16	15.994915	7.976
^{17}O	0.038	8	9	17	16.999131	7.751
^{18}O	0.200	8	10	18	17.999160	7.767
^{19}F	100.0	9	10	19	18.998403	7.779
^{20}Ne	90.48	10	10	20	19.992435	8.032
^{21}Ne	0.27	10	11	21	20.993843	7.972
^{22}Ne	9.25	10	12	22	21.991383	8.081
^{23}Na	100.0	11	12	23	22.989770	8.112
^{24}Mg	78.99	12	12	24	23.985042	8.261
^{25}Mg	10.00	12	13	25	24.985837	8.223
^{26}Mg	11.01	12	14	26	25.982593	8.334
^{27}Al	100.0	13	14	27	26.981538	8.331
^{28}Si	92.23	14	14	28	27.976927	8.448
^{29}Si	4.67	14	15	29	28.976495	8.449
^{30}Si	3.10	14	16	30	29.973770	8.521

Continued on next page

TABLE A3.3   A Selection of Stable Isotopes[a] (Continued)

Isotope ^{A}X	Natural Abundance (%)	Atomic Number Z	Neutron Number N	Mass Number A	Atomic Mass (amu)	Binding Energy per Nucleon (MeV)[b]
^{31}P	100.0	15	16	31	30.973761	8.481
^{32}S	95.02	16	16	32	31.972070	8.493
^{33}S	0.75	16	17	33	32.971456	8.498
^{34}S	4.21	16	18	34	33.967866	8.584
^{36}S	0.02	16	20	36	35.967080	8.575
^{35}Cl	75.77	17	18	35	34.968852	8.520
^{37}Cl	24.23	17	20	37	36.965903	8.570
^{36}Ar	0.337	18	18	36	35.967545	8.520
^{38}Ar	0.063	18	20	38	37.962732	8.614
^{40}Ar	99.600	18	22	40	39.962384	8.595
^{39}K	93.258	19	20	39	38.963707	8.557
^{41}K	6.730	19	22	41	40.961825	8.576
^{40}Ca	96.941	20	20	40	39.962591	8.551
^{42}Ca	0.647	20	22	42	41.958618	8.617
^{43}Ca	0.135	20	23	43	42.958766	8.601
^{44}Ca	2.086	20	24	44	43.955480	8.658
^{46}Ca	0.004	20	26	46	45.953689	8.669
^{48}Ca	0.187	20	28	48	47.952533	8.666
^{45}Sc	100.0	21	24	45	44.955910	8.619
^{46}Ti	8.0	22	24	46	45.952629	8.656
^{47}Ti	7.3	22	25	47	46.951764	8.661
^{48}Ti	73.8	22	26	48	47.947947	8.723
^{49}Ti	5.5	22	27	49	48.947871	8.711
^{50}Ti	5.4	22	28	50	49.944792	8.756
^{51}V	99.750	23	28	51	50.943962	8.742
^{50}Cr	4.345	24	26	50	49.946046	8.701
^{52}Cr	83.789	24	28	52	51.940509	8.776
^{53}Cr	9.501	24	29	53	52.940651	8.760
^{54}Cr	2.365	24	30	54	53.938882	8.778
^{55}Mn	100.0	25	30	55	54.938049	8.765
^{54}Fe	5.9	26	28	54	53.939612	8.736
^{56}Fe	91.72	26	30	56	55.934939	8.790
^{57}Fe	2.1	26	31	57	56.935396	8.770
^{58}Fe	0.28	26	32	58	57.933277	8.792
^{59}Co	100.0	27	32	59	58.933200	8.768
^{204}Pb	1.4	82	122	204	203.973020	7.880
^{206}Pb	24.1	82	124	206	205.974440	7.875
^{207}Pb	22.1	82	125	207	206.975872	7.870
^{208}Pb	52.4	82	126	208	207.976627	7.868
^{209}Bi	100.0	83	126	209	208.980380	7.848

[a]Selection is complete through cobalt-59. Where natural abundances do not add to 100%, the differences are made up by radioactive isotopes with exceedingly long half-lives: potassium-40 (0.0117%, $t_{1/2} = 1.3 \times 10^{9}$ yr); vanadium-50 (0.250%, $t_{1/2} > 1.4 \times 10^{17}$ yr).

[b]1 MeV (mega electron-volt) = 1.6022×10^{-13} J.

TABLE A3.4 A Selection of Radioactive Isotopes

Isotope ^{A}X	Decay Mode[a]	Half-Life $t_{1/2}$	Atomic Number Z	Neutron Number N	Mass Number A	Atomic Mass (amu)	Binding Energy per Nucleon (MeV)[b]
^{3}H	β^-	12.3 yr	1	2	3	3.01605	2.827
^{8}Be	α	$\sim 7 \times 10^{-17}$ s	4	4	8	8.005305	7.062
^{14}C	β^-	5.7×10^3 yr	6	8	14	14.003241	7.520
^{22}Na	β^+	2.6 yr	11	11	22	21.994434	7.916
^{24}Na	β^-	15.0 hr	11	13	24	23.990961	8.064
^{32}P	β^-	14.3 d	15	17	32	31.973907	8.464
^{35}S	β^-	87.2 d	16	19	35	34.969031	8.538
^{59}Fe	β^-	44.5 d	26	33	59	58.934877	8.755
^{60}Co	β^-	5.3 yr	27	33	60	59.933819	8.747
^{90}Sr	β^-	29.1 yr	38	52	90	89.907738	8.696
^{99}Tc	β^-	2.1×10^5 yr	43	56	99	98.906524	8.611
^{109}Cd	EC	462 d	48	61	109	108.904953	8.539
^{125}I	EC	59.4 d	53	72	125	124.904620	8.450
^{131}I	β^-	8.04 d	53	78	131	130.906114	8.422
^{137}Cs	β^-	30.3 yr	55	82	137	136.907073	8.389
^{222}Rn	α	3.82 d	86	136	222	222.017570	7.695
^{226}Ra	α	1600 yr	88	138	226	226.025402	7.662
^{232}Th	α	1.4×10^{10} yr	90	142	232	232.038054	7.615
^{235}U	α	7.0×10^8 yr	92	143	235	235.043924	7.591
^{238}U	α	4.5×10^9 yr	92	146	238	238.050784	7.570
^{239}Pu	α	2.4×10^4 yr	94	145	239	239.052157	7.560

[a]Modes of decay include alpha emission (α), beta emission (β^-), positron emission (β^+), and electron capture (EC).
[b]1 MeV (mega electron-volt) = 1.6022×10^{-13} J.

Chemical Bonds and Thermodynamic Data

TABLE A4.1	Average Lengths and Energies of Covalent Bonds		
Atom	Bond	Bond Length (pm)	Bond Energy (kJ/mol)
H	H—H	75	436
	H—F	92	567
	H—Cl	127	431
	H—Br	141	366
	H—I	161	299
C	C—C	154	348
	C=C	134	614
	C≡C	120	839
	C—H	110	413
	C—N	147	293
	C=N	127	615
	C≡N	116	891
	C—O	143	358
	C=O	123	743[a]
	C≡O	113	1072
	C—F	133	485
	C—Cl	177	328
	C—Br	179	276
	C—I	215	238
N	N—N	147	163
	N=N	124	418
	N≡N	110	945
	N—H	104	391
	N—O	136	201
	N=O	122	607
	N≡O	106	678
O	O—O	148	146
	O=O	121	498
	O—H	96	463
S	S—O	151	265
	S=O	143	523
	S—S	204	266
	S—H	134	347
F	F—F	143	155
Cl	Cl—Cl	200	243
Br	Br—Br	228	193
I	I—I	266	151

[a]The bond energy of C=O in CO_2 is 799 kJ/mol.

TABLE A4.2 Critical Temperatures (T_c) and van der Waals Parameters (a, b) of Real Gases

Gas[a]	Molar Mass (g/mol)	T_c (K)	a ($L^2 \cdot atm/mol^2$)	b (L/mol)
H_2O	18.015	647.14	5.46	0.0305
Br_2	159.808	588	9.75	0.0591
CCl_3F	137.367	471.2	14.68	0.1111
Cl_2	70.906	416.9	6.343	0.0542
CO_2	44.010	304.14	3.59	0.0427
Kr	83.798	209.41	2.325	0.0396
CH_4	16.043	190.53	2.25	0.0428
O_2	31.999	154.59	1.36	0.0318
Ar	39.948	150.87	1.34	0.0322
F_2	37.997	144.13	1.171	0.0290
CO	28.010	132.91	1.45	0.0395
N_2	28.013	126.21	1.39	0.0391
H_2	2.016	32.97	0.244	0.0266
He	4.003	5.19	0.0341	0.0237

[a]Listed in descending order of critical temperature.

TABLE A4.3 Thermodynamic Properties at 25°C

Substance[a,b]	Molar Mass (g/mol)	ΔH_f° (kJ/mol)	S° [J/(mol · K)]	ΔG_f° (kJ/mol)
ELEMENTS AND MONATOMIC IONS				
$Ag^+(aq)$	107.87	105.6	72.7	77.1
$Ag(g)$	107.87	284.9	173.0	246.0
$Ag(s)$	107.87	0.0	42.6	0.0
$Al^{3+}(aq)$	26.982	−531	−321.7	−485
$Al(g)$	26.982	330.0	164.6	289.4
$Al(s)$	26.982	0.0	28.3	0.0
$Al(\ell)$	26.982	10.6	39.6	−1.2
$Ar(g)$	39.948	0.0	154.8	0.0
$Au(g)$	196.97	366.1	180.5	326.3
$Au(s)$	196.97	0.0	47.4	0.0
$B(g)$	10.811	565.0	153.4	521.0
$B(s)$	10.811	0.0	5.9	0.0
$Ba^{2+}(aq)$	137.33	−537.6	9.6	−560.8
$Ba(g)$	137.33	180.0	170.2	146.0
$Ba(s)$	137.33	0.0	62.8	0.0
$Be(g)$	9.0122	324.0	136.3	286.6
$Be(s)$	9.0122	0.0	9.5	0.0
$Br^-(aq)$	79.904	−121.6	82.4	−104.0
$Br(g)$	79.904	111.9	175.0	82.4
$Br_2(g)$	159.808	30.9	245.5	3.1
$Br_2(\ell)$	159.808	0.0	152.2	0.0
$C(g)$	12.011	716.7	158.1	671.3

Continued on next page

TABLE A4.3	Thermodynamic Properties at 25°C (Continued)			
Substance[a,b]	Molar Mass (g/mol)	ΔH_f° (kJ/mol)	S° [J/(mol · K)]	ΔG_f° (kJ/mol)
C(s, diamond)	12.011	1.9	2.4	2.9
C(s, graphite)	12.011	0.0	5.7	0.0
$Ca^{2+}(aq)$	40.078	−542.8	−55.3	−553.6
Ca(g)	40.078	177.8	154.9	144.0
Ca(s)	40.078	0.0	41.6	0.0
$Cl^{-}(aq)$	35.453	−167.2	56.5	−131.2
Cl(g)	35.453	121.3	165.2	105.3
$Cl_2(g)$	70.906	0.0	223.0	0.0
$Co^{2+}(aq)$	58.933	−58.2	−113	−54.4
$Co^{3+}(aq)$	58.933	92	−305	134
Co(g)	58.933	424.7	179.5	380.3
Co(s)	58.933	0.0	30.0	0.0
Cr(g)	51.996	396.6	174.5	351.8
Cr(s)	51.996	0.0	23.8	0.0
$Cs^{+}(aq)$	132.91	−258.3	133.1	−292.0
Cs(g)	132.91	76.5	175.6	49.6
Cs(s)	132.91	0.0	85.2	0.0
$Cu^{+}(aq)$	63.546	71.7	40.6	50.0
$Cu^{2+}(aq)$	63.546	64.8	−99.6	65.5
Cu(g)	63.546	337.4	166.4	297.7
Cu(s)	63.546	0.0	33.2	0.0
$F^{-}(aq)$	18.998	−332.6	−13.8	−278.8
F(g)	18.998	79.4	158.8	62.3
$F_2(g)$	37.997	0.0	202.8	0.0
$Fe^{2+}(aq)$	55.845	−89.1	−137.7	−78.9
$Fe^{3+}(aq)$	55.845	−48.5	−315.9	−4.7
Fe(g)	55.845	416.3	180.5	370.7
Fe(s)	55.845	0.0	27.3	0.0
$H^{+}(aq)$	1.0079	0.0	0.0	0.0
H(g)	1.0079	218.0	114.7	203.3
$H_2(g)$	2.0158	0.0	130.6	0.0
He(g)	4.0026	0.0	126.2	0.0
$Hg_2^{2+}(aq)$	401.18	172.4	84.5	153.5
$Hg^{2+}(aq)$	200.59	171.1	−32.2	164.4
Hg(g)	200.59	61.4	175.0	31.8
Hg(ℓ)	200.59	0.0	75.9	0.0
$I^{-}(aq)$	126.90	−55.2	111.3	−51.6
I(g)	126.90	106.8	180.8	70.2
$I_2(g)$	253.81	62.4	260.7	19.3
$I_2(s)$	253.81	0.0	116.1	0.0
$K^{+}(aq)$	39.098	−252.4	102.5	−283.3
K(g)	39.098	89.0	160.3	60.5

TABLE A4.3	Thermodynamic Properties at 25°C *(Continued)*			
Substance[a,b]	Molar Mass (g/mol)	ΔH_f° (kJ/mol)	S° [J/(mol · K)]	ΔG_f° (kJ/mol)
K(s)	39.098	0.0	64.7	0.0
$Li^+(aq)$	6.941	−278.5	13.4	−293.3
Li(g)	6.941	159.3	138.8	126.6
$Li^+(g)$	6.941	685.7	133.0	648.5
Li(s)	6.941	0.0	29.1	0.0
$Mg^{2+}(aq)$	24.305	−466.9	−138.1	−454.8
Mg(g)	24.305	147.1	148.6	112.5
Mg(s)	24.305	0.0	32.7	0.0
$Mn^{2+}(aq)$	54.938	−220.8	−73.6	−228.1
Mn(g)	54.938	280.7	173.7	238.5
Mn(s)	54.938	0.0	32.0	0.0
N(g)	14.007	472.7	153.3	455.5
$N_2(g)$	28.013	0.0	191.5	0.0
$Na^+(aq)$	22.990	−240.1	59.0	−261.9
Na(g)	22.990	107.5	153.7	77.0
$Na^+(g)$	22.990	609.3	148.0	574.3
Na(s)	22.990	0.0	51.3	0.0
Ne(g)	20.180	0.0	146.3	0.0
$Ni^{2+}(aq)$	58.693	−54.0	−128.9	−45.6
Ni(g)	58.693	429.7	182.2	384.5
Ni(s)	58.693	0.0	29.9	0.0
O(g)	15.999	249.2	161.1	231.7
$O_2(g)$	31.999	0.0	205.0	0.0
P(g)	30.974	314.6	163.1	278.3
$P_4(s,$ red)	123.895	−17.6	22.8	−12.1
$P_4(s,$ white)	123.895	0.0	41.1	0.0
$Pb^{2+}(aq)$	207.2	−1.7	10.5	−24.4
Pb(g)	207.2	195.2	162.2	175.4
Pb(s)	207.2	0.0	64.8	0.0
$Rb^+(aq)$	85.468	−251.2	121.5	−284.0
Rb(g)	85.468	80.9	170.1	53.1
Rb(s)	85.468	0.0	76.8	0.0
S(g)	32.065	277.2	167.8	236.7
$S_8(g)$	256.520	102.3	430.2	49.1
$S_8(s)$	256.520	0.0	32.1	0.0
Sc(g)	44.956	377.8	174.8	336.0
Si(g)	28.086	450.0	168.0	405.5
Si(s)	28.086	0.0	18.8	0.0
Sn(g)	118.71	301.2	168.5	266.2
Sn(s, gray)	118.71	−2.1	44.1	0.1
Sn(s, white)	118.71	0.0	51.2	0.0
$Sr^{2+}(aq)$	87.62	−545.8	−32.6	−559.5
Sr(g)	87.62	164.4	164.6	130.9
Sr(s)	87.62	0.0	52.3	0.0

Continued on next page

TABLE A4.3 Thermodynamic Properties at 25°C (Continued)

Substance[a,b]	Molar Mass (g/mol)	ΔH_f° (kJ/mol)	S° [J/(mol · K)]	ΔG_f° (kJ/mol)
Ti(g)	47.867	473.0	180.3	428.4
Ti(s)	47.867	0.0	30.7	0.0
V(g)	50.942	514.2	182.2	468.5
V(s)	50.942	0.0	28.9	0.0
W(s)	183.84	0.0	32.6	0.0
Zn^{2+}(aq)	65.38	−153.9	−112.1	−147.1
Zn(g)	65.38	130.4	161.0	94.8
Zn(s)	65.38	0.0	41.6	0.0
POLYATOMIC IONS				
CH_3COO^-(aq)	59.045	−486.0	86.6	−369.3
CO_3^{2-}(aq)	60.009	−677.1	−56.9	−527.8
$C_2O_4^{2-}$(aq)	88.020	−825.1	45.6	−673.9
CrO_4^{2-}(aq)	115.994	−881.2	50.2	−727.8
$Cr_2O_7^{2-}$(aq)	215.988	−1490.3	261.9	−1301.1
$HCOO^-$(aq)	45.018	−425.6	92	−351.0
HCO_3^-(aq)	61.017	−692.0	91.2	−586.8
HSO_4^-(aq)	97.072	−887.3	131.8	−755.9
MnO_4^-(aq)	118.936	−541.4	191.2	−447.2
NH_4^+(aq)	18.038	−132.5	113.4	−79.3
NO_3^-(aq)	62.005	−205.0	146.4	−108.7
OH^-(aq)	17.007	−230.0	−10.8	−157.2
PO_4^{3-}(aq)	94.971	−1277.4	−222	−1018.7
SO_4^{2-}(aq)	96.064	−909.3	20.1	−744.5
INORGANIC COMPOUNDS				
AgCl(s)	143.32	−127.1	96.2	−109.8
AgI(s)	234.77	−61.8	115.5	−66.2
$AgNO_3$(s)	169.87	−124.4	140.9	−33.4
Al_2O_3(s)	101.961	−1675.7	50.9	−1582.3
B_2H_6(g)	27.669	35.0	232.0	86.6
B_2O_3(s)	69.622	−1263.6	54.0	−1184.1
$BaCO_3$(s)	197.34	−1216.3	112.1	−1137.6
$BaSO_4$(s)	233.39	−1473.2	132.2	−1362.2
$CaCO_3$(s)	100.087	−1206.9	92.9	−1128.8
$CaCl_2$(s)	110.984	−795.4	108.4	−748.8
CaF_2(s)	78.075	−1228.0	68.5	−1175.6
CaO(s)	56.077	−634.9	38.1	−603.3
$Ca(OH)_2$(s)	74.093	−985.2	83.4	−897.5
$CaSO_4$(s)	136.142	−1434.5	106.5	−1322.0
CO(g)	28.010	−110.5	197.7	−137.2
CO_2(g)	44.010	−393.5	213.8	−394.4
CO_2(aq)	44.010	−412.9	121.3	−386.2
CS_2(g)	76.143	115.3	237.8	65.1
CS_2(ℓ)	76.143	87.9	151.0	63.6

TABLE A4.3   Thermodynamic Properties at 25°C (Continued)

Substance[a,b]	Molar Mass (g/mol)	ΔH_f° (kJ/mol)	S° [J/(mol · K)]	ΔG_f° (kJ/mol)
$CsCl(s)$	168.358	−443.0	101.2	−414.6
$CuSO_4(s)$	159.610	−771.4	109.2	−662.2
$FeCl_2(s)$	126.750	−341.8	118.0	−302.3
$FeCl_3(s)$	162.203	−399.5	142.3	−334.0
$FeO(s)$	71.844	−271.9	60.8	−255.2
$Fe_2O_3(s)$	159.688	−824.2	87.4	−742.2
$HBr(g)$	80.912	−36.3	198.7	−53.4
$HCl(g)$	36.461	−92.3	186.9	−95.3
$HF(g)$	20.006	−273.3	173.8	−275.4
$HI(g)$	127.912	26.5	206.6	1.7
$HNO_3(g)$	63.013	−135.1	266.4	−74.7
$HNO_3(\ell)$	63.013	−174.1	155.6	−80.7
$HNO_3(aq)$	63.013	−206.6	146.0	−110.5
$HgCl_2(s)$	271.50	−224.3	146.0	−178.6
$Hg_2Cl_2(s)$	472.09	−265.4	191.6	−210.7
$H_2O(g)$	18.015	−241.8	188.8	−228.6
$H_2O(\ell)$	18.015	−285.8	69.9	−237.2
$H_2S(g)$	34.082	−20.17	205.6	−33.01
$H_2O_2(g)$	34.015	−136.3	232.7	−105.6
$H_2O_2(\ell)$	34.015	−187.8	109.6	−120.4
$H_2SO_4(\ell)$	98.079	−814.0	156.9	−690.0
$H_2SO_4(aq)$	98.079	−909.2	20.1	−744.5
$KBr(s)$	119.002	−393.8	95.9	−380.7
$KCl(s)$	74.551	−436.5	82.6	−408.5
$KHCO_3(s)$	100.115	−963.2	115.5	−863.6
$K_2CO_3(s)$	138.205	−1151.0	155.5	−1063.5
$LiBr(s)$	86.845	−351.2	74.3	−342.0
$LiCl(s)$	42.394	−408.6	59.3	384.4
$Li_2CO_3(s)$	73.891	−1215.9	90.4	−1132.1
$MgCl_2(s)$	95.211	−641.3	89.6	591.8
$Mg(OH)_2(s)$	58.320	−924.5	63.2	−833.5
$MgSO_4(s)$	120.369	−1284.9	91.6	−1170.6
$MnO_2(s)$	86.937	−520.0	53.1	−465.1
$NaCH_3OO(s)$	82.034	−708.8	123.0	−607.2
$NaBr(s)$	102.894	−361.1	86.82	−349.0
$NaCl(s)$	58.443	−411.2	72.1	−384.2
$NaCl(g)$	58.443	−181.4	229.8	−201.3
$Na_2CO_3(s)$	105.989	−1130.7	135.0	−1044.4
$NaHCO_3(s)$	84.007	−950.8	101.7	−851.0
$NaNO_3(s)$	84.995	−467.9	116.5	−367.0
$NaOH(s)$	39.997	−425.6	64.5	−379.5
$Na_2SO_4(s)$	142.043	−1387.1	149.6	−1270.2
$NF_3(g)$	71.002	−132.1	260.8	−90.6

Continued on next page

TABLE A4.3 Thermodynamic Properties at 25°C (Continued)

Substance[a,b]	Molar Mass (g/mol)	ΔH_f° (kJ/mol)	S° [J/(mol · K)]	ΔG_f° (kJ/mol)
$NH_3(aq)$	17.031	−80.3	111.3	−26.50
$NH_3(g)$	17.031	−46.1	192.5	−16.5
$NH_4Cl(s)$	53.491	−314.4	94.6	−203.0
$NH_4NO_3(s)$	80.043	−365.6	151.1	−183.9
$N_2H_4(g)$	32.045	95.35	238.5	159.4
$N_2H_4(\ell)$	32.045	50.63	121.52	149.3
$NiCl_2(s)$	129.60	−305.3	97.7	−259.0
$NiO(s)$	74.60	−239.7	38.0	−211.7
$NO(g)$	30.006	90.3	210.7	86.6
$NO_2(g)$	46.006	33.2	240.0	51.3
$N_2O(g)$	44.013	82.1	219.9	104.2
$N_2O_4(g)$	92.011	9.2	304.2	97.8
$NOCl(g)$	65.459	51.7	261.7	66.1
$O_3(g)$	47.998	142.7	238.8	163.2
$PCl_3(g)$	137.33	−288.07	311.7	−269.6
$PCl_3(\ell)$	137.33	−319.6	217	−272.4
$PF_5(g)$	125.96	−1594.4	300.8	−1520.7
$PH_3(g)$	33.998	5.4	210.2	13.4
$PbCl_2(s)$	278.1	−359.4	136.0	−314.1
$PbSO_4(s)$	303.3	−920.0	148.5	−813.0
$SO_2(g)$	64.065	−296.8	248.2	−300.1
$SO_3(g)$	80.064	−395.7	256.8	−371.1
$ZnCl_2(s)$	136.30	−415.1	111.5	−369.4
$ZnO(s)$	81.37	−348.0	43.9	−318.2
$ZnSO_4(s)$	161.45	−982.8	110.5	−871.5
ORGANIC COMPOUNDS				
$CCl_4(g)$	153.823	−102.9	309.7	−60.6
$CCl_4(\ell)$	153.823	−135.4	216.4	−65.3
$CH_4(g)$	16.043	−74.8	186.2	−50.8
$CH_3COOH(g)$	60.053	−432.8	282.5	−374.5
$CH_3COOH(\ell)$	60.053	−484.5	159.8	−389.9
$CH_3OH(g)$	32.042	−200.7	239.9	−162.0
$CH_3OH(\ell)$	32.042	−238.7	126.8	−166.4
$C_2H_2(g)$	26.038	226.7	200.8	209.2
$C_2H_4(g)$	28.054	52.4	219.5	68.1
$C_2H_6(g)$	30.070	−84.67	229.5	−32.9
$CH_3CH_2OH(g)$	46.069	−235.1	282.6	−168.6
$CH_3CH_2OH(\ell)$	46.069	−277.7	160.7	−174.9
$CH_3CHO(g)$	44.053	−166	266	−133.7
$C_3H_8(g)$	44.097	−103.8	269.9	−23.5
$n\text{-}CH_3(CH_2)_2CH_3(g)^c$	58.123	−125.6	310.0	−15.7
$n\text{-}CH_3(CH_2)_2CH_3(\ell)^c$	58.123	−147.6	231.0	−15.0
$CH_3COCH_3(\ell)$	58.079	−248.4	199.8	

TABLE A4.3   Thermodynamic Properties at 25°C (Continued)

Substance[a,b]	Molar Mass (g/mol)	ΔH_f° (kJ/mol)	S° [J/(mol · K)]	ΔG_f° (kJ/mol)
$CH_3COCH_3(g)$	58.079	−217.1	295.3	−152.7
$CH_3(CH_2)_2CH_2OH(\ell)$	74.122	−327.3	225.8	
$(CH_3CH_2)_2O(\ell)$	74.122	−279.6	172.4	
$(CH_3CH_2)_2O(g)$	74.122	−252.1	342.7	
$(CH_3)_2C{=}C(CH_3)_2(\ell)$	84.161	66.6	362.6	−69.2
$(CH_3)_2NH(\ell)$	45.084	−43.9	182.3	
$(CH_3)_2NH(g)$	45.084	−18.5	273.1	
$(C_2H_5)_2NH(\ell)$	73.138	−103.3		
$(C_2H_5)_2NH(g)$	73.138	−71.4		
$(CH_3)_3N(\ell)$	59.111	−46.0	208.5	
$(CH_3)_3N(g)$	59.111	−23.6	287.1	
$(CH_3CH_2)_3N(\ell)$	101.191	−134.3		
$(CH_3CH_2)_3N(g)$	101.191	−95.8		
$C_6H_6(g)$	78.114	82.9	269.2	129.7
$C_6H_6(\ell)$	78.114	49.0	172.9	124.5
$C_6H_{12}O_6(s)$	180.158	−1274.4	212.1	−910.1
$n\text{-}C_8H_{18}(\ell)^c$	114.231	−249.9	361.1	6.4
$n\text{-}C_8H_{18}(g)$	114.231	−208.6	466.7	16.4
$C_{12}H_{22}O_{11}(s)$	342.300	−2221.7	360.2	−1543.8
$HCOOH(\ell)$	46.026	−424.7	129.0	−361.4

[a]Substances are arranged alphabetically by chemical formula within each class: (1) elements and monatomic ions; (2) polyatomic ions; (3) inorganic compounds (including CO and CO_2); (4) organic compounds (hydrocarbon-based).

[b]Symbols denote standard enthalpy of formation (ΔH_f°), standard third-law entropy (S°), and standard Gibbs free energy of formation (ΔG_f°). Entropies in aqueous solution are referred to $S^\circ[H^+(aq)] = 0$, not to absolute zero.

[c]The symbol n denotes the "normal" unbranched alkane.

TABLE A4.4   Vapor Pressure of Water as a Function of Temperature

T (°C)	P (torr)
0.0	4.579
10.0	9.209
20.0	17.535
25.0	23.756
30.0	31.824
40.0	55.324
60.0	149.4
70.0	233.7
90.0	525.8
100	760.0
105	906.0

Equilibrium Constants

TABLE A5.1 Ionization Constants of Selected Acids at 25°C

Acid	Step	Aqueous Equilibrium[a]	K_a	pK_a
acetic	1	$CH_3COOH(aq) \rightleftharpoons H^+(aq) + CH_3COO^-(aq)$	1.76×10^{-5}	4.75
arsenic	1	$H_3AsO_4(aq) \rightleftharpoons H^+(aq) + H_2AsO_4^-(aq)$	5.5×10^{-3}	2.26
	2	$H_2AsO_4^-(aq) \rightleftharpoons H^+(aq) + AsO_4^{2-}(aq)$	1.7×10^{-7}	6.77
	3	$HAsO_4^{2-}(aq) \rightleftharpoons H^+(aq) + AsO_4^{3-}(aq)$	5.1×10^{-12}	11.29
ascorbic	1	$H_2C_6H_6O_6(aq) \rightleftharpoons H^+(aq) + HC_6H_6O_6^-(aq)$	9.1×10^{-5}	4.04
	2	$HC_6H_6O_6^-(aq) \rightleftharpoons H^+(aq) + C_6H_6O_6^{2-}(aq)$	5×10^{-12}	11.3
benzoic	1	$C_6H_5COOH(aq) \rightleftharpoons H^+(aq) + C_6H_5COO^-(aq)$	6.25×10^{-5}	4.20
boric	1	$H_3BO_3(aq) \rightleftharpoons H^+(aq) + H_2BO_3^-(aq)$	5.4×10^{-10}	9.27
	2	$H_2BO_3^-(aq) \rightleftharpoons H^+(aq) + HBO_3^{2-}(aq)$	$<10^{-14}$	>14
bromoacetic	1	$CH_2BrCOOH(aq) \rightleftharpoons H^+(aq) + CH_2BrCOO^-(aq)$	2.0×10^{-3}	2.70
butanoic	1	$CH_3CH_2CH_2COOH(aq) \rightleftharpoons$ $H^+(aq) + CH_3CH_2CH_2COO^-(aq)$	1.5×10^{-5}	4.82
carbonic	1	$H_2CO_3(aq) \rightleftharpoons H^+(aq) + HCO_3^-(aq)$	4.3×10^{-7}	6.37
	2	$HCO_3^-(aq) \rightleftharpoons H^+(aq) + CO_3^{2-}(aq)$	4.7×10^{-11}	10.33
chloric	1	$HClO_3(aq) \rightleftharpoons H^+(aq) + ClO_3^-(aq)$	~ 1	~ 0
chloroacetic	1	$CH_2ClCOOH(aq) \rightleftharpoons H^+(aq) + CH_2ClCOO^-(aq)$	1.4×10^{-3}	2.85
chlorous	1	$HClO_2(aq) \rightleftharpoons H^+(aq) + ClO_2^-(aq)$	1.1×10^{-2}	1.96
citric	1	$HOC(CH_2)_2(COOH)_3(aq) \rightleftharpoons$ $H^+(aq) + HOC(CH_2)_2(COOH)_2COO^-(aq)$	7.4×10^{-4}	3.13
	2	$HOC(CH_2)_2(COOH)_2COO^-(aq) \rightleftharpoons$ $H^+(aq) + HOC(CH_2)_2(COOH)(COO^-)_2(aq)$	1.7×10^{-5}	4.77
	3	$HOC(CH_2)_2(COOH)(COO^-)_2(aq) \rightleftharpoons$ $H^+(aq) + HOC(CH_2)_2(COO^-)_3(aq)$	4.0×10^{-7}	6.40
dichloroacetic	1	$CHCl_2COOH(aq) \rightleftharpoons H^+(aq) + CHCl_2COO^-(aq)$	5.5×10^{-2}	1.26
ethanol	1	$CH_3CH_2OH(aq) \rightleftharpoons H^+(aq) + CH_3CH_2O^-(aq)$	1.3×10^{-16}	15.9
fluoroacetic	1	$CH_2FCOOH(aq) \rightleftharpoons H^+(aq) + CH_2FCOO^-(aq)$	2.6×10^{-3}	2.59
formic	1	$HCOOH(aq) \rightleftharpoons H^+(aq) + HCOO^-(aq)$	1.77×10^{-4}	3.75
germanic	1	$H_2GeO_3(aq) \rightleftharpoons H^+(aq) + HGeO_3^-(aq)$	9.8×10^{-10}	9.01
	2	$HGeO_3^-(aq) \rightleftharpoons H^+(aq) + GeO_3^{2-}(aq)$	5×10^{-13}	12.3
hydr(o)azoic	1	$HN_3(aq) \rightleftharpoons H^+(aq) + N_3^-(aq)$	1.9×10^{-5}	4.72

TABLE A5.1		Ionization Constants of Selected Acids at 25°C *(Continued)*		
Acid	**Step**	**Aqueous Equilibrium**[a]	K_a	pK_a
hydrobromic	1	$HBr(aq) \rightleftharpoons H^+(aq) + Br^-(aq)$	$\gg 1$ (strong)	<0
hydrochloric	1	$HCl(aq) \rightleftharpoons H^+(aq) + Cl^-(aq)$	$\gg 1$ (strong)	<0
hydrocyanic	1	$HCN(aq) \rightleftharpoons H^+(aq) + CN^-(aq)$	6.2×10^{-10}	9.21
hydrofluoric	1	$HF(aq) \rightleftharpoons H^+(aq) + F^-(aq)$	6.8×10^{-4}	3.17
hydr(o)iodic	1	$HI(aq) \rightleftharpoons H^+(aq) + I^-(aq)$	$\gg 1$ (strong)	<0
hydrosulfuric	1	$H_2S(aq) \rightleftharpoons H^+(aq) + HS^-(aq)$	8.9×10^{-8}	7.05
	2	$HS^-(aq) \rightleftharpoons H^+(aq) + S^{2-}(aq)$	$\sim 10^{-19}$	~19
hypobromous	1	$HBrO(aq) \rightleftharpoons H^+(aq) + BrO^-(aq)$	2.3×10^{-9}	8.64
hypochlorous	1	$HClO(aq) \rightleftharpoons H^+(aq) + ClO^-(aq)$	2.9×10^{-8}	7.54
hypoiodous	1	$HIO(aq) \rightleftharpoons H^+(aq) + IO^-(aq)$	2.3×10^{-11}	10.64
iodic	1	$HIO_3(aq) \rightleftharpoons H^+(aq) + IO_3^-(aq)$	1.7×10^{-1}	0.77
iodoacetic	1	$CH_2ICOOH(aq) \rightleftharpoons H^+(aq) + CH_2ICOO^-(aq)$	7.6×10^{-4}	3.12
lactic	1	$CH_3CHOHCOOH(aq) \rightleftharpoons$ $H^+(aq) + CH_3CHOHCOO^-(aq)$	1.4×10^{-4}	3.85
maleic (*cis*-butenedioic)	1	$HOOCCH{=}CHCOOH(aq) \rightleftharpoons$ $H^+(aq) + HOOCCH{=}CHCOO^-(aq)$	1.2×10^{-2}	1.92
	2	$HOOCCH{=}CHCOO^-(aq) \rightleftharpoons$ $H^+(aq) + {}^-OOCCH{=}CHCOO^-(aq)$	4.7×10^{-7}	6.33
malonic	1	$HOOCCH_2COOH(aq) \rightleftharpoons$ $H^+(aq) + HOOCCH_2COO^-(aq)$	1.5×10^{-3}	2.82
	2	$HOOCCH_2COO^-(aq) \rightleftharpoons$ $H^+(aq) + {}^-OOCCH_2COO^-(aq)$	2.0×10^{-6}	5.70
nitric	1	$HNO_3(aq) \rightleftharpoons H^+(aq) + NO_3^-(aq)$	$\gg 1$ (strong)	<0
nitrous	1	$HNO_2(aq) \rightleftharpoons H^+(aq) + NO_2^-(aq)$	4.0×10^{-4}	3.40
oxalic	1	$HOOCCOOH(aq) \rightleftharpoons H^+(aq) + HOOCCOO^-(aq)$	5.9×10^{-2}	1.23
	2	$HOOCCOO^-(aq) \rightleftharpoons H^+(aq) + {}^-OOCCOO^-(aq)$	6.4×10^{-5}	4.19
perchloric	1	$HClO_4(aq) \rightleftharpoons H^+(aq) + ClO_4^-(aq)$	$\gg 1$ (strong)	<0
periodic	1	$HIO_4(aq) \rightleftharpoons H^+(aq) + IO_4^-(aq)$	2.3×10^{-2}	1.64
phenol	1	$C_6H_5OH(aq) \rightleftharpoons H^+(aq) + C_6H_5O^-(aq)$	1.3×10^{-10}	9.89
phosphoric	1	$H_3PO_4(aq) \rightleftharpoons H^+(aq) + H_2PO_4^-(aq)$	6.9×10^{-3}	2.16
	2	$H_2PO_4^-(aq) \rightleftharpoons H^+(aq) + HPO_4^{2-}(aq)$	6.4×10^{-8}	7.19
	3	$HPO_4^{2-}(aq) \rightleftharpoons H^+(aq) + PO_4^{3-}(aq)$	4.8×10^{-13}	12.32
propanoic	1	$CH_3CH_2COOH(aq) \rightleftharpoons$ $H^+(aq) + CH_3CH_2COO^-(aq)$	1.4×10^{-5}	4.85
pyruvic	1	$CH_3C(O)COOH(aq) \rightleftharpoons$ $H^+(aq) + CH_3C(O)COO^-(aq)$	2.8×10^{-3}	2.55
sulfuric	1	$H_2SO_4(aq) \rightleftharpoons H^+(aq) + HSO_4^-(aq)$	$\gg 1$ (strong)	<0
	2	$HSO_4^-(aq) \rightleftharpoons H^+(aq) + SO_4^{2-}(aq)$	1.2×10^{-2}	1.92
sulfurous	1	$H_2SO_3(aq) \rightleftharpoons H^+(aq) + HSO_3^-(aq)$	1.7×10^{-2}	1.77
	2	$HSO_3^-(aq) \rightleftharpoons H^+(aq) + SO_3^{2-}(aq)$	6.2×10^{-8}	7.21

Continued on next page

TABLE A5.1 Ionization Constants of Selected Acids at 25°C (Continued)

Acid	Step	Aqueous Equilibrium[a]	K_a	pK_a
thiocyanic	1	$HSCN(aq) \rightleftharpoons H^+(aq) + SCN^-(aq)$	$\gg 1$ (strong)	<0
trichloroacetic	1	$CCl_3COOH(aq) \rightleftharpoons H^+(aq) + CCl_3COO^-(aq)$	2.3×10^{-1}	0.64
trifluoroacetic	1	$CF_3COOH(aq) \rightleftharpoons H^+(aq) + CF_3COO^-(aq)$	5.9×10^{-1}	0.23
water	1	$H_2O(aq) \rightleftharpoons H^+(aq) + OH^-(aq)$	1.0×10^{-14}	14.00

[a]The formulas of most carboxylic acids are written in an RCOOH format to highlight their molecular structures.

TABLE A5.2 Acid Ionization Constants of Hydrated Metal Ions at 25°C

Free Ion	Hydrated Ion	K_a
Fe^{3+}	$Fe(H_2O)_6^{3+}$	3×10^{-3}
Sn^{2+}	$Sn(H_2O)_6^{2+}$	4×10^{-4}
Cr^{3+}	$Cr(H_2O)_6^{3+}$	1×10^{-4}
Al^{3+}	$Al(H_2O)_6^{3+}$	1×10^{-5}
Cu^{2+}	$Cu(H_2O)_6^{2+}$	3×10^{-8}
Pb^{2+}	$Pb(H_2O)_6^{2+}$	3×10^{-8}
Zn^{2+}	$Zn(H_2O)_6^{2+}$	1×10^{-9}
Co^{2+}	$Co(H_2O)_6^{2+}$	2×10^{-10}
Ni^{2+}	$Ni(H_2O)_6^{2+}$	1×10^{-10}

TABLE A5.3 Ionization Constants of Selected Bases at 25°C

Base	Aqueous Equilibrium	K_b	pK_b
ammonia	$NH_3(aq) + H_2O(\ell) \rightleftharpoons NH_4^+(aq) + OH^-(aq)$	1.76×10^{-5}	4.75
aniline	$C_6H_5NH_2(aq) + H_2O(\ell) \rightleftharpoons C_6H_5NH_3^+(aq) + OH^-(aq)$	4.0×10^{-10}	9.40
diethylamine	$(CH_3CH_2)_2NH(aq) + H_2O(\ell) \rightleftharpoons (CH_3CH_2)_2NH_2^+(aq) + OH^-(aq)$	8.6×10^{-4}	3.07
dimethylamine	$(CH_3)_2NH(aq) + H_2O(\ell) \rightleftharpoons (CH_3)_2NH_2^+(aq) + OH^-(aq)$	5.9×10^{-4}	3.23
methylamine	$CH_3NH_2(aq) + H_2O(\ell) \rightleftharpoons CH_3NH_3^+(aq) + OH^-(aq)$	4.4×10^{-4}	3.36
nicotine (1)		1.0×10^{-6}	6.0
nicotine (2)		1.3×10^{-11}	10.9
pyridine	$C_5H_5N(aq) + H_2O(\ell) \rightleftharpoons C_5H_5NH^+(aq) + OH^-(aq)$	1.7×10^{-9}	8.77
quinine (1)		3.3×10^{-6}	5.48
quinine (2)		1.4×10^{-10}	9.9
urea	$H_2NCONH_2(aq) + H_2O(\ell) \rightleftharpoons H_2NCONH_3^+(aq) + OH^-(aq)$	1.3×10^{-14}	13.9

TABLE A5.4		Solubility-Product Constants at 25°C	
Cation	**Anion**	**Heterogeneous Equilibrium**[a]	K_{sp}
aluminum	hydroxide	$Al(OH)_3(s) \rightleftharpoons Al^{3+}(aq) + 3\,OH^-(aq)$	1.9×10^{-33}
	phosphate	$AlPO_4(s) \rightleftharpoons Al^{3+}(aq) + PO_4^{3-}(aq)$	9.8×10^{-21}
barium	carbonate	$BaCO_3(s) \rightleftharpoons Ba^{2+}(aq) + CO_3^{2-}(aq)$	2.6×10^{-9}
	fluoride	$BaF_2(s) \rightleftharpoons Ba^{2+}(aq) + 2\,F^-(aq)$	1.0×10^{-6}
	sulfate	$BaSO_4(s) \rightleftharpoons Ba^{2+}(aq) + SO_4^{2-}(aq)$	9.1×10^{-11}
calcium	carbonate	$CaCO_3(s) \rightleftharpoons Ca^{2+}(aq) + CO_3^{2-}(aq)$	5.0×10^{-9}
	fluoride	$CaF_2(s) \rightleftharpoons Ca^{2+}(aq) + 2\,F^-(aq)$	3.9×10^{-11}
	hydroxide	$Ca(OH)_2(s) \rightleftharpoons Ca^{2+}(aq) + 2\,OH^-(aq)$	4.7×10^{-6}
	phosphate	$Ca_3(PO_4)_2(s) \rightleftharpoons 3\,Ca^{2+}(aq) + 2\,PO_4^{3-}(aq)$	2.1×10^{-33}
	sulfate	$CaSO_4(s) \rightleftharpoons Ca^{2+}(aq) + SO_4^{2-}(aq)$	7.1×10^{-5}
cobalt(II)	carbonate	$CoCO_3(s) \rightleftharpoons Co^{2+}(aq) + CO_3^{2-}(aq)$	1.0×10^{-10}
	phosphate	$Co_3(PO_4)_2(s) \rightleftharpoons 3\,Co^{2+}(aq) + 2\,PO_4^{3-}(aq)$	2.1×10^{-35}
	sulfide	$CoS(s) \rightleftharpoons Co^{2+}(aq) + S^{2-}(aq)$	4×10^{-21}
copper(I)	bromide	$CuBr(s) \rightleftharpoons Cu^+(aq) + Br^-(aq)$	6.3×10^{-9}
	chloride	$CuCl(s) \rightleftharpoons Cu^+(aq) + Cl^-(aq)$	1.0×10^{-6}
	iodide	$CuI(s) \rightleftharpoons Cu^+(aq) + I^-(aq)$	1.3×10^{-12}
copper(II)	phosphate	$Cu_3(PO_4)_2(s) \rightleftharpoons 3\,Cu^{2+}(aq) + 2\,PO_4^{3-}(aq)$	1.4×10^{-37}
	hydroxide	$Cu(OH)_2(s) \rightleftharpoons Cu^{2+}(aq) + 2\,OH^-(aq)$	4.8×10^{-20}
iron(II)	carbonate	$FeCO_3(s) \rightleftharpoons Fe^{2+}(aq) + CO_3^{2-}(aq)$	3.1×10^{-11}
	fluoride	$FeF_2(s) \rightleftharpoons Fe^{2+}(aq) + 2\,F^-(aq)$	2.4×10^{-6}
	hydroxide	$Fe(OH)_2(s) \rightleftharpoons Fe^{2+}(aq) + 2\,OH^-(aq)$	4.9×10^{-17}
	sulfide	$FeS(s) \rightleftharpoons Fe^{2+}(aq) + S^{2-}(aq)$	6.3×10^{-18}
lead	bromide	$PbBr_2(s) \rightleftharpoons Pb^{2+}(aq) + 2\,Br^-(aq)$	6.6×10^{-6}
	carbonate	$PbCO_3(s) \rightleftharpoons Pb^{2+}(aq) + CO_3^{2-}(aq)$	1.5×10^{-13}
	chloride	$PbCl_2(s) \rightleftharpoons Pb^{2+}(aq) + 2\,Cl^-(aq)$	1.6×10^{-5}
	fluoride	$PbF_2(s) \rightleftharpoons Pb^{2+}(aq) + 2\,F^-(aq)$	3.2×10^{-8}
	iodide	$PbI_2(s) \rightleftharpoons Pb^{2+}(aq) + 2\,I^-(aq)$	8.5×10^{-9}
	sulfate	$PbSO_4(s) \rightleftharpoons Pb^{2+}(aq) + SO_4^{2-}(aq)$	1.8×10^{-8}
lithium	carbonate	$Li_2CO_3(s) \rightleftharpoons 2\,Li^+(aq) + CO_3^{2-}(aq)$	8.2×10^{-4}
magnesium	carbonate	$MgCO_3(s) \rightleftharpoons Mg^{2+}(aq) + CO_3^{2-}(aq)$	6.8×10^{-6}
	fluoride	$MgF_2(s) \rightleftharpoons Mg^{2+}(aq) + 2\,F^-(aq)$	6.5×10^{-9}
	hydroxide	$Mg(OH)_2(s) \rightleftharpoons Mg^{2+}(aq) + 2\,OH^-(aq)$	5.6×10^{-12}
manganese(II)	carbonate	$MnCO_3(s) \rightleftharpoons Mn^{2+}(aq) + CO_3^{2-}(aq)$	2.2×10^{-11}
	hydroxide	$Mn(OH)_2(s) \rightleftharpoons Mn^{2+}(aq) + 2\,OH^-(aq)$	5.6×10^{-12}
mercury(I)	bromide	$Hg_2Br_2(s) \rightleftharpoons Hg_2^{2+}(aq) + 2\,Br^-(aq)$	6.4×10^{-23}
	carbonate	$Hg_2CO_3(s) \rightleftharpoons Hg_2^{2+}(aq) + CO_3^{2-}(aq)$	3.7×10^{-17}
	chloride	$Hg_2Cl_2(s) \rightleftharpoons Hg_2^{2+}(aq) + 2\,Cl^-(aq)$	1.5×10^{-18}
	iodide	$Hg_2I_2(s) \rightleftharpoons Hg_2^{2+}(aq) + 2\,I^-(aq)$	5.3×10^{-29}
	sulfate	$Hg_2SO_4(s) \rightleftharpoons Hg_2^{2+}(aq) + SO_4^{2-}(aq)$	8.0×10^{-7}
mercury(II)	hydroxide	$Hg(OH)_2(s) \rightleftharpoons Hg^{2+}(aq) + 2\,OH^-(aq)$	3.1×10^{-26}
	iodide	$HgI_2(s) \rightleftharpoons Hg^{2+}(aq) + 2\,I^-(aq)$	2.8×10^{-29}
nickel(II)	carbonate	$NiCO_3(s) \rightleftharpoons Ni^{2+}(aq) + CO_3^{2-}(aq)$	1.4×10^{-7}
	phosphate	$Ni_3(PO_4)_2(s) \rightleftharpoons 3\,Ni^{2+}(aq) + 2\,PO_4^{3-}(aq)$	4.7×10^{-32}
	sulfide	$NiS(s) \rightleftharpoons Ni^{2+}(aq) + S^{2-}(aq)$	1×10^{-24}
silver	bromide	$AgBr(s) \rightleftharpoons Ag^+(aq) + Br^-(aq)$	5.4×10^{-13}
	carbonate	$Ag_2CO_3(s) \rightleftharpoons 2\,Ag^+(aq) + CO_3^{2-}(aq)$	8.5×10^{-12}
	chloride	$AgCl(s) \rightleftharpoons Ag^+(aq) + Cl^-(aq)$	1.8×10^{-10}
	chromate	$Ag_2CrO_4(s) \rightleftharpoons 2\,Ag^+(aq) + CrO_4^{2-}(aq)$	1.1×10^{-12}
	hydroxide	$AgOH(s) \rightleftharpoons Ag^+(aq) + OH^-(aq)$	1.52×10^{-8}
	iodide	$AgI(s) \rightleftharpoons Ag^+(aq) + I^-(aq)$	8.3×10^{-17}
	phosphate	$Ag_3PO_4(s) \rightleftharpoons 3\,Ag^+(aq) + PO_4^{3-}(aq)$	8.9×10^{-17}
	sulfate	$Ag_2SO_4(s) \rightleftharpoons 2\,Ag^+(aq) + SO_4^{2-}(aq)$	1.2×10^{-5}
	sulfide	$Ag_2S(s) \rightleftharpoons 2\,Ag^+(aq) + S^{2-}(aq)$	1.6×10^{-49}
strontium	carbonate	$SrCO_3(s) \rightleftharpoons Sr^{2+}(aq) + CO_3^{2-}(aq)$	5.6×10^{-10}
	fluoride	$SrF_2(s) \rightleftharpoons Sr^{2+}(aq) + 2\,F^-(aq)$	4.3×10^{-9}
	sulfate	$SrSO_4(s) \rightleftharpoons Sr^{2+}(aq) + SO_4^{2-}(aq)$	3.4×10^{-7}
zinc	carbonate	$ZnCO_3(s) \rightleftharpoons Zn^{2+}(aq) + CO_3^{2-}(aq)$	1.2×10^{-10}
	hydroxide	$Zn(OH)_2(s) \rightleftharpoons Zn^{2+}(aq) + 2\,OH^-(aq)$	3.0×10^{-16}

[a]Equilibrium is between solid phase and aqueous solution.

TABLE A5.5   Formation Constants of Complex Ions at 25°C

Complex Ion	Aqueous Equilibrium	K_f
$[Ag(NH_3)_2]^+$	$Ag^+(aq) + 2\,NH_3(aq) \rightleftharpoons Ag(NH_3)_2^+(aq)$	1.7×10^7
$[AgCl_2]^-$	$Ag^+(aq) + 2\,Cl^-(aq) \rightleftharpoons AgCl_2^-(aq)$	2.5×10^5
$[Ag(CN)_2]^-$	$Ag^+(aq) + 2\,CN^-(aq) \rightleftharpoons Ag(CN)_2^-(aq)$	1.0×10^{21}
$[Ag(S_2O_3)_2]^{3-}$	$Ag^+(aq) + 2\,S_2O_3^{2-}(aq) \rightleftharpoons Ag(S_2O_3)_2^{3-}(aq)$	4.7×10^{13}
$[AlF_6]^{3-}$	$Al^{3+}(aq) + 6\,F^-(aq) \rightleftharpoons AlF_6^{3-}(aq)$	4.0×10^{19}
$[Al(OH)_4]^-$	$Al^{3+}(aq) + 4\,OH^-(aq) \rightleftharpoons Al(OH)_4^-(aq)$	7.7×10^{33}
$[Au(CN)_2]^-$	$Au^+(aq) + 2\,CN^-(aq) \rightleftharpoons Au(CN)_2^-(aq)$	2.0×10^{38}
$[Co(NH_3)_6]^{2+}$	$Co^{2+}(aq) + 6\,NH_3(aq) \rightleftharpoons Co(NH_3)_6^{2+}(aq)$	7.7×10^4
$[Co(NH_3)_6]^{3+}$	$Co^{3+}(aq) + 6\,NH_3(aq) \rightleftharpoons Co(NH_3)_6^{3+}(aq)$	5.0×10^{31}
$[Co(en)_3^{2+}]$	$Co^{2+}(aq) + 3\,en(aq) \rightleftharpoons Co(en)_3^{2+}(aq)$	8.7×10^{13}
$[Co(C_2O_4)_3]^{4-}$	$Co^{2+}(aq) + 3\,C_2O_4^{2-}(aq) \rightleftharpoons Co(C_2O_4)_3^{4-}(aq)$	4.5×10^6
$[Cu(NH_3)_4]^{2+}$	$Cu^{2+}(aq) + 4\,NH_3(aq) \rightleftharpoons Cu(NH_3)_4^{2+}(aq)$	5.0×10^{13}
$[Cu(en)_2^{2+}]$	$Cu^{2+}(aq) + 2\,en(aq) \rightleftharpoons Cu(en)_2^{2+}(aq)$	3.2×10^{19}
$[Cu(CN)_4]^{2-}$	$Cu^{2+}(aq) + 4\,CN^-(aq) \rightleftharpoons Cu(CN)_4^{2-}(aq)$	1.0×10^{25}
$[Cu(C_2O_4)_2]^{2-}$	$Cu^{2+}(aq) + 2\,C_2O_4^{2-}(aq) \rightleftharpoons Cu(C_2O_4)_2^{2-}(aq)$	1.7×10^{10}
$[Fe(C_2O_4)_3]^{4-}$	$Fe^{2+}(aq) + 3\,C_2O_4^{2-}(aq) \rightleftharpoons Fe(C_2O_4)_3^{4-}(aq)$	6×10^6
$[Fe(C_2O_4)_3]^{3-}$	$Fe^{3+}(aq) + 3\,C_2O_4^{2-}(aq) \rightleftharpoons Fe(C_2O_4)_3^{3-}(aq)$	3.3×10^{20}
$[HgCl_4]^{2-}$	$Hg^{2+}(aq) + 4\,Cl^-(aq) \rightleftharpoons HgCl_4^{2-}(aq)$	1.2×10^{15}
$[Ni(NH_3)_6]^{2+}$	$Ni^{2+}(aq) + 6\,NH_3(aq) \rightleftharpoons Ni(NH_3)_6^{2+}(aq)$	5.5×10^8
$[PbCl_4]^{2-}$	$Pb^{2+}(aq) + 4\,Cl^-(aq) \rightleftharpoons PbCl_4^{2-}(aq)$	2.5×10^1
$[Zn(NH_3)_4]^{2+}$	$Zn^{2+}(aq) + 4\,NH_3(aq) \rightleftharpoons Zn(NH_3)_4^{2+}(aq)$	2.9×10^9
$[Zn(OH)_4]^{2-}$	$Zn^{2+}(aq) + 4\,OH^-(aq) \rightleftharpoons Zn(OH)_4^{2-}(aq)$	2.8×10^{15}

Standard Reduction Potentials

TABLE A6.1	Standard Reduction Potentials at 25°C		
Half-Reaction		**n**	**E° (V)**
$F_2(g) + 2\,e^- \rightarrow 2\,F^-(aq)$		2	2.866
$H_2N_2O_2(s) + 2\,H^+(aq) + 2\,e^- \rightarrow N_2(g) + 2\,H_2O(\ell)$		2	2.65
$O(g) + 2\,H^+(aq) + 2\,e^- \rightarrow H_2O(\ell)$		2	2.421
$Cu^{3+}(aq) + e^- \rightarrow Cu^{2+}(aq)$		1	2.4
$XeO_3(s) + 6\,H^+(aq) + 6\,e^- \rightarrow Xe(g) + 3\,H_2O(\ell)$		6	2.10
$O_3(g) + 2\,H^+(aq) + 2\,e^- \rightarrow O_2(g) + H_2O(\ell)$		2	2.076
$OH(g) + e^- \rightarrow OH^-(aq)$		1	2.02
$Co^{3+}(aq) + e^- \rightarrow Co^{2+}(aq)$		1	1.92
$H_2O_2(\ell) + 2\,H^+(aq) + 2\,e^- \rightarrow 2\,H_2O(\ell)$		2	1.776
$N_2O(g) + 2\,H^+(aq) + 2\,e^- \rightarrow N_2(g) + H_2O(\ell)$		2	1.766
$Ce(OH)^{3+}(aq) + H^+(aq) + e^- \rightarrow Ce^{3+}(aq) + H_2O(\ell)$		1	1.70
$Au^+(aq) + e^- \rightarrow Au(s)$		1	1.692
$PbO_2(s) + SO_4^{2-}(aq) + 4\,H^+(aq) + 2\,e^- \rightarrow PbSO_4(s) + 2\,H_2O(\ell)$		2	1.6913
$PbO_2(s) + HSO_4^-(aq) + 3\,H^+(aq) + 2\,e^- \rightarrow PbSO_4(s) + 2\,H_2O(\ell)$		2	1.685
$MnO_4^-(aq) + 4\,H^+(aq) + 3\,e^- \rightarrow MnO_2(s) + 2\,H_2O(\ell)$		3	1.673
$NiO_2(s) + 4\,H^+(aq) + 2\,e^- \rightarrow Ni^{2+}(aq) + 2\,H_2O(\ell)$		2	1.678
$HClO(\ell) + H^+(aq) + e^- \rightarrow \frac{1}{2}\,Cl_2(g) + H_2O(aq)$		1	1.63
$Ce^{4+}(aq) + e^- \rightarrow Ce^{3+}(aq)$		1	1.61
$Mn^{3+}(aq) + e^- \rightarrow Mn^{2+}(aq)$		1	1.542
$MnO_4^-(aq) + 8\,H^+(aq) + 5\,e^- \rightarrow Mn^{2+}(aq) + 4\,H_2O(\ell)$		5	1.507
$BrO_3^-(aq) + 6\,H^+(aq) + 5\,e^- \rightarrow \frac{1}{2}\,Br_2(\ell) + 3\,H_2O(\ell)$		5	1.52
$ClO_3^-(aq) + 6\,H^+(aq) + 5\,e^- \rightarrow \frac{1}{2}\,Cl_2(g) + 3\,H_2O(\ell)$		5	1.47
$PbO_2(s) + 4\,H^+(aq) + 2\,e^- \rightarrow Pb^{2+}(aq) + 2\,H_2O(\ell)$		2	1.455
$Au^{3+}(aq) + 3\,e^- \rightarrow Au(s)$		3	1.40
$Cl_2(g) + 2\,e^- \rightarrow 2\,Cl^-(aq)$		2	1.3583
$Cr_2O_7^{2-}(aq) + 14\,H^+(aq) + 6\,e^- \rightarrow 2\,Cr^{3+}(aq) + 7\,H_2O(\ell)$		6	1.33
$2\,NiO(OH)(s) + 2\,H_2O(\ell) + 2\,e^- \rightarrow 2\,Ni(OH)_2(s) + 2\,OH^-(aq)$		2	1.32
$MnO_2(s) + 4\,H^+(aq) + 2\,e^- \rightarrow Mn^{2+}(aq) + 2\,H_2O(\ell)$		2	1.23

Continued on next page

TABLE A6.1 **Standard Reduction Potentials at 25°C** *(Continued)*

Half-Reaction	n	$E°$ (V)
$O_2(g) + 4\,H^+(aq) + 4\,e^- \rightarrow 2\,H_2O(\ell)$	4	1.229
$IO_3^-(aq) + 6\,H^+(aq) + 5\,e^- \rightarrow \frac{1}{2}\,I_2(s) + 3\,H_2O(\ell)$	5	1.195
$IO_3^-(aq) + 6\,H^+(aq) + 6\,e^- \rightarrow I^-(aq) + 3\,H_2O(\ell)$	6	1.085
$Br_2(\ell) + 2\,e^- \rightarrow 2\,Br^-(aq)$	2	1.066
$HNO_2(\ell) + H^+(aq) + e^- \rightarrow NO(g) + H_2O(\ell)$	1	1.00
$VO_2^+(aq) + 2\,H^+(aq) + e^- \rightarrow VO^{2+}(aq) + H_2O(\ell)$	1	1.00
$NO_3^-(aq) + 4\,H^+(aq) + 3\,e^- \rightarrow NO(g) + 2\,H_2O(\ell)$	3	0.96
$2\,Hg^{2+}(aq) + 2\,e^- \rightarrow Hg_2^{2+}(aq)$	2	0.92
$ClO^-(aq) + H_2O(\ell) + 2\,e^- \rightarrow Cl^-(aq) + 2\,OH^-(aq)$	2	0.89
$HO_2^-(aq) + H_2O(\ell) + 2\,e^- \rightarrow 3\,OH^-(aq)$	2	0.88
$Hg^{2+}(aq) + 2\,e^- \rightarrow Hg(\ell)$	2	0.851
$Ag^+(aq) + e^- \rightarrow Ag(s)$	1	0.7996
$Hg_2^{2+}(aq) + 2\,e^- \rightarrow 2\,Hg(\ell)$	2	0.7973
$Fe^{3+}(aq) + e^- \rightarrow Fe^{2+}(aq)$	1	0.770
$PtCl_4^{2-}(aq) + 2\,e^- \rightarrow Pt(s) + 4\,Cl^-(aq)$	2	0.73
$O_2(g) + 2\,H^+(aq) + 2\,e^- \rightarrow H_2O_2(\ell)$	2	0.68
$MnO_4^-(aq) + 2\,H_2O(\ell) + 3\,e^- \rightarrow MnO_2(s) + 4\,OH^-(aq)$	3	0.59
$H_3AsO_4(s) + 2\,H^+(aq) + 2\,e^- \rightarrow H_3AsO_3(aq) + H_2O(\ell)$	2	0.559
$I_2(s) + 2\,e^- \rightarrow 2\,I^-(aq)$	2	0.5355
$Cu^+(aq) + e^- \rightarrow Cu(s)$	1	0.521
$H_2SO_3(\ell) + 4\,H^+(aq) + 4\,e^- \rightarrow S(s) + 3\,H_2O(\ell)$	4	0.449
$Ag_2CrO_4(s) + 2\,e^- \rightarrow 2\,Ag(s) + CrO_4^{2-}(aq)$	2	0.4470
$O_2(g) + 2\,H_2O(\ell) + 4\,e^- \rightarrow 4\,OH^-(aq)$	4	0.401
$Fe(CN)_6^{3-}(aq) + e^- \rightarrow Fe(CN)_6^{4-}(aq)$	1	0.36
$Ag_2O(s) + H_2O(\ell) + 2\,e^- \rightarrow 2\,Ag(s) + 2\,OH^-(aq)$	2	0.342
$Cu^{2+}(aq) + 2\,e^- \rightarrow Cu(s)$	2	0.342
$BiO^+(aq) + 2\,H^+(aq) + 3\,e^- \rightarrow Bi(s) + H_2O(\ell)$	3	0.32
$AgCl(s) + e^- \rightarrow Ag(s) + Cl^-(aq)$	1	0.2223
$HSO_4^-(aq) + 3\,H^+(aq) + 2\,e^- \rightarrow H_2SO_3(\ell) + H_2O(\ell)$	2	0.17
$Sn^{4+}(aq) + 2\,e^- \rightarrow Sn^{2+}(aq)$	2	0.154
$Cu^{2+}(aq) + e^- \rightarrow Cu^+(aq)$	1	0.153
$2\,MnO_2(s) + H_2O(\ell) + 2\,e^- \rightarrow Mn_2O_3(s) + 2\,OH^-(aq)$	2	0.15
$S(s) + 2\,H^+(aq) + 2\,e^- \rightarrow H_2S(g)$	2	0.141
$HgO(s) + H_2O(\ell) + 2\,e^- \rightarrow Hg(\ell) + 2\,OH^-(aq)$	2	0.0977

TABLE A6.1 **Standard Reduction Potentials at 25°C** *(Continued)*

Half-Reaction	n	$E°$ (V)
$AgBr(s) + e^- \rightarrow Ag(s) + Br^-(aq)$	1	0.095
$Ag(S_2O_3)_2^{3-}(aq) + e^- \rightarrow Ag(s) + 2\ S_2O_3^{2-}(aq)$	1	0.01
$NO_3^-(aq) + H_2O(\ell) + 2\ e^- \rightarrow NO_2^-(aq) + 2\ OH^-(aq)$	2	0.01
$2\ H^+(aq) + 2\ e^- \rightarrow H_2(g)$	2	0.000
$Pb^{2+}(aq) + 2\ e^- \rightarrow Pb(s)$	2	−0.126
$CrO_4^{2-}(aq) + 4\ H_2O(\ell) + 3\ e^- \rightarrow Cr(OH)_3(s) + 5\ OH^-(aq)$	3	−0.13
$Sn^{2+}(aq) + 2\ e^- \rightarrow Sn(s)$	2	−0.136
$AgI(s) + e^- \rightarrow Ag(s) + I^-(aq)$	1	−0.1522
$CuI(s) + e^- \rightarrow Cu(s) + I^-(aq)$	1	−0.185
$N_2(g) + 5\ H^+(aq) + 4\ e^- \rightarrow N_2H_5^+(aq)$	4	−0.23
$Ni^{2+}(aq) + 2\ e^- \rightarrow Ni(s)$	2	−0.257
$PbSO_4(s) + H^+(aq) + 2\ e^- \rightarrow Pb(s) + HSO_4^-(aq)$	2	−0.356
$Co^{2+}(aq) + 2\ e^- \rightarrow Co(s)$	2	−0.277
$Ag(CN)_2^-(aq) + e^- \rightarrow Ag(s) + 2\ CN^-(aq)$	1	−0.31
$Cd^{2+}(aq) + 2\ e^- \rightarrow Cd(s)$	2	−0.403
$Cd(OH)_2(s) + 2\ e^- \rightarrow Cd(s) + 2\ OH^-(aq)$	2	−0.403
$Cr^{3+}(aq) + e^- \rightarrow Cr^{2+}(aq)$	1	−0.41
$Fe^{2+}(aq) + 2\ e^- \rightarrow Fe(s)$	2	−0.447
$2\ CO_2(g) + 2\ H^+(aq) + 2\ e^- \rightarrow H_2C_2O_4(s)$	2	−0.49
$Ni(OH)_2(s) + 2\ e^- \rightarrow Ni(s) + 2\ OH^-(aq)$	2	−0.72
$Cr^{3+}(aq) + 3\ e^- \rightarrow Cr(s)$	3	−0.74
$Zn^{2+}(aq) + 2\ e^- \rightarrow Zn(s)$	2	−0.762
$2\ H_2O(\ell) + 2\ e^- \rightarrow H_2(g) + 2\ OH^-(aq)$	2	−0.828
$SO_4^{2-}(aq) + H_2O(\ell) + 2\ e^- \rightarrow SO_3^{2-}(aq) + 2\ OH^-(aq)$	2	−0.92
$N_2(g) + 4\ H_2O(\ell) + 4\ e^- \rightarrow 4\ OH^-(aq) + N_2H_4(\ell)$	4	−1.16
$Mn^{2+}(aq) + 2\ e^- \rightarrow Mn(s)$	2	−1.185
$Zn(OH)_2(s) + 2\ e^- \rightarrow Zn(s) + 2\ OH^-(aq)$	2	−1.249
$ZnO(s) + H_2O(\ell) + 2\ e^- \rightarrow Zn(s) + 2\ OH^-(aq)$	2	−1.25
$Al^{3+}(aq) + 3\ e^- \rightarrow Al(s)$	3	−1.662
$Mg^{2+}(aq) + 2\ e^- \rightarrow Mg(s)$	2	−2.37
$Na^+(aq) + e^- \rightarrow Na(s)$	1	−2.71
$Ca^{2+}(aq) + 2\ e^- \rightarrow Ca(s)$	2	−2.868
$Ba^{2+}(aq) + 2\ e^- \rightarrow Ba(s)$	2	−2.912
$K^+(aq) + e^- \rightarrow K(s)$	1	−2.95
$Li^+(aq) + e^- \rightarrow Li(s)$	1	−3.05

Naming Organic Compounds

While organic chemistry was becoming established as a discipline within chemistry, many compounds were given trivial names that are still commonly used and recognized. We refer to many of these compounds by their nonsystematic names throughout this book, and their names and structures are listed in Table A7.1.

TABLE A7.1	Organic Compounds and Their Commonly Used Nonsystematic Names	
Name	**Formula**	**Structure**
ethylene	C_2H_4	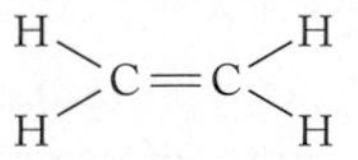 H₂C=CH₂ (shown as H, H on C=C, H, H)
acetylene	C_2H_2	$HC\equiv CH$
benzene	C_6H_6	 (benzene ring)
toluene	$C_6H_5CH_3$	(benzene ring with CH_3)
ethyl alcohol	CH_3CH_2OH	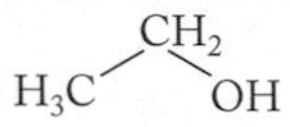 H_3C — CH_2 — OH
acetone	CH_3COCH_3	H_3C — C(=O) — CH_3
acetic acid	CH_3COOH	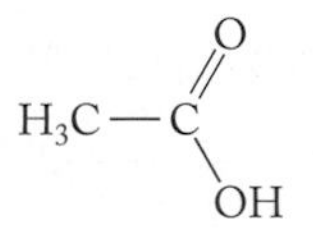 H_3C — C(=O) — OH
formaldehyde	CH_2O	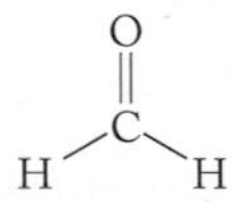 H — C(=O) — H

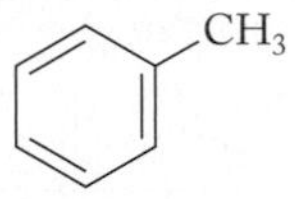

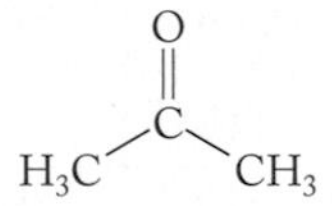

The International Union of Pure and Applied Chemistry (IUPAC) has proposed a set of rules for the systematic naming of organic compounds. The basic principles for naming alkanes, alkenes, and alkynes are presented in Chapter 19. These rules are summarized here and extended to include compounds containing other functional groups. When naming compounds or drawing structures based on names, we need to keep in mind that the IUPAC system of nomenclature is based on two fundamental ideas: (1) the name of a compound must indicate how the carbon atoms in the skeleton are bonded together, and (2) the name must identify the location of any functional groups in the molecule.

Alkanes

Table A7.2 contains the prefixes used for carbon chains ranging in size from C_1 to C_{20} and gives the names for compounds consisting of unbranched chains. The name of a compound consists of a prefix identifying the number of carbons in the chain and a suffix defining the type of hydrocarbon. The suffix *-ane* indicates that the compounds are alkanes and that all carbon–carbon bonds are single bonds.

TABLE A7.2		Prefixes for Naming Carbon Chains			
Prefix	Example	Name	Prefix	Example	Name
meth	CH_4	methane	undec	$C_{11}H_{24}$	undecane
eth	C_2H_6	ethane	dodec	$C_{12}H_{26}$	dodecane
pro	C_3H_8	propane	tridec	$C_{13}H_{28}$	tridecane
but	C_4H_{10}	butane	tetradec	$C_{14}H_{30}$	tetradecane
pent	C_5H_{12}	pentane	pentadec	$C_{15}H_{32}$	pentadecane
hex	C_6H_{14}	hexane	hexadec	$C_{16}H_{34}$	hexadecane
hept	C_7H_{16}	heptane	heptadec	$C_{17}H_{36}$	heptadecane
oct	C_8H_{18}	octane	octadec	$C_{18}H_{38}$	octadecane
non	C_9H_{20}	nonane	nonadec	$C_{19}H_{40}$	nonadecane
dec	$C_{10}H_{22}$	decane	eicos	$C_{20}H_{42}$	eicosane

Branched-Chain Alkanes

The alkane drawn here is used to illustrate each step in the naming rules:

$$CH_3$$
$$|$$
$$CH_3CH_2CHCH_2CHCHCH_2CH_2CH_3$$
$$|\qquad\qquad|$$
$$CH_3\qquad\quad CH_2CH_3$$

1. **Identify and name the longest continuous carbon chain.**

$$\boxed{CH_3CH_2CHCH_2CHCHCH_2CH_2CH_3}\ \text{Nonane}$$
$$|\qquad\qquad|$$
$$CH_3\qquad\quad CH_2CH_3$$

2. **Identify the groups attached to this chain and name them.** Names of substituent groups consist of the prefix from Table A7.2 that identifies the length of the group and the suffix *-yl* that identifies it as an alkyl group.

methyl-

$$\boxed{CH_3}$$
$$CH_3CH_2CHCH_2CHCHCH_2CH_2CH_3$$
$$|\qquad\qquad|$$
$$\boxed{CH_3}\qquad\boxed{CH_2CH_3}$$
$$\text{methyl-}\qquad\text{ethyl-}$$

3. **Number the carbon atoms in the longest chain,** starting at the end nearest a substituent group. Doing this identifies the points of attachment of the alkyl groups with the lowest possible numbers.

methyl-

$$\boxed{CH_3}$$
$$\overset{1}{C}H_3\overset{2}{C}H_2\overset{3}{C}H\overset{4}{C}H_2\overset{5}{C}H\overset{6}{C}H\overset{7}{C}H_2\overset{8}{C}H_2\overset{9}{C}H_3$$
$$|\qquad\qquad|$$
$$\boxed{CH_3}\qquad\boxed{CH_2CH_3}$$
$$\text{methyl-}\qquad\text{ethyl-}$$

4. **Designate the location and identity of each substituent group with a number,** followed by a hyphen, and its name.

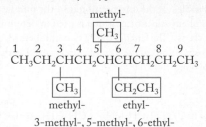

3-methyl-, 5-methyl-, 6-ethyl-

5. **Put together the complete name by listing the substituent groups in alphabetical order.** If more than one of a given type of substituent group is present, prefixes *di-*, *tri-*, *tetra-*, and so forth are appended to the names, but these numerical prefixes are not considered when determining the alphabetical order. The name of the last substituent group is written together with the name identifying the longest carbon chain.

$$CH_3$$
$$|$$
$$CH_3CH_2CHCH_2CHCHCH_2CH_2CH_3$$
$$|\qquad\qquad|$$
$$CH_3\qquad\quad CH_2CH_3$$

6-Ethyl-3,5-dimethylnonane

Cycloalkanes

The simplest examples of this class of compounds consist of one unsubstituted ring of carbon atoms. The IUPAC names of these compounds consist of the prefix *cyclo-* followed by the parent name from Table A7.2 to indicate the number of carbon atoms in the ring. As an illustration, the names, formulas, and line structures of the first three cycloalkanes in the homologous series are

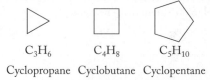

C_3H_6	C_4H_8	C_5H_{10}
Cyclopropane	Cyclobutane	Cyclopentane

Alkenes and Alkynes

Alkenes have carbon–carbon double bonds and alkynes have carbon–carbon triple bonds as functional groups. The names of these types of compounds consist of (1) a parent name that identifies the longest carbon chain that includes the double or triple bond, (2) a suffix that identifies the class of compound, and (3) names of any substituent groups attached to the longest carbon chain. The suffix *-ene* identifies an alkene; *-yne* identifies an alkyne.

The alkene and alkyne drawn here are used to illustrate each step in the naming rules:

$$CH_3\qquad\qquad\qquad CH_3$$
$$|\qquad\qquad\qquad\qquad\ |$$
$$CH_3CHCH=CHCH_2CH_2CH_3\qquad CH_3CHC\equiv CCH_2CH_2CH_3$$

1. **To determine the parent name,** identify the longest chain that contains the unsaturation. Name the parent **compound** with the prefix that defines the number of carbons in that chain and the suffix that identifies the class of compound.

$$CH_3\qquad\qquad\qquad\qquad CH_3$$
$$|\qquad\qquad\qquad\qquad\qquad |$$
$$\boxed{CH_3CHCH=CHCH_2CH_2CH_3}\qquad\boxed{CH_3CHC\equiv CCH_2CH_2CH_3}$$
$$\qquad\text{Heptene}\qquad\qquad\qquad\qquad\text{Heptyne}$$

2. **Number the parent chain from the end nearest the unsaturation so that the first carbon in the double or triple bond has the lowest number possible.** (If the unsaturation is in the middle of a chain, the location of any substituent group is used to determine where the numbering starts.) The smaller of the two numbers identifying the carbon atoms involved in the unsaturation is used as the locator of the multiple bond.

$$\underset{\text{3-Heptene}}{\overset{1\ \ 2\ \ 3\ \ \ \ 4\ \ 5\ \ 6\ \ 7}{CH_3CHCH=CHCH_2CH_2CH_3}} \qquad \underset{\text{3-Heptyne}}{\overset{1\ \ 2\ \ 3\ \ \ 4\ 5\ \ 6\ \ 7}{CH_3CHC\equiv CCH_2CH_2CH_3}}$$

3. **Stereoisomers of alkenes are named by writing *cis-* or *trans-* before the number identifying the location of the double bond.** Chapter 19 in the text addresses naming stereoisomers.

4. **The rules for naming substituted alkanes are followed to name and locate any other groups on the chain.**

$$\underset{\text{2-Methyl-3-heptene}}{\overset{CH_3}{\underset{|}{CH_3CHCH=CHCH_2CH_2CH_3}}} \qquad \underset{\text{2-Methyl-3-heptyne}}{\overset{CH_3}{\underset{|}{CH_3CHC\equiv CCH_2CH_2CH_3}}}$$

Halogens attached to an alkane, alkene, or alkyne are named as fluoro- (F–), chloro- (Cl–), bromo- (Br–), or iodo- (I–) and are located by using the same numbering system described for alkyl groups.

Benzene Derivatives

Naming compounds containing substituted benzene rings is less systematic than naming hydrocarbons. Many compounds have common names that are incorporated into accepted names, but for simple substituted benzene rings, the following rules may be applied.

1. **For monosubstituted benzene rings,** a prefix identifying the group is appended to the parent name benzene:

Chlorobenzene   Nitrobenzene   Ethylbenzene

2. **For disubstituted benzene rings,** three isomers are possible. The relative position of the substituent groups is indicated by numbers in IUPAC nomenclature, but the set of prefixes shown are very commonly used as well:

IUPAC:
1,2-Dichlorobenzene   1,3-Dichlorobenzene   1,4-Dichlorobenzene

Common:
ortho-Dichlorobenzene   *meta*-Dichlorobenzene   *para*-Dichlorobenzene
o-Dichlorobenzene   *m*-Dichlorobenzene   *p*-Dichlorobenzene

3. **When three or more groups are attached to a benzene ring,** the lowest possible numbers are assigned to locate the groups with respect to each other.

1,2,3-Trichlorobenzene   1,2,4-Trichlorobenzene   1,2,3,5-Tetrachlorobenzene
(NOTE: Not 1,3,4-trichlorobenzene; and not 1,3,4,5-tetrachlorobenzene.)

Hydrocarbons Containing Other Functional Groups

The same basic principles developed for naming alkanes apply to naming hydrocarbons with functional groups other than alkyl groups The name must identify the carbon skeleton, locate the functional group, and contain a suffix that defines the class of compound. The following examples give the suffixes for some common functional groups; when suffixes are used, they replace the final -*e* in the name of the parent alkane. Other functional groups may be identified by including the name of the class of compounds in the name of the molecule.

Alcohols: Suffix -*ol*

$$CH_3CH_2CH_2OH \qquad \underset{|}{\overset{}{CH_3CHCH_3}} \qquad \underset{|}{\overset{}{CH_3CH_2CH_2CHCH_3}}$$
$$\qquad\qquad\qquad OH \qquad\qquad\qquad OH$$

1-Propanol   2-Propanol   2-Pentanol

Aldehydes: Suffix -*al*

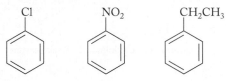

IUPAC: Methanal   Ethanal
Common: Formaldehyde   Acetaldehyde

Because the aldehyde group can only be on a terminal carbon, no number is necessary to locate it on the carbon chain.

Ketones: Suffix -*one*   The location of the carbonyl is given by a number, and the chain is numbered so that the carbonyl carbon has the lowest possible value. Many ketones also have common names generated by identifying the hydrocarbon groups on both sides of the carbonyl group.

IUPAC: Propan-2-one   Butan-2-one
Common: Acetone   Methyl ethyl ketone

Carboxylic Acids: Suffix -*oic acid*   The carboxylic acid group is by definition carbon 1, so no number identifying its location is included in the name.

 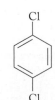

IUPAC: Ethanoic acid   *trans*-2-Butenoic acid
Common: Acetic acid

Salts of Carboxylic Acids Salts are named with the cation first, followed by the anion name of the acid from which *-ic acid* is dropped and the suffix *-ate* is added. The sodium salt of acetic acid is sodium acetate.

Acetic acid   Acetate ion   Sodium acetate

Esters Esters are viewed as derivatives of carboxylic acids. They are named in a manner analogous to that of salts. The alkyl group comes first followed by the name of the carboxylate anion.

Alkyl Carboxylate   Ethyl acetate

Amides Amides are also derivatives of carboxylic acids. They are named by replacing *-ic acid* (of the common names) or *-oic acid* of the IUPAC names with *-amide.*

Parent acid -amide   Acetamide
                            (ethanamide)

Ethers Ethers are frequently named by naming the two groups attached to the oxygen and following those names by the word *ether*.

$$CH_3OCH_3 \qquad CH_3CH_2OCH_2CH_3$$

Dimethyl ether   Diethyl ether

Amines Aliphatic amines are usually named by listing the group or groups attached to the nitrogen and then appending *-amine* as a suffix. They may also be named by prefixing *amino-* to the name of the parent chain.

Methylamine   Ethylmethylamine   2-Aminoethanol

This brief summary will enable you to understand the names of organic compounds used in this book. IUPAC rules are much more extensive than this and can be applied to all varieties of carbon compounds including those with multiple functional groups. It is important to recognize that the rules of systematic nomenclature do not necessarily lead to a unique name for each compound, but they do always lead to an unambiguous one. Furthermore, common names are still used frequently in organic chemistry because the systematic alternatives do not improve communication. Remember that the main purpose of chemical nomenclature is to identify a chemical species by means of written or spoken words. Anyone who reads or hears the name should be able to deduce the structure and thereby the identity of the compound.

A

absolute temperature   Temperature expressed on the Kelvin scale whose zero value is absolute zero, the lowest possible temperature.

absolute zero (0 K)   Zero point on Kelvin temperature scale; theoretically the lowest temperature possible.

accessible microstate   A unique arrangement of the positions and momenta of the particles in a thermodynamic system.

accuracy   The agreement between an experimental value and the true value.

activated complex   A short-lived species formed in a chemical reaction.

activation energy (E_a)   The minimum energy molecules need to react when they collide.

active site   The location on an enzyme where a reactive substance binds.

addition polymer   Macromolecule prepared by adding monomers to a growing polymer chain.

addition reaction   A reaction in which two molecules couple together and form one product.

alcohol   An organic compound whose molecular structure includes a hydroxyl group bonded to a carbon atom that is not bonded to any other functional group(s).

aldehyde   Organic compound containing a carbonyl group bonded to one R group and one hydrogen; its general formula is RCHO.

alkali metal   An element in group 1 of the periodic table.

alkaline earth metal   An element in group 2 of the periodic table.

alkane   A hydrocarbon in which each carbon atom is bonded to four other atoms.

alkene   Hydrocarbon containing one or more carbon–carbon double bonds.

alkyne   Hydrocarbon containing one or more carbon–carbon triple bonds.

allotropes   Different molecular forms of the same element, such as oxygen (O_2) and ozone (O_3).

alloy   A blend of a host metal and one or more other elements, which may or may not be metals, that are added to change the properties of the host metal.

alpha (α) decay   A nuclear reaction in which an unstable nuclide spontaneously emits an alpha particle.

α helix   A coil in a protein chain's secondary structure.

alpha (α) particle   A radioactive emission with a charge of 2+ and a mass equivalent to that of a helium nucleus.

amide   Organic compound in which the –OH of a carboxylic acid group is replaced by $-NH_2$, –NHR, or $-NR_2$, where R can be any organic group.

amine   Organic compound that contains a group with the general formula RNH_2, R_2NH, or R_3N, where R is any organic subgroup.

amino acid   A molecule that contains at least one amine group and one carboxylic acid group; in an *α-amino acid*, the two groups are attached to the same (α) carbon atom.

Amontons's law   The pressure of a gas is proportional to its absolute temperature if its volume does not change.

amphiprotic   Describes a substance that can behave as either a proton acceptor or a proton donor.

analyte   The substance whose concentration is to be determined in a chemical analysis.

angular   (also called *bent*) Molecular geometry about a central atom with a steric number of 3 and one lone pair or a steric number of 4 and two lone pairs.

angular momentum quantum number (ℓ)   An integer having any value from 0 to $(n-1)$ that defines the shape of an orbital.

anion   A negatively charged ion.

anode   An electrode at which an oxidation half-reaction (loss of electrons) takes place.

antibonding orbital   Term in MO theory describing regions of electron density in a molecule that destabilize the molecule because they do not increase the electron density between nuclear centers.

antimatter   Particles that are the charge opposites of normal subatomic particles.

aromatic compound   A cyclic, planar compound with delocalized π electrons above and below the plane of the molecule.

Arrhenius equation   Relates the rate constant of a reaction to absolute temperature (T), the activation energy of the reaction (E_a), and the frequency factor (A).

atom The smallest particle of an element that retains the chemical characteristics of the element.

atomic absorption spectra Characteristic patterns of dark lines produced when an external source of radiation passes through free, gaseous atoms.

atomic emission spectra Characteristic patterns of bright lines produced when atoms are vaporized in high-temperature flames or electrical discharges.

atomic mass unit (amu) Unit used to express the relative masses of atoms and subatomic particles that is exactly 1/12 the mass of 1 atom of carbon with 6 protons and 6 neutrons in its nucleus.

atomic number (Z) The number of protons in the nucleus of an atom.

aufbau principle The method of building electron configurations of atoms by adding one electron at a time as atomic number increases across the rows of the periodic table.

autoionization The process that produces equal and very small concentrations of H_3O^+ and OH^- ions in pure water.

average atomic mass The weighted average of masses of all isotopes of an element, calculated by multiplying the natural abundance of each isotope by its mass in atomic mass units and then summing the products.

Avogadro's law The volume of a gas at constant temperature and pressure is proportional to the quantity (number of moles) of the gas.

Avogadro's number (N_A) The number of carbon atoms in exactly 12 grams of the carbon-12 isotope; $N_A = 6.0221 \times 10^{23}$. It is the number of particles in one mole.

B

band gap (E_g) The energy gap between the valence and conduction bands.

band theory An extension of molecular orbital theory that describes bonding in solids.

barometer An instrument that measures atmospheric pressure.

becquerel (Bq) The SI unit of radioactivity; one becquerel equals one decay event per second.

belt of stability The region on the graph of number of neutrons versus number of protons that includes all stable nuclei.

bent (also called *angular*) Molecular geometry about a central atom with a steric number of 3 and one lone pair or a steric number of 4 and two lone pairs.

beta (β) decay The process by which a neutron in a neuron-rich nucleus decays into a proton and a β particle.

beta (β) particle A radioactive emission that is a high-energy electron.

β-pleated sheet A puckered two-dimensional array of protein strands held together by hydrogen bonds.

bimolecular step A step in a reaction mechanism involving a collision between two molecules.

binding energy (BE) The energy that holds the nucleons together in a nucleus.

biocatalysis The strategy of using enzymes to catalyze reactions on a large scale.

biomass The sum total of the mass of organic matter in any given ecological system.

biomolecule An organic molecule present naturally in a living system.

body-centered cubic (bcc) unit cell A cell with atoms at the eight corners of a cube and at the center of the cell.

bomb calorimeter A constant-volume device used to measure the energy released during a combustion reaction.

bond angle The angle (in degrees) defined by lines joining the centers of two atoms to a third atom to which they are chemically bonded.

bond dipole Separation of electrical charge created when atoms with different electronegativities form a covalent bond.

bond energy The energy needed to break 1 mole of a particular covalent bond in a molecule or polyatomic ion in the gas phase; also called *bond strength*.

bond length The distance between the nuclear centers of two atoms joined together in a bond.

bond order The number of bonds between atoms: 1 for a single bond, 2 for a double bond, and 3 for a triple bond.

bonding capacity The number of covalent bonds an atom forms to have an octet of electrons in its valence shell.

bonding orbital Term in MO theory describing regions of increased electron density between nuclear centers that serve to hold atoms together in molecules.

bonding pair A pair of electrons shared between two atoms.

Born–Haber cycle A series of steps with corresponding enthalpy changes that describes the formation of an ionic solid from its constituent elements.

Boyle's law The volume of a gas at constant temperature is inversely proportional to its pressure.

Bragg equation Relates the angle of diffraction (2θ) of X-rays to the spacing (d) between the layers of ions or atoms in a crystal: $n\lambda = 2d \sin \theta$.

branched-chain hydrocarbon An organic molecule in which the chain of carbon atoms is not linear.

breeder reactor A nuclear reactor in which fissionable material is produced during normal reactor operation.

Brønsted–Lowry acid A proton donor.

Brønsted–Lowry base A proton acceptor.

Brønsted–Lowry model Defines acids as H^+ ion donors and bases as H^+ ion acceptors.

buffer capacity The quantity of acid or base that a pH buffer can neutralize while maintaining its pH within a desired range.

C

calorimeter A device used to measure the absorption or release of energy by a physical change or chemical process.

calorimeter constant ($C_{calorimeter}$) The heat capacity of a calorimeter.

calorimetry The experimental determination of the quantity of energy transferred during a physical change or chemical process.

capillary action The rise of a liquid in a narrow tube as a result of adhesive forces between the liquid and the tube and cohesive forces within the liquid.

carbohydrate An organic molecule with the generic formula $C_x(H_2O)_y$.

carbonyl group A functional group that consists of a carbon atom with a double bond to an oxygen atom.

carboxylic acid A compound containing the –COOH functional group.

catalyst A substance added to a reaction that increases the rate of the reaction but is not consumed in the process.

cathode An electrode at which a reduction half-reaction (gain of electrons) takes place.

cathode rays Streams of electrons emitted by the cathode in a partially evacuated tube.

cation A positively charged ion.

cell diagram Symbols that show how the components of an electrochemical cell are connected.

cell potential (E_{cell}) The electromotive force with which an electrochemical cell can pump electrons through an external circuit.

ceramic A solid inorganic compound or mixture that has been transformed into a harder, more heat-resistant material by heating.

chain reaction A self-sustaining series of fission reactions in which the neutrons released when nuclei split apart initiate additional fission events and sustain the reaction.

Charles's law The volume of a gas at constant pressure is directly proportional to its absolute temperature.

chelate effect The greater affinity of metal ions for polydentate ligands than for monodentate ligands.

chelation The interaction of a metal with a polydentate ligand (chelating agent); pairs of electrons on one molecule of the ligand occupy two or more coordination sites on the central metal.

chemical bond A force that holds two atoms in a molecule or compound together.

chemical energy Potential energy stored in chemical bonds.

chemical equation A description of the identities and proportions of reactants (substances consumed during a chemical reaction) and products (substances formed).

chemical equilibrium A dynamic process in which the concentrations of reactants and products remain constant over time and the rate of a reaction in the forward direction matches its rate in the reverse direction.

chemical formula A notation for representing the elemental composition of a pure substance using the symbols of the elements; subscripts indicate the relative number of atoms of each element in the substance.

chemical kinetics The study of the rates of change of concentrations of substances involved in chemical reactions.

chemical property A property of a substance that can be observed only by reacting the substance chemically to form another substance.

chemical reaction The conversion of one or more substances into one or more different substances.

chemistry The study of the composition, structure, and properties of matter and of the energy consumed or given off when matter undergoes a change.

chiral Describes a molecule that is not superimposable on its mirror image.

chromatography A process involving stationary and mobile phases for separating a mixture of substances based on their different affinities for the two types of phases.

cis isomer (also called Z *isomer*) Molecule with two like groups (such as two R groups or two hydrogen atoms) on the same side of the molecule.

Clausius–Clapeyron equation Relates the vapor pressure of a substance at different temperatures to its heat of vaporization.

codon A three-nucleotide sequence that codes for a specific amino acid.

colligative properties Characteristics of solutions that depend on the concentration and not the identity of particles dissolved in the solvent.

combination reaction A reaction in which two or more substances form a single product.

combined gas law The ratio PV/T for a given quantity of gas is a constant.

combustion A rapid reaction between fuel and oxygen that produces energy.

combustion analysis A laboratory procedure in which a substance is burned completely in oxygen to produce known compounds whose masses are used to determine the composition of the original material.

common-ion effect The shift in the position of an equilibrium caused by the addition of an ion taking part in the reaction.

complex ion An ionic species consisting of a metal ion bonded to one or more Lewis bases.

compound A pure substance that is composed of two or more elements linked together in fixed proportions and that can be broken down into those elements by some chemical process.

condensation polymer Macromolecule formed by the reaction of monomers yielding a polymer and water or another small molecule as products of the reaction.

condensation reaction Two molecules combining to form a larger molecule and a small molecule (typically water).

conduction band An unoccupied band higher in energy than a valence band in which electrons are free to migrate.

conjugate acid Formed when a Brønsted–Lowry base accepts a H^+ ion.

conjugate acid–base pair A Brønsted–Lowry acid and base differing from each other only by the presence or absence of a H^+ ion: acid $\rightleftharpoons$ conjugate base + H^+.

conjugate base Formed when a Brønsted–Lowry acid donates a H^+ ion.

constitutional isomer One of a set of compounds with the same molecular formula but different connections between the atoms in their molecules; also called *structural isomer*.

conversion factor Fraction in which the numerator is equivalent to the denominator but is expressed in different units, making the fraction equivalent to 1.

coordinate bond Formed when one anion or molecule donates a pair of electrons to another ion or molecule to form a covalent bond.

coordination compound Made up of at least one complex ion.

coordination number Identifies the number of electron pairs surrounding a metal ion in a complex.

copolymer A macromolecule formed from the chemical combination of two different monomers.

core electrons Electrons in the filled, inner shells in an atom or ion that are not involved in chemical reactions.

counter ion An ion whose charge balances the charge of a complex ion in a coordination compound.

covalent bond A bond created by two atoms sharing one or more pairs of electrons.

covalent network solid A solid consisting of atoms held together by extended arrays of covalent bonds.

critical mass The minimum quantity of fissionable material needed to sustain a chain reaction.

critical point A specific temperature and pressure at which the liquid and gas phases of a substance have the same density and are indistinguishable from each other.

critical temperature (T_c) The temperature below which a material becomes a superconductor.

crystal field splitting The separation of a set of *d* orbitals into subsets with different energies as a result of interactions between electrons in those orbitals and lone pairs of electrons in ligands.

crystal field splitting energy (Δ) The difference in energy between subsets of *d* orbitals split by interactions in a crystal field.

crystal lattice An ordered three-dimensional array of particles.

crystal structure An ordered arrangement in three-dimensional space of the particles (atoms, ions, or molecules) that make up a crystalline solid.

crystalline solid A solid made of an ordered array of atoms, ions, or molecules.

cubic closest-packed (ccp) A crystal structure composed of face-centered cubic unit cells and layers of particles having an *abcabc . . .* stacking pattern.

curie (Ci) Non-SI unit of radioactivity; 1 Ci = 3.70×10^{10} decay events per second.

cycloalkane A ring-containing alkane with the general formula C_nH_{2n}.

D

dalton (Da) A unit of mass equal to 1 atomic mass unit.

Dalton's law of partial pressures The total pressure of a mixture of gases is the sum of the partial pressures of all the gases in the mixture.

degree of ionization The ratio of the quantity of a substance that is ionized to the concentration of the substance before ionization; when expressed as a percentage, also called *percent ionization*.

density (d) The ratio of the mass (*m*) of an object to its volume (*V*).

deposition Transformation of a vapor (gas) directly into a solid.

diamagnetic Describes a substance with no unpaired electrons that is weakly repelled by a magnetic field.

diffusion The spread of one substance (usually a gas or liquid) through another.

dilution The process of lowering the concentration of a solution by adding more solvent.

dipole moment (μ) A measure of the degree to which a molecule aligns itself in an applied electric field; a quantitative expression of the polarity of a molecule.

dipole–dipole interaction An attraction between regions of polar molecules that have partial charges of opposite sign.

dipole–induced dipole interaction An attraction between a polar molecule and the oppositely charged pole it induces in another molecule.

distillation A process using evaporation and condensation to separate a mixture of substances with different volatilities.

double bond A bond that results when two atoms share two pairs of electrons.

E

E isomer (also called *trans isomer*) Molecule with two like groups (such as two R groups or two hydrogen atoms) on opposite sides of the molecule.

effective nuclear charge (Z_{eff}) The attraction toward the nucleus experienced by an electron in an atom; the positive charge on the nucleus reduced by the extent to which other electrons in the atom shield the electron from the nucleus.

effusion The process by which a gas escapes from its container through a tiny hole into a region of lower pressure.

electrochemical cell An apparatus that converts chemical energy into electrical work or electrical work into chemical energy.

electrochemistry The branch of chemistry that examines the transformations between chemical and electrical energy.

electrode A solid electrical conductor that is used to make contact with a solution or other nonmetallic component of an electrical circuit.

electrolysis A process in which electrical energy is used to drive a nonspontaneous chemical reaction.

electrolyte A material that conducts electricity because it contains free ions; ionic solutions and molten salts are examples of electrolytes.

electrolytic cell A device in which an external source of electrical energy does work on a chemical system, turning reactant(s) into higher energy product(s).

electromagnetic radiation Any form of radiant energy in the electromagnetic spectrum.

electromagnetic spectrum A continuous range of radiant energy that includes gamma rays, X-rays, ultraviolet radiation, visible light, infrared radiation, and radio waves.

electromotive force (emf) Also called voltage, the force pushing electrons through an electrical circuit.

electron A subatomic particle that has a relative charge of 1– and essentially zero mass.

electron affinity (EA) The energy change that occurs when 1 mole of electrons combines with 1 mole of atoms or ions in the gas phase.

electron capture A nuclear reaction in which a neutron-poor nucleus draws in one of its surrounding electrons, which transforms a proton in the nucleus into a neutron.

electron configuration The distribution of electrons among the orbitals of an atom or ion.

electron transition Movement of an electron between energy levels.

electronegativity A relative measure of the ability of an atom in a bond to attract electrons to itself when bonded to another atom.

electron-group geometry The three-dimensional arrangement of bonding pairs and lone pairs of electrons about a central atom.

electrostatic potential energy (E_{el}) The energy a charged particle has because of its position relative to another charged particle; it is directly proportional to the product of the charges of the particles and inversely proportional to the distance between them; also called *coulombic attraction*.

element A pure substance that cannot be separated into simpler substances by any chemical process.

elementary step A molecular-level view of a single process taking place in a chemical reaction.

empirical formula A formula showing the smallest whole-number ratio of elements in a compound.

enantiomer One of a pair of optical isomers of a compound.

end point The point in a titration when a color change or other signal indicates that enough titrant has been added to react with all of the analyte in the sample.

endothermic process One in which energy—usually in the form of heat—flows from the surroundings into the system.

energy The ability to do work.

energy profile Graph showing the changes in potential energy for a reaction as a function of the progress of the reaction from reactants to products.

enthalpy (H) The sum of the internal energy and the pressure–volume product of a system; $H = E + PV$.

enthalpy change (ΔH) The energy absorbed by the reactants (endothermic reaction) or the energy given off by the products (exothermic reaction) for a reaction carried out at constant pressure.

enthalpy of fusion (ΔH_{fus}) The energy required to convert 1 mole of a solid substance at its melting point into the liquid state; also called *heat of fusion*.

enthalpy of reaction (ΔH_{rxn}) The enthalpy change that accompanies a chemical reaction; also called *heat of reaction*.

enthalpy of solution ($\Delta H_{solution}$) The overall change in enthalpy that occurs when a solute is dissolved in a solvent; also called *heat of solution*.

enthalpy of vaporization (ΔH_{vap}) The energy required to convert 1 mole of a liquid substance at its boiling point into the vapor state; also called *heat of vaporization*.

entropy (S) A measure of how dispersed the energy in a system is at a specific temperature.

enzyme   A protein that catalyzes a reaction.

equilibrium constant (K)   The value of the ratio of concentration (or partial pressure) terms in the equilibrium constant expression at a specific temperature.

equilibrium constant expression   The ratio of the equilibrium concentrations or partial pressures of products to reactants, each term raised to a power equal to the coefficient of that substance in the balanced chemical equation for the reaction.

equivalence point   The point in a titration when just enough titrant has been added to react with all of the analyte in the sample.

essential amino acid   Any of the 8 amino acids that make up peptides and proteins but are not synthesized in the human body and must be obtained through the food we eat.

ester   Organic compound in which the –OH of a carboxylic acid group is replaced by –OR, where R can be any organic group.

ether   Organic compound with the general formula R—O—R′, where R and R′ are any alkyl group or aromatic ring; the R and R′ groups may be the same.

excited state   Any energy state above the ground state.

exothermic process   One in which energy—usually in the form of heat—flows from a system into its surroundings.

extensive property   A property that varies with the amount of substance present.

F

face-centered cubic (fcc) unit cell   An array of closest-packed particles that has eight of the particles at the corners of a cube and six of them at the centers of each face of the cube.

family   (also called *group*) All the elements in a column of the periodic table.

Faraday constant (F)   The magnitude of electric charge in 1 mole of electrons. Its value to three significant figures is 9.65×10^4 C/mol.

fat   Solid triglyceride containing primarily saturated fatty acids.

filtration   A process for separating solid particles from a liquid or gaseous sample by passing the sample through a porous material that retains the solid particles.

first law of thermodynamics   The principle that the energy gained or lost by a system must equal the energy lost or gained by the surroundings.

formal charge (FC)   Value calculated for an atom in a molecule or polyatomic ion by determining the difference between the number of valence electrons in the free atom and the sum of lone-pair electrons plus half of the electrons in the atom's bonding pairs.

formation constant (K_f)   Equilibrium constant describing the formation of a metal complex from a free metal ion and its ligands.

formation reaction   A reaction in which 1 mole of a substance is formed from its component elements in their standard states.

formula mass   The mass in amu of one formula unit of an ionic compound.

formula unit   The smallest electrically neutral unit of an ionic compound.

fractional distillation   A method of separating a mixture of compounds on the basis of their different boiling points.

Fraunhofer lines   A set of dark lines in the otherwise continuous solar spectrum.

free energy   A measure of the maximum amount of work a thermodynamic system can perform.

free radical   An odd-electron atom, ion, or molecule.

frequency (ν)   The number of crests of a wave that pass a stationary point of reference per second.

frequency factor (A)   The product of the frequency of molecular collisions and a factor that expresses the probability that the orientation of the molecules is appropriate for a reaction to occur.

fuel cell   A voltaic cell based on the oxidation of a continuously supplied fuel. The reaction is the equivalent of combustion, but chemical energy is converted directly into electrical energy.

fuel density   The quantity of energy released during the complete combustion of a particular volume of a liquid fuel.

fuel value   The quantity of energy released during the complete combustion of 1 g of a substance.

functional group   A group of atoms in a molecule that is largely responsible for the physical and chemical properties of a molecular compound.

G

galvanic cell   (also called *voltaic cell*) An electrochemical cell in which chemical energy is transformed into electrical work by a spontaneous redox reaction.

gas   A form of matter that has neither definite volume nor shape, and that expands to fill its container; also called *vapor*.

Geiger counter   A portable device for determining nuclear radiation levels by measuring how much the radiation ionizes the gas in a sealed detector.

Gibbs free energy (G)   The maximum amount of energy released by a process occurring at constant temperature and pressure that is available to do useful work.

glyceride   Lipid consisting of esters formed between fatty acids and the alcohol glycerol.

glycolysis   A series of reactions that converts glucose into pyruvate; a major anaerobic (no oxygen required) pathway for the metabolism of glucose in the cells of almost all living organisms.

glycosidic bond   A C—O—C bond between sugar molecules.

Graham's law of effusion   The rate of effusion of a gas is inversely proportional to the square root of its molar mass.

gray (Gy)   The SI unit of absorbed radiation; 1 Gy = 1 J/kg of tissue.

ground state   The most stable, lowest energy state of a particle.

group   (also called *family*) All the elements in a column of the periodic table.

H

half-life ($t_{1/2}$)   The time interval during which the concentration of a reactant decreases by half in the course of a chemical reaction.

half-reaction   One of the two halves of a redox reaction; one half-reaction is the oxidation component, and the other is the reduction component.

halogen   An element in group 17 of the periodic table.

heat   A flow of energy from one object or place to another due to differences in the temperatures of the objects or places.

heat capacity (C_P)   The energy required to raise the temperature of an object 1°C at constant pressure.

Heisenberg uncertainty principle   The principle that one cannot simultaneously know the exact position and the exact momentum of an electron.

Henderson–Hasselbalch equation   Used to calculate the pH of a solution in which the concentrations of an acid and conjugate base are known.

Henry's law The concentration of a sparingly soluble gas in a liquid is proportional to the partial pressure of the gas.

hertz (Hz) The SI unit of frequency with units of reciprocal seconds: $1\ Hz = 1\ s^{-1} = 1$ cycle per second (cps).

Hess's law The principle that the heat of reaction ΔH_{rxn} for a process that is the sum of two or more reactions is equal to the sum of the ΔH_{rxn} values of the constituent reactions; also called *Hess's law of constant heat of summation*.

heteroatom Any atom other than a carbon, hydrogen, or metal atom in an organic compound.

heterogeneous catalyst A catalyst in a different phase than the reactants.

heterogeneous equilibrium Involves reactants and products in more than one phase.

heterogeneous mixture A mixture in which the components are not distributed uniformly, so that the mixture contains distinct regions of different compositions.

heteropolymer A polymer made of three or more different monomer units.

hexagonal closest-packed (hcp) A crystal lattice in which the layers of atoms or ions have an *ababab* . . . stacking pattern.

hexagonal unit cell An array of 9 closest-packed particles that are the repeating unit in a hexagonal closest-packed crystal.

homogeneous catalyst A catalyst in the same phase as the reactants.

homogeneous equilibrium Involves reactants and products in the same phase.

homogeneous mixture A mixture in which the components are distributed uniformly throughout and have no visible boundaries or regions.

homologous series A set of related organic compounds that differ from one another by the number of common subgroups, such as $-CH_2-$, in their molecular structures.

homopolymer A polymer composed of only one kind of monomer unit.

Hund's rule The lowest energy electron configuration of an atom has the maximum number of unpaired electrons, all of which have the same spin, in degenerate orbitals.

hybrid atomic orbital In valence bond theory, one of a set of equivalent orbitals about an atom created when specific atomic orbitals are mixed.

hybridization In valence bond theory, the mixing of atomic orbitals to generate new sets of orbitals that then are available to form covalent bonds with other atoms.

hydrocarbon An organic compound whose molecules contain only carbon and hydrogen atoms.

hydrogen bond The strongest dipole–dipole interaction, which occurs between a hydrogen atom bonded to a N, O, or F atom and another N, O, or F atom.

hydrogenation The reaction of an unsaturated hydrocarbon with hydrogen.

hydronium ion (H_3O^+) An H^+ ion plus a water molecule, H_2O; the form in which the hydrogen ion is found in an aqueous solution.

hydrophilic Describes a "water-loving" or attractive interaction between a solute and water that promotes water solubility.

hydrophobic Describes a "water-fearing" or repulsive interaction between a solute and water that diminishes water solubility.

hydroxyl group A functional group that consists of an oxygen atom with a single bond to a hydrogen atom.

hypothesis A tentative and testable explanation for an observation or a series of observations.

I

ideal gas A gas whose behavior is predicted by the linear relations defined by the combined gas law.

ideal gas equation (also called *ideal gas law*) The pressure, volume, number of moles, and temperature of an ideal gas are related by the equation $PV = nRT$, where R is the universal gas constant.

ideal gas law (also called *ideal gas equation*) The pressure, volume, number of moles, and temperature of an ideal gas are related by the equation $PV = nRT$, where R is the universal gas constant.

ideal solution One that obeys Raoult's law.

immiscible liquids Combinations of liquids that are incapable of mixing with, or dissolving in, each other.

inhibitor A compound that diminishes or destroys the ability of an enzyme to catalyze a reaction.

inner coordination sphere The ligands that are bound directly to a metal via coordinate bonds.

integrated rate law A mathematical expression that describes the change in concentration of a reactant in a chemical reaction with time.

intensive property A property that is independent of the amount of substance present.

intermediate A species produced in one step of a reaction and consumed in a subsequent step.

internal energy (E) The sum of all the kinetic and potential energies of all of the components of a system.

interstitial alloy An alloy in which the nonhost atoms occupy spaces between atoms of the host.

ion An atom or molecule that has a positive or negative charge.

ion exchange A process in which one ion is displaced by another.

ion pair A cluster formed when a cation and an anion associate with each other in solution.

ion–dipole interaction An attractive force between an ion and a molecule that has a permanent dipole.

ionic bond A bond resulting from the electrostatic attraction of a cation for an anion.

ionic solid A solid consisting of monatomic or polyatomic ions held together by ionic bonds.

ionization energy (IE) The amount of energy needed to remove 1 mole of electrons from 1 mole of ground-state atoms or ions in the gas phase.

ionizing radiation High-energy products of radioactive decay that can ionize molecules.

isoelectronic Describes atoms or ions that have identical electron configurations.

isomer One of a group of compounds having the same chemical formula but different molecular structures.

isotopes Atoms of an element containing the same number of protons but different numbers of neutrons.

J

joule (J) The SI unit of energy, equivalent to $1\ kg \cdot (m/s)^2$.

K

Kekulé structure A structure using lines to show all of the bonds in a covalently bonded molecule, but not showing lone pairs on the atoms.

kelvin (K) The SI unit of temperature.

ketone An organic compound that contains a carbonyl group bonded to two other carbon atoms.

kinetic energy (KE) The energy of an object in motion due to its mass (m) and its speed (u): $KE = \frac{1}{2}mu^2$.

kinetic molecular theory (KMT) A model that explains the behavior of gases based on the motion of the particles that make them up.

L

lattice energy (U) The energy released when 1 mole of an ionic compound forms from its free ions in the gas phase.

law of conservation of energy The principle that energy cannot be created or destroyed, but can be changed from one form to another.

law of conservation of mass The principle that the sum of the masses of the reactants in a chemical reaction is equal to the sum of the masses of the products.

law of constant composition The principle that all samples of a particular compound always contain the same elements combined in the same proportions.

law of definite proportions The principle that compounds always contain the same proportions of their component elements; equivalent to the *law of constant composition*.

law of mass action The ratio of the concentrations or partial pressures of products to reactants at equilibrium has a characteristic value at a given temperature when each term is raised to a power equal to the coefficient of that substance in the balanced chemical equation for the reaction.

law of multiple proportions The principle that, when two masses of one element react with a given mass of another element to form two different compounds, the two masses of the first element have a ratio of two small whole numbers.

Le Châtelier's principle A system at equilibrium responds to a stress in such a way that it relieves that stress.

leveling effect The observation that strong acids all have the same strength in water and are completely converted into solutions of H_3O^+ ions; strong bases are likewise leveled in water and are completely converted into solutions of OH^- ions.

Lewis acid A substance that *accepts* a lone pair of electrons in a chemical reaction.

Lewis base A substance that *donates* a lone pair of electrons in a chemical reaction.

Lewis structure A two-dimensional representation of the bonds and lone pairs of valence electrons in an ionic or molecular compound.

Lewis symbol The chemical symbol for an element surrounded by one or more dots representing valence electrons; also called *Lewis dot symbol*.

ligand A Lewis base bonded to the central metal ion of a complex ion.

limiting reactant A reactant that is consumed completely in a chemical reaction. The amount of product formed depends on the amount of the limiting reactant available.

linear Molecular geometry about a central atom with a steric number of 2 and no lone pairs of electrons.

lipid A class of water-insoluble, oily organic compounds that are common structural materials in cells.

lipid bilayer A double layer of molecules whose polar head groups interact with water molecules and whose nonpolar tails interact with each other.

liquid A form of matter that occupies a definite volume but flows to assume the shape of its containers.

London dispersion force An intermolecular force between nonpolar molecules caused by the presence of temporary dipoles in the molecules.

lone pair A pair of electrons that is not shared.

M

macrocyclic ligand A ring containing multiple electron-pair donors that bind to a metal ion.

magnetic quantum number (m_ℓ) Defines the orientation of an orbital in space; an integer that may have any value from $-\ell$ to $+\ell$, where ℓ is the angular momentum quantum number.

main group elements (also called *representative elements*) The elements in groups 1, 2, and 13 through 18 of the periodic table.

manometer An instrument for measuring the pressure exerted by a gas.

mass (m) The property that defines the quantity of matter in an object.

mass action expression Equivalent to the equilibrium constant expression, but applied to reaction mixtures that may or may not be at equilibrium.

mass defect (Δm) The difference between the mass of a stable nucleus and the masses of the individual nucleons that comprise it.

mass number (A) The number of nucleons in an atom.

mass spectrum A plot of the number of ions (y axis) that are produced in a mass spectrometer and then separated based on their mass-to-charge (m/z) ratios (x axis).

matter Anything that has mass and occupies space.

matter wave The wave associated with any moving particle.

mean free path The average distance that a particle can travel through air or any gas before colliding with another particle.

meniscus The concave or convex surface of a liquid.

messenger RNA (mRNA) The form of RNA that carries the code for synthesizing proteins from DNA to the site of protein synthesis in a cell.

metallic bond A bond consisting of the nuclei of metal atoms surrounded by a "sea" of shared electrons.

metalloids (also called *semimetals*) Elements that tend to have the physical properties of metals and the chemical properties of nonmetals.

metals Elements that are typically shiny, malleable, ductile solids that conduct heat and electricity well and tend to form positive ions.

meter (m) The standard unit of length, named after the Greek *metron*, which means "measure"; equivalent to 39.37 inches.

methanogenic bacteria Bacteria using simple organic compounds and hydrogen for energy; their respiration produces methane, carbon dioxide, and water, depending on the compounds they consume.

methyl group ($-CH_3$), a structural unit that can make only one bond.

methylene group ($-CH_2-$), a structural unit that can make two bonds.

miscible Liquids that are mutually soluble in any proportion.

mixture A combination of pure substances in variable proportions in which the individual substances retain their chemical identities and can be separated from one another by a physical process.

molality (m) Concentration expressed as the number of moles of solute per kilogram of solvent.

molar heat capacity ($c_{P,n}$) The energy required to raise the temperature of 1 mole of a substance by 1°C at constant pressure.

molar mass ($\mathcal{M}$) The mass of 1 mole of a substance.

molarity (M) The number of moles of solute per liter of solution: $M = n/V$; also called *molar concentration*.

mole (mol) An amount of a substance that contains Avogadro's number ($N_A = 6.0221 \times 10^{23}$) of particles (atoms, ions, molecules, or formula units).

mole fraction (x_i) The ratio of the number of moles of a particular component i in a mixture to the total number of moles in the mixture.

molecular equation   A balanced equation that describes a reaction in solution in which the reactants are written as undissociated molecules.

molecular formula   A chemical formula that shows how many atoms of each element are in one molecule of a pure substance.

molecular geometry   The three-dimensional arrangement of the atoms in a molecule.

molecular ion (M^+)   An ion formed in a mass spectrometer when a molecule loses an electron.

molecular mass   The mass in amu of one molecule of a molecular compound.

molecular orbital   A region of characteristic shape and energy where electrons in a molecule are located.

molecular orbital diagram   In MO theory, an energy-level diagram showing the relative energies and electron occupancy of the molecular orbitals for a molecule.

molecular orbital (MO) theory   A bonding theory based on the mixing of atomic orbitals of similar shapes and energies to form molecular orbitals that belong to the molecule as a whole.

molecular recognition   The process by which molecules interact with other molecules to produce a biological effect.

molecular solid   A solid formed by neutral, covalently bonded molecules held together by intermolecular attractive forces.

molecularity   The number of ions, atoms, or molecules involved in an elementary step in a reaction.

molecule   A collection of atoms chemically bonded together.

monodentate ligand   A species that forms only a single coordinate bond to a metal ion in a complex.

monomer   A small molecule that bonds with others to form polymers.

monoprotic acid   Has one ionizable hydrogen atom per molecule.

monosaccharide   A single-sugar unit and the simplest carbohydrate.

N

natural abundance   The proportion of a particular isotope, usually expressed as a percentage, relative to all the isotopes of that element in a natural sample.

Nernst equation   An equation relating the potential of a cell (or half-cell) reaction to its standard potential ($E°$) and to the concentrations of its reactants and products.

net ionic equation   A balanced equation that describes the actual reaction taking place in solution; it is obtained by eliminating the spectator ions from the total ionic equation.

neutralization reaction   A reaction that takes place when an acid reacts with a base and produces a solution of a salt in water.

neutron   An electrically neutral (uncharged) subatomic particle with a mass number of 1.

neutron capture   The absorption of a neutron by a nucleus.

noble gases   The elements in group 18 of the periodic table.

node   A location in a standing wave that experiences no displacement.

nonelectrolyte   A molecular substance that does not dissociate into ions when it dissolves in water.

nonmetals   Elements with properties opposite those of metals, including poor conductivity of heat and electricity.

nonpolar covalent bond   A bond characterized by an even distribution of charge; electrons in the bonds are shared equally by the two atoms; pure covalent bonds give rise to nonpolar diatomic molecules.

normal boiling point   The temperature at which the vapor pressure of a liquid equals 1 atm (760 torr).

normal (straight-chain) hydrocarbon   A hydrocarbon in which the carbon atoms are bonded together in one continuous line. Linear alkane chains have a methyl group at each end with methylene groups connecting them.

n-type semiconductor   A semiconductor containing an electron-rich dopant.

nuclear chemistry   The study of reactions that involve changes in the nuclei of atoms.

nuclear fission   A nuclear reaction in which the nucleus of an element splits into two lighter nuclei. The process is usually accompanied by the release of one or more neutrons and energy.

nuclear fusion   A nuclear reaction in which subatomic particles or atomic nuclei collide with each other at very high speeds and fuse together, forming more massive nuclei and releasing energy.

nucleic acid   One of a family of large molecules, which includes deoxyribonucleic acid (DNA) and ribonucleic acid (RNA), that stores the genetic blueprint of an organism and controls the production of proteins.

nucleon   A proton or neutron in a nucleus.

nucleosynthesis   The natural formation of nuclei as a result of fusion and other nuclear processes.

nucleotide   A monomer unit from which nucleic acids are made.

nucleus   (of an atom) The positively charged center of an atom that contains nearly all the atom's mass.

nuclide   A specific isotope of an element.

O

octahedral   Molecular geometry about a central atom with a steric number of 6 and no lone pairs of electrons, in which all six sites are equivalent.

octet rule   Atoms of main group elements make bonds by gaining, losing, or sharing electrons to achieve a valence shell containing 8 electrons, or four electron pairs.

oil   Liquid triglyceride containing primarily unsaturated fatty acids.

orbital diagram   Depiction of the arrangement of electrons in an atom or ion using boxes to represent orbitals.

orbitals   Defined by the square of the wave function (ψ^2); regions in an atom where the probability of finding an electron is high.

ore   A mineral that contains one or more metals valuable enough to be mined.

organic chemistry   The study of compounds containing C—C and/or C—H bonds.

organic compounds   Compounds that contain carbon, hydrogen, and sometimes other elements including oxygen, nitrogen, sulfur, and a halogen.

osmosis   The flow of a fluid through a semipermeable membrane to balance the concentration of solutes in solutions on the two sides of the membrane. The flow of solvent molecules proceeds from the more dilute solution into the more concentrated one.

osmotic pressure (Π)   The pressure applied across a semipermeable membrane to stop the flow of water from the compartment containing pure solvent or a less concentrated solution to the compartment containing a more concentrated solution.

overall reaction order   The sum of the exponents of the concentration terms in the rate law.

overlap   A term in valence bond theory describing bonds arising from two orbitals on different atoms that occupy the same region of space.

oxidation   A chemical change in which an element loses electrons; the oxidation number of the element increases.

oxidation number (O.N.)   (also called *oxidation state*) A positive or negative number based on the number of electrons that an atom gains or loses when it forms an ion, or that it shares when it forms a covalent bond with an atom of another element.

oxidation state   (also called *oxidation number [O.N.]*) A positive or negative number based on the number of electrons that an atom gains or loses when it forms an ion, or that it shares when it forms a covalent bond with an atom of another element.

oxidizing agent   A reactant that accepts electrons from another in a redox reaction, thereby oxidizing the other reactant; the oxidizing agent is reduced in the reaction.

oxoacid   An acid whose molecules contain oxygen atoms, ionizable hydrogen atoms, and atoms of another element.

oxoanion   A polyatomic anion that contains at least one nonoxygen central atom bonded to one or more oxygen atoms.

P

packing efficiency   Percentage of the total volume of a unit cell occupied by the spheres.

paramagnetic   Describes a substance with unpaired electrons that is attracted to a magnetic field.

partial pressure   The contribution to the total pressure made by a component in a mixture of gases.

Pauli exclusion principle   No two electrons in an atom can have the same set of four quantum numbers.

peptide   A compound of two or more amino acids joined by peptide bonds. Small peptides containing up to 20 amino acids are *oligopeptides*; and the term *polypeptide* is used for chains longer than 20 amino acids but shorter than proteins.

peptide bond   The result of a condensation reaction between the carboxylic acid group of one amino acid and the amine group of another.

percent composition   The composition of a compound expressed in terms of the percentage by mass of each element in the compound.

percent ionization   The ratio of the quantity of a substance that is ionized to the concentration of the substance before ionization, expressed as a percentage.

percent yield   The ratio, expressed as a percentage, of the actual yield of a chemical reaction to the theoretical yield.

period   All the elements in a row of the periodic table.

periodic table of the elements   A chart of the elements in order of their atomic numbers and in a pattern based on their physical and chemical properties.

permanent dipole   Permanent separation of electrical charge in a molecule due to unequal distributions of bonding and/or lone pairs of electrons.

pH   The negative logarithm of the hydrogen ion concentration in an aqueous solution.

pH buffer   A solution that resists changes in pH when acids or bases are added to it; typically a solution of a weak acid and its conjugate base.

pH indicator   A water-soluble weak organic acid that changes color as pH changes.

phase diagram   A graphical representation of the dependence of the stabilities of the physical states of a substance on temperature and pressure.

phospholipid   A molecule of glycerol with two fatty acid chains and one polar group containing a phosphate; phospholipids are major constituents of cell membranes.

phosphorylation   A reaction resulting in the addition of a phosphate group to an organic molecule.

photochemical smog   A mixture of gases formed in the lower atmosphere when sunlight interacts with compounds produced in internal combustion engines and other pollutants.

photoelectric effect   The release of electrons from a material as a result of electromagnetic radiation striking it.

photon   A quantum of electromagnetic radiation.

physical process   A transformation of a sample of matter, such as a change in its physical state, that does not alter the chemical identity of any substance in the sample.

physical property   A property of a substance that can be observed without changing the substance into another substance.

pi (π) bond   A covalent bond in which electron density is greatest above and below the bonding axis.

pi (π) molecular orbital   In MO theory, an orbital formed by the mixing of atomic orbitals oriented above and below, or in front of and behind, the bonding axis; electrons in π orbitals form π bonds.

Planck constant (h)   The proportionality constant between the energy and frequency of electromagnetic radiation expressed in $E = hv$; $h = 6.626 \times 10^{-34}$ J $\cdot$ s.

pOH   The negative logarithm of the hydroxide ion concentration in an aqueous solution.

polar covalent bond   A bond resulting from unequal sharing of bonding pairs of electrons between atoms.

polarizability   The relative ease with which the electron cloud in a molecule, ion, or atom can be distorted, inducing a temporary dipole.

polyatomic ion   A charged group of more than one kind of atom joined together by covalent bonds.

polydentate ligand   A species that can form more than one coordinate bond per molecule.

polymer   A very large molecule with high molar mass; the root word *meros* is Greek for "part" or "unit," so *polymer* literally means "many units."

polyprotic acid   Has two or more ionizable hydrogen atoms per molecule.

polysaccharide   A polymer of monosaccharides.

porphyrin   A type of tetradentate macrocyclic ligand.

positron   A particle with the mass of an electron but with a positive charge.

positron emission   The spontaneous emission of a positron from a neutron-poor nucleus.

potential energy (PE)   The energy stored in an object because of its position or composition.

precipitate   A solid product formed from a reaction in solution.

precision   The extent to which repeated measurements of the same variable agree.

pressure (P)   The ratio of a force to the surface area over which the force is applied.

pressure–volume (P–V) work   The work associated with the expansion or compression of a gas.

primary (1°) structure   The sequence in which the amino acid monomers occur in a protein chain.

principal quantum number (n)   A positive integer describing the relative size and energy of an atomic orbital or group of orbitals in an atom.

product   Substance formed as a result of a chemical reaction.

protein   Biological polymer made of amino acids.

proton   A subatomic particle, present in the nucleus of an atom, that has a relative charge of 1+ and a mass number of 1.

pseudo-first-order A reaction in which all the reactants but one are present at such high concentrations that they do not decrease significantly during the course of the reaction, so that reaction rate is controlled by the concentration of the limiting reactant.

p-type semiconductor A semiconductor containing an electron-poor dopant.

pure substance Matter that cannot be separated into simpler matter by a physical process.

Q

quantized Having values restricted to whole-number multiples of a specific base value.

quantum (plural *quanta*) The smallest discrete quantity of a particular form of energy.

quantum mechanics (also called *wave mechanics*) A mathematical description of the wavelike behavior of electrons and other particles.

quantum number One of four related numbers that specify the energy, shape, and orientation of orbitals in an atom and the spin orientation of electrons in the orbitals.

quantum theory A model based on the idea that energy is absorbed and emitted in discrete quantities of energy called quanta.

quarks Elementary particles that combine to form neutrons and protons.

quaternary (4°) structure The larger structure functioning as a single unit that results when two or more proteins associate.

R

R Symbol in a general formula standing for an organic group that has one available bond; it is used to indicate the variable part of a molecule so that the focus is placed on the functional group.

racemic mixture A sample containing equal amounts of both enantiomers of a compound.

radioactive decay The spontaneous disintegration of unstable particles accompanied by the release of radiation.

radioactivity The spontaneous emission of high-energy radiation and particles by materials.

radiocarbon dating A method for establishing the age of a carbon-containing object by measuring the amount of radioactive carbon-14 remaining in the object.

radiometric dating A method for determining the age of an object based on the quantity of a radioactive nuclide and/or the products of its decay that the object contains.

radionuclide A radioactive (unstable) nuclide.

random coil An irregular or rapidly changing part of the secondary structure of a protein.

Raoult's law The vapor pressure of a solution is the sum of the vapor pressures of the volatile components of the solution, which are each the product of the vapor pressure of the pure component and its mole fraction in the solution.

rate constant The proportionality constant that relates the rate of a reaction to the concentrations of reactants.

rate law An equation that defines the experimentally determined relation between the concentration of reactants in a chemical reaction and the rate of that reaction.

rate-determining step The slowest step in a multistep chemical reaction.

reactant Substance consumed during a chemical reaction.

reaction mechanism A set of steps that describes how a reaction occurs at the molecular level; the mechanism must be consistent with the rate law for the reaction.

reaction order An experimentally determined number defining the dependence of the reaction rate on the concentration of a reactant.

reaction quotient (Q) The numerical value of the mass action expression for *any values* of the concentrations (or partial pressures) of reactants and products; at equilibrium, $Q = K$.

reaction rate How rapidly a reaction occurs; it is related to rates of change in the concentrations of reactants and products over time.

reducing agent A reactant that donates electrons to another in a redox reaction, thereby reducing the other reactant; the reducing agent is oxidized in the reaction.

reduction A chemical change in which an element gains electrons; the oxidation number of the element decreases.

relative biological effectiveness (RBE) A factor that accounts for the differences in physical damage caused by different types of radiation.

replication The process by which one double-stranded DNA forms two new DNA molecules, each one containing one strand from the original molecule and one new strand.

representative elements (also called *main group elements*) The elements in groups 1, 2, and 13 through 18 of the periodic table.

resonance Characteristic of electron distributions when two or more equivalent Lewis structures can be drawn for one compound.

resonance stabilization The stability of a molecular structure due to delocalization of its electrons.

resonance structure One of two or more Lewis structures with the same arrangement of atoms but different arrangements of bonding pairs of electrons.

reverse osmosis (RO) A purification process in which solvent is forced through semipermeable membranes, leaving dissolved impurities behind.

reversible process A process that happens so slowly that an incremental change can be reversed by another tiny change, restoring the original state of the system with no net flow of energy between the system and its surroundings.

root-mean-square speed (u_{rms}) The square root of the average of the squared speeds of all the particles in a population of gas particles; a particle possessing the average kinetic energy moves at this speed.

S

salt The product of a neutralization reaction; it is made up of the cation of the base in the reaction and the anion of the acid.

saturated hydrocarbon An alkane.

saturated solution A solution that contains the maximum concentration of a solute possible at a given temperature.

Schrödinger wave equation A description of how the electron matter wave varies with location and time around the nucleus of a hydrogen atom.

scientific law A concise and generally applicable statement of a fundamental scientific principle.

scientific method An approach to acquiring knowledge based on observation of phenomena, development of a testable hypothesis, and additional experiments that test the validity of the hypothesis.

scientific theory A general explanation of widely observed phenomena that has been extensively tested.

scintillation counter An instrument that determines the level of radioactivity in samples by measuring the intensity of light emitted by phosphors in contact with the samples.

second law of thermodynamics The principle that the total entropy of the universe increases in any spontaneous process.

secondary (2°) structure The pattern of arrangement of segments of a protein chain.

seesaw Molecular geometry about a central atom with a steric number of 5 and one lone pair of electrons in an equatorial position; the atoms occupy two axial sites and two equatorial sites.

semiconductor A semimetal (metalloid) with electrical conductivity between that of metals and insulators that can be chemically altered to increase its electrical conductivity.

semimetals (also called *metalloids*) Elements that tend to have the physical properties of metals and the chemical properties of nonmetals.

sievert (Sv) SI unit used to express the amount of biological damage caused by ionizing radiation.

sigma (σ) bond A covalent bond in which the highest electron density lies between the two atoms along the bond axis.

sigma (σ) molecular orbital In MO theory, the orbital that results in the highest electron density between the two bonded atoms.

significant figures All the certain digits in a measured value plus one estimated digit. The greater the number of significant figures, the greater the certainty with which the value is known.

simple cubic (sc) unit cell A cell with atoms only at the eight corners of a cube.

single bond A bond that results when two atoms share one pair of electrons.

solid A form of matter that has a definite shape and volume.

solubility The maximum quantity of a substance that can dissolve in a given volume of solution.

solubility product, K_{sp} (also called *solubility-product constant*) An equilibrium constant that describes the formation of a saturated solution of a slightly soluble salt.

solubility-product constant (also called *solubility product*, K_{sp}) An equilibrium constant that describes the formation of a saturated solution of a slightly soluble salt.

solute Any component in a solution other than the solvent. A solution may contain one or more solutes.

solution Another name for a *homogeneous mixture*. Solutions are often liquids, but they may also be solids or gases.

solvent The component of a solution that is present in the largest number of moles.

sp hybrid orbitals Two hybrid orbitals on opposite sides of the hybridized atom formed by mixing one *s* and one *p* orbital.

sp^2 hybrid orbitals Three hybrid orbitals in a trigonal planar orientation formed by mixing one *s* and two *p* orbitals.

sp^3 hybrid orbitals A set of four hybrid orbitals with a tetrahedral orientation produced by mixing one *s* and three *p* atomic orbitals.

sp^3d hybrid orbitals Five equivalent hybrid orbitals with lobes pointing toward the vertices of a trigonal bipyramid that form by mixing one *s* orbital, three *p* orbitals, and one *d* orbital from the same shell.

sp^3d^2 hybrid orbitals Six equivalent hybrid orbitals pointing toward the vertices of an octahedron that form by mixing one *s* orbital, three *p* orbitals, and two *d* orbitals from the same shell.

specific heat (c_P) The energy required to raise the temperature of 1 g of a substance by 1°C at constant pressure.

spectator ion An ion that is unchanged by a chemical reaction.

spectrochemical series A list of ligands rank-ordered by their ability to split the energies of the *d* orbitals of transition metal ions.

sphere of hydration The cluster of water molecules surrounding an ion in aqueous solution.

spin quantum number (m_s) Either $+\frac{1}{2}$ or $-\frac{1}{2}$, indicating the spin orientation of an electron.

spontaneous process A process that proceeds without outside intervention.

square planar Molecular geometry about a central atom with a steric number of 6 and two lone pairs of electrons that occupy axial sites; the atoms occupy four equatorial positions.

square pyramidal Molecular geometry about a central atom with a steric number of 6 and one lone pair of electrons; as typically drawn, the atoms occupy four equatorial and one axial site.

standard atmosphere (atm) The average pressure at sea level on Earth.

standard cell potential ($E°_{cell}$) A measure of how forcefully an electrochemical cell, in which all reactants and products are in their standard states, can pump electrons through an external circuit.

standard conditions In thermodynamics: a pressure of 1 atm (~1 bar) and some specified temperature, assumed to be 25°C unless otherwise stated; for solutions, a concentration of 1 *M* is specified.

standard enthalpy of formation ($\Delta H°_f$) The enthalpy change of a formation reaction; also called *standard heat of formation*.

standard enthalpy of reaction ($\Delta H°_{rxn}$) The enthalpy change associated with a reaction that takes place under standard conditions; also called *standard heat of reaction*.

standard free energy of formation ($\Delta G°_f$) The change in free energy associated with the formation of 1 mole of a compound in its standard state from its elements in their standard states.

standard hydrogen electrode (SHE) A reference electrode based on the half-reaction $2\,H^+(aq) + 2\,e^- \rightarrow H_2(g)$ that produces a standard electrode potential defined to be 0.000 V.

standard molar entropy ($S°$) The absolute entropy of 1 mole of a substance in its standard state.

standard reduction potential ($E°$) The potential of a reduction half-reaction in which all reactants and products are in their standard states at 25°C.

standard solution A solution of known concentration that is used in chemical analysis.

standard state The most stable form of a substance under 1 atm pressure and some specified temperature (usually 25.0°C).

standard temperature and pressure (STP) 0°C and 1 bar as defined by IUPAC; 0°C and 1 atm are commonly used in the United States.

standing wave A wave confined to a given space with a wavelength λ related to the length *L* of the space by $L = n(\lambda/2)$, where *n* is a whole number.

state function A property of an entity based solely on its chemical or physical state or both, but not on how it achieved that state.

stereoisomerism Isomerism created by differences in the orientations of the bonds between atoms in molecules.

stereoisomers Molecules with the same formulas and the same connectivities between their atoms, but with different spatial arrangements of their atoms.

steric number (SN) The sum of the number of atoms bonded to a central atom plus the number of lone pairs of electrons on the central atom.

stoichiometry The mole ratios among the reactants and products in a chemical reaction.

strong acid An acid that completely dissociates into ions in aqueous solution.

strong base A base that completely dissociates into ions in aqueous solution.

strong electrolyte An ionic substance that dissociates completely when it dissolves in water.

strong nuclear force The fundamental force of nature that keeps quarks together in subatomic particles and nucleons together in atomic nuclei.

structural formula A representation of a molecule that uses short lines between the symbols of elements to show chemical bonds between atoms.

subatomic particles The neutrons, protons, and electrons in an atom.

sublimation Transformation of a solid directly into a vapor (gas).

substitutional alloy An alloy in which atoms of the nonhost metal replace host atoms in the crystal lattice.

substrate The reactant that binds to the active site in an enzyme-catalyzed reaction.

superconductor A material that has zero resistance to the flow of electric current.

supercritical fluid The state of a substance that is above the temperature and pressure at the critical point, where the liquid and vapor phases are indistinguishable.

supersaturated solution A solution that contains more than the maximum quantity of solute predicted to be soluble in a given volume of solution at a given temperature.

surface tension The ability of the surface of a liquid to resist an external force.

surroundings Everything in a thermochemical study that is not part of the system.

system The part of the universe that is the focus of a thermochemical study.

T

temporary dipole The separation of charge produced in an atom or molecule by a momentary uneven distribution of electrons; also called *induced dipole*.

termolecular step A step in a reaction mechanism involving a collision among three molecules.

tertiary (3°) structure The three-dimensional, biologically active structure of the protein that arises because of interactions between R groups on amino acids.

tetrahedral Molecular geometry about a central atom with a steric number of 4 and no lone pairs of electrons.

theoretical yield The maximum amount of product possible in a chemical reaction for given quantities of reactants; also called *stoichiometric yield*.

thermal energy The portion of the total internal energy of a system that is proportional to its absolute temperature.

thermochemical equation The chemical equation of a reaction that includes the change in enthalpy that accompanies the reaction.

thermochemistry The study of the changes in energy that accompany chemical reactions.

thermodynamics The study of energy and its transformations.

third law of thermodynamics The principle that the entropy of a perfect crystal is zero at absolute zero.

threshold frequency (ν_0) The minimum frequency of light required to produce the photoelectric effect.

titrant The standard solution added to the sample in a titration.

titration An analytical method for determining the concentration of a solute in a sample by reacting the solute with a solution of known concentration.

trans isomer (also called E *isomer*) Molecule with two like groups (such as two R groups or two hydrogen atoms) on opposite sides of the molecule.

transcription The process of copying the information in DNA to RNA.

transfer RNA (tRNA) The form of RNA that delivers amino acids, one at a time, to polypeptide chains being assembled by the ribosome–mRNA complex.

transition metals The elements in groups 3 through 12 of the periodic table.

transition state A high-energy state between reactants and products in a chemical reaction.

translation The process of assembling proteins from the information encoded in RNA.

tricarboxylic acid (TCA) cycle A series of reactions that continue the oxidation of pyruvate formed in glycolysis.

trigonal bipyramidal Molecular geometry about a central atom with a steric number of 5 and no lone pairs of electrons, in which three atoms occupy equatorial sites and two other atoms occupy axial sites above and below the equatorial plane.

trigonal planar Molecular geometry about a central atom with a steric number of 3 and no lone pairs of electrons.

trigonal pyramidal Molecular geometry about a central atom with a steric number of 4 and one lone pair of electrons.

triple bond A bond that results when two atoms share three pairs of electrons.

triple point The temperature and pressure where all three phases of a substance coexist. Freezing and melting, boiling and liquefaction, and sublimation and deposition all proceed at the same rate, so no net change takes place in the system.

T-shaped Molecular geometry about a central atom with a steric number of 5 and two lone pairs of electrons that occupy equatorial positions; the atoms occupy two axial sites and one equatorial site.

U

unimolecular step A step in a reaction mechanism involving only one molecule on the reactant side.

unit cell The basic repeating unit of the arrangement of atoms, ions, or molecules in a crystalline solid.

universal gas constant The constant R in the ideal gas equation; its value and units depend on the units used for the variables in the equation.

unsaturated hydrocarbon An alkene or alkyne.

unsaturated solution A solution that contains less than the maximum quantity of solute predicted to be soluble in a given volume of solution at a given temperature.

V

valence The capacity of the atoms of an element to form chemical bonds.

valence band A band of orbitals that are filled or partially filled by valence electrons.

valence bond theory A quantum mechanics–based theory of bonding that assumes covalent bonds form when half-filled orbitals on different atoms overlap or occupy the same region in space.

valence electrons   Electrons in the outermost occupied shell of an atom having the most influence on the atom's chemical behavior.

valence shell   The outermost occupied shell of an atom.

valence-shell electron-pair repulsion theory (VSEPR)   A model predicting the arrangement of valence electron pairs around a central atom that minimizes their mutual repulsion to produce the lowest energy orientations.

van der Waals equation   An equation describing how the pressure, volume, and temperature of a quantity of a real gas are related; it includes terms that account for the incompressibility of gas particles and interactions between them.

van der Waals force   Any interaction between neutral atoms and molecules including hydrogen bonds, other dipole–dipole interactions, and London dispersion forces; the term does not apply to interactions involving ions.

van 't Hoff factor (*i*)   The ratio of the concentration of solute particles in a solution to the concentration of particles that would be there if the solute did not dissociate.

vapor pressure   The pressure exerted by a gas in equilibrium with its liquid phase at a given temperature.

vinyl group   The subgroup $CH_2{=}CH-$.

vinyl polymer   One of the family of polymers formed from monomers containing the subgroup $CH_2{=}CH-$.

viscosity   The measure of the resistance to flow of a liquid.

volatile   Having a significant vapor pressure at a given temperature.

volatility   A measure of how readily a substance vaporizes.

voltaic cell   (also called *galvanic cell*) An electrochemical cell in which chemical energy is transformed into electrical work by a spontaneous redox reaction.

W

wave function (ψ)   A solution to the Schrödinger wave equation.

wave mechanics   (also called *quantum mechanics*) A mathematical description of the wavelike behavior of electrons and other particles.

wavelength (λ)   The distance from crest to crest or trough to trough on a wave.

weak acid   An acid that only partially dissociates in aqueous solutions.

weak base   A base that only partially dissociates in aqueous solutions.

weak electrolyte   A substance that only partly dissociates into ions when it dissolves in water.

work   The energy required to move an object through a given distance.

work function (Φ)   The amount of energy needed to dislodge an electron from the surface of a material.

X

X-ray diffraction (XRD)   A technique for determining the arrangement of atoms or ions in a crystal by analyzing the pattern that results when X-rays are scattered after bombarding the crystal.

Z

Z isomer   (also called *cis isomer*) Molecule with two like groups (such as two R groups or two hydrogen atoms) on the same side of the molecule.

zeolites   Natural crystalline minerals or synthetic materials consisting of three-dimensional networks of channels that contain sodium or other 1+ cations.

zwitterion   A molecule that has both positively and negatively charged groups in its structure.

Answers to Concept Tests and Practice Exercises

Chapter 1

Concept Tests

p. 6 (a) Sublimation; (b) deposition

p. 7 Kinetic energy increases by a factor of 4.

p. 9 1:1

p. 12 (top) (a) Filtration; (b) chromatography; (c) distillation

p. 12 (bottom) b and d

p. 13 (a) Chemical; (b) physical; (c) physical; (d) chemical

p. 16 CH_4O or CH_3OH

p. 22 3, 3, 4

p. 23 Fails to disprove: The density value matches that of gold, but the density alone does not prove conclusively that it *is* gold.

p. 24 (a) 14.7; (b) We only know the elapsed time to the closest minute.

Practice Exercises

1.1. (a) 1.130 g/cm^3; (b) yes

1.2. Statistics a and e are exact numbers; statistics b, c, and d have inherent uncertainty.

1.3. 0.324 km; 3.24×10^4 cm

1.4. (a) 3.8439×10^5 km; (b) 1.2822 s

1.5. Store A

1.6. $K_{low} = 40$ K and $K_{high} = 396$ K; $°F_{low} = -387°F$ and $°F_{high} = 253°F$

Chapter 2

Concept Tests

p. 44 Top plate

p. 46 The alpha particle reacts with the electron.

p. 50 The concept of atomic number was unknown in the mid-19th century.

p. 57 It represents the number of objects (tissues) per box. This is analogous having Avogadro's number of particles in a mole.

p. 60 One gram of silver (Ag)

Practice Exercises

2.1. (a) $^{56}_{26}Fe$; (b) $^{15}_{7}N$; (c) $^{37}_{17}Cl$; (d) $^{39}_{19}K$

2.2. (a) $^{39}_{19}K^+$; (b) $^{24}_{12}Mg^{2+}$; (c) $^{35}_{17}Cl^-$

2.3. (a) As, arsenic; (b) Ca, calcium; (c) Hg, mercury

2.4. $^{107}Ag = 51.5\%$; $^{109}Ag = 48.5\%$

2.5. 1.5×10^{10} atoms

2.6. 2.80 g Si

2.7. 0.0541 mole C or 3.26×10^{22} atoms C

2.8. 180.15 g/mol

2.9. 1.81×10^{-3} moles, 1.09×10^{21} molecules

2.10. $^{239}_{94}Pu + ^4_2He \rightarrow ^{240}_{96}Cm + 3\,^1_0n$

Chapter 3

Concept Tests

p. 80 Higher frequency

p. 82 c

p. 83 b and d

p. 90 $c > a > b > d$

p. 94 The vibrations of the strings have wavelengths that are equal to $2L/n$, where n is a whole number.

p. 106 ns^2np^5

p. 107 Ultraviolet

p. 113 Half-filled $6s$ and $4f$ orbitals have lower energy.

p. 119 Because the valence electrons are farther from the nucleus and shielded from it by more inner-shell electrons

p. 120 The magnitude of IE and EA both increase with increasing Z across a row (except for group 18). EA values do not display clear trends within groups whereas IE values decrease with increasing Z.

Practice Exercises

3.1. $\lambda = 3.30$ m

3.2. $\lambda = 1.99$ nm or 1.99×10^{-9} m

3.3. $\lambda = 2.62 \times 10^{-7}$ m or 262 nm

3.4. 375.0 nm; Balmer may just have been able to see this line as it is very close to violet.

3.5. Prediction: Less energy is required to remove an electron from the hydrogen atom in the $n = 3$ state versus that for a hydrogen atom in the $n = 1$ state. Calculated value: 2.420×10^{-19} J

3.6. $\lambda = 3.3 \times 10^{-10}$ m

3.7. $\Delta x \geq 7 \times 10^{-11}$ m

3.8. For $n = 5$ there are $n^2 = 25$ orbitals: $5s, 5p, 5d, 5f,$ and $5g$

3.9.

n	ℓ	m_ℓ	m_s
3	1	−1	$+\frac{1}{2}$
3	1	0	$+\frac{1}{2}$
3	1	1	$+\frac{1}{2}$

3.10. $Co = [Ar]3d^74s^2$

3.11. $K^+ = [Ar]$; $I^- = [Kr]4d^{10}5s^25p^6 = [Xe]$; $Ba^{2+} = [Xe]$; $S^{2-} = [Ne]3s^23p^6 = [Ar]$; $Al^{3+} = [Ne]$; K^+ and S^{2-} are isoelectronic with Ar.

3.12. $Mn^{3+} = [Ar]3d^4$; $Mn^{4+} = [Ar]3d^3$

3.13. (a) $Na^+ < F^- < Cl^-$; (b) $Al^{3+} < Mg^{2+} < P^{3-}$

3.14. $Ne > Ca > Cs$

Chapter 4

Concept Tests

p. 135 (a) $NaCl > KCl > RbCl$; (b) $MgO > CaO > KF$

p. 144 Hydrofluoric acid

p. 146 $BC = 18 - GN$

p. 154 $\longleftrightarrow$

$:C \equiv O:$

$\delta+ \quad \delta-$

p. 156 Not active because they are symmetric stretches

p. 162 $FC = -1$

p. 170 P ($Z = 15$) can expand its octet, N ($Z = 7$) cannot.

Practice Exercises

4.1. -3.25×10^{-18} J

4.2. (a) $SrCl_2$; (b) MgO; (c) NaF; (d) $CaBr_2$

4.3. (a) $MnCl_2$; (b) MnO_2

4.4. (a) $Sr(NO_3)_2$; (b) $KHCO_3$

4.5. (a) Calcium phosphate; (b) magnesium perchlorate; (c) lithium nitrite; (d) sodium hypochlorite; (e) potassium permanganate

4.6. (a) Tetraphosphorus decaoxide; (b) carbon monoxide; (c) nitrogen trichloride

4.7. (a) Hydrobromous acid; (b) bromous acid; (c) carbonic acid

4.8. Mg^{2+} $2\left[:\ddot{\underset{\cdot\cdot}{F}}:\right]^-$

4.9.

4.10.

4.11. $:\ddot{O}=C=\ddot{O}:$

4.12. Be and Cl; $\Delta\chi = 1.5$, polar covalent

4.13.

4.14.

4.15.

4.16.

4.17.

The resonance structure on the left has all-zero formal charges and should contribute more to bonding than the structure on the right. Therefore, the length of the terminal N–O bond should be about 122 pm, and the N–O bond in the N–O–H group should be about 136 pm long.

Chapter 5

Concept Tests

p. 190 Less: 2, 3 SN value; expanded: 5, 6; octet: 4

p. 196 Slightly smaller

p. 200 Because the difference in electronegativity between H and S is less than between H and O

p. 206 No p orbitals are available.

p. 211 a and c, because the double bonds are conjugated

p. 216 If the muscarine came from mushrooms, it would be one enantiomer and its solution would be optically active. Because the solution did not rotate the plane of polarized light, the muscarine present was a racemic mixture, which had to be the product of a laboratory synthesis. The coroner concluded that the victim was poisoned by someone who had access to synthetic muscarine.

p. 223 (a) Be_2 and Ne_2; (b) B_2

p. 226 No, bond order is lower because one more electron is in an antibonding orbital and one fewer is in a bonding orbital.

Practice Exercises

5.1. Tetrahedral;

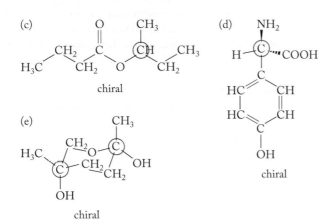

5.2. Largest bond angle: N_2O; smallest bond angle: NO_2^-
5.3. Tetrahedral, bond angles ~109.5°
5.4. No
5.5. CCl_4 and PH_3
5.6. sp^3d^2
5.7. (a)

achiral

(b)

achiral

(c)

chiral

(d)

chiral

(e)

chiral

5.8. H_2^+ may exist; its calculated bond order is 0.5.
5.9. The bond order increases on the addition of an electron for Be_2 to Be_2^-, B_2 to B_2^-, C_2 to C_2^-, and Ne_2 to Ne_2^-.
5.10.

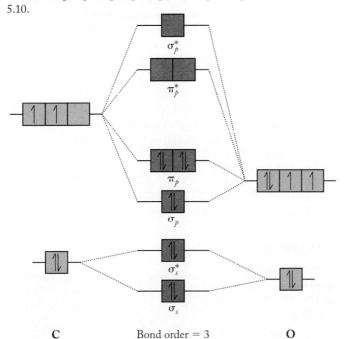

C   Bond order = 3   O

Chapter 6
Concept Tests

p. 241 CCl_4 is larger and experiences larger dispersion forces.
p. 249 The intermolecular interactions accounting for the huge difference in boiling points are hydrogen bonds: N_2H_4 can form H bonds with each other; C_2H_6 can't, it has only weaker London dispersion forces. It's harder for N_2H_4 to overcome attractions and enter gas phase.
p. 252 Normal alcohols: most soluble $C_3 > C_4 > C_5 > C_6$ (least soluble)
p. 255 A weighted wire can pass through a block of ice without melting it because pressure locally increases, melts ice near wire, and allows wire to pass but coldness leads to refreezing above wire.
p. 258 No, the adhesive forces are weak.

Practice Exercises

6.1. To enter the vapor phase from the liquid phase (to boil), ethylene glycol would break two hydrogen bonds compared to isopropanol's one hydrogen bond. Therefore, ethylene glycol has a higher boiling point.
6.2. In overproof rum, the solvent is water (1.36 mol) and the solute is ethanol (1.29 mol).
6.3. At 25 atm pressure and −100°C, the sample of CO_2 is a solid. As the temperature is increased to about −50°C the solid melts into a liquid, and as the temperature is raised further (to about −20°C) the liquid CO_2 boils to form a gas. At 20°C, as the pressure is raised from 25 to 100 atm the gaseous CO_2 condenses into a liquid.

Chapter 7
Concept Tests

p. 272 Yes, mass of Mg + O_2 consumed equals mass of MgO produced. No, mass of Mg consumed is not equal to mass of O_2 consumed because reactants are consumed in 2:1 mole ratio, which may not necessarily be the same as the mass ratio owing to the different molar masses of Mg (24 g/mol) and O_2 (32 g/mol).
p. 287 (a) C_2H_4 and (b) $C_{20}H_{40}$ have the same empirical formula (CH_2); (c) C_2H_2 and (d) C_6H_6 have the same empirical formula (CH).
p. 289 All three are molecular formulas but only (b) C_3H_8O (isopropanol) is also an empirical formula.
p. 291 Molecular formula
p. 295 A chemical reaction in which a reactant is present in excess is solvation (dissolving salt in water) or hydrolysis (ionization of acetic acid in water).

Practice Exercises

7.1. $P_4O_{10}(s) + 6\ H_2O(\ell) \rightarrow 4\ H_3PO_4(\ell)$
7.2. $2\ C_4H_{10}(g) + 13\ O_2(g) \rightarrow 8\ CO_2(g) + 10\ H_2O(\ell)$
7.3. 3.30 g CO_2 produced
7.4. Percent composition of tetracycline, $C_{22}H_{24}N_2O_8$: C = 59.45%; H = 5.443%; N = 6.305%; O = 28.80%
7.5. Cu_2S
7.6. $FeCr_2O_4$
7.7. Empirical formula: P_2O_5; molecular formula: P_4O_{10}
7.8. $C_8H_8O_3$
7.9. The fuel−oxygen mixture is fuel-rich.
7.10. Percent yield of CH_4 = 75.0%

Chapter 8

Concept Tests

p. 313 $b > a > c > d$ (b is the most concentrated solution

p. 318 $a < d < c < e < b$ (a is the most dilute solution)

p. 322 $EDTA^{4-} < HEDTA^{3-} < H_2EDTA^{2-} < H_3EDTA^-$
$< H_4EDTA$ (H_4EDTA is the strongest acid)

p. 329 React aqueous lead(II) nitrate with the stoichiometric amount of aqueous potassium dichromate. Stir the mixture for a minute or so, let it stand for 10 min, filter off the yellow $PbCr_2O_7$ precipitate, wash it with water in the filter, and allow the washed solid to air-dry overnight.

p. 332 The oxygen content of Fe_3O_4 changes when it reacts with oxygen to form Fe_2O_3. In Fe_3O_4 there are 4 O for 3 Fe, or 1.33:1, while the O to Fe ratio in Fe_2O_3 is 3:2 or 1.5:1. Thus the oxygen content increases in this reaction and iron is oxidized.

p. 337 Reactions a and b are redox reactions. In reaction a, the Br in Br_2 (O.N. = 0) is reduced to Br^-, so Br_2 is the oxidizing agent and Sn^{2+} is the reducing agent. In reaction b, the F in F_2 (O.N. = 0) is reduced (O.N. in HF is −1), so F_2 is the oxidizing agent. The O in H_2O has O.N. = −2 but the O in O_2 has O.N. = 0, so H_2O is the reducing agent.

Practice Exercises

8.1. 1.88 M $MgCl_2$

8.2. 0.109 M K^+

8.3. 4.48 g $NaCH_3CH(OH)CO_2$

8.4. $V_{initial} = 1.25$ mL

8.5. Molecular: $H_3PO_4(aq) + 3\ NaOH(aq) \rightarrow$
$3\ H_2O(\ell) + Na_3PO_4(aq)$
Total ionic: $3\ H^+(aq) + PO_4^{3-}(aq) + 3\ Na^+(aq) + 3\ OH^-(aq) \rightarrow$
$3\ H_2O(\ell) + 3\ Na^+(aq) + PO_4^{3-}(aq)$
Net ionic: $3\ H^+(aq) + 3\ OH^-(aq) \rightarrow 3\ H_2O(\ell)$, or
$H^+(aq) + OH^-(aq) \rightarrow H_2O(\ell)$

8.6. a. No precipitate forms.
b. Hg_2Cl_2 precipitates from the mixture. The net ionic equation is $Hg_2^{2+}(aq) + 2\ Cl^-(aq) \rightarrow Hg_2Cl_2(s)$.

8.7. 0.174 g HgS

8.8. 26.6 mmol/L SO_4^{2-}

8.9. (a) +4; (b) +1; (c) +5

8.10. Oxygen is reduced and is the oxidizing agent; SO_2 is oxidized and is the reducing agent.

8.11. $2\ Fe(s) + 3\ Pd^{2+}(aq) \rightarrow 2\ Fe^{3+}(aq) + 3\ Pd(s)$

8.12. $3\ HO_2^-(aq) + H_2O(\ell) + 2\ MnO_4^-(aq) \rightarrow$
$2\ MnO_2(s) + 3\ O_2(g) + 5\ OH^-(aq)$

8.13. The lemon juice is 0.292 M $C_6H_8O_7$; 100 mL of juice contains 5.62 g $C_6H_8O_7$.

Chapter 9

Concept Tests

p. 366 a. Cooling pot of water with tight lid is a closed system; cooling pot of water is an open system.
b. The pot of water without the lid cools faster because it can exchange matter with the surroundings and also exchanges energy faster without the lid.

p. 373 Water has the highest heat of vaporization in Table 9.2 because H_2O experiences the strongest intermolecular forces of attraction, namely, multiple hydrogen bonds among neighbors; it takes a lot of energy for molecules to overcome intermolecular forces and enter the gas phase.

p. 382 Equation 9.9 gives the relationship between specific heat and heat, mass, and temperature: $c_P = q/m\Delta T$.
a. Slow transfer of the metal would make the initial temperature less than 100°C so ΔT would be smaller (less negative) and that would increase c_P.
b. Drops of water adhering to the metal would make the mass greater than the mass of the metal and that would decrease c_P.
c. Ignoring the thermal mass of the thermometer would not affect c_P.
d. Heat transfer from the system would lower the final temperature below the expected value so that ΔT would be larger (more negative) and that would decrease c_P.

p. 383 Combustion of 2 mol of H_2 gives $\Delta H_{rxn} = 572$ kJ

p. 397 Fuel densities (kJ/mL) of C_3 to C_{10} alkanes increase with number of carbons and molar mass even though fuel values (kJ/g) decrease because fuel densities will vary with the density of the fuel (g/mL). The more g/mL a fuel has, the more kJ/mL it will have. We expect densities of alkanes to increase with molar mass as London forces become stronger among larger molecules.

p. 399 Combustion of 1 g of glucose releases less energy than combustion of 1 g of sucrose because the carbon content of sucrose (whose formula has one less H_2O than two molecules of glucose) is slightly higher than that of glucose (sucrose is 144.12/342.30 or 42% C; glucose is 72.06/180.16 or 40% C).

Practice Exercises

9.1. a. The match is the system, $q < 0$, and the process is exothermic.
b. The wax is the system, $q < 0$, and the process is exothermic.
c. The dry ice (CO_2) is the system, $q > 0$, and the process (sublimation) is endothermic

9.2. $w = 1.56 \times 10^7$ L · atm $= 1.58 \times 10^9$ J

9.3. $\Delta E = 32$ J

9.4. 1.13×10^3 g or 1.13 kg

9.5. 0.0°C

9.6. +68.8 kJ/mol

9.7. When 0.500 g of the hydrocarbon mixture is burned, the energy released is 24.6 kJ. When 1.000 g of the hydrocarbon mixture is burned, the energy released is 49.2 kJ.

9.8. $2\ CH_4(g) + 3\ O_2(g) \rightarrow 2\ CO(g) + 4\ H_2O(g)$
$\Delta H_{comb} = -1038$ kJ
$\underline{2\ CO(g) + O_2(g) \rightarrow 2\ CO_2(g) \qquad \Delta H_{comb} = -566\ \text{kJ}}$
$2\ CH_4(g) + 4\ O_2(g) \rightarrow 2\ CO_2(g) + 4\ H_2O(g)$
$2 \times \Delta H_{comb} = -1604$ kJ
$\Delta H_{comb} = -802$ kJ

9.9. a. $Ca(s) + C(s) + \frac{3}{2} O_2(g) \rightarrow CaCO_3(s)$
b. $2\ C(s) + 2\ H_2(g) + O_2(g) \rightarrow CH_3COOH(\ell)$
c. $K(s) + Mn(s) + 2\ O_2(g) \rightarrow KMnO_4(s)$

9.10. $\Delta H°_{rxn} = -41.2$ kJ

9.11. $\Delta H_{rxn} = -36$ kJ

9.12. 1.1×10^3 L of kerosene per second

9.13. $C_{calorimeter} = 11.2$ kJ/°C

Chapter 10

Concept Tests

p. 414 c

p. 424 d

p. 428 c

p. 435   (a) i; (b) iv
p. 441   Yes
p. 446   c
p. 448   The van der Waals pressure correction factor a of SO_2 is nearly 2 times that for $CO_2(g)$ because bent, polar SO_2 experiences collectively greater van der Waals forces, including dipole–dipole interactions, than linear CO_2

Practice Exercises

10.1.   Ar
10.2. $\Delta h = 144$ mmHg
10.3.   220% increase in volume
10.4. $V_1/V_2 = 0.622$
10.5.   38 psi
10.6. $V_2 = 2.5 \times 10^3$ L
10.7.   (a) 9.41×10^5 g He; (b) 6.77×10^6 g air $- 9.4 \times 10^5$ g He $= 5.8 \times 10^6$ g lift
10.8.   The O_2 balloon in air will sink to the floor.
10.9. $\mathcal{M} = 44.0$ g/mol, CO_2
10.10. 17 g NaN_3
10.11. $P_{O_2} = 5.24 \times 10^{-2}$ atm, outside air must be compressed 3.82 times
10.12. 2.2×10^{-3} g H_2
10.13. 4.6 g CO_2
10.14. $u_{rms,He} = 1.36 \times 10^3$ m/s, or 2.65 times faster than N_2
10.15. $P_{ideal} = 250$ atm for tank of CH_4 ($d = 275$ g/L) at 25°C; $P_{real} = 210$ atm

Chapter 11
Concept Tests

p. 473   Gasoline
p. 475   Dimethyl ether
p. 478   c
p. 480   Since the vapor pressures of pure solvents and solutions depend not on amount of substance but only on the identity of substance, P_{vap}(solvent) and P_{vap}(solution) are intensive properties.
p. 482   The solution is mostly water, which has a density of 1 kg/L.
p. 492   Toward the KCl

Practice Exercises

11.1. $U = -3792$ kJ
11.2. $\Delta H_{vap} = 28.4$ kJ/mol
11.3.   1.4
11.4. $P_{solution} = 70$ torr
11.5.   0.840 m
11.6.   4.4 m
11.7.   109.3°C
11.8.   b
11.9. $d > b > a > c > e$ (d has highest melting point, lowest i)
11.10. 26.7 atm
11.11. 27 atm
11.12. 180 g/mol
11.13. 6.40×10^4 g/mol

Chapter 12
Concept Tests

p. 508   a, c, and d
p. 512   Prediction: (c) 10^6; $W = 4.47 \times 10^3$

p. 520 $\Delta S_{univ} > 0$
p. 523   a. Yes, it is spontaneous ($\Delta G < 0$).
         b. No, spontaneity tells you nothing about the rate of reaction.
p. 525   (top) The reaction is very slow at room temperature and pressure.
p. 525   (middle) No, $\Delta G°_{rxn}$ values for combustion of three C_8 alkanes are not the same because each compound has a different $\Delta G°_f$ value; reaction stoichiometry is the same for all three, so the other values in Equation 12.12 for calculating $\Delta G°_{rxn}$ do not change.
p. 527

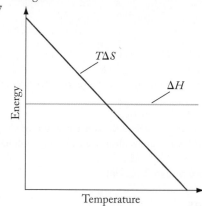

Practice Exercises

12.1.   (a) Increases; (b) decreases
12.2. $a > b > d > c$ (benzene has lowest $S°$)
12.3.   Prediction: S_{sys} decreases; $\Delta S°_{rxn} = -242.6$ J/K
12.4.   a. ΔS_{rxn} is expected to be negative.
         b. $\Delta S°_{rxn} = -326.4$ J/K
         c. $\Delta H°_{rxn} = -571.6$ kJ
         d. The reaction is spontaneous at 298 K and 1 bar.
12.5. $\Delta G°_{rxn} = 142.2$ kJ
12.6.   The reaction is spontaneous at high temperatures.
12.7. $\Delta G°_{rxn} = -196.0$ kJ

Chapter 13
Concept Tests

p. 544   (a) Positive ($\Delta S°_{rxn} > 0$); (b) same number of moles of gaseous reactants and products
p. 549   d
p. 554   (a) 1; (b) 0
p. 565   Plot k' as $[O_3]_0$; slope $= k$
p. 572   c
p. 574 n
p. 578   Similar rate laws indicate similar reaction mechanisms.

Practice Exercises

13.1.   CO; the rate of consumption of CO is twice that of O_2. The rate of change of $[CO_2]$ (a product) is -2 times the rate of change of $[O_2]$ (a reactant).
13.2. 1.2×10^{-6} M/s
13.3.   Rate $= k[NO][NO_3]$; $k = 1.57 \times 10^{10}$ M^{-1} s^{-1}
13.4.   The decomposition of H_2O_2 is first order in H_2O_2; $k = 8.30 \times 10^{-4}$ s^{-1}
13.5. $k = 2.5 \times 10^{-2}$ day^{-1}
13.6.   The reaction is second order in $[NO_2]$; $k = 0.757$ M^{-1} s^{-1}
13.7. $t_{1/2} = 9.23 \times 10^{-3}$ s

13.8. The pseudo-first-order rate constant is $k' = 6.11 \times 10^{-4}\ \mu s^{-1}$. The second-order rate constant is $k = 7.2 \times 10^6\ M^{-1}\ \mu s^{-1}$ or $7.2 \times 10^{12}\ M^{-1}\ s^{-1}$.

13.9. 6.9 kJ/mol

13.10. Rate $= k_{overall}[A]^2[B]^2$

13.11. Because none of the rate laws match the experimental rate law, this proposed mechanism cannot be valid.

13.12. Yes, NO_2 acts as a catalyst in the reaction.

Chapter 14

Concept Tests

p. 605 3

p. 606 b. $[CO_2] = [H_2] > [CO] = [H_2O]$

p. 618 Zone a in Figure 14.5 where $Q < K$

p. 622 There is no term for liquid water because its concentration is considered constant.

p. 623 a. The value of Q will decrease.

b. After equilibrium is reached the concentrations of CO_2, H_2O, and CO will all have decreased and $[H_2]$ will have increased.

p. 635 (e) There is more B than A present.

Practice Exercises

14.1. $K_c = \dfrac{[CO][H_2]^3}{[CH_4][H_2O]}$ $\quad K_p = \dfrac{(P_{CO})(P_{H_2})^3}{(P_{CH_4})(P_{H_2O})}$

14.2. $K_c = [CH_3OH]/[CO][H_2]^2 = 2.9 \times 10^2$

14.3. $K_p = 2.7 \times 10^4$

14.4. $K_p = 4.5 \times 10^{-12}$

14.5. $K_c = 0.32$

14.6. $K_{c,overall} = 1.7 \times 10^2$

14.7. This reaction is not at equilibrium and proceeds to the right.

14.8. a. $K_p = \dfrac{(P_{CO})^2}{(P_{CO_2})}$

b. $K_p = \dfrac{(P_{CO})}{(P_{CO_2})(P_{H_2})}$

14.9. a. When the reaction is cooled and water vapor condenses, one product is removed from the reaction mixture and the equilibrium shifts to the right, forming more SO_2.

b. When SO_2 gas dissolves in liquid water as it condenses, products are removed and the equilibrium shifts to the right, forming more products.

c. When O_2 is added, the concentration of one reactant increases and the equilibrium shifts to the right, forming more products.

14.10. Increasing the pressure shifts the equilibrium in the reaction to the products, the side of the reaction that has the fewest moles of gas.

14.11. The value of K for the endothermic reaction increases with increasing reaction temperature.

14.12. $P_{HI} = 0.156$ atm

14.13. $[NO_2] = 0.058\ M$ and $[N_2O_4] = 0.016\ M$ at equilibrium

14.14. $K_p = 7.80 \times 10^2$

14.15. $K_{p,298\ K} = 2.95 \times 10^{-37}$ and $K_{p,2000\ K} = 9.20 \times 10^{-13}$

Chapter 15

Concept Tests

p. 663 $H_3PO_4 > H_3AsO_4 > H_3SbO_4 \approx H_3BiO_4$

p. 665 (a) T; (b) T; (c) T; (d) T; (e) T

p. 671 Most: C; least: A

p. 673 C and D

p. 676 No for phosphoric acid, but the second ionization step of citric acid will influence pH

p. 689 A 1:1 mixture of NaH_2PO_4 and Na_2HPO_4 would have a $pH = pK_a = 7.21$. If the $[Na_2HPO_4]$ is slightly greater than $[NaH_2PO_4]$ then the pH would be exactly 7.41.

p. 693 There would be too small a change in pH at the equivalence point to detect it precisely.

p. 696 It takes less titrant to reach the first equivalence point than the second because neutralization of the CO_3^{2-} produces more HCO_3^-, which adds to the HCO_3^- present initially in the sample to make the second plateau wider than the first.

p. 700 If $[H^+]$ decreases, less F^- combines with H^+ to form HF. Therefore, the solubility of CaF_2 decreases.

Practice Exercises

15.1. $CH_3COOH(aq) + H_2O(\ell) \rightarrow CH_3COO^-(aq) + H_3O^+(aq)$

acid $\qquad$ base $\qquad$ conjugate base $\quad$ conjugate acid

15.2. pH = 2.160

15.3. $[H^+] = 2 \times 10^{-12}\ M$ and $[OH^-] = 5 \times 10^{-3}\ M$

15.4. pH = 2.23; percent ionization = 12%; $K_a = 7.9 \times 10^{-4}$

15.5. pH = 11.97

15.6. pH = 5.63

15.7. $SO_4^{2-}(aq) + H_2O(\ell) \rightleftharpoons HSO_4^-(aq) + OH^-(aq)$

15.8. pH = 9.08

15.9. pH = 4.01

15.10. pH = 4.75. There is essentially no change in the pH.

15.11. a. $[Base] = 6.51 \times 10^{-2}\ M$, $pK_a = 9.2$ so $pK_b = 4.8$.

b. The pK_b of ammonia is 4.75 so the unknown base might be ammonia (answer depends on exactly where one reads the graph).

15.12. $S = 2.6 \times 10^{-3}\ M$

15.13. $S = 3.3 \times 10^{-6}\ M$

15.14. Yes

15.15. a. Yes, both BaF_2 and CaF_2 are slightly soluble.

b. Yes, Ba^{2+} and Ca^{2+} ions in solution can be completely separated by selective precipitation with F^-.

Chapter 16

Concept Tests

p. 721 Yes, N donates a pair of electrons to an empty valence-shell orbital on B to form the N—B bond.

p. 723 CN^- ions occupy the inner sphere of Fe^{3+} with Na^+ as the counter ions. $Na_3[Fe(CN)_6]$ would have the same conductivity as orange $[Co(NH_3)_6]Cl_3$.

p. 726 Tetrachloroplatinate(II)

p. 732 $\Delta H < 0$; $\Delta S \approx 0$

p. 737 Green

p. 739 a. CN^- is a stronger field ligand.

b. Ru^{2+} ions are larger than Fe^{2+} and their $4d$ electrons interact more with ligand lone pairs than do the $3d$ electrons of Fe^{2+}.

p. 742 No for square planar, yes for tetrahedral

p. 749 Yes, $Al(OH)_3(s) + 3\ H^+(aq) \rightarrow Al^{3+}(aq) + 3\ H_2O(\ell)$

$\qquad Al(OH)_3(s) + OH^-(aq) \rightarrow Al(OH)_4^-(aq)$

Or $\qquad Al(OH)_3(s) \rightleftharpoons Al(OH)_2^+(aq) + OH^-(aq)$

$\qquad Al(OH)_3(s) + H_2O(\ell) \rightleftharpoons Al(OH)_4^-(aq) + H^+(aq)$

Practice Exercises

16.1. CaO is the Lewis base, CO_2 is the Lewis acid.

16.2. $[Ru(NH_3)_4Cl_2]^+$ is the complex ion, chloride is the counter ion, and ruthenium is present as Ru^{3+}.

16.3. (a) Ligands: NH_3, counterions: Cl^-; tetraamminezinc(II) chloride

(b) Ligands: NH_3 and H_2O, counterions: NO_2^-; tetraamminediaquacobalt(II) nitrite

16.4. Four ($-OH$ can also coordinate, either as neutral or deprotonated)

16.5. Only d^7 Ni^{3+} (c) can have either a high-spin or a low-spin configuration; none of the ions are diamagnetic.

16.6.

cis-Diammine-*trans*-dibromo(ethylenediamine)cobalt(III),
cis-diammine-*cis*-dibromo(ethylenediamine)cobalt(III), and
trans-diammine-*cis*-dibromo(ethylenediamine)cobalt(III)

16.7. 1.6×10^{-8} M

Chapter 17

Concept Tests

p. 765 No because Zn and Cu have slightly different molar masses.

p. 768 Only a is spontaneous.

p. 773 $\Delta G_{cell} > 0$, $E_{cell} < 0$

p. 776 0.257 V

p. 781 (a) watt-minute; (b) watt-hour or kilowatt-hour; (c) kilowatt-hour

p. 787 anode: oxygen
cathode: hydrogen

p. 789 CO_3^{2-} migrates toward anode (−)

Practice Exercises

17.1. $2 NO_2^-(aq) + O_2(g) \rightarrow 2 NO_3^-(aq)$

17.2. Oxidation: $HNO_2(aq) + H_2O(\ell) \rightarrow$
$NO_3^-(aq) + 3 H^+(aq) + 2 e^-$
Reduction: $MnO_4^-(aq) + 8 H^+(aq) + 5 e^- \rightarrow$
$Mn^{2+}(aq) + 4 H_2O(\ell)$
Overall: $5 HNO_2(aq) + 2 MnO_4^-(aq) + H^+(aq) \rightarrow$
$5 NO_3^-(aq) + Mn^{2+}(aq) + 3 H_2O(\ell)$

17.3. The balanced redox reaction is
$3 Cu^{2+}(aq) + 2 Al(s) \rightarrow 3 Cu(s) + 2 Al^{3+}(aq)$.
The cell diagram is $Al(s) \mid Al^{3+}(aq) \parallel Cu^{2+}(aq) \mid Cu(s)$.

17.4. The net ionic equation is $Cd(s) + 2 NiO(OH)(s) + 2 H_2O(\ell) \rightarrow$
$Cd(OH)_2(s) + 2 Ni(OH)_2(s)$. $E°_{cell} = 1.72$ V

17.5. $\Delta G°_{cell} = -290$ kJ

17.6. $E_{cell} = 1.64$ V

17.7. $K = 1.8 \times 10^{62}$

17.8. 130 g Mg

17.9. 1.5 g Pb

Chapter 18

Concept Tests

p. 803 Bands from filled Mg $3s$ orbitals overlap with bands from empty $3p$ orbitals.

p. 804 Selenium

p. 808 A crystal lattice extends indefinitely in three dimensions. A unit cell is the smallest repeating piece of the lattice.

p. 816 Li/Al

p. 820 Ice is a molecular solid in which the molecules are linked by a network of hydrogen bonds.

p. 826 $x = y = 1.2$ or $Mg_{1.2}Al_{1.2}(Si_2O_5)(OH)_4$

Practice Exercises

18.1. 128 pm

18.2. For silver, $d = 10.57$ g/cm^3, close to the 10.50 g/mL value for the density of silver from Appendix 3; for gold, $d = 19.41$ g/cm^3, close to the 19.3 g/mL value for the density of gold from Appendix 3

18.3. Gold forms substitutional alloys with both silver and copper.

18.4. 101 pm

18.5. 2.16 g/cm^3

18.6. $d = 412$ pm

Chapter 19

Concept Tests

p. 848 Dispersion forces

p. 849 No

p. 863 e

p. 869 The π bonds are localized in the linear molecule.

p. 872 Amphetamine is a primary amine, adrenaline is a secondary amine, Benadryl and pargyline are tertiary amines.

p. 876 MTBE > diethyl ether > ethanol > methanol

p. 880 Zingerone: ether, aromatic rings, $-OH$; carvone: C=C; cinnamaldehyde: aromatic ring, C=C

Practice Exercises

19.1. Yes, A absorbs 2 moles H_2/mole A, B absorbs 3 moles H_2/mole B. The product is heptane, CH_3—CH_2—CH_2—CH_2—CH_2—CH_2—CH_3, in both cases.

19.2. *n*-Hexane: ; *n*-heptane: $CH_3(CH_2)_5CH_3$

19.3. (a) 7 carbon atoms; (b) 9 carbon atoms

19.4. (a) Constitutional isomers; (b) different compounds; (c) constitutional isomers

19.5.

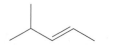 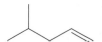

2-Methyl-1-pentene   2-Methyl-2-pentene

trans-4-Methyl-2-pentene   *cis*-4-Methyl-2-pentene   4-Methyl-1-pentene

19.6. The carbon skeleton of the monomer is

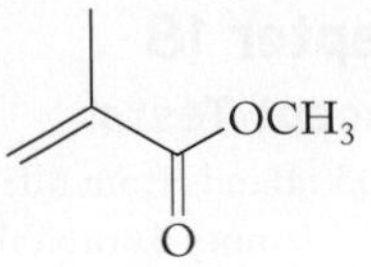

The condensed structure of the monomer is
$H_2C{=}C(CH_3)C(O)OCH_3$.

19.7. Butane by 3.7 kJ/mL

19.8. London dispersion and dipole–induced dipole interactions

19.9.

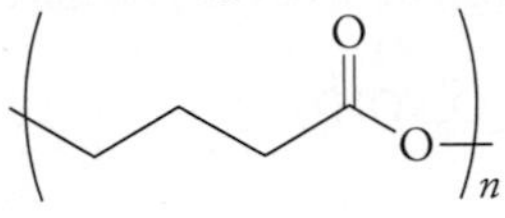

19.10. The polar fibers of cotton and polyester repel very nonpolar greases and oils but attract water molecules so perspiration wicks out of the gloves to cool the skin.

19.11. The carbon skeleton structures of the monomers are

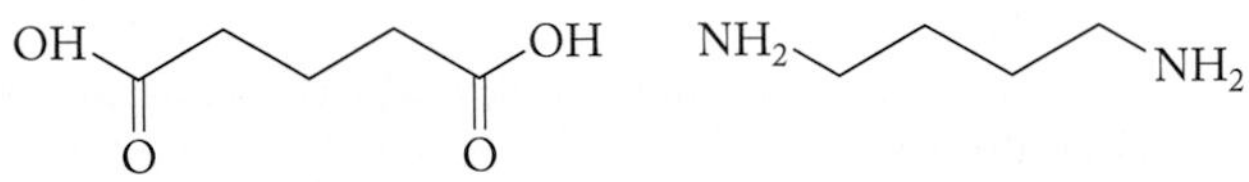

The repeating unit in the polymer is

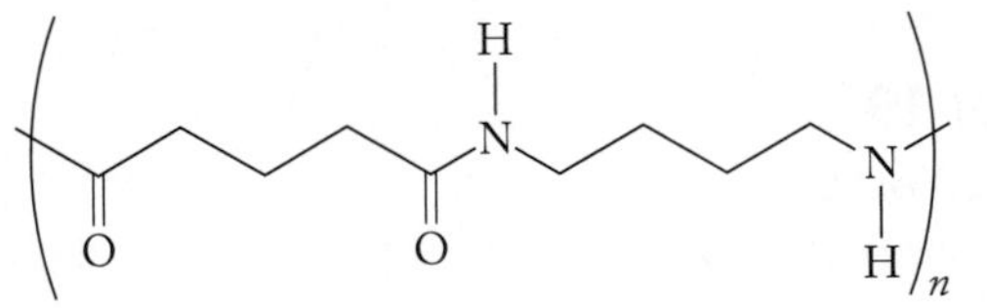

Chapter 20

Concept Tests

p. 906 sp^3; 109.5°

p. 913 Amino acids with nonpolar R groups

p. 921 a

p. 925 Olestra is a much larger molecule with many more C—C and C—H bonds compared to a triglyceride so on a per mole basis it would give off more energy.

p. 931 No

Practice Exercises

20.1.

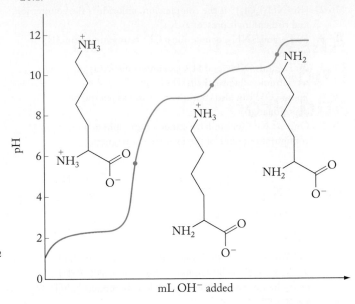

20.2. Six

20.3. Six

20.4. (a) GCCATGGCTA
     (b) AATTCGGCGATC

Chapter 21

Concept Tests

p. 946 Neutron

p. 953 The n–p ratio for stable isotopes increases with atomic number. For heavy isotopes it is about 1.5 to 1. Losing α particles increases this ratio, creating neutron-rich nuclides that undergo β decay.

p. 957 To fuse nuclei their coulombic repulsion must be overcome, which requires they collide at very high velocities, which require very high temperatures. Man-made fusion has not been carried out except in a hydrogen bomb. One cannot make a fusion reactor until the process itself can be safely carried out.

p. 969 Increasing fossil (C-14 depleted) CO_2 in the air will reduce the $^{14}C/^{12}C$ ratio in the air and in plants.

Practice Exercises

21.1. 1.753×10^9 kJ/mol

21.2. β decay to ^{28}Si

21.3. 1.4%

21.4. $A = 3.7 \times 10^7$ Bq; $A = 1.0$ mCi

21.5. 1.4×10^{11} X-rays

21.6. 4950 years old

Answers to Selected End-of-Chapter Questions and Problems

Chapter 1

1.1. a. A pure compound in the gas phase.
   b. A mixture of blue element atoms and red element atoms: blue atoms are in the gas phase, red spheres are in the liquid phase.

1.3. b

1.5. CH_2O_2

1.7. In both liquid and ice, the water molecules are touching each other. In ice, however, the water molecules arrange themselves in a rigid hexagonal arrangement; in the liquid, the molecules can move around each other and there is no long-range structure to their arrangement.

1.9. At the triple point, the gas's particles have the greatest motion and the solid's particles have the least motion.

1.11. The snow sublimed directly into the gas phase.

1.13. Energy is the ability to do work; work is force times distance; energy makes work possible.

1.15. a, b, c

1.17. 13 times as much

1.19. A Snickers bar (b) and an uncooked hamburger (d)

1.21. Orange juice (with pulp)

1.23. One chemical property of gold is its resistance to corrosion (oxidation). Gold's physical properties include its density, color, melting temperature, and electrical and thermal conductivity.

1.25. We can distinguish between table sugar, water, and oxygen by examining their physical states (sugar is a solid, water is a liquid, and oxygen is a gas) and by their densities, melting points, and boiling points.

1.27. Density, melting point, thermal and electrical conductivity, softness (a–d) are all physical properties, whereas tarnishing and reaction with water (e and f) are both chemical properties.

1.29. Extensive properties will change with the size of the sample and therefore cannot be used to identify a substance.

1.31. To form a hypothesis we need at least one observation, experiment, or idea (from examining nature).

1.33. Yes

1.35. *Theory* in normal conversation is someone's idea or opinion or speculation that can be changed.

1.37. SI units can be easily converted into a larger or smaller unit by multiplying or dividing by multiples of 10. English units are based on other number multiples and thus are more complicated to manipulate.

1.39. 93.2%

1.41. 2.5 mi

1.43. 1330 Cal

1.45. 4.1×10^{13} km

1.47. 2.0 hr

1.49. 4.0 m/s

1.51. 23 g

1.53. 19.0 mL

1.55. 26.5 g; 0.0265 kg

1.57. 58.0 cm^3

1.59. 73.8 mL

1.61. 5.1 g/cm^3

1.63. Yes

1.65. 0.28 cm^3

1.67. Precise: techniques 1 and 3, accurate: techniques 2 and 3, precise and accurate: technique 3; ranges: technique 1, 2 mg; technique 2, 10 mg; technique 3, 2 mg

1.69. a, c, d, f

1.71. a. 17.4

b. 1×10^{-13}

c. 5.70×10^{-23}

d. 3.58×10^{-3}

1.73. Because the Celsius scale is based on the freezing and boiling points of a common liquid—water.

1.75. Zero is the lowest possible temperature.

1.77. $-269.0°C$

1.79. 285.4 K; 54.1°F

1.81. $-89.2°C$; 183.9 K

1.83. The T_c for $YBa_2Cu_3O_7$ is already expressed in kelvin, $T_c = 93.0$ K. The T_c of Nb_3Ge converted to K is 23.2 K. The T_c of $HgBa_2CaCu_2O_6$ converted to K is 127.0 K. The superconductor with the highest T_c is $HgBa_2CaCu_2O_6$.

1.85. 0.031 mg/L

1.87. Both mixtures a and b react so that there is neither sodium nor chlorine left over.

1.89. Day 11

1.91. 16 times more administered than prescribed

1.93. (a) All three; (b) none

Chapter 2

2.1. The elements shaded purple (H, hydrogen) and dark blue (He, helium)

2.3. Purple (H, hydrogen)

2.5. a. Yellow (Cl, chlorine) and red (Ne, neon)

b. Red (Ne, neon)

c. Blue (Na, sodium), green (Au, gold), and orange (Lr, lawrencium)

2.7. Red arrow, alpha; green arrow, beta

2.9. a. Blue (K) and green (Ag)

b. Gray (Mg)

c. Yellow (Sc)

d. Purple (I)

e. Red (O)

2.11. Rutherford concluded that the positive charge in the atom could not be spread out (the pudding) in the atom but must result from a concentration of charge in the center of the atom (the nucleus). Most of the particles were deflected only slightly or passed directly through the gold foil; so he reasoned that the nucleus must be small compared to the size of the entire atom. The negatively charged electrons do not deflect the particles, and Rutherford reasoned that the electrons took up the remainder of the space of the atom outside the nucleus.

2.13. The fact that cathode rays were deflected by a magnetic field indicated that the rays were streams of charged particles.

2.15. Through the alpha decay of the radioactive uranium ore and its products

2.17. Greater than 1

2.19. Hydrogen (^1H)

2.21.

Atom	Mass Number	Atomic Number = Number of Protons	Number of Neutrons = Mass Number – Atomic Number	Number of Electrons = Number of Protons
(a) ^{14}C	14	6	8	6
(b) ^{59}Fe	59	26	33	26
(c) ^{90}Sr	90	38	52	38
(d) ^{210}Pb	210	82	128	82

2.23.

Symbol	^{23}Na	^{89}Y	^{118}Sn	^{197}Au
Number of Protons	11	39	50	79
Number of Neutrons	12	50	68	118
Number of Electrons	11	39	50	79
Mass Number	23	89	118	197

2.25.

Symbol	$^{37}Cl^-$	$^{23}Na^+$	$^{81}Br^-$	$^{226}Ra^{2+}$
Number of Protons	17	11	35	88
Number of Neutrons	20	12	46	138
Number of Electrons	18	10	36	86
Mass Number	37	23	81	226

2.27. Group 2, RO; group 3, R_2O_3; group 4, RO_2

2.29. Mendeleev based his groups on chemical reactivity. No compounds of the noble gases existed to indicate their presence as a group.

2.31. C, N, and O

2.33. a. Palladium (Pd)

b. Rhodium (Rh)

c. Platinum (Pt)

2.35. Three (Na, Mg, and Al)

2.37. A *weighted average* takes into account the proportion of each value in the group of values to be averaged.

2.39. $(m_X + m_Y)/2$

2.41. ^{51}V

2.43. (a) ^{11}B; (b) ^7Li; (c) ^{14}N

2.45. 63.55 amu

2.47. Yes

2.49. 47.95 amu

2.51. a. CaF_2, 78.074 amu

b. Na_2S, 78.045 amu

c. Cr_2O_3, 151.989 amu

2.53. (a) 1; (b) 3; (c) 6; (d) 6

2.55. (e) $CH_4 <$ (d) $NH_3 <$ (a) CO $<$ (c) $CO_2 <$ (b) Cl_2

2.57. A dozen is too small a unit to express the very large number of atoms, ions, or molecules present in laboratory quantities such as a mole.

2.59. (a) 7.3×10^{-10} mol Ne; (b) 7.0×10^{-11} mol CH_4; (c) 4.2×10^{-12} mol O_3; (d) 8.1×10^{-15} mol NO_2

2.61. (a) 1 mol; (b) 2 mol; (c) 1 mol; (d) 3 mol

2.63. 10.3 g

2.65. a. 7.53×10^{22} atoms

b. 7.53×10^{22} atoms

c. 1.51×10^{23} atoms

d. 2.26×10^{23} atoms

2.67. (a) Both contain the same; (b) N_2O_4; (c) CO_2

2.69. (a) 3.00 mol; (b) 4.50 mol; (c) 1.50 mol

2.71. a. 64.063 g/mol
b. 47.997 g/mol
c. 44.009 g/mol
d. 108.009 g/mol

2.73. a. 152.148 g/mol
b. 164.203 g/mol
c. 148.204 g/mol
d. 132.161 g/mol

2.75. 41.63 mol

2.77. 0.25 mol; 10 g

2.79. (a) NO; (b) CO_2; (c) O_2

2.81. 0.752 mol

2.83. Diamond

2.85. First formed (d) quark, last formed (a) deuteron

2.87. Energy is released in fusion processes when product has $Z < 26$ (Fe) because binding energy per nucleon is relatively high but drops off after Fe.

2.89. Because for fusion of helium, we have to force together two nuclei of higher positive charge and higher temperatures are required to overcome the higher repulsion.

2.91. The expanding universe was cooling and expanding and therefore could not support the high temperatures and densities needed for fusion to produce elements heavier than lithium.

2.93. The effect of β decay on the neutron-to-proton ratio is to decrease it as a neutron is transformed into a proton plus β particle.

2.95. a. $^{16}_{8}O$
b. $^{24}_{12}Mg$
c. $^{36}_{18}Ar$

2.97. a. $^{99}_{43}Tc$
b. $^{121}_{51}Sb$
c. $^{110}_{80}Hg$

2.99. The transformation of ^{137}I to ^{137}Xe can be balanced as $^{137}_{53}I \rightarrow {}^{137}_{54}Xe + {}^{0}_{-1}\beta$.
The transformation of ^{137}Xe to ^{137}Cs can be balanced as $^{137}_{54}Xe \rightarrow {}^{137}_{53}Cs + {}^{0}_{-1}\beta$.
Both of the nuclear reactions involve β emission.

2.101. $^{209}_{83}Bi + 2\,^{1}_{0}n \rightarrow {}^{211}_{85}At + 2\,^{0}_{-1}\beta$

2.103. a. $^{32}_{15}P$
b. $4\,^{1}_{0}n$
c. $2\,^{1}_{1}H$
d. $^{125}_{54}Xe,\ ^{0}_{1}\beta$

2.105. a. $^{0}_{-1}\beta$
b. $^{122}_{53}I$
c. $^{10}_{5}B$
d. $^{67}_{30}Zn$

2.107. a. Electrons
b. The negatively charged electrons were attracted to the positively charged plate as the electrons passed through the electric field.
c. If the polarities of the plates were switched, the electron would still be deflected toward the positively charged plate, which would now be at the opposite side of the screen.

2.109. 10.00% ^{25}Mg and 11.01% ^{26}Mg

2.111. $^{11}_{5}B + {}^{1}_{0}n \rightarrow {}^{12}_{5}B \rightarrow {}^{8}_{3}Li + {}^{4}_{2}He$
$^{11}_{5}B + {}^{1}_{0}n \rightarrow {}^{12}_{5}B \rightarrow {}^{12}_{6}C + {}^{0}_{-1}\beta$

2.113. $^{208}_{82}Pb + {}^{62}_{28}Ni \rightarrow {}^{269}_{110}Ds + {}^{1}_{0}n$

2.115. (a) 0.7580 mol C; (b) 4.565×10^{23} atoms C

Chapter 3

3.1. a. Purple (Na), red (Cr), and orange (Au)
b. Blue (Ne)
c. Orange (Au)
d. Red (Cr)
e. Blue (Ne) and green (Cl)

3.3. a. Green (Cl)
b. Purple (Na)

3.5. Blue (Rb), green (Sr), and orange (Y)

3.7. c

3.9. a

3.11. All these forms of light have perpendicular, oscillating electric and magnetic fields that travel together through space.

3.13. The lead shield must protect the parts of your body that might be exposed to X-rays but are not being imaged. Lead is a very high density metal with many electrons, which interact with X-rays and absorb nearly all the X-rays before they can reach your body.

3.15. It still emits infrared radiation.

3.17. $4.87 \times 10^{14}\ s^{-1}$

3.19. (a) 3.32 m; (b) 3.14 m; (c) 3.04 m; (d) 2.79 m

3.21. The radio station has the lower frequency.

3.23. 8.317 min

3.25. The absorption spectrum consists of dark lines at wavelengths specific to that element. The emission spectrum has bright lines on a dark background with the lines appearing at the exact same wavelengths as the dark lines in the absorption spectrum.

3.27. Because each element shows distinctive and unique absorption and emission lines, the bright emission lines observed for the pure elements could be matched to the many dark absorption lines in the spectrum of sunlight. This approach can be used to deduce the sun's elemental composition.

3.29. The quantum is the smallest indivisible amount of radiant energy that an atom can absorb or emit.

3.31. From Figure 3.11 we see that tungsten as a blackbody metal will emit in the visible range at 1000 K and that it will glow red-orange.

3.33. 6.62×10^{-19} J

3.35. b

3.37. 6.93×10^{-19} J

3.39. No

3.41. Potassium; 8.04×10^{5} m/s

3.43. 3.17×10^{18} photons/s

3.45. The Rydberg equation is more general then the Balmer equation; the Balmer equation is equivalent to the Rydberg when $n_1 = 2$.

3.47. It is the difference between n levels that determines emission energy.

3.49. a

3.51. No

3.53. At $n = 7$, the wavelength of the electron's transition ($n = 7$ to $n = 2$) has moved out of the visible region.

3.55. 1875 nm; infrared

3.57. No, because for hydrogen the transition is in the ultraviolet region and as Z increases the wavelength will shorten further.

3.59. 72.9 nm

3.61. In the de Broglie equation, λ is the wavelength the particle of mass m exhibits as it travels at speed u, where h is Planck's constant. This equation states that (1) any moving particle has wavelike properties because a wavelength can be calculated through the equation, and (2) the wavelength of the particle is inversely related to its momentum (mass multiplied by velocity).

3.63. No

3.65. b and c

3.67. a. 10.8 nm
  b. 0.180 nm
  c. 8.2×10^{-37} nm
  d. 3.7×10^{-54} nm

3.69. $\Delta x \geq 1.3 \times 10^{-13}$ m

3.71. The Bohr model orbit showed the quantized nature of the electron in the atom as a particle moving around the nucleus in concentric orbits. In quantum theory, an orbital is a region of space where the probability of finding the electron is high. The electron is not viewed as a particle, but as a wave, and it is not confined to a clearly defined orbit; rather, we refer to the probability of the electron being at various locations around the nucleus.

3.73. Three: n, ℓ, and m_ℓ.

3.75. (a) 1; (b) 4; (c) 9; (d) 16; (e) 25

3.77. 3, 2, 1, 0

3.79. (a) $2s$; (b) $3p$; (c) $4d$; (d) $1s$

3.81. (a) 2; (b) 2; (c) 10; (d) 2

3.83. b

3.85. Degenerate orbitals have the same energy and are indistinguishable from each other.

3.87. As we start from an argon core of electrons, we move to potassium and calcium, which are located in the s block on the periodic table. It is not until Sc, Ti, V, etc., that we begin to fill electrons into the $3d$ shell.

3.89. (c) $3s$ < (a) $3d$ < (d) $4p$ < (b) $7f$

3.91. Li^+: $1s^2$ or [He]
  Ca: $[Ar]4s^2$
  F^-: $[He]2s^22p^6$ or [Ne]
  Mg^{2+}: $[He]2s^22p^6$ or [Ne]
  Al^{3+}: $[He]2s^22p^6$ or [Ne]

3.93. K: $[Ar]4s^1$
  K^+: [Ar]
  Ba: $[Xe]6s^2$
  Ti^{4+}: $[Ne]3s^23p^6$ or [Ar]
  Ni: $[Ar]4s^23d^8$

3.95. Ra: $[Rn]7s^2$
  I: $[Kr] 5s^24d^{10}5p^5$
  In: $[Kr] 5s^24d^{10}5p^1$
  Mn: $[Ar] 4s^23d^5$
  Mn^{2+}: $[Ar]3d^5$

3.97. (a) 3; (b) 2; (c) 0; (d) 0

3.99. Ti, two unpaired electrons.

3.101. Cl^-, no unpaired electrons

3.103. a and d

3.105. $5p$, yes

3.107. If electrons do not repel each other as much in Na^+ as they do in Na, they will have lower energy and be, on average, closer to the nucleus, resulting in a smaller size. When electrons are added to an atom (Cl), the e^-–e^- repulsion increases, so the electrons have higher energy and they will be, on average, farther from the nucleus, thereby creating a larger size species (Cl^-).

3.109. Rb. The size of atoms increases down a group because electrons have been added to higher n levels.

3.111. a. As the atomic number increases down a group, electrons are added to higher n levels, leading to a decrease in ionization energy.
  b. As the atomic number increases across a period, the effective nuclear charge increases. This means that the ionization energy increases across a period of elements.

3.113. Fluorine, with a higher nuclear charge, exerts a higher Z_{eff} on the $2p$ electrons than boron, resulting in higher ionization energy.

3.115. Sr

3.117. No, sodium has a negative (favorable) electron affinity but is never found as an anion in nature. It is always a cation in salts such as NaCl and Na_2CO_3.

3.119. The electron is added farther away from the nucleus as we descend the halogens because the atoms get larger. This means that the electron affinities become more positive (less negative).

3.121. a. -1.11×10^{-26} J
  b. 17.9 m
  c. A radio telescope

3.123. a, c, and d

3.125. Yes. It is generally observed that as Z increases so does the IE_2. However, Ge's second IE_2 is lower than Ga's because to ionize the second electron in Ga, we need to remove an electron from a lower energy $4s$ orbital. Also, Br's IE_2 is lower than Se's because the electron pairing ($4p^4$) in one of the p orbitals for the Br^+ ion lowers its IE_2 slightly.

3.127. a. Sn^{2+}: $[Kr]4d^{10}5s^2$
  Sn^{4+}: $[Kr]4d^{10}$
  Mg^{2+}: $[He]2s^22p^6$ or [Ne]
  b. Cadmium has the same electron configuration as Sn^{2+} and neon has the same electron configuration as Mg^{2+}.
  c. Cd^{2+}

3.129. a. Ne, 5.76; Ar, 6.76
  b. The outermost electron in argon is a $3p$ electron that is mostly shielded by the electrons in the $n = 2$ level (10 electrons) and the $n = 1$ level (2 electrons), whereas the outermost electron in neon is a $2p$ electron that is shielded only by the electrons in the $n = 1$ level (2 electrons).

3.131. When we think of the electron as a wave, we can envision the node between the two lobes as a wave of zero amplitude and the p orbital as a standing wave.

Chapter 4

4.1. a. Group 1 (red)
  b. Group 14 (blue)
  c. Group 16 (purple)

4.3. Mg^{2+}

4.5. Group 14 (blue, carbon)

4.7. Lithium (red) and fluorine (lilac)

4.9. b

4.11. The arrangement of the atoms in two of the structures is S—O—S and in the other two structures it is S—S—O. Because the arrangement of atoms differs, they are not resonance structures. Also, for each arrangement, the structures do not show a different arrangement of electrons on the atoms; only the bonds are drawn bent, not straight. The "bent form" and "linear form" are not resonance forms of each other if the numbers of lone pairs and bonding pairs of electrons on each atom are the same.

4.13. a

4.15. Fluorine (purple) and oxygen (light blue)

4.17. Group 17 (blue)

4.19. Yes

4.21. Yes, for hydrogen and helium

4.23. No

4.25. -6.94×10^{-18} J

4.27. TiO_2

4.29. $CsBr < KBr < SrBr_2$

4.31. Roman numerals indicate the charge on the transition metal cation.

4.33. $XO_2{}^{2-}$

4.35. a. NO_3, nitrogen trioxide
 b. N_2O_5, dinitrogen pentoxide
 c. N_2O_4, dinitrogen tetroxide
 d. NO_2, nitrogen dioxide
 e. N_2O_3, dinitrogen trioxide
 f. NO, nitrogen monoxide
 g. N_2O, dinitrogen monoxide
 h. N_4O, tetranitrogen monoxide

4.37. a. Na_2S, sodium sulfide
 b. $SrCl_2$, strontium chloride
 c. Al_2O_3, aluminum oxide
 d. LiH, lithium hydride

4.39. a. cobalt(II) oxide
 b. cobalt(III) oxide
 c. cobalt(IV) oxide

4.41. a. BrO^-
 b. $SO_4{}^{2-}$
 c. $IO_3{}^-$
 d. $NO_2{}^-$

4.43. a. nickel(II) carbonate
 b. sodium cyanide
 c. lithium hydrogen carbonate
 d. calcium hypochlorite

4.45. a. hydrofluoric acid
 b. bromic acid
 c. H_3PO_4
 d. HNO_2

4.47. a. sodium oxide
 b. sodium sulfide
 c. sodium sulfate
 d. sodium nitrate
 e. sodium nitrite

4.49. a. K_2S
 b. K_2Se
 c. Rb_2SO_4
 d. $RbNO_2$
 e. $MgSO_4$

4.51. a. manganese(II) sulfide
 b. vanadium(II) nitride
 c. chromium(III) sulfate
 d. cobalt(II) nitrate
 e. iron(III) oxide

4.53. b. Na_2SO_3

4.55. In the diatomic molecule XY shown here

$$:\ddot{X}:\ddot{Y}:$$

Lewis counts 6 e⁻ in 3 lone pairs on both X and Y. He also counts the 2 e⁻ shared between X and Y separately (2 e⁻ for X and 2 e⁻ for Y). However, there are not 4 e⁻ being shared, only 2 e⁻. It seems that the Lewis counting scheme counts the shared electrons twice.

4.57. For the H—O—H bonding pattern, the oxygen of the central atom forms bonds to the two hydrogen atoms. This uses 4 of the 8 e⁻ leaving 4 e⁻ left over for the two lone pairs. Each hydrogen atom has a duet of electrons, so the lone pairs reside on oxygen and form an octet on oxygen.

$$H-\ddot{\underset{\cdot\cdot}{O}}-H$$

For H—H—O bonding, the two covalent bonds again use 4 of the 8 e⁻, leaving 4 e⁻ for two lone pairs. If these are placed on the oxygen atom as shown here,

$$H-H-\ddot{\underset{\cdot\cdot}{O}}$$

oxygen does not complete its octet and the central hydrogen atom has 4 e⁻, not a duet. This structure would violate the Lewis structure formalism.

4.59. $Cs\cdot \quad \cdot Ba\cdot \quad \cdot\dot{Al}\cdot$

4.61. $Na^+ \quad [\cdot In\cdot]^+ \quad Ca^{2+} \quad \left[:\ddot{S}:\right]^{2-}$

4.63. I^- and Ca^{2+}

4.65. (a) 8; (b) 8; (c) 8; (d) 10

4.67. a. $:C\equiv O:$

 b. $\ddot{O}=\ddot{O}$

 c. $\left[:\ddot{Cl}-\ddot{O}:\right]^-$

 d. $[:C\equiv N:]^-$

4.69. a., b., c., d. (Lewis structures)

4.71. a., b., c., d., e. (Lewis structures)

4.73. a. $[:\ddot{O}-\ddot{C}l-\ddot{O}:]^-$ c. $[:O=C-\ddot{O}-H]^-$ with $:\ddot{O}:$ below C

b. $[:\ddot{O}-S-\ddot{O}:]^{2-}$ with $:\ddot{O}:$ below S

4.75.

H—C—C—C—C—$\ddot{S}$—H (each C bearing H above and below)

H—$\ddot{S}$—H

4.77. $:\ddot{C}l-\ddot{C}l-\ddot{O}:$

$[:\ddot{O}-\ddot{C}l-\ddot{O}:]^-$ with $:\ddot{O}:$ above Cl

4.79. If there is an electronegativity difference of 2.0 or greater, the bond between the atoms is ionic; below 2.0, the bond is covalent.

4.81. The size of an atom is the result of the nucleus pulling on the electrons. The higher the nuclear charge, the stronger the pull on the electrons within a given valence shell. This is why the size of atoms generally decreases across a period. A small atom will form a shorter bond with another atom and the electrons in the bond will feel a strong pull from the nucleus of a smaller atom since the bonding electrons will be "closer" to the nucleus. This stronger pull results in a higher electronegativity for smaller atoms.

4.83. A polar covalent bond is one in which the electrons are shared, but not equally, by the atoms.

4.85. The polar bonds and the atoms with the greater electronegativity (underlined) are $\underline{C}$—Se, C—$\underline{O}$, $\underline{N}$—H, and $\underline{C}$—H.

4.87. Binary compounds of (b) C and O and (c) Al and Cl have polar covalent bonds. The binary compound of (d) Ca and O has ionic bonds.

4.89. Like the panes of glass in a greenhouse, the greenhouse gases in the atmosphere are transparent to visible light. Once the visible light warms the surface of the Earth and is reemitted as infrared (lower energy) light the greenhouse gases absorb the infrared light, in the same way that the panes of glass do not allow the heat from inside the greenhouse to escape.

4.91. The nitrogen–oxygen bond would be expected to absorb IR radiation on stretching.

4.93. CO does absorb IR radiation because stretching its linear bond gives rise to a fluctuating electric field.

4.95. Infrared radiation with its longer wavelengths and lower energy than UV radiation cause chemical bonds only to stretch and bend, but not to break.

4.97. More

4.99. Resonance occurs when two or more valid Lewis structures may be drawn for a molecular species. The true structure of the species is a hybrid of the structures drawn.

4.101. A molecule or ion shows resonance when there is more than one correct Lewis structure, that is, when the electrons in the correct Lewis structure may be distributed in more than one way. Often, when the central atom has both a single and a double bond resonance is possible.

4.103. Either N–O bond in the NO_2 structure could be double-bonded, and the formal charges for each structure are identical,

so there is more than one correct Lewis structure and NO_2 will exhibit resonance.

$:\overset{-1}{\ddot{O}}-\overset{+1}{N}=\overset{+0}{\ddot{O}:} \longleftrightarrow :\overset{+0}{\ddot{O}}=\overset{+1}{N}-\overset{-1}{\ddot{O}:}$

The resonance forms of CO_2 show that one is dominant (the one in which all formal charges are zero) and so the other forms contribute little to the true structure of CO_2.

$:\overset{+0}{\ddot{O}}=\overset{+0}{C}=\overset{+0}{\ddot{O}:} \longleftrightarrow :\overset{+1}{O}\equiv\overset{+0}{C}-\overset{-1}{\ddot{O}:} \longleftrightarrow :\overset{-1}{\ddot{O}}-\overset{+0}{C}\equiv\overset{+1}{O:}$

4.105.

H—C=C—H ring with H—C=C—H (cyclobutadiene) $\longleftrightarrow$ resonance form

4.107. N_2O_2:

$\ddot{O}=\ddot{N}-\ddot{N}=\ddot{O} \longleftrightarrow :O\equiv N-\ddot{N}-\ddot{O}: \longleftrightarrow \ddot{O}=N=\ddot{N}-\ddot{O}: \longleftrightarrow$

$:\ddot{O}-N\equiv N-\ddot{O}: \longleftrightarrow :\ddot{O}-\ddot{N}-N\equiv O: \longleftrightarrow :\ddot{O}-\ddot{N}=N=\ddot{O}$

N_2O_3:

$:\ddot{O}=\ddot{N}-N\overset{\ddot{O}:}{\underset{\ddot{O}:}{}} \longleftrightarrow :\ddot{O}=\ddot{N}-N\overset{\ddot{O}:}{\underset{\ddot{O}:}{}} \longleftrightarrow$

$:\ddot{O}=N=N\overset{\ddot{O}:}{\underset{\ddot{O}:}{}} \longleftrightarrow :O\equiv N-N\overset{\ddot{O}:}{\underset{\ddot{O}:}{}}$

4.109.

$H-C\equiv N-\ddot{O}: \longleftrightarrow H-\ddot{C}-N\equiv O: \longleftrightarrow H-\ddot{C}=\ddot{N}=O:$

4.111.

[Resonance structures with N and O atoms] $\longleftrightarrow$ [Resonance structures]

$\updownarrow$ $\updownarrow$

[Resonance structures] $\longleftrightarrow$ [Resonance structures]

4.113. The best possible structure for a molecule judging by formal charges is the structure in which the formal charges are minimized and the negative formal charges are on the most electronegative atoms in the structure.

4.115. No

4.117. $\overset{0}{H}-\overset{+1}{N}\equiv\overset{-1}{C}:$ $\overset{0}{H}-\overset{0}{C}\equiv\overset{0}{N}:$

The formal charges are zero for all the atoms in HCN, whereas in HNC the carbon atom, with a lower electronegativity than N, has a -1 formal charge.

4.119. $\overset{0}{H}\diagdown\overset{+1}{N}=\overset{0}{C}=\overset{-1}{\ddot{N}:}$ over $\overset{0}{H}\diagup$ $\longleftrightarrow$ $\overset{0}{H}\diagdown\overset{0}{\ddot{N}}-\overset{0}{C}\equiv\overset{0}{N}:$ over $\overset{0}{H}\diagup$

The preferred structure is the one with the C triple bonded to N.

4.121. $:\overset{-2}{N}-\overset{+2}{O}\equiv\overset{0}{N}: \longleftrightarrow :\overset{-1}{N}=\overset{+2}{O}=\overset{-1}{\ddot{N}:} \longleftrightarrow :\overset{0}{N}\equiv\overset{+2}{O}-\overset{-2}{\ddot{N}:}$

Because oxygen is more electronegative than nitrogen, none of these structures is likely to be stable because the formal charge on O is positive.

4.123. a.

$$H-\overset{\overset{\displaystyle H}{|}}{\underset{\underset{\displaystyle H}{|}}{C}}-N\overset{O}{\underset{O}{\diagup}} \longleftrightarrow H-\overset{\overset{\displaystyle H}{|}}{\underset{\underset{\displaystyle H}{|}}{C}}-N\overset{O}{\underset{O}{\diagup}}$$

b.

$$:\overset{-1}{C}\equiv\overset{+1}{N}-\overset{+1}{N}\overset{O}{\underset{\underset{-1}{O}}{\diagup}}^{0} \longleftrightarrow :\overset{-1}{C}\equiv\overset{+1}{N}-\overset{+1}{N}\overset{O^{-1}}{\underset{\underset{0}{O}}{\diagup}}$$

(1)   (2)

$$:\overset{0}{N}\equiv\overset{0}{C}-\overset{+1}{N}\overset{O}{\underset{\underset{-1}{O}}{\diagup}}^{0} \longleftrightarrow :\overset{0}{N}\equiv\overset{0}{C}-\overset{+1}{N}\overset{O^{-1}}{\underset{\underset{0}{O}}{\diagup}}$$

(3)   (4)

Formal charges are minimized in structures 3 and 4, so they are preferred.

c. No

4.125. Yes

4.127. In order for the atom to accommodate more than 8 e⁻ in covalently bonded molecules, it would require the use of orbitals beyond s and p. The d orbitals are not available to the small elements in the second period.

4.129. (a) SF_6, (b) SF_5, and (c) SF_4

4.131. (a) 12; (b) 8; (c) 12; (d) 10

4.133.

$$\overset{\overset{\displaystyle :O:^{-1}}{|}}{\underset{\underset{\displaystyle :F:}{|}}{\overset{0}{F}-\overset{+1}{N}-\overset{0}{F}}}\qquad\overset{\overset{\displaystyle :O:^{0}}{||}}{\underset{\underset{\displaystyle :F:}{|}}{\overset{0}{F}-\overset{0}{P}-\overset{0}{F}}}$$

In POF_3 there is a double bond and no formal charges; in NOF_3 there are only single bonds and formal charges are present on N and O.

4.135.

$$\overset{:F:}{\underset{:F:}{\overset{|}{Se}}}\quad\left[\overset{:F:}{\underset{:F:}{\overset{|}{Se}}}\right]^{-}$$

In both structures Se has more than 8 valence electrons.

4.137.

$$\overset{:O:^{0}}{\underset{:O:^{0}}{\overset{}{Cl}}}-\overset{}{Cl}:^{0}$$

The central chlorine atom has an expanded octet.

4.139. (c) ClO_4, (d) ClO_3, and (e) ClO_2

4.141. (a) S; (b) N; (c) C; (d) O

4.143. d

4.145. No

4.147. The nitrogen–oxygen bond in N_2O_4 has a bond order of 1.5 due to four equivalent resonance forms:

$$\overset{:O:}{\underset{:O:}{\overset{}{}}}N-N\overset{O:}{\underset{O:}{}} \longleftrightarrow \overset{:O:}{\underset{:O:}{\overset{}{}}}N-N\overset{O:}{\underset{O:}{}} \longleftrightarrow$$

$$\overset{:O:}{\underset{:O:}{\overset{}{}}}N-N\overset{O:}{\underset{O:}{}} \longleftrightarrow \overset{:O:}{\underset{:O:}{\overset{}{}}}N-N\overset{O:}{\underset{O:}{}}$$

The nitrogen–oxygen bond in N_2O has a bond order of 1.5 due to resonance between three resonance forms (where the last resonance structure shown does not significantly contribute

to the structure of the molecule because of the buildup of too much formal charge):

$$:\overset{-1}{N}=\overset{+1}{N}=\overset{+0}{O}: \longleftrightarrow :\overset{+0}{N}\equiv\overset{+1}{N}-\overset{-1}{O}: \longleftrightarrow :\overset{-2}{N}-\overset{+1}{N}\equiv\overset{+1}{O}:$$

Therefore, owing to resonance, N_2O_4 and N_2O are expected to have nearly equal bond lengths.

4.149. $NO^+ < NO_2^- < NO_3^-$

4.151. $NO_3^- < NO_2^- < NO^+$

4.153. (a) ·Be·   (b) ·Ȧl·   (c) ·Ċ·   (d) He:

4.155. $:\overset{0}{S}=\overset{0}{C}=\overset{0}{S}: \qquad :\overset{0}{S}=\overset{+2}{S}=\overset{-2}{C}:$

The preferred structure for carbon disulfide has C as the central atom.

4.157.

$$\overset{\overset{\displaystyle :O:}{||}}{\underset{\underset{}{\overset{}{}}}{\overset{}{}}}\\ :Cl\overset{C}{\diagup\diagdown}Cl:$$

4.159.

a. $:\overset{+1}{O}\equiv\overset{0}{C}-\overset{-1}{N}-\overset{-1}{N}-\overset{0}{C}\equiv\overset{+1}{O}: \longleftrightarrow :\overset{0}{O}=\overset{0}{C}=\overset{0}{N}-\overset{0}{N}=\overset{0}{C}=\overset{0}{O}: \longleftrightarrow$

$:\overset{0}{O}=\overset{-1}{C}-\overset{+1}{N}\equiv\overset{+1}{N}-\overset{-1}{C}=\overset{0}{O}: \longleftrightarrow :\overset{0}{O}-\overset{-1}{C}\equiv\overset{+1}{N}-\overset{+1}{N}\equiv\overset{-1}{C}-\overset{0}{O}: \longleftrightarrow$

$:\overset{0}{O}=\overset{0}{C}=\overset{0}{N}-\overset{+1}{N}\equiv\overset{0}{C}-\overset{-1}{O}: \longleftrightarrow :\overset{-1}{O}-\overset{0}{C}\equiv\overset{+1}{N}-\overset{0}{N}=\overset{0}{C}=\overset{0}{O}:$

b. $:\overset{}{Br}-\overset{}{N}=\overset{}{O}:$

$:\overset{+1}{O}\equiv\overset{0}{C}-\overset{0}{N}=\overset{0}{N}-\overset{-1}{O}: \longleftrightarrow :\overset{0}{O}=\overset{-1}{C}-\overset{0}{N}=\overset{+1}{N}=\overset{0}{O}: \longleftrightarrow$

$:\overset{-1}{O}-\overset{-1}{C}\equiv\overset{0}{N}-\overset{+1}{N}\equiv\overset{+1}{O}:$

c. $:\overset{+1}{O}\equiv\overset{0}{C}-\overset{-1}{N}-\overset{}{C}-\overset{-1}{N}-\overset{0}{C}\equiv\overset{+1}{O}: \longleftrightarrow :\overset{0}{O}=\overset{0}{C}=\overset{0}{N}-\overset{}{C}-\overset{0}{N}=\overset{0}{C}=\overset{0}{O}: \longleftrightarrow$

$:\overset{0}{O}=\overset{0}{C}=\overset{0}{N}-\overset{}{C}=\overset{+1}{N}=\overset{0}{C}=\overset{0}{O}: \longleftrightarrow :\overset{-1}{O}-\overset{0}{C}\equiv\overset{+1}{N}-\overset{}{C}-\overset{+1}{N}\equiv\overset{0}{C}-\overset{-1}{O}: \longleftrightarrow$

$:\overset{-1}{O}-\overset{0}{C}\equiv\overset{+1}{N}-\overset{}{C}-\overset{0}{N}=\overset{0}{C}=\overset{0}{O}: \longleftrightarrow :\overset{0}{O}=\overset{0}{C}=\overset{0}{N}-\overset{}{C}-\overset{+1}{N}\equiv\overset{0}{C}-\overset{-1}{O}:$

4.161. For Cl_2O_6 with a Cl—Cl bond

$$:\overset{0}{O}=\overset{0}{Cl}-\overset{0}{Cl}=\overset{0}{O}:$$
with O atoms above and below

For Cl_2O_6 with a Cl—O—Cl bond

$$:\overset{0}{O}=\overset{0}{Cl}-\overset{0}{O}-\overset{0}{Cl}=\overset{0}{O}:$$
with O atoms above and below

4.163. a. ·C≡N:; the more likely structure for cyanogen is the one that contains the C—C bond.

b. It would be expected that oxalic acid would retain the C—C bond from the cyanogen from which it is formed in the reaction of cyanogen with water. This is consistent with the structure for cyanogen predicted by formal charge analysis.

4.165.

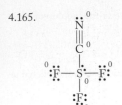

4.167.

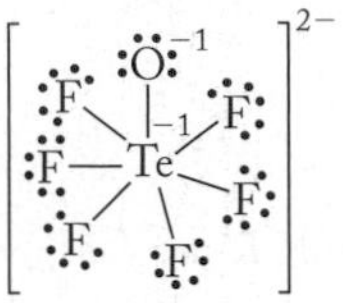

4.169.

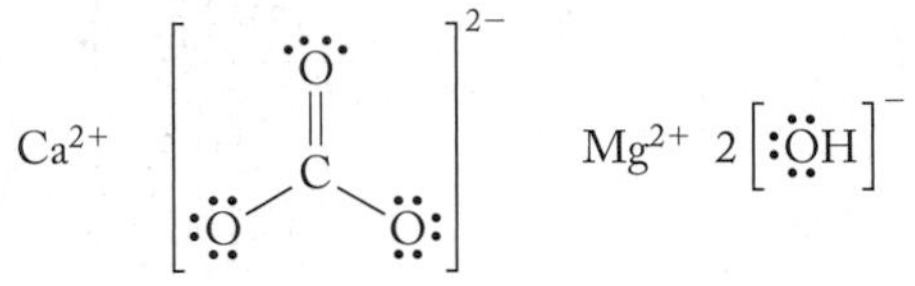

Ca²⁺ [structure] ²⁻   Mg²⁺ 2[:ÖH]⁻

4.171. a, b.

:N≡N—N=N: ⁻¹ ↔ :N=N=N=N: ⁻¹ ↔

N=N—N≡N:

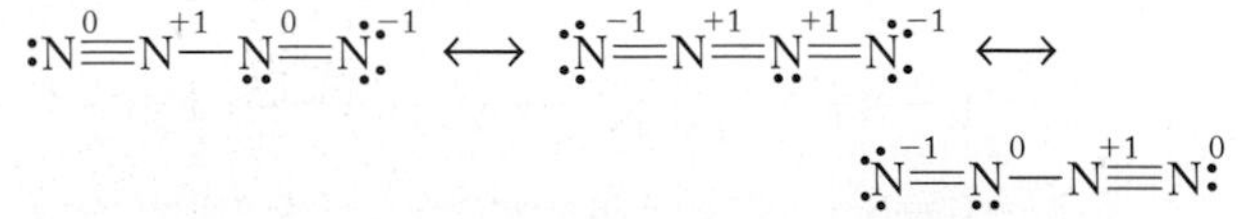

The middle structure has the most nonzero formal charges separated over three bond lengths, so this one is least preferred. The first and last resonance structures are preferred and are indistinguishable from each other.

c.

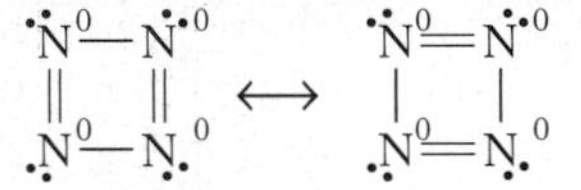

4.173. b and c

4.175. a, b.

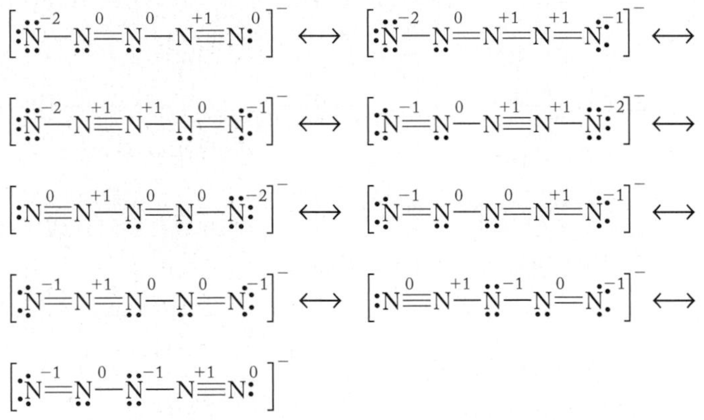

The structures that contribute most have the lowest formal charges (last four structures shown).

c. N₃⁻ has the Lewis structures

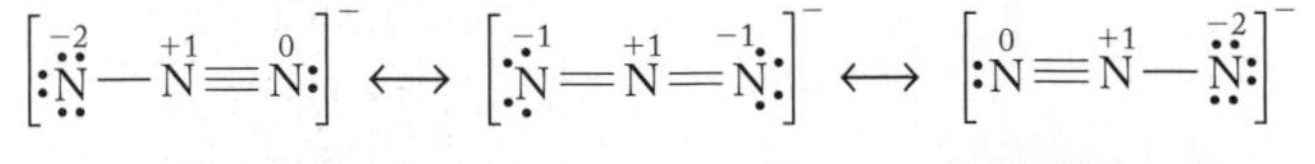

From these resonance structures we see that each bond is predicted to be of double bond character in N₃⁻. Therefore, in N₅⁻ there are two longer N–N bonds than in N₃⁻. N₃⁻ has the higher average bond order.

4.177.

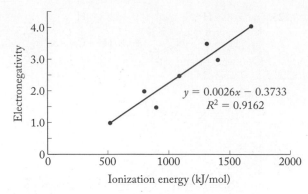

Using the equation for the best-fit line where x = the ionization energy of neon gives a value of y (electronegativity) of neon: $y = 5.0$

4.179. a. Isoelectronic means that the two species have the same number of electrons.

b–d.

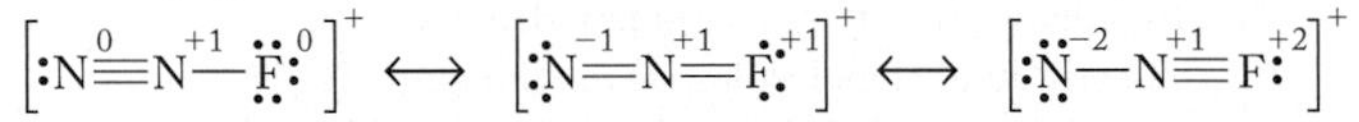

The central nitrogen atom in all the resonance structures always carries a +1 formal charge. The second and third resonance forms shown are unacceptable because they have greater than the minimal formal charges on the atoms.

e. Yes, the fluorine could be the central atom in the molecule, but this would place significant positive formal charge on the fluorine atom (the most electronegative element). These structures are unlikely:

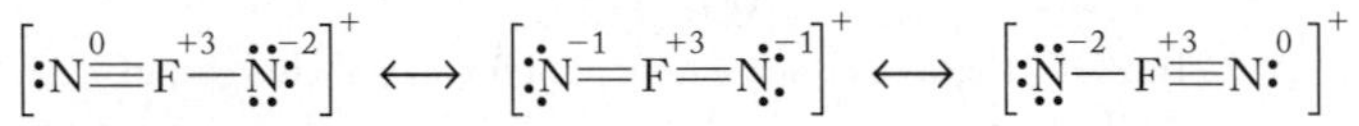

Chapter 5

5.1. a

5.3. N_2F_2 and NCCN are planar; there are no delocalized π electrons in any of these molecules.

5.5. More

5.7. Yes, there is a carbon in the ring structure connected to four different groups, so the compound is chiral.

5.9. The axial F–Re–axial F bond angle is 180°. The axial F–Re–equatorial F angle is 90°. The equatorial F–Re–equatorial F bonds are all 72°.

5.11. Because the electrons take up most of the space in the atom and because the nucleus is located in the center of the electron cloud, the electron clouds repel each other before the nuclei get close enough to each other.

5.13. Because they both have the same steric number around the central nitrogen atom.

5.15. Because the lone pair feels attraction from only one nucleus, it is less confined than bonding pairs and therefore occupies more space around the central N atom in ammonia.

5.17. The seesaw geometry has only two lone pair–bond pair interactions at 90° (compared to trigonal pyramidal's three), so it has lower energy.

5.19. (c) tetrahedral < (a) trigonal planar < (b) linear

5.21. a and c

5.23. a. Tetrahedral
b. Trigonal pyramidal
c. Bent
d. Tetrahedral

5.25. a. Tetrahedral
b. Trigonal planar
c. Bent
d. Square pyramidal

5.27. a. Tetrahedral
b. Tetrahedral
c. Trigonal planar
d. Linear

5.29. O_3 and SO_2

5.31. SCN^- and CNO^-

5.33. $\ddot{S}=\ddot{S}=\ddot{O}$ Bent

$\ddot{O}=\ddot{S}=\ddot{S}=\ddot{O}$ Bent at each S atom

or

$\ddot{O}=\ddot{S}=\ddot{S}$ Trigonal planar

5.35. Pentagonal planar

5.37.

The geometry around the P atom in Sarin is tetrahedral.

5.39. A polar bond is only between two atoms in a molecule. Molecular polarity takes into account all the individual bond polarities and the geometry of the molecule. A polar molecule has a permanent, measurable dipole moment.

5.41. Yes

5.43. Polar molecules are (b) $CHCl_3$, (d) H_2S, and (e) SO_2. Nonpolar molecules are (a) CCl_4 and (c) CO_2.

5.45. All of the molecules (a–c) are polar.

5.47. (a) $CBrF_3$; (b) CHF_2Cl

5.49. $COCl_2 < COBr_2 < COI_2$

5.51. Atomic orbitals must have similar size (energy) and the same phase to form hybrid orbitals.

5.53. All are sp^2 hybridized.

5.55. Both Lewis structures of N_2F_2 have sp^2 hybridized orbitals on N. Each F atom is sp^3 hybridized. In acetylene, C_2H_2, the carbon atoms are sp hybridized.

5.57.
CO_2	NO_2	O_3	ClO_2
sp	sp^2	sp^2	sp^3

5.59.

Tetrahedral molecular geometry. At first glance this would mean that the hybridization would be assigned as sp^3. However,

notice that the Cl forms three π bonds to three of the oxygen atoms. This requires that three of the p orbitals on Cl not be involved in the hybridization so that it can form parallel π bonds. Therefore, Cl must use low-lying d orbitals in place of the p orbitals for sd^3 hybridization to form the 4 σ bonds to oxygen.

5.61. $H—\ddot{Ar}—\ddot{F}:$ sp^3d hybridized

5.63. Yes

5.65. a. 109.5°
b. sp^3
c.

d. Trigonal bipyramidal

5.67. Yes

5.69. Yes, in resonance structures the electron distribution is blurred across all the resonance forms, which, in essence, defines the delocalization of electrons.

5.71. One N atom has trigonal pyramidal geometry. The other N atom has trigonal planar geometry. No, the hybridization of both N atoms is not the same.

5.73. Both the S and N atoms have SN = 4 for an electron-pair geometry of tetrahedral. The presence of a lone pair on N gives this atom trigonal pyramidal geometry and the nitrogen atom is sp^3 hybridized. The steric number for S is also 4, which, at first glance, would also mean that the hybridization would be assigned as sp^3. However, notice that the S forms two π bonds to two of the oxygen atoms. This requires that two of the p orbitals on S not be involved in the hybridization so that it can form parallel π bonds. Therefore, S must use two low-lying d orbitals in place of two of the p orbitals for spd^2 hybridization to form the four bonds σ to oxygen and nitrogen.

5.75. a, c, and d

5.77. No, sp hybridized carbon is linear with only two bonds to each carbon and to be chiral the carbon must be bonded to four different groups.

5.79. Homogeneous

5.81. a and c

5.83. a

5.85.

Saccharin

Sodium cyclamate

Aspartame

5.87.

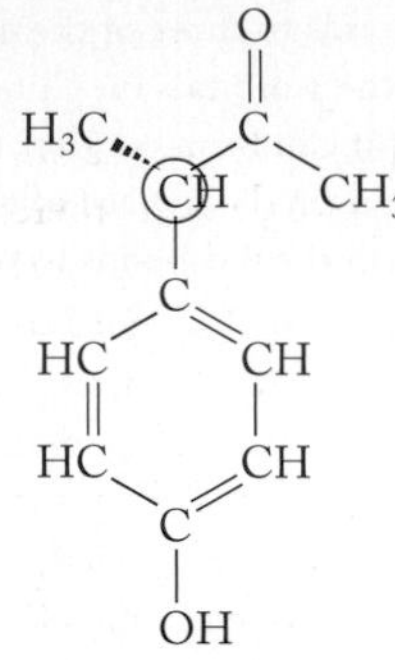

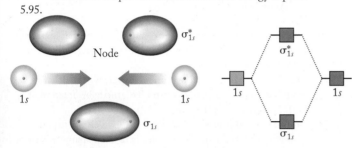

5.89. Molecular orbital theory

5.91. No

5.93. No. The overlap of $1s$ and $2s$ orbitals is not as efficient as $1s$–$1s$ or $2s$–$2s$ overlaps. The match in size and energy is poor.

5.95.

5.97. N_2^+ BO = 2.5
O_2^+ BO = 2.5
C_2^+ BO = 1.5
Br_2^{2-} BO = 0
All species with nonzero bond order (N_2^+, O_2^+, and C_2^+) are expected to exist.

5.99. The paramagnetic species with one or more unpaired electrons are (a) N_2^+, (b) O_2^+, and (c) C_2^+.

5.101. The species with electrons in π^* orbitals are (b) N_2^{2-}, (c) O_2^{2-}, and (d) Br_2^{2-}.

5.103. (a) B_2 and (b) C_2

5.105. No

5.107.

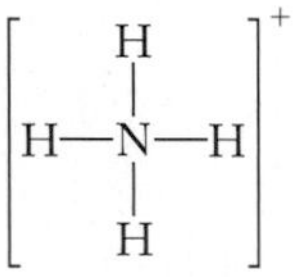

Tetrahedral

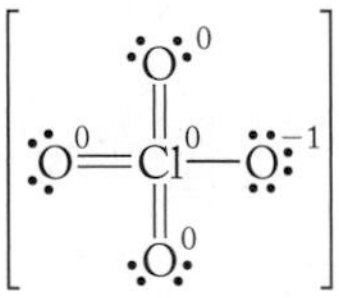

Tetrahedral

5.109.

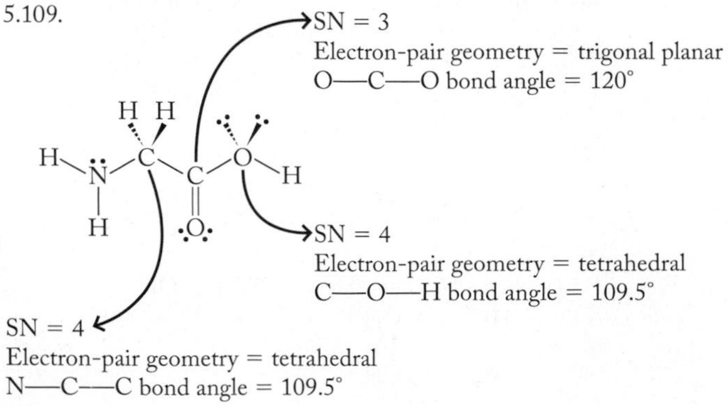

→ SN = 3
Electron-pair geometry = trigonal planar
O—C—O bond angle = 120°

→ SN = 4
Electron-pair geometry = tetrahedral
C—O—H bond angle = 109.5°

SN = 4 ←
Electron-pair geometry = tetrahedral
N—C—C bond angle = 109.5°

5.111. a. No, neither of the two molecules is linear.
b. Cl—O—Cl—O would have a permanent dipole.

5.113. a. $\left[\overset{..}{\underset{..}{Cl}} = \overset{..}{\underset{.}{O}} \right]^+$
b. BO = 2

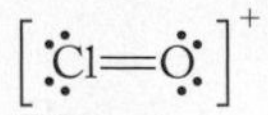

5.115.

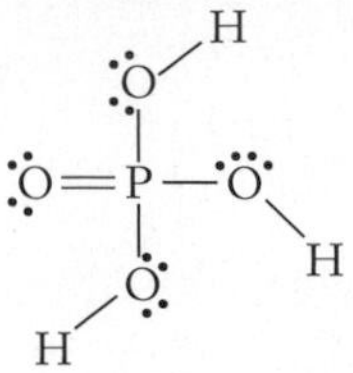

Molecular geometry around P = tetrahedral

5.117. a. :B—B≡C═Ö: Both B atoms have incomplete octets.
Molecular geometry = linear

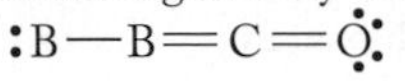

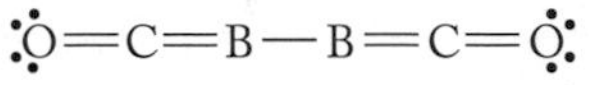

:Ö═C═B—B═C═Ö: Both B atoms have incomplete octets.
Molecular geometry = linear

b. 180°

5.119. a, b.

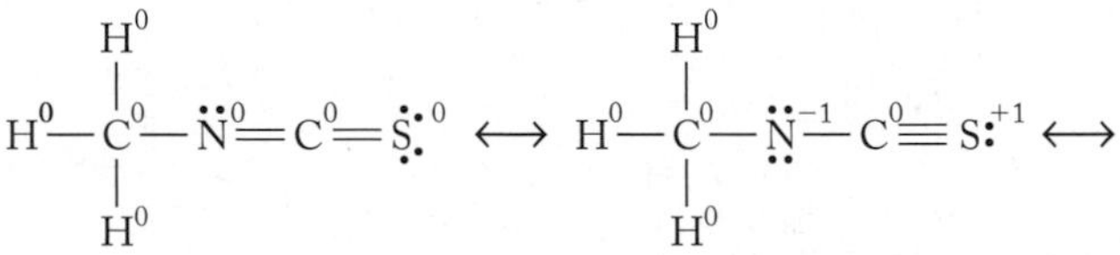

(1)   (2)

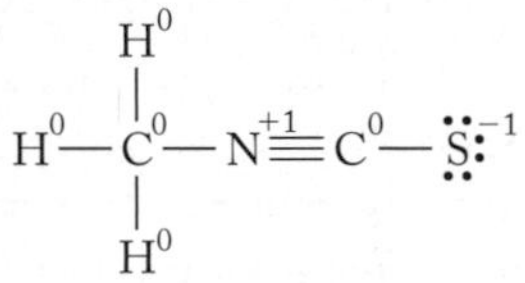

(3)

Structure 1 is likely to contribute the most to bonding. The methyl (CH_3) carbon is tetrahedral.
c. The isothiocyanate (NCS) carbon is linear.

5.121. Yes

5.123. BO = 0

5.125. N_2O_5 and N_2O_3; N_2O_2 depending on its actual structure.

5.127. BO for O_2^{2-} = 1 and is consistent with the Lewis structure; BO for O_2^- = 1.5 and is not consistent with the Lewis structure.

5.129. sp^3

5.131. The ozone molecule is polar because, although the oxygen–oxygen bonds themselves are nonpolar, the lone pair has its own "pull" on the electrons in the molecule. Also, the π bonds between the oxygen atoms places slightly more electron density on the terminal O atoms and makes the "nonpolar O—O bond" actually polar.

Chapter 6

6.1. a

6.3. solid

6.5. solid to liquid to gas

6.7. no

6.9. liquid to solid

6.11. Because there is less molecular surface area on a branched alkane and, therefore, lower total intermolecular forces.

6.13. a. CCl_4
    b. C_3H_8
    c. CS_2

6.15. diesel oil

6.17. The water molecule is oriented around an anion so as to point the partially positive hydrogen atoms toward the anion.

6.19. Because of the full positive or negative charge on the ion, the ion–dipole interaction is stronger than the dipole–dipole interaction.

6.21. The charge buildup on H (partially positive) and the electronegative element (partially negative) means that the X—H bond is polar. It is still a dipole–dipole interaction except that its strength is noticeably higher than other dipole–dipole interactions.

6.23. CH_3F is a polar molecule and therefore has stronger intermolecular forces than the nonpolar molecules of CH_4, which have only the weak dispersion forces. As it takes more energy to overcome strong intermolecular forces, CH_3F has a higher melting point than CH_4.

6.25. The H in methane has just a single bond to the relatively low electronegativity C atom and, therefore, the carbon–hydrogen bond is not polar enough to exhibit hydrogen bonding. In methanol, however, one of the H atoms is bonded to oxygen, which is second to fluorine in electronegativity. It is this H that shows hydrogen bonding in methanol.

6.27. (b) CF_2Cl_2 and (d) $CFCl_3$

6.29. Cl^-

6.31. Miscible solutes and solvents dissolve completely in each other; an insoluble solute does not dissolve at all.

6.33. Hydrophilic substances dissolve in water. Hydrophobic substances do not dissolve, or are immiscible, in water.

6.35. a. $CHCl_3$
    b. CH_3OH
    c. NaF
    d. BaF_2

6.37. (b) KI

6.39. (c) CH_3OCH_3

6.41. In sublimation the solid does not first liquefy before evaporating. In evaporation a liquid becomes a gas.

6.43. If you are along the equilibrium line in a phase diagram, the two phases that border that line are stable and coexist at that pressure–temperature combination.

6.45. a. Solid phase
    b. Gas phase

6.47. Yes

6.49. Reduce the temperature from 25°C to 0.01°C, then reduce the pressure from 1 atm to 0.006 atm.

6.51. The water vaporizes from liquid to gas.

6.53. −57°C

6.55. a. Liquid
    b. Liquid
    c. Gas

6.57. A needle floats on water but not on methanol because of the high surface tension of water. This is because water can hydrogen bond through two O—H bonds with other water molecules whereas methanol has only one O—H bond through which to form strong hydrogen bonds.

6.59. The expansion of water in the pipes on freezing may create sufficient pressure on the wall of the pipes to cause them to burst.

6.61. The cohesive forces in mercury are stronger than the adhesive forces of the mercury to the glass.

6.63. Molecules in the bulk liquid are "pulled" by all the other liquid molecules surrounding them and they are, therefore, "suspended" in the bulk liquid. Molecules on the surface of a liquid, however, are only pulled by molecules under and beside them, creating a tight film of molecules on the surface.

6.65. Although the dispersion forces between methanol molecules are greater than those between water molecules because methanol has more electrons and greater molar mass, water can form two hydrogen bonds compared to methanol's one hydrogen bond. This greater number of stronger interactions between water molecules raises the boiling point of water above that of methanol.

6.67. The greater dispersion forces of CH_2Cl_2 add to the dipole–dipole interactions to give stronger intermolecular forces between the CH_2Cl_2 molecules compared to those of CH_2F_2 molecules. Also, the molar mass of CH_2Cl_2 is higher than that of CH_2F_2, so it takes more energy to vaporize.

6.69. Increases

6.71. Contracts

6.73. b

Chapter 7

7.1. empirical, NO_2; molecular, N_2O_4

7.3. a. $4\,X(g) + 4\,Y(g) \rightarrow 4\,XY(g)$
    b. $4\,X(g) + 4\,Y(g) \rightarrow 4\,XY(s)$
    c. $4\,X(g) + 4\,Y(g) \rightarrow 2\,XY_2(g) + 2\,X(g)$
    d. $4\,X_2(g) + 4\,Y_2(g) \rightarrow 8\,XY(g)$

7.5. b and d

7.7. 5.00 g

7.9. reactant B

7.11. $2x$

7.13. 1.5 times more

7.15. No

7.17. 2 mol

7.19. a. $N_2(g) + O_2(g) \rightarrow 2\,NO(g)$
    b. $2\,N_2(g) + O_2(g) \rightarrow 2\,N_2O(g)$
    c. $NO(g) + NO_3(g) \rightarrow 2\,NO_2(g)$
    d. $4\,NO(g) + O_2(g) + 2\,H_2O(\ell) \rightarrow 4\,HNO_2(\ell)$

7.21. a. $N_2O_5(g) + Na(s) \rightarrow NaNO_3(s) + NO_2(g)$
    b. $N_2O_4(g) + H_2O(\ell) \rightarrow HNO_3(aq) + HNO_2(aq)$
    c. $3\,NO(g) \rightarrow N_2O(g) + NO_2(g)$

7.23. a. $C_3H_8(g) + 5\,O_2(g) \rightarrow 3\,CO_2(g) + 4\,H_2O(g)$
    b. $2\,C_4H_{10}(g) + 13\,O_2(g) \rightarrow 8\,CO_2(g) + 10\,H_2O(g)$
    c. $2\,C_6H_6(\ell) + 15\,O_2(g) \rightarrow 12\,CO_2(g) + 6\,H_2O(g)$
    d. $2\,C_8H_{18}(\ell) + 25\,O_2(g) \rightarrow 16\,CO_2(g) + 18\,H_2O(g)$

7.25. a. $C_2H_4(g) + 3\,O_2(g) \rightarrow 2\,CO_2(g) + 2\,H_2O(g)$
    b. $2\,C_3H_6(g) + 9\,O_2(g) \rightarrow 6\,CO_2(g) + 6\,H_2O(g)$
    c. $2\,C_4H_{10}(g) + 13\,O_2(g) \rightarrow 8\,CO_2(g) + 10\,H_2O(g)$
    d. $C_4H_8(g) + 6\,O_2(g) \rightarrow 4\,CO_2(g) + 4\,H_2O(g)$

7.27. a. $2\,SO_2(g) + O_2(g) \rightarrow 2\,SO_3(g)$
    b. $2\,H_2S(g) + 3\,O_2(g) \rightarrow 2\,SO_2(g) + 2\,H_2O(g)$
    c. $16\,H_2S(g) + 8\,SO_2(g) \rightarrow 3\,S_8(s) + 16\,H_2O(g)$

7.29. No, because the mole ratio of CO_2 to C_2H_6 (2:1) is the same in both.

7.31. (a) 4.5×10^{11} mol C; (b) 2.0×10^{10} kg CO_2

7.33. a. $2\,NaHCO_3(s) \rightarrow CO_2(g) + H_2O(g) + Na_2CO_3(s)$
    b. 6.55 g CO_2

7.35. 1.17 kg

7.37. 29 g O_2

7.39. (a) 1.48 kg; (b) 1.11 kg

7.41. 346 g

7.43. An empirical formula shows the lowest whole-number ratio of atoms in a substance. A molecular formula shows the actual numbers of each kind of atom that compose one molecule of the substance.

7.45. No

7.47. (a) 74.19% Na, 25.81% O; (b) 57.48% Na, 40.00% O, 2.52% H; (c) 27.37% Na, 1.20% H, 14.30% C, 57.13% O; (d) 43.38% Na, 11.33% C, 45.28% O

7.49. Pyrene, $C_{16}H_{10}$

7.51. NO, N_2O_3, and NO_2

7.53. No

7.55. Ti_6Al_4V

7.57. (a) P_2O_5; (b) P_4O_{10}

7.59. $Mg_2Si_2H_4O_9$

7.61. $CuCl_2O_8$

7.63. The excess of oxygen is required in combustion analysis to ensure the complete reaction of the hydrogen and carbon to form water and carbon dioxide.

7.65. Yes

7.67. $C_3H_6O_2$

7.69. empirical, C_2H_3; molecular, $C_{20}H_{30}$

7.71. (c) Less than the sum of the masses of Fe and S to start

7.73. Theoretical yield is the greatest amount of a product possible from a reaction and assumes that the reaction goes to 100% completion. The percent yield is the observed experimental yield divided by the theoretical yield and multiplied by 100.

7.75. Reactions do not always go to completion because the reaction may be slow or may have, for a portion of the reaction, yielded different products than expected.

7.77. 3 cups

7.79. 87% efficient; 0.60 metric tons escaped

7.81. a. $NH_3(g) + HCl(g) \rightarrow NH_4Cl(s)$
  b. HCl
  c. 7.3 g
  d. 0.7 g

7.83. a. $C(s) + H_2O(g) \rightarrow CO(g) + H_2(g)$
  b. 61%

7.85. a. $C_6H_{12}O_6(aq) \rightarrow 2\ C_2H_5OH(\ell) + 2\ CO_2(g)$
  b. 77.1%

7.87. a. calcium triphosphate hydroxide
  b. 39.89%
  c. decreases slightly

7.89. 99%

7.91. (a) 45 g; (b) 15 g; (c) 1.67 cm

7.93. (a) $a = 1$, $b = 3$, charge on U is 6+; (b) $c = 3$, $d = 8$, charge on U is 5.33+; (c) $x = 2$, $y = 2$, $z = 6$

7.95. a. No
  b. $C_5H_{10}O_5(s) + 5\ O_2(g) \rightarrow 5\ CO_2(g) + 5\ H_2O(\ell)$
    $2\ C_7H_{12}O_7(s) + 13\ O_2(g) \rightarrow 14\ CO_2(g) + 12\ H_2O(\ell)$

7.97. a. FeS, Fe^{2+} and S^{2-}; FeS_2, Fe^{4+} and S^{2-}
  b. FeS is iron(II) sulfide, FeS_2 is iron(IV) sulfide [actually, this compound is Fe^{2+} with S_2^{2-} and is named iron(II) persulfide]
  c. 0.26 g HCO_2H

7.99. a. $3\ FeO(s) + H_2O(\ell) \rightarrow Fe_3O_4(s) + H_2(g)$
  b. $12\ FeO(s) + 2\ H_2O(\ell) + CO_2(aq) \rightarrow 4\ Fe_3O_4(s) + CH_4(g)$

7.101. (a) 55 mol ethanol; (b) 110 mol

7.103. Re

7.105. 82.4%

7.107. a. $2\ SO_2(g) + 2\ CaO(s) + O_2(g) \rightarrow 2\ CaSO_4(s)$
  b. 2.13

7.109. Mg_2SiO_4

Chapter 8

8.1. Yellow

8.3. Blue, Na^+; green, SO_4^{2-}

8.5. From +2 to +4

8.7. (a) $N_2O_5 >$ (b) $N_2O_4 =$ (e) $NO_2 >$ (d) $N_2O_3 >$ (c) NO

8.9. a. $2\ H^+(aq) + SO_4^{2-}(aq) + Ba^{2+}(aq) + 2\ OH^-(aq) \rightarrow$ $BaSO_4(s) + 2\ H_2O(\ell)$
  b. c

8.11. a. 5.6 M $BaCl_2$
  b. 1.00 M Na_2CO_3
  c. 1.30 M $C_6H_{12}O_6$
  d. 5.92 M KNO_3

8.13. a. 0.29 M
  b. 0.23 M
  c. 1.76 M
  d. 0.084 M

8.15. a. 11.7 g NaCl
  b. 4.99 g $CuSO_4$
  c. 6.41 g CH_3OH

8.17. 2.72 g

8.19. a. 9.6×10^{-3} mol
  b. 7.80×10^{-4} mol
  c. 8.8×10^{-2} mol
  d. 4.22 mol

8.21. Orchard sample: 3.4×10^{-4} mmol/L
  Residential area sample: 5.6×10^{-5} mmol/L
  After storm sample: 3.2×10^{-2} mmol/L

8.23. 1.5×10^{-4} M

8.25. $AgNO_3$, $Fe(NO_3)_2$; $6\ H_2O$, and $Ca(OH)_2$

8.27. 4.57×10^{-2} M Mg^{2+}

8.29. a. The final concentration after diluting will be 1.81×10^{-2} M Na^+.
  b. The final concentration after diluting will be 2.7×10^{-1} mM LiCl.
  c. The final concentration after diluting will be 1.28×10^{-2} mM Zn^{2+}.

8.31. 1.95 M

8.33. The concentration of the adult-strength medication is 1.8 mg/mL; 23 mL is needed to prepare the child-strength cough syrup.

8.35. Table salt produces Na^+ and Cl^- ions in solution when it dissolves. Sugar does not dissociate into ions because it is not a salt. Ions are required to conduct electricity.

8.37. The lack of ions in methanol means that the liquid is nonconductive. Molten NaOH, however, has freely moving Na^+ and OH^- ions, which can conduct electricity.

8.39. In order of decreasing conductivity, 1.0 M Na_2SO_4 (c) $>$ 1.2 M KCl (b) $>$ 1.0 M NaCl (a) $>$ 0.75 M LiCl (d)

8.41. a. 0.025 M
  b. 0.050 M
  c. 0.075 M

8.43. An acid donates H^+ to another species (a base) in solution.

8.45. Strong acids include HCl, HNO_3, $HClO_4$, H_2SO_4, HI, HBr; weak acids include CH_3COOH, $HCOOH$, HF, H_3PO_4.

8.47. A base accepts H^+ from another species (an acid) in solution.

8.49. Strong bases include $NaOH$, KOH, $CsOH$, $LiOH$, $RbOH$, $Ba(OH)_2$, $Sr(OH)_2$, $Ca(OH)_2$; weak bases include NH_3, CH_3NH_2, C_5H_5N.

8.51. a. Ionic and net ionic equation: $2\,H^+(aq) + SO_4^{2-}(aq) + Ca^{2+}(aq) + 2\,OH^-(aq) \rightarrow CaSO_4(s) + 2\,H_2O(\ell)$
The acid is H_2SO_4; the base is $Ca(OH)_2$.
b. Ionic and net ionic equation: $PbCO_3(s) + 2\,H^+(aq) + SO_4^{2-}(aq) \rightarrow PbSO_4(s) + CO_2(g) + H_2O(\ell)$
$PbCO_3$ is the base; sulfuric acid is the acid.
c. Ionic equation: $Ca^{2+}(aq) + 2\,OH^-(aq) + 2\,CH_3COOH(aq) \rightarrow Ca^{2+}(aq) + 2\,CH_3COO^-(aq) + 2\,H_2O(\ell)$
Calcium is a spectator ion. $Ca(OH)_2$ is the base; CH_3COOH is the acid.
Net ionic equation: $OH^-(aq) + CH_3COOH(aq) \rightarrow CH_3COO^-(aq) + H_2O(\ell)$

8.53. a. Molecular equation:
$Mg(OH)_2(s) + H_2SO_4(aq) \rightarrow MgSO_4(aq) + 2\,H_2O(\ell)$
Net ionic equation:
$Mg(OH)_2(s) + 2\,H^+(aq) \rightarrow Mg^{2+}(aq) + 2\,H_2O(\ell)$
b. Molecular equation:
$MgCO_3(s) + 2\,HCl(aq) \rightarrow MgCl_2(aq) + H_2CO_3(aq)$
Net ionic equation:
$MgCO_3(s) + 2\,H^+(aq) \rightarrow Mg^{2+}(aq) + H_2O(\ell) + CO_2(g)$
c. Molecular equation: $NH_3(g) + HCl(g) \rightarrow NH_4Cl(s)$
This is also the net ionic equation.

8.55. $PbCO_3(s) + 2\,H^+(aq) \rightarrow Pb^{2+}(aq) + CO_2(g) + H_2O(\ell)$;
$Pb(OH)_2(s) + 2\,H^+(aq) \rightarrow Pb^{2+}(aq) + 2\,H_2O(\ell)$

8.57. A saturated solution contains the maximum concentration of a solute. A supersaturated solution *temporarily* contains *more* than the maximum concentration of a solute at a given temperature.

8.59. A precipitation reaction occurs when two solutions are mixed to form an insoluble compound.

8.61. A saturated solution may not be a concentrated solution if the solute is only sparingly or slightly soluble in the solution. In that case, the solution is a saturated dilute solution.

8.63. (a) Barium sulfate is insoluble; (e) lead hydroxide is insoluble; (f) calcium phosphate is insoluble.

8.65. a. Balanced reaction:
$Pb(NO_3)_2(aq) + Na_2SO_4(aq) \rightarrow PbSO_4(s) + 2\,NaNO_3(aq)$
Net ionic equation: $Pb^{2+}(aq) + SO_4^{2-}(aq) \rightarrow PbSO_4(s)$
b. No precipitation reaction occurs.
c. Balanced reaction:
$FeCl_2(aq) + Na_2S(aq) \rightarrow FeS(s) + 2\,NaCl(aq)$
Net ionic equation: $Fe^{2+}(aq) + S^{2-}(aq) \rightarrow FeS(s)$
d. Balanced reaction:
$MgSO_4(aq) + BaCl_2(aq) \rightarrow MgCl_2(aq) + BaSO_4(s)$
Net ionic equation: $Ba^{2+}(aq) + SO_4^{2-}(aq) \rightarrow BaSO_4(s)$

8.67. $CaCO_3$

8.69. 2.11×10^{-2} g

8.71. 5.4×10^{-2} g

8.73. 130 kg

8.75. The number of electrons gained or lost is directly related to the change in oxidation number of a species.

8.77. (a) -1; (b) $+1$; (c) -2; (d) -3

8.79. Silver

8.81. As n increases the oxidation state of carbon increases.

8.83. (a) $+1$; (b) $+5$; (c) $+7$

8.85. 0 mol

8.87. 0.25 mol

8.89. a. Reactants
SiO_2: $Si = +4$, $O = -2$
Fe_3O_4: $Fe = +8/3$, $O = -2$
Products
Fe_2SiO_4: $Fe = +2$, $Si = +4$, $O = -2$
O_2: $O = 0$
Oxygen is oxidized (O^{2-} to O_2) and iron is reduced (Fe^{3+} to Fe^{2+}).
b. Reactants
SiO_2: $Si = +4$, $O = -2$
Fe: $Fe = 0$
O_2: $O = 0$
Products
Fe_2SiO_4: $Fe = +2$, $Si = +4$, $O = -2$
Iron is oxidized (Fe^0 to Fe^{2+}) and oxygen is reduced (O_2 to O^{2-}).
c. Reactants
FeO: $Fe = +2$, $O = -2$
O_2: $O = 0$
H_2O: $H = +1$, $O = -2$
Products
$Fe(OH)_3$: $Fe = +3$, $O = -2$, $H = +1$
Iron is oxidized (Fe^{2+} to Fe^{3+}) and oxygen is reduced (O_2 to O^{2-}).

8.91. (a) 0.25 mol; (b) 0.17 mol

8.93. $NH_4^+(aq) + 2\,O_2(g) \rightarrow NO_3^+(aq) + 2\,H^+(aq) + H_2O(\ell)$

8.95. $2\,Fe(OH)_2^+(aq) + Mn^{2+}(aq) \rightarrow 2\,Fe^{2+}(aq) + 2\,H_2O(\ell) + MnO_2(s)$

8.97. $2\,H_2O(\ell) + 4\,Ag(s) + 8\,CN^-(aq) + O_2(g) \rightarrow 4\,Ag(CN)_2^-(aq) + 4\,OH^-(aq)$

8.99. a. $2\,ClO_3^-(aq) + SO_2(g) \rightarrow 2\,ClO_2(g) + SO_4^{2-}(aq)$
b. $4\,H^+(aq) + 2\,ClO_3^-(aq) + 2\,Cl^-(aq) \rightarrow 2\,ClO_2(g) + 2\,H_2O(\ell) + Cl_2(g)$
c. $2\,ClO_3^-(aq) + Cl_2(g) \rightarrow 2\,ClO_2(g) + 2\,Cl^-(aq) + O_2(g)$

8.101. a. 5.00 mL
b. 31.5 mL
c. 21.5 mL

8.103. 500 mL

8.105. 556 mM, 19.2 g/kg

8.107. To deionize water, cations such as Na^+ and Ca^{2+} are exchanged for H^+ at cation-exchange sites. Anions such as Cl^- and SO_4^{2-} are exchanged for OH^- at the anion-exchange sites. The released ions (H^+ and OH^-) at these sites combine to form H_2O.

8.109. K^+ would act just like Na^+, but it is less abundant and more expensive.

8.111. $7.98 \times 10^{-4}\ M\ SO_4^{2-}$

8.113. a. 11.7 M
b. 42.7 mL
c. 1.72 kg

8.115. a. $2\,OH^-(aq) + 2\,H_2O(\ell) + 3\,S_2O_4^{2-}(aq) + 2\,CrO_4^{2-}(aq) \rightarrow 6\,SO_3^{2-}(aq) + 2\,Cr(OH)_3(s)$
b. Sulfur is oxidized; chromium is reduced.
c. Oxidizing agent = CrO_4^{2-}; reducing agent = $S_2O_4^{2-}$
d. 38.7 g

8.117. a. Balanced equation: $4 Ag(s) + 2 H_2S(g) + O_2(g) \rightarrow$
$2 Ag_2S(s) + 2 H_2O(\ell)$

   b. $3 Ag_2S(s) + 3 H_2O(\ell) + Al(s) \rightarrow$
$6 Ag(s) + 3 HS^-(g) + Al(OH)_3(s)$

8.119. a. H_3PO_4; phosphoric acid

   b. H_2SeO_3; selenous acid

   c. H_3BO_3; boric acid

8.121. $2 H^+(aq) + ClO^-(aq) + 2 I^-(aq) \rightarrow Cl^-(aq) + H_2O(\ell) + I_2(aq)$
$I_2(aq) + 2 S_2O_3^{2-}(aq) \rightarrow 2 I^-(aq) + S_4O_6^{2-}(aq)$

8.123. a. $NaClO_4$, NH_4ClO_4

   b. 427 kg

   c. 2.80×10^{10} gal

   d. The MA lab

8.125. a. $3 CH_2O \rightarrow CO_2 + C_2H_5OH$

   b. $C_2H_5OH + O_2 \rightarrow HC_2H_3O_2 + H_2O$

   c. CH_2O: $C = 0$

   CO_2: $C = +4$

   C_2H_5OH: $C = -4$ over two carbon atoms, so oxidation
   number on each carbon $= -2$

   $HC_2H_3O_2$: $C = 0$

   d. 66.7 g acetic acid

8.127. a. The first reaction is a redox reaction; eight electrons are
   transferred.

   b. $2 H^+(aq) + SO_4^{2-}(aq) + CaCO_3(s) \rightarrow$
   $CaSO_4(s) + H_2O(\ell) + CO_2(g)$

   c. $SO_4^{2-}(aq) + CaCO_3(s) \rightarrow CaSO_4(s) + CO_3^{2-}(aq)$

8.129. c and d

Chapter 9

9.1. Internal energy increases

9.3. Highest fuel value, (c) C_5H_{12}; lowest fuel value, (b) C_6H_{12}

9.5. q is negative, w is positive, ΔE is negative; 67% yield

9.7. a. $2 SO_2(g) + O_2(g) \rightarrow 2 SO_3(g)$

   b. -98.9 kJ/mol

   c. Heat flows out from the reaction mixture.

9.9. Energy makes work possible.

9.11. The value of a state function is independent of the path; only
   the initial and final values are important.

9.13. Potential energy is locked up in the bonds of the acetylene. The
   gas also has potential energy to do P–V work.

9.15. The internal energy of the gas sample can be increased by
   raising the temperature or by increasing the pressure through
   compression.

9.17. P–V work has energy units because work is done by expending
   energy and it is equivalent to force $\times$ distance.

9.19. a. Exothermic

   b. Endothermic

   c. Exothermic

9.21. Energy is absorbed from the surroundings. Thus, q increases
   and therefore ΔE increases.

9.23. $w = -0.500$ L $\cdot$ atm $= -50.7$ J

9.25. a. 50.0 J

   b. 6.3 kJ

   c. -1.23×10^4 kJ

9.27. -276 kJ

9.29. b; w is negative

9.31. A change in enthalpy is the sum of the change of internal
   energy and the product of the system's pressure and change in
   volume.

9.33. If the system transfers energy to the surroundings its energy
   will be less after the process than at the start of the process.

9.35. Negative

9.37. Positive

9.39. Negative

9.41. Specific heat is specified for a gram of the substance. Heat
   capacity does not take into account how much of a substance
   there is; it is defined for a given object.

9.43. Because to vaporize water you have to completely break
   the strong intermolecular hydrogen bonds between water
   molecules, not just loosen them which occurs during melting

9.45. Water's high heat capacity compared to air means that water
   carries away more energy from the engine for every Celsius
   degree rise in temperature, so water is a good choice to cool
   automobile engines.

9.47. 29.3 kJ

9.49.

9.51. 96.5 g

9.53. $-47.5°C$

9.55. To know how much energy (generated or absorbed by
   the system) is required to change the temperature of the
   surroundings (the calorimeter) in order to calculate the heat
   capacity or final temperature of the system in an experiment

9.57. Yes

9.59. 8.044 kJ/°C

9.61. 5129 kJ/mol is produced

9.63. 3.027°C

9.65. -464 kJ

9.67. When we apply Hess's law all the heat is accounted for in the
   reaction; energy is neither created nor destroyed when using
   Hess's law.

9.69. -297 kJ/mol

9.71. 28.0 kJ/mol

9.73. -103 kJ

9.75. Because CO_2 formation is energetically more favored

9.77. No, because to make ozone from oxygen you would have to
   break bonds and so these are not energetically equal.

9.79. We must account for all the bonds that break and all the bonds
   that form in the reaction. In order to do so we must have a
   balanced chemical reaction.

9.81. When all the reactants and products are gases we have to
   only consider the bond energies and not the energy of the
   intermolecular forces that would be present in the liquid or
   solid state.

9.83. a and d

9.85. -252.9 kJ

9.87. -35.9 kJ

9.89. -7198 kJ

9.91. a. 862 kJ
     b. 98 kJ
     c. 93 kJ
9.93. 554 kJ less energy
9.95. −667 kJ
9.97. The energy per gram a fuel releases on burning
9.99. (a) CH_4; (b) H_2
9.101. 201 kg
9.103. a. 48.99 kJ/g
      b. 4.90×10^4 kJ
      c. 5.97 g
9.105. We are given that ΔE for this process is less than q absorbed $(\Delta E < q)$. If this is the case, for the equality $\Delta E = q + w$ to be maintained, then w must be negative. A negative value of w must mean that work is done by the system on the surroundings.
9.107. a. $2\,NaOH(aq) + H_2SO_4(aq) \rightarrow 2\,H_2O(\ell) + Na_2SO_4(aq)$
      b. No
      c. −114 kJ/mol H_2SO_4
9.109. 26.0°C
9.111. a. −1255.5 kJ/mol
      b. 48.219 kJ/g
9.113. Exothermic
9.115. a. $CH_3OH(g) + N_2(g) \rightarrow HCN(g) + NH_3(g) + \frac{1}{2}O_2(g)$
      b. A reactant
      c. 307 kJ
9.117. −1398 kJ
9.119. 67.8 kJ/mol
9.121. 31.5 kJ
9.123. Hydrogen

Chapter 10

10.1. c
10.3. There is no effect.
10.5. a. The pressure will double.
     b. The frequency of collisions will double.
     c. The most probable speed does not change.
10.7. (a) a; (b) b; (c) b
10.9. a
10.11. SO_2, curve 1; curve 2, propane
10.13. a. 2. Mt. Everest
      b. 1. San Diego
      c. 3. Gold mine in South Africa
10.15. Line 2 (red)
10.17. Line 1
10.19. b
10.21. The speed of a particle in a gas that has the average kinetic energy of all the molecules of the sample
10.23. No, only temperature affects the root-mean-square speed.
10.25. The rank order in terms of increasing root-mean-square speed is $N_2O_5 < N_2O_4 < NO_2 < NO$.
10.27. 32.3 g/mol
10.29. 18.2 g/mol
10.31. Hydrogen
10.33. Force is the product of the mass of an object and the acceleration due to gravity. Pressure uses force in its definition: it is the force an object exerts over a given area.
10.35. The ethanol barometer
10.37. A sharpened blade has a smaller area over which the force is distributed compared to a dull blade.

10.39. a. 0.020 atm
      b. 0.739 atm
10.41. 914 g, 3584 N/m² or 3.54×10^{-2} atm
10.43. a. 814.6 mmHg
      b. 1.072 atm
      c. 1086 mbar
10.45. As temperature increases the speed of gas particles increases and they will more often collide with the container walls and with more force, resulting in an increase in pressure.
10.47. The balloonist should decrease the temperature.
10.49. 1.45 atm
10.51. a. 3.06 atm
      b. 21 m
10.53. 6.79 L
10.55. 4.2 L
10.57. They are essentially the same.
10.59. a. No change
      b. Decrease to 1/4 the original volume
      c. Increase of 17%
10.61. 144 L
10.63. 6.6 atm
10.65. STP is defined as 1 atm and 0°C (273 K); $V = 22.4$ L
10.67. 0.67 mol
10.69. 0.788 atm
10.71. 1%
10.73. 4.20×10^4 g
10.75. a. 1.00 mol
      b. Helium
10.77. The densities of different gases are not necessarily the same for a particular temperature and pressure.
10.79. Density (a) increases with increasing pressure and (b) increases with decreasing temperature.
10.81. a. 9.08 g/L
      b. In the basement
10.83. SO_2
10.85. CO
10.87. a. 0.20 mol
      b. 51 g
10.89. 6.2×10^2 g
10.91. 589 g
10.93. The pressure that a particular gas individually contributes to the total pressure
10.95. 0.20
10.97. $P_{total} = 31$ atm
      $P(N_2) = 22$ atm
      $P(H_2) = 6.2$ atm
      $P(CH_4) = 3.1$ atm
10.99. 0.0190 mol
10.101. a. Greater than
       b. Less than
       c. Greater than
10.103. 67% more
10.105. 650 mmHg
10.107. 13%
10.109. The reaction of CO_2 with water changes dissolved CO_2 into carbonic acid. Once this occurs, more CO_2 can dissolve in water.
10.111. London dispersion–dipole
10.113. 3.7×10^{-2} mol/(L · atm)
10.115. a. 2.74×10^{-3} M
       b. 2.34×10^{-2} M

10.117. 681 m/s

10.119. 509 m/s

10.121. Br

10.123. At low temperatures the gas particles move more slowly, and their collisions become inelastic; they stick together due to the weak attractive forces between them. The particles, therefore, do not act separately to contribute to the pressure in the container, and the pressure is lower than would be expected by the ideal gas law. Also, the gas particles take up real volume in the container and as the pressure increases the volume of the particles takes up a greater volume of the free space in the container. This has the effect of raising the pressure–volume product above what we would expect from the ideal gas law (in a plot of PV/RT versus P).

10.125. Since b is a measure of the volume that the gas particles occupy, b increases as the sizes of the particles increase.

10.127. Because Ar has more electrons and will have stronger London forces between its atoms than He does

10.129. H_2

10.131. a. $P = 1603$ atm
b. $P = 597$ atm

10.133. a. 70.33 mL
b. No, because when gases are compressed they heat up.

10.135. 0.25 m/s

10.137. 18 kg

10.139. a. 1.4 L
b. 7.8 g/L

10.141. a. CH_3CO_2H and $(CH_3)_3N$
b. HCl and $(CH_3)_3N$
c. No

10.143. $P_{H_2} = 110$ kPa
$P_H = 20$ kPa
$P_{CH_4} = 3$ kPa

10.145. 2.7 atm

10.147. a. 38.4 L N_2
b. 192 L CO_2
c. 1.74 g/L

10.149. 1.68×10^3 g

Chapter 11

11.1. a

11.3. a. NaCl
b. $MgCl_2$
c. $C_6H_{12}O_6$
d. K_3PO_4

11.5. $bp_X \approx 20°C$ and $bp_Y \approx 38°C$; Y has stronger intermolecular forces

11.7. Solution A must be the more concentrated solution because solvent flows through the membrane from the least to the most concentrated side.

11.9. As temperature increases the molecules in the liquid gain kinetic energy and so more of the molecules have sufficient energy to enter into the gas phase, thus increasing the vapor pressure.

11.11. As intermolecular forces increase in strength, the vapor pressure decreases.

11.13. -717 kJ/mol

11.15. (a) $CH_3CH_2OH <$ (b) $CH_3OCH_3 <$ (c) $CH_3CH_2CH_3$

11.17. 41.0 kJ/mol

11.19. For isooctane, 0.105 atm or 80.0 torr
For tetramethylbutane, 0.0483 atm or 36.7 torr

11.21. The components of crude oil can be separated by fractional distillation, which uses differences in boiling points of the compounds.

11.23. C_5H_{12}

11.25. 60 torr

11.27. The vapor pressure of water in the compartment of pure water is greater than that in the seawater compartment. These conditions never change because the seawater will always contain dissolved salts and so the transfer process of water to the seawater compartment continues.

11.29. Molarity is the moles of the solute in one liter of solution. Molality is the moles of solute in one kilogram of solvent.

11.31. The greater the concentration of dissolved solutes in water (as present in seawater), the lower the freezing point of the water. The presence of nonvolatile solutes shifts the solid–liquid line on the phase diagram to lower temperature.

11.33. A strong electrolyte completely dissociates in the solvent. This dissociation yields two or more particles in solution from one dissolved solute particle. This results in greater changes in the melting and boiling points compared to that of a solute that does not dissociate.

11.35. Because ions of opposite charge associate with each other in solution

11.37. $\chi_{water} = 0.70$; $P_{soln} = 17$ torr

11.39. a. 0.58 m
b. 0.18 m
c. 1.12 m

11.41. a. 307 g
b. 86.8 g
c. 28.8 g

11.43. 6.5×10^{-5} m NH_3, 8.7×10^{-6} m NO_2^-, 2.195×10^{-2} m NO_3^-

11.45. 3.1°C

11.47. 2.52×10^{-2} m

11.49. $-1.89°C$

11.51. 0.5 m $CaCl_2$

11.53. 0.0100 m $Ca(NO_3)_2$

11.55. (a) 0.06 m $FeCl_3 <$ (b) 0.10 m $MgCl_2 <$ (c) 0.20 m KCl

11.57. A semipermeable membrane is a boundary between two solutions through which some molecules may pass through but others cannot. Usually, small molecules may pass through but large molecules are excluded.

11.59. Solvent flows across a semipermeable membrane from the more dilute solution side to the more concentrated solution side to balance the concentration of solutes on both sides of the membrane.

11.61. Reverse osmosis transfers solvent across a semipermeable membrane from a region of higher solute concentration to a region of lower solute concentration. Because reverse osmosis goes against the natural flow of solvent across the membrane, the key component needed is a pump to apply pressure to the more concentrated side of the membrane.

11.63. a. From side A to side B
b. From side B to side A
c. From side A to side B

11.65. a. 57.5 atm
     b. 0.682 atm
     c. 52.9 atm
     d. 46.5 atm

11.67. a. $2.75 \times 10^{-2}\ M$
     b. $1.11 \times 10^{-3}\ M$
     c. $1.00 \times 10^{-2}\ M$

11.69. False, the molarity of the NaCl solution would be greater by 1.5 times than the molarity of $CaCl_2$

11.71. a. Osmotic pressure increases
     b. Freezing point decreases
     c. Boiling point increases

11.73. 94.1 g/mol

11.75. Molar mass = 164 g/mol. The molecular formula of eugenol is $C_{10}H_{12}O_2$.

11.77. Yes

11.79. For 0.0935 m NH_4Cl, $i = 1.85$
     For 0.0378 m $(NH_4)_2SO_4$, $i = 2.46$

11.81. 2.2 atm

11.83. 4270 g/mol

Chapter 12

12.1. Tire with blue gas on the right has greater pressure; tire with blue gas on the right has greater entropy.

12.3. No, the distribution of gases is not affected by gravity because the particles have sufficient energy to move throughout the container.

12.5. Spontaneous at low temperature

12.7. Condensation, freezing, and deposition

12.9. The sign is reversed.

12.11. Eight microstates; the most likely microstates have sums of +1 and −1.

12.13. 4.47×10^3

12.15. b < a < c

12.17. (d) $Cr(NO_3)_3$

12.19. a. $S_8(g)$
     b. $S_2(g)$
     c. $O_3(g)$
     d. 1 g of $O_2(g)$

12.21. Fullerenes

12.23. a. $CH_4(g) < CF_4(g) < CCl_4(g)$
     b. $CH_2O(g) < CH_3CHO(g) < CH_3CH_2CHO(g)$
     c. $HF(g) < H_2O(g) < NH_3(g)$

12.25. ΔS_{sys} is positive; ΔS_{surr} is negative.

12.27. a and b

12.29. ΔS_{sys} is negative; ΔS_{univ} is positive.

12.31. ΔS_{surr} must be more positive than 66.0 J/K

12.33. ΔS_{rxn} is positive.

12.35. $\Delta S^\circ_{rxn} < 0$ because the separated ions or molecules in solution are being organized into a solid.

12.37. a. 24.9 J/K
     b. −146.4 J/K
     c. −73.2 J/K
     d. −175.8 J/K

12.39. 218.9 J/(mol · K)

12.41. When ΔG is positive the reaction is nonspontaneous; when ΔG is negative the reaction is spontaneous

12.43. The magnitude of ΔH is generally greater than the magnitude of $T\Delta S$; therefore, an exothermic reaction with negative ΔH is often dominant in determining spontaneity.

12.45. ΔS is positive, ΔH is positive, ΔG is negative

12.47. a and d

12.49. For NaBr, −18 kJ/mol; for NaI, −29 kJ/mol

12.51. $\Delta G^\circ_{rxn} = 91.4$ kJ

12.53. −35.3 kJ

12.55. −90.3 kJ

12.57. No, if ΔS_{rxn} were positive the exothermic reaction would be spontaneous at all temperatures.

12.59. 981.3 K

12.61. $\Delta H^\circ = 51.5$ kJ/mol; $\Delta S^\circ = 123.1$ J/(mol · K); 418 K or 145°C

12.63. a. Spontaneous only at low temperatures (i)
     b. Spontaneous only at low temperatures (i)
     c. Spontaneous all temperatures (iii)

12.65. The spontaneous reaction must have a more negative or less positive free energy than the nonspontaneous reaction.

12.67. ATP has the important function of energy storage and transfer. The cell must make this important molecule and so the overall process must be spontaneous. Even if some of the glycolysis steps are nonspontaneous, at least some glycolysis steps must be spontaneous in order to produce ATP.

12.69. a. 142.2 kJ
     b. −28.6 kJ
     c. $CH_4(g) + 2\,H_2O(g) \rightarrow CO_2(g) + 4\,H_2(g)$
     d. 113.6 kJ; nonspontaneous

12.71. 83.5 J/(mol · K)

12.73. $T = 618$ K

12.75. −630.4 kJ; spontaneous

12.77. $T_b \approx 294$ K

12.79. 9.58 J/(mol · K)

12.81. 0.805 J/(mol · K)

12.83. There are more atoms in $CaCO_3$ than in CaO, so the S° is more positive; 1099 K

12.85. a. ΔH is negative, ΔS is positive
     b. The reverse reaction would have $+\Delta H$ and $-\Delta S$ and, therefore, would never be spontaneous.

Chapter 13

13.1. The [N_2O] is represented by the green line and [O_2] is represented by the red line.

13.3. b

13.5. (a) 1; (b) 5; (c) 2; (d) 4; (e) 3

13.7. b

13.9. c

13.11. b

13.13. Nitrogen (light blue)

13.15. Palladium (blue) and platinum (orange)

13.17. The presence of NO_2 in the atmosphere and ample sunlight allows the O atoms to react with O_2 to generate O_3. The reactant NO_2 is present in the atmosphere due to automobile exhausts, which build up during the day. The buildup of O_3 lags behind until later in the day, until [NO_2] increases and the sunlight becomes stronger as midday approaches.

13.19. In the evening the sunlight (and UV radiation) is less intense, so the photochemical breakdown of NO_2 does not occur to as great an extent as after the morning rush hour.

13.21. −114.2 kJ

13.23. a. $2 N_2(g) + O_2(g) \rightarrow 2 N_2O(g)$
  b. $2 N_2(g) + 5 O_2(g) \rightarrow 2 N_2O_5(g)$

13.25. The average rate is the rate averaged over a defined time interval, whereas the instantaneous rate is the rate at a specific moment.

13.27. As the reaction proceeds, the concentrations of the reactants decrease. Because most reactions depend on the availability (i.e., concentration) of reactants to proceed, the decrease in reactant concentrations lowers the reaction rate.

13.29. a. The rates are the same.
  b. The rate of formation of NO_2^- and of H^+ is two-thirds the rate of consumption of O_2.
  c. The rate of consumption of NH_3 is two-thirds the rate of consumption of O_2.

13.31. a. $Rate = -\dfrac{\Delta[H_2O_2]}{\Delta t} = \dfrac{1}{2}\dfrac{\Delta[OH]}{\Delta t}$

  b. $Rate = -\dfrac{\Delta[ClO]}{\Delta t} = -\dfrac{\Delta[O_2]}{\Delta t} = \dfrac{\Delta[ClO_3]}{\Delta t}$

  c. $Rate = -\dfrac{\Delta[N_2O_5]}{\Delta t} = -\dfrac{\Delta[H_2O]}{\Delta t} = \dfrac{1}{2}\dfrac{\Delta[HNO_3]}{\Delta t}$

13.33. a. $Rate = \dfrac{\Delta[CO_2]}{\Delta t} = -\dfrac{2}{3}\dfrac{\Delta[CO]}{\Delta t}$

  b. $Rate = \dfrac{\Delta[COS]}{\Delta t} = -\dfrac{\Delta[SO_2]}{\Delta t}$

  c. $Rate = \dfrac{\Delta[CO]}{\Delta t} = 3\dfrac{\Delta[SO_2]}{\Delta t}$

13.35. a. 1.2×10^7 M/s
  b. 2.9×10^4 M/s

13.37. Between 0 and 100 μs: 1.4×10^{-5} M/μs
  Between 200 and 300 μs: 5.5×10^{-6} M/μs

13.39. For the change in concentration of ClO versus time we obtain the following plot:

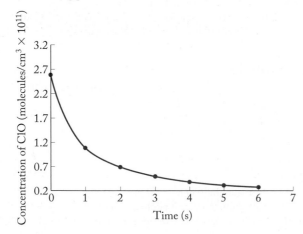

The instantaneous rate at 1 s is 8.28×10^{10} molecules · cm^{-3} · s^{-1}. For the change in concentration of Cl_2O_2 versus time we obtain the following plot:

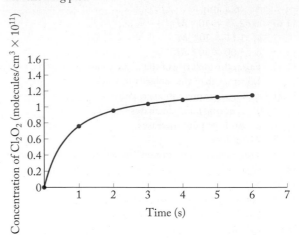

The instantaneous rate at 1 s is 4.13×10^{10} molecules · cm^{-3} · s^{-1}.

13.41. Yes

13.43. Yes

13.45. The half-life will be halved.

13.47. a. First order in both A and B, and second order overall
  b. Second order in A, first order in B, and third order overall
  c. First order in A, third order in B, and fourth order overall

13.49. a. $Rate = k[O][NO_2]$; k units $= M^{-1} s^{-1}$
  b. $Rate = k[NO]^2[Cl_2]$; k units $= M^{-2} s^{-1}$
  c. $Rate = k[CHCl_3][Cl_2]^{1/2}$; k units $= M^{-1/2} s^{-1}$
  d. $Rate = k[O_3]^2[O]^{-1}$; k units $= s^{-1}$

13.51. a. $Rate = k[BrO]$
  b. $Rate = k[BrO]^2$
  c. $Rate = k[BrO]$
  d. $Rate = k[BrO]^0 = k$

13.53. We need to determine the change in the rate when only [NO] or [ClO] is changed.

13.55. a. $Rate = k[NO_2][O_3]$
  b. 4.9×10^{-11} M/s
  c. 4.9×10^{-11} M/s
  d. The rate doubles.

13.57. c

13.59. $Rate = k[NO][NO_2]$

13.61. $Rate = k[ClO_2][OH^-]$; $k = 14 M^{-1} s^{-1}$

13.63. $Rate = k[NO]^2[H_2]$; $k = 6.32 M^{-2} s^{-1}$

13.65. $0.32 \mu M^{-1} \cdot min^{-1}$

13.67. $Rate = k[NH_3]$; $k = 0.0030 s^{-1}$

13.69. $0.051 M$, 89%

13.71. a. $Rate = k[N_2O]$
  b. 4

13.73. a. $Rate = k[^{32}P]$
  b. $0.0485 day^{-1}$
  c. 14.3 days

13.75. $k = 5.40 \times 10^{-12}$ cm^3 molecule^{-1} s^{-1}; $t_{1/2} = 0.712$ s

13.77. Rate $= k[C_{12}H_{22}O_{11}][H_2O] = k'[C_{12}H_{22}O_{11}]$;
$k' = 6.21 \times 10^{-5}$ s^{-1}

13.79. The larger the activation energy, the slower the reaction.

13.81. When the energy of the products is lower than the energy of the reactants.

13.83. An increase in temperature increases the frequency and the kinetic energy at which the reactants collide. This speeds up the reaction. The order of the reaction is unaffected.

13.85. The reaction with the larger activation energy (150 kJ/mol)

13.87. $E_a = 17.1$ kJ/mol; $A = 1.002$ cm^3/molecule $\cdot$ s

13.89. a. $E_a = 314$ kJ/mol
b. $A = 5.03 \times 10^{10}$ M^{-1} s^{-1}
c. $k = 1.06 \times 10^{-44}$ $M^{-1/2}$ s^{-1}

13.91. $E_a = 39.1$ kJ/mol; $A = 1.27 \times 10^{12}$ M^{-1} s^{-1}

13.93. No

13.95. No, because they have different rate laws

13.97. Pseudo-first-order kinetics occurs when one of the reactants is in sufficiently high concentration that its concentration does not change appreciably over the course of the reaction.

13.99.

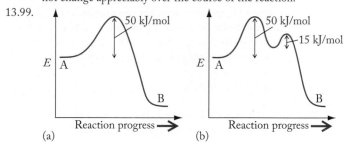

(a)   (b)

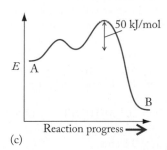

(c)

13.101. a. Rate $= k[SO_2Cl_2]$; unimolecular
b. Rate $= k[NO_2][CO]$; bimolecular
c. Rate $= k[NO_2]^2$; bimolecular

13.103. $N_2O_5(g) + O(g) \rightarrow 2\,NO_2(g) + O_2(g)$

13.105. The second step

13.107. The first step

13.109. Photochemical decomposition: a; thermal decomposition: b or c

13.111. Yes

13.113. Yes

13.115. Because the catalyst itself is not involved in the rate-limiting step

13.117. NO is the catalyst.

13.119. The reaction of O_3 with Cl has the larger rate constant.

13.121. When the concentration of a reactant (O_2 for the combustion reaction) increases, the rate of combustion increases.

13.123. The bodily reactions that use O_2 are slower at colder temperatures.

13.125. Yes, we could use other times, not just $t = 0$, as long as the rate of the reverse reaction is still much slower than the forward reaction.

13.127. In this plot $1/[X] - 1/[X]_0$ divided by $t - t_0$ is the slope of the line that corresponds to k, the reaction rate constant.

13.129. Zero order means that a reactant is not involved in a particular step in a reaction. However, the reactants in an elementary step ARE involved in that step.

13.131. The rate of consumption of O_3 is the same as the rate of formation of N_2O_5 and O_2 and one-half the rate of consumption of NO_2.

13.133. $k = 3.6 \times 10^{-4}$ s^{-1}; Rate $= (3.6 \times 10^{-4}$ s$^{-1})[N_2O_5]$

13.135. a. Yes
b. $E_a = 62.5$ kJ/mol
c. Rate $= 1.2 \times 10^{-12}$ M/s
d. At 10°C (283 K), $k = 21$ M^{-1} s^{-1}; at 35°C (308 K), $k = 1.8 \times 10^2$ M^{-1} s^{-1}

13.137. a. Rate $= k[Na(H_2O)_6^+]$
b. Neither

13.139. a. Rate $= k[NO][ONOO^-]$; $k = 1.30 \times 10^{-3}$ M^{-1} s^{-1}
b. The first Lewis structure is preferred:

$$\left[\overset{0}{\ddot{\text{O}}}=\overset{0}{\ddot{\text{N}}}-\overset{0}{\ddot{\text{O}}}-\overset{-1}{\ddot{\text{O}}}\colon\right]^- \longleftrightarrow \left[\colon\overset{-1}{\ddot{\text{O}}}-\overset{-1}{\ddot{\text{N}}}=\overset{+1}{\text{O}}-\overset{-1}{\ddot{\text{O}}}\colon\right]^- \longleftrightarrow$$

$$\left[\colon\overset{-1}{\ddot{\text{O}}}-\overset{-1}{\ddot{\text{N}}}-\overset{+1}{\text{O}}=\overset{0}{\ddot{\text{O}}}\right]^-$$

c. -55 kJ

13.141. a. Second order
b. No

13.143. a. Rate $= k[NH_2][NO]$
b. 1.2×10^9 M^{-1} s^{-1}

Chapter 14

14.1. Reaction C $\rightleftharpoons$ D has the larger k_f, the smaller k_r, and the larger K_c.

14.3. (a) A $\rightleftharpoons$ B; (b) 2.0

14.5. The reaction is endothermic. As temperature increases, K increases, indicating that more products form at higher temperatures.

14.7. No, because at 20 µs the concentrations of A and B are still changing.

14.9. Any process in which something is removed and replaced immediately would be an example. A common example is that of an unopened soda bottle in which the dissolved carbon dioxide is continually entering the gas phase at the same rate at which undissolved carbon dioxide enters the liquid phase in a dynamic equilibrium process.

14.11. Greater than 1

14.13.

Molar Mass	Compound	How Present
28	$^{14}N_2$	Originally present
29	$^{15}N^{14}N$	From decomposition of $^{15}N^{14}NO$
30	$^{15}N_2$	From decomposition of $^{15}N_2O$
32	O_2	Originally present
44	$^{14}N_2O$	From combination of $^{14}N_2$ and O_2
45	$^{15}N^{14}NO$	From combination of $^{15}N^{14}N$ and O_2
46	$^{15}N_2O$	Originally present

14.15. 0.333

14.17. When $\Delta n = 0$; when the number of moles of gaseous products equals the number of moles of gaseous reactants

14.19. a. $K_c = \dfrac{[N_2O_4]}{[N_2][O_2]^2}$ and $K_p = \dfrac{(P_{N_2O_4})}{(P_{N_2})(P_{O_2})^2}$

b. $K_c = \dfrac{[NO_2][N_2O]}{[NO]^3}$ and $K_p = \dfrac{(P_{NO_2})(P_{N_2O})}{(P_{NO})^3}$

c. $K_c = \dfrac{[N_2]^2[O_2]}{[N_2O]^2}$ and $K_p = \dfrac{(P_{N_2})^2(P_{O_2})}{(P_{N_2O})^2}$

14.21. 0.50

14.23. 0.068

14.25. 0.50

14.27. 1.5

14.29. 780

14.31. 0.0583

14.33. b and c

14.35. 0.10

14.37. When scaling the coefficients of a reaction up or down the new value of the equilibrium constant is the first K raised to the power of the scaling constant.

14.39. 11.0

14.41. $K_{c,forward} = \dfrac{[NO_2]^2}{[NO][NO_3]}$; $K_{c,reverse} = \dfrac{[NO][NO_3]}{[NO_2]^2}$;

$K_{c,reverse} = \dfrac{1}{K_{c,forward}}$

14.43. $K_c = \dfrac{[SO_3]}{[SO_2][O_2]^{1/2}}$; $K'_c = \dfrac{[SO_3]^2}{[SO_2]^2[O_2]}$; $K'_c = (K_c)^2$

14.45. a. 0.049

b. 420

c. 20

14.47. 7.4

14.49. The reaction quotient Q is the ratio of the concentrations or partial pressures of the products of a reaction raised to their stoichiometric coefficients to the concentrations of reactants raised to their stoichiometric coefficients. The reaction quotient has the same form as the equilibrium constant K expression, but the reaction is not necessarily at equilibrium.

14.51. The system is at equilibrium.

14.53. No, $Q < K$ so the reaction proceeds to the right to reach equilibrium.

14.55. Mixture a is at equilibrium.

14.57. $Q > K$, so the reaction will proceed to the left.

14.59. a

14.61. $K_c = [Cu^{2+}][S^{2-}]$

14.63. The concentrations of pure solids ($CaCO_3$ and CaO) do not change during the reaction and so they do not appear in the equilibrium constant expression.

14.65. No

14.67. As the concentration of O_2 increases the reaction shifts to the right and the CO on the hemoglobin is displaced.

14.69. According to Le Châtelier's principle an increase in the partial pressure (or concentration) of O_2 above the water shifts the equilibrium to the right so that more oxygen becomes dissolved in the water. This is consistent with Henry's law.

14.71. b and d

14.73. a. Increasing the concentration of the reactant O_3 shifts the equilibrium to the right, increasing the concentration of the product O_2 and decreasing the concentration of O_3.

b. Increasing the concentration of the product O_2 shifts the equilibrium to the left, increasing the concentration of the reactant O_3 and decreasing the concentration of O_2.

c. Decreasing the volume of the reaction to 1/10 its original volume shifts the equilibrium to the left, increasing the concentration of the reactant O_3 and decreasing the concentration of O_2.

14.75. The equilibrium shifts to the left.

14.77. a

14.79. When K is very small the amount of reactants that are transformed into products may be so small that the equilibrium concentrations of the reactants is approximately equal to the initial concentrations. This means that we can make an approximation in the K expression to make our calculations easier. When there are no products (Z) present, we also know that the reaction proceeds to the right to achieve equilibrium.

14.81. a. $P_{Cl_5} = 0.024$ atm, $P_{PCl_3} = 1.036$ atm, $P_{Cl_2} = 0.536$ atm

b. The partial pressure of PCl_3 decreases and the partial pressure of PCl_5 increases.

14.83. $[H_2O] = [Cl_2O] = 3.76 \times 10^{-3}\ M$; $[HOCl] = 1.13 \times 10^{-3}\ M$

14.85. 9×10^5

14.87. a. $P_{CO} = 2.4$ atm, $P_{CO_2} = 3.8$ atm

14.89. a. $P_{NO} = 0.272$ atm; $P_{NO_2} = 7.98 \times 10^{-3}$ atm

b. $P_T = 0.416$ atm

14.91. a. $P_{O_2} = 0.17$ atm, $P_{N_2} = 0.75$ atm, $P_{NO} = 0.080$ atm

14.93. 5.75 M

14.95. a. $P_{CO} = P_{Cl_2} = 0.258$ atm, $P_{COCl_2} = 0.00680$ atm

14.97. $[CO] = [H_2O] = 0.031\ M$, $[CO_2] = [H_2] = 0.069\ M$

14.99. $[NO] = 0.027\ M$, $[NO_2] = 0.28\ M$, $[N_2O_3] = 0.098\ M$

14.101. Yes

14.103. To the right

14.105. c

14.107. a. -197.8 kJ

b. 7.4×10^{24}

c. -142 kJ/mol from K_P; -142.0 kJ/mol from Appendix 4

14.109. 3.0×10^{-20}

14.111. Exothermic

14.113. Exothermic

14.115. 1.3×10^{-31}

14.117. -115 kJ

14.119. The reaction is endothermic:

At 1500 K, $K_p = 5.5 \times 10^{-11}$
At 2500 K, $K_p = 4.0 \times 10^{-3}$
At 3000 K, $K_p = 0.40$

This reaction does not favor products even at very high temperature, so this is not a viable source of CO and is not a remedy to decrease CO_2 as a contributor to global warming. Also, the process produces poisonous CO gas.

14.121. 9×10^{-22} M

14.123. $K_{p,25°C} = 3.15 \times 10^{-59}$; $K_{p,500°C} = 5.14 \times 10^{26}$

14.125. $P_{SO_2} = 9.2 \times 10^{-74}$ atm

Chapter 15

15.1. Red line

15.3. The blue titration curve represents the titration of a 1 M solution of strong acid. The red titration curve represents the titration of a 1 M solution of weak acid.

15.5. The indicator with a pK_a of 9.0

15.7. The red titration curve represents the titration of Na_2CO_3; the blue titration curve represents the titration of $NaHCO_3$.

15.9. HBr is the acid; H_2O is the base.

15.11. OH^- is the base; H_2O is the acid.

15.13. a. HNO_3 is the acid; NaOH is the base.
b. HCl is the acid; $CaCO_3$ is the base.
c. HCN is the acid; NH_3 is the base.

15.15. NO_2^-; OCl^-; $H_2PO_4^-$; NH_2^-

15.17. 1.50 M

15.19. 0.160 M

15.21. Dissolve 70.0 g of NaOH(s) in water and dilute to a total volume of 2.50 L.

15.23. Sulfur is more electronegative than selenium. The higher electronegativity on the sulfur atom stabilizes the anion HSO_4^- more than the anion $HSeO_4^-$.

15.25. a. H_2SO_3
b. H_2SeO_4

15.27. Because the pH function is a −log function, as $[H^+]$ increases, the value of $-\log[H^+]$ decreases.

15.29. When $[H^+]$ is greater than 1 M

15.31. a. pH = 7.462; pOH = 6.538; basic
b. pH = 4.70; pOH = 9.30; acidic
c. pH = 7.15; pOH = 6.85; basic
d. pH = 10.932; pOH = 3.068; basic

15.33. 0.810

15.35. pOH = 1.347; pH = 12.653

15.37. 6.95

15.39. 2.3

15.41. In order of largest K_a (strongest acid) to smallest K_a (weakest acid): $HNO_2 > CH_3COOH > HClO > HCN$

15.43. $NaNO_2$ is soluble in water, separating into Na^+ and NO_2^- ions, each in 1.0 M concentration for a total ion concentration of 2.0 M. HNO_2, however, only weakly dissociates in water and so produces just slightly greater than 1.0 M ions in solution. $NaNO_2$, therefore, with more dissolved ions in solution, is a better conductor of electricity.

15.45. $K_a = \dfrac{[H^+][F^-]}{[H]}$

15.47. (a) Water; (b) water

15.49. H_2O is the acid, CH_3NH_2 is the base.

15.51. 8.91×10^{-4}

15.53. 1.63%; $K_a = 6.74 \times 10^{-5}$

15.55. 2.49

15.57. 2.3 times

15.59. 10.77

15.61. 9.36

15.63. 3%

15.65. With each successive ionization, it becomes more difficult to remove H^+ from a species that is more negatively charged.

15.67. 0.51

15.69. 2.80

15.71. 9.50

15.73. 10.27

15.75. Increase

15.77. Ammonium nitrate

15.79. The citric acid in the lemon juice neutralizes the volatile trimethylamine to make a nonvolatile dissolved salt.

15.81. 3.32

15.83. 7.35

15.85. A solution of acetic acid and acetate ions can neutralize additions of acid or base. However, a solution of HCl and NaCl has no acid-neutralizing power because the Cl^- ion is too weak a base.

15.87. a. 4.453
b. 4.484

15.89. pH = 12.34; pOH = 1.65

15.91. 0.064

15.93. 9.25

15.95. 3.0 mL

15.97. pH = 3.50 before adding HCl; pH = 3.42 after adding HCl

15.99. The weak acid titration curve has an initial pH that is higher (less acidic) than that of an equimolar solution of a strong acid (lower pH, more acidic). The pH at the equivalence point in the titration of a strong acid is 7.00 whereas the pH at the equivalence point for a weak acid is basic.

15.101. No

15.103. After 10.0 mL of OH^- has been added, pH = 4.754; after 20.0 mL of OH^- has been added, pH = 8.750; after 30.0 mL of OH^- has been added, pH = 12.356

15.105. 0.02559 M

15.107. 25.0 mL and 25.0 mL more

15.109. 4.44

15.111.

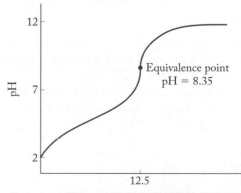

Equivalence point
pH = 8.35

Volume of 1.00 M NaOH (mL)

15.113.

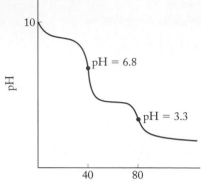

Volume of 0.100 M HCl (mL)

15.115. Molar solubility is the quantity (moles) of substance that dissolves in a liter of solution. The solubility product is the equilibrium constant for the dissolution of a substance.

15.117. Sr^{2+}

15.119. Endothermic

15.121. Acidic substances react with the OH^- released on dissolution of hydroxyapatite. The equilibrium is shifted to the right, dissolving more hydroxyapatite.

15.123. 1.08×10^{-10}

15.125. $[Cu^+] = [Cl^-] = 1.01 \times 10^{-3} M$

15.127. 9.96×10^{-6} g/mL

15.129. 10.091

15.131. d

15.133. No

15.135. Yes

15.137. (a) SO_4^{2-}; (b) $1.34 \times 10^{-4} M$

15.139. The hydrogen bonds between some of the water molecules must break and re-form around the species CH_3NH_2. Also, the amine hydrolyzes and forms $CH_3NH_3^+$ and OH^-, resulting in ion–dipole forces between these ions and the surrounding water molecules.

15.141. Buffers should have their pK_a values close to the pH of the target buffer.
pH 3.00 buffer: citric acid/citrate
pH 5.00 buffer: ascorbic acid/ascorbate
pH 7.00 buffer: dihydrogen phosphate/hydrogen phosphate
pH 12.00 buffer: hydrogen phosphate/phosphate

15.143. Subsequent additions of HCO_3^- react with water to form bicarbonate's conjugate acid (H_2CO_3) and its conjugate base (CO_3^{2-}) in the same proportions as the first addition, so pH does not change.

15.145. 1.4×10^9 L

15.147. The structure on the right in Figure P15.147 is the acid form because it has one more H^+ ion than the structure on the left.

15.149. (a) $K = 2.9 \times 10^{-4}$; (b) $4.4 \times 10^{-4} M$

15.151. a.

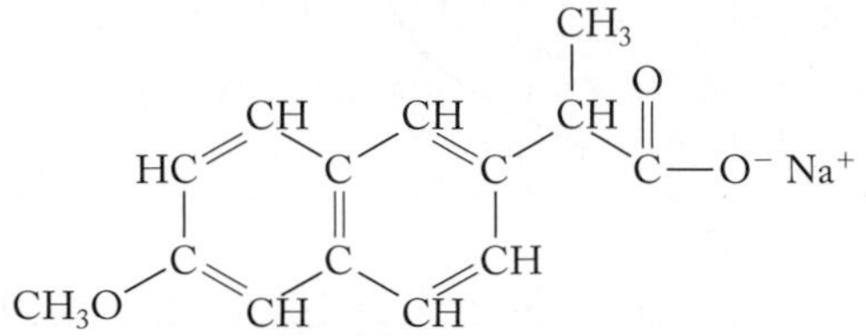

b. A solution of Naproxen should be basic because the anion in Naproxen is the conjugate base of a weak acid, and is, therefore, a weak base.

c. Naproxen is an ionic compound and likely to be more soluble in water than the acid form, which is composed of molecules with a large, hydrophobic middle section that contains two fused aromatic rings.

Chapter 16

16.1. Chromium (green) and cobalt (yellow)

16.3. Zinc (blue)

16.5. 4

16.7. a. $[Co(CN)_6]^{3-}$
b. $[CoF_6]^{3-}$
c. $[Co(NH_3)_6]^{3+}$

16.9. a. identical
b. isomers
c. identical

16.11. Yes, because it may donate an electron pair but not accept a proton.

16.13. BF_3 can accept electron pairs but has no H atoms to donate to be a Brønsted–Lowry acid.

16.15.

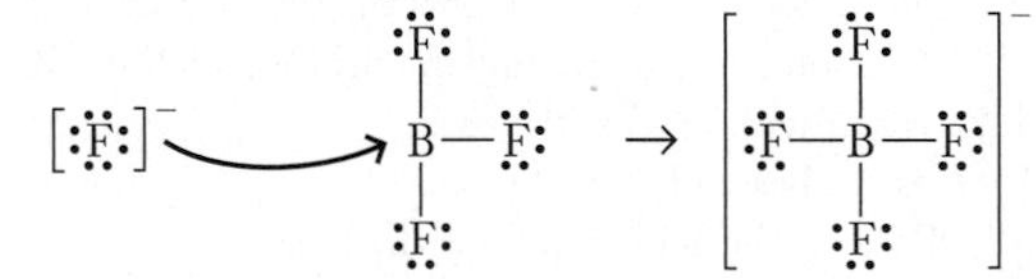

Lewis base   Lewis acid

16.17.

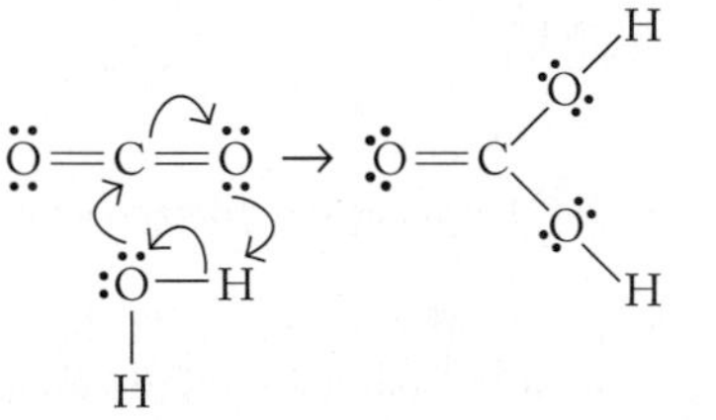

CO_2 and H_2O in this reaction act as both Lewis acids and Lewis bases.

16.19.

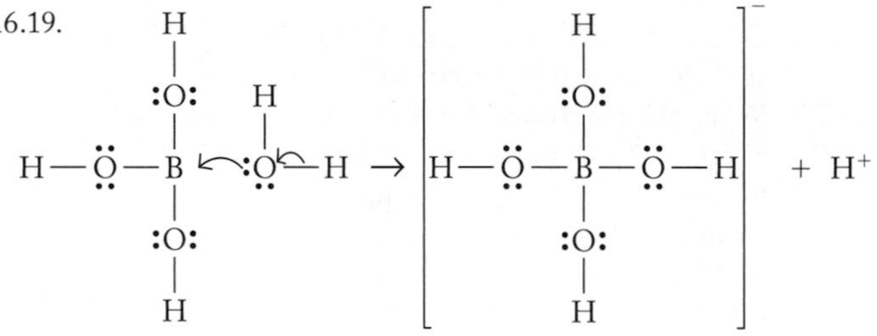

$B(OH)_3$ is the Lewis acid and H_2O is the Lewis base.

16.21. Water

16.23. Water

16.25. Na^+

16.27. $[Pt(NH_3)_6]Cl_4$, $[Pt(NH_3)_5Cl]Cl_3$, $[Pt(NH_3)_4Cl_2]Cl_2$, $[Pt(NH_3)_3Cl_3]Cl$, $[Pt(NH_3)_2Cl_4]$

16.29. Hexaammineplatinum(IV) chloride, pentaamminechloroplatinum(IV) chloride, tetraamminedichloroplatinum(IV) chloride, triamminetrichloroplatinum(IV) chloride, diamminetetrachloroplatinum(IV)

16.31. a. Hexaamminechromium(III)
b. Hexaaquacobalt(III)
c. Pentaamminechloroiron(III)

16.33. a. Tetrabromocobaltate(II)
 b. Aquatrihydroxozincate(II)
 c. Pentacyanonickelate(II)

16.35. a. Ethylenediaminezinc(II) sulfate
 b. Pentaammineaquanickel(II) chloride
 c. Potassium hexacyanoferrate(II)

16.37. A sequestering agent is a multidentate ligand that separates metal ions from other substances so that they can no longer react. Properties that make a sequestering agent effective include strong bonds formed between the metal and the ligand and large formation constants.

16.39. As pH increases the chelating ability increases because OH^- removes the H on the carboxylic acid groups, providing an additional site for binding to the metal cation.

16.41. When the transition metals bond to ligands the d orbitals split in energy. If there is a d to d transition possible for the ion, the compound is likely to be colored.

16.43. The repulsions due to the ligands in a square planar crystal field are highest for the d_{xy} orbital and so it is raised in energy because this orbital lies in the plane of the ligands.

16.45. The yellow solution contains (b) $Cr(NH_3)_6^{3+}$. The violet solution contains (a) $Cr(H_2O)_6^{3+}$.

16.47. Colorless

16.49. $NiCl_4^{2-}$

16.51. The magnitude of the crystal field splitting energy compared to the pairing energy of the electrons in a lower energy d orbital

16.53. Fe^{2+} has 4 unpaired electrons.
 Cu^{2+} has 1 unpaired electron.
 Co^{2+} has 3 unpaired electrons.
 Mn^{3+} has 4 unpaired electrons.

16.55. Cr^{3+}

16.57. a. Mn^{4+} in MnO_2; 2 Mn^{3+} and 1 Mn^{2+} in Mn_3O_4
 b. Both low-spin and high-spin configurations are possible in Mn_3O_4 (d^4 and d^5) but not in MnO_2 (d^3).

16.59. Paramagnetic

16.61. For an octahedral geometry cis- means that two ligands are side by side and have a 90° bond angle between them. Ligands that are $trans$- to each other have a 180° bond angle between them.

16.63. At least two different ligands

16.65. Yes

16.67.

$$\begin{bmatrix} Br & Cl \\ & Cu & \\ Br & Cl \end{bmatrix}^{2-} \quad \begin{bmatrix} Cl & Br \\ & Cu & \\ Br & Cl \end{bmatrix}^{2-}$$

 Cis   Trans

 No, neither isomer is chiral

16.69. Because the Cu^{2+} ion is fully complexed by the EDTA

16.71. Ag^+ forms a soluble complex with NH_3, removing Ag^+ from solution and shifting the equilibrium for the dissolution of AgCl to the right.

16.73. (a) $4.0 \times 10^{-3}\ M$; (b) $6.3 \times 10^{-10}\ M$

16.75. $2.6 \times 10^{-15}\ M$

16.77. $2 \times 10^{-3}\ M$

16.79. b and d

16.81. The solution will become more acidic.

16.83. In basic solution: $Cr(OH)_3(s) + OH^-(aq) \rightleftharpoons Cr(OH)_4^-(aq)$
 In acidic solution: $Cr(OH)_3(s) + 3\ H^+(aq) \rightleftharpoons$
 $Cr^{3+}(aq) + 3\ H_2O(\ell)$

16.85. $Al(OH)_3$ reacts with OH^- in solution to form soluble $Al(OH)_4^-$. The other ions do not form this type of soluble complex ion.

16.87. 2.65

16.89. 1.80

16.91.

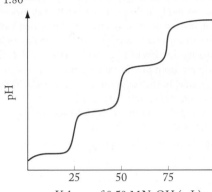

 Volume of 0.50 M NaOH (mL)

16.93. 4×10^{-13}

16.95. The yellow complex containing Co^{3+} in aqueous ammonia has the larger Δ_o.

16.97. Ag^{2+} has 9 d electrons, leaving an unpaired electron in the $d_{x^2-y^2}$ orbital to make it paramagnetic. Ag^{3+} has 8 d electrons and Ag^+ has 10 d electrons. Both have all electrons paired, so those silver ions are diamagnetic.

16.99. To longer wavelengths

Chapter 17

17.1. Because of the careful layering, each half-cell has its metal in contact with its cation solution. The solutions are not mixing, but nevertheless the layers allow the ions needed to balance the charge in each half-cell to pass.

17.3. Ag is the cathode; Pt in the SHE is the anode; electrons flow from the SHE to Ag.

17.5. Blue line

17.7. a. $2\ H_2O(\ell) + 2\ e^- \rightarrow H_2(g) + 2\ OH^-(aq)$
 $E°_{cathode} = -0.8277\ V$
 $2\ H_2O(\ell) \rightarrow O_2(g) + 4\ H^+(aq) + 4\ e^-$ $E°_{anode} = 1.229\ V$
 b. Increases the conductivity of the solution

17.9. A half-reaction is one couple of the redox reaction, either the reduction reaction or the oxidation reaction.

17.11. A wire can only pass electrons through it, not ions.

17.13. a. $Br_2(\ell) + 2\ e^- \rightarrow 2\ Br^-(aq)$, reduction
 b. $Pb(s) + 2\ Cl^-(aq) \rightarrow PbCl_2(s) + 2\ e^-$, oxidation
 c. $O_3(g) + 2\ H^+(aq) + 2\ e^- \rightarrow O_2(g) + H_2O(\ell)$, reduction
 d. $H_2S(g) \rightarrow S(s) + 2\ H^+(aq) + 2\ e^-$, oxidation

17.15. $2\ Fe_3O_4(s) + H_2O(\ell) \rightarrow 3\ Fe_2O_3(s) + 2\ H^+(aq) + 2\ e^-$

17.17. a. $Pb^{2+}(aq) + 2\ e^- \rightarrow Pb(s)$ cathode
 $Zn(s) \rightarrow Zn^{2+}(aq) + 2\ e^-$ anode
 b. $Pb^{2+}(aq) + Zn(s) \rightarrow Zn^{2+}(aq) + Pb(s)$
 c. $Zn(s)\ |\ Zn^{2+}(aq)\ ||\ Pb^{2+}(aq)\ |\ Pb(s)$

17.19. a. $MnO_4^-(aq) + 2\ H_2O(\ell) + 3\ e^- \rightarrow MnO_2(s) + 4\ OH^-(aq)$
 cathode
 $Cd(s) + 2\ OH^-(aq) \rightarrow Cd(OH)_2(s) + 2\ e^-$ anode
 b. $2\ MnO_4^-(aq) + 4\ H_2O(\ell) + 3\ Cd(s) \rightarrow$
 $2\ MnO_2(s) + 3\ Cd(OH)_2(s) + 2\ OH^-(aq)$
 c. $Cd(s)\ |\ Cd(OH)_2(s)\ ||\ MnO_4^-(aq)\ |\ MnO_2(s)\ |\ Pt(s)$

17.21. a. 6

b. FeO_4^{2-} has Fe^{6+}, Fe_2O_3 has Fe^{3+}, Zn has Zn^0, ZnO and ZnO_2^{2-} have Zn^{2+}

c. $Zn(s) \mid ZnO(s) \mid ZnO_2^{2-}(aq) \parallel FeO_4^{2-}(aq) \mid Fe_2O_3(s) \mid Pt(s)$

17.23. We have to assume that $E°_{oxidation}$(anode) represents the potential of the anode half-reaction written *as an oxidation* half-reaction instead of the usual way as a reduction half-reaction. Reversing a half-reaction changes the sign of its $E°$ value. Changing the sign of $E°_{anode}$ in Equation 17.1 explains why the − sign in front of it in that equation is a + sign in front of $E°_{oxidation}$(anode) in the equation in this problem. Therefore, the two equations are equivalent.

17.25. a. $Hg^{2+}(aq) + 2\,e^- \rightarrow Hg(\ell)$ with $Co(s) \rightarrow Co^{2+}(aq) + 2\,e^-$; $E°_{cell} = 1.128$ V

b. $Hg^{2+}(aq) + 2\,e^- \rightarrow Hg(\ell)$ with $Cu(s) \rightarrow Cu^{2+}(aq) + 2\,e^-$; $E°_{cell} = 0.509$ V

17.27. No

17.29. Less than 1.10 V

17.31. a.

Anode: $\quad Zn(s) \rightarrow Zn^{2+}(aq) + 2\,e^- \quad E°_{anode} = -0.7618$ V

Cathode: $Hg^{2+}(aq) + 2\,e^- \rightarrow Hg(\ell) \quad E°_{cathode} = 0.851$ V

$Zn(s) + Hg^{2+}(aq) \rightarrow Zn^{2}(aq) + Hg(\ell)$

$E°_{cell} = E°_{cathode} - E°_{anode} = 1.613$ V

b.

Anode: $\quad Zn(s) + 2\,OH^-(aq) \rightarrow ZnO(s) + H_2O(\ell) + 2\,e^-$

$E°_{anode} = -1.25$ V

Cathode: $Ag_2O(s) + H_2O(\ell) + 2\,e^- \rightarrow 2\,Ag(s) + 2\,OH^-(aq)$

$E°_{cathode} = 0.342$ V

$Zn(s) + Ag_2O(s) \rightarrow ZnO(s) + 2\,Ag(s)$

$E°_{cell} = E°_{cathode} - E°_{anode} = 1.59$ V

c.

Anode: $2 \times [Ni(s) + 2\,OH^-(aq) \rightarrow Ni(OH)_2(s) + 2\,e^-]$

$E°_{anode} = -0.72$ V

Cathode: $O_2(g) + 2\,H_2O(\ell) + 4\,e^- \rightarrow 4\,OH^-(aq)$

$E°_{cathode} = 0.401$ V

$2\,Ni(s) + O_2(g) + 2\,H_2O(\ell) \rightarrow Ni(OH)_2(s)$

$E°_{cell} = E°_{cathode} - E°_{anode} = 1.12$ V

17.33. Because the voltaic cell does work on the surroundings

17.35. a. $\Delta G° = -34.5$ kJ; $\Delta E°_{cell} = 0.358$ V

b. $\Delta G° = 2.9$ kJ; $\Delta E°_{cell} = -0.030$ V

17.37. −290 kJ

17.39. −116 kJ

17.41. a. $\Delta E°_{cell} = -0.478$ V; $\Delta G° = 92.2$ kJ

b. $\Delta E°_{cell} = 0.548$ V; $\Delta G° = -97.4$ kJ

17.43. It is an inert electrode, a surface on which electron transfer can occur.

17.45. Voltage of a battery (a voltaic cell) is governed by the Nernst equation: $E_{cell} = E°_{cell} - \dfrac{RT}{nF} \ln Q$. As a battery discharges, the value of Q, the reaction quotient, changes: $Q = \dfrac{[\text{products}]^x}{[\text{reactants}]^y}$.

At the start of the reaction, Q is very small because [reactants] >> [products]. As the reaction proceeds [products] grows and Q increases but does not increase significantly until significant amounts of products form, that is, when the battery is nearly discharged.

17.47. 1.27 V

17.49. 8.56×10^{19}

17.51. −0.414 V

17.53. $E_{cell} = 1.54$ V; E_{cell} will decrease

17.55. a. 0.62 V

b. 0.61 V

17.57. a. 0.349 V

b. 1.02×10^{57}

17.59. c and f

17.61. $Al–O_2$

17.63. $Li–MnO_2$

17.65. In a voltaic cell, the electrons are produced at the anode so a negative (−) charge builds up there; in an electrolytic cell, electrons are being forced onto the cathode so that it builds up negative (−) charge. The flow of electrons in the outside circuit is reversed in an electrolytic cell compared to the flow in a voltaic cell.

17.67. Br_2

17.69. More negative

17.71. 6.8 g

17.73. +3

17.75. 18.0 minutes

17.77. a. 5.78×10^{-3} L

b. No, some Cl_2 and Br_2 would be produced.

17.79. −0.270 V

17.81. A hybrid vehicle uses a relatively inexpensive fuel (gasoline) in the internal combustion engine and has good fuel economy, but it still gives off emissions. A fuel-cell vehicle does not give off emissions (the reaction produces H_2O) but requires a more expensive and explosive fuel (hydrogen); moreover, current battery technologies incorporate materials that are still very expensive and bulky.

17.83. Electric engines are more efficient by converting more of the energy into motion instead of losing it as heat.

17.85. a. $\overset{-4+1}{CH_4}(g) + \overset{+1}{H_2O}(g) \rightarrow \overset{+2}{CO}(g) + 3\,\overset{0}{H_2}(g)$

$\overset{+2}{CO}(g) + \overset{+1}{H_2O}(g) \rightarrow \overset{+1}{H_2}(g) + \overset{+4}{CO_2}(g)$

b. For the reaction of CH_4 with H_2O, $\Delta G°_{rxn} = 142.2$ kJ. For the reaction of CO with H_2O, $\Delta G°_{rxn} = -28.6$ kJ. For the overall reaction, $\Delta G°_{overall} = \Delta G°_{rxn_1} + \Delta G°_{rxn_2} = 113.6$ kJ.

17.87. All the values in Appendix 6 would increase by 0.8277 V.

17.89. $O_2(g) + 2\,H_2O(\ell) + 2\,Zn(s) + 4\,OH^-(aq) \rightarrow 2\,Zn(OH)_4^{2-}(aq)$

17.91. a. $NiO(OH)(s) + TiZr_2H(s) \rightarrow TiZr_2(s) + Ni(OH)_2(s)$

b. 1.32 V

17.93. The cell consists of a tin anode which is immersed in a solution of 0.025 $M\ Sn^{2+}(aq)$. This half-cell is connected via an external wire circuit and a salt bridge to a second half-cell consisting of a silver cathode immersed in a solution of 0.010 $M\ Ag^+(aq)$.

17.95. a. −0.87 V

b. Mo_3S_4: Mo = +2.67; $MgMoS_4$: Mo = +2

c. Mg^{2+} is added to the electrolyte to better carry the charge in the cell. This cation is produced at the anode and consumed at the cathode.

17.97. a. In K_2MnF_6: K = +1, Mn = +4, F = −1

In SbF_5: Sb = +5, F = −1

In $KSbF_6$: K = +1, Sb = +5, F = −1

In MnF_3: Mn = +3, F = −1

In F_2: F = 0

This is a one-electron process.

b. −656 kJ

c. 6.80 V

d. Too low

e. In H_2: H = 0
   In F_2: F = 0
   In KF: K = +1, F = −1
   In KHF_2: K = +1, H = +1, F = −1
   This is a two-electron process.

17.99. a. Reducing agent is Mg(s)
   b. Element reduced is U
   c. $E^\circ_{cathode} > -2.37$ V

17.101. a. Cathode
   b. No. Mg^{2+}, with a higher positive charge, has a lower (less negative) reduction potential than Na^+.
   c. No
   d. H_2 and O_2

Chapter 18

18.1. b and d are crystalline; a and c are amorphous.

18.3. The chemical formula is A_4B_4 or AB.

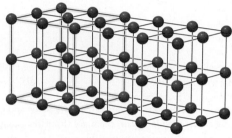

18.5. 3 A atoms, 1 B atom

18.7. The chemical formula is AB_3X.

18.9. 3.81 g/cm³

18.11. $MgAl_2O_4$

18.13. MS

18.15. Cs (blue) and Sr (purple)

18.17. MgB_2

18.19. Application of an electrical potential across a metal causes its mobile valence electrons to move toward the positive potential.

18.21. Ionic bonds are stronger than metallic bonds.

18.23. Yes

18.25. Groups 2 and 12

18.27. Phosphorus gives silicon a higher conductivity because it has one more valence electron making an n-type semiconductor.

18.29. Group 14

18.31. a. n-type
   b.

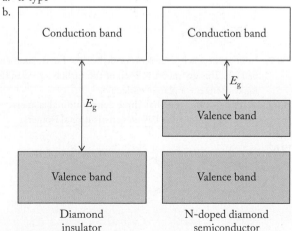

Diamond insulator

N-doped diamond semiconductor

   c. 4.68×10^{-19} J

18.33. InN

18.35. Cubic closest-packed structures have an *abcabc* pattern and hexagonal closest-packed structures have an *abab* pattern.

18.37. Body-centered cubic

18.39. These structural forms are not allotropes because iron is not molecular.

18.41. For bcc $\ell = \frac{4r}{\sqrt{3}}$; for fcc $\ell = \frac{4r}{\sqrt{2}}$

18.43. 104.2 pm

12.45. 513 pm

18.47. c

18.49. No

18.51. Because the metallic bonding between Cu and Ag in the alloy is weaker due to a mismatch of their atomic sizes

18.53. Tungsten is the host and carbon occupies the holes.

18.55. (a) XY_3; (b) YX_3

18.57. Octahedral holes

18.59. Substitutional alloy

18.61. Yes

18.63. (a) AB; (b) A_2B; (c) AB

18.65. One-fifth

18.67. Yes, the alloy is more dense.

18.69. Each S atom has a bent geometry due to sp^3 hybridization, and therefore the ring is not flat.

18.71. The ring of BN atoms is flat due to sp^2 hybridization of the B and N atoms.

18.73. a. 2 O^{2-} ions and 4 H^+ ions
   b. $\left[:\ddot{O}: \right]^{2-}$ 2 H^+

18.75. 109.5°

18.77. K^+ is large and so does not fit well into the octahedral holes of the fcc lattice.

18.79. The radius of Cl^- is 181 pm and the radius of Cs^+ is 170 pm and so their radii are very similar. The Cs^+ ion at the center of Figure P18.79 occupies the center of the cubic cell, so CsCl could be viewed as a body-centered cubic structure when taking into account the ions' slight difference in size. However, if we look at the ions as roughly equal in size, the unit cell becomes two interpenetrating simple cubic unit cells.

18.81. No, the Cl^- radius is so much larger than that of Na^+ that the Na^+ would not be closest-packed.

18.83. Less than

18.85. $MgFe_2O_4$

18.87. UO_2

18.89. (a) Octahedral; (b) half

18.91. This rock salt arrangement is more dense than the sphalerite arrangement because in sphalerite the lattice of S^{2-} ions must expand to accommodate the Cd^{2+} ions.

18.93. 5.25 g/cm³

18.95. 421 pm

18.97. Ceramics are (b) thermal insulators. (a) Ductility, (c) electrical conductivity, and (d) malleability describe metals.

18.99. $Mg_3(Si_2O_5)(OH)_4$

18.101. $2\ KAlSi_3O_8(s) + 2\ H_2O(\ell) + CO_2(g) \rightarrow Al_2(Si_2O_5)(OH)_4(s) + 4\ SiO_2(s) + K_2CO_3(aq)$; this is not a redox reaction.

18.103. a. $3\ CaAl_2Si_2O_8(s) \rightarrow Ca_3Al_2(SiO_4)_3(s) + 2\ Al_2SiO_5(s) + SiO_2(s)$
   b. In anorthite, the silicate anion is $Si_2O_8^{8-}$.
   In grossular, the silicate anion is SiO_4^{4-}.
   In kyanite, the silicate anion is SiO_5^{6-}.

18.105. Cubic holes can accommodate Ba^{2+}. Octahedral holes can accommodate Ti^{4+}.

18.107. An amorphous solid has no regular, repeating lattice to diffract X-rays.

18.109. X-rays have wavelengths of the order of the separation of atoms in crystals. Microwaves have wavelengths too long to be diffracted by crystal lattices.

18.111. If a crystallographer uses a shorter λ wavelength, the data set can be collected over a smaller scanning range.

18.113. Halite

18.115. The values of n are 2 ($\theta = 6.99°$) and 3 ($\theta = 10.62°$). The average lattice spacing is $d = 582$ pm.

18.117. 4.76°

18.119. XYZ_3

18.121. 33.5%

18.123. a. 139 pm
   b. 9.96 g/cm³
   c. 3375 unit cells, 6750 Mo atoms

18.125. a. 7.53 g/cm³
   b. 3.42 g/cm³
   c. 3.36 g/cm³

18.127. AuZn

18.129. 52.4%

18.131. Substitutional alloy

18.133. A cluster of three simple cubic unit cells with an Al atom in the center of one of them is consistent with the formula (Cu_3Al).

Chapter 19

19.1. a. One degree of unsaturation
   b. Two degrees of unsaturation
   c. No degrees of unsaturation
   d. Three degrees of unsaturation

19.3. Pine oil and oil of celery

19.5. b and d

19.7. Benzyl acetate contains an ester group; carvone contains two alkene groups and a ketone group; cinnamaldehyde contains an alkene and an aldehyde group.

19.9.
$$\left[\begin{array}{c} CH_3 \\ | \\ -Si-O- \\ | \\ CH_3 \end{array} \right]_n$$

19.11. For *cis*-polyisoprene:

For *trans*-polyisoprene:

19.13. 8

19.15. An *sp* hybridized carbon atom can form two double bonds or one triple bond; an *sp²* hybridized carbon atom can form one double bond; an *sp³* hybridized carbon atom can form only single bonds.

19.17. No

19.19. Amine, alcohol, ether, aldehyde, ketone, carboxylic acid, ester, and amide

19.21. Sample A because of its higher average molar mass and greater number of intermolecular forces

19.23. Yes

19.25. *sp³*

19.27. The structure of cyclohexane shows that C atoms are *sp³* hybridized with bond angles of 109.5°. It cannot be a planar molecule.

19.29. No

19.31. No

19.33.

Pentane   2-Methylbutane   2,2-Dimethylpropane

19.35. (a) 2,3-Dimethylhexane, (c and d) 2-methylheptane

19.37. a. C_8H_{18}
   b. C_9H_{20}
   c. C_8H_{18}
   d. C_8H_{18}
   e. C_9H_{20}

19.39. $C_3H_8 < C_8H_{16} < C_{14}H_{30}$

19.41. Hexane

19.43. No

19.45. When the double bond is "terminal" (occurs at the end or beginning of the carbon chain) there are three like groups (H) so no cis and trans isomers are possible.

19.47. The C=C double bond outside of the ring does not show cis–trans isomerism because there are not two dissimilar groups on the terminal carbon atom. The C=C double bond in the ring of carbon atoms is cis in the structure of carvone. This bond cannot be trans or the ring of 6 carbon atoms would not be possible.

19.49. Ethylene has a C=C bond with which HBr is reactive but polyethylene has only saturated C—C bonds that do not react with HBr.

19.51. −124 kJ

19.53. Isomer a is trans, *E*; b is cis, *Z*.

19.55. 681.2 kJ, endothermic

19.57. −174.30 kJ

19.59.

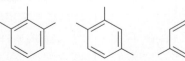

19.61. In benzene, each C atom is *sp²* hybridized with bond angles of 120°. This geometry at each of the carbon atoms in the ring makes benzene a planar molecule.

19.63. Tetramethylbenzene has three constitutional isomers; pentamethylbenzene has no constitutional isomers.

19.65. Yes

19.67.

19.69. Fuel value for 1 mole benzene = 41.83 kJ/g
Fuel value for 3 moles ethylene = 50.30 kJ/g
One mole benzene has a lower energy content than 3 moles ethylene.

19.71. Methylamine has a smaller nonpolar hydrocarbon chain compared to *n*-butylamine and so it is more soluble in water.

19.73.

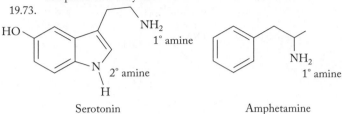

Serotonin   Amphetamine

19.75. −138.7 kJ

19.77. a, b, and c

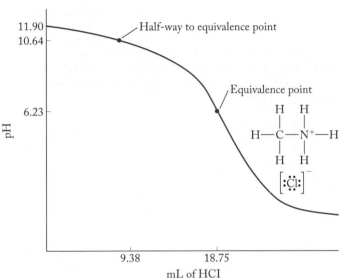

19.79. The more oxygenated the fuel, the lower the fuel value.

19.81. Ethers have lower boiling points compared to alcohols because they have weaker dipole–dipole forces compared to the alcohols, which have hydrogen bonding between the molecules.

19.83. Evaporation of ethanol from the skin is an endothermic process (phase change from liquid to vapor). The heat transfers from the skin to the ethanol so the skin feels cold.

19.85. a and d are alcohols, b and c are ethers; b < c < d < a

19.87. Fuel value for diethyl ether = 36.74 kJ/g
Fuel value for *n*-butanol = 36.10 kJ/g
Diethyl ether has a slightly higher fuel value.

19.89. Fuel value for methanol = 22.67 kJ/g
Fuel value for ethanol = 29.67 kJ/g
Yes, the answer supports the prediction made in Problem 19.80.

19.91. Both carboxylic acids and aldehydes have polar functional groups. Carboxylic acids, however, are more soluble in water because they form strong hydrogen bonds with water.

19.93. Yes

19.95. No

19.97. Structure a because all of the formal charges are zero

19.99. An amide includes a carbonyl (C=O) as part of its functional group in addition to the −NH$_2$ group.

19.101. a, b, and d

19.103. b

19.105.

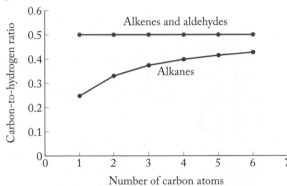

The plot of C:H ratio versus number of C atoms for aldehydes correlates exactly to that of alkenes and poorly to that of alkanes.

19.107. a. Pineapples

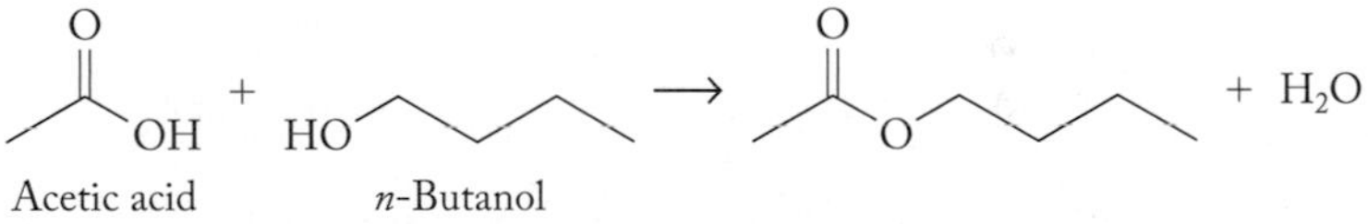

Acetic acid   *n*-Butanol

b. Bananas

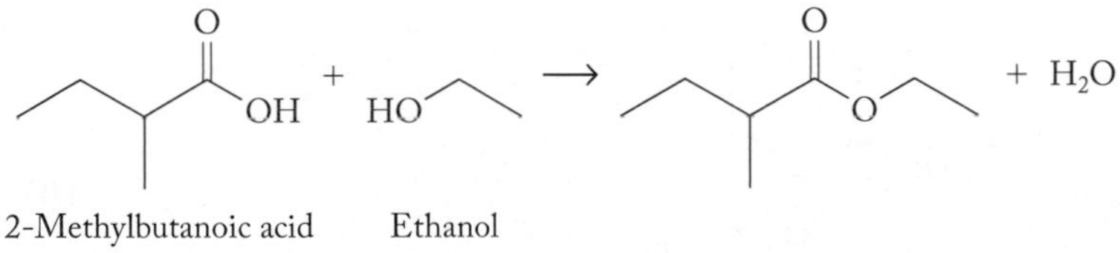

2-Methylbutanoic acid   Ethanol

c. Apples

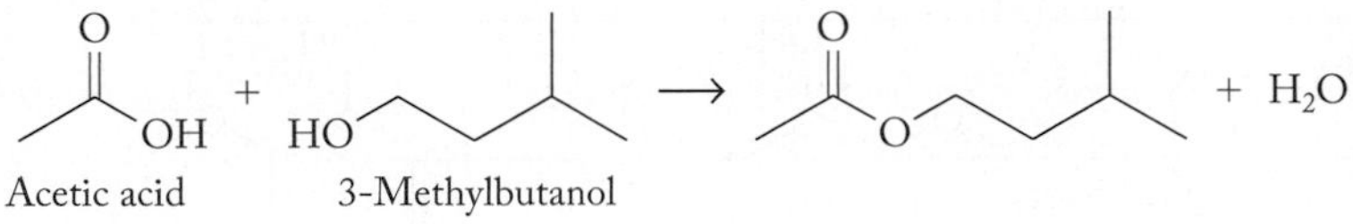

Acetic acid   3-Methylbutanol

19.109. Nicotine's highlighted N atom is in a tertiary amine group; valium's highlighted N atom is in an amide group.

19.111. Fuel value for formaldehyde = 19.00 kJ/g
Fuel value for formic acid = 6.531 kJ/g
Formaldehyde has a significantly higher fuel value than formic acid.

19.113. For reaction 1, ΔH°_{rxn} = 17.5 kJ
For reaction 2, ΔH°_{rxn} = 312.1 kJ

19.115. (a) 4; (b) 6; (c) 8

19.117. a. Condensation; methanol
b. Because of the presence of the six-membered ring, Kodel might be better able to accept nonpolar organic dyes.

19.119. No; enantiomer and optically active can describe the same chiral molecule, but achiral cannot.

19.121. Yes. If R contains a chiral center the cis or trans isomer of RCH=CHR would have optical isomers.

19.123.

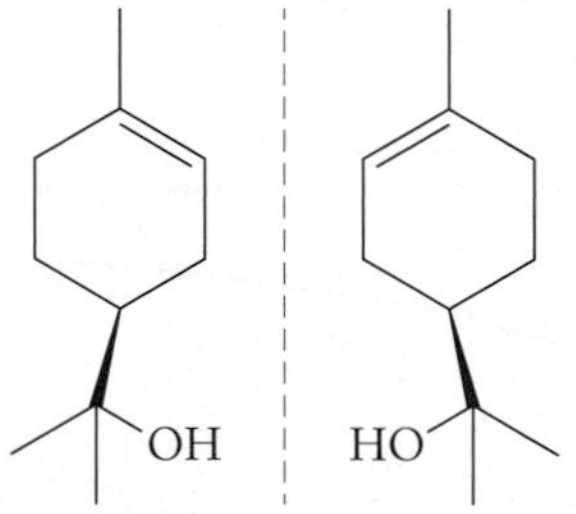

19.125. None

19.127. 2.52 g methanol; 3.45 g carbon dioxide

19.129. Methane and decane, both being nonpolar, have dispersion forces between their molecules. Dissolving methane in decane then is facile. When methane dissolves in water the dipole–induced dipole interactions it forms with water molecules are not strong enough to break the hydrogen bonds of water molecules. Therefore, methane is more soluble in decane than in water.

19.131. a. Trans
b.

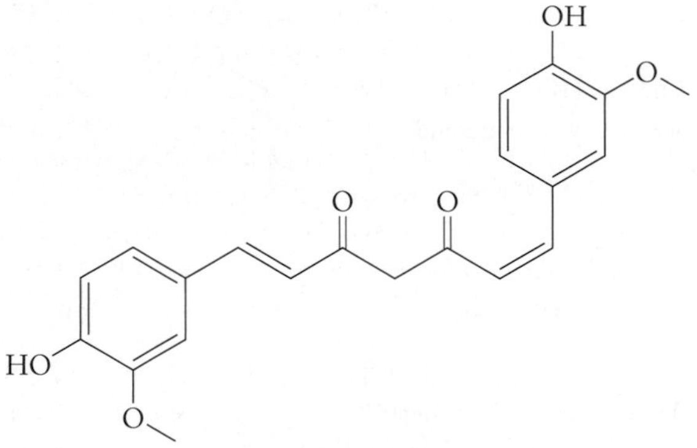

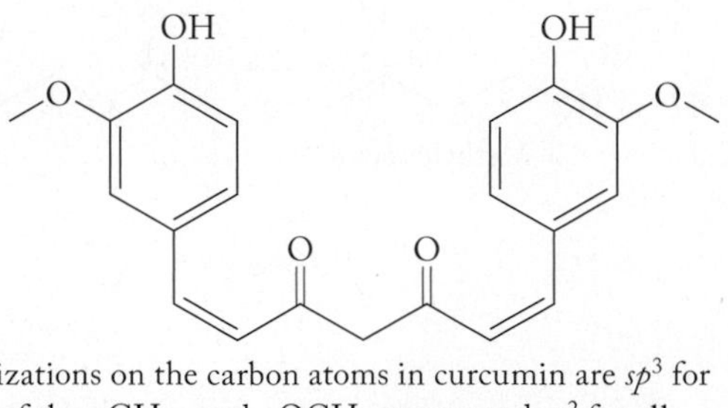

c. The hybridizations on the carbon atoms in curcumin are sp^3 for the carbon of the −CH$_2$− and −OCH$_3$ groups and sp^2 for all other carbon atoms.

19.133.

For polymer a

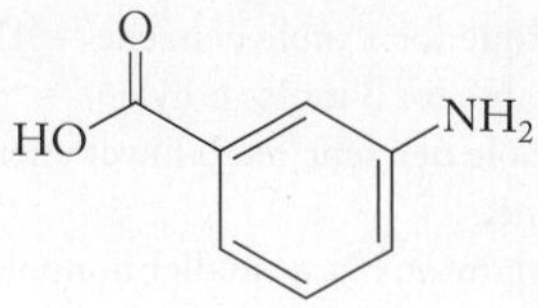

For polymer b

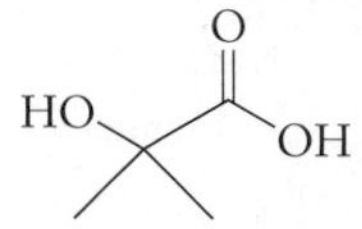

19.135.

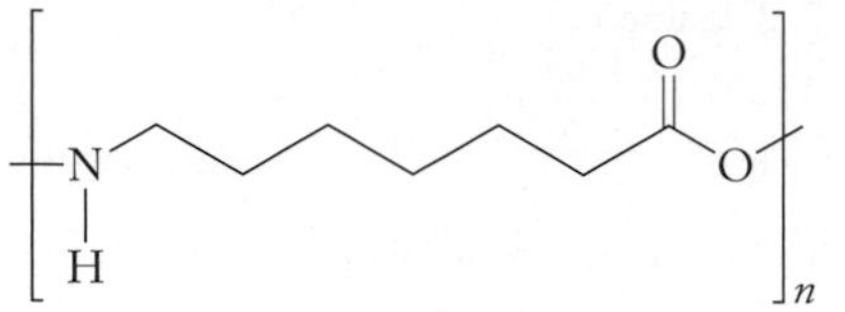

In this polymer there is one monomeric repeating unit with 7 carbon atoms because it is prepared from the difunctional H$_2$N(CH$_2$)$_6$COOH monomer. In nylon-6 the polymer also has a single monomeric unit but with 6 carbon atoms.

19.137. Cross-linking increases the strength, hardness, melting (softening) point, and the chemical resistance of a polymer.

19.139.

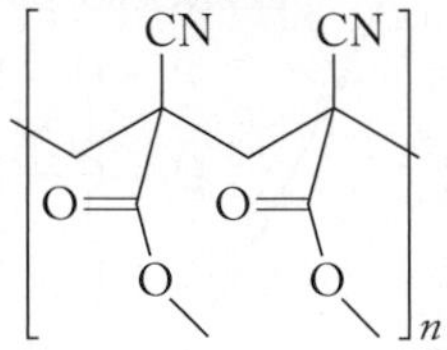

19.141. a.

For piperine

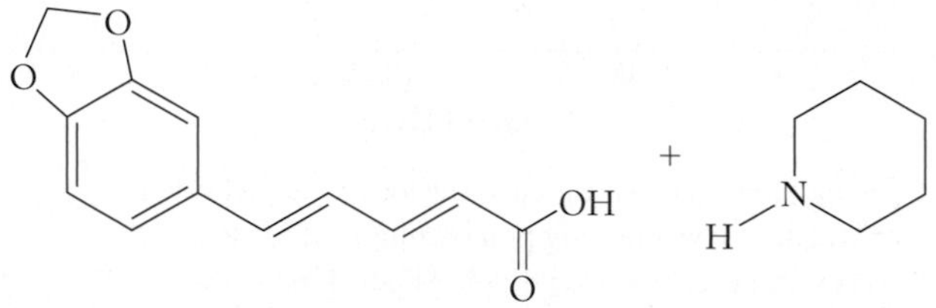

For capsaicin

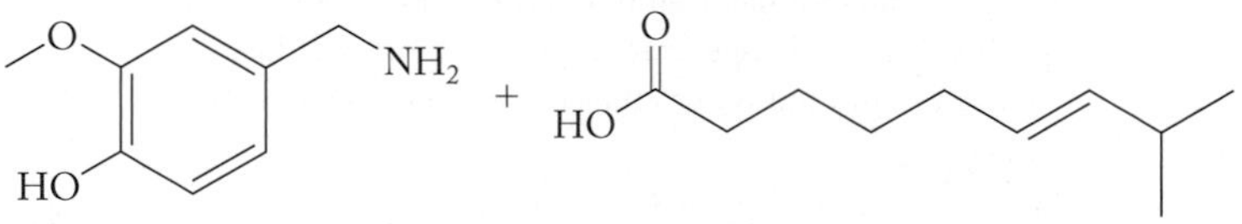

b. All C=C bonds are trans.
c. In piperine there are two ether groups (in the five-membered ring) and the amide group. In capsaicin there is an ether group, an alcohol group, and the amide group.

Chapter 20

20.1.

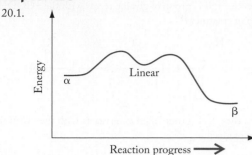

20.3. a. Palmitic acid and stearic acid
b. Cis

20.5. Tyrosine, glycine, glycine, phenylalanine, and methionine

20.7. Trans fats exhibit geometric isomerism around the C=C bond where similar groups on the two carbon atoms are situated on opposite sides of the double bond. Structures a and c contain trans fats.

20.9. Sucrose. The difference in the structures is that in sucralose, three −OH groups on sucrose have been replaced by Cl atoms. Being derived from sucrose implies that the sugar is natural, but the presence of Cl atoms on sugars is not natural.

20.11. Decreases

20.13. The "α" refers to the single carbon atom in amino acids to which both −NH$_2$ and −COOH groups are bonded.

20.15. D- and L- refer to how the four groups on a chiral carbon are oriented.

20.17. a and c

20.19. Most amino acids are zwitterions at pH ≈ 7.4 because the amino group will be protonated and the carboxylic acid group will be deprotonated, giving RC$\overset{+}{N}$H$_3$COO$^−$.

20.21. a.

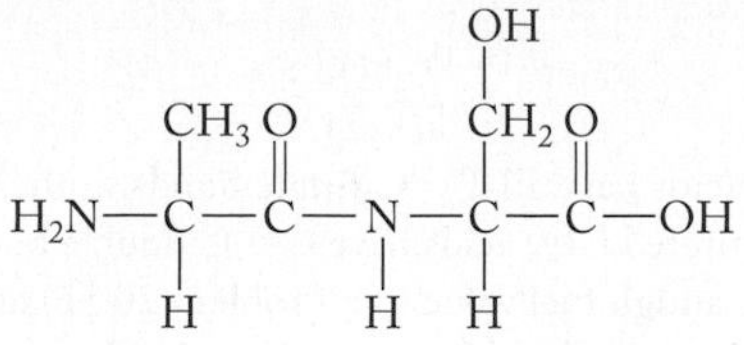

$$H_2N-\underset{\underset{H}{|}}{\overset{\overset{CH_3}{|}}{C}}-\overset{O}{\overset{||}{C}}-\underset{\underset{H}{|}}{\overset{}{N}}-\underset{\underset{H}{|}}{\overset{\overset{CH_2}{|}\overset{|}{OH}}{C}}-\overset{O}{\overset{||}{C}}-OH$$

b.

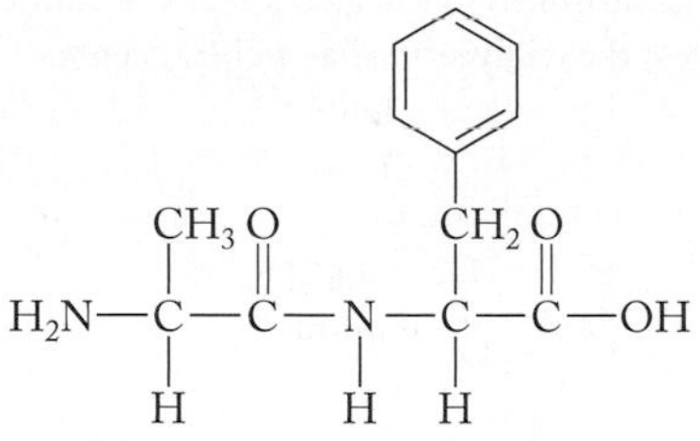

c.

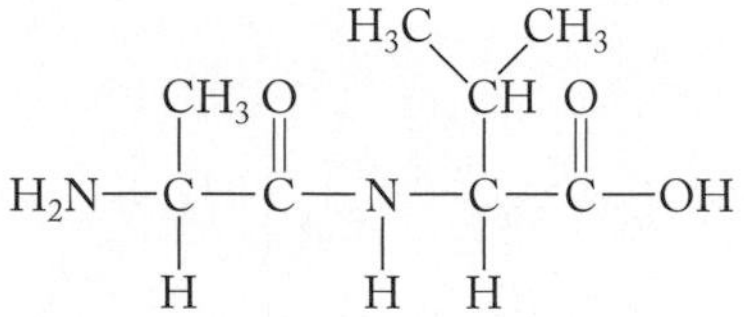

20.23. a. Alanine + glycine
b. Leucine + leucine
c. Tyrosine + phenylalanine

20.25. NH$_3$

20.27. The primary structure

20.29. Tertiary structure

20.31. In this model the binding site changes slightly when the substrate binds with the active site so that it fits the transition state better for the reaction.

20.33. The substrate and the active site must fit each other so if the molecular lock and key do not match then the reaction cannot be catalyzed at that site.

20.35. Lysine contains two amino groups, one of which is on a long carbon tail. This can react with the carboxylic acid on the carbon tail of glutamic acid to form a bridge.

20.37. Starch has α-glycosidic bonds and cellulose has β-glycosidic bonds. Starch coils into granules and cellulose forms linear molecules.

20.39. No

20.41. The bonding in fructose and glucose is nearly the same.

20.43. To calculate the free-energy change for a two-step process we need only to sum the individual ΔG values for each reaction.

20.45.

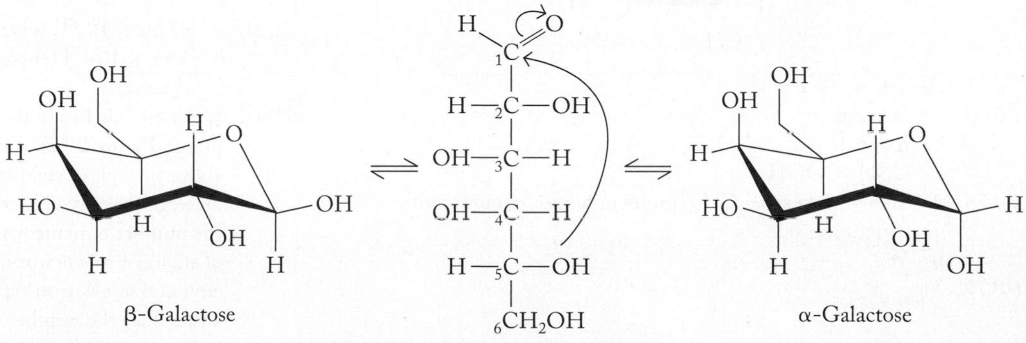

β-Galactose ⇌ ⇌ α-Galactose

20.47. c

20.49. a

20.51. b

20.53. -16.4 kJ

20.55. Saturated fatty acids have all C—C single bonds in their structure; unsaturated fatty acids have C=C double bonds.

20.57. Fatty acids have a high fuel value (see Problem 20.56) and eating sticks of butter affords Arctic explorers with more energy per gram of food compared to carbohydrates or proteins.

20.59. If the two fatty acids linked to the glycerol at C-1 and C-3 are different then, yes, the triglyceride has a chiral center.

20.61. b and c

20.63.

(a) Glycerol with octanoic acid

(b) Glycerol with decanoic acid

(c) Glycerol with dodecanoic acid

20.65. Phosphate group, a five-carbon sugar, and a nitrogen base; the backbone of DNA is composed of alternating sugar residues and phosphate groups.

20.67. Hydrogen bonds

20.69.

20.71. A-G-C-C-A-T

20.73. a. Sucrose
b. Esters
c. $C_{15}H_{31}COOH$

20.75. a. There is an extra $-CH_2-$ group in homocysteine's sulfur-containing side chain.
b. Yes

20.77. Yes

20.79. a. No
b. When the $-NH_2$ group of glycine reacts with the $-COOH$ group of creatine:

When the $-NH_2$ group of creatine reacts with the $-COOH$ group of glycine:

20.81. Glutamic acid, cysteine, and glycine

20.83. Yes. Because there is no difference in the number of C—C, C—H, C=O, C—O, or N—H bonds between the two compounds, we expect on the basis of average bond energies that the fuel values of leucine and isoleucine should be identical. Isoleucine might have a lower fuel value because the CH_3 group is closer to the COOH and NH_2 groups, and this difference in shape must contribute to the slightly different fuel values.

Chapter 21

21.1. Purple (lithium)

21.3. Red (radium)

21.5. a

21.7. Blue line (b)

21.9. Process 1 represents fission; process 2 represents fusion

21.11. Electron = β particle = positron < proton < neutron < deuteron < α particle

21.13. Antihydrogen has the same mass as hydrogen but its nucleus has a negative charge with a positively charged electron. It contains the antiproton in the nucleus and a positron in place of the electron.

21.15. Fusion in the sun has more steps than the primordial synthesis did. Fusion in the sun produces beta particles along with alpha particles (helium) while primordial nucleosynthesis produced deuterium that was later used along with neutrons as fuel for the process.

21.17. $\Delta E = 3.01 \times 10^{-10}$ J; $\lambda = 1.32 \times 10^{-15}$ m or 1.32×10^{-6} nm

21.19. a. 4.37×10^{-12} J
b. 6.80×10^{-12} J
c. 2.69×10^{-12} J
d. 1.60×10^{-12} J

21.21. a. -7.66×10^{-13} J (energy is released in this reaction)
b. 3.96×10^{-13} J (energy will have to be provided for this reaction)

21.23. If the nuclide lies in the belt of stability (green dots on the plot in Figure 21.2), it is not radioactive and is stable. If it lies above the belt of stability, then it is neutron-rich and tends to undergo β decay to increase the number of protons and reduce the number of neutrons in its nucleus. If it lies below the belt of stability, it is neutron-poor and tends to undergo positron emission or electron capture to increase the number of neutrons and reduce the number of protons in its nucleus.

21.25. Alpha decay increases the neutron-to-proton ratio to produce less stable isotopes, which can then be made more stable through β emission to decrease the neutron-to-proton ratio.

21.27. Both of these processes are β decays.

21.29. Greater than 1

21.31. $^{26}_{13}\text{Al} \rightarrow ^{0}_{1}\beta + ^{26}_{12}\text{Mg}$

21.33. a. Electron capture or positron emission
   b. Electron capture or positron emission
   c. This isotope is stable.

21.35. ^{56}Co has 27 protons and 29 neutrons and is neutron-poor; it may undergo electron capture or positron emission. ^{44}Ti has 22 protons and 22 neutrons and is neutron-poor; it may undergo electron capture or positron emission.

21.37. 25%

21.39. 2.2 days

21.41. 131 years from 2011 or in the year 2142

21.43. Control rods made of boron or cadmium are used to absorb the excess neutrons to control the rate of energy release.

21.45. The neutron-to-proton ratio for heavy nuclei is high and when the nuclide undergoes fission to form smaller nuclides, it must emit neutrons because the fission products require a lower neutron-to-proton ratio for stability.

21.47. a. $^{138}_{52}\text{Te}$
   b. $^{133}_{51}\text{Sb}$
   c. $^{143}_{55}\text{Cs}$

21.49. a. $^{103}_{39}\text{Y}$
   b. $^{130}_{48}\text{Cd}$
   c. $^{138}_{52}\text{Te}$

21.51. The level of radioactivity is the amount of radioactive particles present in a given instant of time. The dose is the accumulation of exposure over a length of time.

21.53. When radon-222 decays to polonium-218 while in the lungs, the ^{218}Po, a reactive solid that is chemically similar to oxygen, lodges in the lung tissue where it continues to emit α radiation. Alpha radiation is one of the most damaging kinds of radiation when in contact with biological tissues. The result of exposure to high levels of radon is an increased risk for lung cancer.

21.55. 5 μSv = 5 μGy; 250 μJ

21.57. a. $^{90}_{38}\text{Sr} \rightarrow ^{0}_{-1}\beta + ^{90}_{39}\text{Y}$
   b. 3.28×10^8 atoms ^{90}Sr
   c. Strontium-90 is found in milk and not other foods because it is chemically similar to calcium and milk is rich in calcium.

21.59. a. 0.15 decays/s
   b. 7.0×10^4

21.61. a. The half-life should be long enough to effect treatment of the cancerous cells but not so long as to cause damage to healthy tissues.
   b. Because α radiation does not penetrate far beyond a tumor, the α decay mode is best.
   c. Products should be nonradioactive, if possible, or have short half-lives and be able to be flushed from the body by normal cellular and biological processes.

21.63. a. Positron emission or electron capture
   b. Positron emission or electron capture
   c. Positron emission or electron capture

21.65. 74.3 days

21.67. Yes

21.69. 136 min

21.71. a. $^{10}_{5}\text{B} + ^{1}_{0}\text{n} \rightarrow ^{7}_{3}\text{Li} + ^{4}_{2}\alpha$
   b. 4.43×10^{-13} J
   c. Alpha particles have a high RBE and they do not penetrate into healthy tissue if the radionuclide is placed inside a tumor.

21.73. After 8.726 half-lives the ratio of ^{14}C present to that originally in an artifact is $N_t/N_0 = 0.50^{8.726} = 0.00236$ or 0.236%. This is too little to detect.

21.75. After 0.00023 half-lives the ratio of ^{40}K present to that originally in a sample is $N_t/N_0 = 0.50^{0.00023} = 0.9998$ or 99.98%. This level is just when we can detect the difference in amounts of ^{40}K.

21.77. 35%

21.79. 85%

21.81. 36,640 yr

21.83. a. Besides releasing a large amount of energy to power the starship *Enterprise*, hydrogen is an abundant fuel in the universe and therefore could easily react with any antihydrogen produced.
   b. Antihydrogen would react with all matter so it would be difficult to contain except maybe with a magnetic or energy field.

21.85. The energy released in the fusion reaction is $\Delta E = 9.91 \times 10^{-13}$ J/nucleon. The energy released in the fission reaction is $\Delta E = 1.4 \times 10^{-13}$ J/nucleon. On a per nucleon basis, the fusion reaction generates more energy.

21.87. a. Because the ^{241}Am ionizes the air and the smoke detector registers a change in current, a smoke detector resembles a Geiger counter in its operation.
   b. $t = \dfrac{433 \text{ yr}}{0.693} \ln \dfrac{1}{100} = 2877$ yr
   c. Smoke detectors are safe to handle because the ^{241}Am is an α emitter and α particles do not travel more than a few inches in air and cannot penetrate the first layer of skin.

21.89. a. $^{249}_{98}\text{Cf} + ^{48}_{20}\text{Ca} \rightarrow ^{294}_{118}\text{Uuo} + 3\ ^{1}_{0}\text{n}$
   b. $^{290}_{116}\text{Uuh}$
   c. $^{286}_{114}\text{Uuq}$
   d. $^{282}_{112}\text{Cn}$
   e. Because ^{294}Uuo is a member of the noble gas family, it has chemical and physical properties similar to naturally occurring radon.

21.91. ^{210}Pb

21.93. a. $^{64}_{28}\text{Ni} + ^{124}_{50}\text{Sn} \rightarrow ^{188}_{78}\text{Pt}$
   b. $^{196}_{78}\text{Pt}$

21.95. 3.35

21.97. a. $^{40}_{19}\text{K} + ^{40}_{18}\text{Ar} \rightarrow ^{0}_{1}\beta$
   b. Because the half-life of ^{40}K is so much longer than that of ^{14}C

21.99. 118 s

Chapter 1

Page 3: Gianni Dagli Orti/The Art Archive at Art Resource, NY; p. 5 (all): David Wrobel/Visuals Unlimited; p. 6 (icicles): Podisu/Dreamstime; (ice cubes): Arenacreative/Dreamstime; (beaker): Dorling Kindersley/Getty Images; (glass) Shutterstock; (freezer) Katherine Fawssett/Getty Images; (pond): Erik Page Photography/Getty Images; p. 7 (top): Tony Garcia/Getty Images; (bottom): Christian Charisius/DPA/Landov; p. 8 (salad dressing): iStockphoto; (ice): Daniel Smith/Corbis; (vinegar): NRH Photography; (gold): Photographer's Choice/Punchstock; p. 9 (a): iStockphoto; (b): Andy Clarke/Photo Researchers; (bottom): Lester V. Bergman/Corbis; p. 10: Aquacone courtesy Solar Solutions, Inc.; p. 11 (top left): GeoEye; (inset): Markus Geisen/The Natural History Museum, London; (bottom, left): © 2009 Richard Megna/Fundamental Photographs; p. 12: (both): © 2009 Richard Megna/Fundamental Photographs; p. 13 (top): Mason Morfit; (bottom): IBM/Photo Researchers; p. 20 (both): Photo courtesy of A&D Weighing, San Jose, CA. www.andweighing.com; p. 23: © 2009 Richard Megna/Fundamental Photographs; p. 25 (top): Shutterstock; (center): Shutterstock; (bottom): Canadian Press Images; p. 27: AP Photo; p. 28: Courtesy Chris Joosen/White Mountain National Forest; p. 33: Aquacone courtesy Solar Solutions, Inc.; p. 35: Rob Tringali/MLB Photos via Getty Images; p. 36: Brian Vance, Motor Trend; p. 38: Brian Hartshorn/Alamy; p. 39: Shutterstock.

Chapter 2

Page 41: Michael Hoch/© CERN; p. 42: The Royal Institution, London/Bridgeman Art Library; p. 45: Courtesy Jacob Lewis Bourjaily; p. 55 (top): AP Photo; (bottom): Dirk Wiersma/Science Photo Library; p. 56: Richard Megna/Fundamental Photographs; p. 58 (top): Rona Tuccillo; p. (bottom): Vladimir Mucibabic/Dreamstime; p. 60: moodboard/Corbis; p. 65: NASA/HST/J. Morse/K. Davidson; p. 66 (top): X-ray: NASA/CXC/ASU/J. Hester et al.; Optical: NASA/ESA/ASU/J. Hester & A. Loll; Infrared: NASA/JPL-Caltech/Univ. Minn./R. Gehrz; (bottom): Corbis; p. 69: Richard Megna/Fundamental Photographs; p. 69: X-ray: NASA/CXC/ASU/J. Hester et al.; Optical: NASA/ESA/ASU/J. Hester & A. Loll; Infrared: NASA/JPL-Caltech/Univ. Minn./R. Gehrz; p. 75: Smithsonian Institution/Corbis.

Chapter 3

Page 77: Stocktrek Images/Richard Roscoe/Getty Images; p. 78: (radioactive symbol): Digital Vision/Getty Images; (TSA scanner X-ray): imagebroker/Jochen Tack/Newscom; (rainbow): Andrew Holt/Photographer's Choice/Getty Images; (infrared soldier): Jason T. Bailey/UPI/Landov; (tower): Shutterstock; p. 80: Universal History Archive/Getty Images; p. 81 (top): Physics Dept., Imperial College/Science Photo Library; (middle, a–c): Richard Megna/Fundamental Photographs; p. 82 (all): © 1984 Richard Megna, Fundamental Photographs, NYC; p. 83 (top): Science Source/Photo Researchers; (bottom): James Leynse/Corbis; p. 85: (both): U.S. Army NVESD; p. 94: Lebrecht Music and Arts Photo Library/Alamy; p. 107 (top): David Taylor/Photo Researchers; (bottom): Dorling Kindersley/Getty Images; p. 121: Shutterstock; p. 122: Richard Megna/Fundamental Photographs.

Chapter 4

Page 133: iStockphoto; p. 137: David Wrobel/Visuals Unlimited; p. 138: Charles D. Winters/Photo Researchers; p. 153: Roman Samokhin/Dreamstime.com; p. 157: W. Perry Conway/Corbis; p. 164: Bob Rowan/Progressive Image/Corbis; p. 180 (3): NRH Photography; p. 184: NRH Photography.

Chapter 5

Page 187: Darren Robb/Alamy; p. 190 (all): © 2007 Richard Megna—Fundamental Photographs; p. 210: Clive Freeman, The Royal Institution/Photo Researchers; p. 216 (both): Phil Degginger/www.color-pic.com; p. 217: Shutterstock; p. 222: © Yoav Levy/Phototake; p. 229: Clive Freeman, The Royal Institution/Photo Researchers; p. 230: © Yoav Levy/Phototake; p. 232: Mark Garlick/SPL/Getty Images.

Chapter 6

Page 239: Shutterstock; p. 242: © 2012 Richard Megna, Fundamental Photographs; p. 252: Goo Gone® is a premier brand of The Homax Group, Inc.; p. 254: Dr. Ryan Maue, WeatherBell Analytics; p. 255: © 1998 Richard Megna/Fundamental Photographs; p. 257: Martin Shields/PhotoResearchers; p. 257: Jeff Daly/Visuals Unlimited; p. 258 (top): Sinclair Stammers/Science Photo Library/Photo Researchers; (bottom): Phil Degginger/www.color-pic.com; p. 259: iStockphoto; p. 261: Jeff Daly/Visuals Unlimited.

Chapter 7

Page 267: Paul A. Souders/Corbis; p. 270: Jon Bower Pollution/Alamy; p. 271 (left): © 1994 NYC Parks Photo Archive/Fundamental Photographs; (right): © 1994 Kristen Brochmann/Fundamental Photographs; p. 276: Shutterstock; p. 285: Mark A. Schneider/Photo Researchers; p. 287: Stephen Earle, PhD, Geology Department, Malaspina University College, Naimo, Canada; p. 288: PJF/Alamy; p. 301 (left): Shutterstock; (right): Paul A. Souders/Corbis.

Chapter 8

Page 311: NASA/NOAA/GSFC/Suomi NPP/VIIRS/Norman Kuring; p. 314: Leigh Smith Images/Alamy; p. 318 (left): Photo courtesy of A&D Weighing, San Jose, CA. www.andweighing.com; (right): © 2009 Richard Megna, Fundamental Photographs; p. 319 (a–e): © 2009 Richard Megna—Fundamental Photographs; p. 321 (both): © 1994 Richard Megna/Fundamental Photographs; p. 327 (both): Richard Megna/Fundamental Photographs; p. 331 (top right): Javier Trueba/MSF/Photo Researchers; (bottom, left, all): © 1990 Richard Megna/Fundamental Photographs; (bottom, right): Richard Thom/Visuals Unlimited; p. 332: Istockphoto; p. 335 (all): Phil Degginger/www.color-pic.com; p. 337: NASA/JPL-Caltech; p. 338 (both): Peticolas/Megna/Fundamental Photographs; p. 339: Johnbell/Dreamstime.com; p. 341: Bill Ross/Corbis; p. 342 (top, both): Wetlands Field Manual/Courtesy USDA; (bottom, all): © 2009 Richard Megna, Fundamental Photographs; p. 345 (both): Richard Megna/Fundamental Photographs; p. 348 (top): Joel Arem/Photo Researchers; (bottom, left): © 2002 Richard Megna, Fundamental Photographs; (bottom, right): TUMS and the shape of the TUMS bottle are trademarks of GlaxoSmithKline. The image was provided courtesy of GlaxoSmithKline; p. 349: © 2009 Richard Megna/Fundamental Photographs; p. 350 (both): Peticolas/Megna/Fundamental Photographs; p. 355: P. Rona/NOAA; p. 357 (both): Courtesy Thomas Gilbert; p. 358: Courtesy Richard Sugarek/Environmental Protection Agency; p. 359: Joel Arem/Photo Researchers.

Chapter 9

Page 361: Shutterstock; p. 362: Richard and Ellen Thane/Getty Images; p. 363: Stocktrek Images, Inc./Alamy; p. 367: Hans Neleman/Getty Images; p. 370: Reuters/Corbis; p. 371: AP Photo; p. 374: © Jon Gnass/Gnass Photo Images; p. 377: Sprokop/Dreamstime.com; p. 397: AP Photo; p. 399 (all): Courtesy of Alcoa, Inc; p. 406: Shutterstock.

Chapter 10

Page 413: Istockphoto; p. 414 (top): British Antarctic Survey/Science Source/Photo Researchers; (center): Shutterstock; p. 416: Istockphoto; p. 419 (top): Galen Rowell/Corbis; (center): Hubert Stadler/Corbis; (bottom): Bill Ross/Corbis; p. 421: (top): Sam Ogden/Photo Researchers; (bottom): Stan Pritchard/Alamy; p. 423 (top): Gianni Dagli Orti/Corbis; (bottom): Istockphoto; p. 433: © 2001 Richard Megna, Fundamental Photographs; p. 434 (top): Istockphoto; (center): Richard Anthony/

Minden Pictures; (bottom): Thierry Orban/Sygma/Corbis; p. 435 (all): © 2009 Richard Megna, Fundamental Photographs; p. 438: Shutterstock; p. 439: Art Directors & TRIP/Alamy; p. 441: © 2008 Richard Megna, Fundamental Photographs; p. 443: Phil Degginger/Color-Pic; p. 444: © Elifranssens/Dreamstime.com; p. 449 (top): Jason Butcher/Getty Images; (bottom): Stephen B. Goodwin/Shutterstock.com; p. 450: Nicholas Burningham/Dreamstime.com; p. 451: Art Directors & TRIP/Alamy; p. 452: © 2008 Richard Megna, Fundamental Photographs; p. 456 (left): Imageshop/Corbis; (right): Jonathan Blair/Corbis; p. 458: F. Jack Jackson/Alamy; p. 460: Transtock/Corbis.

Chapter 11

Page 463: Istockphoto; p. 464: DigitalVues/Alamy; p. 465 (all): Courtesy Tom Gilbert; p. 466 (top): Charles D. Winters/Science Photo Library/Photo Researchers; (bottom): Paul Whitehill/Science Photo Library/Photo Researchers; p. 472: © Valentino Visentini/Dreamstime.com; p. 474: Martyn F. Chillmaid/Science Source/Photo Researchers; p. 491 (left): David M. Phillips/Science Source/Photo Researchers; (center): David M. Phillips/Science Source/Photo Researchers; (right): SPL/Science Source/Photo Researchers; p. 497: David M. Phillips/Science Source/Photo Researchers; p. 500 (left): G. Flayols/PhotoCuisine/Corbis; (right): Mark Bolton/Corbis; p. 501: OSH/Alamy.

Chapter 12

Page 505: iStockphoto; p. 506: Phil Degginger/Alamy; p. 507: © 1987 Richard Megna, Fundamental Photographs; p. 520: B. & C. Alexander/Photo Researchers; p. 529 (left): Hans Reinhard; (right): SCIMAT/Science Photo Library/Photo Researchers.

Chapter 13

Page 543: Chris Detrick/Salt Lake Tribune; p. 544: Courtesy Bob Burkhart; p. 579: Image courtesy the TOMS science team and Scientific Visualization Studio, NASA/Goddard Space Flight Center; p. 582: Clive Streeter/Getty Images; p. 591: Lawrence Migdale/Photo Researchers; p. 597: Andrew Lambert Photography/Science Photo Library/Photo Researchers.

Chapter 14

Page 601: Phil Degginger/Alamy; p. 604 (top): Robert Landau/Corbis; (bottom): © 1983 Chip Clark—Fundamental Photographs; p. 620: Morev Valery/ITAR-TASS/Landov; p. 621: Thermo Scientific Model 48000/Courtesy of Thermofisher; p. 623: (a–c): © 2010 Richard Megna/Fundamental Photographs; p. 627(both): Richard Megna/Fundamental Photographs; p. 643 (both): Richard Megna/Fundamental Photographs.

Chapter 15

Page 655: JTB Photo Communications/Age Fotostock; p. 690 (a–c): Courtesy Thomas Gilbert; p. 691: © 1994 Richard Megna, Fundamental Photographs; p. 708: Richard Megna, Fundamental Photographs.

Chapter 16

Page 717: Dmitry Kushch/Shutterstock; p. 721: Photo by Albris, May 2011; http://creativecommons.org/licenses/by-sa/3.0/deed.en; p. 723: Image copyright © The Metropolitan Museum of Art. Image source: Art Resource, NY; p. 731 (both) © 1994 Richard Megna, Fundamental Photographs; p. 732: © 1997 Richard Megna, Fundamental Photographs; p.734: © 1990 Richard Megna, Fundamental Photographs; p. 736 © 1994 Richard Megna, Fundamental Photographs; p. 737: Vaughan Fleming/Photo Researchers; p. 741 (both): Phil Degginger/www.color-pic.com;

NOTE: Material in figures or tables is indicated by *italic* page numbers. Footnotes are indicated by n after the page number.

ATOMIC COLOR PALETTE

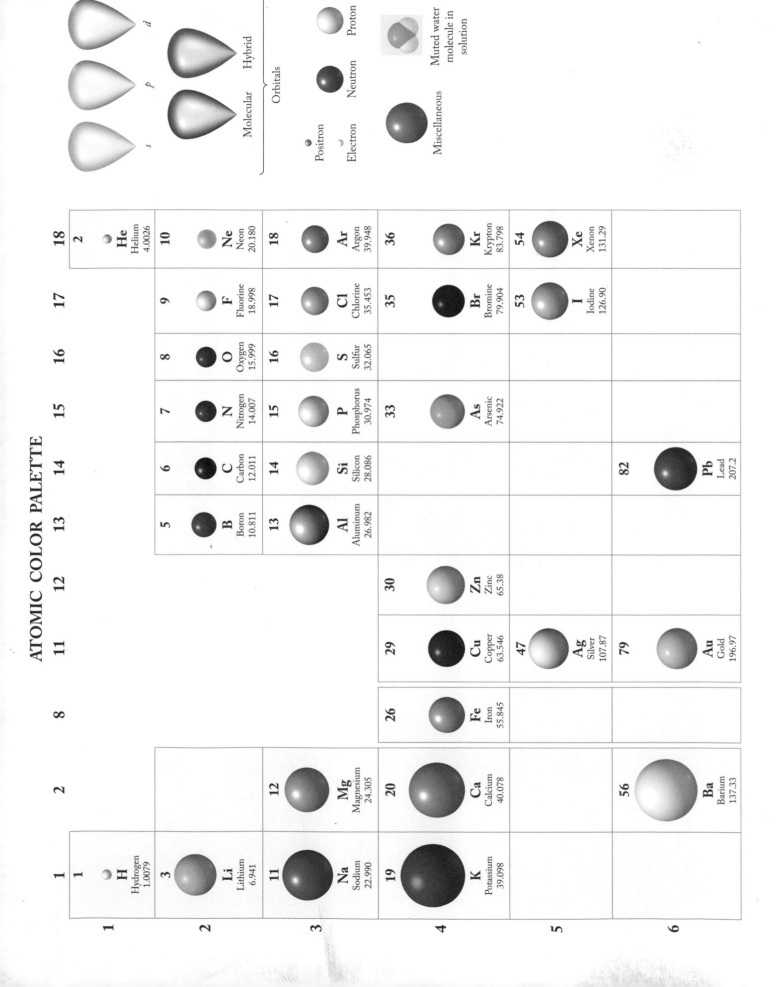

Orbitals
- d
- p
- s

Molecular Orbitals
- Hybrid

Proton
Neutron
Electron
Positron

Muted water molecule in solution

Miscellaneous

	1	2	8	11	12	13	14	15	16	17	18
1	1 **H** Hydrogen 1.0079										2 **He** Helium 4.0026
2	3 **Li** Lithium 6.941	12 **Mg** Magnesium 24.305				5 **B** Boron 10.811	6 **C** Carbon 12.011	7 **N** Nitrogen 14.007	8 **O** Oxygen 15.999	9 **F** Fluorine 18.998	10 **Ne** Neon 20.180
3	11 **Na** Sodium 22.990					13 **Al** Aluminum 26.982	14 **Si** Silicon 28.086	15 **P** Phosphorus 30.974	16 **S** Sulfur 32.065	17 **Cl** Chlorine 35.453	18 **Ar** Argon 39.948
4	19 **K** Potassium 39.098	20 **Ca** Calcium 40.078	26 **Fe** Iron 55.845	29 **Cu** Copper 63.546	30 **Zn** Zinc 65.38			33 **As** Arsenic 74.922		35 **Br** Bromine 79.904	36 **Kr** Krypton 83.798
5				47 **Ag** Silver 107.87						53 **I** Iodine 126.90	54 **Xe** Xenon 131.29
6		56 **Ba** Barium 137.33		79 **Au** Gold 196.97			82 **Pb** Lead 207.2				